thermophysical properties research literature retrieval guide

1900–1980

volume 5
oxide mixtures and minerals

Editors
J. F. CHANEY
V. RAMDAS
C. R. RODRIGUEZ
M. H. WU

thermophysical properties research literature retrieval guide

1900–1980

A Comprehensive Compilation of Scientific and Technical Literature by the Center for Information and Numerical Data Analysis and Synthesis (CINDAS), Purdue University

Volume 1. *Elements*

Volume 2. *Inorganic Compounds*

Volume 3. *Organic Compounds and Polymeric Materials*

Volume 4. *Alloys, Intermetallic Compounds, and Cermets*

Volume 5. *Oxide Mixtures and Minerals*

Volume 6. *Mixtures and Solutions*

Volume 7. *Coatings, Systems, Composites, Foods, Animal and Vegetable Products*

See inside front and back covers for CONDENSED MATERIALS INDEX to the SEVEN-VOLUME RETRIEVAL GUIDE.

New literature on thermophysical properties is being constantly accumulated at CINDAS. Contact CINDAS and use its interim updating search services for the most current scientific information

thermophysical properties research literature retrieval guide

1900–1980

volume 5
oxide mixtures and minerals

Editors

J. F. CHANEY

V. RAMDAS

C. R. RODRIGUEZ

M. H. WU

PART A. MATERIALS DIRECTORY
PART B. SEARCH PARAMETERS
PART C. BIBLIOGRAPHY
PART D. AUTHOR INDEX

IFI/PLENUM · NEW YORK · WASHINGTON · LONDON

Library of Congress Cataloging in Publication Data

Main entry under title:

Thermophysical properties research literature retrieval guide, 1900-1980.

Rev. ed. of: Thermophysical properties research literature retrieval guide. 1967.

Includes bibliographes and indexes.

Contents: v. 1. Elements—v. 2. Inorganic compounds—v. 3. Organic compounds and polymeric materials—[etc.]

1. Materials—Thermal properties—Bibliography. I. Chaney, J. F. (James F.)

Z5853.M38T48 [TA418.52] 016.6201'1296 81-15776

ISBN 0-306-67225-1 (v. 5) AACR2

IFI/Plenum Data Company
A Division of Plenum Publishing Corporation
233 Spring Street, New York, N.Y. 10013

Printed in the United States of America

CONDENSED INSTRUCTIONS ON THE USE OF VOLUME 5*

EXAMPLE

Obtain references on the thermal diffusivity of Corning 7900 glass.

STEP 1

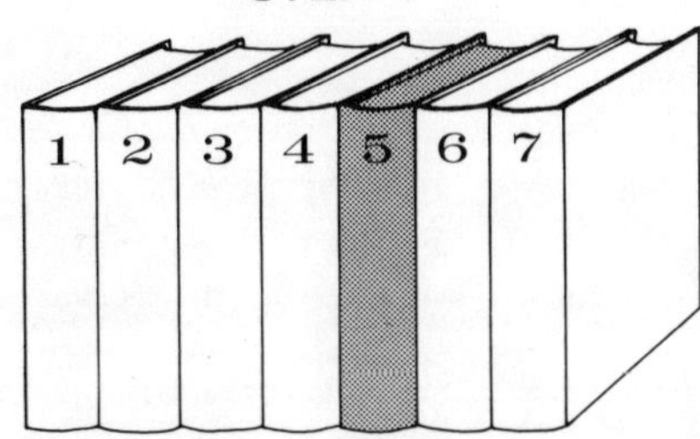

a. **Select** volume containing material of interest.

b. Scan the *Table of Contents;* note also that **instructions** on use are given at the beginning of each part.

a. Glasses are indexed in *Volume 5* (Oxide Mixtures and Minerals).

STEP 2

a. Locate **material name** in Part A, the *Materials Directory* (see example). Note the **properties** listed for the material.

b. Record the seven-digit **substance number.**

a. On page A7, Part A (the *Materials Directory*) locate the following entry:

Substance Name	Property	Number
Corning 7900 Glass $SiO_2 > B_2O_3$	deghijnr	503-0537
see also, Corning Vycor Glass		503-0020

The property of interest d (thermal diffusivity) is listed for the material.

b. The seven-digit substance number for Corning 7900 glass is 503-0537.

STEP 3

a. Select **property chapter** of interest, and locate the seven-digit **substance number** (obtained from Part A, the *Materials Directory*).

b. Use the five **search parameters** (physical state, subject, language, temperature, and year) to locate and record TPRC numbers of interest.

a. Locate property chapter Thermal Diffusivity (Chapter 2, page B25). The seven-digit substance number is located on page B25, column 1.

b. Locate the following entries:

Substance Number	Phys. State	Subject	Language	Temperature	Year	TPRC Number
503-0537	S	D	E	N	1957	19864
0537	S	D	E	N	1962	27922
0537	S	S	E	N	1973	70678

STEP 4

a. Locate **TPRC numbers** of interest (recorded from Part B, the *Search Parameters*) to obtain complete **bibliographic citation** information of documents for the material and properties of interest.

a. Page C26, column 1, has the following citation for TPRC number 19864:

19864 TRANSPARENT MATERIALS FOR AIRCRAFT ENCLOSURES.
WADC-UNIVERSITY OF DAYTON JOINT CONFERENCE.
WITTMAN, R. E.
WRIGHT AIR DEVELOPMENT CENTER
1-209, 1957.
(WADC-TR 57-421, AD 142021)

The citations for TPRC numbers 27922 and 70678 are given on pages C41, column 2, and C168, column 2, respectively.

NOTE

a. Locate author's name in the *Author Index* to obtain TPRC numbers of all documents written by this author and reported in this volume.

a. To obtain TPRC numbers of articles published by the author R. E. Wittman, turn to page D48, column 1, and locate the following entry:

WITTMAN R E 19864 29300

The complete bibliographic citations for the above numbers are given in Part C of this volume.

*For additional details on the use of this volume, see the introductory remarks for Parts A, B, C, and D, scan the *Table of Contents,* and also note the *Condensed Materials Index* on the opposite page.

CONTENTS OF VOLUME 5

*For earlier coding see Thermal Radiative Properties.

PREFACE

The phenomenal growth of science and technology has brought about a universal appreciation of the fact that present limitations in many technical developments are often a direct result of the paucity of knowledge on material properties. Engineering developments in the years ahead will be closely linked to the research that is done today to contribute to a better understanding of the properties of matter, of which thermophysical properties constitute a major segment.

While research on the properties of materials continues, adequate steps are not being taken to ensure that this invaluable body of information is coordinated, synthesized, organized, and disseminated to the ultimate user, namely, the individual scientist and engineer.

It is generally agreed that the present level of research support on thermophysical properties of matter falls short of existing needs and anticipated future demands; but what is even more disturbing is the fact that engineering groups across the nation are using no more than a fraction of the information already available, either because it is in a form not directly useful to them or, often, because its existence is not generally known. As a result, such information remains buried in the world's scientific literature. The repercussions of this latter condition are indeed serious since it leads to unintentional duplication of research effort with the resultant waste of time and scientific manpower.

In conjunction with its research activities, CINDAS screens the world's literature and collects published information on a wide range of materials in the field of thermophysics. This information concerns data, theoretical estimation methods, and experimental measurement techniques. Technical papers come from journals, abstracting services, reports, doctoral dissertations, masters theses, and many other sources. Full evaluation and analysis of the collected raw data are needed before publications on recommended values can be prepared. Such an effort is obviously time consuming and expensive, and therefore this critical evaluation is currently performed at a rather modest funding level. The end result is that much of the available world literature is not being processed and distilled.

As a complementary effort to its Data Table Series, CINDAS published in 1967 a work entitled *Thermophysical Properties Research Literature Retrieval Guide.* This three-book work reported 33,700 references on seven thermophysical property groups and about 45,000 materials. This Basic Edition systematically covered the world's unclassified literature published essentially between 1920 and mid-1964, in many instances going much earlier. Then, in 1973, Supplement I to the Basic Edition was published, with 26,000 additional references on sixteen thermophysical properties of 20,000 materials, covering the years mid-1964 to 1970.

In 1979, CINDAS published Supplement II to the Basic Edition, reporting over 15,000 new references on fourteen thermophysical properties of 12,674 substances. It essentially covered the years 1971 to 1977, going beyond these dates in a few instances.

In the period between 1967 and 1980, a number of improvements and other editorial changes were introduced in the organization of the Retrieval Guide, thus creating certain inconsistencies in presentation between the Basic Edition, Supplement I, and Supplement II. Furthermore, the user was forced to perform repetitive searches in each of the three publications in order to obtain complete coverage. Hence, it was thought desirable to merge the 1967 Basic Edition and the two Supplements of 1973 and 1977 as well as add new citations which would update the new volume to 1980.

This merged and enlarged new Basic Edition of the *Thermophysical Properties Research Literature Retrieval Guide* covers the world literature on fourteen thermophysical properties over the period 1900 to 1980. It should be a welcome reference work to all those who have not been acquainted with the earlier versions of this work. Even those who have been using the earlier Basic Edition and its Supplements will find the new Basic Edition more convenient to use and more complete in coverage.

The new Basic Edition of the Retrieval Guide, consisting of over 4800 pages, comprises seven volumes, each covering a specialized or related group of materials. Thus, a specialist group may procure one or more of the set's volumes while a central library would be interested in the complete set. Some statistical data on the seven volumes are presented below:

Volume Designation and Number	*Number of Substances**	*Number of References*	*Number of Pages*
Volume 1—Elements	340	15,884	801
Volume 2—Inorganic Compounds	6,032	19,407	1,092
Volume 3—Organic Compounds and Polymeric Materials	6,377	8,307	628
Volume 4—Alloys, Intermetallic Compounds, and Cermets	12,803	7,839	734
Volume 5—Oxide Mixtures and Minerals	4,605	6,443	411
Volume 6—Mixtures and Solutions	5,201	8,128	496
Volume 7—Coatings, Systems, Composites, Foods, Animal and Vegetable Products	8,980	9,200	639
Totals:	44,338	75,208	4,801

*An additional 8.000 synonyms and trade names have been incorporated to assist the user in identifying the material or substance of interest.

It is sincerely hoped that this new Basic Edition: 1900-1980 for the *Thermophysical Properties Research Literature Retrieval Guide* will constitute a permanent and valuable contribution to science and technology as well as to scientific documentation. These volumes, and those to follow, should prove to be an invaluable source of information to every scientist and engineer, with a scope of knowledge humanly impossible to master for any one individual or any group of individuals. Perhaps even more important, it is hoped that a wealth of information, heretofore unknown, will have been made available to many, including the specialist. The CINDAS staff is most anxious to receive comments, suggestions, and criticism from all users of these volumes. All communications will be gratefully received. Specific information concerning CINDAS' operations, services, publications, and research activities can be obtained by communicating with the Director.

The preparation of these volumes was made possible through the collective financial support of a large number of governmental, industrial, and nonprofit research organizations. Their interest and support are gratefully acknowledged. I wish to specifically acknowledge the major support received from the Department of Defense in connection with the operation of the Thermophysical and Electronic Properties Information Analysis Center (TEPIAC), which has made this work possible.

In closing, I wish to acknowledge the individual and collective accomplishments of CINDAS' Scientific Documentation Division's supporting staff: Mrs. M. C. Day, Coder; Mrs. M. R. Troyer, Documentalist; Mrs. F. L. Frist and Mrs. G. M. Halstead, Computer Operations; Mrs. S. J. Creamer and Mrs. J. A. Brittingham, Library. Special thanks are due to Mr. T. M. Putnam for many helpful suggestions and assistance. Thanks are also extended to the staff of R. R. Donnelley and Sons, Chicago, Illinois, for their cooperation and the computer-assisted phototypesetting of this new Retrieval Guide.

Y. S. Touloukian
Director, Center for Information and Numerical Data Analysis and Synthesis
Distinguished Atkins Professor of Engineering
Purdue University

Part A
MATERIALS DIRECTORY

PART A

USE OF MATERIALS DIRECTORY

I. ORGANIZATION AND GENERAL CONSIDERATIONS

The organization of thermophysical properties information at CINDAS is by substance/material. A *Condensed Materials Index* to the seven-volume Retrieval Guide is located on the inside front and back covers of this volume.

In order to index the world literature the classification system must be systematic and yet flexible. Therefore, CINDAS has adopted a classification scheme which arranges materials into logical groups that have closely related chemical composition or into compatible groups by their physical form and/or by their use and application.

Within this system of classification by substance name, there exists a class of pertinent information for which no specific substance name would be appropriate, e.g., "Theory of the Thermal Conductivity of Gases" or "New Technique for the Viscosity Measurement of Liquids." Literature covering this class is not reported in this series but is available at CINDAS. A special computer search and retrieval can be made upon request.

The index to materials in this volume is divided into oxide mixtures and minerals. Synonyms, trade names, and cross-references have been incorporated in the *Materials Directory*. Cross-references are preceded by the words "see also." The user should browse in the general region where the material is likely to be found and look at each cross-reference for additional information and possibly additional cross-references. An example on the use of this volume is given on the inside front and back covers.

The user of this index should not be concerned with the structure of the seven-digit number associated with each material. This number uniquely identifies a material, and its structure is of significance only to CINDAS.

II. DEFINITIONS AND LIMITATIONS USED IN MATERIAL CLASSIFICATION

For the effective use of the *Materials Directory* of this volume certain definitions and limitations of terms accepted by CINDAS should be understood. These are listed below:

1. *Metallic Elements*—The elements which are internationally accepted as metallic (all elements *not* listed in 2 below).

2. *Nonmetallic Elements*—The elements which are internationally accepted as nonmetallic:

Argon	Chlorine	Krypton	Selenium
Arsenic	Deuterium	Neon	Silicon
Astatine	Fluorine	Nitrogen	Sulfur
Boron	Hydrogen	Oxygen	Tellurium
Bromine	Helium	Phosphorus	Tritium
Carbon	Iodine	Radon	Xenon

3. *Impurities*—For the purposes of classification, CINDAS defines the "impurity" limits as follows:

   Elements. Total of impurities must be <0.5 percent and individual impurity <0.2 percent by weight. A metallic element having impurities equal to or in excess of these limits is considered an alloy whereas a nonmetallic element is considered a mixture.

   Compounds. Total of impurities must be <5.0 percent and individual impurity <2.0 percent by weight. A compound with impurities equal to or in excess of these limits is considered a mixture.

4. *Doping*—Doped materials have intentionally added impurities and are indicated by the physical state code letter D in Part B of this volume.

5. *Oxide Mixtures*—An oxide mixture is a glass, ceramic, glass ceramic, slag, scale, etc., consisting of only simple oxides (binary oxides such as Al_2O_3, Na_2O, SiO_2, CaO, etc.).

   Oxide mixtures are named by first placing the single oxide components ≥ 2 weight percent in decreasing weight order. Next, a name is determined by the largest component. The four largest components are kept and the word "Other" is added if there is a fifth component. A descriptor of glass, glass ceramic, ceramic, slag, porcelain, or scale is used where appropriate. Brand or code name oxide mixtures and general entries such as flint glass, soda lime glass, and window glass are also used. Compositions are given if known.

   Much of the older coding of thermophysical properties is located under Oxide Mixture, where the components which have a formula are listed in alphabetical order followed by other words arranged alphabetically. The components are connected by plus signs. The naming does not indicate how much of a component is present.

6. *Brand Names*—Brand name materials are listed separately with cross-references to the generic name. In the earlier coding of TPRC documents, brand name materials were listed under the generic name. It is essential to retrieve information under the generic name for possible additional documents on a brand name material.

7. *Minerals, Rocks, and Celestial Bodies*—Minerals and rocks are the naturally occurring substances listed with generally accepted names. Composition for minerals are taken mainly from *Glossary of Geology*, American Geological Institute.

III. EXCLUSIONS TO MATERIAL AND PROPERTY COVERAGE

While CINDAS attempts to cover the world research literature on all matter for the fourteen thermophysical properties it monitors, for reasons of scientific and technical rationale and practical expediency it has become necessary to put a number of constraints on certain classes of materials in the coverage of specific properties.

1. GENERAL CONSTRAINTS

a. Data reported under unsteady-state and nonequilibrium conditions are excluded.
b. Patents and patent applications are excluded.
c. Unidentifiable product catalogs and promotional literature are excluded.
d. Data reported in arbitrary units or relative ratio without specification of reference used are excluded.
e. Data on solutions of undefined character and all polymeric mixtures are excluded.

2. SPECIFIC CONSTRAINTS RELATIVE TO CERTAIN PROPERTIES

Property	*Constraint*
a. Viscosity	Data for slags and scales, ceramics, glasses, polymers, oils, liquid mixtures containing one or more organic compounds, solids, and suspensions are excluded. However, the viscosity of vegetable oils, foods, and biological materials is included.
b. Emittance and reflectance	Data outside the wavelength range of 10 to 2×10^5 cm^{-1} or 0.05 to 1000 μm or 1.24×10^{-3} to 25 eV are excluded.
c. Absorptance, transmittance, and absorptance-to-emittance ratio	Data outside the same wavelength range as above and data for liquids, gases, organic compounds, and inorganic complexes are excluded.
d. Thermal linear and volumetric expansion	Data for all liquids, gases, and irreversible processes are excluded.

IV. THERMAL RADIATIVE PROPERTIES

In the earlier coding of TPRC documents, the properties Emittance, Reflectance, Absorptance, Transmittance, and Absorptance to Emittance Ratio were grouped under Thermal Radiative Properties. The user should consult this chapter in addition to the specific property chapter to obtain possible additional references.

V. USEFUL REFERENCES FOR MATERIALS IDENTIFICATION

The latest editions of the following selected references have been used by CINDAS in the classification and identification of materials:

1. *Chemical Synonyms and Trade Names*, W. Gardner and E. I. Cooke, CRC Press, Cleveland, Ohio.
2. *The Condensed Chemical Dictionary*, G. G. Hawley, Van Nostrand Reinhold Co., New York, New York.
3. *Handbook of Material Trade Names*, O. T. Zimmerman and I. Lavine, Industrial Research Service, Inc., Dover, New Hampshire.
4. *Webster's New International Dictionary, Unabridged*, G. & C. Merriam Co., Springfield, Massachusetts.
5. *A Concise Encyclopedia of Metallurgy*, A. D. Merriman, American Elsevier Publishing Co., Inc., New York, New York.
6. *Glossary of Geology*, M. Gary, R. McAfee, Jr., and C. L. Wolf, American Geological Institute, Falls Church, Virginia.
7. *Materials Handbook*, G. S. Brady and H. R. Clauser, McGraw-Hill Book Co., New York, New York.

VOLUME 5. OXIDE MIXTURES AND MINERALS

Property: a. Thermal conductivity; **b.** Accommodation coefficient; **c.** Thermal contact resistance; **d.** Thermal diffusivity; **e.** Specific heat; **f.** Viscosity; **g.** Emittance; **h.** Reflectance; **i.** Absorptance; **j.** Transmittance; **k.** a_s/ϵ_t ratio; **l.** Prandtl number; **n.** Thermal linear expansion coefficient; **o.** Thermal volumetric expansion coefficient; **r.** Thermal radiative properties (g,h,i,j,k)

Substance Name	*Property*	*Number*
Aluminum Oxide Mixture: -(Cont)-		
$Al_2O_3 > NiO$	g	503-2774
$Al_2O_3 > NiO > TiO_2$ Ceramic	g	503-2505
$Al_2O_3 >$ Other A D 85 (Coors Ceramics Co.)	gj	503-2770
$Al_2O_3 >$ Other A D 94 (Coors Ceramics Co.)	gj	503-2771
$Al_2O_3 >$ Other Alsimag 614 Ceramic	n	503-0531
$Al_2O_3 >$ Other Wesgo Al -500 Ceramic	n	503-2671
$Al_2O_3 >$ Other Wesgo Al -600 S Ceramic	n	503-2670
$Al_2O_3 > Rb_2O$	e	503-1975
$Al_2O_3 > SiO_2$	adegh	503-0631
$Al_2O_3 > SiO_2$ Ceramic	aden	503-2000
$Al_2O_3 > SiO_2$ Korterit 90 Ceramic	a	503-2310
$Al_2O_3 > SiO_2 > B_2O_3 > Na_2O$ Glass	adeno	503-1811
$Al_2O_3 > SiO_2 > Fe_2O_3$	a	503-2704
$Al_2O_3 > SiO_2 > Fe_2O_3$ Ceramic	a	503-2484
$Al_2O_3 > SiO_2 > Fe_2O_3 > TiO_2$ Ceramic	an	503-2807
$Al_2O_3 > SiO_2 > Fe_2O_3 > TiO_2 >$ Other Ceramic	an	503-2433
$Al_2O_3 > SiO_2 > MgO$	n	503-0650
$Al_2O_3 > SiO_2 > MgO > CaO$ Ceramic	aen	503-1827
$Al_2O_3 > SiO_2 > MgO > Cr_2O_3 >$ Other Slag	ade	503-1990
$Al_2O_3 > SiO_2 > MgO > Fe_2O_3 >$ Other Ceramic	aen	503-2380
$Al_2O_3 > SiO_2 >$ Other Mullite Refractory	aden	503-1125
$Al_2O_3 > SiO_2 > P_2O_5$ Ceramic	n	503-2425
$Al_2O_3 > SiO_2 > TiO_2$ Ceramic	ae	503-2002
$Al_2O_3 > SiO_2 > TiO_2 > Fe_2O_3$ Ceramic	an	503-2434
$Al_2O_3 > ThO_2 > BeO$	n	503-2881
$Al_2O_3 > TiO_2$ Ceramic	dgn	503-1388
$Al_2O_3 > TiO_2 > BaO > ZrO_2 >$ Other Ceramic	n	503-2980
$Al_2O_3 > TiO_2 > CaO > MgO$ Ceramic	a	503-2498
$Al_2O_3 > TiO_2 > CaO > MgO >$ Other Ceramic	a	503-2499
$Al_2O_3 > TiO_2 > CaO > SiO_2 >$ Other Ceramic	a	503-2497
$Al_2O_3 > TiO_2 > Cr_2O_3$ Ceramic	n	503-2981
$Al_2O_3 > TiO_2 > MgO > CaO >$ Other Ceramic	a	503-2496
$Al_2O_3 > TiO_2 > MnO_2$ Porcelain	a	503-2455
$Al_2O_3 > TiO_2 > Na_2O > CaO >$ Other Glass	a	503-1937
$Al_2O_3 > TiO_2 > SiO_2$	a	503-2705
$Al_2O_3 > TiO_2 > SiO_2$ Ceramic	a	503-2468
$Al_2O_3 > V_2O_3$	e	503-1886
$Al_2O_3 > ZnO > CoO$	h	503-1784
$Al_2O_3 > ZnO > Cr_2O_3 > CoO$	h	503-1800
$Al_2O_3 > ZrO_2$	a	503-2683
$Al_2O_3 > ZrO_2$ Ceramic	a	503-2878
$Al_2O_3 > ZrO_2 > SiO_2$ Ceramic	ae	503-2001
$Al_2O_3 > ZrO_2 > SiO_2 > Na_2O$ Ceramic	a	503-2295
Aluminum Oxide Porcelain	aden	503-1219
see also: Aluminum Oxide Ceramic		503-0026
see also: Aluminum Oxide Refractory		503-1173
Aluminum Oxide Refractory	adegjnr	503-1173
see also: Aluminum Oxide Porcelain		503-1219
Aluminum Oxide Slag	r	503-1364
Aluminum Phosphate Ceramic	n	503-1343
Aluminum Phosphate Glass	o	503-1033
Alundum Refractory	ag	503-1184
American Optical A o-0835 Glass	n	503-1034
American Optical A o-1053 Glass	j	503-1035
American Optical Colored Filter Glass	j	503-1963
American Optical E O D839 Glass	n	503-1036
Andesite Glaze	n	503-1327
Anorthite Ceramic	n	503-1312
Anorthosite Ceramic	n	503-1297
Antimony Oxide Mixture:		———
$Sb_2O_3 > K_2O > Al_2O_3 > Na_2O >$ Other Glass	r	503-2410
$Sb_2O_3 > P_2O_5$ Glass	n	503-1879
$Sb_2O_3 > SiO_2 > K_2O > Na_2O >$ Other Glass	r	503-2450
Antireflection Glass	j	503-2307
A o-0835 American Optical Glass	n	503-1034
A o-1053 American Optical Glass	j	503-1035
A R; Glass	g	503-1434
A S-2, Russian Enamel	h	503-1295
A S 023 Glass Ceramic, Russian	a	503-2366
A S 370 Glass Ceramic, Russian	a	503-2367
A S 418 Glass Ceramic, Russian	a	503-2368
Asbestos Brick	a	503-1200
Ash + Tripoli, Ceramic	a	503-2336
see also: Silicon Oxide Mixture Ceramic $SiO_2 > CaO > Fe_2O_3 > Al_2O_3 >$ Other		503-2337
B 1463, Borosilicate Crown Glass $SiO_2 > Na_2O > B_2O_3 > K_2O >$ Other	ae	503-2103
B 2156, Crown Glass $SiO_2 > Na_2O > CaO > MgO$	e	503-2111
B 3556, Crown Glass	e	503-0977
B 3860, Light Barium Crown Glass $SiO_2 > Na_2O > B_2O_3 > BaO >$ Other	ae	503-2108
B 4446, Extra Light Flint Glass $SiO_2 > Na_2O > PbO$	e	503-2110
B 9060, Dense Barium Crown Glass $SiO_2 > BaO > B_2O_3 > Al_2O_3$	ae	503-2109
Baer And Stroud B S-39 B Glass	en	503-0704
Ba K 1 G 12, Schott Glass	j	503-0996
Ball Clay + Zirconium Oxide Refractory	n	503-1348
Ball Clay Refractory	nr	503-1190
Barium - Soda Glass	r	503-1085
Barium Cordierite Ceramic	n	503-2977
Barium Crown, Dense Glass	ae	503-2114
see also: Bausch And Lomb D B C 1, Dense Barium Crown Glass		503-2986
see also: Bausch And Lomb D B C 3, Dense Barium Crown Glass		503-2987
Barium Crown Glass	aejnor	503-0818
Barium Feldspar Ts M 3 Ceramic	a	503-1240
Barium Flint, Light Glass	ae	503-0291
Barium Flint Glass	jn	503-0584
see also: Bausch And Lomb B F 1, Barium Flint Glass		503-2990
Barium Oxide Mixture:		———
$BaO > B_2O_3$ Glass	n	503-0110
$BaO > B_2O_3 > SiO_2$ Glass	n	503-0416
$BaO > B_2O_3 > SiO_2 > Al_2O_3$ Glass	n	503-1895
$BaO > Cr_2O_3 > SiO_2 > B_2O_3 >$ Other Ceramic	g	503-2874
$BaO > Cr_2O_3 > SiO_2 > B_2O_3 >$ Other Porcelain	h	503-2817
$BaO > K_2O > Na_2O > B_2O_3$ Glass	j	503-2555
$BaO > MgO > SiO_2$ Ceramic	n	503-2902
$BaO > SiO_2 > Al_2O_3$	en	503-2904
$BaO > SiO_2 > B_2O_3$ Glass	n	503-0417
$BaO > SiO_2 > B_2O_3 > CdO$ Glass	n	503-2271
$BaO > SiO_2 > B_2O_3 > La_2O_3$ Glass	n	503-2269
$BaO > SiO_2 > B_2O_3 > ThO_2$ Glass	n	503-2267
$BaO > SiO_2 > B_2O_3 > ZnO >$ Other	n	503-2312
$BaO > SiO_2 > B_2O_3 > ZrO_2$ Glass	n	503-2272
$BaO > SiO_2 > CeO_2 > SnO_2 >$ Other Ceramic	g	503-2869
$BaO > SiO_2 > CeO_2 > SnO_2 >$ Other Porcelain	g	503-2810
$BaO > SiO_2 > CeO_2 > ZrO_2 >$ Other Ceramic	g	503-2870
$BaO > SiO_2 > CeO_2 > ZrO_2 >$ Other Porcelain	h	503-2814
$BaO > SiO_2 > Cr_2O_3 > Fe_2O_3 >$ Other Ceramic	g	503-2862
$BaO > SiO_2 > Cr_2O_3 > Fe_2O_3 >$ Other Porcelain	h	503-2818
$BaO > SiO_2 > Fe_2O_3 > CoO >$ Other Ceramic	g	503-2863
$BaO > SiO_2 > La_2O_3 > B_2O_3$ Glass	n	503-2270
$BaO > SiO_2 > SnO_2 > CeO_2 >$ Other Ceramic	g	503-2871
$BaO > SiO_2 > SnO_2 > CeO_2 >$ Other Porcelain	g	503-2811
$BaO > SiO_2 > ThO_2 > B_2O_3$ Glass	n	503-2268
$BaO > SiO_2 > TiO_2$ Glass	n	503-0361
$BaO > SiO_2 > ZnO > Al_2O_3 >$ Other Glass	j	503-2126
$BaO > SiO_2 > ZnO > B_2O_3$ Glass	a	503-2413
$BaO > SiO_2 > ZrO_2 > CeO_2 >$ Other Ceramic	g	503-2872
$BaO > SrO$	a	503-0125
$BaO > SrO > ZrO$	a	503-2706
$BaO > TiO_2$	n	503-2914
$BaO > TiO_2 > SrO$	n	503-2913
Barium Silicate Glass	j	503-1040
Bausch And Lomb B F 1, Barium Flint Glass	n	503-2990
see also: Barium Flint Glass		503-0584
Bausch And Lomb B S C 1, Borosilicate Crown Glass	n	503-2983
see also: Borosilicate Crown Glass		503-0001
Bausch And Lomb B S C 2, Borosilicate Crown Glass	n	503-2984
see also: Borosilicate Crown Glass		503-0001

Property: **a.** Thermal conductivity; **b.** Accommodation coefficient; **c.** Thermal contact resistance; **d.** Thermal diffusivity; **e.** Specific heat; **f.** Viscosity; **g.** Emittance; **h.** Reflectance; **i.** Absorptance; **j.** Transmittance; **k.** α_s/ϵ_t ratio; **l.** Prandtl number; **n.** Thermal linear expansion coefficient; **o.** Thermal volumetric expansion coefficient; **r.** Thermal radiative properties (g,h,i,j,k)

Substance Name	Property	Number
Bausch And Lomb C 1, Crown Glass	n	503-2982
see also: Crown Glass		503-0024
Bausch And Lomb C F 1, Crown Flint Glass	n	503-2991
see also: Crown Flint Glass		503-0976
Bausch And Lomb Colored Filter Glass	j	503-1964
Bausch And Lomb D B C 1, Dense Barium Crown Glass	n	503-2986
see also: Barium Crown, Dense Glass		503-2114
Bausch And Lomb D B C 3, Dense Barium Crown Glass	n	503-2987
see also: Barium Crown, Dense Glass		503-2114
Bausch And Lomb D F 2, Dense Flint Glass	n	503-2988
see also: Flint, Dense Glass		503-0771
Bausch And Lomb E D F 3, Extra Dense Flint Glass	n	503-2989
see also: Flint, Extra Dense Glass		503-0815
Bausch And Lomb Glass	jn	503-1041
Bausch And Lomb L B C 2, Light Barium Crown Class	n	503-2985
Bausch And Lomb O 2235, Optical Flint Glass	jn	503-2088
Bausch And Lomb S 8688, Optical Crown Glass	jn	503-2087
Bauxite Refractory	an	503-1197
Beryllium Oxide Mixture:		———
$BeO > Al_2O_3$ Ceramic	a	503-2175
$BeO > Al_2O_3 > MgO$	n	503-2883
$BeO > Al_2O_3 > SiO_2 > CaO$ Ceramic	a	503-2667
$BeO > Al_2O_3 > ThO_2$	an	503-2707
$BeO > Al_2O_3 > ThO_2 > MgO$	a	503-2714
$BeO > Al_2O_3 > ZrO_2$	a	503-2708
$BeO > Al_2O_3 > ZrO_2 > MgO$	a	503-2718
$BeO > MgO$	an	503-2189
$BeO > MgO$ Ceramic	a	503-2664
$BeO > MgO > Al_2O_3$	n	503-2882
$BeO > MgO > Al_2O_3 > ThO_2$	a	503-2709
$BeO > MgO > Al_2O_3 > ZrO_2$	a	503-2712
$BeO > MgO > ThO_2 > Al_2O_3$	a	503-2710
$BeO > MgO > ZrO_2 > Al_2O_3$	a	503-2711
$BeO > SiO_2$ Ceramic	a	503-2665
$BeO > ThO_2 > Al_2O_3$	an	503-2715
$BeO > ThO_2 > Al_2O_3 > MgO$	a	503-2713
$BeO > TiO_2$	an	503-1869
$BeO > UO_2$	a	503-2684
$BeO > ZrO_2$ Ceramic	a	503-2666
$BeO > ZrO_2 > Al_2O_3$	a	503-2717
$BeO > ZrO_2 > Al_2O_3 > MgO$	a	503-2716
$BeO > ZrO_2 > MgO > Al_2O_3$	a	503-2719
Beryllium Oxide Refractory	adenr	503-1180
Beryllium Phosphate Glass	j	503-1043
Beryl Porcelain	an	503-1292
B F 1; Bausch And Lomb Barium Flint Glass	n	503-2990
see also: Barium Flint Glass		503-0584
B F 25 Glass, Russian	n	503-0719
B G 1; Schott Glass $SiO_2 > K_2O > CaO > B_2O_3$	j	503-2044
B G 1, Optical Industries Glass	h	503-1449
B G 1 Glass + Pyrex Glass	h	503-1450
B G 2; Schott Glass $SiO_2 > B_2O_3 > K_2O > PbO$	j	503-2045
B G 9; Schott Glass $SiO_2 > Na_2O > CaO > K_2O$	j	503-2046
B G 12; Schott Glass $SiO_2 > K_2O > ZnO > Na_2O$	j	503-2047
B G 16; Schott Glass $SiO_2 > Na_2O > CaO > K_2O >$ Other	j	503-2048
B G 20; Schott Glass $SiO_2 >$ Lanthan- Didymoxyd $> Na_2O >$ Other	j	503-2049
B G 24; Schott Glass	hjr	503-1567
B G 26; Schott Glass	j	503-1588
B G 28; Schott Glass $SiO_2 > K_2O > ZnO > Na_2O$	j	503-2050
B K 4 Glass, Russian	e	503-1402
B K 7; Schott Glass	adhj	503-0080
B K 7 G; Schott Glass; Cerium Doped	j	503-2875
B K 7 G 25, Schott Glass	adejn	503-0997
B K 8; Schott Glass $SiO_2 > B_2O_3 > K_2O > Na_2O >$ Other	jn	503-1570
B K 10 Glass, Russian	n	503-2284
Blast Furnace Brick	d	503-2765
see also: Silicon Oxide Mixture Ceramic $SiO_2 > Al_2O_3$		503-2437
Bone China	n	503-1345
see also: Silicon Oxide Mixture Porcelain $SiO_2 > CaO > Al_2O_3 > P_2O_5$		503-1801
see also: Silicon Oxide Mixture Porcelain $SiO_2 > CaO > P_2O_5 > Al_2O_3$		503-1802
Borate Flint Glass	ajn	503-1063
Borate Glass	jn	503-0083
Boron Chromium Glass	a	503-1118
Boron Oxide Mixture:		———
$B_2O_3 > Al_2O_3 > K_2O$ Glass	j	503-2543
$B_2O_3 > Al_2O_3 > K_2O > SiO_2 >$ Other Glass	n	503-1889
$B_2O_3 > Al_2O_3 > Li_2O$ Glass	n	503-2933
$B_2O_3 > Al_2O_3 > Li_2O > SiO_2 >$ Other Glass	n	503-2652
$B_2O_3 > Al_2O_3 > Na_2O$ Glass	n	503-2941
$B_2O_3 > Al_2O_3 > PbO > Na_2O$ Glass	j	503-2340
$B_2O_3 > Al_2O_3 > ZnO > PbO >$ Other Glass	a	503-2399
$B_2O_3 > BaO$ Glass	n	503-0396
$B_2O_3 > BaO > SiO_2$ Glass	n	503-1900
$B_2O_3 > BaO > SiO_2 > Al_2O_3$ Glass	n	503-1619
$B_2O_3 > CaO$ Glass	jn	503-0179
$B_2O_3 > Gd_2O_3 > La_2O_3 > CaO$ Glass	j	503-2548
$B_2O_3 > Gd_2O_3 > La_2O_3 > ZnO$ Glass	j	503-2551
$B_2O_3 > Gd_2O_3 > La_2O_3 > ZnO >$ Other Glass	j	503-2394
$B_2O_3 > Gd_2O_3 > ZnO > La_2O_3$ Glass	j	503-2552
$B_2O_3 > GeO > Li_2O > ZnO >$ Other H R 12 B G Glass	jn	503-1864
$B_2O_3 > GeO_2 > Al_2O_3 > ZnO >$ Other Glass	j	503-2384
$B_2O_3 > GeO_2 > Li_2O > SiO_2 >$ Other Glass	n	503-2656
$B_2O_3 > GeO_2 > Li_2O > ZnO >$ Other Glass	n	503-2655
$B_2O_3 > GeO_2 > SiO_2 > Ta_2O_5 >$ Other Glass	n	503-2648
$B_2O_3 > GeO_2 > Ta_2O_5 > Al_2O_3 >$ Other Glass	n	503-2647
$B_2O_3 > GeO_2 > Ta_2O_5 > Li_2O >$ Other Glass	n	503-2650
$B_2O_3 > GeO_2 > Ta_2O_5 > SiO_2 >$ Other Glass	n	503-2649
$B_2O_3 > GeO_2 > Ta_2O_5 > ZnO >$ Other Glass	jn	503-2386
$B_2O_3 > GeO_2 > ZnO > Ta_2O_5 >$ Other Glass	jn	503-2646
$B_2O_3 > GeO_2 > ZrO > SiO_2 >$ Other Glass	jn	503-2387
$B_2O_3 > K_2O$ Glass	an	503-0214
$B_2O_3 > K_2O > Al_2O_3 > Fe_2O_3 >$ Other Glass	j	503-1834
$B_2O_3 > K_2O > SiO_2$ Glass	n	503-1903
$B_2O_3 > K_2O > SiO_2 > Al_2O_3$ Glass	n	503-1907
$B_2O_3 > K_2O > SiO_2 > CaO$ Glass	n	503-1899
$B_2O_3 > K_2O > SiO_2 > Na_2O$ Glass	n	503-1887
$B_2O_3 > K_2O > SiO_2 > Na_2O >$ Other Glass	n	503-1888
$B_2O_3 > K_2O_3$	a	503-2681
$B_2O_3 > La_2O_3 > Gd_2O_3 > ZnO$ Glass	j	503-2550
$B_2O_3 > La_2O_3 > Gd_2O_3 > ZnO >$ Other Glass	j	503-2545
$B_2O_3 > La_2O_3 > PbO > Na_2O$ Glass	j	503-2342
$B_2O_3 > La_2O_3 > Yb_2O_3 > Gd_2O_3$ Glass	j	503-2393
$B_2O_3 > La_2O_3 > ZnO > CaO$ Glass	j	503-2547
$B_2O_3 > La_2O_3 > ZnO > Gd_2O_3$ Glass	j	503-2549
$B_2O_3 > La_2O_3 > ZnO > Gd_2O_3 >$ Other Glass	j	503-2546
$B_2O_3 > Li_2O$ Glass	n	503-0393
$B_2O_3 > Li_2O > Al_2O_3$ Glass	n	503-2934
$B_2O_3 > Li_2O > BeO$ Glass	n	503-2935
$B_2O_3 > Li_2O > BeO$ Lindeman X - Ray Transmitting Glass	n	503-2944
$B_2O_3 > Li_2O > MgO$ Glass	n	503-2936
$B_2O_3 > Li_2O > SiO_2$ Glass	n	503-1743
$B_2O_3 > Li_2O > Ta_2O_5 > ZnO >$ Other Glass	n	503-2653
$B_2O_3 > MgO > Al_2O_3$ Glass	n	503-2938
$B_2O_3 > MgO > BeO$ Glass	n	503-2939
$B_2O_3 > MgO > Li_2O$ Glass	n	503-2937
$B_2O_3 > MgO > Na_2O$ Glass	n	503-0392
$B_2O_3 > Na_2O$	a	503-2680

Property: **a.** Thermal conductivity; **b.** Accommodation coefficient; **c.** Thermal contact resistance; **d.** Thermal diffusivity; **e.** Specific heat; **f.** Viscosity; **g.** Emittance; **h.** Reflectance; **i.** Absorptance; **j.** Transmittance; **k.** α_s/ϵ_t ratio; **l.** Prandtl number; **n.** Thermal linear expansion coefficient; **o.** Thermal volumetric expansion coefficient; **r.** Thermal radiative properties (g,h,i,j,k)

Property: a. Thermal conductivity; **b.** Accommodation coefficient; **c.** Thermal contact resistance; **d.** Thermal diffusivity; **e.** Specific heat; **f.** Viscosity; **g.** Emittance; **h.** Reflectance; **i.** Absorptance; **j.** Transmittance; **k.** a_s/ϵ_t ratio; **l.** Prandtl number; **n.** Thermal linear expansion coefficient; **o.** Thermal volumetric expansion coefficient; **r.** Thermal radiative properties (g,h,i,j,k)

Substance Name	Property	Number
B S C – 2, Borosilicate Crown Glass	h	503-1829
B S C 1; Bausch And Lomb Borosilicate Crown Glass	n	503-2983
see also: Borosilicate Crown Glass		503-0001
B S C 2; Bausch And Lomb Borosilicate Crown Glass	n	503-2984
see also: Borosilicate Crown Glass		503-0001
C 1; Bausch And Lomb Crown Glass	n	503-2982
see also: Crown Glass		503-0024
C 2336, Dense Flint Glass $SiO_2 > PbO > Na_2O > K_2O$	ae	503-2112
C 6733, Dense Flint Glass $PbO > SiO_2 > K_2O > CaO$	e	503-2113
Calcia Stabilized Zirconia Ceramic $ZrO_2 > CaO$	adghin	503-0284
see Zirconium Oxide Mixture		
Calcine Ceramic	n	503-1329
Calcite Ceramic	ae	503-1156
Calcium Aluminate Glass	j	503-0800
Calcium Oxide Mixture:		———
$CaO > Al_2O_3$ Slag	d	503-1932
$CaO > Al_2O_3 > BaO > MgO >$ Other Glass	j	503-0297
$CaO > Al_2O_3 > BaO > PbO >$ Other Glass	j	503-2388
$CaO > Al_2O_3 > MgO > SiO_2$ Glass	jn	503-0476
$CaO > Al_2O_3 > PbO > BaO >$ Other Glass	j	503-2645
$CaO > Al_2O_3 > SiO_2$ Glass	j	503-2772
$CaO > B_2O_3$ Glass	j	503-2341
$CaO > FeO > MgO > SiO_2 >$ Other Slag	g	503-2016
$CaO > FeO > SiO_2$ Slag	e	503-2025
$CaO > FeO > SiO_2 > Al_2O_3 >$ Other Slag	g	503-2010
$CaO > FeO > SiO_2 > MgO >$ Other Slag	g	503-2015
$CaO > FeO > SiO_2 > MnO >$ Other Slag	i	503-2019
$CaO > FeO > SiO_2 > ZnO$ Slag	e	503-2026
$CaO > MgO$ Ceramic	no	503-2800
$CaO > MgO > FeO > SiO_2 >$ Other Slag	g	503-2014
$CaO > MgO > SiO_2$ Ceramic	an	503-2490
$CaO > MgO > SiO_2 > Al_2O_3 >$ Other Ceramic	an	503-2514
$CaO > MgO > SiO_2 > Fe_2O_3 >$ Other Ceramic	e	503-2466
$CaO > Rb_2O > SiO_2$ Glass	n	503-2857
$CaO > SiO_2$ Slag	j	503-1985
$CaO > SiO_2 > Al_2O_3$ Slag	e	503-2028
$CaO > SiO_2 > Al_2O_3 > FeO$ Slag	g	503-2011
$CaO > SiO_2 > Al_2O_3 > FeO >$ Other Slag	h	503-1991
$CaO > SiO_2 > Al_2O_3 > Fe_2O_3 >$ Other Slag	adejno	503-1987
$CaO > SiO_2 > Al_2O_3 > MgO$ Slag	adej	503-0884
$CaO > SiO_2 > Al_2O_3 > MgO >$ Other Slag	ade	503-2017
$CaO > SiO_2 > Al_2O_3 > MnO$ Slag	adeg	503-1989
$CaO > SiO_2 > Al_2O_3 > TiO_2 >$ Other Slag	ade	503-1988
$CaO > SiO_2 > FeO$ Slag	e	503-2022
$CaO > SiO_2 > FeO > Al_2O_3$ Slag	g	503-2012
$CaO > SiO_2 > FeO > Cr_2O_3 >$ Other Slag	g	503-2009
$CaO > SiO_2 > FeO > MgO >$ Other Slag	e	503-2029
$CaO > SiO_2 > FeO > MnO >$ Other Slag	i	503-2018
$CaO > SiO_2 > FeO > ZnO$ Slag	e	503-2023
$CaO > SiO_2 > Fe_2O_3 > Al_2O_3 >$ Other Scale	a	503-1996
$CaO > SiO_2 > MgO > FeO >$ Other Slag	eg	503-2013
$CaO > SiO_2 > MnO > Al_2O_3$ Slag	g	503-2008
$CaO > SiO_2 > Rb_2O$ Glass	n	503-2856
$CaO > SiO_2 > ZnO$ Slag	e	503-2024
$CaO > SiO_2 > ZnO > FeO$ Slag	e	503-2027
$CaO > TiO_2$	n	503-2884
Calorex Glass	r	503-1055
Carbon Glass	r	503-1126
Carborundum	aeghjnr	503-1194
Celkate T 21 $SiO_2 > MgO$	d	503-2762
Cellular Glass	aden	503-2726
Cenco Glass	j	503-1413
Ceramic	adeghijnor	503-1142
Ceramic:		———
Alumina	aegjn	503-0026
see also: Aluminum Oxide Porcelain		503-1219
Aluminum Oxide	aegjn	503-0026
see also: Aluminum Oxide Porcelain		503-1219
Aluminum Oxide + Steatite	a	503-1271
Aluminum Phosphate	n	503-1343
Anorthite	n	503-1312
Anorthosite	n	503-1297

Substance Name	Property	Number
Ceramic: -(Cont)-		
Ash + Tripoli,	a	503-2336
see also: Silicon Oxide Mixture Ceramic		503-2337
$SiO_2 > CaO > Fe_2O_3 > Al_2O_3 >$ Other		
Barium Cordierite	n	503-2977
Bone China	n	503-1345
see also: Silicon Oxide Mixture Porcelain		503-1801
$SiO_2 > CaO > Al_2O_3 > P_2O_5$		
see also: Silicon Oxide Mixture Porcelain		503-1802
$SiO_2 > CaO > P_2O_5 > Al_2O_3$		
Calcine	n	503-1329
Calcite	ae	503-1156
Cast	a	503-1333
Chlorite + Cordierite	n	503-1341
Clay	adeghn	503-1185
Clay + Flint + Feldspar,	n	503-2202
Cooking Ware	n	503-1290
Cordierite	agn	503-0461
see also: Silicon Oxide Mixture Ceramic		503-2480
$SiO_2 > Al_2O_3 > MgO$		
Cutting Tool	d	503-2325
see also: C C T 707 (Carborundum) Al_2O_3		102-0767
see also: C O – 6 (Kennametal Corp.) Al_2O_3		102-0868
see also: V R – 97 (V R Wesson Corp.) Al_2O_3		102-0869
Diaspore	a	503-2669
Diopside	n	503-1300
Earthenware	g	503-2190
Electrical	aer	503-1220
Electrical Insulating	n	503-1291
Feldspar	dn	503-1296
Ferrochrome Slag + Quartzite + Water Glass,	a	503-2330
Fireclay Refractory	adeghnr	503-1147
see also: Grog		503-1321
Forsterite	an	503-0544
see also: Alsimag 243 (American Lava Corp.)		119-0248
$2MgO . SiO_2$		
Foundry Mold	ade	503-1331
Glazed	dn	503-1309
Greenware	den	503-1310
Hotel Whiteware	n	503-2276
Illite + Kaolinite	n	503-1337
Kaolinite + Potassium Acetate	n	503-1338
Kaolinite + Sodium Acetate	n	503-1339
Lead Cordierite	n	503-2976
Lithium Oxide	an	503-1154
Low Lead	n	503-1320
Mechanical	ae	503-1221
Mica	aen	503-1187
Nephelite Syenite	n	503-1298
Niobate	a	503-2371
Nuclear	aen	503-1222
Nuclear Fuel	ad	503-1267
Optical	aejn	503-2274
Oven Whiteware	n	503-2275
Petalite	an	503-1128
see also: Silicon Oxide Mixture Glass		503-0481
$SiO_2 > Al_2O_3 > Li_2O$		
Porous	adn	503-1226
Pyroelectric	e	503-1344
Pyrophyllite	n	503-1318
Quartz Feldspar + Tripoli,	a	503-2334
see also: Silicon Oxide Mixture Ceramic		503-2335
$SiO_2 > Al_2O_3 > SO_3 > Fe_2O_3$		
Radio Equipment	a	503-1141
Sand	a	503-1223
Schist	n	503-1325
Silicon Oxide	aden	503-1933
Spark Plugs	a	503-1165
Steatite	aen	503-0765
see also: Steatite Porcelain		503-1183
Stoneware	aen	503-1218
Tableware	n	503-1287
Technical	an	503-1302
Trachyte	n	503-1323

Property: **a.** Thermal conductivity; **b.** Accommodation coefficient; **c.** Thermal contact resistance; **d.** Thermal diffusivity; **e.** Specific heat; **f.** Viscosity; **g.** Emittance; **h.** Reflectance; **i.** Absorptance; **j.** Transmittance; **k.** α_s/ϵ_t ratio; **l.** Prandtl number; **n.** Thermal linear expansion coefficient; **o.** Thermal volumetric expansion coefficient; **r.** Thermal radiative properties (g,h,i,j,k)

Substance Name	Property	Number
Ceramic: -(Cont)-		
Tremolite	n	503-1299
Tripoli	a	503-2332
see also: Silicon Oxide Mixture Ceramic $SiO_2 > Al_2O_3 > Fe_2O_3$		503-2333
Unfired	den	503-1310
Wollastonite	an	503-1471
Zinc Oxide	a	503-1322
UO_2 > Other Uranium Oxide Mixture	a	503-2668
Ceramic, Russian:		———
M G Sh	a	503-2374
Cerium Oxide Mixture:		———
$CeO_2 > BaO > SiO_2 > MgO$ > Other Ceramic	g	503-2873
$CeO_2 > BaO > SiO_2 > MgO$ > Other Porcelain	h	503-2813
$CeO_2 > MgO$	a	503-2685
$CeO_2 > UO_2$	a	503-2686
Cer Vit; Owens Illinois Glass Ceramic	adehjn	503-1828
Cesium Oxide Mixture:		———
$Cs_2O > SiO_2$ Glass	n	503-0353
$Cs_2O > SiO_2 > Na_2O$ Glass	n	503-0254
C F 1; Bausch And Lomb Crown Flint Glass	n	503-2991
see also: Crown Flint Glass		503-0976
Chamotte Refractory	adeghnr	503-1147
see also: Grog		503-1321
Chance - Pilkington Colored Filter Glass	j	503-1965
Chance - Pilkington D E D F 927210 $PbO > SiO_2$	j	503-2116
see also: Flint, Double Extra Dense Glass		503-2120
Chance - Pilkington E D F 700303 $PbO > SiO_2 > K_2O$	j	503-2115
see also: Flint, Extra Dense Glass		503-0815
Chance - Pilkington O W 7 $PbO > SiO_2 > K_2O$	j	503-2117
Chance - Pilkington O W 8 $PbO > SiO_2$	j	503-2118
Chance - Pilkington O W 10 Glass	j	503-1464
Chance - Pilkington R S F Glass	j	503-1463
Chance Brothers Colored Filter	r	503-1251
Chance O Gr3 Glass	j	503-1430
Chance O Y 13 Glass	j	503-1512
Chance O Y 21 Glass	j	503-1513
Charge Layer	a	503-1379
Charge Layer:		———
Blast Furnace	e	503-1366
Ferromanganese	a	503-1370
Iron, Furnace	a	503-1375
Lead, Blast Furnace	ade	503-1374
Silicomanganese	a	503-1371
Steel, Furnace	a	503-1376
Chemcor:		———
Corning 0317 Glass	hijn	503-2347
Corning L E M	j	503-2679
L E M	j	503-2679
China Clay Refractory	adeghnr	503-1162
Chinaware	an	503-1812
Chlorite + Cordierite Ceramic	n	503-1341
Chrome + Forsterite Refractory	a	503-1315
Chrome + Magnesite Refractory	aeghnr	503-1131
Chrome + Magnesium Oxide Refractory	a	503-1351
Chrome + Magnetite Refractory	n	503-1319
Chrome + Silicon Oxide Refractory	a	503-1238
Chrome Refractory	an	503-1262
Chromite + Forsterite Brick	a	503-2139
Chromite + Magnesite Refractory	an	503-1248
see also: Aluminum Oxide + Chromium Oxide + Iron Oxide + Magnesium Oxide + Phosphorus + Silicon Oxide, Chromite - Magnesite Refractory $Al_2O_3 + Cr_2O_3 + Fe_2O_3 + MgO + P + SiO_2$		513-3579
Chromite + Magnesium Oxide Refractory	a	503-1326
Chromite + Sillimanite Refractory	n	503-1301
Chromite Refractory	aegn	503-1139
Chromium Glass	r	503-1070
Chromium Oxide Mixture:		———
$Cr_2O_3 > Al_2O_3$	gno	503-0580
$Cr_2O_3 > Al_2O_3 > Fe_2O_3 > MgO$ > Other Ceramic	adeno	503-2478
$Cr_2O_3 > Al_2O_3 > MgO > Fe_2O_3$ > Other Ceramic	e	503-2464

Substance Name	Property	Number
Chromium Oxide Mixture: -(Cont)-		
$Cr_2O_3 > BaO > SiO_2 > CoO$ > Other Ceramic	g	503-2864
$Cr_2O_3 > BaO > SiO_2 > NiO$ > Other Ceramic	g	503-2865
$Cr_2O_3 > Co_2O_3 > PbO > SiO_2$ > Other Porcelain	g	503-2815
$Cr_2O_3 > Fe_2O_3 > SiO_2 > B_2O_3$ > Other Porcelain	g	503-2826
$Cr_2O_3 > MgO > Al_2O_3$	n	503-2899
$Cr_2O_3 > MgO > Fe_2O_3 > SiO_2$ > Other	a	503-2721
$Cr_2O_3 > SiO_2 > Al_2O_3 > MgO$ Ceramic	g	503-2859
$Cr_2O_3 > SiO_2 > MgO$	g	503-2860
$Cr_2O_3 > TiO_2$ Ceramic	an	503-2308
$Cr_2O_3 > V_2O_3$	e	503-1885
$Cr_2O_3 > ZnO$	h	503-1777
$Cr_2O_3 > ZnO > Al_2O_3$	h	503-1779
$Cr_2O_3 > ZnO > Al_2O_3 > CoO$	h	503-1783
$Cr_2O_3 > ZnO > Al_2O_3 > TiO_2$ > Other	h	503-1781
$Cr_2O_3 > ZnO > CoO$	h	503-1780
$Cr_2O_3 > ZnO > TiO_2$	h	503-1778
$Cr_2O_3 > ZnO > TiO_2 > CoO$	h	503-1782
Clay + Flint + Feldspar, Ceramic	n	503-2202
Clay - Diatomite Brick	a	503-1268
Clay - Grog - Sand Brick	n	503-1289
Clay - Grog Brick	ae	503-1186
Clay - Lime Brick	r	503-1212
Clay - Shale Brick	n	503-1307
Clay Brick	adehinr	503-1136
Clay Ceramic	adeghn	503-1185
Clinker, Portland Cement	eh	503-1340
Clinker Brick	e	503-1495
Coal Slag	a	503-1363
Cobalt Glass	ajr	503-1095
Cobalt Oxide Mixture:		———
$CoO > CuO$	e	503-2900
$CoO > NiO$	e	503-0210
$Co_2O_3 > Cr_2O_3 > PbO > SiO_2$ > Other Porcelain	g	503-2816
$Co_2O_3 > NiO > SiO_2 > B_2O_3$ > Other Porcelain	g	503-2820
$Co_2O_3 > SiO_2 > B_2O_3 > CaO$ > Other Porcelain	g	503-2819
Colored Glass Filter	jr	503-1252
Coors 4811 Porcelain	a	503-2979
Coors A D 85 Al_2O_3 > Other	gj	503-2770
Coors A D 94 Al_2O_3 > Other	gj	503-2771
Copper Glass	r	503-0787
Copper Oxide Glass	j	503-2134
Cordierite Ceramic	agn	503-0461
see also: Silicon Oxide Mixture Ceramic $SiO_2 > Al_2O_3 > MgO$		503-2480
Cordierite Glass Ceramic	d	503-1825
Cordierite Refractory	n	503-1346
Corex Glass	jr	503-1061
Corion, Corning Glass	jn	503-2877
Corning 1 -59 Glass	j	503-1407
Corning 1 -69 Glass	j	503-1401
Corning 5 -59 Glass	j	503-2507
Corning 7 -54 Glass	j	503-0971
Corning 024 Glass	n	503-1427
Corning 0080 Glass $SiO_2 > Na_2O > CaO > MgO$	adn	503-0262
Corning 0120 Glass $SiO_2 > PbO > K_2O > Na_2O$ > Other	n	503-0316
Corning 0160 Glass	egjnr	503-0018
Corning 0211 Borosilicate Glass	adejn	503-0200
see also: Glass Slide		503-0746
Corning 0317 Glass	hijn	503-2347
Corning 0580 Glass	j	503-0241
Corning 774 Glass $SiO_2 > B_2O_3 > K_2O > Na_2O$ > Other	adeghjno	503-1466
Corning 0790 Vycor Glass $SiO_2 > B_2O_3$	jn	503-0863
see also: Corning Vycor Glass		503-0020
Corning 0791 Vycor Glass $SiO_2 > B_2O_3$	j	503-2794
Corning 0800 Soda Lime Glass	n	503-0721
Corning 1717 Glass $SiO_2 > Al_2O_3 > BaO > CaO$	n	503-0927

Property: a. Thermal conductivity; **b.** Accommodation coefficient; **c.** Thermal contact resistance; **d.** Thermal diffusivity; **e.** Specific heat; **f.** Viscosity; **g.** Emittance; **h.** Reflectance; **i.** Absorptance; **j.** Transmittance; **k.** α_s/ϵ_t ratio; **l.** Prandtl number; **n.** Thermal linear expansion coefficient; **o.** Thermal volumetric expansion coefficient; **r.** Thermal radiative properties (g,h,i,j,k)

Substance Name	Property	Number
Corning 1720 Glass	jn	503-0736
$SiO_2 > Al_2O_3 > MgO > CaO >$ Other		
Corning 1722 Glass	d	503-0928
Corning 1723 Glass	adeghijnr	503-0604
$SiO_2 > Al_2O_3 > CaO > MgO >$ Other		
Corning 1762 Glass	n	503-0929
Corning 2403 Glass	j	503-0930
Corning 2408 Glass	j	503-0931
Corning 2412 Glass	r	503-0932
Corning 2418 Glass	j	503-0933
Corning 2434 Glass	j	503-0934
Corning 3304 Glass	j	503-0935
Corning 3307 Glass	j	503-0936
Corning 3384 Glass	j	503-0937
Corning 3387 Glass	j	503-0938
Corning 3480 Glass	j	503-0939
Corning 3482 Glass	jr	503-0940
Corning 3962 Glass	j	503-0941
Corning 4010 Glass	j	503-0942
Corning 4015 Glass	j	503-0943
Corning 4060 Glass	j	503-0944
Corning 4084 Glass	j	503-0945
Corning 4305 Glass	j	503-0946
Corning 4605 Glass	j	503-0947
Corning 4784 Glass	j	503-0948
Corning 5045 Glass	j	503-0949
Corning 5070 Glass	j	503-0950
Corning 5120 Glass	h	503-0951
Corning 5330 Glass	hj	503-0108
Corning 5551 Glass	j	503-0952
Corning 5562 Glass	j	503-0953
Corning 5850 Glass	r	503-0954
Corning 7052 Glass	n	503-0723
$SiO_2 > B_2O_3 > Al_2O_3 > K_2O >$ Other		
Corning 7059 Glass	adejn	503-0126
Corning 7160 Glass	n	503-0955
Corning 7560 Glass $SiO_2 > Na_2O > ZnO > CaO$	e	503-2415
Corning 7570 Glass $PbO > B_2O_3 > Al_2O_3 > SiO_2$	n	503-0304
Corning 7740 Pyrex Glass	adeghijnr	503-0216
$SiO_2 > B_2O_3 > Na_2O > Al_2O_3$		
see also: Phoenix Glass		503-1049
Corning 7900 Glass $SiO_2 > B_2O_3$	deghijnr	503-0537
see also: Corning Vycor Glass		503-0020
Corning 7905 Glass	egjnr	503-0974
see also: Corning Vycor Glass		503-0020
Corning 7910 Glass	jr	503-0975
Corning 7913 Glass $SiO_2 > B_2O_3$	hij	503-0305
Corning 7930 Glass	jn	503-0957
Corning 7971 U L E Glass $SiO_2 > TiO_2$	aden	503-0562
Corning 8263 Glass	n	503-0958
Corning 8325 Glass	dr	503-0959
$SiO_2 > BaO > ZrO_2 > Na_2O >$ Other		
Corning 8361 Glass	r	503-0960
Corning 8362 Glass	d	503-2247
$SiO_2 > PbO > K_2O > Na_2O >$ Other		
Corning 8363 Glass $PbO > B_2O_3 > SiO_2 > Al_2O_3$	adjr	503-0309
Corning 8365 Glass	j	503-0961
Corning 8370 Glass	d	503-2768
$SiO_2 > B_2O_3 > Na_2O > K_2O >$ Other		
Corning 8392 Glass	r	503-0962
Corning 8606 Glass Ceramic	r	503-0656
Corning 8608 Glass Ceramic	r	503-1144
Corning 9363 Glass	j	503-0963
Corning 9454 Glass Ceramic	n	503-1336
Corning 9606 Glass Ceramic	adeghijnr	503-0659
Corning 9608 Glass Ceramic	adeghijnr	503-0660
Corning 9623 Glass Ceramic	n	503-1264
Corning 9658 Glass Ceramic	deno	503-0964
Corning 9690 Glass Ceramic	adegnr	503-0452
Corning 9692 Glass Ceramic	n	503-1263
Corning 9700 Glass	ej	503-0648
$SiO_2 > B_2O_3 > Na_2O > Al_2O_3$		
Corning 9751 Glass	r	503-0965
Corning 9752 Glass	eghjnr	503-0966

Substance Name	Property	Number
Corning 9753 Glass $Al_2O_3 > CaO > SiO_2$	aeghijn	503-0967
Corning 9754 Glass	aegj	503-0968
Corning 9788 Glass	j	503-0969
Corning 9830 Glass	h	503-0970
Corning 9863 Glass	j	503-0971
Corning Aklo Glass	r	503-1027
Corning Cercor 9690 Glass Ceramic	adegnr	503-0452
Corning Color Filter Glass	j	503-0089
Corning Corion Glass	jn	503-2877
Corning C S 1-60 Glass	j	503-1087
Corning C S 1-64 Glass	hj	503-0108
Corning C S 9-53 Glass	j	503-2293
Corning C S 9-54 Glass	j	503-2294
Corning Fotoceram	a	503-0539
Corning G 5 Lead Glass	n	503-1536
Corning G 8 Lime Glass	n	503-1537
Corning G 12 Glass	n	503-1535
Corning G 986 A Glass	j	503-1442
Corning Glass	adejnr	503-1110
Corning Heat Transmitting Glass	jr	503-1107
Corning H Glass $SiO_2 > B_2O_3 > Na_2O > Al_2O_3$	adeghijnr	503-0216
see also: Phoenix Glass		503-1049
Corning Infrared Glass	j	503-1421
Corning L E M Chemcor	j	503-2679
Corning Noviweld Glass	jr	503-1116
Corning Photochromic Glass	j	503-1420
Corning Pyroceram	adeghijnr	503-0667
see also: Glass Ceramic		503-0066
see also: Sitall		503-2370
Corning U L E Glass $SiO_2 > TiO_2$	aden	503-0562
Corning Vycor Glass	adehjnor	503-0020
see also: Corning 7900 Glass $SiO_2 > B_2O_3$		503-0537
see also: Corning 0790 Vycor Glass $SiO_2 > B_2O_3$		503-0863
see also: Corning 7905 Glass		503-0974
Cortran 9753 Glass $Al_2O_3 > CaO > SiO_2$	aeghijn	503-0967
Corundum + Sillimanite Brick	a	503-2137
Corundum Brick	a	503-2138
Crolite No. 7 Ceramic	n	503-1293
Crown B 3556 Glass	e	503-0977
Crown Flint Glass	j	503-0976
see also: Bausch And Lomb C F 1, Crown Flint Glass		503-2991
Crown Glass	aehjnor	503-0024
see also: 7904; Crown Glass		503-1124
see also: Bausch And Lomb C 1, Crown Glass		503-2982
C S 1-60, Corning Glass	j	503-1087
C S 1-64, Corning Glass	hj	503-0108
C S Z; Calcia Stabilized Zirconia Ceramic $ZrO_2 > CaO$	adghin	503-0284
see Zirconium Oxide Mixture		
Cupola Brick	d	503-2764
see also: Silicon Oxide Mixture Ceramic		503-2437
$SiO_2 > Al_2O_3$		
Cyanite + Sillimanite Refractory	n	503-1273
Cyanite Refractory	n	503-1272
D B C 1; Bausch And Lomb Dense Barium Crown Glass	n	503-2986
see also: Barium Crown, Dense Glass		503-2114
D B C 3; Bausch And Lomb Dense Barium Crown Glass	n	503-2987
see also: Barium Crown, Dense Glass		503-2114
D E D F 927210, Chance - Pilkington $PbO > SiO_2$	j	503-2116
see also: Flint, Double Extra Dense Glass		503-2120
Deutsche Spiegelglas Colored Filter Glass	j	503-1966
D F 2; Bausch And Lomb Dense Flint Glass	n	503-2988
see also: Flint, Dense Glass		503-0771
Diabase Glass	a	503-0979
Diaspore Ceramic	a	503-2669
Diatomite, Brick	aden	503-1152
see also: Silicon Oxide Refractory		503-1133
Didymium, Schott Glass	j	503-2064
Didymium Glass	hjr	503-0033
see also: S R M 2009 (N B S), Didymium Glass Filter		503-2796
see also: S R M 2010 (N B S), Didymium Glass Filter		503-2797
see also: S R M 2013 (N B S), Didymium Glass Filter		503-2798
see also: S R M 2014 (N B S), Didymium Glass Filter		503-2799
Dinasochromite Refractory	ae	503-1148

Property: a. Thermal conductivity; **b.** Accommodation coefficient; **c.** Thermal contact resistance; **d.** Thermal diffusivity; **e.** Specific heat; **f.** Viscosity; **g.** Emittance; **h.** Reflectance; **i.** Absorptance; **j.** Transmittance; **k.** α_s/ϵ_t ratio; **l.** Prandtl number; **n.** Thermal linear expansion coefficient; **o.** Thermal volumetric expansion coefficient; **r.** Thermal radiative properties (g,h,i,j,k)

Property: **a.** Thermal conductivity; **b.** Accommodation coefficient; **c.** Thermal contact resistance; **d.** Thermal diffusivity; **e.** Specific heat; **f.** Viscosity; **g.** Emittance; **h.** Reflectance; **i.** Absorptance; **j.** Transmittance; **k.** a_s/ϵ_t ratio; **l.** Prandtl number; **n.** Thermal linear expansion coefficient; **o.** Thermal volumetric expansion coefficient; **r.** Thermal radiative properties (g,h,i,j,k)

Substance Name	Property	Number
F K 14 Glass, Russian	ade	503-1483
Flint, Dense Glass	e	503-0771
see also: Bausch And Lomb D F 2, Dense Flint Glass		503-2988
Flint, Double Extra Dense Glass	j	503-2120
see also: Chance - Pilkington D E D F 927210		503-2116
$PbO > SiO_2$		
Flint, Extra Dense Glass	ej	503-0815
see also: Chance - Pilkington E D F 700303		503-2115
$PbO > SiO_2 > K_2O$		
see also: Bausch And Lomb E D F 3, Extra Dense Flint Glass		503-2989
Flint, Extra Light Glass	e	503-0295
Flint, Optical Glass	hr	503-1080
Flint Glass	aeghjnor	503-0270
see also: Schott, Flint Glass		503-1456
see also: Hoya, Flint Glass		503-1457
Flint O 165 Glass	a	503-1101
Foamglas (Pittsburgh Corning Corp.)	n	503-2727
see also: Soda Lime Silicate, Cellular Glass		503-2725
Foam Glass	aden	503-2726
Foamsil-28 (Pittsburgh Corning Corp.)	n	503-2728
see also: Borosilicate, Cellular Glass		503-2724
Foamsil-70 (Pittsburgh Corning Corp.)	n	503-2729
Forsterite Ceramic	an	503-0544
see also: Alsimag 243 (American Lava Corp.)		119-0248
$2MgO.SiO_2$		
Fotoceram, Corning	a	503-0539
Fourcault Glass	ahj	503-1097
F S 1 Glass, Russian	j	503-1837
F S K V 4, Schott Glass	n	503-1426
Fuel:		———
Ceramic Nuclear Fuel	ad	503-1267
Oxide Reactor	a	503-1317
Fusilain	n	503-1305
G 5, Corning Lead Glass	n	503-1536
G 8, Corning Lime Glass	n	503-1537
G 12, Corning Glass	n	503-1535
G 20, Borosilicate Glass	n	503-1412
see also: Borosilicate Glass		503-0782
G 702 P Glass	nr	503-1071
G 986 A, Corning Glass	j	503-1442
Gabbro-norite Glass Ceramic	n	503-2179
Gadolinium Oxide Mixture:		———
$Gd_2O_3 > B_2O_3 > ZnO$ Glass	j	503-2556
$Gd_2O_3 > B_2O_3 > ZnO > La_2O_3$ Glass	j	503-2557
$Gd_2O_3 > Sm_2O_3$	a	503-2687
$Gd_2O_3 > Tb_2O_3$	en	503-2887
G E - 72673 Glass	j	503-1865
G E-1 T L Glass	jn	503-1862
$SiO_2 > B_2O_3 > Li_2O > La_2O_3 >$ Other		
G E-101673 Glass	j	503-1866
Germanate Glass	nr	503-0569
Germanium Oxide Mixture:		———
$GeO_2 > BaO > K_2O > TiO_2 >$ Other Glass	j	503-2101
$GeO_2 > BaO > ZnO > PbO >$ Other Glass	j	503-2098
$GeO_2 > Bi_2O_3$	j	503-0734
$GeO_2 > CaO > K_2O > PbO >$ Other Glass	j	503-2097
$GeO_2 > P_2O_5$ Glass	n	503-1884
$GeO_2 > P_2O_5 > B_2O_3 > SiO_2$	n	503-2786
$GeO_2 > P_2O_5 > SiO_2$ Glass	n	503-2785
$GeO_2 > PbO > K_2O > Na_2O >$ Other Glass	j	503-2099
$GeO_2 > PbO > K_2O > TiO_2 >$ Other Glass	j	503-2102
$GeO_2 > PbO > TiO_2 > K_2O >$ Other Glass	j	503-2100
G G 4; Schott Glass	j	503-1508
G G 5; Schott Glass $SiO_2 > K_2O > CaO > B_2O_3$	j	503-2054
G G 7; Schott Glass	j	503-1509
G G 11; Schott Glass $SiO_2 > K_2O > CaO > B_2O_3$	j	503-2055
G G 13; Schott Glass	j	503-2056
$SiO_2 > PbO > K_2O > Na_2O >$ Other		
G G 14; Schott Glass $SiO_2 > K_2O > ZnO > B_2O_3$	j	503-2057
G G 16; Schott Glass	j	503-2058
G G 17; Schott Glass	j	503-1510
G G 375 G 34; Schott Glass	j	503-1511
G G 495; Schott Glass	j	503-1821

Substance Name	Property	Number
Glass	adeghijnor	503-0082
Glass:		———
Alkali Borosilicate	r	503-0706
Alkali Lead Silicate	n	503-0546
Alkaline Earth Silicate	j	503-0203
Alkali Silicate	ij	503-1028
Alkali Tin Silicate	n	503-2229
Aluminate	j	503-1029
Aluminoborosilicate	adn	503-0599
Aluminosilicate	jnr	503-0739
Aluminum Lanthanum Lead Borate	jn	503-1032
Aluminum Oxide - Soda Lime	e	503-1076
Aluminum Phosphate	o	503-1033
Antireflection	j	503-2307
Apparatus	n	503-1030
Automobile	h	503-1037
Barium - Soda	r	503-1085
Barium Crown	aejnor	503-0818
Barium Crown, Dense	ae	503-2114
see also: Bausch And Lomb D B C 1, Dense Barium Crown Glass		503-2986
see also: Bausch And Lomb D B C 3, Dense Barium Crown Glass		503-2987
Barium Flint	jn	503-0584
see also: Bausch And Lomb B F 1, Barium Flint Glass		503-2990
Barium Flint, Light	ae	503-0291
Barium Silicate	j	503-1040
Bausch And Lomb	jn	503-1041
Beryllium Phosphate	j	503-1043
Bonding	n	503-1044
Borate	jn	503-0083
Borate Flint	ajn	503-1063
Boron Chromium	a	503-1118
Borosilicate	adegjnr	503-0782
see also: Supremax Glass		503-1112
see also: G 20, Borosilicate Glass		503-1412
see also: S R M 731 (N B S), Borosilicate Glass		503-1454
see also: Pittsburgh 3235 Glass		503-1547
$SiO_2 > B_2O_3 > Na_2O > Al_2O_3$		
see also: Nuclear Waste Borosilicate Glass		503-2350
Borosilicate, Cellular	n	503-2724
see also: Foamsil-28 (Pittsburgh Corning Corp.)		503-2728
Borosilicate Crown	aehjno	503-0001
see also: 6556, Borosilicate Crown Glass		503-1047
see also: Silicon Oxide Mixture Glass		503-2921
$SiO_2 > B_2O_3 > Na_2O > K_2O >$ Other		
see also: Bausch And Lomb B S C 1, Borosilicate Crown Glass		503-2983
see also: Bausch And Lomb B S C 2, Borosilicate Crown Glass		503-2984
Bottle	ae	503-1075
Calcium Aluminate	j	503-0800
Carbon	r	503-1126
Cellular	aden	503-2726
Chalkboard	r	503-1122
Chromium	r	503-1070
Cobalt	ajr	503-1095
Colored	adhijr	503-1250
Colored Glass Filter	jr	503-1252
Container	eghjn	503-1081
Copper	r	503-0787
Copper Oxide	j	503-2134
Crown	aehjnor	503-0024
see also: 7904; Crown Glass		503-1124
see also: Bausch And Lomb C 1, Crown Glass		503-2982
Crown Flint	j	503-0976
see also: Bausch And Lomb C F 1, Crown Flint Glass		503-2991
Crystal	aen	503-0978
Diabase	a	503-0979
Didymium	hjr	503-0033
see also: S R M 2009 (N B S), Didymium Glass Filter		503-2796
see also: S R M 2010 (N B S), Didymium Glass Filter		503-2797
see also: S R M 2013 (N B S), Didymium Glass Filter		503-2798
see also: S R M 2014 (N B S), Didymium Glass Filter		503-2799

Property: **a.** Thermal conductivity; **b.** Accommodation coefficient; **c.** Thermal contact resistance; **d.** Thermal diffusivity; **e.** Specific heat; **f.** Viscosity; **g.** Emittance; **h.** Reflectance; **i.** Absorptance; **j.** Transmittance; **k.** a_s/ϵ_t ratio; **l.** Prandtl number; **n.** Thermal linear expansion coefficient; **o.** Thermal volumetric expansion coefficient; **r.** Thermal radiative properties (g,h,i,j,k)

Property: a. Thermal conductivity; **b.** Accommodation coefficient; **c.** Thermal contact resistance; **d.** Thermal diffusivity; **e.** Specific heat; **f.** Viscosity; **g.** Emittance; **h.** Reflectance; **i.** Absorptance; **j.** Transmittance; **k.** α_s/ϵ_t ratio; **l.** Prandtl number; **n.** Thermal linear expansion coefficient; **o.** Thermal volumetric expansion coefficient; **r.** Thermal radiative properties (g,h,i,j,k)

Substance Name	Property	Number
Glass: -(Cont)-		
Window	adeghijnr	503-0025
see also: Silicon Oxide Mixture Glass		503-0376
$Si O_2 > Na_2 O > Ca O > Mg O$		
see also: Air Traffic Control Tower Glass		535-1708
see also: Glass, Window - Silver Glass - Ag		535-2352
see also: Glass, Window - Gold Glass - Au		535-2353
see also: Glass, Window - Iron Glass - Fe		535-2354
see also: Glass, Window - Nickel Glass - Ni		535-2355
see also: Copper - Glass, Window Cu - Glass		535-2356
see also: Aluminum - Glass, Window Al - Glass		535-2357
see also: Chromium - Glass, Window Cr - Glass		535-2358
X-ray Plate	j	503-1452
X-ray Shield Glass (High Iron Content)	ar	503-1048
Zinc	r	503-1255
Zinc Crown	n	503-1395
Zinc Silicate Crown	o	503-1392
Glass, Russian:		———
3c-5k	n	503-1918
3 S 4	n	503-1571
3 S 5	n	503-1572
3 S 9	n	503-1573
7 I V-23 Glass Ceramic, Russian	n	503-1936
13 V	jn	503-1010
23;	e	503-1058
95 B K	n	503-1011
97 B K	n	503-1012
108;	n	503-1013
316	aden	503-1766
369 V	n	503-1014
A 10	n	503-2251
A S 023 Glass Ceramic, Russian	a	503-2366
A S 370 Glass Ceramic, Russian	a	503-2367
A S 418 Glass Ceramic, Russian	a	503-2368
B F 25	n	503-0719
B K 4	e	503-1402
B K 10	n	503-2284
Colored Filter	j	503-1973
F 1	adejr	503-1083
see also: Silicon Oxide Mixture Glass		503-0442
$Si O_2 > Pb O > K_2 O$		
F 8	ade	503-1480
F 18	ade	503-1481
F K 1	ade	503-1482
F K 14	ade	503-1483
F S 1	j	503-1837
I V-23 Glass Ceramic, Russian	n	503-1935
K 2	n	503-1550
K 8, Borosilicate Crown	adeghjn	503-1405
K 17	n	503-1551
K 20	n	503-1552
K 208	no	503-1553
K 515	ade	503-1554
K F 4	ade	503-1015
K F 6, Optical	dn	503-1432
K R L;	j	503-1017
K S 6	o	503-1393
K S 19	j	503-1018
K S P	g	503-1019
L F 7	n	503-1488
L F 9	ade	503-1489
L F 10	ade	503-1490
L F 11	n	503-1491
L K 1	ade	503-1561
L K 5	aj	503-1562
L K 6, Light Borosilicate	aden	503-1563
M 19	n	503-1020
M 519 Glass Ceramic, Russian	n	503-2250
M C 20 Opal	h	503-2346
M K R 1	no	503-1021
M S 14 Opal	ghjr	503-1114
N S 6	j	503-1519
N S 7	j	503-1520
N S 8	j	503-1521

Substance Name	Property	Number
Glass, Russian: -(Cont)-		
N S 9	j	503-1522
N S 10	j	503-1523
N S 11	j	503-1524
N S 55/6	n	503-2348
O F 2	ade	503-0868
O F 4, Flint	ade	503-1478
O F 5, Flint	e	503-2289
O K 3	ade	503-1451
P E S 5	no	503-2287
P S 7	j	503-1455
S 48-1	n	503-1575
S 48-3	n	503-1576
S 54-1	n	503-1577
S 87-1	n	503-1578
S 88-1	n	503-1579
S 88-1v	n	503-1580
see also: Silicon Oxide Mixture Glass		503-1835
$Si O_2 > Ba O > K_2 O > Al_2 O_3$		
S 89-1	n	503-1581
S 89-2	n	503-2253
S 90-1	n	503-1582
S E T 1	n	503-1022
Sitall	a	503-2370
see also: Glass Ceramic		503-0066
see also: Corning Pyroceram		503-0667
S O 115 M Glass Ceramic, Russian	a	503-2369
S T K 3, Superheavy Crown	e	503-2290
T 0254	n	503-1484
T 0257	n	503-1485
T B F;	hj	503-0479
T B F 25	ade	503-1023
T F 1	adejn	503-0463
see also: Lead Oxide Mixture Glass		503-2119
$Pb O > Si O_2 > K_2 O$		
T F 3	jn	503-0464
see also: Lead Oxide Mixture Glass		503-2119
$Pb O > Si O_2 > K_2 O$		
T F 5	jn	503-0465
T F 7	j	503-2291
T F 8	n	503-1492
T F 10	aden	503-0466
T F 11	ade	503-1493
T F 12	ade	503-1494
T F 101	n	503-2286
T F K 1	ade	503-1024
T K 4	n	503-1555
T K 14	den	503-1556
T K 16	aden	503-1557
T K 20	n	503-0237
T K 21	aden	503-1559
T K 25	n	503-2285
T K 2621	ade	503-1560
T R L 10	n	503-1025
T S M 500	n	503-1525
T S M 501	n	503-1526
T S M 502	n	503-1527
T S M 504	n	503-2288
U F S 1	j	503-1836
V 073-1	n	503-1486
V 080-1	n	503-1487
Vertical-drawing	j	503-1117
V O 1	n	503-1499
V O 2	n	503-1500
V O 3	n	503-1501
V O 5	n	503-1502
V O 6	n	503-1503
V S 1	ejn	503-1504
V S 2	jn	503-1505
V S 6	jn	503-1506
V S 73-1	n	503-2292
V S 80-1	n	503-1507
V V;	hijno	503-1026
Z S 4	n	503-1571

Property: **a.** Thermal conductivity; **b.** Accommodation coefficient; **c.** Thermal contact resistance; **d.** Thermal diffusivity; **e.** Specific heat; **f.** Viscosity; **g.** Emittance; **h.** Reflectance; **i.** Absorptance; **j.** Transmittance; **k.** α_s/ϵ_t ratio; **l.** Prandtl number; **n.** Thermal linear expansion coefficient; **o.** Thermal volumetric expansion coefficient; **r.** Thermal radiative properties (g,h,i,j,k)

Property: **a.** Thermal conductivity; **b.** Accommodation coefficient; **c.** Thermal contact resistance; **d.** Thermal diffusivity; **e.** Specific heat; **f.** Viscosity; **g.** Emittance; **h.** Reflectance; **i.** Absorptance; **j.** Transmittance; **k.** a_s/ϵ_t ratio; **l.** Prandtl number; **n.** Thermal linear expansion coefficient; **o.** Thermal volumetric expansion coefficient; **r.** Thermal radiative properties (g,h,i,j,k)

Substance Name	Property	Number
Kz F S 4, Schott Glass	j	503-1004
Ladle Brick	d	503-2763
see also: Silicon Oxide Mixture Ceramic		503-2438
$SiO_2 > Al_2O_3 > TiO_2$		
Lanthanum Borate Glass	j	503-1460
Lanthanum Crown Glass	jno	503-0290
Lanthanum Lead Crown Glass	jn	503-1417
Lanthanum Lead Silicate Glass	jn	503-1418
Lanthanum Oxide Mixture:		———
$La_2O_3 > Al_2O_3 > B_2O_3 > SiO_2$ Glass	j	503-2571
$La_2O_3 > Al_2O_3 > B_2O_3 > SiO_2 >$ Other Glass	j	503-2395
$La_2O_3 > B_2O_3$ Glass	j	503-2563
$La_2O_3 > B_2O_3 > Al_2O_3 > SiO_2$ Glass	j	503-2559
$La_2O_3 > B_2O_3 > BaO$ Glass	j	503-2560
$La_2O_3 > B_2O_3 > CaO > BaO$ Glass	j	503-2565
$La_2O_3 > B_2O_3 > Lu_2O_3$ Glass	j	503-2564
$La_2O_3 > B_2O_3 > Y_2O_3$ Glass	j	503-2561
$La_2O_3 > B_2O_3 > Y_2O_3 > CaO >$ Other Glass	j	503-2397
$La_2O_3 > B_2O_3 > Yb_2O_3$ Glass	j	503-2562
$La_2O_3 > B_2O_3 > Yb_2O_3 > CaO >$ Other Glass	j	503-2566
$La_2O_3 > B_2O_3 > Yb_2O_3 > ZnO >$ Other Glass	j	503-2392
$La_2O_3 > B_2O_3 > ZnO$ Glass	j	503-2570
$La_2O_3 > B_2O_3 > ZnO > Gd_2O_3$ Glass	j	503-2569
$La_2O_3 > B_2O_3 > ZnO > Lu_2O_3$ Glass	j	503-2567
$La_2O_3 > B_2O_3 > ZnO > Sm_2O_3$ Glass	j	503-2558
$La_2O_3 > SiO_2 > Al_2O_3 > B_2O_3 >$ Other Glass	j	503-2396
$La_2O_3 > Ta_2O_{13} > ThO_2 > B_2O_3$ Glass	j	503-2512
$La_2O_3 > ThO_2 > B_2O_3 > BaO >$ Other Glass	j	503-2511
$La_2O_3 > WO_3$	n	503-1877
Laser Glass	adejn	503-1953
L B C 2; Bausch And Lomb Light Barium Crown Class	n	503-2985
Lead - Soda Glass	r	503-1086
Lead Borate Glass	j	503-1497
Lead Cordierite Ceramic	n	503-2976
Lead Crystal Glass	a	503-1098
Lead Glass	aeghjr	503-0545
Lead Oxide Mixture:		———
$PbO > B_2O_3$ Glass	aden	503-0849
$PbO > B_2O_3 > Al_2O_3 > SiO_2$ Corning 7570 Glass	n	503-0304
$PbO > B_2O_3 > Fe_2O_3$ Glass	ade	503-2365
$PbO > B_2O_3 > Na_2O$ Glass	j	503-2128
$PbO > B_2O_3 > SiO_2$ Glass	n	503-0833
$PbO > B_2O_3 > SiO_2 > Al_2O_3$ Corning 8363 Glass	adjr	503-0309
$PbO > B_2O_3 > SiO_2 > Al_2O_3$ Glass	d	503-2767
$PbO > B_2O_3 > SiO_2 > CaO >$ Other Glass	n	503-1774
$PbO > B_2O_3 > SiO_2 > Na_2O$ Glass	n	503-1872
$PbO > B_2O_3 > SiO_2 > SO_3$ Glass	n	503-1873
$PbO > B_2O_3 > ZnO > CuO >$ Other Glass	n	503-2676
$PbO > B_2O_3 > ZnO > TiO_2 >$ Other Glass	n	503-2677
$PbO > GeO_2 > Bi_2O_3$ Glass	jn	503-1753
$PbO > GeO_2 > HfO_2$ Glass	jn	503-1754
$PbO > GeO_2 > Nb_2O_5$ Glass	jn	503-1750
$PbO > GeO_2 > Sb_2O_3$ Glass	jn	503-1751
$PbO > GeO_2 > Ta_2O_5$ Glass	jn	503-1752
$PbO > GeO_2 > TiO_2$ Glass	jn	503-0173
$PbO > GeO_2 > WO_3$ Glass	jn	503-1756
$PbO > GeO_2 > Y_2O_3$ Glass	j	503-1757
$PbO > GeO_2 > ZnO$ Glass	jn	503-1755
$PbO > P_2O_5 > Al_2O_3 > K_2O >$ Other Ceramic	n	503-2429
$PbO > P_2O_5 > K_2O > Al_2O_5 >$ Other Ceramic	n	503-2805
$PbO > SiO_2$ Chance - Pilkington D E D F 927210	j	503-2116
see also: Flint, Double Extra Dense Glass		503-2120
$PbO > SiO_2$ Chance - Pilkington O W 8	j	503-2118
$PbO > SiO_2$ Glass	aj	503-0105
$PbO > SiO_2 > Al_2O_3 > B_2O_3$ Glass	n	503-1929
$PbO > SiO_2 > B_2O_3$ Glass	n	503-2258

Substance Name	Property	Number
Lead Oxide Mixture: -(Cont)-		
$PbO > SiO_2 > B_2O_3 > SO_3$ Glass	n	503-1876
$PbO > SiO_2 > CeO_2 > ZnO >$ Other Porcelain	g	503-2812
$PbO > SiO_2 > K_2O$	a	503-0515
$PbO > SiO_2 > K_2O$ Chance - Pilkington E D F 700303	j	503-2115
see also: Flint, Extra Dense Glass		503-0815
$PbO > SiO_2 > K_2O$ Chance - Pilkington O W 7	j	503-2117
$PbO > SiO_2 > K_2O$ Glass	ajn	503-2119
see also: T F 1 Glass, Russian		503-0463
see also: T F 3 Glass, Russian		503-0464
$PbO > SiO_2 > K_2O > CaO$ C 6733, Dense Flint Glass	e	503-2113
$PbO > SiO_2 > K_2O > CaO$ Glass	ade	503-2107
$PbO > SiO_2 > K_2O > Na_2O$ Glass	j	503-2508
$PbO > SiO_2 > MnO$ Glass	j	503-1748
$PbO > SiO_2 > Na_2O$ Ceramic	n	503-2842
$PbO > SiO_2 > Na_2O$ Ceramic Cement	n	503-2281
$PbO > SiO_2 > Na_2O$ Glass	j	503-0516
$PbO > SiO_2 > Na_2O > Al_2O_3 >$ Other Ceramic	n	503-2844
$PbO > SiO_2 > Na_2O > K_2O$ Glass	ade	503-2404
$PbO > SiO_2 > NiO > B_2O_3$ Glass	n	503-2259
$PbO > SiO_2 > SO_3$ Glass	n	503-1871
$PbO > SiO_2 > SO_3 > B_2O_3$ Glass	n	503-1875
$PbO > SiO_2 > SO_3 > Na_2O$ Glass	n	503-1874
$PbO > SiO_2 > TiO_2 > K_2O >$ Other Glass	j	503-2083
$PbO > SiO_2 > ZrO_2 > TiO_2 >$ Other Glass	n	503-2391
$PbO > Ta_2O_3 > B_2O_3 > Al_2O_3 >$ Other Glass	j	503-2509
$PbO > ZnO > B_2O_3 > BaO$ Glass	n	503-2678
$PbO > ZnO > B_2O_3 > SiO_2$ Glass	n	503-2675
$PbO > ZnO > B_2O_3 > TiO_2 >$ Other Glass	n	503-2674
Lead Silicate Glass	ajnr	503-0567
Lead Slag	e	503-1354
Leeds And Northrup Absorbing Glass	j	503-2339
Leeds And Northrup Filter	r	503-1258
L E M Chemcor	j	503-2679
L F 7 Glass, Russian	n	503-1488
L F 9 Glass, Russian	ade	503-1489
L F 10 Glass, Russian	ade	503-1490
L F 11 Glass, Russian	n	503-1491
Libbey Owens Ford 81 E-19778 Glass	ghjr	503-1094
see also: Electroconducting Glass		503-2793
Libbey Owens Ford 547-26 Glass	ghjr	503-1093
see also: Electroconducting Glass		503-2793
Libbey Owens Ford 9330 Glass	ae	503-1067
$SiO_2 > Na_2O > CaO > MgO$		
Libbey Owens Ford P B-19195 Glass	ghjr	503-1092
see also: Electroconducting Glass		503-2793
Libbey Owens Glass	aj	503-1084
Lime + Potassium Glass	n	503-1396
Lime Brick	en	503-1270
Lime Glass	an	503-1064
Lime Silicate Brick	a	503-1166
Lindeman X - Ray Transmitting Glass	n	503-2944
$B_2O_3 > Li_2O > BeO$		
Liquid Glass	aehij	503-1411
Lithium Oxide Ceramic	an	503-1154
Lithium Phosphate Glass	a	503-1415
L K 1 Glass, Russian	ade	503-1561
L K 5 Glass, Russian	aj	503-1562
L K 6, Light Borosilicate Glass, Russian	aden	503-1563
Low Iron Glass	jn	503-2803
M 19 Glass, Russian	n	503-1020
M 519 Glass Ceramic, Russian	n	503-2250
Macor Glass Ceramic	deno	503-0964
Magmalox Brick	a	503-2140
Magnesite Refractory	adeginr	503-1146
see also: Magnesium Oxide Mixture Ceramic		503-2486
$MgO > SiO_2 > Al_2O_3$		
see also: Calcium Oxide + Iron Oxide + Magnesium Oxide + Phosphorus + Silicon Oxide, Magnesite Refractory		513-3578
$CaO + Fe_2O_3 + MgO + P + SiO_2$		

Property: a. Thermal conductivity; **b.** Accommodation coefficient; **c.** Thermal contact resistance; **d.** Thermal diffusivity; **e.** Specific heat; **f.** Viscosity; **g.** Emittance; **h.** Reflectance; **i.** Absorptance; **j.** Transmittance; **k.** a_s/ϵ_t ratio; **l.** Prandtl number; **n.** Thermal linear expansion coefficient; **o.** Thermal volumetric expansion coefficient; **r.** Thermal radiative properties (g,h,i,j,k)

Substance Name	Property	Number
Magnesium Oxide Mixture:		——
$MgO > Al_2O_3$	d	503-2227
$MgO > Al_2O_3 > BeO$	n	503-2888
$MgO > Al_2O_3 > CaO$ Ceramic	e	503-2517
$MgO > Al_2O_3 > CaO > SiO_2$ Ceramic	a	503-2373
$MgO > Al_2O_3 > Fe_2O_3$ Ceramic	a	503-2493
$MgO > Al_2O_3 > Fe_2O_3 > CaO >$ Other Ceramic	a	503-2494
$MgO > Al_2O_3 > Fe_2O_3 > Cr_2O_3 >$ Other Ceramic	a	503-2495
$MgO > Al_2O_3 > Fe_2O_3 > SiO_2 >$ Other	a	503-2886
$MgO > Al_2O_3 > SiO_2$	n	503-1808
$MgO > BeO$	an	503-2086
$MgO > CaO$ Ceramic	no	503-2801
$MgO > CaO > Al_2O_3$ Ceramic	e	503-2516
$MgO > CaO > Fe_2O_3$	a	503-3000
$MgO >$ Clay	a	503-2688
$MgO > CoO$	j	503-0373
$MgO > CoO > ZnO$	j	503-2722
$MgO > Cr_2O_3$	n	503-2919
$MgO > Cr_2O_3 > Al_2O_3 > Fe_2O_3 >$ Other Ceramic	an	503-2427
$MgO > Cr_2O_3 > Fe_2O_3 > Al_2O_3$ M Kh V P Ceramic, Russian	d	503-2660
$MgO > Cr_2O_3 > Fe_2O_3 > Al_2O_3 >$ Other Ceramic	aen	503-2428
$MgO > Cr_2O_3 > Fe_2O_3 > FeO >$ Other Ceramic	n	503-2597
$MgO > Cr_2O_3 > Fe_2O_3 > SiO_2 >$ Other Ceramic	an	503-2586
$MgO > Cr_2O_3 > SiO_2 > Fe_2O_3 >$ Other Ceramic	an	503-2489
$MgO > FeO > CaO$ Ceramic	i	503-2444
$MgO > FeO > CaO > SiO_2 >$ Other Ceramic	i	503-2445
$MgO > Fe_2O_3$	h	503-2172
$MgO > Fe_2O_3$ Ceramic	e	503-2518
$MgO > Fe_2O_3 > Al_2O_3 > SiO_2 >$ Other Ceramic	ae	503-2492
$MgO > Fe_2O_3 > CaO$	a	503-2885
$MgO > Fe_2O_3 > CaO > SiO_2$	a	503-2731
$MgO > Fe_2O_3 > Cr_2O_3 > Al_2O_3 >$ Other Ceramic	e	503-2520
$MgO > Fe_2O_3 > Cr_2O_3 > SiO_2$ Ceramic	e	503-2519
$MgO > Fe_2O_3 > SiO_2 > Al_2O_3 >$ Other Ceramic	a	503-2472
$MgO > Fe_2O_3 > SiO_2 > CaO$ Ceramic	ae	503-2453
$MgO > MnO > FeO > CaO$ Ceramic	i	503-2446
$MgO > MnO_2$	n	503-0594
$MgO > Na_2O > SiO_2$ Glass	n	503-1921
$MgO > NiO$	an	503-0793
$MgO > SiO_2$	an	503-0768
$MgO > SiO_2$ Ceramic	aen	503-2463
$MgO > SiO_2 > Al_2O_3$	n	503-1809
$MgO > SiO_2 > Al_2O_3$ Ceramic	a	503-2486
see also: Magnesite Refractory		503-1146
$MgO > SiO_2 > Al_2O_3 > CaO$ Ceramic	a	503-2806
$MgO > SiO_2 > Al_2O_3 > FeO >$ Other Ceramic	a	503-2470
$MgO > SiO_2 > CaO$ Ceramic	a	503-2659
$MgO > SiO_2 > CaO > Al_2O_3 >$ Other Ceramic	a	503-2471
$MgO > SiO_2 > CaO > Fe_2O_3$	a	503-2732
$MgO > SiO_2 > Fe_2O_3$ Ceramic	a	503-2485
$MgO > SiO_2 > Fe_2O_3 > Al_2O_3$ Ceramic	e	503-2521
$MgO > SiO_2 > Fe_2O_3 > Al_2O_3 >$ Other Ceramic	a	503-2469
$MgO > SiO_2 > PbO$ Porcelain	n	503-2441
$MgO > SnO_2$	a	503-2689
$MgO > TiO_2$	n	503-0211
$MgO > UO_2$	a	503-2690
$MgO > Y_2O_3$ Ceramic	n	503-2672
$MgO > ZnO$	a	503-2691
$MgO > ZnO > B_2O_3 > SiO_2 >$ Other Glass	n	503-2963
$MgO > ZnO > CoO$	j	503-2723
Magnesium Oxide Refractory	adeghnr	503-1189
Magnetite Refractory	a	503-1196
Manganese Concentrate	a	503-1372
Manganese Glass	r	503-1096

Substance Name	Property	Number
Manganese Oxide Mixture:		——
$MnO > MgO$ Glass	n	503-1764
$MnO > SiO_2 > Al_2O_3 > CaO >$ Other Ceramic	adeno	503-2479
$MnO > TiO_2$	h	503-2777
$MnO_2 > Fe_2O_3 > SiO_2 > B_2O_3 >$ Other Porcelain	g	503-2828
$MnO_2 > MnO > SiO_2 > CaO$ Slag	ae	503-2007
$MnO_2 > SiO_2 > Co_2O_3 > B_2O_3 >$ Other Porcelain	g	503-2829
$Mn_2O_3 > Al_2O_3$	a	503-2692
$Mn_2O_3 > BaO > SiO_2 > CoO >$ Other Ceramic	g	503-2868
$Mn_2O_3 > MgO$	a	503-2693
$Mn_2O_3 > SiO_2$	a	503-2694
Marl Brick	a	503-1247
Marquardt Mass Refractory	en	503-1241
M C 20 Opal Glass, Russian	h	503-2346
M G Sh Ceramic, Russian	a	503-2374
Mica Brick	a	503-1170
Mica Ceramic	aen	503-1187
Mica Glass Ceramic	d	503-1824
Microcline Glass	e	503-1065
Mineral Wool	a	503-1179
Mirror Glass	ae	503-2500
M Kh V P Ceramic, Russian $MgO > Cr_2O_3 > Fe_2O_3 > Al_2O_3$	d	503-2660
M K R 1 Glass, Russian	no	503-1021
Moler Brick	an	503-1193
Molochite Refractory	n	503-1330
Molybdenum Oxide Mixture:		——
$MoO_3 > SiO_2$	h	503-2778
Monofrax M Brick	a	503-2141
M S 14 Opal Glass, Russian	ghjr	503-1114
M S 20 Glass, Russian	h	503-1823
Mullite Brick	anr	503-1208
Mullite Refractory $Al_2O_3 > SiO_2 >$ Other	aden	503-1125
N 51, Kimble Glass $SiO_2 > B_2O_3 > K_2O > Na_2O >$ Other	gh	503-2782
N 51 A, Kimble Glass	ghj	503-2783
see also: Silicon Oxide Mixture Glass $SiO_2 > B_2O_3 > Na_2O > Al_2O_3 >$ Other		503-1697
N B S, F 467 Glass	j	503-1414
N B S, I R 1 Glass $SiO_2 > PbO > Na_2O > Cr_2O_3$	r	503-2249
N B S, I R 5 Glass	j	503-1461
N B S 2101 Glass	j	503-0939
N B S 2102 Glass	j	503-0936
N B S 2103 Glass	j	503-0942
N B S 2104 Glass	j	503-0952
N B S 2105 Glass	j	503-0949
N D Y; Neodymium Doped Yttralox Ceramic $Y_2O_3 > ThO_2$	aen	503-0003
Neophan, Schott Glass	j	503-2063
Nephelite Syenite Ceramic	n	503-1298
N G 1; Schott Glass $SiO_2 > B_2O_3 > K_2O > Fe_2O_3 >$ Other	j	503-2061
N G 4, Schott Glass	j	503-1005
Nickel Glass	nr	503-1054
Nickel Oxide Mixture:		——
$NiO > Al_2O_3$	g	503-2776
$NiO > Co_2O_3 > SiO_2 > B_2O_3 >$ Other Porcelain	g	503-2821
$NiO > Fe_2O_3 > SiO_2 > B_2O_3 >$ Other Porcelain	g	503-2827
$NiO > MgO$	a	503-2085
$NiO > MgO$ Glass	n	503-1765
Niobate Ceramic	a	503-2371
Niobium Oxide Mixture:		——
$NbO_2 > TiO_2$	n	503-0103
$Nb_2O_5 > Al_2O_5$	n	503-2901
$Nb_2O_5 > HfO_2$	n	503-1857
$Nb_2O_5 > MgO$	n	503-2912
$Nb_2O_5 > TiO_2$	n	503-0749
$Nb_2O_5 > ZrO_2$	n	503-1858
Nonex Glass	nr	503-1071
Nonlead Glass	r	503-1256

Property: a. Thermal conductivity; **b.** Accommodation coefficient; **c.** Thermal contact resistance; **d.** Thermal diffusivity; **e.** Specific heat; **f.** Viscosity; **g.** Emittance; **h.** Reflectance; **i.** Absorptance; **j.** Transmittance; **k.** α_s/ϵ_t ratio; **l.** Prandtl number; **n.** Thermal linear expansion coefficient; **o.** Thermal volumetric expansion coefficient; **r.** Thermal radiative properties (g,h,i,j,k)

Substance Name	Property	Number
Noviweld, Corning Glass	jr	503-1116
N S 6 Glass, Russian	j	503-1519
N S 7 Glass, Russian	j	503-1520
N S 8 Glass, Russian	j	503-1521
N S 9 Glass, Russian	j	503-1522
N S 10 Glass, Russian	j	503-1523
N S 11 Glass, Russian	j	503-1524
N S 55/6 Glass, Russian	n	503-2348
Nuclear Waste:		——
Borosilicate Glass	an	503-2350
see also: Borosilicate Glass		503-0782
Glass	a	503-2349
Glass Ceramic	n	503-2351
see also: Glass Ceramic		503-0066
Phosphate Glass	an	503-2352
see also: Phosphate Glass		503-0096
O 165, Flint Glass	a	503-1101
O 2235, Bausch And Lomb Optical Flint Glass	jn	503-2088
Obsidian Glass	a	503-1090
O F 2 Glass, Russian	ade	503-0868
O F 4, Flint Glass, Russian	ade	503-1478
O F 5, Flint Glass, Russian	e	503-2289
O G 1, Schott Glass	r	503-1006
O G 4; Schott Glass $SiO_2 > K_2O > ZnO > B_2O_3$	j	503-2059
O G 550; Schott Glass	j	503-1803
O G 590; Schott Glass	j	503-2876
O Gr3, Chance Glass	j	503-1430
Ohara Colored Filter Glass	j	503-1969
Ohara Glass	jn	503-1446
O K 3 Glass, Russian	ade	503-1451
Opal Glass	ehr	503-1077
Optical Ceramic	aejn	503-2274
Optical Glass	ahjnr	503-0731
Optical Industries, B G 1 Glass	h	503-1449
O Q 3, Kimble Glass	j	503-1530
O Q 4, Kimble Glass	j	503-1531
Osira H Z 6 X Glass	r	503-1108
Osram 712 I I Glass	r	503-1056
Oven Whiteware	n	503-2275
O W 7, Chance - Pilkington $PbO > SiO_2 > K_2O$	j	503-2117
O W 8, Chance - Pilkington $PbO > SiO_2$	j	503-2118
O W 10, Chance - Pilkington Glass	j	503-1464
Owens Illinois Cer Vit Glass Ceramic	adehjn	503-1828
Oxide Mixture	aen	503-1202
Oxide Mixture:		——
$Ag_2O + Al_2O_3 + Li_2O + P_2O_5 + PbO$ Glass	j	503-8887
$Ag_2O + B_2O_3$ Glass	jn	503-8526
$Ag_2O + K_2O + Nb_2O_5$ Ceramic	n	503-9713
$Ag_2O + MoO_3 + P_2O_5$ Glass	n	503-8884
$Ag_2O + MoO_3 + P_2O_5 + PdO$ Glass	n	503-8507
$Al_2O_3 + B_2O_3$	n	503-9250
$Al_2O_3 + B_2O_3$ Glass	jo	503-9146
$Al_2O_3 + B_2O_3$ + Bauxite Refractory	o	503-8254
$Al_2O_3 + B_2O_3 + BaO$ Glass	eno	503-8746
$Al_2O_3 + B_2O_3 + BaO + CdO + SiO_2$ Glass	n	503-8891
$Al_2O_3 + B_2O_3 + BaO + Fe_2O_3 + SiO_2$ Glass	r	503-9040
$Al_2O_3 + B_2O_3 + BaO + K_2O + Na_2O$ Glass	n	503-8587
$Al_2O_3 + B_2O_3 + BaO + K_2O + SiO_2$ Glass	n	503-8860
$Al_2O_3 + B_2O_3 + BaO + K_2O + SiO_2$ Glaze	n	503-9374
$Al_2O_3 + B_2O_3 + BaO + La_2O_3 + SiO_2$ Glass	jn	503-8892
$Al_2O_3 + B_2O_3 + BaO + Na_2O$ Glass	an	503-8838
$Al_2O_3 + B_2O_3 + BaO + Na_2O + SiO_2$ Enamel	n	503-9386
$Al_2O_3 + B_2O_3 + BaO + Na_2O + SiO_2$ Glass	nr	503-8991
$Al_2O_3 + B_2O_3 + BaO + PbO + SiO_2 + ZnO$ Glass	n	503-9149
$Al_2O_3 + B_2O_3 + BaO + SiO_2$	n	503-9571
$Al_2O_3 + B_2O_3 + BaO + SiO_2$ Glass	no	503-8894

Substance Name	Property	Number
Oxide Mixture: -(Cont)-		
$Al_2O_3 + B_2O_3 + BaO + SiO_2 + ThO_2$ Glass	n	503-8893
$Al_2O_3 + B_2O_3 + BaO + SiO_2 + ZnO$ Glass	an	503-8833
$Al_2O_3 + B_2O_3 + BaO + SiO_2 + ZrO_2$ Glass	n	503-8890
$Al_2O_3 + B_2O_3 + BaO + Y_2O_3 + ZrO_2$	n	503-9572
$Al_2O_3 + B_2O_3 + Bi_2O_3 + PbO + SiO_2$ Glaze	n	503-9584
$Al_2O_3 + B_2O_3 + CaO$ Glass	n	503-8748
$Al_2O_3 + B_2O_3 + CaO + FeO + SiO_2$ Glaze	n	503-9438
$Al_2O_3 + B_2O_3 + CaO + K_2O + SiO_2$ Glaze	n	503-9587
$Al_2O_3 + B_2O_3 + CaO + La_2O_3 + ZrO_2$ Glass	j	503-8693
$Al_2O_3 + B_2O_3 + CaO + MgO + SiO_2$ Glass	aejn	503-8882
$Al_2O_3 + B_2O_3 + CaO + MgO + SiO_2$ Glaze	hn	503-9588
$Al_2O_3 + B_2O_3 + CaO + MgO + SiO_2 + ZnO$ Glaze	n	503-9255
$Al_2O_3 + B_2O_3 + CaO + Na_2O + SiO_2$ Glass	o	503-9025
$Al_2O_3 + B_2O_3 + CaO + Na_2O + SiO_2$ Glaze	n	503-9324
$Al_2O_3 + B_2O_3 + CaO + P_2O_5 + SiO_2$ Glass	j	503-8691
$Al_2O_3 + B_2O_3 + CaO + PbO$ Glaze	n	503-9411
$Al_2O_3 + B_2O_3 + CaO + PbO + SiO_2$ Ceramic	n	503-9586
$Al_2O_3 + B_2O_3 + CaO + PbO + SiO_2$ Glass	n	503-8709
$Al_2O_3 + B_2O_3 + CaO + SiO_2$ Glass	jn	503-9078
$Al_2O_3 + B_2O_3 + CaO + SiO_2$ Glaze	n	503-9288
$Al_2O_3 + B_2O_3 + CaO + SiO_2$ + Other Glaze	n	503-8200
$Al_2O_3 + B_2O_3 + CaO + SiO_2 + TiO_2$ Glaze	n	503-9287
$Al_2O_3 + B_2O_3 + CaO + SiO_2 + ZnO$ Glaze	n	503-9260
$Al_2O_3 + B_2O_3 + CaO + SiO_2 + ZrO_2$ Glaze	n	503-9286
$Al_2O_3 + B_2O_3 + CdO$ Glass	n	503-8744
$Al_2O_3 + B_2O_3 + CoO + Na_2O + SiO_2$ Enamel	n	503-9578
$Al_2O_3 + B_2O_3 + Cr_2O_3 + K_2O + SiO_2$ Glass	n	503-8855
$Al_2O_3 + B_2O_3 + Cr_2O_3 + MnO$ Pigment	h	503-9419
$Al_2O_3 + B_2O_3 + CuO + K_2O + SiO_2$ Glass	n	503-8854
$Al_2O_3 + B_2O_3 + Fe_2O_3 + K_2O + SiO_2$ Glass	n	503-8856
$Al_2O_3 + B_2O_3 + Fe_2O_3 + K_2O + SiO_2$ Glaze	n	503-9259
$Al_2O_3 + B_2O_3 + GeO_2$ Glass	j	503-8432
$Al_2O_3 + B_2O_3 + K_2O$ Glass	jn	503-8927
$Al_2O_3 + B_2O_3 + K_2O + MgO + P_2O_5$ Glass	a	503-8834
$Al_2O_3 + B_2O_3 + K_2O + MgO + SiO_2$ Glaze	n	503-9376
$Al_2O_3 + B_2O_3 + K_2O + Na_2O + SiO_2$ Ceramic	n	503-9686
$Al_2O_3 + B_2O_3 + K_2O + Na_2O + SiO_2$ Glass	eno	503-8943
$Al_2O_3 + B_2O_3 + K_2O + PbO + SiO_2$ Glass	an	503-8837
$Al_2O_3 + B_2O_3 + K_2O + PbO + SiO_2$ Glaze	n	503-9380
$Al_2O_3 + B_2O_3 + K_2O + SiO_2$ Glass	n	503-8545
$Al_2O_3 + B_2O_3 + K_2O + SiO_2 + ZnO$ Glaze	n	503-9375
$Al_2O_3 + B_2O_3 + La_2O_3 + PbO + SiO_2$ Glass	j	503-8812
$Al_2O_3 + B_2O_3 + La_2O_3 + SiO_2$ Glass	j	503-8697
$Al_2O_3 + B_2O_3 + La_2O_3 + SiO_2 + ZnO$ Glass	j	503-8694
$Al_2O_3 + B_2O_3 + Li_2O$ Glass	n	503-8584
$Al_2O_3 + B_2O_3 + Li_2O + Na_2O + SiO_2$ Flux	n	503-9289
$Al_2O_3 + B_2O_3 + Li_2O + Na_2O + SiO_2$ Glass	n	503-8586
$Al_2O_3 + B_2O_3 + Li_2O + SiO_2$ Glass	n	503-8585
$Al_2O_3 + B_2O_3 + Li_2O + SiO_2$ + Clay Ceramic	n	503-8310
$Al_2O_3 + B_2O_3 + MgO$ Glass	n	503-8749

Property: a. Thermal conductivity; **b.** Accommodation coefficient; **c.** Thermal contact resistance; **d.** Thermal diffusivity; **e.** Specific heat; **f.** Viscosity; **g.** Emittance; **h.** Reflectance; **i.** Absorptance; **j.** Transmittance; **k.** α_s/ϵ_t ratio; **l.** Prandtl number; **n.** Thermal linear expansion coefficient; **o.** Thermal volumetric expansion coefficient; **r.** Thermal radiative properties (g,h,i,j,k)

Substance Name	Property	Number
Oxide Mixture: -(Cont)-		
$Al_2O_3 + B_2O_3 + MgO + Na_2O + P_2O_5$ Glass Frit	a	503-8603
$Al_2O_3 + B_2O_3 + MgO + Na_2O + SiO_2$ Ceramic	n	503-8355
$Al_2O_3 + B_2O_3 + MgO + Na_2O + SiO_2$ Glaze	n	503-9373
$Al_2O_3 + B_2O_3 + MgO + PbO + SiO_2$ Glaze	n	503-9396
$Al_2O_3 + B_2O_3 + MgO + SiO_2$ Ceramic	h	503-9874
$Al_2O_3 + B_2O_3 + MgO + SiO_2$ Glass	n	503-8869
$Al_2O_3 + B_2O_3 + MnO$ Pigment	h	503-9420
$Al_2O_3 + B_2O_3 + Na_2O$ Glass	aeno	503-9175
$Al_2O_3 + B_2O_3 + Na_2O + P_2O_5 + SiO_2$ Glass Frit	a	503-8604
$Al_2O_3 + B_2O_3 + Na_2O + PbO + SiO_2$ Glass	n	503-8614
$Al_2O_3 + B_2O_3 + Na_2O + PbO + SiO_2$ Glaze	n	503-9585
$Al_2O_3 + B_2O_3 + Na_2O + SiO_2$ Ceramic	ahjn	503-9378
$Al_2O_3 + B_2O_3 + Na_2O + SiO_2$ Glass	aejno	503-9160
$Al_2O_3 + B_2O_3 + Na_2O + SiO_2 + SrO$ Frit	j	503-9247
$Al_2O_3 + B_2O_3 + Na_2O + SiO_2 + TiO_2$ Glaze Frit, Fayence	n	503-9664
$Al_2O_3 + B_2O_3 + Na_2O + SiO_2 + ZnO$ Glass	n	503-8726
$Al_2O_3 + B_2O_3 + Na_2O + SiO_2 + ZnO$ Glaze	n	503-9371
$Al_2O_3 + B_2O_3 + Na_2O + SiO_2 + ZrO_2$ Enamel	n	503-9221
$Al_2O_3 + B_2O_3 + NiO + P_2O_5 + R_2O$ Glass	ar	503-9056
$Al_2O_3 + B_2O_3 + P_2O_5$	n	503-9487
$Al_2O_3 + B_2O_3 + P_2O_5$ Glass	n	503-8750
$Al_2O_3 + B_2O_3 + P_2O_5 + ZnO + R_2O$ Glass	ar	503-9057
$Al_2O_3 + B_2O_3 + PbO + SiO_2$ Glass	aeno	503-9099
$Al_2O_3 + B_2O_3 + PbO + SiO_2$ Glass Ceramic	n	503-9521
$Al_2O_3 + B_2O_3 + PbO + SiO_2 + TiO_2$ Glass	n	503-8642
$Al_2O_3 + B_2O_3 + PbO + SiO_2 + ZnO$ Glass	n	503-8708
$Al_2O_3 + B_2O_3 + PbO + SiO_2 + ZrO_2$ Glass	n	503-8643
$Al_2O_3 + B_2O_3 + PbO + Ta_2O_5 + ZnO$ Glass	j	503-8658
$Al_2O_3 + B_2O_3 + PbO + ZnO$ Glass	a	503-8516
$Al_2O_3 + B_2O_3 + Rb_2O + SiO_2$ Glass	jn	503-8707
$Al_2O_3 + B_2O_3 + SiO_2$ Ceramic	an	503-9430
$Al_2O_3 + B_2O_3 + SiO_2$ Glass	no	503-8899
$Al_2O_3 + B_2O_3 + SiO_2$ + Clay + Feldspar Porcelain	n	503-8314
$Al_2O_3 + B_2O_3 + SiO_2 + SrO$ Glaze	n	503-9249
$Al_2O_3 + B_2O_3 + SiO_2 + SrO + ZnO$ Glaze	n	503-9268
$Al_2O_3 + B_2O_3 + SiO_2 + Ta_2O_5$ Glass	n	503-8425
$Al_2O_3 + B_2O_3 + SiO_2 + ZnO$ Glass Ceramic	j	503-9776
$Al_2O_3 + B_2O_3 + SiO_2 + ZnO + ZrO_2$ Glaze	n	503-9267
$Al_2O_3 + B_2O_3 + SrO$ Glass	n	503-8747
$Al_2O_3 + B_2O_3 + ZnO$ Glass	n	503-8745
Al_2O_3 + Bauxite Refractory	o	503-8255
$Al_2O_3 + BaO$	e	503-9641
$Al_2O_3 + BaO + BeO + Bi_2O_3 + CaO$ Yellow Glass	jr	503-8972
$Al_2O_3 + BaO + BeO + CaO$ Glass	j	503-8809
$Al_2O_3 + BaO + BeO + CaO + PbO$ Glass	j	503-8808
$Al_2O_3 + BaO + Bi_2O_3 + PbO + SiO_2$ Glaze	n	503-9581
$Al_2O_3 + BaO + CaO$ Glass	j	503-8816
$Al_2O_3 + BaO + CaO + Fe_2O_3 + Na_2O$ Glass	j	503-8825
$Al_2O_3 + BaO + CaO + GeO_2$ Glass	j	503-8830
$Al_2O_3 + BaO + CaO + GeO_2 + MgO$ Glass	j	503-8847
$Al_2O_3 + BaO + CaO + GeO_2 + Na_2O$ Glass	j	503-8821
$Al_2O_3 + BaO + CaO + GeO_2 + ZnO$ Glass	a	503-8527
$Al_2O_3 + BaO + CaO + K_2O + Na_2O + SiO_2$ Glaze	n	503-9210
$Al_2O_3 + BaO + CaO + K_2O + SiO_2$ Glaze	n	503-9392
Oxide Mixture: -(Cont)-		
$Al_2O_3 + BaO + CaO + La_2O_3$ Glass	j	503-8811
$Al_2O_3 + BaO + CaO + MgO + Na_2O + SiO_2$ Glass	n	503-8599
$Al_2O_3 + BaO + CaO + MgO + PbO$	n	503-9569
$Al_2O_3 + BaO + CaO + MgO + SiO_2$ Ceramic	n	503-9205
$Al_2O_3 + BaO + CaO + MgO + SiO_2$ Glass	jn	503-8848
$Al_2O_3 + BaO + CaO + MgO + SiO_2 + SrO$ Ceramic	n	503-9204
$Al_2O_3 + BaO + CaO + Na_2O + SiO_2$ Glass	n	503-8757
$Al_2O_3 + BaO + CaO + Na_2O + SnO_2$ Glass	j	503-8829
$Al_2O_3 + BaO + CaO + PbO + SiO_2$ Glaze	n	503-9583
$Al_2O_3 + BaO + CaO + SiO_2$ Glaze	n	503-9816
$Al_2O_3 + BaO + CaO + SiO_2 + R_2O$ Glass	r	503-9069
$Al_2O_3 + BaO + CaO + SiO_2 + ZrO_2$ Glass	n	503-8597
$Al_2O_3 + BaO + K_2O + MgO + Na_2O + SiO_2$ Glaze	n	503-9251
$Al_2O_3 + BaO + K_2O + Na_2O + SiO_2$ Glass	jn	503-8878
$Al_2O_3 + BaO + K_2O + Na_2O + SiO_2$ Glaze	n	503-9377
$Al_2O_3 + BaO + K_2O + P_2O_5$ Glass	i	503-9142
$Al_2O_3 + BaO + K_2O + PbO + SiO_2$ Glaze	n	503-9382
$Al_2O_3 + BaO + K_2O + SiO_2$ Glass	n	503-8592
$Al_2O_3 + BaO + K_2O + SiO_2 + ZnO$ Glaze	n	503-9354
$Al_2O_3 + BaO + La_2O_3 + SiO_2$ Glass	j	503-8455
$Al_2O_3 + BaO + La_2O_3 + TeO_2$ Glass	r	503-9072
$Al_2O_3 + BaO + MgO + SiO_2$ Ceramic	n	503-9734
$Al_2O_3 + BaO + MgO + SiO_2 + ZnO$ Glaze	n	503-9370
$Al_2O_3 + BaO + MgO + SiO_2 + ZrO_2$ Glass	n	503-8595
$Al_2O_3 + BaO + Na_2O + PbO + SiO_2 + ZnO$ Glass	j	503-8769
$Al_2O_3 + BaO + Na_2O + SiO_2$ Ceramic	n	503-9831
$Al_2O_3 + BaO + Na_2O + SiO_2$ Glass	n	503-8589
$Al_2O_3 + BaO + Na_2O + SiO_2 + TiO_2$ Glass Ceramic, Nephelite	n	503-9946
$Al_2O_3 + BaO + P_2O_5$ Glass	aejo	503-8520
$Al_2O_3 + BaO + P_2O_5 + SiO_2$ Glass	j	503-8714
$Al_2O_3 + BaO + P_2O_5 + SiO_2 + TiO_2$ Glass	j	503-8715
$Al_2O_3 + BaO + P_2O_5 + SiO_2 + WO_3$ Glass	j	503-8713
$Al_2O_3 + BaO + PbO + SiO_2 + TiO_2$ Glass	jn	503-8442
$Al_2O_3 + BaO + Rb_2O + Ta_2O_5$ Glass	j	503-8471
$Al_2O_3 + BaO + SiO_2$	n	503-9660
$Al_2O_3 + BaO + SiO_2$ Glass	gijn	503-8734
$Al_2O_3 + BaO + SiO_2 + TiO_2$ Glass	jno	503-8954
$Al_2O_3 + BaO + ZrO_2$ Refractory	n	503-9494
$Al_2O_3 + BeO$	aeno	503-9956
$Al_2O_3 + BeO + CaO$ Glass	j	503-8817
$Al_2O_3 + BeO + CaO + MgO + SiO_2$ Ceramic	a	503-9785
$Al_2O_3 + BeO + CaO + ThO_2$ Porcelain	a	503-9186
$Al_2O_3 + BeO + CaO + ZrO_2$ Ceramic	an	503-9954
$Al_2O_3 + BeO + MgO$ Ceramic	aeno	503-9983
$Al_2O_3 + BeO + MgO + SiO_2$ Glass	n	503-8543
$Al_2O_3 + BeO + MgO + ThO_2$ Ceramic	a	503-9928
$Al_2O_3 + BeO + MgO + ZrO_2$ Ceramic	a	503-9953
$Al_2O_3 + BeO + SiO_2$ Ceramic	ano	503-9829
$Al_2O_3 + BeO + SiO_2 + ThO_2$ Porcelain	a	503-9184
$Al_2O_3 + BeO + SrO + ThO_2$ Porcelain	a	503-9185
$Al_2O_3 + BeO + SrO + ZrO_2$ Porcelain	a	503-9183
$Al_2O_3 + BeO + ThO_2$ Ceramic	an	503-9952
$Al_2O_3 + BeO + ThO_2 + TiO_2$ Ceramic	an	503-9332
$Al_2O_3 + BeO + TiO_2$ Ceramic	ano	503-9961
$Al_2O_3 + BeO + TiO_2 + ZrO_2$ Porcelain	a	503-9182
$Al_2O_3 + BeO + ZrO_2$ Ceramic	an	503-9982
$Al_2O_3 + Bi_2O_3 + CaO$ Glass	j	503-8810
$Al_2O_3 + Bi_2O_3 + Li_2O$ Glass	r	503-9020
$Al_2O_3 + Bi_2O_3 + PbO + SiO_2$ Glass	a	503-8486
$Al_2O_3 + CaO$	aegjn	503-9960
$Al_2O_3 + CaO$ Colored Glass	aejn	503-8822
$Al_2O_3 + CaO + Cr_2O_3$ Refractory	n	503-9358
$Al_2O_3 + CaO + Cr_2O_3 + FeO + MgO + SiO_2$ Refractory	er	503-9885
$Al_2O_3 + CaO + Cr_2O_3 + Li_2O + SiO_2$ Glass	n	503-8538

Property: a. Thermal conductivity; **b.** Accommodation coefficient; **c.** Thermal contact resistance; **d.** Thermal diffusivity; **e.** Specific heat; **f.** Viscosity; **g.** Emittance; **h.** Reflectance; **i.** Absorptance; **j.** Transmittance; **k.** α_s/ϵ_t ratio; **l.** Prandtl number; **n.** Thermal linear expansion coefficient; **o.** Thermal volumetric expansion coefficient; **r.** Thermal radiative properties (g,h,i,j,k)

Substance Name	Property	Number
Oxide Mixture: -(Cont)-		
$Al_2O_3 + CaO + Cr_2O_3 + MgO + SiO_2$ Ceramic	an	503-9720
$Al_2O_3 + CaO + Cr_2O_3 + MgO + SiO_2$ Glass	n	503-8541
$Al_2O_3 + CaO + Cr_2O_3 + Na_2O + SiO_2$ Glass	nr	503-9067
$Al_2O_3 + CaO + Cr_2O_3 + SiO_2$ Glass Ceramic, Sintox	n	503-9645
$Al_2O_3 + CaO + Cr_2O_3 + ZrO_2$ Refractory	n	503-9360
$Al_2O_3 + CaO + CuO + Na_2O + SiO_2$ Glass	n	503-8861
$Al_2O_3 + CaO + FeO + Fe_2O_3 + MgO + SiO_2$ Ceramic, Iron Ore	ad	503-8392
$Al_2O_3 + CaO + FeO + Fe_2O_3 + MgO + SiO_2$ Refractory, Forsterite	a	503-9793
$Al_2O_3 + CaO + FeO + Fe_2O_3 + SiO_2$ Glass	n	503-8706
$Al_2O_3 + CaO + FeO + Fe_2O_3 + SiO_2$ Slag	gi	503-9988
$Al_2O_3 + CaO + FeO + MgO + MnO + SiO_2$ Refractory, Magnesite	r	503-9886
$Al_2O_3 + CaO + FeO + MgO + SiO_2$ Slag	g	503-9996
$Al_2O_3 + CaO + FeO + MgO + SiO_2 + TiO_2$ Converter Lining	an	503-9533
$Al_2O_3 + CaO + FeO + SiO_2$ Ceramic	r	503-9888
$Al_2O_3 + CaO + FeO + SiO_2$ Slag	a	503-9993
$Al_2O_3 + CaO + FeO + SiO_2 + R_2O$ Glaze	n	503-8280
$Al_2O_3 + CaO + FeO + SiO_2 + ZnO$	a	503-9740
$Al_2O_3 + CaO + Fe_2O_3$ Glass	jr	503-8990
$Al_2O_3 + CaO + Fe_2O_3 + K_2O + MgO + SiO_2$	a	503-9768
$Al_2O_3 + CaO + Fe_2O_3 + K_2O + Na_2O$ Glass	jn	503-8826
$Al_2O_3 + CaO + Fe_2O_3 + K_2O + SiO_2$	ah	503-9684
$Al_2O_3 + CaO + Fe_2O_3 + MgO + MnO + SiO_2$	ade	503-9792
$Al_2O_3 + CaO + Fe_2O_3 + MgO + SiO_2$	adeno	503-9809
$Al_2O_3 + CaO + Fe_2O_3 + MgO + SiO_2$ Glass	n	503-8896
$Al_2O_3 + CaO + Fe_2O_3 + MgO + SiO_2$ Refractory	a	503-8390
$Al_2O_3 + CaO + Fe_2O_3 + MnO + SiO_2$ Ceramic	adeno	503-9773
$Al_2O_3 + CaO + Fe_2O_3 + Na_2O + SiO_2$	n	503-9642
$Al_2O_3 + CaO + Fe_2O_3 + Na_2O + SiO_2$ Boiler Deposit	ag	503-9994
$Al_2O_3 + CaO + Fe_2O_3 + Na_2O + SiO_2$ Charge, Slime	n	503-9984
$Al_2O_3 + CaO + Fe_2O_3 + Na_2O + SiO_2$ Glass	nr	503-9141
$Al_2O_3 + CaO + Fe_2O_3 + SiO_2$	adnor	503-9835
$Al_2O_3 + CaO + Fe_2O_3 + SiO_2$ Glass	r	503-8989
$Al_2O_3 + CaO + Fe_2O_3 + SiO_2 + R_2O$ Refractory	an	503-8370
$Al_2O_3 + CaO + Fe_2O_3 + SiO_2 + TiO_2$	adn	503-9688
$Al_2O_3 + CaO + GeO_2$ Glass	j	503-8805
$Al_2O_3 + CaO + GeO_2 + MgO + ZnO$ Glass	j	503-8849
$Al_2O_3 + CaO + HfO_2 + ZrO_2$ Ceramic	a	503-9895
$Al_2O_3 + CaO + K_2O + Li_2O + SiO_2$ Glass	n	503-8540
$Al_2O_3 + CaO + K_2O + Li_2O + SiO_2$ Glaze	n	503-9294
$Al_2O_3 + CaO + K_2O + MgO + Na_2O + SiO_2$ Glaze	n	503-9935
$Al_2O_3 + CaO + K_2O + MgO + SiO_2$ Glaze	n	503-9445
$Al_2O_3 + CaO + K_2O + MgO + SiO_2 + ZnO$ Glaze	n	503-9443
$Al_2O_3 + CaO + K_2O + MnO + SiO_2$ Glaze	n	503-9573
$Al_2O_3 + CaO + K_2O + Na_2O$ Glass	jn	503-8824
$Al_2O_3 + CaO + K_2O + Na_2O + SiO_2$ Ceramic	n	503-9936
$Al_2O_3 + CaO + K_2O + Na_2O + SiO_2$ Glass	aejn	503-8876
$Al_2O_3 + CaO + K_2O + PbO + SiO_2$ Glaze	n	503-9389
$Al_2O_3 + CaO + K_2O + SiO_2$ Glaze	n	503-9444
$Al_2O_3 + CaO + K_2O + SiO_2 + SrO$ Glaze	n	503-9334
$Al_2O_3 + CaO + K_2O + SiO_2 + ZnO$ Glaze	n	503-9372
$Al_2O_3 + CaO + Li_2O + MgO + SiO_2$ Glass	n	503-8985
$Al_2O_3 + CaO + Li_2O + MgO + SiO_2 + TiO_2$ Glass	n	503-8942
$Al_2O_3 + CaO + Li_2O + SiO_2$ Ceramic	n	503-9399
$Al_2O_3 + CaO + Li_2O + SiO_2$ Glass	n	503-8524
$Al_2O_3 + CaO + MgO$	aen	503-9978
$Al_2O_3 + CaO + MgO + MnO + SiO_2$ Glass	n	503-8895
Oxide Mixture: -(Cont)-		
$Al_2O_3 + CaO + MgO + MnO + SiO_2$ Glaze	n	503-9442
$Al_2O_3 + CaO + MgO + Na_2O + SiO_2$ Ceramic	n	503-9367
$Al_2O_3 + CaO + MgO + Na_2O + SiO_2$ Glass	adjnr	503-9042
$Al_2O_3 + CaO + MgO + NdO + SiO_2$ Glass	j	503-8441
$Al_2O_3 + CaO + MgO + P_2O_5 + SiO_2$ Glass	n	503-8542
$Al_2O_3 + CaO + MgO + SiO_2$	ano	503-9902
$Al_2O_3 + CaO + MgO + SiO_2$ Glass	jn	503-9166
$Al_2O_3 + CaO + MgO + SiO_2 + RO$ Porcelain	n	503-8296
$Al_2O_3 + CaO + MgO + SiO_2 + R_2O$ Glaze	n	503-8261
$Al_2O_3 + CaO + MgO + SiO_2 + SmO$ Glass	j	503-8440
$Al_2O_3 + CaO + MgO + SiO_2 + TiO_2$ Glass	aen	503-8874
$Al_2O_3 + CaO + MgO + SiO_2 + TiO_2$ Slag	ade	503-9995
$Al_2O_3 + CaO + MgO + SiO_2 + TiO_2 + ZrO_2$ Glass	n	503-8537
$Al_2O_3 + CaO + MgO + SiO_2 + ZnO$ Glass	jn	503-8850
$Al_2O_3 + CaO + MgO + SiO_2 + ZrO_2$ Glass	n	503-8598
$Al_2O_3 + CaO + MnO + Na_2O + SiO_2$ Glass	n	503-8947
$Al_2O_3 + CaO + Na_2O$ Glass	jn	503-8510
$Al_2O_3 + CaO + Na_2O + P_2O_5$ Glass	n	503-8602
$Al_2O_3 + CaO + Na_2O + PbO + SiO_2$ Glaze	n	503-9390
$Al_2O_3 + CaO + Na_2O + SiO_2$	an	503-9850
$Al_2O_3 + CaO + Na_2O + SiO_2$ Glass	ghjnor	503-9159
$Al_2O_3 + CaO + Na_2O + SiO_2 + SrO$ Glaze	n	503-9556
$Al_2O_3 + CaO + Na_2O + SiO_2 + TiO_2$ Frit	n	503-9665
$Al_2O_3 + CaO + Na_2O + SiO_2 + TiO_2$ Glass	n	503-8662
$Al_2O_3 + CaO + Na_2O + SiO_2 + ZnO$ Ceramic	no	503-9555
$Al_2O_3 + CaO + Na_2O + SiO_2 + ZnO$ Glass	hn	503-9092
$Al_2O_3 + CaO + Na_2O + SiO_2 + ZrO_2$ Glass	n	503-8596
$Al_2O_3 + CaO + P_2O_5$ Glass	j	503-8523
$Al_2O_3 + CaO + P_2O_5 + SiO_2$ Bone China	n	503-9187
$Al_2O_3 + CaO + PbO + SiO_2$ Glaze	n	503-9412
$Al_2O_3 + CaO + PbO + SiO_2 + SrO$ Glaze	n	503-9333
$Al_2O_3 + CaO + PbO + SiO_2 + ZnO$ Glaze	n	503-9391
$Al_2O_3 + CaO + SiO_2$	aden	503-9980
$Al_2O_3 + CaO + SiO_2$ Glass	jn	503-8889
$Al_2O_3 + CaO + SiO_2$ Slag	e	503-9999
$Al_2O_3 + CaO + SiO_2$ + Borocalcite Refractory	n	503-8263
$Al_2O_3 + CaO + SiO_2 + SrO + ZrO_2$ Glaze	o	503-9432
$Al_2O_3 + CaO + SiO_2 + TiO_2$	en	503-9401
$Al_2O_3 + CaO + SiO_2 + TiO_2$ Glass	n	503-8928
$Al_2O_3 + CaO + SiO_2 + TiO_2 + ZrO_2$ Ceramic	a	503-9450
$Al_2O_3 + CaO + SiO_2 + TiO_2 + ZrO_2$ Glass	n	503-8535
$Al_2O_3 + CaO + SiO_2 + ZnO + ZrO_2$ Glaze	o	503-9434
$Al_2O_3 + CaO + SiO_2 + ZrO_2$ Ceramic	a	503-9449
$Al_2O_3 + CaO + SnO_2$ Glass	j	503-8845
$Al_2O_3 + CaO + SrO$ Glass	j	503-8522
$Al_2O_3 + CaO + TiO_2$ Ceramic	n	503-9576
$Al_2O_3 + CaO + ZrO_2$ Refractory	n	503-9362
$Al_2O_3 + CeO_2$ Ceramic, Cerium Oxide	a	503-9963
$Al_2O_3 + CeO_2 + TiO_2$ Ceramic	n	503-9779
Al_2O_3 + Clay Refractory, Aluminum Oxide	ain	503-8334
Al_2O_3 + Clay + Feldspar + Sand Porcelain	n	503-8342
Al_2O_3 + Clay + Steatite Ceramic	n	503-8337
$Al_2O_3 + CoO$ Ceramic	dr	503-9744
$Al_2O_3 + CoO + Cr_2O_3$ Spinel	n	503-9226
$Al_2O_3 + CoO + Cr_2O_3 + TiO_2 + ZnO$ Ceramic	h	503-9188
$Al_2O_3 + CoO + Cr_2O_3 + ZnO$ Ceramic	h	503-9191
$Al_2O_3 + CoO + TiO_2 + ZnO$ Ceramic	h	503-9189
$Al_2O_3 + CoO + ZnO$ Ceramic	h	503-9190
$Al_2O_3 + Co_3O_4$ Porcelain	n	503-9312
$Al_2O_3 + Co_3O_4 + MgO + SiO_2$ Porcelain	n	503-9526
$Al_2O_3 + CrO_2 + P_2O_5$	n	503-9488
$Al_2O_3 + Cr_2O_3$	agnor	503-9748
$Al_2O_3 + Cr_2O_3 + FeO$ Refractory	an	503-9868
$Al_2O_3 + Cr_2O_3 + FeO + MgO$ Refractory	n	503-9272
$Al_2O_3 + Cr_2O_3 + FeO + MgO + SiO_2$ Ceramic	n	503-9957
$Al_2O_3 + Cr_2O_3 + Fe_2O_3 + MgO$ Refractory	n	503-9738
$Al_2O_3 + Cr_2O_3 + Fe_2O_3 + MgO + SiO_2$	aen	503-9898
$Al_2O_3 + Cr_2O_3 + Fe_2O_3 + MgO + TiO_2$ Ceramic	an	503-9326

Property: **a.** Thermal conductivity; **b.** Accommodation coefficient; **c.** Thermal contact resistance; **d.** Thermal diffusivity; **e.** Specific heat; **f.** Viscosity; **g.** Emittance; **h.** Reflectance; **i.** Absorptance; **j.** Transmittance; **k.** α_s/ϵ_t ratio; **l.** Prandtl number; **n.** Thermal linear expansion coefficient; **o.** Thermal volumetric expansion coefficient; **r.** Thermal radiative properties (g,h,i,j,k)

Substance Name	Property	Number
Oxide Mixture: -(Cont)-		
$Al_2O_3 + Cr_2O_3 + MgO$ Ceramic	an	503-9826
$Al_2O_3 + Cr_2O_3 + MgO + SiO_2$	n	503-9284
$Al_2O_3 + Cr_2O_3 + MgO + ZnO$ Ceramic	n	503-9855
$Al_2O_3 + Cr_2O_3 + MnO$ Pigment	h	503-9418
$Al_2O_3 + Cr_2O_3 + NiO$ Spinel	n	503-9227
$Al_2O_3 + Cr_2O_3 + NiO_2$ Ceramic	r	503-9872
$Al_2O_3 + Cr_2O_3 + P_2O_5$	n	503-9486
$Al_2O_3 + Cr_2O_3 + P_2O_5 + SiO_2$ Ceramic	a	503-9802
$Al_2O_3 + Cr_2O_3 + SiO_2$	aen	503-9704
$Al_2O_3 + Cr_2O_3 + TiO_2$ Ceramic	n	503-9949
$Al_2O_3 + Cr_2O_3 + TiO_2 + ZnO$ Ceramic	h	503-9194
$Al_2O_3 + Cr_2O_3 + ZnO$ Ceramic	h	503-9195
$Al_2O_3 + CuO + FeO + K_2O + SiO_2$ Tektite Glass	ad	503-8566
$Al_2O_3 + Dy_2O_3$ Ceramic	n	503-9552
Al_2O_3 + Feldspar	r	503-8384
Al_2O_3 + Feldspar + Kaolin + Nepheline Syenite Ceramic	n	503-8194
$Al_2O_3 + FeO$ Refractory, Spinel	n	503-9553
$Al_2O_3 + FeO + Fe_2O_3$ Refractory	n	503-9276
$Al_2O_3 + FeO + Fe_2O_3 + MgO$ Refractory	n	503-9274
$Al_2O_3 + FeO + Fe_2O_3 + SiO_2$ Refractory	a	503-9846
$Al_2O_3 + FeO + MgO$ Ceramic	an	503-9772
$Al_2O_3 + FeO + SiO_2$ Refractory, Fireclay	r	503-9887
$Al_2O_3 + Fe_2O_3$	nr	503-9745
$Al_2O_3 + Fe_2O_3 + K_2O + MgO + SiO_2$ Porcelain, Refractory	e	503-9775
$Al_2O_3 + Fe_2O_3 + K_2O + MnO + SiO_2$ Ceramic, Stone Ware	n	503-9669
$Al_2O_3 + Fe_2O_3 + K_2O + Na_2O + SiO_2$ Refractory	an	503-9771
$Al_2O_3 + Fe_2O_3 + K_2O + SiO_2 + TiO_2$	n	503-9625
$Al_2O_3 + Fe_2O_3 + Li_2O + SiO_2$	n	503-9506
$Al_2O_3 + Fe_2O_3 + Li_2O + SiO_2 + R_2O$ Semiporcelain	n	503-8320
$Al_2O_3 + Fe_2O_3 + MgO$	aen	503-9694
$Al_2O_3 + Fe_2O_3 + MgO + SiO_2$ Refractory	an	503-9762
$Al_2O_3 + Fe_2O_3 + MgO + SiO_2 + R_2O$ Refractory	a	503-8369
$Al_2O_3 + Fe_2O_3 + MgO + SiO_2 + TiO_2$	n	503-9593
$Al_2O_3 + Fe_2O_3 + MgO + SiO_2 + TiO_2$ Slag	ade	503-9989
$Al_2O_3 + Fe_2O_3 + MnO + SiO_2 + RO$ Glass	r	503-8414
$Al_2O_3 + Fe_2O_3 + Na_2O + SiO_2$	aden	503-9859
$Al_2O_3 + Fe_2O_3 + Na_2O + SiO_2 + TiO_2$	jo	503-9355
$Al_2O_3 + Fe_2O_3 + SiO_2$	adeno	503-9836
$Al_2O_3 + Fe_2O_3 + SiO_2$ Glass	j	503-8479
$Al_2O_3 + Fe_2O_3 + SiO_2$ Refractory, Fireclay	a	503-8372
$Al_2O_3 + Fe_2O_3 + SiO_2$ + Dolomite + Gypsum Refractory	ae	503-8395
$Al_2O_3 + Fe_2O_3 + SiO_2 + R_2O$ Glass	an	503-8418
$Al_2O_3 + Fe_2O_3 + SiO_2 + R_2O$ Refractory	n	503-8245
$Al_2O_3 + Fe_2O_3 + SiO_2 + TiO_2$	adeno	503-9815
$Al_2O_3 + Fe_2O_3 + SiO_2 + TiO_2$ Concentrate, Titanium	ade	503-9986
$Al_2O_3 + Fe_2O_3 + SiO_2 + TiO_2$ Refractory, Fireclay	a	503-8373
$Al_2O_3 + Fe_2O_3 + SiO_2 + TiO_2 + R_2O$ Refractory	aden	503-8374
$Al_2O_3 + Fe_2O_3 + SiO_2 + TiO_2 + ZrO_2$	no	503-9538
$Al_2O_3 + Fe_2O_3 + SiO_2 + ZrO + R_2O$ Refractory	n	503-8394
$Al_2O_3 + Fe_2O_3 + SiO_2 + ZrO_2$ Ceramic	an	503-9596
$Al_2O_3 + Fe_2O_3 + TiO_2$ Ceramic	no	503-9896
$Al_2O_3 + Ga_2O_3$	n	503-9563
$Al_2O_3 + Ga_2O_3 + Na_2O + SiO_2$ Glass	ijn	503-8873
$Al_2O_3 + GeO_2$ Glass	r	503-9060
$Al_2O_3 + GeO_2 + K_2O + Na_2O + Sb_2O_3$ Glass	r	503-8949
$Al_2O_3 + GeO_2 + Li_2O$	n	503-9403
$Al_2O_3 + GeO_2 + Li_2O$ Glass	n	503-8664
$Al_2O_3 + GeO_2 + Li_2O + SiO_2$	n	503-9477

Substance Name	Property	Number
Oxide Mixture: -(Cont)-		
$Al_2O_3 + GeO_2 + Li_2O + SiO_2$ + Clay Ceramic	n	503-8309
$Al_2O_3 + GeO_2 + Li_2O + TiO_2$ Glass	jn	503-8663
$Al_2O_3 + GeO_2 + PbO$ Glass	j	503-8463
Al_2O_3 + Glass Ceramic	n	503-8268
$Al_2O_3 + K_2O$ + Bauxite Refractory	o	503-8252
$Al_2O_3 + K_2O + Li_2O + MgO + P_2O_5$ Glass	n	503-8929
$Al_2O_3 + K_2O + Li_2O + MgO + SiO_2$	n	503-9451
$Al_2O_3 + K_2O + Li_2O + P_2O_5 + SiO_2$ Glass	n	503-8539
$Al_2O_3 + K_2O + Li_2O + SiO_2$	n	503-9504
$Al_2O_3 + K_2O + Li_2O + SiO_2 + ZnO$ Glass	aen	503-8754
$Al_2O_3 + K_2O + Li_2O + SiO_2 + ZrO_2$ Glass Ceramic	n	503-9780
$Al_2O_3 + K_2O + MgO + Na_2O + SiO_2$ Glaze	n	503-9507
$Al_2O_3 + K_2O + MgO + PbO + SiO_2$ Glaze	n	503-9387
$Al_2O_3 + K_2O + MgO + SiO_2 + TiO_2$ Glass Ceramic	n	503-9781
$Al_2O_3 + K_2O + MgO + SiO_2 + ZrO_2$ Glass Ceramic	n	503-9777
$Al_2O_3 + K_2O + Na_2O$	n	503-9408
$Al_2O_3 + K_2O + Na_2O + PbO + SiO_2$ Glass	hjn	503-8941
$Al_2O_3 + K_2O + Na_2O + PbO + SiO_2$ Glaze	n	503-9388
$Al_2O_3 + K_2O + Na_2O + Sb_2O_3$ Glass	jr	503-9015
$Al_2O_3 + K_2O + Na_2O + Sb_2O_3 + SiO_2$ Glass	r	503-9061
$Al_2O_3 + K_2O + Na_2O + SiO_2$	n	503-9839
$Al_2O_3 + K_2O + Na_2O + SiO_2$ Ceramic, Feldspar	an	503-8363
$Al_2O_3 + K_2O + Na_2O + SiO_2$ Glass	no	503-9135
$Al_2O_3 + K_2O + Na_2O + SiO_2$ + Clay Ceramic	n	503-8232
$Al_2O_3 + K_2O + Na_2O + SiO_2 + SrO$ Glass	j	503-8879
$Al_2O_3 + K_2O + Na_2O + SiO_2 + TiO_2$ Ceramic	no	503-9965
$Al_2O_3 + K_2O + Na_2O + SiO_2 + ZnO$ Glass	jn	503-8875
$Al_2O_3 + K_2O + Na_2O + SiO_2 + ZrO_2$ Glass	jn	503-8877
$Al_2O_3 + K_2O + P_2O_5$ Glass	aeo	503-8517
$Al_2O_3 + K_2O + PbO + SiO_2 + SrO$ Glaze	n	503-9335
$Al_2O_3 + K_2O + PbO + SiO_2 + ZnO$ Glaze	n	503-9381
$Al_2O_3 + K_2O + SiO_2$	en	503-9671
$Al_2O_3 + K_2O + SiO_2$ Ceramic	a	503-8364
$Al_2O_3 + K_2O + SiO_2$ Glass	n	503-9137
Al_2O_3 + Kaolin	n	503-8308
Al_2O_3 + Kaolin + Nephelite Syenite Porcelain	n	503-8359
$Al_2O_3 + La_2O_3 + SiO_2 + TiO_2$ Glass	j	503-8582
$Al_2O_3 + Li_2O$	an	503-9407
$Al_2O_3 + Li_2O + MgO$ Ceramic	n	503-9201
$Al_2O_3 + Li_2O + MgO + SiO_2$ Ceramic	n	503-9429
$Al_2O_3 + Li_2O + MgO + SiO_2$ Glass	n	503-8767
$Al_2O_3 + Li_2O + MgO + SiO_2$ + Kaolin Ceramic	n	503-8344
$Al_2O_3 + Li_2O + MgO + SiO_2 + ZnO$ Glass	n	503-8762
$Al_2O_3 + Li_2O + MgO + SiO_2 + ZrO_2$ Glass Ceramic	n	503-9921
$Al_2O_3 + Li_2O + Na_2O + SiO_2$	n	503-9505
$Al_2O_3 + Li_2O + Na_2O + SiO_2 + TiO_2$ Glass	n	503-8770
$Al_2O_3 + Li_2O + Na_2O + SiO_2 + TiO_2$ Glass Ceramic	n	503-9441
$Al_2O_3 + Li_2O + P_2O_5 + SiO_2$ Glass	n	503-8863
$Al_2O_3 + Li_2O + PbO + SiO_2$ Glaze	n	503-9398
$Al_2O_3 + Li_2O + Sb_2O_3$ Glass	r	503-9021
$Al_2O_3 + Li_2O + SiO_2$	adehijno	503-9869
$Al_2O_3 + Li_2O + SiO_2$ Glass	aeijno	503-9068
$Al_2O_3 + Li_2O + SiO_2$ + Clay Ceramic	n	503-8313
$Al_2O_3 + Li_2O + SiO_2$ + Kaolin Ceramic	n	503-8346
$Al_2O_3 + Li_2O + SiO_2 + SnO_2$	n	503-9659
$Al_2O_3 + Li_2O + SiO_2 + SrO + ZrO_2$ Ceramic	a	503-9925
$Al_2O_3 + Li_2O + SiO_2$ + Steatite Ceramic	n	503-8285
$Al_2O_3 + Li_2O + SiO_2 + TiO_2$	n	503-9658
$Al_2O_3 + Li_2O + SiO_2 + TiO_2$ + Clay Ceramic	n	503-8312

Property: **a.** Thermal conductivity; **b.** Accommodation coefficient; **c.** Thermal contact resistance; **d.** Thermal diffusivity; **e.** Specific heat; **f.** Viscosity; **g.** Emittance; **h.** Reflectance; **i.** Absorptance; **j.** Transmittance; **k.** α_s/ϵ_t ratio; **l.** Prandtl number; **n.** Thermal linear expansion coefficient; **o.** Thermal volumetric expansion coefficient; **r.** Thermal radiative properties (g,h,i,j,k)

Substance Name	Property	Number
Oxide Mixture: -(Cont)-		
$Al_2O_3 + Li_2O + SiO_2 + ZnO$ Glass	n	503-8764
$Al_2O_3 + Li_2O + SiO_2 + ZnO + ZrO_2$ Glass Ceramic	n	503-9910
$Al_2O_3 + Li_2O + SiO_2 + ZrO_2$ Ceramic	n	503-9400
$Al_2O_3 + Li_2O + SiO_2 + ZrO_2$ Glass	a	503-8659
$Al_2O_3 + Li_2O + SrO$ Ceramic, Strontium Oxide	a	503-9970
$Al_2O_3 + MgO$	adeghjnor	503-9750
$Al_2O_3 + MgO + $ Clay Ceramic	n	503-8336
$Al_2O_3 + MgO + MnO + SiO_2$ Porcelain	n	503-9528
$Al_2O_3 + MgO + Na_2O + P_2O_5$ Glass	n	503-9125
$Al_2O_3 + MgO + Na_2O + SiO_2$ Ceramic	an	503-9834
$Al_2O_3 + MgO + Na_2O + SiO_2 + TiO_2$ Glass Ceramic, Nephelite	n	503-9942
$Al_2O_3 + MgO + Ni_2O_3 + SiO_2$ Porcelain	n	503-9527
$Al_2O_3 + MgO + P_2O_5$ Glass	hjn	503-8979
$Al_2O_3 + MgO + P_2O_5 + SiO_2$	n	503-9558
$Al_2O_3 + MgO + SiO_2$	aden	503-9979
$Al_2O_3 + MgO + SiO_2$ Glass	jn	503-8888
$Al_2O_3 + MgO + SiO_2 + TiO_2$ Ceramic	n	503-9900
$Al_2O_3 + MgO + SiO_2 + TiO_2$ Glass	aen	503-8940
$Al_2O_3 + MgO + SiO_2 + TiO_2 + R_2O$ Ceramic	n	503-8350
$Al_2O_3 + MgO + SiO_2 + TiO_2 + ZrO_2$ Glass	n	503-8765
$Al_2O_3 + MgO + SiO_2 + ZnO$ Glass	n	503-8763
$Al_2O_3 + MgO + SiO_2 + ZnO$ Porcelain	n	503-9523
$Al_2O_3 + MgO + SiO_2 + ZrO_2$ Ceramic	n	503-9913
$Al_2O_3 + MgO + Ta_2O_5$	n	503-9231
$Al_2O_3 + MgO + TiO_2$ Ceramic	n	503-9579
$Al_2O_3 + MgO + ZrO_2$ Refractory	nr	503-9725
$Al_2O_3 + MnO$	ao	503-9510
$Al_2O_3 + MnO + Mn_2O_3$ Ceramic	a	503-9908
$Al_2O_3 + MnO + P_2O_5$ Glass	n	503-8609
$Al_2O_3 + MnO + SiO_2$	n	503-9648
$Al_2O_3 + MnO + SiO_2 + RO$ Glass	r	503-8415
$Al_2O_3 + MnO + TiO_2$ Ceramic	an	503-9806
$Al_2O_3 + MnO + V_2O_3$	n	503-9462
$Al_2O_3 + MnO_2$ Porcelain	n	503-9311
$Al_2O_3 + MnO_2 + TiO_2$ Ceramic	an	503-9876
$Al_2O_3 + Mn_2O_3$ Ceramic	a	503-9973
$Al_2O_3 + Mn_3O_4 + SiO_2$ Ceramic	n	503-9501
$Al_2O_3 + Mn_3O_4 + TiO_2$ Ceramic	n	503-9502
$Al_2O_3 + $ Mullite Porcelain	n	503-8341
$Al_2O_3 + Na_2O$	a	503-9889
$Al_2O_3 + Na_2O + $ Bauxite Refractory	o	503-8253
$Al_2O_3 + Na_2O + $ Clay Ceramic	n	503-8238
$Al_2O_3 + Na_2O + P_2O_5$ Glass	n	503-8999
$Al_2O_3 + Na_2O + P_2O_5 + SiO_2$ Glass	n	503-9116
$Al_2O_3 + Na_2O + P_2O_5 + V_2O_5$ Glass	r	503-8924
$Al_2O_3 + Na_2O + PbO + SiO_2$ Glaze	n	503-9379
$Al_2O_3 + Na_2O + PbO + SiO_2 + ZnO$ Glass	a	503-8832
$Al_2O_3 + Na_2O + PbO + SiO_2 + ZnO$ Glaze	n	503-9369
$Al_2O_3 + Na_2O + Sb_2O_3$	r	503-8948
$Al_2O_3 + Na_2O + SiO_2$	aejn	503-9901
$Al_2O_3 + Na_2O + SiO_2$ Glass	ahjno	503-9158
$Al_2O_3 + Na_2O + SiO_2 + $ Clay Ceramic	n	503-8251
$Al_2O_3 + Na_2O + SiO_2 + $ Clay $+$ Pyrophyllite Ceramic	n	503-8231
$Al_2O_3 + Na_2O + SiO_2 + SrO$ Glass	no	503-9176
$Al_2O_3 + Na_2O + SiO_2 + TiO_2$ Glass	n	503-8590
$Al_2O_3 + Na_2O + SiO_2 + TiO_2$ Glass Ceramic	n	503-9966
$Al_2O_3 + Na_2O + SiO_2 + ZnO$ Ceramic	n	503-9833
$Al_2O_3 + Na_2O + SiO_2 + ZnO$ Ceramic Adhesive	n	503-9661
$Al_2O_3 + Na_2O + SiO_2 + ZnO + $ Clay Ceramic	o	503-8283
$Al_2O_3 + Nb_2O_5$	n	503-9657
$Al_2O_3 + Nd_2O_3 + SiO_2$ Lilac Glass	j	503-9066
$Al_2O_3 + NiO$	gnr	503-9747
$Al_2O_3 + P_2O_5$	aen	503-9456
$Al_2O_3 + P_2O_5$ Glass	j	503-8521
$Al_2O_3 + P_2O_5 + $ Feldspar	n	503-8319
$Al_2O_3 + P_2O_5 + $ Kaolin Ceramic	n	503-8227
$Al_2O_3 + P_2O_5 + Sb_2O_3 + ZnO$ Glass	j	503-8401
$Al_2O_3 + P_2O_5 + SiO_2$ Ceramic	n	503-9522
$Al_2O_3 + P_2O_5 + SiO_2$ Glass	j	503-8718
$Al_2O_3 + P_2O_5 + SiO_2 + $ Color Oxide Ceramic	a	503-8377
Oxide Mixture: -(Cont)-		
$Al_2O_3 + P_2O_5 + SiO_2 + $ Diatomite Ceramic	a	503-8376
$Al_2O_3 + P_2O_5 + SiO_2 + TiO_2$ Glass	j	503-8717
$Al_2O_3 + P_2O_5 + SiO_2 + TiO_2 + WO_3$ Glass	j	503-8712
$Al_2O_3 + P_2O_5 + TiO_2$	an	503-9611
$Al_2O_3 + P_2O_5 + V_2O_5$	n	503-9484
$Al_2O_3 + P_2O_5 + WO_3$	n	503-9485
$Al_2O_3 + PbO$	n	503-9597
$Al_2O_3 + PbO + SiO_2$ Ceramic	n	503-9612
$Al_2O_3 + PbO + SiO_2$ Glass	nr	503-8911
$Al_2O_3 + PbO + SiO_2 + TiO_2$	r	503-9796
$Al_2O_3 + PbO + SiO_2 + TiO_2$ Glass	n	503-8637
$Al_2O_3 + PbO + SiO_2 + TiO_2 + ZrO_2$ Glass	n	503-8636
$Al_2O_3 + PbO + SiO_2 + ZnO$ Ceramic	n	503-9436
$Al_2O_3 + PbO + SiO_2 + ZrO_2$ Glass	n	503-8638
$Al_2O_3 + PbO + TeO_2$ Glass	r	503-9018
$Al_2O_3 + Pr_2O_3 + SiO_2$ Green Glass	j	503-9048
$Al_2O_3 + R_2O$ Refractory	an	503-8381
$Al_2O_3 + SiO_2$	adeghijnor	503-9916
$Al_2O_3 + SiO_2$ Glass	jn	503-8735
$Al_2O_3 + SiO_2$ Refractory	ad	503-8371
$Al_2O_3 + SiO_2 + $ Ballclay $+$ Feldspar $+$ Kaolin Chinaware	n	503-8295
$Al_2O_3 + SiO_2 + $ Biotite Refractory	n	503-8265
$Al_2O_3 + SiO_2 + $ Borocalcite Refractory	n	503-8262
$Al_2O_3 + SiO_2 + $ Clay Refractory	ghn	503-8228
$Al_2O_3 + SiO_2 + $ Clay $+$ Feldspar Porcelain	n	503-8315
$Al_2O_3 + SiO_2 + $ Dolomite Refractory	n	503-8305
$Al_2O_3 + SiO_2 + $ Mica Refractory	n	503-8264
$Al_2O_3 + SiO_2 + RO + R_2O$ Porcelain	ae	503-8389
$Al_2O_3 + SiO_2 + R_2O$ Refractory	an	503-8385
$Al_2O_3 + SiO_2 + SrO$	n	503-9557
$Al_2O_3 + SiO_2 + TiO_2$	aeno	503-9760
$Al_2O_3 + SiO_2 + TiO_2$ Electrocorundum	n	503-9409
$Al_2O_3 + SiO_2 + TiO_2 + $ Clay Ceramic	no	503-8282
$Al_2O_3 + SiO_2 + TiO_2 + $ Grog $+ R_2O$ Ceramic	n	503-8352
$Al_2O_3 + SiO_2 + TiO_2 + $ Mica	n	503-8256
$Al_2O_3 + SiO_2 + TiO_2 + R_2O$ Ceramic	an	503-8391
$Al_2O_3 + SiO_2 + TiO_2 + R_2O$ Refractory	an	503-8379
$Al_2O_3 + SiO_2 + TiO_2 + ZnO$	r	503-9797
$Al_2O_3 + SiO_2 + TiO_2 + ZnO + ZrO_2$ Glass	n	503-8766
$Al_2O_3 + SiO_2 + TiO_2 + ZrO_2$ Ceramic	n	503-9621
$Al_2O_3 + SiO_2 + ZnO + ZrO_2$ Zirconium Spinel	i	503-9842
$Al_2O_3 + SiO_2 + ZrO_2$	ahno	503-9736
$Al_2O_3 + SiO_2 + ZrO_2 + $ Ball Clay $+$ Kaolin Ceramic	dn	503-8243
$Al_2O_3 + SnO_2 + TiO_2$ Ceramic	n	503-9786
$Al_2O_3 + SrO$	en	503-9647
$Al_2O_3 + TeO_2$ Glass	jn	503-8456
$Al_2O_3 + ThO_2 + TiO_2$ Ceramic	n	503-9764
$Al_2O_3 + TiO_2$	agno	503-9811
$Al_2O_3 + TiO_2 + $ Clay Ceramic	n	503-8230
$Al_2O_3 + TiO_2 + $ Feldspar $+$ Kaolin $+$ Petalite Ceramic	n	503-8326
$Al_2O_3 + TiO_2 + $ Kaolin $+$ Petalite Ceramic	n	503-8325
$Al_2O_3 + TiO_2 + $ Petalite Ceramic	n	503-8327
$Al_2O_3 + TiO_2 + ZrO_2$ Ceramic	n	503-9575
$Al_2O_3 + V_2O_3$	aj	503-9733
$Al_2O_3 + Y_2O_3$	gn	503-9628
$Al_2O_3 + ZnO$	adeghn	503-9633
$Al_2O_3 + ZrO_2$	adghn	503-9838
$As_2O_3 + B_2O_3 + K_2O$ Glass	r	503-8901
$As_2O_3 + B_2O_3 + Li_2O$ Glass	r	503-8900
$As_2O_3 + B_2O_3 + Na_2O$ Glass	r	503-8902
$As_2O_3 + B_2O_3 + Na_2O + PbO + SiO_2$ Enamel	o	503-9410
$As_2O_3 + BaO + La_2O_3 + TeO_2$ Glass	r	503-9071
$As_2O_3 + CaO + K_2O + PbO + SiO_2$ Glass	o	503-8647
$As_2O_3 + Cr_2O_3 + Na_2O + SiO_2$ Glass	j	503-8925
$As_2O_3 + Cr_2O_3 + SiO_2$ Glass	j	503-8926
$As_2O_3 + K_2O + Sb_2O_3$ Glass	r	503-9019
$B_2O_3 + $ Andesite Glaze	n	503-8206
$B_2O_3 + BaO$	ano	503-9929

Property: a. Thermal conductivity; **b.** Accommodation coefficient; **c.** Thermal contact resistance; **d.** Thermal diffusivity; **e.** Specific heat; **f.** Viscosity; **g.** Emittance; **h.** Reflectance; **i.** Absorptance; **j.** Transmittance; **k.** α_s/ϵ_t ratio; **l.** Prandtl number; **n.** Thermal linear expansion coefficient; **o.** Thermal volumetric expansion coefficient; **r.** Thermal radiative properties (g,h,i,j,k)

Substance Name	Property	Number
Oxide Mixture: -(Cont)-		
$B_2O_3 + BaO$ Glass	jnr	503-9133
$B_2O_3 + BaO + Bi_2O_3 + PbO + SiO_2$ Glaze	n	503-9582
$B_2O_3 + BaO + Bi_2O_3 + PbO + TiO_2$ Glass	n	503-8489
$B_2O_3 + BaO + Bi_2O_3 + TiO_2$ Glass	n	503-8506
$B_2O_3 + BaO + Bi_2O_3 + TiO_2 + ZrO_2$ Glass	n	503-8504
$B_2O_3 + BaO + CaO + Gd_2O_3 + La_2O_3$ Glass	j	503-8675
$B_2O_3 + BaO + CaO + La_2O_3$ Glass	j	503-8703
$B_2O_3 + BaO + CaO + La_2O_3 + Lu_2O_3$ Glass	j	503-8683
$B_2O_3 + BaO + CaO + La_2O_3 + SiO_2$ Glass	j	503-8672
$B_2O_3 + BaO + CaO + La_2O_3 + Yb_2O_3$ Glass	j	503-8674
$B_2O_3 + BaO + CaO + La_2O_3 + ZrO_2$ Glass	j	503-8687
$B_2O_3 + BaO + CdO + La_2O_3$ Glass	j	503-8773
$B_2O_3 + BaO + CdO + La_2O_3 + SiO_2$ Glass	j	503-8796
$B_2O_3 + BaO + CdO +$ Other Glass	a	503-8402
$B_2O_3 + BaO + CoO + Na_2O + SiO_2$ Enamel	n	503-9531
$B_2O_3 + BaO + FeO + NiO + SiO_2$ Ceramic Adhesive	n	503-9667
$B_2O_3 + BaO + Fe_2O_3 + Y_2O_3$	a	503-9199
$B_2O_3 + BaO + Gd_2O_3$ Glass	j	503-8679
$B_2O_3 + BaO + Gd_2O_3 + La_2O_3 + Y_2O_3$ Glass	j	503-8681
$B_2O_3 + BaO + K_2O + Li_2O + SiO_2$ Flux	n	503-9291
$B_2O_3 + BaO + K_2O + Na_2O + SiO_2$ Glass	adjno	503-9091
$B_2O_3 + BaO + K_2O + Na_2O + SiO_2 + ZnO$ Glass	n	503-8563
$B_2O_3 + BaO + K_2O + SiO_2 + ZnO$ Glass	n	503-9094
$B_2O_3 + BaO + La_2O_3$ Glass	j	503-8815
$B_2O_3 + BaO + La_2O_3 + PbO$ Glass	j	503-8772
$B_2O_3 + BaO + La_2O_3 + SiO_2$ Glass	j	503-8696
$B_2O_3 + BaO + La_2O_3 + SiO_2 + SrO$ Glass	j	503-8795
$B_2O_3 + BaO + La_2O_3 + SiO_2 + ZnO$ Glass	j	503-8819
$B_2O_3 + BaO + La_2O_3 + SiO_2 + ZrO_2$ Glass	j	503-8689
$B_2O_3 + BaO + La_2O_3 + Sm_2O_3$ Glass	r	503-8994
$B_2O_3 + BaO + La_2O_3 + SrO$ Glass	j	503-8776
$B_2O_3 + BaO + La_2O_3 + SrO + ThO_2$ Glass	j	503-8959
$B_2O_3 + BaO + La_2O_3 + Ta_2O_5$ Glass	j	503-8801
$B_2O_3 + BaO + La_2O_3 + Ta_2O_5 + ThO_2$ Glass	j	503-8967
$B_2O_3 + BaO + La_2O_3 + Yb_2O_3$ Glass	r	503-8996
$B_2O_3 + BaO + La_2O_3 + ZnO$ Glass	j	503-8780
$B_2O_3 + BaO + Na_2O + Sb_2O_3 + SiO_2$ Enamel	n	503-9383
$B_2O_3 + BaO + Na_2O + SiO_2 + SnO_2$ Enamel	n	503-9384
$B_2O_3 + BaO + Na_2O + SiO_2 + TiO_2$ Enamel	n	503-9944
$B_2O_3 + BaO + Na_2O + SiO_2 + ZnO$ Glass	j	503-8571
$B_2O_3 + BaO + Na_2O + SiO_2 + ZrO_2$ Enamel	n	503-9385
$B_2O_3 + BaO + PbO + SiO_2 + ZnO$ Glass	jn	503-8886
$B_2O_3 + BaO + PbO + TiO_2$ Glass	n	503-8490
$B_2O_3 + BaO + SiO_2$ Glass	aejno	503-8742
$B_2O_3 + BaO + SiO_2 + TiO_2$ Glass	jn	503-8738
$B_2O_3 + BaO + SiO_2 + TiO_2 + Ti_2O_3$ Glass	j	503-8721
$B_2O_3 + BaO + SiO_2 + Ti_2O_3$ Glass	j	503-8722
$B_2O_3 + BaO + SiO_2 + ZnO$ Glass	aeo	503-8515
$B_2O_3 + BeO + CoO + SiO_2 + TiO_2$ Glass	n	503-8898
$B_2O_3 + BeO + Li_2O$ Glass	r	503-9098
$B_2O_3 + BeO + Na_2O$ Glass	n	503-8858
$B_2O_3 + BeO + Na_2O + SiO_2$ Glass	n	503-8857
$B_2O_3 + Bi_2O_3$ Glass	n	503-8613
$B_2O_3 + Bi_2O_3 + CuO + PbO + ZnO$ Glass	n	503-8737
$B_2O_3 + Bi_2O_3 + K_2O + Na_2O + SiO_2$ Flux	n	503-9290
$B_2O_3 + Bi_2O_3 + La_2O_3 + Ta_2O_5$ Glass	j	503-8799
$B_2O_3 + Bi_2O_3 + PbO + ZnO$ Glass	n	503-8740
$B_2O_3 + CaO$	an	503-9918
$B_2O_3 + CaO$ Glass	aijr	503-9148
$B_2O_3 + CaO + Gd_2O_3$ Glass	j	503-8680
$B_2O_3 + CaO + Gd_2O_3 + La_2O_3$ Glass	j	503-8684

Substance Name	Property	Number
Oxide Mixture: -(Cont)-		
$B_2O_3 + CaO + Gd_2O_3 + La_2O_3 + ZnO$ Glass	j	503-8676
$B_2O_3 + CaO + K_2O + Na_2O + PbO + SiO_2$ Glass	n	503-8559
$B_2O_3 + CaO + K_2O + Na_2O + SiO_2$ Glass	ejno	503-8870
$B_2O_3 + CaO + La_2O_3$ Glass	j	503-8701
$B_2O_3 + CaO + La_2O_3 + MgO + ZrO_2$ Glass	j	503-8686
$B_2O_3 + CaO + La_2O_3 + SiO_2 + ZrO_2$ Glass	j	503-8692
$B_2O_3 + CaO + La_2O_3 + Yb_2O_3 + ZnO$ Glass	j	503-8677
$B_2O_3 + CaO + La_2O_3 + ZnO$ Glass	j	503-8699
$B_2O_3 + CaO + La_2O_3 + ZnO + ZrO_2$ Glass	j	503-8690
$B_2O_3 + CaO + La_2O_3 + ZrO_2$ Glass	j	503-8698
$B_2O_3 + CaO + MgO + Na_2O + SiO_2$ Glass	n	503-8565
$B_2O_3 + CaO + MgO + PbO + SiO_2$ Glaze	n	503-9397
$B_2O_3 + CaO + MgO + SiO_2$	an	503-9710
$B_2O_3 + CaO + Na_2O + SiO_2$ Glass	no	503-9107
$B_2O_3 + CaO + Na_2O + SiO_2 + TiO_2$ Glass	n	503-8955
$B_2O_3 + CaO + Na_2O + SiO_2 + ZrO_2$ Enamel	hn	503-9222
$B_2O_3 + CaO + PbO$ Glaze	n	503-9413
$B_2O_3 + CaO + PbO + SiO_2$ Glaze	no	503-9414
$B_2O_3 + CaO + SiO_2 + ZnO + ZrO_2$ Glaze	n	503-9254
$B_2O_3 + CdO$ Glass	jn	503-8950
$B_2O_3 + CdO + La_2O_3$ Glass	jn	503-8952
$B_2O_3 + CdO + La_2O_3 + PbO$ Glass	j	503-8774
$B_2O_3 + CdO + La_2O_3 + SiO_2$ Glass	j	503-8951
$B_2O_3 + CdO + La_2O_3 + SrO$ Glass	j	503-8778
$B_2O_3 + CdO + La_2O_3 + Ta_2O_5$ Glass	j	503-8798
$B_2O_3 + CdO + La_2O_3 + ZnO$ Glass	j	503-8782
$B_2O_3 + CdO + PbO$ Glass	j	503-8790
$B_2O_3 + CeO_2 + Na_2O + SiO_2$ Enamel	n	503-9303
$B_2O_3 + CeO_2 + Na_2O + SiO_2$ Glass	n	503-9001
$B_2O_3 + Co_3O_4 + Na_2O$ Colored Glass	r	503-8964
$B_2O_3 + Cr_2O_3 + MnO$ Pigment	h	503-9417
$B_2O_3 + Cr_2O_3 + Na_2O$ Green Glass	r	503-8963
$B_2O_3 + Cs_2O$ Glass	nr	503-9082
$B_2O_3 + CuO + Na_2O$ Colored Glass	r	503-8966
$B_2O_3 + CuO + PbO + TiO_2 + ZnO$ Glass	n	503-8594
$B_2O_3 + CuO + PbO + ZnO$ Glass	n	503-8741
$B_2O_3 +$ Dolomite $+$ Perlite Glaze	n	503-8209
$B_2O_3 + Fe_2O_3 + Na_2O$ Glass	r	503-8998
$B_2O_3 + Fe_2O_3 + Na_2O + SiO_2 + U_3O_8$ Glass	a	503-8660
$B_2O_3 +$ Fireclay $+$ Grog Refractory	n	503-8299
$B_2O_3 + Ga_2O_3 + Na_2O + SiO_2$ Glass	ijn	503-8872
$B_2O_3 + Gd_2O_3 + La_2O_3 + ZnO$ Glass	j	503-8678
$B_2O_3 + Gd_2O_3 + ZnO$ Glass	j	503-8670
$B_2O_3 + GeO_2$ Glass	n	503-8880
$B_2O_3 + GeO_2 + K_2O$ Glass	n	503-8511
$B_2O_3 + GeO_2 + Li_2O$ Glass	n	503-8512
$B_2O_3 + GeO_2 + Li_2O + SiO_2 + Ta_2O_5$ Glass	n	503-8430
$B_2O_3 + GeO_2 + Li_2O + SiO_2 + ZnO$ Glass	n	503-8426
$B_2O_3 + GeO_2 + Li_2O + Ta_2O_5$ Glass	n	503-8429
$B_2O_3 + GeO_2 + Li_2O + ZnO$ Glass	n	503-8433
$B_2O_3 + GeO_2 + Na_2O$ Glass	no	503-8881
$B_2O_3 + GeO_2 + PbO$ Glass	j	503-8467
$B_2O_3 + GeO_2 + Ta_2O_5 + ZnO$ Glass	jn	503-8437
$B_2O_3 +$ Grog $+$ Kaolin Refractory	n	503-8303
$B_2O_3 + HfO_2 + La_2O_3 + Ta_2O_5$ Glass	j	503-8797
$B_2O_3 + K_2O$	a	503-9926
$B_2O_3 + K_2O$ Glass	ijnr	503-9134
$B_2O_3 + K_2O + La_2O_3 + Na_2O + SiO_2$ Glass	j	503-8671
$B_2O_3 + K_2O + Na_2O + PbO + SiO_2$ Ceramic	an	503-9508
$B_2O_3 + K_2O + Na_2O + PbO + SiO_2$ Glass	n	503-9153
$B_2O_3 + K_2O + Na_2O + SiO_2$ Glass	ano	503-9102
$B_2O_3 + K_2O + Na_2O + SiO_2 + TiO_2$ Enamel	n	503-9577
$B_2O_3 + K_2O + Na_2O + SiO_2 + ZnO$ Glass	n	503-8578
$B_2O_3 + K_2O + P_2O_5 + ZnO$ Glass	n	503-8533
$B_2O_3 + K_2O + PbO$ Glass	j	503-8705

Property: a. Thermal conductivity; **b.** Accommodation coefficient; **c.** Thermal contact resistance; **d.** Thermal diffusivity; **e.** Specific heat; **f.** Viscosity; **g.** Emittance; **h.** Reflectance; **i.** Absorptance; **j.** Transmittance; **k.** α_s/ϵ_t ratio; **l.** Prandtl number; **n.** Thermal linear expansion coefficient; **o.** Thermal volumetric expansion coefficient; **r.** Thermal radiative properties (g,h,i,j,k)

Substance Name	Property	Number
Oxide Mixture: -(Cont)-		
$B_2O_3 + K_2O + PbO + SiO_2$ Glass	o	503-8685
$B_2O_3 + K_2O + SiO_2$ Glass	jn	503-8752
$B_2O_3 + K_2O + V_2O_5$ Brown Glass	r	503-8921
$B_2O_3 + La_2O_3$ Glass	jn	503-9080
$B_2O_3 + La_2O_3 + Lu_2O_3 + ZnO$ Glass	j	503-8682
$B_2O_3 + La_2O_3 + Nb_2O_5 + Ta_2O_5$ Glass	j	503-8800
$B_2O_3 + La_2O_3 + NiO + Ta_2O_5$ Glass	j	503-8791
$B_2O_3 + La_2O_3 + PbO + SrO$ Glass	j	503-8777
$B_2O_3 + La_2O_3 + PbO + Ta_2O_5$ Glass	j	503-8802
$B_2O_3 + La_2O_3 + PbO + ZnO$ Glass	j	503-8781
$B_2O_3 + La_2O_3 + SiO_2 + Ta_2O_5$ Glass	j	503-8803
$B_2O_3 + La_2O_3 + SrO + Ta_2O_5$ Glass	j	503-8775
$B_2O_3 + La_2O_3 + SrO + ZnO$ Glass	j	503-8783
$B_2O_3 + La_2O_3 + Ta_2O_5$ Glass	j	503-8804
$B_2O_3 + La_2O_3 + Ta_2O_5 + ThO_2$ Glass	j	503-8958
$B_2O_3 + La_2O_3 + Ta_2O_5 + ZnO$ Glass	j	503-8779
$B_2O_3 + La_2O_3 + Y_2O_3$ Glass	j	503-8700
$B_2O_3 + La_2O_3 + Yb_2O_3$ Glass	j	503-8702
$B_2O_3 + La_2O_3 + ZnO$ Glass	jn	503-8704
$B_2O_3 + Li_2O$ Ceramic	a	503-9972
$B_2O_3 + Li_2O$ Glass	aeijnr	503-9147
$B_2O_3 + Li_2O + Na_2O + SiO_2 + ZrO_2$ Glass	ho	503-8530
$B_2O_3 + Li_2O + P_2O_5 + ZnO$ Glass	n	503-8532
$B_2O_3 + Li_2O + SiO_2$ Glass	no	503-8628
$B_2O_3 + Li_2O + SiO_2 + ZnO$ Glass	n	503-8427
$B_2O_3 + Li_2O + V_2O_5$ Brown Glass	r	503-8920
$B_2O_3 + MgO$	n	503-9446
$B_2O_3 + MgO + Na_2O + SiO_2 + ZnO$ Glass	n	503-8580
$B_2O_3 + Mn_2O_3 + R_2O$ Glass	j	503-8409
$B_2O_3 + Mn_3O_4 + Na_2O$ Brown Glass	r	503-8962
$B_2O_3 + MoO_3 + PbO + SiO_2$ Glass	n	503-8493
$B_2O_3 + Na_2O$	a	503-9934
$B_2O_3 + Na_2O$ Glass	deijnr	503-9169
$B_2O_3 + Na_2O + NiO$ Colored Glass	r	503-8965
$B_2O_3 + Na_2O + P_2O_5 + ZnO$ Glass	n	503-8534
$B_2O_3 + Na_2O + PbO$ Glass	e	503-9065
$B_2O_3 + Na_2O + PbO + SiO_2$	n	503-9848
$B_2O_3 + Na_2O + PbO + SiO_2 + ZnO$ Flux	n	503-9292
$B_2O_3 + Na_2O + Sb_2O_3 + SiO_2 + ZnO$ Enamel	hn	503-9271
$B_2O_3 + Na_2O + SiO_2$	hn	503-9662
$B_2O_3 + Na_2O + SiO_2$ Glass	aeijnor	503-9171
$B_2O_3 + Na_2O + SiO_2 + TiO_2$ Glass	jn	503-9151
$B_2O_3 + Na_2O + SiO_2 + TiO_2 + ZrO_2$ Enamel	n	503-9943
$B_2O_3 + Na_2O + SiO_2 + V_2O_5$ Brown Glass	r	503-8922
$B_2O_3 + Na_2O + SiO_2 + ZnO$ Glass	n	503-9172
$B_2O_3 + Na_2O + SiO_2 + ZnO + ZrO_2$ Enamel	o	503-9310
$B_2O_3 + Na_2O + SiO_2 + ZrO_2$ Glass	ho	503-8531
$B_2O_3 + Na_2O + SiO_2 + ZrO_2$ Glaze	hn	503-9285
$B_2O_3 + Na_2O + Tb_2O_3$ Glass	r	503-8995
$B_2O_3 + Na_2O + TiO_2$ Glass	r	503-8960
$B_2O_3 + Na_2O + V_2O_4$ Green Glass	h	503-9073
$B_2O_3 + Na_2O + V_2O_5$ Colored Glass	r	503-8961
$B_2O_3 + P_2O_5 + PbO$ Glass	n	503-8743
$B_2O_3 + P_2O_5 + ZnO$ Glass	j	503-9054
$B_2O_3 + PbO$	jo	503-9562
$B_2O_3 + PbO$ Glass	dhn	503-9023
$B_2O_3 + PbO + SiO_2$	no	503-9428
$B_2O_3 + PbO + SiO_2$ Glass	ehnr	503-9074
$B_2O_3 + PbO + SiO_2 + TiO_2$ Glass	n	503-8645
$B_2O_3 + PbO + SiO_2 + TiO_2 + ZnO$ Glass	n	503-8739
$B_2O_3 + PbO + SiO_2 + TiO_2 + ZrO_2$ Glass	n	503-8644
$B_2O_3 + PbO + SiO_2 + ZnO$ Glass	n	503-8761
$B_2O_3 + PbO + SiO_2 + ZrO_2$ Glass	n	503-8646
$B_2O_3 + PbO + TeO_2$ Glass	n	503-8574
$B_2O_3 + PbO + TiO_2$ Glass	n	503-8635
$B_2O_3 + PbO + TiO_2 + ZnO$ Glass	n	503-8593
$B_2O_3 + PbO + ZnO$ Glass	n	503-9055
B_2O_3 + Perlite Glaze	n	503-8211
$B_2O_3 + Rb_2O$ Glass	n	503-9083
$B_2O_3 + Rb_2O + SiO_2$ Glass	jn	503-8723

Substance Name	Property	Number
Oxide Mixture: -(Cont)-		
$B_2O_3 + Sb_2O_3$ Glass	n	503-8862
$B_2O_3 + SiO_2$	jn	503-9948
$B_2O_3 + SiO_2$ Glass	aejnor	503-9181
$B_2O_3 + SiO_2$ + Clay Porcelain	n	503-8273
$B_2O_3 + SiO_2$ + Kaolin Ceramic	n	503-8248
$B_2O_3 + SiO_2$ + Other Glass	a	503-8406
$B_2O_3 + SiO_2 + R_2O$ Glass	ae	503-8416
$B_2O_3 + SiO_2 + Ta_2O_5 + ZnO$ Glass	n	503-8431
$B_2O_3 + SiO_2 + Tl_2O$ Glass	n	503-8583
$B_2O_3 + SiO_2 + ZnO$	n	503-9364
$B_2O_3 + SiO_2 + ZnO$ Glass	n	503-8728
$B_2O_3 + SrO$	n	503-9927
$B_2O_3 + SrO$ Glass	jnr	503-9132
B_2O_3 + Trachyte Glaze	n	503-8204
$B_2O_3 + V_2O_4$ Yellow Glass	h	503-9138
$B_2O_3 + ZnO$ Glass	jno	503-9081
$BaO + BaO_2 + La_2O_3 + SiO_2 + ZnO$ Glass	j	503-8806
$BaO + BeO$	n	503-9649
$BaO + BeO + PbO + SiO_2$ Glass	j	503-8784
$BaO + BeO + SiO_2$	j	503-9619
$BaO + Bi_2O_3 + CaO + PbO + SiO_2$ Glaze	n	503-9580
$BaO + Bi_2O_3 + SiO_2$ Glass	a	503-8485
$BaO + Bi_2O_3 + TiO_2$ Glass	jn	503-8505
$BaO + Bi_2O_3 + TiO_2 + ZrO_2$ Glass	n	503-8503
$BaO + CaO$	n	503-9366
$BaO + CaO + Cr_2O_3 + SiO_2 + ZnO$ Ceramic	n	503-9614
$BaO + CaO + K_2O + Na_2O + SiO_2$ Glass	aegno	503-9178
$BaO + CaO + MgO + Na_2O + SiO_2$ Glass	n	503-8756
$BaO + CaO + MgO + SiO_2 + ZnO$ Enamel	n	503-9313
$BaO + CaO + MoO_3 + SiO_2 + ZnO$ Enamel	n	503-9316
$BaO + CaO + Na_2O + SiO_2$ Glass	jn	503-9113
$BaO + CaO + NiO + SrO$ Ceramic, Oxide Cathode	a	503-9696
$BaO + CaO + SiO_2$	jn	503-9820
$BaO + CaO + SiO_2 + TiO_2 + ZrO_2$ Ceramic	n	503-9613
$BaO + CaO + SiO_2 + ZnO$ Glaze	n	503-9209
$BaO + CaO + SiO_2 + ZrO_2$ Ceramic	n	503-9440
$BaO + CaO + SrO$	a	503-9697
$BaO + CaO + SrO + TiO_2$	ae	503-9626
$BaO + CaO + TiO_2$ Ceramic	ade	503-9812
$BaO + CaO + TiO_2 + ZrO_2$ + Clay	n	503-2835
BaO + Clay + Steatite	n	503-2831
$BaO + Cr_2O_3 + Na_2O + Nb_2O_5$	n	503-9228
$BaO + Cr_2O_3 + SiO_2$ Ceramic	n	503-9224
$BaO + Fe_2O_3$ Ceramic	hn	503-9503
$BaO + Fe_2O_3 + SiO_2$	n	503-9616
$BaO + Gd_2O_3 + Na_2O + Nb_2O_5$	n	503-9229
$BaO + GeO_2 + K_2O$ Glass	j	503-8494
$BaO + GeO_2 + Na_2O$ Glass	j	503-8491
$BaO + Ge_2O + SrO + TiO_2 + ZrO_2$ Glass	j	503-8827
$BaO + K_2O + Na_2O + Nd_2O_3 + SiO_2$ Glass	jn	503-8956
$BaO + K_2O + Na_2O + PbO + SiO_2$ Glass	n	503-8588
$BaO + K_2O + Na_2O + SiO_2$ Glass	eno	503-9124
$BaO + K_2O + Na_2O + SiO_2 + ZnO$ Glass	n	503-8736
$BaO + K_2O + Nd_2O_3 + SiO_2$ Glass	nr	503-8992
$BaO + K_2O + PbO + SiO_2$ Glass	o	503-8519
$BaO + K_2O + PbO + SiO_2 + ZnO$ Glass	an	503-9097
$BaO + K_2O + SiO_2$ Glass	aeo	503-9123
$BaO + K_2O + SiO_2 + ZnO$ Glass	ao	503-8839
$BaO + La_2O_3 + Li_2O + SiO_2 + ZnO$ Glass	j	503-8807
$BaO + La_2O_3 + Nd_2O_3 + SiO_2$ Glass	r	503-8993
$BaO + La_2O_3 + PbO + SiO_2 + ZnO$ Glass	j	503-8814
$BaO + La_2O_3 + SiO_2$ Glass	jnr	503-9064
$BaO + La_2O_3 + SiO_2 + SrO + TiO_2$ Glass	j	503-8450
$BaO + La_2O_3 + SiO_2 + SrO + ZnO$ Glass	j	503-8813
$BaO + La_2O_3 + SiO_2 + Ta_2O_5$ Glass	jr	503-9063
$BaO + La_2O_3 + SiO_2 + TiO_2$ Glass	jn	503-8452
$BaO + La_2O_3 + SiO_2 + TiO_2 + ZnO$ Glass	h	503-8997
$BaO + Li_2O + SiO_2$	n	503-9617
$BaO + MgO$ Refractory	n	503-9495
$BaO + MgO + SiO_2$	jn	503-9620
$BaO + MgO + SiO_2$ Porcelain	a	503-8396
$BaO + MgO + TiO_2 + ZrO_2$ Ceramic	a	503-9801
$BaO + MnO + Nb_2O_5 + TiO_2$ Ceramic	a	503-9765

Property: a. Thermal conductivity; **b.** Accommodation coefficient; **c.** Thermal contact resistance; **d.** Thermal diffusivity; **e.** Specific heat; **f.** Viscosity; **g.** Emittance; **h.** Reflectance; **i.** Absorptance; **j.** Transmittance; **k.** α_s/ϵ_t ratio; **l.** Prandtl number; **n.** Thermal linear expansion coefficient; **o.** Thermal volumetric expansion coefficient; **r.** Thermal radiative properties (g,h,i,j,k)

Substance Name	Property	Number
Oxide Mixture: -(Cont)-		
$BaO + MoO_3 + SiO_2 + SrO + ZnO$ Enamel	n	503-9314
$BaO + MoO_3 + SiO_2 + TiO_2 + ZnO$ Enamel	n	503-9315
$BaO + Na_2O + Nb_2O_5$	n	503-9230
$BaO + Na_2O + SiO_2$ Glass	ajnr	503-9157
$BaO + Na_2O + SiO_2 + SnO_2$ Glass	n	503-8550
$BaO + Na_2O + SiO_2 + TiO_2$ Glass	n	503-8549
$BaO + Na_2O + SiO_2 + ZnO$ Glass	n	503-8551
$BaO + Nb_2O_5 + PbO + TiO_2 + ZnO$ Ceramic	j	503-9263
$BaO + Nb_2O_5 + SiO_2$ Glass	j	503-8459
$BaO + Nb_2O_5 + SrO$	j	503-9207
$BaO + NiO + SiO_2$ Ceramic	n	503-9223
$BaO + P_2O_5$	ae	503-9308
$BaO + P_2O_5$ Glass	ajn	503-8818
$BaO + P_2O_5 + V_2O_5$ Green Glass	r	503-8923
$BaO + PbO + SiO_2$ Glass	anr	503-8914
$BaO + PbO + TeO_2$ Glass	n	503-8575
$BaO + PbO + TiO_2 + ZrO_2$ Ceramic	an	503-9865
$BaO + Rb_2O + SiO_2$ Glass	jn	503-8725
$BaO + SO_3 + TiO_2$ Glass	r	503-9070
$BaO + SiO_2$	ehj	503-9917
$BaO + SiO_2$ Glass	jn	503-8842
$BaO + SiO_2 + SnO_2$ Glaze	n	503-9208
$BaO + SiO_2 + SrO$	j	503-9699
$BaO + SiO_2 + SrO + TiO_2$ Glass	j	503-8451
$BaO + SiO_2 + SrO + TiO_2 + ZrO_2$ Glass	j	503-8828
$BaO + SiO_2 + Ta_2O_3$ Glass	jn	503-8461
$BaO + SiO_2 + Ta_2O_5$ Glass	j	503-8454
$BaO + SiO_2 + TiO_2$	n	503-9618
$BaO + SiO_2 + TiO_2$ Glass	ahjnr	503-8987
$BaO + SiO_2 + ZnO$ Glass	jn	503-8460
$BaO + SiO_2 + ZrO_2$	in	503-9701
$BaO + SnO_2$ Ceramic	n	503-9635
$BaO + SnO_2 + TiO_2$	a	503-9357
$BaO + SnO_2 + TiO_2 + ZrO_2$ Ceramic	a	503-9393
$BaO + SrO$ Ceramic, Oxide Cathode	ae	503-9950
$BaO + SrO + TiO_2$	dejn	503-9951
$BaO + TeO_2$ Glass	j	503-8453
$BaO + TiO_2$ Ceramic	adehn	503-9824
$BaO + TiO_2 + ZrO_2$	a	503-9356
$BaO + UO_2$	a	503-9737
$BaO + ZrO_2$	en	503-9511
BeO + Bentonite	no	503-8322
$BeO + CaO$ Ceramic	a	503-9860
$BeO + CaO + Na_2O + SiO_2$ Glass	n	503-9114
$BeO + CaO + ZrO_2$ Refractory	n	503-9491
$BeO + CeO_2 + ZrO_2$ Refractory	n	503-9489
$BeO + Cr_2O_3$ Refractory	n	503-9493
$BeO + GeO_2 + Na_2O$ Glass	j	503-8492
$BeO + K_2O + Na_2O + SiO_2$ Glass	n	503-8631
$BeO + Li_2O + SiO_2$ Glass	r	503-8412
$BeO + MgO$ Ceramic	aenor	503-9955
$BeO + MgO + ThO_2$ Refractory, Beryllium Oxide	a	503-9964
$BeO + MgO + ZrO_2$ Refractory	an	503-9958
$BeO + Na_2O + SiO_2$ Glass	n	503-9155
$BeO + P_2O_5$ Glass	n	503-8611
$BeO + P_2O_5 + V_2O_5$ Glass	r	503-9005
$BeO + SiO_2$	no	503-9610
$BeO + SiO_2$ + Clay + Feldspar Porcelain	n	503-8316
$BeO + ThO_2 + UO_2$	n	503-9609
$BeO + ThO_2 + UO_2 + Y_2O_3$	n	503-9608
$BeO + TiO_2$	n	503-9650
$BeO + ZnO$	j	503-9685
$BeO + ZrO_2$	n	503-9726
$BiO + Fe_2O_3 + Nb_2O_3 + PbO$ Ceramic	n	503-9328
$BiO_2 + ZrO_2$	n	503-9788
$Bi_2O_3 + CaO + La_2O_3 + MnO$ Ceramic	e	503-9673
$Bi_2O_3 + CaO + MnO$ Refractory, Perovskite Manganite	e	503-9672
$Bi_2O_3 + GeO_2$	j	503-9220
$Bi_2O_3 + GeO_2$ Glass	n	503-8476
$Bi_2O_3 + GeO_2 + SiO_2$	j	503-9218
$Bi_2O_3 + K_2O + Na_2O + SiO_2$ Glass	n	503-8615
$Bi_2O_3 + Mn_2O_3 + PbO + TiO_2$	no	503-9305
Oxide Mixture: -(Cont)-		
$Bi_2O_3 + MoO_3 + TiO_2$	n	503-9302
$Bi_2O_3 + P_2O_5$ Glass	n	503-9093
$Bi_2O_3 + P_2O_5 + V_2O_5$ Glass	n	503-8771
$Bi_2O_3 + PbO$ Ceramic	hj	503-9465
$Bi_2O_3 + PbO$ Glass	a	503-8488
$Bi_2O_3 + PbO + RO_2$ I R Transmitting Glass	jn	503-8403
$Bi_2O_3 + PbO + SiO_2$ Glass	a	503-8487
$Bi_2O_3 + PbO + WO_3$ Glass	j	503-8469
$Bi_2O_3 + SiO_2$	j	503-9219
$Bi_2O_3 + SiO_2$ Glass	an	503-8612
$Bi_2O_3 + TiO_2$	h	503-9629
$Bi_2O_3 + TiO_2$ Glass	j	503-8468
$Bi_2O_3 + V_2O_5$	i	503-9256
CaO + Ball Clay + Feldspar + Kaolin + Tripoli Ceramic	n	503-8258
$CaO + CdO$	h	503-9524
$CaO + CeO_2 + ZrO_2$ Refractory, Zirconium Oxide	gr	503-9717
$CaO + CoO + Na_2O + SiO_2$ Glass	r	503-9111
$CaO + Co_3O_4 + Na_2O + SiO_2$ Glass	r	503-9028
$CaO + Cr_2O_3$ Refractory	n	503-9359
$CaO + Cr_2O_3 + FeO + MgO + SiO_2$ Ceramic	n	503-9790
$CaO + Cr_2O_3 + Fe_2O_3 + MgO$ Ceramic	n	503-9252
$CaO + Cr_2O_3 + Fe_2O_3 + MgO + SiO_2$ Ceramic	aen	503-9431
$CaO + Cr_2O_3 + MgO + SiO_2$ Refractory	an	503-9483
$CaO + Cr_2O_3 + Na_2O + SiO_2$ Glass	jr	503-9027
$CaO + Cr_2O_3 + ZrO_2$ Refractory	n	503-9361
$CaO + CuO + Na_2O + SiO_2$ Glass	nr	503-9109
$CaO + FeO + Fe_2O_3 + MgO + SiO_2$ Ceramic	o	503-9711
$CaO + FeO + Fe_2O_3 + Na_2O + SiO_2$ Enamel	n	503-9879
$CaO + FeO + Fe_2O_3 + SiO_2$	n	503-9827
$CaO + FeO + Fe_2O_3 + SiO_2$ Concentrate	ade	503-9985
$CaO + FeO + MgO + SiO_2$	n	503-9623
$CaO + FeO + MgO + SiO_2$ Slag	ae	503-9998
$CaO + FeO + SiO_2$	an	503-9742
$CaO + FeO + SiO_2$ Slag	ae	503-9997
$CaO + FeO + SiO_2 + ZnO$	a	503-9741
$CaO + FeO + TiO_2 + ZrO_2$	j	503-9233
$CaO + FeO + ZrO_2$	j	503-9234
$CaO + Fe_2O_3 + MgO$ Refractory, Magnesite	ad	503-9784
$CaO + Fe_2O_3 + MgO + SiO_2$ Ceramic	an	503-9541
$CaO + Fe_2O_3 + MnO + Na_2O + SiO_2$ Glass	g	503-8871
$CaO + Fe_2O_3 + Na_2O + SiO_2$ Enamel	n	503-9880
$CaO + Fe_2O_3 + Na_2O + SiO_2$ Glass	jnr	503-9029
$CaO + Fe_2O_3 + P_2O_5$ Glass	j	503-8720
$CaO + Fe_2O_3 + SiO_2$	dn	503-9682
$CaO + Fe_2O_3 + SiO_2 + TiO_2 + ZrO_2$ Ceramic	a	503-9923
$CaO + Fe_2O_3 + V_2O_5 + Y_2O_3$ Ferrite - Garnet Mixture	a	503-9264
$CaO + GeO_2 + SiO_2$	j	503-9211
$CaO + HfO_2$ Refractory, Hafnium Oxide	n	503-9640
$CaO + HfO_2 + TiO_2 + ZrO_2$	n	503-9426
$CaO + HfO_2 + ZrO_2$	aeghn	503-9891
$CaO + K_2O + MgO + Na_2O + SiO_2$ Colored Glass	j	503-8867
$CaO + K_2O + MgO + Na_2O + SiO_2 + R_2O_3$ Ceramic, Steatite	r	503-8387
$CaO + K_2O + Na_2O + PbO + SiO_2$ Glass	ejn	503-9058
$CaO + K_2O + Na_2O + SiO_2$ Glass	agjno	503-9115
$CaO + K_2O + Na_2O + SiO_2 + SnO_2$ Pigment	h	503-9421
$CaO + K_2O + Na_2O + SiO_2 + ZnO$ Glass	a	503-8711
$CaO + K_2O + PbO + SiO_2$ Colored Glass	n	503-8606
$CaO + K_2O + PbO + SiO_2$ Glass	no	503-8605
$CaO + K_2O + SiO_2$ Glass	aegno	503-9168
$CaO + K_2O + SiO_2 + R_2O_3$ Glass	n	503-8399
$CaO + K_2O + SiO_2 + ZnO + ZrO_2$ Glass	jn	503-8424
$CaO + Li_2O + SiO_2$ Glass	n	503-8727
$CaO + Li_2O + SiO_2 + SnO_2$ Pigment	h	503-9423
$CaO + MgO$	aegnr	503-9808
$CaO + MgO + Na_2O + PbO + SiO_2$ Glass	n	503-8572
$CaO + MgO + Na_2O + SiO_2$ Glass	nr	503-9161
$CaO + MgO + Na_2O + SiO_2 + R_2O_3$ Glass	n	503-8420
$CaO + MgO + Na_2O + SiO_2 + ZnO$ Glaze	n	503-9368
$CaO + MgO + Na_2O + SiO_2 + ZnO + ZrO_2$ Glass	j	503-8423

Property: **a.** Thermal conductivity; **b.** Accommodation coefficient; **c.** Thermal contact resistance; **d.** Thermal diffusivity; **e.** Specific heat; **f.** Viscosity; **g.** Emittance; **h.** Reflectance; **i.** Absorptance; **j.** Transmittance; **k.** a_s/ϵ_t ratio; **l.** Prandtl number; **n.** Thermal linear expansion coefficient; **o.** Thermal volumetric expansion coefficient; **r.** Thermal radiative properties (g,h,i,j,k)

Substance Name	Property	Number
Oxide Mixture: -(Cont)-		
$CaO + MgO + SiO_2$	no	503-9981
$CaO + MgO + SiO_2$ Glass	jno	503-9177
$CaO + MgO + ZrO_2$	n	503-9592
$CaO + MnO + Na_2O + SiO_2$ Glass	n	503-9112
$CaO + Na_2O$ Glass	n	503-8661
$CaO + Na_2O + Nd_2O_3 + SiO_2$ Glass	jr	503-9032
$CaO + Na_2O + NiO + SiO_2$ Glass	nr	503-9110
$CaO + Na_2O + Ni_2O_3 + SiO_2$ Glass	r	503-9030
$CaO + Na_2O + P_2O_5$ Glass	n	503-8601
$CaO + Na_2O + P_2O_5 + SiO_2$ Glass	n	503-8548
$CaO + Na_2O + PbO + SiO_2$ Glass	n	503-8398
$CaO + Na_2O + PbO + SiO_2 + SrO$ Glass	n	503-8557
$CaO + Na_2O + Pr_6O_{11} + SiO_2$ Glass	r	503-9033
$CaO + Na_2O + Sb_2O_3 + SiO_2$ Glass	n	503-8547
$CaO + Na_2O + SiO_2$ Glass	aeghijnor	503-9173
$CaO + Na_2O + SiO_2 + R_2O_3$ Glass	e	503-8419
$CaO + Na_2O + SiO_2 + Sm_2O_3$ Glass	r	503-9031
$CaO + Na_2O + SiO_2 + SnO_2$ Glass	n	503-8552
$CaO + Na_2O + SiO_2 + SnO_2$ Pigment	h	503-9422
$CaO + Na_2O + SiO_2 + SrO + TiO_2$ Glass	hj	503-8445
$CaO + Na_2O + SiO_2 + SrO + ZrO_2$ Glaze	o	503-9433
$CaO + Na_2O + SiO_2 + TiO_2$ Glass	no	503-9106
$CaO + Na_2O + SiO_2 + TiO_2 + ZnO$ Glass	j	503-8446
$CaO + Na_2O + SiO_2 + TiO_2 + ZnO + ZrO_2$ Glass	j	503-8447
$CaO + Na_2O + SiO_2 + UO_2$ Glass	n	503-8546
$CaO + Na_2O + SiO_2 + ZnO$ Glass	aeno	503-9108
$CaO + Na_2O + SiO_2 + ZnO + ZrO_2$ Enamel	o	503-9309
$CaO + Na_2O + SiO_2 + ZnO + ZrO_2$ Glass	jn	503-8448
$CaO + Na_2O + SiO_2 + ZrO_2$ Glass	n	503-9105
$CaO + NiO$	j	503-9261
$CaO + P_2O_5$ Glass	ijn	503-9143
$CaO + P_2O_5 + V_2O_5$ Glass	r	503-9006
$CaO + P_2O_5 + ZrO_2$ Ceramic	aen	503-9363
$CaO + PuO_2 + ZrO_2$	aden	503-9854
$CaO + Rb_2O + SiO_2$ Glass	n	503-8724
$CaO + SiO_2$	ehnr	503-9977
$CaO + SiO_2$ Glass	eijr	503-9084
$CaO + SiO_2$ + Ball Clay + Feldspar + Kaolin Ceramic	n	503-8259
$CaO + SiO_2$ + Clay + Feldspar	n	503-2830
$CaO + SiO_2$ + Kaolin Refractory	r	503-8382
$CaO + SiO_2 + SnO_2$ Pigment	h	503-9424
$CaO + SiO_2 + TiO_2$	n	503-9622
$CaO + SiO_2 + TiO_2 + ZrO_2$ Ceramic	e	503-9924
$CaO + SiO_2 + TiO_2 + ZrO_2$ + Ball Clay Ceramic	n	503-8328
$CaO + SiO_2 + ZrO_2$	in	503-9706
$CaO + SiO_2 + ZrO_2$ + Clay Ceramic	an	503-8338
$CaO + ThO_2$	j	503-9482
$CaO + ThO_2 + ZrO_2$ Refractory	n	503-9490
$CaO + TiO_2$ Ceramic	n	503-9532
$CaO + TiO_2 + ZrO_2$ Refractory	ad	503-9463
$CaO + UO_2$	a	503-9783
$CaO + UO_2 + ZrO_2$	aden	503-9712
$CaO + ZrO_2$	adeghnor	503-9922
$CdO + CoO + MnO_2 + PbO + WO_3$	n	503-9304
$CdO + Cr_2O_3 + Fe_2O_3$	an	503-9499
$CdO + Fe_2O_3 + NiO$ Ceramic, Ferrate	a	503-9627
$CdO + P_2O_5 + V_2O_5$ Glass	r	503-9009
$CdO + PbO$ Ceramic	i	503-9466
$CdO + PbO + SiO_2$ Reduced Glass	r	503-8910
$CdO + PbO_2$ Ceramic	i	503-9467
$CdO + SrO$	h	503-9278
$CeO_2 + Ce_2O_3$	e	503-9810
$CeO_2 + Dy_2O_3$ Ceramic	n	503-9551
$CeO_2 + HfO_2 + SrO$	no	503-9198
$CeO_2 + LaO_2 + ZrO_2$	r	503-9730
$CeO_2 + MgO$ Ceramic, Cerium Oxide	aor	503-9962
$CeO_2 + PbO$ Ceramic, Cerium Oxide	a	503-9959
$CeO_2 + SrO$	n	503-9480
$CeO_2 + ThO_2$	agr	503-9883
$CeO_2 + UO_2$	a	503-9933
$CeO_2 + V_2O_4$	n	503-9404

Substance Name	Property	Number
Oxide Mixture: -(Cont)-		
$CeO_2 + Y_2O_3 + ZrO_2$ Refractory, Zirconium Oxide	r	503-9715
$CeO_2 + ZrO_2$	dnor	503-9731
$Ce_2O_3 + ZrO_2$ Refractory, Zirconium Oxide	n	503-9932
$CoO + Cr_2O_3$ Pigment	h	503-9215
$CoO + Cr_2O_3 + NiO$ Spinel	n	503-9225
$CoO + Cr_2O_3 + Sb_2O_3 + TiO_2$ Pigment	h	503-9217
$CoO + Cr_2O_3 + TiO_2$ Pigment	h	503-9216
$CoO + Cr_2O_3 + TiO_2 + ZnO$ Ceramic	h	503-9193
$CoO + Cr_2O_3 + ZnO$ Ceramic	h	503-9427
$CoO + FeO + Fe_2O_3$ Oxide Magnet	e	503-9930
$CoO + K_2O + P_2O_5$ Glass	i	503-8482
$CoO + MgO$	hjn	503-9851
$CoO + MnO$	e	503-9911
$CoO + Na_2O + P_2O_5$ Glass	ij	503-8481
$CoO + Na_2O + SiO_2$ Glass	n	503-9128
$CoO + NiO$	e	503-9912
$CoO + P_2O_5$ Glass	ei	503-8483
$CoO + PbO + SiO_2$ Reduced Glass	r	503-8907
$CoO + TiO$	a	503-9213
$CoO + TiO_2$ Ceramic	n	503-9636
$CoO + TiO_2 + ZnO$ Ceramic	h	503-9192
Co_2O_3 + Andesite Glaze	n	503-8208
Co_2O_3 + Andesite + Porcelain Glaze	n	503-8207
$Co_3O_4 + Na_2O + SiO_2$ Blue Glass	r	503-8970
$CrO + MgO + SiO_2 + R_2O$ Ceramic	n	503-8274
$CrO_3 + PbO + SiO_2$	j	503-9296
Cr_2O_3 + Dolomite Refractory	n	503-8304
$Cr_2O_3 + FeO$ Refractory, Spinel	an	503-9554
$Cr_2O_3 + FeO + Fe_2O_3$ Refractory	n	503-9277
$Cr_2O_3 + FeO + Fe_2O_3 + MgO$ Refractory	n	503-9275
$Cr_2O_3 + FeO + MgO$ Refractory	n	503-9273
$Cr_2O_3 + Fe_2O_3 + MgO$	aen	503-9695
$Cr_2O_3 + Fe_2O_3 + MgO + SiO_2$ Ceramic	an	503-9297
$Cr_2O_3 + Fe_2O_3 + NiO$	an	503-9498
$Cr_2O_3 + K_2O + Na_2O + PbO + SiO_2$ Glass	n	503-8853
$Cr_2O_3 + Li_2O + MgO$ Ceramic	n	503-9864
$Cr_2O_3 + MgO$	ahnr	503-9837
$Cr_2O_3 + MgO + SiO_2$	n	503-9693
$Cr_2O_3 + Na_2O + SiO_2$ Green Glass	r	503-8969
$Cr_2O_3 + NiO$	h	503-9281
$Cr_2O_3 + P_2O_5 + Sb_2O_3$ Glass	n	503-8986
$Cr_2O_3 + P_2O_5 + ZrO_2$ Refractory	n	503-9492
Cr_2O_3 + Sillimanite Refractory	n	503-8244
$Cr_2O_3 + SiO_2$	agn	503-9795
$Cr_2O_3 + SiO_2$ Glass	jr	503-9034
$Cr_2O_3 + TiO_2$	hno	503-9543
$Cr_2O_3 + TiO_2 + ZnO$ Ceramic	h	503-9196
$Cr_2O_3 + Y_2O_3$	ghr	503-9749
$Cr_2O_3 + ZnO$	hi	503-9206
$Cr_2O_3 + ZrO_2$ Refractory	eijn	503-9536
$Cs_2O + GeO_2$ Glass	n	503-8475
$Cs_2O + GeO_2 + Na_2O$ Glass	n	503-8472
$Cs_2O + K_2O + Na_2O + SiO_2$ Glass	n	503-8632
$Cs_2O + Li_2O + SiO_2$ Glass	o	503-8732
$Cs_2O + Na_2O + SiO_2$ Glass	o	503-8730
$Cs_2O + PbO + SiO_2$ Reduced Glass	r	503-8903
$Cs_2O + SiO_2$ Glass	j	503-9075
$CuO + Fe_2O_3 + NiO + ZnO$	n	503-9262
$CuO + Fe_2O_3 + ZnO$	n	503-9325
$CuO + K_2O + Na_2O + PbO + SiO_2$ Glass	n	503-8852
$CuO + MgO$	h	503-9529
$CuO + MgO + Na_2O + SiO_2$ Glass	n	503-9130
$CuO + Na_2O + NiO + PbO + SiO_2$ Glass	a	503-9047
$CuO + Na_2O + PbO + SiO_2$ Glass	a	503-9046
$CuO + Na_2O + SiO_2$ Colored Glass	nr	503-9129
$CuO + P_2O_5$	n	503-9460
$CuO + P_2O_5$ Glass	n	503-8608
$CuO + P_2O_5 + TiO_2$	n	503-9457
$CuO + TeO + V_2O_5$ Glass	n	503-8189
$CuO + ZnO$	n	503-9473
$CuO + ZrO_2$	n	503-9615
$Cu_2O + K_2O + PbO + SiO_2$ Glass	o	503-8668

Property: **a.** Thermal conductivity; **b.** Accommodation coefficient; **c.** Thermal contact resistance; **d.** Thermal diffusivity; **e.** Specific heat; **f.** Viscosity; **g.** Emittance; **h.** Reflectance; **i.** Absorptance; **j.** Transmittance; **k.** α_s/ϵ_t ratio; **l.** Prandtl number; **n.** Thermal linear expansion coefficient; **o.** Thermal volumetric expansion coefficient; **r.** Thermal radiative properties (g,h,i,j,k)

Substance Name	Property	Number
Oxide Mixture: -(Cont)-		
$Dy_2O_3 + Nb_2O_5$ Ceramic	n	503-9550
$Dy_2O_3 + UO_2$ Ceramic	n	503-9549
$Dy_2O_3 + ZrO_2$ Ceramic	n	503-9548
$Er_2O_3 + Fe_2O_3 + MgO$	n	503-9461
$Er_2O_3 + K_2O + Na_2O + SiO_2$ Glass	n	503-8624
$Er_2O_3 + SiO_2$ + Other Glass	j	503-8411
$Er_2O_3 + ThO_2$	r	503-9840
$Er_2O_3 + Yb_2O_3$	i	503-9437
$Er_2O_3 + ZrO_2$	n	503-9537
$EuO + Eu_2O_3$	a	503-9646
$FeO + Fe_2O_3$	n	503-9828
$FeO + Fe_2O_3$ Slag, Iron	e	503-9991
$FeO + Fe_2O_3 + MgO$ Refractory, Spinel	an	503-9680
$FeO + Fe_2O_3 + Na_2O + SiO_2$ Glass	nr	503-9096
$FeO + Fe_2O_3 + SiO_2$ Glass	aer	503-9180
$FeO + Fe_2O_3 + SiO_2$ Slag, Iron	e	503-9990
$FeO + Fe_2O_3 + ZnO$ Refractory, Spinel	a	503-9679
$FeO + K_2O + SiO_2$	an	503-9453
$FeO + MgO$	ajn	503-9624
$FeO + MgO + SiO_2$	n	503-9689
$FeO + MnO$	j	503-9894
$FeO + Na_2O + SiO_2$ Ceramic Adhesive	n	503-9670
$FeO + Na_2O + SiO_2$ Glass	n	503-9126
$FeO + P_2O_5$ Ceramic	n	503-9454
$FeO + SiO_2$	en	503-9919
$FeO + TiO_2$ + Other Titanium – Ilmenite Concentrate	ad	503-8196
$Fe_2O_3 + Fe_3O_4$	er	503-9746
$Fe_2O_3 + Fe_3O_4 + MgO$	a	503-9561
$Fe_2O_3 + Fe_3O_4 + ZnO$	a	503-9560
$Fe_2O_3 + K_2O + SiO_2$ Glass	j	503-8509
$Fe_2O_3 + K_2O + SiO_2 + TiO_2$ Glass	j	503-8508
$Fe_2O_3 + La_2O_3$ Ceramic	a	503-9971
$Fe_2O_3 + MgO$	aenr	503-9728
$Fe_2O_3 + MgO + NiO$	an	503-9517
$Fe_2O_3 + MgO + SiO_2$ Refractory, Forsterite	n	503-9253
$Fe_2O_3 + MnO$	an	503-9516
$Fe_2O_3 + MnO + NiO$	an	503-9515
$Fe_2O_3 + MnO + ZnO$ Ceramic	hn	503-9595
$Fe_2O_3 + MoO_3 + SrO$	j	503-9674
$Fe_2O_3 + Na_2O + SO_3$	ag	503-9512
$Fe_2O_3 + Na_2O + SiO_2$	e	503-9756
$Fe_2O_3 + Na_2O + SiO_2$ Glass	ajn	503-9139
$Fe_2O_3 + Nb_2O_5 + TiO_2$	j	503-9248
$Fe_2O_3 + NiO$	aen	503-9518
$Fe_2O_3 + NiO + ZnO$	aen	503-9666
$Fe_2O_3 + NiO + ZnO$ Ferramic E, Ferromagnetic Spinel	e	503-9878
$Fe_2O_3 + P_2O_5$	n	503-9455
$Fe_2O_3 + P_2O_5$ Glass	e	503-8514
$Fe_2O_3 + P_2O_5 + V_2O_5$ Glass	r	503-9004
$Fe_2O_3 + PbO + SiO_2$ Glass	jr	503-8913
$Fe_2O_3 + PbO + SiO_2 + ZnO$ Lead Concentrate	e	503-8306
$Fe_2O_3 + PbO + Ta_2O_5$	n	503-9604
$Fe_2O_3 + PbO + WO_3$	n	503-9605
$Fe_2O_3 + SiO_2$	a	503-9634
$Fe_2O_3 + SiO_2$ + Kaolin Ceramic	n	503-8249
$Fe_2O_3 + V_2O_3$	n	503-9236
$Fe_2O_3 + Y_2O_3$ Ceramic	h	503-9416
$Fe_2O_3 + ZnO$	ae	503-9514
$Fe_3O_4 + MgO$	r	503-9782
$Ga_2O_3 + K_2O + Na_2O + SiO_2$ Glass	n	503-8623
$Ga_2O_3 + Li_2O + MgO$ Ceramic	n	503-9862
$Ga_2O_3 + Li_2O + SiO_2$ Glass	no	503-8868
$Ga_2O_3 + Li_2O + SiO_2 + TiO_2$ Glass	n	503-8513
$Ga_2O_3 + Na_2O + SiO_2$ Glass	ijn	503-9118
$Ga_2O_3 + NiO$	h	503-9282
$Ga_2O_3 + TiO_2$	n	503-9565
$Ga_2O_3 + TiO_2 + VO_2$	n	503-9564
$Ga_2O_3 + TiO_2 + V_2O_5$	n	503-9474
$Gd_2O_3 + Sm_2O_3$ Refractory	a	503-9890
$Gd_2O_3 + UO_2$	n	503-9448
$GeO_2 + K_2O$	e	503-9435
$GeO_2 + K_2O$ Glass	n	503-8980

Substance Name	Property	Number
Oxide Mixture: -(Cont)-		
$GeO_2 + K_2O + Na_2O$ Glass	n	503-8474
$GeO_2 + K_2O + Na_2O + SiO_2$ Glass	n	503-8619
$GeO_2 + La_2O_3 + PbO$ Glass	j	503-8464
$GeO_2 + MgO + ZnO$ Ceramic	n	503-9570
$GeO_2 + MnO + PbO$ Glass	j	503-8462
$GeO_2 + Na_2O$ Glass	eijn	503-9039
$GeO_2 + Na_2O + PbO + TiO_2$ Glass	j	503-8844
$GeO_2 + Na_2O + PbO + TiO_2 + ZrO_2$ Glass	j	503-8843
$GeO_2 + Na_2O + Rb_2O$ Glass	n	503-8473
$GeO_2 + P_2O_5$	n	503-9336
$GeO_2 + P_2O_5$ Glass	n	503-9012
$GeO_2 + P_2O_5 + PbO$	n	503-9603
$GeO_2 + P_2O_5 + V_2O_5$	n	503-9870
$GeO_2 + P_2O_5 + V_2O_5$ Glass	n	503-9011
$GeO_2 + PbO$	hjn	503-9598
$GeO_2 + PbO$ Glass	hjnr	503-8977
$GeO_2 + PbO + SiO_2$ Glass	n	503-8478
$GeO_2 + PbO + TiO_2$ Glass	jr	503-9016
$GeO_2 + PbO + V_2O_5$	n	503-9601
$GeO_2 + Pb_3O_4$ Glass	j	503-8422
$GeO_2 + RO + RO_2$ IR Transmitting Glass	jn	503-8404
$GeO_2 + Rb_2O$ Glass	n	503-8976
$GeO_2 + SiO_2$	jn	503-9594
$GeO_2 + SiO_2$ Glass	jn	503-8865
$GeO_2 + SnO_2$ Ceramic	n	503-9853
$GeO_2 + Ta_2O_5$	h	503-9283
$GeO_2 + Ta_2O_5 + ZnO$	n	503-9348
$GeO_2 + TiO_2$	n	503-9852
$GeO_2 + V_2O_5$ Glass	n	503-9013
$GeO_2 + ZnO$	j	503-9202
$HfO_2 + K_2O + Na_2O + SiO_2$ Glass	n	503-8620
$HfO_2 + La_2O_3 + SO_4 + ZrO_2$	n	503-9698
$HfO_2 + MgO$	n	503-9591
$HfO_2 + MgO + SO_4 + ZrO_2$	n	503-9687
$HfO_2 + MgO + SiO_2 + ZrO_2$ Ceramic	d	503-9721
$HfO_2 + Nb_2O_5 + PbO$ Ceramic	e	503-9804
$HfO_2 + SO_4 + ZrO_2$	n	503-9703
$HfO_2 + SiO_2$	n	503-9743
$HfO_2 + SrO$	n	503-9519
$HfO_2 + SrO + ZrO_2$	n	503-9520
$HfO_2 + Ta_2O_5$	n	503-9590
$HfO_2 + Ta_2O_5 + ZrO_2$	n	503-9607
$HfO_2 + TiO_2$	en	503-9652
$HfO_2 + TiO_2 + ZrO_2$	n	503-9651
$HfO_2 + Y_2O_3$	an	503-9678
$HfO_2 + Y_2O_3 + ZrO_2$	gnr	503-9714
$HfO_2 + ZrO_2$	aejn	503-9789
$Ho_2O_3 + Na_2O + SiO_2$ Glass	j	503-8502
$In_2O_3 + K_2O + Na_2O + SiO_2$ Glass	n	503-8622
$In_2O_3 + Na_2O + SiO_2 + SrO$ Glass	n	503-8758
$In_2O_3 + Nb_2O_5 + PbO + TiO_2$ Ceramic	n	503-9496
$K_2O + La_2O_3 + Na_2O + SiO_2$ Glass	n	503-8625
$K_2O + La_2O_3 + SiO_2$	hi	503-9331
$K_2O + Li_2O + Na_2O + P_2O_5$ Glass	n	503-8897
$K_2O + Li_2O + Na_2O + PbO + SiO_2$ Enamel	o	503-9317
$K_2O + Li_2O + Na_2O + SiO_2$ Glass	n	503-8634
$K_2O + Li_2O + Na_2O + SiO_2 + SrO + TiO_2$ Enamel	o	503-9318
$K_2O + Li_2O + Na_2O + SiO_2 + TiO_2$ Enamel	o	503-9319
$K_2O + Li_2O + Na_2O + TiO_2$ Glass	n	503-8466
$K_2O + Li_2O + P_2O_5 + SiO_2 + ZnO$ Glass Ceramic	n	503-9439
$K_2O + Li_2O + P_2O_5 + SiO_2 + ZnO$ + Other Glass Ceramic	n	503-8281
$K_2O + Li_2O + SiO_2$ Glass	hno	503-8859
$K_2O + Li_2O + SiO_2 + ZnO$ Glass	aen	503-8755
$K_2O + MgO + Na_2O + PbO + SiO_2$ Glass	j	503-8785
$K_2O + MgO + Na_2O + SiO_2$ Glass	n	503-9122
$K_2O + MgO + SiO_2$ Glass	jn	503-9121
$K_2O + MnO + PbO + SiO_2$ Glass	o	503-8669
$K_2O + MoO_3 + P_2O_5$ Glass	n	503-8885
$K_2O + Na_2O + Nb_2O_5$	h	503-9470
$K_2O + Na_2O + Nb_2O_5 + SiO_2$ Glass	n	503-8617

Property: a. Thermal conductivity; **b.** Accommodation coefficient; **c.** Thermal contact resistance; **d.** Thermal diffusivity; **e.** Specific heat; **f.** Viscosity; **g.** Emittance; **h.** Reflectance; **i.** Absorptance; **j.** Transmittance; **k.** a_s/ϵ_t ratio; **l.** Prandtl number; **n.** Thermal linear expansion coefficient; **o.** Thermal volumetric expansion coefficient; **r.** Thermal radiative properties (g,h,i,j,k)

Substance Name	Property	Number
Oxide Mixture: -(Cont)-		
$K_2O + Na_2O + Nd_2O_3 + SiO_2$ Glass	n	503-8525
$K_2O + Na_2O + PbO + Sb_2O_3 + SiO_2$ Glass	a	503-8710
$K_2O + Na_2O + PbO + SiO_2$ Glass	hjnor	503-9101
$K_2O + Na_2O + PbO + SiO_2 + SrO$ Glass	n	503-8558
$K_2O + Na_2O + PbO + SiO_2 + TiO_2$ Glass	j	503-8840
$K_2O + Na_2O + PbO + SiO_2 + ZnO$ Glass	jn	503-8841
$K_2O + Na_2O + Rb_2O + SiO_2$ Glass	n	503-8633
$K_2O + Na_2O + Sb_2O_3 + SiO_2$ Glass	n	503-8616
$K_2O + Na_2O + Sc_2O_3 + SiO_2$ Glass	n	503-8627
$K_2O + Na_2O + SiO_2$ Glass	ehjno	503-9167
$K_2O + Na_2O + SiO_2 + R_2O$ Ceramic	n	503-8277
$K_2O + Na_2O + SiO_2 + SnO_2$ Glass	n	503-8618
$K_2O + Na_2O + SiO_2 + SrO$ Glass	n	503-8630
$K_2O + Na_2O + SiO_2 + Ta_2O_5$ Glass	jn	503-8787
$K_2O + Na_2O + SiO_2 + TiO_2$ Glass	n	503-8751
$K_2O + Na_2O + SiO_2 + Tl_2O_3$ Glass	n	503-8621
$K_2O + Na_2O + SiO_2 + Y_2O_3$ Glass	n	503-8626
$K_2O + Na_2O + SiO_2 + ZnO$ Glass	no	503-8629
$K_2O + P_2O_5$	a	503-9306
$K_2O + P_2O_5 + ZnO$ Glass	j	503-8544
$K_2O + PbO + Sb_2O_3 + SiO_2$ Glass	o	503-8667
$K_2O + PbO + SiO_2$ Glass	aehjnor	503-9095
$K_2O + Rb_2O + SiO_2$ Glass	o	503-8729
$K_2O + SiO_2$	ehj	503-9940
$K_2O + SiO_2$ Glass	aehjnor	503-9179
$K_2O + SiO_2$ + Fireclay Ceramic	n	503-8291
$K_2O + SiO_2 + SrO$ Glass	r	503-9014
$K_2O + SiO_2 + V_2O_5$ Brown Glass	r	503-8917
$K_2O + SiO_2 + ZnO$	hi	503-9330
$K_2O + SiO_2 + ZnO$ Glass	aeno	503-8518
$K_2O + TeO_2$ Glass	n	503-8568
$K_2O + TiO_2$	n	503-9200
$LaO + PbO + SnO + TiO_2 + ZrO_2$ Ceramic	e	503-9676
$LaO + PbO + TiO_2 + ZrO_2$ Ceramic	j	503-9299
$La_2O_3 + Na_2O + SiO_2$ Glass	jn	503-8864
$La_2O_3 + Na_2O + SiO_2 + SrO + TiO_2$ Glass	j	503-8444
$La_2O_3 + Na_2O + SiO_2 + TiO_2$ Glass	j	503-8443
$La_2O_3 + Na_2O + SiO_2 + ZnO + ZrO_2$ Glass	j	503-8449
$La_2O_3 + Na_2O + SiO_2 + ZrO_2$ Glass	n	503-8866
$La_2O_3 + NiO$	ah	503-9969
$La_2O_3 + Pr_2O_3$	h	503-9280
$La_2O_3 + SiO_2$	n	503-9509
$La_2O_3 + ThO_2$	r	503-9844
$La_2O_3 + TiO_2$ Ceramic	a	503-9967
$La_2O_3 + UO_2$	a	503-9727
$La_2O_3 + Y_2O_3$	r	503-9825
$Li_2O + MgO$	n	503-9452
$Li_2O + MgO + Na_2O$	a	503-9394
$Li_2O + MgO + SiO_2$ Glass	en	503-8753
$Li_2O + Na_2O + Nb_2O_5$	n	503-9300
$Li_2O + Na_2O + SiO_2$ Glass	ho	503-8733
$Li_2O + Na_2O + SiO_2 + TiO_2$ Glass	n	503-8591
$Li_2O + P_2O_5$	a	503-9307
$Li_2O + P_2O_5$ Glass	j	503-8529
$Li_2O + PbO + SiO_2$ Glass	jr	503-8905
$Li_2O + PbO + TeO_2$ Glass	n	503-8577
$Li_2O + Sb_2O_3$ Glass	r	503-9010
$Li_2O + SiO_2$	e	503-9941
$Li_2O + SiO_2$ Glass	ehjnor	503-9154
$Li_2O + SiO_2 + SnO_2$ Glass	n	503-8665
$Li_2O + SiO_2 + SrO + TiO_2 + ZrO_2$ Ceramic	a	503-9905
$Li_2O + SiO_2 + SrO + ZnO + ZrO_2$ Ceramic	a	503-9904
$Li_2O + SiO_2 + SrO + ZrO_2$ Ceramic	a	503-9968
$Li_2O + SiO_2 + Ta_2O_5$ Glass	j	503-8789
$Li_2O + SiO_2 + V_2O_5$ Brown Glass	r	503-8916
$Li_2O + SiO_2 + ZnO$ Glass	n	503-8484
$Li_2O + SiO_2 + ZnO$ Glass Ceramic	n	503-9203
$Li_2O + SiO_2 + ZrO_2$ Glass	n	503-8480
$Li_2O + TeO_2$ Glass	n	503-8570
$Li_2O + TiO_2$	n	503-9327
$Li_2O + ZnO$	n	503-9472
MgO + Ball Clay Refractory	an	503-8397
MgO + Fireclay + Grog Refractory	n	503-8298

Substance Name	Property	Number
Oxide Mixture: -(Cont)-		
MgO + Grog + Kaolin Refractory	n	503-8302
$MgO + MnO$	nr	503-9739
$MgO + MnO + Mn_2O_3$ Ceramic	a	503-9906
$MgO + Na_2O + SiO_2$ Glass	ajn	503-9156
$MgO + Na_2O + SiO_2 + ZrO_2$ Glass	n	503-8579
$MgO + Nb_2O_5$	n	503-9655
$MgO + Nb_2O_5 + P_2O_5$	n	503-9344
$MgO + Nd_2O_3$ Ceramic	a	503-9975
$MgO + NiO$	anr	503-9892
$MgO + NiO + ZrO_2$ Refractory, Zirconium Oxide	r	503-9724
$MgO + P_2O_5$ Ceramic	aen	503-9458
$MgO + P_2O_5$ Glass	n	503-8823
$MgO + PbO$	h	503-9530
$MgO + PbO + TiO_2 + W_2O_5 + ZrO_2$ Ceramic	n	503-8340
$MgO + PbO + WO_3$	n	503-8321
$MgO + SiO_2$	adehnr	503-9830
$MgO + SiO_2$ Ceramic	e	503-8190
$MgO + SiO_2 + R_2O$ Ceramic	n	503-8275
$MgO + SiO_2 + ZnO$	an	503-9681
$MgO + SiO_2 + ZrO_2$	hi	503-9702
$MgO + SiO_2 + ZrO_2$ + Clay Ceramic	n	503-8339
$MgO + SnO_2$ Ceramic	a	503-9945
$MgO + SnO_2 + ZnO$ Ceramic	a	503-9937
$MgO + TiO_2$	an	503-9920
$MgO + TiO_2 + ZrO_2$	n	503-9447
$MgO + V_2O_3$	n	503-9237
$MgO + V_2O_5$	h	503-9758
$MgO + Y_2O_3$	n	503-9567
$MgO + ZnO$	aghjnr	503-9882
$MgO + ZrO_2$	adnr	503-9974
$MnO + Mn_2O_3$ Ceramic	a	503-9909
$MnO + Mn_2O_3 + SiO_2$ Ceramic	a	503-9907
$MnO + Na_2O + SiO_2$ Glass	n	503-9127
$MnO + Nb_2O_5 + P_2O_5$	n	503-9343
$MnO + P_2O_5$ Glass	n	503-8610
$MnO + P_2O_5 + TiO_2$	n	503-9197
$MnO + SiO_2$	n	503-9914
$MnO + TiO$	a	503-9212
$MnO + TiO_2$	h	503-9705
$MnO + ZnO$	h	503-9643
$MnO_2 + Mn_2O_3$	a	503-9751
$MnO_2 + PbO + SiO_2$ Reduced Glass	r	503-8912
$MnO_2 + ThO_2$	r	503-9856
$MnO_2 + TiO_2$	no	503-9545
$Mn_2O_3 + SiO_2$ Ceramic	a	503-9976
$Mn_3O_4 + Na_2O + SiO_2$ Purple Glass	r	503-8968
$MoO_3 + PbO$	n	503-9599
$MoO_3 + SiO_2$	h	503-9708
$Na_2O + Nb_2O_5$ Ceramic	n	503-9691
$Na_2O + Nb_2O_5 + PbO + TiO_2$ Ceramic	n	503-9723
$Na_2O + Nb_2O_5 + SiO_2$ Glass	h	503-8465
$Na_2O + NiO + PbO + SiO_2 + TiO_2$ Glass	a	503-9045
$Na_2O + NiO + SiO_2$ Brown Glass	nr	503-8971
$Na_2O + P_2O_5$	aen	503-9402
$Na_2O + P_2O_5$ Glass	n	503-9145
$Na_2O + P_2O_5 + V_2O_5$ Green Glass	r	503-8918
$Na_2O + P_2O_5 + V_2O_5 + ZnO$ Colored Glass	r	503-8919
$Na_2O + PbO + Sb_2O_3 + SiO_2 + ZnO$ Glass	n	503-8562
$Na_2O + PbO + SiO_2$	en	503-9757
$Na_2O + PbO + SiO_2$ Glass	aenor	503-9163
$Na_2O + PbO + SiO_2 + SrO$ Glass	jn	503-8788
$Na_2O + PbO + SiO_2 + TiO_2$ Glass	a	503-9044
$Na_2O + PbO + SiO_2 + ZnO$ Glass	an	503-8836
$Na_2O + PbO + TeO_2$ Glass	n	503-8576
$Na_2O + PbO + TiO_2$	no	503-9232
$Na_2O + Rb_2O + SiO_2$ Glass	o	503-8731
$Na_2O + SO_3$	ag	503-9513
$Na_2O + SiO_2$	ehjn	503-9938
$Na_2O + SiO_2$ Glass	aehijnor	503-9174
$Na_2O + SiO_2 + RO + R_2O_3$ Glass	r	503-8417
$Na_2O + SiO_2 + SnO_2$ Glass	jn	503-8666
$Na_2O + SiO_2 + SrO$ Glass	jn	503-9140
$Na_2O + SiO_2 + TiO_2$ Glass	jnr	503-9120

Property: a. Thermal conductivity; **b.** Accommodation coefficient; **c.** Thermal contact resistance; **d.** Thermal diffusivity; **e.** Specific heat; **f.** Viscosity; **g.** Emittance; **h.** Reflectance; **i.** Absorptance; **j.** Transmittance; **k.** a_s/ϵ_t ratio; **l.** Prandtl number; **n.** Thermal linear expansion coefficient; **o.** Thermal volumetric expansion coefficient; **r.** Thermal radiative properties (g,h,i,j,k)

Substance Name	Property	Number
Oxide Mixture: -(Cont)-		
$Na_2O + SiO_2 + UO_3$ Glass	i	503-8759
$Na_2O + SiO_2 + V_2O_5$ Colored Glass	r	503-8407
$Na_2O + SiO_2 + V_2O_5$ Glass	jr	503-9117
$Na_2O + SiO_2 + WO_3$ Refractory	n	503-9395
$Na_2O + SiO_2 + ZnO$ Glass	aejno	503-9131
$Na_2O + SiO_2 + ZrO_2$ Glass	n	503-9119
$Na_2O + TeO_2$ Glass	jn	503-8569
$Nb_2O_5 + P_2O_5$	n	503-9341
$Nb_2O_5 + P_2O_5 + Ta_2O_5 + ZnO$	n	503-9345
$Nb_2O_5 + P_2O_5 + ZnO$	n	503-9346
$Nb_2O_5 + PbO + TiO_2 + ZrO_2$	an	503-9574
$Nb_2O_5 + Ta_2O_5$	n	503-9347
$Nb_2O_5 + Ta_2O_5 + V_2O_5$	n	503-9338
$Nb_2O_5 + TiO_2$	no	503-9654
$Nb_2O_5 + UO_2$	a	503-9814
$Nb_2O_5 + V_2O_5$	n	503-9406
$Nb_2O_5 + ZnC$	n	503-9350
$Nb_2O_5 + ZrO_2$	n	503-9656
$Nd_2O_3 + SiO_2$ + Other Glass	j	503-8410
$Nd_2O_3 + ThO_2$	r	503-9841
$Nd_2O_3 + ZrO_2$	ajn	503-9481
$NiO + PbO + SiO_2$ Glass	r	503-8908
$NiO + Sb_2O_3 + TiO_2$	h	503-9632
$NiO + ThO_2$	r	503-9858
$NiO + TiO$	a	503-9214
$NiO + TiO_2$	hn	503-9631
$NiO + ZnO$ Ceramic	n	503-9861
P_2O_5 + Clay + Perlite Ceramic	a	503-8201
$P_2O_5 + PbO$ Glass	jn	503-9022
$P_2O_5 + PbO + V_2O_5$ Glass	r	503-9003
$P_2O_5 + Sb_2O_3 + V_2O_5$ Glass	r	503-9007
$P_2O_5 + SiO_2$	n	503-9337
$P_2O_5 + SiO_2$ + Kaolin Ceramic	n	503-8250
$P_2O_5 + SiO_2 + TiO_2$ Glass	j	503-8716
$P_2O_5 + SrO$ Glass	n	503-8607
$P_2O_5 + SrO_2 + V_2O_5$	n	503-9877
$P_2O_5 + Ta_2O_5$	n	503-9340
$P_2O_5 + ThO_2$ Ceramic	n	503-9301
$P_2O_5 + ThO_2 + ZrO_2$ Ceramic	n	503-9719
$P_2O_5 + TiO_2$	n	503-9459
$P_2O_5 + TiO_2 + WO_3$ Glass	j	503-8719
$P_2O_5 + Tl_2O$ Glass	n	503-8851
$P_2O_5 + UO_2$ Ceramic	n	503-9653
$P_2O_5 + V_2O_5$	n	503-9342
$P_2O_5 + V_2O_5$ Glass	jn	503-9136
$P_2O_5 + V_2O_5 + ZnO$	n	503-9351
$P_2O_5 + V_2O_5 + ZnO$ Colored Glass	r	503-9008
$P_2O_5 + WO_3 + ZrO_2$	n	503-9479
$P_2O_5 + ZnO$	n	503-9353
$P_2O_5 + ZnO$ Glass	j	503-8528
$P_2O_5 + ZrO_2$ Refractory	n	503-9535
$PbO + PbO_2$	de	503-9675
$PbO + Rb_2O + SiO_2$ Glass	nr	503-8904
$PbO + SiO_2$	ehjn	503-9893
$PbO + SiO_2$ Glass	aehjnor	503-9152
$PbO + SiO_2 + RO$ I R Transmitting Glass	jn	503-8405
$PbO + SiO_2 + TiO_2$ Glass	jr	503-8906
$PbO + SiO_2 + TiO_2 + ZrO_2$ Glass	n	503-8640
$PbO + SiO_2 + V_2O_5$	n	503-9602
$PbO + SiO_2 + V_2O_5$ Glass	r	503-8909
$PbO + SiO_2 + ZnO$ Reduced Glass	r	503-8915
$PbO + SiO_2 + ZrO_2$ Glass	n	503-8641
$PbO + SnO + TiO_2 + ZrO_2$ Ceramic	e	503-9677
$PbO + SnO_2$ Glass	j	503-8470
$PbO + SnO_2 + TiO_2$ Ceramic	e	503-9897
$PbO + SnO_2 + TiO_2 + ZrO_2$ Ceramic	e	503-9803
$PbO + SrO$	ij	503-9469
$PbO + SrO + TiO_2$ Ceramic	e	503-9947
$PbO + TeO_2 + WO_3$ Glass	n	503-8573
$PbO + TiO_2$	ahn	503-9822
$PbO + TiO_2 + ZrO_2$ Ceramic	aden	503-9709
$PbO + WO_3$	n	503-9600
$PbO + ZrO_2$ Ceramic	a	503-9769
Oxide Mixture: -(Cont)-		
$PbO_2 + SrO$ Ceramic	ijn	503-9468
$Pb_3O_4 + SiO_2$	j	503-9295
$Pm_2O_3 + Sm_2O_3$	adn	503-9534
$Pr_2O_3 + TiO_2$	h	503-9279
$Pr_4O_7 + ThO_2$	r	503-9847
$PuO_2 + ThO_2$	aden	503-9843
$PuO_2 + UO_2$	aden	503-9819
$PuO_2 + UO_2 + ZrO_2$	n	503-9663
$Rb_2O + SiO_2$ Glass	nor	503-9043
$Rb_2O + TeO_2$ Glass	n	503-8567
$Rb_2O_5 + ZrO_2$	r	503-9729
$Sb_2O_3 + TiO_2$	hno	503-9630
$Sb_2O_3 + V_2O_5$	i	503-9257
$Sc_2O_3 + TiO_2$	n	503-9476
$Sc_2O_3 + TiO_2 + VO_2$	n	503-9566
$Sc_2O_3 + TiO_2 + V_2O_5$	n	503-9475
$Sc_2O_3 + Ti_2O_3$	j	503-9265
$Sc_2O_3 + V_2O_3$	j	503-9266
SiO_2 + Ball Clay + Dolomite + Feldspar + Kaolin Refractory	n	503-8246
SiO_2 + Ball Clay + Feldspar Ceramic	n	503-8292
SiO_2 + Ball Clay + Feldspar + Kaolin Ceramic	n	503-8272
SiO_2 + Ball Clay + Feldspar + Kaolin + Pyrophyllite Refractory	n	503-8247
SiO_2 + Ball Clay + Feldspar + Kaolin + Steatite Ceramic	n	503-8257
SiO_2 + Ball Clay + Feldspar + Kaolin + Sylenite Ceramic	n	503-8329
SiO_2 + Ball Clay + Kaolin + Nephelite Syenite Sanitary Porcelain	n	503-8284
SiO_2 + Ball Clay + Kaolin + Nephelite Syenite Tile	n	503-8288
SiO_2 + Ball Clay + Kaolin + Pyrophyllite + Steatite Ceramic	n	503-8294
SiO_2 + Ball Clay + Kaolin + Steatite Ceramic	n	503-8278
SiO_2 + Bentonite	an	503-8318
SiO_2 + Bentonite + Grog + Kaolin Tile	an	503-8214
SiO_2 + Clay Ceramic	n	503-8366
SiO_2 + Clay + Feldspar Refractory	n	503-8317
SiO_2 + Clay + Feldspar + Frit Enamel	no	503-8269
SiO_2 + Clay + Feldspar + Kaolin + Nephelite Syenite Sanitary Porcelain	n	503-8279
SiO_2 + Clay + Frit Enamel	no	503-8270
SiO_2 + Clay + Grog + Kaolin Tile	an	503-8215
SiO_2 + Dolomite + Feldspar Glaze	n	503-8212
SiO_2 + Dolomite + Feldspar + Kaolin Glaze	n	503-8213
SiO_2 + Dolomite + Glass + Perlite Glaze	n	503-8210
SiO_2 + Feldspar	n	503-8332
SiO_2 + Feldspar + Kaolin Refractory	an	503-8195
SiO_2 + Feldspar + Kaolin + Spodumene Ceramic	jn	503-8236
SiO_2 + Feldspar + Kaolin + Steatite Ceramic	n	503-8234
SiO_2 + Glass	h	503-8202
SiO_2 + Glass + Mullite Glaze	n	503-8271
SiO_2 + Kaolin	n	503-8307
SiO_2 + Kaolin + Nephelite Syenite Porcelain	n	503-8358
SiO_2 + Magnesite Refractory	e	503-8354
$SiO_2 + R_2O$ Ceramic	n	503-8276
$SiO_2 + SrO$	eh	503-9939
$SiO_2 + SrO$ Glass	j	503-8846
$SiO_2 + ThO_2$	n	503-9735
$SiO_2 + TiO_2$	ano	503-9915
$SiO_2 + TiO_2$ Glass	n	503-8988
$SiO_2 + TiO_2 + ZrO_2$ Glass	n	503-8536
$SiO_2 + Tl_2O$ Glass	h	503-8581
$SiO_2 + UO_2$	adn	503-9644
$SiO_2 + VO_2$ Glass	r	503-8933
$SiO_2 + V_2O_5$ Glass	r	503-9000
$SiO_2 + Yb_2O_3$ + Other Glass	j	503-8413
$SiO_2 + ZnO$	n	503-9365
$SiO_2 + ZnO$ + Chalk + Feldspar Glaze	j	503-8219
$SiO_2 + ZnO$ + Clay + Feldspar Glaze	j	503-8222
$SiO_2 + ZnO$ + Dolomite + Feldspar Glaze	j	503-8221
$SiO_2 + ZnO$ + Feldspar + Marble Glaze	j	503-8220
$SiO_2 + ZnO + ZrO_2$	i	503-9899

Property: a. Thermal conductivity; **b.** Accommodation coefficient; **c.** Thermal contact resistance; **d.** Thermal diffusivity; **e.** Specific heat; **f.** Viscosity; **g.** Emittance; **h.** Reflectance; **i.** Absorptance; **j.** Transmittance; **k.** α_s/ϵ_t ratio; **l.** Prandtl number; **n.** Thermal linear expansion coefficient; **o.** Thermal volumetric expansion coefficient; **r.** Thermal radiative properties (g,h,i,j,k)

Substance Name	Property	Number
Oxide Mixture: -(Cont)-		
$SiO_2 + ZrO_2$	aeghnor	503-9791
$SiO_2 + ZrO_2$ + Clay Refractory	n	503-8333
$SiO_2 + ZrO_2$ + Clay + Feldspar Ceramic	n	503-8335
$SiO_2 + ZrO_2$ + Glass Ceramic	n	503-8267
$Sm_2O_3 + ZrO_2$	n	503-9293
$SnO_2 + TiO_2$	no	503-9544
$SnO_2 + V_2O_5$	no	503-9542
$SnO_2 + ZnO$ Ceramic	an	503-9732
$SrO + TiO_2$ Ceramic	agjn	503-9823
$SrO + V_2O_5$	n	503-9405
$SrO + ZrO_2$	egn	503-9589
$SrO + ZrO_2$ Refractory	r	503-8356
$Ta_2O_5 + TeO_2$ Glass	r	503-9017
$Ta_2O_5 + V_2O_5$	n	503-9339
$Ta_2O_5 + ZnO$	n	503-9349
$Ta_2O_5 + ZrO_2$	gn	503-9752
$TeO_2 + V_2O_5$ Glass	r	503-8978
$TeO_2 + ZnO$ Glass	j	503-8883
$ThO_2 + Tm_2O_3$	n	503-9668
$ThO_2 + UO_2$	adenr	503-9718
$ThO_2 + UO_2$ Refractory	a	503-8393
$ThO_2 + Y_2O_3$	j	503-9238
$ThO_2 + ZrO_2$	jn	503-9606
TiO_2 + Chalk Ceramic	an	503-8223
$TiO_2 + UO_2$	n	503-9329
$TiO_2 + VO_2$	n	503-9471
$TiO_2 + V_2O_5$	no	503-9547
$TiO_2 + WO_3$	no	503-9546
$TiO_2 + WO_3 + ZrO_2$	n	503-9478
$TiO_2 + ZnO$	hn	503-9845
$TiO_2 + ZrO_2$	aghno	503-9707
$TiO_2 + ZrO_2$ + Bentonite Ceramic, Tikond	an	503-8224
$Ti_2O_3 + V_2O_3$	e	503-9298
$Tm_2O_3 + Yb_2O_3$	a	503-9415
$UO + Y_2O_3$ Refractory	a	503-8383
$UO_2 + Y_2O_3$	a	503-9857
$UO_2 + ZrO_2$	aden	503-9813
$V_2O_3 + ZrO_2$	n	503-9235
$V_2O_5 + ZnO$	n	503-9352
$WO_3 + ZnO$ Ceramic	a	503-9931
$Y_2O_3 + ZrO_2$ Refractory, Zirconium Oxide	adgjnr	503-9716
ZnO + Fireclay + Grog Refractory	n	503-8297
ZnO + Grog + Kaolin Refractory	n	503-8301
$ZnO + ZrO_2$	n	503-9787
ZrO_2 + Glass Ceramic	n	503-8266
Oxide Reactor Fuel	a	503-1317
O Y 13; Chance Glass	j	503-1512
O Y 21; Chance Glass	j	503-1513
Parra - Mantois Colored Filter Glass	j	503-1970
P B-19195, Libbey Owens Ford Glass	ghjr	503-1092
see also: Electroconducting Glass		503-2793
Periclase - Spinellide Brick	ae	503-1335
Periclase Spinel Refractory	en	503-1201
see also: Aluminum Oxide + Calcium Oxide + Chromium Oxide + Iron Oxide + Magnesium Oxide + Phosphorus + Silicon Oxide, Periclase Spinel Refractory $Al_2O_3 + CaO + Cr_2O_3 + Fe_2O_3 + MgO + P + SiO_2$		513-3580
Perlite Refractory	ad	503-1178
P E S 5 Glass, Russian	no	503-2287
Petalite Ceramic	an	503-1128
see also: Silicon Oxide Mixture Glass $SiO_2 > Al_2O_3 > Li_2O$		503-0481
Phoenix Glass	a	503-1049
see also: Corning 7740 Pyrex Glass $SiO_2 > B_2O_3 > Na_2O > Al_2O_3$		503-0216
Phosphate Crown Glass	a	503-0819
Phosphate Glass	ejnr	503-0096
see also: Nuclear Waste Phosphate Glass		503-2352
Phosphate Turbid Glass	r	503-2457
Phosphorus Oxide Mixture:		———
$P_2O_3 > BaO > La_2O_3$ Glass	j	503-2572
Phosphorus Oxide Mixture: -(Cont)-		
$P_2O_5 > Al_2O_3 > GeO_2 > SiO_2$	n	503-2788
$P_2O_5 > B_2O_3 > Al_2O_3 > BaO >$ Other Glass	j	503-2585
$P_2O_5 > B_2O_3 > Al_2O_3 > K_2O$ Glass	j	503-2583
$P_2O_5 > B_2O_3 > Al_2O_3 > K_2O >$ Other Glass	j	503-2584
$P_2O_5 > B_2O_3 > Al_2O_3 > Li_2O >$ Other Glass	j	503-2582
$P_2O_5 > B_2O_3 > Al_2O_3 > Na_2O$ Glass	j	503-2580
$P_2O_5 > B_2O_3 > Al_2O_3 > Na_2O >$ Other Glass	j	503-2581
$P_2O_5 > B_2O_3 > Al_2O_3 > ZrO_2 >$ Other Glass	j	503-2389
$P_2O_5 > BaO > UO_3 > Na_2O$ Schott V G 3 Glass	j	503-2051
$P_2O_5 > CaO$ Glass	j	503-0735
$P_2O_5 > CuO > CaO$ Glass	j	503-1775
$P_2O_5 > GeO_2$ Glass	n	503-1883
$P_2O_5 > GeO_2 > SiO_2 > Al_2O_3$	n	503-2787
$P_2O_5 > K_2O > Al_2O_3$ Glass	j	503-1831
$P_2O_5 > Li_2O$ Glass	n	503-0808
$P_2O_5 > Na_2O$ Glass	in	503-0809
$P_2O_5 > Na_2O > Al_2O_3$ Glass	j	503-2344
$P_2O_5 > Na_2O > K_2O > Al_2O_3 >$ Other Glass	j	503-1832
$P_2O_5 > Na_2O > PbO$ Glass	j	503-2129
$P_2O_5 > PbO > Na_2O$ Glass	j	503-2130
$P_2O_5 > Sb_2O_3$ Glass	n	503-1880
$P_2O_5 > SiO_2$ Glass	n	503-1882
Pilkington D E D F (See also Chance - Pilkington) $PbO > SiO_2$	j	503-2116
see also: Flint, Double Extra Dense Glass		503-2120
Pittsburgh 2043 Glass	j	503-1400
Pittsburgh 3235 Glass $SiO_2 > B_2O_3 > Na_2O > Al_2O_3$	aeghijr	503-1547
see also: Borosilicate Glass		503-0782
Pittsburgh 6695 Glass	n	503-1391
Pittsburgh 6788 Glass	r	503-1548
Pittsburgh Glass	ar	503-1260
Pittsburgh Plate Colored Filter Glass	j	503-1971
Pittsburgh Solex 2808 Glass $SiO_2 > Na_2O > CaO > MgO$	adeno	503-1062
Pittsburgh Solex Glass	j	503-1425
Pittsburgh Solex S Glass $SiO_2 > Na_2O > CaO > MgO >$ Other	adeno	503-1072
Plutonium Oxide Mixture:		———
$PuO_2 > UO_2$	a	503-2163
Polaroid Glass	j	503-1406
Porcelain	adeghnr	503-0801
see also: Silicon Oxide Mixture Porcelain $SiO_2 > Al_2O_3 > K_2O$		503-0866
Porcelain:		———
Alumina	aden	503-1219
see also: Aluminum Oxide Ceramic		503-0026
see also: Aluminum Oxide Refractory		503-1173
Aluminum Oxide	aden	503-1219
see also: Aluminum Oxide Ceramic		503-0026
see also: Aluminum Oxide Refractory		503-1173
Berlin	aen	503-1228
Beryl	an	503-1292
Chemical Technical	n	503-1314
Chinaware	an	503-1812
Electrical	aen	503-1467
Glazed	anr	503-1277
Hard	a	503-1134
Hotel Whiteware	n	503-2276
Lithia	n	503-0541
Meissner	n	503-1229
Oven Whiteware	n	503-2275
Rashing Rings	a	503-1276
Sanitary Ware	n	503-1306
Semiporcelain	n	503-1347
Sericite	a	503-1243

Property: a. Thermal conductivity; **b.** Accommodation coefficient; **c.** Thermal contact resistance; **d.** Thermal diffusivity; **e.** Specific heat; **f.** Viscosity; **g.** Emittance; **h.** Reflectance; **i.** Absorptance; **j.** Transmittance; **k.** α_s/ϵ_t ratio; **l.** Prandtl number; **n.** Thermal linear expansion coefficient; **o.** Thermal volumetric expansion coefficient; **r.** Thermal radiative properties (g,h,i,j,k)

Property: **a.** Thermal conductivity; **b.** Accommodation coefficient; **c.** Thermal contact resistance; **d.** Thermal diffusivity; **e.** Specific heat; **f.** Viscosity; **g.** Emittance; **h.** Reflectance; **i.** Absorptance; **j.** Transmittance; **k.** α_s/ϵ_t ratio; **l.** Prandtl number; **n.** Thermal linear expansion coefficient; **o.** Thermal volumetric expansion coefficient; **r.** Thermal radiative properties (g,h,i,j,k)

Substance Name	Property	Number
Rubidium Oxide Mixture: -(Cont)-		
$Rb_2O > SiO_2$ Glass	n	503-2855
$Rb_2O > SiO_2 > B_2O_3$ Glass	n	503-2852
$Rb_2O > SiO_2 > CaO$ Glass	n	503-2854
$Rb_2O > SiO_2 > Na_2O$ Glass	n	503-1960
S; Solex Glass $SiO_2 > Na_2O > CaO > MgO >$ Other	adeno	503-1072
S 48-1 Glass, Russian	n	503-1575
S 48-3 Glass, Russian	n	503-1576
S 54-1 Glass, Russian	n	503-1577
S 87-1 Glass, Russian	n	503-1578
S 88-1 Glass, Russian	n	503-1579
S 88-1v Glass, Russian	n	503-1580
see also: Silicon Oxide Mixture Glass $SiO_2 > BaO > K_2O > Al_2O_3$		503-1835
S 89-1 Glass, Russian	n	503-1581
S 89-2 Glass, Russian	n	503-2253
S 90-1 Glass, Russian	n	503-1582
S 8688, Bausch And Lomb Optical Crown Glass	jn	503-2087
Sagger Mix Refractory	aen	503-1242
Saint Gobain Glass	ao	503-1100
Sand-Lime Brick	air	503-1151
Sandstone Refractory	jn	503-1324
Sanitary Ware Porcelain	n	503-1306
Sapphirin Glass	r	503-1120
Sauereisen Ceramic Cement	agnr	503-1529
Scale:		——
Nuclear Reactor	a	503-1361
Steel	adeg	503-1378
Schist Ceramic	n	503-1325
Schott, Colored Filter Glass	j	503-1968
Schott, Flint Glass	an	503-1456
see also: Flint Glass		503-0270
Schott, Glass	aejnr	503-0239
Schott, Optical Glass	j	503-0984
Schott 016 I I I Glass	nor	503-0985
Schott 059 I I I Glass	n	503-0986
see also: Glass Thermometer		503-1111
Schott 102 I I I Glass	no	503-0987
Schott 121 I I I Glass	no	503-0988
Schott 431 I I I Glass	j	503-2133
Schott 444 I I I Raughglass	j	503-2135
Schott 1565 I I I Glass	n	503-0989
Schott 1801 E I I I Glass	n	503-0990
Schott 2745 Glass	hjr	503-0991
Schott 2954 I I I Glass	n	503-0992
Schott 3815 Glass	h	503-0993
Schott 4512 Glass	jr	503-0994
Schott 8563 Glass	n	503-0995
Schott Ba K 1 G 12 Glass	j	503-0996
Schott B G 1 Glass $SiO_2 > K_2O > CaO > B_2O_3$	j	503-2044
Schott B G 2 Glass $SiO_2 > B_2O_3 > K_2O > PbO$	j	503-2045
Schott B G 9 Glass $SiO_2 > Na_2O > CaO > K_2O$	j	503-2046
Schott B G 12 Glass $SiO_2 > K_2O > ZnO > Na_2O$	j	503-2047
Schott B G 16 Glass $SiO_2 > Na_2O > CaO > K_2O >$ Other	j	503-2048
Schott B G 20 Glass $SiO_2 >$ Lanthan-Didymoxyd $> Na_2O >$ Other	j	503-2049
Schott B G 24 Glass	hjr	503-1567
Schott B G 26 Glass	j	503-1588
Schott B G 28 Glass $SiO_2 > K_2O > ZnO > Na_2O$	j	503-2050
Schott B K 7, Glass	adhj	503-0080
Schott B K 7 G 25 Glass	adejn	503-0997
Schott B K 7 G Glass; Cerium Doped	j	503-2875
Schott B K 8 Glass $SiO_2 > B_2O_3 > K_2O > Na_2O >$ Other	jn	503-1570
Schott Didymium Glass	j	503-2064
Schott F 6 G 40 Glass	j	503-0999
Schott F 3086 Glass	j	503-2132
Schott F G 9 Glass	j	503-1000
Schott F Glass	j	503-0998
Schott F K 1 Glass	j	503-1001
Schott F S K V 4 Glass	n	503-1426
Schott G G 4 Glass	j	503-1508

Substance Name	Property	Number
Schott G G 5 Glass $SiO_2 > K_2O > CaO > B_2O_3$	j	503-2054
Schott G G 7 Glass	j	503-1509
Schott G G 11 Glass $SiO_2 > K_2O > CaO > B_2O_3$	j	503-2055
Schott G G 13 Glass $SiO_2 > PbO > K_2O > Na_2O >$ Other	j	503-2056
Schott G G 14 Glass $SiO_2 > K_2O > ZnO > B_2O_3$	j	503-2057
Schott G G 16 Glass	j	503-2058
Schott G G 17 Glass	j	503-1510
Schott G G 375 G 34 Glass	j	503-1511
Schott G G 495 Glass	j	503-1821
Schott I R G 7 Glass	j	503-1002
Schott K 5 Glass	j	503-1003
Schott K F 7 Glass	n	503-0783
Schott K G 1 Glass	j	503-1473
Schott K G 2 Glass	j	503-1474
Schott K G 3 Glass	j	503-1475
Schott K G 4 Glass	j	503-1476
Schott Kz F S 4 Glass	j	503-1004
Schott Neophan Glass	j	503-2063
Schott N G 1 Glass $SiO_2 > B_2O_3 > K_2O > Fe_2O_3 >$ Other	j	503-2061
Schott N G 4 Glass	j	503-1005
Schott O G 1 Glass	r	503-1006
Schott O G 4 Glass $SiO_2 > K_2O > ZnO > B_2O_3$	j	503-2059
Schott O G 550 Glass	j	503-1803
Schott O G 590 Glass	j	503-2876
Schott Raughglass 444 I I I	j	503-2135
Schott R G 2 Glass	r	503-1568
Schott R G 3 Glass $SiO_2 > B_2O_3 > Na_2O > Al_2O_3$	j	503-2060
Schott R G 7 Glass	j	503-1586
Schott R G 8 Glass	jr	503-1007
Schott R G 10 Glass	j	503-1569
Schott S F 1 G 7 Glass	j	503-1516
Schott S F 2 Glass	j	503-1514
Schott S F 8 G 7 Glass	j	503-1517
Schott S F 11 Glass	j	503-0231
Schott S F 14 Glass	j	503-0159
Schott S F 19 G 7 Glass	j	503-1518
Schott S F S 1 Glass	j	503-0855
Schott S F S 2 Glass	j	503-1515
Schott S K 4 G 13 Glass	j	503-1008
Schott U G 1 Glass $SiO_2 > ZnO > K_2O > Ni_2O_3 >$ Other	j	503-2041
Schott U G 2 Glass	jr	503-1564
Schott U G 3 Glass	j	503-1587
Schott U G 5 Glass	r	503-1565
Schott U G 6 Glass $SiO_2 > K_2O > PbO > MnO >$ Other	j	503-2042
Schott U G 7 Glass $SiO_2 > Na_2O > B_2O_3 > BaO >$ Other	j	503-2043
Schott U G 11 Glass	r	503-1566
Schott V G 3 Glass $P_2O_5 > BaO > UO_3 > Na_2O$	j	503-2051
Schott V G 4 Glass $SiO_2 > PbO > B_2O_3 > Na_2O >$ Other	j	503-2052
Schott V G 11 Glass $SiO_2 > PbO > K_2O > Na_2O >$ Other	j	503-2053
Schott W 3 G Glass	jr	503-1539
Schott W G 1 Glass	jr	503-0983
Schott W G 2 Glass	jr	503-1538
Schott W G 4 Glass	r	503-1540
Schott W G 5 Glass	hjr	503-1541
Schott W G 6 Glass	hjr	503-1542
Schott W G 7 Glass	jr	503-1543
Schott W G 8 Glass	jr	503-1544
Schott W G 9 G 9 Glass	j	503-1546
Schott W G 10 Glass	j	503-1545
Schott Zerodur Glass	n	503-1009
Selenium Glass	jr	503-1254
Semiporcelain	n	503-1347
Servo A 137 Glass	i	503-1532
Servo A 147 Glass	i	503-1533
Servo A 198 Glass	i	503-1534
S E T 1 Glass, Russian	n	503-1022
Sewer Brick	n	503-1352

Property: a. Thermal conductivity; **b.** Accommodation coefficient; **c.** Thermal contact resistance; **d.** Thermal diffusivity; **e.** Specific heat; **f.** Viscosity; **g.** Emittance; **h.** Reflectance; **i.** Absorptance; **j.** Transmittance; **k.** α_s/ϵ_t ratio; **l.** Prandtl number; **n.** Thermal linear expansion coefficient; **o.** Thermal volumetric expansion coefficient; **r.** Thermal radiative properties (g,h,i,j,k)

Substance Name	Property	Number
S F 1 G 7; Schott Glass	j	503-1516
S F 2; Schott Glass	j	503-1514
S F 8 G 7; Schott Glass	j	503-1517
S F 11; Schott Glass	j	503-0231
S F 14; Schott Glass	j	503-0159
S F 19 G 7; Schott Glass	j	503-1518
S F S 1; Schott Glass	j	503-0855
S F S 2; Schott Glass	j	503-1515
S Glass	n	503-0590
see also: Silicon Oxide Mixture $SiO_2 > Al_2O_3 > MgO$		503-0676
Shale Brick	ahn	503-1246
Sibor Glass	a	503-1099
Sierracote No. 3 Glass	j	503-1585
Silica Glass	aeghjnor	503-1059
see also: Silicon Oxide, Fused SiO_2		122-0520
Silicate Glass	ehjn	503-0090
Silicate Of Soda Glass	aehij	503-1411
Silicon Oxide Ceramic	aden	503-1933
Silicon Oxide Mixture:		———
A 10 Glass, Russian	n	503-2251
Corning 9606 Glass Ceramic	adeghijnr	503-0659
M 519 Glass Ceramic, Russian	n	503-2250
$SiO_2 > Al_2O_3$	adegh	503-0240
$SiO_2 > Al_2O_3$ Ceramic	aden	503-2437
see also: Cupola Brick		503-2764
see also: Blast Furnace Brick		503-2765
$SiO_2 > Al_2O_3$ Glass	n	503-2327
$SiO_2 > Al_2O_3$ Porcelain	n	503-2574
$SiO_2 > Al_2O_3 > B_2O_3 > CaO >$ Other Glass	n	503-2950
$SiO_2 > Al_2O_3 > B_2O_3 > K_2O >$ Other Glass	n	503-1892
$SiO_2 > Al_2O_3 > B_2O_3 > Li_2O >$ Other Glass	n	503-2949
$SiO_2 > Al_2O_3 > B_2O_3 > MgO >$ Other Glass	n	503-2633
$SiO_2 > Al_2O_3 > B_2O_3 > Na_2O$ Glass	adeno	503-2142
$SiO_2 > Al_2O_3 > B_2O_3 > TiO_2 >$ Other Glass	n	503-2390
$SiO_2 > Al_2O_3 > BaO$ Porcelain	n	503-2577
$SiO_2 > Al_2O_3 > BaO > CaO$ Corning 1717 Glass	n	503-0927
$SiO_2 > Al_2O_3 > CaO$	e	503-1941
$SiO_2 > Al_2O_3 > CaO$ Ceramic	a	503-2630
$SiO_2 > Al_2O_3 > CaO$ Porcelain	n	503-2576
$SiO_2 > Al_2O_3 > CaO$ Slag	ag	503-2032
$SiO_2 > Al_2O_3 > CaO > B_2O_3$ Glass	n	503-1923
$SiO_2 > Al_2O_3 > CaO > FeO >$ Other	d	503-2759
$SiO_2 > Al_2O_3 > CaO > FeO >$ Other Glass	aden	503-1815
$SiO_2 > Al_2O_3 > CaO > Fe_2O_3$ Slag	ag	503-2033
$SiO_2 > Al_2O_3 > CaO > Fe_2O_3 >$ Other Ceramic	n	503-2592
$SiO_2 > Al_2O_3 > CaO > K_2O$ Ceramic	n	503-2640
$SiO_2 > Al_2O_3 > CaO > MgO >$ Other	d	503-2758
$SiO_2 > Al_2O_3 > CaO > MgO >$ Other Ceramic	n	503-2536
$SiO_2 > Al_2O_3 > CaO > MgO >$ Other Corning 1723 Glass	adeghijnr	503-0604
$SiO_2 > Al_2O_3 > CaO > MgO >$ Other Glass	aden	503-2590
$SiO_2 > Al_2O_3 > CaO > Na_2O$	n	503-2905
$SiO_2 > Al_2O_3 > CaO > Na_2O$ Slag	ag	503-2034
$SiO_2 > Al_2O_3 > FeO > CaO$ Slag	g	503-2459
$SiO_2 > Al_2O_3 > FeO > CaO >$ Other Glass	e	503-1816
$SiO_2 > Al_2O_3 > FeO > CuO >$ Other Glass	d	503-2769
$SiO_2 > Al_2O_3 > FeO > Fe_2O_3 >$ Other Slag	a	503-2476
$SiO_2 > Al_2O_3 > Fe_2O_3$	a	503-2737
$SiO_2 > Al_2O_3 > Fe_2O_3$ Ceramic	a	503-2333
see also: Brick		503-0503
see also: Tripoli Ceramic		503-2332
$SiO_2 > Al_2O_3 > Fe_2O_3 > CaO >$ Other	aden	503-2432
$SiO_2 > Al_2O_3 > Fe_2O_3 > CaO >$ Other Ceramic	a	503-2439
$SiO_2 > Al_2O_3 > Fe_2O_3 > FeO >$ Other	d	503-2757
$SiO_2 > Al_2O_3 > Fe_2O_3 > FeO >$ Other Glass	aden	503-2589
$SiO_2 > Al_2O_3 > Fe_2O_3 > MgO >$ Other Ceramic	a	503-2736

Substance Name	Property	Number
Silicon Oxide Mixture: -(Cont)-		
$SiO_2 > Al_2O_3 > Fe_2O_3 > Na_2O$ Ceramic	n	503-2593
$SiO_2 > Al_2O_3 > Fe_2O_3 > R_2O$ Ceramic	e	503-2462
$SiO_2 > Al_2O_3 > K_2O$ Ceramic	an	503-1387
$SiO_2 > Al_2O_3 > K_2O$ Porcelain	n	503-0866
see also: Porcelain		503-0801
$SiO_2 > Al_2O_3 > K_2O > Fe_2O_3$ Ceramic	a	503-2491
$SiO_2 > Al_2O_3 > K_2O > Li_2O$	n	503-2906
$SiO_2 > Al_2O_3 > K_2O > MgO >$ Other Porcelain	n	503-2431
$SiO_2 > Al_2O_3 > K_2O > Na_2O$ Ceramic	an	503-2861
$SiO_2 > Al_2O_3 > K_2O > Na_2O$ Glass	e	503-1814
see also: Perlite		521-0073
$SiO_2 > Al_2O_3 > K_2O > Na_2O$ Porcelain	n	503-1592
$SiO_2 > Al_2O_3 > K_2O > Na_2O >$ Other Glass	e	503-1817
$SiO_2 > Al_2O_3 > Li_2O$ Ceramic	an	503-1386
$SiO_2 > Al_2O_3 > Li_2O$ Glass	n	503-0481
see also: Petalite Ceramic		503-1128
$SiO_2 > Al_2O_3 > Li_2O$ Glass Ceramic	j	503-2773
$SiO_2 > Al_2O_3 > Li_2O > B_2O_3 >$ Other Glass	n	503-2636
$SiO_2 > Al_2O_3 > Li_2O > MgO >$ Other Glass	n	503-2637
$SiO_2 > Al_2O_3 > Li_2O > TiO_2$ Glass	ijn	503-1654
$SiO_2 > Al_2O_3 > Li_2O > TiO_2 >$ Other Glass	n	503-2638
$SiO_2 > Al_2O_3 > MgO$	n	503-0676
see also: S Glass		503-0590
see also: Cordierite $(Mg, Fe)_2 Al_4 Si_5 O_{18}$		521-0443
$SiO_2 > Al_2O_3 > MgO$ Ceramic	a	503-2480
see also: Cordierite Ceramic		503-0461
$SiO_2 > Al_2O_3 > MgO$ Glass	jn	503-2795
$SiO_2 > Al_2O_3 > MgO$ Porcelain	n	503-2575
$SiO_2 > Al_2O_3 > MgO > B_2O_3$ Glass	n	503-1924
$SiO_2 > Al_2O_3 > MgO > B_2O_3 >$ Other Glass	n	503-2159
$SiO_2 > Al_2O_3 > MgO > CaO >$ Other Corning 1720 Glass	jn	503-0736
$SiO_2 > Al_2O_3 > MgO > Fe_2O_3$ Ceramic	a	503-2481
$SiO_2 > Al_2O_3 > MgO > Fe_2O_3 >$ Other Ceramic	e	503-2467
$SiO_2 > Al_2O_3 > MgO > K_2O >$ Other Ceramic	n	503-2537
$SiO_2 > Al_2O_3 > MgO > Li_2O >$ Other Glass	n	503-2634
$SiO_2 > Al_2O_3 > MgO > Na_2O >$ Other Ceramic	n	503-2535
$SiO_2 > Al_2O_3 > MgO > TiO_2 >$ Other Glass	n	503-2635
$SiO_2 > Al_2O_3 > MnO > K_2O$ Ceramic	n	503-2836
$SiO_2 > Al_2O_3 > Na_2O$ Ceramic	n	503-2908
$SiO_2 > Al_2O_3 > Na_2O$ Porcelain	n	503-2578
$SiO_2 > Al_2O_3 > Na_2O > B_2O_3 >$ Other Glass	r	503-2409
$SiO_2 > Al_2O_3 > Na_2O > CaO >$ Other Ceramic	n	503-2909
$SiO_2 > Al_2O_3 > Na_2O > FeO >$ Other Glass	e	503-1818
$SiO_2 > Al_2O_3 > Na_2O > K_2O$	n	503-2907
$SiO_2 > Al_2O_3 > Na_2O > K_2O$ Porcelain	n	503-2579
$SiO_2 > Al_2O_3 > SO_3 > Fe_2O_3$ Ceramic	a	503-2335
see also: Quartz Feldspar + Tripoli, Ceramic		503-2334
$SiO_2 > Al_2O_3 > TiO_2$ Ceramic	ade	503-2438
see also: Ladle Brick		503-2763
$SiO_2 > Al_2O_3 > TiO_2 > B_2O_3 >$ Other Glass	n	503-2631
$SiO_2 > Al_2O_3 > TiO_2 > Fe_2O_3$ Ceramic	ad	503-2735
$SiO_2 > Al_2O_3 > TiO_2 > K_2O$ Ceramic	a	503-2522
$SiO_2 > Al_2O_3 > TiO_2 > MgO >$ Other Glass	n	503-2632
$SiO_2 > Al_2O_3 > ZnO$ Glass	n	503-2588
$SiO_2 > Al_2O_3 > ZnO > CaO >$ Other Ceramic	n	503-2534
$SiO_2 > Al_2O_3 > ZnO > MgO >$ Other Glass	n	503-2643
$SiO_2 > Al_2O_3 > ZnO > Na_2O >$ Other Glass	a	503-2411
$SiO_2 > Al_2O_3 > ZrO_2 > CaO >$ Other Ceramic	n	503-2542
$SiO_2 > Al_2O_3 > ZrO_2 > MgO >$ Other Ceramic	n	503-2541
$SiO_2 > Al_2O_3 > ZrO_2 > ZnO >$ Other Ceramic	n	503-2540
$SiO_2 > B_2O_3$ Corning 0790 Vycor Glass	jn	503-0863
see also: Corning Vycor Glass		503-0020
$SiO_2 > B_2O_3$ Corning 0791 Vycor Glass	j	503-2794
$SiO_2 > B_2O_3$ Corning 7900 Glass	deghijnr	503-0537
see also: Corning Vycor Glass		503-0020
$SiO_2 > B_2O_3$ Corning 7913 Glass	hij	503-0305
$SiO_2 > B_2O_3$ Glass	dghjn	503-0019
$SiO_2 > B_2O_3 > Al_2O_3$ Glass	n	503-1911

Property: a. Thermal conductivity; **b.** Accommodation coefficient; **c.** Thermal contact resistance; **d.** Thermal diffusivity; **e.** Specific heat; **f.** Viscosity; **g.** Emittance; **h.** Reflectance; **i.** Absorptance; **j.** Transmittance; **k.** α_s/ϵ_t ratio; **l.** Prandtl number; **n.** Thermal linear expansion coefficient; **o.** Thermal volumetric expansion coefficient; **r.** Thermal radiative properties (g,h,i,j,k)

Substance Name	Property	Number
Silicon Oxide Mixture: -(Cont)-		
$SiO_2 > B_2O_3 > Al_2O_3 > K_2O$ Glass	n	503-1914
$SiO_2 > B_2O_3 > Al_2O_3 > K_2O >$ Other Ceramic	n	503-2641
$SiO_2 > B_2O_3 > Al_2O_3 > K_2O >$ Other Corning 7052 Glass	n	503-0723
$SiO_2 > B_2O_3 > Al_2O_3 > K_2O >$ Other Glass	n	503-1891
$SiO_2 > B_2O_3 > Al_2O_3 > Na_2O$ Glass	n	503-1930
$SiO_2 > B_2O_3 > Al_2O_3 > Na_2O >$ Other Glass	jn	503-1694
$SiO_2 > B_2O_3 > CaO > K_2O >$ Other Glass	n	503-2527
$SiO_2 > B_2O_3 > CaO > Na_2O >$ Other Glass	n	503-2526
$SiO_2 > B_2O_3 > K_2O$ Glass	n	503-0180
$SiO_2 > B_2O_3 > K_2O > Al_2O_3$ Glass	n	503-1917
$SiO_2 > B_2O_3 > K_2O > Al_2O_3 >$ Other Glass	n	503-2947
$SiO_2 > B_2O_3 > K_2O > CaO$ Glass	n	503-1893
$SiO_2 > B_2O_3 > K_2O > Fe_2O_3 >$ Other Glass	j	503-2081
$SiO_2 > B_2O_3 > K_2O > Fe_2O_3 >$ Other Schott N G 1 Glass	j	503-2061
$SiO_2 > B_2O_3 > K_2O > Li_2O >$ Other Glass	n	503-2946
$SiO_2 > B_2O_3 > K_2O > Na_2O$ Glass	n	503-0897
$SiO_2 > B_2O_3 > K_2O > Na_2O >$ Other Glass	ghjn	503-1890
$SiO_2 > B_2O_3 > K_2O > Na_2O >$ Other Kimble N 51 Glass	gh	503-2782
$SiO_2 > B_2O_3 > K_2O > Na_2O >$ Other Pyrex 774 Glass	adeghjno	503-1466
$SiO_2 > B_2O_3 > K_2O > Na_2O >$ Other Schott B K 8 Glass	jn	503-1570
$SiO_2 > B_2O_3 > K_2O > PbO$ Glass	j	503-2070
$SiO_2 > B_2O_3 > K_2O > PbO$ Schott B G 2 Glass	j	503-2045
$SiO_2 > B_2O_3 > K_2O > ZnO$ Glass	j	503-2082
$SiO_2 > B_2O_3 > Li_2O$ Glass	n	503-0612
$SiO_2 > B_2O_3 > Li_2O > Al_2O_3 >$ Other Glass	jn	503-2383
$SiO_2 > B_2O_3 > Li_2O > La_2O_3 >$ Other G E-1 T L Glass	jn	503-1862
$SiO_2 > B_2O_3 > Li_2O > La_2O_3 >$ Other Glass	jn	503-2385
$SiO_2 > B_2O_3 > Na_2O$ Glass	n	503-0767
$SiO_2 > B_2O_3 > Na_2O > Al_2O_3$ Corning 7740 Pyrex Glass	adeghijnr	503-0216
see also: Phoenix Glass		503-1049
$SiO_2 > B_2O_3 > Na_2O > Al_2O_3$ Corning 9700 Glass	ej	503-0648
$SiO_2 > B_2O_3 > Na_2O > Al_2O_3$ Glass	adejn	503-1912
$SiO_2 > B_2O_3 > Na_2O > Al_2O_3$ Pittsburgh 3235 Glass	aeghijr	503-1547
see also: Borosilicate Glass		503-0782
$SiO_2 > B_2O_3 > Na_2O > Al_2O_3$ Schott R G 3 Glass	j	503-2060
$SiO_2 > B_2O_3 > Na_2O > Al_2O_3 >$ Other Glass	nr	503-1697
see also: Kimble N 51 A Glass		503-2783
$SiO_2 > B_2O_3 > Na_2O > CaO >$ Other Glass	r	503-2406
$SiO_2 > B_2O_3 > Na_2O > K_2O$ Glass	adej	503-0879
see also: Pyrex Glass		503-0031
$SiO_2 > B_2O_3 > Na_2O > K_2O >$ Other Corning 8370 Glass	d	503-2768
$SiO_2 > B_2O_3 > Na_2O > K_2O >$ Other Glass	dghn	503-2921
see also: Borosilicate Crown Glass		503-0001
$SiO_2 > B_2O_3 > Na_2O > PbO >$ Other Glass	r	503-2408
$SiO_2 > B_2O_3 > Na_2O > Sb_2O_3 >$ Other Ceramic	hn	503-2591
$SiO_2 > B_2O_3 > Na_2O > ZnO >$ Other Glass	nr	503-2407
$SiO_2 > B_2O_3 > Na_2O > ZrO_2 >$ Other Ceramic	n	503-2599
$SiO_2 > B_2O_3 > PbO > CaO >$ Other Glass	n	503-1773
$SiO_2 > B_2O_3 > PbO > Na_2O >$ Other Glass	j	503-2077
$SiO_2 > B_2O_3 > Rb_2O$ Glass	n	503-2848
$SiO_2 > B_2O_3 > Rb_2O > Al_2O_3$ Glass	j	503-2573
Silicon Oxide Mixture: -(Cont)-		
$SiO_2 > B_2O_3 > Sb_2O_3$	n	503-2296
$SiO_2 > B_2O_3 > ZnO > Na_2O >$ Other Glass	a	503-2403
$SiO_2 > BaO > Al_2O_3 > B_2O_3$ Glass	n	503-1927
$SiO_2 > BaO > B_2O_3 > Al_2O_3$ B 9060, Dense Barium Crown Glass	ae	503-2109
$SiO_2 > BaO > B_2O_3 > Al_2O_3$ Glass	ae	503-1610
$SiO_2 > BaO > B_2O_3 > NiO >$ Other Ceramic Cement	n	503-2282
$SiO_2 > BaO > K_2O > Al_2O_3$ Glass	n	503-1835
see also: S 88 -1v Glass, Russian		503-1580
$SiO_2 > BaO > K_2O > ZnO >$ Other Glass	a	503-2412
$SiO_2 > BaO > MgO$ Ceramic	n	503-2834
$SiO_2 > BaO > Na_2O$ Glass	n	503-0424
$SiO_2 > BaO > Na_2O > TiO_2 >$ Other Ceramic	n	503-2617
$SiO_2 > BaO > TiO_2 > Na_2O >$ Other Ceramic	n	503-2618
$SiO_2 > BaO > ZrO_2 > Na_2O >$ Other Corning 8325 Glass	dr	503-0959
$SiO_2 > CaO$	a	503-2265
$SiO_2 > CaO$ Ceramic	adeno	503-2461
$SiO_2 > CaO$ Glass	a	503-1472
$SiO_2 > CaO > Al_2O_3$ Ceramic	a	503-2482
$SiO_2 > CaO > Al_2O_3$ Glass	an	503-0473
$SiO_2 > CaO > Al_2O_3 > B_2O_3$ Glass	n	503-1715
$SiO_2 > CaO > Al_2O_3 > B_2O_3 >$ Other Glass	n	503-1713
$SiO_2 > CaO > Al_2O_3 > Fe_2O_3$ Glass	ae	503-1716
$SiO_2 > CaO > Al_2O_3 > Fe_2O_3 >$ Other Glass	e	503-1984
$SiO_2 > CaO > Al_2O_3 > MgO >$ Other	d	503-2760
$SiO_2 > CaO > Al_2O_3 > MgO >$ Other Glass	aden	503-2303
$SiO_2 > CaO > Al_2O_3 > P_2O_5$ Porcelain	n	503-1801
see also: Bone China		503-1345
$SiO_2 > CaO > Al_2O_3 > TiO_2 >$ Other Glass Ceramic	n	503-2304
$SiO_2 > CaO > BaO$ Ceramic	n	503-2903
$SiO_2 > CaO > FeO$ Slag	e	503-1999
$SiO_2 > CaO > FeO > ZnO$ Slag	e	503-1992
$SiO_2 > CaO > Fe_2O_3 > Al_2O_3 >$ Other Ceramic	a	503-2337
see also: Ash + Tripoli, Ceramic		503-2336
$SiO_2 > CaO > K_2O > Na_2O$ Glass	aghj	503-1741
$SiO_2 > CaO > K_2O > Na_2O >$ Other Glass	a	503-0633
$SiO_2 > CaO > Li_2O$ Glass	n	503-2920
$SiO_2 > CaO > Na_2O$ Glass	adn	503-0423
$SiO_2 > CaO > Na_2O > K_2O >$ Other Glass	gh	503-1718
$SiO_2 > CaO > P_2O_5 > Al_2O_3$ Porcelain	n	503-1802
see also: Bone China		503-1345
$SiO_2 > CaO > ZnO$ Slag	e	503-2021
$SiO_2 > CaO > ZnO > FeO$ Slag	e	503-2020
$SiO_2 > CdO > Al_2O_3 > B_2O_3$ Glass	n	503-1928
$SiO_2 > CoO > Rb_2O$ Glass	n	503-2849
$SiO_2 > Cr_2O_3 > Al_2O_3 > MgO >$ Other Ceramic	a	503-2488
$SiO_2 > CuO > Na_2O$ Glass	n	503-2926
$SiO_2 > FeO > Al_2O_3 > CaO$ Slag	a	503-2477
$SiO_2 > FeO > Al_2O_3 > CaO >$ Other Slag	ag	503-2458
$SiO_2 > FeO > CaO$ Slag	e	503-1997
$SiO_2 > FeO > Fe_2O_3 > ZnO >$ Other Slag	e	503-2006
$SiO_2 > FeO > MnO > Al_2O_3 >$ Other Slag	an	503-2030
$SiO_2 > FeO > MnO > TiO_2 >$ Other Slag	an	503-1993
$SiO_2 > FeO > Na_2O$ Ceramic	n	503-2841
$SiO_2 > FeO > Na_2O$ Ceramic Cement	n	503-2283
$SiO_2 > FeO > Na_2O > CaO >$ Other Glass	g	503-2460
$SiO_2 > FeO > Na_2O > Fe_2O_3$ Glass	e	503-1819
$SiO_2 > Fe_2O_3$	ah	503-2191
$SiO_2 > Fe_2O_3 > CoO > Cr_2O_3 >$ Other Porcelain	g	503-2822
$SiO_2 > Fe_2O_3 > Na_2O > FeO$ Glass	n	503-2531
$SiO_2 > K_2O$ Glass	hn	503-0663
$SiO_2 > K_2O > Al_2O_3 > B_2O_3$ Glass	n	503-1915
$SiO_2 > K_2O > B_2O_3$ Glass	n	503-0178
$SiO_2 > K_2O > B_2O_3 > Al_2O_3$ Glass	n	503-1916
$SiO_2 > K_2O > B_2O_3 > Al_2O_3 >$ Other Glass	n	503-2948
$SiO_2 > K_2O > B_2O_3 > Na_2O$ Glass	an	503-0899

Property: **a.** Thermal conductivity; **b.** Accommodation coefficient; **c.** Thermal contact resistance; **d.** Thermal diffusivity; **e.** Specific heat; **f.** Viscosity; **g.** Emittance; **h.** Reflectance; **i.** Absorptance; **j.** Transmittance; **k.** α_s/ϵ_t ratio; **l.** Prandtl number; **n.** Thermal linear expansion coefficient; **o.** Thermal volumetric expansion coefficient; **r.** Thermal radiative properties (g,h,i,j,k)

Substance Name	Property	Number
Silicon Oxide Mixture: -(Cont)-		
$SiO_2 > K_2O > BaO > Na_2O >$ Other Glass	j	503-2625
$SiO_2 > K_2O > BaO > Nd_2O_3 >$ Other Glass	j	503-2626
$SiO_2 > K_2O > CaO$ Glass	j	503-0554
$SiO_2 > K_2O > CaO > B_2O_3$ Glass	jn	503-1894
$SiO_2 > K_2O > CaO > B_2O_3$ Schott B G 1 Glass	j	503-2044
$SiO_2 > K_2O > CaO > B_2O_3$ Schott G G 5 Glass	j	503-2054
$SiO_2 > K_2O > CaO > B_2O_3$ Schott G G 11 Glass	j	503-2055
$SiO_2 > K_2O > CaO > B_2O_3 >$ Other Glass	r	503-2474
$SiO_2 > K_2O > CaO > MgO$ Glass	j	503-2523
$SiO_2 > K_2O > CaO > Na_2O$ Glass	n	503-2170
$SiO_2 > K_2O > CaO > P_2O_5 >$ Other Glass	r	503-2473
$SiO_2 > K_2O > Na_2O$ Glass	n	503-0482
$SiO_2 > K_2O > Na_2O > BaO >$ Other Glass	j	503-2629
$SiO_2 > K_2O > Na_2O > CaO >$ Other Glass	a	503-2447
$SiO_2 > K_2O > Na_2O > Nd_2O_3 >$ Other Glass	j	503-2624
$SiO_2 > K_2O > Nd_2O_3 > BaO >$ Other Glass	j	503-2627
$SiO_2 > K_2O > Nd_2O_3 > Na_2O >$ Other Glass	j	503-2628
$SiO_2 > K_2O > P_2O_5 > CaO >$ Other Glass	r	503-2414
$SiO_2 > K_2O > PbO > MnO >$ Other Schott U G 6 Glass	j	503-2042
$SiO_2 > K_2O > PbO > MnO_2 >$ Other Glass	j	503-2067
$SiO_2 > K_2O > ZnO > B_2O_3$ Glass	j	503-2080
$SiO_2 > K_2O > ZnO > B_2O_3$ Schott G G 14 Glass	j	503-2057
$SiO_2 > K_2O > ZnO > B_2O_3$ Schott O G 4 Glass	j	503-2059
$SiO_2 > K_2O > ZnO > Na_2O$ Glass	j	503-2072
$SiO_2 > K_2O > ZnO > Na_2O$ Schott B G 12 Glass	j	503-2047
$SiO_2 > K_2O > ZnO > Na_2O$ Schott B G 28 Glass	j	503-2050
$SiO_2 > K_2O > ZnO > Na_2O >$ Other Glass	j	503-2075
$SiO_2 >$ Lanthan- Didymoxyd $> Na_2O >$ Other Schott B G 20 Glass	j	503-2049
$SiO_2 > Li_2O$ Ceramic	n	503-2975
$SiO_2 > Li_2O$ Glass	hn	503-0588
$SiO_2 > Li_2O > Al_2O_3$ Ceramic	n	503-2838
$SiO_2 > Li_2O > Al_2O_3$ Glass	n	503-0431
$SiO_2 > Li_2O > Al_2O_3 > GeO_2$ Ceramic	n	503-2837
$SiO_2 > Li_2O > B_2O_3$ Glass	n	503-1745
$SiO_2 > Li_2O > CaO$ Glass	n	503-2169
$SiO_2 > MgO$	d	503-0769
$SiO_2 > MgO$ Celkate T 21	d	503-2762
$SiO_2 > MgO > Al_2O_3$	n	503-0409
$SiO_2 > MgO > Al_2O_3$ Glass	n	503-2929
$SiO_2 > MgO > Al_2O_3 > B_2O_3$ Glass	n	503-2952
$SiO_2 > MgO > Al_2O_3 > B_2O_3 >$ Other Glass	n	503-2953
$SiO_2 > MgO > Al_2O_3 > Fe_2O_3 >$ Other Ceramic	a	503-2456
$SiO_2 > MgO > Al_2O_3 > ZnO$ Glass	n	503-2954
$SiO_2 > MgO > B_2O_3$ Glass	n	503-2278
$SiO_2 > MgO > BaO$ Ceramic	n	503-2833
$SiO_2 > MgO > CaO > Al_2O_3 >$ Other Glass	n	503-2594
$SiO_2 > MgO > Li_2O > Ce_2O_3$ H P, Glass	r	503-2248
$SiO_2 > Mn_3O_4 > Na_2O > Fe_2O_3 >$ Other Glass	e	503-2299
$SiO_2 > MoO_3$	h	503-2779
$SiO_2 > Na_2O$ Glass	hjn	503-0585
$SiO_2 > Na_2O > Al_2O_3$ Glass	an	503-0008
$SiO_2 > Na_2O > Al_2O_3 > B_2O_3$ Glass	n	503-1913
$SiO_2 > Na_2O > Al_2O_3 > CaO >$ Other Ceramic	n	503-2607
$SiO_2 > Na_2O > Al_2O_3 > CaO >$ Other Glass	ae	503-0555
$SiO_2 > Na_2O > Al_2O_3 > SrO$ Glass	n	503-2927
$SiO_2 > Na_2O > Al_2O_3 > TiO_2 >$ Other Ceramic	n	503-2608
$SiO_2 > Na_2O > Al_2O_3 > ZnO >$ Other Ceramic	n	503-2840
$SiO_2 > Na_2O > Al_2O_3 > ZnO >$ Other Ceramic Cement	en	503-2279
$SiO_2 > Na_2O > Al_2O_3 > ZrO_2 >$ Other Ceramic	n	503-2609

Substance Name	Property	Number
Silicon Oxide Mixture: -(Cont)-		
$SiO_2 > Na_2O > B_2O_3$ Glass	n	503-0526
$SiO_2 > Na_2O > B_2O_3 > Al_2O_3$ Glass	n	503-2951
$SiO_2 > Na_2O > B_2O_3 > Al_2O_3 >$ Other Glass	j	503-2524
$SiO_2 > Na_2O > B_2O_3 > BaO >$ Other B 3860, Light Barium Crown Glass	ae	503-2108
$SiO_2 > Na_2O > B_2O_3 > BaO >$ Other Glass	aej	503-2068
$SiO_2 > Na_2O > B_2O_3 > BaO >$ Other Schott U G 7 Glass	j	503-2043
$SiO_2 > Na_2O > B_2O_3 > K_2O >$ Other B 1463, Borosilicate Crown Glass	ae	503-2103
$SiO_2 > Na_2O > B_2O_3 > K_2O >$ Other Glass	ae	503-2104
$SiO_2 > Na_2O > B_2O_3 > Sb_2O_3 >$ Other Ceramic	hn	503-2422
$SiO_2 > Na_2O > BaO$ Glass	n	503-2530
$SiO_2 > Na_2O > BaO > Al_2O_3$ Glass	n	503-2922
$SiO_2 > Na_2O > BaO > B_2O_3 >$ Other Glass	j	503-2069
$SiO_2 > Na_2O > CaO$	n	503-2416
$SiO_2 > Na_2O > CaO$ Glass	aejn	503-0556
see also: Glass Plate		503-1050
$SiO_2 > Na_2O > CaO > Al_2O_3$ Glass	aegn	503-1723
$SiO_2 > Na_2O > CaO > Al_2O_3 >$ Other Glass	adn	503-1721
$SiO_2 > Na_2O > CaO > B_2O_3 >$ Other Glass	a	503-1722
$SiO_2 > Na_2O > CaO > CoO$ Glass	a	503-2405
$SiO_2 > Na_2O > CaO > K_2O$	j	503-2122
$SiO_2 > Na_2O > CaO > K_2O$ Schott B G 9 Glass	j	503-2046
$SiO_2 > Na_2O > CaO > K_2O >$ Other Ceramic	n	503-2424
$SiO_2 > Na_2O > CaO > K_2O >$ Other Glass	jn	503-2071
$SiO_2 > Na_2O > CaO > K_2O >$ Other Schott B G 16 Glass	j	503-2048
$SiO_2 > Na_2O > CaO > Li_2O$ Glass	j	503-2131
$SiO_2 > Na_2O > CaO > MgO$ B 2156, Crown Glass	e	503-2111
$SiO_2 > Na_2O > CaO > MgO$ Corning 0080 Glass	adn	503-0262
$SiO_2 > Na_2O > CaO > MgO$ Glass	adegjnr	503-0376
see also: Window Glass		503-0025
$SiO_2 > Na_2O > CaO > MgO$ Libbey Owens Ford 9330 Glass	ae	503-1067
$SiO_2 > Na_2O > CaO > MgO$ Solex 2808 Glass	adeno	503-1062
$SiO_2 > Na_2O > CaO > MgO >$ Other Glass	aer	503-1720
$SiO_2 > Na_2O > CaO > MgO >$ Other Solex S Glass	adeno	503-1072
$SiO_2 > Na_2O > CaO > MnO$ Glass	n	503-2532
$SiO_2 > Na_2O > CaO > MnO >$ Other Glass	n	503-2533
$SiO_2 > Na_2O > CaO > Ni_2O_3 >$ Other Glass	j	503-2066
$SiO_2 > Na_2O > CaO > TiO_2 >$ Other Ceramic	n	503-2606
$SiO_2 > Na_2O > CaO > ZnO >$ Other Glass	n	503-2297
$SiO_2 > Na_2O > CaO > ZrO_2 >$ Other Ceramic	n	503-2610
$SiO_2 > Na_2O > CuO > Al_2O_3 >$ Other Ceramic	n	503-2605
$SiO_2 > Na_2O > CuO > MgO$ Glass	n	503-2925
$SiO_2 > Na_2O > Fe_2O_3 > CaO$ Glass	a	503-2400
$SiO_2 > Na_2O > Fe_2O_3 > CaO >$ Other Glass	a	503-2449
$SiO_2 > Na_2O > Fe_2O_3 > MgO >$ Other Glass	a	503-2401
$SiO_2 > Na_2O > K_2O > CaO >$ Other Glass	ghn	503-1709
$SiO_2 > Na_2O > La_2O_3 > Pr_2O_3 >$ Other Glass	j	503-2074
$SiO_2 > Na_2O > Li_2O > CaO >$ Other Glass	n	503-2328
$SiO_2 > Na_2O > MgO$ Glass	n	503-2923
$SiO_2 > Na_2O > MgO > CuO$ Glass	n	503-2924
$SiO_2 > Na_2O > PbO$ B 4446, Extra Light Flint Glass	e	503-2110
$SiO_2 > Na_2O > PbO$ Glass	ej	503-2106
$SiO_2 > Na_2O > PbO > K_2O >$ Other Glass	j	503-2123
$SiO_2 > Na_2O > PbO > Sb_2O_3 >$ Other Glass	n	503-2525
$SiO_2 > Na_2O > PbO > ZnO$ Glass	n	503-2127
$SiO_2 > Na_2O > SrO$ Glass	n	503-2928
$SiO_2 > Na_2O > SrO > Al_2O_3$ Glass	n	503-0288
$SiO_2 > Na_2O > TiO_2 > Al_2O_3 >$ Other Ceramic	n	503-2611
$SiO_2 > Na_2O > TiO_2 > CaO >$ Other Ceramic	n	503-2612
$SiO_2 > Na_2O > TiO_2 > ZrO_2 >$ Other Ceramic	n	503-2613
$SiO_2 > Na_2O > ZnO$ Glass	n	503-2529
$SiO_2 > Na_2O > ZnO > CaO$ Corning 7560 Glass	e	503-2415
$SiO_2 > Na_2O > ZnO > K_2O >$ Other Glass	r	503-2475

Property: a. Thermal conductivity; **b.** Accommodation coefficient; **c.** Thermal contact resistance; **d.** Thermal diffusivity; **e.** Specific heat; **f.** Viscosity; **g.** Emittance; **h.** Reflectance; **i.** Absorptance; **j.** Transmittance; **k.** α_s/ϵ_t ratio; **l.** Prandtl number; **n.** Thermal linear expansion coefficient; **o.** Thermal volumetric expansion coefficient; **r.** Thermal radiative properties (g,h,i,j,k)

Property: **a.** Thermal conductivity; **b.** Accommodation coefficient; **c.** Thermal contact resistance; **d.** Thermal diffusivity; **e.** Specific heat; **f.** Viscosity; **g.** Emittance; **h.** Reflectance; **i.** Absorptance; **j.** Transmittance; **k.** α_s/ϵ_t ratio; **l.** Prandtl number; **n.** Thermal linear expansion coefficient; **o.** Thermal volumetric expansion coefficient; **r.** Thermal radiative properties (g,h,i,j,k)

Substance Name	Property	Number
S R M 2014 (N B S), Didymium Glass Filter	j	503-2799
see also: Didymium Glass		503-0033
Steatite Brick	n	503-1349
Steatite Ceramic	aen	503-0765
see also: Steatite Porcelain		503-1183
Steatite Porcelain	an	503-1183
see also: Steatite Ceramic		503-0765
S T K 3, Superheavy Crown Glass, Russian	e	503-2290
Stoneware	aen	503-1218
Stoneware, Sillimanite	a	503-1230
Strontium Oxide Mixture:		———
$SrO > Al_2O_3 > Li_2O$	a	503-2738
$SrO > Al_2O_3 > ZrO > SiO_2 >$ Other	a	503-2739
$SrO > SiO_2$	j	503-1986
$SrO > TiO_2 > ZrO > SiO_2 >$ Other	a	503-2743
$SrO > ZnO > ZrO > SiO_2 >$ Other	a	503-2741
$SrO > ZrO > Al_2O_3 > SiO_2 >$ Other	a	503-2740
$SrO > ZrO > ZnO > SiO_2 >$ Other	a	503-2742
Sulfur Glass	r	503-1088
Supramica 620 B B Ceramic	an	503-1280
Supremax Glass	en	503-1112
see also: Borosilicate Glass		503-0782
T 21; Celkate $SiO_2 > MgO$	d	503-2762
T 0254 Glass, Russian	n	503-1484
T 0257 Glass, Russian	n	503-1485
Tableware	n	503-1287
Tantalum Oxide Mixture:		———
$Ta_2O_3 > La_2O_3 > ThO_2 > B_2O_3 >$ Other Glass	j	503-2510
$Ta_2O_5 > Al_2O_3$	n	503-1847
$Ta_2O_5 > HfO_2$	n	503-1844
$Ta_2O_5 > HfO_2 > Al_2O_3$	n	503-1854
$Ta_2O_5 > HfO_2 > TiO_2$	n	503-1852
$Ta_2O_5 > HfO_2 > Y_2O_3$	n	503-1856
$Ta_2O_5 > MgO$	n	503-1848
$Ta_2O_5 > NiO$	n	503-1849
$Ta_2O_5 > SnO_2$	n	503-1845
$Ta_2O_5 > TiO_2$	n	503-1846
$Ta_2O_5 > TiO_2 > Al_2O_3$	n	503-1853
$Ta_2O_5 > WO_3$	n	503-1843
$Ta_2O_5 > WO_3 > HfO_2$	n	503-1851
$Ta_2O_5 > Y_2O_3 > TiO_2$	n	503-1855
$Ta_2O_5 > ZrO_2$	n	503-1850
T B F; Glass, Russian	hj	503-0479
T B F 25 Glass, Russian	ade	503-1023
Tellurite Glass	jn	503-1445
Tellurite Glass $TeO_2 > WO_3 > PbO$	j	503-1974
Tellurium Glass	h	503-1444
Tellurium Oxide Mixture:		———
$TeO_2 > MoO_3$ Glass	n	503-2970
$TeO_2 > WO_3$ Glass	n	503-2971
$TeO_2 > WO_3 > PbO$ Tellurite Glass	j	503-1974
T F 1 Glass, Russian	adejn	503-0463
see also: Lead Oxide Mixture Glass $PbO > SiO_2 > K_2O$		503-2119
T F 3 Glass, Russian	jn	503-0464
see also: Lead Oxide Mixture Glass $PbO > SiO_2 > K_2O$		503-2119
T F 5 Glass, Russian	jn	503-0465
T F 7 Glass, Russian	j	503-2291
T F 8 Glass, Russian	n	503-1492
T F 10 Glass, Russian	aden	503-0466
T F 11 Glass, Russian	ade	503-1493
T F 12 Glass, Russian	ade	503-1494
T F 101 Glass, Russian	n	503-2286
T F K 1 Glass, Russian	ade	503-1024
Thermolux Plate Glass	a	503-1053
Thorium Oxide Mixture:		———
$ThO_2 > BeO$	a	503-2162
$ThO_2 > CeO_2$	a	503-2164
$ThO_2 > UO_2$	aeo	503-0559
Thorium Oxide Refractory	anr	503-1188
Thuringian Glass	aeno	503-1057
Tin Oxide Mixture:		———
$SnO_2 > MgO$	a	503-2695

Substance Name	Property	Number
Tin Oxide Mixture: –(Cont)–		
$SnO_2 > MgO > ZnO$	a	503-2744
$SnO_2 > Sb_2O_3$	j	503-1749
$SnO_2 > ZnO$	a	503-2696
$SnO_2 > ZnO > MgO$	a	503-2745
Titanium Enamel	n	503-1159
Titanium Oxide Enamel	r	503-1168
see also: Titanium Oxide Refractory		503-1198
Titanium Oxide Glass	r	503-1119
Titanium Oxide Mixture:		———
$TiO_2 > Al_2O_3$	g	503-0759
$TiO_2 > Al_2O_3 > SiO_2 > CaO >$ Other Slag	ade	503-2809
$TiO_2 > Al_2O_3 > SiO_2 > Fe_2O_3 >$ Other Slag	ade	503-2440
$TiO_2 > Al_2O_3 > SiO_2 > MgO >$ Other Slag	ade	503-2808
$TiO_2 > BaO$	n	503-2918
$TiO_2 > BaO > BeO$	n	503-2915
$TiO_2 > BaO > MgO$	n	503-2916
$TiO_2 > BeO$	an	503-1870
$TiO_2 > MgO$	n	503-0403
$TiO_2 > MgO > BeO > SiO_2 >$ Other	n	503-2891
$TiO_2 > MnO$	h	503-2780
$TiO_2 > MnO_2$	no	503-2188
$TiO_2 > Nb_2O_5$	no	503-2186
$TiO_2 > Sb_2O_3$	no	503-2185
$TiO_2 > SiO_2$	no	503-1944
$TiO_2 > SiO_2 > Al_2O_3 > FeO >$ Other Slag	an	503-2031
$TiO_2 > SiO_2 > GeO_2$	n	503-2890
$TiO_2 > SnO_2$	no	503-1867
$TiO_2 > SrO > BaO$	n	503-2917
$TiO_2 > V_2O_5$	no	503-2184
$TiO_2 > WO_3$	no	503-2187
$TiO_2 > ZnO$	h	503-1977
$TiO_2 > ZrO_2$	no	503-0213
$Ti_2O_3 > V_2O_3$	e	503-0751
Titanium Oxide Refractory	an	503-1198
see also: Titanium Oxide Enamel		503-1168
Titanium Slag	a	503-1365
T K 4 Glass, Russian	n	503-1555
T K 14 Glass, Russian	den	503-1556
T K 16 Glass, Russian	aden	503-1557
T K 20 Glass, Russian	n	503-0237
T K 21 Glass, Russian	aden	503-1559
T K 25 Glass, Russian	n	503-2285
T K 2621 Glass, Russian	ade	503-1560
Toshiba Colored Filter Glass	j	503-1972
Trachyte Ceramic	n	503-1323
Tremolite Ceramic	n	503-1299
Tripoli Ceramic	a	503-2332
see also: Silicon Oxide Mixture Ceramic $SiO_2 > Al_2O_3 > Fe_2O_3$		503-2333
T R L 10 Glass, Russian	n	503-1025
Ts M 3, Barium Feldspar Ceramic	a	503-1240
T S M 500 Glass, Russian	n	503-1525
T S M 501 Glass, Russian	n	503-1526
T S M 502 Glass, Russian	n	503-1527
T S M 504 Glass, Russian	n	503-2288
Tungsten Oxide Mixture	e	503-1227
Tungsten Oxide Mixture:		———
$WO_3 > ZnO$	a	503-2697
U F S 1 Glass, Russian	j	503-1836
U G 1; Schott Glass $SiO_2 > ZnO > K_2O > Ni_2O_3 >$ Other	j	503-2041
U G 2; Schott Glass	jr	503-1564
U G 3; Schott Glass	j	503-1587
U G 5; Schott Glass	r	503-1565
U G 6; Schott Glass $SiO_2 > K_2O > PbO > MnO >$ Other	j	503-2042
U G 7; Schott Glass $SiO_2 > Na_2O > B_2O_3 > BaO >$ Other	j	503-2043
U G 11; Schott Glass	r	503-1566
U L E Glass $SiO_2 > TiO_2$	aden	503-0562
Unfired Ceramic	den	503-1310
Uranium Glass	hnr	503-1121

Property: **a.** Thermal conductivity; **b.** Accommodation coefficient; **c.** Thermal contact resistance; **d.** Thermal diffusivity; **e.** Specific heat; **f.** Viscosity; **g.** Emittance; **h.** Reflectance; **i.** Absorptance; **j.** Transmittance; **k.** α_s/ϵ_t ratio; **l.** Prandtl number; **n.** Thermal linear expansion coefficient; **o.** Thermal volumetric expansion coefficient; **r.** Thermal radiative properties (g,h,i,j,k)

Property: a. Thermal conductivity; **b.** Accommodation coefficient; **c.** Thermal contact resistance; **d.** Thermal diffusivity; **e.** Specific heat; **f.** Viscosity; **g.** Emittance; **h.** Reflectance; **i.** Absorptance; **j.** Transmittance; **k.** α_s/ϵ_t ratio; **l.** Prandtl number; **n.** Thermal linear expansion coefficient; **o.** Thermal volumetric expansion coefficient; **r.** Thermal radiative properties (g,h,i,j,k)

Substance Name	Property	Number
Zirconium Silicate Glaze	n	503-1266
Zirconium Silicate Porcelain	an	503-1468
Zircon Porcelain	an	503-1468
Z K N 7 Glass	dhj	503-1459
Z S 4 Glass, Russian	n	503-1571
Z S 5 Glass, Russian	n	503-1572
Z S 6 Glass, Russian	r	503-1257
Z S 9 Glass, Russian	n	503-1573
Z S 11 Glass, Russian	o	503-1574
Z S 112 Glass, Russian	o	503-1394

MINERALS, CELESTIAL BODIES, AND ROCKS CLASS 521

Substance Name	Property	Number
Acmite $Na\,Fe(Si\,O_3)_2$	ahn	521-0947
see also: Iron Sodium Silicate $Na\,Fe(Si\,O_3)_2$		110-0495
Acmite- Trachyte Pyroxene	f	521-0224
Actinolite $Ca_2(Mg, Fe)_5\,Si_8\,O_{22}(O\,H)_2$	ahijn	521-0950
see also: Asbestos		521-0055
Adamellite Granite	a	521-1490
Adamite $Zn_2\,As\,O_4\,O\,H$	j	521-1189
Adularia $K\,Al\,Si_3\,O_8$	aehnor	521-0060
see also: Potassium Aluminum Silicate $K\,Al\,Si_3\,O_8$		102-0173
Aegirite $Na\,Fe(Si\,O_3)_2$	ahn	521-0947
see also: Iron Sodium Silicate $Na\,Fe(Si\,O_3)_2$		110-0495
Aegirite Riebeckite Granite	n	521-1258
Aeschynite $(Ce, Ca, Fe, Th)(Ti, Nb)_2(O, O\,H)_6$	h	521-1340
Afwillite $Ca_3\,Si_2\,O_4(O\,H)_6$	n	521-1602
Agate $Si\,O_2$	j	521-1223
see also: Silicon Oxide $Si\,O_2$		122-0009
Agatized Wood	r	521-0474
Agglomerate	h	521-0997
see also: Volcanic Breccia		521-1034
Ahlfeldite $(Ni, Co)\,Se\,O_3$	j	521-1369
Aikinite $Pb\,Cu\,Bi\,S_3$	h	521-1573
Akibara, Japan Coal	j	521-1126
Alabama Talc $Mg_3\,Si_4\,O_{10}(O\,H)_2$	adehijnr	521-0116
Alabandite $Mn\,S$	hj	521-0863
see also: Manganese Sulfide $Mn\,S$		119-0043
Alabastine	a	521-0102
Alalite $Ca\,Mg(Si\,O_3)_2$	aefhijno	521-0036
see also: Calcium Magnesium Silicate $Ca\,Mg\,Si_2\,O_6$		106-0940
see also: Chrome Diopside		521-1593
Albertite	h	521-1129
Albite $Na\,Al\,Si_3\,O_8$	adefghijnor	521-0057
see also: Sodium Aluminosilicate $Na\,Al\,Si_3\,O_8$		102-0218
Albite:		———
Cleavelandite	h	521-0985
Albite - Anorthite $Na\,Al\,Si_3\,O_8 - Ca\,Al_2\,Si_2\,O_8$	d	521-1549
Albitite	a	521-0353
Aleurite	aeh	521-0599
Aleurolite	a	521-1246
Alfeldite $(Ni, Co)\,Se\,O_3$	j	521-1369
Alkali Feldspar	ad	521-0913
see also: Aluminum Potassium Silicate		102-0878
Alkali Feldspar:		———
Adularia $K\,Al\,Si_3\,O_8$	aehnor	521-0060
see also: Potassium Aluminum Silicate $K\,Al\,Si_3\,O_8$		102-0173
Albite $Na\,Al\,Si_3\,O_8$	adefghijnor	521-0057
see also: Sodium Aluminosilicate $Na\,Al\,Si_3\,O_8$		102-0218
Analbite $(Na, K)\,Al\,Si_3\,O_8$	n	521-1613
see also: Anorthoclase $(Na, K)\,Al\,Si_3\,O_8$		521-1047
Anorthoclase $(Na, K)\,Al\,Si_3\,O_8$	egh	521-1047
see also: Analbite $(Na, K)\,Al\,Si_3\,O_8$		521-1613
Microcline $K\,Al\,Si_3\,O_8$	aehjnor	521-0392
see also: Potassium Aluminum Silicate $K\,Al\,Si_3\,O_8$		102-0173
Microcline - Perthite	a	521-1458
Moonstone	ijnr	521-0457
Orthoclase $K\,Al\,Si_3\,O_8$	adefhijnor	521-0026
see also: Potassium Aluminum Silicate $K\,Al\,Si_3\,O_8$		102-0173
Perthite	fhn	521-0153

Substance Name	Property	Number
Alkali Feldspar: -(Cont)-		
Plagioclase $(Na, Ca)\,Al\,(Si, Al)\,Si_2\,O_8$	adeh	521-0495
Potassium Feldspar $K\,Al\,Si_3\,O_8$	adfgn	521-0170
see also: Potassium Aluminum Silicate $K\,Al\,Si_3\,O_8$		102-0173
Sanidine $K\,Al\,Si_3\,O_8$	aehno	521-1045
see also: Potassium Aluminum Silicate $K\,Al\,Si_3\,O_8$		102-0173
Sodium Feldspar	fn	521-0332
Allanite $(Ce, Ca, Y)(Al, Fe)_3(Si\,O_4)_3\,O\,H$	h	521-1395
Alluvial Soil	adh	521-1345
Almandine $Fe_3\,Al_2(Si\,O_4)_3$	ahjnor	521-0412
Almandite $Fe_3\,Al_2(Si\,O_4)_3$	ahjnor	521-0412
Altered Rock	h	521-1631
Alum $K\,Al(S\,O_4)_2$	hj	521-1471
see also: Aluminum Potassium Sulfate $K\,Al(S\,O_4)_2$		102-0024
Alumina, Natural:		———
Corundum, Natural $Al_2\,O_3$	adeghijnor	521-0077
see also: Aluminum Oxide $Al_2\,O_3$		102-0002
Ruby, Natural $Al_2\,O_3$	adeghijnr	521-0191
see also: Aluminum Oxide $Al_2\,O_3$		102-0002
Sapphire, Natural $Al_2\,O_3$	adeghjnr	521-0064
see also: Aluminum Oxide $Al_2\,O_3$		102-0002
Aluminum Beryllium Cesium Borate Hydroxide, Natural $Cs\,Al_4\,Be_4\,B_{11}\,O_{25}(O\,H)_4$	j	521-0834
Aluminum Beryllium Sodium Silicate Chloride, Natural $Na_4\,Be\,Al\,Si_4\,O_{12}\,Cl$	hn	521-1437
Aluminum Borate Silicate, Natural $Al_7(B\,O_3)(Si\,O_4)_3\,O_3$	ehjn	521-0911
Aluminum Calcium Iron Magnesium Silicate, Natural $(Ca, Mg, Fe)(Si, Al)\,O_3$	h	521-1397
Aluminum Calcium Potassium Sodium Silicate, Natural:		———
Mordenite $(Ca,(Na, K)_2)_4\,Al_8\,Si_{40}\,O_{96}$	ij	521-1606
Aluminum Calcium Silicate, Natural $Ca\,Al_2\,Si_3\,O_{10}$	j	521-0781
Aluminum Calcium Silicate, Natural $Ca\,Al_2\,Si_4\,O_{12}$	ah	521-0489
Aluminum Calcium Silicate, Natural $Ca_2\,Al_4\,Si_8\,O_{24}$	e	521-0488
Aluminum Calcium Silicate Carbonate, Natural $3\,Ca\,Al_2\,Si_2\,O_8\,.\,Ca\,C\,O_3$	a	521-0961
Aluminum Calcium Silicate Hydroxide, Natural $Ca\,Al_4\,Si_2\,O_{10}(O\,H)_2$	j	521-1331
Aluminum Calcium Silicate Hydroxide, Natural $Ca_2\,Al_2\,Si_3\,O_{10}(O\,H)_2$	aj	521-0968
Aluminum Calcium Silicate Hydroxide, Natural:		———
Clinozoisite $Ca_2\,Al_3\,Si_3\,O_{12}\,O\,H$	aehij	521-1131
Zoisite $Ca_2\,Al_3\,Si_3\,O_{12}\,O\,H$	aehj	521-0941
Aluminum Calcium Sodium Fluoride, Natural $Na\,Ca\,Al\,F_6$	j	521-1474
Aluminum Calcium Sodium Silicate, Natural $(Na, Ca)\,Al\,(Si, Al)\,Si_2\,O_8$	adeh	521-0495
Aluminum Calcium Sodium Silicate, Natural $Na\,Ca_2\,Al_5\,Si_{13}\,O_{36}$	ahj	521-0779
Aluminum Calcium Sodium Silicate, Natural $(Na, Ca)_5\,Al_6(Al, Si)_4\,Si_{26}\,O_{72}$	j	521-0778
Aluminum Calcium Sodium Silicate Sulfate, Natural $(Na, Ca)_8(Al, Si)_{12}\,O_{24}(S, S\,O_4)$	hj	521-1351
see also: Lapis Lazuli		521-1350
Aluminum Carbonate, Natural $Al_2[C_6(C\,O\,O)_6]$	j	521-0782
Aluminum Copper Phosphate Hydroxide, Natural $Cu\,Al_6(P\,O_4)_4(O\,H)_8$	hj	521-1285
Aluminum Fluorosilicate Hydroxide, Natural $Al_2\,Si\,O_4\,(F, O\,H)_2$	aehijnor	521-0082
Aluminum Iron Garnet $Fe_3\,Al_2(Si\,O_4)_3$	ahjnor	521-0412
Aluminum Iron Magnesium Phosphate Hydroxide, Natural $(Mg, Fe)\,Al_2(P\,O_4)_2(O\,H)_2$	hj	521-1283
Aluminum Iron Magnesium Potassium Silicate Hydroxide, Natural $K(Mg, Fe)_3(Al, Fe)\,Si_3\,O_{10}(O\,H)_2$	adehjnr	521-0430
Aluminum Iron Magnesium Silicate, Natural $(Mg, Fe)_3\,Al_2\,(Si\,O_4)_3$	aehno	521-0411
Aluminum Iron Magnesium Silicate Hydroxide, Natural $(Mg, Fe, Al)_6(Si, Al)_4\,O_{10}(O\,H)_8$	j	521-0177
Aluminum Iron Magnesium Silicate Hydroxide, Natural $Na_2(Mg, Fe)_3\,Al_2\,Si_8\,O_{22}(O\,H)_2$	ah	521-1137
Aluminum Iron Magnesium Silicate Hydroxide, Natural Mica $(Mg, Fe, Al)_3\,(Al, Si)_4\,O_{10}\,(O\,H)_2$	adehjn	521-0020

Property: a. Thermal conductivity; **b.** Accommodation coefficient; **c.** Thermal contact resistance; **d.** Thermal diffusivity; **e.** Specific heat; **f.** Viscosity; **g.** Emittance; **h.** Reflectance; **i.** Absorptance; **j.** Transmittance; **k.** α_s/ϵ_t ratio; **l.** Prandtl number; **n.** Thermal linear expansion coefficient; **o.** Thermal volumetric expansion coefficient; **r.** Thermal radiative properties (g,h,i,j,k)

Substance Name	*Property*	*Number*
Aluminum Iron Magnesium Zinc Oxide, Natural $Zn_n Fe_x Mg_y Al_2 O_4$	n	521-1429
Aluminum Iron Silicate, Natural $Fe_3 Al_2(Si O_4)_3$	ahjnor	521-0412
Aluminum Iron Silicate Hydroxide, Natural $Fe_2 Al_4 Si_2 O_{10}(O H)_4$	j	521-1480
Aluminum Iron Sodium Silicate, Natural $Na (Al, Fe) Si_2 O_6$	adehij	521-0058
Aluminum Lithium Manganese Oxide Hydroxide, Natural $(Al, Li) Mn O_2(O H)_2$	j	521-1176
Aluminum Lithium Potassium Silicate, Natural $(Li, K) Al Si_3 O_8$	n	521-1653
Aluminum Lithium Silicate, Natural $Li Al Si_3 O_8$	n	521-1652
Aluminum Lithium Silicate, Natural $Li Al Si_4 O_{10}$	ahijn	521-0772
Aluminum Lithium Sodium Fluoride, Natural $Na_3 Li_3 Al_2 F_{12}$	j	521-1293
Aluminum Magnesium Borate, Natural $Mg Al(B O_4)$	j	521-1164
Aluminum Magnesium Carbonate Hydroxide, Natural $Mg_6 Al_2 C O_3(O H)_{16}$	hjr	521-0554
Aluminum Magnesium Silicate Hydroxide, Natural $(Mg, Al)_2 Si_4 O_{10}(O H)$	aeij	521-0182
Aluminum Magnesium Silicate Soil	a	521-0864
Aluminum Manganese Silicate, Natural $Mn_3 Al_2(Si O_4)_3$	ahjno	521-0957
Aluminum Mellitate, Hydrated $Al_2[C_6(C O O)_6]$	j	521-0782
Aluminum Phosphate Hydroxide, Natural $Al_3(P O_4)_2(O H)_3$	hj	521-1286
Aluminum Potassium Silicate, Natural $K Al Si O_4$	e	521-0801
Aluminum Potassium Sodium Silicate, Natural $(Na, K) (Al, Si)_2 O_4$	aeghijno	521-0059
see also: Sodium Aluminosilicate $Na_2 O . Al_2 O_3 . 2 Si O_2$		102-0953
see also: Carnegieite $Na_2 O . Al_2 O_3 . 2 Si O_2$		102-0954
Aluminum Potassium Sodium Silicate, Natural:		——
Analbite $(Na, K) Al Si_3 O_8$	n	521-1613
see also: Anorthoclase $(Na, K) Al Si_3 O_8$		521-1047
Anorthoclase $(Na, K) Al Si_3 O_8$	egh	521-1047
see also: Analbite $(Na, K) Al Si_3 O_8$		521-1613
Aluminum Silicate, Natural $3 Al_2 O_3 . 2 Si O_2 + Si O_2$	hir	521-0516
Aluminum Silicate Hydroxide, Natural $Al_2 Si_2 O_5(O H)_4$	ehijn	521-0506
see also: Aluminum Silicate Hydroxide $Al_2 Si_2 O_5(O H)_4$		102-0871
Aluminum Sodium Arsenate Fluoride, Natural $Na Al As O_4 F$	j	521-1299
Aluminum Sodium Fluoride, Natural $5 Na F . 3 Al F_3$	r	521-0469
Aluminum Sodium Hydrogen Silicate, Natural $H Na_2 Al_2(Si O_4)_3$	j	521-1479
Aluminum Sodium Silicate, Natural $Na_2 Al_2 Si_3 O_{10}$	ahij	521-0780
Aluminum Sodium Silicate Chloride, Natural:		——
Sodalite $Na_4 Al_3 Si_3 O_{12} Cl$	ahjn	521-0970
Aluminum Sodium Silicate Sulfate, Natural $Na_8 Al_6 Si_6 O_{24} S O_4$	n	521-1108
Alumite $K Al_3(S O_4)_2(O H)_6$	aehij	521-0341
see also: Aluminum Potassium Sulfate $K_2 O . 3 Al_2 O_3 . 4 S O_3$		102-0201
Alum Stone $K Al_3(S O_4)_2(O H)_6$	aehij	521-0341
see also: Aluminum Potassium Sulfate $K_2 O . 3 Al_2 O_3 . 4 S O_3$		102-0201
Alunite $K Al_3(S O_4)_2(O H)_6$	aehij	521-0341
see also: Aluminum Potassium Sulfate $K_2 O . 3 Al_2 O_3 . 4 S O_3$		102-0201
Alunite:		——
Jarosite $K Fe_3 (S O_4)_2(O H)_6$	h	521-1493
Amazonite	hn	521-0984
Amazon Stone	hn	521-0984
Amber	hjn	521-0933
Amber Mica $K(Mg, Fe)_3 Al Si_3 O_{10}(O H, F)_2$	adefhijn	521-0241
see also: Aluminum Magnesium Potassium Silicate Fluoride $K Mg_3 Al Si_3 O_{10} F_2$		102-0262
see also: Aluminum Germanium Magnesium Potassium Oxide Fluoride $K Mg_3 Al Ge_3 O_{10} F_2$		102-0962
see also: Aluminum Germanium Magnesium Potassium Oxide Fluoride, Nonstoichiometric $K Mg_3 Al_n Ge_4 O_{10} F_2$		102-0963

Substance Name	*Property*	*Number*
Amblygonite $(Li, Na) Al P O_4(F, O H)$	ah	521-1149
Ambroid	hjn	521-0933
Amethyst $Si O_2$	hj	521-0682
see also: Silicon Oxide $Si O_2$		122-0009
Ampangabeite $(Y, Ce, U, Ca, Fe, Pb, Th)(Nb, Ta, Ti, Sn)_2 O_6$	eh	521-0297
Amphibole:		——
Actinolite $Ca_2(Mg, Fe)_5 Si_8 O_{22}(O H)_2$	ahijn	521-0950
see also: Asbestos		521-0055
Anthophyllite $(Mg, Fe)_7 Si_8 O_{22}(O H)_2$	aehij	521-1134
Barkevikite	a	521-1136
Cummingtonite $(Fe, Mg)_7 Si_8 O_{22}(O H)_2$	ah	521-1135
Gedrite $(Mg, Fe)_2 Mg Al_2 Si_2 O_6$	e	521-1441
Glaucophane $Na_2(Mg, Fe)_3 Al_2 Si_8 O_{22}(O H)_2$	ah	521-1137
Grunerite $Fe_7 Si_8 O_{22}(O H)_2$	a	521-0949
Holmquistite $(Na, K, Ca) Li(Mg, Fe)_3 Al_2 Si_8 O_{22}(O H)_2$	h	521-1401
Hornblende $Ca_2 Na(Mg, Fe)_4(Al, Fe, Ti) Si_8 O_{22} (O, O H)_2$	aeghijnor	521-0951
Hornblende Andesite	n	521-0011
Magnesioriebeckite	a	521-1138
Nephrite	h	521-1400
Richterite $(Na, K)_2(Mg, Mn, Ca)_6 Si_8 O_{22}(O H)_2$	ah	521-1139
Riebeckite $Na_2(Fe, Mg)_5 Si_8 O_{22}(O H)_2$	ah	521-0952
Smaragdite	h	521-1399
Tremolite	aeghij	521-0109
Uralite	h	521-1398
Amphibole Biotite Schist	ade	521-1413
Amphibole Schist	a	521-0282
Amphibolic Gneiss	ade	521-0564
Amphibolic With Serpentine Gneiss	ade	521-0570
Amphibolite	adefhn	521-0274
Amphibolite - Biotite	a	521-1489
Amphibolite Basalt	ah	521-0848
Amphigene $K Al Si_2 O_6$	aeno	521-0062
see also: Potassium Aluminosilicate $K Al(Si O_3)_2$		102-0257
see also: Potassium Aluminosilicate $K_2 O . Al_2 O_3 . 4 Si O_2$		102-0957
Analbite $(Na, K) Al Si_3 O_8$	n	521-1613
see also: Anorthoclase $(Na, K) Al Si_3 O_8$		521-1047
Analcime Zeolite $Na Al Si_2 O_6$	adej	521-0041
see also: Sodium Aluminosilicate $Na Al(Si O_3)_2$		102-0266
Analcite Zeolite $Na Al Si_2 O_6$	adej	521-0041
see also: Sodium Aluminosilicate $Na Al(Si O_3)_2$		102-0266
Anatase, Natural $Ti O_2$	ehj	521-0094
see also: Titanium Oxide $Ti O_2$		122-0005
see also: Rutile, Natural $Ti O_2$		521-0092
see also: Brookite $Ti O_2$		521-1484
Andalusite $Al_2 Si O_5$	aehijno	521-0032
see also: Aluminum Silicate $Al_2 Si O_5$		102-0331
Andalusite Basalt	f	521-0269
Andalusite Biotite Schist	ad	521-0840
Andesine	ehno	521-0859
Andesite	aefghn	521-0330
Andesite:		——
Hypersthene	fg	521-0228
Old Red Sandstone Lava	f	521-0227
Olivine	ade	521-0377
Pyroxene	ade	521-0383
Sanukite	ade	521-0378
Andesite Basalt	adefg	521-0027
Andesite Porphyry	h	521-1069
Andorite $Pb Ag Sb_3 S_6$	h	521-1203
Andradite $Ca_3 Fe_2(Si O_4)_3$	ahijno	521-0958
see also: Calcium Iron Silicate $Ca_3 Fe_2(Si O_4)_3$		106-1096
Andradite:		——
Melanite	i	521-1386
Anglesite $Pb S O_4$	hjn	521-0786
see also: Lead Sulfate $Pb S O_4$		122-0021

Property: **a.** Thermal conductivity; **b.** Accommodation coefficient; **c.** Thermal contact resistance; **d.** Thermal diffusivity; **e.** Specific heat; **f.** Viscosity; **g.** Emittance; **h.** Reflectance; **i.** Absorptance; **j.** Transmittance; **k.** α_s/ϵ_t ratio; **l.** Prandtl number; **n.** Thermal linear expansion coefficient; **o.** Thermal volumetric expansion coefficient; **r.** Thermal radiative properties (g,h,i,j,k)

Substance Name	Property	Number
Anhydrite $CaSO_4$	adehj	521-0251
see also: Calcium Sulfate $CaSO_4$		106-0051
see also: Gypsum $CaSO_4$		521-0083
Anhydrite - Halite	a	521-1621
Anhydrite Sand	g	521-0906
Ankerite	aeh	521-0409
Annabergite $(Ni, Co)_3(AsO_4)_2$	h	521-1281
Anorthite, Natural $CaAl_2(SiO_4)_2$	aefhijno	521-0035
see also: Calcium Aluminum Silicate $CaAl_2(SiO_4)_2$		102-0125
Anorthoclase $(Na, K)AlSi_3O_8$	egh	521-1047
see also: Analbite $(Na, K)AlSi_3O_8$		521-1613
Anorthosite	aegno	521-0348
Anthophyllite $(Mg, Fe)_7Si_8O_{22}(OH)_2$	aehij	521-1134
Anthracite Coal	adehnr	521-0024
Anthraloxite	h	521-1130
Antigorite $Mg_3Si_2O_5(OH)_4$	ij	521-1605
Antimonite Sb_2S_3	ahjr	521-0095
see also: Antimony Sulfide Sb_2S_3		126-0014
Antimony, Natural Sb	h	521-1426
see also: Antimony Sb		101-0062
Antimony - Westerwaldite Fe(As Sb)	h	521-1448
Antimony Bismuth Silver Sulfide, Natural $Ag(Sb, Bi)S_2$	h	521-1199
Antimony Blende Sb_2S_2O	h	521-1173
Antimony Cobalt Nickel Sulfide, Natural (Co, Ni) Sb S	hj	521-0929
Antimony Copper Lead Sulfide, Natural $CuPb_{13}Sb_7S_{24}$	h	521-1561
Antimony Copper Telluride $Cu_2Sb(Te)$	h	521-1438
Antimony Glance Sb_2S_3	ahjr	521-0095
see also: Antimony Sulfide Sb_2S_3		126-0014
Antimony Hydroxide, Natural Sb_3O_6OH	j	521-1374
Antimony Iron Arsenide, Natural Fe(As Sb)	h	521-1448
Antimony Iron Arsenide, Natural $Fe(As, Sb)_2$	h	521-1447
Antimony Iron Oxide, Natural $FeSb_2O_6$	j	521-1296
Antimony Iron Sulfide, Natural Fe Sb S	h	521-1352
Antimony Lead Chlorate, Natural $PbSbO_2Cl$	n	521-1428
Antimony Lead Hydroxide, Natural $Pb_2Sb_2O_6(O, OH)$	j	521-1375
Antimony Lead Silver Sulfide, Natural $PbAgSb_3S_6$	h	521-1203
Antimony Lead Sulfide, Natural:		———
Boulangerite $Pb_5Sb_4S_{11}$	j	521-1180
Heteromorphite $Pb_7Sb_8S_{19}$	j	521-1178
Plagionite $Pb_5Sb_8S_{17}$	j	521-1177
Semseyite $Pb_9Sb_8S_{21}$	j	521-1179
Zinkenite $Pb_6Sb_{14}S_{27}$	h	521-1222
Antimony Nickel Sulfide, Natural Ni Sb S	hj	521-0928
Antimony Nickel Sulfide, Natural $Ni_9Sb_2S_8$	h	521-1547
Antimony Niobium Tantalum Oxide, Natural $Sb(Ta, Nb)O_4$	e	521-1231
Antimony Palladium Platinum Intermetallic, Natural $(Pt, Pd)_4Sb_3$	h	521-1439
Antimony Silver Sulfide, Natural Ag_5SbS_4	h	521-1204
Antimony Sulfate, Natural Sb_2S_2O	h	521-1173
Apatite	aehij	521-1463
Apatite:		———
Chlorapatite $Ca_5(PO_4)_3Cl$	ah	521-1461
see also: Calcium Phosphate Chloride $Ca_5(PO_4)_3Cl$		106-1088
Fluorapatite $Ca_5(PO_4)_3F$	aeghjn	521-0091
see also: Calcium Fluoride Phosphate $Ca_5(PO_4)_3F$		106-1314
Hydroxyapatite $Ca_5(PO_4)_3OH$	eh	521-0090
Mimetite $Pb_5(AsO_4)_3Cl$	hj	521-1284
see also: Lead Arsenate Chloride $Pb_5(AsO_4)_3Cl$		102-0436
Pyromorphite $Pb_5(PO_4)_3Cl$	h	521-0533
see also: Lead Phosphate Chloride $Pb_5(PO_4)_3Cl$		106-1661
Svabite $Ca_5(AsO_4)_3F$	j	521-1298
Vanadinite $Pb_5(VO_4)_3Cl$	j	521-1102
Apatite - Calcite	j	521-1569
Apatite - Forsterite - Magnetite Ore	ad	521-1322
Apatite - Quartz	j	521-1570
Aplite	adeg	521-0925
Aplite, Augite Porphyrite	g	521-0893
Aplite, Granite	g	521-0892
Aplite, Plagiomicrocline	ade	521-1334
Apophyllite $KCa_4Si_8O_{20}(F, OH)$	ahj	521-0967

Substance Name	Property	Number
Aquamarine	hj	521-1116
Aragonite $CaCO_3$	aeghijno	521-0105
see also: Calcium Carbonate $CaCO_3$		106-0005
see also: Calcite $CaCO_3$		106-2645
see also: Calcite, Natural $CaCO_3$		521-0075
see also: Chalk $CaCO_3$		521-0183
Aragonite:		———
Cerussite $PbCO_3$	hij	521-0626
see also: Lead Carbonate $PbCO_3$		106-0496
Strontianite $SrCO_3$	aehj	521-0030
see also: Strontium Carbonate $SrCO_3$		106-0758
Witherite $BaCO_3$	aehj	521-0195
see also: Barium Carbonate $BaCO_3$		104-0233
Aramayoite $Ag(Sb, Bi)S_2$	h	521-1199
Argentite Ag_2S	h	521-0754
see also: Silver Sulfide Ag_2S		102-0120
Argillaceous Sand	ad	521-1604
Argilla Clay	adefhjnr	521-0135
see also: Kaolinite, Mineral $Al_2Si_2O_5(OH)_4$		521-0174
see also: Kaolin Board		661-0444
see also: Kaolin Paper		661-0484
Argillite	aen	521-1010
Arite	h	521-1191
Arkose	h	521-1499
Armalcolite $(Mg, Fe)Ti_2O_5$	h	521-1599
see also: Pseudobrookite Fe_2TiO_5		521-0101
Arsenic, Natural As	hj	521-1309
see also: Arsenic As		101-0061
Arsenical Nickel Ni As	hj	521-0751
see also: Nickel Arsenide Ni As		102-0612
Arsenical Pyrite $FeS_2 . FeAs_2$	hj	521-0744
Arsenopyrite $FeS_2 . FeAs_2$	hj	521-0744
Artinite $Mg_2CO_3(OH)_2$	e	521-1229
Asbestos	adeghjn	521-0055
see also: Serpentine $(Mg, Fe)_3Si_2O_5(OH)_4$		521-0236
see also: Chrysotile $Mg_3Si_2O_5(OH)_4$		521-0799
see also: Actinolite $Ca_2(Mg, Fe)_5Si_8O_{22}(OH)_2$		521-0950
Asbestos:		———
Actinolite $Ca_2(Mg, Fe)_5Si_8O_{22}(OH)_2$	ahijn	521-0950
see also: Asbestos		521-0055
Anthophyllite $(Mg, Fe)_7Si_8O_{22}(OH)_2$	aehij	521-1134
Chrysotile $Mg_3Si_2O_5(OH)_4$	eijn	521-0799
see also: Asbestos		521-0055
see also: Serpentine $(Mg, Fe)_3Si_2O_5(OH)_4$		521-0236
Tremolite	aeghij	521-0109
Ascharite $MgBO_2(OH)$	ade	521-1645
Asphaltite	h	521-1110
see also: Asphalt		671-0015
Asteroid	h	521-1226
Asthenosphere	f	521-0641
Atacamite $Cu_2Cl(OH)_3$	h	521-1280
Attapulgite $(Mg, Al)_2Si_4O_{10}(OH)$	aeij	521-0182
Augen Gneiss	a	521-1551
Augite	h	521-1592
see also: Augite $(Ca, Na)(Mg, Fe, Al)(Si, Al)_2O_6$		521-0794
Augite $(Ca, Na)(Mg, Fe, Al)(Si, Al)_2O_6$	aghijno	521-0794
see also: Augite		521-1592
Augite:		———
Fassaite $(Ca, Mg, Fe)(Si, Al)O_3$	h	521-1397
Omphacite	j	521-1460
Augite Diorite	g	521-0898
Augite Diorite Gabbro	g	521-0900
Augite Granite	ad	521-1209
Augite Porphyry	ade	521-0373
Augite Syenite	an	521-0266
Aurichalcite $(Zn, Cu)_5(CO_3)_2(OH)_6$	h	521-1349
Australia Continent	ad	521-0836
Australite Tektite	j	521-1415
Aventurine SiO_2	h	521-1082
see also: Silicon Oxide SiO_2		122-0009
Axinite $(Ca, Mn, Fe)_3Al_2BSi_4O_{15}OH$	ahj	521-0946

Property: a. Thermal conductivity; **b.** Accommodation coefficient; **c.** Thermal contact resistance; **d.** Thermal diffusivity; **e.** Specific heat; **f.** Viscosity; **g.** Emittance; **h.** Reflectance; **i.** Absorptance; **j.** Transmittance; **k.** α_s/ϵ_t ratio; **l.** Prandtl number; **n.** Thermal linear expansion coefficient; **o.** Thermal volumetric expansion coefficient; **r.** Thermal radiative properties (g,h,i,j,k)

Substance Name	Property	Number
Azurite $Cu_3(CO_3)_2(OH)_2$	ehjno	521-0396
Baikalite	no	521-0860
Baikalite Hedenbergite	no	521-0860
Barite $BaSO_4$	aehijnor	521-0223
see also: Barium Sulfate $BaSO_4$		104-0025
Barium Feldspar $BaAl_2Si_2O_8$	aehn	521-0021
see also: Aluminum Barium Silicate $BaAl_2(SiO_4)_2$		102-0982
see also: Paracelsian $BaAl_2Si_2O_8$		521-1300
Barium Lead Sulfate, Natural $BaPb(SO_4)_2$	ij	521-1607
Barium Manganese Hydroxide Oxide, Natural $BaMn_9O_{16}(OH)_4$	h	521-1015
Barium Manganese Oxide, Natural $BaMn_8O_{16}$	h	521-1024
Barium Manganese Sodium Titanium Borate Silicate, Natural $Na_2BaMnTiB_2Si_6O_{20}$	j	521-1065
Barium Sodium Titanium Borate Silicate, Natural $BaNa_4Ti_2B_2Si_{10}O_{30}$	j	521-1063
Barium Titanium Silicate, Natural $BaTiSi_3O_9$	hj	521-1376
Barium Uranyl Hydroxide Selenate, Natural $Ba(UO_2)_3(SeO_3)_2(OH)_4$	j	521-1371
Barkevikite	a	521-1136
Barytes $BaSO_4$	aehijnor	521-0223
see also: Barium Sulfate $BaSO_4$		104-0025
Basalt	adefghjn	521-0123
Basalt:		———
Amphibolite	ah	521-0848
Andalusite	f	521-0269
Andesite	adefg	521-0027
Chlorite	a	521-0849
Nepheline	fg	521-0031
Olivine	adefgh	521-0028
Olivine, Carboniferous Lava	f	521-0232
Olivine - Pyroxene	h	521-1430
Olivine Melanephelinite	f	521-1237
Olivine Tholeiite	f	521-1487
Plagioclase	g	521-0901
Quartz	g	521-0890
Scoriaceous	h	521-0979
Tephrite	n	521-0678
Tholeiite	adfhn	521-0978
Trachydolerite	ad	521-1326
Trachydolerite Olivine Andesite	f	521-0231
Basalt Glass	an	521-1422
Basalt Glass Moon Rock	h	521-1596
Basalt Moon Rock	adeh	521-1392
Basalt Porphyry	n	521-1262
Basement Rock	o	521-1503
Bauxite	aehjno	521-0217
see also: Boehmite $AlO(OH)$		521-0302
see also: Diaspore $AlO(OH)$		521-0429
Bed Rock	a	521-1609
Beidellite	n	521-0945
Bellidoite Cu_2Se	h	521-1435
see also: Copper Selenide Cu_2Se		106-1879
Benitoite $BaTiSi_3O_9$	hj	521-1376
Benjaminite $Pb_2(Cu,Ag)_2Bi_4S_9$	h	521-1450
Bentonite, Calcium Clay	n	521-1056
Bentonite, Sodium Clay	e	521-0876
Bentonite Clay	adehijn	521-0111
Berlinite $AlPO_4$	ej	521-0389
see also: Aluminum Phosphate $AlPO_4$		102-0210
Bernstein	hjn	521-0933
Berryite $Pb_2(Cu,Ag)_3Bi_5S_{11}$	hn	521-0838
Beryl $Be_3Al_2Si_6O_{18}$	aehijnor	521-0130
see also: Beryllium Aluminosilicate $Be_3Al_2(SiO_3)_6$		102-0389
Beryl:		———
Aquamarine	hj	521-1116
Emerald	hjn	521-1117
Beryl - Quartz	h	521-0986
Beryllium Borate	hj	521-0696
Beryllium Calcium Manganese Silicate, Natural $Be(Ca,Mn)SiO_4$	j	521-1301

Substance Name	Property	Number
Berzelianite Cu_2Se	h	521-1435
see also: Copper Selenide Cu_2Se		106-1879
Berzeliite $(Mg,Mn)_2(Ca,Na)_3(AsO_4)_3$	j	521-1292
Betafite $(Ca,Na,U)_2(Nb,Ta)_2O_6(O,OH)$	h	521-1342
Betafite - Priazovite - Pyrochlore	h	521-1575
Bifeldspathic Granite	ade	521-1241
Bindheimite $Pb_2Sb_2O_6(O,OH)$	j	521-1375
Biotite $K(Mg,Fe)_3(Al,Fe)Si_3O_{10}(OH)_2$	adehjnr	521-0430
Biotite:		———
Lepidomelane	a	521-1141
Biotite - Granodiorite	a	521-1623
Biotite Chlorite Schist	ade	521-1411
Biotite Feldspar Garnet Quartz Schist	ade	521-1406
Biotite Feldspar Garnet Schist	adn	521-1405
Biotite Feldspar Quartz Schist	h	521-1328
Biotite Garnet Muscovite Schist	ade	521-1412
Biotite Garnet Schist	ade	521-1404
Biotite Granite Gneiss	n	521-1260
Biotite Granodiorite	a	521-1488
Biotite Granodiorite Gneiss	a	521-1627
Biotite Hornblende Granite	adn	521-1210
Biotite Mica Gneiss	a	521-0833
Biotite Muscovite Granite	n	521-1257
Biotite Muscovite Granite Gneiss	n	521-1259
Biotite Plagioclase Gneiss	no	521-0982
Biotite Schist	ade	521-1403
Biotite Sillimanite Gneiss	no	521-0983
Bismuth Copper Lead Silver Sulfide, Natural $Pb_2(Cu,Ag)_2Bi_4S_9$	h	521-1450
Bismuth Copper Lead Silver Sulfide, Natural $Pb_2(Cu,Ag)_3Bi_5S_{11}$	hn	521-0838
Bismuth Copper Lead Sulfide, Natural $PbCuBiS_3$	h	521-1573
Bismuth Copper Sulfide, Natural Cu_3BiS_3	h	521-1200
Bismuth Glance Bi_2S_3	h	521-0753
see also: Bismuth Sulfide Bi_2S_3		104-0174
Bismuthinite Bi_2S_3	h	521-0753
see also: Bismuth Sulfide Bi_2S_3		104-0174
Bismuth Lead Sulfide, Natural $PbBi_2S_4$	h	521-1452
Bismuth Lead Sulfide, Natural $Pb_3Bi_2S_6$	h	521-1456
Bismuth Lead Sulfide, Natural $Pb_6Bi_2S_9$	h	521-1453
Bismuth Lead Telluride, Natural $(Bi,Pb)_3Te_4$	h	521-1552
Bismuth Silver Sulfide, Natural $AgBiS_2$	h	521-1006
Bismuth Silver Sulfide, Natural $AgBi_3S_5$	h	521-1449
Bismuth Tungsten Oxide, Natural Bi_2WO_6	j	521-1100
Bitter Spar $CaMg(CO_3)_2$	adefghjnor	521-0084
see also: Calcium Magnesium Carbonate $CaMg(CO_3)_2$		106-0795
Bitumen, Natural	hj	521-1107
Bitumen, Natural:		———
Asphaltite	h	521-1110
see also: Asphalt		671-0015
Bituminous Coal	adefhjr	521-0249
Bituminous Shale	adej	521-0115
see also: Kerogen		521-1630
Bituminous Vitrinite Coal	h	521-1338
Bixbyite $(Mn,Fe)_2O_3$	h	521-0966
Black Earth Soil	adhr	521-0384
Black Jack $(Zn,Fe)S$	aehjnr	521-0086
see also: Zinc Sulfide ZnS		126-0005
see also: Wurtzite $(Zn,Fe)S$		521-1303
Black Lead C	aehn	521-0878
see also: Graphite C		101-0001
Black Loam Soil	h	521-0676
Black Tin SnO_2	hjnor	521-0470
see also: Tin Oxide SnO_2		122-0053
Blanc Fixe $BaSO_4$	aehijnor	521-0223
see also: Barium Sulfate $BaSO_4$		104-0025
Blende $(Zn,Fe)S$	aehjnr	521-0086
see also: Zinc Sulfide ZnS		126-0005
see also: Wurtzite $(Zn,Fe)S$		521-1303
Bloedite $Na_2Mg(SO_4)_2$	hj	521-1476
Blomstrandine $(Y,Ca,Th)(Ti,Nb)_2O_6$	h	521-1577
Blomstrandite $(Ca,Na,U)_2(Nb,Ta)_2O_6(O,OH)$	h	521-1342
Blue Copper Carbonate $Cu_3(CO_3)_2(OH)_2$	ehjno	521-0396

Property: a. Thermal conductivity; **b.** Accommodation coefficient; **c.** Thermal contact resistance; **d.** Thermal diffusivity; **e.** Specific heat; **f.** Viscosity; **g.** Emittance; **h.** Reflectance; **i.** Absorptance; **j.** Transmittance; **k.** α_s/ϵ_t ratio; **l.** Prandtl number; **n.** Thermal linear expansion coefficient; **o.** Thermal volumetric expansion coefficient; **r.** Thermal radiative properties (g,h,i,j,k)

Substance Name	Property	Number
Boehmite $AlO(OH)$	eijn	521-0302
see also: Aluminum Oxide Al_2O_3		102-0002
see also: Aluminum Hydroxide Oxide $AlOOH$		102-0427
see also: Bauxite		521-0217
see also: Diaspore $AlO(OH)$		521-0429
Bog	r	521-0541
Boghead Coal	adefhjr	521-0249
Bohdanowiczite $AgBiSe_2$	h	521-1213
see also: Silver Bismuth Selenide $AgBiSe_2$		102-0583
Bohemian Garnet $(Mg, Fe)_3Al_2(SiO_4)_3$	aehno	521-0411
Bologna Stone $BaSO_4$	aehijnor	521-0223
see also: Barium Sulfate $BaSO_4$		104-0025
Bolus Alba Clay	adefhjnr	521-0135
see also: Kaolinite, Mineral $Al_2Si_2O_5(OH)_4$		521-0174
see also: Kaolin Board		661-0444
see also: Kaolin Paper		661-0484
Bonaccordite Ni_2FeBO_5	h	521-1600
Borax $Na_2B_4O_7$	ah	521-1287
see also: Sodium Borate $Na_2B_4O_7$		104-0035
Borehole Sample	ade	521-1424
Bornite Cu_5FeS_4	h	521-0758
see also: Copper Iron Sulfide Cu_5FeS_4		106-1686
Bornite - Digenite	h	521-1084
Boulangerite $Pb_5Sb_4S_{11}$	j	521-1180
Boulder	adefghnor	521-0046
Boulder Clay	a	521-0107
Brackebuschite $Pb_4MnFe(VO_4)_4$	j	521-1101
Brannerite $(U, Ca, Ce)(Ti, Fe)_2O_6$	h	521-1214
Braunite $3Mn_2O_3 . MnSiO_3$	h	521-0992
Bravoite $(Ni, Fe)S_2$	h	521-0710
see also: Iron Nickel Sulfide, Nonstoichiometric $NiFeS_2$		110-0370
Breccia Moon Rock	adehjno	521-1357
Breithauptite $NiSb$	h	521-1196
see also: Antimony Nickel $NiSb$		135-0448
Briquet Coal	adef	521-0180
Britholite $(Ca, Ce)_5(SiO_4)_x(PO_4)_y(OH)_nF_z$	i	521-1271
Bronzite Pyroxene	ahno	521-0791
Bronzitite	a	521-0350
Brookite TiO_2	j	521-1484
see also: Titanium Oxide. TiO_2		122-0005
see also: Rutile, Natural TiO_2		521-0092
see also: Anatase, Natural TiO_2		521-0094
Brown Coal	adeh	521-0246
Brown Hematite	eh	521-0414
Brown Iron Ore	eh	521-0414
Brown Soil	h	521-1087
Brucite $Mg(OH)_2$	aehjn	521-0093
see also: Magnesium Hydroxide $Mg(OH)_2$		112-0017
Brushite $CaHPO_4$	j	521-1185
see also: Calcium Phosphate $CaHPO_4$		106-0568
Buhrstone	e	521-0511
see also: Sandstone		521-0196
Bunsenite NiO	j	521-1294
see also: Nickel Oxide NiO		120-0001
Burnt Clay	agh	521-0735
Burrstone	e	521-0511
see also: Sandstone		521-0196
Bytownite	ahno	521-0969
Bytownite Gabbro	ho	521-0975
Calamine (Eng) $ZnCO_3$	ahj	521-0611
see also: Zinc Carbonate $ZnCO_3$		106-0495
Calcareous Clay	adefn	521-0371
Calcareous Rock	n	521-1560
see also: Limestone		521-0085
see also: Chalk $CaCO_3$		521-0183
Calcareous Soil	ad	521-1048
see also: Limestone		521-0085
Calcedony SiO_2	aehijnr	521-0222
see also: Silicon Oxide SiO_2		122-0009
Calciclase $CaAl_2(SiO_4)_2$	aefhijno	521-0035
see also: Calcium Aluminum Silicate $CaAl_2(SiO_4)_2$		102-0125

Substance Name	Property	Number
Calcite, Natural $CaCO_3$	adeghijnor	521-0075
see also: Calcium Carbonate $CaCO_3$		106-0005
see also: Aragonite $CaCO_3$		521-0105
Calcite - Magnetite Ore	ad	521-1327
Calcium Aluminum Silicate, Natural $CaAl_2Si_4O_{12}$	ah	521-0489
Calcium Aluminum Silicate, Natural $Ca_2Al_4Si_8O_{24}$	e	521-0488
Calcium Arsenate Fluoride, Natural $Ca_5(AsO_4)_3F$	j	521-1298
Calcium Borate, Natural:		———
Colemanite $Ca_2B_6O_{11}$	ehj	521-0563
Meyerhofferite $Ca_2B_6O_{11}$	r	521-0342
Calcium Borate Hydroxide, Natural $Mg_3B_{11}O_{15}(OH)_9$	j	521-1170
Calcium Boride Hydroxide, Natural $Ca_3B_4(OH)_{18}$	j	521-1169
Calcium Borosilicate, Natural $CaB_2(SiO_4)_2$	ehj	521-0720
Calcium Cerium Iron Titanium Uranium Oxide, Natural $(U, Ca, Ce)(Ti, Fe)_2O_6$	h	521-1214
Calcium Cerium Lanthanum Borate Silicate, Natural $(Ce, La, Ca)BSiO_5$	j	521-1062
Calcium Feldspar	efn	521-0344
Calcium Feldspar:		———
Anorthite, Natural $CaAl_2(SiO_4)_2$	aefhijno	521-0035
see also: Calcium Aluminum Silicate $CaAl_2(SiO_4)_2$		102-0125
Calcium Hydrogen Arsenate, Natural $CaHAsO_4$	j	521-1186
Calcium Iron Magnesium Silicate, Natural $(Mg, Fe, Ca)(Mg, Fe)Si_2O_6$	h	521-0792
Calcium Iron Silicate, Natural $CaFeSi_2O_6$	hjno	521-0793
Calcium Iron Silicate Hydroxide, Natural $CaFe_3(SiO_4)_2(OH)$	h	521-1550
see also: Epidote $Ca_2(Al, Fe)_3Si_3O_{12}OH$		521-0942
Calcium Larsenite $(Ca, Pb)ZnSiO_4$	hj	521-0987
Calcium Magnesium Arsenate Fluoride, Natural $CaMgAsO_4F$	j	521-1291
Calcium Magnesium Borate, Natural $CaMgB_6O_{11}$	e	521-1647
Calcium Magnesium Carbonate, Natural $CaMg_3(CO_3)_4$	e	521-1228
Calcium Magnesium Manganese Sodium Arsenate, Natural $(Mg, Mn)_2(Ca, Na)_3(AsO_4)_3$	j	521-1292
Calcium Magnesium Phosphate Fluoride, Natural $CaMgPO_4F$	j	521-1290
Calcium Phosphate Hydroxide, Natural $Ca_5(PO_4)_3OH$	eh	521-0090
Calcium Potassium Silicate Hydroxide Fluoride, Natural $KCa_4Si_8O_{20}(F, OH)$	ahj	521-0967
Calcium Silicate Hydroxide, Natural $Ca_3Si_2O_4(OH)_6$	n	521-1602
Calcium Silicide Carbonate Sulfate Hydroxide, Natural $Ca_3Si(OH)_6(CO_3)SO_4$	j	521-1477
Calcium Sodium Borate, Natural $NaCaB_5O_9$	h	521-1289
Calcium Sodium Carbonate, Natural $Na_2Ca(CO_3)_2$	h	521-1347
Calcium Sodium Feldspar	ahjnor	521-0552
Calcium Sodium Silicate Hydroxide, Natural $NaCa_2Si_3O_8OH$	ahj	521-0948
Calcium Sodium Sulfate, Natural $Na_2Ca(SO_4)_2$	hj	521-0784
Calcspar $CaCO_3$	adeghijnor	521-0075
see also: Calcium Carbonate $CaCO_3$		106-0005
see also: Aragonite $CaCO_3$		521-0105
Calkybeborosilite	j	521-1064
Cancrinite $(Na, Na, Ca)_4(AlSiO_4)_6CO_3$	hjno	521-0930
Cannel Coal	a	521-0632
Cape Ruby $(Mg, Fe)_3Al_2(SiO_4)_3$	aehno	521-0411
Carbonaceous Shale	adehjnr	521-0070
Carbonate - Silica Sand	h	521-1156
Carbonate Rock	a	521-1118
Carbonate Sand	h	521-1158
Carbonatite, Calcitic	ad	521-1323
Carbonizing Coal	adehjn	521-0660
Carboperlite	a	521-0711
Carbuncle $Fe_3Al_2(SiO_4)_3$	ahjnor	521-0412
Carnallite $KCl . MgCl_2$	e	521-0256
see also: Magnesium Potassium Chloride $MgCl_2 . KCl$		106-0236
Carnelian SiO_2	h	521-1105
see also: Silicon Oxide SiO_2		122-0009
Carnotite $K(UO_2)_2(VO_4)_2$	h	521-1282
Carrollite $Cu(Co, Ni)_2S_4$	h	521-0826

Property: **a.** Thermal conductivity; **b.** Accommodation coefficient; **c.** Thermal contact resistance; **d.** Thermal diffusivity; **e.** Specific heat; **f.** Viscosity; **g.** Emittance; **h.** Reflectance; **i.** Absorptance; **j.** Transmittance; **k.** α_s/ϵ_t ratio; **l.** Prandtl number; **n.** Thermal linear expansion coefficient; **o.** Thermal volumetric expansion coefficient; **r.** Thermal radiative properties (g,h,i,j,k)

Substance Name	Property	Number
Cassiterite SnO_2	hjnor	521-0470
see also: Tin Oxide SnO_2		122-0053
Catlinite	a	521-0482
Cawk $BaSO_4$	aehijnor	521-0223
see also: Barium Sulfate $BaSO_4$		104-0025
Celestine $SrSO_4$	ahjn	521-0785
see also: Strontium Sulfate $SrSO_4$		122-0132
Celestite $SrSO_4$	ahjn	521-0785
see also: Strontium Sulfate $SrSO_4$		122-0132
Celsian Feldspar $BaAl_2Si_2O_8$	aehn	521-0021
see also: Aluminum Barium Silicate $BaAl_2(SiO_4)_2$		102-0982
see also: Paracelsian $BaAl_2Si_2O_8$		521-1300
Ceramic Clay	n	521-0915
Cerium Lanthanum Neodymium Thorium Silicate Phosphate, Natural (Ce, La, Nd, Th)(SiO_4PO_4)	aehj	521-0216
Cerolite	e	521-1172
Cerussite $PbCO_3$	hij	521-0626
see also: Lead Carbonate $PbCO_3$		106-0496
Ceylonite	o	521-0990
see also: Picotite		521-1162
Chabasite $CaAl_2Si_4O_{12}$	ah	521-0489
Chabazite $CaAl_2Si_4O_{12}$	ah	521-0489
Chalcedonite SiO_2	aehijnr	521-0222
see also: Silicon Oxide SiO_2		122-0009
Chalcedony SiO_2	aehijnr	521-0222
see also: Silicon Oxide SiO_2		122-0009
Chalcedony:		———
Agate SiO_2	j	521-1223
see also: Silicon Oxide SiO_2		122-0009
Carnelian SiO_2	h	521-1105
see also: Silicon Oxide SiO_2		122-0009
Flint, Mineral SiO_2	ahn	521-1465
see also: Silicon Oxide SiO_2		122-0009
Hornstone SiO_2	nor	521-0629
see also: Silicon Oxide SiO_2		122-0009
Prase SiO_2	j	521-1224
see also: Silicon Oxide SiO_2		122-0009
Chalcocite Cu_2S	h	521-0756
see also: Copper Sulfide Cu_2S		106-0104
see also: Djurleite Cu_2S		521-1436
Chalcomenite $CuSeO_3$	j	521-1370
see also: Copper Selenite $CuSeO_3$		106-1374
Chalcophanite $(Zn, Mn, Fe)Mn_2O_5$	hj	521-1022
Chalcopyrite $CuFeS_2$	ahj	521-0099
see also: Copper Iron Sulfide $CuFeS_2$		106-0791
Chalcostibite $CuSbS_2$	h	521-1197
see also: Antimony Copper Sulfide $CuSbS_2$		106-0505
Chalk $CaCO_3$	adeg	521-0183
see also: Calcium Carbonate $CaCO_3$		106-0005
see also: Aragonite $CaCO_3$		521-0105
see also: Calcareous Rock		521-1560
Chernozem Soil	adhr	521-0384
Chert	agjnr	521-0534
see also: Flint SiO_2		521-0481
see also: Jasper		521-1566
Chert - Limestone	h	521-1500
Chervetite $Pb_2V_2O_7$	h	521-1379
Chessy Copper $Cu_3(CO_3)_2(OH)_2$	ehjno	521-0396
Chessylite $Cu_3(CO_3)_2(OH)_2$	ehjno	521-0396
Chestnut-brown Earth Soil	h	521-0693
China Clay	adefhjnr	521-0135
see also: Kaolinite, Mineral $Al_2Si_2O_5(OH)_4$		521-0174
see also: Kaolin Board		661-0444
see also: Kaolin Paper		661-0484
Chiolite $5NaF.3AlF_3$	r	521-0469
Chlorapatite $Ca_5(PO_4)_3Cl$	ah	521-1461
see also: Calcium Phosphate Chloride $Ca_5(PO_4)_3Cl$		106-1088
Chlorite $(Mg, Fe)_6AlSi_3O_{10}(OH)_8$	ahijn	521-0912
Chlorite:		———
Clinochlore $(Mg, Fe, Al)_3(Si, Al)_2O_5(OH)_4$	j	521-1481
Penninite $(Mg, Fe, Al)_6(Si, Al)_4O_{10}(OH)_8$	j	521-0177
Ripidolite $(Mg, Fe)_9Al_6Si_5O_{20}(OH)_{16}$	n	521-0818
see also: Mica		521-0108
Chlorite - Muscovite	j	521-1378
Chlorite Basalt	a	521-0849
Chlorite Mica Quartz Rock	a	521-0841
Chlorite Muscovite Schist	ade	521-1410
Chlorite Schist	adehn	521-0296
Chloritoid $Fe_2Al_4Si_2O_{10}(OH)_4$	j	521-1480
Chondrite	aghik	521-0887
see also: Meteorite		521-0239
Chripobul	h	521-1391
Chrome Diopside	h	521-1593
see also: Diopside $CaMg(SiO_3)_2$		521-0036
Chrome Iron Ore $(Fe, Mg)(Cr, Al)_2O_4$	adehijn	521-0366
see also: Magnesiochromite $(Mg, Fe)(Cr, Al)_2O_4$		521-0963
Chrome Ore	n	521-0870
Chrome Sand	n	521-0870
Chromite $(Fe, Mg)(Cr, Al)_2O_4$	adehijn	521-0366
see also: Magnesiochromite $(Mg, Fe)(Cr, Al)_2O_4$		521-0963
Chromite - Magnesite	n	521-1113
Chromite Sand	a	521-1618
Chrysoberyl $BeAl_2O_4$	hn	521-1216
see also: Aluminum Beryllium Oxide $BeAl_2O_4$		102-0085
Chrysocolla $Cu_2H_2Si_2O_5(OH)_4$	h	521-1396
Chrysolite $(Mg, Fe)_2SiO_4$	adeghijnor	521-0258
Chrysoprase SiO_2	r	521-0628
see also: Silicon Oxide SiO_2		122-0009
Chrysotile $Mg_3Si_2O_5(OH)_4$	eijn	521-0799
see also: Asbestos		521-0055
see also: Serpentine $(Mg, Fe)_3Si_2O_5(OH)_4$		521-0236
Churchite YPO_4	j	521-1184
see also: Yttrium Phosphate YPO_4		122-0296
Cinnabar HgS	hj	521-0749
see also: Mercury Sulfide HgS		112-0402
Cinnamon Stone	a	521-1071
see also: Zircon $ZrSiO_4$		521-0250
Citrine SiO_2	hj	521-0684
see also: Silicon Oxide SiO_2		122-0009
Clay	adj	521-1571
Clay:		———
Bentonite	adehijn	521-0111
Bentonite, Calcium	n	521-1056
Bentonite, Sodium	e	521-0876
Burnt	agh	521-0735
Calcareous	adefn	521-0371
Ceramic	n	521-0915
Colloidal	adehijn	521-0111
Dixie	ade	521-0215
Fireclay	adefgjn	521-0393
Fuller's Earth	aj	521-0572
Kaolin,	adefhjnr	521-0135
see also: Kaolinite, Mineral $Al_2Si_2O_5(OH)_4$		521-0174
see also: Kaolin Board		661-0444
see also: Kaolin Paper		661-0484
Kaolin, Sodium	e	521-0707
Kaolinite, Sodium	e	521-0707
Keramzit	adefhjnr	521-0040
Leda	a	521-0630
see also: Clay Soil		521-0645
Loess	ej	521-0798
Marl	adefn	521-0371
Micaceous	n	521-0856
Molding	a	521-0441
Red	hj	521-0677
Sodium	e	521-0708
Wilkanite	adehijn	521-0111
Yellow Sandstone	h	521-1651
Clay; Mineral:		———
Attapulgite $(Mg, Al)_2Si_4O_{10}(OH)$	aeij	521-0182
Beidellite	n	521-0945
Dickite	ehijn	521-0505
see also: Aluminum Silicate $Al_2Si_2O_7$		102-0866
see also: Kaolinite, Mineral $Al_2Si_2O_5(OH)_4$		521-0174

Property: **a.** Thermal conductivity; **b.** Accommodation coefficient; **c.** Thermal contact resistance; **d.** Thermal diffusivity; **e.** Specific heat; **f.** Viscosity; **g.** Emittance; **h.** Reflectance; **i.** Absorptance; **j.** Transmittance; **k.** a_s/ϵ_t ratio; **l.** Prandtl number; **n.** Thermal linear expansion coefficient; **o.** Thermal volumetric expansion coefficient; **r.** Thermal radiative properties (g,h,i,j,k)

Substance Name	Property	Number
Clay; Mineral: -(Cont)-		
Halloysite $Al_2 Si_2 O_5(OH)_4$	ehijn	521-0506
see also: Aluminum Silicate Hydroxide $Al_2 Si_2 O_5(OH)_4$		102-0871
Hectorite	ij	521-1536
Illite	ehijn	521-0765
Kaolinite, Mineral $Al_2 Si_2 O_5(OH)_4$	adehijn	521-0174
see also: Aluminum Silicate $Al_2 Si_2 O_7$		102-0866
see also: Aluminum Silicate Hydroxide $Al_2 Si_2 O_5(OH)_4$		102-0871
see also: Kaolin,		521-0135
see also: Dickite		521-0505
Montmorillonite	ehijn	521-0764
Montmorillonite, Aluminum	e	521-0706
Montmorillonite, Calcium	e	521-0705
Montmorillonite, Lithium	ei	521-0729
Montmorillonite, Potassium	e	521-0689
Montmorillonite, Sodium	ei	521-0730
Nontronite	hj	521-1329
Sepiolite $Mg_4 (Si_2 O_5)_3(OH)_2$	e	521-1171
see also: Magnesium Silicate $Mg_2 Si_3 O_8$		119-0065
Clay, Brown Ironstone	eh	521-0414
Clay Loam Soil	r	521-0544
Clay Soil	adehj	521-0645
see also: Leda Clay		521-0630
Clay Soil With Ground Water	g	521-1654
Claystone	ah	521-0709
Clay With Gravel And Sand Soil	aden	521-1255
Cleavelandite	h	521-0985
Clevelandite	h	521-0985
Cliff	hr	521-0546
Clinkstone	ghn	521-1011
Clinochlore $(Mg, Fe, Al)_3(Si, Al)_2 O_5(OH)_4$	j	521-1481
Clinoenstatite $Mg Si O_3$	n	521-1054
see also: Magnesium Silicate $Mg Si O_3$		119-0009
Clinopyroxene	no	521-1339
Clinopyroxene:		———
Acmite $Na Fe(Si O_3)_2$	ahn	521-0947
see also: Iron Sodium Silicate $Na Fe(Si O_3)_2$		110-0495
Augite $(Ca, Na)(Mg, Fe, Al)(Si, Al)_2 O_6$	aghijno	521-0794
see also: Augite		521-1592
Chrome Diopside	h	521-1593
see also: Diopside $Ca Mg(Si O_3)_2$		521-0036
Clinoenstatite $Mg Si O_3$	n	521-1054
see also: Magnesium Silicate $Mg Si O_3$		119-0009
Diopside $Ca Mg(Si O_3)_2$	aefhijno	521-0036
see also: Calcium Magnesium Silicate $Ca Mg Si_2 O_6$		106-0940
see also: Chrome Diopside		521-1593
Hedenbergite $Ca Fe Si_2 O_6$	hjno	521-0793
Jadeite $Na (Al, Fe) Si_2 O_6$	adehij	521-0058
Kunzite	j	521-1486
Omphacite	j	521-1460
Pigeonite $(Mg, Fe, Ca)(Mg, Fe) Si_2 O_6$	h	521-0792
Salite $Ca(Mg, Fe) Si_2 O_6$	h	521-1578
Spodumene $Li Al Si_2 O_6$	aehn	521-0346
see also: Lithium Aluminosilicate $Li Al(Si O_3)_2$		102-0269
Clinopyroxene Moon Rock	hj	521-1059
Clinozoisite $Ca_2 Al_3 Si_3 O_{12} OH$	aehij	521-1131
Coal	adefhjnr	521-0069
Coal:		———
Akibara, Japan	j	521-1126
Anthracite	adehnr	521-0024
Bituminous	adefhjr	521-0249
Bituminous Vitrinite	h	521-1338
Boghead	adefhjr	521-0249
Briquet	adef	521-0180
Brown	adeh	521-0246
Cannel	a	521-0632
Carbonizing	adehjn	521-0660
Coking	adehjn	521-0660
Fat	ej	521-0740
Fusain	e	521-0056

Substance Name	Property	Number
Coal: -(Cont)-		
Fusinite	h	521-1119
Gas	adefhjr	521-0249
Hard	adehnr	521-0024
Jet	hj	521-0796
Lean	ej	521-0739
Leiptinite	h	521-1120
Lignite	adeh	521-0065
Long Flame	ej	521-0817
Metamorphosed	h	521-1208
Miike, Japan	fj	521-0019
Mineral Charcoal	e	521-0056
Mother Of	e	521-0056
Nut	adefhjr	521-0249
Sea	adefhjr	521-0249
Semifusinite	h	521-0816
Semivitrinite	h	521-1121
Soft	adefhjr	521-0249
Subbituminous	aden	521-1114
Vitrinite	h	521-0738
Coal, Gasified	a	521-1557
Coal - Iron Ore	ade	521-1572
Coal Concentrate	h	521-1002
Cobalt Arsenide Sulfide, Natural $Co As S$	eh	521-0043
Cobalt Copper Nickel Sulfide, Natural $Cu(Co, Ni)_2 S_4$	h	521-0826
Cobalt Glance $Co As S$	eh	521-0043
Cobalt Iron Selenide, Natural $(Co, Fe) Se_2$	h	521-1614
Cobaltite $Co As S$	eh	521-0043
Cobalt Nickel Arsenate, Natural $(Ni, Co)_3(As O_4)_2$	h	521-1281
Cobalt Nickel Arsenide, Natural $(Co, Ni) As_2$	h	521-0752
Cobalt Nickel Arsenide, Natural:		———
Skutterudite $(Co, Ni) As_3$	hj	521-1308
Skutterudite, Smaltite $(Co, Ni) As_3$	h	521-1433
Cobalt Nickel Arsenide Ore	h	521-1073
Cobalt Nickel Selenate, Natural $(Ni, Co) Se O_3$	j	521-1369
Cobaltomenite $Co Se O_3$	j	521-1368
Cobalt Selenate, Natural $Co Se O_3$	j	521-1368
Cobblestone	adefghnor	521-0046
Coesite, Natural $Si O_2$	aej	521-0823
see also: Silicon Oxide $Si O_2$		122-0009
see also: Quartz, Natural $Si O_2$		521-0732
Coffinite $U(Si O_4)_x (OH)_y$	h	521-0851
Coking Coal	adehjn	521-0660
Colemanite $Ca_2 B_6 O_{11}$	ehj	521-0563
Colloidal Clay	adehijn	521-0111
Columbite $(Fe, Mn)(Nb, Ta)_2 O_6$	h	521-1574
see also: Tantalite $(Fe, Mn)(Nb, Ta)_2 O_6$		521-0750
Comet	gh	521-1219
Conglomerate	aehr	521-0525
Cooperite $(Pt, Pd) S$	h	521-0884
Copper, Natural Cu	a	521-1639
Copper Arsenate Hydroxide, Natural $Cu_2 As O_4 OH$	j	521-1188
Copper Arsenide Sulfide, Natural $(Cu, Fe)_{12} As_4 S_{13}$	h	521-0880
see also: Fahlore		521-0748
see also: Tetrahedrite $(Cu, Fe)_{12} Sb_4 S_{13}$		521-1205
Copper Carbonate Hydroxide, Natural $Cu_2 C O_3(OH)_2$	hij	521-1227
Copper Carbonate Hydroxide, Natural $Cu_3(C O_3)_2(OH)_2$	ehjno	521-0396
Copper Gallium Germanium Iron Arsenide Sulfide, Natural $Cu_3(Ge, Ga, Fe)(S, As)_4$	h	521-0882
Copper Germanium Iron Zinc Arsenide Sulfide, Natural $Cu_3(Fe, Ge, Zn)(S, As)_4$	h	521-0881
Copper Glance $Cu_2 S$	h	521-0756
see also: Copper Sulfide $Cu_2 S$		106-0104
see also: Djurleite $Cu_2 S$		521-1436
Copper Hydrogen Silicate, Natural $Cu_2 H_2 Si_2 O_5(OH)_4$	h	521-1396
Copper Hydroxide Chloride, Natural $Cu_2 Cl(OH)_3$	h	521-1280
Copper Iron Sulfide, Natural $Cu_3 Fe S_4$	h	521-1636
see also: Nukundamite $Cu_5 Fe S_6$		521-1635
Copper Iron Sulfide, Natural $Cu_5 Fe S_6$	h	521-1635
see also: Idaite $Cu_3 Fe S_4$		521-1636
Copper Iron Tin Sulfide, Natural $Cu_2 Fe Sn S_4$	h	521-1637
Copper Iron Tin Sulfide, Natural $Cu_7 Fe_2 Sn S_{10}$	h	521-1638

Property: a. Thermal conductivity; **b.** Accommodation coefficient; **c.** Thermal contact resistance; **d.** Thermal diffusivity; **e.** Specific heat; **f.** Viscosity; **g.** Emittance; **h.** Reflectance; **i.** Absorptance; **j.** Transmittance; **k.** α_s/ϵ_t ratio; **l.** Prandtl number; **n.** Thermal linear expansion coefficient; **o.** Thermal volumetric expansion coefficient; **r.** Thermal radiative properties (g,h,i,j,k)

Substance Name	Property	Number
Copper Lead Uranyl Hydroxide Selenate, Natural	j	521-1373
$Pb_2Cu_5(UO_2)_2(SeO_3)_6(OH)_6$		
Copper Lead Vanadium Zinc Hydroxide Oxide, Natural:		———
Mottramite $Pb(Cu, Zn)(VO_4)(OH)$	j	521-1103
Zincdescloizite $Pb(Zn, Cu)(VO_4)(OH)$	j	521-1104
Copper Manganese Oxide, Natural $CuMnO_2$	h	521-1021
Copper Phosphate Hydroxide, Natural Cu_2PO_4OH	j	521-1187
Copper Pyrites $CuFeS_2$	ahj	521-0099
see also: Copper Iron Sulfide $CuFeS_2$		106-0791
Copper Sulfide, Natural Cu_9S_5	h	521-1083
Copper Sulfide Ore	ade	521-1074
Copper Uranyl Hydroxide Selenate, Natural	j	521-1372
$Cu(UO_2)_3(SeO_3)_3(OH)_2$		
Copper Uranyl Hydroxide Sulfate, Natural	e	521-0284
$Cu(UO_2)_2(SO_4)_2(OH)_2$		
Copper Vanadium Sulfide, Natural Cu_3VS_4	h	521-0747
Copper Zinc Carbonate Hydroxide, Natural	h	521-1349
$(Zn, Cu)_5(CO_3)_2(OH)_6$		
Coral, White	h	521-1649
Coral Limestone	j	521-1563
Coral Sand	h	521-1036
Cordierite $(Mg, Fe)_2Al_4Si_5O_{18}$	aehijno	521-0443
see also: Magnesium Aluminosilicate		102-0143
$Mg_2Al_4Si_5O_{18}$		
see also: Silicon Oxide Mixture		503-0676
$SiO_2 > Al_2O_3 > MgO$		
Coronadite $PbMn_8O_{16}$	h	521-1023
Corundum, Natural Al_2O_3	adeghijnor	521-0077
see also: Aluminum Oxide Al_2O_3		102-0002
Corundum, Natural:		———
Ruby, Natural Al_2O_3	adeghijnr	521-0191
see also: Aluminum Oxide Al_2O_3		102-0002
Sapphire, Natural Al_2O_3	adeghjnr	521-0064
see also: Aluminum Oxide Al_2O_3		102-0002
Cosalite $Pb_2Si_2S_5$	h	521-1451
Covelline CuS	eh	521-0877
see also: Copper Sulfide CuS		106-0023
Covellite CuS	eh	521-0877
see also: Copper Sulfide CuS		106-0023
Crednerite $CuMnO_2$	h	521-1021
Cristobalite, Natural SiO_2	aehijnr	521-0254
see also: Silicon Oxide SiO_2		122-0009
see also: Tridymite SiO_2		122-0646
see also: Tridymite, Natural SiO_2		521-0255
see also: Quartz, Natural SiO_2		521-0732
Crocoicite $PbCrO_4$	hj	521-0854
see also: Lead Chromate $PbCrO_4$		106-0678
Crocoisite $PbCrO_4$	hj	521-0854
see also: Lead Chromate $PbCrO_4$		106-0678
Crocoite $PbCrO_4$	hj	521-0854
see also: Lead Chromate $PbCrO_4$		106-0678
Cryolite Na_3AlF_6	efhjr	521-0385
see also: Aluminum Sodium Fluoride Na_3AlF_6		102-0016
Cryolithionite $Na_3Li_3Al_2F_{12}$	j	521-1293
Cryophyllite	j	521-1091
Cryptomelane KMn_8O_{16}	h	521-1025
Cumberlandite	a	521-1218
Cummingtonite $(Fe, Mg)_7Si_8O_{22}(OH)_2$	ah	521-1135
Cuprite Cu_2O	h	521-0644
see also: Copper Oxide Cu_2O		106-0127
Cuprodescloizite $Pb(Cu, Zn)(VO_4)(OH)$	j	521-1103
Cuprostibite $Cu_2Sb(Te)$	h	521-1438
Cyanite Al_2OSiO_4	aehijnor	521-0033
see also: Aluminum Silicate Al_2SiO_5		102-0331
Dacite	afghn	521-0243
Dacite, Alkaline	f	521-0244
Danburite $CaB_2(SiO_4)_2$	ehj	521-0720
Datolite $CaBSiO_4(OH)$	adehj	521-0783
D E; Diatomaceous Earth	adeghijnr	521-0015
Demesmaekerite	j	521-1373
$Pb_2Cu_5(UO_2)_2(SeO_3)_6(OH)_6$		
Descloizite $Pb(Zn, Cu)(VO_4)(OH)$	j	521-1104
Desert	ghr	521-0238

Substance Name	Property	Number
Desert Soil	hr	521-0451
Desmine $NaCa_2Al_5Si_{13}O_{36}$	ahj	521-0779
Deuterium Faujasite	j	521-0681
Deweylite	hj	521-1472
Diabase	adefghno	521-0308
see also: Ophite		521-0760
Dialogite $MnCO_3$	ahjn	521-0614
see also: Manganese Carbonate $MnCO_3$		106-0497
Diamond, Blue C	j	521-0776
Diamond, Natural C	adehijnor	521-0063
see also: Carbon C		101-0069
Diaspore $AlO(OH)$	ehijnr	521-0429
see also: Aluminum Oxide Al_2O_3		102-0002
see also: Bauxite		521-0217
see also: Boehmite $AlO(OH)$		521-0302
Diatomaceous Earth	adeghijnr	521-0015
Diatomite	adeghijnr	521-0015
Diatomite, Soil	j	521-1388
Dichroite $(Mg, Fe)_2Al_4Si_5O_{18}$	aehijno	521-0443
see also: Magnesium Aluminosilicate		102-0143
$Mg_2Al_4Si_5O_{18}$		
see also: Silicon Oxide Mixture		503-0676
$SiO_2 > Al_2O_3 > MgO$		
Dickite	ehijn	521-0505
see also: Aluminum Silicate $Al_2Si_2O_7$		102-0866
see also: Kaolinite, Mineral $Al_2Si_2O_5(OH)_4$		521-0174
Digenite Cu_9S_5	h	521-1083
Dike (Dyke)	a	521-0492
Diopside $CaMg(SiO_3)_2$	aefhijno	521-0036
see also: Calcium Magnesium Silicate $CaMgSi_2O_6$		106-0940
see also: Chrome Diopside		521-1593
Dioptase $CuSiO_2(OH)_2$	ei	521-0637
see also: Copper Silicate $CuSiO_3$		106-0796
Diorite	adeghino	521-0293
Diorite, Epidotized	f	521-0197
Diorite, Hornblende	ade	521-1211
Diorite, Silicified	a	521-0273
Diorite Porphyry	ade	521-1245
Disthene Al_2OSiO_4	aehijnor	521-0033
see also: Aluminum Silicate Al_2SiO_5		102-0331
Dixie Clay	ade	521-0215
Djurleite Cu_2S	h	521-1436
see also: Copper Sulfide Cu_2S		106-0104
see also: Chalcocite Cu_2S		521-0756
Dolerite	adefghno	521-0308
see also: Ophite		521-0760
Dolerite, Olivine	f	521-0233
Dolomite $CaMg(CO_3)_2$	adefghjnor	521-0084
see also: Calcium Magnesium Carbonate $CaMg(CO_3)_2$		106-0795
Dolomite – Quartzite	a	521-1446
Dolomitic Limestone	aj	521-1444
Dolostone	ae	521-0475
Domeykite Cu_3As	j	521-1310
Dry Bone $ZnCO_3$	ahj	521-0611
see also: Zinc Carbonate $ZnCO_3$		106-0495
Dufrenoysite $Pb_2As_2S_5$	j	521-1181
Dumortierite $Al_7(BO_3)(SiO_4)_3O_3$	ehjn	521-0911
Dunite	adghnr	521-0352
Dunite, Serpentinized	h	521-1642
Dunite Moon Rock	h	521-1393
Durangite $NaAlAsO_4F$	j	521-1299
Dyscrasite Ag_3Sb	h	521-1427
see also: Antimony Silver Ag_3Sb		135-0031
Earth	a	521-1633
Earth Core	adfn	521-0797
Earth Crust	aer	521-0054
Earth Mantle	aefn	521-0918
Earth Surface	dghir	521-0423
Eclogite	ano	521-0718
Elaeolite $(Na, K)(Al, Si)_2O_4$	aeghijno	521-0059
see also: Sodium Aluminosilicate		102-0953
$Na_2O \cdot Al_2O_3 \cdot 2SiO_2$		
see also: Carnegieite $Na_2O \cdot Al_2O_3 \cdot 2SiO_2$		102-0954

Property: a. Thermal conductivity; b. Accommodation coefficient; c. Thermal contact resistance; d. Thermal diffusivity; e. Specific heat; f. Viscosity; g. Emittance; h. Reflectance; i. Absorptance; j. Transmittance; k. α_s/ϵ_t ratio; l. Prandtl number; n. Thermal linear expansion coefficient; o. Thermal volumetric expansion coefficient; r. Thermal radiative properties (g,h,i,j,k)

Substance Name	Property	Number
Ellsworthite $(Ca, Na, U)_2(Nb, Ta)_2 O_6(O, OH)$	h	521-1342
Emerald	hjn	521-1117
Emplectite $Cu Bi S_2$	h	521-1198
see also: Bismuth Copper Sulfide $Cu Bi S_2$		104-0176
Enargite $Cu_3 As S_4$	h	521-1221
see also: Copper Arsenide Sulfide $Cu_3 As S_4$		102-0219
Enstatite Pyroxene $Mg Si O_3$	adhij	521-0769
see also: Magnesium Silicate $Mg Si O_3$		119-0009
Epidote $Ca_2(Al, Fe)_3 Si_3 O_{12} OH$	aehijno	521-0942
see also: Ilvaite $Ca Fe_3(Si O_4)_2(OH)$		521-1550
Epidote:		———
Allanite $(Ce, Ca, Y)(Al, Fe)_3(Si O_4)_3 OH$	h	521-1395
Clinozoisite $Ca_2 Al_3 Si_3 O_{12} OH$	aehij	521-1131
Piemontite $Ca_2 (Al, Mn, Fe)_3 Si_3 O_{12} OH$	j	521-1459
Tanzanite	hj	521-1115
Zoisite $Ca_2 Al_3 Si_3 O_{12} OH$	aehj	521-0941
Epidote Schist	ade	521-0380
Eskimoite	h	521-1454
Esperite $(Ca, Pb) Zn Si O_4$	hj	521-0987
Essonite	a	521-1071
see also: Zircon $Zr Si O_4$		521-0250
Estancia Playa Soil	ag	521-0866
Ethiops Mineral $Hg S$	hj	521-0749
see also: Mercury Sulfide $Hg S$		112-0402
Eucryptite $Li Al Si O_4$	n	521-1629
see also: Lithium Aluminosilicate $Li Al Si O_4$		102-0332
Euxenite $(Y, Ca, Ce, U, Th)(Nb, Ta)_2 O_6$	h	521-1343
Fahlerz	h	521-0748
see also: Tennantite $(Cu, Fe)_{12} As_4 S_{13}$		521-0880
see also: Tetrahedrite $(Cu, Fe)_{12} Sb_4 S_{13}$		521-1205
Fahlore	h	521-0748
see also: Tennantite $(Cu, Fe)_{12} As_4 S_{13}$		521-0880
see also: Tetrahedrite $(Cu, Fe)_{12} Sb_4 S_{13}$		521-1205
False Topaz $Si O_2$	hj	521-0684
see also: Silicon Oxide $Si O_2$		122-0009
Fassaite $(Ca, Mg, Fe)(Si, Al) O_3$	h	521-1397
Fat Coal	ej	521-0740
Fayalite $Fe_2 Si O_4$	aehjno	521-0261
see also: Iron Silicate $Fe_2 Si O_4$		110-0051
see also: Fayalite,		551-3076
Fayalite:		———
Hortonolite $(Fe, Mg, Mn)_2 Si O_4$	ah	521-0790
Feldspar	adefghijno	521-0394
Feldspar:		———
Adularia $K Al Si_3 O_8$	aehnor	521-0060
see also: Potassium Aluminum Silicate $K Al Si_3 O_8$		102-0173
Albite $Na Al Si_3 O_8$	adefghijnor	521-0057
see also: Sodium Aluminosilicate $Na Al Si_3 O_8$		102-0218
Albite - Anorthite $Na Al Si_3 O_8 - Ca Al_2 Si_2 O_8$	d	521-1549
Alkali Feldspar	ad	521-0913
see also: Aluminum Potassium Silicate		102-0878
Amazonite	hn	521-0984
Analbite $(Na, K) Al Si_3 O_8$	n	521-1613
see also: Anorthoclase $(Na, K) Al Si_3 O_8$		521-1047
Andesine	ehno	521-0859
Anorthite, Natural $Ca Al_2(Si O_4)_2$	aefhijno	521-0035
see also: Calcium Aluminum Silicate $Ca Al_2(Si O_4)_2$		102-0125
Anorthoclase $(Na, K) Al Si_3 O_8$	egh	521-1047
see also: Analbite $(Na, K) Al Si_3 O_8$		521-1613
Barium $Ba Al_2 Si_2 O_8$	aehn	521-0021
see also: Aluminum Barium Silicate $Ba Al_2(Si O_4)_2$		102-0982
see also: Paracelsian $Ba Al_2 Si_2 O_8$		521-1300
Bytownite	ahno	521-0969
Calcium	efn	521-0344
Celsian $Ba Al_2 Si_2 O_8$	aehn	521-0021
see also: Aluminum Barium Silicate $Ba Al_2(Si O_4)_2$		102-0982
see also: Paracelsian $Ba Al_2 Si_2 O_8$		521-1300
Cleavelandite	h	521-0985
Ferriferous Orthoclase	e	521-0061

Substance Name	Property	Number
Feldspar: -(Cont)-		
Labradorite	ahjnor	521-0471
Lime	efn	521-0344
Lithium Feldspar $Li Al Si_3 O_8$	n	521-1652
Lithium Potassium Feldspar $(Li, K) Al Si_3 O_8$	n	521-1653
Microcline $K Al Si_3 O_8$	aehjnor	521-0392
see also: Potassium Aluminum Silicate $K Al Si_3 O_8$		102-0173
Microcline - Perthite	a	521-1458
Moonstone	ijnr	521-0457
Orthoclase $K Al Si_3 O_8$	adefhijnor	521-0026
see also: Potassium Aluminum Silicate $K Al Si_3 O_8$		102-0173
Paracelsian $Ba Al_2 Si_2 O_8$	j	521-1300
see also: Aluminum Barium Silicate $Ba Al_2(Si O_4)_2$		102-0982
see also: Celsian Feldspar $Ba Al_2 Si_2 O_8$		521-0021
Perthite	fhn	521-0153
Plagioclase $(Na, Ca) Al (Si, Al) Si_2 O_8$	adeh	521-0495
Potassium Feldspar $K Al Si_3 O_8$	adfgn	521-0170
see also: Potassium Aluminum Silicate $K Al Si_3 O_8$		102-0173
Sanidine $K Al Si_3 O_8$	aehno	521-1045
see also: Potassium Aluminum Silicate $K Al Si_3 O_8$		102-0173
Sodium Feldspar	fn	521-0332
Feldspathic Mica Gneiss	a	521-0832
Feldspathic Pegmatite	a	521-0830
Feldspathoid:		———
Cancrinite $(Na, Na, Ca)_4(Al Si O_4)_6 C O_3$	hjno	521-0930
Hackmanite	j	521-1385
Hauyne $(Na, Ca)_8(Al_6 Si_6 O_{24})(S O_4)_x S_y$	hn	521-1109
Lazurite $(Na, Ca)_8(Al, Si)_{12} O_{24}(S, S O_4)$	hj	521-1351
see also: Lapis Lazuli		521-1350
Leucite $K Al Si_2 O_6$	aeno	521-0062
see also: Potassium Aluminosilicate $K Al(Si O_3)_2$		102-0257
see also: Potassium Aluminosilicate $K_2 O . Al_2 O_3 . 4 Si O_2$		102-0957
Melilite $Ca_2 Al_2 Si O_7$	ah	521-0943
see also: Calcium Aluminum Silicate $Ca_2 Al_2 Si O_7$		102-0092
see also: Gehlenite $Ca_2 Al_2 Si O_7$		521-1640
Nepheline $(Na, K) (Al, Si)_2 O_4$	aeghijno	521-0059
see also: Sodium Aluminosilicate $Na_2 O . Al_2 O_3 . 2 Si O_2$		102-0953
see also: Carnegieite $Na_2 O . Al_2 O_3 . 2 Si O_2$		102-0954
Nosean $Na_8 Al_6 Si_6 O_{24} S O_4$	n	521-1108
Sodalite $Na_4 Al_3 Si_3 O_{12} Cl$	ahjn	521-0970
Felsite Porphyry	hno	521-0976
Felspar	adefghijno	521-0394
Feolite	e	521-1443
Ferberite $Fe W O_4$	hj	521-1077
see also: Iron Tungsten Oxide $Fe W O_4$		110-0338
Ferriferous Orthoclase	e	521-0061
Ferrimolybdenite	j	521-1098
Ferroferrichrompicotite $Fe(Fe, Cr, Al)_2 O_4$	h	521-0937
Ferroferripicotite $Fe(Fe, Al)_2 O_4$	h	521-0938
Ferrophosphore Rock	a	521-0763
Ferropicotite	h	521-0759
Ferroplatinum Ore	h	521-1086
Ferroselite $Fe Se_2$	h	521-0874
see also: Iron Selenide $Fe Se_2$		110-0498
Ferrosilite $Fe Si O_3$	ah	521-1133
Fibrolite $Al_2 Si O_5$	aehijnor	521-0034
see also: Aluminum Silicate $Al_2 Si O_5$		102-0331
Fireclay	adefgjn	521-0393
Flagstone	hr	521-0550
Flagstone, Grey	r	521-0634
Flagstone, White	r	521-0633
Flint $Si O_2$	aehn	521-0481
see also: Chert		521-0534

Property: a. Thermal conductivity; **b.** Accommodation coefficient; **c.** Thermal contact resistance; **d.** Thermal diffusivity; **e.** Specific heat; **f.** Viscosity; **g.** Emittance; **h.** Reflectance; **i.** Absorptance; **j.** Transmittance; **k.** α_s/ϵ_t ratio; **l.** Prandtl number; **n.** Thermal linear expansion coefficient; **o.** Thermal volumetric expansion coefficient; **r.** Thermal radiative properties (g,h,i,j,k)

Substance Name	Property	Number
Flint, Mineral SiO_2	ahn	521-1465
see also: Silicon Oxide SiO_2		122-0009
Flint Clay	adefgjn	521-0393
Flood Plain, Crust Clay	r	521-0450
Flood Plain, Gravel	r	521-0448
Florspar CaF_2	aehjnr	521-0096
see also: Calcium Fluoride CaF_2		106-0026
Flow	a	521-0286
Fluoborite $Mg_3BO_3(F,OH)_3$	j	521-1168
Fluorapatite $Ca_5(PO_4)_3F$	aeghjn	521-0091
see also: Calcium Fluoride Phosphate $Ca_5(PO_4)_3F$		106-1314
Fluorite CaF_2	aehjnr	521-0096
see also: Calcium Fluoride CaF_2		106-0026
Fluorspar CaF_2	aehjnr	521-0096
see also: Calcium Fluoride CaF_2		106-0026
Fool's Gold FeS_2	adehijnor	521-0025
see also: Iron Sulfide FeS_2		110-0010
see also: Marcasite FeS_2		521-1304
Forsterite Mg_2SiO_4	adehijno	521-0022
see also: Magnesium Silicate Mg_2SiO_4		119-0004
see also: Forsterite,		551-3194
Forsterite - Apatite - Magnetite Ore	ad	521-1322
Fossil Resin:		———
Amber	hjn	521-0933
Retinite	h	521-1111
Franklinite $(Zn,Mn,Fe)(Fe,Mn)_2O_4$	h	521-1017
see also: Iron Zinc Oxide $ZnFe_2O_4$		110-0029
French Chalk $Mg_3Si_4O_{10}(OH)_2$	adehijnr	521-0116
Fuller's Earth	aj	521-0572
Fusain Coal	e	521-0056
Fusinite Coal	h	521-1119
Gabbro	adefghjno	521-0264
Gabbro:		———
Augite Diorite	g	521-0900
Basalt Dolerite Interbedded	ad	521-0843
Bytownite	ho	521-0975
Garnet	g	521-0899
Hornblende	adeo	521-0281
Hypersthene	n	521-1263
Norite	ano	521-0292
Olivine	ag	521-0904
Porphyrite Dikes Interbedded	ad	521-0844
Pyroxene	a	521-1247
Scapolite	dn	521-1655
Gabbro - Diabase	aden	521-1335
Gahnite $ZnAl_2O_4$	ahno	521-0962
see also: Aluminum Zinc Oxide $ZnAl_2O_4$		102-0185
Gahnite, Ferroan $Zn_nFe_xMg_yAl_2O_4$	n	521-1429
Galena PbS	aehjnr	521-0192
see also: Lead Sulfide PbS		123-0003
Galenite PbS	aehjnr	521-0192
see also: Lead Sulfide PbS		123-0003
Galenobismutite $PbBi_2S_4$	h	521-1452
Ganister	adehjno	521-0298
Garnet, Natural	adeghijno	521-0003
Garnet, Natural:		———
Almandine $Fe_3Al_2(SiO_4)_3$	ahjnor	521-0412
Andradite $Ca_3Fe_2(SiO_4)_3$	ahijno	521-0958
see also: Calcium Iron Silicate $Ca_3Fe_2(SiO_4)_3$		106-1096
Essonite	a	521-1071
see also: Zircon $ZrSiO_4$		521-0250
Grossular $Ca_3Al_2(SiO_4)_3$	aehjno	521-0959
see also: Aluminum Calcium Silicate $Ca_3Al_2(SiO_4)_3$		102-0797
Kimzeyite $Ca_3(Zr,Ti)_2(Al,Si)_3O_{12}$	j	521-1336
Melanite	i	521-1386
Pyrope $(Mg,Fe)_3Al_2(SiO_4)_3$	aehno	521-0411
Rhodolite	a	521-1072
Schorlomite $Ca_3(Fe,Ti)_2(Si,Ti)_3O_{12}$	hij	521-1337
Spessartine $Mn_3Al_2(SiO_4)_3$	ahjno	521-0957
Garnet, Natural: -(Cont)-		
Uvarovite $Ca_3Cr_2(SiO_4)_3$	h	521-1419
see also: Calcium Chromium Silicate $Ca_3Cr_2Si_3O_{12}$		106-2005
Garnet Gabbro	g	521-0899
Gas Coal	adefhjr	521-0249
Gaylussite $Na_2Ca(CO_3)_2$	h	521-1347
Gedrite $(Mg,Fe)_2MgAl_2Si_2O_6$	e	521-1441
Gehlenite $Ca_2Al_2SiO_7$	e	521-1640
see also: Calcium Aluminum Silicate $Ca_2Al_2SiO_7$		102-0092
see also: Melilite $Ca_2Al_2SiO_7$		521-0943
Geikielite $MgTiO_3$	eh	521-0098
see also: Magnesium Titanium Oxide $MgTiO_3$		119-0021
Genkinite $(Pt,Pd)_4Sb_3$	h	521-1439
Germanite $Cu_3(Ge,Ga,Fe)(S,As)_4$	h	521-0882
Gersdorffite $NiAsS$	h	521-1190
Getchellite $AsSbS_3$	j	521-1312
see also: Antimony Arsenide Sulfide $SbAsS_3$		102-0145
Giant Granite	adeghn	521-0762
Gibbsite $Al(OH)_3$	aehijn	521-0303
see also: Aluminum Oxide Al_2O_3		102-0002
Giimenite $Ba(UO_2)_3(SeO_3)_2(OH)_4$	j	521-1371
Gilders Whiting $CaCO_3$	adeg	521-0183
see also: Calcium Carbonate $CaCO_3$		106-0005
see also: Aragonite $CaCO_3$		521-0105
see also: Calcareous Rock		521-1560
Gilpinite $Cu(UO_2)_2(SO_4)_2(OH)_2$	e	521-0284
Gismondine $CaAl_2(SiO_4)_2$	j	521-1475
see also: Calcium Aluminum Silicate $CaAl_2(SiO_4)_2$		102-0125
see also: Anorthite $CaAl_2(SiO_4)_2$		102-0796
Gismondite $CaAl_2(SiO_4)_2$	j	521-1475
see also: Calcium Aluminum Silicate $CaAl_2(SiO_4)_2$		102-0125
see also: Anorthite $CaAl_2(SiO_4)_2$		102-0796
Glacial Rock	ade	521-1358
Glacial Till	a	521-0107
Glauberite $Na_2Ca(SO_4)_2$	hj	521-0784
Glauber's Salt $Na_2SO_4 . 10H_2O$	d	521-0731
see also: Sodium Sulfate Na_2SO_4		120-0016
Glauconite $(K,Na)(Al,Fe,Mg)_2(Al,Si)_4O_{10}(OH)_2$	a	521-1140
Glaucophane $Na_2(Mg,Fe)_3Al_2Si_8O_{22}(OH)_2$	ah	521-1137
Gneiss	adehn	521-0349
Gneiss:		———
Amphibolic	ade	521-0564
Amphibolic With Serpentine	ade	521-0570
Augen	a	521-1551
Biotite Granite	n	521-1260
Biotite Granodiorite	a	521-1627
Biotite Mica	a	521-0833
Biotite Muscovite Granite	n	521-1259
Biotite Plagioclase	no	521-0982
Biotite Sillimanite	no	521-0983
Feldspathic Mica	a	521-0832
Granite	adeg	521-0894
Granite, Biotite	n	521-1260
Granite, Biotite Muscovite	n	521-1259
Granodiorite	a	521-1626
Hornblende Gabbro	gh	521-0902
Leucocratic Granodiorite	a	521-1625
Leucocratic Monzogranite	a	521-1624
Sillimanite	a	521-0831
Sillimanite Garnet	ah	521-0835
Gneiss - Schist	a	521-1445
Goethite $FeOOH$	aehjn	521-0604
see also: Iron Hydroperoxide $FeOOH$		110-0480
Gossan	h	521-1390
Grandidierite $(Mg,Fe)Al_3BO_4SiO_4O$	j	521-0868
Granite	adefghijnor	521-0072
Granite:		———
Adamellite	a	521-1490
Aegirite Riebeckite	n	521-1258
Albite	e	521-0476
Augite	ad	521-1209
Bifeldspathic	ade	521-1241

Property: **a.** Thermal conductivity; **b.** Accommodation coefficient; **c.** Thermal contact resistance; **d.** Thermal diffusivity; **e.** Specific heat; **f.** Viscosity; **g.** Emittance; **h.** Reflectance; **i.** Absorptance; **j.** Transmittance; **k.** α_s/ϵ_t ratio; **l.** Prandtl number; **n.** Thermal linear expansion coefficient; **o.** Thermal volumetric expansion coefficient; **r.** Thermal radiative properties (g,h,i,j,k)

Substance Name	Property	Number
Granite: -(Cont)-		
Biotite	adehno	521-0364
Biotite Hornblende	adn	521-1210
Biotite Muscovite	n	521-1257
Hornblende	adehn	521-0363
Leucocratic	ade	521-1238
Muscovite	n	521-1261
Pyroxene	ad	521-1239
Quartz Monzonite	a	521-1490
Two Mica	ade	521-0365
Granite, Adamellite Gneiss	a	521-1622
Granite, Biotite Gneiss	n	521-1260
Granite, Biotite Muscovite Gneiss	n	521-1259
Granite, Graphic	adeghn	521-0762
Granite, Quartz Monzonite Gneiss	a	521-1622
Granite - Syenite	a	521-1144
Granite Gneiss	adeg	521-0894
Granite Porphyry	ade	521-0671
Granodiorite	adefghno	521-0242
Granodiorite, Biotite	a	521-1488
Granodiorite, Leucocratic	ade	521-1243
Granodiorite Gneiss	a	521-1626
Granodiorite Porphyry	ade	521-1244
Granulite	d	521-1556
Graphite, Natural C	aehn	521-0878
see also: Graphite C		101-0001
Gratonite $Pb_9 As_4 S_{15}$	j	521-1182
Gravel	adeghnor	521-0001
Gravel - Sand	ano	521-0944
Gravelite	no	521-0889
Gray Antimony $Sb_2 S_3$	ahjr	521-0095
see also: Antimony Sulfide $Sb_2 S_3$		126-0014
Gray Copper Ore	h	521-0748
see also: Tennantite $(Cu, Fe)_{12} As_4 S_{13}$		521-0880
see also: Tetrahedrite $(Cu, Fe)_{12} Sb_4 S_{13}$		521-1205
Graywacke	eh	521-0478
Gray With Gravel Soil	h	521-1501
Green Glass Moon Rock	h	521-1597
Greenland Spar $Na_3 Al F_6$	efhjr	521-0385
see also: Aluminum Sodium Fluoride $Na_3 Al F_6$		102-0016
Green Lead Ore $Pb_5 (P O_4)_3 Cl$	h	521-0533
see also: Lead Phosphate Chloride $Pb_5(P O_4)_3 Cl$		106-1661
Greenschist	a	521-0275
Greenstone	adehn	521-0370
Greenstone, Breccia	an	521-0569
Greenstone, Hornblende Plagioclase	ad	521-0842
Greigite $Fe_3 S_4$	hi	521-1220
see also: Iron Sulfide $Fe_3 S_4$		110-0009
Greisen, Quartz - Muscovitic	ae	521-0715
Greisen, Quartz - Muscovitic With Potassium Feldspar	e	521-0716
Griphite $(Na, Al, Ca, Fe)_6 Mn_4(P O_4)_5(O H)_4$	a	521-1150
Grit	adeghjnor	521-0110
Gritrock	no	521-0889
Gritstone	no	521-0889
Grossular $Ca_3 Al_2(Si O_4)_3$	aehjno	521-0959
see also: Aluminum Calcium Silicate $Ca_3 Al_2(Si O_4)_3$		102-0797
Grossularite $Ca_3 Al_2(Si O_4)_3$	aehjno	521-0959
see also: Aluminum Calcium Silicate $Ca_3 Al_2(Si O_4)_3$		102-0797
Groutite $H Mn O_2$	h	521-1016
see also: Manganite $Mn O(O H)$		521-0042
Grunerite $Fe_7 Si_8 O_{22}(O H)_2$	a	521-0949
Gudmundite $Fe Sb S$	h	521-1352
Guhr	adeghijnr	521-0015
Guilleminite $Ba(U O_2)_3(Se O_3)_2(O H)_4$	j	521-1371
Gymnite	hj	521-1472
Gyproc Wool $Ca S O_4$	adeghijnor	521-0083
see also: Calcium Sulfate $Ca S O_4$		106-0051
see also: Anhydrite $Ca S O_4$		521-0251
see also: Plaster Board		661-0019
see also: Plaster Of Paris		661-0573

Substance Name	Property	Number
Gypsum $Ca S O_4$	adeghijnor	521-0083
see also: Calcium Sulfate $Ca S O_4$		106-0051
see also: Anhydrite $Ca S O_4$		521-0251
see also: Plaster Board		661-0019
see also: Plaster Of Paris		661-0573
Gypsum:		———
Selenite	aehj	521-1148
see also: Calcium Sulfate $Ca S O_4$		106-0051
Gypsum Sand	agh	521-0613
Hackmanite	j	521-1385
Halite $Na Cl$	adehjnor	521-0007
see also: Sodium Chloride $Na Cl$		106-0024
see also: Sea Salt		521-1122
Halloysite $Al_2 Si_2 O_5(O H)_4$	ehijn	521-0506
see also: Aluminum Silicate Hydroxide $Al_2 Si_2 O_5(O H)_4$		102-0871
Hambergite	hj	521-0696
Hard Coal	adehnr	521-0024
Harzburgite	h	521-1641
Harzburgite, Serpentinized	h	521-1643
Hauerite $Mn S_2$	eh	521-0981
see also: Manganese Sulfide $Mn S_2$		119-0051
Hausmannite $Mn_3 O_4$	h	521-1019
see also: Manganese Oxide $Mn_3 O_4$		119-0026
Hauyne $(Na, Ca)_8(Al_6 Si_6 O_{24})(S O_4)_x S_y$	hn	521-1109
Hauynite $(Na, Ca)_8(Al_6 Si_6 O_{24})(S O_4)_x S_y$	hn	521-1109
Heavy Spar $Ba S O_4$	aehijnor	521-0223
see also: Barium Sulfate $Ba S O_4$		104-0025
Heazlewoodite $Ni_3 S_2$	j	521-1305
see also: Nickel Sulfide $Ni_3 S_2$		120-0198
Hectorite	ij	521-1536
Hedenbergite $Ca Fe Si_2 O_6$	hjno	521-0793
Hematite $Fe_2 O_3$	adehjno	521-0104
see also: Iron Oxide $Fe_2 O_3$		110-0011
Hematite:		———
Martite $Fe O . 3 Fe_2 O_3$	h	521-1432
Hematite, Black (Romanechite) $Ba Mn_9 O_{16}(O H)_4$	h	521-1015
Hematite, Oolitic	h	521-1634
Hematite - Magnetite - Quartz Mineral	no	521-0886
Hematite - Silica Sand	h	521-1159
Hematite Concentrate	e	521-1268
Hemimorphite $Zn C O_3$	ahj	521-0611
see also: Zinc Carbonate $Zn C O_3$		106-0495
Hercynite:		———
Ferroferrichrompicotite $Fe(Fe, Cr, Al)_2 O_4$	h	521-0937
Ferroferripicotite $Fe(Fe, Al)_2 O_4$	h	521-0938
Picotite	h	521-1162
see also: Magnesiochromite $(Mg, Fe)(Cr, Al)_2 O_4$		521-0963
see also: Ceylonite		521-0990
Hessonite	a	521-1071
see also: Zircon $Zr Si O_4$		521-0250
Hetaerolite $Zn Mn_2 O_4$	h	521-1020
Heteromorphite $Pb_7 Sb_8 S_{19}$	j	521-1178
Heulandite $(Na, Ca)_5 Al_6(Al, Si)_4 Si_{26} O_{72}$	j	521-0778
Heyrovskyite $Pb_6 Bi_2 S_9$	h	521-1453
Hill	h	521-0692
Hokutolite $Ba Pb (S O_4)_2$	ij	521-1607
Hollandite $Ba Mn_8 O_{16}$	h	521-1024
Holmquistite $(Na, K, Ca) Li(Mg, Fe)_3 Al_2 Si_8 O_{22}(O H)_2$	h	521-1401
Hornblende $Ca_2 Na(Mg, Fe)_4(Al, Fe, Ti) Si_8 O_{22} (O, O H)_2$	aeghijnor	521-0951
Hornblende Andesite	n	521-0011
Hornblende Diorite	ade	521-1211
Hornblende Gabbro	adeo	521-0281
Hornblende Gabbro Gneiss	gh	521-0902
Hornblende Plagioclase Greenstone	ad	521-0842
Hornblende Schist	h	521-1383
Hornstone $Si O_2$	nor	521-0629
see also: Silicon Oxide $Si O_2$		122-0009
Hortonolite $(Fe, Mg, Mn)_2 Si O_4$	ah	521-0790

Property: **a.** Thermal conductivity; **b.** Accommodation coefficient; **c.** Thermal contact resistance; **d.** Thermal diffusivity; **e.** Specific heat; **f.** Viscosity; **g.** Emittance; **h.** Reflectance; **i.** Absorptance; **j.** Transmittance; **k.** α_s/ϵ_t ratio; **l.** Prandtl number; **n.** Thermal linear expansion coefficient; **o.** Thermal volumetric expansion coefficient; **r.** Thermal radiative properties (g,h,i,j,k)

Substance Name	Property	Number
Huebnerite $MnWO_4$	hj	521-1076
see also: Manganese Tungsten Oxide $MnWO_4$		119-0053
Humus Soil	adeh	521-0530
Huntite $CaMg_3(CO_3)_4$	e	521-1228
Hurzburgite	a	521-1007
Hyacinth $ZrSiO_4$	aeghijnor	521-0250
see also: Zirconium Silicate $ZrSiO_4$		122-0007
see also: Essonite		521-1071
Hyalite SiO_2	aij	521-1128
see also: Silicon Oxide SiO_2		122-0009
Hyalosiderite	a	521-0955
Hydrargillite $Al(OH)_3$	aehijn	521-0303
see also: Aluminum Oxide Al_2O_3		102-0002
Hydroboracite $CaMgB_6O_{11}$	e	521-1647
Hydrogen Faujasite	j	521-0680
Hydromagnesite $Mg_4(OH)_2(CO_3)_3$	e	521-1230
Hydronephelite $HNa_2Al_2(SiO_4)_3$	j	521-1479
Hydrotalcite $Mg_6Al_2CO_3(OH)_{16}$	hjr	521-0554
Hydroxyapatite $Ca_5(PO_4)_3OH$	eh	521-0090
Hydroxylapatite $Ca_5(PO_4)_3OH$	eh	521-0090
Hypersthene $(Mg,Fe)SiO_3$	ahij	521-0351
Hypersthene, Iron $FeSiO_3$	ah	521-1133
Hypersthene Andesite	fg	521-0228
Hypersthene Gabbro	n	521-1263
Iceland Spar $CaCO_3$	adeghijnor	521-0075
see also: Calcium Carbonate $CaCO_3$		106-0005
see also: Aragonite $CaCO_3$		521-0105
Ice Spar Na_3AlF_6	efhjr	521-0385
see also: Aluminum Sodium Fluoride Na_3AlF_6		102-0016
Icestone Na_3AlF_6	efhjr	521-0385
see also: Aluminum Sodium Fluoride Na_3AlF_6		102-0016
Idaite Cu_3FeS_4	h	521-1636
see also: Nukundamite Cu_5FeS_6		521-1635
Idocrase $Ca_{10}Mg_2Al_4(SiO_4)_5(Si_2O_7)_2(OH)_4$	ahij	521-0939
Igneous Rock	ahn	521-0999
Ijolite	ad	521-1316
Ijolite:		———
Urtite	adn	521-1057
Illite	ehijn	521-0765
Ilmenite $FeO.TiO_2$	aehj	521-0133
see also: Iron Titanium Oxide $FeTiO_3$		110-0005
Ilmenorutile $(Ti,Nb,Fe)_3O_6$	h	521-1576
Ilvaite $CaFe_3(SiO_4)_2(OH)$	h	521-1550
see also: Epidote $Ca_2(Al,Fe)_3Si_3O_{12}OH$		521-0942
Infusorial Earth	adeghijnr	521-0015
Intermetallic:		———
Lead Palladium Intermetallic, Natural $PbPd_3$	h	521-0872
Interplanetary Matter	af	521-0624
Interstellar Dust	h	521-1421
Io	h	521-1161
Iolite $(Mg,Fe)_2Al_4Si_5O_{18}$	aehijno	521-0443
see also: Magnesium Aluminosilicate $Mg_2Al_4Si_5O_{18}$		102-0143
see also: Silicon Oxide Mixture $SiO_2 > Al_2O_3 > MgO$		503-0676
Iron Agglomerate	ade	521-0581
Iron Arsenide Sulfide, Natural $FeS_2 . FeAs_2$	hj	521-0744
Iron Formation	e	521-0479
Iron Lead Manganese Vanadium Oxide, Natural $Pb_4MnFe(VO_4)_4$	j	521-1101
Iron Lithium Manganese Phosphate, Natural $Li(Fe,Mn)PO_4$	a	521-0964
Iron Magnesium Borate, Natural $(Mg,Fe)_2FeBO_5$	ade	521-1646
Iron Magnesium Manganese Phosphate, Natural $(Fe,Mg,Mn)_3(PO_4)_2$	e	521-0777
Iron Magnesium Manganese Silicate, Natural $(Fe,Mg,Mn)_2SiO_4$	ah	521-0790
Iron Magnesium Oxide, Natural $(Fe,Mg)Fe_2O_4$	adehjn	521-0067
see also: Iron Oxide Fe_3O_4		110-0025
Iron Magnesium Silicate, Natural $(Mg,Fe)SiO_3$	ahij	521-0351
Iron Magnesium Silicate, Natural $(Mg,Fe)_2SiO_4$	adeghijnor	521-0258

Substance Name	Property	Number
Iron Magnesium Silicate Hydroxide, Natural:		———
Anthophyllite $(Mg,Fe)_7Si_8O_{22}(OH)_2$	aehij	521-1134
Cummingtonite $(Fe,Mg)_7Si_8O_{22}(OH)_2$	ah	521-1135
Iron Manganese Niobium Tantalum Oxide, Natural $(Fe,Mn)(Nb,Ta)_2O_6$	h	521-0750
see also: Tapiolite $Fe(Ta,Nb)_2O_6$		521-1297
see also: Columbite $(Fe,Mn)(Nb,Ta)_2O_6$		521-1574
Iron Manganese Niobium Tantalum Oxide, Natural $(Fe,Mn)(Nb,Ta)_2O_6$	h	521-1574
see also: Tantalite $(Fe,Mn)(Nb,Ta)_2O_6$		521-0750
Iron Manganese Oxide, Natural $(Mn,Fe)_2O_3$	h	521-0966
Iron Manganese Tungsten Oxide, Natural $(Fe,Mn)WO_4$	ehj	521-0555
Iron Manganese Zinc Oxide, Natural $(Zn,Mn,Fe)Mn_2O_5$	hj	521-1022
Iron Nickel Borate, Natural Ni_2FeBO_5	h	521-1600
Iron Nickel Sulfide, Natural $(Fe,Ni)_9S_8$	h	521-0827
Iron Niobium Tantalum Oxide, Natural $Fe(Ta,Nb)_2O_6$	hj	521-1297
see also: Tantalite $(Fe,Mn)(Nb,Ta)_2O_6$		521-0750
Iron Niobium Titanium Oxide, Natural $(Ti,Nb,Fe)_3O_6$	h	521-1576
Iron Ore	adeh	521-0382
Iron Ore, Taconite	ade	521-0567
Iron Oxide, Natural $FeO.3Fe_2O_3$	h	521-1432
Iron Oxide, Red Fe_2O_3	adehjno	521-0104
see also: Iron Oxide Fe_2O_3		110-0011
Iron Potassium Cyanide, Natural $K_4Fe(CN)_6$	j	521-1473
see also: Potassium Cyanoferrate $K_4Fe(CN)_6$		106-0577
Iron Potassium Sulfate Hydroxide, Natural $KFe_3(SO_4)_2(OH)_6$	h	521-1493
Iron Pyrites:		———
Marcasite FeS_2	j	521-1304
see also: Iron Sulfide FeS_2		110-0010
see also: Pyrite FeS_2		521-0025
Pyrite FeS_2	adehijnor	521-0025
see also: Iron Sulfide FeS_2		110-0010
see also: Marcasite FeS_2		521-1304
Iron Silicate, Natural $FeSiO_3$	ah	521-1133
Iron Silicate Hydroxide, Natural $Fe_7Si_8O_{22}(OH)_2$	a	521-0949
Iron Spar $FeCO_3$	aehijo	521-0566
see also: Iron Carbonate $FeCO_3$		106-0498
Iron Thallium Sulfide, Natural $TlFe_2S_3$	h	521-1468
Iserine $FeO.TiO_2$	aehj	521-0133
see also: Iron Titanium Oxide $FeTiO_3$		110-0005
Isinglass $KAl_2(AlSi_3)O_{10}(OH)_2$	adeghijknr	521-0240
see also: Aluminum Potassium Silicate Hydroxide $KAl_2(AlSi_3O_{10})(OH)_2$		102-0798
see also: Potassium Aluminum Silicate $KAl_3Si_3O_{11}$		102-0966
Isokite $CaMgPO_4F$	j	521-1290
Isolantite	a	521-1145
Jacinth $ZrSiO_4$	aeghijnor	521-0250
see also: Zirconium Silicate $ZrSiO_4$		122-0007
see also: Essonite		521-1071
Jacobsite $(Mn,Fe,Mg)(Fe,Mn)_2O_4$	h	521-1018
see also: Iron Manganese Oxide $MnFe_2O_4$		110-0187
Jadeite $Na(Al,Fe)Si_2O_6$	adehij	521-0058
Jargon $ZrSiO_4$	aeghijnor	521-0250
see also: Zirconium Silicate $ZrSiO_4$		122-0007
see also: Essonite		521-1071
Jarosite $KFe_3(SO_4)_2(OH)_6$	h	521-1493
Jasper	j	521-1566
see also: Chert		521-0534
Jasper – Limestone	j	521-1565
Jet Coal	hj	521-0796
Joaquinite $NaBa_2Ce_2Fe(Ti,Nb)_2Si_8O_{26}(OH,F)_2$	i	521-1548
Johannite $Cu(UO_2)_2(SO_4)_2(OH)_2$	e	521-0284
Jordanite $(Pb,Tl)_{13}As_7S_{23}$	j	521-1183
Jupiter Satellite	h	521-1161
Jupiter Satellite Surface, Synthetic	h	521-1417
Jupiter Surface	hjr	521-0424

Property: **a.** Thermal conductivity; **b.** Accommodation coefficient; **c.** Thermal contact resistance; **d.** Thermal diffusivity; **e.** Specific heat; **f.** Viscosity; **g.** Emittance; **h.** Reflectance; **i.** Absorptance; **j.** Transmittance; **k.** α_s/ϵ_t ratio; **l.** Prandtl number; **n.** Thermal linear expansion coefficient; **o.** Thermal volumetric expansion coefficient; **r.** Thermal radiative properties (g,h,i,j,k)

Substance Name	Property	Number
Kaliophilite $K Al Si O_4$	e	521-0801
Kalkibeborosilit	j	521-1064
Kanaekanite	j	521-1067
Kankar	j	521-1562
see also: Limestone		521-0085
Kaolin,	adefhjnr	521-0135
see also: Kaolinite, Mineral $Al_2 Si_2 O_5(O H)_4$		521-0174
see also: Kaolin Board		661-0444
see also: Kaolin Paper		661-0484
Kaolin, Mineral $Al_2 Si_2 O_5(O H)_4$	adehijn	521-0174
see also: Aluminum Silicate $Al_2 Si_2 O_7$		102-0866
see also: Aluminum Silicate Hydroxide $Al_2 Si_2 O_5(O H)_4$		102-0871
see also: Kaolin,		521-0135
see also: Dickite		521-0505
Kaolin, Sodium Clay	e	521-0707
Kaolinite, Mineral $Al_2 Si_2 O_5(O H)_4$	adehijn	521-0174
see also: Aluminum Silicate $Al_2 Si_2 O_7$		102-0866
see also: Aluminum Silicate Hydroxide $Al_2 Si_2 O_5(O H)_4$		102-0871
see also: Kaolin,		521-0135
see also: Dickite		521-0505
Kaolinite, Sodium Clay	e	521-0707
Keramzit	adefhjnr	521-0040
Kermesite $Sb_2 S_2 O$	h	521-1173
Kernite $Na_2 B_4 O_7$	h	521-1288
see also: Sodium Borate $Na_2 B_4 O_7$		104-0035
Kerogen	h	521-1630
see also: Oil Shale		521-0115
Kerolite	e	521-1172
Kersantite Lamprophyre	f	521-0229
Kesterite $Cu_2(Zn, Fe) Sn S_4$	hj	521-1366
Kharkerite	j	521-1167
Kieselguhr	adeghijnr	521-0015
Kieserite $Mg S O_4$	hj	521-1483
see also: Magnesium Sulfate $Mg S O_4$		119-0003
Kimzeyite $Ca_3(Zr, Ti)_2(Al, Si)_3 O_{12}$	j	521-1336
Koechlinite $Bi_2 Mo O_6$	j	521-1095
see also: Bismuth Molybdenum Oxide $Bi_2 Mo O_6$		104-0562
Kornerupine $Mg_3 Al_6(Si, B, Al)_5 O_{21}(O H)$	j	521-0869
Kotoite $Mg_3(B O_3)_2$	ej	521-1163
see also: Magnesium Borate $Mg_3 B_2 O_6$		104-0368
Kreittonite $Zn Al_2 O_4$	ahno	521-0962
see also: Aluminum Zinc Oxide $Zn Al_2 O_4$		102-0185
Kryolith $Na_3 Al F_6$	efhjr	521-0385
see also: Aluminum Sodium Fluoride $Na_3 Al F_6$		102-0016
Kunzite	j	521-1486
Kupferkies $Cu Fe S_2$	ahj	521-0099
see also: Copper Iron Sulfide $Cu Fe S_2$		106-0791
Kupfernickel $Ni As$	hj	521-0751
see also: Nickel Arsenide $Ni As$		102-0612
Kyanite $Al_2 O Si O_4$	aehijnor	521-0033
see also: Aluminum Silicate $Al_2 Si O_5$		102-0331
Labradorite	ahjnor	521-0471
Lake Bed	gh	521-0767
Lamprophyre	a	521-0789
Lamprophyre:		———
Kersantite	f	521-0229
Monchiquite	g	521-0903
Lapilli	adefghno	521-0053
Lapis Lazuli	j	521-1350
see also: Lazurite $(Na, Ca)_8(Al, Si)_{12} O_{24}(S, S O_4)$		521-1351
Larsenite, Calcium $(Ca, Pb) Zn Si O_4$	hj	521-0987
Larvikite	g	521-0972
Laterite Soil	ahj	521-0865
Latite	g	521-1278
Laurite $Ru S_2$	h	521-0971
Lava	adefghno	521-0053
Lava, Amygdaloidal	ae	521-0468
Lava, Glassy	f	521-0226
Lava, Highly Crystalline	f	521-0234
Lavite $Mg_3 Si_4 O_{10}(O H)_2$	adehijnr	521-0116

Substance Name	Property	Number
Lavroffite Pyroxene	g	521-0908
Lavrovite Pyroxene	g	521-0908
Lawsonite $Ca Al_2 Si_2 O_7(O H)_2$	e	521-0487
see also: Calcium Aluminum Silicate $Ca Al_2(Si O_4)_2$		102-0125
Lazulite $(Mg, Fe) Al_2(P O_4)_2(O H)_2$	hj	521-1283
Lazurite $(Na, Ca)_8(Al, Si)_{12} O_{24}(S, S O_4)$	hj	521-1351
see also: Lapis Lazuli		521-1350
Lead Arsenide Sulfide, Natural:		———
Dufrenoysite $Pb_2 As_2 S_5$	j	521-1181
Gratonite $Pb_9 As_4 S_{15}$	j	521-1182
Lead Carbonate Chloride, Natural $Pb_2 Cl_2 C O_3$	h	521-1348
Lead Glance $Pb S$	aehjnr	521-0192
see also: Lead Sulfide $Pb S$		123-0003
Lead Manganese Oxide, Natural $Pb Mn_8 O_{16}$	h	521-1023
Lead Palladium Intermetallic, Natural $Pb Pd_3$	h	521-0872
Lead Selenate, Natural $Pb Se O_3$	j	521-1365
see also: Lead Selenite $Pb Se O_3$		122-0509
Lead Selenate Sulfate, Natural $Pb_2 Se O_4 S O_4$	j	521-1367
Lead Silicide Sulfide, Natural $Pb_2 Si_2 S_5$	h	521-1451
Lead Spar $Pb S O_4$	hjn	521-0786
see also: Lead Sulfate $Pb S O_4$		122-0021
Lead Thallium Arsenide Sulfide, Natural $(Pb, Tl)_{13} As_7 S_{23}$	j	521-1183
Lead Vanadium Oxide, Natural $Pb_2 V_2 O_7$	h	521-1379
Lead Vitriol $Pb S O_4$	hjn	521-0786
see also: Lead Sulfate $Pb S O_4$		122-0021
Lean Coal	ej	521-0739
Lechatelierite $Si O_2$	adefghijnor	521-0732
see also: Quartz $Si O_2$		122-0002
see also: Silicon Oxide $Si O_2$		122-0009
see also: Stishovite $Si O_2$		122-0609
see also: Cristobalite $Si O_2$		122-0628
see also: Coesite $Si O_2$		122-0629
see also: Tridymite $Si O_2$		122-0646
see also: Cristobalite, Natural $Si O_2$		521-0254
see also: Tridymite, Natural $Si O_2$		521-0255
see also: Coesite, Natural $Si O_2$		521-0823
see also: Stishovite, Natural $Si O_2$		521-0824
see also: Quartz Fiber Felt		661-0331
Leda Clay	a	521-0630
see also: Clay Soil		521-0645
Leikosfenit $Ba Na_4 Ti_2 B_2 Si_{10} O_{30}$	j	521-1063
Leiptinite Coal	h	521-1120
Leonhardite $Ca_2 Al_4 Si_8 O_{24}$	e	521-0488
Lepidocrocite $Fe O(O H)$	j	521-1175
see also: Iron Hydroperoxide $Fe O O H$		110-0480
Lepidolite Mica $K(Li, Al)_3(Si, Al)_4 O_{10}(F, O H)$	ahj	521-0965
see also: Polylithionite $K Li_2 Al Si_4 O_{10}(F, O H)_2$		521-1092
Lepidomelane	a	521-1141
Leucite $K Al Si_2 O_6$	aeno	521-0062
see also: Potassium Aluminosilicate $K Al(Si O_3)_2$		102-0257
see also: Potassium Aluminosilicate $K_2 O . Al_2 O_3 . 4 Si O_2$		102-0957
Leucochalcite $Cu_2 As O_4 O H$	j	521-1188
Leucocratic Granite	ade	521-1238
Leucocratic Granodiorite Gneiss	a	521-1625
Leucocratic Monzogranite Gneiss	a	521-1624
Leucopyrite	ehj	521-1306
see also: Rammelsbergite $Ni As_2$		521-1307
Leucosapphire $Al_2 O_3$	adeghijnor	521-0077
see also: Aluminum Oxide $Al_2 O_3$		102-0002
Leucosphenite $Ba Na_4 Ti_2 B_2 Si_{10} O_{30}$	j	521-1063
Lherzolite	a	521-1008
Libethenite $Cu_2 P O_4 O H$	j	521-1187
Lignite Coal	adeh	521-0065
Lillianite $Pb_3 Bi_2 S_6$	h	521-1456
Lillianite - Gustavite	h	521-1457
Limburgite	g	521-0905
Lime - Soda Feldspar	ahjnor	521-0552
Lime Feldspar	efn	521-0344
Lime Sand	a	521-1361

Property: a. Thermal conductivity; **b.** Accommodation coefficient; **c.** Thermal contact resistance; **d.** Thermal diffusivity; **e.** Specific heat; **f.** Viscosity; **g.** Emittance; **h.** Reflectance; **i.** Absorptance; **j.** Transmittance; **k.** α_s/ϵ_t ratio; **l.** Prandtl number; **n.** Thermal linear expansion coefficient; **o.** Thermal volumetric expansion coefficient; **r.** Thermal radiative properties (g,h,i,j,k)

Substance Name	*Property*	*Number*
Limestone	adehijnor	521-0085
see also: Calcareous Soil		521-1048
see also: Calcareous Rock		521-1560
see also: Kankar		521-1562
Limestone:		——
Chalk $CaCO_3$	adeg	521-0183
see also: Calcium Carbonate $CaCO_3$		106-0005
see also: Aragonite $CaCO_3$		521-0105
see also: Calcareous Rock		521-1560
Coral Limestone	j	521-1563
Dolomitic	aj	521-1444
Siliceous	j	521-1564
Limestone - Mud	h	521-1495
Limestone - Sandstone - Shale Breccia	a	521-0741
Limestone - Shale	a	521-1469
Limonite	eh	521-0414
Linnaeite $(Co, Ni)_3S_4$	h	521-1192
Linnaeite:		——
Carrollite $Cu(Co, Ni)_2S_4$	h	521-0826
Polydymite Ni_3S_4	h	521-1195
Siegenite $(Co, Ni)_3S_4$	h	521-1193
Liparite	adefghjn	521-0272
Lithic Soil	h	521-1070
Lithiophorite $(Al, Li)MnO_2(OH)_2$	j	521-1176
Lithium Feldspar $LiAlSi_3O_8$	n	521-1652
Lithium Magnesium Potassium Silicate Fluoride, Natural $KLiMg_2Si_4O_{10}F_2$	n	521-1355
Lithium Potassium Feldspar $(Li, K)AlSi_3O_8$	n	521-1653
Liver Ore HgS	hj	521-0749
see also: Mercury Sulfide HgS		112-0402
Lizardite $Mg_3Si_2O_5(OH)_4$	a	521-1143
Loam Soil	adehjnor	521-0186
Loam With Gravel And Sand Soil	a	521-0013
Loam With Gravel Soil	a	521-0497
Loamy Sand Soil	j	521-1598
Lochaline Sand SiO_2	adeghnor	521-0010
see also: Silicon Oxide SiO_2		122-0009
Lodestone $(Fe, Mg)Fe_2O_4$	adehjn	521-0067
see also: Iron Oxide Fe_3O_4		110-0025
Loellingite	ehj	521-1306
see also: Rammelsbergite $NiAs_2$		521-1307
Loess Clay	ej	521-0798
Lollingite	ehj	521-1306
see also: Rammelsbergite $NiAs_2$		521-1307
Long Flame Coal	ej	521-0817
Loparite $(Ce, Na, Ca)_2(Ti, Nb)_2O_6$	e	521-1442
Ludlamite $(Fe, Mg, Mn)_3(PO_4)_2$	e	521-0777
Ludwigite $(Mg, Fe)_2FeBO_5$	ade	521-1646
Lunar Rock (See Moon Rock)	adehjno	521-0922
Lussatite SiO_2	ij	521-1420
see also: Silicon Oxide SiO_2		122-0009
Lyutsonite	h	521-1207
Magma	adef	521-0145
Magnesia, Natural MgO	aegjn	521-0132
see also: Magnesium Oxide MgO		119-0002
Magnesia Mica $K(Mg, Fe)_3AlSi_3O_{10}(OH, F)_2$	adefhijn	521-0241
see also: Aluminum Magnesium Potassium Silicate Fluoride $KMg_3AlSi_3O_{10}F_2$		102-0262
see also: Aluminum Germanium Magnesium Potassium Oxide Fluoride $KMg_3AlGe_3O_{10}F_2$		102-0962
see also: Aluminum Germanium Magnesium Potassium Oxide Fluoride, Nonstoichiometric $KMg_3Al_nGe_4O_{10}F_2$		102-0963
Magnesian Soil	a	521-0864
Magnesiochromite $(Mg, Fe)(Cr, Al)_2O_4$	an	521-0963
see also: Chromium Magnesium Oxide $MgCr_2O_4$		106-0136
see also: Chromite $(Fe, Mg)(Cr, Al)_2O_4$		521-0366
see also: Picotite		521-1162
Magnesioferrite $(Mg, Fe)Fe_2O_4$	r	521-0670
see also: Iron Magnesium Oxide, Nonstoichiometric $MgFe_2O_4$		110-0229
Magnesioriebeckite	a	521-1138

Substance Name	*Property*	*Number*
Magnesite $MgCO_3$	adehjn	521-0006
see also: Magnesium Carbonate $MgCO_3$		106-0015
Magnesium Aluminum Carbonate Hydroxide, Natural $Mg_6Al_2CO_3(OH)_{16}$	hjr	521-0554
Magnesium Borate Hydroxide Fluoride, Natural:		——
Fluoborite $Mg_3BO_3(F, OH)_3$	j	521-1168
Magnesium Carbonate Hydroxide, Natural $Mg_5(CO_3)_4(OH)_2$	j	521-1363
see also: Tetrakis(Carbonato)dihydroxypentamagnesium $4MgCO_3 . Mg(OH)_2$		218-0004
Magnesium Carbonate Hydroxide, Natural:		——
Artinite $Mg_2CO_3(OH)_2$	e	521-1229
Hydromagnesite $Mg_4(OH)_2(CO_3)_3$	e	521-1230
Magnesium Silicate, Natural $H_2Mg_3Si_4O_{12}$	e	521-0519
Magnesium Silicate Hydroxide, Natural:		——
Antigorite $Mg_3Si_2O_5(OH)_4$	ij	521-1605
Chrysotile $Mg_3Si_2O_5(OH)_4$	eijn	521-0799
see also: Asbestos		521-0055
see also: Serpentine $(Mg, Fe)_3Si_2O_5(OH)_4$		521-0236
Lizardite $Mg_3Si_2O_5(OH)_4$	a	521-1143
Magnesium Sodium Sulfate, Natural $Na_2Mg(SO_4)_2$	hj	521-1476
Magnetic Pyrite FeS	ahj	521-0746
see also: Iron Sulfide FeS		110-0007
see also: Iron Sulfide, Nonstoichiometric FeS		110-0295
Magnetite $(Fe, Mg)Fe_2O_4$	adehjn	521-0067
see also: Iron Oxide Fe_3O_4		110-0025
Magnetite:		——
Franklinite $(Zn, Mn, Fe)(Fe, Mn)_2O_4$	h	521-1017
see also: Iron Zinc Oxide $ZnFe_2O_4$		110-0029
Jacobsite $(Mn, Fe, Mg)(Fe, Mn)_2O_4$	h	521-1018
see also: Iron Manganese Oxide $MnFe_2O_4$		110-0187
Magnesioferrite $(Mg, Fe)Fe_2O_4$	r	521-0670
see also: Iron Magnesium Oxide, Nonstoichiometric $MgFe_2O_4$		110-0229
Trevorite $NiFe_2O_4$	h	521-1360
see also: Iron Nickel Oxide $NiFe_2O_4$		110-0028
Magnetite - Apatite - Forsterite Ore	ad	521-1322
Magnetite - Calcite Ore	ad	521-1327
Magnetite - Lime	d	521-0663
Magnetite - Limestone	d	521-0664
Magnetite - Mica, Biotite - Quartz	n	521-0988
Magnetite - Staffelite Ore	ad	521-1321
Magnochromite $(Mg, Fe)(Cr, Al)_2O_4$	an	521-0963
see also: Chromium Magnesium Oxide $MgCr_2O_4$		106-0136
see also: Chromite $(Fe, Mg)(Cr, Al)_2O_4$		521-0366
see also: Picotite		521-1162
Malachite $Cu_2CO_3(OH)_2$	hij	521-1227
Malachite, Blue $Cu_3(CO_3)_2(OH)_2$	ehjno	521-0396
Malacolite $CaMg(SiO_3)_2$	aefhijno	521-0036
see also: Calcium Magnesium Silicate $CaMgSi_2O_6$		106-0940
see also: Chrome Diopside		521-1593
Manganblende MnS	hj	521-0863
see also: Manganese Sulfide MnS		119-0043
Manganese Hydrogen Oxide, Natural $HMnO_2$	h	521-1016
see also: Manganite $MnO(OH)$		521-0042
Manganese Hydroxide, Natural $Mn(OH)_2$	h	521-1014
Manganese Hydroxide Oxide, Natural $Mn(O, OH)_2$	h	521-1013
Manganese Ore	a	521-1215
γ- Manganese Oxide $Mn(O, OH)_2$	h	521-1013
Manganese Potassium Oxide, Natural KMn_8O_{16}	h	521-1025
Manganese Silicate, Natural $3Mn_2O_3 . MnSiO_3$	h	521-0992
Manganese Spar $MnCO_3$	ahjn	521-0614
see also: Manganese Carbonate $MnCO_3$		106-0497
Manganese Spar $MnSiO_3$	hij	521-1043
see also: Manganese Silicate $MnSiO_3$		119-0020
Manganese Zinc Oxide, Natural $ZnMn_2O_4$	h	521-1020
Manganite $MnO(OH)$	ehj	521-0042
see also: Groutite $HMnO_2$		521-1016
Manganosite MnO	h	521-0954
see also: Manganese Oxide MnO		119-0008
Manganotantalite $(Mn, Fe)(Ta, Nb)_2O_6$	h	521-1005
Marble	adeghinor	521-0103
Marble, Black	an	521-0483

Property: a. Thermal conductivity; **b.** Accommodation coefficient; **c.** Thermal contact resistance; **d.** Thermal diffusivity; **e.** Specific heat; **f.** Viscosity; **g.** Emittance; **h.** Reflectance; **i.** Absorptance; **j.** Transmittance; **k.** a_s/ϵ_t ratio; **l.** Prandtl number; **n.** Thermal linear expansion coefficient; **o.** Thermal volumetric expansion coefficient; **r.** Thermal radiative properties (g,h,i,j,k)

Substance Name	Property	Number
Marble, Brown	a	521-0485
Marble, White	adehnr	521-0432
Marcasite $Fe S_2$	j	521-1304
see also: Iron Sulfide $Fe S_2$		110-0010
see also: Pyrite $Fe S_2$		521-0025
Margarite $Ca Al_4 Si_2 O_{10}(O H)_2$	j	521-1331
Marl, Dolomitized	f	521-0499
Marl, Siliceous	a	521-1364
Marl Clay	adefn	521-0371
Marmatite	h	521-0953
Mars:		———
Core	df	521-1124
Soil	ahj	521-0923
Surface	aeghir	521-0413
Surface, Synthetic	h	521-0719
Marshalite Mica	ade	521-1094
Marsh Mud	h	521-0931
Marthozite $Cu(U O_2)_3(Se O_3)_3(O H)_2$	j	521-1372
Martite $Fe O . 3 Fe_2 O_3$	h	521-1432
Martozite $Cu(U O_2)_3(Se O_3)_3(O H)_2$	j	521-1372
Matildite $Ag Bi S_2$	h	521-1006
Mawsonite $Cu_7 Fe_2 Sn S_{10}$	h	521-1638
Mechernichite $(Ni, Fe) S_2$	h	521-0710
see also: Iron Nickel Sulfide, Nonstoichiometric $Ni Fe S_2$		110-0370
Meerschaum $Mg_4 (Si_2 O_5)_3(O H)_2$	e	521-1171
see also: Magnesium Silicate $Mg_2 Si_3 O_8$		119-0065
Meionite $3 Ca Al_2 Si_2 O_8 . Ca C O_3$	a	521-0961
Melanite	i	521-1386
Melilite $Ca_2 Al_2 Si O_7$	ah	521-0943
see also: Calcium Aluminum Silicate $Ca_2 Al_2 Si O_7$		102-0092
see also: Gehlenite $Ca_2 Al_2 Si O_7$		521-1640
Melinose $Pb Mo O_4$	hj	521-0736
see also: Lead Molybdenum Oxide $Pb Mo O_4$		119-0073
Mellite $Al_2[C_6(C O O)_6]$	j	521-0782
Menaccanite $Fe O . Ti O_2$	aehj	521-0133
see also: Iron Titanium Oxide $Fe Ti O_3$		110-0005
Meneghinite $Cu Pb_{13} Sb_7 S_{24}$	h	521-1561
Mercury Core	f	521-1603
Mercury Soil	h	521-1558
Mercury Surface	ghr	521-0421
Merwinite $Ca_3 Mg(Si O_4)_2$	n	521-0862
see also: Calcium Magnesium Silicate $Ca_3 Mg Si_2 O_8$		106-0832
Metabentonite	ij	521-1537
Metakaolin	e	521-0267
Metamorphic Rock	adehn	521-0825
Metamorphosed Coal	h	521-1208
Metapyrophyllite	e	521-1346
see also: Pyrophyllite $Al Si_2 O_5(O H)$		521-0362
Meteorite	adefghnr	521-0239
see also: Chondrite		521-0887
Meyerhofferite $Ca_2 B_6 O_{11}$	r	521-0342
Miargyrite $Ag Sb S_2$	h	521-1202
see also: Antimony Silver Sulfide $Ag Sb S_2$		102-0131
Mica	adefghjnr	521-0108
see also: Germanium Magnesium Potassium Oxide Fluoride, Nonstoichiometric $K Mg_3 Ge_4 O_{10} F_2$		110-0752
see also: Ripidolite $(Mg, Fe)_9 Al_6 Si_5 O_{20}(O H)_{16}$		521-0818
Mica:		———
Aluminum Iron Magnesium Silicate Hydroxide, Natural $(Mg, Fe, Al)_3 (Al, Si)_4 O_{10} (O H)_2$	adehjn	521-0020
Biotite $K(Mg, Fe)_3(Al, Fe) Si_3 O_{10}(O H)_2$	adehjnr	521-0430
Chloritoid $Fe_2 Al_4 Si_2 O_{10}(O H)_4$	j	521-1480
Cryophyllite	j	521-1091
Glauconite $(K, Na)(Al, Fe, Mg)_2(Al, Si)_4 O_{10}(O H)_2$	a	521-1140
Illite	ehijn	521-0765
Lepidolite $K(Li, Al)_3(Si, Al)_4 O_{10}(F, O H)$	ahj	521-0965
see also: Polylithionite $K Li_2 Al Si_4 O_{10}(F, O H)_2$		521-1092
Lepidomelane	a	521-1141
Margarite $Ca Al_4 Si_2 O_{10}(O H)_2$	j	521-1331
Marshalite	ade	521-1094
Mica: -(Cont)-		
Muscovite $K Al_2 (Al Si_3) O_{10} (O H)_2$	adeghijknr	521-0240
see also: Aluminum Potassium Silicate Hydroxide $K Al_2(Al Si_3 O_{10})(O H)_2$		102-0798
see also: Potassium Aluminum Silicate $K Al_3 Si_3 O_{11}$		102-0966
Muscovite, Green	ajr	521-0686
Muscovite, Ruby	ahjnr	521-0687
Phlogopite $K(Mg, Fe)_3 Al Si_3 O_{10}(O H, F)_2$	adefhijn	521-0241
see also: Aluminum Magnesium Potassium Silicate Fluoride $K Mg_3 Al Si_3 O_{10} F_2$		102-0262
see also: Aluminum Germanium Magnesium Potassium Oxide Fluoride $K Mg_3 Al Ge_3 O_{10} F_2$		102-0962
see also: Aluminum Germanium Magnesium Potassium Oxide Fluoride, Nonstoichiometric $K Mg_3 Al_n Ge_4 O_{10} F_2$		102-0963
Pinite	no	521-1032
Polylithionite $K Li_2 Al Si_4 O_{10}(F, O H)_2$	j	521-1092
see also: Lepidolite Mica $K(Li, Al)_3(Si, Al)_4 O_{10}(F, O H)$		521-0965
Taeniolite $K Li Mg_2 Si_4 O_{10} F_2$	n	521-1355
Vermiculite $(Mg, Fe, Al)_3 (Al, Si)_4 O_{10} (O H)_2$	adehjn	521-0020
Zinnwaldite $K_2(Li, Fe, Al)_6(Si, Al)_8 O_{20}(O H, F)_4$	ahn	521-0819
Micaceous Clay	n	521-0856
Mica Quartz Schist	ah	521-1147
Mica Schist	j	521-1568
Microcline $K Al Si_3 O_8$	aehjnor	521-0392
see also: Potassium Aluminum Silicate $K Al Si_3 O_8$		102-0173
Microcline:		———
Amazonite	hn	521-0984
Microcline - Perthite	a	521-1458
Microlite $(Na, Ca)_2(Ta, Nb)_2 O_6(O, O H, F)$	ah	521-1003
Microsyenite	e	521-1267
Migmatite	ade	521-1408
Miike, Japan Coal	fj	521-0019
Millstone	e	521-0511
see also: Sandstone		521-0196
Mimetesite $Pb_5(As O_4)_3 Cl$	hj	521-1284
see also: Lead Arsenate Chloride $Pb_5(As O_4)_3 Cl$		102-0436
Mimetite $Pb_5(As O_4)_3 Cl$	hj	521-1284
see also: Lead Arsenate Chloride $Pb_5(As O_4)_3 Cl$		102-0436
Mineral, General	adeghijnor	521-0299
Mineral Charcoal Coal	e	521-0056
Mirabilite $Na_2 S O_4 . 10 H_2 O$	d	521-0731
see also: Sodium Sulfate $Na_2 S O_4$		120-0016
Mispickel $Fe S_2 . Fe As_2$	hj	521-0744
Moldavite Tektite	fj	521-0237
Molding Clay	a	521-0441
Molding Sand	aden	521-0500
Molochite $3 Al_2 O_3 . 2 Si O_2 + Si O_2$	hir	521-0516
Molten Earth	adef	521-0145
Molybdenite $Mo S_2$	adehj	521-0675
see also: Molybdenum Sulfide $Mo S_2$		119-0001
Molybdenoscheelite	j	521-1099
Molybdenum Glance $Mo S_2$	adehj	521-0675
see also: Molybdenum Sulfide $Mo S_2$		119-0001
Molybdomenite $Pb Se O_3$	j	521-1365
see also: Lead Selenite $Pb Se O_3$		122-0509
Monazite $(Ce, La, Nd, Th)(Si O_4 P O_4)$	aehj	521-0216
Monchiquite Lamprophyre	g	521-0903
Monopyroxene	no	521-1339
Monticellite $Ca Mg Si O_4$	ahn	521-0861
see also: Calcium Magnesium Silicate $Ca Mg Si O_4$		106-0939
Montmorillonite	ehijn	521-0764
Montmorillonite:		———
Beidellite	n	521-0945
Hectorite	ij	521-1536
Metabentonite	ij	521-1537
Montmorillonite, Aluminum	e	521-0706
Montmorillonite, Calcium	e	521-0705
Montmorillonite, Lithium	ei	521-0729
Montmorillonite, Potassium	e	521-0689

Property: a. Thermal conductivity; **b.** Accommodation coefficient; **c.** Thermal contact resistance; **d.** Thermal diffusivity; **e.** Specific heat; **f.** Viscosity; **g.** Emittance; **h.** Reflectance; **i.** Absorptance; **j.** Transmittance; **k.** α_s/ϵ_t ratio; **l.** Prandtl number; **n.** Thermal linear expansion coefficient; **o.** Thermal volumetric expansion coefficient; **r.** Thermal radiative properties (g,h,i,j,k)

Substance Name	Property	Number
Montmorillonite: -(Cont)-		
Montmorillonite, Sodium	ei	521-0730
Nontronite	hj	521-1329
Montmorillonite, Aluminum	e	521-0706
Montmorillonite, Calcium	e	521-0705
Montmorillonite, Lithium	ei	521-0729
Montmorillonite, Potassium	e	521-0689
Montmorillonite, Sodium	ei	521-0730
Monzonite	adghn	521-0924
Moon:		———
Armalcolite $(Mg, Fe) Ti_2 O_5$	h	521-1599
see also: Pseudobrookite $Fe_2 Ti O_5$		521-0101
Augite	h	521-1592
see also: Augite $(Ca, Na)(Mg, Fe, Al)(Si, Al)_2 O_6$		521-0794
Core	df	521-1125
Dust	h	521-1632
Soil	adefgh	521-0917
Soil, Synthetic	adh	521-1038
Surface	adeghr	521-0259
Moon Rock	adehjno	521-0922
Moon Rock:		———
Basalt	adeh	521-1392
Basalt Glass	h	521-1596
Breccia	adehjno	521-1357
Clinopyroxene	hj	521-1059
Dunite	h	521-1393
Green Glass	h	521-1597
Obsidian	h	521-1587
Olivine	h	521-1393
Pigeonite	j	521-1272
Pyroxene	j	521-1249
Red Glass	h	521-1594
Regolith	ahj	521-1384
Synthetic	ah	521-1001
Titanium Iron Oxide	h	521-0921
Titanochromite	h	521-0920
Ulvospinel	h	521-0919
Moonstone	ijnr	521-0457
Mordenite $(Ca,(Na, K)_2)_4 Al_8 Si_{40} O_{96}$	ij	521-1606
Morion	aej	521-0517
Mother Of Coal	e	521-0056
Mottramite $Pb(Cu, Zn)(V O_4)(O H)$	j	521-1103
Mountain	r	521-0520
Mountain, Snowcapped	r	521-0480
Muck Soil	adeh	521-0530
Mud, Dried	r	521-0357
Mudstone	aeh	521-0599
Mullite $Al_6 Si_2 O_{13}$	aeghn	521-1418
see also: Aluminum Silicate $Al_6 Si_2 O_{13}$		102-0021
Muscovite $K Al_2 (Al Si_3) O_{10} (O H)_2$	adeghijknr	521-0240
see also: Aluminum Potassium Silicate Hydroxide $K Al_2(Al Si_3 O_{10})(O H)_2$		102-0798
see also: Potassium Aluminum Silicate $K Al_3 Si_3 O_{11}$		102-0966
Muscovite, Green Mica	ajr	521-0686
Muscovite, Ruby Mica	ahjnr	521-0687
Muscovite Granite	n	521-1261
Nadorite $Pb Sb O_2 Cl$	n	521-1428
Native Antimony Sb	h	521-1426
see also: Antimony Sb		101-0062
Natroborocalcite $Na Ca B_5 O_9$	h	521-1289
Natrolite $Na_2 Al_2 Si_3 O_{10}$	ahij	521-0780
Nepheline $(Na, K) (Al, Si)_2 O_4$	aeghijno	521-0059
see also: Sodium Aluminosilicate $Na_2 O . Al_2 O_3 . 2 Si O_2$		102-0953
see also: Carnegieite $Na_2 O . Al_2 O_3 . 2 Si O_2$		102-0954
Nepheline Basalt	fg	521-0031
Nepheline Syenite	adehn	521-0477
Nephelite $(Na, K) (Al, Si)_2 O_4$	aeghijno	521-0059
see also: Sodium Aluminosilicate $Na_2 O . Al_2 O_3 . 2 Si O_2$		102-0953
see also: Carnegieite $Na_2 O . Al_2 O_3 . 2 Si O_2$		102-0954
Nephelite Syenite	adehn	521-0477

Substance Name	Property	Number
Nephrite	h	521-1400
Neptune Surface	r	521-0527
Niccolite $Ni As$	hj	521-0751
see also: Nickel Arsenide $Ni As$		102-0612
Nickel Arsenide, Natural:		———
Pararammelsbergite $Ni As_2$	h	521-1381
see also: Rammelsbergite $Ni As_2$		521-1307
Rammelsbergite $Ni As_2$	hj	521-1307
see also: Loellingite		521-1306
see also: Pararammelsbergite $Ni As_2$		521-1381
Nickel Arsenide Sulfide, Natural $Ni As S$	h	521-1190
Nickeline $Ni As$	hj	521-0751
see also: Nickel Arsenide $Ni As$		102-0612
Nickel Pyrite $(Ni, Fe) S_2$	h	521-0710
see also: Iron Nickel Sulfide, Nonstoichiometric $Ni Fe S_2$		110-0370
Nickel Sulfide, Natural $Ni_3 S_4$	h	521-1195
Nigrine $Ti O_2$	aehijno	521-0092
see also: Titanium Oxide $Ti O_2$		122-0005
see also: Anatase, Natural $Ti O_2$		521-0094
see also: Brookite $Ti O_2$		521-1484
Nitrate Soil	h	521-0674
Nontronite	hj	521-1329
Nordmarkite Syenite	e	521-0484
Norite Gabbro	ano	521-0292
Nosean $Na_8 Al_6 Si_6 O_{24} S O_4$	n	521-1108
Noselite $Na_8 Al_6 Si_6 O_{24} S O_4$	n	521-1108
Novaculite	a	521-1127
Nsutite $Mn(O, O H)_2$	h	521-1013
Nukundamite $Cu_5 Fe S_6$	h	521-1635
see also: Idaite $Cu_3 Fe S_4$		521-1636
Nut Coal	adefhjr	521-0249
Obsidian	adefghjnor	521-0307
Obsidian Moon Rock	h	521-1587
Ocean Floor	a	521-0812
Octahedrite $Ti O_2$	ehj	521-0094
see also: Titanium Oxide $Ti O_2$		122-0005
see also: Rutile, Natural $Ti O_2$		521-0092
see also: Brookite $Ti O_2$		521-1484
Oil Shale	adej	521-0115
see also: Kerogen		521-1630
Oil Well	ad	521-0837
Oligoclase	ahjnor	521-0552
Olivenite $Cu_2 As O_4 O H$	j	521-1188
Olivine $(Mg, Fe)_2 Si O_4$	adeghijnor	521-0258
Olivine:		———
Fayalite $Fe_2 Si O_4$	aehjno	521-0261
see also: Iron Silicate $Fe_2 Si O_4$		110-0051
see also: Fayalite,		551-3076
Forsterite $Mg_2 Si O_4$	adehijno	521-0022
see also: Magnesium Silicate $Mg_2 Si O_4$		119-0004
see also: Forsterite,		551-3194
Hortonolite $(Fe, Mg, Mn)_2 Si O_4$	ah	521-0790
Hyalosiderite	a	521-0955
Peridot	h	521-1394
Tephroite $Mn_2 Si O_4$	ah	521-0956
see also: Manganese Silicate $Mn_2 Si O_4$		119-0052
Olivine, Carboniferous Lava Basalt	f	521-0232
Olivine – Pyroxene Basalt	h	521-1430
Olivine Andesite	ade	521-0377
Olivine Basalt	adefgh	521-0028
Olivine Gabbro	ag	521-0904
Olivine Melanephelinite Basalt	f	521-1237
Olivine Moon Rock	h	521-1393
Olivine Rock	adghnr	521-0352
Olivine Sand	a	521-1619
Olivine Tholeiite Basalt	f	521-1487
Olivinorite	n	521-1256
Olsacherite $Pb_2 Se O_4 S O_4$	j	521-1367
Olshanskyite $Ca_3 B_4(O H)_{18}$	j	521-1169
Olzakherite $Pb_2 Se O_4 S O_4$	j	521-1367
Omphacite	j	521-1460

Property: **a.** Thermal conductivity; **b.** Accommodation coefficient; **c.** Thermal contact resistance; **d.** Thermal diffusivity; **e.** Specific heat; **f.** Viscosity; **g.** Emittance; **h.** Reflectance; **i.** Absorptance; **j.** Transmittance; **k.** α_s/ϵ_t ratio; **l.** Prandtl number; **n.** Thermal linear expansion coefficient; **o.** Thermal volumetric expansion coefficient; **r.** Thermal radiative properties (g,h,i,j,k)

Substance Name	Property	Number
Opal SiO_2	eghijnr	521-0496
see also: Silicon Oxide SiO_2		122-0009
Opal:		———
Hyalite SiO_2	aij	521-1128
see also: Silicon Oxide SiO_2		122-0009
Lussatite SiO_2	ij	521-1420
see also: Silicon Oxide SiO_2		122-0009
Opalized Wood	r	521-0474
Ophite	a	521-0760
see also: Diabase		521-0308
Ore	adeh	521-1089
Ore:		———
Anglesite $PbSO_4$	hjn	521-0786
see also: Lead Sulfate $PbSO_4$		122-0021
Apatite - Forsterite - Magnetite Ore	ad	521-1322
Argentite Ag_2S	h	521-0754
see also: Silver Sulfide Ag_2S		102-0120
Arsenopyrite $FeS_2 . FeAs_2$	hj	521-0744
Azurite $Cu_3(CO_3)_2(OH)_2$	ehjno	521-0396
Barite $BaSO_4$	aehijnor	521-0223
see also: Barium Sulfate $BaSO_4$		104-0025
Borax $Na_2B_4O_7$	ah	521-1287
see also: Sodium Borate $Na_2B_4O_7$		104-0035
Bornite Cu_5FeS_4	h	521-0758
see also: Copper Iron Sulfide Cu_5FeS_4		106-1686
Braunite $3Mn_2O_3 . MnSiO_3$	h	521-0992
Calcite - Magnetite Ore	ad	521-1327
Carnotite $K(UO_2)_2(VO_4)_2$	h	521-1282
Cassiterite SnO_2	hjnor	521-0470
see also: Tin Oxide SnO_2		122-0053
Celestite $SrSO_4$	ahjn	521-0785
see also: Strontium Sulfate $SrSO_4$		122-0132
Cerussite $PbCO_3$	hij	521-0626
see also: Lead Carbonate $PbCO_3$		106-0496
Chalcocite Cu_2S	h	521-0756
see also: Copper Sulfide Cu_2S		106-0104
see also: Djurleite Cu_2S		521-1436
Chalcopyrite $CuFeS_2$	ahj	521-0099
see also: Copper Iron Sulfide $CuFeS_2$		106-0791
Chrome Ore	n	521-0870
Chromite $(Fe, Mg)(Cr, Al)_2O_4$	adehijn	521-0366
see also: Magnesiochromite $(Mg, Fe)(Cr, Al)_2O_4$		521-0963
Cinnabar HgS	hj	521-0749
see also: Mercury Sulfide HgS		112-0402
Cobaltite $CoAsS$	eh	521-0043
Cobalt Nickel Arsenide Ore	h	521-1073
Columbite $(Fe, Mn)(Nb, Ta)_2O_6$	h	521-1574
see also: Tantalite $(Fe, Mn)(Nb, Ta)_2O_6$		521-0750
Copper Sulfide Ore	ade	521-1074
Covellite CuS	eh	521-0877
see also: Copper Sulfide CuS		106-0023
Crocoite $PbCrO_4$	hj	521-0854
see also: Lead Chromate $PbCrO_4$		106-0678
Cuprite Cu_2O	h	521-0644
see also: Copper Oxide Cu_2O		106-0127
Enargite Cu_3AsS_4	h	521-1221
see also: Copper Arsenide Sulfide Cu_3AsS_4		102-0219
Fahlore	h	521-0748
see also: Tennantite $(Cu, Fe)_{12}As_4S_{13}$		521-0880
see also: Tetrahedrite $(Cu, Fe)_{12}Sb_4S_{13}$		521-1205
Ferroplatinum Ore	h	521-1086
Fluorite CaF_2	aehjnr	521-0096
see also: Calcium Fluoride CaF_2		106-0026
Franklinite $(Zn, Mn, Fe)(Fe, Mn)_2O_4$	h	521-1017
see also: Iron Zinc Oxide $ZnFe_2O_4$		110-0029
Galena PbS	aehjnr	521-0192
see also: Lead Sulfide PbS		123-0003
Hematite Fe_2O_3	adehjno	521-0104
see also: Iron Oxide Fe_2O_3		110-0011
Huebnerite $MnWO_4$	hj	521-1076
see also: Manganese Tungsten Oxide $MnWO_4$		119-0053
Ilmenite $FeO . TiO_2$	aehj	521-0133
see also: Iron Titanium Oxide $FeTiO_3$		110-0005

Substance Name	Property	Number
Ore: -(Cont)-		
Iron Ore	adeh	521-0382
Iron Ore, Taconite	ade	521-0567
Limonite	eh	521-0414
Magnesite $MgCO_3$	adehjn	521-0006
see also: Magnesium Carbonate $MgCO_3$		106-0015
Magnetite $(Fe, Mg)Fe_2O_4$	adehjn	521-0067
see also: Iron Oxide Fe_3O_4		110-0025
Malachite $Cu_2CO_3(OH)_2$	hij	521-1227
Manganese Ore	a	521-1215
Manganite $MnO(OH)$	ehj	521-0042
see also: Groutite $HMnO_2$		521-1016
Martite $FeO . 3Fe_2O_3$	h	521-1432
Mimetite $Pb_5(AsO_4)_3Cl$	hj	521-1284
see also: Lead Arsenate Chloride $Pb_5(AsO_4)_3Cl$		102-0436
Molybdenite MoS_2	adehj	521-0675
see also: Molybdenum Sulfide MoS_2		119-0001
Niccolite $NiAs$	hj	521-0751
see also: Nickel Arsenide $NiAs$		102-0612
Pentlandite $(Fe, Ni)_9S_8$	h	521-0827
Powellite $CaMoO_4$	j	521-1079
see also: Calcium Molybdenum Oxide $CaMoO_4$		106-0768
Proustite Ag_3AsS_3	hjn	521-0755
see also: Silver Arsenide Sulfide Ag_3AsS_3		102-0408
Pyrargyrite Ag_3SbS_3	h	521-1201
see also: Antimony Silver Sulfide Ag_3SbS_3		102-0296
Pyrolusite MnO_2	h	521-1012
see also: Manganese Oxide MnO_2		119-0010
Pyromorphite $Pb_5(PO_4)_3Cl$	h	521-0533
see also: Lead Phosphate Chloride $Pb_5(PO_4)_3Cl$		106-1661
Pyrrhotite FeS	ahj	521-0746
see also: Iron Sulfide FeS		110-0007
see also: Iron Sulfide, Nonstoichiometric FeS		110-0295
Rhodochrosite $MnCO_3$	ahjn	521-0614
see also: Manganese Carbonate $MnCO_3$		106-0497
Romanechite $BaMn_9O_{16}(OH)_4$	h	521-1015
Rutile, Natural TiO_2	aehijno	521-0092
see also: Titanium Oxide TiO_2		122-0005
see also: Anatase, Natural TiO_2		521-0094
see also: Brookite TiO_2		521-1484
Salt Ore $KCl + MgCl_2 + MgSO_4 + NaCl$	a	521-0565
Scheelite $CaWO_4$	hij	521-0737
see also: Calcium Tungsten Oxide $CaWO_4$		106-0610
Siderite $FeCO_3$	aehijo	521-0566
see also: Iron Carbonate $FeCO_3$		106-0498
Skutterudite $(Co, Ni)As_3$	hj	521-1308
Skutterudite, Smaltite $(Co, Ni)As_3$	h	521-1433
Smithsonite $ZnCO_3$	ahj	521-0611
see also: Zinc Carbonate $ZnCO_3$		106-0495
Sphalerite $(Zn, Fe)S$	aehjnr	521-0086
see also: Zinc Sulfide ZnS		126-0005
see also: Wurtzite $(Zn, Fe)S$		521-1303
Spodumene $LiAlSi_2O_6$	aehn	521-0346
see also: Lithium Aluminosilicate $LiAl(SiO_3)_2$		102-0269
Staffelite - Magnetite Ore	ad	521-1321
Stibnite Sb_2S_3	ahjr	521-0095
see also: Antimony Sulfide Sb_2S_3		126-0014
Sylvite KCl	aehjr	521-0087
see also: Potassium Chloride KCl		106-0006
Tantalite $(Fe, Mn)(Nb, Ta)_2O_6$	h	521-0750
see also: Tapiolite $Fe(Ta, Nb)_2O_6$		521-1297
see also: Columbite $(Fe, Mn)(Nb, Ta)_2O_6$		521-1574
Tapiolite $Fe(Ta, Nb)_2O_6$	hj	521-1297
see also: Tantalite $(Fe, Mn)(Nb, Ta)_2O_6$		521-0750
Tennantite $(Cu, Fe)_{12}As_4S_{13}$	h	521-0880
see also: Fahlore		521-0748
see also: Tetrahedrite $(Cu, Fe)_{12}Sb_4S_{13}$		521-1205

Property: **a.** Thermal conductivity; **b.** Accommodation coefficient; **c.** Thermal contact resistance; **d.** Thermal diffusivity; **e.** Specific heat; **f.** Viscosity; **g.** Emittance; **h.** Reflectance; **i.** Absorptance; **j.** Transmittance; **k.** α_s/ϵ_t ratio; **l.** Prandtl number; **n.** Thermal linear expansion coefficient; **o.** Thermal volumetric expansion coefficient; **r.** Thermal radiative properties (g,h,i,j,k)

Substance Name	Property	Number
Ore: -(Cont)-		
Tetrahedrite $(Cu, Fe)_{12} Sb_4 S_{13}$	h	521-1205
see also: Antimony Copper Iron Sulfide, Nonstoichiometric $(Cu, Fe)_{12} Sb_4 S_{13}$		106-1802
see also: Antimony Copper Sulfide, Nonstoichiometric $Cu_{12} Sb_4 S_{13}$		106-2441
see also: Antimony Copper Sulfide, Nonstoichiometric $Cu_{14} Sb_4 S_{13}$		106-2442
see also: Antimony Copper Sulfide, Nonstoichiometric $Cu_{13} Sb_4 S_{13}$		106-2443
see also: Antimony Copper Sulfide $Cu_{13} Sb_4 S_{13}$		106-2444
see also: Fahlore		521-0748
see also: Tennantite $(Cu, Fe)_{12} As_4 S_{13}$		521-0880
Uraninite UO_2	hj	521-1353
see also: Uranium Oxide UO_2		122-0008
Uranium	h	521-1650
Vanadinite $Pb_5(VO_4)_3 Cl$	j	521-1102
Willemite $Zn_2 SiO_4$	ehjr	521-0650
see also: Zinc Silicate $Zn_2 SiO_4$		122-0027
Wolframite $(Fe, Mn) WO_4$	ehj	521-0555
Wulfenite $PbMoO_4$	hj	521-0736
see also: Lead Molybdenum Oxide $PbMoO_4$		119-0073
Zincite $(Zn, Mn) O$	hno	521-0879
see also: Zinc Oxide ZnO		122-0023
Zircon $ZrSiO_4$	aeghijnor	521-0250
see also: Zirconium Silicate $ZrSiO_4$		122-0007
see also: Essonite		521-1071
Organic Soil	adeh	521-0530
Orgueil	h	521-1502
Orpiment $As_2 S_3$	hj	521-0757
see also: Arsenic Sulfide $As_2 S_3$		102-0086
Orthoclase $KAlSi_3 O_8$	adefhijnor	521-0026
see also: Potassium Aluminum Silicate $KAlSi_3 O_8$		102-0173
Orthoclase:		——
Adularia $KAlSi_3 O_8$	aehnor	521-0060
see also: Potassium Aluminum Silicate $KAlSi_3 O_8$		102-0173
Ferriferous	e	521-0061
Moonstone	ijnr	521-0457
Sanidine $KAlSi_3 O_8$	aehno	521-1045
see also: Potassium Aluminum Silicate $KAlSi_3 O_8$		102-0173
Orthopyroxene	h	521-1601
Orthopyroxene:		——
Bronzite Pyroxene	ahno	521-0791
Enstatite Pyroxene $MgSiO_3$	adhij	521-0769
see also: Magnesium Silicate $MgSiO_3$		119-0009
Ferrosilite $FeSiO_3$	ah	521-1133
Hypersthene $(Mg, Fe) SiO_3$	ahij	521-0351
Ottawa Sand SiO_2	adeghnor	521-0010
see also: Silicon Oxide SiO_2		122-0009
Palladium Platinum Sulfide, Natural $(Pt, Pd) S$	h	521-0884
Palladium Selenide, Natural $Pd_{17} Se_{15}$	h	521-1492
Palladseite $Pd_{17} Se_{15}$	h	521-1492
Palygorskite $(Mg, Al)_2 Si_4 O_{10}(OH)$	aeij	521-0182
Pantellerite	f	521-1235
Paracelsian $BaAl_2 Si_2 O_8$	j	521-1300
see also: Aluminum Barium Silicate $BaAl_2(SiO_4)_2$		102-0982
see also: Celsian Feldspar $BaAl_2 Si_2 O_8$		521-0021
Pararammelsbergite $NiAs_2$	h	521-1381
see also: Rammelsbergite $NiAs_2$		521-1307
Paris White $CaCO_3$	adeg	521-0183
see also: Calcium Carbonate $CaCO_3$		106-0005
see also: Aragonite $CaCO_3$		521-0105
see also: Calcareous Rock		521-1560
Paris Whiting $CaCO_3$	adeg	521-0183
see also: Calcium Carbonate $CaCO_3$		106-0005
see also: Aragonite $CaCO_3$		521-0105
see also: Calcareous Rock		521-1560
Pavonite $AgBi_3 S_5$	h	521-1449
Peacock Ore $Cu_5 FeS_4$	h	521-0758
see also: Copper Iron Sulfide $Cu_5 FeS_4$		106-1686

Substance Name	Property	Number
Pearl Spar $CaMg(CO_3)_2$	adefghjnor	521-0084
see also: Calcium Magnesium Carbonate $CaMg(CO_3)_2$		106-0795
Peat	adefg	521-0066
Pebble	adefghnor	521-0046
Pectolite $NaCa_2 Si_3 O_8 OH$	ahj	521-0948
Pedalfer Soil	h	521-1152
Pedocal Soil	h	521-1157
Pegmatite	adeghn	521-0762
Pegmatite, Feldspathic	a	521-0830
Pegmatite Quartz	a	521-0743
Penninite $(Mg, Fe, Al)_6(Si, Al)_4 O_{10}(OH)_8$	j	521-0177
Pentlandite $(Fe, Ni)_9 S_8$	h	521-0827
Periclase MgO	aegjn	521-0132
see also: Magnesium Oxide MgO		119-0002
Periclase – Spinel	n	521-1234
Peridot	h	521-1394
Peridotite	adeghjno	521-0852
Peridotite:		——
Dunite	adghnr	521-0352
Harzburgite	h	521-1641
Lherzolite	a	521-1008
Peridotite, Serpentinized	ad	521-0277
Perlite	adefgjnor	521-0073
see also: Silicon Oxide Mixture Glass $SiO_2 > Al_2 O_3 > K_2 O > Na_2 O$		503-1814
Permafrost Soil	ae	521-1359
Perovskite $CaO \cdot TiO_2$	eh	521-0097
see also: Calcium Titanium Oxide $CaTiO_3$		106-0218
Perrierite $(Ca, Ce, Th)_4(Mg, Fe)_2(Ti, Fe)_3 Si_4 O_{22}$	j	521-1009
Perthite	fhn	521-0153
Petalite $LiAlSi_4 O_{10}$	ahijn	521-0772
Petrified Wood	r	521-0474
Petroleum Basin Sediment	h	521-1106
Petroleum Stratum	a	521-1264
Pharmacolite $CaHAsO_4$	j	521-1186
Phenacite $Be_2 SiO_4$	ehjn	521-0858
see also: Beryllium Silicate $Be_2 SiO_4$		104-0142
Phenakite $Be_2 SiO_4$	ehjn	521-0858
see also: Beryllium Silicate $Be_2 SiO_4$		104-0142
Phlogopite $K(Mg, Fe)_3 AlSi_3 O_{10}(OH, F)_2$	adefhijn	521-0241
see also: Aluminum Magnesium Potassium Silicate Fluoride $KMg_3 AlSi_3 O_{10} F_2$		102-0262
see also: Aluminum Germanium Magnesium Potassium Oxide Fluoride $KMg_3 AlGe_3 O_{10} F_2$		102-0962
see also: Aluminum Germanium Magnesium Potassium Oxide Fluoride, Nonstoichiometric $KMg_3 Al_n Ge_4 O_{10} F_2$		102-0963
Phonolite	ghn	521-1011
Phosgenite $Pb_2 Cl_2 CO_3$	h	521-1348
Phosphate Fertilizer	f	521-0262
Phosphate Rock	h	521-1055
Phyllite	hn	521-1035
Picotite	h	521-1162
see also: Magnesiochromite $(Mg, Fe)(Cr, Al)_2 O_4$		521-0963
see also: Ceylonite		521-0990
Picotpaulite $TlFe_2 S_3$	h	521-1468
Picrite	en	521-1540
Picroilmenite	h	521-1270
Piedmontite Schist	ade	521-0379
Piemontite $Ca_2 (Al, Mn, Fe)_3 Si_3 O_{12} OH$	j	521-1459
Pigeonite $(Mg, Fe, Ca)(Mg, Fe) Si_2 O_6$	h	521-0792
Pigeonite Moon Rock	j	521-1272
Pinite	no	521-1032
Pipestone	a	521-0482
Pitchstone	no	521-1029
Plagioclase $(Na, Ca) Al (Si, Al) Si_2 O_8$	adeh	521-0495
Plagioclase:		——
Albite $NaAlSi_3 O_8$	adefghijnor	521-0057
see also: Sodium Aluminosilicate $NaAlSi_3 O_8$		102-0218
Albite – Anorthite $NaAlSi_3 O_8 - CaAl_2 Si_2 O_8$	d	521-1549
Andesine	ehno	521-0859

Property: **a.** Thermal conductivity; **b.** Accommodation coefficient; **c.** Thermal contact resistance; **d.** Thermal diffusivity; **e.** Specific heat; **f.** Viscosity; **g.** Emittance; **h.** Reflectance; **i.** Absorptance; **j.** Transmittance; **k.** a_s/ϵ_t ratio; **l.** Prandtl number; **n.** Thermal linear expansion coefficient; **o.** Thermal volumetric expansion coefficient; **r.** Thermal radiative properties (g,h,i,j,k)

Substance Name	Property	Number
Plagioclase: -(Cont)-		
Anorthite, Natural $CaAl_2(SiO_4)_2$	aefhijno	521-0035
see also: Calcium Aluminum Silicate $CaAl_2(SiO_4)_2$		102-0125
Bytownite	ahno	521-0969
Labradorite	ahjnor	521-0471
Oligoclase	ahjnor	521-0552
Plagioclase Basalt	g	521-0901
Plagiogranite	adeh	521-0713
Plagiogranite, Greisenized	e	521-0714
Plagiogranite, Leucocratic	ad	521-1242
Plagiomicrocline	ade	521-1334
Plagionite $Pb_5Sb_8S_{17}$	j	521-1177
Plagiopegmatite	ade	521-1333
Planet	h	521-1356
Platinum, Natural Pt	h	521-1085
see also: Platinum Pt		101-0029
Playa Soil	h	521-1165
Pleonaste	o	521-0990
see also: Picotite		521-1162
Pluto Surface	r	521-0528
Podzol Soil	ahr	521-0542
Polycrase $(Y, Ca, Ce, U, Th)(Ti, Nb, Ta)_2O_6$	h	521-1341
Polydymite Ni_3S_4	h	521-1195
Polylithionite $KLi_2AlSi_4O_{10}(F, OH)_2$	j	521-1092
see also: Lepidolite Mica $K(Li, Al)_3(Si, Al)_4O_{10}(F, OH)$		521-0965
Polyxene Pt	h	521-1085
see also: Platinum Pt		101-0029
Porcelain Clay	adefhjnr	521-0135
see also: Kaolinite, Mineral $Al_2Si_2O_5(OH)_4$		521-0174
see also: Kaolin Board		661-0444
see also: Kaolin Paper		661-0484
Porphyrite	adeno	521-0375
Porphyry	adeno	521-0375
Porphyry:		———
Andesite	h	521-1069
Augite	ade	521-0373
Basalt	n	521-1262
Diorite	ade	521-1245
Felsite	hno	521-0976
Granite	ade	521-0671
Granodiorite	ade	521-1244
Latite	g	521-1278
Pyroxene	g	521-1559
Quartz	ade	521-1212
Quartz - Feldspar	a	521-0455
Quartz Monzonite	ag	521-0907
Rhyolite	a	521-1217
Portlandite $Ca(OH)_2$	j	521-1174
see also: Calcium Hydroxide $Ca(OH)_2$		106-0042
Potash Alum $KAl(SO_4)_2$	hj	521-1471
see also: Aluminum Potassium Sulfate $KAl(SO_4)_2$		102-0024
Potassium Cyanoferrate, Natural $K_4Fe(CN)_6$	j	521-1473
see also: Potassium Cyanoferrate $K_4Fe(CN)_6$		106-0577
Potassium Feldspar $KAlSi_3O_8$	adfgn	521-0170
see also: Potassium Aluminum Silicate $KAlSi_3O_8$		102-0173
Potassium Feldspar:		———
Adularia $KAlSi_3O_8$	aehnor	521-0060
see also: Potassium Aluminum Silicate $KAlSi_3O_8$		102-0173
Microcline $KAlSi_3O_8$	aehjnor	521-0392
see also: Potassium Aluminum Silicate $KAlSi_3O_8$		102-0173
Orthoclase $KAlSi_3O_8$	adefhijnor	521-0026
see also: Potassium Aluminum Silicate $KAlSi_3O_8$		102-0173
Sanidine $KAlSi_3O_8$	aehno	521-1045
see also: Potassium Aluminum Silicate $KAlSi_3O_8$		102-0173
Potassium Mica $KAl_2(AlSi_3)O_{10}(OH)_2$	adeghijknr	521-0240
see also: Aluminum Potassium Silicate Hydroxide $KAl_2(AlSi_3O_{10})(OH)_2$		102-0798
see also: Potassium Aluminum Silicate $KAl_3Si_3O_{11}$		102-0966

Substance Name	Property	Number
Potassium Uranium Vanadium Oxide, Natural $K(UO_2)_2(VO_4)_2$	h	521-1282
Potter's Flint SiO_2	ahn	521-1465
see also: Silicon Oxide SiO_2		122-0009
Powellite $CaMoO_4$	j	521-1079
see also: Calcium Molybdenum Oxide $CaMoO_4$		106-0768
Prairie Soil	h	521-1050
Prase SiO_2	j	521-1224
see also: Silicon Oxide SiO_2		122-0009
Prehnite $Ca_2Al_2Si_3O_{10}(OH)_2$	aj	521-0968
Preobrazhenskite $Mg_3B_{11}O_{15}(OH)_9$	j	521-1170
Priorite $(Y, Ca, Th)(Ti, Nb)_2O_6$	h	521-1577
Prochlorite $(Mg, Fe)_9Al_6Si_5O_{20}(OH)_{16}$	n	521-0818
see also: Mica		521-0108
Proustite Ag_3AsS_3	hjn	521-0755
see also: Silver Arsenide Sulfide Ag_3AsS_3		102-0408
Pseudobrookite Fe_2TiO_5	e	521-0101
see also: Armalcolite $(Mg, Fe)Ti_2O_5$		521-1599
Psilomelane $BaMn_9O_{16}(OH)_4$	h	521-1015
Pulsar	a	521-1232
Pumice	adeghjn	521-0004
Pyrargyrite Ag_3SbS_3	h	521-1201
see also: Antimony Silver Sulfide Ag_3SbS_3		102-0296
Pyrite FeS_2	adehijnor	521-0025
see also: Iron Sulfide FeS_2		110-0010
see also: Marcasite FeS_2		521-1304
Pyrite Breccia	no	521-0885
Pyritic Andalusite Schist	ad	521-0847
Pyritic Slate Schist	ad	521-0845
Pyrochlore $(Na, Ca)_2(Nb, Ta)_2O_6(OH, F)$	h	521-1004
Pyrochlore:		———
Betafite $(Ca, Na, U)_2(Nb, Ta)_2O_6(O, OH)$	h	521-1342
Microlite $(Na, Ca)_2(Ta, Nb)_2O_6(O, OH, F)$	ah	521-1003
Pyrochroite $Mn(OH)_2$	h	521-1014
Pyrolusite MnO_2	h	521-1012
see also: Manganese Oxide MnO_2		119-0010
Pyromorphite $Pb_5(PO_4)_3Cl$	h	521-0533
see also: Lead Phosphate Chloride $Pb_5(PO_4)_3Cl$		106-1661
Pyrope $(Mg, Fe)_3Al_2(SiO_4)_3$	aehno	521-0411
Pyrophyllite $AlSi_2O_5(OH)$	adehijno	521-0362
see also: Aluminum Silicate Hydroxide $AlSi_2O_5(OH)$		102-0799
see also: Metapyrophyllite		521-1346
Pyroxene	aeho	521-0494
Pyroxene:		———
Acmite $NaFe(SiO_3)_2$	ahn	521-0947
see also: Iron Sodium Silicate $NaFe(SiO_3)_2$		110-0495
Acmite-Trachyte	f	521-0224
Andesite	ade	521-0383
Augite $(Ca, Na)(Mg, Fe, Al)(Si, Al)_2O_6$	aghijno	521-0794
see also: Augite		521-1592
Baikalite Hedenbergite	no	521-0860
Bronzite	ahno	521-0791
Chrome Diopside	h	521-1593
see also: Diopside $CaMg(SiO_3)_2$		521-0036
Clinoenstatite $MgSiO_3$	n	521-1054
see also: Magnesium Silicate $MgSiO_3$		119-0009
Clinopyroxene	no	521-1339
Diopside $CaMg(SiO_3)_2$	aefhijno	521-0036
see also: Calcium Magnesium Silicate $CaMgSi_2O_6$		106-0940
see also: Chrome Diopside		521-1593
Enstatite $MgSiO_3$	adhij	521-0769
see also: Magnesium Silicate $MgSiO_3$		119-0009
Ferrosilite $FeSiO_3$	ah	521-1133
Hedenbergite $CaFeSi_2O_6$	hjno	521-0793
Hypersthene $(Mg, Fe)SiO_3$	ahij	521-0351
Jadeite $Na(Al, Fe)Si_2O_6$	adehij	521-0058
Kunzite	j	521-1486
Lavroffite	g	521-0908
Lavrovite	g	521-0908
Omphacite	j	521-1460
Orthopyroxene	h	521-1601
Pigeonite $(Mg, Fe, Ca)(Mg, Fe)Si_2O_6$	h	521-0792

Property: **a.** Thermal conductivity; **b.** Accommodation coefficient; **c.** Thermal contact resistance; **d.** Thermal diffusivity; **e.** Specific heat; **f.** Viscosity; **g.** Emittance; **h.** Reflectance; **i.** Absorptance; **j.** Transmittance; **k.** α_s/ϵ_t ratio; **l.** Prandtl number; **n.** Thermal linear expansion coefficient; **o.** Thermal volumetric expansion coefficient; **r.** Thermal radiative properties (g,h,i,j,k)

Substance Name	Property	Number
Pyroxene: -(Cont)-		
Rhodonite $Mn\,Si\,O_3$	hij	521-1043
see also: Manganese Silicate $Mn\,Si\,O_3$		119-0020
Salite $Ca(Mg, Fe)\,Si_2\,O_6$	h	521-1578
Spodumene $Li\,Al\,Si_2\,O_6$	aehn	521-0346
see also: Lithium Aluminosilicate $Li\,Al(Si\,O_3)_2$		102-0269
Pyroxene Gabbro	a	521-1247
Pyroxene Granite	ad	521-1239
Pyroxene Moon Rock	j	521-1249
Pyroxene Porphyry	g	521-1559
Pyroxenite	adgh	521-0356
Pyroxenite:		———
Bronzitite	a	521-0350
Pyroxenite, Serpentinized	h	521-1644
Pyrrhotine $Fe\,S$	ahj	521-0746
see also: Iron Sulfide $Fe\,S$		110-0007
see also: Iron Sulfide, Nonstoichiometric $Fe\,S$		110-0295
Pyrrhotite $Fe\,S$	ahj	521-0746
see also: Iron Sulfide $Fe\,S$		110-0007
see also: Iron Sulfide, Nonstoichiometric $Fe\,S$		110-0295
Pyrrhotite:		———
Troilite $Fe\,S$	h	521-1273
see also: Iron Sulfide $Fe\,S$		110-0007
Quartz, Milky	h	521-1081
Quartz, Natural $Si\,O_2$	adefghijnor	521-0732
see also: Quartz $Si\,O_2$		122-0002
see also: Silicon Oxide $Si\,O_2$		122-0009
see also: Stishovite $Si\,O_2$		122-0609
see also: Cristobalite $Si\,O_2$		122-0628
see also: Coesite $Si\,O_2$		122-0629
see also: Tridymite $Si\,O_2$		122-0646
see also: Cristobalite, Natural $Si\,O_2$		521-0254
see also: Tridymite, Natural $Si\,O_2$		521-0255
see also: Coesite, Natural $Si\,O_2$		521-0823
see also: Stishovite, Natural $Si\,O_2$		521-0824
see also: Quartz Fiber Felt		661-0331
Quartz, Natural:		———
Agate $Si\,O_2$	j	521-1223
see also: Silicon Oxide $Si\,O_2$		122-0009
Amethyst $Si\,O_2$	hj	521-0682
see also: Silicon Oxide $Si\,O_2$		122-0009
Aventurine $Si\,O_2$	h	521-1082
see also: Silicon Oxide $Si\,O_2$		122-0009
Carnelian $Si\,O_2$	h	521-1105
see also: Silicon Oxide $Si\,O_2$		122-0009
Chalcedony $Si\,O_2$	aehijnr	521-0222
see also: Silicon Oxide $Si\,O_2$		122-0009
Chrysoprase $Si\,O_2$	r	521-0628
see also: Silicon Oxide $Si\,O_2$		122-0009
Citrine $Si\,O_2$	hj	521-0684
see also: Silicon Oxide $Si\,O_2$		122-0009
Flint, Mineral $Si\,O_2$	ahn	521-1465
see also: Silicon Oxide $Si\,O_2$		122-0009
Hornstone $Si\,O_2$	nor	521-0629
see also: Silicon Oxide $Si\,O_2$		122-0009
Hyalite $Si\,O_2$	aij	521-1128
see also: Silicon Oxide $Si\,O_2$		122-0009
Lussatite $Si\,O_2$	ij	521-1420
see also: Silicon Oxide $Si\,O_2$		122-0009
Opal $Si\,O_2$	eghijnr	521-0496
see also: Silicon Oxide $Si\,O_2$		122-0009
Prase $Si\,O_2$	j	521-1224
see also: Silicon Oxide $Si\,O_2$		122-0009
Quartz, Milky	h	521-1081
Quartz, Rose	e	521-0395
Quartz, Smoky	aej	521-0517
Tiger's Eye $Si\,O_2$	r	521-0473
see also: Silicon Oxide $Si\,O_2$		122-0009
Quartz, Rose	e	521-0395
Quartz, Smoky	aej	521-0517
Quartz - Feldspar Porphyry	a	521-0455
Quartz Andesite	afghn	521-0243

Substance Name	Property	Number
Quartz Basalt	g	521-0890
Quartz Breccia	a	521-0276
Quartz Diorite	adeg	521-0279
Quartz Dolerite	a	521-0850
Quartz Gabbro, Metamorphosed	a	521-0280
Quartz Glass, Natural $Si\,O_2$	adefghijnor	521-0732
see also: Quartz $Si\,O_2$		122-0002
see also: Silicon Oxide $Si\,O_2$		122-0009
see also: Stishovite $Si\,O_2$		122-0609
see also: Cristobalite $Si\,O_2$		122-0628
see also: Coesite $Si\,O_2$		122-0629
see also: Tridymite $Si\,O_2$		122-0646
see also: Cristobalite, Natural $Si\,O_2$		521-0254
see also: Tridymite, Natural $Si\,O_2$		521-0255
see also: Coesite, Natural $Si\,O_2$		521-0823
see also: Stishovite, Natural $Si\,O_2$		521-0824
see also: Quartz Fiber Felt		661-0331
Quartzite	adehjno	521-0298
Quartzite - Sandstone	a	521-1112
Quartz Monzonite	aegno	521-0354
Quartz Monzonite Granite	a	521-1490
Quartz Monzonite Porphyry	ag	521-0907
Quartzose Rock	a	521-1344
Quartz Porphyry	ade	521-1212
Quartz Rock	adehjno	521-0298
Quartz Schist	a	521-0263
Quartz Syenite	gn	521-0896
Quicksand	adeghjnor	521-0110
Rain Forest, Tropical Soil	ahj	521-0865
Rammelsbergite $Ni\,As_2$	hj	521-1307
see also: Loellingite		521-1306
see also: Pararammelsbergite $Ni\,As_2$		521-1381
Ravine	r	521-0522
Realgar $As\,S$	j	521-1311
see also: Arsenic Sulfide $As\,S$		102-0264
Red Clay	hj	521-0677
Red Glass Moon Rock	h	521-1594
Red Iron Ore $Fe_2\,O_3$	adehjno	521-0104
see also: Iron Oxide $Fe_2\,O_3$		110-0011
Red Silica - Calcite Sand	h	521-1153
Red Soil	h	521-1498
Reedmergnerite $Na\,B\,Si_3\,O_8$	j	521-1061
Regolith Moon Rock	ahj	521-1384
Renierite $Cu_3(Fe, Ge, Zn)(S, As)_4$	h	521-0881
Retinite	h	521-1111
Rhodizite $Cs\,Al_4\,Be_4\,B_{11}\,O_{25}(O\,H)_4$	j	521-0834
Rhodochrosite $Mn\,C\,O_3$	ahjn	521-0614
see also: Manganese Carbonate $Mn\,C\,O_3$		106-0497
Rhodolite	a	521-1072
Rhodonite $Mn\,Si\,O_3$	hij	521-1043
see also: Manganese Silicate $Mn\,Si\,O_3$		119-0020
Rhyolite	adefghjn	521-0272
Rhyolite, Gray	h	521-0973
Rhyolite, Pink	h	521-0974
Rhyolite Breccia	a	521-0271
Rhyolite Porphyry	a	521-1217
Rhyolite Tuff	ahn	521-0828
Richterite $(Na, K)_2(Mg, Mn, Ca)_6\,Si_8\,O_{22}(O\,H)_2$	ah	521-1139
Ridmerdzhnerit $Na\,B\,Si_3\,O_8$	j	521-1061
Riebeckite $Na_2(Fe, Mg)_5\,Si_8\,O_{22}(O\,H)_2$	ah	521-0952
Ripidolite $(Mg, Fe)_9\,Al_6\,Si_5\,O_{20}(O\,H)_{16}$	n	521-0818
see also: Mica		521-0108
Ristschorrite	ad	521-1324
Riverbank	hr	521-0523
Rock	adefghnor	521-0046
Rock:		———
Altered	h	521-1631
Basement	o	521-1503
Bed	a	521-1609
Calcareous	n	521-1560
see also: Limestone		521-0085
see also: Chalk $Ca\,C\,O_3$		521-0183

Property: a. Thermal conductivity; **b.** Accommodation coefficient; **c.** Thermal contact resistance; **d.** Thermal diffusivity; **e.** Specific heat; **f.** Viscosity; **g.** Emittance; **h.** Reflectance; **i.** Absorptance; **j.** Transmittance; **k.** α_s/ϵ_t ratio; **l.** Prandtl number; **n.** Thermal linear expansion coefficient; **o.** Thermal volumetric expansion coefficient; **r.** Thermal radiative properties (g,h,i,j,k)

Substance Name	Property	Number
Rock: -(Cont)-		
Carbonate	a	521-1118
Chlorite Mica Quartz	a	521-0841
Glacial	ade	521-1358
Igneous	ahn	521-0999
Metamorphic	adehn	521-0825
Phosphate	h	521-1055
Quartzose	a	521-1344
Sandy	ah	521-1615
Schistose	h	521-1440
Sedimentary	adehn	521-1028
Terrigenous	a	521-0717
Trap	aeno	521-0761
Volcanic	a	521-1434
Rock - Shell - Silica Sand	h	521-1154
Rock - Silica Sand	h	521-1155
Rock Crystal SiO_2	adefghijnor	521-0732
see also: Quartz SiO_2		122-0002
see also: Silicon Oxide SiO_2		122-0009
see also: Stishovite SiO_2		122-0609
see also: Cristobalite SiO_2		122-0628
see also: Coesite SiO_2		122-0629
see also: Tridymite SiO_2		122-0646
see also: Cristobalite, Natural SiO_2		521-0254
see also: Tridymite, Natural SiO_2		521-0255
see also: Coesite, Natural SiO_2		521-0823
see also: Stishovite, Natural SiO_2		521-0824
see also: Quartz Fiber Felt		661-0331
Rock Salt $NaCl$	adehjnor	521-0007
see also: Sodium Chloride $NaCl$		106-0024
see also: Sea Salt		521-1122
Rock Strata	a	521-0049
Rock Strata, Metamorphic	ade	521-1414
Romanechite $BaMn_9O_{16}(OH)_4$	h	521-1015
Ruby, Natural Al_2O_3	adeghijnr	521-0191
see also: Aluminum Oxide Al_2O_3		102-0002
Rucklidgite $(Bi,Pb)_3Te_4$	h	521-1552
Russellite Bi_2WO_6	j	521-1100
Ruthenium, Natural Ru	h	521-1591
Rutile, Natural TiO_2	aehijno	521-0092
see also: Titanium Oxide TiO_2		122-0005
see also: Anatase, Natural TiO_2		521-0094
see also: Brookite TiO_2		521-1484
Safflorite - Rammelsbergite $(Co,Ni)As_2$	h	521-0752
Sal Ammoniac NH_4Cl	n	521-0873
see also: Ammonium Chloride NH_4Cl		106-0045
Saline Playa Soil	ag	521-0866
Salite $Ca(Mg,Fe)Si_2O_6$	h	521-1578
Salmiac NH_4Cl	n	521-0873
see also: Ammonium Chloride NH_4Cl		106-0045
Salt $NaCl$	adehjnor	521-0007
see also: Sodium Chloride $NaCl$		106-0024
see also: Sea Salt		521-1122
Salt, Sea	eh	521-1122
see also: Sodium Chloride $NaCl$		106-0024
see also: Halite $NaCl$		521-0007
Salt Flat	j	521-0980
Salt Marsh Soil	hr	521-0548
Salt Ore $KCl + MgCl_2 + MgSO_4 + NaCl$	a	521-0565
Samarskite $(Y,Ce,U,Ca,Fe,Pb,Th)(Nb,Ta,Ti,Sn)_2O_6$	eh	521-0297
Sand	adeghjnor	521-0110
Sand:		——
Anhydrite	g	521-0906
Argillaceous	ad	521-1604
Carbonate	h	521-1158
Carbonate - Silica	h	521-1156
Chromite	a	521-1618
Coral	h	521-1036
Gray	a	521-0927
Gypsum	agh	521-0613
Hematite - Silica	h	521-1159
Lime	a	521-1361
Sand: -(Cont)-		
Lochaline SiO_2	adeghnor	521-0010
see also: Silicon Oxide SiO_2		122-0009
Molding	aden	521-0500
Olivine	a	521-1619
Ottawa SiO_2	adeghnor	521-0010
see also: Silicon Oxide SiO_2		122-0009
Red	a	521-0926
Red Silica - Calcite	h	521-1153
Rock - Shell - Silica	h	521-1154
Rock - Silica	h	521-1155
Silica SiO_2	adeghnor	521-0010
see also: Silicon Oxide SiO_2		122-0009
Silty	ade	521-1467
Silver	a	521-0551
Slag	ah	521-1617
Titanium Beach	n	521-0685
see also: Iron Titanium Oxide $FeTiO_3$		110-0005
White	h	521-1648
Yellow	h	521-1088
Zirconium	ade	521-1093
Sand - Shale	a	521-1546
Sand Soil	adehn	521-0442
Sandstone	adefghijnor	521-0196
see also: Buhrstone		521-0511
Sandstone:		——
Arkose	h	521-1499
Chloritic	ade	521-1409
Gray	h	521-0694
Green	h	521-1496
Limonitic	h	521-1494
Micaceous	a	521-0397
Quartzitic	a	521-0347
Red	ah	521-0688
Yellow	h	521-1497
Sandstone - Shale	a	521-1470
Sandstone - Slate, Interbedded	ad	521-0839
Sand With Pebbles	h	521-0690
Sandy Clay Soil	adejn	521-0190
Sandy Loam Soil	adeghjr	521-0368
Sandy Rock	ah	521-1615
Sandy Silt Soil	ad	521-0822
Sanidine $KAlSi_3O_8$	aehno	521-1045
see also: Potassium Aluminum Silicate $KAlSi_3O_8$		102-0173
Sanukite Andesite	ade	521-0378
Sapphire, Natural Al_2O_3	adeghjnr	521-0064
see also: Aluminum Oxide Al_2O_3		102-0002
Sapphirine $(Mg,Fe)_{15}(Al,Fe)_{34}Si_7O_{80}$	ej	521-0875
Sapropel	f	521-0391
Sapropelite Shale	j	521-1090
Saturn Rings	h	521-1058
Saturn Satellite	h	521-1330
Saturn Surface	hr	521-0425
Scapolite:		——
Meionite $3CaAl_2Si_2O_8 . CaCO_3$	a	521-0961
Wernerite	aeh	521-0960
Scapolite Gabbro	dn	521-1655
Schalstein	adehn	521-0370
Scheelite $CaWO_4$	hij	521-0737
see also: Calcium Tungsten Oxide $CaWO_4$		106-0610
Schist	adegn	521-0270
Schist:		——
Amphibole	a	521-0282
Amphibole Biotite	ade	521-1413
Andalusite Biotite	ad	521-0840
Biotite	ade	521-1403
Biotite Chlorite	ade	521-1411
Biotite Feldspar Garnet	adn	521-1405
Biotite Feldspar Garnet Quartz	ade	521-1406
Biotite Feldspar Quartz	h	521-1328
Biotite Garnet	ade	521-1404
Biotite Garnet Muscovite	ade	521-1412
Chlorite	adehn	521-0296
Chlorite Muscovite	ade	521-1410

Property: a. Thermal conductivity; **b.** Accommodation coefficient; **c.** Thermal contact resistance; **d.** Thermal diffusivity; **e.** Specific heat; **f.** Viscosity; **g.** Emittance; **h.** Reflectance; **i.** Absorptance; **j.** Transmittance; **k.** α_s/ϵ_t ratio; **l.** Prandtl number; **n.** Thermal linear expansion coefficient; **o.** Thermal volumetric expansion coefficient; **r.** Thermal radiative properties (g,h,i,j,k)

Substance Name	Property	Number
Schist: -(Cont)-		
Epidote	ade	521-0380
Greenschist	a	521-0275
Hornblende	h	521-1383
Mica	j	521-1568
Mica Quartz	ah	521-1147
Piedmontite	ade	521-0379
Pyritic Andalusite	ad	521-0847
Pyritic Slate	ad	521-0845
Quartz	a	521-0263
Silicified	a	521-0295
Sillimanite	ade	521-1407
Talc	ade	521-0381
Tremolite	h	521-1382
Schistose Rock	h	521-1440
Schorlomite $Ca_3(Fe, Ti)_2(Si, Ti)_3 O_{12}$	hij	521-1337
Schreyerite $V_2 Ti_3 O_9$	h	521-1491
Scolecite $Ca Al_2 Si_3 O_{10}$	j	521-0781
Scoria	adefghno	521-0053
Scoria, Volcanic	ade	521-0376
Scoriaceous Basalt	h	521-0979
Sea Coal	adefhjr	521-0249
Seacoast	h	521-1431
Sea Mud	a	521-0766
Sea Salt	eh	521-1122
see also: Sodium Chloride $Na Cl$		106-0024
see also: Halite $Na Cl$		521-0007
Sea Sediment	ade	521-0253
Sedimentary Rock	adehn	521-1028
Seinaejokite $Fe(As, Sb)_2$	h	521-1447
Selenite	aehj	521-1148
see also: Calcium Sulfate $Ca S O_4$		106-0051
Selenium, Natural Se	h	521-1485
see also: Selenium Se		101-0030
Sellaite $Mg F_2$	j	521-1295
see also: Magnesium Fluoride $Mg F_2$		110-0023
Semifusinite Coal	h	521-0816
Semivitrinite Coal	h	521-1121
Semseyite $Pb_9 Sb_8 S_{21}$	j	521-1179
Senamontite $Sb_2 O_3$	jo	521-0991
see also: Antimony Oxide $Sb_2 O_3$		122-0019
Senarmontite $Sb_2 O_3$	jo	521-0991
see also: Antimony Oxide $Sb_2 O_3$		122-0019
Sepiolite $Mg_4 (Si_2 O_5)_3(O H)_2$	e	521-1171
see also: Magnesium Silicate $Mg_2 Si_3 O_8$		119-0065
Serendibite $Ca_4(Mg, Fe, Al)_6(Al, Fe)_9(Si, Al)_6 B_3 O_{40}$	j	521-1166
Sericite $K Al_2 (Al Si_3) O_{10} (O H)_2$	adeghijknr	521-0240
see also: Aluminum Potassium Silicate Hydroxide $K Al_2(Al Si_3 O_{10})(O H)_2$		102-0798
see also: Potassium Aluminum Silicate $K Al_3 Si_3 O_{11}$		102-0966
Serpentine $(Mg, Fe)_3 Si_2 O_5(O H)_4$	adeghijn	521-0236
see also: Asbestos		521-0055
see also: Chrysotile $Mg_3 Si_2 O_5(O H)_4$		521-0799
Serpentine:		———
Antigorite $Mg_3 Si_2 O_5(O H)_4$	ij	521-1605
Chrysotile $Mg_3 Si_2 O_5(O H)_4$	eijn	521-0799
see also: Asbestos		521-0055
see also: Serpentine $(Mg, Fe)_3 Si_2 O_5(O H)_4$		521-0236
Lizardite $Mg_3 Si_2 O_5(O H)_4$	a	521-1143
Serpentine Asbestos $Mg_3 Si_2 O_5(O H)_4$	eijn	521-0799
see also: Asbestos		521-0055
see also: Serpentine $(Mg, Fe)_3 Si_2 O_5(O H)_4$		521-0236
Serpentine Rock	ah	521-1389
Serpentine With Garnets	ade	521-0502
Serpentinite	ah	521-1389
Serpentinized Dunite	h	521-1642
Serpentinized Harzburgite	h	521-1643
Serpentinized Peridotite	ad	521-0277
Serpentinized Pyroxenite	h	521-1644
Shale	adehjnr	521-0070

Substance Name	Property	Number
Shale:		———
Carbonaceous Shale	adehjnr	521-0070
Oil Shale	adej	521-0115
see also: Kerogen		521-1630
Sapropelite	j	521-1090
Siderite $Fe C O_3$	aehijo	521-0566
see also: Iron Carbonate $Fe C O_3$		106-0498
Siegenite $(Co, Ni)_3 S_4$	h	521-1193
Silex $Si O_2$	aehn	521-0481
see also: Chert		521-0534
Silica, Bentonite Bonded	a	521-0257
Silica, Natural:		———
Agate $Si O_2$	j	521-1223
see also: Silicon Oxide $Si O_2$		122-0009
Amethyst $Si O_2$	hj	521-0682
see also: Silicon Oxide $Si O_2$		122-0009
Aventurine $Si O_2$	h	521-1082
see also: Silicon Oxide $Si O_2$		122-0009
Carnelian $Si O_2$	h	521-1105
see also: Silicon Oxide $Si O_2$		122-0009
Chalcedony $Si O_2$	aehijnr	521-0222
see also: Silicon Oxide $Si O_2$		122-0009
Chrysoprase $Si O_2$	r	521-0628
see also: Silicon Oxide $Si O_2$		122-0009
Citrine $Si O_2$	hj	521-0684
see also: Silicon Oxide $Si O_2$		122-0009
Coesite, Natural $Si O_2$	aej	521-0823
see also: Silicon Oxide $Si O_2$		122-0009
see also: Quartz, Natural $Si O_2$		521-0732
Cristobalite, Natural $Si O_2$	aehijnr	521-0254
see also: Silicon Oxide $Si O_2$		122-0009
see also: Tridymite $Si O_2$		122-0646
see also: Tridymite, Natural $Si O_2$		521-0255
see also: Quartz, Natural $Si O_2$		521-0732
Flint, Mineral $Si O_2$	ahn	521-1465
see also: Silicon Oxide $Si O_2$		122-0009
Hornstone $Si O_2$	nor	521-0629
see also: Silicon Oxide $Si O_2$		122-0009
Hyalite $Si O_2$	aij	521-1128
see also: Silicon Oxide $Si O_2$		122-0009
Lussatite $Si O_2$	ij	521-1420
see also: Silicon Oxide $Si O_2$		122-0009
Opal $Si O_2$	eghijnr	521-0496
see also: Silicon Oxide $Si O_2$		122-0009
Prase $Si O_2$	j	521-1224
see also: Silicon Oxide $Si O_2$		122-0009
Quartz, Natural $Si O_2$	adefghijnor	521-0732
see also: Quartz $Si O_2$		122-0002
see also: Silicon Oxide $Si O_2$		122-0009
see also: Stishovite $Si O_2$		122-0609
see also: Cristobalite $Si O_2$		122-0628
see also: Coesite $Si O_2$		122-0629
see also: Tridymite $Si O_2$		122-0646
see also: Cristobalite, Natural $Si O_2$		521-0254
see also: Tridymite, Natural $Si O_2$		521-0255
see also: Coesite, Natural $Si O_2$		521-0823
see also: Stishovite, Natural $Si O_2$		521-0824
see also: Quartz Fiber Felt		661-0331
Stishovite, Natural $Si O_2$	eno	521-0824
see also: Silicon Oxide $Si O_2$		122-0009
see also: Quartz, Natural $Si O_2$		521-0732
Tiger's Eye $Si O_2$	r	521-0473
see also: Silicon Oxide $Si O_2$		122-0009
Tridymite, Natural $Si O_2$	ejn	521-0255
see also: Silicon Oxide $Si O_2$		122-0009
see also: Cristobalite $Si O_2$		122-0628
see also: Cristobalite, Natural $Si O_2$		521-0254
see also: Quartz, Natural $Si O_2$		521-0732
Silica Sand $Si O_2$	adeghnor	521-0010
see also: Silicon Oxide $Si O_2$		122-0009
Silicate Rock:		———
Chert	agjnr	521-0534
see also: Flint $Si O_2$		521-0481
see also: Jasper		521-1566

Property: a. Thermal conductivity; **b.** Accommodation coefficient; **c.** Thermal contact resistance; **d.** Thermal diffusivity; **e.** Specific heat; **f.** Viscosity; **g.** Emittance; **h.** Reflectance; **i.** Absorptance; **j.** Transmittance; **k.** a_s/ϵ_t ratio; **l.** Prandtl number; **n.** Thermal linear expansion coefficient; **o.** Thermal volumetric expansion coefficient; **r.** Thermal radiative properties (g,h,i,j,k)

Substance Name	Property	Number
Silicate Rock: -(Cont)-		
Diatomite	adeghijnr	521-0015
Flint $Si\,O_2$	aehn	521-0481
see also: Chert		521-0534
Quartzite	adehjno	521-0298
Sand	adeghjnor	521-0110
Siliceous Earth	adeghijnr	521-0015
Siliceous Limestone	j	521-1564
Siliceous Sinter	ade	521-0372
Silicified Schist	a	521-0295
Silicified Wood	r	521-0474
Sillimanite $Al_2\,Si\,O_5$	aehijnor	521-0034
see also: Aluminum Silicate $Al_2\,Si\,O_5$		102-0331
Sillimanite:		———
Andalusite $Al_2\,Si\,O_5$	aehijno	521-0032
see also: Aluminum Silicate $Al_2\,Si\,O_5$		102-0331
Dumortierite $Al_7(B\,O_3)(Si\,O_4)_3\,O_3$	ehjn	521-0911
Kyanite $Al_2\,O\,Si\,O_4$	aehijnor	521-0033
see also: Aluminum Silicate $Al_2\,Si\,O_5$		102-0331
Mullite $Al_6\,Si_2\,O_{13}$	aeghn	521-1418
see also: Aluminum Silicate $Al_6\,Si_2\,O_{13}$		102-0021
Topaz $Al_2\,Si\,O_4\,(F,O\,H)_2$	aehijnor	521-0082
Sillimanite Garnet Gneiss	ah	521-0835
Sillimanite Gneiss	a	521-0831
Sillimanite Schist	ade	521-1407
Silt Soil	adehjr	521-0402
Siltstone	adehn	521-0727
Silty Clay Loam Soil	ah	521-0768
Silty Clay Soil	adehn	521-0612
Silty Loam Soil	adeh	521-0643
Silver Glance $Ag_2\,S$	h	521-0754
see also: Silver Sulfide $Ag_2\,S$		102-0120
Silver Sand	a	521-0551
Sinhalite $Mg\,Al(B\,O_4)$	j	521-1164
Sinter, Siliceous	ade	521-0372
Skarn	e	521-1160
Skutterudite $(Co,Ni)\,As_3$	hj	521-1308
Skutterudite, Smaltite $(Co,Ni)\,As_3$	h	521-1433
Slag Sand	ah	521-1617
Slate	adeghinor	521-0181
Slate, Gray	h	521-0728
Slate, Mauve	r	521-0652
Slate, Micaceous	ad	521-0846
Smaragdite	h	521-1399
Smithsonite $Zn\,C\,O_3$	ahj	521-0611
see also: Zinc Carbonate $Zn\,C\,O_3$		106-0495
Smythite $Fe_3\,S_4$	h	521-1254
see also: Iron Sulfide $Fe_3\,S_4$		110-0009
Soapstone $Mg_3\,Si_4\,O_{10}(O\,H)_2$	adehijnr	521-0116
Sodalite $Na_4\,Al_3\,Si_3\,O_{12}\,Cl$	ahjn	521-0970
Sodalite:		———
Hackmanite	j	521-1385
Hauyne $(Na,Ca)_8(Al_6\,Si_6\,O_{24})(S\,O_4)_x\,S_y$	hn	521-1109
Lazurite $(Na,Ca)_8(Al,Si)_{12}\,O_{24}(S,S\,O_4)$	hj	521-1351
see also: Lapis Lazuli		521-1350
Nosean $Na_8\,Al_6\,Si_6\,O_{24}\,S\,O_4$	n	521-1108
Sodium Aluminum Fluoride, Natural $5\,Na\,F\,.\,3\,Al\,F_3$	r	521-0469
Sodium A Zeolite $Na\,Al\,Si_2\,O_6$	adej	521-0041
see also: Sodium Aluminosilicate $Na\,Al(Si\,O_3)_2$		102-0266
Sodium Borate Silicate, Natural $Na\,B\,Si_3\,O_8$	j	521-1061
Sodium Calcium A Zeolite $(Na,Ca)_5\,Al_6(Al,Si)_4\,Si_{26}\,O_{72}$	j	521-0778
Sodium Clay	e	521-0708
Sodium Feldspar	fn	521-0332
Sodium Feldspar:		———
Albite $Na\,Al\,Si_3\,O_8$	adefghijnor	521-0057
see also: Sodium Aluminosilicate $Na\,Al\,Si_3\,O_8$		102-0218
Soft Coal	adefhjr	521-0249
Sogdianite $(K,Na)_2\,Li_2(Li,Fe,Al)_2\,Zr\,Si_{12}\,O_{30}$	j	521-1066
Sogdianovite $(K,Na)_2\,Li_2(Li,Fe,Al)_2\,Zr\,Si_{12}\,O_{30}$	j	521-1066
Soil	adefghjnr	521-0009

Substance Name	Property	Number
Soil:		———
Alluvial	adh	521-1345
Aluminum Magnesium Silicate	a	521-0864
Black Earth	adhr	521-0384
Black Loam	h	521-0676
Brown	h	521-1087
Calcareous	ad	521-1048
see also: Limestone		521-0085
Chernozem	adhr	521-0384
Chestnut-brown Earth	h	521-0693
Clay	adehj	521-0645
see also: Leda Clay		521-0630
Clay Loam	r	521-0544
Clay Soil With Ground Water	g	521-1654
Clay With Gravel And Sand	aden	521-1255
Desert	hr	521-0451
Diatomite,	j	521-1388
Estancia Playa	ag	521-0866
Gray With Gravel	h	521-1501
Humus	adeh	521-0530
Laterite	ahj	521-0865
Lithic	h	521-1070
Loam	adehjnor	521-0186
Loam With Gravel	a	521-0497
Loam With Gravel And Sand	a	521-0013
Loamy Sand	j	521-1598
Magnesian	a	521-0864
Muck	adeh	521-0530
Nitrate	h	521-0674
Organic	adeh	521-0530
Pedalfer	h	521-1152
Pedocal	h	521-1157
Permafrost	ae	521-1359
Playa	h	521-1165
Podzol	ahr	521-0542
Prairie	h	521-1050
Rain Forest, Tropical	ahj	521-0865
Red	h	521-1498
Saline Playa	ag	521-0866
Salt Marsh	hr	521-0548
Sand	adehn	521-0442
Sandy Clay	adejn	521-0190
Sandy Loam	adeghjr	521-0368
Sandy Silt	ad	521-0822
Silt	adehjr	521-0402
Silty Clay	adehn	521-0612
Silty Clay Loam	ah	521-0768
Silty Loam	adeh	521-0643
Tripoli	j	521-1387
Yellow	h	521-0934
Sorbent, Natural	j	521-0867
Sovite	ad	521-1323
Spar $Ca\,C\,O_3$	adeghijnor	521-0075
see also: Calcium Carbonate $Ca\,C\,O_3$		106-0005
see also: Aragonite $Ca\,C\,O_3$		521-0105
Spartalite $(Zn,Mn)\,O$	hno	521-0879
see also: Zinc Oxide $Zn\,O$		122-0023
Spathic Iron $Fe\,C\,O_3$	aehijo	521-0566
see also: Iron Carbonate $Fe\,C\,O_3$		106-0498
Specular Iron $Fe_2\,O_3$	adehjno	521-0104
see also: Iron Oxide $Fe_2\,O_3$		110-0011
Spessartine $Mn_3\,Al_2(Si\,O_4)_3$	ahjno	521-0957
Spessartite $Mn_3\,Al_2(Si\,O_4)_3$	ahjno	521-0957
Sphalerite $(Zn,Fe)\,S$	aehjnr	521-0086
see also: Zinc Sulfide $Zn\,S$		126-0005
see also: Wurtzite $(Zn,Fe)\,S$		521-1303
Sphalerite:		———
Marmatite	h	521-0953
Sphene $Ca\,Ti\,Si\,O_5$	aehjr	521-0089
see also: Calcium Titanium Silicate $Ca\,Ti\,Si\,O_5$		106-1805
Spinel:		———
Ceylonite	o	521-0990
see also: Picotite		521-1162

Property: a. Thermal conductivity; **b.** Accommodation coefficient; **c.** Thermal contact resistance; **d.** Thermal diffusivity; **e.** Specific heat; **f.** Viscosity; **g.** Emittance; **h.** Reflectance; **i.** Absorptance; **j.** Transmittance; **k.** α_s/ϵ_t ratio; **l.** Prandtl number; **n.** Thermal linear expansion coefficient; **o.** Thermal volumetric expansion coefficient; **r.** Thermal radiative properties (g,h,i,j,k)

Substance Name	Property	Number
Spinel: -(Cont)-		
Chromite $(Fe, Mg)(Cr, Al)_2 O_4$	adehijn	521-0366
see also: Magnesiochromite $(Mg, Fe)(Cr, Al)_2 O_4$		521-0963
Ferroferrichrompicotite $Fe(Fe, Cr, Al)_2 O_4$	h	521-0937
Ferroferripicotite $Fe(Fe, Al)_2 O_4$	h	521-0938
Ferropicotite	h	521-0759
Franklinite $(Zn, Mn, Fe)(Fe, Mn)_2 O_4$	h	521-1017
see also: Iron Zinc Oxide $Zn Fe_2 O_4$		110-0029
Gahnite $Zn Al_2 O_4$	ahno	521-0962
see also: Aluminum Zinc Oxide $Zn Al_2 O_4$		102-0185
Gahnite, Ferroan $Zn_n Fe_x Mg_y Al_2 O_4$	n	521-1429
Jacobsite $(Mn, Fe, Mg)(Fe, Mn)_2 O_4$	h	521-1018
see also: Iron Manganese Oxide $Mn Fe_2 O_4$		110-0187
Magnesiochromite $(Mg, Fe)(Cr, Al)_2 O_4$	an	521-0963
see also: Chromium Magnesium Oxide $Mg Cr_2 O_4$		106-0136
see also: Chromite $(Fe, Mg)(Cr, Al)_2 O_4$		521-0366
see also: Picotite		521-1162
Magnesioferrite $(Mg, Fe) Fe_2 O_4$	r	521-0670
see also: Iron Magnesium Oxide, Nonstoichiometric $Mg Fe_2 O_4$		110-0229
Magnetite $(Fe, Mg) Fe_2 O_4$	adehjn	521-0067
see also: Iron Oxide $Fe_3 O_4$		110-0025
Picotite	h	521-1162
see also: Magnesiochromite $(Mg, Fe)(Cr, Al)_2 O_4$		521-0963
see also: Ceylonite		521-0990
Trevorite $Ni Fe_2 O_4$	h	521-1360
see also: Iron Nickel Oxide $Ni Fe_2 O_4$		110-0028
Spinel, Natural $Mg Al_2 O_4$	adeghnor	521-0648
see also: Aluminum Magnesium Oxide $Mg Al_2 O_4$		102-0013
Spodumene $Li Al Si_2 O_6$	aehn	521-0346
see also: Lithium Aluminosilicate $Li Al(Si O_3)_2$		102-0269
Spodumene:		———
Kunzite	j	521-1486
Staffelite - Magnetite Ore	ad	521-1321
Stannite $Cu_2 Fe Sn S_4$	h	521-1637
Star	afh	521-1332
Staurolite $(Fe, Mg)_2 Al_{10} Si_4 O_{23} O H$	ahij	521-0940
Steatite, Chloritic	a	521-0647
Steatite, Natural $Mg_3 Si_4 O_{10}(O H)_2$	adehijnr	521-0116
Stellar Matter	ae	521-1075
Stellar System	a	521-0770
Stephanite $Ag_5 Sb S_4$	h	521-1204
Steppe	r	521-0521
Stibiconite $Sb_3 O_6 O H$	j	521-1374
Stibiolyutsonite	h	521-1206
Stibiotantalite $Sb(Ta, Nb) O_4$	e	521-1231
Stibnite $Sb_2 S_3$	ahjr	521-0095
see also: Antimony Sulfide $Sb_2 S_3$		126-0014
Stilbite $Na Ca_2 Al_5 Si_{13} O_{36}$	ahj	521-0779
Stillwellite $(Ce, La, Ca) B Si O_5$	j	521-1062
Stilpnomelane $K(Fe, Mg, Al)_3 Si_4 O_{10}(O H)_2$	a	521-1142
Stiluelit $(Ce, La, Ca) B Si O_5$	j	521-1062
Stishovite, Natural $Si O_2$	eno	521-0824
see also: Silicon Oxide $Si O_2$		122-0009
see also: Quartz, Natural $Si O_2$		521-0732
Stolzite $Pb W O_4$	j	521-1078
see also: Lead Tungsten Oxide $Pb W O_4$		122-0054
Stone	adefghnor	521-0046
Stratosphere	r	521-0467
Strontianite $Sr C O_3$	aehj	521-0030
see also: Strontium Carbonate $Sr C O_3$		106-0758
Stumpflite	h	521-1402
Subbituminous Coal	aden	521-1114
Succinite	hjn	521-0933
Sulfur, Natural S_8	ahn	521-1377
see also: Sulfur, Octatomic S_8		101-0045
Sulvanite $Cu_3 V S_4$	h	521-0747
Sundtite $Pb Ag Sb_3 S_6$	h	521-1203
Svabite $Ca_5(As O_4)_3 F$	j	521-1298
Sychnodymite $Cu(Co, Ni)_2 S_4$	h	521-0826
Syenite	adefghjno	521-0184

Substance Name	Property	Number
Syenite:		———
Nepheline	adehn	521-0477
Nephelite	adehn	521-0477
Nordmarkite	e	521-0484
Syenite Porphyry	ag	521-0265
Syenodiorite	adghn	521-0924
Sylvine $K Cl$	aehjr	521-0087
see also: Potassium Chloride $K Cl$		106-0006
Sylvite $K Cl$	aehjr	521-0087
see also: Potassium Chloride $K Cl$		106-0006
Szaibelyite $Mg B O_2(O H)$	ade	521-1645
Table Salt $Na Cl$	adehjnor	521-0007
see also: Sodium Chloride $Na Cl$		106-0024
see also: Sea Salt		521-1122
Taconite	ade	521-0567
Taeniolite $K Li Mg_2 Si_4 O_{10} F_2$	n	521-1355
Talc $Mg_3 Si_4 O_{10}(O H)_2$	adehijnr	521-0116
Talc, Fibrous	aeghij	521-0109
Talc Schist	ade	521-0381
Tantalite $(Fe, Mn)(Nb, Ta)_2 O_6$	h	521-0750
see also: Tapiolite $Fe(Ta, Nb)_2 O_6$		521-1297
see also: Columbite $(Fe, Mn)(Nb, Ta)_2 O_6$		521-1574
Tantalite:		———
Manganotantalite $(Mn, Fe)(Ta, Nb)_2 O_6$	h	521-1005
Tanzanite	hj	521-1115
Tapiolite $Fe(Ta, Nb)_2 O_6$	hj	521-1297
see also: Tantalite $(Fe, Mn)(Nb, Ta)_2 O_6$		521-0750
Tectosphere	f	521-0641
Tektite	aefghj	521-0225
Tektite:		———
Australite	j	521-1415
Moldavite	fj	521-0237
Tellurobismuthite $Bi_2 Te_3$	h	521-1553
see also: Bismuth Telluride $Bi_2 Te_3$		104-0731
Tennantite $(Cu, Fe)_{12} As_4 S_{13}$	h	521-0880
see also: Fahlore		521-0748
see also: Tetrahedrite $(Cu, Fe)_{12} Sb_4 S_{13}$		521-1205
Tephra	h	521-1595
Tephrite, Nephelite-leucite	f	521-0230
Tephrite Basalt	n	521-0678
Tephroite $Mn_2 Si O_4$	ah	521-0956
see also: Manganese Silicate $Mn_2 Si O_4$		119-0052
Terra Alba Clay	adefhjnr	521-0135
see also: Kaolinite, Mineral $Al_2 Si_2 O_5(O H)_4$		521-0174
see also: Kaolin Board		661-0444
see also: Kaolin Paper		661-0484
Terrain	r	521-0343
Terrain, Dry Mat - Fire Burn	r	521-0447
Terrain, Moss - Fire Burn	r	521-0449
Terra Rosa $Fe_2 O_3$	adehjno	521-0104
see also: Iron Oxide $Fe_2 O_3$		110-0011
Terrigenous Rock	a	521-0717
Teschenite	f	521-0306
Tetradymite $Bi_2 Te_2 S$	hj	521-0745
see also: Bismuth Telluride Sulfide $Bi_2 Te_2 S$		104-0191
Tetrahedrite $(Cu, Fe)_{12} Sb_4 S_{13}$	h	521-1205
see also: Antimony Copper Iron Sulfide, Nonstoichiometric $(Cu, Fe)_{12} Sb_4 S_{13}$		106-1802
see also: Antimony Copper Sulfide, Nonstoichiometric $Cu_{12} Sb_4 S_{13}$		106-2441
see also: Antimony Copper Sulfide, Nonstoichiometric $Cu_{14} Sb_4 S_{13}$		106-2442
see also: Antimony Copper Sulfide, Nonstoichiometric $Cu_{13} Sb_4 S_{13}$		106-2443
see also: Antimony Copper Sulfide $Cu_{13} Sb_4 S_{13}$		106-2444
see also: Fahlore		521-0748
see also: Tennantite $(Cu, Fe)_{12} As_4 S_{13}$		521-0880
Thaumasite $Ca_3 Si(O H)_6(C O_3) S O_4$	j	521-1477
Thenardite $Na_2 S O_4$	j	521-1482
see also: Sodium Sulfate $Na_2 S O_4$		120-0016
Tholeiite	adfhn	521-0978
Thomsenolite $Na Ca Al F_6$	j	521-1474

Property: a. Thermal conductivity; **b.** Accommodation coefficient; **c.** Thermal contact resistance; **d.** Thermal diffusivity; **e.** Specific heat; **f.** Viscosity; **g.** Emittance; **h.** Reflectance; **i.** Absorptance; **j.** Transmittance; **k.** α_s/ϵ_t ratio; **l.** Prandtl number; **n.** Thermal linear expansion coefficient; **o.** Thermal volumetric expansion coefficient; **r.** Thermal radiative properties (g,h,i,j,k)

Substance Name	*Property*	*Number*
Thorite $ThSiO_4$	e	521-0294
see also: Thorium Silicate $ThO_2 . SiO_2$		122-0402
Thortveitite $(Sc,Y)_2Si_2O_7$	j	521-0910
see also: Scandium Silicate $Sc_2Si_2O_7$		122-0380
Tienshanite $Na_2BaMnTiB_2Si_6O_{20}$	j	521-1065
Tiff $BaSO_4$	aehijnor	521-0223
see also: Barium Sulfate $BaSO_4$		104-0025
Tiger's Eye SiO_2	r	521-0473
see also: Silicon Oxide SiO_2		122-0009
Tilasite $CaMgAsO_4F$	j	521-1291
Till	a	521-0107
Tinstone SnO_2	hjnor	521-0470
see also: Tin Oxide SnO_2		122-0053
Titan	h	521-1330
Titanic Iron Ore $FeO . TiO_2$	aehj	521-0133
see also: Iron Titanium Oxide $FeTiO_3$		110-0005
Titanite $CaTiSiO_5$	aehjr	521-0089
see also: Calcium Titanium Silicate $CaTiSiO_5$		106-1805
Titanium Beach Sand	n	521-0685
see also: Iron Titanium Oxide $FeTiO_3$		110-0005
Titanium Iron Oxide Moon Rock	h	521-0921
Titanium Vanadium Oxide, Natural $V_2Ti_3O_9$	h	521-1491
Titanochromite Moon Rock	h	521-0920
Titanomagnetite	eh	521-0029
Tonalite	adeg	521-0279
Topaz $Al_2SiO_4(F,OH)_2$	aehijnor	521-0082
Tourmaline	ahijr	521-0131
Trachydolerite	ad	521-1326
Trachydolerite Olivine Andesite Basalt	f	521-0231
Trachyte	ghjn	521-0895
Trap	aeno	521-0761
Tremolite	aeghij	521-0109
Tremolite Schist	h	521-1382
Trevorite $NiFe_2O_4$	h	521-1360
see also: Iron Nickel Oxide $NiFe_2O_4$		110-0028
Tridymite, Natural SiO_2	ejn	521-0255
see also: Silicon Oxide SiO_2		122-0009
see also: Cristobalite SiO_2		122-0628
see also: Cristobalite, Natural SiO_2		521-0254
see also: Quartz, Natural SiO_2		521-0732
Trimerite $Be(Ca,Mn)SiO_4$	j	521-1301
Triphylite $Li(Fe,Mn)PO_4$	a	521-0964
Tripoli Soil	j	521-1387
Tripolite	adeghijnr	521-0015
Tripuhyite $FeSb_2O_6$	j	521-1296
Troilite FeS	h	521-1273
see also: Iron Sulfide FeS		110-0007
Trondhjemite	a	521-0742
Tucekite $Ni_9Sb_2S_8$	h	521-1547
Tuff	adefghnr	521-0369
see also: Volcanic Ash		521-1068
Tuff, Carboniferous	a	521-0287
Tuffaceous Stone	ade	521-0169
Tuff Glass	a	521-1425
Tugtupite $Na_4BeAlSi_4O_{12}Cl$	hn	521-1437
Tundra	r	521-0549
Tungstic Ochre WO_3	j	521-1096
see also: Tungsten Oxide WO_3		122-0012
Tungstite WO_3	j	521-1096
see also: Tungsten Oxide WO_3		122-0012
Turquoise $CuAl_6(PO_4)_4(OH)_8$	hj	521-1285
Tyanshanit $Na_2BaMnTiB_2Si_6O_{20}$	j	521-1065
Ulexite $NaCaB_5O_9$	h	521-1289
Ullmannite $NiSbS$	hj	521-0928
Ulvospinel Moon Rock	h	521-0919
Umohoite $(UO_2)MoO_4$	j	521-1097
Uralite	h	521-1398
Uraninite UO_2	hj	521-1353
see also: Uranium Oxide UO_2		122-0008
Uranium Mineral	j	521-1354
Uranium Ore	h	521-1650
Uranium Silicate Hydroxide, Natural $U(SiO_4)_x(OH)_y$	h	521-0851

Substance Name	*Property*	*Number*
Uranus Surface	r	521-0526
Urtite	adn	521-1057
Uvarovite $Ca_3Cr_2(SiO_4)_3$	h	521-1419
see also: Calcium Chromium Silicate $Ca_3Cr_2Si_3O_{12}$		106-2005
Vanadinite $Pb_5(VO_4)_3Cl$	j	521-1102
Variegated Copper Ore Cu_5FeS_4	h	521-0758
see also: Copper Iron Sulfide Cu_5FeS_4		106-1686
Variscite $AlPO_4$	hj	521-1478
see also: Aluminum Phosphate $AlPO_4$		102-0210
Venus Core	df	521-1123
Venus Surface	hr	521-0422
Vermiculite Mica $(Mg,Fe,Al)_3(Al,Si)_4O_{10}(OH)_2$	adehjn	521-0020
Vermilion HgS	hj	521-0749
see also: Mercury Sulfide HgS		112-0402
Vesicular Basalt	adefghno	521-0053
Vesuvianite $Ca_{10}Mg_2Al_4(SiO_4)_5(Si_2O_7)_2(OH)_4$	ahij	521-0939
Vikingite	h	521-1455
Vitrinite Coal	h	521-0738
Vitrophyre, Hypersthene	g	521-0897
Vivianite $Fe_3(PO_4)_2$	hj	521-0472
see also: Iron Phosphate $Fe_3(Po_4)_2$		110-0289
Volcanic Ash	hn	521-1068
see also: Tuff		521-0369
Volcanic Breccia	hn	521-1034
see also: Agglomerate		521-0997
Volcanic Glass:		——
Obsidian	adefghjnor	521-0307
Pitchstone	no	521-1029
Volcanic Rock	a	521-1434
Volcanic Slag	adefghno	521-0053
Wavellite $Al_3(PO_4)_2(OH)_3$	hj	521-1286
Weinschenkite YPO_4	j	521-1184
see also: Yttrium Phosphate YPO_4		122-0296
Wernerite	aeh	521-0960
White Clay	adefhjnr	521-0135
see also: Kaolinite, Mineral $Al_2Si_2O_5(OH)_4$		521-0174
see also: Kaolin Board		661-0444
see also: Kaolin Paper		661-0484
White Coral	h	521-1649
White Iron Pyrites FeS_2	j	521-1304
see also: Iron Sulfide FeS_2		110-0010
see also: Pyrite FeS_2		521-0025
White Mica $KAl_2(AlSi_3)O_{10}(OH)_2$	adeghijknr	521-0240
see also: Aluminum Potassium Silicate Hydroxide $KAl_2(AlSi_3O_{10})(OH)_2$		102-0798
see also: Potassium Aluminum Silicate $KAl_3Si_3O_{11}$		102-0966
White Nickel $NiAs_2$	hj	521-1307
see also: Loellingite		521-1306
see also: Pararammelsbergite $NiAs_2$		521-1381
White Sand	h	521-1648
Whiting $CaCO_3$	adeg	521-0183
see also: Calcium Carbonate $CaCO_3$		106-0005
see also: Aragonite $CaCO_3$		521-0105
see also: Calcareous Rock		521-1560
Wilhemite Zn_2SiO_4	ehjr	521-0650
see also: Zinc Silicate Zn_2SiO_4		122-0027
Wilkanite Clay	adehijn	521-0111
Willemite Zn_2SiO_4	ehjr	521-0650
see also: Zinc Silicate Zn_2SiO_4		122-0027
Willyamite $(Co,Ni)SbS$	hj	521-0929
Witherite $BaCO_3$	aehj	521-0195
see also: Barium Carbonate $BaCO_3$		104-0233
Wittichenite Cu_3BiS_3	h	521-1200
Wolfbergite $CuSbS_2$	h	521-1197
see also: Antimony Copper Sulfide $CuSbS_2$		106-0505
Wolframite $(Fe,Mn)WO_4$	ehj	521-0555

Property: a. Thermal conductivity; **b.** Accommodation coefficient; **c.** Thermal contact resistance; **d.** Thermal diffusivity; **e.** Specific heat; **f.** Viscosity; **g.** Emittance; **h.** Reflectance; **i.** Absorptance; **j.** Transmittance; **k.** a_s/ϵ_t ratio; **l.** Prandtl number; **n.** Thermal linear expansion coefficient; **o.** Thermal volumetric expansion coefficient; **r.** Thermal radiative properties (g,h,i,j,k)

Substance Name	Property	Number
Wolframite:		———
Ferberite $Fe W O_4$	hj	521-1077
see also: Iron Tungsten Oxide $Fe W O_4$		110-0338
Huebnerite $Mn W O_4$	hj	521-1076
see also: Manganese Tungsten Oxide $Mn W O_4$		119-0053
Wollastonite $Ca Si O_3$	aehjnr	521-0514
see also: Calcium Silicate $Ca Si O_3$		106-0392
Wood:		———
Agatized	r	521-0474
Opalized	r	521-0474
Petrified	r	521-0474
Silicified	r	521-0474
Wulfenite $Pb Mo O_4$	hj	521-0736
see also: Lead Molybdenum Oxide $Pb Mo O_4$		119-0073
Wurtzite $(Zn, Fe) S$	j	521-1303
see also: Zinc Sulfide $Zn S$		126-0005
see also: Sphalerite $(Zn, Fe) S$		521-0086
Xenotime $Y P O_4$	hj	521-0445
see also: Yttrium Phosphate $Y P O_4$		122-0296
Yellow Copper $Cu Fe S_2$	ahj	521-0099
see also: Copper Iron Sulfide $Cu Fe S_2$		106-0791
Yellow Lead Spar $Pb Mo O_4$	hj	521-0736
see also: Lead Molybdenum Oxide $Pb Mo O_4$		119-0073
Yellow Sand	h	521-1088
Yellow Sandstone Clay	h	521-1651
Zeolite	adeh	521-0771
Zeolite:		———
Analcime $Na Al Si_2 O_6$	adej	521-0041
see also: Sodium Aluminosilicate $Na Al(Si O_3)_2$		102-0266
Analcite $Na Al Si_2 O_6$	adej	521-0041
see also: Sodium Aluminosilicate $Na Al(Si O_3)_2$		102-0266
Chabazite $Ca Al_2 Si_4 O_{12}$	ah	521-0489
Deuterium Faujasite	j	521-0681
Gismondite $Ca Al_2(Si O_4)_2$	j	521-1475
see also: Calcium Aluminum Silicate $Ca Al_2(Si O_4)_2$		102-0125
see also: Anorthite $Ca Al_2(Si O_4)_2$		102-0796
Heulandite $(Na, Ca)_5 Al_6(Al, Si)_4 Si_{26} O_{72}$	j	521-0778
Hydrogen Faujasite	j	521-0680
Leonhardite $Ca_2 Al_4 Si_8 O_{24}$	e	521-0488
Mordenite $(Ca,(Na, K)_2)_4 Al_8 Si_{40} O_{96}$	ij	521-1606
Natrolite $Na_2 Al_2 Si_3 O_{10}$	ahij	521-0780
Scolecite $Ca Al_2 Si_3 O_{10}$	j	521-0781
Sodium A $Na Al Si_2 O_6$	adej	521-0041
see also: Sodium Aluminosilicate $Na Al(Si O_3)_2$		102-0266
Stilbite $Na Ca_2 Al_5 Si_{13} O_{36}$	ahj	521-0779
Zeolite Y, Modified:		———
Deuterium Faujasite	j	521-0681
Hydrogen Faujasite	j	521-0680
Zhozeite	h	521-0883
Zinc Arsenate Hydroxide, Natural $Zn_2 As O_4 O H$	j	521-1189
Zinc Blende $(Zn, Fe) S$	aehjnr	521-0086
see also: Zinc Sulfide $Zn S$		126-0005
see also: Wurtzite $(Zn, Fe) S$		521-1303
Zincdescloizite $Pb(Zn, Cu)(V O_4)(O H)$	j	521-1104
Zincite $(Zn, Mn) O$	hno	521-0879
see also: Zinc Oxide $Zn O$		122-0023
Zinc Spar $Zn C O_3$	ahj	521-0611
see also: Zinc Carbonate $Zn C O_3$		106-0495
Zinkenite $Pb_6 Sb_{14} S_{27}$	h	521-1222
Zinnwaldite:		———
Cryophyllite	j	521-1091
Zinnwaldite Mica $K_2(Li, Fe, Al)_6(Si, Al)_8 O_{20}(O H, F)_4$	ahn	521-0819
Zircon $Zr Si O_4$	aeghijnor	521-0250
see also: Zirconium Silicate $Zr Si O_4$		122-0007
see also: Essonite		521-1071
Zirconium Sand	ade	521-1093
Zoisite $Ca_2 Al_3 Si_3 O_{12} O H$	aehj	521-0941
Zoisite, Blue	hj	521-1115
Zone Of Mobility	f	521-0641
Zone Of Weakness	f	521-0641
Zopfstein $H_2 Mg_3 Si_4 O_{12}$	e	521-0519

Property: a. Thermal conductivity; **b.** Accommodation coefficient; **c.** Thermal contact resistance; **d.** Thermal diffusivity; **e.** Specific heat; **f.** Viscosity; **g.** Emittance; **h.** Reflectance; **i.** Absorptance; **j.** Transmittance; **k.** a_s/ϵ_t ratio; **l.** Prandtl number; **n.** Thermal linear expansion coefficient; **o.** Thermal volumetric expansion coefficient; **r.** Thermal radiative properties (g,h,i,j,k)

Part B
SEARCH PARAMETERS

PART B

USE OF SEARCH PARAMETERS

This part is arranged in 12 chapters, each representing a specific property. Within each chapter, the seven-digit substance numbers are listed in numerical order followed by physical state within each substance, subject, the language, the temperature range, and the year. The last column gives the TPRC serial number pertaining to the document retrieved. The code designations for the search parameters need not be memorized since they are interpreted at the bottom of each page of this section.

The substance numbers link this portion of the volume with the *Materials Directory* (Part A), and the TPRC numbers provide the link to the *Bibliography* (Part C). The procedure for the retrieval of references in literature search is as follows:

1. Locate the desired substance in Part A, the *Materials Directory*. Check if the desired property is listed, then record the seven-digit substance number.
2. In Part B (this section) locate the property chapter and the seven-digit substance number of interest.
3. Use as many of the five search parameters (physical state, subject, language, temperature, and year) as desired to locate and record TPRC numbers.
4. Obtain the complete bibliographic citations for the recorded TPRC numbers in Part C, the *Bibliography*.

Example: Obtain references on the transmittance of Aragonite ($CaCO_3$).

1. Aragonite, with cross-references to minerals and compounds with similar composition, is listed in Part A (the *Materials Directory*) on page A38, column 2. Transmittance, Property *j* is listed for the material, and the seven-digit substance number is 521-0105.
2. In this section locate Chapter 8 (Transmittance), page B59. Locate substance number 521-0105, page B64, column 2.
3. For the substance number 521-0105, the following entries are listed:

Substance Number	Phys. State	Sub-ject	Lan-guage	Temper-ature	Year	TPRC Number
521-0105	P	D	F	N	1967	42313
0105	P	D	E	N	1978	94862
0105	S	D	E	N	1950	58879
0105	S	D	R	N	1972	69083
0105	S	D	J	N	1978	99886

(*Note:* The code letters of the parameters are interpreted in the footnote below.)

4. Obtain complete bibliographic citations for the TPRC numbers in Part C (the *Bibliography*).

Note: In the earlier coding transmittance was grouped under Thermal Radiative Properties. The user should also retrieve TPRC numbers for 521-0105 in Chapter 12 for possible documents on the transmittance of aragonite.

Phys. State: C, Superconductive; D, Doped; E, Expanded; F, Fibrous or Whisker; G, Gas; I, Ionized or Plasma; L, Liquid; M, Multiphase; P, Powder or Fine Particle; S, Solid; T, Thick or Thin Film.

Subject: D, Data; E, Experiment; G, General (Data + Theory + Experiment); S, Survey (Review, Compendium, etc.); T, Theory.

Language: C, Czech; D, Dutch; E, English; F, French; G, German; I, Italian; J, Japanese; O, Other Languages; P, Polish; R, Russian; S, Spanish.

Temperature: F, Full Range (Low + Normal + High); L, Low (O to 75 K) + Overlap into Normal; N, Normal (75 to 1273 K); H, High (above 1273 K) + Overlap into Normal; blank, Not Coded.

Chapter 1 Thermal Conductivity

Substance Number	Phys. State	Subject	Language	Temperature	Year	TPRC Number
503-0001	S	D	G		1911	6708
0001	S	D	F	N	1959	18033
0001	S	D	G	N	1951	21301
0003	S	S	R	N	1973	79859
0003	S	S	E	N	1973	79860
0008	S	D	G	N	1894	48020
0012	S	G	E	L	1962	26836
0012	S	D	E	N	1975	82363
0020	M	D	E	H	1960	18644
0020	S	D	E		1954	6565
0020	S	D	E	N	1960	10947
0020	S	S	E	N	1960	24931
0020	S	S	E	H	1970	61233
0020	S	S	E	N	1970	61233
0020	S	D	E	L	1974	74374
0020	S	E	E	L	1974	74374
0024	S	D	E	N	1892	16421
0024	S	G	E	L	1962	26836
0024	S	G	E	N	1966	46250
0024	S	D	E	L	1977	90706
0025	L	D	R	N	1958	16447
0025	L	D	G	H	1959	17892
0025	L	D	E	H	1961	18317
0025	L	T	E	H	1961	18317
0025	L	D	E	N	1958	28537
0025	L	D	E	N	1962	29103
0025	L	D	E	N	1969	44511
0025	L	T	E	N	1969	44511
0025	S	G	G		1952	145
0025	S	D	E		1947	6356
0025	S	T	E		1947	6356
0025	S	D	E		1958	7695
0025	S	G	E		1956	9943
0025	S	D	R	N	1958	16447
0025	S	D	G	N	1959	17892
0025	S	D	E	N	1928	22494
0025	S	S	E	N	1960	24931
0025	S	D	E	N	1958	28537
0025	S	D	E	N	1962	29103
0025	S	G	R	N	1962	31437
0025	S	D	E	N	1968	52209
0025	S	D	J	N	1968	53139
0025	S	E	J	N	1968	53139
0025	S	S	E	H	1970	61233
0025	S	S	E	N	1970	61233
0026	S	D	E	N	1975	82040
0026	S	T	E	N	1975	82040
0026	S	S	P	H	1976	92927
0026	S	S	E	N	1976	94271
0031	S	D	E		1949	798
0031	S	D	E		1943	1287
0031	S	D	E		1958	6939
0031	S	D	E		1953	7102
0031	S	D	E	N	1954	10100
0031	S	D	E	N	1940	11000
0031	S	D	E	N	1955	15754
0031	S	D	E	N	1955	15755
0031	S	S	E	L	1960	16701
0031	S	S	E	N	1960	16701
0031	S	D	E	N	1960	17320
0031	S	D	F	N	1931	21006
0031	S	D	G	N	1951	21301
0031	S	D	E	N	1958	23521
0031	S	D	E	N	1924	24551
0031	S	S	E	N	1960	24931
0031	S	D	E	N	1963	25562
0031	S	D	O	N	1949	28661
0031	S	D	E	N	1940	28780
0031	S	T	E	N	1940	28780
0031	S	E	E	N	1963	29811
0031	S	G	E	N	1963	31783
0031	S	D	E	N	1941	31862
0031	S	G	E	L	1963	32571
0031	S	D	E	N	1964	35390
0031	S	D	E	N	1968	37491
0031	S	D	E	L	1966	42676
0031	S	D	E	N	1968	53519
0031	S	D	E	N	1967	58953
0031	S	S	E	N	1970	61233
0031	S	D	E	L	1971	75183
0031	S	T	E	L	1971	75183
0031	S	D	E	N	1976	85580
0047	S	S	E	H	1970	61233
0047	S	S	E	N	1970	61233
0047	S	D	R	H	1972	70092
0047	S	D	R	N	1972	70092
0055	S	D	E	L	1976	88203
0055	S	D	E	N	1976	88203
0055	S	E	E	L	1976	88203
0055	S	E	E	N	1976	88203
0066	E	D	R	N	1966	59500
0066	E	D	E	N	1966	59501
0066	S	D	E	N	1963	49933
0066	S	D	E	N	1971	68241
0066	S	D	R	N	1972	70625

Substance Number	Phys. State	Subject	Language	Temperature	Year	TPRC Number
503-0066	S	D	E	N	1972	70626
0066	S	D	E	L	1975	85499
0066	S	D	E	N	1975	85499
0080	S	D	G	N	1965	34763
0082	E	D	E		1951	810
0082	E	D	E		1950	1097
0082	E	D	E		1944	1470
0082	E	D	G		1954	1630
0082	E	T	G		1954	1630
0082	E	D	G		1950	6215
0082	E	T	G		1950	6215
0082	E	D	E		1952	6243
0082	E	D	E		1945	6308
0082	E	D	E		1946	6321
0082	E	D	E		1954	6325
0082	E	D	E		1935	6329
0082	E	D	E		1945	8940
0082	E	D	E	N	1957	10271
0082	E	D	E	N	1946	12715
0082	E	D	E	N	1960	15664
0082	E	D	E	L	1959	17069
0082	E	D	E	N	1958	17617
0082	E	T	E	N	1958	17617
0082	E	D	E	N	1952	24201
0082	E	D	E	N	1950	24328
0082	E	D	E	N	1963	30287
0082	E	D	E	N	1960	30362
0082	E	D	E	N	1964	37058
0082	E	D	E	N	1961	39897
0082	E	T	E	N	1961	39897
0082	E	D	E	N	1943	40228
0082	E	D	E	N	1966	42331
0082	E	D	E	N	1966	42708
0082	E	D	E	N	1957	45699
0082	E	D	R	N	1968	51025
0082	E	D	G	N	1967	51692
0082	E	S	E	N	1948	54088
0082	E	D	E	N	1969	54423
0082	E	D	R	N	1969	56318
0082	E	D	G	N	1970	60237
0082	E	D	I	N	1970	63776
0082	E	E	I	N	1970	63776
0082	E	D	E	N	1972	66207
0082	E	D	E	N	1971	69687
0082	E	D	E	N	1972	94710
0082	E	E	E	N	1972	94710
0082	F	D	G		1950	1628
0082	F	G	E		1952	1939
0082	F	D	E		1942	5258
0082	F	D	E		1945	6308
0082	F	D	E		1956	6324
0082	F	D	E		1951	6357
0082	F	D	E		1947	6359
0082	F	D	E		1949	6929
0082	F	D	E		1954	7020
0082	F	D	E		1953	7099
0082	F	D	E		1954	7336
0082	F	D	E		1942	7694
0082	F	D	E		1953	8233
0082	F	G	E		1956	9943
0082	F	D	E	N	1955	10184
0082	F	D	E	N	1955	10206
0082	F	D	E	N	1941	11001
0082	F	D	E	N	1960	11047
0082	F	D	E	N	1960	11373
0082	F	D	E	N	1948	12133
0082	F	D	E	N	1945	12661
0082	F	D	E	N	1959	13954
0082	F	D	E	N	1960	16520
0082	F	D	F	N	1935	16570
0082	F	D	F	N	1955	20452
0082	F	D	E	N	1953	20540
0082	F	E	F	L	1952	21185
0082	F	D	R	N	1929	22225
0082	F	D	E	N	1953	24364
0082	F	D	R	N	1960	24772
0082	F	D	F	N	1956	24968
0082	F	D	F	N	1956	24969
0082	F	D	G	N	1935	25920
0082	F	D	E	N	1941	28215
0082	F	D	E	N	1952	28781
0082	F	D	E	N	1930	28819
0082	F	D	E	N	1960	30362
0082	F	D	G	N	1951	31815
0082	F	S	G	N	1949	31871
0082	F	D	R	N	1956	32425
0082	F	D	R	N	1965	32857
0082	F	D	E	N	1962	32865
0082	F	D	E	N	1965	32866
0082	F	D	R	N	1965	34599
0082	F	D	E	N	1964	36262
0082	F	D	E	N	1964	36327
0082	F	D	E	N	1949	37072
0082	F	D	R	N	1964	37200
0082	F	T	R	N	1964	37200

Substance Number	Phys. State	Subject	Language	Temperature	Year	TPRC Number
503-0082	F	D	E	N	1963	37349
0082	F	D	E	N	1964	37450
0082	F	T	E	N	1964	37450
0082	F	D	E	N	1968	37491
0082	F	G	E	N	1961	38320
0082	F	D	E	N	1955	39857
0082	F	D	E	N	1943	40228
0082	F	D	E	N	1968	41208
0082	F	D	E	N	1966	41219
0082	F	D	I	N	1966	41292
0082	F	D	E	N	1949	42306
0082	F	D	E	N	1966	42331
0082	F	D	R	N	1959	42504
0082	F	D	E	N	1960	42824
0082	F	D	I	N	1965	43137
0082	F	D	E	N	1959	44092
0082	F	D	E	N	1966	44554
0082	F	G	E	N	1966	44681
0082	F	D	F	N	1966	44906
0082	F	E	F	N	1966	44906
0082	F	D	E	N	1967	44986
0082	F	D	E	N	1966	44988
0082	F	D	I	N	1966	44990
0082	F	G	F	N	1966	45257
0082	F	S	E	L	1966	46217
0082	F	D	E	N	1965	47289
0082	F	D	E	N	1938	48291
0082	F	D	E	N	1967	49153
0082	F	D	E	N	1969	49616
0082	F	T	E	N	1969	49616
0082	F	D	E	N	1968	50638
0082	F	D	R	N	1968	51010
0082	F	G	E	N	1968	52206
0082	F	G	E	N	1967	52232
0082	F	G	S	N	1967	52232
0082	F	D	E	N	1969	52402
0082	F	T	E	N	1969	52402
0082	F	D	E	N	1969	53585
0082	F	D	E	N	1969	53597
0082	F	D	E	N	1969	54423
0082	F	D	E	N	1961	56385
0082	F	G	R	N	1969	56887
0082	F	D	I	N	1970	60900
0082	F	D	E	N	1970	61192
0082	F	D	R	N	1968	61210
0082	F	T	R	N	1968	61210
0082	F	D	E	N	1970	63135
0082	F	T	E	N	1970	63135
0082	F	D	E	N	1971	65087
0082	F	E	E	N	1971	65087
0082	F	D	R	N	1971	65427
0082	F	T	R	N	1971	65427
0082	F	D	E	N	1971	65523
0082	F	E	E	N	1971	65523
0082	F	D	R	N	1972	68895
0082	F	T	R	N	1972	68895
0082	F	D	R	N	1972	69051
0082	F	T	R	N	1972	69051
0082	F	D	E	N	1972	69052
0082	F	T	E	N	1972	69052
0082	F	G	R	N	1972	69818
0082	F	G	E	N	1972	69819
0082	F	D	E	N	1971	73365
0082	F	T	E	N	1971	73365
0082	F	D	R	N	1971	73730
0082	F	D	E	N	1971	73731
0082	F	D	E	N	1974	74724
0082	F	D	E	N	1974	75009
0082	F	T	E	N	1974	75009
0082	F	D	E	N	1972	76957
0082	L	G	R		1951	1121
0082	L	D	R		1940	3698
0082	L	T	E		1950	5195
0082	L	D	E		1898	9285
0082	L	D	G	N	1949	11204
0082	L	S	E	N	1961	18315
0082	L	S	E	N	1961	18316
0082	L	D	E	H	1952	21446
0082	L	T	E	H	1952	21446
0082	L	E	G	H	1959	24403
0082	L	G	E	H	1965	30858
0082	L	G	J	N	1966	42659
0082	L	T	R	N	1966	50712
0082	L	D	J	N	1972	67784
0082	L	E	J	N	1972	67784
0082	M	D	E	N	1974	75085
0082	M	E	E	N	1974	75085
0082	M	D	R	N	1973	77334
0082	M	E	R	N	1973	77334
0082	P	G	E		1933	7644
0082	P	D	G	N	1959	17892
0082	P	G	E	N	1962	22559
0082	P	D	E	N	1962	29103
0082	P	D	E	N	1962	29551
0082	P	D	E	N	1964	31231

Phys. State: **C.** Superconductive; **D.** Doped; **E.** Expanded; **F.** Fibrous or Whisker; **G.** Gas; **I.** Ionized or Plasma; **L.** Liquid; **M.** Multiphase; **P.** Powder or Fine Particle; **S.** Solid; **T.** Thick or Thin Film

Subject: **D.** Data; **E.** Experiment; **G.** General (Data + Theory + Experiment); **S.** Survey (Review, Compendium etc.); **T.** Theory

Language: **C.** Czech; **D.** Dutch; **E.** English; **F.** French; **G.** German; **I.** Italian; **J.** Japanese; **O.** Other Languages; **P.** Polish; **R.** Russian; **S.** Spanish

Temperature: **F.** Full Range (Low + Normal + High); **L.** Low (0 to 75K) + Overlap into Normal; **N.** Normal (75 to 1273K); **H.** High (above 1273K) + Overlap into Normal; Blank = Not Coded

Substance Number	Phys. State	Subject	Language	Temperature	Year	TPRC Number
503-0082	P	D	J	N	1963	36246
0082	P	D	E	N	1964	38646
0082	P	D	E	N	1966	39869
0082	P	D	E	N	1966	41217
0082	P	D	E	N	1960	42824
0082	P	D	E	N	1964	45653
0082	P	D	E	N	1964	45753
0082	P	D	E	N	1967	49193
0082	P	D	E	N	1966	49738
0082	P	D	E	N	1950	51604
0082	P	S	E	N	1968	51855
0082	P	D	E	N	1968	51904
0082	P	G	E	N	1969	52300
0082	P	D	E	N	1969	53610
0082	P	G	E	N	1968	56537
0082	P	D	R	N	1969	57572
0082	P	D	E	N	1970	57632
0082	P	D	E	N	1971	64041
0082	P	D	R	N	1970	65443
0082	P	E	R	N	1970	65443
0082	P	D	F	N	1972	66103
0082	P	E	F	N	1972	66103
0082	P	G	E	N	1974	75084
0082	P	D	E	N	1973	76783
0082	P	T	E	N	1973	76783
0082	P	G	R	N	1973	77332
0082	P	D	E	N	1974	78827
0082	P	E	E	N	1974	78827
0082	P	D	J	N	1977	94212
0082	P	E	J	N	1977	94212
0082	S	D	E		1949	26
0082	S	E	G		1953	853
0082	S	T	E		1949	2704
0082	S	D	E		1949	7477
0082	S	T	E		1949	7477
0082	S	G	E		1958	9471
0082	S	D	E	L	1955	10187
0082	S	D	E	L	1955	10206
0082	S	D	E	N	1936	11107
0082	S	D	G	N	1933	11172
0082	S	D	R	N	1934	15457
0082	S	T	R	N	1934	15457
0082	S	D	E	N	1935	15503
0082	S	D	G	N	1898	15516
0082	S	D	G	L	1911	15576
0082	S	D	G	N	1953	16064
0082	S	D	E	L	1959	16875
0082	S	D	E	L	1956	16895
0082	S	D	E	N	1960	18628
0082	S	T	E	N	1960	18628
0082	S	D	E	N	1961	18657
0082	S	T	E	N	1961	18657
0082	S	S	G	F	1961	20059
0082	S	D	E	H	1925	20074
0082	S	G	E	N	1962	20435
0082	S	D	I	N	1899	20655
0082	S	S	E	N	1961	20964
0082	S	S	G	N	1923	21761
0082	S	D	E	N	1928	22494
0082	S	D	E	N	1921	22633
0082	S	E	G	N	1928	22654
0082	S	S	E	L	1953	23568
0082	S	D	E	N	1961	24021
0082	S	S	E	N	1956	24971
0082	S	D	E	N	1963	25061
0082	S	G	E	N	1896	25164
0082	S	D	E	N	1960	25797
0082	S	D	J	N	1961	27093
0082	S	D	E	N	1961	27117
0082	S	S	E	H	1961	28014
0082	S	T	E	L	1960	28074
0082	S	D	G	N	1904	28731
0082	S	E	G	L	1961	28803
0082	S	G	E	N	1963	28956
0082	S	S	E	L	1961	29185
0082	S	D	E	N	1951	29285
0082	S	G	E	N	1963	30146
0082	S	G	E	N	1872	31896
0082	S	D	E	N	1957	31948
0082	S	S	E	N	1939	32125
0082	S	S	E	N	1963	32162
0082	S	S	R	N	1962	33745
0082	S	S	E	N	1966	33746
0082	S	D	I	N	1949	35068
0082	S	D	E	N	1968	37601
0082	S	E	E	N	1968	37601
0082	S	S	E	F	1969	37682
0082	S	D	E	N	1955	39857
0082	S	S	S	N	1965	41303
0082	S	S	E	N	1966	42284
0082	S	D	E	N	1960	42824
0082	S	T	E	L	1965	43040
0082	S	D	E	N	1919	43637
0082	S	D	E	N	1965	44628
0082	S	E	E	N	1965	44628
0082	S	G	J	N	1967	46251
0082	S	G	E	N	1968	46252
0082	S	S	E	N	1967	48182
0082	S	D	E	N	1960	49210
0082	S	S	E	L	1968	49258
0082	S	S	D	N	1968	49783

Substance Number	Phys. State	Subject	Language	Temperature	Year	TPRC Number
503-0082	S	D	E	N	1959	50688
0082	S	S	E	H	1962	50779
0082	S	D	E	N	1967	53206
0082	S	E	E	N	1967	53206
0082	S	D	E	N	1969	53587
0082	S	D	E	N	1969	53610
0082	S	S	E	L	1967	56366
0082	S	S	E	F	1970	61233
0082	S	T	E	N	1972	65794
0082	S	S	R	N	1971	67587
0082	S	D	J	N	1972	67784
0082	S	E	J	N	1972	67784
0082	S	D	E	N	1972	68674
0082	S	E	E	N	1972	68674
0082	S	D	E	N	1973	70941
0082	S	D	E	N	1974	72742
0082	S	T	O	N	1974	75943
0082	S	D	E	N	1975	75982
0082	S	D	E	N	1975	76427
0082	S	D	E	N	1960	77341
0082	S	D	E	L	1974	82998
0082	S	E	E	L	1974	82998
0082	S	D	E	L	1975	85499
0082	S	D	E	N	1975	85499
0082	S	D	E	N	1976	85580
0082	S	T	E	N	1976	85580
0082	S	T	E	N	1975	88857
0082	S	D	R	N	1977	90126
0082	S	D	R	N	1978	93254
0082	S	D	E	N	1978	94579
0082	S	D	E	N	1977	94999
0082	S	T	E	N	1800	96284
0082	S	D	E	N	1978	98018
0105	S	D	E	N	1960	15618
0105	S	G	E	N	1963	32163
0105	S	S	E	N	1970	61233
0125	S	S	E	N	1970	61233
0126	S	D	E	N	1973	71894
0199	S	D	R	N	1973	80006
0200	S	D	E	L	1975	80989
0200	S	E	E	L	1975	80989
0214	L	D	R	H	1976	88484
0214	L	D	R	N	1976	88484
0214	L	D	E	H	1976	89373
0214	L	D	E	N	1976	89373
0216	M	D	E	N	1960	10052
0216	S	D	E		1952	6567
0216	S	D	E	N	1960	10052
0216	S	D	E	N	1960	16712
0216	S	E	E	N	1960	18009
0216	S	D	E	N	1940	20780
0216	S	S	E	N	1957	24970
0216	S	D	E	N	1964	25668
0216	S	S	E	L	1962	26069
0216	S	D	E	N	1962	26684
0216	S	D	E	N	1961	27117
0216	S	D	E	N	1961	29152
0216	S	D	E	N	1963	32772
0216	S	S	E	N	1963	32772
0216	S	D	E	N	1965	36419
0216	S	D	E	N	1966	36468
0216	S	D	E	N	1965	36738
0216	S	D	E	N	1965	38860
0216	S	D	E	L	1968	47910
0216	S	D	E	N	1963	49933
0216	S	D	E	N	1961	53006
0216	S	G	E	N	1969	53595
0216	S	D	E	N	1966	57665
0216	S	D	E	N	1970	61233
0216	S	S	E	N	1970	61233
0216	S	D	E	L	1971	63375
0216	S	G	E	L	1971	66152
0216	S	D	E	N	1955	68323
0216	S	D	E	H	1969	77371
0216	S	E	E	H	1969	77371
0216	S	S	E	N	1976	87488
0216	S	D	E	N	1978	98927
0216	S	E	E	N	1978	98927
0239	S	S	E	N	1920	22280
0240	P	D	E	L	1975	82557
0240	P	D	E	N	1975	82557
0240	P	E	E	L	1975	82557
0240	P	E	E	N	1975	82557
0262	M	D	E	N	1960	10052
0262	M	D	E	N	1953	10164
0262	S	D	E	N	1960	10052
0262	S	D	E	N	1954	10192
0262	S	S	E	N	1960	24931
0262	S	S	E	N	1970	61233
0270	L	D	E	N	1961	18321
0270	S	D	G		1911	6708
0270	S	D	E	N	1892	16421
0270	S	D	G	N	1951	21301
0270	S	G	R	N	1949	31625
0270	S	D	E	N	1975	76616
0280	S	G	E	N	1963	32163
0284	S	D	J	N	1957	23772
0284	S	S	E	H	1960	24931
0284	S	S	E	N	1960	24931
0284	S	S	E	H	1970	61233
0284	S	S	E	N	1970	61233

Substance Number	Phys. State	Subject	Language	Temperature	Year	TPRC Number
503-0284	S	D	E	N	1976	85644
0284	S	D	R	H	1976	87278
0284	S	D	R	N	1976	87278
0284	S	D	E	H	1976	89160
0284	S	D	E	N	1976	89160
0291	S	D	F	N	1959	18033
0309	S	D	E	N	1963	49933
0343	S	S	E	N	1975	80596
0376	M	D	E	N	1960	10052
0376	S	D	R	H	1947	622
0376	S	D	R	N	1947	622
0376	S	E	R	H	1947	622
0376	S	D	E	N	1960	10052
0376	S	D	E	N	1954	10192
0376	S	S	E	N	1960	24931
0376	S	D	E	H	1962	29111
0376	S	D	E	N	1962	29111
0376	S	E	E	H	1962	29111
0376	S	G	R	N	1962	31437
0376	S	G	E	N	1963	32163
0376	S	S	E	N	1970	61233
0423	S	D	R	H	1947	622
0423	S	D	R	H	1947	622
0423	S	E	R	H	1947	622
0423	S	D	E	H	1962	29111
0423	S	D	E	N	1962	29111
0423	S	E	E	H	1962	29111
0442	S	D	E	N	1960	15618
0442	S	G	E	N	1963	32163
0442	S	S	E	N	1970	61233
0452	S	S	E	N	1963	32162
0452	S	D	E	N	1962	47240
0452	S	D	E	N	1963	58889
0461	M	D	E	N	1955	28067
0461	S	D	E		1958	6939
0461	S	D	E		1953	7102
0461	S	D	E		1954	7104
0461	S	D	E		1953	9404
0461	S	D	E	N	1949	10272
0461	S	D	E	N	1955	15754
0461	S	D	E	N	1955	15755
0461	S	S	E	N	1960	24931
0461	S	D	E	N	1949	30225
0461	S	D	E	N	1955	30255
0461	S	D	E	N	1950	31538
0461	S	D	E	N	1951	43395
0461	S	D	E	N	1954	44149
0461	S	D	O	N	1974	76179
0463	S	D	R	N	1971	67588
0463	S	D	R	N	1975	85563
0463	S	E	R	N	1975	85563
0463	S	D	E	N	1975	85564
0463	S	E	E	N	1975	85564
0466	S	D	R	N	1971	67588
0473	S	S	E	H	1970	61233
0473	S	S	E	N	1970	61233
0503	S	S	E	N	1960	24931
0515	S	S	E	N	1970	61233
0539	S	G	E	L	1962	26836
0539	S	D	E	L	1965	43041
0544	M	D	R	H	1960	51562
0544	P	D	R	N	1963	30309
0544	S	D	G	H	1957	20977
0544	S	S	E	N	1960	24931
0544	S	D	E	H	1962	27640
0544	S	D	R	H	1960	31072
0544	S	D	R	N	1942	31909
0544	S	D	R	N	1963	32426
0544	S	D	G	H	1964	37519
0544	S	E	G	H	1964	37519
0544	S	D	F	H	1961	38063
0544	S	D	R	H	1967	48335
0544	S	D	E	H	1967	48336
0544	S	S	E	N	1935	60445
0544	S	D	G	H	1971	66895
0545	S	D	E		1958	6939
0545	S	D	E		1953	7102
0545	S	D	E	N	1955	15754
0545	S	D	E	N	1955	15755
0545	S	D	E	N	1958	23521
0545	S	G	R	N	1962	31437
0545	S	D	E	N	1957	35437
0545	S	D	E	N	1967	40398
0545	S	D	E	N	1968	53178
0555	M	D	E	N	1960	10052
0555	S	D	E	N	1960	10052
0555	S	D	E	N	1954	10192
0555	S	S	E	N	1970	61233
0556	L	D	G	H	1959	17892
0556	L	D	E	H	1962	29103
0556	S	D	G	N	1959	17892
0556	S	S	E	N	1960	24931
0556	S	D	E	N	1962	29103
0556	S	G	E	N	1963	32163
0556	S	S	E	N	1970	61233
0559	S	D	E	N	1960	10052
0559	S	S	E	N	1970	61233
0559	S	S	E	H	1975	80596
0559	S	S	E	N	1975	80596
0559	S	S	E	N	1977	94788
0559	S	D	E	H	1978	98161

Phys. State: **C.** Superconductive; **D.** Doped; **E.** Expanded; **F.** Fibrous or Whisker; **G.** Gas; **I.** Ionized or Plasma; **L.** Liquid; **M.** Multiphase; **P.** Powder or Fine Particle; **S.** Solid; **T.** Thick or Thin Film

Subject: **D.** Data; **E.** Experiment; **G.** General (Data + Theory + Experiment); **S.** Survey (Review, Compendium etc.); **T.** Theory

Language: **C.** Czech; **D.** Dutch; **E.** English; **F.** French; **G.** German; **I.** Italian; **J.** Japanese; **O.** Other Languages; **P.** Polish; **R.** Russian; **S.** Spanish

Temperature: **F.** Full Range (Low + Normal + High); **L.** Low (0 to 75K) + Overlap into Normal; **N.** Normal (75 to 1273K); **H.** High (above 1273K) + Overlap into Normal; Blank = Not Coded

Substance Number	Phys. State	Subject	Language	Temperature	Year	TPRC Number
503-0559	S	D	E	N	1978	98161
0559	S	T	E	H	1978	98161
0559	S	T	E	N	1978	98161
0562	S	S	E	N	1979	97587
0566	S	D	E	N	1955	20062
0566	S	D	E	N	1963	37530
0566	S	S	E	N	1970	61233
0567	S	S	E	N	1960	24931
0589	S	S	E	N	1968	48754
0599	S	D	R	N	1970	61257
0599	S	D	E	N	1970	67223
0600	D	D	R	L	1970	59436
0600	D	D	E	L	1970	59437
0600	S	D	R	L	1970	59436
0600	S	D	E	L	1970	59437
0604	S	D	E		1958	7695
0631	F	D	E	N	1977	91785
0631	S	D	E	N	1977	91785
0633	L	D	E	N	1961	18321
0651	S	S	E	N	1960	24931
0652	S	D	E	H	1960	10052
0652	S	D	E	N	1960	10052
0652	S	S	E	N	1970	61233
0659	S	D	E		1959	8100
0659	S	D	E	N	1959	22183
0659	S	D	E	H	1964	25668
0659	S	D	E	H	1963	26844
0659	S	D	E	N	1961	29152
0659	S	D	E	N	1961	29925
0659	S	S	E	N	1963	32162
0659	S	G	E	H	1963	32772
0659	S	D	E	N	1958	33214
0659	S	D	E	N	1965	36419
0659	S	D	E	N	1966	36468
0659	S	D	E	N	1964	37328
0659	S	D	E	N	1962	40297
0659	S	S	E	H	1964	45435
0659	S	D	E	N	1967	46877
0659	S	D	E	N	1963	49933
0659	S	D	E	N	1966	57665
0659	S	G	E	N	1970	60590
0659	S	D	E	H	1970	60612
0659	S	D	E	H	1970	61233
0659	S	D	E	N	1970	61233
0659	S	S	E	F	1970	61233
0659	S	D	E	H	1974	72699
0659	S	E	E	H	1974	72699
0659	S	D	E	H	1973	72866
0659	S	E	E	H	1973	72866
0659	S	D	E	N	1972	78327
0659	S	E	E	N	1972	78327
0659	S	S	E	H	1972	78327
0659	S	S	E	N	1972	78327
0659	S	S	E	N	1976	83746
0659	S	S	E	H	1975	84691
0659	S	S	E	N	1975	84691
0659	S	D	E	N	1976	88185
0659	S	E	E	N	1976	88185
0659	S	S	E	N	1977	93643
0659	S	D	E	H	1977	94200
0659	S	D	E	N	1977	94200
0659	S	S	E	F	1977	94200
0659	S	D	E	N	1978	98927
0659	S	E	E	N	1978	98927
0660	S	D	E		1959	8100
0660	S	D	E	N	1961	29925
0660	S	D	E	N	1958	33214
0667	S	E	J	H	1962	24315
0667	S	S	E	N	1961	25075
0667	S	G	E	L	1962	26836
0667	S	S	E	N	1964	30525
0667	S	D	E	N	1964	35390
0675	S	D	E	H	1976	92035
0731	D	D	E	N	1966	40547
0731	S	D	E	N	1947	15591
0765	M	D	E	N	1955	28067
0765	S	D	E		1958	6939
0765	S	D	E		1953	7102
0765	S	D	E		1954	7104
0765	S	D	E		1953	9404
0765	S	D	E	N	1949	10272
0765	S	D	G	N	1933	11172
0765	S	D	E	N	1955	15754
0765	S	D	E	N	1955	15755
0765	S	D	R	N	1956	19823
0765	S	S	E	N	1960	24931
0765	S	D	E	N	1963	25061
0765	S	S	G	N	1928	28680
0765	S	D	E	N	1949	30225
0765	S	D	E	N	1955	30255
0765	S	D	E	N	1951	43395
0765	S	D	E	N	1954	44149
0765	S	D	E	N	1960	52136
0765	S	D	R	N	1970	62314
0768	S	S	E	N	1960	24931
0768	S	S	E	H	1970	61233
0768	S	S	E	N	1970	61233
0782	P	S	E	N	1970	61233
0782	S	S	E	L	1960	16701
0782	S	S	E	N	1960	16701
0782	S	S	E	N	1960	24931

Substance Number	Phys. State	Subject	Language	Temperature	Year	TPRC Number
503-0782	S	D	E	N	1962	29551
0782	S	G	R	N	1949	31625
0782	S	D	E	N	1958	33214
0782	S	D	E	N	1965	34765
0782	S	D	E	N	1966	48383
0782	S	S	E	N	1970	61233
0782	S	D	E	N	1971	68241
0782	S	D	E	L	1973	72882
0782	S	E	E	L	1973	72882
0782	S	D	E	L	1974	78202
0782	S	S	E	N	1975	78777
0782	S	G	E	L	1975	86706
0793	S	D	E	H	1960	10052
0793	S	D	E	N	1960	10052
0793	S	S	E	H	1960	24931
0793	S	S	E	N	1960	24931
0793	S	S	E	H	1970	61233
0793	S	S	E	N	1970	61233
0801	M	D	E	N	1976	87714
0801	M	E	E	N	1976	87714
0801	S	D	E		1958	6939
0801	S	D	E		1953	7102
0801	S	D	E		1953	9404
0801	S	G	E		1956	9943
0801	S	D	E	N	1949	10272
0801	S	D	E	N	1936	11107
0801	S	D	G	N	1933	11172
0801	S	D	G	N	1938	13489
0801	S	D	F	N	1827	16209
0801	S	D	E	N	1955	20062
0801	S	D	E	N	1960	20514
0801	S	S	E	N	1960	24931
0801	S	D	E	N	1955	25109
0801	S	D	G	N	1923	25148
0801	S	G	E	N	1896	25164
0801	S	S	G	N	1928	28680
0801	S	D	E	N	1950	31538
0801	S	D	E	N	1957	31948
0801	S	D	E	N	1968	36068
0801	S	D	G	N	1963	37025
0801	S	D	E	N	1965	37045
0801	S	D	E	N	1951	43395
0801	S	D	E	N	1954	44149
0801	S	D	R	N	1967	47303
0801	S	D	E	N	1968	47358
0801	S	D	R	N	1968	50457
0801	S	D	R	N	1968	53478
0801	S	S	E	H	1970	61233
0801	S	S	E	N	1970	61233
0801	S	D	R	N	1969	62953
0801	S	D	E	N	1969	62954
0801	S	D	E	N	1968	67345
0801	S	D	R	N	1976	94572
0803	L	D	R	H	1976	88484
0803	L	D	R	N	1976	88484
0803	L	D	E	H	1976	89373
0803	L	D	E	N	1976	89373
0818	D	D	E	N	1966	40547
0818	S	D	J	N	1966	41630
0819	S	D	G		1911	6708
0819	S	D	G	N	1951	21301
0822	S	D	E	N	1914	16313
0822	S	D	O	N	1949	28661
0822	S	D	E	N	1967	40398
0822	S	D	J	N	1967	50592
0822	S	E	J	N	1967	50592
0822	S	G	E	N	1972	66811
0822	S	G	E	N	1973	76238
0849	S	D	E	N	1979	97889
0868	S	D	R	N	1971	67588
0879	S	D	F	N	1959	25490
0879	S	T	F	N	1959	25490
0884	S	D	R	N	1977	94615
0899	S	D	E	N	1960	15618
0899	S	G	E	N	1963	32163
0899	S	S	E	N	1970	61233
0967	S	D	E	N	1969	78264
0967	S	E	E	N	1969	78264
0968	S	D	E	L	1971	63375
0968	S	D	E	N	1969	78264
0968	S	E	E	N	1969	78264
0978	L	D	G	H	1959	17892
0978	L	D	E	H	1962	29103
0979	S	D	E	N	1940	20780
0979	S	D	E	N	1940	28780
0979	S	T	E	N	1940	28780
0997	S	S	E	N	1976	88411
1015	S	D	R	N	1971	67588
1023	S	D	R	N	1971	67588
1024	S	D	R	N	1971	67588
1048	S	G	G		1952	145
1049	S	D	E		1951	1411
1049	S	S	E	L	1960	24931
1049	S	S	E	N	1960	24931
1049	S	S	E	L	1962	26069
1050	L	D	R	N	1958	16447
1050	L	D	G	H	1959	17892
1050	L	D	E	N	1958	28537
1050	L	D	E	H	1962	29103
1050	S	D	E		1954	6565
1050	S	G	E		1951	6940

Substance Number	Phys. State	Subject	Language	Temperature	Year	TPRC Number
503-1050	S	G	G		1881	8295
1050	S	D	E	N	1960	10947
1050	S	D	E	N	1955	15754
1050	S	D	E	N	1955	15755
1050	S	D	R	N	1958	16447
1050	S	D	E	N	1920	22166
1050	S	D	G	N	1928	22654
1050	S	D	E	N	1919	23435
1050	S	D	E	N	1958	23521
1050	S	S	E	L	1960	24931
1050	S	S	E	N	1960	24931
1050	S	D	E	N	1955	25109
1050	S	D	E	N	1962	26684
1050	S	D	E	N	1958	28537
1050	S	D	E	N	1931	31895
1050	S	D	E	N	1912	31897
1050	S	D	G	N	1965	34763
1050	S	D	E	N	1957	35437
1050	S	D	E	N	1953	48387
1050	S	S	E	N	1970	61233
1050	S	D	E	H	1971	69284
1050	S	E	E	H	1971	69284
1050	S	D	E	N	1919	73665
1052	S	D	G		1942	3688
1052	S	D	E		1958	6939
1052	S	D	E		1953	7102
1052	S	D	E		1942	7694
1053	S	D	G		1942	3688
1057	E	D	G	N	1965	53105
1057	E	D	E	N	1967	53106
1057	S	D	E		1945	5402
1057	S	S	E	L	1970	61233
1059	S	D	E	N	1955	15754
1059	S	D	E	N	1955	15755
1059	S	S	E	L	1960	16701
1059	S	S	E	N	1960	16701
1059	S	S	E	L	1960	24931
1059	S	S	E	N	1960	24931
1059	S	D	E	N	1958	33214
1059	S	D	E	N	1957	35437
1059	S	D	E	N	1965	36144
1059	S	G	F	L	1966	44519
1059	S	D	E	H	1963	77272
1059	S	E	E	H	1963	77272
1062	S	D	E		1954	6565
1062	S	D	E	N	1960	10947
1062	S	S	E	N	1960	24931
1062	S	S	E	N	1970	61233
1063	S	D	G		1911	6708
1063	S	D	G	N	1951	21301
1064	S	D	E		1958	6939
1064	S	D	E		1953	7102
1067	S	D	E		1958	7695
1072	S	D	E		1954	6565
1072	S	D	E	N	1960	10947
1072	S	S	E	N	1960	24931
1072	S	S	E	N	1970	61233
1075	L	D	G	H	1959	17892
1075	L	D	E	H	1962	29103
1075	S	D	G	N	1959	17892
1075	S	D	E	N	1962	29103
1078	E	D	E	N	1955	40671
1078	E	T	E	N	1955	40671
1078	S	D	R	L	1970	59436
1078	S	D	E	L	1970	59437
1083	S	D	R	N	1971	67588
1084	L	D	G	H	1959	17892
1084	L	D	E	H	1962	29103
1090	S	D	E	N	1940	20780
1090	S	D	E	N	1940	28780
1090	S	T	E	N	1940	28780
1095	L	D	G	H	1959	17892
1095	L	D	E	H	1962	29103
1097	L	D	G	H	1959	17892
1097	L	D	E	H	1962	29103
1097	S	D	R	H	1947	622
1097	S	D	R	N	1947	622
1097	S	D	E	H	1962	29111
1097	S	D	E	N	1962	29111
1098	L	D	G	H	1959	17892
1098	L	D	E	H	1962	29103
1099	S	D	F	N	1931	21006
1100	S	D	F	N	1931	21006
1101	S	D	G	N	1951	21301
1102	L	D	G	H	1959	17892
1102	L	D	E	H	1962	29103
1102	S	D	G	N	1959	17892
1102	S	D	E	N	1962	29103
1103	S	D	E	N	1958	23657
1110	S	S	E	N	1961	25075
1115	L	D	R	N	1958	16447
1115	L	D	E	N	1958	28537
1115	S	D	R	N	1958	16447
1115	S	D	E	N	1958	28537
1118	S	D	G	N	1936	31927
1125	M	D	E	N	1964	32835
1125	S	D	E		1958	6939
1125	S	D	E		1953	7102
1125	S	D	E		1954	7104
1125	S	D	E	N	1958	10458
1125	S	D	E	H	1954	10481

Phys. State: **C.** Superconductive; **D.** Doped; **E.** Expanded; **F.** Fibrous or Whisker; **G.** Gas; **I.** Ionized or Plasma; **L.** Liquid; **M.** Multiphase; **P.** Powder or Fine Particle; **S.** Solid; **T.** Thick or Thin Film

Subject: **D.** Data; **E.** Experiment; **G.** General (Data + Theory + Experiment); **S.** Survey (Review, Compendium etc.); **T.** Theory

Language: **C.** Czech; **D.** Dutch; **E.** English; **F.** French; **G.** German; **I.** Italian; **J.** Japanese; **O.** Other Languages; **P.** Polish; **R.** Russian; **S.** Spanish

Temperature: **F.** Full Range (Low + Normal + High); **L.** Low (0 to 75K) + Overlap into Normal; **N.** Normal (75 to 1273K); **H.** High (above 1273K) + Overlap into Normal; Blank = Not Coded

Substance Number	Phys. State	Subject	Language	Temperature	Year	TPRC Number
503-1125	S	T	E	H	1954	10481
1125	S	D	E	H	1954	10482
1125	S	T	E	H	1954	10482
1125	S	D	E	N	1959	13944
1125	S	D	E	N	1960	16639
1125	S	D	E	H	1951	18353
1125	S	D	E	N	1955	20062
1125	S	D	G	H	1961	24603
1125	S	D	E	N	1949	30225
1125	S	D	E	N	1950	31538
1125	S	D	E	H	1969	55769
1125	S	S	E	N	1970	61233
1125	S	D	E	H	1971	69284
1125	S	E	E	H	1971	69284
1125	S	D	E	N	1973	77957
1125	S	E	E	N	1973	77957
1128	S	D	E		1958	6939
1128	S	D	E		1953	7102
1128	S	D	E	N	1950	31538
1128	S	S	E	N	1970	61233
1129	S	D	E		1958	6939
1129	S	D	E		1953	7102
1129	S	D	E		1954	7104
1129	S	D	E	N	1949	30225
1130	S	D	G	N	1912	20505
1130	S	S	G	N	1928	28680
1131	M	D	E		1957	646
1131	S	D	R		1955	757
1131	S	D	E		1945	1719
1131	S	D	E		1956	9171
1131	S	D	E	N	1958	10458
1131	S	D	R	H	1958	19079
1131	S	D	R	N	1956	19823
1131	S	D	R	H	1961	24125
1131	S	D	R	N	1961	24125
1131	S	D	E	H	1962	27640
1131	S	D	G	H	1962	27691
1131	S	D	E	N	1938	29136
1131	S	D	R	N	1963	35746
1131	S	E	R	N	1963	35746
1131	S	D	G	H	1964	37519
1131	S	E	G	H	1964	37519
1131	S	D	R	N	1966	46591
1131	S	D	E	N	1966	46592
1131	S	D	R	N	1966	46593
1131	S	D	E	N	1966	46594
1131	S	D	R	H	1967	48335
1131	S	D	E	H	1967	48336
1132	M	D	R	N	1940	48511
1132	M	D	R	H	1960	51562
1132	S	D	R		1949	5009
1132	S	D	R	H	1959	17217
1132	S	D	R	N	1956	19823
1132	S	D	G	N	1920	22016
1132	S	D	G	N	1930	23249
1132	S	S	G	N	1928	28680
1132	S	D	R	H	1960	31072
1132	S	D	R	N	1942	31909
1132	S	D	G	H	1914	31910
1132	S	D	R	N	1963	35746
1132	S	E	R	N	1963	35746
1132	S	D	R	N	1968	58479
1132	S	T	R	N	1968	58479
1132	S	D	R	H	1971	64346
1132	S	E	R	H	1971	64346
1132	S	D	E	H	1971	64851
1132	S	E	E	H	1971	64851
1132	S	D	R	H	1973	70496
1132	S	D	E	H	1973	74223
1133	E	D	R	N	1977	90713
1133	E	D	E	N	1977	95009
1133	M	D	E		1957	646
1133	S	D	J		1957	220
1133	S	D	E		1942	1264
1133	S	T	E		1942	1264
1133	S	D	I		1942	1716
1133	S	D	E		1945	1719
1133	S	D	G		1949	1972
1133	S	D	E		1943	3610
1133	S	D	E		1941	8097
1133	S	T	E		1941	8097.
1133	S	D	E		1956	9171
1133	S	D	E	N	1952	10321
1133	S	D	E	N	1958	10458
1133	S	D	E	N	1941	11001
1133	S	D	E	H	1936	12641
1133	S	D	E	N	1937	13366
1133	S	D	E	H	1937	13367
1133	S	D	G	H	1935	14283
1133	S	D	E	N	1935	15252
1133	S	D	E	H	1960	15664
1133	S	D	E	N	1960	15664
1133	S	D	G	N	1932	16913
1133	S	S	E	N	1925	19806
1133	S	D	G	N	1929	20582
1133	S	D	E	N	1924	20589
1133	S	D	E	H	1927	20929
1133	S	D	E	H	1927	21490
1133	S	D	G	N	1920	22016
1133	S	S	G	H	1919	22023
1133	S	D	F	H	1924	22489

Substance Number	Phys. State	Subject	Language	Temperature	Year	TPRC Number
503-1133	S	D	G	N	1930	23249
1133	S	D	E	N	1962	24680
1133	S	D	R	N	1960	24772
1133	S	S	E	N	1960	24931
1133	S	D	G	N	1923	25148
1133	S	D	E	H	1962	27640
1133	S	D	G	H	1962	27817
1133	S	S	S	H	1955	28264
1133	S	S	G	H	1928	28680
1133	S	S	G	N	1928	28680
1133	S	D	F	N	1909	28745
1133	S	D	J	N	1931	28757
1133	S	D	E	N	1938	29136
1133	S	D	E	N	1920	30669
1133	S	D	E	N	1896	30723
1133	S	E	E	N	1896	30723
1133	S	G	E	N	1916	31915
1133	S	D	E	N	1964	36805
1133	S	E	E	N	1964	36805
1133	S	D	G	H	1964	37519
1133	S	E	G	H	1964	37519
1133	S	D	R	N	1971	37743
1133	S	D	R	N	1966	46591
1133	S	D	E	N	1966	46592
1133	S	D	R	H	1967	48335
1133	S	D	E	H	1967	48336
1133	S	D	G	H	1967	51075
1133	S	D	G	N	1967	51075
1133	S	D	E	N	1968	53178
1133	S	S	G	N	1968	54351
1133	S	D	F	H	1969	55618
1133	S	D	F	H	1968	57448
1133	S	S	E	N	1935	60445
1133	S	D	E	N	1936	60447
1133	S	S	E	N	1970	61233
1133	S	D	E	H	1971	69284
1133	S	E	E	H	1971	69284
1133	S	D	E	H	1974	73503
1133	S	E	E	H	1974	73503
1134	S	D	I		1951	6752
1134	S	T	I		1951	6752
1135	E	D	E	H	1964	30443
1135	E	D	E	N	1962	30934
1135	E	D	E	H	1964	33320
1135	E	D	E	N	1966	44941
1135	L	D	E	N	1947	48002
1135	M	D	E	N	1958	10458
1135	P	D	R	H	1961	26561
1135	P	D	E	H	1961	27771
1135	P	D	E	N	1966	42987
1135	P	E	E	N	1966	42987
1135	S	D	J		1957	220
1135	S	D	E		1951	799
1135	S	D	E		1951	800
1135	S	D	E		1952	1458
1135	S	D	E		1958	6939
1135	S	D	E	H	1959	10759
1135	S	D	E	H	1952	10851
1135	S	D	E	H	1951	16818
1135	S	D	E	N	1950	16820
1135	S	D	E	N	1951	16821
1135	S	D	E	N	1955	20062
1135	S	D	E	H	1927	21490
1135	S	D	G	N	1930	23249
1135	S	D	J	N	1957	23772
1135	S	D	E	N	1958	25574
1135	S	D	G	H	1962	27817
1135	S	S	G	H	1928	28680
1135	S	S	G	N	1928	28680
1135	S	D	E	N	1965	36144
1135	S	D	F	H	1965	42421
1135	S	E	F	H	1965	42421
1135	S	D	E	H	1964	43422
1135	S	S	R	H	1963	43656
1135	S	D	E	N	1954	44149
1135	S	D	R	H	1967	47147
1136	M	G	G	N	1940	12997
1136	M	G	G	N	1964	37646
1136	M	D	E	N	1967	43052
1136	M	D	R	N	1967	45418
1136	M	D	G	N	1958	51235
1136	M	G	E	N	1970	60608
1136	P	G	E	N	1896	25164
1136	P	D	R	N	1931	31857
1136	S	D	R		1949	801
1136	S	D	E		1951	810
1136	S	D	G		1942	1070
1136	S	D	G		1949	1972
1136	S	D	R		1947	2473
1136	S	G	R		1948	4492
1136	S	G	G		1925	6706
1136	S	D	E		1957	9238
1136	S	D	E	N	1966	11559
1136	S	G	E	N	1949	12033
1136	S	D	F	N	1953	12592
1136	S	G	G	N	1940	12997
1136	S	D	G	N	1960	14148
1136	S	D	E	N	1960	15640
1136	S	D	I	N	1954	16072
1136	S	D	G	N	1956	16078
1136	S	D	G	N	1959	16208

Substance Number	Phys. State	Subject	Language	Temperature	Year	TPRC Number
503-1136	S	D	R	N	1958	16447
1136	S	G	G	N	1957	20061
1136	S	D	E	N	1938	20064
1136	S	D	E	N	1944	20137
1136	S	D	I	N	1955	20975
1136	S	E	E	H	1950	21333
1136	S	G	E	N	1924	21969
1136	S	D	G	N	1920	22016
1136	S	E	E	N	1923	23248
1136	S	D	O	N	1952	23260
1136	S	G	F	H	1955	23912
1136	S	G	E	N	1955	24261
1136	S	D	G	N	1923	25148
1136	S	D	G	N	1962	26150
1136	S	D	E	N	1964	26653
1136	S	D	E	N	1958	28537
1136	S	S	G	N	1928	28680
1136	S	D	E	N	1963	30132
1136	S	D	E	N	1920	30669
1136	S	G	G	N	1941	31923
1136	S	S	E	N	1939	32125
1136	S	D	R	N	1951	32674
1136	S	D	E	N	1964	37058
1136	S	G	G	N	1964	37646
1136	S	D	E	H	1909	39859
1136	S	D	E	N	1966	42030
1136	S	S	E	N	1966	42030
1136	S	D	E	N	1919	43637
1136	S	D	O	N	1967	46158
1136	S	G	O	H	1967	46347
1136	S	D	E	N	1967	48531
1136	S	D	G	N	1958	51235
1136	S	D	O	N	1966	53698
1136	S	G	R	N	1972	67963
1136	S	D	E	N	1972	68595
1137	M	S	E		1942	1064
1137	M	T	E		1949	1082
1137	M	G	E		1958	8136
1137	S	D	E		1954	108
1137	S	E	E		1954	191
1137	S	D	E		1956	243
1137	S	D	R		1945	398
1137	S	T	R		1945	398
1137	S	E	E		1951	1114
1137	S	D	E		1950	1151
1137	S	E	E		1944	1930
1137	S	E	E		1952	1967
1137	S	D	E		1942	3671
1137	S	G	E		1953	3910
1137	S	S	E		1937	6750
1137	S	D	E		1934	7181
1137	S	D	E		1950	8911
1137	S	D	E		1955	9439
1137	S	E	E	H	1940	14573
1137	S	D	R	N	1935	14987
1137	S	D	E	H	1963	20507
1137	S	E	E	N	1921	22633
1137	S	S	E	H	1922	33208
1137	S	E	E	H	1936	33219
1137	S	S	E	H	1926	33546
1137	S	D	G	H	1967	45067
1137	S	D	E	H	1971	69284
1137	S	E	E	H	1971	69284
1139	S	D	E		1942	1264
1139	S	T	E		1942	1264
1139	S	D	E		1945	1719
1139	S	D	R		1932	6277
1139	S	D	F		1934	6279
1139	S	D	E		1935	8241
1139	S	D	E	N	1935	15252
1139	S	D	G	N	1932	16913
1139	S	D	E	H	1963	20507
1139	S	D	E	H	1927	20929
1139	S	D	E	H	1927	21490
1139	S	D	E	N	1918	24767
1139	S	S	G	N	1928	28680
1139	S	D	F	N	1909	28745
1139	S	D	E	N	1931	28757
1139	S	D	E	H	1909	39859
1139	S	D	G	H	1967	51075
1139	S	D	G	N	1967	51075
1139	S	D	O	N	1967	68514
1139	S	D	O	H	1971	71516
1140	S	D	G		1957	74
1141	S	D	E	N	1963	23263
1142	F	D	E	N	1959	25424
1142	M	E	O	N	1958	20915
1142	S	E	E		1954	110
1142	S	E	E		1954	196
1142	S	E	E	N	1952	247
1142	S	D	E		1955	1095
1142	S	T	E		1955	1095
1142	S	D	E		1952	1115
1142	S	S	G		1954	1399
1142	S	E	J		1956	1991
1142	S	E	G		1957	6619
1142	S	D	E		1958	6939
1142	S	T	E		1958	7156
1142	S	D	R		1953	8876
1142	S	E	E		1959	9102
1142	S	E	E	H	1953	10083

Phys. State: **C.** Superconductive; **D.** Doped; **E.** Expanded; **F.** Fibrous or Whisker; **G.** Gas; **I.** Ionized or Plasma; **L.** Liquid; **M.** Multiphase; **P.** Powder or Fine Particle; **S.** Solid; **T.** Thick or Thin Film
Subject: **D.** Data; **E.** Experiment; **G.** General (Data + Theory + Experiment); **S.** Survey (Review, Compendium etc.); **T.** Theory
Language: **C.** Czech; **D.** Dutch; **E.** English; **F.** French; **G.** German; **I.** Italian; **J.** Japanese; **O.** Other Languages; **P.** Polish; **R.** Russian; **S.** Spanish
Temperature: **F.** Full Range (Low + Normal + High); **L.** Low (0 to 75K) + Overlap into Normal; **N.** Normal (75 to 1273K); **H.** High (above 1273K) + Overlap into Normal; Blank = Not Coded

Substance Number	Phys. State	Sub-ject	Lan-guage	Temper-ature	Year	TPRC Number
503-1142	S	T	E	H	1953	10083
1142	S	E	E	N	1955	10127
1142	S	S	E	N	1958	10681
1142	S	E	E	N	1959	10714
1142	S	E	E	H	1952	10851
1142	S	E	E	H	1956	10923
1142	S	E	E	N	1959	15634
1142	S	E	G	N	1959	17275
1142	S	T	E	N	1958	17622
1142	S	T	E	N	1960	18632
1142	S	E	E	N	1961	19138
1142	S	S	E	N	1958	19768
1142	S	E	E	N	1948	20581
1142	S	D	E	H	1961	20637
1142	S	S	E	N	1961	20740
1142	S	E	E	N	1961	20774
1142	S	S	E	H	1961	20947
1142	S	S	E	N	1961	20964
1142	S	D	G	N	1926	21109
1142	S	S	E	N	1961	23520
1142	S	S	E	N	1961	23527
1142	S	D	E	N	1960	24322
1142	S	S	E	H	1960	24931
1142	S	S	E	N	1956	24958
1142	S	S	E	N	1961	24962
1142	S	E	E	H	1962	25194
1142	S	S	E	N	1962	25194
1142	S	S	G	N	1961	26370
1142	S	T	G	N	1961	26371
1142	S	S	E	N	1928	28673
1142	S	S	G	H	1928	28680
1142	S	S	E	H	1960	29246
1142	S	S	E	N	1956	29362
1142	S	E	G	N	1963	29768
1142	S	E	E	N	1955	30255
1142	S	S	E	N	1928	31217
1142	S	E	E	H	1954	32611
1142	S	S	R	N	1962	33745
1142	S	S	E	N	1966	33746
1142	S	D	E	H	1965	36139
1142	S	G	R	N	1964	36313
1142	S	S	E	F	1969	37682
1142	S	E	E	N	1963	37822
1142	S	E	E	H	1963	40384
1142	S	S	E	H	1962	40432
1142	S	S	E	H	1966	42284
1142	S	D	R	H	1965	43324
1142	S	S	E	H	1955	43596
1142	S	S	E	H	1961	44863
1142	S	S	E	F	1969	45627
1142	S	D	E	N	1966	45649
1142	S	T	E	N	1966	45649
1142	S	D	R	N	1965	47287
1142	S	T	R	N	1965	47287
1142	S	D	E	N	1965	47288
1142	S	T	E	N	1965	47288
1142	S	D	O	N	1965	47373
1142	S	S	D	N	1968	49783
1142	S	T	E	N	1968	51473
1142	S	S	E	H	1951	52106
1142	S	D	R	N	1968	53478
1142	S	E	R	H	1969	59605
1142	S	E	E	H	1969	59606
1142	S	S	E	N	1970	61233
1142	S	D	E	N	1968	67345
1142	S	D	R	N	1973	70445
1142	S	T	R	N	1973	70445
1142	S	D	E	N	1973	71677
1142	S	T	E	N	1973	71677
1142	S	T	R	N	1972	73845
1142	S	T	E	N	1972	73846
1142	S	D	P	N	1975	83281
1142	S	E	P	N	1975	83281
1142	S	D	E	L	1975	85499
1142	S	D	E	N	1975	85499
1142	S	D	E	N	1977	91785
1142	S	T	E	N	1979	99309
1145	P	S	E	N	1939	32125
1145	S	E	E		1954	109
1145	S	D	J		1957	220
1145	S	S	R		1945	398
1145	S	E	R		1947	617
1145	S	E	S		1955	747
1145	S	T	G		1955	761
1145	S	D	F		1950	806
1145	S	D	E		1945	1719
1145	S	D	E		1949	1922
1145	S	E	E		1952	1967
1145	S	D	G		1949	1972
1145	S	G	R		1948	5008
1145	S	D	E		1942	5258
1145	S	S	F		1937	6481
1145	S	E	R		1957	6657
1145	S	E	R		1957	7095
1145	S	T	E		1935	7185
1145	S	E	I		1947	7515
1145	S	D	E		1943	7757
1145	S	T	E		1941	8097
1145	S	D	E		1937	8244
1145	S	D	E	N	1952	10321
1145	S	S	E	H	1958	10681
503-1145	S	E	E	H	1961	10721
1145	S	S	E	H	1952	10851
1145	S	D	F	N	1953	12592
1145	S	E	E	H	1937	13367
1145	S	E	F	H	1934	14061
1145	S	E	G	N	1938	14291
1145	S	E	G	N	1939	14762
1145	S	E	E	N	1942	16279
1145	S	E	R	H	1956	17406
1145	S	E	E	H	1958	17407
1145	S	E	E	H	1960	19485
1145	S	E	F	N	1933	19612
1145	S	E	E	N	1926	19815
1145	S	E	E	H	1954	20345
1145	S	S	E	H	1958	20631
1145	S	E	I	N	1939	20657
1145	S	S	G	H	1923	21104
1145	S	E	F	N	1927	21563
1145	S	E	R	N	1928	21758
1145	S	D	F	H	1924	22489
1145	S	E	G	N	1930	23249
1145	S	S	E	N	1922	23628
1145	S	E	J	N	1951	23634
1145	S	S	E	H	1949	25466
1145	S	T	R	N	1960	25545
1145	S	E	G	N	1965	26779
1145	S	E	E	H	1936	27140
1145	S	E	E	N	1947	27189
1145	S	E	G	N	1962	27715
1145	S	E	E	H	1961	27836
1145	S	T	E	N	1963	28160
1145	S	D	F	N	1909	28745
1145	S	D	E	N	1931	28757
1145	S	S	G	H	1931	28786
1145	S	S	E	H	1961	29615
1145	S	E	E	N	1961	29826
1145	S	E	E	N	1961	29828
1145	S	E	E	N	1896	30723
1145	S	E	E	H	1923	30988
1145	S	S	E	H	1923	30988
1145	S	E	R	H	1961	31073
1145	S	S	E	N	1917	31611
1145	S	E	E	N	1915	31868
1145	S	S	D	H	1952	31870
1145	S	S	R	H	1942	31909
1145	S	S	R	N	1939	31942
1145	S	S	G	N	1930	33613
1145	S	S	E	N	1910	35241
1145	S	E	E	N	1964	36431
1145	S	D	G	N	1964	37617
1145	S	S	E	H	1964	37819
1145	S	S	E	H	1932	39845
1145	S	D	E	H	1909	39859
1145	S	D	E	N	1929	40826
1145	S	E	E	H	1947	41141
1145	S	E	G	H	1966	41276
1145	S	E	G	N	1966	41277
1145	S	D	G	H	1966	42406
1145	S	E	G	H	1966	42406
1145	S	D	G	N	1964	43101
1145	S	S	R	H	1966	44058
1145	S	S	E	H	1966	44059
1145	S	G	F	H	1969	44549
1145	S	S	E	N	1924	44568
1145	S	D	G	H	1967	51075
1145	S	D	G	N	1967	51075
1145	S	T	G	H	1963	51236
1145	S	S	E	H	1967	51671
1145	S	S	E	N	1967	51672
1145	S	S	E	H	1956	52789
1145	S	S	E	H	1963	52812
1145	S	S	E	H	1969	55310
1145	S	E	J	N	1968	56147
1145	S	S	R	H	1966	60815
1145	S	S	E	H	1970	60816
1145	S	D	R	N	1970	61168
1145	S	D	E	N	1970	61169
1145	S	D	R	H	1970	61318
1145	S	D	E	H	1970	61319
1145	S	E	F	H	1970	61396
1145	S	D	R	N	1970	62538
1145	S	D	E	N	1970	67377
1145	S	E	C	H	1973	71322
1146	M	D	E		1957	646
1146	M	D	R	H	1960	51562
1146	P	D	E	N	1905	30693
1146	S	D	J		1957	220
1146	S	D	R		1955	757
1146	S	D	R		1950	838
1146	S	D	E		1942	1264
1146	S	T	E		1942	1264
1146	S	D	E		1945	1719
1146	S	D	O		1947	1958
1146	S	D	G		1949	1972
1146	S	D	G		1942	3391
1146	S	G	R		1941	4892
1146	S	D	E		1935	8241
1146	S	D	E	N	1958	10458
1146	S	D	E	H	1961	10721
1146	S	D	G	H	1935	14283
1146	S	D	E	F	1933	14416
503-1146	S	D	J	N	1938	15065
1146	S	D	E	N	1935	15252
1146	S	D	R	H	1959	17217
1146	S	S	E	N	1925	19806
1146	S	D	R	N	1956	19823
1146	S	D	E	H	1927	20929
1146	S	D	E	H	1927	21490
1146	S	D	G	N	1920	22016
1146	S	D	F	H	1924	22489
1146	S	D	E	N	1961	23066
1146	S	D	G	N	1930	23249
1146	S	D	G	N	1918	24323
1146	S	D	E	N	1961	24326
1146	S	D	E	N	1962	24680
1146	S	D	E	N	1918	24767
1146	S	S	E	N	1960	24931
1146	S	D	G	N	1923	25148
1146	S	D	G	N	1965	26779
1146	S	D	E	H	1962	27640
1146	S	D	G	N	1962	27715
1146	S	D	G	N	1962	27817
1146	S	S	G	H	1928	28680
1146	S	D	F	N	1909	28745
1146	S	D	E	N	1931	28757
1146	S	D	E	N	1938	29136
1146	S	D	E	N	1920	30669
1146	S	D	E	N	1896	30723
1146	S	E	E	N	1896	30723
1146	S	D	R	H	1960	31072
1146	S	D	R	N	1942	31909
1146	S	D	G	H	1914	31910
1146	S	D	R	N	1963	35746
1146	S	E	R	N	1963	35746
1146	S	D	G	H	1964	37519
1146	S	E	G	H	1964	37519
1146	S	D	R	H	1962	38012
1146	S	D	R	N	1966	46591
1146	S	D	E	N	1966	46592
1146	S	D	R	H	1967	48335
1146	S	D	E	H	1967	48336
1146	S	D	G	H	1967	51075
1146	S	D	G	N	1967	51075
1146	S	S	G	N	1968	54351
1146	S	S	E	N	1935	60445
1146	S	S	E	N	1970	61233
1146	S	D	R	N	1970	62687
1146	S	G	R	N	1971	65014
1146	S	D	R	H	1971	74368
1146	S	D	E	H	1971	74369
1146	S	D	R	N	1974	78988
1146	S	D	E	N	1974	83663
1146	S	D	R	H	1977	90440
1146	S	D	R	N	1977	90440
1146	S	D	R	N	1978	93218
1146	S	D	E	H	1977	95993
1146	S	D	E	N	1977	95993
1146	S	D	E	N	1978	98789
1147	E	D	E	N	1964	32835
1147	E	D	R	N	1965	38402
1147	E	G	R	H	1965	42535
1147	E	G	E	H	1966	43468
1147	E	D	E	N	1968	47499
1147	E	T	E	N	1968	47499
1147	E	D	R	N	1967	49687
1147	E	T	R	N	1967	49687
1147	E	G	R	N	1971	65014
1147	E	D	E	N	1974	74696
1147	E	T	E	N	1974	74696
1147	M	G	G	N	1952	19030
1147	M	D	E	N	1964	32835
1147	M	D	R	N	1972	71835
1147	M	D	R	N	1971	73732
1147	M	D	E	N	1971	73733
1147	M	D	E	N	1976	85570
1147	M	T	E	N	1976	85570
1147	P	D	E	N	1905	30693
1147	P	D	G	H	1964	37519
1147	P	E	G	H	1964	37519
1147	P	D	E	N	1905	41683
1147	S	D	R		1950	838
1147	S	D	E		1942	1264
1147	S	T	E		1942	1264
1147	S	D	E		1943	3610
1147	S	D	G	N	1934	6342
1147	S	D	E		1934	7180
1147	S	D	E		1934	7181
1147	S	D	E		1941	8097
1147	S	T	E		1941	8097
1147	S	D	E		1935	8241
1147	S	D	E		1956	9247
1147	S	D	E	N	1952	10321
1147	S	D	E	N	1958	10458
1147	S	D	E	H	1961	10721
1147	S	D	E	N	1960	10950
1147	S	D	E	N	1941	11001
1147	S	D	E	N	1945	11116
1147	S	D	E	H	1936	12641
1147	S	D	E	N	1948	12708
1147	S	D	G	H	1935	14283
1147	S	D	E	N	1935	15252
1147	S	D	E	H	1960	15664

Phys. State: **C.** Superconductive; **D.** Doped; **E.** Expanded; **F.** Fibrous or Whisker; **G.** Gas; **I.** Ionized or Plasma; **L.** Liquid; **M.** Multiphase; **P.** Powder or Fine Particle; **S.** Solid; **T.** Thick or Thin Film

Subject: **D.** Data; **E.** Experiment; **G.** General (Data + Theory + Experiment); **S.** Survey (Review, Compendium etc.); **T.** Theory

Language: **C.** Czech; **D.** Dutch; **E.** English; **F.** French; **G.** German; **I.** Italian; **J.** Japanese; **O.** Other Languages; **P.** Polish; **R.** Russian; **S.** Spanish

Temperature: **F.** Full Range (Low + Normal + High); **L.** Low (0 to 75K) + Overlap into Normal; **N.** Normal (75 to 1273K); **H.** High (above 1273K) + Overlap into Normal; Blank = Not Coded

Substance Number	Phys. State	Subject	Language	Temperature	Year	TPRC Number
503-1147	S	D	E	N	1914	16313
1147	S	D	E	N	1950	16816
1147	S	D	E	N	1951	16818
1147	S	D	E	N	1951	16821
1147	S	G	E	N	1949	16826
1147	S	D	G	N	1932	16913
1147	S	D	R	H	1959	17217
1147	S	S	E	N	1925	19806
1147	S	D	R	N	1956	19823
1147	S	D	E	N	1955	20062
1147	S	D	E	N	1938	20064
1147	S	D	E	H	1954	20345
1147	S	D	G	N	1912	20505
1147	S	D	E	H	1963	20507
1147	S	D	E	N	1960	20514
1147	S	D	G	N	1929	20582
1147	S	D	E	N	1924	20589
1147	S	D	E	N	1928	20652
1147	S	D	E	N	1909	20765
1147	S	D	G	H	1957	20977
1147	S	D	J	N	1958	21141
1147	S	G	J	N	1959	21142
1147	S	D	E	N	1950	21333
1147	S	D	E	H	1927	21490
1147	S	D	E	N	1939	21536
1147	S	G	E	N	1924	21969
1147	S	D	G	N	1920	22016
1147	S	S	G	H	1919	22023
1147	S	D	E	H	1927	22689
1147	S	D	E	N	1961	23066
1147	S	D	G	N	1930	23249
1147	S	D	G	N	1918	24323
1147	S	E	G	N	1908	24397
1147	S	D	E	N	1962	24680
1147	S	D	E	N	1960	24761
1147	S	D	E	N	1918	24767
1147	S	D	R	N	1960	24772
1147	S	S	E	N	1960	24931
1147	S	D	G	N	1923	25148
1147	S	D	E	H	1959	25424
1147	S	D	J	N	1960	26221
1147	S	T	J	N	1960	26221
1147	S	D	O	N	1934	26674
1147	S	G	E	H	1936	27140
1147	S	G	E	N	1936	27140
1147	S	D	E	N	1962	27283
1147	S	D	E	N	1962	27498
1147	S	D	E	H	1962	27640
1147	S	D	G	H	1962	27817
1147	S	S	G	H	1928	28680
1147	S	D	J	N	1931	28757
1147	S	D	E	N	1938	29136
1147	S	D	E	N	1920	30669
1147	S	D	E	N	1912	31897
1147	S	D	R	N	1942	31909
1147	S	D	G	H	1914	31910
1147	S	G	E	N	1916	31915
1147	S	D	E	N	1954	32611
1147	S	D	E	N	1966	34401
1147	S	D	R	N	1963	35746
1147	S	E	R	N	1963	35746
1147	S	D	F	H	1964	36132
1147	S	E	F	H	1964	36132
1147	S	D	O	N	1964	36898
1147	S	D	E	N	1964	37068
1147	S	D	G	H	1964	37519
1147	S	E	G	H	1964	37519
1147	S	D	G	H	1964	37617
1147	S	D	F	H	1961	38063
1147	S	D	G	H	1964	38128
1147	S	D	E	H	1941	38840
1147	S	E	E	N	1947	39130
1147	S	S	E	F	1932	39856
1147	S	D	E	N	1965	40717
1147	S	D	E	N	1963	41911
1147	S	D	R	N	1963	41912
1147	S	D	J	H	1964	43236
1147	S	D	E	N	1924	43414
1147	S	D	E	N	1959	43731
1147	S	D	R	N	1966	44488
1147	S	D	G	H	1967	45067
1147	S	D	E	H	1968	46475
1147	S	D	R	H	1966	46591
1147	S	D	E	N	1966	46592
1147	S	D	E	N	1966	46599
1147	S	D	R	H	1967	48335
1147	S	D	E	H	1967	48336
1147	S	D	G	H	1967	51075
1147	S	D	G	N	1967	51075
1147	S	D	G	H	1969	53723
1147	S	S	G	N	1968	54351
1147	S	D	R	H	1968	54773
1147	S	D	E	H	1968	54774
1147	S	D	R	N	1969	54917
1147	S	D	E	N	1969	57266
1147	S	D	R	N	1969	61220
1147	S	T	R	N	1969	61220
1147	S	S	E	H	1970	61233
1147	S	S	E	N	1970	61233
1147	S	D	R	N	1970	61607
1147	S	D	E	N	1970	61608

Substance Number	Phys. State	Subject	Language	Temperature	Year	TPRC Number
503-1147	S	D	R	N	1970	62687
1147	S	D	R	N	1971	66439
1147	S	D	E	N	1971	66440
1147	S	D	E	H	1971	69284
1147	S	E	E	H	1971	69284
1147	S	D	E	N	1974	74696
1147	S	T	E	N	1974	74696
1147	S	D	E	N	1964	78045
1147	S	E	E	N	1964	78045
1147	S	D	E	N	1964	78047
1147	S	E	E	N	1964	78047
1147	S	D	E	H	1973	78263
1147	S	S	E	H	1976	90192
1147	S	S	E	N	1976	90192
1148	M	D	R		1957	8791
1149	S	E	E		1957	945
1149	S	D	G		1949	1972
1149	S	D	E		1950	8911
1149	S	D	E		1955	9027
1149	S	D	E	N	1957	10271
1149	S	D	E	N	1952	10321
1149	S	D	E	N	1960	10950
1149	S	D	E	N	1941	11001
1149	S	D	G	N	1956	12170
1149	S	D	R	N	1937	14196
1149	S	E	G	N	1936	14870
1149	S	D	G	N	1938	26667
1149	S	G	E	H	1936	27140
1149	S	D	E	N	1962	27498
1149	S	D	E	H	1962	27640
1149	S	D	E	H	1961	28858
1149	S	D	E	N	1920	30669
1149	S	D	E	H	1934	31858
1149	S	D	E	N	1965	35875
1149	S	D	E	N	1963	37822
1149	S	D	E	H	1931	39843
1149	S	D	E	N	1966	46439
1149	S	D	E	H	1969	55769
1149	S	D	E	H	1971	69284
1149	S	E	E	H	1971	69284
1149	S	G	E	N	1974	74606
1151	S	D	G		1942	1070
1151	S	D	G		1958	9894
1151	S	D	G	N	1912	20505
1151	S	D	E	N	1960	20529
1151	S	D	G	H	1957	20977
1151	S	D	G	N	1920	22016
1151	S	D	O	N	1952	23260
1152	M	G	G	N	1952	19030
1152	M	G	R	N	1941	23645
1152	P	D	E	N	1932	31912
1152	S	D	E	N	1941	8097
1152	S	D	O		1956	9767
1152	S	D	E	N	1952	10321
1152	S	D	G	N	1956	12170
1152	S	D	G	N	1938	13489
1152	S	D	E	N	1934	13903
1152	S	D	E	N	1938	14860
1152	S	D	E	H	1961	16157
1152	S	D	E	N	1932	16349
1152	S	D	E	N	1941	17280
1152	S	D	G	N	1912	20505
1152	S	D	G	N	1929	20582
1152	S	D	E	N	1928	20652
1152	S	G	J	N	1959	21142
1152	S	D	E	N	1950	21333
1152	S	D	E	N	1927	21490
1152	S	D	G	N	1920	22016
1152	S	D	E	N	1920	22166
1152	S	D	E	N	1928	23444
1152	S	D	R	N	1941	23645
1152	S	D	O	N	1934	26674
1152	S	S	G	N	1928	28680
1152	S	D	E	N	1938	29136
1152	S	D	E	N	1920	30669
1152	S	D	E	N	1917	31611
1152	S	D	G	N	1951	31815
1152	S	D	E	N	1932	31912
1152	S	G	E	N	1916	31915
1152	S	D	R	N	1956	32423
1152	S	D	R	N	1955	32646
1152	S	D	E	N	1963	37822
1152	S	D	E	H	1909	39859
1152	S	D	E	N	1943	40228
1152	S	D	R	N	1971	66439
1152	S	D	E	N	1971	66440
1152	S	D	E	N	1973	71892
1152	S	D	E	H	1974	73503
1152	S	E	E	H	1974	73503
1152	S	D	E	N	1919	73665
1153	M	D	E		1957	646
1153	S	D	E		1945	1719
1153	S	D	E	N	1941	8097
1153	S	D	E		1953	9404
1153	S	D	E	N	1958	10458
1153	S	D	G	N	1933	11172
1153	S	D	G	H	1935	14283
1153	S	D	G	N	1932	16913
1153	S	D	G	H	1957	20977
1153	S	D	E	N	1950	21333
1153	S	D	E	H	1961	27783

Substance Number	Phys. State	Subject	Language	Temperature	Year	TPRC Number
503-1153	S	S	G	N	1928	28680
1153	S	D	G	H	1964	37519
1153	S	E	G	H	1964	37519
1153	S	D	E	N	1951	43395
1153	S	D	E	N	1954	44149
1153	S	G	F	H	1969	44549
1153	S	D	E	N	1966	46439
1153	S	S	G	N	1968	54351
1153	S	D	F	H	1969	55618
1153	S	D	F	H	1968	57448
1154	S	D	E		1953	9404
1154	S	D	E	N	1954	44149
1155	S	D	E		1958	8103
1155	S	D	E	N	1935	12720
1156	S	D	G		1947	1822
1157	S	D	E		1935	7182
1158	S	D	E	N	1936	11107
1158	S	G	J	N	1959	21142
1158	S	D	J	N	1951	23634
1158	S	D	J	N	1931	28757
1158	S	D	E	N	1964	37541
1160	S	D	R		1950	838
1160	S	D	E	N	1961	23066
1161	S	D	G		1942	1070
1162	E	D	G	N	1969	56882
1162	E	E	G	N	1969	56882
1162	S	D	E		1943	3610
1162	S	D	E	H	1937	13367
1162	S	D	G	N	1963	37025
1162	S	S	E	H	1970	61233
1162	S	S	E	N	1970	61233
1162	S	S	E	H	1976	90192
1162	S	S	E	N	1976	90192
1163	S	D	E		1935	7182
1163	S	D	R	N	1966	46591
1163	S	D	E	N	1966	46592
1164	S	D	G		1942	1070
1164	S	D	J	N	1931	28757
1164	S	D	G	H	1964	37519
1164	S	E	G	H	1964	37519
1165	S	D	E	N	1940	31902
1166	S	D	E	H	1961	16157
1170	S	D	E	N	1957	10271
1170	S	G	E	N	1958	17623
1170	S	D	E	N	1938	29136
1172	M	G	G	N	1940	12997
1172	P	G	G	N	1940	12997
1172	S	G	G	N	1940	12997
1172	S	D	O	N	1952	23260
1172	S	S	R	H	1936	31941
1172	S	D	G	H	1964	37519
1172	S	E	G	H	1964	37519
1172	S	D	G	H	1967	51075
1172	S	D	G	N	1967	51075
1173	E	D	E	N	1962	30934
1173	E	D	R	N	1963	32427
1173	M	D	E		1957	646
1173	M	D	R		1951	1969
1173	M	D	E	N	1960	13346
1173	M	D	O	N	1958	20915
1173	M	D	E	N	1970	58368
1173	M	T	E	N	1970	58368
1173	P	D	E	N	1966	42987
1173	P	E	E	N	1966	42987
1173	S	D	E		1954	108
1173	S	D	E		1951	1114
1173	S	D	E		1945	1719
1173	S	D	G		1941	4721
1173	S	D	E		1958	6939
1173	S	D	E	N	1941	8097
1173	S	G	E	N	1960	10013
1173	S	D	E	N	1955	10127
1173	S	D	E	N	1949	10272
1173	S	D	E	N	1956	10518
1173	S	D	E	N	1935	15252
1173	S	D	G	N	1932	16913
1173	S	D	R	N	1959	17216
1173	S	D	E	H	1963	20507
1173	S	D	E	H	1927	20929
1173	S	D	E	N	1950	21333
1173	S	D	E	H	1927	21490
1173	S	D	E	H	1961	23579
1173	S	D	R	H	1961	24125
1173	S	D	R	N	1961	24125
1173	S	D	E	N	1961	24326
1173	S	G	G	H	1959	24395
1173	S	D	G	H	1961	24603
1173	S	D	E	N	1962	24680
1173	S	D	G	N	1923	25148
1173	S	D	E	N	1958	25574
1173	S	D	G	N	1965	26779
1173	S	D	R	N	1962	26930
1173	S	D	E	N	1962	27498
1173	S	D	E	H	1962	27640
1173	S	D	G	N	1962	27715
1173	S	D	E	H	1961	27783
1173	S	D	G	H	1962	27817
1173	S	D	E	N	1957	28630
1173	S	S	G	H	1928	28680
1173	S	D	E	N	1949	30225
1173	S	D	E	H	1964	30443

Phys. State: **C.** Superconductive; **D.** Doped; **E.** Expanded; **F.** Fibrous or Whisker; **G.** Gas; **I.** Ionized or Plasma; **L.** Liquid; **M.** Multiphase; **P.** Powder or Fine Particle; **S.** Solid; **T.** Thick or Thin Film

Subject: **D.** Data; **E.** Experiment; **G.** General (Data + Theory + Experiment); **S.** Survey (Review, Compendium etc.); **T.** Theory

Language: **C.** Czech; **D.** Dutch; **E.** English; **F.** French; **G.** German; **I.** Italian; **J.** Japanese; **O.** Other Languages; **P.** Polish; **R.** Russian; **S.** Spanish

Temperature: **F.** Full Range (Low + Normal + High); **L.** Low (0 to 75K) + Overlap into Normal; **N.** Normal (75 to 1273K); **H.** High (above 1273K) + Overlap into Normal; Blank = Not Coded

Substance Number	Phys. State	Subject	Language	Temperature	Year	TPRC Number
503-1173	S	G	E	N	1958	31786
1173	S	D	E	N	1963	32128
1173	S	D	R	N	1963	32427
1173	S	D	E	N	1958	33214
1173	S	D	E	N	1965	35403
1173	S	D	R	N	1963	35746
1173	S	E	R	N	1963	35746
1173	S	D	R	H	1964	36008
1173	S	D	R	H	1971	36180
1173	S	T	R	H	1971	36180
1173	S	D	G	H	1964	37519
1173	S	E	G	H	1964	37519
1173	S	D	E	N	1963	37530
1173	S	D	F	H	1961	38063
1173	S	S	E	F	1966	42294
1173	S	D	F	H	1965	42421
1173	S	E	F	H	1965	42421
1173	S	D	E	N	1951	43395
1173	S	D	E	H	1964	43422
1173	S	D	E	N	1954	44149
1173	S	D	G	H	1967	45067
1173	S	D	E	H	1968	46475
1173	S	D	E	N	1952	48341
1173	S	D	G	H	1967	51075
1173	S	D	G	N	1967	51075
1173	S	D	E	N	1960	52136
1173	S	S	G	N	1968	54351
1173	S	D	E	H	1969	55769
1173	S	D	P	H	1969	56331
1173	S	S	E	H	1970	56454
1173	S	D	F	H	1970	60726
1173	S	E	F	H	1970	60726
1173	S	D	R	N	1970	60833
1173	S	D	E	N	1970	60834
1173	S	D	R	H	1970	61531
1173	S	D	E	H	1970	61532
1173	S	D	R	N	1970	62314
1173	S	D	R	N	1971	62758
1173	S	E	R	N	1971	62758
1173	S	D	E	H	1964	63069
1173	S	D	R	H	1971	64822
1173	S	D	E	H	1971	64823
1173	S	D	R	H	1971	64846
1173	S	D	E	H	1971	64847
1173	S	D	G	N	1972	67454
1173	S	E	G	N	1972	67454
1173	S	D	R	N	1971	68805
1173	S	D	R	N	1971	68807
1173	S	T	R	N	1971	68807
1173	S	D	E	H	1971	69284
1173	S	E	E	H	1971	69284
1173	S	D	O	H	1971	71516
1173	S	D	E	L	1973	72882
1173	S	E	E	L	1973	72882
1173	S	D	E	H	1974	73503
1173	S	E	E	H	1974	73503
1173	S	D	E	N	1974	74549
1173	S	D	E	N	1974	75087
1173	S	D	R	N	1974	78988
1173	S	D	E	N	1974	83663
1173	S	S	E	H	1976	87488
1173	S	S	E	N	1976	87488
1173	S	S	E	H	1976	90192
1173	S	S	E	N	1976	90192
1175	S	D	E	H	1961	16157
1175	S	D	E	N	1950	21333
1175	S	S	E	N	1976	90192
1177	S	D	E	N	1941	11001
1177	S	D	E	N	1960	20529
1177	S	D	E	N	1928	20652
1177	S	D	E	N	1927	21490
1177	S	D	O	N	1952	23260
1177	S	D	R	N	1960	24772
1177	S	D	J	N	1931	28757
1177	S	G	R	N	1971	66294
1177	S	G	E	N	1971	70931
1178	S	D	E	H	1961	16157
1178	S.	D	E	N	1964	36552
1178	S	D	R	N	1967	54757
1179	F	S	E	N	1970	61233
1179	S	D	E	H	1961	16157
1179	S	D	G	H	1964	37519
1179	S	E	G	H	1964	37519
1179	S	S	E	N	1976	90192
1180	D	D	E	N	1966	44329
1180	L	D	E	N	1947	48002
1180	M	G	E	H	1963	31430
1180	S	D	E	N	1949	10272
1180	S	D	E	N	1949	30225
1180	S	D	E	H	1964	30443
1180	S	D	E	N	1963	32128
1180	S	D	E	N	1951	43395
1180	S	D	E	N	1954	44149
1180	S	D	E	N	1966	44329
1180	S	D	E	N	1968	53502
1180	S	D	E	N	1970	77556
1181	S	D	E	N	1920	30669
1183	S	D	J	N	1938	13636
1184	M	D	E	N	1963	28710
1184	M	G	E	N	1963	30401
1184	S	D	E	H	1960	15664

Substance Number	Phys. State	Subject	Language	Temperature	Year	TPRC Number
503-1184	S	D	E	N	1966	43774
1185	S	D	G	N	1960	25525
1185	S	D	R	N	1951	32674
1185	S	D	G	N	1963	37025
1185	S	D	E	N	1963	37822
1185	S	D	E	N	1967	42955
1185	S	D	R	N	1967	45153
1186	S	E	E	H	1940	14575
1187	S	D	E		1956	6524
1187	S	T	E		1956	6524
1187	S	D	E		1958	6939
1187	S	D	E		1953	7102
1187	S	D	E		1957	7288
1187	S	D	G	N	1933	11172
1187	S	D	G	N	1937	13487
1187	S	D	E	N	1955	15754
1187	S	D	E	N	1955	15755
1187	S	D	E	N	1963	25061
1187	S	D	E	N	1962	29551
1187	S	D	E	N	1949	30225
1187	S	D	E	N	1957	31948
1188	L	D	E	N	1947	48002
1189	E	D	R	N	1964	37572
1189	E	D	E	N	1968	47920
1189	E	D	R	N	1968	49597
1189	P	S	G	N	1928	28680
1189	P	D	E	N	1966	42987
1189	P	E	E	N	1966	42987
1189	S	D	E	N	1924	20589
1189	S	S	G	N	1928	28680
1189	S	D	R	N	1964	37572
1189	S	D	F	H	1961	38063
1189	S	D	F	H	1965	42421
1189	S	E	F	H	1965	42421
1189	S	D	F	H	1969	55618
1189	S	D	F	H	1968	57448
1189	S	D	E	H	1974	73503
1189	S	E	E	H	1974	73503
1189	S	D	R	H	1974	76473
1189	S	D	E	H	1974	76474
1193	S	D	O	N	1952	23260
1193	S	D	E	N	1969	57606
1194	P	G	E		1933	7644
1194	P	G	E	N	1962	22559
1194	P	D	G	N	1929	22931
1194	P	D	E	N	1962	29551
1194	P	D	E	N	1905	30693
1194	P	G	E	N	1909	31920
1194	P	D	E	N	1905	41683
1194	P	D	E	N	1969	53610
1194	P	G	E	N	1974	75084
1194	P	G	R	N	1973	77332
1194	S	D	R		1950	838
1194	S	D	E		1942	1264
1194	S	T	E		1942	1264
1194	S	D	E	H	1927	21490
1194	S	D	E	N	1961	23066
1194	S	D	E	N	1963	25061
1194	S	D	F	N	1909	28745
1194	S	D	E	N	1924	43414
1195	S	D	I	N	1954	16072
1195	S	D	G	N	1920	22016
1195	S	G	F	H	1955	23912
1195	S	D	I	N	1958	23960
1196	S	D	R	H	1959	17217
1197	S	D	E		1942	1264
1197	S	T	E		1942	1264
1197	S	D	G	N	1932	16913
1197	S	D	F	N	1909	28745
1198	M	D	E	N	1955	28067
1198	S	D	E	N	1949	30225
1198	S	D	E	N	1955	30255
1199	M	D	E	N	1955	28067
1199	S	D	E	N	1955	30255
1200	S	D	E	N	1938	29136
1202	S	S	E	H	1961	18230
1208	M	D	E	N	1964	32835
1208	S	D	G	H	1957	20977
1208	S	D	E	N	1950	21333
1208	S	S	E	N	1960	24931
1208	S	D	E	H	1962	27640
1208	S	D	E	H	1961	27783
1208	S	D	G	H	1967	51075
1208	S	D	G	N	1967	51075
1208	S	D	E	H	1974	73503
1208	S	E	E	H	1974	73503
1209	S	E	E	H	1961	29826
1213	S	D	G	N	1963	33448
1213	S	D	E	N	1965	33449
1214	S	D	G	N	1912	20505
1214	S	D	R	N	1929	22225
1216	S	D	E	H	1963	20507
1218	S	S	E	N	1941	10273
1218	S	D	G	N	1933	11172
1218	S	D	G	H	1957	20977
1218	S	D	G	N	1923	25148
1218	S	D	E	H	1909	39859
1219	S	D	E		1958	6939
1219	S	D	E		1953	7036
1219	S	D	E		1953	7102
1219	S	D	E	H	1960	10052

Substance Number	Phys. State	Subject	Language	Temperature	Year	TPRC Number
503-1219	S	D	E	H	1960	20600
1219	S	D	E	N	1963	23121
1219	S	S	E	N	1960	24931
1219	S	D	E	N	1958	28870
1219	S	D	E	N	1952	48341
1219	S	S	E	N	1970	61233
1219	S	S	P	H	1976	92927
1220	S	S	E	H	1958	20631
1221	S	S	E	H	1958	20631
1222	S	S	E	H	1958	20631
1222	S	E	G	H	1962	27691
1223	S	D	E	N	1961	32869
1224	S	D	R	N	1928	21758
1224	S	D	G	N	1920	22016
1224	S	D	R	N	1929	22225
1224	S	D	O	N	1952	23260
1224	S	E	G	N	1908	24397
1224	S	D	R	N	1960	24772
1224	S	S	G	N	1928	28680
1226	M	T	G	N	1970	59479
1226	S	D	R	N	1960	24772
1226	S	G	R	N	1965	43098
1228	S	D	G	N	1923	25148
1230	S	D	G	N	1923	25148
1236	S	D	E	H	1961	27783
1237	P	D	C	H	1974	80491
1237	P	D	C	N	1974	80491
1237	S	D	R	N	1962	26930
1238	S	D	G	H	1962	27817
1240	P	D	R	N	1963	30309
1242	S	D	E	N	1918	24767
1243	S	D	J		1955	17
1244	M	D	R	N	1940	48511
1245	S	D	E	N	1966	42030
1245	S	S	E	N	1966	42030
1245	S	S	E	N	1976	90192
1246	S	D	E	N	1966	42030
1246	S	S	E	N	1966	42030
1247	S	D	E	N	1966	42030
1247	S	S	E	N	1966	42030
1248	S	D	R	H	1968	44358
1248	S	D	E	H	1968	46849
1248	S	D	R	H	1967	48335
1248	S	D	E	H	1967	48336
1248	S	D	R	N	1968	50671
1248	S	D	E	N	1968	50672
1248	S	D	G	H	1967	51075
1248	S	D	G	N	1967	51075
1250	L	D	G	H	1959	17892
1250	L	D	E	H	1962	29103
1250	S	G	R	N	1949	31625
1260	L	D	E	N	1972	66008
1260	L	E	E	N	1972	66008
1262	S	D	E	H	1968	46475
1262	S	D	O	H	1971	71516
1262	S	S	E	H	1976	90192
1262	S	S	E	N	1976	90192
1267	M	D	E	H	1978	97131
1267	M	D	E	N	1978	97131
1267	M	T	E	H	1978	97131
1267	M	T	E	N	1978	97131
1267	M	D	E	H	1978	97168
1267	M	D	E	N	1978	97168
1267	M	T	E	H	1978	97168
1267	M	T	E	N	1978	97168
1267	S	E	G	H	1967	45394
1267	S	E	G	H	1966	46223
1267	S	S	J	H	1967	51199
1267	S	T	E	H	1970	59268
1268	M	D	I	N	1966	44990
1268	S	D	I	N	1966	41292
1268	S	D	E	N	1969	53585
1271	S	D	E	N	1969	53587
1276	S	D	E	N	1960	25797
1277	S	D	E	N	1940	31902
1280	S	D	E	N	1963	58889
1281	S	D	E	N	1955	20062
1281	S	D	E	H	1971	69284
1281	S	E	E	H	1971	69284
1285	S	S	F	N	1975	94378
1286	S	D	R	N	1969	58848
1286	S	D	E	N	1969	58849
1288	S	D	F	H	1961	38063
1292	S	S	E	N	1960	24931
1294	S	S	E	N	1970	61233
1302	S	S	E	N	1947	60371
1304	S	D	F	H	1964	36132
1304	S	E	F	H	1964	36132
1315	S	D	G	H	1964	37519
1315	S	E	G	H	1964	37519
1316	S	D	G	H	1964	37519
1316	S	E	G	H	1964	37519
1317	S	T	G	N	1970	61891
1321	P	D	R	N	1971	65687
1322	S	D	E	L	1972	68190
1326	S	D	E	H	1971	69284
1326	S	E	E	H	1971	69284
1331	S	D	R	N	1972	67971
1332	S	D	R	N	1972	68003
1333	S	D	R	N	1970	61898
1333	S	D	E	N	1970	72171

Phys. State: **C.** Superconductive; **D.** Doped; **E.** Expanded; **F.** Fibrous or Whisker; **G.** Gas; **I.** Ionized or Plasma; **L.** Liquid; **M.** Multiphase; **P.** Powder or Fine Particle; **S.** Solid; **T.** Thick or Thin Film

Subject: **D.** Data; **E.** Experiment; **G.** General (Data + Theory + Experiment); **S.** Survey (Review, Compendium etc.); **T.** Theory

Language: **C.** Czech; **D.** Dutch; **E.** English; **F.** French; **G.** German; **I.** Italian; **J.** Japanese; **O.** Other Languages; **P.** Polish; **R.** Russian; **S.** Spanish

Temperature: **F.** Full Range (Low + Normal + High); **L.** Low (0 to 75K) + Overlap into Normal; **N.** Normal (75 to 1273K); **H.** High (above 1273K) + Overlap into Normal; Blank = Not Coded

Substance Number	Phys. State	Subject	Language	Temperature	Year	TPRC Number
503-1335	S	D	R	H	1971	74368
1335	S	D	E	H	1971	74369
1342	S	D	E	H	1974	73503
1342	S	E	E	H	1974	73503
1350	S	S	E	N	1960	24931
1350	S	D	E	N	1896	30723
1350	S	E	E	N	1896	30723
1351	S	D	E	H	1974	73503
1351	S	E	E	H	1974	73503
1353	M	D	E	N	1967	48632
1356	S	S	G	N	1961	28087
1356	S	S	E	N	1962	28088
1357	E	D	E		1956	4737
1357	E	D	E		1954	7336
1357	F	D	I		1950	76
1357	F	D	E		1942	1264
1357	F	T	E		1942	1264
1357	F	D	G		1950	1628
1357	F	D	E		1952	5069
1357	F	D	E		1942	5258
1357	F	D	E		1953	6231
1357	F	D	E		1947	6283
1357	F	D	E		1945	6308
1357	F	D	E		1952	6309
1357	F	D	E		1956	6324
1357	F	D	E		1951	6357
1357	F	D	E		1947	6359
1357	F	D	E		1954	6695
1357	F	D	E		1949	6929
1357	F	D	E		1954	7020
1357	F	D	E		1954	7336
1357	F	D	E		1956	7348
1357	F	D	E		1953	8233
1357	F	D	E	N	1941	11001
1357	F	G	E	N	1949	12033
1357	F	D	E	N	1937	14744
1357	F	D	I	N	1954	16072
1357	F	D	E	N	1935	16241
1357	F	D	E	N	1940	16263
1357	F	D	E	N	1938	16306
1357	F	D	E	L	1948	19642
1357	F	D	E	N	1920	20414
1357	F	D	F	N	1955	20452
1357	F	D	I	N	1950	20956
1357	F	D	E	N	1927	21490
1357	F	D	E	N	1919	21929
1357	F	D	R	N	1960	24772
1357	F	D	R	N	1962	27694
1357	F	D	E	N	1941	28215
1357	F	D	R	N	1939	28414
1357	F	D	E	N	1952	28781
1357	F	D	E	N	1912	28785
1357	F	D	G	L	1960	28817
1357	F	D	E	N	1930	28819
1357	F	D	D	N	1959	29375
1357	F	D	G	N	1951	31815
1357	F	D	R	N	1931	31857
1357	F	S	G	N	1949	31871
1357	F	D	E	N	1932	31912
1357	F	D	R	N	1956	32423
1357	F	D	R	N	1956	32425
1357	F	D	R	N	1955	32646
1357	F	D	E	N	1962	32865
1357	F	D	R	N	1964	38609
1357	F	D	E	N	1964	38610
1357	F	D	E	N	1953	39860
1357	F	D	E	N	1933	39866
1357	F	D	E	N	1925	39868
1357	F	D	E	N	1921	40025
1357	F	D	E	N	1943	40228
1357	F	D	I	N	1966	41292
1357	F	D	R	N	1959	42504
1357	F	D	I	N	1965	43137
1357	F	D	I	N	1966	44990
1357	F	D	G	N	1968	46358
1357	F	D	E	N	1961	47930
1357	F	D	E	N	1899	48324
1357	F	D	R	N	1968	51010
1357	F	D	E	N	1969	53585
1357	F	D	E	N	1969	54423
1357	F	D	J	N	1969	55345
1357	F	D	E	N	1953	55981
1357	F	D	I	N	1970	60900
1357	F	D	E	N	1971	65523
1357	F	E	E	N	1971	65523
1357	F	G	R	N	1972	69818
1357	F	G	E	N	1972	69819
1357	L	E	G	H	1966	41478
1357	L	D	E	N	1971	61518
1357	P	D	E	N	1952	20071
1357	S	D	E	N	1937	14744
1357	S	D	E	N	1963	40741
1357	S	E	G	H	1966	41478
1357	S	D	E	N	1971	61518
1359	S	D	R		1954	8877
1360	P	D	E	N	1947	25437
1360	P	D	E	N	1968	53178
1360	P	D	E	N	1945	61037
1360	S	D	R	N	1960	20043
1360	S	D	G	N	1920	22016
1360	S	S	E	N	1970	61233

Substance Number	Phys. State	Subject	Language	Temperature	Year	TPRC Number
503-1360	S	D	R	N	1977	94615
1361	S	D	E	N	1960	16712
1361	S	D	E	N	1972	70652
1361	S	E	E	N	1972	70652
1362	S	G	R	N	1958	21016
1362	S	G	E	N	1959	21017
1363	S	D	G	N	1920	22016
1365	S	D	E	N	1963	31481
1365	S	T	E	N	1963	31481
1368	L	D	R	H	1973	72892
1369	P	D	R	N	1971	65687
1370	M	D	R	N	1968	56196
1370	M	D	E	N	1972	65531
1371	M	D	R	N	1968	56196
1371	M	D	E	N	1972	65531
1372	M	D	R	N	1968	56196
1372	M	D	E	N	1972	65531
1373	M	D	R	N	1968	56196
1373	M	D	E	N	1972	65531
1374	S	D	R	N	1970	74070
1374	S	E	R	N	1970	74070
1374	S	D	E	N	1970	74071
1374	S	E	E	N	1970	74071
1375	S	D	R	N	1970	73689
1375	S	E	R	N	1970	73689
1375	S	D	E	N	1970	73690
1375	S	E	E	N	1970	73690
1376	S	D	R	N	1970	73689
1376	S	E	R	N	1970	73689
1376	S	D	E	N	1970	73690
1376	S	E	E	N	1970	73690
1377	S	D	R	N	1973	71503
1377	S	E	R	N	1973	71503
1378	S	D	R	N	1974	74804
1378	S	E	R	N	1974	74804
1379	M	D	R	N	1960	18058
1386	S	S	E	N	1960	24931
1387	S	D	J	N	1959	23128
1387	S	D	G	N	1962	25181
1387	S	D	G	N	1930	33613
1405	S	D	R	N	1971	67588
1405	S	S	R	N	1973	79859
1405	S	S	E	N	1973	79860
1411	L	D	R	H	1961	51623
1411	L	T	R	H	1961	51623
1411	L	D	E	H	1972	66064
1411	L	T	E	H	1972	66064
1411	S	D	R	H	1961	51623
1411	S	T	R	H	1961	51623
1411	S	D	E	N	1972	66064
1411	S	T	E	N	1972	66064
1415	S	D	E	L	1965	43041
1429	S	G	E	N	1968	53618
1443	S	D	R	N	1974	97029
1451	S	D	R	N	1971	67588
1456	S	D	E	N	1975	76616
1457	S	D	E	N	1975	76616
1466	S	D	E	N	1960	10947
1466	S	E	E	N	1960	10947
1466	S	S	E	N	1960	24931
1467	M	D	G		1942	3135
1467	S	D	E		1954	2428
1467	S	D	E	H	1952	10851
1467	S	S	E	H	1960	24931
1467	S	S	E	N	1960	24931
1467	S	D	E	N	1957	28630
1467	S	S	E	H	1970	61233
1467	S	S	E	N	1970	61233
1467	S	D	E	N	1973	73921
1467	S	T	E	N	1973	73921
1468	S	D	E		1947	120
1468	S	D	E		1950	1132
1468	S	D	E		1953	7102
1468	S	D	E		1953	9404
1468	S	D	E	N	1957	23593
1468	S	S	E	N	1960	24931
1471	E	D	E	N	1965	40717
1471	S	D	E		1958	6939
1471	S	D	E		1953	7102
1471	S	E	E		1955	7333
1471	S	D	E		1953	9404
1471	S	D	E		1952	9410
1471	S	D	E	N	1954	44149
1472	L	D	G	H	1959	17892
1472	L	D	E	H	1962	29103
1472	S	D	G	N	1959	17892
1472	S	D	E	N	1962	29103
1478	S	D	R	N	1971	67588
1480	S	D	R	N	1971	67588
1481	S	D	R	N	1971	67588
1482	S	D	R	N	1971	67588
1483	S	D	R	N	1971	67588
1489	S	D	R	N	1971	67588
1490	S	D	R	N	1971	67588
1493	S	D	R	N	1971	67588
1494	S	D	R	N	1971	67588
1528	S	D	E		1958	6939
1528	S	D	E		1953	7102
1528	S	D	E	N	1952	48341
1529	S	D	E	N	1961	46148
1529	S	D	E	N	1969	52470

Substance Number	Phys. State	Subject	Language	Temperature	Year	TPRC Number
503-1529	S	T	E	N	1969	52470
1547	S	D	E		1958	7695
1554	S	D	R	N	1971	67588
1557	S	D	R	N	1971	67588
1559	S	D	R	N	1971	67588
1560	S	D	R	N	1971	67588
1561	S	D	R	N	1971	67588
1562	S	D	R	N	1975	85563
1562	S	E	R	N	1975	85563
1562	S	D	E	N	1975	85564
1562	S	E	E	N	1975	85564
1563	S	D	R	N	1971	67588
1610	S	D	F	N	1959	18033
1716	S	G	R	N	1962	31437
1720	S	D	E	L	1955	9398
1720	S	S	E	L	1970	61233
1721	M	D	E	N	1960	10052
1721	S	D	E	N	1960	10052
1721	S	D	E	N	1954	10192
1721	S	S	E	N	1960	24931
1721	S	S	E	N	1970	61233
1722	S	D	E	N	1960	15618
1722	S	G	E	N	1963	32163
1722	S	S	E	N	1970	61233
1723	L	G	R	N	1958	21016
1723	L	G	E	N	1959	21017
1723	S	G	R	N	1958	21016
1723	S	G	E	N	1959	21017
1723	S	S	E	N	1960	24931
1724	S	D	E	N	1960	15618
1724	S	G	E	N	1963	32163
1724	S	S	E	N	1970	61233
1741	L	D	E	N	1961	18321
1766	S	D	R	N	1976	88314
1766	S	D	E	N	1976	91214
1804	S	D	E	N	1976	94580
1805	S	D	E	N	1977	92443
1810	S	D	J	N	1977	94558
1811	S	D	J	N	1977	94558
1812	S	T	E	N	1800	96284
1815	S	D	R	N	1967	48703
1815	S	D	E	N	1967	49415
1826	S	D	E	H	1971	94768
1826	S	D	E	N	1971	94768
1826	S	S	E	N	1977	94788
1826	S	D	E	H	1978	97131
1826	S	D	E	N	1978	97131
1826	S	T	E	H	1978	97131
1826	S	T	E	N	1978	97131
1826	S	D	E	H	1978	97168
1826	S	D	E	N	1978	97168
1826	S	T	E	H	1978	97168
1826	S	T	E	N	1978	97168
1827	S	S	E	N	1976	94271
1828	S	D	E	N	1967	35986
1828	S	S	E	N	1976	88411
1828	S	S	E	N	1979	97587
1869	D	D	E	N	1966	78023
1869	S	D	E	N	1966	78023
1870	D	D	E	N	1966	78023
1870	D	D	E	N	1967	78350
1870	S	D	E	N	1966	78023
1912	M	D	E	N	1960	10052
1912	S	D	E	N	1960	10052
1912	S	D	E	N	1960	15618
1912	S	S	E	N	1960	24931
1912	S	G	E	N	1963	32163
1912	S	D	E	N	1970	61233
1912	S	S	E	N	1970	61233
1931	S	D	R	N	1969	58148
1933	S	D	G	N	1964	37617
1933	S	D	R	N	1969	58233
1933	S	D	E	N	1975	82040
1933	S	T	E	N	1975	82040
1934	S	D	R	N	1968	58479
1934	S	T	R	N	1968	58479
1937	L	D	R	H	1966	50710
1937	L	D	R	N	1966	50710
1938	L	D	R	H	1976	88484
1938	L	D	R	N	1976	88484
1938	L	D	E	H	1976	89373
1938	L	D	E	N	1976	89373
1939	L	D	R	H	1976	88484
1939	L	D	R	N	1976	88484
1939	L	D	E	H	1976	89373
1939	L	D	E	N	1976	89373
1945	S	S	E	H	1970	61233
1945	S	S	E	N	1970	61233
1950	S	D	R	H	1976	89429
1950	S	D	R	N	1976	89429
1950	S	D	E	H	1976	91205
1950	S	D	E	N	1976	91205
1951	S	D	R	H	1976	89429
1951	S	D	R	N	1976	89429
1951	S	D	E	H	1976	91205
1951	S	D	E	N	1976	91205
1952	S	D	E	N	1957	23779
1953	D	D	E	N	1971	61349
1953	D	D	E	N	1973	70944
1954	M	D	E		1952	6991
1954	P	D	E	N	1963	30163

Phys. State: **C.** Superconductive; **D.** Doped; **E.** Expanded; **F.** Fibrous or Whisker; **G.** Gas; **I.** Ionized or Plasma; **L.** Liquid; **M.** Multiphase; **P.** Powder or Fine Particle; **S.** Solid; **T.** Thick or Thin Film

Subject: **D.** Data; **E.** Experiment; **G.** General (Data + Theory + Experiment); **S.** Survey (Review, Compendium etc.); **T.** Theory

Language: **C.** Czech; **D.** Dutch; **E.** English; **F.** French; **G.** German; **I.** Italian; **J.** Japanese; **O.** Other Languages; **P.** Polish; **R.** Russian; **S.** Spanish

Temperature: **F.** Full Range (Low + Normal + High); **L.** Low (0 to 75K) + Overlap into Normal; **N.** Normal (75 to 1273K); **H.** High (above 1273K) + Overlap into Normal; Blank = Not Coded

Substance Number	Phys. State	Subject	Language	Temperature	Year	TPRC Number
503-1955	M	D	E		1952	6991
1955	M	S	E	N	1954	16648
1955	M	D	E	N	1958	19419
1955	M	D	G	N	1962	28811
1955	M	D	G	H	1963	30125
1955	M	D	E	N	1962	32716
1955	S	D	E	N	1961	16746
1955	S	D	E	N	1947	16842
1955	S	S	E	N	1960	24931
1955	S	D	E	N	1962	25950
1955	S	D	E	N	1964	37438
1955	S	D	E	N	1960	45513
1955	S	D	E	H	1964	45610
1955	S	D	E	N	1960	47102
1955	S	S	E	N	1970	61233
1956	S	D	E	L	1976	88203
1956	S	D	E	N	1976	88203
1956	S	E	E	L	1976	88203
1956	S	E	E	N	1976	88203
1957	S	D	E	L	1976	88203
1957	S	D	E	N	1976	88203
1957	S	E	E	L	1976	88203
1957	S	E	E	N	1976	88203
1958	S	D	E	L	1976	88203
1958	S	D	E	N	1976	88203
1958	S	E	E	L	1976	88203
1958	S	E	E	N	1976	88203
1961	S	D	E	N	1977	93507
1981	S	S	E	N	1970	61233
1987	S	D	E	N	1921	22633
1988	S	D	R	N	1977	94615
1989	S	D	E	N	1921	22633
1990	S	D	E	N	1921	22633
1993	S	D	J	H	1958	23478
1993	S	D	J	N	1958	23478
1994	P	G	E	H	1962	29477
1994	P	G	R	H	1961	29478
1996	S	D	R	N	1965	34986
1996	S	T	R	N	1965	34986
1996	S	D	E	N	1965	34987
1996	S	T	E	N	1965	34987
2000	S	D	G	H	1929	22687
2000	S	D	G	N	1929	22687
2000	S	D	G	H	1930	22688
2000	S	D	G	N	1930	22688
2000	S	D	R	H	1961	24125
2000	S	D	R	N	1961	24125
2000	S	S	E	H	1960	24931
2000	S	S	E	N	1960	24931
2000	S	D	E	N	1968	49179
2000	S	S	E	N	1970	61233
2000	S	D	R	H	1978	93881
2000	S	D	R	N	1978	93881
2000	S	D	E	H	1978	95017
2000	S	D	E	N	1978	95017
2001	S	D	E	N	1968	49179
2002	M	G	E	N	1963	30401
2002	S	D	E	N	1924	20589
2002	S	D	J	N	1951	23634
2002	S	S	E	H	1960	24931
2002	S	S	E	N	1960	24931
2002	S	D	E	N	1968	49179
2002	S	S	E	N	1970	61233
2003	S	D	G	H	1966	41479
2003	S	D	G	N	1966	41479
2007	P	D	R	H	1970	63632
2007	P	D	R	N	1970	63632
2017	S	D	R	N	1977	94615
2030	S	D	J	N	1958	23478
2031	S	D	J	N	1958	23478
2032	S	D	E	N	1969	56664
2033	S	D	E	N	1969	56664
2034	S	D	E	N	1969	56664
2035	S	D	G	H	1966	41479
2035	S	D	G	N	1966	41479
2036	S	D	G	H	1966	41479
2036	S	D	G	N	1966	41479
2037	S	D	G	H	1966	41479
2037	S	D	G	N	1966	41479
2038	S	D	G	H	1966	41479
2038	S	D	G	N	1966	41479
2039	S	D	G	H	1966	41479
2039	S	D	G	N	1966	41479
2068	S	D	F	N	1959	18033
2085	S	D	E	H	1960	10052
2085	S	D	E	N	1960	10052
2085	S	S	E	H	1960	24931
2085	S	S	E	N	1960	24931
2086	S	D	E	H	1960	10052
2086	S	D	E	N	1960	10052
2086	S	S	E	N	1960	24931
2086	S	S	E	N	1970	61233
2103	S	D	F	N	1959	18033
2104	S	D	F	N	1959	18033
2105	S	D	F	N	1959	18033
2107	S	D	F	N	1959	25490
2107	S	T	F	N	1959	25490
2108	S	D	F	N	1959	18033
2109	S	D	F	N	1959	18033
2112	S	D	F	N	1959	18033
2114	S	D	F	N	1959	18033

Substance Number	Phys. State	Subject	Language	Temperature	Year	TPRC Number
503-2119	S	D	E	N	1960	15618
2119	S	G	E	N	1963	32163
2136	S	D	G	H	1967	51075
2136	S	D	G	N	1967	51075
2137	S	D	G	H	1967	51075
2137	S	D	G	N	1967	51075
2138	S	D	G	H	1967	51075
2138	S	D	G	N	1967	51075
2139	S	D	G	H	1967	51075
2139	S	D	G	N	1967	51075
2140	S	D	G	H	1967	51075
2140	S	D	G	N	1967	51075
2141	S	D	G	H	1967	51075
2141	S	D	G	N	1967	51075
2142	S	D	J	N	1977	94558
2160	P	D	E	N	1963	45477
2160	S	D	E	N	1964	52271
2160	S	S	E	N	1970	61233
2160	S	S	E	N	1975	80596
2162	S	S	E	N	1975	80596
2163	S	S	E	N	1975	80596
2164	S	S	E	N	1975	80596
2175	S	D	R	N	1975	80319
2175	S	D	E	N	1975	81987
2189	S	S	E	N	1960	24931
2191	S	S	E	H	1970	61233
2191	S	S	E	N	1970	61233
2265	S	S	E	N	1970	61233
2274	S	S	R	N	1973	79859
2274	S	S	E	N	1973	79860
2295	S	D	E	N	1950	1459
2303	S	D	R	N	1967	48703
2303	S	D	E	N	1967	49415
2308	S	D	R	H	1977	91732
2308	S	D	R	N	1977	91732
2308	S	D	E	H	1977	96251
2308	S	D	E	N	1977	96251
2309	D	D	R	H	1977	91732
2309	D	D	R	N	1977	91732
2309	D	D	E	H	1977	96251
2309	D	D	E	N	1977	96251
2309	S	D	R	H	1961	24125
2309	S	D	R	N	1961	24125
2310	S	D	R	H	1977	91732
2310	S	D	R	N	1977	91732
2310	S	D	E	H	1977	96251
2310	S	D	E	N	1977	96251
2330	S	D	R	H	1976	89883
2330	S	D	R	N	1976	89883
2330	S	D	E	H	1976	89884
2330	S	D	E	N	1976	89884
2332	S	D	R	N	1975	82103
2333	S	D	G	H	1960	20925
2333	S	D	J	N	1958	21141
2333	S	D	G	N	1930	33613
2333	S	S	E	N	1970	61233
2333	S	D	R	N	1975	82103
2333	S	D	E	N	1975	82185
2334	S	D	R	N	1975	82103
2335	S	D	R	N	1975	82103
2336	S	D	R	N	1975	82103
2337	S	D	R	N	1975	82103
2349	S	S	E	N	1978	99742
2350	S	D	E	N	1979	98803
2350	S	E	E	N	1979	98803
2352	S	D	E	N	1979	98803
2352	S	E	E	N	1979	98803
2365	S	D	E	N	1979	97889
2366	S	D	R	N	1978	93254
2366	S	D	E	N	1978	98018
2367	S	D	R	N	1978	93254
2367	S	D	E	N	1978	98018
2368	S	D	R	N	1978	93254
2368	S	D	E	N	1978	98018
2369	S	D	R	N	1978	93254
2369	S	D	E	N	1978	98018
2370	S	D	R	N	1978	93254
2370	S	D	E	N	1978	98018
2371	S	D	E	N	1979	98240
2371	S	E	E	N	1979	98240
2373	S	D	R	N	1978	93218
2373	S	D	E	N	1978	98789
2374	S	D	R	N	1978	93218
2374	S	D	E	N	1978	98789
2379	S	D	G	H	1932	14367
2379	S	T	G	H	1932	14367
2379	S	D	E	H	1927	20929
2379	S	D	G	H	1929	22687
2379	S	D	G	N	1929	22687
2379	S	D	G	H	1930	22688
2379	S	D	G	N	1930	22688
2380	S	D	R	N	1960	19081
2380	S	D	E	N	1960	32629
2381	S	D	R	H	1958	19078
2399	S	D	G	N	1894	48020
2400	S	G	E	N	1963	32163
2401	S	G	E	N	1963	32163
2402	S	G	E	N	1963	32163
2403	S	G	E	N	1963	32163
2404	S	D	F	N	1959	25490
2404	S	T	F	N	1959	25490

Substance Number	Phys. State	Subject	Language	Temperature	Year	TPRC Number
503-2405	L	G	R	N	1958	21016
2405	L	G	E	N	1959	21017
2405	S	G	R	N	1958	21016
2405	S	G	E	N	1959	21017
2411	S	D	E	N	1960	15618
2411	S	G	E	N	1963	32163
2412	S	D	E	N	1960	15618
2412	S	G	E	N	1963	32163
2412	S	S	E	N	1970	61233
2413	S	D	E	N	1960	15618
2413	S	G	E	N	1963	32163
2413	S	S	E	N	1970	61233
2427	S	S	E	N	1960	24931
2428	S	S	E	N	1960	24931
2432	S	D	R	N	1967	48703
2432	S	D	E	N	1967	49415
2433	S	D	E	N	1961	24763
2434	S	D	J	N	1951	23634
2434	S	D	E	N	1961	24763
2434	S	D	E	H	1962	31014
2434	S	D	E	N	1962	31014
2436	S	D	J	N	1951	23634
2437	S	D	E	N	1952	247
2437	S	E	E	N	1952	247
2437	S	D	E	N	1924	20589
2437	S	D	J	N	1958	21141
2437	S	D	G	H	1929	22687
2437	S	D	G	N	1929	22687
2437	S	D	J	N	1951	23634
2437	S	D	G	N	1962	25181
2437	S	D	G	N	1930	33613
2437	S	S	E	H	1970	61233
2437	S	S	E	N	1970	61233
2437	S	D	E	N	1975	82185
2438	M	G	G	N	1963	32596
2438	S	D	E	N	1924	20589
2438	S	D	J	N	1951	23634
2438	S	S	E	H	1970	61233
2438	S	S	E	N	1970	61233
2439	S	D	G	N	1956	12170
2440	S	D	R	N	1967	54757
2447	S	D	E	N	1960	15618
2447	S	G	E	N	1963	32163
2447	S	S	E	N	1970	61233
2449	S	G	E	N	1963	32163
2453	S	D	G	N	1932	18962
2453	S	D	G	H	1929	22687
2453	S	D	G	N	1929	22687
2453	S	D	G	H	1930	22688
2453	S	D	G	N	1930	22688
2454	S	D	J	H	1960	19008
2455	S	D	J	H	1960	19008
2456	S	D	R	H	1958	19078
2458	L	G	R	H	1958	21016
2458	L	G	E	H	1959	21017
2458	S	G	R	H	1958	21016
2458	S	G	R	N	1958	21016
2458	S	G	E	H	1959	21017
2458	S	G	E	N	1959	21017
2461	P	D	E	N	1921	22633
2461	S	D	E	N	1921	22633
2461	S	D	G	H	1930	22688
2461	S	D	G	N	1930	22688
2461	S	D	J	N	1951	23634
2461	S	S	E	H	1960	24931
2461	S	S	E	N	1960	24931
2461	S	D	G	N	1930	33613
2463	S	S	E	N	1960	24931
2468	S	D	E	N	1924	20589
2469	S	D	R	H	1959	20705
2470	S	D	R	H	1959	20705
2471	S	D	R	H	1959	20705
2472	S	D	R	H	1959	20705
2476	L	G	R	H	1958	21016
2476	L	G	E	H	1959	21017
2476	S	G	R	H	1958	21016
2476	S	G	R	N	1958	21016
2476	S	G	E	H	1959	21017
2476	S	G	E	N	1959	21017
2477	L	G	R	H	1958	21016
2477	L	G	E	H	1959	21017
2477	S	G	R	H	1958	21016
2477	S	G	R	N	1958	21016
2477	S	G	E	H	1959	21017
2477	S	G	E	N	1959	21017
2478	P	D	E	N	1921	22633
2478	S	D	E	N	1921	22633
2479	S	D	E	N	1921	22633
2480	S	D	G	H	1929	22687
2480	S	D	G	N	1929	22687
2481	S	D	G	H	1929	22687
2481	S	D	G	N	1929	22687
2482	S	D	G	H	1929	22687
2482	S	D	G	N	1929	22687
2482	S	D	G	H	1930	22688
2482	S	D	G	N	1930	22688
2484	S	D	G	H	1930	22688
2484	S	D	G	N	1930	22688
2484	S	D	J	N	1951	23634
2485	S	D	J	N	1951	23634
2485	S	D	G	H	1962	27692

Phys. State: **C.** Superconductive; **D.** Doped; **E.** Expanded; **F.** Fibrous or Whisker; **G.** Gas; **I.** Ionized or Plasma; **L.** Liquid; **M.** Multiphase; **P.** Powder or Fine Particle; **S.** Solid; **T.** Thick or Thin Film
Subject: **D.** Data; **E.** Experiment; **G.** General (Data + Theory + Experiment); **S.** Survey (Review, Compendium etc.); **T.** Theory
Language: **C.** Czech; **D.** Dutch; **E.** English; **F.** French; **G.** German; **I.** Italian; **J.** Japanese; **O.** Other Languages; **P.** Polish; **R.** Russian; **S.** Spanish
Temperature: **F.** Full Range (Low + Normal + High); **L.** Low (0 to 75K) + Overlap into Normal; **N.** Normal (75 to 1273K); **H.** High (above 1273K) + Overlap into Normal; Blank = Not Coded

Substance Number	Phys. State	Subject	Language	Temperature	Year	TPRC Number
503-2485	S	D	G	N	1962	27692
2485	S	D	R	H	1968	54686
2485	S	D	E	H	1968	54687
2486	S	S	E	N	1970	61233
2488	S	D	R	H	1961	24125
2488	S	D	R	N	1961	24125
2489	S	D	E	N	1961	24763
2489	S	D	E	H	1962	31014
2489	S	D	E	N	1962	31014
2490	S	D	E	N	1961	24763
2491	S	D	G	N	1962	25181
2492	S	D	G	H	1962	27692
2492	S	D	G	N	1962	27692
2493	S	D	G	H	1962	27692
2493	S	D	G	N	1962	27692
2494	S	D	G	H	1962	27692
2494	S	D	G	N	1962	27692
2495	S	D	G	H	1962	27692
2495	S	D	G	N	1962	27692
2496	S	D	R	H	1956	28492
2496	S	D	R	N	1956	28492
2497	S	D	R	H	1956	28492
2497	S	D	R	N	1956	28492
2498	S	D	R	H	1956	28492
2498	S	D	R	N	1956	28492
2499	S	D	R	H	1956	28492
2499	S	D	R	N	1956	28492
2500	S	D	R	H	1947	622
2500	S	D	R	N	1947	622
2500	S	D	E	H	1962	29111
2500	S	D	E	N	1962	29111
2500	S	G	R	N	1962	31437
2514	S	D	E	H	1962	31014
2514	S	D	E	N	1962	31014
2522	M	G	G	N	1963	32596
2586	S	D	R	N	1966	46593
2586	S	D	E	N	1966	46594
2589	S	D	R	N	1967	48703
2589	S	D	E	N	1967	49415
2590	S	D	R	N	1967	48703
2590	S	D	E	N	1967	49415
2630	S	D	R	H	1970	60839
2630	S	D	R	N	1970	60839
2630	S	D	E	H	1970	60840
2630	S	D	E	N	1970	60840
2630	S	D	E	N	1975	82185
2659	S	D	R	H	1977	90440
2659	S	D	R	N	1977	90440
2659	S	D	E	H	1977	95993
2659	S	D	E	N	1977	95993
2664	S	D	R	N	1975	80319
2664	S	D	E	N	1975	81987
2665	S	D	R	N	1975	80319
2665	S	D	E	N	1975	81987
2666	S	D	R	N	1975	80319
2666	S	D	E	N	1975	81987
2667	S	D	R	N	1975	80319
2667	S	D	E	N	1975	81987
2668	S	D	E	N	1975	82040
2668	S	T	E	N	1975	82040
2669	S	D	E	N	1975	82040
2669	S	T	E	N	1975	82040
2680	S	D	R	N	1973	80006
2681	S	D	R	N	1973	80006
2682	S	S	E	N	1970	61233
2683	S	S	E	N	1970	61233
2684	S	S	E	N	1960	24931
2684	S	S	E	N	1970	61233
2685	S	S	E	N	1970	61233
2686	S	S	E	N	1970	61233
2687	S	S	E	N	1970	61233
2688	S	S	E	N	1970	61233
2689	S	S	E	N	1960	24931
2689	S	S	E	N	1970	61233
2690	S	S	E	N	1960	24931
2690	S	S	E	N	1970	61233
2691	S	S	E	N	1960	24931
2691	S	S	E	N	1970	61233
2692	S	S	E	N	1970	61233
2693	S	S	E	N	1970	61233
2694	S	S	E	N	1970	61233
2695	S	S	E	N	1960	24931
2695	S	S	E	N	1970	61233
2696	S	S	E	N	1970	61233
2697	S	S	E	N	1970	61233
2698	S	S	E	N	1970	61233
2699	S	S	E	N	1970	61233
2700	S	S	E	N	1970	61233
2701	S	S	E	H	1970	61233
2701	S	S	E	N	1970	61233
2702	S	S	E	N	1960	24931
2702	S	S	E	N	1970	61233
2703	S	S	E	N	1970	61233
2704	S	S	E	N	1970	61233
2705	S	S	E	N	1970	61233
2706	S	S	E	N	1970	61233
2707	S	S	E	N	1960	24931
2707	S	S	E	N	1970	61233
2708	S	S	E	N	1960	24931
2708	S	S	E	N	1970	61233
2709	S	S	E	N	1960	24931

Substance Number	Phys. State	Subject	Language	Temperature	Year	TPRC Number
503-2709	S	S	E	N	1970	61233
2710	S	S	E	N	1960	24931
2710	S	S	E	N	1970	61233
2711	S	S	E	N	1960	24931
2711	S	S	E	N	1970	61233
2712	S	S	E	N	1960	24931
2712	S	S	E	N	1970	61233
2713	S	S	E	N	1960	24931
2713	S	S	E	N	1970	61233
2714	S	S	E	N	1960	24931
2714	S	S	E	N	1970	61233
2715	S	S	E	N	1960	24931
2715	S	S	E	N	1970	61233
2716	S	S	E	N	1960	24931
2716	S	S	E	N	1970	61233
2717	S	S	E	N	1960	24931
2717	S	S	E	N	1970	61233
2718	S	S	E	N	1960	24931
2718	S	S	E	N	1970	61233
2719	S	S	E	N	1970	61233
2721	S	S	E	N	1970	61233
2726	E	D	G	N	1955	1125
2726	E	E	G	N	1955	1125
2726	E	S	E	N	1970	61233
2731	S	S	E	H	1970	61233
2731	S	S	E	N	1970	61233
2732	S	S	E	N	1970	61233
2733	S	S	E	N	1970	61233
2734	S	S	E	N	1970	61233
2735	S	S	E	H	1970	61233
2735	S	S	E	N	1970	61233
2736	S	S	E	N	1970	61233
2737	S	S	E	H	1970	61233
2737	S	S	E	N	1970	61233
2738	S	S	E	N	1970	61233
2739	S	S	E	N	1970	61233
2740	S	S	E	N	1970	61233
2741	S	S	E	N	1970	61233
2742	S	S	E	N	1970	61233
2743	S	S	E	N	1970	61233
2744	S	S	E	N	1970	61233
2745	S	S	E	N	1970	61233
2746	S	S	E	N	1970	61233
2747	S	S	E	N	1970	61233
2748	S	S	E	H	1970	61233
2748	S	S	E	N	1970	61233
2749	S	S	E	H	1970	61233
2749	S	S	E	N	1970	61233
2750	S	S	E	H	1970	61233
2750	S	S	E	N	1970	61233
2806	S	D	J	N	1951	23634
2807	S	D	E	H	1962	31014
2807	S	D	E	N	1962	31014
2808	S	D	R	N	1967	54757
2809	S	D	R	N	1967	54757
2861	S	D	J	N	1959	23128
2878	S	S	E	N	1960	24931
2885	S	S	E	H	1960	24931
2885	S	S	E	N	1960	24931
2886	S	S	E	H	1960	24931
2886	S	S	E	N	1960	24931
2979	S	S	E	N	1960	24931
3000	S	S	E	H	1970	61233
3000	S	S	E	N	1970	61233
8195	S	D	G	N	1947	1822
8196	S	D	R	N	1967	54757
8201	S	D	R	N	1972	68003
8214	S	D	R	N	1969	54745
8214	S	D	E	N	1969	66786
8215	S	D	R	N	1969	54745
8215	S	D	E	N	1969	66786
8223	S	D	R	N	1970	62314
8224	S	D	R	N	1970	62314
8318	S	D	E	N	1970	63010
8318	S	E	E	N	1970	63010
8334	L	D	E	N	1947	48002
8334	P	D	E	N	1957	18607
8334	S	D	R	H	1968	54773
8334	S	D	E	H	1968	54774
8334	S	D	R	N	1970	62314
8338	S	S	E	N	1960	24931
8363	S	D	J	N	1959	23128
8364	S	D	J	N	1959	23128
8369	S	D	I	N	1939	20657
8369	S	D	I	N	1955	20975
8370	S	D	I	N	1939	20657
8370	S	D	I	N	1955	20975
8371	S	D	I	N	1939	20657
8371	S	D	E	H	1927	20929
8371	S	D	I	N	1955	20975
8371	S	D	G	N	1918	24323
8371	S	D	E	N	1915	25767
8371	S	G	E	H	1915	31868
8372	S	D	E	H	1927	20929
8373	S	D	E	H	1927	20929
8374	S	D	G	N	1930	33613
8374	S	D	R	H	1969	58852
8374	S	D	E	H	1969	58853
8376	S	D	E	N	1960	24824
8376	S	G	E	N	1959	31244
8377	S	D	E	N	1960	24824

Substance Number	Phys. State	Subject	Language	Temperature	Year	TPRC Number
503-8377	S	G	E	N	1959	31244
8379	S	D	R	N	1966	46587
8379	S	D	E	N	1966	46588
8381	S	D	E	H	1965	37802
8383	M	D	E	H	1964	26233
8383	S	D	E	H	1965	38404
8383	S	D	E	H	1964	49249
8385	S	D	E	H	1965	37802
8389	S	D	E	N	1959	13944
8390	S	D	G	N	1930	33613
8391	S	D	E	N	1940	30989
8392	P	G	E	H	1962	29477
8392	P	G	R	H	1961	29478
8392	S	G	E	H	1962	29477
8392	S	G	R	H	1961	29478
8393	S	D	E	H	1960	10052
8395	S	D	G		1949	1972
8395	S	D	R		1945	8886
8396	S	D	J	N	1960	20659
8397	S	D	E		1958	6939
8402	S	D	J	N	1966	41630
8406	F	D	R	N	1967	46271
8406	F	D	E	N	1967	46272
8416	E	D	E	N	1963	25157
8418	S	D	G	H	1932	14367
8418	S	T	G	H	1932	14367
8485	S	D	E	N	1965	34089
8486	S	D	E	N	1965	34089
8487	S	D	E	N	1965	34089
8488	S	D	E	N	1965	34089
8515	S	D	G	N	1893	50995
8515	S	T	G	N	1893	50995
8516	S	D	G	N	1893	50995
8516	S	T	G	N	1893	50995
8517	S	D	G	N	1893	50995
8517	S	T	G	N	1893	50995
8518	S	D	G	N	1893	50995
8518	S	T	G	N	1893	50995
8520	S	D	G	N	1893	50995
8520	S	T	G	N	1893	50995
8527	S	G	E	L	1971	66152
8566	S	D	E	N	1963	49933
8603	P	D	F	N	1968	37370
8604	P	D	F	N	1968	37370
8612	S	D	E	N	1965	34089
8659	S	D	R	N	1964	36398
8659	S	D	E	N	1964	36399
8660	S	D	E	N	1964	36017
8710	S	D	E	N	1960	49210
8711	S	D	E	N	1960	49210
8742	S	D	G	N	1893	50995
8742	S	T	G	N	1893	50995
8754	S	D	E	N	1963	37530
8755	S	D	E	N	1963	37530
8818	S	D	J	N	1966	41630
8822	S	D	J	N	1966	41630
8822	S	D	E	N	1960	76856
8832	S	D	G	N	1894	48020
8833	S	D	G	N	1894	48020
8833	S	D	E	N	1960	49210
8834	S	D	G	N	1894	48020
8836	S	D	G	N	1894	48020
8837	S	D	G	N	1894	48020
8838	S	D	G	N	1894	48020
8839	S	D	G	N	1894	48020
8874	S	D	R	N	1972	70625
8874	S	D	E	N	1972	70626
8876	S	D	O		1947	1413
8882	S	D	F	N	1970	58498
8914	S	D	E	N	1965	34089
8940	S	D	R	N	1972	70625
8940	S	D	E	N	1972	70626
8987	S	D	J	N	1966	41630
9042	L	G	E	H	1959	29365
9044	S	D	E	N	1956	19964
9045	S	D	E	N	1956	19964
9046	S	D	E	N	1956	19964
9047	S	D	E	N	1956	19964
9056	L	G	G	H	1959	24403
9057	L	G	G	H	1959	24403
9068	S	D	E	N	1963	37530
9091	S	D	G	N	1894	48020
9091	S	D	E	N	1963	49933
9095	L	D	O	N	1973	74859
9095	S	D	G	N	1928	20567
9095	S	D	G	N	1894	48020
9095	S	D	E	N	1960	49210
9095	S	D	O	N	1973	74859
9097	S	D	E	N	1960	49210
9099	S	D	G	N	1893	50995
9099	S	T	G	N	1893	50995
9102	S	D	E	N	1960	49210
9108	S	D	G	N	1893	50995
9108	S	T	G	N	1893	50995
9115	S	D	G	N	1894	48020
9115	S	D	G	N	1969	55722
9123	S	D	G	N	1893	50995
9123	S	T	G	N	1893	50995
9131	S	D	G	N	1928	20567
9131	S	D	G	N	1894	48020
9131	S	D	G	N	1893	50995

Phys. State: **C.** Superconductive; **D.** Doped; **E.** Expanded; **F.** Fibrous or Whisker; **G.** Gas; **I.** Ionized or Plasma; **L.** Liquid; **M.** Multiphase; **P.** Powder or Fine Particle; **S.** Solid; **T.** Thick or Thin Film
Subject: **D.** Data; **E.** Experiment; **G.** General (Data + Theory + Experiment); **S.** Survey (Review, Compendium etc.); **T.** Theory
Language: **C.** Czech; **D.** Dutch; **E.** English; **F.** French; **G.** German; **I.** Italian; **J.** Japanese; **O.** Other Languages; **P.** Polish; **R.** Russian; **S.** Spanish
Temperature: **F.** Full Range (Low + Normal + High); **L.** Low (0 to 75K) + Overlap into Normal; **N.** Normal (75 to 1273K); **H.** High (above 1273K) + Overlap into Normal; Blank = Not Coded

Substance Number	Phys. State	Subject	Language	Temperature	Year	TPRC Number
503-9131	S	T	G	N	1893	50995
9139	S	D	G	N	1928	20567
9147	S	D	E		1953	7102
9147	S	D	E	N	1952	48341
9148	S	D	E		1958	6939
9148	S	D	E		1953	7102
9148	S	D	E	N	1952	48341
9152	S	D	E	N	1965	34089
9152	S	D	J	N	1966	41630
9152	S	D	G	N	1894	48020
9152	S	D	E	N	1960	49210
9152	S	D	G	N	1893	50995
9152	S	T	G	N	1893	50995
9156	S	D	G	N	1928	20567
9157	S	D	G	N	1928	20567
9158	S	D	G	N	1928	20567
9158	S	D	G	N	1893	50995
9158	S	T	G	N	1893	50995
9160	S	G	E	N	1932	16179
9160	S	D	G	N	1894	48020
9163	S	D	E		1951	810
9163	S	D	I		1951	6752
9163	S	T	I		1951	6752
9163	S	D	G	N	1928	20567
9163	S	D	G	N	1893	50995
9163	S	T	G	N	1893	50995
9168	S	D	G	N	1893	50995
9168	S	T	G	N	1893	50995
9168	S	D	G	N	1970	66394
9168	S	E	G	N	1970	66394
9171	S	D	G	N	1928	20567
9171	S	D	G	N	1893	50995
9171	S	T	G	N	1893	50995
9173	S	D	E		1943	1287
9173	S	D	E	N	1955	20062
9173	S	D	G	N	1928	20567
9173	S	D	G	N	1893	50995
9173	S	T	G	N	1893	50995
9173	S	D	G	N	1970	66394
9173	S	E	G	N	1970	66394
9173	S	D	E	L	1972	67775
9174	S	G	G		1925	6706
9174	S	G	E		1956	9943
9174	S	D	E	L	1972	67569
9175	S	D	G	N	1893	50995
9175	S	T	G	N	1893	50995
9178	S	D	G	N	1969	55722
9179	S	D	G	N	1893	50995
9179	S	T	G	N	1893	50995
9180	S	D	E		1954	206
9181	S	G	G		1952	145
9181	S	D	E		1958	6939
9181	S	D	E		1953	7102
9181	S	D	E		1957	7288
9181	S	G	E		1956	9943
9181	S	D	J	N	1966	41630
9182	S	D	E	N	1946	3081
9183	S	D	E	N	1946	3081
9184	S	D	E	N	1946	3081
9185	S	D	E	N	1946	3081
9186	S	D	E	N	1946	3081
9199	L	D	R	H	1973	59198
9199	L	D	E	H	1970	62295
9212	S	D	R	N	1973	72260
9212	S	D	E	N	1973	72261
9213	S	D	R	N	1973	72260
9213	S	D	E	N	1973	72261
9214	S	D	R	N	1973	72260
9214	S	D	E	N	1973	72261
9264	S	D	R	N	1971	64268
9264	S	D	E	N	1971	72874
9297	S	D	R	N	1963	35746
9297	S	E	R	N	1963	35746
9306	S	D	E	N	1971	62030
9307	S	D	E	N	1971	62030
9308	S	D	E	N	1971	62030
9326	S	D	R	N	1963	35746
9326	S	E	R	N	1963	35746
9332	S	D	E	N	1946	3081
9356	S	D	R	N	1965	34098
9356	S	D	E	N	1965	35752
9357	S	D	R	N	1965	34098
9357	S	D	E	N	1965	35752
9357	S	D	R	N	1971	67665
9357	S	T	R	N	1971	67665
9363	S	D	E	N	1971	62030
9378	F	D	E	N	1955	23881
9378	F	D	E	H	1961	28858
9378	F	D	E	H	1964	45639
9393	S	D	R	N	1965	34098
9393	S	D	E	N	1965	35752
9393	S	D	E	N	1967	36035
9393	S	D	R	N	1967	48739
9394	S	D	E	N	1970	62655
9394	S	D	E	N	1970	62656
9402	S	D	E	N	1968	53133
9407	S	D	E	L	1976	88203
9407	S	D	E	N	1976	88203
9407	S	E	E	L	1976	88203
9407	S	E	E	N	1976	88203
9415	S	D	E	N	1970	61783

Substance Number	Phys. State	Subject	Language	Temperature	Year	TPRC Number
503-9430	F	D	E	N	1972	71449
9431	S	D	R	H	1970	59471
9431	S	D	E	H	1970	61167
9449	S	D	R	H	1964	38013
9450	S	D	R	H	1964	38013
9453	S	D	R	H	1969	58854
9453	S	D	E	H	1969	58855
9456	S	D	R	H	1969	57263
9456	S	D	E	H	1969	57264
9456	S	D	E	N	1971	62030
9458	S	D	E	N	1971	62030
9463	S	D	R	H	1968	56210
9463	S	D	E	H	1969	56605
9481	F	D	E	N	1963	37349
9483	S	D	R	H	1969	57271
9483	S	D	E	H	1969	57272
9498	S	D	R	N	1966	44790
9498	S	D	R	N	1968	55793
9498	S	T	R	N	1968	55793
9498	S	D	E	N	1969	56586
9498	S	T	E	N	1969	56586
9499	S	D	R	N	1966	44790
9499	S	D	R	N	1968	55793
9499	S	T	R	N	1968	55793
9499	S	D	E	N	1969	56586
9499	S	T	E	N	1969	56586
9508	S	D	O	N	1968	53266
9510	S	D	R	N	1966	44598
9510	S	D	E	N	1969	51032
9512	S	D	E	H	1966	41579
9513	S	D	E	H	1966	41579
9514	S	D	R	N	1966	49684
9514	S	T	R	N	1966	49684
9515	S	D	R	N	1966	49684
9515	S	T	R	N	1966	49684
9515	S	D	R	N	1966	51043
9515	S	T	R	N	1966	51043
9515	S	D	R	N	1969	57098
9515	S	D	E	N	1969	57099
9516	S	D	R	N	1966	49684
9516	S	T	R	N	1966	49684
9517	S	D	R	N	1966	49684
9517	S	T	R	N	1966	49684
9517	S	D	R	N	1966	51043
9517	S	T	R	N	1966	51043
9517	S	D	R	N	1969	57098
9517	S	D	E	N	1969	57099
9518	S	D	R	N	1966	49684
9518	S	T	R	N	1966	49684
9533	S	D	G	H	1966	45267
9534	S	D	E	N	1968	50485
9541	S	D	R	H	1969	57587
9554	S	D	O	H	1971	71516
9560	S	D	R	N	1968	53442
9560	S	D	E	N	1968	54102
9561	S	D	R	N	1968	53442
9561	S	D	E	N	1968	54102
9574	S	D	E	N	1959	24048
9596	S	D	E	N	1929	40826
9611	S	D	R	H	1970	60837
9611	S	D	E	H	1970	60838
9624	S	D	R	N	1969	59079
9624	S	T	R	N	1969	59079
9624	S	D	E	N	1969	59080
9624	S	T	E	N	1969	59080
9626	S	D	E	N	1967	46877
9626	S	D	E	N	1966	53877
9627	S	D	R	N	1965	43289
9627	S	T	R	N	1965	43289
9633	S	D	O	N	1966	43203
9634	S	D	E	N	1939	47960
9644	S	D	G	N	1968	36520
9644	S	D	E	N	1967	46294
9644	S	T	E	N	1967	46294
9644	S	D	G	N	1969	52808
9646	S	G	E	L	1967	47036
9646	S	D	E	L	1968	53424
9646	S	T	E	L	1968	53424
9666	S	D	E	L	1967	48890
9666	S	T	E	L	1967	48890
9666	S	D	R	N	1966	49684
9666	S	T	R	N	1966	49684
9666	S	D	R	N	1968	55793
9666	S	T	R	N	1968	55793
9666	S	D	E	N	1969	56586
9666	S	T	E	N	1969	56586
9678	S	D	R	N	1970	61097
9678	S	D	E	N	1970	67215
9679	S	D	R	N	1967	47476
9679	S	D	E	N	1968	49405
9680	S	D	R	N	1967	47476
9680	S	D	E	N	1968	49405
9681	S	D	J	N	1960	20659
9684	S	D	E	H	1909	39859
9688	S	D	R	N	1966	46597
9688	S	D	E	N	1966	46598
9694	S	D	R	N	1968	55793
9694	S	T	R	N	1968	55793
9694	S	D	E	N	1969	56586
9694	S	T	E	N	1969	56586
9694	S	D	R	N	1969	57098

Substance Number	Phys. State	Subject	Language	Temperature	Year	TPRC Number
503-9694	S	D	E	N	1969	57099
9695	S	D	R	N	1963	35746
9695	S	E	R	N	1963	35746
9695	S	D	R	N	1966	44790
9695	S	D	R	N	1968	55793
9695	S	T	R	N	1968	55793
9695	S	D	E	N	1969	56586
9695	S	T	E	N	1969	56586
9696	S	D	R	H	1966	42537
9696	S	D	E	N	1966	53877
9697	P	D	E	N	1969	56578
9697	P	D	R	N	1968	61217
9697	P	E	R	N	1968	61217
9697	S	D	R	H	1966	42537
9697	S	D	E	N	1967	46877
9697	S	D	R	N	1969	53653
9704	F	G	E	H	1967	43744
9704	F	D	E	H	1965	45121
9704	F	E	E	H	1965	45121
9704	F	D	E	H	1967	46648
9704	F	D	E	N	1967	46648
9704	F	D	G	N	1973	75986
9704	F	T	G	N	1973	75986
9704	S	D	E	H	1966	39358
9704	S	G	E	H	1967	43744
9704	S	D	E	H	1967	46648
9704	S	D	E	H	1966	47098
9704	S	D	E	H	1966	49761
9707	S	D	R	H	1970	61170
9707	S	D	E	H	1970	61171
9709	S	D	E	N	1961	23542
9710	S	D	J	N	1960	20659
9712	S	D	E	H	1966	35917
9712	S	D	E	H	1967	39997
9712	S	D	E	H	1965	45297
9712	S	G	E	H	1967	46952
9712	S	D	E	N	1962	48904
9712	S	D	E	H	1968	54697
9712	S	D	E	N	1968	54697
9712	S	D	E	N	1968	54698
9716	S	D	E	H	1961	29231
9716	S	D	E	H	1964	49249
9716	S	D	E	N	1966	49739
9716	S	D	F	H	1970	60595
9716	S	D	F	H	1970	61396
9716	S	E	F	H	1970	61396
9718	M	D	E	H	1963	22176
9718	M	T	E	H	1963	22176
9718	S	D	E	H	1963	33409
9718	S	D	E	H	1965	42731
9718	S	G	E	H	1967	46952
9718	S	D	E	N	1969	52844
9718	S	D	E	H	1969	53591
9718	S	D	E	N	1969	54478
9718	S	D	E	H	1969	55204
9718	S	D	E	H	1969	56069
9718	S	D	E	H	1968	57194
9718	S	D	F	H	1968	57457
9718	S	D	E	H	1970	60316
9718	S	D	E	N	1970	61913
9718	S	T	E	N	1970	61913
9718	S	D	E	H	1970	65535
9718	S	D	E	H	1972	71915
9718	S	T	E	H	1972	71915
9718	S	S	S	H	1973	74958
9720	S	D	E	N	1960	28670
9727	M	D	E	N	1962	25950
9727	S	D	E	N	1962	26552
9728	S	D	R	N	1964	35347
9728	S	D	E	N	1965	35348
9728	S	D	R	H	1965	42464
9728	S	D	R	N	1966	49684
9728	S	T	R	N	1966	49684
9728	S	D	R	L	1971	64309
9728	S	D	E	L	1971	64310
9732	S	D	E	N	1954	25427
9733	S	D	E	L	1962	23210
9733	S	D	F	L	1962	27738
9736	S	D	E	H	1962	27640
9736	S	D	R	H	1968	53620
9736	S	D	E	H	1968	53621
9737	S	E	E	N	1962	25588
9740	S	D	R	N	1961	25827
9741	S	D	R	N	1961	25827
9742	S	D	R	N	1961	25827
9748	S	D	E	N	1960	10052
9748	S	D	E	N	1950	14673
9748	S	D	E	H	1960	18644
9748	S	D	E	L	1962	23210
9748	S	D	F	L	1962	27738
9748	S	D	R	N	1966	44369
9750	M	D	E	N	1958	10458
9750	S	D	E	N	1946	3081
9750	S	D	E		1935	8241
9750	S	D	E	N	1959	13944
9750	S	D	E	N	1955	15754
9750	S	D	E	N	1955	15755
9750	S	D	E	N	1960	16036
9750	S	D	E	N	1960	16639
9750	S	D	E	N	1955	20062
9750	S	G	E	N	1961	29253

Phys. State: **C.** Superconductive; **D.** Doped; **E.** Expanded; **F.** Fibrous or Whisker; **G.** Gas; **I.** Ionized or Plasma; **L.** Liquid; **M.** Multiphase; **P.** Powder or Fine Particle; **S.** Solid; **T.** Thick or Thin Film

Subject: **D.** Data; **E.** Experiment; **G.** General (Data + Theory + Experiment); **S.** Survey (Review, Compendium etc.); **T.** Theory

Language: **C.** Czech; **D.** Dutch; **E.** English; **F.** French; **G.** German; **I.** Italian; **J.** Japanese; **O.** Other Languages; **P.** Polish; **R.** Russian; **S.** Spanish

Temperature: **F.** Full Range (Low + Normal + High); **L.** Low (0 to 75K) + Overlap into Normal; **N.** Normal (75 to 1273K); **H.** High (above 1273K) + Overlap into Normal; Blank = Not Coded

Substance Number	Phys. State	Subject	Language	Temperature	Year	TPRC Number
503-9750	S	D	E	N	1949	30225
9750	S	D	R	N	1965	34784
9750	S	D	E	N	1966	34785
9750	S	D	F	H	1965	42421
9750	S	E	F	H	1965	42421
9750	S	D	E	N	1959	43731
9750	S	D	R	N	1966	44598
9750	S	D	E	N	1969	45285
9750	S	D	E	N	1969	51032
9750	S	D	E	N	1955	52252
9750	S	D	E	N	1970	56455
9750	S	D	R	N	1969	59079
9750	S	T	R	N	1969	59079
9750	S	D	E	N	1969	59080
9750	S	T	E	N	1969	59080
9750	S	D	R	H	1971	64844
9750	S	D	E	H	1971	64845
9751	S	D	E	N	1961	24937
9751	S	D	R	N	1959	24938
9760	S	D	R	H	1969	58852
9760	S	D	E	H	1969	58853
9760	S	D	R	N	1971	64345
9760	S	D	E	N	1971	64850
9762	S	D	G	H	1932	14367
9762	S	T	G	H	1932	14367
9762	S	D	E	H	1909	39859
9762	S	D	R	N	1966	46597
9762	S	D	E	N	1966	46598
9765	S	D	R	N	1961	23796
9765	S	D	E	N	1961	25307
9768	S	D	G	N	1918	24323
9769	M	D	E	N	1960	22625
9771	S	D	G	N	1930	33613
9771	S	D	E	H	1909	39859
9771	S	D	R	N	1968	47851
9771	S	D	E	N	1968	47852
9772	S	G	E	N	1961	29253
9773	P	D	E	N	1921	22633
9773	S	D	E	N	1921	22633
9783	M	D	E	N	1962	25950
9783	S	D	E	N	1960	20510
9784	S	D	E	H	1921	20009
9785	S	D	E	N	1963	20313
9789	F	D	E	N	1963	37349
9791	E	D	E	N	1969	37570
9791	E	D	E	N	1969	55028
9791	E	D	E	N	1968	57191
9791	S	D	E	N	1949	10272
9791	S	D	E	N	1959	10757
9791	S	D	E	N	1955	15754
9791	S	D	E	N	1955	15755
9791	S	D	E	N	1960	16639
9791	S	D	G	N	1961	24393
9791	S	D	E	N	1962	24680
9791	S	D	E	N	1968	48434
9792	S	D	E	N	1921	22633
9793	P	D	R	H	1961	26561
9793	P	D	E	H	1961	27771
9793	S	D	R	N	1958	19538
9795	E	D	E	N	1969	37570
9795	E	D	E	N	1969	55028
9795	E	D	E	N	1968	57191
9795	S	D	E	N	1939	47960
9795	S	D	E	N	1968	48434
9801	S	D	E	N	1950	31538
9802	S	D	E	N	1960	24824
9802	S	G	E	N	1959	31244
9806	M	D	E	N	1958	31788
9808	S	D	R	H	1971	66445
9808	S	D	E	H	1971	66446
9809	M	D	G	N	1962	27579
9809	S	D	E		1959	7696
9809	S	D	G	H	1932	14367
9809	S	T	G	H	1932	14367
9809	S	D	E	N	1921	22633
9809	S	D	R	H	1961	24125
9809	S	D	E	N	1915	25767
9809	S	D	E	N	1910	28772
9809	S	D	G	N	1962	30999
9809	S	G	E	H	1915	31868
9809	S	D	E	H	1909	39859
9809	S	D	R	H	1970	59471
9809	S	D	E	H	1970	61167
9811	S	D	R	N	1966	44598
9811	S	D	E	N	1969	51032
9812	M	D	E	N	1962	29549
9812	S	D	E	N	1963	29780
9813	S	D	E	H	1964	29398
9813	S	T	E	H	1964	29398
9813	S	D	E	N	1963	31964
9813	S	D	E	H	1966	35917
9813	S	D	E	N	1964	37438
9813	S	D	E	N	1963	37609
9813	S	D	E	H	1967	46878
9813	S	G	E	H	1967	46952
9813	S	D	E	N	1960	47102
9814	M	D	E	N	1959	19992
9814	M	D	E	N	1959	19993
9814	S	D	E	N	1960	20510
9814	S	D	E	N	1962	25950
9815	S	D	E		1942	1264

Substance Number	Phys. State	Subject	Language	Temperature	Year	TPRC Number
503-9815	S	D	E		1959	7696
9815	S	G	E	N	1953	10855
9815	S	D	G	N	1942	29774
9815	S	D	E	N	1929	40826
9815	S	G	R	H	1965	42535
9815	S	G	E	H	1966	43468
9815	S	D	R	N	1966	46577
9815	S	D	E	N	1966	46578
9815	S	D	R	H	1969	58852
9815	S	D	E	H	1969	58853
9819	M	D	E	N	1963	33507
9819	P	D	E	H	1964	33692
9819	P	G	E	H	1967	49986
9819	S	S	E	N	1965	35565
9819	S	S	E	N	1966	35566
9819	S	D	E	H	1965	35740
9819	S	D	E	N	1967	43539
9819	S	D	E	N	1967	46259
9819	S	T	E	N	1967	46259
9819	S	D	E	N	1966	49737
9819	S	D	E	H	1959	52112
9819	S	D	E	H	1969	53573
9819	S	D	F	H	1968	54964
9819	S	D	E	H	1968	55211
9819	S	T	E	H	1968	55211
9819	S	D	E	H	1969	56741
9819	S	D	E	H	1969	56968
9819	S	D	F	H	1968	57456
9819	S	D	J	H	1968	60568
9819	S	D	E	H	1969	60569
9819	S	D	E	H	1970	65138
9819	S	D	E	H	1971	65186
9819	S	D	E	H	1971	68234
9819	S	E	E	H	1971	68234
9819	S	D	E	N	1972	73900
9819	S	G	F	H	1972	74283
9819	S	S	S	H	1973	74958
9822	M	D	E	N	1960	22625
9823	M	D	E	N	1960	22625
9823	S	D	J	N	1958	18522
9823	S	D	E	L	1965	36785
9824	M	D	E	N	1960	22625
9824	M	D	E	N	1962	29549
9824	S	D	J	N	1958	18522
9824	S	D	E	N	1961	23542
9824	S	D	E	N	1963	29780
9824	S	D	E	N	1950	31538
9824	S	D	E	L	1965	36785
9824	S	D	E	N	1964	37434
9824	S	D	R	N	1971	67665
9824	S	T	R	N	1971	67665
9826	S	D	E	N	1962	24680
9826	S	D	R	H	1969	57271
9826	S	D	E	H	1969	57272
9829	S	D	E	N	1961	16746
9829	S	D	J	N	1963	37566
9830	M	G	E	H	1955	10045
9830	S	D	E		1935	8241
9830	S	D	E	H	1955	10110
9830	S	D	E	N	1958	10682
9830	S	D	E	N	1959	13944
9830	S	D	E	N	1955	15754
9830	S	D	E	N	1955	15755
9830	S	D	E	N	1955	16823
9830	S	D	J	N	1960	20659
9830	S	G	E	N	1958	31786
9830	S	D	R	N	1963	35746
9830	S	E	R	N	1963	35746
9834	S	D	R	N	1971	66449
9834	S	D	E	N	1971	66450
9835	F	D	R	N	1967	60936
9835	F	D	E	N	1967	60937
9835	P	D	R	H	1973	74048
9835	S	D	I	H	1955	20975
9835	S	D	I	H	1946	21703
9835	S	D	G	N	1918	24323
9835	S	G	E	H	1915	31868
9835	S	D	R	H	1969	58542
9836	E	D	R	H	1961	42503
9836	S	D	E	H	1921	20009
9836	S	D	E	N	1964	36805
9836	S	E	E	N	1964	36805
9836	S	D	E	H	1909	39859
9836	S	D	E	N	1929	40826
9836	S	D	R	N	1969	55846
9836	S	T	R	N	1969	55846
9837	S	D	R	H	1965	42464
9837	S	D	R	N	1966	44369
9837	S	D	R	N	1969	59079
9837	S	T	R	N	1969	59079
9837	S	D	E	N	1969	59080
9837	S	T	E	N	1969	59080
9838	M	G	E	H	1955	10045
9838	M	D	E	N	1963	31136
9838	M	T	E	N	1963	31136
9838	S	D	E	H	1960	10052
9838	S	D	E	H	1927	22689
9838	S	D	R	H	1971	64846
9838	S	D	E	H	1971	64847
9838	S	D	E	H	1974	73503
9838	S	E	E	H	1974	73503

Substance Number	Phys. State	Subject	Language	Temperature	Year	TPRC Number
503-9843	S	D	E	N	1968	54196
9843	S	D	E	H	1969	55199
9843	S	D	E	H	1968	56785
9846	E	D	E	N	1966	40459
9850	S	D	E	N	1950	14673
9854	S	D	E	N	1968	54196
9857	M	D	E	H	1959	19992
9857	M	D	E	N	1959	19993
9857	M	D	E	N	1962	31182
9857	S	D	E	N	1960	14113
9857	S	D	E	N	1960	20510
9857	S	D	E	N	1962	25950
9859	M	D	E	N	1960	13346
9859	S	D	O	N	1972	71321
9860	E	D	E	H	1962	30774
9860	E	D	R	H	1962	30774
9860	E	E	E	H	1962	30774
9865	S	D	F	N	1968	48178
9868	S	D	E	N	1940	30989
9868	S	D	O	H	1971	71516
9869	S	D	E	N	1973	70757
9876	S	D	J	N	1960	17203
9882	S	D	E		1953	7067
9883	M	D	E	N	1963	25157
9883	S	S	E	H	1959	10831
9883	S	D	E	H	1962	29508
9889	S	D	E	L	1976	88203
9889	S	D	E	N	1976	88203
9889	S	E	E	L	1976	88203
9889	S	E	E	N	1976	88203
9890	S	D	E	H	1955	10110
9891	S	D	E	N	1959	13944
9891	S	D	R	N	1966	44598
9891	S	D	E	N	1969	51032
9891	S	D	E	H	1960	52260
9892	S	D	E	H	1960	10052
9892	S	D	E	H	1954	10481
9895	S	D	E	H	1965	35041
9898	S	D	E	N	1965	35041
9898	S	D	R	H	1970	59471
9898	S	D	E	H	1970	61167
9901	P	D	E		1952	6309
9901	S	D	E		1949	6929
9901	S	D	E	N	1940	30989
9902	S	D	E	N	1960	28670
9902	S	D	E	N	1964	36805
9902	S	E	E	N	1964	36805
9904	S	D	E		1953	9404
9905	S	D	E		1953	9404
9906	S	D	E		1952	9410
9907	S	D	E		1952	9410
9908	S	D	E		1952	9410
9909	S	D	E		1952	9410
9915	E	D	E	N	1969	37570
9915	E	D	E	N	1969	55028
9915	E	D	E	N	1968	57191
9915	S	D	E	N	1968	48434
9916	E	D	R	N	1971	64345
9916	E	D	E	N	1971	64850
9916	F	D	E	H	1965	36420
9916	F	D	E	H	1964	45639
9916	F	D	G	N	1973	75986
9916	F	T	G	N	1973	75986
9916	P	D	E	N	1957	45699
9916	P	D	E	N	1968	47499
9916	P	T	E	N	1968	47499
9916	S	D	E	N	1959	10757
9916	S	D	J	N	1949	14665
9916	S	D	E	N	1958	17648
9916	S	D	E	H	1921	20009
9916	S	D	E	N	1926	21670
9916	S	D	R	H	1961	24126
9916	S	D	E	N	1960	24761
9916	S	T	E	N	1951	29285
9916	S	D	R	N	1963	35746
9916	S	E	R	N	1963	35746
9916	S	D	E	N	1964	36805
9916	S	E	E	N	1964	36805
9916	S	D	E	H	1965	45219
9916	S	E	E	H	1965	45219
9916	S	D	E	H	1959	47022
9916	S	D	G	H	1969	53723
9916	S	S	E	H	1970	56454
9916	S	D	R	N	1970	60872
9916	S	D	R	N	1971	64345
9916	S	D	E	N	1971	64431
9916	S	D	R	H	1971	64846
9916	S	D	E	H	1971	64847
9916	S	D	E	N	1971	64850
9916	S	D	R	N	1971	68807
9916	S	T	R	N	1971	68807
9916	S	D	O	N	1972	71321
9916	S	D	R	N	1972	71631
9916	S	T	R	N	1972	71631
9916	S	D	E	N	1974	73503
9916	S	E	E	N	1974	73503
9918	S	D	E	N	1952	48341
9920	S	D	E		1952	6991
9922	E	D	E	H	1965	43489
9922	F	D	E	H	1965	36420
9922	L	D	E	N	1947	48002

Phys. State: **C.** Superconductive; **D.** Doped; **E.** Expanded; **F.** Fibrous or Whisker; **G.** Gas; **I.** Ionized or Plasma; **L.** Liquid; **M.** Multiphase; **P.** Powder or Fine Particle; **S.** Solid; **T.** Thick or Thin Film

Subject: **D.** Data; **E.** Experiment; **G.** General (Data + Theory + Experiment); **S.** Survey (Review, Compendium etc.); **T.** Theory

Language: **C.** Czech; **D.** Dutch; **E.** English; **F.** French; **G.** German; **I.** Italian; **J.** Japanese; **O.** Other Languages; **P.** Polish; **R.** Russian; **S.** Spanish

Temperature: **F.** Full Range (Low + Normal + High); **L.** Low (0 to 75K) + Overlap into Normal; **N.** Normal (75 to 1273K); **H.** High (above 1273K) + Overlap into Normal; Blank = Not Coded

Substance Number	Phys. State	Subject	Language	Temperature	Year	TPRC Number
503-9922	M	D	F	H	1964	28674
9922	M	D	R	H	1961	31073
9922	P	D	F	H	1964	28674
9922	P	D	F	H	1965	35705
9922	S	D	E		1959	7696
9922	S	S	E	H	1959	10831
9922	S	E	E	N	1961	19822
9922	S	D	E	H	1962	23378
9922	S	D	E	H	1964	26233
9922	S	D	F	H	1965	35705
9922	S	D	E	H	1965	38404
9922	S	D	E	H	1967	41352
9922	S	D	F	H	1965	42421
9922	S	E	F	H	1965	42421
9922	S	D	E	H	1965	46471
9922	S	D	F	H	1966	46819
9922	S	D	E	H	1963	48451
9922	S	D	E	H	1964	49249
9922	S	D	F	H	1970	60595
9922	S	D	F	H	1970	61396
9922	S	E	F	H	1970	61396
9922	S	D	R	H	1971	64844
9922	S	D	E	H	1971	64845
9922	S	D	E	H	1965	77294
9923	S	D	F		1959	7696
9925	S	D	E		1953	7036
9926	L	D	R	N	1973	75201
9926	S	D	R	N	1973	75201
9928	S	D	E		1950	1180
9928	S	D	E	N	1946	3081
9928	S	D	E		1959	7696
9928	S	D	E	N	1961	16746
9929	L	D	R	H	1970	59198
9929	L	D	R	H	1970	60944
9929	L	D	E	H	1970	60945
9929	L	D	E	H	1970	62295
9931	S	D	R		1955	790
9931	S	D	E	N	1958	14540
9933	M	D	E	N	1959	19992
9933	M	D	E	N	1962	25950
9933	S	D	E		1958	8195
9933	S	D	E	N	1962	26552
9933	S	D	E	N	1962	48904
9933	S	D	E	N	1968	54697
9933	S	D	E	N	1968	54698
9933	S	D	E	N	1970	61913
9933	S	T	E	N	1970	61913
9934	L	D	R	N	1973	75201
9934	S	D	R	N	1973	75201
9937	S	D	E		1953	7067
9945	S	D	E		1953	7036
9945	S	D	E		1953	7067
9950	S	D	E		1955	1639
9950	S	D	R	N	1966	41922
9950	S	D	R	H	1966	42537
9950	S	D	R	N	1969	53653
9950	S	D	E	N	1966	53877
9952	S	D	E		1950	1180
9952	S	D	E	N	1946	3081
9952	S	D	E		1958	6939
9952	S	D	E		1959	7696
9952	S	D	E	N	1961	16746
9953	S	D	E		1950	1180
9953	S	D	E	N	1946	3081
9953	S	D	E		1958	6939
9953	S	D	E		1959	7696
9953	S	D	E	N	1961	16746
9954	S	D	E	N	1946	3081
9954	S	D	E		1958	6939
9955	S	D	E	N	1946	3081
9955	S	D	E	H	1960	10052
9955	S	D	E	H	1954	10483
9955	S	D	E	H	1960	16022
9955	S	E	E	H	1960	16022
9955	S	D	E	H	1961	16746
9955	S	D	E	H	1966	43496
9956	S	D	E	N	1963	20313
9958	S	D	E		1950	1180
9958	S	D	E		1958	6939
9958	S	D	E		1953	7102
9959	S	D	E		1958	6939
9959	S	D	E		1953	7102
9960	S	D	E		1958	6939
9960	S	D	E		1953	7102
9960	S	D	E	N	1958	30547
9960	S	D	E	H	1967	44823
9960	S	D	R	H	1969	55280
9960	S	D	E	H	1969	57270
9961	S	D	E		1953	7102
9962	S	D	E		1958	6939
9962	S	D	E		1953	7102
9962	S	D	E	N	1952	48341
9963	S	D	E		1958	6939
9963	S	D	E		1953	7102
9963	S	D	E	N	1967	39186
9963	S	D	E	N	1952	48341
9964	S	D	E		1958	6939
9964	S	D	E		1953	7102
9967	S	D	E		1958	6939
9967	S	D	E		1953	7102
9968	S	D	E		1958	6939

Substance Number	Phys. State	Subject	Language	Temperature	Year	TPRC Number
503-9968	S	D	E		1953	7102
9969	S	D	E		1958	6939
9969	S	D	E		1953	7102
9970	S	D	E		1958	6939
9970	S	D	E		1953	7102
9970	S	D	E		1953	9404
9971	S	D	E		1958	6939
9971	S	D	E		1953	7102
9972	S	D	E		1958	6939
9972	S	D	E		1953	7102
9972	S	D	E	N	1952	48341
9973	S	D	E		1958	6939
9973	S	D	E		1953	7102
9974	E	D	E	H	1965	39261
9974	E	D	E	H	1965	43489
9974	S	D	E		1958	6939
9974	S	D	E		1953	7102
9974	S	D	E	N	1964	34296
9974	S	G	E	H	1964	48050
9975	S	D	E		1958	6939
9975	S	D	E		1953	7102
9976	S	D	E		1958	6939
9976	S	D	E		1953	7102
9978	S	D	E	H	1927	20929
9978	S	D	E	N	1958	30547
9978	S	D	R	H	1969	58854
9978	S	D	E	H	1969	58855
9979	S	D	E		1958	6939
9979	S	D	E		1953	7102
9979	S	D	R	H	1961	24125
9979	S	D	E	N	1962	24680
9979	S	S	E	N	1960	24931
9979	S	D	E	N	1962	29551
9979	S	D	J	N	1969	61940
9979	S	E	J	N	1969	61940
9979	S	D	G	N	1960	62279
9979	S	D	G	H	1971	66895
9980	S	D	G	N	1930	10317
9980	S	D	E	N	1960	15664
9980	S	D	E	H	1921	20009
9980	S	D	E	H	1927	22690
9980	S	D	G	H	1929	28664
9980	S	D	E	N	1964	36805
9980	S	E	E	N	1964	36805
9980	S	D	E	H	1909	39859
9980	S	D	R	N	1964	43252
9980	S	D	R	N	1970	62314
9980	S	D	E	N	1971	63275
9980	S	D	R	N	1971	64345
9980	S	D	E	N	1971	64850
9980	S	D	O	N	1972	71321
9982	S	D	E	N	1946	3081
9982	S	D	E		1958	6939
9982	S	D	E		1953	7102
9982	S	D	E		1959	7696
9982	S	D	E	N	1961	16746
9983	S	D	E	N	1946	3081
9983	S	D	E		1958	6939
9983	S	D	E		1953	7102
9983	S	D	E	H	1966	43496
9985	S	D	E	H	1973	74558
9985	S	T	E	H	1973	74558
9986	S	D	R	N	1967	54757
9989	S	D	R	N	1967	54757
9993	L	G	E	N	1959	21017
9993	S	G	E	N	1959	21017
9994	P	D	E	H	1966	41579
9994	S	D	E	H	1966	41579
9995	S	D	R	N	1967	54757
9997	L	S	E	H	1955	23533
9998	S	D	R	H	1976	89429
9998	S	D	R	N	1976	89429
9998	S	D	E	H	1976	91205
9998	S	D	E	N	1976	91205
521-0001	M	D	E	N	1907	31019
0001	M	D	E	N	1943	40228
0001	M	D	E	N	1966	42708
0001	P	D	E	N	1968	53178
0001	P	D	E	N	1949	61706
0001	S	G	R		1951	223
0001	S	D	R		1949	1154
0001	S	D	R	N	1934	15129
0001	S	D	G	N	1920	22016
0001	S	D	G	N	1910	28694
0001	S	D	E	N	1969	56053
0001	S	D	E	N	1949	61706
0003	S	D	E		1954	7104
0003	S	D	E	N	1968	47311
0003	S	S	E	N	1970	61233
0004	E	D	E	N	1965	42374
0004	M	D	E	N	1935	14829
0004	P	D	G	N	1910	28694
0004	P	D	E	N	1964	38646
0004	P	D	E	N	1966	41217
0004	P	D	E	N	1965	44628
0004	P	E	E	N	1965	44628
0004	P	D	E	N	1964	45753
0004	P	D	E	N	1967	46040
0004	P	D	E	N	1967	46042
0004	P	G	E	N	1966	46250

Substance Number	Phys. State	Subject	Language	Temperature	Year	TPRC Number
521-0004	P	D	E	N	1967	49193
0004	P	D	E	N	1966	49738
0004	P	D	E	N	1969	63002
0004	P	E	E	N	1969	63002
0004	S	D	G		1942	1070
0004	S	D	E	N	1935	14829
0004	S	D	R	N	1934	15129
0004	S	D	R	H	1934	15138
0004	S	D	G	N	1912	20505
0004	S	D	E	N	1932	31912
0004	S	D	R	N	1965	39987
0004	S	D	E	N	1965	39988
0004	S	D	E	N	1964	45753
0004	S	G	E	N	1966	46250
0006	P	D	R	N	1969	57572
0006	P	D	E	N	1971	62371
0006	P	D	E	N	1971	64041
0006	S	D	E	N	1921	22633
0006	S	D	R	N	1964	38184
0006	S	D	E	N	1969	55729
0006	S	D	E	L	1969	55762
0006	S	D	E	N	1973	73516
0007	P	D	E	N	1971	62371
0007	S	D	E		1952	6787
0007	S	S	E		1952	6787
0007	S	T	E		1952	6787
0007	S	G	G	N	1928	15610
0007	S	D	E	N	1960	15929
0007	S	D	E	N	1892	16421
0007	S	D	E	N	1940	20780
0007	S	D	E	N	1939	21536
0007	S	D	G	N	1931	22802
0007	S	D	E	N	1924	24551
0007	S	D	E	N	1963	28116
0007	S	G	E	N	1964	28649
0007	S	D	F	N	1895	28730
0007	S	D	G	N	1884	28778
0007	S	E	E	N	1963	29811
0007	S	D	E	N	1966	43774
0007	S	G	E	L	1961	46324
0007	S	G	E	N	1960	51286
0007	S	S	E	N	1970	61233
0007	S	D	E	N	1976	89727
0007	S	D	E	N	1977	89894
0007	S	E	E	N	1977	89894
0007	S	S	E	N	1976	90192
0007	S	D	E	N	1956	91695
0007	S	S	E	N	1960	94260
0007	S	D	E	N	1978	94672
0007	S	E	E	N	1978	94672
0007	S	D	E	N	1978	94673
0007	S	D	E	N	1978	95858
0007	S	D	E	N	1974	98070
0007	S	D	E	N	1979	98173
0007	S	D	E	N	1979	98785
0009	M	D	E		1939	4561
0009	M	E	E	N	1959	31893
0009	M	G	E	N	1959	31893
0009	M	S	E	N	1959	31893
0009	M	D	R	N	1958	41905
0009	M	D	E	N	1966	42695
0009	M	E	R	N	1959	43002
0009	M	E	F	N	1959	43003
0009	M	T	E	N	1967	46332
0009	M	D	E	N	1943	46820
0009	M	T	E	N	1943	46820
0009	M	T	R	N	1966	48085
0009	M	S	E	N	1976	90192
0009	M	D	E	N	1948	94419
0009	M	E	E	N	1948	94419
0009	M	D	F	N	1978	96877
0009	P	D	E		1952	6309
0009	P	D	E	N	1921	22633
0009	P	E	E	N	1909	31920
0009	P	S	E	N	1966	42284
0009	P	D	E	N	1968	53178
0009	P	E	E	N	1969	53598
0009	P	E	E	N	1969	57235
0009	P	D	E	N	1945	61037
0009	P	E	E	N	1949	61706
0009	P	D	E	N	1970	61712
0009	P	T	E	N	1970	61712
0009	P	D	J	N	1976	87993
0009	P	D	E	N	1948	94419
0009	P	E	E	N	1948	94419
0009	S	G	E		1950	182
0009	S	D	E		1954	183
0009	S	D	E		1956	184
0009	S	T	E		1956	185
0009	S	E	F		1941	1720
0009	S	G	E		1942	5200
0009	S	D	E		1950	6205
0009	S	T	E		1950	6205
0009	S	G	E		1951	6227
0009	S	D	E		1952	6252
0009	S	T	E		1952	6252
0009	S	E	E		1958	6478
0009	S	T	E		1957	6938
0009	S	E	E		1951	7346
0009	S	D	E		1957	7350
0009	S	D	E		1950	7365

Phys. State: **C.** Superconductive; **D.** Doped; **E.** Expanded; **F.** Fibrous or Whisker; **G.** Gas; **I.** Ionized or Plasma; **L.** Liquid; **M.** Multiphase; **P.** Powder or Fine Particle; **S.** Solid; **T.** Thick or Thin Film

Subject: **D.** Data; **E.** Experiment; **G.** General (Data + Theory + Experiment); **S.** Survey (Review, Compendium etc.); **T.** Theory

Language: **C.** Czech; **D.** Dutch; **E.** English; **F.** French; **G.** German; **I.** Italian; **J.** Japanese; **O.** Other Languages; **P.** Polish; **R.** Russian; **S.** Spanish

Temperature: **F.** Full Range (Low + Normal + High); **L.** Low (0 to 75K) + Overlap into Normal; **N.** Normal (75 to 1273K); **H.** High (above 1273K) + Overlap into Normal; Blank = Not Coded

Substance Number	Phys. State	Subject	Language	Temperature	Year	TPRC Number
521-0009	S	D	E		1948	7368
0009	S	E	E		1948	7369
0009	S	G	E		1949	8230
0009	S	G	E		1957	8476
0009	S	D	E		1955	9232
0009	S	D	E		1958	9266
0009	S	G	E	N	1954	10097
0009	S	E	E	N	1960	10917
0009	S	D	E	N	1966	11559
0009	S	G	E	N	1938	14188
0009	S	D	E	N	1960	15640
0009	S	T	E	N	1958	15736
0009	S	D	G	N	1956	16078
0009	S	S	E	N	1959	16322
0009	S	D	E	N	1953	16343
0009	S	T	E	N	1953	16343
0009	S	E	E	N	1955	16344
0009	S	E	E	N	1955	18237
0009	S	E	E	N	1950	21290
0009	S	D	G	N	1920	22016
0009	S	D	E	N	1963	24010
0009	S	G	E	N	1961	24978
0009	S	G	E	N	1896	25164
0009	S	T	E	N	1956	25369
0009	S	G	E	N	1905	25461
0009	S	D	E	N	1951	25557
0009	S	D	E	N	1951	25559
0009	S	E	E	N	1922	26034
0009	S	G	E	N	1960	26766
0009	S	D	E	N	1947	28799
0009	S	E	G	N	1932	31875
0009	S	D	E	N	1966	39355
0009	S	D	E	N	1964	39737
0009	S	G	E	N	1952	41004
0009	S	E	R	N	1959	43002
0009	S	E	F	N	1959	43003
0009	S	E	E	N	1965	44633
0009	S	T	R	N	1956	46091
0009	S	T	E	N	1960	46092
0009	S	D	E	N	1943	46820
0009	S	T	E	N	1943	46820
0009	S	D	E	N	1963	48879
0009	S	D	E	N	1963	48881
0009	S	D	E	N	1970	57988
0009	S	S	E	N	1970	61233
0009	S	D	E	N	1970	61537
0009	S	E	E	N	1949	61706
0009	S	D	E	N	1970	61712
0009	S	T	E	N	1970	61712
0009	S	G	E	N	1970	65069
0009	S	S	R	N	1957	68049
0009	S	S	E	N	1971	68050
0009	S	D	J	N	1976	87993
0009	S	S	E	N	1976	90192
0009	S	D	J	N	1975	98110
0009	S	E	J	N	1975	98110
0010	M	D	E	N	1907	31019
0010	M	G	E	N	1959	31893
0010	M	G	E	N	1959	31906
0010	M	G	E	N	1909	31920
0010	M	D	E	N	1948	94419
0010	M	E	E	N	1948	94419
0010	P	D	E	N	1947	25437
0010	P	S	G	N	1928	28680
0010	P	D	E	N	1961	29364
0010	P	D	E	N	1965	36861
0010	P	D	E	N	1962	39516
0010	P	D	R	N	1965	42532
0010	P	D	E	N	1966	42709
0010	P	D	E	N	1964	45653
0010	P	D	E	N	1968	47499
0010	P	T	E	N	1968	47499
0010	P	D	E	H	1969	53598
0010	P	D	E	N	1969	53610
0010	P	D	R	N	1969	56232
0010	P	D	E	H	1969	57235
0010	P	D	E	N	1949	61706
0010	P	D	R	N	1971	65687
0010	P	D	E	N	1969	69506
0010	P	D	E	N	1973	72674
0010	P	E	E	N	1973	72674
0010	P	G	E	N	1974	82019
0010	P	D	J	N	1977	94212
0010	P	E	J	N	1977	94212
0010	P	D	E	N	1948	94419
0010	P	E	E	N	1948	94419
0010	S	E	J		1953	192
0010	S	T	E		1956	742
0010	S	D	R		1946	1734
0010	S	G	E		1949	1945
0010	S	D	E	N	1924	25145
0010	S	D	E	N	1962	29258
0010	S	D	E	N	1962	29491
0010	S	D	I	N	1963	36904
0010	S	D	E	N	1948	40389
0010	S	D	R	N	1966	44953
0010	S	D	E	N	1967	44954
0010	S	T	R	N	1956	49765
0010	S	S	E	N	1970	61233
0010	S	D	E	N	1949	61706
0010	S	D	R	N	1971	63757

Substance Number	Phys. State	Subject	Language	Temperature	Year	TPRC Number
521-0010	S	E	R	N	1971	63757
0010	S	D	E	N	1971	67155
0010	S	E	E	N	1971	67155
0010	S	G	E	N	1974	82019
0013	S	D	E	N	1943	40228
0015	E	D	J	N	1973	72129
0015	E	T	J	N	1973	72129
0015	L	D	R		1954	7704
0015	M	D	E		1955	7356
0015	M	D	E		1955	9738
0015	M	G	E	N	1956	48292
0015	P	D	E		1954	6325
0015	P	G	E		1955	6609
0015	P	G	E		1955	6624
0015	P	D	E	N	1957	10271
0015	P	D	E	L	1959	17069
0015	P	D	E	L	1956	19362
0015	P	D	E	N	1920	20414
0015	P	D	E	N	1959	20933
0015	P	D	G	N	1929	22931
0015	P	D	R	N	1941	23645
0015	P	D	F	N	1910	24647
0015	P	D	O	N	1931	28679
0015	P	D	E	N	1963	29585
0015	P	D	E	N	1905	30693
0015	P	D	G	N	1962	31088
0015	P	D	G	N	1951	31815
0015	P	D	E	N	1941	31862
0015	P	D	E	N	1932	31912
0015	P	D	E	N	1925	39868
0015	P	D	E	N	1921	40025
0015	P	D	E	N	1943	40228
0015	P	D	E	N	1905	41683
0015	P	D	E	N	1957	45699
0015	P	D	E	N	1968	46484
0015	P	D	R	N	1968	49504
0015	P	E	R	N	1968	49504
0015	P	D	G	N	1908	50697
0015	P	E	G	N	1908	50697
0015	P	D	E	N	1968	52102
0015	P	D	E	N	1968	53178
0015	P	D	E	N	1961	53596
0015	P	E	E	N	1961	53596
0015	P	D	E	N	1969	53598
0015	P	D	E	N	1961	56385
0015	P	D	E	N	1969	57235
0015	P	D	E	N	1968	65123
0015	P	E	E	N	1968	65123
0015	P	D	E	N	1974	73843
0015	P	E	E	N	1974	73843
0015	S	D	E		1951	817
0015	S	D	E		1942	1264
0015	S	T	E		1942	1264
0015	S	D	E		1945	1719
0015	S	D	G		1940	3421
0015	S	D	E		1954	7129
0015	S	D	E		1959	7688
0015	S	D	E		1953	8233
0015	S	D	G	N	1930	10317
0015	S	D	E	N	1958	10458
0015	S	D	E	N	1941	11001
0015	S	D	G	N	1953	16064
0015	S	D	E	N	1940	16263
0015	S	D	E	N	1928	20652
0015	S	D	E	H	1927	21490
0015	S	D	E	N	1919	21929
0015	S	D	R	N	1929	22225
0015	S	D	F	H	1924	22489
0015	S	D	E	H	1926	22638
0015	S	D	E	N	1919	23435
0015	S	T	E	N	1919	23435
0015	S	D	G	N	1935	25920
0015	S	D	G	H	1929	28664
0015	S	S	G	N	1928	28680
0015	S	D	F	N	1909	28745
0015	S	D	G	N	1930	33613
0015	S	D	E	H	1931	39843
0015	S	D	E	N	1925	39868
0015	S	D	E	N	1943	40228
0015	S	D	E	N	1924	43414
0015	S	D	E	N	1916	48034
0015	S	G	E	N	1956	48292
0015	S	D	E	N	1969	55257
0015	S	D	E	N	1974	77908
0015	S	E	E	N	1974	77908
0020	E	D	E		1949	6929
0020	E	D	R	N	1962	27694
0020	E	D	G	N	1967	51692
0020	E	D	E	N	1971	69687
0020	E	D	E	N	1974	73843
0020	E	E	E	N	1974	73843
0020	M	D	R	N	1941	23645
0020	P	D	E	N	1957	10271
0020	P	D	E	L	1948	19642
0020	P	D	I	N	1950	20956
0020	P	S	E	N	1960	24931
0020	P	D	E	N	1966	47174
0020	P	D	R	N	1971	65687
0020	S	D	J		1956	59
0020	S	D	I		1950	76
0020	S	D	E		1942	5258

Substance Number	Phys. State	Subject	Language	Temperature	Year	TPRC Number
521-0020	S	D	E		1951	6237
0020	S	D	E		1954	7336
0020	S	D	E		1942	7694
0020	S	D	E	N	1957	10271
0020	S	D	E	N	1941	11001
0020	S	D	E	N	1960	15640
0020	S	E	E	N	1958	20140
0020	S	S	E	N	1960	24931
0020	S	D	E	N	1950	40447
0020	S	D	G	N	1964	43101
0020	S	D	E	N	1969	55257
0020	S	S	E	N	1970	61233
0020	S	D	E	N	1973	71780
0021	P	D	E	N	1971	62371
0022	P	D	E	N	1971	62371
0022	S	D	E		1953	7102
0022	S	D	E	N	1958	10458
0022	S	D	E	L	1958	10723
0022	S	D	E	N	1958	23657
0022	S	D	R	H	1961	24125
0022	S	D	R	N	1961	24125
0022	S	S	E	N	1960	24931
0022	S	D	G	N	1965	26779
0022	S	D	G	N	1962	27715
0022	S	D	E	H	1937	28449
0022	S	D	E	N	1957	35437
0022	S	D	R	N	1964	38184
0022	S	D	G	H	1968	55499
0022	S	E	G	H	1968	55499
0022	S	D	E	N	1969	55729
0022	S	D	R	H	1969	58786
0022	S	D	E	H	1972	68734
0022	S	E	E	H	1972	68734
0022	S	G	E	N	1971	74012
0022	S	D	E	H	1975	83595
0022	S	D	E	N	1975	83595
0022	S	S	E	N	1977	92134
0024	M	D	E	N	1957	25524
0024	P	G	C	N	1970	62537
0024	S	D	G		1939	9693
0024	S	D	G	N	1958	21268
0024	S	D	R	N	1960	21288
0024	S	D	E	N	1960	25356
0024	S	D	G	N	1940	29870
0024	S	D	G	N	1939	31888
0024	S	D	E	N	1963	35602
0024	S	D	R	N	1968	51815
0025	P	D	E	N	1971	62371
0025	S	G	G	N	1928	15610
0025	S	D	E	N	1923	21665
0025	S	D	R	N	1966	45831
0025	S	D	E	N	1967	45832
0025	S	D	E	N	1969	55729
0025	S	D	E	N	1979	97680
0026	P	D	E	N	1971	62371
0026	S	G	G	N	1928	15610
0026	S	D	E	N	1969	55729
0026	S	D	P	N	1971	63225
0026	S	D	E	N	1965	88950
0026	S	S	E	N	1977	92134
0027	P	D	R	N	1971	69768
0027	P	D	E	N	1972	69769
0028	P	D	E	N	1964	38646
0028	P	D	E	N	1962	39516
0028	P	D	E	N	1965	42374
0028	P	D	E	N	1964	45753
0028	P	G	E	N	1963	49023
0028	P	D	E	N	1969	54820
0028	S	D	E	N	1963	24360
0028	S	D	E	N	1962	29491
0028	S	D	E	N	1973	70421
0028	S	E	E	N	1973	70421
0030	P	D	E	N	1971	62371
0030	S	D	E	N	1969	55729
0032	P	D	E	N	1971	62371
0032	S	G	G	N	1928	15610
0032	S	D	E	N	1969	55729
0033	P	D	E	N	1971	62371
0033	S	D	E		1958	6939
0033	S	D	E		1954	7104
0033	S	D	E	N	1969	55729
0034	F	D	G		1950	6215
0034	F	T	G		1950	6215
0034	P	D	E	N	1971	62371
0034	S	D	G	H	1932	14367
0034	S	T	G	H	1932	14367
0034	S	D	E	N	1969	55729
0034	S	S	E	N	1970	61233
0034	S	G	E	N	1971	74012
0035	P	D	E	N	1971	62371
0035	S	D	E	N	1969	55729
0035	S	D	E	N	1965	88950
0036	P	D	E	N	1971	62371
0036	S	D	E	N	1963	24342
0036	S	T	E	N	1963	24342
0036	S	D	E	H	1967	34350
0036	S	D	E	H	1967	34861
0036	S	D	E	N	1969	55729
0036	S	G	E	H	1970	58983
0040	E	D	E		1954	7336
0040	E	D	R	N	1965	39987

Phys. State: **C.** Superconductive; **D.** Doped; **E.** Expanded; **F.** Fibrous or Whisker; **G.** Gas; **I.** Ionized or Plasma; **L.** Liquid; **M.** Multiphase; **P.** Powder or Fine Particle; **S.** Solid; **T.** Thick or Thin Film

Subject: **D.** Data; **E.** Experiment; **G.** General (Data + Theory + Experiment); **S.** Survey (Review, Compendium etc.); **T.** Theory

Language: **C.** Czech; **D.** Dutch; **E.** English; **F.** French; **G.** German; **I.** Italian; **J.** Japanese; **O.** Other Languages; **P.** Polish; **R.** Russian; **S.** Spanish

Temperature: **F.** Full Range (Low + Normal + High); **L.** Low (0 to 75K) + Overlap into Normal; **N.** Normal (75 to 1273K); **H.** High (above 1273K) + Overlap into Normal; Blank = Not Coded

Substance Number	Phys. State	Subject	Language	Temperature	Year	TPRC Number
521-0040	E	D	E	N	1965	39988
0040	E	D	R	N	1966	46566
0040	E	D	E	N	1966	46567
0040	M	D	E		1948	620
0040	M	D	G		1950	1448
0040	M	D	R	N	1960	20247
0040	M	D	E	N	1952	21187
0040	M	T	E	N	1952	21187
0040	M	D	E	N	1964	42307
0040	M	S	E	N	1976	90192
0040	P	D	E		1956	8514
0040	P	D	R	N	1960	51563
0040	S	D	E		1956	226
0040	S	D	E		1952	268
0040	S	D	R		1952	785
0040	S	D	R		1946	1734
0040	S	D	E		1953	6231
0040	S	D	E		1949	6929
0040	S	D	R		1953	8876
0040	S	D	E		1956	9825
0040	S	S	E	N	1958	10667
0040	S	G	E	N	1938	14188
0040	S	D	F	N	1827	16209
0040	S	D	E	N	1942	16287
0040	S	D	E	N	1952	20918
0040	S	D	E	N	1939	21536
0040	S	D	E	N	1921	22633
0040	S	D	E	N	1951	25557
0040	S	D	E	N	1951	25559
0040	S	D	E	H	1960	31874
0040	S	D	E	N	1967	50428
0040	S	D	E	N	1969	54631
0040	S	D	E	N	1962	55247
0040	S	D	R	N	1968	58386
0040	S	G	E	N	1970	65069
0041	P	D	E	N	1971	62371
0041	S	G	G	N	1928	15610
0041	S	D	E	N	1979	97680
0046	M	S	E	N	1964	45753
0046	M	D	R	N	1969	54587
0046	M	G	E	N	1976	87714
0046	M	D	E	N	1947	89613
0046	M	E	E	N	1947	89613
0046	M	D	E	N	1973	89619
0046	P	G	E	N	1958	19701
0046	P	D	F	N	1965	42373
0046	P	T	F	N	1965	42373
0046	P	G	E	N	1971	74012
0046	P	D	E	N	1968	74222
0046	S	E	E		1957	63
0046	S	D	E		1957	9238
0046	S	D	E	N	1966	11559
0046	S	D	G	N	1956	16078
0046	S	D	E	N	1955	16345
0046	S	D	R	N	1930	21451
0046	S	E	E	N	1921	22633
0046	S	E	E	N	1959	23900
0046	S	D	E	N	1878	25165
0046	S	E	E	N	1941	31862
0046	S	S	G	N	1924	33616
0046	S	E	E	N	1967	35605
0046	S	S	E	N	1965	41112
0046	S	G	E	N	1970	41374
0046	S	S	E	N	1966	42284
0046	S	E	E	N	1964	43773
0046	S	E	E	N	1966	43774
0046	S	G	E	N	1964	44867
0046	S	D	E	N	1956	44868
0046	S	S	E	N	1967	44944
0046	S	D	G	N	1968	54725
0046	S	E	G	N	1968	54725
0046	S	D	E	N	1967	54790
0046	S	D	E	N	1969	56491
0046	S	T	E	N	1969	56491
0046	S	T	E	N	1969	56570
0046	S	D	R	N	1968	58386
0046	S	D	R	H	1969	58631
0046	S	T	R	N	1968	61216
0046	S	S	E	N	1970	61233
0046	S	D	E	N	1970	63519
0046	S	E	S	N	1971	63639
0046	S	E	E	N	1971	64422
0046	S	G	E	N	1971	65601
0046	S	T	R	N	1968	67269
0046	S	T	E	N	1968	67270
0046	S	T	E	N	1974	73781
0046	S	S	E	N	1971	74002
0046	S	D	E	N	1972	74003
0046	S	D	E	N	1971	74013
0046	S	E	E	N	1970	74014
0046	S	D	E	N	1976	87316
0046	S	E	F	N	1976	88190
0046	S	D	E	N	1947	89613
0046	S	E	E	N	1947	89613
0046	S	D	E	N	1977	90680
0046	S	D	E	N	1976	91238
0046	S	D	E	N	1956	91695
0046	S	D	E	N	1977	92136
0046	S	S	E	N	1970	92931
0046	S	E	E	N	1976	92952
0046	S	T	E	N	1979	98173
521-0046	S	D	E	N	1979	99992
0049	S	D	E		1956	8150
0049	S	D	R	N	1967	58388
0049	S	T	R	N	1967	58388
0053	P	D	E	N	1964	45753
0053	S	D	E	N	1921	22633
0053	S	G	E	N	1939	28787
0053	S	D	E	N	1954	35118
0053	S	D	I	N	1963	36904
0053	S	D	E	N	1951	40424
0053	S	D	E	N	1966	43774
0053	S	D	E	N	1964	45753
0053	S	D	G	N	1968	54724
0053	S	E	G	N	1968	54724
0053	S	D	E	N	1972	64938
0053	S	E	E	N	1972	64938
0054	M	D	E	N	1970	61541
0054	S	E	E	N	1947	16341
0054	S	S	E	N	1947	16341
0054	S	D	E	N	1963	24342
0054	S	T	E	N	1963	24342
0054	S	D	R	H	1958	25156
0054	S	T	R	H	1958	25156
0054	S	T	R	N	1967	58387
0055	F	D	R	N	1941	23645
0055	F	D	E	N	1968	36576
0055	F	T	E	N	1968	36576
0055	F	D	I	N	1966	41292
0055	F	D	E	N	1960	42824
0055	F	D	I	N	1966	44990
0055	F	D	R	N	1968	46500
0055	F	T	R	N	1968	46500
0055	F	D	R	N	1960	51563
0055	F	D	E	N	1969	53585
0055	F	S	E	N	1970	61233
0055	L	D	R	N	1970	61258
0055	L	E	R	N	1970	61258
0055	L	D	E	N	1970	67224
0055	L	E	E	N	1970	67224
0055	M	D	R	N	1941	23645
0055	M	D	E	N	1899	48324
0055	P	D	R	N	1960	51563
0055	S	D	E		1951	810
0055	S	D	E		1942	1264
0055	S	D	E		1954	7336
0055	S	D	E		1953	8233
0055	S	G	E		1956	9943
0055	S	D	E	N	1941	11001
0055	S	D	G	N	1933	11172
0055	S	D	E	N	1940	16263
0055	S	D	E	N	1938	16306
0055	S	D	E	N	1892	16421
0055	S	D	F	N	1935	16570
0055	S	D	I	N	1955	20975
0055	S	D	E	N	1927	21490
0055	S	D	E	N	1919	21929
0055	S	D	G	N	1920	22016
0055	S	D	E	N	1921	22633
0055	S	D	E	N	1947	25437
0055	S	D	G	N	1910	28694
0055	S	D	E	N	1930	28819
0055	S	D	I	N	1949	35068
0055	S	D	E	N	1943	40228
0055	S	D	R	N	1966	48545
0055	S	E	R	N	1966	48545
0055	S	D	I	N	1975	80464
0055	S	E	I	N	1975	80464
0057	P	D	E	N	1971	62371
0057	S	D	E	N	1969	55729
0057	S	D	E	N	1965	88950
0057	S	S	E	N	1977	92134
0057	S	D	E	N	1979	97680
0058	P	D	E	N	1971	62371
0058	S	D	E	N	1968	47311
0058	S	D	E	N	1969	55729
0059	P	D	E	N	1971	62371
0059	S	D	E	N	1969	55729
0060	P	D	E	N	1971	62371
0062	P	D	E	N	1971	62371
0062	S	D	E	N	1969	55729
0062	S	G	E	N	1971	74012
0063	D	D	E	N	1973	74495
0063	D	E	E	N	1973	74495
0063	S	D	E		1957	64
0063	S	T	E		1957	64
0063	S	D	R		1954	179
0063	S	D	E		1953	266
0063	S	G	E		1956	1161
0063	S	D	R		1952	1390
0063	S	T	R		1952	1390
0063	S	D	R		1942	1395
0063	S	T	R		1942	1395
0063	S	G	E		1953	6702
0063	S	G	E		1951	7200
0063	S	T	E		1942	9357
0063	S	D	E		1956	9433
0063	S	D	E		1958	9878
0063	S	T	E		1958	9878
0063	S	D	E	L	1962	10561
0063	S	T	E	L	1962	10561
0063	S	S	E	H	1956	10581
521-0063	S	T	F	N	1959	10911
0063	S	D	E	L	1938	11913
0063	S	D	G	L	1911	15576
0063	S	G	G	N	1928	15610
0063	S	S	E	N	1960	16704
0063	S	D	E	L	1922	21008
0063	S	T	E	L	1922	21008
0063	S	D	G	N	1927	21664
0063	S	D	E	N	1961	23683
0063	S	D	G	L	1954	28157
0063	S	T	G	L	1954	28157
0063	S	D	E	L	1963	28158
0063	S	T	E	L	1963	28158
0063	S	D	E	N	1928	33185
0063	S	D	E	N	1965	33801
0063	S	T	E	N	1965	33801
0063	S	T	E	L	1966	33967
0063	S	D	G	N	1954	43592
0063	S	T	G	N	1954	43592
0063	S	D	E	L	1967	44768
0063	S	D	R	N	1968	51066
0063	S	G	E	L	1965	55170
0063	S	D	E	L	1970	61233
0063	S	D	E	N	1970	61233
0063	S	S	E	L	1970	61233
0063	S	S	E	N	1970	61233
0063	S	D	E	L	1972	66323
0063	S	D	E	N	1972	66323
0063	S	T	E	L	1972	66323
0063	S	T	E	N	1972	66323
0063	S	D	E	L	1973	70467
0063	S	D	E	L	1972	71398
0063	S	T	E	L	1972	71398
0063	S	D	E	N	1973	74495
0063	S	E	E	N	1973	74495
0063	S	D	E	N	1975	77237
0063	S	E	E	N	1975	77237
0063	S	D	E	L	1974	79241
0063	S	D	E	N	1974	79241
0063	S	S	E	L	1974	79241
0063	S	S	E	N	1974	79241
0063	S	D	E	N	1975	81297
0063	S	D	E	L	1976	85815
0063	S	D	E	N	1976	85815
0063	S	D	E	N	1977	87185
0063	S	S	E	N	1977	87185
0063	S	S	E	N	1976	89725
0063	S	D	R	N	1977	90120
0063	S	E	R	N	1977	90120
0063	S	D	E	N	1978	90953
0063	S	D	E	L	1977	91646
0063	S	D	E	N	1977	91646
0063	S	D	R	N	1977	91920
0063	S	D	E	N	1977	92059
0063	S	E	E	N	1977	92059
0063	S	D	E	L	1976	92918
0063	S	D	E	N	1976	92918
0063	S	D	E	N	1977	95298
0063	S	D	E	L	1978	96083
0063	S	D	E	N	1979	98122
0063	S	D	E	N	1980	100115
0063	S	S	E	N	1980	100400
0064	S	D	E		1957	64
0064	S	T	E		1957	64
0064	S	D	E		1955	1108
0064	S	D	E		1943	1287
0064	S	D	E		1952	1964
0064	S	G	E		1953	6702
0064	S	D	E		1953	7036
0064	S	D	E		1953	7102
0064	S	D	E		1954	7104
0064	S	G	E		1951	7200
0064	S	D	E		1959	7688
0064	S	D	E		1959	7696
0064	S	D	E		1935	8241
0064	S	E	E		1952	8818
0064	S	D	E	N	1960	10052
0064	S	D	E	H	1955	10101
0064	S	D	E	L	1955	10187
0064	S	D	E	L	1955	10206
0064	S	D	E	H	1958	10703
0064	S	D	E	N	1960	13432
0064	S	D	E	N	1959	13944
0064	S	D	E	L	1960	16756
0064	S	D	E	L	1952	21176
0064	S	D	E	N	1952	21176
0064	S	D	E	N	1951	25435
0064	S	D	E	L	1962	27279
0064	S	D	O	N	1949	28661
0064	S	T	E	L	1966	33967
0064	S	D	E	N	1965	37045
0064	S	S	E	N	1964	37618
0064	S	D	E	N	1968	49390
0064	S	G	E	L	1965	55170
0064	S	D	E	L	1971	61376
0064	S	D	G	L	1969	64980
0064	S	E	G	L	1969	64980
0064	S	D	E	L	1971	66159
0064	S	D	E	N	1972	68602
0064	S	E	E	N	1972	68602
0064	S	D	E	N	1973	68985

Phys. State: **C.** Superconductive; **D.** Doped; **E.** Expanded; **F.** Fibrous or Whisker; **G.** Gas; **I.** Ionized or Plasma; **L.** Liquid; **M.** Multiphase; **P.** Powder or Fine Particle; **S.** Solid; **T.** Thick or Thin Film
Subject: **D.** Data; **E.** Experiment; **G.** General (Data + Theory + Experiment); **S.** Survey (Review, Compendium etc.); **T.** Theory
Language: **C.** Czech; **D.** Dutch; **E.** English; **F.** French; **G.** German; **I.** Italian; **J.** Japanese; **O.** Other Languages; **P.** Polish; **R.** Russian; **S.** Spanish
Temperature: **F.** Full Range (Low + Normal + High); **L.** Low (0 to 75K) + Overlap into Normal; **N.** Normal (75 to 1273K); **H.** High (above 1273K) + Overlap into Normal; Blank = Not Coded

Substance Number	Phys. State	Subject	Language	Temperature	Year	TPRC Number
521-0064	S	D	E	L	1972	69949
0064	S	S	E	N	1959	85032
0065	M	D	E	N	1979	99744
0065	S	D	G		1942	3699
0065	S	D	R	N	1960	21288
0065	S	D	E	N	1960	25356
0065	S	G	E	N	1969	59226
0065	S	D	E	N	1979	99744
0065	S	D	E	N	1979	100246
0066	M	D	R	N	1960	24247
0066	M	D	E	N	1960	25361
0066	M	G	R	N	1948	31914
0066	M	D	R	N	1968	51025
0066	M	D	R	N	1969	56318
0066	M	D	E	N	1949	61706
0066	P	D	E	N	1968	53178
0066	P	D	E	N	1949	61706
0066	S	D	R		1952	1992
0066	S	D	R		1939	9699
0066	S	D	G	N	1920	22016
0066	S	D	E	N	1922	39850
0066	S	D	E	N	1969	56053
0066	S	D	R	N	1967	60317
0066	S	D	E	N	1970	60318
0066	S	D	E	N	1949	61706
0066	S	D	E	N	1972	66207
0067	M	D	R	N	1973	72486
0067	M	D	E	N	1973	74596
0067	P	D	E	N	1971	62371
0067	S	D	E		1954	7104
0067	S	D	R	L	1962	20285
0067	S	D	E	L	1963	20413
0067	S	G	R	N	1962	27608
0067	S	G	E	N	1963	32761
0067	S	D	G	N	1905	38729
0067	S	D	E	N	1969	55729
0067	S	D	E	N	1971	63145
0067	S	E	E	N	1971	63145
0067	S	D	R	N	1973	76173
0069	M	D	E	N	1955	40671
0069	M	T	E	N	1955	40671
0069	P	D	E		1952	164
0069	P	D	R	N	1937	15019
0069	P	D	E	N	1920	22166
0069	P	D	E	N	1919	23435
0069	P	T	E	N	1919	23435
0069	P	D	E	N	1934	25152
0069	P	D	R	N	1964	35710
0069	P	D	E	N	1968	53178
0069	P	D	R	N	1969	56898
0069	P	T	R	N	1969	56898
0069	P	D	R	N	1971	65746
0069	P	T	R	N	1971	65746
0069	P	D	J	N	1970	67704
0069	P	E	J	N	1970	67704
0069	P	D	E	N	1973	76783
0069	P	T	E	N	1973	76783
0069	S	D	G		1952	73
0069	S	T	G		1952	73
0069	S	D	S		1950	1120
0069	S	G	G		1943	4512
0069	S	G	G		1925	6706
0069	S	S	G		1948	6748
0069	S	D	R		1957	8974
0069	S	D	G		1939	9693
0069	S	D	E	N	1936	11107
0069	S	D	R	N	1960	11725
0069	S	D	E	N	1862	16210
0069	S	D	O	N	1958	17338
0069	S	D	E	N	1959	18202
0069	S	D	E	N	1961	18657
0069	S	T	E	N	1961	18657
0069	S	D	R	N	1959	18710
0069	S	D	G	N	1959	20041
0069	S	D	E	N	1962	20611
0069	S	D	E	N	1960	20660
0069	S	D	R	N	1960	21288
0069	S	D	E	N	1923	21665
0069	S	D	E	N	1863	22527
0069	S	D	E	N	1921	22633
0069	S	D	R	N	1961	23771
0069	S	D	R	N	1961	24992
0069	S	D	E	N	1960	25356
0069	S	D	G	N	1904	28731
0069	S	D	E	N	1951	28792
0069	S	S	R	N	1958	30882
0069	S	S	E	N	1961	30884
0069	S	D	E	N	1956	38510
0069	S	D	E	N	1966	42432
0069	S	D	R	N	1968	49919
0069	S	T	R	N	1968	49919
0069	S	S	E	N	1970	56648
0069	S	D	R	N	1969	56898
0069	S	T	R	N	1969	56898
0069	S	D	R	N	1969	58119
0069	S	D	J	N	1969	59043
0069	S	E	J	N	1969	59043
0069	S	S	E	H	1970	61233
0069	S	S	E	N	1970	61233
0069	S	D	E	N	1968	65121
0069	S	T	E	N	1968	65121
521-0069	S	D	J	N	1970	67704
0069	S	E	J	N	1970	67704
0069	S	D	E	N	1968	70866
0069	S	D	E	N	1978	94682
0070	M	D	E	N	1941	43835
0070	M	D	R	N	1969	54587
0070	M	D	E	N	1976	87714
0070	M	E	E	N	1976	87714
0070	M	D	E	N	1973	89619
0070	M	S	E	N	1976	90192
0070	M	D	E	N	1973	92930
0070	M	E	E	N	1973	92930
0070	M	D	E	N	1976	92954
0070	S	D	E	N	1958	12174
0070	S	D	E	N	1939	21536
0070	S	D	E	N	1923	21665
0070	S	D	E	N	1921	22633
0070	S	G	E	N	1939	28787
0070	S	D	E	N	1951	28792
0070	S	D	E	N	1951	40424
0070	S	D	E	N	1966	43774
0070	S	D	E	N	1941	43835
0070	S	D	E	N	1963	48294
0070	S	G	E	N	1960	51286
0070	S	D	J	N	1969	59044
0070	S	E	J	N	1969	59044
0070	S	G	E	N	1971	74012
0070	S	D	E	N	1967	89035
0070	S	E	E	N	1967	89035
0070	S	D	E	N	1963	89614
0070	S	D	E	N	1969	89618
0070	S	S	E	N	1976	90192
0070	S	D	E	N	1956	91695
0070	S	D	E	N	1973	92930
0070	S	E	E	N	1973	92930
0070	S	D	E	N	1978	94680
0070	S	E	E	N	1978	94680
0070	S	D	E	N	1978	98172
0070	S	E	E	N	1978	98172
0070	S	D	E	N	1979	99928
0070	S	D	E	N	1979	100616
0072	L	D	E	H	1973	70719
0072	M	D	R	N	1967	46172
0072	M	D	E	N	1965	89617
0072	M	D	E	N	1973	89619
0072	M	D	E	N	1948	94419
0072	M	E	E	N	1948	94419
0072	M	D	E	N	1978	96710
0072	P	D	R	N	1966	41480
0072	P	T	R	N	1966	41480
0072	P	D	F	N	1965	42373
0072	P	T	F	N	1965	42373
0072	P	D	E	N	1964	45753
0072	P	D	E	H	1969	53598
0072	P	D	E	H	1969	57235
0072	P	D	E	N	1966	59491
0072	P	T	E	N	1966	59491
0072	P	D	E	N	1949	61706
0072	P	D	E	N	1977	94057
0072	P	D	E	N	1948	94419
0072	P	E	E	N	1948	94419
0072	S	D	E		1951	810
0072	S	D	E	N	1933	16320
0072	S	D	E	N	1940	20780
0072	S	D	E	N	1921	22633
0072	S	G	E	N	1959	23900
0072	S	D	E	N	1947	25437
0072	S	D	E	N	1964	25860
0072	S	D	E	N	1963	28116
0072	S	G	E	N	1940	28780
0072	S	G	E	N	1939	28787
0072	S	G	E	N	1912	31883
0072	S	G	E	N	1914	31884
0072	S	D	E	N	1965	35554
0072	S	D	E	N	1965	40782
0072	S	T	E	N	1965	40782
0072	S	D	E	N	1966	43774
0072	S	D	E	N	1956	44868
0072	S	D	E	N	1964	45753
0072	S	D	E	N	1966	54110
0072	S	T	E	N	1966	54110
0072	S	D	G	N	1968	54724
0072	S	E	G	N	1968	54724
0072	S	D	G	H	1968	55499
0072	S	E	G	H	1968	55499
0072	S	D	R	N	1968	57555
0072	S	D	R	H	1969	58786
0072	S	S	E	N	1970	61233
0072	S	D	E	N	1949	61706
0072	S	D	P	N	1971	63225
0072	S	D	F	N	1971	64435
0072	S	D	E	N	1972	64938
0072	S	E	E	N	1972	64938
0072	S	D	R	N	1971	67298
0072	S	D	E	N	1972	67570
0072	S	D	E	N	1972	67571
0072	S	T	E	N	1972	67571
0072	S	D	R	N	1972	71302
0072	S	D	E	N	1974	73939
0072	S	T	E	N	1974	73939
0072	S	D	R	N	1969	74008
521-0072	S	D	E	N	1969	74009
0072	S	G	E	N	1971	74012
0072	S	D	E	N	1971	74013
0072	S	D	E	N	1972	78340
0072	S	E	E	N	1972	78340
0072	S	D	E	N	1975	82018
0072	S	D	F	N	1976	88190
0072	S	D	E	N	1965	89617
0072	S	D	E	N	1977	92134
0072	S	S	E	N	1977	92134
0072	S	D	E	N	1956	92919
0072	S	D	E	N	1977	93606
0072	S	G	E	N	1955	94431
0072	S	D	R	N	1976	94572
0072	S	D	E	N	1978	94674
0072	S	E	E	N	1978	94674
0072	S	D	J	N	1978	96703
0072	S	E	J	N	1978	96703
0072	S	D	E	N	1979	98105
0072	S	D	E	N	1980	100244
0072	S	E	E	N	1980	100244
0072	S	D	E	N	1979	100606
0073	E	D	E	N	1966	41219
0073	E	D	R	N	1965	41987
0073	E	D	E	N	1965	42456
0073	E	T	E	N	1965	42456
0073	E	D	E	N	1965	42457
0073	E	T	E	N	1965	42457
0073	E	D	E	N	1967	49193
0073	E	D	R	N	1969	56893
0073	E	D	E	N	1972	66055
0073	E	D	E	N	1974	73843
0073	E	E	E	N	1974	73843
0073	M	D	E	L	1956	19362
0073	M	D	E	N	1962	27842
0073	M	G	E	N	1962	27842
0073	M	T	E	N	1962	27842
0073	M	D	R	N	1965	41987
0073	M	D	R	N	1959	42504
0073	P	D	E		1955	9739
0073	P	D	R	N	1939	13474
0073	P	D	E	N	1959	13954
0073	P	D	E	L	1959	17069
0073	P	D	E	L	1956	19362
0073	P	D	E	N	1959	20933
0073	P	G	E	N	1962	27842
0073	P	D	G	N	1960	28816
0073	P	D	E	N	1960	28824
0073	P	G	E	N	1963	33467
0073	P	D	E	N	1965	35535
0073	P	D	E	N	1968	37491
0073	P	D	E	N	1964	38646
0073	P	D	E	N	1965	42374
0073	P	D	R	N	1959	42504
0073	P	D	E	N	1957	45699
0073	P	D	E	N	1964	45753
0073	P	D	E	L	1968	47499
0073	P	D	E	N	1968	47499
0073	P	T	E	L	1968	47499
0073	P	T	E	N	1968	47499
0073	P	D	E	N	1961	47935
0073	P	D	R	N	1966	48083
0073	P	D	E	N	1967	49193
0073	P	D	R	N	1967	49687
0073	P	T	R	N	1967	49687
0073	P	D	R	N	1968	51661
0073	P	T	R	N	1968	51661
0073	P	D	E	N	1968	51662
0073	P	T	E	N	1968	51662
0073	P	D	R	N	1967	55858
0073	P	D	E	N	1961	56385
0073	P	G	E	N	1969	59176
0073	P	D	E	N	1968	60177
0073	P	D	E	N	1967	63347
0073	P	D	R	N	1971	65687
0073	P	D	E	N	1961	83559
0073	P	E	E	N	1961	83559
0073	S	E	E		1955	7333
0073	S	D	R	N	1939	13474
0073	S	D	E	N	1959	16875
0073	S	D	E	N	1956	16895
0073	S	D	E	N	1921	22633
0073	S	D	R	N	1962	27694
0073	S	D	E	N	1962	29846
0073	S	D	R	N	1965	39987
0073	S	D	R	N	1965	39988
0073	S	D	E	N	1965	42374
0073	S	S	E	N	1970	61233
0075	P	D	E	N	1971	62371
0075	P	D	E	N	1973	76783
0075	P	T	E	N	1973	76783
0075	S	G	G		1925	7028
0075	S	D	R	N	1934	15457
0075	S	T	R	N	1934	15457
0075	S	G	G	N	1928	15610
0075	S	D	E	N	1892	16421
0075	S	D	E	N	1940	20780
0075	S	D	G	N	1927	21664
0075	S	S	G	N	1928	28680
0075	S	D	G	N	1884	28778
0075	S	D	E	N	1940	28780

Phys. State: **C.** Superconductive; **D.** Doped; **E.** Expanded; **F.** Fibrous or Whisker; **G.** Gas; **I.** Ionized or Plasma; **L.** Liquid; **M.** Multiphase; **P.** Powder or Fine Particle; **S.** Solid; **T.** Thick or Thin Film

Subject: **D.** Data; **E.** Experiment; **G.** General (Data + Theory + Experiment); **S.** Survey (Review, Compendium etc.); **T.** Theory

Language: **C.** Czech; **D.** Dutch; **E.** English; **F.** French; **G.** German; **I.** Italian; **J.** Japanese; **O.** Other Languages; **P.** Polish; **R.** Russian; **S.** Spanish

Temperature: **F.** Full Range (Low + Normal + High); **L.** Low (0 to 75K) + Overlap into Normal; **N.** Normal (75 to 1273K); **H.** High (above 1273K) + Overlap into Normal; Blank = Not Coded

Substance Number	Phys. State	Subject	Language	Temperature	Year	TPRC Number
521-0075	S	D	E	N	1928	33185
0075	S	E	F	N	1847	45048
0075	S	D	E	N	1969	55729
0075	S	D	E	L	1969	55762
0075	S	S	E	N	1970	61233
0075	S	D	E	N	1973	73516
0075	S	S	E	N	1976	90192
0075	S	S	E	N	1977	92134
0075	S	D	E	N	1979	97680
0077	E	D	G	N	1960	25175
0077	P	D	F	N	1910	24647
0077	P	D	R	N	1969	57572
0077	P	D	E	N	1971	64041
0077	S	D	E		1943	1287
0077	S	G	E		1951	1411
0077	S	D	E		1953	6200
0077	S	D	E		1958	6939
0077	S	D	E		1952	6990
0077	S	D	E		1959	7696
0077	S	D	G	H	1932	14367
0077	S	T	G	H	1932	14367
0077	S	D	G	N	1936	14889
0077	S	T	G	N	1936	14889
0077	S	S	E	N	1960	24931
0077	S	D	F	N	1965	42373
0077	S	T	F	N	1965	42373
0077	S	S	E	N	1970	61233
0082	P	D	E	N	1971	62371
0082	S	D	E		1943	1287
0082	S	D	E		1958	6939
0082	S	D	E		1954	7104
0082	S	G	G	N	1928	15610
0082	S	D	E	N	1969	55729
0082	S	S	E	N	1970	61233
0082	S	D	R	N	1971	64741
0082	S	D	E	N	1971	64742
0082	S	D	E	N	1973	68985
0082	S	G	E	N	1971	74012
0083	F	D	E		1949	6929
0083	P	D	E	N	1971	62371
0083	S	D	E		1954	7336
0083	S	D	F	N	1941	13299
0083	S	D	E	N	1933	13898
0083	S	D	J	N	1959	19561
0083	S	G	R	N	1958	20245
0083	S	D	E	N	1920	20414
0083	S	G	E	N	1962	20436
0083	S	D	G	N	1912	20505
0083	S	D	F	N	1895	28730
0083	S	D	G	N	1904	28731
0083	S	D	E	N	1927	39852
0083	S	D	E	N	1955	40671
0083	S	T	E	N	1955	40671
0083	S	E	F	N	1847	45048
0083	S	D	E	N	1966	45502
0083	S	D	E	N	1969	55729
0083	S	G	R	N	1971	62997
0083	S	G	E	N	1971	72195
0084	M	G	E		1954	1963
0084	M	D	E	N	1976	87714
0084	M	E	E	N	1976	87714
0084	P	G	E		1954	1963
0084	P	D	E	N	1971	62371
0084	S	D	O		1947	1958
0084	S	D	E	N	1940	20780
0084	S	D	E	N	1921	22633
0084	S	D	E	N	1963	28116
0084	S	G	E	N	1939	28787
0084	S	E	E	N	1963	29811
0084	S	D	E	N	1954	35118
0084	S	D	F	N	1965	42373
0084	S	T	F	N	1965	42373
0084	S	D	E	N	1969	54631
0084	S	D	G	N	1968	55392
0084	S	T	G	N	1968	55392
0084	S	D	E	N	1969	55729
0084	S	D	E	L	1969	55762
0084	S	S	E	N	1970	61233
0084	S	D	E	N	1969	61242
0084	S	D	I	N	1969	61242
0084	S	D	E	N	1973	73516
0084	S	D	E	N	1969	89618
0084	S	S	E	N	1976	90192
0084	S	D	E	N	1956	91695
0084	S	D	E	N	1977	92134
0084	S	S	E	N	1977	92134
0084	S	D	R	N	1976	94572
0084	S	D	E	N	1979	97680
0085	F	D	E	N	1937	14744
0085	M	D	E	N	1960	17349
0085	M	T	E	N	1960	17349
0085	M	D	E	N	1956	27825
0085	M	D	E	N	1941	43835
0085	M	D	E	N	1976	85570
0085	M	T	E	N	1976	85570
0085	M	D	E	N	1976	87714
0085	M	E	E	N	1976	87714
0085	M	D	E	N	1945	89612
0085	M	D	E	N	1976	92954
0085	P	D	G	H	1959	17892
0085	P	D	E	H	1962	29103
521-0085	P	D	E	H	1969	53598
0085	S	D	F	N	1941	13299
0085	S	G	J	N	1940	16239
0085	S	D	E	N	1951	16342
0085	S	D	R	N	1960	20043
0085	S	D	E	N	1940	20780
0085	S	D	E	N	1939	21536
0085	S	D	E	N	1921	22633
0085	S	D	E	N	1963	24010
0085	S	D	E	N	1961	24326
0085	S	D	E	N	1924	24551
0085	S	D	R	N	1960	24772
0085	S	D	E	N	1940	25372
0085	S	D	E	N	1947	25437
0085	S	D	E	N	1960	25797
0085	S	D	E	N	1964	25860
0085	S	G	R	N	1962	27608
0085	S	D	E	N	1956	27825
0085	S	D	E	N	1940	27917
0085	S	D	E	N	1963	28116
0085	S	S	G	N	1928	28680
0085	S	D	E	N	1940	28780
0085	S	D	E	N	1951	28792
0085	S	G	E	N	1962	29477
0085	S	G	R	N	1961	29478
0085	S	G	E	N	1912	31883
0085	S	D	E	N	1931	31895
0085	S	G	E	N	1963	32761
0085	S	D	E	N	1964	35413
0085	S	D	E	N	1966	41593
0085	S	D	E	N	1965	43156
0085	S	G	E	N	1964	43773
0085	S	D	E	N	1966	43774
0085	S	D	E	N	1941	43835
0085	S	D	E	N	1966	45502
0085	S	D	E	N	1963	48294
0085	S	D	E	N	1958	48883
0085	S	D	E	N	1967	50428
0085	S	G	E	N	1960	51286
0085	S	D	E	N	1968	53178
0085	S	T	E	N	1968	53178
0085	S	D	E	N	1969	54631
0085	S	D	G	N	1968	55392
0085	S	T	G	N	1968	55392
0085	S	D	E	N	1971	61009
0085	S	T	E	N	1971	61009
0085	S	S	E	H	1970	61233
0085	S	S	E	N	1970	61233
0085	S	D	E	N	1969	61242
0085	S	D	I	N	1969	61242
0085	S	D	E	N	1972	64938
0085	S	E	E	N	1972	64938
0085	S	D	E	N	1973	73516
0085	S	D	R	N	1969	74008
0085	S	D	E	N	1969	74009
0085	S	G	E	N	1971	74012
0085	S	D	E	N	1973	77418
0085	S	D	E	N	1975	82040
0085	S	T	E	N	1975	82040
0085	S	D	F	N	1976	88190
0085	S	D	E	N	1945	89612
0085	S	D	E	N	1963	89614
0085	S	D	E	N	1969	89618
0085	S	S	E	N	1976	90192
0085	S	D	O	N	1976	90397
0085	S	E	O	N	1976	90397
0085	S	D	E	N	1977	92134
0085	S	S	E	N	1977	92134
0085	S	S	E	N	1970	92968
0085	S	D	I	N	1970	92968
0085	S	D	E	N	1977	93606
0085	S	D	E	N	1980	100244
0085	S	E	E	N	1980	100244
0086	M	D	F		1954	7703
0086	P	D	E	N	1971	62371
0086	S	G	G	N	1928	15610
0086	S	D	G	N	1927	21664
0086	S	D	E	N	1928	33185
0086	S	D	E	L	1964	45417
0086	S	T	E	L	1964	45417
0086	S	D	E	N	1969	55729
0087	S	D	G	L	1911	15576
0087	S	G	G	N	1928	15610
0089	P	D	E	N	1971	62371
0089	S	D	E	N	1969	55729
0091	P	D	E	N	1971	62371
0091	S	D	E	N	1969	55729
0092	P	D	E	N	1971	62371
0092	S	D	E		1958	6939
0092	S	D	E		1953	7036
0092	S	D	E		1953	7102
0092	S	D	E		1954	7104
0092	S	E	E		1952	8818
0092	S	D	E	H	1955	10110
0092	S	D	E	N	1956	10518
0092	S	D	G	N	1936	14889
0092	S	T	G	N	1936	14889
0092	S	D	E	N	1951	25435
0092	S	D	E	N	1969	55729
0093	P	D	E	N	1971	62371
0093	S	D	E	N	1969	55729
521-0095	S	G	F		1906	6720
0096	P	D	E	N	1971	62371
0096	S	D	E		1958	6939
0096	S	D	E		1953	7102
0096	S	G	G	N	1928	15610
0096	S	D	G	N	1927	21664
0096	S	D	E	N	1928	33185
0099	P	D	E	N	1971	62371
0099	S	D	R	N	1960	27562
0099	S	D	E	N	1969	55729
0102	S	D	E		1949	6929
0103	G	G	G		1881	8295
0103	M	D	F	N	1963	36058
0103	M	D	E	N	1941	43835
0103	P	D	E		1953	768
0103	P	S	E	N	1970	61233
0103	S	D	E		1951	810
0103	S	D	G	N	1955	1125
0103	S	E	G	N	1955	1125
0103	S	D	G		1911	6708
0103	S	G	G		1925	7028
0103	S	G	E		1900	9389
0103	S	D	E	N	1936	11107
0103	S	D	G	N	1933	11172
0103	S	D	R	N	1934	15457
0103	S	T	R	N	1934	15457
0103	S	G	G	N	1928	15610
0103	S	D	G	N	1959	16208
0103	S	D	F	N	1827	16209
0103	S	D	E	N	1892	16421
0103	S	G	G	N	1957	20061
0103	S	D	E	N	1940	20780
0103	S	D	E	N	1921	21495
0103	S	D	E	N	1921	22633
0103	S	D	E	N	1921	25099
0103	S	D	E	N	1947	25437
0103	S	D	F	N	1895	28730
0103	S	D	E	N	1940	28780
0103	S	G	E	N	1872	31896
0103	S	G	E	N	1898	35675
0103	S	D	E	N	1966	41593
0103	S	D	E	N	1966	42030
0103	S	D	F	N	1965	42373
0103	S	T	F	N	1965	42373
0103	S	D	E	N	1960	42824
0103	S	D	E	N	1941	43835
0103	S	D	E	N	1966	45502
0103	S	D	E	N	1900	48245
0103	S	D	G	N	1968	54724
0103	S	E	G	N	1968	54724
0103	S	D	F	N	1970	57876
0103	S	S	E	N	1970	61233
0103	S	D	E	N	1972	64938
0103	S	E	E	N	1972	64938
0103	S	D	F	N	1976	88190
0103	S	S	E	N	1976	90192
0103	S	D	O	N	1976	90397
0103	S	E	O	N	1976	90397
0103	S	S	E	N	1977	92134
0103	S	D	E	N	1977	93606
0103	S	D	R	N	1976	94572
0104	P	D	E		1953	768
0104	P	D	E	N	1971	62371
0104	S	D	E		1954	7104
0104	S	D	G	N	1905	38729
0104	S	D	E	N	1969	55729
0104	S	D	E	N	1971	63145
0104	S	E	E	N	1971	63145
0105	P	D	E	N	1971	62371
0105	S	D	E		1949	6929
0105	S	D	E	N	1969	55729
0107	P	D	E		1958	6939
0107	S	D	G	N	1967	70791
0107	S	E	G	N	1967	70791
0107	S	D	E	N	1973	70792
0107	S	E	E	N	1973	70792
0108	E	D	E		1951	810
0108	E	D	E	N	1959	20933
0108	E	D	E	N	1938	29136
0108	E	D	E	N	1943	40228
0108	M	D	R	N	1964	38189
0108	M	D	E	N	1941	43835
0108	P	D	E	N	1957	10271
0108	P	D	E	N	1921	22633
0108	P	D	R	N	1931	31857
0108	P	D	R	N	1968	49504
0108	P	E	R	N	1968	49504
0108	P	D	E	N	1968	65123
0108	P	E	E	N	1968	65123
0108	S	D	E		1942	5258
0108	S	D	E		1949	6929
0108	S	D	E		1923	7648
0108	S	G	E		1956	9943
0108	S	D	G	N	1933	11172
0108	S	D	E	N	1960	14171
0108	S	D	E	N	1935	14959
0108	S	D	R	N	1959	15835
0108	S	D	E	N	1892	16421
0108	S	D	E	N	1930	23196
0108	S	D	E	N	1919	23435
0108	S	D	E	N	1955	25109

Phys. State: **C.** Superconductive; **D.** Doped; **E.** Expanded; **F.** Fibrous or Whisker; **G.** Gas; **I.** Ionized or Plasma; **L.** Liquid; **M.** Multiphase; **P.** Powder or Fine Particle; **S.** Solid; **T.** Thick or Thin Film
Subject: **D.** Data; **E.** Experiment; **G.** General (Data + Theory + Experiment); **S.** Survey (Review, Compendium etc.); **T.** Theory
Language: **C.** Czech; **D.** Dutch; **E.** English; **F.** French; **G.** German; **I.** Italian; **J.** Japanese; **O.** Other Languages; **P.** Polish; **R.** Russian; **S.** Spanish
Temperature: **F.** Full Range (Low + Normal + High); **L.** Low (0 to 75K) + Overlap into Normal; **N.** Normal (75 to 1273K); **H.** High (above 1273K) + Overlap into Normal; Blank = Not Coded

Substance Number	Phys. State	Subject	Language	Temperature	Year	TPRC Number
521-0108	S	D	E	H	1960	31874
0108	S	D	E	N	1957	35437
0108	S	D	E	N	1965	37045
0108	S	D	E	N	1968	37491
0108	S	D	R	N	1964	38189
0108	S	D	E	N	1941	43835
0108	S	D	E	N	1968	47768
0108	S	D	E	N	1973	70941
0108	S	E	E	N	1975	89330
0109	P	D	E	N	1971	62371
0109	S	D	E	N	1969	55729
0110	M	D	E		1941	4438
0110	M	D	E	N	1960	17349
0110	M	T	E	N	1960	17349
0110	M	G	G	N	1952	19030
0110	M	D	E	N	1952	21187
0110	M	T	E	N	1952	21187
0110	M	G	E	N	1961	25813
0110	M	D	E	N	1962	29528
0110	M	D	E	N	1907	31019
0110	M	G	E	N	1959	31893
0110	M	G	E	N	1909	31920
0110	M	D	E	N	1965	36377
0110	M	D	E	N	1966	42695
0110	M	G	E	N	1966	43424
0110	M	D	F	N	1967	45088
0110	M	D	E	N	1899	48324
0110	M	D	G	N	1958	51235
0110	M	D	R	N	1959	51567
0110	M	E	R	N	1959	51567
0110	M	G	E	N	1973	76238
0110	M	D	R	N	1970	92946
0110	M	E	R	N	1970	92946
0110	P	D	E		1942	8797
0110	P	D	R		1957	9500
0110	P	D	G	H	1959	17892
0110	P	G	R	N	1958	20245
0110	P	G	E	N	1962	20436
0110	P	D	E	N	1921	22633
0110	P	D	G	N	1929	22931
0110	P	D	F	N	1910	24647
0110	P	G	E	N	1896	25164
0110	P	D	E	N	1947	25437
0110	P	G	E	N	1961	25813
0110	P	D	O	N	1931	28679
0110	P	D	G	N	1910	28694
0110	P	G	E	N	1961	29059
0110	P	D	E	H	1962	29103
0110	P	D	E	N	1961	29364
0110	P	D	E	N	1962	29786
0110	P	G	R	N	1939	29889
0110	P	D	E	N	1905	30693
0110	P	G	E	N	1872	31896
0110	P	G	E	N	1959	31906
0110	P	G	E	N	1909	31920
0110	P	D	R	N	1956	32405
0110	P	D	I	N	1963	32627
0110	P	D	R	N	1964	35710
0110	P	D	E	N	1965	36377
0110	P	D	E	N	1962	39516
0110	P	D	E	N	1905	41683
0110	P	D	E	N	1964	42307
0110	P	D	E	N	1966	42709
0110	P	D	R	N	1967	46172
0110	P	D	G	N	1958	51235
0110	P	G	E	N	1968	51330
0110	P	D	R	N	1960	51563
0110	P	D	E	N	1968	53178
0110	P	D	E	H	1969	53598
0110	P	D	E	N	1969	55156
0110	P	D	E	H	1969	57235
0110	P	D	E	N	1945	61037
0110	P	D	E	N	1949	61706
0110	P	D	F	N	1965	62182
0110	P	D	E	N	1973	70628
0110	P	D	R	N	1970	73669
0110	P	D	E	N	1970	73670
0110	P	D	E	N	1973	76783
0110	P	T	E	N	1973	76783
0110	P	D	E	N	1960	77341
0110	P	S	E	N	1976	90192
0110	S	G	E		1956	742
0110	S	D	E		1942	1264
0110	S	T	E		1942	1264
0110	S	D	R		1946	1734
0110	S	D	E		1942	7694
0110	S	E	E	N	1953	10123
0110	S	S	E	N	1958	10667
0110	S	D	E	N	1966	11559
0110	S	G	E	N	1949	12033
0110	S	D	G	N	1960	14148
0110	S	G	E	N	1938	14188
0110	S	D	E	N	1960	15640
0110	S	D	E	N	1958	15736
0110	S	D	G	N	1956	16078
0110	S	G	E	N	1960	16203
0110	S	G	G	N	1950	23243
0110	S	D	R	N	1960	24204
0110	S	G	E	N	1955	24261
0110	S	D	E	N	1924	25145
0110	S	G	E	N	1905	25461

Substance Number	Phys. State	Subject	Language	Temperature	Year	TPRC Number
521-0110	S	D	E	N	1922	26034
0110	S	D	E	N	1951	28792
0110	S	D	E	N	1965	39028
0110	S	G	E	N	1966	39168
0110	S	D	E	N	1966	39355
0110	S	D	E	N	1953	39860
0110	S	D	E	N	1964	40682
0110	S	D	E	N	1964	40900
0110	S	G	E	N	1966	43424
0110	S	D	F	N	1967	45088
0110	S	D	E	N	1966	45843
0110	S	G	R	N	1956	49765
0110	S	D	E	N	1953	55981
0110	S	S	E	N	1970	61233
0110	S	D	E	N	1949	61706
0110	S	D	S	N	1971	63639
0110	S	E	S	N	1971	63639
0110	S	D	E	N	1971	65828
0110	S	E	E	N	1971	65828
0110	S	D	E	N	1975	75982
0110	S	D	E	N	1956	91695
0111	M	D	O	N	1977	99357
0111	M	E	O	N	1977	99357
0111	P	D	R	N	1964	38488
0111	S	G	R	N	1960	24338
0111	S	D	E	N	1959	25275
0111	S	D	R	N	1959	25276
0111	S	D	R	N	1964	38488
0115	S	D	E	N	1966	44533
0115	S	S	R	N	1966	47360
0115	S	D	R	N	1966	48082
0115	S	D	E	N	1968	51892
0115	S	G	E	N	1969	58484
0115	S	D	E	N	1975	80507
0115	S	E	E	N	1975	80507
0115	S	D	E	N	1977	93608
0115	S	E	E	N	1977	93608
0115	S	T	E	N	1979	98525
0115	S	D	E	N	1978	98927
0115	S	E	E	N	1978	98927
0116	L	D	R	N	1970	61258
0116	L	E	R	N	1970	61258
0116	L	D	E	N	1970	67224
0116	L	E	E	N	1970	67224
0116	M	S	E	N	1976	90192
0116	P	D	E	N	1959	10898
0116	P	D	E	N	1959	20933
0116	P	G	E	N	1962	22559
0116	P	D	E	N	1962	29551
0116	P	D	E	N	1957	45699
0116	P	D	R	N	1968	49504
0116	P	E	R	N	1968	49504
0116	P	D	E	N	1969	53610
0116	P	D	E	N	1971	62371
0116	P	D	E	N	1968	65123
0116	P	E	E	N	1968	65123
0116	S	D	E		1954	7104
0116	S	D	E		1957	9974
0116	S	D	E	N	1923	10331
0116	S	D	E	N	1960	18662
0116	S	D	E	N	1920	22166
0116	S	D	E	N	1919	23435
0116	S	D	E	N	1924	24551
0116	S	D	E	N	1963	25061
0116	S	D	E	N	1955	25109
0116	S	D	O	N	1949	28661
0116	S	S	G	N	1928	28680
0116	S	D	E	N	1962	29551
0116	S	D	E	N	1950	31538
0116	S	D	E	N	1967	35150
0116	S	D	E	N	1957	35437
0116	S	D	E	N	1965	37045
0116	S	D	E	N	1966	42107
0116	S	D	E	N	1969	55729
0116	S	S	E	N	1970	61233
0116	S	D	E	N	1919	73665
0123	E	D	E	N	1965	42374
0123	F	D	E	N	1970	38913
0123	F	T	E	N	1970	38913
0123	F	D	I	N	1966	41292
0123	F	D	I	N	1966	44990
0123	F	D	E	N	1969	53585
0123	F	D	R	N	1970	62437
0123	F	T	R	N	1970	62437
0123	F	G	R	N	1972	69818
0123	F	G	E	N	1972	69819
0123	L	D	E	H	1970	38148
0123	L	D	E	H	1973	70719
0123	M	D	E	N	1976	87966
0123	P	D	E	N	1961	29364
0123	P	D	E	N	1964	35550
0123	P	D	E	N	1968	37601
0123	P	E	E	N	1968	37601
0123	P	D	E	N	1964	38646
0123	P	D	E	N	1962	39516
0123	P	D	F	N	1965	42373
0123	P	T	F	N	1965	42373
0123	P	D	E	N	1964	45753
0123	P	G	E	N	1966	46250
0123	P	G	E	N	1963	49023
0123	P	D	E	N	1966	49738

Substance Number	Phys. State	Subject	Language	Temperature	Year	TPRC Number
521-0123	P	D	E	N	1969	54820
0123	P	D	E	N	1969	56491
0123	P	T	E	N	1969	56491
0123	P	D	E	N	1970	59561
0123	P	D	E	N	1969	63002
0123	P	E	E	N	1969	63002
0123	P	D	E	N	1971	64053
0123	P	E	E	N	1971	64053
0123	P	D	R	N	1971	69768
0123	P	D	E	N	1972	69769
0123	P	D	E	N	1973	73810
0123	P	D	E	N	1970	74010
0123	P	E	E	N	1970	74010
0123	P	D	E	N	1973	76783
0123	P	T	E	N	1973	76783
0123	S	D	G		1950	1628
0123	S	D	G	N	1933	11172
0123	S	D	E	N	1921	22633
0123	S	D	E	N	1924	24551
0123	S	D	E	N	1963	28116
0123	S	D	F	N	1895	28730
0123	S	D	G	N	1904	28731
0123	S	G	E	N	1914	31884
0123	S	D	E	N	1965	35554
0123	S	D	E	H	1970	38148
0123	S	D	R	N	1965	44434
0123	S	D	E	N	1964	44969
0123	S	D	E	N	1964	45753
0123	S	G	E	N	1966	46250
0123	S	D	G	N	1968	55392
0123	S	T	G	N	1968	55392
0123	S	S	E	N	1970	61233
0123	S	D	P	N	1971	63225
0123	S	D	E	N	1972	67570
0123	S	D	E	N	1973	76783
0123	S	T	E	N	1973	76783
0123	S	D	E	N	1972	78340
0123	S	E	E	N	1972	78340
0123	S	D	R	N	1975	83260
0123	S	D	E	N	1969	89618
0123	S	D	E	N	1976	91244
0123	S	D	E	N	1977	92134
0123	S	S	E	N	1977	92134
0123	S	D	E	N	1975	92233
0123	S	E	E	N	1975	92233
0123	S	D	E	N	1977	93606
0123	S	D	E	N	1978	98902
0123	S	D	E	N	1978	99991
0123	S	D	E	N	1980	100244
0123	S	E	E	N	1980	100244
0130	P	D	E	N	1971	62371
0130	S	D	E		1943	1287
0130	S	D	E		1958	6939
0130	S	D	E		1953	7102
0130	S	D	E		1954	7104
0130	S	D	E		1953	9404
0130	S	G	G	N	1928	15610
0130	S	D	G	N	1927	21664
0130	S	S	E	N	1960	24931
0130	S	D	E	N	1928	33185
0130	S	D	E	N	1969	55729
0130	S	S	E	N	1970	61233
0130	S	D	E	N	1973	68985
0130	S	G	E	N	1971	74012
0131	P	D	E	N	1971	62371
0131	S	D	E		1943	1287
0131	S	G	G	N	1928	15610
0131	S	D	G	N	1927	21664
0131	S	D	E	N	1928	33185
0131	S	D	E	N	1969	55729
0131	S	S	E	N	1970	61233
0131	S	D	E	N	1973	68985
0132	P	G	E		1958	8133
0132	P	G	E	N	1960	19745
0132	S	D	E		1943	1287
0132	S	D	E		1958	6939
0132	S	D	E		1953	7036
0132	S	D	E		1953	7102
0132	S	D	G	H	1932	14367
0132	S	T	G	H	1932	14367
0133	P	D	E	N	1971	62371
0133	S	D	E	N	1948	40389
0133	S	D	E	N	1969	55729
0135	P	D	E	N	1921	22633
0135	P	G	G	N	1963	30734
0135	P	D	E	N	1969	53610
0135	P	D	R	N	1971	65687
0135	S	D	R	N	1959	15835
0135	S	D	R	N	1958	19650
0135	S	D	E	H	1961	27783
0145	L	D	S	N	1970	92963
0145	S	E	G		1941	2523
0145	S	D	S	N	1970	92963
0169	S	D	R	N	1960	20281
0170	M	D	E	N	1948	94419
0170	M	E	E	N	1948	94419
0170	P	D	E	N	1948	94419
0170	P	E	E	N	1948	94419
0170	S	D	G	N	1927	21664
0170	S	D	E	N	1928	33185
0174	F	D	R	H	1966	42004

Phys. State: **C.** Superconductive; **D.** Doped; **E.** Expanded; **F.** Fibrous or Whisker; **G.** Gas; **I.** Ionized or Plasma; **L.** Liquid; **M.** Multiphase; **P.** Powder or Fine Particle; **S.** Solid; **T.** Thick or Thin Film

Subject: **D.** Data; **E.** Experiment; **G.** General (Data + Theory + Experiment); **S.** Survey (Review, Compendium etc.); **T.** Theory

Language: **C.** Czech; **D.** Dutch; **E.** English; **F.** French; **G.** German; **I.** Italian; **J.** Japanese; **O.** Other Languages; **P.** Polish; **R.** Russian; **S.** Spanish

Temperature: **F.** Full Range (Low + Normal + High); **L.** Low (0 to 75K) + Overlap into Normal; **N.** Normal (75 to 1273K); **H.** High (above 1273K) + Overlap into Normal; Blank = Not Coded

Substance Number	Phys. State	Subject	Language	Temperature	Year	TPRC Number
521-0174	F	D	E	H	1966	43896
0174	F	G	R	N	1972	69818
0174	F	G	E	N	1972	69819
0174	S	D	E	N	1973	71892
0180	S	D	E	N	1959	18202
0181	M	D	E	N	1941	43835
0181	M	S	E	N	1976	90192
0181	S	D	E	N	1936	11107
0181	S	D	G	N	1933	11172
0181	S	D	E	N	1892	16421
0181	S	D	E	N	1960	17320
0181	S	D	E	N	1940	20780
0181	S	G	E	N	1896	25164
0181	S	D	E	N	1951	25559
0181	S	D	E	N	1940	27917
0181	S	D	E	N	1931	31895
0181	S	G	E	N	1872	31896
0181	S	D	E	N	1966	42030
0181	S	D	E	N	1941	43835
0181	S	D	E	N	1964	44968
0181	S	D	G	N	1968	55392
0181	S	T	G	N	1968	55392
0181	S	D	R	N	1968	58386
0181	S	S	E	N	1970	61233
0181	S	S	E	N	1976	90192
0181	S	D	E	N	1977	92134
0181	S	S	E	N	1977	92134
0181	S	D	R	N	1976	94572
0182	S	G	E	N	1954	12405
0183	E	D	F	N	1953	12592
0183	L	D	R	N	1970	61258
0183	L	E	R	N	1970	61258
0183	L	D	E	N	1970	67224
0183	L	E	E	N	1970	67224
0183	M	D	E	N	1976	92954
0183	P	G	E	N	1954	12405
0183	S	D	F	N	1965	42373
0183	S	T	F	N	1965	42373
0183	S	D	R	N	1968	58386
0184	S	D	E	N	1951	16342
0184	S	D	E	N	1966	45502
0184	S	D	E	N	1952	51540
0184	S	S	E	N	1977	92134
0184	S	D	E	N	1978	94674
0184	S	E	E	N	1978	94674
0186	P	D	E	N	1921	22633
0186	S	S	E	N	1958	10667
0186	S	D	E	N	1976	87028
0190	S	D	E	N	1924	25145
0190	S	D	E	N	1922	26034
0190	S	D	E	N	1965	44633
0190	S	E	E	N	1965	44633
0190	S	S	E	N	1970	61233
0190	S	D	E	N	1973	72674
0190	S	E	E	N	1973	72674
0191	S	D	E	N	1960	10052
0191	S	D	E	L	1952	21176
0191	S	D	E	N	1952	21176
0191	S	D	E	L	1962	27279
0191	S	D	E	L	1963	27317
0191	S	T	E	L	1963	27317
0191	S	D	E	L	1965	35028
0191	S	D	E	N	1973	70944
0192	P	D	G	N	1913	15603
0192	P	D	E	N	1971	62371
0192	S	D	G	N	1913	15603
0192	S	D	R	N	1960	27562
0192	S	D	E	N	1965	34423
0192	S	D	R	N	1966	44372
0195	P	D	E	N	1971	62371
0195	S	D	E	N	1969	55729
0196	E	D	E	N	1974	74696
0196	E	T	E	N	1974	74696
0196	E	D	E	N	1977	90289
0196	L	D	E	H	1973	70719
0196	M	D	E	N	1960	17349
0196	M	T	E	N	1960	17349
0196	M	D	E	N	1960	19024
0196	M	D	E	N	1961	20265
0196	M	D	E	N	1961	25737
0196	M	D	E	N	1956	27825
0196	M	G	E	N	1962	32632
0196	M	D	E	N	1964	32876
0196	M	T	E	N	1964	32876
0196	M	D	E	N	1961	39858
0196	M	D	E	N	1941	43835
0196	M	D	E	N	1965	44339
0196	M	D	E	N	1968	48921
0196	M	T	E	N	1968	48921
0196	M	D	E	N	1968	51693
0196	M	D	E	N	1968	51754
0196	M	T	E	N	1968	51754
0196	M	D	R	N	1969	54587
0196	M	D	E	N	1970	58368
0196	M	T	E	N	1970	58368
0196	M	D	E	N	1968	74222
0196	M	D	E	N	1976	85570
0196	M	T	E	N	1976	85570
0196	M	D	E	N	1973	89620
0196	M	D	E	N	1977	91242
0196	M	G	E	N	1973	92930
521-0196	M	D	E	N	1977	93609
0196	M	S	E	N	1977	93609
0196	M	T	E	N	1977	93609
0196	P	D	E	N	1968	74222
0196	S	D	G		1952	1444
0196	S	D	E	N	1958	12174
0196	S	D	E	N	1933	16320
0196	S	T	E	N	1944	16642
0196	S	D	E	N	1961	20265
0196	S	D	G	N	1912	20505
0196	S	D	E	N	1939	21536
0196	S	D	G	N	1920	22016
0196	S	D	E	N	1921	22633
0196	S	D	E	N	1963	24010
0196	S	D	E	N	1940	25372
0196	S	D	E	N	1947	25437
0196	S	D	E	N	1962	27283
0196	S	D	E	N	1956	27825
0196	S	D	G	N	1904	28731
0196	S	D	F	N	1909	28745
0196	S	E	E	N	1963	29811
0196	S	D	E	N	1964	35413
0196	S	D	J	N	1961	39818
0196	S	D	E	N	1951	40424
0196	S	D	E	N	1966	41593
0196	S	S	E	N	1966	42284
0196	S	D	E	N	1965	43156
0196	S	G	E	N	1964	43773
0196	S	D	E	N	1966	43774
0196	S	D	E	N	1941	43835
0196	S	D	E	N	1965	44339
0196	S	D	E	N	1963	44970
0196	S	D	E	N	1966	45502
0196	S	D	E	N	1963	48294
0196	S	D	E	N	1968	48921
0196	S	T	E	N	1968	48921
0196	S	D	E	N	1967	50428
0196	S	G	E	N	1960	51286
0196	S	D	E	N	1968	51693
0196	S	D	E	N	1969	54631
0196	S	D	G	N	1968	55392
0196	S	T	G	N	1968	55392
0196	S	S	E	N	1970	61233
0196	S	D	S	N	1971	63639
0196	S	E	S	N	1971	63639
0196	S	D	E	N	1972	64938
0196	S	E	E	N	1972	64938
0196	S	G	E	N	1971	74012
0196	S	D	E	N	1974	74696
0196	S	T	E	N	1974	74696
0196	S	D	E	N	1973	77418
0196	S	D	E	N	1975	82040
0196	S	T	E	N	1975	82040
0196	S	D	F	N	1976	88190
0196	S	D	E	N	1963	89614
0196	S	D	E	N	1973	89620
0196	S	S	E	N	1976	90192
0196	S	D	E	N	1977	91242
0196	S	G	E	N	1973	92930
0196	S	D	E	N	1977	93606
0196	S	D	E	N	1980	100244
0196	S	E	E	N	1980	100244
0215	S	D	E	N	1923	10331
0216	P	D	E	N	1971	62371
0216	S	D	E	N	1969	55729
0217	S	D	G	N	1930	10317
0217	S	S	G	N	1928	28680
0217	S	D	E	H	1931	39843
0217	S	D	F	N	1965	42373
0217	S	T	F	N	1965	42373
0222	S	S	S	H	1955	28264
0223	P	D	E	N	1971	62371
0223	S	D	E	N	1969	55729
0225	P	D	E	N	1964	38646
0225	P	D	E	N	1965	42374
0225	P	D	E	N	1964	45753
0225	P	G	E	N	1963	49023
0236	M	S	E	N	1976	90192
0236	P	D	F	N	1965	42373
0236	P	T	F	N	1965	42373
0236	P	G	F	N	1966	45257
0236	P	D	E	N	1971	62371
0236	S	D	G	N	1933	11172
0236	S	D	E	N	1951	16342
0236	S	D	E	N	1921	22633
0236	S	D	E	N	1940	25372
0236	S	D	G	N	1904	28731
0236	S	D	E	N	1964	45753
0236	S	D	E	N	1969	55729
0239	S	D	R	N	1974	76265
0240	M	S	E	N	1976	90192
0240	P	D	E	N	1971	62371
0240	S	D	E	N	1937	13404
0240	S	D	E	N	1937	13405
0240	S	D	E	N	1921	22633
0240	S	D	E	N	1930	23196
0240	S	D	G	N	1968	51994
0240	S	D	E	N	1969	55729
0240	S	D	R	N	1972	68113
0240	S	D	E	N	1972	68114
0240	S	D	E	N	1974	77865
521-0240	S	E	E	N	1974	77865
0240	S	D	E	N	1975	81256
0240	S	E	E	N	1975	81256
0240	S	D	E	L	1976	86877
0240	S	E	E	L	1976	86877
0240	S	D	E	L	1977	89460
0240	S	D	E	N	1977	89460
0240	S	E	E	L	1977	89460
0240	S	E	E	N	1977	89460
0240	S	S	E	N	1977	92134
0241	P	D	E	N	1971	62371
0241	S	D	E	N	1937	13404
0241	S	D	E	N	1937	13405
0241	S	D	E	N	1930	23196
0241	S	D	E	N	1943	40228
0241	S	D	G	N	1968	51994
0241	S	D	E	N	1969	55729
0241	S	S	E	N	1970	61233
0241	S	D	G	N	1971	63749
0241	S	E	G	N	1971	63749
0241	S	D	R	N	1972	68113
0241	S	D	E	N	1972	68114
0241	S	D	E	N	1974	77865
0241	S	E	E	N	1974	77865
0241	S	D	E	N	1975	81256
0241	S	E	E	N	1975	81256
0241	S	D	E	L	1977	89460
0241	S	D	E	N	1977	89460
0241	S	E	E	L	1977	89460
0241	S	E	E	N	1977	89460
0241	S	S	E	N	1977	92134
0242	L	D	E	H	1973	70719
0242	P	D	E	N	1964	38646
0242	P	D	E	N	1965	42374
0242	P	G	E	N	1963	49023
0242	S	D	R	N	1960	20043
0242	S	D	E	N	1964	44969
0242	S	D	E	N	1964	45753
0242	S	D	G	N	1968	54724
0242	S	E	G	N	1968	54724
0242	S	D	E	N	1972	64938
0242	S	E	E	N	1972	64938
0242	S	D	R	N	1971	67298
0242	S	D	E	N	1972	67570
0242	S	D	E	N	1974	73939
0242	S	E	E	N	1974	73939
0242	S	D	E	N	1957	91693
0242	S	D	E	N	1977	91966
0242	S	T	E	N	1977	91966
0242	S	D	E	N	1977	93606
0242	S	D	E	N	1978	94674
0242	S	E	E	N	1978	94674
0242	S	D	E	N	1979	98105
0242	S	D	E	N	1980	100244
0242	S	E	E	N	1980	100244
0243	S	D	E	N	1974	73939
0243	S	E	E	N	1974	73939
0246	P	D	R	N	1964	35710
0246	S	D	G	N	1937	13892
0249	M	D	E	N	1957	25524
0249	S	D	G	N	1940	29870
0249	S	D	G	N	1939	31888
0249	S	D	R	N	1968	51815
0249	S	D	R	N	1968	54490
0250	P	G	F	N	1966	45257
0250	P	D	E	N	1971	62371
0250	S	D	E	N	1959	13944
0250	S	D	G	N	1936	14889
0250	S	T	G	N	1936	14889
0250	S	D	E	H	1960	15874
0250	S	D	E	H	1951	18353
0250	S	S	E	N	1960	24931
0250	S	D	E	N	1965	37045
0250	S	D	E	N	1969	55729
0250	S	G	E	N	1971	74012
0251	P	D	E	N	1971	62371
0251	S	D	G	N	1936	14889
0251	S	T	G	N	1936	14889
0251	S	D	F	N	1895	28730
0251	S	D	E	N	1969	54631
0251	S	D	G	N	1968	55392
0251	S	T	G	N	1968	55392
0251	S	D	E	N	1969	55729
0251	S	D	E	N	1977	89894
0251	S	E	E	N	1977	89894
0251	S	D	E	N	1956	91695
0251	S	D	E	N	1978	94672
0251	S	E	E	N	1978	94672
0251	S	D	E	N	1974	98070
0253	M	D	E	N	1960	16286
0253	M	D	E	N	1976	89094
0253	M	D	E	N	1964	89616
0253	M	S	E	N	1976	90192
0253	M	D	E	N	1977	92038
0253	P	D	E	N	1970	61725
0253	P	D	E	N	1971	65225
0253	P	D	E	N	1972	68859
0253	P	D	E	N	1973	72792
0253	P	D	E	N	1968	74222
0253	P	D	E	N	1972	77763
0253	S	D	E	N	1959	14812

Phys. State: **C.** Superconductive; **D.** Doped; **E.** Expanded; **F.** Fibrous or Whisker; **G.** Gas; **I.** Ionized or Plasma; **L.** Liquid; **M.** Multiphase; **P.** Powder or Fine Particle; **S.** Solid; **T.** Thick or Thin Film
Subject: **D.** Data; **E.** Experiment; **G.** General (Data + Theory + Experiment); **S.** Survey (Review, Compendium etc.); **T.** Theory
Language: **C.** Czech; **D.** Dutch; **E.** English; **F.** French; **G.** German; **I.** Italian; **J.** Japanese; **O.** Other Languages; **P.** Polish; **R.** Russian; **S.** Spanish
Temperature: **F.** Full Range (Low + Normal + High); **L.** Low (0 to 75K) + Overlap into Normal; **N.** Normal (75 to 1273K); **H.** High (above 1273K) + Overlap into Normal; Blank = Not Coded

Substance Number	Phys. State	Subject	Language	Temperature	Year	TPRC Number
521-0253	S	D	E	N	1954	16331
0253	S	D	E	N	1968	48263
0253	S	D	E	N	1969	54631
0253	S	D	E	N	1975	92233
0253	S	E	E	N	1975	92233
0254	P	D	E	N	1972	68349
0254	S	D	R	H	1960	18939
0254	S	D	E	H	1960	19727
0257	S	D	E	H	1960	15874
0258	P	D	E	N	1964	31231
0258	P	D	G	H	1964	37519
0258	P	E	G	H	1964	37519
0258	P	D	G	H	1971	66895
0258	S	D	E	H	1960	15874
0258	S	D	E	N	1963	24342
0258	S	T	E	N	1963	24342
0258	S	G	E	N	1940	28780
0258	S	D	E	H	1967	34350
0258	S	D	R	H	1966	41414
0258	S	D	E	N	1968	47311
0258	S	D	G	H	1968	55499
0258	S	E	G	H	1968	55499
0258	S	D	E	N	1969	55729
0258	S	D	R	H	1969	58786
0258	S	G	E	H	1970	58983
0258	S	D	E	H	1972	64938
0258	S	E	E	H	1972	64938
0258	S	D	G	H	1971	66895
0258	S	D	E	N	1972	67571
0258	S	T	E	N	1972	67571
0258	S	D	E	H	1972	68734
0258	S	E	E	H	1972	68734
0258	S	S	E	N	1977	92134
0258	S	D	E	H	1970	94446
0258	S	D	E	N	1970	94446
0258	S	T	E	H	1970	94446
0258	S	T	E	N	1970	94446
0258	S	D	E	N	1978	95858
0259	P	S	E	N	1969	41316
0259	P	T	E	N	1965	44465
0259	S	D	R	N	1955	15774
0259	S	T	R	N	1955	15774
0259	S	D	E	N	1959	15778
0259	S	T	E	N	1959	15778
0259	S	D	E	N	1960	20866
0259	S	T	E	N	1960	20866
0259	S	D	E	N	1960	20867
0259	S	T	E	N	1960	20867
0259	S	D	E	L	1962	28041
0259	S	T	E	L	1962	28041
0259	S	G	E	N	1962	29137
0259	S	D	E	N	1967	40657
0259	S	S	E	N	1969	41316
0259	S	S	E	N	1966	45535
0259	S	D	E	N	1968	50610
0259	S	T	E	N	1968	50610
0259	S	S	E	L	1970	59124
0259	S	S	E	N	1971	68503
0259	S	S	E	N	1973	73308
0261	P	D	E	N	1971	62371
0261	S	D	E	N	1969	55729
0261	S	S	E	N	1977	92134
0263	S	D	E	N	1933	16320
0263	S	D	E	N	1951	16342
0264	L	D	E	H	1973	70719
0264	M	D	E	N	1941	43835
0264	M	D	E	N	1976	87966
0264	S	D	E	N	1933	16320
0264	S	D	E	N	1940	20780
0264	S	D	E	N	1964	25860
0264	S	G	E	N	1940	28780
0264	S	D	E	N	1941	43835
0264	S	D	E	N	1964	44969
0264	S	D	E	N	1966	45502
0264	S	D	G	N	1968	54724
0264	S	E	G	N	1968	54724
0264	S	S	E	N	1970	61233
0264	S	D	E	N	1972	64938
0264	S	E	E	N	1972	64938
0264	S	D	E	N	1972	67570
0264	S	D	E	N	1972	67571
0264	S	T	E	N	1972	67571
0264	S	D	E	N	1974	73939
0264	S	E	E	N	1974	73939
0264	S	D	F	N	1976	88190
0264	S	D	E	N	1977	92134
0264	S	S	E	N	1977	92134
0264	S	D	R	N	1976	94572
0264	S	D	E	N	1978	94674
0264	S	E	E	N	1978	94674
0265	S	D	E	N	1951	16342
0265	S	D	E	N	1952	51540
0265	S	D	E	N	1972	67570
0265	S	D	E	N	1972	67571
0265	S	T	E	N	1972	67571
0266	S	D	E	N	1951	16342
0266	S	D	E	N	1940	20780
0266	S	D	E	N	1952	51540
0270	S	D	E	N	1951	16342
0270	S	D	R	N	1971	67298
0270	S	D	F	N	1976	88190
521-0270	S	D	E	N	1964	89615
0270	S	S	E	N	1976	90192
0270	S	D	E	N	1956	92919
0271	S	D	E	N	1951	16342
0272	L	D	E	H	1973	70719
0272	S	D	E	N	1951	16342
0272	S	D	E	N	1921	22633
0272	S	D	E	N	1964	45753
0272	S	D	E	N	1973	89620
0272	S	D	E	N	1977	90746
0273	S	D	E	N	1951	16342
0274	S	D	E	N	1951	16342
0274	S	D	E	N	1921	22633
0274	S	D	E	N	1964	25860
0274	S	D	E	N	1964	44966
0274	S	D	E	N	1963	44970
0274	S	D	E	N	1966	45502
0274	S	D	R	N	1971	67298
0274	S	D	R	N	1972	71302
0274	S	D	R	N	1967	74004
0274	S	D	E	N	1967	74005
0274	S	D	O	N	1976	90397
0274	S	E	O	N	1976	90397
0274	S	D	E	N	1978	94674
0274	S	E	E	N	1978	94674
0274	S	D	E	N	1978	94680
0274	S	E	E	N	1978	94680
0274	S	D	E	N	1979	97505
0275	S	D	E	N	1951	16342
0276	S	D	E	N	1951	16342
0277	M	D	E	N	1976	87966
0277	S	D	E	N	1951	16342
0279	P	D	E	N	1947	25437
0279	P	D	E	N	1945	61037
0279	S	D	E	N	1951	16342
0279	S	D	E	N	1940	20780
0279	S	G	E	N	1940	28780
0279	S	S	E	N	1970	61233
0279	S	D	E	N	1974	73939
0279	S	T	E	N	1974	73939
0279	S	D	E	N	1971	74013
0280	S	D	E	N	1951	16342
0281	S	D	E	N	1951	16342
0281	S	D	E	N	1921	22633
0282	S	D	E	N	1951	16342
0286	S	D	E	N	1951	16342
0287	S	D	E	N	1951	16342
0292	S	D	E	N	1951	16342
0292	S	G	E	N	1941	31862
0293	S	D	E	N	1951	16342
0293	S	D	E	N	1921	22633
0293	S	D	G	N	1968	54724
0293	S	E	G	N	1968	54724
0293	S	D	E	N	1972	64938
0293	S	E	E	N	1972	64938
0293	S	D	R	N	1971	67298
0293	S	D	E	N	1971	74013
0293	S	D	E	N	1977	90746
0293	S	D	R	N	1976	94572
0295	S	D	E	N	1951	16342
0296	S	D	E	N	1951	16342
0296	S	D	R	N	1972	71302
0298	M	D	R	N	1968	56196
0298	M	D	E	N	1972	65531
0298	M	D	E	N	1973	89620
0298	P	D	E	N	1921	22633
0298	S	D	E	N	1966	11559
0298	S	D	G	N	1956	16078
0298	S	D	E	N	1921	22633
0298	S	D	R	H	1961	24125
0298	S	D	R	N	1961	24125
0298	S	D	G	H	1962	27817
0298	S	G	E	N	1939	28787
0298	S	D	E	N	1931	31895
0298	S	D	E	N	1954	35118
0298	S	D	E	N	1965	35554
0298	S	D	E	N	1951	40424
0298	S	D	F	N	1965	42373
0298	S	T	F	N	1965	42373
0298	S	D	R	N	1965	44434
0298	S	D	E	N	1963	44970
0298	S	D	E	N	1966	45502
0298	S	D	R	N	1969	74008
0298	S	D	E	N	1969	74009
0298	S	G	E	N	1971	74012
0298	S	D	E	N	1972	78340
0298	S	E	E	N	1972	78340
0298	S	D	E	N	1977	90289
0298	S	D	O	N	1976	90397
0298	S	E	O	N	1976	90397
0298	S	S	E	N	1977	92134
0298	S	D	E	N	1977	93606
0298	S	D	E	N	1980	100244
0298	S	E	E	N	1980	100244
0299	P	D	E	N	1920	20414
0299	P	D	E	N	1921	40025
0299	P	D	E	N	1916	48034
0299	S	S	G	H	1928	28680
0299	S	S	E	N	1928	31217
0299	S	D	E	N	1957	31948
0299	S	S	G	N	1924	33616
521-0299	S	S	E	F	1969	37682
0299	S	S	E	N	1966	42284
0299	S	D	E	N	1970	57604
0299	S	T	E	N	1970	57604
0299	S	T	R	N	1972	66234
0299	S	S	P	N	1975	84618
0303	P	D	E	N	1971	62371
0303	S	D	E	N	1969	55729
0307	L	D	E	H	1970	38148
0307	S	D	E	H	1970	38148
0307	S	D	E	N	1964	45753
0307	S	D	G	H	1968	55499
0307	S	E	G	H	1968	55499
0307	S	D	R	H	1969	58786
0307	S	D	E	H	1972	64938
0307	S	E	E	H	1972	64938
0307	S	D	E	N	1955	68323
0308	F	G	R	N	1972	69818
0308	F	G	E	N	1972	69819
0308	P	D	E	N	1969	59671
0308	S	D	E	N	1940	20780
0308	S	G	E	N	1940	28780
0308	S	D	E	N	1967	36055
0308	S	D	E	N	1964	37463
0308	S	D	E	N	1951	40424
0308	S	D	E	N	1964	44969
0308	S	D	E	N	1960	44970
0308	S	D	R	N	1967	45103
0308	S	D	G	N	1968	54724
0308	S	E	G	N	1968	54724
0308	S	D	G	H	1968	55499
0308	S	E	G	H	1968	55499
0308	S	D	P	N	1971	63225
0308	S	D	E	H	1972	64938
0308	S	D	E	N	1972	64938
0308	S	E	E	H	1972	64938
0308	S	E	E	N	1972	64938
0308	S	D	E	N	1972	67570
0308	S	D	E	N	1974	73939
0308	S	T	E	N	1974	73939
0308	S	D	R	N	1967	74004
0308	S	D	E	N	1967	74005
0308	S	D	E	N	1977	90289
0308	S	D	E	N	1977	92134
0308	S	S	E	N	1977	92134
0308	S	D	R	N	1976	94572
0330	L	D	E	H	1970	38148
0330	S	D	E	H	1970	38148
0330	S	D	E	N	1977	90746
0341	S	G	J	N	1940	16239
0346	P	D	E	N	1971	62371
0346	S	D	E	N	1960	20514
0346	S	D	E	N	1969	55729
0346	S	S	E	N	1970	61233
0346	S	G	E	N	1971	74012
0347	M	D	E	N	1973	89620
0347	M	D	E	N	1976	92954
0347	M	S	E	N	1977	93609
0347	S	D	E	N	1940	20780
0347	S	D	G	N	1968	55392
0347	S	T	G	N	1968	55392
0347	S	G	E	N	1971	74012
0347	S	D	E	N	1975	82040
0347	S	T	E	N	1975	82040
0347	S	D	E	N	1973	89620
0347	S	S	E	N	1976	90192
0347	S	D	E	N	1977	92134
0347	S	S	E	N	1977	92134
0347	S	D	E	N	1976	92954
0348	M	D	E	N	1941	43835
0348	S	D	E	N	1940	20780
0348	S	G	E	N	1940	28780
0348	S	D	E	N	1941	43835
0348	S	D	E	N	1966	45502
0348	S	D	E	N	1972	67570
0348	S	D	E	N	1972	67571
0348	S	T	E	N	1972	67571
0348	S	D	E	N	1974	73939
0348	S	T	E	N	1974	73939
0348	S	D	E	N	1977	92134
0348	S	S	E	N	1977	92134
0349	M	D	E	N	1965	89617
0349	S	D	E	N	1940	20780
0349	S	D	E	N	1921	22633
0349	S	D	E	N	1964	25860
0349	S	D	F	N	1965	42373
0349	S	T	F	N	1965	42373
0349	S	D	E	N	1963	44970
0349	S	D	E	N	1966	45502
0349	S	D	G	N	1968	55392
0349	S	T	G	N	1968	55392
0349	S	D	R	N	1971	67298
0349	S	D	E	N	1972	68206
0349	S	D	E	N	1974	73939
0349	S	T	E	N	1974	73939
0349	S	D	E	N	1971	74013
0349	S	D	E	N	1965	89617
0349	S	S	E	N	1976	90192
0349	S	D	E	N	1956	92919
0349	S	D	E	N	1979	97505
0350	S	D	E	N	1940	20780

Phys. State: **C.** Superconductive; **D.** Doped; **E.** Expanded; **F.** Fibrous or Whisker; **G.** Gas; **I.** Ionized or Plasma; **L.** Liquid; **M.** Multiphase; **P.** Powder or Fine Particle; **S.** Solid; **T.** Thick or Thin Film

Subject: **D.** Data; **E.** Experiment; **G.** General (Data + Theory + Experiment); **S.** Survey (Review, Compendium etc.); **T.** Theory

Language: **C.** Czech; **D.** Dutch; **E.** English; **F.** French; **G.** German; **I.** Italian; **J.** Japanese; **O.** Other Languages; **P.** Polish; **R.** Russian; **S.** Spanish

Temperature: **F.** Full Range (Low + Normal + High); **L.** Low (0 to 75K) + Overlap into Normal; **N.** Normal (75 to 1273K); **H.** High (above 1273K) + Overlap into Normal; Blank = Not Coded

Substance Number	Phys. State	Subject	Language	Temperature	Year	TPRC Number	Substance Number	Phys. State	Subject	Language	Temperature	Year	TPRC Number	Substance Number	Phys. State	Subject	Language	Temperature	Year	TPRC Number
521-0350	S	G	E	N	1940	28780	521-0375	S	D	G	N	1968	54724	521-0442	S	G	E	N	1966	43424
0350	S	S	E	N	1977	92134	0375	S	E	G	N	1968	54724	0442	S	D	E	N	1965	44633
0351	S	D	E	N	1940	20780	0375	S	D	P	N	1971	63225	0442	S	E	E	N	1965	44633
0351	S	G	E	N	1940	28780	0375	S	D	E	N	1972	64938	0442	S	D	E	N	1953	55981
0352	S	D	E	N	1940	20780	0375	S	E	E	N	1972	64938	0442	S	D	R	N	1967	60317
0352	S	D	E	N	1967	35150	0375	S	D	E	N	1957	91693	0442	S	D	E	N	1970	60318
0352	S	D	E	N	1966	42107	0376	S	D	E	N	1921	22633	0443	P	D	E	N	1971	62371
0352	S	D	E	H	1966	59827	0377	S	D	E	N	1921	22633	0443	S	D	E	N	1955	25109
0352	S	D	E	N	1972	67570	0378	S	D	E	N	1921	22633	0443	S	D	E	N	1957	35437
0352	S	D	E	N	1972	67571	0379	S	D	E	N	1921	22633	0443	S	D	E	N	1969	55729
0352	S	T	E	N	1972	67571	0380	S	D	E	N	1921	22633	0443	S	S	E	N	1970	61233
0352	S	D	E	H	1972	68734	0381	S	D	E	N	1921	22633	0455	S	G	E	N	1939	28787
0352	S	E	E	H	1972	68734	0382	P	G	E	H	1962	29477	0468	S	D	E	N	1951	40424
0352	S	D	E	N	1974	73939	0382	P	G	R	H	1961	29478	0471	P	D	E	N	1971	62371
0352	S	E	E	N	1974	73939	0382	P	G	R	H	1962	29903	0471	S	D	E	N	1969	55729
0352	S	D	E	N	1978	95858	0382	P	D	R	N	1959	51567	0475	S	E	E	N	1963	29811
0353	S	D	E	N	1940	20780	0382	P	E	R	N	1959	51567	0477	S	D	E	N	1972	67570
0353	S	G	E	N	1940	28780	0382	S	G	E	H	1962	29477	0477	S	D	E	N	1972	68206
0354	S	D	E	N	1940	20780	0382	S	G	R	H	1961	29478	0477	S	D	R	N	1967	74004
0354	S	G	E	N	1940	28780	0382	S	G	R	H	1962	29903	0477	S	D	E	N	1967	74005
0354	S	D	E	N	1972	67570	0382	S	D	G	N	1905	38729	0481	S	D	F	N	1965	42373
0354	S	D	E	N	1974	73939	0382	S	D	F	N	1965	42373	0481	S	T	F	N	1965	42373
0354	S	T	E	N	1974	73939	0382	S	T	F	N	1965	42373	0482	S	D	E	N	1924	24551
0354	S	D	E	N	1971	74013	0382	S	D	E	N	1964	44968	0482	S	S	G	N	1928	28680
0354	S	D	E	N	1979	100605	0382	S	D	E	N	1960	44970	0483	S	D	E	N	1940	27917
0354	S	E	E	N	1979	100605	0382	S	D	R	H	1976	84491	0483	S	G	E	N	1872	31896
0356	S	D	E	N	1940	20780	0382	S	D	R	N	1976	84491	0485	S	D	E	N	1940	27917
0356	S	D	R	N	1967	74004	0383	S	D	E	N	1921	22633	0489	P	D	E	N	1971	62371
0356	S	D	E	N	1967	74005	0384	M	D	R	N	1960	20247	0492	S	G	E	N	1939	28787
0362	M	S	E	N	1976	90192	0384	P	D	R	N	1971	67531	0492	S	D	E	N	1954	35118
0362	P	D	E	N	1971	62371	0384	P	D	R	N	1959	69694	0494	S	G	E	N	1940	28780
0362	S	D	E	N	1921	22633	0384	P	D	E	N	1972	69695	0494	S	D	E	N	1972	67571
0362	S	G	E	N	1955	32862	0384	S	D	R	N	1959	69694	0494	S	T	E	N	1972	67571
0362	S	D	E	N	1969	55729	0384	S	D	E	N	1972	69695	0494	S	S	E	N	1977	92134
0362	S	D	E	N	1972	64938	0392	P	D	E	N	1971	62371	0495	S	G	E	N	1940	28780
0362	S	E	E	N	1972	64938	0392	S	D	E	N	1966	45502	0495	S	D	P	N	1971	63225
0362	S	G	E	N	1971	70966	0392	S	D	E	N	1969	55729	0495	S	D	E	N	1972	67571
0362	S	S	E	N	1976	90192	0392	S	D	E	N	1965	88950	0495	S	T	E	N	1972	67571
0362	S	D	J	N	1977	90890	0393	P	D	R	N	1964	38488	0497	P	D	G	N	1910	28694
0363	S	D	E	N	1921	22633	0393	P	D	R	N	1960	51563	0500	P	G	E	N	1964	40403
0364	S	D	E	N	1921	22633	0393	P	D	R	N	1969	57572	0500	P	D	E	N	1966	42967
0364	S	D	E	N	1965	40782	0393	P	D	E	N	1971	64041	0500	P	D	R	N	1971	65687
0364	S	T	E	N	1965	40782	0393	P	D	R	N	1971	65687	0500	P	D	R	N	1971	65854
0364	S	D	G	N	1968	54724	0393	S	D	E	H	1961	27783	0500	P	D	E	N	1971	73366
0364	S	E	G	N	1968	54724	0393	S	D	R	N	1964	38488	0500	S	D	O	H	1957	17692
0364	S	D	E	N	1972	64938	0393	S	S	E	N	1970	61233	0500	S	D	E	N	1937	22568
0364	S	E	E	N	1972	64938	0394	P	D	E	N	1949	61706	0500	S	G	R	N	1960	24338
0365	S	D	E	N	1921	22633	0394	S	S	G	N	1928	28680	0502	S	D	G	N	1904	28731
0366	P	D	E	N	1971	62371	0394	S	D	E	N	1965	40782	0514	F	D	E	N	1964	78047
0366	S	D	E	N	1921	22633	0394	S	T	E	N	1965	40782	0514	F	E	E	N	1964	78047
0366	S	D	E	H	1937	28449	0394	S	D	E	N	1949	61706	0514	P	D	E	N	1971	62371
0366	S	D	E	H	1964	28591	0394	S	D	R	N	1971	67298	0514	S	D	E	N	1969	55729
0366	S	D	R	H	1963	30257	0397	S	D	E	N	1939	21536	0514	S	S	E	N	1970	61233
0366	S	D	R	H	1965	42464	0402	M	D	E	N	1966	42708	0517	S	G	E	L	1958	24015
0368	M	D	R	N	1960	20247	0402	P	G	E	N	1965	44150	0517	S	D	E	N	1978	95858
0368	M	G	E	N	1909	31920	0402	S	D	E	N	1939	21536	0525	S	D	E	N	1951	40424
0368	M	D	E	N	1948	94419	0402	S	D	E	N	1969	56053	0530	M	G	E	N	1909	31920
0368	M	E	E	N	1948	94419	0409	S	D	E	N	1923	21665	0530	P	D	E		1952	6309
0368	P	G	E	N	1909	31920	0411	P	D	E	N	1971	62371	0530	P	G	E	N	1909	31920
0368	P	D	E	N	1945	61037	0411	S	D	E	N	1963	24342	0530	S	G	G	N	1950	23243
0368	P	D	E	N	1949	61706	0411	S	T	E	N	1963	24342	0530	S	D	E	N	1953	39860
0368	P	D	E	N	1948	94419	0411	S	D	E	N	1969	55729	0530	S	D	E	N	1953	55981
0368	P	E	E	N	1948	94419	0412	P	D	E	N	1971	62371	0530	S	D	E	N	1969	56053
0368	S	G	E	N	1970	58782	0412	S	D	E	N	1963	24342	0530	S	D	R	N	1967	60317
0368	S	D	R	N	1967	60317	0412	S	T	E	N	1963	24342	0530	S	D	E	N	1970	60318
0368	S	D	E	N	1970	60318	0412	S	D	E	N	1969	55729	0534	P	D	E	N	1971	62371
0368	S	D	E	N	1949	61706	0412	S	D	E	L	1971	62719	0534	S	G	E	N	1939	28787
0368	S	G	E	N	1974	82019	0412	S	D	E	L	1971	67372	0542	P	D	E	H	1969	53598
0369	S	D	R	N	1930	21451	0412	S	G	E	N	1971	74012	0542	P	D	E	H	1969	57235
0369	S	D	E	N	1921	22633	0413	S	D	E	N	1963	24342	0551	M	S	E	N	1976	90192
0369	S	D	R	N	1960	24772	0413	S	T	E	N	1963	24342	0551	P	G	E	N	1962	22559
0369	S	D	E	N	1963	28116	0430	M	S	E	N	1976	90192	0551	S	D	E	N	1963	25061
0369	S	D	E	N	1964	45753	0430	P	D	E	N	1971	62371	0551	S	D	E	N	1951	25559
0369	S	G	E	N	1960	51286	0430	S	D	E	N	1969	55729	0551	S	D	E	N	1962	29551
0369	S	D	G	N	1968	54724	0430	S	D	R	N	1971	67298	0551	S	D	E	N	1943	40228
0369	S	E	G	N	1968	54724	0430	S	D	E	N	1974	77865	0552	P	D	E	N	1971	62371
0369	S	S	E	H	1970	61233	0430	S	E	E	N	1974	77865	0552	S	D	E	H	1967	34350
0369	S	S	E	N	1970	61233	0430	S	D	E	N	1975	81256	0552	S	D	E	H	1967	34861
0369	S	D	E	N	1972	64938	0430	S	E	E	N	1975	81256	0552	S	D	E	N	1969	55729
0369	S	E	E	N	1972	64938	0432	S	D	R	N	1960	24772	0552	S	G	E	H	1970	58983
0369	S	G	E	N	1971	74012	0432	S	D	E	N	1940	27917	0552	S	G	E	N	1971	74012
0369	S	D	E	N	1975	82018	0432	S	D	G	N	1904	28731	0552	S	D	E	N	1965	88950
0370	S	D	E	N	1921	22633	0441	S	G	R	N	1960	24338	0564	S	D	E	N	1964	25860
0370	S	D	E	N	1964	25860	0442	M	D	E	N	1907	31019	0565	M	D	G	N	1962	30126
0370	S	D	E	N	1966	45502	0442	M	G	F	N	1965	39350	0566	P	D	E	N	1971	62371
0371	M	D	E	N	1977	91242	0442	M	G	R	N	1959	43002	0566	S	D	E	N	1969	55729
0371	S	D	E	N	1939	21536	0442	M	G	F	N	1959	43003	0567	S	D	E	N	1964	25860
0371	S	D	E	N	1921	22633	0442	M	G	E	N	1966	43424	0567	S	D	E	N	1965	35554
0371	S	D	E	N	1967	50428	0442	M	D	E	H	1970	61713	0569	S	D	E	N	1964	25860
0371	S	D	E	N	1973	77418	0442	M	T	E	H	1970	61713	0570	S	D	E	N	1964	25860
0371	S	S	E	N	1976	90192	0442	P	D	E	N	1967	47508	0572	S	D	E	N	1920	20414
0371	S	D	E	N	1977	91242	0442	P	D	R	N	1970	73679	0572	S	D	E	N	1921	40025
0372	S	D	E	N	1921	22633	0442	P	D	E	N	1970	73680	0581	M	G	R	H	1962	29903
0373	S	D	E	N	1921	22633	0442	S	D	R	N	1959	25367	0599	M	D	E	N	1973	89620
0375	M	D	E	N	1973	89620	0442	S	T	R	N	1959	25367	0599	S	D	G	N	1968	55392
0375	P	D	F	N	1965	42373	0442	S	D	R	N	1958	29482	0599	S	T	G	N	1968	55392
0375	P	T	F	N	1965	42373	0442	S	D	E	N	1961	29483	0599	S	D	E	N	1973	89620
0375	S	D	E	N	1921	22633	0442	S	G	F	N	1965	39350	0604	P	D	E	N	1971	62371
0375	S	D	E	N	1964	44968	0442	S	D	E	N	1953	39860	0604	S	D	E	N	1969	55729
0375	S	D	E	N	1964	44969	0442	S	G	R	N	1959	43002	0611	P	D	E	N	1971	62371
0375	S	D	E	N	1952	51540	0442	S	G	F	N	1959	43003	0612	P	D	E	N	1965	44633

Phys. State: **C.** Superconductive; **D.** Doped; **E.** Expanded; **F.** Fibrous or Whisker; **G.** Gas; **I.** Ionized or Plasma; **L.** Liquid; **M.** Multiphase; **P.** Powder or Fine Particle; **S.** Solid; **T.** Thick or Thin Film

Subject: **D.** Data; **E.** Experiment; **G.** General (Data + Theory + Experiment); **S.** Survey (Review, Compendium etc.); **T.** Theory

Language: **C.** Czech; **D.** Dutch; **E.** English; **F.** French; **G.** German; **I.** Italian; **J.** Japanese; **O.** Other Languages; **P.** Polish; **R.** Russian; **S.** Spanish

Temperature: **F.** Full Range (Low + Normal + High); **L.** Low (0 to 75K) + Overlap into Normal; **N.** Normal (75 to 1273K); **H.** High (above 1273K) + Overlap into Normal; Blank = Not Coded

Substance Number	Phys. State	Subject	Language	Temperature	Year	TPRC Number
521-0612	P	E	E	N	1965	44633
0612	P	D	E	N	1949	61706
0612	S	D	E	N	1965	44633
0612	S	E	E	N	1965	44633
0612	S	G	E	N	1970	65069
0613	P	D	E	N	1968	47316
0613	P	T	E	N	1968	47316
0614	P	D	E	N	1971	62371
0614	S	D	E	N	1969	55729
0624	G	S	E	F	1962	31279
0624	G	T	E	H	1973	73315
0624	G	T	E	H	1973	73316
0630	M	G	E	N	1959	31907
0630	P	G	E	N	1959	31907
0630	S	E	E	N	1978	94662
0632	S	D	G	N	1939	31888
0643	M	G	E	N	1909	31920
0643	M	G	E	N	1966	43424
0643	M	D	E	N	1948	94419
0643	M	E	E	N	1948	94419
0643	P	G	E	N	1909	31920
0643	P	D	E	N	1969	52361
0643	P	T	E	N	1969	52361
0643	P	D	E	N	1945	61037
0643	P	D	E	N	1949	61706
0643	P	D	E	N	1948	94419
0643	P	E	E	N	1948	94419
0643	S	D	E	N	1965	39028
0643	S	G	E	N	1966	43424
0643	S	D	E	N	1966	45843
0643	S	D	E	N	1949	61706
0645	M	G	E	N	1909	31920
0645	M	G	R	N	1959	43002
0645	M	G	F	N	1959	43003
0645	M	D	E	H	1970	61713
0645	M	T	E	H	1970	61713
0645	M	D	E	N	1948	94419
0645	M	E	E	N	1948	94419
0645	P	G	E	N	1909	31920
0645	P	D	E	N	1967	47508
0645	P	D	E	H	1969	53598
0645	P	D	E	H	1969	57235
0645	P	D	E	N	1945	61037
0645	P	D	E	N	1949	61706
0645	P	D	E	N	1948	94419
0645	P	E	E	N	1948	94419
0645	S	D	R	N	1959	25367
0645	S	T	R	N	1959	25367
0645	S	D	E	N	1965	39028
0645	S	D	E	N	1943	40228
0645	S	G	R	N	1959	43002
0645	S	G	F	N	1959	43003
0645	S	D	E	N	1968	46410
0645	S	D	E	N	1949	61706
0645	S	D	R	N	1970	73679
0645	S	D	E	N	1970	73680
0647	P	D	R	N	1931	31857
0648	P	D	E	N	1971	62371
0648	S	S	E	N	1960	24931
0648	S	D	E	H	1937	28449
0648	S	D	E	L	1964	33123
0648	S	D	E	N	1957	35437
0648	S	D	E	N	1965	37045
0648	S	D	E	N	1969	55729
0648	S	S	E	L	1970	61233
0648	S	S	E	N	1970	61233
0660	P	D	E	N	1944	59826
0660	S	D	O	H	1966	44322
0671	S	D	E	N	1921	22633
0671	S	D	G	N	1968	54724
0671	S	E	G	N	1968	54724
0671	S	D	E	N	1972	64938
0671	S	E	E	N	1972	64938
0671	S	D	R	N	1972	71302
0675	P	D	R	N	1966	39422
0686	S	D	G	N	1971	63749
0686	S	E	G	N	1971	63749
0687	S	D	G	N	1971	63749
0687	S	E	G	N	1971	63749
0688	S	D	E	N	1978	94680
0688	S	E	E	N	1978	94680
0709	M	D	E	N	1977	91242
0709	S	D	E	N	1967	50428
0709	S	D	E	N	1973	77418
0709	S	D	E	N	1977	91242
0711	S	D	R	N	1967	54554
0713	S	D	G	N	1968	54724
0713	S	E	G	N	1968	54724
0713	S	D	R	N	1976	94572
0715	P	D	E	N	1977	94057
0717	S	D	R	N	1968	58386
0718	S	D	E	N	1967	35150
0718	S	D	E	N	1967	36055
0718	S	D	R	N	1967	45103
0718	S	D	G	H	1968	55499
0718	S	E	G	H	1968	55499
0718	S	D	E	H	1972	64938
0718	S	E	E	H	1972	64938
0727	E	D	E	N	1977	90289
0727	M	D	E	N	1977	91242
0727	S	D	E	N	1958	12174

Substance Number	Phys. State	Subject	Language	Temperature	Year	TPRC Number
521-0727	S	D	E	N	1967	50428
0727	S	G	E	N	1971	74012
0727	S	D	E	N	1973	77418
0727	S	D	E	N	1977	91242
0732	L	D	E		1957	9238
0732	L	D	E	N	1953	10164
0732	L	D	E	H	1958	10703
0732	L	D	E	N	1955	20062
0732	L	S	E	N	1926	20886
0732	L	D	G	N	1923	25148
0732	L	T	E	H	1963	26274
0732	L	T	R	H	1963	26275
0732	L	D	E	N	1964	36391
0732	L	D	R	N	1964	38075
0732	M	D	G	N	1955	16070
0732	M	D	G	N	1959	22957
0732	M	D	E	N	1907	31019
0732	M	G	E	N	1959	31893
0732	M	D	E	N	1974	75083
0732	M	T	E	N	1974	75083
0732	M	D	R	N	1973	77333
0732	M	T	R	N	1973	77333
0732	M	D	E	N	1948	94419
0732	M	E	E	N	1948	94419
0732	P	D	R		1958	8506
0732	P	D	E		1958	9254
0732	P	E	E		1958	9254
0732	P	D	G	N	1955	16070
0732	P	D	G	N	1929	22931
0732	P	D	F	N	1910	24647
0732	P	D	E	N	1934	25152
0732	P	D	E	N	1912	28785
0732	P	D	E	N	1905	30693
0732	P	G	G	N	1963	30734
0732	P	D	E	N	1964	31231
0732	P	G	E	N	1959	31906
0732	P	D	R	N	1964	38488
0732	P	D	F	N	1965	42373
0732	P	T	F	N	1965	42373
0732	P	D	R	N	1961	50701
0732	P	E	R	N	1961	50701
0732	P	D	R	N	1959	51567
0732	P	E	R	N	1959	51567
0732	P	D	R	N	1969	57572
0732	P	D	E	N	1945	61037
0732	P	D	E	N	1949	61706
0732	P	D	E	N	1971	62371
0732	P	D	E	N	1971	64041
0732	P	G	E	N	1974	75084
0732	P	D	E	N	1973	76783
0732	P	T	E	N	1973	76783
0732	P	G	R	N	1973	77332
0732	P	D	E	N	1948	94419
0732	P	E	E	N	1948	94419
0732	S	D	E		1949	56
0732	S	D	E		1957	63
0732	S	D	E		1957	64
0732	S	T	E		1957	64
0732	S	D	E		1952	107
0732	S	D	E		1943	1287
0732	S	G	E		1951	1410
0732	S	G	E		1951	1411
0732	S	D	E		1953	6200
0732	S	D	E		1954	6565
0732	S	G	E		1953	6702
0732	S	D	G		1911	6708
0732	S	G	F		1906	6720
0732	S	D	E		1958	6939
0732	S	D	E		1959	7013
0732	S	G	G		1925	7028
0732	S	D	E		1953	7036
0732	S	D	E		1953	7102
0732	S	D	E		1954	7104
0732	S	G	E		1951	7200
0732	S	D	E		1957	7288
0732	S	D	E		1923	7648
0732	S	D	E		1959	7688
0732	S	D	E		1957	8116
0732	S	D	F	L	1936	8245
0732	S	D	E		1938	8327
0732	S	T	E		1938	8327
0732	S	D	R		1957	8991
0732	S	T	R		1957	8991
0732	S	D	E		1957	9238
0732	S	D	E		1936	9293
0732	S	D	E		1938	9294
0732	S	D	E		1952	9410
0732	S	D	E	N	1936	11107
0732	S	D	G	N	1933	11172
0732	S	D	E	N	1959	13944
0732	S	D	G	H	1932	14367
0732	S	T	G	H	1932	14367
0732	S	D	R	N	1934	15457
0732	S	T	R	N	1934	15457
0732	S	D	G	L	1911	15576
0732	S	G	G	N	1928	15610
0732	S	D	E	N	1892	16421
0732	S	D	G	N	1928	16451
0732	S	T	G	N	1928	16451
0732	S	D	R	H	1960	18939
0732	S	D	E	H	1960	19727

Substance Number	Phys. State	Subject	Language	Temperature	Year	TPRC Number
521-0732	S	D	R	N	1960	20183
0732	S	D	E	H	1961	20530
0732	S	D	E	H	1960	20600
0732	S	D	E	N	1940	20780
0732	S	D	E	L	1952	21176
0732	S	D	E	N	1952	21176
0732	S	S	E	L	1952	21184
0732	S	D	G	N	1951	21301
0732	S	D	E	N	1958	23521
0732	S	D	E	L	1958	23940
0732	S	G	E	L	1958	24015
0732	S	G	R	N	1960	24338
0732	S	D	R	N	1961	24510
0732	S	S	E	N	1957	24970
0732	S	D	E	N	1960	24979
0732	S	D	E	N	1963	25061
0732	S	D	E	N	1955	25109
0732	S	D	G	N	1923	25148
0732	S	D	R	N	1957	25257
0732	S	T	R	N	1957	25257
0732	S	D	E	L	1950	25453
0732	S	D	E	N	1960	26060
0732	S	T	E	H	1963	26274
0732	S	T	R	H	1963	26275
0732	S	D	E	N	1962	26684
0732	S	D	E	N	1961	27117
0732	S	S	E	L	1960	28074
0732	S	S	S	H	1955	28264
0732	S	D	E	H	1937	28449
0732	S	D	O	N	1949	28661
0732	S	S	G	N	1928	28680
0732	S	D	F	N	1895	28730
0732	S	D	G	N	1884	28778
0732	S	D	E	N	1940	28780
0732	S	D	E	N	1962	29551
0732	S	E	E	N	1963	29811
0732	S	D	R	N	1962	30911
0732	S	D	G	N	1935	31878
0732	S	G	E	N	1872	31896
0732	S	S	E	N	1963	32162
0732	S	D	J	N	1964	33029
0732	S	D	E	N	1962	33174
0732	S	D	E	N	1928	33185
0732	S	D	R	N	1961	33393
0732	S	E	E	N	1961	33393
0732	S	D	R	N	1964	38184
0732	S	D	R	N	1964	38488
0732	S	D	E	N	1964	40753
0732	S	D	E	N	1965	40782
0732	S	T	E	N	1965	40782
0732	S	G	E	N	1970	41374
0732	S	G	F	N	1965	42720
0732	S	G	E	N	1964	43773
0732	S	D	R	N	1965	44434
0732	S	D	E	N	1956	44868
0732	S	E	F	N	1847	45048
0732	S	G	F	N	1966	45483
0732	S	D	E	N	1968	47311
0732	S	D	E	N	1968	51931
0732	S	D	E	N	1969	54631
0732	S	D	E	N	1969	55729
0732	S	D	E	L	1968	56698
0732	S	T	E	L	1968	56698
0732	S	G	E	N	1969	58909
0732	S	D	E	N	1949	61706
0732	S	D	P	N	1971	63225
0732	S	D	E	N	1970	63519
0732	S	D	E	N	1971	64422
0732	S	E	E	N	1971	64422
0732	S	G	E	L	1971	66152
0732	S	G	E	N	1971	74012
0732	S	G	E	N	1970	74014
0732	S	D	E	N	1976	85837
0732	S	D	E	N	1976	87714
0732	S	E	E	N	1976	87714
0732	S	S	E	N	1977	92134
0732	S	D	R	N	1976	94572
0732	S	D	E	N	1978	95858
0732	S	D	E	L	1979	97008
0732	S	D	E	N	1979	97680
0735	P	D	R	N	1960	51563
0741	S	D	E	N	1963	48294
0742	S	D	E	N	1966	45502
0743	S	D	E	N	1966	45502
0746	P	D	E	N	1971	62371
0760	P	D	F	N	1965	42373
0760	P	T	F	N	1965	42373
0761	M	D	E	N	1948	94419
0761	M	E	E	N	1948	94419
0761	P	D	F	N	1965	42373
0761	P	T	F	N	1965	42373
0761	P	D	E	N	1945	61037
0761	P	D	E	N	1949	61706
0761	P	D	E	N	1948	94419
0761	P	E	E	N	1948	94419
0761	S	D	E	N	1949	61706
0762	P	D	F	N	1965	42373
0762	P	T	F	N	1965	42373
0762	S	D	E	N	1963	44970
0762	S	D	R	N	1972	71302
0762	S	G	E	N	1955	94431

Phys. State: **C.** Superconductive; **D.** Doped; **E.** Expanded; **F.** Fibrous or Whisker; **G.** Gas; **I.** Ionized or Plasma; **L.** Liquid; **M.** Multiphase; **P.** Powder or Fine Particle; **S.** Solid; **T.** Thick or Thin Film

Subject: **D.** Data; **E.** Experiment; **G.** General (Data + Theory + Experiment); **S.** Survey (Review, Compendium etc.); **T.** Theory

Language: **C.** Czech; **D.** Dutch; **E.** English; **F.** French; **G.** German; **I.** Italian; **J.** Japanese; **O.** Other Languages; **P.** Polish; **R.** Russian; **S.** Spanish

Temperature: **F.** Full Range (Low + Normal + High); **L.** Low (0 to 75K) + Overlap into Normal; **N.** Normal (75 to 1273K); **H.** High (above 1273K) + Overlap into Normal; Blank = Not Coded

Substance Number	Phys. State	Subject	Language	Temperature	Year	TPRC Number
521-0763	S	D	F	N	1965	42373
0763	S	T	F	N	1965	42373
0766	L	D	I	N	1966	42018
0766	L	E	I	N	1966	42018
0768	M	D	E	N	1948	94419
0768	M	E	E	N	1948	94419
0768	P	D	E	N	1949	61706
0768	P	D	E	N	1948	94419
0768	P	E	E	N	1948	94419
0768	S	D	E	N	1949	61706
0769	P	D	E	N	1971	62371
0769	S	D	E	N	1969	55729
0769	S	D	E	H	1972	68734
0769	S	E	E	H	1972	68734
0770	G	T	E	H	1970	59690
0771	P	D	E	N	1977	94194
0771	S	D	R	N	1969	59417
0771	S	D	E	N	1969	59418
0771	S	D	E	N	1977	94194
0772	P	D	E	N	1971	62371
0779	P	D	E	N	1971	62371
0780	P	D	E	N	1971	62371
0783	P	D	E	N	1971	62371
0783	S	D	E	N	1969	55729
0783	S	D	R	N	1978	95686
0783	S	D	E	N	1978	98588
0785	P	D	E	N	1971	62371
0785	S	D	E	N	1969	55729
0789	S	D	E	N	1952	51540
0789	S	D	E	N	1972	67570
0790	P	D	E	N	1971	62371
0791	P	D	E	N	1971	62371
0791	S	D	E	N	1969	55729
0791	S	S	E	N	1977	92134
0794	P	D	E	N	1971	62371
0794	S	D	E	N	1969	55729
0797	S	G	E	N	1964	44867
0797	S	D	E	N	1956	44868
0812	M	D	R	N	1969	54785
0812	M	D	E	N	1969	54786
0812	M	D	E	N	1970	61541
0812	S	D	E	N	1966	47059
0812	S	E	E	N	1968	59573
0812	S	D	E	N	1971	68244
0819	P	D	E	N	1971	62371
0822	P	D	E	N	1967	47508
0823	S	D	E	N	1978	95858
0825	S	D	E	N	1964	44969
0825	S	D	E	N	1965	46855
0825	S	D	R	N	1971	67298
0825	S	D	O	N	1976	90397
0825	S	E	O	N	1976	90397
0825	S	D	E	N	1977	90745
0828	S	D	E	N	1977	92038
0830	S	D	E	N	1963	44970
0831	S	D	E	N	1963	44970
0832	S	D	E	N	1963	44970
0833	S	D	E	N	1963	44970
0835	S	D	E	N	1963	44970
0836	S	D	E	N	1964	44966
0836	S	D	E	N	1964	44968
0837	S	D	E	N	1964	44968
0839	S	D	E	N	1964	44968
0840	S	D	E	N	1964	44968
0841	S	D	E	N	1962	44967
0842	S	D	E	N	1964	44969
0843	S	D	E	N	1964	44969
0844	S	D	E	N	1964	44969
0845	S	D	E	N	1964	44966
0846	S	D	E	N	1964	44966
0847	S	D	E	N	1964	44966
0848	S	D	E	N	1964	44969
0849	S	D	E	N	1964	44969
0850	S	D	E	N	1964	44969
0852	S	D	R	N	1971	63961
0852	S	D	E	N	1971	63962
0852	S	D	E	N	1973	74813
0852	S	D	R	N	1974	76265
0861	S	D	E	N	1969	55729
0864	P	D	E	H	1969	53598
0864	P	D	E	H	1969	57235
0865	P	D	E	H	1969	53598
0865	P	D	E	H	1969	57235
0866	P	D	E	H	1969	53598
0866	P	D	E	H	1969	57235
0878	E	D	F	H	1977	92686
0878	E	D	F	N	1977	92686
0878	P	D	R	N	1964	38488
0878	S	S	E	L	1960	16701
0878	S	S	E	N	1960	16701
0878	S	D	R	H	1964	36402
0878	S	D	E	H	1964	36403
0878	S	D	R	N	1964	38488
0878	S	S	E	L	1970	61233
0878	S	S	E	N	1970	61233
0878	S	D	J	H	1972	67883
0878	S	E	J	H	1972	67883
0878	S	S	E	N	1974	79241
0878	S	S	F	N	1975	94378
0887	P	D	E	N	1964	38646
0887	P	D	E	N	1964	45753

Substance Number	Phys. State	Subject	Language	Temperature	Year	TPRC Number
521-0887	P	G	E	N	1963	49023
0894	S	D	R	N	1969	58631
0894	S	D	R	N	1972	71302
0894	S	D	E	N	1977	90745
0894	S	D	R	N	1976	94572
0904	S	D	E	N	1972	67570
0904	S	D	E	N	1972	67571
0904	S	T	E	N	1972	67571
0907	S	D	E	N	1972	67570
0907	S	D	E	N	1972	67571
0907	S	T	E	N	1972	67571
0912	M	S	E	N	1976	90192
0912	P	D	E	N	1971	62371
0913	P	D	E	N	1971	62371
0917	P	T	E	N	1968	52865
0917	P	D	E	N	1970	59239
0917	P	D	E	N	1970	59561
0917	P	D	E	N	1970	60704
0917	P	D	E	N	1971	64053
0917	P	E	E	N	1971	64053
0917	P	D	R	N	1971	64989
0917	P	D	E	N	1972	66830
0917	P	D	E	N	1971	68482
0917	P	D	E	N	1972	68483
0917	P	D	R	N	1971	69768
0917	P	D	E	N	1972	69769
0917	P	D	E	N	1972	70284
0917	P	D	E	N	1972	70817
0917	P	D	E	N	1973	70823
0917	P	D	E	N	1973	72168
0917	P	T	E	N	1973	72168
0917	P	D	E	N	1973	72169
0917	P	T	E	N	1973	73331
0917	P	D	E	N	1973	74265
0917	P	T	E	N	1973	74265
0917	P	D	E	N	1974	74722
0917	P	D	E	N	1973	76783
0917	P	T	E	N	1973	76783
0917	P	T	R	N	1974	77616
0917	P	D	E	N	1974	78683
0917	P	D	R	N	1974	80433
0917	P	T	R	N	1974	80433
0917	P	D	E	N	1975	80434
0917	P	T	E	N	1975	80434
0917	P	D	E	N	1974	82575
0917	P	S	E	N	1977	93573
0917	S	D	E	N	1965	34393
0917	S	S	E	N	1967	44944
0917	S	E	E	N	1967	44946
0917	S	S	E	N	1970	61561
0917	S	S	E	N	1972	68902
0917	S	T	E	N	1973	73331
0918	S	D	R	N	1969	54785
0918	S	D	E	N	1969	54786
0918	S	S	E	N	1969	56231
0918	S	T	R	N	1967	58387
0918	S	D	E	H	1972	68734
0918	S	T	E	H	1972	68734
0918	S	D	E	N	1973	74813
0918	S	S	E	N	1972	77778
0918	S	D	R	N	1975	80843
0918	S	T	R	N	1975	80843
0922	P	D	E	N	1970	56105
0922	P	D	E	N	1970	56106
0922	S	D	E	N	1970	56106
0922	S	D	E	N	1970	56107
0922	S	D	E	L	1970	61558
0922	S	D	E	N	1972	70282
0922	S	D	E	N	1973	70421
0922	S	E	E	N	1973	70421
0922	S	D	E	L	1970	74011
0923	S	D	E	N	1965	35897
0923	S	T	E	N	1965	35897
0924	S	D	E	N	1978	94674
0924	S	E	E	N	1978	94674
0925	S	D	R	N	1971	67298
0926	P	D	E	N	1968	47316
0926	P	T	E	N	1968	47316
0927	P	D	E	N	1968	47316
0927	P	T	E	N	1968	47316
0939	P	D	E	N	1971	62371
0939	S	D	E	N	1969	55729
0940	P	D	E	N	1971	62371
0940	S	D	E	N	1969	55729
0941	P	D	E	N	1971	62371
0941	S	D	E	N	1969	55729
0942	P	D	E	N	1971	62371
0942	S	D	E	N	1969	55729
0943	P	D	E	N	1971	62371
0943	S	D	E	N	1969	55729
0944	M	D	E	N	1948	94419
0944	M	E	E	N	1948	94419
0944	P	D	E	N	1945	61037
0944	P	D	E	N	1948	94419
0944	P	E	E	N	1948	94419
0946	P	D	E	N	1971	62371
0946	S	D	E	N	1969	55729
0947	P	D	E	N	1971	62371
0947	S	D	E	N	1969	55729
0948	P	D	E	N	1971	62371
0948	S	D	E	N	1969	55729

Substance Number	Phys. State	Subject	Language	Temperature	Year	TPRC Number
521-0949	P	D	E	N	1971	62371
0949	S	D	E	N	1969	55729
0950	P	D	E	N	1971	62371
0950	S	D	E	N	1969	55729
0951	P	D	E	N	1964	31231
0951	P	D	E	N	1971	62371
0951	S	D	E	N	1969	55729
0952	P	D	E	N	1971	62371
0952	S	D	E	N	1969	55729
0955	S	D	E	N	1969	55729
0956	P	D	E	N	1971	62371
0956	S	D	E	N	1969	55729
0956	S	G	E	N	1971	74012
0957	P	D	E	N	1971	62371
0957	S	D	E	N	1969	55729
0957	S	D	E	L	1971	62719
0957	S	D	E	L	1971	67372
0958	P	D	E	N	1971	62371
0958	S	D	E	N	1969	55729
0959	P	D	E	N	1971	62371
0959	S	D	E	N	1969	55729
0959	S	D	E	L	1971	62719
0959	S	D	E	L	1971	67372
0959	S	S	E	N	1976	88841
0960	P	D	E	N	1971	62371
0960	S	D	E	N	1969	55729
0961	P	D	E	N	1971	62371
0961	S	D	E	N	1969	55729
0962	P	D	E	N	1971	62371
0962	S	D	E	N	1969	55729
0963	P	D	E	N	1971	62371
0963	S	D	E	N	1969	55729
0964	P	D	E	N	1971	62371
0964	S	D	E	N	1969	55729
0965	P	D	E	N	1971	62371
0965	S	D	E	N	1969	55729
0967	P	D	E	N	1971	62371
0967	S	D	E	N	1969	55729
0968	P	D	E	N	1971	62371
0968	S	D	E	N	1969	55729
0969	P	D	E	N	1971	62371
0969	S	D	E	N	1969	55729
0969	S	G	E	N	1971	74012
0969	S	D	E	N	1965	88950
0970	P	D	E	N	1971	62371
0970	P	D	E	N	1977	94194
0970	S	D	E	N	1969	55729
0970	S	G	E	N	1971	74012
0978	S	D	E	N	1974	73939
0978	S	E	E	N	1974	73939
0999	S	D	E	N	1968	74019
0999	S	E	E	N	1968	74019
1001	L	D	E	H	1970	38148
1001	S	D	E	H	1970	38148
1003	S	D	R	N	1959	51567
1003	S	E	R	N	1959	51567
1007	S	D	E	H	1966	59827
1008	S	D	E	H	1966	59827
1010	S	D	E	N	1972	64938
1010	S	E	E	N	1972	64938
1010	S	D	E	N	1979	99928
1028	M	D	E	N	1976	85570
1028	M	T	E	N	1976	85570
1028	S	D	E	N	1972	64938
1028	S	E	E	N	1972	64938
1028	S	D	R	N	1971	67298
1028	S	D	E	N	1976	91244
1038	P	D	E	N	1969	59671
1045	P	D	E	N	1971	62371
1048	P	D	E	H	1969	57235
1048	S	D	E	N	1924	25145
1057	S	D	R	N	1967	74004
1057	S	D	E	N	1967	74005
1071	S	D	E	L	1971	62719
1071	S	D	E	L	1971	67372
1072	S	D	E	L	1971	62719
1072	S	D	E	L	1971	67372
1074	S	D	R	N	1969	56214
1074	S	E	R	N	1969	56214
1074	S	D	E	N	1969	67256
1074	S	E	E	N	1969	67256
1075	G	T	E	N	1969	59763
1075	G	T	E	H	1968	65226
1075	G	T	E	H	1973	72038
1075	G	T	E	H	1973	73309
1075	S	T	E	H	1976	85206
1089	D	D	R	N	1971	63801
1089	D	E	R	N	1971	63801
1093	P	D	R	N	1971	65687
1093	P	D	J	N	1977	94212
1093	P	E	J	N	1977	94212
1094	P	D	R	N	1971	65687
1112	S	D	R	N	1969	74008
1112	S	D	E	N	1969	74009
1114	P	D	E	N	1974	77420
1118	S	D	E	N	1951	16342
1118	S	D	E	N	1971	64422
1118	S	E	E	N	1971	64422
1118	S	D	E	N	1968	74019
1118	S	E	E	N	1968	74019
1127	P	D	E	N	1971	62371

Phys. State: **C.** Superconductive; **D.** Doped; **E.** Expanded; **F.** Fibrous or Whisker; **G.** Gas; **I.** Ionized or Plasma; **L.** Liquid; **M.** Multiphase; **P.** Powder or Fine Particle; **S.** Solid; **T.** Thick or Thin Film

Subject: **D.** Data; **E.** Experiment; **G.** General (Data + Theory + Experiment); **S.** Survey (Review, Compendium etc.); **T.** Theory

Language: **C.** Czech; **D.** Dutch; **E.** English; **F.** French; **G.** German; **I.** Italian; **J.** Japanese; **O.** Other Languages; **P.** Polish; **R.** Russian; **S.** Spanish

Temperature: **F.** Full Range (Low + Normal + High); **L.** Low (0 to 75K) + Overlap into Normal; **N.** Normal (75 to 1273K); **H.** High (above 1273K) + Overlap into Normal; Blank = Not Coded

Substance Number	Phys. State	Subject	Language	Temperature	Year	TPRC Number
521-1128	P	D	E	N	1971	62371
1131	P	D	E	N	1971	62371
1133	P	D	E	N	1971	62371
1134	P	D	E	N	1971	62371
1135	P	D	E	N	1971	62371
1136	P	D	E	N	1971	62371
1137	P	D	E	N	1971	62371
1138	P	D	E	N	1971	62371
1139	P	D	E	N	1971	62371
1140	M	S	E	N	1976	90192
1140	P	D	E	N	1971	62371
1140	S	D	E	N	1937	22568
1141	P	D	E	N	1971	62371
1142	P	D	E	N	1971	62371
1143	P	D	E	N	1971	62371
1144	S	D	R	N	1969	74008
1144	S	D	E	N	1969	74009
1145	S	G	E	N	1971	74012
1147	S	D	E	N	1971	74013
1148	P	D	E	N	1971	62371
1149	P	D	E	N	1971	62371
1150	P	D	E	N	1971	62371
1209	S	D	G	N	1968	54724
1209	S	E	G	N	1968	54724
1210	S	D	G	N	1968	54724
1210	S	E	G	N	1968	54724
1210	S	D	E	N	1972	64938
1210	S	E	E	N	1972	64938
1211	S	D	G	N	1968	54724
1211	S	E	G	N	1968	54724
1211	S	D	E	N	1972	64938
1211	S	E	E	N	1972	64938
1211	S	D	R	N	1971	67298
1212	S	D	G	N	1968	54724
1212	S	E	G	N	1968	54724
1212	S	D	E	N	1972	64938
1212	S	E	E	N	1972	64938
1212	S	D	E	N	1956	92919
1215	S	D	R	N	1968	56266
1215	S	E	R	N	1968	56266
1215	S	D	E	N	1971	64050
1215	S	E	E	N	1971	64050
1217	S	D	E	N	1972	67570
1218	S	D	E	N	1972	67570
1218	S	D	E	N	1972	67571
1218	S	T	E	N	1972	67571
1232	G	T	E	H	1972	70325
1238	S	D	E	N	1972	64938
1238	S	E	E	N	1972	64938
1238	S	D	E	N	1977	92134
1238	S	S	E	N	1977	92134
1239	S	D	E	N	1972	64938
1239	S	E	E	N	1972	64938
1241	S	D	E	N	1972	64938
1241	S	E	E	N	1972	64938
1242	S	D	E	N	1972	64938
1242	S	E	E	N	1972	64938
1243	S	D	E	N	1972	64938
1243	S	E	E	N	1972	64938
1244	S	D	E	N	1972	64938
1244	S	E	E	N	1972	64938
1245	S	D	E	N	1972	64938
1245	S	E	E	N	1972	64938
1246	S	D	E	N	1972	64938
1246	S	E	E	N	1972	64938
1247	S	D	E	H	1972	64938
1247	S	E	E	H	1972	64938
1247	S	S	E	N	1977	92134
1255	S	D	E	N	1965	44633
1255	S	E	E	N	1965	44633
1264	M	T	R	N	1971	38919
1287	L	D	E	N	1975	81682
1287	S	D	E	N	1975	81682
1316	S	D	R	N	1967	74004
1316	S	D	E	N	1967	74005
1321	S	D	R	N	1967	74004
1321	S	D	E	N	1967	74005
1322	S	D	R	N	1967	74004
1322	S	D	E	N	1967	74005
1323	S	D	R	N	1967	74004
1323	S	D	E	N	1967	74005
1324	S	D	R	N	1967	74004
1324	S	D	E	N	1967	74005
1326	S	D	R	N	1967	74004
1326	S	D	E	N	1967	74005
1327	S	D	R	N	1967	74004
1327	S	D	E	N	1967	74005
1332	G	T	R	N	1972	73600
1332	G	T	E	N	1973	73601
1332	G	T	E	H	1979	99817
1333	S	D	R	N	1971	67298
1334	S	D	R	N	1971	67298
1335	S	D	R	N	1971	67298
1335	S	D	R	H	1970	91230
1335	S	D	R	N	1970	91230
1335	S	D	E	H	1970	91231
1335	S	D	E	N	1970	91231
1335	S	D	R	N	1976	94572
1344	S	D	E	N	1968	74019
1344	S	E	E	N	1968	74019
1345	S	D	R	N	1970	73679

Substance Number	Phys. State	Subject	Language	Temperature	Year	TPRC Number
521-1345	S	D	E	N	1970	73680
1357	S	D	E	N	1970	89122
1358	S	D	R	N	1969	73753
1358	S	D	E	N	1969	73754
1359	S	D	R	N	1969	73763
1359	S	D	E	N	1969	73764
1361	M	D	E	N	1974	75083
1361	M	T	E	N	1974	75083
1361	M	D	R	N	1973	77333
1361	M	T	R	N	1973	77333
1364	S	D	E	H	1974	75441
1364	S	E	E	H	1974	75441
1377	S	D	E	N	1977	93213
1384	S	D	E	N	1972	68901
1384	S	E	E	N	1972	68901
1389	S	D	E	N	1974	73939
1389	S	E	E	N	1974	73939
1392	S	D	E	N	1972	68902
1392	S	D	E	N	1970	89122
1403	S	D	R	N	1972	71302
1404	S	D	R	N	1972	71302
1405	S	D	R	N	1972	71302
1406	S	D	R	N	1972	71302
1407	S	D	R	N	1972	71302
1408	S	D	R	N	1972	71302
1409	S	D	R	N	1972	71302
1410	S	D	R	N	1972	71302
1411	S	D	R	N	1972	71302
1412	S	D	R	N	1972	71302
1413	S	D	R	N	1972	71302
1414	S	D	R	N	1972	71302
1418	S	S	E	N	1960	24931
1422	S	D	E	N	1975	82018
1424	M	D	O	N	1973	81213
1424	M	D	E	N	1976	87714
1424	M	E	E	N	1976	87714
1424	M	D	E	N	1973	89619
1424	M	D	E	N	1977	91242
1424	M	D	E	N	1977	91687
1424	M	D	E	N	1956	91695
1424	M	D	E	N	1975	91972
1424	M	D	E	N	1977	91973
1424	S	D	O	N	1973	81213
1424	S	D	E	N	1976	87316
1424	S	D	E	N	1964	89615
1424	S	D	E	N	1977	90290
1424	S	D	E	N	1977	90680
1424	S	D	E	N	1977	90681
1424	S	D	E	N	1976	90683
1424	S	D	E	N	1977	90746
1424	S	D	E	N	1977	91242
1424	S	D	E	N	1977	91966
1424	S	T	E	N	1977	91966
1424	S	D	E	N	1977	92136
1424	S	S	E	N	1970	92931
1424	S	D	E	N	1978	94672
1424	S	E	E	N	1978	94672
1424	S	D	E	N	1978	94680
1424	S	E	E	N	1978	94680
1424	S	D	E	N	1978	95474
1424	S	D	E	N	1979	97505
1424	S	D	E	N	1979	99992
1424	S	D	E	N	1979	100606
1424	S	D	E	N	1979	100616
1425	S	D	E	N	1975	82018
1434	S	D	E	N	1977	90681
1434	S	D	E	N	1976	91244
1444	S	D	E	N	1956	92919
1445	S	D	E	N	1956	92919
1446	S	D	E	N	1956	92919
1458	S	D	E	N	1965	88950
1461	P	D	E	N	1971	62371
1461	S	D	E	N	1969	55729
1463	S	D	F	N	1906	6720
1465	P	D	E	N	1971	62371
1467	S	D	E	N	1958	12174
1469	S	D	E	N	1963	89614
1469	S	D	E	N	1969	89618
1470	S	D	E	N	1963	89614
1488	S	D	E	N	1978	94680
1488	S	E	E	N	1978	94680
1488	S	D	E	N	1979	97505
1489	S	D	E	N	1978	94680
1489	S	E	E	N	1978	94680
1489	S	D	E	N	1979	97505
1490	S	D	E	N	1978	94680
1490	S	E	E	N	1978	94680
1490	S	D	E	N	1979	97505
1546	S	D	E	N	1956	91695
1551	S	D	O	N	1976	90397
1551	S	E	O	N	1976	90397
1557	S	D	R	N	1935	14269
1571	M	D	E	N	1863	7575
1571	M	E	E	N	1863	7575
1571	M	D	G	N	1861	10036
1571	M	E	G	N	1861	10036
1571	S	S	E	N	1970	61233
1572	M	D	R	N	1971	44543
1572	M	E	R	N	1971	44543
1572	M	D	E	N	1971	75014
1572	M	E	E	N	1971	75014

Substance Number	Phys. State	Subject	Language	Temperature	Year	TPRC Number
521-1604	P	D	E	N	1863	7575
1604	P	E	E	N	1863	7575
1604	P	D	G	N	1861	10036
1604	P	E	G	N	1861	10036
1609	M	D	E	N	1978	96710
1615	M	D	R	N	1970	92949
1615	S	D	R	N	1970	92949
1617	P	D	J	N	1977	94212
1617	P	E	J	N	1977	94212
1618	P	D	J	N	1977	94212
1618	P	E	J	N	1977	94212
1619	P	D	J	N	1977	94212
1619	P	E	J	N	1977	94212
1621	S	D	E	N	1978	94672
1621	S	E	E	N	1978	94672
1622	S	D	E	N	1978	94680
1622	S	E	E	N	1978	94680
1623	S	D	E	N	1978	94680
1623	S	E	E	N	1978	94680
1624	S	D	E	N	1978	94680
1624	S	E	E	N	1978	94680
1625	S	D	E	N	1978	94680
1625	S	E	E	N	1978	94680
1626	S	D	E	N	1978	94680
1626	S	E	E	N	1978	94680
1627	S	D	E	N	1978	94680
1627	S	E	E	N	1978	94680
1633	S	S	E	N	1970	61233
1639	S	S	E	L	1974	79241
1639	S	S	E	N	1974	79241
1645	S	D	R	N	1978	95686
1645	S	D	E	N	1978	98588
1646	S	D	R	N	1978	95686
1646	S	D	E	N	1978	98588

Phys. State: **C.** Superconductive; **D.** Doped; **E.** Expanded; **F.** Fibrous or Whisker; **G.** Gas; **I.** Ionized or Plasma; **L.** Liquid; **M.** Multiphase; **P.** Powder or Fine Particle; **S.** Solid; **T.** Thick or Thin Film
Subject: **D.** Data; **E.** Experiment; **G.** General (Data + Theory + Experiment); **S.** Survey (Review, Compendium etc.); **T.** Theory
Language: **C.** Czech; **D.** Dutch; **E.** English; **F.** French; **G.** German; **I.** Italian; **J.** Japanese; **O.** Other Languages; **P.** Polish; **R.** Russian; **S.** Spanish
Temperature: **F.** Full Range (Low + Normal + High); **L.** Low (0 to 75K) + Overlap into Normal; **N.** Normal (75 to 1273K); **H.** High (above 1273K) + Overlap into Normal; Blank = Not Coded

Chapter 2 Thermal Diffusivity

Substance Number	Phys. State	Subject	Language	Temperature	Year	TPRC Number
503-0012	S	D	E	N	1957	19864
0012	S	S	E	N	1973	70678
0012	S	D	E	N	1975	82363
0019	S	S	E	N	1973	70678
0020	S	D	E	N	1960	10947
0025	L	G	E	H	1961	18317
0025	S	D	J	N	1968	53139
0025	S	E	J	N	1968	53139
0031	S	D	E	N	1964	35542
0031	S	D	E	N	1966	42356
0047	S	S	E	H	1973	70678
0047	S	D	E	N	1976	86042
0047	S	E	E	N	1976	86042
0066	S	D	E	N	1963	49933
0066	S	D	E	N	1971	68241
0066	S	S	E	N	1973	70678
0080	S	G	E	N	1972	78256
0082	E	D	E		1946	6321
0082	E	D	E	N	1964	37058
0082	E	D	R	N	1968	51025
0082	E	S	E	N	1948	54088
0082	E	D	R	N	1969	56318
0082	E	D	E	N	1972	66207
0082	E	S	E	N	1973	70678
0082	F	D	R	N	1960	24772
0082	F	D	E	N	1942	25423
0082	F	D	E	N	1960	42824
0082	M	D	E	N	1974	75085
0082	M	E	E	N	1974	75085
0082	M	D	R	N	1973	77334
0082	M	E	R	N	1973	77334
0082	P	D	F	N	1972	66103
0082	P	E	F	N	1972	66103
0082	S	D	E	N	1921	22633
0082	S	S	E	N	1956	24971
0082	S	D	E	N	1961	27117
0082	S	T	E	N	1961	27117
0082	S	E	E	H	1962	27922
0082	S	D	G	N	1904	28731
0082	S	T	R	N	1962	31437
0082	S	D	O	N	1932	40805
0082	S	D	E	N	1966	40806
0082	S	D	E	N	1960	42824
0082	S	G	J	N	1967	46251
0082	S	G	E	N	1968	46252
0082	S	D	F	H	1969	56148
0082	S	D	E	N	1972	68674
0082	S	E	E	N	1972	68674
0082	S	D	E	N	1974	72742
0082	S	D	E	N	1975	75982
0082	S	D	F	N	1968	76887
0082	S	D	E	N	1976	85868
0126	S	D	E	N	1973	71894
0200	S	D	E	L	1975	80989
0200	S	E	E	L	1975	80989
0216	S	D	E	N	1962	26684
0216	S	D	E	N	1961	27117
0216	S	T	E	N	1961	27117
0216	S	D	E	N	1962	27922
0216	S	D	E	N	1965	36738
0216	S	D	E	L	1968	47910
0216	S	D	E	N	1963	49933
0216	S	S	E	N	1973	70678
0240	P	S	E	L	1973	70678
0240	P	S	E	N	1973	70678
0240	P	D	E	L	1975	82557
0240	P	D	E	N	1975	82557
0240	P	E	E	L	1975	82557
0240	P	E	E	N	1975	82557
0262	S	D	E	N	1962	27922
0262	S	S	E	N	1973	70678
0284	S	S	E	H	1973	70678
0309	S	D	E	N	1963	49933
0309	S	S	E	N	1973	70678
0376	S	S	E	N	1973	70678
0423	L	D	E	H	1951	3911
0423	L	D	E	N	1951	3911
0423	L	E	E	H	1951	3911
0423	L	E	E	N	1951	3911
0452	S	D	E	N	1963	24384
0452	S	D	E	H	1963	49930
0452	S	S	E	H	1973	70678
0452	S	S	E	N	1973	70678
0463	S	D	R	N	1971	67588
0466	S	D	R	N	1971	67588
0537	S	D	E	N	1957	19864
0537	S	D	E	N	1962	27922
0537	S	S	E	N	1973	70678
0562	S	D	F	N	1968	76887
0562	S	S	E	N	1979	97587
0599	S	D	R	N	1970	61257
0599	S	D	E	N	1970	67223
0604	S	D	E	N	1962	27922
0604	S	S	E	N	1973	70678
0631	F	D	E	N	1977	91785

Substance Number	Phys. State	Subject	Language	Temperature	Year	TPRC Number
503-0631	P	S	E	L	1973	70678
0631	P	S	E	N	1973	70678
0631	S	D	E	N	1977	91785
0652	S	D	E	H	1975	81946
0652	S	D	E	N	1975	81946
0652	S	D	E	H	1976	88872
0652	S	D	E	N	1976	88872
0659	S	D	E	N	1962	27922
0659	S	D	E	N	1963	49897
0659	S	D	E	N	1963	49933
0659	S	D	E	N	1968	50841
0659	S	S	E	H	1973	70678
0659	S	S	E	N	1973	70678
0659	S	S	E	N	1976	83746
0659	S	D	E	N	1976	88185
0659	S	E	E	N	1976	88185
0659	S	D	E	N	1976	88872
0659	S	D	E	H	1977	94200
0659	S	D	E	N	1977	94200
0659	S	S	E	H	1977	94200
0659	S	S	E	N	1977	94200
0660	S	D	E	N	1963	24384
0660	S	D	E	N	1962	27922
0660	S	D	E	H	1963	49930
0660	S	S	E	H	1973	70678
0660	S	S	E	N	1973	70678
0660	S	D	E	H	1976	88872
0660	S	D	E	N	1976	88872
0667	S	S	E	N	1961	25075
0667	S	S	E	N	1964	30525
0769	P	S	E	N	1973	70678
0782	S	D	E	N	1968	54081
0782	S	D	E	N	1971	68241
0782	S	S	E	N	1975	78777
0801	S	D	O	N	1932	40805
0801	S	D	E	N	1966	40806
0801	S	D	R	N	1976	94572
0822	S	G	E	N	1972	66811
0849	S	D	E	N	1979	97889
0868	S	D	R	N	1971	67588
0879	S	D	F	N	1959	25490
0879	S	T	F	N	1959	25490
0884	S	D	R	N	1977	94615
0928	S	D	E	N	1957	19864
0959	S	D	E	N	1962	27922
0959	S	S	E	N	1973	70678
0964	S	D	E	N	1977	90324
0964	S	D	E	N	1977	91785
0997	S	S	E	N	1976	88411
1015	S	D	R	N	1971	67588
1023	S	D	R	N	1971	67588
1024	S	D	R	N	1971	67588
1050	S	D	E	N	1960	10947
1050	S	D	E	N	1962	26684
1050	S	S	E	N	1973	70678
1050	S	G	E	N	1972	78256
1062	S	D	E	N	1960	10947
1072	S	D	E	N	1960	10947
1078	S	G	E	L	1972	76522
1083	S	D	R	N	1971	67588
1110	S	S	E	N	1961	25075
1110	S	D	E	N	1976	85868
1125	S	D	E		1955	6941
1125	S	E	E		1955	6941
1132	S	D	R	H	1971	38298
1132	S	D	R	H	1971	64346
1132	S	E	R	H	1971	64346
1132	S	D	E	H	1971	64851
1132	S	E	E	H	1971	64851
1133	S	E	E		1951	1114
1133	S	D	R	N	1960	24772
1135	S	D	R	H	1971	36028
1135	S	E	R	H	1971	36028
1136	S	G	E	N	1949	12033
1136	S	D	E	N	1964	37058
1136	S	G	R	N	1972	67963
1136	S	D	E	N	1972	68595
1137	S	G	E		1953	3910
1142	S	D	R		1953	8876
1142	S	E	E	H	1956	10923
1142	S	E	E	F	1957	15658
1142	S	D	E	H	1961	20637
1142	S	S	E	N	1961	20740
1142	S	D	E	N	1963	24384
1142	S	S	E	H	1960	24931
1142	S	E	E	H	1962	27922
1142	S	E	G	N	1963	29768
1142	S	D	E	H	1965	36139
1142	S	G	R	N	1964	36313
1142	S	D	R	H	1965	43324
1142	S	D	R	N	1965	47287
1142	S	T	R	N	1965	47287
1142	S	D	E	N	1965	47288
1142	S	T	E	N	1965	47288
1142	S	D	E	H	1976	88872

Substance Number	Phys. State	Subject	Language	Temperature	Year	TPRC Number
503-1142	S	D	E	N	1976	88872
1142	S	D	E	N	1977	91785
1145	M	E	E	N	1963	31136
1145	S	D	E		1951	1114
1145	S	E	E		1955	9140
1145	S	T	E	H	1958	10933
1145	S	D	I	N	1950	16411
1145	S	E	I	N	1952	20957
1145	S	E	F	H	1960	23879
1145	S	E	E	H	1962	27922
1145	S	E	E	N	1960	29334
1145	S	E	R	H	1969	56452
1146	S	G	R	N	1963	30775
1146	S	G	E	N	1963	31046
1146	S	D	R	H	1969	58850
1146	S	D	E	H	1969	58851
1147	E	D	E	N	1965	38402
1147	S	D	J	N	1958	17473
1147	S	G	J	N	1959	21142
1147	S	D	G	N	1961	24393
1147	S	D	R	N	1960	24772
1147	S	G	R	N	1963	30775
1147	S	G	E	N	1963	31046
1147	S	D	E	N	1966	34401
1147	S	D	R	N	1966	41857
1147	S	D	E	N	1966	51215
1147	S	D	E	N	1969	56579
1147	S	D	R	N	1969	61220
1147	S	T	R	N	1969	61220
1147	S	S	E	N	1973	70678
1149	S	E	E		1957	945
1149	S	G	E	N	1974	74606
1152	S	G	J	N	1959	21142
1152	S	D	R	N	1956	32423
1152	S	D	E	N	1973	71892
1158	S	G	J	N	1959	21142
1162	F	S	E	H	1973	70678
1162	F	S	E	N	1973	70678
1162	S	D	S	N	1956	29007
1173	S	D	G	N	1961	24393
1173	S	D	R	N	1971	62758
1173	S	E	R	N	1971	62758
1177	S	D	R	N	1960	24772
1177	S	G	R	N	1971	66294
1177	S	G	E	N	1971	70931
1178	S	D	R	N	1967	54757
1180	S	D	E	N	1970	77556
1182	M	D	G	N	1963	30732
1182	S	D	G	N	1962	30040
1185	S	D	E	N	1967	42955
1185	S	D	R	N	1967	45153
1189	S	D	R	H	1969	58850
1189	S	D	E	H	1969	58851
1219	S	D	E	H	1960	20600
1219	S	D	E	N	1963	23121
1219	S	D	E	N	1958	28870
1224	S	D	R	N	1960	24772
1226	S	D	R	N	1960	24772
1226	S	G	R	N	1965	43098
1250	S	D	F	H	1969	56148
1267	S	T	E	H	1970	59268
1296	S	D	E	N	1934	43126
1309	S	D	R	N	1970	64246
1309	S	T	R	N	1970	64246
1310	M	D	R	N	1970	64246
1310	M	T	R	N	1970	64246
1321	P	D	R	N	1971	65687
1331	S	D	R	N	1972	67971
1357	E	D	E		1956	4737
1357	F	G	E	N	1949	12033
1357	F	D	R	N	1960	24772
1357	F	D	E	N	1942	25423
1357	F	D	R	N	1956	32423
1360	P	D	E	N	1947	25437
1360	S	D	R	N	1977	94615
1368	L	D	R	H	1973	72892
1369	P	D	R	N	1971	65687
1374	S	D	R	N	1970	74070
1374	S	E	R	N	1970	74070
1374	S	D	E	N	1970	74071
1374	S	E	E	N	1970	74071
1378	S	D	R	N	1974	74804
1378	S	E	R	N	1974	74804
1388	S	D	E	H	1975	81946
1388	S	D	E	N	1975	81946
1405	S	D	R	N	1971	67588
1429	S	G	E	N	1968	53618
1432	S	D	E	N	1969	40513
1432	S	D	O	N	1966	41369
1451	S	D	R	N	1971	67588
1459	S	G	E	N	1972	78256
1466	S	D	E	N	1960	10947
1466	S	E	E	N	1960	10947
1478	S	D	R	N	1971	67588
1480	S	D	R	N	1971	67588

Phys. State: **C.** Superconductive; **D.** Doped; **E.** Expanded; **F.** Fibrous or Whisker; **G.** Gas; **I.** Ionized or Plasma; **L.** Liquid; **M.** Multiphase; **P.** Powder or Fine Particle; **S.** Solid; **T.** Thick or Thin Film
Subject: **D.** Data; **E.** Experiment; **G.** General (Data + Theory + Experiment); **S.** Survey (Review, Compendium etc.); **T.** Theory
Language: **C.** Czech; **D.** Dutch; **E.** English; **F.** French; **G.** German; **I.** Italian; **J.** Japanese; **O.** Other Languages; **P.** Polish; **R.** Russian; **S.** Spanish
Temperature: **F.** Full Range (Low + Normal + High); **L.** Low (0 to 75K) + Overlap into Normal; **N.** Normal (75 to 1273K); **H.** High (above 1273K) + Overlap into Normal; Blank = Not Coded

Substance Number	Phys. State	Subject	Language	Temperature	Year	TPRC Number
503-1481	S	D	R	N	1971	67588
1482	S	D	R	N	1971	67588
1483	S	D	R	N	1971	67588
1489	S	D	R	N	1971	67588
1490	S	D	R	N	1971	67588
1493	S	D	R	N	1971	67588
1494	S	D	R	N	1971	67588
1554	S	D	R	N	1971	67588
1556	S	D	E	N	1969	40513
1556	S	D	R	N	1966	41369
1557	S	D	R	N	1971	67588
1559	S	D	E	N	1969	40513
1559	S	D	R	N	1966	41369
1559	S	D	R	N	1971	67588
1560	S	D	R	N	1971	67588
1561	S	D	R	N	1971	67588
1563	S	D	R	N	1971	67588
1721	L	D	E	H	1951	3911
1721	L	D	E	N	1951	3911
1721	L	E	E	H	1951	3911
1721	L	E	E	N	1951	3911
1724	S	S	E	N	1973	70678
1766	S	D	R	N	1976	88314
1766	S	D	E	N	1976	91214
1810	S	D	J	N	1977	94558
1810	S	E	J	N	1977	94558
1811	S	D	J	N	1977	94558
1811	S	E	J	N	1977	94558
1815	S	D	R	N	1967	48703
1815	S	D	E	N	1967	49415
1824	S	D	E	N	1977	91785
1825	S	D	E	N	1977	91785
1826	P	S	E	H	1973	70678
1826	P	S	E	N	1973	70678
1826	S	S	E	H	1973	70678
1826	S	S	E	N	1973	70678
1828	S	D	E	N	1967	35986
1828	S	D	F	N	1968	76887
1828	S	S	E	N	1976	88411
1828	S	S	E	N	1979	97587
1912	S	S	E	N	1973	70678
1932	S	D	R	N	1969	58148
1933	S	D	R	N	1969	58233
1945	S	S	E	H	1973	70678
1945	S	S	E	N	1973	70678
1950	S	D	R	H	1976	89429
1950	S	D	R	N	1976	89429
1950	S	E	R	H	1976	89429
1950	S	E	R	N	1976	89429
1950	S	D	E	H	1976	91205
1950	S	D	E	N	1976	91205
1950	S	E	E	H	1976	91205
1950	S	E	E	N	1976	91205
1951	S	D	R	H	1976	89429
1951	S	D	R	N	1976	89429
1951	S	E	R	H	1976	89429
1951	S	E	R	N	1976	89429
1951	S	D	E	H	1976	91205
1951	S	D	E	N	1976	91205
1951	S	E	E	H	1976	91205
1951	S	E	E	N	1976	91205
1953	D	D	E	N	1973	70944
1961	S	D	E	N	1977	93507
1987	S	D	E	N	1921	22633
1988	S	D	R	N	1977	94615
1989	S	D	E	N	1921	22633
1990	S	D	E	N	1921	22633
1994	P	D	E	H	1962	29477
1994	P	D	R	H	1961	29478
2000	S	S	E	H	1973	70678
2000	S	S	E	N	1973	70678
2017	S	D	R	N	1977	94615
2107	S	D	F	N	1959	25490
2107	S	T	F	N	1959	25490
2142	S	D	J	N	1977	94558
2142	S	E	J	N	1977	94558
2160	S	D	E	H	1964	52271
2160	S	D	E	N	1964	52271
2227	S	S	E	N	1973	70678
2247	S	D	E	N	1962	27922
2247	S	S	E	N	1973	70678
2303	S	D	R	N	1967	48703
2303	S	D	E	N	1967	49415
2325	S	D	E	H	1975	81946
2325	S	D	E	N	1975	81946
2365	S	D	E	N	1979	97889
2404	S	D	F	N	1959	25490
2404	S	T	F	N	1959	25490
2432	S	D	R	N	1967	48703
2432	S	D	E	N	1967	49415
2432	S	S	E	N	1973	70678
2437	S	S	E	H	1973	70678
2437	S	S	E	N	1973	70678
2438	S	S	E	H	1973	70678
2438	S	S	E	N	1973	70678
2440	S	D	R	N	1967	54757
2461	S	D	E	N	1921	22633
2478	S	D	E	N	1921	22633
2479	S	D	E	N	1921	22633
2589	S	D	R	N	1967	48703
2589	S	D	E	N	1967	49415

Substance Number	Phys. State	Subject	Language	Temperature	Year	TPRC Number
503-2590	S	D	R	N	1967	48703
2590	S	D	E	N	1967	49415
2590	S	S	E	H	1973	70678
2660	S	D	R	H	1977	90440
2660	S	D	R	N	1977	90440
2660	S	D	E	H	1977	95993
2660	S	D	E	N	1977	95993
2726	E	D	G	N	1955	1125
2726	E	E	G	N	1955	1125
2735	S	S	E	H	1973	70678
2735	S	S	E	N	1973	70678
2752	S	S	E	H	1973	70678
2752	S	S	E	N	1973	70678
2757	S	S	E	N	1973	70678
2758	S	S	E	N	1973	70678
2759	S	S	E	N	1973	70678
2760	S	S	E	N	1973	70678
2762	P	S	E	N	1973	70678
2763	S	S	E	H	1973	70678
2763	S	S	E	N	1973	70678
2764	S	S	E	H	1973	70678
2764	S	S	E	N	1973	70678
2765	S	S	E	H	1973	70678
2765	S	S	E	N	1973	70678
2767	S	S	E	N	1973	70678
2768	S	S	E	N	1973	70678
2769	S	S	E	N	1973	70678
2808	S	D	R	N	1967	54757
2809	S	D	R	N	1967	54757
2921	S	S	E	N	1973	70678
8196	S	D	R	N	1967	54757
8243	S	D	E	N	1934	43126
8371	S	D	E	H	1915	31868
8371	S	T	E	H	1915	31868
8374	S	D	R	H	1970	59046
8374	S	D	E	H	1970	60836
8392	P	D	E	H	1962	29477
8392	P	D	R	H	1961	29478
8392	S	D	E	H	1962	29477
8392	S	D	R	H	1961	29478
8566	S	D	E	N	1963	49933
9023	S	D	E	N	1974	76145
9042	L	G	E	H	1959	29365
9042	S	D	F	H	1969	56148
9091	S	D	E	N	1963	49933
9169	S	D	E	N	1974	76145
9463	S	D	R	H	1969	56452
9534	S	D	E	H	1968	46044
9534	S	D	E	H	1968	50485
9633	S	D	O	N	1966	43203
9644	S	D	G	N	1968	36520
9644	S	D	G	N	1969	52808
9644	S	D	G	N	1968	54534
9644	S	E	G	N	1968	54534
9644	S	D	F	N	1970	57871
9644	S	T	F	N	1970	57871
9644	S	D	E	N	1970	61774
9675	S	S	E	N	1973	70678
9682	S	D	R	H	1969	55514
9682	S	D	E	H	1969	57273
9688	S	D	R	H	1969	55514
9688	S	D	E	H	1969	57273
9688	S	D	R	H	1970	59046
9688	S	D	E	H	1970	60836
9709	S	D	E	N	1961	23542
9712	S	D	E	H	1966	35917
9716	S	D	F	H	1970	60595
9718	S	D	E	H	1967	52120
9718	S	D	E	H	1969	53591
9718	S	D	E	N	1969	54478
9718	S	D	E	H	1968	57194
9718	S	D	F	H	1968	57457
9718	S	D	E	H	1970	65535
9718	S	D	E	H	1972	71915
9718	S	T	E	H	1972	71915
9718	S	D	E	H	1972	73901
9718	S	S	S	H	1973	74958
9721	S	D	E	H	1960	16593
9731	M	D	E	N	1963	31136
9744	S	D	E	H	1963	49930
9750	S	D	E	H	1962	27922
9773	P	D	E	N	1921	22633
9773	S	D	E	N	1921	22633
9784	S	D	E	H	1921	20009
9792	S	D	E	N	1921	22633
9809	S	D	E	N	1921	22633
9809	S	D	E	H	1915	31868
9809	S	T	E	H	1915	31868
9812	M	D	E	H	1962	29549
9812	S	D	E	N	1963	29780
9813	S	D	E	H	1966	35917
9815	S	D	R	H	1970	59046
9815	S	D	E	H	1970	60836
9819	S	D	E	H	1969	53573
9819	S	D	F	H	1968	54964
9819	S	D	E	H	1968	55211
9819	S	T	E	H	1968	55211
9819	S	D	E	H	1969	56741
9819	S	D	F	H	1968	57456
9819	S	D	J	H	1968	60568
9819	S	D	E	H	1969	60569

Substance Number	Phys. State	Subject	Language	Temperature	Year	TPRC Number
503-9819	S	G	F	H	1972	74283
9824	M	D	E	H	1962	29549
9824	S	D	E	N	1961	23542
9824	S	D	E	N	1963	29780
9830	P	G	E	N	1968	55639
9830	P	G	E	N	1969	59176
9835	S	D	E	H	1915	31868
9835	S	T	E	H	1915	31868
9836	S	D	E	H	1921	20009
9836	S	D	R	N	1969	55846
9836	S	T	R	N	1969	55846
9838	M	D	E	N	1963	31136
9843	S	D	E	N	1968	54196
9854	S	D	E	N	1968	54196
9859	S	D	O	N	1972	71321
9869	S	D	E	N	1973	70757
9916	S	D	E	H	1921	20009
9916	S	D	E	N	1926	21670
9916	S	D	R	H	1966	46573
9916	S	D	E	H	1966	46574
9916	S	D	O	N	1972	71321
9922	M	D	E	N	1963	31136
9922	S	D	F	H	1970	60595
9951	S	D	F	N	1966	39275
9951	S	S	F	N	1968	54531
9974	M	D	E	N	1963	31136
9974	S	D	E	H	1960	16593
9974	S	D	E	H	1963	24367
9974	S	D	E	N	1964	34296
9974	S	G	E	H	1964	48050
9979	S	D	G	N	1960	62279
9980	S	D	E	H	1921	20009
9980	S	D	R	N	1964	43252
9980	S	D	O	N	1972	71321
9985	S	D	E	H	1973	74558
9985	S	T	E	H	1973	74558
9986	S	D	R	N	1967	54757
9989	S	D	R	N	1967	54757
9995	S	D	R	N	1967	54757
521-0001	S	D	E	N	1969	56053
0003	S	D	E	N	1968	47311
0004	L	S	E	N	1969	54305
0004	P	S	E	N	1969	54305
0006	S	D	E	N	1921	22633
0006	S	D	R	N	1964	38184
0007	S	D	G	N	1967	45388
0007	S	G	E	N	1960	51286
0007	S	D	E	N	1976	89727
0007	S	D	E	N	1978	94673
0009	M	E	R	N	1959	43002
0009	M	E	F	N	1959	43003
0009	M	D	E	N	1943	46820
0009	M	T	E	N	1943	46820
0009	M	D	E	N	1974	78688
0009	M	S	E	N	1976	90192
0009	M	D	E	N	1913	94424
0009	M	T	E	N	1913	94424
0009	P	D	E	N	1921	22633
0009	P	S	E	N	1969	55244
0009	P	E	E	N	1973	72674
0009	P	D	E	N	1974	82320
0009	P	E	E	N	1974	82320
0009	S	G	E		1951	6227
0009	S	D	E	N	1862	16210
0009	S	D	E	N	1953	16343
0009	S	D	E	N	1863	22527
0009	S	G	R	N	1958	25142
0009	S	D	E	N	1951	25557
0009	S	E	R	N	1959	43002
0009	S	E	F	N	1959	43003
0009	S	D	E	N	1966	46035
0009	S	D	E	N	1943	46820
0009	S	T	E	N	1943	46820
0009	S	S	R	N	1957	68049
0009	S	S	E	N	1971	68050
0009	S	S	E	N	1973	70678
0009	S	S	E	N	1976	90192
0009	S	D	J	N	1975	98110
0009	S	E	J	N	1975	98110
0010	P	D	E	N	1947	25437
0010	P	D	R	N	1961	27540
0010	P	D	E	N	1965	36861
0010	P	D	E	N	1962	39516
0010	P	D	R	N	1965	42532
0010	P	D	R	N	1971	65687
0010	P	D	E	N	1973	72674
0010	P	E	E	N	1973	72674
0010	S	G	E		1949	1945
0010	S	D	E	N	1924	25145
0010	S	D	E	N	1962	29491
0010	S	D	R	N	1971	63757
0010	S	E	R	N	1971	63757
0010	S	D	E	N	1971	67155
0010	S	E	E	N	1971	67155
0015	P	D	R	N	1968	49504
0015	P	E	R	N	1968	49504
0015	P	D	E	N	1968	65123
0015	P	E	E	N	1968	65123
0015	S	D	E		1955	9738
0015	S	T	E		1955	9738

Phys. State: **C.** Superconductive; **D.** Doped; **E.** Expanded; **F.** Fibrous or Whisker; **G.** Gas; **I.** Ionized or Plasma; **L.** Liquid; **M.** Multiphase; **P.** Powder or Fine Particle; **S.** Solid; **T.** Thick or Thin Film

Subject: **D.** Data; **E.** Experiment; **G.** General (Data + Theory + Experiment); **S.** Survey (Review, Compendium etc.); **T.** Theory

Language: **C.** Czech; **D.** Dutch; **E.** English; **F.** French; **G.** German; **I.** Italian; **J.** Japanese; **O.** Other Languages; **P.** Polish; **R.** Russian; **S.** Spanish

Temperature: **F.** Full Range (Low + Normal + High); **L.** Low (0 to 75K) + Overlap into Normal; **N.** Normal (75 to 1273K); **H.** High (above 1273K) + Overlap into Normal; Blank = Not Coded

Substance Number	Phys. State	Subject	Language	Temperature	Year	TPRC Number
521-0015	S	D	E	H	1926	22638
0020	P	D	R	N	1971	65687
0020	S	D	E	N	1942	25423
0020	S	D	E	N	1973	71780
0022	S	D	R	N	1964	38184
0022	S	D	E	H	1972	68734
0022	S	E	E	H	1972	68734
0022	S	D	E	H	1975	83595
0022	S	D	E	N	1975	83595
0024	P	G	C	N	1970	65248
0024	S	D	R	N	1960	21288
0024	S	D	E	N	1960	25356
0024	S	D	G	N	1940	29870
0024	S	D	R	N	1968	51815
0024	S	S	E	N	1973	70678
0025	S	D	E	N	1979	97680
0026	S	D	P	N	1971	63225
0027	P	D	R	N	1971	69768
0027	P	D	E	N	1972	69769
0028	P	D	E	N	1962	39516
0028	S	D	E	N	1963	24360
0028	S	D	E	N	1962	29491
0028	S	D	E	N	1973	70421
0028	S	E	E	N	1973	70421
0040	M	D	R	N	1960	20247
0040	P	D	E		1956	8514
0040	S	D	E		1956	226
0040	S	D	E		1952	268
0040	S	D	R		1952	785
0040	S	D	R		1953	8876
0040	S	D	E		1956	9825
0040	S	S	E	N	1958	10667
0040	S	D	E	N	1921	22633
0040	S	G	E	N	1958	42981
0041	S	D	E	N	1979	97680
0046	M	D	R	N	1969	54587
0046	S	E	E	N	1951	23406
0046	S	S	G	N	1924	33616
0046	S	E	R	N	1965	38199
0046	S	G	E	N	1970	41374
0046	S	E	E	N	1969	55839
0046	S	T	E	N	1969	56570
0046	S	D	R	N	1968	58386
0046	S	D	R	H	1969	58631
0046	S	T	R	N	1968	61216
0046	S	E	E	N	1971	64422
0046	S	T	R	N	1968	67269
0046	S	T	E	N	1968	67270
0046	S	E	E	N	1973	72674
0046	S	T	E	N	1974	73781
0046	S	S	E	N	1971	74002
0046	S	E	E	N	1970	74014
0046	S	D	E	N	1976	91238
0053	S	D	E	N	1921	22633
0053	S	D	E	N	1951	40424
0053	S	D	G	N	1968	54724
0053	S	E	G	N	1968	54724
0053	S	D	E	N	1972	64938
0053	S	E	E	N	1972	64938
0055	S	D	E	N	1921	22633
0057	S	D	E	N	1979	97680
0058	S	D	E	N	1968	47311
0063	S	D	E	L	1969	53600
0063	S	D	E	L	1973	70678
0063	S	D	E	N	1973	70678
0064	S	D	E	N	1961	29043
0064	S	D	G	L	1969	64980
0064	S	E	G	L	1969	64980
0065	M	D	E	N	1979	99744
0065	S	D	R	N	1960	21288
0065	S	D	E	N	1960	25356
0065	S	G	E	N	1969	59226
0065	S	D	E	N	1979	100246
0065	S	E	E	N	1979	100246
0066	M	D	R	N	1960	24247
0066	M	D	E	N	1960	25361
0066	M	D	R	N	1968	51025
0066	M	D	R	N	1969	56318
0066	S	D	E	N	1969	56053
0066	S	D	E	N	1972	66207
0067	S	G	R	N	1962	27608
0067	S	D	R	H	1963	32198
0067	S	G	E	N	1963	32761
0069	P	D	R	N	1964	35710
0069	P	D	R	N	1969	56898
0069	P	T	R	N	1969	56898
0069	S	D	R	N	1960	11725
0069	S	D	E	N	1862	16210
0069	S	D	R	N	1959	18710
0069	S	D	E	N	1962	20611
0069	S	D	E	N	1960	20660
0069	S	D	R	N	1960	21288
0069	S	D	E	N	1863	22527
0069	S	D	E	N	1921	22633
0069	S	D	R	N	1961	23771
0069	S	D	E	N	1960	25356
0069	S	D	G	N	1904	28731
0069	S	S	R	N	1958	30882
0069	S	S	E	N	1961	30884
0069	S	D	R	N	1957	32417
0069	S	G	R	N	1957	32675
521-0069	S	D	R	N	1964	48353
0069	S	S	E	N	1970	56648
0069	S	D	R	N	1969	56898
0069	S	T	R	N	1969	56898
0069	S	D	R	N	1969	58119
0069	S	D	G	N	1970	59789
0069	S	E	G	N	1970	59789
0069	S	D	E	N	1968	70866
0070	M	D	R	N	1969	54587
0070	S	D	E	N	1958	12174
0070	S	D	E	N	1921	22633
0070	S	D	R	N	1961	25832
0070	S	D	E	N	1951	40424
0070	S	G	E	N	1960	51286
0070	S	D	E	N	1969	56809
0072	S	D	E	N	1862	16210
0072	S	D	E	N	1863	22527
0072	S	D	E	N	1921	22633
0072	S	D	E	N	1964	25860
0072	S	D	G	N	1968	54724
0072	S	E	G	N	1968	54724
0072	S	D	P	N	1971	63225
0072	S	D	E	N	1972	64938
0072	S	E	E	N	1972	64938
0072	S	D	R	N	1971	65570
0072	S	E	R	N	1971	65570
0072	S	D	R	N	1971	67298
0072	S	D	R	N	1972	71302
0072	S	D	E	H	1971	73275
0072	S	E	E	H	1971	73275
0072	S	D	E	N	1972	78340
0072	S	E	E	N	1972	78340
0072	S	E	E	N	1976	90853
0072	S	D	E	N	1976	91645
0072	S	D	E	N	1977	93606
0072	S	E	E	N	1977	93606
0072	S	D	R	N	1976	94572
0072	S	D	E	N	1978	94675
0072	S	E	E	N	1978	94675
0072	S	D	J	N	1978	96703
0072	S	E	J	N	1978	96703
0072	S	D	E	N	1979	100606
0073	E	D	R	N	1969	56893
0073	E	D	E	N	1972	66055
0073	P	D	R	N	1966	48083
0073	P	D	R	N	1967	55858
0073	P	G	E	N	1969	59176
0073	P	D	E	N	1967	63347
0073	P	D	R	N	1971	65687
0073	S	D	E	N	1921	22633
0075	S	S	E	N	1977	92134
0075	S	D	E	N	1979	97680
0077	S	T	E		1955	6930
0077	S	D	E		1955	6942
0077	S	D	E	N	1968	47311
0077	S	S	E	N	1973	70678
0083	S	D	I	N	1950	16411
0083	S	G	R	N	1958	20245
0083	S	G	E	N	1962	20436
0083	S	D	G	N	1904	28731
0083	S	D	O	N	1932	40805
0083	S	D	E	N	1966	40806
0083	S	G	R	N	1971	62997
0083	S	G	E	N	1971	72195
0084	M	G	E		1954	1963
0084	P	G	E		1954	1963
0084	S	D	E	N	1921	22633
0084	S	D	R	N	1976	94572
0084	S	D	E	N	1979	97680
0085	L	D	E	N	1951	23406
0085	S	D	E	N	1921	22633
0085	S	D	E	N	1961	24001
0085	S	D	R	N	1960	24772
0085	S	D	E	N	1940	25372
0085	S	D	E	N	1964	25860
0085	S	D	E	N	1962	29477
0085	S	D	R	N	1961	29478
0085	S	D	E	H	1958	48883
0085	S	G	E	N	1960	51286
0085	S	D	E	N	1971	61009
0085	S	T	E	N	1971	61009
0085	S	D	E	N	1972	64938
0085	S	E	E	N	1972	64938
0085	S	D	E	N	1977	93606
0085	S	E	E	N	1977	93606
0103	S	D	G	N	1955	1125
0103	S	E	G	N	1955	1125
0103	S	D	E	N	1921	21495
0103	S	D	E	N	1921	22633
0103	S	D	O	N	1932	40805
0103	S	D	E	N	1966	40806
0103	S	D	E	N	1960	42824
0103	S	D	E	N	1900	48245
0103	S	D	G	N	1968	54724
0103	S	E	G	N	1968	54724
0103	S	D	F	N	1970	57876
0103	S	S	E	N	1973	70678
0103	S	D	E	N	1977	93606
0103	S	E	E	N	1977	93606
0103	S	D	R	N	1976	94572
0104	S	D	R	H	1963	32198
521-0108	P	D	E	N	1921	22633
0108	P	D	R	N	1968	49504
0108	P	E	R	N	1968	49504
0108	P	D	E	N	1968	65123
0108	P	E	E	N	1968	65123
0108	S	D	E	N	1960	14171
0110	M	D	F	N	1967	45088
0110	P	G	R	N	1958	20245
0110	P	G	E	N	1962	20436
0110	P	D	E	N	1921	22633
0110	P	D	E	N	1947	25437
0110	P	G	R	N	1939	29889
0110	P	D	E	N	1962	39516
0110	P	D	F	N	1965	62182
0110	P	D	R	N	1970	73669
0110	P	D	E	N	1970	73670
0110	P	D	E	N	1974	82320
0110	P	E	E	N	1974	82320
0110	S	S	E	N	1958	10667
0110	S	G	E	N	1949	12033
0110	S	D	R	N	1960	24204
0110	S	D	E	N	1924	25145
0110	S	D	E	N	1965	39028
0110	S	D	F	N	1967	45088
0110	S	D	E	N	1975	75982
0111	S	D	E	N	1959	25275
0111	S	D	R	N	1959	25276
0115	S	S	R	N	1966	47360
0115	S	D	R	N	1966	48082
0115	S	D	E	N	1968	51892
0115	S	D	E	N	1975	80507
0115	S	D	E	N	1977	93608
0115	S	E	E	N	1977	93608
0115	S	D	E	N	1979	98525
0115	S	T	E	N	1979	98525
0115	S	D	E	N	1980	100317
0115	S	T	E	N	1980	100317
0116	P	D	R	N	1968	49504
0116	P	E	R	N	1968	49504
0116	P	D	E	N	1968	65123
0116	P	E	E	N	1968	65123
0116	S	D	E		1950	5155
0116	S	D	E	N	1923	10331
0116	S	D	G	N	1968	54534
0116	S	E	G	N	1968	54534
0123	L	D	E	H	1970	59685
0123	P	D	E	N	1962	39516
0123	P	S	E	N	1969	54305
0123	P	D	R	N	1971	69768
0123	P	D	E	N	1972	69769
0123	S	D	E	N	1921	22633
0123	S	D	G	N	1904	28731
0123	S	S	E	N	1969	54305
0123	S	E	E	N	1969	55839
0123	S	D	E	N	1970	59685
0123	S	D	R	N	1971	62710
0123	S	D	P	N	1971	63225
0123	S	D	E	N	1971	69441
0123	S	D	E	N	1972	78340
0123	S	E	E	N	1972	78340
0123	S	D	R	N	1975	83260
0123	S	D	E	N	1977	93606
0123	S	E	E	N	1977	93606
0123	S	D	E	N	1978	98902
0135	P	D	E	N	1921	22633
0135	P	D	R	N	1971	65687
0135	S	G	F	N	1967	45089
0145	L	D	S	N	1970	92963
0145	S	D	S	N	1970	92963
0169	S	D	R	N	1960	20281
0170	S	D	E	N	1979	97680
0174	F	D	R	H	1966	42004
0174	F	D	E	H	1966	43896
0174	S	D	E	N	1973	71892
0180	S	D	R	N	1959	21047
0181	S	D	R	N	1964	44968
0181	S	D	R	N	1976	94572
0183	S	E	R	N	1956	54084
0184	S	D	E	N	1978	94675
0184	S	E	E	N	1978	94675
0186	P	D	E	N	1921	22633
0186	S	S	E	N	1958	10667
0190	S	D	E	N	1924	25145
0190	S	D	E	N	1965	44633
0190	S	E	E	N	1965	44633
0190	S	D	E	N	1973	72674
0190	S	E	E	N	1973	72674
0191	S	D	E	N	1973	70944
0196	M	D	R	N	1969	54587
0196	S	D	E	N	1958	12174
0196	S	D	E	N	1862	16210
0196	S	D	E	N	1863	22527
0196	S	D	E	N	1921	22633
0196	S	D	E	N	1961	24001
0196	S	D	E	N	1940	25372
0196	S	D	G	N	1904	28731
0196	S	D	E	N	1951	40424
0196	S	D	E	N	1958	48883
0196	S	G	E	N	1960	51286
0196	S	E	E	N	1976	90853
0196	S	D	E	N	1976	91645

Phys. State: **C.** Superconductive; **D.** Doped; **E.** Expanded; **F.** Fibrous or Whisker; **G.** Gas; **I.** Ionized or Plasma; **L.** Liquid; **M.** Multiphase; **P.** Powder or Fine Particle; **S.** Solid; **T.** Thick or Thin Film

Subject: **D.** Data; **E.** Experiment; **G.** General (Data + Theory + Experiment); **S.** Survey (Review, Compendium etc.); **T.** Theory

Language: **C.** Czech; **D.** Dutch; **E.** English; **F.** French; **G.** German; **I.** Italian; **J.** Japanese; **O.** Other Languages; **P.** Polish; **R.** Russian; **S.** Spanish

Temperature: **F.** Full Range (Low + Normal + High); **L.** Low (0 to 75K) + Overlap into Normal; **N.** Normal (75 to 1273K); **H.** High (above 1273K) + Overlap into Normal; Blank = Not Coded

Substance Number	Phys. State	Subject	Language	Temperature	Year	TPRC Number
521-0196	S	D	E	N	1976	92953
0196	S	E	E	N	1976	92953
0196	S	T	E	N	1976	92954
0196	S	D	E	N	1977	93606
0196	S	E	E	N	1977	93606
0215	S	D	E	N	1923	10331
0215	S	S	E	N	1973	70678
0236	S	D	E	N	1862	16210
0236	S	D	E	N	1863	22527
0236	S	D	E	N	1921	22633
0236	S	D	E	N	1940	25372
0236	S	D	G	N	1904	28731
0239	S	D	R	N	1974	76265
0240	S	D	E	N	1921	22633
0241	S	S	E	N	1973	70678
0242	S	D	E	N	1964	44969
0242	S	D	G	N	1968	54724
0242	S	E	G	N	1968	54724
0242	S	D	E	N	1972	64938
0242	S	E	E	N	1972	64938
0242	S	D	R	N	1971	67298
0242	S	D	E	N	1974	78860
0242	S	E	E	N	1974	78860
0242	S	D	E	N	1977	91966
0242	S	D	E	N	1977	93606
0242	S	E	E	N	1977	93606
0242	S	D	E	N	1978	94675
0242	S	E	E	N	1978	94675
0246	P	D	R	N	1964	35710
0246	S	S	E	N	1973	70678
0249	S	D	G	N	1940	29870
0249	S	D	R	N	1968	51815
0249	S	D	R	N	1968	54490
0249	S	S	E	N	1973	70678
0251	S	E	E	N	1976	90853
0251	S	D	E	N	1976	92953
0251	S	E	E	N	1976	92953
0253	M	D	E	N	1976	89094
0253	S	D	E	N	1959	14812
0258	S	D	E	N	1968	47311
0258	S	D	E	H	1972	68734
0258	S	E	E	H	1972	68734
0258	S	S	E	N	1973	70678
0258	S	S	E	N	1977	92134
0259	S	E	E	N	1967	43123
0264	S	D	G	N	1968	54724
0264	S	E	G	N	1968	54724
0264	S	D	E	N	1972	64938
0264	S	E	E	N	1972	64938
0264	S	D	E	N	1974	78860
0264	S	E	E	N	1974	78860
0264	S	D	R	N	1976	94572
0264	S	D	E	N	1978	94675
0264	S	E	E	N	1978	94675
0270	S	D	R	N	1971	67298
0272	S	D	E	N	1921	22633
0272	S	D	E	N	1974	78860
0272	S	E	E	N	1974	78860
0274	S	D	E	N	1921	22633
0274	S	D	E	N	1964	44966
0274	S	D	R	N	1971	67298
0274	S	D	R	N	1972	71302
0274	S	D	R	N	1967	74004
0274	S	D	E	N	1967	74005
0274	S	D	E	N	1978	94675
0274	S	E	E	N	1978	94675
0277	S	D	E	N	1969	55839
0279	P	D	E	N	1947	25437
0281	S	D	E	N	1921	22633
0293	S	D	E	N	1921	22633
0293	S	D	G	N	1968	54724
0293	S	E	G	N	1968	54724
0293	S	D	E	N	1972	64938
0293	S	E	E	N	1972	64938
0293	S	D	R	N	1971	67298
0293	S	D	R	N	1976	94572
0296	S	D	R	N	1972	71302
0298	P	D	E	N	1921	22633
0298	S	D	E	N	1921	22633
0298	S	D	E	N	1951	40424
0298	S	D	E	N	1972	78340
0298	S	E	E	N	1972	78340
0298	S	D	E	N	1977	93606
0298	S	E	E	N	1977	93606
0299	S	S	G	N	1924	33616
0299	S	E	E	N	1969	55839
0307	S	D	E	N	1974	78860
0307	S	E	E	N	1974	78860
0308	S	D	E	N	1964	37463
0308	S	D	E	N	1951	40424
0308	S	D	G	N	1968	54724
0308	S	E	G	N	1968	54724
0308	S	D	P	N	1971	63225
0308	S	D	E	N	1972	64938
0308	S	E	E	N	1972	64938
0308	S	D	R	N	1967	74004
0308	S	D	E	N	1967	74005
0308	S	D	R	N	1976	94572
0349	S	D	E	N	1921	22633
0349	S	D	E	N	1964	25860
0349	S	D	R	N	1971	67298

Substance Number	Phys. State	Subject	Language	Temperature	Year	TPRC Number
521-0352	S	D	E	H	1972	68734
0352	S	E	E	H	1972	68734
0352	S	D	E	N	1974	78860
0352	S	E	E	N	1974	78860
0356	S	D	R	N	1967	74004
0356	S	D	E	N	1967	74005
0362	S	D	E	N	1921	22633
0362	S	D	E	N	1955	32862
0362	S	D	J	N	1977	90890
0362	S	E	J	N	1977	90890
0363	S	D	E	N	1921	22633
0364	S	D	E	N	1921	22633
0364	S	D	G	N	1968	54724
0364	S	E	G	N	1968	54724
0364	S	D	E	N	1972	64938
0364	S	E	E	N	1972	64938
0365	S	D	E	N	1921	22633
0366	S	D	E	N	1921	22633
0368	M	D	R	N	1960	20247
0368	P	S	E	N	1973	70678
0368	S	G	E	N	1970	58782
0369	S	D	E	N	1921	22633
0369	S	D	R	N	1960	24772
0369	S	G	E	N	1960	51286
0369	S	D	G	N	1968	54724
0369	S	E	G	N	1968	54724
0369	S	D	E	N	1972	64938
0369	S	E	E	N	1972	64938
0370	S	D	E	N	1921	22633
0371	S	D	E	N	1921	22633
0372	S	D	E	N	1921	22633
0373	S	D	E	N	1921	22633
0375	S	D	E	N	1921	22633
0375	S	D	E	N	1964	44968
0375	S	D	G	N	1968	54724
0375	S	E	G	N	1968	54724
0375	S	D	P	N	1971	63225
0375	S	D	E	N	1972	64938
0375	S	E	E	N	1972	64938
0376	S	D	E	N	1921	22633
0377	S	D	E	N	1921	22633
0378	S	D	E	N	1921	22633
0379	S	D	E	N	1921	22633
0380	S	D	E	N	1921	22633
0381	S	D	E	N	1921	22633
0382	P	D	E	H	1962	29477
0382	P	D	R	H	1961	29478
0382	P	G	R	H	1962	29903
0382	S	D	E	H	1962	29477
0382	S	D	R	H	1961	29478
0382	S	G	R	H	1962	29903
0382	S	D	R	H	1963	32198
0382	S	D	E	H	1964	44968
0382	S	D	R	H	1976	84491
0382	S	D	R	N	1976	84491
0383	S	D	E	N	1921	22633
0384	M	D	R	N	1960	20247
0384	P	D	R	N	1971	67531
0384	P	D	R	N	1959	69694
0384	P	D	E	N	1972	69695
0384	S	D	R	N	1959	69694
0384	S	D	E	N	1972	69695
0393	P	D	R	N	1971	65687
0394	S	D	R	N	1971	67298
0402	S	D	E	N	1969	56053
0423	S	E	E	N	1965	38994
0430	S	D	R	N	1971	67298
0432	S	D	R	N	1960	24772
0432	S	D	G	N	1904	28731
0442	M	G	F	N	1965	39350
0442	M	G	R	N	1959	43002
0442	M	G	F	N	1959	43003
0442	P	D	E	N	1967	47508
0442	P	D	R	N	1970	73679
0442	P	D	E	N	1970	73680
0442	S	D	R	N	1959	25367
0442	S	T	R	N	1959	25367
0442	S	G	F	N	1965	39350
0442	S	G	R	N	1959	43002
0442	S	G	F	N	1959	43003
0442	S	D	E	N	1965	44633
0442	S	E	E	N	1965	44633
0477	S	D	R	N	1967	74004
0477	S	D	E	N	1967	74005
0495	S	D	P	N	1971	63225
0495	S	D	E	N	1979	97680
0500	P	D	G	N	1959	28721
0500	P	D	G	H	1971	65158
0500	P	E	G	H	1971	65158
0500	P	D	R	N	1971	65687
0500	P	D	E	N	1975	84944
0500	S	D	R	N	1963	49400
0500	S	D	E	N	1967	49401
0502	S	D	G	N	1904	28731
0530	S	D	E	N	1969	56053
0564	S	D	E	N	1964	25860
0567	S	D	E	N	1964	25860
0570	S	D	E	N	1964	25860
0581	M	G	R	H	1962	29903
0612	S	D	E	N	1965	44633
0612	S	E	E	N	1965	44633

Substance Number	Phys. State	Subject	Language	Temperature	Year	TPRC Number
521-0643	S	D	E	N	1965	39028
0645	M	G	R	N	1959	43002
0645	M	G	F	N	1959	43003
0645	P	D	E	N	1967	47508
0645	S	D	R	N	1959	25367
0645	S	T	R	N	1959	25367
0645	S	D	E	N	1965	39028
0645	S	G	R	N	1959	43002
0645	S	G	F	N	1959	43003
0645	S	D	R	N	1970	73679
0645	S	D	E	N	1970	73680
0648	S	D	E	N	1968	47311
0660	S	D	G	N	1970	59789
0660	S	E	G	N	1970	59789
0663	S	D	R	H	1963	32198
0664	S	D	R	H	1963	32198
0671	S	D	E	N	1921	22633
0671	S	D	G	N	1968	54724
0671	S	E	G	N	1968	54724
0671	S	D	E	N	1972	64938
0671	S	E	E	N	1972	64938
0671	S	D	R	N	1972	71302
0675	P	D	R	N	1966	39422
0713	S	D	G	N	1968	54724
0713	S	E	G	N	1968	54724
0713	S	D	R	N	1971	67298
0713	S	D	R	N	1976	94572
0727	S	D	E	N	1958	12174
0731	S	D	R	N	1950	43829
0732	P	D	R	N	1961	50701
0732	P	E	R	N	1961	50701
0732	P	D	R	N	1973	75951
0732	P	E	R	N	1973	75951
0732	S	D	E		1951	85
0732	S	D	E	H	1960	20600
0732	S	D	E	N	1962	26684
0732	S	D	E	N	1961	27117
0732	S	T	E	N	1961	27117
0732	S	D	R	N	1961	33393
0732	S	E	E	N	1961	33393
0732	S	D	R	N	1964	38184
0732	S	G	E	N	1970	41374
0732	S	G	F	N	1965	42720
0732	S	D	E	N	1968	47311
0732	S	D	E	N	1969	55839
0732	S	D	P	N	1971	63225
0732	S	S	E	N	1973	70678
0732	S	G	E	N	1970	74014
0732	S	S	E	N	1977	92134
0732	S	D	R	N	1976	94572
0732	S	D	E	N	1979	97680
0762	S	D	R	N	1972	71302
0769	S	D	E	H	1972	68734
0769	S	E	E	H	1972	68734
0771	S	D	R	N	1969	59417
0771	S	D	E	N	1969	59418
0783	S	D	R	N	1978	95686
0783	S	D	E	N	1978	98588
0797	L	D	E	H	1969	64934
0797	L	T	E	H	1969	64934
0822	P	D	E	N	1967	47508
0825	S	D	E	N	1964	44969
0825	S	D	R	N	1971	67298
0836	S	D	E	N	1964	44966
0836	S	D	E	N	1964	44968
0837	S	D	E	N	1964	44968
0839	S	D	E	N	1964	44968
0840	S	D	E	N	1964	44968
0842	S	D	E	N	1964	44969
0843	S	D	E	N	1964	44969
0844	S	D	E	N	1964	44969
0845	S	D	E	N	1964	44966
0846	S	D	E	N	1964	44966
0847	S	D	E	N	1964	44966
0852	S	D	R	N	1974	76265
0894	S	D	R	N	1969	58631
0894	S	D	R	N	1972	71302
0894	S	D	R	N	1976	94572
0913	S	D	E	N	1968	47311
0917	P	D	E	N	1971	64053
0917	P	E	E	N	1971	64053
0917	P	D	E	N	1972	66830
0917	P	D	R	N	1971	69768
0917	P	D	E	N	1972	69769
0917	P	D	E	N	1973	70823
0917	P	D	E	N	1974	74722
0917	P	D	E	N	1974	78683
0917	P	D	R	N	1974	80433
0917	P	D	E	N	1975	80434
0917	P	D	E	N	1974	82575
0922	S	D	E	N	1970	56107
0922	S	S	E	N	1972	68902
0922	S	D	E	N	1972	70282
0922	S	D	E	N	1973	70421
0922	S	E	E	N	1973	70421
0924	S	D	E	N	1978	94675
0924	S	E	E	N	1978	94675
0925	S	D	R	N	1971	67298
0978	S	D	E	N	1974	78860
0978	S	E	E	N	1974	78860
1028	S	D	R	N	1971	67298

Phys. State: **C.** Superconductive; **D.** Doped; **E.** Expanded; **F.** Fibrous or Whisker; **G.** Gas; **I.** Ionized or Plasma; **L.** Liquid; **M.** Multiphase; **P.** Powder or Fine Particle; **S.** Solid; **T.** Thick or Thin Film

Subject: **D.** Data; **E.** Experiment; **G.** General (Data + Theory + Experiment); **S.** Survey (Review, Compendium etc.); **T.** Theory

Language: **C.** Czech; **D.** Dutch; **E.** English; **F.** French; **G.** German; **I.** Italian; **J.** Japanese; **O.** Other Languages; **P.** Polish; **R.** Russian; **S.** Spanish

Temperature: **F.** Full Range (Low + Normal + High); **L.** Low (0 to 75K) + Overlap into Normal; **N.** Normal (75 to 1273K); **H.** High (above 1273K) + Overlap into Normal; Blank = Not Coded

Substance Number	Phys. State	Subject	Language	Temperature	Year	TPRC Number
521-1038	P	D	E	N	1971	64504
1038	P	T	E	N	1971	64504
1048	S	D	E	N	1924	25145
1057	S	D	R	N	1967	74004
1057	S	D	E	N	1967	74005
1074	S	D	R	N	1969	56214
1074	S	E	R	N	1969	56214
1074	S	D	E	N	1969	67256
1074	S	E	E	N	1969	67256
1089	D	D	R	N	1971	63801
1089	D	E	R	N	1971	63801
1093	P	D	R	N	1971	65687
1094	P	D	R	N	1971	65687
1114	P	D	E	N	1974	77420
1123	L	D	E	H	1969	64934
1123	L	T	E	H	1969	64934
1124	L	D	E	H	1969	64934
1124	L	T	E	H	1969	64934
1125	L	D	E	H	1969	64934
1125	L	T	E	H	1969	64934
1209	S	D	G	N	1968	54724
1209	S	E	G	N	1968	54724
1210	S	D	G	N	1968	54724
1210	S	E	G	N	1968	54724
1211	S	D	G	N	1968	54724
1211	S	E	G	N	1968	54724
1211	S	D	E	N	1972	64938
1211	S	E	E	N	1972	64938
1211	S	D	R	N	1971	67298
1212	S	D	G	N	1968	54724
1212	S	E	G	N	1968	54724
1212	S	D	E	N	1972	64938
1212	S	E	E	N	1972	64938
1238	S	D	E	N	1972	64938
1238	S	E	E	N	1972	64938
1239	S	D	E	N	1972	64938
1239	S	E	E	N	1972	64938
1241	S	D	E	N	1972	64938
1241	S	E	E	N	1972	64938
1242	S	D	E	N	1972	64938
1242	S	E	E	N	1972	64938
1243	S	D	E	N	1972	64938
1243	S	E	E	N	1972	64938
1244	S	D	E	N	1972	64938
1244	S	E	E	N	1972	64938
1245	S	D	E	N	1972	64938
1245	S	E	E	N	1972	64938
1255	S	D	E	N	1965	44633
1255	S	E	E	N	1965	44633
1316	S	D	R	N	1967	74004
1316	S	D	E	N	1967	74005
1321	S	D	R	N	1967	74004
1321	S	D	E	N	1967	74005
1322	S	D	R	N	1967	74004
1322	S	D	E	N	1967	74005
1323	S	D	R	N	1967	74004
1323	S	D	E	N	1967	74005
1324	S	D	R	N	1967	74004
1324	S	D	E	N	1967	74005
1326	S	D	R	N	1967	74004
1326	S	D	E	N	1967	74005
1327	S	D	R	N	1967	74004
1327	S	D	E	N	1967	74005
1333	S	D	R	N	1971	67298
1334	S	D	R	N	1971	67298
1335	S	D	R	N	1971	67298
1335	S	D	R	N	1976	94572
1345	S	D	R	N	1970	73679
1345	S	D	E	N	1970	73680
1357	E	D	E	N	1976	89124
1357	E	E	E	N	1976	89124
1357	S	D	E	N	1970	89122
1357	S	E	E	N	1970	89122
1357	S	D	E	N	1976	89124
1357	S	E	E	N	1976	89124
1358	S	D	R	N	1969	73753
1358	S	D	E	N	1969	73754
1392	E	D	E	N	1975	89123
1392	P	D	E	N	1976	89124
1392	P	E	E	N	1976	89124
1392	S	D	E	N	1970	89122
1392	S	E	E	N	1970	89122
1392	S	D	E	N	1975	89123
1392	S	D	E	N	1976	89124
1392	S	E	E	N	1976	89124
1403	S	D	R	N	1972	71302
1404	S	D	R	N	1972	71302
1405	S	D	R	N	1972	71302
1406	S	D	R	N	1972	71302
1407	S	D	R	N	1972	71302
1408	S	D	R	N	1972	71302
1409	S	D	R	N	1972	71302
1410	S	D	R	N	1972	71302
1411	S	D	R	N	1972	71302
1412	S	D	R	N	1972	71302
1413	S	D	R	N	1972	71302
1414	S	D	R	N	1972	71302
1424	M	D	E	N	1973	89620
1424	M	D	E	N	1977	91973
1424	M	T	E	N	1977	91973
1424	S	D	E	N	1977	91966
521-1424	S	D	E	N	1979	100606
1467	S	D	E	N	1958	12174
1549	S	D	E	N	1979	97680
1556	S	D	E	N	1978	95156
1571	M	D	E	N	1863	7575
1571	M	E	E	N	1863	7575
1571	M	D	G	N	1861	10036
1571	M	E	G	N	1861	10036
1572	M	D	R	N	1971	44543
1572	M	E	R	N	1971	44543
1572	M	D	E	N	1971	75014
1572	M	E	E	N	1971	75014
1604	P	D	E	N	1863	7575
1604	P	E	E	N	1863	7575
1604	P	D	G	N	1861	10036
1604	P	E	G	N	1861	10036
1645	S	D	R	N	1978	95686
1645	S	D	E	N	1978	98588
1646	S	D	R	N	1978	95686
1646	S	D	E	N	1978	98588
1655	S	D	E	N	1978	94675
1655	S	E	E	N	1978	94675

Phys. State: **C.** Superconductive; **D.** Doped; **E.** Expanded; **F.** Fibrous or Whisker; **G.** Gas; **I.** Ionized or Plasma; **L.** Liquid; **M.** Multiphase; **P.** Powder or Fine Particle; **S.** Solid; **T.** Thick or Thin Film

Subject: **D.** Data; **E.** Experiment; **G.** General (Data + Theory + Experiment); **S.** Survey (Review, Compendium etc.); **T.** Theory

Language: **C.** Czech; **D.** Dutch; **E.** English; **F.** French; **G.** German; **I.** Italian; **J.** Japanese; **O.** Other Languages; **P.** Polish; **R.** Russian; **S.** Spanish

Temperature: **F.** Full Range (Low + Normal + High); **L.** Low (0 to 75K) + Overlap into Normal; **N.** Normal (75 to 1273K); **H.** High (above 1273K) + Overlap into Normal; Blank = Not Coded

Chapter 3 Specific Heat

Substance Number	Phys. State	Subject	Language	Temperature	Year	TPRC Number
503-0001	S	G	F	N	1959	18033
0001	S	S	E	N	1960	24931
0001	S	D	O	N	1970	59018
0001	S	D	E	N	1971	63983
0003	S	S	R	N	1973	79859
0003	S	S	E	N	1973	79860
0012	S	S	E	N	1970	61236
0018	S	D	E	N	1960	76856
0020	S	D	E		1954	6565
0020	S	D	E	N	1960	10947
0020	S	S	E	N	1960	24931
0020	S	D	E	N	1968	48550
0024	S	G	F	N	1959	18033
0024	S	D	G	N	1951	21301
0024	S	D	O	N	1970	59018
0024	S	D	E	N	1971	63983
0025	S	D	E		1948	1557
0025	S	D	E		1947	6356
0025	S	T	E		1947	6356
0025	S	G	R	N	1962	31437
0025	S	D	R	H	1967	49598
0025	S	D	J	N	1968	53139
0025	S	E	J	N	1968	53139
0026	S	S	E	N	1976	94271
0031	S	D	E		1958	8182
0031	S	T	E		1958	8182
0031	S	S	E	L	1960	16701
0031	S	D	G	N	1951	21301
0031	S	D	E	N	1930	21724
0031	S	D	E	N	1927	23204
0031	S	D	E	N	1963	37530
0031	S	D	E	N	1964	37606
0031	S	D	E	N	1970	59444
0031	S	S	E	N	1970	61236
0047	S	D	E	L	1980	99672
0055	S	D	E	L	1977	88621
0055	S	D	E	L	1977	91994
0066	S	D	E	N	1971	68241
0066	S	D	R	N	1972	70625
0066	S	D	E	N	1972	70626
0066	S	D	E	L	1975	85499
0066	S	D	E	N	1975	85499
0082	E	D	E		1950	1097
0082	E	D	E		1946	6321
0082	E	D	E	N	1960	15664
0082	E	D	E	N	1950	24328
0082	E	D	E	N	1964	37058
0082	E	S	E	N	1948	54088
0082	F	D	E	N	1959	16094
0082	F	D	E	N	1950	20019
0082	F	D	E	N	1942	25423
0082	F	D	E	N	1942	28712
0082	F	D	E	N	1960	42824
0082	F	D	G	N	1967	46159
0082	F	D	G	N	1968	49680
0082	F	D	G	L	1970	58456
0082	F	E	G	L	1970	58456
0082	F	S	E	N	1954	73367
0082	F	D	E	N	1947	74339
0082	F	E	E	N	1947	74339
0082	F	D	E	N	1947	76486
0082	F	E	E	N	1947	76486
0082	F	D	E	N	1972	76957
0082	F	D	E	N	1967	77941
0082	G	D	G	N	1900	25402
0082	L	D	E		1954	588
0082	L	T	E		1951	1378
0082	L	E	E		1941	4627
0082	L	D	E	H	1936	14872
0082	L	T	E	H	1936	14872
0082	L	S	F	H	1957	25488
0082	L	T	R	N	1964	32851
0082	L	T	E	N	1968	47535
0082	M	T	E		1954	1713
0082	M	D	E	N	1974	75085
0082	M	E	E	N	1974	75085
0082	M	D	R	N	1973	77334
0082	M	E	R	N	1973	77334
0082	S	E	E		1956	426
0082	S	D	J		1953	1282
0082	S	T	E		1951	1378
0082	S	D	E		1958	8182
0082	S	T	E		1958	8182
0082	S	D	G	N	1933	11172
0082	S	D	E	F	1936	14872
0082	S	T	E	F	1936	14872
0082	S	S	E	N	1936	16524
0082	S	E	G	N	1959	17394
0082	S	E	E	N	1960	17395
0082	S	S	G	N	1923	21761
0082	S	D	G	N	1911	22607
0082	S	D	E	N	1921	22633
0082	S	D	E	N	1963	24550
0082	S	S	E	N	1956	24971
0082	S	D	E	N	1919	25193
503-0082	S	S	F	H	1957	25488
0082	S	T	G	N	1962	26483
0082	S	D	E	N	1961	27117
0082	S	T	E	N	1962	27989
0082	S	S	E	H	1961	28014
0082	S	D	G	N	1904	28731
0082	S	D	E	N	1951	29285
0082	S	S	R	L	1959	31165
0082	S	S	E	L	1963	31166
0082	S	D	E	N	1957	31948
0082	S	D	E	N	1963	32128
0082	S	S	E	N	1963	32162
0082	S	S	R	N	1962	33745
0082	S	S	E	N	1966	33746
0082	S	D	O	N	1965	35665
0082	S	T	O	N	1965	35665
0082	S	S	E	F	1969	37682
0082	S	E	E	L	1964	40697
0082	S	D	E	N	1960	42824
0082	S	G	J	N	1967	46251
0082	S	G	E	N	1968	46252
0082	S	S	E	L	1968	49258
0082	S	S	D	N	1968	49783
0082	S	D	E	N	1968	50083
0082	S	D	O	H	1966	55890
0082	S	D	O	N	1966	55890
0082	S	T	E	N	1972	65794
0082	S	T	E	L	1972	65856
0082	S	S	R	N	1971	67587
0082	S	D	J	N	1972	67784
0082	S	E	J	N	1972	67784
0082	S	D	E	N	1972	68674
0082	S	E	E	N	1972	68674
0082	S	D	E	N	1973	70941
0082	S	T	O	N	1974	75943
0082	S	D	E	N	1975	76427
0082	S	D	O	N	1974	78659
0082	S	T	O	N	1974	78659
0082	S	D	E	L	1974	82998
0082	S	E	E	L	1974	82998
0082	S	D	E	L	1975	85499
0082	S	D	E	N	1975	85499
0082	S	T	E	N	1975	88857
0082	S	S	E	N	1979	99434
0090	D	D	R	L	1972	75950
0096	S	D	J	N	1956	19593
0126	S	D	E	N	1973	71894
0200	S	D	E	L	1975	80989
0200	S	E	E	L	1975	80989
0210	S	S	E	N	1960	24931
0216	S	D	E		1954	6565
0216	S	D	E	N	1962	26684
0216	S	D	E	N	1961	27117
0216	S	D	E	N	1965	36738
0216	S	D	E	N	1964	38131
0216	S	D	E	L	1968	47910
0216	S	D	E	N	1968	49113
0216	S	E	E	N	1968	49113
0216	S	D	E	L	1971	63375
0216	S	G	E	L	1971	66152
0216	S	D	E	L	1973	69939
0216	S	D	E	L	1973	71899
0239	S	D	G	L	1970	58456
0239	S	E	G	L	1970	58456
0239	S	D	E	N	1913	94422
0240	P	D	E	N	1975	82557
0270	S	D	F		1955	695
0270	S	D	G	N	1951	21301
0270	S	D	O	N	1970	59018
0270	S	D	E	N	1971	63983
0291	S	G	F	N	1959	18033
0295	S	G	F	N	1959	18033
0376	L	G	G		1953	448
0376	S	G	G		1953	448
0376	S	G	F	N	1959	18033
0376	S	S	E	H	1960	24931
0376	S	S	E	N	1960	24931
0376	S	D	E	H	1946	28355
0376	S	D	E	N	1946	28355
0376	S	E	E	H	1946	28355
0376	S	E	E	N	1946	28355
0376	S	G	R	N	1962	31437
0376	S	S	E	N	1970	61236
0452	S	S	E	N	1963	32162
0452	S	D	E	N	1962	47240
0452	S	D	E	N	1963	58889
0463	S	D	R	N	1971	67588
0466	S	D	R	N	1971	67588
0537	S	S	E	N	1970	61236
0545	S	D	E	H	1934	15070
0545	S	G	R	N	1962	31437
0545	S	D	E	N	1967	40398
0555	L	G	O	H	1959	23751
0555	S	G	O	H	1959	23751
0556	L	G	G	H	1953	448
503-0556	S	G	G	N	1953	448
0556	S	S	E	H	1960	24931
0556	S	S	E	N	1960	24931
0559	L	S	E	H	1977	94788
0559	S	S	E	H	1977	94788
0559	S	S	E	N	1977	94788
0562	S	G	E	N	1972	71416
0562	S	S	E	N	1979	97587
0566	S	D	E	N	1963	37530
0566	S	S	E	N	1970	61236
0604	S	D	E		1958	7695
0604	S	S	E	N	1970	61236
0631	S	S	E	N	1976	90192
0648	S	D	E	L	1971	63375
0648	S	G	E	L	1971	66152
0648	S	D	E	L	1973	71899
0659	S	D	E		1959	8100
0659	S	D	E	N	1959	22183
0659	S	D	E	N	1961	29925
0659	S	D	E	N	1958	33214
0659	S	D	E	N	1963	49933
0659	S	S	E	N	1970	61236
0659	S	S	E	N	1976	83746
0659	S	D	E	H	1977	94200
0659	S	D	E	N	1977	94200
0659	S	S	E	N	1977	94200
0660	S	D	E		1959	8100
0660	S	D	E	N	1963	24384
0660	S	D	E	N	1961	29925
0660	S	D	E	N	1958	33214
0660	S	S	E	N	1970	61236
0667	S	S	E	N	1961	25075
0667	S	S	E	N	1964	30525
0704	S	S	R	N	1973	79859
0704	S	S	E	N	1973	79860
0751	S	D	E	L	1975	88798
0751	S	E	E	L	1975	88798
0765	S	D	G	N	1933	11172
0771	S	G	F	N	1959	18033
0782	L	D	E	N	1975	86303
0782	S	D	E	H	1934	15070
0782	S	D	E	N	1944	28248
0782	S	D	E	N	1958	33214
0782	S	S	E	L	1970	61236
0782	S	D	E	N	1971	68241
0782	S	D	R	N	1972	69632
0782	S	D	E	N	1973	69633
0782	S	S	E	N	1975	78777
0801	S	D	G	N	1933	11172
0801	S	S	E	N	1936	16524
0801	S	D	G	H	1926	21459
0801	S	S	G	H	1928	28680
0801	S	D	E	N	1957	31948
0801	S	D	R	N	1967	47303
0801	S	D	E	N	1968	47358
0801	S	D	J	N	1969	58137
0801	S	E	J	N	1969	58137
0815	S	S	E	N	1960	24931
0818	S	S	E	N	1960	24931
0822	S	D	E	N	1967	40398
0822	S	D	E	N	1914	41121
0822	S	G	E	N	1972	66811
0822	S	G	E	N	1973	76238
0849	S	D	E	N	1979	97889
0861	S	D	E	L	1975	81116
0868	S	D	R	N	1971	67588
0879	S	G	F	N	1959	25490
0884	L	G	R		1940	8860
0884	L	D	E	H	1928	22639
0884	S	D	E	H	1928	22639
0964	S	D	E	L	1980	100311
0966	S	D	E	N	1960	99523
0967	S	D	E	N	1969	78264
0967	S	E	E	N	1969	78264
0968	S	D	E	N	1969	78264
0968	S	E	E	N	1969	78264
0974	S	D	E	N	1960	76856
0977	S	G	F	N	1959	18033
0978	S	D	G	N	1951	21301
0981	S	D	G	L	1970	58456
0981	S	E	G	L	1970	58456
0997	S	S	E	N	1976	88411
1015	S	D	R	N	1971	67588
1023	S	D	R	N	1971	67588
1024	S	D	R	N	1971	67588
1050	S	D	E		1954	6565
1050	S	D	E	N	1960	10947
1050	S	D	E	H	1934	15070
1050	S	D	G	N	1951	21301
1050	S	S	E	H	1960	24931
1050	S	S	E	H	1960	24931
1050	S	D	E	N	1962	26684
1050	S	D	E	H	1946	28355
1050	S	D	E	N	1946	28355
1050	S	E	E	H	1946	28355

Phys. State: **C.** Superconductive; **D.** Doped; **E.** Expanded; **F.** Fibrous or Whisker; **G.** Gas; **I.** Ionized or Plasma; **L.** Liquid; **M.** Multiphase; **P.** Powder or Fine Particle; **S.** Solid; **T.** Thick or Thin Film

Subject: **D.** Data; **E.** Experiment; **G.** General (Data + Theory + Experiment); **S.** Survey (Review, Compendium etc.); **T.** Theory

Language: **C.** Czech; **D.** Dutch; **E.** English; **F.** French; **G.** German; **I.** Italian; **J.** Japanese; **O.** Other Languages; **P.** Polish; **R.** Russian; **S.** Spanish

Temperature: **F.** Full Range (Low + Normal + High); **L.** Low (0 to 75K) + Overlap into Normal; **N.** Normal (75 to 1273K); **H.** High (above 1273K) + Overlap into Normal; Blank = Not Coded

Substance Number	Phys. State	Subject	Language	Temperature	Year	TPRC Number
503-1050	S	E	E	N	1946	28355
1050	S	S	E	N	1970	61236
1057	S	D	G	N	1951	21301
1057	S	G	G	N	1911	29642
1058	S	D	R	N	1957	9440
1058	S	D	E	N	1957	17360
1059	L	T	E	N	1953	19384
1059	M	G	E		1933	8112
1059	P	S	E	N	1970	61236
1059	S	T	R		1954	674
1059	S	T	E	N	1953	19384
1059	S	S	E	H	1960	24931
1059	S	S	E	N	1960	24931
1059	S	D	E	N	1958	33214
1059	S	D	E	N	1965	36144
1059	S	D	E	H	1919	43370
1059	S	D	E	N	1977	90682
1060	S	D	E	H	1934	15070
1062	S	D	E		1954	6565
1062	S	D	E	N	1960	10947
1062	S	S	E	N	1960	24931
1062	S	S	E	N	1970	61236
1065	S	D	E		1958	8182
1065	S	T	E		1958	8182
1065	S	D	E	H	1919	43370
1067	S	D	E		1958	7695
1067	S	S	E	N	1970	61236
1072	S	D	E		1954	6565
1072	S	D	E	N	1960	10947
1072	S	S	E	H	1960	24931
1072	S	S	E	N	1960	24931
1072	S	D	E	H	1946	28355
1072	S	D	E	N	1946	28355
1072	S	E	E	H	1946	28355
1072	S	E	E	N	1946	28355
1072	S	S	E	N	1970	61236
1075	S	D	E	H	1934	15070
1076	S	D	E	H	1934	15070
1077	S	D	E	H	1934	15070
1078	S	G	E	L	1972	76522
1081	L	D	E	N	1961	18322
1083	S	D	R	N	1971	67588
1102	S	D	G	N	1951	21301
1110	S	S	E	N	1961	25075
1112	S	S	E	N	1960	24931
1112	S	D	G	N	1941	30260
1125	S	D	E		1935	8241
1125	S	D	E	N	1959	13944
1125	S	S	E	N	1976	90192
1130	S	D	F	N	1954	23317
1131	S	D	R		1955	757
1131	S	D	R	N	1939	48342
1131	S	D	E	N	1969	59294
1131	S	E	E	N	1969	59294
1132	S	D	R	H	1934	16471
1133	S	D	E		1945	1719
1133	S	D	R		1947	1942
1133	S	D	G		1949	1972
1133	S	S	E	N	1925	19806
1133	S	S	G	H	1919	22023
1133	S	D	G	N	1930	23249
1133	S	D	G	N	1923	25148
1133	S	D	E	N	1961	25173
1133	S	D	E	N	1962	28005
1133	S	S	S	H	1955	28264
1133	S	S	G	H	1928	28680
1133	S	D	E	H	1926	33546
1133	S	D	E	H	1924	43414
1133	S	D	E	N	1969	59294
1133	S	E	E	N	1969	59294
1135	M	D	E	N	1958	10458
1135	S	S	G	H	1928	28680
1135	S	S	R	H	1963	43656
1136	M	D	G	N	1958	51235
1136	S	D	R		1949	801
1136	S	D	G		1949	1972
1136	S	G	R		1948	4492
1136	S	S	E	N	1958	10667
1136	S	G	E	N	1949	12033
1136	S	G	E	N	1955	24261
1136	S	D	G	N	1923	25148
1136	S	D	E	N	1962	28005
1136	S	D	E	N	1963	30132
1136	S	D	E	N	1964	37058
1136	S	D	G	N	1958	51235
1136	S	D	E	N	1969	59294
1136	S	E	E	N	1969	59294
1136	S	D	E	N	1972	68595
1137	S	D	R		1945	398
1137	S	T	R		1945	398
1137	S	E	E	N	1921	22633
1138	S	D	G	H	1926	21459
1139	S	D	E		1935	8241
1139	S	D	G	N	1930	23249
1139	S	D	E	N	1918	24767
1139	S	S	G	N	1928	28680
1139	S	D	R	N	1939	48342
1142	S	E	E		1955	901
1142	S	D	R		1953	8876
1142	S	E	E		1959	9102
1142	S	D	E	H	1961	20637
503-1142	S	S	E	N	1961	20740
1142	S	D	E	N	1960	24322
1142	S	S	E	H	1960	24931
1142	S	S	G	H	1928	28680
1142	S	S	E	N	1928	31217
1142	S	E	E	N	1963	31998
1142	S	D	E	H	1965	36139
1142	S	S	E	F	1969	37682
1142	S	D	R	H	1965	43324
1142	S	S	D	N	1968	49783
1142	S	S	E	H	1926	59969
1142	S	D	E	L	1975	85499
1142	S	D	E	N	1975	85499
1145	S	E	J		1953	1045
1145	S	D	E		1948	1557
1145	S	D	E		1945	1719
1145	S	D	G		1949	1972
1145	S	D	E		1943	7757
1145	S	E	E	H	1960	19485
1145	S	D	G	N	1926	19685
1145	S	E	E	N	1962	20006
1145	S	S	E	H	1958	20631
1145	S	E	G	N	1930	23249
1145	S	E	E	N	1960	29334
1145	S	S	E	H	1961	29615
1145	S	E	R	H	1965	32972
1145	S	E	E	H	1965	33031
1145	S	D	E	N	1965	35579
1145	S	S	E	H	1964	37819
1145	S	S	E	N	1932	39845
1145	S	S	E	N	1924	44568
1145	S	S	E	N	1967	51672
1145	S	S	E	H	1956	52789
1145	S	S	E	H	1963	52812
1145	S	S	E	H	1969	55310
1145	S	D	R	H	1970	61318
1145	S	D	E	H	1970	61319
1145	S	D	R	N	1974	75244
1146	S	D	R		1955	757
1146	S	D	E		1945	1719
1146	S	D	G		1949	1972
1146	S	D	J	N	1938	15065
1146	S	S	E	N	1925	19806
1146	S	D	G	N	1930	23249
1146	S	D	E	N	1918	24767
1146	S	D	G	N	1923	25148
1146	S	D	E	N	1962	28005
1146	S	S	G	N	1928	28680
1146	S	D	E	H	1926	33546
1146	S	D	R	N	1939	48342
1146	S	D	E	N	1969	59294
1146	S	E	E	N	1969	59294
1146	S	D	R	H	1971	74368
1146	S	D	E	H	1971	74369
1147	S	D	G	N	1934	6342
1147	S	D	E	N	1945	11116
1147	S	D	R	H	1934	16471
1147	S	S	E	N	1925	19806
1147	S	D	G	H	1926	21459
1147	S	S	G	H	1919	22023
1147	S	D	G	N	1930	23249
1147	S	D	E	N	1918	24767
1147	S	D	G	N	1923	25148
1147	S	D	E	N	1962	28005
1147	S	D	E	H	1924	43414
1147	S	D	R	N	1939	48342
1147	S	D	E	N	1969	59294
1147	S	E	E	N	1969	59294
1147	S	S	E	H	1976	90192
1147	S	S	E	N	1976	90192
1147	S	S	E	N	1976	92625
1148	M	D	R		1957	8791
1149	S	D	G		1949	1972
1149	S	G	E	N	1974	74606
1152	S	D	E	N	1973	71892
1153	S	D	E		1945	1719
1153	S	D	E	N	1949	10272
1153	S	D	G	N	1933	11172
1156	S	D	G		1947	1822
1162	S	D	E	N	1970	57984
1162	S	D	E	N	1970	67511
1164	S	D	G		1943	1911
1173	E	D	E	N	1960	15664
1173	M	D	E	N	1960	13346
1173	S	D	E	N	1962	28005
1173	S	D	E	N	1963	32128
1173	S	D	E	N	1958	33214
1173	S	D	E	H	1926	33546
1173	S	D	E	N	1963	37530
1173	S	D	E	N	1969	59294
1173	S	E	E	N	1969	59294
1173	S	D	R	H	1970	61531
1173	S	D	E	H	1970	61532
1173	S	D	R	H	1971	64822
1173	S	D	E	H	1971	64823
1173	S	D	E	N	1974	75087
1177	S	D	F	N	1954	23317
1177	S	S	G	N	1928	28680
1177	S	G	R	N	1971	66294
1177	S	G	E	N	1971	70931
1180	S	D	E	N	1963	32128
503-1185	S	D	E	N	1967	42955
1185	S	D	I	N	1967	45153
1185	S	G	E	N	1967	46930
1186	S	E	E	H	1940	14575
1187	S	D	E		1956	6524
1187	S	T	E		1956	6524
1187	S	D	E	N	1957	31948
1189	S	D	E	N	1966	44390
1189	S	E	E	N	1966	44390
1189	S	D	E	N	1969	59294
1189	S	E	E	N	1969	59294
1189	S	S	E	H	1976	90192
1189	S	S	E	N	1976	90192
1194	P	D	E	N	1909	31920
1194	P	G	E	N	1969	49097
1194	S	S	G	N	1928	28680
1201	S	D	E	N	1962	28005
1201	S	D	E	N	1969	59294
1201	S	E	E	N	1969	59294
1202	L	E	G	H	1927	22655
1202	S	E	G	H	1927	22655
1218	S	D	G	N	1933	11172
1218	S	D	G	H	1926	21459
1218	S	D	G	N	1923	25148
1219	S	D	E	H	1960	20600
1219	S	D	E	N	1963	23121
1220	S	S	E	H	1958	20631
1221	S	S	E	H	1958	20631
1222	S	S	E	H	1958	20631
1227	S	D	D	H	1927	22700
1228	S	D	G	N	1920	24372
1228	S	D	G	N	1923	25148
1241	S	D	G	N	1923	25148
1242	S	D	E	N	1918	24767
1270	S	S	E	N	1976	90192
1310	M	T	R	N	1970	64246
1321	P	D	R	N	1971	65687
1331	S	D	R	N	1972	67971
1335	S	D	R	H	1971	74368
1335	S	D	E	H	1971	74369
1340	S	D	R	N	1974	75244
1344	S	S	E	N	1974	78899
1354	F	D	E	N	1942	28712
1356	S	S	G	N	1961	28087
1356	S	S	E	N	1962	28088
1357	F	G	E	N	1949	12033
1357	F	D	E	N	1921	22401
1357	F	D	E	N	1942	25423
1357	L	D	G	N	1933	13923
1357	L	T	G	N	1933	13923
1357	L	E	G	H	1927	22655
1357	L	D	E	N	1971	61518
1357	S	E	G	H	1927	22655
1357	S	D	F	N	1954	23317
1357	S	D	E	N	1965	35579
1360	L	G	R	H	1940	8860
1360	L	G	R	N	1940	8860
1360	S	D	G	H	1935	16474
1360	S	D	R	N	1974	75244
1366	L	D	R	H	1966	34910
1366	L	T	R	H	1966	34910
1366	S	D	R	H	1966	34910
1366	S	T	R	H	1966	34910
1367	L	D	R	H	1965	43143
1367	L	T	R	H	1965	43143
1369	P	D	R	N	1971	65687
1374	S	D	R	N	1970	74070
1374	S	E	R	N	1970	74070
1374	S	D	E	N	1970	74071
1374	S	E	E	N	1970	74071
1378	S	D	R	N	1974	74804
1378	S	E	R	N	1974	74804
1402	D	D	R	H	1967	49598
1402	S	D	R	H	1967	49598
1405	S	D	O	N	1970	59018
1405	S	D	E	N	1971	63983
1405	S	D	R	N	1971	67588
1405	S	S	R	N	1973	79859
1405	S	S	E	N	1973	79860
1411	S	D	R	L	1960	36659
1411	S	E	R	L	1960	36659
1451	S	D	R	N	1971	67588
1466	S	D	E	N	1960	10947
1466	S	E	E	N	1960	10947
1466	S	S	E	N	1960	24931
1466	S	S	E	N	1970	61236
1467	S	D	E	N	1973	73921
1467	S	T	E	N	1973	73921
1478	S	D	O	N	1970	59018
1478	S	D	E	N	1971	63983
1478	S	D	R	N	1971	67588
1480	S	D	R	N	1971	67588
1481	S	D	R	N	1971	67588
1482	S	D	R	N	1971	67588
1483	S	D	R	N	1971	67588
1489	S	D	R	N	1971	67588
1490	S	D	R	N	1971	67588
1493	S	D	R	N	1971	67588
1494	S	D	R	N	1971	67588
1495	S	D	G	N	1923	25148
1504	S	D	O	N	1970	59018

Phys. State: **C.** Superconductive; **D.** Doped; **E.** Expanded; **F.** Fibrous or Whisker; **G.** Gas; **I.** Ionized or Plasma; **L.** Liquid; **M.** Multiphase; **P.** Powder or Fine Particle; **S.** Solid; **T.** Thick or Thin Film

Subject: **D.** Data; **E.** Experiment; **G.** General (Data + Theory + Experiment); **S.** Survey (Review, Compendium etc.); **T.** Theory

Language: **C.** Czech; **D.** Dutch; **E.** English; **F.** French; **G.** German; **I.** Italian; **J.** Japanese; **O.** Other Languages; **P.** Polish; **R.** Russian; **S.** Spanish

Temperature: **F.** Full Range (Low + Normal + High); **L.** Low (0 to 75K) + Overlap into Normal; **N.** Normal (75 to 1273K); **H.** High (above 1273K) + Overlap into Normal; Blank = Not Coded

Substance Number	Phys. State	Subject	Language	Temperature	Year	TPRC Number
503-1504	S	D	E	N	1971	63983
1547	S	D	E		1958	7695
1547	S	S	E	N	1970	61236
1554	S	D	R	N	1971	67588
1556	S	D	R	H	1967	49598
1557	S	D	R	H	1967	49598
1557	S	D	R	N	1971	67588
1559	S	D	R	N	1971	67588
1560	S	D	R	N	1971	67588
1561	S	D	R	N	1971	67588
1563	S	D	O	N	1970	59018
1563	S	D	E	N	1971	63983
1563	S	D	E	N	1968	67357
1563	S	D	R	N	1971	67588
1610	S	G	F	N	1959	18033
1716	S	G	R	N	1962	31437
1720	L	G	O	H	1959	23751
1720	S	G	O	N	1959	23751
1723	L	G	G	H	1953	448
1723	L	G	G	H	1954	28474
1723	S	G	G	N	1953	448
1723	S	S	E	H	1960	24931
1723	S	S	E	N	1960	24931
1723	S	D	G	N	1954	28474
1723	S	G	G	H	1954	28474
1766	S	D	R	N	1976	88314
1766	S	D	E	N	1976	91214
1768	S	D	R	H	1976	88372
1768	S	D	R	N	1976	88372
1768	S	D	E	H	1976	90594
1768	S	D	E	N	1976	90594
1806	S	D	E	L	1975	81116
1810	S	D	J	N	1977	94558
1811	S	D	J	N	1977	94558
1814	S	S	E	F	1970	61236
1814	S	D	E	H	1977	90682
1814	S	D	E	N	1977	90682
1815	S	D	R	N	1967	48703
1815	S	D	E	N	1967	49415
1815	S	D	E	N	1977	90682
1816	S	D	E	N	1977	90682
1817	S	D	E	N	1977	90682
1818	L	D	E	H	1977	90682
1818	S	D	E	H	1977	90682
1818	S	D	E	N	1977	90682
1819	S	D	E	N	1977	90682
1820	S	D	E	N	1977	90682
1826	L	S	E	H	1977	94788
1826	S	S	E	H	1977	94788
1826	S	S	E	N	1977	94788
1827	S	S	E	N	1976	94271
1828	S	D	E	N	1967	35986
1828	S	S	E	N	1976	88411
1828	S	S	E	N	1979	97587
1885	S	D	E	L	1975	88798
1885	S	E	E	L	1975	88798
1886	S	D	E	L	1975	88798
1886	S	E	E	L	1975	88798
1912	S	D	G	N	1963	31475
1933	S	D	R	N	1969	58233
1941	S	S	E	L	1960	24931
1941	S	S	E	N	1960	24931
1950	S	D	R	N	1976	89429
1950	S	E	R	N	1976	89429
1950	S	D	E	N	1976	91205
1950	S	E	E	N	1976	91205
1951	S	D	R	N	1976	89429
1951	S	E	R	N	1976	89429
1951	S	D	E	N	1976	91205
1951	S	E	E	N	1976	91205
1953	D	D	E	N	1973	70944
1955	S	D	E	H	1964	45610
1956	S	D	E	L	1977	88621
1956	S	D	E	L	1977	91994
1957	S	D	E	L	1977	88621
1957	S	D	E	L	1977	91994
1958	S	D	E	L	1977	88621
1958	S	D	E	L	1977	91994
1961	S	D	E	N	1977	93507
1975	S	D	E	L	1977	88621
1983	L	G	G		1953	448
1983	S	G	G		1953	448
1984	P	D	G		1943	1910
1984	S	D	G		1943	1910
1987	S	D	E	N	1921	22633
1988	S	D	R	N	1974	74868
1988	S	T	R	N	1974	74868
1989	S	D	E	N	1921	22633
1990	S	D	E	N	1921	22633
1992	L	D	R	H	1959	17782
1995	L	G	R	H	1966	33720
1995	S	G	R	H	1966	33720
1997	L	D	R	H	1966	42066
1997	S	D	R	H	1966	42066
1998	L	D	R	H	1966	42066
1998	S	D	R	H	1966	42066
1999	L	D	R	H	1959	17782
1999	L	D	R	H	1966	42066
1999	S	D	R	H	1966	42066
2000	S	D	E	H	1962	20006
2000	S	D	E	N	1962	20006

Substance Number	Phys. State	Subject	Language	Temperature	Year	TPRC Number
503-2000	S	D	F	H	1963	31125
2000	S	D	F	N	1963	31125
2001	S	D	F	H	1963	31125
2001	S	D	F	N	1963	31125
2002	S	D	F	H	1963	31125
2002	S	D	F	N	1963	31125
2004	S	D	R	H	1968	53327
2005	S	D	R	N	1968	53327
2006	S	D	R	N	1968	53327
2007	P	D	R	H	1970	63632
2007	P	D	R	N	1970	63632
2013	L	D	E	H	1928	22639
2013	S	D	E	H	1928	22639
2017	S	D	R	N	1974	74868
2017	S	T	R	N	1974	74868
2020	L	D	R	H	1959	17782
2021	L	D	R	H	1959	17782
2022	L	D	R	H	1959	17782
2023	L	D	R	H	1959	17782
2024	L	D	R	H	1959	17782
2025	L	D	R	H	1959	17782
2026	L	D	R	H	1959	17782
2027	L	D	R	H	1959	17782
2028	L	D	E	H	1928	22639
2028	S	D	E	H	1928	22639
2029	L	D	E	H	1928	22639
2029	S	D	E	H	1928	22639
2040	S	D	R	N	1968	53327
2068	S	G	F	N	1959	18033
2103	S	G	F	N	1959	18033
2104	S	G	F	N	1959	18033
2105	S	G	F	N	1959	18033
2106	S	G	F	N	1959	18033
2107	S	G	F	N	1959	18033
2107	S	G	F	N	1959	25490
2108	S	G	F	N	1959	18033
2109	S	G	F	N	1959	18033
2110	S	G	F	N	1959	18033
2111	S	G	F	N	1959	18033
2112	S	G	F	N	1959	18033
2113	S	G	F	N	1959	18033
2114	S	G	F	N	1959	18033
2114	S	S	E	N	1960	24931
2142	S	D	J	N	1977	94558
2197	S	D	E	N	1979	97659
2274	S	S	R	N	1973	79859
2274	S	S	E	N	1973	79860
2279	L	G	G	H	1954	28474
2279	S	D	G	N	1954	28474
2279	S	G	G	H	1954	28474
2289	S	D	O	N	1970	59018
2289	S	D	E	N	1971	63983
2290	S	D	O	N	1970	59018
2290	S	D	E	N	1971	63983
2298	L	G	G	H	1954	28474
2298	S	D	G	N	1954	28474
2298	S	G	G	H	1954	28474
2299	L	G	G	H	1954	28474
2299	S	D	G	N	1954	28474
2299	S	G	G	H	1954	28474
2303	S	D	R	N	1967	48703
2303	S	D	E	N	1967	49415
2309	S	D	E	H	1962	20006
2309	S	D	E	N	1962	20006
2309	S	D	F	H	1963	31125
2309	S	D	F	N	1963	31125
2365	S	D	E	N	1979	97889
2380	S	D	R	H	1960	19081
2380	S	D	E	H	1960	32629
2404	S	G	F	N	1959	25490
2415	S	D	E	L	1968	48157
2428	S	D	F	H	1963	31125
2428	S	D	F	N	1963	31125
2432	S	D	R	N	1967	48703
2432	S	D	E	N	1967	49415
2437	S	D	F	H	1962	20006
2437	S	D	F	N	1962	20006
2438	S	D	F	H	1963	31125
2438	S	D	F	N	1963	31125
2440	S	D	R	N	1967	54757
2453	S	D	E	H	1962	20006
2453	S	D	E	N	1962	20006
2461	P	D	E	N	1921	22633
2461	S	D	F	H	1962	20006
2461	S	D	F	N	1962	20006
2461	S	D	E	N	1921	22633
2461	S	D	F	H	1963	31125
2461	S	D	F	N	1963	31125
2462	S	D	F	H	1962	20006
2462	S	D	E	N	1962	20006
2463	S	D	E	H	1962	20006
2463	S	D	E	N	1962	20006
2463	S	D	F	H	1963	31125
2463	S	D	F	N	1963	31125
2464	S	D	E	H	1962	20006
2464	S	D	E	N	1962	20006
2466	S	D	E	H	1962	20006
2466	S	D	E	N	1962	20006
2467	S	D	E	H	1962	20006
2467	S	D	E	N	1962	20006
2478	P	D	E	N	1921	22633

Substance Number	Phys. State	Subject	Language	Temperature	Year	TPRC Number
503-2478	S	D	E	N	1921	22633
2479	S	D	E	N	1921	22633
2487	L	G	O	H	1959	23751
2487	S	G	O	H	1959	23751
2492	S	D	F	H	1963	31125
2492	S	D	F	N	1963	31125
2500	S	G	R	N	1962	31437
2513	S	D	F	H	1963	31125
2513	S	D	F	N	1963	31125
2516	S	D	F	H	1963	31125
2516	S	D	F	N	1963	31125
2517	S	D	F	H	1963	31125
2517	S	D	F	N	1963	31125
2518	S	D	F	H	1963	31125
2518	S	D	F	N	1963	31125
2519	S	D	F	H	1963	31125
2519	S	D	F	N	1963	31125
2520	S	D	F	H	1963	31125
2520	S	D	F	N	1963	31125
2521	S	D	F	H	1963	31125
2521	S	D	F	N	1963	31125
2589	S	D	R	N	1967	48703
2589	S	D	E	N	1967	49415
2590	S	D	R	N	1967	48703
2590	S	D	E	N	1967	49415
2726	E	D	G	N	1955	1125
2726	E	E	G	N	1955	1125
2751	S	S	E	L	1970	61236
2751	S	S	E	N	1970	61236
2808	S	D	R	N	1967	54757
2809	S	D	R	N	1967	54757
2887	S	S	E	N	1960	24931
2900	S	S	E	N	1960	24931
2904	S	S	E	N	1960	24931
8190	M	D	R	N	1970	64246
8190	M	T	R	N	1970	64246
8306	S	D	R	N	1964	38203
8306	S	D	E	N	1967	49388
8354	S	D	R	N	1939	48342
8374	S	D	R	H	1969	58852
8374	S	D	E	H	1969	58853
8389	S	D	E	N	1959	13944
8389	S	D	R	F	1962	37343
8389	S	D	E	F	1964	37344
8395	S	D	G		1949	1972
8416	E	D	E	N	1963	25157
8419	S	D	E	H	1934	15070
8483	S	D	E	L	1972	73134
8514	S	D	E	L	1972	73134
8515	S	D	G	N	1893	50995
8515	S	T	G	N	1893	50995
8517	S	D	G	N	1893	50995
8517	S	T	G	N	1893	50995
8518	S	D	G	N	1893	50995
8518	S	T	G	N	1893	50995
8520	S	D	G	N	1893	50995
8520	S	T	G	N	1893	50995
8742	S	D	G	N	1893	50995
8742	S	T	G	N	1893	50995
8746	S	D	E	N	1969	54502
8746	S	T	E	N	1969	54502
8753	S	D	E	N	1963	37530
8754	S	D	E	N	1963	37530
8755	S	D	E	N	1963	37530
8822	S	D	E	N	1960	76856
8870	L	D	G	H	1969	54931
8870	L	T	G	H	1969	54931
8870	S	D	G	N	1969	54931
8870	S	T	G	N	1969	54931
8874	S	D	R	N	1972	70625
8874	S	D	E	N	1972	70626
8876	S	D	O		1947	1413
8882	S	D	F	N	1970	58498
8940	S	D	R	N	1972	70625
8940	S	D	E	N	1972	70626
8943	S	D	R	N	1962	39649
8943	S	D	E	N	1965	39650
9039	S	D	E	L	1973	72936
9058	S	G	F	N	1959	18033
9065	M	D	E	N	1960	23278
9068	S	D	E	N	1963	37530
9068	S	D	E	L	1974	73785
9074	L	G	G	N	1963	30270
9074	S	G	G	N	1963	30270
9084	S	G	E	N	1942	14570
9095	S	G	E	N	1942	14570
9099	S	D	G	N	1893	50995
9099	S	T	G	N	1893	50995
9108	S	D	G	N	1893	50995
9108	S	T	G	N	1893	50995
9123	S	D	G	N	1893	50995
9123	S	T	G	N	1893	50995
9124	S	G	G	H	1972	68513
9131	S	D	G	N	1893	50995
9131	S	T	G	N	1893	50995
9147	S	D	E		1958	6939
9152	S	G	E	N	1942	14570
9152	S	D	G	N	1893	50995
9152	S	T	G	N	1893	50995
9154	S	D	E	N	1967	50646
9154	S	E	E	N	1967	50646

Phys. State: **C.** Superconductive; **D.** Doped; **E.** Expanded; **F.** Fibrous or Whisker; **G.** Gas; **I.** Ionized or Plasma; **L.** Liquid; **M.** Multiphase; **P.** Powder or Fine Particle; **S.** Solid; **T.** Thick or Thin Film

Subject: **D.** Data; **E.** Experiment; **G.** General (Data + Theory + Experiment); **S.** Survey (Review, Compendium etc.); **T.** Theory

Language: **C.** Czech; **D.** Dutch; **E.** English; **F.** French; **G.** German; **I.** Italian; **J.** Japanese; **O.** Other Languages; **P.** Polish; **R.** Russian; **S.** Spanish

Temperature: **F.** Full Range (Low + Normal + High); **L.** Low (0 to 75K) + Overlap into Normal; **N.** Normal (75 to 1273K); **H.** High (above 1273K) + Overlap into Normal; Blank = Not Coded

Substance Number	Phys. State	Subject	Language	Temperature	Year	TPRC Number
503-9154	S	G	J	N	1973	73007
9160	S	D	R	N	1962	39649
9160	S	D	E	N	1965	39650
9163	L	G	G	N	1963	30270
9163	S	G	G	N	1963	30270
9163	S	D	G	N	1893	50995
9163	S	T	G	N	1893	50995
9167	S	G	E	N	1942	14570
9168	S	G	E	N	1942	14570
9169	S	D	E	N	1963	33523
9169	S	D	R	L	1967	34293
9169	S	D	E	L	1967	34294
9169	S	D	R	L	1966	43351
9169	S	D	R	L	1968	53391
9169	S	D	E	N	1971	62098
9171	S	G	E	N	1942	14570
9171	S	G	R	L	1959	31165
9171	S	G	E	L	1963	31166
9171	S	D	G	N	1893	50995
9171	S	T	G	N	1893	50995
9173	L	D	G		1955	1520
9173	S	D	G		1955	1520
9173	S	D	D	N	1948	10394
9173	S	T	D	N	1948	10394
9173	S	G	E	N	1942	14570
9173	S	D	G	N	1893	50995
9173	S	T	G	N	1893	50995
9173	S	G	G	H	1972	68513
9174	L	D	G		1955	1520
9174	S	D	R		1953	274
9174	S	D	R		1954	1277
9174	S	T	R		1954	1277
9174	S	D	G		1955	1520
9174	S	G	E	N	1942	14570
9174	S	D	R	L	1967	34293
9174	S	D	E	L	1967	34294
9174	S	D	E	L	1964	40697
9174	S	D	E	N	1968	49113
9174	S	E	E	N	1968	49113
9174	S	D	E	N	1967	50646
9174	S	E	E	N	1967	50646
9174	S	D	E	L	1972	67569
9174	S	G	G	H	1972	68513
9174	S	D	E	L	1973	72936
9174	S	G	J	N	1973	73007
9175	S	D	G	N	1893	50995
9175	S	T	G	N	1893	50995
9178	S	G	G		1957	311
9178	S	D	F		1955	695
9178	S	G	E	H	1958	15534
9179	L	D	G		1955	1520
9179	S	D	R		1953	274
9179	S	D	G		1955	1520
9179	S	D	E		1958	8182
9179	S	T	E		1958	8182
9179	S	G	E	N	1942	14570
9179	S	D	E	N	1967	50646
9179	S	E	E	N	1967	50646
9179	S	D	G	N	1893	50995
9179	S	T	G	N	1893	50995
9179	S	G	J	N	1973	73007
9180	S	D	J		1953	1282
9181	S	D	F		1955	695
9181	S	D	R		1954	904
9181	S	G	R		1956	1324
9181	S	G	E		1952	1798
9181	S	G	E	N	1942	14570
9298	S	D	E	L	1971	64004
9298	S	D	E	L	1972	65865
9298	S	D	E	L	1973	70202
9298	S	D	E	L	1972	71391
9308	S	D	E	N	1971	62030
9363	S	D	E	N	1971	62030
9401	S	D	E	N	1960	15664
9402	S	D	E	N	1971	62030
9431	S	D	R	H	1970	59471
9431	S	D	E	H	1970	61167
9435	S	D	R	L	1970	60073
9435	S	D	E	L	1970	60074
9456	S	D	E	N	1971	62030
9458	S	D	E	N	1971	62030
9511	S	D	E	L	1966	40494
9514	S	D	E	N	1968	37038
9514	S	D	R	N	1968	51368
9518	S	D	E	N	1968	37038
9518	S	D	R	N	1968	51368
9536	S	D	O	N	1965	44879
9589	S	D	E	L	1966	40494
9626	S	D	O		1956	9045
9633	S	D	O	N	1966	43203
9641	S	D	R	N	1967	45866
9647	L	D	R	H	1971	62094
9647	S	D	R	N	1971	62094
9652	S	G	E	N	1971	77938
9666	S	D	E	N	1968	37038
9666	S	D	R	N	1968	51368
9671	S	D	E	N	1967	48518
9672	S	D	E	N	1968	50339
9673	S	D	E	N	1968	50339
9675	S	D	E	L	1968	47507
9676	S	D	E	N	1968	48155

Substance Number	Phys. State	Subject	Language	Temperature	Year	TPRC Number
503-9677	S	D	E	N	1968	48155
9694	S	D	R	N	1965	42567
9695	S	D	R	N	1965	42567
9704	S	D	E	H	1967	43744
9709	S	D	E	N	1961	23542
9709	S	D	E	N	1968	48155
9709	S	D	E	N	1953	58921
9712	S	D	E	H	1966	35917
9712	S	D	E	H	1967	39997
9712	S	D	E	H	1965	45297
9718	S	D	E	H	1967	52120
9718	S	D	E	H	1972	71915
9718	S	T	E	H	1972	71915
9728	S	G	R	N	1962	31431
9746	S	D	R	N	1974	75244
9750	M	D	E	N	1958	10458
9750	S	D	E		1935	8241
9750	S	D	E	N	1959	13944
9750	S	D	G	N	1961	24393
9750	S	D	E	N	1959	43731
9750	S	D	E	N	1959	47023
9750	S	D	E	N	1955	52252
9756	S	D	E	N	1958	20590
9757	L	G	R	H	1961	24514
9757	L	G	R	N	1962	27612
9760	S	D	R	H	1969	58852
9760	S	D	E	H	1969	58853
9773	P	D	E	N	1921	22633
9773	S	D	E	N	1921	22633
9775	S	D	F	H	1928	22457
9789	S	D	R	H	1966	44062
9789	S	D	E	H	1966	44063
9791	S	D	E		1935	8241
9791	S	D	G	N	1961	24393
9791	S	T	E	H	1962	26008
9792	S	D	E	N	1921	22633
9803	S	G	E	N	1963	31998
9804	D	G	E	N	1963	31998
9804	S	G	E	N	1963	31998
9808	S	D	E	L	1969	56014
9809	S	D	F	H	1928	22457
9809	S	D	E	H	1921	22633
9809	S	D	E	N	1941	28353
9809	S	D	R	H	1970	59471
9809	S	D	E	H	1970	61167
9810	S	G	E	H	1963	31604
9812	M	D	E	N	1962	29549
9812	M	T	E	N	1962	29549
9813	S	D	E	H	1966	35917
9813	S	D	E	H	1967	46878
9815	S	D	R	H	1965	42535
9815	S	D	E	H	1966	43468
9815	S	D	R	H	1969	58852
9815	S	D	E	H	1969	58853
9819	S	D	F	H	1973	71478
9819	S	E	F	H	1973	71478
9819	S	D	E	H	1971	75182
9819	S	T	E	H	1971	75182
9824	M	D	E	N	1962	29549
9824	M	T	E	N	1962	29549
9824	S	D	E	N	1961	23542
9830	G	D	E	H	1967	49525
9830	M	D	E	H	1967	49525
9830	S	D	E		1935	8241
9830	S	D	E	N	1959	13944
9836	S	D	R	N	1969	55846
9836	S	T	R	N	1969	55846
9843	S	D	E	N	1968	54196
9854	S	D	E	H	1968	54196
9859	M	D	E	N	1960	13346
9869	S	D	E	N	1973	70757
9878	S	D	E	L	1957	10586
9878	S	T	E	L	1957	10586
9878	S	D	E	L	1957	10587
9878	S	T	E	L	1957	10587
9885	P	D	E	N	1941	28353
9885	S	D	E	N	1941	28353
9891	S	D	E	N	1959	13944
9891	S	D	R	N	1966	44598
9891	S	D	E	N	1969	51032
9893	L	G	R	H	1961	24514
9893	L	D	R	H	1970	75449
9893	L	D	E	H	1970	75450
9893	S	D	R	H	1970	75449
9893	S	D	E	H	1970	75450
9897	S	G	E		1955	716
9898	S	D	R	H	1970	59471
9898	S	D	E	H	1970	61167
9901	P	D	E		1952	6309
9911	S	D	G		1958	9728
9911	S	D	F	N	1967	54354
9912	S	D	F		1954	1213
9912	S	D	F	N	1967	54354
9916	L	D	G	N	1958	19586
9916	S	D	G	N	1961	24393
9916	S	D	E	N	1951	29285
9916	S	D	E	N	1954	29302
9916	S	E	E	N	1954	29302
9916	S	D	E	N	1959	47022
9916	S	D	G	H	1967	50122
9917	S	D	E	L	1966	40494

Substance Number	Phys. State	Subject	Language	Temperature	Year	TPRC Number
503-9919	L	D	R	H	1959	18100
9919	L	D	R	N	1959	19842
9919	S	D	R	N	1959	19842
9922	S	D	E	H	1962	26012
9922	S	D	E	L	1966	40494
9922	S	D	E	H	1965	46471
9924	S	D	E		1959	7696
9930	S	G	E		1950	1198
9938	L	G	R	H	1961	24514
9938	S	G	J	N	1973	73007
9939	S	D	E	L	1966	40494
9940	S	G	J	N	1973	73007
9941	S	G	J	N	1973	73007
9947	S	D	J		1953	612
9947	S	D	E		1955	717
9950	S	E	E	N	1951	26388
9951	S	D	E		1950	1504
9951	S	D	R	N	1969	55502
9951	S	T	R	N	1969	55502
9955	S	D	E	F	1966	43496
9956	S	D	E	L	1964	38152
9956	S	D	E	L	1966	39351
9960	L	D	G	N	1958	19586
9960	S	D	E	N	1958	30547
9960	S	D	E	H	1967	44823
9977	L	D	G	N	1958	19586
9977	S	D	G	N	1971	63748
9977	S	T	G	N	1971	63748
9978	S	D	E	N	1958	30547
9979	S	D	G	N	1960	62279
9980	S	D	E	N	1960	15664
9980	S	D	R	N	1964	43252
9983	S	D	E	F	1966	43496
9985	P	D	R	H	1964	35493
9985	S	D	R	H	1963	37067
9985	S	D	E	H	1973	74558
9985	S	T	E	H	1973	74558
9986	S	D	R	N	1967	54757
9989	S	D	R	N	1967	54757
9990	P	D	R	H	1964	35493
9990	S	D	R	H	1974	79147
9991	S	D	R	H	1971	61642
9991	S	E	R	H	1971	61642
9991	S	D	E	H	1972	66202
9991	S	E	E	H	1972	66202
9991	S	D	E	H	1973	74558
9991	S	T	E	H	1973	74558
9995	S	D	R	N	1967	54757
9997	L	D	R	H	1938	13824
9997	L	D	R	H	1968	45107
9997	S	D	R	H	1968	45107
9998	S	D	R	N	1976	89429
9998	S	E	R	N	1976	89429
9998	S	D	E	N	1976	91205
9998	S	E	E	N	1976	91205
9999	L	D	R	H	1940	28585
521-0001	P	D	E	N	1949	61706
0001	S	S	E	N	1958	10667
0003	S	D	G	N	1922	22647
0003	S	D	E	N	1979	97863
0004	E	D	E	N	1965	42374
0006	S	D	E	N	1921	22633
0006	S	S	E	N	1976	92625
0007	S	S	G		1933	6798
0007	S	D	E		1949	9411
0007	S	T	E		1949	9411
0007	S	S	E		1952	9922
0007	S	D	E	L	1944	12210
0007	S	D	E	N	1960	15929
0007	S	D	E	N	1961	19379
0007	S	T	E	N	1961	19379
0007	S	D	E	N	1961	20864
0007	S	T	E	N	1961	20864
0007	S	D	G	N	1912	20865
0007	S	T	G	N	1912	20865
0007	S	D	G	N	1920	22785
0007	S	T	G	N	1920	22785
0007	S	D	F	N	1895	28730
0007	S	D	R	N	1972	66602
0007	S	D	E	N	1972	69793
0007	S	D	E	N	1976	89727
0007	S	D	E	N	1978	94673
0007	S	D	E	N	1979	98785
0009	M	E	R	N	1959	43002
0009	M	E	F	N	1959	43003
0009	M	D	E	N	1943	46820
0009	M	T	E	N	1943	46820
0009	M	T	E	H	1970	61713
0009	P	D	E	N	1921	22633
0009	P	S	E	N	1969	55244
0009	S	D	E		1954	183
0009	S	D	E		1952	652
0009	S	G	E		1949	8230
0009	S	D	E		1948	8236
0009	S	D	E		1948	8237
0009	S	T	E	N	1958	15736
0009	S	D	E	N	1953	16343
0009	S	D	E	N	1961	24910
0009	S	D	E	N	1951	25557
0009	S	S	G	H	1928	28680

Phys. State: **C.** Superconductive; **D.** Doped; **E.** Expanded; **F.** Fibrous or Whisker; **G.** Gas; **I.** Ionized or Plasma; **L.** Liquid; **M.** Multiphase; **P.** Powder or Fine Particle; **S.** Solid; **T.** Thick or Thin Film
Subject: **D.** Data; **E.** Experiment; **G.** General (Data + Theory + Experiment); **S.** Survey (Review, Compendium etc.); **T.** Theory
Language: **C.** Czech; **D.** Dutch; **E.** English; **F.** French; **G.** German; **I.** Italian; **J.** Japanese; **O.** Other Languages; **P.** Polish; **R.** Russian; **S.** Spanish
Temperature: **F.** Full Range (Low + Normal + High); **L.** Low (0 to 75K) + Overlap into Normal; **N.** Normal (75 to 1273K); **H.** High (above 1273K) + Overlap into Normal; Blank = Not Coded

Substance Number	Phys. State	Subject	Language	Temperature	Year	TPRC Number
521-0009	S	E	G	N	1932	31875
0009	S	E	R	N	1959	43002
0009	S	E	F	N	1959	43003
0009	S	E	E	N	1965	44633
0009	S	D	E	N	1943	46820
0009	S	T	E	N	1943	46820
0009	S	T	E	H	1970	61713
0009	S	S	R	N	1957	68049
0009	S	S	E	N	1971	68050
0009	S	D	J	N	1975	98110
0009	S	E	J	N	1975	98110
0010	M	G	E	N	1962	30997
0010	M	D	E	N	1909	31920
0010	M	D	E	N	1949	61706
0010	P	D	E	N	1949	61706
0010	P	D	R	N	1971	65687
0010	P	D	E	N	1971	67445
0010	S	D	E	N	1962	29491
0010	S	D	E	N	1948	40389
0015	P	D	R	N	1968	49504
0015	P	E	R	N	1968	49504
0015	P	D	E	N	1968	65123
0015	P	E	E	N	1968	65123
0015	S	D	J	N	1940	13228
0015	S	D	R	N	1934	16918
0015	S	D	J	N	1956	19593
0015	S	D	E	N	1921	22401
0015	S	D	E	H	1926	22638
0020	P	D	R	N	1971	65687
0020	S	D	E	N	1942	25423
0020	S	D	E	N	1973	71780
0021	S	S	E	N	1960	24931
0021	S	D	E	N	1951	29539
0022	S	D	E	N	1937	28449
0022	S	D	E	N	1971	66172
0022	S	T	E	N	1971	66172
0024	S	D	E	N	1932	14353
0024	S	D	R	N	1960	21288
0024	S	D	E	N	1960	25356
0024	S	D	G	N	1940	29870
0024	S	D	R	N	1965	42576
0024	S	D	R	N	1965	42663
0024	S	D	E	N	1965	49232
0024	S	D	R	N	1968	51815
0024	S	D	R	N	1968	53054
0024	S	D	R	N	1970	59042
0024	S	D	E	N	1970	67687
0024	S	D	R	N	1973	71862
0025	S	S	E		1952	9922
0025	S	S	E	N	1959	10076
0025	S	D	E	N	1923	21665
0025	S	D	G	N	1920	22785
0025	S	T	G	N	1920	22785
0025	S	D	O	L	1964	35713
0025	S	T	O	L	1964	35713
0025	S	D	E	N	1950	39752
0025	S	S	E	L	1970	61236
0025	S	S	E	N	1970	61236
0025	S	D	E	N	1969	63778
0026	S	D	G	N	1922	22647
0026	S	D	P	N	1971	63225
0027	P	D	R	N	1971	69768
0027	P	D	E	N	1972	69769
0028	P	D	E	N	1962	39516
0028	S	D	E	N	1962	29491
0029	S	D	E		1955	359
0029	S	D	E		1953	474
0029	S	D	E		1956	7001
0030	S	D	R		1947	1805
0030	S	T	R		1947	1805
0030	S	D	E	L	1934	10348
0030	S	S	E	L	1970	61236
0030	S	S	E	N	1970	61236
0032	P	D	R	H	1969	56752
0032	P	D	E	H	1969	70864
0032	S	D	E		1950	408
0032	S	D	E		1935	8241
0032	S	D	G	L	1926	22861
0032	S	S	E	L	1960	24931
0032	S	S	E	N	1960	24931
0032	S	D	E	N	1937	28449
0032	S	D	E	H	1964	32979
0032	S	T	E	H	1964	32979
0032	S	S	E	F	1970	61236
0032	S	D	E	N	1966	99938
0033	P	D	R	H	1964	37217
0033	S	D	E		1950	408
0033	S	D	F	N	1930	21511
0033	S	T	F	N	1930	21511
0033	S	D	G	L	1926	22861
0033	S	S	E	L	1960	24931
0033	S	S	E	N	1960	24931
0033	S	D	E	H	1964	32979
0033	S	T	E	H	1964	32979
0033	S	D	R	H	1964	41484
0033	S	E	R	H	1964	41484
0033	S	G	R	H	1965	42492
0033	S	S	E	F	1970	61236
0033	S	D	E	N	1966	99938
0034	S	D	E		1950	408
0034	S	D	E		1935	8241

Substance Number	Phys. State	Subject	Language	Temperature	Year	TPRC Number
521-0034	S	D	G	H	1926	21459
0034	S	D	G	L	1926	22861
0034	S	S	E	L	1960	24931
0034	S	S	E	N	1960	24931
0034	S	D	E	N	1937	28449
0034	S	S	G	H	1928	28680
0034	S	D	E	H	1964	32979
0034	S	T	E	H	1964	32979
0034	S	S	E	F	1970	61236
0034	S	D	R	H	1972	66891
0034	S	E	R	H	1972	66891
0034	S	D	E	N	1966	99938
0035	S	D	E		1957	8941
0035	S	D	E	L	1961	26327
0035	S	D	E	H	1919	43370
0035	S	S	E	L	1970	61236
0035	S	S	E	N	1970	61236
0036	S	D	E	H	1919	43370
0036	S	D	E	N	1971	66172
0036	S	T	E	N	1971	66172
0040	P	D	E		1956	8514
0040	P	D	R	N	1969	55535
0040	S	D	E		1956	226
0040	S	G	E		1954	669
0040	S	D	E		1956	9825
0040	S	S	E	N	1958	10667
0040	S	S	E	N	1927	20950
0040	S	D	G	H	1926	21459
0040	S	D	E	N	1921	22633
0040	S	D	E	N	1961	24910
0040	S	S	G	N	1928	28680
0040	S	D	E	N	1966	44390
0040	S	E	E	N	1966	44390
0040	S	D	R	N	1968	60916
0041	S	D	E		1955	435
0041	S	D	E	L	1961	26327
0041	S	D	E	N	1968	49403
0041	S	S	E	L	1970	61236
0041	S	S	E	N	1970	61236
0042	S	D	J		1954	566
0042	S	T	J		1954	566
0043	S	D	J		1954	566
0043	S	T	J		1954	566
0046	M	D	R	N	1969	54587
0046	S	E	E	N	1921	22633
0046	S	T	E	N	1974	73781
0046	S	E	E	N	1970	74014
0046	S	D	E	N	1976	91238
0053	S	D	J		1954	606
0053	S	D	E	N	1921	22633
0053	S	D	F	N	1954	23317
0053	S	D	E	N	1951	40424
0053	S	D	G	N	1968	54724
0053	S	E	G	N	1968	54724
0053	S	D	E	N	1972	64938
0053	S	E	E	N	1972	64938
0054	S	T	R		1954	559
0054	S	D	E	N	1963	24342
0054	S	T	E	N	1963	24342
0055	F	D	G	N	1968	49680
0055	F	S	E	N	1954	73367
0055	F	D	E	N	1972	78278
0055	F	E	E	N	1972	78278
0055	S	D	E	L	1958	10723
0055	S	D	E	N	1921	22633
0055	S	D	F	N	1954	23317
0055	S	D	E	H	1962	77274
0056	S	D	G		1956	615
0057	M	G	E	N	1962	30997
0057	S	D	E		1953	710
0057	S	D	E	H	1919	43370
0057	S	D	E	N	1971	66172
0057	S	T	E	N	1971	66172
0057	S	D	E	L	1974	85507
0057	S	D	E	N	1974	85507
0057	S	S	E	N	1977	92134
0058	S	D	E		1953	710
0058	S	D	E	N	1971	66172
0058	S	T	E	N	1971	66172
0059	S	G	E		1954	669
0059	S	D	E		1953	710
0060	S	D	E		1953	710
0060	S	D	E	N	1971	66172
0060	S	T	E	N	1971	66172
0061	S	D	E		1953	710
0062	S	D	E		1953	710
0062	S	D	E	H	1968	49403
0063	S	G	E		1956	22
0063	S	G	E		1953	882
0063	S	D	E		1953	917
0063	S	G	E		1956	1321
0063	S	D	E		1948	1359
0063	S	T	E		1948	1359
0063	S	G	E		1944	1863
0063	S	D	E		1952	2122
0063	S	G	E		1951	3546
0063	S	G	E		1954	7457
0063	S	G	E		1946	9350
0063	S	D	E		1958	9820
0063	S	S	E		1952	9922
0063	S	S	E	N	1959	10076

Substance Number	Phys. State	Subject	Language	Temperature	Year	TPRC Number
521-0063	S	D	D	N	1948	10394
0063	S	T	D	N	1948	10394
0063	S	G	E	L	1958	10669
0063	S	D	E	N	1938	11621
0063	S	D	E	N	1941	14482
0063	S	T	E	N	1941	14482
0063	S	S	E	L	1960	16701
0063	S	S	E	N	1960	16701
0063	S	S	E	N	1960	16704
0063	S	D	E	N	1961	17062
0063	S	T	E	N	1961	17062
0063	S	S	E	L	1952	17353
0063	S	D	E	N	1959	19236
0063	S	T	E	N	1959	19236
0063	S	D	E	L	1957	19365
0063	S	T	E	L	1957	19365
0063	S	D	E	N	1961	19621
0063	S	D	E	N	1959	20784
0063	S	T	E	N	1959	20784
0063	S	D	G	N	1923	21398
0063	S	D	G	N	1926	21404
0063	S	D	G	N	1928	21663
0063	S	T	G	N	1928	21663
0063	S	D	E	N	1931	22467
0063	S	T	E	N	1931	22467
0063	S	D	G	H	1922	23176
0063	S	T	G	H	1922	23176
0063	S	D	E	L	1958	23953
0063	S	D	E	L	1962	26731
0063	S	T	E	L	1962	26731
0063	S	D	E	N	1962	27469
0063	S	D	R	H	1965	28567
0063	S	T	R	H	1965	28567
0063	S	D	E	H	1965	28597
0063	S	T	E	H	1965	28597
0063	S	D	R	L	1962	29020
0063	S	T	R	L	1962	29020
0063	S	D	E	L	1963	29021
0063	S	T	E	L	1963	29021
0063	S	G	G	L	1911	29642
0063	S	G	E	N	1961	32553
0063	S	S	E	L	1963	33277
0063	S	D	R	N	1963	33544
0063	S	T	R	N	1963	33544
0063	S	D	E	N	1963	33545
0063	S	T	E	N	1963	33545
0063	S	T	E	L	1966	39188
0063	S	D	E	L	1913	40240
0063	S	D	E	L	1957	40722
0063	S	D	E	N	1966	41195
0063	S	E	E	N	1966	41195
0063	S	S	E	L	1968	53495
0063	S	D	E	L	1969	56036
0063	S	T	E	L	1969	56036
0063	S	D	R	L	1969	56996
0063	S	T	R	L	1969	56996
0063	S	S	E	L	1970	61236
0063	S	S	E	N	1970	61236
0063	S	D	E	L	1971	66020
0063	S	T	E	L	1971	66020
0063	S	D	R	N	1976	91482
0064	S	D	E		1959	6453
0064	S	E	E		1959	6453
0064	S	D	E		1959	7688
0064	S	D	E		1958	7689
0064	S	D	E	N	1957	10583
0064	S	D	E	H	1958	10703
0064	S	T	E	H	1958	10703
0064	S	D	E	N	1959	13944
0064	S	D	G	N	1930	23249
0064	S	D	E	N	1963	24559
0064	S	D	E	H	1962	24856
0064	S	S	E	H	1960	24931
0064	S	S	E	N	1960	24931
0064	S	D	E	N	1963	25199
0064	S	T	E	H	1962	26008
0064	S	G	E	L	1962	26393
0064	S	D	E	N	1962	27641
0064	S	D	E	N	1965	36419
0064	S	D	E	N	1967	49389
0064	S	S	E	L	1970	61236
0064	S	S	E	N	1970	61236
0064	S	D	G	L	1969	64980
0064	S	E	G	L	1969	64980
0064	S	D	F	N	1971	65621
0064	S	E	F	N	1971	65621
0064	S	S	E	N	1959	85032
0065	S	D	R	N	1960	21288
0065	S	D	E	N	1960	25356
0065	S	G	E	N	1969	59226
0065	S	D	E	N	1968	60192
0065	S	D	R	N	1970	60745
0065	S	D	E	N	1979	99744
0066	G	D	R	N	1974	75361
0066	G	T	R	N	1974	75361
0066	M	D	R	N	1960	24247
0066	M	D	E	N	1960	25361
0066	S	S	E	N	1958	10667
0066	S	D	I	N	1961	24368
0066	S	T	I	N	1961	24368
0066	S	D	R	N	1966	51577

Phys. State: **C.** Superconductive; **D.** Doped; **E.** Expanded; **F.** Fibrous or Whisker; **G.** Gas; **I.** Ionized or Plasma; **L.** Liquid; **M.** Multiphase; **P.** Powder or Fine Particle; **S.** Solid; **T.** Thick or Thin Film

Subject: **D.** Data; **E.** Experiment; **G.** General (Data + Theory + Experiment); **S.** Survey (Review, Compendium etc.); **T.** Theory

Language: **C.** Czech; **D.** Dutch; **E.** English; **F.** French; **G.** German; **I.** Italian; **J.** Japanese; **O.** Other Languages; **P.** Polish; **R.** Russian; **S.** Spanish

Temperature: **F.** Full Range (Low + Normal + High); **L.** Low (0 to 75K) + Overlap into Normal; **N.** Normal (75 to 1273K); **H.** High (above 1273K) + Overlap into Normal; Blank = Not Coded

Substance Number	Phys. State	Subject	Language	Temperature	Year	TPRC Number
521-0066	S	D	R	N	1967	60317
0066	S	D	E	N	1970	60318
0067	S	D	E		1950	975
0067	S	D	G	H	1933	16107
0067	S	D	R	N	1956	16240
0067	S	D	G	N	1929	22766
0067	S	D	E	L	1963	30875
0067	S	T	E	L	1963	30875
0067	S	D	R	N	1965	34590
0067	S	D	E	N	1969	37300
0067	S	D	F	N	1917	48024
0067	S	D	R	N	1968	53688
0067	S	D	E	L	1965	55946
0067	S	D	E	N	1950	55957
0067	S	D	E	L	1969	59025
0067	S	D	E	N	1969	59294
0067	S	E	E	N	1969	59294
0067	S	S	E	L	1970	61236
0067	S	D	E	N	1975	82970
0067	S	D	E	L	1976	85637
0067	S	D	E	N	1976	85637
0069	G	D	R	N	1969	55272
0069	P	D	E	N	1934	25152
0069	S	D	J		1954	324
0069	S	D	G		1956	984
0069	S	S	E		1949	7618
0069	S	D	R	N	1960	11725
0069	S	T	E	N	1957	14689
0069	S	S	E	N	1936	16524
0069	S	D	O	N	1958	17338
0069	S	D	E	N	1959	18202
0069	S	D	R	N	1959	18710
0069	S	D	E	N	1962	20611
0069	S	D	E	N	1960	20660
0069	S	D	R	N	1960	21288
0069	S	D	E	N	1923	21665
0069	S	D	E	N	1921	22633
0069	S	S	E	N	1923	23239
0069	S	D	E	N	1960	25356
0069	S	D	G	N	1904	28731
0069	S	D	R	N	1958	30882
0069	S	S	R	N	1958	30882
0069	S	D	E	N	1961	30884
0069	S	S	E	N	1961	30884
0069	S	S	R	N	1965	34116
0069	S	D	E	N	1965	35579
0069	S	D	I	H	1966	41943
0069	S	T	I	H	1966	41943
0069	S	D	R	N	1968	49919
0069	S	T	R	N	1968	49919
0069	S	D	R	N	1968	53054
0069	S	D	E	N	1968	60192
0069	S	D	E	N	1968	65121
0069	S	T	E	N	1968	65121
0069	S	D	R	N	1972	70478
0069	S	D	E	N	1968	70866
0069	S	D	E	N	1972	70989
0069	S	D	R	N	1973	71345
0069	S	D	E	N	1913	94422
0070	M	D	R	N	1969	54587
0070	S	D	S		1949	1509
0070	S	D	E		1947	1823
0070	S	D	E		1951	1889
0070	S	D	E	N	1958	12174
0070	S	E	R	N	1959	19849
0070	S	D	R	N	1959	19856
0070	S	D	E	N	1923	21665
0070	S	D	E	N	1921	22633
0070	S	S	G	H	1928	28680
0070	S	D	E	N	1951	40424
0070	S	D	R	N	1968	60916
0070	S	D	E	N	1978	98172
0070	S	D	E	N	1979	99928
0072	P	D	E	N	1949	61706
0072	S	S	E	N	1958	10667
0072	S	D	J	N	1956	19558
0072	S	D	E	N	1921	22633
0072	S	D	F	N	1954	23317
0072	S	D	E	N	1964	25860
0072	S	D	E	N	1962	28005
0072	S	G	E	N	1914	31884
0072	S	D	R	N	1964	41484
0072	S	E	R	N	1964	41484
0072	S	G	R	N	1965	42492
0072	S	D	R	N	1965	44434
0072	S	D	R	N	1967	45144
0072	S	D	G	N	1968	54724
0072	S	E	G	N	1968	54724
0072	S	D	E	N	1969	59294
0072	S	E	E	N	1969	59294
0072	S	D	P	N	1971	63225
0072	S	D	E	N	1972	64938
0072	S	E	E	N	1972	64938
0072	S	D	R	N	1971	67298
0072	S	D	E	N	1967	70862
0072	S	D	R	N	1972	71302
0072	S	D	R	N	1969	74008
0072	S	D	E	N	1969	74009
0072	S	D	E	N	1972	78340
0072	S	E	E	N	1972	78340
0072	S	D	E	N	1977	93606

Substance Number	Phys. State	Subject	Language	Temperature	Year	TPRC Number
521-0072	S	D	R	N	1976	94572
0072	S	D	J	N	1978	96703
0072	S	E	J	N	1978	96703
0073	E	D	R	H	1971	75495
0073	E	D	E	H	1971	75496
0073	P	D	E	N	1965	42374
0073	P	D	R	N	1971	65687
0073	S	D	E	N	1921	22633
0073	S	D	E	L	1948	30254
0073	S	D	E	N	1965	42374
0073	S	S	E	F	1970	61236
0073	S	D	R	H	1971	75495
0073	S	D	E	H	1971	75496
0075	S	D	I		1951	1247
0075	S	T	I		1951	1247
0075	S	D	E		1951	1835
0075	S	T	E		1951	1835
0075	S	G	J		1951	2025
0075	S	S	E	N	1959	10076
0075	S	D	E	L	1934	10348
0075	S	D	G	L	1935	11805
0075	S	D	E	N	1932	14353
0075	S	D	E	N	1925	21790
0075	S	D	O	L	1964	35713
0075	S	T	O	L	1964	35713
0075	S	D	E	L	1969	55399
0075	S	S	E	L	1970	61236
0075	S	S	E	N	1970	61236
0075	S	S	E	N	1977	92134
0077	M	G	E	N	1962	30997
0077	S	D	R		1955	507
0077	S	D	G	L	1935	11805
0077	S	S	E	N	1936	16524
0077	S	D	O	H	1958	17821
0077	S	T	O	H	1958	17821
0077	S	D	G	H	1926	21459
0077	S	D	R	H	1962	27546
0077	S	T	R	H	1962	27546
0082	S	D	R	N	1971	64741
0082	S	D	E	N	1971	64742
0083	S	D	E		1948	1481
0083	S	G	R	N	1958	20245
0083	S	G	E	N	1962	20436
0083	S	D	F	N	1895	28730
0083	S	D	G	N	1904	28731
0083	S	S	E	L	1970	61236
0083	S	S	E	N	1970	61236
0083	S	G	R	N	1971	62997
0083	S	G	E	N	1971	72195
0084	L	D	G	N	1958	19586
0084	S	D	E		1951	1503
0084	S	D	E		1935	8241
0084	S	D	E	N	1921	22633
0084	S	D	E	N	1937	28449
0084	S	G	E	L	1963	30496
0084	S	S	E	L	1970	61236
0084	S	S	E	N	1970	61236
0084	S	D	R	N	1976	94572
0085	M	G	E	N	1962	30997
0085	S	D	E		1951	1503
0085	S	S	E	N	1958	10667
0085	S	G	J	N	1940	16239
0085	S	D	G	H	1935	16474
0085	S	D	E	N	1921	22633
0085	S	G	E	N	1961	23287
0085	S	D	F	N	1954	23317
0085	S	D	E	N	1940	25372
0085	S	D	E	N	1964	25860
0085	S	D	E	N	1962	28005
0085	S	D	E	N	1969	59294
0085	S	E	E	N	1969	59294
0085	S	D	R	N	1968	60916
0085	S	D	E	N	1971	61009
0085	S	T	E	N	1971	61009
0085	S	S	E	N	1970	61236
0085	S	D	E	N	1971	61734
0085	S	D	R	N	1969	74008
0085	S	D	E	N	1969	74009
0085	S	D	E	N	1977	93606
0086	D	D	E	L	1977	89480
0086	S	S	G		1933	6798
0086	S	G	G	N	1936	43371
0087	S	S	G		1933	6798
0087	S	S	E		1952	9922
0087	S	D	E	N	1961	20864
0087	S	T	E	N	1961	20864
0087	S	D	G	N	1912	20865
0087	S	T	G	N	1912	20865
0087	S	D	G	N	1920	22785
0087	S	T	G	N	1920	22785
0089	S	D	E		1954	518
0089	S	D	E		1956	7001
0089	S	D	R	N	1975	91435
0089	S	E	R	N	1975	91435
0090	S	D	E		1950	1476
0091	S	D	E	L	1951	1383
0091	S	D	E	N	1951	1383
0091	S	D	E		1951	1383
0091	S	D	J	N	1956	19593
0092	S	D	G		1956	887
0092	S	D	E		1956	7001

Substance Number	Phys. State	Subject	Language	Temperature	Year	TPRC Number
521-0092	S	S	E	N	1959	10076
0092	S	D	E	N	1971	66172
0092	S	T	E	N	1971	66172
0093	S	D	R		1950	1340
0093	S	T	R		1950	1340
0094	S	D	G		1956	887
0094	S	D	E		1956	7001
0096	S	S	G		1933	6798
0096	S	S	E		1952	9922
0096	S	S	E	N	1959	10076
0096	S	D	E	N	1959	19385
0096	S	T	E	N	1959	19385
0096	S	D	J	N	1956	19593
0096	S	D	G	N	1920	22785
0096	S	T	G	N	1920	22785
0096	S	D	O	L	1964	35713
0096	S	T	O	L	1964	35713
0096	S	D	E	N	1950	55957
0096	S	S	E	L	1970	61236
0096	S	S	E	N	1970	61236
0097	S	D	E		1956	7001
0098	S	D	E		1956	7001
0101	S	D	E		1956	7001
0103	S	D	G	N	1955	1125
0103	S	E	G	N	1955	1125
0103	S	D	G	N	1933	11172
0103	S	D	E	N	1921	22633
0103	S	D	G	N	1922	22647
0103	S	D	F	N	1954	23317
0103	S	D	F	N	1895	28730
0103	S	D	R	N	1965	34590
0103	S	D	E	N	1960	42824
0103	S	D	E	N	1900	48245
0103	S	D	F	N	1970	57876
0103	S	D	E	N	1971	61734
0103	S	D	E	N	1977	93606
0103	S	D	E	N	1913	94422
0103	S	D	R	N	1976	94572
0104	M	G	E	N	1962	30997
0104	S	D	G	N	1929	22766
0104	S	G	E	N	1961	23287
0104	S	D	E	N	1969	59294
0104	S	E	E	N	1969	59294
0105	P	D	G	L	1916	27515
0105	S	G	J		1951	2025
0105	S	S	E	N	1959	10076
0105	S	D	E	L	1934	10348
0105	S	D	E	L	1969	55399
0108	M	D	R	N	1964	38189
0108	P	D	E	N	1921	22633
0108	P	D	R	N	1968	49504
0108	P	E	R	N	1968	49504
0108	P	D	E	N	1968	65123
0108	P	E	E	N	1968	65123
0108	S	D	G	N	1933	11172
0108	S	D	R	N	1964	38189
0108	S	D	E	N	1973	70941
0109	P	D	E	L	1963	30497
0109	S	S	E	L	1970	61236
0109	S	S	E	N	1970	61236
0110	M	D	E	N	1909	31920
0110	M	D	G	N	1958	51235
0110	M	G	E	N	1973	76238
0110	P	G	R	N	1958	20245
0110	P	G	E	N	1962	20436
0110	P	D	E	N	1921	22633
0110	P	D	E	N	1909	31920
0110	P	D	G	N	1958	51235
0110	P	D	E	N	1949	61706
0110	P	D	F	N	1965	62182
0110	P	D	E	N	1973	70628
0110	P	D	R	N	1970	73669
0110	P	D	E	N	1970	73670
0110	P	G	E	N	1973	76238
0110	S	S	E	N	1958	10667
0110	S	G	E	N	1949	12033
0110	S	D	E	N	1932	14353
0110	S	T	E	N	1957	14689
0110	S	D	E	N	1960	14795
0110	S	D	E	N	1958	15736
0110	S	D	G	H	1926	21459
0110	S	D	R	N	1960	24204
0110	S	G	E	N	1955	24261
0110	S	D	E	N	1961	24910
0110	S	S	G	N	1928	28680
0111	P	D	E	N	1966	43796
0111	P	T	E	N	1966	43796
0111	S	D	E	N	1959	25275
0111	S	D	R	N	1959	25276
0115	S	D	R	N	1966	42469
0115	S	S	R	N	1966	47360
0115	S	D	R	N	1966	47082
0116	P	D	E	L	1963	30497
0116	P	D	R	N	1968	49504
0116	P	E	R	N	1968	49504
0116	P	D	E	N	1968	65123
0116	P	E	E	N	1968	65123
0116	S	D	E	N	1923	10331
0116	S	D	R	N	1934	16918
0116	S	S	G	N	1928	28680
0116	S	D	R	N	1967	47305

Phys. State: **C.** Superconductive; **D.** Doped; **E.** Expanded; **F.** Fibrous or Whisker; **G.** Gas; **I.** Ionized or Plasma; **L.** Liquid; **M.** Multiphase; **P.** Powder or Fine Particle; **S.** Solid; **T.** Thick or Thin Film
Subject: **D.** Data; **E.** Experiment; **G.** General (Data + Theory + Experiment); **S.** Survey (Review, Compendium etc.); **T.** Theory
Language: **C.** Czech; **D.** Dutch; **E.** English; **F.** French; **G.** German; **I.** Italian; **J.** Japanese; **O.** Other Languages; **P.** Polish; **R.** Russian; **S.** Spanish
Temperature: **F.** Full Range (Low + Normal + High); **L.** Low (0 to 75K) + Overlap into Normal; **N.** Normal (75 to 1273K); **H.** High (above 1273K) + Overlap into Normal; Blank = Not Coded

Substance Number	Phys. State	Subject	Language	Temperature	Year	TPRC Number
521-0116	S	D	R	N	1967	50602
0116	S	D	E	N	1969	59294
0116	S	E	E	N	1969	59294
0116	S	S	E	L	1970	61236
0116	S	S	E	N	1970	61236
0116	S	D	E	N	1967	70863
0123	E	D	E	N	1965	42374
0123	P	D	E	N	1962	39516
0123	P	D	R	N	1971	69768
0123	P	D	E	N	1972	69769
0123	S	S	E	N	1958	10667
0123	S	D	G	N	1933	11172
0123	S	D	E	N	1921	22633
0123	S	D	F	N	1954	23317
0123	S	D	E	N	1962	28005
0123	S	D	F	N	1895	28730
0123	S	D	G	N	1904	28731
0123	S	G	E	N	1914	31884
0123	S	D	R	N	1965	44434
0123	S	D	R	N	1967	45144
0123	S	D	E	N	1964	45753
0123	S	D	E	N	1969	59294
0123	S	E	E	N	1969	59294
0123	S	D	E	N	1971	61734
0123	S	D	P	N	1971	63225
0123	S	D	E	N	1967	70862
0123	S	D	E	N	1972	78340
0123	S	E	E	N	1972	78340
0123	S	D	R	N	1975	83260
0123	S	D	E	N	1977	93606
0123	S	D	E	N	1978	98902
0123	S	D	E	N	1978	99991
0130	S	S	E	N	1959	10076
0132	S	D	E	H	1958	10703
0132	S	T	E	H	1958	10703
0133	S	D	E		1946	1888
0133	S	D	E		1956	7001
0133	S	S	E	H	1960	24931
0133	S	S	E	N	1960	24931
0133	S	D	E	N	1865	30410
0133	S	D	E	N	1948	40389
0135	P	D	E	N	1921	22633
0135	P	D	E	H	1967	47943
0135	P	D	R	N	1969	55535
0135	P	D	R	N	1971	65687
0135	P	D	R	N	1973	75274
0135	P	E	R	N	1973	75274
0135	S	D	G	H	1935	16474
0135	S	S	E	N	1936	16524
0135	S	D	R	N	1934	16918
0135	S	D	R	N	1958	19650
0135	S	D	G	H	1926	21459
0135	S	S	G	H	1928	28680
0135	S	D	E	N	1966	44390
0135	S	E	E	N	1966	44390
0135	S	G	F	N	1967	45089
0145	L	D	S	N	1970	92963
0145	S	D	S	N	1970	92963
0169	S	D	R	N	1960	20281
0174	P	D	R	N	1969	56752
0174	P	D	E	N	1969	70864
0174	S	D	F	N	1930	21511
0174	S	T	F	N	1930	21511
0174	S	D	E	L	1961	26326
0174	S	S	G	H	1928	28680
0174	S	S	E	L	1970	61236
0174	S	S	E	N	1970	61236
0174	S	D	E	N	1973	71892
0180	S	D	E	N	1959	18202
0181	S	D	G	N	1933	11172
0181	S	D	R	N	1976	94572
0182	S	G	E	N	1954	12405
0182	S	D	R	N	1967	50602
0183	P	G	E	N	1954	12405
0183	S	D	R	N	1968	60916
0184	S	D	E	N	1962	28005
0184	S	D	E	N	1969	59294
0184	S	E	E	N	1969	59294
0186	M	G	E	N	1962	30997
0186	P	D	E	N	1921	22633
0186	S	S	E	N	1958	10667
0190	P	D	E	N	1965	44633
0190	P	E	E	N	1965	44633
0190	S	D	E	N	1965	44633
0190	S	E	E	N	1965	44633
0191	S	D	E	N	1973	70944
0192	S	S	E	N	1959	10076
0192	S	D	E	L	1918	20920
0192	S	S	E	L	1970	61236
0192	S	S	E	N	1970	61236
0195	S	D	F	N	1908	10282
0195	S	D	E	L	1934	10348
0195	S	D	R	H	1910	10454
0195	S	S	E	L	1970	61236
0195	S	S	E	N	1970	61236
0196	M	D	R	N	1969	54587
0196	S	S	E	N	1958	10667
0196	S	D	E	N	1958	12174
0196	S	D	E	N	1921	22633
0196	S	D	E	N	1940	25372
0196	S	D	E	N	1962	28005
521-0196	S	D	G	N	1904	28731
0196	S	D	E	N	1951	40424
0196	S	D	E	N	1969	59294
0196	S	E	E	N	1969	59294
0196	S	D	E	N	1977	93606
0215	S	D	E	N	1923	10331
0216	S	D	J	N	1934	10365
0216	S	D	E	N	1935	16537
0217	S	D	F	N	1930	21511
0217	S	T	F	N	1930	21511
0222	S	D	R	N	1910	10454
0222	S	D	G	H	1926	21459
0222	S	S	S	H	1955	28264
0222	S	S	G	H	1928	28680
0222	S	D	G	H	1921	30241
0223	S	D	R	H	1910	10454
0225	P	D	E	N	1965	42374
0236	S	D	G	N	1933	11172
0236	S	D	E	N	1921	22633
0236	S	D	E	N	1940	25372
0236	S	D	G	N	1904	28731
0236	S	D	R	N	1966	42491
0236	S	D	R	N	1967	45144
0236	S	D	E	N	1964	45753
0236	S	D	E	L	1967	49237
0236	S	D	R	N	1967	50602
0236	S	D	E	N	1969	59294
0236	S	E	E	N	1969	59294
0236	S	D	E	N	1967	70862
0239	S	D	R	N	1974	76265
0240	P	D	E	L	1963	27814
0240	P	D	E	N	1968	52038
0240	S	D	E	N	1921	22633
0240	S	D	E	N	1964	32980
0240	S	T	E	N	1964	32980
0240	S	S	E	L	1970	61236
0240	S	S	E	N	1970	61236
0240	S	D	R	N	1973	71882
0240	S	D	E	L	1976	89267
0240	S	D	E	N	1976	89267
0240	S	E	E	L	1976	89267
0240	S	E	E	N	1976	89267
0240	S	S	E	N	1977	92134
0241	S	D	E	L	1959	17122
0241	S	D	R	N	1973	71882
0242	P	D	E	N	1965	42374
0242	S	D	E	N	1962	28005
0242	S	D	R	N	1965	34088
0242	S	D	E	N	1964	45753
0242	S	D	G	N	1968	54724
0242	S	E	G	N	1968	54724
0242	S	D	E	N	1969	59294
0242	S	E	E	N	1969	59294
0242	S	D	E	N	1971	61734
0242	S	D	E	N	1972	64938
0242	S	E	E	N	1972	64938
0242	S	D	R	N	1971	67298
0242	S	D	E	N	1977	93606
0246	S	D	G	N	1927	19249
0246	S	D	E	N	1962	28005
0246	S	G	G	N	1966	44294
0246	S	D	G	N	1966	44295
0246	S	D	R	N	1968	53054
0246	S	D	R	N	1968	54490
0246	S	D	R	N	1973	71862
0249	S	D	E	N	1932	14353
0249	S	D	D	N	1931	23230
0249	S	D	G	N	1940	29870
0249	S	D	R	N	1965	42663
0249	S	D	E	N	1965	49232
0249	S	D	R	N	1968	53054
0249	S	D	R	N	1968	54490
0249	S	D	R	N	1970	59042
0249	S	D	E	N	1970	67687
0249	S	D	R	N	1973	71862
0249	S	D	E	N	1913	94422
0250	S	D	E	N	1959	13944
0251	S	D	F	N	1895	28730
0251	S	S	E	L	1970	61236
0251	S	S	E	N	1970	61236
0253	M	D	E	N	1976	89094
0253	S	D	R	N	1970	68596
0254	P	D	G	L	1922	21397
0254	S	D	E	L	1936	15366
0254	S	D	G	H	1926	21459
0254	S	D	E	L	1956	26210
0254	S	S	S	H	1955	28264
0254	S	S	G	H	1928	28680
0254	S	D	E	H	1941	30160
0254	S	T	E	H	1941	30160
0254	S	D	G	H	1921	30241
0254	S	T	E	L	1963	32466
0254	S	D	E	H	1919	43370
0254	S	D	E	N	1966	99938
0255	S	D	E	L	1936	15366
0255	S	D	E	H	1941	30160
0255	S	T	E	H	1941	30160
0255	L	D	E	N	1966	99938
0256	L	G	R	N	1935	15471
0256	L	G	E	N	1963	31234
0256	S	G	R	N	1935	15471
521-0256	S	G	E	N	1963	31234
0258	P	D	E	N	1965	42374
0258	S	D	E	N	1865	30410
0258	S	D	E	N	1969	59294
0258	S	E	E	N	1969	59294
0258	S	S	E	N	1977	92134
0258	S	S	E	N	1976	92625
0258	S	D	E	N	1979	97863
0259	P	S	E	N	1969	41316
0259	S	D	E	N	1960	20866
0259	S	T	E	N	1960	20866
0259	S	D	E	N	1960	20867
0259	S	T	E	N	1960	20867
0259	S	G	E	N	1962	29137
0259	S	S	E	N	1969	41316
0259	S	D	E	N	1968	50610
0259	S	T	E	N	1968	50610
0259	S	S	E	L	1970	59124
0259	S	S	E	N	1971	68503
0261	S	D	G	H	1933	16107
0261	S	D	G	N	1929	22766
0264	S	D	E	N	1969	59294
0264	S	E	E	N	1969	59294
0264	S	D	R	N	1976	94572
0267	S	D	G	H	1935	16474
0267	S	D	E	N	1970	57984
0267	S	D	E	N	1970	67511
0267	S	D	R	H	1973	71847
0267	S	T	R	H	1973	71847
0270	S	D	R	N	1971	67298
0272	S	D	E	N	1921	22633
0272	S	D	E	N	1962	28005
0272	S	D	E	N	1964	45753
0272	S	D	E	N	1969	59294
0272	S	E	E	N	1969	59294
0274	S	D	E	N	1921	22633
0274	S	D	E	N	1969	59294
0274	S	E	E	N	1969	59294
0274	S	D	R	N	1971	67298
0274	S	D	R	N	1972	71302
0279	S	D	J	N	1956	19558
0281	S	D	E	N	1921	22633
0284	S	D	E	N	1935	16537
0293	S	D	E	N	1921	22633
0293	S	D	R	N	1971	67298
0293	S	D	R	N	1976	94572
0294	S	D	E	N	1935	16537
0296	S	D	R	N	1965	34590
0296	S	D	R	N	1972	71302
0297	S	D	E	N	1935	16537
0298	P	D	E	N	1921	22633
0298	S	D	E	N	1921	22633
0298	S	D	E	N	1962	28005
0298	S	D	E	N	1951	40424
0298	S	D	R	N	1965	44434
0298	S	D	E	N	1969	59294
0298	S	E	E	N	1969	59294
0298	S	D	E	N	1971	61734
0298	S	D	R	N	1969	74008
0298	S	D	E	N	1969	74009
0298	S	D	E	N	1972	78340
0298	S	E	E	N	1972	78340
0298	S	D	E	N	1977	93606
0299	S	S	E	N	1936	16524
0299	S	S	G	H	1928	28680
0299	S	S	E	N	1928	31217
0299	S	D	E	N	1957	31948
0299	S	S	E	F	1969	37682
0299	S	T	R	L	1969	57010
0299	S	D	E	N	1970	57604
0299	S	T	E	N	1970	57604
0299	S	S	R	N	1970	61897
0299	S	S	P	N	1975	84618
0302	S	D	E	L	1959	17036
0302	S	D	O	H	1958	17821
0302	S	T	O	H	1958	17821
0302	S	D	E	N	1966	99938
0303	S	D	E	L	1959	17036
0303	S	D	O	H	1958	17821
0303	S	T	O	H	1958	17821
0303	S	D	E	N	1966	99938
0307	S	D	E	N	1964	45753
0308	S	D	E	N	1964	37463
0308	S	D	E	N	1951	40424
0308	S	D	P	N	1971	63225
0308	S	D	E	L	1971	63851
0308	S	D	R	N	1976	94572
0330	S	D	J	N	1956	19558
0341	S	G	J	N	1940	16239
0344	S	S	E	L	1960	24931
0344	S	S	E	N	1960	24931
0346	S	D	E	L	1967	47495
0348	S	D	E	N	1962	28005
0348	S	D	E	N	1969	59294
0348	S	E	E	N	1969	59294
0349	S	D	E	N	1921	22633
0349	S	D	E	N	1964	25860
0349	S	D	R	N	1971	67298
0354	S	D	E	N	1962	28005
0354	S	D	E	N	1969	59294
0354	S	E	E	N	1969	59294

Phys. State: **C.** Superconductive; **D.** Doped; **E.** Expanded; **F.** Fibrous or Whisker; **G.** Gas; **I.** Ionized or Plasma; **L.** Liquid; **M.** Multiphase; **P.** Powder or Fine Particle; **S.** Solid; **T.** Thick or Thin Film

Subject: **D.** Data; **E.** Experiment; **G.** General (Data + Theory + Experiment); **S.** Survey (Review, Compendium etc.); **T.** Theory

Language: **C.** Czech; **D.** Dutch; **E.** English; **F.** French; **G.** German; **I.** Italian; **J.** Japanese; **O.** Other Languages; **P.** Polish; **R.** Russian; **S.** Spanish

Temperature: **F.** Full Range (Low + Normal + High); **L.** Low (0 to 75K) + Overlap into Normal; **N.** Normal (75 to 1273K); **H.** High (above 1273K) + Overlap into Normal; Blank = Not Coded

Substance Number	Phys. State	Subject	Language	Temperature	Year	TPRC Number
521-0354	S	D	E	N	1971	61734
0362	S	D	E	N	1921	22633
0362	S	D	E	N	1955	32862
0362	S	D	R	H	1967	48397
0362	S	D	E	L	1970	57611
0362	S	D	E	H	1967	70865
0362	S	D	R	N	1973	71847
0362	S	T	R	N	1973	71847
0362	S	D	E	L	1976	89267
0362	S	D	E	N	1976	89267
0362	S	E	E	L	1976	89267
0362	S	E	E	N	1976	89267
0362	S	D	J	N	1977	90890
0363	S	D	E	N	1921	22633
0364	S	D	E	N	1921	22633
0365	S	D	E	N	1921	22633
0366	S	D	E	N	1921	22633
0366	S	D	E	N	1937	28449
0366	S	D	E	N	1865	30410
0368	M	G	E	N	1962	30997
0368	M	D	E	N	1909	31920
0368	P	D	E	N	1909	31920
0368	S	G	E	N	1970	58782
0369	S	D	E	N	1921	22633
0369	S	D	E	N	1964	45753
0369	S	D	G	N	1968	54724
0369	S	E	G	N	1968	54724
0369	S	D	E	N	1972	64938
0369	S	E	E	N	1972	64938
0370	S	D	E	N	1921	22633
0371	S	D	E	N	1921	22633
0372	S	D	E	N	1921	22633
0373	S	D	E	N	1921	22633
0375	S	D	E	N	1921	22633
0375	S	D	G	N	1968	54724
0375	S	E	G	N	1968	54724
0375	S	D	P	N	1971	63225
0375	S	D	R	N	1969	74008
0375	S	D	E	N	1969	74009
0376	S	D	E	N	1921	22633
0377	S	D	E	N	1921	22633
0378	S	D	E	N	1921	22633
0379	S	D	E	N	1921	22633
0380	S	D	E	N	1921	22633
0381	S	D	E	N	1921	22633
0382	P	D	E	N	1930	33182
0382	P	T	E	N	1930	33182
0382	P	D	R	H	1964	35493
0382	S	G	R	H	1962	27593
0382	S	G	E	H	1963	27761
0382	S	D	R	H	1963	37067
0382	S	D	R	H	1971	64967
0382	S	D	E	H	1971	64968
0382	S	D	R	H	1976	84491
0382	S	D	R	N	1976	84491
0383	S	D	E	N	1921	22633
0385	S	D	G	N	1929	22766
0385	S	D	E	H	1957	36165
0389	S	D	E	L	1960	23718
0389	S	D	E	N	1966	99938
0392	M	G	E	N	1962	30997
0392	S	D	E	H	1919	43370
0392	S	D	E	N	1971	66172
0392	S	T	E	N	1971	66172
0392	S	D	E	L	1974	85507
0392	S	D	E	N	1974	85507
0393	P	D	R	N	1971	65687
0393	S	D	G	H	1926	21459
0393	S	S	G	H	1928	28680
0394	P	D	E	N	1949	61706
0394	S	D	G	H	1926	21459
0394	S	S	G	H	1928	28680
0394	S	D	R	N	1971	67298
0395	S	D	G	H	1926	21459
0395	S	S	G	H	1928	28680
0396	S	D	E	L	1959	23934
0402	S	D	R	N	1973	71862
0409	S	D	E	N	1923	21665
0411	S	D	E	N	1976	91691
0411	S	D	E	N	1976	91692
0411	S	D	R	N	1979	99173
0413	S	D	E	N	1963	24342
0413	S	T	E	N	1963	24342
0414	S	G	E	N	1961	23287
0429	S	D	E		1935	8241
0429	S	D	E	L	1961	26326
0429	S	D	E	N	1937	28449
0429	S	D	E	N	1966	99938
0430	S	D	R	N	1971	67298
0430	S	D	R	N	1973	71882
0432	S	D	G	N	1904	28731
0442	M	G	F	N	1965	39350
0442	M	G	R	N	1959	43002
0442	M	G	F	N	1959	43003
0442	P	D	E	N	1965	44633
0442	P	E	E	N	1965	44633
0442	S	D	R	N	1959	25367
0442	S	T	R	N	1959	25367
0442	S	G	R	N	1959	43002
0442	S	G	F	N	1959	43003
0442	S	D	E	N	1965	44633
521-0442	S	E	E	N	1965	44633
0443	S	S	E	L	1970	61236
0443	S	S	E	N	1970	61236
0468	S	D	E	N	1951	40424
0475	M	G	E	N	1962	30997
0475	S	D	E	N	1962	28005
0475	S	D	E	N	1969	59294
0475	S	E	E	N	1969	59294
0476	S	D	E	N	1962	28005
0476	S	D	E	N	1969	59294
0476	S	E	E	N	1969	59294
0477	S	D	E	N	1962	28005
0477	S	D	E	N	1969	59294
0477	S	E	E	N	1969	59294
0478	S	D	E	N	1962	28005
0479	S	D	E	N	1962	28005
0479	S	D	E	N	1969	59294
0479	S	E	E	N	1969	59294
0481	S	D	F	N	1954	23317
0484	S	D	E	N	1962	28005
0484	S	D	E	N	1969	59294
0484	S	E	E	N	1969	59294
0487	S	D	E	L	1961	26327
0487	S	S	E	L	1970	61236
0487	S	S	E	N	1970	61236
0488	S	D	E	L	1961	26327
0488	S	S	E	L	1970	61236
0488	S	S	E	N	1970	61236
0494	S	D	E	H	1919	43370
0494	S	D	E	N	1979	97863
0495	S	D	P	N	1971	63225
0496	S	S	S	H	1955	28264
0500	P	D	R	N	1971	65687
0500	S	D	O	H	1957	17692
0502	S	D	G	N	1904	28731
0505	S	D	E	L	1961	26326
0505	S	S	E	L	1970	61236
0505	S	S	E	N	1970	61236
0506	S	D	E	L	1961	26326
0506	S	S	E	L	1970	61236
0506	S	S	E	N	1970	61236
0506	S	D	E	N	1966	99938
0511	S	D	F	N	1954	23317
0514	S	S	E	H	1970	61236
0514	S	S	E	N	1970	61236
0517	S	D	E	L	1978	94314
0519	S	D	G	N	1922	22647
0525	S	D	E	N	1951	40424
0530	M	D	E	N	1909	31920
0530	P	D	E	N	1909	31920
0530	S	D	E	N	1961	24910
0530	S	D	R	N	1967	60317
0530	S	D	E	N	1970	60318
0555	S	D	E	N	1865	30410
0563	S	S	E	H	1960	24931
0563	S	S	E	N	1960	24931
0564	S	D	E	N	1964	25860
0566	S	D	G	N	1929	22766
0566	S	D	E	N	1865	30410
0567	S	D	E	N	1964	25860
0570	S	D	E	N	1964	25860
0581	S	G	R	H	1962	27593
0581	S	G	E	H	1963	27761
0599	S	D	E	N	1951	40424
0604	S	D	E	L	1970	57611
0612	M	G	E	N	1962	30997
0612	P	D	E	N	1965	44633
0612	P	E	E	N	1965	44633
0612	S	D	E	N	1965	44633
0612	S	E	E	N	1965	44633
0637	S	G	E	L	1963	43286
0643	M	D	E	N	1909	31920
0643	P	D	E	N	1909	31920
0643	P	D	E	N	1949	61706
0645	M	G	E	N	1962	30997
0645	M	D	E	N	1909	31920
0645	M	G	R	N	1959	43002
0645	M	G	F	N	1959	43003
0645	P	D	E	N	1909	31920
0645	S	D	R	N	1959	25367
0645	S	T	R	N	1959	25367
0645	S	G	R	N	1959	43002
0645	S	G	F	N	1959	43003
0645	S	D	E	N	1968	46410
0648	S	S	E	L	1960	24931
0648	S	S	E	N	1960	24931
0648	S	D	E	N	1937	28449
0648	S	S	G	H	1928	28680
0648	S	S	G	N	1928	28680
0648	S	D	E	N	1971	66172
0648	S	T	E	N	1971	66172
0650	S	S	E	L	1970	61236
0650	S	S	E	N	1970	61236
0660	S	D	D	N	1931	23230
0660	S	D	E	N	1965	35579
0660	S	D	R	N	1965	42663
0660	S	D	O	H	1966	44322
0660	S	D	E	N	1965	49232
0660	S	D	R	N	1970	59042
0660	S	D	E	N	1970	67687
0660	S	D	R	N	1973	71862
521-0671	S	D	E	N	1921	22633
0671	S	D	G	N	1968	54724
0671	S	E	G	N	1968	54724
0671	S	D	E	N	1972	64938
0671	S	E	E	N	1972	64938
0671	S	D	R	N	1972	71302
0675	P	D	R	N	1966	39422
0689	P	D	E	N	1964	36076
0689	S	D	E	N	1964	45713
0705	P	D	E	N	1964	36076
0705	S	D	E	N	1964	45713
0706	P	D	E	N	1964	36076
0706	S	D	E	N	1964	45713
0707	P	D	E	N	1964	36076
0707	S	D	E	N	1964	45713
0708	P	D	E	N	1964	36076
0708	S	D	E	N	1964	45713
0713	S	D	R	N	1965	34088
0713	S	D	G	N	1968	54724
0713	S	E	G	N	1968	54724
0713	S	D	R	N	1971	67298
0713	S	D	R	N	1976	94572
0714	S	D	R	N	1965	34088
0715	S	D	R	N	1965	34088
0716	S	D	R	N	1965	34088
0720	S	D	R	L	1977	92523
0720	S	D	R	N	1977	92523
0720	S	D	R	N	1977	92524
0720	S	E	R	N	1977	92524
0727	S	D	E	N	1958	12174
0729	P	D	E	N	1964	36076
0729	S	D	E	N	1964	45713
0730	P	D	E	N	1964	36076
0730	S	D	E	N	1964	45713
0730	S	D	E	N	1967	49308
0732	G	G	E		1959	8282
0732	L	D	F		1946	2046
0732	L	G	E		1959	8282
0732	L	T	E	H	1939	13369
0732	L	S	E	N	1926	20886
0732	L	D	G	N	1923	25148
0732	L	T	E	H	1963	26274
0732	L	T	R	H	1963	26275
0732	P	D	R		1958	8506
0732	P	D	E		1958	9254
0732	P	E	E		1958	9254
0732	P	D	G	L	1922	21397
0732	P	S	E	N	1960	24931
0732	P	D	E	N	1934	25152
0732	P	D	R	N	1961	50701
0732	P	E	R	N	1961	50701
0732	P	D	E	N	1949	61706
0732	S	D	E		1957	296
0732	S	D	E		1956	341
0732	S	T	E		1956	341
0732	S	G	E		1953	500
0732	S	D	R		1953	619
0732	S	D	E		1955	718
0732	S	T	E		1955	718
0732	S	D	G		1956	887
0732	S	D	R		1954	904
0732	S	G	G		1936	6314
0732	S	D	E		1954	6565
0732	S	S	G		1933	6798
0732	S	D	E		1959	7688
0732	S	D	E		1935	8241
0732	S	G	E		1959	8282
0732	S	D	E		1957	9207
0732	S	S	E	N	1959	10076
0732	S	D	R	N	1910	10454
0732	S	G	E	N	1951	10853
0732	S	D	G	N	1933	11172
0732	S	T	E	L	1939	13369
0732	S	D	E	N	1959	13944
0732	S	D	E	N	1932	14353
0732	S	D	E	N	1941	14484
0732	S	T	E	N	1941	14484
0732	S	D	E	N	1958	14916
0732	S	D	E	L	1936	15366
0732	S	D	E	N	1954	15501
0732	S	D	G	H	1960	20600
0732	S	D	G	N	1920	21393
0732	S	T	G	N	1920	21393
0732	S	D	G	H	1926	21459
0732	S	D	G	N	1924	21691
0732	S	D	G	N	1922	22647
0732	S	D	G	N	1930	22749
0732	S	T	G	N	1930	22749
0732	S	D	G	N	1929	22766
0732	S	S	E	N	1960	24931
0732	S	D	G	N	1923	25148
0732	S	D	E	N	1919	25193
0732	S	D	R	N	1960	25502
0732	S	D	E	L	1956	26210
0732	S	T	E	H	1963	26274
0732	S	T	R	H	1963	26275
0732	S	D	E	N	1962	26684
0732	S	D	E	N	1961	27117
0732	S	G	E	N	1966	27375
0732	S	D	E	N	1962	28005
0732	S	S	S	H	1955	28264

Phys. State: **C.** Superconductive; **D.** Doped; **E.** Expanded; **F.** Fibrous or Whisker; **G.** Gas; **I.** Ionized or Plasma; **L.** Liquid; **M.** Multiphase; **P.** Powder or Fine Particle; **S.** Solid; **T.** Thick or Thin Film
Subject: **D.** Data; **E.** Experiment; **G.** General (Data + Theory + Experiment); **S.** Survey (Review, Compendium etc.); **T.** Theory
Language: **C.** Czech; **D.** Dutch; **E.** English; **F.** French; **G.** German; **I.** Italian; **J.** Japanese; **O.** Other Languages; **P.** Polish; **R.** Russian; **S.** Spanish
Temperature: **F.** Full Range (Low + Normal + High); **L.** Low (0 to 75K) + Overlap into Normal; **N.** Normal (75 to 1273K); **H.** High (above 1273K) + Overlap into Normal; Blank = Not Coded

Substance Number	Phys. State	Subject	Language	Temperature	Year	TPRC Number
521-0732	S	D	E	N	1937	28449
0732	S	D	E	N	1957	28453
0732	S	D	R	H	1965	28567
0732	S	T	R	H	1965	28567
0732	S	D	E	H	1965	28597
0732	S	T	E	H	1965	28597
0732	S	S	G	N	1928	28680
0732	S	D	F	N	1895	28730
0732	S	G	G	L	1911	29642
0732	S	D	E	H	1941	30160
0732	S	T	E	H	1941	30160
0732	S	D	G	H	1921	30241
0732	S	D	E	N	1865	30410
0732	S	G	G	N	1913	30553
0732	S	D	G	L	1926	31607
0732	S	G	G	H	1918	32447
0732	S	T	E	L	1963	32466
0732	S	D	E	N	1960	32705
0732	S	D	R	H	1964	37217
0732	S	D	E	L	1965	38865
0732	S	G	E	N	1970	41374
0732	S	D	R	H	1964	41484
0732	S	E	R	H	1964	41484
0732	S	G	R	H	1965	42492
0732	S	D	E	H	1919	43370
0732	S	D	R	N	1965	44434
0732	S	D	E	N	1968	52896
0732	S	S	E	N	1968	55124
0732	S	D	E	N	1969	59294
0732	S	E	E	N	1969	59294
0732	S	S	E	L	1970	61236
0732	S	S	E	N	1970	61236
0732	S	D	E	N	1971	62185
0732	S	D	P	N	1971	63225
0732	S	G	E	L	1971	66152
0732	S	G	E	N	1970	74014
0732	S	S	E	N	1977	92134
0732	S	D	R	N	1976	94572
0739	S	D	R	N	1965	42663
0739	S	D	E	N	1965	49232
0739	S	D	R	N	1970	59042
0739	S	D	E	N	1970	67687
0739	S	D	R	N	1973	71862
0740	S	D	R	N	1965	42663
0740	S	D	R	N	1970	59042
0740	S	D	E	N	1970	67687
0740	S	D	R	N	1973	71862
0761	P	D	E	N	1949	61706
0762	S	D	R	N	1972	71302
0764	P	D	G	N	1962	37853
0764	P	E	G	N	1962	37853
0764	P	D	R	N	1969	55535
0765	S	D	E	N	1966	44390
0765	S	E	E	N	1966	44390
0765	S	D	E	L	1976	89267
0765	S	D	E	N	1976	89267
0765	S	E	E	L	1976	89267
0765	S	E	E	N	1976	89267
0771	P	D	R	N	1967	51171
0771	S	D	J	N	1975	94691
0777	S	D	E	L	1970	57997
0783	S	D	R	L	1977	92523
0783	S	D	R	N	1977	92523
0783	S	D	R	N	1977	92524
0783	S	E	R	N	1977	92524
0783	S	D	R	N	1978	95686
0783	S	D	E	N	1978	98588
0798	S	D	R	N	1968	60916
0799	F	D	E	N	1979	97928
0799	S	D	E	L	1967	49237
0801	S	D	E	H	1968	49403
0817	S	D	E	N	1965	49232
0817	S	D	R	N	1970	59042
0817	S	D	E	N	1970	67687
0817	S	D	R	N	1973	71862
0823	S	D	E	L	1967	47086
0823	S	D	E	N	1971	62185
0823	S	D	E	N	1971	66172
0823	S	T	E	N	1971	66172
0824	S	D	E	L	1967	47086
0824	S	D	E	N	1971	66172
0824	S	T	E	N	1971	66172
0825	S	D	R	N	1971	67298
0852	S	D	R	N	1967	45144
0852	S	D	E	N	1967	70862
0852	S	D	E	N	1973	74813
0852	S	D	R	N	1974	76265
0858	S	G	E	L	1939	11861
0858	S	G	E	N	1939	11861
0858	S	S	E	L	1970	61236
0858	S	S	E	N	1970	61236
0859	S	D	E	H	1919	43370
0875	S	D	R	N	1975	80733
0875	S	D	E	N	1975	91263
0876	S	D	E	N	1966	44535
0877	S	S	E	L	1970	61236
0877	S	S	E	N	1970	61236
0878	D	S	E	L	1973	71483
0878	D	S	E	L	1979	98944
0878	P	S	E	L	1973	71483
0878	S	S	E	N	1960	16704

Substance Number	Phys. State	Subject	Language	Temperature	Year	TPRC Number
521-0878	S	D	E	L	1964	45094
0878	S	D	R	H	1969	60888
0878	S	S	E	L	1970	61236
0878	S	S	E	N	1970	61236
0878	S	D	R	N	1970	63211
0878	S	S	E	L	1973	71483
0878	S	S	E	N	1973	71483
0878	S	D	E	L	1957	76640
0878	S	E	E	L	1957	76640
0878	S	D	E	L	1978	97784
0878	S	S	E	L	1979	98944
0894	S	D	R	N	1969	58631
0894	S	D	R	N	1972	71302
0894	S	D	R	N	1976	94572
0911	S	D	E	N	1969	59294
0911	S	E	E	N	1969	59294
0917	P	D	R	N	1971	64989
0917	P	D	R	N	1971	69768
0917	P	D	E	N	1972	69769
0917	P	D	R	N	1974	80433
0917	P	D	E	N	1975	80434
0917	S	S	E	N	1967	44944
0917	S	E	E	N	1967	44946
0917	S	D	E	N	1970	56109
0917	S	D	R	N	1971	66494
0917	S	D	E	N	1971	66495
0918	S	D	E	N	1973	74813
0918	S	S	E	N	1972	77778
0922	S	D	E	N	1970	56106
0922	S	D	E	N	1970	56107
0922	S	D	E	N	1970	56109
0922	S	D	E	L	1970	61558
0922	S	S	E	L	1972	68902
0922	S	D	E	L	1970	74011
0925	S	D	R	N	1971	67298
0941	S	D	R	N	1974	76135
0941	S	D	E	N	1974	91262
0942	S	D	R	N	1974	76135
0942	S	D	E	N	1974	91262
0951	S	D	E	H	1919	43370
0951	S	D	E	N	1969	59294
0951	S	E	E	N	1969	59294
0959	S	D	R	H	1972	66891
0959	S	E	R	H	1972	66891
0959	S	D	E	N	1979	96889
0959	S	D	R	L	1979	99173
0959	S	D	R	N	1979	99173
0960	S	D	G	N	1922	22647
0981	S	D	E	L	1970	57963
1010	S	D	E	N	1979	99928
1028	S	D	R	N	1971	67298
1045	S	D	E	L	1974	85507
1045	S	D	E	N	1974	85507
1047	S	D	E	L	1974	85507
1047	S	D	E	N	1974	85507
1074	S	D	R	N	1969	56214
1074	S	E	R	N	1969	56214
1074	S	D	E	N	1969	67256
1074	S	E	E	N	1969	67256
1075	G	D	E	H	1969	42040
1075	G	T	E	H	1969	42040
1089	D	D	R	N	1971	63801
1089	D	E	R	N	1971	63801
1093	P	D	R	N	1971	65687
1094	P	D	R	N	1971	65687
1114	P	D	E	N	1974	77420
1122	S	D	E	N	1970	57380
1131	S	D	R	N	1974	76135
1131	S	D	E	N	1974	91262
1134	S	D	R	N	1967	50602
1148	S	S	E	L	1970	61236
1148	S	S	E	N	1970	61236
1160	S	D	R	N	1965	34590
1163	S	D	R	N	1978	95686
1163	S	D	E	N	1978	98588
1171	S	D	R	N	1967	50602
1172	S	D	R	N	1967	50602
1211	S	D	R	N	1971	67298
1212	S	D	G	N	1968	54724
1212	S	E	G	N	1968	54724
1212	S	D	E	N	1972	64938
1212	S	E	E	N	1972	64938
1228	S	D	E	L	1972	69520
1229	S	D	E	L	1972	69520
1230	S	D	E	L	1972	69521
1231	S	D	R	N	1972	68080
1231	S	D	E	N	1972	68081
1238	S	D	E	N	1972	64938
1238	S	E	E	N	1972	64938
1241	S	D	E	N	1972	64938
1241	S	E	E	N	1972	64938
1243	S	D	E	N	1972	64938
1243	S	E	E	N	1972	64938
1244	S	D	E	N	1972	64938
1244	S	E	E	N	1972	64938
1245	S	D	E	N	1972	64938
1245	S	E	E	N	1972	64938
1255	P	D	E	N	1965	44633
1255	P	E	E	N	1965	44633
1255	S	D	E	N	1965	44633
1255	S	E	E	N	1965	44633

Substance Number	Phys. State	Subject	Language	Temperature	Year	TPRC Number
521-1267	S	D	E	N	1969	59294
1267	S	E	E	N	1969	59294
1268	S	D	E	N	1969	59294
1268	S	E	E	N	1969	59294
1306	S	D	E	N	1969	63778
1333	S	D	R	N	1971	67298
1334	S	D	R	N	1971	67298
1335	S	D	R	N	1971	67298
1335	S	D	R	H	1970	91230
1335	S	D	R	N	1970	91230
1335	S	D	E	H	1970	91231
1335	S	D	E	N	1970	91231
1335	S	D	R	N	1976	94572
1346	S	D	R	H	1973	71847
1346	S	T	R	H	1973	71847
1357	S	D	E	N	1970	89122
1358	S	D	R	N	1969	73753
1358	S	D	E	N	1969	73754
1359	S	D	R	N	1969	73763
1359	S	D	E	N	1969	73764
1392	S	D	E	N	1970	89122
1403	S	D	R	N	1972	71302
1404	S	D	R	N	1972	71302
1406	S	D	R	N	1972	71302
1407	S	D	R	N	1972	71302
1408	S	D	R	N	1972	71302
1409	S	D	R	N	1972	71302
1410	S	D	R	N	1972	71302
1411	S	D	R	N	1972	71302
1412	S	D	R	N	1972	71302
1413	S	D	R	N	1972	71302
1414	S	D	R	N	1972	71302
1418	S	D	E	N	1966	99938
1424	S	D	E	N	1976	85313
1424	S	D	E	N	1976	91690
1441	S	D	E	N	1976	85313
1441	S	D	E	N	1976	91690
1442	S	D	R	N	1975	91435
1442	S	E	R	N	1975	91435
1443	S	S	E	N	1976	92625
1463	M	G	E	N	1962	30997
1467	S	D	E	N	1958	12174
1540	S	D	E	N	1978	98199
1572	M	D	R	N	1971	44543
1572	M	E	R	N	1971	44543
1572	M	D	R	H	1971	62563
1572	M	T	R	H	1971	62563
1572	M	D	E	H	1971	75013
1572	M	T	E	H	1971	75013
1572	M	D	E	N	1971	75014
1572	M	E	E	N	1971	75014
1640	S	S	E	L	1970	61236
1640	S	S	E	N	1970	61236
1645	S	D	R	N	1978	95686
1645	S	D	E	N	1978	98588
1646	S	D	R	N	1978	95686
1646	S	D	E	N	1978	98588
1647	S	D	R	N	1978	95686
1647	S	D	E	N	1978	98588

Phys. State: **C.** Superconductive; **D.** Doped; **E.** Expanded; **F.** Fibrous or Whisker; **G.** Gas; **I.** Ionized or Plasma; **L.** Liquid; **M.** Multiphase; **P.** Powder or Fine Particle; **S.** Solid; **T.** Thick or Thin Film

Subject: **D.** Data; **E.** Experiment; **G.** General (Data + Theory + Experiment); **S.** Survey (Review, Compendium etc.); **T.** Theory

Language: **C.** Czech; **D.** Dutch; **E.** English; **F.** French; **G.** German; **I.** Italian; **J.** Japanese; **O.** Other Languages; **P.** Polish; **R.** Russian; **S.** Spanish

Temperature: **F.** Full Range (Low + Normal + High); **L.** Low (0 to 75K) + Overlap into Normal; **N.** Normal (75 to 1273K); **H.** High (above 1273K) + Overlap into Normal; Blank = Not Coded

Chapter 4 Viscosity

Substance Number	Phys. State	Subject	Language	Temperature	Year	TPRC Number
521-0009	M	D	E	N	1953	13720
0019	P	D	J		1958	9789
0026	L	S	E		1934	9970
0026	L	G	E	H	1939	10288
0026	L	D	E	H	1960	17346
0027	L	S	E		1934	9970
0027	L	D	E	H	1936	14067
0028	L	S	E		1934	9970
0028	L	D	G	N	1957	11205
0028	L	D	E	H	1934	16499
0028	L	D	E	H	1973	70488
0031	L	S	E		1934	9970
0031	L	D	E	H	1934	16499
0035	L	S	E		1934	9970
0035	L	G	E	H	1939	10288
0035	L	D	E	H	1960	17346
0036	L	S	E		1934	9970
0036	L	G	E	H	1939	10288
0040	L	G	R		1945	2094
0040	L	G	R	H	1959	16489
0040	L	D	R	H	1960	17278
0040	L	D	R	H	1939	28724
0040	L	G	E	H	1960	32967
0040	M	T	E		1947	2089
0040	M	D	O		1939	3300
0040	M	D	R	N	1934	15805
0040	M	T	R	H	1958	16452
0046	L	D	G	H	1935	15692
0046	L	S	E	H	1966	42284
0053	L	T	E		1950	4688
0053	L	D	E		1954	5926
0053	L	D	E	N	1932	14411
0053	L	D	J	N	1951	14655
0053	L	T	J	N	1951	14655
0053	L	D	E	N	1936	15420
0053	L	D	F	H	1959	24186
0053	L	D	E	N	1955	28151
0053	L	D	E	H	1967	40652
0053	L	D	E	H	1968	52311
0053	L	D	E	H	1970	58201
0053	L	D	E	H	1973	70497
0057	L	S	E		1934	9970
0057	L	G	E	H	1939	10288
0057	L	D	E	H	1960	17346
0066	L	D	G	N	1937	11777
0066	L	D	R	N	1937	11885
0066	L	D	R	N	1937	11886
0066	L	D	R	N	1937	11887
0066	M	T	R		1949	2507
0066	M	D	G	N	1935	15693
0066	M	D	R	N	1946	20985
0069	M	D	J		1951	3118
0069	M	D	E		1946	8882
0069	M	G	G	N	1956	12879
0069	M	D	R	N	1950	29210
0069	P	D	E	N	1938	14907
0069	P	D	E	N	1958	17994
0072	L	D	E	H	1973	70719
0072	L	E	E	H	1973	70719
0073	L	G	R	H	1962	31079
0073	L	D	R	H	1966	44611
0084	L	D	R	H	1939	28724
0108	L	D	J		1952	2172
0123	L	G	R		1956	6002
0123	L	D	G	N	1957	11205
0123	L	D	E	H	1936	14067
0123	L	D	J	N	1934	15850
0123	L	D	R	H	1934	16925
0123	L	D	O	H	1962	27719
0123	L	D	R	N	1967	44836
0123	L	D	R	H	1968	54552
0123	L	D	R	H	1969	55142
0123	L	D	O	H	1970	57536
0123	L	D	E	H	1971	59464
0123	L	D	E	H	1968	60513
0123	L	D	E	H	1973	70719
0123	L	E	E	H	1973	70719
0123	L	D	E	H	1975	82018
0135	L	G	R		1945	2094
0135	L	D	R		1950	6185
0145	L	D	I	H	1968	56973
0145	L	D	E	H	1976	88455
0145	L	E	E	H	1976	88455
0153	L	D	E	H	1960	17346
0153	M	D	J	H	1935	14263
0170	L	D	E	H	1959	20249
0180	M	D	E	N	1955	12036
0184	L	D	E	H	1950	11540
0196	L	D	E	H	1973	70719
0196	L	E	E	H	1973	70719
0197	L	D	E	H	1946	10291
0224	L	D	G	N	1957	11205
0225	L	D	E	H	1939	13347
0226	L	D	J	N	1953	13281
0227	L	D	G	N	1957	11205
521-0227	L	D	E	H	1936	14067
0227	L	D	R	H	1934	16925
0228	L	D	G	N	1957	11205
0229	L	D	G	N	1957	11205
0230	L	D	G	N	1957	11205
0231	L	D	G	N	1957	11205
0232	L	D	G	N	1957	11205
0233	L	D	G	N	1957	11205
0234	L	D	J	N	1953	13281
0237	L	D	E	H	1939	13347
0239	L	D	E	H	1939	13347
0241	L	D	J	H	1962	33108
0242	L	D	E	H	1973	70719
0242	L	E	E	H	1973	70719
0243	L	D	E	H	1936	14067
0244	L	D	E	H	1936	14067
0249	L	D	E	N	1958	17866
0249	M	D	E	N	1958	17084
0262	L	D	J	H	1953	16133
0264	L	D	R	N	1967	44836
0264	L	D	O	H	1970	57536
0264	L	D	E	H	1971	59464
0264	L	D	E	H	1973	70719
0264	L	E	E	H	1973	70719
0269	L	D	E	H	1934	16499
0272	L	D	E	H	1936	14067
0272	L	G	E	N	1963	32896
0272	L	D	E	H	1973	70719
0272	L	E	E	H	1973	70719
0274	L	D	R	H	1968	54552
0274	L	D	E	H	1968	60513
0306	L	D	R	H	1934	16925
0307	L	D	R	H	1934	16925
0307	L	G	E	H	1963	32894
0308	L	D	R	H	1934	16925
0308	L	D	O	H	1962	27719
0308	L	D	R	H	1968	54552
0308	L	D	E	H	1968	60513
0330	L	D	E	H	1973	70488
0330	L	D	E	H	1976	88455
0330	L	E	E	H	1976	88455
0330	L	D	E	H	1978	98666
0332	L	D	E	H	1959	20249
0344	L	D	E	H	1959	20249
0369	L	D	E	H	1975	82018
0371	L	D	R	H	1939	28724
0385	L	G	R	H	1961	27609
0385	L	G	O	N	1961	27888
0391	M	D	R	N	1946	20985
0393	L	D	R	H	1966	42041
0393	L	E	R	H	1966	42041
0394	L	D	J		1957	2932
0499	L	D	R	H	1939	28724
0624	G	S	E	F	1962	31279
0641	L	D	R	N	1966	41349
0641	L	T	R	N	1966	41349
0641	L	D	E	N	1967	41350
0641	L	T	E	N	1967	41350
0732	L	G	E		1956	3880
0732	L	D	E		1940	3956
0732	L	D	E	H	1955	12071
0732	L	D	E	H	1960	17229
0732	L	D	E	H	1958	25498
0732	L	T	E	H	1963	26274
0732	L	T	R	H	1963	26275
0732	L	D	E	H	1963	30807
0732	L	T	E	H	1963	30807
0732	L	D	E	H	1962	31216
0732	L	D	F	H	1964	32113
0732	M	D	R		1941	3420
0732	M	D	E		1949	3989
0732	M	D	E	H	1936	12875
0732	M	D	E	H	1962	26777
0732	M	S	E	H	1962	26777
0797	L	D	E	N	1972	44792
0797	L	T	E	N	1972	44792
0797	L	D	E	N	1966	47203
0797	L	D	E	H	1969	64934
0797	L	T	E	H	1969	64934
0917	L	D	E	H	1972	70285
0917	L	E	E	H	1972	70285
0918	L	D	E	H	1968	52804
0918	L	T	E	H	1968	52804
0918	L	D	E	N	1969	54213
0918	L	T	E	N	1969	54213
0918	L	D	E	N	1969	54984
0918	L	T	E	N	1969	54984
0918	L	T	E	N	1972	73335
0918	L	S	E	N	1972	77778
0978	L	D	E	H	1973	70488
1123	L	D	E	H	1969	64934
1123	L	T	E	H	1969	64934
1124	L	D	E	H	1969	64934
1124	L	T	E	H	1969	64934
1125	L	D	E	H	1969	64934
521-1125	L	T	E	H	1969	64934
1125	L	D	E	N	1973	73330
1235	L	D	E	H	1973	70488
1237	L	D	E	H	1973	70488
1332	G	T	E	H	1979	99817
1487	L	D	E	H	1976	88455
1487	L	E	E	H	1976	88455
1603	L	T	E	H	1977	90372

Phys. State: **C.** Superconductive; **D.** Doped; **E.** Expanded; **F.** Fibrous or Whisker; **G.** Gas; **I.** Ionized or Plasma; **L.** Liquid; **M.** Multiphase; **P.** Powder or Fine Particle; **S.** Solid; **T.** Thick or Thin Film

Subject: **D.** Data; **E.** Experiment; **G.** General (Data + Theory + Experiment); **S.** Survey (Review, Compendium etc.); **T.** Theory

Language: **C.** Czech; **D.** Dutch; **E.** English; **F.** French; **G.** German; **I.** Italian; **J.** Japanese; **O.** Other Languages; **P.** Polish; **R.** Russian; **S.** Spanish

Temperature: **F.** Full Range (Low + Normal + High); **L.** Low (0 to 75K) + Overlap into Normal; **N.** Normal (75 to 1273K); **H.** High (above 1273K) + Overlap into Normal; Blank = Not Coded

Chapter 5 Emittance

Substance Number	Phys. State	Subject	Language	Temperature	Year	TPRC Number
503-0012	S	S	E	N	1960	24931
0012	S	S	E	L	1972	66579
0012	S	S	E	N	1972	66579
0018	S	D	E	N	1960	76856
0019	S	S	E	N	1972	66579
0025	S	D	E	N	1968	50191
0025	S	D	E	N	1977	97520
0026	S	D	E	N	1972	66579
0066	S	D	R	N	1968	50642
0082	F	D	E	N	1966	40324
0082	F	D	E	N	1958	44273
0082	F	T	E	N	1958	44273
0082	S	D	G	N	1935	20382
0082	S	E	G	N	1935	20382
0082	S	G	E	N	1964	35856
0082	S	D	G	N	1964	36717
0082	S	T	G	N	1964	36717
0082	S	S	E	N	1930	40619
0082	S	D	E	N	1965	42764
0082	S	D	E	N	1965	42781
0082	S	D	E	N	1958	44273
0082	S	T	E	N	1958	44273
0082	S	D	E	N	1960	50317
0082	S	S	E	N	1959	52098
0082	S	D	E	N	1968	53616
0082	S	S	E	N	1951	98632
0082	S	S	E	H	1979	98998
0216	S	S	E	N	1960	24931
0216	S	D	E	N	1965	36738
0216	S	S	E	N	1972	66579
0240	S	S	E	N	1972	66579
0270	S	S	E	N	1972	66579
0284	S	S	E	H	1960	24931
0284	S	S	E	N	1960	24931
0284	S	S	E	F	1972	66579
0284	T	S	E	H	1972	70629
0376	D	D	R	N	1958	19646
0376	D	D	E	N	1959	21015
0376	S	D	R	N	1958	19646
0376	S	D	E	N	1959	21015
0452	S	S	E	N	1972	66579
0461	S	G	R	H	1971	64649
0461	S	G	E	H	1971	64650
0503	S	S	E	N	1951	98632
0537	S	S	E	N	1960	24931
0537	S	S	E	N	1972	66579
0545	S	S	E	N	1960	24931
0580	S	S	E	N	1972	66579
0600	S	D	R	N	1968	50642
0604	S	S	E	N	1960	24931
0604	S	S	E	N	1972	66579
0631	S	S	E	N	1972	66579
0645	T	S	E	H	1972	70629
0645	T	S	E	N	1972	70629
0651	T	S	E	H	1972	70629
0651	T	S	E	N	1972	70629
0652	S	S	E	N	1972	66579
0659	S	S	E	H	1960	24932
0659	S	S	E	H	1972	66579
0659	S	S	E	N	1972	66579
0659	S	S	E	H	1976	83746
0659	S	S	E	N	1976	83746
0660	S	S	E	H	1960	24932
0660	S	S	E	N	1960	24932
0660	S	S	E	H	1972	66579
0660	S	S	E	N	1972	66579
0660	S	S	E	H	1976	83746
0660	S	S	E	N	1976	83746
0667	S	D	E	N	1959	77381
0667	S	E	E	N	1959	77381
0677	T	S	E	N	1960	24932
0677	T	D	E	N	1979	97884
0759	T	S	E	N	1972	70629
0782	S	D	E	N	1964	39347
0801	S	S	E	N	1951	98632
0966	S	D	E	N	1960	15886
0966	S	S	E	N	1972	66579
0967	S	D	E	N	1969	78264
0967	S	E	E	N	1969	78264
0967	S	D	E	N	1976	83746
0967	S	S	E	N	1976	83746
0968	S	D	E	N	1969	78264
0968	S	E	E	N	1969	78264
0974	S	D	E	N	1960	76856
1019	S	D	E	N	1972	71455
1019	S	E	E	N	1972	71455
1050	S	G	E	N	1964	35856
1050	S	D	E	N	1965	42764
1059	D	D	R	N	1972	68812
1059	D	D	E	N	1972	77029
1059	S	D	E	N	1965	36144
1059	S	D	R	N	1972	68812
1059	S	D	E	N	1972	77029
1069	S	S	E	N	1960	24931
1069	S	S	E	N	1972	66579

Substance Number	Phys. State	Subject	Language	Temperature	Year	TPRC Number
503-1081	S	S	E	N	1960	24931
1081	S	S	E	N	1972	66579
1092	S	S	E	N	1960	24932
1092	S	S	E	L	1972	66579
1092	S	S	E	N	1972	66579
1093	S	S	E	N	1972	66579
1094	S	S	E	N	1960	24932
1094	S	S	E	N	1972	66579
1114	S	D	R	N	1968	50642
1131	S	S	E	H	1960	24931
1131	S	S	E	N	1960	24931
1131	S	D	R	H	1972	67899
1131	S	E	R	H	1972	67899
1133	S	S	E	H	1960	24931
1133	S	S	E	N	1960	24931
1133	S	D	F	H	1949	77511
1133	S	D	E	H	1941	90219
1133	S	D	E	N	1941	90219
1135	E	D	E	N	1966	44941
1135	S	D	E	N	1965	36144
1135	S	D	E	H	1965	42165
1139	P	D	R	N	1970	73866
1139	P	D	E	N	1970	73867
1142	E	D	R	H	1967	56880
1142	E	D	R	N	1967	56880
1142	S	D	E	N	1965	36491
1142	S	T	E	N	1965	36491
1142	S	E	E	H	1965	36510
1142	S	D	E	N	1970	37273
1142	S	T	E	N	1970	37273
1142	S	G	E	N	1970	59273
1145	E	D	R	H	1967	56880
1145	E	D	R	N	1967	56880
1145	S	S	E	H	1964	37819
1145	S	S	E	N	1924	44568
1145	S	S	E	N	1959	52098
1145	S	S	E	H	1969	55310
1146	S	D	R	H	1972	67899
1146	S	E	R	H	1972	67899
1147	P	D	R	N	1970	73866
1147	P	D	E	N	1970	73867
1147	S	D	E	H	1965	42165
1147	S	D	R	H	1973	72276
1147	S	T	R	H	1973	72276
1147	S	D	E	H	1973	72277
1147	S	T	E	H	1973	72277
1147	S	D	E	N	1941	90219
1149	S	G	E	N	1974	74606
1153	S	D	F	H	1971	65283
1153	S	E	F	H	1971	65283
1162	S	D	E	H	1965	42165
1173	S	D	E	H	1965	42165
1173	S	G	E	N	1967	46276
1173	S	D	G	N	1969	55722
1173	S	D	G	N	1970	66394
1173	S	E	G	N	1970	66394
1173	S	D	F	H	1949	77511
1173	S	S	E	N	1951	98632
1177	S	D	R	N	1964	34769
1177	S	T	R	N	1964	34769
1177	S	D	E	N	1965	34770
1177	S	T	E	N	1965	34770
1177	S	S	E	N	1975	78777
1184	S	D	E	N	1957	39841
1185	S	D	E	N	1969	59615
1185	S	E	E	N	1969	59615
1189	S	D	E	H	1965	42165
1194	P	D	E	N	1965	34137
1194	P	D	R	N	1970	73866
1194	P	D	E	N	1970	73867
1294	S	S	E	N	1972	70629
1294	S	S	E	N	1951	98632
1357	L	D	E	N	1971	61518
1378	S	S	E	N	1961	17039
1388	S	D	E	N	1962	29555
1388	T	S	E	N	1972	70629
1405	S	D	R	N	1968	50642
1434	S	D	G	N	1969	54365
1466	S	S	E	N	1960	24931
1466	S	S	E	H	1972	66579
1466	S	S	E	N	1972	66579
1529	S	D	E	N	1959	36167
1547	S	S	E	N	1960	24931
1547	S	S	E	L	1972	66579
1547	S	S	E	N	1972	66579
1709	S	S	E	N	1960	24931
1718	S	S	E	N	1960	24931
1723	L	D	R	H	1958	19646
1723	L	D	E	H	1959	21015
1723	S	D	R	N	1958	19646
1723	S	D	E	N	1959	21015
1741	S	S	E	N	1972	66579
1826	S	S	E	H	1977	94788
1826	S	S	E	N	1977	94788
1890	S	S	E	H	1972	66579

Substance Number	Phys. State	Subject	Language	Temperature	Year	TPRC Number
503-1890	S	S	E	N	1972	66579
1945	S	S	E	H	1960	24931
1945	S	S	E	N	1960	24931
1945	S	S	E	F	1972	66579
1989	L	D	R	H	1970	38969
2008	L	D	R	H	1970	38969
2009	L	D	R	H	1970	38969
2010	L	D	R	H	1970	38969
2011	L	D	R	H	1970	38969
2012	L	D	R	H	1970	38969
2013	L	D	R	H	1970	38969
2014	L	D	R	H	1970	38969
2015	L	D	R	H	1970	38969
2016	L	D	R	H	1970	38969
2032	S	D	E	N	1969	56664
2033	S	D	E	N	1969	56664
2034	S	D	E	N	1969	56664
2190	S	S	E	N	1951	98632
2309	T	S	E	H	1972	70629
2309	T	S	E	N	1972	70629
2452	S	D	E	H	1961	18648
2458	L	D	R	H	1958	19646
2458	L	D	E	H	1959	21015
2458	S	D	R	N	1958	19646
2458	S	D	E	N	1959	21015
2459	S	D	R	N	1958	19646
2459	S	D	E	N	1959	21015
2460	L	D	R	H	1958	19646
2460	L	D	E	H	1959	21015
2460	S	D	R	N	1958	19646
2460	S	D	E	N	1959	21015
2501	T	D	E	N	1962	29555
2502	T	D	E	N	1962	29555
2503	T	D	E	N	1962	29555
2504	T	D	E	N	1962	29555
2505	T	D	E	N	1962	29555
2770	S	S	E	H	1972	66579
2770	S	S	E	N	1972	66579
2770	S	S	E	H	1975	81954
2770	S	S	E	N	1975	81954
2770	S	S	E	H	1976	83746
2770	S	S	E	N	1976	83746
2771	S	S	E	H	1976	83746
2771	S	S	E	N	1976	83746
2774	S	S	E	N	1972	66579
2776	S	S	E	N	1972	66579
2781	S	S	E	N	1972	66579
2782	S	S	E	N	1972	66579
2783	S	S	E	N	1960	24931
2793	S	S	E	N	1960	24932
2793	S	S	E	N	1972	66579
2810	T	S	E	N	1972	70629
2811	T	S	E	N	1972	70629
2812	T	S	E	N	1972	70629
2815	T	S	E	N	1972	70629
2816	T	S	E	N	1972	70629
2819	T	S	E	N	1972	70629
2820	T	S	E	N	1972	70629
2821	T	S	E	N	1972	70629
2822	T	S	E	N	1972	70629
2823	T	S	E	N	1972	70629
2824	T	S	E	N	1972	70629
2825	T	S	E	N	1972	70629
2826	T	S	E	N	1972	70629
2827	T	S	E	N	1972	70629
2828	T	S	E	N	1972	70629
2829	T	S	E	N	1972	70629
2859	T	S	E	H	1972	70629
2859	T	S	E	N	1972	70629
2860	S	S	E	H	1972	70629
2860	S	S	E	N	1972	70629
2862	T	S	E	N	1961	17039
2863	T	S	E	N	1961	17039
2864	T	S	E	N	1961	17039
2865	T	S	E	N	1961	17039
2866	T	S	E	N	1961	17039
2867	T	S	E	N	1961	17039
2868	T	S	E	N	1961	17039
2869	T	S	E	N	1961	17039
2870	T	S	E	N	1961	17039
2871	T	S	E	N	1961	17039
2872	T	S	E	N	1961	17039
2873	T	S	E	N	1961	17039
2874	T	S	E	N	1961	17039
2921	S	S	E	N	1960	24931
8228	S	S	E	H	1960	24931
8228	S	S	E	N	1960	24931
8734	S	D	E	N	1966	43484
8871	L	D	E	H	1966	41579
8871	S	D	E	H	1966	41579
9115	S	D	G	N	1969	55722
9159	L	D	E	H	1966	41579
9159	S	D	E	H	1966	41579
9168	S	D	G	N	1970	66394
9168	S	E	G	N	1970	66394

Phys. State: C. Superconductive; D. Doped; E. Expanded; F. Fibrous or Whisker; G. Gas; I. Ionized or Plasma; L. Liquid; M. Multiphase; P. Powder or Fine Particle; S. Solid; T. Thick or Thin Film
Subject: D. Data; E. Experiment; G. General (Data + Theory + Experiment); S. Survey (Review, Compendium etc.); T. Theory
Language: C. Czech; D. Dutch; E. English; F. French; G. German; I. Italian; J. Japanese; O. Other Languages; P. Polish; R. Russian; S. Spanish
Temperature: F. Full Range (Low + Normal + High); L. Low (0 to 75K) + Overlap into Normal; N. Normal (75 to 1273K); H. High (above 1273K) + Overlap into Normal; Blank = Not Coded

Substance Number	Phys. State	Subject	Language	Temperature	Year	TPRC Number
503-9168	S	G	G	N	1970	66395
9173	P	G	E	N	1971	64971
9173	S	D	E	N	1960	40738
9173	S	D	G	N	1970	66394
9173	S	E	G	N	1970	66394
9178	S	D	G	N	1969	55722
9512	S	D	E	H	1966	41579
9513	S	D	E	H	1966	41579
9589	S	D	E	H	1969	52936
9628	S	D	J	H	1967	53697
9633	P	G	G	H	1932	40644
9707	S	D	E	H	1967	48048
9714	S	D	E	H	1966	48361
9716	S	D	F	H	1970	60595
9716	S	D	R	H	1969	61089
9716	S	E	R	H	1969	61089
9716	S	D	R	H	1969	61090
9716	S	E	R	H	1969	61090
9716	S	D	E	H	1971	67864
9716	S	E	E	H	1971	67864
9716	S	D	E	H	1971	67865
9716	S	E	E	H	1971	67865
9717	S	D	E	H	1966	48361
9747	S	D	E	N	1965	34835
9747	S	T	E	N	1965	34835
9747	S	D	E	N	1966	48364
9748	P	G	G	H	1932	40644
9748	S	D	E	N	1965	34835
9748	S	T	E	N	1965	34835
9748	S	G	G	H	1934	46996
9748	S	D	E	N	1966	48364
9748	S	D	G	N	1970	66394
9748	S	E	G	N	1970	66394
9749	S	D	E	N	1965	34835
9749	S	T	E	N	1965	34835
9749	S	D	E	N	1966	48364
9750	P	G	G	H	1932	40644
9750	S	D	E	N	1965	34835
9750	S	T	E	N	1965	34835
9750	S	D	E	N	1965	36510
9750	S	D	E	N	1959	47023
9750	S	D	G	N	1970	66394
9750	S	E	G	N	1970	66394
9752	S	S	E	N	1972	66579
9791	S	D	E	N	1965	34835
9791	S	T	E	N	1965	34835
9791	S	D	E	N	1965	36510
9795	S	D	E	H	1964	47248
9808	P	G	G	H	1932	40644
9811	T	D	E	N	1962	29555
9823	S	G	E	N	1964	37424
9838	L	D	J	H	1965	36598
9882	S	G	G	H	1934	46996
9883	S	G	G	H	1934	46996
9891	S	D	E	N	1965	36510
9916	P	G	G	H	1932	40644
9916	P	G	E	N	1971	64971
9916	P	D	E	N	1973	71769
9916	S	D	E	N	1965	34835
9916	S	T	E	N	1965	34835
9916	S	D	E	N	1965	36510
9916	S	D	E	N	1966	48364
9916	S	D	F	H	1971	65283
9916	S	E	F	H	1971	65283
9916	S	D	F	H	1949	77511
9922	L	D	J	H	1965	36598
9922	S	D	E	N	1965	34835
9922	S	T	E	N	1965	34835
9922	S	D	E	H	1965	36510
9922	S	D	E	H	1965	36511
9922	S	D	E	H	1965	37899
9922	S	D	E	H	1965	46471
9922	S	D	E	H	1966	48361
9960	P	G	G	H	1932	40644
9960	S	D	E	H	1967	44823
9988	L	D	R	H	1969	54727
9988	S	G	E	H	1969	56348
9994	P	D	E	H	1966	41579
9994	S	D	E	H	1966	41579
9996	L	D	R	H	1969	54727
521-0001	S	D	E	N	1965	42764
0003	P	G	E	N	1971	64971
0003	P	D	E	N	1973	71769
0004	P	D	E	N	1970	61715
0004	P	D	E	N	1970	61726
0004	S	D	E	N	1965	35595
0004	S	D	E	N	1964	37490
0004	S	D	E	N	1966	43498
0004	S	D	E	N	1967	54965
0004	S	D	E	N	1972	71749
0009	P	D	R	N	1964	34769
0009	P	T	R	N	1964	34769
0009	P	D	E	N	1965	34770
0009	P	T	E	N	1965	34770
0009	P	S	E	N	1968	53122
0009	P	D	E	N	1970	61710
0009	S	D	E	N	1969	34391
0009	S	D	R	N	1969	46008
0009	S	D	E	N	1967	53986
0009	S	D	E	N	1977	96185

Substance Number	Phys. State	Subject	Language	Temperature	Year	TPRC Number
521-0010	S	D	E	N	1964	37490
0015	P	G	E	H	1965	35669
0027	L	D	R	H	1978	96807
0027	L	D	R	N	1978	96807
0027	L	D	E	H	1978	98853
0027	L	D	E	N	1978	98853
0028	S	D	E	N	1966	43498
0031	S	D	E	N	1964	37490
0046	S	D	E	N	1965	35595
0046	S	D	E	N	1967	47283
0046	S	S	E	N	1971	69328
0053	P	D	E	N	1968	49290
0055	F	D	E	N	1972	78278
0055	F	E	E	N	1972	78278
0057	S	D	E	N	1967	47282
0059	S	D	E	N	1964	37490
0064	S	D	E	N	1964	37490
0064	S	D	E	L	1966	40450
0064	S	G	E	L	1966	48360
0064	S	S	E	N	1958	85031
0064	S	D	E	N	1959	85032
0066	S	D	E	N	1969	34391
0066	S	D	R	N	1969	46008
0072	P	D	E	N	1964	37490
0072	P	D	E	N	1970	61726
0072	P	D	E	N	1973	77753
0072	S	D	E	N	1965	34138
0072	S	D	E	N	1969	34391
0072	S	D	E	N	1965	35414
0072	S	D	E	N	1965	35595
0072	S	G	E	N	1964	35856
0072	S	D	E	N	1964	37490
0072	S	D	E	N	1965	42764
0072	S	D	R	N	1969	46008
0072	S	D	E	N	1972	71749
0073	S	D	E	N	1972	71749
0075	S	D	E	N	1965	34138
0075	S	D	E	N	1965	35414
0075	S	G	E	N	1964	35856
0075	S	D	E	N	1964	37490
0075	S	D	E	N	1965	42764
0075	S	D	E	N	1971	62718
0075	S	D	E	N	1972	67055
0075	S	T	E	N	1972	67055
0077	P	G	E	N	1971	64971
0077	S	D	G	N	1935	20382
0077	S	E	G	N	1935	20382
0083	S	D	E	N	1958	44273
0083	S	T	E	N	1958	44273
0084	S	D	E	N	1965	34138
0084	S	D	E	N	1969	34391
0084	S	D	E	N	1965	35414
0084	S	G	E	N	1964	35856
0084	S	D	E	N	1964	37490
0084	S	D	E	N	1965	42764
0084	S	D	R	N	1969	46008
0091	S	D	E	H	1908	23456
0103	S	D	E	N	1965	34138
0103	S	D	E	N	1965	35414
0103	S	G	E	N	1964	35856
0103	S	D	E	N	1965	42764
0103	S	D	F	N	1976	90381
0103	S	T	F	N	1976	90381
0103	S	S	E	N	1951	98632
0105	S	D	E	N	1964	37490
0108	S	D	E	N	1958	44273
0108	S	T	E	N	1958	44273
0109	S	D	E	N	1967	47282
0110	P	D	R	N	1964	34769
0110	P	T	R	N	1964	34769
0110	P	D	E	N	1965	34770
0110	P	T	E	N	1965	34770
0110	P	D	E	N	1964	37490
0110	P	D	E	N	1972	77037
0110	S	D	E	N	1965	34138
0110	S	D	E	N	1969	34391
0110	S	D	E	N	1965	35414
0110	S	G	E	N	1964	35856
0110	S	D	E	N	1965	42764
0110	S	D	E	N	1966	43498
0110	S	D	R	N	1969	46008
0110	S	D	E	N	1967	52863
0123	L	D	R	H	1978	96807
0123	L	D	R	N	1978	96807
0123	L	D	E	H	1978	98853
0123	L	D	E	N	1978	98853
0123	P	D	E	N	1964	38646
0123	P	D	E	N	1969	53645
0123	P	T	E	N	1969	53645
0123	P	D	E	N	1969	54820
0123	P	D	E	N	1970	61715
0123	P	D	E	N	1970	61726
0123	P	D	E	N	1973	77753
0123	S	D	E	N	1965	34138
0123	S	D	E	N	1969	34391
0123	S	D	E	N	1965	35414
0123	S	D	E	N	1965	35595
0123	S	G	E	N	1964	35856
0123	S	D	E	N	1964	37490
0123	S	D	E	N	1965	42764
0123	S	D	R	N	1969	46008

Substance Number	Phys. State	Subject	Language	Temperature	Year	TPRC Number
521-0123	S	D	E	N	1969	53645
0123	S	T	E	N	1969	53645
0123	S	D	E	N	1969	54820
0123	S	D	E	N	1967	54965
0123	S	D	E	N	1972	71749
0132	S	S	E	N	1958	85031
0170	S	D	E	N	1964	37490
0181	S	D	E	N	1972	77037
0183	P	D	E	N	1965	34137
0184	P	D	E	N	1970	61715
0184	P	D	E	N	1970	61726
0184	S	D	E	N	1964	37490
0184	S	D	E	N	1972	71749
0191	S	E	R	N	1967	47158
0191	S	E	E	N	1967	47159
0196	S	D	E	N	1965	35414
0196	S	G	E	N	1964	35856
0196	S	D	E	N	1965	42764
0225	S	D	E	N	1965	35595
0225	S	D	E	N	1964	37490
0228	S	D	E	N	1964	37490
0236	P	D	E	N	1970	61715
0236	P	D	E	N	1970	61726
0236	S	D	E	N	1964	37490
0236	S	D	E	N	1972	77037
0236	S	S	E	N	1951	98632
0238	M	D	E	N	1967	47506
0239	S	D	E	N	1965	35595
0239	S	D	E	N	1967	47283
0240	S	D	E	N	1967	47261
0242	P	D	E	N	1964	38646
0242	S	D	E	N	1972	71749
0243	S	D	E	N	1964	37490
0250	S	S	E	N	1972	66579
0258	P	D	E	N	1970	61726
0258	P	D	E	N	1972	64451
0258	S	D	E	N	1967	47282
0259	S	D	E	N	1969	54243
0259	S	D	E	N	1970	59098
0259	S	S	E	N	1970	59662
0259	S	D	E	N	1971	68434
0259	S	D	E	N	1972	69104
0259	S	S	E	N	1973	73308
0264	P	D	E	N	1970	61715
0264	P	D	E	N	1970	61726
0264	S	D	E	N	1972	71749
0265	S	D	E	N	1965	35595
0270	S	D	E	N	1964	37490
0272	P	D	E	N	1973	77753
0272	S	D	E	N	1966	43498
0272	S	D	E	N	1967	54965
0272	S	D	E	N	1972	71749
0279	S	D	E	N	1964	37490
0293	S	D	E	N	1965	35595
0293	S	D	E	N	1972	71749
0299	S	S	E	N	1968	53122
0299	S	S	E	N	1971	69328
0307	P	D	E	N	1964	38646
0307	S	D	E	N	1965	34138
0307	S	D	E	N	1969	34391
0307	S	D	E	N	1965	35414
0307	S	D	E	N	1965	35595
0307	S	G	E	N	1964	35856
0307	S	D	E	N	1964	37490
0307	S	D	E	N	1965	42764
0307	S	D	E	N	1966	43498
0307	S	D	R	N	1969	46008
0308	S	D	E	N	1964	37490
0308	S	D	E	N	1972	71749
0330	P	D	E	N	1964	38646
0330	S	D	E	N	1964	37490
0330	S	D	E	N	1967	54965
0330	S	D	E	N	1972	71749
0348	S	D	E	N	1972	71749
0352	P	D	E	N	1964	37490
0352	S	D	E	N	1965	34138
0352	S	D	E	N	1969	34391
0352	S	D	E	N	1965	35414
0352	S	D	E	N	1965	35595
0352	S	G	E	N	1964	35856
0352	S	D	E	N	1964	37490
0352	S	D	E	N	1965	42764
0352	S	D	R	N	1969	46008
0354	S	D	E	N	1966	43498
0354	S	D	E	N	1967	54965
0356	S	D	E	N	1972	71749
0368	M	D	E	N	1970	61710
0368	P	D	E	N	1970	61710
0369	S	D	E	N	1965	35595
0369	S	D	E	N	1964	37490
0369	S	D	E	N	1966	43498
0369	S	D	E	N	1967	54965
0393	S	D	G	N	1935	20382
0393	S	E	G	N	1935	20382
0394	S	D	E	N	1965	34138
0394	S	D	E	N	1969	34391
0394	S	D	E	N	1965	35414
0394	S	D	E	N	1965	35595
0394	S	G	E	N	1964	35856
0394	S	D	E	N	1965	42764
0394	S	D	R	N	1969	46008

Phys. State: **C.** Superconductive; **D.** Doped; **E.** Expanded; **F.** Fibrous or Whisker; **G.** Gas; **I.** Ionized or Plasma; **L.** Liquid; **M.** Multiphase; **P.** Powder or Fine Particle; **S.** Solid; **T.** Thick or Thin Film

Subject: **D.** Data; **E.** Experiment; **G.** General (Data + Theory + Experiment); **S.** Survey (Review, Compendium etc.); **T.** Theory

Language: **C.** Czech; **D.** Dutch; **E.** English; **F.** French; **G.** German; **I.** Italian; **J.** Japanese; **O.** Other Languages; **P.** Polish; **R.** Russian; **S.** Spanish

Temperature: **F.** Full Range (Low + Normal + High); **L.** Low (0 to 75K) + Overlap into Normal; **N.** Normal (75 to 1273K); **H.** High (above 1273K) + Overlap into Normal; Blank = Not Coded

Substance Number	Phys. State	Sub-ject	Lan-guage	Temper-ature	Year	TPRC Number
521-0394	S	D	E	N	1967	47283
0413	S	D	E	N	1970	61193
0421	S	D	E	N	1974	79799
0421	S	T	E	N	1974	79799
0423	S	G	E	N	1964	35856
0423	S	D	E	N	1965	42764
0423	S	D	E	N	1967	46040
0423	S	S	R	N	1969	57577
0496	S	D	E	N	1964	37490
0534	S	D	E	N	1964	37490
0613	S	D	E	N	1968	49669
0648	S	S	E	N	1972	66579
0732	P	D	E	N	1964	37490
0732	P	D	E	N	1964	38646
0732	P	D	E	N	1969	53645
0732	P	T	E	N	1969	53645
0732	P	D	E	N	1972	64451
0732	P	G	E	N	1971	64971
0732	S	D	E	N	1969	34391
0732	S	D	E	N	1965	35414
0732	S	D	E	N	1965	35595
0732	S	D	E	N	1964	35796
0732	S	G	E	N	1964	35856
0732	S	D	E	N	1959	36167
0732	S	D	E	N	1964	37490
0732	S	D	E	N	1965	42764
0732	S	D	R	N	1969	46008
0732	S	D	E	N	1967	47282
0732	S	D	E	N	1967	47283
0732	S	D	E	N	1969	53645
0732	S	T	E	N	1969	53645
0732	S	D	E	N	1972	67055
0732	S	T	E	N	1972	67055
0732	S	D	E	N	1978	100061
0735	S	D	G	N	1935	20382
0735	S	E	G	N	1935	20382
0762	S	D	E	N	1964	37490
0767	P	D	E	N	1968	49290
0794	P	D	E	N	1964	38646
0852	P	D	E	N	1970	61726
0852	P	D	E	N	1973	77753
0852	S	D	E	N	1965	35595
0852	S	D	E	N	1964	37490
0852	S	D	E	N	1972	71749
0866	M	D	E	N	1970	61710
0866	P	D	E	N	1970	61710
0887	P	D	E	N	1964	38646
0887	S	D	E	N	1964	37490
0887	S	D	E	N	1965	39333
0890	S	D	E	N	1964	37490
0892	S	D	E	N	1964	37490
0893	S	D	E	N	1964	37490
0894	S	D	E	N	1964	37490
0895	P	D	E	N	1973	77753
0895	S	D	E	N	1964	37490
0895	S	D	E	N	1972	71749
0896	S	D	E	N	1964	37490
0897	S	D	E	N	1964	37490
0898	S	D	E	N	1964	37490
0899	S	D	E	N	1964	37490
0900	S	D	E	N	1964	37490
0901	S	D	E	N	1964	37490
0902	S	D	E	N	1964	37490
0903	S	D	E	N	1964	37490
0904	S	D	E	N	1964	37490
0905	S	D	E	N	1964	37490
0906	S	D	E	N	1964	37490
0907	S	D	E	N	1964	37490
0908	S	D	E	N	1965	35414
0917	P	D	E	N	1972	66830
0917	P	D	E	N	1972	68484
0917	P	E	E	N	1972	68484
0917	P	D	E	N	1972	68911
0917	P	E	E	N	1972	68911
0917	P	D	E	N	1972	70817
0917	P	D	E	N	1973	73329
0917	P	D	E	N	1974	78683
0917	S	D	E	N	1972	65323
0917	S	E	E	N	1972	65323
0917	S	D	E	N	1972	71749
0924	S	D	E	N	1965	35595
0924	S	D	E	N	1972	71749
0925	S	D	E	N	1965	35595
0951	P	D	E	N	1969	53645
0951	P	T	E	N	1969	53645
0951	P	D	E	N	1970	59099
0951	P	D	E	N	1970	61715
0951	P	D	E	N	1970	61726
0951	P	D	E	N	1972	64451
0951	S	S	E	N	1970	59662
0972	S	G	E	N	1964	35856
0972	S	D	E	N	1965	42764
1011	S	D	E	N	1972	71749
1047	P	D	E	N	1972	64451
1219	M	D	E	N	1971	63703
1219	M	T	E	N	1971	63703
1278	S	D	E	N	1972	71749
1418	S	S	E	H	1960	24931
1418	S	S	E	N	1960	24931
1418	S	S	E	N	1972	66579
1559	L	D	R	H	1978	96807
521-1559	L	D	R	N	1978	96807
1559	L	D	E	H	1978	98853
1559	L	D	E	N	1978	98853
1654	M	D	E	N	1977	96185

Phys. State: **C.** Superconductive; **D.** Doped; **E.** Expanded; **F.** Fibrous or Whisker; **G.** Gas; **I.** Ionized or Plasma; **L.** Liquid; **M.** Multiphase; **P.** Powder or Fine Particle; **S.** Solid; **T.** Thick or Thin Film
Subject: **D.** Data; **E.** Experiment; **G.** General (Data + Theory + Experiment); **S.** Survey (Review, Compendium etc.); **T.** Theory
Language: **C.** Czech; **D.** Dutch; **E.** English; **F.** French; **G.** German; **I.** Italian; **J.** Japanese; **O.** Other Languages; **P.** Polish; **R.** Russian; **S.** Spanish
Temperature: **F.** Full Range (Low + Normal + High); **L.** Low (0 to 75K) + Overlap into Normal; **N.** Normal (75 to 1273K); **H.** High (above 1273K) + Overlap into Normal; Blank = Not Coded

Chapter 6 Reflectance

Substance Number	Phys. State	Subject	Language	Temperature	Year	TPRC Number
503-0001	S	D	E	N	1976	88047
0012	S	S	E	N	1960	24931
0012	S	D	E	N	1965	36144
0012	S	D	E	N	1971	66243
0012	S	S	E	N	1972	66579
0012	S	S	E	N	1980	100404
0019	S	S	E	N	1960	24931
0019	S	S	E	N	1972	66579
0020	S	S	E	N	1960	24931
0022	S	D	E	N	1972	77770
0024	S	D	E	N	1961	45736
0024	S	T	E	N	1961	45736
0024	S	D	F	N	1960	47249
0025	S	D	R	N	1965	35610
0025	S	D	F	N	1965	38592
0025	S	D	E	N	1966	40908
0025	S	D	R	N	1967	48320
0025	S	D	E	N	1977	97519
0031	S	D	F	H	1966	39292
0031	S	D	E	N	1967	42891
0033	P	D	E	N	1971	62547
0033	P	D	E	N	1975	76621
0033	P	T	E	N	1975	76621
0033	S	D	E	N	1930	73996
0033	S	E	E	N	1930	73996
0047	S	D	R	N	1976	88951
0047	S	D	E	N	1976	88952
0080	S	G	E	N	1972	78256
0080	S	D	E	N	1977	88941
0082	F	D	E	N	1968	52892
0082	F	D	E	N	1965	54294
0082	F	T	E	N	1965	54294
0082	P	D	E	N	1970	56104
0082	P	G	E	N	1971	63502
0082	P	D	E	N	1972	69102
0082	P	E	E	N	1972	69102
0082	P	S	E	N	1972	71106
0082	S	D	E	N	1964	34325
0082	S	D	E	N	1957	39842
0082	S	D	E	N	1966	40528
0082	S	D	E	N	1966	44875
0082	S	D	E	N	1961	45929
0082	S	T	E	N	1961	45929
0082	S	D	E	N	1966	52900
0082	S	D	E	N	1965	54294
0082	S	T	E	N	1965	54294
0082	S	D	E	N	1953	55041
0082	S	D	I	N	1969	55533
0082	S	T	I	N	1969	55533
0082	S	D	E	N	1969	57031
0082	S	D	E	N	1971	60669
0082	S	T	E	N	1971	60669
0082	S	D	E	N	1971	61361
0082	S	G	E	N	1971	61369
0082	S	D	E	L	1971	61457
0082	S	E	E	L	1971	61457
0082	S	D	E	N	1970	61942
0082	S	D	E	N	1972	67441
0082	S	D	R	N	1971	67728
0082	S	T	R	N	1971	67728
0082	S	D	E	N	1972	69253
0082	S	E	E	N	1972	69253
0082	S	S	E	N	1972	70629
0082	S	G	E	N	1973	72581
0082	S	D	E	N	1972	76803
0082	S	D	E	N	1971	77338
0082	S	D	E	N	1967	77362
0082	S	G	E	N	1972	77502
0082	S	D	E	N	1976	88022
0082	S	T	E	N	1976	88022
0082	S	S	G	N	1977	92758
0082	S	D	E	N	1919	95768
0082	S	D	E	N	1978	99888
0082	S	E	E	N	1978	99888
0082	S	D	E	N	1980	100601
0090	S	D	R	N	1972	69642
0090	S	D	E	N	1973	69643
0108	P	G	E	N	1973	71027
0108	P	G	E	N	1973	73838
0108	S	G	E	N	1973	71027
0108	S	G	E	N	1973	73838
0216	S	S	E	N	1960	24931
0216	S	S	E	N	1972	66579
0216	S	D	E	N	1977	90330
0240	S	S	E	N	1972	66579
0270	S	D	G	N	1898	46995
0270	S	D	F	N	1960	47249
0270	S	S	E	N	1972	66579
0280	P	G	E	N	1976	83577
0284	S	S	E	H	1960	24931
0284	S	S	E	N	1960	24931
0284	S	S	E	N	1972	66579
0305	S	D	E	N	1971	66243
0479	S	D	R	N	1974	83581
0479	S	D	E	N	1974	83582

Substance Number	Phys. State	Subject	Language	Temperature	Year	TPRC Number
503-0503	S	D	E	N	1969	77603
0503	S	E	E	N	1969	77603
0537	S	S	E	N	1960	24931
0537	S	S	E	N	1972	66579
0545	S	S	E	N	1960	24931
0585	S	D	E	N	1972	77769
0585	S	D	E	N	1972	77770
0588	S	D	E	N	1972	77769
0588	S	D	E	N	1972	77770
0604	S	S	E	N	1960	24931
0604	S	D	E	N	1971	66243
0604	S	S	E	N	1972	66579
0631	S	S	E	N	1972	66579
0659	S	S	E	N	1960	24932
0659	S	S	E	N	1972	66579
0659	S	D	E	N	1976	83746
0659	S	S	E	N	1976	83746
0660	S	S	E	N	1960	24932
0660	S	S	E	N	1972	66579
0660	S	S	E	N	1976	83746
0662	S	D	E	N	1972	77770
0663	S	D	E	N	1972	77770
0663	S	D	J	N	1976	90666
0667	S	D	E	N	1959	77381
0667	S	E	E	N	1959	77381
0677	T	S	E	N	1960	24932
0677	T	D	E	N	1978	96544
0677	T	D	E	N	1979	97884
0729	P	D	E	N	1977	92103
0731	S	D	R	N	1968	55537
0731	S	E	R	N	1968	55537
0746	S	D	G	N	1970	58419
0746	S	S	E	N	1972	70629
0801	S	D	E	N	1966	34814
0801	S	G	E	N	1928	40591
0801	S	D	E	N	1906	96758
0951	P	D	E	N	1964	34833
0966	S	D	E	N	1960	15886
0966	S	S	E	N	1972	66579
0967	S	D	E	N	1976	83746
0967	S	S	E	N	1976	83746
0970	P	D	E	N	1964	34833
0970	P	D	E	N	1975	76621
0970	P	T	E	N	1975	76621
0970	S	D	E	N	1964	34833
0991	S	D	E	N	1906	96758
0993	S	G	E	N	1928	40591
1026	S	D	R	N	1968	55318
1026	S	D	E	N	1972	70720
1037	S	D	R	N	1967	48320
1050	S	D	E	N	1961	33844
1050	S	D	G	N	1898	46995
1050	S	D	F	N	1969	56731
1050	S	E	F	N	1969	56731
1050	S	D	E	N	1970	63038
1050	S	D	E	N	1972	68251
1050	S	E	E	N	1972	68251
1050	S	D	E	N	1972	68853
1050	S	G	E	N	1972	78256
1050	S	D	E	N	1906	96758
1059	S	S	E	N	1960	24931
1059	S	D	E	N	1977	92471
1059	S	D	E	N	1908	96759
1069	S	S	E	N	1960	24931
1069	S	S	E	N	1972	66579
1077	S	D	G	N	1967	51716
1077	S	D	E	N	1969	58395
1077	S	S	E	N	1972	66579
1077	S	D	E	N	1972	67054
1077	S	T	E	N	1972	67054
1077	S	D	R	N	1974	76274
1077	S	D	E	N	1974	77217
1080	S	S	E	N	1960	24931
1081	S	S	E	N	1960	24931
1081	S	S	E	N	1972	66579
1092	S	S	E	N	1960	24932
1092	S	S	E	N	1972	66579
1093	S	S	E	N	1972	66579
1094	S	S	E	N	1960	24932
1094	S	S	E	N	1972	66579
1097	S	D	R	N	1968	55318
1097	S	D	E	N	1972	70720
1104	S	D	E	N	1963	31237
1104	S	D	E	N	1962	36117
1104	S	D	E	N	1965	36348
1104	S	D	E	N	1950	46403
1104	S	G	E	N	1968	47917
1104	S	D	E	N	1968	51403
1104	S	G	G	N	1960	51581
1104	S	D	E	N	1955	78310
1104	S	E	E	N	1955	78310
1104	S	G	E	N	1977	90398
1113	S	D	R	N	1968	50914
1113	S	D	E	N	1968	50915
1113	S	D	E	N	1966	52897

Substance Number	Phys. State	Subject	Language	Temperature	Year	TPRC Number
503-1113	S	D	R	N	1970	59391
1113	S	D	E	N	1970	59392
1114	S	D	R	N	1965	35073
1114	S	D	E	N	1965	36833
1114	S	D	R	N	1966	43360
1114	S	D	R	N	1966	44356
1114	S	D	E	N	1966	60334
1114	S	D	R	N	1971	64907
1114	S	T	R	N	1971	64907
1114	S	D	E	N	1971	64908
1114	S	T	E	N	1971	64908
1121	S	D	E	N	1906	96758
1131	S	S	E	H	1960	24931
1131	S	S	E	N	1960	24931
1133	S	S	E	H	1960	24931
1133	S	S	E	N	1960	24931
1133	S	D	R	N	1974	76274
1133	S	D	E	N	1974	77217
1136	S	D	E	N	1969	52594
1136	S	D	E	N	1967	72670
1136	S	E	E	N	1967	72670
1136	S	D	E	N	1969	77603
1136	S	E	E	N	1969	77603
1142	S	D	E	N	1965	36491
1142	S	T	E	N	1965	36491
1142	S	D	E	N	1967	46650
1142	S	E	E	N	1973	69347
1145	S	D	E	N	1970	60580
1147	S	D	E	N	1963	40420
1147	S	D	G	N	1965	43102
1147	S	D	E	N	1972	68251
1147	S	E	E	N	1972	68251
1147	S	D	E	N	1972	78254
1147	S	E	E	N	1972	78254
1162	S	S	E	H	1972	66579
1177	S	D	E	N	1966	33879
1177	S	D	R	N	1964	34769
1177	S	T	R	N	1964	34769
1177	S	D	E	N	1965	34770
1177	S	T	E	N	1965	34770
1177	S	D	E	N	1964	36100
1177	S	D	E	N	1967	46650
1185	S	D	E	N	1964	37534
1189	S	D	E	N	1970	59646
1189	S	T	E	N	1970	59646
1194	M	D	E	N	1966	41373
1194	P	D	E	N	1954	78248
1194	S	D	E	N	1965	42064
1194	S	D	E	N	1906	96758
1194	S	D	E	N	1908	96759
1246	S	D	E	N	1967	72670
1246	S	E	E	N	1967	72670
1250	D	D	R	N	1966	41087
1250	D	D	E	N	1966	41088
1250	P	G	E	N	1971	63502
1250	S	D	R	N	1966	41087
1250	S	D	E	N	1966	41088
1250	S	D	E	N	1945	46837
1250	S	D	R	N	1968	53880
1250	S	D	E	N	1968	53881
1250	S	D	E	N	1969	54711
1250	S	D	E	N	1972	67054
1250	S	T	E	N	1972	67054
1261	S	D	R	N	1969	56964
1261	S	D	E	N	1971	63018
1278	S	D	E	N	1964	36100
1294	S	S	E	N	1972	70629
1295	S	D	R	N	1970	59460
1295	S	E	R	N	1970	59460
1295	S	D	E	N	1970	60046
1295	S	T	E	N	1970	60046
1295	S	D	E	N	1970	71940
1295	S	E	E	N	1970	71940
1313	D	D	E	N	1971	61329
1313	S	G	E	N	1965	34683
1313	S	D	E	N	1967	39212
1328	S	D	R	N	1968	50006
1328	S	D	E	N	1972	66504
1340	S	D	R	N	1973	75634
1340	S	D	E	N	1973	75635
1390	S	D	E	N	1966	33854
1390	S	D	E	N	1977	92045
1397	S	D	G	N	1952	43597
1405	S	D	R	N	1967	48320
1405	S	D	R	N	1976	87213
1405	S	E	R	N	1976	87213
1405	S	D	E	N	1976	87214
1405	S	E	E	N	1976	87214
1411	L	D	E	N	1908	96759
1440	S	G	G	N	1966	45348
1444	S	D	R	N	1968	55537
1444	S	E	R	N	1968	55537
1447	S	D	E	N	1972	68853
1449	P	D	E	N	1972	70035
1450	P	D	E	N	1972	70035

Phys. State: **C.** Superconductive; **D.** Doped; **E.** Expanded; **F.** Fibrous or Whisker; **G.** Gas; **I.** Ionized or Plasma; **L.** Liquid; **M.** Multiphase; **P.** Powder or Fine Particle; **S.** Solid; **T.** Thick or Thin Film
Subject: **D.** Data; **E.** Experiment; **G.** General (Data + Theory + Experiment); **S.** Survey (Review, Compendium etc.); **T.** Theory
Language: **C.** Czech; **D.** Dutch; **E.** English; **F.** French; **G.** German; **I.** Italian; **J.** Japanese; **O.** Other Languages; **P.** Polish; **R.** Russian; **S.** Spanish
Temperature: **F.** Full Range (Low + Normal + High); **L.** Low (0 to 75K) + Overlap into Normal; **N.** Normal (75 to 1273K); **H.** High (above 1273K) + Overlap into Normal; Blank = Not Coded

Substance Number	Phys. State	Subject	Language	Temperature	Year	TPRC Number
503-1459	S	G	E	N	1972	78256
1466	S	S	E	H	1972	66579
1466	S	S	E	N	1972	66579
1541	S	D	G	N	1970	58419
1542	S	D	G	N	1970	58419
1547	S	S	E	N	1960	24931
1547	S	S	E	N	1972	66579
1567	P	D	G	N	1966	44476
1567	P	T	G	N	1966	44476
1583	S	D	E	N	1906	96758
1709	S	S	E	N	1960	24931
1718	S	S	E	N	1960	24931
1741	S	S	E	N	1972	66579
1777	S	D	J	N	1976	88782
1778	S	D	J	N	1976	88782
1779	S	D	J	N	1976	88782
1780	S	D	J	N	1976	88782
1781	S	D	J	N	1976	88782
1782	S	D	J	N	1976	88782
1783	S	D	J	N	1976	88782
1784	S	D	J	N	1976	88782
1785	S	D	J	N	1976	88782
1786	S	D	J	N	1976	88782
1787	S	D	J	N	1976	88782
1788	S	D	J	N	1976	88782
1789	S	D	J	N	1976	88782
1790	S	D	J	N	1976	88782
1791	S	D	J	N	1976	88782
1792	S	D	J	N	1976	88782
1793	S	D	J	N	1976	88782
1794	S	D	J	N	1976	88782
1795	S	D	J	N	1976	88782
1796	S	D	J	N	1976	88782
1797	S	D	J	N	1976	88782
1798	S	D	J	N	1976	88782
1799	S	D	J	N	1976	88782
1800	S	D	J	N	1976	88782
1823	S	D	R	N	1976	87604
1823	S	D	E	N	1976	91194
1828	S	D	E	N	1974	75232
1829	S	D	E	N	1976	88047
1829	S	D	E	N	1975	99652
1829	S	E	E	N	1975	99652
1890	S	S	E	H	1972	66579
1890	S	S	E	N	1972	66579
1945	S	S	E	H	1960	24931
1945	S	S	E	N	1960	24931
1945	S	S	E	N	1972	66579
1977	P	G	E	N	1976	83577
1978	P	G	E	N	1976	83577
1979	P	G	E	N	1976	83577
1980	P	G	E	N	1976	83577
1981	P	G	E	N	1976	83577
1982	P	G	E	N	1976	83577
1991	L	D	R	H	1972	75511
1991	L	D	E	H	1972	75512
1991	S	D	R	N	1972	75511
1991	S	D	E	N	1972	75512
2172	S	D	E	N	1976	96156
2191	S	D	E	N	1976	96156
2273	S	D	E	N	1976	93227
2273	S	T	E	N	1976	93227
2346	S	G	E	N	1977	90398
2347	S	D	E	N	1971	66243
2421	S	D	R	N	1966	49853
2422	S	D	R	N	1966	49853
2591	S	D	R	N	1966	49853
2777	S	S	E	N	1972	66579
2778	S	S	E	N	1972	66579
2779	S	S	E	N	1972	66579
2780	S	S	E	N	1972	66579
2782	S	S	E	N	1972	66579
2783	S	S	E	N	1960	24931
2784	S	S	E	N	1972	66579
2793	S	S	E	N	1960	24932
2793	S	S	E	N	1972	66579
2813	T	S	E	N	1972	70629
2814	T	S	E	N	1972	70629
2817	T	S	E	N	1972	70629
2818	T	S	E	N	1972	70629
2858	S	S	E	N	1972	70629
2921	S	S	E	N	1960	24931
8202	S	D	E	N	1972	68980
8202	S	T	E	N	1972	68980
8228	S	S	E	H	1960	24931
8228	S	S	E	N	1960	24931
8445	S	D	R	N	1967	47639
8465	S	D	R	N	1970	60925
8465	S	D	E	N	1970	76917
8530	S	D	R	N	1969	56254
8531	S	D	R	N	1967	47639
8531	S	D	R	N	1969	56254
8581	S	D	R	N	1970	60158
8581	S	D	E	N	1970	60159
8733	S	D	E	N	1968	76986
8859	S	D	E	N	1968	76986
8941	S	D	E	N	1974	74602
8977	S	D	R	N	1971	64677
8977	S	D	E	N	1971	64678
8979	S	D	E	N	1966	33878
8987	S	D	E	N	1966	33878

Substance Number	Phys. State	Subject	Language	Temperature	Year	TPRC Number
503-8997	S	D	E	N	1966	33878
9023	S	D	R	N	1972	70002
9023	S	D	E	N	1972	77025
9073	S	D	E	N	1967	40813
9074	D	D	E	N	1971	61737
9074	S	D	E	N	1971	61737
9092	S	D	E	N	1960	52687
9095	S	D	R	N	1967	46969
9095	S	D	E	N	1967	46970
9101	S	D	E	N	1965	42588
9138	S	D	E	N	1967	40813
9152	S	D	E	N	1966	33878
9152	S	D	R	N	1968	34183
9152	S	D	R	N	1965	38598
9152	S	D	E	N	1965	38599
9152	S	D	E	N	1965	42588
9152	S	D	R	N	1968	52757
9152	S	D	E	N	1965	61203
9154	S	D	R	N	1963	47474
9154	S	D	F	N	1965	47475
9154	S	D	R	H	1969	58575
9154	S	D	E	H	1969	58576
9154	S	D	R	N	1971	63673
9154	S	D	E	N	1971	63674
9154	S	D	E	N	1971	65024
9154	S	D	E	N	1973	70906
9154	S	D	E	N	1973	71366
9154	S	D	E	N	1968	76986
9154	S	D	E	N	1973	77755
9158	S	D	E	N	1971	65024
9159	S	D	E	N	1960	52687
9167	S	D	E	N	1968	76986
9173	P	G	E	N	1971	64971
9173	S	D	E	N	1960	40738
9173	S	D	E	N	1971	65024
9174	S	D	F	N	1970	35674
9174	S	D	R	N	1968	55426
9174	S	D	E	N	1968	55427
9174	S	D	E	N	1971	65024
9174	S	D	E	N	1973	70906
9174	S	D	E	N	1968	76986
9174	S	D	E	N	1973	77755
9179	S	D	R	N	1965	61201
9179	S	D	E	N	1968	76986
9188	S	D	J	N	1976	88782
9189	S	D	J	N	1976	88782
9190	S	D	J	N	1976	88782
9191	S	D	J	N	1976	88782
9192	S	D	J	N	1976	88782
9193	S	D	J	N	1976	88782
9194	S	D	J	N	1976	88782
9195	S	D	J	N	1976	88782
9196	S	D	J	N	1976	88782
9206	S	D	J	N	1976	88782
9215	S	D	R	N	1966	49147
9216	S	D	R	N	1966	49147
9217	S	D	R	N	1966	49147
9222	S	D	R	N	1967	47639
9271	S	D	R	N	1966	49853
9278	S	D	E	N	1972	73299
9279	S	D	R	N	1971	64816
9279	S	D	E	N	1971	64817
9280	S	D	R	N	1971	64816
9280	S	D	E	N	1971	64817
9281	S	D	R	N	1971	64816
9281	S	D	E	N	1971	64817
9282	S	D	R	N	1971	64816
9282	S	D	E	N	1971	64817
9283	S	D	R	N	1971	64814
9283	S	D	E	N	1971	64815
9285	S	D	R	N	1969	56255
9330	P	D	E	N	1969	59196
9330	S	D	E	N	1970	59647
9331	P	D	E	N	1969	59196
9331	S	D	E	N	1970	59647
9378	F	D	E	N	1966	39951
9378	F	G	E	N	1968	47917
9378	F	D	E	N	1969	52594
9416	S	D	E	N	1964	38224
9417	S	D	R	N	1970	61002
9417	S	D	E	N	1970	61003
9418	S	D	R	N	1970	61002
9418	S	D	E	N	1970	61003
9419	S	D	R	N	1970	61002
9419	S	D	E	N	1970	61003
9420	S	D	R	N	1970	61002
9420	S	D	E	N	1970	61003
9421	S	D	R	N	1970	60986
9421	S	D	E	N	1970	60987
9422	S	D	R	N	1970	60986
9422	S	D	E	N	1970	60987
9423	S	D	R	N	1970	60986
9423	S	D	E	N	1970	60987
9424	S	D	R	N	1970	60986
9424	S	D	E	N	1970	60987
9427	S	D	E	N	1965	36067
9427	S	D	R	N	1965	38884
9465	P	D	E	N	1969	37364
9470	S	D	E	N	1968	52870
9503	S	D	E	N	1964	38224
9524	S	D	G	N	1969	55624

Substance Number	Phys. State	Subject	Language	Temperature	Year	TPRC Number
503-9529	P	D	E	N	1965	37046
9530	P	D	E	N	1965	37046
9543	S	D	R	N	1966	49147
9588	S	D	R	N	1967	45346
9595	S	D	E	N	1964	38224
9598	S	D	R	N	1969	57136
9598	S	D	E	N	1969	57137
9598	S	D	R	N	1969	58350
9598	S	D	E	N	1969	76913
9629	P	D	R	N	1964	43251
9630	P	D	R	N	1964	43251
9631	P	D	R	N	1964	43251
9631	S	D	R	N	1970	66304
9631	S	D	E	N	1970	66305
9632	P	D	R	N	1964	43251
9632	S	D	R	N	1970	66304
9632	S	D	E	N	1970	66305
9633	P	D	E	N	1965	37046
9643	S	D	E	N	1965	46537
9662	S	D	R	N	1969	56255
9684	S	D	E	N	1970	75466
9702	P	D	E	N	1961	41149
9705	M	D	E	N	1966	40351
9705	P	D	E	N	1965	35907
9705	P	D	E	N	1967	40230
9705	S	D	E	N	1966	48364
9707	P	D	E	N	1967	40230
9707	P	D	E	N	1966	40351
9708	M	D	E	N	1966	40351
9708	P	D	E	N	1965	35907
9708	P	D	E	N	1967	40230
9708	S	D	E	N	1966	48364
9736	P	D	E	N	1961	41149
9749	S	S	E	N	1972	66579
9750	P	D	E	N	1964	50030
9750	S	D	G	N	1974	75213
9758	P	D	E	N	1965	43415
9758	P	T	E	N	1965	43415
9791	S	D	E	N	1965	36510
9822	S	D	E	N	1970	59445
9824	S	D	R	N	1967	48465
9824	S	D	E	N	1968	51222
9824	S	D	E	N	1967	67144
9830	P	D	E	N	1962	42292
9837	S	D	E	N	1970	57964
9838	M	D	E	N	1966	40351
9838	P	D	E	N	1965	35907
9838	P	D	E	N	1967	40230
9838	P	D	E	N	1968	51997
9838	S	D	E	N	1966	48364
9845	P	D	E	N	1964	34819
9851	S	D	E	N	1974	74526
9869	D	D	E	N	1973	71365
9869	P	D	E	N	1964	50030
9869	S	D	E	N	1973	71365
9874	S	D	R	N	1965	44133
9874	S	D	E	N	1965	44134
9882	P	D	E	N	1965	37046
9891	P	D	E	N	1964	37398
9891	S	D	E	N	1965	36510
9893	S	D	R	N	1965	61203
9916	P	D	E	N	1967	40230
9916	P	D	E	N	1966	40351
9916	P	D	E	N	1964	50030
9916	P	D	E	N	1968	51997
9916	P	G	E	N	1971	64971
9916	P	D	E	N	1973	71769
9916	S	D	E	N	1966	33924
9917	S	D	E	N	1953	41918
9922	S	D	E	N	1965	36510
9922	S	D	E	H	1965	36511
9922	S	D	E	H	1965	37899
9938	S	D	E	N	1953	41918
9939	S	D	E	N	1953	41918
9940	P	D	E	N	1962	42292
9940	S	D	E	N	1953	41918
9969	S	D	R	N	1971	64816
9969	S	D	E	N	1971	64817
9977	S	D	E	N	1953	41918
521-0001	P	G	E	N	1965	35027
0001	S	D	E	N	1965	39092
0001	S	D	E	N	1970	60103
0001	S	G	E	N	1971	65946
0003	P	G	E	N	1971	64971
0003	P	D	E	N	1973	71769
0003	P	D	E	N	1976	85954
0003	S	D	R	N	1967	46404
0003	S	D	E	N	1976	85517
0003	S	E	E	N	1976	85517
0004	P	D	E	N	1967	46432
0004	S	D	E	N	1967	38580
0004	S	D	R	N	1965	42693
0004	S	D	E	N	1966	42694
0006	P	D	E	N	1954	78248
0006	S	D	F	N	1970	59429
0006	S	D	E	N	1971	66376
0006	S	D	E	N	1906	96758
0006	S	D	E	N	1908	96759
0007	S	D	E	N	1966	33881
0007	S	D	E	N	1966	34538

Phys. State: **C.** Superconductive; **D.** Doped; **E.** Expanded; **F.** Fibrous or Whisker; **G.** Gas; **I.** Ionized or Plasma; **L.** Liquid; **M.** Multiphase; **P.** Powder or Fine Particle; **S.** Solid; **T.** Thick or Thin Film

Subject: **D.** Data; **E.** Experiment; **G.** General (Data + Theory + Experiment); **S.** Survey (Review, Compendium etc.); **T.** Theory

Language: **C.** Czech; **D.** Dutch; **E.** English; **F.** French; **G.** German; **I.** Italian; **J.** Japanese; **O.** Other Languages; **P.** Polish; **R.** Russian; **S.** Spanish

Temperature: **F.** Full Range (Low + Normal + High); **L.** Low (0 to 75K) + Overlap into Normal; **N.** Normal (75 to 1273K); **H.** High (above 1273K) + Overlap into Normal; Blank = Not Coded

Substance Number	Phys. State	Subject	Language	Temperature	Year	TPRC Number
521-0007	S	D	E	N	1964	37966
0007	S	D	E	N	1965	40480
0007	S	D	E	N	1966	43477
0007	S	D	G	N	1898	46995
0007	S	D	E	N	1956	51285
0007	S	D	E	N	1964	57860
0007	S	S	E	N	1972	66579
0007	S	D	E	N	1972	71762
0007	S	D	E	N	1974	75330
0007	S	D	E	N	1919	95768
0009	M	D	E	N	1966	33879
0009	M	D	R	N	1960	45899
0009	M	D	E	N	1965	45900
0009	M	D	E	N	1974	78688
0009	P	D	E	N	1966	33879
0009	P	D	E	N	1966	33881
0009	P	D	E	N	1966	34538
0009	P	D	R	N	1964	34769
0009	P	T	R	N	1964	34769
0009	P	D	E	N	1965	34770
0009	P	T	E	N	1965	34770
0009	P	D	E	N	1966	44932
0009	P	E	E	N	1974	76329
0009	S	S	E	N	1966	34136
0009	S	D	E	N	1964	36100
0009	S	D	E	N	1965	39092
0009	S	D	E	N	1965	40480
0009	S	D	E	N	1967	46650
0009	S	D	E	N	1945	51284
0009	S	D	E	N	1962	51415
0009	S	D	R	N	1965	53622
0009	S	D	E	N	1969	53623
0009	S	S	E	N	1968	55083
0009	S	S	R	N	1955	55424
0009	S	S	E	N	1969	55425
0009	S	D	E	N	1968	56553
0009	S	D	E	N	1972	68251
0009	S	E	E	N	1972	68251
0009	S	D	E	N	1972	68816
0009	S	S	E	N	1971	69328
0009	S	D	E	N	1972	69340
0009	S	D	E	N	1965	70139
0009	S	E	E	N	1965	70139
0009	S	G	E	N	1968	76273
0009	S	D	E	N	1974	76371
0009	S	T	E	N	1974	76371
0009	S	E	E	N	1974	77746
0009	S	D	E	N	1973	77757
0009	S	T	E	N	1973	77757
0009	S	D	E	N	1976	96156
0010	P	D	E	N	1966	33881
0010	P	D	E	N	1966	34538
0010	P	D	E	N	1966	39007
0010	P	D	E	N	1972	64449
0010	P	S	E	N	1972	66579
0010	P	D	E	N	1954	78248
0010	S	D	E	N	1965	40480
0015	P	D	E	N	1965	36525
0015	P	D	E	N	1962	42292
0020	S	D	G	N	1965	43102
0021	P	D	E	N	1973	73529
0022	P	D	E	N	1968	51709
0022	P	D	E	N	1970	65200
0022	P	D	E	N	1975	83895
0022	S	D	E	N	1966	48370
0022	S	D	F	N	1973	72371
0024	P	D	E	N	1973	71846
0024	S	D	R	N	1965	42494
0024	S	D	R	N	1970	65698
0024	S	D	E	N	1960	93151
0025	M	D	E	N	1966	41373
0025	P	D	E	N	1954	78248
0025	S	D	E	N	1965	35672
0025	S	D	E	N	1966	41709
0025	S	E	E	N	1966	41709
0025	S	D	O	N	1964	41980
0025	S	D	E	N	1965	42064
0025	S	D	R	N	1964	42175
0025	S	D	E	N	1967	45890
0025	S	D	E	N	1967	46425
0025	S	D	E	N	1968	49893
0025	S	D	E	N	1968	52945
0025	S	D	G	N	1969	55278
0025	S	D	E	N	1968	58992
0025	S	D	E	N	1970	61942
0025	S	D	R	N	1970	63189
0025	S	T	R	N	1970	63189
0025	S	D	G	N	1971	64343
0025	S	D	E	N	1973	75868
0025	S	E	E	N	1973	75868
0025	S	D	R	N	1972	76266
0025	S	D	R	N	1969	77844
0025	S	D	E	N	1974	81307
0025	S	E	E	N	1974	81307
0025	S	D	J	N	1974	81726
0025	S	D	E	N	1960	90396
0025	S	E	E	N	1960	90396
0025	S	D	E	N	1906	96758
0026	P	D	E	N	1969	57189
0026	P	D	E	N	1970	65200
0026	P	D	E	N	1973	73529

Substance Number	Phys. State	Subject	Language	Temperature	Year	TPRC Number
521-0028	S	D	E	N	1965	39312
0028	S	D	E	L	1970	62676
0028	S	D	E	N	1976	87489
0029	S	G	F	N	1967	49675
0030	S	D	E	N	1971	66376
0030	S	D	F	N	1973	71786
0030	S	D	E	N	1908	96759
0032	P	D	E	N	1970	65200
0033	P	D	E	N	1973	73529
0033	S	D	E	N	1966	43477
0033	S	D	F	N	1964	57860
0033	S	D	F	N	1970	59429
0033	S	D	E	N	1908	96759
0034	P	D	E	N	1973	73529
0035	P	D	E	N	1973	73529
0035	P	D	E	N	1975	83895
0035	S	D	E	N	1978	99431
0036	P	D	E	N	1968	51709
0036	P	D	E	N	1970	65200
0036	P	D	E	N	1973	73529
0036	P	D	E	N	1975	83895
0036	S	D	E	H	1967	34861
0040	D	D	E	N	1969	59185
0040	L	D	E	N	1969	59185
0040	M	D	E	N	1966	33881
0040	P	D	E	N	1966	33879
0040	P	D	E	N	1968	50190
0040	P	D	E	N	1969	59185
0040	S	D	E	N	1969	35412
0042	S	D	E	N	1965	35672
0043	S	D	R	N	1965	41710
0043	S	D	R	N	1964	42175
0043	S	D	E	N	1973	75868
0043	S	E	E	N	1973	75868
0043	S	D	E	N	1974	81307
0043	S	E	E	N	1974	81307
0043	S	D	E	N	1960	90396
0043	S	E	E	N	1960	90396
0046	M	D	E	N	1966	33879
0046	S	D	E	N	1966	33879
0046	S	D	E	N	1967	47283
0046	S	S	E	N	1971	69328
0046	S	D	E	N	1972	69340
0046	S	E	E	N	1974	76329
0046	S	E	E	N	1974	77746
0046	S	D	E	N	1976	96156
0053	P	D	E	N	1969	59106
0053	P	T	E	N	1969	59106
0053	S	D	E	N	1967	38580
0053	S	D	R	N	1965	42693
0053	S	D	E	N	1966	42694
0053	S	D	E	N	1967	46650
0053	S	D	E	N	1956	51285
0055	S	D	E	N	1966	43477
0055	S	D	E	N	1972	68251
0055	S	E	E	N	1972	68251
0057	P	D	E	N	1969	57189
0057	P	D	E	N	1970	65200
0057	P	D	E	N	1973	73529
0057	P	D	E	N	1975	83895
0057	S	D	E	N	1967	46042
0057	S	D	E	L	1970	62676
0057	S	D	E	N	1906	96758
0057	S	D	E	N	1978	99431
0058	P	D	E	N	1973	73529
0059	P	D	E	N	1970	65200
0060	S	D	E	N	1906	96758
0063	P	D	R	N	1972	74170
0063	P	D	E	N	1972	74171
0063	S	D	E	N	1966	34177
0063	S	D	E	N	1967	35143
0063	S	T	E	N	1965	35246
0063	S	D	E	N	1964	37442
0063	S	D	E	N	1967	43501
0063	S	D	E	N	1967	44768
0063	S	D	E	N	1945	46837
0063	S	D	R	N	1968	55537
0063	S	E	R	N	1968	55537
0063	S	D	E	N	1973	75868
0063	S	E	E	N	1973	75868
0063	S	D	E	N	1975	83960
0063	S	D	E	N	1960	93151
0064	S	D	E	L	1965	36486
0064	S	D	E	N	1967	40853
0064	S	D	E	N	1945	46837
0064	S	D	E	N	1973	68990
0064	S	D	E	N	1972	69102
0064	S	E	E	N	1972	69102
0064	S	D	E	N	1967	77362
0065	S	D	E	N	1971	95916
0067	P	D	E	N	1965	34090
0067	P	D	E	N	1970	49757
0067	P	D	E	N	1975	81230
0067	P	D	E	N	1975	83895
0067	S	D	R	N	1964	42175
0067	S	D	E	N	1967	46425
0067	S	D	E	N	1970	61942
0067	S	D	E	N	1971	63066
0067	S	D	E	N	1971	63739
0067	S	D	E	N	1973	71795
0067	S	D	E	N	1973	75868

Substance Number	Phys. State	Subject	Language	Temperature	Year	TPRC Number
521-0067	S	E	E	N	1973	75868
0067	S	D	E	N	1974	81307
0067	S	E	E	N	1974	81307
0067	S	D	E	N	1960	90396
0067	S	E	E	N	1960	90396
0067	S	D	E	N	1908	96759
0069	P	D	E	N	1957	51268
0069	S	D	E	N	1965	35594
0069	S	D	R	N	1965	41690
0069	S	E	R	N	1965	41690
0069	S	D	E	N	1966	42338
0069	S	D	R	N	1965	42494
0069	S	D	R	N	1965	42542
0069	S	D	R	N	1965	42543
0069	S	D	E	N	1966	44297
0069	S	D	E	N	1967	45106
0069	S	D	E	N	1968	49087
0069	S	D	G	N	1968	49510
0069	S	D	E	N	1968	49785
0069	S	D	G	N	1968	50847
0069	S	T	G	N	1968	50847
0069	S	D	R	N	1968	55537
0069	S	E	R	N	1968	55537
0069	S	D	R	N	1968	55694
0069	S	D	R	N	1969	58120
0069	S	D	E	N	1970	59296
0069	S	D	E	N	1970	61941
0069	S	D	O	N	1972	69533
0069	S	D	C	N	1972	71198
0069	S	T	C	N	1972	71198
0069	S	S	E	N	1975	85290
0069	S	D	E	N	1968	88413
0070	S	D	E	N	1966	33879
0070	S	D	E	N	1967	46650
0070	S	D	E	N	1974	78685
0070	S	D	E	N	1976	89182
0070	S	D	E	N	1976	96156
0072	P	D	E	N	1966	33774
0072	P	D	E	N	1965	42772
0072	P	D	E	N	1967	46432
0072	S	D	E	N	1966	33774
0072	S	D	E	N	1966	33879
0072	S	D	E	N	1965	39312
0072	S	D	E	N	1965	42772
0072	S	D	E	N	1967	44403
0072	S	D	E	N	1967	46650
0072	S	D	E	N	1956	51285
0072	S	D	E	N	1972	66640
0072	S	D	E	N	1970	67859
0072	S	D	E	N	1970	67860
0072	S	D	E	N	1974	78685
0072	S	D	E	N	1906	96758
0075	P	D	E	L	1970	62676
0075	P	D	E	N	1954	78248
0075	P	D	E	N	1975	83895
0075	P	D	E	N	1979	99163
0075	S	G	G	N	1967	34931
0075	S	D	E	N	1966	43477
0075	S	D	F	N	1967	44753
0075	S	D	E	N	1967	47283
0075	S	D	R	N	1969	57308
0075	S	D	E	N	1970	57309
0075	S	D	E	N	1970	57768
0075	S	D	E	N	1964	57860
0075	S	D	R	N	1970	59430
0075	S	D	E	N	1970	59431
0075	S	D	E	N	1971	62145
0075	S	D	E	L	1970	62676
0075	S	D	E	N	1971	62718
0075	S	D	E	N	1971	64502
0075	S	D	E	N	1971	66376
0075	S	S	E	N	1972	66579
0075	S	D	E	N	1973	71147
0075	S	T	E	N	1973	71147
0075	S	D	E	N	1973	72167
0075	S	D	R	N	1973	72646
0075	S	D	E	N	1974	72647
0075	S	D	R	N	1973	73651
0075	S	D	E	N	1973	73652
0075	S	D	E	N	1977	95448
0075	S	D	E	N	1919	95768
0075	S	D	E	N	1906	96758
0075	S	D	E	N	1908	96759
0077	P	G	E	N	1971	64971
0077	P	D	E	N	1972	69102
0077	P	E	E	N	1972	69102
0077	P	S	E	N	1972	71106
0077	S	D	E	N	1966	43477
0077	S	D	E	N	1964	57860
0077	S	D	R	N	1969	61200
0077	S	D	E	N	1971	63739
0077	S	D	E	N	1908	96759
0082	P	D	E	N	1973	73529
0082	P	D	E	N	1954	78248
0082	S	D	E	N	1965	42979
0082	S	D	E	N	1966	43477
0082	S	D	E	N	1945	46837
0082	S	D	E	N	1964	57860
0082	S	D	F	N	1970	59429
0082	S	D	G	N	1971	64274
0082	S	D	G	N	1971	66369

Phys. State: **C.** Superconductive; **D.** Doped; **E.** Expanded; **F.** Fibrous or Whisker; **G.** Gas; **I.** Ionized or Plasma; **L.** Liquid; **M.** Multiphase; **P.** Powder or Fine Particle; **S.** Solid; **T.** Thick or Thin Film

Subject: **D.** Data; **E.** Experiment; **G.** General (Data + Theory + Experiment); **S.** Survey (Review, Compendium etc.); **T.** Theory

Language: **C.** Czech; **D.** Dutch; **E.** English; **F.** French; **G.** German; **I.** Italian; **J.** Japanese; **O.** Other Languages; **P.** Polish; **R.** Russian; **S.** Spanish

Temperature: **F.** Full Range (Low + Normal + High); **L.** Low (0 to 75K) + Overlap into Normal; **N.** Normal (75 to 1273K); **H.** High (above 1273K) + Overlap into Normal; Blank = Not Coded

Substance Number	Phys. State	Sub-ject	Lan-guage	Temper-ature	Year	TPRC Number
521-0082	S	D	F	N	1973	72371
0082	S	D	R	N	1975	80566
0082	S	D	E	N	1975	86759
0082	S	D	E	N	1908	96759
0083	M	D	E	N	1970	59127
0083	P	D	E	N	1979	99163
0083	S	D	G	N	1965	43102
0083	S	D	E	N	1966	43477
0083	S	D	E	N	1964	57860
0083	S	D	E	N	1970	59127
0083	S	S	E	N	1972	66579
0083	S	D	E	N	1976	96156
0084	P	D	E	N	1975	83895
0084	S	D	F	N	1967	40913
0084	S	D	E	N	1966	43477
0084	S	D	E	N	1967	47283
0084	S	D	E	N	1964	57860
0084	S	D	F	N	1970	59429
0084	S	D	F	N	1970	59793
0084	S	D	E	N	1971	66376
0084	S	D	E	N	1970	67860
0084	S	D	E	N	1908	96759
0085	P	D	E	N	1966	39007
0085	P	D	E	N	1966	44932
0085	S	D	E	N	1966	33879
0085	S	D	E	N	1965	34262
0085	S	D	E	N	1969	35412
0085	S	D	E	N	1967	46650
0085	S	D	E	N	1967	54966
0085	S	D	E	N	1964	57860
0085	S	D	E	N	1970	60103
0085	S	D	E	N	1970	67860
0085	S	D	E	N	1972	68251
0085	S	E	E	N	1972	68251
0085	S	D	E	N	1974	78685
0085	S	D	E	N	1976	89182
0085	S	D	E	N	1978	96156
0086	S	D	E	N	1965	35672
0086	S	D	O	N	1964	41980
0086	S	D	R	N	1964	42175
0086	S	D	E	N	1966	43477
0086	S	D	R	N	1960	44197
0086	S	D	F	N	1960	44198
0086	S	D	E	N	1967	46425
0086	S	D	R	N	1968	55537
0086	S	E	R	N	1968	55537
0086	S	D	E	N	1969	56819
0086	S	D	E	N	1964	57860
0086	S	D	E	N	1970	60195
0086	S	D	G	N	1972	70031
0086	S	E	G	N	1972	70031
0086	S	D	E	N	1973	75868
0086	S	E	E	N	1973	75868
0086	S	D	E	N	1973	85285
0086	S	D	E	N	1960	90396
0086	S	E	E	N	1960	90396
0086	S	D	E	N	1971	92956
0086	S	D	E	N	1906	96758
0087	S	D	G	N	1898	46995
0087	S	D	E	N	1972	71762
0089	P	D	E	N	1973	73529
0089	P	D	E	N	1954	78248
0090	S	D	F	N	1963	32110
0090	S	D	E	N	1964	37156
0090	S	D	E	N	1972	71762
0091	S	D	E	N	1906	22517
0091	S	D	F	N	1963	32110
0091	S	D	F	N	1964	37156
0091	S	D	E	N	1968	51750
0091	S	D	E	L	1970	62676
0091	S	D	E	N	1972	71762
0091	S	D	E	N	1906	96758
0092	P	D	E	N	1977	92103
0092	S	D	R	N	1964	42175
0092	S	D	E	N	1971	63739
0092	S	D	R	N	1973	72652
0092	S	D	E	N	1974	72653
0092	S	D	E	N	1908	96759
0093	S	D	E	N	1971	63739
0094	P	D	E	N	1975	83895
0094	S	D	F	N	1973	75296
0095	S	D	R	N	1964	42175
0095	S	D	O	N	1962	45488
0095	S	D	E	N	1931	47008
0095	S	D	G	N	1970	65144
0095	S	D	G	N	1971	67135
0095	S	E	G	N	1971	67135
0095	S	D	E	N	1973	74443
0095	S	D	E	N	1906	96758
0096	P	D	E	N	1954	78248
0096	S	D	E	N	1966	43477
0096	S	D	G	N	1898	46995
0096	S	D	E	N	1964	57860
0096	S	D	E	N	1972	71762
0096	S	D	E	N	1919	95768
0097	P	D	E	N	1977	92103
0098	S	G	F	N	1967	49675
0099	S	D	O	N	1964	41980
0099	S	D	R	N	1964	42175
0099	S	D	R	N	1960	44197
0099	S	D	F	N	1960	44198

Substance Number	Phys. State	Sub-ject	Lan-guage	Temper-ature	Year	TPRC Number
521-0099	S	D	E	N	1973	75868
0099	S	E	E	N	1973	75868
0099	S	D	R	N	1972	76266
0099	S	D	E	N	1974	81307
0099	S	E	E	N	1974	81307
0099	S	D	E	N	1960	90396
0099	S	E	E	N	1960	90396
0103	S	D	E	N	1967	46650
0103	S	D	E	N	1974	78685
0103	S	D	E	N	1976	89183
0104	P	D	E	N	1965	34090
0104	P	D	E	N	1965	50543
0104	P	S	E	N	1972	66579
0104	P	D	E	N	1975	83895
0104	S	D	E	N	1969	35541
0104	S	D	G	N	1965	35678
0104	S	D	E	N	1965	42064
0104	S	D	R	N	1964	42175
0104	S	D	E	N	1967	46425
0104	S	G	F	N	1967	49675
0104	S	D	E	N	1965	50543
0104	S	D	R	N	1965	53622
0104	S	D	E	N	1969	53623
0104	S	D	F	N	1970	59429
0104	S	D	E	N	1971	63739
0104	S	D	R	N	1971	64719
0104	S	D	E	N	1971	64720
0104	S	S	E	N	1972	66579
0104	S	D	E	N	1973	71795
0104	S	D	R	N	1973	73659
0104	S	D	E	N	1973	73660
0104	S	D	E	N	1973	75868
0104	S	E	E	N	1973	75868
0104	S	D	E	N	1974	81307
0104	S	E	E	N	1974	81307
0104	S	D	E	N	1960	90396
0104	S	E	E	N	1960	90396
0104	S	D	E	N	1977	93760
0104	S	D	E	N	1978	96156
0104	S	D	E	N	1908	96759
0104	S	D	E	N	1979	99162
0105	P	D	E	N	1975	83895
0105	S	D	E	N	1967	47283
0105	S	D	R	N	1969	57308
0105	S	D	E	N	1970	57309
0105	S	D	E	L	1970	62676
0105	S	D	E	N	1908	96759
0108	P	D	E	N	1954	78248
0108	S	D	G	N	1898	46995
0108	S	D	E	N	1931	47008
0109	P	D	E	N	1970	65200
0109	P	D	E	N	1973	73529
0110	D	D	E	N	1969	59185
0110	L	D	E	N	1969	59185
0110	M	D	E	N	1969	59185
0110	M	D	E	N	1974	78687
0110	M	E	E	N	1974	78687
0110	M	D	E	N	1977	93594
0110	M	E	E	N	1977	93594
0110	P	D	E	N	1966	33879
0110	P	D	R	N	1964	34769
0110	P	T	R	N	1964	34769
0110	P	D	E	N	1965	34770
0110	P	T	E	N	1965	34770
0110	P	G	E	N	1965	35027
0110	P	D	E	N	1966	39007
0110	P	D	E	N	1966	48369
0110	P	D	E	N	1969	59185
0110	P	D	E	N	1970	60103
0110	P	D	E	N	1971	61868
0110	P	D	E	N	1970	77522
0110	P	D	E	N	1977	93594
0110	P	E	E	N	1977	93594
0110	S	D	E	N	1965	34262
0110	S	D	E	N	1969	35412
0110	S	D	E	N	1964	36100
0110	S	D	E	N	1965	40480
0110	S	D	R	N	1965	42693
0110	S	D	E	N	1966	42694
0110	S	D	E	N	1956	51285
0110	S	D	E	N	1968	52046
0110	S	D	E	N	1967	52863
0110	S	D	R	N	1965	53622
0110	S	D	E	N	1969	53623
0110	S	D	E	N	1968	56553
0110	S	D	E	N	1972	68251
0110	S	E	E	N	1972	68251
0110	S	D	E	N	1974	77316
0111	S	D	E	N	1972	66863
0111	S	D	E	N	1965	77521
0111	S	D	E	N	1978	96156
0116	P	D	E	N	1965	36525
0116	P	D	E	N	1962	42292
0116	P	D	E	N	1967	47385
0116	P	D	E	N	1964	50030
0116	P	D	E	N	1970	65200
0116	P	D	E	N	1973	73529
0116	S	D	E	N	1968	49251
0116	S	G	R	N	1969	58230
0116	S	D	E	N	1970	60580
0116	S	S	E	N	1972	66579

Substance Number	Phys. State	Sub-ject	Lan-guage	Temper-ature	Year	TPRC Number
521-0116	S	D	E	N	1906	96758
0123	P	D	E	N	1966	33774
0123	P	D	E	N	1966	41239
0123	P	D	E	N	1965	42772
0123	P	D	E	N	1967	46432
0123	P	D	E	N	1968	51709
0123	P	D	E	N	1969	55600
0123	P	D	E	N	1969	57189
0123	P	D	E	N	1969	59106
0123	P	T	E	N	1969	59106
0123	P	D	E	N	1968	59961
0123	P	S	E	N	1972	66579
0123	P	D	E	N	1971	68875
0123	P	D	E	N	1972	70141
0123	P	E	E	N	1972	70141
0123	S	D	E	N	1966	33774
0123	S	D	E	N	1965	42772
0123	S	D	E	N	1966	43477
0123	S	D	E	N	1967	46650
0123	S	D	E	N	1968	49669
0123	S	D	E	N	1967	59569
0123	S	D	E	L	1970	62676
0123	S	D	E	N	1972	66640
0123	S	D	E	N	1973	72869
0123	S	D	E	N	1973	73153
0130	P	D	E	N	1970	65200
0130	S	D	E	N	1945	46837
0130	S	D	G	N	1971	64274
0130	S	D	F	N	1972	65129
0130	S	D	G	N	1971	66369
0130	S	D	E	H	1972	67713
0130	S	D	F	N	1973	72371
0130	S	D	E	N	1908	96759
0131	P	D	E	N	1973	73529
0131	P	D	E	N	1954	78248
0131	S	D	E	N	1945	46837
0131	S	D	F	N	1970	54251
0131	S	D	F	N	1970	59429
0131	S	D	E	N	1906	96758
0133	P	D	E	N	1970	49757
0133	P	D	E	N	1977	92103
0133	S	D	G	N	1965	35678
0133	S	D	R	N	1964	42175
0133	S	G	F	N	1967	49675
0133	S	D	E	N	1970	56100
0133	S	D	E	N	1970	56101
0133	S	D	E	L	1970	62676
0133	S	D	E	N	1971	63739
0133	S	D	R	N	1972	70800
0133	S	D	E	N	1973	75868
0133	S	E	E	N	1973	75868
0133	S	D	E	N	1960	90396
0133	S	E	E	N	1960	90396
0135	P	D	E	N	1965	36525
0135	P	D	E	N	1962	42292
0135	P	D	E	N	1973	73529
0135	S	S	E	N	1972	66579
0135	S	D	E	N	1972	66852
0135	S	D	E	N	1965	77521
0153	P	D	E	N	1973	73529
0153	P	D	E	N	1975	83895
0174	D	D	R	N	1977	91530
0174	P	D	E	N	1964	36837
0174	P	D	E	N	1970	65200
0174	P	D	E	N	1973	71872
0174	P	D	E	N	1954	78248
0174	P	D	E	N	1979	99163
0174	S	D	E	N	1979	99162
0181	S	D	G	N	1965	43102
0181	S	D	E	N	1976	89183
0184	S	D	E	N	1964	35550
0184	S	D	E	N	1965	39312
0186	M	D	E	N	1974	78687
0186	M	E	E	N	1974	78687
0186	P	D	E	N	1966	39007
0191	S	D	E	N	1966	34852
0191	S	D	E	N	1965	35028
0191	S	D	E	N	1965	36324
0191	S	D	F	N	1966	39281
0191	S	D	E	N	1967	42891
0191	S	D	F	N	1966	44924
0191	S	D	E	N	1945	46837
0192	S	D	R	N	1966	41130
0192	S	D	O	N	1964	41980
0192	S	D	O	N	1966	42467
0192	S	D	R	N	1960	44197
0192	S	D	F	N	1960	44198
0192	S	D	E	N	1967	46425
0192	S	D	E	N	1931	47008
0192	S	D	R	N	1969	56759
0192	S	D	R	N	1969	59480
0192	S	D	E	N	1970	61942
0192	S	S	E	N	1972	66579
0192	S	D	E	N	1973	75868
0192	S	E	E	N	1973	75868
0192	S	D	R	N	1969	77844
0192	S	D	E	N	1960	90396
0192	S	E	E	N	1960	90396
0192	S	D	E	N	1971	92956
0192	S	D	E	N	1978	94616
0192	S	D	E	N	1906	96758

Phys. State: **C.** Superconductive; **D.** Doped; **E.** Expanded; **F.** Fibrous or Whisker; **G.** Gas; **I.** Ionized or Plasma; **L.** Liquid; **M.** Multiphase; **P.** Powder or Fine Particle; **S.** Solid; **T.** Thick or Thin Film
Subject: **D.** Data; **E.** Experiment; **G.** General (Data + Theory + Experiment); **S.** Survey (Review, Compendium etc.); **T.** Theory
Language: **C.** Czech; **D.** Dutch; **E.** English; **F.** French; **G.** German; **I.** Italian; **J.** Japanese; **O.** Other Languages; **P.** Polish; **R.** Russian; **S.** Spanish
Temperature: **F.** Full Range (Low + Normal + High); **L.** Low (0 to 75K) + Overlap into Normal; **N.** Normal (75 to 1273K); **H.** High (above 1273K) + Overlap into Normal; Blank = Not Coded

Substance Number	Phys. State	Subject	Language	Temperature	Year	TPRC Number
521-0195	S	D	E	L	1970	62676
0195	S	D	E	N	1971	66376
0195	S	D	F	N	1973	71786
0195	S	D	E	N	1908	96759
0196	S	D	E	N	1969	35412
0196	S	D	E	N	1976	89182
0216	S	D	E	N	1966	43477
0216	S	D	E	N	1964	57860
0216	S	D	E	N	1972	71762
0217	P	D	E	N	1954	78248
0222	P	D	F	N	1971	65152
0222	S	D	F	N	1971	65152
0223	P	D	E	N	1964	50030
0223	S	D	E	N	1906	96758
0225	P	D	E	N	1966	33774
0225	P	D	E	N	1966	41239
0225	P	D	E	N	1965	42772
0225	P	D	E	N	1975	83895
0225	S	D	E	N	1966	33774
0225	S	D	E	N	1967	40853
0225	S	D	E	N	1965	42772
0225	S	D	E	N	1964	57860
0236	P	D	E	N	1966	33774
0236	P	D	E	N	1965	42772
0236	P	D	E	N	1970	65200
0236	P	D	E	N	1973	73529
0236	S	D	E	N	1966	33774
0236	S	D	E	N	1969	35412
0236	S	D	E	N	1965	42772
0236	S	D	E	N	1966	43477
0236	S	D	E	N	1967	44403
0236	S	D	E	N	1965	50543
0236	S	D	E	L	1970	62676
0236	S	S	E	N	1972	66579
0236	S	D	E	N	1906	96758
0238	M	D	E	N	1967	47506
0238	P	D	E	N	1966	33879
0238	P	D	E	N	1966	33880
0238	P	D	E	N	1969	54960
0238	S	D	E	N	1956	51285
0239	P	D	E	N	1968	59961
0239	P	D	E	N	1973	71987
0239	S	D	E	N	1967	47283
0239	S	D	E	L	1966	48370
0240	P	D	E	N	1970	65200
0240	P	D	E	N	1973	73529
0240	P	D	E	N	1975	83895
0240	P	D	E	N	1979	99163
0240	S	D	E	N	1967	47261
0240	S	D	R	N	1969	50057
0240	S	T	R	N	1969	50057
0240	S	D	E	N	1969	50058
0240	S	T	E	N	1969	50058
0240	S	D	E	N	1906	96758
0240	S	D	E	N	1979	99162
0241	P	D	E	N	1973	73529
0241	P	D	E	N	1975	83895
0241	S	D	E	N	1972	73127
0241	S	D	E	N	1979	99162
0242	P	D	E	N	1969	57189
0242	S	D	E	N	1965	39312
0242	S	D	E	N	1966	43477
0243	S	D	E	N	1965	39312
0246	S	D	R	N	1965	42494
0246	S	D	G	N	1967	50888
0249	S	E	G	N	1968	48311
0249	S	D	G	N	1967	50888
0249	S	D	R	N	1968	50914
0249	S	D	E	N	1968	50915
0249	S	D	E	N	1967	56197
0250	S	D	E	N	1945	46837
0250	S	S	E	N	1972	66579
0250	S	D	F	N	1973	72371
0250	S	D	E	H	1973	72419
0250	S	D	E	N	1908	96759
0251	S	D	F	N	1970	59429
0251	S	D	E	N	1906	96758
0254	S	D	E	N	1955	41955
0254	S	D	E	N	1967	43012
0258	P	D	E	N	1973	73529
0258	P	D	E	H	1967	34861
0258	S	G	E	L	1962	40665
0258	S	D	E	N	1966	48370
0258	S	D	E	N	1965	50543
0258	S	G	E	H	1970	58983
0258	S	D	F	N	1970	59429
0258	S	S	E	N	1972	66579
0259	M	D	E	N	1966	59540
0259	S	D	E	N	1965	39092
0259	S	D	E	N	1966	41239
0259	S	D	R	N	1965	42693
0259	S	D	E	N	1966	42694
0259	S	D	E	N	1969	55404
0259	S	D	E	N	1969	58200
0259	S	D	E	N	1968	59961
0259	S	D	E	N	1972	65340
0259	S	D	E	N	1971	68875
0259	S	D	E	N	1972	69103
0259	S	D	E	N	1973	70043
0259	S	D	E	N	1972	70141
0259	S	E	E	N	1972	70141
521-0259	S	D	E	N	1972	70324
0259	S	D	E	N	1973	73151
0259	S	D	R	N	1974	78950
0259	S	D	E	N	1974	79915
0259	S	E	E	N	1974	79915
0259	S	D	E	N	1975	86304
0259	S	D	E	N	1977	90178
0259	S	D	E	N	1977	92103
0259	S	D	E	N	1975	92421
0259	S	D	E	N	1906	96758
0259	S	D	E	N	1979	98303
0261	P	D	E	N	1968	51709
0261	P	D	E	N	1970	65200
0261	P	S	E	N	1972	66579
0261	P	D	E	N	1973	73529
0261	P	D	E	N	1975	83895
0261	P	D	E	N	1978	98168
0261	S	D	E	N	1966	48370
0261	S	S	E	N	1972	66579
0261	S	D	E	N	1976	85517
0261	S	E	E	N	1976	85517
0261	S	D	E	N	1978	98168
0264	P	D	E	N	1967	46432
0264	P	S	E	N	1972	66579
0264	S	D	E	N	1965	39312
0264	S	D	E	N	1967	44403
0272	P	D	E	N	1967	46432
0272	S	D	E	N	1965	39312
0272	S	D	E	N	1966	43477
0272	S	D	E	N	1965	50543
0272	S	D	E	N	1967	54966
0274	S	D	E	N	1970	67859
0274	S	G	E	N	1974	74262
0293	P	D	E	N	1969	57189
0293	S	D	E	N	1969	35412
0293	S	D	E	N	1965	39312
0293	S	D	E	N	1972	66640
0293	S	D	E	N	1974	78685
0296	S	D	E	N	1976	89183
0297	S	D	R	N	1975	91464
0298	S	D	E	N	1964	36100
0298	S	D	E	N	1964	57860
0298	S	D	E	N	1974	78685
0298	S	D	E	N	1976	89183
0299	P	D	E	N	1972	68980
0299	P	T	E	N	1972	68980
0299	S	S	E	N	1966	34136
0299	S	D	E	N	1965	46419
0299	S	S	E	N	1968	53122
0299	S	D	R	N	1969	61103
0299	S	S	E	N	1971	66306
0299	S	S	R	N	1971	68576
0299	S	S	E	N	1971	69328
0299	S	D	E	N	1972	69340
0299	S	T	E	N	1978	96140
0303	S	D	E	N	1971	63739
0307	P	D	E	N	1967	46432
0307	P	D	E	N	1969	57189
0307	P	S	E	N	1972	66579
0307	P	D	E	N	1972	70141
0307	P	E	E	N	1972	70141
0307	S	D	E	N	1964	35550
0307	S	D	E	N	1966	43477
0307	S	D	F	N	1971	65152
0307	S	D	E	N	1973	73153
0308	P	D	E	N	1978	99103
0308	S	D	E	N	1974	78685
0330	P	D	E	N	1969	57189
0330	S	D	E	N	1965	39312
0330	S	D	E	N	1967	54966
0330	S	D	E	L	1970	62676
0330	S	D	E	N	1973	73153
0341	P	D	E	N	1979	99163
0341	S	D	E	N	1978	96156
0341	S	D	E	N	1979	99162
0346	P	D	E	N	1973	73529
0346	S	D	F	N	1973	72371
0346	S	D	E	N	1973	72419
0346	S	D	E	N	1908	96759
0349	S	D	E	N	1976	89183
0351	P	D	E	N	1968	51709
0351	P	D	E	N	1970	65200
0351	P	D	E	N	1975	83895
0352	P	D	E	N	1966	33774
0352	P	D	E	N	1966	41239
0352	P	D	E	N	1965	42772
0352	P	D	E	N	1967	46432
0352	S	D	E	N	1966	33774
0352	S	D	E	N	1964	35550
0352	S	D	E	N	1965	39312
0352	S	D	E	N	1965	42772
0352	S	D	E	N	1974	78685
0352	S	D	E	N	1979	97563
0356	S	D	E	N	1964	35550
0356	S	D	E	N	1974	78685
0362	P	D	E	N	1973	73529
0362	P	D	E	N	1979	99163
0362	S	D	E	N	1979	99162
0363	P	D	E	N	1969	57189
0364	P	D	E	N	1969	57189
0364	S	D	E	N	1974	78685
521-0366	P	G	F	N	1972	68386
0366	P	D	E	N	1979	97563
0366	S	D	R	N	1964	42175
0366	S	D	E	N	1967	46425
0366	S	D	F	N	1970	59429
0366	S	D	R	N	1971	64719
0366	S	D	E	N	1971	64720
0366	S	D	R	N	1972	66892
0366	S	G	F	N	1972	68386
0366	S	D	E	N	1973	75868
0366	S	E	E	N	1973	75868
0366	S	D	E	N	1908	96759
0366	S	D	E	N	1979	97563
0368	M	D	E	N	1966	33879
0368	M	D	E	N	1966	39008
0368	P	D	E	N	1966	33879
0368	P	D	E	N	1966	48369
0368	P	D	E	N	1968	50190
0368	S	D	E	N	1967	46650
0368	S	D	E	N	1965	77521
0369	P	D	E	N	1966	33774
0369	P	D	E	N	1965	42772
0369	S	D	E	N	1966	33774
0369	S	D	R	N	1965	42693
0369	S	D	E	N	1966	42694
0369	S	D	E	N	1965	42772
0369	S	D	E	N	1966	43477
0369	S	D	E	L	1970	62676
0370	S	D	E	N	1967	54966
0382	P	D	R	N	1972	74170
0382	P	D	E	N	1972	74171
0382	S	D	E	N	1973	71795
0384	M	D	E	N	1966	33879
0384	P	D	E	N	1966	33879
0384	P	D	E	N	1972	64449
0384	P	D	E	N	1970	77522
0385	S	S	E	N	1972	66579
0385	S	D	E	N	1972	71762
0392	P	D	E	N	1973	73529
0392	P	D	E	N	1975	83895
0392	S	D	R	N	1968	56285
0392	S	D	E	N	1906	96758
0394	P	D	E	N	1964	50030
0394	S	D	E	N	1967	46650
0396	S	D	E	N	1971	66376
0396	S	D	F	N	1973	71786
0396	S	D	E	N	1908	96759
0402	P	D	E	N	1966	33879
0409	S	D	F	N	1973	71786
0411	S	D	E	N	1975	85218
0412	P	D	F	N	1967	34629
0412	P	G	F	N	1972	68386
0412	P	D	E	N	1973	73529
0412	P	D	E	N	1954	78248
0412	S	D	F	N	1967	34629
0412	S	D	F	N	1970	59429
0412	S	D	E	N	1975	85218
0413	M	D	E	N	1966	59535
0413	M	D	E	N	1966	59540
0413	P	D	E	N	1966	45538
0413	P	D	E	N	1967	52919
0413	S	D	E	N	1965	38998
0413	S	D	E	N	1965	42776
0413	S	T	E	N	1965	42776
0413	S	D	R	N	1965	53622
0413	S	D	E	N	1969	53623
0413	S	D	E	N	1969	55600
0413	S	D	E	N	1971	64871
0413	S	D	R	N	1971	64993
0413	S	D	E	N	1972	66640
0413	S	T	R	N	1972	68884
0413	S	T	E	N	1972	68885
0413	S	D	E	N	1971	69194
0413	S	D	E	N	1972	69810
0413	S	G	E	N	1973	70824
0413	S	D	E	N	1971	70955
0413	S	T	E	N	1971	70955
0413	S	D	E	N	1971	70976
0413	S	E	E	N	1971	70976
0413	S	D	E	N	1972	72549
0413	S	D	E	N	1975	86711
0413	S	D	E	N	1978	100061
0414	P	D	E	N	1966	39007
0414	P	D	E	N	1965	50543
0414	P	D	E	N	1967	52919
0414	P	D	R	N	1965	53622
0414	P	D	E	N	1969	53623
0414	P	S	E	N	1972	66579
0414	S	D	E	N	1965	35623
0414	S	D	E	N	1967	38580
0414	S	D	R	N	1964	42175
0414	S	D	E	N	1966	48369
0414	S	D	E	N	1968	49893
0414	S	D	E	N	1965	50543
0414	S	D	R	N	1965	53622
0414	S	D	E	N	1969	53623
0414	S	D	E	N	1971	63739
0414	S	S	E	N	1972	66579
0414	S	D	E	N	1978	96156
0421	S	D	E	N	1972	69895
0421	S	D	E	N	1972	70324

Phys. State: **C.** Superconductive; **D.** Doped; **E.** Expanded; **F.** Fibrous or Whisker; **G.** Gas; **I.** Ionized or Plasma; **L.** Liquid; **M.** Multiphase; **P.** Powder or Fine Particle; **S.** Solid; **T.** Thick or Thin Film
Subject: **D.** Data; **E.** Experiment; **G.** General (Data + Theory + Experiment); **S.** Survey (Review, Compendium etc.); **T.** Theory
Language: **C.** Czech; **D.** Dutch; **E.** English; **F.** French; **G.** German; **I.** Italian; **J.** Japanese; **O.** Other Languages; **P.** Polish; **R.** Russian; **S.** Spanish
Temperature: **F.** Full Range (Low + Normal + High); **L.** Low (0 to 75K) + Overlap into Normal; **N.** Normal (75 to 1273K); **H.** High (above 1273K) + Overlap into Normal; Blank = Not Coded

Substance Number	Phys. State	Subject	Language	Temperature	Year	TPRC Number
521-0421	S	D	E	N	1976	86771
0421	S	D	E	N	1977	92103
0422	M	D	R	N	1973	72910
0422	M	T	R	N	1973	72910
0422	M	D	R	N	1972	73320
0422	S	D	E	N	1966	40479
0422	S	T	E	N	1966	40479
0422	S	D	E	N	1966	47643
0422	S	E	E	N	1966	47643
0422	S	D	E	N	1972	69810
0422	S	D	E	N	1972	70303
0423	M	D	E	N	1967	45739
0423	S	D	E	N	1965	34834
0423	S	D	E	N	1965	34848
0423	S	D	E	N	1967	47238
0423	S	D	E	N	1971	64445
0423	S	E	E	N	1971	64445
0423	S	G	E	N	1971	65946
0423	S	D	E	N	1972	68850
0423	S	E	E	N	1972	68850
0423	S	T	F	N	1972	69066
0423	S	D	E	N	1972	77765
0423	S	D	E	N	1975	87491
0423	S	T	E	N	1975	87491
0423	S	D	E	N	1979	99698
0423	S	E	E	N	1979	99698
0424	G	D	R	N	1971	68472
0424	G	E	R	N	1971	68472
0424	M	D	E	N	1965	42389
0424	S	D	E	N	1966	40479
0424	S	T	E	N	1966	40479
0424	S	D	E	N	1966	47643
0424	S	E	E	N	1966	47643
0424	S	D	E	N	1972	69810
0424	S	D	E	N	1973	70311
0424	S	E	E	N	1973	70311
0424	S	D	E	N	1974	76412
0424	S	D	E	N	1975	83343
0424	S	D	E	N	1978	95930
0425	G	D	R	N	1971	68473
0425	S	D	E	N	1972	69810
0425	S	D	E	N	1973	70822
0425	S	D	R	N	1973	70912
0425	S	D	E	N	1971	70976
0425	S	E	E	N	1971	70976
0425	S	D	E	N	1973	71376
0425	S	D	E	N	1973	73943
0425	S	D	E	N	1974	76412
0429	S	D	E	N	1971	63739
0429	S	D	E	N	1906	96758
0429	S	D	E	N	1979	99162
0430	P	D	E	N	1969	57189
0430	P	D	E	N	1970	65200
0430	P	D	E	N	1975	83895
0430	S	D	E	N	1966	39134
0430	S	D	E	N	1972	73127
0430	S	D	E	N	1906	96758
0432	S	D	E	N	1976	89183
0442	M	D	E	N	1966	33879
0442	P	D	E	N	1966	33879
0442	P	S	E	N	1972	66579
0442	P	D	E	N	1968	76340
0442	S	D	E	N	1962	42139
0442	S	D	E	N	1967	46650
0442	S	D	E	N	1972	77038
0442	S	D	E	N	1974	77316
0442	S	D	E	N	1976	96156
0443	P	D	E	N	1973	73529
0445	S	D	F	N	1970	59429
0451	G	D	E	N	1970	57798
0451	M	D	E	N	1970	59127
0451	S	D	E	N	1967	46650
0451	S	D	E	N	1970	59127
0470	S	D	E	N	1966	40528
0470	S	D	R	N	1964	42175
0470	S	D	E	N	1966	43477
0470	S	D	E	N	1967	46425
0470	S	D	E	N	1964	57860
0470	S	T	E	N	1968	59170
0470	S	D	E	N	1971	63739
0471	P	D	E	N	1969	57189
0471	P	D	E	N	1970	65200
0471	P	D	E	N	1973	73529
0471	P	D	E	N	1975	83895
0471	S	D	E	N	1968	49893
0471	S	D	E	N	1978	99431
0472	S	D	F	N	1970	59429
0472	S	D	E	N	1972	71762
0477	P	D	E	N	1964	50030
0478	S	D	E	N	1974	78685
0481	S	D	E	N	1967	46650
0489	P	D	E	N	1970	65200
0494	S	D	E	N	1976	85517
0494	S	E	E	N	1976	85517
0495	P	D	E	N	1970	65200
0495	S	D	R	N	1968	56285
0496	S	D	F	N	1971	65152
0505	P	D	E	N	1973	73529
0505	S	D	E	N	1972	66852
0506	P	D	E	N	1973	73529
0506	S	D	E	N	1972	66852

Substance Number	Phys. State	Subject	Language	Temperature	Year	TPRC Number
521-0514	P	D	E	N	1965	36525
0514	P	D	E	N	1962	42292
0514	S	S	E	N	1972	66579
0516	P	D	E	N	1965	36525
0516	P	D	E	N	1962	42292
0523	M	D	E	N	1966	33879
0525	S	D	E	N	1966	33879
0530	S	D	E	N	1967	38580
0533	S	D	F	N	1970	59429
0542	M	D	E	N	1966	33879
0542	P	D	E	N	1966	33879
0546	S	D	E	N	1966	33879
0548	P	D	E	N	1966	33879
0550	S	D	E	N	1966	33879
0552	P	D	E	N	1970	65200
0552	P	D	E	N	1973	73529
0552	P	D	E	N	1975	83895
0552	S	D	E	H	1967	34861
0552	S	D	E	N	1966	48370
0552	S	S	E	N	1972	66579
0552	S	D	E	N	1978	99431
0554	S	D	E	N	1906	96758
0555	S	D	R	N	1964	42175
0555	S	D	R	N	1971	65250
0563	S	D	F	N	1970	59429
0563	S	D	E	N	1972	71762
0563	S	D	E	N	1906	96758
0566	P	D	E	N	1975	83895
0566	S	D	F	N	1967	40913
0566	S	D	E	L	1970	62676
0566	S	D	E	N	1971	66376
0566	S	D	E	N	1973	75868
0566	S	E	E	N	1973	75868
0566	S	D	E	N	1908	96759
0599	S	D	E	N	1978	96156
0604	P	D	E	N	1965	34090
0604	P	D	E	N	1965	50543
0604	P	S	E	N	1972	66579
0604	P	D	E	N	1975	83895
0604	S	D	E	N	1969	35541
0604	S	D	E	N	1965	50543
0604	S	D	E	N	1971	63739
0604	S	S	E	N	1972	66579
0604	S	D	E	N	1973	71795
0604	S	D	E	N	1973	75868
0604	S	E	E	N	1973	75868
0604	S	D	E	N	1974	81307
0604	S	E	E	N	1974	81307
0604	S	D	E	N	1978	96156
0604	S	D	E	N	1979	99162
0611	S	D	F	N	1967	40913
0611	S	D	E	N	1966	43477
0611	S	D	F	N	1970	59429
0611	S	D	E	N	1971	66376
0611	S	D	E	N	1908	96759
0612	P	D	E	N	1973	71872
0612	S	D	E	N	1965	77521
0613	P	D	E	N	1966	33881
0613	P	D	E	N	1966	34538
0613	P	D	E	N	1966	39007
0613	P	D	E	N	1966	48369
0613	P	S	E	N	1972	66579
0613	P	D	E	N	1973	72869
0613	P	D	E	N	1970	77522
0613	P	D	E	N	1968	77523
0613	P	E	E	N	1968	77523
0613	S	D	E	N	1965	40480
0613	S	D	E	N	1968	49669
0614	P	G	E	N	1973	71027
0614	P	G	E	N	1973	73838
0614	S	D	F	N	1967	40913
0614	S	D	E	N	1971	63136
0614	S	D	E	N	1971	66376
0614	S	G	E	N	1973	71027
0614	S	D	F	N	1973	71786
0614	S	G	E	N	1973	73838
0626	S	D	F	N	1967	40913
0626	S	D	E	N	1908	96759
0643	P	D	E	N	1973	71872
0643	S	D	E	N	1967	52855
0643	S	D	E	N	1965	77521
0644	S	D	R	N	1964	42175
0644	S	D	E	N	1967	46425
0644	S	D	E	N	1971	63739
0645	P	D	E	N	1970	77522
0645	S	D	E	N	1967	46650
0645	S	D	E	N	1972	77038
0645	S	D	E	N	1976	96156
0648	P	D	E	N	1965	36525
0648	P	D	E	N	1975	83895
0648	S	D	G	N	1964	36713
0648	S	E	G	N	1964	36713
0650	S	D	E	N	1966	43477
0650	S	D	E	N	1964	57860
0650	S	D	F	N	1970	59429
0650	S	D	E	N	1906	96758
0660	S	D	R	N	1968	55694
0674	P	D	E	N	1966	34538
0674	S	D	E	N	1965	40480
0675	S	D	E	N	1966	41373
0675	S	D	E	N	1971	62141

Substance Number	Phys. State	Subject	Language	Temperature	Year	TPRC Number
521-0675	S	D	E	N	1973	71849
0675	S	D	E	N	1908	96759
0676	P	D	E	N	1966	39007
0676	P	D	E	N	1966	48369
0676	S	D	E	N	1965	34262
0676	S	D	E	N	1967	46650
0677	P	D	E	N	1966	39007
0677	S	D	E	N	1970	60103
0682	S	D	E	N	1945	46837
0684	S	D	E	N	1945	46837
0687	S	D	E	N	1972	73127
0688	P	D	E	N	1966	33774
0688	P	D	E	N	1965	42772
0688	S	D	E	N	1966	33774
0688	S	D	E	N	1966	33879
0688	S	D	E	N	1965	42772
0688	S	D	E	N	1974	78685
0688	S	D	E	N	1976	89182
0688	S	D	E	N	1976	96156
0690	M	D	E	N	1966	33879
0692	P	D	E	N	1966	33879
0693	P	D	E	N	1966	33879
0694	S	D	E	N	1966	33879
0694	S	D	E	N	1976	96156
0696	S	D	F	N	1970	59429
0709	S	D	E	N	1978	96156
0710	S	D	G	N	1969	55278
0710	S	D	E	N	1970	60195
0713	S	D	R	N	1971	67298
0719	M	D	E	N	1969	55600
0719	P	D	E	N	1966	45538
0719	P	D	E	N	1970	61071
0719	S	D	O	N	1971	64601
0720	P	D	E	N	1970	65200
0720	S	D	F	N	1970	59429
0720	S	D	G	N	1971	64274
0727	S	D	E	N	1976	89182
0727	S	D	E	N	1976	96156
0728	S	D	E	N	1969	52594
0728	S	D	E	N	1976	89183
0732	D	D	R	N	1970	61281
0732	D	D	E	N	1971	61282
0732	P	D	E	N	1965	42772
0732	P	D	E	N	1969	57189
0732	P	D	G	N	1971	63552
0732	P	E	G	N	1971	63552
0732	P	G	E	N	1971	64971
0732	P	D	F	N	1971	65152
0732	P	D	E	N	1972	69102
0732	P	E	E	N	1972	69102
0732	P	D	E	N	1972	70141
0732	P	E	E	N	1972	70141
0732	P	S	E	N	1972	71106
0732	P	D	E	N	1954	78248
0732	S	D	E	N	1969	33888
0732	S	T	E	N	1969	33888
0732	S	D	R	N	1969	33889
0732	S	T	R	N	1969	33889
0732	S	D	E	N	1964	35550
0732	S	D	E	N	1964	35856
0732	S	D	E	N	1963	40420
0732	S	G	E	L	1962	40665
0732	S	D	E	N	1955	41955
0732	S	D	E	N	1965	42064
0732	S	D	E	N	1965	42764
0732	S	D	E	N	1966	43477
0732	S	D	G	N	1898	46995
0732	S	D	E	L	1966	48370
0732	S	D	E	N	1969	55822
0732	S	D	R	N	1970	63800
0732	S	D	R	N	1969	77128
0732	S	D	E	N	1969	77129
0732	S	G	F	N	1973	78785
0732	S	D	E	N	1919	95768
0732	S	D	E	N	1906	96758
0732	S	D	E	N	1908	96759
0735	S	D	G	N	1965	43102
0736	S	D	E	N	1965	42979
0736	S	D	E	N	1967	49233
0736	S	S	E	N	1972	66579
0736	S	D	E	N	1908	96759
0737	S	D	E	N	1965	42979
0737	S	D	E	N	1966	43477
0737	S	D	E	N	1964	57860
0737	S	D	F	N	1970	59429
0737	S	D	E	N	1908	96759
0738	P	D	E	N	1971	62451
0738	P	D	E	N	1971	65035
0738	P	E	E	N	1971	65035
0738	P	D	E	N	1973	71846
0738	S	D	R	N	1965	41690
0738	S	D	R	N	1965	42494
0738	S	D	R	N	1965	42542
0738	S	D	R	N	1965	42543
0738	S	D	R	N	1965	43084
0738	S	S	R	N	1967	48541
0738	S	D	E	N	1968	49087
0738	S	D	G	N	1968	49510
0738	S	D	E	N	1967	50619
0738	S	D	G	N	1968	50847
0738	S	T	G	N	1968	50847

Phys. State: **C.** Superconductive; **D.** Doped; **E.** Expanded; **F.** Fibrous or Whisker; **G.** Gas; **I.** Ionized or Plasma; **L.** Liquid; **M.** Multiphase; **P.** Powder or Fine Particle; **S.** Solid; **T.** Thick or Thin Film
Subject: **D.** Data; **E.** Experiment; **G.** General (Data + Theory + Experiment); **S.** Survey (Review, Compendium etc.); **T.** Theory
Language: **C.** Czech; **D.** Dutch; **E.** English; **F.** French; **G.** German; **I.** Italian; **J.** Japanese; **O.** Other Languages; **P.** Polish; **R.** Russian; **S.** Spanish
Temperature: **F.** Full Range (Low + Normal + High); **L.** Low (0 to 75K) + Overlap into Normal; **N.** Normal (75 to 1273K); **H.** High (above 1273K) + Overlap into Normal; Blank = Not Coded

Substance Number	Phys. State	Subject	Language	Temperature	Year	TPRC Number
521-0738	S	D	R	N	1968	55537
0738	S	E	R	N	1968	55537
0738	S	D	R	N	1968	55692
0738	S	D	R	N	1968	55693
0738	S	D	R	N	1968	55694
0738	S	D	G	N	1970	57539
0738	S	D	R	N	1970	59304
0738	S	D	E	N	1972	66213
0738	S	D	E	N	1972	66214
0738	S	D	E	N	1972	69537
0738	S	T	R	N	1976	91436
0738	S	D	E	N	1976	91694
0738	S	D	R	N	1975	92746
0738	S	D	E	N	1960	93150
0738	S	E	E	N	1960	93150
0738	S	D	E	N	1960	93151
0738	S	D	E	N	1961	93152
0744	S	D	R	N	1964	42175
0744	S	D	G	N	1971	64343
0744	S	D	R	N	1973	72392
0744	S	D	E	N	1973	75868
0744	S	E	E	N	1973	75868
0744	S	D	E	N	1974	81307
0744	S	E	E	N	1974	81307
0744	S	D	E	N	1960	90396
0744	S	E	E	N	1960	90396
0745	S	D	R	N	1964	42175
0746	S	D	R	N	1964	42175
0746	S	D	E	N	1965	42964
0746	S	D	E	N	1971	63771
0746	S	D	E	N	1973	71157
0746	S	D	E	N	1973	75868
0746	S	E	E	N	1973	75868
0746	S	D	E	N	1974	81307
0746	S	E	E	N	1974	81307
0746	S	D	E	N	1960	90396
0746	S	E	E	N	1960	90396
0746	S	D	E	N	1908	96759
0747	S	D	R	N	1964	42175
0748	S	D	R	N	1964	42175
0749	S	D	R	N	1964	42175
0749	S	D	F	N	1972	66730
0749	S	D	R	N	1971	66906
0749	S	D	R	N	1977	88573
0749	S	D	E	N	1977	92560
0750	S	D	R	N	1964	42175
0750	S	D	R	N	1966	44608
0751	S	D	R	N	1965	41710
0751	S	D	R	N	1964	42175
0751	S	D	E	N	1973	75868
0751	S	E	E	N	1973	75868
0751	S	D	E	N	1974	81307
0751	S	E	E	N	1974	81307
0752	S	D	R	N	1964	42175
0753	S	D	R	N	1964	42175
0753	S	D	E	N	1973	74443
0753	S	D	O	N	1976	91485
0753	S	D	E	N	1978	94616
0754	S	D	R	N	1964	42175
0754	S	D	E	N	1978	94616
0755	S	D	R	N	1965	41710
0755	S	D	R	N	1964	42175
0756	S	D	R	N	1964	42175
0756	S	D	O	N	1965	43202
0756	S	E	O	N	1965	43202
0756	S	D	E	N	1908	96759
0757	S	D	R	N	1964	42175
0757	S	D	E	N	1971	62140
0757	S	D	E	N	1973	75868
0757	S	E	E	N	1973	75868
0757	S	D	E	N	1975	77670
0757	S	D	E	N	1974	81307
0757	S	E	E	N	1974	81307
0758	S	D	R	N	1964	42175
0758	S	D	R	N	1960	44197
0758	S	D	F	N	1960	44198
0758	S	D	R	N	1969	59194
0758	S	D	E	N	1970	61942
0758	S	D	R	N	1971	63609
0758	S	D	R	N	1971	64869
0758	S	D	E	N	1973	75868
0758	S	E	E	N	1973	75868
0758	S	D	E	N	1974	81307
0758	S	E	E	N	1974	81307
0759	S	D	F	N	1970	59429
0762	S	D	E	N	1974	78685
0764	D	D	E	N	1978	96156
0764	P	D	E	N	1970	65200
0764	P	D	E	N	1973	73529
0764	P	D	E	N	1975	83895
0764	P	D	E	N	1979	99163
0764	S	D	E	N	1972	66852
0764	S	D	E	N	1978	96156
0764	S	D	E	N	1979	99162
0765	P	D	E	N	1973	71872
0767	S	D	R	N	1964	42175
0768	P	D	E	N	1968	50190
0769	P	D	E	N	1968	51709
0769	P	D	E	N	1975	83895
0769	S	D	E	N	1966	48370
0769	S	S	E	N	1972	66579

Substance Number	Phys. State	Subject	Language	Temperature	Year	TPRC Number
521-0769	S	D	E	N	1906	96758
0771	S	T	E	N	1968	49567
0772	P	D	E	N	1964	50030
0779	P	D	E	N	1973	73529
0780	P	D	E	N	1973	73529
0780	S	D	E	N	1906	96758
0783	P	D	E	N	1973	73529
0783	S	D	G	N	1971	64274
0783	S	D	G	N	1971	66369
0783	S	D	E	N	1906	96758
0784	S	D	F	N	1970	59429
0784	S	D	E	N	1906	96758
0785	S	D	F	N	1970	59429
0785	S	D	E	N	1906	96758
0786	S	D	F	N	1970	59429
0790	P	D	E	N	1968	51709
0790	P	D	E	N	1975	83895
0790	S	D	E	N	1976	85517
0790	S	E	E	N	1976	85517
0791	P	D	E	N	1968	51709
0791	P	D	E	N	1970	65200
0792	P	D	E	N	1968	51709
0792	P	D	E	N	1973	73529
0792	P	D	E	N	1975	83895
0792	P	D	E	N	1978	99103
0793	P	D	E	N	1968	51709
0793	P	D	E	N	1970	65200
0794	P	D	E	N	1968	51709
0794	P	D	E	N	1969	57189
0794	P	D	E	N	1970	65200
0794	P	D	E	N	1973	73529
0794	P	D	E	N	1975	83895
0794	P	D	E	N	1978	99103
0794	S	D	E	L	1970	62676
0794	S	D	E	N	1978	98168
0796	S	D	E	N	1968	46105
0816	S	D	R	N	1965	42494
0816	S	D	E	N	1968	49087
0816	S	D	R	N	1968	55693
0816	S	D	E	N	1972	66213
0819	S	D	E	N	1966	43477
0819	S	D	E	N	1964	57860
0825	S	D	E	N	1976	89183
0826	S	D	R	N	1965	41710
0826	S	D	E	N	1967	46425
0827	S	D	E	N	1967	46425
0827	S	D	E	N	1973	75868
0827	S	E	E	N	1973	75868
0827	S	D	E	N	1974	81307
0827	S	E	E	N	1974	81307
0828	P	D	E	N	1967	46432
0828	P	S	E	N	1972	66579
0835	S	D	E	N	1974	78685
0835	S	D	E	N	1976	89183
0838	S	D	E	N	1966	47072
0838	S	D	E	N	1978	94616
0848	S	D	E	N	1976	87489
0851	S	D	O	N	1967	45409
0852	P	D	E	N	1969	57189
0852	S	D	E	N	1965	39312
0852	S	D	E	N	1979	97563
0854	S	D	G	N	1968	52383
0854	S	T	G	N	1968	52383
0858	S	D	G	N	1971	64274
0858	S	D	F	N	1973	72371
0858	S	D	E	H	1973	72822
0859	P	D	E	N	1975	83895
0859	S	D	E	N	1978	99431
0861	P	D	E	N	1973	73529
0863	S	D	E	N	1970	60195
0865	P	D	E	N	1972	64449
0865	P	D	E	N	1970	77522
0872	S	D	E	N	1966	45486
0874	S	D	E	N	1968	52945
0877	S	D	E	N	1965	42966
0877	S	D	F	N	1971	63774
0877	S	E	F	N	1971	63774
0877	S	D	E	N	1971	65030
0877	S	D	E	N	1908	96759
0878	S	S	E	N	1972	66579
0878	S	D	E	N	1960	93151
0878	S	D	E	N	1977	93700
0878	S	D	E	N	1906	96758
0879	S	D	E	N	1931	40580
0879	S	D	E	N	1971	63739
0879	S	D	E	N	1908	96759
0880	S	D	R	N	1965	41710
0880	S	D	R	N	1960	44197
0880	S	D	F	N	1960	44198
0880	S	D	E	N	1973	75868
0880	S	E	E	N	1973	75868
0880	S	D	E	N	1974	81307
0880	S	E	E	N	1974	81307
0880	S	D	F	N	1976	88837
0880	S	D	E	N	1977	91731
0881	S	D	R	N	1960	44197
0881	S	D	F	N	1960	44198
0882	S	D	R	N	1960	44197
0882	S	D	F	N	1960	44198
0883	S	D	R	N	1968	48997
0884	S	D	R	N	1968	49133

Substance Number	Phys. State	Subject	Language	Temperature	Year	TPRC Number
521-0887	P	D	E	N	1966	41239
0887	P	D	E	N	1968	51709
0887	P	D	E	N	1978	98168
0887	S	D	E	N	1978	98168
0895	S	D	E	N	1976	87489
0902	S	D	E	N	1976	89183
0911	P	D	E	N	1970	65200
0912	P	D	E	N	1970	65200
0917	D	D	E	N	1971	34077
0917	P	D	E	N	1971	34077
0917	P	D	E	N	1970	56102
0917	P	D	E	N	1970	56103
0917	P	D	E	N	1970	56104
0917	P	D	E	N	1970	61099
0917	P	S	E	N	1972	66579
0917	P	D	E	N	1972	66830
0917	P	D	E	N	1972	68484
0917	P	D	E	N	1972	68911
0917	P	E	E	N	1972	68911
0917	P	D	E	N	1973	70043
0917	P	D	E	N	1972	70058
0917	P	D	E	N	1972	70060
0917	P	D	E	N	1972	70141
0917	P	E	E	N	1972	70141
0917	P	D	E	N	1972	70817
0917	P	D	E	N	1973	71054
0917	P	D	E	N	1973	73151
0917	P	D	E	N	1973	73307
0917	P	E	E	N	1973	73307
0917	P	D	E	N	1973	73329
0917	P	D	E	N	1973	78661
0917	P	D	E	N	1974	78683
0917	P	D	E	N	1975	86304
0917	P	D	E	N	1974	89708
0917	P	D	E	N	1978	99103
0917	S	D	E	N	1970	61099
0917	S	D	E	N	1976	92161
0917	S	D	E	N	1979	98302
0917	S	D	E	N	1979	98328
0917	S	E	E	N	1979	98328
0919	S	D	E	N	1970	56101
0920	S	D	E	N	1970	56101
0921	S	D	E	N	1970	56101
0922	P	D	E	N	1970	56104
0922	P	D	E	N	1970	56105
0922	P	D	E	N	1970	56108
0922	P	D	E	N	1972	70058
0922	P	D	E	N	1972	70061
0922	P	D	E	N	1971	70142
0922	P	D	E	N	1975	86304
0922	S	D	E	N	1970	56102
0922	S	D	E	N	1970	56104
0922	S	D	E	N	1970	56105
0922	S	D	E	N	1970	56108
0922	S	S	E	N	1972	66579
0922	S	D	E	N	1973	70043
0922	S	D	E	N	1972	70060
0922	S	D	E	N	1972	74269
0923	S	D	O	N	1971	64601
0924	S	D	E	N	1965	39312
0928	S	D	R	N	1965	41710
0928	S	D	E	N	1970	58974
0928	S	D	E	N	1973	75868
0928	S	E	E	N	1973	75868
0929	S	D	R	N	1965	41710
0929	S	D	E	N	1970	58974
0930	P	D	E	N	1973	73529
0931	S	D	E	N	1969	35412
0933	S	D	F	N	1971	62702
0934	S	D	E	N	1969	35412
0934	S	D	E	N	1967	46650
0937	S	D	E	N	1969	59001
0938	S	D	E	N	1969	59001
0939	P	D	E	N	1973	73529
0940	P	D	E	N	1973	73529
0941	P	D	E	N	1973	73529
0942	P	D	E	N	1973	73529
0942	P	D	E	N	1975	83895
0943	S	D	E	N	1978	98168
0946	P	D	E	N	1973	73529
0947	P	D	E	N	1975	83895
0948	S	D	E	N	1906	96758
0950	P	D	E	N	1970	65200
0950	P	D	E	N	1973	73529
0950	P	D	E	N	1975	83895
0951	P	D	E	N	1969	57189
0951	P	D	E	N	1970	65200
0951	P	D	E	N	1973	73529
0951	P	D	E	N	1975	83895
0951	S	D	E	N	1906	96758
0952	P	D	E	N	1973	73529
0953	S	D	E	N	1965	35672
0954	S	D	E	N	1965	35672
0956	P	D	E	N	1973	73529
0957	P	D	E	N	1973	73529
0957	S	D	E	N	1975	85218
0958	P	D	E	N	1973	73529
0958	S	D	E	N	1975	85218
0959	P	G	F	N	1972	68386
0959	P	D	E	N	1973	73529
0959	S	D	F	N	1970	59429

Phys. State: **C.** Superconductive; **D.** Doped; **E.** Expanded; **F.** Fibrous or Whisker; **G.** Gas; **I.** Ionized or Plasma; **L.** Liquid; **M.** Multiphase; **P.** Powder or Fine Particle; **S.** Solid; **T.** Thick or Thin Film

Subject: **D.** Data; **E.** Experiment; **G.** General (Data + Theory + Experiment); **S.** Survey (Review, Compendium etc.); **T.** Theory

Language: **C.** Czech; **D.** Dutch; **E.** English; **F.** French; **G.** German; **I.** Italian; **J.** Japanese; **O.** Other Languages; **P.** Polish; **R.** Russian; **S.** Spanish

Temperature: **F.** Full Range (Low + Normal + High); **L.** Low (0 to 75K) + Overlap into Normal; **N.** Normal (75 to 1273K); **H.** High (above 1273K) + Overlap into Normal; Blank = Not Coded

Substance Number	Phys. State	Subject	Language	Temperature	Year	TPRC Number
521-0959	S	G	F	N	1972	68386
0959	S	D	E	N	1975	85218
0960	P	D	E	N	1973	73529
0962	S	D	F	N	1970	59429
0965	P	D	E	N	1973	73529
0966	S	D	E	N	1965	35672
0967	S	D	E	N	1906	96758
0969	P	D	E	N	1970	65200
0969	P	D	E	N	1975	83895
0969	S	D	E	N	1978	99431
0970	P	G	F	N	1972	68386
0970	P	D	E	N	1973	73529
0970	S	G	F	N	1972	68386
0971	S	D	E	N	1969	55641
0973	P	D	E	N	1969	57189
0974	P	D	E	N	1969	57189
0975	P	D	E	N	1969	57189
0976	P	D	E	N	1969	57189
0976	S	D	E	N	1967	54966
0978	P	D	E	N	1975	83895
0978	S	D	E	N	1965	39312
0979	S	D	E	N	1965	39312
0981	S	D	E	N	1970	60195
0984	S	D	E	N	1966	43477
0984	S	D	E	N	1964	57860
0985	S	D	E	N	1966	43477
0985	S	D	E	N	1964	57860
0986	S	D	E	N	1966	43477
0986	S	D	E	N	1964	57860
0987	S	D	E	N	1964	57860
0992	S	D	E	N	1965	35672
0997	S	D	R	N	1965	42693
0997	S	D	E	N	1966	42694
0999	P	D	E	N	1965	42772
0999	S	D	E	N	1965	42772
1001	P	D	E	N	1969	59106
1001	P	T	E	N	1969	59106
1002	S	D	O	N	1968	49556
1003	S	D	R	N	1966	44608
1004	S	D	R	N	1966	44608
1005	S	D	R	N	1966	44608
1006	S	D	R	N	1969	55648
1006	S	D	E	N	1978	94616
1011	S	D	E	N	1974	78685
1012	S	D	E	N	1965	35672
1012	S	D	E	N	1971	63739
1013	S	D	E	N	1965	35672
1014	S	D	E	N	1965	35672
1015	S	D	E	N	1965	35672
1015	S	D	E	N	1971	63739
1016	S	D	E	N	1965	35672
1017	S	D	E	N	1965	35672
1018	S	D	E	N	1965	35672
1019	S	D	E	N	1965	35672
1020	S	D	E	N	1965	35672
1021	S	D	E	N	1965	35672
1022	S	D	E	N	1965	35672
1023	S	D	E	N	1965	35672
1024	S	D	E	N	1965	35672
1025	S	D	E	N	1965	35672
1028	S	G	F	N	1971	65747
1028	S	D	E	N	1975	80884
1028	S	D	E	N	1976	89182
1034	S	D	E	L	1970	62676
1035	S	D	E	N	1976	89183
1036	M	D	E	N	1970	60103
1036	P	D	E	N	1970	60103
1038	P	D	E	N	1970	65008
1043	P	D	E	N	1973	73529
1043	S	D	E	L	1970	62676
1045	P	D	E	N	1975	83895
1045	S	D	E	L	1970	62676
1047	P	D	E	N	1970	65200
1047	P	D	E	N	1973	73529
1047	S	D	E	L	1970	62676
1050	M	D	E	N	1970	59127
1050	S	D	E	N	1970	59127
1055	S	D	E	N	1967	54966
1058	G	D	E	N	1973	70822
1058	G	D	E	N	1973	73943
1058	S	D	E	N	1970	62513
1059	S	D	E	N	1972	70058
1068	S	D	E	N	1972	66640
1069	S	D	E	N	1972	66640
1070	S	D	E	N	1972	66640
1073	S	D	R	N	1967	46404
1076	S	D	R	N	1971	65250
1077	S	D	R	N	1971	65250
1081	P	D	E	N	1970	65200
1082	P	D	E	N	1970	65200
1083	S	D	R	N	1969	59194
1084	S	D	R	N	1969	59194
1085	S	D	R	N	1970	62413
1086	S	D	R	N	1970	62413
1087	P	D	R	N	1965	68331
1087	P	E	R	N	1965	68331
1087	P	D	E	N	1965	68332
1087	P	E	E	N	1965	68332
1087	S	D	E	N	1967	46650
1088	P	D	R	N	1965	68331
1088	P	E	R	N	1965	68331
521-1088	P	D	E	N	1965	68332
1088	P	E	E	N	1965	68332
1088	S	D	E	N	1967	46650
1089	S	T	R	N	1975	81872
1105	S	D	F	N	1971	65152
1106	S	G	F	N	1971	65747
1107	S	D	F	N	1971	62702
1107	S	G	F	N	1971	65747
1109	P	D	E	N	1973	73529
1110	S	D	F	N	1971	62702
1111	S	D	F	N	1971	62702
1115	S	D	G	N	1971	66369
1116	S	D	G	N	1971	66369
1117	S	D	G	N	1971	66369
1119	S	D	R	N	1965	42494
1119	S	D	R	N	1968	55693
1119	S	D	E	N	1972	66213
1120	S	D	R	N	1965	42494
1120	S	D	R	N	1968	55693
1120	S	D	E	N	1972	66213
1121	S	D	R	N	1965	42494
1121	S	D	R	N	1968	55693
1121	S	D	E	N	1972	66213
1122	S	D	R	N	1968	55693
1122	S	D	E	N	1970	60220
1129	S	D	F	N	1971	62702
1130	S	D	F	N	1971	62702
1131	P	D	E	N	1973	73529
1133	P	D	E	N	1975	83895
1134	P	D	E	N	1973	73529
1135	P	D	E	N	1973	73529
1137	P	D	E	N	1973	73529
1139	P	D	E	N	1973	73529
1147	S	D	E	N	1976	89183
1148	S	D	E	N	1906	96758
1149	S	D	E	N	1972	71762
1152	P	D	E	N	1972	64449
1152	P	D	E	N	1970	77522
1153	P	D	E	N	1972	64449
1153	P	D	E	N	1970	77522
1154	P	D	E	N	1972	64449
1154	P	D	E	N	1970	77522
1155	P	D	E	N	1972	64449
1155	P	D	E	N	1970	77522
1156	P	D	E	N	1972	64449
1156	P	D	E	N	1970	77522
1157	P	D	E	N	1972	64449
1157	P	D	E	N	1970	77522
1158	P	D	E	N	1972	64449
1158	P	D	E	N	1970	77522
1159	P	D	E	N	1972	64449
1159	P	D	E	N	1970	77522
1161	S	D	E	N	1971	64992
1161	S	D	E	N	1972	70287
1161	S	D	E	N	1974	75330
1162	P	G	F	N	1972	68386
1165	S	D	E	N	1973	72869
1173	S	D	F	N	1972	66864
1173	S	D	F	N	1972	69877
1190	S	D	R	N	1965	41710
1190	S	D	E	N	1972	68228
1190	S	D	E	N	1973	75868
1190	S	E	E	N	1973	75868
1190	S	D	E	N	1974	81307
1190	S	E	E	N	1974	81307
1191	S	D	R	N	1965	41710
1192	S	D	R	N	1965	41710
1193	S	D	R	N	1965	41710
1195	S	D	R	N	1965	41710
1196	S	D	R	N	1965	41710
1197	S	D	R	N	1965	41710
1198	S	D	R	N	1965	41710
1198	S	D	O	N	1976	91485
1199	S	D	R	N	1965	41710
1200	S	D	R	N	1965	41710
1201	S	D	R	N	1965	41710
1202	S	D	R	N	1965	41710
1203	S	D	R	N	1965	41710
1204	S	D	R	N	1965	41710
1205	S	D	R	N	1965	41710
1205	S	D	E	N	1973	75868
1205	S	E	E	N	1973	75868
1205	S	D	F	N	1976	88837
1205	S	D	E	N	1960	90396
1205	S	E	E	N	1960	90396
1205	S	D	E	N	1979	99174
1206	S	D	R	N	1965	41710
1207	S	D	R	N	1965	41710
1208	S	D	R	N	1968	55694
1213	S	D	P	N	1969	55680
1214	S	D	R	N	1968	50280
1214	S	D	E	N	1971	64014
1216	S	D	E	N	1971	63739
1219	M	D	E	N	1971	63703
1219	M	T	E	N	1971	63703
1220	S	D	R	N	1972	66874
1221	S	D	F	N	1971	63774
1221	S	E	F	N	1971	63774
1221	S	D	S	N	1971	67278
1221	S	D	E	N	1973	75868
1221	S	E	E	N	1973	75868
521-1221	S	D	E	N	1974	81307
1221	S	E	E	N	1974	81307
1222	S	D	S	N	1971	67277
1226	P	D	E	N	1973	71987
1226	S	D	E	N	1971	66222
1226	S	T	E	N	1971	66222
1226	S	D	E	N	1971	68875
1226	S	T	E	N	1973	73152
1226	S	D	E	N	1974	76745
1227	S	D	E	N	1971	66376
1227	S	D	F	N	1973	71786
1227	S	D	E	N	1908	96759
1254	S	D	E	N	1971	63771
1270	S	D	R	N	1972	70800
1273	S	D	E	N	1973	71157
1273	S	D	R	N	1975	81871
1273	S	D	E	N	1978	98168
1280	S	D	E	N	1972	71762
1281	S	D	E	N	1972	71762
1282	S	D	E	N	1972	71762
1283	S	D	E	N	1972	71762
1284	S	D	E	N	1972	71762
1285	P	D	E	N	1954	78248
1285	S	D	E	N	1972	71762
1286	S	D	E	N	1972	71762
1287	S	D	E	N	1972	71762
1288	S	D	E	N	1972	71762
1289	S	D	E	N	1972	71762
1297	S	D	E	N	1973	80711
1297	S	D	R	N	1975	91464
1306	S	D	E	N	1973	75868
1306	S	E	E	N	1973	75868
1306	S	D	E	N	1974	81307
1306	S	E	E	N	1974	81307
1307	S	D	E	N	1973	75868
1307	S	E	E	N	1973	75868
1307	S	D	R	N	1974	76107
1307	S	D	E	N	1974	81307
1307	S	E	E	N	1974	81307
1308	S	D	E	N	1973	75868
1308	S	E	E	N	1973	75868
1308	S	D	E	N	1974	81307
1308	S	E	E	N	1974	81307
1309	S	D	E	N	1974	81307
1309	S	E	E	N	1974	81307
1328	S	D	E	N	1974	74262
1328	S	T	E	N	1974	74262
1329	P	D	E	N	1973	71872
1329	S	D	E	N	1978	96156
1330	S	D	E	N	1973	73943
1330	S	D	E	N	1974	76412
1330	S	D	E	N	1975	86712
1332	G	T	R	N	1973	73585
1332	G	T	E	N	1973	73586
1337	S	D	E	N	1975	85218
1338	P	D	E	N	1973	71846
1338	S	D	E	N	1977	90886
1340	S	D	E	N	1973	72037
1341	S	D	E	N	1973	72037
1342	S	D	E	N	1973	72037
1343	S	D	E	N	1973	72037
1345	S	D	E	N	1974	83346
1345	S	E	E	N	1974	83346
1345	S	D	E	N	1976	96156
1347	S	D	F	N	1973	71786
1348	S	D	F	N	1973	71786
1349	S	D	F	N	1973	71786
1351	P	D	E	N	1973	73529
1352	S	D	R	N	1973	72392
1352	S	D	R	N	1975	91513
1353	S	D	E	N	1972	68786
1356	S	S	R	N	1964	73825
1356	S	S	E	N	1973	73826
1357	S	S	E	N	1972	66579
1357	S	D	E	N	1972	68900
1357	S	D	E	N	1972	74269
1357	S	D	E	N	1970	91828
1360	S	D	E	N	1968	58152
1366	S	D	E	N	1979	99351
1376	P	D	E	N	1954	78248
1377	P	D	E	N	1954	78248
1377	P	D	E	N	1978	95944
1377	P	E	E	N	1978	95944
1377	S	S	E	N	1972	66579
1377	S	D	E	N	1974	81307
1377	S	E	E	N	1974	81307
1377	S	D	E	N	1906	96758
1379	S	D	F	N	1973	75987
1381	S	D	R	N	1974	76107
1382	S	D	E	N	1974	78685
1382	S	D	E	N	1976	89183
1383	S	D	E	N	1974	78685
1383	S	D	E	N	1976	89183
1384	S	D	E	N	1973	78661
1389	S	D	E	N	1979	97563
1390	S	G	E	N	1974	74262
1391	P	D	E	N	1954	78248
1392	P	D	E	N	1978	99103
1392	S	D	E	N	1972	68900
1392	S	D	E	N	1977	90178
1393	S	D	E	N	1972	68900

Phys. State: **C.** Superconductive; **D.** Doped; **E.** Expanded; **F.** Fibrous or Whisker; **G.** Gas; **I.** Ionized or Plasma; **L.** Liquid; **M.** Multiphase; **P.** Powder or Fine Particle; **S.** Solid; **T.** Thick or Thin Film

Subject: **D.** Data; **E.** Experiment; **G.** General (Data + Theory + Experiment); **S.** Survey (Review, Compendium etc.); **T.** Theory

Language: **C.** Czech; **D.** Dutch; **E.** English; **F.** French; **G.** German; **I.** Italian; **J.** Japanese; **O.** Other Languages; **P.** Polish; **R.** Russian; **S.** Spanish

Temperature: **F.** Full Range (Low + Normal + High); **L.** Low (0 to 75K) + Overlap into Normal; **N.** Normal (75 to 1273K); **H.** High (above 1273K) + Overlap into Normal; Blank = Not Coded

Substance Number	Phys. State	Subject	Language	Temperature	Year	TPRC Number
521-1394	P	D	E	N	1973	73529
1394	P	D	E	N	1978	98168
1394	S	D	E	N	1976	85517
1394	S	E	E	N	1976	85517
1394	S	D	E	N	1978	98168
1395	P	D	E	N	1973	73529
1396	P	D	E	N	1973	73529
1397	P	D	E	N	1973	73529
1397	S	D	E	N	1978	98168
1398	P	D	E	N	1973	73529
1399	P	D	E	N	1973	73529
1400	P	D	E	N	1973	73529
1401	P	D	E	N	1973	73529
1402	S	D	F	N	1972	70913
1417	S	D	E	N	1974	78981
1418	S	S	E	H	1960	24931
1418	S	S	E	N	1960	24931
1419	S	D	E	N	1975	85218
1421	P	D	G	N	1970	64194
1421	P	T	G	N	1970	64194
1426	S	D	R	N	1975	91513
1427	S	D	R	N	1975	91513
1430	S	D	E	N	1976	87489
1431	M	D	E	N	1976	87595
1431	M	T	E	N	1976	87595
1432	S	D	E	N	1973	75868
1432	S	E	E	N	1973	75868
1432	S	D	E	N	1974	81307
1432	S	E	E	N	1974	81307
1433	S	D	E	N	1973	75868
1433	S	E	E	N	1973	75868
1433	S	D	E	N	1974	81307
1433	S	E	E	N	1974	81307
1435	P	D	E	N	1975	80845
1435	S	D	E	N	1975	80845
1436	S	D	E	N	1975	80845
1437	S	D	E	N	1977	91623
1438	S	D	E	N	1975	85747
1439	S	D	E	N	1977	91838
1440	S	D	E	N	1976	89183
1447	S	D	E	N	1977	94376
1448	S	D	E	N	1977	94376
1449	S	D	E	N	1978	94616
1450	S	D	E	N	1978	94616
1451	S	D	E	N	1978	94616
1452	S	D	E	N	1978	94616
1453	S	D	E	N	1978	94616
1454	S	D	E	N	1978	94616
1455	S	D	E	N	1978	94616
1456	S	D	E	N	1978	94616
1457	S	D	E	N	1978	94616
1461	S	D	F	N	1963	32110
1461	S	D	E	N	1964	37156
1461	S	D	E	N	1972	71762
1463	S	D	E	L	1970	62676
1465	P	D	E	N	1954	24602
1468	S	D	F	N	1970	94406
1471	S	D	E	N	1906	96758
1472	S	D	E	N	1906	96758
1476	P	D	E	N	1978	95944
1476	P	E	E	N	1978	95944
1478	S	D	E	N	1906	96758
1483	S	D	E	N	1906	96758
1485	S	D	E	N	1908	96759
1491	S	D	E	N	1978	95500
1492	S	D	E	N	1977	92321
1493	P	D	E	N	1979	99163
1493	S	D	E	N	1978	96156
1493	S	D	E	N	1979	99162
1494	S	D	E	N	1978	96156
1495	S	D	E	N	1978	96156
1496	S	D	E	N	1978	96156
1497	S	D	E	N	1967	46650
1497	S	D	E	N	1978	96156
1498	S	D	E	N	1967	46650
1498	S	D	E	N	1976	96156
1499	S	D	E	N	1976	96156
1500	S	D	E	N	1976	96156
1501	S	D	E	N	1976	96156
1502	P	D	E	N	1978	98168
1502	S	D	E	N	1978	98168
1547	S	D	E	N	1978	96789
1550	S	D	E	N	1960	90396
1550	S	E	E	N	1960	90396
1552	S	D	R	N	1977	91551
1553	S	D	R	N	1977	91551
1558	S	D	E	N	1979	98302
1561	S	D	R	N	1976	94687
1573	S	D	O	N	1976	91485
1574	S	D	R	N	1975	91464
1575	S	D	R	N	1975	91464
1576	S	D	R	N	1975	91464
1577	S	D	R	N	1975	91464
1578	P	D	E	N	1975	83895
1587	P	D	E	N	1975	83895
1591	S	D	E	N	1974	83509
1592	P	D	E	N	1975	83895
1593	P	D	E	N	1975	83895
1594	P	D	E	N	1975	83895
1595	P	D	E	N	1975	83895
1596	P	D	E	N	1975	83895
521-1597	P	D	E	N	1975	83895
1599	S	D	E	N	1974	84280
1600	S	D	E	N	1975	84446
1601	P	D	E	N	1978	99103
1614	S	D	E	N	1974	80708
1615	S	D	E	N	1967	46650
1617	S	D	E	N	1967	46650
1630	S	D	J	N	1979	100083
1631	P	D	E	N	1979	99163
1631	S	D	E	N	1979	99162
1632	P	S	E	N	1972	66579
1634	P	S	E	N	1972	66579
1634	S	S	E	N	1972	66579
1635	S	D	E	N	1979	99278
1636	S	D	E	N	1979	99278
1637	S	D	E	N	1979	99351
1638	S	D	E	N	1979	99351
1641	S	D	E	N	1979	97563
1642	S	D	E	N	1979	97563
1643	S	D	E	N	1979	97563
1644	S	D	E	N	1979	97563
1648	S	D	E	N	1967	46650
1649	S	D	E	N	1967	46650
1650	S	D	E	N	1967	46650
1651	S	D	E	N	1967	46650

Phys. State: **C.** Superconductive; **D.** Doped; **E.** Expanded; **F.** Fibrous or Whisker; **G.** Gas; **I.** Ionized or Plasma; **L.** Liquid; **M.** Multiphase; **P.** Powder or Fine Particle; **S.** Solid; **T.** Thick or Thin Film
Subject: **D.** Data; **E.** Experiment; **G.** General (Data + Theory + Experiment); **S.** Survey (Review, Compendium etc.); **T.** Theory
Language: **C.** Czech; **D.** Dutch; **E.** English; **F.** French; **G.** German; **I.** Italian; **J.** Japanese; **O.** Other Languages; **P.** Polish; **R.** Russian; **S.** Spanish
Temperature: **F.** Full Range (Low + Normal + High); **L.** Low (0 to 75K) + Overlap into Normal; **N.** Normal (75 to 1273K); **H.** High (above 1273K) + Overlap into Normal; Blank = Not Coded

Chapter 7 Absorptance

Substance Number	Phys. State	Sub-ject	Lan-guage	Temper-ature	Year	TPRC Number
503-0012	S	D	E	N	1971	66243
0012	S	S	E	N	1972	66579
0025	S	D	E	N	1969	37823
0025	S	T	E	N	1969	37823
0025	S	D	E	N	1965	40905
0025	S	D	E	N	1966	40907
0025	S	D	E	N	1966	40908
0082	S	S	R	N	1962	33745
0082	S	S	E	N	1966	33746
0082	S	D	E	N	1966	40903
0082	S	D	E	N	1966	40904
0082	S	D	E	N	1963	45668
0082	S	D	I	N	1969	55533
0082	S	T	I	N	1969	55533
0216	S	S	E	N	1972	66579
0280	P	G	E	N	1976	83577
0284	S	S	E	N	1972	66579
0305	S	D	E	N	1971	66243
0537	S	S	E	N	1972	66579
0604	S	D	E	N	1971	66243
0604	S	S	E	N	1972	66579
0651	T	S	E	H	1972	70629
0651	T	S	E	N	1972	70629
0659	S	S	E	N	1972	66579
0660	S	S	E	N	1972	66579
0667	S	S	E	N	1972	66579
0746	S	S	E	N	1972	70629
0809	D	D	J	N	1972	67595
0822	S	D	E	N	1971	66243
0967	S	D	E	N	1976	83746
1026	S	D	R	N	1966	44184
1026	S	D	E	N	1966	44185
1028	D	D	F	N	1969	53706
1050	S	D	E	N	1970	63038
1060	S	D	E	N	1934	97516
1066	S	D	E	N	1968	51936
1133	S	D	G	H	1931	40645
1136	S	S	E	N	1951	98632
1142	S	S	E	N	1969	56355
1146	S	D	G	H	1931	40645
1151	S	S	E	N	1951	98632
1250	S	D	R	N	1973	71976
1250	S	D	E	N	1973	71977
1397	S	D	G	N	1952	43597
1411	S	D	E	N	1970	38064
1532	S	D	E	N	1963	45698
1533	S	D	E	N	1963	45698
1534	S	D	E	N	1963	45698
1547	S	S	E	N	1972	66579
1654	S	S	E	N	1972	66579
1945	S	S	E	N	1972	66579
2018	S	D	G	H	1949	12993
2018	S	D	G	N	1949	12993
2019	S	D	G	H	1949	12993
2019	S	D	G	N	1949	12993
2347	S	D	E	N	1971	66243
2444	S	D	G	H	1949	12993
2444	S	D	G	N	1949	12993
2445	S	D	G	H	1949	12993
2445	S	D	G	N	1949	12993
2446	S	D	G	H	1949	12993
2446	S	D	G	N	1949	12993
8334	S	D	G	H	1931	40645
8481	S	D	R	N	1972	68525
8481	S	D	E	N	1972	70883
8482	S	D	R	N	1972	68525
8482	S	D	E	N	1972	70883
8483	S	D	R	N	1972	68525
8483	S	D	E	N	1972	70883
8734	S	D	E	N	1966	43484
8759	S	D	E	N	1964	44408
8872	S	D	R	N	1965	42546
8873	S	D	R	N	1965	42545
9039	S	D	E	N	1964	37891
9068	S	D	R	N	1966	40976
9068	S	D	E	N	1966	40977
9084	S	D	E	N	1969	55904
9118	S	D	R	N	1965	42545
9134	S	D	E	N	1969	55904
9142	D	D	E	N	1972	68348
9143	S	D	E	N	1969	55904
9147	S	D	E	N	1969	55904
9148	S	D	E	N	1969	55904
9169	S	D	E	N	1969	55904
9171	S	D	R	N	1965	42546
9173	D	D	E	N	1972	67781
9174	D	D	E	N	1970	38064
9174	S	D	R	N	1965	42545
9206	S	D	R	N	1970	75467
9206	S	D	E	N	1970	75468
9256	S	D	O	N	1970	60880
9257	S	D	O	N	1970	60881
9330	P	D	E	N	1969	59196
9330	S	D	E	N	1970	59647
9331	S	D	E	N	1970	59647
503-9437	S	D	E	N	1970	59674
9466	S	D	E	N	1969	37359
9467	S	D	E	N	1969	37359
9468	S	D	E	N	1969	37359
9469	S	D	E	N	1969	37359
9536	S	D	O	N	1965	44879
9701	P	D	E	N	1966	48492
9702	P	D	E	N	1966	48231
9702	P	D	E	N	1966	48492
9706	P	D	E	N	1966	48231
9706	P	D	E	N	1966	48492
9842	P	D	E	N	1964	34819
9869	P	D	E	N	1964	34819
9869	S	D	R	N	1966	40976
9869	S	D	E	N	1966	40977
9869	S	D	R	N	1969	57127
9869	S	D	E	N	1969	57128
9899	P	D	E	N	1966	48231
9899	P	D	E	N	1966	48492
9916	P	D	P	N	1971	65209
9916	S	D	G	H	1931	40645
9916	S	D	E	N	1965	48230
9988	S	G	E	H	1969	56348
521-0003	S	D	J	N	1978	99886
0015	S	D	F	N	1973	71497
0022	P	D	E	N	1966	48492
0022	S	S	E	N	1972	66579
0022	S	D	J	N	1978	99886
0025	S	D	J	N	1974	81726
0026	S	D	J	N	1978	99886
0032	S	D	J	N	1978	99886
0033	S	D	J	N	1978	99886
0034	S	D	J	N	1978	99886
0035	S	D	J	N	1978	99886
0036	S	D	E	N	1966	59530
0057	S	D	J	N	1978	99886
0058	S	D	J	N	1978	99886
0059	S	D	J	N	1978	99886
0063	P	D	R	N	1970	63227
0063	S	D	E	N	1967	41206
0063	S	D	E	N	1967	44768
0063	S	D	E	N	1972	67890
0063	S	D	O	N	1972	73171
0063	S	D	E	N	1976	85439
0072	S	S	E	N	1951	98632
0075	P	D	G	N	1968	52305
0075	P	T	G	N	1968	52305
0075	S	D	J	N	1978	99886
0077	S	D	J	N	1978	99886
0082	S	D	J	N	1978	99886
0083	S	D	J	N	1978	99886
0085	S	S	E	N	1951	98632
0092	S	D	J	N	1978	99886
0103	S	S	E	N	1951	98632
0105	S	D	J	N	1978	99886
0109	S	D	J	N	1978	99886
0111	S	D	F	N	1973	71497
0116	S	D	E	N	1968	49251
0116	S	G	R	N	1969	58230
0130	S	D	J	N	1978	99886
0131	S	D	J	N	1978	99886
0174	S	D	J	N	1978	99886
0181	S	S	E	N	1951	98632
0182	S	D	J	N	1978	99886
0191	S	D	E	N	1966	34852
0191	S	D	E	N	1965	36465
0196	S	S	E	N	1951	98632
0222	S	D	J	N	1978	99886
0223	S	S	E	N	1972	66579
0223	S	G	E	N	1976	85445
0236	S	D	E	N	1965	50543
0240	S	D	J	N	1978	99886
0241	S	D	J	N	1978	99886
0250	S	S	E	N	1972	66579
0254	S	D	O	N	1973	75883
0254	S	D	J	N	1978	99886
0258	S	D	E	N	1966	59530
0258	S	D	E	N	1977	98673
0258	S	D	J	N	1978	99886
0293	S	D	E	N	1965	35595
0299	S	S	E	N	1968	53122
0302	S	D	J	N	1978	99886
0303	S	D	J	N	1978	99886
0341	S	D	J	N	1978	99886
0351	S	D	J	N	1978	99886
0362	S	D	J	N	1978	99886
0366	S	D	J	N	1978	99886
0394	S	S	E	N	1951	98632
0413	S	D	E	N	1970	61193
0423	S	D	E	N	1967	47238
0429	S	D	J	N	1978	99886
0443	S	D	J	N	1978	99886
0457	S	D	J	N	1978	99886
0496	S	D	E	N	1974	85093
521-0505	S	D	J	N	1978	99886
0506	S	D	J	N	1978	99886
0516	P	D	E	N	1964	34819
0566	S	D	J	N	1978	99886
0626	S	D	J	N	1978	99886
0637	P	D	E	N	1967	40931
0729	P	D	R	N	1967	46073
0729	P	D	E	N	1967	46074
0730	P	D	R	N	1967	46073
0730	P	D	E	N	1967	46074
0732	S	D	E	N	1970	37023
0732	S	D	E	N	1968	51936
0732	S	D	J	N	1978	99886
0737	S	D	J	N	1978	99886
0764	S	D	J	N	1978	99886
0765	S	D	J	N	1978	99886
0769	P	D	E	N	1966	48492
0769	S	D	E	N	1966	59530
0769	S	S	E	N	1972	66579
0772	S	D	J	N	1978	99886
0780	S	D	J	N	1978	99886
0794	S	D	J	N	1978	99886
0799	S	D	J	N	1978	99886
0887	S	D	E	N	1965	39333
0912	S	D	E	N	1966	59530
0939	S	D	J	N	1978	99886
0940	S	D	J	N	1978	99886
0942	S	D	E	N	1966	59530
0942	S	D	J	N	1978	99886
0950	S	D	E	N	1966	59530
0951	P	D	E	N	1969	53645
0951	P	T	E	N	1969	53645
0951	P	D	E	N	1970	59099
0951	S	S	E	N	1970	59662
0951	S	D	J	N	1978	99886
0958	S	D	E	N	1970	77520
1043	S	D	E	N	1974	76328
1128	S	D	E	N	1974	85093
1131	S	D	J	N	1978	99886
1134	S	D	J	N	1978	99886
1220	S	D	R	N	1972	66874
1227	S	D	J	N	1978	99886
1271	S	D	O	N	1972	69881
1337	S	D	E	N	1970	77520
1386	S	D	E	N	1970	77520
1420	S	D	E	N	1974	85093
1463	S	D	J	N	1978	99886
1536	S	D	J	N	1978	99886
1537	S	D	J	N	1978	99886
1548	P	D	E	N	1975	83863
1605	S	D	J	N	1978	99886
1606	S	D	J	N	1978	99886
1607	S	D	J	N	1978	99886

Phys. State: **C.** Superconductive; **D.** Doped; **E.** Expanded; **F.** Fibrous or Whisker; **G.** Gas; **I.** Ionized or Plasma; **L.** Liquid; **M.** Multiphase; **P.** Powder or Fine Particle; **S.** Solid; **T.** Thick or Thin Film

Subject: **D.** Data; **E.** Experiment; **G.** General (Data + Theory + Experiment); **S.** Survey (Review, Compendium etc.); **T.** Theory

Language: **C.** Czech; **D.** Dutch; **E.** English; **F.** French; **G.** German; **I.** Italian; **J.** Japanese; **O.** Other Languages; **P.** Polish; **R.** Russian; **S.** Spanish

Temperature: **F.** Full Range (Low + Normal + High); **L.** Low (0 to 75K) + Overlap into Normal; **N.** Normal (75 to 1273K); **H.** High (above 1273K) + Overlap into Normal; Blank = Not Coded

Chapter 8 Transmittance

Substance Number	Phys. State	Subject	Language	Temperature	Year	TPRC Number
503-0001	D	D	E	N	1965	38556
0001	S	D	E	N	1963	30814
0001	S	D	E	N	1965	38556
0001	S	D	E	N	1971	68241
0012	S	D	E	N	1965	36144
0012	S	D	E	N	1965	37896
0012	S	D	E	N	1971	64206
0012	S	E	E	N	1971	64206
0012	S	D	E	N	1971	66243
0012	S	S	E	N	1972	66579
0012	S	D	R	N	1971	68616
0012	S	D	E	N	1977	92721
0012	S	D	E	N	1977	93343
0012	S	S	E	N	1980	100404
0018	S	D	E	N	1965	38674
0018	S	S	E	N	1972	66579
0018	S	D	E	N	1957	76819
0018	S	E	E	N	1957	76819
0018	S	D	E	N	1960	76856
0019	S	S	E	N	1960	24931
0019	S	S	E	N	1972	66579
0020	S	D	E	N	1969	59714
0020	S	D	E	N	1957	76819
0020	S	E	E	N	1957	76819
0022	S	D	E	N	1963	30814
0024	P	D	P	N	1971	65759
0024	P	E	P	N	1971	65759
0024	S	D	F	N	1960	47249
0024	S	D	R	N	1968	50789
0024	S	D	E	N	1968	67356
0024	S	D	E	N	1971	67411
0024	S	D	E	N	1972	68344
0025	S	D	R	N	1965	35610
0025	S	D	E	N	1969	37823
0025	S	T	E	N	1969	37823
0025	S	D	E	N	1965	38592
0025	S	D	E	N	1965	40905
0025	S	D	E	N	1966	40907
0025	S	D	E	N	1966	40908
0025	S	D	E	N	1966	40910
0025	S	D	E	N	1967	45052
0025	S	D	E	N	1966	49395
0025	S	D	E	N	1928	72863
0025	S	E	E	N	1928	72863
0025	S	D	P	N	1975	83208
0025	S	E	P	N	1975	83208
0025	S	D	E	N	1976	87238
0025	S	S	I	N	1974	92925
0025	S	D	E	N	1977	97519
0026	S	S	E	N	1972	66579
0026	S	D	R	N	1976	87322
0026	S	D	E	N	1976	89153
0031	S	D	E	N	1967	47761
0031	S	D	E	N	1930	51282
0031	S	D	E	N	1975	76210
0031	S	D	E	N	1951	76810
0031	S	D	E	N	1957	76819
0031	S	E	E	N	1957	76819
0031	S	D	E	N	1959	77381
0031	S	E	E	N	1959	77381
0033	S	D	E	N	1971	62547
0033	S	D	E	N	1975	77166
0033	S	E	E	N	1975	77166
0033	S	D	E	N	1979	99709
0047	S	D	R	N	1976	88951
0047	S	D	E	N	1976	88952
0080	S	D	E	N	1971	66243
0080	S	G	E	N	1972	78256
0080	S	D	E	N	1972	78269
0082	F	G	E	N	1965	36758
0082	F	D	E	N	1968	49921
0082	F	E	E	N	1968	49921
0082	F	D	E	N	1965	54294
0082	F	T	E	N	1965	54294
0082	F	D	R	N	1969	57183
0082	F	D	E	N	1969	57184
0082	F	D	E	N	1968	65125
0082	F	E	E	N	1968	65125
0082	S	S	R	N	1962	33745
0082	S	S	E	N	1966	33746
0082	S	D	E	N	1964	35357
0082	S	D	E	N	1969	37122
0082	S	E	E	N	1965	38861
0082	S	D	E	N	1966	40324
0082	S	D	E	N	1950	40784
0082	S	D	E	N	1955	40789
0082	S	D	E	N	1966	40903
0082	S	D	E	N	1966	40904
0082	S	D	I	N	1966	42019
0082	S	E	I	N	1966	42019
0082	S	D	E	N	1967	42742
0082	S	D	G	N	1898	46995
0082	S	D	E	N	1931	47008
0082	S	D	O	N	1961	49805
0082	S	D	E	N	1968	49971

Substance Number	Phys. State	Subject	Language	Temperature	Year	TPRC Number
503-0082	S	D	E	N	1965	54294
0082	S	T	E	N	1965	54294
0082	S	D	I	N	1969	55533
0082	S	T	I	N	1969	55533
0082	S	D	E	N	1971	60669
0082	S	T	E	N	1971	60669
0082	S	D	E	N	1934	68062
0082	S	D	E	N	1972	68853
0082	S	G	E	N	1973	72581
0082	S	D	E	N	1974	74727
0082	S	T	E	N	1974	74727
0082	S	T	E	N	1974	74933
0082	S	D	E	N	1953	76853
0082	S	E	E	N	1953	76853
0082	S	S	F	N	1974	77206
0082	S	S	E	N	1955	77330
0082	S	D	E	N	1964	77530
0082	S	E	E	N	1964	77530
0082	S	D	E	N	1974	77552
0082	S	D	E	N	1972	78249
0082	S	E	E	N	1972	78249
0082	S	S	G	N	1967	78275
0082	S	S	E	N	1972	78276
0082	S	S	F	N	1962	78359
0082	S	D	E	N	1974	79014
0082	S	D	E	N	1906	96757
0082	S	D	E	N	1908	96760
0082	S	S	E	N	1979	99434
0082	S	D	E	N	1980	100601
0083	D	D	E	N	1965	34147
0083	D	T	E	N	1965	34147
0089	S	D	E	N	1963	39738
0089	S	D	E	N	1962	39896
0089	S	D	E	N	1967	44551
0089	S	S	E	N	1977	89235
0089	S	D	E	N	1941	90225
0090	D	D	E	N	1965	34147
0090	D	T	E	N	1965	34147
0090	D	D	P	N	1975	83208
0090	D	E	P	N	1975	83208
0090	S	D	R	N	1966	63162
0090	S	D	E	N	1966	63163
0090	S	D	R	N	1972	69642
0090	S	D	E	N	1973	69643
0090	S	D	E	N	1976	93353
0096	D	D	E	N	1965	34147
0096	D	T	E	N	1965	34147
0096	D	D	E	N	1951	76810
0105	S	D	E	N	1961	18367
0108	P	G	E	N	1973	71027
0108	P	G	E	N	1973	73838
0108	S	G	E	N	1973	71027
0108	S	G	E	N	1973	73838
0112	S	D	G	N	1969	63268
0126	S	D	E	N	1976	85565
0126	S	D	E	N	1976	85566
0126	S	D	E	N	1977	90271
0159	S	D	E	N	1972	78269
0173	S	D	E	N	1977	91585
0179	S	D	E	N	1968	40352
0200	S	D	E	N	1974	77552
0203	S	D	E	N	1973	70702
0216	S	D	E	N	1965	36738
0216	S	S	E	N	1972	66579
0231	S	D	E	N	1972	78269
0239	S	D	E	N	1967	44427
0241	D	D	E	N	1967	38576
0270	D	D	E	N	1965	38556
0270	P	D	P	N	1971	65759
0270	P	E	P	N	1971	65759
0270	S	D	E	N	1963	30814
0270	S	D	E	N	1965	38556
0270	S	D	E	N	1962	42530
0270	S	D	F	N	1960	47249
0270	S	D	R	N	1968	50789
0270	S	D	P	N	1971	65759
0270	S	E	P	N	1971	65759
0270	S	S	E	N	1972	66579
0270	S	D	E	N	1968	67356
0270	S	D	E	N	1972	68344
0290	S	D	E	N	1966	40468
0290	S	D	E	N	1968	47870
0297	S	D	E	N	1956	77317
0305	S	D	E	N	1971	66243
0309	S	D	E	N	1965	37896
0309	S	D	E	N	1965	38674
0309	S	S	E	N	1972	66579
0373	S	D	E	N	1979	100051
0376	S	D	E	N	1961	18367
0376	S	D	F	N	1961	32688
0442	S	D	E	N	1959	18428
0463	S	D	E	N	1953	15126
0463	S	D	R	N	1953	20572
0464	S	D	E	N	1953	15126
0464	S	D	R	N	1953	20572

Substance Number	Phys. State	Subject	Language	Temperature	Year	TPRC Number
503-0465	S	D	E	N	1953	15126
0465	S	D	R	N	1953	20572
0476	S	D	E	N	1958	15895
0479	S	D	R	N	1974	83581
0479	S	D	E	N	1974	83582
0516	D	D	E	N	1958	18615
0537	S	D	E	N	1965	36738
0537	S	S	E	N	1972	66579
0545	S	D	E	N	1951	76810
0554	D	D	E	N	1959	18621
0556	D	D	E	N	1959	18621
0556	D	D	E	N	1936	22584
0556	D	D	E	N	1976	93353
0566	F	D	E	N	1976	88395
0566	F	T	E	N	1976	88395
0567	S	D	E	N	1966	40468
0567	S	D	E	N	1968	47870
0567	S	D	E	N	1957	48109
0584	S	D	R	N	1968	50789
0584	S	D	E	N	1968	67356
0585	D	S	E	N	1972	66579
0585	D	D	E	N	1976	85230
0604	S	D	E	N	1965	36144
0604	S	D	E	N	1969	57597
0604	S	D	E	N	1971	66243
0604	S	S	E	N	1972	66579
0648	S	D	E	N	1972	68284
0659	S	G	E	N	1966	39365
0659	S	D	E	N	1976	83746
0660	S	S	E	N	1960	24932
0660	S	S	E	N	1972	66579
0660	S	D	E	N	1976	83746
0667	S	S	E	N	1972	66579
0667	S	S	E	N	1976	83746
0677	T	D	E	N	1978	94818
0677	T	D	E	N	1978	96544
0677	T	D	E	N	1979	97392
0677	T	D	E	N	1979	97884
0677	T	D	E	N	1979	98629
0731	D	D	E	N	1966	40547
0731	S	D	E	N	1965	43033
0731	S	T	E	N	1965	43033
0731	S	D	R	N	1968	50789
0731	S	D	E	N	1968	67356
0734	S	D	E	N	1949	4320
0735	S	D	E	N	1949	4320
0735	S	D	E	N	1974	77445
0736	S	D	E	N	1949	4320
0739	S	D	E	N	1962	24553
0746	S	D	E	N	1969	37122
0746	S	D	E	N	1930	51282
0746	S	D	E	N	1969	57221
0746	S	D	E	N	1957	76819
0746	S	E	E	N	1957	76819
0782	D	D	E	N	1965	34147
0782	D	T	E	N	1965	34147
0782	D	D	E	N	1966	39012
0782	D	D	G	N	1966	42502
0782	D	D	E	N	1969	56119
0782	S	D	E	N	1965	37896
0782	S	D	E	N	1966	39012
0782	S	D	E	N	1964	39347
0782	S	D	G	N	1966	42502
0782	S	D	E	N	1950	58879
0800	S	S	E	N	1972	66579
0800	S	D	E	N	1963	76855
0800	S	E	E	N	1963	76855
0803	S	D	E	N	1963	30814
0803	S	D	R	N	1974	79531
0803	S	D	E	N	1974	83743
0815	S	D	E	N	1961	18367
0818	D	D	E	N	1966	40547
0818	P	D	P	N	1971	65759
0818	P	E	P	N	1971	65759
0818	S	D	R	N	1968	50789
0818	S	D	P	N	1971	65759
0818	S	E	P	N	1971	65759
0818	S	D	E	N	1968	67356
0818	S	D	E	N	1951	76810
0822	S	D	E	N	1971	66243
0855	S	D	E	N	1967	44427
0863	S	S	E	H	1960	24931
0863	S	S	E	N	1960	24931
0863	S	S	E	N	1972	66579
0879	S	D	E	N	1968	40352
0884	S	D	R	H	1967	49346
0884	S	D	E	H	1967	49347
0930	S	D	E	N	1968	48473
0931	S	D	E	N	1968	48473
0933	S	D	E	N	1968	48473
0934	S	D	E	N	1968	48473
0935	S	D	E	N	1968	48473
0936	S	D	E	N	1965	36348
0936	S	D	E	N	1969	56119
0936	S	D	E	N	1977	91831

Phys. State: **C.** Superconductive; **D.** Doped; **E.** Expanded; **F.** Fibrous or Whisker; **G.** Gas; **I.** Ionized or Plasma; **L.** Liquid; **M.** Multiphase; **P.** Powder or Fine Particle; **S.** Solid; **T.** Thick or Thin Film

Subject: **D.** Data; **E.** Experiment; **G.** General (Data + Theory + Experiment); **S.** Survey (Review, Compendium etc.); **T.** Theory

Language: **C.** Czech; **D.** Dutch; **E.** English; **F.** French; **G.** German; **I.** Italian; **J.** Japanese; **O.** Other Languages; **P.** Polish; **R.** Russian; **S.** Spanish

Temperature: **F.** Full Range (Low + Normal + High); **L.** Low (0 to 75K) + Overlap into Normal; **N.** Normal (75 to 1273K); **H.** High (above 1273K) + Overlap into Normal; Blank = Not Coded

Substance Number	Phys. State	Subject	Language	Temperature	Year	TPRC Number
503-0937	S	D	E	N	1968	48473
0938	S	D	E	N	1968	48473
0939	S	D	E	N	1965	36348
0939	S	D	E	N	1977	91831
0940	S	D	E	N	1968	48473
0941	S	D	E	N	1969	56119
0942	S	D	E	N	1965	36348
0942	S	D	E	N	1977	91831
0943	S	D	E	N	1968	48473
0944	S	D	E	N	1968	48473
0945	S	D	E	N	1968	48473
0946	S	D	E	N	1968	48473
0947	S	D	E	N	1977	93343
0948	S	D	E	N	1969	56119
0949	S	D	E	N	1965	36348
0949	S	D	E	N	1977	91831
0950	S	D	E	N	1968	48473
0952	S	D	E	N	1965	36348
0952	S	D	E	N	1968	48473
0952	S	D	E	N	1977	91831
0953	S	D	E	N	1968	48473
0957	M	D	E	N	1970	35888
0957	M	T	E	N	1970	35888
0957	S	D	E	N	1968	41236
0957	S	D	E	N	1969	43716
0961	S	D	E	N	1965	37896
0963	D	D	E	N	1969	38662
0963	S	D	E	N	1969	38662
0966	S	D	E	N	1960	15886
0967	S	D	E	N	1969	78264
0967	S	E	E	N	1969	78264
0967	S	D	E	N	1976	83746
0967	S	S	E	N	1976	83746
0968	S	D	E	N	1969	78264
0968	S	E	E	N	1969	78264
0969	S	D	E	N	1969	56119
0971	S	D	E	N	1964	32047
0971	S	D	E	N	1966	39011
0971	S	D	E	N	1965	39259
0971	S	D	E	N	1965	43840
0971	S	D	E	N	1969	59196
0971	S	D	E	N	1969	59714
0971	S	S	E	N	1972	66579
0971	S	D	E	N	1974	72733
0974	S	D	E	N	1963	76855
0974	S	E	E	N	1963	76855
0974	S	D	E	N	1960	76856
0975	S	S	E	N	1972	66579
0976	S	D	R	N	1968	50789
0976	S	D	E	N	1968	67356
0983	S	D	E	N	1949	4320
0983	S	D	E	N	1957	76819
0983	S	E	E	N	1957	76819
0983	S	T	E	N	1957	76819
0984	S	S	G	N	1967	78275
0984	S	S	E	N	1972	78276
0991	S	D	E	N	1908	96760
0994	S	D	E	N	1941	90225
0996	D	D	E	N	1976	86147
0997	D	D	E	N	1976	86147
0998	F	D	E	N	1971	61511
0998	F	E	E	N	1971	61511
0998	S	D	E	N	1971	61511
0998	S	E	E	N	1971	61511
0999	D	D	E	N	1976	86147
1000	S	D	E	N	1969	56119
1001	S	D	E	N	1972	78269
1002	F	D	E	N	1971	61511
1002	F	E	E	N	1971	61511
1002	S	D	E	N	1971	61511
1002	S	E	E	N	1971	61511
1003	S	D	E	N	1967	44427
1003	S	S	E	N	1972	66579
1004	S	D	E	N	1967	44427
1004	S	S	E	N	1972	66579
1005	S	D	E	N	1972	68742
1005	S	E	E	N	1972	68742
1007	S	D	E	N	1949	4320
1008	D	D	E	N	1976	86147
1010	S	D	R	N	1966	44184
1010	S	D	E	N	1966	44185
1017	S	D	E	N	1965	38334
1017	S	D	R	N	1965	38846
1018	S	D	R	N	1968	51767
1018	S	D	E	N	1968	51768
1026	S	D	R	N	1968	55318
1026	S	D	E	N	1972	70720
1028	S	D	E	N	1973	70702
1029	S	D	E	N	1973	77754
1032	S	D	E	N	1966	40468
1032	S	D	E	N	1968	47870
1035	S	D	E	N	1967	44427
1035	S	S	E	N	1972	66579
1040	D	D	R	N	1966	63162
1040	D	D	E	N	1966	63163
1041	S	D	E	N	1967	42742
1041	S	D	E	N	1967	44427
1041	S	D	E	N	1958	48105
1043	D	D	E	N	1965	34147
1043	D	T	E	N	1965	34147
1050	S	D	E	N	1961	18367
503-1050	S	D	E	N	1970	63038
1050	S	D	E	N	1972	68853
1050	S	D	E	N	1975	76210
1050	S	G	E	N	1972	78256
1059	D	D	R	N	1971	68297
1059	S	D	E	N	1965	37896
1059	S	D	E	N	1970	59141
1059	S	D	E	N	1957	76819
1059	S	E	E	N	1957	76819
1061	S	D	E	N	1930	51282
1063	S	D	E	N	1951	76810
1069	S	S	E	N	1972	66579
1073	S	D	E	N	1972	68853
1078	S	D	E	N	1966	40530
1078	S	D	R	N	1971	64731
1078	S	D	E	N	1971	64732
1081	S	S	E	N	1972	66579
1083	S	D	E	N	1953	15126
1083	S	D	R	N	1953	20572
1084	S	D	E	N	1967	47094
1087	S	D	E	N	1975	77166
1087	S	E	E	N	1975	77166
1092	S	S	E	N	1972	66579
1093	S	S	E	N	1972	66579
1094	S	S	E	N	1972	66579
1095	S	D	E	N	1976	93353
1095	S	D	E	N	1908	96760
1097	S	D	R	N	1968	55318
1097	S	D	E	N	1972	70720
1105	S	D	E	N	1930	51282
1107	S	D	E	N	1941	90225
1110	S	D	E	N	1968	51936
1110	S	S	G	N	1967	78275
1110	S	S	E	N	1972	78276
1113	S	D	R	N	1970	59391
1113	S	D	E	N	1970	59392
1114	S	D	R	N	1971	64907
1114	S	T	R	N	1971	64907
1114	S	D	E	N	1971	64908
1114	S	T	E	N	1971	64908
1116	S	D	E	N	1941	90225
1117	S	D	R	N	1966	44184
1117	S	D	E	N	1966	44185
1133	S	D	R	H	1968	59345
1133	S	D	E	H	1968	59346
1133	S	D	P	N	1970	65195
1142	S	D	E	N	1970	37273
1142	S	T	E	N	1970	37273
1173	P	D	E	N	1978	94862
1177	P	D	E	N	1978	94862
1194	S	D	E	N	1908	96760
1250	D	D	E	N	1962	42530
1250	S	D	E	N	1962	42530
1250	S	D	O	N	1961	49805
1250	S	D	E	N	1955	51500
1250	S	D	E	N	1971	64206
1250	S	E	E	N	1971	64206
1250	S	D	E	N	1908	96760
1252	S	D	E	N	1970	60103
1252	S	S	E	N	1977	89235
1254	S	D	E	N	1976	85197
1259	S	D	E	N	1915	47965
1313	D	D	R	N	1972	70260
1313	D	D	E	N	1972	72358
1313	D	D	E	N	1955	77324
1324	S	D	P	N	1970	65195
1358	S	D	R	N	1972	67137
1358	S	D	E	N	1972	69791
1360	P	D	E	N	1978	94862
1369	P	D	E	N	1978	94862
1390	S	D	E	N	1977	92045
1397	S	D	G	N	1952	43597
1399	S	D	O	N	1963	41521
1400	S	D	E	N	1968	47524
1401	S	D	E	N	1968	49583
1405	D	D	R	N	1967	46583
1405	D	D	E	N	1967	46584
1405	S	D	R	N	1967	46583
1405	S	D	E	N	1967	46584
1406	D	D	E	N	1968	47909
1407	S	D	E	N	1968	49583
1408	S	D	E	N	1930	51282
1408	S	D	E	N	1973	69660
1409	S	D	E	N	1930	51282
1410	S	D	E	N	1930	51282
1411	D	D	R	N	1968	59375
1411	D	D	E	N	1968	59376
1411	D	D	E	N	1973	77754
1411	S	S	E	N	1972	66579
1411	S	D	E	N	1973	77754
1413	S	D	E	N	1968	51936
1414	S	D	E	N	1958	48105
1417	S	D	E	N	1966	40468
1417	S	D	E	N	1968	47870
1418	S	D	E	N	1966	40468
1418	S	D	E	N	1968	47870
1420	S	D	E	N	1967	47094
1421	S	D	E	N	1967	47094
1425	S	D	E	N	1965	37896
1428	S	D	O	N	1965	34645
1428	S	D	C	N	1971	63750
503-1428	S	D	E	N	1928	72863
1428	S	E	E	N	1928	72863
1429	S	D	E	N	1964	37430
1429	S	D	E	N	1968	49971
1430	S	D	E	N	1969	56119
1435	S	D	R	N	1966	44624
1439	S	D	R	N	1971	64483
1439	S	T	R	N	1971	64483
1439	S	D	E	N	1971	64484
1439	S	T	E	N	1971	64484
1439	S	S	G	N	1977	92758
1440	S	G	G	N	1966	45348
1440	S	D	C	N	1971	63750
1442	S	D	E	N	1928	72863
1442	S	E	E	N	1928	72863
1443	D	D	E	N	1971	64140
1443	D	D	E	N	1972	69035
1445	S	D	E	N	1971	63695
1446	S	D	E	N	1971	67411
1447	S	D	E	N	1972	68853
1452	S	D	E	N	1951	76810
1453	S	D	E	N	1951	76810
1455	S	D	R	N	1974	76069
1455	S	E	R	N	1974	76069
1455	S	D	E	N	1974	76440
1455	S	E	E	N	1974	76440
1459	S	G	E	N	1972	78256
1460	D	D	E	N	1975	77119
1460	S	D	E	N	1975	77119
1461	S	D	E	N	1953	77327
1462	S	S	R	N	1970	77553
1462	S	S	E	N	1974	77554
1463	D	D	E	N	1976	86147
1464	D	D	E	N	1976	86147
1466	S	S	E	N	1972	66579
1473	S	D	E	N	1977	93343
1474	S	D	E	N	1977	93343
1475	S	D	E	N	1977	93343
1476	S	D	E	N	1977	93343
1497	S	D	E	N	1972	64464
1504	S	D	R	N	1968	53472
1504	S	D	E	N	1968	67357
1505	S	D	R	N	1968	53472
1505	S	D	E	N	1968	67357
1506	S	D	R	N	1968	53472
1506	S	D	E	N	1968	67357
1508	S	D	E	N	1969	56119
1509	S	D	E	N	1969	56119
1510	S	D	E	N	1969	56119
1511	D	D	E	N	1976	86147
1512	S	D	E	N	1969	56119
1513	S	D	E	N	1969	56119
1514	S	D	E	N	1967	44427
1514	S	S	E	N	1972	66579
1515	S	D	E	N	1967	44427
1516	S	D	E	N	1976	86147
1517	S	D	E	N	1976	86147
1518	D	D	E	N	1976	86147
1518	S	D	E	N	1976	86147
1519	S	D	R	N	1968	63102
1519	S	T	R	N	1968	63102
1519	S	D	E	N	1968	63103
1519	S	T	E	N	1968	63103
1520	S	D	R	N	1968	63102
1520	S	T	R	N	1968	63102
1520	S	D	E	N	1968	63103
1520	S	T	E	N	1968	63103
1521	S	D	R	N	1968	63102
1521	S	T	R	N	1968	63102
1521	S	D	E	N	1968	63103
1521	S	T	E	N	1968	63103
1522	S	D	R	N	1968	63102
1522	S	T	R	N	1968	63102
1522	S	D	E	N	1968	63103
1522	S	T	E	N	1968	63103
1523	S	D	R	N	1968	63102
1523	S	T	R	N	1968	63102
1523	S	D	E	N	1968	63103
1523	S	T	E	N	1968	63103
1524	S	D	R	N	1968	63102
1524	S	T	R	N	1968	63102
1524	S	D	E	N	1968	63103
1524	S	T	E	N	1968	63103
1530	S	D	E	N	1968	49971
1531	S	D	E	N	1968	49971
1538	S	D	E	N	1957	76819
1538	S	T	E	N	1957	76819
1539	S	D	E	N	1957	76819
1539	S	T	E	N	1957	76819
1541	S	S	E	N	1965	45481
1542	S	S	E	N	1965	45481
1543	S	S	E	N	1965	45481
1543	S	D	E	N	1975	76210
1544	S	D	E	N	1949	4320
1544	S	S	E	N	1965	45481
1545	S	S	E	N	1965	45481
1546	D	D	E	N	1976	86147
1546	S	D	E	N	1976	86147
1547	S	S	E	N	1972	66579
1562	S	D	R	N	1966	44184
1562	S	D	E	N	1966	44185

Phys. State: **C.** Superconductive; **D.** Doped; **E.** Expanded; **F.** Fibrous or Whisker; **G.** Gas; **I.** Ionized or Plasma; **L.** Liquid; **M.** Multiphase; **P.** Powder or Fine Particle; **S.** Solid; **T.** Thick or Thin Film

Subject: **D.** Data; **E.** Experiment; **G.** General (Data + Theory + Experiment); **S.** Survey (Review, Compendium etc.); **T.** Theory

Language: **C.** Czech; **D.** Dutch; **E.** English; **F.** French; **G.** German; **I.** Italian; **J.** Japanese; **O.** Other Languages; **P.** Polish; **R.** Russian; **S.** Spanish

Temperature: **F.** Full Range (Low + Normal + High); **L.** Low (0 to 75K) + Overlap into Normal; **N.** Normal (75 to 1273K); **H.** High (above 1273K) + Overlap into Normal; Blank = Not Coded

Substance Number	Phys. State	Subject	Language	Temperature	Year	TPRC Number
503-1564	S	D	E	N	1949	4320
1567	S	D	G	N	1966	44476
1567	S	T	G	N	1966	44476
1569	S	D	E	N	1949	4320
1569	S	T	E	N	1967	47364
1570	S	D	E	N	1967	44427
1570	S	S	E	N	1972	66579
1585	S	D	E	N	1958	40374
1586	S	D	E	N	1949	4320
1586	S	D	E	N	1957	76819
1586	S	T	E	N	1957	76819
1587	S	D	E	N	1949	4320
1587	S	D	E	N	1957	76819
1587	S	T	E	N	1957	76819
1588	S	D	E	N	1957	76819
1588	S	T	E	N	1957	76819
1654	S	S	E	N	1972	66579
1694	S	D	G	N	1963	32867
1694	S	D	E	N	1974	77552
1724	S	D	E	N	1949	4320
1741	S	S	E	N	1972	66579
1748	S	D	E	N	1949	4320
1749	S	D	J	N	1977	94196
1749	T	D	J	N	1978	95041
1750	S	D	E	N	1977	91585
1751	S	D	E	N	1977	91585
1752	S	D	E	N	1977	91585
1753	S	D	E	N	1977	91585
1754	S	D	E	N	1977	91585
1755	S	D	E	N	1977	91585
1756	S	D	E	N	1977	91585
1757	S	D	E	N	1977	91585
1775	S	D	E	N	1974	77445
1803	S	D	E	N	1974	75444
1821	S	D	E	N	1978	92257
1828	S	D	E	N	1968	49293
1830	D	D	E	N	1976	85232
1831	D	D	E	N	1976	85232
1832	D	D	E	N	1976	85232
1834	S	D	R	N	1975	87131
1834	S	D	E	N	1976	87132
1836	S	D	R	N	1976	89575
1836	S	T	R	N	1976	89575
1836	S	D	E	N	1976	89576
1836	S	T	E	N	1976	89576
1837	S	D	R	N	1976	89575
1837	S	T	R	N	1976	89575
1837	S	D	E	N	1976	89576
1837	S	T	E	N	1976	89576
1862	S	D	E	N	1974	88682
1864	S	D	E	N	1974	88682
1865	S	D	E	N	1974	88682
1866	S	D	E	N	1974	88682
1878	S	S	G	N	1977	92758
1890	S	S	E	N	1972	66579
1894	S	D	E	N	1949	4320
1912	S	D	E	N	1949	4320
1912	S	D	E	N	1963	30814
1953	D	D	E	N	1973	70703
1963	S	S	E	N	1977	89235
1964	S	S	E	N	1977	89235
1965	S	S	E	N	1977	89235
1966	S	S	E	N	1977	89235
1967	S	S	E	N	1977	89235
1968	S	S	E	N	1977	89235
1969	S	S	E	N	1977	89235
1970	S	S	E	N	1977	89235
1971	S	S	E	N	1977	89235
1972	S	S	E	N	1977	89235
1973	S	S	E	N	1977	89235
1974	D	D	E	N	1974	89398
1974	D	D	E	N	1977	90258
1985	S	D	R	H	1967	49346
1985	S	D	E	H	1967	49347
1986	S	D	R	H	1967	49346
1986	S	D	E	H	1967	49347
1987	S	D	R	H	1967	49346
1987	S	D	E	H	1967	49347
2041	S	D	E	N	1949	4320
2042	S	D	E	N	1949	4320
2043	S	D	E	N	1949	4320
2044	S	D	E	N	1949	4320
2045	S	D	E	N	1949	4320
2046	S	D	E	N	1949	4320
2047	S	D	E	N	1949	4320
2048	S	D	E	N	1949	4320
2049	S	D	E	N	1949	4320
2050	S	D	E	N	1949	4320
2051	S	D	E	N	1949	4320
2052	S	D	E	N	1949	4320
2053	S	D	E	N	1949	4320
2054	S	D	E	N	1949	4320
2055	S	D	E	N	1949	4320
2056	S	D	E	N	1949	4320
2057	S	D	E	N	1949	4320
2058	S	D	E	N	1949	4320
2059	S	D	E	N	1949	4320
2060	S	D	E	N	1949	4320
2061	S	D	E	N	1949	4320
2063	S	D	E	N	1949	4320
2064	S	D	E	N	1949	4320

Substance Number	Phys. State	Subject	Language	Temperature	Year	TPRC Number
503-2065	S	D	E	N	1949	4320
2066	S	D	E	N	1949	4320
2067	S	D	E	N	1949	4320
2068	S	D	E	N	1949	4320
2069	S	D	E	N	1949	4320
2070	S	D	E	N	1949	4320
2071	S	D	E	N	1949	4320
2072	S	D	E	N	1949	4320
2073	S	D	E	N	1949	4320
2074	S	D	E	N	1949	4320
2075	S	D	E	N	1949	4320
2076	S	D	E	N	1949	4320
2077	S	D	E	N	1949	4320
2078	S	D	E	N	1949	4320
2079	S	D	E	N	1949	4320
2080	S	D	E	N	1949	4320
2081	S	D	E	N	1949	4320
2082	S	D	E	N	1949	4320
2083	S	D	E	N	1949	4320
2087	S	D	E	N	1958	15895
2088	S	D	E	N	1958	15895
2089	D	D	E	N	1959	18622
2089	S	D	E	N	1958	15895
2089	S	D	E	N	1959	18622
2090	S	D	E	N	1958	15895
2091	D	D	E	N	1959	18622
2091	S	D	E	N	1958	15895
2091	S	D	E	N	1959	18622
2092	S	D	E	N	1958	15895
2093	S	D	E	N	1958	15895
2094	S	D	E	N	1958	15895
2095	S	D	E	N	1958	15895
2097	S	D	E	N	1958	15895
2098	S	D	E	N	1958	15895
2099	S	D	E	N	1958	15895
2100	S	D	E	N	1958	15895
2101	S	D	E	N	1958	15895
2102	S	D	E	N	1958	15895
2106	D	D	E	N	1958	18615
2106	S	D	E	N	1958	18615
2115	S	D	E	N	1961	18367
2116	S	D	E	N	1961	18367
2117	S	D	E	N	1961	18367
2118	S	D	E	N	1961	18367
2119	S	D	E	N	1961	18367
2119	S	D	E	N	1959	18428
2120	S	D	E	N	1961	18367
2121	S	D	E	N	1959	18428
2122	S	D	E	N	1959	18428
2123	S	D	E	N	1959	18428
2124	S	D	E	N	1959	18428
2125	S	D	E	N	1959	18428
2126	S	D	E	N	1959	18428
2128	D	D	E	N	1958	18615
2128	S	D	E	N	1958	18615
2129	D	D	E	N	1958	18615
2129	S	D	E	N	1958	18615
2130	D	D	E	N	1958	18615
2131	D	D	E	N	1959	18621
2132	S	D	E	N	1908	96760
2133	S	D	E	N	1908	96760
2134	S	D	E	N	1908	96760
2135	S	D	E	N	1908	96760
2273	S	D	E	N	1976	93227
2273	S	T	E	N	1976	93227
2274	S	S	R	N	1973	79859
2274	S	S	E	N	1973	79860
2291	S	D	E	N	1953	15126
2291	S	D	R	N	1953	20572
2293	S	D	E	N	1964	32047
2294	S	D	E	N	1964	32047
2305	P	D	E	N	1978	94862
2307	S	D	E	N	1977	92045
2331	S	D	E	N	1941	90225
2338	S	D	E	N	1941	90225
2339	S	D	E	N	1941	90225
2340	D	D	E	N	1977	90258
2341	D	D	E	N	1977	90258
2342	D	D	E	N	1977	90258
2343	D	D	E	N	1977	90258
2344	D	D	E	N	1977	90258
2345	S	D	J	N	1977	90383
2347	S	D	E	N	1971	66243
2383	S	D	E	N	1974	77552
2384	S	D	E	N	1974	77552
2385	S	D	E	N	1974	77552
2386	S	D	E	N	1974	77552
2387	D	D	E	N	1974	77552
2387	S	D	E	N	1974	77552
2388	S	D	E	N	1956	77317
2389	S	D	S	N	1966	45057
2392	S	D	E	N	1968	40352
2393	S	D	E	N	1968	40352
2394	S	D	E	N	1968	40352
2395	S	D	E	N	1968	40352
2395	S	D	E	N	1968	49971
2396	S	D	E	N	1968	49971
2397	S	D	E	N	1968	49971
2398	S	D	E	N	1964	37430
2506	S	D	E	N	1963	30814
2507	S	D	E	N	1963	30814

Substance Number	Phys. State	Subject	Language	Temperature	Year	TPRC Number
503-2508	S	D	E	N	1963	30814
2509	S	D	E	N	1963	30814
2510	S	D	E	N	1963	30814
2511	S	D	E	N	1963	30814
2512	S	D	E	N	1963	30814
2523	S	D	F	N	1961	32688
2524	S	D	G	N	1963	32867
2543	S	D	E	N	1968	40352
2544	S	D	E	N	1968	40352
2545	S	D	E	N	1968	40352
2546	S	D	E	N	1968	40352
2547	S	D	E	N	1968	40352
2548	S	D	E	N	1968	40352
2549	S	D	E	N	1968	40352
2550	S	D	E	N	1968	40352
2551	S	D	E	N	1968	40352
2552	S	D	E	N	1968	40352
2553	S	D	E	N	1968	40352
2554	S	D	E	N	1968	40352
2555	S	D	E	N	1968	40352
2556	S	D	E	N	1968	40352
2557	S	D	E	N	1968	40352
2558	S	D	E	N	1968	40352
2559	S	D	E	N	1968	40352
2560	S	D	E	N	1968	40352
2561	S	D	E	N	1968	40352
2562	S	D	E	N	1968	40352
2563	S	D	E	N	1968	40352
2564	S	D	E	N	1968	40352
2565	S	D	E	N	1968	40352
2566	S	D	E	N	1968	40352
2567	S	D	E	N	1968	40352
2569	S	D	E	N	1968	40352
2570	D	D	E	N	1968	40352
2570	S	D	E	N	1968	40352
2571	S	D	E	N	1968	40352
2572	S	D	E	N	1968	40352
2573	S	D	E	N	1968	40352
2580	S	D	S	N	1966	45057
2581	S	D	S	N	1966	45057
2582	S	D	S	N	1966	45057
2583	S	D	S	N	1966	45057
2584	S	D	S	N	1966	45057
2585	S	D	S	N	1966	45057
2624	S	D	E	N	1970	58799
2625	S	D	E	N	1970	58799
2626	S	D	E	N	1970	58799
2627	S	D	E	N	1970	58799
2628	S	D	E	N	1970	58799
2629	S	D	E	N	1970	58799
2645	S	D	E	N	1956	77317
2646	S	D	E	N	1974	77552
2679	S	D	E	N	1967	47094
2722	S	D	E	N	1979	100051
2723	S	D	E	N	1979	100051
2770	S	S	E	N	1976	83746
2771	S	S	E	N	1976	83746
2772	S	S	E	N	1976	83746
2773	S	S	E	N	1976	83746
2783	S	S	E	N	1972	66579
2784	S	S	E	N	1972	66579
2789	S	S	E	N	1972	66579
2790	S	S	E	N	1972	66579
2791	S	S	E	N	1972	66579
2792	S	S	E	N	1972	66579
2793	S	S	E	N	1972	66579
2794	S	S	E	N	1972	66579
2795	S	S	E	N	1972	66579
2796	S	D	E	N	1979	99709
2797	S	D	E	N	1979	99709
2798	S	D	E	N	1979	99709
2799	S	D	E	N	1979	99709
2802	S	S	E	N	1979	99434
2803	S	S	E	N	1979	99434
2875	D	D	E	N	1979	97880
2876	S	D	E	N	1979	97880
2877	S	D	E	N	1979	97880
8219	S	D	R	N	1969	56249
8220	S	D	R	N	1969	56249
8221	S	D	R	N	1969	56249
8222	S	D	R	N	1969	56249
8236	D	D	E	N	1973	71248
8401	D	D	E	N	1968	47842
8403	S	D	J	N	1962	40485
8403	S	D	E	N	1962	40486
8404	S	D	J	N	1962	40485
8404	S	D	E	N	1962	40486
8405	S	D	J	N	1962	40485
8405	S	D	E	N	1962	40486
8409	S	D	E	N	1966	39761
8410	S	D	E	N	1966	39603
8410	S	D	E	N	1969	54897
8411	S	D	E	N	1966	39603
8413	S	D	E	N	1966	39603
8422	S	D	E	N	1955	77328
8423	S	D	E	N	1955	77328
8424	S	D	E	N	1955	77328
8432	S	D	E	N	1974	77552
8437	D	D	E	N	1974	77552
8440	S	D	R	H	1974	76628
8441	S	D	R	H	1974	76628

Phys. State: **C.** Superconductive; **D.** Doped; **E.** Expanded; **F.** Fibrous or Whisker; **G.** Gas; **I.** Ionized or Plasma; **L.** Liquid; **M.** Multiphase; **P.** Powder or Fine Particle; **S.** Solid; **T.** Thick or Thin Film
Subject: **D.** Data; **E.** Experiment; **G.** General (Data + Theory + Experiment); **S.** Survey (Review, Compendium etc.); **T.** Theory
Language: **C.** Czech; **D.** Dutch; **E.** English; **F.** French; **G.** German; **I.** Italian; **J.** Japanese; **O.** Other Languages; **P.** Polish; **R.** Russian; **S.** Spanish
Temperature: **F.** Full Range (Low + Normal + High); **L.** Low (0 to 75K) + Overlap into Normal; **N.** Normal (75 to 1273K); **H.** High (above 1273K) + Overlap into Normal; Blank = Not Coded

Substance Number	Phys. State	Subject	Language	Temperature	Year	TPRC Number
503-8442	S	D	E	N	1953	77327
8443	S	D	E	N	1954	77326
8444	S	D	E	N	1954	77326
8445	S	D	E	N	1954	77326
8446	S	D	E	N	1954	77326
8447	S	D	E	N	1954	77326
8448	S	D	E	N	1954	77326
8448	S	D	E	N	1955	77328
8449	S	D	E	N	1954	77326
8450	S	D	E	N	1954	77326
8451	S	D	E	N	1954	77326
8452	S	D	E	N	1954	77326
8452	S	D	E	N	1953	77327
8452	S	D	E	N	1955	77328
8453	S	D	E	N	1955	77324
8454	S	D	E	N	1957	77323
8455	S	D	E	N	1957	77323
8456	S	D	E	N	1956	77322
8456	S	D	E	N	1955	77324
8459	S	D	E	N	1973	76990
8460	S	D	E	N	1973	76990
8461	S	D	E	N	1973	76990
8462	S	D	E	N	1967	78266
8463	S	D	E	N	1967	78266
8464	S	D	E	N	1967	78266
8467	S	D	E	N	1956	76865
8468	S	D	E	N	1956	76857
8469	S	D	E	N	1956	76857
8470	S	D	E	N	1956	76857
8471	S	D	E	N	1956	76857
8479	D	D	E	N	1972	73840
8479	S	D	E	N	1972	73840
8481	S	D	R	N	1972	68525
8481	S	D	E	N	1972	70883
8491	S	D	R	N	1971	66247
8492	S	D	R	N	1971	66247
8494	D	D	E	N	1973	72771
8494	S	D	E	N	1973	72771
8502	S	D	I	N	1966	45643
8505	S	D	E	N	1956	76857
8508	S	D	R	N	1971	64818
8508	S	D	E	N	1971	64819
8509	S	D	R	N	1971	64818
8509	S	D	E	N	1971	64819
8510	S	D	R	N	1971	64711
8510	S	D	E	N	1971	64712
8520	S	D	R	N	1971	64691
8520	S	D	E	N	1971	64692
8521	S	D	R	N	1971	64691
8521	S	D	E	N	1971	64692
8522	S	D	R	N	1971	64691
8522	S	D	E	N	1971	64692
8523	S	D	R	N	1971	64691
8523	S	D	E	N	1971	64692
8526	S	D	E	N	1971	63042
8528	D	D	R	N	1967	63242
8528	D	D	E	N	1967	63243
8529	D	D	R	N	1967	63242
8529	D	D	E	N	1967	63243
8544	S	D	R	N	1970	61440
8544	S	D	E	N	1970	61441
8569	S	D	R	N	1972	70260
8569	S	D	E	N	1972	72358
8571	D	D	R	N	1970	60982
8571	D	D	E	N	1970	60983
8582	S	D	R	N	1970	60144
8582	S	D	E	N	1970	60145
8658	S	D	E	N	1962	39896
8663	S	D	R	N	1969	57179
8663	S	D	E	N	1969	57180
8666	S	D	R	N	1969	61597
8666	S	D	E	N	1969	61598
8670	S	D	E	N	1968	49971
8671	S	D	E	N	1968	49971
8672	S	D	E	N	1968	49971
8674	S	D	E	N	1968	49971
8674	S	D	E	N	1969	54273
8675	S	D	E	N	1968	49971
8675	S	D	E	N	1969	54273
8676	S	D	E	N	1968	49971
8676	S	D	E	N	1969	54273
8677	S	D	E	N	1968	49971
8677	S	D	E	N	1969	54273
8678	S	D	E	N	1968	49971
8678	S	D	E	N	1969	54273
8679	S	D	E	N	1968	49971
8680	S	D	E	N	1968	49971
8681	S	D	E	N	1968	49971
8682	S	D	E	N	1968	49971
8683	S	D	E	N	1968	49971
8684	S	D	E	N	1968	49971
8686	S	D	E	N	1968	49971
8687	S	D	E	N	1968	49971
8689	S	D	E	N	1968	49971
8690	S	D	E	N	1968	49971
8691	S	D	E	N	1968	49971
8692	S	D	E	N	1968	49971
8693	S	D	E	N	1968	49971
8694	S	D	E	N	1968	49971
8696	S	D	E	N	1968	49971
8697	S	D	E	N	1968	49971
503-8698	S	D	E	N	1968	49971
8699	S	D	E	N	1968	49971
8699	S	D	E	N	1971	67411
8700	S	D	E	N	1968	49971
8701	S	D	E	N	1968	49971
8702	S	D	E	N	1968	49971
8703	S	D	E	N	1968	49971
8703	S	D	E	N	1969	54273
8704	D	D	E	N	1971	67411
8704	S	D	E	N	1968	49971
8704	S	D	E	N	1969	54273
8704	S	D	E	N	1971	67411
8705	S	D	R	N	1943	51635
8705	S	D	E	N	1943	51636
8705	S	D	F	N	1943	51636
8707	S	D	E	N	1969	54273
8712	S	D	R	N	1967	49692
8713	S	D	R	N	1967	49692
8714	S	D	R	N	1967	49692
8715	S	D	R	N	1967	49692
8716	S	D	R	N	1967	49692
8717	S	D	R	N	1967	49692
8718	D	D	R	N	1969	57161
8718	D	D	E	N	1969	57162
8718	S	D	R	N	1967	49692
8718	S	D	R	N	1969	57161
8718	S	D	E	N	1969	57162
8719	S	D	R	N	1967	49692
8720	S	D	E	N	1969	41419
8720	S	T	E	N	1969	41419
8721	S	D	E	N	1969	41333
8722	S	D	E	N	1969	41333
8723	S	D	E	N	1971	67411
8725	D	D	E	L	1971	61835
8734	S	D	E	N	1973	76990
8734	S	D	E	N	1957	77323
8735	S	D	R	N	1972	67739
8735	S	D	E	N	1972	76930
8738	S	D	E	N	1969	41333
8742	S	D	E	N	1969	41333
8752	D	D	E	N	1973	76072
8769	S	D	R	N	1967	45797
8772	S	D	E	N	1968	47870
8773	S	D	E	N	1968	47870
8774	S	D	E	N	1968	47870
8775	S	D	E	N	1968	47870
8776	S	D	E	N	1968	47870
8777	S	D	E	N	1968	47870
8778	S	D	E	N	1968	47870
8779	S	D	E	N	1968	47870
8780	S	D	E	N	1968	47870
8781	S	D	E	N	1968	47870
8782	S	D	E	N	1968	47870
8783	S	D	E	N	1968	47870
8784	S	D	E	N	1955	39835
8785	S	D	E	N	1968	47870
8787	S	D	E	N	1968	47870
8788	S	D	E	N	1968	47870
8789	S	D	E	N	1968	47870
8790	S	D	E	N	1968	47870
8791	S	D	E	N	1968	47870
8795	S	D	E	N	1968	47870
8796	S	D	E	N	1968	47870
8797	S	D	E	N	1968	47870
8798	S	D	E	N	1968	47870
8799	S	D	E	N	1968	47870
8800	S	D	E	N	1968	47870
8801	S	D	E	N	1968	47870
8802	S	D	E	N	1968	47870
8803	S	D	E	N	1968	47870
8804	S	D	E	N	1968	47870
8805	S	D	E	N	1955	39835
8806	S	D	E	N	1968	47870
8807	S	D	E	N	1968	47870
8808	S	D	E	N	1955	39835
8809	S	D	E	N	1955	39835
8810	S	D	E	N	1955	39835
8811	S	D	E	N	1955	39835
8812	S	D	E	N	1968	47870
8813	S	D	E	N	1968	47870
8814	S	D	E	N	1968	47870
8815	S	D	E	N	1968	47870
8816	S	D	E	N	1955	39835
8817	S	D	E	N	1955	39835
8818	D	D	R	N	1967	63242
8818	D	D	E	N	1967	63243
8818	S	D	R	N	1943	51635
8818	S	D	E	N	1943	51636
8818	S	D	F	N	1943	51636
8819	S	D	E	N	1967	42742
8819	S	D	E	N	1968	47870
8821	S	D	E	N	1957	48109
8822	D	D	R	N	1971	63871
8822	D	D	E	N	1971	67363
8822	S	D	E	N	1957	76819
8822	S	E	E	N	1957	76819
8822	S	D	E	N	1960	76856
8822	S	D	E	N	1955	77328
8824	D	D	E	N	1957	48109
8825	S	D	E	N	1957	48109
8826	D	D	E	N	1957	48109
503-8827	D	D	E	N	1957	48109
8828	D	D	E	N	1957	48109
8829	D	D	E	N	1957	48109
8829	S	D	E	N	1957	48109
8830	S	D	E	N	1957	48109
8840	S	D	E	N	1958	48105
8841	S	D	E	N	1958	48105
8842	S	D	R	N	1968	55436
8842	S	D	E	N	1968	55437
8842	S	D	E	N	1973	76990
8842	S	D	E	N	1957	77323
8842	S	D	E	N	1954	77326
8843	S	D	E	N	1958	48105
8844	S	D	E	N	1958	48105
8845	S	D	E	N	1958	48105
8846	S	D	R	N	1968	55436
8846	S	D	E	N	1968	55437
8847	S	D	E	N	1958	48105
8848	S	D	E	N	1958	48105
8849	S	D	E	N	1958	48105
8850	S	D	E	N	1958	48105
8864	S	D	E	N	1954	77326
8865	S	D	E	N	1969	54980
8867	D	D	E	N	1968	47426
8867	S	D	P	N	1975	83208
8867	S	E	P	N	1975	83208
8870	S	D	O	N	1961	49805
8872	S	D	R	N	1965	42546
8873	S	D	R	N	1965	42545
8875	D	D	R	N	1966	44206
8875	D	D	E	N	1966	44207
8875	S	D	R	N	1966	44206
8875	S	D	E	N	1966	44207
8876	D	D	R	N	1966	44206
8876	D	D	E	N	1966	44207
8876	D	D	R	N	1970	60960
8876	D	D	E	N	1970	60961
8876	S	D	R	N	1966	44206
8876	S	D	E	N	1966	44207
8876	S	D	E	N	1968	49971
8877	D	D	R	N	1966	44206
8877	D	D	E	N	1966	44207
8877	S	D	R	N	1966	44206
8877	S	D	E	N	1966	44207
8878	D	D	R	N	1966	44206
8878	D	D	E	N	1966	44207
8878	S	D	R	N	1966	44206
8878	S	D	E	N	1966	44207
8879	D	D	R	N	1966	44206
8879	D	D	E	N	1966	44207
8879	S	D	R	N	1966	44206
8879	S	D	E	N	1966	44207
8882	D	D	E	N	1970	37250
8882	S	D	E	N	1970	37250
8883	S	D	E	N	1967	35227
8886	S	D	E	N	1968	47870
8887	S	D	E	N	1967	34653
8888	S	D	R	N	1968	51253
8888	S	D	E	N	1968	51254
8889	D	D	R	N	1970	60960
8889	D	D	E	N	1970	60961
8889	S	D	E	N	1955	39835
8889	S	D	R	N	1972	67739
8889	S	D	E	N	1972	76930
8889	S	D	R	N	1971	78332
8889	S	D	E	N	1972	78333
8892	S	D	E	N	1968	47870
8905	S	D	E	N	1958	48105
8906	S	D	E	N	1955	77324
8906	S	D	E	N	1955	77325
8913	S	D	R	N	1970	75457
8913	S	D	E	N	1970	75458
8925	S	D	E	N	1966	39945
8926	S	D	E	N	1966	39945
8927	D	D	E	N	1969	40481
8927	S	D	E	N	1966	34479
8927	S	D	E	N	1969	54273
8941	D	D	R	N	1966	44206
8941	D	D	E	N	1966	44207
8941	S	D	R	N	1966	44206
8941	S	D	E	N	1966	44207
8950	S	D	E	N	1968	49971
8951	S	D	E	N	1968	47870
8952	S	D	E	N	1968	47870
8954	D	D	E	N	1966	36254
8954	D	D	R	N	1966	47091
8956	S	D	E	N	1966	39603
8956	S	D	F	N	1969	54171
8956	S	D	E	N	1971	62634
8956	S	E	E	N	1971	62634
8958	S	D	E	N	1962	39896
8959	S	D	E	N	1962	39896
8967	S	D	E	N	1962	39896
8972	S	D	E	N	1955	39835
8977	P	D	E	N	1974	76073
8977	S	D	E	N	1965	35043
8977	S	D	E	N	1955	39835
8977	S	D	E	N	1974	76073
8979	S	D	R	N	1971	64691
8979	S	D	E	N	1971	64692
8987	S	D	E	N	1973	76990

Phys. State: **C.** Superconductive; **D.** Doped; **E.** Expanded; **F.** Fibrous or Whisker; **G.** Gas; **I.** Ionized or Plasma; **L.** Liquid; **M.** Multiphase; **P.** Powder or Fine Particle; **S.** Solid; **T.** Thick or Thin Film

Subject: **D.** Data; **E.** Experiment; **G.** General (Data + Theory + Experiment); **S.** Survey (Review, Compendium etc.); **T.** Theory

Language: **C.** Czech; **D.** Dutch; **E.** English; **F.** French; **G.** German; **I.** Italian; **J.** Japanese; **O.** Other Languages; **P.** Polish; **R.** Russian; **S.** Spanish

Temperature: **F.** Full Range (Low + Normal + High); **L.** Low (0 to 75K) + Overlap into Normal; **N.** Normal (75 to 1273K); **H.** High (above 1273K) + Overlap into Normal; Blank = Not Coded

Substance Number	Phys. State	Subject	Language	Temperature	Year	TPRC Number
503-8987	S	D	E	N	1955	77324
8987	S	D	E	N	1954	77326
8987	S	D	E	N	1955	77328
8990	S	D	E	N	1958	48105
9015	S	D	J	N	1962	40485
9015	S	D	E	N	1962	40486
9016	S	D	E	N	1955	77325
9022	S	D	R	N	1943	51635
9022	S	D	E	N	1943	51636
9022	S	D	F	N	1943	51636
9027	S	D	E	N	1966	39945
9029	S	D	G	N	1966	45549
9029	S	D	G	N	1968	50596
9032	D	D	E	N	1963	45668
9032	S	D	E	N	1963	45668
9034	S	D	E	N	1966	39945
9039	S	D	R	N	1971	66247
9042	S	D	E	N	1968	49236
9048	S	D	R	N	1966	40885
9048	S	D	E	N	1966	40886
9054	S	D	E	N	1968	47870
9058	S	D	E	N	1968	47870
9063	S	D	E	N	1957	77323
9064	S	D	E	N	1973	76990
9064	S	D	E	N	1957	77323
9066	S	D	R	N	1966	40885
9066	S	D	E	N	1966	40886
9068	D	D	R	N	1971	64244
9068	D	D	E	N	1971	72666
9068	S	D	R	N	1965	41898
9068	S	D	R	N	1968	51980
9068	S	D	E	N	1968	51981
9068	S	D	E	N	1971	62678
9068	S	D	R	N	1971	64244
9068	S	D	E	N	1971	72666
9075	D	D	R	N	1970	60960
9075	D	D	E	N	1970	60961
9075	S	D	R	N	1970	60960
9075	S	D	E	N	1970	60961
9078	S	D	R	N	1971	78332
9078	S	D	E	N	1972	78333
9080	S	D	E	N	1968	47870
9080	S	D	E	N	1968	49971
9080	S	D	E	N	1971	67411
9081	S	D	E	N	1968	49971
9084	S	D	R	N	1972	67739
9084	S	D	E	N	1972	76930
9091	S	D	E	N	1968	49971
9095	S	D	E	N	1958	48105
9101	S	D	E	N	1962	39896
9101	S	D	O	N	1966	44269
9113	S	D	R	N	1971	65218
9113	S	D	E	N	1971	66790
9115	D	D	O	N	1964	45263
9117	S	D	R	N	1968	51970
9117	S	D	E	N	1968	51971
9118	S	D	R	N	1965	42545
9118	S	D	R	N	1965	42546
9120	S	D	E	N	1974	74816
9121	D	D	E	N	1964	77530
9121	D	E	E	N	1964	77530
9131	S	D	E	N	1970	37340
9132	S	D	E	N	1968	49971
9133	S	D	E	N	1968	49971
9134	S	D	G	N	1964	37241
9134	S	D	G	N	1967	45763
9134	S	D	R	N	1943	51635
9134	S	D	E	N	1943	51636
9134	S	D	F	N	1943	51636
9134	S	D	E	N	1971	67411
9136	S	D	E	N	1973	68992
9139	S	D	R	N	1971	64818
9139	S	D	E	N	1971	64819
9140	S	D	R	N	1968	55436
9140	S	D	E	N	1968	55437
9143	S	D	E	N	1966	34479
9143	S	D	E	N	1969	41419
9143	S	T	E	N	1969	41419
9146	D	D	E	N	1968	50230
9147	S	D	E	N	1958	59534
9147	S	D	E	N	1971	63042
9148	S	D	E	N	1968	49971
9151	S	D	R	N	1970	60142
9151	S	D	E	N	1970	60143
9152	S	D	R	N	1968	34183
9152	S	D	E	N	1955	39835
9152	S	D	R	N	1943	51635
9152	S	D	E	N	1943	51636
9152	S	D	F	N	1943	51636
9152	S	D	E	N	1968	52757
9152	S	D	E	N	1955	77324
9152	S	D	E	N	1955	77325
9154	D	D	R	N	1970	60960
9154	D	D	E	N	1970	60961
9154	S	D	R	N	1970	60960
9154	S	D	E	N	1970	60961
9156	D	D	E	N	1964	77530
9156	D	E	E	N	1964	77530
9157	S	D	R	N	1968	55436
9157	S	D	E	N	1968	55437
9158	S	D	R	N	1965	42545
503-9158	S	D	R	N	1965	42546
9158	S	D	R	N	1972	67510
9158	S	D	E	N	1972	76928
9158	S	D	G	N	1959	76936
9158	S	D	E	N	1964	77530
9158	S	E	E	N	1964	77530
9158	S	D	E	N	1973	77755
9158	S	E	E	N	1973	77755
9159	S	D	E	N	1968	49971
9159	S	D	R	N	1971	64711
9159	S	D	E	N	1971	64712
9160	S	D	E	N	1962	39896
9160	S	D	R	N	1965	42546
9160	S	D	E	N	1968	49971
9160	S	D	E	N	1969	54273
9160	S	D	R	N	1972	67510
9160	S	D	E	N	1972	76928
9166	S	D	E	N	1955	39835
9167	S	D	O	N	1966	44269
9167	S	D	E	N	1971	63697
9169	D	D	E	N	1962	39896
9169	D	D	I	N	1966	45643
9169	D	D	E	N	1972	69035
9169	S	D	E	N	1962	39896
9169	S	D	E	N	1958	59534
9169	S	D	E	N	1971	63042
9169	S	D	E	N	1971	67411
9171	D	D	I	N	1966	45643
9171	S	D	R	N	1965	42546
9171	S	D	J	N	1970	59003
9171	S	D	E	N	1971	67411
9171	S	D	R	N	1972	67510
9171	S	D	E	N	1972	76928
9173	D	D	P	N	1969	56272
9173	D	D	E	N	1970	58318
9173	S	D	G	N	1966	45549
9173	S	D	G	N	1968	50596
9173	S	D	P	N	1969	56272
9173	S	D	R	N	1970	65219
9173	S	D	E	N	1974	75753
9173	S	D	E	N	1973	77755
9173	S	E	E	N	1973	77755
9174	D	D	E	N	1965	34147
9174	D	T	E	N	1965	34147
9174	D	D	E	N	1965	36547
9174	D	D	I	N	1966	45643
9174	D	D	R	N	1968	51970
9174	D	D	E	N	1968	51971
9174	D	D	R	N	1970	60960
9174	D	D	E	N	1970	60961
9174	D	D	E	N	1973	77755
9174	D	E	E	N	1973	77755
9174	S	D	E	N	1965	35972
9174	S	D	E	N	1970	37340
9174	S	D	E	N	1964	37799
9174	S	D	E	N	1970	38064
9174	S	D	E	N	1962	39896
9174	S	D	G	N	1966	41424
9174	S	D	R	N	1965	42545
9174	S	D	R	N	1965	42546
9174	S	D	I	N	1966	45643
9174	S	D	R	N	1943	51635
9174	S	D	E	N	1943	51636
9174	S	D	F	N	1943	51636
9174	S	D	R	N	1968	51970
9174	S	D	E	N	1968	51971
9174	S	D	R	N	1970	60960
9174	S	D	E	N	1970	60961
9174	S	D	E	N	1971	61383
9174	S	D	R	N	1969	61597
9174	S	D	E	N	1969	61598
9174	S	D	R	N	1971	64818
9174	S	D	E	N	1971	64819
9174	S	D	E	N	1971	65024
9174	S	D	E	N	1973	77755
9174	S	E	E	N	1973	77755
9177	S	D	R	H	1974	76628
9179	D	D	I	N	1966	45643
9179	D	D	R	N	1970	60960
9179	D	D	E	N	1970	60961
9179	S	D	R	N	1970	60960
9179	S	D	E	N	1970	60961
9179	S	D	E	N	1973	77755
9179	S	E	E	N	1973	77755
9181	M	D	E	N	1944	47956
9181	S	D	E	N	1967	33849
9181	S	D	E	N	1965	38674
9181	S	D	E	N	1966	39011
9181	S	D	G	N	1966	41425
9181	S	D	E	N	1965	43840
9181	S	D	R	N	1972	67510
9181	S	D	E	N	1972	76928
9202	S	D	R	N	1973	75668
9202	S	D	E	N	1973	75669
9207	S	D	R	N	1973	74621
9207	S	D	E	N	1974	74622
9211	S	D	R	N	1973	71271
9211	S	D	E	N	1973	72305
9218	S	D	E	N	1973	70940
9219	S	D	E	N	1973	70940
9220	S	D	E	N	1973	70940
503-9233	S	D	E	N	1972	69070
9234	S	D	E	N	1972	69070
9238	S	D	E	N	1973	70215
9238	S	G	E	N	1970	77935
9247	P	D	F	N	1970	65176
9248	P	D	F	N	1970	65176
9261	S	D	E	N	1971	61603
9263	S	D	E	N	1972	67398
9265	S	D	R	N	1962	68159
9265	S	D	E	N	1972	68160
9266	S	D	R	N	1962	68159
9266	S	D	E	N	1972	68160
9295	P	D	G	N	1971	64533
9296	P	D	G	N	1971	64533
9299	S	D	F	N	1972	65133
9355	D	D	R	H	1971	75485
9355	D	D	E	H	1971	75486
9355	S	D	R	H	1971	75485
9355	S	D	E	H	1971	75486
9378	F	D	E	N	1966	39951
9465	P	D	E	N	1969	37364
9468	P	D	E	N	1969	37362
9469	P	D	E	N	1969	37362
9481	F	D	E	N	1963	37349
9482	S	D	E	N	1970	56114
9536	F	D	E	N	1963	37349
9562	S	D	E	N	1970	59521
9562	S	T	E	N	1970	59521
9594	S	D	F	N	1969	52558
9598	S	D	R	N	1969	58350
9598	S	D	E	N	1969	76913
9606	F	D	E	N	1963	37349
9619	S	D	J	N	1967	45440
9620	S	D	J	N	1967	45440
9624	S	D	F	N	1966	40087
9674	P	D	E	N	1968	48968
9685	P	D	E	L	1968	49768
9699	S	D	J	N	1967	45440
9716	P	D	E	N	1966	44936
9716	S	D	E	N	1967	35229
9733	D	D	E	N	1964	37868
9733	S	D	E	N	1964	37868
9750	D	D	E	N	1968	49278
9750	D	D	E	N	1971	61512
9750	S	D	E	N	1960	42960
9750	S	D	E	N	1969	45285
9750	S	D	E	N	1969	54057
9776	S	D	R	N	1965	35610
9776	S	D	E	N	1965	38592
9789	F	D	E	N	1963	37349
9820	S	D	J	N	1967	45440
9823	S	G	E	N	1966	39365
9851	D	D	R	N	1973	72437
9851	D	T	R	N	1973	72437
9869	D	D	E	N	1973	71365
9869	S	D	E	N	1969	55290
9869	S	D	R	N	1969	57971
9869	S	D	E	N	1969	57972
9869	S	D	E	N	1973	71365
9882	D	D	E	N	1972	73297
9893	P	D	G	N	1971	64533
9894	S	D	F	N	1966	40087
9901	D	D	R	H	1971	75485
9901	D	D	E	H	1971	75486
9901	S	D	R	H	1971	75485
9901	S	D	E	H	1971	75486
9916	S	D	E	N	1964	59652
9916	S	D	R	N	1971	66447
9916	S	D	E	N	1971	66448
9916	S	D	R	N	1971	75483
9916	S	D	E	N	1971	75484
9917	S	D	J	N	1967	45440
9938	S	G	E	L	1962	40665
9940	S	D	E	N	1967	78266
9948	P	D	E	N	1970	56867
9951	S	D	E	N	1967	43455
9960	S	D	R	N	1973	75648
9960	S	D	E	N	1973	75649
521-0003	P	D	E	N	1973	71252
0003	P	D	E	L	1973	71361
0003	S	D	E	N	1950	58879
0003	S	D	E	N	1906	96757
0003	S	D	J	N	1978	99886
0004	P	D	E	N	1970	61726
0006	P	D	F	N	1967	42313
0006	P	D	E	N	1978	94862
0006	S	D	E	N	1950	58879
0006	S	D	E	N	1973	75300
0006	S	D	E	N	1906	96757
0007	S	D	E	N	1964	37966
0007	S	D	E	N	1940	40581
0007	S	D	G	N	1898	46995
0007	S	S	E	N	1972	66579
0007	S	S	E	N	1960	94260
0007	S	D	E	N	1906	96757
0009	S	D	E	N	1969	57407
0015	S	D	E	N	1950	58879
0019	S	D	E	N	1971	62414
0020	S	D	E	N	1973	70809
0022	S	G	E	N	1966	42701

Phys. State: **C.** Superconductive; **D.** Doped; **E.** Expanded; **F.** Fibrous or Whisker; **G.** Gas; **I.** Ionized or Plasma; **L.** Liquid; **M.** Multiphase; **P.** Powder or Fine Particle; **S.** Solid; **T.** Thick or Thin Film
Subject: **D.** Data; **E.** Experiment; **G.** General (Data + Theory + Experiment); **S.** Survey (Review, Compendium etc.); **T.** Theory
Language: **C.** Czech; **D.** Dutch; **E.** English; **F.** French; **G.** German; **I.** Italian; **J.** Japanese; **O.** Other Languages; **P.** Polish; **R.** Russian; **S.** Spanish
Temperature: **F.** Full Range (Low + Normal + High); **L.** Low (0 to 75K) + Overlap into Normal; **N.** Normal (75 to 1273K); **H.** High (above 1273K) + Overlap into Normal; Blank = Not Coded

Substance Number	Phys. State	Subject	Language	Temperature	Year	TPRC Number
521-0022	S	D	E	N	1970	58748
0022	S	D	J	N	1978	99886
0025	S	D	E	N	1950	58879
0025	S	D	E	N	1969	63778
0025	S	D	R	N	1971	69101
0025	S	D	R	N	1972	69135
0025	S	D	J	N	1974	81726
0026	P	D	E	N	1978	94862
0026	S	D	E	N	1906	96757
0026	S	D	J	N	1978	99886
0030	P	D	F	N	1967	42313
0032	S	D	J	N	1978	99886
0033	P	D	E	N	1978	94862
0033	S	D	J	N	1978	99886
0034	S	D	J	N	1978	99886
0035	P	D	E	N	1973	77753
0035	S	D	E	N	1950	58879
0035	S	D	J	N	1978	99886
0036	S	D	E	H	1967	34861
0036	S	D	E	N	1970	58748
0036	S	G	E	H	1970	58983
0040	S	D	E	N	1967	52863
0041	S	D	E	N	1906	96757
0042	S	D	E	N	1906	51607
0042	S	D	R	N	1971	66289
0042	S	S	E	N	1972	66579
0042	S	D	E	N	1906	96757
0055	P	D	E	N	1978	94862
0055	S	D	R	N	1966	41075
0055	S	D	E	N	1966	41076
0057	P	D	E	N	1968	52907
0057	P	T	E	N	1968	52907
0057	P	D	E	N	1973	77753
0057	S	D	E	N	1950	58879
0057	S	D	E	N	1906	96757
0057	S	D	J	N	1978	99886
0058	S	D	R	N	1974	75189
0058	S	D	E	N	1974	76546
0058	S	D	J	N	1978	99886
0059	S	D	E	N	1950	58879
0059	S	D	J	N	1978	99886
0063	S	D	E	N	1965	35127
0063	S	D	E	N	1967	44768
0063	S	D	E	N	1973	71361
0063	S	D	E	N	1972	72761
0063	S	D	E	N	1978	90953
0063	S	D	E	N	1976	91628
0063	S	E	E	N	1976	91628
0064	S	S	E	H	1960	24931
0064	S	S	E	N	1960	24931
0064	S	D	E	L	1966	33775
0064	S	D	E	N	1966	34965
0064	S	D	E	N	1963	37791
0064	S	D	E	N	1967	44164
0064	S	D	E	N	1968	49666
0064	S	D	E	N	1971	61911
0064	S	D	E	L	1973	71361
0064	S	G	E	H	1957	76816
0064	S	D	E	H	1973	77361
0064	S	E	E	H	1973	77361
0064	S	S	E	N	1958	85031
0064	S	D	E	N	1959	85032
0067	S	D	E	N	1971	63066
0069	S	D	E	N	1970	58434
0069	S	D	J	N	1969	62420
0069	S	D	J	N	1970	62754
0069	S	D	E	N	1972	66890
0069	S	D	E	N	1968	88413
0070	S	D	E	N	1950	58879
0072	P	D	E	N	1970	61726
0072	P	D	E	N	1973	77753
0072	P	D	E	N	1978	94862
0072	S	D	E	N	1950	58879
0073	P	D	E	N	1973	75869
0075	P	D	F	N	1967	42313
0075	P	D	E	N	1969	77342
0075	P	D	E	N	1973	77753
0075	P	D	E	N	1978	94862
0075	S	D	E	N	1967	38654
0075	S	T	E	N	1967	38654
0075	S	D	E	N	1967	44164
0075	S	D	G	N	1898	46995
0075	S	D	E	N	1950	58879
0075	S	D	R	N	1970	59430
0075	S	D	E	N	1970	59431
0075	S	D	E	N	1969	59714
0075	S	D	E	N	1971	62718
0075	S	D	R	N	1973	72646
0075	S	D	E	N	1974	72647
0075	S	D	R	N	1972	72911
0075	S	D	E	N	1963	76818
0075	S	E	E	N	1963	76818
0075	S	D	R	N	1976	88957
0075	S	D	E	N	1976	88958
0075	S	D	E	N	1906	96757
0075	S	D	J	N	1978	99886
0077	D	D	R	N	1972	67791
0077	S	D	R	N	1972	67791
0077	S	D	E	N	1970	77232
0077	S	D	E	N	1968	77836
0077	S	E	E	N	1968	77836
521-0077	S	D	J	N	1978	99886
0082	P	D	E	N	1978	94862
0082	S	D	G	N	1971	64274
0082	S	D	G	N	1971	66369
0082	S	D	J	N	1978	99886
0083	P	D	E	N	1973	77753
0083	P	D	E	N	1978	94862
0083	S	D	G	N	1898	46995
0083	S	D	G	N	1968	50169
0083	S	D	E	N	1906	51607
0083	S	D	E	N	1970	57048
0083	S	D	E	N	1950	58879
0083	S	D	O	N	1971	63599
0083	S	D	J	N	1978	99886
0084	P	D	F	N	1967	42313
0084	P	D	E	N	1978	94862
0084	S	D	E	N	1950	58879
0085	P	D	E	N	1978	94862
0085	S	D	E	N	1950	58879
0086	S	D	R	N	1971	69101
0086	S	D	R	N	1972	69135
0086	S	D	E	N	1906	96757
0086	S	D	E	N	1908	96760
0087	S	D	E	N	1967	45947
0087	S	D	G	N	1898	46995
0089	S	D	R	N	1972	69083
0091	S	D	R	N	1972	69083
0091	S	D	E	N	1972	72814
0091	S	D	E	N	1906	96757
0092	P	D	E	N	1966	33956
0092	S	D	R	N	1972	69083
0092	S	D	R	N	1973	72381
0092	S	D	E	N	1906	96757
0092	S	D	J	N	1978	99886
0093	S	D	E	N	1906	51607
0093	S	D	R	N	1971	66289
0093	S	S	E	N	1972	66579
0093	S	D	E	N	1905	96756
0093	S	D	E	N	1906	96757
0094	S	D	R	N	1973	72381
0095	S	D	O	N	1971	64208
0095	S	D	E	N	1971	68829
0095	S	D	R	N	1971	69101
0095	S	D	E	N	1906	96757
0096	S	D	E	N	1952	46828
0096	S	D	G	N	1898	46995
0096	S	D	E	N	1906	96757
0096	S	D	E	N	1908	96760
0099	S	D	E	N	1969	63778
0099	S	D	R	N	1972	69135
0104	P	D	E	N	1973	77753
0104	S	D	E	N	1950	58879
0104	S	D	R	N	1973	72381
0105	P	D	F	N	1967	42313
0105	P	D	E	N	1978	94862
0105	S	D	E	N	1950	58879
0105	S	D	R	N	1972	69083
0105	S	D	J	N	1978	99886
0108	S	D	E	N	1964	35906
0108	S	D	E	N	1950	40784
0108	S	D	G	N	1898	46995
0108	S	D	E	N	1931	47008
0108	S	D	E	N	1968	53509
0108	S	D	E	N	1961	54099
0108	S	D	E	N	1932	59587
0108	S	S	E	N	1972	66579
0108	S	D	E	N	1976	87344
0109	P	D	E	N	1978	94862
0109	S	D	E	N	1950	58879
0109	S	D	J	N	1978	99886
0110	S	D	E	N	1967	52863
0111	D	D	R	N	1971	77545
0111	D	D	E	N	1971	77546
0111	P	D	R	N	1971	77545
0111	P	D	E	N	1971	77546
0111	P	D	E	N	1978	94862
0111	S	D	O	N	1972	68509
0115	P	D	E	N	1950	58879
0116	P	D	E	N	1978	94862
0116	S	D	E	N	1906	51607
0116	S	D	E	N	1950	58879
0116	S	S	E	N	1972	66579
0116	S	D	E	N	1973	73145
0116	S	D	E	N	1906	96757
0123	P	D	E	N	1970	61726
0123	P	D	E	N	1973	77753
0123	P	D	E	N	1978	94862
0130	S	D	F	N	1970	61903
0130	S	D	G	N	1971	64274
0130	S	D	G	N	1971	66369
0130	S	D	J	N	1978	99886
0131	P	D	E	N	1978	94862
0131	S	D	R	N	1970	62988
0131	S	T	R	N	1970	62988
0131	S	D	E	N	1906	96757
0131	S	D	J	N	1978	99886
0132	S	D	R	N	1972	69083
0132	S	S	E	N	1958	85031
0133	S	D	E	N	1950	58879
0133	S	D	R	N	1973	72381
0135	P	D	E	N	1978	94862
521-0135	S	D	E	N	1950	58879
0174	P	D	E	N	1964	36837
0174	P	D	E	N	1973	77753
0174	S	D	E	N	1973	70809
0174	S	D	R	N	1972	72984
0174	S	D	J	N	1978	99886
0177	S	D	E	N	1906	51607
0177	S	D	E	N	1906	96757
0182	S	D	J	N	1978	99886
0184	P	D	E	N	1970	61726
0186	P	D	E	N	1975	84118
0186	S	D	E	N	1950	58879
0190	P	D	E	N	1975	84118
0191	D	D	R	N	1972	73465
0191	D	D	E	N	1972	73466
0191	D	D	R	N	1973	73637
0191	D	D	E	N	1973	73638
0191	S	D	E	N	1966	34913
0191	S	D	E	N	1966	34965
0191	S	D	E	H	1965	35028
0191	S	D	E	N	1965	36324
0191	S	D	F	L	1965	36326
0191	S	D	E	N	1967	44164
0191	S	G	E	N	1957	76819
0191	S	D	E	N	1968	77836
0191	S	E	E	N	1968	77836
0192	P	S	E	N	1972	66579
0192	S	D	E	N	1950	58879
0192	S	D	O	N	1971	64208
0192	S	D	E	N	1971	68829
0192	S	D	R	N	1971	69101
0192	T	S	E	N	1972	66579
0195	P	D	F	N	1967	42313
0196	S	D	P	N	1970	65195
0216	S	D	E	N	1906	96757
0217	S	D	E	N	1906	51607
0217	S	S	E	N	1972	66579
0217	S	D	E	N	1906	96757
0222	S	D	R	N	1970	68921
0222	S	D	J	N	1978	99886
0223	S	D	E	N	1906	51607
0223	S	D	E	N	1950	58879
0223	S	S	E	N	1972	66579
0223	S	D	R	N	1972	69083
0223	S	D	E	N	1906	96757
0225	S	D	E	N	1967	40853
0225	S	S	E	N	1956	56984
0236	P	D	E	N	1970	61726
0236	S	D	E	N	1965	50543
0236	S	D	E	N	1906	51607
0236	S	S	E	N	1972	66579
0236	S	D	E	N	1906	96757
0237	P	D	E	N	1973	75869
0240	P	D	E	N	1967	47261
0240	S	D	E	N	1906	51607
0240	S	D	E	N	1969	54234
0240	S	D	E	N	1950	58879
0240	S	D	R	N	1970	61231
0240	S	D	R	N	1969	63741
0240	S	D	R	N	1971	65249
0240	S	D	E	N	1973	73145
0240	S	D	E	N	1970	76916
0240	S	D	E	N	1906	96757
0240	S	D	J	N	1978	99886
0240	T	D	E	N	1978	94862
0241	S	D	E	N	1973	70809
0241	S	D	E	N	1972	73127
0241	S	D	E	N	1973	73145
0241	S	D	J	N	1978	99886
0241	T	D	E	N	1978	94862
0249	S	D	R	N	1971	63859
0250	P	D	E	N	1977	90188
0250	S	D	R	N	1972	69083
0250	S	D	E	N	1971	73101
0250	S	D	E	N	1906	96757
0251	S	D	E	N	1906	51607
0251	S	D	E	N	1906	96757
0254	P	D	E	N	1967	43975
0254	S	G	E	N	1962	40665
0254	S	D	E	L	1967	43975
0254	S	D	E	N	1958	59584
0254	S	D	R	N	1970	68921
0254	S	D	J	N	1978	99886
0255	P	D	E	N	1967	43975
0255	S	D	E	N	1958	59584
0255	S	D	R	N	1970	68921
0258	P	D	E	N	1970	61726
0258	P	D	E	N	1973	77753
0258	S	D	E	H	1967	34861
0258	S	G	E	L	1962	40665
0258	S	D	E	N	1968	50275
0258	S	D	E	N	1970	58748
0258	S	D	E	N	1950	58879
0258	S	G	E	H	1970	58983
0258	S	D	E	N	1977	98332
0258	S	D	J	N	1978	99886
0261	P	D	E	N	1966	48370
0261	S	D	E	N	1976	91628
0261	S	E	E	N	1976	91628
0264	P	D	E	N	1970	61726
0272	P	D	E	N	1978	94862

Phys. State: **C.** Superconductive; **D.** Doped; **E.** Expanded; **F.** Fibrous or Whisker; **G.** Gas; **I.** Ionized or Plasma; **L.** Liquid; **M.** Multiphase; **P.** Powder or Fine Particle; **S.** Solid; **T.** Thick or Thin Film
Subject: **D.** Data; **E.** Experiment; **G.** General (Data + Theory + Experiment); **S.** Survey (Review, Compendium etc.); **T.** Theory
Language: **C.** Czech; **D.** Dutch; **E.** English; **F.** French; **G.** German; **I.** Italian; **J.** Japanese; **O.** Other Languages; **P.** Polish; **R.** Russian; **S.** Spanish
Temperature: **F.** Full Range (Low + Normal + High); **L.** Low (0 to 75K) + Overlap into Normal; **N.** Normal (75 to 1273K); **H.** High (above 1273K) + Overlap into Normal; Blank = Not Coded

Substance Number	Phys. State	Subject	Language	Temperature	Year	TPRC Number
521-0298	S	D	R	N	1970	68921
0299	S	D	O	N	1969	60920
0299	S	T	O	N	1969	60920
0299	S	S	E	N	1971	66306
0302	S	D	R	N	1971	66289
0302	S	D	J	N	1978	99886
0303	S	D	E	N	1906	51607
0303	S	D	R	N	1971	66289
0303	S	S	E	N	1972	66579
0303	S	D	E	N	1906	96757
0303	S	D	J	N	1978	99886
0307	P	D	E	N	1973	75869
0341	S	D	J	N	1978	99886
0351	S	D	J	N	1978	99886
0362	S	D	J	N	1978	99886
0366	S	D	J	N	1978	99886
0368	P	D	E	N	1975	84118
0385	P	D	E	N	1978	94862
0385	S	D	E	N	1908	96760
0389	S	D	R	N	1972	69083
0392	P	D	E	N	1978	94862
0392	S	D	E	N	1950	58879
0393	P	D	E	N	1978	94862
0394	P	D	P	N	1971	65759
0394	P	E	P	N	1971	65759
0394	S	D	E	N	1906	96757
0396	S	D	E	N	1906	51607
0396	S	S	E	N	1972	66579
0396	S	D	E	N	1906	96757
0402	S	D	E	N	1967	52863
0412	S	D	E	N	1972	70410
0424	S	D	E	N	1973	70311
0424	S	E	E	N	1973	70311
0429	S	D	E	N	1906	51607
0429	S	D	R	N	1971	66289
0429	S	S	E	N	1972	66579
0429	S	D	E	N	1906	96757
0429	S	D	J	N	1978	99886
0430	S	D	E	N	1966	39134
0430	S	D	E	N	1906	51607
0430	S	D	E	N	1950	58879
0430	S	D	R	N	1969	63741
0430	S	D	R	N	1972	71078
0430	S	D	E	N	1973	73145
0430	S	D	E	N	1906	96757
0430	T	D	E	N	1978	94862
0443	S	D	R	N	1968	51253
0443	S	D	E	N	1968	51254
0443	S	S	E	N	1972	66579
0443	S	D	E	N	1979	97378
0443	S	D	J	N	1978	99886
0445	S	D	R	N	1972	69083
0457	S	D	J	N	1978	99886
0470	S	D	E	N	1906	96757
0471	P	D	E	N	1973	77753
0472	S	D	E	N	1906	96757
0496	S	D	E	N	1906	51607
0496	S	D	E	N	1950	58879
0496	S	D	R	N	1970	68921
0496	S	S	E	N	1976	83746
0496	S	D	E	N	1974	85093
0496	S	D	E	N	1906	96757
0505	S	D	J	N	1978	99886
0506	S	D	E	N	1973	70809
0506	S	D	E	N	1973	72795
0506	S	D	J	N	1978	99886
0514	S	D	E	N	1950	58879
0517	S	D	E	N	1966	34810
0517	S	D	R	N	1973	70840
0517	S	D	E	N	1973	72359
0534	S	D	E	N	1950	58879
0552	S	D	E	H	1967	34861
0552	S	D	E	N	1950	58879
0552	S	D	E	N	1906	96757
0554	S	D	E	N	1906	96757
0555	S	D	R	N	1969	59066
0555	S	D	R	N	1971	65920
0563	S	D	E	N	1906	96757
0566	S	D	E	N	1950	58879
0566	S	D	E	N	1963	76818
0566	S	E	E	N	1963	76818
0566	S	D	E	N	1908	96760
0566	S	D	J	N	1978	99886
0572	S	D	E	N	1950	58879
0604	P	D	E	N	1973	77753
0604	S	D	E	N	1906	51607
0604	S	D	R	N	1971	66289
0604	S	S	E	N	1972	66579
0604	S	D	E	N	1906	96757
0611	P	D	F	N	1967	42313
0611	S	D	E	N	1950	58879
0614	P	D	F	N	1967	42313
0614	S	D	E	N	1950	58879
0614	S	D	E	N	1971	63136
0614	S	G	E	N	1973	71027
0614	S	G	E	N	1973	73838
0626	P	D	F	N	1967	42313
0626	S	D	R	N	1972	69083
0626	S	D	J	N	1978	99886
0645	S	D	E	N	1950	58879
0645	S	D	E	N	1973	70809

Substance Number	Phys. State	Subject	Language	Temperature	Year	TPRC Number
521-0650	S	D	E	N	1906	96757
0660	S	D	R	N	1971	63859
0675	S	D	R	N	1971	69101
0675	S	D	E	N	1908	96760
0677	P	D	E	N	1978	94862
0680	S	D	E	N	1967	43179
0681	S	D	E	N	1967	43179
0682	S	D	E	N	1966	34962
0682	S	D	F	N	1968	50925
0682	S	D	R	N	1970	68921
0684	S	D	E	N	1966	34962
0684	S	D	R	N	1973	70840
0684	S	D	E	N	1973	72359
0686	S	D	E	N	1966	34399
0686	S	D	E	N	1964	35906
0687	S	D	E	N	1966	34399
0687	S	D	E	N	1964	35906
0696	S	D	F	N	1970	59293
0720	S	D	R	N	1970	63213
0720	S	D	R	N	1972	69083
0732	D	D	R	N	1970	61281
0732	D	D	E	N	1971	61282
0732	P	D	E	N	1973	70406
0732	P	D	R	N	1973	73059
0732	P	D	J	N	1973	73927
0732	P	D	E	N	1973	77753
0732	P	D	E	N	1978	94862
0732	S	D	E	N	1963	37791
0732	S	D	E	N	1965	37896
0732	S	G	E	L	1962	40665
0732	S	D	E	N	1964	40756
0732	S	D	E	N	1950	40784
0732	S	G	E	N	1962	40794
0732	S	D	E	L	1967	43975
0732	S	D	G	N	1898	46995
0732	S	D	E	N	1906	51607
0732	S	D	E	N	1961	54099
0732	S	D	E	N	1968	54244
0732	S	D	E	N	1950	58879
0732	S	D	E	N	1958	59584
0732	S	D	E	N	1932	59587
0732	S	D	E	N	1936	59958
0732	S	S	E	N	1972	66579
0732	S	D	R	N	1970	68921
0732	S	D	R	N	1972	69083
0732	S	D	G	N	1973	70590
0732	S	E	G	N	1973	70590
0732	S	D	R	N	1973	70840
0732	S	D	E	N	1973	72359
0732	S	D	R	N	1973	73059
0732	S	D	J	N	1973	73927
0732	S	D	R	N	1973	75615
0732	S	D	E	N	1973	75616
0732	S	D	E	L	1963	76818
0732	S	E	E	L	1963	76818
0732	S	G	E	N	1957	76819
0732	S	D	E	N	1972	78269
0732	S	D	E	N	1906	96757
0732	S	D	J	N	1978	99886
0736	S	D	R	N	1969	59066
0736	S	D	R	N	1971	65920
0736	S	D	E	N	1908	96760
0737	S	D	R	N	1969	59066
0737	S	D	R	N	1971	65920
0737	S	D	J	N	1978	99886
0739	S	D	R	N	1971	63859
0740	S	D	R	N	1970	58991
0740	S	D	R	N	1971	63859
0744	S	D	R	N	1971	69101
0745	S	D	R	N	1971	69101
0746	S	D	R	N	1971	69101
0746	S	D	R	N	1972	69135
0749	S	D	R	N	1971	69101
0751	S	D	R	N	1971	69101
0755	S	D	R	N	1968	48802
0755	S	D	E	N	1968	49364
0757	S	D	O	N	1971	64208
0757	S	D	E	N	1971	68829
0757	S	D	R	N	1971	69101
0764	D	D	E	N	1973	70809
0764	P	D	E	N	1973	77753
0764	S	D	E	N	1950	58879
0764	S	D	E	N	1973	70809
0764	S	D	J	N	1978	99886
0765	S	D	E	N	1950	58879
0765	S	D	O	N	1972	68509
0765	S	D	J	N	1978	99886
0769	S	D	E	N	1906	96757
0772	S	D	J	N	1978	99886
0776	S	D	E	N	1964	50480
0778	S	D	E	N	1906	51607
0778	S	S	E	N	1972	66579
0778	S	D	E	N	1906	96757
0779	S	D	E	N	1906	51607
0779	S	S	E	N	1972	66579
0779	S	D	E	N	1906	96757
0780	S	D	E	N	1906	51607
0780	S	S	E	N	1972	66579
0780	S	D	E	N	1906	96757
0780	S	D	J	N	1978	99886
0781	S	D	E	N	1906	51607

Substance Number	Phys. State	Subject	Language	Temperature	Year	TPRC Number
521-0781	S	S	E	N	1972	66579
0781	S	D	E	N	1906	96757
0782	S	D	E	N	1906	51607
0782	S	D	E	N	1906	96757
0783	S	D	E	N	1906	51607
0783	S	D	G	N	1971	64274
0783	S	D	G	N	1971	66369
0783	S	S	E	N	1972	66579
0783	S	D	E	N	1906	96757
0784	S	D	E	N	1906	51607
0784	S	S	E	N	1972	66579
0784	S	D	E	N	1906	96757
0785	S	D	E	N	1906	51607
0785	S	S	E	N	1972	66579
0785	S	D	R	N	1972	69083
0785	S	D	E	N	1906	96757
0786	S	D	E	N	1906	51607
0786	S	S	E	N	1972	66579
0786	S	D	R	N	1972	69083
0786	S	D	R	N	1973	73173
0786	S	D	E	N	1906	96757
0793	P	D	E	N	1973	77753
0794	S	D	E	N	1950	58879
0794	S	D	J	N	1978	99886
0796	S	D	E	N	1968	46105
0798	S	D	E	N	1950	58879
0799	P	D	E	N	1978	94862
0799	S	D	J	N	1978	99886
0817	S	D	R	N	1971	63859
0823	S	D	E	N	1958	59584
0834	S	D	F	N	1970	59293
0852	P	D	E	N	1970	61726
0852	P	D	E	N	1973	77753
0854	S	D	G	N	1968	52383
0854	S	T	G	N	1968	52383
0854	S	D	R	N	1973	73173
0858	S	D	G	N	1971	64274
0863	S	D	R	N	1971	69101
0865	P	D	E	N	1978	94862
0867	S	D	R	N	1967	49655
0867	S	D	E	N	1967	49656
0868	S	D	F	N	1970	59293
0868	S	D	R	N	1970	62988
0868	S	T	R	N	1970	62988
0869	S	D	F	N	1970	59293
0875	S	D	F	N	1970	59293
0895	P	D	E	N	1973	77753
0910	S	D	F	N	1970	59293
0911	S	D	F	N	1970	59293
0911	S	D	R	N	1970	62988
0911	S	T	R	N	1970	62988
0912	P	D	E	N	1973	77753
0912	S	D	R	N	1970	61231
0912	S	D	E	N	1973	73145
0912	S	D	E	N	1970	76916
0922	P	D	E	N	1972	70061
0922	P	D	E	N	1971	70142
0922	S	D	E	N	1972	74269
0922	S	D	E	N	1973	80784
0923	P	D	E	N	1973	77753
0928	S	D	E	N	1969	54234
0929	S	D	E	N	1969	54234
0930	S	D	E	N	1969	54234
0933	S	D	E	N	1906	96757
0939	S	D	J	N	1978	99886
0940	S	D	J	N	1978	99886
0941	S	D	E	N	1971	61303
0942	S	D	E	N	1906	96757
0942	S	D	J	N	1978	99886
0946	S	D	R	N	1970	62988
0946	S	T	R	N	1970	62988
0948	S	D	E	N	1906	96757
0950	S	D	E	N	1950	58879
0951	P	D	E	N	1969	53645
0951	P	T	E	N	1970	53645
0951	P	D	E	N	1970	59099
0951	P	D	E	N	1970	61726
0951	S	D	E	N	1950	58879
0951	S	S	E	N	1970	59662
0951	S	D	E	N	1906	96757
0951	S	D	J	N	1978	99886
0957	S	D	E	N	1972	70410
0958	S	D	E	N	1972	70410
0958	S	D	E	N	1976	91628
0958	S	E	E	N	1976	91628
0959	S	D	R	N	1972	69083
0965	S	D	E	N	1973	73145
0967	S	D	E	N	1906	96757
0968	S	D	E	N	1906	96757
0970	S	D	R	N	1965	34118
0970	S	D	E	N	1966	60093
0970	S	D	E	N	1964	77530
0970	S	E	E	N	1964	77530
0970	S	D	E	N	1977	90208
0980	S	D	E	N	1969	57407
0987	S	D	R	N	1972	69083
0991	S	D	E	N	1972	75779
1009	S	D	R	N	1970	60144
1009	S	D	E	N	1970	60145
1022	S	D	R	N	1971	66289
1043	S	D	E	N	1974	76328

Phys. State: **C.** Superconductive; **D.** Doped; **E.** Expanded; **F.** Fibrous or Whisker; **G.** Gas; **I.** Ionized or Plasma; **L.** Liquid; **M.** Multiphase; **P.** Powder or Fine Particle; **S.** Solid; **T.** Thick or Thin Film
Subject: **D.** Data; **E.** Experiment; **G.** General (Data + Theory + Experiment); **S.** Survey (Review, Compendium etc.); **T.** Theory
Language: **C.** Czech; **D.** Dutch; **E.** English; **F.** French; **G.** German; **I.** Italian; **J.** Japanese; **O.** Other Languages; **P.** Polish; **R.** Russian; **S.** Spanish
Temperature: **F.** Full Range (Low + Normal + High); **L.** Low (0 to 75K) + Overlap into Normal; **N.** Normal (75 to 1273K); **H.** High (above 1273K) + Overlap into Normal; Blank = Not Coded

Substance Number	Phys. State	Sub- ject	Lan- guage	Temper- ature	Year	TPRC Number
521-1059	S	D	E	N	1972	70057
1059	S	D	E	N	1972	70058
1061	S	D	R	N	1970	63213
1062	S	D	R	N	1970	62988
1062	S	T	R	N	1970	62988
1062	S	D	R	N	1970	63213
1063	S	D	R	N	1970	63213
1064	S	D	R	N	1970	63213
1065	S	D	R	N	1970	63213
1066	S	D	R	N	1970	63213
1067	S	D	R	N	1970	63213
1076	S	D	R	N	1969	59066
1076	S	D	R	N	1971	65920
1077	S	D	R	N	1969	59066
1077	S	D	R	N	1971	65920
1078	S	D	R	N	1969	59066
1078	S	D	R	N	1971	65920
1079	S	D	R	N	1969	59066
1079	S	D	R	N	1971	65920
1090	S	D	R	N	1971	63859
1091	S	D	R	N	1971	65249
1092	S	D	R	N	1971	65249
1095	S	D	R	N	1971	65920
1096	S	D	R	N	1971	65920
1097	S	D	R	N	1971	65920
1098	S	D	R	N	1971	65920
1099	S	D	R	N	1971	65920
1100	S	D	R	N	1971	65920
1101	S	D	R	N	1971	65791
1102	S	D	R	N	1971	65791
1103	S	D	R	N	1971	65791
1104	S	D	R	N	1971	65791
1107	S	D	R	N	1971	64215
1115	S	D	G	N	1971	66369
1116	S	D	G	N	1971	64274
1116	S	D	G	N	1971	66369
1117	S	D	G	N	1971	64274
1117	S	D	G	N	1971	66369
1126	S	D	E	N	1971	62414
1128	S	D	E	N	1974	85093
1131	S	D	J	N	1978	99886
1134	S	D	J	N	1978	99886
1148	S	D	E	N	1905	96756
1148	S	D	E	N	1906	96757
1163	S	D	R	N	1970	62988
1163	S	T	R	N	1970	62988
1164	S	D	O	N	1969	60920
1164	S	T	O	N	1969	60920
1164	S	D	R	N	1970	62988
1164	S	T	R	N	1970	62988
1166	S	D	R	N	1970	62988
1166	S	T	R	N	1970	62988
1167	S	D	R	N	1970	62988
1167	S	T	R	N	1970	62988
1168	S	D	R	N	1970	62988
1168	S	T	R	N	1970	62988
1169	S	D	R	N	1970	62988
1169	S	T	R	N	1970	62988
1170	S	D	R	N	1970	62988
1170	S	T	R	N	1970	62988
1174	S	D	R	N	1971	66289
1175	S	D	R	N	1971	66289
1176	S	D	R	N	1971	66289
1177	S	D	O	N	1971	64208
1177	S	D	E	N	1971	68829
1178	S	D	O	N	1971	64208
1178	S	D	E	N	1971	68829
1179	S	D	O	N	1971	64208
1179	S	D	E	N	1971	68829
1180	S	D	O	N	1971	64208
1180	S	D	E	N	1971	68829
1181	S	D	O	N	1971	64208
1181	S	D	E	N	1971	68829
1182	S	D	O	N	1971	64208
1182	S	D	E	N	1971	68829
1183	S	D	O	N	1971	64208
1183	S	D	E	N	1971	68829
1184	S	D	O	N	1971	63599
1185	S	D	O	N	1971	63599
1186	S	D	O	N	1971	63599
1187	S	D	O	N	1969	60920
1187	S	T	O	N	1969	60920
1188	S	D	O	N	1969	60920
1188	S	T	O	N	1969	60920
1189	S	D	O	N	1969	60920
1189	S	T	O	N	1969	60920
1223	S	D	R	N	1970	68921
1224	S	D	R	N	1970	68921
1227	S	D	J	N	1978	99886
1249	S	D	E	N	1972	70057
1249	S	D	E	N	1976	91628
1249	S	E	E	N	1976	91628
1272	S	D	E	N	1973	71054
1283	S	D	E	N	1906	96757
1284	S	D	R	N	1973	73173
1285	S	D	E	N	1906	96757
1286	S	D	E	N	1906	96757
1290	S	D	R	N	1972	69083
1291	S	D	R	N	1972	69083
1292	S	D	R	N	1972	69083
1293	S	D	R	N	1972	69083
521-1294	S	D	R	N	1972	69083
1295	S	D	R	N	1972	69083
1296	S	D	R	N	1972	69083
1297	S	D	R	N	1972	69083
1298	S	D	R	N	1972	69083
1299	S	D	R	N	1972	69083
1300	S	D	R	N	1972	69083
1301	S	D	R	N	1972	69083
1303	S	D	R	N	1971	69101
1304	S	D	E	N	1969	63778
1304	S	D	R	N	1971	69101
1305	S	D	R	N	1971	69101
1306	S	D	R	N	1971	69101
1307	S	D	R	N	1971	69101
1308	S	D	R	N	1971	69101
1309	S	D	R	N	1971	69101
1310	S	D	R	N	1971	69101
1311	S	D	R	N	1971	69101
1312	S	D	R	N	1971	69101
1329	D	D	E	N	1973	70809
1331	S	D	E	N	1973	73145
1336	S	D	E	N	1972	70410
1337	S	D	E	N	1972	70410
1350	S	D	R	N	1973	72394
1351	S	D	R	N	1973	72394
1353	S	D	R	N	1973	72381
1354	S	D	R	N	1973	72381
1357	S	D	E	N	1973	80784
1363	S	D	E	N	1973	75300
1365	S	D	R	N	1973	73173
1366	S	D	R	N	1973	73173
1367	S	D	R	N	1973	73173
1368	S	D	R	N	1973	73173
1369	S	D	R	N	1973	73173
1370	S	D	R	N	1973	73173
1371	S	D	R	N	1973	73173
1372	S	D	R	N	1973	73173
1373	S	D	R	N	1973	73173
1374	S	D	E	N	1972	75779
1375	S	D	E	N	1972	75779
1376	S	D	G	N	1974	74674
1376	S	T	G	N	1974	74674
1378	S	D	R	N	1970	61231
1378	S	D	E	N	1970	76916
1384	P	D	E	N	1973	80784
1385	S	D	E	N	1964	77530
1385	S	E	E	N	1964	77530
1387	D	D	R	N	1971	77545
1387	D	D	E	N	1971	77546
1387	P	D	R	N	1971	77545
1387	P	D	E	N	1971	77546
1388	D	D	R	N	1971	77545
1388	D	D	E	N	1971	77546
1388	P	D	R	N	1971	77545
1388	P	D	E	N	1971	77546
1415	P	D	E	N	1973	75869
1420	S	D	E	N	1974	85093
1444	P	D	E	N	1978	94862
1459	P	D	E	N	1976	91628
1459	P	E	E	N	1976	91628
1460	S	D	E	N	1976	91628
1460	S	E	E	N	1976	91628
1463	P	D	E	N	1978	94862
1463	S	D	J	N	1978	99886
1471	S	S	E	N	1972	66579
1471	S	D	E	N	1906	96757
1472	S	D	E	N	1906	96757
1473	S	D	E	N	1906	96757
1474	S	D	E	N	1906	96757
1475	S	D	E	N	1906	96757
1476	S	D	E	N	1906	96757
1477	S	D	E	N	1906	96757
1478	S	D	E	N	1906	96757
1479	S	D	E	N	1906	96757
1480	S	D	E	N	1906	96757
1481	S	D	E	N	1906	96757
1482	S	D	E	N	1906	96757
1483	S	D	E	N	1906	96757
1484	S	D	E	N	1906	96757
1486	S	D	E	N	1908	96760
1536	S	D	J	N	1978	99886
1537	S	D	J	N	1978	99886
1562	P	D	E	N	1978	94862
1563	P	D	E	N	1978	94862
1564	P	D	E	N	1978	94862
1565	P	D	E	N	1978	94862
1566	P	D	E	N	1978	94862
1568	T	D	E	N	1978	94862
1569	P	D	E	N	1978	94862
1570	P	D	E	N	1978	94862
1571	P	D	E	N	1978	94862
1598	P	D	E	N	1975	84118
1605	S	D	J	N	1978	99886
1606	S	D	J	N	1978	99886
1607	S	D	J	N	1978	99886

Phys. State: **C.** Superconductive; **D.** Doped; **E.** Expanded; **F.** Fibrous or Whisker; **G.** Gas; **I.** Ionized or Plasma; **L.** Liquid; **M.** Multiphase; **P.** Powder or Fine Particle; **S.** Solid; **T.** Thick or Thin Film
Subject: **D.** Data; **E.** Experiment; **G.** General (Data + Theory + Experiment); **S.** Survey (Review, Compendium etc.); **T.** Theory
Language: **C.** Czech; **D.** Dutch; **E.** English; **F.** French; **G.** German; **I.** Italian; **J.** Japanese; **O.** Other Languages; **P.** Polish; **R.** Russian; **S.** Spanish
Temperature: **F.** Full Range (Low + Normal + High); **L.** Low (0 to 75K) + Overlap into Normal; **N.** Normal (75 to 1273K); **H.** High (above 1273K) + Overlap into Normal; Blank = Not Coded

Chapter 9 Absorptance To Emittance Ratio

Substance Number	Phys. State	Sub-ject	Lan-guage	Temper-ature	Year	TPRC Number
521-0240	S	D	E	N	1967	47261
0887	S	D	E	N	1965	39333

Phys. State: **C.** Superconductive; **D.** Doped; **E.** Expanded; **F.** Fibrous or Whisker; **G.** Gas; **I.** Ionized or Plasma; **L.** Liquid; **M.** Multiphase; **P.** Powder or Fine Particle; **S.** Solid; **T.** Thick or Thin Film
Subject: **D.** Data; **E.** Experiment; **G.** General (Data + Theory + Experiment); **S.** Survey (Review, Compendium etc.); **T.** Theory
Language: **C.** Czech; **D.** Dutch; **E.** English; **F.** French; **G.** German; **I.** Italian; **J.** Japanese; **O.** Other Languages; **P.** Polish; **R.** Russian; **S.** Spanish
Temperature: **F.** Full Range (Low + Normal + High); **L.** Low (0 to 75K) + Overlap into Normal; **N.** Normal (75 to 1273K); **H.** High (above 1273K) + Overlap into Normal; Blank = Not Coded

Chapter 10 Thermal Linear Expansion

Substance Number	Phys. State	Subject	Language	Temperature	Year	TPRC Number
503-0001	D	D	E	N	1967	35230
0001	S	D	F	L	1950	46948
0001	S	D	F	N	1935	59813
0001	S	D	E	N	1946	60008
0001	S	D	E	N	1971	68241
0001	S	S	E	N	1975	88036
0003	S	S	R	N	1973	79859
0003	S	S	E	N	1973	79860
0008	S	D	E	N	1976	85762
0012	S	D	R	N	1958	47220
0012	S	D	E	N	1958	47221
0012	S	D	E	N	1963	47403
0012	S	D	E	N	1971	62397
0012	S	D	E	N	1972	78497
0012	S	S	E	N	1967	89315
0012	S	S	E	N	1979	100149
0018	S	D	E	N	1960	76856
0019	S	S	E	N	1960	24931
0019	S	D	E	N	1930	37544
0019	S	D	E	N	1976	85762
0019	S	S	E	N	1977	88199
0019	S	S	E	N	1967	89315
0020	S	D	E	N	1960	10947
0020	S	E	E	N	1960	10947
0020	S	S	E	N	1960	24931
0020	S	D	D	N	1967	47237
0020	S	G	E	N	1957	76819
0024	S	D	E	N	1971	67411
0025	S	D	O	N	1965	42150
0025	S	D	E	N	1967	45885
0025	S	D	R	N	1967	45886
0025	S	D	D	N	1967	47237
0025	S	S	E	N	1967	89315
0025	S	D	E	N	1978	98331
0026	S	D	E	N	1975	81874
0026	S	S	E	N	1977	88199
0026	S	S	E	N	1967	89315
0026	S	S	P	H	1976	92927
0026	S	S	E	H	1976	94271
0026	S	S	E	N	1976	94271
0031	S	D	E	N	1933	35687
0031	S	D	E	N	1963	37530
0031	S	D	E	N	1951	38163
0031	S	D	E	N	1943	38858
0031	S	D	E	N	1934	43056
0031	S	S	E	L	1961	44227
0031	S	S	E	N	1961	44227
0031	S	D	D	N	1967	47237
0031	S	D	E	N	1926	50877
0031	S	D	E	N	1960	52250
0031	S	D	R	N	1967	53157
0031	S	D	E	L	1963	55962
0031	S	D	E	N	1932	57548
0031	S	D	E	N	1948	58465
0031	S	D	E	N	1951	60526
0031	S	D	E	L	1977	88199
0031	S	D	E	N	1977	88199
0031	S	S	E	L	1977	88199
0031	S	S	E	N	1977	88199
0031	S	D	E	N	1978	98331
0066	S	D	E	N	1963	49644
0066	S	D	R	N	1968	56270
0066	S	D	E	N	1971	68241
0066	S	D	E	N	1972	78497
0066	S	D	R	N	1974	79125
0066	S	D	E	N	1974	79405
0082	E	D	E	N	1946	6321
0082	E	E	E	N	1946	6321
0082	E	S	E	N	1948	54088
0082	F	D	E	N	1964	36084
0082	F	D	E	N	1955	37152
0082	F	D	R	N	1967	46367
0082	F	G	R	N	1967	50070
0082	F	D	E	N	1960	52129
0082	F	G	E	N	1967	64177
0082	F	D	E	N	1972	69344
0082	F	D	E	N	1974	74724
0082	F	D	E	N	1974	74725
0082	F	D	E	N	1960	77144
0082	F	D	E	N	1967	77941
0082	F	D	E	N	1975	84492
0082	F	D	E	N	1978	95849
0082	S	E	E	N	1945	1555
0082	S	S	G	N	1928	28680
0082	S	S	R	N	1962	33745
0082	S	S	E	N	1966	33746
0082	S	D	E	N	1965	34592
0082	S	D	E	N	1964	36084
0082	S	D	E	N	1965	36738
0082	S	S	F	N	1930	38471
0082	S	S	E	H	1950	41921
0082	S	T	E	N	1959	43394
0082	S	D	R	N	1966	44204
0082	S	D	E	N	1966	44205
0082	S	D	E	L	1961	44227
503-0082	S	D	R	N	1966	45198
0082	S	D	O	N	1964	45262
0082	S	D	E	N	1946	45807
0082	S	T	E	N	1946	45807
0082	S	D	R	N	1966	47242
0082	S	D	E	N	1907	49396
0082	S	S	D	N	1968	49783
0082	S	D	E	H	1958	50170
0082	S	S	R	H	1965	52875
0082	S	S	E	H	1966	52876
0082	S	S	R	N	1967	53147
0082	S	S	R	N	1967	53148
0082	S	D	O	N	1960	53304
0082	S	E	R	N	1967	53740
0082	S	D	E	L	1969	54713
0082	S	D	E	N	1969	54728
0082	S	D	R	N	1968	55460
0082	S	D	E	N	1968	55461
0082	S	D	E	N	1935	57409
0082	S	D	E	N	1946	57698
0082	S	D	E	N	1934	57719
0082	S	S	G	N	1941	58735
0082	S	D	G	N	1954	58737
0082	S	E	G	N	1954	58737
0082	S	E	E	N	1928	58935
0082	S	S	E	N	1938	59053
0082	S	S	E	N	1942	59580
0082	S	D	E	N	1956	59583
0082	S	D	E	N	1936	59995
0082	S	D	E	N	1946	60454
0082	S	D	G	N	1963	62289
0082	S	S	E	H	1971	65258
0082	S	D	E	N	1972	67519
0082	S	T	E	N	1972	67519
0082	S	T	E	N	1972	67656
0082	S	T	E	L	1972	68273
0082	S	D	E	N	1950	72108
0082	S	E	E	N	1950	72108
0082	S	D	R	N	1972	75623
0082	S	D	E	N	1972	75624
0082	S	T	O	N	1974	75943
0082	S	D	F	N	1968	76887
0082	S	D	E	N	1969	77506
0082	S	D	G	N	1972	81966
0082	S	S	E	N	1977	88199
0082	S	D	E	N	1977	90241
0082	S	E	E	N	1977	90241
0082	S	D	E	N	1979	100105
0083	S	D	E	N	1976	85762
0090	S	D	J	N	1976	89071
0096	S	D	J	N	1976	89071
0103	S	S	E	N	1960	24931
0110	S	S	E	N	1960	24931
0110	S	S	E	N	1977	88199
0112	S	D	G	N	1967	49008
0112	S	S	G	N	1977	92758
0126	S	D	E	N	1973	71894
0126	S	D	E	N	1976	88416
0126	S	D	E	N	1977	91632
0173	S	D	E	N	1977	91585
0178	S	D	E	N	1976	85762
0179	S	S	E	N	1960	24931
0179	S	S	E	N	1977	88199
0180	S	D	E	N	1976	85762
0200	S	D	E	N	1967	43437
0200	S	D	E	N	1976	88416
0200	S	D	E	N	1977	91632
0211	S	S	E	H	1960	24931
0211	S	S	E	N	1960	24931
0213	S	D	E	N	1960	45778
0214	S	D	E	N	1976	85762
0216	M	D	E	N	1944	47956
0216	S	D	E	N	1955	37152
0216	S	D	E	N	1970	37248
0216	S	D	E	N	1968	47871
0216	S	D	E	L	1952	50208
0216	S	D	E	N	1967	53989
0216	S	D	E	N	1955	68323
0216	S	S	E	N	1971	78841
0216	S	S	E	N	1977	88199
0224	S	D	E	N	1976	85762
0237	S	D	R	N	1968	53472
0237	S	D	E	N	1968	67357
0239	S	D	E	N	1903	35787
0239	S	D	D	N	1903	38441
0239	S	D	E	L	1925	43401
0239	S	D	D	L	1925	43841
0239	S	D	E	N	1969	54036
0239	S	D	E	N	1969	60104
0254	S	D	E	N	1976	87561
0262	S	D	E	N	1956	41939
0262	S	S	E	N	1971	78841
0262	S	S	E	N	1977	88199
0270	S	D	E	N	1967	35230
0270	S	D	F	N	1950	46948
503-0270	S	D	E	N	1939	65139
0270	S	D	E	N	1975	76616
0284	S	D	J	N	1957	23772
0284	S	S	E	H	1960	24931
0284	S	S	E	N	1960	24931
0284	S	D	R	H	1967	49621
0284	S	D	R	N	1976	87278
0284	S	D	E	N	1976	89160
0284	S	D	E	H	1977	94058
0284	S	D	E	N	1977	94058
0288	S	S	E	N	1960	24931
0290	S	D	E	N	1966	40468
0290	S	D	E	N	1968	47870
0304	S	D	E	N	1956	41939
0304	S	S	E	N	1977	88199
0316	S	S	E	N	1971	78841
0353	S	D	E	N	1976	87561
0361	S	S	E	N	1960	24931
0376	S	D	E	N	1934	37974
0376	S	S	E	N	1967	89315
0392	S	S	E	N	1960	24931
0393	S	S	E	N	1960	24931
0396	S	S	E	N	1960	24931
0396	S	S	E	L	1977	88199
0396	S	S	E	N	1977	88199
0403	S	S	E	N	1960	24931
0409	S	D	E	N	1973	92934
0416	S	D	E	N	1976	85762
0417	S	D	C	N	1963	34151
0417	S	D	F	N	1965	46017
0423	S	D	E	N	1934	37974
0424	S	S	E	N	1960	24931
0431	S	D	E	N	1978	97376
0440	S	S	E	N	1960	24931
0440	S	D	E	N	1930	37544
0442	S	D	E	N	1930	37544
0452	S	D	E	N	1962	47240
0452	S	D	E	N	1963	58889
0461	S	S	E	N	1960	24931
0461	S	D	R	N	1960	39500
0461	S	D	E	N	1951	43395
0461	S	D	E	N	1954	44149
0461	S	D	R	N	1966	44216
0461	S	D	E	N	1966	44217
0461	S	D	F	N	1967	46340
0461	S	D	E	N	1968	49276
0461	S	D	R	N	1968	50669
0461	S	D	E	N	1960	52128
0461	S	D	E	N	1952	52749
0461	S	D	R	N	1967	53471
0461	S	D	E	N	1949	56176
0461	S	D	E	N	1967	57594
0461	S	D	E	N	1973	69346
0461	S	S	E	N	1967	89315
0463	S	D	R	N	1966	41499
0463	S	D	R	N	1968	53472
0463	S	D	E	N	1968	67357
0464	S	D	R	N	1968	53472
0464	S	D	E	N	1968	67357
0465	S	D	R	N	1966	41499
0465	S	D	R	N	1968	53472
0465	S	D	R	N	1971	62599
0465	S	D	E	N	1971	66484
0465	S	D	E	N	1968	67357
0466	S	D	R	N	1968	53472
0466	S	D	E	N	1968	67357
0473	S	S	E	N	1960	24931
0476	S	D	E	N	1958	15895
0481	S	S	E	N	1960	24931
0481	S	D	G	N	1967	47477
0481	S	T	G	N	1967	47477
0481	S	D	G	N	1963	62287
0481	S	D	E	N	1973	76782
0481	S	T	E	N	1973	76782
0481	S	D	E	N	1978	97376
0482	S	D	E	N	1976	87561
0526	S	S	E	N	1960	24931
0526	S	D	E	N	1976	85762
0531	S	S	E	N	1977	88199
0537	S	D	E	N	1970	37248
0537	S	S	E	N	1971	78841
0541	S	S	E	N	1959	52678
0541	S	S	E	N	1967	89315
0544	S	D	J	N	1962	41964
0544	S	D	E	N	1954	43336
0544	S	D	E	N	1955	43396
0544	S	D	E	H	1942	43643
0544	S	D	P	N	1967	48452
0544	S	E	P	N	1967	48452
0544	S	D	O	H	1967	49026
0544	S	D	E	F	1963	54415
0544	S	S	E	N	1935	60445
0544	S	S	E	N	1967	89315
0546	S	S	E	N	1967	89315
0556	S	S	E	N	1960	24931

Phys. State: **C.** Superconductive; **D.** Doped; **E.** Expanded; **F.** Fibrous or Whisker; **G.** Gas; **I.** Ionized or Plasma; **L.** Liquid; **M.** Multiphase; **P.** Powder or Fine Particle; **S.** Solid; **T.** Thick or Thin Film
Subject: **D.** Data; **E.** Experiment; **G.** General (Data + Theory + Experiment); **S.** Survey (Review, Compendium etc.); **T.** Theory
Language: **C.** Czech; **D.** Dutch; **E.** English; **F.** French; **G.** German; **I.** Italian; **J.** Japanese; **O.** Other Languages; **P.** Polish; **R.** Russian; **S.** Spanish
Temperature: **F.** Full Range (Low + Normal + High); **L.** Low (0 to 75K) + Overlap into Normal; **N.** Normal (75 to 1273K); **H.** High (above 1273K) + Overlap into Normal; Blank = Not Coded

Substance Number	Phys. State	Subject	Language	Temperature	Year	TPRC Number
503-0556	S	D	E	N	1930	37544
0556	S	D	E	N	1931	37634
0556	S	D	E	N	1934	37974
0556	S	S	E	N	1967	89315
0562	S	D	E	L	1970	63400
0562	S	D	E	N	1972	67282
0562	S	E	E	N	1972	67282
0562	S	G	E	N	1972	71416
0562	S	D	E	L	1973	72981
0562	S	D	E	L	1974	74498
0562	S	D	E	L	1974	74521
0562	S	D	E	N	1974	74522
0562	S	E	E	N	1974	74522
0562	S	D	F	N	1968	76887
0562	S	D	E	L	1976	87010
0562	S	D	E	N	1976	87010
0562	S	D	E	N	1976	87429
0562	S	S	E	N	1977	88199
0562	S	S	E	N	1977	92040
0562	S	S	E	N	1979	97587
0566	S	D	E	N	1955	20062
0566	S	D	E	N	1963	37530
0566	S	D	E	N	1960	52250
0567	S	D	E	N	1966	40468
0567	S	D	E	N	1968	47870
0569	S	D	J	N	1976	89071
0580	S	D	E	N	1960	45778
0580	S	D	R	H	1979	97572
0580	S	D	R	N	1979	97572
0580	S	D	E	H	1979	100641
0580	S	D	E	N	1979	100641
0583	S	D	R	N	1968	56269
0584	D	D	E	N	1967	35230
0585	D	D	E	N	1976	85230
0585	S	S	E	N	1960	24931
0585	S	D	E	N	1930	37544
0585	S	D	E	N	1930	37545
0585	S	D	R	N	1967	53762
0585	S	D	E	N	1976	87561
0585	S	S	E	N	1977	88199
0585	S	D	R	N	1977	89777
0585	S	D	E	N	1977	91150
0588	S	S	E	N	1960	24931
0588	S	D	R	N	1977	89777
0588	S	D	E	N	1977	91150
0588	S	D	E	N	1978	97376
0589	S	S	E	N	1968	48754
0589	S	D	E	N	1979	97010
0590	S	S	E	N	1968	48754
0594	S	D	E	H	1967	77589
0594	S	D	E	N	1967	77589
0599	S	D	R	N	1966	41499
0599	S	S	E	N	1967	89315
0604	S	D	E	N	1970	37248
0604	S	D	E	N	1960	52654
0604	S	D	E	N	1974	74515
0604	S	T	E	N	1974	74515
0604	S	S	E	N	1971	78841
0612	S	S	E	N	1960	24931
0612	S	D	E	N	1976	85762
0650	S	D	G	N	1976	89038
0650	S	D	E	N	1973	92934
0651	S	D	E	N	1973	92934
0652	S	S	E	N	1960	24931
0652	S	D	E	N	1960	45778
0652	S	D	E	N	1964	91250
0652	S	D	R	H	1979	97572
0652	S	D	R	N	1979	97572
0652	S	D	E	H	1979	100641
0652	S	D	E	N	1979	100641
0659	S	S	E	H	1964	45435
0659	S	D	E	N	1968	47422
0659	S	S	E	N	1962	52795
0659	S	S	E	N	1976	83746
0659	S	S	E	L	1977	88199
0659	S	S	E	N	1977	88199
0659	S	D	E	H	1977	94200
0659	S	D	E	N	1977	94200
0659	S	S	E	L	1977	94200
0659	S	S	E	N	1977	94200
0660	S	D	E	N	1962	49968
0660	S	D	E	N	1961	52100
0662	S	D	E	N	1976	87561
0663	S	S	E	N	1960	24931
0663	S	D	E	N	1976	87561
0667	S	D	R	N	1965	49154
0667	S	D	E	N	1968	49155
0667	S	D	R	N	1969	57969
0667	S	D	E	N	1969	57970
0676	S	S	E	N	1977	92163
0676	S	D	E	N	1973	92934
0683	S	D	R	N	1967	47173
0683	S	D	E	N	1969	56358
0704	S	S	R	N	1973	79859
0704	S	S	E	N	1973	79860
0719	S	D	R	N	1968	53472
0719	S	D	R	N	1971	62599
0719	S	D	E	N	1971	66484
0719	S	D	E	N	1968	67357
0721	S	D	E	N	1967	43437
0721	S	S	E	N	1971	78841
503-0723	S	D	E	N	1970	37248
0723	S	D	E	N	1974	74511
0723	S	E	E	N	1974	74511
0731	D	D	E	N	1966	40547
0731	S	D	E	N	1966	39948
0731	S	T	R	N	1966	43027
0731	S	D	E	N	1928	58935
0736	S	S	E	N	1971	78841
0736	S	D	E	N	1976	88416
0736	S	D	E	N	1977	91632
0739	S	S	E	N	1967	89315
0749	S	S	E	N	1977	92163
0765	S	S	E	N	1960	24931
0765	S	D	E	H	1940	36234
0765	S	D	E	N	1951	43395
0765	S	D	E	H	1942	43643
0765	S	D	S	N	1967	44117
0765	S	D	E	N	1954	44149
0765	S	D	E	N	1961	47544
0765	S	D	E	N	1960	52136
0765	S	D	E	N	1952	59512
0765	S	D	E	N	1936	60407
0765	S	D	E	N	1938	60409
0765	S	D	E	N	1946	60454
0765	S	D	R	N	1970	62314
0765	S	S	E	N	1967	89315
0767	S	D	R	N	1968	56202
0767	S	D	E	N	1976	85762
0767	S	S	E	N	1967	89315
0768	S	S	E	H	1960	24931
0768	S	S	E	N	1960	24931
0782	S	D	E	N	1938	36191
0782	S	D	F	N	1947	46297
0782	S	D	E	N	1947	46798
0782	S	D	R	N	1958	47220
0782	S	D	E	N	1958	47221
0782	S	D	E	N	1963	47403
0782	S	D	E	N	1964	49184
0782	S	D	E	N	1967	51989
0782	S	D	E	N	1934	56075
0782	S	D	E	N	1949	59951
0782	S	D	R	N	1970	60136
0782	S	D	E	N	1970	60137
0782	S	D	E	N	1971	68241
0782	S	D	E	N	1972	78497
0782	S	S	E	N	1977	88199
0783	S	S	E	N	1967	89315
0793	S	D	E	H	1967	77589
0793	S	D	E	N	1967	77589
0793	S	D	E	N	1964	91250
0801	S	D	E	N	1955	20062
0801	S	S	G	N	1928	28680
0801	S	D	E	N	1949	35817
0801	S	T	E	N	1949	35817
0801	S	D	G	H	1952	42961
0801	S	D	E	N	1948	43305
0801	S	D	E	N	1951	43395
0801	S	D	E	N	1936	44125
0801	S	D	E	N	1954	44149
0801	S	D	E	N	1936	44296
0801	S	D	E	H	1958	50170
0801	S	D	E	H	1952	52094
0801	S	D	E	N	1959	52680
0801	S	D	R	N	1955	54638
0801	S	D	R	N	1970	60146
0801	S	D	E	N	1970	60147
0801	S	D	E	N	1926	60435
0801	S	D	E	N	1927	60436
0801	S	D	E	N	1933	60786
0801	S	D	G	N	1959	62278
0801	S	D	J	N	1958	62291
0801	S	D	E	N	1955	63074
0801	S	D	G	N	1971	64254
0801	S	D	E	N	1919	71468
0801	S	E	E	N	1919	71468
0801	S	D	E	N	1950	72108
0801	S	E	E	N	1950	72108
0801	S	D	C	N	1974	76085
0801	S	D	E	N	1836	91329
0801	S	E	E	N	1836	91329
0803	S	S	E	N	1960	24931
0803	S	S	E	N	1977	88199
0808	S	D	R	N	1977	89777
0808	S	D	E	N	1977	91150
0809	S	D	R	N	1977	89777
0809	S	D	E	N	1977	91150
0818	D	D	E	N	1966	40547
0818	S	D	E	N	1939	65139
0833	S	D	E	N	1976	91261
0849	S	S	E	N	1960	24931
0859	S	D	R	N	1969	56962
0863	S	S	E	N	1960	24931
0863	S	S	E	N	1977	88199
0866	S	D	E	N	1960	41838
0866	S	D	R	N	1967	46274
0866	S	D	E	N	1967	46275
0871	S	D	E	N	1974	77552
0897	S	D	E	N	1976	85762
0899	S	D	E	N	1930	37544
0927	S	S	E	N	1971	78841
0929	S	D	E	N	1974	74515
503-0929	S	T	E	N	1974	74515
0955	S	D	E	N	1955	68323
0957	S	D	E	N	1970	44781
0958	D	D	E	L	1972	65401
0958	D	E	E	L	1972	65401
0958	D	G	E	N	1971	66142
0964	S	D	E	L	1976	87010
0964	S	D	E	N	1976	87010
0964	S	D	E	L	1980	100311
0964	S	D	E	N	1980	100311
0966	S	D	E	N	1960	99523
0967	S	S	E	N	1971	78841
0967	S	S	E	N	1976	83746
0974	S	D	E	N	1963	76855
0974	S	E	E	N	1963	76855
0974	S	D	E	N	1960	76856
0978	S	D	R	L	1978	96430
0978	S	D	R	N	1978	96430
0978	S	D	E	L	1978	98501
0978	S	D	E	N	1978	98501
0982	S	D	E	N	1949	55925
0985	S	D	E	L	1929	54992
0985	S	D	G	N	1921	55996
0985	S	D	G	L	1928	56000
0985	S	S	E	N	1977	88199
0986	S	D	G	N	1918	40320
0986	S	D	G	N	1907	47659
0986	S	D	G	N	1921	55996
0986	S	D	G	L	1928	56000
0986	S	D	E	N	1949	59951
0986	S	S	E	N	1977	88199
0987	S	D	G	N	1893	60016
0988	S	D	G	N	1893	60016
0989	S	D	G	N	1921	55996
0989	S	G	G	L	1924	55998
0989	S	G	G	N	1924	58086
0990	S	D	G	N	1921	55996
0992	S	D	E	L	1934	36300
0992	S	D	E	L	1934	48880
0992	S	D	E	L	1952	50208
0992	S	D	E	N	1949	59951
0992	S	S	E	L	1977	88199
0992	S	S	E	N	1977	88199
0995	S	D	E	N	1970	59834
0997	S	S	E	N	1976	88411
1009	S	D	E	L	1973	72981
1009	S	D	E	L	1974	74521
1009	S	D	E	N	1974	74522
1009	S	E	E	N	1974	74522
1009	S	D	E	L	1976	87010
1009	S	D	E	N	1976	87010
1009	S	D	E	N	1976	87429
1009	S	D	E	N	1977	88904
1009	S	E	E	N	1977	88904
1009	S	S	E	N	1977	92040
1009	S	S	G	N	1977	92758
1010	S	D	R	N	1969	56888
1011	S	D	R	N	1968	53472
1011	S	D	E	N	1968	67357
1012	S	D	R	N	1968	53472
1012	S	D	E	N	1968	67357
1013	P	D	R	N	1967	53149
1013	S	D	R	N	1967	53149
1014	S	D	R	N	1971	62599
1014	S	D	E	N	1971	66484
1020	S	D	R	N	1967	53153
1020	S	T	R	N	1967	53153
1021	S	D	R	N	1969	56888
1022	S	D	R	N	1969	56888
1025	S	D	R	N	1969	56888
1026	S	D	R	N	1969	56888
1030	S	D	G	N	1967	49008
1032	S	D	E	N	1966	40468
1032	S	D	E	N	1968	47870
1034	S	D	E	N	1970	57787
1036	S	D	E	N	1970	57787
1041	S	D	E	N	1949	55925
1044	S	D	E	N	1928	60440
1050	S	D	E	N	1960	10947
1050	S	E	E	N	1960	10947
1050	S	S	E	N	1960	24931
1050	S	D	D	N	1967	47237
1051	D	D	E	N	1967	35230
1052	S	S	E	N	1967	89315
1054	S	S	E	N	1967	89315
1057	S	D	E	N	1903	35787
1057	S	D	D	N	1903	38441
1059	S	S	G	N	1928	28680
1059	S	D	E	N	1908	40328
1059	S	S	E	L	1961	44227
1059	S	S	E	N	1961	44227
1059	S	D	E	N	1967	45885
1059	S	D	R	N	1967	45886
1059	S	D	E	N	1971	67369
1059	S	E	E	N	1971	67369
1059	S	S	G	N	1977	92758
1060	S	D	G	N	1966	41958
1060	S	T	G	N	1966	41958
1060	S	S	E	N	1967	89315
1062	S	D	E	N	1960	10947
1062	S	E	E	N	1960	10947

Phys. State: **C.** Superconductive; **D.** Doped; **E.** Expanded; **F.** Fibrous or Whisker; **G.** Gas; **I.** Ionized or Plasma; **L.** Liquid; **M.** Multiphase; **P.** Powder or Fine Particle; **S.** Solid; **T.** Thick or Thin Film

Subject: **D.** Data; **E.** Experiment; **G.** General (Data + Theory + Experiment); **S.** Survey (Review, Compendium etc.); **T.** Theory

Language: **C.** Czech; **D.** Dutch; **E.** English; **F.** French; **G.** German; **I.** Italian; **J.** Japanese; **O.** Other Languages; **P.** Polish; **R.** Russian; **S.** Spanish

Temperature: **F.** Full Range (Low + Normal + High); **L.** Low (0 to 75K) + Overlap into Normal; **N.** Normal (75 to 1273K); **H.** High (above 1273K) + Overlap into Normal; Blank = Not Coded

Substance Number	Phys. State	Subject	Language	Temperature	Year	TPRC Number
503-1062	S	S	E	N	1960	24931
1063	S	D	E	N	1971	63137
1064	S	D	E	N	1920	37542
1071	S	D	E	N	1934	43056
1072	S	D	E	N	1960	10947
1072	S	E	E	N	1960	10947
1072	S	S	E	N	1960	24931
1074	S	D	E	N	1941	38475
1081	S	S	O	N	1969	57568
1081	S	S	E	N	1967	89315
1082	S	D	G	N	1930	58957
1110	S	D	E	N	1943	38858
1110	S	D	E	N	1946	43404
1111	S	D	E	N	1934	56075
1111	S	D	E	N	1949	59951
1111	S	S	E	N	1977	88199
1112	S	D	D	N	1967	47237
1112	S	D	G	N	1930	58957
1121	S	D	E	N	1934	43056
1125	P	D	O	H	1972	67719
1125	S	D	E	N	1955	20062
1125	S	D	E	H	1933	55938
1128	S	S	E	N	1960	24931
1128	S	D	E	N	1973	69346
1131	S	D	R	N	1966	41085
1131	S	D	E	N	1966	41086
1131	S	D	E	N	1945	42265
1131	S	D	E	H	1965	44136
1131	S	D	R	H	1966	46593
1131	S	D	E	H	1966	46594
1131	S	D	G	H	1967	48042
1131	S	D	E	H	1943	50711
1131	S	D	E	H	1968	53652
1131	S	D	E	F	1963	54415
1131	S	D	G	H	1969	55539
1131	S	D	R	N	1969	57257
1131	S	D	E	N	1969	57258
1132	D	D	R	N	1968	58259
1132	S	D	R	N	1968	58259
1132	S	D	O	H	1958	62300
1133	D	D	E	H	1931	64936
1133	P	D	O	H	1972	67719
1133	S	D	E	N	1937	13366
1133	S	S	E	H	1960	24931
1133	S	S	E	N	1960	24931
1133	S	S	G	H	1928	28680
1133	S	S	G	N	1928	28680
1133	S	D	E	N	1921	36041
1133	S	D	G	H	1966	36685
1133	S	D	E	H	1966	36690
1133	S	D	E	N	1930	38880
1133	S	D	G	H	1931	39517
1133	S	D	O	H	1959	39712
1133	S	T	O	H	1959	39712
1133	S	D	E	N	1946	41392
1133	S	D	E	N	1930	41738
1133	S	D	G	H	1964	44351
1133	S	D	E	N	1943	45557
1133	S	D	E	N	1967	45885
1133	S	D	R	N	1967	45886
1133	S	D	E	H	1940	47367
1133	S	D	O	H	1967	49026
1133	S	D	O	N	1960	53304
1133	S	D	E	N	1925	54335
1133	S	D	E	F	1963	54415
1133	S	D	E	N	1933	54491
1133	S	D	E	H	1933	55938
1133	S	D	E	H	1917	55983
1133	S	D	E	H	1969	56368
1133	S	D	R	N	1969	58856
1133	S	D	E	N	1969	58857
1133	S	D	G	H	1928	59017
1133	S	D	G	H	1934	59091
1133	S	D	R	H	1968	59345
1133	S	D	E	H	1968	59346
1133	S	D	E	N	1920	60430
1133	S	D	E	N	1934	60444
1133	S	S	E	N	1935	60445
1133	S	D	E	N	1936	60447
1133	S	D	E	N	1943	60559
1133	S	D	J	N	1962	62294
1133	S	D	R	H	1971	64836
1133	S	D	E	H	1971	64837
1133	S	D	E	H	1931	64936
1133	S	D	P	N	1970	65195
1133	S	D	E	N	1974	74511
1133	S	E	E	N	1974	74511
1135	E	D	E	N	1962	52417
1135	S	D	E	N	1955	20062
1135	S	S	G	H	1928	28680
1135	S	D	E	N	1965	36144
1135	S	D	E	H	1964	43422
1135	S	S	R	H	1963	43656
1135	S	D	E	N	1954	44149
1135	S	D	E	N	1947	48002
1135	S	D	R	H	1967	53201
1135	S	D	E	N	1969	54728
1135	S	D	E	H	1970	61027
1135	S	D	E	N	1974	74511
1135	S	E	E	N	1974	74511
1135	S	D	E	H	1971	75168

Substance Number	Phys. State	Subject	Language	Temperature	Year	TPRC Number
503-1136	S	S	G	N	1928	28680
1136	S	D	E	N	1941	36248
1136	S	D	E	N	1934	42447
1136	S	D	E	N	1935	44812
1136	S	D	E	N	1960	45215
1136	S	D	E	N	1958	46791
1136	S	D	E	N	1967	48531
1136	S	D	E	N	1965	50414
1136	S	D	E	H	1925	52689
1136	S	D	E	N	1836	91329
1136	S	E	E	N	1836	91329
1138	S	D	E	N	1949	35817
1138	S	T	E	N	1949	35817
1138	S	D	E	N	1940	42209
1138	S	D	O	N	1967	44894
1138	S	D	G	N	1970	58486
1138	S	D	R	N	1970	60146
1138	S	D	E	N	1970	60147
1138	S	D	G	N	1971	64254
1138	S	D	C	N	1974	76085
1139	S	S	G	H	1928	28680
1139	S	D	F	H	1921	46395
1140	S	D	E	N	1940	46430
1142	S	S	E	N	1960	24931
1142	S	S	R	N	1962	33745
1142	S	S	E	N	1966	33746
1142	S	D	E	N	1953	35277
1142	S	S	E	H	1950	41921
1142	S	S	E	N	1955	43596
1142	S	D	E	N	1967	45072
1142	S	S	E	L	1969	45606
1142	S	S	E	H	1961	47146
1142	S	D	E	N	1941	48788
1142	S	D	E	N	1942	49128
1142	S	S	D	N	1968	49783
1142	S	S	E	H	1951	52106
1142	S	S	E	N	1959	52521
1142	S	D	E	N	1959	52680
1142	S	S	E	N	1961	52792
1142	S	S	E	N	1962	52794
1142	S	D	E	N	1968	53479
1142	S	D	E	N	1961	54086
1142	S	D	E	N	1969	58153
1142	S	D	E	N	1970	58547
1142	S	S	E	H	1926	59969
1142	S	S	E	N	1938	59970
1142	S	D	G	N	1961	62282
1142	S	D	G	N	1963	62289
1142	S	S	E	H	1971	65258
1142	S	S	E	N	1977	88199
1142	S	T	E	N	1979	99309
1143	S	S	G	N	1928	28680
1143	S	D	E	N	1941	48788
1143	S	D	E	N	1959	52680
1143	S	D	G	N	1961	62282
1143	S	D	E	N	1973	69346
1145	S	S	E	H	1964	37819
1145	S	E	G	H	1966	41276
1145	S	S	R	H	1966	44058
1145	S	S	E	H	1966	44059
1145	S	S	E	N	1924	44568
1145	S	D	G	H	1966	45145
1145	S	D	E	N	1940	46430
1145	S	T	G	N	1961	47250
1145	S	S	E	H	1967	51671
1145	S	S	E	H	1967	51672
1145	S	S	E	H	1956	52789
1145	S	S	E	H	1963	52812
1145	S	D	O	N	1960	53304
1145	S	S	E	H	1969	55310
1145	S	D	R	N	1969	57259
1145	S	D	E	N	1969	57260
1145	S	E	F	H	1969	57519
1145	S	D	E	N	1929	60442
1145	S	D	R	H	1970	61318
1145	S	D	E	H	1970	61319
1146	D	D	R	H	1971	64834
1146	D	D	E	H	1971	64835
1146	S	S	G	H	1928	28680
1146	S	D	E	N	1896	30723
1146	S	E	E	N	1896	30723
1146	S	D	E	N	1934	35758
1146	S	D	R	H	1962	38012
1146	S	D	R	N	1966	41085
1146	S	D	E	N	1966	41086
1146	S	D	G	H	1932	41792
1146	S	D	P	N	1967	48452
1146	S	E	P	N	1967	48452
1146	S	D	O	H	1967	49026
1146	S	D	E	H	1943	50711
1146	S	D	G	H	1968	53251
1146	S	D	E	H	1968	53652
1146	S	D	E	F	1963	54415
1146	S	D	E	H	1917	55983
1146	S	D	G	H	1928	59017
1146	S	E	E	N	1935	60445
1146	S	D	R	H	1971	64828
1146	S	D	E	H	1971	64829
1146	S	D	R	H	1971	64834
1146	S	D	E	H	1971	64835
1147	P	D	O	H	1972	67719

Substance Number	Phys. State	Subject	Language	Temperature	Year	TPRC Number
503-1147	S	D	G	N	1934	6342
1147	S	D	E	N	1955	20062
1147	S	S	E	N	1960	24931
1147	S	S	G	N	1928	28680
1147	S	D	E	N	1941	36248
1147	S	D	E	H	1941	38840
1147	S	D	R	N	1966	41085
1147	S	D	E	N	1966	41086
1147	S	D	E	N	1966	41727
1147	S	D	G	N	1932	41792
1147	S	D	E	H	1961	41820
1147	S	D	E	N	1963	41911
1147	S	D	R	N	1963	41912
1147	S	D	E	H	1945	42273
1147	S	D	E	N	1934	42447
1147	S	D	E	H	1958	43298
1147	S	E	E	H	1958	43298
1147	S	D	G	H	1930	47183
1147	S	D	E	H	1940	47367
1147	S	D	G	N	1967	48442
1147	S	D	O	H	1967	49026
1147	S	D	G	H	1962	53241
1147	S	D	O	N	1960	53304
1147	S	D	E	N	1925	54335
1147	S	D	E	F	1963	54415
1147	S	D	R	H	1968	54773
1147	S	D	E	H	1968	54774
1147	S	D	R	N	1969	54917
1147	S	D	E	H	1933	55938
1147	S	D	E	H	1917	55983
1147	S	D	E	N	1969	57266
1147	S	D	F	H	1969	57519
1147	S	E	F	H	1969	57519
1147	S	D	G	H	1928	59017
1147	S	D	E	N	1920	60430
1147	S	D	E	H	1925	60558
1147	S	D	O	H	1958	62300
1147	S	D	E	H	1973	78263
1147	S	D	E	N	1836	91329
1147	S	E	E	N	1836	91329
1149	S	D	E	H	1925	52689
1152	S	S	G	N	1928	28680
1153	P	D	O	H	1972	67719
1153	S	D	E	N	1951	43395
1153	S	D	E	N	1954	44149
1153	S	D	E	N	1925	54416
1153	S	D	E	H	1943	59052
1154	S	S	E	N	1960	24931
1154	S	D	E	N	1954	44149
1159	S	D	R	N	1968	59372
1159	S	D	E	N	1968	59373
1162	P	D	E	N	1959	52680
1162	P	D	O	H	1972	67719
1162	S	S	G	N	1928	28680
1162	S	D	E	H	1940	36234
1162	S	D	E	N	1960	41838
1162	S	D	R	N	1965	41896
1162	S	D	E	N	1964	42037
1162	S	D	E	N	1935	44727
1162	S	D	F	N	1967	50734
1162	S	E	F	N	1967	50734
1162	S	D	E	N	1968	54551
1162	S	D	E	N	1964	54981
1162	S	D	E	N	1920	60430
1162	S	D	G	N	1959	62278
1162	S	D	E	N	1965	63068
1173	P	D	O	H	1972	67719
1173	S	S	G	H	1928	28680
1173	S	D	E	N	1963	37530
1173	S	D	R	N	1952	37567
1173	S	D	E	N	1964	41670
1173	S	S	E	F	1966	42294
1173	S	D	E	N	1951	43395
1173	S	D	E	H	1964	43422
1173	S	D	E	N	1967	43437
1173	S	D	E	N	1954	44149
1173	S	D	R	N	1966	44192
1173	S	D	E	N	1966	44193
1173	S	D	R	N	1966	44216
1173	S	D	E	N	1966	44217
1173	S	D	R	N	1966	44508
1173	S	D	E	N	1935	44812
1173	S	D	R	N	1967	46096
1173	S	D	E	N	1967	46097
1173	S	D	E	H	1947	48002
1173	S	D	O	H	1967	49026
1173	S	D	E	N	1962	49968
1173	S	D	E	N	1960	52136
1173	S	D	E	N	1966	52249
1173	S	E	E	H	1958	52758
1173	S	S	E	N	1962	52795
1173	S	D	R	H	1967	53201
1173	S	D	R	N	1967	53471
1173	S	G	R	N	1967	53739
1173	S	D	E	F	1963	54415
1173	S	D	E	H	1933	55938
1173	S	S	E	H	1970	56454
1173	S	D	E	N	1920	60430
1173	S	D	E	H	1970	61027
1173	S	D	R	H	1970	61316
1173	S	D	E	H	1970	61317

Phys. State: **C.** Superconductive; **D.** Doped; **E.** Expanded; **F.** Fibrous or Whisker; **G.** Gas; **I.** Ionized or Plasma; **L.** Liquid; **M.** Multiphase; **P.** Powder or Fine Particle; **S.** Solid; **T.** Thick or Thin Film

Subject: **D.** Data; **E.** Experiment; **G.** General (Data + Theory + Experiment); **S.** Survey (Review, Compendium etc.); **T.** Theory

Language: **C.** Czech; **D.** Dutch; **E.** English; **F.** French; **G.** German; **I.** Italian; **J.** Japanese; **O.** Other Languages; **P.** Polish; **R.** Russian; **S.** Spanish

Temperature: **F.** Full Range (Low + Normal + High); **L.** Low (0 to 75K) + Overlap into Normal; **N.** Normal (75 to 1273K); **H.** High (above 1273K) + Overlap into Normal; Blank = Not Coded

Substance Number	Phys. State	Subject	Language	Temperature	Year	TPRC Number
503-1173	S	D	R	H	1970	61531
1173	S	D	E	H	1970	61532
1173	S	D	R	N	1970	62314
1173	S	D	E	N	1952	63993
1173	S	D	E	L	1969	63999
1173	S	E	E	L	1969	63999
1173	S	D	R	H	1971	64822
1173	S	D	E	H	1971	64823
1173	S	D	R	H	1971	64846
1173	S	D	E	H	1971	64847
1173	S	D	E	N	1974	74549
1173	S	D	E	N	1974	75087
1173	S	D	E	H	1971	75168
1173	S	D	E	N	1965	77529
1173	S	D	G	N	1976	89014
1177	S	S	G	N	1928	28680
1180	S	D	E	N	1951	43395
1180	S	D	E	N	1954	44149
1180	S	D	E	H	1947	48002
1180	S	D	E	N	1970	77556
1183	S	D	E	N	1948	43305
1185	S	D	E	N	1966	41727
1185	S	D	E	N	1956	43645
1185	S	D	E	N	1935	44639
1185	S	D	E	N	1940	45889
1185	S	D	E	N	1968	49276
1185	S	D	R	N	1968	50669
1185	S	D	F	N	1967	50734
1185	S	E	F	N	1967	50734
1185	S	D	E	N	1958	52543
1185	S	D	E	N	1959	52683
1185	S	D	I	N	1968	52947
1185	S	D	E	N	1944	55475
1185	S	D	R	N	1969	55684
1185	S	D	C	N	1969	58204
1185	S	D	E	N	1970	59727
1185	S	D	E	H	1926	60435
1185	S	D	G	N	1961	62280
1185	S	D	E	N	1971	64044
1185	S	D	E	N	1836	91329
1185	S	E	E	N	1836	91329
1187	S	D	E	H	1940	36234
1188	S	S	G	N	1928	28680
1188	S	D	E	H	1947	48002
1189	E	D	R	N	1964	37572
1189	E	D	E	N	1968	47920
1189	E	D	R	N	1968	49597
1189	S	D	R	N	1964	37572
1189	S	D	R	H	1967	53201
1189	S	D	E	N	1920	60430
1189	S	D	E	H	1970	61027
1189	S	D	R	H	1970	61316
1189	S	D	E	H	1970	61317
1189	S	D	R	N	1971	68490
1189	S	T	R	N	1971	68490
1189	S	D	E	N	1974	74548
1189	S	E	E	N	1974	74548
1189	S	D	R	H	1974	76473
1189	S	D	E	H	1974	76474
1190	S	S	G	N	1928	28680
1190	S	D	E	N	1968	47626
1190	S	D	E	N	1968	48767
1190	S	D	E	N	1920	60430
1193	S	D	E	N	1969	57606
1194	P	D	O	H	1972	67719
1194	S	S	G	N	1928	28680
1194	S	D	E	N	1920	60430
1197	S	D	E	H	1930	44258
1198	S	D	E	N	1941	41723
1199	S	D	R	N	1953	150
1199	S	D	E	N	1960	52128
1201	S	D	R	N	1966	41085
1201	S	D	E	N	1966	41086
1201	S	D	P	N	1967	48452
1201	S	E	P	N	1967	48452
1201	S	D	R	H	1971	64828
1201	S	D	E	H	1971	64829
1201	S	D	R	N	1971	68490
1201	S	T	R	N	1971	68490
1202	S	S	E	N	1950	59515
1208	S	S	G	N	1928	28680
1213	F	D	E	N	1974	77949
1213	F	E	E	N	1974	77949
1213	S	S	E	N	1960	24931
1213	S	D	E	N	1960	52128
1213	S	D	E	N	1968	58396
1213	S	D	R	N	1974	79125
1213	S	D	E	N	1974	79405
1218	S	S	G	N	1928	28680
1218	S	D	G	F	1930	37195
1218	S	D	E	N	1948	43305
1218	S	D	G	N	1927	59560
1218	S	D	E	N	1927	60438
1218	S	D	G	N	1959	62278
1218	S	D	G	N	1971	64254
1219	S	D	R	H	1967	49332
1219	S	D	E	H	1967	49333
1219	S	S	E	N	1967	89315
1219	S	S	P	H	1976	92927
1222	S	D	E	N	1961	48241
1226	S	S	E	N	1967	89315

Substance Number	Phys. State	Subject	Language	Temperature	Year	TPRC Number
503-1228	S	S	G	N	1928	28680
1228	S	D	G	N	1918	40320
1228	S	D	G	N	1907	47659
1228	S	D	G	N	1921	55996
1228	S	D	G	L	1928	56000
1228	S	D	G	H	1930	58957
1229	S	D	G	H	1930	58957
1241	S	D	G	F	1930	37195
1241	S	D	G	N	1931	39517
1241	S	D	G	H	1930	47183
1241	S	D	G	H	1930	58957
1242	S	S	G	N	1928	28680
1242	S	D	E	N	1944	55475
1246	S	D	E	N	1941	36248
1246	S	D	E	N	1934	42447
1248	S	D	R	H	1968	44358
1248	S	D	E	H	1968	46849
1248	S	D	P	N	1967	48452
1248	S	E	P	N	1967	48452
1248	S	D	R	H	1971	64828
1248	S	D	E	H	1971	64829
1262	S	D	J	H	1965	44606
1262	S	D	E	H	1968	53652
1262	S	D	E	F	1963	54415
1263	S	D	E	N	1968	48472
1264	S	D	E	N	1968	48472
1266	S	D	O	N	1967	47181
1269	S	D	G	H	1930	58957
1270	S	D	E	H	1925	52689
1272	S	D	E	N	1926	41254
1272	S	D	E	N	1940	46430
1272	S	D	E	H	1969	54059
1273	S	D	E	H	1969	54059
1277	S	D	O	N	1967	44894
1277	S	D	E	N	1933	60786
1277	S	D	C	N	1974	76085
1277	S	D	J	N	1978	92614
1280	S	D	E	N	1963	58889
1281	S	D	E	N	1955	20062
1281	S	D	G	H	1964	44351
1281	S	D	E	F	1963	54415
1281	S	D	E	H	1933	55938
1282	S	S	E	N	1962	52795
1285	S	D	G	H	1968	53727
1285	S	D	R	H	1970	61529
1285	S	D	E	H	1970	61530
1287	S	D	E	N	1934	43127
1287	S	D	E	N	1941	45181
1287	S	D	E	N	1940	46430
1287	S	D	E	H	1940	47367
1287	S	D	E	N	1942	49821
1287	S	D	E	N	1944	55475
1289	S	D	R	H	1969	57827
1289	S	D	E	H	1969	57828
1290	S	D	O	N	1969	58222
1290	S	D	G	N	1971	64254
1291	S	S	E	N	1942	59580
1292	S	D	E	N	1938	58611
1293	S	D	E	N	1932	57548
1293	S	D	E	N	1931	58531
1294	S	D	E	N	1941	51176
1294	S	D	E	N	1969	55331
1294	S	D	R	N	1956	55416
1294	S	D	R	N	1969	56333
1294	S	T	G	N	1969	56820
1294	S	D	E	N	1935	60446
1294	S	D	E	N	1956	63075
1294	S	D	G	N	1972	81966
1296	S	D	E	N	1931	42029
1296	S	D	E	N	1931	42418
1296	S	D	E	H	1931	42476
1296	S	D	E	N	1934	43126
1296	S	D	E	N	1946	55621
1296	S	D	P	N	1969	55868
1296	S	D	E	N	1970	58547
1296	S	D	E	N	1927	60437
1297	S	D	E	N	1970	58547
1298	S	S	E	N	1960	24931
1298	S	D	E	N	1970	58547
1299	S	D	E	N	1945	59786
1300	S	D	E	N	1945	59786
1301	S	D	E	H	1943	59052
1302	S	S	E	N	1947	60371
1303	S	D	E	N	1937	60406
1303	S	D	E	N	1937	60408
1304	S	D	E	N	1928	60441
1305	S	D	E	N	1928	60439
1306	S	S	G	N	1928	28680
1306	S	D	E	N	1931	42418
1306	S	D	E	N	1939	44506
1306	S	D	G	N	1959	62278
1306	S	D	G	N	1971	64254
1306	S	S	E	N	1967	89315
1307	S	D	E	N	1941	36248
1308	S	D	E	N	1939	45258
1309	S	S	G	N	1928	28680
1309	S	D	G	N	1959	62278
1309	S	D	G	N	1961	62282
1309	S	D	E	N	1973	69346
1310	S	D	E	H	1940	47367
1312	S	D	E	N	1971	61912

Substance Number	Phys. State	Subject	Language	Temperature	Year	TPRC Number
503-1313	D	D	E	N	1955	77324
1313	S	D	E	N	1947	36715
1314	S	D	G	N	1971	64254
1318	S	D	J	H	1965	44605
1319	S	D	J	H	1965	44606
1320	S	D	R	N	1965	46455
1323	S	D	P	N	1969	55868
1324	S	D	P	N	1970	65195
1325	S	D	F	N	1967	50734
1325	S	E	F	N	1967	50734
1327	S	D	R	N	1970	62319
1327	S	D	E	N	1970	66788
1328	S	D	R	N	1968	50006
1328	S	D	E	N	1972	66504
1329	S	D	E	N	1919	71468
1329	S	E	E	N	1919	71468
1330	P	D	O	H	1972	67719
1336	S	D	E	N	1974	74511
1336	S	E	E	N	1974	74511
1337	S	D	E	N	1973	74556
1338	S	D	E	N	1973	74556
1339	S	D	E	N	1973	74556
1341	S	D	E	N	1967	57594
1343	S	D	E	N	1964	77313
1345	S	D	J	H	1977	89029
1345	S	D	J	H	1977	89033
1345	S	D	J	N	1978	92619
1346	S	D	G	N	1976	89038
1347	S	S	G	N	1928	28680
1348	S	S	G	N	1928	28680
1349	S	S	G	N	1928	28680
1350	S	S	G	H	1928	28680
1350	S	D	E	N	1896	30723
1350	S	E	E	N	1896	30723
1352	S	D	E	N	1966	41727
1357	F	D	J	N	1969	55345
1357	S	D	E	N	1964	37777
1357	S	D	E	N	1951	60526
1369	S	D	R	N	1967	58543
1386	S	S	E	N	1960	24931
1387	S	D	J	N	1959	23128
1387	S	S	E	N	1960	24931
1387	S	S	E	N	1977	88199
1388	S	S	E	N	1960	24931
1388	S	D	E	H	1963	52805
1388	S	S	E	N	1977	92163
1391	S	D	E	N	1960	52654
1395	D	D	E	N	1967	35230
1395	S	D	E	N	1969	60104
1396	S	D	E	N	1967	35230
1398	S	G	G	N	1959	62301
1405	S	D	R	N	1966	41499
1405	S	S	R	N	1973	79859
1405	S	S	E	N	1973	79860
1412	S	D	G	N	1967	49008
1417	S	D	E	N	1966	40468
1417	S	D	E	N	1968	47870
1418	S	D	E	N	1966	40468
1418	S	D	E	N	1968	47870
1426	S	D	G	N	1961	60772
1427	S	D	E	N	1954	43335
1429	S	D	E	N	1970	37248
1429	S	D	E	N	1971	67411
1431	S	D	E	N	1949	35817
1431	S	T	E	N	1949	35817
1431	S	D	E	N	1941	40392
1431	S	D	J	N	1955	43250
1431	S	T	J	N	1955	43250
1431	S	D	E	N	1964	44122
1431	S	S	E	N	1949	60524
1432	S	D	E	N	1969	40513
1432	S	D	O	N	1966	41369
1433	S	D	G	N	1966	41958
1433	S	T	G	N	1966	41958
1435	S	D	R	N	1967	53157
1435	S	D	R	N	1967	53471
1435	S	D	E	N	1920	59946
1437	S	D	E	N	1920	59946
1441	S	D	E	N	1970	67661
1441	S	T	E	N	1970	67661
1443	S	D	G	N	1966	49064
1445	S	D	E	N	1971	63695
1446	S	D	E	N	1971	67411
1454	S	D	E	N	1974	74507
1454	S	S	E	N	1977	93644
1454	S	S	E	N	1978	96027
1456	S	D	E	N	1975	76616
1457	S	D	E	N	1975	76616
1462	S	S	R	N	1970	77553
1462	S	S	E	N	1974	77554
1465	S	S	G	N	1928	28680
1466	S	D	E	N	1960	10947
1466	S	E	E	N	1960	10947
1466	S	S	E	L	1960	24931
1466	S	S	E	N	1960	24931
1467	P	D	O	H	1972	67719
1467	S	S	E	N	1960	24931
1467	S	D	E	N	1963	37530
1467	S	D	E	N	1931	42418
1467	S	D	E	N	1948	43305
1467	S	D	E	N	1952	52749

Phys. State: **C.** Superconductive; **D.** Doped; **E.** Expanded; **F.** Fibrous or Whisker; **G.** Gas; **I.** Ionized or Plasma; **L.** Liquid; **M.** Multiphase; **P.** Powder or Fine Particle; **S.** Solid; **T.** Thick or Thin Film
Subject: **D.** Data; **E.** Experiment; **G.** General (Data + Theory + Experiment); **S.** Survey (Review, Compendium etc.); **T.** Theory
Language: **C.** Czech; **D.** Dutch; **E.** English; **F.** French; **G.** German; **I.** Italian; **J.** Japanese; **O.** Other Languages; **P.** Polish; **R.** Russian; **S.** Spanish
Temperature: **F.** Full Range (Low + Normal + High); **L.** Low (0 to 75K) + Overlap into Normal; **N.** Normal (75 to 1273K); **H.** High (above 1273K) + Overlap into Normal; Blank = Not Coded

Substance Number	Phys. State	Subject	Language	Temperature	Year	TPRC Number
503-1467	S	D	E	N	1928	60439
1467	S	D	R	N	1969	61595
1467	S	D	E	N	1969	61596
1467	S	D	G	N	1971	64254
1467	S	S	E	N	1967	89315
1468	S	S	E	N	1960	24931
1468	S	D	E	N	1948	43305
1468	S	D	E	H	1947	48002
1468	S	S	E	N	1967	89315
1471	S	D	E	N	1954	44149
1471	S	D	E	N	1945	59786
1484	S	D	R	N	1971	62599
1484	S	D	E	N	1971	66484
1485	S	D	R	N	1971	62599
1485	S	D	E	N	1971	66484
1486	S	D	R	N	1971	62599
1486	S	D	E	N	1971	66484
1487	S	D	R	N	1971	62599
1487	S	D	E	N	1971	66484
1488	S	D	R	N	1966	41499
1491	S	D	R	N	1966	41499
1492	S	D	R	N	1968	53472
1492	S	D	E	N	1968	67357
1499	S	D	R	N	1968	53472
1499	S	D	E	N	1968	67357
1500	S	D	R	N	1968	53472
1500	S	D	E	N	1968	67357
1501	S	D	R	N	1968	53472
1501	S	D	E	N	1968	67357
1502	S	D	R	N	1968	53472
1502	S	D	E	N	1968	67357
1503	S	D	R	N	1968	53472
1503	S	D	E	N	1968	67357
1504	S	D	R	N	1968	53472
1504	S	D	E	N	1968	67357
1505	S	D	R	N	1968	53472
1505	S	D	E	N	1968	67357
1506	S	D	R	N	1968	53472
1506	S	D	E	N	1968	67357
1507	S	D	R	N	1971	62599
1507	S	D	E	N	1971	66484
1525	S	D	R	N	1969	56888
1526	S	D	R	N	1969	56888
1527	S	D	R	N	1969	56888
1529	S	D	E	N	1961	46148
1535	S	D	E	N	1954	43335
1536	S	D	E	N	1934	43056
1537	S	D	E	N	1934	43056
1550	S	D	R	N	1968	53472
1550	S	D	E	N	1968	67357
1551	S	D	R	N	1968	53472
1551	S	D	R	N	1971	62599
1551	S	D	E	N	1971	66484
1551	S	D	E	N	1968	67357
1552	S	D	R	N	1968	53472
1552	S	D	E	N	1968	67357
1553	S	D	R	N	1969	56888
1555	S	D	R	N	1968	53472
1555	S	D	E	N	1968	67357
1556	S	D	E	N	1969	40513
1556	S	D	R	N	1966	41369
1557	S	D	R	N	1966	41499
1559	S	D	E	N	1969	40513
1559	S	D	R	N	1966	41369
1559	S	D	R	N	1968	53472
1559	S	D	E	N	1968	67357
1563	S	D	R	N	1966	41499
1563	S	D	R	N	1968	53472
1563	S	D	E	N	1968	67357
1570	S	D	E	L	1965	46615
1571	S	D	E	N	1968	38081
1571	S	D	R	N	1968	38898
1571	S	D	R	N	1967	53154
1572	S	D	R	N	1967	47173
1572	S	G	R	N	1967	53739
1572	S	D	E	N	1969	56358
1573	S	D	R	N	1967	53154
1575	S	D	R	N	1967	53153
1575	S	T	R	N	1967	53153
1576	S	D	R	N	1968	50115
1576	S	D	R	N	1967	53149
1576	S	G	R	N	1967	53739
1576	S	D	E	N	1968	57289
1577	S	D	R	N	1971	62599
1577	S	D	E	N	1971	66484
1578	S	D	R	N	1967	53153
1578	S	T	R	N	1967	53153
1579	S	D	R	N	1971	62599
1579	S	D	E	N	1971	66484
1580	S	D	R	N	1968	53472
1580	S	D	E	N	1968	67357
1581	S	D	R	N	1971	62599
1581	S	D	E	N	1971	66484
1582	S	D	R	N	1967	53153
1582	S	T	R	N	1967	53153
1592	S	D	E	N	1939	44506
1619	S	D	E	N	1976	85762
1654	S	D	G	N	1963	62287
1694	S	D	E	N	1974	77552
1697	S	S	E	N	1960	24931
1697	S	D	E	N	1930	37544

Substance Number	Phys. State	Subject	Language	Temperature	Year	TPRC Number
503-1697	S	S	E	N	1967	89315
1709	S	D	R	N	1977	91533
1713	S	S	E	N	1960	24931
1715	S	S	E	N	1960	24931
1721	S	D	E	N	1931	37634
1723	S	D	E	N	1931	37634
1742	S	D	E	N	1976	85762
1743	S	D	E	N	1976	85762
1745	S	S	E	N	1960	24931
1750	S	D	E	N	1977	91585
1751	S	D	E	N	1977	91585
1752	S	D	E	N	1977	91585
1753	S	D	E	N	1977	91585
1754	S	D	E	N	1977	91585
1755	S	D	E	N	1977	91585
1756	S	D	E	N	1977	91585
1764	S	D	E	N	1967	77589
1765	S	D	E	H	1967	77589
1765	S	D	E	N	1967	77589
1765	S	D	E	N	1964	91250
1766	S	D	R	N	1976	88314
1766	S	D	E	N	1976	91214
1769	S	D	E	N	1969	63005
1770	S	D	E	N	1969	63005
1771	S	D	E	N	1969	63005
1772	S	D	E	N	1969	63005
1773	S	D	E	N	1969	63005
1774	S	D	E	N	1969	63005
1801	S	D	J	H	1977	89033
1802	S	D	J	H	1977	89033
1806	S	D	E	N	1974	79928
1807	S	D	E	H	1977	90800
1807	S	D	E	N	1977	90800
1808	S	D	E	N	1973	92934
1809	S	D	E	N	1973	92934
1810	S	D	J	N	1977	94558
1811	S	D	J	N	1977	94558
1812	S	D	J	N	1978	92614
1812	S	D	J	N	1977	92618
1812	S	D	J	N	1978	92619
1813	S	D	J	N	1978	92614
1813	S	D	J	N	1977	92618
1813	S	D	J	N	1978	92619
1815	S	D	R	N	1967	48703
1815	S	D	E	N	1967	49415
1822	S	D	J	N	1978	92614
1827	S	S	E	H	1976	94271
1827	S	S	E	N	1976	94271
1828	S	D	E	N	1967	35986
1828	S	D	E	N	1970	58679
1828	S	D	E	N	1970	59834
1828	S	D	E	N	1969	60601
1828	S	D	E	L	1970	63400
1828	S	D	E	L	1973	72981
1828	S	D	E	L	1974	74521
1828	S	D	E	N	1974	74522
1828	S	E	E	N	1974	74522
1828	S	D	F	N	1968	76887
1828	S	D	E	L	1976	87010
1828	S	D	E	N	1976	87010
1828	S	D	E	N	1976	87429
1828	S	S	E	N	1976	88411
1828	S	D	E	N	1977	88904
1828	S	E	E	N	1977	88904
1828	S	S	E	N	1979	97587
1833	S	S	E	N	1960	24931
1833	S	D	R	H	1974	83593
1833	S	D	R	N	1974	83593
1833	S	D	E	H	1974	83594
1833	S	D	E	N	1974	83594
1835	S	D	R	N	1968	53472
1835	S	D	E	N	1968	67357
1838	S	D	E	H	1976	87952
1838	S	D	E	N	1976	87052
1838	S	D	E	N	1977	92163
1838	S	S	E	N	1977	92163
1838	S	D	E	N	1979	99593
1839	S	S	E	N	1977	92163
1840	S	S	E	N	1977	92163
1841	S	S	E	N	1977	92163
1842	S	S	E	N	1977	92163
1843	S	S	E	N	1977	92163
1844	S	S	E	N	1977	92163
1845	S	S	E	N	1977	92163
1846	S	S	E	N	1977	92163
1847	S	S	E	N	1977	92163
1848	S	S	E	N	1977	92163
1849	S	S	E	N	1977	92163
1850	S	S	E	N	1977	92163
1851	S	S	E	N	1977	92163
1852	S	S	E	N	1977	92163
1853	S	S	E	N	1977	92163
1854	S	S	E	N	1977	92163
1855	S	S	E	N	1977	92163
1856	S	S	E	N	1977	92163
1857	S	S	E	N	1977	92163
1858	S	S	E	N	1960	24931
1858	S	S	E	N	1977	92163
1859	S	S	E	N	1977	92163
1860	S	S	E	H	1960	24931
1860	S	S	E	N	1960	24931

Substance Number	Phys. State	Subject	Language	Temperature	Year	TPRC Number
503-1860	S	S	E	N	1977	92163
1861	S	S	E	N	1977	92163
1862	S	D	E	N	1974	88682
1863	S	D	E	N	1974	88682
1864	S	D	E	N	1974	88682
1867	S	D	E	N	1960	45778
1867	S	D	E	N	1964	91250
1868	S	S	E	N	1975	88036
1868	S	D	E	N	1964	91250
1869	D	D	E	H	1967	78350
1869	D	D	E	N	1967	78350
1870	D	D	E	H	1967	78350
1870	D	D	E	N	1967	78350
1871	S	D	E	N	1976	91261
1872	S	D	E	N	1976	91261
1873	S	D	E	N	1976	91261
1874	S	D	E	N	1976	91261
1875	S	D	E	N	1976	91261
1876	S	D	E	N	1976	91261
1877	S	S	E	H	1977	93565
1879	S	D	J	N	1975	89070
1880	S	D	J	N	1975	89070
1881	S	D	J	N	1976	89071
1882	S	D	J	N	1976	89071
1883	S	D	J	N	1976	89071
1884	S	D	J	N	1976	89071
1884	S	D	R	N	1978	99206
1887	S	D	E	N	1976	85762
1888	S	D	E	N	1976	85762
1889	S	D	E	N	1976	85762
1890	S	D	E	N	1976	85762
1891	S	D	E	N	1976	85762
1892	S	D	E	N	1976	85762
1893	S	D	E	N	1976	85762
1894	S	D	E	N	1976	85762
1895	S	D	E	N	1976	85762
1896	S	D	E	N	1976	85762
1897	S	D	E	N	1976	85762
1898	S	D	E	N	1976	85762
1899	S	D	E	N	1976	85762
1900	S	D	E	N	1976	85762
1901	S	D	E	N	1976	85762
1902	S	D	E	N	1976	85762
1903	S	D	E	N	1976	85762
1904	S	D	E	N	1976	85762
1905	S	D	E	N	1976	85762
1906	S	D	E	N	1976	85762
1907	S	D	E	N	1976	85762
1908	S	D	E	N	1976	85762
1909	S	D	E	N	1976	85762
1910	S	D	E	N	1976	85762
1911	S	D	E	N	1976	85762
1912	S	D	R	N	1968	56202
1912	S	D	E	N	1976	85762
1912	S	S	E	N	1967	89315
1913	S	D	E	N	1976	85762
1914	S	D	E	N	1976	85762
1915	S	D	E	N	1976	85762
1916	S	S	E	N	1960	24931
1916	S	D	E	N	1976	85762
1917	S	D	E	N	1976	85762
1918	S	D	R	N	1969	60905
1919	S	D	R	N	1967	53762
1920	S	D	R	N	1967	53762
1921	S	D	R	N	1967	53762
1922	S	D	R	N	1967	54933
1923	S	D	R	N	1968	56200
1924	S	S	E	N	1960	24931
1924	S	D	R	N	1968	56200
1925	S	D	R	N	1968	56200
1926	S	D	R	N	1968	56200
1927	S	D	R	N	1968	56200
1928	S	D	R	N	1968	56200
1020	S	D	R	N	1968	56200
1930	S	D	R	N	1968	56202
1933	S	S	E	N	1960	24931
1935	S	D	R	N	1969	58996
1936	S	D	R	N	1969	58996
1944	S	S	E	N	1960	24931
1944	S	D	E	N	1960	45778
1945	S	S	E	H	1960	24931
1945	S	S	E	N	1960	24931
1953	D	D	E	N	1971	61349
1954	S	S	E	N	1960	24931
1954	S	D	E	N	1952	52759
1955	S	S	E	N	1960	24931
1955	S	D	E	N	1960	45513
1955	S	D	E	N	1952	52759
1960	S	D	E	N	1976	87561
1961	S	D	E	N	1976	87015
1961	S	D	E	N	1977	93507
1987	S	D	E	N	1921	22633
1987	S	E	E	N	1921	22633
1993	S	D	J	N	1958	23478
2000	S	D	R	H	1970	60843
2000	S	D	R	N	1970	60843
2000	S	D	E	H	1970	60844
2000	S	D	E	N	1970	60844
2000	S	S	E	H	1977	88199
2000	S	S	E	N	1977	88199
2030	S	D	J	N	1958	23478

Phys. State: **C.** Superconductive; **D.** Doped; **E.** Expanded; **F.** Fibrous or Whisker; **G.** Gas; **I.** Ionized or Plasma; **L.** Liquid; **M.** Multiphase; **P.** Powder or Fine Particle; **S.** Solid; **T.** Thick or Thin Film

Subject: **D.** Data; **E.** Experiment; **G.** General (Data + Theory + Experiment); **S.** Survey (Review, Compendium etc.); **T.** Theory

Language: **C.** Czech; **D.** Dutch; **E.** English; **F.** French; **G.** German; **I.** Italian; **J.** Japanese; **O.** Other Languages; **P.** Polish; **R.** Russian; **S.** Spanish

Temperature: **F.** Full Range (Low + Normal + High); **L.** Low (0 to 75K) + Overlap into Normal; **N.** Normal (75 to 1273K); **H.** High (above 1273K) + Overlap into Normal; Blank = Not Coded

Substance Number	Phys. State	Subject	Language	Temperature	Year	TPRC Number
503-2031	S	D	J	N	1958	23478
2071	S	D	E	N	1930	37544
2086	S	S	E	H	1960	24931
2086	S	S	E	N	1960	24931
2086	S	D	E	N	1960	45778
2087	S	D	E	N	1958	15895
2088	S	D	E	N	1958	15895
2089	S	D	E	N	1958	15895
2090	S	D	E	N	1958	15895
2091	S	D	E	N	1958	15895
2092	S	D	E	N	1958	15895
2105	S	S	E	N	1967	89315
2119	S	S	E	N	1967	89315
2127	S	S	G	N	1959	17892
2127	S	S	E	N	1962	29103
2142	S	D	J	N	1977	94558
2154	S	D	E	N	1963	52770
2154	S	D	E	N	1963	52788
2154	S	S	E	H	1977	88199
2154	S	S	E	N	1977	88199
2159	S	D	G	N	1963	62287
2160	S	S	E	H	1960	24931
2160	S	S	E	N	1960	24931
2160	S	D	E	N	1964	52271
2169	S	S	E	N	1960	24931
2170	S	D	E	N	1930	37544
2179	S	D	R	N	1967	55628
2184	S	D	E	N	1960	45778
2185	S	D	E	N	1960	45778
2186	S	D	E	N	1960	45778
2187	S	D	E	N	1960	45778
2188	S	D	E	N	1960	45778
2189	S	D	E	N	1960	45778
2202	S	S	E	N	1967	89315
2229	S	D	R	N	1968	56201
2250	E	D	E	N	1959	41863
2251	S	D	R	N	1967	45799
2253	S	D	R	N	1971	62599
2253	S	D	E	N	1971	66484
2258	S	S	E	N	1960	24931
2258	S	D	E	N	1976	90099
2259	S	D	E	N	1976	90099
2267	S	D	C	N	1963	34151
2267	S	D	F	N	1965	46017
2268	S	D	C	N	1963	34151
2268	S	D	F	N	1965	46017
2269	S	D	C	N	1963	34151
2269	S	D	F	N	1965	46017
2270	S	D	C	N	1963	34151
2270	S	D	F	N	1965	46017
2271	S	D	C	N	1963	34151
2271	S	D	F	N	1965	46017
2272	S	D	C	N	1963	34151
2272	S	D	F	N	1965	46017
2274	S	S	R	N	1973	79859
2274	S	S	E	N	1973	79860
2275	S	S	E	N	1967	89315
2276	S	S	E	N	1967	89315
2277	S	S	E	N	1967	89315
2278	S	S	E	N	1967	89315
2279	S	D	E	N	1962	45217
2280	S	D	E	N	1962	45217
2281	S	D	E	N	1962	45217
2282	S	D	E	N	1962	45217
2283	S	D	E	N	1962	45217
2284	S	D	R	N	1968	53472
2284	S	D	E	N	1968	67357
2285	S	D	R	N	1968	53472
2285	S	D	E	N	1968	67357
2286	S	D	R	N	1968	53472
2286	S	D	E	N	1968	67357
2287	S	D	R	N	1969	56888
2288	S	D	R	N	1969	56888
2292	S	D	R	N	1971	62599
2292	S	D	E	N	1971	66484
2296	E	D	R	N	1965	41976
2297	S	D	R	N	1977	91533
2303	S	D	R	N	1967	48703
2303	S	D	E	N	1967	49415
2303	S	D	E	N	1978	98331
2304	S	D	E	N	1978	98331
2308	S	D	R	H	1977	91732
2308	S	D	R	N	1977	91732
2308	S	D	E	H	1977	96251
2308	S	D	E	N	1977	96251
2309	D	D	R	H	1977	91732
2309	D	D	R	N	1977	91732
2309	D	D	E	H	1977	96251
2309	D	D	E	N	1977	96251
2309	S	S	E	H	1960	24931
2309	S	S	E	N	1960	24931
2309	S	S	E	N	1977	88199
2312	S	D	E	N	1960	16732
2326	S	D	E	N	1978	97376
2327	S	S	E	N	1960	24931
2327	S	D	E	N	1978	97376
2328	S	D	R	N	1977	91533
2347	S	S	E	N	1979	100149
2348	S	D	R	L	1972	67805
2348	S	D	R	N	1972	67805
2348	S	E	R	L	1972	67805

Substance Number	Phys. State	Subject	Language	Temperature	Year	TPRC Number
503-2348	S	E	R	N	1972	67805
2348	S	D	E	L	1972	79601
2348	S	D	E	N	1972	79601
2348	S	E	E	L	1972	79601
2348	S	E	E	N	1972	79601
2350	S	S	E	N	1978	99742
2351	S	S	E	N	1978	99742
2352	S	S	E	N	1978	99742
2380	S	D	R	H	1960	19081
2380	S	D	R	N	1960	19081
2380	S	D	E	H	1960	32629
2380	S	D	E	N	1960	32629
2382	S	D	E	N	1974	77552
2383	S	S	E	N	1960	24931
2385	S	D	E	N	1974	77552
2386	S	D	E	N	1974	77552
2387	S	D	E	N	1974	77552
2390	S	D	G	N	1963	62287
2391	S	D	E	N	1965	40714
2407	S	S	E	N	1960	24931
2416	S	D	E	N	1930	37544
2417	S	D	R	N	1969	56962
2418	S	D	R	N	1969	56962
2419	S	D	R	N	1969	56962
2420	S	D	G	N	1914	66628
2421	S	D	R	N	1966	49853
2422	S	D	R	N	1966	49853
2423	P	D	R	N	1969	58287
2424	P	D	R	N	1969	58287
2425	S	D	R	H	1970	60843
2425	S	D	R	N	1970	60843
2425	S	D	E	H	1970	60844
2426	S	D	O	N	1961	39755
2427	S	D	R	H	1968	54777
2427	S	D	R	N	1968	54777
2427	S	D	E	H	1968	54778
2427	S	D	E	N	1968	54778
2428	S	D	R	H	1968	54777
2428	S	D	R	N	1968	54777
2428	S	D	E	H	1968	54778
2428	S	D	E	N	1968	54778
2429	S	D	R	N	1964	46214
2431	S	D	R	N	1967	46274
2431	S	D	E	N	1967	46275
2432	S	D	R	N	1967	48703
2432	S	D	E	N	1967	49415
2433	S	D	E	N	1961	24763
2434	S	D	E	N	1961	24763
2434	S	D	E	N	1962	31014
2437	S	S	E	N	1960	24931
2437	S	D	R	H	1970	60843
2437	S	D	R	N	1970	60843
2437	S	D	E	H	1970	60844
2437	S	D	E	N	1970	60844
2437	S	S	E	N	1977	88199
2441	S	D	J	N	1960	52431
2461	S	D	E	N	1921	22633
2461	S	E	E	N	1921	22633
2463	S	D	R	H	1968	54777
2463	S	D	R	N	1968	54777
2463	S	D	E	H	1968	54778
2463	S	D	E	N	1968	54778
2478	S	D	E	N	1921	22633
2478	S	E	E	N	1921	22633
2479	S	D	E	N	1921	22633
2479	S	E	E	N	1921	22633
2489	S	D	E	N	1961	24763
2489	S	D	E	N	1962	31014
2490	S	D	E	N	1961	24763
2514	S	D	E	N	1962	31014
2525	S	D	E	N	1930	37544
2526	S	D	E	N	1930	37544
2527	S	D	E	N	1930	37544
2528	S	D	E	N	1930	37544
2529	S	D	E	N	1930	37544
2530	S	D	E	N	1930	37544
2531	S	D	E	N	1930	37544
2532	S	D	E	N	1931	37634
2533	S	D	E	N	1931	37634
2534	S	D	O	N	1961	39755
2535	S	D	O	N	1961	39755
2536	S	D	O	N	1961	39755
2537	S	D	O	N	1961	39755
2538	S	D	O	N	1961	39755
2539	S	D	O	N	1961	39755
2540	S	D	O	N	1961	39755
2541	S	D	O	N	1961	39755
2542	S	D	O	N	1961	39755
2544	S	S	E	N	1960	24931
2544	S	S	E	N	1977	88199
2574	S	D	E	N	1960	41838
2575	S	D	E	N	1960	41838
2576	S	D	E	N	1960	41838
2577	S	D	E	N	1960	41838
2578	S	D	E	N	1960	41838
2579	S	D	E	N	1939	44506
2586	S	D	R	N	1966	46593
2586	S	D	E	N	1966	46594
2586	S	D	R	H	1968	54777
2586	S	D	R	N	1968	54777
2586	S	D	E	H	1968	54778

Substance Number	Phys. State	Subject	Language	Temperature	Year	TPRC Number
503-2586	S	D	E	N	1968	54778
2588	S	D	G	N	1967	47477
2588	S	T	G	N	1967	47477
2589	S	D	R	N	1967	48703
2589	S	D	E	N	1967	49415
2590	S	D	R	N	1967	48703
2590	S	D	E	N	1967	49415
2591	S	D	R	N	1966	49853
2592	S	D	I	N	1968	52947
2593	S	D	I	N	1968	52947
2594	S	D	R	N	1967	53156
2597	S	D	R	H	1968	54777
2597	S	D	R	N	1968	54777
2597	S	D	E	H	1968	54778
2597	S	D	E	N	1968	54778
2598	S	D	R	N	1968	56892
2598	S	D	E	N	1971	64048
2599	S	D	R	N	1968	56892
2599	S	D	E	N	1971	64048
2600	S	D	R	N	1969	56962
2601	S	D	R	N	1969	56962
2602	S	D	R	N	1969	56962
2603	S	D	R	N	1969	56962
2604	S	D	R	N	1969	56962
2605	P	D	R	N	1969	58287
2606	P	D	R	N	1969	58287
2607	P	D	R	N	1969	58287
2608	P	D	R	N	1969	58287
2609	P	D	R	N	1969	58287
2610	P	D	R	N	1969	58287
2611	P	D	R	N	1969	58287
2612	P	D	R	N	1969	58287
2613	P	D	R	N	1969	58287
2614	P	D	R	N	1969	58287
2615	P	D	R	N	1969	58287
2616	P	D	R	N	1969	58287
2617	P	D	R	N	1969	58287
2618	P	D	R	N	1969	58287
2619	P	D	R	N	1969	58287
2620	P	D	R	N	1969	58287
2621	P	D	R	N	1969	58287
2622	P	D	R	N	1969	58287
2623	P	D	R	N	1969	58287
2631	S	D	G	N	1963	62287
2632	S	D	G	N	1963	62287
2633	S	D	G	N	1963	62287
2634	S	D	G	N	1963	62287
2635	S	D	G	N	1963	62287
2636	S	D	G	N	1963	62287
2637	S	D	G	N	1963	62287
2638	S	D	G	N	1963	62287
2640	S	D	G	N	1914	66628
2641	S	D	G	N	1914	66628
2642	S	D	G	N	1914	66628
2643	S	D	E	N	1973	76782
2643	S	T	E	N	1973	76782
2646	S	D	E	N	1974	77552
2647	S	D	E	N	1974	77552
2648	S	D	E	N	1974	77552
2649	S	D	E	N	1974	77552
2650	S	D	E	N	1974	77552
2651	S	D	E	N	1974	77552
2652	S	D	E	N	1974	77552
2653	S	D	E	N	1974	77552
2654	S	D	E	N	1974	77552
2655	S	D	E	N	1974	77552
2656	S	D	E	N	1974	77552
2657	S	D	E	N	1974	77552
2658	S	D	E	N	1974	77552
2663	P	D	P	N	1969	56205
2670	S	D	E	N	1976	90099
2671	S	D	E	N	1974	79605
2672	S	D	R	H	1976	84581
2672	S	D	E	H	1976	91876
2673	S	D	R	H	1976	84581
2673	S	D	E	H	1976	91876
2674	S	D	E	N	1976	83382
2675	D	D	E	N	1976	83382
2675	S	D	E	N	1976	83382
2676	D	D	E	N	1976	83382
2677	S	D	E	N	1976	83382
2678	D	D	E	N	1976	83382
2678	S	D	E	N	1976	83382
2707	S	S	E	N	1960	24931
2715	S	S	E	N	1960	24931
2724	E	S	E	N	1979	100149
2725	E	S	E	N	1979	100149
2726	E	S	E	N	1979	100149
2727	E	S	E	N	1979	100149
2728	E	S	E	N	1979	100149
2729	E	S	E	N	1979	100149
2785	S	D	R	N	1978	99206
2786	S	D	R	N	1979	99207
2787	S	D	R	N	1979	99207
2788	S	D	R	N	1979	99207
2795	S	S	E	N	1960	24931
2800	S	D	E	N	1979	99265
2801	S	D	E	N	1979	99265
2802	S	S	E	N	1979	99434
2803	S	S	E	N	1979	99434
2805	S	D	R	N	1964	46214

Phys. State: **C.** Superconductive; **D.** Doped; **E.** Expanded; **F.** Fibrous or Whisker; **G.** Gas; **I.** Ionized or Plasma; **L.** Liquid; **M.** Multiphase; **P.** Powder or Fine Particle; **S.** Solid; **T.** Thick or Thin Film

Subject: **D.** Data; **E.** Experiment; **G.** General (Data + Theory + Experiment); **S.** Survey (Review, Compendium etc.); **T.** Theory

Language: **C.** Czech; **D.** Dutch; **E.** English; **F.** French; **G.** German; **I.** Italian; **J.** Japanese; **O.** Other Languages; **P.** Polish; **R.** Russian; **S.** Spanish

Temperature: **F.** Full Range (Low + Normal + High); **L.** Low (0 to 75K) + Overlap into Normal; **N.** Normal (75 to 1273K); **H.** High (above 1273K) + Overlap into Normal; Blank = Not Coded

Substance Number	Phys. State	Sub-ject	Lan-guage	Temper-ature	Year	TPRC Number
503-2807	S	D	E	N	1962	31014
2830	S	S	E	N	1977	88199
2831	S	S	E	N	1977	88199
2832	S	S	E	N	1977	88199
2833	S	S	E	N	1960	24931
2833	S	S	E	N	1977	88199
2834	S	S	E	N	1977	88199
2835	S	S	E	N	1977	88199
2836	S	S	E	N	1977	88199
2837	S	S	E	N	1977	88199
2838	S	S	E	N	1977	88199
2839	S	S	E	N	1977	88199
2840	S	S	E	N	1977	88199
2841	S	S	E	N	1977	88199
2842	S	S	E	N	1977	88199
2843	S	S	E	N	1977	88199
2844	S	S	E	N	1977	88199
2845	S	S	E	N	1977	88199
2846	S	S	E	N	1977	88199
2847	S	S	E	N	1977	88199
2848	S	S	E	N	1977	88199
2849	S	S	E	N	1977	88199
2850	S	S	E	N	1977	88199
2851	S	S	E	N	1977	88199
2852	S	S	E	N	1977	88199
2853	S	S	E	N	1977	88199
2854	S	S	E	N	1977	88199
2855	S	S	E	N	1977	88199
2856	S	S	E	N	1977	88199
2857	S	S	E	N	1977	88199
2861	S	D	J	N	1959	23128
2861	S	S	E	N	1960	24931
2877	S	D	R	N	1978	97669
2877	S	D	E	N	1978	98577
2879	S	S	E	N	1960	24931
2880	S	S	E	N	1960	24931
2881	S	S	E	N	1960	24931
2882	S	S	E	N	1960	24931
2883	S	S	E	N	1960	24931
2884	S	S	E	H	1960	24931
2884	S	S	E	N	1960	24931
2887	S	S	E	N	1960	24931
2888	S	S	E	N	1960	24931
2889	S	S	E	N	1960	24931
2890	S	S	E	N	1960	24931
2891	S	S	E	N	1960	24931
2892	S	S	E	H	1960	24931
2892	S	S	E	N	1960	24931
2893	S	S	E	N	1960	24931
2894	S	S	E	H	1960	24931
2894	S	S	E	N	1960	24931
2895	S	S	E	H	1960	24931
2895	S	S	E	N	1960	24931
2896	S	S	E	N	1960	24931
2897	S	S	E	H	1960	24931
2897	S	S	E	N	1960	24931
2898	S	S	E	H	1960	24931
2898	S	S	E	N	1960	24931
2899	S	S	E	H	1960	24931
2899	S	S	E	N	1960	24931
2901	S	S	E	N	1960	24931
2902	S	S	E	N	1960	24931
2903	S	S	E	N	1960	24931
2904	S	S	E	N	1960	24931
2905	S	S	E	N	1960	24931
2906	S	S	E	N	1960	24931
2907	S	S	E	N	1960	24931
2908	S	S	E	N	1960	24931
2909	S	S	E	N	1960	24931
2910	S	S	E	N	1960	24931
2911	S	S	E	N	1960	24931
2912	S	S	E	N	1960	24931
2913	S	S	E	N	1960	24931
2914	S	S	E	N	1960	24931
2915	S	S	E	N	1960	24931
2916	S	S	E	N	1960	24931
2917	S	S	E	N	1960	24931
2918	S	S	E	N	1960	24931
2919	S	S	E	H	1960	24931
2919	S	S	E	N	1960	24931
2920	S	S	E	N	1960	24931
2921	S	S	E	N	1960	24931
2922	S	S	E	N	1960	24931
2923	S	S	E	N	1960	24931
2924	S	S	E	N	1960	24931
2925	S	S	E	N	1960	24931
2926	S	S	E	N	1960	24931
2927	S	S	E	N	1960	24931
2928	S	S	E	N	1960	24931
2929	S	S	E	N	1960	24931
2931	S	S	E	N	1960	24931
2932	S	S	E	N	1960	24931
2933	S	S	E	N	1960	24931
2934	S	S	E	N	1960	24931
2935	S	S	E	N	1960	24931
2936	S	S	E	N	1960	24931
2937	S	S	E	N	1960	24931
2938	S	S	E	N	1960	24931
2939	S	S	E	N	1960	24931
2940	S	S	E	N	1960	24931
2941	S	S	E	N	1960	24931
503-2942	S	S	E	N	1960	24931
2943	S	S	E	N	1960	24931
2944	S	S	E	N	1960	24931
2945	S	S	E	N	1960	24931
2946	S	S	E	N	1960	24931
2947	S	S	E	N	1960	24931
2948	S	S	E	N	1960	24931
2949	S	S	E	N	1960	24931
2950	S	S	E	N	1960	24931
2951	S	S	E	N	1960	24931
2952	S	S	E	N	1960	24931
2953	S	S	E	N	1960	24931
2954	S	S	E	N	1960	24931
2955	S	S	E	N	1960	24931
2956	S	S	E	N	1960	24931
2957	S	S	E	N	1960	24931
2958	S	S	E	N	1960	24931
2959	S	S	E	N	1960	24931
2960	S	S	E	N	1960	24931
2961	S	S	E	N	1960	24931
2962	S	S	E	N	1960	24931
2963	S	S	E	N	1960	24931
2964	S	S	E	N	1960	24931
2965	S	S	E	N	1960	24931
2966	S	S	E	N	1960	24931
2967	S	S	E	N	1960	24931
2968	S	S	E	N	1960	24931
2969	S	S	E	N	1960	24931
2970	S	S	E	N	1960	24931
2971	S	S	E	N	1960	24931
2972	S	S	E	N	1960	24931
2973	S	S	E	N	1960	24931
2974	S	S	E	N	1960	24931
2975	S	S	E	N	1960	24931
2976	S	S	E	N	1960	24931
2977	S	S	E	N	1960	24931
2980	S	S	E	N	1960	24931
2981	S	S	E	N	1960	24931
2982	S	S	E	N	1961	44227
2983	S	S	E	N	1961	44227
2984	S	S	E	N	1961	44227
2985	S	S	E	N	1961	44227
2986	S	S	E	N	1961	44227
2987	S	S	E	N	1961	44227
2988	S	S	E	N	1961	44227
2989	S	S	E	N	1961	44227
2990	S	S	E	N	1961	44227
2991	S	S	E	N	1961	44227
2992	S	S	E	N	1961	44227
2993	S	S	E	N	1961	44227
2994	S	S	E	N	1961	44227
2995	S	S	E	N	1961	44227
8189	S	D	O	N	1977	92525
8194	S	D	E	N	1970	56159
8195	S	S	E	N	1977	88199
8200	S	D	E	N	1970	67661
8200	S	T	E	N	1970	67661
8204	S	D	R	N	1970	62319
8204	S	D	E	N	1970	66788
8206	S	D	R	N	1970	62319
8206	S	D	E	N	1970	66788
8207	S	D	R	N	1970	62319
8207	S	D	E	N	1970	66788
8208	S	D	R	N	1970	62319
8208	S	D	E	N	1970	66788
8209	S	D	R	N	1970	62319
8209	S	D	E	N	1970	66788
8210	S	D	R	N	1970	62319
8210	S	D	E	N	1970	66788
8211	S	D	R	N	1970	62319
8211	S	D	E	N	1970	66788
8212	S	D	R	N	1970	60753
8212	S	D	E	N	1970	66787
8213	S	D	R	N	1970	60753
8213	S	D	E	N	1970	66787
8214	S	D	R	N	1969	54745
8214	S	D	E	N	1969	66786
8215	S	D	R	N	1969	54745
8215	S	D	E	N	1969	66786
8223	S	D	R	N	1970	62314
8224	S	D	R	N	1970	62314
8227	S	D	R	H	1971	64840
8227	S	D	E	H	1971	64841
8228	S	D	R	H	1971	64830
8228	S	D	E	H	1971	64831
8228	S	D	R	H	1974	75192
8228	S	D	E	H	1974	76872
8230	S	S	E	N	1960	24931
8230	S	D	E	N	1959	44960
8231	S	D	E	N	1942	49821
8232	S	D	E	N	1942	49821
8234	S	D	G	N	1961	62283
8234	S	T	G	N	1961	62283
8236	S	D	G	N	1961	62281
8238	S	D	E	N	1940	45889
8243	S	D	E	N	1934	43126
8244	S	D	E	H	1943	59052
8245	S	D	E	N	1928	60525
8246	S	D	E	N	1941	60451
8247	S	D	E	N	1941	60451
8248	S	D	E	N	1939	60450
503-8249	S	D	E	N	1939	60450
8250	S	D	E	N	1939	60450
8251	S	D	E	N	1939	60449
8256	S	D	E	N	1932	34144
8257	S	D	E	N	1937	44341
8257	S	D	E	N	1940	44640
8257	S	D	E	N	1941	50775
8258	S	D	E	N	1941	50775
8259	S	D	E	N	1941	50775
8261	S	D	R	N	1970	60972
8261	S	D	E	N	1970	60973
8262	S	D	E	N	1925	60560
8263	S	D	E	N	1925	60560
8264	S	D	E	N	1925	60560
8265	S	D	E	N	1925	60560
8266	S	D	R	N	1967	53463
8267	S	D	R	N	1967	53463
8268	S	D	R	N	1967	53463
8269	S	D	E	N	1936	60006
8270	S	D	E	N	1936	60006
8271	S	D	R	N	1970	60146
8271	S	D	E	N	1970	60147
8272	S	D	E	N	1939	44506
8272	S	D	E	N	1940	44640
8272	S	D	E	N	1946	47340
8272	S	D	E	N	1949	56188
8272	S	D	E	N	1924	59968
8272	S	D	E	N	1941	60451
8273	S	D	E	N	1943	59806
8274	S	D	E	H	1935	43312
8275	S	D	E	H	1935	43312
8276	S	D	E	H	1935	43312
8277	S	D	E	H	1935	43312
8278	S	D	E	N	1937	44341
8278	S	D	E	N	1940	44640
8279	S	D	E	N	1939	44506
8280	S	D	R	N	1969	56890
8281	S	D	R	N	1969	58286
8282	S	S	E	N	1960	24931
8284	S	D	E	N	1939	44506
8284	S	D	E	N	1940	44640
8285	S	D	E	N	1952	59512
8288	S	D	E	N	1940	44640
8291	S	D	E	H	1968	53124
8292	S	D	E	N	1942	45416
8294	S	D	E	N	1942	45416
8295	S	D	E	N	1946	47340
8296	S	D	E	N	1947	47612
8297	S	D	E	N	1936	60515
8298	S	D	E	N	1936	60515
8299	S	D	E	N	1936	60515
8301	S	D	E	N	1936	60515
8302	S	D	E	N	1936	60515
8303	S	D	E	N	1936	60515
8304	S	D	E	H	1946	41574
8305	S	D	E	H	1946	41574
8307	P	D	F	N	1962	53233
8307	S	D	E	N	1964	42037
8308	P	D	F	N	1962	53233
8309	S	D	E	N	1960	52568
8310	S	D	E	N	1960	52568
8312	S	D	E	N	1959	46144
8312	S	D	E	N	1960	52568
8312	S	S	E	N	1977	88199
8313	D	D	E	N	1955	46143
8313	S	D	E	N	1955	46143
8313	S	D	E	N	1959	46144
8313	S	D	E	N	1960	52568
8313	S	D	E	N	1952	59512
8313	S	S	E	N	1977	88199
8314	S	D	R	N	1959	46428
8314	S	D	E	N	1959	52164
8315	S	D	R	N	1959	46428
8315	S	D	E	N	1959	52164
8316	S	D	R	N	1959	46428
8316	S	D	E	N	1959	52164
8317	S	D	E	N	1939	45359
8317	S	D	R	N	1959	46428
8317	S	D	E	N	1959	52164
8317	S	D	E	N	1973	69346
8318	S	D	E	N	1958	46791
8319	S	D	E	N	1964	37509
8320	S	D	R	N	1958	41706
8320	S	D	E	N	1958	45585
8321	S	D	E	N	1963	45352
8322	S	D	E	H	1965	39834
8325	S	D	E	N	1955	52405
8326	S	D	E	N	1955	52405
8327	S	D	E	N	1955	52405
8327	S	S	E	N	1977	88199
8328	S	D	E	N	1962	52418
8328	S	S	E	N	1977	88199
8329	S	D	E	N	1959	52279
8332	S	D	E	N	1931	42418
8332	S	D	R	H	1962	43170
8333	S	D	E	H	1947	48002
8334	S	D	E	H	1956	46847
8334	S	D	E	H	1947	48002
8334	S	D	R	H	1968	54773
8334	S	D	E	H	1968	54774
8334	S	D	E	N	1926	60435

Phys. State: **C.** Superconductive; **D.** Doped; **E.** Expanded; **F.** Fibrous or Whisker; **G.** Gas; **I.** Ionized or Plasma; **L.** Liquid; **M.** Multiphase; **P.** Powder or Fine Particle; **S.** Solid; **T.** Thick or Thin Film

Subject: **D.** Data; **E.** Experiment; **G.** General (Data + Theory + Experiment); **S.** Survey (Review, Compendium etc.); **T.** Theory

Language: **C.** Czech; **D.** Dutch; **E.** English; **F.** French; **G.** German; **I.** Italian; **J.** Japanese; **O.** Other Languages; **P.** Polish; **R.** Russian; **S.** Spanish

Temperature: **F.** Full Range (Low + Normal + High); **L.** Low (0 to 75K) + Overlap into Normal; **N.** Normal (75 to 1273K); **H.** High (above 1273K) + Overlap into Normal; Blank = Not Coded

Substance Number	Phys. State	Subject	Language	Temperature	Year	TPRC Number
503-8334	S	D	R	N	1970	62314
8334	S	D	R	H	1974	75192
8334	S	D	E	H	1974	76872
8335	S	D	E	N	1950	47941
8335	S	S	E	N	1977	88199
8336	S	D	E	N	1960	52665
8336	S	S	E	N	1977	88199
8337	S	D	E	N	1960	52665
8337	S	S	E	N	1977	88199
8338	S	D	E	N	1960	52665
8338	S	D	E	N	1951	59510
8338	S	S	E	N	1977	88199
8339	S	D	E	N	1960	52665
8339	S	S	E	N	1977	88199
8340	S	D	R	N	1967	47846
8340	S	D	E	N	1968	47847
8341	S	D	R	H	1967	49332
8341	S	D	E	H	1967	49333
8342	S	D	R	H	1967	49332
8342	S	D	E	H	1967	49333
8344	S	D	E	N	1958	52670
8346	S	D	E	N	1958	52670
8346	S	D	E	N	1964	54981
8350	S	D	R	N	1967	45798
8350	S	D	E	N	1967	48177
8352	S	D	R	H	1967	46801
8352	S	D	E	H	1967	46802
8355	S	D	E	N	1961	52519
8355	S	D	E	N	1962	52662
8355	S	T	E	N	1962	52662
8358	S	D	E	N	1967	43864
8358	S	D	E	N	1968	49871
8358	S	S	E	N	1977	88199
8359	S	D	E	N	1967	43864
8359	S	S	E	N	1977	88199
8363	S	D	J	N	1959	23128
8366	S	D	E	N	1966	45006
8366	S	D	E	H	1958	46791
8366	S	D	E	N	1920	60430
8366	S	D	E	N	1939	60449
8366	S	D	R	H	1971	64830
8366	S	D	E	H	1971	64831
8370	S	D	R	N	1958	41706
8370	S	D	E	N	1958	45585
8374	S	D	G	H	1931	39517
8374	S	D	R	H	1967	46608
8374	S	D	E	H	1967	46609
8374	S	D	R	H	1967	46801
8374	S	D	E	H	1967	46802
8374	S	D	R	H	1969	58852
8374	S	D	E	H	1969	58853
8374	S	D	E	N	1928	60525
8379	S	D	R	N	1966	46587
8379	S	D	E	N	1966	46588
8379	S	D	R	H	1967	46608
8379	S	D	E	H	1967	46609
8381	S	D	E	H	1965	37802
8385	S	D	E	H	1965	37802
8385	S	D	R	N	1970	60972
8385	S	D	E	N	1970	60973
8391	S	D	E	H	1925	52689
8394	S	D	G	H	1931	39517
8397	S	D	E	N	1968	48767
8398	S	D	E	N	1941	37976
8399	S	D	G	N	1931	61780
8403	S	D	J	N	1962	40485
8403	S	D	E	N	1962	40486
8404	S	D	J	N	1962	40485
8404	S	D	E	N	1962	40486
8405	S	D	J	N	1962	40485
8405	S	D	E	N	1962	40486
8418	S	D	R	N	1967	48543
8420	S	D	E	N	1969	52549
8424	S	D	E	N	1955	77328
8425	S	D	E	N	1974	77552
8426	S	D	E	N	1974	77552
8427	S	D	E	N	1974	77552
8429	S	D	E	N	1974	77552
8430	S	D	E	N	1974	77552
8431	S	D	E	N	1974	77552
8433	S	D	E	N	1974	77552
8437	S	D	E	N	1974	77552
8442	S	D	E	N	1953	77327
8448	S	D	E	N	1955	77328
8452	S	D	E	N	1953	77327
8456	S	D	E	N	1955	77324
8460	S	D	E	N	1973	76990
8461	S	D	E	N	1973	76990
8466	S	D	R	N	1974	76046
8466	S	D	E	N	1974	76906
8472	S	D	E	N	1975	76620
8473	S	D	E	N	1975	76620
8474	S	D	E	N	1975	76620
8475	S	D	E	N	1975	76620
8476	S	D	E	N	1974	76565
8478	S	D	E	N	1974	74535
8480	S	D	R	N	1969	55800
8484	S	D	R	N	1972	68304
8484	S	D	E	N	1972	70606
8489	S	D	R	N	1969	56246
8489	S	D	E	N	1972	66144

Substance Number	Phys. State	Subject	Language	Temperature	Year	TPRC Number
503-8490	S	D	R	N	1969	56246
8490	S	D	E	N	1972	66144
8493	S	D	R	N	1969	53252
8493	S	D	E	N	1969	66785
8503	S	D	R	N	1969	56247
8504	S	D	R	N	1969	56247
8505	S	D	R	N	1969	56247
8506	S	D	R	N	1965	46408
8506	S	D	R	N	1969	56247
8507	F	D	E	N	1971	61382
8510	S	D	R	N	1971	64711
8510	S	D	E	N	1971	64712
8511	S	D	R	H	1971	65109
8511	S	D	E	H	1971	65110
8512	S	D	R	H	1971	65109
8512	S	D	E	H	1971	65110
8513	S	D	R	H	1971	64703
8513	S	D	E	H	1971	64704
8518	S	D	R	N	1972	68304
8518	S	D	E	N	1972	70606
8524	S	D	E	N	1971	62634
8524	S	E	E	N	1971	62634
8525	S	D	E	N	1971	62634
8525	S	E	E	N	1971	62634
8526	S	D	E	N	1971	63042
8532	S	D	R	N	1967	49829
8532	S	D	E	N	1971	63808
8533	S	D	R	N	1967	49829
8533	S	D	E	N	1971	63808
8534	S	D	R	N	1967	49829
8534	S	D	E	N	1971	63808
8535	S	D	R	N	1963	59462
8535	S	D	E	N	1971	61556
8536	S	D	R	N	1963	59462
8536	S	D	E	N	1971	61556
8537	S	D	R	N	1963	59462
8537	S	D	E	N	1971	61556
8538	S	D	J	N	1961	62298
8539	S	D	J	N	1961	62297
8540	D	D	J	N	1960	62296
8540	S	D	J	N	1960	62296
8541	S	D	R	N	1963	59315
8541	S	D	E	N	1971	59660
8542	S	D	R	N	1963	59315
8542	S	D	E	N	1971	59660
8543	F	D	E	N	1968	54821
8545	S	D	E	N	1970	59667
8546	S	D	R	N	1953	54459
8546	S	D	E	N	1953	63073
8547	S	D	R	N	1953	54459
8547	S	D	E	N	1953	63073
8548	S	D	R	N	1953	54459
8548	S	D	E	N	1953	63073
8549	S	D	R	N	1953	54459
8549	S	D	E	N	1953	63073
8550	S	D	R	N	1953	54459
8550	S	D	E	N	1953	63073
8551	S	D	R	N	1953	54459
8551	S	D	E	N	1953	63073
8552	S	D	R	N	1953	54459
8552	S	D	E	N	1953	63073
8557	S	D	E	N	1941	37976
8558	S	D	E	N	1941	37976
8559	S	D	E	N	1941	37976
8562	S	D	E	N	1933	37748
8563	S	D	E	N	1934	37975
8565	D	D	E	N	1968	53539
8565	S	D	E	N	1968	53539
8567	S	D	R	N	1970	60998
8567	S	D	E	N	1970	60999
8568	S	D	R	N	1970	60998
8568	S	D	E	N	1970	60999
8569	S	D	R	N	1970	60998
8569	S	D	E	N	1970	60999
8570	S	D	R	N	1970	60998
8570	S	D	E	N	1970	60999
8572	S	D	E	N	1942	60535
8573	S	D	E	N	1952	59877
8574	S	D	E	N	1952	59877
8575	S	D	E	N	1952	59877
8576	S	D	E	N	1952	59877
8577	S	D	E	N	1952	59877
8578	S	D	G	N	1958	58457
8578	S	T	G	N	1958	58457
8579	S	D	G	N	1958	58457
8579	S	T	G	N	1958	58457
8580	S	D	E	N	1920	59946
8583	S	D	E	N	1968	52968
8584	S	D	E	N	1935	60005
8585	S	D	E	N	1935	60005
8585	S	D	R	N	1958	62285
8586	S	D	E	N	1935	60005
8587	S	D	E	N	1941	60370
8588	S	D	E	N	1941	40392
8588	S	D	G	N	1958	58457
8588	S	T	G	N	1958	58457
8588	S	D	E	N	1931	60003
8589	S	D	R	N	1953	54459
8589	S	D	E	N	1941	60370
8589	S	D	E	N	1953	63073
8590	S	D	R	N	1969	56889

Substance Number	Phys. State	Subject	Language	Temperature	Year	TPRC Number
503-8591	S	D	R	N	1969	56889
8592	S	D	E	N	1950	59514
8593	S	D	R	N	1967	48459
8593	S	D	E	N	1967	59338
8594	S	D	R	N	1967	48459
8594	S	D	E	N	1967	59338
8595	S	D	R	N	1969	58217
8596	S	D	R	N	1969	58217
8597	S	D	R	N	1969	58217
8598	S	D	R	N	1969	58217
8598	S	D	R	N	1963	59315
8598	S	D	E	N	1971	59660
8599	S	D	R	N	1969	58217
8601	S	D	E	N	1970	58694
8602	S	D	E	N	1970	58694
8605	S	D	E	N	1928	58924
8606	S	D	E	N	1928	58924
8607	S	D	R	N	1967	51836
8607	S	D	R	N	1970	61701
8607	S	D	E	N	1970	61702
8608	S	D	R	N	1967	51836
8609	S	D	R	N	1967	51836
8610	S	D	R	N	1967	51836
8611	S	D	R	N	1967	51836
8612	S	D	R	N	1967	54573
8613	S	D	R	N	1967	54573
8614	S	D	E	N	1941	40392
8614	S	D	G	N	1958	58457
8614	S	T	G	N	1958	58457
8615	S	D	R	N	1969	58579
8615	S	D	E	N	1969	58580
8616	S	D	R	N	1969	58579
8616	S	D	E	N	1969	58580
8617	S	D	R	N	1969	58579
8617	S	D	E	N	1969	58580
8618	S	D	R	N	1969	58579
8618	S	D	E	N	1969	58580
8619	S	D	R	N	1969	58579
8619	S	D	E	N	1969	58580
8620	S	D	R	N	1969	58579
8620	S	D	E	N	1969	58580
8621	S	D	R	N	1969	58579
8621	S	D	E	N	1969	58580
8622	S	D	R	N	1969	58579
8622	S	D	E	N	1969	58580
8623	S	D	R	N	1969	58579
8623	S	D	E	N	1969	58580
8624	S	D	R	N	1969	58579
8624	S	D	E	N	1969	58580
8625	S	D	R	N	1969	58579
8625	S	D	E	N	1969	58580
8626	S	D	R	N	1969	58579
8626	S	D	E	N	1969	58580
8627	S	D	R	N	1969	58579
8627	S	D	E	N	1969	58580
8628	S	D	E	N	1954	38130
8629	S	D	R	N	1969	58579
8629	S	D	E	N	1969	58580
8630	S	D	R	N	1969	58579
8630	S	D	E	N	1969	58580
8631	S	D	R	N	1969	58579
8631	S	D	E	N	1969	58580
8632	S	D	R	N	1969	58579
8632	S	D	E	N	1969	58580
8633	S	D	R	N	1969	58579
8633	S	D	E	N	1969	58580
8634	S	D	R	N	1969	58579
8634	S	D	E	N	1969	58580
8635	S	D	J	N	1968	56166
8636	S	D	E	N	1965	40714
8637	S	D	E	N	1965	40714
8638	S	D	E	N	1965	40714
8640	S	D	E	N	1965	40714
8641	S	D	E	N	1965	40714
8642	S	D	E	N	1965	40714
8643	S	D	E	N	1965	40714
8644	S	D	E	N	1965	40714
8645	S	D	E	N	1965	40714
8645	S	D	R	N	1969	53252
8645	S	D	E	N	1969	66785
8646	S	D	E	N	1965	40714
8661	S	D	R	N	1967	57203
8661	S	D	E	N	1969	57204
8662	S	D	R	N	1964	44592
8662	S	D	R	N	1969	54916
8663	S	D	R	N	1969	57179
8663	S	D	E	N	1969	57180
8664	S	D	R	N	1969	57179
8664	S	D	E	N	1969	57180
8665	S	D	R	N	1969	54435
8665	S	D	E	N	1969	57079
8666	S	D	R	N	1969	54435
8666	S	D	E	N	1969	57079
8704	S	D	E	N	1969	54273
8704	S	D	E	N	1971	67411
8706	S	D	E	N	1969	54350
8707	S	D	E	N	1969	54273
8708	S	D	E	N	1956	41939
8709	S	D	E	N	1969	38662
8723	S	D	E	N	1961	52524
8723	S	D	E	N	1971	67411

Phys. State: **C.** Superconductive; **D.** Doped; **E.** Expanded; **F.** Fibrous or Whisker; **G.** Gas; **I.** Ionized or Plasma; **L.** Liquid; **M.** Multiphase; **P.** Powder or Fine Particle; **S.** Solid; **T.** Thick or Thin Film
Subject: **D.** Data; **E.** Experiment; **G.** General (Data + Theory + Experiment); **S.** Survey (Review, Compendium etc.); **T.** Theory
Language: **C.** Czech; **D.** Dutch; **E.** English; **F.** French; **G.** German; **I.** Italian; **J.** Japanese; **O.** Other Languages; **P.** Polish; **R.** Russian; **S.** Spanish
Temperature: **F.** Full Range (Low + Normal + High); **L.** Low (0 to 75K) + Overlap into Normal; **N.** Normal (75 to 1273K); **H.** High (above 1273K) + Overlap into Normal; Blank = Not Coded

Substance Number	Phys. State	Subject	Language	Temperature	Year	TPRC Number
503-8724	S	D	E	N	1961	52524
8725	S	D	E	N	1959	52523
8726	S	D	E	N	1920	59946
8727	S	D	E	N	1954	52479
8728	S	D	E	N	1959	41697
8728	S	D	G	N	1966	41958
8728	S	T	G	N	1966	41958
8728	S	D	O	N	1974	76092
8728	S	T	O	N	1974	76092
8728	S	D	E	N	1974	77552
8734	S	D	E	H	1969	52550
8734	S	D	G	N	1969	58212
8734	S	D	E	N	1950	59514
8734	S	D	E	N	1973	76990
8735	S	D	E	N	1967	50145
8735	S	D	E	H	1969	52550
8735	S	D	G	N	1969	58212
8735	S	D	E	N	1937	60007
8736	S	D	R	N	1966	43026
8736	S	D	R	N	1967	53152
8736	S	T	R	N	1967	53152
8737	S	D	J	N	1966	45054
8738	S	D	R	N	1968	58288
8739	S	D	J	N	1966	45054
8740	S	D	J	N	1966	45054
8741	S	D	J	N	1966	45054
8742	S	D	R	N	1968	58288
8742	S	D	R	N	1963	59225
8742	S	D	E	N	1971	61709
8742	S	D	R	N	1972	67819
8742	S	D	E	N	1972	70605
8743	S	D	E	N	1959	47140
8744	S	D	E	N	1961	52253
8745	S	D	E	N	1961	52253
8746	S	D	E	N	1961	52253
8746	S	D	E	N	1969	54502
8746	S	T	E	N	1969	54502
8746	S	D	E	N	1970	58727
8746	S	D	E	N	1970	60282
8746	S	T	E	N	1970	60282
8747	S	D	E	N	1961	52253
8748	S	D	E	N	1962	41729
8748	S	D	E	N	1961	52253
8749	S	D	E	N	1961	52253
8750	S	D	E	N	1960	52250
8751	S	D	R	N	1966	43026
8751	S	D	R	N	1967	53152
8751	S	T	R	N	1967	53152
8751	S	D	R	N	1969	56889
8752	S	D	R	N	1966	43026
8752	S	D	R	N	1967	53152
8752	S	T	R	N	1967	53152
8752	S	D	E	N	1971	67411
8752	S	D	E	N	1970	67661
8752	S	T	E	N	1970	67661
8753	S	D	E	N	1963	37530
8753	S	D	E	N	1951	60011
8754	S	D	E	N	1963	37530
8755	S	D	E	N	1963	37530
8755	S	D	O	N	1968	53267
8756	S	D	G	N	1958	58457
8756	S	T	G	N	1958	58457
8757	S	D	R	N	1969	58217
8757	S	D	E	N	1931	60003
8757	S	D	E	N	1941	60370
8758	S	D	R	N	1964	46218
8761	S	D	O	N	1963	44532
8762	S	D	G	N	1967	47477
8762	S	T	G	N	1967	47477
8762	S	D	E	N	1973	76782
8762	S	T	E	N	1973	76782
8763	S	D	G	N	1967	47477
8763	S	T	G	N	1967	47477
8763	S	D	E	N	1973	76782
8763	S	T	E	N	1973	76782
8764	S	D	G	N	1967	47477
8764	S	T	G	N	1967	47477
8764	S	D	E	N	1973	76782
8764	S	T	E	N	1973	76782
8765	S	D	G	N	1967	47477
8765	S	T	G	N	1967	47477
8765	S	D	E	N	1973	76782
8765	S	T	E	N	1973	76782
8766	S	D	G	N	1967	47477
8766	S	T	G	N	1967	47477
8766	S	D	E	N	1973	76782
8766	S	T	E	N	1973	76782
8767	S	D	E	N	1966	43514
8767	S	D	G	N	1967	47477
8767	S	T	G	N	1967	47477
8767	S	D	E	N	1968	52688
8767	S	D	R	N	1969	55146
8767	S	D	E	N	1973	76782
8767	S	T	E	N	1973	76782
8770	S	D	R	N	1968	50198
8770	S	D	E	N	1968	50199
8771	S	D	R	N	1968	50061
8771	S	D	E	N	1968	51782
8787	S	D	R	N	1969	58579
8787	S	D	E	N	1969	58580
8788	S	D	E	N	1952	60012

Substance Number	Phys. State	Subject	Language	Temperature	Year	TPRC Number
503-8818	S	D	R	N	1967	51836
8818	S	D	R	N	1970	61701
8818	S	D	E	N	1970	61702
8822	S	D	E	N	1973	72582
8822	S	D	E	N	1960	76856
8823	S	D	R	N	1967	51836
8824	D	D	E	N	1957	48109
8826	D	D	E	N	1957	48109
8833	S	D	E	N	1964	43407
8833	S	D	G	N	1958	58457
8833	S	T	G	N	1958	58457
8836	S	D	G	N	1958	58457
8836	S	T	G	N	1958	58457
8837	S	D	G	N	1958	58457
8837	S	T	G	N	1958	58457
8838	S	D	E	N	1941	60370
8841	S	D	G	N	1958	58457
8841	S	T	G	N	1958	58457
8842	S	D	R	N	1967	47640
8842	S	D	R	N	1972	67819
8842	S	D	E	N	1972	70605
8842	S	D	E	N	1973	76990
8842	S	D	E	N	1957	77323
8848	S	D	E	N	1968	52688
8848	S	D	E	N	1967	52854
8848	S	D	R	N	1969	58217
8848	S	D	G	N	1958	58457
8848	S	T	G	N	1958	58457
8850	S	D	E	N	1968	52688
8851	S	D	R	N	1967	51836
8851	S	D	R	N	1968	55446
8851	S	D	E	N	1968	55447
8852	S	D	J	N	1967	50020
8853	S	D	J	N	1967	50020
8854	S	D	J	N	1967	50020
8855	S	D	J	N	1967	50020
8856	S	D	J	N	1967	50020
8857	S	D	R	N	1967	48659
8858	S	D	R	N	1967	48659
8859	S	D	R	N	1967	48659
8860	S	D	J	N	1967	50020
8861	S	D	J	N	1967	50020
8862	S	D	J	N	1969	55538
8863	S	D	J	N	1961	47229
8863	S	D	J	N	1961	62297
8864	S	D	R	N	1966	41504
8865	S	D	E	N	1968	49199
8865	S	D	E	N	1974	74535
8866	S	D	R	N	1967	47685
8866	S	D	E	N	1967	47686
8868	S	D	E	N	1968	38338
8868	S	D	R	N	1962	42232
8868	S	D	R	N	1968	50051
8868	S	D	R	N	1968	59359
8868	S	D	E	N	1968	59360
8868	S	D	R	H	1971	64703
8868	S	D	E	H	1971	64704
8869	S	D	E	N	1968	52688
8870	S	D	G	N	1958	58457
8870	S	T	G	N	1958	58457
8872	S	D	R	N	1965	42546
8873	S	D	R	N	1965	42545
8874	S	D	R	N	1964	44592
8874	S	D	E	N	1968	52688
8874	S	D	R	N	1969	56248
8874	S	D	R	N	1963	59315
8874	S	D	R	N	1963	59462
8874	S	D	E	N	1971	59660
8874	S	D	E	N	1971	61556
8875	S	D	G	N	1958	58457
8875	S	T	G	N	1958	58457
8876	S	G	F	N	1928	42820
8876	S	D	R	N	1953	53130
8876	S	D	G	N	1958	58457
8876	S	T	G	N	1958	58457
8876	S	D	F	N	1926	59810
8876	S	D	E	N	1953	63072
8877	S	D	R	N	1953	53130
8877	S	D	E	N	1953	63072
8878	S	D	R	N	1953	53130
8878	S	D	E	N	1950	59514
8878	S	D	E	N	1953	63072
8880	S	D	J	N	1965	44607
8880	S	D	E	N	1967	46030
8880	S	D	J	N	1969	55538
8880	S	D	E	N	1974	76037
8881	S	D	J	N	1965	44607
8881	S	D	E	N	1967	46030
8881	S	D	R	H	1971	65109
8881	S	D	E	H	1971	65110
8882	S	D	E	N	1968	52688
8882	S	D	G	N	1958	58457
8882	S	T	G	N	1958	58457
8882	S	D	F	N	1970	58498
8884	S	D	E	N	1967	35226
8885	S	D	E	N	1967	35226
8886	S	D	E	N	1967	44019
8888	D	D	R	N	1967	45268
8888	D	D	R	N	1967	77590
8888	D	D	E	N	1974	77591
8888	S	D	R	N	1966	42125

Substance Number	Phys. State	Subject	Language	Temperature	Year	TPRC Number
503-8888	S	T	R	N	1967	48707
8888	S	D	E	N	1968	52688
8888	S	D	E	N	1951	60011
8888	S	D	E	N	1970	67661
8888	S	T	E	N	1970	67661
8888	S	D	R	N	1967	77590
8888	S	D	E	N	1974	77591
8889	S	D	G	N	1968	54930
8889	S	D	R	N	1969	55287
8889	S	D	E	N	1972	65505
8890	S	D	O	N	1963	34151
8891	S	D	O	N	1963	34151
8892	S	D	O	N	1963	34151
8893	S	D	O	N	1963	34151
8894	S	D	O	N	1963	34151
8894	S	D	R	N	1966	43026
8894	S	D	R	N	1967	53152
8894	S	T	R	N	1967	53152
8895	S	D	E	N	1968	52688
8896	S	D	E	N	1968	52688
8897	S	D	E	N	1966	49818
8898	S	D	R	N	1961	52569
8898	S	D	E	N	1961	52570
8899	S	D	R	N	1961	52569
8899	S	D	E	N	1961	52570
8899	S	D	G	N	1969	58212
8904	S	D	E	N	1959	52523
8911	S	D	E	N	1965	40714
8914	S	T	R	N	1967	48707
8927	S	D	E	N	1969	54273
8928	S	D	E	N	1953	52564
8928	S	D	R	N	1963	59462
8928	S	D	E	N	1971	61556
8929	S	D	R	N	1966	44218
8929	S	D	E	N	1966	44219
8940	D	D	R	H	1969	55886
8940	S	D	E	N	1953	52564
8940	S	D	R	H	1969	55886
8941	S	D	G	N	1958	58457
8941	S	T	G	N	1958	58457
8942	S	D	E	N	1953	52564
8943	S	D	E	N	1964	43407
8943	S	D	E	N	1964	44122
8943	S	D	E	N	1935	57681
8943	S	D	G	N	1958	58457
8943	S	T	G	N	1958	58457
8943	S	D	R	N	1958	62285
8947	S	D	R	N	1966	44188
8947	S	D	E	N	1966	44189
8950	S	G	E	N	1970	45628
8952	S	G	E	N	1970	45628
8954	S	D	E	H	1966	36253
8954	S	D	E	H	1966	36254
8954	S	D	R	H	1966	47091
8954	S	D	R	H	1966	49015
8955	S	D	R	N	1969	54916
8956	S	D	E	N	1971	62634
8956	S	E	E	N	1971	62634
8971	S	D	R	N	1953	54459
8971	S	D	E	N	1953	63073
8976	S	D	E	N	1975	76620
8977	S	D	E	N	1974	74535
8977	S	D	E	N	1974	74536
8979	S	D	E	N	1973	72582
8980	S	D	E	N	1968	51221
8980	S	D	R	H	1971	65109
8980	S	D	E	H	1971	65110
8980	S	D	E	N	1975	76620
8985	S	D	E	N	1968	52688
8986	S	D	R	N	1967	49689
8986	S	D	R	N	1969	56245
8987	S	D	R	N	1973	76990
8988	S	D	E	N	1967	50145
8988	S	D	R	N	1970	60940
8988	S	D	E	N	1970	60941
8988	S	D	E	N	1972	71410
8988	S	T	E	N	1972	71410
8988	S	D	E	L	1972	76886
8991	S	D	E	N	1955	37152
8991	S	D	E	N	1964	43407
8991	S	D	E	N	1941	47074
8991	S	D	E	N	1941	55942
8992	S	D	E	N	1971	62634
8992	S	E	E	N	1971	62634
8999	S	D	R	N	1967	49691
9001	S	D	E	N	1965	36875
9011	S	D	E	N	1965	38742
9012	S	D	E	N	1965	38742
9013	S	D	E	N	1965	38742
9022	S	D	E	N	1959	47140
9022	S	D	R	N	1967	51836
9023	S	G	F	N	1928	42820
9023	S	D	E	N	1959	47140
9029	S	D	G	N	1966	45549
9029	S	D	G	N	1968	50596
9039	S	D	E	N	1967	46030
9039	S	D	R	H	1971	65109
9039	S	D	E	H	1971	65110
9039	S	D	E	N	1974	75123
9039	S	D	E	N	1975	76620
9042	S	D	R	N	1966	44194

Phys. State: **C.** Superconductive; **D.** Doped; **E.** Expanded; **F.** Fibrous or Whisker; **G.** Gas; **I.** Ionized or Plasma; **L.** Liquid; **M.** Multiphase; **P.** Powder or Fine Particle; **S.** Solid; **T.** Thick or Thin Film
Subject: **D.** Data; **E.** Experiment; **G.** General (Data + Theory + Experiment); **S.** Survey (Review, Compendium etc.); **T.** Theory
Language: **C.** Czech; **D.** Dutch; **E.** English; **F.** French; **G.** German; **I.** Italian; **J.** Japanese; **O.** Other Languages; **P.** Polish; **R.** Russian; **S.** Spanish
Temperature: **F.** Full Range (Low + Normal + High); **L.** Low (0 to 75K) + Overlap into Normal; **N.** Normal (75 to 1273K); **H.** High (above 1273K) + Overlap into Normal; Blank = Not Coded

Substance Number	Phys. State	Subject	Language	Temperature	Year	TPRC Number
503-9042	S	D	E	N	1966	44195
9042	S	D	E	N	1968	52688
9042	S	D	O	N	1968	54559
9042	S	D	R	N	1969	58217
9042	S	D	G	N	1958	58457
9042	S	T	G	N	1958	58457
9042	S	D	E	L	1964	58875
9043	S	D	E	N	1961	52524
9055	S	D	G	N	1966	41958
9055	S	T	G	N	1966	41958
9055	S	D	J	N	1966	45054
9058	S	D	E	N	1941	37976
9058	S	D	G	N	1958	58457
9058	S	T	G	N	1958	58457
9058	S	D	E	N	1920	59946
9058	S	D	E	N	1941	60370
9064	S	D	E	N	1973	76990
9067	S	D	J	N	1967	50020
9068	S	D	E	N	1963	37530
9068	S	D	R	N	1965	41898
9068	S	D	J	N	1961	47229
9068	S	D	E	H	1968	51220
9068	S	D	R	N	1955	54638
9068	S	D	G	N	1958	58457
9068	S	T	G	N	1958	58457
9068	S	D	E	N	1951	60010
9068	S	D	E	N	1971	62075
9068	S	D	E	N	1971	62678
9068	S	D	E	N	1955	63074
9068	S	D	E	N	1970	67661
9068	S	T	E	N	1970	67661
9074	S	D	E	N	1938	36061
9074	S	D	E	N	1965	40714
9074	S	D	R	N	1969	57112
9074	S	D	E	N	1969	57113
9078	S	D	E	N	1953	52564
9080	S	G	E	N	1970	45628
9081	D	D	R	N	1966	42052
9081	D	D	R	N	1966	42053
9081	S	D	E	N	1959	41697
9081	S	D	R	N	1966	42052
9081	S	D	R	N	1966	42053
9081	S	D	E	N	1961	52253
9082	S	D	E	N	1969	55763
9083	S	D	E	N	1969	55763
9091	S	D	E	N	1934	37975
9091	S	D	E	N	1967	44019
9091	S	D	G	N	1958	58457
9091	S	T	G	N	1958	58457
9091	S	D	E	N	1920	59946
9092	S	D	G	N	1958	58457
9092	S	T	G	N	1958	58457
9093	S	D	E	N	1970	37232
9094	S	D	R	N	1966	43026
9094	S	D	R	N	1967	53152
9094	S	T	R	N	1967	53152
9094	S	D	G	N	1958	58457
9094	S	T	G	N	1958	58457
9094	S	D	E	N	1920	59946
9094	S	D	G	N	1931	61780
9095	D	D	E	N	1928	58924
9095	S	D	E	N	1936	35781
9095	S	D	E	N	1967	44019
9095	S	D	E	N	1961	47390
9095	S	D	E	N	1963	47403
9095	S	D	G	N	1958	58457
9095	S	T	G	N	1958	58457
9095	S	D	E	N	1928	58924
9095	S	D	R	N	1970	62576
9095	S	D	E	N	1970	67361
9095	S	D	E	N	1970	67661
9095	S	T	E	N	1970	67661
9096	S	D	G	N	1958	58457
9096	S	T	G	N	1958	58457
9097	S	D	G	N	1958	58457
9097	S	T	G	N	1958	58457
9097	S	D	E	N	1920	59946
9099	S	D	E	N	1938	36061
9099	S	D	E	N	1969	38662
9099	S	D	E	N	1965	40714
9099	S	D	E	N	1956	41939
9101	S	D	E	N	1934	37975
9101	S	D	E	N	1941	37976
9101	S	D	J	N	1967	50020
9101	S	D	G	N	1958	58457
9101	S	T	G	N	1958	58457
9101	S	D	R	N	1969	58579
9101	S	D	E	N	1969	58580
9101	S	D	E	N	1931	60003
9101	S	D	E	N	1942	60535
9102	S	D	E	N	1960	52283
9102	S	D	G	N	1969	58212
9102	S	D	R	N	1969	58579
9102	S	D	E	N	1969	58580
9105	S	D	R	N	1953	54459
9105	S	D	E	N	1953	63073
9106	S	D	R	N	1953	54459
9106	S	D	E	N	1953	63073
9107	S	D	R	N	1953	54459
9107	S	D	G	N	1969	58212
9107	S	D	E	N	1953	63073

Substance Number	Phys. State	Subject	Language	Temperature	Year	TPRC Number
503-9108	S	D	R	N	1970	62576
9108	S	D	E	N	1970	67361
9108	S	D	E	N	1970	67661
9108	S	T	E	N	1970	67661
9109	S	D	R	N	1953	54459
9109	S	D	E	N	1953	63073
9110	S	D	R	N	1953	54459
9110	S	D	E	N	1953	63073
9112	S	D	E	N	1965	42031
9113	S	D	R	N	1971	65218
9113	S	D	E	N	1971	66790
9114	S	D	R	N	1967	48659
9115	S	D	D	N	1967	47237
9115	S	D	R	N	1967	53151
9115	S	D	G	N	1958	58457
9115	S	T	G	N	1958	58457
9115	S	D	R	N	1969	58579
9115	S	D	E	N	1969	58580
9116	S	D	R	N	1967	49691
9118	S	D	R	N	1965	42545
9118	S	D	R	N	1965	42546
9119	S	D	R	N	1967	47685
9119	S	D	E	N	1967	47686
9119	S	D	G	N	1958	58457
9119	S	T	G	N	1958	58457
9119	S	D	E	N	1927	60002
9119	S	D	E	N	1961	77266
9120	S	D	R	N	1969	56889
9120	S	D	G	N	1958	58457
9120	S	T	G	N	1958	58457
9120	S	D	E	N	1961	77266
9121	S	D	J	N	1943	59830
9122	S	D	R	N	1967	53471
9122	S	D	J	N	1943	59830
9124	S	D	R	N	1969	58579
9124	S	D	E	N	1969	58580
9125	S	D	R	N	1966	44218
9125	S	D	E	N	1966	44219
9126	S	D	R	N	1953	54459
9126	S	D	E	N	1953	63073
9127	S	D	R	N	1953	54459
9127	S	D	E	N	1953	63073
9128	S	D	R	N	1953	54459
9128	S	D	E	N	1953	63073
9129	S	D	E	N	1952	60012
9130	S	D	E	N	1952	60012
9131	D	D	E	N	1973	71060
9131	S	D	E	N	1970	37340
9131	S	D	E	N	1967	45618
9131	S	D	G	N	1958	58457
9131	S	T	G	N	1958	58457
9131	S	D	E	N	1929	60001
9131	S	D	E	N	1927	60002
9131	S	D	E	N	1970	67661
9131	S	T	E	N	1970	67661
9131	S	D	R	N	1972	68304
9131	S	D	E	N	1972	70606
9132	S	D	E	N	1961	52253
9133	S	D	E	N	1961	52253
9133	S	D	E	N	1973	72582
9134	S	D	G	N	1964	37241
9134	S	D	J	N	1965	44607
9134	S	D	E	N	1969	55763
9134	S	D	R	H	1971	65109
9134	S	D	E	H	1971	65110
9134	S	D	E	N	1971	67411
9135	S	D	E	H	1967	40429
9135	S	D	R	N	1953	53130
9135	S	D	R	N	1969	58579
9135	S	D	E	N	1969	58580
9135	S	D	E	N	1953	63072
9135	S	D	J	N	1972	66292
9135	S	T	J	N	1972	66292
9136	S	D	E	N	1965	38742
9136	S	D	E	N	1969	54023
9137	S	D	E	H	1967	40429
9137	S	D	J	N	1972	66292
9137	S	T	J	N	1972	66292
9139	S	D	R	N	1953	54459
9139	S	D	G	N	1958	58457
9139	S	T	G	N	1958	58457
9139	S	D	E	N	1953	63073
9140	S	D	R	N	1964	37755
9140	S	D	E	N	1944	55268
9140	S	D	E	N	1952	60012
9141	S	D	J	N	1967	50020
9143	S	D	R	N	1967	51836
9145	S	G	F	N	1928	42820
9147	S	D	J	N	1965	44607
9147	S	D	E	N	1969	55763
9147	S	D	E	N	1971	63042
9147	S	D	R	H	1971	65109
9147	S	D	E	H	1971	65110
9149	S	D	E	N	1934	37975
9151	S	D	R	N	1969	56889
9152	D	D	E	N	1973	71060
9152	S	D	E	N	1934	35768
9152	S	D	E	N	1965	40714
9152	S	D	E	N	1961	47390
9152	S	D	E	N	1963	47403
9152	S	D	E	N	1973	71060

Substance Number	Phys. State	Subject	Language	Temperature	Year	TPRC Number
503-9152	S	D	E	N	1974	74535
9153	S	D	E	N	1941	37976
9153	S	D	E	N	1967	44019
9153	S	D	E	N	1935	57681
9153	S	D	G	N	1958	58457
9153	S	T	G	N	1958	58457
9154	S	D	E	N	1956	37172
9154	S	D	E	N	1966	41584
9154	S	D	R	N	1967	48623
9154	S	D	E	N	1967	48624
9154	S	D	R	N	1967	48659
9154	S	D	R	N	1969	54435
9154	S	D	R	N	1969	55800
9154	S	D	E	N	1969	57079
9154	S	D	E	N	1972	64957
9155	S	D	R	N	1967	48659
9155	S	D	G	N	1958	58457
9155	S	T	G	N	1958	58457
9156	P	D	E	N	1968	52243
9156	S	D	E	N	1920	37542
9156	S	D	E	N	1968	52243
9156	S	D	G	N	1958	58457
9156	S	T	G	N	1958	58457
9156	S	D	J	N	1943	59830
9156	S	D	E	N	1927	60002
9156	S	D	E	N	1952	60012
9157	S	D	R	N	1964	37755
9157	S	D	R	N	1953	54459
9157	S	D	G	N	1969	58212
9157	S	D	G	N	1958	58457
9157	S	T	G	N	1958	58457
9157	S	D	J	N	1942	59828
9157	S	D	E	N	1929	60001
9157	S	D	E	N	1927	60002
9157	S	D	E	N	1952	60012
9157	S	D	E	N	1941	60370
9157	S	D	E	N	1953	63073
9158	S	D	E	H	1967	40429
9158	S	D	R	N	1965	42545
9158	S	D	R	N	1965	42546
9158	S	D	J	N	1955	43250
9158	S	T	J	N	1955	43250
9158	S	D	R	N	1967	49691
9158	S	D	E	N	1970	56113
9158	S	D	G	N	1958	58457
9158	S	T	G	N	1958	58457
9158	S	D	F	N	1934	59812
9158	S	D	E	N	1935	59948
9158	S	D	E	N	1927	60002
9158	S	D	J	N	1972	66292
9158	S	T	J	N	1972	66292
9159	S	D	E	N	1933	37748
9159	S	D	J	N	1955	43250
9159	S	T	J	N	1955	43250
9159	S	D	J	N	1967	50020
9159	S	D	R	N	1953	54459
9159	S	D	E	N	1931	60003
9159	S	D	E	N	1953	63073
9159	S	D	R	N	1971	64711
9159	S	D	E	N	1971	64712
9160	S	D	R	N	1965	42546
9160	S	G	F	N	1928	42820
9160	S	D	E	N	1941	47074
9160	S	D	E	N	1969	54273
9160	S	D	E	N	1941	55942
9160	S	D	G	N	1958	58457
9160	S	T	G	N	1958	58457
9160	S	D	E	L	1964	58875
9160	S	D	E	N	1920	59946
9160	S	D	E	N	1931	60003
9160	S	D	R	N	1958	62285
9161	S	D	G	N	1958	58457
9161	S	T	G	N	1958	58457
9161	S	D	E	N	1920	59946
9161	S	D	E	N	1941	60370
9161	S	D	E	N	1942	60535
9163	S	D	G	N	1958	58457
9163	S	T	G	N	1958	58457
9163	S	D	E	N	1927	60002
9163	S	D	E	N	1952	60012
9166	S	D	D	N	1967	47237
9166	S	D	R	N	1967	49690
9166	S	D	E	N	1953	52564
9166	S	D	E	N	1968	52688
9166	S	D	G	N	1968	54930
9166	S	D	R	N	1969	55287
9166	S	D	R	N	1969	56248
9166	S	D	R	N	1969	57969
9166	S	D	E	N	1969	57970
9166	S	D	R	N	1963	59315
9166	S	D	E	N	1971	59660
9166	S	D	E	N	1972	65505
9166	S	D	E	N	1970	67661
9166	S	T	E	N	1970	67661
9167	S	D	R	N	1969	58579
9167	S	D	E	N	1969	58580
9168	S	D	E	N	1934	37975
9168	S	D	R	N	1967	53151
9168	S	D	G	N	1958	58457
9168	S	T	G	N	1958	58457
9169	S	D	E	N	1934	40799

Phys. State: **C.** Superconductive; **D.** Doped; **E.** Expanded; **F.** Fibrous or Whisker; **G.** Gas; **I.** Ionized or Plasma; **L.** Liquid; **M.** Multiphase; **P.** Powder or Fine Particle; **S.** Solid; **T.** Thick or Thin Film

Subject: **D.** Data; **E.** Experiment; **G.** General (Data + Theory + Experiment); **S.** Survey (Review, Compendium etc.); **T.** Theory

Language: **C.** Czech; **D.** Dutch; **E.** English; **F.** French; **G.** German; **I.** Italian; **J.** Japanese; **O.** Other Languages; **P.** Polish; **R.** Russian; **S.** Spanish

Temperature: **F.** Full Range (Low + Normal + High); **L.** Low (0 to 75K) + Overlap into Normal; **N.** Normal (75 to 1273K); **H.** High (above 1273K) + Overlap into Normal; Blank = Not Coded

Substance Number	Phys. State	Sub-ject	Lan-guage	Temper-ature	Year	TPRC Number
503-9169	S	G	F	N	1928	42820
9169	S	D	J	N	1965	44607
9169	S	D	G	N	1967	45763
9169	S	D	E	N	1967	46030
9169	S	D	E	N	1969	55007
9169	S	D	E	N	1969	55763
9169	S	D	E	N	1971	63042
9169	S	D	R	H	1971	65109
9169	S	D	E	H	1971	65110
9171	S	D	E	N	1954	38130
9171	S	D	E	N	1934	40799
9171	S	D	R	N	1965	42546
9171	S	G	F	N	1928	42820
9171	S	D	J	N	1955	43250
9171	S	T	J	N	1955	43250
9171	S	D	R	N	1967	48659
9171	S	D	R	N	1968	51983
9171	S	D	E	N	1968	51984
9171	S	D	E	N	1969	55711
9171	S	D	E	N	1970	56113
9171	S	D	E	N	1970	57256
9171	S	D	G	N	1969	58212
9171	S	D	G	N	1958	58457
9171	S	T	G	N	1958	58457
9171	S	D	J	N	1970	59003
9171	S	D	E	N	1927	60002
9171	S	D	E	N	1971	67411
9171	S	D	E	N	1973	69333
9171	S	D	E	N	1970	77005
9172	S	D	G	N	1969	58212
9172	S	D	O	N	1974	76092
9172	S	T	O	N	1974	76092
9173	D	D	P	N	1969	56272
9173	D	D	E	N	1928	58924
9173	S	D	E	N	1933	37748
9173	S	D	R	N	1964	37755
9173	S	D	E	N	1934	37825
9173	S	D	C	N	1964	41897
9173	S	T	C	N	1964	41897
9173	S	D	E	N	1965	42031
9173	S	D	J	N	1955	43250
9173	S	T	J	N	1955	43250
9173	S	D	G	N	1966	45549
9173	S	D	R	N	1967	48659
9173	S	D	G	N	1968	50596
9173	S	D	E	N	1954	52479
9173	S	D	R	N	1967	53151
9173	S	D	R	N	1967	53155
9173	S	D	R	N	1953	54459
9173	S	D	E	N	1944	55268
9173	S	D	E	N	1965	55966
9173	S	D	P	N	1969	56272
9173	S	D	G	N	1958	58457
9173	S	T	G	N	1958	58457
9173	S	D	E	N	1928	58924
9173	S	D	E	N	1930	59875
9173	S	D	E	N	1920	59946
9173	S	D	E	N	1934	59947
9173	S	D	E	N	1927	60002
9173	S	D	E	N	1952	60012
9173	S	D	E	N	1941	60370
9173	S	D	G	N	1931	61780
9173	S	D	R	N	1970	62576
9173	S	D	E	N	1953	63073
9173	S	D	E	N	1971	63238
9173	S	T	E	N	1971	63238
9173	S	D	E	N	1970	67361
9173	S	D	O	N	1973	73520
9173	S	T	O	N	1973	73520
9174	D	D	E	N	1956	37574
9174	D	D	R	N	1956	60736
9174	P	D	E	N	1968	52243
9174	S	D	E	N	1956	37172
9174	S	D	E	N	1970	37340
9174	S	D	E	N	1956	37574
9174	S	D	E	N	1933	37748
9174	S	D	E	N	1951	38312
9174	S	T	E	N	1951	38312
9174	S	D	E	N	1934	40799
9174	S	D	R	N	1966	41504
9174	S	D	C	N	1964	41897
9174	S	T	C	N	1964	41897
9174	S	D	R	N	1965	42545
9174	S	D	R	N	1965	42546
9174	S	G	F	N	1928	42820
9174	S	D	E	N	1967	45618
9174	S	D	E	N	1968	47538
9174	S	D	E	N	1968	50611
9174	S	D	E	N	1968	52243
9174	S	D	R	N	1969	54435
9174	S	D	R	N	1969	56889
9174	S	D	E	N	1969	57079
9174	S	D	G	N	1969	58212
9174	S	D	G	N	1958	58457
9174	S	T	G	N	1958	58457
9174	S	D	F	N	1926	59811
9174	S	D	F	N	1934	59812
9174	S	D	E	N	1934	59947
9174	S	D	E	N	1935	59948
9174	S	D	E	N	1929	60001
9174	S	D	E	N	1927	60002

Substance Number	Phys. State	Sub-ject	Lan-guage	Temper-ature	Year	TPRC Number
503-9174	S	D	R	N	1956	60736
9174	S	D	E	N	1971	63238
9174	S	T	E	N	1971	63238
9174	S	D	E	N	1970	67661
9174	S	T	E	N	1970	67661
9174	S	D	E	N	1970	77005
9175	S	D	E	N	1969	55007
9176	S	D	E	N	1969	52843
9177	S	D	G	N	1968	54930
9178	S	D	E	N	1931	60003
9179	S	D	E	N	1956	37172
9179	S	D	R	N	1967	48659
9181	S	D	G	N	1966	41425
9181	S	G	F	N	1928	42820
9181	S	D	E	N	1941	47074
9181	S	D	D	N	1967	47237
9181	S	D	E	N	1967	50145
9181	S	D	J	N	1969	55538
9181	S	D	E	N	1941	55942
9181	S	D	G	N	1969	58212
9181	S	D	E	L	1964	58875
9181	S	D	E	N	1942	59949
9181	S	D	E	N	1972	67214
9181	S	D	E	N	1971	67411
9181	S	D	E	N	1970	67661
9181	S	T	E	N	1970	67661
9181	S	D	E	N	1973	69333
9181	S	D	E	N	1970	77005
9187	S	D	J	H	1977	89033
9197	S	D	E	N	1965	77596
9198	S	D	R	H	1974	83589
9198	S	D	R	N	1974	83589
9198	S	D	E	H	1974	83590
9198	S	D	E	N	1974	83590
9200	F	D	E	N	1974	77949
9200	F	E	E	N	1974	77949
9201	S	D	G	H	1959	77314
9201	S	D	E	H	1963	77315
9203	S	D	R	N	1973	75640
9203	S	D	E	N	1973	75641
9204	S	D	E	H	1963	77282
9205	S	D	E	N	1960	41838
9205	S	D	E	H	1963	77282
9205	S	S	E	N	1977	88199
9208	S	D	E	N	1971	73893
9209	S	D	E	N	1971	73893
9210	S	D	E	N	1971	73893
9221	S	D	R	N	1967	47638
9222	S	D	R	N	1967	47639
9222	S	D	R	N	1969	56321
9223	S	D	R	N	1967	47640
9224	S	D	R	N	1967	47640
9225	S	D	E	H	1971	61176
9225	S	E	E	H	1971	61176
9226	S	D	E	H	1971	61176
9226	S	E	E	H	1971	61176
9227	S	D	E	H	1971	61176
9227	S	E	E	H	1971	61176
9228	S	D	E	N	1971	70964
9228	S	E	E	N	1971	70964
9229	S	D	E	N	1971	70964
9229	S	E	E	N	1971	70964
9230	D	D	E	N	1971	70964
9230	D	E	E	N	1971	70964
9230	S	D	E	N	1971	70964
9230	S	E	E	N	1971	70964
9231	S	D	E	N	1973	71232
9232	S	D	E	N	1973	71231
9235	D	D	E	N	1973	69622
9236	D	D	E	N	1973	69622
9237	D	D	E	N	1973	69622
9249	S	D	R	N	1971	63644
9249	S	D	E	N	1971	66789
9250	S	D	E	H	1942	64933
9251	S	D	E	N	1971	73893
9252	S	D	P	N	1967	48452
9252	S	E	P	N	1967	48452
9253	S	D	P	N	1967	48452
9253	S	E	P	N	1967	48452
9254	S	D	O	N	1969	57518
9255	S	D	O	N	1969	57518
9259	S	D	G	N	1914	66628
9260	S	D	R	N	1969	56322
9262	S	D	E	N	1972	66636
9267	S	D	R	N	1965	46455
9268	S	D	R	N	1965	46455
9271	S	D	R	N	1966	49853
9272	S	D	J	H	1965	44606
9273	S	D	J	H	1965	44606
9274	S	D	J	H	1965	44606
9275	S	D	J	H	1965	44606
9276	S	D	J	H	1965	44606
9277	S	D	J	H	1965	44606
9284	S	D	R	H	1971	66437
9284	S	D	E	H	1971	66438
9285	S	D	R	N	1967	47638
9285	S	D	R	N	1967	47639
9285	S	D	R	N	1969	56255
9286	S	D	R	N	1969	56251
9287	S	D	R	N	1969	56251
9288	S	D	R	N	1969	56251

Substance Number	Phys. State	Sub-ject	Lan-guage	Temper-ature	Year	TPRC Number
503-9288	S	D	R	N	1969	56322
9288	S	D	R	N	1970	63601
9289	S	D	R	N	1969	56253
9290	S	D	R	N	1969	56253
9291	S	D	R	N	1969	56253
9292	S	D	R	N	1969	56253
9293	S	D	R	H	1971	66125
9293	S	D	E	H	1971	66126
9294	S	D	G	N	1971	64254
9297	S	D	P	N	1967	48452
9297	S	E	P	N	1967	48452
9297	S	D	R	H	1971	56437
9297	S	D	E	H	1971	66438
9300	S	D	E	N	1971	64462
9301	S	D	E	N	1971	62073
9301	S	D	E	N	1971	63043
9302	S	D	R	N	1971	64915
9302	S	D	E	N	1971	64916
9303	D	D	E	N	1971	64013
9303	S	D	E	N	1971	64013
9304	S	D	R	N	1965	54787
9304	S	D	E	N	1965	54788
9305	S	D	R	N	1969	64018
9305	S	D	E	N	1969	64019
9311	S	D	J	H	1962	62293
9312	S	D	J	H	1962	62293
9313	S	D	R	N	1969	56326
9314	S	D	R	N	1969	56326
9315	S	D	R	N	1969	56326
9316	S	D	R	N	1969	56326
9324	S	D	R	N	1969	56321
9324	S	D	R	N	1964	58769
9324	S	D	E	N	1971	62788
9325	S	D	R	N	1956	43654
9325	S	D	R	N	1950	62309
9326	S	D	E	N	1937	42208
9327	S	D	E	N	1970	59974
9327	S	D	E	N	1971	63401
9327	S	D	E	N	1970	77532
9327	S	D	E	N	1970	77562
9328	S	D	E	N	1967	36019
9328	S	D	R	N	1967	46353
9329	S	D	E	N	1970	59655
9332	S	D	E	H	1949	55856
9333	S	D	E	N	1944	55268
9334	S	D	E	N	1944	55268
9335	S	D	E	N	1944	55268
9336	S	D	E	N	1964	34704
9337	S	D	E	N	1964	34704
9338	S	D	E	N	1964	34704
9339	S	D	E	N	1964	34704
9340	S	D	E	N	1964	34704
9341	S	D	E	N	1964	34704
9342	S	D	E	N	1964	34704
9343	S	D	E	N	1964	34704
9344	S	D	E	N	1964	34704
9345	S	D	E	N	1964	34704
9346	S	D	E	N	1964	34704
9347	S	D	E	N	1964	34704
9347	S	D	E	N	1971	67369
9347	S	E	E	N	1971	67369
9348	S	D	E	N	1964	34704
9349	S	D	E	N	1964	34704
9350	S	D	E	N	1964	34704
9351	S	D	E	N	1964	34704
9352	S	D	E	N	1964	34704
9353	S	D	E	N	1964	34704
9354	S	D	G	N	1927	59560
9358	S	D	E	H	1943	50711
9359	S	D	E	H	1943	50711
9360	S	D	E	H	1943	50711
9361	S	D	E	H	1943	50711
9362	S	D	G	N	1966	48685
9362	S	D	E	H	1943	50711
9363	S	D	E	H	1943	50711
9364	S	D	E	N	1932	35483
9365	S	D	E	N	1932	34703
9366	S	D	E	N	1949	37123
9367	S	D	E	N	1930	41725
9367	S	D	R	N	1964	58769
9367	S	D	E	N	1971	62788
9368	S	D	E	N	1930	41725
9369	S	D	E	N	1930	41725
9370	S	D	E	N	1930	41725
9370	S	D	G	N	1927	59560
9371	S	D	E	N	1930	41725
9372	S	D	E	N	1930	41725
9373	S	D	E	N	1930	41725
9374	S	D	E	N	1930	41725
9374	S	D	G	N	1927	59560
9374	S	D	E	N	1970	67661
9374	S	T	E	N	1970	67661
9375	S	D	E	N	1930	41725
9376	S	D	E	N	1930	41725
9377	S	D	E	N	1930	41725
9378	S	D	E	N	1930	41725
9379	S	D	E	N	1930	41725
9380	S	D	E	N	1930	41725
9380	S	D	G	N	1927	59560
9381	S	D	E	N	1930	41725
9382	S	D	E	N	1930	41725

Phys. State: **C.** Superconductive; **D.** Doped; **E.** Expanded; **F.** Fibrous or Whisker; **G.** Gas; **I.** Ionized or Plasma; **L.** Liquid; **M.** Multiphase; **P.** Powder or Fine Particle; **S.** Solid; **T.** Thick or Thin Film
Subject: **D.** Data; **E.** Experiment; **G.** General (Data + Theory + Experiment); **S.** Survey (Review, Compendium etc.); **T.** Theory
Language: **C.** Czech; **D.** Dutch; **E.** English; **F.** French; **G.** German; **I.** Italian; **J.** Japanese; **O.** Other Languages; **P.** Polish; **R.** Russian; **S.** Spanish
Temperature: **F.** Full Range (Low + Normal + High); **L.** Low (0 to 75K) + Overlap into Normal; **N.** Normal (75 to 1273K); **H.** High (above 1273K) + Overlap into Normal; Blank = Not Coded

Substance Number	Phys. State	Subject	Language	Temperature	Year	TPRC Number
503-9383	S	D	E	N	1935	44430
9384	S	D	E	N	1935	44430
9385	S	D	E	N	1935	44430
9386	S	D	E	N	1935	44430
9387	S	D	E	N	1930	41725
9388	S	D	E	N	1930	41725
9388	S	D	G	N	1927	59560
9389	S	D	E	N	1930	41725
9389	S	D	E	N	1944	55268
9389	S	D	G	N	1927	59560
9390	S	D	E	N	1930	41725
9391	S	D	E	N	1930	41725
9391	S	D	E	N	1926	60434
9392	S	D	G	N	1927	59560
9392	S	D	E	N	1926	60434
9395	S	D	E	N	1946	50114
9396	S	D	E	N	1949	56188
9397	S	D	E	N	1949	56188
9398	S	D	S	N	1964	59722
9399	S	D	J	N	1955	60412
9399	S	D	E	N	1973	70701
9400	S	D	J	N	1955	60412
9401	S	D	J	N	1955	60412
9402	S	D	E	N	1968	53133
9403	S	D	E	N	1969	55290
9404	S	D	E	N	1955	60460
9405	S	D	E	N	1955	60460
9406	S	D	E	N	1964	34704
9406	S	D	E	N	1955	60460
9407	S	D	E	N	1951	60456
9407	S	D	G	H	1959	77314
9407	S	D	E	H	1963	77315
9408	S	D	E	N	1938	60448
9409	D	D	G	N	1960	34127
9411	S	D	G	N	1914	59966
9412	S	D	G	N	1914	59966
9412	S	D	E	N	1970	67661
9412	S	T	E	N	1970	67661
9413	S	D	G	N	1914	59966
9414	S	D	G	N	1914	59966
9426	S	D	R	H	1970	60841
9426	S	D	E	H	1970	60842
9428	S	D	R	N	1969	56253
9428	S	D	G	N	1914	59966
9429	S	D	E	H	1953	7036
9429	S	D	J	N	1955	60412
9429	S	D	E	N	1956	60562
9429	S	S	E	N	1977	88199
9430	S	D	E	N	1968	54814
9430	S	D	E	N	1925	60560
9431	S	D	R	H	1970	61172
9431	S	D	E	H	1970	61173
9436	S	D	E	N	1952	59988
9438	S	D	R	N	1969	56890
9439	S	D	R	N	1969	58286
9440	S	D	E	N	1951	59510
9441	S	D	O	N	1969	57020
9442	S	D	G	N	1970	58486
9443	S	D	G	N	1970	58486
9444	S	D	E	N	1930	41725
9444	S	D	E	N	1936	44125
9444	S	D	E	N	1939	45359
9444	S	D	G	N	1970	58486
9444	S	D	G	N	1971	64254
9445	S	D	E	N	1930	41725
9445	S	D	G	N	1970	58486
9445	S	D	G	N	1927	59560
9445	S	D	G	N	1971	64254
9446	S	D	R	H	1969	57817
9446	S	D	E	H	1969	57818
9447	S	D	E	H	1970	58343
9448	S	D	E	N	1969	58271
9448	S	T	E	N	1969	58271
9451	S	D	R	N	1969	55387
9451	S	D	G	N	1971	64254
9452	S	D	E	N	1968	53097
9453	S	D	R	H	1969	58854
9453	S	D	E	H	1969	58855
9454	S	D	R	N	1969	57149
9454	S	D	E	N	1969	57150
9455	S	D	R	N	1968	56794
9455	S	D	R	N	1969	57149
9455	S	D	E	N	1969	57150
9456	S	D	R	N	1968	56794
9456	S	D	R	N	1969	57149
9456	S	D	E	N	1969	57150
9456	S	D	R	H	1969	57263
9456	S	D	E	H	1969	57264
9456	S	D	E	H	1972	66639
9457	S	D	R	N	1968	56794
9457	S	D	R	N	1969	57149
9457	S	D	E	N	1969	57150
9458	S	D	R	N	1968	56794
9458	S	D	R	N	1969	57149
9458	S	D	E	N	1969	57150
9459	S	D	R	N	1968	56794
9459	S	D	R	N	1969	57149
9459	S	D	E	N	1969	57150
9460	S	D	R	N	1968	56794
9460	S	D	R	N	1969	57149
9460	S	D	E	N	1969	57150

Substance Number	Phys. State	Subject	Language	Temperature	Year	TPRC Number
503-9461	S	D	E	H	1970	56455
9462	S	D	R	H	1969	57086
9462	S	D	E	H	1969	57087
9468	S	D	E	N	1969	37362
9471	S	D	E	N	1967	52868
9472	S	D	E	N	1967	52868
9473	S	D	E	N	1967	52868
9474	S	D	E	N	1967	52868
9475	S	D	E	N	1967	52868
9476	S	D	E	N	1967	52868
9477	S	D	E	N	1964	52525
9478	S	D	E	N	1968	50499
9479	S	D	E	N	1968	50499
9480	S	D	R	N	1965	51130
9480	S	D	E	N	1967	51131
9481	S	D	R	H	1965	51130
9481	S	D	E	H	1967	51131
9481	S	D	R	H	1971	66125
9481	S	D	E	H	1971	66126
9481	S	D	R	H	1971	66453
9481	S	D	E	H	1971	66454
9483	S	D	G	N	1968	51350
9483	S	D	R	H	1969	57271
9483	S	D	E	H	1969	57272
9484	S	D	E	N	1964	37509
9485	S	D	E	N	1964	37509
9486	S	D	E	N	1964	37509
9486	S	D	E	H	1972	66639
9486	S	D	R	N	1972	67684
9486	S	D	E	N	1972	70601
9487	S	D	E	N	1964	37509
9488	S	D	E	N	1964	37509
9489	S	D	E	H	1947	41534
9490	S	D	E	H	1947	41534
9491	S	D	E	H	1947	41534
9492	S	D	E	N	1946	41574
9493	S	D	E	N	1946	41574
9494	S	D	E	H	1946	41574
9495	S	D	E	N	1948	37113
9495	S	D	E	H	1946	41574
9496	S	D	E	N	1969	38523
9498	S	D	R	N	1968	55793
9498	S	T	R	N	1968	55793
9498	S	D	E	N	1969	56586
9498	S	T	E	N	1969	56586
9499	S	D	R	N	1968	55793
9499	S	T	R	N	1968	55793
9499	S	D	E	N	1969	56586
9499	S	T	E	N	1969	56586
9501	S	D	E	N	1967	53283
9502	S	D	E	N	1967	53283
9503	S	D	E	N	1957	43625
9504	S	D	E	N	1964	45842
9504	S	D	R	N	1969	55387
9505	S	D	E	N	1964	45842
9506	S	D	E	N	1964	45842
9507	S	D	E	N	1930	41725
9508	S	D	O	N	1968	53266
9508	S	D	E	N	1949	56188
9509	S	D	R	N	1963	39194
9511	S	D	E	H	1946	41574
9515	S	D	E	N	1968	37106
9515	S	D	R	N	1968	50768
9515	S	D	R	N	1966	51043
9515	S	T	R	N	1966	51043
9515	S	D	R	N	1968	55793
9515	S	T	R	N	1968	55793
9515	S	D	E	N	1969	56586
9515	S	T	E	N	1969	56586
9515	S	D	R	N	1969	57098
9515	S	D	E	N	1969	57099
9516	S	D	E	N	1968	37106
9516	S	D	R	N	1968	50768
9517	S	D	E	N	1968	37106
9517	S	D	R	N	1968	50768
9517	S	D	R	N	1966	51043
9517	S	T	R	N	1966	51043
9517	S	D	R	N	1968	55793
9517	S	T	R	N	1968	55793
9517	S	D	E	N	1969	56586
9517	S	T	E	N	1969	56586
9517	S	D	R	N	1969	57098
9517	S	D	E	N	1969	57099
9518	S	D	E	N	1968	37106
9518	S	D	R	N	1968	50768
9519	S	D	R	H	1968	55434
9519	S	D	E	H	1968	55435
9520	S	D	R	H	1968	55434
9520	S	D	E	H	1968	55435
9521	S	D	R	H	1968	55442
9521	S	D	E	N	1968	55443
9522	S	D	G	N	1962	53242
9523	S	D	J	N	1960	52431
9526	S	D	J	N	1960	52431
9527	S	D	J	N	1960	52431
9528	S	D	J	N	1960	52431
9531	S	D	G	N	1968	50065
9532	S	D	E	N	1949	37123
9532	S	D	E	H	1946	55185
9533	S	D	G	N	1966	45267
9534	S	D	E	N	1968	50485

Substance Number	Phys. State	Subject	Language	Temperature	Year	TPRC Number
503-9535	S	D	E	H	1946	41574
9535	S	D	E	N	1971	63043
9535	S	D	R	N	1968	63328
9535	S	D	E	N	1968	63329
9536	S	D	E	N	1946	41574
9536	S	D	O	N	1965	44879
9537	S	D	E	H	1970	58729
9538	P	D	R	H	1968	54779
9538	P	D	E	H	1968	54780
9541	S	D	E	H	1942	43643
9541	S	D	G	H	1968	53251
9541	S	D	G	H	1969	56623
9542	S	D	E	N	1962	52662
9542	S	T	E	N	1962	52662
9542	S	D	E	N	1962	52694
9543	S	D	E	N	1965	52586
9543	S	D	E	N	1962	52662
9543	S	T	E	N	1962	52662
9543	S	D	E	N	1962	52694
9544	S	D	E	N	1965	52586
9544	S	D	E	N	1962	52662
9544	S	T	E	N	1962	52662
9544	S	D	E	N	1962	52694
9545	S	D	E	N	1962	52662
9545	S	T	E	N	1962	52662
9545	S	D	E	N	1962	52694
9546	S	D	E	N	1962	52662
9546	S	T	E	N	1962	52662
9546	S	D	E	N	1962	52694
9547	S	D	E	N	1962	52662
9547	S	T	E	N	1962	52662
9547	S	D	E	N	1962	52694
9548	S	D	E	N	1960	52693
9548	S	D	R	H	1971	66125
9548	S	D	E	H	1971	66126
9549	S	D	E	N	1960	52693
9550	S	D	E	N	1960	52693
9551	S	D	E	N	1960	52693
9552	S	D	E	N	1960	52693
9553	S	D	E	N	1932	52690
9554	S	D	E	N	1932	52690
9555	S	D	R	N	1959	52538
9555	S	D	E	N	1959	52539
9555	S	D	E	N	1962	52661
9556	S	D	R	N	1959	52538
9556	S	D	E	N	1959	52539
9557	S	D	E	N	1964	41670
9558	S	D	E	N	1964	37509
9563	S	D	E	N	1965	52586
9563	S	D	E	N	1969	53775
9563	S	D	E	N	1964	77265
9564	S	D	E	N	1967	52868
9564	S	D	E	N	1969	53775
9565	S	D	E	N	1967	52868
9565	S	D	E	N	1969	53775
9566	S	D	E	N	1967	52868
9566	S	D	E	N	1969	53775
9567	S	D	E	H	1964	52699
9569	S	D	E	N	1971	73893
9570	S	D	E	N	1964	46491
9571	S	D	E	N	1970	58727
9572	S	D	E	N	1970	58727
9573	S	D	R	N	1966	41970
9574	S	D	E	N	1963	45380
9575	S	D	E	N	1953	52563
9575	S	D	E	H	1963	52805
9576	S	D	E	H	1963	52805
9577	S	D	R	N	1967	45269
9577	S	D	E	N	1967	46273
9578	S	D	R	N	1967	45269
9578	S	D	E	N	1967	46273
9579	S	D	E	H	1963	52805
9580	S	D	E	N	1968	47117
9581	S	D	E	N	1968	47117
9582	S	D	E	N	1968	47117
9583	S	D	E	N	1930	41725
9583	S	D	E	N	1968	47117
9584	S	D	E	N	1968	47117
9585	S	D	E	N	1930	41725
9585	S	D	E	N	1968	46742
9586	S	D	G	N	1954	36744
9586	S	T	G	N	1954	36744
9586	S	D	E	N	1930	41725
9586	S	D	E	N	1968	46742
9586	S	D	G	N	1927	59560
9586	S	D	G	N	1914	59966
9586	S	D	E	N	1952	59988
9586	S	D	E	N	1926	60434
9587	S	D	R	N	1966	40891
9587	S	D	E	N	1966	40892
9587	S	D	E	N	1930	41725
9587	S	D	G	N	1927	59560
9587	S	D	R	N	1970	63601
9588	S	D	R	N	1967	45346
9588	S	D	R	N	1969	56321
9588	S	D	R	N	1970	63601
9589	S	D	R	H	1968	55434
9589	S	D	E	H	1968	55435
9590	S	D	E	H	1958	52565
9590	S	D	E	H	1957	52778
9591	S	D	E	H	1958	52565

Phys. State: **C.** Superconductive; **D.** Doped; **E.** Expanded; **F.** Fibrous or Whisker; **G.** Gas; **I.** Ionized or Plasma; **L.** Liquid; **M.** Multiphase; **P.** Powder or Fine Particle; **S.** Solid; **T.** Thick or Thin Film
Subject: **D.** Data; **E.** Experiment; **G.** General (Data + Theory + Experiment); **S.** Survey (Review, Compendium etc.); **T.** Theory
Language: **C.** Czech; **D.** Dutch; **E.** English; **F.** French; **G.** German; **I.** Italian; **J.** Japanese; **O.** Other Languages; **P.** Polish; **R.** Russian; **S.** Spanish
Temperature: **F.** Full Range (Low + Normal + High); **L.** Low (0 to 75K) + Overlap into Normal; **N.** Normal (75 to 1273K); **H.** High (above 1273K) + Overlap into Normal; Blank = Not Coded

Substance Number	Phys. State	Subject	Language	Temperature	Year	TPRC Number
503-9591	S	D	E	H	1957	52778
9592	S	D	E	H	1947	41534
9592	S	D	E	H	1963	52481
9593	S	D	E	N	1930	41738
9593	S	D	E	N	1961	43328
9593	S	D	E	H	1925	52689
9593	S	D	O	H	1968	54560
9593	S	D	R	N	1969	58856
9593	S	D	E	N	1969	58857
9593	S	D	E	N	1929	60442
9593	S	D	E	N	1928	60525
9594	S	D	E	H	1961	52522
9595	S	D	E	N	1957	43625
9596	S	D	E	H	1925	52689
9597	S	D	E	N	1963	45352
9598	S	D	E	N	1963	45352
9599	S	D	E	N	1963	45352
9600	S	D	E	N	1963	45352
9601	S	D	E	N	1963	45352
9602	S	D	E	N	1963	45352
9603	S	D	E	N	1963	45352
9604	S	D	E	N	1963	45352
9605	S	D	E	N	1963	45352
9606	S	D	E	N	1941	36711
9606	S	D	E	H	1947	41534
9606	S	D	J	H	1957	43650
9606	S	D	E	H	1957	45056
9606	S	D	E	N	1935	47033
9606	S	D	E	H	1965	52286
9606	S	D	E	N	1926	55980
9607	S	D	E	H	1963	52285
9608	S	D	E	H	1965	39834
9608	S	D	E	H	1964	52768
9609	S	D	E	H	1965	39834
9609	S	D	E	H	1964	52768
9609	S	D	E	N	1941	60452
9610	S	D	E	H	1965	39834
9610	S	D	E	N	1949	43559
9611	S	D	E	N	1958	52287
9611	S	D	R	H	1970	60837
9611	S	D	E	H	1970	60838
9612	S	S	E	N	1960	24931
9612	S	D	E	N	1955	52405
9612	S	D	G	N	1927	59560
9613	S	D	E	N	1962	52418
9614	S	D	E	N	1963	52276
9614	S	D	R	N	1969	56326
9615	S	D	R	H	1968	50788
9615	S	D	E	H	1968	52347
9616	S	D	R	N	1967	47640
9617	S	D	E	N	1968	58396
9618	S	D	R	N	1967	47640
9620	S	S	E	N	1960	24931
9620	S	D	E	N	1955	52405
9620	S	S	E	N	1977	88199
9621	S	D	E	N	1959	44960
9621	S	D	E	H	1925	52689
9621	S	D	J	N	1955	60412
9622	S	D	R	N	1966	49052
9623	S	D	G	H	1931	39517
9623	S	D	E	N	1946	52256
9624	S	D	E	N	1946	52256
9625	S	D	E	N	1962	52684
9628	S	D	E	H	1962	46010
9630	S	D	E	N	1962	52662
9630	S	T	E	N	1962	52662
9630	S	D	E	N	1962	52694
9631	S	D	E	N	1967	52587
9633	S	D	E	N	1932	34703
9635	S	D	E	N	1967	52587
9636	S	D	E	N	1967	52587
9640	S	D	E	H	1954	46789
9642	S	D	E	N	1958	47988
9644	S	D	F	N	1969	54268
9644	S	D	E	H	1970	58186
9645	S	D	E	N	1967	52055
9645	S	D	E	H	1964	77283
9647	S	D	E	N	1959	52652
9648	S	D	E	N	1970	56161
9649	S	D	E	N	1951	52135
9650	S	D	E	N	1951	52135
9651	S	D	E	N	1960	52128
9652	S	D	E	N	1960	52128
9652	S	D	R	H	1974	75178
9652	S	D	E	H	1974	76900
9652	S	G	E	N	1971	77933
9653	S	D	E	N	1959	52652
9654	S	D	E	N	1952	47530
9654	S	D	E	N	1962	52662
9654	S	T	E	N	1962	52662
9654	S	D	E	N	1962	52694
9655	S	D	E	N	1952	47530
9656	S	D	E	N	1952	47530
9656	S	D	E	N	1952	52749
9657	S	D	E	N	1952	47530
9658	S	D	E	N	1965	51723
9658	S	D	O	N	1968	54558
9658	S	D	R	N	1969	54919
9658	S	D	R	N	1968	55442
9658	S	D	E	N	1968	55443
9658	S	D	R	N	1965	60876

Substance Number	Phys. State	Subject	Language	Temperature	Year	TPRC Number
503-9658	S	S	E	N	1977	88199
9659	S	D	E	N	1952	59512
9660	S	D	E	N	1932	34703
9660	S	D	E	N	1964	41670
9660	S	D	R	N	1967	47640
9660	S	D	E	N	1962	48993
9660	S	D	E	N	1932	57549
9661	D	D	E	N	1962	52661
9662	S	D	E	N	1932	35483
9662	S	D	E	N	1962	52661
9662	S	D	R	N	1969	56255
9662	S	D	E	N	1971	64013
9663	S	D	E	H	1967	48428
9664	S	D	R	N	1967	48447
9665	S	D	R	N	1967	48447
9666	S	D	E	N	1957	43625
9666	S	D	R	N	1956	43654
9666	S	D	R	N	1968	55793
9666	S	T	R	N	1968	55793
9666	S	D	E	N	1969	56586
9666	S	T	E	N	1969	56586
9666	S	D	R	N	1950	62309
9667	S	D	E	N	1962	52661
9668	S	D	G	H	1967	50309
9669	S	D	I	N	1967	48686
9670	S	D	E	N	1962	52661
9671	S	D	E	N	1933	42478
9671	S	D	E	H	1967	46164
9671	S	D	R	H	1968	49613
9671	S	D	R	H	1958	52116
9671	S	D	E	H	1958	52117
9671	S	D	R	N	1955	54638
9671	S	D	E	N	1943	60453
9671	S	D	E	N	1925	60560
9671	S	D	C	N	1971	63012
9671	S	T	C	N	1971	63012
9671	S	D	E	N	1955	63074
9678	S	D	E	H	1964	37378
9678	S	D	E	H	1966	43748
9678	S	D	E	H	1964	45353
9678	S	D	F	H	1967	49826
9678	S	D	E	H	1963	52285
9678	S	D	E	H	1965	52286
9678	S	D	E	H	1969	54983
9678	S	D	F	H	1969	57614
9678	S	E	F	H	1969	57614
9678	S	D	E	H	1971	65538
9678	S	D	E	H	1973	69867
9678	S	D	R	N	1972	71864
9680	S	D	J	H	1965	44606
9681	S	D	E	N	1964	46491
9681	S	D	E	N	1968	48014
9682	S	D	E	N	1971	44828
9682	S	D	R	N	1931	60851
9686	S	D	E	N	1930	41725
9686	S	D	G	N	1968	50065
9686	S	D	G	N	1927	59560
9687	S	D	E	N	1961	52574
9688	S	D	G	H	1931	39517
9688	S	D	E	N	1957	46503
9688	S	D	E	N	1929	60442
9689	S	D	E	N	1946	52256
9691	S	D	E	N	1968	51217
9693	S	D	E	H	1967	43379
9693	S	D	G	N	1968	51350
9694	S	D	E	N	1968	37106
9694	S	D	J	H	1965	44606
9694	S	D	R	N	1968	50768
9694	S	D	R	N	1968	55793
9694	S	T	R	N	1968	55793
9694	S	D	E	N	1969	56586
9694	S	T	E	N	1969	56586
9694	S	D	R	N	1969	57098
9694	S	D	E	N	1969	57099
9695	S	D	J	H	1965	44606
9695	S	D	P	N	1967	48452
9695	S	E	P	N	1967	48452
9695	S	D	R	N	1968	55793
9695	S	T	R	N	1968	55793
9695	S	D	E	N	1969	56586
9695	S	T	E	N	1969	56586
9698	S	D	E	N	1961	52574
9701	S	D	R	N	1967	47640
9703	S	D	E	N	1961	52574
9704	S	D	E	N	1963	55332
9706	S	D	F	H	1966	46819
9707	S	D	E	H	1956	40763
9707	S	D	E	H	1959	44960
9707	S	D	E	H	1958	52565
9707	S	D	E	H	1958	52575
9707	S	D	E	N	1962	52662
9707	S	T	E	N	1962	52662
9707	S	D	E	N	1962	52694
9707	S	D	E	H	1954	60458
9709	S	D	E	N	1953	58921
9710	S	D	R	N	1969	56251
9710	S	D	R	H	1969	57817
9710	S	D	E	H	1969	57818
9712	S	D	E	H	1967	39997
9712	S	D	E	H	1969	53771
9713	P	D	E	N	1967	46031

Substance Number	Phys. State	Subject	Language	Temperature	Year	TPRC Number
503-9714	S	D	E	H	1963	52285
9714	S	D	E	H	1965	52286
9716	S	D	E	H	1961	29231
9716	S	D	E	N	1967	35229
9716	S	D	E	H	1964	37378
9716	S	D	E	H	1966	44936
9716	S	D	E	H	1945	45785
9716	S	D	R	H	1965	51130
9716	S	D	E	H	1967	51131
9716	S	D	R	H	1968	52350
9716	S	D	E	H	1968	52351
9716	S	D	E	H	1964	52429
9716	S	D	E	H	1963	52481
9716	S	D	E	N	1969	54502
9716	S	T	E	N	1969	54502
9716	S	D	R	H	1968	55422
9716	S	D	E	H	1968	55423
9716	S	D	R	N	1968	55432
9716	S	D	E	N	1968	55433
9716	S	D	G	H	1926	55999
9716	S	D	E	N	1970	58727
9716	S	D	R	H	1971	64848
9716	S	D	E	H	1971	64849
9716	S	D	R	N	1972	71864
9716	S	D	E	N	1973	76994
9718	S	D	E	H	1967	52120
9718	S	D	E	L	1974	74420
9719	S	D	E	N	1963	43911
9719	S	D	E	N	1971	63043
9720	S	D	G	H	1968	53251
9720	S	D	E	N	1967	53283
9723	S	D	E	N	1963	43909
9725	S	D	G	N	1966	48685
9725	S	D	E	H	1942	49491
9726	S	D	E	H	1947	41534
9726	S	D	F	H	1949	53232
9728	S	D	E	N	1968	37106
9728	S	D	R	H	1962	38443
9728	S	D	E	H	1962	41856
9728	S	D	P	N	1967	48452
9728	S	E	P	N	1967	48452
9728	S	D	R	N	1968	50768
9728	S	D	E	N	1932	52690
9731	S	D	E	N	1965	38236
9731	S	D	E	H	1956	40763
9731	S	D	E	H	1947	41534
9731	S	D	J	H	1965	44604
9731	S	D	R	H	1966	45561
9731	S	D	E	H	1945	45785
9731	S	D	R	H	1968	47019
9731	S	D	E	H	1968	47020
9731	S	D	E	N	1950	47674
9731	S	D	E	H	1967	51545
9731	S	D	R	H	1968	52350
9731	S	D	E	H	1968	52351
9731	S	D	E	H	1963	52481
9731	S	D	E	H	1958	52575
9731	S	D	F	H	1949	53232
9731	S	D	R	N	1969	54441
9731	S	D	R	H	1969	54518
9731	S	D	E	N	1969	57104
9731	S	D	E	H	1969	57265
9731	S	D	E	H	1956	77264
9732	S	D	E	N	1954	25427
9734	S	S	E	N	1960	24931
9734	S	D	E	N	1967	43453
9734	S	D	J	N	1971	64876
9734	S	S	E	N	1977	88199
9735	S	D	E	H	1958	52565
9735	S	D	E	H	1959	52652
9736	S	G	C	H	1968	49012
9736	S	D	E	H	1925	52689
9736	S	D	R	N	1967	53463
9736	S	D	E	H	1971	75168
9738	S	D	R	N	1966	46568
9738	S	D	E	N	1966	46569
9738	S	D	P	N	1967	48452
9738	S	E	P	N	1967	48452
9739	S	D	E	H	1968	53097
9742	S	D	E	N	1945	55985
9743	S	D	E	H	1958	52565
9745	S	D	R	N	1968	56794
9745	S	D	J	H	1962	62293
9747	S	D	E	H	1963	52698
9748	S	D	E	N	1966	34768
9748	S	D	E	H	1954	52567
9748	S	D	E	N	1965	52586
9748	S	D	E	N	1962	52662
9748	S	T	E	N	1962	52662
9748	S	D	E	N	1959	52667
9748	S	D	E	N	1962	52694
9748	S	D	E	N	1967	52868
9748	S	D	E	N	1969	53775
9748	S	D	R	N	1956	56394
9748	S	D	E	N	1970	59655
9748	S	D	J	H	1962	62293
9750	S	D	E	N	1946	3081
9750	S	D	E	N	1932	34703
9750	S	D	E	N	1966	34768
9750	S	D	E	N	1968	37106
9750	S	D	R	H	1962	38443

Phys. State: **C.** Superconductive; **D.** Doped; **E.** Expanded; **F.** Fibrous or Whisker; **G.** Gas; **I.** Ionized or Plasma; **L.** Liquid; **M.** Multiphase; **P.** Powder or Fine Particle; **S.** Solid; **T.** Thick or Thin Film

Subject: **D.** Data; **E.** Experiment; **G.** General (Data + Theory + Experiment); **S.** Survey (Review, Compendium etc.); **T.** Theory

Language: **C.** Czech; **D.** Dutch; **E.** English; **F.** French; **G.** German; **I.** Italian; **J.** Japanese; **O.** Other Languages; **P.** Polish; **R.** Russian; **S.** Spanish

Temperature: **F.** Full Range (Low + Normal + High); **L.** Low (0 to 75K) + Overlap into Normal; **N.** Normal (75 to 1273K); **H.** High (above 1273K) + Overlap into Normal; Blank = Not Coded

Substance Number	Phys. State	Subject	Language	Temperature	Year	TPRC Number
503-9750	S	D	E	N	1967	40427
9750	S	D	E	N	1964	41670
9750	S	D	E	H	1962	41856
9750	S	D	E	N	1960	42960
9750	S	D	E	N	1959	43731
9750	S	D	E	H	1962	46010
9750	S	D	E	N	1964	49184
9750	S	D	E	H	1942	49491
9750	S	D	R	H	1967	50196
9750	S	D	E	H	1967	50197
9750	S	D	R	N	1968	50768
9750	S	D	E	N	1955	52252
9750	S	D	E	N	1961	52519
9750	S	D	E	N	1965	52586
9750	S	D	E	N	1960	52653
9750	S	D	E	N	1961	52656
9750	S	D	E	N	1962	52662
9750	S	T	E	N	1962	52662
9750	S	D	E	N	1932	52690
9750	S	D	R	H	1967	53201
9750	S	D	E	N	1969	54946
9750	S	D	E	N	1970	56455
9750	S	D	R	H	1969	57271
9750	S	D	E	H	1969	57272
9750	S	D	E	H	1963	57515
9750	S	D	R	H	1969	59723
9750	S	D	F	H	1931	59929
9750	S	D	E	H	1970	61027
9750	S	D	R	H	1970	61316
9750	S	D	E	H	1970	61317
9750	S	D	E	N	1971	61687
9750	S	D	J	H	1962	62293
9750	S	D	R	H	1971	64834
9750	S	D	E	H	1971	64835
9750	S	D	R	H	1971	64844
9750	S	D	E	H	1971	64845
9750	S	D	E	N	1964	77265
9752	S	D	E	H	1958	52565
9757	S	D	E	N	1962	52661
9760	S	D	G	N	1960	34127
9760	S	D	E	N	1932	34144
9760	S	D	E	N	1930	41738
9760	S	D	R	N	1966	42998
9760	S	D	E	N	1959	44960
9760	S	D	E	N	1966	46572
9760	S	D	R	H	1966	46595
9760	S	D	E	H	1966	46596
9760	S	D	E	N	1953	52563
9760	S	D	G	N	1962	53242
9760	S	D	E	N	1968	54551
9760	S	D	E	N	1963	55332
9760	S	D	R	N	1969	58856
9760	S	D	E	N	1969	58857
9760	S	D	E	N	1943	60453
9760	S	D	E	N	1928	60525
9762	S	D	R	H	1969	58542
9764	S	D	E	N	1953	52563
9771	E	D	R	N	1970	60650
9771	E	D	E	N	1970	60651
9771	S	D	R	N	1968	47851
9771	S	D	E	N	1968	47852
9772	S	D	J	H	1965	44606
9773	S	D	E	N	1921	22633
9773	S	E	E	N	1921	22633
9777	S	D	E	N	1967	42910
9779	S	D	E	N	1953	52563
9780	S	D	E	N	1967	42910
9781	S	D	E	N	1967	42910
9786	S	D	E	N	1953	52563
9787	S	D	E	N	1965	38236
9788	S	D	E	N	1965	38236
9789	S	D	E	H	1965	36927
9789	S	D	E	H	1965	52286
9789	S	D	E	N	1961	52574
9790	S	D	R	H	1967	46604
9790	S	D	E	H	1967	46605
9791	S	D	E	H	1954	9032
9791	S	D	E	H	1946	41574
9791	S	D	E	H	1945	45785
9791	S	G	E	H	1964	48050
9791	S	D	R	H	1968	50076
9791	S	D	E	H	1968	52086
9791	S	D	E	H	1960	52577
9791	S	D	E	F	1963	54415
9791	S	D	E	H	1933	55938
9791	S	D	E	N	1926	55980
9795	S	D	E	N	1960	52576
9806	S	D	E	N	1962	48993
9806	S	D	E	N	1958	52287
9806	S	D	E	N	1953	52563
9808	S	D	R	H	1971	66445
9808	S	D	E	H	1971	66446
9809	S	D	E	N	1921	22633
9809	S	E	E	N	1921	22633
9809	S	D	G	H	1931	39517
9809	S	D	F	H	1921	46395
9809	S	D	R	H	1966	46579
9809	S	D	E	H	1966	46580
9809	S	D	E	N	1958	47988
9809	S	D	R	N	1969	57257
9809	S	D	E	N	1969	57258

Substance Number	Phys. State	Subject	Language	Temperature	Year	TPRC Number
503-9809	S	D	S	N	1950	58630
9809	S	D	R	H	1970	61172
9809	S	D	E	H	1970	61173
9811	D	D	E	N	1959	44960
9811	D	D	E	N	1953	52563
9811	P	D	G	N	1969	54743
9811	S	D	E	N	1954	9032
9811	S	D	E	N	1965	41301
9811	S	D	E	H	1959	44960
9811	S	D	E	N	1953	52563
9811	S	D	E	H	1925	52689
9811	S	D	E	N	1962	52780
9811	S	D	R	H	1967	53201
9811	S	D	E	N	1952	57977
9811	S	D	E	H	1970	61027
9811	S	D	J	H	1962	62293
9813	S	D	E	H	1963	37609
9813	S	D	J	H	1957	43650
9813	S	D	E	H	1967	46878
9813	S	D	E	L	1974	74420
9815	S	D	G	H	1931	39517
9815	S	D	E	N	1930	41738
9815	S	D	E	N	1937	42208
9815	S	D	R	N	1966	46564
9815	S	D	E	N	1966	46565
9815	S	D	R	N	1966	46577
9815	S	D	E	N	1966	46578
9815	S	D	E	N	1962	52684
9815	S	D	E	N	1968	54551
9815	S	D	R	H	1969	58852
9815	S	D	E	H	1969	58853
9815	S	D	E	N	1929	60442
9815	S	D	E	N	1943	60453
9815	S	D	E	N	1928	60525
9815	S	D	G	H	1960	62303
9816	S	D	E	N	1971	73893
9819	P	D	E	H	1965	51007
9819	S	S	E	N	1966	35566
9819	S	D	E	H	1967	47497
9819	S	D	E	H	1965	48431
9819	S	D	E	N	1962	58238
9819	S	D	J	H	1968	60568
9819	S	D	E	H	1969	60569
9820	S	S	E	N	1960	24931
9822	S	D	E	N	1962	45272
9823	S	D	E	N	1947	47957
9823	S	D	E	H	1946	55185
9824	S	D	E	N	1947	47957
9824	S	D	E	N	1960	52665
9824	S	D	E	H	1946	55185
9824	S	S	E	N	1950	59448
9824	S	D	E	N	1971	63401
9824	S	D	E	N	1970	77562
9826	S	D	E	N	1966	34768
9826	S	D	E	H	1946	41574
9826	S	D	E	N	1967	42909
9826	S	D	J	H	1965	44606
9826	S	D	E	N	1965	52586
9826	S	D	R	H	1969	57271
9826	S	D	E	H	1969	57272
9826	S	D	R	H	1971	66437
9826	S	D	E	H	1971	66438
9827	S	D	E	N	1943	60559
9828	S	D	E	N	1932	52690
9829	S	D	J	H	1963	37566
9830	S	D	E	N	1932	34703
9830	S	D	E	N	1966	34768
9830	S	D	E	H	1946	41574
9830	S	D	J	H	1965	44606
9830	S	D	R	H	1967	47694
9830	S	D	E	H	1967	47695
9830	S	D	G	N	1968	51350
9830	S	D	E	N	1946	52256
9830	S	D	E	F	1963	54415
9830	S	D	R	N	1969	58856
9830	S	D	E	N	1969	58857
9831	S	D	E	N	1930	41725
9833	S	D	E	N	1930	41725
9834	S	D	E	N	1930	41725
9834	S	D	R	N	1971	66449
9834	S	D	E	N	1971	66450
9835	S	D	E	N	1921	22633
9835	S	E	E	N	1921	22633
9835	S	D	E	N	1971	44828
9835	S	D	F	H	1921	46395
9835	S	D	E	N	1946	47666
9835	S	D	E	N	1971	48695
9835	S	D	R	N	1966	49052
9835	S	D	E	N	1946	50114
9835	S	D	R	H	1968	52348
9835	S	D	E	H	1968	52349
9835	S	D	R	H	1969	58542
9835	S	D	E	N	1921	60557
9835	S	D	R	N	1931	60851
9835	S	D	R	N	1931	60852
9836	S	D	E	N	1921	22633
9836	S	E	E	N	1921	22633
9836	S	D	G	H	1931	39517
9836	S	D	E	H	1925	52689
9836	S	D	E	N	1963	55332
9836	S	D	S	N	1950	58630

Substance Number	Phys. State	Subject	Language	Temperature	Year	TPRC Number
503-9836	S	D	E	N	1943	60453
9836	S	D	E	N	1928	60525
9836	S	D	E	N	1925	60560
9837	S	D	E	N	1966	34768
9837	S	D	E	H	1946	41574
9837	S	D	E	N	1967	45099
9837	S	D	G	N	1968	51350
9837	S	D	E	N	1965	52586
9837	S	D	E	N	1932	52690
9837	S	D	F	H	1931	59929
9837	S	D	R	H	1971	66437
9837	S	D	E	H	1971	66438
9838	S	D	E	N	1946	41574
9838	S	G	E	H	1964	48050
9838	S	D	G	N	1966	48685
9838	S	D	R	H	1971	64846
9838	S	D	E	H	1971	64847
9839	S	D	E	N	1930	41725
9839	S	D	E	N	1960	41838
9839	S	D	E	N	1939	45053
9839	S	D	E	N	1940	46430
9839	S	D	R	N	1968	49613
9839	S	D	O	N	1969	55274
9839	S	D	E	N	1941	60452
9843	S	D	E	N	1968	54196
9845	S	D	E	N	1970	59974
9845	S	D	E	N	1971	63401
9845	S	D	E	N	1970	77562
9848	S	D	R	N	1969	56253
9848	S	D	G	N	1927	59560
9850	S	D	E	N	1930	41725
9850	S	D	E	N	1941	60452
9851	S	D	E	N	1966	34768
9851	S	D	E	N	1965	52586
9851	S	D	E	N	1964	77265
9852	S	D	E	N	1966	34768
9852	S	D	E	N	1967	52868
9852	S	D	E	N	1969	53775
9853	S	D	E	N	1966	34768
9853	S	D	E	N	1969	53775
9854	S	D	E	N	1968	54196
9855	S	D	E	N	1966	34768
9859	S	D	E	N	1932	34144
9859	S	D	O	H	1972	71321
9861	S	D	E	N	1966	34768
9862	S	D	E	N	1966	34768
9864	S	D	E	N	1966	34768
9865	S	D	E	N	1967	34027
9868	S	D	J	H	1965	44606
9869	S	D	E	H	1953	7036
9869	S	D	E	N	1952	41699
9869	S	D	R	N	1965	41896
9869	S	D	E	N	1954	42277
9869	S	G	E	N	1959	43564
9869	S	D	O	N	1966	44266
9869	S	D	J	N	1966	45644
9869	S	D	R	N	1967	46267
9869	S	D	E	N	1967	46268
9869	S	D	E	N	1952	48341
9869	S	D	E	N	1965	51723
9869	S	D	E	N	1961	52507
9869	S	D	E	N	1958	52682
9869	S	D	E	N	1962	52780
9869	S	D	R	N	1955	54638
9869	S	D	R	N	1969	54919
9869	S	D	E	N	1964	54981
9869	S	D	E	N	1951	56215
9869	S	T	E	N	1951	56215
9869	S	D	E	N	1952	56237
9869	S	D	E	N	1953	58115
9869	S	D	E	N	1954	58145
9869	S	D	O	N	1970	58998
9869	S	D	S	N	1964	59722
9869	S	D	J	N	1955	60412
9869	S	D	E	N	1951	60456
9869	S	D	R	N	1965	60876
9869	S	D	G	N	1961	62281
9869	S	D	O	N	1960	62302
9869	S	D	E	N	1965	63068
9869	S	D	E	N	1955	63074
9869	S	D	G	H	1959	77314
9869	S	D	E	H	1963	77315
9869	S	D	R	N	1974	78699
9869	S	E	R	N	1974	78699
9870	S	D	E	N	1969	54896
9876	S	D	E	N	1952	59511
9877	S	D	E	N	1969	54896
9879	S	D	G	N	1966	45549
9879	S	D	G	N	1968	50596
9880	S	D	G	N	1966	45549
9880	S	D	G	N	1968	50596
9882	S	D	E	N	1966	34768
9882	S	D	E	N	1969	53775
9891	S	D	E	N	1968	49277
9891	S	D	R	N	1968	50670
9892	S	D	E	N	1966	34768
9892	S	D	E	N	1965	52586
9892	S	D	E	N	1968	53097
9893	S	D	E	N	1965	46298
9893	S	D	R	N	1969	56253
9893	S	D	G	N	1927	59560

Phys. State: **C.** Superconductive; **D.** Doped; **E.** Expanded; **F.** Fibrous or Whisker; **G.** Gas; **I.** Ionized or Plasma; **L.** Liquid; **M.** Multiphase; **P.** Powder or Fine Particle; **S.** Solid; **T.** Thick or Thin Film

Subject: **D.** Data; **E.** Experiment; **G.** General (Data + Theory + Experiment); **S.** Survey (Review, Compendium etc.); **T.** Theory

Language: **C.** Czech; **D.** Dutch; **E.** English; **F.** French; **G.** German; **I.** Italian; **J.** Japanese; **O.** Other Languages; **P.** Polish; **R.** Russian; **S.** Spanish

Temperature: **F.** Full Range (Low + Normal + High); **L.** Low (0 to 75K) + Overlap into Normal; **N.** Normal (75 to 1273K); **H.** High (above 1273K) + Overlap into Normal; Blank = Not Coded

Substance Number	Phys. State	Sub-ject	Lan-guage	Temper-ature	Year	TPRC Number
503-9896	S	D	E	N	1953	52563
9898	S	D	G	H	1931	39517
9898	S	D	E	N	1937	42208
9898	S	D	E	N	1957	46503
9898	S	D	R	H	1970	61172
9898	S	D	E	H	1970	61173
9898	S	D	R	H	1971	66437
9898	S	D	E	H	1971	66438
9900	S	D	E	N	1967	42910
9900	S	D	R	N	1967	49336
9900	S	D	E	N	1967	49337
9901	S	D	E	N	1932	34144
9901	S	D	E	N	1930	41725
9901	S	D	F	H	1921	46395
9901	S	D	R	N	1968	49613
9901	S	D	R	N	1955	54638
9901	S	D	E	N	1925	60560
9901	S	D	E	N	1955	63074
9902	S	D	E	N	1967	53283
9902	S	D	S	N	1950	58630
9902	S	D	J	N	1955	60412
9910	S	D	E	N	1967	42910
9913	S	D	E	N	1967	42910
9913	S	D	E	N	1949	55889
9914	S	D	E	N	1970	56161
9915	D	D	E	N	1951	41923
9915	S	D	E	N	1951	41923
9915	S	D	R	N	1966	46575
9915	S	D	E	N	1966	46576
9915	S	D	E	N	1962	52662
9915	S	T	E	N	1962	52662
9915	S	D	E	N	1962	52694
9916	D	D	E	H	1971	62381
9916	P	D	G	H	1962	53241
9916	S	D	E	N	1932	34144
9916	S	D	E	N	1932	34703
9916	S	D	E	H	1928	36277
9916	S	D	G	H	1931	39517
9916	S	D	E	N	1964	41670
9916	S	D	E	N	1930	41738
9916	S	D	E	N	1955	43573
9916	S	D	E	N	1959	44960
9916	S	D	F	H	1921	46395
9916	S	D	E	H	1959	47022
9916	S	D	J	N	1967	47150
9916	S	D	E	N	1946	47302
9916	S	D	E	N	1946	47666
9916	S	D	R	H	1967	50196
9916	S	D	E	H	1967	50197
9916	S	D	S	H	1967	50986
9916	S	D	R	H	1958	52116
9916	S	D	E	H	1958	52117
9916	S	D	J	N	1960	52431
9916	S	D	E	H	1960	52577
9916	S	D	E	N	1954	52681
9916	S	D	E	N	1958	52682
9916	S	D	E	H	1925	52689
9916	S	D	G	H	1962	53241
9916	S	D	E	N	1967	53283
9916	S	D	E	F	1963	54415
9916	S	S	E	H	1970	56454
9916	S	D	G	H	1966	59707
9916	S	D	F	H	1931	59929
9916	S	D	E	N	1928	60441
9916	S	D	E	N	1928	60525
9916	S	D	E	H	1971	62381
9916	S	D	C	N	1971	63012
9916	S	T	C	N	1971	63012
9916	S	D	J	H	1971	63549
9916	S	D	R	H	1971	64836
9916	S	D	E	H	1971	64837
9916	S	D	R	H	1971	64840
9916	S	D	E	H	1971	64841
9916	S	D	R	H	1971	64846
9916	S	D	E	H	1971	64847
9916	S	D	O	H	1972	71321
9916	S	D	R	H	1974	75192
9916	S	D	C	N	1974	76085
9916	S	D	E	H	1974	76872
9918	S	D	E	N	1956	52534
9919	S	D	E	N	1946	52256
9920	S	D	E	N	1948	37113
9920	S	D	R	H	1962	38443
9920	S	D	E	H	1946	41574
9920	S	D	E	H	1962	41856
9920	S	D	E	N	1959	44960
9920	S	D	E	H	1946	55185
9920	S	D	E	N	1970	59974
9920	S	D	E	N	1971	63401
9920	S	D	E	N	1970	77562
9921	S	D	E	H	1967	42910
9922	S	D	G	H	1966	41276
9922	S	E	G	H	1966	41276
9922	S	D	E	H	1947	41534
9922	S	D	E	H	1946	41574
9922	S	D	E	N	1949	41902
9922	S	D	R	H	1963	42333
9922	S	D	F	H	1963	42334
9922	S	D	F	H	1966	44131
9922	S	E	F	H	1966	44131
9922	S	D	G	H	1964	44351

Substance Number	Phys. State	Sub-ject	Lan-guage	Temper-ature	Year	TPRC Number
503-9922	S	D	E	H	1968	44778
9922	S	D	E	H	1968	44799
9922	S	D	C	H	1967	45261
9922	S	E	C	H	1967	45261
9922	S	D	R	H	1966	45561
9922	S	D	E	H	1945	45785
9922	S	D	E	H	1965	46471
9922	S	D	E	H	1954	46789
9922	S	D	E	H	1947	48002
9922	S	D	R	H	1967	48619
9922	S	D	E	H	1967	48620
9922	S	D	E	H	1943	50711
9922	S	D	R	H	1965	51130
9922	S	D	E	H	1967	51131
9922	S	D	E	H	1967	51545
9922	S	D	R	H	1968	52350
9922	S	D	E	H	1968	52351
9922	S	D	E	H	1963	52481
9922	S	D	R	H	1968	53021
9922	S	D	R	H	1968	53023
9922	S	D	R	H	1967	53201
9922	S	D	F	H	1949	53232
9922	S	D	E	F	1963	54415
9922	S	D	R	N	1969	54441
9922	S	D	R	H	1968	55422
9922	S	D	E	H	1968	55423
9922	S	D	E	N	1969	57104
9922	S	D	F	H	1969	57614
9922	S	E	F	H	1969	57614
9922	S	D	R	H	1969	60884
9922	S	D	E	H	1970	61027
9922	S	D	R	H	1970	61316
9922	S	D	E	H	1970	61317
9922	S	D	R	H	1971	64844
9922	S	D	E	H	1971	64845
9927	S	D	E	N	1956	52534
9929	S	D	E	N	1956	52534
9932	S	D	R	H	1966	40883
9932	S	D	E	H	1966	40884
9932	S	D	R	H	1966	46570
9932	S	D	E	H	1966	46571
9935	S	D	E	N	1936	44125
9936	S	D	E	N	1930	41725
9936	S	D	E	N	1961	43326
9936	S	D	E	N	1941	60452
9938	S	D	E	N	1932	35483
9938	S	D	E	N	1921	60557
9942	S	D	E	N	1967	40693
9943	S	D	G	N	1968	50065
9944	S	D	E	N	1935	44430
9946	S	D	E	N	1967	40693
9946	S	D	E	N	1968	46742
9948	S	D	E	N	1932	35483
9948	S	D	R	H	1971	64836
9948	S	D	E	H	1971	64837
9949	S	D	E	N	1953	52563
9951	S	D	E	N	1950	1504
9951	S	D	E	N	1947	47957
9952	S	D	E	N	1946	3081
9954	S	D	E	H	1949	55856
9955	S	D	E	H	1965	39834
9955	S	D	E	H	1946	41574
9955	S	D	E	H	1966	43496
9956	S	D	E	N	1946	3081
9956	S	D	E	H	1965	39834
9957	S	D	G	H	1931	39517
9957	S	D	E	H	1974	75284
9958	S	D	E	H	1947	41534
9958	S	D	E	H	1949	55856
9960	S	D	E	N	1964	41670
9960	S	D	E	H	1967	44823
9960	S	D	E	H	1963	52805
9960	S	D	R	N	1969	54437
9960	S	D	R	H	1969	55280
9960	S	D	E	N	1969	57096
9960	S	D	E	H	1969	57270
9960	S	D	R	H	1969	59723
9960	S	D	E	N	1956	59993
9960	S	D	R	N	1971	66443
9960	S	D	E	N	1971	66444
9961	S	D	E	N	1952	55944
9965	S	D	E	N	1967	40693
9966	S	D	E	N	1967	40693
9966	S	D	E	N	1968	46742
9974	S	D	E	N	1964	34296
9974	S	D	E	N	1965	38236
9974	S	D	E	H	1947	41534
9974	S	D	E	H	1946	41574
9974	S	D	R	H	1963	42333
9974	S	D	F	H	1963	42334
9974	S	D	E	H	1968	44778
9974	S	D	E	H	1945	45785
9974	S	G	E	H	1964	48050
9974	S	D	E	H	1963	52481
9974	S	D	E	H	1954	52650
9974	S	D	R	H	1968	53021
9974	S	D	R	H	1967	53201
9974	S	D	F	H	1949	53232
9974	S	D	F	H	1931	59929
9974	S	D	E	H	1970	61027
9974	S	D	E	H	1974	75122

Substance Number	Phys. State	Sub-ject	Lan-guage	Temper-ature	Year	TPRC Number
503-9977	S	D	E	N	1932	34703
9977	S	D	E	N	1921	36041
9977	S	D	E	N	1943	45557
9977	S	D	E	N	1946	47666
9977	S	D	E	N	1946	50114
9977	S	D	E	N	1961	52793
9978	S	D	E	H	1943	55984
9978	S	D	E	N	1970	56161
9978	S	D	R	H	1969	58854
9978	S	D	E	H	1969	58855
9979	S	S	E	N	1960	24931
9979	S	D	E	N	1932	34703
9979	S	D	P	N	1967	48452
9979	S	E	P	N	1967	48452
9979	S	D	I	H	1967	49047
9979	S	D	J	N	1960	52431
9979	S	D	E	N	1967	53283
9979	S	D	R	N	1955	54638
9979	S	D	E	N	1949	55889
9979	S	D	E	N	1970	56161
9979	S	D	E	N	1932	57549
9979	S	D	G	N	1970	58487
9979	S	D	R	H	1969	58542
9979	S	D	E	N	1952	60457
9979	S	D	E	N	1954	60459
9979	S	D	E	N	1956	60562
9979	S	D	E	N	1932	60793
9979	S	D	G	N	1960	62279
9979	S	D	J	N	1958	62292
9979	S	D	C	N	1971	63012
9979	S	T	C	N	1971	63012
9979	S	D	E	N	1955	63074
9980	S	D	E	N	1932	34703
9980	S	D	G	H	1966	36685
9980	S	D	E	H	1966	36690
9980	S	D	G	N	1954	36744
9980	S	T	G	N	1954	36744
9980	S	D	G	H	1931	39517
9980	S	D	E	N	1964	41670
9980	S	D	E	N	1946	47666
9980	S	D	E	N	1967	53283
9980	S	D	E	N	1970	56161
9980	S	D	E	N	1932	57549
9980	S	D	G	H	1934	59091
9980	S	D	E	N	1921	60557
9980	S	D	E	N	1925	60560
9980	S	D	R	N	1970	62314
9980	S	D	C	N	1971	63012
9980	S	T	C	N	1971	63012
9980	S	D	E	N	1971	63275
9980	S	D	J	H	1971	63549
9980	S	D	O	H	1972	71321
9980	S	D	C	N	1974	76085
9981	S	D	E	H	1938	39867
9981	S	D	E	H	1942	43643
9981	S	D	J	H	1965	44606
9981	S	D	G	N	1968	51350
9981	S	D	E	H	1925	52689
9981	S	D	G	H	1968	53251
9981	S	D	E	N	1970	56161
9981	S	D	R	H	1970	60841
9981	S	D	E	H	1970	60842
9982	S	D	E	N	1946	3081
9983	S	D	E	N	1946	3081
9983	S	D	E	H	1966	43496
9984	S	D	R	H	1970	74074
9984	S	D	E	H	1970	74075
521-0001	S	D	E	N	1951	60526
0003	S	D	E	N	1967	53490
0003	S	T	E	N	1967	53490
0003	S	D	E	N	1970	60221
0004	S	D	E	N	1972	67289
0004	S	E	E	N	1972	67289
0006	P	D	I	N	1953	94412
0006	S	S	G	N	1928	28680
0006	S	D	P	N	1967	48452
0006	S	E	P	N	1967	48452
0006	S	D	E	H	1933	55938
0006	S	D	E	N	1944	59950
0006	S	D	E	H	1932	60468
0007	S	D	F	N	1867	45463
0007	S	D	E	N	1955	55920
0007	S	D	E	N	1961	55934
0007	S	D	G	N	1937	56410
0007	S	T	R	N	1969	57561
0007	S	D	F	N	1934	59784
0007	S	S	E	L	1977	88199
0007	S	S	E	N	1977	88199
0007	S	D	D	N	1924	90913
0007	S	S	E	N	1960	94260
0007	S	D	E	N	1962	94434
0009	S	E	E	N	1965	44633
0010	P	D	E	N	1951	60526
0010	P	D	G	N	1963	62306
0010	S	D	E	N	1960	41838
0011	S	D	E	H	1927	34128
0015	S	S	G	H	1928	28680
0020	S	D	R	N	1969	57140
0020	S	D	E	N	1969	57141
0021	S	S	E	N	1960	24931

Phys. State: **C.** Superconductive; **D.** Doped; **E.** Expanded; **F.** Fibrous or Whisker; **G.** Gas; **I.** Ionized or Plasma; **L.** Liquid; **M.** Multiphase; **P.** Powder or Fine Particle; **S.** Solid; **T.** Thick or Thin Film

Subject: **D.** Data; **E.** Experiment; **G.** General (Data + Theory + Experiment); **S.** Survey (Review, Compendium etc.); **T.** Theory

Language: **C.** Czech; **D.** Dutch; **E.** English; **F.** French; **G.** German; **I.** Italian; **J.** Japanese; **O.** Other Languages; **P.** Polish; **R.** Russian; **S.** Spanish

Temperature: **F.** Full Range (Low + Normal + High); **L.** Low (0 to 75K) + Overlap into Normal; **N.** Normal (75 to 1273K); **H.** High (above 1273K) + Overlap into Normal; Blank = Not Coded

Substance Number	Phys. State	Subject	Language	Temperature	Year	TPRC Number
521-0022	S	D	P	N	1967	48452
0022	S	E	P	N	1967	48452
0022	S	D	E	N	1932	57549
0022	S	D	R	H	1969	59405
0022	S	D	E	H	1969	59406
0022	S	D	E	N	1976	87315
0022	T	D	E	N	1976	87315
0024	S	D	F	N	1869	59537
0024	S	D	O	N	1971	65626
0025	S	G	G	N	1915	37174
0025	S	D	G	N	1918	37184
0025	S	T	G	N	1918	37184
0025	S	D	E	N	1965	38527
0025	S	D	F	N	1868	45692
0025	S	D	E	N	1964	51146
0025	S	D	R	N	1968	54585
0025	S	D	E	N	1955	55920
0025	S	D	E	N	1949	60337
0025	S	D	E	N	1951	60341
0026	S	D	E	H	1925	38313
0026	S	D	F	N	1868	56483
0032	P	D	E	N	1979	97914
0032	S	D	E	H	1933	55938
0033	S	S	G	N	1928	28680
0033	S	D	E	N	1927	36274
0033	S	D	E	H	1933	55938
0034	P	D	E	N	1967	48521
0034	P	E	E	N	1967	48521
0035	S	D	E	N	1933	43623
0035	S	D	E	H	1971	62575
0035	S	D	E	N	1974	91243
0035	S	T	E	N	1974	91243
0036	S	D	E	N	1933	43624
0036	S	D	E	N	1973	74326
0040	M	D	E	N	1969	52318
0040	P	D	E	N	1965	50414
0040	P	D	E	N	1969	52318
0040	P	D	F	N	1962	53233
0040	P	D	F	H	1960	53243
0040	P	D	P	N	1969	56205
0040	P	D	F	H	1969	57519
0040	P	E	F	H	1969	57519
0040	P	D	S	N	1964	59708
0040	S	S	G	N	1928	28680
0040	S	D	E	N	1936	44168
0040	S	D	E	N	1940	45889
0040	S	D	F	N	1967	50734
0040	S	E	F	N	1967	50734
0040	S	D	G	N	1958	53305
0040	S	D	E	H	1933	55938
0040	S	D	E	N	1970	59727
0040	S	D	G	N	1961	62280
0046	S	D	E	H	1956	40763
0046	S	S	E	N	1966	42284
0046	S	E	E	N	1970	60293
0046	S	E	E	N	1972	68377
0046	S	D	E	N	1836	91329
0046	S	E	E	N	1836	91329
0053	S	D	R	N	1969	57574
0053	S	T	R	N	1969	57574
0053	S	D	E	N	1972	67289
0053	S	E	E	N	1972	67289
0055	F	D	E	N	1972	78278
0055	F	E	E	N	1972	78278
0055	P	D	E	N	1965	46637
0055	S	D	E	H	1942	43643
0055	S	D	O	N	1967	48527
0057	S	D	E	N	1933	43623
0057	S	D	E	H	1967	45743
0057	S	D	E	N	1974	91243
0057	S	T	E	N	1974	91243
0057	S	D	E	N	1979	100063
0059	S	D	E	N	1933	38381
0059	S	D	R	N	1965	44807
0060	S	D	E	H	1925	38313
0060	S	D	O	N	1960	53304
0060	S	D	F	N	1868	56483
0060	S	D	E	N	1979	100063
0062	P	D	E	N	1968	51345
0063	P	S	E	H	1960	16704
0063	P	S	E	N	1960	16704
0063	P	D	E	L	1971	66170
0063	P	E	E	L	1971	66170
0063	P	S	E	F	1977	88199
0063	S	S	R	N	1954	1093
0063	S	S	E	N	1960	16704
0063	S	D	G	N	1918	37184
0063	S	T	G	N	1918	37184
0063	S	D	E	L	1966	39188
0063	S	T	E	L	1966	39188
0063	S	D	E	N	1951	41971
0063	S	S	E	L	1961	44227
0063	S	S	E	N	1961	44227
0063	S	D	E	N	1957	52532
0063	S	D	R	L	1960	54097
0063	S	D	E	L	1961	54098
0063	S	D	R	N	1970	55768
0063	S	D	F	N	1869	59537
0063	S	D	E	H	1975	76618
0063	S	D	E	F	1977	88199
0063	S	S	E	F	1977	88199

Substance Number	Phys. State	Subject	Language	Temperature	Year	TPRC Number
521-0063	S	D	R	N	1977	93719
0063	S	T	R	N	1977	93719
0063	S	D	E	N	1977	93720
0063	S	T	E	N	1977	93720
0063	S	S	E	N	1980	100400
0064	S	S	E	N	1964	37618
0064	S	D	E	H	1956	40763
0064	S	D	E	L	1964	43411
0064	S	D	E	N	1931	43558
0064	S	D	E	N	1960	46880
0064	S	D	E	H	1962	47946
0064	S	D	E	H	1966	49755
0064	S	D	E	L	1968	52205
0064	S	D	E	H	1961	52754
0064	S	S	E	N	1962	52795
0064	S	D	R	N	1967	53738
0064	S	E	R	N	1967	53738
0064	S	D	E	F	1963	54415
0064	S	D	R	N	1969	58624
0064	S	D	E	N	1972	66040
0064	S	E	E	N	1972	66040
0064	S	D	E	N	1974	74511
0064	S	E	E	N	1974	74511
0064	S	S	E	N	1958	85031
0064	S	S	E	N	1959	85032
0067	P	D	E	N	1965	44135
0067	P	D	R	H	1972	69712
0067	P	D	E	H	1972	72179
0067	S	D	E	N	1945	42265
0067	S	D	E	N	1950	55957
0067	S	D	E	N	1932	55974
0067	S	D	E	H	1946	58939
0067	S	D	F	N	1921	59932
0067	S	S	E	N	1977	88199
0069	S	D	E	N	1966	42432
0069	S	D	F	N	1869	59537
0069	S	D	O	N	1971	65626
0070	P	D	E	N	1965	50414
0070	S	D	E	N	1978	98172
0070	S	D	E	N	1979	99928
0072	S	D	E	N	1939	38901
0072	S	D	E	N	1953	43480
0072	S	D	R	N	1968	54585
0072	S	D	E	N	1944	59950
0072	S	D	E	N	1951	60526
0072	S	D	E	N	1970	61727
0072	S	D	E	N	1972	68377
0072	S	E	E	N	1972	68377
0072	S	D	E	N	1974	74504
0072	S	E	E	N	1974	74504
0072	S	D	E	N	1972	78340
0072	S	E	E	N	1972	78340
0072	S	S	E	N	1976	90192
0072	S	D	E	N	1836	91329
0072	S	E	E	N	1836	91329
0072	S	D	E	N	1977	91696
0072	S	D	O	N	1978	94675
0073	S	D	O	N	1960	53304
0075	S	D	E	N	1929	34568
0075	S	D	E	N	1934	38439
0075	S	G	G	N	1936	43371
0075	S	D	F	N	1868	45692
0075	S	D	E	N	1943	55409
0075	S	D	F	N	1934	55725
0075	S	D	E	N	1940	55947
0075	S	D	E	N	1949	59820
0075	S	D	E	N	1944	59950
0075	S	D	E	N	1955	60343
0075	S	D	E	N	1933	60348
0077	S	D	R	N	1971	34689
0077	S	E	R	N	1971	34689
0077	S	D	R	N	1966	41750
0077	S	D	E	N	1966	41751
0077	S	D	F	N	1866	44793
0077	S	D	E	H	1962	47946
0077	S	D	E	H	1961	52754
0077	S	D	E	H	1968	52786
0077	S	E	E	H	1968	52786
0077	S	D	R	N	1967	53738
0077	S	E	R	N	1967	53738
0077	S	D	E	N	1933	60348
0077	S	D	E	N	1972	66040
0077	S	E	E	N	1972	66040
0077	S	D	G	N	1976	89038
0082	S	D	F	N	1868	56483
0082	S	D	E	N	1929	60116
0082	S	D	R	N	1974	74406
0083	S	D	O	N	1967	48527
0083	S	D	F	N	1868	56483
0084	P	D	J	N	1965	50714
0084	S	D	E	N	1939	38901
0084	S	D	E	H	1946	41574
0084	S	D	E	N	1967	46291
0084	S	D	O	N	1967	48527
0084	S	D	O	N	1960	53304
0084	S	D	E	N	1944	59950
0084	S	D	E	N	1951	60526
0085	M	D	E	N	1936	38383
0085	P	D	I	N	1953	94412
0085	S	D	E	N	1939	38901
0085	S	D	E	H	1946	41574

Substance Number	Phys. State	Subject	Language	Temperature	Year	TPRC Number
521-0085	S	D	E	N	1953	43480
0085	S	D	E	N	1967	46291
0085	S	D	O	N	1967	48527
0085	S	D	R	N	1968	54585
0085	S	D	E	N	1944	59950
0085	S	D	E	N	1951	60526
0085	S	D	E	N	1970	61727
0085	S	D	G	H	1961	62304
0085	S	D	G	H	1962	62305
0085	S	D	E	N	1919	71468
0085	S	E	E	N	1919	71468
0085	S	D	E	N	1937	71469
0085	S	D	E	N	1950	72108
0085	S	E	E	N	1950	72108
0085	S	S	E	N	1976	90192
0085	S	D	I	N	1953	94412
0086	D	D	E	L	1977	89480
0086	S	G	G	N	1936	43371
0086	S	D	E	N	1962	91702
0091	S	D	E	N	1970	101370
0092	P	D	E	N	1970	56672
0092	P	S	E	N	1977	88199
0092	S	D	E	H	1956	40763
0092	S	D	F	N	1866	44793
0093	S	D	E	N	1933	60348
0096	S	D	E	N	1959	19385
0096	S	T	E	N	1959	19385
0096	S	G	G	N	1915	37174
0096	S	D	G	N	1918	37184
0096	S	T	G	N	1918	37184
0096	S	D	F	N	1868	45692
0096	S	D	R	L	1967	51195
0096	S	D	R	N	1967	51195
0096	S	D	E	N	1950	55957
0096	S	D	E	N	1949	60337
0096	S	S	E	N	1977	88199
0103	S	D	E	N	1953	43480
0103	S	D	R	N	1968	54585
0103	S	D	E	N	1950	57504
0103	S	D	E	N	1949	59820
0103	S	D	E	N	1951	60526
0103	S	D	E	N	1972	68377
0103	S	E	E	N	1972	68377
0103	S	D	E	N	1919	71468
0103	S	E	E	N	1919	71468
0103	S	D	E	N	1937	71469
0103	S	D	E	N	1948	71916
0103	S	D	E	N	1950	72108
0103	S	E	E	N	1950	72108
0103	S	S	E	N	1976	90192
0103	S	D	I	N	1953	94412
0104	P	D	E	N	1965	44135
0104	S	D	F	N	1866	44793
0104	S	D	J	H	1967	49049
0104	S	D	E	N	1950	60339
0104	S	S	E	H	1977	88199
0104	S	S	E	N	1977	88199
0105	P	D	I	N	1953	94412
0105	S	D	E	N	1934	38439
0105	S	G	G	N	1936	43371
0105	S	D	F	N	1868	56483
0108	S	S	E	N	1960	24931
0108	S	D	G	N	1952	42961
0108	S	D	E	N	1952	52094
0108	S	D	G	L	1928	56000
0108	S	D	G	N	1961	62280
0110	P	D	E	N	1951	60526
0110	S	S	G	N	1928	28680
0111	S	D	E	H	1941	42172
0111	S	D	O	N	1960	53304
0111	S	D	G	N	1961	62280
0116	P	D	E	N	1965	46637
0116	P	D	S	N	1964	59708
0116	S	S	G	N	1928	28680
0116	S	D	E	H	1966	41544
0116	S	D	G	N	1937	43381
0116	S	D	O	N	1967	48527
0116	S	D	E	N	1959	52680
0116	S	D	E	N	1963	58889
0123	S	D	E	N	1939	38901
0123	S	D	E	N	1936	39337
0123	S	D	E	N	1953	43480
0123	S	D	E	N	1944	59950
0123	S	D	E	N	1972	78340
0123	S	E	E	N	1972	78340
0123	S	S	E	N	1976	90192
0123	S	D	E	N	1978	99991
0123	S	D	E	N	1979	100720
0130	P	D	E	N	1976	94823
0130	S	S	E	N	1960	24931
0130	S	D	E	N	1932	34703
0130	S	D	F	N	1866	44793
0130	S	D	F	N	1868	45692
0130	S	D	G	L	1939	48022
0130	S	D	E	N	1952	52749
0130	S	D	E	N	1932	57549
0130	S	D	E	N	1972	67443
0132	S	D	E	H	1933	55938
0132	S	D	E	N	1933	60348
0135	P	D	P	N	1966	45597
0135	P	E	P	N	1966	45597

Phys. State: **C.** Superconductive; **D.** Doped; **E.** Expanded; **F.** Fibrous or Whisker; **G.** Gas; **I.** Ionized or Plasma; **L.** Liquid; **M.** Multiphase; **P.** Powder or Fine Particle; **S.** Solid; **T.** Thick or Thin Film
Subject: **D.** Data; **E.** Experiment; **G.** General (Data + Theory + Experiment); **S.** Survey (Review, Compendium etc.); **T.** Theory
Language: **C.** Czech; **D.** Dutch; **E.** English; **F.** French; **G.** German; **I.** Italian; **J.** Japanese; **O.** Other Languages; **P.** Polish; **R.** Russian; **S.** Spanish
Temperature: **F.** Full Range (Low + Normal + High); **L.** Low (0 to 75K) + Overlap into Normal; **N.** Normal (75 to 1273K); **H.** High (above 1273K) + Overlap into Normal; Blank = Not Coded

Substance Number	Phys. State	Subject	Language	Temperature	Year	TPRC Number
521-0135	P	D	E	N	1965	50414
0135	P	D	F	N	1962	53233
0135	P	D	F	H	1960	53243
0135	P	D	F	H	1969	57519
0135	P	E	F	H	1969	57519
0135	P	D	S	N	1964	59708
0135	S	S	G	N	1928	28680
0135	S	D	F	N	1920	37779
0135	S	D	E	N	1931	42029
0135	S	D	G	N	1952	42961
0135	S	D	E	N	1936	44168
0135	S	D	F	N	1967	50734
0135	S	E	F	N	1967	50734
0135	S	D	E	N	1952	52094
0135	S	D	E	N	1959	52680
0135	S	D	O	N	1960	53304
0135	S	D	E	H	1933	55938
0135	S	D	E	H	1938	59921
0135	S	D	E	N	1927	60437
0135	S	D	G	N	1961	62280
0135	S	D	E	H	1979	100206
0135	S	D	E	N	1979	100206
0153	S	D	E	H	1925	38313
0170	S	S	E	N	1960	24931
0170	S	D	E	N	1939	45053
0174	P	D	E	N	1965	46637
0174	S	S	G	N	1928	28680
0174	S	D	G	N	1931	42592
0174	S	D	E	N	1941	60451
0174	S	D	E	N	1973	74556
0174	S	D	I	N	1977	98301
0181	S	D	R	N	1968	54585
0181	S	D	E	N	1972	68377
0181	S	E	E	N	1972	68377
0181	S	S	E	N	1976	90192
0181	S	D	E	N	1836	91329
0181	S	E	E	N	1836	91329
0184	S	D	R	N	1968	54585
0184	S	D	E	N	1978	94675
0186	P	D	R	H	1969	57588
0190	S	D	E	N	1965	44633
0190	S	E	E	N	1965	44633
0191	S	G	E	N	1967	45061
0191	S	D	R	N	1967	53738
0191	S	E	R	N	1967	53738
0191	S	D	E	N	1972	66040
0191	S	E	E	N	1972	66040
0192	S	D	F	N	1868	45692
0192	S	D	E	N	1951	60341
0196	M	D	E	N	1936	38383
0196	S	D	E	N	1939	38901
0196	S	D	E	N	1953	43480
0196	S	D	E	N	1946	47302
0196	S	D	G	N	1961	53239
0196	S	D	E	N	1925	54335
0196	S	D	E	N	1944	59950
0196	S	D	E	N	1970	61727
0196	S	D	P	N	1970	65195
0196	S	D	E	N	1972	68377
0196	S	E	E	N	1972	68377
0196	S	D	E	N	1937	71469
0196	S	D	E	N	1950	72108
0196	S	E	E	N	1950	72108
0196	S	S	E	N	1976	90192
0196	S	D	E	N	1836	91329
0196	S	E	E	N	1836	91329
0217	S	S	G	N	1928	28680
0217	S	D	E	H	1930	36288
0217	S	D	E	H	1933	55938
0222	S	D	E	N	1953	43480
0222	S	D	E	N	1944	59950
0223	S	D	S	N	1967	49900
0223	S	D	E	N	1951	60340
0223	S	D	G	H	1962	62305
0236	S	D	O	N	1967	48527
0236	S	D	E	N	1944	59950
0236	S	D	E	N	1972	67289
0236	S	E	E	N	1972	67289
0239	S	D	R	N	1974	76265
0239	S	D	R	N	1974	77626
0240	P	D	E	N	1965	46637
0240	P	D	S	N	1964	59708
0240	S	S	E	N	1960	24931
0240	S	D	E	N	1965	41300
0240	S	D	E	N	1945	48237
0240	S	D	O	N	1960	53304
0241	P	D	E	N	1965	46637
0241	S	S	E	N	1960	24931
0241	S	D	E	H	1965	41300
0241	S	D	E	N	1945	48237
0241	S	D	E	N	1933	60348
0242	S	D	R	N	1968	54585
0242	S	D	E	N	1944	59950
0242	S	D	E	N	1970	60293
0242	S	D	E	N	1972	67289
0242	S	E	E	N	1972	67289
0242	S	D	E	N	1978	94675
0243	S	D	E	N	1944	59950
0243	S	D	E	N	1972	67289
0243	S	E	E	N	1972	67289
0250	S	S	E	H	1960	24931

Substance Number	Phys. State	Subject	Language	Temperature	Year	TPRC Number
521-0250	S	S	E	N	1960	24931
0250	S	S	E	H	1958	43298
0250	S	E	E	H	1958	43298
0250	S	D	E	N	1931	43558
0250	S	D	F	N	1868	45692
0250	S	D	E	H	1945	45785
0250	S	D	R	H	1976	87277
0250	S	D	R	N	1976	87277
0250	S	D	E	H	1976	89152
0250	S	D	E	N	1976	89152
0254	S	S	G	H	1928	28680
0254	S	D	E	N	1933	60348
0254	S	D	G	N	1963	62306
0254	S	D	E	L	1976	87010
0254	S	D	E	N	1976	87010
0254	S	D	E	H	1977	91331
0254	S	D	E	N	1977	91331
0255	S	S	G	N	1928	28680
0255	S	D	E	N	1954	43565
0258	S	D	E	N	1934	35758
0258	S	D	E	H	1972	76999
0258	S	T	E	H	1972	76999
0261	S	D	E	N	1965	42608
0264	S	D	E	N	1953	43480
0264	S	D	R	N	1968	54585
0264	S	D	R	H	1969	58631
0264	S	D	E	N	1972	67289
0264	S	E	E	N	1972	67289
0264	S	D	E	N	1950	72108
0264	S	E	E	N	1950	72108
0264	S	D	E	H	1972	76999
0264	S	T	E	H	1972	76999
0264	S	S	E	N	1976	90192
0264	S	D	E	N	1977	91696
0264	S	D	E	N	1978	94675
0266	S	D	E	N	1944	59950
0270	S	D	E	N	1953	43480
0270	S	D	F	N	1967	50734
0270	S	E	F	N	1967	50734
0272	S	D	E	N	1953	43480
0272	S	D	E	N	1972	67289
0272	S	E	E	N	1972	67289
0272	S	S	E	N	1976	90192
0274	S	D	E	N	1927	34129
0274	S	D	E	N	1978	94675
0292	S	D	R	N	1968	54585
0292	S	D	E	N	1972	68377
0292	S	E	E	N	1972	68377
0293	S	D	E	N	1953	43480
0293	S	D	R	N	1968	54585
0293	S	D	E	N	1944	59950
0293	S	S	E	N	1976	90192
0296	S	D	E	N	1944	59950
0298	P	D	E	N	1967	48521
0298	P	E	E	N	1967	48521
0298	S	S	G	N	1928	28680
0298	S	D	E	N	1939	38901
0298	S	D	E	N	1953	43480
0298	S	D	G	H	1967	45495
0298	S	D	G	N	1961	53239
0298	S	D	R	N	1968	54585
0298	S	D	R	H	1969	58631
0298	S	D	E	N	1944	59950
0298	S	D	E	N	1970	60293
0298	S	D	E	N	1921	60557
0298	S	D	E	N	1937	71469
0298	S	D	E	N	1972	78340
0298	S	E	E	N	1972	78340
0298	S	S	E	N	1976	90192
0299	S	E	E	N	1949	59820
0299	S	S	E	H	1971	65258
0299	S	S	E	N	1974	76112
0302	S	D	F	N	1972	44166
0303	S	D	E	H	1968	48982
0303	S	D	E	N	1933	60348
0307	S	D	E	N	1944	59950
0307	S	D	E	N	1970	60293
0307	S	D	E	N	1972	67289
0307	S	E	E	N	1972	67289
0307	S	D	E	N	1955	68323
0308	S	D	E	N	1972	70283
0308	S	S	E	N	1976	90192
0330	S	D	E	N	1953	43480
0330	S	D	E	N	1944	59950
0330	S	S	E	N	1976	90192
0332	S	D	E	H	1926	38360
0344	S	S	E	N	1960	24931
0346	S	S	E	N	1960	24931
0346	S	D	E	N	1964	45842
0346	S	D	S	N	1967	49679
0346	S	D	E	N	1948	60516
0348	S	D	R	N	1968	54585
0349	S	D	E	N	1953	43480
0352	S	D	R	H	1969	59405
0352	S	D	E	H	1969	59406
0352	S	D	E	N	1972	67289
0352	S	E	E	N	1972	67289
0354	S	D	E	N	1950	72108
0354	S	E	E	N	1950	72108
0354	S	D	E	N	1977	91696
0362	P	D	E	N	1965	46637

Substance Number	Phys. State	Subject	Language	Temperature	Year	TPRC Number
521-0362	P	D	E	N	1971	98069
0362	S	D	E	N	1936	44168
0362	S	D	J	H	1965	44605
0362	S	D	E	N	1940	46430
0362	S	D	E	N	1926	58417
0362	S	D	E	H	1979	100206
0362	S	D	E	N	1979	100206
0363	S	D	E	N	1950	72108
0363	S	E	E	N	1950	72108
0364	S	D	R	N	1968	54585
0364	S	D	E	N	1950	72108
0364	S	E	E	N	1950	72108
0366	S	S	G	N	1928	28680
0366	S	D	E	N	1931	43558
0366	S	D	E	H	1966	54245
0366	S	E	E	H	1966	54245
0369	S	D	O	N	1960	53304
0369	S	D	E	N	1969	55322
0369	S	D	E	N	1972	67289
0369	S	E	E	N	1972	67289
0370	S	D	E	N	1944	59950
0370	S	D	E	N	1836	91329
0370	S	E	E	N	1836	91329
0371	S	D	G	H	1962	62305
0375	S	D	R	N	1968	54585
0392	S	D	E	N	1944	59950
0392	S	D	E	N	1974	91243
0392	S	T	E	N	1974	91243
0392	S	D	E	N	1979	100063
0393	P	D	F	N	1962	53233
0393	S	D	E	N	1940	46430
0393	S	D	E	N	1929	60442
0394	S	S	G	N	1928	28680
0394	S	D	E	N	1931	42029
0394	S	D	E	N	1931	42418
0394	S	D	G	N	1952	42961
0394	S	D	E	N	1953	43480
0394	S	D	E	N	1936	44168
0394	S	D	E	N	1952	52094
0394	S	D	P	N	1969	55868
0394	S	D	F	N	1868	56483
0396	S	D	F	N	1868	56483
0411	S	D	E	N	1965	43391
0412	S	D	E	N	1965	43391
0429	S	S	G	N	1928	28680
0429	S	D	E	H	1930	36288
0429	S	D	E	H	1933	55938
0430	S	S	E	N	1960	24931
0430	S	D	E	N	1934	38438
0430	S	D	E	N	1965	41300
0430	S	D	E	N	1945	48237
0432	S	D	E	N	1836	91329
0432	S	E	E	N	1836	91329
0442	S	D	E	N	1965	44633
0442	S	E	E	N	1965	44633
0443	S	D	E	N	1952	52749
0443	S	D	E	N	1932	57549
0443	S	D	G	N	1976	89038
0457	S	D	E	H	1925	38313
0470	S	D	F	N	1866	44793
0471	S	D	E	H	1966	42366
0477	S	S	E	N	1960	24931
0477	S	D	E	N	1939	45053
0481	S	S	E	N	1960	24931
0481	S	S	G	N	1928	28680
0481	S	D	E	N	1936	44168
0481	S	D	E	N	1944	59950
0483	S	D	E	N	1836	91329
0483	S	E	E	N	1836	91329
0496	S	D	E	N	1944	59950
0496	S	D	E	L	1976	87010
0496	S	D	E	N	1976	87010
0500	P	D	R	H	1968	50178
0500	P	E	R	H	1968	50178
0500	P	D	E	H	1972	65520
0500	P	E	E	H	1972	65520
0500	S	D	E	H	1941	42172
0500	S	D	O	N	1965	45264
0500	S	D	E	H	1967	47464
0500	S	D	E	H	1935	60425
0505	P	D	E	N	1965	46637
0506	P	D	E	N	1965	46637
0506	P	D	S	N	1964	59708
0506	S	D	E	H	1938	59921
0506	S	D	G	N	1961	62280
0514	S	D	E	N	1932	57549
0514	S	D	E	N	1976	85755
0514	S	D	E	N	1978	95052
0534	S	D	E	N	1939	38901
0534	S	D	E	N	1953	43480
0534	S	D	E	N	1944	59950
0552	S	D	E	N	1933	43623
0552	S	D	E	N	1944	59950
0552	S	D	E	N	1974	91243
0552	S	T	E	N	1974	91243
0569	S	D	E	N	1944	59950
0604	P	D	E	N	1965	44135
0612	P	D	E	N	1965	44633
0612	P	E	E	N	1965	44633
0612	S	D	E	N	1965	44633
0612	S	E	E	N	1965	44633

Phys. State: **C.** Superconductive; **D.** Doped; **E.** Expanded; **F.** Fibrous or Whisker; **G.** Gas; **I.** Ionized or Plasma; **L.** Liquid; **M.** Multiphase; **P.** Powder or Fine Particle; **S.** Solid; **T.** Thick or Thin Film
Subject: **D.** Data; **E.** Experiment; **G.** General (Data + Theory + Experiment); **S.** Survey (Review, Compendium etc.); **T.** Theory
Language: **C.** Czech; **D.** Dutch; **E.** English; **F.** French; **G.** German; **I.** Italian; **J.** Japanese; **O.** Other Languages; **P.** Polish; **R.** Russian; **S.** Spanish
Temperature: **F.** Full Range (Low + Normal + High); **L.** Low (0 to 75K) + Overlap into Normal; **N.** Normal (75 to 1273K); **H.** High (above 1273K) + Overlap into Normal; Blank = Not Coded

Substance Number	Phys. State	Subject	Language	Temperature	Year	TPRC Number
521-0614	P	S	E	N	1977	88199
0629	S	D	R	N	1968	54585
0648	S	D	E	N	1933	60348
0648	S	D	E	N	1975	82684
0648	T	D	E	N	1975	82684
0660	S	D	O	H	1966	44322
0678	S	D	R	N	1966	44186
0678	S	D	E	N	1966	44187
0685	S	D	E	H	1969	54983
0687	S	D	E	N	1969	51732
0718	S	D	R	N	1968	54585
0727	S	D	E	N	1953	43480
0732	P	D	E	N	1967	50243
0732	P	D	G	N	1963	62306
0732	S	S	E	N	1960	24931
0732	S	S	G	N	1928	28680
0732	S	D	R	N	1971	34689
0732	S	E	R	N	1971	34689
0732	S	D	E	N	1965	36419
0732	S	D	F	N	1920	37779
0732	S	D	E	H	1925	38313
0732	S	D	E	L	1965	38865
0732	S	D	G	N	1918	40320
0732	S	D	E	N	1941	41810
0732	S	E	E	N	1941	41810
0732	S	D	G	N	1965	41873
0732	S	S	G	N	1965	41873
0732	S	D	G	N	1952	42961
0732	S	D	E	N	1953	43480
0732	S	D	E	N	1933	43627
0732	S	D	F	N	1866	44793
0732	S	D	F	N	1868	45692
0732	S	D	O	N	1967	48527
0732	S	D	E	N	1907	49396
0732	S	D	E	N	1967	50145
0732	S	D	E	N	1926	50877
0732	S	D	E	N	1952	52094
0732	S	D	E	H	1956	52425
0732	S	D	E	N	1968	52896
0732	S	D	G	N	1961	53239
0732	S	D	O	N	1960	53304
0732	S	D	R	N	1967	53469
0732	S	D	E	N	1967	53490
0732	S	T	E	N	1967	53490
0732	S	D	R	N	1967	53738
0732	S	E	R	N	1967	53738
0732	S	D	E	H	1966	54245
0732	S	E	E	H	1966	54245
0732	S	D	R	N	1968	54585
0732	S	D	J	N	1969	54751
0732	S	S	E	N	1968	55124
0732	S	D	E	N	1969	55736
0732	S	D	G	N	1916	56408
0732	S	D	E	N	1941	56479
0732	S	G	E	N	1958	56540
0732	S	D	E	N	1924	58024
0732	S	D	E	N	1968	58396
0732	S	D	R	N	1966	58999
0732	S	D	F	N	1957	59213
0732	S	D	E	N	1944	59950
0732	S	D	R	N	1970	60799
0732	S	D	G	N	1963	62306
0732	S	D	E	N	1972	65498
0732	S	D	E	N	1972	66040
0732	S	E	E	N	1972	66040
0732	S	G	E	N	1972	71416
0732	S	G	E	N	1957	76819
0732	S	D	J	N	1973	79311
0732	S	S	E	L	1977	88199
0732	S	S	E	N	1977	88199
0755	S	D	E	N	1968	51017
0761	S	D	E	N	1951	60526
0762	S	D	R	H	1969	58631
0764	P	D	S	N	1964	59708
0764	S	D	O	N	1960	53304
0765	P	D	S	N	1964	59708
0765	S	S	E	N	1960	24931
0765	S	D	O	N	1960	53304
0765	S	D	G	N	1961	62280
0765	S	D	E	N	1973	74556
0772	S	D	E	N	1951	56215
0772	S	T	E	N	1951	56215
0772	S	D	E	N	1948	60516
0785	S	D	S	N	1967	49900
0786	S	D	S	N	1967	49900
0791	P	D	E	N	1972	68735
0793	S	D	R	N	1968	54585
0794	S	D	E	N	1934	38438
0794	S	D	F	N	1868	56483
0797	E	D	E	N	1979	99957
0799	F	D	E	N	1979	97928
0818	S	S	E	N	1960	24931
0818	S	D	E	N	1945	48237
0819	S	S	E	N	1960	24931
0819	S	D	E	N	1945	48237
0824	S	D	E	N	1969	53775
0824	S	D	E	N	1974	76074
0825	S	D	E	N	1936	45621
0828	S	D	E	N	1953	43480
0838	S	D	E	N	1971	63198
0852	S	D	R	N	1968	54585
521-0852	S	D	E	N	1973	74813
0852	S	D	R	N	1974	76265
0852	S	D	R	N	1974	77626
0856	P	D	S	N	1964	59708
0856	S	D	G	N	1952	42961
0856	S	D	E	N	1952	52094
0856	S	D	E	H	1938	59921
0858	S	D	E	N	1949	43559
0859	S	D	E	N	1933	43623
0859	S	D	E	N	1974	91243
0859	S	T	E	N	1974	91243
0860	S	D	E	N	1933	43624
0861	S	D	E	N	1961	52793
0862	S	D	E	N	1961	52793
0870	S	D	E	N	1930	36298
0870	S	D	E	N	1945	42265
0870	S	D	E	N	1930	44259
0870	S	D	J	H	1965	44606
0870	S	D	E	H	1932	52690
0870	S	D	E	H	1933	55938
0873	S	G	G	N	1936	43371
0873	S	D	F	N	1867	45463
0878	E	D	F	N	1977	92686
0878	P	S	E	N	1973	71483
0878	P	D	E	N	1973	79806
0878	S	S	E	N	1960	16704
0878	S	D	R	N	1966	45522
0878	S	D	E	N	1945	56414
0878	S	D	E	H	1970	60729
0878	S	D	E	H	1970	64009
0878	S	E	E	H	1970	64009
0878	S	S	E	F	1973	71483
0878	S	D	E	H	1974	77861
0878	S	D	E	N	1974	77861
0878	S	S	E	H	1977	88199
0878	S	S	E	N	1977	88199
0878	S	D	E	N	1962	91326
0878	S	E	E	N	1962	91326
0878	S	D	E	N	1963	91328
0878	S	D	R	N	1977	91438
0878	S	D	E	H	1969	97338
0878	S	D	E	N	1969	97338
0879	S	D	F	N	1866	44793
0885	S	D	R	N	1968	54585
0886	S	D	R	N	1968	54585
0889	S	D	R	N	1968	54585
0895	S	D	P	N	1969	55868
0895	S	D	E	N	1944	59950
0896	S	D	E	N	1944	59950
0911	S	D	E	H	1933	55938
0912	P	D	E	N	1965	46637
0915	S	S	G	N	1928	28680
0915	S	D	E	N	1959	52683
0915	S	D	E	N	1927	60437
0918	S	D	E	N	1973	74813
0918	S	S	E	N	1972	77778
0922	S	D	E	N	1972	70283
0924	S	D	E	N	1944	59950
0924	S	D	E	N	1978	94675
0930	S	D	E	H	1933	38381
0933	S	D	E	N	1939	38474
0942	S	D	F	N	1868	56483
0944	S	D	E	N	1951	60526
0945	S	D	E	H	1938	59921
0947	S	D	S	N	1967	49679
0950	S	D	S	N	1967	49679
0951	S	D	E	H	1927	34128
0951	S	D	E	N	1927	34129
0951	S	D	E	N	1934	38410
0951	S	D	S	N	1967	49679
0951	S	D	F	N	1868	56483
0957	S	D	E	N	1965	43391
0958	S	D	E	N	1965	43391
0959	S	D	E	N	1965	43391
0959	S	D	E	N	1956	43499
0962	S	D	E	N	1975	82684
0963	S	D	E	N	1975	82684
0963	T	D	E	N	1975	82684
0969	S	D	E	N	1944	59950
0969	S	D	E	N	1974	91243
0969	S	T	E	N	1974	91243
0970	S	D	E	N	1968	50443
0976	S	D	E	N	1944	59950
0976	S	D	E	N	1977	91696
0978	S	D	E	N	1972	67289
0978	S	E	E	N	1972	67289
0982	S	D	R	N	1968	54585
0983	S	D	R	N	1968	54585
0984	S	D	E	H	1925	38313
0988	S	D	E	H	1946	58939
0999	S	D	E	N	1936	45621
1010	S	D	E	N	1953	43480
1010	S	D	E	N	1944	59950
1010	S	D	E	N	1979	99928
1011	S	D	E	N	1953	43480
1028	S	D	E	N	1936	45621
1029	S	D	E	N	1944	59950
1032	S	D	E	N	1940	46430
1034	S	D	E	N	1944	59950
1035	S	D	E	N	1944	59950
1045	S	D	E	N	1974	91243
521-1045	S	T	E	N	1974	91243
1045	S	D	E	N	1979	100063
1054	S	D	E	N	1932	57549
1056	S	D	G	N	1961	62280
1057	S	D	R	H	1969	58631
1068	S	D	E	N	1969	55765
1108	S	D	E	N	1968	50443
1109	S	D	E	N	1968	50443
1113	S	D	P	N	1967	48452
1113	S	E	P	N	1967	48452
1114	S	D	O	N	1971	65626
1117	S	D	E	N	1972	67443
1210	S	D	E	N	1950	72108
1210	S	E	E	N	1950	72108
1216	S	S	E	N	1960	24931
1234	S	D	P	N	1967	48452
1234	S	E	P	N	1967	48452
1255	S	D	E	N	1965	44633
1255	S	E	E	N	1965	44633
1256	S	D	E	N	1950	72108
1256	S	E	E	N	1950	72108
1257	S	D	E	N	1950	72108
1257	S	E	E	N	1950	72108
1258	S	D	E	N	1950	72108
1258	S	E	E	N	1950	72108
1259	S	D	E	N	1950	72108
1259	S	E	E	N	1950	72108
1260	S	D	E	N	1950	72108
1260	S	E	E	N	1950	72108
1261	S	D	E	N	1950	72108
1261	S	E	E	N	1950	72108
1262	S	D	E	N	1950	72108
1262	S	E	E	N	1950	72108
1263	S	D	E	N	1950	72108
1263	S	E	E	N	1950	72108
1335	S	D	R	H	1970	91230
1335	S	D	R	N	1970	91230
1335	S	D	E	H	1970	91231
1335	S	D	E	N	1970	91231
1339	S	D	E	N	1973	71982
1355	S	D	J	N	1973	72447
1357	S	D	E	N	1972	70283
1377	S	D	E	N	1977	93213
1405	S	D	R	N	1972	71302
1418	S	S	E	N	1960	24931
1418	S	S	G	H	1928	28680
1418	S	D	G	N	1976	89038
1422	S	D	E	N	1975	82018
1428	S	D	E	N	1977	92180
1429	T	D	E	N	1975	82684
1437	S	D	E	N	1977	91236
1465	S	S	E	N	1960	24931
1465	S	D	E	N	1921	60557
1540	S	D	E	N	1978	98199
1560	P	D	I	N	1953	94412
1560	S	D	I	N	1953	94412
1602	S	D	E	N	1953	95767
1613	S	D	E	N	1979	100063
1629	S	S	E	N	1960	24931
1629	S	D	J	N	1973	79311
1652	S	S	E	N	1960	24931
1653	S	S	E	N	1960	24931
1655	S	D	E	N	1978	94675

Phys. State: **C.** Superconductive; **D.** Doped; **E.** Expanded; **F.** Fibrous or Whisker; **G.** Gas; **I.** Ionized or Plasma; **L.** Liquid; **M.** Multiphase; **P.** Powder or Fine Particle; **S.** Solid; **T.** Thick or Thin Film

Subject: **D.** Data; **E.** Experiment; **G.** General (Data + Theory + Experiment); **S.** Survey (Review, Compendium etc.); **T.** Theory

Language: **C.** Czech; **D.** Dutch; **E.** English; **F.** French; **G.** German; **I.** Italian; **J.** Japanese; **O.** Other Languages; **P.** Polish; **R.** Russian; **S.** Spanish

Temperature: **F.** Full Range (Low + Normal + High); **L.** Low (0 to 75K) + Overlap into Normal; **N.** Normal (75 to 1273K); **H.** High (above 1273K) + Overlap into Normal; Blank = Not Coded

Chapter 11 Thermal Volumetric Expansion

Substance Number	Phys. State	Subject	Language	Temperature	Year	TPRC Number
503-0001	S	D	E	N	1960	37965
0020	S	D	E	N	1951	6940
0020	S	D	E	N	1960	10947
0020	S	E	E	N	1960	10947
0024	S	D	E	N	1960	37965
0031	S	D	E	N	1933	35687
0031	S	D	E	N	1951	60526
0066	S	D	R	H	1968	51967
0066	S	D	E	H	1968	51968
0082	F	D	E	N	1960	52129
0082	F	D	E	N	1958	52783
0082	F	T	E	N	1958	52783
0082	F	D	E	N	1960	77144
0082	S	D	G	N	1911	22607
0082	S	E	G	N	1911	22607
0082	S	D	E	N	1963	24550
0082	S	E	E	N	1963	24550
0082	S	S	E	N	1966	42284
0082	S	T	E	N	1959	43394
0082	S	D	E	N	1916	47677
0082	S	S	G	N	1941	58735
0082	S	S	E	N	1938	59053
0082	S	G	F	N	1875	59926
0082	S	T	R	N	1973	75835
0082	S	T	E	N	1973	75836
0112	S	D	E	N	1960	37965
0213	S	D	E	N	1960	45778
0270	S	D	E	N	1960	37965
0290	S	D	E	N	1960	37965
0559	S	S	E	H	1977	94788
0559	S	S	E	N	1977	94788
0580	S	D	E	N	1960	45778
0652	S	D	E	N	1960	45778
0652	S	D	E	N	1964	91250
0818	S	D	E	N	1960	37965
0964	S	D	E	N	1980	100311
0985	S	D	E	N	1929	54992
0987	S	D	G	N	1893	60016
0988	S	D	G	N	1893	60016
1021	S	D	R	N	1967	48173
1021	S	D	E	N	1967	48174
1026	S	D	R	N	1969	56888
1033	S	D	E	N	1960	37965
1050	S	D	E	N	1951	6940
1050	S	D	E	N	1960	10947
1050	S	E	E	N	1960	10947
1057	P	D	G	N	1930	58390
1057	S	D	G	N	1930	58390
1059	S	D	E	N	1960	37965
1059	S	D	E	H	1960	56656
1062	S	D	E	N	1951	6940
1062	S	D	E	N	1960	10947
1062	S	E	E	N	1960	10947
1072	S	D	E	N	1951	6940
1072	S	D	E	N	1960	10947
1072	S	E	E	N	1960	10947
1100	S	D	F	N	1866	44793
1133	S	D	E	N	1944	55526
1142	S	S	E	L	1969	45606
1142	S	S	E	H	1971	65258
1143	S	D	E	N	1941	48788
1294	S	D	E	N	1937	44921
1294	S	D	E	N	1956	59992
1311	S	D	E	N	1942	50623
1313	S	D	E	N	1941	49059
1392	S	D	E	N	1960	37965
1393	S	D	R	N	1967	48173
1393	S	D	E	N	1967	48174
1394	S	D	R	N	1967	48173
1394	S	D	E	N	1967	48174
1466	S	D	E	N	1951	6940
1466	S	D	E	N	1960	10947
1466	S	E	E	N	1960	10947
1553	S	D	R	N	1969	56888
1574	S	D	R	N	1967	48173
1574	S	D	E	N	1967	48174
1806	S	D	E	N	1974	79928
1810	S	D	J	N	1977	94558
1811	S	D	J	N	1977	94558
1826	S	S	E	H	1977	94788
1826	S	S	E	N	1977	94788
1867	S	D	E	N	1960	45778
1867	S	D	E	N	1964	91250
1944	S	D	E	N	1960	45778
1987	S	D	E	N	1921	22633
1987	S	E	E	N	1921	22633
2142	S	D	J	N	1977	94558
2184	S	D	E	N	1960	45778
2185	S	D	E	N	1960	45778
2186	S	D	E	N	1960	45778
2187	S	D	E	N	1960	45778
2188	S	D	E	N	1960	45778
2287	S	D	R	N	1969	56888
2461	S	D	E	N	1921	22633
2461	S	E	E	N	1921	22633
503-2478	S	D	E	N	1921	22633
2478	S	E	E	N	1921	22633
2479	S	D	E	N	1921	22633
2479	S	E	E	N	1921	22633
2800	S	D	E	N	1979	99265
2801	S	D	E	N	1979	99265
8252	S	D	E	H	1961	47675
8253	S	D	E	H	1961	47675
8254	S	D	E	H	1961	47675
8255	S	D	E	H	1961	47675
8269	S	D	E	N	1936	60006
8270	S	D	E	N	1936	60006
8282	S	D	E	N	1950	59516
8283	S	D	E	N	1950	59516
8322	S	D	E	H	1965	39834
8515	S	D	G	N	1893	50995
8515	S	T	G	N	1893	50995
8517	S	D	G	N	1893	50995
8517	S	T	G	N	1893	50995
8518	S	D	G	N	1893	50995
8518	S	T	G	N	1893	50995
8519	S	D	G	N	1893	50995
8519	S	T	G	N	1893	50995
8520	S	D	G	N	1893	50995
8520	S	T	G	N	1893	50995
8530	S	D	R	N	1969	56254
8531	S	D	R	N	1969	56254
8605	S	D	G	N	1911	59933
8628	S	D	E	N	1954	38130
8629	S	D	G	N	1893	50995
8629	S	T	G	N	1893	50995
8647	S	D	G	N	1911	59933
8667	S	D	G	N	1911	59933
8668	S	D	G	N	1911	59933
8669	S	D	G	N	1911	59933
8685	S	D	G	N	1911	59933
8729	S	D	E	N	1969	52935
8730	S	D	E	N	1969	52935
8731	S	D	E	N	1969	52935
8732	S	D	E	N	1969	52935
8733	S	D	E	N	1969	52935
8742	S	D	G	N	1893	50995
8742	S	T	G	N	1893	50995
8746	S	D	E	N	1969	54502
8746	S	T	E	N	1969	54502
8839	S	D	G	N	1893	50995
8839	S	T	G	N	1893	50995
8859	S	D	E	N	1969	52935
8868	S	D	E	N	1968	38338
8868	S	D	R	N	1968	50051
8870	S	D	G	N	1969	54931
8870	S	T	G	N	1969	54931
8881	S	D	E	N	1968	47537
8894	S	D	F	N	1949	42450
8899	S	D	C	N	1969	55528
8943	S	D	G	N	1911	59965
8954	S	D	E	H	1966	36254
8954	S	D	R	H	1966	47091
9025	S	D	E	N	1950	60009
9043	S	D	E	N	1969	52935
9068	S	D	R	N	1965	41898
9068	S	D	E	N	1971	62678
9081	S	D	G	N	1893	50995
9081	S	T	G	N	1893	50995
9091	S	D	F	N	1949	42450
9095	S	D	G	N	1893	50995
9095	S	T	G	N	1893	50995
9095	S	D	G	N	1911	59933
9099	S	D	G	N	1893	50995
9099	S	T	G	N	1893	50995
9101	S	D	F	N	1949	42450
9101	S	D	G	N	1893	50995
9101	S	T	G	N	1893	50995
9101	S	D	G	N	1911	59933
9102	S	D	G	N	1893	50995
9102	S	T	G	N	1893	50995
9106	S	D	E	N	1965	42435
9107	S	D	E	N	1965	42435
9108	S	D	G	N	1893	50995
9108	S	T	G	N	1893	50995
9115	S	D	G	N	1893	50995
9115	S	T	G	N	1893	50995
9115	S	D	G	N	1911	59933
9123	S	D	G	N	1893	50995
9123	S	T	G	N	1893	50995
9124	S	G	G	H	1972	68513
9131	S	D	G	N	1893	50995
9131	S	T	G	N	1893	50995
9135	S	D	G	N	1893	50995
9135	S	T	G	N	1893	50995
9146	S	D	G	N	1893	50995
9146	S	T	G	N	1893	50995
9152	S	D	G	N	1893	50995
9152	S	T	G	N	1893	50995
9154	S	D	E	N	1969	52935
503-9154	S	D	E	H	1960	56656
9158	S	D	E	N	1968	47539
9159	S	D	E	N	1965	42435
9160	D	D	G	N	1911	59965
9160	S	D	G	N	1911	59965
9163	S	D	G	N	1893	50995
9163	S	T	G	N	1893	50995
9163	S	D	G	N	1911	59965
9167	S	D	G	N	1893	50995
9167	S	T	G	N	1893	50995
9167	S	D	E	N	1969	52935
9168	S	D	G	N	1893	50995
9168	S	T	G	N	1893	50995
9171	S	D	E	N	1954	38130
9171	S	D	E	N	1968	47537
9171	S	D	G	N	1893	50995
9171	S	T	G	N	1893	50995
9171	S	D	C	N	1969	55468
9173	S	D	E	N	1965	42435
9173	S	D	G	N	1893	50995
9173	S	T	G	N	1893	50995
9173	S	D	E	N	1965	55966
9173	S	G	G	H	1972	68513
9174	S	D	E	N	1968	47539
9174	S	D	G	N	1893	50995
9174	S	T	G	N	1893	50995
9174	S	D	E	N	1969	52935
9174	S	D	E	H	1960	56656
9174	S	G	F	N	1929	59925
9174	S	G	G	H	1972	68513
9175	S	D	G	N	1893	50995
9175	S	T	G	N	1893	50995
9176	S	D	E	N	1968	47539
9177	S	D	E	N	1968	47539
9178	S	D	G	N	1969	41799
9179	S	D	E	N	1968	47539
9179	S	D	G	N	1893	50995
9179	S	T	G	N	1893	50995
9179	S	D	E	N	1969	52935
9179	S	D	E	H	1960	56656
9181	S	D	C	N	1969	55468
9181	S	D	C	N	1969	55528
9198	S	D	R	H	1974	83589
9198	S	D	R	N	1974	83589
9198	S	D	E	H	1974	83590
9198	S	D	E	N	1974	83590
9232	S	D	E	N	1973	71231
9305	S	D	R	N	1969	64018
9305	S	D	E	N	1969	64019
9309	S	D	R	N	1969	56324
9310	S	D	R	N	1969	56324
9317	S	D	R	N	1969	56824
9318	S	D	R	N	1969	56824
9319	S	D	R	N	1969	56824
9355	S	D	G	H	1960	59565
9410	S	D	G	N	1911	59965
9414	S	D	G	N	1914	59966
9428	S	D	G	N	1914	59966
9432	S	D	E	N	1955	59941
9433	S	D	E	N	1955	59941
9434	S	D	E	N	1955	59941
9510	S	D	E	H	1957	39699
9538	P	D	R	H	1968	54779
9538	P	D	E	H	1968	54780
9542	S	D	E	N	1962	52694
9543	S	D	E	N	1962	52694
9544	S	D	E	N	1962	52694
9545	S	D	E	N	1962	52694
9546	S	D	E	N	1962	52694
9547	S	D	E	N	1962	52694
9555	S	D	E	N	1955	59941
9562	S	D	E	N	1970	58728
9610	S	D	E	H	1965	39834
9630	S	D	E	N	1962	52694
9654	S	D	E	N	1962	52694
9707	S	D	E	N	1962	52694
9711	S	D	R	N	1968	54781
9711	S	D	E	N	1968	54782
9731	S	D	R	N	1969	54441
9731	S	D	E	N	1969	57104
9736	S	G	C	H	1968	49012
9748	S	D	E	N	1962	52694
9750	S	D	E	H	1959	47023
9760	S	D	G	H	1960	59565
9773	S	D	E	N	1921	22633
9773	S	E	E	N	1921	22633
9791	S	D	R	H	1968	52350
9791	S	D	R	H	1968	52351
9809	S	D	E	N	1921	22633
9809	S	E	E	N	1921	22633
9809	S	D	S	N	1950	58630
9811	S	D	E	N	1952	57977
9815	S	D	G	H	1960	59565
9829	S	D	E	N	1958	52777
9835	S	D	E	N	1921	22633

Phys. State: **C.** Superconductive; **D.** Doped; **E.** Expanded; **F.** Fibrous or Whisker; **G.** Gas; **I.** Ionized or Plasma; **L.** Liquid; **M.** Multiphase; **P.** Powder or Fine Particle; **S.** Solid; **T.** Thick or Thin Film

Subject: **D.** Data; **E.** Experiment; **G.** General (Data + Theory + Experiment); **S.** Survey (Review, Compendium etc.); **T.** Theory

Language: **C.** Czech; **D.** Dutch; **E.** English; **F.** French; **G.** German; **I.** Italian; **J.** Japanese; **O.** Other Languages; **P.** Polish; **R.** Russian; **S.** Spanish

Temperature: **F.** Full Range (Low + Normal + High); **L.** Low (0 to 75K) + Overlap into Normal; **N.** Normal (75 to 1273K); **H.** High (above 1273K) + Overlap into Normal; Blank = Not Coded

Substance Number	Phys. State	Subject	Language	Temperature	Year	TPRC Number
503-9835	S	E	E	N	1921	22633
9836	S	D	E	N	1921	22633
9836	S	E	E	N	1921	22633
9836	S	D	S	N	1950	58630
9869	S	D	E	N	1958	52777
9869	S	D	R	N	1974	78699
9869	S	E	R	N	1974	78699
9896	S	D	R	H	1968	54775
9896	S	D	E	H	1968	54776
9902	S	D	S	N	1950	58630
9915	S	D	E	N	1962	52694
9916	D	D	R	H	1971	64842
9916	D	D	E	H	1971	64843
9916	S	D	S	H	1967	50986
9916	S	D	G	H	1960	59565
9916	S	D	R	H	1971	64842
9916	S	D	E	H	1971	64843
9922	P	D	R	H	1968	52350
9922	P	D	E	H	1968	52351
9922	S	D	R	H	1968	52350
9922	S	D	E	H	1968	52351
9922	S	D	R	N	1969	54441
9922	S	D	E	N	1969	57104
9922	S	D	R	H	1969	57261
9922	S	D	E	H	1969	57262
9929	S	D	E	N	1970	58728
9955	S	D	E	H	1965	39834
9955	S	D	E	H	1966	43496
9956	S	D	E	H	1965	39834
9961	S	D	E	N	1952	55944
9962	S	D	E	N	1951	60526
9965	S	D	G	H	1960	59565
9981	S	D	E	N	1958	52777
9983	S	D	E	H	1966	43496
521-0001	S	D	E	N	1951	60526
0003	S	D	E	N	1979	97863
0007	S	T	R	N	1969	57561
0007	S	D	E	N	1932	59589
0007	S	D	D	N	1924	90913
0007	S	S	E	N	1960	94260
0007	S	D	R	N	1958	94485
0007	S	D	E	N	1958	94486
0010	P	D	E	N	1951	60526
0022	S	D	E	N	1976	87315
0022	T	D	E	N	1976	87315
0025	S	D	R	N	1968	54585
0026	S	D	F	N	1868	56483
0032	P	D	E	H	1961	52509
0032	P	D	E	N	1979	97914
0033	P	D	E	H	1961	52509
0034	P	D	E	H	1961	52509
0035	S	D	E	N	1933	43623
0035	S	D	E	H	1971	62575
0035	S	D	E	N	1974	91243
0035	S	T	E	N	1974	91243
0036	S	D	E	N	1934	38438
0036	S	D	E	N	1933	43624
0046	S	S	E	N	1966	42284
0053	S	D	R	N	1969	57574
0053	S	T	R	N	1969	57574
0053	S	D	E	N	1929	58609
0057	S	D	E	N	1933	43623
0057	S	D	E	H	1967	45743
0057	S	D	E	N	1974	91243
0057	S	T	E	N	1974	91243
0059	S	D	E	N	1933	38381
0059	S	D	R	N	1965	44807
0060	S	D	F	N	1868	56483
0062	P	D	E	N	1968	51345
0063	S	D	F	N	1866	44793
0063	S	D	E	N	1957	52532
0063	S	D	E	N	1932	59589
0063	S	D	E	N	1976	90502
0063	S	T	E	N	1977	92117
0063	S	D	E	N	1946	98462
0063	S	E	E	N	1946	98462
0072	S	D	R	N	1968	54585
0072	S	D	E	N	1951	60526
0072	S	D	E	N	1977	91696
0072	S	D	E	N	1978	96401
0072	S	D	E	N	1979	98173
0072	S	T	E	N	1979	98173
0073	S	D	E	N	1929	58609
0075	S	D	E	N	1929	34568
0075	S	D	E	N	1950	57504
0075	S	D	G	N	1930	58390
0075	S	D	E	N	1932	59589
0075	S	D	E	N	1949	59820
0075	S	D	E	N	1955	60343
0077	S	D	F	N	1866	44793
0082	S	D	E	N	1929	60116
0083	S	D	F	N	1868	56483
0084	S	D	E	N	1951	60526
0085	S	D	R	N	1968	54585
0085	S	D	E	N	1951	60526
0085	S	D	E	N	1964	70710
0085	S	E	E	N	1964	70710
0085	S	D	E	N	1979	98173
0085	S	T	E	N	1979	98173
0092	P	D	E	N	1970	56672
521-0092	S	D	F	N	1866	44793
0103	S	D	R	N	1968	54585
0103	S	D	E	N	1950	57504
0103	S	D	E	N	1951	59551
0103	S	D	E	N	1949	59820
0103	S	D	E	N	1951	60526
0103	S	D	E	N	1979	98173
0103	S	T	E	N	1979	98173
0104	S	D	F	N	1866	44793
0104	S	D	E	N	1950	60339
0105	S	D	E	N	1934	38439
0110	P	D	E	N	1951	60526
0130	S	D	F	N	1866	44793
0181	S	D	R	N	1968	54585
0184	S	D	R	N	1968	54585
0186	P	D	R	H	1969	57588
0196	S	D	E	N	1979	98173
0196	S	T	E	N	1979	98173
0217	S	D	E	H	1961	47675
0223	S	D	G	N	1934	60768
0242	S	D	R	N	1968	54585
0250	S	D	R	H	1976	87277
0250	S	D	R	N	1976	87277
0250	S	D	E	H	1976	89152
0250	S	D	E	N	1976	89152
0258	S	D	E	N	1979	97863
0261	S	D	E	N	1965	42608
0261	S	D	E	N	1969	94447
0264	S	D	R	N	1968	54585
0264	S	D	E	N	1977	91696
0264	S	D	E	N	1979	98173
0264	S	T	E	N	1979	98173
0281	S	D	E	N	1979	98173
0281	S	T	E	N	1979	98173
0292	S	D	R	N	1968	54585
0293	S	D	R	N	1968	54585
0298	S	G	E	N	1968	44400
0298	S	D	R	N	1968	54585
0299	S	T	G	N	1920	22785
0299	S	E	E	N	1949	59820
0307	S	D	E	N	1929	58609
0308	S	D	E	N	1972	70283
0348	S	D	R	N	1968	54585
0354	S	D	E	N	1977	91696
0362	S	D	E	N	1936	44168
0364	S	D	R	N	1968	54585
0364	S	D	E	N	1979	98173
0364	S	T	E	N	1979	98173
0375	S	D	R	N	1968	54585
0392	S	D	E	N	1974	91243
0392	S	T	E	N	1974	91243
0394	S	D	E	N	1936	60006
0396	S	D	F	N	1868	56483
0411	S	D	E	N	1965	43391
0412	S	D	E	N	1965	43391
0443	S	D	E	N	1979	97378
0470	S	D	F	N	1866	44793
0470	S	D	G	N	1934	60768
0471	S	D	E	H	1966	42366
0494	S	D	E	N	1979	97863
0552	S	D	E	N	1933	43623
0552	S	D	E	N	1974	91243
0552	S	T	E	N	1974	91243
0566	S	D	G	N	1934	60768
0629	S	D	R	N	1968	54585
0648	S	D	F	N	1866	44793
0718	S	D	R	N	1968	54585
0732	S	D	F	N	1866	44793
0732	S	D	R	N	1968	54585
0732	S	D	E	N	1977	91331
0761	S	D	E	N	1951	60526
0791	P	D	E	N	1972	68735
0793	S	D	R	N	1968	54585
0794	S	D	E	N	1934	38438
0794	S	D	F	N	1868	56483
0824	S	D	E	N	1969	53775
0824	S	T	E	N	1972	67440
0824	S	D	E	N	1974	76074
0852	S	D	R	N	1968	54585
0859	S	D	E	N	1933	43623
0859	S	D	E	N	1974	91243
0859	S	T	E	N	1974	91243
0860	S	D	E	N	1933	43624
0879	S	D	F	N	1866	44793
0885	S	D	R	N	1968	54585
0886	S	D	R	N	1968	54585
0889	S	D	R	N	1968	54585
0922	S	D	E	N	1972	70283
0930	S	D	E	H	1933	38381
0942	S	D	F	N	1868	56483
0944	S	D	E	N	1951	60526
0951	S	D	E	N	1934	38410
0951	S	D	E	N	1934	38438
0951	S	D	F	N	1868	56483
0957	S	D	E	N	1965	43391
0958	S	D	E	N	1965	43391
0959	S	D	E	N	1965	43391
0962	S	D	F	N	1866	44793
0969	S	D	E	N	1974	91243
0969	S	T	E	N	1974	91243
0975	S	D	E	N	1979	98173
521-0975	S	T	E	N	1979	98173
0976	S	D	E	N	1977	91696
0982	S	D	R	N	1968	54585
0983	S	D	R	N	1968	54585
0990	S	D	F	N	1866	44793
0991	S	D	F	N	1866	44793
1029	S	D	E	N	1929	58609
1032	S	D	E	N	1940	46430
1045	S	D	E	N	1974	91243
1045	S	T	E	N	1974	91243
1339	S	D	E	N	1973	71982
1357	S	D	E	N	1972	70283
1503	S	D	E	N	1979	98173
1503	S	T	E	N	1979	98173

Phys. State: **C.** Superconductive; **D.** Doped; **E.** Expanded; **F.** Fibrous or Whisker; **G.** Gas; **I.** Ionized or Plasma; **L.** Liquid; **M.** Multiphase; **P.** Powder or Fine Particle; **S.** Solid; **T.** Thick or Thin Film

Subject: **D.** Data; **E.** Experiment; **G.** General (Data + Theory + Experiment); **S.** Survey (Review, Compendium etc.); **T.** Theory

Language: **C.** Czech; **D.** Dutch; **E.** English; **F.** French; **G.** German; **I.** Italian; **J.** Japanese; **O.** Other Languages; **P.** Polish; **R.** Russian; **S.** Spanish

Temperature: **F.** Full Range (Low + Normal + High); **L.** Low (0 to 75K) + Overlap into Normal; **N.** Normal (75 to 1273K); **H.** High (above 1273K) + Overlap into Normal; Blank = Not Coded

Chapter 12 Thermal Radiative Properties

Substance Number	Phys. State	Sub-ject	Lan-guage	Temper-ature	Year	TPRC Number
503-0012	D	D	E	N	1917	17754
0012	M	D	E	F	1959	10060
0012	S	D	E		1959	7696
0012	S	D	E	F	1959	10060
0012	S	D	E	N	1963	30621
0012	S	E	E	N	1963	30621
0018	S	D	E	N	1964	20810
0018	S	D	E	N	1959	24081
0020	F	D	E	N	1962	29611
0020	M	D	E	H	1960	18644
0020	S	D	E		1951	6456
0020	S	G	E		1954	6596
0020	S	D	E		1951	9560
0020	S	D	E	N	1958	20771
0020	S	D	E	N	1962	33458
0024	S	D	E	N	1953	15126
0024	S	D	R	N	1953	20572
0024	S	D	E	N	1960	29300
0025	L	G	E	N	1961	18319
0025	S	G	G		1952	145
0025	S	E	E		1954	4038
0025	S	G	E		1956	8408
0025	S	G	E	N	1961	18319
0025	S	D	E	N	1958	18613
0025	S	G	E	N	1958	20502
0025	S	D	E	N	1957	20833
0025	S	D	E	N	1955	24082
0025	S	D	R	N	1957	24221
0025	S	D	E	N	1955	29200
0031	F	D	E	N	1962	29611
0031	S	D	E		1951	6456
0031	S	G	E		1954	6596
0031	S	D	E	H	1961	24893
0031	S	D	E	N	1955	26325
0031	S	D	E	N	1964	27035
0031	S	D	E	N	1962	29643
0031	S	G	E	N	1939	31171
0031	S	D	E	N	1951	31931
0033	S	G	E	N	1954	22147
0082	D	G	E	N	1964	22704
0082	D	G	R	N	1964	22705
0082	D	D	E	N	1963	23289
0082	D	E	E	N	1963	23289
0082	E	D	E	N	1955	24082
0082	F	D	E	N	1961	23361
0082	G	T	E	N	1950	20468
0082	G	G	E	H	1961	29352
0082	L	S	E	N	1961	18316
0082	L	S	E	N	1960	20650
0082	L	D	E	H	1952	21446
0082	L	T	E	H	1952	21446
0082	S	D	E		1958	7023
0082	S	T	E		1956	8408
0082	S	E	E		1946	8831
0082	S	D	E	N	1959	13954
0082	S	G	E	N	1961	15251
0082	S	G	E	N	1961	16961
0082	S	D	E	N	1948	17044
0082	S	D	O	N	1958	17573
0082	S	S	E	N	1961	18316
0082	S	D	G	N	1935	20382
0082	S	D	G	N	1898	20396
0082	S	S	E	N	1960	20650
0082	S	S	G	N	1923	21761
0082	S	S	R	N	1931	22058
0082	S	D	E	N	1961	22990
0082	S	D	E	N	1962	23258
0082	S	D	E	N	1963	23291
0082	S	S	E	N	1960	23528
0082	S	D	E	H	1913	23759
0082	S	D	G	N	1889	24145
0082	S	D	E	N	1961	24746
0082	S	D	E	N	1961	24893
0082	S	S	E	N	1956	24971
0082	S	G	E	N	1963	25235
0082	S	D	E	N	1962	25723
0082	S	D	E	N	1954	26253
0082	S	E	E	N	1954	26253
0082	S	D	E	N	1963	27886
0082	S	S	E	H	1961	28014
0082	S	D	E	N	1941	31731
0082	S	G	E	N	1959	32429
0082	S	D	E	N	1927	33283
0082	S	D	E	N	1942	33298
0082	S	D	E	N	1962	33388
0082	S	D	E	N	1962	33460
0082	S	T	E	N	1962	33460
0096	S	D	E	N	1963	23289
0096	S	D	E	N	1951	24552
0216	M	D	E	F	1959	10060
0216	S	D	E		1959	7696
0216	S	D	E		1951	9560
0216	S	D	E	F	1959	10060
0216	S	D	E	N	1961	29705
0216	S	G	E	N	1950	31153

Substance Number	Phys. State	Sub-ject	Lan-guage	Temper-ature	Year	TPRC Number
503-0216	S	D	E	N	1963	32772
0239	S	G	E	H	1920	31310
0270	L	D	E	H	1961	18320
0270	S	D	E		1951	9560
0270	S	D	E	N	1905	27306
0270	S	T	E	N	1905	27306
0270	S	G	E	N	1956	33315
0309	S	D	E	N	1964	20810
0376	S	D	E	N	1960	18638
0452	M	D	E	N	1964	29570
0452	S	G	E	L	1963	26738
0537	M	D	E	F	1959	10060
0537	P	D	E	N	1963	28755
0537	S	D	E		1959	7696
0537	S	D	E	F	1959	10060
0545	D	D	E	N	1917	17754
0545	D	D	E	N	1916	28205
0545	S	D	E	H	1961	24893
0545	S	D	E	N	1955	26325
0545	S	G	E	N	1939	31171
0567	S	D	J	N	1963	33004
0569	S	D	E	N	1963	22963
0569	S	D	E	N	1959	24081
0604	M	D	E	F	1959	10060
0604	S	D	E		1959	7696
0604	S	D	E	F	1959	10060
0640	D	D	E	N	1962	27280
0656	S	D	E		1959	7696
0659	M	D	E	F	1959	10060
0659	S	D	E	F	1959	10060
0659	S	G	E	L	1963	26738
0659	S	D	E	H	1964	29570
0659	S	D	E	N	1963	30602
0660	M	D	E	F	1959	10060
0660	S	D	E	F	1959	10060
0660	S	D	E	N	1961	16221
0660	S	D	E	H	1960	18630
0660	S	G	E	L	1963	26738
0660	S	D	E	H	1964	29570
0667	S	D	E	N	1958	20771
0667	S	S	E	N	1961	25075
0667	S	S	E	N	1964	30525
0667	S	D	G	N	1964	31343
0667	S	D	E	N	1964	31344
0706	E	D	E	N	1963	32319
0731	S	D	G	N	1958	19095
0731	S	D	E	N	1964	32047
0739	S	D	E	N	1962	24553
0782	D	D	E	N	1960	18637
0782	D	D	E	N	1916	28205
0782	D	D	E	N	1920	28418
0782	M	D	E	F	1959	10060
0782	S	D	E	F	1959	10060
0782	S	D	E	N	1960	18634
0782	S	D	E	N	1963	30621
0782	S	E	E	N	1963	30621
0786	S	D	E	N	1955	24082
0787	S	D	E	N	1917	17754
0787	S	D	E	N	1958	21327
0787	S	D	E	N	1955	24082
0801	S	D	E	N	1963	23291
0801	S	D	E	H	1908	23456
0801	S	D	E	H	1913	23759
0801	S	D	E	N	1953	26316
0801	S	D	R	N	1947	26342
0801	S	D	G	N	1941	31926
0801	S	G	E	N	1954	32388
0801	S	D	G	H	1911	33218
0818	S	D	E	N	1955	26325
0822	S	D	E	H	1961	24893
0822	S	G	E	N	1939	31171
0932	S	G	E	N	1954	22147
0940	S	G	E	N	1954	22147
0954	S	G	E	N	1961	19818
0954	S	D	E	N	1920	28418
0959	S	D	E	N	1959	24081
0960	F	D	E	N	1962	29611
0960	F	E	E	N	1962	29611
0962	S	D	E	N	1960	18635
0965	S	D	E	N	1959	24081
0966	S	D	E	N	1960	10967
0966	S	D	E	N	1960	15886
0974	S	D	E	N	1964	20810
0974	S	D	E	N	1959	24081
0974	S	D	E	N	1961	29705
0975	S	D	E	N	1961	29705
0983	S	S	G	N	1957	20221
0985	L	G	G	H	1959	17922
0991	S	G	E	H	1920	31310
0994	S	G	E	H	1920	31310
1006	S	D	E	N	1960	17182
1007	S	D	E	N	1960	17182
1027	S	D	E	N	1955	24082
1047	S	D	F	N	1929	33623
1048	S	G	G		1952	145

Substance Number	Phys. State	Sub-ject	Lan-guage	Temper-ature	Year	TPRC Number
503-1050	S	D	E	N	1955	24082
1050	S	D	E	N	1955	26325
1050	S	D	E	N	1905	27306
1050	S	T	E	N	1905	27306
1050	S	D	E	N	1961	27388
1050	S	D	E	N	1962	27848
1050	S	D	E	N	1960	29596
1050	S	G	E	N	1954	32388
1050	S	D	E	N	1939	33152
1050	S	G	E	N	1942	33296
1054	L	D	E		1955	4043
1055	L	D	E		1955	4043
1056	S	S	G	N	1957	20221
1059	S	D	E	N	1963	22963
1059	S	G	E	N	1961	25994
1059	S	D	E	N	1960	29300
1060	S	D	E	H	1917	27822
1061	S	D	E		1947	9353
1061	S	D	E	N	1955	24082
1066	S	D	E	H	1908	23456
1069	S	D	E		1951	9560
1070	S	D	E		1951	9560
1070	S	D	E	N	1917	17754
1070	S	D	E	N	1955	24082
1071	S	D	E		1947	9353
1073	S	D	E	N	1960	29596
1073	S	D	E	N	1963	30621
1073	S	E	E	N	1963	30621
1077	S	D	E		1956	9450
1077	S	D	G	N	1935	16087
1077	S	D	E	N	1953	26316
1077	S	D	R	N	1947	26342
1080	S	D	E	N	1953	14835
1080	S	D	E	N	1953	14986
1080	S	D	E	N	1953	15126
1080	S	D	R	N	1953	20572
1080	S	D	R	N	1953	23580
1080	S	D	R	N	1953	23581
1080	S	D	R	N	1953	23582
1083	S	D	R	N	1963	30784
1083	S	D	E	N	1963	31295
1083	S	E	E	N	1963	31295
1085	D	D	E	N	1917	17754
1086	D	D	E	N	1917	17754
1088	S	D	E	N	1917	17754
1089	S	D	E	N	1917	17754
1089	S	D	E	N	1910	33140
1092	M	D	E	F	1959	10060
1092	S	D	E		1959	7696
1092	S	D	E	F	1959	10060
1093	M	D	E	F	1959	10060
1093	S	D	E	F	1959	10060
1094	M	D	E	F	1959	10060
1094	S	D	E	F	1959	10060
1095	S	D	E	N	1917	17754
1095	S	G	E	N	1954	22147
1095	S	G	E	N	1964	22704
1095	S	G	R	N	1964	22705
1095	S	D	E	H	1908	23456
1095	S	D	E	N	1910	33140
1096	S	D	E	N	1917	17754
1104	S	D	E	N	1963	22272
1105	S	D	E	N	1955	24082
1107	S	D	E	N	1955	24082
1108	S	D	E	H	1961	24893
1108	S	G	E	N	1939	31171
1109	S	D	E	H	1961	24893
1109	S	G	E	N	1939	31171
1110	S	D	E	N	1956	24972
1110	S	S	E	N	1961	25075
1110	S	G	E	H	1920	31310
1113	S	G	E	H	1920	24246
1114	S	D	E	H	1962	31121
1116	S	G	E	H	1920	31310
1119	S	S	E	N	1963	32581
1119	S	S	E	N	1963	33171
1120	S	D	E	N	1910	33140
1121	S	D	E	N	1910	33140
1122	S	D	E	N	1942	33298
1123	S	D	E	N	1965	33451
1123	S	D	E	N	1965	33453
1124	S	D	F	N	1929	33623
1126	S	D	E	N	1917	17754
1131	M	D	R		1941	4632
1131	S	D	G	N	1962	27821
1131	S	D	F	N	1963	31427
1132	M	D	R		1941	4632
1133	S	D	F	H	1958	19583
1133	S	S	E	N	1925	19806
1133	S	D	G	H	1927	21108
1133	S	D	F	H	1963	31427
1133	S	D	E	H	1963	32477
1135	M	D	F	H	1963	31427
1135	S	D	E		1957	6740
1135	S	D	F	H	1963	31427

Phys. State: **C.** Superconductive; **D.** Doped; **E.** Expanded; **F.** Fibrous or Whisker; **G.** Gas; **I.** Ionized or Plasma; **L.** Liquid; **M.** Multiphase; **P.** Powder or Fine Particle; **S.** Solid; **T.** Thick or Thin Film
Subject: **D.** Data; **E.** Experiment; **G.** General (Data + Theory + Experiment); **S.** Survey (Review, Compendium etc.); **T.** Theory
Language: **C.** Czech; **D.** Dutch; **E.** English; **F.** French; **G.** German; **I.** Italian; **J.** Japanese; **O.** Other Languages; **P.** Polish; **R.** Russian; **S.** Spanish
Temperature: **F.** Full Range (Low + Normal + High); **L.** Low (0 to 75K) + Overlap into Normal; **N.** Normal (75 to 1273K); **H.** High (above 1273K) + Overlap into Normal; Blank = Not Coded

Substance Number	Phys. State	Sub-ject	Lan-guage	Temper-ature	Year	TPRC Number
503-1136	L	S	E	N	1960	23528
1136	M	D	R		1941	4632
1136	S	S	E	H	1939	22514
1136	S	D	E	N	1912	22520
1136	S	D	R	N	1957	24221
1136	S	D	E	N	1953	26316
1136	S	D	R	N	1947	26342
1136	S	G	E	N	1931	33165
1142	S	D	E	H	1959	15666
1142	S	E	G	H	1960	20138
1142	S	S	E	N	1961	23527
1142	S	S	E	N	1960	23528
1142	S	S	E	F	1962	24748
1142	S	E	E	N	1960	24831
1142	S	S	E	H	1960	24931
1142	S	S	E	F	1962	25558
1142	S	S	E	N	1928	31217
1144	S	D	E		1959	7696
1145	S	D	E		1955	4013
1145	S	T	E		1955	4013
1145	S	T	E		1955	4200
1145	S	E	E	H	1955	10101
1145	S	E	E	H	1954	10483
1145	S	S	E	H	1958	10681
1145	S	S	E	H	1960	18000
1146	M	D	R		1941	4632
1146	S	D	J	H	1937	14107
1146	S	S	E	N	1925	19806
1147	M	D	R		1941	4632
1147	M	D	E	N	1962	25127
1147	S	S	E	N	1925	19806
1147	S	D	G	H	1927	21108
1147	S	D	E	N	1963	23291
1147	S	D	G	N	1941	31926
1151	S	G	E	N	1931	33165
1153	M	D	E		1953	4000
1153	S	D	F	N	1963	31427
1155	S	D	F	H	1963	31427
1162	S	D	E	H	1958	10458
1162	S	D	F	N	1963	31427
1168	S	D	E	N	1959	28666
1173	M	G	E	N	1939	31171
1173	S	D	E		1959	7696
1173	S	D	E	H	1960	18630
1173	S	D	F	H	1958	19583
1173	S	D	E	N	1958	20771
1173	S	D	E	H	1959	29099
1173	S	D	E	H	1964	29570
1173	S	D	F	H	1954	31836
1173	S	D	E	N	1964	32045
1173	S	D	E	H	1963	32477
1177	S	D	E	N	1963	23291
1177	S	D	G	N	1941	31926
1177	S	G	E	N	1931	33165
1180	S	D	E	H	1964	29570
1188	M	G	E	N	1939	31171
1189	M	G	E	N	1939	31171
1190	S	D	E	H	1961	24893
1190	S	G	E	N	1939	31171
1194	S	D	E	N	1915	20292
1208	S	D	F	N	1963	31427
1212	S	G	E	N	1931	33165
1220	S	D	R	N	1957	24221
1250	D	D	E	N	1955	24082
1250	D	D	E	N	1910	33140
1250	L	D	E	H	1961	18320
1250	S	D	E		1951	9560
1250	S	D	E	N	1958	21327
1250	S	G	E	N	1954	22147
1250	S	D	E	N	1960	29596
1250	S	G	E	N	1963	30604
1250	S	G	R	N	1949	31625
1250	S	D	E	N	1910	33140
1250	S	D	F	N	1929	33623
1251	S	D	E	N	1955	24082
1252	S	D	E	N	1955	24082
1253	S	D	E	N	1955	24082
1254	S	D	E	N	1917	17754
1254	S	D	E	N	1958	21327
1254	S	G	E	N	1954	22147
1255	D	D	E	N	1916	28205
1255	D	D	E	N	1920	28418
1256	D	D	E	N	1916	28205
1256	D	D	E	N	1920	28418
1257	P	G	E	N	1956	28573
1257	P	G	R	N	1956	33222
1258	S	G	E	H	1920	31310
1260	S	D	E	N	1956	24972
1277	S	D	R	N	1957	24221
1278	S	G	E	N	1931	33165
1357	L	D	J	H	1937	14107
1359	S	D	R		1954	8877
1364	S	D	G	N	1962	27821
1529	S	D	E	N	1961	28924
1529	S	E	E	N	1961	28924
1538	S	S	G	N	1957	20221
1539	S	S	G	N	1957	20221
1540	S	S	G	N	1957	20221
1541	S	S	G	N	1957	20221
1542	S	S	G	N	1957	20221
1543	S	S	G	N	1957	20221

Substance Number	Phys. State	Sub-ject	Lan-guage	Temper-ature	Year	TPRC Number
503-1544	S	S	G	N	1957	20221
1547	S	D	E		1959	7696
1548	S	D	E	N	1962	22720
1564	S	D	E	N	1964	32047
1565	S	D	E	N	1964	32047
1566	S	D	E	N	1964	32047
1567	S	D	E	N	1964	32047
1568	S	D	E	N	1960	17182
1697	S	D	G	N	1959	19587
1720	S	D	E	N	1960	18638
2248	S	D	E	N	1961	30293
2249	S	G	E	N	1950	31153
2406	S	D	G	N	1959	19587
2407	S	D	G	N	1959	19587
2408	S	D	G	N	1959	19587
2409	S	D	G	N	1959	19587
2410	S	D	E	N	1960	18635
2414	S	D	R	N	1959	20917
2450	S	D	E	N	1960	18635
2451	S	D	E	N	1960	18635
2451	S	D	E	N	1960	18638
2457	S	D	G	N	1959	19587
2473	S	D	R	N	1959	20917
2474	S	D	R	N	1959	20917
2475	S	D	R	N	1959	20917
8356	S	D	E	H	1965	31115
8382	S	D	F	H	1954	31836
8384	S	D	E	N	1910	23480
8387	S	D	E	N	1958	18668
8407	S	D	J	H	1961	26224
8412	S	G	E	N	1950	31153
8414	S	D	R	N	1962	26143
8415	S	D	R	N	1962	26143
8417	S	D	E	N	1963	33172
8900	S	D	E	N	1966	27407
8901	S	D	E	N	1966	27407
8902	S	D	E	N	1966	27407
8903	S	D	E	N	1963	20683
8903	S	D	R	N	1962	31150
8904	S	D	E	N	1963	20683
8904	S	D	R	N	1962	31150
8905	S	D	E	N	1963	20683
8905	S	D	R	N	1962	31150
8906	S	D	E	N	1963	20683
8906	S	D	R	N	1962	31150
8907	S	D	E	N	1963	20683
8907	S	D	R	N	1962	31150
8908	S	D	E	N	1963	20683
8908	S	D	R	N	1962	31150
8909	S	D	E	N	1963	20683
8909	S	D	R	N	1962	31150
8910	S	D	E	N	1963	20683
8910	S	D	R	N	1962	31150
8911	S	D	E	N	1963	20683
8911	S	D	R	N	1962	31150
8912	S	D	E	N	1963	20683
8912	S	D	R	N	1962	31150
8913	S	D	E	N	1963	20683
8913	S	D	R	N	1962	31150
8914	S	D	E	N	1963	20683
8914	S	D	R	N	1962	31150
8915	S	D	E	N	1963	20683
8915	S	D	R	N	1962	31150
8916	S	D	J	H	1961	26224
8917	S	D	J	H	1961	26224
8918	S	D	J	H	1961	26224
8919	S	D	J	H	1961	26224
8920	S	D	J	H	1961	26224
8921	S	D	J	H	1961	26224
8922	S	D	J	H	1961	26224
8923	S	D	J	H	1961	26224
8924	S	D	J	H	1961	26224
8933	L	T	E	H	1962	32616
8948	S	D	J	N	1962	31464
8949	S	D	J	N	1962	31464
8960	S	D	E	N	1962	30804
8961	S	D	J	H	1961	26224
8961	S	D	E	N	1962	30804
8962	S	D	E	N	1962	30804
8963	S	D	E	N	1962	30804
8964	S	D	E	N	1962	30804
8965	S	D	E	N	1962	30804
8966	S	D	E	N	1962	30804
8968	S	D	E	N	1962	30804
8969	S	D	E	N	1962	30804
8970	S	D	E	N	1962	30804
8971	S	D	E	N	1962	30804
8972	S	D	E	N	1952	30239
8977	S	D	E	N	1963	29914
8978	S	D	E	N	1963	29914
8987	S	D	E	N	1959	18624
8989	S	D	E	N	1957	28311
8990	S	D	E	N	1957	28311
8991	S	D	E		1951	9560
8992	S	D	E	N	1962	27280
8993	S	D	E	N	1962	27280
8994	S	D	E	N	1962	27280
8995	S	D	E	N	1962	27280
8996	S	D	E	N	1962	27280
8998	S	D	E	N	1960	28804
9000	S	D	E	N	1962	26777

Substance Number	Phys. State	Sub-ject	Lan-guage	Temper-ature	Year	TPRC Number
503-9003	S	D	R	N	1960	25795
9003	S	D	E	N	1961	29003
9004	S	D	R	N	1960	25795
9004	S	D	E	N	1961	29003
9005	S	D	R	N	1960	25795
9005	S	D	E	N	1961	29003
9006	S	D	R	N	1960	25795
9006	S	D	E	N	1961	29003
9007	S	D	R	N	1960	25795
9007	S	D	E	N	1961	29003
9008	S	D	R	N	1960	25795
9008	S	D	J	H	1961	26224
9008	S	D	E	N	1961	29003
9009	S	D	R	N	1960	25795
9009	S	D	E	N	1961	29003
9010	S	D	E	N	1963	29914
9014	S	D	E	N	1962	25985
9014	S	D	G	N	1959	30334
9015	S	D	E	N	1963	29914
9016	S	D	E	N	1959	24712
9016	S	D	E	N	1963	29914
9017	S	D	E	N	1959	24712
9017	S	D	E	N	1963	29914
9018	S	D	E	N	1959	24712
9018	S	D	E	N	1963	29914
9019	S	D	E	N	1959	24712
9019	S	D	E	N	1963	29914
9020	S	D	E	N	1959	24712
9020	S	D	E	N	1963	29914
9021	S	D	E	N	1959	24712
9021	S	D	E	N	1963	29914
9027	S	D	E	N	1955	24082
9028	S	D	E	N	1955	24082
9029	S	D	E	N	1955	24082
9030	S	D	E	N	1955	24082
9031	S	D	E	N	1955	24082
9032	S	D	E	N	1955	24082
9033	S	D	E	N	1955	24082
9034	S	G	E	H	1963	28056
9040	L	G	G	H	1959	17922
9042	L	G	G	H	1959	17922
9042	L	G	E	H	1959	29365
9042	M	D	E	H	1925	21960
9042	M	S	E	H	1925	21960
9042	S	D	E	N	1963	21323
9043	L	G	G	H	1959	17922
9056	S	D	G	N	1959	24403
9057	S	D	G	N	1959	24403
9060	S	D	E	N	1960	18636
9061	S	D	J	N	1962	31464
9063	S	D	E	N	1959	18624
9064	S	D	E	N	1959	18624
9067	L	D	E	H	1961	18318
9069	S	D	E	N	1959	17924
9070	S	D	E	N	1959	17924
9071	S	D	E	N	1959	17924
9072	S	D	E	N	1959	17924
9074	S	D	E	N	1963	20683
9074	S	D	R	N	1962	31150
9082	S	D	E	N	1955	26325
9084	S	D	E	H	1961	16377
9095	S	D	E	N	1963	20683
9095	S	D	R	N	1962	31150
9096	S	D	G	N	1931	14371
9098	S	G	G	N	1936	13340
9101	L	G	G	H	1959	17922
9109	S	D	E	N	1955	24082
9110	S	D	E	N	1955	24082
9111	S	D	E	N	1955	24082
9117	S	D	J	H	1961	26224
9117	S	D	E	N	1962	30804
9120	S	D	E	N	1962	30804
9129	S	D	E	N	1962	30804
9132	S	D	E	N	1955	26325
9133	S	D	E	N	1955	26325
9133	S	D	G	N	1962	27639
9134	S	D	E	N	1955	26325
9141	L	D	E	H	1961	18318
9147	S	D	E	N	1955	26325
9148	S	D	E	N	1955	26325
9152	S	D	E	N	1953	15126
9152	S	D	R	N	1953	20572
9152	S	D	E	N	1963	20683
9152	S	D	R	N	1962	31150
9154	L	G	G	H	1959	17922
9154	S	D	E	N	1959	17925
9154	S	D	E	N	1955	26325
9154	S	G	E	N	1950	31153
9154	S	D	E	N	1963	33172
9157	S	D	G	N	1962	27646
9157	S	G	E	N	1950	31153
9159	S	D	E	N	1962	25985
9159	S	D	G	N	1959	30334
9161	L	G	G	H	1959	17922
9163	S	D	E	N	1963	20683
9163	S	D	R	N	1962	31150
9169	S	D	E	N	1955	26325
9169	S	D	E	N	1966	27407
9169	S	D	E	N	1962	30804
9171	S	D	E	N	1955	26325
9173	S	D	E	N	1955	24082

Phys. State: **C.** Superconductive; **D.** Doped; **E.** Expanded; **F.** Fibrous or Whisker; **G.** Gas; **I.** Ionized or Plasma; **L.** Liquid; **M.** Multiphase; **P.** Powder or Fine Particle; **S.** Solid; **T.** Thick or Thin Film

Subject: **D.** Data; **E.** Experiment; **G.** General (Data + Theory + Experiment); **S.** Survey (Review, Compendium etc.); **T.** Theory

Language: **C.** Czech; **D.** Dutch; **E.** English; **F.** French; **G.** German; **I.** Italian; **J.** Japanese; **O.** Other Languages; **P.** Polish; **R.** Russian; **S.** Spanish

Temperature: **F.** Full Range (Low + Normal + High); **L.** Low (0 to 75K) + Overlap into Normal; **N.** Normal (75 to 1273K); **H.** High (above 1273K) + Overlap into Normal; Blank = Not Coded

Substance Number	Phys. State	Subject	Language	Temperature	Year	TPRC Number
503-9173	S	D	E	N	1962	25985
9173	S	D	G	N	1959	30334
9174	L	G	G	H	1959	17922
9174	S	D	E	N	1959	17925
9174	S	D	E	N	1955	26325
9174	S	D	E	N	1962	30804
9174	S	D	E	H	1963	31735
9174	S	D	R	H	1960	31736
9174	S	D	E	N	1963	33172
9179	L	G	G	H	1959	17922
9179	S	D	E	N	1953	14586
9179	S	D	E	N	1959	17925
9179	S	D	R	N	1963	23584
9179	S	D	E	N	1955	26325
9179	S	G	E	N	1950	31153
9179	S	D	E	N	1963	33172
9180	L	D	R	H	1958	19646
9180	L	D	E	H	1959	21015
9180	S	D	R	H	1958	19646
9180	S	D	E	H	1959	21015
9181	S	G	G		1952	145
9714	S	D	E	H	1965	31115
9715	L	D	E	N	1962	28922
9715	S	D	E	N	1962	28922
9716	L	D	E	N	1962	28922
9716	S	D	E	H	1911	23691
9716	S	D	E	N	1962	28922
9717	L	D	E	N	1962	28922
9717	S	D	E	N	1962	28922
9717	S	D	E	H	1965	31115
9718	S	D	E	H	1918	31087
9724	S	D	E	H	1965	31115
9725	S	D	E	H	1965	31115
9728	S	D	E	H	1965	31115
9729	S	D	E	N	1963	27886
9730	S	D	E	N	1963	27886
9731	S	D	E	N	1963	27886
9739	S	D	E	H	1965	31115
9744	M	D	E	H	1963	23145
9745	M	D	E	H	1963	23145
9745	S	D	F	N	1963	31427
9746	M	D	E	N	1962	25127
9747	M	D	E	N	1962	25127
9748	M	D	E	H	1963	23145
9748	M	D	E	N	1962	25127
9748	P	D	G	H	1932	10242
9748	S	D	E	H	1960	18644
9748	S	D	F	N	1963	31427
9748	S	G	G	H	1932	31928
9749	M	D	E	N	1962	25127
9750	M	D	E	H	1964	29570
9782	P	E	J	N	1958	25858
9791	S	D	F	N	1963	31427
9796	P	D	E	N	1964	32760
9797	P	D	E	N	1964	32760
9808	S	D	E	N	1964	32045
9825	S	D	E	N	1961	24582
9830	S	D	E	N	1958	20771
9835	S	D	F	H	1958	19583
9835	S	D	E	H	1963	32477
9837	M	D	E	H	1963	22261
9837	S	D	E	H	1965	31115
9840	S	D	E	H	1918	31087
9841	S	D	E	H	1918	31087
9844	S	D	E	H	1918	31087
9847	S	D	E	H	1918	31087
9856	S	D	E	H	1918	31087
9858	S	D	E	H	1918	31087
9872	E	D	E	H	1963	30641
9872	E	E	E	H	1963	30641
9882	P	D	G	H	1932	10242
9883	P	D	G	H	1932	10242
9883	S	D	G	H	1906	20643
9883	S	D	E	H	1908	23456
9883	S	D	E	H	1910	23480
9883	S	D	E	H	1911	23691
9883	S	D	G	H	1905	26365
9883	S	D	E	H	1918	31087
9883	S	G	G	H	1915	31809
9883	S	D	G	H	1905	33184
9883	S	D	G	H	1911	33218
9885	S	D	G	N	1949	12993
9886	S	D	G	N	1949	12993
9887	S	D	G	N	1949	12993
9888	S	D	G	H	1949	12993
9892	S	D	E	H	1965	31115
9916	M	D	E	N	1962	25127
9916	S	D	E	N	1910	23480
9922	L	D	E	N	1962	28922
9922	S	D	E		1959	7696
9922	S	D	E	N	1962	28922
9922	S	D	E	H	1963	30638
9922	S	E	E	H	1963	30638
9922	S	D	E	H	1965	31115
9922	S	D	E	N	1964	32045
9922	S	G	E	H	1962	33360
9955	S	D	E	H	1960	23333
9962	S	D	E	H	1965	31115
9974	S	D	E		1959	7696
9974	S	D	E	H	1964	29570
9974	S	D	E	H	1965	31115

Substance Number	Phys. State	Subject	Language	Temperature	Year	TPRC Number
503-9974	S	D	G	H	1911	33218
9977	L	D	R	N	1952	14318
521-0001	S	D	E	N	1963	23291
0001	S	D	R	N	1957	24221
0001	S	D	E	N	1963	27519
0007	S	G	G	N	1930	33626
0009	G	D	E	N	1961	24943
0009	S	D	E		1956	7351
0009	S	S	E	N	1958	10667
0009	S	D	E	N	1963	23291
0009	S	D	E	N	1953	26316
0009	S	D	R	N	1947	26342
0009	S	D	E	N	1961	28944
0010	P	D	E	N	1954	24602
0015	P	D	E	N	1963	28755
0015	S	D	E	N	1962	33458
0024	S	D	E	N	1939	33181
0025	S	D	G	N	1933	20376
0025	S	D	E	N	1951	30221
0026	S	D	E	N	1906	22517
0033	S	D	E	N	1910	23480
0034	S	D	E	N	1957	18609
0040	S	G	E		1954	4834
0040	S	D	R	N	1957	24221
0040	S	D	E	N	1953	26316
0040	S	D	R	N	1947	26342
0046	S	S	E	N	1956	10594
0046	S	D	E	N	1953	26316
0046	S	D	R	N	1947	26342
0046	S	D	E	N	1963	27886
0046	S	D	E	N	1961	28944
0054	S	S	E	N	1956	10594
0054	S	D	E	H	1957	20260
0054	S	T	E	H	1957	20260
0054	S	D	E	N	1958	24363
0057	S	D	E	N	1906	22517
0057	S	D	E	H	1908	23456
0060	S	D	E	H	1908	23456
0063	S	D	E		1956	9433
0063	S	D	E	N	1960	20847
0063	S	D	E	N	1963	26640
0063	S	D	E	N	1962	26839
0063	S	D	E	N	1962	29155
0063	S	D	E	N	1964	32864
0063	S	D	E	N	1955	33156
0063	S	T	E	N	1955	33156
0063	S	D	E	N	1954	33158
0064	S	D	E	H	1958	10703
0064	S	D	E	N	1964	20810
0064	S	D	E	N	1960	20847
0064	S	D	E	N	1951	24552
0064	S	D	E	N	1960	24960
0064	S	D	E	H	1964	25207
0064	S	D	E	N	1963	26640
0064	S	D	E	N	1962	28460
0064	S	G	E	N	1961	29193
0064	S	D	E	L	1963	29377
0064	S	D	E	N	1963	30100
0064	S	D	G	H	1964	31343
0064	S	D	E	H	1964	31344
0069	S	D	R	N	1957	24221
0070	S	D	E	N	1953	26316
0070	S	D	R	N	1947	26342
0072	S	D	E	N	1962	21660
0072	S	T	E	N	1962	21660
0072	S	D	E	N	1962	21932
0072	S	D	E	N	1963	21998
0072	S	T	E	N	1963	21998
0072	S	D	E	N	1906	22517
0072	S	D	E	N	1963	23291
0072	S	D	E	N	1953	26316
0072	S	D	R	N	1947	26342
0073	P	D	E	N	1962	33475
0075	S	D	E	N	1895	30195
0075	S	D	E	N	1964	32047
0075	S	D	E	N	1927	33283
0077	S	D	R	N	1966	30598
0077	S	D	E	N	1966	30661
0082	S	D	E	H	1908	23456
0082	S	D	E	N	1910	23480
0082	S	D	F	N	1963	32110
0083	P	D	E	N	1940	3025
0083	P	D	E	H	1908	23456
0083	S	D	E	N	1956	12328
0083	S	D	R	N	1956	12328
0083	S	T	E	N	1956	12328
0083	S	T	R	N	1956	12328
0083	S	D	R	N	1956	14939
0083	S	T	R	N	1956	14939
0083	S	D	R	N	1957	24221
0083	S	D	E	N	1953	26316
0083	S	D	R	N	1947	26342
0084	S	D	E		1955	4200
0085	S	D	E	N	1912	22520
0085	S	D	E	N	1953	26316
0085	S	D	R	N	1947	26342
0086	S	D	G	N	1932	33228
0087	S	G	G	N	1930	33626
0089	S	D	J	N	1961	27213
0089	S	T	J	N	1961	27213

Substance Number	Phys. State	Subject	Language	Temperature	Year	TPRC Number
521-0095	S	D	G	N	1932	33228
0096	S	D	E	H	1913	23759
0096	S	D	E	N	1910	33140
0103	P	D	E	H	1908	23456
0103	S	D	R	N	1957	24221
0108	S	D	E		1951	4515
0110	S	S	E	N	1956	10594
0110	S	D	E	N	1957	20833
0110	S	D	R	N	1957	24221
0110	S	D	E	N	1953	26316
0110	S	D	R	N	1947	26342
0110	S	D	E	N	1928	33285
0116	P	D	E	N	1940	3025
0116	P	D	E	N	1910	23480
0116	P	D	E	N	1963	28755
0116	S	D	G		1955	9046
0116	S	D	E	N	1906	22517
0116	S	D	E	N	1910	23480
0130	S	D	E	H	1908	23456
0131	S	D	E	N	1895	30195
0135	P	D	E	N	1956	10306
0135	P	D	E	N	1963	28755
0135	S	D	E	N	1956	10306
0135	S	D	E	H	1958	10458
0135	S	D	E	N	1961	23525
0135	S	D	E	N	1922	23559
0135	S	D	G	N	1961	23874
0135	S	D	E	N	1962	33458
0181	S	D	E	N	1912	22520
0181	S	G	E	N	1931	33165
0181	S	D	E	N	1942	33298
0186	S	D	E	N	1963	23291
0186	S	D	E	N	1963	27519
0191	S	G	E	N	1954	22147
0191	S	D	E	L	1961	29206
0191	S	D	R	N	1966	30598
0191	S	D	E	N	1966	30661
0191	S	G	E	N	1963	32489
0192	S	D	G	N	1932	33228
0196	S	D	E	N	1953	26316
0196	S	D	R	N	1947	26342
0222	P	D	E	N	1954	24602
0222	S	D	G	N	1959	16088
0222	S	D	G	N	1921	31233
0223	S	D	E	N	1906	22517
0238	S	S	E	N	1956	10594
0238	S	D	E	N	1953	26316
0238	S	D	R	N	1947	26342
0238	S	D	E	N	1963	27850
0238	S	D	E	N	1961	28944
0239	S	D	E	N	1962	21660
0239	S	T	E	N	1962	21660
0239	S	D	E	N	1962	21932
0239	S	D	E	N	1963	21998
0239	S	T	E	N	1963	21998
0240	S	D	E	N	1906	22517
0249	S	D	E	N	1939	33181
0250	P	D	E	N	1964	32760
0250	S	D	E	N	1910	23480
0254	P	D	E	N	1956	31610
0254	P	D	R	N	1956	31853
0254	S	D	E	N	1956	31610
0254	S	D	R	N	1956	31853
0258	P	D	E	N	1954	24602
0259	S	D	E	N	1960	20866
0259	S	T	E	N	1960	20866
0259	S	D	E	N	1960	20867
0259	S	T	E	N	1960	20867
0259	S	D	E	N	1961	24744
0259	S	D	R	N	1952	24745
0299	S	S	R	N	1959	23083
0299	S	S	E	N	1928	31217
0307	S	D	E	N	1962	21660
0307	S	T	E	N	1962	21660
0307	S	D	E	N	1962	21932
0307	S	D	E	N	1963	21998
0307	S	T	E	N	1963	21998
0342	S	D	G	N	1962	26958
0343	S	G	E	N	1960	16740
0352	S	D	E	N	1962	21660
0352	S	T	E	N	1962	21660
0352	S	D	E	N	1962	21932
0352	S	D	E	N	1963	21998
0352	S	T	E	N	1963	21998
0357	S	D	E	N	1957	20833
0368	S	D	E	N	1953	26316
0368	S	D	R	N	1947	26342
0368	S	D	E	N	1956	28999
0369	S	D	E	N	1962	25391
0369	S	D	E	N	1963	27519
0384	S	D	E	N	1953	26316
0384	S	D	R	N	1947	26342
0385	S	D	E	N	1963	22963
0385	S	D	E	N	1962	25029
0392	S	D	E	N	1906	22517
0402	S	D	E	N	1953	26316
0402	S	D	R	N	1947	26342
0412	S	D	E	N	1963	22963
0413	S	D	E	N	1961	24893
0413	S	D	F	N	1962	27944
0421	S	D	E	N	1961	24893

Phys. State: **C.** Superconductive; **D.** Doped; **E.** Expanded; **F.** Fibrous or Whisker; **G.** Gas; **I.** Ionized or Plasma; **L.** Liquid; **M.** Multiphase; **P.** Powder or Fine Particle; **S.** Solid; **T.** Thick or Thin Film
Subject: **D.** Data; **E.** Experiment; **G.** General (Data + Theory + Experiment); **S.** Survey (Review, Compendium etc.); **T.** Theory
Language: **C.** Czech; **D.** Dutch; **E.** English; **F.** French; **G.** German; **I.** Italian; **J.** Japanese; **O.** Other Languages; **P.** Polish; **R.** Russian; **S.** Spanish
Temperature: **F.** Full Range (Low + Normal + High); **L.** Low (0 to 75K) + Overlap into Normal; **N.** Normal (75 to 1273K); **H.** High (above 1273K) + Overlap into Normal; Blank = Not Coded

Substance Number	Phys. State	Subject	Language	Temperature	Year	TPRC Number
521-0422	S	D	E	N	1961	24893
0423	L	T	E	N	1961	29355
0423	S	D	E	H	1964	23397
0423	S	T	E	H	1964	23397
0423	S	D	E	N	1961	24893
0423	S	T	E	N	1961	28900
0423	S	D	E	N	1961	28944
0423	S	T	E	N	1961	29355
0424	S	D	E	N	1961	24893
0425	S	D	E	N	1961	24893
0429	S	D	E	N	1906	22517
0429	S	D	E	N	1910	23480
0430	S	D	E	N	1906	22517
0432	S	D	E	N	1912	22520
0447	S	D	E	N	1963	27519
0448	S	D	E	N	1963	27519
0449	S	D	E	N	1963	27519
0450	S	D	E	N	1963	27519
0451	S	D	E	N	1962	25391
0451	S	D	E	N	1963	27519
0457	S	D	E	N	1963	27886
0467	G	D	E	N	1963	26746
0467	G	T	E	N	1963	26746
0469	S	D	E	N	1963	22963
0470	S	D	E	N	1963	22963
0471	S	D	E	N	1963	22963
0473	S	D	E	N	1963	22963
0474	S	D	E	N	1963	22963
0480	S	D	E	N	1961	28944
0496	P	D	E	N	1954	24602
0496	S	D	G	N	1921	31233
0514	P	D	E	N	1963	28755
0516	P	D	E	N	1963	28755
0516	S	D	E	N	1962	33458
0520	S	D	E	N	1953	26316
0520	S	D	R	N	1947	26342
0521	S	D	E	N	1953	26316
0521	S	D	R	N	1947	26342
0522	S	D	E	N	1953	26316
0522	S	D	R	N	1947	26342
0523	S	D	E	N	1953	26316
0523	S	D	R	N	1947	26342
0525	S	D	E	N	1953	26316
0525	S	D	R	N	1947	26342
0526	S	D	E	N	1961	24893
0527	S	D	E	N	1961	24893
0528	S	D	E	N	1961	24893
0534	S	D	E	N	1963	22963
0541	S	D	E	N	1953	26316
0541	S	D	R	N	1947	26342
0542	S	D	E	N	1953	26316
0542	S	D	R	N	1947	26342
0544	S	D	E	N	1953	26316
0544	S	D	R	N	1947	26342
0546	S	D	E	N	1953	26316
0546	S	D	R	N	1947	26342
0548	S	D	E	N	1953	26316
0548	S	D	R	N	1947	26342
0549	S	D	E	N	1953	26316
0549	S	D	R	N	1947	26342
0550	S	D	E	N	1953	26316
0550	S	D	R	N	1947	26342
0552	S	D	E	H	1908	23456
0552	S	D	E	N	1910	23480
0554	S	D	E	N	1910	23480
0628	S	D	G	N	1921	31233
0629	S	D	G	N	1921	31233
0633	S	D	G	N	1941	31926
0634	S	D	G	N	1941	31926
0648	S	D	E	N	1951	24552
0648	S	D	E	N	1964	33123
0648	S	D	F	N	1966	33568
0650	S	S	E	N	1941	33164
0652	S	G	E	N	1931	33165
0670	M	D	R	N	1964	33084
0686	S	G	E	N	1963	31759
0686	S	D	E	N	1963	33279
0687	S	D	E	N	1963	33279
0732	L	D	E		1947	9353
0732	L	D	E		1940	9354
0732	P	D	R	N	1938	13387
0732	P	D	E	N	1954	24602
0732	P	D	E	N	1956	31610
0732	P	D	R	N	1956	31853
0732	S	D	E		1951	4515
0732	S	G	E		1959	9395
0732	S	D	E	H	1958	10703
0732	S	D	G	N	1959	16088
0732	S	G	E	N	1961	16961
0732	S	D	E	N	1958	18425
0732	S	D	E	H	1961	19478
0732	S	G	E	N	1961	19818
0732	S	S	G	N	1957	20221
0732	S	D	E	N	1964	20810
0732	S	D	E	N	1957	20833
0732	S	D	E	N	1963	21998
0732	S	T	E	N	1963	21998
0732	S	D	E	N	1906	22517
0732	S	D	E	N	1912	22520
0732	S	D	G	N	1957	22979
0732	S	D	E	N	1961	22990
521-0732	S	S	E	N	1960	23528
0732	S	D	E	H	1913	23759
0732	S	D	E	N	1955	24082
0732	S	D	E	N	1915	24339
0732	S	D	E	N	1957	24431
0732	S	D	E	H	1961	24893
0732	S	D	E	H	1960	25230
0732	S	D	E	N	1962	25723
0732	S	G	R	N	1960	25839
0732	S	D	E	N	1961	25906
0732	S	D	E	N	1962	26087
0732	S	D	E	N	1955	26325
0732	S	D	E	N	1963	26640
0732	S	D	E	N	1962	26777
0732	S	D	E	N	1961	26790
0732	S	D	E	N	1964	27035
0732	S	D	E	N	1964	27141
0732	S	D	E	N	1963	27345
0732	S	D	E	L	1963	27957
0732	S	D	E	N	1963	27983
0732	S	D	E	H	1959	28099
0732	S	G	G	N	1912	28159
0732	S	D	E	N	1961	28913
0732	S	D	E	L	1963	29377
0732	S	D	E	N	1961	29429
0732	S	D	E	N	1961	29595
0732	S	D	E	N	1963	30100
0732	S	D	E	N	1895	30195
0732	S	D	E	N	1962	30789
0732	S	D	E	N	1963	30889
0732	S	D	E	L	1953	31084
0732	S	D	O	N	1963	31129
0732	S	D	E	N	1933	31149
0732	S	G	E	N	1950	31153
0732	S	G	E	N	1939	31171
0732	S	D	G	N	1921	31233
0732	S	D	R	N	1962	31296
0732	S	D	G	N	1962	31564
0732	S	D	E	N	1956	31610
0732	S	D	E	N	1941	31731
0732	S	D	E	N	1963	31733
0732	S	D	R	N	1960	31734
0732	S	D	E	L	1963	31749
0732	S	D	R	N	1956	31853
0732	S	D	E	N	1964	32047
0732	S	D	E	N	1963	32051
0732	S	D	E	N	1963	32170
0732	S	D	R	N	1963	32433
0732	S	D	E	N	1930	33142
0732	S	D	G	N	1950	33230
0732	S	D	E	N	1927	33283
0732	S	D	E	N	1928	33285
0732	S	D	E	N	1948	33304
0732	S	G	E	N	1956	33315
0951	S	D	E	N	1906	22517

Phys. State: **C.** Superconductive; **D.** Doped; **E.** Expanded; **F.** Fibrous or Whisker; **G.** Gas; **I.** Ionized or Plasma; **L.** Liquid; **M.** Multiphase; **P.** Powder or Fine Particle; **S.** Solid; **T.** Thick or Thin Film

Subject: **D.** Data; **E.** Experiment; **G.** General (Data + Theory + Experiment); **S.** Survey (Review, Compendium etc.); **T.** Theory

Language: **C.** Czech; **D.** Dutch; **E.** English; **F.** French; **G.** German; **I.** Italian; **J.** Japanese; **O.** Other Languages; **P.** Polish; **R.** Russian; **S.** Spanish

Temperature: **F.** Full Range (Low + Normal + High); **L.** Low (0 to 75K) + Overlap into Normal; **N.** Normal (75 to 1273K); **H.** High (above 1273K) + Overlap into Normal; Blank = Not Coded

Part C
BIBLIOGRAPHY

PART C

USE OF THE BIBLIOGRAPHY

A. GENERAL CONSIDERATIONS

Because of the wide variety of literature sources (i.e., journal publications, dissertations, government and industrial reports, books, etc.), three specific formats for the citations are used in the *Bibliography*. CINDAS' procedures are described below:

1. Titles reported in the *Bibliography* are taken from an abstract or from the original document. In the case of translated titles, discrepancies often exist between various sources. In general, CINDAS makes no special effort to check the accuracy of titles.
2. The names of authors reported in abstracts, and even in original publications, are at times misspelled and/or incomplete. CINDAS attempts to report the correct names of the authors.
3. The names of scientific and technical journals are abbreviated according to the notations used by the abstracting journals. In those cases where a journal name is not applicable, the name of the publisher, symposium, or disseminating agency is entered in place of the journal name, depending upon the reference work.
4. Foreign language documents and their English translations, if any, are reported in the *Bibliography*. The translation refers to the original document by the words "For original article, see T . . ."; the original article refers to the translation by "For English translation, see T. . . ."
5. Subscripts are either spelled out or written on the line. Examples: $C_P \equiv$ specific heat at constant pressure; $Si_3Fe_5 \equiv$ Si(3) Fe(5); etc.
6. Superscripts are spelled out. Examples: $V^2 \equiv$ V squared; $\sqrt{2} \equiv$ square root of two; etc.

B. ABBREVIATIONS AND ACRONYMS USED IN BIBLIOGRAPHIC CITATIONS

AAAS. . . . . .American Association for the Advancement of Science
AAIE. . . . . .American Association of Industrial Engineers
AD. . . . . .Prefix Catalog Codes for Defense Documentation Center (DDC) (Formerly ASTIA)
AEC. . . . . .Atomic Energy Commission
AEDC. . . . . .Arnold Engineering Development Center
AELRDL. . . . . .Army Electronics Research and Development Laboratory
AERDL. . . . . . .Army Engineering Research and Development Laboratories
AERE. . . . . .Atomic Energy Research Establishment (Great Britain)
AF. . . . . .Air Force
AFB. . . . . .Air Force Base
AFCRC. . . . . . .Air Force Cambridge Research Center
AFML. . . . . .Air Force Materials Laboratory
AFOSR. . . . . . . .Air Force Office of Scientific Research
AFWL. . . . . .Air Force Weapons Laboratory
AGARD. . . . . . .Advisory Group for Aeronautical Research and Development
AIAA. . . . . .American Institute of Aeronautics and Astronautics (Formerly ARS & IAS)
AIAE. . . . . .Associate Institute of Automobile Engineering
AICE. . . . . .American Institute of Chemical Engineers
AIEE. . . . . .American Institute of Electrical Engineers (Now IEEE)
AIIE. . . . . .American Institute of Industrial Engineers
AIME. . . . . .American Institute of Mining and Metallurgical Engineers
AIP. . . . . .American Institute of Physics
AISI. . . . . .American Iron and Steel Institute
AM. . . . . .Applied Mechanics Reviews
AMRA. . . . . . . .Army Materials Research Agency
ANL. . . . . .Argonne National Laboratory
AOA. . . . . .American Ordnance Association
APDA. . . . . .Atomic Power Development Association, Inc.
API. . . . . .American Petroleum Institute
ARC. . . . . .Aeronautical Research Council
ARDE. . . . . .Armament Research and Development Establishment
ARF. . . . . .Armour Research Foundation
ARL. . . . . .Aeronautical Research Laboratories
ARS. . . . . .American Rocket Society (Now AIAA)
ASHRAE. . . . . .American Society of Heating, Refrigeration and Air Conditioning Engineers
ASM. . . . . .American Society for Metals
ASME. . . . . .American Society of Mechanical Engineers
ASR. . . . . .American Society of Rocketry
ASSE. . . . . .American Society of Safety Engineers
ASTE. . . . . .American Society of Tool Engineers
ASTIA. . . . . . . .Armed Services Technical Information Agency (Now DDC)
ASTM. . . . . .American Society for Testing and Materials
ASTME. . . . . . .American Society of Tool and Manufacturing Engineers
BC. . . . . .British Ceramic Society
BISI. . . . . .British Iron and Steel Industry
BM. . . . . .Bureau of Mines
BMI. . . . . .Battelle Memorial Institute
BNL. . . . . .Brookhaven National Laboratories
BR. . . . . .Battelle Technical Review
BRL. . . . . .Ballistic Research Laboratories
BU ORD. . . . . .Bureau of Ordnance
BU WEPS. . . . .Bureau of Weapons
CDAP. . . . . .Idaho National Engineering Laboratories

CEA.......... Commissariat à l'Energie Atomique (France)
CERN......... Counseil Européen pour la Recherche Nucléaire
CFSTI......... Clearinghouse for Federal Scientific and Technical Information (Now NTIS)
CINDAS–EPIC........ Center for Information and Numerical Data Analysis and Synthesis—Electronic Properties Information Center
CINDAS–TPRC....... Center for Information and Numerical Data Analysis and Synthesis—Thermophysical Properties Research Center
CNEN......... Comitato Nazionale per l'Energia Nucleare (National Nuclear Energy Commission)
CNDR......... Consiglio Nazionale delle Ricerche (National Research Council of Italy)
CNES......... Centre National d'Etudes Spatiales (National Center for Space Studies)
COO.......... Chicago Operations Office, Atomic Energy Commission
COSATI....... Committee on Scientific and Technical Information (Formerly COSI)
COSI.......... Committee on Scientific Information (Now COSATI)
CPIA.......... Chemical Propulsion Information Agency
CRC.......... Communications Research Center (Canada)
CRREL........ Cold Regions Research and Engineering Laboratory
DASA......... Defense Atomic Support Agency
DDC.......... Defense Documentation Center (Formerly ASTIA)
DDR&E....... Directorate of Defense Research and Engineering
DESY......... Deutsches Elektron Synchrotron (W. Germany)
DFULR....... Deutsche Luft- und Raumfahrt Forschungsberichte (W. Germany)
DMIC......... Defense Metals Informaton Center
DOD.......... Department of Defense
DOE.......... Department of Energy
DOFL......... Diamond Ordnance Fuze Laboratory (HDL)
DPGR......... Dugway Proving Ground
DRL.......... Defense Research Laboratory
ECOM........ Electronics Command, U.S. Army
EPIC.......... Electronic Properties Information Center
ERDA......... Energy Research and Development Administration
EURATOM.... European Atomic Energy Community
FAI........... Federation Aeronautique Internationale
FID........... Federation Internationale de Documentation
FTD........... Foreign Technology Division
GRA.......... Government Research Announcements (Formerly UGAR)
HDL.......... Harry Diamond Laboratory
IAA........... International Academy of Astronautics
IAS........... Institute of Aerospace Science (Now AIAA)
IEE........... Institute of Electrical Engineers (Great Britain)
IEEE.......... Institute of Electrical and Electronic Engineers (Formerly AIEE & IRE)
IITRI.......... Illinois Institute of Technology Research Institute
IPST.......... Israel Program for Scientific Translation, Ltd.
IRE........... Institute of Radio Engineers (Now IEEE)
ISA........... Instrument Society of America
JAERI......... Japan Atomic Energy Research Institute
JANAF........ Joint Army–Navy–Air Force
JINR.......... Joint Institute for Nuclear Research (USSR)
JPL........... Jet Propulsion Laboratory
KAPL......... Knolls Atomic Power Plant
LA............ Los Alamos Scientific Laboratories
LBL........... Lawrence Berkeley Laboratory, University of California
LRL........... Livermore Research Laboratory, University of California
ML–TDR...... Materials Laboratory, Technical Division Report, Wright-Patterson Air Force Base
MR............ ASM Review of Current Metal Literature (in Metals Review)
NADC......... Naval Air Development Center
NAS........... National Academy of Sciences
NASA......... National Aeronautics and Space Administration
NASC......... National Aeronautics and Space Council
NAS/NRC..... National Academy of Sciences—National Research Council
NATO......... North Atlantic Treaty Organization
NBS........... National Bureau of Standards
NDAC......... National Defense Advisory Committee
NDRC......... National Defense Research Council
NOTS......... Naval Ordnance Test Station
NPL........... National Physical Laboratory (Great Britain)
NRC.......... National Research Council (Canada)
NRD.......... Naval Research and Development (Now ONR)
NRL.......... Naval Research Laboratory
NS............ Nuclear Science Abstracts
NSF........... National Science Foundation
NTIS.......... National Technical Information Service (Formerly CFSTI)
NUREG....... Nuclear Regulatory Commission
OAR.......... Office of Aerospace Research
ONERA....... Office National d'Etudes et de Recherches Aerospatiales (France)
ONR.......... Office of Naval Research (Formerly NRD)
ORNL......... Oak Ridge National Laboratory
OSI........... Office of Scientific Information
OSR........... Office of Scientific Research
OTS........... Office of Technical Services
PB............ Office of Technical Services and Office of Publication Board
PRL........... Properties Research Laboratory, Purdue University
RADC......... Rome Air Development Center, Griffiss Air Force Base
RAE........... Royal Aircraft Establishment (Great Britain)
REIC.......... Radiation Effects Information Center
RI............. Radievyi Institute (USSR)
RIA........... Rock Island Arsenal
RM............ ASM Review of Metal Literature
RPL........... Rocket Propulsion Laboratory
RR............ U.S. Government Research Reports (OTS)
SERI.......... Solar Energy Research Institute
SIPRE......... Snow, Ice and Permafrost Research Establishment
SORI.......... Southern Research Institute
SRDE......... Signal Research and Development Establishment
SRI............ Stanford Research Institute
TACOM....... Army Tank Automotive Command
TDVI.......... Atomic Energy of Canada, Limited
TID........... Technical Information Service USAEC
TML.......... Titanium Metallurgical Laboratory
TPRC......... Thermophysical Properties Research Center
TREE......... Idaho National Engineering Laboratories
TT............ Technical Translations (OTS)
UCRL......... University of California Radiation Laboratory
UGAR......... United States Government Announcements Reports (Now GRA)
USAEC........ U.S. Atomic Energy Commission
USBM......... U.S. Bureau of Mines
USDA......... U.S. Department of Agriculture
USGS......... U.S. Geological Survey
USJRPS....... U.S. Joint Publication Research Service
USL........... Underwater Sound Laboratory
WADC........ Wright Air Development Center
WADD........ Wright Air Development Division
WRAC........ Willow Run Aeronautical Center

BIBLIOGRAPHY

TPRC Number | *Bibliographic Citation*

17 PROPERTIES OF SERICITE FROM THE STANDPOINT OF CERAMICS
SHIRAKI, Y.
YOGYO KYOKAI SHI
63 633–40 1955 CA 51 10022

22 THE SPECIFIC HEATS OF CRYSTALS. II. THE CASE OF DIAMOND.
RAMAN C V
PROC INDIAN ACAD SCI
44 A 160–4 1956 CA 51 6264

26 EXPERIMENTS ON MAGNETIC THERMOMETRY AND HEAT CONDUCTIVITY OF CHROMIUM POTASSIUM ALUM AND GLASS AT LOW TEMPERATURES
BIJL D
PHYSICA
14 684–93 1949 CA 43 7316

56 THERMAL CONDUCTIVITY OF GLASSES AT LOW TEMPERATURES
BERMAN R
PHYS REV
76 315–16 1949 CA 43 8786

59 EFFECTS OF FIRING ON THE PHYSICAL PROPERTIES OF EXPANDED VERMICULITE
TANAKA, M. TSUBAKI, T. UESHIMA, T.
KAMIIKE, O.
J CERAM ASSOC JAPAN
64 83–9 1956 CA 51 12459

63 A STEADY-STATE METHOD FOR THE RAPID MEASUREMENT OF THE THERMAL CONDUCTIVITY OF ROCKS
BECK A
J SCI INSTR
34 186–9 1957 CA 51 11773

64 EFFECT OF ISOTOPES ON LOW-TEMPERATURE THERMAL CONDUCTIVITY
SLACK, G. A.
PHYS REV
105 829–31 1957 CA 51 11786

73 THERMAL CONDUCTIVITY OF THE CHARGE IN COKE OVENS AND GAS WORKS
PIEPER PAUL
BERGBAU ARCH
13 62–73 1952 CA 47 5663

74 PROPERTIES OF MATERIALS USED IN LINING SULFITE DIGESTERS. II.
WESTER AXEL AGGERYD BARBRO
DAS PAPIER
11 98–104 1957 CA 51 15122

76 THERMAL CONDUCTIVITY AT MINUS 180 DEGREES.
CODEGONE, C.
RICERCA SCI
20 68–70 1950 CA 45 4543

85 NOTES ON THE ELECTRICAL AND THERMAL CONDUCTIVITIES OF GRAPHITE AND AMORPHOUS CARBON
MIZUSHIMA, S. OKADA, J.
PHYS REV
82 94–5 1951 CA 45 4989

107 THERMAL CONDUCTIVITY MEASUREMENTS ON SILICA
COLOSKY BENJAMIN P
AM CERAM SOC BULL
31 465–6 1952 CA 47 842

108 A COOPERATIVE TEST ON THERMAL CONDUCTIVITY
WATSON A F CLEMENTS J F VYSE J
TRANS BRIT CERAM SOC
53 156–64 1954 CA 48 7864

109 A NEW THERMAL CONDUCTIVITY APPARATUS FOR REFRACTORY MATERIALS
CLEMENTS J F VYSE J
TRANS BRIT CERAM SOC
53 134–55 1954 CA 48 7867

110 IMPROVED METHOD OF MEASURING THERMAL CONDUCTIVITY OF DENSE CERAMICS
RUH, E.
J AM CERAM SOC
37 224–9 1954 CA 48 7868

120 CHARACTERISTICS OF ZIRCON PORCELAIN
RUSSELL, R., JR. MOHR, W. C.
J AM CERAM SOC
30 32–5 1947 CA 41 2221

145 HEAT CONDUCTION IN GLASS AT HIGH TEMPERATURES
GEFFCKEN I W
GLASTECH BER
25 392–6 1952 CA 47 5081
(FOR ENGLISH TRANSLATION SEE T20435)

150 THERMAL RESISTIVITY OF LI2O.AL2O3.4SIO2 CERAMIC MATERIALS.
FINKELSHTEIN I D FRID E I
VOPROSY PETROG I MINERAL AKAD NAUK S S S R
2 381–5 1953 CA 49 12802

164 THERMAL CONDUCTIVITY OF POWDERED COAL
TANAKA, M. YAMASAKI, T.
J MINING INST JAPAN
68 255–9 1952 CA 48 3661

179 CERTAIN REGULARITIES IN THE VALUES OF THE THERMAL CONDUCTIVITIES OF SEMICONDUCTORS
IOFFE A V IOFFE A F
DOKLADY AKAD NAUK S S S R
97 821–2 1954 CA 49 3639

182 THE MEASUREMENT OF THE THERMAL CONDUCTIVITY OF SOILS AND THE CALCULATION OF OIL TEMPERATURES IN BURIED PIPELINES
BRUMMAGE K G
J INST PETROLEUM
36 575–92 1950 CA 45 1330

183 TEMPERATURES AND THERMAL CONDUCTIVITY OF THE GROUND ALONG THE ALIGNMENT OF THE 30-INCH PIPELINE KIRKUK-BANIAS IN RELATION TO OIL TEMPERATURES
MURRAY T B WHALLEY W C R
J INST PETROLEUM
40 372–400 1954 CA 49 8587

184 THERMAL CONDUCTIVITY OF SOIL
WEBB JOHN
NATURE
178 1074–5 1956 CA 51 3895

185 THERMAL CONDUCTIVITY OF SOIL
DE VRIES D A
NATURE
178 1074 1956 CA 51 3894

191 NOTE ON THE RELATIONSHIP BETWEEN BULK DENSITY AND THERMAL CONDUCTIVITY IN REFRACTORY INSULATING BRICKS
COWLING K W ELLIOTT A HALE W T
TRANS BRIT CERAM SOC
53 461–73 1954 CA 48 13191

192 A STUDY OF THE THERMAL CONDUCTIVITY OF THE SILICA SAND FOR STEEL CASTING
MIKASHIMA, H. NAKAO, Y.
TETSU-TO-HAGANE
39 1229–33 1953 CA 49 7477

196 THERMAL CONDUCTIVITY. I. CONCEPTS OF MEASUREMENT AND FACTORS AFFECTING THERMAL CONDUCTIVITY OF CERAMIC MATERIALS.
KINGERY W D MCQUARRIE MALCOLM
J AM CERAM SOC
37 67–72 1954 CA 48 7786

206 HEAT TRANSMISSION THROUGH MOLTEN GLASS
KUNUGI, M. TANAKA, S. SHIMIZU, S.
KITAMURA, K.
J CERAM ASSOC JAPAN
62 780–4 1954 CA 49 6557

220 MEASUREMENT OF THERMAL CONDUCTIVITY OF CERAMIC MATERIALS
SUZUKI, H. KUWAYAMA, N. YAMAUCHI, T.
YOGYO KYOKAI SHI
65 206–11 1957 CA 52 1571

223 COEFFICIENTS OF THERMAL CONDUCTIVITY IN A GRANULAR STRATUM
AEROV M E UMNIK N N
ZHUR TEKH FIZ
21 1351–63 1951 CA 47 7836

226 TRANSIENT HEAT FLOW DETERMINATION OF THE THERMAL PROPERTIES OF CLAY
PENROD E B PRIVON G R
INDONESIAN J NAT SCI
112 196–206 1956 CA 52 681

243 A CORRELATION OF DATA OBTAINED FROM TESTS ON REFRACTORY INSULATING MATERIALS
GRADY J G MATTHEWS S
CAN CERAM SOC
25 33–8 1956 CA 51 3954

247 MEASUREMENT OF THERMAL CONDUCTIVTY AT HIGH TEMPERATURES
RASI ANTONIO
RICERCA SCI
22 46–55 1952 CA 49 3638

266 THE THERMAL CONDUCTIVITY OF DIAMOND AT LOW TEMPERATURES
BERMAN R SIMON F E ZIMAN J M
PROC ROY SOC
220 A 171–83 1953 CA 48 5587

TPRC Number	Bibliographic Citation
268	**THERMAL CONDUCTIVITY AND THERMAL DIFFUSIVITY OF DEAERATED CLAYS** BUDNIKOV P P ALPEROVICH I A J APPL CHEM U S S R 25 665-73 1952 CA 48 5454 (ENGLISH TRANSLATION OF ZH. PRIKL. KHIM., 25, 582-91, 1952; FOR ORIGINAL SEE T785)
274	**MEASURING TRUE HEAT CAPACITY OF GLASSES, OF SILVER IODIDE, AND OF AGI-KI EUTECTIC MELT IN THE TEMPERATURE RANGE OF 50-300 DEGREES.** BENDEROVICH A M VYATKIN A M TRUDY SIBIR FIZ-TEKH INST TOMSK UNIV 32 203-9 1953 CA 50 11093
296	**CALCULATION OF THE HEAT CAPACITY OF ALPHA QUARTZ AND VITREOUS SILICA FROM SPECTROSCOPIC DATA** LORD R C MORROW J C J CHEM PHYS 26 230-2 1957 CA 51 7827
311	**MEAN SPECIFIC HEATS OF INDUSTRIALLY IMPORTANT GLASSES. III. SPECIFIC HEAT OF A BARIUM GLASS.** HARTMANN, H. KIESSLING, K. H. GLASTECH BER 30 186-8 1957 CA 51 10859 (FOR ENGLISH TRANSLATION SEE T15534)
324	**EXOTHERMIC PHENOMENON DURING CARBONIZATION OF JAPANESE COAL AND ITS SPECIFIC HEAT** IWASAKI, T. NOMURA, S. KON, Y. J FUEL SOC JAPAN 33 249-55 1954 CA 48 11757
341	**ALPHA TO BETA TRANSFORMATION OF QUARTZ** SILVERMAN S M J CHEM PHYS 25 1081 1956 CA 51 4112
359	**HIGH-TEMPERATURE HEAT CONTENTS OF SOME TITANATES OF ALUMINUM, IRON, AND ZINC.** BONNICKSON K R J AM CHEM SOC 77 2152-4 1955 CA 49 15433
398	**HEAT ACCUMULATION IN THE LINING OF PERIODICALLY OPERATING FURNACES** IVANTSOV G P TRUDY URAL IND INST IM S M KIROVA 20 61-97 1945 CA 40 3944
408	**HEAT CAPACITIES AT LOW TEMPERATURES AND ENTROPIES AT 298.16 K. OF ANDALUSITE, KYANITE, AND SILLIMANITE.** TODD S S J AM CHEM SOC 72 4742-3 1950 CA 45 3702
426	**METHOD FOR THE CONTINUOUS MEASUREMENT OF THE HIGH-TEMPERATURE HEAT CONTENT OF GLASSES AND OF FUSED SALTS** TASHIRO MEGUMI GLASS IND 37 549-52 1956 CA 51 9226
435	**LOW-TEMPERATURE HEAT CAPACITY AND ENTROPY AT 298.16 K. OF ANALCITE.** KING E G J AM CHEM SOC 77 2192-3 1955 CA 49 15432
448	**AVERAGE SPECIFIC HEATS OF INDUSTRIAL GLASSES** HARTMANN H BRAND H GLASTECH BER 26 29-33 1953 CA 47 5081
474	**HEAT CAPACITIES AT LOW TEMPERATURES AND ENTROPIES AT 298.16 K. OF TITANOMAGNETITE AND FERRIC TITANATE.** TODD S S KING E G J AM CHEM SOC 75 4547-9 1953 CA 47 11935
500	**SECOND-ORDER TRANSITIONS AND CRITICAL PHENOMENA** SCOTT ROBERT L J CHEM PHYS 21 209-11 1953 CA 47 3641
507	**THE HEAT CONTENT OF ALPHA-ALUMINUM OXIDE /CORUNDUM/ MODIFICATION AT HIGH TEMPERATURE** RODIGINA E N GOMELSKII K Z ZHUR FIZ KHIM 29 1105-12 1955 CA 51 833
518	**LOW-TEMPERATURE HEAT CAPACITY, ENTROPY AT 298.16 K, AND HIGH-TEMPERATURE HEAT CONTENT OF SPHENE (CALCIUM, TITANIUM, SILICON OXIDE).** KING, E. G. ORR, R. L. BONNIOKSON, K. R. J. AM. CHEM. SOC. 76, 4320-1, 1954.
559	**POSSIBILITIES OF THE ABSORPTION OF SOLAR ENERGY IN THE CRYSTALLINE MATERIAL OF THE EARTH** LEBEDEV V I IZVEST AKAD NAUK S S S R SER GEOL 4 50-74 1954 CA 49 8755
566	**MAGNETIC PROPERTIES OF MANGANITE, /LA, CA/MNO3, AND COBALTITE, /LA, SR/COO3. II. HEAT CAPACITY.** KISHI HISASHI BUSSEIRON KENKYU 79 1-9 1954 CA 49 5906
588	**PHYSICS OF THE GLASSY STATE. I. CONSTITUTION AND STRUCTURE.** CONDON E U AM J PHYS 22 43-53 1954 CA 48 4786
606	**THE 1950-1951 ERUPTIONS OF MT. MIHARA, OSHIMA VOLCANO, SEVEN IZU ISLANDS, JAPAN. I. THE 1950 ERUPTION.** TSUYA, H. MORIMOTO, R. OSSAKA, J. BULL EARTHQUAKE RESEARCH INST TOKYO UNIV 32 35-66 1954 CA 48 11995
612	**THE DIELECTRIC SUBSTANCES HAVING HIGH PERMITTIVITY. I. SOLID SOLUTIONS OF TITANATES. 2.** NOMURA, S. ANDO, R. REPT INST SCI TECHNOL UNIV TOKYO 7 1-5 1953 CA 47 8438
615	**DECIDING THE PROBLEM OF FORMATION OF FUSAIN ON THE BASIS OF ITS SPECIFIC HEAT** TERRES, E. DAHNE, H. NANDI, B. SCHEIDEL, C. TRAPPE, K. BRENNSTOFF-CHEM 37 269-77 1956 CA 51 1581
617	**APPARATUS FOR THE DETERMINATION OF THE HEAT CONDUCTIVITY OF REFRACTORIES** DUDAVSKII I E ZAVODSKAYA LAB 13 710-15 1947 CA 43 3584
619	**A VACUUM, ADIABATIC CALORIMETER AND SOME NEW DATA ON THE BETA IN EQUILIBRIUM WITH ALPHA TRANSITION FOR QUARTZ.** SINELNIKOV N N DOKLADY AKAD NAUK S S S R 92 369-72 1953 CA 49 5099 (FOR ENGLISH TRANSLATION SEE T15501)
620	**39TH REPORT OF THE REFRACTORY MATERIALS JOINT COMMITTEE—1947-48.** CLEMENTS J F VYSE J RIGBY G R GAS RESEARCH BOARD 41 54-65 1948 CA 43 3582
622	**HEAT CONDUCTIVITY OF GLASS MASS** GINZBURG D B STEKOL I KERAM PROM 4 7 9-11 1947 CA 43 3579 (FOR ENGLISH TRANSLATION SEE T29111)
646	**THERMAL CONDUCTIVITY OF SOME REFRACTORY MATERIALS** CLEMENTS J F VYSE J TRANS BRIT CERAM SOC 56 296-308 1957 CA 51 17129
652	**SOME THERMODYNAMIC PROPERTIES OF SOIL MOISTURE** ROBINS J S SOIL SCI 74 127-39 1952 CA 47 12714
669	**DETN OF SPECIFIC HEATS AND HEATS OF REACTION OF CLAY MINERALS** ALLISON E B SILICATES IND 19 363-73 1954 CA 49 7954
674	**QUANTUM THEORY OF THE SPECIFIC HEAT AND THE STRUCTURE OF SILICATE GLASSES** TARASOV V V STEKLO I KERAM 11 2 6-9 1954 CA 49 7823
695	**TEMPERATURE DEPENDENCE OF THE SPECIFIC HEAT OF GLASSES** PRODHOMME, M. COMPT REND 240 180-1 1955 CA 49 7208
710	**THERMODYNAMIC PROPERTIES OF SODIUM ALUMINUM AND POTASSIUM ALUMINUM SILICATES** KELLEY, K. K. TODD, S. S. ORR, R. L. KING, E. G. BONNICKSON, K. R. U. S. BUREAU OF MINES 1-21 1953 CA 47 5237 (USBM REPT INVEST 4955)
716	**DIELECTRIC PROPERTIES OF TITANATES CONTAINING SN PLUS 4 IONS.** NOMURA, S. J PHYS SOC JAPAN 10 112-19 1955 CA 49 9987
717	**THE DIELECTRIC PROPERTIES OF LEAD-STRONTIUM TITANATE** NOMURA, S. SAWADA, S. J PHYS SOC JAPAN 10 108-11 1955 CA 49 9987

TPRC Number	Bibliographic Citation
718	**LOW-TEMPERATURE SPECIFIC HEATS OF VITREOUS AND CRYSTALLINE SILICA** DANK M BARBER S W J CHEM PHYS 23 597-8 1955 CA 49 10038
742	**DETERMINATION OF THE TEMPERATURE FUNCTION FOR THE COEFFICIENT OF THERMAL CONDUCTIVITY.** VULIS, L. A. POTSELUIKO, V. A. SOVIET PHYS. TECH. PHYS. 1 (1), 70-7, 1956. (FOR ENGLISH TRANSLATION SEE T49765)
747	**APPARATUS FOR TESTING RESISTANCE OF REFRACTORIES TO HEAT UNDER PRESSURE.** BLANCO, E. P. INST. HIERRC Y. ACERO 8, 370-3PP., 1955.
757	**HEAT CONDUCTIVITY OF BASIC REFRACTORIES.** KOLECHKOVA, A. F. GONCHAROV, V. V. OGNEUPCRY 20, 39-44PP., 1955.
761	**THE IMPROVEMENT OF THE HEAT CONDUCTIVITY OF REFRACTORY BRICKS FOR SPECIAL APPLICATIONS BY REINFORCING.** NEUMANN, J. SPRECHSAAL 88, 469-71, 1955.
768	**MEASUREMENT OF THERMAL CONDUCTIVITY OF POWDERS.** MARATHE, M. N. TENDOLKAR, G. S. TRANS INDIAN INST. CHEM. ENGRS. 6, 90-104, 1953.
785	**THERMAL CONDUCTIVITY AND THERMAL DIFFUSIVITY OF DEAERATED CLAYS** BUDNIKOV P P ALPEROVICH I A ZHUR PRIKLAD KHIM 25 582-91 1952 CA 47 9583 (FOR ENGLISH TRANSLATION SEE T268)
790	**THE PREPARATION OF SEMICONDUCTING CERAMICS ON AN WO3 BASE, AND THE INVESTIGATION OF SOME OF THEIR ELECTRICAL AND THERMAL PROPERTIES.** SKANAVI G I KASHTANOVA A M ZHUR TEKH FIZ 25 1883-92 1955 CA 50 2139 (FOR ENGLISH TRANSLATION SEE T14540)
798	**THE EFFECT OF AN ELECTRIC FIELD ON THE THERMAL CONDUCTIVITY OF GLASS** BERRY CLIFFORD E J CHEM PHYS 17 1355-6 1949 CA 44 4645
799	**PROPERTIES AND USES OF ZIRCONIA PRODUCTS** WHITTEMORE OSGOOD J BRICK AND CLAY RECORD 118 3 58-9 1951 CA 45 4419
800	**DEVELOPMENT OF FUSED STABILIZED ZIRCONIA** KISTLER, S. S. BRICK AND CLAY RECORD 118 3 57-8 1951 CA 45 4419
801	**HEAT CONDUCTIVITY OF GROG- AND LIGHTWEIGHT REFRACTORIES** KOLECHKOVA A F GONCHAROV V V OGNEUPORY 14 445-53 1949 CA 44 2722
806	**REFRACTORY CONCRETES IN THE CONSTRUCTION AND REPAIR OF FURNACES AND CHIMNEYS** CROZEL MARCEL REV MATERIAUX CONSTRUCTION TRAV PUBL C 412 16-20 1950 CA 44 4222
810	**EXPANDED RUBBER AS A THERMAL-INSULATING MATERIAL** COOPER A TRANS INST RUBBER IND 27 84-100 1951 CA 45 6413
817	**EFFECTIVE THERMAL CONDUCTIVITY OF GRANULAR SOLIDS THROUGH WHICH GASES ARE FLOWING** HOUGEN, J. O. PIRET, E. L. CHEM ENG PROGRESS 47 295-303 1951 CA 45 6437
838	**DETERMINATION OF THE COEFFICIENT OF HEAT CONDUCTIVITY OF CERAMIC MATERIALS** PLOTNIKOV L A ZAVODSKAYA LAB 16 1136-9 1950 CA 47 4572 (FOR ENGLISH TRANSLATION SEE T23066)
853	**THE DETERMINATION OF THE HEAT CONDUCTIVITY OF POORLY CONDUCTING MATERIALS** HOCK LOTHAR KESSLER ALBERT H CHEM BER 86 1166-70 1953 CA 48 1791
882	**SPECIFIC HEAT OF DIAMOND AT LOW TEMPERATURES** DE SORBO, W. J CHEM PHYS 21 876-80 1953 CA 47 7879
887	**THE SPECIFIC HEAT OF RUTILE AND ANATASE** LIETZ JOACHIM HAMBURGER BEITR ANGEW MINERAL U KRISTALLPHYS 1 229-38 1956 CA 51 6308
901	**NEW TYPE OF CALORIMETER FOR THE MEASUREMENT OF TRUE HEAT CAPACITY OF CERAMIC MATERIALS** SKOGEN, H. S. RUTGERS UNIVERSITY, NEW BRUNSWICK, N. J., PH. D. THESIS 1-147, 1955. (UNIV MICROFILMS 14058)
904	**CHEMISTRY OF GLASS-FORMING MATERIALS OF HIGH FUSION POINT ON THE BASIS OF HEAT CAPACITY** MYULLER R L ZHUR FIZ KHIM 28 1831-6 1954 CA 50 2133
917	**SPECIFIC HEAT OF DIAMOND AT LOW TEMPERATURES** BERMAN R POULTER J J CHEM PHYS 21 1906-7 1953 CA 48 1130
945	**DETERMINATION OF THE THERMAL CONDUCTIVITY OF INSULATING BRICKS** HIMSWORTH F R TRANS BRIT CERAM SOC 56 345-55 1957 CA 51 18522
975	**RESEARCH ON THE SPECIFIC HEAT AND THE HALL EFFECT OF MAGNETITE** OKAMURA, T. TORIZUKA, Y. SCIENCE REPTS RESEARCH INSTS TOHOKU UNIV 2 A 352-60 1950 CA 45 6443
984	**DECIDING THE QUESTION OF THE FORMATION OF FIBROUS COAL ON THE BASIS OF ITS SPECIFIC HEAT** TERRES, E. DAHNE, H. NANDI, B. SCHEIDEL, C. TRAPPE, K. BRENNSTOFF-CHEM 37 342-7 1956 CA 51 1584
1045	**MEASUREMENT OF SPECIFIC HEAT OF REFRACTORIES AT HIGH TEMPERATURES. I.** KIYOURA RAISAKU ITO YOSHITAKA J CERAM ASSOC JAPAN 61 369-72 1953 CA 48 2340
1064	**HEAT TRANSFER IN INSULATING MATERIALS AND THE INFLUENCE OF CONDUCTIONG AND RADIATION** SEIFFERT K CHEM APP 29 145-9 1942 CA 38 900
1070	**THE THERMAL PROTECTION AND THE BEHAVIOR AGAINST HUMIDITY OF WALLS OF METALLURGICAL BRICK AND METALLURGICAL VITRIFIED BRICK** CAMMERER J S STAHL U EISEN 62 503-10 1942 CA 38 1335
1082	**HEAT TRANSFER IN REFRACTORY INSULATING MATERIALS. I. TEXTURE AND INSULATING POWER.** BARRETT L R TRANS BRIT CERAM SOC 48 235-62 1949 CA 44 296
1093	**THE RELATION OF THE THERMAL RESISTANCE OF CRYSTALS TO THEIR COEFFICIENT OF LINEAR EXPANSION** ZHUZE V P DOKLADY AKAD NAUK S S S R 99 711-14 1954 CA 49 13724
1095	**THERMAL CONDUCTIVITY. XII. TEMPERATURE DEPENDENCE OF CONDUCTIVITY FOR SINGLE-PHASE CERAMICS.** KINGERY W D J AM CERAM SOC 38 251-5 1955 CA 49 14288
1097	**THERMAL INSULATION IN REFINERIES** STONE JOHN F PETROLEUM ENGR 22 1 C11-16 1950 CA 49 14312
1108	**THERMAL CONDUCTION IN ARTIFICIAL SAPPHIRE CRYSTALS AT LOW TEMPERATURES. I. NEARLY PERFECT CRYSTALS.** BERMAN R FOSTER E L PROC ROY SOC /LONDON/ 231 A 130-44 1955 CA 49 14409
1114	**42ND REPORT OF THE REFRACTORY MATERIAL JOINT COMMITTEE. 1950-51.** RHEAD T F E GAS RESEARCH BOARD GRB 62 1-37 1951 CA 46 3724

TPRC Number	Bibliographic Citation
1115	**HEAT CONDUCTIVITY OF ALUMINUM OXIDE AT HIGH TEMPERATURES** SHULMAN A R FEDOROV V N SHEPSENVOL M A ZHUR TEKH FIZ 22 1271–80 1952 CA 49 10721
1120	**APPLICATION OF SPECIAL TESTS TO THE PREPARATION OF MIXTURES SUBSTITUTED FOR NATURAL COALS IN THE COAL-CARBONIZATION INDUSTRY.** FE FRANCISCO PINTADO COMBUSTIBLES /ZARAGOZA/ 10 304–34 1950 CA 46 1232
1121	**THERMAL CONDUCTIVITY OF GLASSMELT** RODNIKOVA V V STEKLO I KERAM 8 4 9–12 1951 CA 46 2766 (FOR ENGLISH TRANSLATION SEE T30858)
1125	**SIMPLE, SHORT METHOD FOR SIMULTANEOUS DETERMINATION OF HEAT CONDUCTIVITY, HEAT CAPACITY, AND HEAT PENETRATION OF SOLIDS.** KRISCHER O ESDORN H VDI. FORSCHUNGSHEFT 450 B, 28–39, 1955.
1132	**THERMALLY RESISTANT WHITEWARE** SMOKE E J J AM CERAM SOC 33 174–7 1950 CA 44 6092
1151	**HEAT TRANSFER IN REFRACTORY INSULATING MATERIALS. II. STUDIES OF A HEAT-DISSIPATION TEST FOR INSULATING MATERIALS.** BARRETT L R VYSE J GREEN A T TRANS BRIT CERAM SOC 49 95–121 1950 CA 44 7035
1154	**PROTECTIVE COVERS FOR HEAT INSULATIONS** YANKELEV L F ELEK STANTSII 20 11 29–31 1949 CA 44 5076
1161	**THE THERMAL CONDUCTIVITY OF DIELECTRIC CRYSTALS. THE EFFECT OF ISOTOPES.** BERMAN R FOSTER E L PROC ROY SOC /LONDON/ 237 A 344–54 1956 CA 51 15192
1180	**THERMAL CONDUCTIVITY OF BODIES OF HIGH BERYLLIUM OXIDE CONTENT** SCHOLES, W. A. J AM CERAM SOC 33 111–17 1950 CA 44 5557
1198	**THE MAGNETIC-FIELD COOLING EFFECT OF A METAL OXIDE MAGNET. I.** SUGIURA ICHIRO OYO BUTSURI 19 198–203 1950 CA 46 3343
1213	**AN ANOMALY IN THE HEAT CAPACITIES OF COO AND OF COO-NIO AND COO-CUO SOLID SOLUTIONS.** ASSAYAG, G. BIZETTE, H. COMPT REND 239 238–40 1954 CA 49 5915
1247	**EXTERNAL OSCILLATIONS AND THE SPECIFIC HEAT OF CALCITE** GIULOTTO L LOINGER A NUOVO CIMENTO 8 475–86 1951 CA 45 8307
1264	**HEAT TRANSFER AND THERMAL RESISTANCE. I.** HASSID N J J SOC CHEM IND 61 63–5 1942 CA 37 1902
1277	**CHAIN AND LAYER STRUCTURES IN CRYSTALS AND GLASSES AND THE QUANTUM THEORY OF HEAT CAPACITY** TARASOV V V TRUDY INST KRIST AKAD NAUK SSSR 10 136–48 1954 CA 50 1387
1282	**MEASUREMENT OF THE SPECIFIC HEAT OF GLASS AT HIGH TEMPERATURES** KANAI, E. AKEYOSHI, K. REPTS RESEARCH LAB ASAHI GLASS CO 3 9–18 1953 CA 50 5254
1287	**THERMAL CONDUCTIVITY OF NONMETALLIC SINGLE CRYSTALS** KNAPP W J J AM CERAM SOC 26 48–55 1943 CA 37 2239
1321	**SPECIFIC HEATS OF CRYSTALS. III.** RAMAN C V PROC INDIAN ACAD SCI 44 A 367–74 1956 CA 51 13552
1324	**THE STRUCTURE OF MICROHETEROGENEOUS GLASS AND GLASSY B2O3 DETERMINED BY SPECIFIC HEAT AT LOW TEMPERATURES** TARASOV V V STROGANOV E F TRUDY MOSKOV KHIM-TEKHNOL INST IM D I MENDELEEVA 21 26–33 1956 CA 51 13552
1340	**THEORY OF THE HEAT CAPACITY OF CHAIN AND LAYER STRUCTURES** TARASOV V V ZHUR FIZ KHIM 24 111–28 1950 CA 44 4742
1359	**LATTICE VIBRATIONS AND SPECIFIC HEAT OF DIAMOND** HOUSTON W V Z NATURFORSCH 3 A 607–11 1948 CA 44 6226
1378	**EFFECT OF COMPOSITION AND TEMPERATURE ON THE SPECIFIC HEAT OF GLASS** SHARP D E GINTHER L B J AM CERAM SOC 34 260–71 1951 CA 45 10527
1383	**THERMODYNAMIC PROPERTIES OF FLUORAPATITE, 15 TO 1600 K.** EGAN, E. P., JR. WAKEFIELD, Z. T. ELMORE, K. L. J AM CHEM SOC 73 5581–2 1951 CA 46 2389
1390	**ESTIMATION OF THE HEAT CONDUCTIVITY OF SEMICONDUCTORS** IOFFE A F DOKLADY AKAD NAUK S S S R 87 369–72 1952 CA 47 6208
1395	**THERMAL CONDUCTIVITY OF DIELECTRICS AT TEMPERATURES LOWER THAN THE DEBYE TEMPERATURE** POMERANCHUK I J EXPTL THEORET PHYS /U S S R/ 12 245–63 1942 CA 37 2972
1399	**CERAMIC MATERIALS FOR USE AT TEMPERATURES ABOVE 1500 DEGREES** WILKENDORF E CHEM-ING-TECH 26 533–8 1954 CA 49 581
1410	**THERMAL CONDUCTIVITY OF DIELECTRIC SOLIDS AT LOW TEMPERATURES** KLEMENS P G PROC ROY SOC /LONDON/ 208 A 108–33 1951 CA 46 5385
1411	**THERMAL CONDUCTIVITIES OF SOME DIELECTRIC SOLIDS AT LOW TEMPERATURES** BERMAN R PROC ROY SOC /LONDON/ 208 A 90–108 1951 CA 46 5385
1413	**KAVALIER GLASS** VOLF M B CZECHOSLOVAK GLASS REV 2 1 12–15 1947 CA 46 5802
1444	**MEASURING OF HEAT CONDUCTIVITY TO FIND THE BEST SHAPES OF CALCAREOUS SANDSTONES WITH AIR CHANNELS** CAMMERER J S TONIND-ZTG 76 99–101 1952 CA 46 7304
1448	**COMPOSITION OF SAGGER BODIES** KEMPCKE E PFAFFENBERGER K BER DEUT KERAM GES 27 24–31 1950 CA 44 8080
1458	**FUSED STABILIZED ZIRCONIA AND REFRACTORIES** WHITTEMORE O J JR MARSHALL D W J AM CERAM SOC 35 85–9 1952 CA 46 6354
1459	**FUSION-CAST REFRACTORIES. I.** MOORE H CERAMICS 2 405–10 462–7 1950 CA 46 6355
1470	**FOAM GLASS—AN INSULATING AND FLOATING MATERIAL.** KITAIGORODSKII I I KESHISHYAN T N STEKOLNAYA I KERAM PROM 6 4–5 1944 CA 42 5190
1476	**HIGH-TEMPERATURE HEAT CONTENT OF HYDROXYAPATITE** EGAN, E. P., JR. WAKEFIELD, Z. T. ELMORE, K. L. J AM CHEM SOC 72 2418–21 1950 CA 44 8757
1481	**DISSOCIATION OF GYPSUM** LEVY J P REV MATERIAUX CONSTRUCTION TRAV PUBL 388 12–14 1948 CA 44 9240
1503	**SPECIFIC HEAT AND HEAT OF CALCINATION OF LIMESTONE AND DOLOMITE** AZBE, V. J. ROCK PRODUCTS 54 1 122–3 1951 CA 45 3141

TPRC Number | *Bibliographic Citation*

1504 SPECIFIC HEAT AND THERMAL EXPANSION OF CERAMIC WARE CONTAINING BARIUM STRONTIUM TITANATE
SAWADA, S. NOMURA, S.
J PHYS SOC JAPAN
5 231–2 1950 CA 45 3139

1509 THE MEAN SPECIFIC HEAT OF PYROBITUMINOUS SHALE
JORDAN IVO
ANAIS ASSOC QUIM BRASIL
8 155–65 1949 CA 45 849

1520 SPECIFIC HEAT OF GLASSES
SCHWIETE H E ZIEGLER G
GLASTECH BER
28 137–46 1955 CA 49 15194

1555 NEW APPARATUS FOR DETERMINATION OF COEFFICIENT OF EXPANSION, TRANSFORMATION POINT, AND SOFTENING POINT OF GLASS.
LINDROTH STIG
BULL AM CERAM SOC
24 450–1 1945 CA 40 1295

1557 A VERSATILE CALORIMETER FOR SPECIFIC–HEAT DETERMINATIONS
HILL HARRY BELL RAYMOND M
ASTM BULL
151 88–91 1948 CA 42 4802

1628 PROPERTIES OF INORGANIC FIBERS, THEIR CHEMICAL AND PHYSICAL BEHAVIOR AS INSULATING MATERIALS FOR SOUND AND HEAT.
BRAUN G
BAUWIRTSCHAFT
47 3–10 1950 CA 47 4523

1630 FOAM GLASS
HUBSCHER MARTIN
SILIKATTECH
5 243–7 1954 CA 49 579

1639 HEAT TRANSFER THROUGH OXIDE–CATHODE MATERIALS
PENGELLY A E
BRIT J APPL PHYS
6 18–20 1955 CA 49 6713

1713 VARIATION OF THE PHYSICAL PROPERTIES OF GLASS WITH TIME IN THE TRANSFORMATION RANGE
DOUGLAS R W
G E C JOURNAL
31 187–96 1954 CA 49 578

1716 NEW THERMAL INSULATING MATERIALS FROM THE ASHES OF RICE CHAFF
SERNAGIOTTO DI CASAVECCHIA E
CHIM IND AGR BIOL
18 508–14 1942 CA 38 6425

1719 THE PROPERTIES OF REFRACTORY MATERIALS AND THEIR SIGNIFICANCE TO FUEL ECONOMY
GREEN A T DODD A E
J INST FUEL
18 74–9 90 1945 CA 39 1745

1720 APPARATUS FOR DETERMINATION OF THE THERMAL CONDUCTIVITY OF SOILS
ROMANOVSKY VSEVOLOD
COMPT REND
213 584–6 1941 CA 38 6029

1734 METHOD OF INVESTIGATION OF THE THERMAL PROPERTIES OF INSULATORS OF DISPERSE STRUCTURE
CHUDNOVSKII A F
J TECH PHYS /U S S R/
16 231–42 1946 CA 41 235
(FOR ENGLISH TRANSLATION SEE T29059)

1798 PROPERTIES AND STRUCTURES OF VITREOUS AND CRYSTALLINE BORON OXIDE
FAJANS KASIMIR BARBER STEPHEN W
J AM CHEM SOC
74 2761–8 1952 CA 46 9374

1805 DISSOCIATION OF STRONTIANITE AT HIGH TEMPERATURES
KAPUSTINSKII A F STAKHANOVA M S
BULL ACAD SCI URSS CLASSE SCI CHIM
11–17 1947 CA 42 4027

1822 PORCELAIN AND SPECIAL CERAMIC MASSES AS TECHNICAL MATERIALS
NAUMANN O
DIE TECHNIK
2 385–92 1947 CA 42 1036

1823 SPECIFIC HEAT OF COLORADO OIL SHALES
SHAW, R. J.
U. S. BUREAU OF MINES
1–9 1947 CA 42 1043
(USBM REPT INVEST 4151)

1835 EXTERNAL OSCILLATIONS AND SPECIFIC HEAT OF CALCITE
GIULOTTO L LOINGER A
J CHEM PHYS
19 1316–17 1951 CA 46 4900

1863 THE LATTICE SPECTRUM AND SPECIFIC HEAT OF DIAMOND
DAYAL, B.
PROC INDIAN ACAD SCI
19 A 224–30 1944 CA 39 459

1888 THERMODYNAMIC PROPERTIES OF ILMENITE AND SELECTIVE REDUCTION OF IRON IN ILMENITE
SHOMATE, C. H. NAYLOR, B. F. BOERICKE, F. S.
U. S. BUREAU OF MINES
1–19 1946 CA 40 6035
(USBM REPT INVEST 3864)

1889 GREEN RIVER OIL SHALES AND PRODUCTS. THEIR CHARACTER, PROCESSING REQUIREMENTS, AND UTILIZATION.
THORNE H M MURPHY W I R STANFIELD K E
BALL J S HORNE J W
OIL SHALE AND CANNEL COAL 2ND CONF
2 301–44 1951 CA 46 11649

1910 HEATS OF DEVITRIFICATION
ROTH W A
GLASTECH BER
21 14–15 1943 CA 39 1802

1911 CUPOLA SLAGS
GLASER O ROLL F
GIESSEREI
30 37–43 1943 CA 39 1830

1922 40TH REPORT OF THE REFRACTORY MATERIALS JOINT COMMITTEE. 1948–49. II. INVESTIGATIONS ON REFRACTORY INSULATING MATERIALS.
CLEMENTS J F VYSE J RIGBY G R
GAS RESEARCH BOARD COPYRIGHT PUBL
4–5, 27–48, 1949.
(GRB 53)

1930 HEAT TRANSFER IN REFRACTORY INSULATING MATERIALS. THERMAL CONDUCTIVITY AND PERMEABILITY AS RELATED TO DIRECTIONAL PROPERTIES.
BARRETT L R VYSE J GREEN A T
GAS RESEARCH BOARD COMMUN
19 82–8 1944 CA 42 6504

1939 HEAT TRANSFER BY GAS CONDUCTION AND RADIATION IN FIBROUS INSULATION
VERSCHOOR, J. D. GREEBLER, P.
TRANS AM SOC MECH ENGRS
74 961–8 1952 CA 46 8426

1942 UTILIZATION OF QUARTZ PELITES IN THE PRODUCTION OF SILICA AND SEMIACID REFRACTORIES
BRON V A
OGNEUPORY
12 113–24 1947 CA 42 1403

1945 THE THERMAL PROPERTIES AND CHILLING POWER OF SOME NON–METALLIC MOLD MATERIALS
RUDDLE R W MINCHER A L
J INST METALS
76 43–90 1949 CA 44 1384

1958 STABILIZED DOLOMITE
DE KEYSER W
REV UNIVERSELLE MINES
90 605–9 1947 CA 42 1715

1963 THERMAL DIFFUSIVITIES AND CONDUCTIVITIES OF GREENSALT, GREENSALT–MAGNESIUM BLENDS, DOLOMITE, AND SLAG.
BEATTY, K. O., JR.
NATIONAL LEAD OF CINCINNATI, OHIO
3–21, 1954.
(FMPC 471)

1964 THE THERMAL CONDUCTIVITY OF SYNTHETIC SAPPHIRE
WEEKS, J. L. SEIFERT, R. L.
J AM CERAM SOC
35 15 1952 CA 46 2365

1967 IMPROVED DESIGN OF APPARATUS FOR MEASURING THERMAL CONDUCTIVITY OF REFRACTORIES AND INSULATION AT HIGH TEMPERATURES
DUPLIN V J JR FITZSIMMONS E S
J AM CERAM SOC
35 226–9 1952 CA 46 10568

1969 USE OF TECHNICAL ALUMINA FOR HIGH–ALUMINA SHAPES
ZHIKHAREVICH S A KRUSHEL L E
OGNEUPORY
16 119–27 1951 CA 45 9235

1972 THE MOST IMPORTANT REFRACTORY CERAMIC MATERIALS AND THE NECESSITY OF THE MANUFACTURE OF LIGHT REFRACTORY BRICKS FOR INDUSTRIAL FURNACES
HOFMANN G
DIE TECHNIK
4 353–6 1949 CA 44 295

1991 AN IMPROVED APPARATUS FOR MEASURING THERMAL CONDUCTIVITY OF HIGHLY CONDUCTIVE HARD CERAMICS
SUZUKE, H. KUWAYAMA, N. YAMAUCHI, T.
YAMAUCHI TOSHIYOSHI
YOGYO KYOKAI SHI
64 161–6 1956 CA 51 14335

TPRC Number	Bibliographic Citation
1992	THERMAL MOISTURE CONDUCTIVITY IN PEAT KLYUCHAREV A E SHALYGINA V S TORFYANAYA PROM 29 5 25-6 1952 CA 46 8829
2025	HEAT CAPACITIES OF INORGANIC SUBSTANCES AT HIGH TEMPERATURES. III. HEAT CAPACITY OF SYNTHETIC CALCITE. KOBAYASHI, K. SCIENCE REPTS TOHOKU UNIV FIRST SER 35 103-18 1951 CA 46 4348
2046	THE PHYSICAL BASES OF THE ELECTRIC MELTING OF GLASS. PEYCHES, I. REV GEN ELEC 55 143-51 1946 CA 40 4185
2089	TESTING OF CLAYS FOR PAPER MACHINE AND BRUSH COATING LESTER IVAN L LYONS S C PAPER TRADE J 125 17 49-55 1947 CA 42 364
2094	VISCOSITY OF CERAMIC MATERIALS AT HIGH TEMPERATURE SOLOMIN N V J TECH PHYS /U S S R/ 15 862-72 1945 CA 40 6232
2122	THE SPECIFIC HEATS OF GERMANIUM AND GRAY TIN HILL R W PARKINSON D H PHIL MAG 43 309-16 1952 CA 46 10842
2172	A TRIAL PREPARATION OF A SINGLE CRYSTAL OF SYNTHETIC MICA BY THE KYROULOS METHOD DAIMON, N. J CERAM ASSOC JAPAN 60 179-80 1952 CA 46 8340
2428	X. DATA FOR SEVERAL PURE OXIDE MATERIALS CORRECTED TO ZERO POROSITY. KINGERY W D FRANCL J COBLE R L VASILOS T J AM CERAM SOC 37 107-10 1954 CA 48 7786
2473	HEAT-INSULATING REFRACTORIES GLEBOV S V EKON TOPLIVA ZA 4 3 9-15 1947 CA 43 9405
2507	VISCOUS-PLASTIC FLOW OF A PEAT MASS HAVING VARIABLE VALUES OF VISCOSITY AND LIMITING SHEARING STRESS VOLAROVICH M P SHCHIPANOV P K KOLLOID ZHUR 11 384-9 1949 CA 44 2822
2523	NEW PROTECTION FOR IRON COOLING COILS-COMPLETELY PERMEABLE TO HEAT. COMBINED MAGNO-ATRAMENT PROCESS. WESLY W KORROSION U METALLSCHUTZ 17 188-90 1941 CA 35 7921
2704	INTERPRETATION OF THE THERMAL CONDUCTIVITY OF GLASSES KITTEL, C. PHYS REV 75 972-4 1949 CA 43 4070
2932	VISCOSITY OF BODIES OF THE FELDSPAR-QUARTZ SYSTEM WITH SPECIAL REFERENCE TO HIGH FELDSPAR CONTENT HAMANO, K. YOGYO KYOKAI SHI 65 1-8 1957 CA 51 15914
3025	METALLIC REFLECTION BY COMPRESSED CRYSTALLINE POWDERS SANDERSON J A J OPTICAL SOC AM 30 566-7 1940 CA 35 948
3081	STUDIES OF BINARY AND TERNARY COMBINATIONS OF MAGNESIA, CALCIA, BARIA, BERYLLIA, ALUMINA, THORIA, AND ZIRCONIA IN RELATION TO THEIR USE AS PORCELAINS. GELLER R F YAVORSKY P J STEIERMAN B L CREAMER A S J RESEARCH NATL BUR STANDARDS 36 277-312 1946 CA 40 5217
3118	RHEOLOGY OF COAL INOUE, K. REPT FUEL RESEARCH INST 66 1951 CA 46 7731
3135	THE INFLUENCE OF THE CHEMICAL COMPOSITION OF GLAZES AND THE TENSION BETWEEN GLAZE AND BODY ON SOME PHYSICAL PROPERTIES OF HARD PORCELAIN NEUBAUER, F. SPRECHSAAL 75, 397-9, 417-20, 1942. 75, 433-6, 451-4, 1942. 75, 473-5, 1942.
3300	INVESTIGATION OF THE COAGULATION OF HEAVY GLACIAL CLAY. VUORINEN, J. MAATALOUSKOELAITOKSEN MAATUKIMUSOSASTO, (AGROGEOL. JULKAISUJA) 50 5-105 1939 CA 35 5232
3391	SPECIAL MAGNESITE BRICKS FOR METALLURGICAL FURNACES KALPERS H WIEN CHEM-ZTG 45 235-7 1942 CA 38 2459
3420	VISCOSITY OF FUSED-SILICA GLASS. SOLOMIN, N. V. AKAD. NAUK SSSR OTDEL. TEKH. NAUK INST. MASHINOVEDENIYA, SOVESHCHANIE PO VYAZKOSTI ZHIDKOSTEI I KOLLOID. RASTVOROV 1 317-25 1941 CA 40 3319
3421	A NEW METHOD FOR DETERMINING THE SPECIFIC HEATS OF GASES AND VAPORS BENNEWITZ K SCHULZE O Z PHYSIK CHEM 186 A 299-313 1940 CA 35 1693
3546	EVALUATION OF THE SPECIFIC HEAT OF DIAMOND FROM ITS RAMAN FREQUENCIES KRISHNAMURTI D PROC INDIAN ACAD SCI 34 A 121-6 1951 CA 46 10896
3610	MEASUREMENTS OF THE THERMAL CONDUCTIVITY OF FIRECLAY REFRACTORIES PATTON T C NORTON C L JR J AM CERAM SOC 26 350-8 1943 CA 37 6831
3671	APPARATUS FOR MEASURING THERMAL CONDUCTIVITY OF REFRACTORIES NORTON C L JR J AM CERAM SOC 25 451-9 1942 CA 37 734
3688	THE HEAT CONDUCTIVITY OF PLATE GLASS HUBLER A SPRECHSAAL 75 37-8 1942 CA 37 6100
3698	THE THERMAL CONDUCTIVITY OF GLASS CHARGES GUTOP V G STEKOLNAYA PROM 11 24-7 1940 CA 37 6419
3699	THERMAL CONDUCTIVITY, SPECIFIC HEAT AND THERMOMETRIC CONDUCTIVITY OF COKE AND LIGNITE. FRITZ W KNEESE H FEUERUNGSTECH 30 273-9 1942 CA 37 5848
3880	THE RELATION BETWEEN THE VISCOSITY OF SILICATE GLASSES IN THE RANGE 10 TO THE THIRD POWER TO 10 TO THE SEVENTH POWER POISES AND THEIR CHEMICAL COMPOSITION. OKHOTIN M V J APPL CHEM USSR 29 1387-91 1956 (ENGLISH TRANSLATION OF ZH. PRIKL. KHIM., 29, 1287-92, 1956; FOR ORIGINAL SEE T6020)
3910	DEVELOPMENT OF INSULATING BRICKS FOR FURNACE CONSTRUCTION OLIVER H CERAMICS SYMPOSIUM 592-636 1953 CA 48 7270
3911	A METHOD FOR THE MEASUREMENT OF THERMAL DIFFUSIVITY OF MOLTEN GLASS VAN ZEE A F BABCOCK C L J AM CERAM SOC 34 244-50 1951 CA 45 9233
3956	VISCOSITY AND STRUCTURE OF MOLTEN QUARTZ GLASS SOLOMIN N V J PHYS CHEM /U.S.S.R./ 14 235-43 1940 CA 36 3917 (FOR ENGLISH TRANSLATION SEE T31216)
3989	SYSTEMATIC STUDY OF EFFECT OF OXIDE CONSTITUENTS ON VISCOSITY OF SILICATE GLASSES AT ANNEALING TEMPERATURES POOLE, J. P. GENSAMER, M. J AM CERAM SOC 32 220-9 1949 CA 43 7654
4000	METHOD OF MEASURING FLAME RADIATION AND EMISSIVITY CONTINUOUSLY IN OPEN-HEARTH FURNACES MAYORCAS R THOMAS D J IRON STEEL INST /LONDON/ 175 359 1953 CA 48 1218
4013	THERMAL EMISSIVITY OF REFRACTORIES AND ITS IMPORTANCE FOR GLASS-FURNACE OPERATION EITEL WILLIAM GLASS IND 36 193-7 228 1955 CA 49 14285
4038	EFFECTS OF RECENT KNOWLEDGE OF ATOMIC CONSTANTS AND OF HUMIDITY ON THE CALIBRATIONS OF THE NATIONAL BUREAU OF STANDARDS THERMAL-RADIATION STANDARDS STAIR, R. JOHNSTON, R. G. J RESEARCH NATL BUR STANDARDS 53 211-15 1954 CA 49 5954

TPRC Number	Bibliographic Citation
4043	THE INFRARED TRANSMISSION OF GLASS IN THE RANGE ROOM TEMPERATURE TO 1400 DEGREES GROVE F J JELLYMAN P E J SOC GLASS TECHNOL 39 3-15T 1955 CA 49 7208
4200	THE TOTAL EMISSIVITY OF SOME REFRACTORY MATERIALS ABOVE 900 DEGREES PATTERSON J R TRANS BRIT CERAM SOC 54 698-705 1955 CA 50 6007
4320	THE SPECTRAL-TRANSMISSIVE CHARACTERISTICS OF SOME GERMAN GLASSES STAIR, R. GLAZE, F. W. BALL, J. J. GLASS IND 30 331-5 354 1949 CA 44 295
4438	THE UNSTEADY FLOW OF HEAT THROUGH POROUS MATERIALS HARBERT WM D CAIN D C HUNTINGTON R L IND ENG CHEM 33 257-63 1941 CA 35 3488
4492	HEAT CHARACTERISTICS OF LIGHT-WEIGHT REFRACTORIES FROM THE PODOLSK REFRACTORY WORKS KOLECHKOVA A F GONCHAROV V V OGNEUPORY 13 401-7 1948 CA 43 5563
4512	A GENERAL SURVEY OF THE HEAT-AND TEMPERATURE-CONDUCTIVITY OF COAL FRITZ, W. FORSCH GEBIETE INGENIEURW 14 1 1-10 1943 CA 38 3448
4515	SPECTROSCOPY IN INFRARED BY RELECTION AND ITS USE FOR HIGHLY ABSORBING SUBSTANCES SIMON, I. J OPTICAL SOC AM 41 336-45 1951 CA 45 6056
4561	THERMAL CONDUCTIVITIES IN MOIST SOILS SMITH W O SOIL SCI SOC AM PROC 4 32-40 1939 CA 35 244
4627	MANNER OF MOLECULAR GROUPING IN GLASS YOSHIDA USABURO MEM COLL SCI KYOTO IMP UNIV 23 A 233-44 1941 CA 42 7592
4632	INVESTIGATION OF THE RELATIVE LIGHT-ABSORPTION ABILITIES /DEGREE OF BLACKNESS/ OF REFRACTORIES. GLINKOV M A GLINKOVA O A TRUDY URAL IND INST IM S M KIROVA 17 97-114 1941 CA 40 3862
4688	THE ERUPTION OF HEKLA 1947-1948. IV. 5. THE BASIC MECHANISM OF VOLCANIC ERUPTIONS AND THE ULTIMATE CAUSES OF VOLCANISM. EINARSSON TRAUSTI VISINDAFELAG ISLENDINGA 1-30 1950 CA 48 1214
4721	THE PROPERTIES OF CORUNDUM BRICK WITH SPECIAL REFERENCE TO HEAT TRANSFER LASCH HANS KERAM RUNDSCHAU 49 469-73 1941 CA 37 3893
4737	THERMAL CONDUCTIVITY OF BUILDING MATERIALS. METHODS OF DETERMINATION AND RESULTS. PRATT A W BALL J M E J INST HEATING VENTILATING ENGRS 24 201-26 1956 CA 52 2368
4834	THERMAL EMISSIVITY OF REFRACTORY MATERIALS AT HIGH TEMPERATURES MICHAUD M SILICATES IND 19 243-50 1954 CA 48 13190
4892	THERMAL CONDUCTIVITY OF MAGNESITE BRICK LASCH HANS TONIND-ZTG 65 421-3 433-4 1941 CA 37 4867
5008	TEMPERATURE CONDITIONS DURING THE USE OF BRICK IN COKING CHAMBERS KUSTOV B I VOLOSHIN A I OGNEUPORY 13 79-81 1948 CA 43 5926
5009	COKE-OVEN DINAS WITH MANGANESE BINDER SMELYANSKII I S TSIGLER V D OGNEUPORY 14 1 9-21 1949 CA 43 5916
5069	UTILIZATION OF SUGAR-INDUSTRY WASTE, MINERAL WOOL FROM FILTER PRESS MUD. DARUVALLA D N KAMATH V G J SCI AND IND RESEARCH /INDIA/ 11 A 412-13 1952 CA 47 3539
5155	THERMAL DIFFUSIVITY OF REFRACTORY OXIDES FITZSIMMONS E S J AM CERAM SOC 33 327-32 1950 CA 45 1743
5195	THE CONSTITUTION OF GLASS MORIYA, T. JAPAN SCI REV SER I ENG SCI 1 4 61-6 1950 CA 46 2765
5200	THE THERMAL CONDUCTIVITY OF DRY SOIL. SMITH, W. O. SOIL SCI. 53, 435-59, 1942.
5258	THERMAL CONDUCTIVITY OF SOME INDUSTRIAL MATERIALS GRIFFITHS, E. POWELL, R. W. HICKMAN, M. J. J INST FUEL 15 107-20 1942 CA 36 5579
5402	HEAT CONDUCTIVITY OF GLASS AT 1.3 K. KEESOM P H PHYSICA 11 4 339-42 1945 CA 40 3972
5926	ELASTIC AND VISCOUS PROPERTIES OF VOLCANIC ROCKS. IV. EFFECT OF THERMAL HISTORY ON VISCOSITY OF OOSIMA LAVAS. SAKUMA SHUZO BULL EARTHQUAKE RESEARCH INST TOKYO UNIV 32 215-29 1954 CA 48 12634
6002	PHYSICOCHEMICAL PROPERTIES OF FUSED VITREOUS AND RECRYSTALLIZED BASALTS. II. STUDY OF VISCOSITY OF THE FUSED SUPERCOOLED BASALT AT THE INTERVAL OF SOFTNESS. ABRAMYAN, A. V. IZVEST. AKAD. NAUK ARMYAN SSR FIZ.-MAT. ESTESTVEN I TEKH. NAUKI 9 8 17-23 1956 CA 51 7090
6185	INVESTIGATION OF DEFORMATION OF REFRACTORY MATERIALS UNDER LOAD AND OF EFFECTIVE VISCOSITY AT TEMPERATURES UP TO 2800 DEGREES SOLOMIN N V OGNEUPORY 15 183-8 1950 CA 45 1318
6200	APPARATUS FOR THE MEASUREMENT OF THE THERMAL CONDUCTIVITY OF SOLIDS. WEEKS, J. L. SEIFERT, R. L. REV SCI INSTRUMENTS 24 10 1054-1057 1953 MA 21 894
6205	THE THERMAL CONDUCTIVITY OF SOILS. GEMANT, A. J APPD PHYS 21 750-2 1950 RA 7 117
6215	DIE PHYSIKALISHC-TECHNISCHEN EIGENSCHAFTEN DER GEBRAUCHLICHEN KALTESCHUTZSTOFFE //PHYSICAL PROPERTIES OF SOME USUAL INSULATING MATERIALS FOR COLD STORES//. CAMMERER J S KALTETECHNIK 2 270-3 1950 RA 7 13
6227	SOIL TEMPERATURE, MOISTURE CONTENT AND THERMAL PROPERTIES—TENNESSEE VALLEY AREA, 1949, 1950 AND 1951. CARTER CLYDE L UNIV TENN ENGG EXP STA BUL NO 15 1-85 1951 RA 8 115
6231	THERMAL CONDUCTIVITY OF MOIST MATERIALS AND ITS MEASUREMENT. JESPERSEN H B J INSTN HTG AND VENTG ENGR 21 157-74 1953 RA 9 193
6237	MEASUREMENTS OF THE THERMAL CONDUCTIVITY OF SOME AUSTRALIAN POROUS AND FIBROUS MATERIALS. STONEY A J M GRANT G R M PROC 8TH INT CONG REFRIGN 253-9 1951 RA 8 69
6243	A STUDY OF A TRANSIENT HEAT METHOD FOR MEASURING THERMAL CONDUCTIVITY. DEUSTACHIO D SCHREINER R E HTG PIPING AND AIR CONDG ASHVE J SECT 24 6 113-7 1952 RA 7 174
6252	HOW TO COMPUTE THERMAL SOIL CONDUCTIVITIES. GEMANT, A. HTG PIPING AND AIR CONDG 24 1 122-3 1952 RA 7 117
6277	CHROMITE. ITS PROPERTIES AND USES AS A REFRACTORY. MAMYKIN P URALSKIY TEHNIK /URAL TECHNOLOGIST/ 4 21-29 1932 MA 1 262
6279	CHROMIUM-BASE REFRACTORIES. DERIBERE, M REV MAT CONSTR TRAV PUBL 297 101B-104B 1934 MA 1 623

TPRC Number	Bibliographic Citation
6283	**FACTORS THAT INFLUENCE THERMAL CONDUCTIVITY OF POROUS-TYPE INSULATING MATERIALS.** BUTTERWORTH A V CHEM ENG PROG TRANS SECT 43 597-600 1947 RA 3 125
6308	**THERMAL CONDUCTIVITY OF INSULATING MATERIALS AT LOW MEAN TEMPERATURES.** ROWLEY, F. B. JORDAN, R. C. LANDER, R. M. REFRIG ENG 50 541-4 1945 RA 1 80
6309	**DEVELOPMENT OF THE THERMAL CONDUCTIVITY PROBE.** HOOPER F C CHANG S C HTG PIPING AND AIR CONDG ASHVE J SECT 24 10 125-9 1952 RA 8 69
6314	**MEASUREMENTS OF THE TRUE SPECIFIC HEATS OF SILVER, NICKEL, BETA-BRASS, QUARTZ CRYSTAL, AND QUARTZ GLASS BETWEEN PLUS 50 C. AND 700 C. BY AN IMPROVED METHOD.** MOSER, H. PHYSIKAL Z 37 21 737-753 1936 MA 4 178 (FOR ENGLISH TRANSLATION SEE T27375)
6321	**THERMAL CONDUCTIVITY, EXPANSION AND SPECIFIC HEAT OF INSULATORS AT EXTREMELY LOW TEMPERATURES.** WILKES, G. B. REFRIG ENG 52 37-42 1946 RA 1 199
6324	**MEASUREMENT OF THE THERMAL CONDUCTIVITY OF SAMPLES OF THERMAL INSULATING MATERIALS AND OF INSULATION IN SITU BY THE HEATED PROBE METHOD.** MANN G FORSYTH F G E MOD REFRIGN 59 188-91 1956 RA 11 142
6325	**THERMAL CONDUCTIVITY OF COMMERCIAL INSULATIONS AT LOW TEMPERATURE.** VERSCHOOR J D REFRIGG ENGG 62 9 35-7 98 1954 RA 10 69
6329	**THE EXACT MEASUREMENT OF THE SPECIFIC HEATS OF SOLID SUBSTANCES AT HIGHER TEMPERATURES. XIX.—THE SPECIFIC HEATS OF ZINC, MAGNESIUM, AND THEIR BINARY ALLOY MGZN2.** POPPEMA T J JAEGER F M PROC K AKAD WET AMSTERDAM 38 510-520 1935 MA 2 561
6342	**GAS PERMEABILITY, COMPRESSIVE STRENGTH, HEAT CONDUCTIVITY, THERMAL EXPANSION, HEAT CAPACITY, AND TENDENCY TO SPALLING OF REFRACTORY BRICKS IN RELATION TO THE POROSITY.** HERBST, H. FEUERUNGSTECHNIK 22 115-116 1934 MA 2 115
6356	**THE THERMAL PROPERTIES OF BUILDING MATERIALS USED IN HEAT FLOW CALCULATIONS.** NOTTAGE H B ASHVE RES BULL 53 2 1-35 1947 RA 3 259
6357	**GAS IS AN IMPORTANT FACTOR IN THE THERMAL CONDUCTIVITY OF MOST INSULATING MATERIALS.** ROWLEY, F. B. JORDAN, R. C. LUND, C. E. LANDER, R. M. HTG PIPING AND AIR CONDG ASHVE J SECT 23 12 103-9 1951 RA 7 123
6359	**LOW MEAN TEMPERATURE THERMAL CONDUCTIVITY STUDIES.** ROWLEY, F. B. JORDAN, R. C. LANDER, R. M. REFRIG ENG 53 35-9 1947 RA 2 80
6453	**SPECIFIC HEAT OF MATERIALS. FROM THERMODYNAMIC AND TRANPSORT PROPERTIES OF GASES, LIQUIDS AND SOLIDS.** LANG J I ASME SYMP ON THERMAL PROPERTIES 405-414 1959
6456	**MEASUREMENT OF THE TOTAL NORMAL EMISSIVITY OF MATERIALS.** WILKES, G. B. ARMY MISSLE COMMAND, REDSTONE ARSENAL 37PP., 1951. (ATI 117902)
6478	**ON THE CYLINDRICAL PROBE OF MEASURING THERMAL CONDUCTIVITY WITH SPECIAL REFERENCE TO SOILS. I. EXTENSION OF THEORY AND DISCUSSION OF PROBE CHARACTERISTICS.** DE VRIES D A PECK A J AUSTRAL J PHYS 11 2 255-271 1958 AM 12 4093
6481	**NOTE ON THREE YEARS OF RESEARCH ON THE HEAT CONDUCTIVITY OF REFRACTORY MATERIALS IN THE LABORATORY OF THE COMPAGNIE GENERALE DE CONSTRUCTION DE FOURS.** CASSAN HENRY CHALEUR ET IND 18 202 55-72 1937 MA 5 353

TPRC Number	Bibliographic Citation
6524	**THE ROLE OF MATERIAL THERMAL PROPERTIES IN BUILD-UP OF MISSILE SKIN TEMPERATURES.** TAYLOR, W. C. MISSILE SYSTEMS DIV. LOCKHEED AIRCRAFT CORPORATION 6PP., 1956. (AD 105757)
6565	**THE EXPERIMENTAL MEASUREMENT OF THERMAL CONDUCTIVITIES, SPECIFIC HEATS, AND DENSITIES OF METALLIC, TRANSPARENT, AND PROTECTIVE MATERIALS. PART III.** LUCKS, C. F. MATOLICH, J. VAN VELZOR, J. A. BATTELLE MEMORIAL INSTITUTE, WRIGHT AIR DEVELOPMENT CENTER 71PP., 1954. (AF TR 6145, AD 95406)
6567	**THE EXPERIMENTAL MEASUREMENT OF THERMAL CONDUCTIVITIES, SPECIFIC HEATS, AND DENSITIES OF METALLIC, TRANSPARENT, AND PROTECTIVE MATERIALS. PART II.** LUCKS, C. F. BING, G. F. MATOLICH, J. DEEM, H. W. THOMPSON, H. B. BATTELLE MEMORIAL INSTITUTE, WRIGHT AIR DEVELOPMENT CENTER 32PP., 1952. (AF TR 6145, AD 95239)
6596	**TOTAL NORMAL EMISSIVITIES AND SOLAR ABSORPTIVITIES OF MATERIALS.** WILKES, G. B. MASSACHUSETTS INSTITUTE OF TECHNOLOGY, CAMBRIDGE, WRIGHT AIR DEVELOPMENT CENTER 94PP., 1954. (WADC TR 54-42, AD 88066)
6609	**THE INVESTIGATION OF THERMAL AND ELECTRICAL PROPERTIES OF METALS AT HIGH TEMPERATURES.** SAWYER, R. B. LEHIGH UNIVERSITY, BETHLEHEM, PA., INST. RESEARCH 38PP., 1955. (AD 81977)
6619	**COMPARATIVE METHOD FOR DETERMINING THE THERMAL CONDUCTIVITY OF CERAMIC MATERIALS.** KLASSE F HEINZ A HEIN J BERICHTE DER DEUTSCHEN KERAMISCHEN GESELLSCHAFT 34 183-189 1957 BR 6 14126 (FOR ENGLISH TRANSLATION SEE T15634)
6624	**THE THERMAL AND ELECTRICAL CONDUCTIVITIES OF OF LEAD-BISMUTH ALLOYS.** CLIFFORD, J. M. LEHIGH UNIV., BETHLEHEM, PA., INST. RESEARCH 78PP., 1955. (AD 75584)
6657	**METHOD OF MEASUREMENT OF THE HEAT CONDUCTIVITY COEFFICIENT OF REFRACTORY MATERIALS AT HIGH TEMPERATURES.** PUSTOVALOV V V ZAVODSKAIA LABORATORIIA 23 9 1093-1094 1957 BR 7 201
6695	**THERMAL CONDUCTIVITY OF METAL COMPOSITES.** MASTERS, E. B. GENERAL ELECTRIC, AIRCRAFT NUCLEAR PROPULSION PROJECT 4PP., 1954. (APEX 186, AD 63476)
6702	**THE THERMAL CONDUCTIVITY OF DIELECTRIC SOLIDS AT LOW TEMPERATURES.** BERMAN R ADVANCES IN PHYSICS 2 103-140 1953
6706	**DAS THERMISCHE LEITVERMOGEN VON DRAHTEN UND STABEN.** BARRATT T WINTER R M ANN PHYSIK 77 9 1-15 1925
6708	**UBER DIE TEMPERATURABHANGIGKEIT DER WARMELEITFAHIGKEIT FESTER NICHTMETALLE.** EUCKEN A ANN PHYSIK 34 2 185-221 1911
6720	**RECHERCHES SUR LES CONDUCTIBILITES THERMIQUE ET ELECTRIQUE DES PHASES CRISTALLINES ANISTROPES.** JAEGER F M ARCH SCI PHYS ET NAT 22 240-256 1906
6740	**STUDIES RELATING TO THE REACTION BETWEEN ZIRCONIUM AND WATER AT HIGH TEMPERATURES.** LEMMON, A. W., JR. BATTELLE MEMORIAL INSTITUTE 1-114, 1957. (BMI 1154)
6748	**THERMODYNAMICS OF SOLIDS. THERMAL CONDUCTIVITY.** MEISSNER, W. SCHUBERT, G. U. PHYSICS OF SOLIDS 212-220, 1948.

TPRC Number	Bibliographic Citation
6750	INSULATING REFRACTORIES. THEIR MANUFACTURE AND PHYSICAL PROPERTIES. BOLE G A TRANS AMER FOUND ASSOC 45 261-292 1937 MA 5 622
6752	COEFFICIENTS OF THERMAL CONDUCTIVITY OF MATERIALS USED IN THE CONSTRUCTION OF HEAT-EXCHANGERS. DELLE CANNE GIUSEPPE CHIM E IND 33 1 11-15 1951 MA 20 392
6787	AN INVESTIGATION OF THE VALIDITY OF THE WIEDEMANN-FRANZ-LORENZ LAW. LINDE J O ARKIV FYSIK 4 6 541-554 1952 MA 21 415
6798	ON THE CALCULATION OF THE SPECIFIC HEAT OF SOLIDS. HONNEFELDER K Z PHYSIKAL CHEM 21 B 53-64 1933 MA 1 118
6929	THERMAL CONDUCTIVITY TESTS AND RESULTS. ALLCUT E A J INST HEATING VENTILATING ENG 17 151-195 1949
6930	THERMAL SHOCK ANALYSIS OF SPHERICAL SHAPES. CRANDALL W B GING J J AM CERAM SOC 38 1 44-54 1955
6938	CRITERIA FOR THERMAL INSULATION FOR USE ON UNDERGROUND PIPING. DEUSTACHIO, D. AMERICAN SOCIETY FOR TESTING AND MATERIALS 81-6 1957 (ASTM SPEC TECH PUBL 217)
6939	THERMAL PROPERTIES OF CERAMICS. SMOKE, E. J. KOENIG, J. H. ENG. RES. BULL. 40 1-53 1958
6940	THE EXPERIMENTAL MEASUREMENT OF THERMAL CONDUCTIVITIES, SPECIFIC HEATS, AND DENSITIES OF METALLIC, TRANSPARENT, AND PROTECTIVE MATERIALS. PART I. LUCKS, C. F. THOMPSON, H. B. SMITH, A. R. CURRY, F. P. DEEM, H. W. BING, F. G. U.S. AIR FORCE TECHNICAL REPORT 1-127, 1951. (AFTR 6145 PT I, ATI 117715)
6941	ADAPTATION OF A MODIFIED ANGSTROM METHOD TO THE MEASUREMENT OF THERMAL DIFFUSIVITY OF NON-METALLIC MATERIALS. SOXMAN, E. J. STUDY OF HEAT TRANSFER OF CERAMIC MATERIALS. 1-36 1955
6942	INFRARED TRANSMISSION AND RADIATION CONDUCTIVITY OF ALUMINA FROM ROOM TEMPERATURE TO 1800 C. MONROE, J. E., JR. STUDY OF HEAT TRANSFER OF CERAMIC MATERIALS 38-86 1955
6990	APPARATUS FOR THE MEASUREMENT OF THE THERMAL CONDUCTIVITY OF SOLIDS. WEEKS, J. L. SEIFERT, R. L. ARGONNE NATIONAL LABORATORY 1-14, 1952. (ANL 4938, AD 1929)
6991	THERMAL CONDUCTIVITY DATA FOR SOME NUCLEAR FUELS. MC CREIGHT, L. R. KNOLLS ATOMIC POWER LAB., SCHENECTADY, N. Y. 1-19, 1952. (TIO 10062)
7001	PROPERTIES OF TITANIUM COMPOUNDS AND RELATED SUBSTANCES. ROSSINI, F. D. COWIE, P. A. ELLISON, F. D. BROWNE, C. C. OFFICE OF NAVAL RESEARCH 1-448 1956 (ONR RPT ACR 17)
7013	THERMAL CONDUCTIVITIES OF FUSED AND CRYSTALLINE QUARTZ. RATCLIFFE E H BRIT J APPL PHYS 10 22-5 1959
7020	GAS IS AN IMPORTANT FACTOR IN THE THERMAL CONDUCTIVITY OF MOST INSULATING MATERIALS. PART II. LANDER R M HEATING PIPING AIR CONDITIONING 26 12 121-6 1954
7023	THE THERMAL RADIATION CHARACTERISTICS OF SOLID MATERIALS - A REVIEW. BLAU HENRY H JR MILES JOHN L ASHMAN LELAND E OFFICE OF TECHNICAL SERVICE 1-86, 1958. (AD 146833)
7028	DIE WARMELEITFAHIGKEIT FESTER KORPER BEI TEIFEN TEMPERATUREN. EUCKEN A Z TECH PHYSIK 6 689-94 1925
7036	I. DEVELOPMENT OF CERAMIC BODIES WITH HIGH THERMAL CONDUCTIVITY. SNYDER, N. H. SMOKE, E. J. WISELY, H. R. RUH, E. BAUER, H. L. ILLYN, A. V. NIEMCZURA, S. W. N. J. CERAM. RES. STATION, RUTGERS, UNIVERSITY 1-33, 1953. (AD 5552)
7067	I. DEVELOPMENT OF CERAMIC BODIES WITH HIGH THERMAL CONDUCTIVITY. SMOKE, E. J. SNYDER, N. H. WISELY, H. R. RUH, E. ILLYN, A. V. EICHBAUM, B. R. RUTGERS UNIVERSITY 1-25, 1953. (AD 19833)
7095	MEASURING THE HEAT CONDUCTIVITY OF REFRACTORY MATERIALS IN VACUUM. SHAKHTIN D M VISHNEVSKII I I ZAVODSKAIA LABORATORIIA 23 8 927-929 1957 BR 6 17260
7099	THERMAL CONDUCTIVITY OF HEAT INSULATING MATERIALS. POWERS, R. W. JOHNSTON, H. L. HANSEN, R. H. ZIEGLER, J. B. CRYOGENIC LAB., OHIO STATE UNIV. 1-29, 1953. (TR-254-16, AD 27569)
7102	I. DEVELOPMENT OF CERAMIC BODIES WITH HIGH THERMAL CONDUCTIVITY. SMOKE, E. J. WISELY, H. R. RUH, E. ILLYN, A. V. EICHBAUM, B. R. N. J. CERAM. RES. STATION, RUTGERS UNIV. 1-35, 1953. (AD 29335)
7104	I. DEVELOPMENT OF CERAMIC BODIES WITH HIGH THERMAL CONDUCTIVITY. SMOKE, E. J. ILLYN, A. V. EICHBAUM, B. R. N. J. CERAM. RES. STATION, RUTGERS UNIV. 1-24, 1954. (AD 30849)
7129	THERMAL AND ELECTRICAL CONDUCTIVITIES OF AISI C1010 STEEL IN THE RANGE FROM 25 C TO 800 C. RAEZER, S. D. OFFICE OF ORDNANCE RESEARCH, DURHAM, N. C. 1-36, 1954. (AD 49544)
7156	DEVELOPMENT OF CERAMIC INSULATING MATERIALS FOR HIGH-TEMPERATURE USE. KINGERY W D KLEIN J D MCQUARRIE M C ASME TRANSACTIONS 80 3 705-710 1958 BR 7 9142
7180	THE THERMAL CONDUCTIVITY OF REFRACTORIES. WILKES, G. B. J AMER CERAM SOC 17 173-177 1934 MA 2 443
7181	A METHOD OF MEASURING THERMAL CONDUCTIVITY //OF REFRACTORY MATERIALS// AT FURNACE TERMPERATURES. WEINLAND, C. E. J AMER CERAM SOC 17 194-202 1934 MA 2 443
7182	AN APPARATUS FOR MEASURING THE THERMAL CONDUCTIVITY OF REFRACTORIES AT HIGH TEMPERATURES. FINCK J L J AMER CERAM SOC 18 6-12 1935 MA 2 444
7185	THE RELIABILITY OF MEASUREMENTS OF THE THERMAL CONDUCTIVITY OF REFRACTORY BRICK. AUSTIN J B PIERCE R H H JR J AMER CERAM SOC 18 48-54 1935 MA 2 628
7200	THERMAL CONDUCTIVITY OF DIELECTRIC CRYSTALS. THE UMKLAPP PROCESS. BERMAN R SIMON F E WILKS J NATURE 168 277-280 1951

TPRC Number	Bibliographic Citation
7288	**EXPERIMENTS USING A SIMPLE THERMAL COMPARATOR FOR MEASUREMENT OF THERMAL CONDUCTIVITY, SURFACE ROUGHNESS AND THICKNESS OF FOILS OR OF SURFACE DEPOSITS.** POWELL R W JOURNAL OF SCIENTIFIC INSTRUMENTS 34 485–492 1957 BR 7 5992
7333	**THE MEASUREMENT OF THE THERMAL CONDUCTIVITY OF HEAT INSULATORS AT LOW TEMPERATURES. PAPER PRESENTED BEFORE 9TH INT CONG REFRIGN, AG–S 1955** HICKMAN M J RATCLIFFE E H INT INST REFRIGN /IIF/ BUL 35 4 794 1955 RA 10 166
7336	**SYMPOSIUM ON HEATING AND VENTILATING OF INDUSTRIAL BUILDINGS. III. FACTORY INSULATION.** BRODERICK A H T J INSTN HTG AND VENTG ENGR 21 429–41 1954 RA 10 69
7346	**THERMAL CHARACTERISTCS OF SOILS FOR GROUND COIL DESIGN.** PARKERSON WILLIAM HTG AND VENTG 48 10 83–6 1951 RA 7 117
7348	**COLD STORAGE PLANT INSULATION IN GERMANY.** SEIFFERT, K. WORLD REFRIGN 7 663–6 1956 RA 12 99
7350	**EMISSIVITY OF ICE, SNOW, AND FROZEN GROUND.** DUNKLE R V GIER J T BEVANS J T REFRIGG ENGG 65 4 33–5 89 1957 RA 12 85
7351	**EMISSIVITY OF ICE, SNOW, AND FROZEN GROUND. PAPER PRESENTED BEFORE ASRE, N 1956.** DUNKLE R V GIER J T BEVANS J T REFRIGG ENGG 64 11 56 1956 RA 12 1
7356	**VARIATION OF THERMAL CONDUCTIVITY WITH TEMPERATURE OF INSULATING POWDERS. PAPER PRESENTED BEFORE ASRE, D 1955.** GLASER, P. E. KAYAN, C. F. REFRIGG ENGG 63 12 41 1955 RA 11 22
7365	**THERMAL CONDUCTIVITY OF SOILS FOR DESIGN OF HEAT PUMP INSTALLATIONS.** SMITH, G. S. YAMAUCHI, T. HEAT PIPING AND AIR COND ASHVE J SECT 22 7 129–35 1950 RA 5 185
7368	**THE THERMAL CONDUCTIVITY OF SOILS.** KERSTEN, M. S. PROC HGY RES BD 28 391–409 1948 RA 5 127
7369	**APPARATUS FOR MEASURING THERMAL CONDUCTIVITY OF SOIL.** KERSTEN, M. S. PROC 2ND INT CONF SOIL MECH AND FOUNDN ENG 3 162–5 1948 RA 5 128
7457	**THE VIBRATIONAL SPECTRUM AND THE SPECIFIC HEAT OF GERMANIUM AND SILICON.** HSIEH YU–CHANG JOURNAL OF CHEMICAL PHYSICS 22 306–311 1954 RM 11 171–P
7477	**SOME MEASUREMENTS OF HEAT FLOW ALONG TECHNICAL MATERIALS IN THE REGION 4 TO 20 K.** WILKINSON, K. R. WILKS, J. SCI. INSTR. AND PHYS. IN IND. 26 19–20 1949 RM 6 3A–32
7515	**DETERMINATION OF THE COEFFICIENT OF INTERNAL THERMAL CONDUCTIVITY.** MACOLA M LA METALLURGIA ITALIANA 39 12–14 1947 RM 5 17–101
7575	**NEW METHOD OF DETERMINING THE THERMAL CONDUCTIVITY OF BODIES.** ANGSTROM A J PHIL MAG 25 130–142 1863 (ENGLISH TRANSLATION OF ANN. PHYSIK., 114 (2), 513–30, 1861; FOR ORIGINAL SEE T 10036)
7618	**GERMAN WORK ON THE PHYSICS OF SOLIDS.** ADAMS A M SANDOR A J BULL BRIT COAL UTILISATION RESEARCH ASSOC 13 3 73–84 1949 MA 17 903
7644	**CONDUCTION OF HEAT IN POWDERS.** KANNULUIK W G MARTIN L H PROC ROY SOC /LONDON/ 141 A 144–58 1933
7648	**THE MEASUREMENT OF THERMAL CONDUCTIVITY.** GRIFFITHS, E. KAYE, G. W. C. PROC ROY SOC /LONDON/ 104 A 71–98 1923
7688	**OPTICAL MATERIALS FOR INFRARED INSTRUMENTATION.** BALLARD, S. S. MC CARTHY, K. A. WOLFE, W. L. INFRARED INFOR. – ANALYSIS CENTER, UNIV. OF MICHIGAN 1–115, 1959. (IRIA RPT NO. 2389–11–S)
7689	**MEASUREMENTS OF THERMAL PROPERTIES.** FIELDHOUSE, I. B. HEDGE, J. C. LANG, J. I. WRIGHT AIR DEVELOPMENT CENTER 1–79, 1958. (WADC–TR–58–274, AD 206892, PB 151583)
7694	**EFFECT OF SURFACE RESISTANCE ON THERMAL CONDUCTIVITY BY THE HOT PLATE METHOD.** LANDER ROBERT UNIV MINN INST TECHNOL ENG EXP STA TECH PAPER 40 1–13 1942
7695	**MEASUREMENT OF SOME THERMAL PROPERTIES OF THREE GLASSES.** MELONAS, J. V. COVINGTON, P. C. PEARS, C. D. WRIGHT AIR DEVELOPMENT CENTER 1–15, 1958. (WADC TR 58–129, AD 155816)
7696	**THERMOPHYSICAL PROPERTIES OF SOLID MATERIALS.** GOLDSMITH, A. WATERMAN, T. E. WRIGHT AIR DEVELOPMENT CENTER 1–416, 1959. (WADC TR 58–476, AD 207905)
7703	**THE APPARENT THERMAL CONDUCTIVITY OF CERTAIN SINTERED ZINC BLENDES.** DECROLY CLAUD INDUSTRIE CHIM BELGE 20 476–80 1955 CA 50 9969
7704	**THE DETERMINATION OF THE COEFFICIENT OF THERMAL CONDUCTIVITY OF GAS–SYNTHESIS CATALYSTS.** BLYUDOV, A. P. TR. VSES. NAUCH. ISSLED. INST. ISKUSSTV. ZHIDK. TOPL. I GAZA 6 85–9 1954 CA 51 9278
7757	**SPHERICAL FURNACE CALORIMETER FOR DIRECT MEASUREMENT OF SPECIFIC HEAT AND THERMAL CONDUCTIVITY.** WINCKLER JOHN R J AM CERAM SOC 26 339–49 1943 CA 37 6831
8097	**DETERMINATION OF THE THERMAL CONDUCTIVITY OF REFRACTORY MATERIALS.** GRIFFITHS, E. CHALLONER, A. R. TRANS BRIT CERAM SOC 40 40–53 1941 CA 35 4559
8100	**CATALYZED CRYSTALLIZATION OF GLASS IN THEORY AND PRACTICE.** STOOKEY S D IND ENG CHEM 51 805–808 1959
8103	**A BASIC INSULATING REFRACTORY.** HARRIS, H. M. KELLY, H. J. AM CERAM SOC BULL 37 7 307–311 1958
8112	**CRITICAL TEMPERATURES IN SILICATE GLASSES.** LITTLETON J T IND ENG CHEM 25 748–767 1933
8116	**NATURE OF RADIATION DAMAGE IN CRYSTALLINE SOLIDS.** CRAWFORD J H JR AM CERAM SOC BULL 36 3 95–8 1957
8133	**THERMAL CONDUCTIVITY AND DIELECTRIC STRENGTH OF PERICLASE INSULATION.** KARPINSKI J M HASSELMAN D P H TERVO R FETTERLEY G H AM CERAM SOC BULL 37 7 329–333 1958
8136	**EFFECTS OF HIGH–CONDUCTIVITY GASES ON THE THERMAL CONDUCTIVITY OF INSULATING REFRACTORY CONCRETE.** WYGANT J F CROWLEY M S J AM CERAM SOC 41 5 183–8 1958
8150	**THE MEASUREMENT OF THE THERMAL CONDUCTIVITIES OF ROCKS BY OBSERVATIONS IN BOREHOLES.** BECK A JAEGER J C NEWSTEAD G AUSTR J PHYS 9 2 286–296 1956 AM 10 549
8182	**NOTE ON CALCULATION OF EFFECT OF TEMPERATURE AND COMPOSITION ON SPECIFIC HEAT OF GLASS.** MOORE JOHN SHARP D E AMERICAN CERAMIC SOCIETY JOURNAL 41 461–463 1958 BR 8 1281

TPRC Number	Bibliographic Citation
8195	PROPERTIES OF URANIA-CERIA BODIES FOR NUCLEAR FUEL APPLICATION. PLOETZ G L MUCCIGROSSO A T KRYSTYNIAK C W KNOLLS ATOMIC POWER LAB., SCHENECTADY, N. Y. 1-24, 1958. (KAPL-1918)
8230	THERMAL PROPERTIES OF SOILS. KERSTEN, M. S. UNIV MINN INST TECHNOL ENG EXP STA BUL 28 1-227 1949 RA 5 127-3
8233	INSULATION GLOSSARY. AUTHOR ANON. WORLD REFRIG. 375-80, 389, 1953.
8236	SPECIFIC HEAT TESTS ON SOILS. KERSTEN, M. S. PROC 2ND INT CONF SOIL MECH AND FOUNDN ENG 3 158-62 1948 RA 5 128
8237	THERMAL CONDUCTIVITY OF SOIL. KERSTEN, M. S. PROC 2ND INT CONF SOIL MECH AND FOUNDN ENG 3 12-5 1948 RA 5 128
8241	PROPERTIES OF REFRACTORY MATERIALS NOT INCLUDING FIRECLAYS. PHELPS S M AMER REFRACT INST TECH BULL 57 1-5 1935 MA 2 722
8244	THE COEFFICIENTS OF THERMAL CONDUCTIVITY OF INSULATING MATERIALS AND REFRACTORIES. AUBERT ALBERT CHALEUR ET IND 18 202 87-90 1937 MA 5 353
8245	ON THERMAL CONDUCTIVITY AT LOW TEMPERATURES. DE HAAS, W. J. BIERMASZ, TH. CONGR. INT. FROID COMM. INT. RAPPORTS ET COMMUNIC. 241, 1-13, 1937.
8282	THERMODYNAMIC PROPERTIES OF COMBUSTION PRODUCTS. SINKE, G. C. WRIGHT AIR DEVELOPMENT CENTER 153PP., 1959. (AD 214587)
8295	III. EINIGE VERSUCHE UBER DIE WARMELEITUNG. CHRISTIANSEN C ANN PHYSIK 14 3 23-33 1881
8327	NOTE ON THE CONDUCTION OF HEAT IN CRYSTALS. CASIMIR H B G COMMUN KAMERLINGH ONNES LAB UNIV LEIDEN SUPPL 23 B 85 1-6 1938
8408	THE EMISSIVITY OF TRANSPARENT MATERIALS. GARDON R J AMER CERAM SOC 39 8 278-87 1956 AM 10 1245
8476	A PROBE FOR MEASUREMENT OF THERMAL CONDUCTIVITY OF FROZEN SOILS IN PLACE. LACHENBRUCH A H TRANS AMER GEOPHYS UN 38 5 691-697 1957 AM 11 1363
8506	THE ADIABATIC CALORIMETER. AN INSTRUMENT FOR THE SIMULTANEOUS DETERMINATION OF HEAT CAPACITY AND THERMAL CONDUCTIVITY. SINELNIKOV N N FILIPOVITCH V N ZHUR TEKHN FIZIKI 28 1 218-221 1958 MA 26 98 (FOR ENGLISH TRANSLATION SEE T9254)
8514	TRANSIENT HEAT FLOW DETERMINATION OF THE THERMAL PROPERTIES OF CLAY. PENROD E B ASME ANN MEET NEW YORK N Y 1-14 1956 AM 10 3434 (ASME ANNUAL MEETING PAP. 56-A-91)
8791	THE PRODUCTIONG OF DINASO-CHROMITE. KAINARSKII I S PINDRIK B E BOVKUN S S SIDORENKO YU P CHUDNOVSKII A M OGNEUPORY 22 529-33 1957 CA 52 7642
8797	THE THERMAL AND ELECTRICAL PROPERTIES OF BITUMASTIC COMPOUNDS CONTAINING QUARTZ SAND. JACKSON WILLIS PHIL MAG 33 81-9 1942 CA 36 3582
8818	THERMAL CONDUCTIVITY OF SAPPHIRE AND RUTILE AS A FUNCTION OF TEMPERATURE. MC CARTHY, K. A. BALLARD, S. S. DOERNER, E. C. PHYS REV 88 153 1952 CA 48 7943
8831	NOTE ON THE MEASUREMENT OF GLASS ABSORPTION. SMITH T PROC PHYS SOC /LONDON/ 58 472-5 1946 CA 40 6989
8860	THE HEAT CAPACITY OF BLAST-FURNACE SLAGS AT HIGH TEMPERATURES. VOSKOBOINIKOV V G TEORIYA PRAKT MET 12 10 3-5 1940 CA 35 5428
8876	THERMAL PARAMETERS OF CERAMIC MATERIALS. IOFFE YU M TRUDY VSESOYUZ NAUCH ISSLEDOVATEL INST STROITEL KERAM 8 68-98 1953 CA 49 2041
8877	DETERMINATION OF THE SPECIFIC GRAVITY AND HEAT CONDUCTIVITY OF FLOTATION PYRITE CINDERS. LYAPUSTINA E M TARASOVA V P TRUDY URAL NAUCH ISSLEDOVATEL KHIM INST 1 225-9 1954 CA 52 4431
8882	CARBONIZING PROPERTIES OF EAGLE-BED COAL FROM PROSPECT SHAFT, CARBON, KANAWHA COUNTY, W. VA. REYNOLDS D A DAVIS J D ODE W H BREWER R E BIRGE G W US BUR MINES TECH PAPER 691 1-43 1946 CA 40 7562 (U.S. BUR. MINES-TP-691)
8886	MANUFACTURE OF LIGHTWEIGHT REFRACTORIES BY A CHEMICAL METHOD. GLEBOV, S. V. MELNIKOV, F. I. VSESOYUZ. GOSUDARST INST. NAUCH-ISSLEDOVATEL I PROEKT. RABOT OGNEUPOR PROM., INST. OGNEUPOR., LEGKOV OGNEUPORY 40-69 1945 CA 43 8631
8911	HEAT TRANSFER IN REFRACTORY INSULATING MATERIALS. III. THERMAL CONDUCTIVITY AND PERMEABILITY AS RELATED DIRECTIONAL PROPERTIES. BARRETT L R VYSE J GREEN A T TRANS BRIT CERAM SOC 49 122-8 1950 CA 44 7035
8940	PROPERTIES OF SOME EXPANDED PLASTICS AND OTHER LOW-DENSITY MATERIALS. AXILROD BENJ M KOENIG EVELYN NATIONAL ADVISORY COMMITTEE OF AERONAUTICS 1-26 1945 CA 40 7697 (NACA-TN 991)
8941	LOW-TEMPERATURE HEAT CAPACITIES AND ENTROPIES AT 298.15 K. OF SOME CRYSTALLINE SILICATES CONTAINING CALCIUM. KING E G J AM CHEM SOC 79 5437-8 1957 CA 52 5955
8974	UNDERGROUND COAL GASIFICATION. EFFECT OF PRELIMINARY HEATING ON THE HEAT CONDUCTIVITY OF MOSCOW COAL AND OIL-SHALE LUMPS. CHANGES IN THE EFFECTIVE HEAT CONDUCTIVITY OF LUMPS OF MOSCOW COAL AND OIL SHALE DURING HEATING. FARBEROV I L AVDONINA E S YUREVSKAYA N P TRUDY INST GORYUCHIKH ISKOPAEMYKH AKAD NAUK S S S R 7 94-102 1957 CA 52 9561
8991	METHODS OF CALCULATING THERMAL CONDUCTION COEFFICIENTS OF CAPILLARY-POROUS MATERIALS. BABANOV A A SOVIET PHYS TECH PHYS 2 476-84 1957 CA 52 9741 (ENGLISH TRANSLATION OF ZH. TEKH. FIZ., 27, (3), 532-42, 1957; FOR ORIGINAL SEE T25257)
9027	NOTE ON THE COMPARISON OF INSULATING BRICK FROM THE STANDPOINT OF THEIR USEFULNESS. FRANCOIS J CHALEUR + IND 36 357-64 1955 JA 40 40
9032	REFRACTORY MATERIALS FOR USE IN HIGH-TEMPERATURE AREAS OF AIRCRAFT. PART 2. OXIDE-BASE AND CARBIDE-BASE MATERIALS. THIELKE, N. R. WRIGHT AIR DEVELOPMENT CENTER 1-50, 1954. (WADC TR 53-9, AD 48907)
9045	THE SPECIFIC HEAT OF SEIGNETTOELECTRIC //FERROELECTRIC// CERAMICS. DANKOVA J CZECH J PHYS 6 4 407-8 1956 SA 60 5285
9046	THE DIRECTIONAL DISTRIBUTION AT REFLECTION OF HEAT RADIATION AND ITS INFLUENCE ON HEAT TRANSMISSION. MUNCH B UNIVERSITY OF ZURICH, PH. D. THESIS 1-89, 1955. (FOR ENGLISH TRANSLATION SEE T49251)

TPRC Number	Bibliographic Citation
9102	PROPERTY MEASUREMENTS AT HIGH TEMPERATURES. FACTORS AFFECTING AND METHODS OF MEASURING MATERIAL PROPERTIES AT TEMPERATURES ABOVE 1400 C /2550 F/. KINGERY W D JOHN WILEY AND SONS INC 1-416 1959
9140	BEHAVIOR OF RAFRACTORY MATERIALS UNDER THERMAL STRESS. MASSIMILLA L BRACALE S METALLURGIA ITAL 47 511-18 1955 JA 40 39
9171	RUSSIAN OPEN-HEARTH FURNACES. SEWELL B IRON AND STEEL /LONDON/ 29 427-28 1956 JA 40 11
9207	PHASE TRANSITIONS AND CRITICAL PHENOMENA IN ANISOTROPIC PHASES. SEMENCHENKO V K KRISTALLOGRAFIYA 2 1 145-52 1957 SA 60 9657
9232	AN EVALUATION OF TWO RAPID METHODS OF ASSESSING THE THERMAL RESISTIVITY OF SOIL. MAKOWSKI M W MOCHLINSKI K PROC. INSTN. ELECT. ENGRS. 103 A 453-70 1955 SA 60 918
9238	THE APPLICATION OF A TRANSIENT METHOD OF THE MEASUREMENT OF THE THERMAL CONDUCTIVITY OF ROCKS AND BUILDING MATERIALS. GAFNER G BRIT J APPL PHYS 8 10 393-397 1957 AM 11 1781
9247	EXPERIMENTAL STUDY OF THE THERMAL CONDUCTIVITY OF SOME EGYPTIAN FIRE-CLAY INSULATING BRICK. SHERIFF I I BAKR M Y SILICATES INDS 21 6 287-89 1956 JA 40 205
9254	ADIABATIC CALORIMETER—AN APPARATUS FOR SIMULTANEOUS DETERMINATION OF HEAT CAPACITY AND THERMAL CONDUCTIVITY. SINELHIKOV N N FILIPOVICH V N SOVIET PHYS -TECH PHYS 3 1 193-196 1958 AM 12 2029 (ENGLISH TRANSLATION OF ZH. TEKHN. FIZ. 28 (1), 218-21, 1958; FOR ORIGINAL SEE T8506)
9266	ON THE CYLINDRICAL PROBE METHOD OF MEASURING THERMAL CONDUCTIVITY WITH SPECIAL REFERENCE TO SOILS. II. ANALYSIS OF MOISTURE EFFECTS. DE VRIES D A PECK A J AUSTRAL J PHYS 11 3 409-423 1958 AM 12 4094
9285	X. ON THE THERMAL CONDUCTIVITIES OF SINGLE AND MIXED SOLIDS AND LIQUIDS AND THEIR VARIATION WITH TEMPERATURE. LEES, C. H. PHIL TRANS ROY SOC LONDON 191 A 399-440 1898
9293	THE THERMAL CONDUCTIVITY OF QUARTZ AT LOW TEMPERATURES. DE HAAS W J BIERMASZ TH PHYSICA 2 3 673-682 1936
9294	THE DEPENDENCE ON THICKNESS OF THE THERMAL RESISTANCE OF CRYSTALS AT LOW TEMPERATURES. DE HAAS W J BIERMASZ TH PHYSICA 5 7 619-624 1938
9350	THE PHYSICAL PROPERTIES OF A SERIES OF STEELS. PART II. AWBERY J H CHALLONER A R PALLISTER P R POWELL R W J IRON STEEL INST /LONDON/ 154 2 83-111 1946
9353	THE TOTAL EMISSIVITY OF VARIOUS MATERIALS AT 100-500 C. BARNES B T FORSYTHE W E ADAMS E Q J OPT SOC AM 37 10 804-7 1947
9354	TOTAL EMISSIVITY OF VARIOUS MATERIALS. BARNES B T ADAMS E Q FORSYTHE W E J OPT SOC AM 30 269 1940
9357	ON THE THERMAL CONDUCTIVITY OF DIELECTRICS AT TEMPERATURES LOWER THAN THAT OF DEBYE. POMERANSHUK I J PHYS U S S R 6 237-50 1942
9389	XXIII. ON THE THERMAL CONDUCTIVITES OF MIXTURES AND OF THEIR CONSTITUENTS. LEES, C. H. PHIL MAG 49 5 286-293 1900
9395	EFFECT OF THERMAL MOTION ON THE X-RAY REFLECTIVITY OF QUARTZ. BERREMAN D W CHANG TETSE J APPL PHYS 30 7 963-6 1959
9398	THE THERMAL CONDUCTIVITY OF SOME TECHNICAL MATERIALS AT LOW TEMPERATURES. BERMAN R FOSTER E L ROSENBERG H M BRIT J APPL PHYS 6 5 181-2 1955
9404	DEVELOPMENT OF CERAMIC BODIES WITH HIGH THERMAL CONDUCTIVITY. KOENIG, J. H. N. J. CERAM. RES. STA., RUTGERS UNIV., SIGNAL CORPS 1-117, 1953. (AD 13154)
9410	CERAMIC BODIES WITH HIGH THERMAL CONDUCTIVITY. KOENIG, J. H. N. J. CERAM. RES. STA., RUTGERS, UNIV., SIGNAL CORPS 1-98, 1952. (ATI 163 519)
9411	LIQUID HYDROGEN PROPELLANT FOR AIRCRAFT AND ROCKETS JOHNSTON, H. L. RESEARCH FOUNDATION, OHIO STATE UNIV., USAF 1-15, 1949. (ATI 58 926)
9433	SOME PHYSICAL PROPERTIES OF DIAMONDS. CHAMPION F C ADVANCES IN PHYS 5 383-411 1956 SA 60 7502
9439	THERMAL CHARACTERISTICS OF INSULATING REFRACTORIES. HALM LOUISE LAPOUJADE PAULETTE BULL SOC FRANC CERAM 29 3-19 1955 JA 40 163
9440	SPECIFIC HEAT OF CHAIN STRUCTURES AT LOW TEMPERATURES. SOCHAVA I V TRAPEZNIKOVA O N DOKL AKAD NAUK S S S R 113 4 784-6 1957 SA 60 8571 (FOR ENGLISH TRANSLATION SEE T17360)
9450	COMPENSATED OPAL WORKING STANDARD EQUIVALENT TO MAGNESIUM OXIDE. BILLMEYER FRED W JR J OPT SOC AM 46 1 72 1956 JA 40 251
9471	SANDWICH METHOD IN CONSTRUCTION OF AIRPLANES AND IN OTHER INDUSTRIES. PT. 2. NOTON B R ALUMINIUM 34 522-529 1958 MR 31 630-P
9500	MEASUREMENTS OF THE THERMAL CONDUCTIVITY OF A SUSPENDED LAYER. BONDAREVA A K DOKLADY AKAD NAUK S S S R 115 768-70 1957 CA 52 12467
9560	THERMAL RADIATION CHARACTERISTICS OF SOME GLASSES. MC MAHON, H. O. J AM CERAM SOC 34 91-6 1951 CA 45 4899
9693	THE THERMAL CONDUCTIVITY OF COAL AND COKE. FRITZ W DIEMKE H FEUERUNGSTECH 27 129-36 1939 CA 33 7076
9699	PEAT AS AN INSULATING MATERIAL. WOWK JOZEF PRZEGLAD CHEM 3 569-71 1939 CA 35 4871
9728	SPECIFIC HEAT /AT CONSTANT PRESSURE/ OF ANTIFERROMAGNETIC MIXED CRYSTALS MNO-COO. HOLZL JOSEF Z PHYSIK 151 220-32 1958 CA 52 14312
9738	A TRANSIENT METHOD OF MEASURING THERMAL PROPERTIES OF INSULATING MATERIALS. KAYAN C F GLASER P E CONGR INTERN FROID 9E PARIS 1955 COMPT REND TRAV COMM 1 2051-62 1955 CA 52 15017
9739	THE MEASUREMENTS OF THE THERMAL CONDUCTIVITY OF HEAT INSULATORS AT LOW TEMPERATURES. HICKMAN M J RATCLIFFE E H CONGR INTERN FROID 9E PARIS 1955 COMPT REND TRAV COMM 1 2149-52 1955 CA 52 15017

TPRC Number	Bibliographic Citation

9767 **ADAPTATION OF NATIVE DIATOMITE FOR THE PRODUCTION OF CERAMIC THERMAL INSULATION MATERIALS.**
KARPACZ JERZY
PRACE INST PRZEMYSL SZKLA I CERAMIKI
3 27–39 1956 CA 52 15864

9789 **MILD OXIDATION OF MIIKE COAL WITH AIR. I. CHANGE OF PROPERTIES IN OXIDATION PROCESS.**
NAKAGOME, T. YOSHIDA, S. MATSUURA, T.
SUGIMURA, H.
NENRYO KYOKAISHI
37 361–7 1958 CA 52 15875

9820 **THE HEAT CAPACITY OF DIAMOND BETWEEN 12.8 AND 277 K.**
DESNOYERS J E MORRISON J A
PHIL MAG
3 8 42–8 1958 CA 52 16854

9825 **AN EXPERIMENTAL DETERMINATION OF THE THERMAL PROPERTIES OF CLAY.**
PENROD, E. B. PRIVON, G. T.
PRASANNA, K. V.
KENTUCKY UNIV ENG EXPT STA BULL
10 3 1–76 1956 CA 52 17579

9878 **THERMAL RESISTIVITY OF POINT DEFECTS.**
TOXEN, A. M.
PHYS REV
110 585–6 1958

9894 **DETERMINATION OF THE DEGREE OF MOISTURE IN BUILDING CONSTRUCTION.**
VOS B H ERKELENS H J
ALLGEM WARMETECH
8 10 211–217 1958

9922 **PROPERTIES OF METALS AT LOW TEMPERATURES.**
MACDONALD D K C
PROGR IN METAL PHYS
3 42–75 1952

9943 **THERMAL CONDUCTIVITY OF DIELECTRIC SOLIDS.**
ARNOLD RUDOLPH J DARLING LEROY A
SPERRY ENG REV
9 24–31 1956

9970 **VISCOSITY DATA FOR SILICATE MELTS.**
BOWEN, N. L.
TRANS. AM. GEOPHYS. UNION
1934.

9974 **THERMAL CONDUCTIVITIES OF GASES, METALS, AND LIQUID METALS.**
PRZYBYCIEN, W. M. LINDE, D. W.
U.S. ATOMIC ENERGY COMMISSION, KNOLLS ATOMIC POWER LAB., SCHENECTADY, N. Y.
1–24 1957
(KAPL–M–WMP–1)

10013 **APPARATUS FOR MEASURING THERMAL CONDUCTIVITY OF CERAMIC AND METALLIC MATERIALS TO 1200 C.**
SUTTON W H
J AM CERAM SOC
43 2 81–6 1960 JA 43 68

10036 **A NEW METHOD OF DETERMINING THE THERMAL CONDUCTIVITY OF SOLIDS.**
ANGSTROM A J
ANN PHYSIK
114 2 513–30 1861
(FOR ENGLISH TRANSLATION SEE T7575)

10045 **THE MEASUREMENT OF THERMAL CONDUCTIVITY OF REFRACTORY MATERIALS.**
KINGERY, W. D. NORTON, F. H.
MASSACHUSETTS INSTITUTE OF TECHNOLOGY, CAMBRIDGE
16PP., 1955.
(NYO–6451, AD 80699)

10052 **THERMAL CONDUCTIVITY. XIV. CONDUCTIVITY OF MULTI-COMPONENT SYSTEMS.**
KINGERY, W. D.
J. AM. CERAM. SOC.
42 (12), 617–27, 1959.
(43 (12), 669, 1960)

10060 **DETERMINATION OF EMISSIVITY AND REFLECTIVITY DATA ON AIRCRAFT STRUCTURAL MATERIALS. PT. 3 TECHNIQUES FOR MEASUREMENT**
OLSON, O. H. MORRIS, J. C.
WRIGHT AIR DEVELOPMENT CENTER
96PP., 1959.
(WADC TR–56–222 (PT 3), AD 239302)

10076 **THE STATE OF A SOLID BODY.**
GRUNEISEN, E.
NATIONAL AERONAUTICS AND SPACE ADMINISTRATION
1–76 1959
(ENGLISH TRANSLATION OF HANDBOOK DER PHYS. 10, 1–52, 1926; FOR ORIGINAL T26036)

10083 **THE MEASUREMENT OF THERMAL CONDUCTIVITY OF REFRACTORY MATERIALS.**
NORTON, F. H. KINGERY, W. D.
MC QUARRIE, M. C. ADAMS, M. LOEB, A. L.
FRANCL, J.
MASSACHUSETTS INSTITUTE OF TECHNOLOGY, CAMBRIDGE
124PP., 1953.
(NYO–3643, AD–13940)

10097 **THE DEVELOPMENT AND OPERATION OF AN INSTRUMENT FOR THE DIRECT MEASUREMENT OF HEAT FLUX IN THE SOIL.**
STALEY, R. C. MORSE, B. E. GERHARDT, J. R.
BUETTNER, K.
UNIVERSITY OF TEXAS, AUSTIN, TEXAS
1954.
(AFCPC TN 54–280, AD 50 795)

10100 **THE MEASUREMENT OF THERMAL CONDUCTIVITY OF REFRACTORY MATERIALS.**
KINGERY, W. D. NORTON, F. H.
MASSACHUSETTS INSTITUTE OF TECHNOLOGY, CAMBRIDGE
6PP., 1954.
(NYO–6446, AD 53808)

10101 **THE MEASUREMENT OF THERMAL CONDUCTIVITY OF REFRACTORY MATERIALS.**
KINGERY, W. D. NORTON, F. H.
MASSACHUSETTS INSTITUTE OF TECHNOLOGY, CAMBRIDGE
16PP., 1955.
(NYC–6447, AD 55595)

10110 **THE MEASUREMENT OF THERMAL CONDUCTIVITY OF REFRACTORY MATERIALS.**
KINGERY, W. D. NORTON, F. H.
MASSACHUSETTS INSTITUTE OF TECHNOLOGY, CAMBRIDGE
16PP., 1955.
(NYO–6450, AD 73173)

10123 **TRANSPORT PROPERTIES OF LIQUIDS. FROM PROJECT SQUID SEMI-ANNUAL PROGRESS REPORT, APRIL 1, 1953.**
JOHNSON, E. F. SHEFFY, W. J. PARISOT, P. E.
PRINCETON UNIVERSITY, NEW JERSEY
19–21, 1953.
(AD 10591)

10127 **DEVELOPMENT OF CERAMIC BODIES WITH HIGH THERMAL CONDUCTIVITY.**
SMOKE, E. J. ILLYN, A. V. EICHBAUM, B. R.
SNYDER, N. H. LASS, G. NUSSBAUM, T.
NEW JERSEY CERAMIC RES. STA., RUTGERS UNIVERSITY, NEW BRUNSWICK
62PP., 1955.
(AD 74093)

10164 **THE MEASUREMENT OF THERMAL CONDUCTIVITY OF REFRACTORY MATERIALS.**
NORTON, F. H. KINGERY, W. D.
MASSACHUSETTS INSTITUTE OF TECHNOLOGY, CAMBRIDGE
8PP., 1953.
(NYO 3646, AD 16492)

10184 **VACUUM-POWDER INSULATION.**
REYNOLDS M M BROWN J D FULK M M PARK O E
CURTIS, G. W.
PROC. CRYOGENIC ENGR. CONF.
142–50, 1955.
(NBS RPT 3517, AD 125047)

10187 **THERMAL CONDUCTIVITY OF SOLIDS AT LOW TEMPERATURES.**
POWELL, R. L. COFFIN, D. O.
PROC. CRYOGENIC ENGR. CONF.
189–93, 1955.
(NBS REPT 3517, AD 125047)

10192 **THE MEASUREMENT OF THERMAL CONDUCTIVITY OF REFRACTORY MATERIALS.**
NORTON, F. H. KINGERY, W. D.
MASSACHUSETTS INSTITUTE OF TECHNOLOGY, CAMBRIDGE
7PP., 1954.
(NYO–6441, AD 27184)

10206 **PROCEEDINGS OF THE 1954 CRYOGENIC ENGINEERING CONFERENCE.**
HANSON, W. B.
NATIONAL BUREAU OF STANDARDS
1–278, 1955.
(NBS RPT 3517, AD 125047)

10242 **THE RADIATION OF INCANDESCENT OXIDES AND OXIDE MIXTURES IN THE VISIBLE SPECTRUM.**
HOPPE H
ANN PHYSIK
15 709–28 1932

10271 **THERMAL CONDUCTIVITY OF SOME CERAMIC HEAT-INSULATING MATERIALS.**
BISHUI B M PRASAD J
CENTRAL GLASS AND CERAM RESEARCH INST BULL /INDIA/
4 144–9 1957 CA 52 7644

10272 **THERMAL ENDURANCE OF SOME VITRIFIED INDUSTRIAL COMPOSITIONS.**
SMOKE E J
CERAM AGE
54 148–9 1949 CA 43 9403

TPRC Number | *Bibliographic Citation*

10273 **IMPROVING THERMAL SHOCK RESISTANCE OF CHEMICAL STONE-WARE BODIES.**
KNIZEK JAN OTTO
CERAM IND
37 74 76 1941 CA 36 875

10282 **SPECIFIC HEATS OF BARYTA, WITHERITE, AND FUSED LIME.**
LATSCHENKO P N
COMPT REND
147 58-61 1908

10288 **CALCULATION OF THE ACTIVATION ENERGY AND HEAT OF FUSION OF FELDSPARS AND PYROXENES FROM VISCOSITY MEASUREMENTS.**
VOLAROVICH M P
COMPT REND ACAD SCI URSS
24 938-41 1939 CA 34 3976

10291 **FACTORS INFLUENCING THE PROCESS OF SLAG WOOL PRODUCTION.**
ZHILIN A I
COMPT REND ACAD SCI URSS
53 339-41 1946 CA 41 2823

10306 **CHEMICAL BENEFICIATION OF WHITE CLAYS.**
CANNING R G
DEPT MINES S AUSTRALIA MINING REV
104 41-4 1957 CA 52 13209

10317 **ON THE CERAMIC PROPERTIES OF CALCIUM-ALUMINIUM SILICATES AND OTHER REFRACTORIES.**
BERL, E. LOBLEIN, F.
FORSCHUNGSARBEITEN
325 1-28 1930 MA 1 35

10321 **THERMAL CONDUCTIVITY OF SOME REFRACTORY MATERIALS AFTER USE.**
CLEMENTS J F VYSE J
GAS COUNCIL RESEARCH COMMUN
GC6 41-50 1952 JA 40 63

10331 **THERMAL PROPERTIES OF VARIOUS PIGMENTS AND OF RUBBER.**
WILLIAMS IRA
IND ENG CHEM
15 154-7 1923

10348 **THE HEAT CAPACITIES AT LOW TEMPERATURES OF THE ALKALINE EARTH CARBONATES.**
ANDERSON, C. T.
J AM CHEM SOC
56 340-2 1934 CA 28 1596

10365 **SPECIFIC HEATS OF SOME LANTHANUM AND SCANDIUM SALTS AND OF MONAZITE.**
TURSKA E
ROCZNIKI CHEM
14 760-3 1934 CA 29 6131

10394 **CALCULATIONS OF CHEMICAL EQUILIBRIUM. IV. DETERMINATION OF SPECIFIC HEAT.**
DHONT M JUNGERS J C
MEDEDEL VLAAM CHEM VER
10 9 193-206 1948 CA 43 3275

10454 **SPECIFIC HEATS OF BARYTES, WITHERITE, FUSED LIME, QUARTZ AND CHALCEDONY AT HIGH TEMPERATURES.**
LASCHENKO N P
ZHUR RUSS FIZ KHIM OBSHCHESTVA
42 1604-14 1910

10458 **CERAMICS AND GLASS AND THEIR APPLICATION IN MODERN AERONAUTICS.**
ZAGAR L
NATIONAL AERONAUTICS AND SPACE ADMINISTRATION
1-15, 1958.
(AGARD RPT 179, AD 21821,)

10481 **THE MEASUREMENT OF THERMAL CONDUCTIVITY OF REFRACTORY MATERIALS.**
NORTON, F. H. KINGFRY, W. D.
MASSACHUSETTS INSTITUTE OF TECHNOLOGY, CAMBRIDGE
1-15, 1954.
(NYO 6442, AD 40872)

10482 **THE MEASUREMENT OF THERMAL CONDUCTIVITY OF REFRACTORY MATERIALS.**
KINGERY, W. D. NORTON, F. H.
MASSACHUSETTS INSTITUTE OF TECHNOLOGY, CAMBRIDGE
1-16, 1954.
(NYO-6444, AD 40873)

10483 **THE MEASUREMENT OF THERMAL CONDUCTIVITY OF REFRACTORY MATERIALS.**
NORTON, F. H. KINGERY, W. D.
MASSACHUSETTS INSTITUTE OF TECHNOLOGY, CAMBRIDGE
1-8, 1954.
(NYO-6445, AD 40874)

10518 **DEVELOPMENT OF CERAMIC BODIES WITH HIGH THERMAL CONDUCTIVITY.**
KOENIG, J. H. SMOKE, E. J. ILLYN, A. V.
EICHBAUM, B. R. SNYDER, N. H.
KOECHLEIN, P. J. LASS, G. RODKIN, D. L.
SCHATZ, W. J. STROMENGER, D. R.
N. J. CERAM. RES. STA., RUTGERS, UNIVERSITY
1-64, 1956.
(AD 104826)

10561 **LATTICE THERMAL CONDUCTIVITY AT LOW TEMPERATURES.**
AGRAWAL, B. K. VERMA, G. S.
PHYS REV /USA/
126 1 24-9 1962 SA 65 10341

10581 **THERMAL CONDUCTIVITY IN ELEMENTAL SEMICONDUCTORS AND SOLID ARGON.**
WHITE G K WOODS S B MACDONALD D K C
BULL INST INTERN FROID ANNEXE
2 91-5 1956 CA 52 886

10583 **PHYSICAL PROPERTIES OF HIGH TEMPERATURE MATERIALS. PT. 1. NEW APPARATUS FOR THE PRECISE MEASUREMENT OF HEAT CONTENT AND HEAT CAPACITY FROM 0 TO 1500 C.**
DOUGLAS, T. B. PAYNE, W. H.
WRIGHT AIR DEVELOPMENT CENTER
1-35, 1957.
(WADC TR 57-374(PT 1), AD 142119)

10586 **LOW TEMPERATURE HEAT CAPACITY AND THERMODYNAMIC PROPERTIES OF ZINC FERRITE. FROM MEASUREMENT AND INTERPRETATION OF THERMAL AND MAGNETIC DATA ON NICKEL - ZINC FERRITES.**
WESTRUM, E. F., JR. GRIMES, D. M.
MICHIGAN UNIV., ANN ARBOR, ENGINEERING RES. INST.
18-35, 1957.
(AD 135955)

10587 **MEASUREMENT AND INTERPRETATION OF THERMAL AND MAGNETIC DATA ON NICKEL-ZINC FERRITES.**
GRIMES, D. M. WESTRUM, E. F., JR.
LEGVOLD, S.
MICHIGAN UNIV., ANN ARBOR, ENGINEERING RES. INST.
1-37, 1957.
(AD 135955, PB 140340)

10594 **HEAT SOURCES AND SINKS AT THE EARTHS SURFACE.**
HOUGHTON, D. M.
GREAT BRITAIN AIR MINISTRY METEOROLOGICAL RES. COMM.
1-13, 1956.
(MRP 1005, AD 141023)

10667 **GROUND TEMPERATURE. VOLUME I.**
CHANG, J. H.
HARVARD UNIV., BLUE HILL METEOROLOGICAL OBSERVATORY, MILTON, MA.
299PP., 1958.
(AD 202752)

10669 **I. THE ATOMIC HEAT OF DIAMOND FROM 11 TO 200 K. II. CALORIMETER RESPONSE. III. PRESSURE REGULATORS.**
BURK, D. L. FRIEDBERG, S. A.
CARNEGIE INST. OF TECH., PITTSBURGH
48PP., 1958.
(AD 203311)

10681 **PROPERTIES OF REFRACTORY MATERIALS. COLLECTED DATA AND REFERENCES.**
BRADSHAW, W. G. MATTHEWS, C. O.
LOCKHEED AIRCRAFT CORP., PALO ALTO, CA.
108PP., 1958.
(LMSD-2466, AD 205452)

10682 **METAL-CERAMIC LAMINATES.**
FRANCIS R K BROWN R MCNAMARA E P
TINKLEPAUGH J R
N. Y. STATE COLLEGE OF CERAM., ALFRED UNIV.
69PP., 1958.
(WADC TR 58-600, AD 205549)

10703 **SPECIAL REPORT ON DESIGN DATA FOR IR DOME MATERIALS.**
KRAUSHAAR, R.
ACF INDUSTRIES, INC., AVION DIV., PARAMUS, N. J.
29PP., 1958.
(AVION RPT 1070, AD 208300)

10714 **GRAIN SIZE EFFECTS ON THE THERMAL CONDUCTIVITY OF CERAMIC OXIDES.**
FRANCIS R K TRUESDALE R S TINKLEPAUGH J R
N. Y. COLLEGE OF CERAMICS, ALFRED UNIVERSITY
13PP., 1959.
(AD 211888)

10721 **SHEAR RATE DEPENDENCE OF THE VISCOSITY OF WHOLE BLOOD AND PLASMA.**
WELLS, R. E., JR. MERRILL, E. W.
SCIENCE
133 763-4 1961

10723 **SOME THERMAL PROPERTIES OF DURESTOS TYPE MATERIALS.**
BISHOP, P. H. H. ROGERS, K. F.
GREAT BRITAIN ROYAL AIRCRAFT ESTABLISHMENT, FARNBOROUGH HANTS ENGLAND
23PP., 1958.
(RAE TECH NOTE CHEM 1333, AD 212601)

TPRC Number	Bibliographic Citation
10757	**GRAIN SIZE EFFECTS ON THE THERMAL CONDUCTIVITY OF CERAMIC OXIDES. PROGR REPT NO. 10.** TRUESDALE, R. S. SWICA, J. J. WRIGHT AIR DEVELOPMENT CENTER 13PP., 1959. (AD 225763)
10759	**CRITERIA FOR THE COMPARISON OF THE NECESSARY WEIGHT OF STRUCTURAL MATERIALS, HEAT SINK MATERIALS AND THERMAL INSULATIONS FOR USE IN MISSILES.** PERRY, H. A. U.S. NAVAL BUREAU OF ORDNANCE 10PP., 1959. (NAVORD RPT 5754, AD 226266)
10831	**METHODS OF PURIFICATION OF METALS AND INTERMETALLIC COMPOUNDS.** SUSMAN, S. ARMOUR RES. FOUNDATION 1-77, 1959. (WADC TR 59-303, AD 231363)
10851	**THE MEASUREMENT OF THERMAL CONDUCTIVITY OF REFRACTORY MATERIALS.** NORTON, F. H. KINGERY, W. D. COLUMBIA UNIVERSITY 1-52, 1952. (NYO-601)
10853	**SOME CALORIMETRIC STUDIES OF THE METALS AND CHLORIDES OF CERIUM AND NEODYMIUM.** SPEDDING, F. H. MILLER, C. F. IOWA STATE UNIVERSITY, AMES 1-131, 1951. (ISC-167)
10855	**THE MEASUREMENT OF THERMAL PROPERTIES OF NONMETALLIC MATERIALS AT HIGH TEMPERATURES.** FIMCH, E. B. BEATTY, K. O., JR. N. CAROLINA STATE COLLEGE, RALEIGH 1-11, 1953. (ORO-87)
10898	**THE INTERNAL INSULATION OF REACTOR PRESSURE TUBES. 3. HEAT TRANSFER STUDIES OF LAYERS OF COMPRESSED POWDERS.** CHARLESWORTH, D. H. ATOMIC ENERGY OF CANADA, LTD. 1-24, 1959. (CRCE-834, AECL-828, AD 218621)
10911	**ON THE CALCULATION OF THE THERMAL CONDUCTIVITY OF THE DISORDERED STRUCTURES. NOTE.** TAVERNIER, J. WYART, J. COMPAGNIE GENERALE DE TELEGRAPHIE SANS FIL, FRANCE 1-5, 1959. (AD 229775)
10917	**AN ANALYSIS OF METHODS AND EQUIPMENTS FOR MEASURING HEAT FLOW THROUGH THE SOIL** NICKERSON, R. J. UMUR, A. U.S. ARMY SIGNAL RESEARCH AND DEVELOPMENT LAB., FT. MONMOUTH, N. J. 1-14, 1960. (AD 237049)
10923	**THE MEASUREMENT OF THERMAL PROPERTIES OF NONMETALLIC MATERIALS AT ELEVATED TEMPERATURES.** BEATTY, K. O., JR. ARMSTRONG, A. A. N. C. STATE COLLEGE, RALEIGH 1-62, 1956. (ORO-170)
10933	**HIGH VISCOSITY REFRACTORY FIBERS.** DRUMMOND, W. W. BJORKSTEN RESEARCH LABS., INC., MADISON, WI. 1-48, 1958. (AD 210948)
10947	**THERMAL PROPERTIES OF SIX GLASSES AND TWO GRAPHITES.** LUCKS C F DEEM H W WOOD W D AM CERAM SOC BULL 39 313-9 1960 CA 54 15877
10950	**HOT WIRE METHOD FOR RAPID DETERMINATION OF THERMAL CONDUCTIVITY.** HAUPIN W E BULL AMER CERAM SOC 39 3 139-41 1960 JA 43 119
10967	**LOW TEMPERATURE STUDIES OF INFRARED TRANSPARENT MATERIALS.** MERGERIAN, D. OLSON, O. H. WEIGANDT, A. ARMOUR RES. FOUNDATION, ILLINOIS 1-8, 1960. (ARF-1159, AD 235025)
11000	**HEAT-TRANSFER COEFFICIENTS IN GLASS EXCHANGERS.** THOMPSON T J FOUST ALAN S CHEM AND MET ENG 47 410-14 1940 CA 34 5321
11001	**THERMAL INSULATION FOR INDUSTRY.** SAGINOR S V CHEM AND MET ENG 48 82-6 1941 CA 35 1894
11047	**PERFORMANCE OF HEAT INSULATING MATERIALS DOWN TO 20 K.** JOHNSTON H L HOOD C B JR BIGELEISAN J POWERS R W ZIEGLER J B PAPER FROM ADVANCES IN CRYOGENIC ENGINEERING 1 212-15 1960 RM 17 274-P
11107	**THE THERMAL CONDUCTIVITIES OF GLASS, CHILLED GLASS, QUARTZ, FUSED QUARTZ /TRANSPARENT/, BAKELITE, RUBBER, COAL, ISOLIT, PORCELAIN, SLATE AND MARBLE.** NUKIYAMA SHIRO TRANS SOC MECH ENGRS JAPAN 2 344-5 1936 CA 30 7020
11116	**PERMEABLE REFRACTORIES FOR FURNACES.** GUNN D C ENGINEERS DIGEST /AMERICAN EDITION/ 2 184 1945 RM 2 17-21
11172	**SEVERAL MECHANICAL AND THERMAL PROPERTIES OF ELECTRIC INSULATOR SUBSTANCES.** RETZOW ULRICH Z TECH PHYSIK 14 424-8 1933 CA 28 235
11204	**THE TERNARY AND QUATERNARY SYSTEMS ALKALI OXIDE-LIME-SILICA-CARBON DIOXIDE. II.** KROGER C GLASTECH BER 22 86-93 248-61 331-8 1949 CA 44 9788
11205	**VISCOSITY OF ROCK AND SILICATE MELTS.** EULER ROBERT WINKLER HELMUT G F GLASTECH BER 30 325-32 1957 CA 52 1869
11373	**VACUUM POWDER INSULATION.** REYNOLDS M M BROWN J D FULK M M PARK V E CURTIS G W ADVANCES IN CRYOGENIC ENGINEERING 1 216-23 1960 RM 17 275-P
11540	**RAW MATERIALS FOR THE PRODUCTION OF MINERAL FIBERS.** ZHILIN A I J APPLIED CHEM / U S S R/ 23 1235-42 1950 CA 47 2398
11559	**THERMAL CONDUCTION IN MOIST POROUS SUBSTANCES OF VARIOUS STRUCTURES.** KRISCHER O ESDORN H SPECIAL LIBRARY ASSOCIATION 1-30, 1966. (ENGLISH TRANSLATION OF FORSCH. GEBIETE INGENIEURW., 22 (1), 1-8, 1956; FOR ORIGINAL SEE T16078) (TT-66-13022)
11621	**THE HEAT CAPACITY OF DIAMOND FROM 70 TO 300 K.** PITZER, K. S. J CHEM PHYS 6 68-70 1938 CA 32 2418
11725	**THERMAL PROPERTIES OF HIGHLY CARBONIZED POLYMER MATERIALS.** ZAMOLUYEV V K PLASTICHESKIE MASSY 8 46-8 1960 (FOR ENGLISH TRANSLATION SEE T20660)
11777	**THE VISCOSITY AND PLASTICITY OF DISPERSE SYSTEMS. XI. COMPARISON OF THREE METHODS FOR DETERMINING THE PLASTIC-VISCOUS PROPERTIES OF PEAT.** KULAKOV N N KOLLOID-Z 80 204-12 1937 CA 31 7725 (FOR ENGLISH TRANSLATION SEE T11885)
11805	**SPECIFIC HEATS AT LOW TEMPERATURES.** SIMON F SWAIN R C Z PHYSIK CHEM 28 B 189-98 1935 CA 29 3588
11861	**THE SPECIFIC HEATS AT LOW TEMPERATURES OF BERYLLIUM OXIDE AND BERYLLIUM ORTHOSILICATE /PHENACITE/.** KELLEY K K J AM CHEM SOC 61 1217-8 1939 CA 33 5274
11885	**A COMPARISON OF THREE METHODS FOR THE DETERMINATION OF PLASTICITY-VISCOSITY PROPERTIES OF PEAT.** KULAKOV N N COLLOID J /U S S R/ 3 3 217-29 1937 CA 32 5280 (ENGLISH TRANSLATION OF KOLLOID. Z., 80, 204-12, 1937; FOR ORIGINAL SEE T11777)
11886	**METHOD OF DETERMINING VISCOSITY OF PEAT PULP AND OTHER DISPERSED SYSTEMS.** SKRYABIN A K COLLOID J /U S S R/ 3 3 209-16 1937 CA 32 5279

TPRC Number	Bibliographic Citation
11887	VISCOSITY AND PLASTICITY OF DISPERSE SYSTEMS. IV. EFFECT OF VARIOUS FACTORS ON THE PLASTIC-VISCOUS PROPERTIES OF PEAT MASSES. VOLAROVICH M P KULAKOV N N SAMARINA K I COLLOID J /U S S R/ 3 163-8 1937 CA 31 8316
11913	THERMAL CONDUCTIVITY OF DIAMOND AND POTASSIUM CHLORIDE. DE HAAS W J BIERMASZ TH PHYSICA 5 47-53 1938 CA 32 2821
12033	THE THERNAL DIFFUSIVITY OF SOME POOR CONDUCTORS. BILLINGTON N S J SCI INSTR 26 20-3 1949 RA 5 193-8
12036	AN APPARATUS FOR MEASURING THE VISCOELASTIC PROPERTIES OF COAL IN THE COURSE OF CARBONIZATION. FITZGERALD D J SCI INSTR 32 359-61 1955 CA 50 5271
12071	ATOMISTIC INTERPRETATION OF THE EFFECT OF THE COMPOSITION ON THE VISCOSITY OF GLASS. MARBOE, E. C. WEYL, W. A. J SOC GLASS TECHNOL 39 16-36T 1955 CA 49 7208
12133	LOW TEMPERATURE INSULATION. !I. GLASS FIBER INSULATION AT LIQUID AIR TEMPERATURES. PALMER, B. M. TAYLOR, R. B. CHEM ENG PROG 44 652-4 1948 RA 3 271-2
12170	UTILIZATION OF COAL ASHES. ROTTER W BRENNSTOFF-WARME-KRAFT 8 584-7 1956 CA 51 5379
12174	SOME THERMAL CHARACTERISTICS OF POROUS ROCKS. SOMERTON W H J PETROL TECHNOL 10 5 61-4 1958 AM 12 1598
12210	THE EVALUATION OF THE SPECIFIC HEAT OF ROCK SALT BY THE NEW CRYSTAL DYNAMICS. DAYAL, B. PROC INDIAN ACAD SCI 19 A 182-7 1944 CA 39 852
12328	SPECTROPHOTOMETRIC METHOD OF INVESTIGATING DISPERSION AND ABSORPTION BY SOLIDS. LISITSA M P SOVIET PHYS DOKLADY 1 716-8 1956 CA 52 6931 (ENGLISH TRANSLATION OF DOKLADY AKAD. NAUK SSSR 111, 803-5, 1956; FOR ORIGINAL SEE T14939)
12405	HEAT-TRANSFER PROPERTIES OF LIQUID-SOLID SUSPENSIONS. ORR, C., JR. DALLAVALLE, J. M. CHEM ENG PROGR SYMPOSIUM SER 50 9 29-45 1954 CA 48 5563
12592	THERMAL CONDUCTIVITY IN FURNACE CONSTRUCTION MATERIALS. RASI ANTONIO CHALEUR ET INDUSTRIE 34 334 117-23 1953 RM 10 423-P
12641	DETERMINATION OF THERMAL CONDUCTIVITY OF REFRACTORIES. NICHOLLS P BULL AM CERAM SOC 15 37-51 1936 CA 30 2339
12661	GLASS FIBERS FOR INSULATION. AUTHOR ANON. HEAT AND VENT 42 (3), 59-61, 1945.
12708	EFFECT OF GAS ATMOSPHERES ON FURNACE HEAT LOSSES. NONKEN G C STEEL PROCESSING 34 377-81 1948 RM 5 17-73
12715	CELLULAR PLASTIC HEAT INSULATORS. HICKMAN M J MOD REFRIG 49 89-90 1946 RA 1 200-9
12720	INSULATING REFRACTORIES. HEPBURN W M J AM CERAM SOC 18 13-7 1935 MA 2 443
12875	DETERMINATION OF THE VISCOSITY OF QUARTZ GLASS WITHIN THE SOFTENING RANGE. VOLAROVICH M P LEONTEVA A A J SOC GLASS TECH 20 139-43 1936 CA 30 8544
12879	THE MECHANISM OF COKING IN THE PLASTIC REGION. FITZGERALD D BRENNSTOFF-CHEM 37 199-201 1956 CA 50 14211
12993	THE RADIATING PROPERTIES OF REFRACTORY BRICKS AND SLAGS AND THEIR EFFECT ON HEAT TRANSFER. NAESER, G. PEPPERHOFF, W. STAHL UND EISEN 69 325-8 1949 RM 6 17-59
12997	HEAT TRANSFER AND VAPOR DIFFUSION IN MOIST MATERIALS. KRISCHER O ROHNALTER H FORSCH GEBIETE INGENIEURW 11 B 402 1940 CA 35 6487
13228	MECHANISM OF CEMENT FORMATION IN ROTARY KILN. VII. CHANGE IN APPEARANCE AND SPECIFIC HEAT OF RAW MEAL IN ROTARY KILN. YOSII T J SOC CHEM IND JAPAN IND CHEM SECT 43 252-5 1940 CA 35 1961
13281	ELASTIC AND VISCOUS PROPERTIES OF VOLCANIC ROCKS AT HIGH TEMPERATURES. III. OOSIMA LAVA. SAKUMA SHUZO BULL EARTHQUAKE RESEARCH INST UNIV TOKYO 31 291-303 1953 CA 48 6927
13299	THE EFFECT OF HEAT ON THE TRANSFORMATION OF HYDRAULIC BINDERS. LEPINGLE MARCEL INST TECH BATIMENT TRAV PUBL CIRC 7 1-25 1941 CA 38 3441
13340	PHYSICAL PROPERTIES OF LITHIUM-BERYLLIUM-BORATE GETAN GLASS. HERTSRIKEN S D TECH PHYS U S S R 3 336-49 1936 CA 31 4782
13346	HYDRATED, CALCINED, TABULAR ALUMINAS AND CALCIUM ALUMINATE CEMENT. ALUMINUM COMPANY OF AMERICA, PITTSBURGH, PA. ALCOA PRODUCT DATA SECT GA2A 1-20 1960
13347	VISCOSITY OF METEORITES. VOLAROVICH M P LEONTEVA A A COMPT REND ACAD SCI URSS 22 589-91 1939 CA 33 7699
13366	COMPARISON OF THE THERMAL CONDUCTIVITY AND THERMAL EXPANSION OF COKE-OVEN LINERS MADE FROM EASTERN AND WESTERN QUARTZITE. AUSTIN, J. B. PIERCE, R. H. H., JR. LUNDBERG, W. O. J AM CERAM SOC 20 363-7 1937 CA 32 1885
13367	IMPROVED APPARATUS FOR MEASURING THERMAL CONDUCTIVITY OF REFRACTORIES AT HIGH TEMPERATURES. FINCK J L J AM CERAM SOC 20 378-82 1937 CA 32 1885
13369	VISCOSITY AND ELECTRICAL RESISTIVITY AND THEIR BEARING ON THE NATURE OF GLASS. TAYLOR, N. W. J AM CERAM SOC 22 1-8 1939 CA 33 2295
13387	BINARY METHOD OF MEASURING EMISSIVITY. ALEKSANDROV B P KURTENER A V J TECH PHYS /U S S R/ 8 528-39 1938 CA 32 6539
13404	THE VARIATION WITH TEMPERATURE OF THE THERMAL CONDUCTIVITY AND THE X-RAY STRUCTURE OF SOME MICAS. I. THE THERMAL CONDUCTIVITY UP TO 600 DEGREES. POWELL, R. W. GRIFFITHS, E. PROC ROY SOC /LONDON/ 163 A 189-98 1937 CA 32 3226
13405	THE VARIATION WITH TEMPERATURE OF THE THERMAL CONDUCTIVITY AND THE X-RAY STRUCTURE OF SOME MICAS. II. THE X-RAY EXAMINATION OF THE STRUCTURE. WOOD W A PROC ROY SOC /LONDON/ 163 A 199-204 1937 CA 32 3226
13432	ALUMINA PROPERTIES. TECHNICAL PAPER NO. 10. SECOND REVISION. NEWSOME J H HEISER H W RUSSELL A S STUMPF H C ASTIA 1-88, 1960. (AD 250599)
13474	THERMAL CONDUCTIVITY OF PLASTIC MATERIALS. KANAVETS I F LEBEDEV A I ORG CHEM IND /USSR/ 6 170-4 1939 CA 33 6985

TPRC Number	Bibliographic Citation
13487	THE THERMAL CONDUCTIVITY OF ARTIFICAL MATERIALS /SYNTHETIC RESINS/. ERK S KELLER A POLTZ H PHYSIK Z 38 394-402 1937 CA 31 5666
13489	THE THERMAL CONDUCTIVITY OF PORCELAIN AND MOLDED DIATOMACEOUS-EARTH BRICK UP TO 800 DEGREES. KOCH W PHYSIK Z 39 431-6 1938 CA 32 7232
13636	TALC PORCELAIN. XIII. THERMAL CONDUCTIVITY. KONDO, S. SUZUKI, S. I. J. SOC. CHEM. IND. JAPAN 41 35 1938 CA 32 4295
13720	ELECTROCHEMICAL PROPERTIES OF HYDROGEN CLAYS FROM SEVERAL INDIAN SOILS IN RELATION TO THEIR MINERALOGICAL MAKEUP. ROY B B DAS S C SOIL SCI 76 97-105 1953 CA 48 12349
13824	HEAT CAPACITY OF SILICA-LIME-FERROUS OXIDE SLAGS. SELIVANOV B P SHPEIZMAN V M METALLURG 13 2 26-34 1938 CA 32 9005
13892	THE HEAT CONDUCTIVITY OF BROWN COAL AS AFFECTED BY WATER CONTENT. KEGEL K MATSCHAK H FEUERUNGS-TECH 25 213-7 1937 CA 32 2713
13898	THE THERMAL CONDUCTIVITIES OF VARIOUS INSULATORS AT ROOM TEMPERATURE. NIVEN C D CAN J RESEARCH 9 146-52 1933 CA 28 3150
13903	THERMAL CONDUCTIVITY OF DIATOMITE AT MODERATELY HIGH TEMPERATURES. NIVEN C D CAN J RESEARCH 11 249-53 1934 CA 28 7455
13923	COMPILATION OF IMPORTANT SPECIFIC HEATS FOR METALLURGICAL CALCULATIONS. SCHWARZ C ARCH EISENHUTTENW 7 281-92 1933 CA 28 79
13944	OXIDES FOR HIGH-TEMPERATURE APPLICATIONS. KINGERY W D INTERNATIONAL SYMPOSIUM ON HIGH TEMPERATURE TECHNOLOGY 108-33, 1959. (AD 249297)
13954	MATERIALS AND TECHNIQUES FOR THERMAL TRANSFER AND ACCOMMODATION. KING HARRY A INTERNATIONAL SYMPOSIUM ON HIGH TEMPERATURE TECHNOLOGY 202-234, 1959. (AD 249297)
14061	TWELVE YEARS OF RESEARCH ON THERMAL CONDUCTIVITY OF REFRACTORY PRODUCTS AT ELEVATED TEMPERATURES. GRANGER ALBERT CERAM VERRERIE EMAILLERIE 2 341-4 1934 CA 28 6963
14067	A STUDY OF THE VISCOSITY OF MOLTEN LAVAS FROM MOUNT ALAGHEZ. VOLAROVICH M P TOLSTOI D M KORCHEMKIN L I COMPT REND ACAD SCI URSS /NS/ 1 333-6 1936 CA 30 5535
14107	THE EMISSIVITY OF SUBSTANCES IN STEEL PLANTS. UMINO, S. TETSU-TO-HAGANE 23 644-55 1937 CA 32 4498
14113	EFFECT OF 4 MOLE PER CENT Y2O3 ON THE THERMAL CONDUCTIVITY OF UO2. ROBERTSON J A L BAIN A S RIDAL A AT ENERGY CAN LTD CHALK RIVER ONT 1-11, 1660. (AECL-1037)
14148	THE HEAT CONDUCTIVITY OF BUILDING MATERIALS IN DEPENDENCE ON WATER IN THE MATERIAL. CAMPAN T I SIMIONESCU A ANGHELACHE D ALLGEM WARMETECH 9 144-7 1960 CA 55 2053
14171	THERMAL CONDUCTION IN MICA ALONG THE PLANES OF CLEAVAGE. GOLDSMID H J BOWLEY A E NATURE /LONDON/ 187 864-5 1960 SA 63 19482
14188	THE THERMAL CONDUCTIVITY OF DRY SOILS OF CERTAIN OF THE GREAT SOIL GROUPS. SMITH W O BYERS H G SOIL SCI SOC AM PROC 3 13-9 1938 CA 33 5107
14196	APPARATUS FOR DETERMINING THERMAL CONDUCTIVITY OF HEAT-INSULATING MATERIALS AT HIGH TEMPERATURE. CHIZHNYAKOV ZITEIN-TECH SO 27-1881 1-6 1937 CA 32 8623
14263	VISCOSITY PHENOMENA OF THE SYSTEM KALSI308—NAALSI308 AND OF PERTHITE AT HIGH TEMPERATURES. KANI, K. PROC IMP ACAD /TOKYO/ 11 334-6 1935 CA 30 2068
14269	HEAT CONDUCTANCE OF A LAYER /OF COAL/. KOLESNIKOV P T PODZEMNAYA GAZIFIKATZIYA UGLEI 2 9-11 1935 CA 30 1206
14283	MEASURING THE THERMAL CONDUCTIVITY OF REFRACTORY MATERIALS AT HIGH TEMPERATURES. SALMANG H FRANK H SPRECHSAAL 68 225-8 1935 CA 30 3188
14291	RECORDING HEAT AND TEMPERATURE CONDUCTIVITY AS FUNCTIONS OF TEMPERATURE. GRUNWALD E ELEKTROWARME 8 219-22 1938 CA 32 8908
14318	REFRACTOMETRIC METHOD OF INVESTIGATION OF MELTS. STARK B V SHASHKOV YU M DOKLADY AKAD NAUK SSSR 85 125-8 1952 CA 46 9367
14353	PHYSICAL PROPERTIES OF PENNSYLVANIA ANTHRACITE AND SOME RELATED MATERIALS. MYER J LELAND AM INST MINING MET ENGRS TECH PUB 482 1-19 1932 CA 27 2279
14367	THE HEAT CONDUCTIVITIES OF CERAMIC REFRACTORY MATERIALS. CALCULATIONS OF HEAT CONDUCTIVITY FROM THE CONSTITUENTS. EUCKEN A FORSCH GEBIETE INGENIEURW 3 B 353 1-16 1932 CA 27 2269
14371	SODIUM SILICATE GLASSES CONTAINING FERROUS AND FERRIC OXIDES. AN INVESTIGATION OF PHYSICAL PROPERTIES. ANDRESEN-KRAFT C GLASTECH BER 9 577-97 1931 CA 26 6085
14411	VISCOSITY PROBLEMS IN IGNEOUS ROCKS. BALK ROBERT J RHEOL 3 461-78 1932 CA 27 686
14416	THERMAL CONDUCTIVITY OF MAGNESITE BRICK. WILKES, G. B. J AM CERAM SOC 16 125-30 1933 CA 27 2005
14482	THE THERMAL ENERGY OF CRYSTALLINE SOLIDS, BASIC THEORY. DIAMOND. ANAND V B PROC INDIAN ACAD SCI 14 A 484-91 1941 CA 36 3421
14484	THE THERMAL ENERGY OF CRYSTALLINE SOLIDS, BASIC THEORY. QUARTZ. NORRIS R PROC INDIAN ACAD SCI 14 A 499-505 1941 CA 36 3421
14540	THE PRODUCTION OF SEMICONDUCTING CERAMICS BASED ON WO3 AND A STUDY OF SOME OF THEIR ELECTRICAL AND THERMAL PROPERTIES. SKANAVI G I KASHTANOVA A M SPECIAL LIBRARY ASSOCIATION TRANS. CENTER 1-11, 1958. (ENGLISH TRANSLATION OF ZH. TEKH. FIZ., 25, (11), 1883-92, 1955; FOR ORIGINAL SEE T790) (SLA-59-10140, CTS-300)
14570	SPECIFIC HEAT OF OXIDE GLASSES. ABE TOSIO PROC PHYS MATH SOC JAPAN 24 455-98 1942 CA 41 6126
14573	REFRACTORY INSULATING MATERIALS. THE EVALUATION OF THE PROPERTIES OF HIGH-TEMPERATURE INSULATING MATERIALS. I. A DISCUSSION OF TESTING METHODS. BARRETT L R CLEWS F H GREEN A T GAS RESEARCH BOARD 2 45-59 1940 CA 35 4167

TPRC Number	Bibliographic Citation
14575	**REFRACTORY INSULATING MATERIALS. II. DRY-PRESSED GROGGED-CLAY BRICKS WITH A SLAG-RESISTANT FACE.** BARRETT L R CLEMENTS J F GREEN A T GAS RESEARCH BOARD 2 23-37 1940 CA 35 4167
14586	**REFLECTION AND TRANSMISSION SPECTRA OF POTASSIUM - SILICATE GLASSES IN THE INFRARED.** FLORINSKAYA V A PECHENKINA R S NATIONAL SCIENCE FOUNDATION, WASHINGTON, D. C. 1-4, 1953. (NSF-TR-108) (ENGLISH TRANSLATION OF DOKL. AKAD. NAUK, SSSR 91, 59-62, 1953; FOR ORIGINAL SEE T23584)
14655	**THE 1950-51 ERUPTION OF MIHARA-YAMA.** MINAKAMI, T. J. GEOGRAPHY (TOKYO) 60, 117-27, 1951.
14665	**THERMAL-ANALYSIS STUDIES ON REACTION VELOCITY IN THE SOLID STATE. IV. FORMATION OF VARIOUS SILICATES.** SEGAWA KIYOSHI J JAPNA CERAM ASSOC 57 1-3 1949 CA 45 7419
14673	**GLASS-REFRACTORIES SYMPOSIUM. III. PHYSICAL PROPERTIES OF FUSED CAST REFRACTORIES.** MCMULLEN J C THOMPSON A P AM CERAM SOC BULL 29 12-5 1950 CA 44 2717
14689	**ESTIMATE ENGINEERING PROPERTIES. USE AVAILABLE DATA TO ESTIMATE HEAT CAPACITIES.** GAMBILL W C CHEM ENG 64 6 243-8 1957 CA 51 16009
14744	**THERMAL CONDUCTIVITY OF BUILDING MATERIALS.** ROWLEY, F. B. ALGREN, A. UNIV MINN ENG EXPT STA BULL 12 1-134 1937 CA 32 4747
14762	**A NEW PROCEDURE FOR DETERMINING HEAT CONDUCTIVITY AT HIGH TEMPERATURES.** DINGER C KIND A SCHUTZ W DIETZEL A BER DEUT KERAM GES 20 347-62 1939 CA 33 9567
14795	**THE SPECIFIC HEAT OF CRYSTALLINE QUARTZ BETWEEN 2 AND 4 K.** JONES, G. H. S. HALLETT, A. C. H. CAN J PHYS 38 696-700 1960 CA 54 14913
14812	**THE MEASUREMENTS OF THERMAL CONDUCTIVITY OF DEEP-SEA SEDIMENTS BY A NEEDLE-PROBE METHOD.** VON HERZEN R MAXWELL A E J. OF GEOPHYSICAL RESEARCH 64, 1557-63, 1959. (AD 229299)
14829	**THE THERMAL CONDUCTIVITY OF INSULATING MATERIALS AT LOW TEMPERATURES IN ATMOSPHERES CONTAINING CARBON DIOXIDE.** GRIFFITHS, E. AWBERY, J. H. POWELL, R. W. DEPT SCI IND RESEARCH /BRIT/ REPT FOOD INVEST BOARD 226-7 1936 CA 32 4244
14835	**REFLECTION SPECTRA OF COMPOUND SILICATE GLASSES IN THE INFRARED BEFORE AND AFTER THERMAL TREATMENT.** FLORINSKAYA V A NATIONAL SCIENCE FOUNDATION, WASHINGTON, D. C. 1-5, 1953. (ENGLISH TRANSLATION OF DOKL. AKAD. NAUK SSSR 90, 1011-4, 1953; FOR ORIGINAL SEE T23582) (NSF-TR-148)
14860	**NOTE ON THE RELIABILITY OF THERMAL CONDUCTIVITY MEASUREMENTS FOR INSULATING MATERIALS.** OLIVER H TRANS CERAM SOC 37 49-59 1938 CA 32 3509
14870	**THERMAL CONDUCTIVITY OF MODERN CERAMIC INSULATING MATERIAL.** STEGER W STEMAG-NACHR 14 1936 CA 31 1567
14872	**CALCULATION OF THE SPECIFIC HEAT OF VITREOUS SILICA AND COMPONENTS OF GLASSES AS FUNCTIONS OF THE TEMPERATURE.** THURET ANDRE J SOC GLASS TECH 20 680-4 1936 CA 31 3651 (ENGLISH TRANSLATION OF COMP. REND 202, 1368-70, 1936; FOR ORIGINAL SEE T13894)
14889	**THERMAL CONDUCTIVITY IN RELATION TO CRYSTAL STRUCTURE.** WOOSTER W A Z KRIST 95 138-49 1936 CA 31 1671
14907	**DETERMINATION OF THE VISCOSITY OF COAL DURING THE PLASTIC PERIOD.** KUSHNIREVICH N R COKE AND CHEM /USSR/ 10 24-6 1938 CA 33 8384
14916	**THERMOCHEMISTRY AND THERMODYNAMIC PROPERTIES OF SUBSTANCES.** PAUL M A ANN REV PHYS CHEM 9 1-26 1958 CA 52 17840
14939	**SPECTROPHOTOMETRIC METHOD OF INVESTIGATING DISPERSION AND ABSORPTION BY SOLIDS.** LISITSA M P DOKLADY AKAD NAUK SSSR 111 803-5 1956 (FOR ENGLISH TRANSLATION SEE T12328)
14959	**FUNDAMENTAL PHYSICAL PROPERTIES OF LAC. II. THERMAL PROPERTIES.** VERMAN LAL C LONDON SHELLAC RESEARCH BUR TECH PAPER 4 1-20 1935 CA 30 307
14986	**REFLECTION SPECTRA OF ORDINARY AND DEVITRIFIED LEAD GLASSES IN THE INFRARED.** FLORINSKAYA V A NATIONAL SCIENCE FOUNDATION, WASHINGTON, D. C. 1-5, 1953. (ENGLISH TRANSLATION OF DOKL. AKAD. NAUK SSSR, 89, 261-4, 1953; FOR ORIGINAL SEE T23580) (NSF-TR-50)
14987	**DETERMINATION OF THE HEAT CONDUCTION IN INSULATING MATERIALS.** STALHANE, B. TID VARME- VENTILATIONS-SANITETSTEK 6 5-9 1935 CA 30 187
15019	**THE COEFFICIENTS OF HEAT CONDUCTIVITY FOR POWDERED SUBSTANCES.** HORAK, Z. TECH OBZOR 45 68-71 85-9 1937 CA 33 461
15065	**THE THERMOPHYSICAL PROPERTIES OF MAGNESITE BRICK.** MITA MASAAKI J JAPAN CERAM ASSOC 46 83-91 1938 CA 33 9566
15070	**DETERMINATION OF MEAN SPECIFIC HEATS AT HIGH TEMPERATURES OF SOME COMMERCIAL GLASSES.** PARMELEE, C. W. BADGER, A. E. UNIV ILL ENG EXPT STA BULL 271 1-24 1934 CA 29 3124
15126	**TRANSMISSION SPECTRA OF THIN SILICATE GLASS FILMS IN THE INFRARED.** FLORINSKAYA V A PECHENKINA R S NATIONAL SCIENCE FOUNDATION, WASHINGTON, D. C. 1-4, 1953. (ENGLISH TRANSLATION OF DOKL. AKAD. NAUK SSSR 89, 37-40, 1953; FOR ORIGINAL SEE T20572) (NSF-TR-30)
15129	**THE DETERMINATION OF THE HEAT CONDUCTIVITIES OF SPONGY VARIETIES OF PUMICE STONE.** DANILOV V I SIROTENKO D YA BER UKRAIN WISS FORSCH-INST PHYSIK CHEM 3 153-6 1934 CA 29 2434
15138	**THE DETERMINATION OF THE HEAT CONDUCTIVITIES OF A FEW VARIETIES OF PUMICE CONCRETE.** DANILOV V I EGOROV K E BER UKRAIN WISS FORSCH-INST PHYSIK CHEM 3 157-62 1934 CA 29 2434
15251	**FRUSTRATED TOTAL REFLECTION - ITS APPLICATION TO PROXIMITY PROBLEMS IN METROLOGY.** YOUNG T R ROTHROCK BRUCE NATL BUR STANDARDS TECH NEWS BULL 45 7 110-12 1961
15252	**THERMAL CONDUCTIVITY OF REFRACTORIES UNDER OPERATING CONDITIONS.** HEILMAN R H BRADLEY R S J AM CERAM SOC 18 43-8 1935 CA 29 3478
15366	**THE HEAT CAPACITIES OF QUARTZ, CRISTOBALITE AND TRIDYMITE AT LOW TEMPERATURES.** ANDERSON, C. T. J AM CHEM SOC 58 568-70 1936 CA 30 3709
15420	**THE SIMULTANEOUS MEASUREMENT OF VISCOSITY AND ELECTRICAL CONDUCTIVITY OF SOME FUSED SILICATES AT TEMPERATURES UP TO 1400 DEGREES.** VOLAROVICH M P TOLSTOI D M J SOC GLASS TECH 20 54-60 1936 CA 30 4072

TPRC Number	Bibliographic Citation
15457	**THE HEAT CONDUCTIVITY OF SOLID HEAT INSULATORS.** PREDVODITELEV A S J EXPTL THEORET PHYS USSR 4 813-37 1934 CA 29 7780
15471	**SPECIFIC HEAT AND HEAT OF FUSION OF CHLORIDES AND FLUORIDES.** LYASHENKO V S METALLURG 10 11 85-98 1935 CA 30 6278 (FOR ENGLISH TRANSLATION SEE T31234)
15501	**A VACUUM ADIABATIC CALORIMETER AND SOME NEW DATA OF THE BETA IN EQUILIBRIUM WITH A MINUS TRANSFORMATION OF QUARTZ.** SINELNIKOV N N SPECIAL LIBRARY ASSOC. TRANS. CTR. 1-7, 1954. (ENGLISH TRANSLATION OF DOKL. AKAD. NAUK SSSR, 92 (2), 369-72, 1953; FOR ORIGINAL SEE T619) (SLA-61-13676)
15503	**A NEW THERMAL CONDUCTIVITY APPARATUS.** FITCH A L AM PHYS TEACHER 3 135-6 1935
15516	**DETERMINATION OF THE RELATIVE THERMAL CONDUCTIVITIES ACCORDING TO THE ISOTHERMAL LINES METHOD.** VOIGHT W ANN PHYSIK 64 95-100 1898
15534	**CONTRIBUTION TO THE KNOWLEDGE OF THE MEAN SPECIFIC HEATS OF SOME TECHNICALLY IMPORTANT GLASSES. III. THE SPECIFIC HEAT OF A BARIUM GLASS.** HARTMANN H KIESSLING K H SPECIAL LIBRARY ASSOC., TRANS. CTR. 1-10, 1958. (ENGLISH TRANSLATION OF GLASTECH. BER. 30 (5), 186-8, 1957; FOR ORIGINAL SEE T311) (SLA-59-10607)
15576	**THERMAL CONDUCTIVITY OF SOME CRYSTALS AT LOW TEMPERATURES.** EUCKEN, A. PHYSIK. Z. 12, 1005-8, 1911. (VERHANDL. DEUT. PHYSIK. GES.) (13, 839-35, 1911) (14, 9PP., 1912)
15591	**A MODIFIED FITCH THERMAL CONDUCTIVITY APPARATUS.** HARVALIK, Z. V. REV SCI INSTR 18 11 815-7 1947
15603	**CONTRIBUTIONS TO THE KNOWLEDGE OF THERMAL AND ELECTRICAL PROPERTIES OF PRESSED POWDERS FROM ANTIMONY, BISMUTH AND GALENA.** GEHLHOFF G NEUMEIER F VERHANDL DEUT PHYSIK GES 15 1069-81 1913
15610	**NEW MEASUREMENTS OF THE HEAT CONDUCTIVITY OF SOLID CRYSTALLINE SUBSTANCES AT 0 DEGREES AND MINUS 190 DEGREES.** EUCKEN A HUHN G Z PHYSIK CHEM 134 193-219 1928
15618	**PRELIMINARY MEASUREMENTS TO DETERMINE THE EFFECT OF COMPOSITION ON THE THERMAL CONDUCTIVITY OF GLASS.** RATCLIFFE E H PHYS AND CHEM GLASSES 1 3 103-4 1960 JA 43 254
15634	**COMPARISON METHOD FOR THE DETERMINATION OF THE HEAT CONDUCTIVITY OF CERAMIC MATERIALS.** KLASSE F HEINZ A HEIN J U. S. ATOMIC ENERGY COMMISSION 1-27 1959 (ENGLISH TRANSLATION OF BER. DEUT. KERIM. GES. 34 (6), 183-9, 1957; FOR ORIGINAL SEE T6619) (AEC-TR-3751)
15640	**A RAPID IN SITU METHOD FOR DETERMINING THERMAL CONDUCTIVITIES.** GUPTA M L J SCI IND RESEARCH /INDIA/ 19 B 7 240-3 1960
15658	**THERMAL DIFFUSIVITY AND CONDUCTIVITY OF POROUS ZIRCONIUM OXIDE AT HIGH TEMPERATURES.** FLIEGER, H. W., JR. GINNINGS, D. C. NATIONAL BUREAU OF STANDARDS 1-26 1957 (NBS RPT 5642)
15664	**REFRACTORY INORGANIC MATERIALS FOR STRUCTURAL APPLICATIONS.** PEARL, H. A. NOWAK, J. M. CONTI, J. C. URODE, R. J. BELL AIRCRAFT CORP., BUFFALO, N. Y. 1-142, 1960. (WADC TR 59-432, AD 236052)
15666	**DEVELOPMENT AND EVALUATION OF INSULATING TYPE CERAMIC COATINGS.** BLOCKER E W HAUCK C A SKLAREW S LEVY A V MARQUARDT AIRCRAFT CO., VAN NUYS, CA. 1-71, 1959. (WADD TR 59-102, AD 231732)
15692	**THE VISCOSITY AND PLASTICITY OF DISPERSE SYSTEMS. IV. THE PLASTIC, VISCOUS PROPERTIES OF MOLTEN SLAGS AND ROCKS.** VOLAROVICH M P KOLLOID-Z 71 159-65 1935 CA 29 7744
15693	**THE VISCOSITY AND PLASTICITY OF DISPERSE SYSTEMS. V. THE PLASTIC-VISCOUS PROPERTIES OF PEAT.** VOLAROVICH M P KULAKOV N N ROMANSKII A N KOLLOID-Z 71 267-74 1935 CA 29 7744
15736	**THE THERMAL BEHAVIOUR OF SOILS.** DE VRIES D A ARID ZONE RESEARCH ON CLIMATOLOGY AND MICROCLIMATOLOGY PROC CANBERRA SYMP 109-15 1958
15754	**RADIATION EFFECTS ON THERMAL CONDUCTIVITY OF CERAMICS.** SISMAN, O. BOPP, C. D. TOWNS, R. L. OAK RIDGE NATIONAL LAB., TENNESSEE 33-5, 1955. (ORNL-1852)
15755	**SOLID STATE DIVISION SEMIANNUAL PROGRESS REPORT FOR PERIOD ENDING FEBRUARY 28, 1955.** HOWE, J. T. OAK RIDGE NATIONAL LAB., TENNESSEE 1-61, 1955. (ORNL-1852)
15774	**DETERMINATION OF CERTAIN PROPERTIES OF SURFACE LAYERS OF THE MOON FROM ITS RADIO WAVE EMISSION AT A WAVELENGTH OF 3.2 CM.** TROITSKIY V S ZELINSKAYA M P ASTRON ZHUR 32 6 550-4 1955 (FOR ENGLISH TRANSLATION SEE T15778)
15778	**DETERMINATION OF CERTAIN PROPERTIES OF SURFACE LAYERS OF THE MOON FROM ITS RADIO WAVE EMISSION AT A WAVELENGTH OF 3.2 CM.** TROITSKIY V S ZELINSKAYA M P SPECIAL LIBRARY ASSOC., TRANSLATIONS CENTER 1-8, 1959. (ENGLISH TRANSLATION OF ASTRON. ZH. 32 (6), 550-4, 1955; FOR ORIGINAL SEE T15774) (SLA-59-21684)
15805	**A NEW RESISTANCE VISCOMETER FOR CERAMIC SLIPS AND CEMENT SLURRY.** ENDELL K FENDIUS H KERAM RUNDSCHAU 42 459-61 1934 CA 28 7455
15835	**DETERMINATION OF COEFFICIENTS OF TEMPERATURE CONDUCTIVITY AND THERMAL CONDUCTIVITY OF SOLIDS AND LIQUIDS.** VOLKENSHTEIN V S MEDVEDEV N N INZHENER-FIZ ZHUR AKAD NAUK BELORUS SSR 2 10 26-32 1959 CA 54 10489
15850	**MEASUREMENTS OF THE VISCOSITY OF BASALT GLASS AT HIGH TEMPERATURES. I.** KANI, K. PROC IMP ACAD TOKYO 10 29-32 1934 CA 28 2962
15874	**THERMAL PROPERTIES OF MOULDING MATERIALS.** RAMACHANDRAN A INDIAN INSTITUTE OF METALS TRANSACTIONS 13 39-51 1960 RM 17 467-E
15886	**ELEVATED TEMPERATURE STUDIES OF INFRARED TRANSPARENT MATERIALS.** ARMOUR RESEARCH FOUNDATION, ILLINOIS ARMOUR RESEARCH FOUNDATION, ILLINOIS 1-4, 1960. (ARF 1159-3, AD 239151)
15895	**AN INVESTIGATION OF INFRARED TRANSMITTING MATERIALS.** KREIDL N J HAFNER H C HENSLER J R WEIDEL R A LETTER E C BAUSCH AND LOMB OPTICAL CO., ROCHESTER, N. Y. 1-266, 1958. (WADC TR 55-500(PT. 2), AD 203786)
15929	**A THERMAL PROBLEM ASSOCIATED WITH UNDERGROUND STORAGE OF RADIOACTIVE WASTES.** CROWELL, J. PARKER, F. L. OFFICE OF NAVAL RESEARCH 1-72, 1960. (ORNL-3002)

TPRC Number	Bibliographic Citation
16022	**THERMAL CONDUCTIVITY AND THERMAL EXPANSION OF BERYLLIUM OXIDE AT ELEVATED TEMPERATURES.** TAYLOR, R. E. NORTH AMERICAN AVIATION, INC., DOWNEY, CALIFORNIA 1-19 1960 CA 55 1117 (NAA-SR-4905)
16036	**GRAIN SIZE EFFECTS ON THE THERMAL CONDUCTIVITY OF CERAMIC OXIDES. PROGRESS REPT. NO. 11.** TRUESDALE R S SWICA J J TINKLEPAUGH J R WRIGHT AIR DEVELOPMENT DIVISION 1-67, 1960. (AD 240278)
16064	**MEASURING INSTRUMENT FOR THE HEAT PENETRATION NUMBER OF INSULATING MATERIALS.** ELSER KARL MITT INST THERMODYNAMIK U VERBRENNUNGSMOTORENBAU 14 89-96 1953
16070	**ABOUT THE HEAT INSULATION OF FINE-GRAINED POWDERS IN DILUTED GASES.** LEIDENFROST W VDI-ZEITSCHRIFT 97 34 1235-42 1955
16072	**A FAST METHOD TO EVALUATE THE THERMAL CONDUCTIVITY OF INSULATORS.** CODEGONE, C. RICERCA SCI 24 12 2623-7 1954
16078	**HEAT TRANSFER IN DAMP, POROUS MATERIALS OF DIFFERENT COMPOSITION.** KRISCHER O ESDORN H FORSCH GEBIETE INGENIEURW 22 1 1-8 1956 (FOR ENGLISH TRANSLATION SEE T11559)
16087	**INVESTIGATION OF LIGHT-SCATTERING GLASSES FOR ILLUMINATION PURPOSES.** VAN DER HELD E F M MINNAERT M PHYSICA 2 8 769-84 1935
16088	**LATTICE DEFECTS IN CRYSTALLINE AND VITREOUS SIO2.** STEVELS J M GLASTECH BER 32 307-13 1959
16094	**STRONTIUM 90 POWER PROJECT.** MARTIN MARIETTA, CORP., BALTIMORE, MD. U.S. ATOMIC ENERGY COMMISSION 1-72 1959 (MND-SR-1674)
16107	**HEAT CAPACITY OF SOME METALS, ALLOYS AND SLAG FORMERS AT TEMPERATURES UP TO 1200 DEGREES.** ESSER H AVERDIECK R GRASS W ARCH EISENHUTTENW 6 289-92 1933 CA 27 1598
16133	**PHYSICAL PROPERTIES OF FUSED PHOSPHATE FERTILIZER AT HIGH TEMPERATURE. IV. FLUIDITY OF CALCIUM MAGNESIUM PHOSPHATE MELT.** KIYOURA RAISAKU SATA TOSHIYUKI J CHEM SOC JAPAN IND CHEM SECT 56 748-50 1953 CA 48 13147
16157	**BLOCK INSULATION FOR STOVES AND BUSTLE PIPES.** EUSNER G R SHAPLAND J T BULL AM CERAM SOC 40 7 439-44 1961
16179	**TEMPERATURE VARIATION OF THE THERMAL CONDUCTIVITY OF PYREX GLASS.** STEPHENS R W B PHIL MAG 14 897-914 1932 CA 27 3865
16203	**INFLUENCE OF PORE SIZE ON SOME PROPERTIES OF FOUNDRY SAND.** GITTUS J H IRON AND STEEL 33 429-33 1960 RM 17 536-E
16208	**MEASUREMENT OF THERMAL CONDUCTIVITY OF POOR CONDUCTORS BY THE CONTACT METHOD.** CAMPAN T I ALLGEM WARMETECH 9 7 141-4 1959
16209	**THE CONDUCTIVITY OF THE PRINCIPAL METALS AND SOME EARTHY SUBSTANCES.** DESPRETZ C ANN CHIM PHYS 36 422-6 1827
16210	**EXPERIMENTS ON THE CALORIFIC CONDUCTIBILITY OF SOLIDS.** NEUMANN M F ANN CHIM PHYS 66 3 183-7 1862 (FOR ENGLISH TRANSLATION SEE T22527)
16221	**EMITTANCE OF PYROCERAM 9608.** WEIGANDT A OLSON O H J AM CERAM SOC 44 12 632 1961
16239	**THERMAL CONDUCTIVITY OF ALUMITE.** ASADA, Y. BULL INST PHYS CHEM RESEARCH /TOKYO/ 19 45-59 1940
16240	**VACUUM APPARATUS FOR STUDYING THE HEAT CAPACITY OF METALS AT HIGH TEMPERATURES.** LAZAREV, A. I. DYNKOV, B. N. E. ISSLEDOVANIYA V OBLASTI TEPLOVYKH IZMERENII SNORNIK 5-20 1956 CA 54 9381
16241	**THE THERMAL CONDUCTIVITY OF SUNDRY MATERIALS.** NIVEN C D CAN J RESEARCH 13 A 16-18 1935
16263	**HEAT INSULATION.** EMERY E T G ENG AND BOILER HOUSE REV 53 469-74 1940
16279	**MEASURING THERMAL CONDUCTIVITY OF REFRACTORIES.** NORTON C L IND HEATING 9 1532-4 1942
16286	**THE THERMAL CONDUCTIVITIES OF OCEAN SEDIMENTS.** RATCLIFFE E H J GEOPHYS RESEARCH 65 5 1535-41 1960
16287	**THE MEASUREMENT OF THE THERMAL CONDUCTIVITY OF MATERIALS USED IN BUILDING CONSTRUCTION.** GRIFFITHS, E. J INST HEATING VENTILATING ENGRS 10 106-38 1942 CA 37 6837
16306	**HEAT INSULATING MATERIALS.** GRIFFITHS, E. J SCI INSTR 15 117-21 1938
16313	**THERMAL CONDUCTIVITY. PART II. THERMAL CONDUCTIVITY OF BADLY-CONDUCTING SOLIDS.** BARRATT, T. PROC PHYS SOC /LONDON/ 27 81-93 1914
16320	**A METHOD FOR THE DETERMINATION OF THE THERMAL CONDUCTIVITIES OF ROCKS.** NANCARROW H A PROC PHYS SOC /LONDON/ 45 447-61 1933
16322	**METHODS AND EQUIPMENTS FOR MEASURING SOIL TEMPERATURE AND HEAT FLOW THROUGH THE SOIL. A LITERATURE SURVEY.** NICKERSON, R. J. CHOI, H. Y. MARCUS, S. MASSACHUSETTS INSTITUTE OF TECHNOLOGY, CAMBRIDGE 1-15, 1959. (AD 215290)
16331	**THE FLOW OF HEAT THROUGH THE FLOOR OF THE ATLANTIC OCEAN.** BULLARD SIR EDWARD PROC ROY SOC /LONDON/ 222 A 408-29 1954
16341	**CRUSTAL STRUCTURE AND SURFACE HEAT LOW NEAR THE COLORADO FRONT RANGE.** BIRCH, F. TRANS AM GEOPHYS UNION 28 5 792-7 1947
16342	**TERRESTRIAL HEAT FLOW IN ONTARIO AND QUEBEC.** MISENER A D THOMPSON L G D UFFEN R J TRANS AM GEOPHYS UNION 32 5 729-38 1951
16343	**ON THE THERMAL CONDUCTIVITY OF SOIL, WITH SPECIAL REFERENCE TO THAT OF FROZEN SOIL.** HIGASHI AKIRA TRANS AM GEOPHYS UNION 34 5 737-48 1953
16344	**A SMALL PORTABLE METER FOR SOIL HEAT CONDUCTIVITY AND ITS USE IN THE O NEILL TEST.** BUETTNER K TRANS AM GEOPHYS UNION 36 5 827-30 1955
16345	**HEAT FLOW AND DEPTH OF PERMAFROST AT RESOLUTE BAY, CORNWALLIS ISLAND, N. W. T., CANADA.** MISENER A D TRANS AM GEOPHYS UNION 36 6 1055-60 1955
16349	**THERMAL INSULATION.** GRIFFITHS, E. TRANS INST CHEM ENGRS /LONDON/ 10 35-44 1932

TPRC Number	Bibliographic Citation
16377	**HEAT FLOW IN SODA-LIME GLASS.** TOROK JULIUS J BULL AM CERAM SOC 40 8 488-92 1961
16411	**METHOD FOR THE DETERMINATION OF THERMAL DIFFUSIVITY OF BAD THERMAL CONDUCTORS.** MATTAROLO L NUOVO CIMENTO 7 5 809-15 1950
16421	**ON THE THERMAL CONDUCTIVITY OF CRYSTALS AND OTHER BAD CONDUCTORS.** LEES, C. H. PHIL TRANS ROY SOC LONDON 183 481-509 1892
16447	**THE DETERMINATION OF THE COEFFICIENTS OF THERMAL CONDUCTIVITY OF SOME GLASSES.** RODNIKOVA V V STEKLO I KERAM 15 6 20-1 1958 CA 52 16712 (FOR ENGLISH TRANSLATION SEE T28537)
16451	**THERMAL CONDUCTIVITY OF METALS AND NONMETALS.** EUCKEN A PHYSIK Z 29 563-6 1928 CA 22 4339
16452	**THE PHYSICAL BASES OF THE PROCESS OF INFLATING LOW-FUSING CLAYS AND FOAM GLASSES.** CHERNYAK YA N STEKLO I KERAM 15 10 25-8 1958 CA 53 2561
16471	**DETERMINATION OF HEAT CAPACITY OF REFRACTORY MATERIALS AT HIGH TEMPERATURES.** LOZINSKII N GERMAN S ZAVODSKAYA LAB 3 831-8 1934 CA 29 1221
16474	**SPECIFIC HEAT OF RAW //CEMENT// MIXES.** SCHWIETE H E VON GRONOW H E ZEMENT 24 197-9 1935 CA 30 6911
16489	**THE USE OF THE TORSION METHOD FOR DETERMINING THE VISCOSITY OF LOW-FUSING CLAYS.** PAVLOV V F STEKLO I KERAM 16 5 26-31 1959 CA 53 16495 (FOR ENGLISH TRANSLATION SEE T32967)
16499	**MEASUREMENT OF THE VISCOSITY OF BASALT GALSS AT HIGH TEMPERATURES. II.** KANI, K. PROC IMP ACAD /TOKYO/ 10 79-82 1934 CA 28 3545
16520	**A PRELIMINARY INVESTIGATION INTO CONVENTIONAL GLASS-FIBRE INSULATION APPLIED IN SPIRAL FORM TO RESIST MECHANICAL LOADS.** BISHOP, P. H. H. GT BRIT. ROYAL AIRCRAFT ESTABLISHMENT, FARNBOROUGH HANTS ENGLAND 1-13, 1960. (RAE TECH NOTE CHEM. 1366, AD 243810)
16524	**NEW TABLES OF SPECIFIC HEATS.** VETTER HERMAN REFRIGERATING ENG 31 174 1936 CA 30 5865
16537	**SPECIFIC HEAT OF MINERALS AND OF RARE EARTH SALTS.** SWIETOSLAWSKI W SALCEWICZ J USAKIEWICZ J ZMACZYNSKI A ZLOTOWSKI J ROCZNIKI CHEM 15 12-14 1935 CA 29 4252
16570	**THE THERMAL CONDUCTIVITY OF MATERIALS.** ROUX A REV GEN FROID 16 98-105 1935 CA 30 7245
16593	**PHYSICAL PROPERTIES OF HIGH TEMPERATURE MATERIALS. PART 5. THERMAL DIFFUSIVITY OF MAGNESIA-STABILIZED ZIRCONIUM OXIDE AT HIGH TEMPERATURES.** FLIEGER, H. W., JR. KNUDSEN, F. P. GINNINGS, D. C. WRIGHT AIR DEVELOPMENT DIVISION 1-14, 1960. (WADC TR 57-374(PT 5), AD 249385)
16639	**PARAMETRIC STUDIES OF METAL FIBER REINFORCED CERAMIC COMPOSITE MATERIALS.** STOWELL, E. Z. LIU, T. S. BUREAU OF NAVAL WEAPONS 1-17, 1960. (AD 247327)
16642	**THERMAL CONDUCTIVITY OF A SEMI-INFINITE SLAB WHEN THE FLOW TO AND FROM THE SLAB CROSSES THE SURFACE NEAR THE EDGE.** MURRAY, F. H. US ATOMIC ENERGY COMMISSION 1-4, 1944. (AECD-2965, CP-1486, A-2069)
16648	**THERMAL CONDUCTIVITY OF REACTOR FUEL ELEMENT MATERIALS.** WESTPHAL, R. C. WESTINGHOUSE ELECTRIC CORP., ATOMIC POWER DIV., PITTSBURG 1-4, 1954. (WAPD-MM-538, AECD-3864)
16701	**A COMPENDIUM OF THE PROPERTIES OF MATERIALS AT LOW TEMPERATURE /PHASE I/. PART II. PROPERTIES OF SOLIDS.** JOHNSON, V. J. WRIGHT AIR DEVELOPMENT DIVISION 1-333, 1960. (WADD TR 60-56(PT 2), AD 249786)
16704	**THERMOPHYSICAL PROPERTIES OF SOLID MATERIALS. VOLUME I. ELEMENTS. /MELTING TEMPERATURE ABOVE 1000 F/.** GOLDSMITH, A. WATERMAN, T. E. HIRSCHHORN, H. J. WRIGHT AIR DEVELOPMENT DIVISION 730PP., 1960. (WADC TR 58-476(VOL 1), AD 247193)
16712	**WATER CHEMISTRY AND FUEL ELEMENT SCALE IN EBWR.** BREDEN, C. R. CHARAK, I. LEYSE, R. H. ARGONNE NATIONAL LABORATORY 1-105, 1960. (ANL-6136)
16732	**DEVELOPMENT AND EVALUATION OF INSULATING TYPE CERAMIC COATINGS.** LEGGETT H JOHNSON R L BLOCKER E W WEISERT E D WRIGHT AIR DEVELOPMENT DIVISION 1-172, 1960. (WADC TR 59-102(PT 2), AD 248843L)
16740	**APPARENT TEMPERATURES OF SMOOTH AND ROUGH TERRAIN.** CHEN, S. N. C. OHIO STATE UNIV. RESEARCH FOUNDATION 1-28, 1960. (RPT 898-8, AD 251436)
16746	**MARITIME GAS-COOLED REACTOR PROGRAM. THERMAL AND MECHANICAL PROPERTIES OF BERYLLIA CERAMICS.** ILLINOIS INSTITUTE OF TECHNOLOGY, CHICAGO, ILL. GENERAL DYNAMICS CORP. 1-34, 1961. (GA-1906)
16756	**EFFECTS OF IRRADIATION ON THE THERMAL CONDUCTIVITY OF SYNTHETIC SAPPHIRE.** BERMAN R FOSTER E L SCHNEIDMESSER B TIRMIZI S M A J APPL PHYS /USA/ 31 12 2156-9 1960 SA 64 1005
16816	**PROGRESS REPORT FOR JANUARY 1 TO MARCH 31, 1950.** NORTON, F. H. NEW YORK OPERATIONS, AEC 1-22, 1950. (NYO-594)
16818	**THE MEASUREMENT OF THERMAL CONDUCTIVITY OF REFRACTORY MATERIALS.** NORTON, F. H. NEW YORK OPERATIONS, AEC 1-10, 1951. (NYO-598)
16820	**PROGRESS REPORT FOR JULY 1 - SEPTEMBER 30, 1950.** NORTON F H KINGERY W D FELLOWS D M ADAMS M MC QUARRIE M C COBLE R L NEW YORK OPERATIONS, AEC 1-9, 1950. (NYO-596)
16821	**THE MEASUREMENT OF THERMAL CONDUCTIVITY OF REFRACTORY MATERIALS.** NORTON, F. H. KINGERY, W. D. FELLOWS, D. M. ADAMS, M. MC QUARRIE, M. C. COBLE, R. L. FRANCL, J. VASILOS, T. ANDERSON, H. H. NEW YORK OPERATIONS, AEC 1-15, 1951. (NYO-597)
16823	**THE MEASUREMENT OF THERMAL CONDUCTIVITY OF REFRACTORY MATERIALS.** NORTON, F. H. KINGERY, W. D. NEW YORK OPERATIONS, AEC 1-16, 1955. (NYO-6449)

TPRC Number	Bibliographic Citation

16826 **PROGRESS REPORT FOR OCTOBER 1 - DECEMBER 31, 1949.**
NORTON F H FELLOWS D M ADAMS M
MC QUARRIE, M. FULLERTON, C. P.
NEW YORK OPERATIONS, AEC
1-43, 1949.
(NYOO-96)

16842 **PROGRESS ON HIGH TEMPERATURE THERMAL CONDUCTIVITY MEASUREMENTS.**
HUNTER, L. P.
MONSANTO CHEMICAL CO., CLINTON LABS., KNOXVILLE, TN.
1-63, 1947.
(MONM-442)

16875 **CRYOGENIC DATA BOOK.**
CHELTON, D. B. MANN, D. B.
WRIGHT AIR DEVELOPMENT DIVISION
1-144, 1959.
(WADC TR 59-8, PB 151837, AD 208155)

16895 **CRYOGENIC DATA BOOK.**
CHELTON, D. B. MANN, D. B.
UNIVERSITY OF CALIFORNIA RADIATION LABORATORY
1-116 1956
(UCRL-3421)

16913 **THE THERMAL CONDUCTIVITY OF REFRACTORY MATERIALS.**
KANZ ANTON
MITT FORSCH-INST VER STAHLWERKE A-G DORTMUND
2 223-34 1932 CA 27 2547

16918 **HEAT CAPACITY OF MINERAL RAW MATERIALS.**
BAZILEVICH A
MINERAL SUIRE
9 3 48-50 1934 CA 28 4685

16925 **APPLICATION OF VISCOMETRY AND PLASTOMETRY TO PROBLEMS OF APPLIED MINERALOGY.**
VOLAROVICH M P
TRANS INST ECON MINERAL /USSR/
66 1-56 1934 CA 29 4992

16961 **INFRARED SPECTRAL EMISSIVITY OF OPTICAL MATERIALS.**
STIERWALT, D. L.
NAVAL ORDNANCE LAB., CORONA, CA.
1-34, 1961.
(NAVWEPS REPT 7160, NOLC RPT 537, AD 250530)

17036 **B. THERMODYNAMIC PROPERTIES OF LIGHT-ELEMENT COMPOUNDS. CHAPTER 1. THERMAL PROPERTIES OF SOLIDS AND LIQUIDS. FROM PRELIMINARY REPORT ON THE THERMODYNAMIC PROPERTIES OF LITHIUM, BERYLLIUM, MAGNESIUM, ALUMINUM AND THEIR COMPOUNDS WITH HYDROGEN, OXYGEN, NITROGEN. FLUORINE, AND CHLORINE.**
FURUKAWA, G. T. REILLY, M. L.
HENNING, J. M. DOUGLAS, T. B.
VICTOR, A. C. BEAUDOIN, A. R.
NATIONAL BUREAU OF STANDARDS
17-35, 1959.
(NBS RPT 6484, AD 235429)

17039 **THE EMITTANCE OF COATED MATERIALS SUITABLE FOR ELEVATED-TEMPERATURE USE.**
WOOD, W. D. DEEM, H. W. LUCKS, C. F.
BATTELLE MEMORIAL INSTITUTE
1-136, 1961.
(DMIC MEMO 103, PB 171622)

17044 **THE TWO RADIOMETER METHOD FOR THE SIMULTANEOUS DETERMINATION OF EMISSIVITY AND SURFACE TEMPERATURE.**
DUNDLE, R. V. GIER, J. T.
OFFICE OF NAVAL RESEARCH
1-24, 1948.
(ATI 91560)

17062 **ESTIMATION OF SPECIFIC HEATS AT NORMAL TEMPERATURES.**
BROCK F H
ARS J
31 2 265-8 1961 AM 14 3779

17069 **CRYOGENIC INSULATION.**
KROPSCHOT R H
ASHRAE
1 9 48-54 1959 CA 55 848

17084 **FLOW PROPERTIES OF A COKING BITUMINOUS COAL IN THE INITIAL STAGES OF THERMAL SOFTENING.**
FINLAYSON P C MACREA J C
CONF ON SCI USE COAL SHEFFIELD
C15-C17 1958 CA 55 940

17122 **SOME THERMODYNAMIC PROPERTIES OF FLUORPHLOGOPITE MICA.**
KELLEY K K BARANY R KING E G CHRISTENSEN A U
U.S. BUREAU OF MINES
1-16, 1959.
(BM-RI-5436)

17182 **TRANSMISSION OF CUTOFF GLASS FILTERS EMPLOYED IN SOLAR RADIATION RESEARCH. II.**
ANGSTROM A K DRUMMOND A J
J OPT SOC AM
50 974-9 1960 CA 55 14

17203 **ALUMINA PORCELAIN AS A HIGH FREQUENCY INSULATOR. I, EFFECTS OF ALUMINUM OXIDE AS A RAW MATERIAL ON PHYSICAL PROPERTIES.**
KATO, S. OKUDA, H.
NAGOYA KOGYO GIJUTSU SHIKENSHO HOKOKU
9 8 46-52 1960 JA 44 92

17216 **LIGHTWEIGHT REFRACTORIES FROM ALUMINA.**
GUZMAN I YA POLUBOYARINOV D N
OGNEUPORY
24 2 71-9 1959 JA 44 89

17217 **THE DETERMINATION OF THE THERMAL CONDUCTIVITIES OF REFRACTORIES UP TO 1200 DEGREES BY THE STATIC METHOD.**
PUSTOVALOV V V
OGNEUPORY
24 4 180-5 1959 CA 53 13534

17229 **VISCOSITY AND DENSITY OF MOLTEN SILICA AND HIGH SILICA CONTENT GLASSES.**
BACON, J. F. HASAPIS, A. A.
WHOLLEY, J. W., JR.
PHYS AND CHEM GLASSES
1 3 90-8 1960 JA 44 115

17275 **MICROCALORIMETRIC MEASUREMENT OF THERMAL CONDUCTIVITY.**
KROCKEL, O.
SILIKATTECH
10 4 193-5 1959 JA 44 23

17278 **THE EFFECT OF CHANGE OF VISCOSITY IN THE 800-1200 DEGREES INTERVAL ON THE SINTERING AND SWELLING PROPERTIES OF LOW-FUSING CLAYS.**
PAVLOV V F
STEKLO I KERAM
17 3 21-5 1960 CA 54 14612

17280 **THE EFFECT OF INCREASING HEAT TREATMENT ON THE PROPERTIES OF A DIATOMACEOUS INSULATING BRICK.**
OLIVER H RIGBY J S
TRANS BRIT CERAM SOC
40 335-62 1941

17320 **IMPROVEMENT OF THERMAL CONDUCTIVITY OF CHEMICALLY RESISTANT RUBBERS. I. THERMAL-CONDUCTIVITY MEASUREMENTS.**
MOMIN A U SHANKAR U SURYANARAYANA N P
J SCI AND IND RESEARCH /INDIA/
19 A 215-18 1960 CA 54 25931

17338 **DETERMINATION OF THE SPECIFIC HEAT AND THE HEAT CONDUCTIVITY FACTOR OF HUNGARIAN COALS AND COKES.**
WELTNER MARGIT
NEHEZVEGYIPARI KUTATO INTEZET KOZLEMENYEI
1 27-37 1958 CA 53 2576

17346 **FELDSPAR AND ITS INFLUENCE ON THE REACTIONS IN CERAMICS DURING BURNING.**
SUNDIUS NILS
ACTA POLYTECH SCAND
8 7 1-29 1960 CA 54 18918

17349 **HEAT TRANSFER CHARACTERISTICS OF POROUS ROCKS.**
KUNII D SMITH J M
AICHE J
6 1 71-7 1960 AM 13 5980

17353 **THE HEAT CAPACITIES OF THE ELEMENTS BELOW ROOM TEMPERATURE.**
SHIFFMAN C A
U.S. ATOMIC ENERGY COMMISSION
1-70, 1952.
(NP 4945)

17360 **THE SPECIFIC HEAT OF CHAIN STRUCTURES AT LOW TEMPERATURE.**
SOCHAVA I V TRAPEZNIKOVA O N
SOVIET PHYS DOKLADY
2 164-6 1957 CA 53 42
(ENGLISH TRANSLATION OF DOKL. AKAD. NAUK SSSR 113, 784-6, 1957; FOR ORIGINAL SEE T9440)

17394 **DETERMINATION OF HEAT CONTENTS AND OF THE AVERAGE TEMPERATURE OF BOTTLES DURING SHAPING PROCESS.**
UNGER LEOPOLD
GLASTECH BER
32 153-7 1959 CA 53 12611
(FOR ENGLISH TRANSLATION SEE T17395)

17395 **DETERMINATION OF THE HEAT CONTENT AND THE AVERAGE TEMPERATURE OF BOTTLES DURING THE SHAPING PROCESS.**
UNGER LEOPOLD
SPECIAL LIBRARY ASSOC., TRANSLATIONS CENTER
1-20, 1960.
(ENGLISH TRANSLATION OF GLASTECH. BER. 32 (4), 153-7, 1959; FOR ORIGINAL SEE T17394)
(SLA-61-10504)

17406 **APPARATUS FOR DETERMINING THE THERMAL CONDUCTIVITY OF REFRACTORIES.**
SHAKHTIN D M
ZAVODSKAYA LAB
22 7 869-71 1956
(FOR ENGLISH TRANSLATION SEE T17407)

TPRC Number	Bibliographic Citation
17407	**APPARATUS FOR DETERMINING THE THERMAL CONDUCTIVITY OF REFRACTORIES.** SHAKHTIN D M SPECIAL LIBRARY ASSOC., TRANSLATIONS CENTER 1-5, 1958. (ENGLISH TRANSLATION OF ZAVOD. LAB. 22 (7), 869-71, 1956; FOR ORIGINAL SEE T17406) (CTS 432, SLA-59-19612)
17473	**THERMAL DIFFUSIVITY AND COEFFICIENT OF CONVECTIONAL HEAT TRANSFER IN THE LOWER TEMPERATURE RANGE OF THE FIRING OF FIRECLAY BRICK.** TERADA KIYOSHI WAKAMATSU MITSURU WAKABAYASHI KAZUTOSHI YOGYO KYOKAI SHI 66 110-16 1958 CA 52 16714
17573	**RADIATION, CONVECTION AND CONDUCTION COEFFICIENTS IN SOLAR COLLECTORS.** TABOR H BULL RES COUNC ISRAEL 6 C 3 155-76 1958 AM 13 1396
17617	**CALCULATION OF THE THERMAL CONDUCTIVITY OF POROUS MEDIA.** WOODSIDE, W. CAN J PHYS 36 815-23 1958 CA 52 15223
17622	**INFLUENCE OF FIRING TEMPERATURE ON THERMAL CONDUCTIVITY OF CERAMIC MATERIALS.** PRASAD J CENTRAL GLASS AND CERAM RESEARCH INST BULL /INDIA/ 5 111-13 1958 CA 53 8565
17623	**EFFECT OF PORE SIZE ON THE THERMAL CONDUCTIVITY OF HEAT INSULATING MICA BRICK.** PRASAD J CENTRAL GLASS AND CERAM RESEARCH INST BULL /INDIA/ 5 1 29-33 1958 JA 43 88
17648	**THERMAL CONDUCTIVITY OF CATALYST PARTICLES.** SEHR ROBERT A CHEM ENG SCI 9 145-52 1958 CA 54 978
17692	**HIGH-TEMPERATURE PROPERTIES OF SAND TREATED WITH WATER GLASS.** CHOW, Y. S. LEE, S. H. CHI HSIEN KUNG CHENG HSUEH PAO 5 365-92 1957 CA 53 6945
17754	**THE PHYSICAL BASIS OF COLOR-TECHNOLOGY.** LUCKIESH M J FRANKLIN INST 184 73-93 227-50 1917
17782	**PROPERTIES OF MELTS OF THE SYSTEM CAO-FEO-SIO2 IF FEO IS SUBSTITUTED BY ZNO.** CHIZHIKOV D M GULYANITSKAYA Z F SHCHASTLIVYI V P PETROVA R N DOKLADY AKAD NAUK SSSR 129 174-6 1959 CA 54 6227
17821	**CALCINATION OF ALUMINA HYDRATE.** LANYI BELA FEMIPARI KUTATO INTEZET KOZLEMENYEI 2 83-9 1958 CA 53 17656
17866	**KINETICS OF FLUIDITY OF COAL AND THE CHLOROFORM-SOLUBLE EXTRACT.** NADZIAKIEWICZ J FUEL 37 361-4 1958 CA 52 15023
17892	**HEAT CONDUCTANCE OF MELTING GLASS BATCHES.** KROGER, C. ELIGEHAUSEN, H. GLASTECH BER 32 362-73 1959 CA 54 7090 (FOR ENGLISH TRANSLATION SEE T29103)
17922	**WATER CONTENT OF GLASSES. V. DIFFUSION OF WATER IN GLASSES AT HIGH TEMPERATURES.** SCHOLZE, H. MULFINGER, H. O. GLASTECH BER 32 381-6 1959 CA 54 844
17924	**EVALUATION OF SEVERAL INFRARED TRANSMISSIVE GLASSES.** GREKILA R B ROOT H D GLASS IND 40 186-9 212-17 1959 CA 53 14442
17925	**WATER IN THE GLASS STRUCTURE.** SCHOLZE H GLASS IND 40 301-3 1959 CA 53 17458
17994	**VISCOSITY-ITS MEASUREMENT AND IMPORTANCE IN DENSE-MEDIUM CLEANING OF THE FINE SIZES OF COLA.** YANCEY H F GEER M R SOKASKI MICHAEL INTERN COAL PREPARATION CONGR 3RD BRUSSELS-LIEGE F1 583-91 1958 CA 54 23259
18000	**FLAMES AND FURNACES. FORMS OF RADIATION.** THRING M W IRON AND STEEL 33 11-15 1960 RM 17 87-P
18009	**THERMAL CONDUCTIVITY OF SEMI-CONDUCTIVE SOLIDS.. METHOD FOR STEADY-STATE MEASUREMENTS ON SMALL DISC REFERENCE SAMPLES.** FLYNN, D. R. NATIONAL BUREAU OF STANDARDS 1-7, 1960. (NBS RPT 6996, AD 258310)
18033	**SPECIFIC HEAT OF GLASSES. II. DIFFERENTIAL-THERMAL ANALYSIS.** PRODHOMME, M. VERRES ET REFRACTAIRES 13 3-16 1959 CA 53 11783
18058	**THERMAL CONDUCTIVITY OF A CHARGE LAYER CONSISTING OF LARGE PIECES.** MARKOV B L SOLOMENTSEV S L IZVESTIYA VUZ-CHERNAYA METALLURGIYA 3 176-83 1960 RM 17 330-D
18100	**THERMODYNAMIC CHARACTERISTICS OF MOLTEN IRON SILICATES.** ESIN O A BRATCHIKOV S G IZVEST VYSSHIKH UCHEB ZAVEDENII KHIM I KHIM TEKHNOL 2 247-53 1959 CA 53 18800
18202	**MEASUREMENTS OF THE THERMAL PROPERTIES OF CARBONACEOUS MATERIALS.** BATCHELOR J D YAVORSKY P M GORIN EVERETT J CHEM ENG DATA 4 241-6 1959 CA 54 6081
18230	**PARAMETRIC STUDIES OF METAL FIBER REINFORCED CERAMIC COMPOSITE MATERIALS.** LIU, T. S. STOWELL, E. Z. SOUTHWEST RESEARCH INST., SAN ANTONIO, TEXAS 1-73, 1961. (AD 252916)
18237	**EVALUATION OF SOIL HEAT CONDUCTIVITY WITH CYLINDRICAL TEST BODIES.** BUETTNER K TRANS AM GEOPHYS UNION 36 5 831-7 1955
18315	**SYMPOSIUM ON HEAT-TRANSFER PHENOMENA IN GLASS. HEAT-CONDUCTIVITY PROCESSES IN GLASS.** KINGERY W D J AM CERAM SOC 44 7 302-4 1961 JA 44 208
18316	**SYMPOSIUM ON HEAT-TRANSFER PHENOMENA IN GLASS. REVIEW OF RADIANT HEAT TRANSFER IN GLASS.** GARDON, R. J AM CERAM SOC 44 7 305-12 1961 JA 44 208
18317	**SYMPOSIUM ON HEAT-TRANSFER PHENOMENA IN GLASS. EXPERIMENTAL AND THEORETICAL COMPARISON OF RADIATION CONDUCTIVITY PREDICTED BY STEADY-STATE THEORY WITH THAT EFFECTIVE UNDER PERIODIC TEMPERATURE CONDITIONS.** CHARNOCK H J AM CERAM SOC 44 7 313-17 1961 JA 44 208
18318	**SYMPOSIUM ON HEAT-TRANSFER PHENOMENA IN GLASS. SPECTRAL TRANSMISSION OF GLASS AT HIGH TEMPERATURES AND ITS APPLICATION TO HEAT-TRANSFER PROBLEMS.** GROVE F J J AM CERAM SOC 44 7 317-20 1961 JA 44 208
18319	**SYMPOSIUM ON HEAT-TRANSFER PHENOMENA IN GLASS. TEMPERATURE MEASUREMENT OF GLASS BY RADIATION ANALYSIS.** VAN LAETHEM R LEGER L BOFFE M PLUMAT E J AM CERAM SOC 44 7 321-32 1761 JA 44 208
18320	**SYMPOSIUM ON HEAT-TRANSFER PHENOMENA IN GLASS. TOTAL HEAT-TRANSMISSION COEFFICIENTS OF AMBER AND GREEN GLASSES IN TEMPERATURES OF MELTING RANGE.** KRUSZEWSKI S J AM CERAM SOC 44 7 333-9 1961 JA 44 208
18321	**SYMPOSIUM ON HEAT-TRANSFER PHENOMENA IN GLASS. TEMPERATURE DISTRIBUTION AND HEAT FLOW IN GLASS IN BLANK MOLDS OF CONTAINER MACHINES.** TRIER W J AM CERAM SOC 44 7 339-45 1961 JA 44 208
18322	**SYMPOSIUM ON HEAT-TRANSFER PHENOMENA IN GLASS. TRANSFER OF HEAT IN GLASS DURING FORMING.** MCGRAW D A J AM CERAM SOC 44 7 353-63 1961 JA 44 208

TPRC Number	Bibliographic Citation
18353	**THE MEASUREMENT OF THERMAL CONDUCTIVITY OF REFRACTORY MATERIALS.** NORTON, F. H. KINGERY, W. D. MASSACHUSETTS INSTITUTE OF TECHNOLOGY 1-31, 1951. (NYO-599)
18367	**EFFECT OF GAMMA RADIATION ON SPECTRAL TRANSMISSION OF SOME COMMERCIAL GLASSES.** BARKER R S RICHARDSON D A J AM CERAM SOC 44 11 552-60 1961
18425	**TRANSMISSION AND REFLECTANCE OF CESIUM IODIDE IN THE FAR-INFRARED REGION.** PLYLER, E. K. ACQUISTA, N. J OPT SOC AM 48 668-9 1958 CA 53 3878
18428	**REFRACTIVE INDEXES AND TRANSMITTANCES OF SEVERAL OPTICAL GLASSES IN THE INFRARED.** CLEEK GIVEN W VILLA JOHN J HAHNER C H J OPT SOC AM 49 1090-5 1959 CA 54 1069
18522	**THERMAL CONDUCTIVITY OF BARIUM TITANATE CERAMICS.** YOSHIDA, I. NOMURA, S. SAWADA, S. J PHYS SOC JAPAN 13 1550-1 1958 CA 53 12619
18607	**VERIFICATION OF USE OF PEAK AREA FOR QUANTITATIVE DIFFERENTIAL THERMAL ANALYSIS.** DE JOSSELIN DE JONG, G. J AM CERAM SOC 40 2 42-9 1957
18609	**PHASE EQUILIBRIA AT LIQUIDUS TEMPERATURES IN THE SYSTEM IRON OXIDE -AL203 -SI02 IN AIR ATMOSPHERE.** MUAN ARNULF J AM CERAM SOC 40 4 121-33 1957
18613	**CALCULATION OF TEMPERATURE DISTRIBUTIONS IN GLASS PLATES UNDERGOING HEAT-TREATMENT.** GARDON, R. J AM CERAM SOC 41 6 200-9 1958
18615	**COLOR AND LIGHT SCATTERING OF PLATINUM IN SOME LEAD GLASSES.** RYDER R J RINDONE G E J AM CERAM SOC 41 10 415-22 1958
18621	**COLORS AND MAGNETIC PROPERTIES OF IRON IN GLASSES OF VARIOUS TYPES AND THEIR IMPLICATIONS CONCERNING STRUCTURE. I. HIGH ALKALI SILICATE GLASSES.** BISHAY, A. J AM CERAM SOC 42 9 403-7 1959
18622	**THE INFLUENCE OF SOME TRANSITION ELEMENTS ON VISIBLE AND INFRARED TRANSMISSION OF CALCIUM ALUMINATE GLASSES.** WEIDEL, R. A. J AM CERAM SOC 42 9 408-12 1959
18624	**EFFECT OF FLUORIDES ON INFRARED TRANSMITTANCE OF CERTAIN SILICATE GLASSES.** CLEEK, G. W. SCUDERI, T. G. J AM CERAM SOC 42 12 599-603 1959
18628	**EFFECT OF ASPECT RATIO AND VISCOSITY GRADIENTS ON FLOW THROUGH OPEN CHANNELS.** COOPER A R JR J AM CERAM SOC 43 2 97-104 1960
18630	**DETERMINATION OF SPECTRAL EMISSIVITY OF CERAMIC BODIES AT ELEVATED TEMPERATURES.** BLAIR G RICHARD J AM CERAM SOC 43 197-203 1960 CA 54 12529
18632	**APPARENT THERMAL CONDUCTIVITY OF A HETEROGENEOUS CERAMIC BODY UNDER ISOTHERMAL CONDITIONS.** WHITMORE D H J AM CERAM SOC 43 281-2 1960 CA 54 14618
18634	**EFFECT OF WATER CONTENT ON CORROSION OF BOROSILICATE GLASS.** PRIEST D K LEVY A S J AM CERAM SOC 43 7 356-8 1960
18635	**INVESTIGATION OF SOME GLASSES FOR HIGH-LEVEL GAMMA RADIATION DOSIMETERS.** HEDDEN W A KIRCHER J F KING B W J AM CERAM SOC 43 8 413-5 1960
18636	**THERMOLUMINESCENCE OF VITREOUS GERMANIUM OXIDE CONTAINING THE IMPURITY ALUMINUM.** GARINO-CANINA VITTORIO COHEN SABATINO J AM CERAM SOC 43 8 415-6 1960
18637	**APPLICATIONS OF RADIATION EFFECTS IN GLASSES IN LOW- AND HIGH-LEVEL DOSIMETRY.** BLAIR G E J AM CERAM SOC 43 8 426-9 1960
18638	**RADIATION DOSIMETER GLASS.** PAYMAL J BONNAUD M LE CLERC P J AM CERAM SOC 43 8 430-6 1960
18644	**RADIATION ENERGY TRANSFER AND THERMAL CONDUCTIVITY OF CERAMIC OXIDES.** LEE D W KINGERY W D J AM CERAM SOC 43 11 594-607 1960
18648	**STABILITY OF REFRACTORIES IN HYDROGEN-FLUORINE FLAMES.** EBNER M J AM CERAM SOC 44 1 7-12 1961
18657	**ENGINEERING ASPECTS OF SOLID CATALYSTS.** HOUGEN, O. A. IND ENG CHEM 53 509-28 1961
18662	**AERODYNAMIC CLASSIFIED TALCS AND THEIR EFFECT IN A CERAMIC BODY.** FELTON ERNEST WATTERLOND ROBERT BULL AM CERAM SOC 39 12 717-20 1960
18668	**A COMPARATIVE STUDY OF CALIFORNIA AND MONTANA TALCS.** STAFFORD RAY FELTON ERNEST BULL AM CERAM SOC 37 6 274-9 1958
18710	**THERMAL PROPERTIES OF BLACK COALS IN THEIR THERMAL DECOMPOSITION PROCESS.** ZAMOLUEV V K KHIM I TEKHNOL TOPLIV I MASEL 4 7 20-3 1959 CA 54 6084 (FOR ENGLISH TRANSLATION SEE T20611)
18939	**CHANGE IN THERMAL CONDUCTIVITY OF QUARTZ GLASS IN THE PROCESS OF CRYSTALLIZATION.** PUSTOVALOV V V STEKLO I KERAMIKA 17 28-30 1960 (FOR ENGLISH TRANSLATION SEE T19727)
18962	**THE INVESTIGATION OF THE HEAT CONDUCTIVITY OF REFRACTORY MATERIALS WITH PARTICULAR CONSIDERATION OF MAGNESITE BRICKS.** BOETTICHER MARTIN MITT FORSCH-INST VER STAHLWERKE A-G DORTMUND 2 235-48 1932 CA 27 2547
19008	**ALUMINA PORCELAIN AS A HIGH FREQUENCY INSULATOR. I. EFFECTS OF THE KINDS OF ALUMINA ON PHYSICAL PROPERTIES.** KATO, S. OKUDA, H. NAGOYA KOGYO GIJUTSU SHIKENSHO HOKOKU 9 402-8 1960 CA 54 20134
19024	**MOLECULAR EFFECTS IN HEAT CONDUCTION THROUGH POROUS ROCKS.** WOODSIDE W MESSMER J H J GEOPHYS RESEARCH 65 10 3481-5 1960
19030	**THE INFLUENCE OF THE GAS PRESSURE ON THE HEAT CONDUCTIVITY OF INSULATING MATERIALS.** KLING, G. ALLGEM WARME-TECH 3 167-74 1952 CA 47 5581
19078	**LIGHTWEIGHT /REFRACTORIES/ FROM ZIRCONIUM DIOXIDE.** PIRIGOV A A RAKINA V P OGNEUPORY 23 145-50 1958 CA 52 13211
19079	**THE THERMAL CONDUCTIVITIES OF MAGNESITE REFRACTORIES.** PUSTOVALOV V V OGNEUPORY 23 326-8 1958 CA 52 20966
19081	**LIGHT-WEIGHT ALUMINA REFRACTORIES AND THEIR USE.** TSIGLER V D ELTYSHEVA A A PINDRIK B E OGNEUPORY 25 299-307 1960 CA 54 25664 (FOR ENGLISH TRANSLATION SEE T32629)
19095	**TRANSMISSION CURVES OF OPTICAL GLASSES.** TAUERN D OPTIK 15 619-27 1958 CA 53 12613

TPRC Number | Bibliographic Citation

19138 ON MEASUREMENT OF THERMAL CONDUCTIVITY AT HIGH TEMPERATURES.
LAUBITZ M J
CAN J PHYS
39, 1029–39, 1961.
(NRC 6363)

19236 VIBRATION SPECTRA AND SPECIFIC HEATS OF DIAMOND-TYPE LATTICES.
PHILLIPS, J. C.
PHYS REV
113 147–55 1959 CA 53 11931

19249 THE DETERMINATION OF THE SPECIFIC HEAT OF BROWN COAL.
BURCKHARDT G FRITZSCHE A
BRAUNKOHLENARCH
17 20–33 1927 CA 23 499

19362 HEAT TRANSPORT THROUGH POWDERS.
FULK M M DEVEREUX R J SCHRODT J E
PROC CRYOGENIC ENG CONF 2ND BOULDER 1956
163–5 1957 CA 52 17828

19365 THE HEAT CAPACITY OF DIAMOND BETWEEN O AND 1000 K.
RAMAN C V
PROC INDIAN ACAD SCI
46 A 323–32 1957 CA 53 17656

19379 THE SPECTROSCOPIC BEHAVIOUR OF ROCK-SALT AND THE EVALUATION OF ITS SPECIFIC HEAT. IV. SPECIFIC HEAT AND SPECTRAL FREQUENCIES.
RAMAN C V
PROC INDIAN ACAD SCI A
54 5 294–304 1961 SA 65 6261

19384 THE STRUCTURE OF GLASS. THE QUANTUM THEORY OF HEAT CAPACITY AND THE STRUCTURE OF SILICATE GLASSES.
TARASOV V V
STRUCTURE OF GLASS PROC CONF LENINGRAD 1953
49–54 1958 CA 53 17455

19385 SPECIFIC HEAT AND THERMAL EXPANSION OF FLUORSPAR.
GANESAN S SRINIVASAN R
PROC NATL INST SCI INDIA
25 A 139–53 1959 CA 53 19552

19419 EFFECT OF REACTOR IRRADIATION ON THE PHYSICAL PROPERTIES OF BERYLLIUM OXIDE.
GILBREATH J R SIMPSON O C
PROC. UN INTERN. CONF. PEACEFUL USES AT ENERGY, 2ND
5, 367–74, 1958.

19478 INFRARED EMISSION SPECTRA OF QUARTZ.
WENTINK, T., JR. PLANET, W. G., JR.
J OPT SOC AM
51 6 595–600 1961 JA 44 273

19485 THERMAL PROPERTY MEASUREMENTS AT VERY HIGH TEMPERATURES.
RASOR, N. S. MC CLELLAND, J. D.
REV SCI INSTRUMENTS
31 595–604 1960 CA 55 16035

19538 PHYSICOCHEMICAL BASIS OF TECHNOLOGY OF FORSTERITE REFRACTORIES.
BEREZHNOI, A. S.
SBORNIK NAUCH. TRUDOV VSESOYUZ. NAUCH.-ISSLEDOVATEL. INST. OGNEUPOROV
2 5–70 1958 CA 54 12525

19558 SPECIFIC HEAT OF ROCKS AND THE VELOCITY OF ELASTIC WAVES WITHIN THE OUTER LAYER OF THE EARTHS CRUST.
NORITOMI KAZUO
SCI REPTS TOHOKU UNIV
7 5 190–200 1956 CA 53 4040

19561 THERMAL CONDUCTIVITY OF MOLDED PLASTER.
TERAOKA ICHIRO ITO SHUYO
SEKKO TO SEKKAI
39 5–10 1959 CA 53 13719

19583 THERMAL EMISSIVITY OF SIO2—AL2O3 REFRACTORIES IN THE INFRARED SPECTRUM.
ALEGRE R
SILICATES INDS
23 253–60 1958 CA 52 19062
(FOR ENGLISH TRANSLATION SEE T32477)

19586 THERMODYNAMIC INVESTIGATION OF THE SOLID-STATE REACTIONS IN SILICATE SYSTEMS.
MCHEDLOV-PETROSYAN O P BAUBUSHKIN V I
SILIKAT TECH
9 209–12 1958 CA 53 13757

19587 CONSTITUTION AND PROPERTIES OF PHOSPHATE-TURBID GLASSES.
SCHONBORN HERBERT
SILIKAT TECH
10 390–400 1959 CA 54 7093

19593 THE SPECIFIC HEAT AND THE CRYSTAL STRUCTURE OF THE TEETH, BONE, AND MINERAL APATITES BY THE ABSOLUTE SPECIFIC HEATS AND THE THERMAL ANALYSIS METHOD.
KATO, K.
SOGO IGAKU
13 331–5 1956 CA 54 21972

19612 MEASURING THE THERMAL CONDUCTIVITY OF REFRACTORY MATERIALS AND INSULATORS.
LONGCHAMBON L
CERAMIQUE
36 215–17 1933 CA 28 2487

19621 HIGH-TEMPERATURE HEAT CAPACITY OF DIAMOND.
VICTOR ANDREW
NATL BUR STANDARDS /US/ TECH NEWS BULL
45 9 146–7 1961

19642 THE THERMAL CONDUCTIVITY OF SOME INSULATING MATERIALS AT LOW TEMPERATURES.
CHOW C S
PROC PHYS SOC LOND
61 206–16 1948 AM 1 1549

19646 RADIATIVE POWER OF FURNACE SLAGS.
AGABABOV S G
TEPLOENERGETIKA
8 56–60 1958 AM 12 5769
(FOR ENGLISH TRANSLATION SEE T21015)

19650 EXPERIMENTAL INVESTIGATION OF UNSTEADY-STATE HEAT AND MASS TRANSFER DURING PHASE CHANGES AND CHEMICAL TRANSFORMATIONS.
RALKO, A. V.
TEPLO- I MASSOOBMEN V PROTSES. ISPARENIYA, AKAD. NAUK SSSR ENERGET. INST.
57–86 1958 CA 54 23480

19685 THE SPECIFIC HEAT OF OUR /GERMAN/ REFRACTORY BRICK AND ITS DEPENDENCE UPON TEMPERATURE.
MIEHR W IMMKE H KRATZERT I
TONIND ZTG
50 1791–3 1926 CA 21 3113

19701 ON THE MEASUREMENT OF THE THERMAL CONDUCTIVITIES OF ROCKS BY OBSERVATIONS ON A DIVIDED BAR APPARATUS.
BECK, A. E. BECK, J. M.
TRANS AMER GEOPHYS UN
39 6 1111–23 1958 AM 12 5192

19727 CHANGE IN THERMAL CONDUCTIVITY OF QUARTZ GLASS IN THE PROCESS OF CRYSTALLIZATION.
PUSTOVALOV V V
GENERAL ELECTRIC RESEARCH LAB SCHENECTADY N Y
1–7, 1960.
(ENGLISH TRANSLATION OF STEKLO I KERAMIKA 17, 28–30, 1960; FOR ORIGINAL SEE T18939)
(60-RL-/2525M/)

19745 ELECTRICAL AND THERMAL CONDUCTIVITY OF INSULATORS.
FETTERLEY G H
THERMOELECTRICITY PROC CONF 1958
307–19 1960 CA 54 25379

19768 THERMAL CONDUCTIVITY OF CERAMIC MATERIALS.
BISHUI B M PRASAD J
TRANS INDIAN CERAM SOC
17 108–16 1958 CA 54 12529

19806 THE THERMAL PROPERTIES OF REFRACTORY MATERIALS AND A CONSIDERATION OF THE FACTORS INFLUENCING THEM.
GREEN A T
TRANS CERAM SOC /ENGLAND/
25 361–85 1925 CA 21 2542

19815 APPARATUS FOR THE DETERMINATION OF THERMAL CONDUCTIVITIES OF HIGH-TEMPERATURE INSULATION.
HEILMAN R H
TRANS AM INST CHEM ENG
18 283–93 1926 CA 21 3491

19818 COATINGS FOR SOLAR CELLS.
WITUCKI, R. M. LEWIS, A. E.
HOFFMAN ELECTRONICS CORP., SANTA BARBARA, CA.
1–86, 1961.
(AD 258660)

19822 THERMOELECTRIC MATERIALS.
BROOKS, M. H.
NATL. LEAD CO., TITANIUM ALLOY MFG. DIV., NIAGARA FALLS, N. Y.
1–4, 1961.
(AD 258604)

19823 THE THERMAL CONDUCTIVITY OF URAL REFRACTORY MATERIALS.
OGARKOV A F
TRUDY URAL POLITEKH INST SBORNIK
55 5–22 1956 CA 53 15513

19842 THE SPECIFIC HEAT OF FUSED IRON SILICATES.
ESIN O A BRATCHIKOV S G
TRUDY URAL POLITEKH INST IM S M KIROVA
75 243–7 1959 CA 54 19376

19849 SPECIFIC HEAT OF THE MINERAL MATTER AND CARBONIZATION RESIDUES OF BALTIC OIL SHALE.
KOLLEROV, D. K.
TR. VSES. NAUCH.-ISSLED. INST. PO PERERAB. SLANTSEV
7 64–79 1959 CA 54 15901

TPRC Number	Bibliographic Citation
19856	**THERMOPHYSICAL AND PHYSICOCHEMICAL PROPERTIES OF BITUMINOUS SHALE OF THE BALTIC REGION.** KOLLEROV, D. K. TR. VSES. NAUCH-ISSLED. INST. PERERABOTKI I ISPOLZOVAN TOPLIVA 8 35-51 1959 CA 55 2057
19864	**TRANSPARENT MATERIALS FOR AIRCRAFT ENCLOSURES. WADC—UNIVERSITY OF DAYTON JOINT CONFERENCE.** WITTMAN, R. E. WRIGHT AIR DEVELOPMENT CENTER 1-209, 1957. (WADC-TR 57-421, AD 142021)
19964	**DEVELOPMENT OF CERAMIC COATINGS FOR URANIUM.** PLOETZ, G. L. SOWMAN, H. G. KNOLLS ATOMIC POWER LAB. 27PP., 1956. (KAPL-1479)
19992	**HIGH CONDUCTIVITY URANIUM DIOXIDE.** SHAPIRO, H. POWERS, R. M. SYLVANIA - CORNING NUCLEAR CORPORATION 1-43, 1959. (SCNC-294)
19993	**URANIUM DIOXIDE FUEL MATERIALS WITH IMPROVED THERMAL CONDUCTIVITY.** SHAPIRO, H. POWERS, R. M. SYLVANIA - CORNING NUCLEAR COPORATION 1-22, 1959. (SCNC-271)
20006	**THE SPECIFIC HEAT OF SOME REFRACTORY MATERIALS.** CLEMENTS J F TRANS BRIT CERAM SOC 61 452-62 1962 CA 58 1218
20009	**THE THERMAL CONDUCTIVITY OF REFRACTORY MATERIALS AT HIGH TEMPERATURES.** GREEN A T TRANS BRIT CERAM SOC 21 394-414 1921
20019	**A PROPOSED METHOD OF TEST FOR SPECIFIC HEAT OF THERMAL INSULATING MATERIALS.** SPEAR, N. H. AMER SOC TEST MAT BULL 168 79-82 1950 AM 4 2723
20041	**EFFECT OF MOISTURE CONTENT OF COAL ON ITS THERMAL AND ELECTRICAL PROPERTIES.** AGROSKIN A A BERGAKADEMIE 11 7-12 1959 CA 55 11802
20043	**COARSE-POROUS CONCRETE FROM BLAST-FURNACE SLAGS.** SKRAMTAEV B G GEMMERLING G V DOMNICH A I BETON I ZHELEZOBETON 10 439-42 1960 CA 55 13807
20059	**THERMAL CONDUCTIVITY.** FRITZ W POLTZ H GLASHUTTEN-HANDBUCH X-55 1-12 1961
20061	**DETERMINATION OF THE THERMAL CONDUCTIVITY OF POORLY CONDUCTING MATERIALS BY A METHOD OF CONTACT TEMPERATURES.** CAMPAN T I BULL INST POLITEH IASI 3 3-4 215-22 1957 JA 43 270
20062	**FACTORS AFFECTING THERMAL STRESS RESISTANCE OF CERAMIC MATERIALS.** KINGERY W D J AM CERAM SOC 38 1 3-15 1955
20064	**CORRELATION OF ELECTRICAL AND THERMAL PROPERTIES OF BUILDING BRICK.** JOHNSON, J. S. J AM CERAM SOC 21 79-85 1938
20071	**THE TRANSIENT HEAT FLOW METHOD OF DETERMINING THERMAL CONDUCTIVITY. APPLICATION TO INSULATING MATERIALS.** LENTZ C P CANAD J TECHNOL 30 6 153-66 1952 AM 6 283
20074	**FUNDAMENTALS OF HEAT FLOW IN MOLTEN GLASS AND IN WALLS FOR USE AGAINST GLASS.** MC CAULEY, G. V. J AM CERAM SOC 8 493-504 1925
20137	**THE EVALUATION OF THE PROPERTIES OF HIGH-TEMPERATURE INSULATING MATERIALS. PART 5. STUDIES OF A HEAT DISSIPATION TEST FOR INSULATING MATERIALS.** BARRETT L R VYSE J GREEN A T GAS RESEARCH BOARD COMMUN 19 49-73 1944-5
20138	**EMISSIVITY OF TECHNICAL SURFACES AT HIGH TEMPERATURES.** RUDOLPH H H ELEKTROWARME 18 151-7 1960 RM 17 351-P
20140	**THERMAL CONDUCTIVITY OF VERMICULITE.** AUTHOR ANON. ENGINEERING 186, 616P., 1958.
20183	**FUSED QUARTZ AS A MODEL MATERIAL IN THERMAL-CONDUCTIVITY MEASUREMENTS.** DEVYATKOVA E D PETROV A V SMIRNOV I A MOIZHES B YA FIZ TVERDOGO TELA 2 738-46 1960 CA 55 8023 (FOR ENGLISH TRANSLATION SEE T26060)
20221	**INSTRUMENTS FOR THE DETERMINATION OF THE PHYSICAL PROPERTIES OF GLASSES. II.** SCHULZ H GLAS-EMAIL-KERAMO-TECH 8 12 461-4 1957 JA 41 113
20245	**DETERMINATION OF THE THERMAL COEFFICIENTS OF GRANULAR AND SOLID MATERIALS BY THE METHOD OF PLANE THERMAL WAVES.** KRAVCHUK E M INZHENER-FIZ ZH 1 10 29-37 1958 AM 14 3262 (FOR ENGLISH TRANSLATION SEE T20436)
20247	**RELATION BETWEEN THERMOPHYSICAL, RHEOLOGICAL, AND ELECTROPHYSICAL PROPERTIES OF COLLOIDAL CAPILLARY-POROUS BODIES AND THE FORM OF BONDS OF MOISTURE.** MIGLYACHENKO A F INZHENER-FIZ ZHUR AKAD NAUK BELORUS SSR 3 2 25-30 1960 CA 55 3155
20249	**THE INFLUENCE OF ADMIXTURES OF LIME TO CERAMIC BODIES.** SUNDIUS NILS NORDGREN HARRY TRANS BRIT CERAM SOC 58 4 224-30 1959
20260	**RADIATIVE TRANSFER IN THE EARTH S MANTLE.** CLARK, S. P., JR. TRANS AM GEOPHYS UNION 38 6 931-8 1957
20265	**THERMAL CONDUCTIVITY OF POROUS MEDIA. II. CONSOLIDATED ROCKS.** WOODSIDE W MESSMER J H J APPL PHYS /USA/ 32 9 1699-1706 1961 SA 64 20651
20281	**COEFFICIENTS OF TRANSFER OF HEAT BY TUFFACEOUS STONES.** GINDOYAN A G IZVEST AKAD NAUK ARMYAN SSR SER TEKH NAUK 13 2 29-42 1960 CA 55 10838
20285	**THERMAL CONDUCTIVITY OF CRYSTALLINE NATURAL MAGNETITE AT LOW TEMPERATURES.** SEDOV V L SOLOMATINA L V ILCHENKO L N ZH EKSPERIM I TEOR FIZ 43 1125-6 1962 CA 58 7390 (FOR ENGLISH TRANSLATION SEE T20413)
20292	**THE REFLECTING POWER OF METALS IN THE ULTRAVIOLET REGION OF THE SPECTRUM.** HULBURT E O ASTROPHYS J 42 205-30 1915
20313	**THERMAL CONDUCTIVITY OF BERYLLIA CERAMICS FROM -200 TO 150 DEGREES.** BURK MAKSYMILIAN J AM CERAM SOC 46 150-1 1963 CA 59 1117
20345	**APPARATUS FOR THE MEASUREMENT OF THE THERMAL CONDUCTIVITY OF REFRACTORY MATERIALS AT ELEVATED TEMPERATURES.** BRADY J G J CAN CERAM SOC 23 19-30 1954 JA 41 154
20376	**CONTRIBUTION TO THE DETERMINATION OF THE REFLECTIVITY OF METALS IN THE VISIBLE AND ULTRAVIOLET LIGHT.** VON FRAGSTEIN KONRAD ANN PHYSIK 17 1-21 1933
20382	**DIRECTION DISTRIBUTION OF THE THERMAL RADIATION OF SURFACES.** SCHMIDT E ECKERT E FORSCH GEBIETE INGENIEURW 6 175-83 1935

TPRC Number	Bibliographic Citation
20396	**A METHOD TO DETERMINE THE RADIATION IN ABSOLUTE UNITS AND THE RADIATION OF A BLACK BODY BETWEEN 0 AND 100 DEGREES.** KURLBAUM F ANN PHYSIK 65 746-60 1898
20413	**THE HEAT CONDUCTIVITY OF A CRYSTAL OF NATURAL MAGNETITE AT LOW TEMPERATURES.** SEDOV V L SOLOMATINA L V ILCHENKO L N SOVIET PHYSICS—JETP 16 3 794-5 1963 (ENGLISH TRANSLATION OF ZH. EKSP. TEOR. FIZ. 43, 1125-6, 1962; FOR ORIGINAL SEE T20285)
20414	**THE THERMAL CONDUCTIVITY OF HEAT INSULATORS.** VAN DUSEN M S J AM SOC HEATING VENTILATING ENGRS 26 7 625-56 1920
20435	**PROPAGATION OF HEAT IN GLASS AT HIGH TEMPERATURES. PART 1.** GEFFCKEN, W. SPECIAL LIBRARY ASSOC. TRANS. CENTER 1-14, 1962. (ENGLISH TRANSLATION OF GLASTECH. BER. 25 (12), 392-6, 1952; FOR ORIGINAL SEE T145) (SLA-62-16131)
20436	**DETERMINATION OF THE THERMAL COEFFICIENTS OF FRIABLE AND SOLID SUBSTANCES BY THE METHOD OF PLANE THERMAL WAVES.** KRAVCHUK E M NAVAL RESEARCH LABORATORY 1-13, 1962. (ENGLISH TRANSLATION OF INZH. FIZ. ZH. 1 (10), 29-37, 1958; FOR ORIGINAL SEE T20245) (NRL TRANS-885, SLA-62-11609)
20452	**RECENT MEASUREMENTS ON THE HEAT CONDUCTIVITY OF INSULATING MATERIALS AT THE CRYOGENIC LABORATORY OF LOUVAIN. MEASUREMENTS ON THE HEAT TRANSMISSION THROUGH COMPLEX STRUCTURE.** VAN ITTERBEEK A MYNCKE H DE GREVE L CONGR. INTERN. FROID. 2019-28, 1955.
20468	**THERMAL RADIATION FROM PARTIALLY TRANSPARENT REFLECTING BODIES.** MC MAHON H O J OPT SOC AMER 40 6 376-80 1950 AM 4 1354
20502	**INFRARED SPECTRAL EMISSIVITY OF TERRAIN.** FREDRICKSON, W. R. GINSBURG, N. PAULSON, R. SYRACUSE UNIV. DEPT. OF PHYSICS, SYRACUSE, N. Y. 1-153PP., 1958. (WADC-TR-58-229, AD 155 552)
20505	**A TECHNICAL METHOD FOR INVESTIGATING THE THERMAL CONDUCTIVITY OF SLABS OF MATERIAL.** POENSGEN RICHARD VDI ZEITSCHRIFT 56 41 1653-58 1912
20507	**THERMAL CONDUCTIVITY OF REFRACTORY CASTABLES.** RUH, E. RENKEY, A. L. J AM CERAM SOC 46 89-92 1963 CA 58 11080
20510	**THE EFFECTS OF SOLID SOLUTION ADDITIONS ON THE THERMAL CONDUCTIVITY OF UO2.** POWERS, R. M. CAVALLARO, Y. MATHERN, J. P. SYLVANIA-CORNING NUCLEAR CORP., BAYSIDE, N. Y. 1-128, 1960. (SCNC-317)
20514	**MATERIALS FOR ASTRONAUTIC VEHICLES.** MURPHY, A. J. COLLEGE OF AERONAUTICS, CRANFIELD, ENGLAND 1-21, 1960. (COA RPT 139, AD 260616)
20529	**DEVELOPMENT OF A PROBE FOR MEASURING THE COEFFICIENT OF THERMAL CONDUCTIVITY IN BUILDING MATERIALS.** ERKELENS H J J INST HEATING VENTILATING ENGRS 27 281-96 1960
20530	**FUSED SILICA CERAMICS.** WALTON J D CERAM AGE 77 5 52-8 1961
20540	**MEASUREMENT OF THE NET THERMAL CONDUCTIVITY OF FIBROUS SUBSTANCES.** TAKAMURA Y J PHYS SOC JAPAN 8 5 674-6 1953 AM 7 1830
20567	**THE THERMAL CONDUCTIVITY OF GLASSES AS A FUNCTION OF THE CHEMICAL COMPOSITION. III. TESTING MATERIALS.** RUSS ALFRED SPRECHSAAL 61 907-12 1928
20572	**TRANSMISSION SPECTRA OF THIN SILICATE GLASS FILMS IN THE INFRARED.** FLORINSKAYA V A PECHENKINA R S DOKLADY AKAD NAUK SSSR 89 37-40 1953 (FOR ENGLISH TRANSLATION SEE T15126)
20581	**A NEW PRINCIPLE FOR DETERMINATION OF THERMAL CONDUCTIVITY IN SMALL TEST BODIES OF HOMOGENEOUS CERAMIC MATERIALS.** STALHANE, B. IVA /STOCKHOLM/ 19 5 199-200 1948
20582	**NEW TEST METHODS OF THE INSTITUTE FOR RESEARCH IN HEAT INSULATION.** RAISCH E ARCH WARMEWIRTSCH U DAMPFKESSELW 10 369-73 1929
20589	**STUDIES ON THE THERMAL CONDUCTIVITIES OF SOME REFRACTORY MATERIALS.** BUCKNER O S J AM CERAM SOC 7 19-28 1924 CA 18 1040
20590	**CRYSTALLINITY OF SILICA GEL REVEALED IN THE HEAT CAPACITY CURVE AT MODERATELY HIGH TEMPERATURES.** TAKAMURA T KOLLOID Z 157 1 11-16 1958 JA 43 124
20600	**CHARACTERISTICS GOVERNING THE FRICTION AND WEAR BEHAVIOR OF REFRACTORY MATERIALS FOR HIGH TEMPERATURE SEALS AND BEARINGS.** SIBLEY, L. B. MACE, A. E. GRIESER, D. R. ALLEN, C. M. BATTELLE MEMORIAL INSTITUTE 1-47, 1960. (WADD TR 60-54, AD 243897, PB 171010)
20611	**THERMAL PROPERTIES OF COALS UNDERGOING PYROLYSIS.** ZAMOLUEV V K SPECIAL LIBRARY ASSOC., TRANSLATION CENTER 1-7, 1962. (ENGLISH TRANSLATION OF KHIM. I TEKHNOL. TOPLIVA I MASEL 4 (7), 20-3, 1959; FOR ORIGINAL SEE T18710) (OSIR LLM. 3449, SLA-62-24850)
20631	**NEW DEVELOPMENTS IN CERAMICS.** KOENIG, J. H. MAT DESIGN ENG 47 5 121-136 1958 MA 26 1054
20637	**ABLATION, HEAT SINK AND RADIATION.** COLLINS, J. O. SPEIL, S. MATERIALS IN DESIGN ENGINEERING 53 114-8 149-50 1961 RM 18 212-P
20643	**THE EMISSIVITY AND TEMPERATURE OF THE MANTLE /GAS LIGHT/ AT DIFFERENT CERIUM CONTENT.** RUBENS H ANN PHYSIK 20 593-600 1906
20650	**PHYSICAL PROPERTIES AND CONSTITUTION OF LIQUID SLAGS.** BRADBURY B T WILLIAMS D J METALLURGIA 62 235-40 1960 CA 55 8225 63 19-24 1961 CA 55 8225
20652	**METHODS OF MEASURING THE THERMAL CONDUCTIVITY OF INSULATING AND REFRACTORY MATERIALS.** HARTMANN M L WESTMONT O B WEINLAND C E PROC AM SOC TESTING MATERIALS 28 2 820-47 1928
20655	**DETERMINATION OF THERMAL CONDUCTIVITY OF EBONITE AND OF GLASS.** DINA ALBERTO REND IST LOMBARDO SCI /PT 1/ 32 205-12 1899
20657	**AN APPARATUS FOR MEASURING THE THERMAL CONDUCTIVITY OF REFRACTORIES.** CODEGONE, C. RICERCA SCI E RICOSTRUZ 10 701-6 1939
20659	**FORSTERITE PORCELAIN AS A HIGH FREQUENCY INSULATOR. VII. PROPERTIES OF SILICA RICH FORSTERITE PORCELAIN.** HIRAI, M. SUGIURA, M. SANO, S. ISHII, E. NAGOYA KOGYO GIJUTSU SHIKENSHO HOKOKU 9 1 41-8 1960 JA 43 193
20660	**THERMAL AND PHYSICAL PROPERTIES OF HIGHLY CARBONIZED POLYMERS.** ZAMOLUEV V K SOVIET PLASTICS 41-3 1960 (ENGLISH TRANSLATION OF PLASTICHESKIE MASSY (8), 46-8, 1960; FOR ORIGINAL SEE T11725)

TPRC Number *Bibliographic Citation*

20683 **THE INFLUENCE OF CERTAIN OXIDES ON REDUCTION OF LEAD SILICATE GLASSES IN HYDROGEN.**
KITAIGORODSKII I I MENDELEEV D I FAINBERG E A
GRECHANIK L A
GLASS AND CERAM
19 11 645–7 1963
(ENGLISH TRANSLATION OF STEKLO I KERAM. 19, (12), 8–10, 1662; FOR ORIGINAL SEE T31150)

20705 **REFRACTORIES FROM ONOTSK MAGNESITE AND THEIR SERVICE IN THE CHECKERS OF OPEN-HEARTH FURNACES.**
DYACHKOV P N STEPANOVA I A
OGNEUPORY
24 9 403–10 1959 JA 44 165

20740 **SURVEY OF MATERIALS FOR HIGH-TEMPERATURE BEARING AND SLIDING APPLICATIONS.**
AMATEAU, M. F. NICHOLSON, D. W.
GLAESER, W. A.
BATTELLE MEMORIAL INSTITUTE
1–67, 1961.
(DMIC MEMO 106, PB 171625)

20765 **THE THERMAL CONDUCTIVITY OF FIRE-CLAY AT HIGH TEMPERATURES.**
CLEMENT J K EGY W L
UNIV ILLINOIS ENG EXPT STA BULL
36 1–31 1909

20771 **STUDY OF INFRARED SOURCES.**
FINKELSTEIN, I. S.
SERVO CORP., OF AMERICA, NEW HYDE PARK, N. Y.
1–25, 1958.
(RADC-TN-58-326, AD 259853)

20774 **FACTORS AFFECTING THERMAL SHOCK RESISTANCE OF POLYPHASE CERAMIC BODIES.**
SHAFFER, P. T. B. HASSELMAN, D. P. H.
CHABERSKI, A. Z.
CARBORUNDUM CO., NIAGARA FALLS, N. Y.
1–62, 1961.
(WADD TR 60–749(PT 1), AD 258537)

20780 **THE THERMAL CONDUCTIVITY OF ROCKS AND ITS DEPENDENCE UPON TEMPERATURE AND COMPOSITION.**
BIRCH, F. CLARK, H.
AM J SCI
238 529–58 1940

20784 **A HEAT-CAPACITY FUNCTION WITH VELOCITY DISPERSION FOR ISOTROPIC SOLIDS.**
BARBER S W MARTIN B
J PHYS CHEM SOLIDS
9 3–4 198–213 1959 SA 62 8818

20810 **TRANSMITTANCE OF OPTICAL MATERIALS AT HIGH TEMPERATURES IN THE 1- TO 12-MICRON RANGE.**
GILLESPIE, D. T. OLSEN, A. L.
NICHOLS, L. W.
US NAVAL ORDNANCE TEST STATION CHINA LAKE, CALIFORNIA
1–28PP., 1964.
(NAVWEPS REPT. 8558, NOTS-TP-3586, AD 609 036)

20833 **AN ATLAS OF REFLECTIVITIES OF SOME COMMON TYPES OF MATERIALS.**
BELL, E. E.
OHIO STATE UNIVERSITY
1–15, 1957.
(AD 141795)

20847 **FAR ULTRAVIOLET RESEARCH.**
NELSON, J. R. CROCKER, W. C.
CONVAIR DIV. GENERAL DYNAMICS CORP., SAN DIEGO, CA.
III–179–86, 1960.
(ZPH–075, AD 252819)

20864 **VIBRATIONS IN SPACE LATTICES.**
BORN M VON KARMAN THEODOR
GREAT BRITAIN ROYAL AIRCRAFT ESTABLISHMENT
1–27, 1961.
(ENGLISH TRANSLATION OF PHYS. Z. 13, 297–309, 1912; FOR ORIGINAL SEE T20865)
(RAE LIB TRANS 930, AD 253910)

20865 **VIBRATIONS IN SPACE LATTICES.**
BORN, M. VON KARMAN, T.
PHYSIK Z
13 297–309 1912
(FOR ENGLISH TRANSLATION SEE T20864)

20866 **A STUDY OF LUNAR THERMAL EMISSION.**
GIRAUD, A.
RADIATION LAB., UNIV. OF MICHIGAN, ANN ARBOR
1–33, 1960.
(IRIA 03544–1–T, AD 253471)

20867 **ESTIMATION OF THE PHYSICAL CONSTANTS OF THE LUNAR SURFACE.**
BRUNSCHWEG, M. FENSLER, W. E. KNOTT, E.
OLTE, A. SIEGEL, K. M. AHRENS, T. J.
DUNN, J. R. GERHARD, F. B., JR.
KATZ, S. ROSENHOLTZ, J. L.
RADIATION LAB., UNIV. OF MICHIGAN, ANN ARBOR
1–79, 1960.
(IRIA 3544–1–F, PB 162020, AD 253472)

20886 **SOME PROPERTIES OF FUSED QUARTZ AND OTHER FORMS OF SILICON DIOXIDE.**
WATSON H L
J AM CERAM SOC
9 511–34 1926 CA 20 3546

20915 **INSTRUMENT FOR MEASURING THE THERMAL CONDUCTIVITY OF SINTERED REFRACTORIES AT ELEVATED TEMPERATURES.**
PROCHAZKA SVANTE HLAVAC JAN
SILIKATY
2 3 290–5 1958 JA 43 196

20917 **PROPERTIES OF FLUORIDE AND PHOSPHATE OPALINE GLASSES.**
SILVESTROVICH, S. I.
SILIKATY SBORNIK STATEI KHIM. TEKHMOL. SILIKATOV
VSES. KHIM. OBSHCH. MENDELEEVA
1 3–13 1959 CA 55 8800

20918 **A NONSTATIONARY METHOD FOR DETERMINING THERMAL CONDUCTIVITY OF SOIL IN SITU.**
DE VRIES D A
SOIL SCIENCE
73 2 83–9 1952 AM 5 2742

20920 **THE SPECIFIC HEATS AT LOW TEMPERATURES OF SODIUM, POTASSIUM, MAGNESIUM AND CALCIUM METALS, AND OF LEAD SULFIDE.**
EASTMAN E D RODEBUSH W H
J AM CHEM SOC
40 489–500 1918

20925 **EFFECT OF TEXTURE AND KILN ATMOSPHERE ON THE HIGH TEMPERATURE THERMAL CONDUCTIVITY OF CHAMOTTE BRICK.**
ECKHOFF P SCHWIETE H E
SPRECHSAAL
93 19 506–12
20 539–42
21 557–61 1960 JA 45 9

20929 **THE THERMAL CONDUCTIVITY OF SOME REFRACTORIES.**
NORTON F H
J AM CERAM SOC
10 30–52 1927 CA 21 634

20933 **EVACUATED POWDER INSULATION FOR LOW TEMPERATURES.**
FULK M M
PROGR IN CRYOGENICS
1 65–84 1959

20947 **CERAMICS—THERMAL CONDUCTIVITY.**
ARTHUR G
NUCLEAR ENG
6 138–42 1961 CA 55 23964

20950 **SOME THERMAL CHARACTERISTICS OF CLAYS.**
MAC GEE A E
J AM CERAM SOC
10 561–8 1927 CA 21 3435

20956 **THERMAL CONDUCTIVITY MEASUREMENTS TO -180 C.**
CODEGONE C
TERMOTECNICA
4 573–4 1950 AM 4 2709

20957 **ON THE MEASUREMENT OF THERMAL DIFFUSIVITY OF REFRACTORIES.**
GIANNONE A
TERMOTECNICA
6 11 489–92 1952 AM 6 2356

20964 **PROPERTIES AND APPLICATIONS OF GLASSES AND CERAMICS.**
BLIZARD J R
MACHINE DESIGN
33 14 166–7 169 1961 JA 44 275

20975 **METHODS OF MEASUREMENT AND DETERMINATIONS OF HEAT CONDUCTIVITY IN BUILDING MATERIALS.**
CODEGONE C
TERMOTECNICA
9 10 474–90 1955 AM 9 1602

20977 **DEVELOPMENT OF A METHOD FOR THE DETERMINATION OF THE THERMAL CONDUCTIVITY OF RAW AND FIRED PRODUCTS OF THE BRICK AND MINERAL INDUSTRIES AT HIGH TEMPERATURE.**
LEHMANN, H. GATZKE, E.
TONIND-ZTG U KERAM RUNDSCHAU
81 13 217–20
15 249–56 1957 JA 41 98

20985 **QUALITIATIVE CHARACTERISTICS OF SAPROPEL.**
SEMENSKII E P
TORFYANAYA PROM
23 3 19–23 1946

21006 **METHOD FOR MEASURING THE THERMAL CONDUCTIVITY OF GLASS.**
REULOS RENE
REV OPT
10 6 266–72 1931

21008 **A QUANTUM THEORY OF THE THERMAL CONDUCTIVITY IN NON-METALLIC CRYSTALS.**
ENDO, Y.
SCI REPTS TOHOKU IMP UNIV
11 183–93 1922 CA 17 671
(SEE ALSO T23392)

TPRC Number	Bibliographic Citation
21015	EMISSIVITY OF BOILER FURNACE SLAGS. AGABABOV S G SPECIAL LIBRARY ASSOC. TRANSLATION CENTER 1-15, 1959. (ENGLISH TRANSLATION OF TEPLOENERGETIKA 5 (8), 1958; FOR ORIGINAL SEE T19646) (RTS-1052, SLA-59-19813)
21016	THE THERMAL CONDUCTIVITY OF SLAGS IN THE SOLID AND MOLTEN STATES. VARGAFTIK N B OLESHCHUK O N TEPLOENERGETIKA 5 12 79-85 1958 (FOR ENGLISH TRANSLATION SEE T21017)
21017	THE THERMAL CONDUCTIVITY OF SLAGS IN THE SOLID AND MOLTEN STATES. VARGAFTIK N B OLESHCHUK O N SPECIAL LIBRARY ASSOC. TRANSLATION CENTER 1-15, 1959. (ENGLISH TRANSLATION OF TEPLOENERGETIKA 5 (12), 79-85, 1958; FOR ORIGINAL SEE T21017) (RTS-1063, SLA-59-19247)
21047	THERMAL DIFFUSIVITY OF COAL CHARGES AND PRODUCTS OF THEIR THERMAL TREATMENT. KUSAKIN N D SALNIKOV A P BARANOV V I TRUDY INST GORYUCHIKH ISKOPAEMYKH AKAD NAUK SSSR 10 196-200 1959 CA 55 6820
21104	TEXTURE AND HEAT CONDUCTIVITY OF REFRACTORY STONE. JAKOB M Z VER DEUT ING 67 126-7 1923 CA 18 1372
21108	ATTEMPT TO DETERMINE THE RADIATION NUMBERS OF SOLID BODIES. POLAK VIKTOR Z TECH PHYSIK 8 307-12 1927
21109	METHOD FOR THE MEASUREMENT OF THERMAL CONDUCTIVITY OF SOLID SUBSTANCES IN PLATE FORM. JAKOB, M. Z TECH PHYSIK 7 475-81 1926 CA 21 524
21141	THERMAL CONDUCTIVITY OF SOME FIRE-CLAY REFRACTORIES. INAGAKI, K. YOGYO KYOKAI SHI 66 305-10 1958 JA 42 130
21142	MEASUREMENTS OF THERMAL CONSTANTS OF CERAMIC PRODUCTS AND RAW MATERIALS. I. MEASUREMENT OF THERMAL DIFFUSIVITY AND CONDUCTIVITY OF HEAT INSULATORS BY THE METHOD OF PERIODIC HEAT WAVE APPLIED TO A FINITE PLANE. KAWASHIMA, C. SAITO, S. SETAKA, N. YOGYO KYOKAI SHI 67 195-208 1959 JA 43 174
21176	THE THERMAL CONDUCTIVITY OF DISORDERED SOLIDS AT LOW TEMPERATURES. BERMAN R BULL INTERN INST REFRIG ANNEXE 1 13-20 1952
21184	LEIDEN MEASUREMENTS ON THERMAL CONDUCTIVITY SINCE THE SEVENTH INTERNATIONAL CONGRESS OF REFRIGERATION. DE NOBEL J BULL INTERN INST REFRIG ANNEXE 1 89-100 1952
21185	THERMAL CONDUCTIVITY OF GLASS-WOOL AT LOW TEMPERATURE AND LOW PRESSURE OF THE INCLUDED AIR. VAN PAEMEL O BULL INTERN INST REFRIG ANNEXE 1 101-8 1952
21187	THE THERMAL CONDUCTIVITY OF GRANULAR MATERIALS. DE VRIES D A BULL INTERN INST REFRIG ANNEXE 1 115-31 1952
21268	ON THE THERMAL CONDUCTIVITY OF SYNTHETIC COALS. WILD D Z VDI 100 3 95-9 1958 AM 12 4677
21288	THE RELATION OF THERMAL AND MECHANICAL PROPERTIES IN HIGH-CARBON POLYMERS /OBTAINED BY HIGH-TEMPERATURE TREATMENT OF COAL/. ZAMOLUEV V K MUKHANOVA L N TAITS E M DOKLADY AKAD NAUK SSSR 133 1143-5 1960 CA 55 14874 (FOR ENGLISH TRANSLATION SEE T25356)
21290	TRANSIENT HEAT FLOW APPARATUS FOR THE DETERMINATION OF THERMAL CONDUCTIVITIES. HOOPER F C LEPPER F R ASHVE TRANS 56 309-24 1950
21301	ELECTRICALLY HEATED COOLING AND SECONDARY KILNS IN THE GLASS INDUSTRY. HOROWITZ J SPRECHSAAL 84 20 414-15 1951
21323	USE OF LOW-MELTING CLAY IN THE PRODUCTION OF CONTAINERWARE. ZHUNINA L A MIKHLYUKOV E I KUSONSKIY G G FOREIGN TECHNOLOGY DIVISION 21-34, 1963. (ENGLISH TRANSLATION FROM SBORNIK NAUCHNYKH TRUDOV. KHIMIYA I KHIMICHESKIYA TEKHNOLOGIYA SILIKATNYKH MATERIALOV, REDAKTSIONNO-IZDATELSKIY OTDEL BPI IMENI I. V. STALINA (82), MINSK, 86-165, 1960) (FTD-TT-63-688, AD 426815)
21327	RAILWAY COLOR SIGNALS. VAIDYA W M J SCI IND RESEARCH /INDIA/ 17 A 12 489-97 1958 JA 44 238
21333	HIGH-TEMPERATURE INSULATING MATERIALS, THEIR PROPERTIES AND TESTING. CLEMENTS J F INSTN GAS ENGRS PROC OF CONF ON HEAT INSULATION 92-122 1950
21393	SPECIFIC HEAT AND EQUATION OF STATE OF CRYSTALS. FORSTERLING K ANN PHYSIK 61 549-76 1920 CA 14 2576
21397	SPECIFIC HEAT AT LOW TEMPERATURES. SIMON, F. ANN PHYSIK 68 241-80 1922 CA 17 667
21398	THE SPECIFIC HEATS OF CARBON, SILICON, AND SILICON CARBIDE. MAGNUS A ANN PHYSIK 70 303-31 1923 CA 17 2666
21404	MEASUREMENT OF THE SPECIFIC HEATS OF SILVER AND OF DIAMOND AT HIGH TEMPERATURES. MAGNUS A HODLER A ANN PHYSIK 80 808-22 1926 CA 21 1916
21446	TRANSMISSION OF RADIATION THROUGH GLASS IN TANK FURNACES. KELLETT B S J SOC GLASS TECH 36 35 115-23 1952
21451	THERMAL CONDUCTIVITY OF BUILDING MATERIALS. ZHUZE V P AZERBAIDZHANSKOE NEFTYANOE KHOZYAISTVO 2 68-73 1930 CA 24 2571
21459	THE SPECIFIC HEATS AND HEAT VARIATIONS OF CLAYS. COHN, W. M. BER. DEUT. KERAM. GES. 7, 149-55, 1926. (J. SOC. GLASS TECH.) (11 A, 285, 1926)
21490	THERMAL CONDUCTIVITY OF REFRACTORIES. WISHNEVSKY A BLAST FURNACE STEEL PLANT 15 309-10 1927 CA 21 2968
21495	THE THERMAL CONDUCTIVITY AND DIFFUSIVITY OF CONCRETE. CARMAN A P NELSON R A UNIV ILL ENG EXPT STA BULL 122 1-32 1921 CA 16 151
21511	DEVIATION FROM THE LAW OF NEUMANN, JOULE AND KOPP. II. THE APPARENT MOLECULAR HEAT OF WATER IN SOLID COMBINATIONS. KOLOSOVSKII, N. A. SKULSKA, I. M. BULL. SOC. CHIM. 47, 136-47, 1930. (J. RUSS. PHYS. CHEM. SOC.) (62, 427-60, 1930)
21536	TERRESTRIAL HEAT FLOW IN GREAT BRITAIN. BENFIELD A E PROC ROY SOC /LONDON/ 172 A 428-50 1939
21563	METHOD AND APPARATUS FOR DETERMINING THE CONDUCTIVITY OF REFRACTORY MATERIALS. MEKER GEORGES CERAMIQUE 30 445-52 1927 CA 22 4750
21660	INFRARED SPECTRAL ANALYSIS OF THE LUNAR SURFACE FROM AN ORBITING SPACECRAFT. LYON, R. J. P. BURNS, E. A. PROC. SYMP. REMOTE SENSING ENVIRONMENT, 2ND 309-27PP., 1962. (NAS1 N63-14668)

TPRC Number	Bibliographic Citation
21663	**THE SPECIFIC HEAT OF SOLID SUBSTANCES.** VON JUPTNER FEUERUNGSTECH 16 97–102 1928 CA 22 4042
21664	**MEASUREMENTS OF THE THERMAL CONDUCTIVITY OF CRYSTALS AND CRYSTALLINE MATERIAL.** EUCKEN A FORTSCHRITTE MINERAL KRYST PETROG 12 31–2 1927 CA 22 4339 (SEE ALSO T33185)
21665	**THE THERMAL CONDUCTIVITY AND SPECIFIC HEAT OF COAL.** SINNATT F S MACPHERSON H FUEL 3 12–14 1923 CA 18 1044
21670	**TEMPERATURE DIFFUSIVITIES AND THERMAL CONDUCTIVITIES OF SILICA AND FIRECLAY REFRACTORIES.** GREEN A T GAS WORLD /COKING SECT./ 85 112–14 1926 CA 21 2774
21691	**THE SPECIFIC HEAT OF GRAPHITIC CARBON AND OF COKE.** SCHLAPFER P DE BRUNNER PAUL HELVETICA CHIM ACTA 7 31–58 1924 CA 18 2636
21703	**THE MEASUREMENT OF THERMAL CONDUCTIVITY UP TO 1300 C.** CODEGONE, C. RICERCA SCI E RICOSTRUZ 16 3–4 286–89 1946
21724	**SPECIFIC HEAT OF PYREX GLASS FROM 25 TO 175 DEGREES.** DE VRIES THOS IND ENG CHEM 22 617–18 1930 CA 24 3702
21758	**DETERMINATION OF THE COEFFICIENT OF HEAT CONDUCTIVITY.** VLASOV, O. IZVEST. TEPLOTEKH. INST. (TRANS. THERMO-TECH. INST.) 6 32–44 1928 CA 23 4400
21761	**THE PHYSICAL PROPERTIES OF GLASSES.** ECKERT FRITZ JAHRB RADIOAKT ELEKTRONIK 20 93–275 1923 CA 18 1557
21790	**THE THERMODYNAMIC PROPERTIES OF CALCITE AND ARAGONITE.** BACKSTROM H L J J AM CHEM SOC 47 2432–42 1925 CA 19 3414
21929	**THERMAL CONDUCTIVITIES OF INSULATORS IN RELATION TO THE LAGGING OF STEAM PIPES.** THOMAS R J SOC CHEM IND 38 357–60T 1919 CA 14 6
21932	**ERRORS IN THE MEASUREMENT OF THE TEMPERATURE OF THE MOON.** BURNS, E. A. LYON, R. J. P. NATURE /LONDON/ 196 463–4 1962 PA N63–10 641
21960	**THE VISCOSITY OF GLASS.** STOTT V H J SOC GLASS TECH 9 207–25 1925 CA 20 2396
21969	**DETERMINATION OF THERMAL CONDUCTIVITY OF REFRACTORIES.** HERSEY M D BUTZLER E W J WASH ACAD SCI 14 147–51 1924 CA 18 1887
21998	**EVALUATION OF INFRARED SPECTROPHOTOMETRY FOR COMPOSITIONAL ANALYSIS OF LUNAR AND PLANETARY SOILS.** LYON, R. J. P. STANFORD RESEARCH INST., MENLO PARK, CALIFORNIA 1–136 1963 PA N63–10 697 (NASA TN D–1871)
22016	**THE THERMAL CONDUCTIVITY COEFFICIENT OF BUILDING AND INSULATING MATERIALS AND THE DIATHERMANCY OF NEW TYPES OF CONSTRUCTION.** KNOBLAUCH, O. RAISCH, E. REIHER, H. GESUNDHEITS–INGENIEUR 43 52 607–23 1920
22023	**THERMAL CONDUCTIVITY AND SPECIFIC HEAT OF REFRACTORIES EMPLOYED IN THE CONSTRUCTION OF FURNACES AS A BASIS FOR THERMO–TECHNICAL CALCULATIONS.** CZAKO E J GASBEL 62 274–8 1919 CA 14 605
22058	**PROPERTIES OF THE SURFACE OF GLASS.** GREBENSHCHIKOV I V KERAM I STEKLO 7 11 36–41 1931 CA 26 2291
22147	**THE EFFECT OF TEMPERATURE ON THE TRANSMISSION CHARACTERISTICS OF COLORED GLASS FILTERS.** HAFFNER, S. SAMUEL FELTMAN AMMUNITION LABS., PICATINNY ARSENAL, DOVER, N. J. 1–22, 1954. (PA–TR–2069, AD 44542)
22166	**THERMAL CONDUCTIVITY OF INSULATING MATERIALS.** TAYLOR T S MECH ENGINEERING 42 8–10 1920 CA 14 875
22176	**INVESTIGATION OF THO2–UO2 AS A NUCLEAR FUEL.** RAO, S. V. K. ATOMIC ENERGY OF CANADA, LTD, CHALK RIVER, ONTARIO 1–49, 1963. (CRFD–1154, AECL–1785)
22183	**HIGH TEMPERATURE MATCH–BOX ELECTRON TUBES.** WESTINGHOUSE ELECTRIC CORPORATION U.S. ARMY SIGNAL RESEARCH AND DEVELOPMENT LAB., FT. MONMOUTH, N. J. 1–70, 1959. (AD 232770, PB 160825)
22225	**HEAT CONDUCTIVITY OF INSULATING AND BUILDING MATERIALS.** TVERTZUIN V S MILIN V B NEFTYANOE KHOZYAISTVO 17 198–203 1929 CA 24 705
22261	**COMPOSITE MATERIALS RESEARCH AT THE BOEING COMPANY.** SIMPSON, F. H. THE BOEING CO., SEATTLE, WASHINGTON 646–60, 1963. (RTD–TDR–63–4131(VOL III), AD 601266)
22272	**HIGH TEMPERATURE, HIGH EMITTANCE INTERMETALLIC COATINGS. PART 1. EMITTANCE AND REFLECTANCE OF INTERMETALLIC COMPOUNDS.** SCHATZ, E. A. GOLDBERG, D. M. PEARSON, E. G. BURKS, T. L. AERONAUTICAL SYSTEMS DIV., AF MATERIALS LAB., WRIGHT–PATTERSON AFB, OHIO 1–181, 1963. (ASD–TDR–63–657(PT 1))
22280	**THERMAL CONDUCTIVITY OF SOME SOLID INSULATORS.** CLARKE J R PHIL MAG 40 502–4 1920 CA 15 617
22401	**NEW SPECIFIC HEAT APPARATUS.** GRIFFITHS, E. PROC PHYS SOC LONDON 33 355–61 1921 CA 15 3917
22457	**THE CORRECT DETERMINATION OF THE SPECIFIC HEATS OF SOLID MATERIALS AT TEMPERATURES RANGING BETWEEN 0 AND 1625 DEGREES.** JAEGER F M ROSENBOHM E REC TRAV CHIM 47 513–57 1928 CA 24 2368
22467	**ATOMIC HEAT OF CRYSTALLINE SUBSTANCES AT CONSTANT PRESSURE.** ENDO, Y. REPT AERONAUTICAL RESEARCH INST TOKYO IMP UNIV 6 71–82 1931 CA 25 4454
22489	**THE THERMAL CONDUCTIVITY OF REFRACTORY MATERIALS.** GILARD P REV UNIVERSELLE MINES 4 7 34–50 1924 CA 19 205
22494	**THERMAL CONDUCTIVITY OF GLASSES TRANSMITTING ULTRA–VIOLET LIGHT.** KUNERTH, WM. BERKEY, WM. E. SCIENCE 68, 427PP., 1928. (IOWA STATE COLL. J. SCI.) (3 (5), 6PP., 1928)
22514	**THE ERRORS IN GAS TEMPERATURE MEASUREMENT AND THEIR CALCULATION.** FISHENDEN M F SAUNDERS O A J INST FUEL 12 S5–14 1939
22517	**RADIOMETRIC INVESTIGATIONS OF INFRA–RED ABSORPTION AND REFLECTION SPECTRA.** COBLENTZ W W NATL BUR STANDARDS BULL 2 457–78 1906
22520	**THE DIFFUSE REFLECTING POWER OF VARIOUS SUBSTANCES.** COBLENTZ W W NATL BUR STANDARDS BULL 9 283–325 1913

TPRC Number	Bibliographic Citation
22527	IX. EXPERIMENTS ON THE CALORIFIC CONDUCTIBILITY OF SOLIDS. NEUMANN F PHIL MAG 25 63–5 1863 (ENGLISH TRANSLATION OF ANN. CHIM. PHYS., 66 (SER. 3), 183–7, 1862; FOR ORIGINAL SEE T16210)
22559	CHARACTERISTICS OF RADIATION PYROMETERS. BURGESS, G. K. FOOTE, P. D. NATL BUR STANDARDS BULL 12 91–178 1915
22568	HEAT CONDUCTIVITY OF MOLD MATERIALS. DEVELOPMENT OF PLASTIC MOLD COMPOSITIONS. HUDSON F METAL IND /LONDON/ 50 573–5 1937 CA 31 5897
22584	A SERIES OF GLASSES COLORED BY CARBONACEOUS MATTERS. III. THE CHEMICAL COMPOSITION AND THE PHYSICAL PROPERTIES OF GLASSES COLORED BY CARBONACEOUS MATTER. FUWA K J SOC CHEM IND JAPAN /SUPPL BINDING/ 39 373 1936 CA 31 514
22607	DETERMINATION OF THE COEFFICIENT OF THERMAL EXPANSION, SPECIFIC HEAT, AND HEAT OF FUSION OF RUBIDIUM AND OF CADMIUM. DEUSS ELSA VIERTELJAHRESSCHR NATURFORSCH GES ZURICH 56 15–41 1911 (FOR ENGLISH TRANSLATION SEE T24550)
22625	THERMAL CONDUCTION IN FERROELECTRIC CERAMICS. YOSHIDA I J PHYS SOC JAPAN 15 12 2211–19 1960 SA 64 12358
22633	THE DETERMINATION OF THE THERMAL CONDUCTIVITY, SPECIFIC HEAT, DENSITY AND THERMAL EXPANSION OF DIFFERENT ROCKS AND REFRACTORY MATERIALS. TADOKORO Y SCIENCE REPTS TOHOKU IMP UNIV 10 339–410 1921 CA 16 1531
22638	THE EFFECT OF POROSITY UPON THERMAL CONDUCTIVITY, DIFFUSIBILITY AND HEAT CAPACITY AT HIGH TEMPERATURES. TADOKORO, Y. SCIENCE REPTS TOHOKU IMP UNIV 15 567–96 1926 CA 21 1585
22639	THE HEAT CONTENT AND THE SPECIFIC HEAT OF SOME SLAGS AT HIGH TEMPERATURES. UMINO, S. SCIENCE REPTS TOHOKU IMP UNIV 17 5 985–96 1928 CA 23 69
22647	DETERMINATION OF THE SPECIFIC HEATS AND SPECIFIC WEIGHTS OF SEVERAL MINERALS. DANNHOLM ANNA SOC SCI FENNICA COMMENTATIONES PHYSICO–MATH 1 9 1–10 1922 CA 17 1182
22654	THE RELATION OF HEAT CONDUCTIVITY OF GLASS TO THE CHEMICAL COMPOSITION. RUSS ALFRED SPRECHSAAL 61 887–91 1928 CA 23 3549
22655	AN IMPROVED METAL CALORIMETER FOR THE DETERMINATION OF THE SPECIFIC HEAT OF METALS, OXIDES AND SLAGS. GROSSE, W. DINKLER, W. STAHL U EISEN 47 448–53 1927 CA 21 3143
22687	MEASURING THERMAL CONDUCTIVITY OF REFRACTORY MATERIALS AT HIGH TEMPERATURES. EUCKEN A LAUBE H TONIND ZTG 53 1599–1602 1929 CA 24 701
22688	THERMAL CONDUCTIVITY OF REFRACTORY MATERIALS. GOLLA H LAUBE H TONIND–ZTG 54 1411–12 1431–2 1458–60 1930 CA 25 789
22689	COMPARISON OF THE TEMPERATURE DIFFUSIVITIES AND THERMAL CONDUCTIVITIES OF SILICA AND FIRECLAY REFRACTORIES. GREEN A T TRANS. CERAM. SOC. 26 168–76 1927 CA 22 852
22690	THE TEMPERATURE DIFFUSIVITY AND THERMAL CONDUCTIVITY OF SILICA REFRACTORY MATERIAL AT HIGH TEMPERATURES. GREEN A T TRANS. CERAM. SOC. 26 159–67 1927 CA 22 852
22700	ACCURATE DETERMINATION OF SPECIFIC HEATS OF SOLIDS BETWEEN 0 AND 1625 DEGREES. II. SPECIFIC HEATS OF PLATINUM AND OF TUNGSTEN. JAEGER F M ROSENBOHM E VERSLAG AKAD WETENSCHAPPEN AMSTERDAM 36 960–79 1927 CA 22 1086
22704	ROLE OF DIFFUSION IN THE PRODUCTION OF GLASS SPECIMENS OF VARIABLE COMPOSITION. SHELYUBSKII V I RYAZANOV G V SOVIET PHYSICS–DOKLADY 9 5 391–3 1964 (ENGLISH TRANSLATION OF DOKLADY AKAD. NAUK SSSR, 156 (2), 302–5, 1964; FOR ORIGINAL, SEE T022705)
22705	ROLE OF DIFFUSION IN THE PRODUCTION OF GLASS SPECIMENS OF VARIABLE COMPOSITION. SHELYUBSKII V I RYAZANOV G V DOKLADY AKAD NAUK SSSR 156 2 302–5 1964 (FOR ENGLISH TRANSLATION, SEE T022704)
22720	SUPPORTING ENGINEERING STUDIES IN LIQUID METAL SYSTEMS. E. GLASS DOSIMETER. ECKELS, T. W. ARGONNE NATIONAL LAB., 81–4, 1962. (ANL–6477)
22749	GENERAL FORMULA FOR CALCULATING ATOMIC OR MOLECULAR HEAT AS WELL AS THE SPECIFIC HEAT OF ELEMENTS AND THEIR COMPOUNDS IN THE SOLID STATE. MAYDEL I Z ANORG ALLGEM CHEM 186 289–323 1930
22766	MEASURING THE SPECIFIC HEATS OF METALLURGICALLY IMPORTANT MATERIALS THROUGH A LARGE RANGE OF TEMPERATURE WITH THE AID OF TWO NEW TYPES OF CALORIMETERS. ROTH W A BERTRAM W W Z ELEKTROCHEM 35 297–308 1929 CA 23 4876
22785	SPECIFIC HEAT AND THERMAL EXPANSION OF ISOMETRIC CRYSTALS. FORSTERLING K Z PHYSIK 3 9–18 1920 CA 15 333
22802	CONDUCTIVITY MEASUREMENTS ON FORMED AND TEMPERED ROCK SALT CRYSTALS. QUITTNER FRANZ Z PHYSIK 68 796–802 1931 CA 25 5068
22861	INVESTIGATION ON THE SPECIFIC HEAT AT LOW TEMPERATURES. /A/ THE SPECIFIC HEAT OF SODIUM, POTASSIUM, MOLYBDENUM AND PLATINUM. ZEIDLER WALTER Z PHYSIK CHEM 123 383–404 1926 CA 21 2092
22931	METHOD FOR DETERMINING THE THERMAL CONDUCTIVITY OF PULVERIZED SUBSTANCES AT HIGH TEMPERATURES. PIRANI M VON WANGENHEIM Z TECH PHYSIK 10 413–24 1929 CA 24 1004
22957	HEAT TRANSFER IN QUARTZ POWDER—GAS MIXTURES, A METHOD FOR THE DETERMINATION OF THERMAL CONDUCTIVITY OF GASES AT ELEVATED TEMPERATURES. SCHAFER, KL. GRUNDLER, K. H. Z ELEKTROCHEM 63 449–53 1959 CA 53 14669
22963	INFRARED COATING STUDIES. BAUSCH AND LOMB INC., ROCHESTER, N. Y. BAUSCH AND LOMB INC. MATERIALS R AND D LABS., ROCHESTER, N. Y. 1–8, 1963. (AD 296457)
22979	EMISSION OF A HIGH-PRESSURE MERCURY ARC IN THE WHOLE INFRARED. BOHDANSKY JOSEF Z PHYSIK 149 383–96 1957 CA 53 7768
22990	INFRARED SPECTRAL EMISSIVITY OF OPTICAL MATERIALS. STIERWALT, D. L. U.S.N. ORDNANCE LAB., WHSITE OAK, SILVER SPRING, MD. 266–84, 1961. (ONR–9(VOL 1), AD 260239)
23066	DETERMINING THE COEFFICIENT OF THERMAL CONDUCTIVITY OF CERAMIC MATERIALS. PLOTINKOV, L. A. U.S. ATOMIC ENERGY COMMISSION 1–5 1961 (ENGLISH TRANSLATION OF ZAVOD. LAB. 16, 1136–9, 1950; FOR ORIGINAL SEE T838) (AEC–TR–4646)

TPRC Number	Bibliographic Citation
23083	**INFRARED ABSORPTION SPECTRA OF MINERALS.** AKHMANOVA M V USPEKHI KHIM 28 312-35 1959 CA 53 11979
23121	**THERMAL SHOCK ANALYSIS OF SPHERICAL SHAPES. II.** HASSELMAN D P H CRANDALL W B J AM CERAM SOC 46 9 434-7 1963
23128	**THE PROPERTIES OF FIRED SERICITE-FELDSPAR SYSTEM.** SHIRAKI, Y. MATSUMOTO, T. YOGYO KYOKAI SHI 67 75-9 1959 CA 53 18427
23145	**EMITTANCE STUDIES OF VARIOUS HIGH TEMPERATURE MATERIALS AND COATINGS.** SKLAREW, S. RABENSTEINE, A. S. MARQUARDT CORP., VAN NUYS, CA. 1-37, 1963. (PR 281-3Q-1, AD 299417)
23176	**THE THEORY OF SPECIFIC HEAT ADVANCED BY DEBYE.** MEWES RUDOLF Z SAUERSTOFF STICKSTOFF IND 14 93-5 1922 CA 17 3445
23196	**THE THERMAL RESISTIVITY OF SOLID DIELECTRICS.** AUTHOR ANON. J. INST. ELEC. ENG. (LONDON) 68, 1313-55, 1930.
23204	**THE HEAT CAPACITIES OF ACETALDEHYDE AND PARALDEHYDE, AND THE HEAT OF TRANSFORMATION OF ACETALDEHYDE INTO PARALDEHYDE.** COOPER D LEB PROC TRANS NOVA SCOTIAN INST SCI 17 1 82-90 1927 CA 22 531
23210	**THERMAL CONDUCTIVITY OF CRYSTALS CONTAINING MAGNETIC IMPURITIES AT LOW TEMPERATURES AND IN A MAGNETIC FIELD.** DREYFUS B ZADWORNY F J PHYS RADIUM 23 490-3 1962 CA 58 2862
23230	**DETERMINATION OF THE SPECIFIC HEAT OF TAR AND TAR OILS, COKE AND COAL.** PIETERS H A J HET GAS 51 94-5 1931 CA 25 3808
23239	**THE SPECIFIC HEAT OF COAL AND ITS RELATION TO COMPOSITION.** COLES, G. COLLIERY GUARDIAN 126, 973-4, 1923. (J. SOC. CHEM. IND.) (42, 435-9T, 1923) (IRON COAL TRADES REV.) (107, 820, 1923)
23243	**A MEASURING METHOD FOR DETERMINING HEAT CAPACITY OF THE EARTH BY MEANS OF A CONSTANTLY HEATED TEST BODY.** SKEIB G A Z METEOR 4 1/2 32-9 1950 AM 4 3415
23248	**THERMAL CAPACITY OF BRICKS.** TSCHOPLOWITY F BRIT CLAYWORKER 32 91 1923 CA 17 3588
23249	**THE THERMAL QUALITIES OF REFRACTORY MATERIALS AND THE METHODS FOR THEIR DETERMINATION.** COHN, W. M. BRENNSTOFF-U WARMEWIRT 12 77-88 1930 CA 24 5956
23258	**INORGANIC FILMS FOR SOLAR ENERGY ABSORPTION. PART I.** LANGLEY, R. C. ENGELHARD INDUSTRIES, INC., EAST NEWARK, N. J. 1-33, 1962. (ASD-TDR-62-92(PT 1), AD 282008)
23260	**HEAT CONDUCTION OF MOIST BUILDING MATERIALS.** JESPERSEN H B VARME 17 3 41-51 1952 AM 6 279
23263	**NEW NONMETALLIC MATERIALS FOR RADIO EQUIPMENT, MANUAL FOR DESIGNERS.** CHURABO DMITRIY DMITRIYEVICH FOREIGN TECHNOL. DIV. 1-28, 1963. (FTD-TT-62-1873, AD 409066)
23278	**INFLUENCE OF THERMAL TREATMENT ON THE LOW-TEMPERATURE HEAT CAPACITY OF SODIUM BOROSILICATE GLASS.** KHOKOMOV, KH. B. STRUCTURE OF GLASS, PROC. ALL-UNION CONF. GLASSY STATE THIRD LENINGRAD, 2959 465-7 1960 CA 55 8797
23287	**A SIMPLE METHOD OF DETERMINING THE SPECIFIC HEATS OF BLAST-FURNACE RAW MATERIALS.** DHAR P R GUPTA I N MAHAPATRA U P RESEARCH AND IND. 6 6-7 1961 CA 55 23237
23289	**INFRARED OPTICAL MEASUREMENTS.** NATIONAL BUREAU OF STANDARDS NATIONAL BUREAU OF STANDARDS, WASHINGTON, D. C. 1-20, 1963. (NBS REPT 8302, AD 434 237)
23291	**THERMOGRAPHY OF THE HUMAN BODY.** BARNES, R. B. SCIENCE 140 870-7 1963
23317	**MEASURING THE VALUES OF THE SPECIFIC HEATS OF MATERIALS USED IN CONSTRUCTION.** PASCAL M A ANN INST TECH BAT TRAV PUBLICS 7 77 372-85 1954 AM 8 3520
23333	**FABRICATION AND PROPERTIES OF TRANSLUCENT BERYLLIUM OXIDE.** MC CLELLAND, J. D. RICHARDSON, J. H. FRANKLIN, L. R. ATOMICS INTERNATIONAL DIV. OF NORTH AMERICAN AVIATION, INC., CANOGA PARK, CA. 1-13 1960 (NAA-SR-6454)
23361	**TRANSMISSION PROPERTIES OF OPTICAL FIBERS.** POTTER R J J OPT SOC AMER 51 10 1079-89 1961 SA 64 12976
23378	**RESEARCH ON LOW DENSITY THERMAL INSULATION MATERIALS FOR USE ABOVE 3000 F.** STYHR, K. H., JR. HESSINGER, P. S. RHSHKEWITCH, E. NATIONAL BERYLLIA CORP., HASKELL, N. J. 1-32, 1962. (N63-13520)
23397	**THERMAL RADIATION TO A FLAT SURFACE ROTATING ABOUT AN ARBITRARY AXIS IN AN ELLIPTICAL EARTH ORBIT. APPLICATION TO SPIN-STABILIZED SATELLITES.** POWERS, E. I. GODDARD SPACE FLIGHT CENTER, GREENBELT, MD. 1-25, 1964. (NASA-TN-D-2147, N64-17745)
23406	**A TECHNIQUE FOR THE DETERMINATION OF A THERMAL CHARACTERISTIC OF STONE.** FOX R G JR DOLCH W L MEETING, NAT. RES. COUNC. HIGHWAY RES. BOARD, PROC. ANN., 30TH 180-6PP., 1951.
23435	**THERMAL CONDUCTIVITY OF INSULATING AND OTHER MATERIALS.** TAYLOR T S ELEC REV 75 1054-7 1919 CA 14 243
23444	**THERMAL CONDUCTIVITIES OF WALLS, CONCRETES, AND PLASTERS.** GRIFFITHS E DEPT SCI IND RESEARCH BUILDING RESEARCH TECH PAPER 6 1-19 1928 CA 23 2264
23456	**SELECTIVE RADIATION FROM VARIOUS SOLIDS.** COBLENTZ W W BULL NATL BUR STANDARDS 5 159-91 1908
23478	**PHYSICAL PROPERTIES OF WELDING SLAGS.** OKADA K FUKAYA T YOSETSU GAKKAISHI 27 466-72 1958 CA 53 13944
23480	**SELECTIVE RADIATION FROM VARIOUS SOLIDS II.** COBLENTZ W W BULL NATL BUR STANDARDS 6 301-19 1910
23520	**FUNDAMENTALS OF THERMAL CONDUCTIVITY AT HIGH TEMPERATURES.** POWERS, A. E. KMOLLS ATOMIC POWER LABORATORY 1-36, 1961. (KAPL-2143)
23521	**EFFECT OF NUCLEAR RADIATION ON GLASS.** BATTELLE MEMORIAL INSTITUTE, COLUMBUS, OHIO BATTELLE MEMORIAL INST. RADIATION EFFECTS INFO. CTR. 1-24, 1961. (AD 207701)

TPRC Number	Bibliographic Citation

23525 DELAMINATION OF KAOLIN BY HIGH-ENERGY, IONIZING RADIATION.
CORBETT, W. J. BURSON, J. H. III
ENG. EXPT. STA., GEORGIA INST. OF TECHNOLOGY, ATLANTA, GA.
1-61, 1961.
(SRO-57)

23527 A STUDY OF EXTERNAL INSULATING REQUIREMENTS FOR THE EXTENSION OF ALUMINUM ALLOY STRUCTURES TO HOT SUPERSONIC AIRCRAFT.
WEIBE, W.
NATIONAL AERONAUTICAL ESTABLISHMENT, OTTAWA, CANADA
1-30, 1961.
(LR-306, NRC-6365, AD 261262)

23528 THERMAL RADIATION PROPERTIES SURVEY.
GUBAREFF, G. G. JANSSEN, J. E.
TORBORG, R. H.
THERMAL RADIATION PROPERTIES SURVEY
1-293, 1960.

23533 SOME FACTORS INFLUENCING THE PRODUCTION OF MANGANESE FROM OPEN-HEARTH SLAG BY THE TWO-STAGE PROCESS, WITH SPECIAL REFERENCE TO CERTAIN PROPERTIES OF THE SLAGS.
HOSKING H W
AUSTRALASIAN ENGR
70-75 1955 CA 50 743

23542 DEVELOPMENT OF ULTRASONIC WELDING EQUIPMENT FOR REFRACTORY METALS.
JONES J BYRON MAROPIS NICHOLAS
DE PRISCO CARMINE F THOMAS JOHN G DEVINE JANET
AEROPROJECTS, INC., WESTCHESTER, PA.
1-58, 1961.
(ASD PROJ NO TR 7-888, AD 263872)

23559 REFLECTION FACTORS OF INDUSTRIAL PAINTS AND PIGMENTS.
GARDNER HENRY A
TRANS. ILLUM. ENG. SOC.
17 318-22 1922

23568 THERMAL CONDUCTIVITIES OF SOLIDS AT LOW TEMPERATURES.
POWELL, R. L.
NBS CRYOGENIC ENG. LAB.
1-15, 1953.
(NBS TECH MEMO 17, NBS REPT 2484)

23579 ANALYSES OF PRESSURE DROP AND HEAT TRANSFER OF A PEBBLE-BED-STORAGE HEATER FOR A HYPERSONIC WIND TUNNEL.
RANDALL, D. E. BEDFORD, A.
SANDIA CORP., ALBUQUERQUE, N. MEXICO
1-33 1961
(SC-4591(RR))

23580 REFLECTION SPECTRA OF ORDINARY AND DEVITRIFIED LEAD GLASSES IN THE INFRARED.
FLORINSKAYA V A
DOKLADY AKAD NAUK SSSR
89 261-4 1953
(FOR ENGLISH TRANSLATION SEE T14986)

23581 INFRARED ABSORPTION SPECTRUM OF LIQUID WATER IN THE 3200 TO 3600 CM/TO THE MINUS ONE POWER/ REGION.
ANDREYEV S N BALICHEVA T G
DOKLADY AKAD NAUK SSSR
90 149-51 1953
(FOR ENGLISH TRANSLATION SEE T14856)

23582 REFLECTION SPECTRA OF COMPOUND SILICATE GLASSES IN THE INFRARED BEFORE AND AFTER THERMAL TREATMENT.
FLORINSKAYA V A
DOKLADY AKAD NAUK SSSR
90 1011-14 1953
(FOR ENGLISH TRANSLATION SEE T14835)

23584 REFLECTION AND TRANSMISSION SPECTRA OF POTASSIUM-SILICATE GLASSES IN THE INFRARED.
FLORINSKAYA V A PECHENKINA R S
DOKLADY AKAD NAUK SSSR
91 59-62 1953
(FOR ENGLISH TRANSLATION SEE T14586)

23593 ZIRCONIUM BASED ELECTRICAL PORCELAINS OF HIGHER MECHANICAL RESISTANCE.
OLMOS L FERRER
REV CIENC APL /MADRID/
11 301-5 1957 CA 53 18423

23628 THERMAL CONDUCTIVITY OF REFRACTORIES.
HULL W A
CHEM MET ENG
27 538-40 1922 CA 16 4036

23634 MEASUREMENT OF THERMAL CONDUCTIVITY OF REFRACTORIES AT HIGH TEMPERATURES.
OJIRI KOZO
REPTS RESEARCH LAB ASAHI GLASS CO
1 159-68 1951 CA 50 2942

23645 THE EFFECT OF HYDROGEN ON HEAT CONDUCTIVITY OF HEAT-INSULATING MATERIALS.
BUDRIN D V GOLUBEV I I
TRUDY URAL IND INST IM S M KIROVA
17 76-96 1941 CA 40 3861

23657 THERMAL INSULATION FOR NUCLEAR SYSTEMS.
COLLINS C G POMEROY G W
HEATING PIPING AIR CONDITIONING
30 8 129-34 1958

23683 DESIGN IT WITH DIAMONDS.
ASAFF A G
PROD ENG
32 42-5 1961

23691 SELECTIVE RADIATION FROM VARIOUS SUBSTANCES, III.
COBLENTZ W W
BULL NATL BUR STANDARDS
7 2 243-94 1911

23718 LOW-TEMPERATURE HEAT CAPACITY AND ENTROPY OF BERLINITE.
EGAN, E. P., JR. WAKEFIELD, Z. T.
J PHYS CHEM
64 1953-5 1960 CA 55 18276

23751 MEASURING OF MEAN SPECIFIC HEATS OF SOME CZECHOSLOVAK PACKING GLASSES.
KRIVANEK JAROSLAV KRIZ MOJMIR
SILIKATY
3 110-19 1959 CA 55 21513

23759 SELECTIVE RADIATION FROM VARIOUS SUBSTANCES, IV.
COBLENTZ W W
BULL NATL BUR STANDARDS
9 81-115 1913

23771 DETERMINATION OF THE THERMAL CONSTANTS OF COAL AT DIFFERENT BULK DENSITIES.
ZUBILIN I G
ZAVODSKAYA LAB
27 431-3 1961 CA 55 21544

23772 SOME PROPERTIES OF AMERICAN ZIRCONIA BRICKS.
SOMIYA S YAMAUCHI T
J CERAM ASS JAPAN
65 186-90 1957 BC 57 131

23779 THERMAL CONDUCTIVITY OF URANIUM-BEARING MATERIAL UNDER IRRADIATION AT HIGH TEMPERATURES.
HUNTER, L. P.
OAK RIDGE NATIONAL LAB., TENNESSEE
183-210, 1957.
(TID-2501(DEL), CF-48-10,135(DEL))

23796 HEAT CONDUCTANCE OF SOLID SOLUTIONS OF MANGANESE NIOBATE BASED ON BATIO3.
KODZHESPIROV F F
FIZ TVERDOGO TELA
3 781-5 1961 CA 55 24212
(FOR ENGLISH TRANSLATION SEE T25307)

23874 FULLY AUTOMATIC DETERMINATION OF MOISTURE IN SAND AND CERAMIC BODIES. III, DETERMINATION, BY MICROWAVE TECHNIQUES, OF THE MOISTURE CONTENT OF CLAYS IN THE PLASTIC RANGE.
AMRHEIN E M
BER DEUT KERAM GES
38 8 339-41 1961 JA 45 16

23879 METHOD OF MEASUREMENT OF THE THERMAL DIFFUSIVITIES OF REFRACTORY MATERIALS UP TO THE NEIGHBOURHOOD OF THE MELTING POINT.
CABANNES F
J RECH CENT NAT RECH SCI
50 83-9 1960 SA 63 16909

23881 STRUCTURAL DESIGN FOR AERODYNAMIC HEATING. PART I. DESIGN INFORMATION.
DUKES, W. H. SCHNITT, A.
BELL AIRCRAFT CORP., BUFFALO, N. Y.
1 VOL., 1955.
(WADC-TR-55-305(PT. 1), AD 98114)

23900 THE USES OF COMPLETE TEMPERATURE-TIME CURVES FOR DETERMINATION OF THERMAL CONDUCTIVITY WITH PARTICULAR REFERENCE TO ROCKS.
JAEGER J C
AUSTRAL J PHYS
12 3 203-17 1959 SA 63 5202

23912 QUICK MEASUREMENTS OF THE HEAT CONDUCTIVITY AT LOW TEMPERATURES OF INSULATING MATERIALS.
CODEGONE, C.
CONGR. INTERN. FROID. COMPT. REND. TRAV. COMM.
2153-9, 1955.

23934 LOW-TEMPERATURE HEAT CAPACITY OF AZURITE.
FORSTAT H TAYLOR G KING B R
J CHEM PHYS
31 4 929-31 1959 SA 63 1559

23940 THE EFFECT OF FAST NEUTRON BOMBARDMENT ON THE THERMAL CONDUCTIVITY OF SILICA GLASS AT LOW TEMPERATURE.
CRAWFORD J H JR COHEN A F
BULL INST INTERNAT FROID ANNEXE 1
165-72 1958 SA 63 2832

TPRC Number	Bibliographic Citation
23953	**ATOMIC HEAT OF DIAMOND FROM 11 TO 200 K.** BURK D L FRIEDBERG S A PHYS REV 111 5 1275–82 1958 SA 62 1390
23960	**AN APPARATUS FOR THE MEASUREMENT OF THE THERMAL CONDUCTIVITY OF INSULATING MATERIALS OF CYLINDRICAL SHAPE.** NUVOLI A RICERCA SCI 28 10 2095–101 1958 SA 62 3377
24001	**A METHOD OF MEASURING THERMAL DIFFUSIVITIES OF ROCKS AT ELEVATED TEMPERATURES.** SOMERTON W H BOOZER G D A I CH E JOURNAL 7 87–90 1961 CA 56 5787
24010	**WATER MOVEMENT IN POROUS MATERIALS. PART 1. ISOTHERMAL VAPOUR TRANSFER.** ROSE D A BRIT J APPL PHYS 14 256–62 1963
24015	**FUNDAMENTAL STUDIES OF THE PROPERTIES OF NATURAL AND SYNTHETIC QUARTZ CRYSTALS.** KING J C BELL TELEPHONE LABS., INC. 1–50, 1959. (PB 139743)
24021	**ADHERENCE AND THERMAL PERFORMANCE OF ALUMINIUM ENAMELS AND COATINGS.** GELTMAN G L CERAMIC AGE 77 4 73 1961 BC 60 2318
24048	**PIEZOELECTRIC CERAMIC MATERIALS.** BERLINCOURT, D. SANDIA CORP., ALBUQUERQUE, N. M. 1–96, 1959. (SC–4306)
24081	**INFRARED TRANSMITTANCE AT HIGH TEMPERATURE USING A DOUBLE-BEAM SPECTROPHOTOMETER.** PARKER C J NORDBERG M E JR OPTICAL SOCIETY OF AMERICA JOURNAL 49 856–9 1959 BR 8 14735
24082	**THE TRANSMISSION, ABSORPTION AND REFLECTION OF SOLAR RADIATION BY GLASSES.** RAM A PRASAD S N J SCI INDUST RES INDIA 14 A 12 570–84 1955 AM 9 2725
24125	**THE THERMAL CONDUCTIVITIES OF SOME REFRACTORY MATERIALS.** PUSTOVALOV V V OGNEUPORY 26 302–5 1961 CA 55 27825
24126	**THE THERMAL CONDUCTIVITY OF ALUMINUM SILICATE REFRACTORIES.** PUSTOVALOV V V OGNEUPORY 26 362–6 1961 CA 55 27825
24145	**SELECTIVE REFLECTION OF METALS.** RUBENS, H. ANN PHYSIK 37 249–68 1889
24186	**THE RHEOLOGY OF SILICATE ROCKS.** SAUCIER HENRI CAHIER GROUPE FRANC ETUDES RHEOL 4 4 5–9 1959 CA 55 19392
24201	**A STUDY OF A TRANSIENT HEAT METHOD FOR MEASURING THERMAL CONDUCTIVITY.** DEUSTACHIO D SCHREINER R E ASHVE TRANS 58 331–42 1952
24204	**A NEW UNIVERSAL METHOD OF DETERMINING THERMOPHYSICAL CONTANTS.** VERZHINSKAYA A B NOVICHENOK L N INZHENER–FIZ ZH 3 9 65–8 1960 MA 14 5629
24221	**DETERMINATION OF THE EMISSIVITY OF SEVERAL MATERIALS AT NORMAL TEMPERATURE /23–30 DEGREES/.** PAK SSEN–SSU IZVEST AKAD NAUK UZBEK SSR SER FIZ–MAT NAUK 93–102 1957 CA 55 25391
24246	**MEASUREMENT OF REFLECTION FACTORS.** SHARP C H LITTLE W F TRANS ILLUM ENG SOC 15 802–10 1920
24247	**INVESTIGATION OF THE HEAT TRANSPORT BY MOIST PEAT.** VOLAROVICH M P GAMAYUNOV N U CHURAEV N V KOLLOID ZHUR 22 535–42 1960 CA 55 27852 (FOR ENGLISH TRANSLATION SEE T25361)
24261	**THE DETERMINATION OF THERMAL CONDUCTIVITY AND SPECIFIC HEAT BY NONSTEADY HEAT FLOW.** PARSONS J J CONGR INTERN FROID 9E PARIS 1955 COMPT REND TRAV COMM I ET II 2063–71 1955 CA 52 15017
24315	**THERMAL CONDUCTIVITY OF SEMICONDUCTIVE SOLIDS, METHOD FOR STEADY-STATE MEASUREMENTS ON SMALL DISK REFERENCE SAMPLES.** FLYNN, D. R. NATIONAL BUREAU OF STANDARDS 1–41, 1962. (NBS RPT 7740, AD 411157)
24322	**PROPERTIES AND APPLICATIONS OF GLASS–CERAMICS.** STOOKEY S D ENGRS DIG 21 85 1960 BC 60 1436
24323	**THERMAL CONDUCTIVITY OF REFRACTORY BRICKS AT HIGH TEMPERATURES, AND OF STEAM-PIPE COVERINGS AND MASONRY.** VAN RINSUM W Z VER DEUT ING 62 601–5 639–41 1918 CA 14 1421
24326	**THERMAL PROPERTIES OF ELECTRICALLY MELTED MASS CONTAINING DUNITE AS THE PRINCIPAL INGREDIENT.** NIWA, S. YOGYO KYOKAI SHI 69 74–80 1961 CA 56 12558
24328	**HEAT INSULATION. I. HEAT INSULATION IN THE REFRIGERATION INDUSTRY.** RICHARDS S INSTN GAS ENGR PROC OF CONF ON HEAT INSULATION 4–42 1950 RA 8 177–8
24338	**THERMAL CONDUCTIVITY OF MOLDING MATERIALS AS A FUNCTION OF TEMPERATURE CHANGES.** SHEPTUNOV K L IZVEST VYSSHIKH UCHEB ZAVEDENII MASHINOSTROENIE 11 83–90 1960 CA 55 25661
24339	**ABSORPTION, REFLECTION, AND DISPERSION CONSTANTS OF QUARTZ.** COBLENTZ W W BULL NATL BUR STANDARDS 11 471–81 1915
24342	**THE STRUCTURE AND STRENGTH OF THE INNER PLANETS.** MAC DONALD, G. J. F. GODDARD SPACE FLIGHT CENTER, GREENBELT, MD. 1–84 1963 (NASA TR R–173)
24360	**THERMAL PROPERTIES OF A SIMULATED LUNAR MATERIAL IN AIR AND IN VACUUM.** BERNETT E C WOOD H L JAFFE L D MARTENS H E AIAA J 1 6 1402–7 1963
24363	**THE EMISSIVITY OF THE EARTHS SURFACE.** HOUGHTON J T QUART J ROY METEOROL SOC 84 448–50 1958 SA 62 11656
24364	**THE APPARENT THERMAL CONDUCTIVITY OF GLASS WOOL.** BURGOYNE, T. UNITED KINGDOM ATOMIC ENERGY AUTHORITY, WINDSCALE WORKS SELLAFIELD CUMB, ENGLAND 1–3 1953 (R AND DB/W/TN 63)
24367	**THERMAL DIFFUSIVITY MEASUREMENTS ON METALS AND CERAMICS AT HIGH TEMPERATURES.** RUDKIN, R. L. PARKER, W. J. JENKINS, R. J. U.S. RADIOLOGICAL DEFENSE LAB., SAN FRANCISCO, CA. 1–20, 1963. (ASD–TDR–62–24, NASA N63–13636, AD 297836)
24368	**MEASUREMENTS OF HEAT EFFECT ACCOMPANYING HEATING PEAT IN THE TEMPERATURE RANGE 20 TO 220 DEGREES.** SUNNER, S. WADSO, I. ACTA POLYTECH SCAND CHEM MET SER 14 1–40 1961 CA 56 1687
24372	**THE PHYSICAL PROPERTIES OF PORCELAIN.** SINGER, F. ROSENTHAL, E. J. SOC. GLASS TECH. 5, 175–7, 1921. (BER DEUT KERAM GES, 54, 250–1, 1920; SPRECHSAAL, 54, 250–1, 1921)
24384	**THERMAL DIFFUSIVITY MEASUREMENTS ON METALS AND CERAMICS AT HIGH TEMPERATURES.** RUDKIN, R. L. U. S. RADIOLOGICAL DEFENSE LAB., SAN FRANCISCO, CA 1–16, 1963. (ASD–TDR–62–24PPT 2/, NASA–N63–20823, AD–413005)
24393	**CERAMIC HIGH–TEMPERATURE MATERIALS IN ELECTRICAL ENGINEERING.** RICHTER W ELEKTROWARME 19 394–7 1961 BR 11 1493

TPRC Number	Bibliographic Citation
24395	**THE DEVELOPMENT OF ELECTRICALLY-HEATED APPARATUS FOR THE MEASUREMENT OF THE THERMAL CONDUCTIVITY OF REFRACTORIES AT HIGH TEMPERATURES.** SCHWIETE H-E SCHUFFLER A FORSCH-BER LANDES NORDRHEIN-WESTFAL 688 1959 BC 59 1055
24397	**CONTRIBUTION TO THE DETERMINATION OF THE THERMAL CONDUCTIVITY OF FIRE BRICK.** SIMONIS M SPRECHSAAL 41 547-9 1908
24403	**HEAT CONDUCTIVITY AND RADIATION IN GLASSES BETWEEN 100 AND 1,200 C.** ECKHARDT G GLASTECH BER 32 373- 1959 BC 59 628
24431	**STUDIES, RESEARCH AND INVESTIGATIONS OF THE OPTICAL PROPERTIES OF THIN FILMS OF METALS, SEMI-CONDUCTORS AND DIELECTRICS.** HADLEY, L. N., JR. COLORADO STATE UNIV., FT. COLLINS, COLORADO 1-20, 1957. (AD 203183, PB 144057)
24510	**RAPID METHOD FOR DETERMINING THERMAL CONDUCTIVITY OF SEMICONDUCTOR MATERIALS.** KAGANOV M A LISKER I S CHUDNOVSKII A F INZHENER-FIZ ZHUR AKAD NAUK BELORUS SSR 4 3 110-12 1961 CA 56 3002
24514	**THERMOCHEMICAL CHARACTERISTICS OF PBO-SIO2-NA2O MELTS.** TOPORISHCHEV G A ESIN O A BRATCHIKOV S H IZVESTIYA VUZ—TSVETNAYA METALLURGIYA 37-43 1961 RM 18 704-P
24550	**DETERMINATION OF THE COEFFICIENT OF THERMAL EXPANSION, SPECIFIC HEAT, AND HEAT OF FUSION OF RUBIDIUM AND OF CADMIUM.** DEUSS, E. SPECIAL LIBRARY ASSOC. TRANSLATION CENTER 1-40, 1963. (ENGLISH TRANSLATION OF VIERTELJAHRESSCHR. NATURFORSCH. GES. ZURICH, 56, 15-41, 1911; FOR ORIGINAL SEE T22607) (TT-64-10443)
24551	**THE THERMAL CONDUCTIVITY AND COMPRESSIBILITY OF SEVERAL ROCKS UNDER HIGH PRESSURES.** BRIDGMAN P W AM J SCI 7 9 81-102 1924
24552	**THE INFRARED TRANSMISSION OF SOME HF RESISTANT MATERIALS.** GAUNT, J. ATOMIC ENERGY RESEARCH ESTABLISHMENT, HARWELL BERKS, ENGLAND 1-3 1951 (AERE-C-M-120)
24553	**THE EFFECTS OF HIGH VACUUM AND ULTRAVIOLET RADIATION ON METALLIC MATERIALS.** WAHL, N. E. LAPP, R. T. HAAS, F. C. CORNELL AERONAUTICAL LAB., INC., BUFFALO, N. Y. 1-73, 1962. (WADD TR 60-125(PT 2), AD 281799)
24559	**THE THERMOPHYSICAL PROPERTIES OF PLASTIC MATERIALS FROM -50 F TO OVER 700 F.** SOUTHERN RESEARCH INST., BIRMINGHAM, ALA. SOUTHERN RESEARCH INSTITUTE, BIRMINGHAM, ALABAMA 1-83PP., 1963. (5919-1399-XII)
24582	**INFRARED TRANSMITTANCE OF CRYSTALLINE YTTRIUM OXIDE AND RELATED COMPOUNDS.** WICKERSHEIM, K. A. LEFEVER, R. A. J OPT SOC AM 51 1147-8 1961 CA 56 4262
24602	**REFLECTANCE MEASUREMENTS ON SILICEOUS DUSTS AND POWDERS.** DE BOER F NIKLAS W F BRIT J APPL PHYS 5 340-2 1954
24603	**THE THERMAL CONDUCTIVITY OF REFRACTORIES OF THE SYSTEM AL2O3-SIO2 AT 200-1600 DEGREES.** SCHWIETE H E GRANITZKI K E KARSCH K H BER DEUT KERAM GES 38 529-34 1961 CA 57 433
24647	**DENSITY OF THE GAS ON CONDUCTIVITY OF POWDERS OF DIFFERENT MATERIALS.** SMOLUCHOWSKI, M. BULL INTERN ACAD SCI CRACOVIE 3 A 129-53 1910
24680	**THERMAL CONDUCTIVITY OF REFRACTORY BRICK.** RUH, E. MC DOWELL, J. S. J AM CERAM SOC 45 189-95 1962 CA 56 15169
24712	**THE EVALUATION OF SEVERAL NEW INFRARED TRANSMISSIVE GLASS SYSTEMS.** LIN FRANCIS C WESTINGHOUSE ELECTRIC CORP PITTSBURGH PA 1-11 1959 (MATERIALS ENGINEERING PAPER NO. P-5943-3)
24744	**STUDY OF THE REFLECTING POWER OF THE MOON'S SURFACE.** SYTINSKAYA, N. N. SHARONOV, V. V. SPACE TECHNOLOGY LABS. 1-44, 1961. (ENGLISH TRANSLATION OF UCHENYE ZAPISKI LENINGRAD. GOSUDARST. UNIV. IM. A. A. ZHDANOVA, SER. FIZ. NAUK (153), 114- , 1952; FOR ORIGINAL SEE T24745) (STL-TR-61-5110-23, AD 264070)
24745	**STUDY OF THE REFLECTING POWER OF THE MOON'S SURFACE.** SYTINSKAYA, N. N. SHARONOV, V. V. UCH. ZAP. LENINGRAD GOS. UNIV. IM A. A. ZHDANOVA, SER. FIZ. NAUK (153), 114- , 1952. (FOR ENGLISH TRANSLATION SEE T24744)
24746	**THERMAL RADIATIVE CONTROL SURFACES FOR SPACECRAFT.** GAUMER, F. E. MC KELLAR, L. A. LOCKHEED AIRCRAFT CORP., LOCKHEED MISSILES AND SPACE DIVISION, SUNNYVALE, CA. 207PP., 1961. (LMSD-704014, AD 264127)
24748	**THE EMITTANCE OF CERAMICS AND GRAPHITES.** WOOD, W. D. DEEM, H. W. LUCKS, C. F. DEFENSE METALS INFORMATION CENTER, BATTELLE MEMORIAL INSTITUTE, COLUMBUS, OHIO 1-106, 1962. (DMIC MEMO 148, AD-274148)
24761	**THE REFRACTORIES INDUSTRY IN POLAND.** AUTHOR ANON. REFRACT. J. 36, 194, 230, 1960.
24763	**NEW APPROACH TO SOLVE PORTLAND CEMENT KILN HOT ZONE PROBLEMS.** PARNHAM H REFRACTORIES J 37 10 311-12 1961 JA 45 54
24767	**A COMPARATIVE ACCOUNT OF SILICON CARBIDE REFRACTORIES WITH THOSE IN GENERAL USE—GREAT IMPROVEMENT IN THERMAL CONDUCTIVITY, EFFICIENCY AND STRENGTH.** LINBARGER S C CHEM MET ENG 19 489-91 1918
24772	**THERMAL CONDUCTIVITY AND THERMAL DIFFUSIVITY OF BUILDING MATERIALS AND THEIR ROLE IN THE ASSESSMENT OF THE THERMOPHYSICAL PROPERTIES OF PARTITIONS AND EXTERNAL WALLS.** FRANCHUK A U INZHENER-FIZ ZH 3 9 69-73 1960 AM 15 368
24824	**COATINGS FOR SPACE VEHICLES.** RICHMOND, J. C. SYMPOSIUM ON SURFACE EFFECTS ON SPACECRAFT MATERIALS, 1ST 92-116 1960
24831	**HIGH-TEMPERATURE EMISSIVITY MEASUREMENTS AT THE MARTIN COMPANY.** CROMPTON, M. SYMPOSIUM ON SURFACE EFFECTS ON SPACECRAFT MATERIALS, 1ST 212-19 1960
24856	**HIGH TEMPERATURE THERMAL PROPERTY MEASUREMENT TO 5000 F.** NEEL, D. S. PEARS, C. D. ASME SECOND SYMPOSIUM ON THERMOPHYSICAL PROPERTIES SYMPOSIUM ON THERMOPHYSICAL PROPERTIES, ASME SECOND, PROGRESS IN INTERNATIONAL RESEARCH ON THERMODYNAMIC AND TRANSPORT PROPERTIES 500-11, 1962.
24893	**RADIATION HEAT TRANSFER ANALYSIS FOR SPACE VEHICLES.** STEVENSON, J. A. GRAFTON, J. C. AERONAUTICAL SPACE DIVISION 1-426, 1961. (ASD TR-61-119(PT 1), AD 271917)
24910	**SOIL TEMPERATURE AND WEATHER CONDITIONS.** CARSON, J. E. ARGONNE NATIONAL LAB., ARGONNE, ILL. 1-244, 1961. (ANL-6470)
24931	**THERMOPHYSICAL PROPERTIES OF SOLID MATERIALS. VOLUME 3. CERAMICS.** GOLDSMITH, A. HIRSCHHORN, H. J. WATERMAN, T. E. WRIGHT AIR DEVELOPMENT CENTER 1-1162, 1960. (WADC TR 58-476(VOL 3), AD 265597)

TPRC Number	Bibliographic Citation

24932 **THERMOPHYSICAL PROPERTIES OF SOLID MATERIALS. VOLUME IV. CERMETS, INTERMETALLICS, POLYMERICS, AND COMPOSITES.**
GOLDSMITH, A. WATERMAN, T. E.
HIRSCHHORN, H. J.
WRIGHT AIR DEVELOPMENT CENTER
1-798, 1960.
(WADC TR 58-476(VOL 4), AD 265598)

24937 **MANGANESE OXIDE THERMISTORS.**
PERELESHINA, A. P.
SCIE. AND TECHNOL. SECT., AIR INFORMATION DIV.
1-31, 1961.
(ENGLISH TRANSLATION FROM PHYSICOCHEMICAL PROPERTIES OF THERMISTORS MADE OF MANGANESE OXIDES. PROBL. ENERG. SB. AKAD. NAUK SSSR. ENERG. INST. IM. G. M. KRZHIZHANOVSKOGO, MOSKVA, 828-48, 1959; FOR ORIGINAL SEE T24938)
(AID RPT 61-64, AD 257908)

24938 **MANGANESE OXIDE THERMISTORS.**
PERELESHINA, A. P.
PROBL. ENERG. SB. AKAD. NAUK SSSR, ENERG. INST. IM. G. M. KRZHIZHANOVSKOGO
828-48 1959
(FOR ENGLISH TRANSLATION SEE T24937)

24943 **A NEW APPROACH TO THE MEASUREMENT OF TERRESTRIAL RADIATION.**
DRUMMOND, A. J.
AIR FORCE CAMBRIDGE RESEARCH LABORATORY
1-32, 1961.
(AFCRL-828, AD 265139)

24958 **SUMMARY OF DATA ON HIGH TEMPERATURE REACTOR FUEL ELEMENT MATERIALS.**
STRASSER, A.
NUCLEAR DEVELOPMENT CORP. OF AMERICA, WHITE PLAINS
1-180 1956
(NDA-64-103(DEL))

24960 **CONTINUUM INFRARED SPECTRUM OF HIGH TEMPERATURE AIR.**
TAYLOR, R. L.
AVCO-EVERETT RESEARCH LAB., RES. REPT.
1-17, 1960.
(AERL RR-88, AD 243116, PB 150955)

24962 **THE DEVELOPMENT AND TESTING OF HOMOGENEOUS CERAMIC FUELS.**
ZUROMSKY, G. CHERNOCK, W. P.
COMBUSTION ENGINEERING, INC., NUCLEAR DIV.
1-51, 1961.
(CEND-147, NYO-2996)

24968 **THERMAL CONDUCTIVITY MEASUREMENTS ON FIBER-GLASS PANELS, AT DIFFERENT TEMPERATURES, IN VARIOUS ATMOSPHERES.**
BARTOLI R LAINE P
BULL INST INTERN FROID ANNEXE 2
233-40 1956

24969 **INFLUENCE OF THE INCLUDED GASES ON THE THERMAL CONDUCTIVITY OF INSULATING MATERIALS.**
MYNCKE H VAN ITTERBEEK A DE GREVE L
BULL INST INTERN FROID ANNEXE 2
241-9 1956

24970 **THERMAL CONDUCTIVITY OF QUARTZ, FUSED SILICA, AND NO. 7740 GLASS, A LITERATURE SURVEY.**
PARKER, C. J.
CORNING GLASS WORKS CORNING N Y
1-4, 1957.
(R-1238)

24971 **OPTICAL AND THERMAL PROPERTIES OF GLASSES AT HIGH TEMPERATURES.**
SHAND, E. B.
CORNING GLASS WORKS CORNING N Y
1-19, 1956.
(R-845)

24972 **TOTAL EMISSIVITY OF UNIFORMLY HEATED PLATES OF SEVERAL GLASSES FROM 250 TO 700 C.**
PARKER, C. J.
CORNING GLASS WORKS CORNING N Y
1-9, 1956.
(R-851)

24978 **A STUDY OF THERMAL CONDUCTIVITY OF FROZEN AND UNFOROZEN SOILS AND ITS RELATION TO HEAVE.**
KLUS, J. P.
MICHIGAN COLLEGE OF MINING AND TECHNOLOGY, HOUGHTON, M.S. THESIS
1-40 1961

24979 **EVALUATION OF LOW DENSITY, HIGH TEMPERATURE /TO 2000 F/ THERMAL INSULATION.**
HANSEN, A. L.
NORTHROP AIRCRAFT, INC., NORTHROP DIV., HAWTHORNE, CALIFORNIA
1-10, 1660.
(NOR-60-251, AD 270418)

24992 **THE EFFECT OF THE DENSITY OF THE OVEN CHARGE ON ITS THERMOPHYSICAL PROPERTIES.**
ZUBILIN I G
KOKS I KHIM
8 20-1 1961 CA 56 1686

25029 **INFRARED COATING STUDIES.**
MARTIN, T. P.
ARMY ENGINEER R AND D LABS., FT. BELVOIR, VA.
1-25, 1962.
(AD 285022)

25061 **THERMAL CONDUCTIVITY OF CERAMIC MATERIALS AND MEASUREMENTS WITH A NEW FORM OF THERMAL COMPARATOR.**
POWELL, R. W. TYE, R. P.
SPEC. CERAM., PROC. SYMP. BRIT. CERAM. RES. ASSOC. STOKE-ON-TRENT
261-80 1963 BC 62 2641

25075 **MATERIALS HANDBOOK.**
CORNING GLASS WORKS, NEW YORK
CORNING GLASS WORKS, CORNING, N. Y.
1, 1961.

25099 **THERMAL CONDUCTIVITY OF WHITE MARBLE AND NEAT CEMENT.**
NELSON ROY A
PHYS REV
18 113-15 1921 CA 16 3537

25109 **RADIATION STABILITY OF CERAMICS.**
BOPP, C. D. TOWNS, R. L. SISMAN, O.
OAK RIDGE NATIONAL LABORATORY
35-6, 1956.
(ORNL-1945)

25127 **RESEARCH FOR LOW AND HIGH EMITTANCE COATINGS.**
SCHATZ, E. A. MC CANDLESS, L. C.
AERONAUTICAL SPACE DIV, TECHNICAL REPT.
1-92, 1962.
(ASD-TR-62-443, AD 281821)

25142 **DETERMINATION OF THE THERMAL CONDUCTIVITY COEFFICIENT OF SOILS FROM DATA OF OBSERVATIONS UNDER NATURAL CONDITIONS.**
PORKHAEV G V
TRUDY INST MERZLOTOVEDENIIA IM V A OBRUCHEVA
14 80-91 1958

25145 **THERMAL DIFFUSIVITY AND CONDUCTIVITY OF SOME SOIL MATERIALS.**
INGERSOLL L R KOEPP O A
PHYS REV
24 2 92-3 1924

25148 **CERAMIC PRODUCTS AS INDUSTRIAL MATERIALS.**
SINGER, F.
Z VER DEUT INGR
67 24 584-6 1923

25152 **HEAT FLOW THROUGH GRANULATED MATERIAL.**
SCHUMANN T E W VOSS V
FUEL IN SCIENCE AND PRACTICE
8 8 249-56 1934

25156 **THE COEFFICIENT OF THERMAL CONDUCTIVITY IN EARTHS CRUST.**
ZHARKOV V N
IZVEST AKAD NAUK SSSR SER GEOFIZ
23 11 1342-50 1958

25157 **INVESTIGATION OF THE THERMAL PROPERTIES OF HIGH-TEMPERATURE INSULTION MATERIALS.**
WECHSLER, A. E. GLASER, P. E.
ARTHUR D. LITTLE, INC., CAMBRIDGE, MA.
1-171 1963 PA N63-1 1943
(ASD-TDR-63-574)

25164 **ON A SIMPLE APPARATUS FOR DETERMINING THE THERMAL CONDUCTIVITIES OF CEMENTS AND OTHER SUBSTANCES USED IN THE ARTS.**
LEES, C. H. CHORLTON, J. D.
PHIL MAG
41 495-503 1896

25165 **EXPERIMENTS ON THE HEAT-CONDUCTIVITY OF STONE, BASED ON FOURIERS THEORIE DE LA CHALEUR.**
AYRTON W E PERRY JOHN
PHIL MAG
5 5 241-67 1878

25173 **THE THERMAL DIFFUSIVITY OF CHEMICALLY BONDED SILICA BODIES.**
MUEHLEISEN, E. C.
ALFRED UNIV., NEW YORK, M.S. THESIS
1-60 1961

25175 **THE MANUFACTURE, PROPERTIES, AND APPLICATION OF FOAMED CORUNDUM, A HIGHLY POROUS SINTERED ALUMINA.**
HEUBLEIN H KROCKEL O
SILIKATTECHNIK
11 344-6 1960 BC 60 374

TPRC Number	Bibliographic Citation
25181	**THERMAL CONDUCTIVITY AND TEXTURE ANALYSIS OF FIREBRICKS.** SCHWIETE H E WESTMARK HERIBERT SPRECHSAAL 95 1-11 31-5 51-5 1962 CA 56 12557
25193	**THE VARIATION OF THE SPECIFIC HEAT DURING MELTING AND THE HEAT OF FUSION OF SOME METALS.** ITAKA ITIRO SCIENCE REPTS TOHOKU IMP UNIV 8 99-114 1919 CA 16 3797
25194	**RESEARCH ON LOW-DENSITY THERMAL INSULATION MATERIALS FOR USE ABOVE 3000 F.** STYHR, K. H. NATIONAL BERYLLIA CORP., HASKELL, N. J. 1-23, 1962. (NASA N62-15757)
25199	**THE THERMOPHYSICAL PROPERTIES OF PLASTIC MATERIALS FROM -50 F TO OVER 700 F.** ENGELKE, W. T. PEARS, C. D. OGELSBY, S., JR. SOUTHERN RESEARCH INST., BIRMINGHAM, ALABAMA 1-40, 1963. (5783-1399-IX)
25207	**OPTICAL AND INFRARED PROPERTIES OF ALUMINUM OXIDE AT ELEVATED TEMPERATURES.** GRYVNAK, D. A. BURCH, D. E. PHILCO RESEARCH LABS., NEWPORT BEACH, CA. 1-39, 1964. (U-2623, AD 606793)
25230	**WATER IN SILICA GLASS.** MOULSON, A. J. ROBERTS, J. P. TRANS. BRIT. CERAM. SOC. 59 (9), 388-99, 1960. (TRANS. FARADAY SOC., 57, 1208-16, 1961)
25235	**INFRARED COATING STUDIES.** TURNER, A. F. BAUSCH AND LOMB, INC., ROCHESTER, N. Y. 1-12, 1963. (AD 429240)
25257	**METHODS FOR CALCULATION OF THERMAL CONDUCTION COEFFICIENTS OF CAPILLARY-POROUS MATERIALS.** BABANOV A A ZHUR TEKH FIZ 27 3 532-42 1957 (FOR ENGLISH TRANSLATION SEE T8991)
25275	**THERMAL COEFFICIENTS OF CERTAIN UKRANIAN CLAYS AND THEIR DEPENDENCE ON MOISTURE COMBINED WITH THE SOLID PHASE IN DIFFERENT WAYS.** KAZANSKII M F SOVIET PHYS—TECH PHYS 4 2 215-18 1959 (ENGLISH TRANSLATION OF ZHUR. TEKH. FIZ. 29 (2), 247-, 1959; FOR ORIGINAL SEE T25276)
25276	**THERMAL COEFFICIENTS OF CERTAIN UKRANIAN CLAYS AND THEIR DEPENDENCE ON MOISTURE COMBINED WITH THE SOLID PHASE IN DIFFERNT WAYS.** KAZANSKII M F ZHUR TEKH FIZ 29 2 247 1959 (FOR ENGLISH TRANSLATION SEE T25275)
25307	**HEAT CONDUCTIVITY OF BATIO3-BASED SOLID SOLUTIONS OF MANGANESE NIOBATE.** KODZHESPIROV F F SOVIET PHYSICS—SOLID STATE 3 3 567-70 1961 (ENGLISH TRANSLATION OF FIZ. TVERD. TELA 3 (3), 781-5, 1961; FOR ORIGINAL SEE T23796)
25356	**THE THERMOPHYSICAL AND MECHANICAL PROPERTIES OF HIGHLY CARBONIZED POLYMERIC MATERIALS.** ZAMOLUEV V K MUKHANOVA L N TAITS E M PROC ACAD SCI USSR CHEMICAL TECHNOLOGY SECT 133 5 127-9 1960 (ENGLISH TRANSLATION OF DOKLADY AKAD. NAUK SSSR 133, 1143-5, 1960; FOR ORIGINAL SEE T21288)
25361	**INVESTIGATION OF THE HEAT AND MOISTURE CONDUCTION IN PEAT.** VOLAROVICH M P GAMAYUNOV N U CHURAEV N V COLLOID J USSR 22 5 535-41 1960 (ENGLISH TRANSLATION OF KOLLOID. ZH. 22, 535-42, 1960; FOR ORIGINAL SEE T24247)
25367	**CALCULATION OF THE DEPTH OF THAWING OF FROZEN FOUNDATION BEDS.** USHKALOV V P TR. SOVESHCHANIIA PO RATSIONALNYM SPOSOBAM FUNDAMENTOSTROENIIA NA VECHNOMERZLYKH GRUNTAKH VORKUTA 1-16 1959
25369	**THERMAL CONDUCTIVITY OF SOIL.** WEBB JOHN NATURE /LONDON/ 177 989 1956
25372	**A METHOD OF MEASURING THERMAL DIFFUSIVITY AND CONDUCTIVITY OF STONE AND CONCRETE.** THOMSON, W. T. PROC. AM. SOC. TESTING MATERIALS 40, 1073-81, 1940. (KANSAS STATE COLLEGE ENG. EXP. STA. BULL.) (25 (3), 5-13, 1941)
25391	**VARIATION IN REFLECTANCE OF VEGETATION AND SOILS CAUSED BY AN UNDERGROUND NUCLEAR EXPLOSION.** DWORNIK, S. E. ORR, D. G. YOUNG, L. M. DEPT. OF THE ARMY CORPS. OF ENGINEERS, FT. BELVOIR, VA. 77PP., 1962. (AD-292 994)
25402	**UEBER DIE WARMELEITUNGSFAHIGKEIT DER GASE UND IHRE ABHANGIGKEIT VON DER TEMPERATUR /BEI TIEFEN TEMPERATUREN/.** ECKERLEIN P ADALBERT ANN PHYSIK 3 120-54 1900
25423	**THE SPECIFIC HEAT OF THERMAL INSULATING MATERIALS.** WILKES, G. B. WOOD, C. C. HEATING PIPING AIR CONDITIONING 14 6 370-4 1942
25424	**SPECIAL CORROSION RESISTANT REFRACTORIES.** NOBLE, R. W. BROWN, R. W. CORROSION 15 10 92 94 96 98 1959
25427	**PROPERTIES OF A TIN OXIDE-BASE CERAMIC BODY.** QUIRK JOHN HARMAN C G J AM CERAM SOC 37 24-6 1954
25435	**NEW DATA ON THE THERMAL CONDUCTIVITY OF OPTICAL CRYSTALS.** MC CARTHY, K. A. BALLARD, S. S. J OPT SOC AM 41 1062-3 1951
25437	**TESTS FOR THERMAL DIFFUSIVITY OF GRANULAR MATERIALS.** SHANNON, W. L. WELLS, W. A. PROC AM SOC TESTING MATERIALS 47 1044-55 1947
25453	**EFFECT OF NEUTRON IRRADIATION ON THE THERMAL CONDUCTIVITY OF A QUARTZ CRYSTAL AT LOW TEMPERATURE.** BERMAN R KLEMENS P G SIMON F E FRY T M NATURE /LONDON/ 166 864-6 1950
25461	**ON A METHOD OF FINDING THE CONDUCTIVITY FOR HEAT.** NIVEN C PROC ROY SOC /LONDON/ 76 A 34-48 1905
25466	**SOME PHYSICAL PROPERTIES OF PORCELAINS IN THE SYSTEMS MAGNESIA-BERYLLIA-ZIRCONIA AND MAGNESIA-BERYLLIA-THORIA AND THEIR PHASE RELATIONS.** LANG S M MAXWELL L H GELLER R F J RESEARCH NATL BUR STANDARDS 43 429-47 1949
25488	**SPECIFIC HEAT OF GLASSES. I. HISTORICAL.** PRODHOMME M VERRES ET REFRACT 11 287-97 1957 BC 57 1165
25490	**SOME METHODS OF MEASURING THERMAL CONDUCTIVITY OF GLASS.** ACLOQUE, P. VERRES ET REFRACTAIRES 13 (3), 131-42, 1959. (13 (4), 191-203, 1959) (13 (5), 239-52, 1959)
25498	**THE APPARENT VISCOSITY OF REFRACTORY MATERIALS AT HIGH TEMPERATURES.** CHAKLADER A C CARRUTHERS T G ROBERTS A L TRANS. INT. CERAM CONGR. 95-106 1958 BC 59 1926
25502	**ELECTRONIC CALORIMETER.** MITKINA E A ZAVODSKAYA LAB 26 226-9 1960 CA 56 2298 (FOR ENGLISH TRANSLATION SEE T32705)
25524	**THE THERMAL CONDUCTIVITY AND OTHER PROPERTIES OF CARBONACEOUS BRIQUETTES.** BIEN, A. S. PHILLIPS, O. NEWARK COLLEGE OF ENGINEERING, NEWARK, N. J., M.S. THESIS 1-84 1957
25525	**FOAMED KERALITE, A NEW BUILDING MATERIAL.** MALZEW W DIKERMANN N ZIEGELINDUSTRIE 13 234-7 1960 BC 59 2476

TPRC Number	Bibliographic Citation
25545	**THE THERMAL CONDUCTIVITY OF REFRACTORY OXIDES.** VISHNEVSKII I I SB NAUCHN TR UKR NAUCHN-ISSLED INST OGNEUPOROV 3 274-81 1960 (FOR ENGLISH TRANSLATION SEE T28160)
25557	**SOME PRACTICAL APPLICATIONS OF HEAT TRANSFER BETWEEN BURIED OBJECTS AND THE SOIL.** GRIFFITH M V HUTCHINGS E E PROC. GENERAL DISCUSSION ON HEAT TRANSFER 285-7 1951
25558	**HIGH TEMPERATURE MATERIALS. NO. 3 THERMAL RADIATIVE PROPERTIES.** WOOD, W. D. DEEM, H. W. LUCKS, C. F. HANDBOOKS OF HIGH-TEMPERATURE MATERIALS 1-476, 1962. (DMIC RPT 177(VOL 1+2), AD 294346)
25559	**THE THERMAL CONDUCTIVITY OF SOME NONMETALLIC MATERIALS.** GRIFFITHS, E. HICKMAN, M. J. PROC GEN DISCUSSION ON HT TRANSFER 289-90 1951 RA 9 19-7
25562	**THE THERMOPHYSICAL PROPERTIES OF PLASTIC MATERIALS FROM -50 F TO OVER 700 F.** SOUTHERN RESEARCH INST., BIRMINGHAM, ALA. SOUTHERN RESEARCH INST., BIRMINGHAM, AL. 1-114, 1963. (6366-1399-XV)
25574	**PROPERTIES AND APPLICATIONS OF FLAME-SPRAYED REFRACTORY COATINGS.** MANUEL L BULL INST VIT ENAMEL 9 51-4 1958 BC 58 1433
25588	**DEVELOPMENT OF CERAMIC FUEL ELEMENTS FOR HIGH TEMPERATURE REACTORS. FABRICATION OF UO2 AND UO(2)-BEO MIXED PELLETS.** BATTELLE INST., E. V. FRANKFURT, GERMANY U.S. EURATOM JOINT RES. AND DEVELOP. PROGRAM 1-23, 1962. (EURAEC-179)
25668	**THE NATIONAL BUREAU OF STANDARDS HIGH TEMPERATURE ABSOLUTE CUT-BAR APPARATUS FOR DETERMINATION OF THE THERMAL CONDUCTIVITY OF SMALL SOLIDS.** FLYNN, D. R. ROBINSON, H. E. NATIONAL BUREAU OF STANDARDS 1-38, 1964. (NBS REPT 8301, AD 486355)
25723	**BASIC STUDIES ON THE USE AND CONTROL OF SOLAR ENERGY.** EDWARDS, D. K. RODDICK, R. D. UNIV. OF CALIFORNIA, LOS ANGELES 1-103, 1962. (UCB-R-62-27)
25737	**AN INVESTIGATION ON THE THERMAL CONDUCTIVITY OF POROUS MATERIALS AND ITS APPLICATION TO POROUS ROCK.** SUGAWARA A YOSHIZAWA Y AUSTRAL J PHYS 14 4 469-80 1961 AM 16 1724
25767	**THE THERMAL CONDUCTIVITY OF REFRACTORIES.** DUDLEY BOYD JR TRANS AM ELECTROCHEM SOC 27 285-339 1915
25795	**SYNTHESIS AND INVESTIGATION OF VITREOUS OXIDE SEMICONDUCTORS IN SYSTEMS OF THE TYPE VO2.5-PO2.5-RO/SUB X/.** GRECHANIK L A PETROVYKH N V KARPECHENKO V G FIZ TVERDOGO TELA 2 2131-9 1960 CA 56 8145 (FOR ENGLISH TRANSLATION SEE T29003)
25797	**EFFECTIVE THERMAL CONDUCTIVITIES IN PACKED BEDS.** YAGI, S. KUNII, D. WAKAO, N. A I CH E JOURNAL 6 543-6 1960 CA 56 8502
25813	**HEAT TRANSFER CHARACTERISTICS OF POROUS ROCKS. II. THERMAL CONDUCTIVITIES OF UNCONSOLIDATED PARTICLES WITH FLOWING FLUIDS.** KUNII, D. SMITH, J. M. A I CH E JOURNAL 7 29-34 1961 CA 56 9908
25827	**HEAT CONTENT, THERMAL DIFFUSIVITY, AND THERMAL CONDUCTIVITY OF ALLOYS OF THE CALCIUM OXIDE-FERROUS OXIDE, ZINC OXIDE, ALUMINA-SILICA SYSTEM.** CHIZHIKOV, D. M. GULYANITSKAYA, S. F. PETROVA, R. N. IZVEST. AKAD. NAUK SSSR OTDEL. TEKH. NAUK MET. I TOPLIVO 6 37-41 1961 CA 56 7019
25832	**HEAT CONDUCTIVITY OF KUKERSITE /BITUMINOUS SHALE/ DURING HEATING AND THERMAL DECOMPOSITION.** VALDEK R LUTSKOVSKAYA N L EIZEN J IZVEST AKAD NAUK ESTON SSR SER FIZ-MAT I TEKH NAUK 10 3 207-14 1961 CA 56 8994 (FOR ENGLISH TRANSLATION SEE T56809)
25839	**APPARATUS FOR DETERMINING THE COEFFICIENTS OF DIFFUSE REFLECTION IN THE LONGWAVE INFRARED SPECTRAL REGION.** KROPOTKIN M A KOZYREV B P IZVEST LENINGRAD ELEKTROTEKH INST 44 87-99 1960 CA 56 7073
25858	**EFFECT OF PARTICLE-SIZE RATIO OF WHITE POWDER AND BLACK POWDER ON REFLECTIVITY OF THEIR MIXTURES.** KUNO, H. IMAMURA, N. KOGYO KAGAKU ZASSHI 61 813-15 1958 CA 56 3319
25860	**HEAT CONDUCTIVITY DATA FOR SOME SCANDINAVIAN NATURAL STONES.** LORENTZEN, G. BULL INST INTERN FROID ANNEXE 2 45-53 1964
25906	**EFFICIENCIES OF PHOSPHORS FOR SHORT-WAVE ULTRAVIOLET EXCITATION.** BRIL A HOEKSTRA W PHILIPS RESEARCH REPTS 16 356-70 1961 CA 56 9562
25920	**TESTING AND EVALUATION OF HEAT INSULATING MATERIALS.** SCHAIRER W GAS-U WASSERFACH 78 50 939-49 1935
25950	**IMPROVEMENT OF URANIUM OXIDE SINTERING WITH A VIEW TO INCREASING ITS THERMAL CONDUCTIVITY.** PEYSSOU, J. HIRTZ, P. HAUSER, R. EZTAN, S. ATONIQUES FRITTES, CORBEVILLE FRANCE 1-32, 1962. (EURAEC-103)
25985	**THE TRANSMISSION OF GLASSES IN THE NEAR INFRARED AT TEMPERATURES UP TO FREEZING POINT.** SCHOLZE H SPECIAL LIBRARY ASSOC., TRANSLATION CENTER 1-9, 1962. (ENGLISH TRANSLATION OF GLASTECHN. BER. 32K (7), 1-5, 1959; FOR ORIGINAL SEE T30334) (SLA 62-16142)
25994	**WATER IN SILICA GLASS.** MOULSON A J ROBERTS J P TRANS FARADAY SOC /GB/ 57 7 1208-16 1961 SA 65 6378
26008	**THE THERMAL PROPERTIES OF TWENTY-SIX SOLID MATERIALS TO 5000 F OR THEIR DESTRUCTION TEMPERATURES.** PEARS C D WRIGHT PATTERSON AIR FORCE BASE, OHIO 6 1-420, 1962. (ASD-TDR-62-765, AD-298061)
26012	**A CORRELATION OF PROPERTIES FOR VARIOUS FORMULATIONS OF SINTERED ZIRCONIA.** LOWRANCE, D. T. CHANCE VOUGHT CORP., DALLAS, TEXAS 1-95, 1962. (CVC RPT. 2-53420/2R375, AD-273802)
26034	**THE EFFECTS OF MOISTURE ON THE THERMAL CONDUCTIVITY OF SOLIDS, WITH A BIBLIOGRAPHY ON THE HEATING OF CABLES.** SHANKLIN G B J AM INST ELEC ENGRS 41 92-8 1922
26060	**FUSED QUARTZ AS A MODEL MATERIAL IN THERMAL CONDUCTIVITY MEASUREMENTS.** DEVYATKOVA E D PETROV A V SMIRNOV I A MOIZHES B YA SOVIET PHYS—SOLID STATE 2 4 681-8 1960 (ENGLISH TRANSLATION OF FIZ. TVERDOGO TELA, 2 (4), 738-46, 1963; FOR ORIGINAL SEE T20183)
26069	**THERMAL CONDUCTIVITY OF PYREX GLASS. SELECTED VALUES.** CARWILE, L. C. K. HOGE, H. J. QUARTERMASTER RES. AND ENG. CENTER, U. S. ARMY, NATICK, MASS. 10PP., 1962. (REPT TC-1)
26087	**THE RESULTS OF EMITTANCE MEASUREMENTS MADE IN RELATION TO THE THERMAL DESIGN OF EXPLORER SPACECRAFT.** MERRILL, R. SNODDY, W. SCHOCKEN, K. MARSHALL, G. C. GEORGE C MARSHALL SPACE FLIGHT CENTER, HUNTSVILLE 1-147 1962 PA N62-2 342 (NASA TN D-1116)
26143	**DECOLORATION OF MANGANESE GLASSES.** BEZBORODOV M A MAZO E E MELNIK M T ZH PRIKL KHIM 35 1666-70 1962 CA 57 14721

TPRC Number | Bibliographic Citation

26150 EFFECT OF MOISTURE ON THE THERMAL CONDUCTIVITY OF STRUCTURAL AND INSULATING MATERIALS ACCORDING TO THE PRESENT STATE OF RESEARCH.
CAMMERER W F
ZIEGELINDUSTRIE
15 30-3 74-7 1962 BC 61 2173

26210 THE LOW TEMPERATURE HEAT CAPACITY OF NEUTRON IRRADIATED QUARTZ.
WESTRUM E F JR
IV-E CONGRES INTERNATIONAL DU VERRE PARIS
396-9 1956

26221 PHYSICO-MECHANICAL PROPERTIES AND THE THERMAL-SHOCK RESISTANCE OF FIRECLAY CASTING-PIT REFRACTORIES.
SASAKI S
YOGYO KYOKAI SHI
68 283 1960 BC 61 162

26224 STUDIES ON THE COLORATION OF VANADIUM-CONTAINING GLASSES.
SAKAINO T MORIYA T
YOGYO KYOKAI SHI
69 313-21 1961 BC 61 1923

26233 A RADIAL HEAT FLOW APPARATUS FOR HIGH-TEMPERATURE THERMAL CONDUCTIVITY MEASUREMENTS.
FEITH, A. D.
GENERAL ELECTRIC CO., ADVANCED TECHNOLOGY SERVICES
1-25, 1964.
(GEMP-296)

26253 NEW PYROMETERS FOR GLASS AND OTHER SURFACES.
LAND T BARBER R
J SOC GLASS TECHNOL
38 45N-53N 1954

26274 USE OF EXPERIMENTS ON VARYING-RATE ABLATION TO DETERMINE THE THERMAL CONDUCTIVITY AND OTHER PHYSICAL PROPERTIES OF VITREOUS HEAT-SHIELD MATERIALS.
POLEZHAEV YU V
HIGH TEMPERATURE
1 1 27-31 1963
(ENGLISH TRANSLATION OF TEPLOFIZIKA VYSOLOKH TEMPERATUR, 1 (1), 33-8, 1963; FOR ORIGINAL SEE T26275)

26275 USE OF EXPERIMENTS ON VARYING-RATE ABLATION TO DETERMINE THE THERMAL CONDUCTIVITY AND OTHER PHYSICAL PROPERTIES OF VITREOUS HEAT-SHIELD MATERIALS.
POLEZHAEV YU V
TEPLOFIZIKA VYSOKIKH TEMPERATUR
1 1 33-8 1963
(FOR ENGLISH TRANSLATION SEE T26274)

26316 SPECTRAL REFLECTANCE PROPERTIES OF NATURAL FORMATIONS.
KRINOV E L
NATIONAL RESEARCH COUNCIL, CANADA
1-268, 1953.
(ENGLISH TRANSLATION OF SPEKTRALNAIA OTRAZHATELNAIA SPOSOBNOST PRIRODNYKH OBRAZOVANII, LABORATORIIA AEROMETODOV, AKAD. NAUK SSSR, MOSCOW, 271PP., 1947; FOR ORIGINAL SEE T26342)
(NRC-TT-439, AD 113260)

26325 INFRARED REFLECTION SPECTRA OF GLASSES.
JELLYMAN P E PROCTER J P
J SOC GLASS TECHNOL
39 173T-192T 1955

26326 LOW-TEMPERATURE HEAT CAPACITIES AND ENTROPIES AT 298.15 DEGREES OF DIASPORE, KAOLINITE, DICKITE, AND HALLOYSITE.
KING E G WELLER W W
U S BUR MINES REPT INVEST
USBM 5810 1-6 1961 CA 56 4167

26327 LOW-TEMPERATURE HEAT CAPACITIES AND ENTROPIES AT 298.15 K. OF SOME SODIUM- AND CALCIUM-ALUMINUM SILICATES.
KING E G WELLER W W
NATIONAL BUREAU OF STANDARDS
1-8, 1961.
(BM-RI-5855)

26342 SPECTRAL REFLECTANCE PROPERTIES OF NATURAL FORMATIONS.
KRINOV E L
SPECTRAL REFLECTANCE PROPERTIES OF NATURAL FORMATIONS.
1-271 1947
(FOR ENGLISH TRANSLATION SEE T26316)

26365 THE EMISSION SPECTRUM OF WELSBACH BURNERS.
RUBENS H
ANN PHYSIK
18 725-38 1905

26370 THERMAL CONDUCTIVITY OF CERAMIC MATERIALS.
KOLTERMANN, M.
TONIND-ZTG KERAM RUNDSCHAU
85 17 399-407 1961 JA 46 121

26371 THE DETERMINATION BY USE OF ELECTRICAL ANALOGS OF THE HEAT CONDUCTIVITY OF MASSES FORMED BY MIXING COMPONENTS OF DIFFERENT CONDUCTIVITIES.
TRAUSTEL, S. PANDA, J. JACOB, P.
TONIND-ZTG U KERAM RUNDSCHAU
85 565-9 1961 CA 56 14952

26388 EXPERIMENTAL HEAT CONTENTS OF SRO, BAO, CAO, BACO3 AND SRCO3 AT HIGH TEMPERATURES. DISSOCIATION PRESSURES OF BACO3 AND SRCO3.
LANDER J J
J AM CHEM SOC
73 5794-7 1951

26393 LOW-TEMPERATURE ADIABATIC CALORIMETER, AND THE HEAT CAPACITY OF ALPHA-ALUMINA.
EDWARDS J W KINGTON G L
TRANS FARADAY SOC
58 1313-22 1962 CA 58 1956

26483 THERMODYNAMICS IN GLASS INVESTIGATION.
ZAGAR L
SILICATES IND
27 279-83 1962 CA 57 9468

26552 IMPROVING THE SINTERING OF URANIUM OXIDE WITH A VIEW TOWARDS INCREASING ITS THERMAL CONDUCTIVITY.
COMPAGNIE GENERALE DE TELEGRAPHIE
COMPAGNIE GENERALE DE TELEGRAPHIE SANS FIL, PARIS
1-27, 1962.
(ENGLISH TRANSLATION OF A REPORT FROM COMPAGNIE GENERALE DE TELEGRAPHIE SANS FIL, PARIS (FRANCE))
(EURAEC-7 (SUPERSEDES AEC-TIDI 12481))

26561 THERMAL CONDUCTIVITY OF REFRACTORY POWDERS AND SINTERED CERAMICS.
PUSTOVALOV V V
STEKLO I KERAM
18 12 17-19 1961 CA 56 13829
(FOR ENGLISH TRANSLATION SEE T27771)

26640 LONG WAVELENGTH INFRARED DETECTOR RESEARCH AND COMPONENTS MEASUREMENT. PART II. LONG WAVELENGTH INFRARED COMPONENTS MEASUREMENTS.
LANDERS DAVID L KOOZEKANANI SAID BRATT PETER R
RAYTHEON CO., SPENCER LAB, BURLINGTON
1-62, 1963.
(AD 297303)

26653 SOME OBSERVATIONS ON THE VARIATION OF THE THERMAL CONDUCTIVITY OF POROUS INORGANIC SOLIDS WITH MOISTURE CONTENT.
PRATT, A. W.
THERMAL CONDUCTIVITY CONFERENCE
1-10, 1964.
(GT. BRIT. BUILDING RESEARCH SERIES 30, 1-10, 1964)

26667 MEASUREMENT OF THERMAL CONDUCTIVITY OF INSULATING BRICK.
BEUKEN L
WARME U KALTE-TECHNIK
40 11 161-3 1938

26674 DETERMINATION OF THE THERMAL CONDUCTIVITY OF CERAMIC BODIES AT HIGH TEMPERATURE.
STALHANE, B. PYK, S.
TEK TIDSKR
64 48 445-8 1934

26684 ON THIN-FILM HEAT-TRANSFER MEASUREMENTS IN SHOCK TUBES AND SHOCK TUNNELS.
HARTUNIAN R A VARWIG R L
PHYSICS OF FLUIDS
5 2 169-74 1962 AM 15 4812

26731 THE INFRA-RED ABSORPTION BY DIAMOND AND ITS SIGNIFICANCE. X. EVALUATION OF THE SPECIFIC HEAT.
RAMAN C V
PROC INDIAN ACAD SCI
55 A 1 49-61 1962 SA 65 8506

26738 AN APPARATUS FOR THE MEASUREMENT OF TOTAL HEMISPHERICAL EMISSIVITY AND THERMAL CONDUCTIVITY BETWEEN AMBIENT AND LIQUID NITROGEN TEMPERATURES.
HAURY, G. L.
WRIGHT-PATTERSON AFB, OHIO
1-30, 1963.
(ASD-TDR-63-146, AD 411140)

26746 TRANSMITTANCE TABLES FOR SLANT PATHS IN THE STRATOSPHERE.
PLASS, G. N.
AIR FORCE SYSTEMS COMMAND, SPACE SYSTEMS DIVISION
1 VOL, 1963.
(SSD-TDR-62-127 (VOL 5), AD 415207)

26766 SOIL THERMAL CHARACTERISTICS IN RELATION TO UNDERGROUND POWER CABLES. II. SOIL TYPES. IDENTIFICATION AND PHYSICAL PROPERTIES. IV. SOIL THERMAL RESISTIVITY, TYPICAL FIELD VALUES AND CALCULATING FORMULAS.
DEL MAR WILLIAM A BURRELL R W BAUER C A
SINCLAIR W A BULLER F H BENHAM C B
POWER APPARATUS AND SYSTEMS
51 792-856 1960 BR 10 398

TPRC Number	Bibliographic Citation
26777	REFRACTORY BEHAVIOR OF VANADIA-SILICA GLASS. MACKENZIE J D PHYS CHEM GLASSES 3 2 50-3 1962 JA 46 40
26779	THE DETERMINATION OF THE THERMAL CONDUCTIVITY OF REFRACTORIES BY THE HOT-WIRE METHOD. MITTENBUEHLER, A. PURDUE UNIVERSITY, LAFAYETTE 1-19 1965 (ENGLISH TRANSLATION OF BER. DEUT. KERAM. GES., 39 (7), 387-92, 1962; FOR ORIGINAL SEE T27715) (PU TRANS 592, TPRC TRANS 12)
26790	UNIJUNCTION DEVICES MADE FROM RARE EARTH SEMICONDUCTORS. SCLAR, N. NUCLEAR CORP. OF AMERICA, DENVILLE, N. J. 1-86, 1961. (AD 274744)
26836	LOW-TEMPERATURE THERMAL CONDUCTIVITY OF AMORPHOUS SOLIDS. CHANG G K JONES R E PHYS REV 126 2055-8 1962 CA 57 15828
26839	OPTICAL PROPERTIES OF DIAMOND IN THE VACUUM ULTRAVIOLET. PHILIPP H R TAFT E A PHYSICAL REVIEW 127 2 1 159-61 1962 BR 11 8994
26844	THERMAL CONDUCTIVITY REFERENCE STANDARDS. ROBINSON, H. E. NATIONAL BUREAU OF STANDARDS 1-16, 1963. (NBS-R-7835, AD 406059)
26930	HEAT RESISTANCE OF CORUNDUM REFRACTORIES. POLUBOIARINOV D N LUKIN E S OGNEUPORY 27 5 230-5 1962 BR 11 6894
26958	INFRARED ABSORPTION OF MEYERHOFFERITE, A BORON MINERAL, AT 400-5000 CM TO THE MINUS ONE POWER. MOENKE HORST NATURWISSENSCHAFTEN 49 7 1962 CA 57 1763
27035	A STUDY OF RADIATION INDUCED SHOCK WAVES EXTERNAL TO QUARTZ DISCHARGE TUBES AND ASSOCIATED FRACTURE PROBLEMS. ELTON, R. C. NAVAL RESEARCH LAB, WASHINGTON D. C. 1-11, 1964. (NRL-MR-1509, AD 434325)
27093	PHYSICAL PROPERTIES OF POROUS MATERIALS AND THEIR APPLICATIONS. II. THERMAL CONDUCTIVITY OF POWDERS. WAKASHIMA HISAO KANAZAWA DAIGAKU KOGAKUBU KIYO 2 389-95 1961 CA 57 9240
27117	A CORRECTION TO THIN-FILM HEAT TRANSFER MEASUREMENTS. HARTUNIAN, R. A. VARWIG, R. L. AEROSPACE CORP., LOS ANGELES CALIFORNIA 1-46, 1961. (TDR-594/1217-01/TN-2, AD 606036)
27140	THE THERMAL CONDUCTIVITY OF REFRACTORY MATERIALS AT HIGH TEMPERATURE. FRALISH, H. J. OHIO STATE UNIVERSITY, COLUMBUS, PH.D. THESIS 1-73 1936
27141	MEASUREMENT OF RADIATIVE HEAT TRANSFER WITH THIN-FILM RESISTANCE THERMOMETERS. BOGDAN, L. CORNELL AERONAUTICAL LAB INC. 1-39, 1964. (NASA-CR-27)
27189	TENTATIVE METHOD OF TEST FOR THERMAL CONDUCTIVITY OF REFRACTORIES. AUTHOR ANON. AM SOC TESTING MATER ASTM STD 1376-82, 1947. (C-201-45)
27213	SCATTERING COEFFICIENT, REFLECTIVITY AND COVERING POWER OF RECRYSTALLIZED OPAQUE GLAZES. SOGAWA S J CERAM ASSOC JAPAN 69 12 427-33 1961 BC 61 2112
27279	THERMAL CONDUCTIVITY OF SEVERAL OPTICAL MASER MATERIALS. HOLLAND M G J APPL PHYS 33 2910-11 1962 CA 58 13165
27280	LASER MATERIAL STUDY. MAUER, P. B. EASTMAN KODAK CO., ROCHESTER, N. Y. 1-45, 1962. (AD 406936)
27283	EXPERIMENTAL INVESTIGATION ON THE THERMAL CONDUCTIVITY OF CONSOLIDATED POROUS MATERIALS. SUGAWARA A YOSHIZAWA Y J. APPL. PHYS. 33 10 3135-8 1962 JA 46 44
27306	SOME NEW DETERMINATIONS OF THE REFLECTING POWERS OF GLASS AND SILVERED GLASS MIRRORS. CHANT C A ASTROPHYS J 21 211-22 1905
27317	THERMAL RESISTIVITY OF RUBY IN THE OPTICALLY EXCITED STATE. KLEMENS P G APPL PHYS LETTERS 2 81-2 1963 CA 58 12064
27345	TRANSMITTANCE MEASUREMENTS OF OPTICAL MATERIALS AS AFFECTED BY WEDGE ANGLE AND REFRACTIVE INDEX IN THE 2- TO 15-MICRON RANGE. LABAW, K. B. OLSEN, A. L. NICHOLS, L. W. NAVAL ORDNANCE TEST STATION, CHINA LAKE, CALIF. 1-32, 1963. (NAVWEPS 8086, NOTS TP 3116, AD 403988)
27375	MEASUREMENT OF THE TRUE SPECIFIC HEAT OF SILVER, NICKEL, BETA-BRASS, QUARTZ AND QUARTZ GLASS BETWEEN 50 AND 700 C. BY AND IMPROVED METHOD. MOSER H SPECIAL LIBRARY ASSOC. 1-50, 1950. (ENGLISH TRANSLATION OF PHYSIK. Z., 37, 737-53, 1936; FOR ORIGINAL SEE T006314) (G-T-G-65-9, TT-66-61041, AD 631200)
27388	ADDITIONAL COMMENTS ON THE ABSOLUTE CALIBRATION OF THE BRICE-TYPE LIGHT SCATTERING PHOTOMETER. TOMIMATSU YOSHIO PALMER K J J POLYMER SCI 54 159 S21-S25 1961 CA 56 7078
27407	GAMMA-INDUCED ABSORPTION AND STRUCTURAL STUDIES OF ARSENIC BORATE GLASSES. BISHAY A M ARAFA S J AM CERAM SOC 49 8 423-30 1966
27469	HEAT CAPACITY OF DIAMOND AT HIGH TEMPERATURES. VICTOR, A. C. J CHEM PHYS 36 1903-11 1962 CA 57 5367
27498	TESTING PRACTICES IN THE REFRACTORIES INDUSTRY. II. TROUT H E INDUSTR HEAT 29 734-42 1962 BC 61 2431
27515	INVESTIGATIONS ON THE SPECIFIC HEAT AT LOW TEMPERATURES. GUNTHER, P. ANN PHYSIK 51 828-46 1916
27519	REFLECTANCE CURVES OF SOILS, ROCKS, VEGETATION, AND PAVEMENT. ORR, D. G. DWORNIK, S. E. YOUNG, L. M. U S ARMY ENGINEER RESEARCH AND DEVELOPMENT LABS, FT. BELVOIR, VA. 1-64, 1963. (AERDL-RR-1746 RR, AD 410588)
27540	THE DETERMINATION OF COEFFICIENT OF HEAT CONDUCTION AND ULTRA SOUND ABSORPTION FOR MOIST CAPILLARY-POROUS SOLIDS. VENEDIKTOV M V BIBIK V P INZHEN-FIZ ZH /USSR/ 4 11 120-2 1961 SA 65 9429
27546	HEAT CAPACITY AND ENTHALPY OF CORUNDUM IN THE TEMPERATURE INTERVAL FROM 500 TO 2000 DEGREES. CHEKHOVSKOI V YA INZH-FIZ ZH AKAD NAUK BELORUSSK SSR 5 8 62-5 1962 CA 58 71
27562	THERMAL AND ELECTRICAL PROPERTIES OF NATURAL GALENA AND CHALCOPYRITE. ABDULLAEV G B ALIEV G M ANTONOV V B BASHSHALIEV A A KULIEV A Z NASIROV YA N IZV AKAD NAUK AZERB SSR SER FIZ-MAT I TEKHN NAUK 2 49-56 1960 CA 57 1671
27579	PROPERTIES AND BEHAVIOUR OF SINTERED DOLOMITE BRICKS THAT HAVE BEEN FIRED IN A TUNNEL KILN. BISCHOFF F BER DTSCH KERAM GES 39 6 321-5 1962 BC 62 782

TPRC Number	Bibliographic Citation
27593	HEAT CAPACITY OF IRON ORES AND AGGLOMERATES. BRATCHIKOV S G TOPORISHCHEV G A IZV VYSSHIKH UCHEBN ZAVEDENII CHERNAYA MET 5 6 16-20 1962 CA 57 16243 (FOR ENGLISH TRANSLATION SEE T027761)
27608	A METHOD FOR DETERMINING THERMOPHYSICAL PROPERTIES OVER A WIDE TEMPERATURE RANGE. BUDRIN D V YAROSHENKO YU G SUCHKOV V D IZV VYSSHIKH UCHEBN ZAVEDENII PRIBOROSTR 5 1 118-27 1962 CA 57 9525 (FOR ENGLISH TRANSLATION SEE T032761)
27609	VISCOSITY OF THE MOLTEN SYSTEM NA3ALF6-BACL2. MURGULESCU I G ZUCA STEFANIA IZV VYSSHIKH UCHEBN ZAVEDENII TSVETN MET 4 6 33-7 1961 CA 56 14947
27612	THERMOCHEMICAL INVESTIGATION OF MELTS OF PBO-NA2O-SIO2. TOPORISHCHEV G A ESIN O A BRATCHIKOV S G IZV VYSSHIKH UCHEBN ZAVEDENII TSVETN MET 5 1 50-8 1962 CA 57 2925
27639	THE OPTICAL PROPERTIES OF GLASS SYSTEMS. I. THE SPECIFIC EXTINCTION OF LEAD BORATES IN MIXED GLASSES WITH BARIUM BORATE. GEFFCKEN W GLASTECHN BER 35 1 27-36 1962 BC 61 1698
27640	THERMAL PROPERTIES OF GLASS TANK REFRACTORIES. RUK, E. SANDMEYER, K. H. GLASS IND 43 5 252-3 281-2 1962 CA 57 10789
27641	THE THERMOPHYSICAL PROPERTIES OF PLASTIC MATERIALS FROM -50 F TO OVER 700 F. SOUTHERN RESEARCH INST., BIRMINGHAM, ALA. SOUTHERN RESEARCH INST., BIRMINGHAM, ALA. 1-19, 1962. (SRI 5617-1397)
27646	INFLUENCE OF THE WATER-CONTENT OF SILICATE GLASSES ON THEIR TRANSFORMATION AND SOFTENING BEHAVIOUR. MERKER L SCHOLZE H GLASTECHN BER 35 1 37-43 1962 BC 61 1686
27691	APPARATUS FOR THE MEASUREMENT OF THE THERMAL CONDUCTIVITY OF NUCLEAR CERAMICS. MIHAILOVIC Z BER DEUT KERAM GES 39 7 385-6 1962 JA 46 71
27692	THERMAL CONDUCTIVITY OF MAGNESITE BRICK. JESCHKE P SCHWIETE H E BER DEUT KERAM GES 39 7 393-8 1962 JA 46 73
27694	USING PERLITE FOR THERMAL INSULATION. OSTROVSKII A V ELEKTRICHESKIE STANTSII 33 6 32-4 1962 BR 11 8341
27715	MEASUREMENT OF THERMAL CONDUCTIVITY OF REFRACTORIES BY THE HOT-WIRE METHOD. MITTENBUEHLER A BER DEUT KERAM GES 39 7 387-92 1962 JA 46 72 (FOR ENGLISH TRANSLATION SEE T026779)
27719	THE VISCOSITY OF BASALT ROCKS. ALENTEV O O GUDIM L I DOPOVIDI AKAD NAUK UKR RSR 630-3 1962 CA 57 7934
27738	THERMAL CONDUCTION OF DILUTED PARAMAGNETIC SINGLE CRYSTALS AT LOW TEMPERATURE IN A MAGNETIC FIELD. DREYFUS, B. LACAZE, A. ZADWORNY, F. COMPT REND 254 3337-9 1962 CA 57 7922
27761	HEAT CAPACITY OF IRON ORES AND AGGLOMERATES. BRATCHIKOV S G TOPORISHCHEV G A BRITISH IRON AND STEEL INDUSTRY 1-6, 1963. (ENGLISH TRANSLATION OF IZV. VYSSHIKH UCHEBN. ZAVEDENII, CHERNAYA MET., 5 (6), 16-20, 1962; FOR ORIGINAL SEE T027593) (BISI-3424)
27771	THE THERMAL CONDUCTIVITY OF REFRACTORY POWDERS AND SINTERED CERAMICS. PUSTOVALOV V V GLASS AND CERAMICS MOSCOW 18 12 618-20 1961 BC 62 104 (ENGLISH TRANSLATION OF STEKLO I KERAM, 18 (12), 17-19, 1961; FOR ORIGINAL SEE T026561)
27783	THERMAL CONDUCTIVITY OF TANK BLOCK MATERIALS. BUSBY T S GLASS TECHNOL 2 5 209-10 1961 JA 46 12
27814	LOW TEMPERATURE HEAT CAPACITY AND ENTROPY AT 298.15 K OF MUSCOVITE. WELLER W W KING E G U S BUR MINES REPT INVEST 1-4, 1963. (BM-RI-6281)
27817	THERMAL CONDUCTIVITY OF REFRACTORIES ACCORDING TO V. V. PUSTOVALOVS MEASUREMENTS. ZAGAR L BER DEUT KERAM GES 39 7 382-4 1962 CA 57 14726
27821	REFLECTION MEASUREMENTS FOR THE MICROSCOPY OF METALLURGICAL PRODUCTS. TROJER F BERG/HUETTENMAENN MONATSH MONTAN HOCHSCHULE LEOBEN 107 33-9 1962 CA 56 13877
27822	THE EMISSIVE POWER OF TUNGSTEN FOR SHORT WAVE-LENGTHS. HULBURT E O ASTROPHYS J 45 149-63 1917
27825	LABORATORY MEASUREMENTS OF HEAT CONDUCTIVITY OF SEDIMENTARY ROCKS. ZIERFUSS H VAN DER VLIET G BULL AM ASSOC PETROL GEOLOGISTS 40 2475-88 1956
27836	CLASSIFICATION AND METHODS OF SAMPLING AND TESTING OF INSULATING REFRACTORY BRICKS. BRITISH STANDARDS INSTITUTE, LONDON BRIT STAND 2973 1-23 1961 BC 61 155
27842	THERMAL CONDUCTIVITY OF NON-METALLIC MATERIALS. EVEREST, A. GLASER, P. E. WECHSLER, A. E. ARTHUR D. LITTLE INC., CAMBRIDGE, MASS. 1-86, 1962. (N63-16153)
27848	THE BEHAVIOR OF SILVER IN GLASS. I. THE DIFFUSION OF SILVER INTO GLASS FROM A MELT OF SILVER NITRATE. ITO, T. BULL CHEM SOC JAPAN 35 1312-16 1962 CA 57 12113
27850	INFRARED EMISSIVITY OF THE SAHARA FROM TIROS DATA. BUETTNER, K. J. K. KERN, C. D. SCIENCE 142 671-2 1963
27886	INFRARED COATING STUDIES. MARTIN, T. P. MASSO, J. D. TURNER, A. F. LOMB, INC., ROCHESTER, N. Y. 1-9, 1963. (AD 400346)
27888	VISCOSITY OF CRYOLITE. MARGULESCU I G ZUCA STEFANIA ACAD REP POPULARE ROMINE STUDII CERCETARI CHIM 9 1 55-61 1961 CA 57 1576
27917	THERMAL CONDUCTIVITY OF SOME SEDIMENTARY ROCKS. NIVEN C D CAN J RESEARCH 18 A 132-7 1940
27922	METHOD OF MEASUREMENT OF THERMAL DIFFUSIVITY TO 1000 DEGREES. PLUMMER W A CAMPBELL D E COMSTOCK A A J AM CERAM SOC 45 310-16 1962 CA 57 13569
27944	SPECTROPHOTOMETRIC STUDY OF THE REFLECTIVITY OF THE CENTER OF THE MARTIAN DISK AT OPPOSITION. GUERIN PIERRE ANN ASTROPHYS 25 43-9 1962 CA 57 2990
27957	FAR INFRARED SPECTRA OF CRYSTALS AND SOLIDS IN A LARGE RANGE OF TEMPERATURES. LECOMTE, J. HADNI, A. GROUP FOR THE ADVANCEMENT OF SPECTROSCOPIC METHODS GAMS 1-15, 1963. (AFORL-63-326, AD 414528)
27983	HEAT TRANSFER TO A GAS CONTAINING A CLOUD OF PARTICLES. MC ALISTER, A. WARD, H. C. ORR, C., JR. GEORGIA INST. OF TECH. ENGINEERING EXPERIMENT STATION, ATLANTA, GA 1-35, 1963. (NASA-CR-50317, N63-18370)
27989	THEORY OF GLASS STRUCTURE. SMYTH, H. T. MEINKEN, R. H. SKOGEN, H. S. LO, W. C. PETERSON, A. F. BRIDENBAUGH, P. M. U.S. DEPT. COMMERCE OFFICE TECH. SERV. 1-99, 1962. (AD 256599)

TPRC Number	Bibliographic Citation
28005	**NON-METALLIC THERMAL STORAGE MEDIA.** SVIKIS V D CAN DEPT MINES TECH SURV MINES BRANCH RES REPT R-96 1-42 1962 CA 57 13576
28014	**THEORETICAL INVESTIGATIONS OF THE ABLATION OF A GLASS-TYPE HEAT PROTECTION SHIELD OF VARIED MATERIAL PROPERTIES AT THE STAGNATION POINT OF A RE-ENTERING IRBM.** ADAMS, E. W. NATIONAL AERONAUTICS AND SPACE ADMINISTRATION 1-64, 1961. (NASA TN D-564)
28041	**EXPERIMENTAL MEASUREMENT AND THEORETICAL INTERPRETATION OF THERMAL RADIATION FROM THE LUNAR SURFACE.** TYLER WARREN CLAY UNIV. OF ALABAMA, PH.D. THESIS 1-79, 1962. (UMP 62-1351)
28056	**DESIGN AND DEVELOPMENT OF AN E-M WINDOW FOR AIR LIFT REENTRY VEHICLES. INTERIM ENGINEERING REPT. NO. 2, 1 AUG—31 OCT 63.** WALTON J D JR POULOS N E MURPHY C A HARRIS J N WOLF J M GEORGIA INST. OF TECH. ENGINEERING EXPERIMENT STATION ATLANTA 1-78, 1963. (AD 425665)
28067	**REFRACTORY MATERIALS FOR USE IN HIGH TEMPERATURE AREAS OF AIRCRAFT.** THIELKE, N. R. PENNSYLVANIA STATE UNIV., UNIVERSITY PARK 1-88, 1955. (WADC-TR-54-467, AD 88128)
28074	**THERMAL CONDUCTIVITY OF GLASS.** KLEMENS P G NON-CRYST SOLIDS CONF ALFRED NEW YORK 1958 508-30 1960 CA 54 16978
28087	**TEMPERATURE, MATERIAL TRANSFER AND PHYSICAL PROPERTIES OF SLAGS OF UNALLOYED WELDING ELECTRODES.** HUMMITZSCH W SCHWEISSEN UND SCHNEIDEN 13 187-95 1961 (FOR ENGLISH TRANSLATION SEE T028088)
28088	**TEMPERATURE, MATERIAL TRANSFER AND PHYSICAL PROPERTIES OF SLAGS OF UNALLOYED WELDING ELECTRODES.** HUMMITZSCH W BRITISH IRON AND STEEL INDUSTRY 1-29, 1962. (ENGLISH TRANSLATION OF SCHWEISSEN UND SCHNEIDEN, 13, 187-95, 1961; FOR ORIGINAL SEE T028081) (BIST-2617)
28099	**THERMAL PROPERTY DATA REQUIREMENTS FOR NEW WEAPONS SYSTEMS.** WITTEBORT, J. PROCEEDINGS OF THE FLORIDA CONFERENCE ON HIGH-SPEED AERODYNAMICS AND STRUCTURES, GAINESVILLE, FLA., 1, 95-108, 1959. (ARDC-TR-57-46 (VOL. 1), PB 151538, AD 113003)
28116	**HIGH TEMPERATURE THERMAL CONDUCTIVITY OF SIX ROCKS.** STEPHENS, D. R. CALIFORNIA U. LIVERMORE LAWRENCE RADIATION LAB., 1-19, 1963. (UCRL-7605, N64-15399)
28151	**1950 AND 1951 ERUPTIONS OF MIHARA YAMA, O SHIMA VOLCANO, JAPAN.** FOSTER HELEN L MASON ARNOLD C BULL GEOL SOC AMER 66 731-62 1955 CA 49 11517
28157	**HEAT CONDUCTION IN ELECTRICALLY INSULATING CRYSTALS.** LEIBFRIED, G. SCHLOEMANN, E. NACHR. AKAD. WISS. GOETTINGEN, MATH.-PHYSIK. KL. IIA. MATH.-PHYSIK.-CHEM. ABT. 71-93, 1954. (FOR ENGLISH TRANSLATION SEE T28158)
28158	**HEAT CONDUCTION IN ELECTRICALLY INSULATING CRYSTALS.** LEIBFRIED, G. SCHLOEMANN, E. US ATOMIC ENERGY COMMISSION 1-36, 1963. (ENGLISH TRANSLATION OF NACHR. AKAD. WISS. GOETTINGEN, MATH.-PHYSIK. KL. IIA, MATH.-PHYSIK, CHEM. ABT. 71-93, 1954; FOR ORIGINAL SEE T028157) (AEC-TR-5892)
28159	**ON THE ABSORPTION OF LIGHT IN SOLID AND GASEOUS BODIES.** KOENIGSBERGER J KUPFERER K ANN PHYSIK 37 4 600-41 1912
28160	**THE THERMAL CONDUCTIVITY OF REFRACTORY OXIDES.** VISHNEVSKII, I. I. US ATOMIC ENERGY COMMISSION 1-9 1963 NS 17 29331 (ENGLISH TRANSLATION OF SB. NAUCHN. TR., UKR. NAUCHN.-ISSLED. INST. OGNEUPOROV NO. 3, 274-81, 1960; FOR ORIGINAL SEE T025545) (AEC-TR-5872)
28205	**THE EFFECT OF TEMPERATURE UPON THE COEFFICIENT OF ABSORPTION OF CERTAIN GLASSES OF KNOWN COMPOSITION.** GIBSON K S PHYS REV 7 2 2 194-202 1916
28215	**PROPERTIES OF HEAT-INSULATING MATERIALS.** ALLCUT E A HAMLY D H ENG J 24 514-24 1941 CA 36 3582
28248	**PROPERTIES OF PFAUDLER GLASSES. OUTLINE OF CHEMICAL DURABILITY, PHYSICAL PROPERTIES AND THERMAL CONDUCTIVITY.** CHORMANN O I KROHA G F GLASS LINING 14 4 5-7 17 1944 CA 38 3099
28264	**SILICA AS A REFRACTORY MATERIAL.** BLANCO, E. P. INST HIERRO Y ACERO 8 22-49 1955 CA 49 9244
28311	**INFRARED TRANSMITTING GLASSES.** KREIDL N J WEIDEL R A HEFNER H C J OPT SOC AM 47 6 567-8 1957 JA 40 202
28353	**SPECIFIC HEATS OF CHROME AND MAGNESIA REFRACTORIES.** SEIL GILBERT E HECK FRANK G HEILIGMAN H A J AM CERAM SOC 24 204-12 1941 CA 35 5268
28355	**MEASUREMENT OF MEAN SPECIFIC HEAT OF PLATE GLASS AT HIGH TEMPERATURES.** ANDERSON SCOTT J AM CERAM SOC 29 368-70 1946 CA 41 1404
28414	**PHYSICOCHEMICAL PROPERTIES OF CUPOLA-FURNACE SLAG WOOL.** KALTMAN M K KASTNER E P KERAMIKA 10 46-52 1939 CA 34 4484
28418	**THE EFFECT OF TEMPERATURE UPON THE INFRA-RED ABSORPTION OF CERTAIN GLASSES.** GRANTHAM G E PHYS REV 16 6 565-74 1920
28449	**PROPERTIES OF REFRACTORY MATERIALS, EXCEPT FIRECLAY.** PHELPS, S. M. METAL PROGRESS 32 4 383 1937 MA 4 669
28453	**FUSED SILICA.** LUCY CHARLES H MATERIALS AND METHODS 45 2 106-8 1957 JA 40 159
28460	**INFRARED PROPERTIES OF SAPPHIRE AT ELEVATED TEMPERATURES.** OPPENHEIM U P EVEN U J OPT SOC AM 52 1078-9 1962 CA 58 1068
28474	**DETERMINATION OF THE AVERAGE SPECIFIC HEAT OF SOME TECHNICALLY IMPORTANT GLASS. PT. 2** HARTMANN, H. BRAND, H. GLASTECH BER 27 1 12-15 1954
28492	**TITANIUM-ALUMINA SLAGS FOR MANUFACTURE OF REFRACTORY MATERIAL.** DOLKART F Z OGNEUPORY 21 7 300-5 1956 BR 6 3381
28537	**DETERMINATION OF THE REDUCED COEFFICIENTS OF THERMAL CONDUCTIVITY OF CERTAIN GLASSES.** RODNIKOVA V V GLASS CERAM 15 308-10 1960 (ENGLISH TRANSLATION OF STEKLO I KERAM., 15 (6), 20-21, 1958; FOR ORIGINAL SEE T016447)
28567	**AN EMPIRICAL EQUATION FOR THE CALCULATION OF THE ENTHALPY AND HEAT CAPACITY OF CERTAIN SOLIDS.** CHEKHOVSKOI V YA ZHUR FIZ KHIM 39 12 2947- 1965 (FOR ENGLISH TRANSLATION SEE T028597)

TPRC Number	Bibliographic Citation
28573	INVESTIGATION OF DIFFUSE REFLECTION BY MEANS OF POLARIZED LIGHT. PARTS I AND II. IVANOV A P TOPORETS A S SOVIET PHYS – TECH PHYS 1 598–609 1956 (ENGLISH TRANSLATION OF ZH. TEKHN. FIZ., 26 (3), 623–30, 631–35, 1956; FOR ORIGINAL SEE T033222)
28585	THE ISOVISCOUS TEMPERATURES AND THE CONTENT OF HEAT OF BLAST-FURNACE SLAGS AT THESE TEMPERATURES. VOSKOBOINIKOV V G TEORIYA PRAKT MET 12 8 11–16 1940 CA 35 5429
28591	THERMAL CONDUCTIVITY OF CHROMITE SPINELS. VISHNEVSKII I I FRENKEL A S SKRIPAK V N SOVIET PHYSICS–SOLID STATE 5 9 1966–70 1964 (ENGLISH TRANSLATION OF FIZ. TVERD. TELA, 5 (9), 2691–7, 1963; FOR ORIGINAL SEE T030257)
28597	AN EMPIRICAL EQUATION FOR THE CALCULATION OF THE ENTHALPY AND HEAT CAPACITY OF CERTAIN SOLIDS. CHEKHOVSKOI V YA RUSS J PHYS CHEM 39 12 1574–5 1965 (ENGLISH TRANSLATION OF ZHUR. FIZ. KHIM., 39 (12), 2947–, 1965; FOR ORIGINAL SEE T028567)
28630	CERAMICS IN ELECTRON TUBES. GALLET G VIDE 12 65 420–3 1957 JA 40 164
28649	THE THERMAL CONDUCTIVITY OF ROCK SALT. STEPHENS, D. R. MAIMONI, A. LAWRENCE RADIATION LAB., UNIV. OF CALIFORNIA, LIVERMORE 1–14 1964 RR 39 S–7 (UCRL–6894/REV 2)
28661	A NEW METHOD FOR DETERMINATION OF THERMAL CONDUCTIVITY IN SMALL TEST BODIES OF HOMOGENEOUS CERAMIC MATERIALS. II. LUNDQVIST, D. STALHANE, B. IVA /STOCKHOLM/ 20 5 239–44 1949
28664	CONCERNING THE SYSTEM CHALK–ALUMINA–SILICA. BERL, E. LOBLEIN, F. ARCH WARMEWIRTSCH 10 339–42 1929
28666	EFFECT OF RAYLEIGH SCATTERING ON THE COLOR CHARACTERISTICS OF TITANIA ENAMELS. II. SHANNON R D CERAM AGE 73 6 51 1959
28670	DIAMONITE — SPECIALISTS IN TECHNICAL CERAMICS. AUTHOR ANON. CERAM. AGE 76 (1), 24–8, 1960.
28673	HEAT TRANSFER THROUGH CERAMIC MATERIALS. JAKOB, M. CERAM AGE 11 121–30 1928
28674	THERMAL CONDUCTIVITY OF MATERIALS AND REFRACTORY MATERIAL ASSEMBLY USED IN THE JET OF MAGNETOHYDRO-DYNAMIC CONVERSION. DUC, X. N. NEANT, D. SYMPOSIUM INTL. SUR LA PRODUCTION MAGNETO–HYDRODYNAMIQUE D ENERGIE ELECTRIQUE 1213–27, 1964. (CONF–640701–11)
28679	NEW METHOD FOR DETERMINATION OF THERMAL CONDUCTIVITY COEFFICIENTS. STALHANE, B. PYK, S. TEKN TIDSKR 61 28 389–93 1931
28680	THERMAL CONDUCTIVITY, THERMAL EXPANSION, SPECIFIC HEAT AND SOME OTHER THERMAL PROPERTIES OF MINERALS AND CERAMIC BODIES. COHN, W. M. BER DEUT KERAM GES 9 239–99 1928
28694	THERMAL CONDUCTIVITY OF INSULATING AND BUILDING MATERIALS. GROBER HEINRICH Z VER DEUT INGR 54 34 1319–24 1910
28710	DIFFUSION AND HEAT TRANSFER IN POROUS ALUNDUM. HUANG, J. H. SMITH, J. M. IND ENG CHEM FUNDAMENTALS 2 3 189–93 1963 CA 59 4805
28712	THE SPECIFIC HEAT OF THERMAL INSULATING MATERIALS. WILKES, G. B. WOOD, C. O. TRANS A S H V E 48 493–504 1942

TPRC Number	Bibliographic Citation
28721	CONTROL OF BRINELL HARDNESS AND STRUCTURE IN CAST IRON MACHINE TOOL SLIDES. WUELLENWEBER KURT GIESSEREI 46 10 265–74 1959
28724	THE VISCOSITY OF ROCK–WOOL MELTS. KAUFMAN D L KERAMIKA 9 47–54 1939 CA 34 7481
28730	THERMAL CONDUCTIVITY OF ROCKS AND POORLY CONDUCTING BODIES. WEBER ARCH SCI PHYS ET NAT 33 590–1 1895
28731	METHOD TO DETERMINE THE THERMAL CONDUCTIVITY OF POORLY CONDUCTING BODIES IN SPHERE AND CUBE FORM AND ITS PERFORMANCE ON MARBLE, GLASS, SANDSTONE, GYPSUM AND SERPENTINE, BASALT, SULFUR, COAL. HECHT H ANN PHYSIK 14 1008–30 1904
28745	RESEARCH ON CONDUCTIVITY, POROSITY AND PERMEABILITY OF REFRACTORY MATERIAL. WOLOGDINE M S BULL SOC ENCOUR IND NAT 3 5 880–942 1909
28755	STABLE WHITE COATINGS. ZERLAUT, G. A. HARADA, Y. IIT RESEARCH INC., CHICAGO, IL 1–194, 1963. (IITRI–C207–25, NASA CR–52134, N63–23501)
28757	THERMAL CONDUCTIVITY OF REFRACTORY BRICK. KONDO, S. YOSHIDA, H. DAINIPPON YOGYO KYOKAI ZASSHI 39 (466), 657–63, 1931.
28772	MEASUREMENT OF THE THERMAL CONDUCTIVITY OF FIRECLAY. CLEMENT J K EGY W L CHEM MET ENG 8 7 414–16 1910
28778	THE INTERIOR HEAT CONDUCTIVITY PROPERTIES OF QUARTZ, CALCITE AND ROCK SALT. TUCHSCHMID A BEIBL ANN PHYSIK 8 490–2 1884
28780	THE THERMAL CONDUCTIVITY OF ROCKS AND ITS DEPENDENCE UPON TEMPERATURE AND COMPOSITION. PART 2. BIRCH, F. CLARK, H. AM J SCI 238 9 613–35 1940
28781	GAS IS AN IMPORTANT FACTOR IN THE THERMAL CONDUCTIVITY OF MOST INSULATING MATERIALS. ROWLEY F B JORDAN R C LUND C E LANDER R M ASHVE TRANS 58 155–70 1952
28785	THE THERMAL RESISTIVITY OF INSULATING MATERIAL. RANDOLPH C P TRANS AM ELECTROCHEM SOC 21 545–55 1912
28786	STRUCTURE AND THERMAL CONDUCTIVITY OF REFRACTORY BRICK. MAASE E STAHL U EISEN 51 860–1 1931
28787	HEAT FLOW IN SOUTH AFRICA. BULLARD E C PROC ROY SOC /LONDON/ 173 A 474–502 1939
28792	TERRESTRIAL HEAT FLOW IN ENGLAND. BULLARD E C NIBLETT E R ROY ASTRON SOC GEOPHYS SUPPL 6 222–38 1951
28799	TERRESTRIAL HEAT FLOW IN PERSIA. COSTER H P ROY ASTRON SOC GEOPHYS SUPPL 5 131–45 1947
28803	MEASUREMENT OF THERMAL CONDUCTIVITY. FRITZ W POLTZ H GLASHUTTEN–HANDBUCH 1–6, 1961. (Y–55)
28804	A STUDY OF THE MAGNETIC PROPERTIES OF IRON IN RELATION TO ITS COLOURING ACTION IN GLASS. PART 1. IRON IN SODIUM BORATE GLASSES MELTED UNDER OXIDIZING CONDITIONS. PART 2. IRON IN SODIUM BORATE GLASSES MELTED UNDER REDUCING CONDITIONS. BAMFORD C R PHYS CHEM GLASSES 1 5 159–69 1960

TPRC Number	Bibliographic Citation
28811	**MANUFACTURE OF UO2 AND UO2-BEO MIXED PELLETS.** BATTELLE INST., E. V. FRANKFURT, GERMANY BATTELLE INST., E. V. FRANKFURT, GERMANY 1-25, 1962. (EUR-116.D)
28816	**THERMAL CONDUCTIVITY MEASUREMENTS OF INSULATION MATERIALS AT LOW TEMPERATURES WITH THE PLATE APPARATUS.** ACHTZIGER J KALTETECHNIK 12 12 372-5 1960
28817	**MEASUREMENT OF HEAT CONDUCTIVITY OF INSULATION MATERIALS AT LOW TEMPERATURES.** CAMMERER W F KALTETECHNIK 12 4 107-10 1960
28819	**MECHANISM OF HEAT FLOW IN FIBROUS MATERIALS.** FINCK J L J RESEARCH NATL BUR STANDARDS 5 5 973-84 1930
28824	**METAL POWDER ADDITIVES IN EVACUATED-POWDER INSULATION.** HUNTER, B. J. KROPSCHOT, R. H. SCHRODT, J. E. FULK, M. M. PROC. CRYOGENIC ENG. CONF. 5, 146-56, 1960.
28858	**THERMAL PROPERTY MEASUREMENT TECHNIQUES.** CLAYTON, W. A. BOEING CO., SEATTLE, WASH. 1-127, 1961. (D2-80281)
28870	**A STUDY OF REFRACTORY MATERIALS FOR SEAL AND BEARING APPLICATIONS IN AIRCRAFT ACCESSORY UNITS AND ROCKET MOTORS.** SIBLEY L B ALLEN C M ZIELENBACH W J PETERSON C L GOLDTHWAITE W H BATTELLE MEMORIAL INSTITUTE 1-52, 1958. (WADC TR 58-299, AD 203787)
28900	**A SOLAR THERMOELECTRIC GENERATOR SYSTEM STUDY.** FERRAR, J. R. HAMILTON STANDARD DIV. UNITED AIRCRAFT CORP., WINDSOR LOCK, CT 1-266, 1961. (ASD-TR- 61-315, AD 272152)
28913	**INFRARED COATING STUDIES.** MARTIN, T. P. BAUSCH AND LOMB INC., ROCHESTER, N Y 1-9, 1961. (AD 270038)
28922	**PHYSICAL PROPERTIES OF SOME ENGINEERING MATERIALS—UNPUBLISHED DATA FROM COMPANY SPONSORED PROGRAMS.** PETERSON, J. J. CHANCE VOUGHT CORP. AERONATUICS AND MISSILES DIV., DALLAS, TX 1-77, 1962. (CVC RPT 2-53420-2R374, AD 273922)
28924	**JUNO II SUMMARY PROJECT REPORT. VOLUME I EXPLORER SATELLITE.** NATIONAL AERONAUTICS AND SPACE ADMIN., WASHINGTON, D. C. NATIONAL AERONAUTICS AND SPACE ADMIN., WASHINGTON, DC 1-356, 1961. (NASA TN D-608, AD 25999)
28944	**RADIATION FROM PLANET EARTH.** GOLDBERG, I. L ARMY SIGNAL RESEARCH AND DEVELOPMENT LAB, FORT MONMOUTH, N J 1-36, 1961. (USASRDL-TR-2231, AD 266790)
28956	**THERMAL CONDUCTIVITY OF BEDS OF SPHERICAL PARTICLES.** MASAMUNE, S. SMITH, J. M. IND ENG CHEM FUNDAMENTALS 2 136-43 1963 CA 58 10990
28999	**RESEARCH, STUDIES, AND INVESTIGATIONS ON SPECTRAL REFLECTANCE AND ABSORPTION CHARACTERISTICS OF CAMOUFLAGE PAINT MATERIALS AND NATURAL OBJECTS.** KRONSTEIN, M NEW YORK UNIV., N Y 1-68, 1956. (AD 100058)
29003	**PREPARATION AND PROPERTIES OF GLASSY-OXIDE SEMICONDUCTORS OF VO2.5-PO2.5-RO.** GRECHANIK L A PETROVYKH N V KARPECHENKO V G SOVIET PHYS-SOLID STATE 2 9 1908-15 1961 (ENGLISH TRANSLATION OF FIZ. TVERDOGO TELA, 2 (9), 2131-9, 1960; FOR ORIGINAL SEE T025795)
29007	**EVALUATION OF THERMAL DIFFUSIVITY OF REFRACTORY MATERIALS IN NONSTEADY STATE.** OLAVE J ROBREDO REV CIENC APL /MADRID/ 10 334-7 1956 JA 40 253
29020	**CONCERNING THE SIMILARITY BETWEEN THE CHARACTERISTIC FREQUENCY DISPERSION CURVES OF DIAMOND-TYPE CRYSTALS.** KUCHER T I FIZ TVERDOGO TELA 4 9 2385-92 1962 (FOR ENGLISH TRANSLATION SEE T029021)
29021	**CONCERNING THE SIMILARITY BETWEEN THE CHARACTERISTIC FREQUENCY DISPERSION CURVES OF DIAMOND-TYPE CRYSTALS.** KUCHER T I SOVIET PHYSICS-SOLID STATE 4 9 1747-52 1963 (ENGLISH TRANSLATION OF FIZ. TVERDOGO TELA, 4 (9), 2385-92, 1962; FOR ORIGINAL SEE T029020)
29043	**THERMAL DIFFUSIVITY OF SAPPHIRE AND FINE-GRAINED ALUMINA AT LOW TEMPERATURE.** SMITH SIDNEY LOUIS ALFRED UNIV., M.S. THESIS 1-84 1961
29059	**METHODS OF STUDYING THE THERMAL PROPERTIES OF INSULATORS OF DISPERSED STRUCTURE.** CHUDNOVSKII, A. F. ATOMIC ENERGY COMMISSION 1-14, 1961. (ENGLISH TRANSLATION OF ZHUR. TEKH. FIZ., 16 (2), 231-42, 1946; FOR ORIGINAL SEE T001734) (AEC-TR-4772)
29099	**RESEARCH AND DEVELOPMENT SERVICES LEADING TO THE CONTROL OF ELECTRICAL PROPERTIES OF MATERIALS FOR HIGH TEMPERATURE RADOMES.** ATLAS, L. M. ARMOUR RESEARCH FOUNDATION OF ILLINOIS, INST. OF TECH., CHICAGO 1-27, 1959. (WADC-TR-59-300, PB 161423)
29103	**THE THERMAL CONDUCTIVITY OF THE MELTING GLASS BATCH.** KROGER, C. ELIGEHAUSEN, H. SPEICAL LIBRARY ASSOCIATION, TRANSLATIONS CENTER 1-23, 1962. (ENGLISH TRANSLATION OF GLASTECH. BER., 32 (9), 362-73, 1959; FOR ORIGINAL SEE T017892) (SLA 62-14244)
29111	**HEAT CONDUCTIVITY OF THE GLASS MASS.** GINZBERG L B OTS OR SLA 1-10, 1962. (ENGLISH TRANSLATION OF STEKLO I KERAM., 4 (7), 9-11, 1947; FOR ORIGINAL, SEE T000622) (SLA 62-16152)
29136	**THE TESTING AND PROPERTIES OF HEAT INSULATING MATERIALS.** JAY A H LEE L TRANS CERAM SOC 37 4 151-67 1938 MA 6 266
29137	**CHARACTERISTICS OF THE MOONS SURFACE LAYER. AN ANALYSIS OF ITS RADIO EMISSION.** GIRAUD A ASTROPHYS J 135 1 175-86 1962 PA 62-2 57
29152	**THERMAL CONDUCTIVITY OF SEMICONDUCTIVE SOLIDS, METHOD FOR STEADY-STATE MEASUREMENTS ON SMALL DISC REFERENCE SAMPLES.** FLYNN, D. R. ROBINSON, H. E. NATIONAL BUREAU OF STANDARDS 1-36, 1961. (NBS REPT 7135, AD 277034)
29155	**INVESTIGATION OF SEMICONDUCTING PROPERTIES OF TYPE IIB DIAMONDS.** LEIVO, W. J. BELL, M. D. JOHNSON, C. C. KRUMME, J. B. RUSSELL, K. J. STEIN, H. J. WAYLAND, J. H. YOUNG, T. OKLAHOMA STATE UNIV. RESEARCH FOUNDATION, STILLWATER, 1-224, 1962. (AFOSR-2642, AD 277416)
29185	**A COMPENDIUM OF THE PROPERTIES OF MATERIALS AT LOW TEMPERATURE /PHASE 2/. PART 4.** STEWART, R. B. JOHNSON, V. J. NATIONAL BUREAU OF STANDARDS, CRYOGENIC ENGNR. LAB. 1-501, 1961. (WADD TR 60-56(PT 4), AD 272769)
29193	**THE INFRARED HEAT TRANSFER GAUGE.** CAMAC, M. FEINBERG, R. AVCO-EVERETT RESEARCH LAB., EVERETT, MA. 1-95, 1961. (AFCRL-942, AFOSR-1564, AD 275131)

TPRC Number	Bibliographic Citation
29200	**EFFECT OF WINDOW GLASS IN PROTECTING MATERIALS FROM IGNITION BY THERMAL RADIATION.** DOWNS, L. E. BRUCE, H. D. FOREST PRODUCTS LAB., MADISON, WISCONSIN 1-15, 1955. (AFSWP-792, AD 271462)
29206	**FIRST JOINT PROGRESS REPORT OF THE LABORATORIES FOR MOLECULAR SCIENCE AND MOLECULAR ENGINEERING.** MASSACHUSETTS INSTITUTE OF TECHNOLOGY, CAMBRIDGE MASSACHUSETTS INST. OF TECH., CAMBRIDGE 1-204, 1961. (AD 255694)
29210	**VISCOSITY OF COALS IN THE PLASTIC STATE.** TAITS E M TRUDY INST GORYUCH ISKOPAEMYKH AKAD NAUK SSSR 2 32-40 1950 CA 49 4257
29231	**SUMMARY OF THERMAL EXPANSION AND THERMAL CONDUCTIVITY MEASUREMENTS.** FITZSIMMONS, E. S. GENERAL ELECTRIC CO., AIRCRAFT NUCLEAR PROPULSION DEPT., CINCINNATI, OHIO 1-9, 1961. (DC-61-6-4)
29246	**HEAT SINKS, MATERIALS.** PEARCY, M. LOCKHEED AIRCRAFT CORP., MISSILES AND SPACE DIV., SUNNYVALE, CALIFORNIA 1-29, 1960. (SB-60-29, AD 244262)
29253	**INVESTIGATION OF THE STOICHIOMETRIC AND STRUCTURAL VARIABLES WHICH AFFECT CONDUCTIVITY IN FERROMAGNETIC CERAMIC MATERIALS FOR USE IN MAGNETIC CIRCUITS. REPT.** HORN, F. H. SLACK, G. A. ENGELER, W. E. GENERAL ELECTRIC RESEARCH, LAB., SCHENECTADY, N. Y. 1-94, 1961. (ASD-TR 61-284, PB 159966)
29258	**DEVELOPMENT OF A FLUIDIZED BED CALCINATION PROCESS FOR ALUMINUM NITRATE WASTES IN A TWO-FOOT-SQUARE PILOT PLANT CALCINER. PART 2. FACTORS AFFECTING THE INTRAPARTICLE POROSITY OF ALUMINA.** WHEELER, B. R. GRIMMETT, E. S. BUCKHAM, J. A. IDAHO OPERATIONS OFFICE PHILLIPS PETROLEUM CO., ATOMIC ENERGY DIVISION 1-28, 1962. (IDO-14587)
29285	**HEAT TRANSFER AND PRESSURE DROP IN FIXED BEDS OF SPHERICAL AND CYLINDRICAL SOLIDS. PART 1. PRESSURE DROP AND PACKED BED CHARACTERISTICS.** CAMPBELL JOHN M HUNTINGTON R L PETROL REFINER 30 12 127-33 1951
29300	**OPTICALLY TRANSPARENT MATERIALS FOR SPACE FLIGHT VEHICLES.** WITTMAN, R. E. MATERIALS LAB., WRIGHT-PATTERSON AFB, OHIO 1-19, 1960. (WADD TR 60-203, AD 240911)
29302	**THE DESIGN AND CONSTRUCTION OF A BUNSEN ICE CALORIMETER AND HEAT CAPACITIES OF THE SILICA ALUMINA CRACKING CATALYST. PART A. STANDARDIZATION AND DETERMINATION OF HEAT CAPACITIES. PART B. DESIGN AND CONSTRUCTION.** OELKE, W. C. ANUMAN, W. LANE, G. MOORE, R. GRINNELL COLLEGE, IOWA (PT. A+B), 1-8, 1954. (AD 142931)
29334	**THERMAL PROPERTIES OF REFRACTORY MATERIALS.** LEHMAN, G. W. ATOMICS INTERNATIONAL, CANOGA PARK, CA. 1-7, 1960. (AD 246493)
29352	**METHODS FOR PREDICTING INFRARED RADIANCE OF FLAMES BY EXTRAPOLATION FROM LABORATORY MEASUREMENTS.** BABROV, H. J. HENRY, P. M. TOURIN, R. H. WARNER AND SWASEY CO., CONTROL INSTRUMENT DIV., FLUSHING, N. Y. 1-55, 1961. (AFCRL-1070, AD 270832)
29355	**PHYSICAL SIGNIFICANCE OF THE TIROS II RADIATION EXPERIMENT.** HANEL, R. A. WARK, D. Q. GODDARD SPACE FLIGHT CENTER AND U.S. WEATHER BUREAU 1-16, 1961. (NASA TN D-701, AD 268135)
29362	**CERAMICS.** MITCHELL LANE IND ENG CHEM 47 9 1956-61 1955 48 9 1702-9 1956
29364	**THERMAL PROPERTIES OF SIMULATED LUNAR MATERIALS.** BERNETT, E. C. CALIFORNIA INST. OF TECH., PASADENA JET PROPULSION LABORATORY 73 1961 (JPL RS 36-11)
29365	**THERMAL CONDUCTIVITY OF MOLTEN GLASS. A COMPARISON OF THEORY AND EXPERIMENTAL.** GROVE F J CHARNOCK H GLASTECH BER SONDERBAND 32 7 24-8 1959 CA 54 14607
29375	**A NEW INDUSTRIAL THERMAL CONDUCTIVITY PROBE.** HERPOL, G. A. MINNE, R. REVUE M DE MECANIQUE. TIJDSCHRIFT (M) VOOR DE WERKTUIGKUNDE 5 (4), 148-66, 1959.
29377	**AN INVESTIGATION OF THE OPTICAL PROPERTIES OF MATERIALS AT ROOM AND LIQUID HELIUM TEMPERATURES.** ARONSON, J. B. MC LINDEN, H. G. ARTHUR D. LITTLE, INC., CAMBRIDGE, MA. 1-55, 1963. (TP2-84-008(IF//M-RP-T), X64-15469)
29398	**THERMAL CONDUCTIVITY OF UO2-75 M/O ZRO2.** ANDERSON, W. K. GUARE, C. J. LALAK, A. KNOLLS ATOMIC POWER LAB., SCHENECTADY, N. Y. 1-17 1964 NS 18 44140 (KAPL-M-6184)
29429	**ALUMINUM IMPURITIES AND THE 4600-A BAND IN QUARTZ.** WEEKS, R. A. BUTLER, C. T. OAK RIDGE NATIONAL LABORATORY 86-9, 1961. (ORNL-3213)
29477	**STUDY OF THERMAL CONDUCTIVITY OF PELLETS AND CONCENTRATES.** BRATCHIKOV S G HENRY BRUTCHER ALTADENA CALIF 1800 WORDS, 1962. (ENGLISH TRANSLATION OF IZV. VYSSHIKH UCHEBN. ZAVEDENII, CHERN. MET. (6), 157-163, 1961; FOR ORIGINAL SEE T29478) (HB-5264)
29478	**STUDY OF THERMAL CONDUCTIVITY OF PELLETS AND CONCENTRATES.** BRATCHIKOV S G IZV VYSSHIKH UCHEBN ZAVEDENII CHERN MET 6 157-63 1961 (FOR ENGLISH TRANSLATION SEE T29477)
29482	**CERTAIN RESULTS IN MEASURING THERMOPHYSICAL PROPERTIES OF SOIL UNDER NATURAL CONDITIONS.** GORBUNOVA I G DYACHKOVA T V SEROVA N V TRUDY GL GEOFIZ OBSERV 77 79-83 1958 (FOR ENGLISH TRANSLATION SEE T29483)
29483	**CERTAIN RESULTS IN MEASURING THERMOPHYSICAL PROPERTIES OF SOIL UNDER NATURAL CONDITIONS.** GORBUNOVA I G DYACHKOVA T V SEROVA N V SPECIAL LIBRARY ASSOC., TRANSLATIONS CENTER 1-9, 1961. (ENGLISH TRANSLATION OF TR. GL. GEOFIZ. OBSERV. 139 (77), 79-83, 1958; FOR ORIGINAL SEE T29482) (SLA-61-23081)
29491	**THERMAL PROPERTIES OF SIMULATED LUNAR MATERIAL IN AIR AND IN VACUUM.** BERNETT, E. C. WOOD, H. L. JAFFE, L. D. MARTENS, H. E. JET PROPULSION LABORATORY 1-19 1962 (JPL TR 32-368)
29508	**INVESTIGATION OF MATERIALS FOR VACUUM INSULATION UP TO 4000 F.** GLASER, P. E. WECHSLER, A. E. SIMON, I. BERKOWITZ, J. ARTHUR D. LITTLE, INC., CAMBRIDGE, MA. 1-80, 1962. (ASD-TDR-62-88, AD 274742)
29528	**THE THERMAL CONDUCTIVITY OF A FLUIDIZED BED.** BONDAREVA A K TODES O M FOREIGN TECHNOLOGY DIVISION 1-10, 1962. (FTD-TT-62-66, SC-973, AD 271835, 62-19642)
29539	**HIGH-TEMPERATURE MODIFICATION OF BARIUM FELDSPAR.** YOSHIKI, B. MATSUMOTO, K. J AM CERAM SOC 34 283-6 1951
29549	**B2. THERMAL MEASUREMENTS ON FERROELECTRIC BARIUM TITANATE CERAMICS.** GLOWER, D. D. PROC. BLACK HILLS SUMMER CONF. ON TRANSPORT PHENOMENA 73-87, 1962. (AD 289290)

TPRC Number	Bibliographic Citation
29551	**C1. THE USE OF THERMAL COMPARATOR METHODS FOR THE MEASUREMENT OF THERMAL CONDUCTIVITY.** POWELL, R. W. PROC. BLACK HILLS SUMMER CONF. ON TRANSPORT PHENOMENA 95-154, 1962. (AD 289290)
29555	**LARGE CERAMIC RADOME MANUFACTURE, ATTACHMENT AND TESTING. APPENDIX III.** FRYE, H. OKEN, S. THE BOEING CO., SEATTLE, WA. 1-150, 1962. (ASD-TDR-62-967, AD 299300)
29570	**THERMAL RADIATION CHARACTERISTICS OF TRANSPARENT SEMI-TRANSPARENT AND TRANSLUCENT MATERIALS UNDER NON-ISOTHERMAL CONDITIONS.** FOLWEILER, R. C. LEXINGTON LABS., INC., CAMBRIDGE, MA. 1-115, 1964. (ASD-TDR-62-719, AD 600370)
29585	**RADIAL-FLOW APPARATUS FOR DETERMINING THE THERMAL CONDUCTIVITY OF LOOSE-FILL INSULATIONS TO HIGH TEMPERATURES.** FLYNN D R J RES NATL BUR STD 67 C 2 129-37 1963 CA 59 2404
29595	**INTEGRATING SPHERE FOR IMPERFECTLY DIFFUSE SAMPLES.** EDWARDS D K GIER J T NELSON K E RODDICK R D J OPT SOC AM 51 11 1279-88 1961
29596	**SOLAR REFLECTANCE INTEGRATING SPHERE.** DUNKLE R V EDWARDS D K GIER J T BEVANS J T SOLAR ENERGY 4 2 27-39 1960
29611	**RESEARCH AND DEVELOPMENT OF INFRARED FIBER OPTICS.** KAPANY, N. S. OPTICS TECHNOLOGY, INC., BELMONT, CA. 1-75, 1962. (ASD TDR 62-684, AD 290620)
29615	**ADVANCED TECHNOLOGY FOR PROPULSION SYSTEMS. VOLUME 3, REFRACTORY MATERIALS.** ROBINSON, J. W. ROCKETDYNE, CANOGA PARK, CA. 1-143, 1961. (R-3350-3, AD 290247)
29642	**THE ENERGY CONTENT OF SOLID SUBSTANCES.** NERNST W ANN PHYSIK 36 4 395-439 1911
29643	**THE APPLICATION OF PHOTOSENSITIVE GAS COUNTERS TO SCINTILLATION COUNTING.** ELY, R. L., JR. BALLARD, L. F. RESEARCH TRIANGLE INST., ISOTOPE DEVELOP. LAB., DURHAM, N. C. 1-12, 1962. (ORO-490)
29705	**INVESTIGATION OF NEAR ULTRAVIOLET TRANSMITTING LIQUIDS FOR USE AS KERR CELL FLUIDS IN TRANSIENT SPECTROGRAPHIC SHUTTERS.** HAUSER, S. M. MARSHALL, F. R. ELECTRO-OPTICAL SYSTEMS, INC., PASADENA, CA. 1-48, 1961. (ARL-3, PB 153981)
29768	**DETERMINATION OF THE THERMAL CONDUCTIVITY OF CERAMIC MATERIALS BY TEMPERATURE CYCLING.** RUBIN G A BER DEUT KERAM GES 40 1 16-19 1963 CA 59 1117
29774	**INVESTIGATION OF THE PROPERTIES OF CORUNDUM BRICK WITH SPECIAL REFERENCE TO THE THERMAL CONDUCTIVITY.** LASCH HANS TONIND ZTG 66 89-93 1942
29780	**THERMAL STUDIES OF CAXBA1-XTIO3.** GLOWER, D. D. WALLACE, D. C. J PHYS SOC JAPAN 18 679-84 1963 CA 58 13165
29786	**A MICRO METHOD FOR DETERMINING THE THERMAL CONDUCTIVITY OF POWDERS.** BANSAL T D J SCI INDUSTR RES INDIA 21 D 8 285-7 1962 BC 62 879
29811	**AN APPARATUS FOR THE RAPID DETERMINATION OF THE HEAT CONDUCTIVITY OF POOR CONDUCTORS.** ZIERFUSS H J SCI INSTRUM 40 2 69-71 1963 AM 17 2298
29826	**THERMAL CONDUCTIVITY OF PLASTIC REFRACTORIES.** AUTHOR ANON. ASTM STD. (PART V), 343-45, 1961.
29828	**THERMAL CONDUCTIVITY OF CASTABLE REFRACTORIES.** AUTHOR ANON. ASTM STD. (PART V), 397-99, 1961.
29846	**ULTRAFINE SILICON CARBIDE.** AUTHOR ANON. ADVANCED MATERIALS TECHNOLOGY 10-12PP., 1962.
29870	**SPECIFIC HEAT, THERMAL CONDUCTIVITY, AND THERMOMETRIC CONDUCTIVITY OF BITUMINOUS COAL, CHARCOAL AND COKE.** FRITZ W MOSER H FEUERUNGSTECH 28 97-107 1940 CA 35 6764
29889	**A METHOD OF DETERMINING THE THERMAL CHARACTERISTICS OF SURFACE SOILS IN THEIR NATURAL STATE.** BOGOMALOV V Z CHUDNOVSKII A F ZHUR TEKH FIZ 9 14 1325-30 1939
29903	**THERMAL CONDUCTIVITY OF IRON ORES AND AGGLOMERATES.** BRATCHIKOV S G TOPORISHCHEV G A IZV VYSSHIKH UCHEBN ZAVEDENII CHERNAYA MET 5 8 12-17 1962 CA 58 1168
29914	**INFRARED TRANSMISSION OF SEVERAL NON-SILICATE GLASS SYSTEMS.** LIN FRANCIS C GLASS IND 44 19-23 87-91 102 1963 CA 59 294
29925	**CONTROLLED NUCLEATION AND CRYSTALLIZATION LEAD TO VERSATILE NEW GLASS CERAMICS.** STOOKEY S D CHEM ENG NEWS 39 116-25 1961
30040	**THERMAL CONDUCTIVITY AND DIFFUSIVITY OF POROUS SINTER MATERIAL.** RICHTER W ABHANDL DEUT AKAD WISS BERLIN KL MATH PHYSIK TECH 1 99-107 1962 CA 59 1325
30100	**THE REFLECTION AND TRANSMISSION OF INFRARED MATERIALS. I. SPECTRA FROM 2 TO 50 MU. II. BIBLIOGRAPHY.** MC CARTHY, D. E. APPLIED OPT 2 6 591-603 1963 CA 59 8268
30125	**PHYSICAL PROPERTIES OF SINTERED UO2 BODIES AS A FUNCTION OF BEO ADDITIONS.** HARKORT D HONCIA G KLING H G BER DEUT KERAM GES 40 2 123-8 1963 CA 59 2494
30126	**THE HEAT AND MATERIAL EXCHANGE BETWEEN HYGROSCOPIC PACKING MATERIAL AND MINE DAMP.** ANGENEYNDT JAN DERK BERGBAUWISSENSCHAFTEN 9 521-34 1962 CA 59 2409
30132	**TEMPERATURE DISTRIBUTION IN HOMOGENEOUS SLABS DURING FIRE TEST.** HARMATHY T Z TRANS ENGNG INST CANADA 1-21, 1963. (NRCC BR RP 210, NBC 7416, AD 430388)
30146	**THERMOELECTRIC COOLING ELEMENT (= TE) AS THERMAL CONDUCTIVITY METER.** MASUDA, S. MURAKAMI, Y. BULL. JSME 6 22 251-4 1963 SA 66 19044
30160	**THERMODYNAMIC PROPERTIES OF THE CRYSTALLINE FORMS OF SILICA.** MOSESMAN, M. A. PITZER, K. S. J AM CHEM SOC 63 2348-56 1941
30163	**URANIUM COMPOUNDS AS THERMOELECTRIC MATERIALS.** WARREN I H CAN MINING MET BULL 56 288-98 1963 JA 46 282
30195	**ON THE ABSORPTION OF CERTAIN CRYSTALS IN THE INFRARED AS DEPENDENT ON THE DIRECTION OF THE POLARIZATION OF PLANE.** MERRITT E PHYS REV 2 425-41 1895
30221	**OPTICAL CONSTANTS OF GERMANIUM, SILICON, AND PYRITE IN THE INFRARED.** SIMON, I. J OPT SOC AM 41 10 730 1951

TPRC Number	Bibliographic Citation
30225	**LOW LOSS CERAMICS.** SNYDER, N. H. SMOKE, E. J. WISELY, H. R. RUH, E. N. J. CERAMIC RESEARCH STATION, RURGERS UNIVERSITY 1-89, 1949. (AD 89 089, PB 162 945)
30239	**INFRARED TRANSMITTING MATERIALS.** UPTON, L. O. JUPNIK, H. KAVANAGH, A. J. STAMFORD RESEARCH LAB., AMERICAN OPTICAL CO., SOUTHBRIDGE, MA. 1-11, 1952. (AD 463)
30241	**THE DOMAIN OF STABILITY OF GLASS AND CRYSTAL PHASE OF SIO2.** WIETZEL R Z ANORG CHEM 116 71-95 1921
30254	**PERLITE. THERMAL DATA AND ENERGY REQUIRED FOR EXPANSION. REPORT OF INVESTIGATION.** KING, E. G. TODD, S. S. KELLEY, K. K. BUREAU OF MINES, WASHINGTON, D. C. 1-15, 1948. (BMRI 4394)
30255	**THERMAL FRACTURE OF CERAMIC MATERIALS UNDER QUASI-STATIC THERMAL STRESSES /RING TEST/.** BEUSSEM W R BUSH E A J AM CERAMIC SOC 38 27-32 1955
30257	**THERMAL CONDUCTIVITY OF CHROMIUM SPINELS.** VISHNEVSKII, I. I. FRENKEL, A. S. SKRIPAK, V. N. FIZ TVERD TELA 5 9 2691-7 1963 CA 59 13361 (FOR ENGLISH TRANSLATION SEE T28591)
30260	**THE HEAT OF FORMATIONS OF MAGNESIUM TELLURIDE. SPECIFIC HEAT OF TELLURIUM AND FUPREMAXGLAS.** KUBASCHEWSKI O WITTIG F E Z ELEKTROCHEM 47 6 433-8 1941
30270	**THERMOANALYTICAL INVESTIGATIONS OF GLASSY TWO-COMPONENT AND MULTI-COMPONENT SYSTEMS AND CONSIDERATION OF THE PARTIAL MOLAR ENTHALPIES IN THESE SYSTEMS.** HARTMANN H GOLLA G GLASTECHN BER 36 5 179-83 1963 BC 62 2239
30287	**GUARD-DOME THERMAL CONDUCTIVITY APPARATUS-EFFECT OF SMALL TEMPERATURE UNBALANCES ON THE CONDUCTIVITY VALUES.** GUPTA M L RAYCHAUDHURI B C INDIAN J TECH 1 5 183-6 1963 JA 46 313
30293	**GLASS SCINTILLATORS FOR THE DETECTION OF NUCLEAR RADIATIONS. FROM INSTRUMENTS AND MEASUREMENTS.** ANDERSON D G DRACASS J FLANAGAN T P ACADEMIC PRESS PUBLISHERS NEW YORK 616-30 1961 NS 17 10965
30309	**THERMAL CONDUCTIVITY OF POWDERED INORGANIC INSULATORS.** MIMELMAN E L PANOVA YA I INZHEN -FIZ ZH /USSR/ 6 4 83-5 1963 SA 66 25217
30334	**THE TRANSMISSION OF GLASSES IN THE NEAR INFRARED AT TEMPERATURES UP TO FREEZING POINT.** SCHOLZE H GLASTECHN BER 32 K 7 1-5 1959 (FOR ENGLISH TRANSLATION, SEE T025985)
30362	**RIGID FOAM PLASTICS INFORMATION MANUAL.** RESNICK, I. NEW YORK NAVAL SHIPYARD MATERIAL LAB, BROOKLYN 1-79, 1960 (AD 248189)
30401	**HEAT TRANSFER IN POROUS MEDIA WITH KNOWN PORE STRUCTURE /ALUNDUM/.** HUANG, J. H. SMITH, J. M. J CHEM ENG DATA 8 3 437-9 1963 CA 59 8371
30410	**III. INVESTIGATIONS OF THE SPECIFIC HEAT OF SOLID BODIES.** KOPP, H. PHIL TRANS ROY SOC /LONDON/ 155 1 71-202 1865
30443	**RESEARCH ON LOW DENSITY THERMAL INSULATION MATERIALS FOR USE ABOVE 3000 F.** STYHR, K. H. NATIONAL BERYLLIA CORP., HASKELL, N. J. 1-12, 1964. (NASA-CR-56546, N64-23530)
30496	**HEAT CAPACITY FROM 11 TO 300 K., ENTROPY, AND HEAT OF FORMATION OF DOLOMITE.** STOUT, J. W. ROBIE, R. A. J PHYS CHEM 67 11 2248-52 1963 CA 59 14659
30497	**HEAT CAPACITY FROM 12 TO 305 K. AND ENTROPY OF TALC AND TREMOLITE.** ROBIE, R. A. STOUT, J. W. J PHYS CHEM 67 11 2252-6 1963 CA 59 14659
30525	**TECHNOLOGY OF NEW DEVITRIFIED CERAMICS—A LITERATURE REVIEW.** EMRICH, B. R. AF MATERIALS LAB., RESEARCH - TECHNOLOGY DIV., AF SYSTEMS COMMAND, WRIGHT-PATTERSON AIR FORCE BASE 1-149, 1964. (ML-TDR-64-203, AD 608217)
30547	**RESUME OF THERMAL PROPERTY DATA FROM TWENTY-FOUR POTENTIAL RE-ENTRY INSULATION MATERIALS.** GRAY, C. O. ARMY MISSILE COMMAND, REDSTONE ARSENAL, ALABAMA 1-44, 1958. (DSN-TN-6-58, AD 458881)
30553	**MEASUREMENT OF THE SPECIFIC HEAT OF SOLIDS AT HIGH TEMPERATURES.** MAGNUS A PHYSIK Z 14 5-11 1913
30598	**SPECTRAL DISTRIBUTION OF THE LUMINESCENCE YIELD OF RUBY.** MORGENSHTERN Z L NEUSTRUEV V V OPT I SPEKTROSKOPIYA 20 5 837-41 1966 (FOR ENGLISH TRANSLATION SEE T30661)
30602	**AN APPARATUS FOR MEASURING TOTAL HEMISPHERICAL EMITTANCE BETWEEN AMBIENT AND LIQUID NITROGEN TEMPERATURES.** HAURY, G. L. SYMP. ON MEASUREMENT OF THERMAL RADIATION PROPERTIES OF SOLIDS 51-4, 1963. (NASA-SP-31)
30604	**THE REFLECTIVITY OF SOLIDS AT GRAZING ANGLES.** BRANDENBERG, W. M. SYMP. ON MEASUREMENT OF THERMAL RADIATION PROPERTIES OF SOLIDS 75-82, 1963. (NASA-SP-31)
30621	**A SILICON CELL TRANSMISSIVITY-REFLECTIVITY METER FOR USE WITH SOLAR RADIATION.** YELLOTT, J. I. CHAMNESS, L. SYMP. ON MEASUREMENT OF THERMAL RADIATION PROPERTIES OF SOLIDS 269-74, 1963. (NASA-SP-31)
30638	**A TECHNIQUE FOR MEASURING THERMAL RADIATION PROPERTIES OF TRANSLUCENT MATERIALS AT HIGH TEMPERATURE.** COX, R. L. SYMP. ON MEASUREMENT OF THERMAL RADIATION PROPERTIES OF SOLIDS 469-81, 1963. (NASA-SP-31)
30641	**EMITTANCE MEASUREMENT CAPABILITY FOR TEMPERATURES UP TO 3000 F.** KJELBY, A. S. SYMP. ON MEASUREMENT OF THERMAL RADIATION PROPERTIES OF SOLIDS 499-503, 1963. (NASA-SP-31)
30661	**SPECTRAL DISTRIBUTION OF THE LUMINESCENCE YIELD OF RUBY.** MORGENSHTERN Z L NEUSTRUEV V V OPT SPECTRY 20 5 464-6 1966 (ENGLSIH TRANSLATION OF OPT. I SPEKTROSKOPIYA, 20 (5), 837-41, 1966; FOR ORIGINAL SEE T30598)
30669	**HEAT TRANSMISSION OF BRICK AND HIGH TEMPERATURE INSULATING MATERIALS.** HORNING ROY A TRANS AM CERAM SOC 2 865-76 1920
30693	**NOTES ON HEAT INSULATION, PARTICULARLY WITH REGARD TO MATERIALS USED IN FURNACE CONSTRUCTION.** HUTTON R S BEARD J R TRANS FARADAY SOC 1-2 264-70 1905
30723	**LABORATORY-NOTE ON THE HEAT-CONDUCTIVITY, EXPANSION AND FUSIBILITY OF FIRE-BRICK.** PENNOCK J D TRANS AM INST MIN MET ENGRS 26 263-9 1896

TPRC Number	Bibliographic Citation
30732	EVALUATION OF THE THERMAL CONDUCTIVITY AT MODERATE AND HIGH TEMPERATURES. PT. 2. RICHTER, W. LIPPMANN, S. ARNHOLD, A. NEUE HUTTE 8 6 366–70 1963 RM 20 83984P
30734	MEASUREMENT AND THEORETICAL DETERMINATION OF THE THERMAL CONDUCTIVITY OF GOOD-CONDUCTING POWDERS. VON KISS M NEUE TECH 5 247–56 1963 NS 17 27368
30774	SOME PROPERTIES OF POROUS CERAMICS DERIVED FROM BERYLLIUM OXIDE. GUZMAN I YA POLUBOYARINOV D N OGNEUPORY 27 457–62 1962 CA 58 1215
30775	THE RAPID DETERMINATION OF THE COEFFICIENT OF THERMAL CONDUCTIVITY OF REFRACTORY MATERIALS. BUDRIN D V SUCHKOV V D YAROSHENKO YU G OGNEUPORY 28 5 199–206 1963 CA 59 3622 (FOR ENGLISH TRANSLATION SEE T31046)
30784	REFLECTION OF ULTRASOFT X-RAY RADIATION FROM A GLASS AND TITANIUM-COATED SURFACE. LUKIRSKII A P SAVINOV E P OPTIKA I SPEKTROSKOPIYA 14 295–8 1963 CA 58 9756 (FOR ENGLISH TRANSLATION SEE T31295)
30789	THE OPTICAL PROPERTIES AND STRUCTURE OF ZIRCONIUM DIOXIDE FILMS. SVIRIDOVA A I OPT SPECTR /USSR/ 13 236–8 1962 NS 17 8847 (ENGLISH TRANSLATION OF OPTIKA I SPEKTROSKOPIIA 13, 425–8, 1962; FOR ORIGINAL SEE T31296)
30804	THE APPLICATION OF THE LIGAND FIELD THEORY TO COLOURED GLASSES. BAMFORD C R PHYS CHEM GLASSES 3 6 189–202 1962 BC 62 1543
30807	VISCOSITY AND FREE VOLUME OF FUSED BORATES AND SILICATES. KUMAR S PHYS CHEM GLASSES 4 3 106–11 1963 CA 59 8449
30814	EFFECT OF COMPOSITION AND TEMPERATURE ON THE ULTRAVIOLET ABSORPTION OF GLASS. MC SWAIN, B. D. BORRELLI, N. F. SU, G. J. PHYS CHEM GLASSES 4 1 1–10 1963 CA 59 4881
30858	ON THE HEAT-CONDUCTIVITY OF GLASS MASS. RODNIKOVA V V SPECIAL LIBRARY ASSOC., TRANSLATIONS CENTER 1–13, 1965. (ENGLISH TRANSLATION OF STEKLO I KERAM. 8, (4), 9–12, 1951; FOR ORIGINAL SEE T1121) (TT-65-12872)
30875	SPIN-WAVE CONTRIBUTION TO THE HEAT CAPACITY OF MAGNETITE. KENAN R P GLASSER M L MILFORD F J PHYS. REV. 132 1 47–9 1963 SA 66 25200
30882	THERMAL CONSTANTS OF COAL, SEMICOKE AND COKE AND THE HEAT OF CARBONISATION. VOLOSHIN A I KOKS I KHIM 10 20–4 1958 (FOR ENGLISH TRANSLATION SEE T30884)
30884	THERMAL CONSTANTS OF COAL, SEMICOKE AND COKE AND THE HEAT OF CARBONISATION. VOLOSHIN A I SPECIAL LIBRARY ASSOC., TRANSLATIONS CENTER 1–11, 1961. (ENGLISH TRANSLATION OF KOKS I KHIMIYA (10), 20–4, 1958; FOR ORIGINAL SEE T30882) (NCB TRANS A-1853/SEH, TT-62-15264)
30889	OPTICAL PROPERTIES OF LIQUID GERMANIUM. ABRAHAM A TAUC J VELICKY B PHYS STATUS SOLIDI 3 767–72 1963 CA 59 2285
30911	A STUDY OF THE THERMAL CONDUCTIVITY OF SOLIDS IN THE RANGE 80 – 500 K. KAMILOV I K PRIBORY I TEKH EKSPER /USSR/ 3 176–9 1962 SA 66 4831 (FOR ENGLISH TRANSLATION SEE T33174)
30934	THE APPLICATION OF RESIN-IMPREGNATED POROUS CERAMICS TO REENTRY VEHICLE HEAT SHIELDS. STRAUSS, E. L. PROC CONF AERODYN HEATED STRUCT CAMBRIDGE MASS 1961 7–27 1962 CA 58 8747
30988	THE TRANSFER OF HEAT THROUGH REFRACTORIES AND ITS DETERMINATION. WATTS A S KING R M J AM CERAM SOC 6 10 1075–89 1923
30989	USE OF CAST ALUMINA REFRACTORIES IN THE GLASS INDUSTRY. BAUMANN H N JR TURNER A A J AM CERAM SOC 23 11 334–8 1940
30997	SPECIFIC HEAT CAPACITY OF SOILS AND MINERALS AS DETERMINED WITH A RADIATION CALORIMETER. BOWERS S A HANKS R J SOIL SCI 94 6 392–6 1962 BC 62 1443
30999	THERMAL CONDUCTIVITY OF PERICLASE SINGLE CRYSTALS. STARZACHER A RADEX RUNDSCHAU 3 157–63 1962 CA 61 11360
31014	NEW APPROACH TO PORTLAND-CEMENT KILN REFRACTORY PROBLEMS. PARNHAM H REFRACT J 38 8 284–8 1962 BC 62 799
31019	THE HEATING OF COPPER WIRES BY ELECTRIC CURRENTS. KENNELLY A E SHEPARD E R J AM INST ELEC ENGRS 26 969–95 1907
31046	RAPID DETERMINATION OF COEFFICIENTS OF THERMAL DIFFUSIVITY AND THERMAL CONDUCTIVITY OF REFRACTORIES. BUDRIN D V SUCHKOV V D YAROSHENKO YU G REFRACTORIES /MOSCOW/ 5 218–25 1963 BC 63 2181 (ENGLISH TRANSLATION OF OGNEUPORY 28 (5), 199–206, 1963; FOR ORIGINAL SEE T30775)
31072	DETERMINATION OF THERMAL CONDUCTIVITIES OF REFRACTORIES UP TO 1500 DEGREES. PUSTOVALOV V V SB NAUCHN TR UKR NAUCHN-ISSLED INST OGNEUPOROV 3 282–90 1960 CA 58 7702
31073	HIGH-TEMPERATURE MEASUREMENTS /UP TO 2400 DEGREES/ IN VACUUM, OF THE THERMAL CONDUCTIVITIES OF REFRACTORY MATERIALS. PUSTOVALOV V V SB NAUCHN TR UKR NAUCHN-ISSLED INST OGNEUPOROV 5 324–35 1961 CA 59 2493
31079	VISCOSITY AND FUSIBILITY OF SOME VOLCANIC WATER-BEARING GLASSES. YAVITS, I. M. SB. TR. RESP. NAUCHN.-ISSLED. INST. MESTNYKH STROIT. MATERIALOV (25), 54–62, 1962.
31084	STUDY OF THE STRUCTURE OF QUARTZ, CRISTOBALITE, AND VITREOUS SILICA BY REFLECTION IN INFRARED. SIMON I MCMAHON H O J CHEM PHYS 21 23–30 1953
31087	A PHYSICAL STUDY OF THE WELSBACH MANTLE. IVES, H. E. KINGSBURY, E. F. KERRER, E. J FRANKLIN INST 186 585–624 1918
31088	APPARATUS AND TECHNIQUE FOR THE DETERMINATION OF THE THERMAL CONDUCTIVITY OF POWDERS AS AFFECTED BY DENSITY. RODEWALD, H. J. SGUAITAMATTI, B. ILLI, H. SCHWEIZ. ARCH. ANGEW. WISSEN. TECHNIK. 28 (9), 357–61, 1962.
31115	EVALUATION OF THERMAL PROTECTIVE SYSTEMS FOR ADVANCED AEROSPACE VEHICLES, VOL. I. CHRISTIAN W J BLITON J L DALLY J W HEDGE J C HIRSCHHORN, H. J. IIT RESEARCH INST., CHICAGO, IL. 1–259, 1965. (AFML-TDR-64-204(VOL. 1), AD 627139)
31121	SPECTROPHOTOMETRIC AND PHOTOMETRIC CHARACTERISTICS OF THE OPAL GLASS MS 14. VOISHVILLO N A OPTIKA I SPEKTROSKOPIYA 12 443–5 1962 CA 57 4339
31125	MEASUREMENT OF THE MEAN SPECIFIC HEAT OF REFRACTORY MATERIALS. ROBIJN P SILICATES IND 28 5 247–53 1963 NS 18 13779
31129	TEMPERATURE DEPENDENCE OF OPTICAL TRANSMISSION OF MELTED QUARTZ IN THE ULTRAVIOLET SPECTRUM. ABRAHAM A SILHAVA M SILIKATY 7 3 231–3 1963 CA 59 14747

TPRC Number	Bibliographic Citation
31136	**RESEARCH ON THE THERMAL PROPERTIES OF ZIRCONIA.** BURDICK, R. B. HOSKYNS, W. R. PFAUDLER PERMUTIT, INC. 1-72, 1963. (ARL-63-170, AD 420569)
31149	**A BALANCED THERMOCOUPLE AND FILTER METHOD OF ULTRAVIOLET RADIOMETRY, WITH PRACTICAL APPLICATIONS.** COBLENTZ W W STAIR R HOGUE J M J RESEARCH NATL BUR STDS 7 723-49 1933
31150	**THE EFFECT OF CERTAIN OXIDES ON THE HYDROGEN REDUCTION OF LEAD SILICATE GLASSES.** KITAIGORODSKII I I FAINBERG E A GRECHANIK L A MENDELEEV D I STEKLO I KERAM 19 12 8-10 1962 CA 58 13573 (FOR ENGLISH TRANSLATION SEE T20683)
31153	**ABSORPTION OF NEAR-INFRARED ENERGY BY CERTAIN GLASSES.** FLORENCE, J. M. ALLSHOUSE, C. C. GLAZE, F. W. HAHNER, C. H. J RESEARCH NATL BUR STDS 45 121-8 1950
31165	**NEW PROBLEMS IN THE PHYSICS OF GLASS.** TARASOV, V. V. CHERNOPLEKOV, N. A. GOSUDARSTVENNOYE IZDATEL. LITERATURY PO STROITELSTVU ARKHITEKTURE I STROITELNYM MATERIALAM MOSKVA 1-270PP., 1959. (FOR ENGLISH TRANSLATION SEE T31166)
31166	**NEW PROBLEMS IN THE PHYSICS OF GLASS.** TARASOV V V CHERNOPLEKOV N A SPECIAL LIBRARY ASSOC., TRANSLATIONS CENTER 1-225, 1963. 1-225 1963 TT 10 1584 (ENGLISH TRANSLATION OF MONO. NOVYE VOPROSY FIZIKI STEKLA, MOSCOW, 270PP., 1959; FOR ORIGINAL SEE T31165) (TT-63-11004)
31171	**RADIATION PROPERTIES OF DIFFERENT SUBSTANCES WITHIN THE TEMPERATURE RANGE 250 C TO 800 C.** PIRANI M J SCI INSTR 16 372-8 1939
31182	**THE HEAT RATING REQUIRED TO PRODUCE CENTRAL MELTING IN VARIOUS UO2 FUELS. FROM SYMPOSIUM ON RADIATION EFFECTS IN REFRACTORY FUEL COMPOUNDS.** BAIN A S A S T M SPECIAL TECHNICAL PUBLICATION 306 30-46 1962 NS 17 4895
31216	**VISCOSITY OF SILICA.** SOLOMIN N V SPECIAL LIBRARY ASSOC., TRANSLATIONS CENTER 1-9, 1962. (ENGLISH TRANSLATION OF ZH. FIZ. KHIM. 14 (2), 235-43, 1940; FOR ORIGINAL SEE T3956) (TT-62-18318)
31217	**SOME FACTORS INFLUENCING THE THERMAL PROPERTIES OF MINERALS AND PRODUCTS OF THE CERAMIC INDUSTRY.** COHN, W. M. J AM CERAM SOC 11 296-306 1928
31231	**I. THE THERMAL CONDUCTIVITY MEASUREMENTS OF SELECTED SILICATE POWDERS IN VACUUM FROM 150 - 350 K. II. AN INTERPRETATION OF THE MOON'S ECLIPSE AND LUNATION COOLING AS OBSERVED THROUGH THE EARTH'S ATMOSPHERE FROM 8-14 MICRONS.** WATSON, K. CALIFORNIA INST. OF TECHNOL., PH. D. THESIS 1-93 1964
31233	**INFRARED REFLECTIVITY OF SIO2.** SCHAEFER CLEMENS SCHUBERT MARTHA Z PHYSIK 7 313-15 1921
31234	**SPECIFIC HEAT AND HEAT OF FUSION OF VARIOUS CHLORIDES AND FLUORIDES.** LYASHENKO V S SPECIAL LIBRARY ASSOC., TRANSLATIONS CENTER 1-31, 1963. (ENGLISH TRANSLATION OF METALLURG. 10 (11), 85-98, 1935; FOR ORIGINAL SEE T15471) (RT-1517, TT-64-14158)
31237	**EFFECT OF LIGHT UPON WOOL. II. COMPARATIVE REFLECTANCE AS A MEASURE OF COLOR CHANGE.** LAUNER HERBERT F TEXTILE RES J 33 351-6 1963 CA 59 7695
31244	**INVESTIGATION OF ALUMINUM PHOSPHATE COATING FOR THERMAL INSULATION OF AIRFRAMES.** EUBANKS, A. G. MOORE, D. G. NATIONAL BUREAU OF STANDARDS 1-28, 1959. (NASA-TN-D-106, N62-70680)
31279	**KINETIC PROPERTIES OF INTERPLANETARY MATTER.** PARKER E N PLANETARY SPACE SCI 9 461-75 1962
31295	**REFLECTION OF ULTRA-SOFT X-RADIATION FROM GLASS AND TITANATED SURFACES.** LUKIRSKII A P SAVINOV E P OPTICS AND SPECTROSCOPY 14 152-4 1963 (ENGLISH TRANSLATION OF OPTIKA I SPEKTROSKOPIYA 14, 295-8, 1963; FOR ORIGINAL SEE T30784)
31296	**THE OPTICAL PROPERTIES AND STRUCTURE OF ZIRCONIUM DIOXIDE FILMS.** SVIRIDOVA A I OPTIKA I SPEKTROSKOPIYA 13 425-8 1962 (FOR ENGLISH TRANSLATION SEE T30789)
31310	**1919 REPORT OF STANDARDS COMMITTEE ON PYROMETRY.** FORSYTHE W E J OPT SOC AM 4 305-35 1920
31343	**REPORT ON THE BEHAVIOR OF CERAMIC MATERIALS IN THE NEAR INFRARED REGION AT HIGH TEMPERATURES.** KROCKEL O BER DEUT KERAM GESELL 41 7 398-404 1964 (FOR ENGLISH TRANSLATION SEE T31344)
31344	**REPORT ON THE BEHAVIOR OF CERAMIC MATERIALS IN THE NEAR INFRARED REGION AT HIGH TEMPERATURES.** KROCKEL, O. OFFICE OF NAVAL INTELLIGENCE, WASHINGTON, D. C. 1-19, 1964. (ENGLISH TRANSLATION OF BER. DEUT. KERAM. GESELL. 41 (7), 398-404, 1964; FOR ORIGINAL SEE T31343) (ONI-1092, AD-476706)
31427	**EMISSIVITY OF REFRACTORY MATERIALS.** ROBIJN P ANGENOT P VERRES ET REFR 17 1 3-10 1963 BC 62 2016
31430	**ELECTRICAL, MECHANICAL AND THERMAL PROPERTIES OF HOT PRESSED AND SINTERED BERYLLIUM OXIDE.** EMANUELSON, R. C. PRATT AND WHITNEY AIRCRAFT, HARTFORD 1-8 1963 (APR 1048 (SUPPL 1))
31431	**THE TRUE HEAT CAPACITY OF MAGNESIUM FERRITE.** REZNITSKII L A KHOMYAKOV K G VESTN MOSK UNIV SER II KHIM 17 2 50-2 1962 CA 58 2908
31437	**RADIANT HEAT EXCHANGE IN HEATED GLASS AND ITS DEPENDENCE ON CHEMICAL COMPOSITION.** RODNIKOVA V V VOPR RAZVITIYA STEKOLNOI I FARFORO-FAYANSOVOI PROM 70-8 1962 CA 58 4273
31464	**INFRARED ABSORPTION OF ANTIMONATE GLASS.** YOKOTA, R. NAKAI, Y. YOGYO KYOKAI SHI 70 199-202 1962 CA 59 15000
31475	**THIOCARBONATES. V. CARBON SULFIDE DIHYDROSULFIDE SC/SH/2, AND THE SYSTEM H2S-CS2. II. THERMOCHEMISTRY OF SC/SH/2.** GATTOW G KREBS H Z ANORG ALLGEM CHEM 322 113-28 1963 CA 59 3368
31481	**THERMAL CONDUCTIVITY OF SUBLIMATES FORMED IN THE CHLORINATION OF TITANIUM SLAG IN SMELTING.** KHOMYAKOV P P ZHELTOVA V I ADLER YU P NALIMOV V V ZAVODSK LAB 29 330-1 1963 CA 59 279
31538	**LOW LOSS CERAMICS.** SNYDER N H SMOKE E J WISELY H R RUH E NEW JERSEY CERAMIC RESEARCH STATION RUTGERS UNIV 1-90, 1950. (ATI 86345, PB 162947)
31564	**THE OPTICAL CONSTANTS OF QUARTZ IN THE INFRARED AND THEIR TEMPERATURE DEPENDENCE. I.** HAEFELE H G Z PHYSIK 168 530-41 1962 CA 58 1997
31604	**HIGH-TEMPERATURE HEAT CONTENTS AND ENTROPIES OF SESQUIOXIDES OF LUTETIUM, DYSPROSIUM, AND CERIUM.** PANKRATZ L B KELLEY K K U S BUR MINES 1-8, 1963. (BM-RI-6248)

TPRC Number	Bibliographic Citation
31607	**THE QUESTION OF ENTROPY OF AMORPHOUS SUBSTANCES.** SIMON, F. LANGE, F. Z PHYSIK 38 227-36 1926
31610	**REFLECTION AND TRANSMISSION OF 7-24 MU WAVES BY SILICA MODIFICATIONS.** SEVCHENKO N A FLORINSKAYA V A SOVIET PHYS-DOKLADY 1 4 508-11 1956 (ENGLSIH TRANSLATION OF DOKLADY AKAD. NAUK SSSR, 109 (6), 1115-18, 1956; FOR ORIGINAL SEE T31853)
31611	**THE THERMAL CONDUCTIVITY OF MATERIALS EMPLOYED IN FURNACE CONSTRUCTION.** GRIFFITHS, E. TRANS FARADAY SOC 12 193-206 1917
31625	**HEAT TRANSFER BY RADIATION IN HEATED GLASS.** RODNIKOVA V V STEKLO I KERAMIKA 6 3 10-14 1949
31731	**REFLECTING-POWER MEASUREMENTS IN THE SPECTRAL REGION 2000 A. TO 1300 A.** JOHNSON B K PROC PHYS SOC /LONDON/ 53 258-64 1941
31733	**GLASSY STATE.** AVGUSTINIK, A. I. FOREIGH TECH. DIV. AIR FORCE SYSTEMS COMMAND WRIGHT-PATTERSON AIR FORCE BASE OHIO 1-7PP., 1963. (ENGLISH TRANSLATION OF STEKLODBRAZNOYE SOSTOYANIYE, IZDATELSTVD AKADEMII NAUK SSSR, MOSCOW, 115-119PP, 1960; FOR ORIGINAL SEE T031734) (FTD-TT-63-265, AD-409701)
31734	**GLASSY STATE.** AVGUSTINIK, A. I. STEKLOOBRAZNOYE SOSTOYANIYE, IZDATELSTVO AKAD. NAUK. SSSR, MOSCOW 115-19 1960 TA U63-4 74 (FOR ENGLISH TRANSLATION SEE T31733)
31735	**GLASSY STATE (SELECTED ARTICLES).** FLORINSKAYA, V. A. FOREIGN TECH. DIV. AIR FORCE SYSTEMS COMMAND 8-29PP., 1963. (ENGLISH TRANSLATION OF STEKLOOBRAZNOYE SOSTOYANIYE, IZDATELSTVO AKADEMII NAUK SSSR, MOSCOW, 177-94PP., 1960; FOR ORIGINAL SEE T31736)
31736	**GLASSY STATE.** FLORINSKAYA, V. A. STEKOOBRAZNOYE SOSTOYANIYE, IZDATEL'STOV AKAD. NAUK SSSR, MOSCOW 177-94 1960 (FOR ENGLISH TRANSLATION SEE T31735)
31749	**RESEARCH IN THE FAR INFRARED REGION OF THE SPECTRUM.** GINSBURG, N. AIR FORCE CAMBRIDGE RESEARCH LAB 1-72, 1963. (AFCRL-64-73, AD 434770)
31759	**DESIGN AND FABRICATION OF OPTICAL FILTERS FOR LASER FREQUENCY.** HOLMES, S. J. BOWMAN, N. SPECTROLAB INC., SYLMAR, CALIF 1-26, 1963. (AD 427005)
31783	**A PROGRESS REPORT ON 1. A 6500 F FURNACE FOR THERMOPHYSICAL PROPERTY STUDIES. 2. MORE DATA ON SOME HIGH TEMPERATURE STANDARDS FOR THERMAL CONDUCTIVITY. 3. NEW DATA ON ASTM-C-177.** PEARS, C. D. THERMAL CONDUCTIVITY CONF. 453-79, 1963.
31786	**THE BONDING OF CERAMIC-METAL LAMINATES.** STATE UNIV. OF NEW YORK, COLLEGE OF CERAMICS AT ALFRED UNIVERSITY STATE UNIV. OF NEW YORK COLLEGE OF CERAMICS AT ALFRED UNIV. 1-15, 1958. (AD 154872)
31788	**METAL FIBER REINFORCED CERAMICS.** TRUESDALE, R. S. SWICA, J. J. TINKLEPAUGH, J. R. STATE UNIV. OF NEW YORK COLLEGE OF CERAMICS AT ALFRED UNIVERSITY 1-36, 1958. (WADC-TR-58-452, AD 207079)
31809	**DETERMINATION OF THE TRUE TEMPERATURE OF SOLID BODIES FROM THE POINT OF INTERSECTION OF THE LOGARITHMICAL ISOCHROMATES IN THE VISIBLE SPECTRUM.** BENEDICT ELISABETH ANN PHYSIK 47 641-78 1915
31815	**THE PROBLEM OF MEASURING THERMAL CONDUCTIVITY COEFFICIENTS.** SEIFFERT, K. ARCH GES WARMETECH 2 49-55 1951
31836	**STUDY OF THE DIFFUSE REFLECTION FACTOR OF SILICON-ALUMINOUS REFRACTORIES IN THE INFRARED AT HIGH TEMPERATURE.** ALEGRE, R. COMPT REND 238 1209-10 1954
31853	**THE SPECTRUM OF REFLECTION AND TRANSMISSION OF DIFFERENT MODIFICATIONS OF SILICA IN THE WAVELENGTH REGION 7 - 24 MU.** SEVCHENKO N A FLORINSKAYA V A DOKLADY AKAD NAUK SSSR 109 6 1115-18 1956 (FOR ENGLISH TRANSLATION SEE T31610)
31857	**EXPERIMENTS TO DETERMINE THE COEFFICIENT OF THERMAL DIFFUSIVITY AND CONDUCTIVITY OF POWDERS AND FIBROUS INSULATING MATERIALS BY THE METHOD OF REGULAR REGIMEN.** KONDRATEV G M STATSENKO M P IZV. VSES. TEPLOTEKH. INST. IM. FELIKSA DZERZHINSKOGO 7 63-9 1931
31858	**THE APPLICATION OF INSULATING MATERIALS TO INDUSTRIAL FURNACES.** KELLER J D J AM CERAM SOC 17 4 77-87 1934
31862	**A METHOD OF MEASURING THE THERMAL CONDUCTIVITY OF ROCK CORES.** HERSEY J B J APPL PHYS 12 498-501 1941
31868	**THE THERMAL CONDUCTIVITY OF REFRACTORY MATERIALS.** DOUGILL G HODSMAN H J COBB J W J SOC CHEM IND /LONDON/ 34 9 465-70 1915
31870	**THE THERMAL CONDUCTIVITY OF REFRACTORY PRODUCTS.** SALMANG H KLEI 2 23-33 1952
31871	**ARTIFICIAL MINERAL FIBERS FOR THERMAL INSULATION.** SEIFFERT, K. VDI-Z 91 149-53 1949
31874	**VARIATION OF THERMAL CONDUCTIVITY WITH TEMPERATURE FOR A BROAD RANGE OF MATERIALS.** AUTHOR ANON. MATERIALS IN DESIGN ENGINEERING 52 3 177 1960
31875	**AN APPARATUS FOR MEASURING THE HEAT OF TRANSFORMATION IN THE SOIL.** ALBRECHT FRITZ METEROLOGISCHE Z 49 294-9 1932
31878	**INSULATING MATERIALS WITH AUGMENTED THERMAL CONDUCTIVITY.** MEISSNER A NATURWISSENSCHAFTEN 23 3 41-3 1935
31883	**ON THE THERMAL CONDUCTIVITY OF SOME ROCKS AT HIGH TEMPERATURES.** POOLE HORACE H PHIL MAG 24 45-62 1912
31884	**ON THE THERMAL CONDUCTIVITY AND SPECIFIC HEAT OF GRANITE AND BASALT AT HIGH TEMPERATURES.** POOLE HORACE H PHIL MAG 27 58-83 1914
31888	**THE THERMAL CONDUCTIVITY OF NATURAL COAL AND OF COKE.** FRITZ W DIEMKE H PHYSIK Z 40 361-6 1939
31893	**STRUCTURAL AND TEXTURAL INFLUENCES ON THERMAL CONDUCTIVITY OF SOILS.** VAN ROOYEN MARTINUS WINTERKORN HANS F HIGHWAY RESEARCH BOARD PROC NATL RESEARCH COUNCIL 38 576-621 1959
31895	**THE DETERMINATION OF THERMAL CONDUCTIVITY AND ITS TEMPERATURE-VARIATION FOR MEDIUM CONDUCTORS.** ENSOR C R PROC PHYS SOC /LONDON/ 43 581-91 1931

TPRC Number	Bibliographic Citation
31896	**ON THE THERMAL CONDUCTIVITY OF ICE, AND A NEW METHOD OF DETERMINING THE CONDUCTIVITY OF DIFFERENT SUBSTANCES.** FORBES GEORGE PROC ROY SOC EDINBURGH 8 62-8 1872-3
31897	**ON A METHOD OF FINDING THE CONDUCTIVITY FOR HEAT.** NIVEN C GEDDES A E M PROC ROY SOC /LONDON/ 87 A 535-9 1912
31902	**CERAMIC INSULATORS FOR SPARK PLUGS.** RIDDLE FRANK H SAE JOURNAL 46 6 236-42 1940
31906	**HEAT AND MOISTURE TRANSFER IN CLOSED SYSTEMS OF TWO GRANULAR MATERIALS.** WOODSIDE W CLIFFE J B SOIL SCI 87 75-82 1959
31907	**HEAT TRANSFER IN A MOIST CLAY.** WOODSIDE W DE BRUYN C M A SOIL SCI 87 166-73 1959
31909	**HEAT LOSS THROUGH WALLS, LINED WITH REFRACTORY SUBSTITUTES.** IVANTSOV G P STAL 3/4 44-8 1942
31910	**INVESTIGATIONS ON THE THERMAL CONDUCTIVITY OF REFRACTORY BUILDING MATERIALS.** HEYN E STAHL U EISEN 34 20 832-4 1914
31912	**HEAT INSULATION IN COLD STORAGE CONSTRUCTION.** GRIFFITHS, E. STRUCTURAL ENGINEER /LONDON/ 10 202-12 1932
31914	**THE THERMAL CONDUCTIVITY OF PEAT.** KORCHUNOV S S TORFYANAYA PROMYSHLENNOST 2 23-5 1948
31915	**METHODS OF MAKING THERMAL CONDUCTIVITY TESTS, AND THE TRANSMISSION OF HEAT THROUGH BRICKS.** HORNING RAY A TRANS BRITISH CERAM SOC 18 192-202 1916
31920	**HEAT TRANSFERENCE IN SOILS.** PATTEN HARRISON E US DEPT AGR BUR SOILS BULL 59 1-54 1909
31923	**THERMAL CONDUCTIVITY AND VAPOR DIFFUSION IN COLD INSULATORS.** KRISCHER O WARME KALTETECH 43 1-6 1941
31926	**COMPOSITION OF THE REFLECTED HEAT EMISSION FROM CONSTRUCTION AND BUILDING MATERIALS.** SIEBER WERNER Z TECH PHYSIK 22 6 130-5 1941
31927	**THERMAL RESISTIVITY MEASUREMENTS ON INSULATING MATERIALS.** BATSCH H MEISSNER A Z TECH PHYSIK 17 8 283-5 1936
31928	**CRYSTAL EMISSION AND GRAIN BOUNDARY EMISSION OF NON-METALLIC BODIES /OXIDES/.** SKAUPY, F. HOPPE, H. Z TECH PHYSIK 13 226-8 1932
31931	**THE MELTING-POINT-COMPOSITION DIAGRAM OF THE ZIRCONIUM-OXYGEN SYSTEMS.** CUBICCIOTTI, D. J AM CHEM SOC 73 2032-5 1951
31941	**USE OF LIGHT-WEIGHT REFRACTORIES FOR INDUSTRIAL FURNACES.** ELTYSHEV B N VESTNIK METALLOPROMYSHLENNOSTI 16 44-57 1936
31942	**TECHNICAL CONDITIONS FOR LIGHT-WEIGHT REFRACTORIES.** PIROGOV A A ROIZEN A I VESTNIK STANDARTIZATSII 9 9-13 1939
31948	**ELECTRICAL INSULATION MATERIALS.** MONACK A J MATERIALS AND METHODS 45 145-60 1957
31964	**UO2 - ZRO2 STUDIES.** WRIGHT, T. R. FACKELMANN, J. M. KIZER, D. E. KELLER, D. L. BATTELLE MEMORIAL INST., COLUMBUS U1-3 1963 (BMI-1644 (DEL))
31998	**ELECTRICAL AND THERMAL MEASUREMENTS ON FE-AFE PHASE TRANSITIONS IN LEAD HAFNATE TITANATE COMPOSITIONS.** NORTHRIP, J. W. BAUM, G. A. KOHNKE, E. E. SANDIA CORP., ALBURQUERQUE 1-198 1963 NS 18 18458 (SC-DC-64-344)
32045	**SPECTRAL EMITTANCE OF REFRACTORY MATERIALS.** BLAU, H. H., JR. JASPERSE, J. R. APPL OPT 3 2 281-6 1964 CA 60 14022
32047	**TRANSMITTANCE OF SOME OPTICAL MATERIALS FOR USE BETWEEN 1900 AND 3400 A.** PELLICORI S F APPL OPT 3 3 361-6 1964 CA 61 5099
32051	**TRANSMISSION OF SILICON BETWEEN 40 AND 100 MU.** WALLES STEN ARKIV FYSIK 25 4 33-47 1963 CA 60 118
32110	**DETERMINATION OF THE OPTICAL CONSTANTS OF CRYSTALS BY INFRARED REFLECTION.** VINCENT-GEISSE, J. BELORGEOT, C. COMPT REND 257 19 2808-11 1963 CA 60 4890
32113	**VISCOSITY OF MELTED SILICA AND OF LIQUID SLAGS OF THE SIO2-AL2O3 SYSTEM.** ROSSIN, R. BERSAN, J. URBAIN, G. COMPT REND 258 2 562-4 1964 CA 60 8989
32125	**FACTORS INFLUENCING THERMAL CONDUCTIVITY OF NONMETALLIC MATERIALS.** AUSTIN, J. B. SYMPOSIUM ON THERMAL INSULATING MATERIALS 3-67 1939
32128	**HOW GOOD ARE BERYLLIA CERAMICS.** HESSINGER, P. S. ELECTRONICS 36 42 75-6 78 1963 RM 21 88442P
32162	**THERMAL PROPERTIES OF GLASS, CERAMICS, AND GLASS-CERAMICS.** HAGY H E GLASS IND 44 10 563-8 591-4 1963 CA 60 8989
32163	**SURVEY OF MOST PROBABLE VALUES FOR THERMAL CONDUCTIVITIES OF GLASSES BETWEEN ABOUT -150 AND 100 C, INCLUDING NEW DATA ON TWENTY-TWO GLASSES AND A WORKING FORMULA FOR THE CALCULATION OF CONDUCTIVITY FROM COMPOSITION.** RATCLIFFE E H GLASS TECHNOL 4 4 113-28 1963 JA 47 4
32170	**TRANSMISSION OF LONG-WAVELENGTH INFRARED RADIATION /40-200 MU/ BY HEATED CRYSTALLINE AND FUSED QUARTZ.** FILIPPOV O K YAROSLAVSKII N G OPTICS AND SPECTROSC /USA/ 15 4 299-301 1963 SA 67 13053 (ENGLISH TRANSLATION OF OPTIKA I SPEKTROSK, 15 (4) 558-61, 1963; FOR ORIGINAL SEE T32433)
32198	**DETERMINATION OF THERMAL DIFFUSIVITY OF PELLETS.** BEREZHNOI A N INZH-FIZ ZH AKAD NAUK BELORUSSK SSR 6 12 35-9 1963 CA 60 10231
32319	**CHANGES IN INFRARED TRANSMITTANCE OF FLUORINE-CONTAINING POROUS GLASS ON HEATING.** ELMER, T. H. CHAPMAN, I. D. NORDBERG, M. E. J PHYS CHEM 67 10 2219-22 1963 JA 47 84
32388	**REPORT OF INVESTIGATION OF OPTICAL TRANSMITTANCE, REFLECTANCE, AND ABSORPTANCE OF MATERIALS.** BYRNE, R. F. MANCINELL!, L. N. MATERIAL LAB NEW YORK NAVAL SHIPYARD, BROOKLYN 1-39, 1954. (PB 159155)
32405	**DETERMINATION OF THE THERMAL CONDUCTIVITY COEFFICIENT FOR FREE FLOWING MATERIALS BY MALMGREN-SHULEIKIN METHOD.** KOSOV N D VULIS Y A POTSELUIKO V A ZHUR TEKH FIZ 26 84-9 1956
32417	**CHANGE IN THE COEFFICIENT OF THERMAL DIFFUSIVITY OF COALS DURING ISOTHERMIC DECOMPOSITION.** KASATOCHKIN V I ZAMOLUEV V K KOKS I KHIM 6 21-3 1957

TPRC Number	Bibliographic Citation
32423	METHOD OF RAPID DETERMINATION OF HEAT AND TEMPERATURE CONDUCTIVITY COEFFICIENTS WITHOUT TAKING SAMPLES. IANKELEV L F STROIT PREDPRIYATII NEFT PROM 1 5 14-17 1956
32425	HEAT CONDUCTIVITY OF CERTAIN INSULATING MATERIALS AT LOW TEMPERATURES. GOLIAND M M TR LENINGR TEKHNOL INST KHOLODILN PROM 11 83-9 1956
32426	SERVICE TESTS OF LIGHT-WEIGHT REFRACTORIES IN THE LININGS OF TUNNEL KILN LORRIES. TSIGLER V D VINOKUR S B MITROKHINA N S OGNEUPORY 28 11 504-8 1963 CA 60 2617
32427	THE EFFECT OF THE POROSITY AND STRUCTURE OF CORUNDUM REFRACTORIES ON THEIR THERMAL CONDUCTIVITY. DUDEROV I G POLUBOYARINOV D N OGNEUPORY 28 11 518-24 1963 CA 60 3852
32429	OPTICAL CONSTANTS OF EVAPORATED FILMS OF ZINC SULFIDE AND GERMANIUM IN THE INFRARED. HULDT LENNART STAFLIN TORSTEN OPT ACTA 6 1 27-36 1959 CA 60 8755
32433	TRANSMISSION OF LONG-WAVELENGTH INFRARED RADIATION /40-200 MU/ BY HEATED CRYSTALLINE AND FUSED QUARTZ. FILIPPOV O K YAROSLAVSKII N G OPTIKA I SPEKTROSKOPIYA 15 4 558-61 1963 CA 60 2452 (FOR ENGLISH TRANSLATION SEE T32170)
32447	THE TEMPERATURE - HEAT CONTENT CURVES OF THE TECHNICALLY IMPORTANT METALS. WUST F MEUTHEN A DURRER R FORSCH- ARB ING - WES 204 1-63 1918
32466	THE LOW-TEMPERATURE HEAT CAPACITIES OF DIFFERENT FORMS OF SIO2. LEADBETTER A J MORRISON J A PHYS CHEM GLASSES 4 5 188-92 1963 CA 60 1424
32477	EMISSION OF SILICON-ALUMINUM REFRACTORIES IN THE INFRARED SPECTRUM. ALEGRE R SPECIAL LIBRARY ASSOC. 1-23, 1961. 1-24, 1963. (ENGLISH TRANSLATION OF SILICATES IND., 23, 253-60, 1958; FOR ORIGINAL SEE T19583) (TT61-14470, TT63-18694)
32489	INFRARED LATTICE VIBRATIONS AND DIELECTRIC DISPERSION IN CORUNDUM. BARKER A S JR PHYS REV /USA/ 132 4 1474-81 1963 SA 67 4256
32553	THE HEAT CAPACITY OF DIAMOND AT HIGH TEMPERATURES. VICTOR, A. C. UNIV. OF MARYLAND, COLLEGE PARK, M.S. THESIS 1-58 1961
32571	THERMAL CONDUCTIVITY OF SOME AMORPHOUS DIELECTRIC SOLIDS BELOW 1 K. ANDERSON A C REESE W WHEATLEY J C REV SCI INSTRUM /USA/ 34 12 1386-90 1963 SA 67 6758
32581	EFFECTS OF TITANIUM DIOXIDE IN GLASS. I. BEALS M D STRIMPLE J H GLASS IND 44 9 495-501 530-1 1963 BC 63 507
32596	THERMAL CONDUCTIVITY OF MOIST CERAMIC BODIES. SCHULLE W SILIKATTECHNIK 14 3 78-81 1963 BC 63 37
32611	THERMAL CONDUCTIVITY. III, PROLEATE SPHEROIDAL ENVELOPE METHOD. ADAMS MILTON J AM CERAM SOC 37 2 74-9 1954
32616	AN UNUSUALLY REFRACTORY PARTIALLY CRYSTALLINE VANADIA-SILICA GLASS. BROWN S D KISTLER S S SYMP. NUCLEATION CRYSTALLIZATION GLASSES MELTS 35-6 1962 CA 60 2617
32627	AN APPROXIMATE METHOD FOR DETERMINATION OF CONDUCTIVITY IN LOOSE MATERIALS IN TRANSIENT PERIOD. GIAMBELLI G TERMOTECNICA MILAN 17 4 230-4 1963 AM 17 1077
32629	HIGH-ALUMINA LIGHTWEIGHT BRICK AND ITS USE. TSIGLER V D ELTYSHEVA A A PINDRIK B E REFRACTORIES /USSR/ 25 241-7 1960 (ENGLISH TRANSLATION OF OGNEUPORY, 25, 299-307, 1960; FOR ORIGINAL SEE T19081)
32632	HEAT TRANSFER IN POROUS ROCKS THROUGH WHICH SINGLE-PHASE FLUIDS ARE FLOWING. ADIVARAHAN P KUNII D SMITH J M TRANS AIME 225/PT II/ 290-6 1962 CA 60 10437
32646	TEMPERATURE COEFFICIENT OF THERMAL CONDUCTIVITY FOR HEAT INSULATORS. CHERKOV D A ELEKTRICHESKIE STANTSII 26 5 24-5 1955
32674	METHODS OF MEASUREMENT OF THERMAL CONDUCTIVITY OF CERAMIC MATERIALS. ALPEROVICH I A IOFFE IU M STEKLO I KERAMIKA 8 12 21-3 1951
32675	DETERMINING THE THERMAL DIFFUSIVITY OF COAL IN HEATING. AGROSKIN A A MELAMED R IA MIRINGOF N S PODZEMNAYA GAZIFIKATSIYA UGLEI 2 92-5 1957
32688	MIGRATION OF H/PLUS 1 ION/ IONS IN GLASSES. EHRMANN P DE BILLY M ZARZYCKI J VERRES REFRACTAIRES 15 2 63-72 1961 15 3 131-9 1961 JA 47 85
32705	ELECTRONIC CALORIMETER. MITKINA E A INDUSTRIAL LAB /USSR/ 26 2 238-41 1960 (ENGLISH TRANSLATION OF ZAVODSKAYA LAB., 26 (2), 226-9, 1960; FOR ORIGINAL SEE T25502)
32716	FABRICATION OF UO2 AND UO2-BEO MIXED PELLETS. BATTELLE INST., E. V. FRANKFURT, GERMANY BATTELLE INST. E V FRANKFURT GERMANY 1-23, 1962. (EURAEC-317)
32760	PIGMENTED SURFACE COATINGS FOR USE IN THE SPACE ENVIRONMENT. SEARLE, N. Z. HIRT, R. C. DANIEL, J. H., JR. AMERICAN CYANAMID CO., STAMFORD CONN. 1-97, 1964. (ASD-TDR-62-840 PT 2, AD 430529)
32761	METHOD OF DETERMINATION OF THERMOPHYSICAL PROPERTIES IN A WIDE RANGE OF TEMPERATURES. BUDRIN, D. V. YAROSHENKO, YU. G. SUCHKOV, V. D. NEWS OF HIGHER EDUCATIONAL INSTITUTIONS MINISTRY OF HIGHER AND SECONDARY SPECIALIZED EDUCATION USSR, INSTRUMENT BUILDING SERIES 5 (1), 159-70, 1962. (ENGLISH TRANSLATION OF IZV. VYSSHIKH UCHEBN. ZAVEDENII, PRIBOROSTROENIE 5 (1), 118-127, 1962; FOR ORIGINAL SEE T27608)
32772	CURRENT NBS EFFORTS TOWARD DEVELOPMENT OF THERMAL CONDUCTIVITY REFERENCE STANDARDS. FLYNN, D. R. NATIONAL BUREAU OF STANDARDS 1-22, 1963. (NBS REPT 7836, AD 407802)
32835	EFFECT OF VARIOUS ATMOSPHERES ON THERMAL CONDUCTANCE OF REFRACTORIES. YOUNG R C HARTWIG F J NORTON C L J AM CERAM SOC 47 5 205-10 1964 CA 61 4043
32851	RELATION OF ENTROPY OF ACTIVATION OF FLUIDITY, HEAT CAPACITY, AND VALENCE STRUCTURE OF GLASS. NEMILOV S V ZH PRIKL KHIM 37 2 293-300 1964 CA 60 11723
32857	EFFECTIVE HEAT CONDUCTIVITY OF SPUN GLASS IN COMPRESSED GASES. BRAGER N N GOLUBEV I F TEPLOFIZ VYSOKIKH TEMPERATUR 3 5 812-14 1965 (FOR ENGLISH TRANSLATION SEE T32866)
32862	THERMAL CONSTANTS OF PYROPHYLLITE AND THEIR CHANGE ON HEATING. CARTE A E BRIT J APPL PHYS 6 326-8 1955
32864	INTRINSIC EDGE ABSORPTION IN DIAMOND. CLARK C D DEAN P J HARRIS P V ROYAL SOCIETY PROCEEDINGS 277 A 312-29 1964 BR 13 4244

TPRC Number	Bibliographic Citation
32865	**LABORATORY APPARATUS FOR MEASURING THERMAL CONDUCTIVITY OF BULK MATERIALS.** PENNEY W R DEAVER F K ASHRAE J 4 11 64-5 1962 BC 63 124
32866	**EFFECTIVE HEAT CONDUCTIVITY OF SPUN GLASS IN COMPRESSED GASES.** BRAGER N N GOLUBEV I F HIGH TEMPERATURE /USSR/ 3 5 753-5 1965 (ENGLISH TRANSLATION OF TEPLOFIZ. VYSOKIKH TEMPERATUR, 3 (5), 812-4, 1965; FOR ORIGINAL SEE T32857)
32867	**THE INFLUENCE OF THERMAL TREATMENT ON THE ABSORPTION OF A CONTAINER GLASS WHICH INCLUDES IRON AND MANGANESE.** PAOLETTI G SPRECHSAAL 96 20 483-7 1963 BC 63 104
32869	**INSULATING BRICK FROM GANGA SAND.** MISRA M L BHUTA V S INDIAN CERAM 8 1 9-11 1961 JA 47 161
32876	**THERMAL CONDUCTIVITY OF AGGREGATES OF SEVERAL PHASES, INCLUDING POROUS MATERIALS.** BRAILSFORD A D MAGOR K G BRIT J APPL PHYS 15 3 313-19 1964 JA 47 205
32894	**OBSIDIAN-H2O VISCOSITIES AT 1000 AND 2000 BARS IN THE TEMPERATURE RANGE 700 TO 900 C.** SHAW H R J GEOPHYS RES /USA/ 68 23 6337-43 1963 SA 67 11188
32896	**VISCOSITY AND WATER CONTENT OF RHYOLITE GLASS.** FRIEDMAN I LONG W SMITH R L J GEOPHYS RES /USA/ 68 24 6523-35 1963 SA 67 11189
32967	**THE USE OF THE TORSIONAL METHOD FOR DETERMINING THE VISCOSITY OF LOW MELTING CLAYS.** PAVLOV V F GLASS AND CERAMICS 16 5 278-83 1960 (ENGLISH TRANSLATION OF STEKLO I KERAM., 16, (5), 26-31, 1959; FOR ORIGINAL SEE T16489)
32972	**A THEORETICAL BASIS FOR THE MEASUREMENT OF THE TRUE SPECIFIC HEAT OF REFRACTORY MATERIALS WHILE THEY ARE COOLING.** PLATUNOV E S TEPLOFIZ VYSOKIKH TEMPERATUR 3 6 889-94 1965 (FOR ENGLISH TRANSLATION SEE T33031)
32979	**HIGH-TEMPERATURE HEAT CONTENTS AND ENTROPIES OF ANDALUSITE, KYANITE, AND SILLIMANITE.** PANKRATZ L B KELLEY K K U.S. BUREAU OF MINES 1-7, 1964. (BM-RI 6370)
32980	**HIGH-TEMPERATURE HEAT CONTENTS AND ENTROPIES OF MUSCOVITE AND DEHYDRATED MUSCOVITE.** PANKRATZ L B U.S. BUREAU OF MINES 1-6, 1964. (BM-RI 6371)
33004	**THE OPTICAL PROPERTIES OF GLASS SURFACE. II. OPTICAL PROPERTIES OF GLASS ETCHED BY ACIDS.** HASEGAWA, Y. KAWAKUBO, S. BUNKO KENKYU 11 6 245-56 1963 CA 60 14227
33029	**PERMANENT MOLD WASH. II. PROPERTIES OF METAL SALT COATING.** ISOTANI, M. KONDO, Y. TANIGAWA, H. NAGOYA KOGYO GIJUTSU SHIKENSHO HOKOKU 13 53-8 1964 CA 60 14230
33031	**A THEORETICAL BASIS FOR THE MEASUREMENT OF THE TRUE SPECIFIC HEAT OF REFRACTORY MATERIALS WHILE THEY ARE COOLING.** PLATUNOV E S HIGH TEMPERATURE /USSR/ 3 6 827-31 1965 (ENGLISH TRANSLATION OF TEPLOFIZ. VYSOKIKH TEMPERATUR 3 (6), 889-94, 1965; FOR ORIGINAL SEE T32972)
33084	**REFLECTANCE OF MINERALS FROM THE MAGNETITE MAGNESIOFERRITE ISOMORPHIC SERIES.** PONOMAREVA M N PAVLOV N V GEOL RUDN MESTOROZHD 6 1 99-101 1964 CA 60 15230
33108	**VISCOSITY OF SYNTHETIC MICA MELTS.** SUMIYOSHI, Y. NODA, T. KOGYO KAGAKU ZASSHI 65 2 150-4 1962 CA 60 15168
33123	**FEAL2O4-MGAL2O4. GROWTH AND SOME THERMAL, OPTICAL, AND MAGNETIC PROPERTIES OF MIXED SINGLE CRYSTALS.** SLACK, G. A. PHYS REV 134 A 8 1268-80 1964 CA 60 15290
33140	**THE EFFECT OF TEMPERATURE ON FLUORESCENCE AND ABSORPTION.** GIBBS R C PHYS REV 31 4 463-88 1910
33142	**BLACK BODIES IN THE EXTREME INFRA-RED.** CARTWRIGHT C HAWLEY PHYS REV 35 415-20 1930
33152	**REFLECTIVITIES OF EVAPORATED METAL FILMS IN THE NEAR AND FAR ULTRAVIOLET.** SABINE GEORGE B PHYS REV 55 1064-9 1939
33156	**INFRARED LATTICE ABSORPTION IN IONIC AND HOMOPOLAR CRYSTALS.** LAX, M. BURSTEIN, E. PHYS REV 97 1 39-52 1955
33158	**INFRARED LATTICE ABSORPTION BANDS IN GERMANIUM, SILICON, AND DIAMOND.** COLLINS R J FAN H Y PHYS REV 93 4 674-18 1954
33164	**THE ABSORPTION SPECTRA OF ZINCSULPHIDE AND WILLEMITE.** GISOLF J H DE GROOT W KROGER F A PHYSICA 8 7 805-9 1941
33165	**THE REFLECTING POWERS OF ROUGH SURFACES AT SOLAR WAVE-LENGTHS.** BECKETT H E PROC PHYS SOC /LONDON/ 43/PT 3/ 227-41 1931
33171	**EFFECTS OF TITANIUM DIOXIDE IN GLASS. IV.** BEALS M D STRIMPLE J H GLASS IND 44 12 679-83 694 1963 BC 63 1235
33172	**ABSORPTION SPECTRA OF CATIONS IN ALKALI-SILICATES GLASSES OF HIGH ULTRA-VIOLET TRANSMISSION.** SMITH H L COHEN A J PHYS CHEM GLASSES 4 5 173-87 1963 BC 63 1244
33174	**INVESTIGATION OF THE THERMAL CONDUCTIVITY OF SOLIDS IN THE INTERVAL FROM 80 TO 500 K.** KAMILOV I K INSTR EXPTL TECH /USSR/ 3 583-7 1962 (ENGLISH TRANSL. OF PRIBORY I TEKH. EKSPER., USSR, (3), 176-9, 1962; FOR ORIGINAL SEE T30911)
33181	**THE DIRECTIONAL REFLECTION FACTOR OF COAL AND ITS RELATION TO VISION AND ILLUMINATION IN MINES.** SHARPLEY FORBES W PROC S WALES INST ENG 55 158-69 1939
33182	**HEAT TRANSFER FROM A GAS STREAM TO A BED OF BROKEN SOLIDS.** FURNAS C C TRANS AM INST CHEM ENGRS 24 142-93 1930
33184	**THE EMISSION SPECTRUM OF WELSBACH MANTLES.** RUBENS, H. VERHANDL DEUT. PHYS. GES. 7, 346-9, 1905. (VERHANDL GES. DEUT. NATURFORSCHR. ARZTE, LEIPZIG, 2, 44-6, 1906)
33185	**NEW RESULTS CONCERNING THE MEASUREMENT OF THERMAL CONDUCTIVITY OF CRYSTALS AND CRYSTALLINE MATERIALS.** EUCKEN A Z KRISTALLOGRAFIA 66 444-6 1928
33208	**NEED FOR MORE REFRACTORY HEAT INSULATORS. PROPOSED CONDUCTOMETERS FOR MEASURING THERMAL CONDUCTIVITY.** PIKE ROBERT D J AM CERAM SOC 5 554-63 1922
33214	**RECENT DEVELOPMENTS IN GLASS RESEARCH.** SHAVER, W. W. AMERICAN SOCIETY FOR TESTING AND MATERIALS 243 43-8 1958 (ASTM-STIP-243)

TPRC Number	Bibliographic Citation
33218	**ON OPTICAL TEMPERATURE MEASUREMENT.** PIRANI M V VERHANDL DEUT PHYS GES 13 19-25 1911
33219	**APPARATUS FOR THE STUDY OF HEAT-INSULATING MATERIALS FOR USE AT LOW AND HIGH TEMPERATURES** GRIFFITHS, E. TRANS WORLD POWER CONF 1 366-71 1936
33222	**INVESTIGATION OF DIFFUSE REFLECTION BY MEANS OF POLARIZED LIGHT. PARTS I AND II.** IVANOV A P TOPORETS A S ZHUR TEKH FIZ 26 623-30 631-5 1956 (FOR ENGLISH TRANSLATION SEE T28573)
33228	**REFLECTION MEASUREMENTS ON ABSORBENT CRYSTALS, WITH PARTICULAR CONSIDERATION OF MINERAL ORE.** CISSARZ ARNOLD Z KRISTALLOGRAPHIE 82 438-50 1932
33230	**THE REFLECTION SPECTRUM OF FUSED QUARTZ AT 9 MICRONS.** BRUGEL W Z PHYSIK 128 255-9 1950
33277	**DIAMOND. SYNTHESIS AND PROPERTIES.** WEINBERGER, T. WAREKOIS, E. P. MASSACHUSETTS INST. OF TECHNOLOGY 1-17, 1963. (AD 418480)
33279	**DESIGN AND FABRICATION OF OPTICAL FILTERS FOR LASER FREQUENCY.** HOLMES, S. J. BOWMAN, N. SPECTROLAB., SYLMAR, CA. 1-12, 1963. (AD 419554)
33283	**THE SHIFT OF INFRARED REFLECTION MAXIMA WITH TEMPERATURE.** PFUND A H J OPT SOC AM 15 2 69-73 1927
33285	**THE ULTRAVIOLET, VISIBLE AND INFRARED REFLECTIVITIES OF SNOW, SAND AND OTHER SUBSTANCES.** HULBURT E O J OPT SOC AM 17 23-5 1928
33296	**REFLECTION-TRANSMISSION RELATIONSHIPS IN SHEET MATERIALS.** LAUNER HERBERT F J OPT SOC AM 32 84-93 1942
33298	**REFLECTION FACTORS OF SOME MATERIALS USED IN SCHOOL ROOMS.** MOON, P. J OPT SOC AM 32 243-6 1942
33304	**THE TRANSMISSION OF POWDER FILMS IN THE INFRA-RED)** HENRY R L J OPT SOC AM 38 9 775-89 1948
33315	**REFLECTION-REDUCING COATINGS OF MAGNESIUM FLUORIDE AND LITHIUM FLUORIDE IN THE NEAR INFRARED.** JENNESS, J. R., JR. J OPT SOC AM 46 3 157-9 1956
33320	**RESEARCH ON LOW-DENSITY THERMAL INSULATION MATERIALS FOR USE ABOVE 3000 F.** STYHR, K. H. NATIONAL BERYLLIA CORP., HASKELL, N. J. 1-23, 1964. (NASA-CR-53201, N64-17591)
33360	**EMITTANCE OF ZIRCONIA TO 4600 F.** COX, R. L. BOEING CO., SEATTLE, WA. 1-53, 1962. (3-14000/2R19, AD 434003)
33388	**REFLECTIVE COATINGS ON POLYMERIC SUBSTRATES.** BELSER, R. B. CARITHERS, M. D. AERONAUTICAL SYSTEMS DIVISION 1-101, 1962. (ASD-TDR-61-151 PT 2, AD 286836)
33393	**APPARATUS FOR RAPID WIDE-TEMPERATURE THERMOPHYSICAL TESTS OF THERMALLY INSULATING AND SEMICONDUCTING MATERIALS /DYNAMIC ALPHA LAMBDA-CALORIMETER/.** KUREPIN V V PLATUNOV E S IZV VYSSHIKH UCHEBN ZAVEDENII PRIBOROSTR 5 119-26 1961
33409	**METALS AND CERAMICS DIVISION ANNUAL PROGRESS REPORT FOR PERIOD ENDING MAY 31, 1963.** OAK RIDGE NATIONAL LAB., OAK RIDGE, TENN. OAK RIDGE NATIONAL LABORATORY, TENNESSEE 1-268, 1963. (ORNL-3470)
33448	**DEVELOPMENT OF A EUCRYPTITE-GLASS CERAMIC HAVING HIGH THERMAL STRENGTH, AND AN INVESTIGATION OF THE MECHANICAL, THERMAL AND ELECTRICAL PROPERTIES AS A FUNCTION OF THE VITRIFYING AND DEVITRIFYING TEMPERATURES. PT. II. THE INFLUENCE OF DEVITRIFICATION-TIME AND-TEMPERATURE ON THE MECHANICAL, THERMAL AND ELECTRICAL PROPERTIES OF A GLASS-CERAMIC CONTAINING EUCRYPTITE.** BAUM, W. GLASTECH BER /WEST GERMANY/ 36 12 468-81 1963 (FOR ENGLISH TRANSLATION SEE T33449)
33449	**DEVELOPMENT OF A EUCRYPTITE-GLASS CERAMIC HAVING HIGH THERMAL STRENGTH, AND AN INVESTIGATION OF THE MECHANICAL, THERMAL AND ELECTRICAL PROPERTIES AS A FUNCTION OF THE VITRIFYING AND DEVITRIFYING TEMPERATURES. PT. II. THE INFLUENCE OF DEVITRIFICATION-TIME AND-TEMPERATURE ON THE MECHANICAL, THERMAL AND ELECTRICAL PROPERTIES OF A GLASS-CERAMIC CONTAINING EUCRYPTITE.** BAUM, W. SPECIAL LIBRARY ASSOC. 1-40, 1965. (ENGLISH TRANSLATION OF GLASTECH. BER., W. GERMANY, 36 (12), 468-81, 1963; FOR ORIGINAL SEE T33448) (TT65-11743)
33451	**RADIATION MEASUREMENTS OF HOT FLAME GASES WITH THE PHOTOCELL.** MECKE R KEMPTER H Z TECH PHYSIK 21 85-8 1940 (FOR ENGLISH TRANSLATION SEE T33453)
33453	**RADIATION MEASUREMENTS OF HOT FLAME GASES WITH THE PHOTOCELL.** MECKE, R. KEMPTER, H. SPECIAL LIBRARY ASSOC., TRANSLATIONS CENTER 1-7, 1965. (ENGLISH TRANSLATION OF Z. TECH. PHYSIK. 21, 85-88, 1940; FOR ORIGINAL SEE T33451) (TT65-13000)
33458	**HIGHLY REFLECTING COATINGS FOR THE SPACE ENVIRONMENT.** MILLER, C. D. RECHTER, H. L. TOMPKINS, E. H. SYMP. ON EFFECTS OF SPACE ENVIRONMENT ON MATERIALS, 3RD 1-21, 1962. (N63-10917)
33460	**THE EMISSION OF RADIATION BY TRANSPARENT MATERIALS.** GARDON, R. PROC. CONF. ON RADIATIVE TRANSFER FROM SOLID MATERIALS 8-23 1962
33467	**THE DEVELOPMENT OF HIGH PERFORMANCE INSULATION SYSTEMS. APPENDIX B.** LITTLE ARTHUR D, INC., CAMBRIDGE, MASS. LITTLE ARTHUR D, INC., CAMBRIDGE, MASS. 1-97, 1963. (C-62458, AD 410053)
33475	**THE THERMAL PROPERTIES OF POWDER INSULATORS IN THE TEMPERATURE RANGE 300-4 K.** CLINE, D. KROPSCHOT, R. H. PROC. CONF. RADIATIVE TRANSFER FROM SOLID MATERIALS 61-81 1962
33507	**USE OF PLUTONIUM AS A FUEL IN NUCLEAR REACTORS.** CENTRE BELGE DE L'ENERGIE NUCLAIRE CENTRE DETUDE DE LENERGIE NUCLEAIRE ET SOCIETE, BELGE 1-68, 1963. (EURAEC-790)
33523	**PRELIMINARY REPORT ON THE THERMODYNAMIC PROPERTIES OF SELECTED LIGHT-ELEMENT AND SOME RELATED COMPOUNDS.** NATIONAL BUREAU OF STANDARDS NATIONAL BUREAU OF STANDARDS, WASHINGTON D. C. 1-84, 1963. (NBS REPT. 8033, AD-428625)
33544	**NEW FORMULAS FOR CALCULATING CERTAIN PHYSICAL PROPERTIES OF SUBSTANCES.** OSHCHERIN B N POROSHKOVAYA METALLURGIYA 2 6 3-11 1962 (FOR ENGLISH TRANSLATION SEE T33545)

TPRC Number	Bibliographic Citation
33545	NEW FORMULAS FOR CALCULATING CERTAIN PHYSICAL PROPERTIES OF SUBSTANCES. OSHCHERIN B N FOREIGN TECHNOLOGY DIVISION 1-18, 1963. (ENGLISH TRANSLATION OF POROSHKOVAYA METALLURGIYA, 2 (6), 3-11, 1962; FOR ORIGINAL SEE T33544) (AMMETSOC-T-R-410, AD 602191)
33546	THERMAL INSULATION OF ELECTRIC FURNACES. HARTMANN M L WESTMONT O B TRANS AM ELECTROCHEM SOC 53 117-40 1926
33568	OPTICAL SPECTRUM OF THE ION CO /PLUS 2 IONS/ IN A SINGLE CRYSTAL OF SPINEL. DRIFFOED MAURICE RIGNY PAUL COMPT REND 263 B 2 180-3 1966
33613	THERMAL CONDUCTIVITY MEASUREMENTS ON REFRACTORY BUILDING MATERIALS. FRITZ-SCHMIDT MARG GERHLHOFF G GLASTECH BER 8 4 206-28 1930
33616	NUMERICAL SUMMARY OF PHYSICAL AND CHEMICAL PROPERTIES OF MINERALS. C. THE HEAT CONDUCTION IN MINERALS, ROCKS AND THE SYNTHETIC PREPARED MATERIALS OF CORRESPONDING COMPOSITION. SCHULZ, K. FORTSCHR MINERAL(KRISTALLOGRAPHIE PETROGRAPHIE) 9 221-411 1924
33623	DETERMINATION OF REFLECTION FACTORS OF SOME STEEL MIRRORS. CLAVIER J REV OPTIQUE 8 379-91 1929
33626	MEASUREMENTS ON ROCK SALT IN THE INFRARED FOR TESTING THE DISPERSION THEORY. CZERNY M Z PHYSIK 65 600-31 1930
33692	USE OF PLUTONIUM AS FUEL IN NUCLEAR REACTORS. UNITED STATES-EURATOM JOINT RESEARCH AND DEVELOP. UNITED STATES-ERUATOM JOINT RESEARCH AND DEVELOPMENT PROGRAM. 1-63, 1964. (EURAEC-965)
33720	SPEED OF COOLING OF SINTER IN A SINTERING PAN. BRATCHIKOV S G KHUBOROZHKOV I V IZV VYSSHIKH UCHEBN ZAVEDENII CHERN MET 4 32-6 1966
33745	NEW MATERIALS IN ENGINEERING. TROSTYANSKAYA YE B KOLACHEV B A SILVESTROVICH S I NEW MATERIALS IN ENGINEERING. 1-656 1962 (FOR ENGLISH TRANSLATION SEE T033746)
33746	NEW MATERIALS IN ENGINEERING. TROSTYANSKAYA YE B KOLACHEV B A SILVESTROVICH S I FOREIGN TECHNOLOGY DIVISION 1-942, 1966. (ENGLISH TRANSLATION OF NOVYYE MATERIALY V TEKHNIKE (MOSCOW), 656 PP., 1962; FOR ORIGINAL SEE T033745) (FTD-HT-66-269, TT67-60219, N67-20661, AD 643685)
33774	INFRARED REFLECTANCE SPECTRA OF IGNEOUS ROCKS, TUFFS, AND RED SANDSTONE FROM 0.5 TO 22 MICRONS. HOVIS W A JR CALLAHAN W R J OPT SOC AM 56 5 639-43 1966 CA 65 182
33775	SPECTROSCOPY IN THE 5 TO 400 WAVENUMBER REGION WITH THE GRUBB PARSONS INTERFEROMETRIC SPECTROMETER. WHEELER R G HILL J C J OPT SOC AM 56 5 657-65 1966
33801	HEAT TRANSFER IN SOLIDS. BERMAN R BULL INST INTERN FROID ANNEXE 2 23-34 1965
33844	ELLIPSOMETER STUDIES OF SILVER SINGLE CRYSTALS. FROMHOLD, A. T., JR. CORNELL UNIV., PH.D. THESIS 1-187, 1961. (UNIV. MICROFILM NO. 62-159)
33849	THE EFFECTS OF A SIMULATED PROTON SPACE ENVIRONMENT ON THE ULTRAVIOLET TRANSMITTANCE OF OPTICAL MATERIALS BETWEEN 3000 AND 1050 A. SACHER, P. A. GODDARD SPACE FLIGHT CTR., GREENBELT, MD 1-17, 1967. (NASA-TM-X-55911, X-622-67-416, N67-36569)
33854	PHYSICAL AND OPTICAL PROPERTIES OF PROJECTION SCREENS. KLAIBER, R. J. NAVAL TRAINING DEVICE CENTER, ORLANDO, FL 1-45, 1966. (AD 647132)
33878	THE OPTICAL CONSTANTS OF SOME OXIDE GLASSES IN THE STRONG ABSORPTION REGION. CLEEK G W APPL OPT 5 5 771-5 1966
33879	SKY LUMINANCES AND THE DIRECTIONAL LUMINOUS REFLECTANCES OF OBJECTS AND BACKGROUNDS FOR A MODERATELY HIGH SUN. GORDON J I CHURCH P V APPL OPT 5 5 793-801 1966
33880	ATMOSPHERIC PROPERTIES AND REFLECTANCES OF OCEAN WATER AND OTHER SURFACES FOR A LOW SUN. BOILEAU A R GORDON J I APPL OPT 5 5 803-13 1966
33881	OPTIMUM WAVELENGTH INTERVALS FOR SURFACE TEMPERATURE RADIOMETRY. HOVIS W A JR APPL OPT 5 5 815-18 1966
33888	A LUMMER-GEHRKE PLATE-TYPE POLARIZER FOR A NEODYMIUM-GLASS LASER. NESTRIZHENKO YA A OPT SPECTROSC 26 6 542-4 1969 (ENGLISH TRANSLATION OF OPTIKA I SPEKTROSKOPIYA, 26 (6), 1000-5, 1969; FOR ORIGINAL SEE T033889)
33889	A LUMMER-GEHRKE PLATE-TYPE POLARIZER FOR A NEODYMIUM-GLASS LASER. NESTRIZHENKO YA A OPTIKA I SPEKTROSKOPIYA 26 6 1000-5 1969 (FOR ENGLISH TRANSLATION SEE T033888)
33924	COLOR CENTERS AND HYDROGEN-DEUTERIUM EXCHANGE IN GAMMA-IRRADIATED SILICA-ALUMINA CATALYSTS. SHIPMAN G F J PHYS CHEM 70 4 1120-5 1966
33956	INFRARED STUDY OF THE SURFACE OF RUTILE. LEWIS K E PARFITT G D TRANS FARADAY SOC 62 204-14 1966
33967	ULTRASONIC ATTENUATION IN INSULATORS AT ROOM TEMPERATURE. OLIVER D W SLACK G A J APPL PHYS 37 4 1542-8 1966
34027	STRUCTURAL AND DIELECTRIC INVESTIGATION OF THE PBTIO3-BAZRO3 SYSTEM. HARRIS N H TENNERY V J J AM CERAM SOC 50 8 404-7 1967
34077	ALTERATION OF LUNAR OPTICAL PROPERTIES: AGE AND COMPOSITION EFFECTS. ADAMS, J. B. MC CORD, T. B. SCIENCE 171, 567-71, 1971.
34088	ENERGY /HEAT/ ANALYSIS OF GREISENIZATION. DUDAREV A N SOTNIKOV V I GEOL I GEOFIZ AKAD NAUK SSSR SIBIRSK OTD 5 49-58 1965 CA 63 14573
34089	THERMAL CONDUCTIVITIES OF SOME LEAD AND BISMUTH GLASSES. VAN VELDEN, P. F. GLASS TECHNOL. 6 (5), 166-9, 1965.
34090	INFRARED REFLECTIVITY OF IRON OXIDE MINERALS. HOVIS, W. A., JR. ICARUS 4 (4), 425-30, 1965.
34098	THERMAL CONDUCTIVITY OF A BARIUM TITANATE CERAMIC AND SOLID SOLUTIONS BASED ON IT. DIMAROVA E N POPLAVKO YU M IZV AKAD NAUK SSSR SER FIZ 29 6 985-7 1965 CA 63 15587 (FOR ENGLISH TRANSLATION SEE T035752)
34116	HEAT CAPACITY OF COALS. AGROSKIN A A GONCHAROV E I KOKS I KHIM 7 8-13 1965 CA 63 12929

TPRC Number	Bibliographic Citation
34118	**ELECTRICAL AND OPTICAL PROPERTIES OF A SODALITE SINGLE CRYSTAL.** SHALDIN YU V MELNIKOV O K NABATOV V V PISAREVSKII YU V KRISTALLOGRAFIYA 10 4 574-6 1965 CA 63 12443 (FOR ENGLISH TRANSLATION SEE T060093)
34127	**EFFECT OF TITANIUM COMPOUNDS ON THE THERMAL EXPANSION OF STANDARD ELECTRO CORONDUM.** WENZEL K SILIKATTECHN 11 4 164-6 1960
34128	**THE DISSOCIATION-TEMPERATURE OF BROWN HORNBLENDE AND ITS RAPID EXPANSION AT THIS TEMPERATURE.** KOZU S YOSHIKI B SCI REPT TOHOKU UNIV 3 3 2 107-17 1927
34129	**NOTE ON THE STUDY OF THE TRANSFORMATION OF COMMON HORNBLENDE INTO BASALTIC HORNBLENDE AT 750 C.** KOZU S YOSHIKI B KANI K SCI REPT TOHOKU IMP UNIV 3 3 143-59 1927
34136	**RUSSIAN DATA ON SPECTRAL REFLECTANCE OF VEGETATION, SOIL, AND ROCK TYPES.** STEINER D GUTERMANN T ZURICH UNIV U S GOVERNMENT PRINTING OFFICE 1-232, 1966. (AD 614032, N65-33578)
34137	**EMISSIVITY OF GRANULAR SURFACES AT RESONANCE FREQUENCIES.** BLOCK, M. J. PROC. SYMP. REMOTE SENSING ENVIRONMENT, 3RD 435-40, 1965. (AD 614032, N65-33578)
34138	**THE CONSEQUENCES OF TERRESTRIAL SURFACE INFRARED EMISSIVITY.** BUETTNER K J K KERN C D CRONIN J F PROC. SYMP. REMOTE SENSING ENVIRONMENT, 3RD 549-61, 1965. (AD 614032, N65-33586)
34144	**KAOLINS, EFFECT OF FIRING TEMPERATURES ON SOME OF THEIR PHYSICAL PROPERTIES.** HEINDL R A PENDERGAST W L MONG L E J RESEARCH NATIONAL BUREAU STANDARDS 8 199-215 1932 (NBS-RP-410)
34147	**OPTICAL AND SPECTRAL PROPERTIES OF RARE EARTH OXIDES IN INORGANIC GLASSES.** KAN F-H CHIANG C-H TSAI Y-S SCI SINICA /PEKING/ 14 8 1159-70 1965 CA 63 14511
34151	**INFLUENCE OF TH O2, CD O, LA2 O3, AND ZR O2 ON PHYSICOCHEMICAL PROPERTIES OF OPTICAL GLASS.** SUSSER V FANDERLIK I SKLAR KERAM 13 265-8 1963 CA 63 15993 (FOR FRENCH TRANSLATION SEE T046011)
34177	**FINE STRUCTURE IN THE DIRECT ABSORPTION EDGE OF DIAMOND.** ROBERTS R A ROESSLER D M WALKER W C PHYS REV LETTERS 17 6 302-4 1966
34183	**INFRARED-SPECTROSCOPIC INVESTIGATION OF THE VITRIFICATION AND CRYSTALLIZATION OF LEAD GLASS WITH A COMPOSITION CORRESPONDING TO THE COMPOUND PB O.SI O2.** SMIRNOVA E V IZV AKAD NAUK SSSR NEORG MATER 4 7 1124-8 1968 (FOR ENGLISH TRANSLATION SEE T52757)
34262	**A STUDY OF THE REFLECTION AND POLARIZATION CHARACTERISTICS OF SELECTED NATURAL AND ARTIFICIAL SURFACES.** COULSON K L GRAY E L BOURICIUS G M B GENERAL ELECTRIC CO SPACE SCIENCES LAB., PHILADELPHIA 1-150, 1965. (R65SD4, AD 619032, N65-32209)
34293	**A LOW-TEMPERATURE HEAT CAPACITY STUDY OF THE MICROCHEMICAL HETEROGENEITY OF SODIUM SILICATE AND SODIUM BOROSILICATE GLASSES.** TARASOV V V TURDAKIN V A YUNITSKII G A ZHDANOV V M ZH FIZ KHIM 41 2 430- 1967 (FOR ENGLISH TRANSLATION SEE T34294)
34294	**A LOW-TEMPERATURE HEAT CAPACITY STUDY OF THE MICROCHEMICAL HETEROGENEITY OF SODIUM SILICATE AND SODIUM BOROSILICATE GLASSES.** TARASOV V V TURDAKIN V A YUNITSKII G A ZHDANOV V M RUSS J PHYS CHEM 41 2 213-16 1967 (ENGLISH TRANSLATION OF ZH. FIZ. KHIM. 41 (2), 430- , 1967; FOR ORIGINAL SEE T34293)
34296	**EFFECT OF DESTABILIZATION ON THE THERMAL DIFFUSIVITY OF ZIRCONIA-MAGNESIA COMPOSITIONS AT ELEVATED TEMPERATURES.** HOSKYNS, W. R. ALFRED UNIVERSITY, M.S. THESIS 1-74 1964
34325	**POLARIZERS FOR THE EXTREME ULTRAVIOLET.** HUNTER, W. R. US NAVAL RESEARCH LAB., WASHINGTON, D. C. 1-7, 1964. (AD 609535, N65-19059, N65-19061)
34350	**RADIATIVE THERMAL CONDUCTIVITY IN PLANETARY INTERIORS.** ARONSON J R ECKROAD S W EMSLIE A G MC CONNEL R K JR VON THUNA P C NATURE /LONDON/ 216 1096-7 1967
34391	**PROBLEMS OF INFRARED ATMOSPHERIC SPECTROSCOPY RELATED TO THE SATELLITE DETERMINATION OF TEMPERATURE OF UNDERLYING SURFACE.** KONDRATYEV K YA FOREIGN TECHNOLOGY DIVISION 1-19, 1969. (ENGLISH TRANSLATION OF FIZ. ATM. I OKEANA 5 (6), 616-30, 1969; FOR ORIGINAL SEE T46008) (NASA-TT-F-12540, N69-35644)
34393	**ELECTROMAGNETIC AND THERMAL PROPERTIES OF THE MOONS SURFACE.** LAX, B. MASSACHUSETTS INST. OF TECH., CAMBRIDGE, MA. 1-16, 1965. (NASA-CR-69878, N66-16174)
34399	**TRANSMISSION OF INDIAN MICA BETWEEN 300 AND 1,000 MICRONS AND ITS VISUAL CLASSIFICATION.** DHAR R N DAS S R NATURE /LONDON/ 209 185-6 1966
34401	**TRANSIENT METHODS OF MEASURING THERMAL PROPERTIES OF SOLIDS.** DAS M B HOSSAIN M A BRIT J APPL PHYS 17 1 87-97 1966 CA 64 7396
34423	**THERMAL AND ELECTRICAL PROPERTIES OF NATURAL MONO-CRYSTALS OF LEAD SULPHIDE.** FARAG B S SMIRNOV I A YOUSEF Y L PHYSICA 31 11 1673-80 1965 CA 64 99
34479	**OPTICAL SPECTRA OF THALLIUM IN GLASSES AS RELATED TO CRYSTALS AND SOLUTIONS.** GHOSH A K J CHEM PHYS 44 2 535-40 1966
34538	**INFRARED SPECTRAL REFLECTANCE OF SOME COMMON MINERALS.** HOVIS W A JR APPL OPT 5 2 245-8 1966
34568	**CHANGES IN AXIAL RATIO, IN INTERFACIAL ANGLE AND IN VOLUME OF CALCITE, CAUSED BY HEATING.** KOZU S MASUDA M-I UEDA J SCI REPT TOHOKU IMP UNIV 3 3 247-58 1929
34590	**VALUE OF HEAT CAPACITY IN APPRAISING THE ROLE OF THE ENCLOSED ROCKS DURING MINERALIZATION.** DUDAREV, A. N. GEOL. RUDN. MESTOROZHD. 7 (4), 57-62, 1965.
34592	**WATER CONTENT OF CONTAINER GLASSES.** FENSTERMACHER J E LESSER R C RYDER R J GLASS IND 46 9 518-21 1965 CA 64 1781
34599	**HEAT TRANSFER IN A DISPERSE HEAT-INSULATING SYSTEM.** KOSTYLEV V M NABATOV V G INZH FIZ ZH AKAD NAUK BELORUSSK SSR 9 3 877-83 1965 CA 64 1676 (FOR ENGLISH TRANSLATION SEE T47289)
34629	**COMPARISON OF THE INFRARED REFLECTION SPECTRA OF MINERAL POWDERS COMPRESSED TO THOSE OF CORRESPONDING MONOCRYSTALS.** VERGNOUX A-M BRUNEL R COMPT REND 265 B 20 1097-9 1967

TPRC Number	Bibliographic Citation
34645	PHOTOELECTRIC DETERMINATION OF TRANSMITTANCE OF WELDERS GLASSES. WEBER K KEM IND /ZAGREB/ 14 2 69-74 1965 CA 63 17648
34653	RADIOPHOTOLUMINESCENT CENTERS IN SILVER-ACTIVATED PHOSPHATE GLASS. YOKOTA R IMAGAWA H J PHYS SOC JAPAN 23 5 1038-48 1967
34683	PROCEDURES FOR THE PRECISE DETERMINATION OF THERMAL RADIATION PROPERTIES. RICHMOND J C DUNN S T DE WITT D P HAYES, W. D., JR. NATIONAL BUREAU OF STANDARDS 1-62, 1965. (NBS-TN-267, ML-TDR-64-257(PT II), AD 478229, AD 628586, N66-22418)
34689	DILATOMETRIC MEASUREMENTS OF HIGH PRECISION. AMATUNI, A. N. MALIUTINA, T. I. PEVCHENKO, T. I. TEPLOFIZ. SVOISTVA TVERD. TEL. VYS. TEMP. TR. VSES. KONF., 3RD 3-11PP., 1971.
34703	THERMAL EXPANSION OF SOME SILICATES OF ELEMENTS IN GROUP II OF THE PERIODIC SYSTEM. GELLER R F INSLEY H J RESEARCH NATIONAL BUREAU STANDARDS 9 35-46 1932 (NBS-RP-456)
34704	SOLID STATE REACTIONS AND THERMAL EXPANSION OF PHASES IN SYSTEMS CONTAINING ZINC OXIDE AND SELECTED OXIDES OF THE GROUP IV AND V ELEMENTS. BROWN, J. J., JR. PENNSYLVANIA STATE UNIVERSITY, PHILADELPHIA, PH. D. THESIS 1-140, 1964. (UNIV. MICROFILM NO. 65-6725)
34763	DETERMINATION OF THE THERMAL CONDUCTIVITY OF SOLIDS. FRITZ W BODE K H CHEM INGR TECH 37 11 1118-24 1965 CA 64 2768
34765	INTEGRATED CIRCUIT THERMAL STUDY. FINAL REPT. BAUM, J. R. MOTOROLA, INC., SCOTTSDALE, ARIZONA 1-160, 1965. (WF-2990-1, NADC-AE-6516, AD 474696)
34768	CHEMICAL STRENGTHENING OF CERAMIC MATERIALS. KIRCHNER, H. P. GRUVER, R. M. WALKER, R. F. LINDEN LABS., INC., STATE COLLEGE, PA. 1-125, 1966. (AD 634361)
34769	DETERMINING THE EMISSIVITY OF MATERIALS ON THE BASIS OF THEIR INFRARED REFLECTION SPECTRA. KROPOTKIN M A KOZYREV B P INZHEN-FIZ ZHUR /USSR/ 7 9 108-12 1964 (FOR ENGLISH TRANSLATION SEE T34770)
34770	DETERMINING THE EMISSIVITY OF MATERIALS ON THE BASIS OF THEIR INFRARED REFLECTION SPECTRA. KROPOTKIN M A KOZYREV B P SPECIAL LIBRARY ASSOC., TRANSLATIONS CENTER 1-14, 1965. (ENGLISH TRANSLATION OF INZH.-FIZ. ZH., USSR 7, (9), 108-12, 1964; FOR ORIGINAL SEE T34769) (JPRS-R-5536-D, TI-65-63682, FOREIGN TEXT INCL.)
34784	SCATTERING OF PHONONS BY CATION VACANCIES IN SPINELS. VISHNEVSKII I I SKRIPAK V N FIZ TVERDOGO TELA 7 10 2925-9 1965 (FOR ENGLISH TRANSLATION SEE T34785)
34785	SCATTERING OF PHONONS BY CATION VACANCIES IN SPINELS. VISHNEVSKII I I SKRIPAK V N SOVIET PHYSICS-SOLID STATE 7 10 2374-7 1966 (ENGLISH TRANSLATION OF FIZ. TVERD. TELA 7 (10), 2925-9, 1965; FOR ORIGINAL SEE T34784)
34810	THE OPTICAL TRANSMISSION OF SOME QUARTZ SAMPLES IN THE ULTRAVIOLET AND THE 3 MICRON REGION. SAKSENA B D J PHYS CHEM SOLIDS 27 2 465-8 1966 CA 64 15193
34814	OPTICAL REFLECTORS FOR USE IN INTERNAL SAMPLE AQUEOUS CHERENKOV COUNTERS. STRINDEHAG O M REV SCI INSTR 37 3 344-9 1966
34819	DEVELOPMENT OF SPACE-STABLE THERMAL-CONTROL COATINGS /PAINTS WITH LOW SOLAR ABSORPTANCE/EMITTANCE RATIOS/. ZERLAUT, G. A. GILLIGAN, J. E. HARADA, Y. IIT RESEARCH INST., CHICAGO, IL. 1-72, 1964. (NASA-CR-67295, IITRI-C6014-18, N65-35122)
34833	OPTICAL ABSORPTION AND DIFFUSE REFLECTANCE OF POWDERS. JOHNSON P O J APPL PHYS 35 2 334-6 1964
34834	A CORRECTION FOR THE NET RADIOMETER REFLECTION ERROR. ANALYSIS OF MEASURED NET RADIATION VALUES FOR CANADA. FINKE, D. D. WENDLAND, W. M. UNIV. OF WISCONSIN, MADISON, DEPT. OF METEOROLOGY 1-86, 1965. (AD 618721, N65-35196, TR19)
34835	THERMAL RADIATION OF COMPLEX CERAMIC SOLIDS. GRENIS, A. F. ARMY MATERIALS RESEARCH AGENCY 1-19, 1965. (AMRA-TR-65-14, AD 620004, N66-10374)
34848	SPECTRAL REFLECTANCE AND ALBEDO MEASUREMENTS OF THE EARTH FROM HIGH ALTITUDES. BAND, H. E. BLOCK, L. C. AIR FORCE CAMBRIDGE RESEARCH LABS., L. G. HANSCOM FIELD, MA. 1-30, 1965. (AFCRL-65-674, AD 623876, N66-14726)
34852	TOTAL ENERGY AND ENERGY DISTRIBUTION IN A LASER CRYSTAL DUE TO OPTICAL PUMPING, AS CALCULATED BY THE MONTE CARLO METHOD. SKINNER D R TREGELLAS-WILLIAMS J AUSTRALIAN J PHYS 19 1 1-18 1966
34861	THE APPLICATION OF HIGH TEMPERATURE RADIATIVE THERMAL CONDUCTIVITY OF MINERALS AND ROCKS TO A MODEL OF LUNAR VOLCANISM. LITTLE ARTHUR D, INC., CAMBRIDGE, MASS. LITTLE ARTHUR D, INC., CAMBRIDGE, MASS. 1-67, 1967. (NASA-CR-916)
34910	ANALYSIS OF TRANSFER PROCESSES AND RAPIDITY OF CHANGES IN TEMPERATURE WITH REFERENCE TO HEIGHT OF BLAST FURNACE. SUCHKOV V D KITAEV B I IZV VYSSHIKH UCHEBN ZAVEDENII CHERNAYA MET 2 41-7 1966
34913	EFFECT OF IONIZING RADIATION ON RUBY. FORESTIERI, A. F. GRIMES, H. H. LEWIS RESEARCH CENTER, CLEVELAND, OHIO 1-9, 1966. (NASA-TN-D-3379)
34931	THE OPTICAL CONSTANTS OF CALCITE IN THE REGION OF THE FIRST HARMONIC. MATOSSI F HOHLER V Z NATURFORSCH 22 A 10 1516-24 1967
34962	OPTICAL AND PARAMAGNETIC PROPERTIES OF IRON CENTERS IN QUARTZ. LEHMANN G MOORE W J J CHEM PHYS 44 5 1741-5 1966
34965	ULTRAVIOLET ABSORPTION AND EXCITATION SPECTRUM OF RUBY AND SAPPHIRE. LOH E J CHEM PHYS 44 5 1940-5 1966
34986	FOULING OF STEAM BOILER WALLS. GOLOVIN V N TEPLOENERGETIKA 12 4 42-7 1965 (FOR ENGLISH TRANSLATION SEE T34987)
34987	FOULING OF STEAM BOILER WALLS. GOLOVIN V N THERMAL ENGNG 12 4 55-62 1965 (ENGLISH TRANSLATION OF TEPLOENERGETIKA 12 (4), 42-7, 1965; FOR ORIGINAL SEE T34986)
35027	THE INFLUENCE OF LONGWAVE REFLECTIVITY OF NATURAL SURFACES ON THE MEASUREMENTS OF SURFACE TEMPERATURE USING RADIOMETERS. LORENZ, D. U.S. DEPARTMENT OF ARMY 1-51, 1965. (AD 464604)

TPRC Number	Bibliographic Citation
35028	OPTICAL PROPERTIES AND THERMAL CONDUCTIVITY OF ALUMINUM OXIDE. NEUBERGER, M. ELECTRONIC PROPERTIES INFORMATION CENTER, HUGHES AIRCRAFT CO., CULVER CITY, CA. 1-20, 1965. (EPIC-S-6, AD 464823, N65-34038)
35041	COMPARATIVE METHOD AND CHOICE OF STANDARDS FOR THERMAL CONDUCTIVITY DETERMINATIONS. MIRKOVICH V V J AM CERAM SOC 48 8 387-91 1965 CA 63 17179
35043	PHASE RELATIONS AND GLASS FORMATION IN THE SYSTEM PBO-GE O2. PHILLIPS B SCROGER M G J AM CERAM SOC 48 8 398-401 1965
35068	A DYNAMICAL METHOD FOR THE DETERMINATION OF THE THERMAL CONDUCTIVITY OF POOR CONDUCTORS. MERLIN M NUOVO CIMENTO II 6 9 4 283-9 1949
35073	SPECTROPHOTOMETRIC CHARACTERISTICS OF MILKY GLASS MS-14. TOPORETS A S OPT I SPEKTROSKOPIYA 18 4 704-7 1965 CA 63 2720 (FOR ENGLISH TRANSLATION SEE T36833)
35118	HEAT FLOW IN THE TRANSVAAL AND ORANGE FREE STATE. CARTE A E PROC PHYS SOC /LONDON/ 67 B 664-72 1954
35127	OPTICAL ABSORPTION FEATURES ASSOCIATED WITH PARAMAGNETIC NITROGEN IN DIAMOND. DYER H B RAAL F A DU PREEZ L LOUBSER J H N PHIL MAG 11 763-74 1965 JA 48 244
35143	OPTICAL STUDY OF THE ELECTRONIC STRUCTURE OF DIAMOND. ROBERTS R A WALKER W C PHYS REV 161 3 730-5 1967
35150	THERMAL CONDUCTIVITY OF DIELECTRIC SOLIDS AT HIGH PRESSURE. HUGHES D S SAWIN F PHYS REV 161 3 861-3 1967 CA 67 94603
35226	MOLYBDENUM PHOSPHATE GLASSES CONTAINING AG2 O OR K2 O. PROVANCE J D WOOD D C J AM CERAM SOC 50 10 516-20 1967 CA 67 111080
35227	ZINC TELLURITE GLASSES. REDMAN M J CHEN J H J AM CERAM SOC 50 10 523-5 1967
35229	CUBIC PHASE STABILIZATION OF TRANSLUCENT YTTRIA-ZIRCONIA AT VERY LOW TEMPERATURES. MAZDIYASNI K S LYNCH C T SMITH J S II J AM CERAM SOC 50 10 532-7 1967 CA 67 111137
35230	STRESS DISTRIBUTION IN MULTIPHASE SYSTEMS. I. COMPOSITES WITH PLANAR INTERFACES. OEL H J FRECHETTE V D J AM CERAM SOC 50 10 542-9 1967
35241	THE FLOW OF HEAT THROUGH FURNACE WALLS. SNYDER F T AM ELECTROCHEM SOC 18 235-53 1910
35246	HYBRID EXCITONS IN DIAMOND. PHILLIPS J C PHYS REV 139 A 4 1291-4 1965
35277	II. HIGH THERMAL SHOCK CERAMICS. SMOKE E J SNYDER N H WISELY H R RUH E ILLYN A V EICHBAUM B R N. J. CERAM. RES. STATION 27-42, 1953. (AD 19835, AD-19833)
35347	EFFECT OF POINT DEFECTS ON THE THERMAL CONDUCTIVITY OF SUBSTITUTIONAL SOLID SOLUTIONS WITH CATION VACANCIES. VISHNEVSKII I I SUKHAREVSKII B YA FIZ TVERDOGO TELA /USSR/ 6 7 2168-74 1964 SA 68 15061 (FOR ENGLISH TRANSLATION SEE T035348)
35348	EFFECT OF POINT DEFECTS ON THE THERMAL CONDUCTIVITY OF SUBSTITUTIONAL SOLID SOLUTIONS WITH CATION VACANCIES. VISHNEVSKII I I SUKHAREVSKII B YA SOVIET PHYS SOLID STATE /USA/ 6 7 1708-13 1965 (ENGLISH TRANSLATION OF FIZ. TVERDOGO TELA (USSR), 6 (7), 2168-74, 1964; FOR ORIGINAL SEE 035347)
35357	SKIN AND SUBCUTANEOUS TEMPERATURE CHANGES DURING EXPOSURE TO INTENSE THERMAL RADIATION. STOLWIJK J A J HARDY J D JOHN B PIERCE DEFENSE ATOMIC SUPPORT AGENCY, WASHINGTON, D C 1-33, 1964. (DASA-1566, AD 615477)
35390	A THERMAL COMPARATOR APPARATUS FOR THERMAL-CONDUCTIVITY MEASUREMENTS FROM 50 TO 400 C. KOLLIE T G MC ELROY D L GRAVES R S FULKERSON W INTERNAT SYMPOSIUM ON COMPOUNDS OF INTEREST IN NUCLEAR REACTOR TECHNOLOGY 651-76 1964 MA 32 991
35403	THERMAL CONDUCTIVITY OF HIGH CONDUCTIVITY MATERIALS. I. MEASUREMENT. DAY R K AM CERAM SOC BULL 44 8 608-11 1965
35412	A PROGRAM TO ACQUIRE ENVIRONMENTAL FIELD DATA IN FOUR AREAS IN A HUMID SUBTROPIC ENVIRONMENT. WALKER, J. E. CORNELL AERONAUTICAL LAB., INC., BUFFALO, N Y 1-168, 1969. (CAL-VT-2636-0-1, AFCRL-69-0393, AD 698345)
35413	A THERMOELECTRIC DEVICE FOR MEASURING THERMAL CONDUCTIVITY OF ROCK. KHAN A M FATT I J GEOPHYS RES /USA/ 69 20 4414-16 1964 SA 68 13193
35414	THE DETERMINATION OF INFRARED EMISSIVITIES OF TERRESTRIAL SURFACES. BUETTNERK J K KERN C D J GEOPHYS RES /USA/ 70 6 1329-38 1965 SA 68 18354
35437	THE RADIATION STABILITY OF SOME CERAMIC MATERIALS. BOPP, C. D. SISMAN, O. TOWNS, R. L. OAK RIDGE NATIONAL LABORATORY 66-75, 1957. (TID-7530-PT. 1)
35483	A STUDY OF SOME CERAMIC BODIES OF LOW ABSORPTION MATURING AT TEMPERATURES BELOW 1000 C. GELLER R F EVANS D N J RESEARCH NATIONAL BUREAU STANDARDS 9 473-85 1932 (NBS-RP-483)
35493	SPECIFIC HEAT OF KACHKANAR IRON ORE MATERIALS. BRATCHIKOV, S. G. TR. URAL'SK. POLITEKHN. INST. (137), 77-80, 1964.
35535	ON THE THERMAL CONDUCTIVITY OF POWDER INSULATIONS. EVEREST A GLASER P E WECHSLER A E PROC. INTERN. CONGRESS OF REFRIGERATION, 11TH 1 255-63 1965
35541	THE REFLECTANCE SPECTRA OF SOME HETEROGENEOUS MATERIALS. RICHARDSON, J. S. TEXAS TECHNICAL UNIV., LUBBOCK, TEXAS, M.S. THESIS 1-42 1969
35542	THERMAL DIFFUSIVITY OF AMMONIUM PERCHLORATE. ROSSER, W. A. INAMI, S. H. WISE, H. STANFORD RESEARCH INST., MENLO PARK, CA. 1-16, 1964. (AD-464840, AD-614081, N66-39139)
35550	STUDIES OF THE PHYSICAL CHARACTERISTICS OF PROBABLE LUNAR SURFACE MATERIALS. PART II. GLASER, P. E. AIR FORCE CAMBRIDGE RESEARCH LABORATORIES 1-134, 1964. (AFCRL-64-970(PT. 2), AD 613965)
35554	THERMAL CONDUCTIVITY OF ROCK. MEASUREMENT BY THE TRANSIENT LINE SOURCE METHOD. MAROVELLI, R. L. VEITH, K. F. BUREAU OF MINES, WASHINGTON, D, C. 1-19, 1965. (BM-RI-6604, N65-18524)
35565	THE PHYSICAL PROPERTIES OF PLUTONIUM-BEARING OXIDES. PART I. THERMAL CONDUCTIVITY. WARMER, R. J. U.K. ATOMIC ENERGY AUTHROITY, RISLEY, LANCS, ENGLAND 1-12, 1965. (TRG-RPT-586/D(PT. 1), N65-25479)

TPRC Number	Bibliographic Citation
35566	THE PHYSICAL PROPERTIES OF PLUTONIUM-BEARING OXIDES. PART 2. REVIEW OF DATA OTHER THAN THERMAL CONDUCTIVITY. WARMER, R. J. U.K. ATOMIC ENERGY AUTHORITY, RISLEY, LANCS, ENGLAND 1-10, 1965. (TRG-RPT-586/D(PT. 2), N65-25480)
35579	SPECIFIC HEATS AND TOTAL HEAT CONTENTS OF COALS AND RELATED MATERIALS AT ELEVATED TEMPERATURES. KIROV N Y BRIT COAL UTIL RES ASSOC MONTHLY BULL 29 2 33-60 1965 CA 63 2804
35594	REFLECTANCE OF COALS CARBONIZED UNDER PRESSURE. CHANDRA D ECON GEOL 60 3 621-9 1965 CA 63 2775
35595	ANALYSIS OF ROCKS BY SPECTRAL INFRARED EMISSION /8-25 MU/. LYON R J P ECON GEOL 60 4 715-36 1965 CA 63 8046
35602	THE STRUCTURE AND HIGH-TEMPERATURE ELECTRICAL AND THERMAL CONDUCTIVITIES OF SOME CATHODE BLOCK MATERIALS. BEATTIE D D DAY M K B EXTR MET ALUMINIUM 2 387-402 1963 CA 63 6424
35605	RING HEAT SOURCE PROBE FOR RAPID DETERMINATION OF THERMAL CONDUCTIVITY OF ROCKS. SOMERTON W H MOSSAHEBI M REV SCI INSTR 38 10 1368-71 1967
35610	SOME INORGANIC DIELECTRICS IN THE FAR INFRARED REGION. MASHKOVICH M D DEMASHINA A I FIZ TVERD TELA 7 6 1634-8 1965 CA 63 6427 (FOR ENGLISH TRANSLATION SEE T38592)
35623	INFRARED REFLECTIVITY OF FE2 O3, HYDRATED INFLUENCE ON MARTIAN REFLECTION SPECTRA. HOVIS W A JR ICARUS 4 1 41-2 1965 CA 63 6465
35665	CALCULATION OF THE SPECIFIC HEAT OF INDUSTRIAL GLASSES. LYAKH O O KHIM PROM INFORM NAUK-TEKHN ZB 2 30-1 1965 CA 63 5351
35669	RADIATION FROM FINE-PARTICLE CLOUDS IN HIGH-TEMPERATURE COMBUSTION GAS. MILVIDSKII, M. G. EREMEEV, V. V. MEM FAC ENG KYOTO UNIV 27 1 75-88 1965 CA 63 8111
35672	MEASUREMENTS OF SPECTRAL REFLECTIVITY OF MANGANESE OXIDES. NICHOL I PHILLIPS R MINERAL MAG 35 269 200-13 1965 CA 63 5092
35674	ON THE REFLECTION FACTOR OF A BINARY GLASS IN ALL THE INFRARED, FROM ROOM TEMPERATURE AND TO 77 K. CLAUDEL J HADNI A STRIMER P COMPT REND 270 B 1 70-2 1970 CA 72 126952
35675	ON THE THERMAL CONDUCTIVITIES OF CERTAIN POOR CONDUCTORS. I. PEIRCE B O WILLSON R W PROC AM ACAD ARTS SCI 34 1 1-56 1898
35678	OPTICAL PROPERTIES OF HEMATITE AND ILMENITE. GEHLEN, K. V. PILLER, H. NEUES JAHRB. MINERAL., MONATSH. (4), 97-108, 1965.
35687	EFFECT OF HEAT TREATMENT ON THE EXPANSIVITY OF A PYREX GLASS. SAUNDERS J B TOOL A Q J RESEARCH NATIONAL BUREAU STANDARDS 11 799-810 1933
35705	THERMAL CONDUCTIVITY OF STABILIZED ZIRCONIA AT HIGH TEMPERATURES. DUC X N REV HAUTES TEMP REFRACTAIRES 2 1 55-63 1965 JA 50 57
35710	THERMOPHYSICAL COEFFICIENTS OF GROUND COALS. VINOGRADOVA, L. P. RUDN. AEROGAZODINAM. I BEZOPASNOST GORN. RABOT, AKAD. NAUK SSSR, INST. GORN. DELA 201-4, 1964.
35713	HEAT CAPACITY AND EQUATION OF STATE FOR SOLIDS. NOVAK J SB VED PRACI VYSOKA SKOLA CHEM-TECHNOL PARDUBICE 1 79-90 1964 CA 63 6388
35740	IN-PILE THERMAL CONDUCTIVITY OF UO2 AND PU/0.2/ V/0.8/O2 SPECIMENS. BAILY W E HINES D P ZEBROSKI E L TRANS AM NUCL SOC 8 39-40 1965 NS 19 28857
35746	THERMAL CONDUCTIVITY OF INDUSTRIAL REFRACTORIES. GONCHAROV, V. V. KOLECHKOVA, A. F. ZADVORNOVA, E. G. SOLTAN, A. R. TR. VSES. GOS. INST. NAUCHN.- ISSLED. I PROEKTN. RABOT OGNEUPORN. PROM. (35), 26-44, 1963.
35752	THERMAL CONDUCTIVITY OF CERAMIC BARIUM TITANATE AND SOME BARIUM TITANATE BASE SOLID SOLUTIONS. DIMAROVA E N POPLAVKO YU M TVLL ACAD SCI USSR /PHYS SER/ 29 6 985-7 1965 (ENGLISH TRANSLATION OF IZV. AKAD. NAUK SSSR, SER. FIZ. 29 (6), 985-7, 1965; FOR ORIGINAL SEE T34098)
35758	OLIVINE AS A REFRACTORY. HEINDL R A PENDERGAST W L J RESEARCH NATIONAL BUREAU STANDARDS 12 215-22 1934 (NBS-RP-645)
35768	THE SYSTEM PB O-SI O2. GELLER R F CREAMER A S BUNTING E N J RESEARCH NATIONAL BUREAU STANDARDS 13 237-44 1934 (NBS-RP-705)
35781	THE SYSTEM K2 O-PB O-SI O2. GELLER R F BUNTING E N J RESEARCH NATIONAL BUREAU STANDARDS 17 277-89 1936 (NBS-RP-911)
35787	ON THE MEASUREMENT OF VERY LOW TEMPERATURES. V. THE EXPANSION COEFFICIENT OF JENA AND THURINGER GLASS BETWEEN 16 AND -182 C. ONNES H K HEUSE W COMM KAMERLINGH ONNES LAB UNIV LEIDEN 85 1-19 1903 (ENGLISH TRANSLATION OF VERSLAG. GEWONE VERGADER. AFDEL. NATUURK. KONINK. NED. AKAD WETENSCHAPPEN., 212-23, 1903; FOR ORIGINAL SEE T38441)
35796	BASIC INVESTIGATION OF MULTI-LAYER INSULATION SYSTEMS. LITTLE ARTHUR D, INC., CAMBRIDGE, MASS. LITTLE ARTHUR D, INC., CAMBRIDGE, MASS. 1-298, 1964. (ADL-65958-00-04, NASA-CR-54191, N65-23738)
35817	DEVELOPMENT OF A PORCELAIN VACUUM TUBE. MATERIALS. HURSH, R. K. KIRBY, J. J. J AM CERAMIC SOC 32 3 75-80 1949
35856	EVALUATION OF INFRARED EMISSION OF CLOUDS AND GROUND AS MEASURED BY WEATHER SATELLITES. KERN, C. D. WASHINGTON UNIV., SEATTLE, PH.D. THESIS 1-157, 1964. (AD 617417)
35875	HIGH TEMPERATURE THERMOELECTRIC RESEARCH. HENDERSON, C. M. MONSANTO RESEARCH CORP., DAYTON, OHIO 1-88, 1965. (AD 468339)
35888	INTERACTIONS BETWEEN SURFACE HYDROXYL GROUPS AND ADSORBED MOLECULES. II. INFRARED SPECTROSCOPIC STUDY OF BENZENE ADSORPTION. CUSUMANO J A LOW M J D J PHYS CHEM 74 9 1950-6 1970
35897	NOTE ON THERMAL PROPERTIES OF MARS. LEOVY, C. RAND CORP., SANTA MONICA 1-23, 1965. (RM-4551-NASA, NASA-CR-63278, N65-26075)
35906	DESIGN AND FABRICATION OF OPTICAL FILTERS FOR LASER FREQUENCY. HOLMES, S. J. BOWMAN, N. E. TEXTRON ELECTRONICS INC., SYLMAR, CA. 1-64, 1964. (AFAL-TR-64-268, AD 608064)
35907	IMPROVED RADIATOR COATINGS. PART II SCHATZ E A COUNTS C R III ALVAREZ G H HOPPKE M A MATERIALS LAB., WPAFB, OHIO 1-99, 1965. (ML-TDR-64-146 PT. 2, AD 468576)

TPRC Number	Bibliographic Citation
35917	FABRICATION, CHARACTERIZATION, AND THERMAL-PROPERTY MEASUREMENTS OF ZR 02 — BASE FUELS. DEEM, H. W. BATTELLE MEMORIAL INST., COLUMBUS 1-89, 1966. (BMI-1775)
35972	THE STRUCTURE OF SODIUM SILICATE GLASSES AND THEIR FAR INFRARED ABSORPTION SPECTRA. HANNA R J PHYS CHEM 69 11 3846-9 1965
35986	FABRICATION OF A 104-CM MIRROR FROM CER-VIT/SUPERSCRIPT R/ LOW EXPANSION MATERIAL. MONNIER R C APPL OPT 6 8 1437-40 1967
36008	KINETICS OF EXPANDING ALUMINA SUSPENSIONS FOR THE PREPARATION OF LIGHTWEIGHT CORUNDUM. GAODU A N KAINARSKII I S OGNEUPORY 29 6 270-5 1964 CA 61 8040 (FOR ENGLISH TRANSLATION SEE T63069)
36017	THERMAL CONDUCTIVITY OF A BOROSILICATE GLASS SUITABLE FOR FISSION PRODUCT WASTES. AUTY, D. B. GROVER, J. R. COOPER, B. E. ENERGY RES. ESTAB., HARWELL BERKS ENGLAND 1-24, 1964. (AERE-R-4685, N65-21944)
36019	SOME FERROELECTRIC AND ANTIFERROELECTRIC MATERIALS WITH HIGH CURIE POINTS. SMOLENSKII G A KRAINIK N N KHUCHUA N P ISUPOV V A ZHDANOVA V V MUSHTAREEV O M CHERKASHCHENKO A YA BULL ACAD SCI USSR PHYS SER 31 7 1184-8 1967 (EHGLISH TRANSLATION OF IZV. AKAD. NAUK SSSR, SER. FIZ., 31 (7), 1164-7, 1967; FOR ORIGINAL SEE T46353)
36028	METHODS FOR MEASURING COEFFICIENTS OF THERMAL DIFFUSIVITY OF REFRACTORY AND HEAT RESISTANT MATERIALS AT HIGH TEMPERATURES OBTAINED IN THE FOCUS OF SOLAR OVEN. MAVASHEV, IU. Z. BORUKHOV, M. IU. BESPALKO, V. P. TEPLOFIZ. SVOISTVA TVERD TEL VYS. TEMP., TR. VSIS KONF. 46-54PP., 1971.
36035	ON THE INFLUENCE OF A DC FIELD ON THE THERMAL CONDUCTIVITY OF BARIUM TITANATE BASE CERAMIC SPECIMENS AND TRIGLYCINE SULFATE SINGLE CRYSTALS IN THE PHASE TRANSITION REGION. DIMAROVA E N POPLAVKO YU M BULL ACAD SCI USSR PHYS SER 31 11 1885-7 1967 (ENGLISH TRANSLATION OF IZV. AKAD. NAUK SSSR, SER. FIZ., 31 (11), 1842-4, 1967; FOR ORIGINAL SEE T48739)
36041	REVERSIBLE THERMAL EXPANSION OF SILICA. HOULDSWORTH H S COBB J W ENGINEERING 112 576-79 1921
36055	THERMAL CONDUCTIVITY OF ECLOGITE AND DOLERITE AT HIGH TEMPERATURE. MOISEENKO U I SOLOVEVA Z A KUTOLIN V A DOKL ACAD SCIENCES USSR EARTH SCI SECTIONS 173 3 163-5 1967 (ENGLISH TRANSLATION OF DOKL. AKAD. NAUK SSSR, 173 (3), 669-71, 1967; FOR ORIGINAL SEE T45103)
36058	METHODS FOR MEASURING THERMAL CONDUCTIVITY. APPLICATION TO FIBROUS INSULATING MATERIALS. GASQUET R REV THERM 2 23 1245-58 1963 CA 61 13900
36061	SOME SOFT GLAZES OF LOW THERMAL EXPANSION. GELLER R F BUNTING E N CREAMER A S J RESEARCH NATIONAL BUREAU STANDARDS 20 57-66 1938 (NBS-RP-1064)
36067	PRODUCTION OF NEW CHROMIUM PIGMENTS OF THE SPINEL TYPE. TUMANOV S G PYRKOV V P GLASS CERAMICS 22 6 360-2 1965 (ENGLISH TRANSLATION OF STEKLO I KERAM., 22 (6), 2-5, 1965; FOR ORIGINAL SEE T38884)
36068	THERMAL STABILITY OF PORCELAIN. MOROZ I I SARKISOV G G GLASS CERAMICS 25 6 359-62 1968 (ENGLISH TRANSLATION OF STEKLO KERAM., 25 (6), 27-30, 1968; FOR ORIGINAL SEE T50457)
36076	HEAT CAPACITY OF CLAY AND CLAY-WATER MIXTURES. OSTER J D PURDUE UNIV., LAFAYETTE, IND., PH.D. THESIS 1-38, 1964. (UNIV. MICRPFILMS 64-8696)
36084	DEVELOPMENT OF A 1200 F. RADOME. CHASE, V. A. COPELAND, R. L. BRUNSWICK CORP., MARION, VA. 1-155, 1964. (AD 454643)
36100	DIRECTIONAL REFLECTANCE OF CERTAIN MATERIALS IN THE NEAR INFRARED. MARY, D. J. HARRY DIAMOND LABS, U. S. ARMY MATERIAL COMMAND 1-29, 1964. (HDL-TM-64-29, AD 454083, AD 454083)
36117	EMITTANCE AND REFLECTANCE OF INTERMETALLIC COMPOUNDS. SCHATZ, E. A. AMERICAN MACHINE AND FOUNDRY CO., ALEXANDRIA, VA. 1-26, 1962. (ANF-PR-2)
36132	THE THERMAL CONDUCTIVITY AT HIGH TEMPERATURES OF REFRACTORY CLAYS FOR GLASS-POTS. JACQUE, L. DUMEZ, P. BULL. SOC. FRANC. CERAM. (62), 49, 1964.
36139	AUTOMATIC APPARATUS FOR MEASURING PHYSICOTHERMAL CONSTANTS OF CERAMIC MATERIALS. POLUBOYARINOV D N DUDEROV I G INDUSTRIAL LABORATORY 31 11 1761-3 1965 (ENGLISH TRANSLATION OF ZAVODSK. LAB., 31 (11), 1410-12, 1965; FOR ORIGINAL SEE T43324)
36144	ADVANCES IN THE MATERIALS TECHNOLOGY RESULTING FROM THE X-20 PROGRAM. STRATTON W K STACY J T GUNDERSON R D HONEBRINK D A TRIPP J T BRESLICH F N PERKOWSKI W S OTTESTAD D J ARMSTRONG C S LEGAN D J TREPUS G E JR BOEING COMPANY, SEATTLE 1-163, 1965. (AFML-TR-64-396)
36165	HIGH TEMPERATURE HEAT CONTENTS OF CRYOLITE, ANHYDROUS ALUMINUM FLUORIDE AND SODIUM FLUORIDE. O BRIEN C J KELLEY K K J AM CHEM SOC 79 21 5616-18 1957
36167	EMISSIVITY MEASUREMENTS ON SOLAR CELLS. SCHOCKEN, K. ARMY BALLISTIC MISSILE AGENCY, REDSTONE ARSENAL, ALA. 1-14, 1959. (ABMA-DV-TN-16-59, N-82955)
36180	MODEL FOR CALCULATING THERMAL CONDUCTIVITY OF CERAMICS. DULNEV, G. N. ZARICHNIAK, IU. P. LITOVSKII, E. IA. TEPLOFIZ. SROISTVA TVERD. TEL VYS. TEMP., TR. VSIS. KONF. 76-82PP., 1971.
36191	EXPANSION EFFECTS ON THE INVERSION OF SILICA CRYSTALS IN CERTAIN DEVITRIFIED GLASSES. NBS-RP-1153. TOOL A Q SAUNDERS J B J RESEARCH NATIONAL BUREAU STANDARDS 21 773-8 1938
36234	LENGTH CHANGES OF WHITEWARE CLAYS AND BODIES DURING INITIAL HEATING, WITH SUPPLEMENTARY DATA ON MICA. GELLER R F BUNTING E N J RESEARCH NATIONAL BUREAU STANDARDS 25 15-40 1940 (NBS-RP-1311)
36246	PHYSICAL PROPERTIES OF POROUS MATERIALS AND THEIR APPLICATIONS. III. WAKASHIMA H KUWANA K KANAZAWA DAIGAKU KOGAKUBU KIYO 3 2 172-6 1963 CA 62 9808
36248	THERMAL EXPANSION OF CLAY BUILDING BRICKS. CULBERTSON W R J RESEARCH NATIONAL BUREAU STANDARDS 27 197-216 1941 (NBS-RP-1414)
36253	RELATION BETWEEN THE AMOUNT OF CRYSTALLINE PHASE AND THE HEAT TREATMENT REGIME IN SITALL SYSTEMS OF SIO2-AL2O3-BAO-TIO2. BOGDANOVA G S ORLOVA E M ZEVIN L S INORGANIC MATERIALS 2 2 326-8 1966 (ENGLISH TRANSLATION OF IZV. AKAD. NAUK SSSR, NEORGAN. MAT., 2 (2), 380-83, 1966; FOR ORIGINAL SEE T49015)

TPRC Number | Bibliographic Citation

36254 STRUCTURAL TRANSFORMATIONS IN GLASSES OF THE SI O2-AL2 O3-BA O-TI O2 SYSTEM IN THE INITIAL STAGE OF CRYSTALLIZATION.
BOGDANOVA G S ORLOVA E M
INORGANIC MATERIALS
2 3 461-3 1966
(ENGLISH TRANSLATION OF IZV. AKAD. NAUK SSSR, NEOGRAN. MAT., 2 (3), 537-40, 1966; FOR ORIGINAL SEE T47091)

36262 HEAT AND AIR TRANSFER IN COLD STORAGE INSULATION.
LOTZ W A
ASHRAE J
6 4 53 1964

36274 THERMAL EXPANSION OF CLAYS AND SPECIAL REFRACTORIES.
NATIONAL BUREAU OF STANDARDS
NATIONAL BUREAU STANDARDS
(117), 2-3, 1927.
(NBS-117)

36277 THERMAL EXPANSION OF MULLITE FROM 20 TO 1800 C.
NATIONAL BUREAU OF STANDARDS
NATIONAL BUREAU OF STANDARDS, TECHNICAL NEWS BULL.
(139), 158-59, 1928.

36288 THERMAL EXPANSIONS OF FUSED ALUMINA, BAUXITE, AND DIASPORE.
NATIONAL BUREAU OF STANDARDS
NATIONAL BUREAU OF STANDARDS, TECHNICAL NEWS BULL.
(158), 55-56, 1930.

36298 EFFECT OF VARIATIONS IN COMPOSITION ON PROPERTIES OF VITREOUS ENAMELS.
NATIONAL BUREAU OF STANDARDS
NATIONAL BUREAU OF STANDARDS, TECHNICAL NEWS BULL.
(160), 78-79, 1930.

36300 MEASUREMENTS ON THE THERMAL EXPANSION OF JENA THERMOMETER GLASS 2954-III BY THE METHOD OF THE VERTICAL COMPARATOR.
KEESOM W H DOBORZYNSKI D W
PHYSICA
1 2 1085-8 1934
(COMM. KAM. ONNES LAB., LEIDEN. NO. 234A)

36313 ABSOLUTE METHOD OF COMPLEX DETERMINATION OF THERMOPHYSICAL CHARACTERISTICS OF NON-METALLIC MATERIALS.
FRAIMAN YU E
INZH FIZ ZH
7 10 73-9 1964

36324 THE REFLECTION AND TRANSMISSION OF INFRARED MATERIALS. III, SPECTRA FROM 2 TO 50 MICRONS.
MC CARTHY D E
APPL OPT
4 3 317-20 1965 CA 62 14052

36326 ON THE DISSIMILAR BEHAVIOR OF CRYSTALS AND GLASSES IN FAR INFRARED ABSORPTION /40-1500 MU/ AT LIQUID HELIUM TEMPERATURE.
HADNI A CLAUDEL J GERBAUX X MORLOT G
MUNIER J-M
APPL OPT
4 4 487-94 1965 SA 68 18225

36327 THERMAL CONDUCTANCE TESTS ON CABIN WALL INSULATION ASSEMBLIES FOR A SUPERSONIC TRANSPORT AIRCRAFT. SUPERSEDES ARC-27497.
MC NAUGHTAN, I. I. KEENE, P. E.
AERONAUTICAL RESEARCH COUNCIL, LANDON
1-79, 1964.
(RAE-TN-MECH-ENG-405, ARC-CP-910, N67-24037 AD-809896)

36348 PRECISION OF COLOR MEASUREMENT WITH THE GE SPECTROPHOTOMETER. I. ROUTINE INDUSTRIAL PERFORMANCE.
BILLMEYER F W JR
J OPT SOC AM
55 6 707-17 1965

36377 HEAT-TRANSFER CHARACTERISTICS OF PACKED BEDS WITH STAGNANT FLUIDS.
OFUCHI K KUNII D
INTERN J HEAT MASS TRANSFER
8 5 749-57 1965

36391 TEMPERATURE DEPENDENCE OF THE THERMAL AND TEMPERATURE CONDUCTIVITIES OF SOME UNFILLED POLYMERS.
BIL V S AVTOKRATOVA N D
HIGH TEMPERATURE
2 2 169-75 1964
(ENGLISH TRANSLATION OF TEPLOFIZIKA VYSOKIKH TEMPERATUR, 2 (2), 192-8, 1964; FOR ORIGINAL SEE T38075)

36398 VARIATION OF THE HEAT CONDUCTIVITY OF GLASS IN THE SIO2-LI2O-AL2O3-ZRO2 SYSTEM DURING CRYSTALLIZATION.
FONDYMAKIN B I SOLINOV F G
TEPLOFIZIKA VYSOKIKH TEMPERATUR
2 3 397-400 1964
(FOR ENGLISH TRANSLATION SEE T36399)

36399 VARIATION OF THE HEAT CONDUCTIVITY OF GLASS IN THE SIO2-LI2O-AL2O3-ZRO2 SYSTEM DURING CRYSTALLIZATION.
FONDYMAKIN B I SOLINOV F G
HIGH TEMPERATURE
2 3 363-5 1964
(ENGLISH TRANSLATION OF TEPLOFIZIKA VYSOKIKH TEMPERATUR, 2 (3), 397-400, 1964; FOR ORIGINAL SEE T36398)

36402 SOME QUESTIONS ON THE METHOD OF EXPERIMENTALLY STUDYING THE HEAT CONDUCTIVITY OF MATERIALS AT HIGH TEMPERATURES.
BOIKO N V SHPILRAIN E E
TEPLOFIZIKA VYSOKIKH TEMPERATUR
2 4 549-57 1964
(FOR ENGLISH TRANSLATION SEE T36403)

36403 SOME QUESTIONS ON THE METHOD OF EXPERIMENTALLY STUDYING THE HEAT CONDUCTIVITY OF MATERIALS AT HIGH TEMPERATURES.
BOIKO N V SHPILRAIN E E
HIGH TEMPERATURE
2 4 493-500 1964
(ENGLISH TRANSLATION OF TEPLOFIZIKA VYSOKIKH TEMPERATUR, 2 (4), 549-57, 1964; FOR ORIGINAL SEE T36402)

36419 THE THERMAL AND MECHANICAL PROPERTIES OF FIVE ABLATIVE REINFORCED PLASTICS FROM ROOM TEMPERATURE TO 750 F.
PEARS C D ENGELKE W T THORNBURGH J D
SOUTHERN RESEARCH INST., BIRMINGHAM
1-188, 1965.
(AFML-TR-65-133, X-65-17697, AD 465175)

36420 INVESTIGATION AND DEVELOPMENT OF HIGH TEMPERATURE INSULATION SYSTEMS.
WECHSLER, A. E. KRITZ, M. A.
ARTHUR D LITTLE INC., CAMBRIDGE
1-110, 1965.
(AFML-TR-65-138)

36431 APPARATUS FOR RAPID DETERMINATION OF THE THERMAL CONDUCTIVITY OF REFRACTORY BRICK.
JACQUE, L. DUMETZ, P.
IND. CERAM.
(569), I-VPP., VIIIP., 1964.

36465 THE USE OF INTENSE PINCH DISCHARGES FOR LASER ILLUMINATION.
BUSER, R. G. KAINZ, J. J. SULLIVAN, J. J.
U. S. ARMY ELECTRONICS R AND D LABS, FT. MONMOUTH, NJ
121-34, 1965.
(AD 611432)

36468 TECHNIQUE FOR MEASURING THERMAL CONDUCTIVITY OF THERMOELECTRIC MATERIALS.
BIENERT W B TRIMMER D S SKRABEK E A
PROCEEDINGS IEEE/AIAA THERMOELECTRIC SPECIALISTS CONFERENCE
6.1—6.7 1966

36486 FAR-INFRARED SPECTRA OF SOLIDS.
ARONSONM J. R. MC LINDEN, H. G.
SYMP. ON THERMAL RADIATION OF SOLIDS.
29-38, 1965.
(NASA-SP-55, ML-TDR-64-159, N65-26858, AD-629980)

36491 FUNDAMENTALS OF THERMAL RADIATION IN CERAMIC MATERIALS.
COX, R. L.
SYMP. ON THERMAL RADIATION OF SOLIDS
83-101, 1965.
(NASA-SP-55, ML-TDR-64-159, N65-26863, AD-629980)

36510 METHOD AND EQUIPMENT FOR MEASURING THERMAL EMITTANCE OF CERAMIC OXIDES FROM 1200 TO 1800 K.
CLARK, H. E. MOORE, D. G.
SYMP. ON THERMAL RADIATION OF SOLIDS
241-57, 1965.
(NASA-SP-55, ML-TDR-64-159, N65-26882, AD-629980)

36511 HEMISPHERICAL SPECTRAL EMITTANCE OF ABLATION CHARS, CARBON, AND ZIRCONIA.
WILSON, R. G.
SYMP. ON THERMAL RADIATION OF SOLIDS
259-75, 1965.
(NASA-SP-55, ML-TDR-64-159, N65-26883, AD-629980)

36520 THERMAL DIFFUSIVITY OF THE SYSTEMS UO2-SIO2 IN THE TEMPERATURE RANGE 100-800 C.
GLANZ G
J NUCL MATER
27 3 331-4 1968

36525 ULTRAVIOLET IRRADIATION OF WHITE SPACECRAFT COATINGS IN VACUUM.
ZERLAUT, G. A. HARADA, Y. TOMPKINS, E. H.
SYMP. ON THERMAL RADIATION OF SOLIDS
391-420, 1965.
(NASA-SP-55, ML-TDR-64-159, N65-26897, AD-629980)

TPRC Number	Bibliographic Citation
36547	PHOTOTROPY OF REDUCED SILICATE GLASSES CONTAINING THE 570 M-MU COLOR CENTER. SWARTS E L PRESSAU J P J AM CERAM SOC 48 7 333-8 1965
36552	PNEUMATIC PLACEMENT OF REFRACTORY CASTABLES. III. COOK M D COOK C P KING D F BULL AMER CERAM SOC 43 5 380 1964 BC 64 143
36576	DETERMINATION OF THE COEFFICIENT OF THERMAL CONDUCTIVITY. BROVKIN L A INDUSTRIAL LABORATORY 34 6 831-3 1968 (ENGLISH TRANSLATION OF ZAVOD. LAB., 34 (6), 700-2 1968; FOR ORIGINAL SEE T46500)
36598	HIGH TEMPERATURE RESEARCH IN A SOLAR FURNACE. I V. MEASUREMENT OF SPECTRAL EMISSIVITY AND FREEZING POINT OF METAL OXIDES. NOGUCHI, T. MIZUNO, M. KOZUKA, T. YOSHIDA, M. YAMAMOTO, T. NAGOYA KOGYO GIJUTSU SHIKENSHO HOKOKU 14 (1), 28-38, 1965.
36659	THE USE OF LOW-TEMPERATURE CALORIMETRY FOR THE STUDY OF SILICATES. STROGANOV, E. F. MATVEEV, G. M. KHOKONOV, KH. B. SOVREM. METODY ISSLED. SILIKATOV I STROIT. MATERIALOV, VSES. KHIM. OBSHCHESTVO, SB. STATEI 52-66PP., 1960.
36685	INQUIRIES INTO THERMAL EXPANSION AND CREEP RESISTANCE OF SILICA BRICKS. METZGER C SCHWIETE H E GLASTECH BER 39 4 190-202 1966 CA 65 6886 (FOR ENGLISH TRANSLATION SEE T36690)
36690	INQUIRIES INTO THERMAL EXPANSION AND CREEP RESISTANCE OF SILICA BRICKS. METZGER C SCHWIETE H E SPECIAL LIBRARY ASSOC. TT-66-13466 1-31 1966 TT 16-9 68 (ENGLISH TRANSLATION OF GLASTECH. BER., 39 (4), 190-202, 1966; FOR ORIGINAL SEE T36685)
36711	A RESISTOR FURNACE WITH SOME PRELIMINARY RESULTS UP TO 2,000 C. GELLER R F J RESEARCH NATIONAL BUREAU STANDARDS 27 555-65 1941 (NBS-RP-1443)
36713	MEASUREMENT OF THE ABSOLUTE REFLECTIVITY WITH THE INFRARED SPECTROMETER UR 10. VASKO A EXPTL TECH PHYSIK 12 6 392-7 1964 CA 62 14046
36715	CERAMIC COATINGS FOR HIGH-TEMPERATURE PROTECTION OF STEELS. HARRISON W N MOORE D G RICHMOND J C J RESEARCH NATIONAL BUREAU STANDARDS 38 293-307 1947 (NBS-RP-1773)
36717	INFLUENCE OF COLOR CONSTITUENTS ON THE COOLING RATE AND THE TOTAL EMISSIVITY OF GLASSES IN THE WORKING RANGE. WEISS, W. GLASTECH. BER. 37 (12), 553-62, 1964.
36738	PROPERTIES OF SELECTED COMMERCIAL GLASSES. CORNING GLASS WORKS, NEW YORK CORNING GLASS WORKS, MATERIALS HANDBOOK B-83 1-16 1965
36744	EFFECTIVE THERMAL EXPANSION OF WHITEWARE GLAZES. GREINER E GLAS-EMAIL-KERAMO-TECHNIK 5 169-71 1954
36758	FIBER OPTICS. XI. PERFORMANCE IN THE INFRARED REGION. KAPANY N S SIMMS R J J OPT SOC AM 55 8 963-8 1965 CA 63 6480
36785	THERMAL CONDUCTIVITY OF BA TI O3 AND SR TI O3 FROM 4.5 TO 300 K. SUEMUNE Y J PHYS SOC JAPAN 20 1 174-5 1965 SA 68 17592
36805	A COMPARATIVE STUDY OF REFRACTORY BRICKS WITH SPECIAL REFERENCE TO THERMAL CONDUCTIVITY. DUTTA, B. M. RAO, H. V. B. NML TECH. J. 6 (4), 17-22, 1964.

TPRC Number	Bibliographic Citation
36833	SPECTROPHOTOMETRIC CHARACTERISTICS OF MILKY GLASS MS-14. TOPORETS A S OPT SPECTRY /USSR/ 18 4 394-6 1965 (ENGLISH TRANSLATION OF OPT. I SPEKTROSKOPIYA, 18 (4), 704-7, 1965; FOR ORIGINAL SEE T35073)
36837	STUDY OF A SWAT KAOLINITE. SHAH R A NAZ M A NAQVI A A SAFDAR M PAKISTAN J SCI IND RES 7 3 183-7 1964 JA 48 180
36861	THERMOPHYSICAL CHARACTERISTICS OF POLYMER COATINGS. SUKHAREVA L A VORONKOV V A ZUBOV P I J ENG PHYS BSSR 9 2 147-50 1965 (ENGLISH TRANSLATION OF INZHEN.-FIZ. ZH., 9 (2), 211-6, 1965; FOR ORIGINAL SEE T42532)
36875	BEHAVIOR OF INTERLAYERS OF GLASS-TO-TUNGSTEN SEALS. TAKAMORI T TOMOZAWA M J AM CERAM SOC 48 8 405-9 1965
36898	HEAT-INSULATING REFRACTORIES WITH A BULK DENSITY OF 1 G/CM(SUPERSCRIPT 3). TOSHEV, V. VELKOV, I. STROIT. MATER. SILIKAT. PROM. 5 (5), 19-22, 1964.
36904	AN APPARATUS FOR MEASURING THE THERMAL CONDUCTIVITY OF INSULATING MATERIALS. BETTA V TERMOTECNICA MILAN 17 12 739-41 1963 AM 18 3004
36927	THERMAL EXPANSION OF POLYCRYSTALLINE HAFNIUM OXIDE-ZIRCONIUM OXIDE SOLID SOLUTIONS. STANSFIELD O M J AM CERAM SOC 48 8 436-7 1965
37023	STRUCTURE ANALYSIS OF THERMAL OXIDE FILMS OF SILICON BY ELECTRON DIFFRACTION AND INFRARED ABSORPTION. NAGASIMA N JAPANESE J APPLIED PHYS 9 8 879-88 1970
37025	THE HEAT CONDUCTIVITY OF DRYING CERAMIC BODIES. SCHULLE, W. ZIEGELINDUSTRIE 16 (22), 839-44, 1963.
37038	HEAT CAPACITY AT ROOM TEMPERATURE OF FERRITES OF THE SYSTEM NI FE2 O4-ZN FE2 O4. VARAZASHVILI V S LANDIYA N A CHACHANIDZE G D LEZHAVA N G INORGANIC MATERIALS 4 7 1017-20 1968 (ENGLISH TRANSLATION OF IZV. AKAD. NAUK SSSR, NEORG. MATER., 4 (7), 1160-3, 1968; FOR ORIGINAL SEE T51368)
37045	RADIATION DAMAGE IN SOLIDS. A SURVEY. CRAWFORD J H JR BULL AM CERAM SOC 44 12 963-70 1965
37046	ANALYSIS OF SOLID MIXTURES BY DIFFUSE REFLECTANCE MEASUREMENTS. DOYLE W P FORBES F ANAL CHIM ACTA 33 1 108-14 1965 CA 63 9043
37058	THERMOFLASH APPARATUS AND COEFFICIENTS OF HEAT CONDUCTION OF INSULATING AND OTHER BUILDING MATERIALS. SAXENA B K INDIAN J TECHNOL 2 1 27-30 1964 JA 47 198
37067	THE TEMPERATURE DEPENDENCE IN THE CHANGE OF ENTHALPY AND HEAT CAPACITY OF IRON ORES AND AGGLOMERATE. BRATCHIKOV, S. G. IZV. VUZ—CHERN. MET. (12), 35-7, 1963.
37068	VARIABLE-STATE METHODS OF MEASURING THE THERMAL PROPERTIES OF SOLIDS. HARMATHY T Z J APPL PHYS 35 (4), 1190-1200, 1964. (NRCC-8086, REPRINT 221, AD-447244)
37072	HEAT TRANSLFER OF PROPOSED STRUCTURAL JOINTS IN THE ROCKET PACKAGE. JELINEK, D. NORTH AMERICAN AVIATION INC., LOS ANGELES 1-11, 1949. (NAA-49-831)

TPRC Number	Bibliographic Citation
37106	SOME REGULAR FEATURES OF THE THERMAL EXPANSION OF FERRITES WITH THE SPINEL STRUCTURE. BEKKER YA M ZOTOVA G M INORGANIC MATERIALS 4 8 1155-8 1968 (ENGLISH TRANSLATION OF IZV. AKAD. NAUK SSSR, NEORG. MATER., 4 (8), 1319-22, 1968; FOR ORIGINAL SEE T50768)
37113	PROPERTIES OF BARIUM-MAGNESIUM TITANATE DIELECTRICS. SHELTON G R CREAMER A S BUNTING E N J RESEARCH NATIONAL BUREAU STANDSRDS 41 17-26 1948 (NBS-RP-1899)
37122	EXPERIMENTAL DETERMINATION OF THE DIRECTIONAL REFLECTANCE OF DIELECTRIC COATINGS ON METALLIC SUBSTRATES. PIEROWAY, C. S. UNIVERSITY OF OKLAHOMA, NORMAN, M.S. THESIS 1-45 1969
37123	PROPERTIES OF CALCIUM-BARIUM TITANATE DIELECTRICS. BUNTING E N SHELTON G R CREAMER A S J RESEARCH NATIONAL BUREAU STANDARDS 43 237-44 1949 (NBS-RP-2025)
37152	SOME PROPERTIES OF A GLASS USED IN PAPER MANUFACTURE. O LEARY M J HUBBARD D J RESEARCH NATIONAL BUREAU STANDARDS 55 1 1-9 1955 (NBS-RP-2599)
37156	USE OF THIN FILMS FOR THE DETERMINATION OF THE OPTICAL CONSTANTS OF ABSORBENT CRYSTALS IN THE INFRARED. VINCENT-GEISSE J J PHYS /FRANCE/ 25 1 291-4 1964 SA 67 22822
37172	THERMAL EXPANSION OF VINARY ALKALI SILICATE GLASSES. SHERMER H F J RESEARCH NATIONAL BUREAU STANDARDS 57 2 97-101 1956 (NBS-RP-2698)
37174	THE DEPENDENCE OF THE EXPANSION COEFFICIENTS OF SOLIDS ON TEMPERATURE. VALENTINER S WALLOT J ANN PHYSIK 46 4 837-67 1915
37184	THERMAL EXPANSION OF REGULARLY CRYSTALLIZED SOLIDS. GRUNEISEN E ANN PHYSIK 55 4 371-86 1918
37195	STANDARD MATERIAL FOR EXPANSION MEASUREMENTS ON SOLIDS UP TO 1400 C. COHN W M ANN PHYSIK 4 5 493-512 1930
37200	THE THERMAL CONDUCTIVITY OF DISPERSED MATERIALS AT DIFFERENT ATMOSPHERIC PRESSURES. KOSTYLEV V M TEPLOFIZ VYSOKIKH TEMP /USSR/ 2 1 21-8 1964 SA 68 4834 (FOR ENGLISH TRANSLATION SEE T37450)
37217	DETERMINATION OF THE HEAT CAPACITY OF KYANITE AND QUARTZ AT HIGH TEMPERATURE BY THERMAL ANALYSIS. LEONIDOV V YA BARSKII YU P KHITAROV N I GEOKHIMIYA 5 414-19 1964 CA 61 2541
37232	BISMUTH PHOSPHATE GLASSES. TIWARI A N SUBBARAO E C J AM CERAM SOC 53 5 258-61 1970
37241	PERMANENT DENSIFICATION OF GLASSY B2O3 AND POTASSIUM BORATE GLASSES BY HIGH STATIC PRESSURE. POCH W BER DEUTSCH KERAM GESELLSCHAFT 41 2 68-77 1964
37248	DETERMINATION OF AXIAL STRESS IN CLAD GLASS FIBERS. KROHN D A J AM CERAMIC SOC 53 9 505-7 1970
37250	INFRARED ABSORPTION OF OH/-1/ IN E GLASS. NEMEC L GOTZ J J AM CERAMIC SOC 53 9 526 1970
37273	DETERMINATION OF OPTICAL CONSTANTS FOR A SCATTERING AND ABSORBING CERAMIC. PROGELHOF R C THRONE J L J AM CERAM SOC 53 5 262-3 1970
37300	QUANTITATIVE CHARACTERISTICS OF THE SPECIFIC HEAT SINGULARITY IN SECOND-ORDER PHASE-TRANSITION POINTS. VORONEL A V GARBER S R MAMNITSKII V M SOVIET PHYSICS-JETP 28 6 1065-71 1969 (ENGLISH TRANSLATION OF ZH. EKSP. TEOR. FIZ., 55 (6), 2017-30, 1968; FOR ORIGINAL SEE T53688)
37328	THE CURRENT STATUS OF THERMAL CONDUCTIVITY REFERENCE STANDARDS AT THE NATIONAL BUREAU OF STANDARDS. ROBINSON, H. E. FLYNN, D. R. NATIONAL BUREAU OF STANDARDS 1-14, 1964. (NBS-R-8300, AD-439619L)
37340	STRUCTURAL ROLE OF ZINC OXIDE IN GLASSES IN THE SYSTEM NA2O-ZNO-SIO2. HURT J C PHILLIPS C J J AM CERAM SOC 53 5 269-73 1970
37343	RESULTS OF CERTAIN HEAT-ENGINEERING INVESTIGATIONS. YUSUFOVA V D TRUDY ENERG INST AKAD NAUK AZERB SSR 15 180-94 1962 (FOR ENGLISH TRANSLATION SEE T37344)
37344	RESULTS OF CERTAIN HEAT-ENGINEERING INVESTIGATIONS. YUSUFOVA V D FOREIGN TECHNICAL DIV., TRANSLATION DIVISION 1-21, 1964. (ENGLISH TRANSLATION OF TRUDY ENERG. INST., AKAD. ANUK AZERB. SSR, 15 180-94, 1962; FOR ORIGINAL SEE T37343) (FTD-TT-63-922/184, AD-432181)
37349	ZIRCONIA FIBROUS INSULATIONS. ADAMS, J. U. STERRY, J. P. THOMPSON (H I) FIBER GLASS CO., GARDENA, CALIF. 1-78, 1963. (ASD-TDR-63-725, AD-439841)
37359	ELECTRONIC SPECTRA OF THE OXIDES OF LEAD AND OF SOME TERNARY LEAD OXIDE COMPOUNDS. FROM INFRARED TRANSMITTING MATERIALS. KEESTER K L WHITE W B PENNSYLVANIA STATE UNIV. 21-32, 1969. (AD-688933, N69-36716)
37362	SYNTHESIS AND CRYSTALLOGRAPHIC PROPERTIES OF PHASES IN THE SYSTEM SRO-PBO-O. FROM INFRARED TRANSMITTING MATERIALS. KEESTER K L WHITE W B PENNSYLVANIA STATE UNIV. 33-63, 1969. (AD-688933, N69-36716)
37364	INFRARED SPECTRA OF BI2 O3 AND PB/0.6/ BI2 O3. FROM INFRARED TRANSMITTING MATERIALS. DARROW M S WHITE W B PENNSYLVANIA STATE UNIV. 64-80, 1969. (AD-688933, N69-36716)
37370	VITRIFICATION OF THE SOLID FLUORIDE RESIDUE ARISING FROM THE REACTION OF FLUORINE ON IRRADIATED FUEL. BONNIAUD, R. LABE, P. COMMISSARIAT A LENERGIE ATOMIQUE, FONTENAY-AUX-ROSES (FRANCE) 1-19, 1969. (CEA-CONF-1196, N69-29117)
37378	DEVELOPMENT OF FRONTAL SECTION FOR SUPER-ORBITAL, LIFTING, RE-ENTRY VEHICLE. VOL. II. MATERIALS AND COMPOSITE STRUCTURE DEVELOPMENT. LICCIARDELLO M R OHNYSTY B STETSON A R SOLAR, DIV. OF INTERN. HARVESTER CO., SAN DIEGO, CA. 2, 248PP., 1964. (ER-1115-30, FDL-TDR-64-59, AD-442590)
37398	IMPROVED RADIATOR COATINGS. PART I. SCHATZ E A COUNTS C R III BURKS T L AMERICAN MACHINE AND FOUNDRY CO., ALEXANDRIA, VA. 82PP., 1964. (ML-TDR-64-146, AD-442286)
37424	THERMAL RADIATION CHARACTERISTICS OF TRANSPARENT SEMI-TRANSPARENT AND TRANSLUCENT MATERIALS UNDER NONISOTHERMAL CONDITIONS. FOLWEILER, R. C. MALLIO, W. J. LEXINGTON LABS INC., CAMBRIDGE, MASS. 1-130, 1964. (ASD-TDR-62-719(PT. 2), N65-11753, AD-607742)
37430	INFRARED FIBER OPTICS INVESTIGATIONS. KAPANY, N. S. SIMMS, R. J. OPTICS TECHNOLOGY INC., BELMONT, CALIF. 1-84, 1964. (AL-TDR-64-98, N64-27681, AD-601572)
37434	SIMPLIFIED METHOD FOR DETERMINATION OF THERMAL CONDUCTIVITY OF CERAMICS. GLOWER D D AM CERAM SOC BULL 43 11 846-48 1964 JA 48 18

TPRC Number	Bibliographic Citation
37438	IN-PILE EFFECTIVE THERMAL CONDUCTIVITY OF OXIDE FUEL ELEMENTS TO HIGH FISSION DEPLETIONS. DANIEL, R. C. COHEN, I. WESTINGHOUSE ELECTRIC CORP. BETTIS ATOMIC POWER LAB. PITTSBURGH 1-151, 1964. (WAPD-246)
37442	ULTRAVIOLET OPTICAL PROPERTIES OF DIAMOND. WALKER W C OSANTOWSKI J PHYS REV 134 A 1 153-7 1964
37450	THE THERMAL CONDUCTIVITY OF DISPERSED MATERIALS AT DIFFERENT ATMOSPHERIC PRESSURES. KOSTYLEV V M HIGH TEMPERATURE 2 1 15-21 1964 (ENGLISH TRANSLATION OF TEPLOFIZ. VYSOKIKH TEMPERATUR, AKAD. NAUK SSSR, 2 (1), 21-8, 1964; FOR ORIGINAL SEE T37200)
37463	A LINE SOURCE METHOD FOR MEASURING THE THERMAL CONDUCTIVITY AND DIFFUSIVITY OF CYLINDRICAL SPECIMENS OF ROCK AND OTHER POOR CONDUCTORS. JAEGER J C SASS J H BRIT J APPL PHYS 15 10 1187-94 1964 SA 67 29786
37490	EVALUATION OF INFRARED SPECTROPHOTOMETRY FOR COMPOSITIONAL ANALYSIS OF LUNAR AND PLANETARY SOILS. PART II. ROUGH AND POWDERED SURFACES. LYON, R. J. P. STANFORD RESEARCH INSTITUTE, MELNO PARK, CALIF. VOL. 1, 1- , 1964. (NASA-CR-100)
37491	SOLID-CRYOGEN COOLER DESIGN STUDIES AND DEVELOPMENT OF AN EXPERIMENTAL COOLER. GROSS U E JR MANDAL R P LAWSON T W AEROJET-GENERAL CORP., AZUSA, CALIF. 1-211, 1968. (AFFDL-TR-68-1, AGC-3097, AD-834944)
37509	DEVELOPMENT OF A 1200 F RADOME. CHASE, V. A. COPELAND, R. L. BRUNSWICK CORP., MARION, VA. 1-41, 1964. (AD-603730, N64-28569)
37519	APPARATUS FOR MEASURING THE THERMAL CONDUCTIVITY OF BRICKS, GRANULES, AND POWDERS. MITTENBUEHLER, A. BER. DEUT. KERAM. GES. 41 (1), 15-20, 1964.
37530	THE PREPARATION AND PROPERTIES OF GLASS-CERAMICS USING METALLIC PHOSPHATES AS NUCLEATION CATALYSTS. PARTRIDGE G MC MILLAN P W GLASS TECH 4 6 173-82 1963 BC 63 1646
37534	BALL CLAYS. MITCHELL, D. J. BRIT. CERAM. SOC. 1 (2), 303, 1964.
37541	THERMAL CONDUCTIVITY OF UO2, UC, AND UC2. KUBOTA T SEKI Y TAKAHASHI S J NUCL SCI TECHNOL /TOKYO/ 1 93-100 1964 NS 18 34172
37542	THE THERMAL EXPANSION OF MAGNESIA-CONTAINING GLASSES. ENGLISH S TURNER W E S J SOC GLASS TECHNOLOGY 4 115-20 1920
37544	THE THERMAL EXPANSION OF GLASS. PART I. THE GENERAL FORM OF THE EXPANSION CURVE. TURNER W E S WINKS F J SOC GLASS TECHNOLOGY 14 84-109 1930
37545	THE THERMAL EXPANSION OF GLASS. PART II. GLASSES OF THE SERIES SODIUM METASILICATE-SILICA. TURNER W E S WINKS F J SOC GLASS TECHNOLOGY 14 110-26 1930
37566	PROPERTIES OF FORSTERITE - BERYLLIA PORCELAIN AS A HIGH - FREQUENCY INSULATOR. NAKANO, K. SUGIURA, M. OKAWARA, S. NAGOYA KOGYO GIJUTSU SHIKENSHO HOKOKU 12 (12), 583-90, 1963.
37567	ANOMALOUS EXPANSION OF ELECTROCORUNDIUM. FILONENKO, N. E. KUZNETSOVA, O. S. OGNEUPORY 17, 470-4, 1952. (FOR ENGLISH TRANSLATION SEE TPRC 63993)
37570	DEVELOPMENT OF 400 TO 1800 F. FIBROUS-TYPE INSULATION FOR RADIOISOTOPE POWER SYSTEMS. COLLINS J O JAUNARAJS K L REID D R JOHNS-MANVILLE RES. AND ENG. CTR, MANVILLE, N. J. 236PP., 1969. (ALO-2661-12, N70-12852)
37572	SPONGE REFRACTORIES FROM MAGNESIUM OXIDE. GUZMAN I YA SEROVA G A OGNEUPORY 6 281-4 1964 RM 21 05052B
37574	THERMAL EXPANSION OF SODIUM-SILICATE GLASSES CONTAINING FLUORIDES. VARGIN V V KRASOTKINA N I PROC ACAD SCI USSR CHEMICAL TECHNOLOGY 108 6 75-8 1956 (ENGLISH TRANSLATION OF DOKL. AKAD. NAUK, SSSR, 108 (6), 75-8, 1956: FOR ORIGINAL SEE T60736)
37601	PREPARATION FOR AN INVESTIGATION OF THE THERMAL RADIATION CHARACTERISTICS AND THERMAL CONDUCTIVITY OF LUNAR MATERIAL. BIRKEBAK, R. C. CREMERS, C. J. HIGH TEMPERATURE AND THERMAL RADIATION LAB., KENTUCKY UNIV., LEXINGTON, KENTUCKY 44PP., 1968. (N70-41377, NASA-CR-113931)
37606	HEAT OF ADSORPTION OF HYDROGEN ON EVAPORATED FILMS OF TUNGSTEN AND OF NICKEL. BRENNAN D HAYES F H FARADAY SOCIETY TRANSACTIONS 60 495 589-96 1964 RM 21 06564P
37609	UO2-ZR02 STUDIES. WRIGHT T R FACKELMANN J M KIZER D E KELLER D L BATTELLE MEMORIAL INST., COLUMBUS, OHIO 1-6, 1963. (BMI-1655)
37617	THE INFLUENCE OF MOISTURE CONTENT ON THE THERMAL CONDUCTIVITY OF SOME REFRACTORIES AT ROOM TEMPERATURE. SCHULLE, W. ZIEGELINDUSTRIE 17 (3), 77, 1964.
37618	A HALF DECADE OF IMPROVEMENT IN CERAMIC INSULATING MATERIALS. RIGTERINK M D WILLIAMS J C AM CERAM SOC BULL 43 12 894-900 1964
37634	THE INFLUENCE OF MANGANESE OXIDE ON SOME PROPERTIES OF GLASS. CHILDS A A DIMBLEBY V WINKS F TURNER W E S J SOC GLASS TECHNOLOGY 15 172-84 1931
37646	ESTIMATION OF THE EFFECT OF MOISTURE ON THE THERMAL CONDUCTIVITY OF STRUCTURE AND INSULATING MATERIALS. CAMMERER C ACHTZIGER J CHEM-INGR TECH 36 5 493-6 1964
37682	COMPILATION OF THERMAL PROPERTY DATA FOR COMPUTER HEAT-CONDUCTION CALCULATIONS. EDWARDS, A. L. LAWRENCE RADIATION LAB., UNIV. CALIF., LIVERMORE 1-78, 1969. (UCRL-50589, N69-35066)
37743	THERMAL CONDUCTIVITY OF QUARTZ CERAMICS AND QUARTZ GLASS. ROMASHKIN, A. G. TEPLOFIZ. SVOISTVA TVERD. TEL. VYS. TEMP., TR. VSES. KONF. 141-9PP., 1971.
37748	THE BEHAVIOUR OF GLASS WHEN SLOWLY HEATED WITH SPECIAL REFERENCE TO THE THERMAL EXPANSION. SEDDON E TURNER W E S J SOC GLASS TECHNOLOGY 17 324-47 1933
37755	THE FIELD OF GLASS FORMATION AND PROPERTIES OF THE SYSTEM NA2O-SRO-SIO2. DUBROVO S K KASYMOVA S S UZBEKSK KHIM ZH 8 1 14-19 1964 CA 61 4038
37777	THE RELATIONSHIP BETWEEN CHEMICAL COMPOSITION AND PHYSICAL PROPERTIES OF SOME GLAZES IN THE SYSTEM NA2 O-CA O-PB O-AL2 O3-B2 O3-SI O2. ROBERTS G J SALT S ROBERTS W FRANKLIN C E L TRANS BRIT CERAM SOC 63 10 553-602 1964
37779	STUDY OF THE POINTS OF TRANSFORMATION BY A DILATOMETRIC METHOD. BRAESCO P ANNALES DE PHYSIQUE 14 5-75 1920

TPRC Number	Bibliographic Citation
37791	STUDY OF IMPROVED BERYLLIUM OXIDE MATERIALS FOR MICROELECTRONICS TECHNOLOGY. NATIONAL BERYLLIA CORP., HASKELL, N. J. NATIONAL BERYLLIA CORP HASKELL N J 1-60, 1963. (NASA-CR-58527, N64-28843)
37799	INFRARED ABSORPTION SPECTRA OF SODIUM SILICATE GLASSES FROM 4 TO 30 MICRONS. HANNA R SU G-J J AM CERAM SOC 47 12 597-601 1964
37802	A FUSED CAST ALUMINA REFRACTORY. SANDMEYER K H MILLER W A BULL AM CERAM SOC 44 7 541-4 1965
37819	MATERIALS AT HIGH TEMPERATURES. CARPENTER L G BRIT J APPL PHYS 15 8 871-82 1964 RM 21 10255P
37822	APPARATUS FOR MEASURING THERMAL CONDUCTIVITY OF CERAMICS AND METALS AT ELEVATED TEMPERATURES. KAWASHIMA C SAITO S TANAKA S ISHIKAWA I BULL TOKYO INST TECHNOL 53 51-82 1963 AM 17 4759
37823	SOLAR ENERGY REFLECTING GLASS SUN CUT. FURUUCHI S TECHNOCRAT 2 3 46-8 1969
37825	THE THERMAL EXPANSION OF SOME SODA-LIME-SILICA GLASSES. SEDDON E TURNER W E S WINKS F J SOC GLASS TECHNOLOGY 18 5-12 1934
37853	INFLUENCE OF SPECIES, GRAIN SIZE, AND SORBED CATION OF MONTMORILLONITE ON ITS THERMOCHEMICAL BEHAVIOR. SCHWIETE, H. E. GAUGLITZ, R. ACKERMANN, C. FORSCHUNGSBER. LANDES NORDRHEIN-WESTFALEN (1076), 49PP., 1962.
37868	INVESTIGATION OF ABSORPTION AND RE-EMISSION OF OXIDE CRYSTALS CONTAINING DISSIMULAR FOREIGN IONS. SCHNEIDER, W. AIR FORCE CAMBRIDGE RESEARCH LABS. 1-44, 1964. (AFCRL-64-797, AD 606837)
37891	STUDIES IN GERMANIUM OXIDE SYSTEMS. II, PHASE EQUILIBRIA IN THE SYSTEM NA2O-GEO2. MURTHY M K AQUAYO J J AM CERAMIC SOC 47 9 444-7 1964
37896	EFFECTS OF 1.2 AND 0.30 MEV ELECTRONS ON THE OPTICAL TRANSMISSION PROPERTIES OF SEVERAL TRANSPARENT MATERIALS. HAYNES, G. A. MILLER, W. E. LANGLEY RESEARCH CENTER, LANGLEY STA., HAMPTON, VA. 1-32, 1965. (NASA-TN-D-2620, N65-17965)
37899	HEMISPHERICAL SPECTRAL EMITTANCE OF ABLATION CHARS, CARBON, AND ZIRCONIA TO 3700 K. WILSON, R. G. LANGLEY RESEARCH CENTER, LANGLEY STA., HAMPTON, VA. 1-28, 1965. (NASA-TN-D-2704, N65-18607)
37965	A NEW APPROACH TO THE THERMAL CHANGE IN THE REFRACTIVE INDEX OF GLASSES. PRODHOMME L PHYS CHEM GLASSES 1 4 119-22 1960
37966	MEASUREMENT OF THE TOTAL TRANSMITTANCE OF ROCK SALT SAMPLES. DE CORSO S M COLTMAN J W J ENG POWER 86 4 495-6 1964 CA 62 1092
37974	THE INFLUENCE OF BORIC OXIDE ON THE RATE OF MELTING AND ON THE THERMAL EXPANSION AND RESISTANCE TO WEATHERING OF SODA-LIME-SILICA SHEET GLASSES. DIMBLEBY V PARKIN M SEDDON E TURNER W E S WINKS F J SOC GLASS TECHNOLOGY 18 13-18 1934
37975	THE DETERMINATION OF THE TRANSFORMATION POINT OF THE SAME OPTICAL GLASSES IN THREE DIFFERENT LABORATORIES. BERGER E THOMAS M TURNER W E S J SOC GLASS TECHNOLOGY 18 79-87 1934
37976	SOME EXPERIMENTS ON THE SUBSTITUTION OF POTASH IN AN ALKALI-LEAD OXIDE-SILICA GLASS FOR ELECTRIC LAMP PURPOSES. PARTRIDGE J H J SOC GLASS TECHNOLOGY 25 150-8 1941

TPRC Number	Bibliographic Citation
38012	PRODUCTION AND TESTING IN A CONVERTER OF A TRIAL BATCH OF UNFIRED TAR-BONDED MAGNESITE BRICK. PIROGOV A A RAKINA V P GUSEV I D PANKRATOV D I KOISMAN I E KOZIN G N KUKURUZNYAK I S VIT E F OGNEUPORY 27 12 533-9 1962 JA 47 315
38013	LIGHT-WEIGHT HIGH-REFRACTORY ZIRCONIUM DIOXIDE AND ZIRCON. GAODU A N KAINARSKII I S OGNEUPORY 8 380-2 1964 NS 18 44206
38063	BEHAVIOR OF FORSTERITE BRICK IN CEMENT KILNS. BAUMGART W SILICATES IND 26 4 191-5 1961 JA 47 313
38064	RELATION OF DISSOLVED NITROGEN TO PHOTOTROPY OF REDUCED SILICATE GLASSES. SWARTS E L J AMERICAN CERAMIC SOC 53 8 472-5 1970
38075	TEMPERATURE DEPENDENCE OF THERMAL CONDUCTIVITY AND THERMAL DIFFUSIVITY OF SOME UNFILLED POLYMERS. BIL V S AVTOKRATOVA N D TEPLOFIZ VYSOKIKH TEMPERATUR AKAD NAUK SSSR 2 2 192-8 1964 CA 61 9633 (FOR ENGLISH TRANSLATION SEE T36391)
38081	USE OF A SMALL-DISPLACEMENT TRANSDUCER IN VISCOMETRY AND DILATOMETRY. KLYUEV V P MAZURIN O V SAKHOV V B INDUSTRIAL LABORATORY 34 5 735-6 1968 (ENGLISH TRANSLATION OF ZAVOD. LAB. 34 (5), 613-5, 1968; FOR ORIGINAL SEE T38898)
38128	THERMAL CONDUCTIVITY OF FIREBRICKS. KOLTERMANN, M. ELSTNER, I. BER. DEUT. KERAM. GES. 41 (9), 509-11, 1964.
38130	THERMAL EXPANSION OF SOME GLASSES IN THE SYSTEM OF LI2 O-B2 O3-SI O2. HANDWERK J H MC VAY T N OAK RIDGE NATIONAL LAB. 1-22, 1954. (ORO-134)
38131	STUDIES ON THE THERMODYNAMICS AND CONDUCTANCES OF MOLTEN SALTS AND THEIR MIXTURES. II. THE VISCOSITIES, HEATS OF FUSION, AND HEAT CAPACITIES OF LITHIUM CHLORATE AND LITHIUM CHLORATE-LITHIUM NITRATE MIXTURES. CAMPBELL A N NAGARAJAN M K CANAD J CHEM 42 7 1616-26 1964 SA 67 29620
38148	THERMAL CONDUCTIVITY OF LUNAR AND TERRESTRIAL IGNEOUS ROCKS IN THEIR MELTING RANGE. MURASE T MC BIRNEY A R SCIENCE 170 165-7 1970
38152	HEAT CAPACITY AND THERMODYNAMIC PROPERTIES OF BERYLLIUM ALUMINATE /CHRYSOBERYL/, BE O - AL2 O3 FROM 16 TO 380 K. FURUKAWA, G. T. SABA, W. G. J. RES. NATL. BUREAU OF STANDARDS 69 (1), 13-18, 1964. (AFOSR-65-1786, AD 625096)
38163	THE TEMPERATURE VARIATION OF THERMAL EXPANSION AND ELECTRICAL RESISTIVITY OF A BOROSILICATE GLASS. COX S M STIRLING J F KIRBY P L J SOC GLASS TECHNOLOGY 35 103-35 1951
38184	MEASUREMENT OF THE THERMAL CONDUCTIVITY AND THERMAL DIFFUSIVITY COEFFICIENTS OF REFRACTORY MATERIALS BY THE UNSTEADY REGIME METHOD. VISHNEVSKII I I DZYUBENKO M I INZH-FIZ ZH AKAD NAUK BELORUSSK SSR 7 10 45-8 1964 CA 62 3782
38189	ANOMALOUS THERMAL CONDUCTIVITY OF A FILM OF WATER ON MICA CRYSTALS. METSIK M S AIDANOVA O S ISSLED V OBL POVERKHN SIL AKAD NAUK SSSR INST FIZ KHIM SB DOKL NA VTOROI KONF MOSCOE 1962 188-95 1964 CA 62 2262
38199	A SIMPLIFIED METHOD FOR DETERMINING THE THERMAL PROPERTIES OF ROCKS FROM THE CURVE OF THE TEMPERATURE CHANGE. MATAEV G A GARUNOV G A GAIDAROV G M AMIRKHANOV M SH I IZV VYSSH UCHEB ZAVED NEFT GAZ 11 31-3 1965

TPRC Number	Bibliographic Citation
38203	**HEAT CAPACITY OF LEAD CONCENTRATES AND SINTERS AT HIGH TEMPERATURES.** BRATCHIKOV S G IZV VYSSHIKH UCHEBN ZAVEDENII TSVETN MET 7 4 79-82 1964 CA 62 4698 (FOR ENGLISH TRANSLATION SEE T49388)
38224	**INFRARED REFLECTIVITY OF FERRIMAGNETIC OXIDE CERAMICS.** WESTWOOD W D SADLER A G LEWIS D C J CAN CERAM SOC 33 138-46 1964 CA 62 3780
38236	**THERMAL EXPANSION COEFFICIENTS OF ZIRCONATE CERAMICS.** BRANSON D L J AM CERAM SOC 48 8 441 1965
38298	**COEFFICIENT OF THERMAL DIFFUSIVITY OF DINAS REFRACTORY MATERIALS.** LITOVSKII, E. IA. LANDA, IA. A. TEPLOFIZ. SVOISTVA TVERD. TEL. VYS. TEMP., TR. VSES. KONF. 155-65PP., 1971.
38312	**THE RELATIONS OF THERMAL EXPANSION, COMPOSITION AND STRUCTURE OF GLASSES. I. THE SODIUM OXIDE-SILICA GLASSES.** BLAU H H JR J SOC GLASS TECHNOLOGY 35 304-17 1951
38313	**THE THERMAL EXPANSION OF ALKALI-FELDSPARS.** KOZU S SAIKI S SCI REPT TOHOKU UNIV 2 3 205-38 1925
38320	**THERMAL CONDUCTIVITY OF COMPRESSIBLE POROUS MATERIALS.** MC MASTER D G TAPPI 47 12 796-801 1964 CA 63 260
38334	**THE OPTICAL STABILITY OF QUARTZ GLASS TO GAMMA-IRRADIATION CARRIED OUT AT DIFFERENT TEMPERATURES.** BREKHOVSKIKH S M LANDA L M VIKTOROVA YU N SHELYUBSKII V I PROC ACAD SCI USSR CHEM TECHNOL SECT 163 1 127-8 1965 (ENGLISH TRANSLATION OF DOKL. AKAD. NAUK. SSR 163 (1), 164-5, 1965; FOR ORIGINAL SEE T38846)
38338	**CRYSTALLIZATION OF GLASSES IN THE LI2 O-GA2 O3-SI O2 SYSTEM.** DUBROVO S K TSEKHOMSKAYA T S ZASOLOTSKAYA M V PROC ACAD SCI USSR CHEM TECH SECT 180 77-9 1968 (ENGLISH TRANSLATION OF DOKL. AKAD. NAUK SSSR 180 (1), 172-4, 1968; FOR ORIGINAL SEE T50051)
38360	**THE INFLUENCE OF TEMPERATURE ON THE PATH-DIFFERENCE AND ON THE SCHILLERIZATION IN SODA-ORTHOCLASE AND MOONSTONE.** KOZU S MASUDA M SCI REPTS TOHOKU UNIV 3 3 11-12 1926
38381	**OPTICAL AND THERMAL PROPERTIES OF CANCRINITE FROM DODO, KOREA.** KOZU S UEDA J TSURUMI S PROC IMPERIAL ACAD /TOKYO/ 9 1 13-16 1933
38383	**ELASTIC AND THERMAL EXPANSION PROPERTIES OF CONCRETE AS AFFECTED BY SIMILAR PROPERTIES OF THE AGGREGATE.** KOENITZER L H PROC AM SOC TESTING MATERIALS 36 2 393-406 1936
38402	**A METHOD FOR THE SIMULTANEOUS DETERMINATION OF ALL THERMAL PROPERTIES OF POOR HEAT CONDUCTORS OVER THE TEMPERATURE RANGE 80 TO 500 K.** LUIKOV A V VASILEV L L SHASHKOV A G ASME SYMP. ON THERMOPHYSICAL PROPERTIES, 3RD 314-19 1965 CA 63 9085
38404	**THERMAL CONDUCTIVITY OF SEVERAL CERAMIC MATERIALS TO 2500 C.** FEITH A D ASME SYMP. ON THERMOPHYSICAL PROPERTIES, 3RD 328-35 1965 CA 63 9085
38410	**THERMAL EXPANSION OF BASALTIC HORNBLENDE.** KOZU S UEDA J PROC IMPERIAL ACAD /TOKYO/ 10 1 25-8 1934
38438	**THERMAL EXPANSION OF AUGITE.** KOZU S UEDA J PROC IMPERIAL ACAD /TOKYO/ 10 2 87-90 1934
38439	**THERMAL EXPANSION OF ARAGONITE AND ITS ATOMIC DISPLACEMENTS BY TRANSFORMATION INTO CALCITE BETWEEN 450 AND 490 C. IN AIR, I.** KOZU S KANI K PROC IMPERIAL ACAD /TOKYO/ 10 4 222-5 1934
38441	**ON THE MEASUREMENT OF VERY LOW TEMPERATURES. V. THE EXPANSION COEFFICIENT OF JENA AND THURINGER GLASS BETWEEN 16 AND -182 C.** ONNES H K HEUSE W VERSLAG GEWONE VERGADER AFDEL NATUURK KONINK NED AKAD WETENSCHAPPEN 212-23 1903 (FOR ENGLISH TRANSLATION SEE T35787)
38443	**EFFECT OF CERTAIN ADDITIVES ON SINTERING AND PROPERTIES OF SUPERDUTY PERICLASE CERAMICS.** POLUBOYARINOV D N POPILSKIY A YA TUNG-HUA T OGNEUPORY 24 4 178-84 1962 (FOR ENGLISH TRANSLATION SEE T41856)
38471	**ANOMALIES IN THE THERMAL EXPANSION CURVES FOR GLASS.** BELLIERE M REVUE BELGE DES INDUSTRIES VERRIERES 1 262-6 1930-1
38474	**LINEAR COEFFICIENT OF THERMAL EXPANSION OF AMBROID.** YETTER E W REV SCIENTIFIC INSTRUMENTS 10 147 1939
38475	**ALKALI-RESISTANT LABORATORY GLASSWARE.** CORNING GLASS WORKS, NEW YORK REV SCIENTIFIC INSTRUMENTS 12 44 1941
38488	**DEPENDENCE OF THE TRUE (MOLECULAR) HEAT CONDUCTION OF A BODY ON ITS POROSITY.** SHEPTUNOV, K. L. IZV. VYSSHIKH UCHEBN. ZAVEDENII, MASHINOSTR. (3), 170-2, 1964.
38510	**THERMAL CONDUCTIVITY OF SOLID COAL SAMPLES.** GODRIDGE A M FUEL 35 379-81 1956
38523	**PIEZOELECTRIC PROPERTIES AND PHASE TRANSTION OF PB IN/0.5/ NB/0.5/ O3-PB TI O3 SOLID SOLUTION CERAMICS.** KODAMA U OSADA M KUMON O NISHIMOTO T AM CERAM SOC BULL 48 12 1122-4 1969
38527	**THERMAL EXPANSION OF IRON PYRITES.** CHRYSTALL R S B TRANS FARADAY SOC 61 1811-15 1965 SA 50 169
38556	**RADIATION-INDUCED INSTABILITY IN REFRACTIVE PROPERTIES OF SOME OPTICAL GLASSES.** MALITSON, I. H. NATIONAL BUREAU OF STANDARDS, WASHINGTON, D. C. 1-17, 1965. (CONF-651074-1),
38576	**QUANTUM EFFICIENCY OF ND /+3/ IN GLASS, CALCIUM TUNGSTATE, AND YTTRIUM ALUMINUM GARNET.** BRANDEWIE R A TELK C L J OPT SOC AM 57 10 1221-5 1967
38580	**A BIDIRECTIONAL REFLECTANCE ACCESSORY FOR SPECTROSCOPIC MEASUREMENTS.** HUNT G R ROSS H P APPL OPT 6 10 1687-90 1967
38592	**INVESTIGATION OF SOME INORGANIC DIELECTRIC MATERIALS IN THE FAR INFRARED.** MASHKOVICH M D DEMESHINA A I SOVIET PHYSICS-SOLID STATE 7 6 1323-7 1965 (ENGLISH TRANSLATION OF FIZ. TVERD. TELA 7 (6), 1634-8, 1965; FOR ORIGINAL SEE T35610)
38598	**SPECTRA OF SOME TWO-COMPONENT LEAD-SILICATE GLASSES IN THE ULTRAVIOLET REGION.** SMIRNOVA E A DOKLADY AKAD NAUK SSSR 161 3 569-71 1965 (FOR ENGLISH TRANSLATION SEE T38599)
38599	**SPECTRA OF SOME TWO-COMPONENT LEAD-SILICATE GLASSES IN THE ULTRAVIOLET REGION.** SMIRNOVA E V SOVIET PHYSICS-DOKLADY 10 3 247-9 1965 (ENGLISH TRANSLATION OF DOKLADY AKAD. NAUK SSSR 161 (3), 569-71, 1965; FOR ORIGINAL SEE T38598)

TPRC Number	Bibliographic Citation
38609	**A STUDY OF MULTI-LAYER INSULATION.** POVOLOTSKII L V ARKADEV B A TEPLOENERGETIKA 11 1 36-40 1964 (FOR ENGLISH TRANSLATION SEE T38610)
38610	**A STUDY OF MULTI-LAYER INSULATION.** POVOLOTSKII L V ARKADEV B A THERMAL ENGINEERING 11 1 44-8 1964 (ENGLISH TRANSLATION OF TEPLOENERGETIKA 11 (1), 36-40, 1964; FOR ORIGINAL SEE T38609)
38646	**STUDIES OF THE CHARACTERISTICS OF PROBABLE LUNAR SURFACE MATERIALS. PART I.** SALISBURY, J. W. GLASER, P. E. ARTHUR D LITTLE INC., CAMBRIDGE, MASS. 1-309, 1964. (AFCRL-64-970, AD-613018 (PT. 1), N65-26232)
38654	**COLOR CENTER GROWTH CURVES IN CALCITE.** MEDLIN W L J PHYS CHEM SOLIDS 28 9 1725-33 1967
38662	**INFRARED ABSORBING SEALING GLASSES.** HOOGENDOORN H SUNNERS B AM CERAM SOC BULL 48 12 1125-7 1969
38674	**TRANSMITTANCE OF OPTICAL MATERIALS AT HIGH TEMPERATURES IN THE 1-MU TO 12-MU RANGE.** GILLESPIE D T OLSEN A L NICHOLS L W APPL OPT 4 11 1488-93 1965 CA 63 17301
38729	**SOME OF THE PHYSICAL PROPERTIES OF SWEDISH IRON ORES. I. THERMAL CONDUCTIVITY AND MAGNETOSTRICTION.** KRUCKENBERG I ARKIV MATEMATIK ASTRONOMI FYSIK 2 1 1-14 1905
38742	**STRUCTURE AND MECHANISM OF CONDUCTION OF SEMICONDUCTOR GLASSES.** JANAKIRAMA-RAO BH V J AM CERAM SOC 48 6 311-19 1965
38840	**INSULATING FIREBRICK.** JOHNS-MANVILLE CORP REV SCIENTIFIC INSTRUMENTS 12 376 1941
38846	**RADIATION OPTICAL STABILITY OF QUARTZ GLASS WHEN X-RAYED AT VARIOUS TEMPERATURES.** BREKHOVSKIKH S M LANDA L M VIKTOROVA YU N SHELYUBSKII V I DOKL AKAD NAUK SSSR 163 1 164-5 1965 CA 63 9636 (FOR ENGLISH TRANSLATION SEE T38334)
38858	**PYREX BRAND MULTIFORM GLASSWARE.** CORNING GLASS WORKS, NEW YORK REV SCIENTIFIC INSTRUMENTS 14 155 1943
38860	**METHOD FOR THERMAL CONDUCTIVITY MEASUREMENTS ON SOLIDS.** HAACKE G SPITZER D P J SCI INSTR 42 9 702-4 1965 CA 63 7085
38861	**THE ACCURATE MEASUREMENT OF THE SPECTRAL TRANSMISSION OF OPTICAL GLASS.** HARPER D W J SCI INSTR 42 10 746-8 1965 CA 63 11130
38865	**THERMAL PROPERTIES OF SILICA AT LOW TEMPERATURES.** WHITE G K BIRCH J A PHYS CHEM GLASSES 6 3 85-9 1965 JA 48 296
38880	**THERMAL EXPANSION OF SILICA BRICK AND MORTARS.** COLE S S J AM CERAMIC SOC 13 7 437-46 1930
38884	**NEW CHROME PIGMENTS OF THE SPINEL TYPE.** TUMANOV S G PYRKOV V P STEKLO I KERAM 22 6 2-5 1965 CA 63 9642 (FOR ENGLISH TRANSLATION SEE T36067)
38898	**USE OF A SMALL-DISPLACEMENT TRANSDUCER IN VISCOMETRY AND DILATOMETRY.** KLYUEV V P MAZURIN O V SAKHOV V B ZAVOD LAB 34 5 613-5 1968 (FOR ENGLISH TRANSLATION SEE T38081)
38901	**THERMAL VOLUME CHANGE AND ELASTICITY OF AGGREGATES AND THEIR EFFECT ON CONCRETE.** WILLIS T F DE REUS M E PROC AMERICAN SOC TESTING MATERIALS 39 919-28 1939
38913	**DETERMINATION OF THE COEFFICIENT OF THERMAL CONDUCTIVITY IN AN ASYMMETRICAL HEATING MODE.** KAGANER M G IND LAB USSR 36 9 1396-7 1970 (ENGLISH TRANSLATION OF ZAVODSKAYA LAB., 36 (9), 1095-96, 1970; FOR ORIGINAL SEE T62437)
38919	**EQUATION FOR EQUIVALENT COEFFICIENT OF THERMAL CONDUCTIVITY FOR A PETROLEUM STRATUM.** POGORELSKII, A. M. TEPLOFIZ. SVOISTVA TVERD. TEL. VYS. TEMP., TR. VSIS. KONF. 210-16PP., 1971.
38969	**INFLUENCE OF THE CHEMICAL COMPOSITION OF MOLTEN MARTENSITE SLAGS ON ITS RADIATIVE CHARACTERISTICS.** KAIROV E A MASTRYUKOV B S KRIVANDIN V A IZVEST VYSSHIKH UCHEB CHERN MET 13 7 155-8 1970
38994	**EARTH TEMPERATURE AND THERMAL DIFFUSIVITY AT SELECTED STATIONS IN THE UNITED STATES.** KUSUDA, T. ACHENBACH, P. R. NATIONAL BUREAU OF STANDARDS 1-50, 1965. (NBS-R-8972, AD-472916)
38998	**ULTRAVIOLET REFLECTIVITY OF MARS.** EVANS D C SCIENCE 149 969-72 1965 CA 63 17316
39007	**EFFECTS OF REFLECTION PROPERTIES OF NATURAL SURFACES IN AERIAL RECONNAISSANCE.** COULSON K L APPL OPT 5 6 905-17 1966
39008	**OVERCAST SKY LUMINANCES AND DIRECTIONAL LUMINOUS REFLECTANCES OF OBJECTS AND BACKGROUNDS UNDER OVERCAST SKIES.** GORDON J I CHURCH P V APPL OPT 5 6 919-23 1966
39011	**EFFECTS OF A SIMULATED HIGH-ENERGY SPACE ENVIRONMENT ON THE ULTRAVIOLET TRANSMITTANCE OF OPTICAL MATERIALS BETWEEN 1050 ANGSTROMS AND 3000 ANGSTROMS.** HEATH D F SACHER P A APPL OPT 5 6 937-43 1966
39012	**OPTICAL PROPERTIES AND APPLICATIONS OF PHOTOCHROMIC GLASS.** MEGLA G K APPL OPT 5 6 945-60 1966
39028	**THERMAL CONDUCTIVITY AND DIFFUSIVITY OF SOILS AS RELATED TO MOISTURE TENSION AND OTHER PHYSICAL PROPERTIES.** AL-NAKSHABANDI G KOHNKE H AGR METEOROL 2 271-9 1965
39092	**SPECTRA NOTEBOOK. VOLUME I. MATERIAL, TARGET AND BACKGROUND DATA.** WILBURN, D. K. ARMY TANK AUTOMOTIVE CENTER COMPONENTS RESEARCH AND DEVELOPMENT LABS., WARREN, MI. 1-66, 1965. (TR-8863 VOL. 1, AD-475817)
39130	**THERMAL CONDUCTIVITY OF FIRECLAY REFRACTORIES.** AMERICAN SOCIETY FOR TESTING MATERIALS ASTM 427-30, 1947. (ASTM-C-202-47)
39134	**MICACEOUS BIOTITE AS EFFICIENT BREWSTER ANGLE POLARIZER FOR VACUUM ULTRAVIOLET.** ROBIN M B KUEBLER N A PAO Y-H REV SCI INSTR 37 7 922-4 1966
39168	**THE INFLUENCE OF LOW MOISTURE CONTENT ON THE CONDUCTIVITIES OF A GRANULAR SUBSTANCE SUCH AS SAND.** ALBERTS L BOHLMANN M MEIRING G L BRIT J APPL PHYS 17 7 951-5 1966
39186	**REDUCTION BEHAVIOR AND SELECTED PROPERTIES OF ALUMINA-CERIA COMPOSITIONS.** STANISLAW, T. S. GRUMMAN AIRCRAFT ENGINEERING CORP., BETHPAGE, N. Y. 1-23, 1967. (AD-818794)
39188	**THE THERMODYNAMIC AND OPTICAL PROPERTIES OF GERMANIUM, SILICON, DIAMOND AND GALLIUM ARSENIDE.** DOLLING G COWLEY R A PROC PHYS SOC /LONDON/ 88 2 463-94 1966

TPRC Number	Bibliographic Citation
39194	SINTERING CONDITIONS AND PHYSICAL-TECHNICAL PROPERTIES OF LANTANUM SILICATES. ANDREEVA A B KELER E K ZH PRIKL KHIM 36 12 2605-10 1963
39212	THERMAL RADIATION PROPERTY MEASUREMENT TECHNIQUES. DUNN S T GEIST J C MOORE D G CLARK H E RICHARD J C NATIONAL BUREAU OF STANDARDS 1-83, 1967. (NASA-CR-66127, NBS-TN-415, N67-31130)
39259	EFFECTS OF A SIMULATED HIGH ENERGY ELECTRON SPACE ENVIRONMENT ON THE ULTRAVIOLET TRANSMITTANCE OF OPTICAL MATERIALS BETWEEN 1050 AND 3000 A. HEATH, D. F. SACHER, P. A. GODDARD SPACE FLIGHT CENTER, GREENBELT, MD. 1-26, 1965. (NASA-TM-X-55337, N66-11232)
39261	RESEARCH ON THERMAL TRANSFER PHENOMENA. ALLEN, W. C. NATIONAL BERYLLIA CORP., HASKELL, N. J. 1-21, 1965. (NASA-CR-68094, N66-12188)
39275	THERMAL DIFFUSIVITY OF SOLID SOLUTIONS OF BARIUM TITANATE AND STRONTIUM TITANATE. REMOISSENET M GODEFROY L COMPT REND 262 B 1 56-9 1966 CA 64 18509
39281	OPTICAL PROPERTIES OF SINGLE CRYSTALS OF CORUNDUM AND OF RUBIES IN THE EXTREME ULTRAVIOLET. LEMONNIER J-C STEPHAN G ROBIN S COMPT REND 262 B 5 355-8 1966 CA 65 1604
39292	REFLECTION POLARIZERS FOR THE FAR ULTRAVIOLET. STEPHAN G LEMONNIER J-C LE CALVEZ Y ROBIN S COMPT REND 262 B 19 1272-5 1966 CA 65 8189
39312	THE RESPONSE OF SELECTED ROCK SPECIMENS TO VACUUM ULTRAVIOLET LIGHT. DUPEE, W. D. AIR FORCE INST. OF TECH., WPAFB, ONIO, M.S. THESIS 1-148, 1965. (GSF/MECH-65-43, AD-625393)
39333	TEMPERATURES OF METEOROIDS IN SPACE. BUTLER, C. P. NAVAL RADIOLOGICAL REFENSE LAB., SAN FRANCISCO, CALIF 1-27, 1965. (USNRDL-TR-927, AD-626105)
39337	COMPRESSIBILITIES AND ELECTRICAL RESISTANCE UNDER PRESSURE, WITH SPECIAL REFERENCE TO INTERMETALLIC COMPOUNDS. BRIDGMAN P W PROC AMERICAN ACAD ARTS SCI 70 7 285-317 1936
39347	SINTERED CERAMIC COATINGS FOR THERMAL CONTROL. COX, R. L. LING-TEMCO-VOUGHT INC DALLAS TEXAS 1-23, 1964. (REPT NO. 00.369, N65-36213)
39350	SUSCEPTIBILITY OF FROST CRACKING OF SOILS. III. SETTING UP A THERMAL STUDY OF SOILS AND TESTS, WITHOUT FREEZING AND WITH FREEZING, CARRIED OUT ON A WATER-SATURATED SAND MOLD. AGUIRRE-PUENTE, J. LE FUR, B. SZANTO, I. CENTRE NATIONAL DE LA RECHERCHE SCIENTIFIQUE LAB. 1-91, 1965. (REPT.-65-5, N66-14091)
39351	PRELIMINARY REPORT ON THE THERMODYNAMIC PROPERTIES OF SELECTED LIGHT-ELEMENT AND SOME RELATED COMPOUNDS. BECKETT, C. W. DOUGLAS, T. B. NATIONAL BUREAU OF STANDARDS 1-110, 1966. (NBS-REPT-9028, AFOSR-66-1102, AD-485872L)
39355	PROBE CONSTANT FOR IN-SITU DETERMINATION OF THERMAL CONDUCTIVITIES OF SOILS. HUS S T LEE W W PROC 3RD INTERN HEAT TRANSFER CONF 4 81-8 1966 (ASME-HT-122)
39358	THERMAL CONDUCTIVITY MEASUREMENTS OF FIBROUS INSULATIONS UP TO 2500 F. ROLINSKI E J PURCELL G V PROC 3RD INTERN HEAT TRANSFER CONF 4 133-40 1966 (ASME-HT-128)
39365	THERMAL RADIATION CHARACTERISTICS OF TRANSPARENT, SEMI-TRANSPARENT AND TRANSLUCENT MATERIALS UNDER NON-ISOTHERMAL CONDITIONS. HOBBS, H. A. FOLWEILER, R. C. LEXINGTON LABS INC., CAMBRIDGE, MASS. 1-112, 1966. (ASD-TDR-62-719 (PT. 3), N66-37058, AD-635621)
39422	EFFECT OF A HETEROGENEOUS CHEMICAL REACTION ON THE HEAT TRANSFER BETWEEN THE BOILING LAYER AND THE SURFACE. GLINKOV M A BELOUSOV V V IZV VYSSHIKH UCHEBN ZAVEDENII CHERN MET 7 175-9 1966
39500	CORDIERITE CERAMIC- THERMAL SHOCK-RESISTANT REFRACTORY MATERIAL. PANKRATOVA G F POLUBOYARINOV D N ZAIONTS R M OGNEUPORY 25 2 73-6 1960
39516	THERMAL PROPERTIES OF A SIMULATED LUNAR MATERIAL IN AIR AND IN VACUUM. BERNETT E C WOOD H L JAFFE L D MARTENS H E JET PROPULSION LAB., CALIF. INST. OF TECH., PASADENA 23PP., 1962. (JPL-TR-32-369, N63-13767)
39517	REVERSIBLE THERMAL EXPANSION OF REFRACTORIES. KANZ A MITT FORSCHUNGS INST VER STAHLWERKE DORTMUND 2 5 77-96 1931
39603	GLASS LASERS. SNITZER E APPL OPT 5 10 1487-99 1966
39649	A THERMODYNAMIC INVESTIGATION OF MOLYBDENUM DIBROMIDE. SHCHUKAREV S A LENINGRAD UNIV VESTNIK SER FIZ I KHIM 17 4 148-53 1962 (FOR ENGLISH TRANSLATION SEE T39650)
39650	A THERMODYNAMIC INVESTIGATION OF MOLYBDENUM DIBROMIDE. SHCHUKAREV, S. A. BRITISH IRON AND STEEL INDUSTRY LONDON 1-11, 1965. (ENGLISH TRANSLATION OF LENINGRAD. UNIV. VESTNIK SER FIZ. I. KHIM., 17 (4), 148-53, 1962; FOR ORIGINAL SEE T39649) (BISI-4509, NCH-203215)
39699	STRENGTH PROPERTIES OF SINTERED ALUMINA IN RELATION TO POROSITY AND GRAIN SIZE. CUTLER I B J AM CERAM SOC 40 1 20-3 1957
39712	SILICA REFRACTORIES-RELATION BETWEEN THERMAL EXPANSION AND MINERALOGICAL COMPOSITION. STARCZEWSKI M PRACE INST HUTNICZYCH 11 311-17 1959
39737	THERMAL PROPERTIES OF WEST VIRGINIA HIGHWAY MATERIALS. OBLENIS, J. D. WEST VIRGINIA UNIV., MORGANTOWN, M.S. THESIS 1-172 1964
39738	ANISOTROPY OF THE OPTICAL CONSTANTS OF SINGLE CRYSTAL GRAPHITE IN THE ULTRAVIOLET-VISIBLE SPECTRUM. YASINSKY, J. B. UNIVERSITY OF PITTSBURGH, M.S. THESIS 1-49 1963
39752	HIGH-TEMPERATURE HEAT CONTENTS OF MANGANOUS SULFIDE, FERROUS SULFIDE AND PYRITE. COUGHLIN J P J AM CHEM SOC 72 3 5445-7 1950
39755	ZIRCON AS OPACIFIER IN HARD PORCELAIN GLAZES. GULEVICH O SKLAR KERAM 11 32-4 1961 CA 56 12558
39761	MANGANOUS-MANGANIC EQUILIBRIUM IN ALKALI BORATE GLASSES. PAUL A LAHIRI D J AM CERAM SOC 49 10 565-8 1966
39818	EFFECT OF MOISTURE ON THERMAL CONDUCTIVITY OF POROUS MATERIALS. SUGAWARA A J APPL PHYS JAPAN 30 12 899-906 1961

TPRC Number	Bibliographic Citation
39834	**FOURTH ANNUAL REPORT-HIGH-TEMPERATURE MATERIALS AND REACTOR COMPONENT DEVELOPMENT PROGRAMS.** NUCLEAR MATERIALS AND PROPULSION, CINCINNATI, OHIO GENERAL ELECTRIC COMPANY, CINCINNATI, OH 1-237, 1965 (GEMP-334-A VOL I)
39835	**INFRARED TRANSMITTANCE OF SOME CALCIUM ALUMINATE AND GERMANATE GLASSES.** FLORENCE J M GLAZE F W BLACK M H J RESEARCH NAT BUR STANDARDS 55 4 231-50 1955 (NBS-RP-2625)
39841	**STUDY OF THE EFFECTIVE THERMAL EMITTANCE OF CYLINDRICAL CAVITIES.** VOLLMER J J OPT SOC AM 47 10 926-32 1957
39842	**COLORS PRODUCED BY REFLECTION AT GRAZING INCIDENCE FROM ROUGH SURFACES.** MIDDLETON W E K WYSZECKI G J OPT SOC AM 47 11 1020-3 1957
39843	**ALUMINOUS INSULATING MATERIALS RESIST HIGH TEMPERATURES.** BARNITT J B HEILMAN R H CHEM MET ENG 38 390-3 1931
39845	**STRUCTURAL MATERIALS FOR MODERN HEAT TECHNOLOGY.** NORTON F H CHEM MET ENG 39 226-8 1932
39850	**SOME MATERIALS OF LOW THERMAL CONDUCTIVITY.** GRIFFITHS E FARADAY SOC TRANS 18 252-8 1922
39852	**INVESTIGATION OF INSULATED WALLS.** ROWLEY F B TRANS AM SOC MECH ENGRS 49/50 49-55 1927-8
39856	**HEAT LOSS THROUGH FIREBRICK.** DENNY R C COMBUSTION 4 2 26-9 1932
39857	**CHARACTERISTICS OF SOME INSULATIONS FOR LIQUID OXYGEN TRANSFER LINES.** VAN GUNDY D A JACOBS R B PROC 1954 CRYOG ENG CONF 2 156-62 1955
39858	**THERMAL CONDUCTIVITIES OF POROUS ROCKS FILLED WITH STAGNANT FLUID.** KUNII D SMITH J M J SOC PETROL ENGRS 1 1 37-42 1961
39859	**CONDUCTIVITY, POROSITY AND GAS PERMEABILITY OF REFRACTORY MATERIALS.** WOLOGDINE M S QUENEAU A L ELECTROCHEM METALL IND 7 383-9 433-6 1909
39860	**DEVELOPMENT OF THE THERMAL CONDUCTIVITY PROBE.** HOOPER F C CHANG S C AM SOC HEATING VENTILATING ENGRS RES BULL 59 463-72 1953
39866	**THERMAL INSULATION WITH ALUMINUM FOIL.** MASON R B IND ENG CHEM 25 3 245-55 1933
39867	**ELECTRICALLY FUSED MAGNESIA.** WHITE H E J AM CERAM SOC 21 216-29 1938
39868	**THE HEAT-INSULATING EFFICIENCIES OF SOME DIATOMACEOUS EARTH PRODUCTS AND SLAG WOOL.** GREEN A T EDWARDS H GAS J 171 23-7 1925
39869	**THE THERMAL CONDUCTIVITY OF SPHERICAL METAL POWDERS INCLUDING THE EFFECT OF AN OXIDE COATING.** SWIFT D L INTERN J HEAT MASS TRANSFER 9 10 1061-74 1966 CA 66-2 6202
39896	**THE EFFECT OF COMPOSITION AND TEMPERATURE ON THE ULTRAVIOLET ABSORPTION OF GLASS.** MC SWAIN, B. D. UNIV. OF ROCHESTER, ROCHESTER, NY, M.S. THESIS 1-116 1962
39897	**THE APPARENT THERMAL CONDUCTIVITY OF CELLULAR MATERIALS.** HARDEE, H. C., JR. UNIV. OF TEXAS, AUSTIN, M.S. THESIS 1-49 1961
39945	**FIFTY YEARS OF GLASS TECHNOLOGY.** DOUGLAS R W NATURE /LONDON/ 212 769-74 1966
39948	**NONOXIDE CHALCOGENIDE GLASSES AS INFRARED OPTICAL MATERIALS.** HILTON A R APPL OPTICS 5 12 1877-82 1966
39951	**OPTICAL CHARACTERISTICS OF A PROPOSED REFLECTANCE STANDARD.** TRYTTEN G FLOWERS W APPL OPTICS 5 12 1895-7 1966 CA 66-6 24113 CORRECTION.
39987	**ACCELERATED THERMOCOUPLE METHOD OF MEASURING THERMAL CONDUCTIVITY.** YANKELEV L F ROIFE V S INZH FIZ ZH 8 4 511-15 1965 (FOR ENGLISH TRANSLATION SEE T039988)
39988	**ACCELERATED THERMOCOUPLE METHOD OF MEASURING THERMAL CONDUCTIVITY.** YANKELEV L F ROIFE V S J ENG PHYS /BSSR/ 8 4 355-8 1965 (ENGLISH TRANSLATION OF INZH. FIZ. ZH., 8 (4), 511-5, 1965; FOR ORIGINAL SEE T039987)
39997	**THERMAL PROPERTIES OF COMPOSITIONS IN THE SYSTEM UO2-ZRO2-CAO.** KERN R S BEALS R J BULL AM CERAM SOC 46 12 1154-9 1967 CA 68 45260
40025	**THE THERMAL CONDUCTIVTY OF SOME HEAT INSULATORS.** VAN DUSEN, M. S. JOHN HOPKINS UNIV., PH.D. THESIS 1-27 1921
40087	**ABSORPTION SPECTRA IN THE FAR INFRARED OF OXIDES OF MAGNESIUM, IRON, MANGANESE AND THEIR MIXED CRYSTALS /FEO-MGO/ AND /FEO-MNO/.** BREHAT F EVRARD O HADNI A LAMBERT J P COMPT REND 263 B 20 1112-15 1966
40228	**MODERN HEAT INSULATING AND DECKING MATERIALS.** GRIFFITHS E HICKMAN M J TRANSACTIONS NORTH EAST COAST INSTITUTION OF ENGINEERS AND SHIPBUILDERS 59 207-30 1943
40230	**REFLECTANCE OF COMPACTED POWDER MIXTURES.** SCHATZ E A J OPT SOC AM 57 7 941-50 1967 CA 67 58717
40240	**ATOMIC SPECIFIC HEATS BETWEEN THE BOILING POINTS OF LIQUID NITROGEN AND HYDROGEN. I. THE MEAN ATOMIC SPECIFIC HEATS AT 50 DEGREES ABSOLUTE OF THE ELEMENTS A PERIODIC FUNCTION OF THE ATOMIC WEIGHTS.** DEWAR J PROC ROY SOC 89 A 158-69 1913
40297	**METHODS OF PURIFICATION OF METALS AND INTERMETALLIC COMPOUNDS.** BRAMAN, R. S. ARMOUR RES. FOUNDATION, CHICAGO, ILL. 1-48, 1962. (ARF-3133-36, WADD-TR-59-303-OT-3, AD 802970)
40320	**INVESTIGATION ABOUT THE THERMAL EXPANSION OF SOLIDS.** SCHEEL K WISS ABHANDL PHYSIK-TECH REICHSANSTALT 4 49-60 1918
40324	**A MULTI-CHANNEL OPTICAL PYROMETER FOR THE 1500 TO 3000 K RANGE.** PIERSON, A. H. ZEIGLER, G. WEISS, M. BARNES ENGINEERING CO., STAMFORD, CT 1-69, 1966. (AFFDL-TR-65-133, AD-480387)
40328	**FURTHER MEASUREMENTS OF THE COEFFICIENT OF LINEAR EXPANSION AT LOW TEMPERATURES.** DORSEY H G PHYS REV 27 1 1-10 1908
40351	**IMPROVED RADIATOR COATINGS.** SCHATZ, E. A. COUNTS, C. R., III AMERICAN MACHINE AND FOUNDRY CO., ALEXANDRIA, VA 1-95, 1966. (AFML-TDR-146 (PT III), AD-486446L)

TPRC Number	Bibliographic Citation
40352	**FIBER OPTICS WITH EXTENDED ULTRAVIOLET TRANSMISSION.** LI P C ALI M A OLSON O H SCWARTZ M A IIT RES. INST., TECHNOLOGY CTR., CHICAGO, IL. 52PP., 1968. (ECOM-0542-3, AD-678490)
40374	**LIGHT TRANSMISSION CHARACTERISTICS OF F 106B CANOPY WINDOW MATERIALS.** FAULKENBERRY B H GRABER F M KELLER E E GENERAL DYNAMICS CONVAIR DIV., SAN DIEGO, CALIF. 5PP., 1958. (MP-58-422)
40384	**SELECT CERAMIC OXIDE COMPOSITIONS AS HIGH-TEMPERATURE /1600-2100 K./ THERMAL ENERGY STORAGE MATERIALS.** BATUTIS, E. F. GEN. ELECTRIC CO., PHILADELPHIA, PA 1-101, 1963. (NASA-CR-57470, N65-19932)
40389	**HEAT TRANSFER TO GRANULAR MATERIALS.** BRINN M S FRIEDMAN S J GLUCKERT F A PIGFORD R L E I DUPONT DE NEMOURS AND CO., WILMINGTON, DE 192-229, 1948. (NEPA-804-IER-10, AECU-116)
40392	**GLASS-TO-METAL SEALS. II. CONDUCTIVITY IN THE UPPER ATMOSPHERE.** HULL, A. W. BURGER, E. E. NAVIAS, L. J APPLIED PHYS 12 698-707 1941 1-38 1966 PA N66-4-24 4710
40398	**WALL TO FLUID HEAT TRANSFER IN LIQUID FLUIDIZED BEDS.** WASMUND B SMITH J W CAN J CHEM ENG 45 3 156-65 1967
40403	**THERMAL CONDUCTIVITY AND SOME MECHANICAL PROPERTIES OF GREEN SAND MIXTURES.** BHASKARA-RAO, B. OHIO STATE UNIV., COLUMBUS, M.S. THESIS 31-75 1964
40420	**MONOCHROMATIC REFLECTANCE TESTS.** WETMORE, R. A. BOEING CO., SEATTLE, WASH. 1-116, 1963. (BSD-TR-66-140, AD 483037)
40424	**THE THERMAL CONSTANTS OF SOME ROCKS FROM THE ORANGE FREE STATE.** MOSSOP S C GAFNER G J S AFRICAN INST MINING MET 52 61-73 1951
40427	**THERMAL EXPANSION AND ELASTIC PROPERTIES OF TWO-PHASE CERAMIC BODIES.** HUNTER O JR BROWNELL W E J AM CERAM SOC 50 1 19-22 1967 JA 50 58
40429	**THERMAL EXPANSION OF FELDSPAR GLASSES.** VERGANO P J HILL D C UHLMANN D R J AM CERAM SOC 50 1 59-60 1967
40432	**THE THERMAL CONDUCTIVITY OF CERAMIC DIELECTRICS.** KINGERY, W. D. PROGRESS IN CERAMIC SCIENCE. 2 181-235 1962
40447	**THE MEASUREMENT OF THE THERMAL PROPERTIES OF BUILDING MATERIALS - THE THERMAL CONDUCTIVITY OF THIN 12 INCHES SQUARE SAMPLES.** ROUX A J A RICHARDS S J RENNHACKKAMP W M H SOUTH AFRICAN COUNCIL FOR SCIENTIFIC AND INDUS. RES. 28PP., 1950. (DR-6)
40450	**LOW-TEMPERATURE SPECTRAL EMITTANCE MEASUREMENTS.** STIERWALT, D. L. NAVAL ORDNANCE LAB., CORONA, CA 1-17, 1966. (NOLC-667, N66-39494, AD 637742)
40459	**EXPERIMENTAL DETERMINATION OF COEFFICIENT OF THERMAL CONDUCTION OF HEAT-INSULATING MATERIALS BY METHOD OF SELF-SIMULATING CONDITIONS.** VASILYEV G I DEMYANOV YU A KURNAKOV V I MALAKHOV A V RAKHMATULIN KH A RUNYNSKIY A N AIR FORCE SYS. COMMAND, FOREIGN TECH. DIV., WPAFB, OHIO 120-6PP., 1966. (AD-635269, N66-36638)
40468	**FIBER OPTICS WITH HIGH ULTRAVIOLET TRANSMISSION.** MUELLER, A. A. CHICAGO AERIAL INDUSTRIES, INC., BARRINGTON, IL 1-13, 1966. (ECOM-02057, AD-800818)
40479	**ULTRAVIOLET REFLECTIVITY OF VENUS AND JUPITER.** EVANS, D. C. GODDARD SPACE FLIGHT CENTER, GREENBELT, MD 1-27, 1966. (NASA-TM-X-55526, X613-66-172, N66-30349)
40480	**INFRARED REFLECTIVITY OF SOME COMMON MINERALS.** HOVIS, W. A., JR. GODDARD SPACE FLIGHT CENTER, GREENBELT, MD 1-24, 1965. (NASA-TM-X-56482, N66-29335)
40481	**OPTICAL PROPERTIES OF ELECTRON AND HOLE TRAPS IN SILVER-DOPED ALUMINOBORATE GLASS.** GHOSH A K J PHYS CHEM SOLIDS 30 10 2385-94 1969
40485	**INFRARED-TRANSMISSION GLASS.** YOKATO R TOSHIBA REVIEW INTERN ED 529-31 1962 (FOR ENGLISH TRANSLATION SEE T040486)
40486	**INFRARED-TRANSMISSION GLASS.** YOKATO, R. OFFICE OF NAVAL INTELLIGENCE, WASHINGTON, DC 1-7, 1962. (ENGLISH TRANSLATION OF TOSHIBA REVIEW INTERN. ED., 529-31, 1962; FOR ORIGINAL SEE T040485) (AD-482593)
40494	**PRELIMINARY REPORT ON THE THERMODYNAMIC PROPERTIES OF SELECTED LIGHT-ELEMENT AND SOME RELATED COMPOUNDS.** BECKETT, C. W. DOUGLAS, T. B. NATIONAL BUREAU OF STANDARDS, WASHINGTON, DC 1-269, 1966. (NBS REPT 9389, AFOSR-66-1938)
40513	**EFFECT OF OPTICAL GLASS PROPERTIES ON THE COOLING PROCESS.** PRIMENKO V I FOREIGN TECHNOL. DIV. 1-6, 1969. (ENGLISH TRANSLATION OF DOPOV. AKAD. NAUK UKR. RSR, (10), 1321-2, 1966; FOR ORIGINAL SEE T041369) (AD-693850, FTD-HT-23-844-68)
40528	**INFRARED COATING STUDIES.** SULZBACH, F. TURNER, A. F. BAUSCH AND LAMB INC., ROCHESTER, NY 1-33, 1966. (AD 635670)
40530	**MEASUREMENTS OF THE INFRARED SPECTRAL TRANSMITTANCE OF OPTICALLY THICK SAMPLES.** LOW, M. J. D. ABRAMS, L. RUTGERS UNIV., NEW BRUNSWICK, NJ 1-23, 1966. (AD 636-35, N66-37254, TR-2)
40547	**THERMAL OPTIC DISTORTION.** RIEDEL, E. P. BALDWIN, G. D. WESTINGHOUSE ELECTRIC CORP., BALTIMORE, MD 1-113, 1966. (MDE-1198, AD-642540)
40580	**INVESTIGATION IN THE FAR INFRARED.** STRONG J PHYS REV 38 1818-26 1931
40581	**THE INFRA-RED TRANSMISSION OF THIN FILMS OF VARIOUS ORGANIC MATERIALS.** WELLS A J J APP PHYS 11 137-40 1940
40591	**ABSOLUTE METHODS IN REFLECTOMETRY.** MC NICHOLAS H J J RESEARCH OF THE NAT BUR STANDARDS 1 29-73 1928 (NBS-RP-3)
40619	**VACUUM AS AN INSULATOR.** DUEVEL C O JR REFRIG ENG 20 223-8 1930
40644	**TOTAL RADIATION OF SEVERAL OXIDES AND OXIDE MIXTURES.** HILD K MITT KAISER WILHELM INST EISENFORSCH DUSSELDORF 14 5 59-70 1932
40645	**INFRARED RADIATION OF REFRACTORY MATERIALS.** WREDE B MITT KAISER WILHELM INST EISENFORSCH DUSSELDORF 13 131-42 1931
40652	**THICKNESS AND VISCOSITY OF ETNEAN LAVAS.** WALKER G P L NATURE /LONDON/ 213 484-5 1967

TPRC Number	Bibliographic Citation
40657	**THERMAL CONDUCTIVITY OF LUNITE AS DEPENDENT ON TEMPERATURE.** TROITSKY V NATURE /LONDON/ 213 688-9 1967
40665	**A FUNDAMENTAL STUDY OF THE EFFECT OF TEMPERATURE VARIATION ON THE INFRARED TRANSMISSION CHARACTERISTICS OF GLASS.** MILLER, A. R. UNIV. OF ROCHESTER, NY, PH.D. THESIS 1-158, 1962. (UNIV. MICROFILMS 62-6649)
40671	**AN INVESTIGATION OF THE EFFECT OF DENSITY AND WATER VAPOUR CONDENSATION ON THE THERMAL CONDUCTIVITY OF LOW TEMPERATURE INSULATING MATERIALS.** AFIFY, M. Y. M. VICTORIA UNIV. OF MANCHESTER, PH.D. THESIS 1-116 1955
40682	**THE EFFECTS OF GRANULAR SIZE ON THE THERMAL CONDUCTIVITY OF FINE SAND.** YALVAC, F. UNIV. OF MISSOURI, ROLLA, M.S. THESIS 1-45 1964
40693	**CRYSTALLIZATION AND CHEMICAL STRENGTHENING OF NEPHELINE GLASS-CERAMICS.** DUKE D A MAC DOWELL J F KARSTETTER B R J AM CERAM SOC 50 2 67-74 1967
40697	**EXPERIMENTAL DETERMINATION OF LOW TEMPERATURE SPECIFIC HEAT CAPACITY OF AN ALKALI SILICATE GLASS.** SCHAEFER, J. A. UNIV. OF TOLEDO, M.S. THESIS 1-49 1964
40714	**THE DETERMINATION OF THE COEFFICIENT OF THERMAL EXPANSION AND YOUNGS MODULUS OF ELASTICITY OF HIGH LEAD GLASS.** KOENIG, A. R. GEORGIA INST. OF TECH., M.S. THESIS 1-125 1965
40717	**THERMAL CHARACTERISTICS FOR RADAR RELECTING AND ABSORBING INORGANIC FOAMS AND COATING SYSTEMS.** HARKULICH, T. HORIZONS INC., CLEVELAND, OH 1-87, 1965. (AD-478669L)
40722	**THE ATOMIC HEAT OF DIAMOND FROM 11 TO 200 K.** BURK, D. L. CARNEGIE INSTITUTE OF TECHNOLOGY, PH.D. THESIS 1-98 1957
40738	**SPECTRAL EMISSION OF RADIATION BY GLASS.** BEATTIE J R COEN E BRIT J APPL PHYS 11 4 151-7 1960
40741	**EFFECTIVE THERMAL CONDUCTIVITY OF GRANULAR SOLIDS IN ROTARY UNITS.** BRETSZNAJDER S ZIOLKOWSKI D SURDACKI T BULL ACAD POLON SCI SER SCI CHIM 11 12 711-13 1963 CA 60 12903
40753	**AN EXPERIMENTAL STUDY OF THE VARIATION OF THERMAL CONDUCTIVITY WITH TEMPERATURE FOR SINGLE CDS AND ZNS CRYSTALS.** KUTTEMPEROOR, V. Z. UNIV. OF DETROIT, M.S. THESIS 1-99 1964
40756	**IMPROVED ORGANIC COATINGS FOR TEMPERATURE CONTROL IN A SPACE ENVIRONMENT.** HORMANN, H. H. WRIGHT-PATTERSON AIR FORCE BASE, OH 1-31, 1964. (ML-TDR-64-177, AD-453437)
40763	**THERMAL EXPANSION OF CERMET COMPONENTS BY HIGH TEMPERATURE X-RAY DIFFRACTION.** MAUER, F. A. BOLZ, L. H. NBS 1-23, 1956. (NBS-4884)
40782	**COMPUTING THERMAL CONDUCTIVITIES OF ROCKS FROM CHIPS AND CONVENTIONAL SPECIMENS.** BECK J M BECK A E J GEOPHYS RESEARCH 70 20 5227-39 1965
40784	**RECENT INVESTIGATIONS IN THE FAR INFRA-RED.** MC CUBBIN T K SINTON W M J OPT SOC AM 40 8 537-9 1950
40789	**OBSERVATIONS OF SOLAR AND LUNAR RADIATION AT 1.5 MILLIMETERS.** SINTON W M J OPT SOC AM 45 11 975-9 1955
40794	**FAR-INFRARED PROPERTIES OF QUARTZ AND SAPPHIRE.** ROBERTS S COON D D J OPT SOC AM 52 9 1023-9 1962
40799	**IV. A STUDY OF THE SERIES OF GLASSES CONTAINING SODIUM OXIDE, BORIC OXIDE AND SILICA.** GOODING E J TURNER W E S J SOC GLASS TECH 18 69 32-66 1934
40805	**DETERMINATION OF THERMAL DIFFUSIVITY OF A CERAMIC MATERIAL.** PYK S STALHANE B TEK TIDSKR 62 30 285-8 1932 (FOR ENGLISH TRANSLATION SEE T040806)
40806	**DETERMINATION OF THERMAL DIFFUSIVITY OF A CERAMIC MATERIAL.** PYK S STALHANE B SPECIAL LIBRARY ASSOC., TRANSLATIONS CENTER 1-7, 1966. (ENGLISH TRANSLATION OF TEK. TIDSKR., 62 (30), 285-8, 1932; FOR ORIGINAL SEE T040805) (TT-66-12700)
40813	**STUDY OF THE STRUCTURE OF VANADIUM IN SODA-BORIC OXIDE GLASSES.** HECHT H G JOHNSTON T J CHEM PHYS 46 1 23-34 1967
40826	**COOLING REFRACTORIES IN BOILER FURNACES. PT.I. THE THEORY.** SHERMAN R A POWER 70 675-7 1929
40853	**INFRARED REFLECTANCE AND OPTICAL CONSTANTS OF TEKTITES.** PERRY C H WRIGLEY J D JR APPL OPT 6 3 586-7 1967
40883	**INFLUENCE OF THE GAS ATMOSPHERE ON THE REACTION OF ZIRCONIUM DIOXIDE WITH OXIDES OF CERIUM.** LEONOV A I ANDREEVA A B KELER E K IZV AKAD NAUK SSSR NEORGANICHESKIE MATERIALY 2 1 137-44 1966 (FOR ENGLISH TRANSLATION SEE T040884)
40884	**INFLUENCE OF THE GAS ATMOSPHERE ON THE REACTION OF ZIRCONIUM DIOXIDE WITH OXIDES OF CERIUM.** LEONOV A I ANDREEVA A B KELER E K INORGANIC MATERIALS 2 1 115-21 1966 (ENGLISH TRANSLATION OF IZV. AKAD. NAUK SSSR, NEORGANICHESKIE MATERIALY, 2 (1), 137-44, 1966; FOR ORIGINAL SEE T040883)
40885	**SOME SPECTRAL PROPERTIES OF SINGLE CRYSTALS AND GLASSES CONTAINING RARE-EARTH ELEMENTS.** TOROPOV N A BONDAR I A VALTERE A YA IZV AKAD NAUK SSSR NEORGANICHESKIE MATERIALY 2 1 169-72 1966 (FOR ENGLISH TRANSLATION SEE T040886)
40886	**SOME SPECTRAL PROPERTIES OF SINGLE CRYSTALS AND GLASSES CONTAINING RARE-EARTH ELEMENTS.** TOROPOV N A BONDAR I A VALTERE A YA INORGANIC MATERIALS 2 1 144-7 1966 (ENGLISH TRANSLATION OF IZV. AKAD. NAUK SSSR, NEORGANICHESKIE MATERIALY, 2 (1), 169-72, 1966; FOR ORIGINAL SEE T040885)
40891	**FUSIBLE GLAZES BASED ON CALCIUM BORATE.** KRAPIVIN V A DEMIDOVA G A STEKLO I KERAMIKA 23 2 14-15 1966 (FOR ENGLISH TRANSLATION SEE T040892)
40892	**FUSIBLE GLAZES BASED ON CALCIUM BORATE.** KRAPIVIN V A DEMIDOVA G A GLASS AND CERAMICS 23 2 68-9 1966 (ENGLSIH TRANSLATION OF STEKLO I KERAMIKA, 23 (2), 14-5, 1966; FOR ORIGINAL SEE T040891)
40900	**THE EFFECTS OF GRANULAR SIZE ON THE THERMAL CONDUCTIVITY OF SAND.** PATEL, K. M. UNIV. OF MISSOURI, M.S. THESIS 1-43 1964
40903	**SUNPATH DIAGRAMS AND OVERLAYS FOR SOLAR HEAT GAIN CALCULATIONS.** PETHERBRIDGE, P. BUILDING RESEARCH STATION, CURRENT PAPERS RESEARCH SERIES 1-12, 1966. (BRS-NOTE-39)

TPRC Number	Bibliographic Citation
40904	**SUNPATH DIAGRAMS AND OVERLAYS FOR SOLAR HEAT GAIN CALCULATIONS. SUPPLEMENT 1.** PETHERBRIDGE, P. BUILDING RESEARCH STATION, CURRENT PAPERS RESEARCH SERIES SUPPL. 1, 1–12, 1966. (BRS–NOTE–39)
40905	**TRANSMISSION CHARACTERISTICS OF WINDOW–GLASSES AND SUN–CONTROLS.** PETHERBRIDGE, P. BUILDING RESEARCH STATION, GARSTON HERTS, ENGLAND GARSTON HERTS ENGLAND 1–21, 1965. (BRS–NOTE–EN–53–65)
40907	**HEATING EFFECT OF SUNSHINE.** LOUDON A G PETHERBRIDGE P ARCHITECTS JOURNAL 143 138–43 1966 (BRS–NOTE–40)
40908	**INVESTIGATIONS OF SUMMER OVERHEATING AT THE BUILDING RESEARCH STATION, ENGLAND.** LOUDON A G DANTER E BUILDING SCI 1 1 89–94 1966
40910	**THERMAL LOAD ON A VENTILATED ENCLOSURE DUE TO SOLAR HEAT TRANSMISSION THROUGH GLASS.** PRATT A W BALL E F BUILDING SCI 1 237–49 1966
40913	**SIMPLE NEW METHOD FOR DETERMINATION OF THE OPTICAL CONSTANTS OF ABSORBANT CRYSTALS IN THE INFRARED. APPLICATION TO SOME CARBONATES.** VINCENT–GEISSE J TAI N T PINAN–LUCARRE J P J DE PHYSIQUE 28 1 26–30 1967
40931	**MAGNETIC AND OPTICAL PROPERTIES OF DIOPTASE.** NEWNHAM R E SANTORO R P PHYS STAT SOL 19 2 K87–90 1967
40976	**DISTRIBUTION OF INFRARED ABSORPTION BANDS IN THE SPECTRUM OF CRYSTALLISING GLASSES IN THE REGION OF ABSORPTION BY WATER.** TROITSKII O A SHMURAK S Z ZH FIZ KHIM 40 6 1310– 1966 (FOR ENGLISH TRANSLATION SEE T040977)
40977	**DISTRIBUTION OF INFRARED ABSORPTION BANDS IN THE SPECTRUM OF CRYSTALLISING GLASSES IN THE REGION OF ABSORPTION BY WATER.** TROITSKII O A SHMURAK S Z RUSS J PHYS CHEM 40 6 701–2 1966 (ENGLISH TRANSLATION OF ZH. FIZ. KHIM., 40 (6), 1310– , 1966; FOR ORIGINAL SEE T040976)
41004	**RAPID MEASUREMENT OF THE THERMAL RESISTIVITY OF SOIL.** MASON V V KURTZ M POWER APP SYSTEMS 1 570–7 1952
41075	**INVESTIGATION OF THE PRODUCTS OF INTERACTION OF ACTIVE SILICA WITH METAL IONS IN AMMONIACAL SOLUTION.** BULATOV M I ALESKOVSKII V B ZH PRIKL KHIM 39 2 284–8 1966 (FOR ENGLISH TRANSLATION SEE T041076)
41076	**INVESTIGATION OF THE PRODUCTS OF INTERACTION OF ACTIVE SILICA WITH METAL IONS IN AMMONIACAL SOLUTION.** BULATOV M I ALESKOVSKII V B J APPL CHEM /USSR/ 39 2 258–61 1966 (ENGLISH TRANSLATION OF ZH. PRIKL. KHIM., 39 (2), 284–8, 1966; FOR ORIGINAL SEE T041075)
41085	**CRYSTALLIZATION CONDITIONS AND CERTAIN PROPERTIES OF ELECTRICALLY MELTED CORDIERITE CASTINGS.** BUDNIKOV P P SOKHATSKAYA G A KULYGIN I P ZH PRIKL KHIM 39 4 736–43 1966 (FOR ENGLISH TRANSLATION SEE T041086)
41086	**CRYSTALLIZATION CONDITIONS AND CERTAIN PROPERTIES OF ELECTRICALLY MELTED CORDIERITE CASTINGS.** BUDNIKOV P P SOKHATSKAYA G A KULYGIN I P J APPL CHEM /USSR/ 39 4 699–704 1966 (ENGLISH TRANSLATION OF ZH. PRIKL. KHIM., 39 (4), 736–43, 1966; FOR ORIGINAL SEE T041085)
41087	**CAUSES OF THE COLORATION OF RED COPPER–CONTAINING GLASSES.** AVGUSTINIK A I ZHURAVLEV G I VIGDERGAUZ V S ZH PRIKL KHIM 39 4 934–5 1966 (FOR ENGLISH TRANSLATION SEE T041088)

TPRC Number	Bibliographic Citation
41088	**CAUSES OF THE COLORATION OF RED COPPER–CONTAINING GLASSES.** AVGUSTINIK A I ZHURAVLEV G I VIGDERGAUZ V S J APPL CHEM /USSR/ 39 4 876–7 1966 (ENGLISH TRANSLATION OF ZH. PRIKL. KHIM., 39 (4), 934–5, 1966; FOR ORIGINAL SEE T041087)
41112	**TECHNIQUES OF MEASURING HEAT FLOW ON LAND FROM TERRESTRIAL HEAT FLOW. GEOPHYSICAL.** BECK A E AMERICAN GEOPHYSICAL UNION MONOGRAPH NO. 8 24–57 1965
41121	**THE CAPACITY FOR HEAT OF METALS AT LOW TEMPERATURES.** GRIFFITHS E H GRIFFITHS E PHIL TRANS ROY SOC LONDON 214 319–57 1914
41130	**REFLECTANCE, MICROHARDNESS, AND THERMAL ELECTROMOTIVE FORCE OF GALENA IN RELATION TO ISOMORPHOUS SILVER AND BISMUTH.** MOZGOVA N N GEOL RUDN MESTOROZHD 8 3 63–71 1966 CA 65 10337
41141	**THERMAL CONDUCTIVITY OF REFRACTORIES.** AUTHOR ANON. AM. SOC. TESTING MATERIALS 632–638PP., 1947.
41149	**PROGRAM FOR THE DEVELOPMENT OF SPACE RADIATOR SURFACES /SNAP 2 APU/ – PART 1. DEVELOPMENT OF PROTOTYPE SYSTEM.** MILLER, E. CARROLL, W. KLEMM, R. NORTH AMERICAN AVAIATION INC., LOS ANGELES, CA 1–56, 1961. (NA–61–802)
41195	**MEASUREMENT OF SPECIFIC HEAT FUNCTIONS BY DIFFERENTIAL SCANNING CALORIMETRY (DSC).** O'NEILL, M. J. ANAL. CHEM. 38 (10), 1331–6, 1966.
41206	**EXCITATION SPECTRUM OF ALUMINUM ACCEPTORS IN DIAMOND UNDER UNIAXIAL STRESS.** CROWTHER P A DEAN P J SHERMAN W F PHYS REV 154 3 772–85 1967
41208	**AN APPARATUS FOR THERMAL CONDUCTIVITY AT CRYOGENIC TEMPERATURES USING A HEAT FLOW METER.** HOLLINGSWORTH M JR AMERICAN SOC. TESTING MATERIALS 43–51, 1966. (ASTM–STP–411)
41217	**AN IMPROVED GUARDED–COLD–PLATE THERMAL–CONDUCTIVITY APPARATUS.** BLACK I A WECHSLER A E GLASER P E FOUNTAIN J A AMERICAN SOC. TESTING MATERIALS 74–94, 1966. (ASTM–STP–411)
41219	**A CRYOGENIC HEAT–FLOW–METER APPARATUS.** KINZER G R PELANNE C M AMERICAN SOC. TESTING MATERIALS 110–18, 1966. (ASTM–STP–411)
41236	**INFRARED STUDY OF THE ADSORPTION, DESORPTION, AND DEGASSING PROCESSES OF TRIMETHYL–BORATE ON A POROUS HIGH–SILICA GLASS SYSTEM.** SHIMIZU M GESSER H D CANADIAN JOURNAL CHEM 46 22 3517–24 1968
41239	**A COMPARISON OF THE INFRARED SPECTRA OF THE MOON AND SIMULATED LUNAR SURFACE MATERIALS.** WATTSON R B HAPKE B W ASTROPHYS J 144 1 364–8 1966 CA 65 4843
41254	**A STUDY OF MULLITE REFRACTORIES FORMED BY CALCINING CYANITE, THEIR INDUSTRIAL APPLICATION.** FREED M L J AM CERAMIC SOC 9 5 249–56 1926
41276	**INVESTIGATION OF THE MECHANICAL AND THERMAL PROPERTIES OF REFRACTORY MATERIALS AT HIGH TEMPERATURES.** LECRIVAIN, L. BER. DEUT. KERAM. GES. 43 (3), 246–53, 1966.
41277	**ERRORS IN MEASUREMENTS OF THE THERMAL CONDUCTIVITY OF REFRACTORIES RESULTING FROM THE POOR TEMPERATURE DISTRIBUTION IN THE SPECIMEN.** SCHWIETE, H. E. BOEHME, H. J. BER. DEUT. KERAM. GES. 43 (10), 654–9, 1966.

TPRC Number	Bibliographic Citation
41292	AUTOMATIC APPARATUS WITH SQUARE PLATE HEATER FOR DETERMINATION OF THE THERMAL CONDUCTIVITY OF INSULATING MATERIALS USED UP TO 1000 DEGREES. FERRO V SACCHI A CALORE 37 4 135-49 1966 CA 65 13242
41300	THERMAL PROPERTIES OF MICA. I. MANDAL S S ROY S B CENTRAL GLASS CERAM RES INST BULL /INDIA/ 12 3 81-5 1965 CA 65 18291
41301	ALUMINUM TITANATE, FORMATION AND PROPERTIES. BHATTACHARYYA B N SEN S CENTRAL GLASS CERAM RES INST BULL /INDIA/ 12 3 92-103 1965 CA 65 18294
41303	GLASS TECHNOLOGY. ULLMANN, H. H. CERAM. CRISTAL 4 (8), 46-9, 1965.
41316	EFFECTS OF THE LM DESCENT ENGINE EXHAUST ON LUNAR SURFACE TEMPERATURES. WECHSLER, A. E. PROGRESS IN ASTRONAUTICS AND AERONAUTICS 21, 305-17, 1969 (AIAA 3RD THERMOPHYSICS CONFERENCE)
41333	ELECTRICAL RESISTIVITY OF TITANIUM-CONTAINING GLASSES. TAKAMORI T SEO H TASHIRO M J AM CERAM SOC 52 11 582-5 1969
41349	DETERMINATION OF THE DYNAMIC VISCOSITY COEFFICIENT OF THE ASTHENOSPHERE. GRIGORYAN S S KRASS M S LEITES V L USHAKOV S A DOKLADY AKAD NAUK SSSR 170 2 294-6 1966 (FOR ENGLISH TRANSLATION SEE T041350)
41350	DETERMINATION OF THE DYNAMIC VISCOSITY COEFFICIENT OF THE ASTHENOSPHERE. GRIGORYAN S S KRASS M S LEITES V L USHAKOV S A SOVIET PHYSICS-DOKLADY 11 9 782-4 1967 (ENGLISH TRANSLATION OF DOKLADY AKAD. NAUK SSSR, 170 (2), 294-6, 1966; FOR ORIGINAL SEE T041349)
41352	STEADY-STATE TECHNIQUE FOR MEASURING THERMAL CONDUCTIVITY FOR NON-METALLIC MATERIALS TO 2500 K. RAWUKA, A G. ISAKSEN, L. PROC. 5TH OF THE TEMPERATURE MEASUREMENT SOCIETY CONFERENCE II-E-1—15 1967 IAA 7-18 3051
41369	EFFECT OF OPTICAL GLASS PROPERTIES ON THE COOLING PROCESS. PRIMENKO V I DOPOV AKAD NAUK UKR RSR 10 1321-2 1966 CA 66 31705 (FOR ENGLISH TRANSLATION SEE T040513)
41373	REFLECTIVITY MEASUREMENTS ON MOLYBDENITE. LEOW J H ECON GEOL 61 3 598-612 1966 CA 65 6924 CORRECTION.
41374	MEASUREMENT OF THE THERMAL CONDUCTIVITY AND THERMAL DIFFUSIVITY OF NATURAL ROCKS AT LOW TEMPERATURE. BIA, P. COMBARNOUS, M. J. SCI. INSTRUM. 3 (7), 536-40, 1970.
41392	III. SPALLING OF SILICA BRICKS. HOWIE T W TRANS BRIT CERAMIC SOC 45 45-69 1946
41414	TEMPERATURE EFFECT ON THERMAL CONDUCTIVITY OF OLIVINITE. MOISEENKO U I KUTOLIN V A GEOL I GEOFIZ AKAD NAUK SSSR SIBIRSK OTD 4 153-5 1966 CA 65 11999
41419	ROLE OF IRON IN CALCIUM PHOSPHATE GLASSES. BISHAY A M MAKAR L J AM CERAM SOC 52 11 605-9 1969
41424	INVESTIGATIONS OF GLASSES AND MELTS IN THE SYSTEM NA2SIO3-H2O. SCHOLZE H GLIEMEROTH G GLASTECH BER 39 6 279-83 1966 JA 51-6 167
41425	PHYSICOCHEMICAL INVESTIGATIONS IN THE SYSTEM B2O3-SIO2. BRUECKNER R NAVARRO J F GLASTECH BER 39 6 283-93 1966 JA 51-6 168
41478	MEASURING THE THERMAL CONDUCTIVITY OF SOLIDS AT ELEVATED TEMPERATURES. BRAUN, H. B. INTERN. Z. GASWAERME 15 (5), 142-55, 1966.
41479	MEASURING THE THERMAL CONDUCTIVITY OF SOLIDS AT ELEVATED TEMPERATURES UP TO THE MELTING POINT. BRAUN H G INTERN Z GASWAERME 15 7 207-18 1966 CA 66 14592
41480	DEPENDENCE OF CONTACT THERMAL CONDUCTIVITY OF GRANULAR SYSTEMS ON EXTERNAL LOADING. DULNEV G N ZARICHNYAK YU P MURATOVA B L SIGALOVA Z V INZH-FIZ ZH 11 2 202-6 1966 CA 66 30422 (FOR ENGLISH TRANSLATION SEE T059491)
41484	DETERMINATION OF SPECIFIC HEATS OF QUARTZ, KYANITE, AND GRANITE AT HIGH TEMPERATURES. LEONIDOV, V. YA. BARSKII, YU. P. KHITAROV, N. I. ISSLED. PRIR. TEKH. MINERALOOBRAZOV., MATER. SOVESHCH., LVOV. 301-6, 1966.
41499	THERMAL EXPANSION AND STRUCTURE OF GLASS FIBERS. BLOKH K L IZV AKAD NAUK SSSR NEORGAN MATERIALY 2 7 1280-4 1966 JA 51 234
41504	SOME PROPERTIES OF GLASSES IN THE NA2O-LA2O3-SIO2 SYSTEM. DUBROVO S K SHNYPIKOV A D IZV AKAD NAUK SSSR NEORG MATER 2 9 1652-7 1966 JA 51 230
41521	TECHNOLOGICAL PARAMETERS IN TEMPERED SAFETY GLASS MANUFACTURE. LOCSEI B KOCSIS G SOMOGYI A VESZPREMI VEGYIP EGYET KOZLEMEN 6 3 243-50 1963 CA 60 3844
41534	DEVELOPMENT OF ZIRCONIA RESISTANT TO THERMAL SHOCK. CURTIS C E J AM CERAM SOC 30 6 180-96 1947
41544	LATTICE THERMAL CONDUCTIVITY OF DISORDERED ALLOYS OF TERNARY COMPOUND SEMICONDUCTORS CU2/SN,GE//SE,S/3, /AG,PB,SB/TE2, AND /AG,SN,SB/TE2. IRIE T JAPAN J APPL PHYS 5 10 854-9 1966 CA 65 17859
41574	PHYSICAL PROPERTIES OF SOME HIGH-TEMPERATURE REFRACTORY COMPOSITIONS. POLE G R BEINLICH A W JR GILBERT N J AMERICAN CERAMIC SOC 29 8 208-22 1946
41579	FIRESIDE DEPOSITS AND THEIR EFFECT ON HEAT TRANSFER IN A PULVERIZED-FUEL BOILER. I. RADIANT EMITTANCE AND EFFECTIVE THERMAL CONDUCTANCE OF THE DEPOSITS. MULCAHY M F R BOOW J GOARD P R C J INST FUEL 39 308 385-94 1966 CA 66 4160
41584	THE STRUCTURE AND PROPERTIES OF A LITHIUM ZINC SILICATE GLASS-CERAMIC. MC MILLAN, P. W. PHILLIPS, S. V. PARTRIDGE, G. J. MATER. SCI. 1 (3), 269-79, 1966.
41593	THERMAL CONDUCTIVITIES OF PORTLAND CEMENT PASTE, AGGREGATE, AND CONCRETE DOWN TO VERY LOW TEMPERATURES. LENTZ A E MONFORE G E J PCA RES DEVELOP LAB 8 3 27-33 1966 CA 66 31774
41630	MEASUREMENT OF THE THERMAL CONDUCTIVITY OF GLASS IN THE TEMPERATURE RANGE -150 TO 50 C. CHO-JU KU KUEI SUAN YEN HSUEH PAO 5 1 51-4 1966 JA 49 210
41670	EFFECT OF SECONDARY CRYSTALLINE PHASES ON DIELECTRIC LOSSES IN HIGH-ALUMINA BODIES. FLOYD J R J AM CERAMIC SOC 47 11 539-43 1964
41683	BUILDING MATERIALS, CLAYS, MORTARS, AND CEMENTS, HEAT INSULATION. PARTICULARLY WITH REGARD TO MATERIALS USED IN FURNACE CONSTRUCTION. HUTTON R S BEARD J R J SOC CHEM IND /LONDON/ 24 802 1905

TPRC Number	Bibliographic Citation
41690	DETERMINATION OF THE METAMORPHIC GRADE OF COAL FROM THE REFLECTANCE OF VITRINITE AND POSSIBLE USE OF THE METHOD DURING GEOLOGICAL EXPLORATION. STEPANOV, YU. V. MATERIALY PO GEOL. I POLEZN. ISKOP. SEV. VOSTOKA EVROPEISKOI CHASTI SSSR, MIN. GEOL. I OKHRANY NEDR SSSR, SB. STATEI (5), 36–48, 1965.
41697	PROPERTIES OF ZINC BOROSILICATE GLASSES. HAMILTON E H WAXLER R M NIVERT J M JR J RESEARCH NATL BUR STANDARDS 62 2 59–62 1959
41699	A NEW MATERIALS PREVIEW. STARK R E DILKS B H JR MATERIALS AND METHODS 35 98–9 1952
41706	THE EFFECT OF FLUXING ADDITIONS ON THE STRUCTURE AND PROPERTIES OF SEMIPORCELAIN. KOPEIKIN A A STEKLO I KERAM 15 10 18–22 1958 (FOR ENGLISH TRANSLATION SEE T045585)
41709	NEW METHOD FOR THE ABSOLUTE MEASUREMENT OF REFLECTIVITY. BRADSHAW, P. M. D. PHILLIPS, R. SMITH, R. A. MINERAL MAG. 35 (274), 861–6, 1966.
41710	REFLECTIVITY OF SOME SULFIDES AND SIMILAR COMPOUNDS. PONOMAREVA, M. N. MINERALOG. SB., L'VOVSK. GOS. UNIV. 19 (4), 506–8, 1965.
41723	COMPOSITIONS AND PROPERTIES OF SOME HIGH – TITANIA CERAMICS. NAVIAS, L. J. AMER. CERAM. SOC. 24 (5), 148–55, 1941.
41725	THE INFLUENCE OF CHEMICAL COMPOSITION ON THE PHYSICAL PROPERTIES OF GLAZES. HALL F P J AM CERAM SOC 13 182–99 1930
41727	X-RAY DIFFRACTION AND THERMAL EXPANSION PROPERTIES OF CRISTOBALITE-CONTAINING CERAMICS. FOSTER P K HUGHES I R MAC KENZIE K J D NEW ZEALAND J SCI 9 1 249–60 1966 JA 49 291
41729	PROPERTIES OF GLASSES IN THE SYSTEM CAO-B2O3-AL2O3. PART 3. THERMAL EXPANSION IN THE CABAL GLASSES AND ITS CORRELATION WITH STRUCTURE AND ELECTRICAL PROPERTIES. OWEN A E PHYS CHEM GLASSES 3 4 134–8 1962
41738	FIRE CLAYS. SOME FUNDAMENTAL PROPERTIES AT SEVERAL TEMPERATURES. HEINDL R A PENDERGAST W L J AM CERAM SOC 13 725–50 1930
41750	PROPERTIES OF CORUNDUM MICROLITE WHEN USED FOR CONSTRUCTION PURPOSES AT HIGH TEMPERATURES. GOLUBEV B P KHARITONOV F YA KALITIN P P VASILEVA G A SMIRNOV S N TEPLOFIZ VYSOKIKH TEMPERATUR AKAD NAUK SSSR 4 2 202–6 1966 (FOR ENGLISH TRANSLATION SEE T041751)
41751	PROPERTIES OF CORUNDUM MICROLITE WHEN USED FOR CONSTRUCTION PURPOSES AT HIGH TEMPERATURES GOLUBEV B P KHARITONOV F YA KALITIN P P VASILEVA G A SMIRNOV S N HIGH TEMPERATURE 4 2 198–202 1966 (ENGLISH TRANSLATION OF TEPLOFIZ. VYSOKIKH TEMPERATUR, 4 (2), 202–6, 1966; FOR ORIGINAL SEE T041750)
41792	FOR TEMPERATURE CHANGES INSENSITIVE MAGNESITE BRICKS. ENDELL K STAHL UND EISEN 52 31 759–63 1932
41799	CALCULATION OF PHYSICAL AND CHEMICAL PROPERTIES OF GLASSES. WEINMANN G SPRECHSAAL KERAM GLAS EMAIL SIKIKATE 102 23 1050 1969 CA 72 70107
41810	AN INTERFEROMETRIC-DILATOMETER WITH PHOTOGRAPHIC RECORDING. NIC F C MAC NAIR D REV SCI INST 12 66–70 1941
41820	DETERMINATION OF REFRACTORINESS UNDER LOAD AND THERMAL EXPANSION CHARACTERISTICS OF FIRECLAY REFRACTORIES. LOVELL G H B DAVIDSON A M TRANS BRIT CERAMIC SOC 60 10 717–37 1961
41838	THERMAL EXPANSION OF SOME RAW MATERIALS AND PORCELAIN BODIES. RADO P EDGAR J G TRANS 7TH INTERNATIONAL CERAM CONGR 111–20 1960
41856	EFFECT OF CERTAIN ADDITIVES ON SINTERING AND PROPERTIES OF SUPERDUTY PERICLASE CERAMICS. POLUBOYARINOV D N POPILSKIY A YA TUNG-HUA T REFRACTORIES 24 4 137–42 1962 (ENGLISH TRANSLATION OF OGNEUPORY, 24 (4), 178–84, 1962; FOR ORIGINAL SEE T038443)
41857	PROCEDURE AND APPARATUS FOR RAPID DETERMINATION OF THERMAL CONDUCTIVITY OF REFRACTORY MATERIALS OVER A WIDE TEMPERATURE RANGE. ZANEMONETS V F GROSHEVA V M SHCHERBINA O D POROSHKOVAYA MET AKAD NAUK UKR SSR 6 4 34–9 1966 CA 65 5184 (FOR ENGLISH TRANSLATION SEE T051215)
41863	CERTAIN PROPERTIES OF NON-ALKALINE FOAM GLASS. KITAIGORODSKII I I SHIRKEVICH T L GLASS CERAMICS 16 10 533–4 1959 (ENGLISH TRANSLATION OF STEKLO I KERAM., 16 (10), 5–6, 1959; FOR ORIGINAL SEE T017277)
41873	THERMAL EXPANSION OF QUARTZ IN THE RANGE OF THE HIGH-LOW INVERSION. KOHLER K SCHNEIDER G RADEX-RUNDSCHAU 4 584–94 1965
41896	CERAMIC MATERIALS WITH A LOW THERMAL EXPANSION COEFFICIENT. ZINKO E I MEDVEDOVSKAYA E I FOMINA N P STEKLO KERAM 22 5 22–4 1965 (FOR ENGLISH TRANSLATION SEE T063068)
41897	EXPANSION AND THE STRUCTURE OF THE SODA-LIME GLASSES. KNAPP, O. SZKLO I CERAMIKA 15 (9), 240–3, 1964. (FOR ENGLISH TRANSLATION SEE TPRC NO. 63238)
41898	THERMAL EXPANSION AND THE STRUCTURE OF A CRYSTALLIZING GLASS. SORKIN E S STEKLOOB SOSTOYANIE AKAD NAUK SSR VSES KHIM OBSHCHESTVO TR CHETVERTOGO VSES SOVESCH 356–60 1965 (FOR ENGLISH TRANSLATION SEE T062678)
41902	CORRELATION OF PHYSICAL PROPERTIES OF CERAMIC MATERIALS WITH RESISTANCE TO FRACTURE BY THERMAL SHOCK. LIDMAN, W. G. BOBROWSKY, A. R. LEWIS FLIGHT PROPULSION LAB., CLEVELAND, OH 1–15, 1949. (NACA-TN-1918)
41905	EXPERIMENTAL INVESTIGATION OF THE THERMAL CONDUCTIVITY OF SOILS USED IN GREENHOUSES AND HOT BEDS. SHVETS I T SIDNEY V M ROMANYUK L I TRUDY INST TEPLOENERG AKAD NAUK SSSR 14 186–91 1958
41911	ADDING COMBUSTIBLE LIQUIDS TO FIRECLAY BODIES TO REGULATE THE STRUCTURE AND INCREASE SPALLING RESISTANCE. KUKOLEV G V NEMETS I I REFRACTORIES 86–92 1963 (ENGLISH TRANSLATION OF OGNEUPORY, (2), 85–92, 1963; FOR ORIGINAL SEE T041912)
41912	ADDING COMBUSTIBLE LIQUIDS TO FIRECLAY BODIES TO REGULATE THE STRUCTURE AND INCREASE SPALLING RESISTANCE. KUKOLEV G V NEMETS I I OGNEUPORY 2 85–92 1963 (FOR ENGLISH TRANSLATION SEE T041911)
41918	STUDY OF SOME BINARY SILICATE GLASSES BY MEANS OF REFLECTION IN INFRARED. SIMON I MC MAHON H O J AM CERAMIC SOC 36 5 160–4 1953
41921	OBSERVATIONS ON THE THERMAL EXPANSION OF CRYSTALLINE AND GLASSY SUBSTANCES. HUMMEL F A J AMERICAN CERAMIC SOC 33 3 102–7 1950

TPRC Number | **Bibliographic Citation**

41922 MEASUREMENT OF THE THERMAL CONDUCTIVITY OF THE OXIDE COATING AT THE WORKING TEMPERATURE OF THE CATHODE.
MOIZHES B YA PETROV I N SOROKIN O V SHER E M
RADIOTEKHN ELEKTRON
11 9 1674–81 1966 CA 65 17723

41923 REACTIONS IN THE SYSTEM TIO2–SIO2. REVISION OF THE PHASE DIAGRAM.
RICKER R W HUMMEL F A
J AMERICAN CERAMIC SOC
34 9 271–9 1951

41939 SOLDER GLASS SEALING.
DALTON R H
J AMERICAN CERAMIC SOC
39 3 109–12 1956

41943 SPECIFIC HEAT AND TOTAL HEAT CONTENT OF COALS AND RELATED MATERIALS AT HIGH TEMPERATURES.
KIROV N Y
RIV COMBUST
20 4 204–17 1966 CA 65 15108

41955 INFRARED SPECTRA OF SIO2 FROM 400 TO 600 /CM–1/.
REITZEL J
J CHEM PHYS
23 12 2407–9 1955

41958 THERMALLY DEVITRIFYING SOLDER GLASSES.
BROUKAL J
SILIKATTECHNIK
17 8 242–8 1966 CA 66 48863

41964 THE RELATION BETWEEN THE MICROSTRUCTURE AND THE THERMAL EXPANSION OF FORSTERITE PORCELAIN FORSTERITE PORCELAIN AS HIGH FREQUENCY INSULATOR, IV.
SUGIURA M SANO S ISHII E HIRAI M
YOGYO KYOKAI SHI /J CERAM ASSOC JAPAN/
70 4 100–10 1962

41970 DECORATIVE SYENITE GLAZE.
KHIZANISHVILI I G GAPRINDASHVILI G G
SOOBSHCH AKAD NAUK GRUZ SSR
43 2 369–73 1966 CA 65 18298

41971 PRECISION DETERMINATION OF LATTICE PARAMETER, COEFFICIENT OF THERMAL EXPANSION AND ATOMIC WEIGHT OF CARBON IN DIAMOND.
STRAUMANIS M E AKA E Z
J AMERICAN CHEMICAL SOC
73 5643–7 1951

41976 SYNTHESIS OF MATERIALS IN THE SYSTEM SILICON OXIDE - BORON OXIDE - ANTIMONY OXIDE. HIGH-SILICA FOAM MATERIALS.
ARTAMONOVA, N. V. IGNATOVA, V. A.
STEKLO, TR. INST. STEKLA
(3), 1–7, 1965.

41980 METHODS FOR DETERMINING REFLECTIVITY.
VELCHEV V N MADAREVA M SURNEV L
SPISANIE BULGAR GEOL DRUZHESTVO
25 2 169–75 1964 CA 65 10336

41987 DEPENDENCE OF THE HEAT CONDUCTIVITY OF SWELLED PERLITE ON THE INITIAL RAW MATERIAL AND WEIGHT-BY-VOLUME OF THE FINAL PRODUCT.
SPEKTOR B V
STROIT MATERIALY DETALI I IZDELIYA SB
5 24–8 1965 CA 65 19805

42004 INVESTIGATION OF THE THERMAL INSULATING PROPERTIES OF KAOLIN FIBERS.
KRZHIZHANOVSKII R E CHUDNOVSKAYA I I
TEPLOFIZIKA VYSOKIKH TEMPERATUR
4 3 355–9 1966 CA 65 10301
(FOR ENGLISH TRANSLATION SEE T43896)

42018 A RAPID METHOD FOR MEASUREMENT OF LIQUID THERMAL CONDUCTIVITY.
ARRIGONI V
TERMOTECNICA /MILAN/
20 4 197–9 1966 CA 65 6332

42019 METHOD FOR THE MEASURMENT OF TRANSMISSION COEFFICIENTS OF TRANSPARENT GLASSES FOR SOLAR ENERGY.
FANTINI, A.
TERMOTECNICA (MILAN)
20 (8), 469–73, 1966.

42029 INVESTIGATION OF FELDSPAR AND ITS EFFECT IN POTTERY BODIES.
GELLER R F CREAMER A S
J AM CERAM SOC
14 30–71 1931

42030 METHODS OF MEASURING THE THERMAL CONDUCTIVITY OF BRICKS AT ROOM TEMPERATURE.
OWEN V J CLEWS F H
TRANS BRIT CERAM SOC
65 10 559–84 1966 JA 50 50

42031 EFFECTS OF MANGANESE AND IRON OXIDES ON SOME PHYSICAL AND CHEMICAL PROPERTIES OF GLASS.
KUMAR S NATH P
TRANS INDIAN CERAM SOC
24 4 127–33 1965 JA 50 36

42037 THERMAL EXPANSION OF CERAMIC MATERIALS.
SCHUELLER, K. H. LINDL, P.
TRANS. INTERN. CERAM. CONGR., 9TH
139–48, 1964.

42040 SUPERCONDUCTING STATE OF NEUTRON STARS.
ITOH, N.
PROG. THEOR. PHYS.
42 (6), 1478–9, 1969.

42041 EQUIPMENT FOR VISCOSITY DETERMINATION WITH PROGRAMMED TEMPERATURE REGULATION AND CONTINUOUS RECORD OF THE DEFORMATION ANGLE.
PAVLOV, V. F.
TR. GOS. NAUCHN.– ISSLED. INST. STROIT. KERAM.
26, 93–9, 1966.

42052 DEPENDENCE OF SOME PROPERTIES OF GLASSES ON PHYSICAL AND CRYSTALLOCHEMICAL CONSTANTS OF THE ELEMENTS INTRODUCED INTO THE GLASSES.
KITAIGORODSKII, I. I. EGOROVA, L. S.
TR. MOSK. KHIM.– TEKHNOL. INST.
(50), 21–5, 1966.

42053 SYNTHESIS OF BOROSILICATE- BASED GLASSES AND THE STUDY OF SOME OF THEIR PHYSICAL CHEMICAL PROPERTIES.
KITAIGORODSKII, I. I. EGOROVA, L. S.
TR. MOSK. KHIM– TEKHNOL. INST.
(50), 32–5, 1966.

42064 MEASUREMENT OF SPECTRAL REFLECTIVITY WITH THE REICHERT MICROPHOTOMETER.
SINGH D S
BULL INST MINING MET
74 708 901–16 1965 CA 64 9092

42066 VARIATION IN THE HEAT CONTENT OF COPPER SMELTING SLAGS DURING ELECTROTHERMAL FINISHING TREATMENT.
ZAVYALOV M M TIKHONOV A I SMIRNOV V I
TSVETN METAL
39 7 24–8 1966 CA 65 18217

42107 THERMAL CONDUCTIVITY OF DIELECTRIC SOLIDS AT HIGH PRESSURE.
SAWIN, F. C.
UNIV. OF TEXAS, AUSTIN, PH. D. THESIS
1–54, 1966.
(UNIV. MICROFILMS NO. 66–1964)

42125 PROPERTIES AND CRYSTALLIZATION OF CORDIERITE GLASS.
SIRAZHIDDINOV N A
UZBEKSK KHIM ZH
10 4 9–11 1966 CA 66 13692

42139 ENERGY TRANSFER IN JACK PINE AS RELATED TO INFRA-RED REFLECTIVITY OF CONES, BARK, AND NEEDLES.
DE FOREST, S. E.
MICHIGAN COLLEGE OF MINING AND TECHNOLOGY, M. S. THESIS
1–89 1962

42150 INCREASE OF PRODUCTION CAPACITY OF DRAWN SHEET GLASS MANUFACTURE BY CHANGING TECHNOLOGICAL PARAMETERS.
LOCSEI, B. KOCSIS, G. SOMOGYI, A.
VESZPREMI VEGYIP, EGYET. KOZLEMEN.
9 (1), 53–69, 1965.

42165 HIGH EMISSIVITY COATINGS FOR FURNACES.
HOLDEN A F
PROC PORCELAIN ENAMEL INST FORUM
27 226–32 1965 JA 49 232

42172 MEASUREMENT OF FREE EXPANSION OF SAND MIXTURES AT HIGH TEMPERATURES.
EHRHART G W
TRANS AM FOUNDRYMENS ASSOC
49 687–716 1941

42175 REFLECTANCE OF SOME ORE MINERALS IN THIN SECTIONS.
STRONA P A
ZAP VSES MINERALOG OBSHCHESTVA
93 1 69–72 1964 CA 60 13946

42208 A STUDY OF CHROME-ALUMINA BRICKS. PART I.
LYNAM T R REES W J
TRANS BRIT CERAMIC SOC
36 110–32 1937

42209 SOME PHYSICAL EFFECTS OF GLAZES. PART I. THE EFFECT OF THE GLAZE ON THE MECHANICAL STRENGTH OF ELECTRICAL PORCELAIN.
BETTANY C WEBB H W
TRANS BRIT CERAMIC SOC
39 312–35 1940

TPRC Number	Bibliographic Citation
42232	DEPENDENCE OF THE COEFFICIENT OF THERMAL EXPANSION OF GALLOSILICATE AND ALUMINOSILICATE GLASSES ON THEIR COMPOSITION. DUBROVO S K ZH PRIKL KHIM 35 1 47-51 1962 JA 50 66
42265	THE BEHAVIOUR OF CHROME ORES AND CHROME-MAGNESITE BRICKS IN REDUCING ATMOSPHERES. LYNAM T R HOWIE T W CHESTERS J H TRANS BRIT CERAMIC SOC 44 63-7 1945
42273	FURNACE AND APPARATUS FOR CARRYING OUT REFRACTORINESS-UNDER-LOAD TESTS ON FIREBRICKS. SWALLOW H T S TRANS BRIT CERAMIC SOC 44 93-100 1945
42277	THE THERMAL EXPANSION PROPERTIES OF COMPOSITIONS CONTAINING LITHIA, ALUMINA AND SILICA. WHITE R P RIGBY G R TRANS BRIT CERAMIC SOC 53 5 324-34 1954
42284	HANDBOOK OF PHYSICAL CONSTANTS. THERMAL EXPANSION. VISCOSITY. THERMAL CONDUCTIVITY. SKINNER B J CLARK S P JR HANDBOOK OF PHYSICAL CONSTANTS 97 75-96 291-300 459-82 1966 CA 65 4087
42292	STABLE WHITE COATINGS. TOMPKINS, E. H. ARMOUR RESEARCH FOUNDATION, ILLINOIS INST. OF TECHNOLOGY, CHICAGO, ILL. 1-98, 1962. (ARF-3207-5, NASA-CR-50814, N63-19854)
42294	ALUMINA CERAMICS. GITZEN, W. H. OHIO STATE UNIV. RESEARCH FOUNDATION, COLUMBUS, OHIO 1-621, 1966. (AFML-TR-66-13, AD-480064)
42306	THE EFFECT OF MOISTURE ON THE THERMAL CONDUCTIVITY OF INSULATING MATERIALS. BEYERSDORFF, L. E. MICHIGAN COLLEGE OF MINING AND TECHNOLOGY, M. S. THESIS 1-39 1949
42307	AN EVALUATION OF LABORATORY INSTRUMENTATION FOR DETERMINING THE THERMAL CONDUCTIVITIES OF FROZEN AND UNFROZEN SOILS. BOWERS, M. M. MICHIGAN TECHNOLOGICAL UNIVERSITY, M. S. THESIS 1-102 1964
42313	I. EXPERIMENTAL DETERMINATION AND ATTEMPT AT ATTRIBUTION OF THE ACTIVE EXTERNAL VIBRATIONS IN INFRARED IN SOME METALLIC CARBONATES IN THE CRYSTAL STATE. MORANDAT J LORENZELLI V LECOMTE J J DE PHYSIQUE 28 2 152-6 1967
42331	METHOD FOR THE SIMULTANEOUS MEASUREMENTS OF THE THERMAL CONDUCTIVITY OF INSULATING MATERIALS AT DIFFERENT MEAN TEMPERATURES. MYNCKE H VAN ITTERBEEK A BULL INST INT FROID ANNEXE-2 15-22 1966 CA 67 6327
42333	THERMODYNAMICAL CHARACTERISTICS OF CERAMICS COMPOSED OF PURE OXIDES. LUKIN E S POLUBOYARINOV D N OGNEUPORY /USSR/ 28 7 318-23 1963 (FOR ENGLISH TRANSLATION SEE T42234)
42334	THERMODYNAMICAL CHARACTERISTICS OF CERAMICS COMPOSED OF PURE OXIDES. LUKIN E S POLUBOYARINOV D N SPECIAL LIBRARY ASSOC., TRANS. CENTER 1-11, 1963. (FRENCH TRANSLATION OF OGNEUPORY, 28 (7), 318-23, 1963; FOR ORIGINAL SEE T42333) (CNRS-G/R-3568)
42338	NEW METHODS OF COAL PETROGRAPHY IN COAL CHEMISTRY AND COAL TECHNOLOGY. SOOS L ACTA CHIM ACAD SCI HUNG 47 1 67-81 1966 CA 64 10979
42356	THERMAL DIFFUSIVITY OF AMMONIUM PERCHLORATE. ROSSER W A INAMI S H WISE H AIAA /AM INST AERON ASTRONAUT/ J 4 4 663-6 1966 CA 65 3026
42366	PHYSICAL PROPERTIES OF CALCIC LABRADORITE FROM LAKE COUNTY, OREGON. STEWART D B WALKER G W WRIGHT T L FAHEY J J AM MINERALOGIST 51 1/2 177-97 1966 CA 65 496
42373	THEORETICAL AND EXPERIMENTAL RESEARCHES ON THERMAL CONDUCTIVITY OF CONCRETES. MISSENARD F A ANN INST TECH BATIMENT TRAV PUBL 18 949-68 1965 CA 64 15557
42374	THERMAL PROPERTIES OF POSTULATED LUNAR SURFACE MATERIALS. GLASER P E WECHSLER A E GERMELES A E ANN N Y ACAD SCI 123 2 856-70 1965 CA 64 5782
42389	THE REFLECTIVITY OF JUPITER IN THE ULTRAVIOLET. STECHER T P ASTROPHYS J 142 3 1186-90 1965 CA 64 10579
42406	APPARATUS FOR THE RAPID DETERMINATION OF THE THERMAL CONDUCTIVITY OF REFRACTORY MATERIALS. AUERBACH, A. BER. DEUT. KERAM. GES. 43 (1), 19-20, 1966.
42418	AN INVESTIGATION OF THE PROPERTIES OF SOME FELDSPARS. PARMELEE C W MC VAY T N UNIV ILL BULL ENG EXPT STAT BULL 29 9 1-50 1931 (BULLETIN NO. 233)
42421	THE MEASUREMENT OF THERMAL CONDUCTIVITY AT VERY HIGH TEMPERATURES BY THE CYLINDER METHOD. DUMEZ, P. PROVOST, G. LECRIVAIN, L. BULL. SOC. FRANC. CERAM. (69), 89-98, 1965.
42432	MECHANICAL AND THERMAL PROPERTIES OF HEAT-TREATED COALS. HONDA H SANADA Y FURUTA T CARBON 3 4 421-8 1966 CA 65 3610
42435	EFFECT OF SMALL ADDITIONS OF ALUMINA, BORIC OXIDE, AND TITANIA ON THE PROPERTIES OF A SODA-LIME-SILICA GLASS. BHATYE S V CENTRAL GLASS CERAM RES INST BULL /INDIA/ 12 4 111-16 1965 CA 65 3514
42447	SOME FUNDAMENTAL PROPERTIES OF ONTARIO CLAY AND SHALE. MONTGOMERY R J J CANADIAN CERAMIC SOC 3 28-32 1934
42450	EXPANSION OF GLASSES IN THE ZONE OF TRANSFORMATION AND BEYOND THE SOFTENING POINT, DETERMINED DILATOMETRICALLY. GILLOD J VERRES ET REFRACTAIRES 3 4 217-21 1949
42456	THERMAL CONDUCTIVITY OF PERLITE AT LOW TEMPERATURES. ADAMS, L. CRYOG. TECHNOL. 1 (3), 117-8, 1965.
42457	THERMAL CONDUCTIVITY OF EVACUATED PERLITE. ADAMS, L. CRYOG. TECHNOL. 1 (6), 249-51, 1965.
42464	TEMPERATURE HYSTERESIS OF HEAT CONDUCTION IN THE BREAKDOWN OF SOLID SOLUTIONS. VISHNEVSKII I I SKRIPAK V N DOKL AKAD NAUK SSSR 163 2 418-21 1965 CA 64 5783
42467	REFLECTING POWER OF GALENA OF VARIOUS ORIGINS. KNYAZEV G I KUDELYA V K DOPOVIDI AKAD NAUK UKR RSR 4 495-6 1966 CA 65 3562
42469	DETERMINATION OF THE HEAT CAPACITY OF ESTONIAN SCHIST-KUKKER-SITE IN THE PROCESS OF HEATING UP TO 900 DEGREES. AGROSKIN A A GONCHAROV E I EESTI NSV TEADUSTE AKAD TOIMETISED FUUSIK-MAT-JA TEHNIKATEADUSTE SEER 1 94-7 1966 CA 65 3094
42476	COMPARISON OF BODIES CONTAINING BLENDED FELDSPARS AND ONE-MINE FELDSPAR OF SIMILAR COMPOSTION. PARMELEE C W AMBERG C R J AM CERAM SOC 14 309-12 1931
42478	SOME PROPERTIES OF ENGLISH CHINA CLAYS. KLINEFELTER T A MEYER W W VACHUSKA E J J AM CERAM SOC 16 269-76 1933
42491	HEAT CAPACITY OF SERPENTINE. LEONIDOV V YA GEOKHIMIYA 3 348 1966 CA 64 17263

TPRC Number	Bibliographic Citation
42492	**APPLICATION OF THERMODYNAMIC CALCULATIONS TO GEOCHEMISTRY, AND NEW DATA ON THE HEAT CAPACITIES OF QUARTZ, KYANITE, AND GRANITE.** LEONIDOV V YA GEOKHIM ISSLED V OBL POVYSHENNYKH DAVLENII I TEMPERATUR AKAD NAUK SSSR INST GEOKHIM I ANALIT KHIM SB STATEI 191-202 1965 CA 64 12372
42494	**STAGES OF TRANSFORMATION OF COALS AND HOST ROCKS.** AMOSOV, I. I. MALININ, S. I. GEOL. UGLEI SIBIRI I DAL NEGO VOSTOKA, AKAD. NAUK SSSR, SIBIRSK. OTD., INST. GEOL. I GEOFIZ. 5-26, 1965.
42502	**DEGREE OF POLARIZATION OF THE FLUORESCENCE OF BOROSILICATE GLASSES CONTAINING URANIUM.** SCHUSTER E SCHROEDER H GLASTECH BER 39 3 113-18 1966 JA 50 251
42503	**DETERMINING HEAT CONDUCTION OF THERMAL INSULATING MATERIALS UP TO 2000 C.** KHARLOMOV A G TEPLOENERGETIKA 8 3 64-6 1961
42504	**HEAT CONDUCTIVITY OF INSULATION MATERIALS IN VACUO.** KAGANER M G GLEBOVA L I KISLOROD 12 1 13-18 1959
42530	**THE SPECTRAL TRANSMITTANCE OF AMBER GLASS BOTTLES.** HOUGEN, A. G. INTERN. DAIRY CONGR., PROC., 16TH 1, 589-98, 1962.
42532	**THERMOPHYSICAL CHARACTERISTICS OF POLYMER COATINGS.** SUKHAREVA L A VORONKOV V A ZUBOV P I INZHEN-FIZ ZH 9 2 211-16 1965 CA 64 917 (FOR ENGLISH TRANSLATION SEE 36861)
42535	**THERMAL PROPERTIES OF CHAMOTTE CERAMICS WITHIN THE TEMPERATURE RANGE 80 - 1200 K.** VASILEV L L FRAIMAN YU E INZH FIZ ZH AKAD NAUK BELORUSSK SSR 9 6 762-7 1965 CA 64 15540 (FOR ENGLISH TRANSLATION SEE T43468)
42537	**COEFFICIENT OF THERMAL CONDUCTIVITY OF CATHODE OXIDE COATINGS.** CHUDNOVSKII F A INZH-FIZ ZH AKAD NAUK BELORUSSK SSR 10 1 106-14 1966 CA 64 18573 (FOR ENGLISH TRANSLATION SEE T53877)
42542	**REFLECTANCE OF COALS OF DIFFERENT U.S.S.R. BASINS AS AN INDEX OF THEIR METAMORPHISM DEGREE.** DVUZHILNAYA N M PONOMAREVA M N VYRVICH G P ISSLED ISPOLZ I STANDARTIZATSIYA UGLEI DONETSK NAUCHN-ISSLED UGOLN INST 1965 3-19 1965 CA 64 4813
42543	**CERTAIN OPTICAL CONSTANTS OF COALS AS INDEXES OF THEIR METAMORPHISM.** PONOMAREVA M N LIFSHITS M M VYRVICH G P ISSLED ISPOLZ I STANDARTIZATSIYA UGLEI DONETSK NAUCHN-ISSLED UGOLN INST 1965 20-9 1965 CA 64 4815
42545	**PHYSICOCHEMICAL PROPERTIES OF GLASSES IN THE SYSTEM NA2O-AL2O3-GA2O3-SIO2.** DUBROVO S K DANILOVA N P TSEKHOMSKAYA T S ISSLED V OBL KHIM SILIKATOV I OKISLOV AKAD NAUK SSSR SB STATEI 1965 11-17 1965 CA 65 1920
42546	**SOME PROPERTIES OF GLASSES IN THE SYSTEM NA2O-B2O3-GA2O3-SIO2.** DANILOVA N P DUBROVO S K ISSLED V OBL KHIM SILIKATOV I OKISLOV AKAD NAUK SSSR SB STATEI 1965 18-23 1965 CA 65 3515
42567	**HEAT CAPACITY OF SOLID SOLUTIONS OF MAGNESIUM CHROMITE AND ALUMINATE WITH MAGNESIUM FERRITE.** LYUKSHIN V V ZAIONCHKOVSKII YA A IZV AKAD NAUK SSSR NEORGAN MATERIALY 1 9 1602-6 1965 CA 64 5828
42576	**MEASUREMENT OF THE THERMAL CAPACITY OF SOLID FUELS DURING HEATING UP TO 1000 DEGREES.** AGROSKIN A A BARSKII YU P GONCHAROV E I KANAVETS P I IZV VYSSHIKH UCHEBN ZAVEDENII ENERG 8 12 51-7 1965 CA 64 13966
42588	**DETERMINATION OF OPTICAL CONSTANTS OF GLASSES IN THE VACUUM ULTRAVIOLET REGION.** SASAKI T FUKUTANI H ISHIGURO K IZUMITANI T JAPAN J APPL PHYS SUPPL /1964/ 4 1 527-31 1965 CA 64 4756
42592	**CONTRIBUTION TO THE KNOWLEDGE OF KAOLIN.** SCHWARZ R KLOS W ZEIT ANORG ALLGEM CHEM 196 213-20 1931
42608	**MINERALOGY OF FAYALITE, WITH SPECIAL REFERENCE TO ITS THERMAL AND THERMODYNAMICAL PROPERTIES.** SUWA K J EARTH SCI NAGOYA UNIV 12 2 129-46 1965 CA 64 13925
42659	**RADIATIVE HEAT TRANSFER IN MOLTEN GLASS.** SUGIYAMA, S. SHIMIZU, M. MOCHIZUKU, S. KAGAKU KOGAKU 30 (3), 220-5, 1966.
42663	**DETERMINATION OF HEAT CAPACITY OF COAL DURING COKING.** AGROSKIN A A GONCHAROV E I KOKS I KHIM 11 14-19 1965 CA 64 4819
42676	**THERMAL CONDUCTIVITY OF PYREX GLASS. SELECTED VALUES.** CARWILE, L. C. K. HOGE, H. J. ARMY NATICK LABS., PIONEERING RES. DIV., MASS. 1-18, 1966. (TR-66-14-PR, AD-630135)
42693	**EXPERIMENTAL SPECTROPHOTOMETRIC COMPARISON OF THE SURFACE OF THE MOON AND CERTAIN VOLCANIC DEPOSITS.** LEBEDEBA I I UCHENYE ZAPISKI /USSR/ LENINGRAD UNIV 326 99-102 1965 (FOR ENGLISH TRANSLATION SEE T42694)
42694	**EXPERIMENTAL SPECTROPHOTOMETRIC COMPARISON OF THE SURFACE OF THE MOON AND CERTAIN VOLCANIC DEPOSITS.** LEBEDEBA, I. I. AIR FORCE CAMBRIDGE RES. LABS. 9PP., 1966. (ENGLISH TRANSLATION OF UCHENYE ZAPISKI, LENINGRAD UNIVERSITY, (326), 99-102, 1965; FOR ORIGINAL, SEE T42693) (AD-631970)
42695	**THE DEVELOPMENT, FABRICATION, AND EVALUATION OF A DEVICE FOR THE MEASUREMENT OF THERMAL PROPERTIES OF SOIL.** HSU, S. T. KAO, H. S. VIRGINIA POLYTECHNIC INST., BLACKSBURG, VA. 1-250, 1966. (AD-632080)
42701	**SYNTHESIS AND OPTICAL PROPERTIES OF FORSTERITE.** SHANKLAND, T. J. HARVARD UNIV., CAMBRIDGE, MASS. 1-170, 1966. (HP-16, N67-16143, AD-641363)
42708	**DEVELOPMENT OF THERMAL CONDUCTIVITY PROBES FOR SOILS AND INSULATIONS.** WECHSLER, A. E. LITTLE ARTHUR D, INC., CAMBRIDGE, MASS. 1-113, 1966. (CRREL-TR-182, N67-22531, AD-645337)
42709	**HEAT TRANSFER FROM RADIOISOTOPIC HEAT SOURCES BURIED IN THE OCEAN FLOOR.** SCHROCK, V. E. KESAVAN, K. NAVAL RADIOLOGICAL DEFENSE LABS., SAN FRANCISCO, CALIF. 1-55, 1966. (USNRDL-TRC-86, N67-23377, AD-645558)
42720	**METHOD FOR CALORIMETRIC ANALYSIS UNDER TRANSIENT CONDITIONS /OF HEATING/.** BERTHIER, G. COMM. ENERGIE AT. FRANCE 1-99. 1965. (CEA-R-2797)
42731	**IRRADIATION PERFORMANCE ON THORIA-URANIA FUEL MATERIALS.** BABCOCK AND WILCOX COMP., RESEARCH CENTER BABCOCK AND WILCOX CO., NUCLEAR DEVELOPMENT CENTER, LYNCHBURG, VA. 1-279, 1965. (BAW-3376, N66-25089)
42742	**FIBER OPTICS WITH HIGH ULTRAVIOLET TRANSMISSION.** MUELLER, A. A. CHICAGO AERIAL INDUSTRIES, INC., BARRINGTON, ILL. 1-14, 1967. (ECOM-02057-2, AD-807431)
42764	**EVALUATION OF INFRARED EMISSION OF CLOUDS AND GROUND AS MEASURED BY WEATHER SATELLITES. INTERIM REPORT.** KERN, C. D AIR FORCE CAMBRIDGE RES. LABS., BEDFORD, MASS. 1-111, 1965. (AFCRL-65-840, N66-22701, AD-629081)

TPRC Number	Bibliographic Citation
42772	**INFRARED REFLECTRANCE SPECTRA OF IGNEOUS ROCKS, TUFFS AND RED SANDSTONE FROM 0.5 TO 22 MICRONS.** CALLAHAN W R HOVIS W A JR GODDARD SPACE FLIGHT CENTER B83-101, 1965. (NASA-TM-X-55356, N66-23428)
42776	**OBSERVED ULTRAVIOLET REFLECTIVITY OF MARS.** EVANS, D. C. GODDARD SPACE FLIGHT CTR., GREENBELT, MD. 1-18, 1965. (NASA-TM-X-56654, N66-22239)
42781	**DEVELOPMENT OF OPTICAL COATINGS FOR CD S THIN FILM SOLAR CELLS.** SCHAEFER, J. C. HILL, E. R. HARSHAW CHEMICAL COMP., CLEVELAND, OHIO, 1-63. 1965. (NASA-CR-549658 N66-25373)
42820	**CONTRIBUTION TO THE STUDY OF THE VITREOUS STATE AND THE EXPANSION OF GLASSES.** SAMSOEN M ANNALES PHYSIQUE 9 35-127 1928
42824	**FLAT PANEL VACUUM THERMAL INSULATION.** STRONG H M BUNDY F P BOVENKERK H P J APPLIED PHYSICS 31 1 39-50 1960
42891	**OPTICAL PROPERTIES OF INSULATORS IN THE EXTREME ULTRAVIOLET.** STEPHAN G LEMONNIER J-C ROBIN S J OPT SOC AM 57 4 486-92 1967 CA 67 37942
42909	**CHEMICALLY STRENGTHENED, LEACHED ALUMINA AND SPINEL.** KIRCHNER H P GRUVER R M WALKER R E J AM CERAM SOC 50 4 169-73 1967
42910	**CRYSTALLIZATION AND CHEMICAL STRENGTHENING OF STUFFED BETA-QUARTZ GLASS-CERAMICS.** BEALL G H KARSTETTER B R RITTLER H L J AM CERAM SOC 50 4 181-90 1967
42955	**THERMOPHYSICAL PROPERTIES OF A TWO-PHASE DISPERSED SYSTEM SATURATED WITH DIFFERENT LIQUIDS.** VERZHINSKAYA A B VAINBERG V SH J ENGINEERING PHYS 12 1 20-2 1967 (ENGLISH TRANSLATION OF INZH. FIZ. ZH., 12 (1), 38-42, 1967; FOR ORIGINAL SEE T45153)
42960	**FORMATION OF SPINEL BY DIFFUSION OF MG O FROM THE VAPOR INTO ALUMINA.** NAVIAL, S. GENERAL ELECTRIC RESEARCH LAB, SCHENECTADY, N. Y. 1-49, 1960. (60-RL-2544-7)
42961	**DILATOMETRIC INVESTIGATIONS ON PORCELAIN BODIES.** ZWETSCH, A. BER DER DEUT KERAM GESELL 29 3 73-7 1952 (FOR ENGLISH TRANSLATION SEE T52094)
42964	**OPTICS OF HEXAGONAL PYRRHOTINE /FE9 SIO/.** VON GEHLEN K PILLER H MINERAL MAG 35 270 335-46 1965 CA 64 155
42966	**THE OPTICS OF COVELLINE.** VON GEHLEN K PILLER H MINING MAG 113 6 438-45 1965 CA 64 12046
42967	**EFFECT OF VARIATION IN RAMMED SAND DENSITY AND VARIOUS ADDITIVES ON THERMAL CONDUCTIVITY OF GREEN SAND MIXTURES.** RAO B B WILLIAMS D C MOD CASTINGS 49 3 79-88 1966 CA 64 17120
42979	**INFRARED COATING STUDIES. 9TH QUARTERLY REPORT** CHANG, L. BAUSCH AND LOMB INC., ROCHESTER, N. Y. 1-23, 1965. (AD-471653L)
42981	**METHOD OF MEASUREMENT OF THE REAL THERMAL DIFFUSIVITY OF MOIST SOIL.** JACKSON R D KIRKHAM D SOIL SCIENCE SOC AM PROC 22 6 479-82 1958
42987	**HEAT-TRANSFER MEASUREMENTS ON COMPRESSED POWDERS FOR REACTOR PRESSURE-TUBE INSULATION.** LYS, L. A. NEUE TECH. 3 B (3), 111-18, 1966.
42998	**REFRACTORIES WITH HIGH ALUMINA BINDERS.** ZHIKHAREVICH S A GINYAR E A KOZYREVA L A MALCHENKO R S OGNEUPORY 31 4 8-18 1966 CA 65 3527 (FOR ENGLISH TRANSLATION SEE T46572)
43002	**DETERMINATION OF THE THERMOPHYSICAL COEFFICIENTS OF SOILS AND THEIR CALCULATED VALUES.** USHAKOV V P DEFORMATSII GRUNTOV PRI ZAMERZANII /MOSCOW/ MONOGRAPH 52-67 1959 (FOR ENGLISH TRANSLATION SEE T43003)
43003	**DETERMINATION OF THE THERMAL COEFFICIENTS OF SOILS AND VALUES TO REMEMBER FOR THEIR CALCULATIONS.** USHAKOV V P CENTRAL PONTS ET DES CHAUSSUS, PARIS 1-36, 1959. (FRENCH TRANSLATION OF THE MONOGRAPH DEFORMATSII GRUNTOV PRI ZAMERZANII, 52-67, 1959; FOR ORIGINAL SEE T43002) (CNRS-H/R-1093, TT-65-29939)
43012	**THE VIBRATIONAL SPECTRA OF SILICATES. PART I.** GASKELL P H PHYS CHEM GLASSES 8 2 69-80 1967
43026	**TEMPERATURE DEPENDENCE OF THE COEFFICIENT OF EXPANSION AND THE REFRACTIVE INDEX OF GLASS.** SHCHAVELEV O S OPTIKO-MEKHAN PROM 33 1 31-4 1966 CA 64 18639
43027	**DETERMINATION OF THERMAL OPTICAL PROPERTIES OF SILICATE AND BOROSILICATE GLASSES IN A WIDE RANGE OF TEMPERATURES AND WAVE LENGTHS FROM THEIR CHEMICAL COMPOSITIONS.** SHCHAVELEV O S OPTIKO-MEKHAN PROM 33 1 38-42 1966 CA 64 19153
43033	**INFRARED OPTICAL CONSTANTS OF GLASSES.** BLAIN B J DOUGLAS R W PHYS CHEM GLASSES 6 6 233-39 1965 CA 64 6264
43040	**THERMAL CONDUCTIVITY IN VITREOUS SYSTEMS.** KLEMENS P G PHYS. NON-CRYSTALLINE SOLIDS, PROC. INT. CONF. 162-78 1965 CA 64 13884
43041	**DEPENDENCE OF HEAT TREATMENT OF THE THERMAL CONDUCTIVITY OF A CRYSTALLIZING GLASS.** LAUDY J H A PHYS. NON-CRYSTALLINE SOLIDS, PROC. INT. CONF. 189-96 1965 CA 64 13884
43052	**A COMPARISON OF THE THERMAL CONDUCTIVITY OF MOIST CAPILLARY-POROUS MATERIALS AT TEMPERATURES BELOW AND ABOVE 0 C.** PAK N V J ENGINEERING PHYS 12 1 36-7 1967 (ENGLISH TRANSLATION OF INZH. FIZ. ZH. 12 (1), 68-71, 1967; FOR ORIGINAL SEE T45418)
43056	**GLASS-TO-METAL SEALS.** HULL A W BURGER E E PHYSICS 5 384-405 1934
43084	**EFFECT OF OXIDATION ON THE REFLECTANCE OF COAL VITRINITE.** RESHETKO, A. N. PERMITINA, K. S. PODGOTOVA I KOKSOVANIE UGLEI, VOST.-NAUCHN. ISSLED. UGLEKHIM. INST., SB. STATEI (5), 40-7, 1965.
43098	**QUASI-STATIONARY METHOD FOR COMPLEX DETERMINATION OF THE THERMOPHYSICAL CHARACTERISTICS OF BUILDING MATERIALS.** FRAIMAN YU E PROBL STROIT TEPLOFIZ /MINSK VYSSH SHKOLA/ SB 419-26 1965 CA 64 15557
43101	**DETERMINING THE THERMAL CONDUCTIVITY OF CERAMIC AND REFRACTORY MATERIALS.** HARTMANN H ELEKTROWARME 22 11 409-15 1964
43102	**DETERMING THE SPECTRAL RADIATION PROPERTIES /REFLECTING AND ABSORPTIVE POWER/ OF MATERIALS AS A FUNCTION OF TEMPERATURE.** SCHADACH P ELEKTROWARME 23 1 8-15 1965

TPRC Number | *Bibliographic Citation*

43123 **EXPERIMENTERS DESIGN HANDBOOK FOR THE MANNED LUNAR SURFACE PROGRAM.**
FUCHS, R. A.
EXPERIMENTERS DESIGN HANDBOOK FOR THE MANNED LUNAR SURFACE PROGRAM
CHAPTER 7 7-1 - 7-114
CHAPTER 10 10-1 - 10-36 1967

43126 **A STUDY OF THE MINERALOGICAL AND PHYSICAL CHARACTERISTICS OF TWO LITHIA-ZIRCONIA BODIES.**
BARLETT H B THOMAS R R JR
J AM CERAM SOC
17 17-20 1934

43127 **NOTES ON SOME EFFECTS OF THE VARIETY OF FLINT USED ON THE PROPERTIES OF A HOTEL CHINA BODY.**
THOMPSON C L
J AM CERAMIC SOC
17 268-71 1934

43137 **THERMAL CONDUCTIVITY MEASUREMENTS OF INSULATING MATERIALS DOWN TO ABOUT -200 DEGREES.**
FERRO V SACCHI A
RIC SCI REND
8 A 6 1667-79 1965 CA 65 63

43143 **THE DETERMINATION OF HEAT CAPACITIES OF MOLTEN SLAGS.**
BERSHAK V I
SB NAUCHN TR GOS NAUCHN-ISSLED INST TSVETN METAL
23 21-4 1965 CA 64 7733

43156 **A THERMOELECTRIC DEVICE FOR MEASURING THERMAL CONDUCTIVITY OF ROCK.**
KHAN A M FATT I
SOC PETROL ENGRS J
5 2 113-16 1965 CA 64 7901

43170 **RELATION BETWEEN THE SOLUBILITY OF ALUMINA, QUARTZ, AND KAOLIN GROG IN FELDSPATHIC MELTS AND THE FIRING TEMPERATURE OF THE MIXTURES.**
DUDEROV G N SI-TSU C
STEKLO I KERAM
19 7 25-9 1962 JA 50 97

43179 **HIGH-TEMPERATURE INFRARED SPECTROSCOPY OF OLEFINS ADSORBED ON FAUJASITES.**
EBERLY P E JR
J PHYS CHEM
71 6 1717-22 1967

43202 **DETERMINATIONS OF OPTICAL PROPERTIES OF COPPER SULFIDE, SILVER SULFIDE, AND TIN SULFIDE.**
DEZSO, E.
STUDIA UNIV. BABES-BOLYAI, SER. MATH.-PHYS.
(2), 115-21, 1965.

43203 **SOME THERMAL PROBLEMS OF THE SYSTEM ZNO-AL2O3.**
KELEMEN F NEDA A
STUDII CERCETARI FIZ
18 1 13-19 1966 CA 64 18450

43236 **RELATIONS BETWEEN THE THERMAL CONDUCTIVITY OF SILICON CARBIDE STOPPER HEAD AND CORROSION OF THE STOPPER ROD.**
NIWA, S. HAGHIHARA, J.
TAIKABUTSU
15 (77), 338-40, 1964.

43250 **APPLICATION OF NEW FACTORS FOR CALCULATING EXPANSION COEFFICIENTS OF SILICATE AND BOROSILICATE GLASSES.**
TAKAHASHI K
J CERAMIC ASSOC JAPAN
63 707 142-7 1955

43251 **THE SPECTRAL REFLECTION OF COLORED PIGMENTS BASED ON NICKEL TITANATE.**
ALEKSANDROVA L V TROPIN V A SHEINKMAN A I
SHEREMETEV G D
TR CHELYABINSK GOS PED INST
2 165-73 1964 CA 64 11440

43252 **DETERMINATION OF THE THERMOPHYSICAL CHARACTERISTICS OF RAW MATERIAL WITH CONSIDERATION FOR CHEMICAL CONVERSIONS AND HEATING RATE.**
VISHNEVSKII, K. P. KITOV, R. N.
STEPUKHIN, A. S.
TR., GOLOVN. NAUCHN.-ISSLED. INST. TSEMENTN. MASHINOSTR.
(3), 8-12, 1964.

43286 **HEAT CAPACITY OF CU SI O3.H2 O FROM 2 TO 40 K.**
EISENBERG, W. R.
MICHIGAN STATE UNIVERSITY, M. S. THESIS
1-41 1963

43289 **ELECTRICAL, THERMAL, AND SOME MAGNETIC PROPERTIES OF NICKEL-CADMIUM FERRITES.**
CHECHERNIKOV V I SPERANSKII N M MALYSHEV N I
VESTN MOSK UNIV SER III FIZ ASTRON
20 5 45-8 1965 CA 64 140

43298 **A HIGH TEMPERATURE RECORDING DILATOMETER.**
BEALS R J LAUCHNER J H
AM CERAMIC SOC BULL
37 11 486-8 1958

43305 **CERAMICS.**
KOENIG J H
IND ENG CHEM
40 10 1782-5 1948

43312 **TALC IN WHITEWARE BODIES OF THE WALL-TILE TYPE.**
GELLER R F CREAMER A S
J AM CERAM SOC
18 259-69 1935

43324 **AUTOMATIC EQUIPMENT FOR MEASURING THE THERMOPHYSICAL COEFFICIENTS OF CERAMIC MATERIALS.**
POLUBOYARINOV D N DUDEROV I G
ZAVODSK LAB
31 11 1410-12 1965 CA 64 4762
(FOR ENGLISH TRANSLATION SEE T36139)

43326 **HIGH TEMPERATURE MECHANICAL PROPERTIES OF CERAMIC MATERIALS. III. VITRIFIED CHEMICAL AND REFRACTORY PORCELAINS.**
DUNSMORE G K FENSTERMACHER J E HUMMEL F A
AM CERAMIC SOC BULL
40 5 310-13 1961

43328 **CORRELATION BETWEEN PHYSICAL PROPERTIES AND THERMOCHEMICAL REACTIONS IN A MULLITE-CORDIERITE COMPOSITION.**
DAVIS M P JR HACKLER W C
AM CERAMIC SOC BULL
40 6 362-5 1961

43335 **FUSED VACUUM-TIGHT, METAL-TO-CERAMIC, CERAMIC-TO-GLASS METAL-TO-GLASS, AND METAL-TO-MICA SEALING BY POWDERED GLASS TECHNIQUES.**
ANTON N
CERAMIC AGE
63 1 15-19 1954

43336 **ADVANCES IN CERAMICS RELATED TO ELECTRONIC TUBE DEVELOPMENTS**
NAVIAS L
J AM CERAMIC SOC
37 8 329-50 1954

43351 **HEAT CAPACITY OF BORIC ANHYDRIDE AND OF SODIUM BORATE GLASSES AT LOW TEMPERATURES.**
TURDAKIN V A TARASOV V V
ZH NEORGAN KHIM
11 4 931-3 1966 CA 65 454

43360 **DETERMINATION OF OPTICAL PROPERTIES OF OBJECTS WITH ARBITRARY SCATTERING DIAGRAMS ON AN INTEGRAL PHOTOMETER.**
RVACHEV V P SAKHNOVSKII M YU
ZH PRIKL SPEKTROSKOPII AKAD NAUK BELORUSSK SSR
4 2 172-4 1966 CA 65 1608

43370 **SILICATE SPECIFIC HEATS. SECOND SERIES.**
WHITE W P
AM J SCI
47 1-59 1919

43371 **THERMAL EXPANSION OF SOLIDS AT LOW TEMPERATURES /CU, NI, FE, ZINCBLENDE, LIF, CALCITE, ARAGONITE, NH4CL/.**
ADENSTEDT H
ANN PHYSIK
26 5 69-96 1936

43379 **BEHAVIOR OF HIGH-FIRED BASIC BRICK IN SPRUNG ARCHES.**
NEELY J E
BULL AM CERAM SOC
46 6 576-81 1967

43381 **THE SPECIFIC WEIGHT OF LIQUID MAGNESIUM.**
GROTHE H MANGELSDORF C
Z METALLKUNDE
29 352-3 1937

43391 **PHYSICAL PROPERTIES OF END MEMBERS OF THE GARNET GROUP.**
SKINNER B J
AM MINERALOGIST
41 428-36 1956

43394 **AN INTERPRETATION OF THE THERMAL EXPANSION OF THE ALKALI HALIDES AND OF THE STRUCTURAL CHANGES OCCURRING IN GLASS UNDER HIGH PRESSURE.**
WEYL W A
CENTRAL GLASS AND CERAMIC RES INST BULL /INDIA/
6 4 147-74 1959

43395 **THERMAL SHOCK BEHAVIOR OF CERAMIC DIELECTRICS.**
SMOKE E J KOENIG J H
CERAMIC AGE
57 17-19 1951

43396 **SANDWICH-TYPE METAL-TO-CERAMIC VACUUM-TIGHT SEAL.**
PRYSLAK N E
CERAMIC AGE
65 3 21-2 1955

TPRC Number	Bibliographic Citation
43401	**ON THE MEASUREMENT OF VERY LOW TEMPERATURES. XXXII. THE THERMAL EXPANSION OF JENA-GLASS 16-III.** KEESOM W H DE HAAS W J COMMUN KAMERLINGH ONNES LAB UNIV LEIDEN 176-A 1-11 1925 (ENGLISH TRANSLATION OF VERSLAG VAN DE GEWONE VERGADERING DER WISSEN NATUURKUNDIGE AFDEELING DER KON. AKAD. VAN WETTENSCHAPPEN TE AMSTERDAM, DEEL XXXIV, 618-24, 1925; FOR ORIGINAL SEE T43841)
43404	**THEORY AND PRACTICE OF GLASS-METAL SEALS.** MONACK A J GLASS IND 27 8 389-94 408 410 412 414 420 1946
43407	**THERMAL EXPANSION OF BOROSILICATE GLASS IN THE TRANSFORMATION RANGE. II. INFLUENCE OF BORIC OXIDE AND SILICA CONTENT.** OLDFIELD L F GLASS TECHNOLOGY 5 6 224-9 1964
43411	**THERMAL EXPANSION MEASUREMENTS ON SAPPHIRE AND RUBIDIUM IODIDE BELOW 20 K.** SPARKS, P. W. IOWA STATE UNIV., M. S. THESIS 1-67 1964
43414	**SOME PRINCIPLES UNDERLYING THE SUCCESSFUL USE OF METALS AT HIGH TEMPERATURES.** FAHRENWALD F A AMERICAN SOC TESTING MATERIALS PROC 24 310-47 1924
43415	**THEORY AND APPLICATIONS OF DIFFUSE REFLECTANCE SPECTROSCOPY.** COMPANION A L DEVELOP APPL SPECTRY 4 221-34 1965
43422	**SHIELDED CERAMIC COMPOSITE STRUCTURE.** KUMMER, D. L. WRIGHT-PATTERSON AIR FORCE BASE, OHIO 1-42, 1964. (AD-447207)
43424	**THE EFFECTS OF WATER ON THE THERMAL CONDUCTIVITY OF SOIL.** CLOESSNER, J. J. LOUISIANA POLYTECHNIC INST., M. S. THESIS 1-43 1966
43437	**EFFECT OF STRESS ON THE HALL COEFFICIENT OF CHROMIUM FILMS.** LU C-S MILGRAM A A J APPL PHYS 38 5 2038-41 1967
43453	**HIGH-TEMPERATURE ELASTICITY AND EXPANSIVITY OF FORSTERITE AND STEATITE.** SOGA N ANDERSON O L J AM CERAM SOC 50 5 239-42 1967 CA 67 46718
43455	**OPTICAL AND MICROSTRUCTURAL VARIATIONS WITH ELECTRIC FIELD IN BARIUM-STRONTIUM TITANATE CERAMICS.** SANVORDENKER V C J AM CERAM SOC 50 5 261-5 1967
43468	**INVESTIGATION OF THE THERMOPHYSICAL PROPERTIES OF CHAMOTTE CERAMICS IN THE 80-1200 K RANGE.** VASILEV L L FRAIMAN YU E J ENG PHYS 9 6 467-9 1965 (ENGLISH TRANSLATION OF INZH. FIZ. ZH., 9 (6), 762-7, 1965; FOR ORIGINAL SEE T42535)
43477	**REFLECTANCE MEASUREMENTS IN THE 0.6 TO 2.5 MICRON PART OF THE SPECTRUM.** GERHARZ, R. FISCHER, W. A. GEOLOGICAL SURVEY, WASHINGTON, D. C. 1-17, 1966. (NASA-CR-71407, N66-21664)
43480	**THERMAL EXPANSION TESTS ON AGGREGATES, NEAT CEMENTS, AND CONCRETES.** MITCHELL L J AMERICAN SOC TESTING MATERIALS PROC 53 963-75 1953
43484	**LOW SOLAR ABSORPTANCE AND EMITTANCE SURFACES UTILIZING VACUUM DEPOSITED TECHNIQUES.** LOCKHEED MISSILES AND SPACE CO., PALO ALTO, CALIF. LOCKHEED MISSILES AND SPACE CO., PALO ALTO, CALIF. 1-82, 1966. (NASA-CR-73039, N67-17182)
43489	**RESEARCH ON THERMAL TRANSFER PHENOMENA.** ALLEN, W. C. NATL. BERYLLIA CORP., HASKELL, N. J. 1-82, 1965. (NASA-CR-74409, N66-23512)

TPRC Number	Bibliographic Citation
43496	**THERMAL-ENERGY-STORAGE SUPPORTING RESEARCH.** MOAK, D. P. BATTELLE MEMORIAL INST., COLUMBUS, OHIO 1-300, 1966. (NASA-CR-80058, N67-12043)
43498	**INFRARED SPECTRAL EMITTANCE OF ROCKS FROM THE PISGAH CRATER AND MONO CRATERS AREAS, CALIF.** DANIELS, D. L. GEOLOGICAL SURVEY, WASHINGTON, D. C. 1-24, 1966. (NASA-CR-80479, NASA-13, N67-13102)
43499	**THE THERMAL EXPANSION OF A GROSSULARITE GARNET.** WILSON L G AUSTRALIAN J PHYSICS 9 3 403-5 1956
43501	**OPTICAL STUDY OF THE ELECTRONIC BAND STRUCTURE OF DIAMOND.** ROBERTS, R. A. CALIF. UNIV., SANTA BARBARA 1-94, 1967. (NASA-CR-81792, N67-18018)
43514	**THE INFLUENCE OF CONSTITUTION AND MICROSTRUCTURE ON THE TRANSVERSE STRENGTH OF A CRYSTALLIZED GLASS.** DAVIS, R. F. ORDNANCE RES. LAB., PENN. STATE UNIV. 1-90, 1966. (AD 638336)
43539	**MONOXIDE-TYPE COMPOUNDS OF URANIUM AND PLUTONIUM. I, OXYCARBIDES.** TAYLOR K M ANDERSEN J C STRASSER A STAHL D FORBES R L J AM CERAM SOC 50 6 321-5 1967
43558	**THE THERMAL EXPANSION OF SOME REFRACTORY OXIDES.** AUSTIN J B J AM CERAM SOC 14 795-810 1931
43559	**REACTIONS OF BEO AND SIO2, SYNTHESIS AND DECOMPOSITION OF PHENOCITE.** MORGAN R A HUMMEL F A J AM CERAM SOC 32 8 250-5 1949
43564	**THERMAL CONTRACTION OF BETA-EUCRYPTITE /LI2O . 2SIO2/ BY X-RAY AND DILATOMETER METHODS.** GILLERY F H BUSH E A J AM CERAM SOC 42 4 175-7 1959
43565	**COEFFICIENT OF LINEAR THERMAL EXPANSION OF TRIDYMITE.** AUSTIN J B J AM CERAM SOC 76 6019-20 1954
43573	**REVIEW OF HIGH TEMPERATURE METAL-CERAMIC SEALS.** PALMOUR H III J ELECTROCHEM SOC 102 7 160-4C 1955
43592	**HEAT CONDUCTION IN ELECTRIC INSULATED CRYSTALS.** LEIBFRIED G SCHLOMANN E NACHR AKAD WISS GOTTINGEN MATH-PHYS 4 71-93 1954
43596	**CERAMIC FUEL MATERIALS FOR NUCLEAR REACTORS.** JOHNSON J R NUCL. ENG. SCI. CONGR., 1ST 1 71-8 1955
43597	**ON THE MEASUREMENT OF REFLECTION, PERMEABILITY, AND ABSORPTION ON TEST SUBSTANCES OF OPTIONAL FORM IN THE ULBRICHT SPHERE-TYPE PHOTOMETER.** TINGWALDT C P OPTIK 9 7 323-32 1952
43623	**THE THERMAL EXPANSION OF PLAGIOCLASE.** KOZU S UEDA J PROC IMPERIAL ACADEMY /TOKYO/ 9 6 262-4 1933
43624	**THE THERMAL EXPANSION OF DIOPSIDE.** KOZU S UEDA J PROC IMPERIAL ACADEMY /TOKYO/ 9 7 317-19 1933
43625	**THE THERMAL EXPANSION OF FERRITES AT TEMPERATURES NEAR THE CURIE POINT.** BUESSEM W R DORF A PROC METAL POWDER ASSOC 2 196-204 1957
43627	**THE THERMAL EXPANSION OF QUARTZ BY X-RAY MEASUREMENTS.** JAY A H PROC ROY SOC /LONDON/ 142 A 237-47 1933

TPRC Number	Bibliographic Citation
43637	THE THERMAL CONDUCTIVITY OF INSULATING AND OTHER MATERIALS. TAYLOR T S TRANS ASME 41 605–21 1919
43643	THE THERMAL EXPANSION CHARACTERISTICS OF SOME CALCAREOUS AND MAGNESIAN MINERALS. RIGBY G R GREEN A T TRANS BRIT CERAM SOC 41 123–43 1942
43645	BALL CLAYS AND THEIR PROPERTIES. HOLDRIDGE D A TRANS BRIT CERAM SOC 55 369–440 1956
43650	EFFECTS OF UO2 OR THO2 ON THE PROPERTIES OF ZRO2. SOMIYA S YAMAUCHI T SUZUKI H YOGYO KYOKAI SHI /JAPAN/ 65 144–7 1957
43654	THERMAL EXPANSION OF SOLID SOLUTIONS OF FERROMAGNETIC MATERIALS OF THE SYSTEM NICKEL OXIDE–ZINC OXIDE–IRON OXIDE AND COPPER OXIDE–ZINC OXIDE–IRON OXIDE. AVGUSTINIK, A. I. VASILYEV, E. I. ZH. PRIKLAD. KHIMII 29 (6), 1021–4, 1956.
43656	SOME ZIRCONIUM COMPOUNDS AND THE FIELDS OF THEIR APPLICATION. BUDNIKOV, P. P. CHEREPANOV, A. M. ZH. VSES. KHIM. OBSHCHEST. 8 (2), 141–8, 1963.
43716	DUAL INTERACTION OF ANISOLE WITH SURFACE HYDROXYLS. LOW M J D CUSUMANO J A CAN J CHEM 47 20 3906–9 1969
43731	EXTREME TEMPERATURE REFRACTORIES. WHITTEMORE O J JR J CAN CERAM SOC 28 43–8 1959
43744	OPTIMIZATION OF INSULATION AND MECHANICAL SUPPORTS FOR HYPERSONIC AND ENTRY VEHICLES. CROSS, R. I. BLACK, W. E. CONVAIR DIV. OF GENERAL DYNAMICS, SAN DIEGO, CALIF. 1–208, 1967. (AFML–TR–66–414)
43748	PROTECTIVE COATINGS FOR REFRACTORY METALS IN ROCKET ENGINES. RAUSCH, J. J. IIT RESEARCH INST., CHICAGO, ILL. 1–117, 1966. (IITRI–E237–45, NASA–CR–71317, X66–15164)
43773	A THERMOELECTRIC METHOD FOR MEASUREMENT OF STEADY–STATE THERMAL CONDUCTIVITY OF ROCKS. KHAN, A. M. UNIV. OF CALIF., BERKELEY, M. S. THESIS 1–73 1964
43774	THERMAL CONDUCTIVITY OF ROCKS BY A RING SOURCE DEVICE. MOSSAHEBI, M. UNIVERSITY OF CALIFORNIA, BERKELEY, M. S. THESIS 1–77 1966
43796	SOME THERMODYNAMIC RELATIONSHIPS FOR SOILS AT OR BELOW THE FREEZING POINT. I. FREEZING POINT DEPRESSION AND HEAT CAPACITY. LOW, P. F. ANDERSON, D. M. HOEKSTRA, P. COLD REGIONS RES. AND ENG. LAB., HANOVER, N. H. 1–23, 1966. (CRREL–RR–222, AD–649729)
43829	DETERMINATION OF THE COEFFICIENT OF THERMAL DIFFUSIVITY OF ROCKS. SHLAIN I B FILIPPOV M P GORNYI ZHUR 124 10 9–12 1950
43835	THE EFFECTS OF SAMPLE COMPRESSION AND WETTING ON THE THERMAL CONDUCTIVITY OF ROCKS. CLARK H TRANS AMER GEOPHYS UNION 22 543–4 1941
43840	STABLE WHITE COATINGS. ZERLAUT, G. A. GILLIGAN, J. E. HARADA, Y. IIT RES. INST. TECH., CHICAGO, ILL. 1–110, 1965. (IITRI–C–6027, N65–33883)
43841	ON THE MEASUREMENT OF VERY LOW TEMPERATURES. XXXII. THE THERMAL EXPANSION OF JENA–GLASS 16–III. KEESOM W H DE HAAS W J VERSLAG VAN DE GEWONE VERGADERING DER WISSEN NATUURKUNDIGE AFDEELING DER KON. AKAD. VAN WETTENSCHAPPEN TE AMSTERDAM 34 618–24 1925 (FOR ENGLISH TRANSLATION SEE T43401)
43864	COMPARISON OF STRENGTH OF TRIAXIAL PORCELAINS CONTAINING ALUMINA AND SILICA. WARSHAW S I SEIDER R J AM CERAM SOC 50 7 337–43 1967
43896	INVESTIGATION OF THE THERMAL INSULATING PROPERTIES OF KAOLIN FIBERS. KRZHIZHANOVSKII R E CHUDNOVSKAYA I I HIGH TEMPERATURE 4 3 344–7 1966 (ENGLISH TRANSLATION OF TEPLOFIZ. VYS. TEMP., 4 (3), 355–9, 1966; FOR ORIGINAL SEE T42004)
43909	THERMAL EXPANSION AND DIELECTRIC PROPERTIES OF THE NANBO3 RICH END OF THE SYSTEM NANBO3–PBTIO3. CLAUSEN, E. M. UNIVERSITY OF ILLINOIS, M. S. THESIS 1–54 1963
43911	SOLID STATE REACTIONS AND THERMAL EXPANSION IN THE SYSTEMS THO2–P205 AND THO2.P205–ZR02.P205. LAUD, K. R. PENNSYLVANIA STATE UNIVERSITY, M. S. THESIS 1–60 1963
43975	LOW TEMPERATURE FOR INFRARED SPECTRA OF SIO2 POLYMORPHS. PLENDL J N MANSUR L C HADNI A BREHAT F HENRY P MORLOT G NAUDIN F STRIMER P J PHYS CHEM SOLIDS 28 8 1589–97 1967
44019	AN ULTRASONIC STUDY OF SOME OPTICAL GLASSES. INAMURA T JAPANESE J APPL PHYS 6 7 802–7 1967
44058	EXPERIMENTAL METHODS OF DETERMINING THE HEAT RESISTANCE OF REFRACTORY MATERIALS. LANIN A G FEDOROV V B EGOROV V S TEPLOFIZ VYSOKIKH TEMPERATUR 4 6 865–71 1966 (FOR ENGLISH TRANSLATION SEE T44059)
44059	EXPERIMENTAL METHODS OF DETERMINING THE HEAT RESISTANCE OF REFRACTORY MATERIALS. LANIN A G FEDOROV V B EGOROV V S HIGH TEMPERATURE USSR 4 6 799–805 1966 (ENGLISH TRANSLATION OF TEPLOFIZ. VYS. TEMP., 4 (6), 865–71, 1966; FOR ORIGINAL SEE T44058)
44062	ENTHALPY AND SPECIFIC HEAT OF ZIRCONIUM OXIDE IN THE 100–2500 K TEMPERATURE RANGE. KIRILLIN V A SHEINDLIN A E CHEKHOVSKOI V YA ZHUKOVA I A TARASOV V D TEPLOFIZ VYSOKIKH TEMPERATUR 4 6 878–9 1966 CA 66 59545 (FOR ENGLISH TRANSLATION SEE T44063)
44063	ENTHALPY AND SPECIFIC HEAT OF ZIRCONIUM OXIDE IN THE 100–2500 K TEMPERATURE RANGE. KIRILLIN V A SHEINDLIN A E CHEKHOVSKOI V YA ZHUKOVA I A TARASOV V D HIGH TEMPERATURE USSR 4 6 813–4 1966 (ENGLISH TRANSLATION OF TEPLOFIZ. VYS. TEMP., 4 (6), 878–9, 1966; FOR ORIGINAL SEE T44062)
44092	LOW TEMPERATURE INSULATING SYSTEMS. CHRISTIANSEN R M HOLLINGSWORTH R M JR MARSH H N JR PROC CRYOG ENG CONF 171–8 1959
44117	PROPERTIES OF HOT–PRESSED STEATITE. ARROYO E A ALLUE M A S AN REAL SOC ESPAN FIS QUIM 63 B 1 79–84 1967 CA 66 98070
44122	THERMAL EXPANSIONS OF BOROSILICATE GLASS IN THE TRANSFORMATION RANGE. PART I. ANOMALOUS BEHAVIOUR OF TUNGSTEN AND MOLYBDENUM SEALING GLASSES. OLDFIELD L F GLASS TECHNOLOGY 5 4 150–6 1964
44125	INFLUENCE OF GLAZE COMPOSITION ON THE MECHANICAL STRENGTH OF ELECTRICAL PORCELAIN. THIESS L E J AM CERAM SOC 19 70–3 1936
44131	DEVICE FOR MEASURING EXPANSION ABOVE 2000 DEGREES. PROVOST, G. BULL. SOC. FR. CERAM. 73, 39–41, 1966.
44133	ABSORPTION SPECTRA OF AN IRRADIATED SILICATE CERAMIC. BELOUSOV V M IZV VUZ FIZIKA 2 175–7 1965 (FOR ENGLISH TRANSLATION SEE T44134)

TPRC Number	Bibliographic Citation
44134	**ABSORPTION SPECTRA OF AN IRRADIATED SILICATE CERAMIC.** BELOUSOV V M SOVIET PHYS J 2 116-18 1965 (ENGLISH TRANSLATION OF IZV. VYS. UCHEB. ZAVED., FIZ., (2), 175-7, 1965; FOR ORIGINAL SEE T44133)
44135	**THERMAL EXPANSION COEFFICIENTS FOR IRON AND ITS OXIDES FROM X-RAY DIFFRACTION MEASUREMENTS AT ELEVATED TEMPERATURES.** GORTON A T BITSIANES G JOSEPH T L TRANS MET SOC AIME 233 8 1519-25 1965
44136	**DIRECT BONDED BASIC BRICK IN OPEN HEARTH ROOFS.** MIKAMI H M BROWN W E GILBERT V BULL AM CERAM SOC 44 5 433-9 1965
44149	**THERMAL SHOCK AND RELATED PROPERTIES OF DENSE CERAMICS. METAL CARBIDES UP TO 2500 DEGREES.** SMOKE E J CERAMIC AGE 63 6 20-1 38 1954
44150	**DEVELOPMENT AND USE OF THERMAL CONDUCTIVITY PROBES FOR SOILS AND INSULATIONS.** WECHSLER A E KRITZ M A PROC. CONF. THERMAL CONDUCTIVITY, 5TH 5 1-34 1965
44164	**TRANSMITTANCE OF OPTICAL MATERIALS FROM 0.17 MICRONS TO 3.0 MICRONS.** MC CARTHY D E APPL OPT 6 11 1896-8 1967
44166	**INFLUENCE OF THE STATE OF CRYSTALLIZATION ON THE DILATROMETRIC BEHAVIOR OF BOEHMITES.** MURAT, M. CHARBONNIER, M. COMPT. REND. 274 C (3), 221-4, 1972.
44168	**THE USE OF PYROPHYLLITE IN WALL-TILE BODIES.** SPROAT I E J AM CERAM SOC 19 135-42 1936
44184	**USE OF INFRARED HEATING OF GLASS IN INERTIALESS HARDENING FURNACES.** YANISHEVSKII V M VORONKOV E A STEKLO I KERAMIKA 23 3 7-10 1966 (FOR ENGLISH TRANSLATION SEE T44185)
44185	**USE OF INFRARED HEATING OF GLASS IN INERTIALESS HARDENING FURNACES.** YANISHEVSKII V M VORONKOV E A GLASS AND CERAMICS USSR 23 3 119-21 1966 (ENGLISH TRANSLATION OF STEKLO I KERAMIKA, 23 (3), 7-10, 1966; FOR ORIGINAL SEE T44184)
44186	**INVESTIGATING DAUBABINSK TEPHRITE BASALTS FOR MAKING STONE CASTINGS.** SULEIMENOV S T ABDUVALIEV T A SHARAFIEV M SH TROPINA L G STEKLO I KERAMIKA 23 4 23-6 1966 (FOR ENGLISH TRANSLATION SEE T44187)
44187	**INVESTIGATING DAUBABINSK TEPHRITE BASALTS FOR MAKING STONE CASTINGS.** SULEIMENOV S T ABDUVALIEV T A SHARAFIEV M SH TROPINA L G GLASS AND CERAMICS USSR 23 4 192-5 1966 (ENGLISH TRANSLATION OF STEKLO I KERAMIKA, 23 (4), 23-6, 1966; FOR ORIGINAL SEE T44186)
44188	**GLASS COMPOSITIONS BASED ON PERLITE FOR HIGH-VOLTAGE INSULATOR PRODUCTION.** KUTATELADZE K S VERULASHVILI R D STEKLO I KERAMIKA 23 5 19-21 1966 (FOR ENGLISH TRANSLATION SEE T44189)
44189	**GLASS COMPOSITIONS BASED ON PERLITE FOR HIGH-VOLTAGE INSULATOR PRODUCTION.** KUTATELADZE K S VERULASHVILI R D GLASS AND CERAMICS USSR 23 5 241-4 1966 (ENGLISH TRANSLATION OF STEKLO I KERAMIKA, 23 (5), 19-21, 1966; FOR ORIGINAL SEE T44188)
44192	**PROPERTIES OF CORUNDUM MICROLITE MADE BY PRESSING AND HOT CASTING.** KHARITONOV F YA KALITIN P P STEKLO I KERAMIKA 23 5 31-4 1966 (FOR ENGLISH TRANSLATION SEE T44193)
44193	**PROPERTIES OF CORUNDUM MICROLITE MADE BY PRESSING AND HOT CASTING.** KHARITONOV F YA KALITIN P P GLASS AND CERAMICS USSR 23 5 258-61 1966 (ENGLISH TRANSLATION OF STEKLO I KERAMIKA, 23 (5), 31-4, 1966; FOR ORIGINAL SEE T44192)
44194	**EFFECT OF HEATING RATE ON EXPANSION OF GLASS.** ANDRYUKHINA T D STEKLO I KERAMIKA 23 6 10-11 1966 (FOR ENGLISH TRANSLATION SEE T44195)
44195	**EFFECT OF HEATING RATE ON EXPANSION OF GLASS.** ANDRYUKHINA T D GLASS AND CERAMICS USSR 23 6 288-9 1966 (ENGLISH TRANSLATION OF STEKLO I KERAMIKA, 23 (6), 10-11, 1966; FOR ORIGINAL SEE T44194)
44197	**EXPERIMENTAL MEASUREMENT OF THE OPTICAL CONSTANTS OF GERMANITE AND RENIERITE.** LOGINOVA L A INST MINERAL GEOKHIM KRISTALLOKHIM REDK ELEM TRUDY SSSR 4 224-34 1960 (FOR ENGLISH TRANSLATION SEE T44198)
44198	**EXPERIMENTAL MEASUREMENT OF THE OPTICAL CONSTANTS OF GERMANITE AND RENIERITE.** LOGINOVA L A SPECIAL LIBRARY ASSOC., TECH. TRANS. 1-17, 1960. (FRENCH TRANSLATION OF INST. MINERAL. GEOKHIM. KRISTALLOKHIM. REDK. ELEM. TRUDY, SSR, (4), 224-34, 1960; FOR ORIGINAL SEE T44197) (CNRS-H/R-153, TT65-28204)
44204	**STRUCTURAL TRANSFORMATIONS OF GLASS BY HEAT TREATMENT.** ANDRYUKHINA T D STEKLO I KERAM 23 8 11-13 1966 CA 65 15020 (FOR ENGLISH TRANSLTION SEE T44205)
44205	**STRUCTURAL TRANSFORMATIONS DURING HEAT TREATMENT OF GLASS.** ANDRYUKHINA T D GLASS AND CERAMICS USSR 23 8 413-15 1966 (ENGLISH TRANSLATION OF STEKLO I KERAMIKA, 23 (8), 11-13, 1966; FOR ORIGINAL SEE T44204)
44206	**EFFECT OF GYPSUM ON THE LIGHT TRANSMISSION AND OPACIFICATION OF OPAL GLASSES.** NEICH A I TSARITSYN M A STEKLO I KERAMIKA 23 8 16-17 1966 (FOR ENGLISH TRANSLATION SEE T44207)
44207	**EFFECT OF GYPSUM ON THE LIGHT TRANSMISSION AND OPACIFICATION OF OPAL GLASSES.** NEICH A I TSARITSYN M A GLASS AND CERAMICS USSR 23 8 419-20 1966 (ENGLISH TRANSLATION OF STEKLO I KERAMIKA, 23 (8), 16-17, 1966; FOR ORIGINAL SEE T44206)
44216	**NON-ARCING CORDIERITE CERAMICS FOR MAGNETIC SWITCHES.** ZINKO E I PETROVA Z A STEKLO I KERAMIKA 23 10 30-3 1966 (FOR ENGLISH TRANSLATION SEE T44217)
44217	**NON-ARCING CORDIERITE CERAMICS FOR MAGNETIC SWITCHES.** ZINKO E I PETROVA Z A GLASS AND CERAMICS USSR 23 10 550-3 1966 (ENGLISH TRANSLATION OF STEKLO I KERAMIKA, 23 (10), 30-3, 1966; FOR ORIGINAL SEE T44216)
44218	**PHOSPHATE ENAMELS FOR HIGH ALUMINUM ALLOYS.** MIGONADZHIEV A S STEKLO I KERAMIKA 23 12 15-17 1966 (FOR ENGLISH TRANSLATION SEE T44219)
44219	**PHOSPHATE ENAMELS FOR HIGH ALUMINUM ALLOYS.** MIGONADZHIEV A S GLASS AND CERAMICS USSR 23 12 647-9 1966 (ENGLISH TRANSLATION OF STEKLO I KERAMIKA, 23 (12), 15-17, 1966; FOR ORIGINAL SEE T44218)
44227	**THERMAL EXPANSION OF TECHNICAL SOLIDS AT LOW TEMPERATURES. A COMPILATION FROM THE LITERATURE.** CORRUCCINI, R. J. GNIEWEK, J. J. NATIONAL BUREAU OF STANDARDS, WASHINGTON, D. C. 1-22, 1961. (NBS-MONOGRAPH-29)

TPRC Number	Bibliographic Citation
44258	THERMAL EXPANSIONS OF FUSED ALUMINA, BAUXITE, AND DIASPORE. NATIONAL BUREAU OF STANDARDS J FRANKLIN INSTITUTE 210 104-6 1930
44259	THERMAL EXPANSION OF CHROMITE. NATIONAL BUREAU OF STANDARDS J FRANKLIN INSTITUTE 210 382-3 1930
44266	MANUFACTURE OF CERAMIC BODIES WITH A NEGATIVE COEFFICIENT OF LINEAR THERMAL EXPANSION. MOLDVAI A S BUBIK I EPITOANYAG 18 5 165-6 174 1966 CA 65 19810
44269	THE SOLARIZATION OF SODA-POTASH CRYSTAL GLASS BY SUNSHINE AND ITS STABILIZATION. VOBORNIK K EPITOANYAG 18 12 475-7 1966 CA 66 79164
44273	A STUDY ON EMISSIVITIES. SCHOCKEN, K. ARMY BALLISTIC MISSILE AGENCY, REDSTONE, ARSENAL, ALABAMA 1-23, 1958. (DV-TN-64-58, N-83592)
44294	SPECIFIC HEAT AND ITS RELATION TO ANALYTICAL DATA /FOR BROWN COAL/. RAMMLER E SCHMIDT R FREIBERG FORSCHUNGSH A 385 5-38 1966 CA 66 67758
44295	MEASUREMENT OF SPECIFIC HEAT OF BROWN COAL IN THE WORKS LABORATORY. THIELE H FREIBERG FORSCHUNGSH A 385 39-60 1966 CA 66 67760
44296	USE OF SYENITE IN SEMIVITREOUS WARE, 1. /A/ PHYSICAL PROPERTIES OF BISQUES AND /B/ A GLAZE CONTAINING SYENITE. KOENIG C J J AM CERAM SOC 19 295-8 1936
44297	RELATION BETWEEN MOISTURE CONTENT, REFLECTANCE VALUES AND INTERNAL SURFACE AREA OF COAL. HARRISON J A THOMAS J JR FUEL /LONDON/ 45 6 501-3 1966 CA 66 57660
44322	PROPERTIES OF RAMMED CARBON LININGS OF BLAST-FURNACE HEARTHS. DRABINA J HUTN LISTY 21 12 828-32 1966 CA 66 58091
44329	CERAMIC DIELECTRICS WITH RELATIVELY HIGH THERMAL CONDUCTIVITIES. HOLLADAY, C. O. DIVITA, S. IEEE TRANS. PARTS MATER. PACKAG. 2 (4), 110-3, 1966.
44339	THE THERMAL CONDUCTIVITY OF POROUS MEDIA IV SANDSTONES. THE EFFECT OF TEMPERATURE AND SATURATION. MESSMER, J. H. PROC. CONF. THERMAL CONDUCTIVITY, 5TH 1, 1-29, 1965.
44341	TALC IN WHITEWARE. GELLER R F CREAMER A S J AM CERAM SOC 20 137-47 1937
44351	MEASUREMENT OF THE EXPANSION OF REFRACTORIES AT HIGH TEMPERATURE. SCHWIETE H E METZGER C BER DEUT KERAM GES 41 5 334-41 1964
44356	WHITE SURFACE STANDARDS FOR COLORIMETRY AND SPECTROPHOTOMETRY. YUSTOVA E N IVANOVA I S POKROVSKAYA G V IZMER TEKH 8 18-21 1966 CA 66 60482 (FOR ENGLISH TRANSLATION SEE T60334)
44358	EFFECT OF SOME TECHNICAL FACTORS ON THE SPALLING RESISTANCE OF PERICLASE-SPINEL REFRACTORIES. KUKOLEV G V NEMETS I I SHEKHOVTSEVA V A OGNEUPORY 33 5 31-9 1968 (FOR ENGLISH TRANSLATION SEE T46849)
44369	THERMAL CONDUCTIVITY OF SOLID SOLUTIONS MAGNESIUM OXIDE-MAGNESIUM CHROMIUM OXIDE AND ALUMINUM OXIDE-CHROMIUM OXIDE AS A FUNCTION OF TEMPERATURE AND CONCENTRATION. VISHNEVSKII, I. I. SKRIPAK, V. N. IZV. AKAD. NAUK SSSR, NEORG. MATER 2 (10), 1820-6, 1966.
44372	EFFECT OF COPPER TELLURIDE IMPURITY ON THERMOELECTRIC PROPERTIES OF NATURAL GALENA. BARANOVA R KH BERDYEV B M IZV AKAD NAUK TURKM SSR SER FIZ-TEKH KHIM GEOL NAUK 6 119-22 1966 CA 66 80390
44390	DETERMINATION OF APPARENT SPECIFIC HEAT OF CLAY. MAJUMDAR, N. C. STEVENS, T. J. J. AUST. CERAM. SOC. 2 (1), 8-14, 1966.
44400	THERMAL EXPANSION-CONTRACTION OF ASPHALTIC CONCRETE. DARTER, M. I. UNIVERSITY OF UTAH, SALT LAKE, M. S. THESIS 1-95 1968
44403	ULTRAVIOLET REFLECTANCE MEASUREMENTS OF POSSIBLE LUNAR SILICATES. GREENMAN N N BURKIG V W YOUNG J F J GEOPHYS RES 72 4 1355-9 1967 CA 66 70428
44408	VISIBLE AND NEAR-INFRARED ABSOPRTION OF URANIUM IN SODIUM SILICATE GLASS. CARRELL M A WILDER D R J NUCL MATER 13 2 142-51 1964 JA 50 149
44427	GLASS AS A MATERIAL FOR REFRACTIVE SPACE OPTICS. WOODCOCK R F COX M C J SOC MOTION PICT TELEV ENG 76 2 95-9 1967 CA 66 97991
44430	TITANIUM COMPOUNDS AND APPLICATION THEREOF IN VITREOUS ENAMELS. KINZIE C J PLUNKETT J A J AM CERAM SOC 18 117-22 1935
44434	THERMOPHYSICAL PROPERTIES OF STONE LININGS OF INDUSTRIAL REGENERATORS IN AIR DISTILLATION PLANTS. POGOVAYA, I. A. ISHKIN, I. P. KHIM. NEFT. MASHINOSTR. 5, 25-6, 1965.
44465	THERMAL ANALYSIS OF THE LUNAR SURFACE. CHIANG, C. W. PROC. CONF. THERMAL CONDUCTIVITY, 5TH 1, 1-27, 1965.
44476	REFLECTION SPECTRA OF SOLIDS. HYPOTHESIS AND LIMITS OF QUANTITATIVE EVALUATION OF THE KUBELKA-MUNK FUNCTION. KORTUM G OELKRUG D NATURWISSENSCHAFTEN 53 23 600-9 1966 CA 66 89901
44488	APPARATUS FOR MEASUREMENT OF THERMAL CONDUCTIVITY ON SMALL SPECIMENS. VISHNEVSKII I I SKRIPAK V N OGNEUPORY 31 12 13-18 1966 CA 66 59563 (FOR ENGLISH TRANSLATION SEE T46599)
44506	USE OF NEPHELINE SYENITE IN SANITARY PORCELAIN. KOENIG C J J AM CERAM SOC 22 38-46 1939
44508	HIGH PRESSURE /UP TO 18 KILOBARS/ CHAMBER FOR OPERATING AT HELIUM TEMPERATURES. ITSKEVICH E S VORONOVSKII A N GAVRILOV A F SUKHOPAROV V A PRIB TEKH EKSP 11 6 161-4 1966 CA 66 57145 (FOR ENGLISH TRANSLATION SEE T52249)
44511	INTERACTION OF RADIATION AND CONDUCTION IN GLASS. CHUI G K GARDON R J AM CERAM SOC 52 10 548-53 1969 CA 71 116046
44519	MEASUREMENT OF THE DIFFUSIVITY AND THERMAL CONDUCTIVITY OF POLYMERS—VARIATION AS A FUNCTION OF THEIR STRUCTURAL STATE AND TEMPERATURE. BERLOT R PUBL SCI TEC MIN AIR FR 154 102 1966 CA 66 95699
44532	PROPERTIES OF LOW-MELTING GLASSES IN THE ZNO-PBO-B2O3-SIO2 SYSTEM. LYUTSKANOV S KOVACEV R SKLAR KERAM 13 6 152-4 1963 JA 50 66
44533	SOME EFFECTS OF OVERBURDEN PRESSURE ON OIL SHALE DURING UNDERGROUND RETORTING. THOMAS G W SOC PETROL ENG J 6 1 1-8 1966 CA 66 87244
44535	PHASE COMPOSITION OF FROZEN MONTMORILLONITE-WATER MIXTURES FROM HEAT CAPACITY MEASUREMENTS. ANDERSON D M SOIL SCI SOC AMER PROC 30 6 670-5 1966 CA 66 79924

TPRC Number	Bibliographic Citation
44543	**THERMOPHYSICAL PROPERTIES OF A LAYER OF ORE-CARBON PELLETS.** STOROZHEV, YU. I. TELEGIN, A. S. KUDRYAVTSEV, V. S. PCHELKIN, S. A. IZV. VYSSH. UCHEB. ZAVED., CHERN. MET. 14 (10), 30–2, 1971. (FOR ENGLISH TRANSLATION SEE T75014)
44549	**METHOD OF MEASURING THE THERMAL CONDUCTIVITY OF REFRACTORIES AT HIGH TEMPERATURE.** DE SMET A HIGH TEMPERATURES–HIGH PRESSURES 1 2 199–206 1969
44551	**INSTRUMENTATION STUDIES. LXXXVIII. A STUDY OF PHOTOELECTRIC INSTRUMENTS FOR THE MEASUREMENT OF COLOR, REFLECTANCE, AND TRANSMITTANCE. 16. AUTOMATIC COLOR-BRIGHTNESS TESTER.** DEARTH L R SHILLCOX W M VAN DEN AKKER J A TAPPI 50 2 51–8A 1967 CA 66 96387
44554	**PROPERTIES AND USES OF BORON NITRIDE FIBERS.** ECONOMY J ANDERSON R V TEXTILE RES J 36 11 994–1003 1966 CA 66 66594
44568	**SELECTION FACTORS IN CHOOSING REFRACTORIES FOR INDUSTRIAL PLANT.** HOUGEN O A CHEM METALLURGICAL ENGINEERING 30 19 737–41 1924
44592	**ELABORATION OF COMPOSITION OF THE TITANIUM-CONTAINING CHEMICALLY STABLE GLASSES.** ERMOLENKO N N SHAREIKO L V VOP PRIMEN REDK METAL SILIKAT PROM DOKL RESPUB SOVESHCH MINSK 1963 40–5 1964 CA 66 107768
44598	**HEAT CONDUCTIVITY OF PURE OXIDE CERAMICS.** DUDEROV I G POLUBOYARINOV D N VYSOKOOGNEUPOR MATER SB STATEI 92–105 1966 CA 66 107812 (FOR ENGLISH TRANSLATION SEE T51032)
44604	**THERMAL EXPANSION AND ELECTRICAL RESISTANCE OF SINTERED BODIES IN THE SYSTEM ZRO2–CEO2.** SOMIYA S OKIKAWA S SAITO S YOGYO KYOKAI SHI 73 2 25–9 1965 CA 68 24246
44605	**STUDIES ON ROSEKI. I. COMPARISON OF THERMAL PROPERTIES BETWEEN COMPACT AND WET BALL - MILLED ROSEKI (PYROPHYLLITE).** SHIRAKI, Y. INOUE, A. YOGYO KYOKAI SHI 73 (9), 187–95, 1965.
44606	**BURSTING EXPANSION OF THE CHROMIUM BEARING REFRACTORIES.** SOMIYA, I. S. YOGYO KYOKAI SHI 73 (11), 235–44, 1965.
44607	**PROPERTIES AND STRUCTURE OF GLASSES OF THE BORON OXIDE - GERMANIUM OXIDE SYSTEM.** IMAOKA, M. YAMAZAKI, T. YOGYO KYOKAI SHI 73, 264–72, 1965.
44608	**OPTICAL PROPERTIES OF COLUMBITE–TANTALITE AND PYROCHLORE–MICROLITE IN THE REFLECTED LIGHT.** CHERNIK L N ZAP VSES MINERAL OBSHCHEST 95 6 716–19 1966 CA 66 67783
44611	**APPARATUS FOR DETERMINATION OF VISCOSITY OF SILICATE MATERIALS WITHIN THE RANGE OF 10 TO THE EIGHTH POWER TO 10 TO THE FIFTEENTH POWER POISES UNDER GAS PRESSURE.** VARUZHANYAN A A ZAVOD LAB 32 12 1525–7 1966 CA 66 57156
44624	**INVESTIGATION OF THE EFFECT OF NEUTRON- AND GAMMA-IRRADIATION ON THE OPTICAL DENSITY OF BULB GLASSES IN SOME PHOTO-ELECTRONIC MULTIPLIERS.** YUSHKEVICH G F SOSHINA N V ZH PRIKL SPEKTROSK 5 4 486–8 1966 CA 66 68568
44628	**THE IMPROVED GUARDED COLD PLATE THERMAL CONDUCTIVITY APPARATUS.** BLACK, I. A. WECHSLER, A. E. GLASER, P. E. FOUNTAIN, J. A. PROC. CONF. ON THERMAL CONDUCTIVITY, 5TH 2, IV-B-1–IV-B-18, 1965.
44633	**THERMAL PROPERTIES OF SOILS BETWEEN -180 AND 40 C.** MOELLER, C. E. SAUER, H. J., JR. PROC. CONF. ON THERMAL CONDUCTIVITY, 5TH 2, IV-E-1–IV-E-18, 1965.
44639	**PROPERTIES OF SOME OHIO RED-FIRING CLAYS.** MAC GEE A E WHITE W C O KLINEFELTER T A J AM CERAMIC SOC 18 155–62 1935
44640	**USE OF NEPHELINE SYENITE IN FLOOR-TILE AND WALL-TILE BODIES.** KOENIG C J J AM CERAM SOC 23 3 86–91 1940
44681	**MEASUREMENT OF THERMAL CONDUCTANCE OF MULTILAYER AND OTHER INSULATION MATERIALS.** KARP, G. S. FRIED, E. GENERAL ELECTRIC SPACECRAFT DEPT., VALLEY FORGE SPACE TECHNOL. CTR., PHILADELPHIA, PENN. 1–245, 1966. (NASA-CR-78979, N66-39708)
44727	**PROPERTIES OF SOME AMERICAN KAOLINS AND A COMPARISON WITH ENGLISH CHINA CLAYS.** KLINEFELTER T A MEYER W W J AM CERAMIC SOC 18 163–9 1935
44753	**A METHOD TO DETERMINE THE COMPLEX INDICES OF AN UNIAXIAL ABSORBANT CRYSTAL.** STEPHAN G LE CALVEZ Y ROBIN S COMPT REND 265 B 4 281–4 1967
44768	**DIAMOND. /DATA SHEETS/.** NEUBERGER, M. HUGHES AIRCRAFT CO., CULVER CITY, CALIF. 1–146, 1967. (EPIC-DS-153, AD-812827)
44778	**CHANGES IN THE TECHNICAL PROPERTIES OF ZIRCONIA REFRACTORIES AS A FUNCTION OF THE DEGREE OF ZRO2 STABILIZATION.** KLYUCHAROV YA V STRAKHOV V I INORGANIC MATERIALS 4 9 1311–14 1968 (ENGLISH TRANSLATION OF IZV. AKAD. NAUK SSSR, NEORG. MATER., 4 (9), 1502–6, 1968; FOR ORIGINAL SEE T53021)
44781	**LENGTH CHANGE-ADSORPTION RELATIONS FOR THE WATER-POROUS GLASS SYSTEM TO -40 C.** FELDMAN R F CAN J CHEM 48 2 287–97 1970 (NATL. RES. COUNCIL OF CANADA REPT. 11087)
44790	**HEAT CONDUCTIVITY OF NICKEL, MAGNESIUM, AND CADMIUM FERRITE-CHROMITES.** BEKKER YA M FIZ FIZ-KHIM SVOISTVA FERRITOV MATER DOKL VSES SOVESHCH 4TH MINSK 95–102 1966 CA 67 6325
44792	**VISCOSITY OF THE EARTHS CORE.** GANS, R. F. J. GEOPHYS. RES. 77 (2), 360–6, 1972.
44793	**MEMOIR ON THE EXPANSION OF SOLIDS BY HEAT.** FIZEAU M H COMPT RENDUS 62 22 1133–47 1866
44799	**THERMAL EXPANSION OF STABILIZED ZIRCONIUM DIOXIDE IN THE TEMPERATURE INTERVAL 300-2600 K.** SHAKHTIN D M LEVINTOVICH E V PIVOVAR T L ELISEEVA G G INORGANIC MATERIALS 4 9 1401–2 1968 (ENGLISH TRANSALTION OF IZV. AKAD. NAUK SSSR, NEORG. MATER., 4 (9), 1603–4, 1968; FOR ORIGINAL SEE T53023)
44807	**THERMAL EXPANSION OF NEPHELINE.** GOLOVIN A A GEOL I GEOFIZ AKAD NAUK SSSR SIBIRSK OTD 12 135–6 1965 CA 64 19212
44812	**INVESTIGATION OF GLASSHOUSE SUPERSTRUCTURE MATERIALS.** FABIANIC W L J AM CERAM SOC 18 211–19 1935
44823	**REACTIONS OF REFRACTORY MATERIALS WITH THE GAS STREAM IN THE MHD GENERATOR DUCT.** IVANOV A B DOUGLAS AIRCRAFT CO., INC., SANTA MONICA, CALIF. 1–, 1967. (FTD-HT-67-195, SM-74-228)
44828	**LINEAR EXPANSION OF ORDINARY AND BLACK DINAS BRICK DURING HEATING.** BUDNIKOV P P MYULLER V E FOREIGN TECHNOLOGY DIVISION 1–14, 1971. (ENGLISH TRANSLATION OF DOMEZ, (10-11), 24-8, 1961; FOR ORIGINAL SEE T60851) (FTD-MT-24-1389-71)

TPRC Number	Bibliographic Citation

44836 **EFFECT OF VISCOSITY ON PHYSICAL-MECHANICAL PROPERTIES OF COPPER SLAGS.**
KULEMZIN K N ONAEV I A
VESTN AKAD NAUK KAZ SSR
23 3 70-6 1967 CA 67 5091

44863 **REPORT ON THE LITERATURE STUDY FOR THE SELECTION OF A SUITABLE METHOD FOR THE DETERMINATION OF THE THERMAL CONDUCTIVITY OF CERAMIC BODIES ON THE BASIS OF URANIUM OXIDE - BERYLLIUM OXIDE.**
BATTELLE INST., E. V. FRANKFURT, GERMANY
BATTELLE INST., FRANKFURT/MAIN, W. GERMANY
1-29, 1961.
(EURAEC-59)

44867 **GEOTHERMAL MEASUREMENTS IN AUSTRALIA.**
SASS, J. H.
AUSTRALIAN NATIONAL UNIV., CANBERRA, PH. D. THESIS
1-122 1964

44868 **THE MEASUREMENT OF THE FLOW OF HEAT THROUGH THE CRUST OF THE EARTH AND THROUGH ROCKS.**
BECK, A. E.
AUSTRALIAN NATIONAL UNIV., CANBERRA, PH. D. THESIS
1-236 1956

44875 **EVALUATION OF MULTILAYER ANTI-REFLECTION COATINGS FOR INSTRUMENT WEDGES AND COVERGLASSES.**
PARKER, J. W.
U. S. NAVAL AIR DEVELOPMENT CENTER, JOHNSVILLE, PENNSYLVANIA
1-22, 1966.
(NADC-AM-6625, AD-483816L)

44879 **PROPERTIES OF SOLID SOLUTIONS IN THE SYSTEM CHROMIUM OXIDE-ZIRCONIUM OXIDE.**
BECHERESCU, D. IVAN, E. MARX, F.
BUL. STINT. TEH. INST. POLITEH. TIMISOARA
10 (2), 373-87, 1965.

44894 **POSSIBILITY OF DETERMINING THE DEGREE OF INTERACTION BETWEEN THE PORCELAIN BODY AND THE GLAZE.**
MATEJKA J
SKLAR KERAM
17 1 18-21 1967 CA 67 14487

44906 **STUDIES ON THE THERMAL CONDUCTIVITY OF FIBROUS INSULANTS BY MEANS OF A RAPID MEASURING DEVICE FOR THERMAL CONDUCTIVITY UNDER STATIONARY CONDITIONS.**
FOURNIER D ANDRE G KLARSFELD S
BULL INST INT FROID ANNEXE-2
163-74 1966 CA 67 15578

44921 **INTEGRAL EXPANSION OF VITREOUS ENAMELS BETWEEN THE SOFTENING PONT AND ROOM TEMPERATURE.**
ROSENBERG J E LANGERMAN A
J AM CERAM SOC
20 236-44 1937

44924 **OPTICAL PROPERTIES OF VARIOUS MATERIALS IN THE FAR ULTRAVIOLET.**
SIMONE R
METHOD PHYS ANAL
135-43 1966

44932 **RESEARCH ON ATMOSPHERIC OPTICAL RADIATION TRANSMISSION.**
BULLRICH K DE BARY E BLATTNER W EIDEN R
HANEL, G. NOWAK, W.
MAINZ UNIV./WEST GERMANY, METEOROLOGISCH GEOPHYSIKALISCHES INSTITUT
1-68, 1966.
(AFCRL-67-0207, AD-653757)

44936 **DEVELOPMENT OF NEW CERAMIC MATERIALS /ZYTTRITE/ BY THERMAL AND HYDROLYTIC DECOMPOSITION OF METAL ALCOHOLATES.**
MAZDIYASNI, K. S. LYNCH, C. T. SMITH, J. S.
AIR FORCE MATERIALS LAB., WPAFB, OHIO
1-83, 1966.
(AFML-TR-66-418, AD-815055)

44941 **DEVELOPMENT OF ALKALI METAL PEROXIDE AND SUPEROXIDE BLOWN CERAMIC FOAMS.**
BLOCKER, E. W. PAUL, R. D.
UNITED AIRCRAFT CORP., WINDSOR LOCKS, CONN.
1-71, 1966.
(NASA-CR-76406, N66-31363)

44944 **CORRELATION OF MECHANICAL AND THERMAL PROPERTIES OF EXTRATERRESTRIAL MATERIALS.**
HALAJIAN J D REICHMAN J KARAFIATH L L
GRUMMAN AIRCRAFT ENG. CORP., GEO-ASTROPHYSICS SECT., BETHPAGE, N. Y.
1-193, 1967.
(NASA-CR-83895, N67-25958)

44946 **THE IN SITU MEASUREMENT OF LUNAR THERMAL CONDUCTIVITY.**
CLARK, S. P., JR.
YALE UNIV., NEW HAVEN, CONN.
1-28, 1967.
(NASA-CR-84120, N67-26447)

44953 **EFFECT OF FORMATION OIL SATURATION ON THERMAL CONDUCTIVITY AT HIGH TEMPERATURES AND PRESSURES.**
BAIRAMOV E M
DOKL AKAD NAUK AZERB SSR
22 5 45-8 1966
(FOR ENGLISH TRANSLATION SEE T44954)

44954 **EFFECT OF FORMATION OIL SATURATION ON THERMAL CONDUCTIVITY AT HIGH TEMPERATURES AND PRESSURES.**
BAIRAMOV E M
ASSOCIATED TECHNICAL SERVICE
1-4, 1967.
(ENGLISH TRANSLATION OF DOKL. AKAD. NAUK. AZERB. SSR, 22 (5), 45-8, 1966; FOR ORIGINAL SEE T44953)
(ATS-04U100R)

44960 **QUARTERLY PROGRESS REPT. ON THE INVESTIGATION OF REFRACTORY MATERIALS AND APPENDIX.**
MC KINSTRY H A HOCKER C F RICKER R W
U. S. AIR FORCE AIR MATERIAL COMMAND, WPAFB, OHIO
1-61, 1959
(ATI-67891)

44966 **TERRESTRIAL HEAT FLOW IN AUSTRALIA.**
HOWARD L E
J GEOPHYS RESEARCH
69 8 1617-26 1964

44967 **HEAT FLOW AT COBAR, NEW SOUTH WALES.**
LE MARNE A E
J GEOPHYS RESEARCH
67 10 3981-3 1962

44968 **HEAT-FLOW VALUES FROM EASTERN AUSTRALIA.**
SASS J H
J GEOPHYS RESEARCH
69 18 3889-93 1964

44969 **HEAT-FLOW VALUES FROM THE PRECAMBRIAN SHIELD OF WESTERN AUSTRALIA.**
SASS J H
J GEOPHYS RESEARCH
69 2 299-308 1964

44970 **HEAT FLOW AT BROKEN HILL, NEW SOUTH WALES.**
SASS J H LE MARNE A E
GEOPHYSICAL J ROYAL ASTRON SOC
7 4 477-89 1963

44986 **PERFORMANCE OF MULTILAYER INSULATION SYSTEMS FOR TEMPERATURES TO 700 K.**
CUNNINGTON G R JR ZIERMAN C A FUNAI A I
LINDAHN A
LOCKHEED AIRCRAFT CORP., APLO ALTO, CALIF.
1-106, 1967.
(NASA-CR-907)

44988 **THERMAL CONDUCTIVITY MEASUREMENTS OF INSULATING MATERIALS TO -190 C.**
CODEGONE C FERRO V SACCHI A
BULL INST INT FROID ANNEXE-2
23-37 1966

44990 **AN AUTOMATIC APPARATUS WITH A FLAT PLATE FOR THE MEASUREMENT OF THERMAL CONDUCTIVITY OF INSULATING MATERIALS AT HIGH TEMPERATURE /LIMIT AT 1000 C./.**
FERRO V SACCHI A
IL CALORE
4 2-16 1966

45006 **INTERNAL STRESS IN CERAMICS CONTAINING CRISTOBALITE.**
FOSTER P K HUGHES I R
J AM CERAM SOC
49 9 515 1966

45048 **ON THE CONDUCTIVITY OF CRYSTALLINE SUBSTANCES FOR HEAT.**
DE SENARMONT M H
ANN CHIM ET PHYS
21 3 457-70 1847

45052 **THE EFFECT OF ULTRAVIOLET LIGHT ON A TRANSPARENT PLASTIC.**
SAMS W M JR
ARCH DERMATOL
95 2 225-7 1967 CA 67 22522

45053 **SOME FUNDAMENTAL PROPERTIES OF NEPHELINE SYENITE.**
KOENIG C J
J AM CERAMIC SOC
22 35-8 1939

45054 **LOW-MELTING GLASSES FOR PACKAGING OF SEMICONDUCTOR DEVICES.**
ISHIYAMA M MATSUDA T NAGAHARA S SUZUKI Y
ASAHI GARASU KENKYU HOKOKU
16 77-88 1966 CA 67 24950

45056 **PHASE RELATIONSHIPS IN THE SYSTEM ZIRCONIA-THORIA.**
DUWEZ P LOH E
J AM CERAMIC SOC
40 9 321-4 1957

TPRC Number	Bibliographic Citation
45057	**BORIC OXIDE-PHOSPHORIC OXIDE GLASSES. I. TRANSMISSION IN THE SPECTRAL REGION OF 0.2-3.8 MICRONS.** CONDE, C. S. ALVAREZ-ESTRADA, D. BOL. SOC. ESPAN. CERAM. 5 (6), 823-35, 1966.
45061	**THERMAL EXPANSION OF RUTILE FROM 100 TO 700 K.** KIRBY R K J RES NATL BUR STD 71 A 5 363-9 1967 JA 51 118
45067	**THE HOT-WIRE METHOD OF MEASURING THE THERMAL CONDUCTIVITY OF REFRACTORY MATERIALS AND INSULATING BRICKS.** SCHWIETE H E DOHM K D ARCH EISENHUETTENW 38 5 411-14 1967 CA 67 56888
45072	**ALUMINA IMPROVES PROPERTIES OF WHITEWARE BODIES.** FLOYD J R ROYCE D V JR LIPPMAN A JR CERAM IND 88 4 108-11 139 1967 CA 67 24975
45088	**ONE-DIMENSIONAL FREEZING OF WATER-SATURATED SAND.** AGUIRRE-PUENTE J LE FUR B SZANTO I COLLOQ INT CENTRE NAT RECH SCI 1966 160 149-64 1967 CA 67 26550
45089	**FREEZING OF CLAYS.** LE FUR B AGUIRRE-PUENTE J SZANTO I COLLOQ INT CENTRE NAT RECH SCI 1966 160 247-62 1967 CA 67 35874
45094	**THE LOW TEMPERATURE SPECIFIC HEAT OF GRAPHITE AND SEVERAL PURE SUPERCONDUCTORS.** VAN DER HOEVEN, B. J. C., JR. PURDUE UNIVERSITY, W. LAFAYETTE, IND., PH. D. THESIS 1-50, 1964. (UNIVERSITY MICROFILMS NO. 65-2652)
45099	**ELASTIC PROPERTIES AND THERMAL EXPANSION OF BODIES IN THE CHROMIUM/III/ OXIDE-MAGNESIUM OXIDE SYSTEM.** WILLIAMS R M STATE UNIV. NEW YORK, ALFRED, N. Y., PH. D. THESIS 1-111, 1967. (UNIVERSITY MICROFILMS NO. 69-3298)
45103	**THERMAL CONDUCTIVITY OF ECLOGITE AND DOLERITE UNDER HIGH TEMPERATURE.** MOISEENKO U I SOLOVEVA Z A KUTOLIN V A DOKL AKAD NAUK SSSR 173 3 669-71 1967 CA 67 45982 (FOR ENGLISH TRANSLATION SEE T36055)
45106	**TEMPERATURE CONDITIONS AT IGNEOUS CONTACTS WITH CERTAIN PERMIAN COALS OF INDIA.** GHOSH T K ECON GEOL 62 1 109-17 1967 CA 67 23937
45107	**FURNACE WITH MOLYBDENUM DISILICIDE HEATERS.** BERSHAK V I EKSP TEKH METODY VYSOKOTEMP IZMER AKAD NAUK SSSR INST MET 25-8 1966 CA 67 34166
45121	**ELEVATED TEMPERATURE THERMAL CONDUCTIVITY MEASUREMENTS OF FIBROUS INSULATIONS.** ROLINSKI, E. J. PURCELL, G. V. PROC. CONF. ON THERMAL CONDUCTIVITY, 5TH 2, V-B-1-V-B-40, 1965.
45144	**SPECIFIC HEAT OF SOME ROCKS AT HIGH TEMPERATURES.** LEONIDOV V YA GEOKHIMIYA 4 470-2 1967 CA 67 23948 (FOR ENGLISH TRANSLATION SEE T70862)
45145	**EXPANSION BEHAVIOR OF CERAMIC BODIES AFTER DIFFERENT HIGH PREFIRINGS.** SCHMID O GLASTECH BER 39 3 169-73 1966 JA 50 257
45153	**THERMAL PROPERTIES OF TWO-PHASE DISPERSE SYSTEMS SATURATED WITH DIFFERENT LIQUIDS.** VERZHINSKAYA A B VAINBERG V SH INZH-FIZ ZH 12 1 38-42 1967 CA 67 57809 (FOR ENGLISH TRANSLATION SEE T42955)
45181	**NOTE ON MOISTURE EXPANSION OF CERAMIC WHITEWARE IN STORAGE AND IN SERVICE.** GELLER R F CREAMER A S J AM CERAM SOC 24 3 77-9 1941
45198	**PHYSICOCHEMICAL PROPERTIES OF GLASSES OF THE SYSTEM TITANIUM DIOXIDE - SILICON DIOXIDE - PHOSPHORUS PENTOXIDE.** SYRITSKAYA, Z. M. FEIKNER, S. YA. LATV. PSR ZINAT. AKAD. VESTIS, KIM. SER. (5), 515-24, 1966.
45215	**CAUSES OF CHIPPING OF A STIFF-MUD FACING BRICK.** WEST R R FLEISCHER D H HECHT N HOSKYNS W R MUCCIGROSSO A SCHELKER D H J AM CERAMIC SOC 43 12 648-54 1960
45217	**BOND STRENGTH AND ELASTIC PROPERTIES OF CERAMIC ADHESIVES.** THORNTON H R J AMERICAN CERAMIC SOC 45 5 201-9 1962
45219	**THERMAL CONDUCTIVITY OF FIRED GRADE A LAVA.** REDIN, R. D. MAC GREGOR, G. A. PROC. CONF. ON THERMAL CONDUCTIVITY, 5TH 2, V-H-1-V-H-9, 1965.
45257	**THERMAL CONDUCTIVITY OF FINELY DIVIDED AND GRANULAR MATERIALS AND METHODS OF MEASUREMENT IN STABLE CONDITIONS.** DUMEZ P REV GEN THERMIQUE 5 54 561-74 1966 5 55 673-86 1966 CA 67 26554
45258	**CRISTOBALITE FORMATION AS A FACTOR IN FREEZING OF PYROMETRIC CONES.** SMITH P L J AM CERAM SOC 22 6 189-92 1939
45261	**THERMAL EXPANSION OF ZIRCONIA MATERIALS.** TOMSU, F. SILIKATY 11 (2), 129-36, 1967.
45262	**DETERMINATION OF THERMAL EXPANSION OF GLASS BY A MODIFIED PADMOS METHOD.** KOTSMID F DIETRICH W SKLAR KERAM 14 11 303-8 1964 JA 50 251
45263	**STUDY OF CHANGES IN THE SPECTRAL CHARACTERISTICS OF ROSAL TYPE SUNGLASSES IN THE TRANSFORMATION INTERVAL.** FANDERLIK I SKLAR KERAM 14 12 335-9 1964 JA 50 253
45264	**EXPANSION AND DECOMPOSTION PROPERTIES OF SIO2 MIXTURES IN THE PREPARATION OF HEAVY STEEL CASTINGS.** JELINEK P SAIP J SLEVARENSTVI 13 294-7 1965 CA 67 35174
45267	**INVESTIGATIONS ON CHANGES IN THE PROPERTIES OF THE PERMANENT TAR-DOLOMITE LINING OF A BASIC CONVERTER WITH INCREASING TIME IN SERVICE.** LEHMANN H LEERS K-J HARTMANN T STAHL EISEN 86 8 1071-7 1966 JA 50 257
45268	**EFFECT OF MODIFYING ADDITIONS ON GLASS CRYSTALLIZATION.** PAVLUSHKIN N M KHODAKOVSKAYA R YA TIMOFEEVA L K STEKLO KERAM 24 3 11-16 1967 CA 67 35894
45269	**BEADING ENAMELS FOR STEEL.** AZAROV K P GRECHANOVA S B LUKYANOVA A M STEKLO I KERAMIKA 24 6 21-3 1967 CA 67 56908 (FOR ENGLISH TRANSLATION SEE T46273)
45272	**EFFECT OF ADDITIVES ON PROPERTIES OF LEAD TITANATE.** TIEN T-Y CARLSON W G J AM CERAMIC SOC 45 12 567-71 1962
45285	**GROWTH AND CHARACTERIZATION OF SPINEL SINGLE CRYSTALS FOR SUBSTRATE USE IN INTEGRATED ELECTRONICS.** WANG C C J APPL PHYS 40 9 3433-44 1969
45297	**POWER BURST FACILITY FUEL ROD PROTOTYPE OUT-OF-PILE THERMAL TESTS.** BURDG, C. E. ZUROMSKY, G. KOZIOL, J. J. NUCLEAR DIV. COMBUSTION ENGR. INC., WINDSOR, CONN. 1-65, 1965. (CEND-235)
45346	**EFFECT OF COMPOSITION ON PHYSICAL AND CHEMICAL PROPERTIES OF DOLOMITIC OPACIFIED GLAZE ON EARTHENWARE. II. EFFECT OF SILICA CHANGES ON THE PROPERTIES OF OPACIFIED DOLOMITE GLAZES.** AZIMOV I YUNUSOV KH UZB KHIM ZH 11 1 7-12 1967 CA 67 24998
45348	**INHOMOGENEOUS, ABSORBING LAYERS FOR SUN GLASSES.** ANDERS, H. VAK. TECH. 15 (5), 123-6, 1966.

TPRC Number	Bibliographic Citation
45352	**DILATOMETRIC AND X-RAY DATA FOR LEAD COMPOUNDS, II.** ARGYLE J F HUMMEL F A J AM CERAMIC SOC 46 1 10-14 1963
45353	**THERMAL EXPANSION MEASUREMENTS ON THORIA AND HAFNIA TO 4500 F.** OHNYSTY B ROSE F K J AM CERAMIC SOC 47 8 398-400 1964
45359	**STRESSES IN PORCELAIN GLAZES.** JOHNSON A L J AM CERAM SOC 22 11 363-6 1939
45380	**THERMAL EXPANSION AND PYROELECTRICITY IN LEAD TITANATE ZIRCONATE AND BARIUM TITANATE.** COOK W R JR BERLINCOURT D A SCHOLZ F J J APPLIED PHYSICS 34 5 1392-8 1963
45388	**THERMAL IMPULSE METHOD FOR THE DETERMINATION OF THERMAL DIFFUSIVITY.** KELEMEN F ACTA PHYS ACAD SCI HUNG 23 1 111-15 1967 CA 67 76817
45394	**MEASUREMENT OF THERMAL CONDUCTIVITY OF SOLIDS, ESPECIALLY CERAMIC NUCLEAR FUELS AT HIGH TEMPERATURES.** SPIES, J. BER. DEUT. KERAM. GES. 44 (6), 271-8, 1967.
45409	**COFFINITE FROM CARBONATE VEINS.** SOLOVIOVA F I DOPOV AKAD NAUK UKR RSR 29 B 5 423-8 1967 CA 67 83815
45416	**INFLUENCE OF FLINT PARTICLE SIZE ON PERMANENT MOISTURE AND THERMAL EXPANSION IN POROUS EARTHENWARE BODIES.** PALMER W E J AM CERAM SOC 25 14 413-16 1942
45417	**PHONON SCATTERING BY FE /+2/ IMPURITIES AT LOW TEMPERATURES.** SLACK, G. A. PROC CONF. ON THERMAL CONDUCTIVITY, 4TH 4 1-9 1964
45418	**COMPARISON OF THE THERMAL CONDUCTIVITIES OF MOIST CAPILLARY-POROUS MATERIALS AT TEMPERATURES ABOVE AND BELOW 0 DEGREES.** PAK N V INZH-FIZ ZH 12 1 68-71 1967 CA 67 68255 (FOR ENGLISH TRANSLATION SEE T43052)
45435	**PRESENT STATUS OF PYROCERAM CODE 9606 AS A THERMAL CONDUCTIVITY REFERENCE STANDARD.** FLYNN, D. R. ROBINSON, H. E. MARTZ, I. L. PROC. CONF. ON THERMAL CONDUCTIVITY, 4TH 4 1-28 1964
45440	**EFFECTS OF OXIDE ADDITIONS ON THE DENSITY AND SURFACE TENSION OF BARIUM SILICATE MELTS.** IKEDA K SUGINOHARA Y YANAGASE T NIPPON KINZOHU GAKKAISHI 31 4 547-52 1967 CA 67 76483
45463	**ON THE PROPERTY WHICH SILVER IODIDE POSSESSED OF CONTRACTING WITH HEAT AND DILATING WITH COLD.** FIZEAU M H COMPT RENDUS 64 314-23 1867
45477	**IN-PILE EFFECTIVE THERMAL CONDUCTIVITY OF OXIDE FUEL ELEMENTS TO HIGH FISSION DEPLETION.** DANIEL R C COHEN I TRANS AM NUCL SOC 6 2 332-3 1963
45481	**ULTRAVIOLET WINDOWS AND FILTERS FOR THE SPECTRAL REGION BETWEEN 1000 A. AND 3000 A.** BOLDT, G. MAX-PLANCK-INST. PHYS. ASTROPHYS., MUNICH, GER. 1-14, 1965. (ESRO-SM-6, N66-21733)
45483	**MEASUREMENT OF THE DIFFUSIVITY AND THE THERMAL CONDUCTIVITY OF POLYMERS. FUNCTIONAL VARIATION OF THEIR STRUCTURAL STATE AND THE TEMPERATURE.** BERLOT, R. FRANC MINISTERE DE L AIR, PARIS 1-110, 1966. (NT-154, N67-26205)
45486	**NEW PALLADIUM MINERALS FROM NORILSK, WESTERN SIBERIA.** CABRI L J TRAILL R J CAN MINERAL 8 5 541-50 1966 CA 67 83772
45488	**QUANTITATIVE OPTICAL AND SPECTROGRAPHIC STUDY OF STIBNITE FROM CEMERNICA NEAR FOJNICA.** JURKOVIC I GEOL GLAS /SARAJEVO/ 6 13-22 1962 CA 67 23869
45495	**THE TRANSITION BEHAVIOR OF TWO QUARTZITES.** KRUEGER R SALGE H KERAM Z 19 3 163-6 1967 CA 67 35949
45502	**MEASURING THE HEAT CONDUCTIVITY OF NATURAL STONES BY THE THERMAL COMPARATOR METHOD.** LORENTZEN G INTERNATIONAL INST. OF REFRIGERATION 257-68 1966
45513	**REACTOR MATERIALS PROPERTIES.** PAPROCKI S J DICKERSON R F NUCLEONICS 18 11 154-61 1960
45522	**COFFICIENT OF THERMAL EXPANSION OF CARBONACEOUS MATERIALS.** VIRGIL'EV, YU. S. DEEV, A. N. TYRIN, V. A. LUKINA, E. YU. BORISOV, B. N. KONSTR. MATER. OSN. GRAFITA (2), 99-109, 1966.
45535	**MODELS OF THE LUNAR SURFACE INCLUDING TEMPERATURE-DEPENDENT THERMAL PROPERTIES.** LINSKY, L. HARVARD COLLEGE OBSERVATORY, CAMBRIDGE, MASS. 1-59, 1966. (NASA-CR-71554, N66-21679)
45538	**SPECTRAL STUDIES OF MATERIALS POSSIBLY PRESENT ON THE MARTIAN SURFACE.** DRAPER, AL. L. ADAMCIK, J. A. TEXAS TECHNOL. COLLEGE, LUBBOCK, TEXAS 1-11, 1966. (AD-630806)
45549	**CHEMICAL AND PHYSICAL BEHAVIOR OF FERRUGINOUS GLASSES.** PETZOLD A HUEBSCHER M WISS Z HOCHSCH ARCHITEKTUR BAUWESEN WEIMAR 13 1 121-5 1966 CA 65 11934
45557	**MOTTLED OR COLORED SILICA BRICK.** PHELPS S M LIMES R W J AM CERAM SOC 26 11 378-87 1943
45561	**HIGH-TEMPERATURE REFRACTORY MATERIALS BASED ON ZIRCONIA STABILIZED BY RARE-EARTH OXIDES.** KIND N E KOSHUR L T OGNEUPORY /USSR/ 31 10 55-8 1966 (FOR ENGLISH TRANSLATION SEE T51545)
45585	**THE EFFECT OF FLUXING ADDITIONS ON THE STRUCTURE AND PROPERTIES OF SEMIPORCELAIN.** KOPEIKIN A A GLASS AND CERAMICS 15 10 531-6 1958 (ENGLISH TRANSLATION OF STEKLO I KERAM., 15 (10), 18-22, 1958; FOR ORIGINAL SEE T41706)
45597	**DILATOMETRIC PROPERTIES OF KAOLINITIC CLAYS AS A MEASURE OF THEIR MINERALOGICAL COMPOSITION.** BASINSKA-PAMPUCHOWA, S. POL. AKAD. NAUK, ODDZIAL KRAKOWIE, PR. KOM. NAUK TECH., CERAM. (7), 33-102, 1966.
45606	**THERMAL EXPANSION OF CERAMICS.** KIRBY, R. K. MECHANICAL AND THERMAL PROPERTIES OF CERAMICS, PROC. OF A SYMP. 41-61, 1969. (NBS-SP-303, N69-28429)
45610	**THERMAL CONDUCTIVITY AND ENTHALPY OF BEO AND THE BEO-UO2 SYSTEM.** HARRINGTON L C ROWE G H PRATT AND WHITNEY PROC. CONF. ON THERMAL CONDUCTIVITY, 4TH II-E-1—E-30 1964
45618	**AN INVESTIGATION OF THE EFFECTS OF ZINC OXIDE ON THE STRUCTURE OF GLASSES IN THE SYSTEM SODIUM OXIDE ZINC OXIDE SILICON DIOXIDE.** HURT, J. C. RUTGERS STATE UNIV., NEW BRUNSWICK, N. J., PH. D. THESIS 1-224, 1967. (UNIVERSITY MICROFILMS NO. 67-14720)
45621	**THERMAL EXPANSION OF TYPICAL AMERICAN ROCKS.** GRIFFITH J H IOWA ENGINEERING EXPERIMENTAL STATION BULLETIN 35 19 1-36 1936 (RESEARCH BULLETIN NO. 128)

TPRC Number	Bibliographic Citation
45627	**THERMAL CONDUCTIVITY OF CERAMICS.** FLYNN, D. R. PROC. OF A SYMP. ON MECHANICAL AND THERMAL PROPERTIES OF CERAMICS 63–123, 1969. (NBS-SP-303, N69-28430)
45628	**PROPERTIES AND STRUCTURE OF LANTHANUM OXIDE-CADMIUM OXIDE-BORON OXIDE GLASSES.** DENEKA, C. W. RUTGERS UNIVERSITY, PH.D. THESIS 240PP., 1970. (UNIVERSITY MICROFILMS NO. 70-16927, N71-31277)
45639	**THERMAL CONDUCTIVITY OF HIGH TEMPERATURE HETEROGENEOUS INSULATIONS.** WECHSLER, A. E. KRITZ, M. A. GLASER, P. E. PROC. CONF. ON THERMAL CONDUCTIVITY, 4TH 4 1–22 1964
45643	**SPECTRAL TRANSMISSION OF SOME GLASSES WITH HOLMIUM.** PAOLETTI, G. VETRO SILICATI 10 (60), 15–17, 1966.
45644	**EFFECT OF HEAT TREATMENT ON THERMAL EXPANSION PROPERTIES OF LITHIA CERAMICS WITH BETA-EUCRYPTITE COMPOSITION.** IKEDA K MINAGAWA K YOGYO KYOKAI SHI 74 848 109–13 1966 CA 67 93636
45649	**PORES AND THERMAL CONDUCTIVITY.** COOPER, C. F. J. BRIT. CERAM. SOC. 3 (1), 115–23, 1966.
45653	**THERMAL CONDUCTIVITY OF POROUS MEDIA. PACKINGS OF PARTICLES.** MESSMER, J. H. PROC. CONF. ON THERMAL CONDUCTIVITY, 4TH 4 1–12 1964
45668	**GLASS LASER RESEARCH.** MACAVOY T C CHARTERS M L MAURER R D DUMBAUGH W H BORELLI N F CORNING GLASS WORKS, N. Y. N. Y. 1–128, 1963. (AD-429010)
45692	**ON THE EXPANSION OF SOLIDS BY HEAT.** FIZEAU M H COMPT RENDUS 66 21 1005–14 1868
45698	**MEASUREMENT OF THE INFRARED SPECTRAL ABSORPTANCE OF OPTICAL MATERIALS.** STIERWALT D L BERNSTEIN J B KIRK D D APPL OPT 2 1169–73 1963
45699	**PROPERTIES OF MATERIALS AT LOW TEMPERATURES.** CORRUCCINI R J CHEM ENG PROGRESS 53 8 397–402 1957
45713	**HEAT CAPACITIES OF CLAY AND CLAY-WATER MIXTURES.** OSTER J D LOW P F SOIL SCIENCE SOC AM PROC 28 5 605–9 1964
45736	**THE EFFECT OF RADIATION FORCE ON SATELLITES OF CONVEX SHAPE.** HOLL, H. B. GEORGE C MARSHALL SPACE FLIGHT CENTER, HUNTSVILLE, ALABAMA 1–53, 1961. (NASA-TN-D-604, AD-256025)
45739	**EARTH RADIANCE AND REFLECTIVITIES RELATIVE TO THE RESPONSE OF A SILICON DETECTOR.** KRONKE R H RAYMOND F W APPL OPT 6 12 2110–14 1967
45743	**THERMAL EXPANSION OF LOW AND HIGH ALBITE.** STEWART D B VON LIMBACH D AM MINERALOGIST 52 3/4 389–413 1967 JA 51-6 190
45753	**PRESSURE EFFECTS ON POSTULATED LUNAR MATERIALS.** WECHSLER, A. E. GLASER, P. E. PROC. CONF. ON THERMAL CONDUCTIVITY, 4TH 4 1–31 1964
45763	**PROPERTIES AND STRUCTURE OF SODIUM FLUORIDE-BORIC OXIDE GLASSES.** POCH W GLASTECH BER 40 7 261–7 1967 CA 67 111107
45778	**AN INVESTIGATION OF THE THEORETICAL AND PRACTICAL ASPECTS OF THE THERMAL EXPANSION OF CERAMIC MATERIALS.** MERZ, K. M. KIRCHNER, H. P. SMYTH, H. T. CORNELL AERONAUTICAL LAB., INC, BUFFALO, N. Y. 1–72, 1960. (AD-250663, CAL-PI-1273-M-8)
45785	**EFFECTS OF SOME OXIDE ADDITIONS ON THE THERMAL LENGTH CHANGES OF ZIRCONIA.** GELLER R F YAVORSKY P J J RESEARCH NATL BUREAU STANDARDS 35 1 87–110 1945 (NBS RESEARCH PAPER NO. 1662)
45797	**GLASS, TRANSPARENT IN THE INFRARED SPECTRAL REGION.** GAPRINDASHVILI KH I NAKAIDZE D M GOMELAURI V L SOOBSHCH AKAD NAUK GRUZ SSR 47 1 67–70 1967 CA 67 111118
45798	**USE OF BENTONITE AND QUARTZ PORPHYRY IN PRODUCTION OF FACING TILES WITH A SINGLE FIRING.** KUKOLEV G V ORUDZHEVA N T STEKLO I KERAMIKA 24 7 26–30 1967 (FOR ENGLISH TRANSLATION SEE T48177)
45799	**ELECTRICAL PROPERTIES OF IRON-CONTAINING GLASS MANUFACTURED ON THE BASIS OF SILICATE ROCK.** KUTATELADZE K S VERULASHVILI R D STEKLO KERAM 24 8 21–3 1967 CA 67 111114
45807	**THERMAL-EXPANSION STRESSES IN REINFORCED PLASTICS.** TURNER P S J RESEARCH NATL BUREAU STANDARDS 37 4 239–50 1946 (NBS RESEARCH PAPER NO. 1745)
45831	**THERMOELECTRIC PARAMETER MEASUREMENT TECHNIQUE.** AGAYEV YA MIKHAYLOV A R GELIOTEKHNIKA 6 41–6 1966 (FOR ENGLISH TRANSLATION SEE T45832)
45832	**THERMOELECTRIC PARAMETER MEASUREMENT TECHNIQUE.** AGAYEV YA MIKHAYLOV A R SPECIAL LIBRARY ASSOCIATION 29–37, 1967. (ENGLISH TRANSLATION OF GELIOTEKHNIKA, (6), 41–6, 1966; FOR ORIGINAL SEE T45831) (JPRS-41144, TT-67-31785, N67-35470)
45842	**THE THERMAL EXPANSION OF SYNTHETIC BETA-SPODUMENE AS AFFECTED BY ADDITIONS OR IRON, POTASSIUM, AND SODIUM.** MARTIN, L. L. UNIV. OF ILLINOIS, URBANA, M. S. THESIS 1–28 1964
45843	**THE RELATIONSHIP BETWEEN THERMAL CONDUCTIVITY AND POROSITY FOR MAURY SILT LOAM SOIL.** SKAGGS, R. W. UNIV. OF KENTUCKY, LEXINGTON, M. S. THESIS 1–63 1966
45866	**THERMODYNAMIC STUDY OF SOLID PHASE REACTIONS IN A BAO-AL2O3 SYSTEM.** MIKHEEV V N SIROTKIN G D CHERNYSHOVA O V KURSINA V A IZV VYSSH UCHEB ZAVED KHIM KHIM TEKHNOL 10 6 617–19 1967 CA 67 120535
45885	**QUARTZ GLASS CERAMICS.** SOLOMIN, N. V. YAKUBIK, V. V. SEMIRECHENKO, N. A. KRASNOPEROVA, T. E. GLASS CERAM., (USSR) 24 (9), 477–9, 1967. (ENGLISH TRANSLATION OF STEKLO KERAM., 24 (9), 6–8, 1967., FOR ORIGINAL SEE TPRC NO. 45886)
45886	**CERAMICS FROM QUARTZ GLASS.** SOLOMIN, N. V. YAKUBIK, V. V. SEMIRECHENKO, N. A. KRASNOPEROVA, T. E. STEKLO KERAM. 24 (9), 6–8, 1967. (FOR ENGLISH TRANSLATION SEE TPRC NO. 45885)
45889	**SOME CHARACTERISTICS OF ILLINOIS POTTERY CLAY.** HARMAN C G J AM CERAMIC SOC 23 1 26–9 1940
45890	**TWO VARIETIES OF PYRITE FROM THE BASAL REEF OF THE FREE STATE GEDULD MINE, ORANGE FREE STATE GOLD FIELD.** SAAGER R MIHALIK P UNIV WITWATERSRAND JOHANNESBURG ECON GEOL RES UNIT INFORM CIRC 37 1–11 1967 CA 67 119020

TPRC Number	Bibliographic Citation
45899	**DIURNAL AND SEASONAL VARIATIONS OF ALBEDO ON SOME NATURAL SURFACES IN THE ESTONIAN SSR.** TOOMING KH AKAD NAUK ESTONSKOY SSR INST FIZ I ASTRONOMII ISSLED FIZ ATMOSFERY 2 115-63 1960 (FOR ENGLISH TRANSLATION SEE T45900)
45900	**DIURNAL AND SEASONAL VARIATION OF ALBEDO ON SOME NATURAL SURFACES IN THE ESTONIAN SSR.** TOOMING KH AEROSPACE TECHNOLOGY DIVISION 70-109, 1965. (ENGLISH TRANSLATION OF AKAD. NAUK ESTONSKOY SSR, INST. FIZ. ASTRONOMII, ISSLED. FIZ. ATMOSFERY, (2), 115-63, 1960; FOR ORIGINAL SEE T45899) (AD-466477, ATD-T-65-30, N65-29533)
45929	**RELATION BETWEEN SURFACE ROUGHNESS AND SPECULAR REFLECTANCE AT NORMAL INCIDENCE.** BENNETT H E PORTEUS J O J OPT SOC AM 51 2 123-9 1961
45947	**ON MECHANICAL EFFECT OF IONIZING RADIATION IN POTASSIUM CHLORIDE CRYSTALS.** DAMM J Z SUSZYNSKA M J PHYSIQUE SUPPL 28 8/9 C4.168-74 1967
46008	**PROBLEMS OF INFRARED ATMOSPHERIC SPECTROSCOPY RELATED TO THE SATELLITE DETERMINATION OF TEMPERATURE OF UNDERLYING SURFACE.** KONDRATYEV K YA FIZ ATM I OKEANA /MOSCOW/ 5 6 616-30 1969 (FOR ENGLISH TRANSLATION SEE T34391)
46010	**THERMAL EXPANSION IN AIR OF CERAMIC OXIDES TO 2200 C.** HIELSEN, T. LEIPOLD, M. JET PROPULSION LABORATORY, PASADENA, CALIF. 1-25, 1962. (JPL-TR-32-297, N63-11308)
46017	**INFLUENCE OF THORIUM OXIDE, LANTHANUM OXIDE, CADMIUM OXIDE AND ZIRCONIUM OXIDE ON THE PHYSICOCHEMICAL PROPERTIES OF OPTICAL GLASS.** SUSSER, V. FANDERLIK, I. EUROPEAN TRANSLATIONS CENTRE, DOELENSTRAAT, DELFT NETHERLANDS 1-8 PP., 1965. (FRENCH TRANSLATION OF SKLAR A KERAM. (CZECHOSLOVAKIA) 1 (10), 265-68, 1963; FOR ORIGINAL SEE T34151) CNRS-65-H/X-248, TT-65-29484
46030	**STRUCTURE OF GLASSES AND MELTS IN THE NA2O.GEO2.B2O3 SYSTEM.** RIEBLING E F BLASZYK P E SMITH D W J AM CERAM SOC 50 12 641-7 1967
46031	**ELECTRICAL, X-RAY, AND THERMAL EXPANSION STUDIES IN THE SYSTEM KNBO3.AGNBO3.** WEIRAUCH D F TENNERY V J J AM CERAM SOC 50 12 671-3 1967 CA 68 53905
46035	**EFFECTS OF IRRIGATION ON SOIL TEMPERATURE AND THERMAL DIFFUSIVITY.** SHEIKH, G. M. WASHINGTON STATE UNIV., PULLMAN, M. S. THESIS 1-92 1966
46040	**THERMOPHYSICS RESEARCH AT MARSHALL SPACE FLIGHT CENTER.** HELLER, G. B. NATIONAL AERONAUTICS AND SAPCE ADMIN., HUNTSVILLE, ALABAMA 1-30, 1967. (NASA-TM-X-53620, N67-30556)
46042	**LUNAR AND METEOROID PHYSICS RESEARCH AT MARSHALL SPACE FLIGHT CENTER.** HELLER, G. B. DOZIER, J. B., JR. NATIONAL AERONAUTICS AND SPACE ADMIN., HUNTSVILLE, ALABAMA 1-16, 1967. (NASA-TM-X-53620, N67-30611)
46044	**THERMAL DIFFUSIVITY OF /PM-147/203.** GIBBY, R. L. BATTELLE-NORTHWEST PACIFIC NORTHWEST LABS., RICHLAND, WASHINGTON 19-20, 1968. (BNWL-680, N68-36950)
46073	**INVESTIGATION OF THE ADSORPTION OF TRIMETHYLCARBINOL ON MONTMORILLONITE BY THE METHOD OF INFRARED SPECTROSCOPY.** TARASEVICH YU I RADUL N M OVCHARENKO F D DOKL AKAD NAUK SSSR 173 3 615-17 1967 (FOR ENGLISH TRANSLATION SEE T46074)
46074	**INVESTIGATION OF THE ADSORPTION OF TRIMETHYLCARBINOL ON MONTMORILLONITE BY THE METHOD OF INFRARED SPECTROSCOPY.** TARASEVICH YU I RADUL N M OVCHARENKO F D DOKL PHYS CHEM 173 3 230-2 1967 (ENGLISH TRANSLATION OF DOKL. AKAD. NAUK SSSR, 173 (3), 615-7, 1967; FOR ORIGINAL SEE T46073)
46091	**ON CALCULATING THE THERMOMETRIC CONDUCTIVITY COEFFICIENT AND HEAT FLUX IN SOIL ON THE BASIS OF AVERAGE TEMPERATURES.** TSEITIN G KH GLAVNAIA GEOFIZICHESKAIA OBSERVATORIIA IM A I VOEIKOVA TRUDY 60 122 67-79 1956 (FOR ENGLISH TRANSLATION SEE T46092)
46092	**ON CALCULATING THE THERMOMETRIC CONDUCTIVITY COEFFICIENT AND HEAT FLUX IN SOIL ON THE BASIS OF AVERAGE TEMPERATURES.** TSEITIN, G. KG. WEATHER BUREAU, WASHINGTON, D. C. 1-18, 1960. (ENGLISH TRANSLATION OF GLAVNAIA GEOFIZ. OBSERVATORIIA IM. A. I. VOEIKOVA TRUDY, 60 (122), 67-79, 1956; FOR ORIGINAL SEE T46091) (TT-60-15364, AD-660972)
46096	**CORROSION OF CERAMIC INSULATING MATERIALS IN HIGH-TEMPERATURE COOLANTS.** BUDNIKOV P P KHARITONOV F YA ANISIMOVA S V ZH PRIKLAD KHIM 40 3 549-55 1967 (FOR ENGLISH TRANSLATION SEE T46097)
46097	**CORROSION OF CERAMIC INSULATING MATERIALS IN HIGH-TEMPERATURE COOLANTS.** BUDNIKOV P P KHARITONOV F YA ANISIMOVA S V J APPL CHEM USSR 40 3 537-42 1967 (ENGLISH TRANSLATION OF ZH. PRIKL. KHIM., 40 (3), 549-55, 1967; FOR ORIGINAL SEE T46096)
46105	**UTAH JET. A VITRINITE WITH ABERRANT PROPERTIES.** TRAVERSE A KOLVOORD R W SCIENCE 159 302-5 1968
46143	**IMPROVED CERAMICS.** KOENIG, J. H. NEW JERSEY CERAMIC RES. STATION, RUTGERS UNIV. 1-62, 1955. (AD-74091)
46144	**DEVELOPMENT OF CERAMIC DIELECTRICS.** KOENIG, J. H. JEW JERSEY CERAMIC RES. STATION, RUTGERS UNIV. 1-87, 1959. (AD-234321)
46148	**INVESTIGATION OF ULTRA-HIGH TEMPERATURE EMBEDDING COMPOUND ACID PROOF CEMENT NO. 31.** COLLETTI, W. MATERIAL LABS., BROOKLYN, N. Y. 1-30, 1961. (AD-259673L)
46158	**POLYFUNCTIONAL INSULATING BODY ON A KERAMSIT BASE.** TOTH K EPITOANYAG 19 9 349-52 1967 CA 68 5915
46159	**CALORIMETRIC INVESTIGATIONS ON TEXTILE FIBERS. IV. RESULTS.** GOETZE W WINKLER F FASERFORSCH TEXTILTECH 18 8 385-9 1967 CA 68 3725
46164	**A CALCINED ALUMINUM SILICATE FOR REFRACTORY PRODUCTS.** BAILEY R T INTERCERAM /LUEBECK GERMANY/ 16 2 113-15 1967 CA 68 5865
46172	**APPARATUS FOR STUDYING THE EFFECT OF LOAD ON THE THERMAL CONDUCTIVITY OF GRANULAR SYSTEMS IN VACUO.** ZARICHNYAK YU P MURATOVA B L PLATUNOV E S IZV VYSSH UCHEB ZAVED PRIBOROSTR 10 2 105-9 1967 CA 68 7046
46214	**LITHIUM IN GLASSES AND ENAMELS.** RZHEVUSKAYA T L KHODSKII L G MOLOCHKO A P VOP PRIMEN REDK METAL SILIKAT PROM DOKL RESPUB SOVESHCH MINSK 1963 6-10 1964 CA 66 107773
46217	**STORAGE AND HANDLING OF CRYOGENIC FLUIDS.** NORED D L HENNINGS G SINCLAIR D H SMITH G T SMOLAK G R STOFAN A J LEWIS RESEARCH CENTER 125-53, 1966. (NASA-SP-5053)

TPRC Number	Bibliographic Citation
46218	**EFFECT OF INDIUM ON GLASS PROPERTIES.** BOBKOVA N M NEMKOVICH I K VOP PRIMEN REDK METAL SILIKAT PROM DOKL RESPUB SOVESHCH MINSK 1963 37-9 1964 CA 66 107782
46223	**MEASUREMENT OF THE THERMAL CONDUCTIVITY OF SOLIDS ESPECIALLY OF CERAMIC NUCLEAR FUELS AT HIGH TEMPERATURES.** SPIES, J. KERNFORSCHUNGSZENTRUM, KARLSRUHE, W. GERMANY 1-45, 1966. (KFK-415)
46250	**THERMAL CONDUCTIVITY AND DIELECTRIC CONSTANT OF SILICATE MATERIALS.** WECHSLER, A. E. SIMON, I. LITTLE/ARTHUR D/INC., CAMBRIDGE, MASS. 1-133, 1966. (NASA-CR-61495, N68-15849)
46251	**INFRARED RADIATION METHOD OF DETERMINING THERMAL DIFFUSIVITY, HEAT CAPACITY AND THERMAL CONDUCTIVITY OF SOLID MATERIALS.** OGAWA, K. NATIONAL AEROSPACE LABORATORY /TOKYO/ 1-11, 1967. (NAL-TR-128) (FOR ENGLISH TRANSLATION SEE T46252)
46252	**INFRARED RADIATION METHOD OF DETERMINING THERMAL DIFFUSIVITY, HEAT CAPACITY AND THERMAL CONDUCTIVITY OF SOLID MATERIALS.** OGAWA K AZTEC SCHOOL OF LANGUAGES INC RESEARCH AZTEC SCHOOL OF LANGUAGES INC., RES. TRANSLATION DIV., ACTON, MASS. 1-21, 1968. (ENGLISH TRANSLATION OF NAL-TR-128, 1-11, 1967; FOR ORIGINAL SEE T46251) (NASA-TT-F-11453, N68-17901)
46259	**THE THERMAL CONDUCTIVITY OF MULTIPHASE CERAMIC NUCLEAR FUELS.** BIANCHERIA A TRANS AM NUCL SOC 10 1 100-1 1967
46267	**EFFECT OF PARTICLE SIZE DISTRIBUTION ON PROPERTIES AND PHASE COMPOSITION OF SPODUMENE CERAMICS.** BULAVIN I A ZINKO E I MEDVEDOVSKAYA E I FOMINA N P STEKLO I KERAMIKA 24 5 21-3 1967 (FOR ENGLISH TRANSLATION SEE T46268)
46268	**EFFECT OF PARTICLE SIZE DISTRIBUTION ON PROPERTIES AND PHASE COMPOSITION OF SPODUMENE CERAMICS.** BULAVIN I A ZINKO E I MEDVEDOVSKAYA E I FOMINA N P GLASS AND CERAMICS 24 5 253-5 1967 (ENGLISH TRANSLATION OF STEKOL KERAMIKA, 24 (5), 21-3, 1967; FOR ORIGINAL SEE T46267)
46271	**PRODUCING GLASS FIBER BY THE CENTRIFUGAL-BLOW METHOD AND MAKING HEAT AND SOUND INSULATING PRODUCTS FROM IT.** NIGIN E R OLEINIKOVA A N STEKLO I KERAMIKA 24 6 18-21 1967 (FOR ENGLISH TRANSLATION SEE T46272)
46272	**PRODUCING GLASS FIBER BY THE CENTRIFUGAL-BLOW METHOD AND MAKING HEAT AND SOUND INSULATING PRODUCTS FROM IT.** NIGIN E R OLEINIKOVA A N GLASS AND CERAMICS 24 6 312-15 1967 (ENGLISH TRANSLATION OF STEKLO KERAMIKA, 24 (6), 18-21, 1967; FOR ORIGINAL SEE T46271)
46273	**BEADING ENAMELS FOR STEEL.** AZAROV K P GRECHANOVA S B LUKYANOVA A M GLASS AND CERAMICS 24 6 316-18 1967 (ENGLISH TRANSLATION OF STEKLO KERAMIKA, 24 (6), 21-3, 1967; FOR ORIGINAL SEE T45269)
46274	**INCREASING THE MECHANICAL STRENGTH OF PORCELAIN BY SUBSTITUTING CRYSTALLIZING GLASS FOR QUARTZ.** AVGUSTINIK A I SINTSOVA I T STEKLO I KERAMIKA 24 6 26-30 1967 (FOR ENGLISH TRANSLATION SEE T46275)
46275	**INCREASING THE MECHANICAL STRENGTH OF PORCELAIN BY SUBSTITUTING CRYSTALLIZING GLASS FOR QUARTZ.** AVGUSTINIK A I SINTSOVA I T GLASS AND CERAMICS 24 6 322-6 1967 (ENGLISH TRANSLATION OF STEKLO KERAMIKA, 24 (6) 26-30, 1967; FOR ORIGINAL SEE T46274)
46276	**DIRECTIONAL EMITTANCE OF AN ELECTRIC NONCONDUCTOR AS A FUNCTION OF SURFACE ROUGHNESS AND WAVELENGTH.** TORRANCE K E SPARROW E M INTERN J HEAT MASS TRANSFER 10 12 1709-16 1967
46291	**THERMAL EXPANSION OF CERTAIN ILLINOIS LIMESTONES AND DOLOMITES.** HARVEY R D ILL STATE GEOL SURV CIRC NO. 415 1-33 1967 CA 68 4979
46294	**THERMAL CONDUCTIVITY OF UO2-SIO2 VITROCERAMICS IN THE TEMPERATURE RANGE 100 TO 400 K.** LUSCAL N REV ROUM PHYS 12 5 469-77 1967 CA 68 99383
46297	**VARIATION OF GLASS PROPERTIES AS A FUNCTION OF TEMPERATURE AND TIME.** CAHOUR J A CAHIERS PHYSIQUE 29 25-50 1947
46298	**DEVELOPMENT OF INORGANIC REFRACTORY COMPOSITES.** HALLSE, R. L. COLBURN, S. C. BECK, A. P. GENERAL DYNAMICS/POMONA, POMONA, CALIF. 359-69, 1965. (AFML-TR-64-398, AD-477414)
46324	**PHONON SCATTERING IN SODIUM CHLORIDE CONTAINING OXYGEN.** KLEIN M V PHYS REV 122 5 1393-1402 1961
46332	**HEAT TRANSFER THROUGH DISPERSED SYSTEM HAVING IRREGULAR PORE SIZES.** CHAUDHARY D R BHANDARI R C INDIAN J PURE APPL PHYS 5 9 413-16 1967
46340	**THE SYNTHESIS OF CORDIERITE.** JELACIC, C. KACIAN, M. BULL. SOC. FR. CERAM. 75, 5-23, 1967.
46347	**THE MEASUREMENT OF THERMAL CONDUCTIVITY OF INSULATING REFRACTORIES UP TO 1500 DEGREES.** FOLDI T TERENYI G EPITOANYAG 19 8 281-6 1967 CA 68 15664
46353	**SOME FERRO- AND ANTIFERROELECTRICS WITH HIGH CURIE TEMPERATURES.** SMOLENSKII G A KRAINIK N N KHUCHUA N P ISUPOV V A ZHDANOVA V V MUSHTAREEV O M CHERKASHCHENKO A YA IZV AKAD NAUK SSSR SER FIZ 31 7 1164-7 1967 CA 68 16971 (FOR ENGLISH TRANSLATION SEE T36019)
46358	**CONTRIBUTION TO THE MEASUREMENT OF THE THERMAL CONDUCTIVITY OF METALS IN THE RANGE OF 0 TO 500 C.** KUSTER W BODE K H FRITZ W WARME UND STOFFUBERTRAGUNG 1 3 129-39 1968
46367	**FEATURES OF THE THERMAL EXPANSION OF SINTERED OPTICAL GLASS FIBERS.** GOMELSKII M S POLUKHIN V N KHABAROVA E A BABKINA V A OPT-MEKH PROM 34 5 41-7 1967 CA 68 15628
46395	**ON THE EXPANSIVITY OF SEVERAL REFRACTORY MATERIALS AT HIGH TEMPERATURES.** BOGITCH B COMPT RENDUS 173 1358-62 1921
46403	**MEASUREMENTS OF SIXTY-DEGREE SPECULAR GLOSS.** HAMMOND H K III NIMEROFF I J OF RESEARCH NATIONAL BUREAU OF STANDARDS 44 585-9 1950 (NBS-REPT-2105)
46404	**CHANGE IN THE PROPERTIES OF NICKEL - COBALT MINERALS AFTER ROASTING (IN CONNECTION WITH PROBLEMS OF THEIR BENEFICIATION).** NESTEROVA, L. I. OBOGASHCH. RUD 12 (2), 42-3, 1967.
46408	**PHYSICOCHEMICAL AND DIELECTRIC PROPERTIES OF GLASSES IN THE SYSTEM BARIUM TITANATE-BORON OXIDE-BISMUTH OXIDE.** FREIDENFELDS, E. JOKSTA, E. STEINERTE, V. SVINKA, R. UCH. ZAP. RIZHSK. POLITEKH. INST. 16, 321-6, 1965.

TPRC Number	Bibliographic Citation
46410	**RESULTS OF SOIL TEMPERATURE MEASURMENTS AT JOKIOINEN OBSERVATORY.** FINNISH METEOROLOGICAL INSTITUTE, HELSINKI FINNISH METEOROLOGICAL INST., HELSINKI 1-32, 1968. (N69-15013)
46419	**MINERAGRAPHIC ZONES OF REFLECTIVITY AND HARDNESS IN THE STUDY OF ORE MINERALS.** RAO A B NAYAK V K ADUSUMILLI M S ANAIS ACAD BRASIL CIENC 37 2 245-56 1965 CA 68 23741
46425	**CHECKING OF OPTICAL DATA DERIVED FOR POLARIZATION FIGURES.** EALES H V ECON GEOL 62 5 737-8 1967 CA 68 23422
46428	**THE INFLUENCE OF BERYLLIUM OXIDE AND TECHNICAL ALUMINA ADDITIONS ON THE BASIC PROPERTIES OF ELECTRICAL PORCELAIN.** BUDNIKOV P P ZVYAGILSKII A A STEKLO I KERAMIKA 16 7 3-7 1959 (FOR ENGLISH TRANSLATION SEE T52164)
46430	**DEVELOPMENT AND PRELIMINARY STUDIES OF PINITE. A ROCK COMPOSED OF HYDROUS ALUMINUM SILICATES.** PAGE G A RAINE F F SULLIVAN V R J AM CERAM SOC 23 3 71-7 1940
46432	**SPECTRAL REFLECTANCE 0.4 TO 2.0 MICRONS OF SILICATE ROCK POWDERS.** ADAMS J B FILICE A L J GEOPHYS RES 72 22 5705-15 1967 CA 68 25284
46439	**A COMPARATIVE STUDY OF THE ASTM THERMAL CONDUCTIVITY AND THE BLAKELY-COBB APPARATUS AS INSTRUMENTS FOR ROUTINE MEASUREMENTS OF THERMAL CONDUCTIVITY OF REFRACTORIES.** MURTHY H P S GANGADHARAN T K NML /NAT MET LAB/ TECH J 8 2 15-19 1966 CA 68 24285
46455	**EASILY MELTED GLAZES BASED ON ALUMINOBOROSILICATE GLASSES OPACIFIED BY ZIRCONIUM COMPOUNDS.** DUSAUSKAS - DUZ, S. EIDUKS, J. SEDMALIS, U. PAUKSS, P. OZOLINS, J. UCH. ZAP., RIZH. POLITEKH. INST., KHIM. FAK. 16 (9), 275-82, 1965.
46471	**X-20 NOSE CAP MATERIALS SUMMARY REPORT.** SEEGER, J. W. LOWRANCE, D. T. ROGERS, D. C. LTV ASTRONAUTICS DIV., LING-TEMCO-VOUGHT INC., DALLAS, TEXAS. 1021-151, 1965. (AFML-TR-64-398, AD-477414)
46475	**THERMAL CONDUCTIVITY OF CASTABLE REFRACTORIES IN RELATION TO BULK DENSITY.** WALLACE R W CRISS G H BULL AM CERAM SOC 47 2 176-9 1968
46484	**THE EFFECT OF PRESSURE ON THE APPARENT THERMAL CONDUCTIVITY OF A SYSTEM CONTAINING A DIATOMACEOUS POWDER AND AIR.** CHRISTMAN, R. D. UNIVERSITY OF PITTSBURGH, M. S. THESIS 1-38 1968
46491	**DILATOMETER AND X-RAY DATA FOR ZINC COMPOUNDS.I** WENT P-Y BROWN J J HUMMEL F A TRANS BRIT CERAMIC SOC 63 9 501-8 1964
46500	**DETERMINATION OF THE COEFFICIENT OF THERMAL CONDUCTIVITY.** BROVKIN L A ZAVOD LAB 34 6 700-2 1968 (FOR ENGLISH TRANSLATION SEE T36576)
46503	**THE ATTACK ON REFRACTORIES IN THE ROTARY CEMENT KILN. PART I. PHYSICAL AND MINERALOGICAL CHANGES.** BRISBANE S M SEGNIT E R TRANS BRIT CERAMIC SOC 56 5 237-52 1957
46537	**SOLID SOLUTIONS IN THE SYSTEM MNO-ZNO AND THEIR POSSIBLE USE AS PIGMENTS.** WHITE W B MC ILVRIED K E TRANS BRIT CERAM SOC 64 521-30 1965
46564	**PROPERTIES OF GROG OBTAINED BY FIRING CERTAIN CLAYS IN ROTARY FURNACES.** SEGZHDA D P ARZUMANOV M A LEVITAS E G FROLOVA A I DUDAVSKII I E OGNEUPORY 1 5-10 1966 (FOR ENGLISH TRANSLATION SEE T46565)
46565	**PROPERTIES OF GROG OBTAINED BY FIRING CERTAIN CLAYS IN ROTARY FURNACES.** SEGZHDA D P ARZUMANOV M A LEVITAS E G FROLOVA A I DUDAVSKII I E REFRACTORIES 1 6-11 1966 (ENGLISH TRANSLATION OF OGNEUPORY, (1), 5-10, 1966; FOR ORIGINAL SEE T46564)
46566	**METHODS OF PREPARING FOAMED BODY FOR ULTRALIGHTWEIGHT REFRACTORIES.** RABINOVICH M S EREMIN N F OGNEUPORY 3 13-16 1966 (FOR ENGLISH TRANSLATION SEE T46567)
46567	**METHODS OF PREPARING FOAMED BODY FOR ULTRALIGHTWEIGHT REFRACTORIES.** RABINOVICH M S EREMIN N F REFRACTORIES 3 141-4 1966 (ENGLISH TRANSLATION OF OGNEUPORY, (3), 13-6, 1966; FOR ORIGINAL SEE T46566)
46568	**ACTION OF IRON MELTS ON SPINEL-PERICLASE REFRACTORIES.** KLYUCHAROV YA V SUVOROV S A OGNEUPORY 3 34-42 1966 (FOR ENGLISH TRANSLATION SEE T46569)
46569	**ACTION OF IRON MELTS ON SPINEL-PERICLASE REFRACTORIES.** KLYUCHAROV YA V SUVOROV S A REFRACTORIES 3 162-8 1966 (ENGLISH TRANSLATION OF OGNEUPORY, (3), 34-42, 1966; FOR ORIGINAL SEE T46568)
46570	**EFFECT OF GASEOUS ATMOSPHERE ON THE CHEMICAL REACTIONS AND POLYMORPHIC INVERSIONS IN THE SYSTEM ZIRCONIA-CERIA.** LEONOV A I KELER E K ANDREEVA A B OGNEUPORY 3 42-8 1966 (FOR ENGLISH TRANSLATION SEE T46571)
46571	**EFFECT OF GASEOUS ATMOSPHERE ON THE CHEMICAL REACTIONS AND POLYMORPHIC INVERSIONS IN THE SYSTEM ZIRCONIA-CERIA.** LEONOV A I KELER E K ANDREEVA A B REFRACTORIES 3 169-73 1966 (ENGLISH TRANSLATION OF OGNEUPORY, (3), 42-8, 1966; FOR ORIGINAL SEE T46570)
46572	**TECHNOLOGY FOR REFRACTORIES WITH A HIGH-ALUMINA BOND.** ZHIKHAREVICH S A GINYAR E A KOZYREVA L A MALCHENKO R S REFRACTORIES 4 198-207 1966 (ENGLISH TRANSLATION OF OGNEUPORY, (4), 8-18, 1966; FOR ORIGINAL SEE T42998)
46573	**SPEEDING UP THE FIRING OF SILICA IN GAS-CHAMBER FURNACES.** TSIBIN I P SIZOV I D OGNEUPORY 5 29-33 1966 (FOR ENGLISH TRANSLATION SEE T46574)
46574	**SPEEDING UP THE FIRING OF SILICA IN GAS-CHAMBER FURNACES.** TSIBIN I P SIZOV I D REFRACTORIES 5 283-7 1966 (ENGLISH TRANSLATION OF OGNEUPORY, (5), 29-33, 1966; FOR ORIGINAL SEE T46573)
46575	**POLYMORPHIC INVERSIONS OF CRISTOBALITE IN THE PRESENCE OF TITANIA.** SUKHAREVSKII B YA LYSAK S V OGNEUPORY 5 53-8 1966 (FOR ENGLISH TRANSLATION SEE T46576)
46576	**POLYMORPHIC INVERSIONS OF CRISTOBALITE IN THE PRESENCE OF TITANIA.** SUKHAREVSKII B YA LYSAK S V REFRACTORIES 5 307-11 1966 (ENGLISH TRANSLATION OF OGNEUPORY, (5), 53-8, 1966; FOR ORIGINAL SEE T46575)
46577	**REFRACTORIES FROM KAOLINITIC MATERIALS AND HIGH-ALUMINA BOND.** ZHIKHAREVICH S A GINYAR E A KOZYREVA L A MALCHENKO R S DAVYDOV I D RYAZANTSEV V D NOVOKHATKO P I OGNEUPORY 6 11-13 1966 (FOR ENGLISH TRANSLATION SEE T46578)

TPRC Number | Bibliographic Citation

46578 **REFRACTORIES FROM KAOLINITIC MATERIALS AND HIGH-ALUMINA BOND.**
ZHIKHAREVICH S A GINYAR E A KOZYREVA L A
MALCHENKO R S DAVYDOV I D RYAZANTSEV V D
NOVOKHATKO P I
REFRACTORIES
6 325-7 1966
(ENGLISH TRANSLATION OF OGNEUPORY, (6), 11-13, 1966; FOR ORIGINAL SEE T46577)

46579 **SERVICE OF STABILIZED MAGNESITE- DOLOMITE BRICK IN ROTARY CEMENT FURNACES.**
DOLGINA G Z MARKEVICH E P SOKHATSKAYA G A
ORLOV V YA TISHKOVA K S
OGNEUPORY
6 33-6 1966
(FOR ENGLISH TRANSLATION SEE T46580)

46580 **SERVICE OF STABILIZED MAGNESITE-DOLOMITE BRICK IN ROTARY CEMENT FURNACES.**
DOLGINA G Z MARKEVICH E P SOKHATSKAYA G A
ORLOV V YA TISHKOVA K S
REFRACTORIES
6 347-9 1966
(ENGLISH TRANSLATION OF OGNEUPORY, (6), 33-6, 1966; FOR ORIGINAL SEE T46579)

46583 **PROBABILITY OF SPONTANEOUS EMISSION CORRESPONDING TO THE ABSORPTION BAND OF HOLMIUM IN THE 2 MICRON REGION.**
ORLOVA I N GERLOVIN YA I
OPTIKA I SPEKTROSKOPIYA
23 6 988-90 1967
(FOR ENGLISH TRANSLATION SEE T46584)

46584 **PROBABILITY OF SPONTANEOUS EMISSION CORRESPONDING TO THE ABSORPTION BAND OF HOLMIUM IN THE 2 MICRON REGION.**
ORLOVA I N GERLOVIN YA I
OPT SPECTROSC
23 6 539-40 1967
(ENGLISH TRANSLATION OF OPT. SPEKTROSK., 23 (6), 988-90, 1967; FOR ORIGINAL SEE T46583)

46587 **SEMIDRY PRESSING OF SEMIACID REFRACTORIES BASED ON PRIMARY KAOLINS.**
DYGALO M I KOCHETOVA A P
OGNEUPORY
8 8-13 1966
(FOR ENGLISH TRANSLATION SEE T46588)

46588 **SEMIDRY PRESSING OF SEMIACID REFRACTORIES BASED ON PRIMARY KAOLINS.**
DYGALO M I KOCHETOVA A P
REFRACTORIES
8 433-8 1966
(ENGLISH TRANSLATION OF OGNEUPORY, (8), 8-13, 1966; FOR ORIGINAL SEE T46587)

46591 **ANISOTROPY OF THERMAL CONDUCTIVITY OF REFRACTORIES.**
STRELOV K K TSIBIN I P
OGNEUPORY
9 39-43 1966
(FOR ENGLISH TRANSLATION SEE T46592)

46592 **ANISOTROPY OF THERMAL CONDUCTIVITY OF REFRACTORIES.**
STRELOV K K TSIBIN I P
REFRACTORIES
9 527-30 1966
(ENGLISH TRANSLATION OF OGNEUPORY, (9), 39-43, 1966; FOR ORIGINAL SEE T46591)

46593 **USE OF PERICLASE-SPINEL LININGS IN ROTARY CEMENT FURNACES.**
BUDNIKOV P P SOKHATSKAYA G A SHUBIN V I
SIDOCHENKO I M IVANOV N N ZHURBA L G
ZAVGORODNII N S BUDIN I K ZAITSEV K G
MASHEVSKII T A
OGNEUPORY
10 33-40 1966
(FOR ENGLISH TRANSLATION SEE T46594)

46594 **USE OF PERICLASE-SPINEL LININGS IN ROTARY CEMENT FURNACES.**
BUDNIKOV P P SOKHATSKAYA G A SHUBIN V I
SIDOCHENKO I M IVANOV N N ZHURBA L G
ZAVGORODNII N S BUDIN I K ZAITSEV K G
MASHEVSKII T A
REFRACTORIES
10 584-9 1966
(ENGLISH TRANSLATION OF OGNEUPORY, (10), 33-40, 1966; FOR ORIGINAL SEE T46593)

46595 **SOME FEATURES OF THE BEHAVIOR OF KIROVOGRAD CLAY DURING FIRING.**
ZHIKHAREVICH S A GINYAR E A
OGNEUPORY
10 40-50 1966
(FOR ENGLISH TRANSLATION SEE T46596)

46596 **SOME FEATURES OF THE BEHAVIOR OF KIROVOGRAD CLAY DURING FIRING.**
ZHIKHAREVICH S A GINYAR E A
REFRACTORIES
10 590-8 1966
(ENGLISH TRANSLATION OF OGNEUPORY, (10), 40-50, 1966; FOR ORIGINAL SEE T46595)

46597 **REPLACING FIRECLAY INSULANT BL-0.8 BY DIATOMITE IN TUNNEL KILNS AND KILN CAR LININGS.**
TSIBIN I P
OGNEUPORY
10 26-7 1966
(FOR ENGLISH TRANSLATION SEE T46598)

46598 **REPLACING FIRECLAY INSULANT BL-0.8 BY DIATOMITE IN TUNNEL KILNS AND KILN CAR LININGS.**
TSIBIN I P
REFRACTORIES
10 577-8 1966
(ENGLISH TRANSLATION OF OGNEUPORY, (10), 26-7, 1966; FOR ORIGINAL SEE T46597)

46599 **APPARATUS FOR MEASURING THERMAL CONDUCTIVITY OF SMALL SPECIMENS.**
VISHNEVSKII I I SKRIPAK V N
REFRACTORIES
12 694-7 1966
(ENGLISH TRANSLATION OF OGNEUPORY, (12), 13-18, 1966; FOR ORIGINAL SEE T44488)

46604 **DESTRUCTIVE MECHANISM FOR PERICLASE-SPINEL AND MAGNESITE-CHROMITE REFRACTORIES IN OXYGEN CONVERTER LININGS.**
SIMONOV K V PEREPELITSYN V A GOLOV G V
BARANOV V M
OGNEUPORY
1 32-8 1967
(FOR ENGLISH TRANSLATION SEE T46605)

46605 **DESTRUCTIVE MECHANISM FOR PERICLASE-SPINEL AND MAGNESITE-CHROMITE REFRACTORIES IN OXYGEN CONVERTER LININGS.**
SIMONOV K V PEREPELITSYN V A GOLOV G V
BARANOV V M
REFRACTORIES
1 33-40 1967
(ENGLISH TRANSLATION OF OGNEUPORY, (1), 32-8, 1967; FOR ORIGINAL SEE T46604)

46608 **LADLE BRICK FROM SEMIACID CLAYS.**
KARKLIT A K ZEGZHDA V P LEBEDEVA M F
STEPIN R V
OGNEUPORY
2 14-17 1967
(FOR ENGLISH TRANSLATION SEE T46609)

46609 **LADLE BRICK FROM SEMIACID CLAYS.**
KARKLIT A K ZEGZHDA V P LEBEDEVA M F
STEPIN R V
REFRACTORIES
2 84-6 1967
(ENGLISH TRANSLATION OF OGNEUPORY, (2), 14-17, 1967; FOR ORIGINAL SEE T46608)

46615 **THERMAL EXPANSION COEFFICIENT OF BK 8 OPTICAL GLASS BETWEEN 15 AND 300 K.**
LEBESQUE H J M VAN SCHAIK J W KRUIJSHOOP A W
BLAISSE B S
PHYSICA
31 6 967-72 1965

46637 **THERMAL EXPANSION OF CLAY MINERALS.**
MC KINSTRY H A
AM MINERALOGIST
50 1/2 212-22 1965

46648 **INVESTIGATION OF RADIATION AND CONDUCTION HEAT TRANSFER IN FIBROUS HIGH TEMPERATURE INSULATIONS.**
ROLINSKI, E. J. PURCELL, G. V.
WRIGHT-PATTERSON AIR FORCE BASE, OHIO
1-139, 1967.
(AFML-TR-67-251)

46650 **TARGET SIGNATURE ANALYSIS CENTER. DATA COMPILATION. SUPPLEMENTAL DATA SHEETS.**
EARING, D.
WILLOW RUN LABORATORIES UNIV OF MICHIGAN
1-, 1967.
(REPT-7850-9-B AND 8492-12-B/DEL/, DATA SHEETS)

46742 **STRENGTHENING GLASS-CERAMICS BY APPLICATION OF COMPRESSIVE GLAZES.**
DUKE D A MEGLES J E JR MAC DOWELL J F
BOPP H F
J AM CERAM SOC
51 2 98-102 1968

46789 **SOME PROPERTIES OF HAFNIUM OXIDE, HAFNIUM SILICATE, CALCIUM HAFNATE AND HAFNIUM CARBIDE.**
CURTIS, C. E. DONEY, L. M. JOHNSON, J. R.
J AM CERAMIC SOC
37 10 458-65 1954

TPRC Number	Bibliographic Citation
46791	**SYMPOSIUM ON THERMAL SHOCK. SOME GENERAL CONSIDERATIONS ON THERMAL SHOCK.** WHITE J TRANS BRIT CERAMIC SOC 57 10 591–623 1958
46798	**METAL-CERAMIC BRAZED SEALS.** BONDLEY R J ELECTRONICS 20 97–9 1947
46801	**PRODUCTION TECHNOLOGY FOR LADLE BRICK USING SEMIACID CLAYS.** SHCHEGLOV S I KARASIK V A SAZHIN V A OGNEUPORY 5 3–7 1967 (FOR ENGLISH TRANSLATION SEE T46802)
46802	**PRODUCTION TECHNOLOGY FOR LADLE BRICK USING SEMIACID CLAYS.** SHCHEGLOV S I KARASIK V A SAZHIN V A REFRACTORIES 5 272–6 1967 (ENGLISH TRANSLATION OF OGNEUPORY (5), 3–7, 1967; FOR ORIGINAL SEE T46801)
46819	**STUDY OF THE MECHANICAL AND THERMAL CHARACTERISTICS OF REFRACTORIES AT HIGH TEMPERATURES.** LECRIVAIN L REV HATUES TEMP REFRACTAIRES 3 1 27–40 1966
46820	**PREDICTION OF TEMPERATURE-DISTRIBUTION IN FROZEN SOILS.** BERGGREN W P TRANS AM GEOPHYSICAL UNION PART III 71–7 1943
46828	**INFRARED TRANSMISSION LIMITS OF SINGLE CRYSTALS OF SOME FLUORIDES.** JONES D A JONES R V STEVENSON R W H PROC PHYS SOC /LONDON/ 65 B 906–7 1952
46837	**THE IDENTIFICATION OF GEMS.** PFUND A H J OPT SOC AM 35 9 611–14 1945
46847	**THERMAL EXPANSION OF VARIOUS CERAMIC MATERIALS TO 1500 C.** WHITTEMORE O J JR AULT N N J AM CERAM SOC 39 12 443–4 1956
46849	**EFFECT OF SOME TECHNICAL FACTORS ON THE SPALLING RESISTANCE OF PERICLASE-SPINEL REFRACTORIES.** KUKOLEV G V NEMETS I I SHEKHOVTSEVA V A REFRACTORIES 5 292–300 1968 (ENGLISH TRANSLATION OF OGNEUPORY, 33 (5), 31–9, 1968; FOR ORIGINAL SEE T44358)
46855	**SUBSURFACE TEMPERATURE, THERMAL CONDUCTIVITY, AND HEAT FLOW NEAR AIKEN, SOUTH CAROLINA.** DIMENT W H MARINE E W NEIHEISEL J SIPLE G E J GEOPHYS RESEARCH 70 22 5635–44 1965
46877	**THERMAL CONDUCTIVITIES OF SR-90 HEAT SOURCE MATERIALS.** KETCHEN, E. E. MC HENRY, R. E. OAK RIDGE NATIONAL LAB., OAK RIDGE, TENN. 1–22, 1967. (ORNL–TM–1905, CONF–670714–4)
46878	**ZIRCALOY-4, URANIUM DIOXIDE, AND MATERIALS FORMED BY THEIR INTERACTION, A LITERATURE REVIEW WITH EXTRAPOLATION OF PHYSICAL PROPERTIES TO HIGH TEMPERATURES.** HAMMER, R. R. IDAHO NUCLEAR CORP., IDAHO FALLS, IDAHO 1–38, 1967. (IN–1093, N68–13967)
46880	**PHYSICAL PROPERTIES OF THE OXIDE. THE THERMAL EXPANSION OF PUO2 AND SOME OTHER ACTINIDE OXIDES BETWEEN ROOM TEMPERATURE AND 1000 C /5/.** BRETT N H RUSSELL L E PLUTONIUM 397–410 1960
46930	**PHASE COMPOSITION OF FROZEN MONTMORILLONITE-WATER MIXTURES FROM HEAT CAPACITY MEASUREMENTS.** ANDERSON, D. M. COLD REGIONS RES. AND ENGR. LAB., HANOVER, N. H. 1–17, 1967. (CRREL–RR–218, AD–656601)
46948	**THE EVALUATION OF THE EXPANSION COEFFICIENT OF GLASS AT LOW TEMPERATURE.** WINTER-KLEIN A COMPT RENDUS 230 21 1857–8 1950
46952	**THERMAL CONDUCTIVITY OF BULK OXIDE FUELS.** BELLE J BERMAN R M BOURGEOIS W F COHEN I DANIEL R C BETTIS ATOMIC POWER LAB., PITTSBURGH, PAL. 1–99, 1967. (WAPD–TM–586/REV./, N67–35791)
46969	**INVESTIGATION OF OPTICAL PROPERTIES OF SHOCK COMPRESSED LEAD GLASS /ELASTIC-PLASTIC WAVE IN GLASS/.** KORMER S B KRUSHKEVICH G V YUSHKO K B J EXPTL THEORET PHYS /USSR/ 52 6 1478–84 1967 (FOR ENGLISH TRANSLATION SEE T46970)
46970	**INVESTIGATION OF OPTICAL PROPERTIES OF SHOCK COMPRESSED LEAD GLASS /ELASTIC-PLASTIC WAVE IN GLASS.** KORMER S B KRUSHKEVICH G V YUSHKO K B SOVIET PHYSICS JETP 25 6 980–5 1967 (ENGLISH TRANSLATION OF J. EXPTL. THEORET. PHYS., 52 (6), 1478–84, 1967; FOR ORIGINAL SEE T46969)
46995	**THE RESTRAHLEN OF ROCK SALT AND SYLVITE.** RUBENS H ASCHKINASS E ANN PHYSIK 65 6 241–56 1898
46996	**THE TEMPERATURE RADIATION OF INCANDESCENT OXIDES AND OXIDE MIXTURES IN THE ULTRARED SPECTRAL REGION.** RITZOW G ANN PHYSIK 19 5 769–99 1934
47008	**INVESTIGATION IN THE SPECTRAL REGION BETWEEN 20 AND 40 MICRONS.** STRONG J PHYS REV 37 1565–72 1931
47019	**RELATIONSHIP BETWEEN PHASE CHANGES AND ENGINEERING PROPERTIES OF MATERIALS BASED ON THE ZRO2-CEO2 SYSTEM.** KLYUCHAROV YA V STRAKHOV V I OGNEUPORY 33 5 40–3 1968 (FOR ENGLISH TRANSLATION SEE T47020)
47020	**RELATIONSHIP BETWEEN PHASE CHANGES AND ENGINEERING PROPERTIES OF MATERIALS BASED ON THE ZRO2-CEO2 SYSTEM.** KLYUCHAROV YA V STRAKHOV V I REFRACTORIES 5 301–5 1968 (ENGLISH TRANSLATION OF OGNEUPORY, 33 (5), 40–3, 1968; FOR ORIGINAL SEE T47019)
47022	**DISCUSSION 12. FROM SUMMARY OF SECOND HIGH-TEMPERATURE INORGANIC REFRACTORY COATINGS WORKING GROUP MEETING.** TAYLOR, K. M. AIKEN, D. B. WRIGHT-PATTERSON AIR FORCE BASE, OHIO 44–7, 1959. (WADC–TR–59–415, AD–232536)
47023	**DISCUSSION 14. FROM SUMMARY OF SECOND HIGH-TEMPERATURE INORGANIC REFRACTORY COATINGS WORKING GROUP MEETING.** WHEILDON, W. M. WRIGHT-PATTERSON AIR FORCE BASE, OHIO 51–4, 1959. (WADC–TR–59–415, AD–232536)
47033	**SMALL CAST THORIUM OXIDE CRUCIBLES.** RICHARDSON H K J AM CERAM SOC 18 65–9 1935
47036	**THERMAL CONDUCTIVITY MEASUREMENTS OF EUO ACROSS THE CURIE POINT.** CASON, J. L., JR. SOUTH DAKOTA SCHOOL OF MINES AND TECHNOLOGY, RAPID CITY, S. DAKOTA, M. S. THESIS 1–85, 1967. (AD–659181, N68–10715)
47059	**STUDIES OF THE THERMAL STATE OF THE EARTH. THE 19TH PAPER. HEAT-FLOW MEASUREMENTS IN THE NORTHWESTERN PACIFIC.** VACQUIER V UYEDA S YASUI M SCLATER J CORRY C BULL. EARTHQUAKE RES. INSTITUTE 44, 1519–35, 1966. (AD–660518)
47072	**BERRYITE FROM GREENLAND.** KARUP-MOELLER S CAN MINERALOGIST 8 4 414–23 1966 CA 68 31923
47074	**COMPARATIVE TESTS OF CHEMICAL GLASSWARE.** WICHERS E FINN A N CLABAUGH W S IND ENGINEERING CHEM 13 6 419–22 1941

TPRC Number	Bibliographic Citation
47086	**THERMODYNAMICS OF POLYMORPHIC TRANSFORMATIONS IN SILICA. THERMAL PROPERTIES FROM 5 TO 1070 K. AND PRESSURE-TEMPERATURE STABILITY FIELDS FOR COESITE AND STISHOVITE.** HOLM J L KLEPPA O J WESTRUM E F JR GEOCHIM. COSMOCHIM. ACTA 31, 2280-2307, 1967. (AROD-5107.9, AD-667030)
47091	**STRUCTURAL CHANGES IN GLASSES OF THE SIO2-AL2O3-BAO-TIO2 SYSTEM DURING THE INITIAL CRYSTALLIZATION STAGE.** BOGDANOVA G S ORLOVA E M IZV AKAD NAUK SSSR NEORGAN MATERIALY 2 3 537-40 1966 JA 51-2 40 (FOR ENGLISH TRANSLATION SEE T36254)
47094	**MATERIALS EVALUATION FOR SERVICEABILITY OF OPTICAL GLASSES UNDER PROLONGED SPACE CONDITIONS.** HOLLAND, W. R. AVCO ELECTRONICS DIV., AVCO CORP., TULSA, OKLA. 1-73, 1967. (TR-67-G-109-F, NASA-CR-65687, N67-34672)
47098	**THERMOPHYSICAL PROPERTIES OF SOLIDS.** PURCELL, G. V. WRIGHT-PATTERSON AIR FORCE BASE, OHIO 651-76, 1966. (AFML-TR-66-179, AD-804083)
47102	**PRESSURIZED WATER REACTOR /PWR/ PROJECT.** BETTIS ATOMIC POWER LABORATORY PITTSBURGH PA BETTIS ATOMIC POWER LAB., PITTSBURGH, PA. 1-121, 1960. (WAPD-MRP-84)
47117	**GLAZE FOR CERAMIC SUBSTRATES FOR THIN FILMS.** DI MARCELLO F V TREPTOW A W BAKER L A BULL AM CERAM SOC 47 5 511-16 1968
47140	**GLASSES IN THE SYSTEMS CONTAINING LEAD OXIDE, BORIC OXIDE, ALUMINA AND PHOSPHORUS PENTOXIDE.** KUMAR S CENTRAL GLASS CERAM RESEARCH INST BULL /INDIA/ 6 1 13-22 1959
47146	**CERAMICS-PROPERTIES.** ARTHUR G NUCLEAR ENGINEERING 6 253-7 1961
47147	**THERMAL STRESS RESISTANCE OF REFRACTORIES. II. HIGH-TEMPERATURE APPARATUS TO DETERMINE THERMAL CONDUCTIVITY AND PERMISSIBLE TEMPERATURE DROP.** PRANCKEVICIUS G DAUKNYS V JURENAS V LIET TSR MOKSLU AKAD DARB B 3 121-6 1967 CA 68 32837
47150	**CRACKING IN GLAZES OF DOLOMITE EARTHENWARE. I.** KAWAMURA S NISHIMURA T MAEDA T ITO A ASANO K MIZUTA H KUMAZAWA S NAGOYA KOGYO GIJUTSU SHIKENSHO HOKOKU 16 5/6 183-91 1967 CA 68 32824
47158	**ULTRASONIC DEVICE FOR MEASURING THE EMISSIVE POWER OF A LASER.** BONDARENKO A N KRIVOSHCHEKOV G V MARENNIKOV S I PRIBORY I TEKHNIKA EKSPERIMENTA 226-7 1967 IAA 8-5 817 (FOR ENGLISH TRANSLATION SEE T47159)
47159	**ULTRASONIC DEVICE FOR MEASURING THE EMISSIVE POWER OF A LASER.** BONDARENKO A N KRIVOSHCHEKOV G V MARENNIKOV S I INSTR EXP TECH /USSR/ 453-4 1967 (ENGLISH TRANSLATION OF PRIBORY I TEKH. EKSP., 226-27, 1967; FOR ORIGINAL SEE T47158)
47173	**PREPARATION, PROPERTIES, AND UTILIZATION OF ZIRCONIUM-BASED ALLOYS.** PRIDANTSEVA K S SB TR TSENT NAUCH-ISSLED INST CHERN MET 51 139-52 1967 CA 68 32557 (FOR ENGLISH TRANSLATION SEE T56358)
47174	**EFFECT OF CARBON DIOXIDE GAS PRESSURE ON THE THERMAL CONDUCTIVITY OF SOME GRANULAR INSULATIONS.** SANDERSON P D PROC INST MECH ENG /LONDON/ 181 31 57-65 1966-7 NSA 22-4 790
47181	**INVESTIGATION OF CERTAIN TECHNOLOGICAL FACTORS AFFECTING LOW MELTING ZIRCON GLAZES.** BUCHVAROV S KUKUSHEVA M RADKOVA A BUKAREV A STROIT MATER SILIKAT PROM 8 9 28-35 1967 CA 68 32825
47183	**THE DETERMINATION OF THE THERMAL EXPANSION OF SOLID BODIES UP TO 1400 C. WITH THE COMPARATIVE METHOD.** COHN W M Z TECH PHYS 11 4 118-21 1930
47203	**EARTHS VISCOSITY.** ANDERSON D L SCIENCE 151 321-2 1966
47220	**A NEW DILATOMETER DESIGN.** SORKIN E S STEKLO I KERAM 15 10 40-2 1958 (FOR ENGLISH TRANSLATION SEE T47221)
47221	**A NEW DILATOMETER DESIGN.** SORKIN E S GLASS AND CERAMICS 15 558-60 1958 (ENGLISH TRANSLATION OF STEKLO I KERAM., 15 (10), 40-2, 1958; FOR ORIGINAL SEE T47220)
47229	**STUDIES ON FINE GRAINED CRYSTALLINE GLASS-CERAMICS SUBMICROCRYSTALLIZATION OF GLASSES OF THE SYSTEM SIO2-LI2O-AL2O3 BY ADDITION OF P2O5. /IV/.** NAGAOKA K HARA M TANAKA H OSAKA KOGYA GIJUTSU SHIKENSHO KIHO 12 3 292-8 1961
47237	**GLASS USED IN THE MANUFACTURE OF APPARATUS AND INSTRUMENTS.** DE JONG, J. CONSTRUCTIEMATERIALEN 2 (1), 20-2, 1967.
47238	**REFLECTION OF SUNLIGHT TO SPACE AND ABSORPTION BY THE EARTH AND ATMOSPHERE OVER THE UNITED STATES DURING SPRING 1962.** HANSON K J VONDER H T H SUOMI V E MONTHLY WEATHER REVIEW 95 354-62 1967 IAA 8/2/ 256
47240	**CERAMIC HONEYCOMB.** GOSS C L PRODUCT ENGINEERING 33 26 52-5 1962
47242	**HEAT EXPANSION AND SOFTENING TEMPERATURE OF GLASSES IN THE SYSTEM ALUMINUM OXIDE - TITANIUM DIOXIDE - PHOSPHORUS PENTOXIDE.** SYRITSKAYA, Z. M. SHAPOVALOVA, N. F. TR. GOS. NAUCH.-ISSLED. INST. STEKLA (2), 111-17, 1966.
47248	**SLIP-CAST FUSED SILICA.** WALTON, J. D. POULOS, N. E. RES. AND TECHNOL. DIV., WPAFB, OHIO 1-131, 1964. (ML-TDR-64-195)
47249	**TRANSMISSION AND DISPERSION OF SOME GLASSES IN THE INFRA-RED.** VAGNER-DEWULF G VINCENT-GEISSE J TAI N T REV OPT 39 12 578-86 1960
47250	**ON THE MECHANICAL AND THERMAL BEHAVIOUR OF REFRACTORIES. PT. 1. FUNDAMENTAL PHYSICAL PHENOMENA.** SPATH V W BER DEUT KERAM GESELL /WEST GERMANY/ 38 8 351-6 1961
47261	**REFLECTANCE, TRANSMITTANCE, AND THERMAL EMITTANCE OF MICA AS A FUNCTION OF THICKNESS.** BRADFORD A P HASS G HEANEY J B PROGRESS IN ASTRONAUTICS AND AERONAUTICS 20, 3-16, 1967.
47282	**REMOTE THERMAL SENSING.** HOLTER M R LEGAULT R R PROGRESS IN ASTRONAUTICS AND AERONAUTICS 20, 547-66, 1967.
47283	**INFRARED SENSING FROM SPACECRAFT. A GEOLOGICAL INTERPRETATION.** VICKERS R S LYON R J P PROGRESS IN ASTRONAUTICS AND AERONAUTICS 20, 585-607, 1967.
47287	**DETERMINATION OF INTERNAL HEAT AND MASS TRANSFER PARAMETERS USING CHARACTERISTIC FUNCTIONS OF THE THERMODYNAMICS OF IRREVERSIBLE PROCESSES.** TEMKIN A G INZH FIZ ZH AKAD NAUK BELORUSSK SSR 9 3 305-17 1965 (FOR ENGLISH TRANSLATION SEE T47288)
47288	**DETERMINATION OF INTERNAL HEAT AND MASS TRANSFER PARAMETERS USING CHARACTERISTIC FUNCTIONS OF THE THERMODYNAMICS OF IRREVERSIBLE PROCESSES.** TEMKIN A G J ENGINEERING PHYSICS 9 3 203-13 1965 (ENGLISH TRANSLATION OF INZH. FIZ. ZH., AKAD. NAUK BELORUSSK., SSSR, 9 (3), 305-17, 1965; FOR ORIGINAL SEE T47287)

TPRC Number	Bibliographic Citation
47289	**HEAT TRANSFER IN A DISPERSE INSULATOR.** KOSTYLEV V M NABATOV V G J ENGINEERING PHYSICS 9 3 259-63 1965 (ENGLISH TRANSLATION OF INZH. FIZ. ZH., AKAD. NAUK BELORUSSK. SSSR, 9 (3), 377-83, 1965; FOR ORIGINAL SEE T34599)
47302	**PROPERTIES OF SANDSTONES AS A REFRACTORY MATERIAL.** CHEN T HU S-E J AM CERAM SOC 29 7 193-7 1946
47303	**THERMAL CONDUCTIVITY OF AMORPHOUS BODIES, AS ILLUSTRATED BY PORCELAIN.** SHADRICHEV E V SMIRNOV I A TIKHONOV V V FIZ TVERDOGO TELA 9 8 2435-7 1967 CA 68 43838 (FOR ENGLISH TRANSLAITON SEE T47358)
47305	**THERMODYNAMIC PROPERTIES OF TALC.** LEONIDOV V YA KHITAROV N I GEOKHIMIYA 10 1044-9 1967 CA 68 43819 (FOR ENGLISH TRANSLATION SEE T70863)
47311	**THERMAL DIFFUSIVITY MEASUREMENT OF ROCK-FORMING MINERALS FROM 300 TO 110 K.** KANAMORI H FUJII N MIZUTANI H J GEOPHYS RES 73 2 595-605 1968 CA 68 42266
47316	**RESEARCH ON ATMOSPHERIC OPTICAL RADIATION TRANSMISSION.** BULLRICH K BLAETTNER W CONLEY T EIDEN R HAENEL G HEGER K NOWAK W AIR FORCE CAMBRIDGE RES. LAB. 1-163, 1968. (AFCRL-68-0186, N68-30384, AD-670210)
47340	**ALUMINA IN WHITEWARE.** AUSTON C R SCHOFIELD H Z HALDY N L J AM CERAM SOC 29 12 341-54 1946
47358	**THERMAL CONDUCTIVITY OF AMORPHOUS BODIES, AS ILLUSTRATED BY PROCELAIN.** SHADRICHEV E V SMIRNOV I A TIKHONOV V V SOVIET PHYS-SOLID STATE 9 8 1909-11 1968 (ENGLISH TRANSLATION OF FIZ. TVERD. TELA, 9 (8), 2435-7, 1967; FOR ORIGINAL SEE T47303)
47360	**THERMOPHYSICAL PROPERTIES OF KUKERSITE AND THEIR ALTERATION DURING THERMAL DECOMPOSITION OF SHALES.** GUBERGRITS, M. KUIV, K. TERM. PERERAB. SLANTSA-KUKERSITA, INST. KHIM. AKAD. NAUK EST. SSR 118-38, 1966.
47364	**THE MECHANISM OF SPECTRAL ABSORPTION IN GLASSES USED FOR PASSIVE Q-SWITCHES.** DEEG E INTERACTION RADIAT. SOLIDS, PROC. CONF. 607-16 1967 CA 68 42764
47367	**EXPANSION STRAINS IN THE INITIAL FIRING OF CERAMIC WARE.** SIEVERT G E MOFFAT J JR J AM CERAMIC SOC 23 4 122-4 1940
47373	**THERMAL CONDUCTIVITY AND SPECIFIC HEAT MEASURING METHODS APPLICABLE TO SMALL SAMPLES.** VASILIU S ST CERC FIZ 17 9 971-83 1965
47385	**DIFFUSE REFLECTANCE STUDIES OF SOLID-SOLID INTERACTIONS // DRUG-ADJUVANT SYSTEMS/.** BIGHLEY, L. D. UNIV. OF IOWA, IOWA CITY, PH. D. THESIS 1-96, 1967. (UNIVERSITY MICROFILMS NO. 67-2597)
47390	**THERMAL EXPANSION AND OTHER PROPERTIES OF GLASSES IN THE SYSTEM K2O-PBO-SIO2.** BLACHERE, S. T. ALFRED UNIVERSITY, M. S. THESIS 1-41 1961
47403	**THE THERMAL EXPANSION OF LEAD-POTASH-SILICA GLASSES FROM -170 TO 300 C.** SAMDANI, S. G. ALFRED UNIVERSITY, M. S. THESIS 1-33 1963
47422	**FUNCTIONALIZATION OF PYROCERAM 9606 TEST DATA FOR RADOME THERMAL-STRESS ANALYSIS.** TATE, M. B. APPLIED PHYS. LAB., JOHNS HOPKINS UNIV., SILVER SPRING, MD. 1-22, 1968. (ALP-TG-980, N68-27040, AD-688136)
47426	**COLOR CHARACTERISTICS OF ULTRAVIOLET ABSORBING EMERALD GREEN GLASS.** KNUPP R C BERGER D F AM CERAM SOC BULL 47 3 244-7 1968
47464	**THERMAL STABILITY OF CARBON SAND AND OTHER NONSILICA MOLDING MATERIALS.** GENTRY, E. G. BRIT. FOUNDRYMAN 60 (10), 373-9, 1967.
47474	**SPECTRAL REFLECTANCE IN THE INFRARED OF THE PRODUCTS OF CRYSTALLIZATION OF GLASSES OF THE SYSTEM LI2O-SIO2 BETWEEN 7.7 AND 14 MICRONS.** CHERNEVA E F FLORINSKAYA V A PODUSHKO E V ZHUR FIZ KHIMII /USSR/ 37 11 2556-60 1963 (FOR ENGLISH TRANSLATION SEE T47475)
47475	**SPECTRAL REFLECTANCE IN THE INFRARED OF THE PRODUCTS OF CRYSTALLIZATION OF GLASSES OF THE SYSTEM LI2O-SIO2 BETWEEN 7.7 AND 14 MICRONS.** CHERNEVA E F FLORINSKAYA V A PODUSHKO E V CENTRE NATL. RES. SPATIALS, PARIS FRANCE 1-, 1965. (FRENCH TRANSLATION OF ZH. FIZ. KHIM., 37 (11), 2556-60, 1963; FOR ORIGINAL SEE T47474) (H/R-203, TT-65-28254)
47476	**INFLUENCE OF ANION VACANCIES ON THE LATTICE THERMAL CONDUCTIVITY OF COORDINATION CRYSTALS HAVING THE SPINEL STRUCTURE AT LOW TEMPERATURE.** VISHNEVSKII I I SKRIPAK V N FIZ TVERD TELA 9 12 3633-5 1967 CA 68 54237 (FOR ENGLISH TRANSLATION SEE T49405)
47477	**METASTABLE CRYSTALLINE SOLUTIONS WITH QUARTZ STRUCTURE IN THE OXIDE SYSTEM LI2O-MGO-ZNO-AL2O3-SIO2.** PETZOLDT J GLASTECH BER 40 10 385-96 1967 CA 68 52703 (FOR ENGLISH TRANSLATION SEE T76782)
47495	**THERMODYNAMIC PROPERTIES OF THREE LITHIUM-ALUMINUM SILICATES.** PANKRATZ, L. B. WELLER, W. W. BUREAU OF MINES, BERKELEY, CALIF. 1-13, 1967. (BM-RI-7001, N67-34040)
47497	**THE EFFECTS OF STOICHIOMETRY ON THE THERMAL EXPANSION OF 20 WT PERCENT PUO2-UO2 FAST-REACTOR FUEL.** ROTH J HUBERT M E CHERRY J R CALDWELL C S TRANS AMER NUCL SOC 10 457-8 1967 NSA 22/3/502
47499	**THERMAL CONDUCTIVITY OF POROUS SYSTEMS.** LUIKOV A V SHASHKOV A G VASILEV L L FRAIMAN YU E INTERN J HEAT MASS TRANSFER 11 2 117-40 1968 CA 68 72995
47506	**MEASUREMENT OF CLOUD REFLECTANCE PROPERTIES AND THE ATMOSPHERIC ATTENUATION OF SOLAR AND INFRARED ENERGY.** GRIGGS, M. MARGGRAF, W. A. CONVAR DIV., GENERAL DYNAMICS, SAN DIEGO, CALIF. 1-153, 1967. (AFCRL-68-0003, AD-666936)
47507	**THERMODYNAMICS OF THE LEAD STORAGE CELL. THE HEAT CAPACITY AND ENTROPY OF LEAD DIOXIDE FROM 15 TO 318 K** DUISMAN J A GIAUQUE W F J PHYS CHEM 72 2 562-73 1968 CA 68 63352
47508	**SOIL THERMAL CONDUCTIVITY INVESTIGATION FOR UNDERGROUND SHELTERS.** BALDWIN, D. B. PROTECTIVE STRUCTURES DEV. CENTER, FORT BELVOIR, VIRGINIA 1-120, 1967. (PSDC-TR-13, AD-661090)
47524	**ARTIFICIAL LIGHT SOURCES FOR SIMULATING NATURAL DAYLIGHT AND SKYLIGHT.** GRUM F APPL OPT 7 1 183-7 1968
47530	**PROPERTIES OF SOME COLUMBIUM OXIDE-BASIS CERAMICS.** DURBIN, E. A. WAGNER, H. E. HARMAN, C. G. BATTELLE MEMORIAL INST., COLUMBUS, OHIO 1-15, 1952. (BMI-792)
47535	**OXIDE GLASSES IN LIGHT OF THE IDEAL GLASS CONCEPT. I. IDEAL AND NONIDEAL TRANSITIONS, AND DEPARTURES FROM IDEALITY.** ANGELL C A J AM CERAM SOC 51 3 117-24 1968

TPRC Number	Bibliographic Citation
47537	**STRUCTURAL SIMILARITIES BETWEEN A GLASS AND ITS MELT.** RIEBLING E F J AM CERAM SOC 51 3 143-9 1968 CA 68 98270
47538	**STRUCTURE TRANSITION IN ALKALI SILICATE GLASSES.** LACY E D J AM CERAM SOC 51 3 150-7 1968 CA 68 98272
47539	**SUBSTRUCTURES IN SILICATE GLASSES.** BABCOCK C L J AM CERAM SOC 51 3 163-9 1968
47544	**COEFFICIENT OF THERMAL EXPANSION OF STEATITE.** IRWIN G J CLAY PRODS NEWS AND CERAM RECORD 34 5 12-14 1961
47612	**ALKALINE EARTH PORCELAINS POSSESSING LOW DIELECTRIC LOSS.** RIGTERINK M D GRISDALE R O J AM CERAM SOC 30 3 78-81 1947
47626	**DESIGN OF CLAY BODIES FOR CONTROLLED MICROSTRUCTURE. PART I. CONCEPTS AND BODY DESIGN.** ROBINSON G C AM CERAMIC SOC BULLETIN 47 5 477-80 1968
47638	**DEPENDENCE OF CERTAIN PROPERTIES OF ZIRCONIUM ENAMELS ON THEIR COMPOSITION.** VARGIN, V. V. KHEIFETS, V. S. EMAL EMALIROVANIE MET. 49-54, 1967.
47639	**ZIRCONIUM ENAMELS.** LESKOV, A. L. EMAL EMALIROVANIE MET. 60-8, 1967.
47640	**THERMAL EXPANSION OF HEAT-RESISTANT CERAMIC COATINGS.** AZAROV, K. P. DEMCHENKO, N. S. EMAL EMALIROVANIE MET. 285-91, 1967.
47643	**ULTRAVIOLET REFLECTIVITY OF VENUS AND JUPITER.** EVANS, D. C. MOON AND PLANETS, INTERN. SPACE SCI. SYMP., 7TH 135-49, 1967.
47659	**THE EXPANSION OF SOLIDS AT LOW TEMPERATURE.** HENNING F ANN PHYSIK 22 4 631-9 1907
47666	**IV. AN INVESTIGATION INTO THE PROPERTIES OF SILICA BRICKS.** RIGBY G R WHITE R P BOOTH H GREEN A T TRANS BRIT CERAMIC SOC 45 69-109 1946
47674	**PHASE RELATIONSHIPS IN THE SYSTEM ZIRCONIA-CERIA.** DUWEZ P ODELL F J AM CERAM SOC 33 9 274-83 1950
47675	**MECHANISM OF SECONDARY EXPANSION OF HIGH-ALUMINA REFRACTORIES CONTAINING CALCINED BAUXITE.** MC GEE T D DODD C M J AM CERAM SOC 44 6 277-83 1961
47677	**THE DENSITIES AND CUBICAL COEFFICIENTS OF EXPANSION OF THE HALOGEN SALTS OF SODIUM, POTASSIUM, RUBIDIUM AND CESIUM.** BAXTER G P WALLACE C C J AM CHEM SOC 38 259-66 1916
47685	**INFLUENCE OF SMALL ADDITIONS OF LANTHANUM OXIDE ON THE PROPERTIES OF SODIUM ZIRCONIUM SILICATE GLASSES.** DUBROVO S K SHNYPIKOV A D ZHUR PRIKLAD KHIM 40 6 1216-20 1967 (FOR ENGLISH TRANSLATION SEE T47686)
47686	**INFLUENCE OF SMALL ADDITIONS OF LANTHANUM OXIDE ON THE PROPERTIES OF SODIUM ZIRCONIUM SILICATE GLASSES.** DUBROVO S K SHNYPIKOV A D J APPL CHEM USSR 40 6 1174-8 1967 (ENGLISH TRANSLATION OF ZH. PRIKL. KHIM., 40 (6), 1216-20, 1967; FOR ORIGINAL SEE T47685)
47694	**FORMATION OF FORSTERITE BY HEATING OF DUNITE.** BUDNIKOV P P KHOROSHAVIN L B PEREPELITSIN V A USTYANTSEV V M KOSOLAPOVA E P ZHUR PRIKLAD KHIM 40 6 1369-70 1967 (FOR ENGLISH TRANSLATION SEE T47695)
47695	**FORMATION OF FORSTERITE BY HEATING OF DUNITE.** BUDNIKOV P P KHOROSHAVIN L B PEREPELITSIN V A USTYANTSEV V M KOSOLAPOVA E P J APPL CHEM USSR 40 6 1312-13 1967 (ENGLISH TRANSLAITON OF ZH. PRIKL. KHIM., 40 (6), 1369-70, 1967; FOR ORIGINAL SEE T47694)
47761	**VIBRATIONAL EXCITATION OF CARBON MONOXIDE FOLLOWING QUENCHING OF THE A-3-II STATE.** DONOVAN R J HUSAIN D TRANS FARADAY SOC 63 12 2879-87 1967
47768	**HEAT TRANSFER BETWEEN SURFACES IN CONTACT. THE EFFECT OF LOW CONDUCTANCE INTERSTITIAL MATERIALS. PT. 3. COMPARISON OF THE EFFECTIVE THERMAL INSULATION FOR INTERSTITIAL MATERIALS UNDER COMPRESSIVE LOADS. PTS. 4 AND 5. INVESTIGATION OF THERMAL ISOLATION MATERIALS AND THEIR APPLICATION IN FLANGE JOINTS.** SMUDA, P. A. GYOROG, D. A. ARIZONA STATE UNIV., TEMPE ARIZONA 1-82, 1968. 1-68, 1968. (NASA-CR-73244, ME-TR-033-3, N68-31671 NASA-CR-73248, ME-TR-033-5/PTS 4,5/, N68-34080)
47842	**ESR AND OPTICAL ABSORPTION STUDY OF MO /+3/ IN A PHOSPHATE GLASS.** LANDRY R J J CHEM PHYS 48 3 1422-3 1968
47846	**DIELECTRIC POLARIZATION AND THERMAL EXPANSION OF POLARIZED CERAMIC SAMPLES OF THE PB TI O3-PB ZR O3-PB MG/0.5/ O3 SYSTEM.** STOLYPIN YU E ISUPOV V A FIZ TVERD TELA 9 9 2619-24 1967 CA 67 120852 (FOR ENGLISH TRANSLATION SEE T47847)
47847	**DIELECTRIC POLARIZATION AND THERMAL EXPANSION OF POLARIZED CERAMIC SAMPLES OF THE PB TI O3-PB ZR O3-PB MG/0.5/ W/0.5/ O3 SYSTEM.** STOLYPIN YU E ISUPOV V A SOVIET PHYS-SOLID STATE 9 9 2059-63 1968 (ENGLISH TRANSLATION OF FIZ. TVERD. TELA, 9 (9), 2691-24, 1967; FOR ORIGINAL SEE T47846)
47851	**PRODUCTION OF PERLITE-CHAMOTTE INSULATING PRODUCTS.** FAIN I A KAMENETSKII S P RABINOVICH M A GRIGOREV I V OGNEUPORY 1 5-9 1968 (FOR ENGLISH TRANSLATION SEE T47852)
47852	**PRODUCTION OF PERLITE-CHAMOTTE INSULATING PRODUCTS.** FAIN I A KANENETSKII S P RABINOVICH M A GRIGOREV I V REFRACTORIES 1 6-10 1968 (ENGLISH TRANSLATION OF OGNEUPORY, (1), 5-9, 1968; FOR ORIGINAL SEE T47851)
47870	**FIBER OPTICS WITH HIGH ULTRA-VIOLET TRANSMISSION.** DENEKA, C. W. RUTGERS STATE UNIV., N. BRUNSWICK, N. J. 1-46, 1968. (N68-29562)
47871	**INTERFEROMETRIC METHOD FOR MEASURING THE THERMAL EXPANSION OF SMALL SAMPLES OF OPTICAL MATERIALS.** BALLARD S S BROWDER J S KAYLOR H M STREETE J L J OPT SOC AMERICA 58 2 155-9 1968 CA 68 72510
47909	**SELECTIVE POLARIZATION OF LIGHT DUE TO ABSORPTION BY SMALL ELONGATED SILVER PARTICLES IN GLASS.** STOOKEY S D ARAUJO R J APPL OPT 7 5 777-9 1968
47910	**THERMAL CONDUCTIVITY AND HEAT CAPACITY OF 7740 PYREX BELOW 4 K AND IN MAGNETIC FIELDS TO 90 KG.** FISHER R A BRODALE G E HORNUNG E W GIAUQUE W F REV SCI INSTRUM 39 1 108-14 1968 CA 68 54147
47917	**BIDIRECTIONAL REFLECTANCE MEASUREMENTS FOR SATELLITE CALIBRATION TARGET IN THE VISIBLE AND NEAR INFRARED.** SAIEDY F JONES G D APPL OPT 7 3 429-34 1968 CA 68 91553
47920	**MAGNESIA FOAMED CERAMICS.** NOSOVA Z A LYUBINA T M REFRACTORIES 2 113-15 1968 (ENGLISH TRANSLATION OF OGNEUPORY, (2), 50-3, 1968; FOR ORIGINAL SEE T49597)

TPRC Number	Bibliographic Citation
47930	**MEASUREMENT OF THERMAL CONDUCTIVITY AT LOW TEMPERATURES.** VOS B H BULL INTL INST REFRIG ANNEXE 179-90 1961
47935	**SUPERINSULATION FOR THE LARGE SCALE STORAGE AND TRANSPORT OF LIQUEFIED GASES.** DUBS M A DANA L I BULL INTL INST REFRIG ANNEXE 71-83 1961
47941	**EFFECT OF THERMAL SHOCK AND ANNEALING ON CERTAIN PHYSICAL PROPERTIES OF A ZIRCON BODY.** DEADMORE D L CERAMIC AGE 55 17-20 1950
47943	**DIFFERENTIAL THERMAL CALORIMETRIC DETERMINATION OF THE THERMODYNAMIC PROPERTIES OF KAOLINITE.** NICHOLSON, P. S. CALIFORNIA UNIV., BERKELEY, CALIF., PH. D. THESIS 1-53, 1967. (UCRL-17820, N68-14595)
47946	**TRACER DIFFUSION AND SHEAR VISCOSITY IN THE LIQUID-LIQUID CRITICAL REGION.** LEISTER H M ALLEGRA J C ALLEN G F J CHEM PHYS 51 9 3701-8 1969
47956	**PROPERTIES OF SOME VYCOR-BRAND GLASSES.** NORDBERG M E J AM CERAM SOC 27 10 299-305 1944
47957	**PROPERTIES OF BARIUM-STRONTIUM TITANATE DIELECTRICS.** BUNTING E N SHELTON G R CREAMER A S J RESEARCH NATIONAL BUREAU OF STANDARDS 38 337-49 1947
47960	**SILICA AEROGEL. EFFECT OF VARIABLES ON ITS THERMAL CONDUCTIVITY.** WHITE J F IND ENG CHEM 31 7 827-31 1939
47965	**THE EFFECTIVE WAVE-LENGTH OF TRANSMISSION OF RED PYROMETER GLASSES AND OTHER NOTES ON OPTICAL PYROMETRY.** HYDE E P CADY F E FORSYTHE W E ASTROPHYS J 42 294-304 1915
47988	**CONTROL OF REVERSIBLE THERMAL EXPANSION OF PYROPHYLLITE REFRACTORIES BY TALC ADDITIONS.** KENAN, W. M. NORTH CAROLINA STATE COLLEGE 1-19, 1958. (BULLETIN NO. 68)
48002	**PHYSICAL PROPERTIES OF REFRACTORY MATERIALS FOR ROCKET CHAMBER LINERS.** DUWEZ, P. O DELL, F. TAYLOR, J. L. CALIF. INST. OF TECHNOL., JET PROPULSION LAB. 1-70, 1947. (JPL-PR-4-43, ATI-23217)
48014	**HIGH-TEMPERATURE MECHANICAL PROPERTIES OF CERAMIC MATERIALS. V. /ZN/0.65/ MG/0.35//2 SIO4.** BENECKI W T HUMMEL F A J AM CERAM SOC 51 4 220-3 1968 CA 68 107582
48020	**THERMAL CONDUCTIVITY OF SEVERAL COMPOSITE GLASSES.** PAALHORN, O. UNIVERSITY OF JENA, JENA, GERMANY, PH. D. THESIS 1-39 1894
48022	**STUDIES TO THE THERMAL EXPANSION OF SOLIDS AT LOW TEMPERATURES. II. /CR, B-MN, MO, RH, BE, GRAPHITE, TL, ZR, BI, SB, SN AND BERYLL/.** ERFLING H D ANN PHYSIK 34 5 136-60 1939
48024	**CALORIMETRY OF FERRO-MAGNETIC SUBSTANCES.** WEISS P PICCARD A CARRARD A ARCH SCI PHYS ET NAT 43 113-30 1917
48034	**THE TESTING OF THERMAL INSULATORS.** DICKINSON H C VAN DUSEN M S REFRIG ENG 3 2 5-25 1916
48042	**PROPERTIES OF BRICK OF THE MAGNESITE-CHROMITE SERIES.** KONOPICKY K ROUTSCHKA G KOWALCZYK F BER DEUT KERAM GES 44 7 362-5 1967 JA 51 96
48048	**PHASE CHANGES IN SOLIDS MEASURED IN A SOLAR FURNACE— ZRO2-TIO2 SYSTEM.** NOGUCHI T MIZUNO M SOLAR ENERGY 11 1 56-61 1967 JA 51 113
48050	**RESEARCH ON THE THERMAL PROPERTIES OF ZIRCONIA.** BURDICK R B HOSKYNS W R SMITH C F CRANDALL W B MATERIADYNE DIV., PFAUDLER PERMUTIT INC., ALFRED STATION, N. Y. 1-68, 1964. (ARL-64-219, AD-457226)
48082	**THERMOPHYSICAL PROPERTIES OF SLATEBITUMINOUS SHALE.** GUBERGRITZ M YA KUIV K A VOLL M A TEPLO I MASSOPERENOS 7, 73-82, 1966.
48083	**THERMOPHYSICAL PROPERTIES OF PERLITE COMPOUNDS.** MARICHEVSKII I I SPECTOR B V TEPLO I MASSOPERENOS 7, 82-7, 1966.
48085	**A SEMI-EMPIRICAL RELATIONSHIP BETWEEN THE TEMPERATURE AND HUMIDITY OF THE SOIL.** CHUDNOVSKII A F TEPLO I MASSOPERENOS 7, 98-107, 1966.
48105	**INVESTIGATION OF INFRARED TRANSMITTING MATERIALS. PART 3.** KREIDL N J HAFNER H C HENSLER J R WEIDEL R A MATERIALS LAB., WRIGHT-PATTERSON AFB, OHIO 1-132, 1958. (WADC-TR-55-500 / PT 3/, AD-202842)
48109	**AN INVESTIGATION OF INFRARED TRANSMITTING MATERIALS. PART 1.** KREIDL N J HAFNER H C HENSLER J R WEIDEL R A LETTER E C MATERIALS LAB., WRIGHT-PATTERSON AIR FORCE BASE, OHIO 1-150, 1957. (WADC-TR-55-500/PT I/, AD-130904)
48155	**ELECTROCALORIC EFFECTS IN SOME FERROELECTRIC AND ANTIFERROELECTRIC PB/ZR,TI/O3 COMPOUNDS.** THACHER P D J APPL PHYS 39 4 1996-2002 1968
48157	**SPECIFIC HEAT OF CORNING 7560 GLASS BETWEEN 4.5 AND 20 K.** BERG W T J APPL PHYS 39 4 2154 1968
48173	**DEVELOPING THE INDUSTRIAL PROCESS OF LIQUID HARDENING OF HEAT-RESISTANT GLASS.** BOGUSLAVSKII I A PUKHLIK O I RUBINSHIK A M SUDAKOV G V KHALIZEVA O N AKSENOVA E B KOVALENKO N P NEICH A I STEKLO I KERAMIKA 24 8 11-15 1967 (FOR ENGLISH TRANSLATION SEE T48174)
48174	**DEVELOPING THE INDUSTRIAL PROCESS OF LIQUID HARDENING OF HEAT-RESISTANT GLASS.** BOGUSLAVSKII I A PUKHLIK O I RUBINSHIK A M SUDAKOV G V KHALIZEVA O N AKSENOVA E B KOVALENKO N P NEICH A I GLASS AND CERAMICS 24 8 424-7 1967 (ENGLISH TRANSLATION OF STEKLO I KERAMIKA, 24 (8), 11-15, 1967; FOR ORIGINAL SEE T48174)
48177	**USE OF BENTONITE AND QUARTZ PORPHYRY IN PRODUCTION OF FACING TILES WITH A SINGLE FIRING.** KUKOLEV G V ORUDZHEVA N T GLASS AND CERAMICS 24 7 383-6 1967 (ENGLISH TRANSLATION OF STEKLO I KERAMIKA 24 (7), 26-30, 1967; FOR ORIGINAL SEE T45798)
48178	**REALIZATION OF A DEVICE FOR MEASURING THE THERMAL CONDUCTIVITY OF SOLIDS AT LOW TEMPERATURES.** VUILLERMOZ P L PINARD P DAVOINE F REVUE DE PHYSIQUE APPLIQUEE 3 1 11-14 1968
48182	**THERMAL CONDUCTIVITY OF N2 - AR MIXTURES AT HIGH PRESSURES.** PETERSON, J. N. UNIVERSITY OF DELAWARE, M. S. THESIS 1-68 1967
48230	**DEVELOPMENT OF SPACE-STABLE THERMAL-CONTROL COATINGS.** ZERLAUT G A FIRESTONE R F RAZIUNAS V SERWAY R RUBIN G A IIT RESEARCH INSTITUTE, CHICAGO, ILLINOIS 1-42, 1965. (IITRI-U6002-31)
48231	**DEVELOPMENT OF SPACE-STABLE THERMAL-CONTROL COATINGS.** ZERLAUT, G. A. IIT RESEARCH INSTITUTE, CHICAGO, ILLINOIS 1-29, 1966. (IITRI-U6002-42)

TPRC Number	Bibliographic Citation
48237	**SOME PHYSICAL PROPERTIES OF MICA.** HIDNERT P DICKSON G J RESEARCH NATL BUREAU STANDARDS 35 4 309-53 1945 (NBS-RP-1675)
48241	**CERAMICS-THERMAL STRESS RESISTANCE.** EVANS J L NUCLEAR ENGINEERING 6 339-43 1961
48245	**ON THE THERAML DIFFUSIVITIES OF DIFFERENT KINDS OF MARBLE.** PEIRCE B O WILLSON R W PROC AM ACAD ARTS SCIENCES 36 2 13-16 1900
48263	**HEAT FLOW AND WATER TEMPERATURE FLUCTUATIONS IN THE DENMARK STRAIT.** LACHENBRUCH A H MARSHALL B V J GEOPHYSICAL RESEARCH 73 18 5829-42 1968
48291	**NEW INSULATION STUDIES. PART II.** QUEER E R HECHLER F G REFRIG ENG 35 247-52 1938
48292	**VARIATION OF THERMAL CONDUCTIVITY WITH TEMPERATURE OF INSULATING POWDERS.** GLASER P E KAYAN C F REFRIG ENG 64 3 31-6 1956
48294	**HEAT FLOW IN THE SOUTHERN KARROO.** GOUGH D I PROC ROY SOC /LONDON/ 272 A 207-30 1963
48311	**REFLECTIVITY OF BITUMINOUS COAL. I. DESCRIPTION OF PREPARATION OF POLISHED SURFACES REFLECTION APPARATUS, AND CALIBRATION OF THE APPARATUS.** DE VRIES H A W HABETS P J BOKHOVEN C BRENNST-CHEM 49 1 15-21 1968 CA 68 61570
48320	**FEATURES OF SPECTRAL REFLECTIVITY OF GLASS AND QUARTZ IN THE NEAR INFRARED REGION OF THE SPECTRUM.** REKANT N B GELIOTEKH AKAD NAUK UZB SSR 4 38-41 1967 CA 68 64056
48324	**THE CONDUCTIVITY OF HEAT INSULATORS. PART A.** LAMB C G WILSON W G PROC ROYAL SOC 65 283-8 1899
48335	**STUDY OF THE HEAT ENGINEERING PROPERTIES OF CHECKER REFRACTORIES.** FRENKEL A S ANTONOV G I OGNEUPORY 32 8 20-5 1967 CA 67 119868 (FOR ENGLISH TRANSLATION SEE T48336)
48336	**STUDY OF THE HEAT ENGINEERING PROPERTIES OF CHECKER REFRACTORIES.** FRENKEL A S ANTONOV G I REFRACTORIES 8 474-8 1967 (ENGLISH TRANSLATION OF OGNEUPORY (8), 20-25, 1967; FOR ORIGINAL SEE T48335)
48341	**PART 1. DEVELOPMENT OF CERAMIC BODIES WITH HIGH THERMAL CONDUCTIVITY. PART 2. HIGH THERMAL SHOCK CERAMICS.** RUTGERS UNIVERSITY CERAMIC RESEARCH STATION, N. J. NEW JERSEY CERAMIC RESEARCH STATION, RUTGERS UNIV. 1-55, 1952. (AD 167131)
48342	**SPECIFIC HEAT OF REFRACTORIES AT HIGH TEMPERATURES.** LOBANOVA Z OGNEUPORY 1 17-23 1939
48353	**THERMAL DIFFUSIVITY OF COAL SHAPES AND THEIR THERMALLY TREATED PRODUCTS.** KUSAKIN N D SALNIKOV A P BARANOV V I TRUDI IN-TA GEOL I RAZRABOTKI GORYUCHIKH ISKOPAEMYKH AKAD NAUK SSSR 10 196-200 1964
48360	**LOW-TEMPERATURE SPECTRAL EMITTANCE MEASUREMENTS.** STIERWALT D L PROGR ASTRONAUT AERONAUT 18 21-31 1966
48361	**TOTAL NORMAL EMITTANCE MEASUREMENTS TO 2200 C IN AIR.** HEDGE J C PROGR ASTRONAUT AERONAUT 18 33-46 1966 CA 69 14473
48364	**THERMAL RADIATION PROPERTIES OF BINARY MIXTURES.** SCHATZ E A PROGR ASTRONAUT AERONAUT 18 75-100 1966
48369	**EFFECT OF SURFACE PROPERTIES ON PLANETARY ALBEDO.** COULSON K L PROGR ASTRONAUT AERONAUT 18 219-37 1966
48370	**FAR-INFRARED STUDIES OF SILICATE MINERALS.** ARONSON J R MC LINDEN H G PROGR ASTRONAUT AERONAUT 18 291-309 1966
48383	**PERFORMANCE OF MULTILAYER INSULATION SYSTEMS FOR THE 300 TO 800 K TEMPERATURE RANGE.** STREED E R CUNNINGTON G R JR ZIERMAN C A PROGR ASTRONAUT AERONAUT 18 735-71 1966
48387	**AN ACCURATE METHOD FOR THE DETERMINATION OF THE THERMAL CONDUCTIVITY OF INSULATING SOLIDS.** MISCHKE C R FARBER E A AMERICAN SOCIETY OF MECHANICAL ENGINEERS 1-6, 1953. (ASME-53-A-185)
48397	**THERMODYNAMIC CONSTANTS OF PYROPHYLLITE.** FONAREV V I GEOKHIMIYA 12 1505-8 1967 CA 68 72959 (FOR ENGLISH TRANSLATION SEE T70865)
48428	**PREPARATION OF URANIA-PLUTONIA-ZIRCONIA FUELS.** CALDWELL C S PUECHL K H FISHER F D ROTH J PROCEEDINGS INTERNATIONAL CONFERENCE PLUTONIUM, 3RD 654-68 1967 CA 68 74388
48431	**SOME STUDIES OF THE HIGH TEMPERATURE BEHAVIOR OF PLUTONIA AND PLUTONIA-URANIA MIXTURES.** BERGGREN G FORSYTH R S PROCEEDINGS INTERNATIONAL CONFERENCE PLUTONIUM, 3RD 828-34 1967 CA 68 74392
48434	**DEVELOPMENT OF 400 TO 2200 F. FIBROUS-TYPE INSULATIONS FOR RADIOISOTOPE POWER SYSTEMS.** COLLINS J O JUANARAJS K L REID D R JOHNS-MANVILLE RES. AND ENG. CTR., MANVILLE, N. J. 1-56, 1968. (ALO-3633-10)
48442	**REFRACTORY CONCRETE WITH LOW THERMAL EXPANSION.** GUGEL E SPRECHSAAL KERAM GLAS EMAIL SILIKATE 100 21 825-30 1967 CA 68 62411
48447	**PHYSICOCHEMICAL STUDY OF FRITS OF TITANIUM OPACIFIED FAIENCE GLAZES DURING AN ALTERATION OF THE CALCIUM OXIDE CONTENT.** AZIMOV I IRKAKHODZHAEVA A P UZB KHIM ZH 11 6 24-6 1967 CA 68 71830
48451	**THERMAL CONDUCTIVITIES OF SEVERAL METALS AND NON-METALS FROM 200 TO 1300 C. BY THE RADIAL HEAT-FLOW TECHNIQUE.** MOELLER, C. E. WILSON, D. R. PROCEEDINGS OF THE CONFERENCE ON THERMAL CONDUCTIVITY, 3RD 224-51 1963
48452	**THERMAL EXPANSION OF ALKALINE REFRACTORIES.** SZUMAKOWICZ, J. STANIA, B. ZESZ. NAUK. AKAD. GORN.-HUTN. KRAKOWIE, CERAM. 8, 131-44, 1967.
48459	**CRYSTALLIZATION AND THERMAL EXPANSION OF GLASSES IN THE LEAD OXIDE-BORIC OXIDE-ZINC OXIDE SYSTEM.** ZHDANETS N A KHEIFETS V S ZH PRIKL KHIM 40 11 2418-22 1967 CA 68 71764 (FOR ENGLISH TRANSLATION SEE T59338)
48465	**FILTERS, BASED ON BARIUM TITANATE SINGLE CRYSTALS, FOR THE FAR INFRARED REGION.** VERBITSKAYA, T. N. KISLOVSKII, L. D. GALANOV, E. K. SOKOLOVA, L. S. ZH. PRIKL. SPEKTROSK. 7 (2), 276-7, 1967. (FOR ENGLISH TRANSLATION SEE TPRC NO. 67144)
48472	**PRECISION THERMAL EXPANSION MEASUREMENTS ON LOW EXPANSION OPTICAL MATERIALS.** PLUMMER W A HAGY H E APPL OPT 7 5 825-31 1968
48473	**LUMINOUS TRANSMITTANCE, AND CHROMATICITY OF COLORED FILTER GLASSES IN CIE 1964 UNIFORM COLOR SPACE.** WERNER A J APPL OPT 7 5 849-55 1968
48492	**DEVELOPMENT OF SPACE-STABLE THERMAL-CONTROL COATINGS.** ZERLAUT, G. A. RUBIN, G. A. IIT RESEARCH INSTITUTE, CHICAGO, ILLINOIS 1-46, 1966. (IITRI-U6002-36)

TPRC Number	Bibliographic Citation
48511	**METHODS OF PRODUCTION OF LIGHT WEIGHT REFRACTORIES AND HIGHLY RESISTANT PRODUCTS . GENERAL CHARACTERISTICS OF LIGHT WEIGHT REFRACTORIES AND REQUIREMENTS, SET FORTH FOR THEM FROM THE SIDE OF THE CONSUMER.** PIROGOV A A TRUDY I SOVESHCHANIYA PO OGNEUPORNYM MATERIALAM PRI OTDELENII TEKHNICHESKIKH NAUK AKADEMII NAUK 218-30 1940
48518	**PROBLEMS IN THE NONFERROUS INDUSTRIES, AND A NOTE ON ESTIMATION OF HEAT CAPACITY.** KELLOGG H H APPL FUNDAM THERMODYN MET PROCESSES PROC CONF THERMODYN PROPERTIES MATER 1ST 357-66 1967 CA 68 62190
48521	**THERMAL BEHAVIOR OF SILICA AND ITS APPLICATION TO DENTAL INVESTMENTS. II. THE EFFECT OF GRAIN SIZE ON THE THERMAL EXPANSION OF QUARTZ.** JONES, D. W. BRIT. DENT. J. 122, 146-9, 1967.
48527	**A THERMAL STUDY OF THE CERAMIC MINERALS OF TAIWAN.** CHENG, T. Y. LIAW, C. T. KUNG CH'ENG 40 (7), 8-16, 1967.
48531	**EVALUATION OF THE MIANWALI FLINT CLAY DEPOSITS FOR THE MANUFACTURE OF FLINT CLAY BRICKS.** QURESHI M H ARIFF M R SCI IND /KARACHI/ 5 1 86-91 1967 CA 68 71795
48541	**THE REFLECTIVITY OF VITRINITE AND METHODS FOR DETERMINING IT.** KALMYKOV, G. S. PETROLOGIYA UGLEI PARAGENEZ GORYUCH. ISKOP., AKAD. NAUK SSSR, INST. GEOL. RAZRAB. GORYUCH. ISKOP., LAB. PARAGENEZA KAUSTOBIOLITOV 81-126, 1967.
48543	**KAZAKHSTAN VITROPHYRES AS A RAW MATERIAL FOR PRODUCTION OF HIGH QUALITY CHEMICALLY STABLE GLASSES.** SULEIMENOV S T SHARAFIEV M SH ABDUVALIEV T A TR INST GEOL NAUK AKAD NAUK KAZ SSR 24 158-9 1967 CA 68 71760
48545	**DETERMINATION OF THERMAL CONDUCTIVITY IN ELECTRIC INSULATION MATERIALS MADE BY THE ALL-UNION ELECTRICAL INSTITUTE (VEI).** DEREVNINA, T. O. TR. VSES. ELEKTROTEKH. INST. (74), 208-17, 1966.
48550	**STUDIES ON THE THERMODYNAMICS AND CONDUCTANCES OF MOLTEN SALTS AND THEIR MIXTURES. PART VI. CALORIMETRIC STUDIES OF SODIUM CHLORATE AND ITS MIXTURES WITH SODIUM NITRATE.** CAMPBELL A N VAN DER KOUWE E T CAN J CHEM 46 8 1287-91 1968 CA 68 117653
48619	**THE EFFECT OF PHASE COMPOSITION OF ZIRCONIUM DIOXIDE ON SPALLING RESISTANCE.** KARAULOV A G GREBENYUK A A RUDYAK I N IZV AKAD NAUK SSSR NEORG MATER 3 6 1101-3 1967 (FOR ENGLISH TRANSLATION SEE T48620)
48620	**THE EFFECT OF PHASE COMPOSITION OF ZIRCONIUM DIOXIDE ON SPALLING RESISTANCE.** KARAULOV A G GREBENYUK A A RUDYAK I N INORGANIC MATERIALS 3 6 982-4 1967 (ENGLISH TRANSLATION OF IZV. AKAD. NAUK SSSR, NEORG. MATER. 3 (6), 1101-3, 1967; FOR ORIGINAL SEE T48619)
48623	**ELECTRICAL PROPERTIES AND STRUCTURE OF LITHIUM, SODIUM, AND POTASSIUM SILICATE GLASSES.** LEKO V K IZV AKAD NAUK SSSR NEORG MATER 3 7 1224-9 1967 (FOR ENGLISH TRANSLATION SEE T48624)
48624	**ELECTRICAL PROPERTIES AND STRUCTURE OF LITHIUM, SODIUM, AND POTASSIUM SILICATE GLASSES.** LEKO V K INORGANIC MATERIALS 3 7 1079-83 1967 (ENGLISH TRANSLATION OF IZV. AKAD. NAUK SSSR, NEORG. MATER. 3 (7), 1224-9, 1967; FOR ORIGINAL SEE T48623)
48632	**FURNACE SLAG CRUST HEAT CONDUCTIVITY MEASUREMENTS.** BASSA G REMENYI K ACTA TECH ACAD SCI HUNG 59 3/4 331-46 1967 CA 68 88876
48659	**NATURE OF THE DOUBLE ALKALI EFFECT.** ALEINIKOV F K VAITKUS J ZITKEVICIUTE I LIET TSR MOKSLU AKAD DARB B 2 75-88 1967 CA 68 90815
48685	**THEORETICAL AND TECHNICAL VIEWPOINTS CONCERNING NEW GLASSY PLASMA-SPRAYED COATINGS.** KRAUTH, A. MEYER, H. MITT. VER. DEUT. EMAILFACHLEUTE 14 (1), 1-4, 1966.
48686	**CHEMICAL PURIFICATION OF IZUMIYAMA POTTERY STONE AND ITS EFFECTS.** TAKAGI H HAYASHI H NAGOYA KOGYO GIJUTSU SHIKENSHO HOKOKU 16 8 266-71 1967 CA 68 89535
48695	**EXPANSION OF CHASOV-YARSK FIRE BRICK AT HIGH TEMPERATURES.** BUDNIKOV P P MYULLER V E FOREIGN TECHNOLOGY DIVISION 1-13, 1671. (ENGLISH TRANSLATION OF DOMEZ (12), 11-14, 1931; FOR ORIGINAL SEE T60852)
48703	**SOME THERMOPHYSICAL PROPERTIES OF MOLDED CRYSTALLINE GLASS MATERIALS.** BYKOV I I KHAN B KH KLIMENKO V S TEPLOFIZ VYS TEMP 5 6 1005-10 1967 CA 68 89525 (FOR ENGLISH TRANSLATION SEE T49415)
48707	**RELATIONS AMONG COMPOSITION, STRUCTURE, AND SOME PROPERTIES OF GLASSES IN BARIUM OXIDE-LEAD OXIDE-SILICON DIOXIDE AND MAGNESIUM OXIDE-ALUMINUM OXIDE-SILICON DIOXIDE SYSTEMS.** SHELUDYAKOV L N VESTN AKAD NAUK KAZ SSR 23 12 43-8 1967 CA 68 89524
48739	**EFFECT OF A CONSTANT ELECTRIC FIELD ON THE THERMAL CONDUCTIVITY OF CERAMIC SAMPLES OF SOLID SOLUTIONS BASED ON BARIUM TITANATE AND OF A TRIGLYCINE SULFATE SINGLE CRYSTAL IN THE REGION OF PHASE TRANSITIONS.** DIMAROVA E N POPLAVKO YU M IZV AKAD NAUK SSSR SER FIZ 31 11 1842-4 1967 CA 68 99376 (FOR ENGLISH TRANSLATION SEE T36035)
48754	**STRUCTURAL PLASTICS APPLICATIONS HANDBOOK.** JUREVIC, W. G. RITTENHOUSE, J. B. LOCKHEED RESEARCH LAB., PALO ALTO 1-258, 1968. (AFML-TR-67-332)
48767	**DESIGN OF CLAY BODIES FOR CONTROLLED MICROSTRUCTURE. PART II. CLAY-AGGREGATE BODIES.** ROBINSON G C AM CERAM SOC BULL 47 6 548-53 1968
48788	**EFFECT OF GLAZE FILM ON PROPERTIES OF A VITREOUS CHINA BODY.** BELL W C KOENIG J H J AM CERAMIC SOC 24 11 341-8 1941
48802	**PROUSTITE AND PYRARGYRITE AS OPTICAL MATERIALS FOR THE INFRARED.** GUSEVA L M GANEEV I G DRONOV A V REZ I S OPTIKA I SPEKTROSKOPIYA 24 2 298-300 1968 CA 68 109608 (FOR ENGLISH TRANSLATION SEE T49364)
48879	**DEGREE-DAYS AND HEAT CONDUCTION IN SOILS.** SANGER F J PROC. PERMAFROST INTERNATIONAL CONFERENCE 253-62, 1963. (NATL. ACAD. SCI.-NATL. RES. COUNCIL, WASHINGTON, D. C. PUBL. NO. 1287)
48880	**MEASUREMENTS BY THE INTERFEROMETRIC METHOD ON THE THERMAL EXPANSION OF JENA GLASS 2954-III DOWN TO 4 K.** KEESOM W H DOBORZYNSKI D W PHYSICA 1 1089-102 1934 (COMM. KAM. ONNES LAB., LEIDEN, 234B)
48881	**THERMAL PROPERTIES OF FROZEN GROUND.** KERSTEN M S PROC. PERMAFROST INTERNATIONAL CONFERENCE 301-5, 1963. (NATL. ACAD. SCI.-NATL. RES. COUNCIL, WASHINGTON, C. C. PUBL. NO. 1287)
48883	**A METHOD OF MEASURING THE THERMAL DIFFUSIVITY OF CONSOLIDATED MATERIALS AT ELEVATED TEMPERATURES.** BOOZER, G. D. UNIVERSITY OF CALIFORNIA, BERKELEY, M. S. THESIS 1-54 1958
48890	**THE THERMAL CONDUCTIVITY OF A POLYCRYSTALLINE ROD OF NICKEL-ZINC FERRITE AT LIQUID HELIUM TEMPERATURES.** CHARI, M. S. R. NATARAJAN, N. S. PROC. OF THE NUCLEAR PHYSICS AND SOLID STATE PHYSICS SYMPOSIUM 1-3, 1967. (N68-20501, N68-20502)

TPRC Number	Bibliographic Citation
48904	IRRADIATION TESTING OF CERAMIC FUELS. ZUROMSKY, G. CHERNOCK, W. P. NUCLEAR DIVISION COMBUSTION ENGINEERING, INC., WINDSOR, CONNECTICUT 1-188, 1962. (CEND-158, NYO-2998)
48921	HEAT TRANSFER THROUGH A THREE-PHASE POROUS MEDIUM. CHAUDHARY D R BHANDARI R C BRIT J APPL PHYS 1 6 815-17 1968
48968	MAGNETIC AND ELECTRICAL PROPERTIES OF ORDERED PEROVSKITE SR2/FE MO/O6 AND ITS RELATED COMPOUNDS. NAKAGAWA T J PHYS SOC JAPAN 24 4 806-11 1968
48982	COMBINED THERMODILATOMETRIC AND DERIVATOGRAPHIC EXAMINATION OF HYDRARGILLITE AND BARIUM CHLORIDE DIHYDRATE. PAULIK, F. PAULIK, J. ERDEY, L. ANAL. CHIM. ACTA 41 (1), 170-2, 1968.
48993	CERAMIC DIELECTRIC SANDWICH COMPONENTS. AUDA, D. BOEING COMPANY, SEATTLE, WASHINGTON 1-56, 1962. (AD 282903, BOEING DOC. NO. D2-7936-5)
48997	BISMUTH SULFIDE TELLURIDES OF NORTHEAST YAKUTIA. GAMYANIN G N DOKL AKAD NAUK SSSR 178 3 679-82 1968 CA 68 80291
49008	BOROSILICATE GLASSES AND MODERN GLASS APPARATUS CONSTRUCTION. KLEINTEICH R GLAS-EMAIL-KERAMO-TECH 18 12 424-8 1967 CA 68 81026
49012	THERMAL EXPANSION AND VOLUME STABILITY OF BADDELEYITE - CORUNDUM REFRACTORIES. RUZEK, J. SILIKATY 12 (1), 31-41, 1968.
49015	DEPENDENCE OF THE AMOUNT OF CRYSTALLINE PHASE ON THE HEAT-TREATING CONDITIONS OF SITALLS IN THE SIO2-AL2O3-BAO-TIO2 SYSTEM. BOGDANOVA G S ORLOVA E M ZEVIN L S IZV AKAD NAUK SSSR NEORGAN MATERIALY 2 2 380-3 1966 JA 51-6 185 (FOR ENGLISH TRANSLATION SEE T36253)
49023	A LINE HEAT SOURCE METHOD FOR THE DETERMINATION OF THERMAL CONDUCTIVITY OF POWDERS IN VACUUM. WECHSLER, A. E. GLASER, P. E. PROCEEDINGS OF THE CONFERENCE ON THERMAL CONDUCTIVITY, 3RD 613-27 1963
49026	THERMAL EXPANSION OF ROMANIAN REFRACTORIES. POMMER E VEISER I SZABO A METALURGIA /BUCHAREST/ 19 8 436-9 1967 CA 68 81050
49047	EQUILIBRIUM REACTIONS AND DIELECTRIC CHARACTERISTICS OF COMPOSITIONS NEAR THE FIELD OF PRIMARY CRYSTALLIZATION OF CORDIERITE. CINI L SILICATES IND 32 6 220-4 1967 JA 51-6 176
49049	ANOMALOUS THERMAL EXPANSION OF ALPHA-FERRIC OXIDE. SAITO T TOKYO KOGYO SHIKENSHO HOKOKU 62 4 119-22 1967 JA 51-6 183
49052	PREPARATION OF SITALLS BASED ON TYRNAUZ TAILINGS. KITAIGORODSKII, I. I. PETROV, S. V. BEUS, M. D. TR. MOSK. KHIM.- TEKHNOL. INST. (50), 40-6, 1966.
49059	RELATION OF COEFFICIENT OF EXPANSION TO IMPACT RESISTANCE OF PORCELAIN ENAMELS. FELLOWS R L WHEELER P M J AM CERAM SOC 24 11 356-60 1941
49064	INVESTIGATIONS ON SODIUM PHOSPHATE GLASSES. FANDERLIK, M. PALECEK, M. SILIKAT J. 5 (7), 127-30, 1966.
49087	THE REFLECTANCE AND COKING BEHAVIOR OF VITRINITE-SEMIFUSINITE TRANSITION MATERIAL. BENNETT A J R FUEL /LONDON/ 47 1 51-62 1968 CA 68 106700
49097	EFFECTIVE THERMAL CONDUCTIVITY OF COMPOSITE SOLIDS. FIDELLE, T. P., JR. UNIVERSITY OF MASSACHUSETTS, PH.D. THESIS 82PP., 1969. (UNIV. MICROFILM NO. 70-13208, N71-26469)
49113	HEAT CAPACITY OF THREE INORGANIC GLASSES AND SUPER-COOLED LIQUIDS. HAGGERTY, J. S. COOPER, A. R. HEASLEY, J. H. PHYS. CHEM. GLASSES 9 (2), 47-51, 1968.
49128	PROPERTIES OF WARE VERSUS THERMAL HISTORY. II. TESTS ON SEMIVITREOUS CHINA BODIES AND WARE. KOENIG J H J AM CERAMIC SOC 25 2 41-8 1942
49133	X-RAY DIFFRACTION CHARACTERISTICS OF THE NORILSK COOPERITE. ZHURAVLEV N N GENKIN A D STEPANOVA A A ZAP VSES MINERAL OBSHCHEST 97 1 85-8 1968 CA 68 106742
49147	SPECTRAL REFLECTION CAPACITY OF PIGMENTS BASED ON CHROMIUM TITANATE. TROPIN, V. A. FIZ. SVOISTVA STRUKT. NEKOT. ORG. NEORG. VESHCHESTV. 3-8, 1966.
49153	QUARTERLY PROGRESS REPORT IN THERMAL INSULATION STUDY. DE WITT, W. D. REID, R. L. UNION CARBIDE CORPORATION, TONAWANDA, N. Y. 1-115, 1967. (ALO-3632-17)
49154	THE INFLUENCE OF EXPOSURE TO THERMAL NEUTRONS ON SEVERAL PROPERTIES OF HEAT-RESISTANT SITALLS. BREKHOVSKIKH S M GRINSHTEIN YU L IZV AKAD NAUK SSSR NEORGAN MATERIALY 1 6 947-51 1965 (FOR ENGLISH TRANSLATION SEE T49155)
49155	THE INFLUENCE OF EXPOSURE TO THERMAL NEUTRONS ON SEVERAL PROPERTIES OF HEAT-RESISTANT SITALLS. BREKHOVSKIKH S M GRINSHTEIN YU L FOREIGN TECHNOLOGY DIVISION 1-10, 1968. (ENGLISH TRANSLATION OF IZV. AKAD. NAUK SSSR, NEORGAN. MATERIALY 1 (6), 947-51, 1965; FOR ORIGINAL SEE T49154) (FTD-HT-23-1348-67, AD 676175)
49179	CORROSION OF REFRACTORIES BY BLAST FURNACE SLAGS. MILLER W A SHOTT W L AM CERAM SOC BULL 47 7 648-53 1968
49184	THERMAL EXPANSION AND ELASTIC PROPERTIES OF CERAMIC BODIES. HUNTER, O., JR. ALFRED UNIVERSITY, ALFRED, NEW YORK, PH. D. THESIS 1-60, 1964. (UNIV. MICROFILM NO. 64-12268)
49193	THE EFFECT OF POROSITY ON SHEARING RESISTANCE AND THERMAL CONDUCTIVITY FOR AMORPHOUS SOILS IN VACUUM. WATERS, R. H. TEXAS A AND M UNIV., COLLEGE STATION, PH. D. THESIS 1-159, 1967. (UNIV. MICROFILM NO. 67-9816)
49199	NONIDEAL MIXING IN BINARY GEO2-SIO2 GLASSES. RIEBLING E F J AM CERAM SOC 51 7 406-7 1968 CA 69 45750
49210	THE THERMAL COMPARATOR IN NON-DESTRUCTIVE TESTING. POWELL R W TYE R P TECHNIQUES OF NON-DESTRUCTIVE TESTING 175-89 1960
49232	DETERMINATION OF THE SPECIFIC HEAT OF COALS DURING CARBONIZATION. AGROSKIN A A GONCHAROV E I COKE CHEM /USSR/ 11 16-20 1965
49233	INFRARED COATING STUDIES. MAIER, R. L. BAUSCH AND LOMB, INC., ROCHESTER, NEW YORK 1-16, 1967. (AD 664786, N68-20132)
49236	OPTICAL PROPERTIES OF GLASS. TRANSMISSION LOSSES OF EYEPIECES USED IN MINE ENVIRONMENT. REINESS, C. G. MARANO, C. L. BUREAU OF MINES, PITTSBURGH, PA. 1-9, 1968. (BM-RI-7062, N68-13258)

TPRC Number	Bibliographic Citation
49237	**THERMODYNAMIC PROPERTIES OF FORSTERITE AND SERPENTINE.** KING E G BARANY R WELLER W W PANKRATZ L B BUREAU OF MINES, BERKELEY, CALIFORNIA 1-19, 1967. (BM-RI-6962, N67-31801)
49249	**THERMAL CONDUCTIVITY OF SEVERAL CERAMIC MATERIALS TO 2500 C.** FEITH, A. D. NUCLEAR MATERIALS AND PROPULSION OPERATION, CINCINNATI, OHIO 1-17, 1964. (GE-TM-64-10-4, CONF-764-11)
49251	**DIRECTIONAL DISTRIBUTION IN THE REFLECTION OF HEAT RADIATION AND ITS EFFECT ON HEAT TRANSFER.** MUNCH, B. SWISS TECHNICAL COLLEGE (ZURICH), PH. D. THESIS 1-90, 1968. (ENGLISH TRANSLATION OF E. T. H. ZURICH (2434) 89PP., 1955; FOR ORIGINAL SEE T9046) (NASA-TT-F-497, N68-21536)
49258	**THE THERMAL PROPERTIES OF GLASSES AT LOW TEMPERATURES.** LEADBETTER, A. J. PHYS. CHEM. GLASSES 9 (1), 1-13, 1968.
49276	**AUTOMATIC RECORDING OF TEMPERATURE CHARACTERISTICS OF THE TRUE COEFFICIENT OF THERMAL EXPANSION.** MAZO R I REFRACTORIES 1 17-20 1968 (ENGLISH TRANSLATION OF OGNEUPORY (1), 16-19, 1968; FOR ORIGINAL SEE T50669)
49277	**EFFECT OF PHASE COMPOSITION ON THE THERMAL SHOCK RESISTANCE OF ZIRCONIA.** GREBENYUK A A KARAULOV A G DAUKNIS V I PRANTSKYAVICHYUS G A YURENAS V L REFRACTORIES 1 44-51 1968 (ENGLISH TRANSLATION OF OGNEUPORY (1), 42-9, 1968; FOR ORIGINAL SEE T50670)
49278	**VISIBLE AND ULTRA-VIOLET EMISSION AND ABSORPTION SPECTRA OF MG AL2 O4.CR.** LOU F H BALLENTYNE D W G PROC PHYS SOC LONDON SOLID STATE PHYS 1 C 3 608-13 1968
49290	**INFRARED AIRCRAFT SPECTRA OVER DESERT TERRAIN 8.5 MICRONS TO 16 MICRONS.** HOVIS W A JR BLAINE L R CALLAHAN W R APPL OPT 7 6 1137-40 1968
49293	**TRANSMITTANCE OF CER-VIT GLASS-CERAMIC IN THE ULTRAVIOLET, VISIBLE, INFRARED, AND SUBMILLIMETER WAVELENGTH REGIONS.** FULLER R M RATHBUN D G BELL R J APPL OPT 7 6 1243-4 1968
49308	**THE INTERFACE BETWEEN ICE AND SILICATE SURFACES.** ANDERSON, D. M. COLD REGIONS RESEARCH AND ENGINEERING LAB., HANOVER, NEW HAMPSHIRE 1-31, 1967. (CRREL-RR-219, AD 653612)
49332	**THE USE OF ALUMINUM OXIDE IN ELECTRICAL AND RADIO CERAMICS.** BUDNIKOV P P MASLENNIKOVA G N IZV AKAD NAUK SSSR NEORG MATER 3 10 1862-9 1967 (FOR ENGLISH TRANSLATION SEE T49333)
49333	**THE USE OF ALUMINUM OXIDE IN ELECTRICAL AND RADIO CERAMICS.** BUDNIKOV P P MASLENNIKOVA G N INORGANIC MATERIALS 3 10 1622-8 1967 (ENGLISH TRANSLATION OF IZV. AKAD. NAUK SSSR, NEORG. MATER. 3 (10), 1862-9, 1967; FOR ORIGINAL SEE T49332)
49336	**STRUCTURE AND PROPERTIES OF METASTABLE QUARTZITIC PHASES IN SITALLS IN THE SIO2-AL2O3-MGO-TIO2 SYSTEM.** KHODAKOVSKAYA R YA PAVLUSHKIN N M IZV AKAD NAUK SSSR NEORG MATER 3 10 1908-15 1967 (FOR ENGLISH TRANSLATION SEE T49337)
49337	**STRUCTURE AND PROPERTIES OF METASTABLE QUARTZITIC PHASES IN THE SIO2-AL2O3-MGO-TIO2 SYSTEM.** KHODAKOVSKAYA R YA PAVLUSHKIN N M INORGANIC MATERIALS 3 10 1662-8 1967 (ENGLISH TRANSLATION OF IZV. AKAD. NAUK SSSR, NEORG. MATER. 3 (10), 1908-15, 1967; FOR ORIGINAL SEE T49336)
49346	**A STUDY OF THE ION-EXCHANGE PROPERTIES OF ALKALINE EARTH METAL SILICATES, ALUMINATES, AND FERRITES, AND INDUSTRIAL WASTES CONTAINING THESE MATERIALS.** SHEVYAKOV A M FEDOROV N F IZV AKAD NAUK SSSR NEORG MATER 3 11 2088-91 1967 (FOR ENGLISH TRANSLATION SEE T49347)
49347	**A STUDY OF THE ION-EXCHANGE PROPERTIES OF ALKALINE EARTH METAL SILICATES, ALUMINATES, AND FERRITES, AND INDUSTRIAL WASTES CONTAINING THESE MATERIALS.** SHEVYAKOV A M FEDOROV N F INORGANIC MATERIALS 3 11 1815-17 1967 (ENGLISH TRANSLATION OF IZV. AKAD. NAUK SSSR, NEORG. MATER. 3 (11), 2088-91, 1967; FOR ORIGINAL SEE T49346)
49364	**PROUSTITE AND PYRARGYRITE AS OPTICAL MATERIALS FOR THE INFRARED.** GUSEVA L M GANEEV I G DRONOV A V REZ I S OPT SPECTRY 24 2 156-7 1968 (ENGLISH TRANSLATION OF OPTIKA I SPEKTROSKOPIYA 24 (2), 298-300, 1968; FOR ORIGINAL SEE T48802)
49388	**HEAT CAPACITY OF LEAD CONCENTRATES AND SINTERS AT HIGH TEMPERATURES.** BRATCHIKOV, S. G. FOREIGN TECH. DIV., WRIGHT PATTERSON AIR FORCE BASE, OHIO 105-8, 1967. (ENGLISH TRANSLATION OF IZV. VYS. UCHEBN. ZAVED. TSVETN. MET. 7 (4), 79-82, 1964; FOR ORIGINAL SEE T38203) (FTD-MT-65-20, N68-23828, AD-666896)
49389	**DIFFERENTIAL SCANNING CALORIMETRY METHODS IN THE DETERMINATION OF THERMAL PROPERTIES OF EXPLOSIVES.** WILCOX, J. D. AIR FORCE INST. OF TECH., WRIGHT-PATTERSON AIR FORCE BASE, OHIO, M. S. THESIS 1-70, 1967. (AD 818369)
49390	**PHYSICAL CHARACTERIZATION OF ELECTRONIC MATERIALS, DEVICES AND THIN FILMS.** PETERS, E. T. MAN LABS., INC., CAMBRIDGE, MA. 1-44, 1968. (AFCRL-68-0091, AD 667547)
49395	**WINDOW AND WINDOW SCREENS AS MODIFIERS OF THERMAL RADIATION RELEASED IN NUCLEAR DETONATIONS.** BRACCIAVENTI, J. NAVAL APPLIED SCIENCE LAB., BROOKLYN, N. Y. 1-24, 1966. (AD 643019)
49396	**COEFFICIENT OF LINEAR EXPANSION AT LOW TEMPERATURES.** DORSEY H G PHYS REV 25 88-102 1907
49400	**THERMOPHYSICAL PROPERTIES OF MATERIALS.** GALKIN, M. N. TARASUTIN, T. G. PUSHKIN, I. L. TR MOSK AVIATS TEKHNOL INST 58 81-99 1963 (FOR ENGLISH TRANSLATION SEE T49401)
49401	**THERMOPHYSICAL PROPERTIES OF MATERIALS.** GALKIN M N TARASUTIN T G PUSHKIN I L FOREIGN TECH. DIV., WRIGHT-PATTERSON AFB, OHIO 73-90, 1967. (ENGLISH TRANSLATION OF TR. MOSK. AVIATS. TEKHNOL. INST., (58), 81-99, 1963; FOR ORIGINAL SEE T49400) (FTD-MT-64-257, N68-27362, AD-668233)
49403	**HIGH TEMPERATURE HEAT CONTENTS AND ENTROPIES OF DEHYDRATED ANALCITE, KALIOPHILITE, AND LEUCITE.** PANKRATZ, L. B. BUREAU OF MINES, BERKELEY, CA. 1-13, 1968. (BM-RI-7073, N68-14936)
49405	**INFLUENCE OF ANION VACANCIES ON THE LATTICE THERMAL CONDUCTIVITY OF COORDINATION CRYSTALS HAVING THE SPINEL STRUCTURE AT LOW TEMPERATURE.** VISHNEVSKII I I SKRIPAK V N SOVIET PHYSICS-SOLID STATE 9 12 2865-6 1968 (ENGLISH TRANSLATION OF FIZ. TVERD. TELA 9 (12), 3633-5, 1967; FOR ORIGINAL SEE T47476)
49415	**SOME THERMOPHYSICAL PROPERTIES OF CAST VITREOUS-CRYSTALLINE MATERIALS.** BYKOV I I KHAN B KH KLIMENKO V S HIGH TEMPERATURE 5 6 897-901 1967 (ENGLISH TRANSLATION OF TEPLOFIZ. VYS. TEMPERATUR 5 (6), 1005-10, 1967; FOR ORIGINAL SEE T48703)

TPRC Number	Bibliographic Citation
49491	**ZIRCONIA-MAGNESIA SPINEL SYSTEM.** MATHER R F J AM CERAM SOC 25 3 93-6 1942
49504	**DETERMINATION OF THERMAL PROPERTIES OF LOOSE MATERIALS AND THICK BEDS OF VARIOUS MATERIALS.** MEDVEDEV, N. N. INZH. - FIZ. ZH. 14 (2), 329-33, 1968. (FOR ENGLISH TRANSLATION SEE TPRC NO. 65123)
49510	**REFLECTANCE OF BITUMINOUS COAL. III. EFFECT OF TEMPERATURE.** DE VRIES H A W HABETS P J BOKHOVEN C BRENNST CHEM 49 4 105-10 1968 CA 69 4235
49525	**THE THERMODYNAMICS OF THE MAGNESIUM SILICATE/ MAGNESIUM-SILICON-OXYGEN VAPOR SYSTEM.** KRIEGER, F. J. RAND CORP., SANTA MONICA, CALIFORNIA 1-51, 1967. (R-RM-5337-PR, AD 653925)
49556	**OPTICAL PROPERTIES OF COAL CONCENTRATES USED IN THE COKE-CHEMICAL PLANT AT THE KREMIKOVTSI METALLURGICAL COMBINE.** MAKOVSKA M T PIKARD V RUSCHEV D KHIM IND /SOFIA/ 1 5-9 1968 CA 68 116251
49567	**SPECTRA OF SYNTHETIC ZEOLITES CONTAINING TRANSITION METAL IONS-II. NI /+2/ IN TYPE A LINDE MOLECULAR SIEVES.** KLIER K RALEK M J PHYS CHEM SOLIDS 29 6 951-7 1968
49583	**EXPERIMENTS ON FLUCTUATIONS IN OPTICALLY QUENCHED CADMIUM SULFIDE AND CADMIUM SELENIDE CRYSTALS.** HERCZFELD P R VAN VLIET K M PAI M D PHYS STAT SOL 27 2 671-80 1968
49597	**MAGNESIA FOAMED CERAMICS.** NOSOVA Z A LYUBINA T M OGNEUPORY 2 50-3 1968 (FOR ENGLISH TRANSLATION SEE T47920)
49598	**DEPENDENCE OF SPECIFIC HEAT OF BARITE GLASSES ON THEIR CHEMICAL COMPOSITION AND TEMPERATURE.** RODNIKOVA V V SAKHNO V N OPT MEKH PROM 34 4 32-4 1967 CA 69 5783
49613	**THERMAL EXPANSION AND THERMAL STABILITY OF HIGH-VOLTAGE PORCELAIN IN RELATION TO THE COMPOSITION AND DISPERSITY OF THE MATERIALS.** KUKOLEV G V GAIDASH B I KONSTANTINOV E G STEKLO KERAM 25 4 29-31 1968 CA 69 4868
49616	**THERMAL CONDUCTIVITY OF ORDERED FIBROUS SYSTEMS.** DUL'NEV, G. N. ZARICHNYAK, YU. P. MURATOVA, B. L. FOREIGN TECHNOLOGY DIVISION (PT. 1), 85-97, 1969. (ENGLISH TRANSLATION OF TEPLO-MASSOPERENOS, DOKL. VSES. SOVESHCH., 3RD, 7, 82-91, 1968; FOR ORIGINAL SEE T61210) (FTD-HT-23-820-68, AD-698 517)
49621	**SOME PROPERTIES OF ZIRCONIA CERAMICS STABILIZED BY CALCIUM OXIDE.** DEMONIS, I. M. POPIL'SKII, R. YA. SHAPIRO, E. YA. POLUBOYARINOV, D. N. TR. MOSK. KHIM.-TEKHNOL. INST. (55), 146-50, 1967.
49644	**ADVANCES IN INSULATOR-METAL SEALS.** MC MILLAN P W HODGSON B P BULL INST ENGRS /INDIA/ 13 4 1-11 1963
49655	**INFRARED ABSORPTION SPECTRA OF NATURAL SORBENTS.** NADIROV N K NADIROVA G A ZH FIZ KHIM 41 11 2928- 1967 (FOR ENGLISH TRANSLATION SEE T49656)
49656	**INFRARED ABSORPTION SPECTRA OF NATURAL SORBENTS.** NADIROV N K NADIROVA G A RUSS J PHYS CHEM 41 11 1572-4 1967 (ENGLISH TRANSLATION OF ZH. FIZ. KHIM. 41 (11), 2928- , 1967; FOR ORIGINAL SEE T49655)
49666	**LOW SOLAR ABSORPTANCE AND EMITTANCE SURFACES UTILIZING VACUUM DEPOSITED TECHNIQUES.** GREENBERG, S. A. VANCE, D. A. LOCKHEED MISSILES AND SPACE CO., PALO ALTO, CA. 1-83, 1968. (LMSC-4-06-68-1, NASA-CR-73228, N68-25326)
49669	**THERMAL RADIANCE SPECTRA 8 TO 16 MICRONS. NO. 1 WHITE SANDS AND THE MALPAIS /LAVA/ C-47 AIRCRAFT.** HOVIS, W. A., JR. GODDARD SPACE FLIGHT CENTER, GREENBELT, MD. 1-122, 1968. (NASA-TM-X-63375, X-622-68-337, N68-37874)
49675	**THE ILMENITE-GEIKIELITE SERIES.** CERVELLE B FR REP BUR RECH GEOL MINIERES BULL 6 1-26 1967 CA 69 4257
49679	**THERMAL EXPANSION OF PYROXENES AND AMPHIBOLES.** QUADRADO, R. GUTIERREZ, M. BOL. REAL. SOC. ESPAN. HIST. NATUR., SECC. GEOL. 65 (2), 171-9, 1967.
49680	**SPECIFIC HEAT OF FIBERS AND THE DERIVED PARAMETERS.** GOETZE W WINKLER F DEUT TEXTILTECH 18 5 297-303 1968 CA 69 3639
49684	**THERMAL CONDUCTIVITY OF SOLID SOLUTIONS OF NICKEL MAGNESIUM FERRITES.** CHUDNOVSKII A F BEKKER YA M ISSLED NESTATSIONARNOGO TEPLO-MASSOOBMENA INST TEPLO-MASSOOBMENA AKAD NAUK BELORUSS SSR 70-6 1966 CA 69 5817
49687	**THERMAL CONDUCTIVITY OF DRY POROUS SYSTEMS.** VASILEV L L ISSLED TEPLOPROVODNOSTI INST TEPLO-MASSOOBMENA AKAD NAUK BELORUSS SSR 262-72 1967 CA 69 5807
49689	**GLASS FORMATION IN A PHOSPHORUS PENTOXIDE-CHROMIC OXIDE-ANTIMONY TRIOXIDE SYSTEM AND SOME PROPERTIES OF THE PREPARED GLASSES.** DOMBROVSKAYA V K VAIVADS A BERZINS R STEKLOOBRAZNYE SIST MATER AKAD NAUK LATV SSR 105-8 1967 CA 69 4821
49690	**DETERMINATION OF HEAT TREATMENT PARAMETERS DURING THE FORMATION OF MICROCRYSTALLINE GLASS.** YAGLOV V N STEKLOOBRAZNYE SIST MATER AKAD NAUK LATV SSR 135-43 1967 CA 68 116888
49691	**PHYSICOCHEMICAL PROPERTIES OF GLASSES BASED ON A SODIUM OXIDE-ALUMINUM OXIDE-SILICON DIOXIDE-PHOSPHORUS PENTOXIDE SYSTEM.** SEDMALIS U SVINKA V EIDUKS J STEKLOOBRAZNYE SIST MATER AKAD NAUK LATV SSR 169-78 1967 CA 69 4816
49692	**SYNTHESIS OF GLASSES, TRANSPARENT IN THE NEAR INFRARED SPECTRAL REGION.** SYRITSKAYA Z M FEIKNER S YA STEKLOOBRAZNYE SIST MATER AKAD NAUK LATV SSR 235-43 1967 CA 69 4837
49737	**THERMAL CONDUCTIVITY MEASURMENTS ON URANIUM-PLUTONIUM MIXED OXIDES.** SCHMIDT H E RICHTER J VAN DEN BERG M PROC. CONF. THERMAL CONDUCTIVITY, 6TH 527-45 1966
49738	**HEAT TRANSFER BY RADIATION AND CONDUCTION IN EVACUATED POWDERS.** WECHSLER, A. E. PROCEEDINGS OF THE CONFERENCE ON THERMAL CONDUCTIVITY, 6TH 547-69 1966
49739	**HIGH TEMPERATURE THERMAL CONDUCTIVITY AND DIFFUSIVITY STANDARDS-CERAMICS.** WECHSLER, A. E. PROCEEDINGS OF THE CONFERENCE ON THERMAL CONDUCTIVITY, 6TH 603-8 1966
49755	**STANDARDS FOR HIGH TEMPERATURE THERMAL EXPANSION MEASUREMENTS.** CONWAY, J. B. PROCEEDINGS OF THE CONFERENCE ON THERMAL CONDUCTIVITY, 6TH 967-84 1966
49757	**SEARCH FOR MAGNETITE IN LUNAR ROCKS AND FINES.** JEDWAB J HERBOSCH A WOLLAST R NAESSENS G VAN GREEN-PEERS N SCIENCE 167 618-19 1970
49761	**THERMAL AND OPTICAL PROPERTY MEASUREMENTS WITHIN THE AIR FORCE MATERIALS LABORATORY.** PURCELL, G. V. PROCEEDINGS OF THE CONFERENCE ON THERMAL CONDUCTIVITY, 6TH 1127-41 1966

TPRC Number	Bibliographic Citation
49765	**DETERMINATION OF THE TEMPERATURE FUNCTION FOR THE COEFFICIENT OF THERMAL CONDUCTIVITY.** VULIS L A POTSELUIKO V A ZHUR TEKH FIZ 26 76–84 1956 (FOR ENGLISH TRANSLATION SEE T742)
49768	**LOW TEMPERATURE PROPERTIES OF THE RESTRAHLEN POWDER FILTERS IN THE FAR-INFRARED REGION.** SAKAI K NAKAGAWA Y YOSHINAGA H JAPAN J APPL PHYS 7 7 792–3 1968
49783	**PROPERTIES OF CERAMIC MATERIALS.** DEJONG J CONSTRUCTIEMATERIALEN 2 3 10–13 1968 CA 69 12565
49785	**OPTICAL AND CHEMICAL CHARACTERS OF SOME COALS CARBONIZED IN THE LABORATORY.** GHOSH T K ECON GEOL 63 2 182–7 1968 CA 69 12070
49805	**COMPARATIVE PHYSICAL AND CHEMICAL INVESTIGATION OF SOME GLASSES PRODUCED IN HUNGARY FOR USE AS MATERIALS FOR MEDICINE BOTTLES.** HIDVEGI G VIRAGHALMY G GYOGYSZERESZET 5 409–14 1961 CA 56 9725
49818	**THERMAL EXPANSION OF SOME ALKALI PHOSPHATE GLASSES.** MURTHY M K PHYS CHEM GLASSES 7 2 69–70 1966
49821	**LOW-TEMPERATURE VITREOUS BODIES.** KOENIG C J J AM CERAM SOC 25 9 230–6 1942
49826	**STRUCTURAL STUDY IN THE SYSTEM HFO2-Y2O3.** CAILLET M DEPORTES C ROBERT G VITTER G REV INT HAUTES TEMP REFRACT 4 269–71 1967 NSA 22-12 2475
49829	**THE EFFECT OF HALIDES ON THE OPAQUENESS OF ZINC-BOROPHOSPHATE GLASSES.** EIDUKS J PORMANIS I STEKLOOBRAZNYE SIST MATER AKAD NAUK LATV SSR 113–9 1967 CA 69 12580 (FOR ENGLISH TRANSLATION SEE T63808)
49853	**REPLACEMENT OF BORAX BY DANBURITE IN PRIME COATS AND ENAMELS FOR CAST IRON SANITARY WARE.** LOKSHIN, V. YA. CHECHINA, A. F. SB. TR., NAUCH. - ISSLED. INST. SANIT. TEKH. (20), 144–52, 1966.
49871	**X-RAY MEASUREMENT OF STRAIN IN QUARTZ PARTICLES OF WHITEWARE BODIES.** CUCKA P OLIVA R F J AM CERAM SOC 51 8 458–64 1968
49893	**MARTIAN SURFACE MATERIALS. EFFECT OF PARTICLE SIZE ON SPECTRAL BEHAVIOR.** SALISBURY J W HUNT G R SCIENCE 161 365–6 1968
49897	**THE THERMAL DIFFUSIVITY OF PYROCERAM AT HIGH TEMPERATURES.** FLIEGER, H. W., JR. PROCEEDINGS OF THE CONFERENCE ON THERMAL CONDUCTIVITY, 3RD 769–83 1963
49900	**THERMAL EXPANSION OF THE BARITE ISOMORPHOUS GROUP.** GUTIERREZ, M. AMOROS, J. L. BOL. REAL SOC. ESPAN. HIST. NATUR., SECC. GEOL. 65 (1), 57–71, 1967.
49919	**SIMULTANEOUS DETERMINATION OF THE THERMAL PROPERTIES OF DISPERSED MATERIALS DURING HEATING.** DZHAPARIDZE, P. N. LANDAU, I. N. INZH, - FIZ. ZH. 14 (2), 314–21, 1968. (FOR ENGLISH TRANSLATION SEE TPRC NO. 65121)
49921	**TEMPERATURE DEPENDENCE OF INTEGRAL EMISSIVITY OF SOME MATERIALS BELOW 300 DEGREES K.** MIKHALCHENKO, R. S. GERZHIN, A. G. PERSHIN, N. P. INZH. FIZ. ZH. 14 (3), 545–51, 1968. (FOR ENGLISH TRANSLATION SEE TPRC NO. 65125)
49930	**THERMAL DIFFUSIVITY MEASUREMENTS ON CERAMICS AT HIGH TEMPERATURES.** RUDKIN, R. L. PROCEEDINGS OF THE CONFERENCE ON THERMAL CONDUCTIVITY, 3RD 794–808 1963
49933	**A THERMAL DIFFUSIVITY MEASUREMENT TECHNIQUE.** PLUMMER, W. A. PROCEEDINGS OF THE CONFERENCE ON THERMAL CONDUCTIVITY, 3RD 809–28 1963
49968	**SIMPLIFIED THERMAL EXPANSION APPARATUS.** FONTANA E H AM CERAMIC SOC BULL 41 8 516–18 1962
49971	**FIBER OPTICS WITH EXTENDED ULTRAVIOLET TRANSMISSION.** LI P C ALI M A OLSON O H SCHWARTZ M A IIT RESEARCH INST., CHICAGO, ILLINOIS 1–38, 1968. (ECOM-0542-2, IITRI-G-6018-2, N68-30372, AD 670079)
49986	**HEAT TRANSFER INSIDE NUCLEAR FUEL CONSISTING OF COMPACTED U02 AND U02 - PU02 POWDERS.** ANDRIESSEN, H. CENTRE D ETUDE DE L ENERGIE NUCLEAIRE, BRUSSELS, BELGIUM 1–40, 1967. (EURAEC-1998-EUR-3791, N68-28971)
50006	**EFFECT OF A CHANGE IN ALUMINUM OXIDE CONTENT ON THE PHYSICOCHEMICAL PROPERTIES OF DOLOMITIC OPACIFIED GLAZES.** YUNUSOV, KH. AZIMOV, I. TASHKHODZHAEV, S. PARPIEV, N. A. UZB. KHIM. ZH. 12 (1), 14–6, 1968. (FOR ENGLISH TRANSLATION SEE TPRC NO. 66504)
50020	**BONDING MECHANISM IN GLASS-TO-METAL SEALS.** SAMESHIMA Y NISHIYAMA M YOGYO KYOKAI SHI 74 10 301–12 1966 JA 51 242
50030	**DEVELOPMENT OF SPACE-STABLE THERMAL-CONTROL COATINGS /PAINTS WITH LOW SOLAR ABSORPTANCE/EMITTANCE RATIOS./** ZERLAUT, G. A. HARADA, Y. BERMAN, L. U. GEORGE C. MARSHALL SPACE FLIGHT CENTER, HUNTSVILLE, ALABAMA 1–49, 1964. (IITRI-C6014-13)
50051	**GLASS CRYSTALLIZATION IN THE LITHIUM OXIDE-GALLIUM SESQUIOXIDE-SILICON DIOXIDE SYSTEM.** DUBROVO S K TSEKHOMSKAYA T S ZOSOLOTSKAYA M V DOKL AKAD NAUK SSSR 180 1 172–4 1968 CA 69 38416 (FOR ENGLISH TRANSLATION SEE T38338)
50057	**DETERMINATION OF THE OPTICAL CONSTANTS OF STRONGLY ABSORBING SUBSTANCES.** VLASOV A G SMIRNOVA E V OPTIKA I SPEKTROSKOPIYA 26 1 124–6 1969 (FOR ENGLISH TRANSLATION SEE T50058)
50058	**DETERMINATION OF THE OPTICAL CONSTANTS OF STRONGLY ABSORBING SUBSTANCES.** VLASOV A G SMIRNOVA E V OPT SPECTROSC 26 1 67–9 1969 (ENGLISH TRANSLATION OF OPTIKA I SPEKTROSKOPIYA 26 (1), 124–6, 1969; FOR ORIGINAL SEE T50057)
50061	**THE VITREOUS V2O5 - BI2O3 - P2O5 SYSTEM.** MATVEEV M A KHODSKII L G FISYUK G K IZV AKAD NAUK SSSR NEORG MATER 4 1 163–4 1968 (FOR ENGLISH TRANSLATION SEE T51782)
50065	**COMPOSITION OF ENAMEL BATCHES.** ALDINGER R GLAS-EMAIL-KERAMO-TECH 19 4 122–6 1968 CA 69 38449
50070	**MEASUREMENT OF TEMPERATURE COEFFICIENTS OF GLASS-PLASTIC EXPANSION BY THE UNSTEADY STATE METHOD.** MEDVEDEV, N. N. INZH.-FIZ. ZH. 13 (1), 37–43, 1967. (FOR ENGLISH TRANSLATION SEE TPRC NO. 64177)
50076	**DILATOMETRIC STUDIES OF CERAMIC SAMPLES IN THE ZIRCONIUM DIOXIDE-SILICA SYSTEM AT 20–1200 C.** BUDNIKOV P P PASHCHENKO A A STARCHEVSKAYA E A GUMEN V S IZV AKAD NAUK SSSR NEORG MATER 4 5 785–6 1968 CA 69 29810 (FOR ENGLISH TRANSLATION SEE T52086)
50083	**COMPRESSIBILITY AND OTHER THERMODYNAMIC PROPERTIES OF POLYMERS.** SURLAND C C J APPL POLYM SIC 12 6 1423–37 1968 CA 69 28423

TPRC Number	Bibliographic Citation

50114 IVI. THE INFLUENCE OF THE GLASSY BOND ON SOME PROPERTIES OF SILICA REFRACTORIES.
BADGER E H M LEWCOCK W WYLDE J H
TRANS BRIT CERAMIC SOC
45 269-89 1946

50115 EFFECTS OF RUTHENIUM, RHODIUM, AND PALLADIUM ON THE THERMAL EXPANSION OF IRON-NICKEL AND IRON-NICKEL-COBALT ALLOYS.
SOLOVEVA N A YUDKEVICH M I PASTERNAK I I
POGOSOV V Z
METALLOVED TERM OBRAB METAL
4 45-6 1968 CA 69 38151
(FOR ENGLISH TRANSLATION SEE T57289)

50122 THERMODYNAMIC ANALYSIS OF THE FORMATION OF MULLITE FROM THE OXIDES.
MATVEEV G M AGARKOV A S KERBE F
SILIKATTECHNIK
18 10 324-6 1967 CA 69 30803

50145 EFFECTS OF SOME MINOR ADDITIONS ON VITRIFICATION OF AMORPHOUS SILICA.
KUMAR S NAG B B
TRANS INDIAN CERAM SOC
26 6 174-6 1967 CA 69 29811

50169 THE COMBINATION VIBRATION SPECTRUM OF GYPSUM IN THE RANGE FROM 10,000 - 1,200 /CM-1/.
HOHLER V LUTZ H D
Z NATURFORSCH
23 A 5 708-15 1968

50170 A THERMAL EXPANSION APPARATUS WITH A SILICON CARBIDE DILATOMETER FOR TEMPERATURES TO 1500 C.
MARK S D EMANUELSON R C
AM CERAMIC SOC BULL
37 4 193-6 1958

50178 EFFECT OF THERMAL RADIATION ON THE EXPANSION AND CONTRACTION OF MOLDING SANDS.
SZRENIAWSKI, J.
PRZEGL. ODLEW.
18 (2), 41-8, 1968.
(FOR ENGLISH TRANSLATION SEE TPRC NO. 65520)

50190 ELECTROOPTICAL REMOTE SENSING METHODS AS NONDESTRUCTIVE TESTING AND MEASURING TECHNIQUES IN AGRICULTURE.
MYERS V I ALLEN W A
APPL OPT
7 9 1819-38 1968

50191 USE OF CONTIGUOUS OPTICAL FIBERS AS A MEANS OF CARRYING THERMAL INFORMATION FROM WELDS.
BOBO S N CROWLEY A H
APPL OPT
7 9 1839-44 1968

50196 HIGHLY REFRACTORY CERAMICS FROM PURE OXIDES.
POLUBOYARINOV D N
OGNEUPORY
11 19-24 1967
(FOR ENGLISH TRANSLATION SEE T50197)

50197 HIGHLY REFRACTORY CERAMICS FROM PURE OXIDES.
POLUBOYARINOV D N
REFRACTORIES
11 673-8 1967
(ENGLISH TRANSLATION OF OGNEUPORY (11), 19-24, 1967; FOR ORIGINAL SEE T50196)

50198 CRYSTALLIZATION OF SPODUMENE GLASSES FROM GRANITE-BASED MELTS.
KHAN B KH KOSINSKAYA A V
IZV AKAD NAUK SSSR NEORG MATER
4 6 975-9 1968
(FOR ENGLISH TRANSLATION SEE T50199)

50199 CRYSTALLIZATION OF SPODUMENE GLASSES FROM GRANITE-BASED MELTS.
KHAN B KH KOSINSKAYA A V
INORGANIC MATERIALS
4 6 856-60 1968
(ENGLISH TRANSLATION OF IZV. AKAD. NAUK SSSR, NEORG. MATER. 4 (6), 975-9, 1968; FOR ORIGINAL SEE T50198)

50208 LOW TEMPERATURE THERMAL EXPANSION OF VARIOUS MATERIALS.
LAQUER, H. L.
LOS ALAMOS SCIENTIFIC LAB., LOS ALAMOS, NEW MEXICO
1-58, 1952.
(AECD-3706)

50230 THE NATURE OF TRAPPING AND LUMINESCENCE CENTERS DUE TO THALLIUM IN GLASS.
GHOSH A K
J PHYS CHEM SOLIDS
29 8 1387-93 1968

50243 THE THERMAL EXPANSION CHARACTERISTICS OF M3 WO6 AND R2 TI O5 COMPOUNDS.
GRAF, R. H.
MARQUETTE UNIVERSITY, MILWAUKEE, WIS., M. S. THESIS
1-103 1967

50275 BAND GAP OF FORSTERITE.
SHANKLAND T J
SCIENCE
161 51-3 1968 CA 69 47903

50280 PROPERTIES OF BRANNERITE AND THE POSSIBILITY OF USING IT TO DETERMINE ABSOLUTE AGE.
POVILAITIS, M. M. YAKOVLEVSKAYA, T. A.
KNYAZEVA, D. N. BELYAEVA, I. D.
ZAP. VSES. MINERAL. OBSHCHEST.
97 (2), 150-61, 1968.
(FOR ENGLISH TRANSLATION SEE TPRC NO. 64014)

50309 QUASI-BINARY PHASE DIAGRAMS OF THE SYSTEM THULIUM OXIDE-ACTINIDE OXIDE/THO2, UO2, NPO2, PUO2/ BELOW 1700 DEGREES.
LEITNER, L.
INST. RADIOCHEM KERNFORSCHUNGSZENTRUM, KARLSRUHE, GERMANY
1-86, 1967.
(KFK-521)

50317 AN EMISSIVITY STUDY OF VARIOUS MATERIALS.
KELLY, D. P.
UNIVERSITY OF CINCINNATI, M. S. THESIS
1-32 1960

50339 EFFECT OF LATTICE DISTORTIONS ON THE MAGNETIC BEHAVIOUR OF PEROVSKITE-TYPE MANGANITES.
BOKOV V A GRIGORYAN N A BRYZHINA M F
TIKHONOV V V
PHYS STAT SOL
28 2 835-47 1968

50414 INFLUENCE OF LIME ON THE THERMAL EXPANSION OF CLAY.
BELL K E
J CAN CERAM SOC
34 21-7 1965 JA 51 266

50428 COEFFICIENT OF THERMAL CONDUCTIVITY OF SOME SEDIMENTS. ITS DEPENDENCE ON DENSITY AND ON WATER CONTENT OF ROCKS.
CERMAK V
CHEM ERDE
26 4 271-8 1967 CA 69 53610

50443 THERMAL EXPANSION OF THE SODALITE GROUP OF MINERALS.
TAYLOR, D.
MINERAL. MAG.
36 (282), 761-9, 1968.

50457 THERMAL STABILITY OF PORCELAIN.
MOROZ I I SARKISOV G G
STEKLO KERAM
25 6 27-30 1968 CA 69 54033
(FOR ENGLISH TRANSLATION SEE T36068)

50480 PHOTOEFFECTS AND RELATED PROPERTIES OF SEMICONDUCTING DIAMONDS.
JOHNSON C STEIN H YOUNG T WAYLAND J
LEIVO W
J PHYS CHEM SOLIDS
25 827-36 1964

50485 SOME PROPERTY MEASUREMENTS ON PROMETHIUM SESQUIOXIDE.
GIBBY R L CHIKALLA T D NELSON R P
MC NEILLY, C. E.
PACIFIC NORTHWEST LABS., BATTELLE-NORTHWEST, RICHLAND, WASHINGTON
1-30, 1968.
(BNWL-SA-1796, CONF-680408-2, N68-33289)

50499 SUB-SOLIDUS PHASE EQUILIBRIA AND THERMAL EXPANSION PROPERTIES OF THE COMPOUNDS IN THE SYSTEM ZRO2-WO3-P2O5.
MARTINEK, C. A.
PENNSYLVANIA STATE UNIVERSITY, M. S. THESIS
1-57 1968

50543 TOTAL REFLECTION SPECTROPHOTOMETRY AND THERMOGRAVIMETRIC ANALYSIS OF SIMULATED MARTIAN SURFACE MATERIALS.
SAGAN C PHANEUF J P IHNAT M
ICARUS
4 1 43-61 1965

50592 DETERMINATION OF THERMAL CONDUCTIVITY OF SOLIDS BY THE HOT-WIRE METHOD.
UEI, I. FUKUI, M. HAYASHI, K.
ASAHI GARASU KOGYO GIJUTSU SHOREI-KAI KENKYU HOKOKU
13, 363-75, 1967.

50596 EFFECT OF IRON OXIDES ON SOME PROPERTIES OF GLASSES AND ENAMELS.
PETZOLD A HUEBSCHER M
GLAS-EMAIL-KERAMO-TECH
19 5 153-7 1968 CA 69 61247

50602 THERMODYNAMIC ANALYSIS OF THE DECOMPOSITION REACTIONS OF SOME MAGNESIUM HYDRATED SILICATES.
BUDNIKOV, P. P. VED, E. I. BLUDOV, B. F.
IZV. SIB. OTD. AKAD. NAUK SSSR, SER. KHIM. NAUK
(5), 3-7, 1967.

TPRC Number	Bibliographic Citation
50610	INFLUENECE OF THE TEMPERATURE DEPENDENCE OF LUNAR MATERIAL PROPERTIES ON THE SPECTRUM OF THE MOONS RADIO EMISSION. TROITSKIY V S BUROV A B ALYOSHINA T N ICARUS 8 423-33 1968 NSA 22-17 3728
50611	EFFECT OF MICROSTRUCTURE ON THE PHYSICAL PROPERTIES OF GLASSES IN THE SODIUM SILICATE STYSTEM. REDWINE R H FIELD M B J MATER SCI 3 4 380-8 1968 CA 69 61256
50619	CHEMICAL STRUCTURE AND PROPERTIES OF HEAT-TREATED COAL IN THE EARLY STAGES OF CARBONIZATION. X. REFLECTANCE OF COAL. SUGIMURA H SHIWA M NENRYO KYOKAISHI 46 911-18 1967 CA 69 60580
50623	RESISTANCE OF ENAMELS TO THERMAL SHOCK. HOWE E E BOLIN E P J AM CERAM SOC 25 15 463-6 1942
50638	THERMAL CONDUCTIVITY TESTS OF CRYOGENIC INSULATION. BLACK, I. A. GLASER, P. E. PROC. CONF. ON THERMAL CONDUCTIVITY, 2ND 111-31 1962
50642	MEASUREMENT OF THE EMISSIVITY OF SOME MATERIALS HEATED TO 100 DEGREES. VOISHVILLO N A MIRONOVA L R ZH PRIKL SPEKTROSK 8 5 840-3 1968 CA 69 61698
50646	HIGH-TEMPERATURE THERMODYNAMIC PROPERTIES OF SEVERAL BINARY ALKALI SILICATE GLASSES. TAKAHASHI, K. YOSHIO, T. MIURA, Y. HIRAI, T. MEM. SCH. ENG., OKAYAMA UNIV. 2 (1), 55-61, 1967.
50669	AUTOMATIC RECORDING OF TEMPERATURE CHARACTERISTICS OF THE TRUE COEFFICIENT OF THERMAL EXPANSION. MAZO R I OGNEUPORY 1 16-19 1968 (FOR ENGLISH TRANSLATION SEE T49276)
50670	EFFECT OF PHASE COMPOSITION ON THE THERMAL SHOCK RESISTANCE OF ZIRCONIA. GREBENYUK A A KARAULOV A G DAUKNIS V I PRANTSKYAVICHYUS G A YURENAS V L OGNEUPORY 1 42-9 1968 (FOR ENGLISH TRANSLATION SEE T49277)
50671	UNFIRED VOLUME-STABLE MAGNESITE CHROMITE PRODUCTS. UZBERG A I OGNEUPORY 2 4-12 1968 (FOR ENGLISH TRANSLATION SEE T50672)
50672	UNFIRED VOLUME-STABLE MAGNESITE CHROMITE PRODUCTS. UZBERG A I REFRACTORIES 2 65-73 1968 (ENGLISH TRANSLATION OF OGENUPORY, (2), 4-12, 1968; FOR ORIGINAL SEE T50671)
50688	HEAT TRANSFER THROUGH GLASSED STEEL. ACKLEY E J CHEM ENGNG 66 8 181-2 1959
50697	THERMAL CONDUCTIVITY OF THERMAL INSULATORS. NUSSELT, W. Z. VEREINS DEUT. INGENIEURE 52, 906-912, 1908.
50701	INVESTIGATION OF THE THERMAL COEFFICIENTS OF QUARTZ SAND IN RELATION TO TEMPERATURE AND POROSITY. BORISEVICH, V. A. MALYVKEVICH, V. I. INZH. FIZ. ZHUR. 4 (7), 60-3, 1961.
50710	STUDY OF THERMOPHYSICAL AND ELECTRICAL PARAMETERS OF THE TRANSFER OF HEAT AND ELECTRICITY IN MOLTEN GLASS. III. COMPLEX STUDY OF EFFECTIVE COEFFICIENT OF RADIANT-MOLECULAR HEAT CONDUCTIVITY AND VOLUME RESISTIVITY OF MOLTEN GLASS. SEVAST'YANOV, E. F. SOKOLOV, A. A. STEKLO, TRUDY INST. STEKLA (2), 89-94PP., 1966.
50711	NEW REFRACTORY COMPOSITIONS RESISTANT TO MOLTEN ROCK PHOSPHATE. POLE G R BEINLICH A W JR J AM CERAM SOC 26 1 21-37 1943

TPRC Number	Bibliographic Citation
50712	STUDY OF THERMOPHYSICAL AND ELECTRICAL PARAMETERS OF THE TRANSFER OF HEAT AND ELECTRICITY IN MOLTEN GLASS. V. CALCULATION OF THE TEMPERATURE FIELD AND FIELD VELOCITY OF MOLTEN GLASS IN A GLASS MELTING TANK. SEVAST'YANOV, E. F. SOKOLOV, A. A. STEKOL, TRUDY INST. STEKLA (2), 99-104PP., 1966.
50714	HIGH TEMPERATURE PROPERTIES OF DOLOMITE. I, STUDY OF DOLOMITE REFRACTORIES BY HIGH TEMPERATURE X-RAY DIFFRACTION. OHBA H HIRAGUSHI K NAKASHIMA C TAIKABUTSU 17 94 364-71 1965 JA 51 303
50734	CERAMIC CLAYS AND SHALES OF QUEBEC. BRADY, J. G. DEAN, R. S. CAN., DEP. ENERGY, MINES RESOUR., MINES BR., RES. REP. R-188, 114PP., 1967.
50768	THERMAL EXPANSION OF FERRITES WITH A SPINEL STRUCTURE. BEKKER YA M ZOTOVA G M IZV AKAD NAUK SSSR NEORG MATER 4 8 1319-22 1968 CA 69 69856 (FOR ENGLISH TRANSLATION SEE T37106)
50775	SOME FACTORS AFFECTING THE PROPERTIES OF CERAMIC TALCOSE WHITEWARE. GELLER R F CREAMER A S J RESEARCH NAT BUR STANDARDS 26 213-26 1941 (NBS-RP-1371)
50779	ON THE APPARENT THERMAL CONDUCTIVITY OF DIATHERMANOUS MATERIALS. GARDON, R. PROC. CONF. ON THERMAL CONDUCTIVITY, 2ND 167-88 1962
50788	KINETICS OF PHASE TRANSFORMATIONS DURING THE HEATING OF SPECIMENS CONTAINING ZIRCONIUM DIOXIDE AND COPPER OXIDES. BESSONOV A F TAKSIS G A OGNEUPORY 33 5 53-5 1968 CA 69 70420 (FOR ENGLISH TRANSLATION SEE T52347)
50789	TRANSMISSION OF COLORLESS OPTICAL GLASSES IN THE NEAR ULTRAVIOLET SPECTRAL REGION. GUSEVA, V. M. VILCHUR, YA. P. OPT. - MEKH. PROM. 35 (4), 49-53, 1968. (FOR ENGLISH TRANSLATION SEE TPRC NO. 67356)
50841	THE THERMAL DIFFUSIVITY AND THERMAL CONDUCTIVITY OF STOICHIOMETRIC (U(0.8) PU(0.2) O(2)). GIBBY, R. L. PACIFIC NORTHWEST LABS., RICHLAND, WASHINGTON 1-39, 1968. (BNWL-704)
50847	REFLECTIVITY OF BITUMINOUS COAL. II. ANISOTROPY OF REFLECTION. DE VRIES H A W HABETS P J BOKHOVEN C BRENNST-CHEM 49 2 47-52 1968 CA 68 71073
50877	THE MEASURMENT OF COEFFICIENTS OF EXPANSION AT LOW TEMPERATURES. SOME THERMODYNAMIC APPLICATIONS OF EXPANSION DATA. BUFFINGTON R M LATIMER W M J AM CHEM SOC 48 2305-19 1926
50888	COALIFICATION STUDIES ON THE TERTIARY COAL-BEARING SEDIMENTS OF THE UPPER RHINE GRABEN NORTH OF WORMS. TEICHMUELLER M OBERRHEIN GEOL ABHANDL 16 1/2 11-15 1967 CA 69 11978
50914	CHARACTERISTIC REFLECTION CURVES WITH NORMAL IRRADIATION OF A SURFACE. SHCHERBINA D M KIRICHENKO A P ALIEV R S TEPLOFIZ VYS TEMP 6 2 359-63 1968 (FOR ENGLISH TRANSLATION SEE T50915)
50915	CHARACTERISTIC REFLECTION CURVES WITH NORMAL IRRADIATION OF A SURFACE. SHCHERBINA D M KIRICHENKO A P ALIEV R S HIGH TEMP 6 2 351-4 1968 (ENGLISH TRANSLATION OF TEPLOFIZ. VYS. TEMP., 6 (2), 359-63, 1968; FOR ORIGINAL SEE T50914)
50925	OBSERVATION OF AN EFFECT OF PARTIAL POLARIZATION OF LIGHT BY QUARTZ CRYSTALS. REISS R VANBRUGGHE B COMPT REND 267 B 14 643-6 1968

TPRC Number	Bibliographic Citation
50986	THERMAL VOLUMETRIC BEHAVIOR OF BAUXITE-CLAY MIXTURES. PEREIRA, E. TESTA, R. H. ROBREDO, J. BOL. SOC. ESPAN. CERAM. 6 (3), 357-66, 1967.
50995	THERMAL RESISTANCE COEFFICIENTS OF SEVERAL GLASSES AND THEIR DEPENDENCE FROM THE CHEMICAL COMPOSITION. WINKELMANN, A. SCHOTT, O. ANN. PHYS. 51, 730-46, 1893.
51007	THERMAL EXPANSION OF COPRECIPITATED /U, PU/O2 POWDERS BY X-RAY DIFFRACTION TECHNIQUES. ROTH, J. HALTEMAN, E. K. NUCLEAR MATERIALS AND EQUIPMENT CORP., APOLLO, PA. 1-67, 1965. (NUMEC-2389-9)
51010	THERMAL CONDUCTIVITY OF SOME SYNTHETIC FIBER ROVINGS. DOROKHINA YU M IZV VYSSH UCHEB ZAVED TEKHNOL TEKST PROM 3 105-9 1968 CA 69 78345 (FOR ENGLISH TRANSLATION SEE T65523)
51017	THERMAL ANALYSIS OF PROUSTITE. FAKTOR M M HANKS R LEMON T H J INORG NUCL CHEM 30 8 2077-80 1968 CA 69 81096
51025	THERMAL CONDUCTIVITY AND DIFFUSIVITY OF INSULATING MATERIALS. DUDNIK D M STEPANENKO A N KHOLOD TEKH 45 1 27-9 1968 CA 69 81195 (FOR ENGLISH TRANSLATION SEE T66207)
51032	HEAT CONDUCTIVITY OF PURE OXIDE CERAMICS. DUDEROV I G POLUBOYARINOV D N FOREIGN TECHNOLOGY DIVISION 1-16, 1969. (ENGLISH TRANSLATION OF VYSOKOOGNEUPOR MATER., SB. STATEI, 92-105, 1966; FOR ORIGINAL SEE T44598) (FTD-HT-23-1285-68, AD-694833)
51043	CORRELATION BETWEEN THERMAL CONDUCTIVITY AND THERMAL EXPANSION OF FERROSPINELS. BEKKER YA M CHUDNOVSKII A F SB TR AGRON FIZ 13 47-50 1966 CA 69 81190
51066	CALCULATION OF THE THERMAL CONDUCTIVITY OF SOME SEMICONDUCTORS. KHUSAINOVA B N VESTN MOSK UNIV FIZ ASTRON 23 4 117-18 1968 CA 69 81714
51075	EXPERIMENTAL AND THEORETICAL INVESTIGATION OF THE DETERMINATION OF THE THERMAL CONDUCTIVITY OF REFRACTORY BUILDING MATERIALS CONSIDERING THE EFFECTS OF THERMAL STRAINS AND BOUNDARY STRAINS ON HEAT CONDUCTIVITY. SCHWIETE, H. E. BOEHME, H. J. ZIEGELINDUSTRIE 20 (8), 273-81; (9), 331-8; (10), 351-61, 1967.
51130	STUDYING OXIDE CERAMICS BY THE HIGH TEMPERATURE ROENTGENOGRAPHY METHOD. STREKALOVSKII V N BESSONOV A F USTYANTSEV V M BUROV G V TR INST ELEKTROKHIM AKAD NAUK SSSR URALSKII FILIAL 6 123-30 1965 (FOR ENGLISH TRANSLATION SEE T51131)
51131	STUDYING OXIDE CERAMICS BY THE HIGH TEMPERATURE ROENTGENOGRAPHY METHOD. STREKALOVSKII V N BESSONOV A F USTYANTSEV V M BUROV G V FOREIGN TECHNOLOGY DIVISION 1-16, 1967. (ENGLISH TRANSLATION OF TR. INST. ELEKTROKHIM. AKAD. NAUK SSSR, URALSKII FILIAL, (6), 123-30, 1965; FOR ORIGINAL SEE T51130) (FTD-HT-23-688-67, AD-673393)
51146	LATTICE PARAMETERS AND EXPANSION COEFFICIENTS OF FE S2 /NATURAL AND SYNTHETIC, AND OF CO S2. STRAUMANIS M E AMSTUTZ G C CHAN S AMERICAN MINERALOGIST 49 206-12 1964
51171	USE OF ZEOLITES AS GENERATORS OF PRESSURE. MAKARENKOV N N SUKOV YU M IZV VYSSH UCHEB ZAVED MASHINOSTR 5 131-4 1967 CA 69 89930
51176	EFFECT OF SOLUTION OF IRON OXIDE ON COEFFICIENT OF EXPANSION OF ENAMELS. PAQUIN W BULLETIN AMERICAN CERAMIC SOC 20 5 160 1941
51195	THERMAL EXPANSION OF FLUORITE CRYSTALS AT SUBZERO TEMPERATURES. GUREVICH, V. M. SOLODUKHIN, A. V. TR. VSES. NAUCH.-ISSLED. INST. FIZ.-TEKH. RADIOTEKH. IZMER. (92), 114-15, 1967.
51199	THERMAL CONDUCTIVITY OF NUCLEAR FUEL CERAMICS AT HIGH TEMPERATURES. NAITO K YOGYO KYOKAI SHI 75 862 163-74 1967 CA 69 92164
51215	METHOD AND APPARATUS FOR THE RAPID DETERMINATION OF THE THERMAL DIFFUSIVITY OF REFRACTORY MATERIALS OVER A WIDE TEMPERATURE RANGE. ZANEMONETS V F GROSHEVA V M SHCHERBINA O D SOVIET POWDER MET METAL CERAM 4 288-92 1966 JA 51 346 (ENGLISH TRANSLATION OF POROSH. MET. AKAD. NAUK UKR SSR, 6 (4), 34-9, 1966; FOR ORIGINAL SEE T41857)
51217	PROPERTIES OF SODIUM-LITHIUM NIOBATE SOLID SOLUTION CERAMICS WITH SMALL LITHIUM CONCENTRATIONS. NITTA T J AM CERAM SOC 51 11 626-9 1968
51220	THERMAL EXPANSION OF SYNTHETIC BETA-SPODUMENE AND BETA-SPODUMENT-SILICA SOLID SOLUTIONS. OSTERTAG W FISCHER G R WILLIAMS J P J AM CERAM SOC 51 11 651-4 1968
51221	STUDIES IN GERMANIUM OXIDE SYSTEMS. IV. PHASE EQUILIBRIA IN THE SYSTEM K2O-GEO2. MURTHY M K LONG L IP J J AM CERAM SOC 51 11 661-3 1968
51222	ABSOLUTE REFLECTIVE SPECTRA OF CERAMIC BARIUM TITANATE. GOSWAMI A K J AM CERAM SOC 51 11 666 1968
51235	THERMAL CONDUCTIVITY OF SAND, MORTARS AND BRICK DEPENDING ON THE HUMIDITY OF MATERIALS. CAMPAN T SIMIONESCU A ANGHELACHE D BULETINUL INST POLITEH IASI 4 1/2 353-60 1958
51236	THE THERMAL CONDUCTIVITY OF FIRE RESISTANT MATERIALS. JESCHKE P KARSCH K H SCHWIETE H E CHEM INGR TECH 35 8 583-6 1963
51253	STABLE AND METASTABLE PHASES IN THE CRYSTALLISATION OF CORDIERITE GLASS. SIRAZHIDDINOV N A ZH FIZ KHIM 42 1 101- 1968 (FOR ENGLISH TRANSLATION SEE T51254)
51254	STABLE AND METASTABLE PHASES IN THE CRYSTALLISATION OF CORDIERITE GLASS. SIRAZHIDDINOV N A RUSS J PHYS CHEM 42 1 49-52 1968 (ENGLISH TRANSLATION OF ZH. FIZ. KHIM., 42 (1), 101-, 1968; FOR ORIGINAL SEE T51253)
51268	DETERMINATION OF THE MEAN PARTICLE SIZE OF FINE BLACK POWDERS BY THE MEASUREMENT OF REFLECTIVITY. ROSE H E J APPL CHEM 7 244-50 1957
51282	REACTION OF UNTANNED HUMAN SKIN TO ULTRAVIOLET RADIATION. LUCKIESH M HOLLADAY L L TAYLOR A H J OPT SOC AM 20 8 423-32 1930
51284	COLOR OF SOILS. HICKERSON D KELLY K L STULTZ K F J OPTICAL SOC AM 35 4 279-300 1945
51285	SPECTRAL DIFFUSE REFLECTANCE OF DESERT SURFACES. ASHBURN E V WELDON R G J OPT SOC AM 46 8 583-6 1956
51286	THERMAL CHARACTERISTICS OF POROUS ROCKS AT ELEVATED TEMPERATURES. SOMERTON W H BOOZER G D J PETROL TECHNOLOGY 12 77-81 1960
51330	EFFECTIVE THERMAL CONDUCTIVITY OF TWO-PHASE GRANULAR MATERIALS. CHAUDHARY D R BHANDARI R C INDIAN J APPL PHYS 6 6 274-9 1968

TPRC Number	Bibliographic Citation
51345	THERMAL EXPANSION OF THE LEUCITE GROUP OF MINERALS. TAYLOR D HENDERSON C M B AM MINERALOGIST 53 9 1476–89 1968 JA 52-10 362
51350	HIGH-TEMPERATURE MICROSCOPIC STUDIES OF THE MAGNESIUM OXIDE - CALCIUM OXIDE - CHROMIUM OXIDE - SILICON OXIDE SYSTEM. KANCLIR, E. BER. DEUT. KERAM. GES. 45 (7), 362–5, 1968.
51368	HEAT CAPACITY OF FERRITES OF THE NI FE2O4–ZN FE2O4 SYSTEM AT ROOM TEMPERATURE. VARAZASHVILI V S LANDIYA N A CHACHANIDZE G D LEZHAVA N G IZV AKAD NAUK SSSR NEORG MATER 4 7 1160–3 1968 CA 69 100274 (FOR ENGLISH TRANSLATION SEE T37038)
51403	OPTICAL SPHERE PAINT AND A WORKING STANDARD OF REFLECTANCE. GRUM F LUCKEY G W APPL OPT 7 11 2289–94 1968
51415	VARIATION IN REFLECTANCE PROPERTIES OF SOILS AND VEGETATION. ORR, D. G. YOUNG, L. M. WELCH, R. U. S. ARMY ENGR. RES. AND DEVELOPMENT LABS. 73–83, 1962. (AD–289531)
51473	INTRODUCTION. CERAMIC MICROSTRUCTURES. VAN VLACK L H CERAMIC MICROSTRUCTURES 1–21 1968
51500	POTENTIAL APPLICATION OF THE RAPID SCANNING SPECTROPHOTOMETER FOR THE OBJECTIVE VALUATION OF FOOD COLOR. POMERANTZ R GOLDBLITH S A PROCTOR B E FOOD TECHNOLOGY 9 9 478–82 1955
51540	HEAT FLOW AT KIRKLAND LAKE. LEITH T H TRANS AM GEOPHYSICAL UNION 33 3 435–43 1952
51545	HIGH-TEMPERATURE REFRACTORY MATERIALS BASED ON ZIRCONIA STABILIZED BY RARE-EARTH OXIDES. KIND N E KOSHUR L T FOREIGN TECHNOLOGY DIVISION 1–12, 1967. (ENGLISH TRANSLATION OF OGNEUPORY, 31 (10), 55–8, 1966; FOR ORIGINAL SEE T45561) (FTD–HT–67–300, AD–673211)
51562	THE INFLUENCE OF THE DEGREE OF RAREFACTION ON THE EFFECTIVE THERMAL CONDUCTIVITY OF REFRACTORY CERAMICS. PUSTOVALOV V V INZH FIZ ZHUR 3 10 57–9 1960
51563	COEFFICIENTS OF HEAT CONDUCTIVITY OF NONMETALLIC POWDERS AT HIGH TEMPERATURES. KOZAK M I INZH FIZ ZHUR 3 10 110–12 1960
51567	A NEW TYPE OF HORIZONTAL BI CALORIMETER. BEGUNKOVA, A. F. IZV. VYSSHIKH UCHEB. ZAVEDENII PRIBOROSTROENIE (6), 78–88, 1959.
51577	CHANGES OF HEAT CAPACITY OF PEAT IN THE PROCESS OF THERMAL DESTRUCTION. REPRINTSEVA S M KRASNIH S G TORFYANAYA PROM 3 36–9 1966
51581	INFLUENCE OF GLOSSY SURFACES ON REFLECTION MEASUREMENTS. BROCKES A FARBE 9 1/3 53–62 1960
51604	HEAT CONDUCTION BY POWDERS IN VARIOUS GASEOUS ATMOSPHERES AT LOW PRESSURE. PRINS J A SCHENK J SCHRAM A J G L PHYSICA 16 4 379–80 1950
51607	INFRA-RED ABSORPTION AND REFLECTION SPECTRA. COBLENTZ W W PHYS REV 23 2 125–53 1906
51623	USE OF ULTRASONIC DATA IN CALCULATING THE HEAT CONDUCTIVITY OF VITREOUS SUBSTANCES AT HIGH TEMPERATURES. BESSONOV M V PRIMENIE ULTRAAKUSTIKI K ISSLEDOVANIYU VESHCHESTA 13 165–70 1961 (FOR ENGLISH TRANSLATION SEE T66064)
51635	THE RELATION BETWEEN CHEMICAL COMPOSITION AND GLASS TRANSMISSION IN THE INFRARED. GERLOVIN J I DOKLADY AKAD NAUK SSSR 38 5/6 170–2 1943 (FOR ENGLISH TRANSLATION SEE T51636)
51636	THE RELATION BETWEEN CHEMICAL COMPOSITION AND GLASS TRANSMISSION IN THE INFRARED. GERLOVIN J I COMPT RENDUS /DOKLADY/ ACAD SCI URSS 38 5/6 1–4 1943 (ENGLISH AND FRENCH TRANSLATION OF DOKL. AKAD. NAUK SSSR, 38 (5–6), 170–2, 1943; FOR ORIGINAL SEE T51635)
51661	MEASUREMENT OF THE HEAT CONDUCTIVITY COEFFICIENT OF VACUUM-POWDER INSULATION AT HIGH TEMPERATURES. SEREBRYANYI G L ZARUDNYI L B SHORIN S N TEPLOFIZ VYS TEMP 6 3 547–8 1968 (FOR ENGLISH TRANSLATION SEE T51662)
51662	MEASUREMENT OF THE HEAT CONDUCTIVITY COEFFICIENT OF VACUUM-POWDER INSULATION AT HIGH TEMPERATURES. SEREBRYANYI G L ZARUDNYI L B SHORIN S N HIGH TEMP 6 3 523–4 1968 (ENGLISH TRANSLATION OF TEPLOFIZ. VYS. TEMP., 6 (3), 547–8, 1968; FOR ORIGINAL SEE T51661)
51671	OXIDE CERAMICS. BRADSTREET S W HIGH TEMPERATURE MATERIALS AND TECHNOLOGY 235–303 1967
51672	COMMERCIAL OXIDE REFRACTORIES. SHEETS H D SULLIVAN J D HIGH TEMPERATURE MATERIALS AND TECHNOLOGY 304–11 1967
51692	THE DETERMINATION OF THE THERMAL CONDUCTIVITY OF THE SPECIFIC HEAT AND THE DENSITY /GROSS DENSITY/ OF INSULANTS FOR ROCKET TANKS FILLED WITH LIQUID HYDROGEN. PART 1. DESCRIPTION OF MEASURING METHOD. KOGLIN B ZIMNI W F ELDO/CECLES TECH REV. 2 (1), 3–28, 1967. (N67–35022)
51693	EFFECT OF TEMPERATURE AND FLUID SATURATION ON ROCK THERMAL CONDUCTIVITY. VILORIA, G. PENNSYLVANIA STATE UNIVERSITY, M. S. THESIS 1–112 1968
51709	LUNAR AND MARTIAN SURFACES. PETROLOGIC SIGNIFICANCE OF ABSORPTION BANDS IN THE NEAR-INFRARED. ADAMS J B SCIENCE 159 1453–5 1968
51716	MEASUREMENTS OF THE LUMINOUS DENSITY FACTOR ON ARBITRARY REFLECTING SAMPLES. KORTE H SCHMIDT M PHYSIKALISCH TECHNISCHEN BUNDESANSTALT 19 2 1– 1967 (FOR ENGLISH TRANSLATION SEE T58395)
51723	ORDER AND DISORDER IN LITHIUM ALUMINOSILICATES. KONDRATEV YU N INORGANIC MATERIALS 1 8 1275–7 1965 (ENGLISH TRANSLATION OF IZV. AKAD. NAUK SSSR, NEORG. MAT., 1 (8), 1395–8. 1965; FOR ORIGINAL SEE T60876)
51732	THERMAL EXPANSION COEFFICIENTS OF RUBY MUSCOVITE MICA. GOLDSTEIN L POST B J APPL PHYS 40 7 3056–7 1969
51750	RAMAN AND INFRARED STUDIES OF COUPLED PO4/–3/ VIBRATIONS. KRAVITZ L C KINGSLEY J D ELKIN E L J CHEM PHYS 49 10 4600–10 1968
51754	COMPUTATION OF THERMAL CONDUCTIVITY OF SANDSTONES BY CONSIDERING THEM AS UNCONSOLIDATED SANDS. AGRAWAL M P BHANDARI R C INDIAN J PURE APPL PHYS 6 7 388–9 1968

TPRC Number	Bibliographic Citation
51767	**NONLINEAR ABSORPTION IN CDSE.** BONCH-BRUEVICH A M RAZUMOVA T K RUBANOVA G M FIZ TVERD TELA 10 4 1235-6 1968 (FOR ENGLISH TRANSLATION SEE T51768)
51768	**NONLINEAR ABSORPTION IN CDSE.** BONCH-BRUEVICH A M RAZUMOVA T K RUBANOVA G M SOVIET PHYS-SOLID STATE 10 4 981-2 1968 (ENGLISH TRANSLATION OF FIZ. TVERD. TELA, 10 (4), 1235-36, 1968; FOR ORIGINAL SEE T51767)
51782	**THE VITREOUS V205-BI203-P205 SYSTEM.** MATVEEV M A KHODSKII L G FISYUK G K INORGANIC MATERIALS 4 1 135-6 1968 (ENGLISH TRANSLATION OF IZV. AKAD. NAUK SSSR, NEORG. MATER., 4 (1), 163-4, 1968; FOR ORIGINAL SEE T50061)
51815	**COMPLEX DETERMINATION OF THE THERMAL CHARACTERISTICS OF SOLID FUELS DURING HIGH-TEMPERATURE PYROLYSIS.** AGROSKIN A A GONCHAROV E I LOVETSKII L V KHIM TVERD TOPL 4 3-12 1968 CA 69 108467
51836	**INVESTIGATION OF PROPERTIES OF POLYPHOSPATES AND SOME POSSIBILITIES OF THEIR USE.** KUZNETSOV E V CHIZHEVSKAYA I A GERASIMOV V V TRUDY KAZANSK KHIM-TEKHNOL INST IM S M KIROVA 36 420-6 1967
51855	**GLASS BEADS—A STANDARD FOR THE LOW THERMAL CONDUCTIVITY RANGE.** WECHSLER, A. E. PROC. CONF. ON THERMAL CONDUCTIVITY, 7TH 89-96, 1968. (NBS-SP-302)
51892	**THERMAL CONDUCTIVITY AND THERMAL DIFFUSIVITY OF GREEN RIVER OIL SHALE.** TIHEN, S. S. CARPENTER, H. C. SOHNS, H. W. PROC. CONF. ON THERMAL CONDUCTIVITY, 7TH 529-35, 1968. (NBS-SP-302)
51904	**MEASUREMENTS OF THE THERMAL CONDUCTIVITY OF GLASS BEADS IN A VACUUM AT TEMPERATURES FROM 100 TO 500 K.** MERRILL, R. B. PROC. CONF. ON THERMAL CONDUCTIVITY, 7TH 713-20, 1968. (NBS-SP-302)
51931	**THE PRECISE DETERMINATION OF THERMAL CONDUCTIVITY OF PURE FUSED QUARTZ.** SUGAWARA A J APPL PHYS 39 13 5994-7 1968
51936	**FAR INFRARED ABSORPTION IN FUSED QUARTZ AND SOFT GLASS.** BAGDADE W STOLEN R J PHYS CHEM SOLIDS 29 11 2001-8 1968
51967	**DIELECTRIC PERMEABILITY OF SITALS CONTAINING BISMUTH, LEAD, AND CERIUM COMPOUNDS.** BOGDANOVA G S LITVINOV P I KUDRYAKOVA M L IZV AKAD NAUK SSSR NEORG MATER 4 3 417-19 1968 (FOR ENGLISH TRANSLATION SEE T51968)
51968	**DIELECTRIC PERMEABILITY OF SITALS CONTAINING BISMUTH, LEAD, AND CERIUM COMPOUNDS.** BOGDANOVA G S LITVINOV P I KUDRYAKOVA M L INORGANIC MATERIALS 4 3 349-51 1968 (ENGLISH TRANSLATION OF IZV. AKAD. NAUK SSSR, NEORG. MATER., 4 (3), 417-9, 1968; FOR ORIGINAL SEE T51967)
51970	**STUDY OF SODIUM SILICATE GLASSES CONTAINING VANADIUM.** ZELENTSOV V V DENISOV YU V ZELENTSOVA S A BREKHOVSKIKH S M IZV AKAD NAUK SSSR NEORG MATER 4 3 426-30 1968 (FOR ENGLISH TRANSLATION SEE T51971)
51971	**STUDY OF SODIUM SILICATE GLASSES CONTAINING VANADIUM.** ZELENTSOV V V DENISOV YU V ZELENTSOVA S A BREKHOVSKIKH S M INORGANIC MATERIALS 4 3 357-60 1968 (ENGLISH TRANSLATION OF IZV. AKAD. NAUK SSSR, NEORG. MATER., 4 (3), 426-30, 1968; FOR ORIGINAL SEE T51970)
51980	**A STUDY OF THE INITIAL STAGES OF GLASS-CERAMIC FORMATION IN LIGHT-SENSITIVE GLASSES.** BEREZHNOI A I ILCHENKO L N IZV AKAD NAUK SSSR NEORG MATER 4 4 584-9 1968 (FOR ENGLISH TRANSLATION SEE T51981)
51981	**A STUDY OF THE INITIAL STAGES OF GLASS-CERAMIC FORMATION IN LIGHT-SENSITIVE GLASSES.** BEREZHNOI A I ILCHENKO L N INORGANIC MATERIALS 4 4 503-7 1968 (ENGLISH TRANSLATION OF IZV. AKAD. NAUK SSSR, NEORG. MATER., 4 (4), 584-9, 1968; FOR ORIGINAL SEE T51980)
51983	**THE PROBLEM OF QUARTZ-FORMATION IN POROUS GLASSES WITH HIGH SILICA CONTENTS.** MIRONOVA M L ANANICH N I BOTVINKIN O K IZV AKAD NAUK SSSR NEORG MATER 4 4 632-4 1968 (FOR ENGLISH TRANSLATION SEE T51984)
51984	**THE PROBLEM OF QUARTZ-FORMATION IN POROUS GLASSES WITH HIGH SILICA CONTENTS.** MIRONOVA M L ANANICH N I BOTVINKIN O K INORGANIC MATERIALS 4 4 551-3 1968 (ENGLISH TRANSLATION OF IZV. AKAD. NAUK SSSR, NEORG. MATER., 4 (4), 632-4, 1968; FOR ORIGINAL SEE T51983)
51989	**DEVELOPMENTS IN GLASS.** BRAIN, R. C. CHEM. PROCESS. ENG. 48 (2), 85-6, 1967.
51994	**PROPERTIES OF MICA. II.** MUTSCHKE H SILIKATTECHNIK 19 5 160-3 1968 CA 69 89437
51997	**REFLECTANCE VARIABLES OF COMPACTED POWDERS.** SCHATZ E A MOD ASPECTS REFLECTANCE SPECTROSC., PROC. SYMP. 107-31 1968 CA 69 101329
52038	**DEHYDROXYLATION OF MICROCRYSTALLINE MUSCOVITE. KINETICS, MECHANISM AND ENERGY CHANGE.** KODAMA H BRYDON J E TRANS FARADAY SOC 64 11 3112-19 1968
52046	**ENVIRONMENTAL MEASUREMENTS AND INSTRUMENTATION DATA IN SUPPORT OF NAVOCEANO AERIAL COLOR PHOTOGRAPHY TEST.** YOST, E. WENDEROTH, S. ANDERSON, R. SCIENCE ENGR. RES. GROUP, LONG ISLAND UNIV. 1-90, 1968. (SERG-TR-02, AD-671847)
52055	**RESEARCH ON CERAMIC DISCHARGE TUBES FOR GAS LASER.** NASINI M PRATESI R REDAELLI G TECHNOLOGICAL RES. GROUP. PHYSICS DIV. CISE LABS, MILANO 1-40, 1967. (CM-R79-8.7, N68-26016)
52086	**A DILATOMETRIC STUDY OF CERAMIC SPECIMENS IN THE ZRO2-SIO2 SYSTEM AT TEMPERATURES OF 20-1200 C.** BUDNIKOV P P PASHCHENKO A A STARCHEVSKAYA E A GUMEN V S INORGANIC MATERIALS 4 5 688-9 1968 (ENGLISH TRANSLATION OF IZV. AKAD. NAUK SSSR, NEORG. MATER., 4 (5), 785-6, 1968; FOR ORIGINAL SEE T50076)
52094	**DILATOMETRIC INVESTIGATIONS ON PORCELAIN BODIES.** ZWETSCH A SPECIAL LIBRARY ASSOCIATION 1-15, 1952. (ENGLISH TRANSLATION OF BER. DER DEUT. KERAM. GESELL., 29 (3), 73-7, 1952; FOR ORIGINAL SEE T42961) (SLA-62-10530)
52098	**THERMAL PROPERTIES OF SOLIDS. VOLUME II.** WOLFE, G. W. LING-TEMCO-VOUGHT INC DALAS TEXAS 1-24, 1959. (LTV-E9R-12073)
52100	**COEFFICIENT OF THERMAL EXPANSION VS. TEMPERATURE FOR LOW EXPANSION MATERIALS.** EUMARIAN C MATERIALS DESIGN ENGINEERING 53 6 151 1961
52102	**EVALUATION OF AN APPARATUS FOR THE DETERMINATION OF THE THERMAL CONDUCTIVITY OF GOOD CONDUCTING SOLIDS—RESULTS ON ARMCO IRON FROM 50 TO 300 C.** LARSEN, D. C. PURDUE UNIV., W. LAFAYETTE, IND., M. S. THESIS 1-135 1968
52106	**CERAMIC-BASED MATERIALS FOR HIGH TEMPERATURE SERVICE.** BATES, W. A. CALIFORNIA RES. AND DEV. COMP., LIVERMORE, CALIF. 1-12, 1951. (CRD-T2B-20, AECD-3847)

TPRC Number	Bibliographic Citation
52112	**CARBIDE FUEL DEVELOPMENT.** NOLOMEY R LAZERUS S SAPIR J SOFER G STEINMETZ H STRASSER A WEISMAN J DIAL R MC MURTRY C SAULINO F TAYLOR K NUCLEAR DEVELOPMENT ASSOCIATION 1-100, 1959. (NDA-2140-2)
52116	**THE MICROSTRUCTURE OF EARTHENWARE AND SEMI-PORCELAIN FIRED AT 1100-1300 DEGREES.** YAKOVLEVA M E BEZNOSIKOVA A V STEKLO I KERAMIKA 15 9 25-30 1958 (FOR ENGLISH TRANSLATION SEE T52117)
52117	**THE MICROSTRUCTURE OF EARTHENWARE AND SEMI-PORCELAIN FIRED AT 1100-1300 DEGREES.** YAKOVLEVA M E BEZNOSIKOVA A V GLASS AND CERAMICS 15 9 483-90 1958 (ENGLISH TRANSLATION OF STEKLO KERAM., 15 (9), 25-30, 1958; FOR ORIGINAL SEE T52116)
52120	**FABRICATION, CHARACTERIZATION, AND THERMAL-PROPERTY MEASUREMENTS OF THO2-UO2 FUEL MATERIALS. /LWBR DEVELOPMENT PROGRAM/.** SPRINGER J R ELDRIDGE E A GOODYEAR M U WRIGHT T R LAGEDROST J F BATTELLE MEMORIAL INST., COLUMBUS, OHIO 1-140, 1967. (BMI-X-10210, N69-13163)
52128	**CERAMIC AND COMPOSITE CERAMIC-METAL MATERIALS SYSTEMS APPLICABLE TO RE-ENTRY STRUCTURES.** STERRY, W. M. BOEING AIRPLANE COMP., AEROSPACE DIV., SEATTLE, WASHINGTON 333-74, 1960. (WADD-TR-60-58, AD-241597)
52129	**HIGH TEMPERATURE PROPERTIES OF VITREOUS FIBERS AND COATINGS.** EDMUNDS, W. M. OWENS-CORNING FIBERGLAS CORP., GRANVILLE, OHIO 1-65, 1961. (AD-256208)
52135	**PROPERTIES OF BERYLLIUM-BARIUM TITANATE DIELECTRICS.** BUNTING E N SHELTON G R CREAMER A S JAFFE B J RESEARCH NATL BUREAU STANDARDS 47 1 15-24 1951 (NBS-RP-2222)
52136	**DESIGNING WITH HIGH-STRENGTH CERAMICS.** GLOBE-UNION INC., MILWAUKEE, WIS. MACHINE DESIGN 32 22 122-6 1960
52164	**THE INFLUENCE OF BEO AND TECHNICAL ALUMINA ADDITIONS ON THE BASIC PROPERTIES OF ELECTRICAL PORCELAIN.** BUDNIKOV P P ZVYAGILSKII A A GLASS AND CERAMICS 16 362-7 1959 (ENGLISH TRANSLATION OF STEKLO I KERAM., 16 (7), 3-7, 1959; FOR ORIGINAL SEE T46428)
52205	**IMPROVEMENTS IN A THERMAL EXPANSION APPARATUS.** SHEFFIELD, W. F. IOWA STATE UNIVERSITY, M. S. THESIS 1-28 1968
52206	**VARIATIONS OF THE THERMAL CONDUCTIVITY COEFFICIENT FOR FIBROUS INSULATION MATERIALS.** MUMAW, J. R. OHIO STATE UNIVERSITY, M. S. THESIS 1-129 1968
52209	**MEASUREMENT OF THERMAL CONDUCTIVITY OF GLASS AT HIGH TEMPERATURES.** CHEN, K. H. MASSACHUSETTS INSTITUTE OF TECHNOLOGY, M. S. THESIS 1-31 1968
52232	**THERMAL BEHAVIOR OF FIBROUS INSULATING MATERIALS.** GARCIA, A. INSTITUTO EDUARDC TORROJA DE LA CONSTRUCCTION Y DEL CEMENTO, MADRID, SPAIN 1-56, 1967. (N68-28756)
52243	**THERMAL EXPANSION OF SOME GLASSES AND CRYSTALLINE-PHASES IN THE TERNARY SYSTEM NA2O - MGO - SIO2.** RAO, P. S. ALFRED UNIVERSITY, M. S. THESIS 1-85 1968
52249	**HIGH PRESSURE /UP TO 18 KILOBARS/ CHAMBER FOR OPERATING AT HELIUM TEMPERATURE.** ITSKEVICH E S VORONOVSKII A N GAVRILOV A F SUKHOPAROV V A INSTRUM EXP TECH /USSR/ 6 1452-4 1966 (ENGLISH TRANSLATION OF PRIB. TEKH. EKSP., 11 (6), 161-4, 1966; FOR ORIGINAL SEE T44508)
52250	**PROPERTIES OF GLASSES IN THE SYSTEM AL2O3-B2O3-P2O5.** KUMAR S CENTRAL GLASS CERAM RESEARCH INST BULL /INDIA/ 7 3 117-22 1960
52252	**SPECIAL REFRACTORIES FOR USE ABOVE 1700 C.** WHITTEMORE O J JR INDUSTRIAL ENG CHEM 47 12 2510-12 1955
52253	**PROPERTIES OF ALUMINOBORATE GLASSES OF GROUP II METAL OXIDES. I. GLASS FORMATION AND THERMAL EXPANSION.** HIRAYAMA C J AM CERAM SOC 44 12 602-6 1961
52256	**XIII.-THE REVERSIBLE THERMAL EXPANSION AND OTHER PROPERTIES OF SOME MAGNESIAN FERROUS SILICATES.** RIGBY G R LOVELL G H B GREEN A T TRANS BRIT CERAM SOC 45 6 237-50 1946
52260	**DETERMINATION OF THE THERMAL PROPERTIES OF MATERIALS AT ELEVATED TEMPERATURES.** MOELLER, C. E. WILSON, D. R. MIDWEST RESEARCH INSTITUTE 1-95, 1960. (MRI-2059-E)
52271	**STUDIES IN THE UO2-ZRO2 SYSTEM.** WRIGHT, T. R. KIZER, D. E. KELLER, D. L. BATTELLE MEMORIAL INST., COLUMBUS, OHIO 1-34, 1964. (BMI-1689)
52276	**RADIATION-RESISTANT MAGNET WIRE FOR USE IN AIR AND VACUUM AT 850 C.** ANACONDA WIRE AND CABLE COMPANY MAGNET WIRE RESEARCH LABORATORIES MUSKEGON MICHIGAN ANACONDA WIRE AND CABLE COMP., MAGNET WIRE RES. LABS, MUSKEGON, MICHIGAN 1-104, 1963. (ASD-TDR-63-164, AD-414198)
52279	**NONDESTRUCTIVE TESTS FOR CERAMIC, CERMET AND GRAPHITE MATERIALS.** LAUCHNER, J. H. BENNETT, D. G. MORGAN, G. L. AIR RES. AND DEV. COMMAND, WPAFB, OHIO 1-52, 1959. (WADC-TR-59-412, PB-161815)
52283	**AN INVESTIGATION OF METAL-CERAMIC COMPOSITES FOR HIGH-TEMPERATURE APPLICATIONS.** PARIKH, M. M. FISHER, J. I. ARMOUR RES. FOUNDATION, ILL. INST. TECHNOL., CHICAGO, ILL. 1-35, 1960. (ARF-2175-12, AD-238137)
52285	**DEVELOPMENT OF PROTECTIVE COATING FOR TANTALUM BASE ALLOYS.** OHNYSTY, B. STETSON, A. R. AERONAUTICAL SYSTEMS COMMAND, WPAFB, OHIO 1-58, 1963. (RDR-1360-2, AD-429028)
52286	**DEVELOPMENT OF PROTECTIVE COATINGS FOR TANTALUM BASE ALLOYS. TASK II-DEVELOPMENT OF TECHNOLOGY APPLICABLE TO COATINGS USED IN THE 3000 TO 4000 F. TEMPERATURE RANGE.** OHNYSTY B STANSFIELD O M STETSON A R METCALFE A G RES. AND TECHNOL. DIV., WPAFB, OHIO 1-95, 1965. (ML-TDR-64-294, AD-612061)
52287	**METAL REINFORCED CERAMIC RADOME.** ATLAS, L. M. ARMOUR RES. FOUNDATION, ILLINOIS INST. TECHNOLOGY 1-41, 1958. (WADC-TR-58-329, AD-155872)
52300	**THERMAL CONDUCTION THROUGH AN EVACUATED IDEALIZED POWDER OVER THE TEMPERATURE RANGE OF 100 TO 500 K.** MERRILL, R. B. GEORGE C MARSHALL SPACE FLIGHT CTR., HUNTSVILLE, ALA. 1-76, 1969. (NASA-TN-D-5063)
52305	**COLOR-CENTER CONCENTRATION IN PLASTICALLY-DEFORMED CALCITE.** STERZEL W CHORINSKY E Z NATURFORSCH 23 A 7 1052-8 1968
52311	**THE VISCOSITY OF BASALTIC MAGMA. AND ANALYSIS OF FIELD MEASUREMENTS IN MAKAOPUHI LAVA LAKE, HAWAII.** SHAW H R WRIGHT T L PECK D L OKAMURA R AM J SCIENCE 266 4 225-64 1968

TPRC Number	Bibliographic Citation
52318	EFFECTS OF VARYING WATER CONTENTS AND PRESSURES IN THE SEMIDRY COMPACTION OF LOWER KITTANNING CLAY. HUTCHISON C R WILLIAMSON W O BULL AM CERAM SOC 48 2 198–202 1969
52347	KINETICS OF PHASE CHANGES DURING THE HEATING OF SPECIMENS OF ZIRCONIA AND COPPER OXIDES. BESSONOV A F TAKSIS G A REFRACTORIES 5 317–19 1968 (ENGLISH TRANSLATION OF OGNEUPORY, 33 (5), 53–5, 1968; FOR ORIGINAL SEE T50788)
52348	FIRING HEAVY SILICA PRODUCTS IN A TUNNEL KILN. SHIRYAEV S A GUBKO I T SIZOV I D KOSTOMATOV M I NIKOLAEV A M ZHURAVLEVA Z K SATANOVSKII P L OGNEUPORY 33 6 10–19 1968 (FOR ENGLISH TRANSLATION SEE T52349)
52349	FIRING HEAVY SILICA PRODUCTS IN A TUNNEL KILN. SHIRYAEV S A GUBKO I T SIZOV I D KOSTOMATOV M I NIKOLAEV A M ZHURAVLEVA Z K SATANOVSKII P L REFRACTORIES 6 340–8 1968 (ENGLISH TRANSLATION OF OGNEUPORY, 33 (6), 10–19, 1968; FOR ORIGINAL SEE T52348)
52350	INVESTIGATION OF THE SINTERING OF ZIRCONIA. DEGTYAREVA E V KAINARSKII I S ALEKSEENKO L S OGNEUPORY 33 6 33–40 1968 (FOR ENGLISH TRANSLATION SEE T52351)
52351	INVESTIGATION OF THE SINTERING OF ZIRCONIA. DEGTYAREVA E V KAINARSKII I S ALEKSEENKO L S REFRACTORIES 6 362–70 1968 (ENGLISH TRANSLATION OF OGNEUPORY, 33 (6), 33–40, 1968; FOR ORIGINAL SEE T52350)
52361	THERMAL CONDUCTIVITY OF TWO–PHASE POROUS MATERIALS. DRY SOILS. CHAUDHARY D R BHANDARI R C BRIT J APPL PHYS /J PHYS D/ 2 4 609–10 1969
52383	OPTICAL PROPERTIES OF NATURAL AND ARTIFICIAL LEAD CHROMATE SINGLE CRYSTALS. KESSLER F R DANESCHFAR P Z NATURFORSCH 23 A 12 2014–18 1968
52402	MODEL FOR SOLID–PHASE HEAT CONDUCTION IN GLASS FIBER PAD INSULATION. PIKE J N J APPL PHYS 40 2 899–901 1969
52405	IMPROVED CERAMICS. SMOKE, E. J. ILLYN, A. V. EICHBAUM, B. R. NEW JERSEY CERAMIC RESEARCH STATION, RUTGERS UNIV. 1–88, 1955. (AD 74092)
52417	FOAM CERAMICS DEVELOPMENT. SIMPSON, F. H. BOEING AIRPLANE CO., SEATTLE, WASHINGTON 1–54, 1962. (D2–10339, AD 411657)
52418	STUDY OF HIGH TEMPERATURE MATERIALS. SMOKE E J ROODA J MOLONY D A KENNER B PECINA R CHAO C ROME AIR DEVELOPMENT CENTER, GRIFFISS AIR FORCE BASE, NEW YORK 1–142, 1962. (RADC–TDR–62–307, AD 287500)
52425	AN AUTOMATICALLY RECORDING THERMAL–EXPANSION APPARATUS. KELLY H J HARRIS H M J AMER CERAMIC SOC 39 10 344–8 1956
52429	THERMAL EXPANSION OF YTTRIA–STABILIZED ZIRCONIA. /JPL–TR–32–600/ NIELSEN T H LEIPOLD M H J AMER CERAMIC SOC 47 3 155 1964
52431	FORSTERITE AND CORDIERTIE PORCELAINS MADE FROM THE ELECTRICALLY FUSED DUNITE. FUNDAMENTAL STUDIES ON THE RELATION BETWEEN THE STRUCTURE AND THE PROPERTIES OF ELECTRICALLY FUSED DUNITE, PART VII. NIWA S J CERAM ASSOC JAPAN 69 7 224–9 1960
52470	AN EXPERIMENTAL STUDY OF THE VALIDITY OF FOURIERS LAW. FLUMERFELT R W SLATTERY J C A I CH E JOURNAL 15 2 291–2 1969
52479	THERMAL EXPANSION OF SOME GLASSES IN THE SYSTEM LI2O–CAO–SIO2. HANDWERK, J. H. MC VAY, T. N. DEPARTMENT OF CERAMIC TECHNOLOGY, UNIV. OF ALABAMA 1–24, 1954. (ORO–126)
52481	INITIAL TEMPERATURES OF ZIRCONIA PHASE CHANGES AND SOLID–SOLUTION REACTIONS. SMOOT T W RYAN J R J AM CERAMIC SOC 46 12 597–600 1963
52507	HIGH TEMPERATURE CERAMIC STRUCTURES. POULOS N E ELKINS S R MURPHY C A WALTON J D ENGINEERING EXPERIMENT STATION, GEORGIA INST. OF TECHNOLOGY 1–124, 1961. (AD 270805)
52509	MOLAR VOLUMES AND THERMAL EXPANSIONS OF ANDALUSITE, KYANITE, AND SILLIMANITE. SKINNER B J CLARK S P JR APPLEMAN D E AM J SCI 259 9 651–68 1961
52519	AN INVESTIGATION OF THE THEORETICAL AND PRACTICAL ASPECTS OF THE THERMAL EXPANSION OF CERAMIC MATERIALS. SCHEETZ, H. CORNELL AERONAUTICAL LAB., INC., BUFFALO, NEW YORK 1–63, 1961. (CAL–PI–1273–M–11, AD 270178)
52521	A SURVEY OF CERAMIC–TO–METAL BONDING. VAN HOUTEN G R AM CERAMIC SOC BULL 38 6 301–7 1959
52522	POLYMORPHISM AND SUBSOLIDUS EQUILIBRIA IN THE SYSTEM GEO2–TIO2. SARBER J F AM J SCI 259 9 709–18 1961
52523	RUBIDIUM IN LEAD AND BARIUM GLASSES. SIMPSON H E GLASS INDUSTRY 40 411–17 454–7 1959
52524	RUBIDIUM IN LIME AND BORATE GLASSES. SIMPSON H E GLASS INDUSTRY 42 4 189–93 222–3 1961
52525	STUDIES IN LITHIUM OXIDE SYSTEMS. XIII. LI2O.AL2O3.2SIO2–LI2O.AL2O3.2GEO2. TIEN T Y HUMMEL F A J AM CERAMIC SOC 47 11 582–4 1964
52532	THE THERMAL EXPANSIONS OF THORIA, PERICLASE AND DIAMOND. SKINNER B J AM MINERALOGIST 42 39–55 1957
52534	THERMAL EXPANSION OF BINARY ALKALINE–EARTH BORATE GLASSES. SHERMER H F J RESEARCH NATL BUR STANDARDS 56 2 73–9 1956 (NBS–RESEARCH PAPER NO. 2650)
52538	BORON–FREE AND LEAD–FREE POTTERY GLAZES. SIVCHIKOVA M G DAIN F L STEKLO I KERAM 16 7 22–6 1959 (FOR ENGLISH TRANSLATION SEE T52539)
52539	BORON–FREE AND LEAD–FREE POTTERY GLAZES. SIVCHIKOVA M G DAIN F L GLASS AND CERAMICS 16 382–6 1959 (ENGLISH TRANSLATION OF STEKLO I KERAM. 16 (7), 22–6, 1959; FOR ORIGINAL SEE T52538)
52543	3.–EFFECT OF HEAT ON KANDITE–CONTAINING CLAYS. VAUGHAN F TRANS BRIT CERAMIC SOC 57 1 38–68 1958
52549	SINTERING OF GLASS POWDERS DURING CONSTANT RATES OF HEATING. CUTLER I B J AM CERAM SOC 52 1 14–17 1969

TPRC Number	Bibliographic Citation
52550	**IMMISCIBILITY AND CRYSTALLIZATION IN AL2O3-SIO2 GLASSES.** MAC DOWELL J F BEALL G H J AM CERAM SOC 52 1 17-25 1969
52558	**INFRARED SPECTRA AND DISTRIBUTION OF CATIONS IN THE COMPOUND GA AS O4.** PAQUES-LEDENT M-T TARTE P COMPT REND 268 C 3 233-5 1969
52563	**ALUMINUM TITANATE AND RELATED COMPOUNDS.** THIELKE, N. R. MATERIALS LABORATORY, WRIGHT-PATTERSON AFB, OHIO 1-34, 1953. (WADC-TR-53-165)
52564	**SYLVESTER AND COMPANY QUARTERLY REPORT.** SCHOENLAUB, R. A. SYLVESTER AND COMPANY, CLEVELAND, OHIO 1-37, 1953. (AD 222917)
52565	**DEVELOPMENT OF ULTRA REFRACTORY MATERIALS.** SHAFFER, P. T. B. RESEARCH AND DEVELOPMENT DIVISION CARBORUNDUM COMPANY 1-21, 1958. (AD 205942)
52567	**ALUMINA-BASE CERMETS.** SHEVLIN, T. S. HAUCK, C. A. AERONAUTICAL RESEARCH LAB., WRIGHT-PATTERSON AFB, OH. 1-45, 1954. (WADC-TR-54-173(PT. 1), AD 49092, PB-134856)
52568	**DEVELOPMENT OF CERAMIC DIELECTRICS.** SMOKE E J PHILLIPS C J KIM Y S ULRICH D R KASTENBEIN E L SCHNEIDER F WHELAN P BREEN C CERAMIC RES. STATION, RUTGERS STATE UNIV. 1-74, 1960. (AD 287166)
52569	**THE PROTECTION OF CARBON AND GRAPHITE AGAINST OXIDATION AT TEMPERATURES OF 1200 DEGREES.** SAZONOVA M V SITNIKOVA A YA APPEN A A ZHUR PRIKLAD KHIM 34 3 505-12 1961 (FOR ENGLISH TRANSLATION SEE T52570)
52570	**THE PROTECTION OF CARBON AND GRAPHITE AGAINST OXIDATION AT TEMPERATURES OF 1200 DEGREES.** SAZONOVA, M. V. SITNIKOVA, A. YA. APPEM, A. A. WRIGHT-PATTERSON AFB, OHIO 1-13, 1961. (ENGLISH TRANSLATION OF ZH. PRIKL. KHIM., 34 (3), 505-12, 1961; FOR ORIGINAL SEE T52569) (MCL-1285/1 AND 2/, AD 269653)
52574	**DEVELOPMENT OF REINFORCED CERAMIC MATERIAL SYSTEMS.** STEJSKAL L M JENKINS A L GUNDERSON J M BRESLICH F N JR CONA D A STERRY W M BOEING AIRPLANE CO., AERO-SPACE DIV., SEATTLE, WA. 1-173, 1961. (WADD-TR-60-491, AD 259930L)
52575	**ZIRCONIA. ITS CRYSTALLOGRAPHIC POLYMORPHY AND HIGH TEMPERATURE POTENTIALS.** WEBER, B. C. SCHWARTZ, M. A. AERONAUTICAL RESEARCH LAB., WRIGHT-PATTERSON AFB, OH. 1-21, 1958. (WADC-TR-58-646, AD 206908)
52576	**METAL FIBER REINFORCED CERAMICS.** TINKLEPAUGH J R GOSS B R HOSKYNS W R CONNOR J H BUTTON D D COLLEGE OF CERAMICS, ALFRED UNIVERSITY 1-77, 1960. (WADC-TR-58-452(PT. III), PB 171550)
52577	**METAL FIBER REINFORCED CERAMICS.** SWICA J J HOSKYNS W R GOSS B R CONNOR J H TINKLEPAUGH J R COLLEGE OF CERAMICS, ALFRED UNIVERSITY 1-41, 1960. (WADC-TR-58-452(PT. II), PB 161481)
52586	**CHEMICAL STRENGTHENING OF CERAMIC MATERIALS.** KIRCHNER, H. P. GROVER, R. M. LINDEN LABS., INC., STATE COLLEGE, PA 1-68, 1965. (AD 463287)
52587	**CHEMICAL STRENGTHENING OF CERAMIC MATERIALS.** KIRCHNER H P GROVER R M PLATTS D R WALKER R E PENNSYLVANIA STATE UNIV. - TECHNICAL REPORTS 1-143 1967
52594	**NEAR-INFRA-RED DIFFUSE REFLECTIVITIES OF NATURAL AND MAN-MADE MATERIALS.** NICOLLE R L IRVINE J BOWDEN F G BRIT J APPL PHYS /J PHYS D/ 2 2 201-4 1969
52650	**THERMAL EXPANSION OF CERMET COMPONENTS BY HIGH TEMPERATURE X-RAY DIFFRACTION.** MAUER, F. A. BOLZ, L. H. NATIONAL BUREAU OF STANDARDS 1-20, 1954. (NBS-3644, AD 46601)
52652	**AN INVESTIGATION OF THE THEORETICAL AND PRACTICAL ASPECTS OF THE THERMAL EXPANSION OF CERAMIC MATERIALS.** MERZ, K. M. CORNELL AERONAUTICAL LAB., INC., CORNELL UNIV., BUFFALO, NEW YORK 1-18, 1959. (CAL-PI-1273-M-5, AD 230864)
52653	**AN INVESTIGATION OF THE THEORETICAL AND PRACTICAL ASPECTS OF THE THERMAL EXPANSION OF CERAMIC MATERIALS.** MERZ, K. M. CORNELL AERONAUTICAL LAB., INC., CORNELL UNIVERSITY, BUFFALO, N. Y. 1-9, 1960. (CAL-PI-1273-M-6, AD 235034)
52654	**RESEARCH OF TECHNIQUES OF METHODS ON WHICH TO BASE DEVELOPMENT OF HIGH TEMPERATURE GLAZING ATTACHMENTS.** PARTAIN, G. K. PISCOPO, F. A. WILSON, R. E. WRIGHT AIR DEVELOPMENT DIVISION, WRIGHT-PATTERSON AFB, OHIO 1-63, 1960. (WADD-TR-60-11., AD 240357)
52656	**A TITANIUM MATCHING SILICA-FREE CERAMIC.** LESCHEN, J. G. RESEARCH LAB., G. E. CO., SCHENECTADY, N. Y. 1-22, 1961. (AFCRL-298, AD 260363)
52661	**RESEARCH ON ELEVATED TEMPERATURE RESISTANT CERAMIC STRUCTURAL ADHESIVES. PART VI.** FORLANO R J KRUMWIEDE D M BENZEL J F THORNTON H R LEFORT H G WRIGHT AIR DEVELOPMENT CENTER 1-72, 1962. (WADC-TR-55-491(PT. IV), AD 282302)
52662	**INVESTIGATION OF THEORETICAL AND PRACTICAL ASPECTS OF THE THERMAL EXPANSION OF CERAMIC MATERIALS.** KIRCHNER H P SCHEETZ H A BROWN W R SMYTH H T CORNELL AERONAUTICAL LAB. INC., BUFFALO, NEW YORK 1-101, 1962. (CAL-PI-1273-M-12, AD 296188)
52665	**STUDY OF HIGH TEMPERATURE MATERIALS.** SMOKE E J ROODA J SIRKIS M D HOLSBERG P J MOLONY D A CHOA C N. J. CERAMIC RESEARCH STATION, RUTGERS STATE UNIV. 1-35, 1960. (RADC-TR-60-233, AD 248105)
52667	**STRESSES IN POLYCOMPONENT SOLID CERAMICS OR CERMETS.** STRADLEY J G GILES T M SHEVLIN T S EVERHART J O RESEARCH FOUNDATION, OHIO STATE UNIVERSITY 1-11, 1959. (RF-806-4, AD 215330)
52670	**IMPROVED CERAMICS.** SMOKE, E. J. ARLETT, R. H. PHILLIPS, C. J. HALPERN, A. S. N. J. CERAMIC RESEARCH STATION, RUTGERS STATE UNIV. 1-80, 1958. (AD 206803)
52678	**LITHIA IN PORCELAIN ENAMELS.** HUPPERT P A AM CERAM SOC BULL 38 2 57-60 1959
52680	**THE INTERPRETATION OF THERMAL EXPANSION DATA IN RESEARCH AND CONTROL PRACTICES.** FORBES D W A CAN CERAM SOC J 28 77-92 1959
52681	**MECHANISM OF CERAMIC-TO-METAL ADHERERNCE.** PINCUS A G CERAMIC AGE 16-32 1954
52682	**ACCURATE, RAPID MEASUREMENT OF THERMAL EXPANSION.** BAULEKE M P CERAMIC AGE 71 3 24-6 1958
52683	**THERMAL EXPANSION AS A METHOD FOR CHECKING THE COMPOSITION OF CERAMIC CLAYS AND OF STUDYING MINERALOGICAL CHANGES DURING FIRING.** HOLDRIDGE D A CLAY MINERALS BULL 4 22 94-105 1959

TPRC Number	Bibliographic Citation
52684	**THE EFFECT OF FIRING ON SOME PHYSICAL PROPERTIES OF ETRURIA MARLS.** HOLDRIDGE D A CLAY MINERALS BULL 5 28 90-7 1962
52687	**EFFECT OF PARTIAL REPLACEMENTS OF SODA BY LITHIA IN OPAL AND ALABASTER GLASS.** RAUCH H W COMMONS C H JR SILVERMAN A GLASS INDUSTRY 41 261-92 1960
52688	**EFFECTS OF FLUXING INGREDIENTS ON THE PROPERTIES OF MAGNESIUM-CALCIUM-ALUMINO SILICATE GLASSES.** KUMAR S NAG B B INDIAN CERAMIC SOC 21 4 107-14 1962
52689	**THE THERMAL EXPANSION OF REFRACTORIES.** NORTON F H J AM CERAMIC SOC 8 798-815 1925
52690	**SOME PROPERTIES OF CHROME SPINEL.** PARMELEE C W ALLY A J AM CERAMIC SOC 15 213-25 1932
52693	**DYSPROSIUM OXIDE CERAMICS.** PLOETZ G L MUCCIGROSSO A T OSIKA L M JACOBY W R J AM CERAMIC SOC 43 3 154-9 1960
52694	**THERMAL-EXPANSION ANISOTROPY OF OXIDE SOLID SOLUTIONS.** MERZ K M BROWN W R KIRCHNER H P J AM CERAMIC SOC 45 11 531-6 1962
52698	**STRUCTURE OF NICKEL OXIDE CONTAINING ALUMINA.** GILLAM E HOLDEN J P J AM CERAMIC SOC 46 12 601-4 1963
52699	**THERMAL EXPANSION OF YTTRIUM OXIDE AND OF MAGNESIUM OXIDE WITH YTTRIUM OXIDE.** NIELSEN T H LEIPOLD M H J AM CERAMIC SOC 47 5 256 1964
52749	**AN APPRAISAL OF THE SINTERING BEHAVIOR AND THERMAL EXPANSION OF SOME COLUMBATES.** DURBIN, E. A. HARMAN, C. G. BATTELLE MEMORIAL INSTITUTE, COLUMBUS, OHIO 1-12, 1952. (BMI-791)
52754	**THERMAL EXPANSION OF ALPHA ALUMINA.** CAMPBELL, W. J. GRAIN, C. ADVANCES X-RAY ANALYSIS 5, 144- , 1961. (BM-RI-5757)
52757	**INFRARED-SPECTROSCOPIC INVESTIGATION OF THE VITRIFICATION AND CRYSTALLIZATION OF LEAD GLASS WITH A COMPOSITION CORRESPONDING TO THE COMPOUND PB0.SI02.** SMIRNOVA E V INORGANIC MATERIALS 4 7 985-8 1968 (ENGLISH TRANSLATION OF IZV. AKAD. NAUK SSSR, NEORG. MATER. 4 (7), 1124-8, 1968; FOR ORIGINAL SEE T34183)
52758	**THE UNIVERSAL PRECISION DILATOMETER.** CUTLER, G. L. KNOLLS ATOMIC POWER LAB., SCHENECTADY, NEW YORK 1-20, 1958. (KAPL-M-GLC-2)
52759	**THERMAL EXPANSION MEASUREMENTS OF SIX FUEL MATERIALS.** MC CREIGHT, L. R. KNOLLS ATOMIC POWER LAB., SCHENECTADY, NEW YORK 1-9, 1952. (KAPL-M-LRM-7)
52768	**HIGH-TEMPERATURE MATERIALS PROGRAM.** GENERAL ELECTRIC COMPANY GENERAL ELECTRIC CO., NUCLEAR MATERIALS AND PROPULSION OPERATION, CINCINNATI, OHIO 1-38, 1964. (GEMP-36A)
52770	**COMPOSITE CERAMIC RADOME MANUFACTURE BY MOSAIC TECHNIQUES.** FILIPPI, F. J. NARMCO RESEARCH AND DEVELOPMENT, SAN DIEGO, CA. 1-137, 1963. (IR-7-984-/II/, AD 406042)
52777	**AN INVESTIGATION OF THE THEORETICAL AND PRACTICAL ASPECTS OF THE THERMAL EXPANSION OF CERAMIC MATERIALS.** KIRCHNER, H. P. CORNELL AERONAUTICAL LAB., INC., BUFFALO, NEW YORK 1-7, 1958. (CAL-REPT PI-1273-M-1, AD 207202)
52778	**DEVELOPMENT OF ULTRA REFRACTORY MATERIALS.** CLINE, C. F. LEWIS, G. RESEARCH AND DEVELOPMENT DIV., THE CARBORUNDUM CO., NAGARA FALLS, N. Y. 1-18, 1957. (AD 138876)
52780	**ANISOTROPY AND STRENGTH OF CERAMIC BODIES.** BUESSEM, W. R. MC KINSTRY, H. A. PENNSYLVANIA STATE UNIV., UNIVERSITY PARK, PA. 1-39, 1962. (AD 288273)
52783	**INVESTIGATION OF GLASS-METAL COMPOSITE MATERIALS.** WHITEHURST, H. B. WILEY, R. B. OWENS-CORNING FIBERGLAS CORP., NEWARK, OHIO 1-28, 1958. (AD 223057)
52786	**A METHOD FOR THE DETERMINATION OF THE THERMAL EXPANSION OF REFRACTORY MATERIALS FROM 20 - 2300 C.** SHAKHTIN D M LEVINTOVICH E V PIVOVAR T L ELISEEVA G G REFRACTORIES 33 8 517-21 1968 (ENGLISH TRANSLATION OF OGNEUPORY 33 (8), 51-3, 1968; FOR ORIGINAL SEE T51377)
52788	**COMPOSITE CERAMIC RADOME MANUFACTURE BY MOSAIC TECHNIQUES.** FILIPPI, F. J. NARMCO RESEARCH AND DEVELOPMENT, SAN DIEGO, CA. 1-107, 1963. (IR-7-984-(III), ASD-7-984, AD 415269)
52789	**CERAMIC MATERIALS FOR HIGH TEMPERATURES.** LAWRENCE W G NYS COLLEGE OF CERAMICS AT ALFRED UNIVERSITY 21 3 1-78 1956
52792	**INVESTIGATION FOR THE DEVELOPMENT OF CERAMIC BODIES FOR ELECTRON TUBES.** WISELY, H. R. GENERAL ENGINEERING LAB., GENERAL ELECTRIC COMPANY, SCHENECTADY, NEW YORK 1-24, 1961. (AFCRL-74, AD 253795)
52793	**INVESTIGATION FOR THE DEVELOPMENT OF CERAMIC BODIES FOR ELECTRON TUBES.** WISELY, H. R. AIR FORCE CAMBRIDGE RESEARCH LAB., BEDFORD, MA. 1-9, 1961. (AFCRL-344, AD 256938)
52794	**INVESTIGATION FOR THE DEVELOPMENT OF CERAMIC BODIES FOR ELECTRON TUBES.** WISELY, H. R. AIR FORCE CAMBRIDGE RESEARCH LAB., BEDFORD, MA. 1-79, 1962. (AFCRL-62-388, AD 277842)
52795	**DIELECTRIC TO METAL SEAL TECHNOLOGY STUDY.** BERG, M. GRIMM, A. C. MARINARO, F. F. RADIO CORPORATION OF AMERICA, LANCASTER, PA. 1-60, 1962. (RADO-TDR-62-401, AD 296534)
52804	**A FLUID THEORY FOR THE DEEP STRUCTURE OF DIP-SLIP FAULT ZONES.** TURCOTTE D L OXBURGH E R PHYS EARTH PLANET INTERIORS 1 381-6 1968
52805	**REFRACTORY FOR TUNNEL BURNERS OPERATING AT VERY HIGH TEMPERATURES.** ROBERTS, A. L. OTT, T. B. ASTBURY, N. F. GAS COUNCIL RESEARCH COMMUNICATION 15-22 1963
52808	**THERMAL DIFFUSIVITY OF THE URANIUM DIOXIDE-SILICON DIOXIDE SYSTEM WITH VARYING URANIUM DIOXIDE CONCENTRATIONS IN THE 100-800 C. TEMPERATURE RANGE.** GLANZ G J NUCL MATER 29 3 329-33 1969
52812	**REFRACTORY CERAMICS AND INTERMETALLIC COMPOUNDS.** RAMKE W G LATVA J D AEROSPACE ENGR 22 76-84 1963
52843	**SUBSTRUCTURE CLASSIFICATION OF SILICATE GLASSES.** BABCOCK C L J AM CERAM SOC 52 3 151-3 1969

TPRC Number	Bibliographic Citation
52844	**ANNEALING OF IRRADIATION-INDUCED THERMAL CONDUCTIVITY CHANGES IN THO2-1.3 WT PERCENT UO2.** MAC EWAN J R STOUTE R L J AM CERAM SOC 52 3 160-5 1969
52854	**STUDIES OF CERAMIC PROCESSING.** SMOKE E J FLEISCHNER P L JACOBS W G MANGINO D R SCHOOL OF CERAMICS, RUTGERS STATE UNIV., NEW BRUNSWICK, NEW JERSEY 1-90, 1967. (AD 821476)
52855	**TARGET SIGNATURE MEASUREMENTS LABORATORY STUDY.** ANDRYCHUK, D. BENN, M. TEXAS INSTRUMENTS, INC., DALLAS, TEXAS 1-121, 1967. (AD 822204)
52863	**TERRAIN ANALYSIS BY ELECTROMAGNETIC MEANS. REPT. 4. LABORATORY INVESTIGATIONS OF THE INFRARED EMISSIVITY OF SOILS BELOW A WAVELENGTH OF 7.7 MICRONS.** LAVECCHIA N J JR WILLIAMSON A N JR NIKODEM H J U. S. ARMY ENGINEER WATERWAYS EXPERIMENT STATION, VICKSBRUG, MISS. 1-68, 1967. (AD 815453)
52865	**INFRARED EMISSION FROM THE SURFACE OF THE MOON.** WINTER, D. F. BOEING SCIENTIFIC RESEARCH LABS., SEATTLE, WA. 1-78, 1968. (AD 672074)
52868	**STRENGTHENING OXIDES BY REDUCTION OF CRYSTAL ANISOTROPY.** KIRCHNER, H. P. GRUVER, R. M. LINDEN LAB., INC., STATE COLLEGE, PA. 1-65, 1967. (AD 661469)
52870	**INFRARED COATING STUDIES.** BAUSCH AND LOMB INC., ROCHESTER, NEW YORK RESEARCH AND DEVELOPMENT DIVISION, BAUSCH AND LOMB, INC., ROCHESTER, NEW YORK 1-13, 1968. (AD 666787)
52875	**GLASS, GLASS CERAMIC AND SINTER GLASS CERAMIC.** SACK W CHEMIE INGENIEUR TECHNIK 37 1154-65 1965 (FOR ENGLISH TRANSLATION SEE T52876)
52876	**GLASS, GLASS CERAMIC AND SINTER GLASS CERAMIC.** SACK, W. ROYAL AIRCRAFT ESTABLISHMENT, FARNBOROUGH, ENGLAND 1-39, 1966. (ENGLISH TRANSLATION OF CHEMIE INGENIEUR TECHNIK 37, 1154-65, 1965; FOR ORIGINAL SEE T52875)
52892	**BIOLOGICAL METHODS FOR CAMOUFLAGE CONCEALMENT, AND TONEDOWN.** SCAFF, W. L., JR. MACIASR, F. M. BELL, R. T. NORTHROP CORPORATE LABS., HAWTHORNE, CA. 1-69, 1968. (AFAPL-TR-67-153, NCL-67-162, AD 830415)
52896	**ULTRASONIC AND THERMAL STUDIES OF SELECTED PLASTICS LAMINATED MATERIALS, AND METALS.** ASAY, J. R. URZENDOWSKI, S. R. GUENTHER, A. H. AIR FORCE WEAPONS LAB., KIRTLAND AFB, NEW MEXICO 1-266, 1968. (AFWL-TR-67-91, AD 827596)
52897	**NARROW-BAND ULTRAVIOLET FILTER PHOTOGRAPHY.** MANGOLD, V. L. AIR FORCE FLIGHT DYNAMICS LAB., WRIGHT-PATTERSON AFB, OHIO 1-31, 1966. (AFFDL-TR-66-140, AD 808792)
52900	**HIGH PERFORMANCE AIRCRAFT WINDSHIELD DEVELOPMENT PROGRAM.** SHOEMAKER, A. F. CORNING GLASS WORKS, NEW YORK 1-175, 1966. (AFML-TR-66-69, AD 812203)
52907	**OPTICAL EFFECTS UNDERLYING THE ANALYTICAL INFRARED SPECTRA OF SOLID MATERIALS.** PHILLIPPI, C. M. AIR FORCE MATERIALS LAB., WRIGHT-PATTERSON AFB, OHIO 1-36, 1968. (AFML-TR-67-437, AD 834756)
52919	**EFFECTS ON REFLECTION CHARACTERISITCS OF A LIMONITE SURFACE ON RADIATION EMERGING FROM A MODEL MARTIAN ATMOSPHERE.** GRAY E L COULSON K L BOURICIUS G M B MISSILE AND SPACE DIV, GENERAL ELECTRIC CO., PHILADELPHIA, PA. 1-52, 1967. (67-SD-17, AD 822325)
52935	**MECHANICAL RELAXATIONS IN MIXED-ALKALI SILICATE GLASSES. I. RESULTS.** SHELBY J E JR DAY D E J AM CERAM SOC 52 4 169-74 1969
52936	**REACTIONS IN THE SYSTEM ZRO2-SRO.** NOGUCHI T OKUBO T YONEMOCHI O J AM CERAM SOC 52 4 178-81 1969
52945	**REFLECTIVITY AND MICROINDENTATION HARDNESS OF FERROSELITE FROM COLORADO AND NEW MEXICO.** SANTOS E S AMER MINERAL 53 11 2075-7 1968 CA 70 39633
52947	**COMPATIBILITY BETWEEN THE GLAZE AND THE CERAMIC SUPPORT IN MAJOLICA PRODUCTION.** COLLEPARDI, M. ROSSI, G. ANN. CHIM. (ROME) 58 (10), 1014-25, 1968.
52968	**PHYSICOCHEMICAL PROPERTIES OF TL2O-B2O3-SIO2 GLASSES AND THEIR PHASE SEPARATIONS.** KIM K H DAEHAN HWAHAK HWOEJEE 12 2 65-80 1968 CA 70 50074
53006	**COMPARATIVE THERMAL CONDUCTIVITY METHOD FOR SELF-GUARDING DISK SPECIMENS.** LUCKS, C. F. DEEM, H. W. PROCEEDINGS OF THE CONFERENCE ON THERMAL CONDUCTIVITY, 1ST 1-9 1961
53021	**CHANGES IN THE TECHNICAL PROPERTIES OF ZIRCONIUM REFRACTORIES IN RELATION TO THE DEGREE OF STABILIZATION OF ZIRCONIUM DIOXIDE.** KLYUCHAROV YA V STRAKHOV V I IZV AKAD NAUK SSSR NEORG MATER 4 9 1502-6 1968 CA 70 22586 (FOR ENGLISH TRANSLATION SEE T44778)
53023	**THERMAL EXPANSION OF STABILIZED ZIRCONIUM DIOXIDE AT 300-2600 K.** SHAKHTIN D M LEVINTOVICH E V PIVOVAR T L ELISEEVA G G IZV AKAD NAUK SSSR NEORG MATER 4 9 1603-4 1968 CA 70 50104 (FOR ENGLISH TRANSLATION SEE T44799)
53054	**DETERMINING THE EXTENT OF HEAT TREATMENT OF HIGHLY CARBONIZED MATERIALS /COKE, GRAPHITE/.** ZAMOLUEV V K KHIM TVERD TOPL 5 137-40 1968 CA 70 39595
53097	**DIFFUSIONAL PRESTRESSING OF CERAMICS.** BICKELHAUPT, R. E. SOUTHERN RESEARCH INSTITUTE, BIRMINGHAM, ALABAMA 1-102, 1968. (AD 840509)
53105	**OBSERVATIONS ON THE THERMAL CONDUCTIVITY OF FOAM GLASS.** SCHULLE W SILIKAT TECHNIK 16 7 216-8 1965 (FOR ENGLISH TRANSLATION SEE T53106)
53106	**OBSERVATIONS ON THE THERMAL CONDUCTIVITY OF FOAM GLASS.** SCHULLE W FOREIGN TECHNOLOGY DIVISION 1-13, 1967. (ENGLISH TRANSLATION OF SILIKAT. TECHNIK. 16 (7), 216-8, 1965; FOR ORIGINAL SEE T53105) (FTD-HT-23-867-67, AD 835635)
53122	**COMPILATION OF SPECTRAL EMITTANCES OF BACKGROUND AND TARGET CONSTITUENTS IN THE 8 TO 14 MICRON RANGE.** CZARNIK, J. W. LEET, H. P. NAVAL WEAPONS CENTER, CHINA LAKE, CA. 1-188, 1968. (NWC-TP-4624, AD 841158)
53124	**CERAMIC MATERIALS FOR CASTING HIGH STRENGTH STEEL.** WALES, W. WATERVLIET ARSENAL, WATERVLIET, NEW YORK 1-42, 1968. (WVT-6813, AD 831004)
53130	**MICROHARDNESS AND COEFFICIENT OF LINEAR EXPANSION OF CERTAIN GLASSES OF THE TYPE PORCELAIN GLASS PHASES.** AUGUSTINIK A I FEDOROVA E P ZHUR PRIKLAD KHIM 26 9 925-30 1953 (FOR ENGLISH TRANSLATION SEE T63072)
53133	**THERMAL CONDUCTIVITY OF VITREOUS SODIUM PHOSPHATES.** TSUBAKI T HATTORI M TAWAKA M KOLLOID-Z Z POLYMERE 225 2 112-16 1968 JA 52 97

TPRC Number	Bibliographic Citation
53139	THERMOPHYSICAL PROPERTIES OF FIBER-REINFORCED PLASTICS. I. MEASUREMENT OF THERMAL PROPERTIES BY RADIATION HEATING. TAKENAKA, Y. OGAWA, K. KYOKA PURASUCHIKKUSU 14 (4), 191-9, 1968.
53147	METHODS FOR MEASURING THE THERMAL EXPANSION OF GLASSES AND OTHER MATERIALS OF ELECTRONICS TECHNOLOGY. RABINOVICH E M SHELYUBSKII V I METODY IZMER TEPL RASSHIR STEKOL SPAIVAEMYKH NIMI METAL TR VSES SIMP DILATOMETRII 1ST LENINGRAD 1966 41-51 1967 CA 70 6290
53148	REQUIREMENTS FOR MEASUREMENTS AND STANDARDIZATION OF THE THERMAL EXPANSION COEFICIENTS OF GLASSES AND METALS USED IN THE ELECTRONICS INDUSTRY. KUZNETSOVA I N LEVIN A S SOLOVEVA R Z LAVUT E I DULKINA I N METODY IZMER TEPL RASSHIR STEKOL SPAIVAEMYKH NIMI METAL TR VSES SIMP DILATOMETRII 1ST LENINGRAD 1966 75-82 1967 CA 70 51891
53149	THERMAL EXPANSION OF PRECISION ALLOYS USED FOR JOINTS WITH GLASSES. SOLOVEVA N A YUDKEVICH M I PRIDANTSEVA K S PASTERNAK I I METODY IZMER TEPL RASSHIR STEKOL SPAIVAEMYKH NIMI METAL TR VSES SIMP DILATOMETRII 1ST LENINGRAD 1966 83-8 1967 CA 70 22178
53151	THERMAL EXPANSION IN GLASS SYSTEMS STUDIED ON A SAMPLE OF VARIABLE COMPOSITION. SHELYUBSKII V I METODY IZMER TEPL RASSHIR STEKOL SPAIVAEMYKH NIMI METAL TR VSES SIMP DILATOMETRII 1ST LENINGRAD 1966 124-9 1967 CA 70 6292
53152	TEMPERATURE FUNCTION OF THE COEFFICIENT OF EXPANSION OF GLASS AND ITS DEPENDENCE ON CHEMICAL COMPOSITION. SHCHAVELEV O S METODY IZMER TEPL RASSHIR STEKOL SPAIVAEMYKH NIMI METAL TR VSES SIMP DILATOMETRII 1ST LENINGRAD 1966 140-5 1967 CA 70 14100
53153	THERMAL EXPANSION OF GLASSES IN THE TRANSITION REGION. ZHUKHOROVA V M KHEIFETS V S METODY IZMER TEPL RASSHIR STEKOL SPAIVAEMYKH NIMI METAL TR VSES SIMP DILATOMETRII 1ST LENINGRAD 1966 145-51 1967 CA 70 6288
53154	METHODS FOR MEASURING THE COEFFICIENT OF THERMAL EXPANSION OF GLASSES AT TEMPERATURES EXCEEDING THE GLASS POINT. MAZURIN O V METODY IZMER TEPL RASSHIR STEKOL SPAIVAEMYKH NIMI METAL TR VSES SIMP DILATOMETRII 1ST LENINGRAD 1966 151-7 1967 CA 70 40338
53155	EFFECT OF THE CONDITIONS FOR PREPARING A MELT ON THE DILATOMETRIC CURVES. BOBKOVA N M METODY IZMER TEPL RASSHIR STEKOL SPAIVAEMYKH NIMI METAL TR VSES SIMP DILATOMETRII 1ST LENINGRAD 1966 157-60 1967 CA 70 40337
53156	USE OF DILATOMETRIC MEASUREMENTS DURING THE STUDY OF THE CERAMMING OF GLASS. ZHUNINA L A YAGLOV V N METODY IZMER TEPL RASSHIR STEKOL SPAIVAEMYKH NIMI METAL TR VSES SIMP DILATOMETRII 1ST LENINGRAD 1966 167-9 1967 CA 70 6301
53157	EFFECT OF THE THERMAL HISTORY OF SOME ELECTRON TUBE GLASSES ON THE THERMAL EXPANSION. SILVESTROVICH S I FIRSOV V M METODY IZMER TEPL RASSHIR STEKOL SPAIVAEMYKH NIMI METAL TR VSES SIMP DILATOMETRII 1ST LENINGRAD 1966 169-72 1967 CA 70 14097
53178	A FIELD EVALUATION OF CONVENTIONAL PREDICTION EQUATIONS FOR THERMAL CONDUCTIVITY IN HIGHWAY PAVEMENT MATERIALS. HELEY, W. WEST VIRGINIA UNIVERSITY, M. S. THESIS 1-179 1968
53201	THERMAL EXPANSION OF REFRACTORY OXIDES AT HIGH TEMPERATURES. SHAKHTIN C M LEVINTOVICH E V PIVOVAR T L ELISEEVA G G KARAULOV A G RUDYAK I N SB NAUCH TR UKR NAUCH ISSLED INST OGENOUPOROV 10 160-71 1967 CA 70 60407 (FOR ENGLISH TRANSLATION SEE T61027)
53206	THERMAL CONDUCTIVITY OF SILVER TELLURIDE. AKFBI, N. SCI. REP. NIIGATA UNIV. B (1), 12-16, 1967.
53232	CONTRIBUTION TO THE STUDY OF DILATION AT HIGH TEMPERATURE. PIERREY J ANN CHIMIE 4 12 133-95 1949
53233	CONTRIBUTION TO THE DETERMINATION OF CERTAIN CLAYEY MINERALS BY QUANTITIATIVE DIATOMETRIC TESTS. PAQUIN P BULL SOC FRANC CERAM SUPPL 58 1-148 1962
53239	THE EXPANSION OF QUARTZITES AND SANDSTONES WHEN HEATED. KOHLER K RADEX RUNDSCHAU 2 536-41 1961
53241	REHEAT CONTRACTION/EXPANSION OF CERAMIC MATERIALS. REICH H F TONIND ZTG KERAM RUNDSCHAU 86 8 177-81 1962
53242	CONTRIBUTION TO THE THERMAL EXPANSION OF HETEROGENEOUS CERAMIC MATERIALS. CHVATAL TH SCHNITTLER FR TRANS 6TH INTERN CERAMIC CONGR 53-61 1958
53243	EXPANSION BEHAVIOUR AT HIGH TEMPERATURE OF VARIOUS TYPES OF KAOLINS AND KAOLINITIC CLAYS. BAUDRAN A A TRANS 7TH INTERN CERAM CONGR 393-403 1960
53251	TECHNOLOGICAL STUDIES, THERMAL EXPANSION, AND DURABILITY OF IRON-POOR AND IRON-RICH MAGNESITE BRICKS. SCHWIETE H E SCHINNERLING F SPRECHSAAL KERAM GLAS EMAIL SILIKATE 101 22 1000-18 1968 JA 52-9 309
53252	PREPARATION OF CRYSTALLIZED COATINGS ON METAL. PAVLUSHKIN, N. M. ZHURAVLEV, A. K. EGOROVA, L. S. STEKLO KERAM. 26 (1), 11-3, 1969. (FOR ENGLISH TRANSLATION SEE TPRC NO. 66785)
53266	A UNIVERSAL ENAMEL FOR ALUMINUM SHEETS AND FOILS AS WELL AS FOR ALUMINUM-COATED SHEET. NALEPA W SZKLO CERAM 19 8 236-41 1968 CA 70 22602
53267	PROPERTIES OF CRYSTALLIZED GLASSES OF THE SYSTEM POTASSIUM OXIDE-LITHIUM OXIDE-ZINC OXIDE-SILICON OXIDE. SCHLEIFER P KUBACKI W SZKLO CERAM 19 9 258-60 1968 CA 70 60359
53283	PROPERTIES OF HIGH-ALUMINA CERAMICS. POPPER P TRANS. INT. CERAM. CONGR. 10TH 211-22 1967 CA 70 70788
53304	THERMAL EXPANSION STUDIES IN THE SILICATE INDUSTRY. TIBOR T OTTONE F EPITOANYAG 12 425-35 1960
53305	THE ANOMALIES IN THE DILATOMETER CURVE OF UNFIRED CLAYS BELOW AND THROUGH 200 C. ZWETSCH A TRANS 6TH INTERN CERAM CONGRESS 169-82 1958
53327	DETERMINATION OF THE SPECIFIC HEAT AND ENTHALPY OF FUSION OF INDUSTRIAL ZINC-CONTAINING SLAGS. BERSHAK V I SMIRNOVA M N TSVET METAL 41 8 45-9 1968 CA 70 59858
53391	LOW-TEMPERATURE HEAT CAPACITY AND POLYMER STRUCTURE OF SODIUM BORATE GLASSES. TARASOV V V TURDAKIN V A ZH NEORG KHIM 13 10 2651-4 1968 CA 70 14099
53424	THERMAL CONDUCTIVITY OF EUROPIUM OXIDE /EUO/ ACROSS THE CURIE TEMPERATURES. MORRIS R G CASON J L JR HELV PHYS ACTA 41 6/7 1045-51 1968
53442	THERMAL CONDUCTIVITY OF FERRITES-SPINELS AND THEIR SOLID SOLUTIONS. POINT DEFECTS IN FERRITES, DETERMINING ADDITIONAL THERMAL RESISTANCE OF THE LATTICE. VISHNEVSKII I I SKRIPAK V N IZV AKAD NAUK SSSR NEORG MATER 4 11 1989-96 1968 CA 70 61913 (FOR ENGLISH TRANSLATION SEE T54102)

TPRC Number | Bibliographic Citation

53463 **GLASS-CERAMIC COATINGS WITH HIGH DENSITIES PREPARED BY GAS-FLAME SPRAY COATING.**
SVIRSKII L D PIROGOV YU A
KHIM TEKHNOL
9 128-31 1967 CA 70 60387

53469 **METROLOGICAL WORK IN THE AREA OF DILATOMETRY.**
AMATUNI A N
METODY IZMER TEPL RASSHIR STEKOL SPAIVAEMYKH NIMI METAL TR VSES SIMP DILATOMETRII 1ST LENINGRAD 1966
14-21 1967 CA 70 31747

53471 **COEFFICIENTS OF LINEAR EXPANSION OF SOME ELECTRON-TUBE GLASSES.**
RABINOVICH E M FAINBERG E A
METODY IZMER TEPL RASSHIR STEKOL SPAIVAEMYKH NIMI METAL TR VSES SIMP DILATOMETRII 1ST LENINGRAD 1966
184-5 1967 CA 70 60375

53472 **GLASSES FOR FIBER OPTICS.**
POLUKHIN, V. N.
OPT. - MEKH. PROM.
35 (9), 34-8, 1968.
(FOR ENGLISH TRANSLATION SEE TPRC NO. 67357)

53478 **STANDARD MATERIAL FOR MEASURING THERMAL CONDUCTIVITY.**
SHADRICHEV, E. V. SMIRNOV, I. A.
PRIB. TEKH. EKSP.
(5), 218-9, 1968.
(FOR ENGLISH TRANSLATION SEE TPRC NO. 67345)

53479 **TRIAXIAL PORCELAINS—STRENGTH AND MICROSTRUCTURAL RELATIONS.**
WARSHAW S I SEIDER R J
CERAM. MICROSTRUCT. PROC. INT. MATER. SYMP., 3RD
559-71 1968 CA 70 31334

53490 **EQUATION FOR THERMAL EXPANSIVITY IN PLANETARY INTERIORS.**
ANDERSON O L
J GEOPHYSICAL RESEARCH
72 14 3661-8 1967

53495 **HEAT CAPACITIES OF LINEAR HIGH POLYMERS.**
WUNDERLICH, B.
RENSSELAER POLYTECHNIC INSTITUTE, TROY, NEW YORK
1-175, 1968.
(AD 681045, N69-21364)

53502 **EVALUATION OF BERYLLIA CERAMICS FOR MICROWAVE VALVES.**
BRITISH CERAMIC RESEARCH ASSOCIATION
THE BRITISH CERAMIC RESEARCH ASSOCIATION
1-6, 1968.
(AD 846493)

53509 **MULTILAYER FILTERS FOR THE REGION 0.8 TO 100 MICRONS.**
SMITH, S. D. SEELEY, J. S.
AIR FORCE CAMBRIDGE RESEARCH LAB., BEDFORD, MA.
1-198, 1968.
(AFCRL-68-0496(VOLS. 1,2), AD 679565)

53519 **APPROXIMATE MEASUREMENTS OF SOME PHYSICAL PROPERTIES OF CARBON FIBRE REINFORCED POLYMERS, RELEVANT TO THEIR APPLICATIONS IN TRIBOLOGY.**
LANCASTER, J. K.
ROYAL AIRCRAFT ESTABLISHMENT, FARNBOROUGH, ENGLAND
1-34, 1968.
(RAE-TR-68-123(UL), N69-34403, AD 847814)

53539 **EFFECT OF LOW-QUANTITY COMPONENTS ON THE TRANSFORMATION RANGE OF GLASSES.**
WILWERGER F
PROC CONF SILICATE IND 1967
9 533-7 1968 CA 70 70769

53573 **THERMAL DIFFUSIVITY MEASUREMENTS OF HYPOSTOICHIOMETRIC URANIUM DIOXIDE AND URANIUM - PLUTONIUM OXIDE BY A MODULATED ELECTRON BEAM TECHNIQUE.**
SERIZAWA M KANEKO H YOKOUCHI Y KOIZUMI M
PROC. CONF. THERMAL CONDUCTIVITY, 8TH
549-67 1967

53585 **AN AUTOMATIC PLATE APPARATUS FOR MEASUREMENTS OF THERMAL CONDUCTIVITY OF INSULATING MATERIALS AT HIGH TEMPERATURES.**
FERRO, V. SACCHI, A.
PROCEEDING OF THE CONFERENCE ON THERMAL CONDUCTIVITY, 8TH
737-60 1969

53587 **THERMAL COMPARATOR DEVELOPMENTS.**
POWELL R W GROOT H DE WITT D P
PROC. CONF. THERMAL CONDUCTIVITY, 8TH
771-80 1969

53591 **THERMAL DIFFUSIVITY AND CONDUCTIVITY OF SINTERED URANIUM-THORIUM MIXED OXIDES.**
FERRO, C. MORETTI, S. PATIMO, C.
PROCEEDING OF THE CONFERENCE ON THERMAL CONDUCTIVITY, 8TH
815-22 1969

53595 **PRECISE HEAT TRANSFER MEASUREMENT WITH SURFACE THERMOCOUPLES.**
DIXON, W. P.
PROCEEDING OF THE CONFERENCE ON THERMAL CONDUCTIVITY, 8TH
873-85 1969

53596 **CURRENT NBS STEADY-STATE THERMAL CONDUCTIVITY MEASUREMENT METHODS.**
ROBINSON, H. E.
PROCEEDING OF THE CONFERENCE ON THERMAL CONDUCTIVITY, 8TH
11-27, 1961.

53597 **EXPERIMENTS ON THE SEPARATION OF HEAT TRANSFER MECHANISMS IN LOW-DENSITY FIBROUS INSULATION.**
PELANNE, C. M.
PROCEEDING OF THE CONFERENCE ON THERMAL CONDUCTIVITY, 8TH
897-911 1969

53598 **HIGH TEMPERATURE THERMAL CONDUCTIVITY OF SOILS.**
FLYNN, D. R. WATSON, T. W.
PROCEEDING OF THE CONFERENCE ON THERMAL CONDUCTIVITY, 8TH
913-39 1969

53600 **THERMAL DIFFUSIVITY OF THE ELEMENTS.**
HO, C. Y. POWELL, R. W. WU, K. Y.
PROCEEDING OF THE CONFERENCE ON THERMAL CONDUCTIVITY, 8TH
971-98 1969

53610 **THERMAL CONDUCTIVITIES OF TWO-PHASE MEDIA. METHODOLOGY, RESULTS, ESTIMATIONS.**
RATCLIFFE, E. H.
PROCEEDING OF THE CONFERENCE ON THERMAL CONDUCTIVITY, 8TH
1141-7 1969

53616 **A COMPARATIVE STUDY OF THE TOTAL DIRECTIONAL EMITTANCE OF A DIELECTRIC AND A METAL AS A FUNCTION OF SURFACE ROUGHNESS.**
SIEGFRIED, R. G.
UNIVERSITY OF KENTUCKY, M. S. THESIS
1-104 1968

53618 **AN UNSTEADY-STATE METHOD FOR DETERMINING THE THERMAL CONDUCTIVITY OF MATERIALS.**
BULMER, R. H.
UNIVERSITY OF MISSOURI, M. S. THESIS
1-76 1968

53620 **HIGH-ALUMINA INSULATION FROM KYANITE-SILLIMANITE CONCENTRATE.**
TSIGLER V D STOVBUR A V
OGNEUPORY
33 7 56-8 1968
(FOR ENGLISH TRANSLATION SEE T53621)

53621 **HIGH-ALUMINA INSULATION FROM KYANITE-SILLIMANITE CONCENTRATE.**
TSIGLER V D STOVBUR A V
REFRACTORIES
33 7 453-5 1968
(ENGLISH TRANSLATION OF OGNEUPORY 33 (7), 56-8, 1968; FOR ORIGINAL SEE T53620)

53622 **VERIFICATION OF THE HYPOTHESIS OF A MARTIAN LIMONITIC COVER, FROM SPECTROPHOTOMETRIC DATA.**
LEBEDEVA I I
VOPROSY ASTROFIZIKI
99-104 1965
(FOR ENGLISH TRANSLATION SEE T53623)

53623 **VERIFICATION OF THE HYPOTHESIS OF A MARTIAN LIMONITIC COVER, FROM SPECTROPHOTOMETRIC DATA.**
LEBEDEVA I I
NATIONAL AERONAUTICS AND SPACE ADMINISTRATION
85-90, 1966.
(ENGLISH TRANSLATION OF VOPROSY ASTROFIZIKI 99-104, 1965; FOR ORIGINAL SEE T53622)

53645 **INFRARED EMISSIVITIES OF SILICATES. EXPERIMENTAL RESULTS AND A CLOUDY ATMOSPHERE MODEL OF SPECTRAL EMISSION FROM CONDENSED PARTICULATE MEDIUMS.**
CONEL J E
J GEOPHYS RES
74 6 1614-34 1969 CA 70 82640

53652 **HIGH-TEMPERATURE THERMAL EXPANSION OF MAGNESITE AND CHROME-MAGNESITE BRICKS.**
BANERJEE J C BANERJEE S P SIRCAR N R
REV INTERN HAUTES TEMP REFRACTAIRES
5 1 63-8 1968 JA 52 12

53653 **MEASUREMENT OF THE THERMAL CONDUCTIVITY OF THE OXIDE LAYER OF CATHODES USED IN ELECTRONIC EQUIPMENT ACCORDING TO THE TIME FOR ESTABLISHING OF THEIR EMITTER SURFACE TEMPERATURE.**
NILOV O M
RADIOTEKH ELEKTRON
14 1 187-9 1969 CA 70 81705

TPRC Number	Bibliographic Citation
53688	**QUANTITATIVE CHARACTERISTICS OF THE HEAT CAPACITY SINGULARITIES IN SECOND-ORDER PHASE-TRANSITION POINTS** VORONEL A V GARBER S R MAMNITSKII V M ZH EKSP TEOR FIZ 55 6 2017-30 1968 CA 70 81678 (FOR ENGLISH TRANSLATION SEE T37300)
53697	**SYSTEM Y2O3-AL2O3. II, LIQUIDUS CURVE MEASUREMENT.** MIZUNO M NOGUCHI T NAGOYA KOGYO GIJUTSU SHIKENSHO HOKOKU 16 7 210-16 1967 JA 52 159
53698	**THERMAL CONDUCTIVITY OF HOLLOW BRICKS DETERMINED BY ELECTRICAL MODELING.** BALINT, P. BAKOS, J. EPITOANYAG 18 (5), 175-8, 1966.
53706	**EFFECTUAL SECTON OF AMPLIFICATION OF THE TRANSITION AT 1.06 MICRONS BY THE NEODYMIUM /+3/ INSERTED IN A VITREOUS MATRIX.** MICHON M FARCY J-C COMPT REND 268 B 16 1075-8 1969
53723	**THERMAL CONDUCTIVITY OF LIGHTWEIGHT FIRECLAY BRICKS IN VARIOUS GASES.** SCHWIETE, H. E. DOHM, K. D. ARCH. EISENHUETTENW. 40 (1), 41-5, 1969.
53727	**ACID RAMMING MASSES FOR THE LINING OF ELECTRIC INDUCTION FURNACES.** THOENE H MANTELL D GRANITZKI K E GIESSEREI 55 22 686-91 1968 CA 70 22589
53738	**METHOD FOR TESTING DILATOMETERS ON SAMPLE SUBSTANCES TO GUARANTEE THE CONFORMITY OF DILATOMETRIC MEASUREMENTS IN THE SOVIET UNION.** STRELKOV, P. G. LIFANOV, I. I. SHERSTYUKOV, N. G. METODY IZMER. TEPL. RASSHIR. STEKOL SPAIVAEMYKH NIMI METAL., TR. VSES. SIMP. DILATOMETRII, 1ST, LENINGRAD 1966 31-40, 1967. (FOR ENGLISH TRANSLATION SEE TPRC NO. 66040)
53739	**MEASURING THE THERMAL EXPANSION OF MATERIALS FOR JOINTS AND THE REQUIREMENTS OF MODERN DILATOMETERS.** BRAVINSKII V G DUNAEVA N A METODY IZMER TEPL RASSHIR STEKOL SPAIVAEMYKH NIMI METAL TR VSES SIMP DILATOMETRII 1ST LENINGRAD 1966 1 92-6 1967 CA 70 6271
53740	**USE OF SINTERING FOR DETERMINING THE DIFFERENCES IN THE COEFFICIENTS OF EXPANSION OF GLASSES WITH VARIABLE CHEMICAL COMPOSITIONS.** GOMELSKII M S METODY IZMER TEPL RASSHIR STEKOL SPAIVAEMYKH NIMI METAL TR VSES SIMP DILATOMETRII 1ST LENINGRAD 1966 137-40 1967 CA 70 31320
53762	**RELATION BETWEEN SOME SODIUM OXIDE - MAGNESIASILICA GLASSES AND THEIR COMPOSITIONS. VI. THERMAL EXPANSION OF SODIUM MONOXIDE - MAGNESIUM OXIDE - SILICON DIOXIDE GLASSES.** BOTVINKIN, O. K. YAROKER, KH. G. STEKLO (1), 1-18, 1967.
53771	**IRRADIATION TESTING /PULSE CONDITIONS/ OF THE TERNARY FUEL SYSTEM FOR THE POWER BURST FACILITY.** BEALS R J KERN R S J AM CERAM SOC 52 7 355-8 1969
53775	**THERMAL EXPANSION ANISOTROPY OF OXIDES AND OXIDE SOLID SOLUTIONS.** KIRCHNER H P J AM CERAM SOC 52 7 379-86 1969
53877	**INVESTIGATION OF THE HEAT CONDUCTIVITY OF OXIDE CATHODE COATINGS.** CHUDNOVSKII F A J ENG PHYS 10 1 72-8 1966 (ENGLISH TRANSLATION OF INZH.-FIZ. ZH., 10 (1), 106-14, 1966; FOR ORIGINAL SEE T042537)
53880	**SPECULAR REFLECTION FROM A MAT SURFACE WITH COARSE MICROTOPOGRAPHY.** POLYANSKII V K KOVALSKII L V MILOVIDOVA T I OPTIKA I SPEKTROSKOPIYA 25 5 744-7 1968 (FOR ENGLISH TRANSLATION SEE T053881)
53881	**SPECULAR REFLECTION FROM A MAT SURFACE WITH COARSE MICROTOPOGRAPHY.** POLYANSKII V K KOVALSKII L V MILOVIDOVA T I OPT SPECTROSC 25 5 415-16 1968 (ENGLISH TRANSLATION OF OPTIKA I SPEKTROSKOPIYA, 25 (5), 744-7, 1968; FOR ORIGINAL SEE T053880)
53986	**REMOTE MEASUREMENT OF SURFACE TEMPERATURE AND ITS APPLICATION TO ENERGY BALANCE AND EVAPORATION STUDIES OF BARE SOIL SURFACES.** CONAWAY, J. VAN BAVEL, C. H. M. AGRICULTURAL RES. SERVICE, TEMPE, ARIZONA 1-130, 1967. (AD-650286)
53989	**INTERFEROMETRIC METHOD FOR MEASURING THE THERMAL EXPANSION OF SMALL SAMPLES OF OPTICAL MATERIALS. APPENDIX C FROM STUDIES OF OPTICAL MATERIALS.** BALLARD S BROWDER J S KAYLOR H M STREETE J L AIR FORCE CAMBRIDGE RES. LABS. 1-16, 1967. (AFCRL-67-0487, AD-663676)
54023	**ELASTIC AND ANELASTIC BEHAVIOR OF V2O5.P2O5 GLASSES.** FIELD M B J APPL PHYS 40 6 2628-32 1969
54036	**THERMAL COEFFICIENT OF REFRACTIVE INDEX OF OPTICAL GLASSES.** BAAK T J OPT SOC AM 59 7 851-7 1969
54057	**OPTICAL SPECTRA OF 3D TRANSITION METAL IONS IN MGO . 3.5AL2O3 SPINEL.** REED J S KAY H F J AM CERAM SOC 52 6 307-11 1969
54059	**REFRACTORY PROPERTIES OF A FLORIDA KYANITE-SILLIMANITE CONCENTRATE.** VAN DER BECK, R. R. AM CERAM SOC BULL 48 7 703-6 1969
54081	**CHAPTER 2. THERMAL PROPERTIES OF FOODS.** DICKERSON R W JR FREEZING PRESERVATION OF FOODS. 2, 26-51, 1968.
54084	**METHOD FOR DETERMINATION OF THERMAL DIFFUSIVITY AS RELATED TO TEMPERATURE IN ONE EXPERIMENT.** KRAEV O A TEPLOENERGETIKA 4 15-18 1956
54086	**TEST FOR FROST RESISTANCE IN CERAMIC BODIES.** LEHMANN H CERAMICS 12 151 28-34 1961
54088	**CELLULAR-GLASS INSULATION FOR LOW-TEMPERATURE EQUIPMENT.** SANDERS V CHEM ENGINEERING PROGRESS 44 10 804-6 1948
54097	**THERMAL EXPANSION OF DIAMONDS BETWEEN 25 AND 750 K.** NOVIKOVA S I FIZ TVERDOGO TELA 2 7 1617-18 1960 (FOR ENGLISH TRANSLATION SEE T54098)
54098	**THERMAL EXPANSION OF DIAMONDS BETWEEN 25 AND 750 K.** NOVIKOVA S I SOVIET PHYS-SOLID STATE 2 7 1464-5 1961 (ENGLISH TRANSLATION OF FIZ. TVERD. TELA, 2 (7), 1617-18, 1960; FOR ORIGINAL SEE T54097)
54099	**SPECTROSCOPY AT EXTREME INFRA-RED WAVELENGTHS. I. TECHNIQUE.** BLOOR D DEAN T J JONES G O MARTIN D H MAWER P A PERRY C H PROC ROY SOC 260 A 1 510-22 1961
54102	**THE THERMAL CONDUCTIVITY OF FERRITE SPINELS AND THEIR SOLID SOLUTIONS POINT DEFECTS IN FERRITES, DETERMINING THE ADDITIONAL THERMAL RESISTANCE OF THE LATTICE.** VISHNEVSKII I I SKRIPAK V N INORGANIC MATERIALS 4 11 1728-34 1968 (ENGLISH TRANSLATION OF IZV. AKAD. NAUK SSSR, NEORG. MAT., 4 (11), 1989-96, 1968; FOR ORIGINAL SEE T53442)
54110	**EFFECT OF PRESSURE AND SATURATING FLUID ON THE THERMAL CONDUCTIVITY OF COMPACT ROCK.** WALSH J B DECKER E R J GEOPHYSICAL RESEARCH 71 12 3053-61 1966
54171	**SOME CHARACTERISTICS OF LASER EMISSION OBTAINED WITH THE SOLUTION POCL3, SNCL4, ND2O3.** ROUSSEAU S COLLIER F DUBOST H ROOULT G COMPT REND 268 B 23 1501-3 1969

TPRC Number	Bibliographic Citation
54196	THE LASL PLUTONIUM-238 FUEL DEVELOPMENT PROGRAM. LOS ALAMOS SCIENTIFIC LAB., NEW MEXICO LOS ALAMOS SCI. LAB., UNIV. CALIFORNIA 1-48, 1968. (LA-DC-10046, CONF-681022-1, CMB-11-9455, N69-22998)
54213	CONVECTION IN A VARIABLE VISCOSITY FLUID HEATED FROM WITHIN. FOSTER T D J GEOPHYSICAL RESEARCH 74 2 685-93 1969
54234	CALORIMETRIC INVESTIGATION OF FOODSTUFFS. RADIATION. 1. THE SOLAR CYCLE AND SLOWLY VARYING COMPONENTS. PERLICK A FOOD RESEARCH 3 155-60 1938
54243	LUNAR SURFACE STUDIES. MURCRAY, D. G. UNIV. OF DENVER, DEPT. OF PHYSICS, DENVER, COLORADO 1-66, 1969. (AFCRL-69-0089, AD 685729)
54244	SPECTROSCOPY IN THE FAR INFRARED REGION. BELL, E. E. AIR FORCE CAMBRIDGE RESEARCH LABS., BEDFORD, MA. 1-168, 1968. (AFCRL-68-0126, AD 668560)
54245	MATHEMATICAL ANALYSIS OF HEAT CONDUCTION IN A SECTOR OF A HOLLOW CIRCULAR CYLINDER. TORIAN, J. G. MURPHY, V. T. MARTIN CO., ENGINEERING RESEARCH LAB., ORLANDO, FL. 1-138, 1966. (AFFDL-TR-66-16, AD 485307)
54251	INFRARED REFLECTION SPECTRA OF TOURMALINE. VIERNE R BRUNEL R COMPT REND 270 B 7 488-90 1970 CA 72 126951
54268	SHIELDING OF GLASS-CERAMIC NUCLEAR FUEL. I. STABILITY OF THE SYSTEM FUEL-FILM-COOLANT. MAXIM I V J NUCL MATER 31 3 330-8 1969
54273	FIBER OPTICS WITH EXTENDED ULTRAVIOLET TRANSMISSION. BRATSCHUN W R ALI M A PINCUS A G SCHWARTZ M A IIT RESEARCH INST., CHICAGO, IL. 1-46, 1969. (ECOM-0542-4, N69-31375, AD 686338)
54294	DIRECTIONALLY REFLECTIVE COATING STUDY. COX, R. L. LING-TEMCO-VOUGHT, ASTRONAUTICS DIV., DALLAS, TX. 1-151, 1965. (NASA-CR-63772, X65-17370)
54305	THERMAL AND DIELECTRIC PROPERTIES OF A HOMOGENEOUS MOON BASED ON MICROWAVE AND INFRARED TEMPERATURE OBSERVATIONS. CALVERT, T. A. MARSHALL SPACE FLIGHT CENTER, HUNTSVILLE, ALA. 1-43, 1969. (NASA-TM-X-1743, N69-17815)
54335	I.-NOTE ON SOME PROPERTIES OF A SANDSTONE BLOCK, AFTER USE IN A GLASS FURNACE. HOULDSWORTH H S J SOC GLASS TECHNOL 9 2-11 1925
54350	HOW HIGH-IRON GLASS IMPROVES STRUCTURAL CLAY PRODUCTS. WEST R BRICK CLAY REC 154 2 40-3 1969 CA 70 108783
54351	THERMAL CONDUCTIVITY OF REFRACTORIES. KUESTER W BRENNST-WAERME-KRAFT 20 12 567-71 1968 CA 70 108810
54354	ANOMALY OF SPECIFIC HEAT OF ANTIFERROMAGNETIC SOLID SOLUTIONS. BIZETTE H MAINARD R BULL SOC SCI BRETAGNE 42 3/4 209-14 1967 CA 70 101194
54365	MEASUREMENT OF THE EMISSIVITY OF INDUSTRIAL SURFACES BY A SIMPLE TECHNIQUE. DALLMEYER H CHEM-ING-TECH 41 5/6 323-8 1969 CA 70 98308
54415	THERMAL EXPANSION OF REFRACTORY BRICK. WALLACE E W BULLETIN AMERICAN CERAMIC SOC 42 2 52-6 1963
54416	XXII.-SOME PROPERTIES OF SILLIMANITE BRICKS AND KAOLIN-SILLIMANITE MIXTURES. HOULDSWORTH H S J SOC GLASS TECHNOL 9 316-21 1925
54423	APPARATUS FOR THE STUDY OF HEAT- AND MASS-TRANSFER PROPERTIES OF INSULATING MATERIALS AT LOW MEAN TEMPERATURES. AGRAWAL K N VERMA V V INDIAN J TECHNOL 7 1 4-8 1969 CA 70 98212
54435	GLASS FORMATION AND PROPERTIES OF GLASSES OF THE SIO2-R2O-SNO2 SYSTEM. VAKHRAMEEV V I EVSTROPEV K S IZV AKAD NAUK SSSR NEORG MATER 5 1 101-4 1969 CA 70 90336 (FOR ENGLISH TRANSLATION SEE T057079)
54437	THERMAL EXPANSION OF VITREOUS CALCIUM ALUMINATE. YANISHEVSKII V M POPOVA E I KURYLEVA V F IZV AKAD NAUK SSSR NEORG MATER 5 1 183-4 1969 CA 70 90327 (FOR ENGLISH TRANSLATION SEE T057096)
54441	THERMAL EXPANSION OF ZIRCONIUM REFRACTORIES. FILATOV S K FRANK-KAMENETSKII V A ZHURAVINA T A IZV AKAD NAUK SSSR NEORG MATER 5 2 346-51 1969 CA 70 99267 (FOR ENGLISH TRANSLATION SEE T057104)
54459	EFFECT OF CERTAIN COLORS AND OPACIFIERS ON THE COEFFICIENT OF EXPANSION OF SILICATE GLASSES. APPEN A A KAYALOVA S S ZHUR PRIKLAD KHIM 26 11 1127-32 1953 (FOR ENGLISH TRANSLATION SEE T063073)
54478	THERMAL CONDUCTIVITY OF THORIUM DIOXIDE-URANIUM DIOXIDE SYSTEM. MURABAYASHI M NAMBA S TAKAHASHI Y MUKAIBO T J NUCL SCI TECHNOL /TOKYO/ 6 3 128-31 1969 CA 70 100380
54490	THERMAL CHARACTERISTICS OF SOME COALS OF EASTERN RUSSIA. AGROSKIN A A GONCHAROV E I LOVETSKII L V MAKEEV L A GRYAZNOV N S MOCHALOV V V KOKS KHIM 11 1-6 1968 CA 70 108105 (FOR ENGLISH TRANSLATION SEE T70866)
54491	XXV.-THE REVERSIBLE THERMAL EXPANSION OF A SILICA BRICK. SUGDEN J A J SOC GLASS TECHNOL 17 378-83 1933
54502	THERMAL EXPANSION, INTERNAL STRESS, AND STRENGTH OF GLASS-CRYSTAL COMPOSITIES. TUMMALA R R UNIV. ILLINOIS, URBANA, PH.D. THESIS 1-109, 1969. (UM69-15410)
54518	PRODUCTION AND PROPERTIES OF ZIRCONIUM REFRACTORIES STABILIZED BY CERIUM DIOXIDE. ANDREEVA A B LEONOV A I KELER E K OGNEUPORY 34 3 37-44 1969 CA 70 108816 (FOR ENGLISH TRANSLATION SEE T057265)
54531	THERMAL MEASUREMENTS ON FERROELECTRICS. GODEFROY L R REMOISSENET M CHANUSSOT G CHANUSSOT GUY REV GEN ELEC 77 10 929-32 1968 CA 70 91638
54534	DETERMINATION OF THERMAL DIFFUSIVITY OF NUCLEAR FUEL SPECIMENS IN THE COURSE OF THEIR IRRADIATION IN A NUCLEAR REACTOR. GLANZ, G. REV. ROUM. PHYS. 13 (10), 829-35, 1968.
54551	THERMAL EXPANSION OF SILICEOUS KANSAS CLAYS /DAKOTA FORMATION/. BAULEKE M P EDMONDS C S STATE GEOL SURV KANS BULL 191 1 13-15 1968 CA 70 99255
54552	BASALT MELTS FOR MAKING STAPLE FIBERS. DUBROVSKII V A RYCHKO V A BACHILO T M LYSYUK A G STEKLO KERAM 25 12 18-20 1968 CA 70 107382 (FOR ENGLISH TRANSLATION SEE T060513)
54554	THERMAL STABILITY OF CARBOPERLITE. ARKHANGELSKII K P KASHPEROVSKAYA O P STROIT MATER DETALI IZDELIYA 8 57-9 1967 CA 70 99332

TPRC Number	Bibliographic Citation
54558	**PROPERTIES OF A GLASS-CERAMIC IN A SILICON DIOXIDE-ALUMINUM OXIDE-LITHIUM OXIDE-TITANIUM DIOXIDE SYSTEM.** SCHLEIFER P PELCZAR S SZKLO CERAM 19 11 321-4 1968 CA 70 108723
54559	**EFFECT OF THE DIFFUSION OF ALKALI METAL IONS ON THE MECHANICAL STRENGTH OF GLASS.** SCHLEIFER P BUGAJSKI W KUBACKI W SZKLO CERAM 19 12 353-7 1968 CA 70 108766
54560	**SINTERING AND PROPERTIES OF SOME CORDIERITIC MATERIALS.** BUDNIKOV P P SVETLOVA I A KHARITONOV F YA SZKLO CERAM 19 12 367-71 1968 CA 70 108854
54573	**PROPERTIES OF BINARY SYSTEM GLASSES CONTAINING BISMUTH OXIDE.** SHABANOVA E B TR GORK POLITEKH INST 23 4 38-45 1967 CA 70 117660
54585	**THERMAL EXPANSION OF ROCKS AND MINERALS.** ISKANDAROV E UZB GEOL ZH 12 4 22-5 1968 CA 70 89633
54587	**TEMPERATURE DEPENDENCE OF THERMOPHYSICAL PROPERTIES OF MOISTURE-SATURATED ROCKS.** ASTROSHCHENKO, P. P. BOGOMOLOV, YU. G. SHUSHPANOV, A. P. VESTSI AKAD. NAVUK BELARUS. SSR, SER. FIZ.-TEKH. NAVUK (1), 126-9, 1969.
54631	**HEAT CONDUCTIVITY OF SOME CARBONATE ROCKS AND CLAYEY SANDSTONES.** ZIERFUSS, H. BULL. AM. ASSOC. PETROL. GEOLOGISTS 53 (2), 251-60, 1969.
54638	**LINEAR THERMAL EXPANSION OF LITHIUM ALUMINOSILICATES.** AVGUSTINIK A I VASILEV E I ZHUR PRIKLAD KHIM 28 9 939-43 1955 (FOR ENGLISH TRANSLATION SEE T063074)
54686	**REFRACTORIES FOR SAVINSK MAGNESITE.** TIMOFEEV N N STEGANTSEV S A ANOKHINA A D OGNEUPORY 10 14-20 1968 (FOR ENGLISH TRANSLATION SEE T054687)
54687	**REFRACTORIES FROM SAVINSK MAGNESITE.** TIMOFEEV N N STEGANTSEV S A ANOKHINA A D REFRACTORIES 10 621-6 1968 (ENGLISH TRANSLATION OF OGENUPORTY NO. 10, 14-20, 1968; FOR ORIGINAL SEE T054686)
54697	**EX-REACTOR PROPERTY EVALUATIONS. FROM IRRADIATION TESTING OF CERAMIC FUELS.** CHERNOCK W P ZUROMSKY G CHRISTOPHER S KOZIOL J J COMBUSTION ENGINEERING INC, WINDSOR, CT II-1—II-21, 1968. (CEND-2936-335, N69-12878)
54698	**IRRADIATION PROGRAM. FROM IRRADIATION TESTING OF CERAMIC FUELS.** CHERNOCK W P ZUROMSKY G CHRISTOPHER S KOZIOL J. J. COMBUSTION ENGINEERING INC, WINDSOR, CT IV-1—IV-41, 1968. (CEND-2936-335)
54711	**EFFECT OF ILLUMINATING AND VIEWING GEOMETRY ON THE COLOR COORDINATES OF SAMPLES WITH VARIOUS SURFACE TEXTURES.** BILLMEYER F W JR MARCUS R T APPL OPT 8 4 763-8 1969
54713	**LOW TEMPERATURE THERMAL EXPANSION MEASUREMENTS ON OPTICAL MATERIALS.** BROWDER J S BALLARD S S APPL OPT 8 4 793-8 1969
54724	**GEOTHERMAL STUDIES IN SIBERIA.** MOISEENKO, U. I. SOKOLOVA, L. S. FREIBERG. FORSCHUNGSH. 238 C, 79-87, 1968.
54725	**APPARATUS AND METHODS FOR DETERMINATION OF THERMAL CONDUCTIVITY OF ROCKS.** KRESL, M. FREIBERG. FORSCHUNGSH. 238 C, 95-100, 1968.
54727	**CHANGE IN THE OPTICAL CHARACTERISTICS OF MOLTEN OPEN-HEARTH SLAGS.** KAIROV E A KRIVANDIN V A MASTRYUKOV B S IZV VYSSH UCHEB ZAVED CHERN MET 12 1 159-63 1969 CA 71 15316
54728	**STRENGTH OF GLASS-CRYSTAL COMPOSITES.** TUMMALA R R FRIEDBERG A L J AMER CERAM SOC 52 4 228-9 1969 CA 71 15587
54743	**THERMAL EXPANSION OF THE ELEKTRIT A96 ABRASIVE GRAINS.** TURNOVEC, I. SKLAR KERAM. 19 (1), 14-6, 1969.
54745	**EFFECT OF BENTONITE ADDITIVES ON THE PHYSICOMECHANICAL PROPERTIES OF THE PLASTIC CLAY MASS AND ON THE HEAT STABILITY OF FACE TILE DURING FIRING.** KUKOLEV, G. V. ORUDZHEVA, N. T. STEKLO KERAM. 26 (4), 26-9, 1969. (FOR ENGLISH TRANSLATION SEE TPRC NO. 66786)
54751	**AGATE.** IWAI S OSSAKA J MORIKAWA H YOGYO KYOKAI SHI 77 885 172-8 1969 CA 71 14957
54757	**THERMOPHYSICAL PROPERTIES /HEAT REQUIREMENTS, HEAT CAPACITY, THERMAL CONDUCTIVITY, AND THERMAL DIFFUSIVITY/ OF SOME MATERIALS FROM TITANIUM PRODUCTION.** RAFALOVICH I M DENISOVA I A NAUCH TR GOS NAUCH-ISSLED PROEKT INST REDKOMETAL PROM 21 244-52 1967 CA 71 16508
54773	**STOPPER TUBES FOR LADLES USED FOR VACUUM CASTING RIMMED STEELS.** KHOSID G M FEDOROVA E A MATERIKIN YU V ORLOV V A OGNEUPORY 33 11 23-9 1968 (FOR ENGLISH TRANSLATION SEE T054774)
54774	**STOPPER TUBES FOR LADLES USED FOR VACUUM CASTING RIMMED STEELS.** KHOSID G M FEDOROVA E A MATERIKIN YU V ORLOV V A REFRACTORIES 33 11 700-5 1968 (ENGLISH TRANSLATION OF OGNEUPORY, 33 (11), 23-9, 1968; FOR ORIIGINAL SEE T054773)
54775	**SERVICE OF SEMIACID EXPANDING LADLE BRICK.** SHCHEGLOV S I KARASIK V L BARKALOVA T K PILIPCHATIN L D KHILKO M M ALEINIKOV G L CHEVELA L A DUBINA YU G OGNEUPORY 33 12 23-8 1968 (FOR ENGLISH TRANSLATION SEE T054776)
54776	**SERVICE OF SEMIACID EXPANDING LADLE BRICK.** SHCHEGLOV S I KARASIK V L BARKALOVA T K PILIPCHATIN L D KHILKO M M ALEINIKOV G L CHEVELA L A DUBINA YU G REFRACTORIES 33 12 761-5 1968 (ENGLISH TRANSLATION OF OGNEUPORY, 33 (12), 23-8, 1968; FOR ORIGINAL SEE T054775)
54777	**THE MECHANICAL PROPERTIES OF BASIC REFRACTORIES AND THE ROLE OF THERMOPLASTICITY IN THE SERVICE OF REFRACTORY MATERIALS.** BLUVSHTEIN M N KELER E K OGNEUPORY 33 12 28-40 1968 (FOR ENGLISH TRANSLATION SEE T054778)
54778	**THE MECHANICAL PROPERTIES OF BASIC REFRACTORIES AND THE ROLE OF THERMOPLASTICITY IN THE SERVICE OF REFRACTORY MATERIALS.** BLUVSHTEIN M N KELER E K REFRACTORIES 33 12 766-79 1968 (ENGLISH TRANSLATION OF OGNEUPORY, 33 (12), 28-40, 1968; FOR ORIGINAL SEE T054777)
54779	**CHANGES IN KYANITE-SILLIMANITE CONCENTRATE DURING HEATING.** ZHIKHAREVICH S A ZELENSKAYA A T TSYNKINA V M KOCHETOVA A P DRIZHERUK M E OGNEUPORY 33 12 40-6 1968 (FOR ENGLISH TRANSLATION SEE T054780)
54780	**CHANGES IN KYANITE-SILLIMANITE CONCENTRATE DURING HEATING.** ZHIKHAREVICH S A ZELENSKAYA A T TSYNKINA V M KOCHETOVA A P DRIZHERUK M E REFRACTORIES 33 12 780-5 1968 (ENGLISH TRANSLATION OF OGNEUPORY, 33 (12), 40-6, 1968; FOR ORIGINAL SEE T054779)

TPRC Number	Bibliographic Citation
54781	**USING GROG FROM OLD LININGS TO MAKE TAR-BONDED DOLOMITE-MAGNESITE REFRACTORIES.** ALEKSANDROVA T A SERGEEV B I OGNEUPORY 33 12 47-9 1968 (FOR ENGLISH TRANSLATION SEE T054782)
54782	**USING GROG FROM OLD LININGS TO MAKE TAR-BONDED DOLOMITE-MAGNESITE REFRACTORIES.** ALEKSANDROVA T A SERGEEV B I REFRACTORIES 33 12 786-8 1968 (ENGLISH TRANSLATION OF OGNEUPORY, 33 (12), 47-9, 1968; FOR ORIGINAL SEE T054781)
54785	**HEAT FLUX THROUGH THE FLOOR OF THE ARCTIC BASIN IN THE REGION OF THE LOMONOSOV RIDGE.** LIUBIMOVA YE A TOMARO G A ALEKSANDROV A L DOKL AKAD NAUK SSR 184 403-5 1969 (FOR ENGLISH TRANSLATION SEE T054786)
54786	**HEAT FLUX THROUGH THE FLOOR OF THE ARCTIC BASIN IN THE REGION OF THE LOMONOSOV RIDGE.** LIUBIMOVA YE A TOMARO G A ALEKSANDROV A L WRIGHT-PATTERSON AFB, OH AD-689799 1-5 1969 (ENGLISH TRANSLATION OF DOKL. AKAD. NAUK, SSR, 184, 403-50, 1969; FOR ORIGINAL SEE T54785)
54787	**INVESTIGATION OF THE STRUCTURE AND MAGNETIC PROPERTIES OF FERROELECTRIC-FERROMAGNETIC SOLID** ROGINSKAYA YU E VENEVTSEV YU N ZHDANOV G S IZV AKAD NAUK SSSR SER FIZ 29 6 1022 1965 (FOR ENGLISH TRANSLATION SEE T054788)
54788	**INVESTIGATION OF THE STRUCTURE AND MAGNETIC PROPERTIES OF FERROELECTRIC-FERROMAGNETIC SOLID SOLUTIONS IN THE PB-2 CO WO6-CD MN O3 SYSTEM.** ROGINSKAYA YU E VENEVTSEV YU N ZHDANOV G S BULL ACAD SCI USSR /PHYS SER/ 29 6 1021-4 1965 (ENGLISH TRANSLATION OF IZV. AKAD. NAUK SSSR, SER. FIZ., 29 (6), 1022- , 1965; FOR ORIGINAL SEE T054787)
54790	**GEOTHERMAL FIELD OF THE UPPER RHINEGRABEN.** KAPPELMEYER, O. ABH. GEOL. LANDESAMTES BADEN-WUERTTEMBERG (6), 101-3, 1967.
54814	**A REVIEW OF CURRENT REFRACTORY COMPOSITE RESEARCH IN CERAMICS AT IIT RESEARCH INSTITUTE.** BORTZ S A AIR FORCE MATERIALS LAB. 477-514, 1968. (AFML-TR-68-84, AD-838781)
54820	**PROTON BOMBARDMENT-INDUCED TARGET TEMPERATURES AND THERMAL ACCOMMODATION COEFFICIENTS OF ROCK POWDERS. MEASUREMENTS BY INFRARED PYROMETRY.** NASH, D. B. JET PROPULSION LAB CALIFORNIA INST. OF TECHNOLOGY 1-22, 1969. (NASA-CR-1009228 N69-25406, JPL-TM-413)
54821	**MANUFACTURING TECHNOLOGY FOR PRODUCTION OF HIGH-STRENGTH, HIGH-MODULUS, GLASS FIBERS.** LEWIS A RAROGIEWICZ L W NEUMEYER C F DUWEZ P AEROJET-GENERAL CORP., AZUSA, CA 1-173, 1968. (AFML-TR-160, AGC-3543, AD-838901)
54896	**EFFECTS OF VARIOUS DOPING ELEMENTS ON THE TRANSITION TEMPERATURE OF VANADIUM OXIDE SEMICONDUCTORS.** FUTAKI H AOKI M JAPAN J APPL PHYS 8 8 1008-13 1969
54897	**A CONSTRUCTION OF THE HIGH POWER LASER AMPLIFIER USING GLASS AND SELENIUM OXYCHLORIDE DOPED WITH ND /+3/.** SASAKI T YAMANAKA T YAMAGUCHI G YAMANAKA C JAPAN J APPL PHYS 8 8 1037-45 1969
54916	**EFFECT OF SOME FILLERS ON THE THERMAL STABILITY OF GLASS-ENAMEL COATINGS FOR CHEMICAL APPARATUS.** TARASENKO V N KHIM NEFT MASHINOSTR 4 19-20 1969 CA 71 24397
54917	**EFFECTS OF MICROSTRUCTURE ON THE THERMAL STABILITY OF CALCINED GANISTER REFRACTORIES.** NEMETS I I DOBROVOLSKII G B GOGOTSI G A ANTONENKO V M OGNEUPORY 34 4 56-9 1969 CA 71 24419 (FOR ENGLISH TRANSLATION SEE T057266)
54919	**DEPENDENCE OF THE PROPERTIES OF LITHIUM GLASS-CERAMICS ON THE NATURE OF THE MINERALIZER.** BUZHINSKII I M KIRILLOVA I I GURENKO V A OPT-MEKH PROM 36 3 30-4 1969 CA 71 24392
54930	**GLASS FORMATION AND SOME PROPERTIES OF GLASSES OF THE SYSTEM GEHLENITE-AKERMANITE-ANORTHITE.** TOROPOV N A MELNIKOVA O V SILIKATTECHNIK 19 11 341-3 1968 CA 70 108722
54931	**ASPECTS OF THE RANGE OF GLASS PROPERTIES.** WINTER A SILIKATTECHNIK 20 1 10-13 1969 CA 71 24381
54933	**CHEMICAL STABILITY OF ALKALI GLASSES.** ROGOZHIN, YU. V. RODINA, D. V. STEKLO (2), 72-3, 1967.
54946	**PREPARTATION AND PROPERTIES OF DENSE SPINEL CERAMICS IN THE MAGNESIUM ALUMINATE-ALUMINA SYSTEM.** BAILEY J T RUSSELL R JR TRANS BRIT CERAM SOC 68 4 159-64 1969 CA 71 73624
54960	**BIDIRECTIONAL REFLECTANCE MEASUREMENTS FROM AN AIRCRAFT OVER NATURAL EARTH SURFACES.** BRENNAN, B. GODDARD SPACE FLIGHT CENTER, GREENBELT, MD 1-84, 1969. (NASA-TM-X-63564, X-622-69-216, N69-28454)
54964	**THERMAL CONDUCTIVITY OF MIXED OXIDES OF URANIUM AND PLUTONIUM.** VAN CRAEYNEST, J. C. WEIBACHER, J. C. COMMISSARIAT A L ENERGIE ATOMIQUE, FONTENAY-AUX-ROSES, FRANCE 1-14, 1968. (CEA-CONF-681116-2, CEA-CONF-1139, N69-28514)
54965	**ADDITIONAL INFRARED SPECTRAL EMITTANCE MEASUREMENTS OF ROCKS FROM THE MONO CRATERS REGION, CALIFORNIA.** DANIELS, D. L. GEOLOGICAL SURVEY, WASHINGTON, DC 1-12, 1967. (NASA-CR-101457, N69-28378)
54966	**VISIBLE AND ULTRAVIOLET REFLECTANCE AND LUMINESCENCE FROM VARIOUS SAUDI ARABIAN AND INDIANA LIMESTONE ROCKS.** WATTS H V GODLMAN H J NATIONAL AERONAUTICS AND SPACE ADMIN. 1-71, 1967. (NASA-CR-101453)
54980	**INFRARED SPECTRA OF SIO2.GEO2 GLASS.** BORRELLI N F PHYS CHEM GLASSES 10 2 43-5 1969 JA 52 276
54981	**LOW THERMAL EXPANSION COMPOSITIONS IN THE SYSTEMS SPODUMENE-KAOLIN AND PETALITE-KAOLIN.** FISHWICK J H VAN DER BECK R R TALLEY R W BULLETIN AMERICAN CERAMIC SOC 43 11 832-5 1964
54983	**TESTING OF HIGH-EMITTANCE COATINGS.** CLEARY, R. E. AMMANN, C. CLEVELAND OHIO NASA AND CFSTI NASA-CR-1413 1-104 1969
54984	**CONVECTION IN A MANTLE WITH VARIABLE PHYSICAL PROPERTIES.** TURCOTTE D L OXBURGH E R J GEOPHYSICAL RESEARCH 74 6 1458-74 1969
54992	**THERMAL EXPANSION OF JENA GLASS 16 III. COMM. NO. 203A.** KEESOM W H BIJL A PROC KONINKLIJKE AKAD WETENSCHAP AMSTERDAM 32 19 1164-6 1929
55007	**BORIC OXIDE ANOMALY IN NABAL GLASSES.** DE WAAL H PHYS CHEM GLASSES 10 3 101-7 1969 CA 71 33023
55028	**DEVELOPMENT OF 400 TO 2200 F FIBROUS-TYPE INSULATIONS FOR RADIOISOTOPE POWER SYSTEMS.** COLLINS J O JAUNARAJS K L REID D R JOHNS-MANVILLE RESEARCH AND ENGINEERING CTR., MANVILLE, NEW JERSEY 1-31, 1969. (ALO-2661-11, N69-24622)
55041	**EXPERIMENTS CONCERNING INFRARED DIFFUSE REFLECTANCE STANDARDS IN THE RANGE 0.8 TO 20.0 MICRONS.** AGNEW J T MC QUISTAN R B J OPT SOC AMER 43 11 999-1007 1953
55083	**MISSION 73. SUMMARY AND DATA CATALOG.** PASCUCCI, R. F. NORTH, G. W. GEOLOGICAL SURVEY, WASHINGTON, DC 1-296, 1968. (NASA-CR-101466, N69-28163)

TPRC Number	Bibliographic Citation
55124	**HIGH-LOW QUARTZ INVERSION. THERMODYNAMICS OF THE LAMBDA TRANSITION.** KLEMENT W JR COHEN L H J GEOPHYS RES 73 6 2249-59 1968 CA 71 42999
55142	**PARAMETERS FOR THE PRODUCTION OF A BASALT STAPLE FIBER FOR FILTERS IN HYDRAULIC ENGINEERING STRUCTURES.** ROZHANSKII, A. I. KOZLOVSKII, P. P. STEKLO KERAM. 26 (5), 19-21, 1969.
55146	**EFFECT OF MAGNESIUM OXIDE ON FORMATION PROCESSES AND PROPERTIES OF GLASSES BASED ON EUCRYPTITE AND PETALITE.** RAKHMANBEKOV N SIRAZHIDDINOV N A TOROPOV N A UZB KHIM ZH 13 2 83-6 1969 CA 71 41788
55156	**AN INVESTIGATION OF CERTAIN THERMO-PHYSICAL CHARACTERISTICS OF SANDS OF RAJASTHAN DESERT.** CHAUDHARY D R AGRAWAL M P BHANDARI R C INDIAN J PURE APPL PHYS 7 4 252-6 1969
55170	**SOME THERMAL PROPERTIES OF SOLIDS AT LOW TEMPERATURES.** BROCK, J. C. F. OXFORD UNIVERSITY, ENGLAND, PH.D. THESIS 1-107, 1964.
55185	**HIGH DIELECTRIC CONSTANT CERAMICS.** VON HIPPEL A BRECKENRIDGE R G CHESLEY F G TISZA L IND ENG CHEM 38 11 1097-109 1946
55199	**THERMAL CONDUCTIVITY OF THO2-PUO2 UNDER IRRADIATION.** JEFFS, A. T. ATOMIC ENERGY OF CANADA, CHALK RIVER, ONTARIO 1-40, 1969. (AECL-3294, N69-30066)
55204	**IN-CORE STUDY OF FUEL/CLAD INTERACTION AND FUEL CENTRE TEMPERATURE.** KJAERHEIM, G. ROLSTAD, E. INSTITUTT FOR ATOMENERGI, HALDEN, NORWAY 1-36, 1969. (HPR-107)
55211	**EFFECT OF OXYGEN STOICHIOMETRY ON THE THERMAL DIFFUSIVITY AND CONDUCTIVITY OF U/0.75/ PU/0.25/ O(2-X).** GIBBY R L BATTELLE NORTHWEST LABS. 5.7-5.9, 1968. (BNWL-919)
55244	**ON MEASURING PROPERTIES OF SOILS BY THERMAL METHODS WITH SPECIAL REFERENCE TO THE CONTACT METHOD.** STIGTER C J NETH J AGRIC SCI 17 41-9 1969
55247	**THERMAL CONDUCTIVITY OF SATURATED LEDA CLAY.** PENNER E GEOTECHNIQUE 12 2 168-75 1962
55257	**THERMAL PROPERTIES OF COATINGS IN METAL MOLDS FOR CAST IRON.** ISOTANI M KONDO Y HOBO K CAST METALS RES J 5 2 90-2 1969 CA 71 52539
55268	**STRONTIA AND ITS PROPERTIES IN GLAZES.** MC CUTCHEN E S J AM CERAM SOC 27 8 233-8 1944
55272	**DETERMINATION OF THE HEAT OF PYROLYSIS OF COALS.** AGROSKIN A A GONCHAROV E I MAKEEV L A KHIM TVERD TOPL 3 129-33 1969 CA 71 52021
55274	**EFFECT OF KAOLIN-GROG ADDITIVES ON THE LINEAR COEFFICIENT OF THERMAL EXPANSION OF THE GLASS PHASE IN PORCELAIN.** ILCHENKO A I KAGANOVA I V LEGKA PROM 2 43-5 1969 CA 71 53057
55278	**EFFECT OF CHEMICAL COMPOSITION ON THE SPECTRAL REFLECTION FUNCTION AND MICROINDENTATION HARDNESS IN THE IRON SULFIDE /FES2/-NICKEL SULFIDE /NIS2/-COBALT SULFIDE /COS2/ SYSTEM OF NATURAL BRAVOITE CRYSTAL ZONES.** DEMIRSOY S NEUES JAHRB MINERAL MONATSH 7 323-33 1969 CA 71 52076
55280	**DEVELOPMENT OF THE PRODUCTION TECHNOLOGY AND AN EXPERIMENTAL BATCH OF CORUNDUM LIGHTWEIGHT REFRACTORY.** KONETSKII N V MILSHENKO R C OGNEUPORY 34 6 3-6 1969 CA 71 53077 (FOR ENGLISH TRANSLATION SEE T057270)
55287	**SYNTHESIS OF GLASS - CERAMICS MATERIALS FROM BLAST - FURNACE SLAGS OF THE KUZNETSK METALLURGICAL COMBINE.** PAVLUSHKIN, N. M. LOGVINENKO, A. T. PENDYURINA, T. E. IZV. SIB. OTD. AKAD. NAUK S S S R, SER. KHIM. NAUK (2), 148-53, 1969. (FOR ENGLISH TRANSLATION SEE TPRC NO. 65505)
55290	**MECHANISM OF NUCLEATED CRYSTALLIZATION OF GLASSES IN LITHIA-ALUMINA-SILICA AND CORDIERITE SYSTEMS.** BLINOV V A J MATER SCI 4 5 461-8 1969 CA 71 53030
55310	**PROPERTIES OF REFRACTORY MATERIALS.** BURNETT, S. J. ATOMIC ENERGY RESEARCH ESTABLISHMENT, HARWELL, ENGLAND 1-326, 1969. (AERE-R-4657)
55318	**SELECTIVE FILMS ON GLASS, AND THEIR CHARACTERISTICS.** REKANT N B SHEKLEIN A V PREOBRAZOVATELI SOLN ENERG POLUPROV 190-9 1968 CA 71 55230 (FOR ENGLISH TRANSLATION SEE T70720)
55322	**THERMAL EXPANSION OF KANSAS VOLCANIC ASH.** BAULEKE, M. P. HUGH, J. M. STATE GEOL. SURV. KANSAS BULL. PT. 1 (194), 15-6, 1969.
55331	**METHOD FOR THE SIMULTANEOUS MEASUREMENT OF THERMAL EXPANSION, SETTING TEMPERATURE, AND YOUNGS MODULUS OF A VITREOUS ENAMEL IN THE FORM OF A THIN LAYER.** STIRLING J F HARDY L E TRANS BRIT CERAM SOC 68 3 119-24 1969 CA 71 53100
55332	**APPLICATION OF HIGH TEMPERATURE X-RAY DIFFRACTION TO THE INTERPRETATION OF THE X-RAY DIFFRACTION PATTERN OF MULLITE AND THE THERMAL EXPANSION OF MULLITE AND MULLITE SOLID SOLUTIONS.** GODFREY, T. G., JR. CLEMSON COLLEGE, M.S. THESIS 1-71 1963
55345	**MANUFACTURE OF MINERAL WOOL FROM PULVERIZED COAL BURNT ASHES.** NAKAMURA K KITASATO Y TAKAHASHI T SANBONGI K ZAIRYO 18 189 554-60 1969 CA 71 53023
55387	**FINE STRUCTURE AND THERMOMECHANICAL PROPERTIES OF MAGNESIUM-CONTAINING PHOTOSENSITIVE GLASS CERAMICS.** BEREZHNOI A I KRANIKOV A S EROKHOV N A DOKL AKAD NAUK SSSR 186 1 142-5 1969 CA 71 63656
55392	**COMPLEX PETROPHYSICAL STUDIES BY THERMAL CONDUCTIVITY MEASUREMENTS OF ROCK SPECIMENS FROM THE GERMAN DEMOCRATIC REPUBLIC.** HURTIG, E. FREIBERG. FORSCHUNGSH. 238 C, 101-20, 1968.
55399	**HEAT CAPACITY AND ENTROPY OF CALCITE AND ARAGONITE, AND THEIR INTERPRETATION.** STAVELEY, L. A. K. LINFORD, R. G. J. CHEM. THERMODYN. 1 (1), 1-11, 1969.
55404	**RELATIVE SPECTRAL REFLECTIVITY 0.4-1 MICRON OF SELECTED AREAS OF THE LUNAR SURFACE.** MC CORD T B JOHNSON T V J GEOPHYS RES 74 17 4395-401 1969 CA 71 65708
55409	**PHYSICAL PROPERTIES OF TERRAZZO AGGREGATES.** NATIONAL BUREAU OF STANDARDS NBS TECH NEWS BULL 31 6 63-4 1943
55416	**EFFECT OF IRON OXIDE ON THE THERMAL EXPANSION OF BASE ENAMELS OF KITCHEN UTENSILS.** APPEN A A BRESKER R I KUZNETSOBA L A ZHUR PRIKLAD KHIM 29 1753-5 1956 (FOR ENGLISH TRANSLATION SEE T063075)
55422	**CHANGES IN THE PROPERTIES OF ZIRCONIUM-YTTRIUM SOLID SOLUTIONS DURING PROLONGED EXPOSURE TO HIGH TEMPERATURES.** KUZNETSOV A K ZIMINA L A KELER E K IZV AKAD NAUK SSSR NEORG MATER 4 7 1112-17 1968 (FOR ENGLISH TRANSLATION SEE T055423)

TPRC Number	Bibliographic Citation
55423	**CHANGES IN THE PROPERTIES OF ZIRCONIUM-YTTRIUM SOLID SOLUTIONS DURING PROLONGED EXPOSURE TO HIGH TEMPERATURES.** KUZNETSOV A K ZIMINA L A KELER E K INORGANIC MATERIALS 4 7 976-80 1968 (ENGLISH TRANSLATION OF IZV. AKAD. NAUK SSSR, NEORG. MATER., 4 (7), 1112-7, 1968; FOR ORIGINAL SEE T055422)
55424	**USING AERIAL PHOTOGRAPHY IN DIFFERENT SPECTRUM INTERVALS TO STUDY VEGETATION AND SOILS.** VINOFRADOVA A I GEOGRAFICHESKII SBORNIK 7 59-74 1955 (FOR ENGLISH TRANSLATION SEE T055425)
55425	**USING AERIAL PHOTOGRAPHY IN DIFFERENT SPECTRUM INTERVALS TO STUDY VEGETATION AND SOILS.** VINOFRADOVA, A. E. FOREIGN SCIENCE AND TECHNOLOGY CENTER, US ARMY MATERIAL COMMAND, WASHINGTON, DC 1-24, 1969. (ENGLISH TRANSLATION OF GEOGRAFICHESKII SBORNIK, 7, 59-74, 1955; FOR ORIGINAL SEE T055424) (FSTC-HT-23-309-70, AD-693225)
55426	**INFRARED REFLECTANCE SPECTRA OF THE PRODUCTS OF THE CRYSTALLIZATION OF GLASS WITH THE SODIUM DISILICATE COMPOSITION.** DUTOVA K P IZV AKAD NAUK SSSR NEORG MATER 4 8 1296-300 1968 (FOR ENGLISH TRANSLATION SEE T055427)
55427	**INFRARED REFLECTANCE SPECTRA OF THE PRODUCTS OF THE CRYSTALLIZATION OF GLASS WITH THE SODIUM DISILICATE COMPOSITION.** DUTOVA K P INORGANIC MATERIALS 4 8 1136-9 1968 (ENGLISH TRANSLATION OF IZV. AKAD. NAUK SSSR, NEORG. MATER., 4 (8), 1296-300, 1968; FOR ORIGINAL SEE T055426)
55432	**SOME CERAMIC PROPERTIES AND ELECTRICAL CONDUCTIVITY OF ZIRCONIA CERAMICS STABILIZED WITH YTTRIA.** EZERSKII M L KOZLOVA I I POPILSKII R YA DEMONIS I M IZV AKAD NAUK SSSR NEORG MATER 4 9 1599-600 1968 (FOR ENGLISH TRANSLATION SEE T055433)
55433	**SOME CERAMIC PROPERTIES AND ELECTRICAL CONDUCTIVITY OF ZIRCONIA CERAMICS STABILIZED WITH YTTRIA.** EZERSKII M L KOZLOVA I I POPILSKII R YA DEMONIS I M INORGANIC MATERIALS 4 9 1395-7 1968 (ENGLISH TRANSLATION OF IZV. AKAD. NAUK SSSR, NEORG. MATER., 4 (9), 1599-1600, 1968; FOR ORIGINAL SEE T055432)
55434	**SOLID SOLUTIONS OF SR ZR O3 AND SR HF O3.** BEREZHNOI A S BELIK V YA GAVRISH A M GULKO N V IZV AKAD NAUK SSSR NEORG MATER 4 9 1605-6 1968 (FOR ENGLISH TRANSLATION SEE T055435)
55435	**SOLID SOLUTIONS OF SR ZR O(3) AND SR HF O(3).** BEREZHNOI, A. S. BELIK, V. YA. GAVRISH, A. M. GULKO, N. V. INORGANIC MATERIALS 4 9 1403-4 1968 (ENGLISH TRANSLATION OF IZV. AKAD. NAUK SSSR, NEORG. MATER., 4 (9), 1605-6, 1968; FOR ORIGINAL SEE T055434)
55436	**INFRA-RED ABSORPTION SPECTRA OF BARIUM AND STRONTIUM SILICATE GLASSES.** KOLESOVA V A IZV AKAD NAUK SSSR NEORG MATER 4 9 1612-14 1968 (FOR ENGLISH TRANSLATION SEE T055437)
55437	**INFRARED ABSORPTION SPECTRA OF BARIUM AND STRONTIUM SILICATE GLASSES.** KOLESOVA V A INORGANIC MATERIALS 4 9 1410-12 1968 (ENGLISH TRANSLATION OF IZV. AKAD. NAUK SSSR, NEORG. MATER., 4 (9), 1612-4, 1968; FOR ORIGINAL SEE T055436)
55442	**HARDENING OF VITREOCRYSTALLINE MATERIALS BY QUENCHING.** BOGUSLAVSKII I A MARKELOVA M YA IZV AKAD NAUK SSSR NEORG MATER 4 10 1772-6 1968 (FOR ENGLISH TRANSLATION SEE T055443)
55443	**HARDENING OF VITREOCRYSTALLINE MATERIALS BY QUENCHING.** BOGUSLAVSKII I A MARKELOVA M YA INORGANIC MATERIALS 4 10 1544-7 1968 (ENGLISH TRANSLATION OF IZV. AKAD. NAUK SSSR, NEORG. MATER., 4 (10), 1772-6, 1968; FOR ORIGINAL SEE T055442)
55446	**FORMATION AND CHARACTERISTICS OF THE ORIENTED STRUCTURE OF PHOSPHATE GLASSES.** GERASIMOV V V KUZNETSOV-FETISOV L I KUZNETSOV E V IZV AKAD NAUK SSSR NEORG MATER 4 10 1819-21 1968 (FOR ENGLISH TRANSLATION SEE T055447)
55447	**FORMATION AND CHARACTERISTICS OF THE ORIENTED STRUCTURE OF PHOSPHATE GLASSES.** GERASIMOV V V KUZNETSOV-FETISOV L I KUZNETSOV E V INORGANIC MATERIALS 4 10 1589-91 1968 (ENGLISH TRANSLATION OF IZV. AKAD. NAUK SSSR, NEORG. MATER., 4 (10), 1819-21, 1968; FOR ORIGINAL SEE T055446)
55460	**INVESTIGATION OF THE CRYSTALLIZATION OF LOW-ALKALI POLYCOMPONENT GLASSES.** VARGIN V V YASHCHISHIN I N IZV AKAD NAUK SSSR NEORG MATER 4 11 2006-9 1968 (FOR ENGLISH TRANSLATION SEE T055461)
55461	**INVESTIGATION OF THE CRYSTALLIZATION OF LOW-ALKALI POLYCOMPONENT GLASSES.** VARGIN V V YASHCHISHIN I N INORGANIC MATERIALS 4 11 1743-5 1968 (ENGLISH TRANSLATION OF IZV. AKAD. NAUK SSSR, NEORG. MATER., 4 (11), 2006-9, 1968; FOR ORIGINAL SEE T055460)
55468	**BOROSILICATE GLASSES WITH LOW ALKALI CONTENTS.** MAZURIN, O. V. TOTESH, A. S. STRELTSINA, M. V. SKLAR KERAM. 19 (4), 100-1, 1969.
55475	**GLOST KILN REFRACTORIES.** SCHRAMM E J AM CERAM SOC 27 9 282-4 1944
55499	**THERMAL CONDUCTIVITY OF ROCKS AT HIGH TEMPERATURES.** MOISEENKO, U. I. FREIBERG. FORSCHUNGSH. 238 C, 89-94, 1968.
55502	**HEAT CAPACITY OF /BA,SR/TIO3 SOLID SOLUTIONS IN THE REGION OF THE FERROELECTRIC PHASE TRANSITION.** BORMAN K YA STRUKOV B A TARASKIN S A FRITSBERG V YA IZV AKAD NAUK SSSR SER FIZ 33 7 1162-4 1969 CA 71 74920 (FOR ENGLISH TRANSLATION SEE T060362)
55514	**DETERMINATION OF THE COEFFICIENT OF THERMAL CONDUCTIVITY OF DINAS BRICK.** LITOVSKII E YA LANDA YA A OGNEUPORY 34 7 16-21 1969 CA 71 73607 (FOR ENGLISH TRANSLATION SEE T057273)
55526	**VOLUME CHANGE ATTENDING LOW-TO-HIGH INVERSION OF CRISTOBALITE.** BEALS M D ZERFOSS S J AM CERAM SOC 27 10 285-92 1944
55528	**EFFECT OF ALUMINUM OXIDE ON THE PROPERTIES OF BOROSILICATE GLASS WITH A LOW ALKALI CONTENT.** MAZURIN, O. V. TOTESH, A. S. ROSKOVA, G. P. SKLAR KERAM. 19 (5), 131-2, 1969.
55533	**ANALYTICAL METHOD FOR DETERMINATION OF GLASS PROPERTIES.** ATZORI, B. TERMOTECNICA 23 (3), 133-6, 1969.
55535	**MEASUREMENT OF THE HEAT CONTENT OF CEMENT MATERIALS.** VOLKONSKII B V TOLMACHEV G P TSEMENT 5 9-10 1969 CA 71 73686
55537	**MEASUREMENT OF COAL REFLECTANCE.** SARBEEVA, L. I. VOP. METAMORF. UGLEI EPIGENEZA UMESHCHAYUSHCHIKH POROD 50-67, 1968.

TPRC Number	Bibliographic Citation
55538	**PROPERTIES AND STRUCTURE OF GLASSES OF THE BORON SESQUIOXIDE-ANTIMONY SESQUIOXIDE SYSTEM.** IMAOKA M HASEGAWA H SHINDO S YOGYO KYOKAI SHI 77 888 263-71 1969 CA 71 73570
55539	**CHANGES IN A CHROME-MAGNESITE BRICK AFTER SIXTY DAYS ROTARY KILN OPERATION.** DERIE R ZEM-KALK-GIPS 22 6 265-70 1969 CA 71 73599
55600	**MARS. INTERPRETATION OF SPECTRAL REFLECTIVITY OF LIGHT AND DARK REGIONS.** ADAMS J B MC CORD T B J GEOPHYS RES 74 20 4851-6 1969 CA 71 86214
55618	**THERMAL CONDUCTIVITY MEASUREMENT OF REFRACTORIES AT HIGH TEMPERATURE.** DE SMET A REV INT HAUTES TEMP REFRACT 62 2 83-7 1969 CA 71 84179
55621	**USE OF SODA FELDSPAR IN WHITEWARE BODIES.** LOOMIS G A BLACKBURN A R J AM CERAM SOC 29 2 48-57 1946
55624	**RESTSTRAHLEN BANDS AND INTRINSIC ABSORPTION EDGE OF CD/X/ CA/1-X/ O.** FINKENRATH H UHLE N WAIDELICH W Z PHYS 226 5 469-78 1969 CA 71 86320
55628	**GLASS-CERAMIC MATERIAL FROM GABBRO-NITRITE.** SULEIMENOV, S. T. SHARAFIEV, M. SH. ORLOVA, G. V. SB. STATEI ASPIR. SOISKATELEI, MIN. VYSSH. SREDN. SPETS. OBRAZOV. KAZ. SSR, KHIM. KHIM. TEKHNOL. (6), 252-7, 1967.
55639	**THERMAL DIFFUSIVITY OF POWDER INSULATION.** KROPSCHOT, R. H. KNIGHT, B. L. TIMMERHAUS, K. D. ADVAN. CRYOG. ENG. 14, 224-9, 1968.
55641	**ORE MICROSCOPY AND CHEMICAL COMPOSITION OF SOME LAURITES.** LEONARD B F DESBOROUGH G A PAGE N J AMER MINERAL 54 9-10 1328-44 1969 CA 71 93385
55648	**DISCOVERY OF BETA-MATILDITE IN THE EASTERN TRANSBAIKAL REGION.** SAKHAROVA M S DOKL AKAD NAUK SSSR 187 2 418-20 1969 CA 71 93388
55680	**BOHDANOWICZITE (A NATURAL SILVER-BISMUTH SELENIDE).** BANAS, M. OTTEMANN, J. PRZEGL. GEOL. 17 (5), 235-8, 1969.
55684	**PHYSICOCHEMICAL PROPERTIES OF FAIENCE BODIES CONTAINING PERLITE AND DIATOMITE.** KHIZANISHVILI, I. G. SHUSHANISHVILI, A. I. TSANAVA, TS. P. SOOBSHCH. AKAD. NAUK GRUZ. SSR 55 (1), 89-92, 1969. (FOR ENGLISH TRANSLATION SEE TPRC NO. 64044)
55692	**PETROGRAPHIC CHARACTERISTIC OF COALS DURING VARIOUS STAGES OF REGIONAL METAMORPHISM (DONETS AND KUZNETSK BASINS).** KRYLOVA, N. M. SARBEEVA, L. I. VOP. METAMORF. UGLEI EPIGENEZA UMESHCHAYUSHCHIKH POROD 68-87, 1968. (FOR ENGLISH TRANSLATION SEE TPRC NO. 66214)
55693	**REFLECTANCE OF COAL MICROCOMPONENTS IN COALS OF VARIOUS METAMORPHIC GRADES.** SARBEEVA, L. I. KRYLOVA, N. M. VOP. METAMORF. UGLEI EPIGENEZA UMESHCHAYUSHCHIKH 87-106, 1968. (FOR ENGLISH TRANSLATION SEE TPRC NO. 66213)
55694	**PETROGRAPHIC STUDY OF CONTACT-METAMORPHOSED COALS OF THE TUNGUSKA BASIN.** BOGDANOVA, L. A. VOP. METAMORF. UGLEI EPIGENEZA UMESHCHAYUSHCHIKH POROD 205-21, 1968.
55711	**VISCOSITY AND TRANSFORMATION TEMPERATURE OF PHASE-SEPARATED SODIUM BOROSILICATE GLASSES.** MAZURIN O V STRELTSINA M V TOTESH A S PHYS CHEM GLASSES 10 2 63-8 1969 JA 52 277
55722	**TEMPERATURE AND ANGLE DEPENDENCE OF THE TOTAL EMISSIVITY OF POOR HEAT CONDUCTORS IN THE TEMPERATURE RANGE OF -60 TO 250 C.** LOHRENGEL, J. RHEINISCH-WESTFALISIHEN TECHNISCHEN HOCHSCHULE AACHEN, PH.D. THESIS 1-92 1969
55725	**THERMAL EXPANSION OF CALCITE.** WEIGLE J SAINI H HELV PHYS ACTA 7 257-66 1934
55729	**THERMAL CONDUCTIVITY OF ROCK-FORMING MINERALS.** HORAI K SIMMONS G EARTH PLANET SCI LETT 6 5 359-68 1969 CA 71 103914
55736	**THERMAL EXPANSION OF LITHIUM TANTALATE AND LITHIUM NIOBATE SINGLE CRYSTALS.** KIM Y S SMITH R T J APPL PHYS 40 11 4637-41 1969
55762	**NATURAL CARBONATES FOR MAGNETIC COOLING BELOW 1 K.** HSU Y LAMARCHE J L G J LOW TEMP PHYS 1 5 489-512 1969 CA 71 105547
55763	**THERMAL EXPANSION OF ALKALI BORATE GLASSES AND THE BORIC OXIDE ANOMALY.** UHLMANN, D. R. SHAW, R. R. J. NON-CRYST. SOLIDS 1 (5), 347-59, 1969.
55765	**THERMAL EXPANSION OF KANSAS VOLCANIC ASH.** BAULEKE M P HUH J M KANS STATE GEOL SURV BULL 194 1 15-16 1969 CA 71 104764
55768	**X-RAY DIFFRACTION STUDY OF THE THERMAL EXPANSION OF SYNTHETIC AND NATURAL DIAMONDS.** SOKHOR M I VITOLS V KRISTALLOGRAFIYA 14 4 734-5 1969 CA 71 105514
55769	**PRECISION STUDIES OF ASTM THERMAL CONDUCTIVITY TEST FOR REFRACTORIES.** WALLACE R W NORTON C L JR BART R K BRADY J G MATER RES STAND 9 9 27-30 1969 CA 71 104736
55793	**THERMOPHYSICAL CHARACTERISTICS OF FERRITES.** BEKKER YA M CHUDNOVSKII A F TEPLO MASSOPERENOS 7 437-46 1968 CA 71 106974 (FOR ENGLISH TRANSLATION SEE T056586)
55800	**GLASS FORMATION AND CRYSTALLIZATION OF GLASSES IN THE LITHIUM OXIDE-SILICON DIOXIDE-ZIRCONIUM DIOXIDE SYSTEM.** ELLERN, G. A. PAVLUSHKIN, N. M. TR. MOSK. KHIM.-TEKHNOL. INST. (59), 30-5, 1969.
55822	**OPTICAL PROPERTIES OF GEO2 IN THE ULTRAVIOLET REGION.** PAJASOVA L CZECH J PHYS 19 10 1265-70 1969 CA 71 118048
55839	**THERMAL DIFFUSIVITY MEASUREMENT.** KANAMORI H MIZUTANI H FUJII N J PHYS EARTH 17 1 43-53 1969 CA 71 116751
55846	**DETERMINING THE THERMOPHYSICAL PROPERTIES OF HOLLOW CYLINDRICAL REFRACTORY PRODUCTS DURING FIRING.** SEMENENKO A I TAITS N YU STARUN V R RADCHENKO I I KHARCHENKO I G PIMAKHOV D P MET GORNORUD PROM 3 56-9 1969 CA 71 116102
55856	**STRENGTH AND CREEP CHARACTERISTICS OF CERAMIC BODIES AT ELEVATED TEMPERATURES.** BURDICK M D MORELAND R E GELLER R F NATIONAL ADVISORY COMMITTE FOR AERONAUTICS 1-53, 1949. (NACA-TN-1561)
55858	**AN INVESTIGATION OF THE THERMAL DIFFUSIVITY AND THERMAL CONDUCTIVITY OF INSULATING POWDERS AT ATMOSPHERIC PRESSURE AND IN VACUUM BY VARIOUS METHODS.** KAGANER, M. G. SEMENOVA, R. S. INZH. FIZ. ZH. 13 (1), 24-30, 1967. (FOR ENGLISH TRANSLATION SEE TPRC NO. 63347)
55868	**DETERMINATION OF THE COEFFICIENTS OF THERMAL EXPANSION AND OF ELONGATION OF NONPLASTIC CERAMIC CLAYS.** SZYMANSKI, A. SZKLO CERAM. 20 (5), 151-3, 1969.

TPRC Number	Bibliographic Citation
55886	**CRYSTALLIZATION OF GLASSES IN THE SILICON DIOXIDE–ALUMINUM OXIDE–MAGNESIUM OXIDE–TITANIUM DIOXIDE SYSTEM, MODIFIED WITH SMALL ADDITIONS OF VANADIUM PENTOXIDE.** ZUEVA, V. N. PAVLUSHKIN, N. M. KHODAKOVSKAYA, R. YA. TR. MOSK. KHIM.–TEKHNOL. INST. (59), 108–13, 1969.
55889	**DEVELOPMENT OF CORDIERITE BODIES WITH SIERRALITE A NEW CERAMIC MATERIAL.** LAMAR R S J AM CERAM SOC 32 2 65–71 1949
55890	**SPECIFIC HEAT OF GLASSES AS FUNCTION OF THEIR CHEMICAL COMPOSITION AND TEMPERATURE.** RODNIKOVA, V. V. SAKHNO, V. N. VISN. KIIV. POLITEKH. INST., SER. KHIM. MASHINOBUDUV. TEKHNOL. (3), 159–66, 1966.
55904	**ULTRAVIOLET ABSORPTION OF PENTAVALENT VANADIUM IN BINARY ALKALI BORATE GLASSES.** PAUL A RUSIN J M J AM CERAM SOC 52 12 657–60 1969
55920	**THERMAL EXPANSION OF SODIUM AND POTASSIUM CHLORIDES FROM LIQUID–AIR TEMPERATURES TO PLUS 300 C.** SRINIVASAN R INDIAN INST SCI 37 A 232–41 1955
55925	**INDEX OF REFRACTION AND COEFFICIENTS OF EXPANSION OF OPTICAL GLASSES AT LOW TEMPERATURES.** MOLBY F A J OPT SOC AMERICA 39 7 600–11 1949
55934	**THERMAL EXPANSION OF ROCK SALT.** RUBIN T JOHNSTON H L ALTMAN H W J PHYS CHEM 65 65–8 1961
55938	**THE THERMAL EXPANSION OF REFRACTORIES TO 1800 C.** HEINDL R A J RESEARCH NATIONAL BUREAU STANDARDS 10 6 715–35 1933 (NBS–RESEARCH PAPER NO. 562)
55942	**COMPARATIVE TESTS OF CHEMICAL GLASSWARE.** WICHERS E FINN A N CLABAUGH W S J RESEARCH NATIONAL BUREAU STANDARDS 26 537–56 1941 (NBS–RESEARCH PAPER NO. 1394)
55944	**THE SYSTEM BERYLLIA–ALUMINA–TITANIA. PHASE RELATIONS AND GENERAL PHYSICAL PROPERTIES OF THREE–COMPONENT PORCELAINS.** LANG S M FILMORE C L MAXWELL L H J RESEARCH NATIONAL BUREAU STANDARDS 48 4 298–312 1952 (NBS–RESEARCH PAPER NO. 2316)
55946	**THE SPECIFIC HEAT OF MAGNETITE.** DIXON M HOARE F E HOLDEN T M PHYS LETTERS 14 3 184–5 1965
55947	**A DIRECT COMPARISON ON A CRYSTAL OF CALCITE OF THE X–RAY AND OPTICAL INTERFEROMETER METHODS OF DETERMINING LINEAR THERMAL EXPANSION.** AUSTIN J B SAINI H WEIGLE J PIERCE R H H JR PHYS REV 57 931–3 1940
55957	**THERMAL EXPANSION OF CRYSTALS. II. MAGNETITE AND FLUORITE.** SHARMA S S PROC INDIAN ACAD SCI 31 A 261–74 1950
55962	**THERMAL EXPANSION OF SOLIDS AT LOW TEMPERATURE.** WHITE G K PROC INT CONF LOW TEMPERATURE PHYS EIGHTH 394–6 1963
55966	**CONFIGURATIONAL THERMAL EXPANSION OF THREE INORGANIC GLASSES.** HAGGERTY J S COOPER A R JR PROC INT CONF PHYS NON–CRYSTALLINE SOLIDS 436–43 1965
55974	**ON THE TRANSFORMATION MAGNETITE AT A LOW TEMPERATURE.** OKAMURA T SCI REPT TOHOKU IMP UNIV 21 231–41 1932
55980	**THE THERMAL EXPANSION OF SOME FUSED OXIDES USED AS REFRACTORIES.** MERRITT G E TRANS AM ELECTROCHEM SOC 50 165–75 1926
55981	**DEVELOPMENT OF THE THERMAL CONDUCTIVITY PROBE.** HOOPER F C CHANG S C TRANS AM SOC HEATING VENTILATING ENGR 59 463–72 1953
55983	**THE HOT AND COLD SIZES OF FIREBRICKS.** MELLOR J W TRANS CERAMIC SOC /ENGL/ 16 270–3 1917
55984	**THE THERMAL EXPANSION CHARACTERISTICS OF THE CALCIUM ALUMINATES AND CALCIUM FERRITES.** RIGBY G R GREEN A T TRANS BRIT CERAM SOC 42 5 95–103 1943
55985	**THE REVERSIBLE THERMAL EXPANSION AND OTHER PROPERTIES OF SOME CALCIUM FERROUS SILICATES.** RIGBY G R LOVELL G H B GREEN A T TRANS BRIT CERAMIC SOC 44 3 37–52 1945
55996	**THE THERMAL EXPANSION OF SOME MATERIALS. I.** SCHEEL K Z PHYSIK 5 3 167–72 1921
55998	**INVESTIGATION OF METALLIC CRYSTALS. III. THERMAL EXPANSION OF ZINC AND CADMIUM.** GRUNEISEN E GOENS E Z PHYSIK 29 141–56 1924
55999	**AN X–RAY METHOD TO DETERMINE THE THERMAL EXPANSION COEFFICIENT AT HIGH TEMPERATURE.** BECKER K Z PHYSIK 40 37–41 1926
56000	**EXPANSION MEASUREMENTS AT LOW TEMPERATURES.** EBERT H Z PHYSIK 47 712–22 1928
56014	**ANOMALIES IN THE LOW TEMPERATURE HEAT CAPACITIES OF BEO AND MGO, CONTAINING FE /+3/.** GMELIN E J PHYS CHEM SOLIDS 30 12 2789–92 1969
56036	**DETERMINATION OF CRITICAL BEHAVIOUR IN LATTICE STATISTICS FROM SERIES EXPANSIONS II.** THOMPSON C J GUTTMANN A J NINHAM B W J PHYS C /SOLID ST PHYS/ 2 11 1889–99 1969
56053	**PREDICTED DEPTH OF FREEZE OR THAW IN SOILS BY CLIMATOLOGICAL ANALYSIS OF CUMULATIVE HEAT FLOW.** SCOTT, R. F. COLD REGIONS RESEARCH AND ENGINEERING LAB., HANOVER, NH 1–54, 1969. (CRREL–TR–195, AD–696414)
56069	**THE IN–PILE THERMAL CONDUCTIVITY OF SELECTED THORIUM DIOXIDE AND URANIUM DIOXIDE FUELS AT LOW DEPLETIONS /LWBR DEVELOPMENT PROGRAM/.** JACOBS, D. C. BETTIS ATOMIC POWER LAB., PITTSBURGH, PA 1–70, 1969. (WAPD–TM–758, N69–40230)
56075	**AUTOGRAPHIC THERMAL EXPANSION APPARATUS.** SOUDER W HIDNERT P FOX J F J RESEARCH NATL BUREAU STANDARDS 13 4 497–513 1934 (NBS–RESEARCH PAPER NO. 722)
56100	**QUANTITATIVE OPTICAL AND ELECTRON–PROBE STUDIES OF THE OPAQUE PHASES.** SIMPSON P R BOWIE S H U SCIENCE 167 619–21 1970
56101	**OPAQUE MINERALS IN LUNAR SAMPLES.** CAMERON E N SCIENCE 167 623–5 1970
56102	**OPTICAL AND HIGH–FREQUENCY ELECTRICAL PROPERTIES OF THE LUNAR SAMPLE.** GOLD T CAMPBELL M J OLEARY B T SCIENCE 167 707–9 1970
56103	**LUMINESCENCE, ELECTRON PARAMAGNETIC RESONANCE, AND OPTICAL PROPERTIES OF LUNAR MATERIAL.** GEAKE J E DOLLFUS A GARLICK G F J LAMB W WALKER C STEIGMANN G A TITULAER C SCIENCE 167 717–19 1970

TPRC Number	Bibliographic Citation
56104	LUMINESCENCE AND REFLECTANCE OF TRANQUILITY SAMPLES. EFFECTS OF IRRADIATION AND VITRIFICATION. NASH D B CONEL J E GREER R T SCIENCE 167 721-4 1970
56105	THERMAL RADIATION PROPERTIES AND THERMAL CONDUCTIVITY OF LUNAR MATERIAL. BIRKEBAK R C CREMERS C J DAWSON J P SCIENCE 167 724-6 1970
56106	INFRARED AND THERMAL PROPERTIES OF LUNAR ROCK. BASTIN J A CLEGG P E FIELDER G SCIENCE 167 728-30 1970
56107	THERMAL DIFFUSIVITY AND CONDUCTIVITY OF LUNAR MATERIAL. HORAI K SIMMONS G KANAMORI H WONES D SCIENCE 167 730-1 1970
56108	SPECTRAL REFLECTIVITY OF LUNAR SAMPLES. ADAMS J B JONES R L SCIENCE 167 737-9 1970
56109	SPECIFIC HEATS OF LUNAR SURFACE MATERIALS FROM 90 TO 350 K. ROBIE R A HEMINGWAY B WILSON W H SCIENCE 167 749-50 1970
56113	MECHANICAL PROPERTIES AND FRACTURE BEHAVIOR OF CHEMICALLY BONDED COMPOSITES. STETT M A FULRATH R M J AM CERAM SOC 53 1 5-13 1970
56114	FINAL STAGE SINTERING OF THO2. JORGENSEN P J SCHMIDT W G J AM CERAM SOC 53 1 24-7 1970
56119	DESIGN OF GLASS FILTER COMBINATIONS FOR PHOTOMETERS. WRIGHT H SANDERS C L GIGNAC D APPL OPT 8 12 2449-55 1969
56147	SOME PROBLEMS IN TECHNIQUE EMPLOYED FOR MEASURING THERMAL CONDUCTIVITY OF REFRACTORIES. SAITO S TAIKABUTSU 20 374-81 1968 JA 53 110
56148	MEASUREMENT OF TOTAL THERMAL DIFFUSIVITY OF /SOME INDUSTRIAL/ GLASSES UP TO 1400 DEGREES. BRE M KERMABON P VERRES REFRACT 23 1 3-9 1969 CA 72 46918
56159	VITREOUS WHITEWARE BODIES STRENGTHENED BY CALCINED BAUXITE ADDITIONS. FLOYD J R AM CERAM SOC BULL 49 2 172-4 1970 CA 72 82441
56161	SEALING PURE ALUMINA CERAMICS TO METALS. KLOMP J T BOTDEN TH P J AM CERAM SOC BULL 49 2 204-11 1970
56166	CRYSTALLIZATION OF LEAD MONOXIDE-BORON TRIOXIDE-TITANIUM DIOXIDE GLASSES AND THERMAL EXPANSION COEFFICIENTS OF THEIR CRYSTALLIZATION PRODUCTS. SUZUKI Y ICHIMURA N ASAHI GARASU KENKYU HOKOKU 18 2 49-58 1968 CA 71 128081
56176	DENSE CORDIERITE BODIES. GEBLER K A WISELY H R J AM CERAM SOC 32 5 163-65 1949
56188	CORRELATION OF GLAZE-BODY STRESSES WITH THERMAL PROPERTIES OF WHITEWARE BODIES. COOK R L BRUNNER C D J AM CERAM SOC 32 12 401-8 1949
56196	ELECTRICAL RESISTANCE AND THERMAL CONDUCTIVITY OF CHARGES OF CARBON-CONTAINING FERROMANGANESE AND SILICOMANGANESE. NIKOLAISHVILI, G. U. KEKELIDZE, M. A. PROIZVOD. PRIMEN. MARGANTSEVYKH FERROSPLAVOV 37-46, 1968. (FOR ENGLISH TRANSLATION SEE TPRC NO. 65531)
56197	THE RAJHARA SEAM, DALTONGUNJ COALFIELD, BIHAR. DUTT A B KAR S K QUART J GEOL MINING MET SOC INDIA 39 3 147-53 1967 CA 71 127215
56200	PHYSICOCHEMICAL PROPERTIES OF ALKALI-FREE ALUMINUM BOROSILICATE GLASSES. KUZNETSOV, A. I. STEKLO (3), 78-81, 1968.
56201	SYNTHESIS AND PROPERTIES OF ALKALI-TIN SILICATE GLASSES. VAKHRAMEEV, V. I. STEKLO (3), 84-9, 1968.
56202	EFFECT ALUMINUM OXIDE ON THE PROPERTIES OF SLIGHTLY ALKALINE BOROSILICATE GLASSES. ROSKOVA, G. P. STEKLO (3), 89-92, 1968.
56205	THERMAL EXPANSION OF BINDERS USED IN GRINDING WHEELS. SZYMANSKI, A. SZKLO CERAM. 20 (6), 180-2, 1969.
56210	THERMAL CONDUCTIVITY OF STABILIZED ZIRCONIUM DIOXIDE AT HIGH TEMPERATURES. CHEKHOVSKOI V YA BANAEV A M TEPLO MASSOPERENOS 7 591-5 1968 CA 71 128094 (FOR ENGLISH TRANSLATION SEE T056605)
56214	DETERMINATION OF THE THERMOPHYSICAL PROPERTIES OF COPPER SULFIDE ORE AND BRIQUETS FROM FINES. USHAKOV, K. I. FELMAN, R. I. TSVET. METAL. 42 (7), 25-7, 1969. (FOR ENGLISH TRANSLATION SEE TPRC NO. 67256)
56215	CERAMIC COMPOSITIONS HAVING NEGATIVE LINEAR THERMAL EXPANSION. SMOKE E J J AM CERAM SOC 34 3 87-90 1951
56231	HEAT CONDUCTIVITY IN THE MANTLE. CLARK, S. P., JR. GEOPHYS. MONOGR., AMER. GEOPHYS. UNION 13, 622-6, 1969.
56232	HEAT TRANSFER IN A FILLING OF DISPERSED MATERIAL. NIKITIN, V. S. ANTONISHIN, N. V. INZH.-FIZ. ZH. 17 (2), 248-53, 1969. (FOR ENGLISH TRANSLATION SEE TPRC NO. 69506)
56237	SIGNIFICANT ASPECTS OF CERTAIN TERNARY COMPOUNDS AND SOLID SOLUTIONS. HUMMEL F A J AM CERAM SOC 35 3 64-6 1952
56245	PHOSPHORUS PENTOXIDE-ANTIMONY TRIOXIDE-CHROMIC OXIDE AND PHOSPHORUS PENTOXIDE-ANTIMONY TRIOXIDE-VANADIUM PENTOXIDE SYSTEMS. DOMBROVSKAYA V K KONSTANTS Z MILLERS T VAIVADS A NEORG STEKLOVIDNYE POKRYTIYA MATER 101-6 1969
56246	PHYSICOCHEMICAL PROPERTIES OF BARIUM OXIDE-LEAD OXIDE-BORON OXIDE-BISMUTH TRIOXIDE-TITANIUM DIOXIDE SYSTEM GLASSES. FREIDENFELDS, E. JOKSTA, E. ZUSANE, E. POLKA, I. NEORG. STEKLOVIDNYE POKRYTIYA MATER. 107-11, 1969. (FOR ENGLISH TRANSLATION SEE TPRC NO. 66144)
56247	CRYSTALLIZATION OF SOME BARIUM TITANATE-BISMUTH OXIDE-BORON OXIDE SYSTEM GLASSES WITH CALCIUM FLUORIDE AND ZIRCONIUM DIOXIDE ADDITIONS AND THE DIELECTRIC PROPERTIES OF THE RESULTING MATERIALS. FREIDENFELDS, E. JOKSTA, E. VAZA, A. NEORG. STEKLOVIDNYE POKRYTIYA MATER. 119-24, 1969.
56248	EFFECT OF TITANIUM DIOXIDE ON SOME PHYSICOCHEMICAL PROPERTIES OF ALUMINOSILICATE GLASS. OSMANIS A EIDUKS J OKMANIS A DANEBERGS A SVINKA V NEORG STEKLOVIDNYE POKRYTIYA MATER 141-9 1969 CA 72 5856
56249	CHANGE IN THE PHYSICOCHEMICAL PROPERTIES AND STRUCTURE OF A GLAZE ON SUBSTITUTION OF SOME RAW MATERIALS BY OTHERS WHILE STILL PRESERVING THE CHEMICAL COMPOSITION. AVGUSTINIK, A. I. PYZHOVA, A. P. NEORG. STEKLOVIDNYE POKRYTIYA MATER. 153-9, 1969.
56251	BORON-FREE AND LOW-BORON GLAZES FOR CERAMIC FACE TILES. SAKHAROVA, N. A. GOLIK, E. M. NEORG. STEKLOVIDNYE POKRYTIYA MATER. 183-9, 1969.

TPRC Number	Bibliographic Citation
56253	LEAD - FREE FLUXES FOR OVER - GLAZE COLORS. VIZIR, L. A. CHEREPANINA, L. I. NEORG. STEKLOVIDNYE POKRYTIYA MATER. 255-60, 1969.
56254	EFFECT OF REPLACING SODIUM OXIDE BY LITHIUM OXIDE ON THE PROPERTIES OF ZIRCONIUM GLASSES. LESKOV A L NEORG STEKLOVIDNYE POKRYTIYA MATER 351-7 1969 CA 72 5859
56255	EFFECT OF FLUORINE ON THE OPACIFICATION AND OTHER PROPERTIES OF ZIRCONIUM GLASSES AND ENAMELS. LESKOV, A. L. NEORG. STEKLOVIDNYE POKRYTIYA MATER. 359-64, 1969.
56266	THERMAL CONDUCTIVITY OF CHIATURA MANGANESE CONCENTRATES AND THEIR PELLETIZED PRODUCTS. NIKOLAISHVILI, G. U. KEKELIDZE, M. A. PROIZVOD. PRIMEN. MARGANTSEVYKH FERROSPLAVOV 28-36, 1968. (FOR ENGLISH TRANSLATION SEE TPRC NO. 64050)
56269	CRYSTALLIZATION AND SEGREGATION PHENOMENA IN THE SILICON DIOXIDE - TITANIUM DIOXIDE SYSTEM. KOZLOVA, L. N. TARASOV, B. V. CHEPIZHNYI, K. I. STEKLO (3), 6-11, 1968.
56270	GLASS-FORMATION AND CERAMMING REGIONS IN THE SILICON DIOXIDE - CALCIUM OXIDE - MAGNESIUM OXIDE - ALUMINUM OXIDE SYSTEM WITH SOME SODIUM OXIDE, POTASSIUM OXIDE, IRON(III) OXIDE, AND 5 PERCENT FLUORINE IMPURITIES. TRUSHKOV, A. I. STEKLO (3), 57-61, 1968.
56272	NEW GLASSES ABSORBING HARMFUL RADIATION. SCHLEIFER, P. BORCZUCH-LACZKA, M. SZKLO CERAM. 20 (6), 161-5, 1969.
56285	DETERMINATION OF MINERAL REFLECTANCE FOR SELECTING OPTIMUM PHOTOGRAPHIC CONDITIONS DURING GEOLOGICAL PHOTODOCUMENTATION OF MINE WORKS. GOLOVIN G A BESPALOV YU I ZAP LENINGRAD GORN INST 55 2 128-31 1968 CA 72 5043
56318	HEAT-INSULATING PROPERTIES OF MODERN GAS-EXPANDED PLASTICS. DUDNIK D M STEPANENKO A N KHOLOD TEKH TEKHNOL 8 101-4 1969 CA 72 13327
56321	MATT AND SHINY COLORED COATINGS FOR SLAG GLASS-CERAMICS. SAKHAROVA, N. A. GOLIK, E. M. NEORG. STEKLOVIDNYE POKRYTIYA MATER. 191-5, 1969.
56322	PREPARATION OF NONFRITTED MATT AND SEMIMATT GLAZES FROM PLENTIFUL RAW MATERIAL FOR FACE TILE. LYSENKO, I. A. NEORG. STEKLOVIDNYE POKRYTIYA MATER. 197-204, 1969.
56324	EXPERIMENTAL PREPARATION OF ZIRCONIUM-DANBURITE ENAMELS. LESKOV A L RAZLIVANOVA V E AFENKO Z I POSVOLSKAYA L V NEORG STEKLOVIDNYE POKRYTIYA MATER 347-50 1969 CA 72 15286
56326	SYNTHESIS OF HEAT-RESISTANT ENAMEL COATINGS. GORBATENKO V E RATKOVA V P LUKYANOVA Z L NEORG STEKLOVIDNYE POKRYTIYA MATER 437-43 1969 CA 72 15290
56331	THERMAL CONDUCTIVITY OF THERMOCOUPLE SHEATHS USED FOR TEMPERATURE MEASUREMENTS IN FOUNDRIES. NOWAK, J. PR. INST. ODLEW. 19 (1), 29-44, 1969.
56333	PROPERTIES OF INDUSTRIAL ENAMELS FOR CHEMICAL APPARATUS. VARGIN V V ZASUKHINA L Z STEKLO KERAM 26 9 17-19 1969 CA 72 15283
56348	RADIATION PROPERTIES OF SLAG. SMITH, R. A. MASSACHUSETTS INSTITUTE OF TECHNOLOGY, M.S. THESIS 1-127 1969
56355	SNOW ALBEDO MODIFICATION. A REVIEW OF LITERATURE. SLAUGHTER, C. W. COLD REGIONS RESEARCH AND ENGINEERING LAB., HANOVER, NH 1-31, 1969. (CRREL-TR-217, AD-698023)
56358	MANUFACTURE, PROPERTIES AND APPLICATION OF ZIRCONIUM BASE ALLOYS. PRIDANTSEVA, K. S. FOREIGN TECH DIV., WRIGHT-PATTERSON AIR-FORCE BASE, OHIO 21PP., 1969. (ENGLISH TRANSLATION OF SB. TR. TSENT. NAUCH.-ISSLED. INST. CHERN. MET., (51), 139-52, 1967; FOR ORIGINAL SEE T47173) (AD-694 926, FTD-MT-24-457-68)
56366	HANDBOOK OF THERMAL DESIGN DATA FOR MULTILAYER INSULATION SYSTEMS. VOLUME II. COSTON, R. M. LOCKHEED MISSILES AND SPACE CO., SUNNYVALE, CA 1-180, 1967. (LMSC-A84788- VOL 2, NASA-CR-87485, N67-34910)
56368	DIELECTRIC STUDIES OF THE CATION SUBSTITUTED MIXED CRYSTALS OF NIOBATES. WANG, F. F. Y. COLLEGE OF ENGINEERING, STATE UNIV. OF NEW YORK STONY BROOK, NY 1-47, 1969. (NASA-CR-107198, N70-12933)
56385	INSULATION. FROM STUDY OF INTEGRATED CRYOGENIC FUELED POWER GENERATING AND ENVIRONMENTAL CONTROL SYSTEMS. VOLUME II. CRYOGENIC TANKAGE INVESTIGATION. BEECH AIRCRAFT CORPORATION, BOULDER COLORADO BEECH AIRCRAFT CORP., BOULDER DIV., CO 4.1-4.32, 1961. (ASD-TR-61-327 VOL. II, AD-270474)
56394	X-RAY DETERMINATION OF CRYSTALLINE LATTICE PARAMETERS AND COEFFICIENTS OF THERMAL EXPANSION IN LEUCOSAPPHIRE AND RUBY. SHALNIKOVA N A YAKOVLEV Y A KRISTALLOGRAFIYA 1 531-3 1956
56408	EXPANSION MEASUREMENTS ON GOLD AND ON CRYSTALLIZED QUARTZ BETWEEN 18 AND 520 DEGREES BY THE FIZEAU METHOD. MULLER A PHYSIK Z 17 3 29-30 1916
56410	THERMAL CONSTANTS AT HIGH TEMPERATURES. II. THE THERMAL EXPANSION OF ROCK SALT. WALTHER A K WASCHKOWSKY W PH STRELKOV P G PHYS ZEIT SOWJETUNION 12 35-44 1937
56414	THE THERMAL EXPANSION OF GRAPHITE FROM 15 TO 800 C.. PART I. EXPERIMENTAL. NELSON J B RILEY D P PROC PHYS SOC 57 477-85 1945
56452	DETERMINATION OF THE THERMAL DIFFUSIVITY OF MATERIALS AT HIGH TEMPERATURES. MAVASHEV YU Z BORUKHOV M YU GELIOTEKHNIKA 3 36-41 1969 CA 72 25859
56454	ALUMINA AND MULLITE REFRACTORIES AS KILN FURNITURE. MC NAMARA E P JR AM CERAM SOC BULL 49 3 272-5 1970
56455	FERRITE-DIELECTRIC COMPOSITES UTILIZING MGO-MG AL2 O4 MIXTURES TO ACHIEVE THERMAL EXPANSION MATCHED SUBSTRATES. PALADINO A E SNIDER C R AM CERAM SOC BULL 49 3 280-5 1970
56479	LINEAR THERMAL EXPANSION AND INVERSIONS OF QUARTZ, VAR. ROCK CRYSTAL. ROSENHOLTZ J L SMITH D T AM MINERALOGIST 26 103-9 1941
56483	ON THE EXPANSION OF SOLIDS BY HEAT. /2ND MEMORANDUM, PART 2/. FIZEAU M H COMPT REND 66 1072-86 1868
56491	CORRELATION OF MECHANICAL AND THERMAL PROPERTIES OF THE LUNAR SURFACE. HALAJIAN J D REICHMAN J ICARUS 10 179-96 1969
56537	THERMAL CONDUCTION THROUGH AN EVACUATED IDEALIZED POWDER OVER THE TEMPERATURE RANGE OF 100 TO 500 K. MERILL, R. B. BRIGHAM YOUNG UNIVERSITY, PROVO, UTAH, PH. D. THESIS 1-99, 1968. (UNIVERSITY MICROFILMS NO. 69-3006)

TPRC Number	Bibliographic Citation
56540	**THE ANISOTROPIC THERMAL EXPANSION OF BISMUTH.** CAVE, E. F. UNIV. OF MISSOURI, COLUMBIA, PH. D. THESIS 1-96, 1958. (UNIVERSITY MICROFILMS NO. 58-5247)
56553	**DETERMINATION OF THE SPECTRAL COEFFICIENTS OF REFLECTION FOR FRIABLE MATERIALS IN THE 0.7-15 MICROMETERS WAVELENGTH RANGE.** KROPOTKIN M A KOZYREV B P FOREIGN SCIENCE AND TECHNOLOGY CENTER 1-9, 1968. (FSTC-HT-23-256-68, AD-835299)
56570	**THERMOMETRIC DETERMINATION OF THERMOPHYSICAL CHARACTERISTICS.** GERASHCHENKO, O. A. GRISHCHENKO, T. G. PILIPENKO, A. M. FEDOROV, V. G. FOREIGN TECHNOLOGY DIVISION (PT. 2), 280-94, 1969. (ENGLISH TRANSLATION OF TEPLO-MASSOPERENOS, DOKL. VSES. SOVESHCH., 3RD, 7, 261-73, 1968; FOR ORIGINAL SEE T61216) (FTD-HT-23-820-68-PT2, AD-698 518)
56578	**METHODS OF MEASURING THE COEFFICIENT OF THERMAL CONDUCTIVITY OF MATERIALS WITH LOW THERMAL CONDUCTIVITY.** PETROV, I. N. FOREIGN TECHNOLOGY DIVISION (PT. 2), 384-91, 1969. (ENGLISH TRANSLATION OF TEPLO-MASSOPERENOS, DOKL. VSES. SOVESHCH., 3RD, 7, 361-8, 1968; FOR ORIGINAL SEE T61217) (FTD-HT-23-820-68, AD-698 518)
56579	**METHODS OF QUASI-STEADY-STATE REGULAR AND STEADY-STATE REGIMES IN DETERMINING HEAT TRANSFER COEFFICIENTS AT ELEVATED TEMPERATURES.** PETROV-DENISOV V G ZASEDATELEV I B MASLENNIKOV L A FOREIGN TECHNOLOGY DIVISION 392-9, 1969. (FTD-HT-23-820-68/PT. 2/, AD-698518)
56586	**THERMOPHYSICAL CHARACTERISTICS OF FERRITES.** BEKKER YA M CHUDNOVSKII A F FOREIGN TECHNOLOGY DIVISION 470-9, 1969. (ENGLISH TRANSLATION OF TEPLO MASSOPERENOS, 7, 437-46, 1968; FOR ORIGINAL SEE T55793) (FTD-HT-23-820-68/PT. 3/, AD-698519)
56605	**THERMAL CONDUCTIVITY OF STABILIZED ZIRCONIUM DIOXIDE AT HIGH TEMPERATURES.** CHEKHOVSKOI V YA BANAEV A M FOREIGN TECHNOLOGY DIVISION 637-42, 1969. (ENGLISH TRANSLATION OF TEPLO MASSOPERENOS, 7, 591-5, 1968; FOR ORIGINAL SEE T56210) (FTD-HT-23-820-68/PT. 3/, AD-698519)
56623	**PHYSICAL PROPERTIES, THERMAL EXPANSION, AND CREEP BEHAVIOR OF MAGNESITE BRICKS.** SCHWIETE H E SCHINNERLING F ARCH EISENHUETTENW 40 11 881-3 1969 CA 72 35226
56648	**CONCERNING THE TRANSFER OF HEAT IN CERTAIN FOSSIL FUELS.** KIGEL T B TERESHCHENKO V G HEAT TRANSFER-SOVIET RESEARCH 2 1 119-23 1970
56656	**STRUCTURAL AND THERMAL EXPANSIONS IN ALKALI SILICATE BINARY GLASSES.** TILTON L W J AM CERAM SOC 43 1 9-17 1960
56664	**FIRESIDE DEPOSITS AND THEIR EFFECT ON HEAT TRANSFER IN A PULVERIZED-FUEL-FIRED BOILER. III. INFLUENCE OF THE PHYSICAL CHARACTERISTICS OF THE DEPOSIT ON ITS RADIANT EMITTANCE AND EFFECTIVE THERMAL CONDUCTANCE.** BOOW J GOARD P R C J INST FUEL 42 346 412-19 1969 CA 72 22869
56672	**THERMAL EXPANSION OF RUTILE AND ANATASE.** RAO K V K NAIDU S V N IYENGAR L J AMER CERAM SOC 53 3 124-6 1970 JA 53-4 88
56698	**LOW TEMPERATURE THERMAL CONDUCTIVITY OF AMORPHOUS SOLIDS.** DREYFUS B FERNANDES N C MAYNARD R PROC. INT. CONF. ON LOW TEMP. PHYS., 11TH 1, 589-92, 1968.
56731	**REFLECTOMETER DESIGNED FOR SOME MEASUREMENTS IN POLARIZED LIGHT IN THE VACUUM ULTRAVIOLET.** UZAN E MARTIN J DAMANY H REVUE DE PHYSIQUE APPLIQUEE 4 4 507-11 1969
56741	**TRANSPORT PROCESSES IN OXIDE NUCLEAR FUELS.** CHRISTENSEN, J. A. CERAM. NUCL. FUELS, PROC. INT. SYMP. 109-25, 1969.
56752	**DETERMINATION OF SOME THERMODYNAMIC CONSTANTS OF KAOLINITE AND KINETIC PARAMETERS OF ITS DEHYDRATION.** KUSKOV O L KHITAROV N I GEOKHIMIYA 12 1501-6 1969 CA 72 45956 (FOR ENGLISH TRANSLATION SEE T70864)
56759	**REFLECTION SPECTRUM OF NATURAL GALENA AT 77 AND 290 K.** KORSUNSKII M I MURATOV E M IZV AKAD NAUK KAZ SSR SER FIZ-MAT 7 2 44-9 1969 CA 72 49332
56785	**THERMAL CONDUCTIVITY OF THO2-PUO2 DURING IRRADIATION. SEE ALSO TPRC NO. 55199.** JEFFS A T TRANS AMER NUCLEAR SOC 11 497-8 1968 NSA 23-2 439
56794	**DEFORMATIONS DURING HARDENING AND THE COEFFICIENT OF LINEAR EXPANSION OF PHOSPHATE BINDERS.** SYCHEV M M KHIM OSN TEKHNOL PRIMEN FOSFATNYKH SVYAZOK POKRYTII 106-15 1968 CA 72 35357
56809	**ON THE THERMAL CONDUCTIVITY OF SHALE KUKERSIT IN THE PROCESS OF HEATING AND THERMAL DECOMPOSITION.** VALDEK R LUTSKOVSKAYA N L EIZEN J SPECIAL LIBRARY ASSOCIATION, TRANS. CENTER 1-15, 1969. (ENGLISH TRANSLATION OF IZV. AKAD. NAUK ESTON. SSSR, SER. FIZ. MAT. TEKH. NAUK, 10 (3), 207-14, 1961 FOR ORIGINAL SEE T25832) (TT-68-50607)
56819	**PHYSICAL PROPERTIES AND THE CHEMICAL COMPOSITION OF SPHALERITES FROM YUGOSLAVIA.** GRAFENAUER S GORENC B MARINKOVIC V STRMOLE D MAKSIMOVIC Z MINER DEPOSITA 4 3 275-82 1969 CA 72 45893
56820	**EXPANSION COEFFICIENTS OF ENAMEL. RELATION BETWEEN THERMAL EXPANSION, VISCOSITY, TRANSFORMATION POINT, AND STRESSES IN ENAMEL.** DEKKER, P. MITT. VER. DEUT. EMAILFACHLEUTE. 17 (11), 67-73, 1969.
56824	**PREPARATION OF JEWELRY ENAMELS ON ALUMINUM TO BE ANODIZED.** PRONINA L E NEORG STEKLOVIDNYE POKRYTIYA MATER 409-15 1969 CA 72 24166
56867	**REACTIVE SILICA. III. THE FORMATION OF BORON HYDRIDES, AND OTHER REACTIONS, ON THE SURFACE OF BORIA-IMPREGNATED AEROSIL.** MORTERRA C LOW M J D J PHYS CHEM 74 6 1297-302 1970
56880	**SPECTRAL AND INTEGRAL EMISSIVITY OF POROUS REFRACTORY CERAMIC USED IN RADIATION-TYPE GAS BURNERS.** KRIVONOGOV, B. M. SANIT. TEKH. 90-101PP., 1967.
56882	**ROUTINE MEASUREMENTS OF THE THERMAL CONDUCTIVITY COEFFICIENT OF REFRACTORY BUILDING MATERIALS AT ELEVATED TEMPERATURE BY THE HOT-WIRE METHOD.** GAUNITZ, H. SCHULLE, W. LORENZ, H. SILIKATTECHNIK 20 (9), 315, 1969.
56887	**DETERMINATION OF THE COEFFICIENT OF THERMAL CONDUCTIVITY OF FIBERS FROM GLASS AND POLYMER MATERIALS.** SHLENSKII O F GONCHARUK N I GALTSOV V YA STEKLO KERAM 26 9 14-17 1969 CA 72 33102
56888	**THERMAL EXPANSION AND DENSITY OF ANNEALED GLASSES AS DEPENDENT ON THERMAL HISTORY.** BOGUSLAVSKII I A PUKHLIK O I KHALIZEVA O N STEKLO KERAM 26 10 12-15 1969 CA 72 35203
56889	**COEFFICIENT OF THE THERMAL EXPANSION OF TITANIUM-CONTAINING GLASSES AND ENAMELS.** KOMLEVA G P DMITRIEVA V I STEKLO KERAM 26 11 20-2 1969 CA 72 35261
56890	**GLAZES BASED ON BASALTS.** DVORKIN L I GALUSHKO I K STEKLO KERAM 26 11 36-8 1969 CA 72 35209

TPRC Number	Bibliographic Citation
56892	NEW OPAQUE ZIRCONIUM GLAZES FOR STRUCTURAL CERAMICS. EIDUKS, J. DUSAUSKAS-DUZ, S. SEDMALIS, U. PAUKSS, P. STROIT. MATER., DETALI IZDELIYA (10), 44-51, 1968. (FOR ENGLISH TRANSLATION SEE TPRC NO. 64048)
56893	THERMOPHYSICAL PROPERTIES OF HYDROPHOBIZED PERLITE AND MATERIALS BASED ON IT. SPEKTOR, B. V. KRUPA, A. A. STROIT. MATER., DETALI IZDELIYA (11), 81-6, 1969. (FOR ENGLISH TRANSLATION SEE TPRC NO. 66055)
56898	HEAT TRANSFER IN SOME COMBUSTIBLE MINERALS. KIGEL T B TERESHCHENKO V G TEPLOFIZ TEPLOTEKH 15 106-10 1969 CA 72 48317
56962	SYNTHESIS AND STUDY OF STRONTIUM-MAGNESIUM FROSTED GLASSES OPACIFIED WITH CERIUM, ZIRCONIUM, AND TITANIUM COMPOUNDS. AVGUSTINIK, A. I. SHTEINBERG, YU. G. GUBAR, V. N. BELOVA, A. N. ZH. PRIKL. KHIM. (LENINGRAD) 42 (10), 2195-204, 1969.
56964	OPTICAL CHARACTERISTICS OF STRONGLY DIFFUSING GRAY GLASSES. VOISHVILLO, N. A. ZH. PRIKL. SPEKTROSK. 11 (2), 316-9, 1969. (FOR ENGLISH TRANSLATION SEE TPRC NO. 63018)
56968	EFFECT OF DENSITY AND STOICHIOMETRY ON OXIDE FUELS BEHAVIOR. MIKAILOFF, H. VAN CRAEYNEST, J. C. LALLEMENT, R. CERAM. NUCL. FUELS, PROC. INT. SYMP. 126-37, 1969.
56973	VISCOSITY COEFFICIENT OF VILLARRICA VOLCANIC MAGMA. CASERTANO L LUONGO G ASS GEOFIS ITAL ATTI CONV ANNU 17 1 97-108 1968 CA 72 57713
56984	TEKTITES AND THE LOST PLANET. STAIR R SCIENTIFIC MONTHLY 83 1 3-12 1956
56996	VIBRATIONAL FREQUENCIES AND HEAT CAPACITY OF DIAMOND - TYPE CRYSTALS. KUCHER, T. I. KHIM. SVYAZ KRIST. 305-14, 1969.
57010	THERMODYNAMIC POTENTIALS OF WATER-BEARING MINERALS AND CARBONATES. MARAKUSHEV, A. A. PROBL. PETROLOGII GENET. MINERAL. 1, 94-112, 1969.
57020	GLASS-CERAMICS OF HIGH STRENGTH. SCHLEIFER P JARZMIK B KUCHARSKI J ZELAZOWSKI E SZKLO CERAM 20 7/8 197-9 1969 CA 72 58566
57031	INVESTIGATION OF DIELECTRIC OPTICAL COATINGS FOR LASERS. MOYS, B. A. WOOD, R. M. WARR, P. D. GENERAL ELECTRIC COMP., LTD., WEMBLEY, ENGLAND 1-12, 1969. (C.V.D.-RP-4-67, AD-867413)
57048	RADIATION DAMAGE IN GYPSUM. PATEL A R RAJU K S J PHYS CHEM SOLIDS 31 2 331-5 1967
57079	GLASS-FORMATION AND PROPERTIES OF GLASSES IN THE SYSTEM SI O(2)-R(2) O-SN O(2). VAKHRAMEEV, V. I. EVSTROPEV, K. S. INORGANIC MATERIALS 5 1 82-4 1967 (ENGLISH TRANSLATION OF IZV. AKAD. NAUK SSSR, NEORG. MAT., 5 (1), 101-4, 1969; FOR ORIGINAL SEE T54435)
57086	REACTION OF METAL COATINGS OF MO-MN-V2O5 WITH A HIGH-ALUMINA CERAMIC. AVETIKOV V G FADEEVA V I FRENKEL E B IZV AKAD NAUK SSSR NEORG MATER 5 1 54-7 1969 (FOR ENGLISH TRANSLATION SEE T57087)
57087	REACTION OF METAL COATINGS OF MO-MN-V2O5 WITH A HIGH-ALUMINA CERAMIC. AVETIKOV V G FADEEVA V I FRENKEL E B INORGANIC MATERIALS 5 1 44-6 1969 (ENGLISH TRANSLATION OF IZV. AKAD. NAUK SSSR, NEORG. MAT., 5 (1), 54-7, 1969; FOR ORIGINAL SEE T57086)
57096	THERMAL EXPANSION OF CALCIUM ALUMINATE IN THE GLASSY STATE. YANISHEVSKII V M POPOVA E I KURYLEVA V F INORGANIC MATERIALS 5 1 154-5 1969 (ENGLISH TRANSLATION OF IZV. AKAD. NAUK SSSR, NEORG. MAT., 5 (1), 183-4, 1969; FOR ORIGINAL SEE T54437)
57098	THE PROBLEM OF CHEMICAL BONDING IN FERRITES WITH THE SPINEL STRUCTURE. BEKKER YA M BERG I V IZV AKAD NAUK SSSR NEORG MATER 5 1 196-8 1969 (FOR ENGLISH TRANSLATION SEE T57099)
57099	THE PROBLEM OF CHEMICAL BONDING IN FERRITES WITH THE SPINEL STRUCTURE. BEKKER YA M BERG I V INORGANIC MATERIALS 5 1 167-8 1969 (ENGLISH TRANSLATION OF IZV. AKAD. NAUK SSSR, NEORG. MAT., 5 (1), 196-8, 1969; FOR ORIGINAL SEE T57098)
57104	THERMAL EXPANSION OF ZIRCON REFRACTORIES. FILATOV S K FRANK-KAMENETSKII V A ZHURAVINA T A INORGANIC MATERIALS 5 2 286-90 1969 (ENGLISH TRANSLATION OF IZV. AKAD. NAUK SSSR, NEORG. MAT., 5 (2), 346-51, 1969; FOR ORIGINAL SEE T54441)
57112	PECULIARITIES IN THE PHYSICOCHEMICAL PROPERTIES OF LEAD BOROSILICATE GLASSES. PAVLUSHKIN N M ZHURAVLEV A K IZV AKAD NAUK SSSR NEORG MATER 5 3 572-6 1969 (FOR ENGLISH TRANSLATION SEE T57113)
57113	PECULIARITIES IN THE PHYSICOCHEMICAL PROPERTIES OF LEAD BOROSILICATE GLASSES. PAVLUSHKIN N M ZHURAVLEV A K INORGANIC MATERIALS 5 3 481-4 1969 (ENGLISH TRANSLATION OF IZV. AKAD. NAUK SSSR, NEORG. MAT., 5 (3), 572-6, 1969; FOR ORIGINAL SEE T57112)
57127	SYNTHESIS AND IR ABSORPTION SPECTRA OF LOW-TEMPERATURE MODIFICATION OF SPODUMENE. BLINOV V A ROY R IZV AKAD NAUK SSSR NEORG MATER 5 4 811-12 1969 (FOR ENGLISH TRANSLATION SEE T57128)
57128	SYNTHESIS AND IR ABSORPTION SPECTRA OF LOW-TEMPERATURE MODIFICATION OF SPODUMENE. BLINOV V A ROY R INORGANIC MATERIALS 5 4 691-2 1969 (ENGLISH TRANSLATION OF IZV. AKAD. NAUK SSSR, NEORG. MAT., 5 (4), 811-2, 1969; FOR ORIGINAL SEE T57127)
57136	INVESTIGATION ON THE STRUCTURE OF LEAD-GERMANIUM GLASSES BY IR SPECTROSCOPY. MOROZOV V N IZV AKAD NAUK SSSR NEORG MATER 5 5 979-81 1969 (FOR ENGLISH TRANSLATION SEE T57137)
57137	INVESTIGATION ON THE STRUCTURE OF LEAD-GERMANIUM GLASSES BY IR SPECTROSCOPY. MOROZOV V N INORGANIC MATERIALS 5 5 834-5 1969 (ENGLISH TRANSLATION OF IZV. AKAD. NAUK SSSR, NEORG. MAT., 5 (5), 979-81, 1969; FOR ORIGINAL SEE T57136)
57140	PHASE TRANSFORMATIONS AND EXPANSION OF VERMICULITE DURING HEATING. KHVOSTENKOV S I TURKIN A F IZV AKAD NAUK SSSR NEORG MATER 5 6 1121-5 1969 CA 71 75037 (FOR ENGLISH TRANSLATION SEE T57141)
57141	INVESTIGATION OF PHASE CHANGES AND SWELLING OF VERMICULITE DURING HEATING. KHVOSTENKOV S I TURKIN A F INORGANIC MATERIALS 5 6 951-4 1969 (ENGLISH TRANSLATION OF IZV. AKAD. NAUK SSSR, NEORG. MAT., 5 (6), 1121-5, 1969; FOR ORIGINAL SEE T57140)
57149	DEFORMATIONS OF MATERIALS BASED ON PHOSPHATE BINDERS DURING SOLIDIFICATION AND HEATING. SYCHEV M M KOMLEV V G IZV AKAD NAUK SSSR NEORG MATER 5 7 1230-5 1969 CA 71 94459 (FOR ENGLISH TRANSLATION SEE T57150)
57150	THE DEFORMATION OF MATERIALS BASED ON PHOSPHATE BINDING AGENTS DURING HARDENING AND ON HEATING. SYCHEV M M KOMLEV V G INORGANIC MATERIALS 5 7 1046-50 1969 (ENGLISH TRANSLATION OF IZV. AKAD. NAUK SSSR, NEORG. MAT., 5 (7), 1230-5, 1969; FOR ORIGINAL SEE T57149)

TPRC Number	Bibliographic Citation

57161 **A STUDY OF THE ACTION OF IONIZING RADIATION ON GLASSES ACTIVATED BY TRANSITION METALS.**
GALIMOV D G KARAPETYAN G O YUDIN D M
IZV AKAD NAUK SSSR NEORG MATER
5 8 1386-91 1969
(FOR ENGLISH TRANSLATION SEE T57162)

57162 **A STUDY OF THE ACTION OF IONIZING RADIATION ON GLASSES ACTIVATED BY TRANSITION METALS.**
GALIMOV D G KARAPETYAN G O YUDIN D M
INORGANIC MATERIALS
5 8 1184-8 1969
(ENGLISH TRANSLATION OF IZV. AKAD. NAUK SSSR, NEORG. MAT., 5 (8), 1386-91, 1969; FOR ORIGINAL SEE T57161)

57179 **PHASE COMPOSITION AND PROPERTIES OF SITALLS OF THE SYSTEM LI2O-AL2O3-GEO2-TIO2.**
BOGDANOVA G S ORLOVA E M BEZSMERTNAYA Z G
IZV AKAD NAUK SSSR NEORG MATER
5 9 1610-16 1969
(FOR ENGLISH TRANSLATION SEE T57180)

57180 **PHASE COMPOSITION AND PROPERTIES OF SITALLS OF THE SYSTEM LI(2) O-AL(2) O(3)-GE O(2)-TI O(2).**
BOGDANOVA, G. S. ORLOVA, E. M.
BEZSMERTNAYA, Z. G.
INORGANIC MATERIALS
5 9 1363-8 1969
(ENGLISH TRANSLATION OF IZV. AKAD. NAUK SSSR, NEORG. MAT. 5 (9), 1610-6, 1969; FOR ORIGINAL SEE T57179)

57183 **METHODS OF ENHANCING THE CONTRAST TRANSFER FUNCTION OF FIBER-OPTIC COMPONENTS.**
SATTAROV D K TROFIMOVA G YA PERCHERSKAYA K
KORMILITSINA A N
OPTIKA I SPEKTROSKOPIYA
27 1 151-5 1969
(FOR ENGLISH TRANSLATION SEE T57184)

57184 **METHODS OF ENHANCING THE CONTRAST TRANSFER FUNCTION OF FIBER-OPTIC COMPONENTS.**
SATTAROV D K TROFIMOVA G YA PERCHERSKAYA K
KORMILITSINA A N
OPT SPECTRY
27 1 74-6 1969
(ENGLISH TRANSLATION OF OPT. SPEKTROSK., 27 (1), 151-5, 1969; FOR ORIGINAL SEE T57183)

57189 **A STATISTICAL ANALYSIS OF THE REFLECTANCE OF IGNEOUS ROCKS FROM 0.2 TO 2.65 MICRONS.**
ROSS H P ADLER J E M HUNT G R
ICARUS
11, 46-54, 1969.
(AFCRL-70-0049, AD-699704)

57191 **DEVELOPMENT OF 400-2200 F FIBROUS-TYPE INSULATIONS FOR RADIOISOTOPE POWER SYSTEMS.**
COLLINS J O JAUNARAJS K L REID D R
JOHNS-MANVILLE RESEARCH AND ENGR CENTER, MANVILLE, NEW JERSEY
1-65, 1968.
(ALO-3633-10, N69-41277)

57194 **THERMAL PROPERTIES OF BETTIS-FABRICATED THORIA-URANIA FUEL MATERIALS.**
SPRINGER, J. R. LAGEDROST, J. F.
BATTELLE MEMORIAL INST., COLUMBUS, OHIO
1-108, 1968.
(BMI-X-10231)

57203 **ELECTROCHEMICAL METHOD OF INVESTIGATING STRESSES AT THE BOUNDARY OF A GLASS-METAL SOLDERED JOINT.**
DERTEV N K POTAPOV G P
VSESOYUZNYY SIMPOZIUM PO PROBLEME
98-100 1967
(FOR ENGLISH TRANSLATION SEE T57204)

57204 **ELECTROCHEMICAL METHOD OF INVESTIGATING STRESSES AT THE BOUNDARY OF A GLASS-METAL SOLDERED JOINT.**
DERTEV N K POTAPOV G P
FOREIGN TECHNOLOGY DIVISION
1-4, 1969.
(ENGLISH TRANSLATION OF VSESOYUZNYY SIMPOZIUM PO PROBLEME, 98-100, 1967; FOR ORIGINAL SEE T57203)
(FTD-MT-24-22-69, N70-14365, AD-695930)

57221 **TRANSMISSION EFFECTS ON PLASTIC FILMS IRRADIATED WITH ULTRAVIOLET LIGHT, ELECTRONS, AND PROTONS.**
ANAGNOSTOU, E. SPAKOWSKI, A. E.
LEWIS RESEARCH CENTER
1-12, 1969.
(NASA-TM-X-1905, N69-40339)

57235 **MEASUREMENTS OF THE THERMAL CONDUCTIVITY OF SOILS TO HIGH TEMPERATURES.**
FLYNN, D. R. WATSON, T. W.
NATIONAL BUREAU OF STANDARDS, WASHINGTON, D. C.
1-105, 1969.
(SC-CR-3059)

57256 **LIMITATION OF GRIFFITH FLAWS IN GLASS-MATRIX COMPOSITES.**
NIVAS Y FULRATH R M
J AM CERAM SOC
53 4 188-91 1970

57257 **TESTING PERICLASE-BELITE REFRACTORY IN THE SINTER ZONE OF A ROTARY CEMENT KILN.**
ROYAK S M SOKHATSKAYA G A SHUBIN V I
TISHKOVA K S ZAVGORODNII N S BUDIN I K
LAZARENKO N P KOTIK P L
OGNEUPORY
34 2 1-5 1969
(FOR ENGLISH TRANSLATION SEE T57258)

57258 **TESTING PERICLASE-BELITE REFRACTORY IN THE SINTER ZONE OF A ROTARY CEMENT KILN.**
ROYAK S M SOKHATSKAYA G A SHUBIN V I
TISHKOVA K S ZAVGORODNII N S BUDIN I K
LAZARENKO N P KOTIK P L
REFRACTORIES
34 2 71-4 1969
(ENGLISH TRANSLATION OF OGNEUPORY, 34 (2), 1-5, 1969; FOR ORIGINAL SEE T57257)

57259 **CRITERIA OF RESISTANCE TO THERMAL FRACTURE OF INHOMOGENEOUS REFRACTORY MATERIALS.**
KUKOLEV G V NEMETS I I DOBROVOLSKII G B
OGNEUPORY
34 2 41-5 1969
(FOR ENGLISH TRANSLATION SEE T57260)

57260 **CRITERIA OF RESISTANCE TO THERMAL FRACTURE OF INHOMOGENEOUS REFRACTORY MATERIALS.**
KUKOLEV G V NEMETS I I DOBROVOLSKII G B
REFRACTORIES
34 2 112-17 1969
(ENGLISH TRANSLATION OF OGENUPORY, 34 (2), 41-5, 1969; FOR ORIGINAL SEE T57259)

57261 **DENSIFICATION OF RO2 TYPE OXIDES DURING HOT PRESSING.**
BUDNIKOV P P KHARITONOV F YA
OGNEUPORY
34 2 36-40 1969
(FOR ENGLISH TRANSLATION SEE T57262)

57262 **DENSIFICATION OF RO2 TYPE OXIDES DURING HOT PRESSING.**
BUDNIKOV P P KHARITONOV F YA
REFRACTORIES
34 2 107-11 1969
(ENGLISH TRANSLATION OF OGENUPORY, 34 (2), 36-40, 1969; FOR ORIGINAL SEE T57261)

57263 **CORUNDUM WASH FOR REPAIRING CARBON REACTOR LININGS.**
PITAK N V ANSIMOVA T A
OGNEUPORY
34 3 27-33 1969
(FOR ENGLISH TRANSLATION SEE T57264)

57264 **CORUNDUM WASH FOR REPAIRING CARBON REACTOR LININGS.**
PITAK N V ANSIMOVA T A
REFRACTORIES
34 3 164-8 1969
(ENGLISH TRANSLATION OF OGNEUPORY, 34 (3), 27-33, 1969; FOR ORIGINAL SEE T57263)

57265 **PREPARATION AND PROPERTIES OF ZIRCONIUM REFRACTORIES STABILIZED BY CERIUM DIOXIDE.**
ANDREEVA A B LEONOV A I KELER E K
REFRACTORIES
34 3 172-8 1969
(ENGLISH TRANSLATION OF OGNEUPORY, 34 (3), 37-44, 1969; FOR ORIGINAL SEE T54518)

57266 **A STUDY OF THE EFFECT OF STRUCTURE ON THE THERMAL STABILITY OF FIRECALY REFRACTORIES.**
NEMETS I I DOBROVOLSKII G B GOGOTSI G A
ANTONENKO V M
REFRACTORIES
34 4 254-7 1969
(ENGLISH TRANSLATION OF OGNEUPORY, 34 (4), 56-9, 1969; FOR ORIGINAL SEE T54917)

57270 **DEVELOPING A METHOD AND PRODUCING A BATCH OF EXPERIMENTAL CORUNDUM INSULATING BRICK.**
KONETSKII N V MILSHENKO R C
REFRACTORIES
34 6 332-4 1969
(ENGLISH TRANSLATION OF OGENUPORY, 34 (6), 3-6, 1969; FOR ORIGINAL SEE T55280)

57271 **THE PRODUCTION AND SERVICE TESTING OF MAGNESIA WARES FROM RAPNA MAGNESIA.**
IVANOV E V DOLGINA G Z MENDELENKO A K
KOTIK P L ONISHCHENKO P V VINOGRADOV N M
MOLCHANOVA M I
OGNEUPORY
34 7 1-6 1969
(FOR ENGLISH TRANSLATION SEE T57272)

57272 **THE PRODUCTION AND SERVICE TESTING OF MAGNESIA WARES FROM RAPNA MAGNESIA.**
IVANOV E V DOLGINA G Z MENDELENKO A K
KOTIK P L ONISHCHENKO P V VINOGRADOV N M
MOLCHANOVA M I
REFRACTORIES
34 7 397-401 1969
(ENGLISH TRANSLATION OF OGNEUPORY, 34 (7), 1-6, 1969; FOR ORIGINAL SEE T57271)

TPRC Number	Bibliographic Citation
57273	DETERMINING THE TEMPERATURE CONDUCTIVITY COEFFICIENT OF DINAS. LITOVSKII E YA LANDA YA A REFRACTORIES 34 7 410–14 1969 (ENGLISH TRANSLATION OF OGNEUPORY, 34 (7), 16–21, 1969; FOR ORIGINAL SEE T55514)
57289	EFFECT OF RUTHENIUM, RHODIUM, AND PALLADIUM ON THE COEFFICIENT OF THERMAL EXPANSION OF IRON–NICKEL AND IRON–NICKEL–COBALT ALLOYS. SOLOVEVA N A YUDKEVICH M I PASTERNAK I I POGOSOV V Z METAL SCIENCE HEAT TREATMENT METALS 4 293–4 1968 (ENGLISH TRANSLATION OF METALLOVED TERM OBRAB METAL, 4, 45–6, 1968; FOR ORIGINAL SEE T50115)
57308	BAND INTENSITY OF THE EXTRAPLANAR NU–2 VIBRATION OF CO3 /–2/ AND NO3 /–1/ IN CRYSTALS HAVING CALCITE AND ARGONITE STRUCTURE. BELOUSOV M V POGAREV D E SHULTIN A A FIZ TVERD TELA 11 9 2697–9 1969 (FOR ENGLISH TRANSLATION SEE T57309)
57309	BAND INTENSITY OF THE EXTRAPLANAR NU–2 VIBRATION OF CO3 /–5/ AND NO3 /–1/ IN CRYSTALS HAVING CALCITE AND ARGONITE STRUCTURE. BELOUSOV M V POGAREV D E SHULTIN A A SOVIET PHYSICS–SOLID STATE 11 9 2185–6 1970 (ENGLISH TRANSLATION OF FIZ. TVERD. TELA, 11 (9), 2697–9, 1969; FOR ORIGINAL SEE T57308)
57380	HEAT CAPACITIES AND ENTHALPIES OF SEA SALT SOLUTIONS TO 200 C. BROMLEY L A DIAMOND A E SALAMI E WILKINS D G J CHEM ENG DATA 15 2 246–53 1970
57407	SPECTRAL SURVEY OF IRRIGATED REGION CROPS AND SOILS. AGRICULTURAL RESEARCH SERVICE WESLACO TEXAS AGRICULTURAL RESEARCH SERVICE, WESLACO, TEXAS 1–450, 1969. (NASA–CR–107881, N70–17517)
57409	MEASUREMENTS OF THE THERMAL EXPANSION OF CAST AND ROLLED ZINC. JONES H G PROC PHYS SOC 47 263 1117–28 1935
57448	MEASURING METHOD OF THE THERMAL CONDUCTIVITY OF REFRACTORES AT HIGH TEMPERATURE. DE SMET, A. EUROPEAN CONFERENCE ON THERMO–PHYSICAL PROPERTIES OF SOLIDS AT HIGH TEMPERATURES 222–41 1968
57456	MEASUREMENTS OF THERMAL DIFFUSIVITY OF MIXTURES OF URANIUM OXIDE AND THORIUM OXIDE. VAN CRAEYNEST, J. C. WEILBACHER, J. C. EUROPEAN CONFERENCE ON THERMOPHYSICAL PROPERTIES OF SOLIDS AT HIGH TEMPERATURES 393–410 1968
57457	MEASUREMENT OF THE THERMAL DIFFUSIVITY OF MIXTURES OF URANIUM AND THORIUM OXIDES. FERRO C MORETTI S PATIMO C CNEN–C S N EUROPEAN CONFERENCE ON THERMOPHYSICAL PROPERTIES OF SOLIDS AT HIGH TEMPERATURES 411–9 1968
57504	CRESTMORE SKY BLUE MARBLE, ITS LINEAR THERMAL EXPANSION AND COLOR. ROSENHOLTZ J L SMITH D T AMERICAN MINERALOGIST 35 1049–54 1950
57515	MAGNETIC FIELD DEPENDENT PHONON SCATTERING BY MN(+3) IONS IN AL(2) O(3). NIELSEN, T. H. LEIPOLD, M. H. J AM CERAM SOC 46 8 381–7 1963
57518	LOW–MELTING GLAZES WITH LOW THERMAL EXPANSION COEFFICIENTS. GALABUSKAYA, E. A. CHABANOVA, S. BEK, M. BUDIVEL'NI MATER. KONSTR. (4), 5, 1969.
57519	PRINCIPLES AND CONSTRUCTION OF AN ABSOLUTE RECORDING DILATOMETER OPERATING AT HIGH TEMPERATURE IN A CONTROLLED ATMOSPHERE. USE FOR THE STUDY OF THE BEHAVIOR OF CLAY MINERALS. BAUDRAN, A. BULL. SOC. FR. CERAM. (82), 3–48, 1969.
57536	VISCOSITY CHARACTERISTICS OF MELTS OF SOME BASIC ROCKS IN THE UKRAINE. CHERNOVA, V. I. DOPOV. AKAD. NAUK UKR. RSR 32 B (1), 41–5, 1970. (FOR ENGLISH TRANSLATION SEE TPRC NO. 59464)
57539	FREQUENCY DISTRIBUTIONS OF VITRINITE REFLECTANCE VALUES AND VITRINITES DISTINGUISHABLE BY REFLECTANCE. KOCH, J. ERDOEL KOHLE, ERDGAS, PETROCHEM. 23 (1), 2–6, 1970.
57548	COMMENTS ON NEW CERAMIC PRODUCT CLAIMS LOWEST COEFFICIENT OF EXPANSION. WINSHIP W W CERAMIC INDUSTRY 18 38 1932
57549	CERAMIC MATERIAL OF LOW THERMAL EXPANSIVITY. NATIONAL BUREAU OF STANDARDS CERAMIC INDUSTRY 18 100 1932
57555	EFFECTS OF TEMPERATURE AND PRESSURE ON SOME PHYSICAL PROPERTIES OF ROCKS. MOISEENKO, U. I. ISTOMIN, V. E. ALIEVA, M. A. FIZ.–MEKH. SVOISTVA GORN. POROD VERKH. CHASTI ZEMNOI KORY, DOKL. SIMP. 1964 148–52, 1968. (FOR ENGLISH TRANSLATION SEE TPRC NO. 64435)
57561	THERMAL EXPANSION OF SALT AS A POSSIBLE TECTONIC FACTOR. GAVSHIN V M VOLONTEI G M GEOL GEOFIZ 11 146–8 1969 CA 72 81547
57568	PHYSICOCHEMICAL PROPERTIES OF PACKING GLASS. DOBRIN, M. IND. USOARA 16 (5), 302–6, 1969.
57572	EFFECTIVE THERMAL CONDUCTIVITY OF DISPERSED SYSTEMS. KOVENSKII, G. I. ZABRODSKII, S. S. TAMARIN, A. I. ISSLED. PROTSESSOV PERENOSA APP. DISPERSNYMI SIST. 175–9, 1969. (FOR ENGLISH TRANSLATION SEE TPRC NO. 64041)
57574	QUANTITATIVE EVALUATION OF FRACTURING DEGREE AND FRACTURE POROSITY OF ARMENIAN LAVAS. II. TOLOKONNIKOV I S IZV AKAD NAUK ARM SSR NAUKI ZEMLE 22 1 57–65 1969 CA 72 81471
57577	PROBLEMS OF INFRARED ATMOSPHERIC SPECTROSCOPY CONNECTED WITH THE DETERMINATION OF THE UNDERLAYING SURFACE TEMPERATURE. KONDRATEV, K. YA. IZV. AKAD. NAUK SSSR, FIZ. ATMOS. OKEANA 5 (6), 616–30, 1969.
57587	TECHNIQUE FOR DETERMINING THE THERMAL CONDUCTIVITY OF REFRACTORIES OVER A WIDE TEMPERATURE RANGE. LITOVSKII E YA PLATUNOV E S PUCHKELEVICH N A IZV VYSSH UCHEB ZAVED PRIBOROSTR 12 11 124–8 1969 CA 72 92983
57588	DILATOMETRIC METHOD FOR STUDYING LOESSLIKE CLAY LOAMS. KNIGINA G I KATKOVA T F IZV VYSSH UCHEB ZAVED STROIT ARKHITEKT 12 8 87–93 1969 CA 72 82476
57594	USE OF CHLORITE IN LOW EXPANSION CERAMIC BODIES. HAMMER, G. J. AUST. CERM. SOC. 3 (1), 5–9, 1967.
57597	PLANETARY ENVIRONMENT SIMULATION. EROSION AND DUST COATING EFFECTS. ADLON G L RUSERT E L ALLEN T H MCDONNELL–DOUGLAS ASTRONAUTICS CO, ST. LOUIS, MO. 1–39, 1969. (MDC–E0038–VOL.–1, NASA–CR–66878, N70–19860)
57604	EMPIRICAL RELATION BETWEEN THERMAL CONDUCTIVITY AND DEBYE TEMPERATURE FOR SILICATES. HORAI K SIMMONS G J GEOPHYS RES 75 5 978–82 1970 CA 62 81354
57606	MOLAR AND CHIMNEYS. JACKSON, K. R. J. INST. FUEL 4. (344), 350–5, 1969.
57611	LOW–TEMPERATURE HEAT CAPACITIES AND ENTROPIES AT 298.15 K. OF GOETHITE AND PYROPHYLLITE. KING, E. G. WELLER, W. W. METALLURGY RES. CENTER, BUREAU OF MINES, ALBANY, OREGON 1–8, 1970. (BM–RI–7369, PB–190976, N70–24968)

TPRC Number	Bibliographic Citation
57614	**CONDUCTIMETRIC DILATOMETER FOR STUDYING SOLID - PHASE TRANSFORMATIONS AT HIGH TEMPERATURES.** DUCLOT, M. DEPORTES, C. J. THERM. ANAL. 1 (3), 329-37, 1969.
57632	**THERMAL CONDUCTIVITY OF MOIST CAPILLARY-POROUS SYSTEMS.** SHASHKOV, A. G. VASILIEV, L. L. TANAEVA, S. A. PROC. CONF. ON THERMAL CONDUCTIVITY, 9TH 279-87, 1970. (CONF-69-1002)
57665	**TECHNIQUE FOR MEASURING THERMAL CONDUCTIVITY OF T/E MATERIALS.** BIENERT W B TRIMMER D S SKRABEK E A AIAA THERMOPHYSICS CONFERENCE 6 2 1-10 1966
57681	**RECENT DEVELOPMENTS IN METALS SEALING INTO GLASS.** SCOTT H J FRANKLIN INSTITUTE 220 733-54 1935
57698	**GLASS-TO-METAL SEAL DESIGN.** SCOTT W J J SCIENTIFIC INSTRUMENTS 23 9 193-202 1946
57719	**THE EXPANSION CHARACTERISTICS OF SOME COMMON GLASSES AND METALS.** BURGER E E GENERAL ELECTRIC REVIEW 37 2 93-6 1934
57768	**MODIFICATION FOR INCREASED EFFICIENCY OF AN INTRA-CAVITY SECOND HARMONIC NEODYMIUM LASER.** ROSS I N J PHYS D /APPL PHYS/ 3 3 L10-11 1970
57787	**END-REGION ANALYSIS FOR STRESS AND OPTICAL PATHLENGTH CHANGE IN COMPOSITE LASER RODS.** CUFF D W LEON G S LYNCH F D S J APPL PHYS 41 4 1538-48 1970
57798	**ANISOTROPIC REFLECTANCE CHARACTERISTICS OF NATURAL EARTH SURFACES.** BRENNAN B BANDEEN W R APPL OPT 9 2 405-12 1970
57817	**SINTERING MAGNESIA WITH BORON ADDITIVES.** MAMYKIN P S DROZDOVA T A OGNEUPORY 12 41-6 1969 CA 72 70176 (FOR ENGLISH TRANSLATION SEE T57818)
57818	**SINTERING MAGNESIA WITH BORON ADDITIVES.** MAMYKIN P S DROZDOVA T A REFRACTORIES 12 765-8 1969 (ENGLISH TRANSLATION OF OGNEUPORY, (12), 41-6, 1969; FOR ORIGINAL SEE T57817)
57827	**MAKING AND TESTING SEMIACID LADLE BRICK.** PAPAKIN KH M OSTROVSKAYA P S KONOVALOVA T A FINADEEV V F GOLOV G V OGNEUPORY 11 4-6 1969 (FOR ENGLISH TRANSLATION SEE T57828)
57828	**MAKING AND TESTING SEMIACID LADLE BRICK.** PAPAKIN KH M OSTROVSKAYA P S KONOVALOVA T A FINADEEV V F GOLOV G V REFRACTORIES 11 662-4 1969 (ENGLISH TRANSLATION OF OGNEUPORY, (11), 4-6, 1969; FOR ORIGINAL SEE T57827)
57860	**ULTRAVIOLET ABSORPTION AND STIMULATED LUMINESCENCE INVESTIGATIONS. PART 1. ULTRAVIOLET VIDEO IMAGING SYSTEM. PART 2. SPECTRAL DISTRIBUTION OF ULTRAVIOLET STIMULATED LUMINESCENCE. PART 3. MEASUREMENT OF ULTRAVIOLET REFLECTANCE INTERIM REPORT.** DANIELS D L FISCHER W A GAWARECKI S J GERHARZ R HEMPHILL W R MANNED SPACECRAFT CTR., HOUSTON, TEXAS 1-120, 1964. (NASA-TM-X-61721, N69-29390)
57871	**THE INFLUENCE OF IRRADIATION BY FISSION FRAGMENTS ON THE THERMAL DIFFUSIVITY OF VITROCERAMIC MASSES OF THE SYSTEM UO2-SIO2.** GLANZ G RADULESCU O J NUCL MATER 35 1 102-8 1970
57876	**DETERMINATION OF THE THERMAL PARAMETERS OF SOLIDS IN A VARYING REGION.** CHAMPOUSSIN J-C SICARD L COMPT REND 270 B 14 869-71 1970
57963	**MANGANESE DISULFIDE /HAUERITE/ AND MANGANESE DITELLURIDE. THERMAL PROPERTIES FROM 5 TO 350 K. AND ANTI-FERROMAGNETIC TRANSITIONS.** WESTRUM E F JR GROENVOLD F J CHEM PHYS 52 7 3820-6 1970
57964	**APPLICATION OF REFLECTANCE SPECTROSCOPY AND ESR METHOD IN STUDIES ON MGO/CR2O3 SYSTEM.** BIELANSKI A BORONCZYK H DYREK K BULL ACAD POLON SCI SER SCI CHIM 18 1 1-8 1970
57969	**CATALYZED CRYSTALLIZATION OF GLASSES WITH PYROXENE COMPOSITIONS.** TOROPOV N A KHOTIMCHENKO V S IZV AKAD NAUK SSSR NEORG MATER 5 11 1991-4 1969 (FOR ENGLISH TRANSLATION SEE T57970)
57970	**CATALYZED CRYSTALLIZATION OF GLASSES WITH PYROXENE COMPOSITIONS.** TOROPOV N A KHOTIMCHENKO V S INORGANIC MATERIALS 5 11 1694-7 1969 (ENGLISH TRANSLATION OF IZV. AKAD. NAUK SSSR, NEORG. MAT., 5 (11), 1991-4, 1969; FOR ORIGINAL SEE T57969)
57971	**INVESTIGATIONS OF CHEMICAL RESISTANCE OF SITALL IN DEPENDENCE ON VARIOUS METHODS OF THERMAL TREATMENT.** ROGOZHIN YU V ZAIONTS L A EREMEEVA A S IZV AKAD NAUK SSSR NEORG MATER 5 11 2041-3 1969 (FOR ENGLISH TRANSLATION SEE T57972)
57972	**INVESTIGATIONS OF CHEMICAL RESISTANCE OF SITALL IN DEPENDENCE ON VARIOUS METHODS OF THERMAL TREATMENT.** ROGOZHIN YU V ZAIONTS L A EREMEEVA A S INORGANIC MATERIALS 5 11 1742-3 1969 (ENGLISH TRANSLATION OF IZV. AKAD. NAUK SSSR, NEORG. MAT., 5 (11), 2041-3, 1969; FOR ORIGINAL SEE T57971)
57977	**THERMAL EXPANSION HYSTERESIS OF ALUMINUM TITANATE.** BUESSEM W R THIELKE N R SARAKAUSKUS R V CERAMIC AGE 60 5 38-40 1952
57984	**PHASE TRANSITION IN METAKAOLIN AND THE TRUE HEAT CAPACITY OF NEW PHASES.** KUSKOV, O. L. KHITAROV, N. I. GEOKHIMIYA (2), 185-91, 1970. (FOR ENGLISH TRANSLATION SEE TPRC NO. 67511)
57988	**SOIL THERMAL RESISTIVITY MEASURED SIMPLY AND ACCURATELY.** STOLPE J IEEE TRANS POWER APP SYST 89 2 297-304 1970 CA 72 102692
57997	**SPECIFIC HEAT OF AN ISING LINEAR-CHAIN CRYSTAL.** FORSTAT H LOVE N D MC ELEARNEY J N BUTTERWORTH G J PHYS REV 1 B 7 3097-100 1970 (AFOSR-70-2247-TR, AD-710936)
58024	**THE THERMAL EXPANSION OF CRYSTALS OF METALLIC BISMUTH.** ROBERTS J K PROC ROY SOC LONDON 106 385-99 1924
58086	**RESEARCH ON METALLIC CRYSTALS. III. THERMAL EXPANSION OF ZINC AND CADMIUM.** GRUNEISEN E GOENS E WISS ABH PHYS TECH REICH 8 2 43-58 1924
58115	**CERAMIC COMPOSITIONS HAVING NEGATIVE LINEAR THERMAL EXPANSION. PART II.** SMOKE E J CERAMIC AGE 62 1 13-16 19-20 1953
58119	**THERMAL DIFFUSIVITY AND THERMAL CONDUCTIVITY OF SOME DONETS AND KUZNETSK BASIN COALS.** AGROSKIN A A LOVETSKII L V KHIM TVERD TOPL 6 3-10 1969 CA 72 69045
58120	**EFFECT OF CONTACT METAMORPHISM ON THE QUALITY OF THE COAL IN THE SOUTHEASTERN PART OF THE TUNGUSKA BASIN.** STRUGOV A S KHIM TVERD TOPL 6 24-8 1969 CA 72 69036
58137	**MEASUREMENT OF SPECIFIC HEAT OF SOLIDS. MODIFIED MIXING METHOD.** MIYABE, K. KATSUHARA, T. HOSOKAWA, S. TSUBONE, K. KYUSHU KOGYO DAIGAKU KENKYU HOKOKU (19), 25-30, 1969.

TPRC Number	Bibliographic Citation
58145	**THERMAL EXPANSION PROPERTIES OF SOME HIGH SILICA BODIES IN THE SYSTEM LITHIA-ALUMINA-SILICA.** SMOKE E J CERAMIC AGE 64 6 11-15 1954
58148	**THERMAL PROPERTIES OF FURNACE LININGS.** BOVKUN, K. A. YAKOBSON, B. I. KOTOV, K. I. OSHKO, V. P. MET. KOKSOKHIM. (16), 58-62, 1969.
58152	**NICKEL MINERALS FORM BARBERTON, SOUTH AFRICA. FERROAN TREVORITE.** DE WAAL, S. A. HIEMSTRA, S. A. NAT. INST. MET., REPUB. S. AFR., RES. REP. (291), 14PP., 1968.
58153	**DETERMINATION OF THE FIRING TEMPERATURE OF ANCIENT CERAMICS BY MEASUREMENT OF THERMAL EXPANSION.** TITE M S NATURE 222 81 1969 JA 52-10 347
58186	**RELATION BETWEEN VOLUME AND ELECTRICAL CONDUCTIVITY CHANGES IN IRRADIATED UO2-SIO2 SYSTEMS.** COJOCARU L N RADIAT EFF 2 2 133-4 1970 CA 72 60310
58200	**MOON. INFRARED STUDIES OF SURFACE COMPOSITION.** CRUIKSHANK D P SCIENCE 166 215-18 1969 CA 71 107131
58201	**VISCOSITY OF LUNAR LAVAS.** MURASE T MC BIRNEY A R SCIENCE 167 1491-3 1970 CA 72 102736
58204	**CERAMIC MATERIALS FOR WALL TILE PRODUCTION WITHOUT KAOLIN.** LEVCEVA, C. VASILEV, S. SKLAR KERAM. 19 (10), 270-1, 1969.
58212	**CALCULATION OF THE THERMAL EXPANSION OF BOROSILICATE GLASSES.** KNAPP, O. SPRECHSAAL KERAM., GLAS. EMAIL. SILIKATE 102 (23), 1047-8, 1969.
58217	**CHEMICAL LABORATORY GLASS CONTAINING ZIRCONIUM OXIDE.** STOICHEVA V VODENICHAROVA T KARASTOYANOVA E STROIT MATER SILIKAT PROM 10 11 40-3 1969 CA 72 103292
58222	**CERAMIC HEAT-RESISTANT COOKWARE. II.** CHMIEL T SADOWSKI Z SZKLO CERAM 20 10 299-302 1969 CA 72 114607
58230	**INFRARED SPECTROSCOPIC DETERMINATION OF OPTICAL CHARACTERISTICS OF MATERIALS.** SLOBODKIN L S RABINOVICH G D TEPLO-MASSOPERENOS DOKL VSES SOVESHCH 3RD 6 2 3 27-38 1968 CA 72 67425
58233	**POROSITY AND TEMPERATURE DEPENDENCE OF PROPERTIES DETERMINING THE THERMAL STABILITY OF QUARTZ CERAMICS.** GOROBETS, F. T. ROMASHIN, A. G. BORODAI, F. YA. PIVINSKII, YU. E. KOVSHAR, N. K. PRUDNIKOVA, N. I. SAIDOVA, L. L. ILYUKHINA, V. V. TERMOPROCH. MATER. KONSTR. ELEM. (5), 319-25, 1969.
58238	**RESEARCH ON THE THERMAL EXPANSION OF UO2, PUO2 AND UO2 - PUO2. SPECIAL REPORT NO. 6.** LEBLANC J M ANDRIESSEN H U. S. EURATOM JOINT RES. AND DEV. PROGRAM 1-34, 1962. (EURAEC-434)
58259	**INCREASING THE THERMAL STABILITY OF DINAS GREEN REFRACTORIES DURING FIRING.** BICHURINA, A. A. MAMYKIN, P. S. BRON, V. A. KOSTOMAROV, M. I. KHITRO, E. V. TR. VOST. INST. OGNEUPOROV (8), 100-9, 1968.
58271	**METHOD FOR RECONSTRUCTING A THERMAL-EXPANSION GRAPH FROM TWO VALUES OF THE MEAN EXPANSION COEFFICIENT.** YAGGEE, F. L. FOOTE, F. G. ARGONNE NATIONAL LAB., ILLINOIS 1-13, 1969. (ANL-7644, N70-30359)
58286	**CERAMMED COATINGS WITH A HIGH COEFFICIENT OF THERMAL EXPANSION.** PEVZNER B Z APPEN A A ANTONOVA E A ZHAROSTOIKIE TEPLOSTOIKIE POKRYTIYA TR VSES SOVESHCH 4TH 1968 205-10 1969 CA 72 82397
58287	**CHEMICALLY STABLE COATINGS FOR THE PROTECTION OF TITANIUM.** SITNIKOVA A YA APPEN A A ZHAROSTOIKIE TEPLOSTOIKIE POKRYTIYA TR VSES SOVESHCH 4TH 1968 4 211-9 1969 CA 72 82102
58288	**SOME RESULTS OF THE INVESTIGATION OF THE BARIUM OXIDE-SILICON DIOXIDE-BORON OXIDE-TITANIUM DIOXIDE SYSTEM FOR THE PREPARATION OF HEAT-RESISTANT COATINGS.** GORBATENKO, V. E. BOLTENKOVA, N. N. ZHAROSTOIKIE TEPLOSTOIKIE POKRYTIYA, TR. VSES. SOVESHCH., 4TH 1968 242-5, 1969.
58318	**REDUCTION OF IONS IN GLASS BY HYDROGEN.** JOHNSTON W D CHELKO A J J AM CERAM SOC 53 6 295-301 1970
58343	**STABILIZATION OF CUBIC ZIRCONIUM DIOXIDE BY MAGNESIUM OXIDE AND PARTIAL SUBSTITUTION OF TITANIUM, NIOBIUM, OR TANTALUM FOR ZIRCONIUM.** BAYER G J AMER CERAM SOC 53 5 294 1970 CA 73 29578
58350	**INFRARED SPECTRA OF CRYSTALLINE LEAD GERMANATES IN THE REGION OF CONCENTRATIONS CONTAINING 50-100 MOLE PERCENT LEAD OXIDE.** MOROZOV, V. N. ZH. PRIKL. SPEKTROSK. 11 (6), 986-90, 1969. (FOR ENGLISH TRANSLATION SEE T76913)
58368	**A TECHNIQUE FOR PREDICTING THE THERMAL CONDUCTIVITY OF SUSPENSIONS, EMULSIONS AND POROUS MATERIALS.** CHENG S C VACHON R I INTERN J HEAT MASS TRANSFER 13 3 537-46 1970 CA 73 7996
58386	**A TECHNIQUE OF STUDYING THERMAL PROPERTIES OF ROCKS AND CALCULATING HEATFLOW VALUES IN THE CRIMEA.** LEBEDEV T S SAVENKO B YA KORNIETS D V ACAD. SCI. UKR. SSR, GEOPHYSICAL COMMUNICATIONS (24), 61-72, 1968. (N69-37149)
58387	**CRUSTAL THICKNESS AND PECULIARITIES OF CRUSTAL STRUCTURE IN TERMS OF THE PHASE CHARACTER OF THE MOHROVICIC DISCONTINUITY.** RAKHMOVA I SH GOLOVATCH M N DEEP CRUSTAL STRUCTURE 5-22, 1967. (N69-36568)
58388	**GEOTHERMAL CONDITIONS IN THE CRIMEA.** LEBEDEV T S GORDIYENKO V V KUTAS R I DEEP CRUSTAL STRUCTURE 23-8, 1967. (N69-36570)
58390	**MOLECULAR AND ATOMIC VOLUMINA. 24. TECHNICAL EXPERIENCES AT VOLUMETRIC MEASUREMENTS OF LOW TEMPERATURE DENSITIES.** WUNNENBERG E FISCHER W SAPPER A BILTZ W ZEIT PHYS CHEM 151 A 1 1-12 1930
58395	**ON MEASUREMENTS OF THE LIGHT INTENSITY FACTOR ON PROBES OF ANY REFLECTIVITY.** KORTE H SCHMIDT M NATL LENDING LIBRARY BOSTON SPA ENGLAND 1-13, 1969. (ENGLISH TRANSLATION OF LICHTTECHNIK, 19, 135-7A, 1967; FOR ORIGINAL SEE T51716) (NLL-MT-TRANS-6504-/5809.95/, N70-23677)
58396	**PHASE RELATIONS IN THE SYSTEM LI2O-NA2O-AL2O3-SIO2 WITH SPECIAL EMPHASIS ON THE SILICA POLYMORPHS.** BADGER, W. B. PENNSYLVANIA STATE UNIV., UNIVERSITY PARK, PA., PH. D. THESIS 1-284, 1968. (UNIVERSITY MICROFILMS NO. 69-5529)
58417	**NOTES ON AGALMATOLITH, A NEW REFRACTORY MATERIAL.** BURGER O K AM CERAMIC SOC BULL 8 5 343-4 1926
58419	**INFLUENCE OF GLASS PLATES ON THE REFLECTANCE OF VARNISH LAYERS.** HOFFMANN, K. FARBE LACK 76 (3), 249-56, 1970.
58434	**INFRARED SPECTRA OF JAPANESE COAL. ABSORPTION BANDS AT 3030, 2920, AND 1600 CM/-1/.** FUJII S OSAWA Y SUGIMURA H FUEL 49 1 68-75 1970 CA 72 113555

TPRC Number	Bibliographic Citation
58456	**THERMAL PROPERTIES OF FOAMS AND FOAMED HONEYCOMBS IN THE TEMPERATURE RANGE BETWEEN 20 AND 300 K.** ZIMNI, W. F. MEITZNER, K. KAELTETECH. - KLIM. 22 (2), 34-40, 1970.
58457	**RELATION BETWEEN THE STRUCTURAL AND THE PHYSICAL PROPERTIES OF GLASS. T.E. OF GLASS.** NARAY-SZABO I ACTA PHYS ACAD SCI 9 4 403-21 1958
58465	**EXPANSION EFFECTS SHOWN BY SOME PYREX GLASSES.** TOOL A Q SAUNDERS J B J RESEARCH NATIONAL BUREAU STANDARDS 40 67-83 1948 (NBS-RP-1856)
58479	**THERMAL CONDUCTIVITY OF REFRACTORY MATERIALS.** LITOVSKII, E. YA. LANDA, YA. A. TR. VSES. INST. NAUCH.-ISSLED. PROEKT. RAB. OGNEUPOR. PROM. (40), 109-30, 1968.
58484	**DETERMINATION OF LOW-TEMPERATURE THERMAL CONDUCTIVITIES FOR SEVERAL GRADES OF OIL SHALE.** SLADEK, T. A. COLORADO SCHOOL OF MINES, M. S. THESIS 1-51 1969
58486	**THERMAL EXPANSION OF COLORED PORCELAIN GLAZES.** ZAPP F BER DEUT KERAM GES 47 2 101-4 1970 CA 72 124649
58487	**INFLUENCE OF CALCINED AUSTRALIAN TALC ON THE MECHANICAL AND ELECTRICAL PROPERTIES OF A STEATITE BODY.** HARTH R WAGNER M BER DEUT KERAM GES 47 2 116-18 1970 CA 72 124650
58498	**USE OF GLASS FIBERS IN UNDERGROUND PIPING, AND CORROSION PHENOMENA.** BELANGER G INGENIEUR /MONTREAL/ 56 251 21-5 1970 CA 72 122510
58531	**NEW CERAMIC PRODUCT CLAIMS LOWEST COEFFICIENT OF EXPANSION.** AUTHOR ANON. CERAMIC INDUSTRY 17 (6), 476, 1931.
58542	**ALUMINOSILICATE PRODUCTS WITH A CORDIERITE BINDER.** GERASIMOV E LEPKOVA D STROIT MATER SILIKAT PROM 10 11 23-7 1969 CA 72 124655
58543	**HEAT-RESISTANT CONCRETE BASED ON FERROALLOY PRODUCTION SLAGS.** FOMICHEV, N. A. ABYZOV, A. N. ABYZOVA, T. V. STROIT. MATER. BETONY (2), 262-75, 1967.
58547	**VITRIFICATION AND EXPANSILE CHARACTERISTICS OF FELDSPATHIC AND FELSPATHOID FLUXES.** WALKER E G HOLDRIDGE D A TRANS BRIT CERAM SOC 69 1 21-7 1970 CA 72 124635
58575	**STANDARD IR SPECTRA OF THE SIMPLEST SILICATES.** CHERNEVA E F FLORINSKAYA V A IZV AKAD NAUK SSSR NEORG MATER 5 12 2146-50 1969 CA 72 70124 (FOR ENGLISH TRANSLATION SEE T58576)
58576	**STANDARD IR-SPECTRA OF THE SIMPLEST SILICATES.** CHERNEVA E F FLORINSKAYA V A INORGANIC MATERIALS 5 12 1831-4 1969 (ENGLISH TRANSLATION OF IZV. AKAD. NAUK SSSR, NEORG. MAT., 5 (12), 2146-50, 1969; FOR ORIGINAL SEE T58575)
58579	**EFFECT OF VARIOUS OXIDES ON THE COEFFICIENT OF THERMAL EXPANSION AND THE TEMPERATURE COEFFICIENT OF THE REFRACTIVE INDEX OF SILICATE GLASS.** SHCHAVELEV O S BABKINA V A IZV AKAD NAUK SSSR NEORG MATER 5 12 2187-90 1969 CA 72 82414 (FOR ENGLISH TRANSLATION SEE T58580)
58580	**EFFECT OF DIFFERENT OXIDES ON THE COEFFICIENT OF THERMAL EXPANSION AND ON THE TEMPERATURE COEFFICIENT OF THE REFRACTIVE INDEX OF SILICATE GLASS.** SHCHAVELEV O S BABKINA V A INORGANIC MATERIALS 5 12 1867-9 1969 (ENGLISH TRANSLATION OF IZV. AKAD. NAUK SSSR, NEORG. MAT., 5 (12), 2187-90, 1969; FOR ORIGINAL SEE T58579)
58609	**THERMAL STUDIES OF OBSIDIAN, PITCHSTONE AND PERLITE FROM JAPAN.** KOZU S SCI REPTS TOHOKU UNIV 3 225-38 1929
58611	**THE USES OF BERYL IN CERAMICS.** LUKS D W CERAMIC INDUSTRY 30 2 50-2 1938
58624	**THERMOOPTICAL PROPERTIES OF LITHIUM FLUORIDE, BARIUM FLUORIDE, AND LEUCOSAPPHIRE CRYSTALS.** SELEZNEVA A M OPT-MEKH PROM 36 11 30-2 1969 CA 72 138043
58630	**CONSTRUCTION MATERIALS FOR ELECTRIC FURNACES INVESTIGATION OF THEIR THERMAL EXPANSION.** CABELLO A S ANALES DE FISICA Y QUIMICA 46 B 329-34 1950
58631	**ROCKS STUDIED AT HIGH TEMPERATURES.** TURCHANINOV I A MEDVEDEV R V TEPLOFIZ SVOISTVA TVERD TEL VYS TEMP TR VSES KONF 1966 1 442-8 1969 CA 72 135134
58679	**EVALUATION OF A LARGE GLASS-CERAMIC MIRROR BLANK.** STEWART D R DAVIS W L APPLIED OPTICS 9 4 938-41 1970
58694	**PHYSICAL PROPERTIES OF CONTROLLED-SOLUBILITY PHOSPHATE GLASSES.** LUTZ C W J CHEM ENG DATA 15 3 415-16 1970 CA 73 48209
58727	**THERMAL EXPANSION OF COMPOSITES AS AFFECTED BY THE MATRIX.** TUMMALA R R FRIEDBERG A L J AMER CERAM SOC 53 7 376-80 1970 CA 73 58918
58728	**BORON COORDINATION IN PBO.2B2O3 AND BAO.2B2O3 MELTS.** LAIRD J A BERGERON C G J AMER CERAM SOC 53 7 410-12 1970
58729	**STABILIZATION OF ZIRCONIA BY ERBIA.** STEWART R K HUNTER O JR J AMER CERAM SOC 53 7 421-2 1970 CA 73 58906
58735	**CORRELATION BETWEEN EXPANSION COEFFICIENT AND STRUCTURE OF GLASSES.** DIETZEL A GLASTECH BER 19 10 319-25 1941
58737	**STRESS-OPTICAL DETERMINATION OF THE THERMAL EXPANSION COEFFICIENTS OF GLASSES.** SCHWIEGER H FIEDLER H ANN PHYSIK 14 6 64-86 1954
58748	**OPTICAL ABSORPTION MEASUREMENTS ON NATURAL AND SYNTHETIC FERROMAGNESIAN MINERALS SUBJECTED TO HIGH PRESSURES.** PITT G D TOZER D C PHYS EARTH PLANET INTERIORS 2 3 179-88 1970 CA 73 9082
58769	**NEW DECORATIVE GLAZES FOR FAIENCE.** SIVCHIKOVA M G OLEINIK L L KAGANOV I V GLAZURI IKH PROIZVOD I PRIMEN 133-47 1964 (FOR ENGLISH TRANSALTION SEE T62788)
58782	**THERMAL CONDUCTIVITY AND DIFFUSIVITY OF SOIL USING A CYLINDRICAL HEAT SOURCE.** MOENCH A F EVANS D D SOIL SCI SOC AMER PROC 34 3 377-81 1970 CA 73 17502
58786	**THERMAL CONDUCTIVITY AND ELECTRICAL CONDUCTIVITY OF ROCKS AT A HIGH TEMPERATURE.** MOISEENKO U I TEPLOFIZ SVOISTVA TVERD TEL VYS TEMP TR VSES KONF 1966 1 299-302 1969 CA 73 17383
58799	**ENERGY LEVELS AND SPECTRAL BROADENING OF NEODYMIUM IONS IN LASER GLASS.** MANN M M DESHAZER L G J APPLIED PHYSICS 41 7 2951-7 1970

TPRC Number	Bibliographic Citation

58848 **FIRECLAY-PERLITE LIGHTWEIGHT PRODUCTS BASED ON RAW MATERIAL FROM THE LATNENSK DEPOSIT.**
MUKHIN A A MILSHENKO R S GOLDINOVA R N
SOKOLOV A F
OGNEUPORY
10 1-4 1969
(FOR ENGLISH TRANSLATION SEE T58849)

58849 **FIRECLAY-PERLITE LIGHTWEIGHT PRODUCTS BASED ON RAW MATERIAL FROM THE LATNENSK DEPOSIT.**
MUKHIN A A MILSHENKO R S GOLDINOVA R N
SOKOLOV A F
REFRACTORIES
10 599-601 1969
(ENGLISH TRANSLATION OF OGNEUPORY, (10), 1-4, 1969; FOR ORIGINAL SEE T58848)

58850 **DETERMINATION OF THE THERMAL DIFFUSIVITY COEFFICIENT OF MAGNESIA AND MAGNESIOSPINEL REFRACTORIES.**
LITOVSKII E YA LANDA YA A
OGNEUPORY
34 10 12-15 1969 CA 72 35255
(FOR ENGLISH TRANSLATION SEE T58851)

58851 **DETERMINATION OF THE THERMAL DIFFUSIVITY COEFFICIENT OF MAGNESIA AND MAGNESIOSPINEL REFRACTORIES.**
LITOVSKII E YA LANDA YA A
REFRACTORIES
10 611-13 1969
(ENGLISH TRANSLATION OF OGNEUPORY, 34 (10), 12-15, 1969; FOR ORIGINAL SEE T58850)

58852 **PHYSICOCHEMICAL AND THERMODYNAMIC PROPERTIES OF ALUMINOSILICATE PRODUCTS FOR THE BLAST HEATERS OF BLAST FURNACES.**
GINYAR E A MALCHENKO R S PITAK N V
MIKHAILOVA K A
OGNEUPORY
10 31-9 1969
(FOR ENGLISH TRANSLATION SEE T58853)

58853 **PHYSICOCHEMICAL AND THERMODYNAMIC PROPERTIES OF ALUMINOSILICATE PRODUCTS FOR THE BLAST HEATERS OF BLAST FURNACES.**
GINYAR E A MALCHENKO R S PITAK N V
MIKHAILOVA K A
REFRACTORIES
10 630-6 1969
(ENGLISH TRANSLATION OF OGNEUPORY, (10), 31-9, 1969; FOR ORIGINAL SEE T58852)

58854 **PROPERTIES OF PRODUCTS MADE OF FUSED MAGNESITES.**
IGNATOVA T S UZBERG L V TAKSIS G A
DAUKNIS V I YURENAS V L KAZAKYAVICHYUS K A
YANUKENAS V I
OGNEUPORY
10 49-55 1969
(FOR ENGLISH TRANSLATION SEE T58855)

58855 **PROPERTIES OF PRODUCTS MADE OF FUSED MAGNESITES.**
IGNATOVA T S UZBERG L V TAKSIS G A
DAUKNIS V I YURENAS V L KAZAKYAVICHYUS K A
YANUKENAS V I
REFRACTORIES
10 647-52 1969
(ENGLISH TRANSLATION OF OGNEUPORY, (10), 49-55, 1969; FOR ORIGINAL SEE T58854)

58856 **REFRACTORIES AND TECHNOLOGY OF PREPARING CRUCIBLES OF INDUCTION FURNACES IN THE GERMAN DEMOCRATIC REPUBLIC AND THE CZECHOSLOVAK SOCIALIST REPUBLIC.**
KAIBICHEVA M N
OGNEUPORY
10 58-61 1969
(FOR ENGLISH TRANSLATION SEE T58857)

58857 **REFRACTORIES AND TECHNOLOGY OF PREPARING CRUCIBLES OF INDUCTION FURNACES IN THE GERMAN DEMOCRATIC REPUBLIC AND THE CZECHOSLOVAK SOCIALIST REPUBLIC.**
KAIBICHEVA M N
REFRACTORIES
10 655-8 1969
(ENGLISH TRANSLATION OF OGNEUPORY, (10), 58-61, 1969; FOR ORIGINAL SEE T58856)

58875 **THERMAL EXPANSION OF SILICA AT LOW TEMPERATURES.**
WHITE G K
CRYOGENICS
4 2-7 1964

58879 **INFRARED ABSORPTION SPECTRA OF MINERALS AND OTHER INORGANIC COMPOUNDS.**
HUNT J M WISHERD M P BONHAM L C
ANALYTICAL CHEMISTRY
22 12 1478-97 1950

58889 **ANALYSIS OF MECHANICAL, ELECTRICAL, AND THERMAL PROPERTIES OF BERYLLIUM OXIDE AND ALUMINUM OXIDE.**
BELCHER, J. G., JR.
BROWN ENGINEERING COMP., INC.
1-10, 1963.
(AD-439456)

58909 **THERMAL CONDUCTIVITY OF SINGLE CRYSTAL SELENIUM.**
RUNBECK, L. M.
FRANKLIN AND MARSHALL COLLEGE, M. S. THESIS
1-297 1969

58921 **FERROELECTRICITY VERSUS ANTIFERROELECTRICITY IN THE SOLID SOLUTIONS OF PBZRO3 AND PBTIO3.**
SAWAGUCHI E
J PHYS SOC JAPAN
8 615-29 1953

58924 **VII.-A STUDY OF THE CASING OF COLOURLESS BY COBALT BLUE GLASS. PART I. THE THERMAL EXPANSIONS.**
TURNER W E S WINKS F
J SOC GLASS TECHNOLOGY
12 57-74 1928

58935 **EXHIBITION OF AN APPARATUS FOR MEASURING THE THERMAL EXPANSION OF GLASS FROM ROOM TEMPERATURE TO THE SOFTENING POINT.**
WINKS F
PROC PHYS SOC
41 588-91 1928-9

58939 **VII. SOME PROPERTIES OF THE SPINELS ASSOCIATED WITH CHROME ORES.**
RIGBY G R LOVELL G H B GREEN A T
BRIT CERAMIC SOC TRANS
45 4 137-48 1946

58953 **IN VIVO DETERMINATION OF THERMAL CONDUCTIVITY OF BONE USING THE THERMAL COMPARATOR TECHNIQUE.**
VACHON R I WALKER F J WALKER D F NIX G H
PROC 7TH INTERNATIONAL CONFERENCE ON MEDICAL AND BIOLOGICAL ENGINEERING STOCKHOLM
502 1967

58957 **A NEW METHOD TO DETERMINE THE SMALLEST LENGTH CHANGES AND ITS APPLICATION TO MEASURE THE EXPANSION COEFFICIENT OF GLASSES AND CERAMIC MATERIALS.**
THILENIUM R HOLZMANN H
Z ANORG ALLGEM CHEM
189 367-87 1930

58974 **WILLYAMITE REDEFINED.**
CABRI L J HARRIS D C STEWART J M ROWLAND J F
AUSTRALAS INST MINING MET PROC
233 95-100 1970 CA 73 27463

58983 **INFRARED SPECTRA AND RADIATIVE THERMAL CONDUCTIVITY OF MINERALS AT HIGH TEMPERATURES.**
ARONSON, J. R. BELLOTTI, L. H.
ECKROAD, S. W. EMSLIE, A. G.
MC CONNELL, R. K. VON THUNA, P. C.
J GEOPHYS RES
75 17 3443-56 1970 CA 73 27459

58991 **INFRARED ABSORPTION SPECTRA OF SOLID RESIDUES FROM THE THERMAL TREATMENT OF FAT COAL.**
KEKIN N A MIROSHNICHENKO A M
KHIM TVERD TOPL
2 134-40 1970 CA 73 27385

58992 **MINERALOGICAL INVESTIGATION OF PYRITE FROM THE LORAINE GOLD MINE, ORANGE FREE STATE.**
BEUKES, G. J.
NAT. INST. MET., REPUB. S. AFR., RES. REP.
(430), 11PP., 1968.

58996 **PROPERTIES OF GLASS-CERAMICS BEFORE AND AFTER TEMPERING.**
KATAEVA, G. V. GORSHKOVA, Z. S.
SHCHEGLOVA, O. V.
STEKLO
(1), 49-53, 1969.

58998 **GLASS-CERAMIC MATERIAL WITH A LOW THERMAL EXPANSION COEFFICIENT.**
SZWEJDA K
SZKLO CERAM
21 2 36-8 1970 CA 73 28307

58999 **DILATOMETRIC STUDIES OF QUARTZ AND CORUNDUM SINGLE CRYSTALS.**
STRELKOV P G LIFANOV I I SHERSTYUKOV N G
TEPLOFIZ SVOISTVA TVERD TEL VYS TEMP TR VSES KONF 1966
1 81-6 1969 CA 73 29684

59001 **RARE MINERALS OF THE SPINEL GROUP IN ULTRABASITES FROM BULGARIA.**
ZHELYAZKOVA-PANAYOTOVA M IVCHINOVA L ILLIEV Z
TSCHERMAKS MINERAL PETROGR MITT
13 1 1-13 1969 CA 73 27467

59003 **PROPERTIES OF SIO2-BI2O3-NA2O GLASS.**
WATANABE K SUMIYOSHI Y ANBO E
YOGYO KYOKAI SHI
78 897 165-71 1970 CA 73 28319

59017 **MEASUREMENTS OF THE EXPANSION OF REFRACTORY BUILDING MATERIALS TO 1600 C.**
ENDELL K STEGER W
ARCHIV EISENHUTTENWESEN
1 11 721-4 1928

TPRC Number	Bibliographic Citation
59018	CORRELATION OF GLASS HEAT CAPACITY AND DENSITY. RODNIKOVA, V. V. GORODOV, V. S. DOPOV. AKAD. NAUK UKR. RSR, SER. B. 32 (4), 355–7, 1970. (FOR ENGLISH TRANSLATION SEE TPRC NO. 63983)
59025	MAGNETITE HEAT CAPACITY AND THERMODYNAMIC PROPERTIES FROM 5 TO 350 DEGREES K, LOW-TEMPERATURE TRANSITION. WESTRUM, E. F., JR. GROENVOLD, F. J. CHEM. THERMODYN. 1 (6), 543–57, 1969.
59042	SPECIFIC HEAT AND HEAT OF PYROLYSIS OF DONETS BASIN COALS. AGROSKIN, A. A. GONCHAROV, E. I. MAKEEV, L. A. YAKUNIN, V. P. KOKS KHIM. (5), 8–13, 1970. (FOR ENGLISH TRANSLATION SEE TPRC NO. 67687)
59043	MEASUREMENTS OF THERMAL CONDUCTIVITY OF ROCKS BY THE HOT WIRE METHOD. I. ON THE METHOD AND THE THERMAL CONDUCTIVITY OF COALS. EBUCHI, F. UCHINO, K. KYUSHU KOZAN GAKKAI-SHI 37 (1), 28–34, 1969.
59044	MEASUREMENTS OF THERMAL CONDUCTIVITIES OF ROCKS BY THE HOT WIRE METHOD. II. THERMAL CONDUCTIVITIES OF SHALES. EBUCHI, F. UCHINO, K. KYUSHU KOZAN GAKKAI-SHI 37 (3), 88–91, 1969.
59046	THERMAL DIFFUSIVITY OF ALUMINOSILICATE REFRACTORIES IN THE RANGE 200–1600 C. LITOVSKII E YA LANDA YA A MILSHENKO R S OGNEUPORY 35 5 17–19 1970 CA 73 48231 (FOR ENGLISH TRANSLATION SEE T60836)
59052	XII. SILLIMANITE CHROME MIXTURES. CHADEYRON A A REES W J TRANS BRIT CERAMIC SOC 42 163–70 1943
59053	CHAPTER XI. THE COEFFICIENT OF EXPANSION OF GLASS. MOREY G W J AM CHEM SOC 60 262–92 1938
59066	INFRARED SPECTROSCOPIC STUDY OF SYNTHETIC AND NATURAL TUNGSTATES AND MOLYBDATES. PLYUSNINA, I. I. ZAITSEVA, L. A. VESTN. MOSK. UNIV., GEOL. 24 (4), 110–15, 1969.
59079	THERMODYNAMICS OF STRUCTURAL VACANCIES IN OXIDE COMPOUNDS. III. STRUCTURAL VACANCIES IN FERRITE SPINELS. SUKHAREVSKII B YA ALAPIN B G VISHNEVSKII I I ZH FIZ KHIM 43 12 3113– 1969 (FOR ENGLISH TRANSLATION SEE T59080)
59080	THERMODYNAMICS OF STRUCTURAL VACANCIES IN OXIDE COMPOUNDS. III. STRUCTURAL VACANCIES IN FERRITE SPINELS. SUKHAREVSKII B YA ALAPIN B G VISHNEVSKII I I RUSSIAN J PHYS CHEM 43 12 1748–51 1969 (ENGLISH TRANSLATION OF ZH. FIZ. KHIM., 43 (12), 3113– , 1969; FOR ORIGINAL SEE T59079)
59091	HEAT EXPANSION OF SILICA BRICKS. FROMM F ARCHIV EISENHUTTENWESEN 7 7 381–4 1934
59098	INFRARED EMISSIVITY OF LUNAR SURFACE FEATURES. I. BALLOON-BORNE OBSERVATIONS. MURCRAY F H MURCRAY D G WILLIAMS W J JOURNAL OF GEOPHYSICAL RESEARCH 75 2662–9 1970
59099	INFRARED EMISSIVITY OF LUNAR SURFACE FEATURES. II. INTERPRETATION. SALISBURY J W VINCENT R K LOGAN L M HUNT G R JOURNAL OF GEOPHYSICAL RESEARCH 75 2671–82 1970
59106	THE EFFECT OF PULVERIZATION ON THE ALBEDO OF LUNAR ROCKS. MINNAERT M G J ICARUS 11 332–7 1969
59124	REMOTE PROBING OF THE MOON BY INFRARED AND MICROWAVE EMISSIONS AND BY RADAR. HAGFORS T RADIO SCIENCE 5 189–227 1970

TPRC Number	Bibliographic Citation
59127	SOME COMMENTS ON REFLECTANCE MEASUREMENTS OF WET SOILS. PLANET W G REMOTE SENSING OF ENVIRONMENT 1 127–9 1970
59141	LABORATORY EXPERIMENTS TO STUDY SURFACE CONTAMINATION AND DEGRADATION OF OPTICAL COATINGS AND MATERIALS IN SIMULATED SPACE ENVIRONMENTS. HASS G HUNTER W R APPLIED OPTICS 9 9 2101–10 1970 CA 73 93110
59170	ULTRAVIOLET REFLECTIVITY AND BAND STRUCTURE OF STANNIC OXIDE. CLARK, B. P. OKLAHOMA STATE UNIV., STILLWATER, OKLA., PH. D. THESIS 1–108, 1968. (UNIVERSITY MICROFILMS NO. 69–14234)
59176	THERMAL PROPERTIES OF SELECTED EVACUATED POWDER INSULATIONS AT CRYOGENIC TEMPERATURES. KNIGHT B L UNIV. COLORADO, BOULDER, PH. D. THESIS 1–322, 1969. (UNIVERSITY MICROFILMS NO. 69–19554)
59185	SPECTRAL REFLECTANCE QUALITIES OF SOILS WITH VARIOUS SALTS AND MOISTURE LEVELS IN THE ULTRAVIOLET, VISIBLE AND NEAR INFRARED REGIONS OF THE ELECTROMAGNETIC SPECTRUM. ALI, S. S. OKLAHOMA STATE UNIVERSITY, M. S. THESIS 1–103 1969
59194	DEPENDENCE OF REFLECTANCE AND MICROHARDNESS ON THE COMPOSITION OF SOLID SOLUTIONS OF THE DIGENITE - BORNITE SERIES. SLAVSKAYA, A. I. KACHALOVSKAYA, V. M. SHARYBKINA, M. A. RUDNICHENKO, V. E. KHROMOVA, M. M. GEOL. RUD. MESTOROZHD. 11 (5), 121–7, 1969.
59196	STUDY OF ENVIRONMENTAL EFFECTS UPON PARTICULATE RADIATION INDUCED ABSORPTION BANDS IN SPACECRAFT THERMAL CONTROL COATING PIGMENTS. MC CARGO, M. GREENBERG, S. A. BREUCH, R. A. LOCKHEED MISSILES AND SPACE CO., PALO ALTO, CALIF. 1–104, 1969. (NASA–CR–73289, N69–16868)
59198	COMPLEX MEASUREMENT OF THE THERMOPHYSICAL PROPERTIES OF SOLUTION–MELTS INTENDED FOR GROWING FERRITE SINGLE CRYSTALS. DERMAN A S BOGORODSKII O V IZV AKAD NAUK SSSR SER FIZ 34 6 1215–6 1970 CA 73 59772 (FOR ENGLISH TRANSLATION SEE T62295)
59213	FAST NEUTRON IRRADIATION EFFECTS ON CERTAIN PHYSICAL CONSTANTS OF CRYSTAL QUARTZ AND VITREOUS SILICA. MAYER G GIGON J J DE PHYS LE RADIUM 18 2 169–14 1957
59225	PROPERTIES OF GLASSES IN THE SYSTEM SIO2–B2O3–BAO. MAZURENKO V D BELORUS POLITEKH INST IM I V STALINA SBORNIK NAUCH RABOT 14–8 1963 (FOR ENGLISH TRANSLATION SEE T61709)
59226	THERMAL CONDUCTIVITY AND THERMAL DIFFUSIVITY OF NORTH DAKOTA AND MONTANA LIGNITE. DUARTE, J. F. UNIVERSITY OF NORTH DAKOTA, M. S. THESIS 1–49 1969
59239	THERMAL CONDUCTIVITY OF LUNAR FINES FROM APOLLO 11. CREMERS C J PROC SYMP THERMOPHYS PROP 5TH 391–5 1970
59268	ENERGY TRANSPORT IN OXIDE FUEL MATERIALS. CHRISTENSEN, J. A. WADCO CORP., RICHLAND, WASHINGTON 1–6, 1970. (CONF–701033, N71–25379, WHAN–SA–80)
59273	EMITTANCE PROPERTY RESEARCH ON THERMAL SHIELDS FOR SPACE SHUTTLE. REICHMAN, J. GRUMMAN AIRCRAFT ENGR. CORP., BETHPAGE, N. Y. 1–15, 1970. (RM–478, N70–29124)
59293	INFRARED SPECTRA OF SEVEN MINERALS FROM MADAGASCAR. POVARENNYKH A S BULL SOC FR MINERAL CRISTALLOGR 93 2 224–34 1970 CA 73

TPRC Number	Bibliographic Citation
59294	NONMETALLIC THERMAL STORAGE MEDIA FOR BLOCK-TYPE ELECTRIC SPACE HEATERS. SVIKIS, V. D. CAN., MINES BR., RES. REP. R 206, 44 PP., 1969.
59296	REFLECTANCE OF HIMALAYAN COALS. GHOSH T K FUEL 49 2 226-8 1970 CA 73 37403
59304	STATISTICAL RELATIONS BETWEEN THE REFLECTIVITY OF VITRINITE AND THE QUALITY OF COALS FROM THE ALDAN-CHULMAN REGION. FATKULIN, I. YA. GEBLER, I. I. RESHETKO, A. N. KHIM. TVERD. TOPL. (3), 141-3, 1970.
59315	EFFECT OF SOME ADDITIVES ON PROPERTIES OF GLASSES IN THE SYSTEM SIO2-AL2O3-MGO-CAO. ERMOLENKO N N NEMKOVICH I K RAKOV I L BELORUS POLITEKH INST IM I V STALINA SBORNIK NAUCH RABOT 28-32 1963 (FOR ENGLISH TRANSLATION SEE T59660)
59338	CRYSTALLIZATION AND THERMAL EXPANSION OF GLASSES IN THE SYSTEM PBO-B2O3-ZNO. ZHDANETS, N. A. KHEIFETS, V. S. J APPLIED CHEM USSR 40 11 2321-4 1967 (ENGLISH TRANSLATION OF ZH. PRIKL. KHIM., 40 (11), 2418-22, 1967; FOR ORIGINAL SEE T48459)
59345	INVESTIGATION OF SINTERING OF CERAMICS MADE FROM FUSED QUARTZ. BUDNIKOV P P PIVINSKII YU E ZH PRIKL KHIM 41 5 957-64 1968 (FOR ENGLISH TRANSLATION SEE T59346)
59346	INVESTIGATION OF SINTERING OF CERAMICS MADE FROM FUSED QUARTZ. BUDNIKOV P P PIVINSKII YU E J APPLIED CHEM USSR 41 5 910-15 1968 (ENGLISH TRANSLATION OF ZH. PRIKL. KHIM., 41 (5), 957-64, 1968; FOR ORIGINAL SEE T59345)
59359	CRYSTALLIZATION OF GLASS OF THE COMPOSITION LI2O . GA2O3 . 2SIO2. TSEKHOMSKAYA T S DUBROVO S K ZH PRIKL KHIM 41 3 480-7 1968 (FOR ENGLISH TRANSLATION SEE T59360)
59360	CRYSTALLIZATION OF GLASS OF THE COMPOSITION LI2O . GA2O3 . 2SIO2. TSEKHOMSKAYA T S DUBROVO S K J APPLIED CHEM USSR 41 3 468-73 1968 (ENGLISH TRANSLATION OF ZH. PRIKL. KHIM., 41 (3), 480-7, 1968; FOR ORIGINAL SEE T59359)
59372	THE EFFECT OF TEMPERATURE AND TIME OF HEAT TREATMENT ON THE CHEMICAL RESISTANCE OF TITANIUM ENAMELS TO ACIDS. VARGIN V V PEVZNER B Z ZH PRIKL KHIM 41 10 2145-9 1968 (FOR ENGLISH TRANSLATION SEE T59373)
59373	THE EFFECT OF TEMPERATURE AND TIME OF HEAT TREATMENT ON THE CHEMICAL RESISTANCE OF TITANIUM ENAMELS TO ACIDS. VARGIN V V PEVZNER B Z J APPLIED CHEM USSR 41 10 2028-31 1968 (ENGLISH TRANSLATION OF ZH. PRIKL. KHIM., 41 (10), 2145-9, 1968; FOR ORIGINAL SEE T59372)
59375	REACTION OF SOME TRANSITION METALS WITH SODIUM SILICATE MELTS. BANKOVSKAYA I B APPEN A A SAZONOVA M V ZH PRIKL KHIM 41 10 2138-45 1968 (FOR ENGLISH TRANSLATION SEE T59376)
59376	REACTION OF SOME TRANSITION METALS WITH SODIUM SILICATE MELTS. BANKOVSKAYA I B APPEN A A SAZONOVA M V J APPLIED CHEM USSR 41 10 2022-7 1968 (ENGLISH TRANSLATION OF ZH. PRIKL. KHIM., 41 (10), 2138-45, 1968; FOR ORIGINAL SEE T59375)
59391	NONSELECTIVE REFLECTOR MADE OF MILK GLASS. MIKHAILOV O M OPT SPEKTROSK 28 4 801-2 1970 (FOR ENGLISH TRANSLATION SEE T59392)

TPRC Number	Bibliographic Citation
59392	NONSELECTIVE REFLECTOR MADE OF MILK GLASS. MIKHAILOV O M OPTICS SPECTROSCOPY 28 4 429 1970 (ENGLISH TRANSLATION OF OPT. SPEKTROSK., 28 (4), 801-2, 1970; FOR ORIGINAL SEE T59391)
59405	MINERALOGICAL STRUCTURE CHANGES IN DUNITE ON HEATING. BUDNIKOV P P KHOROSHAVIN L B PEREPELITSIN V M USTYANTSEV V M KOSOLAPOVA E P ZH PRIKL KHIM 42 2 260-71 1969 (FOR ENGLISH TRANSLATION SEE T59406)
59406	MINERALOGICAL STRUCTURE CHANGES IN DUNITE ON HEATING. BUDNIKOV P P KHOROSHAVIN L B PEREPELITSIN V M USTYANTSEV V M KOSOLAPOVA E P J APPLIED CHEM USSR 42 2 238-47 1969 (ENGLISH TRANSLATION OF ZH. PRIKL. KHIM., 42 (2), 260-71, 1969; FOR ORIGINAL SEE T59405)
59417	PRESSING OF POROUS SORBENTS AND DETERMINATION OF THEIR THERMAL CHARACTERISTICS. KUATBEKOV M K ROMANKOV P G FROLOV V F ZH PRIKL KHIM 42 4 949-52 1969 (FOR ENGLISH TRANSLATION SEE T59418)
59418	PRESSING OF POROUS SORBENTS AND DETERMINATION OF THEIR THERMAL CHARACTERISTICS. KUATBEKOV M K ROMANKOV P G FROLOV V F J APPLIED CHEM USSR 42 4 907-9 1969 (ENGLISH TRANSLATION OF ZH. PRIKL. KHIM., 42 (4), 949-52, 1969; FOR ORIGINAL SEE T59417)
59429	INFRARED REFLECTION SPECTRA OF SINGLE CRYSTAL AND POWDERED MINERALS. BRUNEL R VIERNE R BULL SOC FR MINERAL CRISTALLOGR 93 3 328-40 1970 CA 73 71665
59430	MANIFESTATION AND CALCULATION OF THE NONANALYTICAL NATURE OF THE CRITICAL FREQUENCIES OF DIPOLE VIBRATIONS. BELOUSOV M V POGAREV D E SHULTIN A A FIZ TVERD TELA 12 4 991-4 1970 CA 73 19935 (FOR ENGLISH TRANSLATION SEE T59431)
59431	MANIFESTATION OF AND ALLOWANCE FOR NONANALYTICITY OF LIMITING FREQUENCIES OF DIPOLE OSCILLATIONS. BELOUSOV M V POGAREV D E SHULTIN A A SOVIET PHYSICS SOLID STATE 12 4 778-80 1970 (ENGLISH TRANSLATION OF FIZ. TVERD. TELA, 12 (4), 991-4, 1970; FOR ORIGINAL SEE T59430)
59436	DIMENSIONAL EFFECT ON GALLIUM IN POROUS GLASS. BOGOMOLOV V N MALKOVICH R SH SMIRNOV I A TIKHONOV V V CHUDNOVSKII F A FIZ TVERD TELA 12 4 1204-7 1970 CA 73 28283 (FOR ENGLISH TRANSLATION SEE T59437)
59437	SIZE EFFECT FOR GALLIUM IN POROUS GLASS. BOGOMOLOV V N MALKOVICH R SH SMIRNOV I A TIKHONOV V V CHUDNOVSKII F A SOVIET PHYSICS-SOLID STATE 12 4 938-40 1970 (ENGLISH TRANSLATION OF FIZ. TVERD. TELA, 12 (4,) 1204-7, 1970; FOR ORIGINAL SEE T59436)
59444	GRUNEISEN PARAMETER FROM THERMAL CONDUCTIVITY MEASUREMENTS UNDER PRESSURE. BARKER R E JR CHEN R Y S J CHEM PHYS 53 7 2616-20 1970
59445	TEMPERATURE-DEPENDENT OPTICAL PHONONS IN LEAD TITANATE. TORNBERG N E PERRY C H J CHEM PHYS 53 7 2946-55 1970
59448	FERROELECTRICITY, DOMAIN STRUCTURE, AND PHASE TRANSITIONS OF BARIUM TITANATE. HIPPEL A REV MODERN PHYSICS 22 3 221-37 1950
59460	TWO-BEAM METHOD FOR DETERMINING THE THERMAL RADIATION OF MATERIALS DISPERSING RADIATION. ILYASOV, S. G. KRASNIKOV, V. V. INZH.-FIZ. ZH. 18 (4), 688-95, 1970. (FOR ENGLISH TRANSLATION SEE TPRC NO. 71940)

TPRC Number	Bibliographic Citation
59462	**PROPERTIES OF GLASSES IN THE SIO2-TIO2-ZRO2-AL2O3-MGO-CAO SYSTEM.** ERMOLENKO N N SHAREIKO L V SAVITSKII S YE GRISHINA N P BELORUS POLITEKH INST IM I V STALINA SBORNIK NAUCH RABOT 48-54 1963 (FOR ENGLISH TRANSLATION SEE T61556)
59464	**VISCOSITY CHARACTERISTICS OF MELTS OF SOME BASIC ROCKS IN THE UKRAINE.** CHERNOVA, V. I. FOREIGN TECHNOLOGY DIVISION 10PP., 1971. (ENGLISH TRANSLATION OF DOPOV. AKAD. NAUK UKR. RSR, 32 B (1), 41-5, 1970., FOR ORIGINAL SEE TPRC NO. 57536) (FTD-HC-23-1523-71)
59471	**THERMAL CONDUCTIVITY AND SPECIFIC HEAT OF PERICLASE-SPINEL AND MAGNESITE-CHROMITE ROOF REFRACTORIES.** SHAKHTIN D M ELISEEVA G G KOVALEV A I ADOLF E M MOZGOVOI A N KOTIK P L OGNEUPORY 7 12-13 1970 (FOR ENGLISH TRANSLATION SEE T61167)
59479	**EFFECT OF HYDROGEN ON THE COEFFICIENT OF THERMAL CONDUCTIVITY OF CERAMIC PRODUCTS.** FUCHS, W. E. VERFAHRENSTECHNIK (MAINZ) 4 (6), 250-2, 1970.
59480	**CHARACTERISTICS OF REFLECTIVITY AND MICROHARDNESS OF GALENAS DEPENDENT ON THE FORM OF SILVER PRESENT IN THEM.** RYABEV, V. L. RYABEVA, E. G. MALAEVA, I. P. TR., TSENT. NAUCH.-ISSLED. GORNORAZVED. INST. (80), 188-95, 1969.
59491	**DEPENDENCE OF THE CONTACT HEAT CONDUCTIVITY OF GRANULAR SYSTEMS ON THE EXTERNAL LOAD.** DULNEV G N ZARICHNYAK YU P MURATOVA B L SIGALOVA Z V J ENGINEERING PHYS 11 2 110-12 1966 (ENGLISH TRANSLATION OF INZH. FIZ. ZH., 11 (2), 202-6, 1966; FOR ORIGINAL SEE T41480)
59500	**VACUUM-TUBE AND TRANSISTOR INSTRUMENTS FOR INVESTIGATING THERMAL CONDUCTIVITY BY MEANS OF A LINEAR HEAT SOURCE.** YANKELEV L F BLUVSHTEIN I M INZH FIZ ZH 11 6 756-60 1966 (FOR ENGLISH TRANSLATION SEE T59501)
59501	**VACUUM-TUBE AND TRANSISTOR INSTRUMENTS FOR INVESTIGATING THERMAL CONDUCTIVITY BY MEANS OF A LINEAR HEAT SOURCE.** YANKELEV L F BLUVSHTEIN I M J ENGINEERING PHYS 11 6 406-9 1966 (ENGLISH TRANSLATION OF INZH. FIZ. ZH., 11 (6), 756-60, 1966; FOR ORIGINAL SEE T59500)
59510	**HIGH THERMAL SHOCK CERAMICS.** RUTGERS UNIVERSITY, CERAMIC RESEARCH STATION, N. J. RUTGERS UNIVERSITY, CERAMIC RESEARCH STATION, N. J. 1-25, 1951. (ATI-123768)
59511	**REFRACTORY MATERIALS FOR USE IN HIGH TEMPERATURE AREAS OF AIRCRAFT.** PENNSYLVANIA STATE COLLEGE, STATE COLLEGE, PA. PENNSYLVANIA STATE COLLEGE, STATE COLLEGE, PA. 1-28, 1952. (ATI-155911)
59512	**HIGH THERMAL SHOCK CERAMICS.** RUTGERS UNIVERSITY, CERAMIC RESEARCH STATION, N. J. RUTGERS UNIVERSITY CERAMIC RESEARCH STATION, N. J. 28-57, 1952. (ATI-187723)
59514	**REFRACTORY PROTECTIVE COATINGS FOR METALS OF AIRCRAFT POWER PLANTS SUBJECTED TO ELEVATED TEMPERATURES.** KOENIG, J. H. HOWER, L. D., JR. SOSMAN, R. B RUTGERS UNIV., NEW BRUNSWICK, N. J. 1-13, 1950. (AF-TR-6071, ATI-80205)
59515	**REFRACTORY MATERIALS FOR USE IN HIGH TEMPERATURE AREAS OF AIRCRAFT.** THIELKE, N. R. HENRY, E. C. PENNSYLVANIA STATE COLLEGE, STATE COLLEGE, PA. 1-47, 1950. (AF-TR-6080, ATI-80745)
59516	**THE EFFECT OF MILL ADDITIONS ON THE THERMAL EXPANSION PROPERTIES OF REFRACTORY CERAMIC TOP COATS.** HICKS R G PLANKENHORN W J BENNETT D G WRIGHT PATTERSON AIR FORCE BASE, DAYTON, OHIO 1-14, 1950. (AF-TR-6094, ATI-90184)
59521	**THEORY AND EXPERIMENTS FOR MULTIELEMENT GRID FILTERS IN A DIELECTRIC.** BELL R J ROMERO H V BLEA J M APPLIED OPTICS 9 10 2350-8 1970
59530	**OPTICAL ABSORPTION SPECTRA OF IRON IN THE ROCK-FORMING SILICATES.** WHITE W B KEESTER K L AM MINERALOGIST 51 774-91 1966
59534	**THE INFRARED SPECTRA OF SOME VITREOUS AND CRYSTALLINE BORATES.** KROGH-MOE J ARKIV KEMI 12 41 475-80 1958
59535	**A SEARCH FOR LIMONITE NEAR-INFRARED SPECTRAL FEATURES ON MARS.** YOUNKIN R L ASTROPHYS J 144 809-18 1966
59537	**TABLES FOR THE THERMAL EXPANSIONS OF DIVERS SIMPLE METALLIC OR NON-METALLIC BODIES, AND OF SOME COMPOUNDS OF HYDROCARBON.** FIZEAU M H COMPT REND 68 20 1125-31 1869
59540	**THE REFLECTIVITY SPECTRUM OF MARS IN THE NEAR-INFRARED.** TULL R G ICARUS 5 505-14 1966
59551	**THE DIRECTIONAL CONCENTRATION OF OPTIC AXES IN YULE MARBLE. A COMPARISON OF THE RESULTS OF PETROFABRIC ANALYSIS AND LINEAR THERMAL EXPANSION.** ROSENHOLTZ J L SMITH D T AMER J SCIENCE 249 5 377-84 1951
59560	**INVESTIGATIONS ON THE THERMAL EXPANSION AND RELAXATION TEMPERATURE OF GLAZES WITH SPECIAL REFERENCE TO ADJUSTMENT OF GLAZES TO BODY.** STEGER W BER DEUT KERAM GESELL 8 24-43 1927
59561	**THERMAL CONDUCTIVITY OF PARTICULATE BASALT AS A FUNCTION OF DENSITY IN SIMULATED LUNAR AND MARTIAN ENVIRONMENTS.** FOUNTAIN J A WEST E A JOURNAL OF GEOPHYSICAL RESEARCH 75 4063-9 1970
59565	**THERMAL EXPANSION AND SHRINKAGE MEASUREMENTS ON SOME FIRE CLAYS AND BODIES.** REICH H F BER DEUT KERAM GESELL 37 3 103-15 1960
59569	**PROTON-IRRADIATION DARKENING OF ROCK POWDERS. CONTAMINATION AND TEMPERATURE EFFECTS, AND APPLICATIONS TO SOLAR-WIND DARKENING OF THE MOON.** NASH D B J GEOPHYSICAL RESEARCH 72 12 3089-104 1967
59573	**INSTRUMENT FOR MEASURING TERRESTRIAL HEAT FLOW THROUGH THE OCEAN FLOOR.** CORRY C DU BOIS C VACQUIER V J MARINE RESEARCH 26 2 165-77 1968
59580	**PHYSICAL PROPERTIES OF HIGH FREQUENCY ELECTRICAL CERAMIC INSULATING PARTS.** GUNZENHAUSER A CERAMIC INDUSTRY 39 11 68-72 1942
59583	**ERRORS IN THE DETERMINATION OF THERMAL EXPANSION.** AUTHOR ANON. SOC. GLASS TECHNOLOGY 40 (194), 70P-82P, 1956.
59584	**INFRARED STUDIES ON POLYMORPHS OF SILICON DIOXIDE AND GERMANIUM DIOXIDE.** LIPPINCOTT E R VAN VALKENBURG A WEIR C E BUNTING E N J RESEARCH NATL BUREAU STANDARDS 61 1 61-70 1958

TPRC Number	Bibliographic Citation
59587	MEASUREMENTS IN THE LONG WAVE-LENGTH INFRARED FROM 20 EPSILON TO 135 EPSILON. BARNES R B PHYS REV 39 4 562-75 1932
59589	A PRECISION COMPARISON OF CALCULATED AND OBSERVED GRATING CONSTANTS OF CRYSTALS. TU Y PHYS REV 40 662-75 1932
59605	AUTOMATIC EQUIPMENT FOR DETERMINING THERMAL CONDUCTIVITY COEFFICIENTS OF CERAMICS. KORTNEV V V DUDEROV I G KOSTYUKOV N S ZAVOD LAB 35 2 242-3 1969 (FOR ENGLISH TRANSLATION SEE T59606)
59606	AUTOMATIC EQUIPMENT FOR DETERMINING THERMAL CONDUCTIVITY COEFFICIENTS OF CERAMICS. KORTNEV V V DUDEROV I G KOSTYUKOV N S IND LAB USSR 35 2 293-4 1969 (ENGLISH TRANSLATION OF ZAVOD. LAB., 35 (2), 242-3, 1969; FOR ORIGINAL SEE T59605)
59615	EMITTANCE MEASUREMENT WITH A DIFFERENTIAL SCANNING CALORIMETER. KHAN, A. M. VILLANOVA UNIVERSITY, M. S. THESIS 1-70 1969
59646	INVESTIGATION OF A MODEL FOR BIDIRECTIONAL REFLECTANCE OF ROUGH SURFACES. TREAT C H WILDIN M W PROGRESS ASTRONAUT AERON 23 77-92 1970
59647	RADIATION-INDUCED ABSORPTION BANDS IN SPACECRAFT THERMAL CONTROL COATING PIGMENTS. MC CARGO M GREENBERG S A DOUGLAS N J PROGRESS ASTRONAUT AERON 23 189-218 1970
59652	STRUCTURE AND PROPERTIES OF AMORPHOUS SILICO-ALUMINAS. I. STRUCTURE FROM X-RAY FLUORESCENCE SPECTROSCOPY AND INFRARED SPECTROSCOPY. LEONARD A SUZUKI S FRIPIAT J J KIMPE C DE J PHYS CHEM 68 9 2608-17 1964
59655	STRENGTHENING OXIDES BY REDUCTION OF CRYSTAL ANISOTROPY. KIRCHNER, H. P. GRUVER, R. M. EWIG, R. A. CERAMIC FINISHING COMP., STATE COLLEGE, PA. 1-51, 1970. (AD-707447)
59660	EFFECT OF SOME ADDITIVES ON PROPERTIES OF GLASSES IN THE SYSTEM SI O(2)-AL(2) O(3)-MG O-CA O ERMOLENKO N N NEMKOVICH I K RAKOV I L FOREIGN TECHNOLOGY DIVISION 1-12, 1971. (ENGLISH TRANSLATION OF BELORUS. POLITEKH. INST. IM. I. V. STALINA, SBORNIK NAUCH. RABOT., 28-32, 1963; FOR ORIGINAL SEE T59315) (FTD-MT-24-1257-71)
59662	COMPOSITION OF THE MOON FROM BALLOON-BORNE MID-INFRARED OBSERVATIONS. SALISBURY J W RADIO SCIENCE 5 2 241-6 1970
59667	FIBER OPTICS WITH EXTENDED ULTRAVIOLET TRANSMISSION. PINCUS, A. G. ALI, M. A. SCHWARTZ, M. A. IIT RESEARCH INST., CHICAGO, ILL. 1-16, 1970. (ECOM-0542-8, AD-710241)
59671	LUNAR SOIL SIMULANT STUDY, PHASE B. PART II. THERMAL CONDUCTIVITY. HUBBARD, J. H. GALL, E. S. KAHN, H. S. OHIO RIVER DIVISION LABS., CINCINNATI, OHIO 1-54, 1969. (ORDL-TR-4-83, AD-709677)
59674	SPECTROSCOPIC PROPERTIES OF SOLID SOLUTIONS OF ERBIUM AND YTTERBIUM OXIDES. GRUSS L SALOMON R E J PHYS CHEM 74 22 3969-72 1970
59685	PROPERTIES OF MOLTEN CERAMICS. BATES J L MC NEILLY C E RASMUSSEN J J PACIFIC NORTHWEST LAB BATTELLE-NORTHWEST, RICHLAND, WASH. 1-18, 1970. (CONF-701206-1, BNWL-SA-3579, N71-27426)
59690	ELECTRICAL AND THERMAL CONDUCTIVITY IN A SUPERDENSE LATTICE. I. HIGH-TEMPERATURE CONDUCTIVITY. SOLINGER A B ASTROPHYS J 161 1 2 553-9 1970 CA 73 81818
59707	CAUSES OF THE DIFFERENT DEGREES OF EXPANSION AND CONTRACTION OF CORUNDIUM-CONTAINING FIRE CLAY BRICK. REICH H F BER DEUT KERAM GESELL 43 1 83-7 1966
59708	IMPORTANCE OF THE DILATOMETRIC METHOD FOR STUDYING AND CHACTERIZING CLAY MATERIALS FOR CERAMIC INDUSTRY. SUAREZ R BOL FAC ING AGRI-MENSURA MONTEVIDEO 8 12 435-60 1964
59714	INVESTIGATION OF ENERGY AND DEPOSITION PROCESSES IN THE UPPER ATMOSPHERE AND THE INTERACTION BETWEEN THE MESOSPHERE AND THE THERMOSPHERE. MC ELANEY, J. H. FARIFIELD UNIV., CONN. 1-14, 1969. (NASA-CR-108227, N70-18960)
59722	EFFECTS OF SUBSTITUTING LEAD OXIDE FOR LITHIUM OXIDE IN CERAMIC GLAZES. FERRANDIS V A NAVARRO J M F BOL SOC ESP CERAM 3 4 377-92 1964
59723	THERMAL EXPANSION OF CORUNDUM, MAGNESIUM OXIDE, AND MAGNESIUM-ALUMINUM SPINEL-CORUNDUM SOLID SOLUTIONS IN A NEUTRAL MEDIUM UP TO 2000-2300 DEGREES. PIVOVAR T L ELISEEVA G G LEVINTOVICH E V TEPLOFIZ SVOISTVA TVERD TEL VYS TEMP TR VSES KONF 1966 1 87-93 1969 CA 73 80833
59727	INTERPRETATION OF THERMAL-EXPANSION DATA FROM CLAY CERAMICS. TITE M S TRANS BRIT CERAM SOC 69 4 183-7 1970 CA 73 80115
59763	THERMAL CONDUCTION BY ELECTRONS IN STELLAR MATTER. HUBBARD, W. B. LAMPE, M. ASTROPHYS. J., SUPPL. SER. 18 (163), 297-346, 1969.
59784	THERMAL EXPANSION OF NATURAL AND SYNTHETIC SODIUM CHLORIDE. SAINI H HELV PHYS ACTA 7 494-500 1934
59786	SOME FIRING AND FIRED PROPERTIES OF NEW YORK STATE DIOPSIDE, WOLLASTONITE, AND A TREMOLITIC MATERIAL. AMBERG C R J AM CERAM SOC 28 5 141-3 1945
59789	EFFECTS OF COKE-PRODUCTION TEMPERATURE AND THE CONTENTS OF WATER AND VOLATILE COMPONENTS IN THE COKE CHARGE ON VARIOUS STAGES OF THE COKING PROCESS. WIESE, H. W. AACHENER BL. AUFBEREITEN-VERKOKEN-BRIKET. 20 (2-3), 124PP., 1970.
59793	OPTICAL PROPERTIES OF DOLOMITE IN THE SCHUMANN REGION. UZAN E DAMANY H CHANDRASEKHARAN V OPT COMMUN 2 1 6-8 1970 CA 73 93138
59806	LOW EXPANSION VITREOUS BODIES ARE RESISTANT TO HEAT SHOCK. BADGER A E CERAMIC INDUSTRY 40 3 49 1943
59810	THE EXPANSION OF COMMERCIAL GLASSES. LE GUILLOU G ALBANY H J SAMSOEN M COMPT REND 182 1384-6 1926
59811	DILATOMETRIC AND THERMAL STUDY OF SODA-SILICA GLASSES. SAMSOEN M COMPT REND 183 285-6 1926
59812	DILATOMETRIC EXAMINATION OF SOME TERNARY SILICON SODIUM-ALUMINUM GLASSES. RENCKER E COMPT REND 199 21 1114-16 1934
59813	VARIATION OF THE EXPANSION COEFFICIENT OF GLASS BY ANNEALING. KLEIN N COMPT REND 200 15 1320-1 1935
59820	LINEAR THERMAL EXPANSION OF CALCITE, VARIOUS, ICELAND SPAR, AND YULE MARBLE. ROSENHOLTZ J L SMITH D T AM MINERALOGIST 34 11 846-54 1949

TPRC Number	Bibliographic Citation
59826	CARBON REFRACTORIES MEET SEVERE TEMPERATURE CONDITIONS. NATIONAL CARBON CORPORATION BRICK CLAY RECORD 105 2 40-2 1944
59827	STUDIES OF THE THERMAL STATE OF THE EARTH. THE 17TH PAPER. VARIATION OF THERMAL CONDUCTIVITY OF ROCKS. PART 2. KAWADA K BULL EARTHQUAKE RESEARCH INSTITUTE TOKYO UNIV 44 1071-91 1966
59828	COEFFICIENT OF THERMAL EXPANSION OF NA2O-BAO-SIO2 GLASSES. YASUHARA N J JAPANESE CERAMIC ASSOC 50 590 61-4 1942
59830	STUDIES ON THE COMPOSITIONS OF GLASS. I. EXPANSION COEFFICIENT OF THE SYSTEM NAIO-K2O-MGO-SIO2. KANEKO T J JAPANESE CERAMIC ASSOC 51 608 487 1943
59834	ULTRAPRECISE MEASUREMENT OF THERMAL COEFFICIENTS OF EXPANSION. JACOBS S F BRADFORD J N BERTHOLD J W III APPLIED OPTICS 9 11 2477-80 1970
59875	THERMAL EXPANSION OF SODA-LIME GLASSES. AUTHOR ANON. NATIONAL BUR. STANDARDS TECH. NEWS BULL. 163, 109PP., 1930.
59877	TELLURITE GLASSES. STANWORTH J E NATURE 169 581-2 1952
59921	THE THERMAL EXPANSION OF SOME CLAY MINERALS. HYSLOP J F MC MURDO A TRANS CERAMIC SOC /ENGLAND/ 37 5 180-2 1938
59925	INVESTIGATIONS ON THE VITREOUS STATE. SAMSOEN M BULL SOC ENCOUR L IND NATIONALE 128 185-204 1929
59926	THE DILATATION, SPECIFIC HEAT OF MELTABLE ALLOYS AND THEIR RELATIONS WITH THE LAW OF THE HEAT CAPACITY OF THE ATOMS OF SIMPLE AND COMPOSITE BODIES. SPRING W BULL SACAD BEAUX ARTS 39 548-602 1875
59929	DILATOMETRIC STUDY OF SOME REFRACTORY OXIDES. GAMBEY F CHAUDRON G CHIMIE ET INDUSTRIE 453-4C 1931
59932	IRREGULARITY OF COEFFICIENT OF EXPANSION ACCOMPANYING THE MAGNETIC TRANSFORMATION IN PYRRHOTITE AND MAGNETITE. CHEVENARD P COMPT REND 172 320-2 1921
59933	A CONTRIBUTION TO THE KNOWLEDGE OF THE EXPANSION COEFFICIENT OF GLASSES. WOLF J SPRECHSAAL 44 43 627-8 1911
59941	ADJUSTMENT OF THERMAL EXPANSION OF CONE 8 GLAZES. VAN GORDON D V SPANGENBERG W C J AM CERAM SOC 38 9 331-5 1955
59946	MEASUREMENTS ON THE THERMAL DILATATION OF GLASS AT HIGH TEMPERATURES. PETERS C G CRAGOE C H J RESEARCH NATL BUREAU STANDARDS 16 449-87 1920 (NBS-SP-393)
59947	THERMAL EXPANSIONS OF SOME SODA-LIME-SILICA GLASSES AS FUNCTIONS OF THE COMPOSITION. SCHMID B C FINN A N YOUNG J C J RESEARCH NATL BUREAU STANDARDS 12 421-28 1934 (NBS-RP-667)
59948	INDEX OF REFRACTION, DENSITY, AND THERMAL EXPANSION OF SOME SODA-ALUMINA-SILICA GLASSES AS FUNCTIONS OF THE COMPOSITION. FAICK C A YOUNG J C HUBBARD D FINN A N J RESEARCH NATL BUREAU STANDARDS 14 2 133-7 1935 (NBS-RP-762)
59949	EXPANSIVITY OF A VYCOR BRAND GLASS. SAUNDERS J B J RESEARCH NATL BUREAU STANDARDS 28 51-5 1942 (NBS-RP-1445)
59950	THERMAL EXPANSION OF CONCRETE AGGREGATE MATERIALS. JOHNSON W H PARSONS W H J RESEARCH NATL BUREAU STANDARDS 32 101-26 1944 (NBS-RP-1578)
59951	EXPANSION EFFECTS OF ANNEALING BOROSILICATE THERMOMETER GLASSES. TOOL A Q SAUNDERS J B J RESEARCH NATL BUREAU STANDARDS 42 2 171-82 1949 (NBS-RP-1960)
59958	THE INFRA-RED ABSORPTION SPECTRA OF QUARTZ AND FUSED SILICA FROM 1 TO 7.5 MICRON. II. EXPERIMENTAL DRUMMOND D G PROC ROY SOC 153 A 328-39 1936
59961	LUNAR SURFACE. COMPOSITION INFERRED FROM OPTICAL PROPERTIES. HAPKE B SCIENCE 159 76-9 1968
59965	THE COEFFICIENT OF EXPANSION OF ENAMELS AND THEIR CHEMICAL COMPOSITION. MAYER M HAVAS B SPRECHSAAL 44 13 188-91 1911 44 13 207-10 220-2 1911
59966	THE COEFFICIENT OF EXPANSION OF GLAZES. RIEKE R STEGER W SPRECHSAAL 47 27 441-3 457-9 1914 47 27 576-8 585-6 1914 47 27 593-5 601-4 1914
59968	EFFECT OF FLINTS ON THE THERMAL EXPANSIONS OF SOME WHITEWARE BODIES. SHEARER W L THE CERAMIST 4 143-65 1924
59969	THE PROPERTIES AND TESTING OF CERAMIC FINISHED PRODUCTS. 3. THERMAL MEASUREMENTS. SCHEEL K THE CERAMIST 8 7 451-65 1926
59970	THE REVERSIBLE THERMAL EXPANSION OF REFRACTORY MATERIALS. RIGBY G R GREEN A T TRANS CERAMIC SOC /ENGLAND/ 37 355-403 1938
59974	MICROWAVE HIGH DIELECTRIC CONSTANT MATERIALS. READEY D W MAGUIRE E A HARTWIG C P MASSE D J RAYTHEON RESEARCH DIVISION, WALTHAM, MASSACHUSETTS 1-29, 1970. (ECOM-0455-4)
59988	COEFFICIENT OF EXPANSION STUDY OF LEAD-ZINC GLAZE. PENCE F K AM CERAMIC SOC BULL 31 12 483-5 1952
59992	THERMAL EXPANSION OF COMMERCIAL DRY PROCESS ENAMELS AND CAST IRONS. DANIELSON R R VAN GORDON D V AM CERAMIC SOC BULL 35 9 347-50 1956
59993	INITIAL THERMAL EXPANSION CHARACTERISTICS OF INSULATION REFRACTORY CONCRETES. CROWLEY M S AM CERAMIC SOC BULL 35 12 465-8 1956
59995	THERMAL EXPANSIONS OF SOME COMMERCIAL GLASSES. BADGER A E TAYLOR R C CERAMIC INDUSTRY 26 6 416-17 1936
60001	THE EFFECTS OF SODA, BARIUM, AND ZINC ON THE ELASTICITY AND THERMAL EXPANSION COEFFICIENTS OF GLASS. FETTEROLF L D PARMELEE C W J AM CERAM SOC 12 3 193-216 1929
60002	RELATIONSHIP BETWEEN CHEMICAL COMPOSITION AND THE THERMAL EXPANSION OF GLASSES. ENGLISH S TURNER W E S J AM CERAMIC SOC 10 8 551-60 1927 CORRECTION.

TPRC Number	Bibliographic Citation
60003	**EXPANSION MEASUREMENTS OF SEVERAL GLASSES BY MEANS OF A SELF-REGISTERING APPARATUS.** COHN W M J AM CERAMIC SOC 14 265-75 1931
60005	**LITHIUM OXIDE AS A CONSTITUENT OF GLASSES-ITS EFFECT ON THERMAL EXPANSION.** NAVIAS L J AM CERAM SOC 18 7 206-10 1935
60006	**EFFECT OF MILL ADDITIONS ON THE THERMAL EXPANSION OF SHEET-IRON GROUND-COAT ENAMELS.** FELLOWS R L HOWE E E J AM CERAM SOC 19 4 109-11 1936
60007	**EFFECT OF HEAT TREATMENT ON THE THERMAL EXPANSION OF SOME SILICA-ALUMINA GLASSES.** THOMPSON C L PARMELEE C W J AM CERAM SOC 20 305-8 1937
60008	**RELATION BETWEEN INELASTIC DEFORMABILITY AND THERMAL EXPANSION OF GLASS IN ITS ANNEALING RANGE.** TOOL A Q J AM CERAM SOC 29 9 240-53 1946
60009	**VOLUME EXPANSION OF GLASS AT HIGH TEMPERATURES.** JOHNSON A G SCHOLES S R SIMPSON H E J AM CERAMIC SOC 33 4 144-7 1950
60010	**THERMAL EXPANSION OF SOME GLASSES IN THE SYSTEM LI2O-AL2O3-SIO2.** BRACKBILL C E MC KINSTRY H A HUMMEL F A J AM CERAM SOC 34 4 107-9 1951
60011	**THERMAL EXPANSION OF SOME GLASSES IN THE SYSTEM MGO-AL2O3-SIO2.** HUMMEL F A REID H W J AM CERAM SOC 34 10 319-21 1951
60012	**THERMAL EXPANSION OF SOME SIMPLE GLASSES.** KARKHANAVALA M D HUMMEL F A J AM CERAM SOC 35 9 215-19 1952
60016	**THE ABBE-FIZEAU DILATOMETER.** PULFRICH C ZEIT INSTRUMENTENKUNDE 13 365-80 401-16 13 437-56 1893
60046	**THE DOUBLE-BEAM METHOD OF DETERMINING THE THERMAL RADIATION PROPERTIES OF DIFFUSING MATERIALS.** ILYASOV S G KRASNIKOV V V HEAT TRANSFER-SOVIET RESEARCH 2 5 68-75 1970
60073	**HEAT CAPACITY OF VITREOUS GERMANIUM DIOXIDE AND POTASSIUM OXIDE-GERMANIUM DIOXIDE GLASSES.** TARASOV V V SOBOLEVA P A ZH FIZ KHIM 44 6 1590- 1970 CA 73 102688 (FOR ENGLISH TRANSLATION SEE T60074)
60074	**HEAT CAPACITY OF VITREOUS GERMANIUM DIOXIDE AND POTASSIUM OXIDE-GERMANIUM DIOXIDE GLASSES.** TARASOV V V SOBOLEVA P A RUSSIAN J PHYS CHEM 44 6 894-5 1970 (ENGLISH TRANSLATION OF ZH. FIZ. KHIM., 44 (6), 1590-, 1970; FOR ORIGINAL SEE T60073)
60093	**ELECTRICAL AND OPTICAL PROPERTIES OF SODALITE SINGLE CRYSTALS.** SHALDIN YU V MELNIKOV O K NABATOV V V PISAREVSKII YU V SOVIET PHYSICS-CRYSTALLOGRAPHY 10 4 484-5 1966 (ENGLISH TRANSLATION OF KRISTALLOGRAFIYA, 10 (4) 574-6, 1965; FOR ORIGINAL SEE T34118)
60103	**INSTRUMENTATION FOR COLOR AERIAL PHOTOGRAPHY.** ROBERTSON, K. D. ARMY ENGINEER TOPOGRAPHIC LABS., FORT BELVOIR, VA. 1-42, 1970. (ETL-RN-70-1, AD-710978)
60104	**OPTICS AT THERMALIZATION.** BAAK T SCHROEDER J B MORSE S E HOOGLAND J FRIEDMAN I LINCOLN H M THE PERKIN-ELMER CORPORATION 1-207, 1969. (AFAL-TR-69-19, AD-850709)
60116	**OPTICAL AND THERMAL STUDIES OF TOPAZ FROM NAEGI, JAPAN.** KOZU S UEDA J SCI REPTS TOHOKU UNIV 3 161-70 1929

TPRC Number	Bibliographic Citation
60136	**ON THE FEASIBILITY OF CALCULATING GLASS SOFTENING POINTS FROM THE MEAN ATOMIZATION ENERGY.** SHUMITSKAYA L F IZV AKAD NAUK SSSR NEORG MATER 6 5 888-92 1970 (FOR ENGLISH TRANSLATION SEE T60137)
60137	**ON THE FEASIBILITY OF CALCULATING GLASS SOFTENING POINTS FROM THE MEAN ATOMIZATION ENERGY.** SHUMITSKAYA L F INORGANIC MATERIALS 6 5 779-82 1970 (ENGLISH TRANSLATION OF IZV. AKAD. NAUK SSSR, NEORG. MATER., 6 (5), 888-92, 1970; FOR ORIGINAL SEE T60136)
60142	**MECHANISM OF COLORATION OF GLASSES BY CERIUM AND TANTALUM IONS.** BOGDANOVA G S ANTONOVA S L DZHURINSKII B F IZV AKAD NAUK SSSR NEORG MATER 6 5 943-8 1970 (FOR ENGLISH TRANSLATION SEE T60143)
60143	**MECHANISM OF COLORATION OF GLASSES BY CERIUM AND TANTALUM IONS.** BOGDANOVA G S ANTONOVA S L DZHURINSKII B F INORGANIC MATERIALS 6 5 825-9 1970 (ENGLISH TRANSLATION OF IZV. AKAD. NAUK SSSR, NEORG. MATER., 6 (5), 943-8, 1970; FOR ORIGINAL SEE T60142)
60144	**PHASE COMPOSITION OF CRYSTALLIZATION PRODUCTS OF GLASSES OF THE SYSTEM SIO2 - AL2O3 - LA2O3 - TIO2.** BOGDANOVA G S KOZELSKAYA L S ZEVIN L S FLANTSBAUM I M BELYAEVA I D IZV AKAD NAUK SSSR NEORG MATER 6 5 957-63 1970 (FOR ENGLISH TRANSLATION SEE T60145)
60145	**PHASE COMPOSITION OF CRYSTALLIZATION PRODUCTS OF GLASSES OF THE SYSTEM SIO2 - AL2O3 - LA2O3 - TIO2.** BOGDANOVA G S KOZELSKAYA L S ZEVIN L S FLANTSBAUM I M BELYAEVA I D INORGANIC MATERIALS 6 5 836-40 1970 (ENGLISH TRANSLATION OF IZV. AKAD. NAUK SSSR, NEORG. MATER., 6 (5), 957-63, 1970; FOR ORIGINAL SEE T60144)
60146	**INTERACTION OF HIGH-VOLTAGE PORCELAIN WITH GLAZE AT VARIOUS TEMPERATURES.** KOCHETKOVA N F MASLENNIKOVA G N IZV AKAD NAUK SSSR NEORG MATER 6 5 963-6 1970 (FOR ENGLISH TRANSLATION SEE T60147)
60147	**INTERACTION OF HIGH-VOLTAGE PORCELAIN WITH GLAZE AT VARIOUS TEMPERATURES.** KOCHETKOVA N F MASLENNIKOVA G N INORGANIC MATERIALS 6 5 841-3 1970 (ENGLISH TRANSLATION OF IZV. AKAD. NAUK SSSR, NEORG. MATER., 6 (5), 963-6, 1970; FOR ORIGINAL SEE T SEE T60146)
60158	**THE CRYSTALLIZATION PROCESS IN GLASS OF THE COMPOSITION TL(2) O . SI O(2).** ZUBAREVA, E. P. TIKHOMIROV, G. P. YAKHKIND, A. K. IZV AKAD NAUK SSSR NEORG MATER 6 5 1021-3 1970 (FOR ENGLISH TRANSLATION SEE T60159)
60159	**THE CRYSTALLIZATION PROCESS IN GLASS OF THE COMPOSITION TL(2) O . SI O(2).** ZUBAREVA, E. P. TIKHOMIROV, G. P. YAKHKIND, A. K. INORGANIC MATERIALS 6 5 898-900 1970 (ENGLISH TRANSLATION OF IZV. AKAD. NAUK SSSR, NEROG. MAT., 6 (5), 1021-3, 1970; FOR ORIGINAL SEE T60158)
60177	**THERMOPHYSICS RESEARCH.** HELLER, G. B. GEORGE C MARSHALL SPACE FLIGHT CENTER, HUNTSVILLE, ALABAMA 143-91, 1968. (NASA-TM-X-62639, N70-18690)
60192	**HEAT CAPACITY OF COAL.** LEE A L AMER CHEM SOC DIV FUEL CHEM PEPR 12 3 19-31 1968 CA 73 122221
60195	**INTERPRETATION OF THE REFLECTIVITY BEHAVIOR OF ORE MINERALS.** BURNS R G VAUGHAN D J AMER MINERAL 55 9 1576-86 1970 CA 73 125375
60220	**DIFFUSE REFLECTANCE AND RELATIVE HARDNESS OF MIXED SALTS.** CHOUDHARI, B. P. SALT. RES. IND. 7 (1), 23-4, 1970.

TPRC Number	Bibliographic Citation
60221	THERMAL EXPANSION OF GARNETS INCLUDED IN DIAMOND. HARRIS J W MILLEDGE H J BARRON T H K MUNN R W J GEOPHYS RES 75 29 5775-92 1970 CA 73 124443
60237	A CONTRIBUTION TO THE KNOWLEDGE OF HEAT CONDUCTIVITY OF POROUS MATERIALS. PART 1. FRITZ W KUSTER W WARME-UND STOFFUBERTRAGUNG 3 3 156-68 1970
60282	THERMAL EXPANSION OF COMPOSITE MATERIALS. TUMMALA R R FRIEDBERG A L J APPL PHYS 41 13 5104-7 1970
60293	EFFECT OF REDUCED PRESSURE ON THERMAL-EXPANSION BEHAVIOR OF ROCKS AND ITS SIGNIFICANCE TO THERMAL FRAGMENTATION. THIRUMALAI K DEMOU S G J APPL PHYS 41 13 5147-51 1970
60316	IN-PILE AND UNIRRADIATED THERMAL CONDUCTIVITY OF A SINGLE-FIRED THO2 + 10 WT. PCT. UO2 /LWBR DEVELOPMENT PROGRAM./ JACOBS, D. C. BETTIS ATOMIC POWER LAB., PITTSBURGH, PA. 1-30, 1970. (WAPD-TM-601, N70-34497)
60317	THE INFLUENCE OF THE SOIL-FORMATION PROCESS ON THE COMPOSITION AND PROPERTIES OF THE DEPOSITS OF THE SEASONALLY FREEZING AND SEASONALLY THAWING LAYERS. MAKSIMOVA L N SBORNIK PO MERZLOTOVEDENIYU 5 7 87-97 1967 (FOR ENGLISH TRANSLATION SEE T60318)
60318	THE INFLUENCE OF THE SOIL-FORMATION PROCESS ON THE COMPOSITION AND PROPERTIES OF THE DEPOSITS OF THE SEASONALLY FREEZING AND SEASONALLY THAWING LAYERS. MAKSIMOVA L N COLD REGIONS RES. AND ENGR. LAB., HANNOVER, N. H. 1-13, 1970. (ENGLISH TRANSLATION OF SBORNIK PO. MERZLOTOVEDENIYU 5 (7), 87-97, 1967; FOR ORIGINAL SEE T60317) (AD-711888)
60334	SPECIMENS OF WHITE SURFACES FOR COLORIMETRY AND SPECTROPHOTOMETRY. YUSTOVA E N IVANOVA I S POKROVSKAYA G V MEASUREMENT TECHNIQUES 15 8 999-1003 1966 (ENGLISH TRANSLATION OF IZMERITELNAYA TEKHNIKA, (8), 18-21, 1966; FOR ORIGINAL SEE T44356)
60337	THERMAL EXPANSION OF FLUORSPAR AND IRON PYRITE. PRESS D C PROC INDIAN ACAD SCI 30 284-94 1949
60339	THERMAL EXPANSION OF CRYSTALS. PART V. HAEMATITE. SHARMA S S PROC INDIAN ACAD SCI 32 285-91 1950
60340	THERMAL EXPANSION OF CRYSTALS. PART VII. BARITE. SHARMA S S PROC INDIAN ACAD SCI 33 283-9 1951
60341	THERMAL EXPANSION OF CRYSTALS. PART VIII. GALENA AND PYRITE. SHARMA S S PROC INDIAN ACAD SCI 34 72-6 1951
60343	THE THERMAL EXPANSION OF CALCITE FROM ROOM TEMPERATURE UP TO 400 C. SRINIVASAN R PROC INDIAN ACAD SCI 42 A 2 81-5 1955
60348	THE THERMAL EXPANSIONS OF CERTAIN CRYSTALS WITH LAYER LATTICES. MEGAW H D PROC ROY SOC LONDON 142 A 198-214 1933
60370	IX. NOTE ON THE EXPANSION CHARACTERISTICS OF GLASSES FOR SODIUM VAPOUR LAMP SEALS. STANWORTH J E J SOC GLASS TECHNOLOGY 25 159-63 1941
60371	PROPERTIES AND USES OF TECHNICAL CERAMICS. THURNAUER H MATERIALS AND METHODS 26 87-92 1947
60406	RELATIVE THERMAL EXPANSION OF SHELL, COATING AND LINING BRICK OF THE CEMENT ROTARY KILN. YOSHII T IWAKIRI I J SOC CHEMICAL INDUSTRY JAPAN 40 5 171B-2B 1937
60407	A STUDY OF TALC PORCELAIN, I-II. KONDO S SUZUKI S J SOC CHEMICAL INDUSTRY JAPAN 39 12 470B-1B 1936
60408	RELATIVE THERMAL EXPANSION COEFFICIENTS OF SHELL, BRICK, COATING AND LINING OF A CEMENT ROTARY KILN, III. YOSHII T J SOC CHEMICAL INDUSTRY JAPAN 40 10 352B-3B 1937
60409	A STUDY OF TALC PROCELAIN, XIV XV. KONDO S SUZUKI S J SOC CHEMICAL INDUSTRY JAPAN 41 6 197B-8B 1938
60412	CERAMIC APPLICATION OF TERNARY SYSTEM, LI2O-AL2O3-SIO2. MATSUMOTO K YOSHIKI B REPORTS RESEARCH LAB ASAHI GLASS CO 5 37-47 1955
60425	THE EXPANSION AND CONTRACTION OF MOLDING SAND AT ELEVATED TEMPERATURES. DIETERT H W VALTIER F TRANS AM FOUNDRYMENS ASSOC 43 107-24 1935
60430	THE REVERSIBLE EXPANSION OF REFRACTORY MATERIALS. HODSMAN H J COBB J W GAS JOURNAL 151 184-7 1920
60434	INTERFEROMETER MEASUREMENTS OF THE THERMAL DILATATION OF GLAZED WARE. MERRITT G E PETERS C G J AM CERAM SOC 9 6 327-42 1926
60435	PROGRESS REPORT ON INVESTIGATION OF SAGGER CLAYS-SOME OBSERVATIONS AS TO THE SIGNIFICANCE OF THEIR THERMAL EXPANSIONS. II. GELLER R F HEINDL R A J AM CERAM SOC 9 9 555-74 1926
60436	THE EFFECT OF CALCINED CYANITE IN PROCELAIN BODIES. MC DOWELL S J VACHUSKA E J J AM CERAM SOC 10 1 64-72 1927
60437	BUREAU OF STANDARDS INVESTIGATION OF FELDSPAR-SECOND PROGRESS REPORT. GELLER R F J AM CERAM SOC 10 411-17 430-3 1927
60438	SOME PHYSICAL PROPERTIES OF CHEMICAL STONEWARE BODIES. MAC GEE A E J AM CERAM SOC 10 8 569-78 1927
60439	THE USE OF FUSED SILICA AS A RAW MATERIAL IN THE MANUFACTURE OF PORCELAIN. WESTMAN A E R J AM CERAM SOC 11 2 82-9 1928
60440	SOME PHYSICAL PROPERTIES OF ARTIFICIAL ALUMINOUS ABRASIVES. PURDY R C MAC GEE A E J AM CERAM SOC 11 3 192-203 1928
60441	SOME PHYSICAL PROPERTIES OF GLASS TANK BLOCK REFRACTORIES. MAC GEE A E J AM CERAM SOC 11 11 858-67 1928
60442	PROGRESS REPORT ON INVESTIGATION OF FIRECLAY BRICK AND THE CLAYS USED IN THEIR PREPARATION. HEINDL R A PENDERGAST W L J AM CERAM SOC 12 10 640-75 1929
60444	RELATION BETWEEN UNCOMBINED QUARTZ AND THERMAL EXPANSION OF CERAMIC BODIES. MORGAN W R J AM CERAM SOC 17 5 117-21 1934
60445	FORSTERITE AND OTHER MAGNESIUM SILICATES AS REFRACTORIES. II. SURVEY OF RAW MATERIALS. BIRCH R E HARVEY F A J AM CERAM SOC 18 6 177-92 1935

TPRC Number	Bibliographic Citation
60446	A DISCUSSION RELATING TO THERMAL EXPANSION METHODS FROM THEORETICAL AND PRACTICAL STANDPOINTS. KINZIE C J COMMONS C H JR J AM CERAM SOC 18 306-8 1935
60447	A COMPARISON OF THE THERMAL EXPANSION OF USED SILICA BRICK FROM AN INSULATED AND AN UNINSULATED OPEN-HEARTH FURNACE ROOF. PIERCE R H H JR AUSTIN J B J AM CERAM SOC 19 10 276-87 1936
60448	LINEAR THERMAL EXPANSION OF BETA-ALUMINA. AUSTIN J B J AM CERAM SOC 21 10 3513 1938
60449	EFFECT OF SODIUM ALUMINATE ON THE THERMAL EXPANSION OF REFRACTORY CLAY BODIES. MORGAN W R J AM CERAM SOC 22 88-90 1939
60450	EFFECT OF BORON OXIDE, PHOSPHOROUS PENTOXIDE, AND FERRIC OXIDE ON THE THERMAL EXPANSION OF A FLORIDA KAOLIN-QUARTZ MIXTURE. THOMPSON C L PARMELEE C W J AM CERAM SOC 22 5 170-2 1939
60451	THERMAL AND MOISTURE EXPANSION OF KAOLINS AND BODIES FIRED TO DIFFERENT TEMPERATURES. THIEMECKE H J AM CERAM SOC 24 2 69-75 1941
60452	THERMAL EXPANSION OF SILICATE FLUXES IN THE CRYSTALLINE AND GLASSY STATES. ORLOWSKI H J KOENIG C J J AM CERAM SOC 24 3 80-4 1941
60453	THERMAL AND MOISTURE EXPANSION OF BALL CLAYS AND BODIES FIRED TO DIFFERENT TEMPERATURES. THIEMECKE H J AM CERAM SOC 26 6 173-9 1943
60454	HERMETICALLY SEALED STEATITE. HAUSNER H H J AM CERAM SOC 29 5 123-8 1946
60456	THERMAL EXPANSION PROPERTIES OF SOME SYNTHETIC LITHIA MINERALS. HUMMEL F A J AM CERAM SOC 34 8 235-9 1951
60457	LOW-EXPANSION CORDIERITE PORCELAINS. BEALS R J COOK R L J AM CERAM SOC 35 2 53-7 1952
60458	THE ZIRCONIA-TITANIA SYSTEM. BROWN F H JR DUWEZ P J AM CERAM SOC 37 3 129-32 1954
60459	REACTION AND FIRED-PROPERTY STUDIES OF CORDIERITE COMPOSITIONS. LAMAR R S WARNER M F J AM CERAM SOC 37 12 602-10 1954
60460	SOME PROPERTIES OF THE OXIDES OF VANADIUM AND THEIR COMPOUNDS. KING B W SUBER L L J AM CERAM SOC 38 9 306-11 1955
60468	THERMAL EXPANSION OF MAGNESITE TO 1800 C. AUTHOR ANON. J. FRANKLIN INST. 740-50, 1932.
60513	BASALT MELTS FOR THE FORMING OF STAPLE FIBER. DUBROSKII V A RYCHKO V A BACHILO T M LYSYUK A G GLASS CERAMICS 25 12 733-5 1968 (ENGLISH TRANSLATION OF STEKLO KERAM., 25 (12), 18-20, 1968; FOR ORIGINAL SEE T54552)
60515	CONTROL OF THE TEMPERATURE OF PLASTIC FLOW AND THE COEFFICIENTS OF ELASTICITY AND THERMAL EXPANSION OF REFRACTORY BODIES. MONTGOMERY R J COUCH E G JR J CANADIAN CERAMIC SOC 5 20-8 1936
60516	THERMAL EXPANSION PROPERTIES OF NATURAL LITHIA MINERALS. HUMMEL F A FOOTE PRINTS 20 2 3-11 1948
60524	THE CONTROL AND APPLICATION OF THE EXPANSION CHARACTERISTICS OF METALS AND ALLOYS. SCHOLEFIELD H H INSTRUMENTS MEASUREMENTS CONF STOCKHOLM 260-7 1949
60525	THE THERMAL EXPANSION OF FIRECLAY BRICKS. WESTMAN A E R UNIV ILLINOIS ENG EXPT STATION BULL 181 1-30 1928
60526	DILATOMETER METHOD FOR DETERMINATION OF THERMAL COEFFICIENT OF EXPANSION OF FINE AND COARSE AGGREGATE. VERBECK G J HASS W E HIGHWAY RES BOARD PROC 30 187-93 1951
60535	LOW EXPANSION ALLOYS FOR GLASS-TO-METAL SEALS. KINGSTON, W. E. TRANS. AM. SOC. METALS 30, 47-67, 1972.
60557	THE REVERSIBLE THERMAL EXPANSION OF SILICA. HOULDSWORTH H S COBB J W TRANS CERAMIC SOC /ENGLAND/ 21 227-76 1921
60558	SOME CHANGES TAKING PLACE IN THE LOW-TEMPERATURE BURNING OF STOURBRIDGE FIRECLAY. II. MOORE C E TRANS CERAMIC SOC /ENGLAND/ 25 127-49 1925
60559	AN INVESTIGATION OF THE PERMANENT EXPANSION OF SILICA PRODUCTS CONTAINING CRISTOBALITE. RIGBY G R DODD A E WHITE R P GREEN A T TRANS BRITISH CERAMIC SOC 42 2 11-16 1943
60560	THE INFLUENCE OF FOREIGN MATTER ON THE THERMAL EXPANSION AND TRANSFORMATION OF SILICA. WOOD J F L HOULDSWORTH H S COBB J W TRANS CERAMIC SOC /ENGLAND/ 25 4 289-303 1925
60562	LOW EXPANSION VITRIFIED BODIES FROM INDIAN TALCS. BHUSHAN B ROY H N TRANS INDIAN CERAMIC SOC 15 2 97-113 1956
60568	PLUTONIUM FUEL DEVELOPMENT, VOLUME 5. PLUTONIUM FUEL DEVELOPMENT LAB., JAPAN POWER REACTOR AND NUCLEAR FUEL DEVELOPMENT CORP., TOKAI JAPAN 216PP., 1968. (FOR ENGLISH TRANSLATION SEE TPRC NO. 60569) (PNC-N-851-69-9005)
60569	PLUTONIUM FUEL DEVELOPMENT, VOLUME 5. PLUTONIUM FUEL DEVELOPMENT LAB., JAPAN POWER REACTOR AND NUCLEAR FUEL DEV. CORP., TOKAI, JAPAN 216PP., 1969. (ENGLISH TRANSLATION OF PNC-N-851-69-9005, 216PP., 1969., FOR ORIGINAL SEE TPRC NO. 60568) (JAPFNR-6, N70-33212)
60580	CORRECTIONS TO PLANCKS FORMULA OF BLACK-BODY RADIATION FOR SMALL CAVITIES AND ITS IMPLICATIONS ON FAR INFRARED STANDARD SOURCES. BALTES H P MURI R KNEUBUHL F K REV INT HAUTES TEMP REFRACT 7 3 192-6 1970
60590	THE DIRECT DETERMINATION OF THERMAL CONDUCTIVITY BY THE FLASH TECHNIQUE. PEGGS I D MILLS R W REV INT HAUTES TEMP REFRACT 7 3 264-7 1970 CA 74 80525
60595	MEASUREMENTS OF THE THERMAL DIFFUSIVITY AND THE TOTAL EMISSIVITY OF SOLIDS BETWEEN 1500 K. AND THEIR MELTING POINT. FAUCHER M CABANNES F ANTHONY A-M PIRIOU B SIMONATO J REV INT HAUTES TEMPER ET REFRACT 7 3 290-7 1970 CA 74 80483
60601	ULTRA-PRECISE MEASUREMENT OF THERMAL EXPANSION COEFFICIENTS. BRADFORD, J. N. UNIVERSITY OF ASIZONA, M.S. THESIS 1-27 1969
60608	HEAT TRANSFER OF CONDENSABLE VAPOUR DIFFUSING THROUGH POROUS MEDIA. KITO M SUGIYAMA S INT J HEAT MASS TRANSFER 13 11 1705-13 1970

TPRC Number	Bibliographic Citation
60612	THERMAL AND MECHANICAL CHARACTERIZATION OF ADVANCED GRAPHITIC MATERIALS. PEARS, C. D. GRAPHITIC MATERIALS FOR ADVANCED RE-ENTRY VECHICLES, PART 1 (PT. 1), 575-726, 1970. (AFML-TR-70-133)
60650	STUDY OF FOAM BRICK SINTERING IN A DILATOMETER. SAAKYAN E R DARVINYAN M V IZV AKAD NAUK SSSR NEORG MATER 6 6 1161-4 1970 (FOR ENGLISH TRANSLATION SEE T60651)
60651	STUDY OF FOAM BRICK SINTERING IN A DILATOMETER. SAAKYAN E R DARVINYAN M V INORGANIC MATERIALS 6 6 1011-13 1970 (ENGLISH TRANSLATION OF IZV. AKAD. NAUK SSSR, NEORG. MATER., 6 (6), 1161-4, 1970; FOR ORIGINAL SEE T60650)
60669	DIFFUSE TRANSMITTANCE OF GLASS. FOSTER, W. H., JR. STEARNS, E. I. J. OPT. SOC. AMER. 61 (1), 60-2, 1971.
60704	THERMAL CONDUCTIVITY OF FINES FROM APOLLO 11. CREMERS, C. J. BIRKEBAK, R. C. DAWSON, J. P. PROC. APOLLO 11 LUNAR SCI. CONF. 3, 2045-50, 1970.
60726	THERMAL CONDUCTIVITY AT VARIOUS ATMOSPHERES AND PRESSURES. PROVOST, G. ZOMBAS, P. KIEHL, J. P. BULL. SOC. FR. CERAM. (87), 49-57, 1970.
60729	LATTICE THERMAL EXPANSION PERPENDICULAR TO THE LAYER PLANES OF PYROCARBONS PRODUCED IN A FLUIDIZED LEAD. PELLEGRINI G CARBON 8 4 497-502 1970 CA 74 7364
60736	THERMAL EXPANSION OF SODIUM-SILICATE GLASSES CONTAINING FLUORIDES. VARGIN V V KRASOTKINA N I DOKL AKAD NAUK SSR 108 6 1133-6 1956 (FOR ENGLISH TRANSLATION SEE T37574)
60745	HEAT CAPACITY OF LIGNITES AND HEAT OF LIGNITE PYROLYSIS. AGROSKIN A A GONCHAROV E I MAKEEV L A KHIM TVERD TOPL 5 118-20 1970 CA 74 5363
60753	EFFECT OF SOME OXIDE ADDITIVES ON THE PROPERTIES OF GLAZES FOR HIGH-VOLTAGE PORCELAIN. MASLENNIKOVA, G. N. KOCHETKOVA, N. F. STEKLO KERAM. 27 (9), 28-30, 1970. (FOR ENGLISH TRANSLATION SEE TPRC NO. 66787)
60768	THE THERMAL EXPANSION OF SOME COMPOUNDS AND ITS ESTIMATION BY MEANS OF THE GRUNEISEN RULE. HULSMANN O BILTZ W ZEIT ANORG ALLGEM CHEM 219 357-66 1934
60772	THE INFLUENCE OF A COPPER ADDITION ON THE EXPANSION IRON-NICKEL-COBALT ALLOYS. RINGPFEIL H HENKEL O NEUE HUTTE 6 5 310-16 1961
60786	EFFECT OF REPEATED HEATINGS ON THE MECHANICAL STRENGTH OF HIGH-TENSION INSULATOR PORCELAINS. GELLER R F BULL AM CERAM SOC 12 18-25 1933
60793	ELECTRICAL REFRACTORY PORCELAIN BODIES CONTAINING MAGNESIUM OXIDE. HIGGINS J M MONTGOMERY R J J CAN CERAMIC SOC 1 59-60 1932
60799	CHANGES IN THE COEFFICIENTS OF LINEAR EXPANSION OF QUARTZES FROM TIN ORE DEPOSITS AS A FUNCTION OF TEMPERATURE. USPENSKAYA, A. B. IZV. VYSSH. UCHEB. ZAVED., GEOL. RAZVED. 13 (7), 26-30, 1970. (FOR ENGLISH TRANSLATION SEE TPRC NO. 65498)
60815	THERMAL CONDUCTIVITY OF REFRACTORY MATERIALS. PUSTOVALOV V V TEPLOPROVODNOST OGNEUPOROV 1-98 1966 (FOR ENGLISH TRANSLATION SEE T60816)
60816	THERMAL CONDUCTIVITY OF REFRACTORY MATERIALS. PUSTOVALOV V V INDIAN NATL. SCI. DOC. CENTRE, NEW DELHI 1-98, 1970. (ENGLISH TRANSLATION OF TEPLOPROVODNOST OGNEUPOROV. IZDATELSTVO METALLURGIYA, 98PP., 1966; FOR ORIGINAL SEE T60815) (NASA-TT-F-516, TT-68-50635, N71-15034)
60833	PRODUCTION OF HIGH-ALUMINA INSULATING REFRACTORIES. RABINOVICH M A GRIGOREV I V KARASEVA M D OGNEUPORY 5 14-16 1970 (FOR ENGLISH TRANSLATION SEE T60834)
60834	PRODUCTION OF HIGH-ALUMINA INSULATING REFRACTORIES. RABINOVICH M A GRIGOREV I V KARASEVA M D REFRACTORIES 5 280-2 1970 (ENGLISH TRANSLATION OF OGNEUPORY, (5), 14-6, 1970; FOR ORIGINAL SEE T60833)
60836	THERMAL DIFFUSIVITY OF ALUMINOSILICATE REFRACTORIES IN THE RANGE 200-1600 C. LITOVSKII E YA LANDA YA A MILSHENKO R S REFRACTORIES 5 284-6 1970 (ENGLISH TRANSLATION OF OGNEUPORY, (5), 17-9, 1970; FOR ORIGINAL SEE T59046)
60837	RAMMING BODIES FOR THE BOTTOMS OF SOAKING PIT FURNACES. PITAK N V ANSIMOVA T A OGNEUPORY 5 32-6 1970 (FOR ENGLISH TRANSLATION SEE T60838)
60838	RAMMING BODIES FOR THE BOTTOMS OF SOAKING PIT FURNACES. PITAK N V ANSIMOVA T A REFRACTORIES 5 299-302 1970 (ENGLISH TRANSLATION OF OGNEUPORY, (5), 32-6, 1970; FOR ORIGINAL SEE T60837)
60839	ANORTHITE INSULATING REFRACTORY. PIROGOV A A RAKINA V P MIRAKYAN M M VOLKOV N V OGNEUPORY 5 36-40 1970 (FOR ENGLISH TRANSLATION SEE T60840)
60840	ANORTHITE INSULATING REFRACTORY. PIROGOV A A RAKINA V P MIRAKYAN M M VOLKOV N V REFRACTORIES 5 303-7 1970 (ENGLISH TRANSLATION OF OGNEUPORY, (5), 36-40 1970; FOR ORIGINAL SEE T60839)
60841	ELECTRIC INSULATING MATERIAL FOR THERMOELECTRODES WORKING IN VACUUM. KAMENETSKII A B GULKO N V GLADKAYA N V OGNEUPORY 5 50-3 1970 (FOR ENGLISH TRANSLATION SEE T60842)
60842	ELECTRIC INSULATING MATERIAL FOR THERMOELECTRODES WORKING IN VACUUM. KAMENETSKII A B GULKO N V GLADKAYA N V REFRACTORIES 5 319-22 1970 (ENGLISH TRANSALTION OF OGNEUPORY, (5), 50-3, 1970; FOR ORIGINAL SEE T60841)
60843	SELECTING STABLE MORTARS FOR LINING INTERMEDIATE LADLES OF CONTINUOUS STEEL CASTING PLANTS. FRIDMAN L YA KHOSID G M ORLOV V A RYAZANTSEV V D NIKOKOSHEV N T OGNEUPORY 6 20-5 1970 (FOR ENGLISH TRANSLATION SEE T60844)
60844	SELECTING STABLE MORTARS FOR LINING INTERMEDIATE LADLES OF CONTINUOUS STEEL CASTING PLANTS FRIDMAN L YA KHOSID G M ORLOV V A RYAZANTSEV V D NIKOKOSHEV N T REFRACTORIES 6 355-9 1970 (ENGLISH TRANSLATION OF OGNEUPORY, (6), 20-5, 1970; FOR ORIGINAL SEE T60843)
60851	LINEAR EXPANSION OF ORDINARY AND BLACK DINAS BRICK DURING HEATING. BUDNIKOV P P MYULLER V E DOMEZ 10 24-8 1931 (FOR ENGLISH TRANSLATION SEE T44828)
60852	EXPANSION OF CHASOV-YARSK FIRE BRICK AT HIGH TEMPERATURES. BUDNIKOV P P MYULLER V E DOMEZ 12 11-4 1931 (FOR ENGLISH TRANSLATION SEE T48695)

TPRC Number	Bibliographic Citation
60872	**CHANGE IN THE MODULUS OF ELASTICITY AND THERMAL EXPANSION RATIO OF PORCELAIN DEPENDENT ON THE INTRODUCTION OF PERLITE.** KHIZANISHVILI, I. G. MAMALADZE, R. A. SOOBSHCH. AKAD. NAUK GRUZ. SSR 58 (3), 597–600, 1970. (FOR ENGLISH TRANSLATION SEE TPRC NO. 64431)
60876	**ORDER AND DISORDER IN LITHIUM ALUMINO-SILICATES.** KONDRATEV YU N IZV AKAD NAUK SSSR NEORG MAT 1 8 1395–8 1965 (FOR ENGLISH TRANSLATION SEE T51723)
60880	**ELECTRIC AND STRUCTURAL PROPERTIES OF THE VANADIUM OXIDE-BISMUTH OXIDE SYSTEM.** STANESCU, L. GOCAN, S. ARDELEAN, I. MAN, S. STUD. UNIV. BABES-BOLYAI, SER. CHEM. 15 (1), 37–41, 1970.
60881	**STRUCTURAL AND ELECTRICAL DATA CONCERNING THE VANADIUM(V) OXIDE-ANTIMONY(III) OXIDE SYSTEM.** GOCAN, S. STANESCU, L. ARDELEAN, I. STUD. UNIV. BABES-BOLYAI, SER. PHYS. 15 (1), 39–44, 1970.
60884	**THERMAL EXPANSION OF FULLY AND PARTIALLY STABILIZED ZIRCONIUM DIOXIDE AT 1000–2300 DEGREES.** SHAKHTIN D M LEVINTOVICH E V ELISEEVA G G KARAULOV A G RUDYAK I N TEPLOFIZ SVOISTVA TVERD TEL VYS TEMP TR VSES KONF 1966 1 94–8 1969 CA 73 134155
60888	**HEAT CAPACITY OF CARBON MATERIALS.** GUTNOVA L B REKOV A I SPIRIDONOV E G TEPLOFIZ SVOISTVA TVERD TEL VYS TEMP TR VSES KONF 1966 1 419–24 1969 CA 73 92295
60900	**GLASS AND MINERAL FIBERS AS THERMAL INSULATORS.** SACCHI, A. TERMOTECNICA 24 (6), 262–8, 1970.
60905	**CHANGE IN THE PROPERTIES OF GLASS AS A RESULT OF EXTERNAL THERMOMECHANICAL ACTION.** SIL'VESTROVICH, S. I. KUTUKOV, A. S. TR. MOSK. KHIM.-TEKHNOL. INST. (63), 21–3, 1969.
60916	**MECHANICAL AND THERMOPHYSICAL PROPERTIES OF RAW MATERIALS.** MALYSHEV, A. P. TSEMENT 34 (2), 5–7, 1968.
60920	**INFRARED SPECTRA AND CRYSTAL CHEMISTRY OF MINERALS.** POVARENNYKH, A. S. VISN. AKAD. NAUK UKR. RSR (8), 24–31, 1969.
60925	**INFRARED AND RAMAN SPECTRA AND STRUCTURE OF SODIUM OXIDE-NIOBIUM PENTOXIDE-SILICON DIOXIDE SYSTEM GLASSES.** SIDOROV, T. A. ZH. PRIKL. SPEKTROSK. 12 (6), 1124–5, 1970. (FOR ENGLISH TRANSLATION SEE TPRC NO. 76917)
60936	**INFRARED DRYING OF MINERAL WOOL PRODUCTS.** KLYUCHAREV A E TIKHONOV N A KARDASHEV G A MIKHAILOV P E INZH FIZ ZH 12 1 99–103 1967 (FOR ENGLISH TRANSLATION SEE T60937)
60937	**INFRARED DRYING OF MINERAL WOOL PRODUCTS.** KLYUCHAREV A E TIKHONOV N A KARDASHEV G A MIKHAILOV P E J ENGINEERING PHYS 12 1 55–8 1967 (ENGLISH TRANSLATION OF INZH. FIZ. ZH., 12 (1), 99–103, 1967; FOR ORIGINAL SEE T60936)
60940	**STRUCTURE OF GLASSES IN THE SILICON DIOXIDE-TITANIUM DIOXIDE SYSTEM.** OSTRIZHKO D G PAVLOVA G A IZV AKAD NAUK SSSR NEORG MATER 6 1 74–7 1970 CA 72 114575 (FOR ENGLISH TRANSLATION SEE T60941)
60941	**STRUCTURE OF GLASSES IN THE SYSTEM SIO2-TIO2.** OSTRIZHKO D G PAVLOVA G A INORGANIC MATERIALS 6 1 58–61 1970 (ENGLISH TRANSLATION OF IZV. AKAD. NAUK SSSR, NEORG. MATER., 6 (1), 74–7, 1970; FOR ORIGINAL SEE T60940)
60944	**MELT TEMPERATURE FIELDS DURING GROWTH OF SINGLE CRYSTALS.** PERILOVA V E BODYACHEVSKII S V AVVAKUMOVA L A DERMAN A S IZV AKAD NAUK SSSR NEORG MATER 6 1 100–3 1970 (FOR ENGLISH TRANSLATION SEE T60945)
60945	**MELT TEMPERATURE FIELDS DURING GROWTH OF SINGLE CRYSTALS.** PERILOVA V E BODYACHEVSKII S V AVVAKUMOVA L A DERMAN A S INORGANIC MATERIALS 6 1 80–2 1970 (ENGLISH TRANSLATION OF IZV. AKAD. NAUK SSSR, NEORG. MATER., 6 (1), 100–3, 1970; FOR ORIGINAL SEE T60944)
60960	**ACTION OF IRRADIATION ON SILICATE GLASS CONTAINING SILVER.** BORGMAN V A IZV AKAD NAUK SSSR NEORG MATER 6 2 336–9 1970 (FOR ENGLISH TRANSLATION SEE T60961)
60961	**ACTION OF IRRADIATION ON SILICATE GLASS CONTAINING SILVER.** BORGMAN V A INORGANIC MATERIALS 6 2 290–3 1970 (ENGLISH TRANSLATION OF IZV. AKAD. NAUK SSSR, NEORG. MATER., 6 (2), 336–9, 1970; FOR ORIGINLA SEE T60960)
60972	**INTERACTION OF HIGH-VOLTAGE PORCELAIN WITH A GLAZE.** MASLENNIKOVA G N KOCHETKOVA N F IZV AKAD NAUK SSSR NEORG MATER 6 3 542–6 1970 (FOR ENGLISH TRANSLATION SEE T60973)
60973	**INTERACTION OF HIGH-VOLTAGE PORCELAIN WITH A GLAZE.** MASLENNIKOVA G N KOCHETKOVA N F INORGANIC MATERIALS 6 3 476–80 1970 (ENGLISH TRANSLATION OF IZV. AKAD. NAUK SSSR, NEORG. MATER., 6 (3), 542–6, 1970; FOR ORIGINAL SEE T60972)
60982	**MECHANISM OF THE COLORATION OF CERIUM-CONTAINING GLASSES.** BOGDANOVA G S DZHURINSKII B F ANTONOVA S L IZV AKAD NAUK SSSR NEORG MATER 6 4 776–80 1970 (FOR ENGLISH TRANSLATION SEE T60983)
60983	**MECHANISM OF THE COLORATION OF CERIUM-CONTAINING GLASSES.** BOGDANOVA G S DZHURINSKII B F ANTONOVA S L INORGANIC MATERIALS 6 4 680–4 1970 (ENGLISH TRANSLATION OF IZV. AKAD. NAUK SSSR, NEORG. MATER., 6 (4), 776–80, 1970; FOR ORIGINAL SEE T60982)
60986	**EFFECT OF ALKALI METAL CATIONS ON FORMATION OF CERAMIC PIGMENTS IN CAO-SNO2-SIO2 SYSTEM.** TUMANOV S G FILIPPOVA E A IZV AKAD NAUK SSSR NEORG MATER 6 4 814–17 1970 (FOR ENGLISH TRANSLATION SEE T60987)
60987	**EFFECT OF ALKALI METAL CATIONS ON FORMATION OF CERAMIC PIGMENTS IN CAO-SNO2-SIO3 SYSTEM.** TUMANOV S G FILIPPOVA E A INORGANIC MATERIALS 6 4 714–16 1970 (ENGLISH TRANSLATION OF IZV. AKAD. NAUK SSSR, NEORG. MATER., 6 (4), 814–7, 1970; FOR ORIGINAL SEE T60986)
60998	**EQUILIBRIUM DIAGRAMS OF ALKALI TELLURITE SYSTEMS AND SOME PROPERTIES OF GLASSES FORMED BY THEM.** YAKHKIND A K MARTYSHCHENKO N S IZV AKAD NAUK SSSR NEORG MATER 6 8 1459–64 1970 (FOR ENGLISH TRANSLATION SEE T60999)
60999	**EQUILIBRIUM DIAGRAMS OF ALKALI TELLURITE SYSTEMS AND SOME PROPERTIES OF GLASSES FORMED BY THEM.** YAKHKIND A K MARTYSHCHENKO N S INORGANIC MATERIALS 6 8 1284–8 1970 (ENGLISH TRANSLATION OF IZV. AKAD. NAUK SSSR, NEORG. MATER., 6 (8), 1459–64; FOR ORIGINAL SEE T60998)
61002	**SYNTHESIS OF SPINEL-TYPE CERAMIC PIGMENTS OF THE SYSTEM MNO - AL2O3 - CR2O3.** TUMANOV S G PYRKOV V P BYSTRIKOV A S IZV AKAD NAUK SSSR NEORG MATER 6 8 1499–502 1970 (FOR ENGLISH TRNASLATION SEE T61003)

TPRC Number	Bibliographic Citation
61003	**SYNTHESIS OF SPINEL-TYPE CERAMIC PIGMENTS OF THE SYSTEM MNO - AL2O3 - CR2O3.** TUMANOV S G PYRKOV V P BYSTRIKOV A S INORGANIC MATERIALS 6 8 1321-3 1970 (ENGLISH TRANSLATION OF IZV. AKAD. NAUK SSSR, NEORG. MATER., 6 (8), 1499-1502, 1970; FOR ORIGINAL SEE T61002)
61009	**PEAK-TIME METHOD FOR MEASURING THERMAL DIFFUSIVITY OF SMALL SOLID SPECIMENS.** HARMATHY, T. Z. AICHE J. 17 (1), 198-201, 1971.
61027	**STUDY OF THE THERMAL EXPANSION OF REFRACTORY OXIDES AT HIGH TEMPERATURES.** SHAKHTIN D M LEVINTOVICH E V PIVOVAR T L ELISEEVA G G KARAULOV A G RUDYAK I N FOREIGN TECHNOLOGY DIVISION 1-17, 1970. (ENGLISH TRANSLATION OF SB. NAUCH. TR., UKR. NAUCH. ISSLED. INST. OGENOUPOROV, 57 (10), 163-71, 1967; FOR ORIGINAL SEE T53201) (FTD-HT-23-343-70)
61037	**FROST INVESTIGATION 1944-1945. APPENDIX 13. REPORT ON LABORATORY TESTS ON FROST PENETRATION AND THERMAL CONDUCTIVITY OF COHESIONLESS SOILS.** ARCTIC CONSTRUCTION AND FROST EFFECTS LAB BOSTON MASS ARCTIC CONSTRUCTION AND FROST EFFECTS LAB., BOSTON 1-45, 1945. (ACFEL-TR-6-APPENDIX-13, AD-712471)
61071	**COMPARISON OF THE SPECTRAL REFLECTIVITY OF MARS WITH OXIDIZED METEORITIC MATERIAL.** GIBSON E K JR ICARUS 13 1 96-9 1970 CA 74 15043
61089	**SPECTRAL EMISSIVITY OF OXIDES AND CONCRETES IN THE VISIBLE REGION OF THE SPECTRUM.** ABRAMOV, A. S. BARYKIN, B. M. ROMANOV, A. I. SPIRIDONOV, E. G. MATER. KANALA MGD MAGNITOGIDRODINAMICHESKII-GENERATORA 119-28, 1969. (FOR ENGLISH TRANSLATION SEE TPRC NO. 67864)
61090	**EMISSIVITY OF OXIDES AND CONCRETES IN THE INFRARED REGION OF THE SPECTRUM.** ABRAMOV, A. S. BARYKIN, B. M. ROMANOV, A. I. MATER. KANALA MGD MAGNITOGIDRODINAMICHESKII- SPIRIDONOV, E. G. 129-34, 1969. (FOR ENGLISH TRANSLATION SEE TPRC NO. 67865)
61097	**THERMAL CONDUCTIVITY OF SOME OXYGEN COMPOUNDS OF ZIRCONIUM AND HAFNIUM.** VISHNEVSKII, I. I. SKRIPAK, V. N. OGNEUPORY 35 (11), 16-8, 1970. (FOR ENGLISH TRANSLATION SEE TPRC NO. 67215)
61099	**REMOTE SENSING OF LUNAR SURFACE MINERALOGY. IMPLICATIONS FROM VISIBLE AND NEAR - INFRARED RELECTIVITY OF APOLLO 11 SAMPLES.** ADAMS, J. B. MC CORD, T. B. PROC. APOLLO 11 (ELEVEN) LUNAR SCI. CONF. 3, 1937-45, 1970.
61103	**MICROSCOPY BY USE OF TRANSMITTED POLARIZED LIGHT (IN THE VISIBLE PART OF THE SPECTRUM).** CHERKASOV, YU. A. SOVREM. METODY MINERAL. ISSLED. 1, 19-77, 1969.
61167	**THERMAL CONDUCTIVITY AND SPECIFIC HEAT OF PERICLASE-SPINEL AND MAGNESITE-CHROMITE ROOF REFRACTORIES.** SHAKHTIN D M ELISEEVA G G KOVALEV A I ADOLF E M MOZGOVOI A N KOTIK P L REFRACTORIES 7 412-14 1967 (ENGLISH TRANSLATION OF OGNEUPORY, (7), 12-3, 1970; FOR ORIGINAL SEE T59471)
61168	**THERMAL SHOCK RESISTANT INSULATING REFRACTORIES.** TSIBIN I P SYREISHCHIKOV YU D OGNEUPORY 7 57 1970 (FOR ENGLISH TRANSLATION SEE T61169)
61169	**THERMAL SHOCK RESISTANT INSULATING REFRACTORIES.** TSIBIN I P SYREISHCHIKOV YU D REFRACTORIES 7 458 1970 (ENGLISH TRANSLATION OF OGNEUPORY, (7), 57, 1970; FOR ORIGINAL SEE T61168)
61170	**EXPERIENCE WITH THE PRODUCTION OF INSULATING REFRACTORIES FROM OXIDES.** KRASOTIN K A MINKOV D B VINOGRADOVA L V MAKAROVA T S LOGACHEVA N S OGNEUPORY 8 5-7 1970 (FOR ENGLISH TRANSLATION SEE T61171)
61171	**EXPERIENCE WITH THE PRODUCTION OF INSULATING REFRACTORIES FROM OXIDES.** KRASOTIN K A MINKOV D B VINOGRADOVA L V MAKAROVA T S LOGACHEVA N S REFRACTORIES 8 470-2 1970 (ENGLISH TRANSLATION OF OGNEUPORY, (8), 5-7, 1970; FOR ORIGINAL SEE T61170)
61172	**THERMAL EXPANSION OF MAGNESIA ROOF REFRACTORIES.** SHAKHTIN D M ELISEEVA G G KOVALEV A I ADOLF E M MOZGOVOI A N KOTIK P L OGNEUPORY 8 12-17 1970 (FOR ENGLISH TRANSLATION SEE T61173)
61173	**THERMAL EXPANSION OF MAGNESIA ROOF REFRACTORIES.** SHAKHTIN D M ELISEEVA G G KOVALEV A I ADOLF E M MOZGOVOI A N KOTIK P L REFRACTORIES 8 478-81 1970 (ENGLISH TRANSLATION OF OGNEUPORY, (8), 12-7 1970; FOR ORIGINAL SEE T61172)
61176	**THERMAL EXPANSION OF SOME NICKEL AND COBALT SPINELS AND THEIR SOLID SOLUTIONS.** ZAPLATYNSKY, I. LEWIS RESEARCH CENTER, CLEVELAND, OHIO 14PP., 1971. (NASA-TN-D-6174)
61192	**THERMAL CONSIDERATIONS OF A LANDED VEHICLE ON THE SURFACE OF MARS.** ROSENBERG M J PROGRESS ASTRONAUT AERON 23 491-514 1970
61193	**MARS LANDER THERMAL CONTROL SYSTEM PARAMETRIC STUDIES.** TRACEY T R MOREY T F PROGRESS ASTRONAUT AERON 23 515-46 1970
61200	**REFLECTION SPECTRA OF LEUCOSAPPHIRE IN THE INFRARED REGION AND THEIR DEPENDENCE ON VARIOUS FACTORS.** DUTOVA, K. P. SPEKTROSK., TR. SIB. SOVESHCH., 4TH 1965 97-9, 1969.
61201	**STRUCTURE OF POTASSIUM SILICATE GLASSES ACCORDING TO THE INFRARED SPECTROSCOPIC DATA AND THE RELATIONS BETWEEN THE INFRARED SPECTRA AND VARIOUS PHYSICOCHEMICAL PROPERTIES.** CHEBOTAREVA, T. E. SPEKTROSK., TR. SIB. SOVESHCH., 4TH 1965 99-101, 1969.
61203	**INFRARED SPECTRA OF CRYSTALLINE AND GLASSY SILICATES OF THE LEAD OXIDE-SILICON DIOXIDE SYSTEM.** SMIRNOVA, E. V. SPEKTROSK., TR. SIB. SOVESHCH., 4TH 1965 212-5, 1969.
61210	**THERMAL CONDUCTIVITY OF ORDERED FIBROUS SYSTEMS.** DUL'NEV, G. N. ZARICHNYAK, YU. P. MURATOVA, B. L. TEPLO- MASSOPERENOS, DOKL. VSES. SOVESHCH., 3RD 7, 82-91, 1968. (FOR ENGLISH TRANSLATION SEE T49616)
61216	**THERMOMETRIC DETERMINATION OF THERMOPHYSICAL CHARACTERISTICS.** GERASHCHENKO, O. A. GRISHCHENKO, T. G. PILIPENKO, A. M. FEDOROV, V. G. TEPLO-MASSOPERENOS, DOKL. VSES. SOVESHCH., 3RD 7, 261-73, 1968. (FOR ENGLISH TRANSLATION SEE T56570)
61217	**MEASUREMENT OF THE THERMAL CONDUCTIVITY OF MATERIALS WITH LOW THERMAL CONDUCTIVITY.** PETROV, I. N. TEPLO-MASSOPERENOS, DOKL. VSES. SOVESHCH., 3RD 7, 361-8, 1968. (FOR ENGLISH TRANSLATION SEE T56578)
61220	**DETERMINATION OF EFFECTIVE THERMOPHYSICAL CHARACTERISTICS OF FIRECLAY CERAMICS IN VARIOUS GAS MEDIA.** FRAIMAN, YU. E. TEPLO-MASSOPERENOS TEPL. SVOISTVA MATER. 70-6, 1969.
61231	**USE OF INFRARED SPECTROSCOPY FOR DETERMINING THE NATURE OF HIDDEN INCLUSIONS IN MICA.** RESHETNIKOVA, A. K. ZH. PRIKL. SPEKTROSK. 12 (4), 749-51, 1970. (FOR ENGLISH TRANSLATION SEE TPRC NO. 76916)

TPRC Number	Bibliographic Citation

61233 **THERMAL CONDUCTIVITY - NONMETALLIC SOLIDS.**
TOULOUKIAN, Y. S. POWELL, R. W. HO, C. Y.
KLEMENS, P. G.
THERMOPHYSICAL PROPERTIES OF MATTER - THE TPRC DATA SERIES
2, 1302 PP., 1970.

61236 **SPECIFIC HEAT - NONMETALLIC SOLIDS.**
TOULOUKIAN, Y. S. BUYCO, E. H.
THERMOPHYSICAL PROPERTIES OF MATTER - THE TPRC DATA SERIES
5, 1737PP., 1970.

61242 **THERMAL CONDUCTIVITY MEASUREMENTS OF SOME APULIAN LIMESTONES BY THE CUT CORE METHOD.**
MONGELLI F
ASS GEOFIS ITAL ATTI CONV ANNU
18 1 137-54 1969 CA 74 24351

61257 **EFFECT OF FILLERS ON THE THERMOPHYSICAL PROPERTIES OF GLASS FIBER - REINFORCED PLASTICS.**
KIRILLOV, V. N. EFIMOV, V. A. KOZIN, V. I.
ABLEKOVA, Z. P. KRASNOV, L. L.
TIKHOMIROVA, R. S.
PLAST. MASSY
(11), 38-40, 1970.
(FOR ENGLISH TRANSLATION SEE TPRC NO. 67223)

61258 **EFFECT OF FILLERS ON THE THERMAL CONDUCTIVITY OF POLYETHYLENE.**
DASHKO, N. M. NOVICHENOK, L. N.
SPORYAGIN, E. A.
PLAST. MASSY
(11), 45-7, 1970.
(FOR ENGLISH TRANSLATION SEE TPRC NO. 67224)

61281 **EPR OF VANADIUM TETRAPOSITIVE IONS IN CRISTOBALITE.**
GRUNIN, V. S.
FIZ. TVERD. TELA
12 (8), 2234-8, 1970.
(FOR ENGLISH TRANSLATION SEE TPRC NO. 61282)

61282 **EPR OF VANADIUM TETRAPOSITIVE IONS IN CRISTOBALITE.**
GRUNIN, V. S.
SOV. PHYS. - SOLID STATE
12 (8), 1785-8, 1971.
(ENGLISH TRANSLATION OF FIZ. TVERD. TELA, 12 (8), 2234-8, 1970; FOR ORIGINAL SEE TPRC NO. 61281)

61303 **ELECTRON PARAMAGNETIC RESONANCE OF VANADIUM, MANGANESE DIPOSITIVE IONS AND FERRIC ION, AND OPTICAL SPECTRA OF VANADIUM TRIPOSITIVE IONS IN BLUE ZOISITE.**
TSANG, T. GHOSE, S.
J. CHEM. PHYS.
54 (3), 856-62, 1971.

61316 **PRODUCTION TECHNOLOGY AND PROPERTIES OF OXIDE REFRACTORIES WITH GRANULAR STRUCTURE.**
POLUBOYARINOV D N BALKEVICH V L LEMESHEV V G
MINKOV D B MAKAROVA T S VINOGRADOVA L V
OGNEUPORY
1 11-14 1970
(FOR ENGLISH TRANSLATION SEE T61317)

61317 **PRODUCTION TECHNOLOGY AND PROPERTIES OF OXIDE REFRACTORIES WITH GRANULAR STRUCTURE.**
POLUBOYARINOV D N BALKEVICH V L LEMESHEV V G
MINKOV D B MAKAROVA T S VINOGRADOVA L V
REFRACTORIES
1 12-14 1970
(ENGLISH TRANSLATION OF OGNEUPORY, (1), 11-4, 1970; FOR ORIGINAL SEE T61316)

61318 **LINING FIREBOXES IN A HIGH-PRESSURE STEAM BOILER.**
KARKLIT A K KRASOTKINA N I PILDISH V G
MALINOVSKII S V
OGNEUPORY
2 18-23 1970
(FOR ENGLISH TRANSLATION SEE T61319)

61319 **LINING FIREBOXES IN A HIGH-PRESSURE STEAM BOILER.**
KARKLIT A K KRASOTKINA N I PILDISH V G
MALINOVSKII S V
REFRACTORIES
2 86-90 1970
(ENGLISH TRANSLATION OF OGNEUPORY, (2), 18-23, 1970; FOR ORIGINAL SEE T61318)

61329 **USE OF SCATTERING THEORY TO INTERPRET OPTICAL DATA FOR ENAMELS.**
EPPLER, R. A.
J. AMER. CERAM. SOC.
54 (2), 116-20, 1971.

61349 **ABSORBING CENTERS IN LASER MATERIALS.**
BENNETT, H. S.
J. APPL. PHYS.
42 (2), 619-30, 1971.

61361 **REFLECTANCE OF ALUMINUM OVERCOATED WITH MAGNESIUM FLUORIDE AND LITHIUM FLUORIDE IN THE WAVELENGTH REGION FROM 1600 TO 300 ANGSTROMS AT VARIOUS ANGLES OF INCIDENCE.**
HUNTER, W. R. OSANTOWSKI, J. F. HASS, G.
APPL. OPT.
10 (3), 540-4, 1971.

61369 **SUPER- AND SUB-SPECULAR MAXIMA IN THE ANGULAR DISTRIBUTION OF POLARIZED RADIATION REFLECTED FROM ROUGHENED DIELECTRIC SURFACES.**
SMITH, A. M. MULLER, P. R.
ARO, INC., ARNOLD AIR FORCE STATION, TENNESSEE
31PP. 1971.
(AEDC-TR-70-286, ARO-VKF-TR-70-286)

61376 **THE THERMAL CONDUCTIVITY OF SAPPHIRE BETWEEN 0.4 AND 4 DEGREES K.**
WOLFMEYER, M. W. DILLINGER, J. R.
PHYS. LETT.
34 A (4), 247-8, 1971.

61382 **ELECTRICALLY CONDUCTING GLASS FIBERS.**
PROVANCE, J. D. HUEBNER, J. S.
J. AMER. CERAM. SOC.
54 (3), 147-51, 1971.

61383 **EFFECT OF WATER CONTENT ON ELECTRICAL RESISTIVITY OF SODIUM OXIDE-SILICON OXIDE GLASSES.**
MARTINSEN, W. E. MC GEE, T. D.
J. AMER. CERAM. SOC.
54 (3), 175-6, 1971.

61396 **METHOD OF MEASURING THERMAL CONDUCTIVITY OF ELECTRICALLY CONDUCTING REFRACTORY MATERIALS.**
MIHAILOVIC, Z. NAUDET, G.
REV. PHYS. APPL.
5 (6), 869-75, 1970.

61440 **ENERGY STATES OF MO /+3/ IN GLASS.**
BOKIN N M KARAPETYAN G O KOLOBKOV V P
MOKEEVA G A KHUDOLEEV A G
OPTIKA I SPEKTROSKOPIYA
29 3 608-10 1970
(FOR ENGLISH TRANSLATION SEE T61441)

61441 **ENERGY STATES OF MO (+3) IN GLASS.**
BOKIN N M KARAPETYAN G O KOLOBKOV V P
MOKEEVA G A KHUDOLEEV A G
OPTICS SPECTROSCOPY
29 3 325-6 1970
(ENGLISH TRANSLATION OF OPT. I SPEKTROSKOPIYA, 29 (3), 608-10, 1970; FOR ORIGINAL SEE T61440)

61457 **RESEARCH PROGRESS AT THE AEROSPACE ENVIRONMENTAL FACILITY- A SUMMARY FOR 1969.**
ARNOLD ENGINEERING DEV. CTR., TULLAHOMA, TENN.
ARNOLD ENGINEERING DEV. CTR., TULLAHOMA, TENN.
102PP., 1971.
(AEDC-TR-70-93, ARO-VKF-TR-70-93)

61511 **ABSORPTION AND SCATTERING LOSSES IN GLASSES AND FIBERS FOR LIGHT GUIDANCE.**
JACOBSEN, A. NEUROTH, N. REITMAYER, F.
J. AMER. CERAM. SOC.
54 (4), 186-7, 1971.

61512 **OPTICAL ABSORPTION SPECTRA OF CHROMIUM TRIPOSITIVE IONS IN MAGNESIUM ALUMINATE SPINELS.**
REED, J. S.
J. AMER. CERAM. SOC.
54 (4), 202-4, 1971.

61518 **THE ROLE OF NATURAL CONVECTION IN LADLES AS AFFECTING TUNDISH TEMPERATURE CONTROL IN CONTINUOUS CASTING.**
SZEKELY, J. CHEN, J. H.
TRANS. MET. SOC. A I M E
2 (4), 1189-92, 1971.

61529 **LINING INDUCTION FURNACES IN WESTERN GERMANY AND OTHER CAPITALIST COUNTRIES.**
KAIBICHEVA M N
OGNEUPORY
35 3 52-8 1970
(FOR ENGLISH TRANSLATION SEE T61530)

61530 **LINING INDUCTION FURNACES IN WESTERN GERMANY AND OTHER CAPITALIST COUNTRIES.**
KAIBICHEVA M N
REFRACTORIES
35 3 189-95 1970
(ENGLISH TRANSLATION OF OGNEUPORY, 3P (3), 52-8, 1970; FOR ORIGINAL SEE T61529)

61531 **CORUNDUM REFRACTORIES.**
KAINARSKII I S DEGTYAREVA E V KABAKOVA I I
OGNEUPORY
35 4 46-53 1970 CA 73 28279
(FOR ENGLISH TRANSLATION SEE T61532)

61532 **CORUNDUM REFRACTORIES.**
KAINARSKII I S DEGTYAREVA E V KABAKOVA I I
REFRACTORIES
35 4 244-51 1970
(ENGLISH TRANSLATION OF OGNEUPORY, 3P (4), 46-53, 1970; FOR ORIGINAL SEE T61531)

61537 **THE MEASUREMENT OF HEAT FLOW IN THE GROUND AND THE THEORY OF HEAT FLUX METERS.**
SCHWERDIFEGER, P.
COLF REGIONS RES. AND ENGR. LAB., HANOVER. N. H.
1-37, 1970.
(CRREL-TR-232, AD-717027)

TPRC Number	Bibliographic Citation
61541	**THE IMPLICATIONS OF TERRESTRIAL HEAT FLOW OBSERVATIONS ON CURRENT TECTONIC AND GEOCHEMICAL MODELS OF THE CRUST AND UPPER MANTLE OF THE EARTH.** SCLATER J G FRANCHETEAU J GEOPHYS J ROY ASTRON SOC 20 509-42 1970
61556	**PROPERTIES OF GLASSES IN THE SIO2 TIO2-ZRO2-AL2O3-MGO-CAO SYSTEM.** ERMOLENKO, N. N. SHAREIKO, L. V. SAVITSKII, S. YE. GRISHINA, N. P. FOREIGN TECHNOLOGY DIVISION 1-13, 1971. (ENGLISH TRANSLATION OF BELORUS. POLITEKH. INST. IM. I. V. STALINE, SBORNIK NAUCH. RABOT, 48-54, 1963; FOR OFIGINAL SEE T59462) (FTD-MT-24-1256-71)
61558	**THE HEAT CAPACITY AND THERMAL CONDUCTIVITY OF APOLLO 11 LUNAR ROCKS 10017 AND 10046 AT LIQUID HELIUM TEMPERATURES.** MORRISON J A NORTON P R JOURNAL OF GEOPHYSICAL RESEARCH 75 6553-7 1970
61561	**THE THERMAL HISTORY OF THE MOON /RESEARCH NOTE/.** CROSS C A THE MOON 2 157-8 1970
61595	**EFFECT OF RADIOACTIVE RADIATION ON ELECTRICAL PORCELAIN.** BUDNIKOV P P KOSTYUKOV N S ANTONOVA N P KOLOTII I I MEDVEDOVSKAYA E I NAIDENOVA G A MOROZ I KH IZV AKAD NAUK SSSR NEORG MATER 5 10 1792-6 1969 CA 72 70144 (FOR ENGLISH TRANSLATION SEE T61596)
61596	**EFFECT OF RADIATION ON ELECTRICAL PORCELAIN.** BUDNIKOV P P KOSTYUKOV N S ANTONOVA N P KOLOTII I I MEDVEDOVSKAYA E I NAIDENOVA G A MOROZ I KH INORGANIC MATERIALS 5 10 1518-21 1969 (ENGLISH TRANSLATION OF IZV. AKAD. NAUK SSSR, NEORG. MATER., 5 (10), 1792-6, 1969; FOR ORIGINAL SEE T61595)
61597	**IR ABSORPTION SPECTRA OF SODIUM-TIN-SILICA GLASSES.** ZORINA M L VAKHRAMEEV V I IZV AKAD NAUK SSSR NEORG MATER 5 10 1834-6 1969 CA 72 70115 (FOR ENGLISH TRANSLATION SEE T61598)
61598	**INFRARED ABSORPTION SPECTRA OF SODIUM-TIN-SILICATE GLASSES.** ZORINA M L VAKHRAMEEV V I INORGANIC MATERIALS 5 10 1563-5 1969 (ENGLISH TRANSLATION OF IZV. AKAD. NAUK SSSR, NEORG. MATER., 5 (10), 1834-6, 1969; FOR ORIGINAL SEE T61597)
61603	**INTERPRETATION OF SOLID SOLUTION HARDENING WITH VIBRATIONAL SPECTRA.** PLENDL, J. N. GIELISSE, P. J. MANSUR, L. C. MITRA, S. S. SMAKULA, A. TARTE, P. C. APPL. OPT. 10 (5), 1129-33, 1971.
61607	**STUDY OF THE ELASTOSTRENGTH PROPERTIES OF INSULATING REFRACTORIES BY THE RESONANCE METHOD.** BLUVSHTEIN M N ZYKOVA Z K RABINOVICH M A KRIGMAN L I OGNEUPORY 35 10 15-19 1970 (FOR ENGLISH TRANSLATION SEE T61608)
61608	**STUDY OF THE ELASTOSTRENGTH PROPERTIES OF INSULATING REFRACTORIES BY THE RESONANCE METHOD.** BLUVSHTEIN M N ZYKOVA Z K RABINOVICH M A KRIGMAN L I REFRACTORIES 35 10 623-6 1970 (ENGLISH TRANSLATION OF OGNEUPORY, 35 (10), 15-9, 1970; FOR ORIGINAL SEE T61607)
61642	**CHANGE IN THE ENTHALPY OF MAGNETITE CONCENTRATES FROM THE SOKOLOVSK-SARBARSK MINING CONCENTRATION COMBINE DURING HEATING IN AN OXIDIZING ENVIRONMENT.** BRATCHIKOV, S. G. YUR'EV, B. P. IZVEST. VYSSH. UCHEB. ZAVED. CHERN. MET. 14 (2), 42-5, 1971. (FOR ENGLISH TRANSLATION SEE TPRC NO. 66202)
61687	**MAGNESIA- RICH MAGNESIUM ALUMINATE SPINEL CERAMICS.** BAILEY, J. T. RUSSELL, R. JR. BULL. AMER. CERAM. SOC. 50 (5), 493-6, 1971.
61701	**PROPERTIES OF GLASSES OF THE SYSTEMS P2O5-SRO AND P2O5-BAO.** SHCHAVELEV O S BABKINA V A IZV AKAD NAUK SSSR NEORG MATER 6 12 2183-6 1970 (FOR ENGLISH TRANSLATION SEE T61702)
61702	**PROPERTIES OF GLASSES OF THE SYSTEMS P2O5-SRO AND P2O5-BAO.** SHCHAVELEV O S BABKINA V A INORGANIC MATERIALS 6 12 1914-17 1970 (ENGLISH TRANSLATION OF IZV. AKAD. NAUK SSSR, NEORG. MATER., 6 (12), 2183-6, 1970; FOR ORIGINAL SEE T61701)
61706	**LABORATORY RESEARCH FOR THE DETERMINATION OF THE THERMAL PROPERTIES OF SOILS.** KERSTEN, M. S. ENGINEERING EXPERIMENT STATION UNIV. OF MINNESOTA 1-227, 1949. (CRREL-TR-23, AD-712516)
61709	**PROPERTIES OF GLASSES IN THE SYSTEM SIO2-B2O3-BAO.** MAZURENKO V D FOREIGN TECHNOLOGY DIVISION 1-10, 1971. (ENGLISH TRANSLATION OF BELORUS. POLITEKH. INST. IM. I. V. STALINA, SBORNIK NAUCH. RABOR, 14-8, 1963; FOR ORIGINAL SEE T59225) (FTD-MT-24-1258-71)
61710	**PASSIVE MICROWAVE MEASUREMENTS OF SNOW, SOILS, AND OCEANOGRAPHIC PHENOMENA.** EDGERTON A T TREXLER D T POE C A STOGRYN A SAKAMOTO S JENKINS J E MEEKS D SOLTIS F SPACE DIVISION AEROJET-GENERAL CORP., EL MONTE, CA. 1-177, 1970. (SD-9016-6, AD-714855)
61712	**METHODS OF ENGINEERING FORECASTING OF DEPTH TO WHICH THE GROUND FREEZES AND THAWS.** PAVLOV, A. V. COLD REGIONS RES. AND ENGR. LAB., HANOVER, N. H. 1-21, 1970. (AD-715027)
61713	**CALCULATION OF THE DEPTH OF THAWING TAKING INTO ACCOUNT THE EXTERNAL HEAT EXCHANGE.** BALOVAYEV, V. T. COLD REGIONS RES. AND ENGR. LAB., HANOVER, N. H. 1-12, 1970. (AD-715069)
61715	**INFRARED EMISSION SPECTRA. ENHANCEMENT OF DIAGNOSTIC FEATURES BY THE LUNAR ENVIRONMENT.** LOGAN L M HUNT G R SCIENCE 169 865-6 1970
61725	**SEA FLOOR GEOTHERMAL MEASUREMENTS FROM VEMA CRUISE 23.** LANGSETH M G JR MALONE I BREGER D LAMONT-DOHERTY GEOLOGICAL OBSERVATORY, PALISADES, N. Y. 1-168, 1970. (AD-718826)
61726	**EMISSION SPECTRA OF PARTICULATE SILICATES UNDER SIMULATED LUNAR CONDITIONS.** LOGAN L M HUNT G R J GEOPHYSICAL RES 75 (32), 6539-48, 1970. (AFCRL-71-0033, AD-718916)
61727	**PHASE COMPOSITION OF PORE WATER IN COLD ROCKS.** MELLOR, M. COLD REGIONS RES. AND ENGR. LAB., HANOVER, N. H. 1-62, 1970. (CRREL-RR-292, AD-719236)
61734	**HEAT CONTENT AND SPECIFIC HEAT OF SIX ROCK TYPES AT TEMPERATURES TO 1000 C.** LINDROTH, D. P. KRAWZA, W. G. BUREAU OF MINES, WASHINGTON, D. C. 1-28, 1971. (PB-199046, BMRI-7503, N71-20613)
61737	**MICROSTRUCTURE AND REFLECTANCE OF A LEAD OXIDE-BORON OXIDE-SILICON OXIDE GLASS WITH CRYSTALLINE OPACIFIER ADDITIONS.** DENNIS, M. D. BRADT, R. C. J. AMER. CERAM. SOC. 54 (5), 232-5, 1971.
61774	**NEW RESULTS IN VITROCERAMIC NUCLEAR FUEL RESEARCH.** ALECU, I. BELEUTA, I. CIOCANESCU, M. COJOCARU, L. N. DRAGOMIR, I. GLANZ, G. GORAN, M. LUNGU, S. MAXIM, I. POPESCU, F. RADULESCU, O. RIBCO, L. RUTTER, I. ACADEMIA R P R, INSTITUTUL DE FIZICA ATOMICA (BUCHAREST) 1-23, 1970. (IFA-MN-11, N71-13795)

TPRC Number	Bibliographic Citation
61780	**INVESTIGATION OF THE VITREOUS STATE BY MEANS OF A DILATOMETER.** KERNER O SALMANG H Z ANORG ALLGEM CHEM 199 235-40 1931
61783	**THERMAL CONDUCTIVITIES OF TM2O3, YB2O3, AND A TM2O3-YB2O3 MIXTURE.** KETCHEN, E. E. OAK RIDGE NATIONAL LAB., TENN. 1-14, 1970. (ORNL-TM-3066, N71-14388)
61835	**EVIDENCE OF SIXFOLD COORDINATION OF NEODYMIUM TRIPOSITIVE IONS IN BARIUM RUBIDIUM SILICATE GLASS.** ROBINSON, C. C. J. CHEM. PHYS. 54 (8), 3572-8, 1971.
61868	**AIRBORNE MEASUREMENTS OF EARTH SURFACE TEMPERATURE (OCEAN AND LAND) IN THE 10-12 MICRONS AND 8-14 MICRONS REGIONS.** WEISS, M. APPL. OPT. 10 (6), 1280-7, 1971.
61891	**THERMAL CONDUCTIVITY OF OXIDE FUELS AS A FUNCTION OF THE (FUEL) POROSITY.** MUELLER, E. M. ATOMWIRT., ATOMTECH. 15 (9-10), 434-5, 1970.
61897	**CORRELATIONS BETWEEN THERMODYNAMIC PROPERTIES AND CRYSTALLOCHEMICAL CHARACTERISTICS OF MINERALS.** MARAKUSHEV, A. A. IDEI E. S. FEDOROVA SOVREM. KRISTALLOGR. MINERAL. 143-78, 1970.
61898	**THERMAL CONDUCTIVITY OF CAST STONE AS A FUNCTION OF ITS STRUCTURE AND TEMPERATURE.** BYKOV, I. I. LEZHENIN, F. F. INZH. FIZ. ZH. 19 (2), 347-9, 1970. (FOR ENGLISH TRANSLATION SEE TPRC NO. 72171)
61903	**FAR INFRARED ABSORPTION SPECTRA OF THREE EMERALDS.** GERBAUX, X. HADNI, A. J. CHIM. PHYS. PHYSICOCHIM. BIOL. 67 (9), 1674-5, 1970.
61911	**THEORETICALLY DENSE (99.9 PERCENT) POLYCRYSTALLINE ALUMINA PREPARED FROM CRYOCHEMICALLY PROCESSED POWDERS.** KIM, Y. S. MONFORTE, F. R. BULL. AMER. CERAM. SOC. 50 (6), 532-5, 1971.
61912	**ANORTHITE CERAMIC DIELECTRICS.** GDULA, R. A. BULL. AMER. CERAM. SOC. 50 (6), 555-7, 1971.
61913	**THERMAL CONDUCTIVITY OF CERAMIC SOLID SOLUTIONS.** MURABAYASHI M J NUCL SCI TECHNOL 7 11 559-63 1970 CA 74 35388
61940	**EFFECTS OF FERRIC OXIDE ON ELECTRICAL AND THERMAL CONDUCTIVITIES OF CORDIERITE BODIES FIRED AT 1300 DEGREES.** INAGAKI, K. SAKAKIBARA, M. TANAKA, Y. AICHI-KEN KOGYO SHIDOSHO HOKOKU (5), 97-102, 1969.
61941	**REFLECTANCE OF SCLEROTIOIDS FROM COALS OF UMARIA, CENTRAL INDIA.** RISHI M K ECON GEOL 65 6 700-5 1970 CA 74 33353
61942	**REFLECTANCE AND MICROHARDNESS OF BORNITE, GALENA, PYRITE, AND MAGNETITE.** HAUSMANN K VON GEHLEN K NEUES JAHRB MINERAL MONATSH 11 498-506 1970 CA 74 33412
62030	**EFFECTS OF DIFFERENT KINDS OF CATIONS ON THERMAL PROPERTIES OF CONDENSED PHOSPHATES.** HATTORI M TSUBAKI T MURATA F TANAKA M YOGYO KYOKAI SHI 79 906 49-55 1971 CA 74 102535
62073	**THE SYSTEM THORIUM OXIDE - PHOSPHORUS OXIDE.** LAUD, K. R. HUMMEL, F. A. J. AMER. CERAM. SOC. 54 (6), 296-8, 1971.
62075	**ELASTIC PROPERTIES OF LITHIUM ALUMINOSILICATE GLASSES AND KEATITE-PHASE GLASS-CERAMICS.** FIELD, M. B. TUCKER, R. W. J. AMER. CERAM. SOC. 54 (6), 309-14, 1971.
62094	**THERMODYNAMIC STUDY OF SOLID- PHASE REACTIONS IN A STRONTIUM OXIDE- ALUMINUM OXIDE SYSTEM.** BROVIKOV, V. N. ORLOV, V. V. MIKHEEV, V. N. IZV. VYSSH. UCHEB. ZAVED., KHIM. KHIM. TECHNOL. 14 (1), 49-52, 1971.
62098	**HEAT CAPACITIES AND THERMAL BEHAVIOR OF ALKALI BORATE GLASSES.** UHLMANN, D. R. KOLBECK, A. G. DE WITTE, D. L. J. NON-CRYST. SOLIDS 5 (5), 426-43, 1971.
62140	**LATTICE VIBRATIONS AND INTERLAYER INTERACTIONS IN CRYSTALLINE ARSENIC SULFIDE AND ARSENIC SELENIDE.** ZALLEN, R. SLADE, M. L. WARD, A. T. PHYS. REV. 3 B (12), 4257-73, 1971.
62141	**INFRARED AND RAMAN STUDIES OF LONG - WAVELENGTH OPTICAL PHONONS IN HEXAGONAL MOLYBDENUM SULFIDE.** WIETING, T. J. VERBLE, J. L. PHYS. REV. 3 B (12), 4286-92, 1971.
62145	**SPLITTING OF THE T O MODE IN CALCITE BY THE POLARIZATION FIELD.** ISHIGAME, M. SATO, T. SAKURAI, T. PHYS. REV. 3 B (12), 4388-91, 1971.
62182	**THERMAL DIFFUSIVITYY OF NON-HOMOGENEOUS MATERIALS.** MATTAROLO L SOVRANO M PROC INTERN CONGRESS OF REFRIGERATION 11TH MUNICH 1963 /PROG REFIRG SCI TECHNOL/ 1 409-13 1965
62185	**A SHOCK WAVE STUDY OF COCONINO SANDSTONE.** SHIPMAN, F. H. GREGSON, V. G. JONES, A. H. GENERAL MOTORS CORP., WARREN, MICH. 46PP., 1971. (NASA-CR-1842)
62278	**TEXTURE STRUCTURE IN CERAMIC BODIES.** ZWETSCH A BER DEUT KERAM GESELL 36 12 388-93 1959
62279	**FORSTERITE CERAMICS WITH THE SAME THERMAL EXPANSION AS TITANIUM.** SONGER E ADOLF C BER DEUT KERAM GESELL 37 1 3-6 1960
62280	**THERMOGRAVIMETRIC INVESTIGATIONS OF CLAYS AND THEIR RELATION TO TECHNICAL PROPERTIES.** HOHLT H G MULLER F ZIMMERMANN K BER DEUT KERAM GES 38 2 58-71 1961
62281	**STRUCTURE AND EXPANSION BEHAVIOUR OF LITHIUM-ALUMINUM-SILICATES.** SAALFELD H BER DEUT KERAM GESELL E V 38 7 281-338 1961
62282	**INFLUENCING THE THERMAL EXPANSION OF A GLAZED BODY BY THE GLAZE.** PAETSCH D BER DEUT KERAM GESELL 38 8 342-4 1961
62283	**INFLUENCE OF THE ADDITIVE OF CRUSHED BISQUE ON THE STRUCTURE AND PROPERTIES OF TALC BODIES.** REISS A MULLER-HESSE H BER DEUT KERAM GES 38 9 420-5 1961
62285	**THERMOSTABLE GLASS PIREKSIL.** KITAIGORODSKII I I BLINOV V A DOKL AKAD NAUK SSSR 118 2 351-3 1958
62287	**PART I. INVESTIGATION OF THE THERMAL EXPANSION OF HETEROGENEOUS DEVITRIFIED MODIFIED EUCRYPTITE GLASS AND THEIR DEVISRIFICATION BEHAVIOUR.** BAUM W GLASSTECHN BER 36 11 444-53 1963
62289	**THE DILATOMETER AND ITS APPLICATION TO CERAMIC PROBLEMS.** METZ A KERAM ZEITSCHRIFT 15 1 32-5 1963
62291	**THE FORSTERITE PORCELAIN AS A HIGH FREQUENCY INSULATOR /III/. PROPERTIES OF THE FORSTERITE PORCELAIN USING SILICIOUS STONE AND SEA-WATER MAGNESIA.** SUGIURA M HIRAI M SANO S ISHII E NAGOYA KOGYO GIJUTSU SHIKENSHO HOKOKU 7 3 65-9 1958

TPRC Number	Bibliographic Citation
62292	**THE FORSTERITE PORCELAIN AS A HIGH FREQUENCY INSULATOR /IV/. THE BODY CONTAINING CORDIERITE.** ISHII E SUGIURA M SANO S HIRAI M NAGOYA KOGYO GIJUTSU SHIKENSHO HOKOKU 7 4 57-62 1958
62293	**ALUMINA PORCELAIN AS A HIGH FREQUENCY INSULATOR /II/ EFFECTS OF METAL OXIDES.** KATO S OKUDA H IGA T OKAWARA S NAGOYA KOGYO GIJUTSU SHIKENSHO HOKOKU 11 8 490-8 1962
62294	**EFFECTS OF THE PARTICLE SIZE OF RAW MATERIALS IN PORCELAIN MANUFACTURING /VIII/. EFFECT OF COMMINUTION OF FELDSPAR AND QUARTZ ON THE PYROPLASTIC DEFORMATION, THERMAL EXPANSION, AND TRANSLUCENCY OF PORCELAIN BODIES.** YAMAMOTO R NAITO R NISHIMURA Y KATO S NAGOYA KOGYO GIJUTSU SHIKENSHO KOKOKU 12 2 106-15 1962
62295	**COMBINED MEASUREMENT OF THE THERMOPHYSICAL PROPERTIES OF FLUXES INTENDED FOR GROWING FERRITE SINGLE CRYSTALS.** DERMAN A S BOGORODSKII O V BULL ACAD SCI USSR /PHYS SER/ 34 6 1081-2 1970 (ENGLISH TRANSLATION OF AZV. AKAD. NAUK SSSR, SER. FIZ., 34 (6), 1215-6, 1970; FOR ORIGINAL SEE T59198)
62296	**STUDIES ON FINE GRAINED CRYSTALLINE GLASS-CERAMICS. CRYSTALLIZATION OF SOME GLASSES CONTAINING CR2O3.II. GLASS-CONVERTED-CERAMICS HAVING LOW THERMAL EXPANSION IN THE SYSTEM LI2O-AL2O3-SIO2-CAO.** HAYAMI R OGURA T TANAKA H OSAKA KOGYO GIJUTSU SHIKENSHO KIHO 11 4 241-5 1960
62297	**STUDIES ON FINE GRAINED CRYSTALLINE GLASS-CERAMICS. STUDIES ON THE SUBMICROCRYSTALLIZATION OF GLASSES OF THE SYSTEM SIO2-LI2O-AL2O3 BY ADDITION OF P2O5 /II/.** NAGAOKA K HARA M TANAKA H OSAKA KOGYO GIJUTSU SHIKENSHO KIHO 12 2 48-53 1961
62298	**STUDIES ON FINE-GRAINED CRYSTALLINE GLASS-CERAMICS. CRYSTALLIZATION OF SOME GLASSES CONTAINING CR2O3. III. CORRELATION BETWEEN CHEMICAL COMPOSTTIONS OF GLASS-CONVERTED-CERAMICS HAVING LOW THERMAL EXPANSION IN SYSTEM LI2O-AL2O3-SIO2-CAO AND THEIR PROPERTIES.** HAYAMI R OGURA T TANAKA H OSAKA KOGYO GIJUTSU SHIKENSHO KIHO 12 2 137-43 1961
62300	**THE COEFFICIENT OF LINEAR EXPANSION OF FIREBRICKS AND SILICA BRICKS.** ENGELTHALER Z ENGELTHALER K SILIKATY 2 227-37 1958
62301	**SUGGESTION FOR THE EVALUATION OF DILATOMETER MEASUREMENTS ON GLASSES.** SCHELINSKI S SILIKAT TECHNIK 10 4 184-7 1959
62302	**LITHIUM CERAMICS-CERAMIC BODIES WITH NEGATIVE HEAT EXPANSION.** TUREK M SKLAR KERAM 10 1 18-20 1960
62303	**TESTING THE VOLUME STABILITY /AFTER EXPANSION OR CONTRACTION/ OF SOME SILICEOUS FIRE-CLAY COKE-OVEN BLOCKS.** REICH H F TONIND-ZEITUNG KERAM RUNDSCHAU 84 2 25-9 1960
62304	**RESULTS WITH THE DTA AND DTA-DS-APPARATUS.** THORMANN P TONIND-ZEITUNG KERAM RUNDSCHAU 85 17 408-10 1961
62305	**THE INVESTIGATION OF RAW MATERIALS WITH THE AID OF COMBINED DILATATOMETRIC AND DTA.** LEHMANN H THORMANN P TONIND-ZEITUNG KERAM RUNDSCHAU 86 24 606-12 1962
62306	**INVESTIGATION OF THE SINTERING OF QUARTZ AND CRISTOBALITE.** DIETZ U SCHULLE W TONIND-ZEITUNG KERAM RUNDSCHAU 87 2 25-8 1963
62309	**INVESTIGATION OF THE THERMAL EXPANSION OF SOLID SOLUTIONS OF FERRO-MAGNETIC MATERIALS OF THE SYSTEMS. NIO-ZNO-FE2O3 AND CUO-ZNO-FE2O3.** AUGUSTINIK A I VASILEV E I ZHUR PRIKLAD KHIM 29 6 941-4 1956
62314	**CERAMIC THERMAL JUNCTIONS USED IN THERMOELECTRIC DEVICES.** ZVYAGINA, E. N. KILIPENKO, V. V. LEBEDEV, V. V. KHOLOD. TEKH. TEKHNOL. (9), 17-23, 1970.
62319	**PHYSICOCHEMICAL PROPERTIES OF GLAZES FROM VOLCANIC ROCKS.** KHIZANISHVILI, I. G. GAPRINDASHVILI, G. G. STEKLO KERAM. 27 (11), 38-40, 1970. (FOR ENGLISH TRANSLATION SEE TPRC NO. 66788)
62371	**THERMAL CONDUCTIVITY OF ROCK-FORMING MINERALS.** HORAI, K. J. GEOPHYS. RES. 76 (5), 1278-308, 1971.
62381	**VOLUME STABILITY OF PHOSPHATE- BONDED HIGH- ALUMINA BRICK.** BAAB, K. A. BLACKWOOD, J. M. BULL. AMER. CERAM. SOC. 50 (7), 607-10, 1971.
62397	**INTERNAL FRACTURE OF GLASS UNDER TRIAXIAL TENSION INDUCED BY THERMAL SHOCK.** KING, C. Y. WEBB, W. W. J. APPL. PHYS. 42 (6), 2386-95, 1971.
62413	**COMPOSITION AND PROPERTIES OF PLATINUM - IRON SERIES MINERALS.** YUSHKO - ZAKHAROVA, O. E. LEBEDEVA, S. I. BYKOV, V. P. SAVON, A. D. DOKL. AKAD. NAUK SSSR 195 (1), 185-7, 1970.
62414	**INFRARED SPECTRA OF JAPANESE COAL.. ABSORPTION BANDS AT 3450 AND 1260 CM-1.** OSAWA, Y. SHIH, J. W. FUEL 50 (1), 53-7, 1971.
62420	**EARLY STAGES OF CARBONIZATION OF COAL. XVII. INFRARED ABSORPTION SPECTRA. 1.** OSAWA Y SUGIMURA H FUJI S NENRYO KYOKAI-SHI 48 505 303-9 1969 CA 74 55975
62437	**DETERMINATION OF THE COEFFICIENT OF THERMAL CONDUCTIVITY IN AN ASYMMETRICAL HEATING MODE.** KAGANER M G ZAVODSKAYA LAB 36 9 1095-6 1970 CA 74 57981 (FOR ENGLISH TRANSLATION SEE T38913)
62451	**DISPERSION OF THE OPTICAL PROPERTIES OF CARBONIZED VITRINITES.** MARSHALL, R. J. MURCHISON, D. G. FUEL 50 (1), 4-22, 1971.
62513	**SATURNS RINGS. SPECTRAL REFLECTIVITY AND COMPOSITIONAL IMPLICATIONS.** LEBOFSKY L A JOHNSON T V MC CORD T B ICARUS 13 2 226-30 1970 CA 74 81245
62537	**THERMAL CONDUCTIVITY OF HARD COALS.** PISTEK, P. KUDELA, V. SB. PR. UVP (22), 248-94, 1970.
62538	**EFFECT OF THERMAL INSULATION OF THE TOP OF STEEL INGOTS ON THEIR SOLIDIFICATION.** SAMOILOVICH, YU. A. KOTLYAREVSKII, E. M. STAL 30 (12), 1086-8, 1970. (FOR ENGLISH TRANSLATION SEE TPRC NO. 67377)
62547	**CONNECTIONS BETWEEN THE BULK ABSORPTION AND DIFFUSE REFLECTANCE SPECTRA OF POWDERED SOLIDS. II.** KARVALY, B. PINTER, F. ACTA PHYS. 29 (2-3), 203-20, 1971.
62563	**DETERMINATION OF THE HEAT CAPACITY OF ORE-CARBON PELLETS IN THE PROCESS OF METALLIZATION.** STOROZHEV, YU. I. TELEGIN, A. S. ZAVARZIN, V. P. IZV. VYSSH. UCHEB. ZAVED., CHERN. MET. 14 (5), 27-9, 1971. (FOR ENGLISH TRANSLATION SEE T75013)
62575	**THERMAL EXPANSION OF ANORTHITE.** CZANK, M. SCHULZ, H. NATURWISSENSCHAFTEN 33, 21, 1937.

TPRC Number	Bibliographic Citation

62576 **UNEVEN NATURE OF A CHANGE IN THE COEFFICIENT OF THERMAL EXPANSION OF SOME GLASSES.**
BOGATYREVA, V. V. GOMELSKII, M. S.
SHAKIROVA, R. N.
OPT. – MEKH. PROM.
37 (11), 48–52, 1970.
(FOR ENGLISH TRANSLATION SEE TPRC NO. 67361)

62599 **DETERMINATION OF THE DIFFERENCE IN THE COEFFICIENTS OF THERMAL EXPANSION OF GLASSES OF VARIOUS COMPOSITIONS.**
GOMEL'SKII, M. S. PLUTALOVA, N. YU.
ZAVOD. LAB.
37 (3), 309–11, 1971.
(FOR ENGLISH TRANSLATION SEE TPRC NO. 66484)

62634 **OPTICAL AND MECHANICAL PROPERTIES OF SOME NEODYMIUM- DOPED LASER GLASSES.**
WAXLER, R. M. CLEEK, G. W. MALITSON, I. H.
DODGE, M. J. HAHN, T. A.
J. RES. NAT. BUR. STAND.
75 A (3), 163–74, 1971.

62655 **THERMAL CONDUCTIVITY OF CARBONATES AND A MIXTURE OF CARBONATES WITH MAGNESIUM OXIDE.**
EGOROV B N REVYAKINA M P
TEPLOFIZ VYS TEMP
8 6 1299–302 1970 CA 74 57987
(FOR ENGLISH TRANSLATION SEE T62656)

62656 **INVESTIGATION OF THE THERMAL CONDUCTIVITY OF CARBONATES IN A MIXTURE OF CARBONATES AND MAGNESIUM OXIDE.**
EGOROV B N REVYAKINA M P
HIGH TEMPERATURE
8 6 1220–3 1970
(ENGLISH TRANSLATION OF TEPLOFIZ. VYS. TEMP., 8 (6), 1299–1302, 1970; FOR ORIGINAL SEE T62655)

62676 **THE STUDY OF THE REFLECTIVITY OF INORGANIC MATERIALS IMPORTANT FOR REMOTE SENSING APPLICATIONS.**
PERRY, C. H. LOWNDES, R. P.
PHYSICS DEPT., NORTHWESTERN UNIV., BOSTON, MASS.
1–118, 1970.
(AFCRL–70–0512, AD–720874)

62678 **THERMAL EXPANSION AND THE STRUCTURE OF A CRYSTALLIZING GLASS.**
SORKIN E S
FOREIGN TECHNOLOGY DIVISION
1–12, 1971.
(ENGLISH TRANSLATION OF STEKLOOB. SOSTOYANIE, AKAD. NAUK SSR, VSES. KHIM. OBSHCHESTOV, TR. CHETVERTOGO VSES. SOVESCH., 356–60, 1965; FOR ORIGINAL SEE T41898)
(FTD–MT–24–1397–71)

62687 **DENSE POLYCRYSTALLINE SILICON CARBIDE AND ITS USE IN THE TECHNIQUE OF HIGH TEMPERATURES.**
GNESIN G G
TUGOPLAVKIE KARBIDY
62–7 1970

62702 **OPTICAL DISTINCTION BETWEEN GELINITE–COLLINITE (COAL) AND BITUMENS–KERABITUMENS.**
ALPERN, B.
C. R. ACAD. SCI.
272 D (13), 1717–20, 1971.

62710 **THERMAL CONDUCTIVITY OF BASALTS AT 300–1200 K.**
PETRUNIN, G. I. YURCHAK, R. P. TKACH, G. F.
IZV. AKAD. NAUK SSSR, FIZ. ZEMLI
(2), 65–8, 1971.
(FOR ENGLISH TRANSLATION SEE TPRC NO. 69441)

62718 **TEMPERATURE DEPENDENCE OF VIBRATIONAL SPECTRA IN CALCITE BY MEANS OF EMISSIVITY MEASUREMENT.**
SAKURAI, T. SATO, T.
PHYS. REV.
4 B (2), 583–91, 1971.

62719 **THERMAL CONDUCTIVITY OF GARNETS AND PHONON SCATTERING BY RARE–EARTH IONS.**
SLACK, G. A. OLIVER, D. W.
PHYS. REV.
4 B (2), 592–609, 1971.

62754 **EARLY STAGE OF COAL CARBONIZATION. XXIII. INFRARED – ABSORPTION SPECTRA. 4. HEAT – TREATED COAL.**
OSAWA, Y. SHIH, J. W. TSUNODA, T.
NENRYO KYOKAI–SHI
49 (524), 898–907, 1970.

62758 **DETERMINATION OF THE THERMAL STABILITY AND THERMOPHYSICAL CHARACTERISTICS OF CORUNDUM MATERIALS.**
GOGOTSI, G. A. KOZDOBA, L. A.
KRIVOSHEI, F. A
PROBL. PROCH.
(1), 92–6, 1971.

62788 **NEW DECORATIVE GLAZES FOR FAIENCE.**
SIVCHIKOVA M G OLEINIK L L KAGANOV I V
FOREIGN TECHNOLOGY DIVISION
1–36, 1971.
(ENGLISH TRANSLATION OF GLAZURI, IKH. PROIZVOD. I OLEINIK., 133–47, 1964; FOR ORIGINAL SEE T58769)
(FTD–MT–71–1621)

62953 **EXPERIMENTAL DETERMINATION OF THE HEAT–CONDUCTIVITY COEFFICIENT FOR NONMETALLIC TUBES /EXCHANGE OF EXPERIENCE/.**
VOLKOV D I DMITRIEV G I ROMANOV V A
TURLAKOV A S
ZAVOD LAB
35 9 1096–7 1969
(FOR ENGLISH TRANSLATION SEE T62954)

62954 **EXPERIMENTAL DETERMINATION OF THE HEAT–CONDUCTIVITY COEFFICIENT FOR NOMETALLIC TUBES /EXCHANGE OF EXPERIENCE/.**
VOLKOV D I DMITRIEV G I ROMANOV V A
TURLAKOV A S
IND LAB USSR
35 9 1320–1 1969
(ENGLISH TRANSLATION OF ZAVOD. LAB., 35 (9), 1096–7, 1969; FOR ORIGINAL SEE T62953)

62988 **RELATION OF MINERAL INFRARED SPECTRA WITH CRYSTALLOCHEMICAL FACTORS.**
POVARENNYKH, A. S.
MINERAL. SB. (LVOV)
24 (1), 12–29, 1970.

62997 **DETERMINATION OF THE THERMOPHYSICAL CHARACTERISTICS OF MATERIALS BY THE SEMENOV METHOD.**
ZHUKOV, N. I.
INZH.–FIZ. ZH.
20 (1), 82–6, 1971.
(FOR ENGLISH TRANSLATION SEE TPRC NO. 72195)

63002 **HEAT TRANSFER BY RADIATION AND CONDUCTION IN EVACUATED POWDERS.**
WECHSLER, A. E.
PROGR. REFRIG. SCI. TECHNOL., PROC. INT. CONGR. REFRIG., 12TH
2, 267–78, 1969.

63005 **DEVELOPMENTS OF LOW–TEMPERATURE GLASSES AND REDUCED COEFFICIENTS OF EXPANSION.**
WILD, A.
PROC., FALL MEET., MATER. EQUIP. WHITE WARES DIV., AMER. CERAM. SOC.
108–13, 1969.

63010 **QUANTITATIVE EVALUATION OF THERMAL CONDUCTIVITY OF MOLD MATERIALS AT ELEVATED TEMPERATURES.**
GUPTA, K. C. KHANDUJA, S. R. GAINDHAR, J. L.
ROORKEE UNIV. RES. J.
12 (1–2), III, 45–54, 1970.

63012 **PORCELAIN GLAZES. I. DETERMINATION OF THERMAL EXPANSION COEFFICIENT OF GLAZES FOR HARD PORCELAIN.**
MATEJKA, J.
SKLAR KERAM.
21 (1), 4–9, 1971.

63018 **OPTICAL CHARACTERISTICS OF STRONGLY DIFFUSING GRAY GLASSES.**
VOISHVILLO, N. A.
FOREIGN TECHNOLOGY DIVISION
12PP., 1971.
(ENGLISH TRANSLATION OF ZH. PRIKL. SPEKTROSK., 11 (2), 316–9, 1969; FOR ORIGINAL SEE T56964)
(FTD–MT–71–1661)

63038 **EFFECT OF INNER SURFACE AIR VELOCITY AND TEMPERATURE UPON HEAT GAIN AND LOSS THROUGH GLASS FENESTRATIONS.**
PENNINGTON C W MC DUFFIE D E JR
ASHRAE TRANS
76 2 190–200 1970

63042 **STRUCTURE AND PROPERTIES OF SILVER BORATE GLASSES.**
BOULOS, E. N. KREIDL, N. J.
J. AMER. CERAM. SOC.
54 (8), 368–75, 1971.

63043 **SUBSOLIDUS RELATIONS IN THE SYSTEM ZIRCONIUM OXIDE–THORIUM OXIDE–PHOSPHORUS OXIDE.**
LAUD, K. R. HUMMEL, F. A.
J. AMER. CERAM. SOC.
54 (8), 407–9, 1971.

63066 **OPTICAL MEASUREMENTS ON MAGNETITE SINGLE CRYSTALS.**
BALBERG, I. PANKOVE, J. I.
PHYS. REV. LETT.
27 (9), 596–9, 1971.

63068 **CERAMIC MATERIALS WITH LOW COEFFICIENT OF LINEAR EXPANSION.**
ZINKO E I MEDVEDOVSKAYA E I FOMINA N P
GLASS CERAMICS
22 5 312–4 1965
(ENGLISH TRANSLATION OF STEKLO KERAM., 22 (5), 22–4, 1965; FOR ORIGINAL SEE T41896)

TPRC Number	Bibliographic Citation
63069	**INVESTIGATION OF THE KINETICS OF BLOATING ALUMINA SLIP FOR MAKING CORUNDUM LIGHTWEIGHT.** GAODU A N KAINARSKII I S REFRACTORIES 6 282-8 1964 (ENGLISH TRANSLATION OF OGNEUPORY, 29 (6), 270-5, 1964; FOR ORIGINAL SEE T36008)
63072	**THE MICROHARDNESS AND COEFFICIENT OF LINEAR EXPANSION OF SEVERAL GLASSES OF THE PORCELAIN GLASS PHASE TYPE.** AVGUSTINIK A E FEDOROVA E P J APPLIED CHEM USSR 26 9 851-5 1953 (ENGLISH TRANSLATION OF ZHUR. PRIKLAD. KHIM., 26 (9), 625-30, 1983; FOR ORIGINAL SEE T53130)
63073	**THE EFFECT OF CERTAIN COLORANTS AND OPACIFIERS ON THE COEFFICIENT OF EXPANSION OF SILICATE GLASSES.** APPEN A A KAYALOVA S S J APPLIED CHEM USSR 26 11 1073-8 1953 (ENGLISH TRANSLATION OF ZHUR. PRIKLAD. KHIM., 26 (11), 1127-32, 1953; FOR ORIGINAL SEE T54459)
63074	**THERMAL EXPANSION OF CERTAIN LITHIUM ALUMINOSILICATES.** AVGUSTINIK A I VASILYEV E I J APPLIED CHEM USSR 28 9 891-4 1955 (ENGLISH TRANSLATION OF ZHUR. PRIKLAD. KHIM., 28 (9), 939-43, 1955; FOR ORIGINAL SEE T54638)
63075	**EFFECT OF IRON OXIDES ON THE THERMAL EXPANSION OF GROUND ENAMELS FOR DISHWARE.** APPEN A A BRESKER R I KUZNETSOVA L A J APPLIED CHEM USSR 29 11 1887-9 1956 (ENGLISH TRANSLATION OF ZHUR. PRIKLAD. KHIM., 29 (11), 1753-55, 1956; FOR ORIGINAL SEE T55416)
63102	**SIMPLIFIED CALCULATION OF THE TOTAL TRANSMISSION COEFFICIENT OF NEUTRAL GLASSES.** RONIS D F ZAVOD LAB 34 11 1348-9 1968 (FOR ENGLISH TRANSLATION SEE T63103)
63103	**SIMPLIFIED CALCULATION OF THE TOTAL TRANSMISSION COEFFICIENT OF NEUTRAL GLASSES.** RONIS D F INDUSTRIAL LABORATORY 34 11 1625-6 1968 (ENGLISH TRANSLATION OF ZAVOD. LAB., 34 (11), 1348-9, 1968; FOR ORIGINAL SEE T63102)
63135	**MARTIAN SOFT LANDER INSULATION STUDY.** WILBERS O J SCHELDEN B J CONTI J C PROGRESS IN ASTRONAUTICS AND AERONAUTICS 24, 630-58, 1971.
63136	**OPTICAL PROPERTIES OF NATURALLY OCCURRING RHODOCHROSITE BETWEEN 0.33 MICROMETER AND 2.5 MICROMETERS.** EGAN, W. G. HILGEMAN, T. APPL. OPT. 10 (9), 2132-6, 1971.
63137	**EXPERIMENTAL DETERMINATION OF THE EFFECT OF TEMPERATURE ON REFRACTIVE INDEX AND OPTICAL PATH LENGTH OF GLASS.** PARKER, C. J. POPOV, W. A. APPL. OPT. 10 (9), 2137-43, 1971.
63145	**THERMAL CONDUCTIVITY OF MAGNETITE AND HEMATITE.** MOLGAARD, J. SMELTZER, W. W. J. APPL. PHYS. 42 (9), 3644-7, 1971.
63162	**DOSIMETERS BASED ON GLASSES WITH OPTICAL DENSITIES VARYING ON IRRADIATION.** BYURGANOVSKAYA G V GVOZDEV E G KHOVANOVICH A I ATOMNAYA ENERGIYA USSR 21 1 38-41 1966 (FOR ENGLISH TRANSLATION SEE T63163)
63163	**DOSIMETERS BASED ON GLASSES WITH OPTICAL DENSITIES VARYING ON IRRADIATION.** BYURGANOVSKAYA G V GVOZDEV E G KHOVANOVICH A I SOVIET ATOMIC ENERGY 21 1 656-8 1966 (ENGLISH TRANSLATION OF ATOMNAYA ENERGIYA, USSR, 21 (1), 38-41, 1966; FOR ORIGINAL SEE T63162)
63189	**REFLECTION SPECTRUM OF PYRITE IN THE 2-65 MICRON RANGE.** KORSUNSKII, M. I. MURATOV, E. M. IZV. AKAD. NAUK KAZ. SSR, SER. FIZ.-MAT. 8 (6), 63-72, 1970.
63198	**THE C - AXIS THERMAL EXPANSION OF CARBONS AND GRAPHITES.** KELLETT, E. A. RICHARDS, B. P. J. APPL. CRYSTALLOGR. 4 (PT. 1), 1-8, 1971.
63211	**THERMODYNAMIC FUNCTIONS OF GRAPHITE AT LOW TEMPERATURES.** VOLGA, V. I. DYMOV, B. K. PUKHLYAKOV, V. P. USOV, V. K. KONSTR. MATER. OSN. GRAFITA (5), 98-104, 1970.
63213	**INFRARED ABSORPTION SPECTRA OF NEW MINERALS FROM CENTRAL ASIA ALKALINE PEGMATITES.** POVARENNYKH, A. S. DUSMATOV, V. D. KONST. SVOISTVA MINER. 4, 3-9, 1970.
63225	**THERMAL PROPERTIES OF SOME MAGMATIC ROCKS.** CHMURA, K. PRZEGL. GEOL. 19 (1), 15-8, 1971.
63227	**GRAPHITIZATION OF MICROPOWDERS OF SYNTHETIC DIAMONDS STUDIES FROM INFRARED ABSORPTION SPECTRA.** MALOGOLOVETS, V. G. GATILOVA, E. G. SIN. ALMAZY 2 (4), 29-32, 1970.
63238	**EXPANSION AND THE STRUCTURE OF THE SODIUM - CALCIUM GLASSES.** KNAPP, O. FOREIGN TECHNOLOGY DIVISION 10PP., 1971. (ENGLISH TRANSLATION OF SZKLO I CERAMIKA, 15 (9), 240-3, 1964; FOR ORGIGINAL SEE TPRC NO. 41897) (FTD-HC-23-1529-71)
63242	**SPECTRAL PROPERTIES OF MANGANESE-ACTIVATED GLASSES.** LUNTER S G KARAPETYAN G O YUDIN D M IZV AKAD NAUK SSSR SER FIZ 31 5 811 1967 (FOR ENGLISH TRANSLATION SEE T63243)
63243	**SPECTRAL PROPERTIES OF MANGANESE-ACTIVATED GLASSES.** LUNTER S G KARAPETYAN G O YUDIN D M BULL ACAD SCI USSR /PHYS SER/ 31 5 817-9 1967 (ENGLISH TRANSLATION OF IZV. AKAD. NAUK SSSR, SER. FIZ., 31 (5), 811- , 1967; FOR ORIGINAL SEE T63242)
63268	**MANUFACTURING AND PROPERTIES OF FILMS OF Y2O3 AND RARE EARTH OXIDES ON GLASS.** FRANK B GROTH R THIN SOLID FILMS 3 41-50 1969
63275	**CALCIUM OXIDE-ALUMINUM OXIDE-SILICON OXIDE COMPOSITION FOR USE AS A SUBSTRATE FOR THIN FILMS.** WILLIAMS, J. C. BULL. AMER. CERAM. SOC. 50 (9), 726-9, 1971.
63328	**CEMENTS FOR STICKING MICA-CERAMICS TO TITANIUM.** BELITSKII M E ABZGILDIN F YU TRESVYATSKII S G POROSH MET USSR 8 5 102-6 1968 (FOR ENGLISH TRANSLATION SEE T63329)
63329	**CEMENTS FOR STICKING MICA-CERAMICS TO TITANIUM.** BELITSKII M E ABZGILDIN F YU TRESVYATSKII S G SOVIET POWDER MET METAL CERAMICS 5 418-21 1968 (ENGLISH TRANSLATION OF POROSH. MET., USSR, 8 (5), 102-6, 1968; FOR ORIGINAL SEE T63328)
63347	**AN INVESTIGATION OF THE THERMAL DIFFUSIVITY AND THERMAL CONDUCTIVITY OF INSULATING POWDERS AT ATMOSPHERIC PRESSURE AND IN VACUUM BY VARIOUS METHODS.** KAGANER, M. G. SEMENOVA, R. S. J. ENG. PHYS. 13 (1), 15-8, 1967. (ENGLISH TRANSLATION OF INZH. FIZ. ZH., 13 (1), 24-30, 1967., FOR ORIGINAL SEE TPRC NO. 55858)
63375	**THERMAL CONDUCTIVITY AND SPECIFIC HEAT OF NONCRYSTALLINE SOLIDS.** ZELLER, R. C. POHL, R. O. PHYS. REV. 4 B (6), 2029-41, 1971.
63400	**OPTICAL MATERIALS STUDY PROGRAM.** GOGGIN, W. R. PAQUIN, R. A. PERKIN-ELMER CORP. OPTICAL OPERATIONS DIV., NORWALK, CN 1-79, 1970. (AD-865842L)
63401	**MICROWAVE HIGH DIELECTRIC CONSTANT MATERIALS.** READEY, D. W. MAGUIRE, E. A. HARTWIG, C. P. MASSE, D. J. RESEARCH DIVISION, RAYTHEON CO., WALTHAM, MASS 129PP., 1971. (AD-725578, ECOM-3455-F)

TPRC Number	Bibliographic Citation
63502	THE DETERMINATION OF LIGHT ABSORPTION IN DIFFUSING MATERIALS BY A PHOTON DIFFUSION MODEL. GATE, L. F. J. PHYS. D 4 (7), 1049-56, 1971.
63519	CRYOGENIC KNOW-HOW AS APPLIED TO INGROUND LIQUID NITROGEN GAS STORAGE. HINCKLEY R B BAKERJIAN B H DRAKE E M J SPACECRAFT ROCKETS 7 8 995-7 1970
63549	THERMAL EXPANSION CHARACTERISTICS OF LIMESTONE-CONTAINING GREEN BODIES FOR EARTHENWARE. IMOTO, Y. YOGYO KYOKAI SHI 79 (909), 156-63, 1971.
63552	ABSOLUTE REFLECTIVITY OF ALKALI HALIDES IN THE SHORT- WAVE ULTRAVIOLET. FASSLER, D. ZIMMERMANN, G. Z. PHYS. CHEM. (LEIPZIG) 246 (1-2), 33-41, 1971.
63599	INFRARED ABSORPTION SPECTRA OF MINERALS ISOSTRUCTURAL WITH GYPSUM. POVARENNYKH, A. S. GEVORK'YAN, S. V. DOPOV. AKAD. NAUK UKR. RSR, 33 B (4), 306-10, 1971.
63601	EFFECT OF POTASSIUM OXIDE ON THE PHYSIOCHEMICAL PROPERTIES OF FAIENCE GLAZES OPACIFIED BY DIOPSIDE CRYSTAL PARTICLES. YUNUSOV, KH. AZIMOV, I. FIZ.-KHIM. ISSLED. GLINISTYKH MINER. SILIKAT. MATER. 101-6, 1970.
63609	MEASUREMENT OF THE REFLECTIVITY AND MICROHARDNESS OF DIFFERENT COLORED BORNITES. KOSYAK, E. A. IZV. AKAD. NAUK KAZ. SSR, SER. FIZ.-MAT. 9 (2), 80-2, 1971.
63632	THERMOPHYSICAL CHARACTERISTICS OF SOME DISPERSED MATERIALS. ILCHENKO, K. D. MET. KOKSOKHIM. (21), 82-5, 1970.
63639	EXPERIMENTAL DETERMINATION OF THERMAL CONDUCTIVITY IN ROCK SAMPLES. RAMOS, R. R. REV. INST. MEX. PETROL. 3 (1), 27-34, 1971.
63644	HEAT-RESISTANT, LOW-ALKALI STRONTIUM GLAZE FOR EARTHENWARE MAJOLICA. SHTEINBERG, YU. G. STEKLO KERAM. (5), 37-8, 1971. (FOR ENGLISH TRANSLATION SEE TPRC NO. 66789)
63673	INTERPRETATION OF INFRARED REFLECTION SPECTRA OF PRODUCTS OF LITHIUM SILICATE GLASS CRYSTALLIZATION. KIND, N. E. TUDOROVSKAYA, N. A. ZH. FIZ. KHIM. 45 (4), 759-63, 1971. (FOR ENGLISH TRANSLATION SEE TPRC NO. 63674)
63674	INTERPRETATION OF THE INFRARED REFLECTION SPECTRA OF THE CRYSTALLIZATION PRODUCTS OF LITHIUM SILICATE GLASSES. KIND, N. E. TUDOROVSKAYA, N. A. RUSS. J. PHYS. CHEM. 45 (4), 424-7, 1971. (ENGLISH TRANSLATION OF ZH. FIZ. KHIM. 45 (4), 759-63, 1971; FOR ORIGINAL SEE TPRC NO. 63673)
63695	TELLURITE GLASS.. A NEW ACOUSTO-OPTIC MATERIAL. YANO, T. FUKUMOTO, A. WATANABE, A. J. APPL. PHYS. 42 (10), 3674-6, 1971.
63697	ULTRAVIOLET-INDUCED TRANSIENT AND STABLE COLOR CENTERS IN SELF-Q-SWITCHING LASER GLASS. LANDRY, R. J. SNITZER, E. BARTRAM, R. H. J. APPL. PHYS. 42 (10), 3827-38, 1971.
63703	NATURE OF PARTICULATE MATTER IN COMETS AS DETERMINED FROM INFRARED OBSERVATIONS. O'DELL, C. R. ASTROPHYS. J. 166 (3), (PT. 1), 675-81, 1971.
63739	VISIBLE AND NEAR-INFRARED SPECTRA OF MINERALS AND ROCKS. III. OXIDES AND HYDROXIDES. HUNT, G. R. SALISBURY, J. W. LENHOFF, C. J. MOD. GEOL. 2 (3), 195-205, 1971. (AD-746820, AFCRL-72-0445)
63741	ABSORPTION SPECTRA OF MUSCOVITE FROM PEGMATITE VEINS OF NORTHERN KARELIA AND THE MAMA-CHUY REGION. VOKHMENTSEV, A. YA. ROMANOV, V. A. KLIMENT'EV, V. N. NOV. ISSLED. GEOL. 6-17, 1969.
63748	STUDY OF THE SYSTEMS TRICALCIUM SILICATE-WATER AND BETA-DICALCIUM SILICATE-WATER IN THE TEMPERATURE INTERVAL BETWEEN 253 AND 373 DEGREES K. MCHEDLOV - PETROSYAN, O. P. CHERNYAVSKII, V. L. SILIKATTECHNIK 22 (2), 45-7, 1971.
63749	MEASUREMENT OF THE THERMAL CONDUCTIVITY OF MICA IN ALL THREE DIRECTIONS OF AXES AS DEPENDENT ON THE TEMPERATURE. RUGE, H. HERRMANN, D. SILIKATTECHNIK 22 (2), 57-9, 1971.
63750	COLORED GLASSES WITH PROTECTIVE EFFECTS. MARES, Z. SKLAR KERAM. 21 (4), 91-2, 1971.
63757	COMPLEX DETERMINATION OF THE TEMPERATURE DEPENDENCE OF THERMOPHYSICAL PROPERTIES OF SUBSTANCES. OSIPOVA, M. N. OSIPOVA, V. A. TEPLOENERGETIKA 18 (6), 84-5, 1971. (FOR ENGLISH TRANSLATION SEE TPRC NO. 67155)
63771	REFLECTANCE AND MICROHARDNESS OF SMYTHITE. NICKEL, E. H. HARRIS, D. C. AMER. MINERAL. 56 (7-8), 1462-7, 1971.
63774	OPTICAL PROPERTIES OF COVELLINE AND ENARGITE IN THE VISIBLE AND NEAR INFRARED. CERVELLE, B. D. LEVY, C. BULL. SOC. FR. MINERAL. CRISTALLOGR. 94 (2), 146-55, 1971.
63776	MEASUREMENT OF THE THERMAL CONDUCTIVITY COEFFICIENT IN EXPANDED MATERIALS. MARANGONI, C. MARTORANA, P. CALORE 41 (1), 42-6, 1970.
63778	INFRARED SPECTROSCOPY OF SOME SULFIDE MINERALS. GILLIESON, A. H. COLLOQ. SPECTROSC. INT., PLENARY CONF. 193-205, 1969.
63800	EFFECT OF TREATMENT ON THE STRUCTURE OF A QUARTZ SURFACE STUDIED BY AN INFRARED-SPECTROSCOPIC METHOD. DIDRIKIL, L. N. SHKLYAR, A. N. KOSHKAREV, E. A. IZV. AKAD. NAUK TADZH. SSR, OTD. FIZ.-MAT. GEOL.-KHIM. NAUK (4), 10-4, 1970.
63801	DETERMINATION OF THERMOPHYSICAL PROPERTIES OF FUEL AND FUEL-ORE MATERIALS. FADEEV, V. V. SILETSKII, V. S. CHERNYKH, V. I. IZV. VYSSH. UCHEB. ZAVED., ENERG. 14 (2), 78-82, 1971.
63808	THE EFFECT OF HALIDES ON THE OPAQUENESS OF ZINC-BOROPHOSPHATE GLASSES. EIDUKS J PORMANIS I FOREIGN TECHNOLOGY DIVISION FTD-MT-24-1384-71 1-19 1971 (ENGLISH TRANSLATION OF STEKLOOBRAZNYE SIST. MATER., AKAD. NAUK LATV. SSR, 113-19, 1967; FOR ORIGINAL SEE T49829) (FTD-MT-24-1384-71)
63851	SPECIFIC HEAT OF PALISADES DIABASE AT LIQUID HELIUM TEMPERATURES AND SOME COMMENTS ON THE VIBRATIONAL SPECTRUM OF COMPLEX STRUCTURES. MORRISON, J. A. NORTON, P. R. J. GEOPHYS. RES. 76 (20), 4993-6, 1971.
63859	OPTICAL DENSITY OF COALS IN THE CONTINUOUS INFRARED-SPECTRAL REGIONS. KEKIN, N. A. MIROSHNICHENKO, A. M. KHIM. TVERD. TOPL. (3), 76-80, 1971.
63871	INFRARED LIGHT FILTER BASED ON ALUMINUM CALCIUM GLASSES WITH TRANSMISSION AT 4.5-5.0 MICRONS. VEINBERG, T. I. LUNKIN, S. P. OPT. - MEKH. PROM. 38 (2), 60-1, 1971. (FOR ENGLISH TRANSLATION SEE TPRC NO. 67363)

TPRC Number	Bibliographic Citation
63961	**APPLICATION OF PERIODIC TEMPERATURE OSCILLATIONS TO HIGH - TEMPERATURE STUDIES OF THE THERMAL PROPERTIES OF DIELECTRICS.** TKACH, G. F. YURCHAK, R. P. TEPLOFIZ. VYS. TEMP. 9 (1), 210-3, 1971. (FOR ENGLISH TRANSLATION SEE TPRC NO. 63962)
63962	**APPLICATION OF PERIODIC TEMPERATURE OSCILLATIONS TO HIGH-TEMPERATURE STUDIES OF THE THERMAL PROPERTIES OF DIELECTRICS.** TKACH, G. F. YURCHAK, R. P. HIGH TEMP. 9 (1), 187-90, 1971. (ENGLISH TRANSLATION OF TEPLOFIZ. VYS. TEMP., 9 (1), 210-3, 1971; FOR ORIGINAL SEE TPRC NO. 63961)
63983	**CORRELATION OF GLASS HEAT CAPACITY AND DENSITY.** RODNIKOVA, V. V. GORODOV, V. S. FOREIGN TECHNOLOGY DIVISION 4PP., 1971. (ENGLISH TRANSLATION OF DOPOV. AKAD. NAUK UKR. RSR, SER. B, 32 (4), 355-7, 1970., FOR ORIGINAL SEE TPRC NO. 59018) (FTD-HC-23-980-71, AD-734638)
63993	**ON THE PROBLEM OF ABNORMAL EXPANSION OF ELECTROCORUNDUM.** FILONENKO, N. E. KUZNETSOVA, O. S. FOREIGN TECHNOLOGY DIVISION 11PP., 1971. (ENGLISH TRANSLATION OF OGNEUPORY, 17, 470-4, 1952; FOR ORIGINAL SEE TPRC NO. 37567) (FTD-HT-23-71-1650)
63999	**DESIGN AND DEVELOPMENT ENGINEERING HANDBOOK ON THERMAL EXPANSION PROPERTIES OF AEROSPACE MATERIALS. AT CRYOGENIC AND ELEVATED TEMPERATURES. (TECHNICAL SUPPORT PACKAGE FOR TECH BRIEF 69-10055).** WILLIAMS, L. R. YOUNG, J. D. SCHMIDT, E. H. ROCKETDYNE DIVISION, ROCKWELL INTERNATIONAL WASHINGTON, D. C. 175PP., 1967. (PB-184 749, R-6981)
64004	**CONTRIBUTION TO THE LOW-TEMPERATURE SPECIFIC HEAT OF VANADIUM-DOPED TITANIUM SESQUIOXIDE BY A ONE DIMENSIONAL ELECTRON GAS.** SJOSTRAND, M. E. KEESOM, P. H. PHYS. REV. LETT. 27 (21), 1434-5, 1971.
64009	**MODIFIED HIGH - TEMPERATURE X-RAY CAMERA FOR DIFFRACTOMETER REFLECTION STUDIES UP TO 2500 DEGREES C (RESEARCH NOTE).** BRAUN, W. WEISENBURGER, S. HIGH TEMP. - HIGH PRESSURES 2 (6), 703-6, 1970.
64013	**THERMAL EXPANSION OF PORCELAIN ENAMELS CONTAINING CERIUM OXIDE.** NEDELJKOVIC, A. I. COOK, R. L. AMER. CERAM. SOC. BULL. 50 (11), 929-32, 1971.
64014	**PROPERTIES OF BRANNERITE AND THE POSSIBILITY OF USING IT TO DETERMINE ABSOLUTE AGE.** POVILAITIS, M. M. YAKOVLEVSKAYA, T. A. KNYAZEVA, D. N. BELYAEVA, I. D. FOREIGN TECHNOLOGY DIVISION 40PP., 1971. (ENGLISH TRANSLATION OF ZAP. VSES. MINERAL. OBSHCHEST., 9M (2), 150-61, 1968; FOR ORIGINAL SEE TPRC NO. 50280) (FTD-MT-24-1623-71)
64018	**FERROELECTRIC AND MAGNETIC PROPERTIES OF /1-X/ BI MNO3-/X/ PB TIO3 SOLID SOLUTIONS.** BOKOV V A GRIGORYAN N A BRYZHINA M F KAZARYAN V S IZV AKAD NAUK SSSR SER FIZ 33 7 1164 1969 (FOR ENGLISH TRANSLATION SEE T64019)
64019	**FERROELECTRIC AND MAGNETIC PROPERTIES OF /1-X/ BI MNO3-/X/ PB TI O3 SOLID SOLUTIONS.** BOKOV V A GRIGORYAN N A BRYZHINA M F KAZARYAN V S BULL ACAD SCI USSR /PHYS SER/ 33 7 1082-4 1969 (ENGLISH TRANSLATION OF IZV. AKAD. NAUK SSSR, SSR. FIZ., 33 (7), 1164- , 1969; FOR ORIGINAL SEE T64018)
64041	**EFFECTIVE THERMAL CONDUCTIVITY OF DISPERSED SYSTEMS.** KOVENSKII, G. I. ZABRODSKII, S. S. TAMARIN, A. I. FOREIGN TECHNOLOGY DIVISION 15PP., 1971. (ENGLISH TRANSLATION OF ISSLED. PROTSESSOV PERENOSA APP. DISPERSNYMI SIST., 175-9, 1969., FOR ORIGINAL SEE TPRC NO. 57572) (FTD-MT-24-1818-71, AD-743686)
64044	**PHYSICOCHEMICAL PROPERTIES OF FAIENCE BODIES CONTAINING PERLITE AND DIATOMITE.** KHIZANISHVILI, I. G. SHUSHANISHVILI, A. I. TSANAVA, TS. P. FOREIGN TECHNOLOGY DIVISION 13PP., 1971. (ENGLISH TRANSLATION OF SOOBSHCH. AKAD. NAUK GRUZ. SSR, 55 (1), 89-92, 1969., FOR ORIGINAL SEE TPRC NO. 55684) (FTD-MT-24-1832-71, AD-740371)
64048	**NEW OPAQUE ZIRCONIUM GLAZES FOR STRUCTURAL CERAMICS.** EIDUKS, J. DUSAUSKAS-DUZ, S. SEDMALIS, U. PAUKSS, P. FOREIGN TECHNOLOGY DIVISION 19PP., 1971. (ENGLISH TRANSLATION OF STROIT. MATER., DETALI IZDELIYA, (10), 44-51, 1968; FOR ORIGINAL SEE TPRC NO. 56892) (FTD-MT-24-1861-71)
64050	**THERMAL CONDUCTIVITY OF CHIATURA MANGANESE CONCENTRATES AND THEIR PELLETIZED PRODUCTS.** NIKOLAISHVILI, G. U. KEKELIDZE, M. A. FOREIGN TECHNOLOGY DIVISION 21PP., 1971. (ENGLISH TRANSLATION OF PROIZVOD. PRIMEN. MARGANTSEVYKH FERROSPLAVOV, 28-36, 1968; FOR ORIGINAL SEE TPRC NO. 56266) (FTD-MT-24-1863-71, AD-739215)
64053	**THERMAL CONDUCTIVITY CELL FOR SMALL POWDERED SAMPLES.** CREMERS, C. J. REV. SCI. INSTRUM. 42 (11), 1694-6, 1971.
64140	**TRANSITION PROBABILITIES OF EUROPIUM IN PHOSPHATE GLASSES.** REISFELD, R. VELAPOLDI, R. A. BOEHM, L. ISH-SHALOM, M. J. PHYS. CHEM. 75 (26), 3980-3, 1971.
64177	**DETERMINATION OF THE COEFFICIENTS OF LINEAR EXPANSION OF FIBERGLAS BY THE UNSTEADY-STATE METHOD.** MEDVEDEV, N. N. J. ENG. PHYS. 13 (1), 22-5, 1967. (ENGLISH TRANSLATION OF INZH. FIZ. ZH., 13 (1), 37-43, 1967., FOR ORIGINAL SEE TPRC NO. 50070)
64194	**THEORETICAL INVESTIGATIONS ON THE INTERSTELLAR DUST. II. OPTICAL PROPERTIES OF SPHERICAL DUST PARTICLES FROM METEORITIC SILICATES AND DIRTY ICE.** DORSCHNER, J. ASTRON. NACHR. (2), 71-8, 1970.
64206	**MEASUREMENT OF SOLAR - OPTICAL PROPERTIES OF GLAZING MATERIALS.** PENNINGTON, C. W. MOORE, G. L. ASHRAE J. 13 (7), 55-8, 1971.
64208	**LONG-WAVE INFRARED SPECTRA OF SOME COMPLEX LEAD SULFIDES.** POVARENNYKH, A. S. SIDORENKO, G. A. SOLNTSEVA, L. S. SOLNTSEV, B. P. DOPOV. AKAD. NAUK UKR. RSR 33 B (7), 596-9, 1971. (FOR ENGLISH TRANSLATION SEE TPRC NO. 68829)
64215	**INFRARED SPECTRA OF BITUMENS FROM ORGANIC SUBSTANCES SCATTERED IN ROCKS.** GLEBOVSKAYA, E. A. MELTSANSKAYA, T. N. SURGOVA, N. Z. IZV. AKAD. NAUK SSSR, SER. GEOL. (5), 110-24, 1971.
64244	**NATURE OF MAXIMUMS IN ULTRAVIOLET ABSORPTION SPECTRA OF SOME GLASSES.** GAVRILOV, M. Z. ZH. PRIKL. SPEKTROSK. 14 (6), 1135-6, 1971. (FOR ENGLISH TRANSLATION SEE TPRC NO. 72666)
64246	**THERMAL CALCULATION OF THE PARTIAL REMOVAL OF A TECHNOLOGICAL BINDER (PLASTICIZER).** LARIONOV, B. A. SLOBODKIN, L. S. TEPLO-MASSOPERENOS APP. DISPERSNYMI SIST. 105-8, 1970.
64254	**THERMAL EXPANSION OF PORCELAIN BODIES AND GLAZES AS A FUNCTION OF FIRING TEMPERATURE.** ZAPP, F. BER. DEUT. KERAM. GES. 48 (8), 337-42, 1971.

TPRC Number	Bibliographic Citation
64268	**THERMAL CONDUCTIVITY OF FERRITE-GARNET SOLID SOLUTIONS.** LYAKAS, M. A. MARKOVIN, P. A. NUROMSKII, A. B. SHALABUTOV, YU. K. SIDOROV, V. G. IZV. VYSSH. UCHEB. ZAVED., FIZ. 14 (7), 113-14, 1971. (FOR ENGLISH TRANSLATION SEE TPRC NO. 72874)
64274	**SPECTRAL REFLECTANCE FOR PERPENDICULAR AND OBLIQUE INCIDENCE OF INFRARED RADIATION AND THE SPECTRAL TRANSMITTANCE FOR PERPENDICULAR INCIDENCE FOR SILICATES.** FRIEDL, W. KRUEGER, H. OPT. ACTA 18 (5), 385-99, 1971.
64309	**PHONON-IMPURITY SCATTERING IN MAGNESIUM OXIDE (IRON DIPOSITIVE ION, IRON TRIPOSITIVE ION) DOPED CRYSTALS.** VISHNEVSKII, I. I. SKRIPAK, V. N. FIZ. TVERD. TELA 13 (5), 1257-62, 1971. (FOR ENGLISH TRANSLATION SEE TPRC NO. 64310)
64310	**PHONON-IMPURITY SCATTERING IN MAGNESIUM OXIDE (IRON DIPOSITIVE ION, IRON TRIPOSITIVE ION) DOPED** VISHNEVSKII, I. I. SKRIPAK, V. N. SOV. PHYS. - SOLID STATE 13 (5), 1053-6, 1971. (ENGLISH TRANSLATION OF FIZ. TVERD. TELA, 13 (5), 1257-62, 1971., FOR ORIGINAL SEE TPRC NO. 64309)
64343	**ARSENIFEROUS PYRITE WITH A BRAVOITE- LIKE STRUCTURE.** BURKART- BAUMANN, I. OTTEMANN, J. MINER. DEPOSITA 6 (2), 148-52, 1971.
64345	**THERMAL CONDUCTIVITY OF LIGHTWEIGHT REFRACTORIES MEASURED BY A STANDARD METHOD AND A CYLINDER METHOD.** VISHNEVSKII, I. I. SKRIPAK, V. N. OGNEUPORY 36 (8), 45-52, 1971. (FOR ENGLISH TRANSLATION SEE TPRC NO. 64850)
64346	**THERMOPHYSICAL PROPERTIES OF DINAS CONCRETE BLOCKS.** TSIBIN, I. P. DYACHKOV, P. N. KURGIN, A. K. TAIGILDINA, E. P. DOMNIKOVA, G. I. KOSOLAPOVA, E. P. OGNEUPORY 36 (8), 56-8, 1971. (FOR ENGLISH TRANSLATION SEE TPRC NO. 64851)
64422	**NEW STEADY-STATE METHOD FOR DETERMINING THERMAL CONDUCTIVITY.** REITER, M. HARTMAN, H. J. GEOPHYS. RES. 76 (29), 7047-51, 1971.
64431	**CHANGE IN THE MODULUS OF ELASTICITY AND THERMAL EXPANSION RATIO OF PORCELAIN DEPENDENT ON THE INTRODUCTION OF PERLITE.** KHIZANISHVILI, I. G. MAMALADZE, R. A. FOREIGN TECHNOLOGY DIVISION 11PP., 1971. (ENGLISH TRANSLATION OF SOOBSHCH. AKAD. NAUK GRUZ. SSR, 58 (3), 597-600, 1970., FOR ORIGINAL SEE TPRC NO. 60872) (FTD-MT-71-1821, AD-747-393, N73-12627)
64435	**EFFECTS OF TEMPERATURE AND PRESSURE ON SOME PHYSICAL PROPERTIES OF ROCKS.** MOISEENKO, U. I. ISTOMIN, V. E. ALIEVA, M. A. FOREIGN TECHNOLOGY DIVISION 17PP., 1971. (ENGLISH TRANSLATION OF FIZ.-MEKH. SVOISTVA GORN. POROD VERKH. CHASTI ZEMNOI KORY, DOKL. SIMP., 148-52, 1968; FOR ORIGINAL SEE TPRC NO. 57555) (FTD-MT-71-1965)
64445	**BIDIRECTIONAL REFLECTANCE OF THE MOONLIT EARTH.** FOWLER, W. B. REED, E. I. BLAMONT, J. E. APPL. OPT. 10 (12), 2657-60, 1971.
64449	**APPLICATION OF CHARACTERISTIC VECTOR ANALYSIS TO THE SPECTRAL ENERGY DISTRIBUTION OF DAYLIGHT AND THE SPECTRAL REFLECTANCE OF AMERICAN SOILS.** CONDIT, H. R. APPL. OPT. 11 (1), 74-86, 1972.
64451	**VARIATION OF SINGLE PARTICLE MID-INFRARED EMISSION SPECTRUM WITH PARTICLE SIZE.** HUNT, G. R. LOGAN, L. M. APPL. OPT. 11 (1), 142-7, 1972. (AD-739791, AFCRL-72-0147)
64462	**X-RAY AND THERMAL EXPANSION STUDY OF NONSTOICHIOMETRIC SODIUM LITHIUM NIOBATE CERAMIC.** NITTA, T. MIYAZAWA, T. J. AMER. CERAM. SOC. 54 (12), 636-7, 1971.
64464	**DETERMINATION OF WATER IN LEAD BORATE GLASSES.** EAGAN, R. J. BERGERON, C. G. J. AMER. CERAM. SOC. 55 (1) 53-4, 1972.
64483	**ACTIVATED PHOTOCHROMIC GLASS AS THREE- DIMENSIONAL HOLOGRAPHIC MEDIA.** ASHCHEULOV, YU. V. SUKHANOV, V. I. OPT. SPEKTROSK. 30 (6), 1148-51, 1971. (FOR ENGLISH TRANSLATION SEE TPRC NO. 64484)
64484	**ACTIVATED PHOTOCHROMIC GLASS AS THREE- DIMENSIONAL HOLOGRAPHIC MEDIA.** ASHCHEULOV, YU. V. SUKHANOV, V. I. OPT. SPECTROS. 30 (6), 612-4, 1971. (ENGLISH TRANSLATION OF OPT. SPEKTROSK. 30 (6), 1148-51, 1971., FOR ORIGINAL SEE TPRC NO. 64483)
64502	**REFLECTION SPECTRA OF CALCITE IN FAR-INFRARED REGION.** ONOMICHI, M. KUDO, K. ARAI, T. J. PHYS. SOC. JAP. 31 (6), 1837, 1971.
64504	**HEAT CONDUCTION CALCULATION FOR A MODEL OF THE SURFACE OF THE MOON.** LIU, M.- K. WILLIAMS, F. A. INT. J. HEAT MASS TRANSFER 14 (11), 1843-51, 1971.
64533	**CHEMICAL NATURE OF COVERING PIGMENTS PRECIPITATED FROM SILICON DIOXIDE. INFRARED SPECTRAL ANALYSIS.** GALLUS- OLENDER, J. BANDROWSKA, C. PLASTE KAUT. 18 (7), 534-7, 1971.
64601	**UPPER SOIL LAYER OF THE SURFACE OF MARS.** BARASHOV, N. P. LUPISHKO, D. F. DOPOV. AKAD. NAUK UKR. RSR 33 B (8), 703-5, 1971.
64649	**MEASUREMENT OF EMISSIVITY OF INORGANIC MATERIALS IN SMALL SAMPLES.** ROMASHIN, A. G. BORZYKH, A. A. TIKHONOV, B. E. VEREVKA, V. G. POTAPOV, YU. M. TEPLOFIZ. VYS. TEMP. 9 (3), 517-21, 1971. (FOR ENGLISH TRANSLATION SEE TPRC NO. 64650)
64650	**MEASUREMENT OF DEGREE OF BLACKNESS ON SMALL SPECIMENS OF INORGANIC MATERIALS.** ROMASHIN, A. G. BORZYKH, A. A. TIKHONOV, B. E. VEREVKA, V. G. POTAPOV, YU. M. HIGH TEMP. 9 (3), 471-4, 1971. (ENGLISH TRANSLATION OF TEPLOFIZ. VYS. TEMP., 9 (3), 517-21, 1971; FOR ORIGINAL SEE TPRC NO. 64649)
64677	**STUDY OF CRYSTALLIZATION OF LEAD - GERMANIUM GLASSES BY INFRARED SPECTROSCOPY AND ELECTRON MICROSCOPY.** MOROZOV, V. N. TIKHOMIROV, G. P. KARTSEVA, L. A. MATSOYAN, B. D. IZV. AKAD. NAUK SSSR, SER. NEORG. MATER. 7 (2), 296-9, 1971. (FOR ENGLISH TRANSLATION SEE TPRC NO. 64678)
64678	**STUDY OF CRYSTALLIZATION OF LEAD - GERMANIUM GLASSES BY INFRARED SPECTROSCOPY AND ELECTRON MICROSCOPY.** MOROZOV, V. N. TIKHOMIROV, G. P. KARTSEVA, L. A. MATSOYAN, B. D. INORG. MAT. (USSR) 7 (2), 259-61, 1971. (ENGLISH TRANSLATION OF IZV. AKAD. NAUK SSSR, SER. NEORG. MATER. 7 (2), 296-9, 1971; FOR ORIGINAL SEE TPRC NO. 64677)
64691	**POINT DEFECTS AND RADIATION DAMAGE IN ALUMINOPHOSPHATE GLASSES.** BERSHOV, L. V. MARTIROSYAN, V. O. ZAVADOVSKAYA, E. K. STARODUBTSEV, V. A. IZV. AKAD. NAUK SSSR, NEORG. MATER. 7 (3), 476-80, 1971. (FOR ENGLISH TRANSLATION SEE TPRC NO. 64692)
64692	**POINT DEFECTS AND RADIATION DAMAGE IN ALUMINOPHOSPHATE GLASSES.** BERSHOV, L. V. MARTIROSYAN, V. O. ZAVADOVSKAYA, E. K. STARODUBTSEV, V. A. INORG. MAT. 7 (3), 417-20, 1971. (ENGLISH TRANSLATION OF IZV. AKAD. NAUK SSSR, NEORG. MATER. 7 (3), 476-80, 1971., FOR ORIGINAL SEE TPRC NO. 64691)

TPRC Number	Bibliographic Citation
64703	**THERMAL EXPANSION OF THE PRODUCTS OF CRYSTALLIZATION OF GLASSES OF THE SYSTEM LITHIUM OXIDE-GALLIUM OXIDE-SILICON OXIDE.** TSEKHOMSKAYA, T. S. DUBROVO, S. K. IZV. AKAD. NAUK SSSR, NEORG. MATER. 7 (4), 665-70, 1971. (FOR ENGLISH TRANSLATION SEE TPRC NO. 64704)
64704	**THERMAL EXPANSION OF THE PRODUCTS OF CRYSTALLIZATION OF GLASSES OF THE SYSTEM LITHIUM OXIDE-GALLIUM OXIDE-SILICON OXIDE.** TSEKHOMSKAYA, T. S. DUBROVO, S. K. INORG. MAT. (USSR) 7 (4), 577-81, 1971. (ENGLISH TRANSLATION OF IZV. AKAD. NAUK SSSR, NEORG. MATER., 7 (4), 665-70, 1971., FOR ORIGINAL SEE TPRC NO. 64703)
64711	**INFLUENCE OF SILICON DIOXIDE ON CERTAIN PROPERTIES OF ALUMINATE AND ALUMINOSILICATE GLASSES.** SOLOMIN, N. V. SHUMITSKAYA, L. F. KUZNETSOVA, L. K. KIM, E. I. IZV. AKAD. NAUK SSSR, NEORG. MATER. 7 (5), 850-2, 1971. (FOR ENGLISH TRANSLATION SEE TPRC NO. 64712)
64712	**INFLUENCE OF SILICON DIOXIDE ON CERTAIN PROPERTIES OF ALUMINATE AND ALUMINOSILICATE GLASSES.** SOLOMIN, N. V. SHUMITSKAYA, L. F. KUZNETSOVA, L. K. KIM, E. I. INORG. MAT. (USSR) 7 (5), 743-5, 1971. (ENGLISH TRANSLATION OF IZV. AKAD. NAUK SSSR, NEORG. MATER., 7 (5), 850-2, 1971., FOR ORIGINAL SEE TPRC NO. 64711)
64719	**BEHAVIOR OF CHROMITE AND HEMATITE DURING NEUTRON BOMBARDMENT.** KORENEVSKII, V. V. KRIVOKONEVA, G. K. PERGAMENSHCHIK, B. K. RYABEVA, E. G. SIDORENKO, G. A. IZV. AKAD. NAUK SSSR, NEORG. MATER. 7 (6), 1040-4, 1971. (FOR ENGLISH TRANSLATION SEE TPRC NO. 64720)
64720	**BEHAVIOR OF CHROMITE AND HEMATITE DURING NEUTRON BOMBARDMENT.** KORENEVSKII, V. V. KRIVOKONEVA, G. K. PERGAMENSHCHIK, B. K. RYABEVA, E. G. SIDORENKO, G. A. INORG. MAT. (USSR) 7 (6), 921-4, 1971. (ENGLISH TRANSLATION OF IZV. AKAD. NAUK SSSR, NEORG. MATER., 7 (6), 1040-4, 1971., FOR ORIGINAL SEE TPRC NO. 64719)
64731	**EXPERIMENTAL STUDIES OF THE SURFACE VIBRATIONS OF IONIC CRYSTALS IN POROUS GLASSES.** ABAEV, M. I. BOGOMOLOV, V. N. BRYKSIN, V. V. KLUSHIN, N. A. FIZ. TVERD. TELA 13 (6), 1578-83, 1971. (FOR ENGLISH TRANSLATION SEE TPRC NO. 64732)
64732	**EXPERIMENTAL STUDIES OF THE SURFACE VIBRATIONS OF IONIC CRYSTALS IN POROUS GLASSES.** ABAEV, M. I. BOGOMOLOV, V. N. BRYKSIN, V. V. KLUSHIN, N. A. SOV. PHYS. - SOLID STATE 13 (6), 1323-7, 1971. (ENGLISH TRANSLATION OF FIZ. TVERD. TELA, 13 (6), 1578-83, 1971; FOR ORIGINAL SEE TPRC NO. 64731)
64741	**THERMAL CONDUCTIVITY AND SPECIFIC HEAT OF TOPAZ.** ROMANOVA, M. V. SMIRNOV, I. A. TIKHONOV, V. V. FIZ. TVERD. TELA 13 (6), 1812-4, 1971. (FOR ENGLISH TRANSLATION SEE TPRC NO. 64742)
64742	**THERMAL CONDUCTIVITY AND SPECIFIC HEAT OF TOPAZ.** ROMANOVA, M. V. SMIRNOV, I. A. TIKHONOV, V. V. SOV. PHYS. - SOLID STATE 13 (6), 1515-6, 1971. (ENGLISH TRANSLATION OF FIZ. TVERD. TELA, 13 (6), 1812-4, 1971., FOR ORIGINAL SEE TPRC NO. 64741)
64814	**CRYSTALLIZATION IN THE SYSTEM TANTALUM OXIDE-GERMANIUM OXIDE.** GUTKINA, N. G. KOZHINA, I. I. SHMATOK, L. K. IZV. AKAD. NAUK SSSR, NEORG. MATER. 7 (8), 1386-8, 1971. (FOR ENGLISH TRANSLATION SEE TPRC NO. 64815)
64815	**CRYSTALLIZATION IN THE SYSTEM TANTALUM OXIDE-GERMANIUM OXIDE.** GUTKINA, N. G. KOZHINA, I. I. SHMATOK, L. K. INORG. MAT. (USSR) 7 (8), 1232-4, 1971. (ENGLISH TRANSLATION OF IZV. AKAD. NAUK SSSR, NEORG. MATER., 7 (8), 1386-8, 1971., FOR ORIGINAL SEE TPRC NO. 64814)
64816	**EFFECTS OF VARIOUS CATIONS ON THE REACTION KINETICS OF THE SYSTEMS LANTHANUM OXIDE-NICKEL OXIDE AND PRASEODYMIUM OXIDE-NICKEL OXIDE.** KNIGA, M. V. RUBINCHIK, YA. S. ZARETSKAYA, R. A. IZV. AKAD. NAUK SSSR, NEORG. MATER. 7 (8), 1389-92, 1971. (FOR ENGLISH TRANSLATION SEE TPRC NO. 64817)
64817	**EFFECTS OF VARIOUS CATIONS ON THE REACTION KINETICS OF THE SYSTEMS LANTHANUM OXIDE-NICKEL OXIDE AND PRASEODYMIUM OXIDE-NICKEL OXIDE.** KNIGA, M. V. RUBINCHIK, YA. S. ZARETSKAYA, R. A. INORG. MAT. (USSR) 7 (8), 1235-7, 1971. (ENGLISH TRANSLATION OF IZV. AKAD. NAUK SSSR, NEORG. MATER., 7 (8), 1389-92, 1971., FOR ORIGINAL SEE TPRC NO. 64816)
64818	**MAGNETIC PROPERTIES OF GLASSES OF THE SYSTEM SODIUM OXIDE-IRON OXIDE-TITANIUM OXIDE-SILICON OXIDE.** STEPANOV, S. A. IZV. AKAD. NAUK SSSR, NEORG. MATER. 7 (8), 1414-6, 1971. (FOR ENGLISH TRANSLATION SEE TPRC NO. 64819)
64819	**MAGNETIC PROPERTIES OF GLASSES OF THE SYSTEM SODIUM OXIDE-IRON OXIDE-TITANIUM OXIDE-SILICON OXIDE.** STEPANOV, S. A. INORG. MAT. (USSR) 7 (8), 1255-7, 1971. (ENGLISH TRANSLATION OF IZV. AKAD. NAUK SSSR, NEORG. MATER., 7 (8), 1414-6, 1971., FOR ORIGINAL SEE TPRC NO. 64818)
64822	**HIGH DENSITY CORUNDUM REFRACTORIES.** KABAKOVA, I. I. DEGTYAREVA, E. V. KAINARSKII, I. S. OGNEUPORY 36 (1), 30-6, 1971. (FOR ENGLISH TRANSLATION SEE TPRC NO. 64823)
64823	**HIGH DENSITY CORUNDUM REFRACTORIES.** KABAKOVA, I. I. DEGTYAREVA, E. V. KAINARSKII, I. S. REFRACTORIES (USSR) 12 (1), 36-41, 1971. (ENGLISH TRANSLATION OF OGNEUPORY, 36 (1), 30-6, 1971; FOR ORIGINAL SEE TPRC NO. 64822)
64828	**MAGNESITE - CHROMITE PRODUCTS FROM HIGH PURITY MAGNESITE AND CHROMITE.** BRON, V. A. KUKURUZOV, A. P. DIESPEROVA, M. I. STEPANOVA, I. A. OGNEUPORY 36 (2), 32-7, 1971. (FOR ENGLISH TRANSLATION SEE TPRC NO. 64829)
64829	**MAGNESITE - CHROMITE PRODUCTS FROM HIGH PURITY MAGNESITE AND CHROMITE.** BRON, V. A. KUKURUZOV, A. P. DIESPEROVA, M. I. STEPANOVA, I. A. REFRACTORIES (USSR) 12 (2), 105-9, 1971. (ENGLISH TRANSLATION OF OGNEUPORY, 36 (2), 32-7, 1971., FOR ORIGINAL SEE TPRC NO. 64828)
64830	**STUDY OF SILICEOUS BODIES FOR MONLITHIC LININGS AND USING THEM IN STEEL LADLES.** FLYAGIN, V. G. RUTMAN, D. S. ISAEV, G. I. POPOV, A. D. TANTSYREV, O. V. AKSYUCHITS, N. N. LARIONOV, E. D. OGNEUPORY 36 (3), 27-31, 1971. (FOR ENGLISH TRANSLATION SEE TPRC NO. 64831)
64831	**STUDY OF SILICEOUS BODIES FOR MONLITHIC LININGS AND USING THEM IN STEEL LADLES.** FLYAGIN, V. G. RUTMAN, D. S. ISAEV, G. I. POPOV, A. D. TANTSYREV, O. V. AKSYUCHITS, N. N. LARIONOV, E. D. REFRACTORIES (USSR) 12 (3), 163-8, 1971. (ENGLISH TRANSLATION OF OGNEUPORY, 36 (3), 27-31, 1971., FOR ORIGINAL SEE TPRC NO. 64830)
64834	**PRODUCTION AND PROPERTIES OF DENSE MAGNESITE REFRACTORIES WITH ENHANCED SPALLING RESISTANCE.** KUKOLEV, G. V. NEMETS, I. I. DOBROVOLSKII, G. B. NESTERTSOV, A. I. OGNEUPORY 36 (3), 43-8, 1971. (FOR ENGLISH TRANSLATION SEE TPRC NO. 64835)
64835	**PRODUCTION AND PROPERTIES OF DENSE MAGNESITE REFRACTORIES WITH ENHANCED SPALLING RESISTANCE.** KUKOLEV, G. V. NEMETS, I. I. DOBROVOLSKII, G. B. NESTERTSOV, A. I. REFRACTORIES (USSR) 12 (3), 180-7, 1971. (ENGLISH TRANSLATION OF OGNEUPORY, 36 (3), 43-8, 1971., FOR ORIGINAL SEE TPRC NO. 64834)

TPRC Number	Bibliographic Citation
64836	**PROPERTIES OF QUARTZ BODIES USED FOR LINING LARGE INDUCTION FURNACES.** KAIBICHEVA, M. N. OGNEUPORY 36 (4), 31–4, 1971. (FOR ENGLISH TRANSLATION SEE TPRC NO. 64837)
64837	**PROPERTIES OF QUARTZ BODIES USED FOR LINING LARGE INDUCTION FURNACES.** KAIBICHEVA, M. N. REFRACTORIES (USSR) 12 (4), 236–40, 1971. (ENGLISH TRANSLATION OF OGNEUPORY, 36 (4), 31–4, 1971., FOR ORIGINAL SEE TPRC NO. 64836)
64840	**HIGH - ALUMINA RAMMING BODIES BASED ON KYANITE - SILLIMANITE CONCENTRATE.** TSEITLIN, L. A. MERKULOVA, E. V. OGNEUPORY 36 (5), 4–10, 1971. (FOR ENGLISH TRANSLATION SEE TPRC NO. 64841)
64841	**HIGH - ALUMINA RAMMING BODIES BASED ON KYANITE - SILLIMANITE CONCENTRATE.** TSEITLIN, L. A. MERKULOVA, E. V. REFRACTORIES (USSR) 12 (5), 278–83, 1971. (ENGLISH TRANSLATION OF OGNEUPORY, 36 (5), 4–10, 1971., FOR ORIGINAL SEE TPRC NO. 64840)
64842	**CHANGES IN THE PROPERTIES OF NOVOSELITS KAOLIN DURING HEATING.** PITAK, N. V. SHULYAK, R. S. OGNEUPORY 36 (5), 40–7, 1971. (FOR ENGLISH TRANSLATION SEE TPRC NO. 64843)
64843	**CHANGES IN THE PROPERTIES OF NOVOSELITS KAOLIN DURING HEATING.** PITAK, N. V. SHULYAK, R. S. REFRACTORIES (USSR) 12 (5), 317–24, 1971. (ENGLISH TRANSLATION OF OGNEUPORY, 36 (5), 40–7, 1971., FOR ORIGINAL SEE TPRC NO. 64842)
64844	**ROLE OF PLASTIC DEFORMATION IN THE THERMAL DESTRUCTION OF REFRACTORIES.** DAUKNIS, V. I. KAZAKYAVICHYUS, K. A. YURENAS, V. L. OGNEUPORY 36 (6), 31–5, 1971. (FOR ENGLISH TRANSLATION SEE TPRC NO. 64845)
64845	**ROLE OF PLASTIC DEFORMATION IN THE THERMAL DESTRUCTION OF REFRACTORIES.** DAUKNIS, V. I. KAZAKYAVICHYUS, K. A. YURENAS, V. L. REFRACTORIES (USSR) 12 (6), 376–81, 1971. (ENGLISH TRANSLATION OF OGNEUPORY, 36 (6), 31–5, 1971., FOR ORIGINAL SEE TPRC NO. 64844)
64846	**CORUNDUM REFRACTORY OF INCREASED THERMAL - SHOCK RESISTANCE.** GAODU, A. N. SHAPIRO, YA. Z. SHAPTALA, V. I. OGNEUPORY 36 (6), 42–5, 1971. (FOR ENGLISH TRANSLATION SEE TPRC NO. 64847)
64847	**CORUNDUM REFRACTORY OF INCREASED THERMAL - SHOCK RESISTANCE.** GAODU, A. N. SHAPIRO, YA. Z. SHAPTALA, V. I. REFRACTORIES (USSR) 12 (6), 389–92, 1971. (ENGLISH TRANSLATION OF OGNEUPORY, 36 (6), 42–5, 1971., FOR ORIGINAL SEE TPRC NO. 64846)
64848	**CERAMICS BASED ON ZIRCONIUM DIOXIDE STABILIZED WITH YTTRIUM CONCENTRATE.** KUZNETSOV, A. K. KRASILNIKOV, M. D. KELER, E. K. OGNEUPORY 36 (6), 45–8, 1971. (FOR ENGLISH TRANSLATION SEE TPRC NO. 64849)
64849	**CERAMICS BASED ON ZIRCONIUM DIOXIDE STABILIZED WITH YTTRIUM CONCENTRATE.** KUZNETSOV, A. K. KRASILNIKOV, M. D. KELER, E. K. REFRACTORIES (USSR) 12 (6), 393–5, 1971. (ENGLISH TRANSLATION OF OGNEUPORY, 36 (6), 45–8, 1971., FOR ORIGINAL SEE TPRC NO. 64848)
64850	**THERMAL CONDUCTIVITY OF INSULATING REFRACTORIES MEASURED BY THE STANDARD METHOD AND BY THE CYLINDER METHOD.** VISHNEVSKII, I. I. SKRIPAK, V. N. REFRACTORIES 12 (8), 523–9, 1971. (ENGLISH TRANSLATION OF OGNEUPORY 36 (8), 45–52, 1971., FOR ORIGINAL SEE TPRC NO. 64345)
64851	**THERMOPHYSICAL PROPERTIES OF DINAS CONCRETE BLOCKS.** TSIBIN, I. P. DYACHKOV, P. N. PURGIN, A. K. TAIGILDINA, E. P. DOMNIKOVA, G. I. KOSOLAPOVA, E. P. REFRACTORIES 12 (8), 534–6, 1971. (ENGLISH TRANSLATION OF OGNEUPORY 36 (8), 56–8, 1971., FOR ORIGINAL SEE TPRC NO. 64346)
64869	**CHANGE IN PHYSICAL PROPERTIES OF BORNITE DURING THE DISSOLUTION OF CHALCOPYRITE IN IT.** SLAVSKAYA, A. I. KACHALOVSKAYA, V. M. SHARYBKINA, M. A. RUDNICHENKO, V. E. GEOL. RUD. MESTOROZHD. 13 (4), 113–15, 1971.
64871	**ASTRONOMICAL INFRARED SPECTROSCOPY WITH A CONES-TYPE INTERFEROMETER. II. MARS, 2500-3500 CM-1.** BEER, R. NORTON, R. H. MARTONCHIK, J. V. ICARUS 15 (1), 1–10, 1971.
64876	**THERMAL EXPANSION OF SOLID SOLUTIONS IN THE FLUORPHLOGOPITE - BARIUM MICA.** KITAJIMA, K. KOGYO KAGAKU ZASSHI 74 (9), 1792–6, 1971.
64907	**STUDY OF SCATTERING MEDIUM WITH LOW ABSORPTION USING THE TWO- PARAMETER APPROXIMATION THEORY AND G. V. ROZENBERGS THEORY. I. USE OF THE TWO PARAMETER APPROXIMATION THEORY.** VOISHVILLO, N. A. OPT. SPEKTROSK. 31 (2), 275–82, 1971. (FOR ENGLISH TRANSLATION SEE TPRC NO. 64908)
64908	**STUDY OF SCATTERING MEDIUM WITH LOW ABSORPTION USING THE TWO- PARAMETER APPROXIMATION THEORY AND G. V. ROZENBERGS THEORY. I. USE OF THE TWO PARAMETER APPROXIMATION THEORY.** VOISHVILLO, N. A. OPT. SPECTROS. 31 (2), 147–51, 1971. (ENGLISH TRANSLATION OF OPT. SPEKTROSK. 31 (2), 275–82, 1971., FOR ORIGINAL SEE TPRC NO. 64907)
64915	**THERMAL AND DILATOMETRIC STUDY OF THE BISMUTH TITANATE - BISMUTH MOLYBDEATE SYSTEM.** SMOLYANINOV, N. P. BOCHKAREVA, O. B. MARENICH, S. S. ARBUZOVA, A. I. ZH. NEORG. KHIM. 16 (8), 2299–302, 1971. (FOR ENGLISH TRANSLATION SEE TPRC NO. 64916)
64916	**THERMOGRAPHIC AND DILATOMETRIC INVESTIGATION OF THE BISMUTH TITANATE-BISMUTH MOLYBDATE SYSTEM.** SMOLYANINOV, N. P. BOCHKAREVA, O. B. MARENICH, S. S. ARBUZOVA, A. I. RUSS. J. INORG. CHEM. 16 (8), 1225–7, 1971. (ENGLISH TRANSLATION OF ZH. NEORG. KHIM., 16 (8), 2299–302, 1971; FOR ORIGINAL SEE TPRC NO. 64915)
64933	**ELECTRIC FURNACE BORALUMINATE.** BAUMANN, H. N., JR. MOORE, C. H., JR. J. AMER. CERAM. SOC. 25 (14), 391–4, 1972.
64934	**STABILITY OF PLANETARY INTERIORS.** SCHUBERT, G. TURCOTTE, D. L. OXBURGH, E. R. GEOPHYS. J. ROY. ASTRON. SOC. 18, 441–60, 1969.
64936	**XXVII.-A FURTHER NOTE ON THE INFLUENCE OF IRON OXIDE ON THE RATE OF QUARTZ INVERSION IN LIME AND LIME-CLAY BONDED SILICA BRICKS.** HUGILL, W. REES, W. J. TRANS. CERAM. SOC. 30 (10), 330–6, 1931.
64938	**ELECTRIC AND THERMAL PROPERTIES OF ROCKS.** MOISEYENKO, U. I. SOKOLOVA, L. S. ISTOMIN, V. YE. NATIONAL AERONAUTICS AND SPACE ADMINISTRATION 59PP., 1972. (NASA-TT-F-671)
64957	**EFFECT OF CRYSTALLIZATION ON THE MECHANICAL PROPERTIES OF LITHIUM OXIDE-SILICON OXIDE GLASS-CERAMICS.** FREIMAN, S. W. HENCH, L. L. J. AMER. CERAM. SOC. 55 (2), 86–90, 1972.
64967	**THE VARIATION/CHANGE IN THE SPECIFIC HEAT OF IRON-ORE PELLETS IN THE PROCESS OF OXIDIZING ROASTING.** BRATCHIKOV, S. G. YUR'YEV, B. P. MAIZEL, G. M. TOPORISHCHEV, I. G. IZV. VYSSH. UCHEB. ZAVED. CHERN. MET. 14 (4), 32–4, 1971. (FOR ENGLISH TRANSLATION SEE TPRC NO. 64968)

TPRC Number	Bibliographic Citation
64968	**THE VARIATION/CHANGE IN THE SPECIFIC HEAT OF IRON-ORE PELLETS IN THE PROCESS OF OXIDIZING ROASTING.** BRATCHIKOV, S. G. YUR'YEV, B. P. MAIZEL, G. M. TOPORISHCHEV, I. G. FOREIGN TECHNOLOGY DIVISION 13PP., 1971. (ENGLISH TRANSLATION OF IZV. VYSSH. UCHEB. ZAVED. CHERN. MET., 1N (4), 32-4, 1971; FOR ORIGINAL SEE TPRC NO. 64967) (FTD-MT-71-1918, AD-748169)
64971	**DEVELOPMENT OF A THEORY OF THE SPECTRAL REFLECTANCE OF MINERALS. PART 2.** ARONSON, J. R. EMSLIE, A. G. ROACH, L. H. STRONG, P. F. VON THUNA, P. C. LITTLE (ARTHUR D.), INC., CAMBRIDGE, MASS. 90PP., 1971. (N71-29925, NASA-CR-115057, ADL-C-72594-TP-2)
64980	**THERMAL CONDUCTIVITY AND SPECIFIC HEAT CAPACITY OF ZINC SULFIDE AND CADMIUM SULFIDE IN THE TEMPERATURE REGION FROM 20 K TO 300 K.** KRUEGER, R. TECHNISCHE UNIVERSITY, BERLIN, PH.D. THESIS 99PP., 1969. (N71-29717)
64989	**THERMOPHYSICAL PROPERTIES OF THE LUNAR MATERIAL PROVIDED BY THE LUNA 16 AUTOMATIC INTERPLANETARY STATION.** AVDUEVSKII, V. S. ANFIMOV, N. A. MAROV, M. L. SHALAEV, S. P. EKONOMOV, A. P. AKAD. NAUK SSSR, DOKL. 196, 801-4, 1971.
64992	**THERMAL INERTIA OF GANYMEDE FROM 20-MICRON ECLIPSE RADIOMETRY.** MORRISON, D. CRUIKSHANK, D. P. MURPHY, R. E. MARTIN, T. Z. BEERY, J. G. SHIPLEY, J. P. ASTROPHYS. J. 167 (PT. 2), L107-11, 1971.
64993	**PHASE DEPENDENCE OF MARTIAN INTEGRAL BRIGHTNESS.** BARABASHOV, N. P. ALEKSANDROV, IU. A. LUPISHKO, T. A. LUPISHKO, D. F. ASTRON. ZH. 48 (3), 581-6, 1971. (FOR ENGLISH TRANSLATION SEE TPRC NO. 69194)
65008	**ALBEDO OF LUNAR SOIL.** SALISBURY, J. W. ICARUS 13, 509-12, 1970.
65014	**APPROXIMATE SOLUTION TO THE INVERSE PROBLEM OF THERMAL CONDUCTIVITY.** ZHEREBYAT'EV, I. F. KOZIN, V. I. LUK'YANOV, A. T. PRILEPSKII, V. N. IZV. AKAD. NAUK KAZ. SSR, SER. FIZ.-MAT. 9 (4), 58-61, 1971.
65024	**VACUUM ULTRAVIOLET ABSORPTION IN ALKALI-DOPED FUSED SILICA AND SILICATE GLASSES.** SIGEL, G. H. JR. J. PHYS. CHEM. SOLIDS 32 (10), 2373-83, 1971.
65030	**IRON - BEARING NORMAL COVELLITE.** CLARK, A. H. SILLITOE, R. H. NEUES JAHRB. MINERAL., MONATSH. 2 (9) 2, 424-8, 1971.
65035	**OPTICAL PROPERTIES OF SOME BOWEN BASIN VITRINITES.** DAVIS, A. QUEENSL. DEP. MINES, GEOL. SURV. QUEENSL. REP. (62), 61-76, 1971.
65069	**THERMAL CONDUCTIVITY OF FROZEN SOILS.** PENNER, E. CAN. J. EARTH SCI. 7 (3), 982-7, 1970.
65087	**THERMAL INSULATION STUDY, PHASE 2.** GRUNET, W. E. MORIHARA, H. NOTARO, F. WEBSTER, D. J. ZAWIERICHA, R. LINDE DIV., UNION CARBIDE CORP., INDIANAPOLIS, IND. 371PP., 1971. (N71-38752, ALO-2832-43)
65109	**PROPERTIES AND STRUCTURE OF MULTI - ALKALI BOROGERMANATE GLASSES.** KRUPKIN, YU. S. EVSTROPEV, K. S. IZV. AKAD. NAUK SSSR, NEORG. MATER. 7 (9), 1591-5, 1971. (FOR ENGLISH TRANSLATION SEE TPRC NO. 65110)
65110	**PROPERTIES AND STRUCTURE OF MULTI - ALKALI BOROGERMANATE GLASSES.** KRUPKIN, YU. S. EVSTROPEV, K. S. INORG. MAT. (USSR) 7 (9), 1410-4, 1971. (ENGLISH TRANSLATION OF IZV. AKAD. NAUK SSSR, NEORG. MATER., 7 (9), 1591-5, 1971., FOR ORIGINAL SEE TPRC NO. 65109)
65121	**SIMULTANEOUS DETERMINATION OF THERMAL CHARACTERISTICS FOR DISPERSE MATERIALS IN THE PROCESS OF HEATING.** DZHAPARIDZE, P. N. LANDAU, I. N. J. ENG. PHYS. 14 (2), 165-70, 1968. (ENGLISH TRANSLATION OF INZH. FIZ. ZH. 14 (2), 314-21, 1968., FOR ORIGINAL SEE TPRC NO. 49919)
65123	**DETERMINATION OF THE THERMOPHYSICAL CHARACTERISTICS OF FREE-FLOWING MATERIALS AND THICK LAYERS OF VARIOUS SUBSTANCES.** MEDVEDEV, N. N. J. ENG. PHYS. 14 (2), 176-9, 1968. (ENGLISH TRANSLATION OF INZH. FIZ. ZH. 14 (2), 329-33, 1968., FOR ORIGINAL SEE TPRC NO. 49504)
65125	**THE TEMPERATURE FUNCTION FOR THE INTEGRAL EMISSIVITY OF CERTAIN MATERIALS BELOW 300 DEGREES K.** MIKHALCHENKO, R. S. GERZHIN, A. G. PERSHIN, N. P. J. ENGINEERING PHYS. 14 (3), 294-7, 1968. (ENGLISH TRANSLATION OF INZH. FIZ. ZH. 14 (3), 545-51, 1968., FOR ORIGINAL SEE TPRC NO. 49921)
65129	**STUDY OF INFRARED REFLECTION SPECTRA OF BERYL IN THE 280-1400 CM(-1) REGION.** GERVAIS, F. PIRIOU, B. COMPT. REND. 274 B (4), 252-5, 1972.
65133	**REGISTRATION OF HOLOGRAPHIC SYSTEMS IN TRANSPARENT FERROELECTRIC CERAMICS.** MICHERON, F. HERMOSIN, A. BISMUTH, G. NICOLAS, J. COMPT. REND. 274 B (5), 361-4, 1972.
65138	**VARIATION IN THE THERMAL CONDUCTIVITY OF AN OXIDE FUEL ELEMENT DURING BURNUP.** KLEYKAMP, H. KERNFORSCHUNGSZENTRUM, KARLSRUKE, W. GERMANY 16PP., 1970. (N71-32154, EURFNR-817, KFK-1245)
65139	**IMPROVED INTERFEROMETRIC PROCEDURE WITH APPLICATION TO EXPANSION MEASUREMENTS.** SAUNDERS, J. B. J. RES. NATL. BUR. STAND. 23, 179-95, 1939.
65144	**INFRARED REFLECTIVITY OF NATURAL ANTIMONY SESQUISULFIDE.** RIEDE, V. ANN. PHYS. (LEIPZIG) 25 (4), 415-6, 1970.
65152	**INFRARED REFLECTION SPECTRA OF THREE TYPES OF SILICEOUS MINERALS.** BRUNEL, R. VERGNOUX, A. M. VIERNE, R. C. R. ACAD. SCI. 272 D (2), 189-92, 1971.
65158	**THERMODIFFUSIVITY OF MOLDING MATERIALS AND SOLIDIFICATION TIMES OF STEEL CASTINGS.** BOEHM, H. J. GIESSEREI 58 (5), 118-21, 1971.
65176	**INFRARED ABSORPTION SPECTRA OF NIOBATES, TANTALATES, AND ANTIMONATES WITH RUTILE STRUCTURE.** ROCCHICCIOLI, C. DUPUIS, T. FRANCK, R. HARMELIN, M. WADIER, C. J. CHIM. PHYS. PHYSICOCHIM. BIOL. 67 (11-2), 2037-44, 1970.
65186	**EFFECT OF PLUTONIUM CONTENT ON THE THERMAL CONDUCTIVITY OF URANIUM OXIDE-PLUTONIUM OXIDE SOLID SOLUTIONS.** GIBBY, R. L. J. NUCL. MATER. 38 (2), 163-77, 1971.
65195	**PROPERTIES OF CRETACEOUS REFRACTORY SANDSTONES OF THE INTRA-SUDETIC TROUGH.** KUBICZ, A. KWART. GEOL. 14 (3), 506-18, 1970.
65200	**VISIBLE AND NEAR - INFRARED SPECTRA OF MINERALS AND ROCKS. I. SILICATE MINERALS.** HUNT, G. R. SALISBURY, J. W. MOD. GEOL. 1 (4), 283-300, 1970.
65209	**EFFECTS OF THE FINENESS OF SILICA-ALUMINA CATALYSTS ON THE SHAPE OF INFRARED ABSORPTION SPECTRA.** WAL, W. PRZEM. CHEM. 50 (10), 650-3, 1971.

TPRC Number	Bibliographic Citation

65218 **CONTINUOUS MELTING OF UVIOL GLASS AND MECHANIZED PRODUCTION OF TUBES FROM IT.**
KOCHETKOVA, G. V. KLEGG, D. I. DYMOV, A. T.
DUBKOVA, A. I. KOVALENKO, I. S.
KELEINIKOV, V. I. SIVKO, A. P.
MOROZOVA, A. N. ZOTOVA, T. M.
STEKLO KERAM.
(9), 22, 1971.
(FOR ENGLISH TRANSLATION SEE TPRC NO. 66790)

65219 **INFRARED - SPECTROSCOPIC STUDY OF THE STRUCTURE OF SODIUM OXIDE - CALCIUM OXIDE - SILICON DIOXIDE SYSTEM GLASSES.**
BOBKOVA, N. M. TIZHOVKA, ZH. S.
STEKLO, SITALLY SILIKAT. MATER.
(1), 12-19, 1970.

65225 **SEA FLOOR GEOTHERMAL MEASUREMENTS FROM VEMA CRUISE 24.**
LANGSETH, G. M., JR. MALONE, I. BREGER, D.
LAMONT-DOHERTY GEOL. OBSERVATORY, PALISADES, N. Y.
452PP., 1971.
(AD-729 682)

65226 **THERMAL CONDUCTION BY ELECTRONS IN STELLAR MATTER.**
HUBBARD, W. B. LAMPE, M.
CALIF. INST. TECHNOL., PASADENA
72PP., 1968.
(AD-729 702)

65248 **THERMODIFFUSIVITY OF GROUND HARD COALS.**
KUDELA, V. PISTEK, P.
SB. PR. UVP
(22), 197-247, 1970.

65249 **INFRARED SPECTRA OF MICA IN THE HYDROXYL VIBRATION**
MELCHAKOVA, L. V. PLYUSNINA, I. I.
SOBOLEV, R. N.
VESTN. MOSK. UNIV., GEOL.
26 (3), 96-100, 1971.

65250 **REFLECTANCE OF WOLFRAMITE GROUP MINERALS AS INFLUENCED BY THEIR CHEMICAL COMPOSITION.**
SHUMSKAYA, N. I.
ZAP. VSES. MINERAL. OBSHCHEST.
100 (5), 629-35, 1971.

65258 **PROBLEM SOLVING (IN THE CERAMIC INDUSTRY) WITH THE DILATOMETER.**
SCHOENLAUB, R. A.
CERAM. AGE
87 (10), 22-3, 26-30, 1971.

65283 **NORMAL EMISSION FACTORS OF DIELECTRIC MATERIALS DETERMINED BETWEEN 1000 AND 1500 DEGREES.**
SACADURA, J. F. O.
REV. INT. HAUTES TEMP. REFRACT.
8 (2), 101-10, 1971.

65323 **TOTAL EMITTANCE OF LUNAR FINES.**
BIRKEBAK, R. C. ABDULKADIR, A.
J. GEOPHYS. RES.
77 (7), 1340-1, 1972.

65340 **LUNAR SPECTRAL TYPES.**
MC CORD, T. B. CHARETTE, M. P.
JOHNSON, T. V. LEBOFSKY, L. A.
PIETERS, C. ADAMS, J. B.
J. GEOPHYS. RES.
77 (8), 1349-59, 1972.

65401 **APPARATUS FOR MEASURING THE THERMAL EXPANSION OF OPTICAL MATERIALS FROM 30 DEGREES C TO 250 DEGREES C.**
EBERSOLE, J. F. BALLARD, S. S.
BROWDER, J. S.
APPL. OPT.
11 (4), 844-8, 1972.

65427 **THERMAL CONDUCTIVITY OF GLASS-FIBER SYSTEMS.**
VNUKOV, S. P. RYABOV, V. A. FEDOSEEV, D. V.
INZH.-FIZ. ZH.
21 (5), 797-803, 1971.
(FOR ENGLISH TRANSLATION SEE T73365)

65443 **THERMAL CONDUCTIVITY OF DISPERSED SYSTEMS.**
TANAEVA, S. A. DOMOROD, L. S.
TEPLO - MASSOOBMEN NIZK. TEMP.
74-81, 1970.

65498 **CHANGES IN THE COEFFICIENTS OF LINEAR EXPANSION OF QUARTZES FROM TIN ORE DEPOSITS AS A FUNCTION OF TEMPERATURE.**
USPENSKAYA, A. B.
FOREIGN TECHNOLOGY DIVISION
6PP., 1972.
(ENGLISH TRANSLATION OF IZV. VYSSH. UCHEB. ZAVED., GEOL. RAZVED., 13 (7), 26-30, 1970; FOR ORIGINAL SEE T60799)
(FTD-HC-23-50-72)

65505 **SYNTHESIS OF GLASS - CERAMICS MATERIALS FROM BLAST FURNACE SLAGS OF THE KUZNETSK METALLURGICAL COMBINE.**
PAVLUSHKIN, N. M. LOGVINENKO, A. T.
PENDYURINA, T. E.
FOREIGN TECHNOLOGY DIVISION
9PP., 1972.
(ENGLISH TRANSLATION OF IZV. SIB. OTD. AKAD. NAUK SSSR, SER. KHIM NAUK, (2), 148-53, 1969; FOR ORIGINAL SEE T55287)
(FTD-HC-23-139-72)

65520 **EFFECT OF THERMAL RADIATION ON THE EXPANSION AND CONTRACTION OF MOLDING SANDS.**
SZRENIAWSKI, J.
FOREIGN TECHNOLOGY DIVISION
17PP., 1972.
(ENGLISH TRANSLATION OF PRZEGL. ODLEW., 18 (2), 41-8, 1968; FOR ORIGINAL SEE T50178)
(AD-744287, FTD-HC-1530-71)

65523 **THERMAL CONDUCTIVITY OF SOME SYNTHETIC FIBER ROVINGS.**
DOROKHINA, YU. M.
FOREIGN TECHNOLOGY DIVISION
10PP., 1971.
(ENGLISH TRANSLATION OF IZV. VYSSH. UCHEB. ZAVED., TEKHNOL. TEKST. PROM., (3), 105-9, 1968., FOR ORIGINAL SEE TPRC NO. 51010)
(FTD-HT-71-1991)

65531 **ELECTRICAL RESISTANCE AND THERMAL CONDUCTIVITY OF CHARGES OF CARBON-CONTAINING FERROMANGANESE AND SILICOMANGANESE.**
NIKOLAISHVILI, G. U. KEKELIDZE, M. A.
FOREIGN TECHNOLOGY DIVISION
11PP., 1972.
(ENGLISH TRANSLATION OF PROIZVOD. PRIMEN. MARGANTSEVYKH. FERROSPLAVOV, 37-46, 1968; FOR ORIGINAL SEE TPRC NO. 56196)
(AD-745231, N72-33553, FTD-HC-23-2106-71)

65535 **THERMAL DIFFUSIVITY MEASUREMENTS OF IRRADIATED OXIDE FUELS (LWBR DEVELOPMENT PROGRAM).**
MATOLICH, J., JR. STORHOK, V. W.
BATTELLE MEMORIAL INST., COLUMBUS, OHIO
52PP., 1970.
(BMI-X-10274)

65538 **X-RAY DIFFRACTION STUDY OF THE YTTRIA-HAFNIA SYSTEM.**
STACY, D. W.
IOWA STATE UNIVERSITY, AMES, IOWA, PH.D. THESIS
84PP., 1971.
(IS-T-425, N72-12721)

65570 **HIGH TEMPERATURE INVESTIGATIONS OF HEAT PROPERTIES OF SOLIDS.**
FILIPPOV, L. P. YURCHAK, R. P.
INZH. FIZ. ZHUR.
21 (3), 561-77, 1971.
(FOR ENGLISH TRANSLATION SEE T73275)

65601 **MEASUREMENT OF THERMAL CONDUCTIVITY IN ROCKS.**
MAY, C. A.
CASE WESTERN RESERVE UNIVERSITY, M. S. THESIS,
72PP., 1971.

65621 **ABSOLUTE DETERMINATION OF SPECIFIC HEATS WITH THE FLUX CALORIMETER.**
TER-MINASSIAN, L. PETIT, J. C. GUYEN-VAN-KIET
CAH. THERM.
(1), (PT. 1), 34-49, 1971.

65626 **THERMAL PROPERTIES OF DOMESTIC COALS.**
PAK, Y. K.
HWAHAK KWA HWAHAK KONGUP
(2), 67-71, 1971.

65687 **THERMOPHYSICAL CHARACTERISTICS OF MOLDING MATERIALS STUDIED BY ROUTINE METHODS.**
NOSKOV, B. A. YANOVER, YA. D.
IZV. VYSSH. UCHEB. ZAVED., MASHINOSTR.
(11), 147-51, 1971.

65698 **CRYSTALLINITY AND REFLECTANCE OF ANTHRACITES.**
POSYLNYI, V. YA.
MATER. MINERAL. PETROGR. NIZHNEGO DONA SEV. KAVKAZA
215-21, 1970.

65746 **NOMOGRAM FOR DETERMINING THE THERMAL CONDUCTIVITY COEFFICIENT OF PULVERIZED COAL.**
STUKALO, V. A.
RAZRAB. MESTOROZHD. POLEZ. ISKOP. (KIEV)
(25) 80-2, 1971.

65747 **PETROGRAPHIC STUDY OF INSOLUBLE ORGANIC SUBSTANCES BY MEASUREMENT OF THEIR REFLECTIVITY. PETROLEUM EXPLORATION AND KNOWLEDGE OF SEDIMENTARY BASINS.**
ROBERT, P.
REV. INST. FR. PETROLE ANN. COMBUST. LIQUIDES
26 (6), 105-35, 1971.

65759 **ANALYSIS OF GLASS RAW MATERIALS AND OPTICAL GLASSES BY INFRARED ABSORPTION SPECTROSCOPY.**
HUBERT, A.
SZKLO CERAM.
22 (11), 321-4, 1971.

TPRC Number	Bibliographic Citation
65791	**OPTICAL SPECTROSCOPY OF VANADIUM IN NATURAL MINERALS. I. OPTICAL ABSORPTION SPECTRA OF NATURAL ORTHOVANADATES.** PLATONOV, A. N. TARASHCHAN, A. N. POVARENNYKH, A. S. ZAKHAROVA, G. M. KONST. SVOISTVA MINER. 5, 92-100, 1971.
65794	**ANOMALOUS LOW - TEMPERATURE THERMAL PROPERTIES OF GLASSES AND SPIN GLASSES.** ANDERSON, P. W. HALPERIN, B. I. VARMA, C. M. PHIL. MAG. 25 (1), 1-9, 1972.
65828	**EVALUATION OF A 20-INCH GUARDED HOT-PLATE THERMAL CONDUCTIVITY APPARATUS, RANGE MINUS 50 TO 250 F.** KAPLER, C. W. COLD REGIONS RESEARCH AND ENGINEERING LAB., HANOVER, N. H. 40PP., 1971. (AD-727668, CRREL-SR-137)
65854	**COEFFICIENTS OF THERMAL CONDUCTIVITY OF THE COATINGS OF METAL FOUNDRY MOLDS.** SEREBRO, V. S. BRENER, A. I. INZH.-FIZ. ZH. 21 (5), 804-6, 1971. (FOR ENGLISH TRANSLATION SEE TPRC NO. 73366)
65856	**ANOMALOUS SPECIFIC HEAT OF GLASSES. ITS TEMPERATURE DEPENDENCE.** ROSENSTOCK, H. B. J. NON - CRYST. SOLIDS 7 (2), 123-6, 1972.
65865	**ANOMALOUS BEHAVIOR OF THE LOW TEMPERATURE SPECIFIC HEAT OF VANADIUM DOPED TITANIUM SESQUIOXIDE.** SJOSTRAND, M. E. KEESOM, P. H. PHYS. LETT. 39 A (2), 147-9, 1972.
65920	**INFRARED SPECTRA OF MINERALS FROM THE GROUPS OF WOLFRAMATES, MOLYBDATES, AND COMPLEX OXIDES CONTAINING TUNGSTEN OR MOLYBDENUM - OXYGEN GROUPS.** YUSHKIN, N. P. BUSHUEVA, E. B. KONST. SVOISTVA MINER. 5, 28-39, 1971.
65946	**SPECTRAL REFLECTIVITY DATA- A PRACTICAL ACQUISITION PROCEDURE.** EGBERT, D. D. UNIVERSITY OF KANSAS, M. S. THESIS 78PP., 1971.
66008	**AN EXPERIMENTAL AND ANALYTICAL STUDY OF RADIATIVE AND CONDUCTIVE HEAT TRANSFER IN MOLTEN GLASS.** ERYOU, N. D. GLICKSMAN, L. R. J. HEAT TRANSFER 94 C (2), 224-30, 1972.
66020	**GENERAL VALENCE FORCE FIELDS AND CRYSTAL VIBRATIONS OF DIAMOND.** SINGH, B. D. DAYAL, B. INDIAN J. PURE APPL. PHYS. 9 (11), 962-8, 1971.
66040	**METHOD FOR TESTING DILATOMETERS ON SAMPLE SUBSTANCES TO GUARANTEE THE CONFORMITY OF DILATOMETRIC MEASUREMENTS IN THE SOVIET UNION.** STRELKOV, P. G. LIFANOV, I. I. SHERSTYUKOV, N. G. FOREIGN TECHNOLOGY DIVISION 13PP., 1972. (ENGLISH TRANSLATION OF METODY IZMER. TEPL. RASSHIR. STEKOL SPAIVAEMYKH NIMI METAL., TR. VSES. SIMP. DILATOMETRII, 1ST, LENINGRAD, 1966, 31-40, 1967., FOR ORIGINAL SEE TPRC NO. 53738) (FTD-HT-2P-569-72, AD-745840)
66055	**THERMOPHYSICAL PROPERTIES OF HYDROPHOBIZED PERLITE AND MATERIALS BASED ON IT.** SPEKTOR, B. V. KRUPA, A. A. FOREIGN TECHNOLOGY DIVISION 14PP., 1972. (ENGLISH TRANSLATION OF STROIT. MATER., DETALI IZDELIYA, (11), 81-6, 1969., FOR ORIGINAL SEE TPRC NO. 56893) (FTD-MT-72-184)
66064	**USE OF ULTRASONIC DATA IN CALCULATING THE HEAT CONDUCTIVITY OF VITREOUS SUBSTANCES AT HIGH TEMPERATURES.** BESSONOV, M. V. FOREIGN TECHNOLOGY DIVISION 7PP., 1972. (ENGLISH TRANSLATION OF PRIMENIE UL'TRAAKUSTIKI K ISSLEDOVANIYU VESHCHESTA, (13), 165-70, 1961., FOR ORIGINAL SEE TPRC NO. 51623) (FTD-HC-23-2108-71, AD-744291)
66103	**SIMULTANEOUS MEASUREMENT OF THE APPARENT THERMAL DIFFUSIVITY AND THERMAL CONDUCTIVITY OF GRANULAR MEDIA BY THE FLASH METHOD.** GERY, M. YEZOU, R. HINZELIN, P. COMPT. REND. 274 B (14), 870-3, 1972.
66125	**STABILITY OF SOLID SOLUTIONS IN THE SYSTEMS ZIRCONIUM OXIDE-LANTHANUM OXIDE.** GLUSHKOVA, V. B. KELER, E. K. SHCHERBAKOVA, L. G. SHCHERBOVSKII, E. YA. IZV. AKAD. NAUK SSSR, NEORG. MATER. 7 (11), 2007-14, 1971. (FOR ENGLISH TRANSLATION SEE TPRC NO. 66126)
66126	**STABILITY OF SOLID SOLUTIONS IN THE SYSTEMS ZIRCONIUM OXIDE - LANTHANUM OXIDE.** GLUSHKOVA, V. B. KELER, E. K. SHCHERBAKOVA, L. G. SHCHERBOVSKII, E. YA. INORG. MAT. (USSR) 7 (11), 1787-93, 1971. (ENGLISH TRANSLATION OF IZV. AKAD. NAUK SSSR, NEORG. MATER., 7 (11), 2007-14, 1971; FOR ORIGINAL SEE TPRC NO. 66125)
66142	**AN APPARATUS FOR MEASURING THE THERMAL EXPANSION OF OPTICAL MATERIALS IN THE RANGE 30 DEGREES C TO 350 DEGREES C.** EBERSOLE, J. F. UNIVERSITY OF FLORIDA, M. S. THESIS, 45PP., 1971.
66144	**PHYSICOCHEMICAL PROPERTIES OF BARIUM OXIDE - LEAD OXIDE - BORON OXIDE - BISMUTH TRIOXIDE - TITANIUM DIOXIDE SYSTEM GLASSES.** FREIDENFELDS, E. JOKSTA, E. ZUSANE, E. POLKA, I. FOREIGN TECHNOLOGY DIVISION 6PP., 1972. (ENGLISH TRANSLATION OF NEORG. STEKLOVIDNYE POKRYTIYA MATER., 107-11, 1969; FOR ORIGINAL SEE TPRC NO. 56246) (FTD-HT-20-512-72, N73-12620, AD-747371)
66152	**HEAT CONDUCTION IN GLASSES.** ZELLER, R. C. CORNELL UNIVERSITY, M. S. THESIS, 86PP., 1971.
66159	**ELECTRON CONTRIBUTION TO THE THERMAL CONDUCTION ACROSS A METAL - SAPPHIRE INTERFACE.** WOLFMEYER, M. W. UNIV. WISCONSIN, PH. D. THESIS 65PP., 1971. (UM-71-20700)
66170	**THERMAL LATTICE EXPANSION OF VARIOUS TYPES OF SOLIDS.** SHAH, J. S. UNIV. MISSOURI, PH. D. THESIS 126PP., 1971. (UM-71-24776)
66172	**SHOCK METAMORPHISM OF THE COCONINO SANDSTONE AT METEOR CRATER, ARIZONA. SPECIFIC HEAT OF SOLIDS OF GEOPHYSICAL INTEREST.** KIEFFER, S. W. CALIFORNIA INST. TECHNOL., PH. D. THESIS 262PP., 1971. (UM-71-27102)
66202	**CHANGE IN THE ENTHALPY OF MAGNETITE CONCENTRATES FROM THE SOKOLOVSK-SARBARSK MINING CONCENTRATION COMBINE DURING HEATING IN AN OXIDIZING ENVIRONMENT.** BRATCHIKOV, S. G. YUR'EV, B. P. FOREIGN TECHNOLOGY DIVISION 5PP., 1972. (ENGLISH TRANSLATION OF IZV. VYSSH. UCHEB. ZAVED. CHERN. MET., 14 (2), 42-5, 1971; FOR ORIGINAL SEE TPRC NO. 61642) (FTD-HC-23-89-72)
66207	**THERMAL CONDUCTIVITY AND DIFFUSIVITY OF INSULATING MATERIALS.** DUDNIK, D. M. STEPANENKO, A. N. FOREIGN TECHNOLOGY DIVISION 5PP., 1972. (ENGLISH TRANSLATION OF KHOLOD. TEKH., 45 (1), 27-9, 1968; FOR ORIGINAL SEE TPRC NO. 51025) (FTD-HC-23-145-72, AD-748748)
66213	**REFLECTANCE OF COAL MICROCOMPONENTS IN COALS OF VARIOUS METAMORPHIC GRADES.** SARBEEVA, L. I. KRYLOVA, N. M. FOREIGN TECHNOLOGY DIVISION 86PP., 1972. (ENGLISH TRANSLATION OF VOP. METAMORF. UGLEI EPIGENEZA UMESHCHAYUSHCHIKH POROD, 87-106, 1968; FOR ORIGINAL SEE TPRC NO. 55693) (FTD-MT-ST-72-763)
66214	**PETROGRAPHIC CHARACTERISTIC OF COALS DURING VARIOUS STAGES OF REGIONAL METAMORPHISM (DONETS AND KUZNETSK BASINS).** KRYLOVA, N. M. SARBEEVA, L. I. FOREIGN TECHNOLOGY DIVISION 58PP., 1972. (ENGLISH TRANSLATION OF VOP METAMORF. UGLEI EPIGENEZA UMESHCHAYUSHCHIKH POROD, 68-87, 1968; FOR ORIGINAL SEE TPRC NO. 55692) (FTD-MT-ST-72-764)

TPRC Number	Bibliographic Citation

66222 **THE POLARIZATION CURVE AND THE ABSOLUTE DIAMETER OF VESTA.**
VEVERKA, J.
ICARUS
15 (1), 11-7, 1971.
(AD-733 579)

66234 **CORRELATION OF MINERAL THERMAL CONDUCTIVITY WITH STRUCTURAL CHARACTERISTICS.**
POVARENNYKH, A. S. PRODAIVODA, G. T.
GEOL. ZH. (RUSS. ED.)
32 (1), 41-7, 1972.

66243 **OPTICAL PROPERTIES OF AEROSPACE TRANSPARENCIES.**
LINFORD, R. M. F. ALLEN, T. H.
SPACE SHUTTLE MATER., NAT. SAMPE
465-71PP., 1971.

66247 **INFRARED SPECTRA OF SODIUM OXIDE-BERYLLIUM OXIDE-GERMANIUM DIOXIDE AND SODIUM OXIDE-BARIUM OXIDE-GERMANIUM DIOXIDE GLASSES.**
AKOPYAN, A. D. TARLAKOV, YU. P.
SHEVYAKOV, A. M. KOSTANYAN, K. A.
ARM. KHIM. ZH.
24 (11), 956-9, 1971.

66289 **REGULARITIES IN THE INFRARED ABSORPTION SPECTRA OF SOME HYDROXIDES AND OXIDE HYDROXIDES.**
BOLDYREV, A. I. EGOROVA, L. N.
POVARENNYKH, A. S.
TR. MINERAL. MUZ., AKAD. NAUK SSSR
(20), 33-9, 1971.

66292 **FORMATION OF LEUCITE AND THERMAL EXPANSION IN SILICON DIOXIDE - ALUMINUM OXIDE - POTASSIUM OXIDE - SODIUM OXIDE GLASSES.**
HOSHIKAWA, T. AKAGI, S.
YOGYO KYOKAI SHI
80 (917), 36-42, 1972.

66294 **SIMULTANEOUS MEASUREMENT OF THE THERMOPHYSICAL PROPERTIES OF INSULATORS.**
FUKS, L. G. SHMANDINA, V. N.
ZAVOD. LAB.
37 (11), 1361-2, 1971.
(FOR ENGLISH TRANSLATION SEE TPRC NO. 70931)

66304 **STRUCTURE OF TITANIUM NICKEL PIGMENTS.**
KLESHCHEV, G. V. SHEYNKMAN, A. I.
GOL'DSHTEYN, L. M. TURLAKOV, V. N.
LAKOKRASOCH. MATER. IKH PRIMEN.
(4), 12-4, 1970.
(FOR ENGLISH TRANSLATION SEE TPRC NO. 66305)

66305 **STRUCTURE OF TITANIUM NICKEL PIGMENTS.**
KLESHCHEV, G. V. SHEYNKMAN, A. I.
GOL'DSHTEYN, L. M. TURLAKOV, V. N.
FOREIGN TECHNOLOGY DIVISION
8PP., 1970.
(ENGLISH TRANSLATION OF LAKOKRASOCH. MATER. IKH. PRIMEN., (4), 12-4, 1970; FOR ORIGINAL SEE TPRC NO. 66304)
(AD-734 908, FTD-HT-23-1139-71)

66306 **SPECTROSCOPIC REMOTE SENSING OF WATER-BEARING MINERALS.**
SALISBURY, J. W. HUNT, G. R.
GEOLOGICAL PROBLEMS LUNAR PLANETARY RESEARCH
25, 25-52, 1971.
(AD-726 302, AFCRL-71-0353)

66323 **THERMAL CONDUCTIVITY OF THE ELEMENTS.**
HO, C. Y. POWELL, R. W. LILEY, P. E.
J. PHYS. CHEM. REF. DATA
1 (2), 279-421, 1972.

66369 **MEASUREMENT OF THE INFRARED ABSORPTION AND REFLECTION SPECTRA OF BERYL, TOPAZ, DATOLITE, AND ZOISITE.**
FRIEDL, W. KRUEGER, H.
OPT. ACTA
18 (12), 913-29, 1971.

66376 **VISIBLE AND NEAR-INFRARED SPECTRA OF MINERALS AND ROCKS. II. CARBONATES.**
HUNT, G. R. SALISBURY, J. W.
MOD. GEOLOG.
2, 23-30, 1971.
(AD-731 101, AFCRL-71-0454)

66394 **MEASUREMENT OF THE ANGLE AND TEMPERATURE DEPENDENCE OF THE TOTAL EMISSIVITY OF GLASSES AND SINTERED OXIDES IN THE REGION OF MINUS 60 TO PLUS 250 DEGREES C.**
LOHRENGEL, J.
GLASTECH. BER.
43, 493-500, 1970.

66395 **DETERMINATION OF THE TOTAL HEMISPHERICAL EMISSIVITY FROM TWO MEASURED VALUES OF THE TOTAL EMISSIVITY OF 0 AND 45 DEGREES.**
LOHRENGEL, J. TINGWALDT, C.
OPTIK
31, 404-9, 1970.

66437 **PERICLASE - SPINEL REFRACTORIES FROM SEAWATER MAGNESIA AND BENEFICIATED CHROMITE.**
ANTONOV, G. I. YANSHINA, A. P.
ZHUBAKOV, S. M. MENZHULINA, F. M.
OGNEUPORY
12 (9), 35-40, 1971.
(FOR ENGLISH TRANSLATION SEE TPRC NO. 66438)

66438 **PERICLASE - SPINEL REFRACTORIES FROM SEAWATER MAGNESIA AND BENEFICIATED CHROMITE.**
ANTONOV, G. I. YANSHINA, A. P.
ZHUBAKOV, S. M. MENZHULINA, F. M.
REFRACTORIES (USSR)
12 (9), 591-5, 1971.
(ENGLISH TRANSLATION OF OGNEUPORY, 12 (9), 35-40, 1971; FOR ORIGINAL SEE TPRC NO. 66437)

66439 **FIBROUS INSULATING REFRACTORIES FOR BLAST FURNACE STOVES.**
GORLOV, YU. P. ZHURBA, V. P.
OGNEUPORY
12 (10), 1-3, 1971.
(FOR ENGLISH TRANSLATION SEE TPRC NO. 66440)

66440 **FIBROUS INSULATING REFRACTORIES FOR BLAST FURNACE STOVES.**
GORLOV, YU. P. ZHURBA, V. P.
REFRACTORIES (USSR)
12 (10), 621-2, 1971.
(ENGLISH TRANSLATION OF OGNEUPORY, 12 (10), 1-3, 1971., FOR ORIGINAL SEE TPRC NO. 66439)

66443 **HIGH - ALUMINA REFRACTORY CEMENTS BASED ON ALUMINUM THERMITE SLAGS.**
ZALDAT, G. I. KUKUI, S. M.
ZALIZOVSKII, E. V. KONDRASHENKO, A. A.
DUBROVIN, A. S. IGNATENKO, G. F.
OGNEUPORY
12 (11), 8-12, 1971.
(FOR ENGLISH TRANSLATION SEE TPRC NO. 66444)

66444 **HIGH - ALUMINA REFRACTORY CEMENTS BASED ON ALUMINUM THERMITE SLAGS.**
ZALDAT, G. I. KUKUI, S. M.
ZALIZOVSKII, E. V. KONDRASHENKO, A. A.
DUBROVIN, A. S. IGNATENKO, G. F.
REFRACTORIES (USSR)
12 (11), 693-7, 1971.
(ENGLISH TRANSLATION OF OGNEUPORY, 12 (11), 8-12, 1971., FOR ORIGINAL SEE TPRC NO. 66443)

66445 **STUDY OF THE THERMAL - SHOCK RESISTANCE OF PERICLASE REFRACTORIES.**
UZBERG, L. V. IGNATOVA, T. S. DAUKNIS, V. I.
YURENAS, V. L. KAZAKYAVICHYUS, K. A.
OGNEUPORY
12 (11), 23-8, 1971.
(FOR ENGLISH TRANSLATION SEE TPRC NO. 66446)

66446 **STUDY OF THE THERMAL - SHOCK RESISTANCE OF PERICLASE REFRACTORIES.**
UZBERG, L. V. IGNATOVA, T. S. DAUKNIS, V. I.
YURENAS, V. L. KAZAKYAVICHYUS, K. A.
REFRACTORIES (USSR)
12 (11), 709-13, 1971.
(ENGLISH TRANSLATION OF OGNEUPORY, 12 (11), 23-8, 1971., FOR ORIGINAL SEE TPRC NO. 66445)

66447 **STUDY OF THE COMPOSITION OF PLASMA MULLITE COATINGS BY INFRARED SPECTROSCOPY.**
KARPINOS, D. M. ZILBERBERG, V. G.
LISTOVNICHAYA, S. P. GROSHEVA, V. M.
OGNEUPORY
12 (11), 47-50, 1971.
(FOR ENGLISH TRANSLATION SEE TPRC NO. 66448)

66448 **STUDY OF THE COMPOSITION OF PLASMA MULLITE COATINGS BY INFRARED SPECTROSCOPY.**
KARPINOS, D. M. ZILBERBERG, V. G.
LISTOVNICHAYA, S. P. GROSHEVA, V. M.
REFRACTORIES (USSR)
12 (11), 735-7, 1971.
(ENGLISH TRANSLATION OF OGNEPORY, 12 (11), 47-50, 1971., FOR ORIGINAL SEE TPRC NO. 66447)

66449 **COMPOSITION AND PROPERTIES OF REFRACTORY CONCRETE BEFORE AND AFTER SERVICE IN BLAST FURNACE TUYERES.**
PRYADKO, V. M. MAGALA, V. S.
OGNEUPORY
12 (12), 22-6, 1971.
(FOR ENGLISH TRANSLATION SEE TPRC NO. 66450)

66450 **COMPOSITION AND PROPERTIES OF REFRACTORY CONCRETE BEFORE AND AFTER SERVICE IN BLAST FURNACE TUYERES.**
PRYADKO, V. M. MAGALA, V. S.
REFRACTORIES (USSR)
12 (12), 776-9, 1971.
(ENGLISH TRANSLATION OF OGNEUPORY, 12 (12), 22-6, 1971., FOR ORIGINAL SEE TPRC NO. 66449)

TPRC Number	Bibliographic Citation

66453 **PHASE COMPOSITION, MICROSTRUCTURE, AND PROPERTIES OF MATERIALS BASED ON THE ZIRCONIUM OXIDE–NEODYMIUM OXIDE SYSTEM.**
STRAKHOV, V. I. KLYUCHAROV, YA. V.
SERGEEV, G. G.
OGNEUPORY
12 (12), 43–7, 1971.
(FOR ENGLISH TRANSLATION SEE TPRC NO. 66454)

66454 **PHASE COMPOSITION, MICROSTRUCTURE, AND PROPERTIES OF MATERIALS BASED ON THE ZIRCONIUM OXIDE–NEODYMIUM OXIDE SYSTEM.**
STRAKHOV, V. I. KLYUCHAROV, YA. V.
SERGEEV, G. G.
REFRACTORIES (USSR)
12 (12), 800–4, 1971.
(ENGLISH TRANSLATION OF OGNEUPORY, 12 (12), 43–7, 1971., FOR ORIGINAL SEE TPRC NO. 66453)

66484 **DETERMINING THE DIFFERENCE OF THE THERMAL EXPANSION COEFFICIENTS OF GLASSES OF DIFFERENT COMPOSITION.**
GOMEL'SKII, M. S. PLUTALOVA, N. YU.
IND. LAB., (USSR)
37 (3), 394–6, 1971.
(ENGLISH TRANSLATION OF ZAVOD. LAB., 37 (3), 309–11, 1971., FOR ORIGINAL SEE TPRC NO. 62599)

66494 **THE WARM JACKET OF THE MOON.**
AVDUYEVSKIY, V. S. EKONOMOV, A. P.
PRIRODA (MOSCOW)
(3), 73–4, 1971.
(FOR ENGLISH TRANSLATION SEE TPRC NO. 66495)

66495 **THE WARM JACKET OF THE MOON.**
AVDUYEVSKIY, V. S. EKONOMOV, A. P.
SCRIPTA TECHNICA, INC., WASHINGTON, D. C.
4PP., 1971.
(ENGLISH TRANSLATION OF PRIRODA (MOSCOW), (3), 73–4, 1971., FOR ORIGINAL SEE TPRC NO. 66494)
(N71–28192, NASA–TT–F–13701, AVAIL. NTIS)

66504 **EFFECT OF A CHANGE IN ALUMINUM OXIDE CONTENT ON THE PHYSICOCHEMICAL PROPERTIES OF DOLOMITIC OPACIFIED GLAZES.**
YUNUSOV, KH. AZIMOV, I. TASHKHODZHAEV, S.
PARPIEV, N. A.
FOREIGN TECHNOLOGY DIVISION
4PP., 1972.
(ENGLISH TRANSLATION OF UZB. KHIM. ZH., 12 (1), 14–6, 1968; FOR ORIGINAL SEE TPRC NO. 50006)
(FTD–HT–23–1339–71)

66579 **THERMAL RADIATIVE PROPERTIES—NONMETALLIC SOLIDS.**
TOULOUKIAN, Y. S. DE WITT, D. P.
THERMOPHYSICAL PROPERTIES OF MATTER – THE TPRC DATA SERIES
8, 1763PP., 1972.

66602 **DETERMINATION OF THE AVERAGE HEAT CAPACITIES OF SALIFEROUS ROCKS AND BRINES.**
YUSOVA, YU. I. ESELEV, I. M.
KARATYGIN, E. P.
ZH. PRIKL. KHIM. (LENINGRAD)
45 (2), 427–8, 1972.
(FOR ENGLISH TRANSLATION SEE TPRC NO. 69793)

66628 **THE THERMAL EXPANSION OF GLAZES I.**
RIEKE, R. STEGER, W.
SPRECHSAAL
47 (29), 476–8, 1914.

66636 **JOINING OF FERRITE – TITANATE CERAMICS BY HOT PRESSING.**
SUEMUNE, Y. IMAMURA, Y.
JAP. J. APPL. PHYS.
11 (5), 641–3, 1972.

66639 **STUDIES IN PHOSPHATE BONDING.**
O'HARA, M. J. OUGA, J. J. SHEETS, H. D., JR.
BULL. AMER. CERAM. SOC.
51 (7), 590–5, 1972.

66640 **SPECTROPHOTMETRIC STUDIES OF THE PHOTOMETRIC FUNCTION, COMPOSITION, AND DISTRIBUTION OF THE SURFACE MATERIALS OF MARS.**
BINDER, A. B. JONES, J. C.
J. GEOPHYS. RES.
77 (17), 3005–20, 1972.

66730 **REFLECTION SPECTRUM AND OPTICAL CONSTANTS OF CINNABAR.**
BARCELO, J. GALTIER, M. MONTANER, A.
COMPT. REND.
274 B (26), 1410–3, 1972.

66785 **PREPARATION OF CRYSTALLIZED COATINGS FOR METALS.**
PAVLUSHKIN, N. M. ZHURAVLEV, A. K.
EGOROVA, L. S.
GLASS CERAM. (USSR)
26 (1), 16–8, 1969.
(ENGLISH TRANSLATION OF STEKLO KERAM., 26 (1), 11–3, 1969., FOR ORIGINAL SEE TPRC NO. 53252)

66786 **EFFECT OF BENTONITE ADDITIVES ON THE PHYSICOMECHANICAL PROPERTIES OF THE PLASTIC CLAY MASS AND ON THE HEAT STABILITY OF FACE TILE DURING FIRING.**
KUKOLEV, G. V. ORUDZHEVA, N. T.
GLASS CERAM. (USSR)
26 (4), 228–31, 1969.
(ENGLISH TRANSLATION OF STEKLO KERAM., 26 (4), 26–9, 1969., FOR ORIGINAL SEE TPRC NO. 54745)

66787 **EFFECT OF ADDITIONS OF CERTAIN OXIDES ON THE PROPERTIES OF GLAZES FOR HIGH–VOLTAGE PORCELAIN.**
MASLENNIKOVA, G. N. KOCHETKOVA, N. F.
GLASS CERAM. (USSR)
27 (9), 543–5, 1970.
(ENGLISH TRANSLATION OF STEKLO KERAM., 27 (9), 28–30, 1970., FOR ORIGINAL SEE TPRC NO. 60753)

66788 **SOME PHYSICAL AND CHEMICAL PROPERTIES OF GLAZES BASED ON VOLCANIC ROCK.**
KHIZANISHVILI, I. G. GAPRINDASHVILI, G. G.
GLASS CERAM., (USSR)
27 (11), 686–8, 1970.
(ENGLISH TRANSLATION OF STEKLO KERAM., 27 (11), 38–40, 1970., FOR ORIGINAL SEE TPRC NO. 62319)

66789 **THERMOSTABLE LOW–ALKALI STRONTIUM GLAZE FOR FAIENCE MAJOLICA.**
SHTEINBERG, YU. G.
GLASS CERAM. (USSR)
28 (5), 316–7, 1971.
(ENGLISH TRANSLATION OF STEKLO KERAM., (5), 37–8, 1971., FOR ORIGINAL SEE TPRC NO. 63644)

66790 **CONTINUOUS MELTING OF UVIOL GLASS AND MECHANIZED PRODUCTION OF TUBES FROM IT.**
KOCHETKOVA, G. V. KLEGG, D. I. DYMOV, A. T.
DUBKOVA, A. I. KOVALENKO, I. S.
KELEINIKOV, V. I. SIVKO, A. P.
MOROZOVA, A. N. ZOTOVA, T. M.
GLASS CERAM. (USSR)
28 (9), 543–4, 1971.
(ENGLISH TRANSLATION OF STEKLO KERAM., (9), 22, 1971., FOR ORIGINAL SEE TPRC NO. 65218)

66811 **A NEW METHOD OF SIMULTANEOUS MEASUREMENT OF SEVERAL THERMAL PROPERTIES BY CONTINUOUS HEATING.**
KATAYAMA, K. OKADA, M. YOSNIDA, T.
HEAT TRANSFER – JAP. RES.
1 (1), 23–32, 1972.

66830 **THERMAL CHARACTERISTICS OF THE LUNAR SURFACE LAYER.**
CREMERS, C. J. BIRKEBAK, R. C. WHITE, J. E.
INT. J. HEAT MASS TRANSFER
15 (5), 1045–55, 1972.

66852 **DIFFUSE RELECTANCE SPECTRA OF SEVERAL CLAY MINERALS.**
LINDBERG, J. D. SNYDER, D. G.
AMER. MINERAL.
57 (3–4), (PT. 1), 485–93, 1972.

66863 **DIFFUSE REFLECTION SPECTRA OF BENTONITES COLORED WITH VARIOUS AROMATIC COMPOUNDS AND RELATED ION–RADICAL SALTS.**
MATSUNAGA, Y.
BULL. CHEM. SOC. JAP.
45 (3), 770–5, 1972.

66864 **OPTICAL PROPERTIES OF AN ABSORBING MONOCLINIC CRYSTAL DETERMINED BY MICROREFLECTOMETRY.**
CERVELLE, B.
BULL. SOC. FR. MINERAL. CRISTALLOGR.
94 (5–6), 486–91, 1972.

66874 **GREIGITE FROM THE YAKUT ASSR, AND ITS OPTICAL PROPERTIES.**
GRUZDEV, V. S. BRYZGALOV, I. A.
CHEPNITSOVA, N. M. SHUMKOVA, N. G.
DOKL. AKAD. NAUK SSSR
202 (4), 915–8, 1972.

66890 **INFRARED SPECTRA OF HEAT–TREATED COAL.**
SHIH, J. W. OSAWA, Y. FUJII, S.
FUEL
51 (2), 153–5, 1972.

66891 **MEASUREMENT OF THE ENTHALPY OF MINERALS BY HIGH–TEMPERATURE MICROCALORIMETRY.**
TOPOR, N. D. KISELEVA, I. A.
MEL'CHAKOVA, L. V.
GEOKHIMIYA
(3), 335–43, 1972.

66892 **RELATION OF OPTICAL PROPERTIES OF NATURAL CHROME–SPINELS TO THEIR COMPOSITION.**
RAKCHEEV, A. D. SMIRNOVA, T. A.
GEOL. RUD. MESTOROZHD.
14 (1), 31–51, 1972.

66895 **TECHNICAL APPLICATION OF OLIVINE.**
BAUMGART, W.
HAUS TECH., ESSEN, VORTRAGSVEROEFF.
(273), 43–6, 1971.

TPRC Number	Bibliographic Citation
66906	OPTICAL PROPERTIES OF MERCURIC SULFIDE, MERCURIC SELENIDE, AND MERCURIC TELLURIDE IN THE FUNDAMENTAL ABSORPTION REGION. SOBOLEV, V. V. DONETSKIKH, V. I. IZV. AKAD. NAUK MOLD. SSR, SER. FIZ.-TEKH. MAT. NAUK (3), 46-51, 1971.
67054	EFFECT OF POLARIZATION ON REFLECTANCE FACTOR MEASUREMENTS. ROBERTSON, A. R. APPL. OPT. 11 (9), 1936-41, 1972.
67055	EMISSION POLARIZATION STUDY ON QUARTZ AND CALCITE. VINCENT, R. K. APPL. OPT. 11 (9), 1942-5, 1972.
67135	QUANTITATIVE STUDY OF ANISOTROPY EFFECTS ON THE (010) PLANE OF STIBNITE USING POLARIZATION MICROSCOPY. KNOSP, H. GAHM, J. PETZOW, G. ZEISS-MITT. FORTSCHR. TECH. OPT. 5 (8), 331-48, 1971.
67137	RELATION OF THE INFRARED SPECTRA OF SLAGS WITH THEIR ACID-BASE PROPERTIES. KURBATOV, G. A. KALMYKOV, V. A. ZH. PRIKL. KHIM. (LENINGRAD) 45 (1), 190-1, 1972. (FOR ENGLISH TRANSLATION SEE TPRC NO. 69791)
67144	THE INFRARED FILTERS BASED ON VARICONDS. VERBITSKAYA, T. N. KISLOVSKII, L. D. GALANOV, E. K. SOKOLOVA, L. S. J. APPL. SPECTROSC., USSR 7 (2), 197, 1967. (ENGLISH TRANSLATION OF ZH. PRIKL. SPEKTROSK., 7 (2), 276-7, 1967., FOR ORIGINAL SEE TPRC NO. 48465)
67155	DETERMINING THE TEMPERATURE DEPENDENCE OF THERMOPHYSICAL PROPERTIES OF MATERIALS. OSIPOVA, M. N. OSIPOVA, V. A. THERM. ENG., USSR 18 (6), 124-7, 1971. (ENGLISH TRANSLATION OF TEPLOENERGETIKA, 18 (6), 84-5, 1971., FOR ORIGINAL SEE TPRC NO. 63757)
67214	DYNAMIC CAMBER MEASUREMENTS OF GLAZED CERAMIC SUBSTRATES. YOUNG, W. S. WILCOX, D. L. AMER. CERAM. SOC. BULL. 51 (9), 672-6, 1972.
67215	THERMAL CONDUCTIVITY OF SOME OXYGEN COMPOUNDS OF ZIRCONIUM AND HAFNIUM. VISHNEVSKII, I. I. SKRIPAK, V. N. REFRACTORIES, USSR 35 (11), 693-4, 1970. (ENGLISH TRANSLATION OF OGNEUPORY, 35 (11), 16-8, 1970., FOR ORIGINAL SEE TPRC NO. 61097)
67223	INFLUENCE OF THE FILLERS ON THE THERMAL PROPERTIES OF GLASS REINFORCED PLASTICS. KIRILLOV, V. N. EFIMOV, V. A. KOZIN, V. I. ABLEKOVA, Z. P. KRASNOV, L. L. TIKHIMIROVA, R. S. SOV. PLAST. (11), 33-5, 1970. (ENGLISH TRANSLATION OF PLAST. MASSY, (11), 38-40, 1970; FOR ORIGINAL SEE TPRC NO. 61257)
67224	INFLUENCE OF FILLERS ON THE THERMAL CONDUCTIVITY OF POLYETHYLENE. DASHKO, N. M. NOVICHENOK, L. N. SPORYAGIN, E. A. SOV. PLAST. (11), 41-3, 1970. (ENGLISH TRANSLATION OF PLAST. MASSY, (11), 45-7, 1970., FOR ORIGINAL SEE TPRC NO. 61258)
67256	DETERMINATION OF THE THERMOPHYSICAL PROPERTIES OF COPPER SULFIDE ORES AND OF BRIQUETS FROM FINES. USHAKOV, K. I. FELMAN, R. I. SOV. J. NON - FERROUS METALS 42 (7), 23-5, 1969. (ENGLISH TRANSLATION OF TSVET. METAL., 42 (7), 25-7, 1969., FOR ORIGINAL SEE TPRC NO. 56214)
67269	THE FORMATION OF MOISTURE AND TEMPERATURE FIELDS IN THE PROCESS OF HEAT- AND MASS-TRANSFER IN HYDROTHERMAL SOLUTIONS WITH ENCLOSING ROCKS. KOCHERGIN, V. N. BALYSHEV, O. A. INZH. - FIZ. ZH. 15 (2), 260-7, 1968. (FOR ENGLISH TRANSLATION SEE TPRC NO. 67270)
67270	THE FORMATION OF MOISTURE AND TEMPERATURE FIELDS IN THE PROCESS OF HEAT- AND MASS-TRANSFER IN HYDROTHERMAL SOLUTIONS WITH ENCLOSING ROCKS. KOCHERGIN, V. N. BALYSHEV, O. A. J. ENG. PHYS., USSR 15 (2), 716-9, 1968. (ENGLISH TRANSLATION OF INZH. - FIZ. ZH., 15 (2), 260-7, 1968., FOR ORIGINAL SEE TPRC NO. 67269)
67277	DETERMINATION AND CALCULATION OF THE OPTICAL CONSTANTS IN A UNIAXIAL OPAQUE CRYSTAL. ZINKENITE. LOPEZ-SOLER, A. BOSCH-FIGUEROA, J. M. ACTA GEOL. HISP. 6 (3), 78-81, 1971.
67278	DETERMINATION AND CALCULATION OF THE OPTICAL CONSTANTS IN A BIAXIAL OPAQUE CRYSTAL. ENARGITE. LOPEZ-SOLER, A. BOSCH-FIGUEROA, J. M. ACTA GEOL. HISP. 6 (3), 82-5, 1971.
67282	THERMAL EXPANSION MEASUREMENTS TO 130 DEGREES BY LASER INTERFEROMETRY. PLUMMER, W. A. AIP CONF. PROC. (3), 36-43, 1972.
67289	THERMAL EXPANSION MEASUREMENTS OF SIMULATED LUNAR ROCKS. GRIFFIN, R. E. DEMOU, S. G. AIP CONF. PROC. (3), 302-11, 1972.
67298	THERMAL PROPERTIES OF PRODUCTS OF REGIONAL METAMORPHISM AND GRANITE FORMATION. DUDAREV, A. N. MIKHALEVA, L. A. FIZ.-KHIM. DIN. PROTSESSOV MAGMAT. RUDOOBRAZOV. 44-83, 1971.
67345	NEW REFERENCE MATERIAL FOR THERMAL - CONDUCTIVITY MEASUREMENTS. SHADRICHEV, E. V. SMIRNOV, I. A. INSTRUM. EXP. TECH., USSR (5), 1249, 1968. (ENGLISH TRANSLATION OF PRIB. TEKH. EKSP., (5), 218-9, 1968., FOR ORIGINAL SEE T P R C NO. 53478)
67356	THE TRANSMISSION OF COLORLESS OPTICAL GLASSES IN THE NEAR ULTRAVIOLET. GUSEVA, V. M. VILCHUR, YA. P. SOV. J. OPT. TECHNOL., USSR 35 (4), 263-6, 1968. (ENGLISH TRANSLATION OF OPT. - MEKH. PROM., 35 (4), 49-53, 1968., FOR ORIGINAL SEE TPRC NO. 50789)
67357	GLASSES FOR FIBER OPTICS. POLUKHIN, V. N. SOV. J. OPT. TECHNOL., USSR 35 (9), 568-71, 1968. (ENGLISH TRANSLATION OF OPT. - MEKH. PROM., 35 (9), 34-8, 1968., FOR ORIGINAL SEE TPRC NO. 53472)
67361	THE DISCONTINUOUS NATURE OF THE VARIATION OF THE THERMAL EXPANSION COEFFICIENT OF SOME GLASSES. BOGATYREVA, V. V. GOMELSKIY, M. S. SHAKIROVA, R. N. SOV. J. OPT. TECHNOL., USSR 37 (11), 741-4, 1970. (ENGLISH TRANSLATION OF OPT. - MEKH. PROM., 37 (11), 48-52, 1970., FOR ORIGINAL SEE TPRC NO. 62576)
67363	AN ALUMOCALCIUM INFRARED FILTER TRANSMITTING IN THE 1.5-5.0 MICRONS REGION. VEYNBERG, T. I. LUNKIN, S. P. SOV. J. OPT. TECHNOL., USSR 38 (2), 119-20, 1971. (ENGLISH TRANSLATION OF OPT. - MEKH. PROM., 38 (2), 60-1, 1971; FOR ORIGINAL SEE TPRC NO. 63871)
67369	THERMAL EXPANSION MEASUREMENTS IN THE NIOBIUM OXIDE-TANTALUM OXIDE SYSTEM. KIDWELL, G. D. RICHARDSON, J. H. WOLTEN, G. M. AEROSPACE CORP., EL SEGUNDO, CALIF. 35PP., 1971. (AD-736026, SAMSO-TR-71-338)
67372	MICROWAVE MEMORY ACOUSTIC CRYSTALS. OLIVER, D. W. YOUNG, J. D. SLACK, G. A. GENERAL ELECTRIC CORP., SCHENECTADY, N. Y. 151PP., 1971. (AD-887024, AFML-TR-71-101)
67377	EFFECT OF HEAT INSULATION OF STEEL INGOT TOPS ON THEIR SOLIDIFICATION. SAMOILOVICH, YU. A. KOTLYAREVSKII, E. M. STAL ENGL., USSR 30 (12), 952-3, 1970. (ENGLISH TRANSLATION OF STAL', 30 (12), 1086-8, 1970; FOR ORIGINAL SEE T62538)
67398	ELECTROOPTIC EFFECTS OF HOT - PRESSED CERAMICS IN THE SYSTEM OF LEAD ZINC NIOBIUM OXIDE. ARIMA, H. MING - CHONG, L. NOMURA, S. JAP. J. APPL. PHYS. 11 (8), 1225-6, 1972.

TPRC Number	Bibliographic Citation
67411	FIBER OPTICS WITH EXTENDED ULTRAVIOLET TRANSMISSION. ALI, M. A. SCHWARTZ, M. A. IIT RESEARCH INST., CHICAGO, ILL. 55PP., 1971. (AD-736514, ECOM-0542-F)
67440	EQUATIONS OF STATE AND PHASE EQUILIBRIA OF STISHOVITE AND A COESITELIKE PHASE FROM SHOCK-WAVE AND OTHER DATA. DAVIES, G. F. J. GEOPHYS. RES. 77 (26), 4920-33, 1972.
67441	OPTICAL PROPERTIES OF EVAPORATED MAGNESIUM OXIDE FILMS IN THE 0.22-8-MICRON WAVELENGTH REGION. BRADFORD, A. P. HASS, G. MC FARLAND, M. APPL. OPT. 11 (10), 2242-4, 1972.
67443	STRUCTURE AND THERMAL EXPANSION OF BERYL. MOROSIN, B. ACTA CRYSTALLOGR., SECT. B 28 (PT. 6), 1899-903, 1972.
67445	HEAT TRANSFER IN A SCREEN-PACKED FLUIDIZED BED. JAIN, S. C. CHEN, B. H. AICHE SYMP. SER. 67 (116), 97-105, 1971.
67454	METHOD FOR THE DETERMINATION OF THE THERMAL CONDUCTIVITY OF SPRAYED LAYERS OF OXIDE CERAMICS. EICHORN, F. METZLER, J. BER. DEUT. KERAM. GES. 49 (4), 112-3, 1972.
67510	STRUCTURAL STUDY OF GLASSES OF SODIUM OXIDE-BORON OXIDE-ALUMINUM OXIDE-SILICON DIOXIDE AND SODIUM OXIDE-BORON OXIDE-ALUMINUM OXIDE-ANTIMONY SESQUIOXIDE-SILICON DIOXIDE SYSTEM BY IR-SPECTROSCOPY, EPR, AND ELECTRON ABSORPTION SPECTRA. GOROBETS, F. T. SIDOROV, T. A. TYUL'KIN, V. A. ZH. PRIKL. SPEKTROSK. 16 (2), 313-19, 1972. (FOR ENGLISH TRANSLATION SEE TPRC NO. 76928)
67511	THE PHASE TRANSITION IN METAKAOLIN AND THE TRUE SPECIFIC HEATS OF THE NEW PHASES. KUSKOV, O. L. KHITAROV, N. I. GEOCHEMISTRY (USSR) (2), 121-6, 1970. (ENGLISH TRANSLATION OF GEOKHIMIYA, (2), 185-91, 1970., FOR ORIGINAL SEE TPRC NO. 57984)
67519	THERMAL EXPANSION OF RIBBON - REINFORCED COMPOSITES. GULATI, S. T. PLUMMER, W. A. AIP CONF. PROC. (3), 257-68, 1972.
67531	THERMOPHYSICAL PROPERTIES OF MOUNTAIN CHERNOZEM SOIL IN THE SOUTHEASTERN PART OF THE GREATER CAUCASUS. MAMEDOV, R. G. MAMEDOV, G. M. DOKL. AKAD. NAUK AZERB. SSR 27 (6), 73-7, 1971.
67569	THERMAL PROPERTIES OF SODIUM SILICATE GLASSES AT LOW TEMPERATURES. KRUEGER, J. PHYS. CHEM. GLASSES 13 (1), 9-13, 1972.
67570	THERMAL CONDUCTIVITY OF NINETEEN IGNEOUS ROCKS. I. APPLICATION OF THE NEEDLE PROBE METHOD TO THE MEASUREMENT OF THE THERMAL CONDUCTIVITY OF ROCK. HORAI, K. BALDRIDGE, S. PHYS. EARTH PLANET. INTERIORS 5 (2), 151-6, 1972.
67571	THERMAL CONDUCTIVITY OF NINETEEN IGNEOUS ROCKS. II. ESTIMATION OF THE THERMAL CONDUCTIVITY OF ROCK FROM THE MINERAL AND CHEMICAL COMPOSITIONS. HORAI, K. BALDRIDGE, S. PHYS. EARTH PLANET. INTERIORS 5 (2), 157-66, 1972.
67587	THERMOPHYSICAL PROPERTIES OF GLASSES. AVAKOVA, I. G. SERGEEV, O. A. TR. METROL. INST. SSSR (129), 13-28, 1971.
67588	THERMOPHYSICAL PROPERTIES OF INDUSTRIAL GLASSES IN THE -50 TO +400 RANGE. GOMELSKII, M. S. KOGAN, G. YA. SERGEEV, O. A. FILATOV, L. I. TR. METROL. INST. SSSR (129), 267-70, 1971.
67595	COMPOSITIONAL AND TEMPERATURE DEPENDENCY OF EMISSION SPECTRA OF EUROPIUM TRIPOSITIVE IONS IN GLASSES. KOMIYAMA, T. UENO, T. YOGYO KYOKAI SHI 80 (6), 227-32, 1972.
67656	ROLE OF TRANSVERSE OXYGEN VIBRATIONS IN THERMAL EXPANSION BEHAVIOR OF GLASSES AND CRYSTALS. SMYTH, H. T. AIP CONF. PROC. (3), 244-56, 1972.
67661	CALCULATION OF THERMAL EXPANSION OF VITREOUS MATERIALS FROM THEIR CHEMICAL COMPOSITION. GUPTA, P. K. KUMAR, S. CERAMICS (CALCUTTA) 10, 9-15, 1970.
67665	THERMAL CONDUCTIVITY OF THE BARIUM TITANIUM TIN OXIDE SYSTEM. DIMAROVA, E. N. GORBOKON, N. V. DIELEKTRIKI (1), 73-6, 1971.
67684	EFFECT OF STRUCTURE ON THE EMISSIVITY OF AN ALUMINOCHROMOPHOSPHATE COATING AT HIGH TEMPERATURES. VERENKOVA, E. M. ZHOROV, G. A. SAMOILOV, A. I. UMANTSEVA, V. M. FROLOV, A. S. IZV. AKAD. NAUK SSSR, NEORG. MATER. 8 (5), 907-10, 1972. (FOR ENGLISH TRANSLATION SEE TPRC NO. 70601)
67687	SPECIFIC HEAT AND HEAT OF PYROLYSIS REACTION FOR SOME DONBAS COALS. AGROSKIN, A. A. GONCHAROV, E. I. MAKEEV, L. A. YAKUNIN, V. P. COKE CHEM., USSR (5), 7-11, 1970. (ENGLISH TRANSLATION OF KOKS KHIM., (5), 8-13, 1970., FOR ORIGINAL SEE TPRC NO. 59042)
67704	THERMAL CONDUCTIVITY OF POWDERED COAL. YANAGIMOTO, T. UCHINP, K. KOMATSU, M. TASHIRO, M. JINTA, H. KYUSHU DAIGAKU KOGAKU SHUHO 43 (5), 688-92, 1970.
67713	ANHARMONICITY OF INFRARED VIBRATION MODES IN BERYL. GERVAIS, F. PIRIOU, B. CABANNES, F. PHYS. STATUS SOLIDI 51 B (2), 701-12, 1972.
67719	THERMAL EXPANSION AND STRENGTH OF MOLDS FABRICATED FROM A SELECT GROUP OF PRECISELY POURABLE REFRACTORY MATERIALS. SEVCIK, F. GABRIEL, J. SLEVARENSTVI 20 (4), 155-8, 1972.
67728	DIFFUSE REFLECTION SPECTRA AS A SOURCE OF INFORMATION ABOUT THE ABSORPTION SPECTRA OF ADSORBED MOLECULES. KOTOV, E. I. TEOR. PRIKL. PROBL. RASSEYAN. SVETA 387-95, 1971. (FOR ENGLISH TRANSLATION SEE T88022)
67739	STRUCTURE OF CALCIUM OXIDE-ALUMINUM OXIDE-SILICON DIOXIDE SYSTEM GLASSES STUDIED BY AN IR SPECTROSCOPIC METHOD. BOBKOVA, N. M. EVMENOVA, I. P. ZH. PRIKL. SPEKTROSK. 16 (3), 494-7, 1972. (FOR ENGLISH TRANSLATION SEE TPRC NO. 76930)
67775	EFFECT OF ELECTRIC AND MAGNETIC FIELDS ON THE THERMAL CONDUCTIVITY OF GLASS AT LOW TEMPERATURES. CHALLIS, L. J. HOOKER, C. N. J. PHYS. 5 C (11), 1153-7, 1972.
67781	OPTICAL CHARACTERISTICS OF TERBIUM TRIPOSITIVE IONS IN SODA GLASS. SHULGIN, B. V. TAYLOR, K. N. R. HOAKSEY, A. HUNT, R. P. J. PHYS. 5 C (13), 1716-26, 1972.
67784	HEAT TRANSFER IN MELTING. NISHIMURA, M. IWATA, T. SUGIYAMA, S. HASATANI, M. KAGAKU KOGAKU 36 (6), 640-7, 1972.
67791	ULTRAVIOLET THRESHOLD OF LEUCOSAPPHIRE CRYSTALS GROWN BY THE CZOCHRALSKI METHOD. MUSTOV, M. I. IVANOV, A. O. OPT.-MEKH. PROM 39 (5), 68, 1972.
67805	COEFFICIENTS OF THERMAL EXPANSION OF GLASS-REINFORCED PLASTICS AND THEIR COMPONENTS AT LOW AND HIGH TEMPERATURES. KRITSUK, A. A. PROBL. PROCH. 4 (5), 98-102, 1972. (FOR ENGLISH TRANSLATION SEE T79601)

TPRC Number	Bibliographic Citation

67819 **OPTICAL STATES, DENSITY, THERMAL EXPANSION COEFFICIENTS, AND REFACTIVE INDEXES OF SILICON DIOXIDE-BORON OXIDE-BARIUM OXIDE-ALUMINUM OXIDE SYSTEM GLASSES.**
SHCHAVELEV, O. S.
ZH. PRIKL. KHIM. (LENINGRAD)
45 (5), 1005-12, 1972.
(FOR ENGLISH TRANSLATION SEE TPRC NO. 70605)

67859 **LINEAR GEOLOGIC STRUCTURE AND MAFIC ROCK DISCRIMINATION AS DETERMINED FROM INFRARED DATA.**
OFFIELD, T. W. ROWAN, L. C. WATSON, R. D.
ANN. EARTH RESOURCES PROGRAM REV.
1, 11-1-11-12PP., 1970.
(N72-12259)

67860 **MULTISPECTRAL ANALYSIS OF LIMESTONE, DOLOMITE, AND GRANITE, MILL CREEK, OKLAHOMA.**
ROWAN, L. C. WATSON, K.
ANN. EARTH RESOURCES PROGRAM REV.
1, 12-1-12-14PP., 1970.
(N72-12260)

67864 **INVESTIGATION OF SPECTRAL EMISSIVITY OF OXIDES AND CONCRETES IN VISIBLE REGION OF SPECTRUM.**
ABRAMOV, A. S. BARYKIN, B. M. ROMANOV, A. I.
SPIRIDONOV, E. G.
JOINT PUBLICATIONS RES. SERVICE, WASHINGTON, D. C.
116-25, 1971.
(ENGLISH TRANSLATION OF MATER. KANALA MGD MAGNITOGIDRODINAMICHESKII-GENERATORA, 119-28, 1969., FOR ORIGINAL SEE TPRC NO. 61089)
(N71-35634)

67865 **STUDY OF EMISSIVITY OF OXIDES AND CONCRETES IN INFRARED REGION OF SPECTRUM.**
ABRAMOV, A. S. BARYKIN, B. M. ROMANOV, A. I.
SPIRIDONOV, E. G.
JOINT PUBLICATIONS RES. SERVICE, WASHINGTON, D.C.
126-31, 1971.
(ENGLISH TRANSLATION OF MATER. KANALA MGD MAGNITOGIDRODINAMICHESKII-GENERATORA, 129-34, 1969., FOR ORIGINAL SEE TPRC NO. 61090)
(N71-35635, AD-938751, FTD-HT-23-1371-71)

67883 **DETERMINATION OF THE THERMAL CONDUCTIVITY OF CARBON MATERIALS AT HIGH TEMPERATURES BY A MODULATED ELECTRON BEAM TECHNIQUE.**
TANAKA, T. SUZUKI, H.
NIPPON GENSHIRYOKU GAKKAISHI
14 (6), 274-7, 1972.

67890 **PROPERTIES OF PROTON-IRRADIATED DIAMOND.**
ANANTHANARAYANAN, K. P. BORER, W. J.
PLENDL, H. S. GIELISSE, P. J.
RADIAT. EFF.
14 (3-4), 245-8, 1972.

67899 **SPECTRAL EMISSIVITY OF MAGNESITE REFRACTORIES.**
KUZNETSOVA, N. P. KRIZANDIN, V. A.
MASTZYUKOV, B. S. SHUTOV, A. P.
IZV. VYSSH. UCHEB. ZAVED., CHERN. MET.
15 (7), 151-3, 1972.

67963 **RAPID DETERMINATION OF THE THERMOPHYSICAL PROPERTIES OF MATERIALS.**
VODOP'YANOVA, N. N. FUKS, L. G.
SHMANDINA, V. N.
IZV. VYSSH. UCHEB. ZAVED., ENERG.
15 (5), 135-7, 1972.

67971 **THERMOPHYSICAL PROPERTIES OF CERAMIC MOLDS.**
ZHARKEVICH, E. G. FEDOROVA, L. A.
ZINOVEV, YU. A. ROSHCHIN, M. I.
VERAKHIN, YU. N.
LITEINOE PROIZVOD.
(6), 16-17, 1972.

68003 **THERMAL AND THERMOPHYSICAL PROPERTIES OF CERAMOPERLITE PHOSPHATE.**
SARTAKOV, YU. A. ZHUKOV, A. V.
TKACHUK, V. P.
STROIT. MATER.
(6), 24-5, 1972.

68049 **CHANGES IN SOIL PROPERTIES ON FREEZING AND THAWING.**
TSYTOVICH, N. A. MARTYNOV, G. A.
MATERIALY PO LAB. ISSLED. MERZLYKH, GRUNTOV, SB.
(3), 84-114, 1957.
(FOR ENGLISH TRANSLATION SEE TPRC NO. 68050)

68050 **CHANGES IN SOIL PROPERTIES ON FREEZING AND THAWING.**
TSYTOVICH, N. A. MARTYNOV, G. A.
COLD REGIONS, RES. AND ENGINEERING LAB., HANOVER, N. H.
31PP., 1971.
(ENGLISH TRANSLATION OF MATERIALY PO LAB. ISSLED. MERZLYKH, GRUNTOV, SB., (3), 84-114, 1957; FOR ORIGINAL SEE TPRC NO. 68049)
(AD-739953, CRREL-TL-329)

68062 **THE RADIATION OF HEAT FROM THE HUMAN BODY. III. THE HUMAN SKIN AS A BLACK - BODY RADIATOR.**
HARDY, J. D.
J. CLIN. INVEST.
13, 615-20, 1934.

68080 **FUNDAMENTAL ABSORPTION AND THERMAL PROPERTIES OF STIBIOTANTALITE.**
NOVIK, V. K. KARYAKINA, N. F.
TIMOSHENKOV, V. A.
FIZ. TVERD. TELA
14 (5), 1503-6, 1972.
(FOR ENGLISH TRANSLATION SEE TPRC NO. 68081)

68081 **INTRINSIC ABSORPTION AND THERMAL PROPERTIES OF STIBIOTANTALITE.**
NOVIK, V. K. KARYAKINA, N. F.
TIMOSHENKOV, V. A.
SOV. PHYS. - SOLID STATE
14 (5), 1288-90, 1972.
(ENGLISH TRANSLATION OF FIZ. TVERD. TELA, 14 (5), 1503-6, 1972., FOR ORIGINAL SEE TPRC NO. 68080)

68113 **THERMAL CONDUCTIVITY AND COEFFICIENT OF LINEAR EXPANSION OF SINGLE CRYSTALS OF SYNTHETIC MICA (FLUOROPHLOGOPITE) AND NATURAL PHLOGOPITE.**
EGOROV, B. N. KONDRATENKOV, V. I.
ANIKIN, I. N.
TEPLOFIZ. VYS. TEMP.
10 (1), 82-6, 1972.
(FOR ENGLISH TRANSLATION SEE TPRC NO. 68114)

68114 **THERMAL CONDUCTIVITY AND COEFFICIENT OF THERMAL EXPANSION OF SINGLE CRYSTALS OF SYNTHETIC MICA (FLUOROPHLOGOPITE) AND OF NATURAL PHLOGOPITE.**
EGOROV, B. N. KONDRATENKOV, V. P.
ANIKIN, I. N.
HIGH TEMP.
10 (1), 68-71, 1972.
(ENGLISH TRANSLATION OF TEPLOFIZ. VYS. TEMP., 10 (1), 82-6, 1972., FOR ORIGINAL SEE TPRC NO. 68113)

68159 **INFRARED SPECTRA OF SOLID SOLUTIONS OF SCANDIUM OXIDE WITH TITANIUM (III) AND VANADIUM (III) OXIDES.**
SHVEIKIN, G. P. BAZUEV, G. V.
PERELYAEV, V. A.
ZH. NEORG. KHIM.
17 (1), 20- , 1972.
(FOR ENGLISH TRANSLATION SEE TPRC NO. 68160)

68160 **INFRARED SPECTRA OF SOLID SOLUTIONS OF SCANDIUM OXIDE WITH TITANIUM (III) AND VANADIUM (III) OXIDES.**
SHVEIKIN, G. P. BAZUEV, G. V.
PERELYAEV, V. A.
RUSS. J. INORG. CHEM.
17 (1), 9-10, 1972.
(ENGLISH TRANSLATION OF ZH. NEORG. KHIM., 17 (1), 20- , 1972., FOR ORIGINAL SEE TPRC NO. 68159)

68190 **THERMAL CONDUCTIVITY OF II-VI COMPOUNDS AND PHONON SCATTERING BY FERROUS DIPOSITIVE ION IMPURITIES.**
SLACK, G. A.
PHYS. REV.
6 B (10), 3791-3800, 1972.

68206 **HEAT FLOW AND SURFACE RADIOACTIVITY AT TWO SITES IN SOUTH GREENLAND.**
SASS, J. H. NIELSEN, B. L. WOLLENBERG, H. A.
MUNROE, R. J.
J. GEOPHYS. RES.
77 (32), 6435-44, 1972.

68228 **REFLECTIVITY OF GERSDORFFITE FROM DOBSINA.**
HALAHYJOVA-ANDRUSOVOVA, G.
GEOL. ZB. (BRATISLAVA)
23 (1), 55-68, 1972.

68234 **THERMAL CONDUCTIVITY OF URANIUM-PLUTONIUM OXIDE.**
LASKIEWICZ, R. A. EVANS, S. K. MELDE, G. F.
BOHABOY, P. E.
GENERAL ELECTRIC COMP., SUNNYVALE, CALIF.
76PP., 1971.
(GEAP-13733)

68241 **NOVEL OPTICAL MATERIALS REPLACE GLASS.**
WALKER, W. W.
MATER. ENG.
73 (5), 34-7, 1971.
(AD-735 593, AFOSR-TR-72-0154, AVAIL. NTIS)

68244 **HEAT FLOW THROUGH THE FLOOR OF CASCADIA BASIN.**
BODVARSSON, G. MESECAR, R. S. KORGEN, B. J.
J. GEOPHYS. RES.
76 (20), 4758-74, 1971.
(AD-745620)

68251 **REFLECTIVITY MEASUREMENTS WITH 10.6 MICROMETER INFRARED RADIATION.**
VAN DAMME, G. E. AMORUSO, M. J.
MC GARVEY, J. W.
ARMY WEAPONS COMMAND, ROCK ISLAND, ILL.
27PP., 1972.
(AD-746238, SWERR-TR-72-39, N73-11482)

68273 **THERMAL EXPANSION OF GLASSES AT LOW TEMPERATURES.**
PAPOULAR, M.
J. PHYS.
5 C (15), 1943-4, 1972.

TPRC Number	Bibliographic Citation
68284	**CHANGES IN ULTRAVIOLET TRANSMISSION OF CORNING CODE 9700 GLASS TUBING.** MORRISON, H. MALESKI, R. PHOTOCHEM. PHOTOBIOL. 16 (2), 145-6, 1972.
68297	**LIGHT TRANSMISSION BY COLORED QUARTZ GLASSES FOR PRODUCTION OF LIGHT FILTERS.** TARASOV, B. V. KOZLOVA, L. N. YUSHANKINA, G. A. STEKLOOBRAZN. SIST. NOV. STEKLA IKH OSN. 11-14, 1971.
68304	**PHYSICOCHEMICAL PROPERTIES AND STRUCTURE OF ALKALINE ZINC SILICATE GLASSES.** VARGIN, V. V. DZHAVUKTSYAN, S. G. MISHEL', V. E. PEVZNER, B. Z. ZH. PRIKL. KHIM. (LENINGRAD) 45 (6), 1187-93, 1972. (FOR ENGLISH TRANSLATION SEE TPRC NO. 70606)
68323	**STUDIES OF OBSIDIAN AS A MATERIAL FOR USE IN MAKING ASTRONOMICAL MIRRORS.** DUNAHAM, T. FUND FOR ASTROPHYSICAL RESEARCH INC, NEW YORK 11PP., 1955. (AD-740689)
68331	**DETERMINATION OF THE SPECTRAL REFLECTION COEFFICIENTS OF FRIABLE MATERIALS IN THE 0.7-15 MICRON WAVELENGTH REGION.** KROPOTKIN, M. A. KOZIREV, B. P. IZV. VYSSH. UCHEB. ZAVED. FIZ. (3), 27-9, 1965. (FOR ENGLISH TRANSLATION SEE TPRC NO. 68332)
68332	**DETERMINATION OF THE SPECTRAL REFLECTION COEFFICIENTS OF FRIABLE MATERIALS IN THE 0.7-15 MICRON WAVELENGTH REGION.** KROPOTKIN, M. A. KOZIREV, B. P. SOV. PHYS. J. (3), 18-9, 1965. (ENGLISH TRANSLATION OF IZV. VYSSH. UCHEB. ZAVED. FIZ., (3), 27-9, 1965., FOR ORIGINAL SEE TPRC NO. 68331)
68344	**PROPOSED 131-CM F/3.5 ACHROMATIC SCHMIDT TELESCOPE.** BUCHROEDER, R. A. APPL. OPT. 11 (12), 2968-71, 1972.
68348	**PHOTOTROPIC GLASSES ACTIVATED BY THALLIUM CHLORIDE.** SAKKA, S. MAC KENZIE, J. D. J. AMER. CERAM. SOC. 55 (11), 553-7, 1972.
68349	**THERMAL CONDUCTIVITY OF CRISTOBALITE.** KUNUGI, M. SOGA, N. SAWA, H. KONISHI, A. J. AMER. CERAM. SOC. 55 (11), 580, 1972.
68377	**APPARATUS FOR DETERMINING THE COEFFICIENT OF THERMAL EXPANSION OF ROCKS, MORTARS, AND CONCRETES.** LOUBSER, P. J. BRYDEN, J. G. MAG. CONCR. RES. 24 (79), 97-100, 1972.
68386	**INFRARED REFLECTION SPECTRA OF COMPRESSED POWDERS.** BRUNEL, R. VIERNE, R. VERGNOUX, A. M. POWDER TECHNOL. 6 (3), 121-31, 1972.
68434	**NON-STELLAR CELESTIAL BACKGROUNDS.** MURCRAY, D. G. MURCRAY, F. H. AMME, R. C. HEWITT, J. G., JR. BARKER, D. B. DENVER UNIV., COLORADO 102PP., 1971. (AD-733738, N72-19891, AFCRL-71-0424)
68472	**ULTRAVIOLET ABSORPTION IN THE CONTINUOUS SPECTRUM OF JUPITER.** KRUGOV, V. D. PHYSICS OF THE MOON AND PLANETS 9-23PP., 1971.
68473	**ENERGY DISTRIBUTION IN THE SHORT-WAVE REGION OF THE SPECTRUM FOR DIFFERENT SEGMENTS OF THE SATURN DISK.** KRUGOV, V. D. PHYSICS OF THE MOON AND PLANETS 23-33., 1971.
68482	**LUNAR SURFACE TEMPERATURES FROM APPOLLO 12.** CREMERS, C. J. BIRKEBAK, R. C. WHITE, J. E. THE MOON 3 (DEC.), 346-351, 1971.
68483	**THERMAL CONDUCTIVITY OF APOLLO 12 FINES AT INTERMEDIATE DENSITY.** CREMERS, C. J. THE MOON 4 (APR.), 88-92, 1972.
68484	**APPOLLO 12 THERMAL RADIATION PROPERTIES.** BIRKEBAK, R. C. THE MOON 4 (APR.), 128-133, 1972.
68490	**HEAT RESISTANCE OF REFRACTORY MATERIALS.** NEMETS, I. I. PROBL. PROCH. 3, 41-3, 1971. (FROM: RUSSIAN)
68503	**THERMAL STUDIES.** BASTIN, J. A. BOWELL, E. L. G. GEOLOGY AND PHYSICS OF THE MOON 143-155, 1971.
68509	**STRUCTURAL CHANGES OF ILLITE AND MONTMORILLONITE STUDIED BY INFRARED SPECTROPHOTOMETRY.** TOROK, I. AGROKEM. TALAJTAN 21 (1-2), 131-8, 1972.
68513	**MEASUREMENT OF ENTHALPY AND CALCULATION OF SPECIFIC HEAT OF GLASSES.** TYDLITAT, V. BLAZEK, A. ENDRYS, J. STANEK, J. GLASTECH. BER. 45 (8), 352-9, 1972.
68514	**THERMAL CONDUCTIVITY OF MATERIALS BASED ON CHROMITE ORE.** DONCHEV, ST. IORDANOV, P. GOD. VISSH. KHIMIKOTEKHNOL. INST., SOFIA 14 (3), 75-81, 1971.
68525	**ABSORPTION SPECTRA OF GLASSES AND MELTS AND VOLUME PROPERTIES OF COBALT METAPHOSPHATE - SODIUM METAPHOSPHATE AND COBALT METAPHOSPHATE - POTASSIUM METAPHOSPHATE.** TANANAEV, I. V. VOSKRESENSKAYA, N. K. DZHURINSKII, B. F. KRIVOVYAZOV, E. L. RAKHIMBEKOVA, KH. M. IZV. AKAD. NAUK SSSR, NEORG. MATER. 8 (8), 1474-7, 1972. (FOR ENGLISH TRANSLATION SEE TPRC NO. 70883)
68576	**A MODERN METHOD FOR REFLECTANCE DETERMINATION IN MINERALS.** KLEINBOK, V. E. METOD. MINERAL. ISSLED. 5-16, 1971.
68595	**PHYSICAL PROPERTIES OF FOAMED PLASTICS AND THEIR DEPENDENCE ON STRUCTURE.** BAXTER, S. JONES, T. T. PLAST. POLYM. 40 (146), 69-76, 1972.
68596	**METALLIC SPHERULES FROM THE SEDIMENTS OF THE NORWEGIAN SEA.** ALEKSEEVA, K. N. FOKA, T. M. PROBL. KOSMOKHIMII METEORITIKI, RASSHIR. YUBILEINYI PLENUM 129-32, 1971.
68602	**THERMAL CONDUCTIVITY OF ALUMINA CERAMICS IN THE TEMPERATURE RANGE 40-400 DEGREES.** GEORGE, W. SPEC. CERAM. (5), 211-23, 1972.
68616	**OPTICAL PROPERTIES OF GLASS WITH AN OXIDE-NICKEL COATING.** PAVLUSHKIN, N. M. PAVLOVA, I. S. TR. MOSK. KHIM.-TEKHNOL. INST. (68), 73-5, 1971.
68674	**THERMOPHYSICAL PROPERTIES OF THERMALLY INSULATING MATERIALS IN THE CRYOGENIC TEMPERATURE REGION.** SHASHKOV, A. G. VASILIEV, L. L. TANAEVA, S. A. DOMOROD, L. S. INT. J. HEAT MASS TRANSFER 15 (12), 2385-90, 1972.
68734	**THERMAL CONDUCTIVITY OF EARTH MATERIALS AT HIGH TEMPERATURES.** SCHATZ, J. F. SIMMONS, G. J. GEOPHYS. RES. 77 (35), 6966-83, 1972.
68735	**LINEAR THERMAL EXPANSION COEFFICIENTS OF ORTHOPYROXENE TO 1000 DEGREES C.** FRISILLO, A. L. BULJAN, S. T. J. GEOPHYS. RES. 77 (35), 7115-7, 1972.
68742	**AN ACCURATE SPECTROPHOTOMETER FOR MEASURING THE TRANSMITTANCE OF SOLID AND LIQUID MATERIALS.** MAVRODINEANU, R. J. RES. NAT. BUR. STAND. 76 A (5), 405-25, 1972.

TPRC Number	Bibliographic Citation
68786	**REFLECTANCE AND MICROINDENTATION HARDNESS VS. CHEMICAL COMPOSITION IN SOME CANADIAN URANINITES.** MORTON, R. D. SASSANO, G. P. NEUES JAHRB. MINERAL., MONATSH. (8), 350–60, 1972.
68805	**PHYSICAL PROPERTIES OF POROUS CORUNDUM CERAMIC.** ADUSHKIN, L. E. GUZMAN, I. YA. TR. MOSK. KHIM.-TEKHNOL. INST. (68), 120–3, 1971.
68807	**RAISING THE THERMAL CONDUCTIVITY OF CORUNDUM CERAMIC** RYAZANOVA, M. YA. MAKAROV, I. A. TR. MOSK. KHIM.-TEKHNOL. INST. (68), 143–6, 1971.
68812	**EMISSIVITY OF DOPED QUARTZ GLASSES.** MEN, A. A. PROKHOROVA, T. I. SETTAROVA, Z. S. ZH. PRIKL. SPEKTROSK. 17 (2), 325–8, 1972. (FOR ENGLISH TRANSLATION SEE TPRC NO. 77029)
68816	**A NOTE ON THE ALBEDO OF SURFACES.** NKEMDIRIM, L. C. J. APPL. METEOROL. 11 (8), 867–74, 1972.
68829	**LONG-WAVE INFRARED SPECTRA OF SOME COMPLEX LEAD SULFIDES.** POVARENNYKH, A. S. SIDORENKO, G. A. SOLNTSEVA, L. S. SOLNTSEV, B. P. FOREIGN TECHNOLOGY DIV., WRIGHT-PATTERSON, AIR FORCE BASE, OHIO 5PP., 1972. (ENGLISH TRANSLATION OF DOPOV. AKAD. NAUK UKR., RSR, SER. B, 33 (7), 596–9, 1971., FOR ORIGINAL SEE TPRC NO. 64208) (AD-749630, FTD-HC-230767-72)
68850	**AIRBORNE MEASUREMENTS OF OPTICAL ATMOSPHERIC PROPERTIES IN SOUTHERN GERMANY.** DUNTLEY, S. Q. JOHNSON, R. W. GORDON, J. I. SCRIPPS INSTITUTION OF OCEANOGRAPHY, SAN DIEGO, CALIF. 221PP., 1972. (AD-747490, AFCRL-72-0255)
68853	**STUDY OF REDUCTION OF GLARE, REFLECTION, HEAT AND NOISE TRANSFER IN AIR TRAFFIC CONTROL TOWER CAB GLASS.** CLINCH, J. M. IIT RESEARCH INST., CHICAGO, ILLINOIS 105PP., 1972. (AD-747069, FAA-RD-72-65)
68859	**SEA FLOOR GEOTHERMAL MEASUREMENTS FROM CONRAD CRUISE 13.** BOOKMAN, C. A. MALONE, I. E. LANGSETH, M. G. LAMONT-DOHERTY GEOLOGICAL OBSERVATORY, PALISADES, N. Y. 279PP., 1972. (AD-749983)
68875	**INFERENCES FROM OPTICAL PROPERTIES CONCERNING THE SURFACE TEXTURE AND COMPOSITION OF ASTEROIDS.** HAPKE, B. UNIV. PITTSBURGH, PITTSBURGH, PA. 67–77, 1971. (NASA-SP-267, N72-25753)
68884	**THE DISTRIBUTION FUNCTION OF THE NORMAL ALBEDO OF THE SURFACE OF MARS.** ALEKSANDROV, IU. V. LUPISHKO, D. F. ASTRON. VESTN. 6 (1), 9–12, 1972. (FOR ENGLISH TRANSLATION SEE TPRC NO. 68885)
68885	**THE DISTRIBUTION FUNCTION OF THE NORMAL ALBEDO OF THE SURFACE OF MARS.** ALEKSANDROV, IU. V. LUPISHKO, D. F. SOLAR SYSTEM RES., (USSR) 6 (1), 7–9, 1972. (ENGLISH TRANSLATION OF ASTRON. VESTN., 6 (1), 9–12, 1972., FOR ORIGINAL SEE TPRC NO. 68884)
68895	**ENERGY TRANSFER IN A MEDIUM OF LOW OPTICAL DENSITY.** KOSTYLEV, V. M. KOMAROVSKAIA, N. V. INZH.-FIZ. ZH. 22, 907–912PP., 1972. (FOR ENGLISH TRANSLATION SEE TPRC NO. 75009)
68900	**INFRARED AND RAMAN SPECTRA OF LUNAR SAMPLES FROM APOLLO 11, 12 AND 14.** PERRY, C. H. AGRAWAL, D. K. ANASTASSAKIS, E. LOWNDES, R. P. RASTOGI, A. TORNBERG, N. E. THE MOON 4 (JUNE-JULY), 315–336, 1972.
68901	**THE APOLLO 15 LUNAR HEAT-FLOW MEASUREMENT.** LANGSETH, M. G., JR. CHUTE, J. L., JR. KEIHM, S. J. CLARK, S. P., JR. WECHSLER, A. E. THE MOON 4 (JUNE-JULY), 390–410, 1972.
68902	**THERMOPHYSICAL PROPERTIES OF LUNAR MATERIAL RETURNED BY APPOLLO MISSIONS.** HORAI, K.-I. FUJII, N. THE MOON 4 (JUNE-JULY), 447–475, 1972.
68911	**SPECTRAL REFLECTANCE AND EMITTANCE OF APOLLO 11 AND 12 LUNAR MATERIAL.** BIRKEBAK, R. C. AIAA J. 10, 1064–7, 1972.
68921	**INFRARED SPECTROSCOPIC STUDY OF CRYPTOCRYSTALLINE VARIETIES OF SILICA.** PLYUSNINA, I. I. MALEEV, M. N. EFIMOVA, M. N. IZV. AKAD. NAUK SSSR, SER. GEOL. (9), 78–83, 1970.
68980	**DEVELOPMENT OF A THEORY OF THE SPECTRAL REFLECTANCE OF MINERALS, PART 3.** ARONSON, J. R. EMSLIE, A. G. ROACH, L. H. SMITH, E. M. VON THUENA, P. C. LITTLE (ARTHUR D.), INC., CAMBRIDGE, MASS. 64PP., 1972. (N72-22738, NASA-CR-115521, ADL-C-72594-PT-3)
68985	**MICROWAVE ULTRASONIC ATTENUATION IN TOPAZ, BERYL, AND TOURMALINE.** LEWIS, M. F. PATTERSON, E. J. APPL. PHYS. 44 (1), 10–3, 1973.
68990	**EPITAXIALLY GROWN AIN AND ITS OPTICAL BAND GAP.** YIM, W. M. STOFKO, E. J. ZANZUCCHI, P. J. PANKOVE, J. I. ETTENBERG, M. GILBERT, S. L. J. APPL. PHYS. 44 (1), 292–6, 1973.
68992	**ABSORPTION TAIL AND FUNDAMENTAL ABSORPTION EDGE IN VANADIUM PHOSPHATE GLASSES.** ANDERSON, G. W. J. APPL. PHYS. 44 (1), 406–9, 1973.
69035	**INORGANIC IONS IN GLASSES AND POLYCRYSTALLINE PELLETS AS FLUORESCENCE STANDARD REFERENCE MATERIALS.** REISFELD, R. J. RES. NAT. BUR. STAND. 76 A (6), 613–35, 1972.
69051	**RADIATIVE HEAT TRANSFER IN FIBERGLASS INSULATION.** VNUKOV, S. P. RYABOV, V. A. FEDOSEEV, D. V. TEPLOFIZ. VYS. TEMP. 10 (3), 666–8, 1972. (FOR ENGLISH TRANSLATION SEE TPRC NO. 69052)
69052	**RADIATIVE HEAT TRANSFER IN FIBERGLASS INSULATION.** VNUKOV, S. P. RYABOV, V. A. FEDOSEEV, D. V. HIGH TEMP. 10 (3), 600–2, 1972. (ENGLISH TRANSLATION OF TEPLOFIZ. VYS. TEMP., 10 (3), 666–8, 1972., FOR ORIGINAL SEE TPRC NO. 69051)
69066	**DETERMINATION OF THE REFLECTING POWER OF A HILLY TERRAIN, KNOWING THE REFLECTIVE POWER OF A FLAT TERRAIN OF THE SAME NATURE.** JAVELOT, M. ANN. TEIECOMMUN. 27, 248–54, 1972.
69070	**CRYSTALLOCHEMICAL BEHAVIOR OF IONS IN GARNET LATTICE. V. STUDY OF DISTRIBUTION OF TITANIUM TETRAPOSITIVE ION AND IRON TRIPOSITIVE ION BY MOESSBAUER AND INFRARED SPECTROSCOPY.** HRICHOVA, R. LIPKA, J. COLLECT. CZECH. CHEM. COMMUN. 37 (10), 3352–5, 1972.
69083	**NATURE OF A SHIFT IN IR ABSORPTION SPECTRAL BANDS OF SOME ISOSTRUCTURAL MINER1LS.** POVARENNYKH, A. S. GEOL. ZH. (RUSS. ED.) 32 (5), 23–30, 1972.
69101	**INFRARED SPECTRA OF SOME SULFIDES AND THEIR ANALOGS OF BINARY COMPOSITION IN THE FAR INFRARED REGION.** POVARENNYKH, A. S. SIDORENKO, G. A. SOLNTSEVA, L. S. SOLNTSEV, B. P. MINERAL. SB. (LOVOV) 25 (4), 306–15, 1971.
69102	**THE PROSPECTS FOR MINERAL ANALYSIS BY REMOTE INFRARED SPECTROSCOPY.** ARONSON, J. R. EMSLIE, A. G. THE MOON 5 (SEPT.), 3–15, 1972.
69103	**SPECTROPHOTOMETRY /0.3 TO 1.1 MICRON/ OF VISITED AND PROPOSED APOLLO LUNAR LANDING SITES.** MC CORD, T. B. CHARETTE, M. P. JOHNSON, T. V. LEBOFSKY, L. A. PIETERS, C. THE MOON 5 (SEPT.), 52–89, 1972.

TPRC Number | Bibliographic Citation

69104 THE SUNLIT LUNAR SURFACE. II – A STUDY OF FAR INFRARED BRIGHTNESS TEMPERATURES.
SAARI, J. M. SHORTHILL, R. W. WINTER, D. F.
THE MOON
5 (SEPT.), 179–199, 1972.

69135 SPHALERITE INFRARED ABSORPTION SPECTRA.
SOLATSEVA, L. S. RYABEVA, E. G.
TR. MINERAL MUZ., AKAD. NAUK SSSR
(21), 201–4, 1972.

69194 PHASE DEPENDENCE OF THE INTEGRAL BRIGHTNESS OF MARS.
BARABASHOV, N. P. ALEKSANDROV, YU. V.
LUPISHKO, T. A. LUPISHKO, D. F.
SOV. ASTRON.–AJ
15 (3), 457–61, 1971.
(ENGLISH TRANSLATION OF ASTRON. ZH., 48 (3), 581–6, 1971., FOR ORIGINAL SEE TPRC NO. 64993)

69253 SINGLE REFLECTION MOVABLE DETECTOR REFLECTOMETER FOR LOW REFLECTION MEASUREMENTS.
TOMAR, M. S. SRIVASTAVA, V. K.
INDIAN J. PURE APPL. PHYS.
10 (7), 573–5, 1972.

69284 DETERMINATION OF THERMAL CONDUCTIVITY OF SOLID MATERIALS BY MEANS OF THE HOT WIRE METHOD.
HAYASHI, K. FUKUI, M. UEI, I.
MEM. FAC. IND. ARTS, KYOTO TECH. UNIV., SCI. TECHNOL.
20, 81–103, 1971.

69328 THE NASA EARTH RESOURCES SPECTRAL INFORMATION SYSTEM: A DATA COMPILATION.
LEEMAN, V. EARING, D. VINCENT, R. K.
LADD, S.
WILLOW RUN LABS., MICHIGAN UNIV., ANN ARBOR
574PP., 1971.
(NASA–CR–115757, N72–28366, WRL–31650–T)

69333 EFFECT OF ALKALI OXIDES ON THE DIFFUSION OF HELIUM IN A SIMPLE BOROSILICATE GLASS.
ALTEMOSE, V. O.
J. AMER. CERAM. SOC.
56 (1), 1–4, 1973.

69340 THE NASA EARTH RESOURCES SPECTRAL INFORMATION SYSTEM: A DATA COMPILATION, FIRST SUPPLEMENT.
LEEMAN, V.
INFRARED AND OPTICS LAB., MICHIGAN UNIV.
159PP., 1972.
(N72–30345, NASA–CR–115756, WRL–31650–69–SUPPL–1)

69344 REINFORCED COMPOSITE CASE, FY70–4.
FOTOPOULOS, C. U.
BENDIX CORP., KANSAS CITY, KANSAS
25PP., 1972.
(N72–33565, BDX–613–289–R5V 8)

69346 SEMICRYSTALLINE GLAZES FOR LOW EXPANSION WHITEWARE BODIES.
O'CONOR, E. F. EPPLER, R. A.
BULL. AMER. CERAM. SOC.
52 (2), 180–4, 1973.

69347 A LASER SPECULAR REFLECTOMETER FOR CERAMIC SURFACE DIAGNOSTICS.
HENSLER, D. SOOS, N. A. HASS, E. H.
FRANKSON, R. W. ROTT, D. A.
BULL. AMER. CERAM. SOC.
52 (2), 191–4, 1973.

69441 TEMPERATURE CONDUCTIVITY OF BASALTS AT TEMPERATURES FROM 300 TO 1200 K.
PETRUNIN, G. I. YURCHAK, R. I. TKACH, G. F.
BULL. ACAD. SCI., USSR, EARTH PHYS.
(2), 126–7, 1971.
(ENGLISH TRANSLATION OF IZV. AKAD. NAUK SSSR, FIZ. ZEMLI, (2), 65–8, 1971., FOR ORIGINAL SEE TPRC NO. 62710)

69506 HEAT TRANSFER IN A LAYER OF DISPERSE MATERIAL.
NIKITIN, V. S. ANTONISHIN, N. V.
J. ENG. PHYS., USSR
17 (2), 948–51, 1969.
(ENGLISH TRANSLATION OF INZH. FIZ. ZH., 17 (2), 248–53, 1969., FOR ORIGINAL SEE TPRC NO. 56232)

69520 HEAT CAPACITIES AT LOW TEMPERATURES AND ENTROPIES AT 298.15 K OF HUNTITE AND ARTINITE.
HEMINGWAY, B. S. ROBIE, R. A.
AMER. MINERAL.
57 (11–2), 1754–67, 1972.

69521 HEAT CAPACITIES AT LOW TEMPERATURES AND ENTROPIES AT 298.15 K OF NESQUEHONITE AND HYDROMAGNESITE.
ROBIE, R. A. HEMINGWAY, B. S.
AMER. MINERAL.
57 (11–2), 1768–81, 1972.

69533 REFLECTIVITY AND METAMORPHISM OF COALS FROM THE LVOV–VOLYN BASIN.
FEDUSHCHAK, M. YU. BYK, S. I.
RADCHENKO, L. M.
DOPOV. AKAD. NAUK UKR. RSR, SER.
34 B (10), 902–4, 1972.

69537 OPTICAL PROPERTIES OF CARBONIZED VITRINITES.
GOODARZI, F. MURCHISON, D. G.
FUEL
51 (4), 322–8, 1972.

69622 METAL–INSULATOR TRANSITIONS IN PURE AND DOPED VANADIUM SESQUIOXIDE.
MC WHAN, D. B. MENTH, A. REMEIKA, J. P.
BRINKMAN, W. F. RICE, T. M.
PHYS. REV.
7 B (5), 1920–31, 1973.

69632 THERMODYNAMIC CHARACTERISTICS OF METALS IN THE FINELY DISPERSED STATE.
BOGOMOLOV, V. N. KLUSHIN, N. A.
ROMANOVA, M. V. SMIRNOV, I. A.
TIKHONOV, V. V.
FIZ. TVERD. TELA
14 (9), 2699–703, 1972.
(FOR ENGLISH TRANSLATION SEE TPRC NO. 69633)

69633 THERMODYNAMIC CHARACTERISTICS OF METALS IN THE FINELY DISPERSED STATE.
BOGOMOLOV, V. N. KLUSHIN, N. A.
ROMANOVA, M. V. SMIRNOV, I. A.
TIKHONOV, V. V.
SOV. PHYS.–SOLID STATE
14 (9), 2330–4, 1973.
(ENGLISH TRANSLATION OF FIZ. TVERD. TELA, 14 (9), 2699–703, 1972., FOR ORIGINAL SEE TPRC NO. 69632)

69642 INFLUENCE OF MECHANICAL STRESSES ON THE HYDROLYSIS OF BONDS ON THE SURFACE OF GLASS.
BERSHTEIN, V. A. MOVCHAN, YU. N.
NIKITIN, V. V.
FIZ. TVERD. TELA
14 (9), 2792–4, 1972.
(FOR ENGLISH TRANSLATION SEE TPRC NO. 69643)

69643 INFLUENCE OF MECHANICAL STRESSES ON THE HYDROLYSIS OF BONDS ON THE SURFACE OF GLASS.
BERSHTEIN, V. A. MOVCHAN, YU. N.
NIKITIN, V. V.
SOV. PHYS.–SOLID STATE
14 (9), 2422–3, 1973.
(ENGLISH TRANSLATION OF FIZ. TVERD. TELA, 14 (9), 2792–4, 1972., FOR ORIGINAL SEE TPRC NO. 69642)

69660 EFFECTS OF STRUCTURE, COMPOSITION, AND STRESS ON THE LASER DAMAGE THRESHOLD OF HOMOGENEOUS AND INHOMOGENEOUS SINGLE FILMS AND MULTILAYERS.
AUSTIN, R. R. MICHAUD, R. GUENTHER, A. H.
PUTMAN, J.
APPL. OPT.
12 (4), 665–76, 1973.

69687 MEASUREMENTS OF HEAT TRANSMISSION IN THERMAL INSULATIONS AT CRYOGENIC TEMPERATURES USING THE GUARDED HOT PLATE METHOD.
TYE, R. P.
PROC. INT. CONGR. REFRIG.
1, 371–8, 1971.

69694 THERMAL PROPERTIES OF THE SOIL AS A FUNCTION OF ITS MOISTURE AND COMPACTNESS.
GUPALO, A. I.
POCHVOVEDENIE
(4), 40–5, 1959.
(FOR ENGLISH TRANSLATION SEE TPRC NO. 69695)

69695 THERMAL PROPERTIES OF THE SOIL AS A FUNCTION OF ITS MOISTURE AND COMPACTNESS.
GUPALO, A. I.
SCIENTIFIC TRANSLATION SERVICE, SANTA BARBARA, CALIF.
13PP., 1972.
(ENGLISH TRANSLATION OF POCHVOVEDENIE, (4), 40–5, 1959; FOR ORIGINAL SEE TPRC NO. 69694)
(N72–28356, NASA–TT–F–14364)

69712 THERMAL EXPANSION OF MAGNETITE.
ARKHAROV, V. I. BOGOSLOVSKII, V. N.
KUZNETSOV, E. N.
IZV. AKAD. NAUK SSSR, NEORG. MATER.
8 (11), 1982–4, 1972.
(FOR ENGLISH TRANSLATION SEE TPRC NO. 72179)

69768 INVESTIGATION OF THE THERMAL PROPERTIES OF LUNAR SOIL AND OF ITS TERRESTRIAL ANALOGS.
DMITRIEV, A. P. DUKHOVSKOI, E. A.
NOVIK, G. YA. PETROCHENKOV, R. G.
DOKL. AKAD. NAUK, SSSR
199 (5), 1036–7, 1971.
(FOR ENGLISH TRANSLATION SEE TPRC NO. 69769)

69769 INVESTIGATION OF THE THERMAL PROPERTIES OF LUNAR SOIL AND OF ITS TERRESTRIAL ANALOGS.
DMITRIEV, A. P. DUKHOVSKOI, E. A.
NOVIK, G. YA. PETROCHENKOV, R. G.
SOV. PHYS.–DOKL.
16 (8), 611–2, 1972.
(ENGLISH TRANSLATION OF DOKL. AKAD. NAUK, SSSR, 199 (5), 1036–7, 1971., FOR ORIGINAL SEE TPRC NO. 69768)

TPRC Number	Bibliographic Citation
69791	RELATION BETWEEN THE INFRARED SPECTRA AND ACID-BASE PROPERTIES OF SLAGS. KURBATOV, G. A. KALMYKOV, V. A. J. APPL. CHEM., USSR 45 (1), 179–80, 1972. (ENGLISH TRANSLATION OF ZH. PRIKL. KHIM., 45 (1), 190–1, 1972; FOR ORIGINAL SEE TPRC NO. 67137)
69793	DETERMINATION OF THE AVERAGE SPECIFIC HEATS OF SALT ROCKS AND BRINES. YUSOVA, YU. I. ESELEV, I. M. KARATYGIN, E. P. J. APPL. CHEM., USSR 45 (2), 425–6, 1972. (ENGLISH TRANSLATION OF ZH. PRIKL. KHIM., 45 (2), 427–8, 1972., FOR ORIGINAL SEE TPRC NO. 66602)
69810	ULTRAVIOLET PHOTOMETRY FROM THE ORBITING ASTRONOMICAL OBSERVATORY. III. OBSERVATIONS OF VENUS, MARS, JUPITER, AND SATURN LONGWARD OF 2000 A. WALLACE, L. CALDWELL, J. J. SAVAGE, B. D. ASTROPHYS. J. 172 (3), (PT. 1), 755–69, 1972.
69818	EFFECTIVE THERMAL CONDUCTIVITY OF FIBROUS INSULATING MATERIAL IN COMPRESSED GASES. BYKOV, V. N. VORONKOV, S. T. DEKHTYAREV, V. L. TOMASHEVA, L. P. POGONTSEV, V. G. TERESHCHENKO, E. N. TEPLOFIZ. VYS. TEMP. 10 (4), 788–94, 1972. (FOR ENGLISH TRANSLATION SEE TPRC NO. 69819)
69819	EXPERIMENTAL RESEARCH ON THE EFFECTIVE THERMAL CONDUCTION OF FIBROUS INSULATION IN COMPRESSED GASES. BYKOV, V. N. VORONKOV, S. T. DEKHTYAREV, V. L. TOMASHEVA, L. P. POGONTSEV, V. G. TERESHCHENKO, E. N. HIGH–TEMP. 10 (4), 707–12, 1972. (ENGLISH TRANSLATION OF TEPLOFIZ. VYS. TEMP., 10 (4), 788–94, 1972., FOR ORIGINAL SEE TPRC NO. 69818)
69867	THERMAL EXPANSION IN THE SYSTEM YTTRIUM OXIDE-HAFNIUM OXIDE. STACY, D. W. WILDER, D. R. J. AMER. CERAM. SOC. 56 (4), 224, 1973.
69877	OPTICAL PROPERTIES OF AN ABSORBING MONOCLINIC CRYSTAL DETERMINED BY MICROREFLECTIVITY MEASUREMENTS. II. CERVELLE, B. BULL. SOC. FR. MINERAL. CRISTALLOGR. 95 (4), 464–9, 1972.
69881	INFRARED SPECTROMETRY OF METAMICT BRITHOLITE. VALTER, A. A. DOPOV. AKAD. NAUK UKR. RSR, SER. B 34 B (10), 870–3, 1972.
69895	MERCURY. INTERPRETATION OF OPTICAL OBSERVATIONS. MC CORD, T. B. ADAMS, J. B. ICARUS 17 (3), 585–8, 1972.
69939	MAGNETOTHERMODYNAMICS OF CERIUM MAGNESIUM NITRATE TETRACOSAHYDRATE. I. HEAT CAPACITY, ENTROPY, MAGNETIC MOMENT FROM 0.5 TO 4.2 K WITH FIELDS TO 90 KG ALONG THE CRYSTAL AXIS. HEAT CAPACITY OF PYREX 7740 GLASS IN FIELDS TO 90 KG. GIAUQUE, W. F. FISHER, R. A. HORNUNG, E. W. BRODALE, G. E. J. CHEM. PHYS. 58 (6), 2621–37, 1973.
69949	LATTICE THERMAL CONDUCTIVITY OF SAPPHIRE IN THE TEMPERATURE RANGE 0.4 TO 4.0 K. DUBEY, K. S. PHYS. STATUS SOLIDI 54 B (2), K139–K141, 1972.
70002	INFRARED REFLECTION SPECTRA OF THE SIMPLEST VITREOUS LEAD BORATES. KURTSINOVSKAYA, R. I. ZH. PRIKL. SPEKTROSK. 17 (5), 834–9, 1972. (FOR ENGLISH TRANSLATION SEE TPRC NO. 77025)
70031	DEPENDENCE OF THE REFLECTANCE OF SPHALERITE ON ITS CHEMICAL COMPOSITION. HABER, M. ACTA GEOL. GEOGR. UNIV. COMENIANAE, GEOL. (22), 169–85, 1972.
70035	APPLICATION OF A PARTICLE MODEL TO SAMPLE DILUTION IN REFLECTANCE SPECTROSCOPY. SIMMONS, E. L. WENDLANDT, W. W. CHEM. ANAL. (WARSAW) 17 (4), 1085–90, 1972.
70043	OPTICAL PROPERTIES OF APOLLO 12 MOON SAMPLES. O'LEARY, B. BRIGGS, F. J. GEOPHYS. RES. 78 (5), 792–7, 1973.
70057	CRYSTAL FIELD SPECTRA OF LUNAR PYROXENES. BURNS, R. G. ABU–EID, R. M. HUGGINS, F. E. PROC. LUNAR SCI. CONF., 3RD 1, 533–43, 1972.
70058	ELECTRONIC SPECTRA OF PYROXENES AND INTERPRETATION OF TELESCOPIC SPECTRAL REFLECTIVITY CURVES OF THE MOON. ADAMS, J. B. MC CORD, T. B. PROC. LUNAR SCI. CONF., 3RD 3, 3021–34, 1972.
70060	FAR INFRARED AND RAMAN SPECTROSCOPIC INVESTIGATIONS OF LUNAR MATERIALS FROM APOLLO 11, 12, 14, AND 15. PERRY, C. H. AGRAWAL, D. K. ANASTASSAKIS, E. LOWNDES, R. P. TORNBERG, N. E. PROC. LUNAR SCI. CONF., 3RD 3, 3077–95, 1972.
70061	REFLECTANCE AND ABSORPTION SPECTRA OF APOLLO 11 AND APOLLO 12 SAMPLES. ANTIPOVA–KARATAEVA, I. I. STACHEEV, YU. I. TARASOV, L. S. PROC. LUNAR SCI. CONF., 3RD 3, 3097–101, 1972.
70092	EXPERIMENTAL STUDY OF THE THERMAL CONDUCTIVITY OF YTTRIUM OXIDE-STABILIZED ZIRCONIUM DIOXIDE IN THE 300–1800 DEGREE RANGE. CHEKHOVSKOI, V. YA. STAVROVSKII, G. I. IVANOV, A. B. TEPLO–MASSOPERENOS, DOKL. VSES. SOVESHCH., 4TH 7, 540–4, 1972.
70139	LASER REFLECTIVITY MEASUREMENTS AT 1.06 MICRONS. ROY, E. L. EMMONS, G. A. ARMY MISSILE COMMAND, REDSTONE ARSENAL, ALA. 19PP., 1965. (AD–368115, AMC–RARE–TR–65–11)
70141	INFRARED REFLECTION, EMISSION, AND ABSORPTION SPECTRA OF REGOLITH FROM THE SEA OF FERTILITY AND ITS SCATTERING COEFFICIENT. AKHMANOVA, M. V. DEMENT'EV, B. V. KARYAKIN, A. V. MARKOV, M. N. PETROV, V. S. PROKHOROV, A. M. SUSHCHINSKII, M. M. SPACE RES. 1 (12), 87–93, 1972.
70142	THE OPTICAL PARAMETERS OF MARE FOECUNDITATIS REGOLITH. ANTIPOVA–KARATAEVA, I. I. STAKHEEV, IU. I. FLORENSKII, K. P. SPACE RESEARCH 1, 95–98PP., 1972.
70202	LOW–TEMPERATURE SPECIFIC HEAT OF METALLIC VANADIUM DOPED TITANIUM SESQUIOXIDE. SJOSTRAND, M. E. KEESOM, P. H. PHYS. REV., B 7 (8), 3558–68, 1973.
70215	FABRICATION OF TRANSPARENT THORIA DOPED YTTRIUM SESQUIOXIDE. GRESKOVICH, C. WOODS, K. N. BULL. AMER. CERAM. SOC. 52 (5), 473–8, 1973.
70260	EFFECT OF WATER ON THE INFRARED TRANSMISSION OF HIGHLY REFRACTIVE TELLURITE GLASSES AND A METHOD FOR WATER DETERMINATION. TATARINTSEV, B. V. YAKHKIND, A. K. OPT.–MEKH. PROM. 39 (10), 72–3, 1972. (FOR ENGLISH TRANSLATION SEE TPRC NO. 72358)
70282	ELASTIC WAVE VELOCITIES AND THERMAL DIFFUSIVITIES OF APOLLO 14 ROCKS. MIZUTANI, H. FUJII, N. HAMANO, Y. OSAKO, M. PROC. LUNAR SCI. CONF., 3RD 2557–2564, 1972.
70283	THERMAL EXPANSION OF APOLLO LUNAR SAMPLES AND FAIRFAX DIABASE. BALDRIDGE, S. MILLER, F. WANG, H. SIMMONS, G. PROC. LUNAR SCI. CONF., 3RD 2599–2609, 1972.
70284	THERMAL CONDUCTIVITY OF APOLLO 14 FINES. CREMERS, C. J. PROC. LUNAR SCI. CONF., 3RD 2611–2617, 1972.
70285	VISCOUS FLOW BEHAVIOR OF LUNAR COMPOSITIONS 14259 AND 14310. CUKIERMAN, M. TUTTS, P. M. UHLMANN, D. R. PROC. LUNAR SCIENCE CONF. 2619–2625, 1972.

TPRC Number **Bibliographic Citation**

70287 GALILEAN SATELLITES - IDENTIFICATION OF WATER FROST.
PILCHER, C. B. MC CORD, T. B.
SCIENCE
178, 1087-9, 1972.

70303 COMPOSITION OF THE VENUS CLOUD TOPS IN LIGHT OF RECENT SPECTROSCOPIC DATA.
LEWIS, J. S.
ASTROPHYS. J.
171 (PT. 2), L75-9, 1972.

70311 PIONEER IMAGING PHOTOPOLARIMETER OPTICAL SYSTEM.
PELLICORI, S. F. RUSSELL, E. E. WATTS, L. A.
APPL. OPT.
12 (6), 1246-58, 1973.

70324 MERCURY: SURFACE COMPOSITION FROM THE REFLECTION SPECTRUM.
MC CORD, T. B. ADAMS, J. B.
SCIENCE
178, 745-7, 1972.

70325 THERMAL CONDUCTIVITY AND HOT MAGNETIC POLES OF PULSARS.
SMOLUCHOWSKI, R.
NATURE (LONDON), PHYS. SCI.
240, 54-6, 1972.

70406 INFRARED INTERNAL REFLECTION SPECTRA OF CRYSTALLINE QUARTZ. I. HYDROXYL GROUPS.
GALLEI, E.
BER. BUNSENGES. PHYS. CHEM.
77 (2), 81-5, 1973.

70410 OPTICAL ABSORPTION SPECTRA OF FERRIC ION IN OCTAHEDRAL AND TETRAHEDRAL SITES IN NATURAL GARNETS.
MANNING, P. G.
CAN. MINERAL.
11 (PT. 4), 826-39, 1972.

70421 THERMAL DIFFUSIVITY OF LUNAR ROCKS UNDER ATMOSPHERIC AND VACUUM CONDITIONS.
FUJII, N. OSAKO, M.
EARTH PLANET. SCI. LETT.
18 (1), 65-71, 1973.

70445 ANALYSIS OF THE EFFECT OF MICROSTRUCTURAL PARAMETERS ON THE THERMAL CONDUCTIVITY OF POROUS CERAMIC MATERIALS.
SOKOLOVA, T. V. LITOVSKII, E. YA.
BUZOVKINA, T. B. BARTENEV, S. S.
IZV. AKAD. NAUK SSSR, NEORG. MATER.
9 (2), 296-300, 1973.
(FOR ENGLISH TRANSLATION SEE TPRC NO. 71677)

70467 NONMETALLIC CRYSTALS WITH HIGH THERMAL CONDUCTIVITY.
SLACK, G. A.
J. PHYS. CHEM. SOLIDS
34 (2), 321-35, 1973.

70478 THERMAL PROPERTIES OF COALS IN THE PLASTIC STATE.
AGROSKIN, A. A. GONCHAROV, E. I.
GRYAZNOV, N. S.
KOKS KHIM.
(9), 1-4, 1972.
(FOR ENGLISH TRANSLATION SEE TPRC NO. 70989)

70488 VISCOSITY OF BASIC MAGMAS AT VARYING PRESSURE.
SCARFE, C. M.
NATURE (LONDON), PHYS. SCI.
241 (109), 101-2, 1973.

70496 THERMAL CONDUCTIVITY OF COKE DINAS BRICK.
STRELOV, K. K. PEREPELITSYN, V. A.
BICHURINA, A. A.
OGNEUPORY
(2), 48-52, 1973.
(FOR ENGLISH TRANSLATION SEE TPRC 74223)

70497 FIELD AND LABORATORY STUDIES OF THE RHEOLOGY OF MOUNT ETNA LAVA.
GAUTHIER, F.
PHIL. TRANS. ROY. SOC. LONDON
274 A (1238), 83-98, 1973.

70590 INFRARED SPECTROSCOPY AT ELEVATED TEMPERATURES. I. CELL, METHOD, AND STUDY OF QUARTZ.
KARGE, H. KLOSE, K.
Z. PHYS. CHEM.
83 (1-4), 92-9, 1973.

70601 EFFECT OF STRUCTURE ON THE EMISSIVITY OF AN ALUMINUM CHROMIUM COATING AT HIGH TEMPERATURES.
VERENKOVA, E. M. ZHOROV, G. A.
SAMOILOV, A. I. UMANTSEVA, V. M.
FROLOV, A. S.
INORG. MAT., USSR
8 (5), 788-90, 1972.
(ENGLISH TRANSLATION OF IZV. AKAD. NAUK SSSR, NEORG. MATER., 8 (5), 907-10, 1972., FOR ORIGINAL SEE TPRC NO. 67684)

70605 OPTICAL CONSTANTS, DENSITIES, COEFFICIENTS OF THERMAL EXPANSION, AND REFRACTIVE INDEXES OF GLASSES IN THE SYSTEM SILICON DIOXIDE - BORON OXIDE - BARIUM OXIDE - ALUMINUM OXIDE.
SHCHAVELEV, O. S.
J. APPL. CHEM., USSR
45 (5), 1033-8, 1972.
(ENGLISH TRANSLATION OF ZH. PRIKL. KHIM., 45 (5), 1005-12, 1972; FOR ORIGINAL SEE T67819)

70606 PHYSICOCHEMICAL PROPERTIES AND STRUCTURE OF ALKALI ZINC SILICATE GLASSES.
VARGIN, V. V. DZHAVUKTSYAN, S. G.
MISHEL', V. E. PEVZNER, B. Z.
J. APPL. CHEM., USSR
45 (6), 1228-33, 1972.
(ENGLISH TRANSLATION OF ZH. PRIKL. KHIM., 45 (6), 1187-93, 1972; FOR ORIGINAL SEE T68304)

70625 THERMOPHYSICAL PROPERTIES OF SOME GLASSES AND GLASS-CERAMICS.
EGOROV, V. N. KONDRATENKOV, V. I.
KILESSO, V. S.
TEPLOFIZ. VYS. TEMP.
10 (5), 1122-3, 1972.
(FOR ENGLISH TRANSLATION SEE TPRC NO. 70626)

70626 THERMOPHYSICAL PROPERTIES OF SOME GLASSES AND GLASS CERAMICS.
EGOROV, V. N. KONDRATENKOV, V. I.
KILESSO, V. S.
HIGH TEMP.
10 (5), 1008-9, 1972.
(ENGLISH TRANSLATION OF TEPLOFIZ. VYS. TEMP., 10 (5), 1122-3, 1972., FOR ORIGINAL SEE TPRC NO. 70625)

70628 CHARACTERISTICS OF VACUUM FREEZE-DRYING.
HAYASHI, Y. KOBAYASHI, N. KATAYAMA, K.
HEAT TRANSFER-JAP. RES.
2 (1), 71-80, 1973.

70629 THERMAL RADIATIVE PROPERTIES. COATINGS
TOULOUKIAN, Y. S. DE WITT, D. P.
HERNICZ, R. S.
THERMOPHYSICAL PROPERTIES OF MATTER
9, 1378PP., 1972.

70652 TEMPERATURE INCREMENTS FROM DEPOSITS ON HEAT TRANSFER SURFACES. THERMAL RESISTIVITY AND THERMAL CONDUCTIVITY OF DEPOSITS OF MAGNETITE, CALCIUM HYDROXY APATITE, HUMUS, AND COPPER OXIDES.
KELEN, T. ARVESEN, J.
AKTIEBOLAGET ATOMENERGI, STOCKHOLM
21PP., 1972.
(AE-459)

70678 THERMAL DIFFUSIVITY.
TOULOUKIAN, Y. S. POWELL, R. W. HO, C. Y.
NICOLAOU, M. C.
THERMOPHYSICAL PROPERTIES OF MATTER
10, 661PP., 1973.

70701 THERMAL EXPANSION COEFFICIENT CRYSTALLINITY RELATIONS IN LITHIUM OXIDE - ALUMINUM OXIDE - SILICON OXIDE GLASS CERAMICS.
RAPP, J. E.
AMER. CERAM. SOC. BULL.
52 (6), 499-500, 504, 1973.

70702 LASERS AND GLASS TECHNOLOGY.
SNITZER, E.
AMER. CERAM. SOC. BULL.
52 (6), 516-25, 1973.

70703 SPATIAL DISTRIBUTION OF INVERSION IN FACE PUMPED NEODYMIUM DOPED GLASS LASER SLABS.
SOURES, J. M. GOLDMAN, L. M. LUBIN, M. J.
APPL. OPT.
12 (5), 927-8, 1973.

70710 LABORATORY THERMAL EXPANSION MEASURING TECHNIQUES APPLIED TO BITUMINOUS CONCRETE.
HOOKS, C. C. GOETZ, W. H.
ARMY ENGINEER WATERWAYS EXPERIMENT STATION
245PP., 1964.
(AD-757419, AEWES-CR-4-102)

70719 DETERMINATION OF ROCK THERMAL PROPERTIES.
BACON, J. F. RUSSELL, S. CARSTENS, J. P.
UNITED AIRCRAFT RESEARCH LABS., EAST HARTFORD, CONN.
56PP., 1973.
(AD-755218, UARL-L911397-4)

70720 SELECTIVE FILMS ON GLASS AND THEIR CHARACTERISTICS.
REKANT, N. B. SHEKLEIN, A. V.
ARMY FOREIGN SCI. TECH. CTR., CHARLOTTESVILLE, VA.
14PP., 1972.
(ENGLISH TRANSLATION OF PREOBRAZOVATELI SOLN. ENERG. POLUPROV., 190-9, 1968; FOR ORIGINAL SEE TPRC NO. 55318)
(AD-755250, FSTC-HT-23-1028-72)

TPRC Number	Bibliographic Citation
70757	**CHARACTERISTIC LAWS GOVERNING CHANGES IN THERMOPHYSICAL PROPERTIES OF TRANSPARENT AND OPAQUE SITALLS.** ROMASHIN, A. G. TIKHONOV, B. E. HEAT TRANSFER-SOV. RES. 5 (2), 183-5, 1973.
70791	**MEASUREMENTS OF THERMAL CONDUCTIVITY IN THE SUBSOIL.** KOITZSCH, R. ZEIT. METEOROL. 19 (11-12), 375-80, 1967. (FOR ENGLISH TRANSLATION SEE TPRC NO. 70792)
70792	**MEASUREMENTS OF THERMAL CONDUCTIVITY IN THE SUBSOIL.** KOITZSCH, R. KANNER (LEO) ASSOCIATES, REDWOOD CITY, CALIF. 14PP., 1973. (ENGLISH TRANSLATION OF ZEIT. METEROL., 19 (11-12), 375-80, 1967; FOR ORIGINAL SEE TPRC NO. 70791) (N73-16446, NASA-TT-F-14664)
70800	**PHYSICAL PROPERTIES OF PICROILMENITE FROM KIMBERLITES AND DEPOSITIONS OF DIFFERENT AGES OF THE ANABARSKY REGION. (WESTERN YAKUT).** SOCHNEVA, E. G. ZAP. VSES. MINERAL. OBSHCHEST. (1), 131-3, 1972.
70809	**OPTICAL ABSORPTION SPECTRA OF CLAY MINERALS.** KARICKHOFF, S. W. BAILEY, G. W. CLAYS CLAY MINER. 21 (1), 59-70, 1973.
70817	**THERMOPHYSICAL PROPERTY MEASUREMTNS ON LUNAR FINES FROM APOLLO MISSIONS.** BIRKEBAK, R. C. CREMERS, C. J. HIGH TEMPERATURE THERMAL RADIATION LAB. KENTUCKY UNIV., LEXINGTON 23PP., 1972. (N73-12888, NASA-CR-12923)
70822	**PHOTOMETRIC PROPERTIES OF SATURNS RINGS.** IRVINE, W. M. LANE, A. P. ICARUS 18 (2), 171-6, 1973.
70823	**THERMOPHYSICAL PROPERTIES OF APOLLO 12 FINES.** CREMERS, C. J. ICARUS 18 (2), 294-303, 1973.
70824	**HIGH ALTITUDE INFRARED SPECTROSCOPIC EVIDENCE FOR BOUND WATER ON MARS.** HOUCK, J. R. POLLACK, J. B. SAGAN, C. SCHAACK, D. DECKER, J. A., JR. ICARUS 18 (3), 470-80, 1973.
70840	**TRANSMISSION AND REFLECTION SPECTRA OF NATURAL AND SYNTHETIC QUARTZ IN THE FAR ULTRAVIOLET REGION.** GERASIMOVA, N. G. KREISKOP, V. N. PANOVA, I. V. SHIN, V. OPT.-MEKH. PROM. 40 (1), 7-11, 1973. (FOR ENGLISH TRANSLATION SEE TPRC NO. 72359)
70862	**HEAT CAPACITIES OF ROCKS AT HIGH TEMPERATURES.** LEONIDOV, V. YA. GEOCHEMISTRY (4), 400-2, 1967. (ENGLISH TRANSLATION OF GEOKHIMIYA, (4), 470-2, 1967; FOR ORIGINAL SEE TPRC NO. 45144)
70863	**NEW DATA ON THE THERMODYNAMIC PROPERTIES OF TALC.** LEONIDOV, V. YA. KHITAROV, N. I. GEOCHEMISTRY (10), 944-9, 1967. (ENGLISH TRANSLATION OF GEOKHIMIYA, (10), 1044-9, 1967; FOR ORIGINAL SEE TPRC NO. 47305)
70864	**THERMODYNAMIC CONSTANTS OF KAOLINITE AND KINETIC PARAMETERS OF KAOLINITE DEHYDRATION.** KUSKOV, O. L. KHITAROV, N. I. GEOCHEMISTRY (12), 1147-51, 1969. (ENGLISH TRANSLATION OF GEOKHIMIYA, (12), 1501-6, 1969; FOR ORIGINAL SEE TPRC NO. 56752)
70865	**THE THERMODYNAMIC CONSTANTS OF PYROPHYLLITE.** FONAREV, V. I. GEOCHEMISTRY (12), 1182-5, 1967. (ENGLISH TRANSLATION OF GEOKHIMIYA, (12), 1505-8, 1967; FOR ORIGINAL SEE TPRC NO. 48397)
70866	**THERMAL PROPERTIES OF SOME EASTERN COALS.** AGROSKIN, A. A. GONCHAROV, E. I. LOVETSKII, L. V. MAKEEV, L. A. GRYAZNOV, N. S. MOCHALOV, V. V. COKE CHEM. (11), 3-8, 1968. (ENGLISH TRANSLATION OF KOKS KHIM, (11), 1-6, 1968; FOR ORIGINAL SEE TPRC NO. 54490)
70883	**ABSORPTION SPECTRA OF GLASSES AND MELTS AND VOLUME PROPERTIES OF COBALT METAPHOSPHATE-SODIUM METAPHOSPHATE AND COBALT METAPHOSPHATE-POTASSIUM METAPHOSPHATE.** TANANAEV, I. V. VOSKRESENSKAYA, N. K. DZHURINSKII, B. F. KRIVOVYAZOV, E. L. RAKHIMBEKOVA, KH. M. INORG. MAT., USSR 8 (8), 1297-1300, 1972. (ENGLISH TRANSLATION OF IZV. AKAD. NAUK SSSR, NEORG. MATER., 8 (8), 1474-7, 1972; FOR ORIGINAL SEE TPRC NO. 68525)
70906	**MECHANISMS OF GLASS CORROSION.** SANDERS, D. M. HENCH, L. L. J. AMER. CERAM. SOC. 56 (7), 373-7, 1973.
70912	**OPTICAL PROPERTIES AND STRUCTURE OF SATURN'S ATMOSPHERE. II. LATITUDINAL VARIATIONS OF ABSORPTION IN THE 0.62 MICRON METHANE BAND AND NEAR-ULTRAVIOLET CHARACTERISTICS OF THE PLANET.** TEIFEL', V. G. USOL'TSEVA, L. A. KHARITONOVA, G. A. ASTRON. ZH. 50 (1), 167-71, 1973. (FOR ENGLISH TRANSLATION SEE TPRC NO. 71376)
70913	**STUMPFLITE, PLATINUM (ANTIMONIDE, BISMUTHIDE), A NEW MINERAL.** JOHAN, Z. PICOT, P. BULL. SOC. FR. MINERAL. CRISTALLOGR. 95 (5), 610-13, 1972.
70931	**SIMULTANEOUS MEASUREMENT OF THE HEAT-TRANSFER PROPERTIES OF INSULATORS.** FUKS, L. G. SHMANDINA, V. N. IND. LAB., USSR 37 (11), 1746-7, 1971. (ENGLISH TRANSLATION OF ZAVOD. LAB., 37 (11), 1361-2, 1971; FOR ORIGINAL SEE TPRC NO. 66294)
70940	**ANOMALOUS ULTRASONIC ATTENUATION IN BISMUTH GERMANIUM OXIDE, BISMUTH SILICON OXIDE AND BISMUTH GERMANIUM SILICON OXIDE.** REHWALD, W. J. APPL. PHYS. 44 (7), 3017-21, 1973.
70941	**PREPARATION OF REPLICATED GERMANIUM FILMS AND THEIR APPLICATIONS TO SENSORS.** ONUMA, Y. SAKAI, Y. J. APPL. PHYS. 44 (7), 3046-51, 1973.
70944	**TRANSIENT THERMAL PROFILE IN OPTICALLY PUMPED LASER RODS.** KOECHNER, W. J. APPL. PHYS. 44 (7), 3162-70, 1973.
70955	**MARTIAN SPECTRAL REFLECTIVITY PROPERTIES FROM MARINER 7 OBSERVATIONS.** CUTTS, J. A. J. CALIFORNIA INST. OF TECH., PASADENA, PH.D. THESIS 102PP., 1971. (UNIV. MICROFULMS NO. 72-470, N72-26709)
70964	**THERMAL EXPANSION OF CATION-SUBSTITUTED BARIUM SODIUM NIOBATES.** MUKHERJEE, J. L. STATE UNIV., NEW YORK, STONY BROOK, N. Y., PH.D. THESIS 127PP., 1971. (UM-72-6388)
70966	**INFLUENCE OF PRESSURE AND CHEMICAL STRUCTURE ON THERMAL CONDUCTIVITY OF POLYMERS.** CHEN, R. Y. UNIVERSITY OF VIRGINIA, PH.D. THESIS 218PP., 1971. (UM-72-7178)
70976	**ULTRAVIOLET OBSERVATIONS OF MARS MADE BY THE ORBITING ASTRONOMICAL OBSERVATORY.** CALDWELL, J. J. WISCONSIN UNIV., MADISON, PH.D. THESIS 136PP., 1971. (UM-71-14129, N72-23874)
70989	**THERMAL PROPERTIES OF COALS IN THE PLASTIC STATE.** AGROSKIN, A. A. GONCHAROV, E. I. GRYAZNOV, N. S. COKE CHEM. (USSR) (9), 3-5, 1972. (ENGLISH TRANSLATION OF KOKS KHIM., (9), 1-4, 1972; FOR ORIGINAL SEE TPRC NO. 70478)
71027	**DETERMINATION OF ABSORPTION AND SCATTERING COEFFICIENTS FOR NONHOMOGENEOUS MEDIA. 2: EXPERIMENT.** EGAN, W. G. HILGEMAN, T. REICHMAN, J. APPL. OPT. 12 (8), 1816-23, 1973.

TPRC Number	Bibliographic Citation
71054	**VISIBLE AND NEAR INFRARED TRANSMISSION AND REFLECTANCE MEASUREMENTS OF THE LUNA 20 SOIL.** ADAMS, J. B. BELL, P. M. CONEL, J. E. MAO, H. K. MC CORD, T. B. NASH, D. B. GEOCHIM. COSMOCHIM. ACTA 37 (4), 731–43, 1973.
71060	**DETERMINATION OF ELASTIC CONSTANTS IN ISOTROPIC SILICATE GLASSES BY BRILLOUIN SCATTERING.** HUANG, Y. Y. HUNT, J. L. STEVENS, J. R. J. APPL. PHYS. 44 (8), 3589–92, 1973.
71078	**USE OF INFRARED SPECTROSCOPY IN THE STUDY OF BIOTITES FROM GRANITIC ROCKS OF THE SARYSUI–TENIZ WATER SHED. (CENTRAL KAZAKHSTAN).** SOBOLEV, R. N. PLYUSNINA, I. I. MEL'CHAKOVA, L. V. IZV. VYSSH. UCHEB. ZAVED., GEOL. RAZVED. 15 (11), 46–51, 1972.
71106	**INFRARED EMISSION AND REFLECTION SPECTROSCOPY OF CONDENSED PHASES.** ARONSON, J. R. PROGR. NUCL. ENERGY 11 (SER. 9), 135–56, 1972.
71147	**INFRARED REFLECTION FROM LONGITIUDINAL MODES IN ANISOTROPIC CRYSTALS.** DECIUS, J. C. FRECH, R. BRUESCH, P. J. CHEM. PHYS. 58 (10), 4056–60, 1973.
71157	**APPLICATION OF RO (O–RAY REFLECTIVITY) AND AR (ANGLE OF APPARENT ROTATION) MEASUREMENTS TO THE STUDY OF PYRRHOTITE AND TROILITE.** CARPENTER, R. H. BAILEY, A. C., JR. AMER. MINERAL. 58 (5–6), 440–3, 1973.
71198	**EVALUATION OF COALS BY REFLECTIVITY.** KOSINA, M. SB. PREDNASEK 50 49–64, 1972.
71231	**FERROELECTRIC–PARAELECTRIC PHASE TRANSITION IN LEAD TITANATE CONTAINING LATTICE DEFECTS.** SHIRASAKI, S.–I. TAKAHASHI, K. KAKEGAWA, K. J. AMER. CERAM. SOC. 56 (8), 430–5, 1973.
71232	**RUTILE–TYPE CRYSTALLINE SOLUTIONS IN THE SYSTEM MAGNESIUM TANTALUM OXIDE – ALUMINUM TANTALUM OXIDE.** PERROTTA, A. J. YOUNG, J. E., JR. J. AMER. CERAM. SOC. 56 (8), 441, 1973.
71248	**CHROMIUM–YTTERBIUM ENERGY TRANSFER IN SILICATE GLASS.** SHARP, E. J. MILLER, J. E. WEBER, M. J. J. APPL. PHYS. 44 (9), 4098–4101, 1973.
71252	**CORRELATIONS BETWEEN THE INFRARED SPECTRUM AND THE COMPOSITION OF GARNETS IN THE PYROPE–ALMANDINE–SPESSARTINE SERIES.** TARTE, P. DELIENS, M. CONTRIB. MINERAL. PETROLOGY 40 (1), 25–37, 1973.
71271	**INFRARED SPECTRA OF SOLID SOLUTIONS IN THE CALCIUM SILICATE – CALCIUM GERMANATE SYSTEM.** LAZAREV, A. N. IGNATEV, I. S. IZV. AKAD. NAUK SSSR, NEORG. MATER. 9 (4), 636–45, 1973. (FOR ENGLISH TRANSLATION SEE TPRC NO. 72305)
71302	**THERMOPHYSICAL ANALYSIS OF THE KUBADRINSK–KURAIKA METAMORPHIC ZONE OF THE MOUNTAINOUS ALTAI REGION.** DUDAREV, A. N. SKURIDIN, V. A. TR. INST. GEOL. GEOFIZ., AKAD. NAUK SSSR, SIB. OTD. (114), 198–217, 1972.
71321	**THERMOPHYSICAL PROPERTIES OF MOLDING MIXTURES AND THEIR INFLUENCE ON CASTING SOLIDIFICATION.** CATANA, V. IATAN, N. DARIE, E. DUMITRU, FL. CERCET. MET., INST. CERCET. MET., BUCHAREST 13, 87–100, 1972.
71322	**MODIFIED METHOD OF KRISHNAN AND JAIN FOR MEASURING THERMAL CONDUCTIVITY IN THE TEMPERATURE RANGE 1000–2200.** PROCHAZKA, V. CESK. CAS. FYS. 23 (3), 280–5, 1973.
71345	**CHANGES IN THE HEAT CAPACITY OF COALS AS A FUNCTION OF THE DEGREE OF METAMORPHISM AND THE AMOUNT OF MINERAL IMPURITIES AT TEMPERATURES UP TO 1000.** GONCHAROV, E. I. DEMIDOV, L. G. KHIM. TVERD. TOPL. (2), 154–7, 1973.
71361	**NEW TECHNIQUES FOR FAR–INFRARED FILTERS.** ARMSTRONG, K. R. LOW, F. J. APPL. OPT. 12 (9), 2007–9, 1973.
71365	**GLASS CERAMIC: A NEW LASER HOST MATERIAL.** MULLER, G. NEUROTH, N. J. APPL. PHYS. 44 (5), 2315–8, 1973.
71366	**ENVIRONMENTAL EFFECTS ON GLASS CORROSION KINETICS.** SANDERS, D. M. HENCH, L. L. BULL. AMER. CERAM. SOC. 52 (9), 662–5, 669, 1973.
71376	**OPTICAL PROPERTIES AND STRUCTURE OF SATURN'S ATMOSPHERE. II. LATITUDINAL VARIATIONS OF ABSORPTION IN THE 0.62 MICRON METHANE BAND AND NEAR–ULTRAVIOLET CHARACTERISTICS OF THE PLANET.** TEIFEL', V. G. USOL'TSEVA, L. A. KHARITONOVA, G. A. SOV. ASTRON.–AJ, USSR 50 (1), 108–11, 1973. (FOR ENGLISH TRANSLATION OF ASTRON. ZH., 50 (1), 167–71, 1973; FOR ORIGINAL SEE TPRC NO. 70912)
71391	**LOW–TEMPERATURE SPECIFIC HEAT OF VANADIUM–DOPED TITANIUM OXIDE.** SJOSTRAND, M. E. C. PURDUE UNIVERSITY, PH.D. THESIS 112PP., 1972. (UM–73–6108)
71398	**LATTICE THERMAL CONDUCTIVITY OF DIAMOND.** TURK, L. A. UNIV. CONNECTICUT, STORRS, CONN., PH.D. THESIS 100PP. 1972. (UM–73–9858)
71410	**THEORETICAL CALCULATION OF FREE ENERGY, LIQUIDUS SHAPE, DENSITY, AND THERMAL EXPANSION OF GLASSES IN THE SILICA–TITANIA SYSTEM.** JAEGER, R. F. RUTGERS STATE UNIV., NEW BRUNSWICK, N. J. PH.D. THESIS 290PP. 1972. (UM–72–27564)
71416	**THERMAL EXPANSION BEHAVIOR OF SILICA AND SILICATES.** CHOW, K. STANFORD UNIVERSITY, PH.D. THESIS 159PP., 1972. (UM–72–30608)
71449	**THERMAL CHARACTERIZATION OF REUSABLE EXTERNAL INSULATION FOR THE SPACE SHUTTLE.** BRAZEL, J. P. TYE, R. P. HIGH TEMP. – HIGH PRESSURES 4 (6), 639–47, 1972.
71455	**MEASUREMENT OF THE EMISSIVITY OF QUARTZ GLASS.** PETROV, V. A. REZNIK, V. YU. HIGH TEMP. – HIGH PRESSURES 4 (6), 687–93, 1972.
71468	**THERMAL EXPANSION OF INSULATING MATERIALS.** SOUDER, W. H. HIDNERT, P. U. S. NAT. BUR. STANDARDS, SCI. PAPERS 15, 387–417, 1919.
71469	**PHYSICAL PROPERTIES OF TYPICAL AMERICAN ROCKS.** GRIFFITH, J. H. IOWA STATE COLL., ENG. EXP. STA. BULL. 35 (36), 56PP., 1937.
71478	**SPECIFIC HEAT OF REFRACTORY COMPOUNDS UP TILL THEIR MELTING POINT BY RAPID HEATING. APPLICATION TO URANIUM AND PLUTONIUM OXIDES.** AFFORTIT, C. BOIVINEAU, J. C. BULL. INFORM. SCI. TECH., COMMIS. ENERG. AT. (FR.) (180), 51–4, 1973.
71483	**THERMAL PROPERTIES OF GRAPHITE.** KELLY, B. T. TAYLOR, R. CHEM. PHYS. CARBON 10, 1–140, 1973.
71497	**INFRARED SPECTRA OF CERTAIN BENTONITES AND DIATOMITES.** LAMBRINO, V. D. DOKL. BOLG. AKAD. NAUK 26 (4), 483–6, 1973.
71503	**THERMAL CONDUCTIVITY OF MATERIALS OF ALUMINUM ELECTROLYTIC CELLS.** VENERAKI, I. E. ROMAN'KO, K. S. URDA, N. N. SEMENOV, V. S. ENERG. ELEK. (2), 52–4, 1973.
71516	**CERAMIC MATERIALS WITH HIGH THERMAL CONDUCTIVITIES.** DONCHEV, S. IORDANOV, P. GOD. VISSH. KHIMIKOTEKHNOL. INST., SOFIA 1, (1), 225–32, 1971.

TPRC Number	Bibliographic Citation
71631	THERMAL CONDUCTIVITY OF REFRACTORY MATERIALS OF THE ALUMINUM OXIDE-SILICON DIOXIDE SYSTEM. LITOVSKII, E. YA. TEPLO-MASSOPERENOS 7, 424-31, 1972.
71677	ANALYSIS OF THE INFLUENCE OF MICROSTRUCTURAL CHARACTERISTICS ON THE THERMAL CONDUCTIVITIES OF POROUS CERAMICS. SOKOLOVA, T. V. LITOVSKII, E. YA. BUZOVKINA, T. B. BARTENEV, S. S. INORG. MAT., USSR 9 (2), 265-8, 1973. (ENGLISH TRANSLATION OF IZV. AKAD. NAUK SSSR, NEORG. MATER., 9 (2), 296-300, 1973; FOR ORIGINAL SEE TPRC NO. 70445)
71749	MIDINFRARED EMISSION SPECTRA OF APOLLO 14 AND 15 SOILS AND REMOTE COMPOSITIONAL MAPPING OF THE MOON. LOGAN, L. M. HUNT, G. R. BALSAMO, S. R. SALISBURY, J. W. PROC. LUNAR SCI. CONF., 3RD 3, 3069-76, 1972. (AD-760848, AFCRL-TR-0299)
71762	VISIBLE AND NEAR-INFRARED SPECTRA OF MINERALS AND ROCKS: V. HALIDES, PHOSPHATES, ARSENATES, VANADATES AND BORATES. HUNT, G. R. SALISBURY, J. W. LENHOFF, C. J. MOD. GEOLOG. 3, 121-32, 1972. (AD-764163, AFCRL-TR-73-0402)
71769	SPECTRAL REFLECTANCE AND EMITTANCE OF PARTICULATE MATERIALS. 2: APPLICATION AND RESULTS. ARONSON, J. R. EMSLIE, A. G. APPL. OPT. 12 (11), 2573-84, 1973.
71780	A STUDY OF THE THERMOPHYSICAL PROPERTIES OF VERMICULITES. VENERALKI, I. E. POLYAKOV, V. F. TOPOLNITSKIY, G. G. PISARENKO, I. M. ROMANKO, K. S. MELNICHENKO, V. S. HEAT TRANSFER-SOV. RES. 5 (5), 165-9, 1973.
71786	INFRARED REFLECTION SPECTRA OF SINGLE-CRYSTAL OR POWEDERED MINERALS. III. CARBONATES. VIERNE, R. BRUNEL, R. BULL. SOC. FR. MINERAL. CRISTALLOGR. 96 (1), 55-62, 1973.
71795	REFLECTION CHARACTERISTICS OF IRON ORES FROM GOA. SAHU, K. C. GURAV, R. P. INDIAN MINERAL 12, 63-6, 1973.
71835	THERMAL CONDUCTIVITY OF POROUS MEDIA IN RELATION TO THE TYPE AND QUANTITY OF FILLER. NOVIKOV, P. A. MIKHNYUK, B. G. SUBACH, V. M. TEPLO-MASSOPERENOS, DOKL. VSES. SOVESHCH. 7, 4TH, 270-4, 1972.
71846	OPTICAL PROPERTIES OF CARBONIZED PREOXIDIZED VITRINITES. GOODARZI, F. MURCHISON, D. G. FUEL 52 (3), 164-7, 1973.
71847	THERMODYNAMIC CONSTANTS OF PRYROPHYLLITE AND ANOMALOUS HEAT CAPACITY OF METAPYROPHYLLITE AND METAKAOLIN. KUSKOV, O. L. GEOKHIMIYA (4), 537-44, 1973.
71849	OPTICAL PROPERTIES AND COLLECTIVE EXCITATIONS IN MOLYBDENUM SULFIDE AND NIOBIUM SELENIDE IN THE 1.7 TO 30 EV RANGE. MARTIN, L. MAMY, R. COUGET, A. RAISIN, C. PHYS. STATUS SOLIDI 58 B (2), 623-7, 1973.
71862	THERMOPHYSICAL CHARACTERISTICS OF SOLID FUEL. AGROSKIN, A. A. KHIM. TVERD. TOPL. (1), 86-91, 19739
71864	PHYSICAL PROPERTIES OF ZIRCONIUM DIOXIDE AND HAFNIUM DIOXIDE SINGLE CRYSTALS. ALEKSANDROV, V. I. LOMONOVA, E. E. MAIER, A. A. OSIKO, V. V. TATARINTSEV, V. M. UDOVENCHIK, V. T. KRATK. SOOBSHCH. FIZ. (11), 3-7, 1972.
71872	SPECTRAL REFLECTANCE OF SELECTED PENNSYLVANIA SOILS. MATHEWS, H. L. CUNNINGHAM, R. L. PETERSEN, G. W. SOIL SCI. SOC. AMER., PROC. 37 (3), 421-4, 1973.
71882	HIGH-TEMPERATURE ENTHALPY AND SPECIFIC HEAT OF MICA GROUP MINERALS MUSCOVIET, PHLOGOPITE, AND BIOTITE. MEL'CHAKOVA, L. V. TOPOR, N. D. VESTN. MOSK. UNIV., GEOL. 28 (1), 102-7, 1973.
71892	COOLING EFFECT OF AN ENDOTHERMIC REACTION ON A SOLID WALL. SATO, A. SUGIYAMA, S. NAKARAI, K. HASATANI, M. HEAT TRANSFER-JAP. RES. 2 (3), 12-21, 1973.
71894	SCRIBING GLASS WITH PULSED AND Q-SWITCHED CARBON DIOXIDE LASER. SAIFI, M. A. PAEK, U-C. BULL. AMER. CERAM. SOC. 52 (11), 838-41, 1973.
71899	LOW-TEMPERATURE SPECIFIC HEAT AND THERMAL CONDUCTIVITY OF NONCRYSTALLINE DIELECTRIC SOLIDS. STEPHENS, R. B. PHYS. REV. 8 B (6), 2896-2905, 1973.
71915	THERMAL CONDUCTIVITY OF POLYCRYSTALLINE THORIA AND THORIA-URANIA SOLID SOLUTIONS. (LWBR DEVELOPMENT PROGRAM). BERMAN, R. M. TULLY, T. S. BELLE, J. GOLDBERG, I. BETTIS AT. POWER LAB., PITTSBURGH, PA. 53PP., 1972. (WAPD-TM-908)
71916	GROWTH OF MARBLE AS A BASIS FOR DURABILITY TESTS. KESSLER, D. W. NAT. BUR. STANDARDS, TECH. NEWS BULL. 32, 28-9, 1948.
71940	DOUBLE-BEAM METHOD OF DETERMINING THE THERMO-RADIATION CHARACTERISTICS OF MATERIALS DISSIPATING RADIATION. ILYASOV, S. G. KRASNIKOV, V. V. J. ENG. PHYS. 18 (4), 472-8, 1970. (ENGLISH TRANSLATION OF INZH.-FIZ. ZH., 18 (4), 688-95, 1970; FOR ORIGINAL SEE TPRC NO. 59460)
71976	MEASUREMENT OF INTEGRAL EMISSIVITY USING A BLACK BODY. SCHCHERBINA, D. M. TEPLOFIZ. VYS. TEMP. 11 (1), 218-21, 1973. (FOR ENGLISH TRANSLATION SEE TPRC NO. 71977)
71977	MEASUREMENT OF THE INTEGRAL DEGREE OF BLACKNESS USING A BLACK BODY. SHCHERBINA, D. M. HIGH TEMP. 11 (1), 196-9, 1973. (ENGLISH TRANSLATION OF TEPLOFIZ. VYS. TEMP., 11 (1), 218-21, 1973; FOR ORIGINAL SEE TPRC NO. 71976)
71982	CLINOPYROXENE LATTICE DEFORMATIONS. ROLES OF CHEMICAL SUBSTITUTION AND TEMPERATURE. OHASHI, Y. BURNHAM, C. W. AMER. MINERAL. 58 (9-10), 843-9, 1973.
71987	COMPARISONS OF METEORITE AND ASTEROID SPECTRAL REFLECTIVITIES. CHAPMAN, C. R. SALISBURY, J. W. ICARUS 19 (4), 507-22, 1973.
72037	VICKERS HARDNESS AND REFLECTANCE DETERMINATIONS FOR METAMICT AB(2)O(6) TYPE RARE EARTH TITANIUM-NIOBIUM-TANTALUM OXIDES. EWING, R. C. AMER. MINERAL. 58 (9-10), 942-4, 1973.
72038	THERMAL CONDUCTIVITY OF PARTIALLY DEGENERATE MAGNETOPLASMAS IN STELLAR CORES. WYLLER, A. A. ASTROPHYS. J. 184 (2), (PT. 1), 517-38, 1973.
72108	THERMAL AND MOISTURE EXPANSION STUDIES OF SOME DOMESTIC GRANITES. HOCKMAN, A. KESSLER, D. W. J. RES. NAT. BUR. STAND. 44, 395-410, 1950.
72129	THERMAL CONDUCTIVITIES OF POROUS SOLIDS. SAEGUSA, T. IIDA, Y. WAKAO, N. KAMATA, K. KAGAKU KOGAKU 37 (8), 811-14, 1973.
72167	LATTICE DYNAMICS OF CALCITE. ONOMICHI, M. SCI. LIGHT (TOKYO) 22 (1), 47-76, 1973.

TPRC Number	Bibliographic Citation
72168	**CONTACT THERMAL CONDUCTIVITY IN LUNAR AGGREGATES.** PILBEAM, C. C. VAISNYS, J. R. J. GEOPHYS. RES. 78, 5233–5236, 1973.
72169	**APOLLO 15 MEASUREMENT OF LUNAR SURFACE BRIGHTNESS TEMPERATURES - THERMAL CONDUCTIVITY OF THE UPPER 1.5 METERS OF REGOLITH.** KEIHM, S. J. PETERS, K. LANGSETH, M. G. EARTH PLANET. SCI. LETT. 19 (3), 337–351, 1973.
72171	**RESEARCH ON THE THERMAL CONDUCTIVITY OF CAST CERAMICS IN RELATION TO ITS STRUCTURE AND TEMPERATURE.** BYKOV, I. I. LEZHENIN, F. F. J. ENG. PHYS. 19 (2), 1042–3, 1970. (ENGLISH TRANSLATION OF INZH. FIZ. ZH., 19 (2), 347–9, 1970; FOR ORIGINAL SEE TPRC NO. 61898)
72179	**THERMAL EXPANSION OF MAGNETITE.** ARKHAROV, V. I. BOGOSLOVSKII, V. N. KUZNETSOV, E. N. INORG. MAT., USSR 8 (11), 1742–3, 1972. (ENGLISH TRANSLATION OF IZV. AKAD. NAUK SSSR, NEORG. MATER., 8 (11), 1982–4, 1972; FOR ORIGINAL SEE TPRC NO. 69712)
72195	**DETERMINATION OF THE THERMOPHYSICAL CHARACTERISTICS OF MATERIALS BY L. A. SEMENOVS METHOD.** ZHUKOV, N. I. J. ENG. PHYS. 20 (1), 60–3, 1971. (ENGLISH TRANSLATION OF INZH. - FIZ. ZH., 20 (1), 82–6, 1971; FOR ORIGINAL SEE TPRC NO. 62997)
72260	**THERMAL CONDUCTIVITY OF TITANIUM MONOXIDE - METAL OXIDE (METAL = NICKEL, COBALT, AND MANGANESE) SOLID SOLUTIONS.** AIVAZOV, M. I. MURANEVICH, A. KH. DOMASHNEV, I. A. TEPLOFIZ. VYS. TEMP. 11 (2), 314–19, 1973. (FOR ENGLISH TRANSLATION SEE TPRC NO. 72261)
72261	**THERMAL CONDUCTIVITY OF TITANIUM MONOXIDE - METAL OXIDE (METAL = NICKEL, COBALT, AND MANGANESE) SOLID SOLUTIONS.** AIVAZOV, M. I. MURANEVICH, A. KH. DOMASHNEV, I. A. HIGH TEMP. 11 (2), 271–5, 1973. (ENGLISH TRANSLATION OF TEPLOFIZ. VYS. TEMP., 11 (2), 314–9, 1973; FOR ORIGINAL SEE TPRC NO. 72260)
72276	**CALCULATION OF RADIATIVE HEAT EXCHANGE IN A CLOSED SYSTEM OF NONGRAY BODIES.** MASTRYUKOV, B. S. TEPLOFIZ. VYS. TEMP. 11 (2), 442–4, 1973. (FOR ENGLISH TRANSLATION SEE TPRC NO. 72277)
72277	**CALCULATION OF RADIATIVE HEAT EXCHANGE IN A CLOSED SYSTEM OF NONGRAY BODIES.** MASTRYUKOV, B. S. HIGH TEMP. 11 (2), 395–8, 1973. (ENGLISH TRANSLATION OF TEPLOFIZ. VYS. TEMP., 11 (2), 442–4, 1973; FOR ORIGINAL SEE TPRC NO. 72276)
72305	**INFRARED SPECTRA OF SOLID SOLUTIONS IN THE CALCIUM SILICATE-CALCIUM GERMANATE SYSTEM.** LAZAREV, A. N. IGNATEV, I. S. INORG. MAT., USSR 9 (4), 574–82, 1973. (ENGLISH TRANSLATION OF IZV. AKAD. NAUK SSSR, NEORG. MATER., 9 (4), 636–45, 1973; FOR ORIGINAL SEE TPRC NO. 71271)
72358	**THE EFFECT OF WATER ON THE INFRARED TRANSMISSION OF HIGHLY REFRACTING TELLURIDE GLASSES AND A METHOD FOR ITS QUANTITATIVE DETERMINATION.** TATARINTSEV, B. V. YAKHKIND, A. K. SOV. J. OPT. TECHNOL. 39 (10), 654–6, 1972. (ENGLISH TRANSLATION OF OPT.-MEKH. PROM., 39 (10), 72–3, 1972; FOR ORIGINAL SEE TPRC NO. 70260)
72359	**TRANSMISSION AND REFLECTION SPECTRA OF NATURAL AND SYNTHETIC QUARTZ IN THE VACUUM ULTRAVIOLET.** GERASIMOVA, N. G. KREYSKOP, V. N. PANOVA, I. V. SHIN, V. SOV. J. OPT. TECHNOL. 40 (1), 4–7, 1973. (ENGLISH TRANSLATION OF OPT.-MEKH. PROM., 40 (1), 7–11, 1973; FOR ORIGINAL SEE TPRC NO. 70840)
72371	**STUDY OF INTERNAL AND EXTERNAL MODES OF SOME SILICATES BY INFRARED REFLECTIVITY.** GERVAIS, F. PIRIOU, B. SERVOIN, J. L. BULL. SOC. FR. MINERAL. CRISTALLOGR. 96 (2), 81–90, 1973.
72381	**METAMICTIC URANIUM MINERALS STUDIED BY AN IR-SPECTROSCOPIC METHOD.** GEVOR'YAN, S. V. RAKOVICH, F. I. POVARENNYKH, A. S. GEOL. ZH. (RUSS. ED.) 33 (4), 65–75, 1973.
72392	**OPTICAL CHARACTERISTICS OF GUDMUNDITE AND ARSENOPYRITE.** CHVILEVA, T. N. ISSLED. OBL. RUD. MINERAL. 61–6, 1973.
72394	**NATURE OF THE BLUE COLOR OF LAZURITE.** SAMOLIVICH, M. I. NOVOZHILOV, A. I. RADYANSKII, V. M. DAVYDCHENKO, A. G. SMIRNOVA, S. A. IZV. AKAD. NAUK SSSR, SER. GEOL. (7), 95–102, 1973.
72419	**ANHARMONICITY IN SILICATE CRYSTALS. TEMPERATURE DEPENDENCE OF A(U) TYPE VIBRATIONAL MODES IN ZIRCONIUM SILICATE AND LITHIUM ALUMINUM SILICATE.** GERVAIS, F. PIRIOU, B. CABANNES, F. J. PHYS. CHEM. SOLIDS 34 (11), 1785–96, 1973.
72437	**OPTICAL ABSORPTION SPECTRA OF COBALT DIPOSITIVE ION IN COBALT MAGNESIUM OXIDE.** MIRONOVA, N. A. ULMANIS, U. LATV. PSR ZINAT. AKAD. VESTIS, FIZ. TEH. ZINAT. SER. (4), 39–44, 1973.
72447	**SWELLING OF SODIUM TAENIOLITE WITH WATER.** KITAJIMA, K. SUGIMORI, K. DAIMON, N. NIPPON KAGAKU KAISHI (10), 1885–92, 1973.
72486	**THERMAL CONDUCTIVITY OF IRON OXIDE DEPOSITS.** RASSOKHIN, N. G. KABANOV, L. P. TEVLIN, S. A. TERSIN, V. A. TEPLOENERGETIKA 20 (9), 12–15, 1973. (FOR ENGLISH TRANSLATION SEE TPRC NO. 74596)
72549	**MARINER 9 ULTRAVIOLET SPECTROMETER EXPERIMENT - PHOTOMETRY AND TOPOGRAPHY OF MARS.** HORD, C. W. BARTH, C. A. STEWART, A. I. LANE, A. L. ICARUS 17, 443–456, 1972.
72581	**REFLECTION CORRECTION FOR HIGH-ACCURACY TRANSMITTANCE MEASUREMENTS ON FILTER GLASSES.** MIELENZ, K. D. MAVRODINEANU, R. J. RES. NAT. BUR. STAND. 77 A (6), 699–703, 1973.
72582	**THE EFFECT OF TEMPERATURE AND PRESSURE ON THE REFRACTIVE INDEX OF SOME OXIDE GLASSES.** WAXLER, R. M. CLEEK, G. W. J. RES. NAT. BUR. STAND. 77 A (6), 755–63, 1973.
72646	**COUPLED AND DECOMPOSING TWO-PHONON STATES AND FERMI RESONANCE IN CALCITE.** BELOUSOV, M. V. POGAREV, D. E. SHULTIN, A. A. FIZ. TVERD. TELA (LENINGRAD) 15 (8), 2553–5, 1973. (FOR ENGLISH TRANSLATION SEE TPRC NO. 72647)
72647	**BOUND AND DISSOCIATING TWO-PHONON STATES AND FERMI RESONANCE IN CALCITE.** BELOUSOV, M. V. POGAREV, D. E. SHULTIN, A. A. SOV. PHYS. SOLID STATE 15 (8), 1701–2, 1974. (ENGLISH TRANSLATION OF FIZ. TVERD. TELA, 15 (8), 2553–5, 1973; FOR ORIGINAL SEE TPRC NO. 72646)
72652	**VIBRATIONAL SPECTRA OF RUTILE.** KNYAZEV, A. S. ZAKHAROV, V. P. MITYUREVA, I. A. POPLAVKO, YU. M. FIZ. TVERD. TELA (LENINGRAD) 15 (8), 2371–7, 1973. (FOR ENGLISH TRANSLATION SEE TPRC NO. 72653)
72653	**VIBRATIONAL SPECTRA OF RUTILE.** KNYAZEV, A. S. ZAKHAROV, V. P. MITYUREVA, I. A. POPLAVKO, YU. M. SOV. PHYS. SOLID STATE 15 (8), 1579–82, 1974. (ENGLISH TRANSLATION OF FIZ. TVERD. TELA, 15 (8), 2371–7, 1973; FOR ORIGINAL SEE TPRC NO. 72652)
72666	**THE NATURE OF THE MAXIMA IN THE UV ABSORPTION SPECTRA OF SOME GLASSES.** GAVRILOV, M. Z. J. APPL. SPECTROSC., USSR 14 (6), 836–7, 1971. (ENGLISH TRANSLATION OF ZH. PRIKL. SPEKTROSK., 14 (6), 1135–6, 1971; FOR ORIGINAL SEE TPRC NO. 64244)

TPRC Number	Bibliographic Citation
72670	**TARGET-SIGNATURE MEASUREMENTS.** RICE, P. J. WILLOW RUN LABS., INST. OF SCIENCE AND TECHNOLOGY, THE UNIV. OF MICHIGAN, ANN ARBOR 272PP., 1967. (WRL-8047-18-P)
72674	**MEASURING THE THERMAL PROPERTIES OF CYLINDRICAL SPECIMENS BY THE USE OF SINUSOIDAL TEMPERATURE WAVES.** HOEKSTRA, P. DELANEY, A. ATKINS, R. COLD REGIONS RESEARCH AND ENGINEERING LAB., HANOVER, N. H. 20PP., 1973. (AD-770425, CRREL-TR-244)
72699	**INCREASING DESIGN ALLOWABLES FOR A GRAPHITE.** STARRETT, H. S. SANDERS, H. G. BROWN, J. R., JR. PEARS, C. D. AIR FORCE MATERIALS LAB., WRIGHT-PATTERSON AIR FORCE BASE, OHIO 404PP., 1974. (AFML-TR-73-300)
72733	**RADIOMETER TO MONITOR LOW LEVELS OF ULTRAVIOLET IRRADIANCE.** GOLDBERG, B. KLEIN, W. H. APPL. OPT. 13 (3), 493-6, 1974.
72742	**REDUCTION IN SWITCHING THRESHOLDS OF CHALCOGENIDE FILMS WITH POLYMERIC COATINGS.** WELBER, B. TAYLOR, R. C. J. APPL. PHYS. 45 (2), 943-5, 1974.
72761	**PULSE IRRADIATION OF DIAMONDS.** BARKATT, A. OGDAN, J. J. PHYS. CHEM. SOLIDS 33 (12), 2217-27, 1972.
72771	**OPTICAL SPECTRA AND RELAXATION OF EUROPIUM TRIPOSITIVE ION IN GERMANATE GLASSES.** REISFELD, R. LIEBLICH, N. J. PHYS. CHEM. SOLIDS 34 (9), 1467-76, 1973.
72792	**IMPLICATIONS OF HEAT FLOW AND BOTTOM WATER TEMPERATURE IN THE EASTERN EQUATORIAL PACIFIC.** VON HERZEN, R. P. ANDERSON, R. N. GEOPHYS. J. ROY. ASTRON. SOC. 26, 427-58, 1972. (AD-751199, WHOI-72-81)
72795	**VISIBLE AND NEAR INFRARED ABSORPTION COEFFICIENTS OF KAOLINITE AND RELATED CLAYS.** LINDBERG, J. D. SMITH, M. S. ARMY ELECTRONICS COMMAND, FORT MONMOUTH, N. J. 17PP., 1973. (AD-770920, ECOM-5522)
72814	**THE DIFFUSION OF CALCIUM, PHOSPHOROUS, AND DEUTEROXIDE-IONS IN FLUORAPATITE.** DEN HARTOG, H. WELCH, D. O. ROYCE, B. S. H. PHYS. STATUS SOLIDI 53 B (1), 201-12, 1972.
72822	**ANHARMONICITY OF INFRARED VIBRATION MODES IN THE NESOSILICATE BERYLLIUM SILICON DIOXIDE.** GERVAIS, F. PIRIOU, B. CABANNES, F. PHYS. STATUS SOLIDI 55 B (1), 143-54, 1973.
72863	**SOME MEASUREMENTS OF THE TRANSMISSION OF ULTRAVIOLET RADIATION THROUGH VARIOUS FABRICS.** COBLENTZ, W. W. STAIR, R. SCHOFFSTALL, C. W. J. RES. NAT. BUR. STAND. 1, 105-24, 1928.
72866	**THERMAL AND MECHANICAL PROPERTY SCREENING OF GRAPHITE MATERIALS FOR ADVANCED RE-ENTRY VEHICLES. VOLUME I. TEST METHODS AND DATA BOOKS A THROUGH E.** LEGG, J. K. BAPAT, S. G. STARRETT, H. S. PEARS, C. D. AIR FORCE MATERIALS LAB., WRIGHT-PATTERSON AIR FORCE BASE, OHIO 240PP., 1973. (AFML-TR-73-151, VOL. I SORI-EAS-73-2300-IXX-184-F
72869	**REFLECTANCE SPECTRA OF GYPSUM SAND FROM THE WHITE SANDS NATIONAL MONUMENT AND BASALT FROM A NEARBY LAVA FLOW.** LINDBERG, J. D. SMITH, M. S. AMER. MINERAL. 58 (11-12), 1062-4, 1973.
72874	**THERMAL CONDUCTIVITY OF FERRITE-GARNET SOLID SOLUTIONS.** LYAKAS, M. A. MARKOVIN, P. A. NUROMSKII, A. B. SHALABUTOV, YU. K. SIDOROV, V. G. SOV. PHYS. J. 14 (7), 963-4, 1971. (ENGLISH TRANSLATION OF IZV. VYSSH. UCHEB. ZAVED., FIZ., 14 (7), 113-4, 1971; FOR ORIGINAL SEE TPRC NO. 64268)
72882	**TRANSMISSION FORCES TO A CRYOSTAT.** GIANESE, P. PORROT, E. REY, G. TAQUET, B. CRYOGENICS 13 (9), 550-3, 1973.
72892	**THERMOPHYSICAL CHARACTERISTICS OF A BOILING OPEN-HEARTH BATH.** GEINEMAN, A. V. IZV. VYSSH. UCHEB. ZAVED., CHERN. MET. (6), 29-32, 1973.
72910	**POSSIBILITY OF CALCULATING THE SPECTRAL ALBEDO OF VENUS IN THE NEAR INFRARED REGION.** KREKOVA, M. M. KREKOV, G. M. TITOV, G. A. FEIGEL'SON, E. M. KOSM. ISSLED. 11 (4), 607-11, 1973.
72911	**SPECTRAL LIGHT TRANSMISSION BY CRYSTALS OF OPTICAL CALCITE.** SHUSTOV, A. V. SHEREMT, I. F. KRISTALLOGR. MINERAL., TR. FEDOROVSKOI YUBILEINOI SESS. 175-7, 1972.
72936	**HEAT CAPACITIES OF A SODA SILICA AND A SODA GERMANIA GLASS BETWEEN 1.5 AND 25 K.** LEADBETTER, A. J. WATERFIELD, C. G. WYCHERLEY, K. E. SOLID STATE COMMUN. 13 (6), 701-3, 1973.
72981	**THERMAL EXPANSION OF REFERENCE MATERIALS. COPPER, SILICA, AND SILICON.** WHITE, G. K. J. PHYS. 6 D (17), 2070-8, 1973.
72984	**INFRARED SPECTRA OF KAOLINITE IN THE REGION OF HYDROXYL STRETCHING VIBRATIONS.** PLYUSNINA, I. I. GRIBINA, I. A. KRISTALLOGR. MINERAL., TR. FEDOROUSKOI YUBILEINOI SESS. 196-9, 1972.
73007	**THERMODYNAMIC QUANTITIES OF ALKALI SILICATES IN THE TEMPERATURE RANGE FROM 25 TO MELTING POINT.** TAKAHASHI, K. YOSHIO, T. YOGYO KYOKAI SHI 81 (12), 524-33, 1973.
73059	**INFRARED SPECTROSCOPIC STUDY OF QUARTZ FROM THE KYZYLALMASAI GOLD ORE DEPOSIT (UZBEK SSR).** RASHIDOVA, G. SH. ZAP. UZB. OTD. VSES. MINERAL. OBSHCHEST. (26), 128-31, 1973.
73101	**THE VIBRATIONAL SPECTRUM OF ZIRCON, (ZIRCONIUM SILICATE).** DAWSON, P. HARGREAVE, M. M. WILKINSON, G. R. J. PHYS., SOLID STATE PHYS. 4 C (2), 240-56, 1971.
73127	**THE OPTICAL PROPERTIES OF MICA IN THE VACUUM ULTRAVIOLET.** DAVIDSON, A. T. VICKERS, A. F. J. PHYS., SOLID STATE PHYS. 5 C (8), 879-87, 1972.
73134	**AMORPHOUS ANTIFERROMAGNETISM IN IRON AND COBALT PHOSPHORUS PENTOXIDE GLASSES.** EGAMI, T. SACLI, O. A. SIMPSON, A. W. TERRY, A. L. WEDGWOOD, F. A. J. PHYS., SOLID STATE PHYS. 5 C (18), L261-L265, 1972.
73145	**OPTICAL VIBRATIONS IN SHEET SILICATES.** LOH, E. J. PHYS., SOLID STATE PHYS. 6 C (6), 1091-1104, 1973.
73151	**MARE HUMORUM: AN INTEGRATED STUDY OF SPECTRAL REFLECTIVITY.** JOHNSON, T. V. PIETERS, C. MC CORD, T. B. ICARUS 19, 224-229, 1973.
73152	**ASTEROID REFLECTIVITIES FROM POLARIZATION CURVES: CALIBRATION OF THE SLOPE-ALBEDO RELATIONSHIP.** VEVERKA, J. NOLAND, M. ICARUS 19, 230-239, 1973.

TPRC Number	Bibliographic Citation
73153	**OPTICAL PROPERTIES OF SOME TERRESTRIAL ROCKS AND GLASSES.** POLLACK, J. B. TOON, O. B. KHARE, B. N. ICARUS 19, 372–389, 1973.
73171	**INFRARED SPECTRA OF DIAMONDS FROM ALTO ARAGUAIA, BRAZIL.** SVISERO, D. P. DE CAMARGO, W. G. R. NAZARIO, G. AN. ACAD. BRASIL. CIENC. 44 (3–4), 441–9, 1972.
73173	**INFRARED SPECTRA OF SELENATES AND SELENITES.** POVARENNYKH, A. S. ONISHCHENKO, L. A. GEOL. ZH. (RUSS. ED.) 33 (6), 98–103, 1973.
73275	**HIGH-TEMPERATURE INVESTIGATIONS OF THE THERMAL PROPERTIES OF SOLIDS.** FILIPPOV, L. P. YURCHAK, R. P. J. ENG. PHYS. 21 (3), 1209–20, 1971. (ENGLISH TRANSLATION OF INZH. FIZ. ZH., 21 (3), 561–77, 1971; FOR ORIGINAL SEE TPRC NO. 65570)
73297	**TRANSIENT EFFECT IN ZINC OXIDE - MAGNESIUM OXIDE: IRON, OXYGEN PHOTOCONDUCTOR.** SINGH, S. SOLID STATE COMMUN. 11 (8), 983–6, 1972.
73299	**RESTSTRAHLEN SPECTRA OF CADMIUM STRONTIUM OXIDE.** FINKENRATH, H. KALINOWSKI, I. UHLE, N. SOLID STATE COMMUN. 11 (9), 1283–5, 1972.
73307	**FAR ULTRAVIOLET REFLECTIVITY OF LUNAR DUST SAMPLES - APOLLO 11, 12, AND 14.** LUCKE, R. L. FASTIE, W. G. HENRY, R. C. ASTRON. J. 78, 263–267, 283, 1973.
73308	**THE MOON AS A PROPOSED RADIOMETRIC STANDARD FOR MICROWAVE AND INFRARED OBSERVATIONS OF EXTENDED SOURCES.** LINSKY, J. L. ASTROPHY. J. SUPPL. SER. 25 (SUPPL. 216), 163–203, 1973.
73309	**THE ELECTRICAL AND THERMAL CONDUCTIVITIES OF STELLAR DEGENERATE MATTER.** KOVETZ, A. SHAVIV, G. ASTRON. ASTROPHY. 28 (2), 315–318, 1973.
73315	**ON THE ROLE OF FLUCTUATIONS IN THE INTERPLANETARY MAGNETIC FIELD ON HEAT CONDUCTION IN THE SOLAR WIND.** CUPERMAN, S. METZIER, N. J. GEOPHYS. RES. 78, 3167–68, 1973.
73316	**SOLAR WIND PROPERTIES AT THE EARTH AS PREDICTED BY THE ONE-FLUID MODEL WITH HELIOCLASSICAL THERMAL ELECTRON CONDUCTIVITY.** DURNEY, B. R. J. GEOPHYS. RES. 78, 7229–7236, 1973.
73320	**SPECTROPHOTOMETRY OF INDIVIDUAL REGIONS OF VENUS.** STARODUBTSEVA, O. M. PHYS. MOON PLANETS 338–45, 1972.
73329	**THERMAL RADIATION PROPERTIES OF APOLLO 14 FINES.** BIRKEBAK, R. C. DAWSON, J. P. THE MOON 6, 93–99, 1973.
73330	**VISCOSITY OF THE MOON. I - AFTER MARE FORMATION. II - DURING MARE FORMATION.** ARKANI-HAMED, J. THE MOON 6, 100–124, 1973.
73331	**THERMAL INSTABILITIES AND THE FORMATION OF LUNAR MARIA.** BASTIN, J. A. THE MOON 8, (SEPT.), 335–339, 1973.
73335	**VISCOUS STRATIFICATION OF THE EARTH AND CONVECTION.** ELSASSER, W. M. PHYSICS OF THE EARTH AND PLANETARY INTERIORS 6 (1–3), 198–204, 1972.
73365	**THERMAL CONDUCTIVITY OF GLASS-FIBER SYSTEMS.** VNUKOV, S. P. RYABOV, V. A. FEDOSEEV, D. V. J. ENG. PHYS. 21 (5), 1350–4, 1971. (ENGLISH TRANSLATION OF INZH. FIZ. ZH., 21 (5), 797–803, 1971; FOR ORIGINAL SEE TPRC NO. 65427)
73366	**THERMAL CONDUCTIVITIES OF METAL CASTING MOLD FACINGS.** SEREBRO, V. S. BRENER, A. I. J. ENG. PHYS. 21 (5), 1355–7, 1971. (ENGLISH TRANSLATION OF INZH. FIZ. ZH., 21 (5), 804–6, 1971; FOR ORIGINAL SEE TPRC NO. 65854)
73367	**THERMAL PROPERTIES.** HARRIS, M. HANDBOOK OF TEXTILE FIBERS 167–82, 1954.
73465	**STIMULATED EMISSION FROM A RUBY LASER EXPOSED TO COBALT-60 GAMMA-RAYS.** BEDILOV, M. R. KHAIDAROV, K. ZH. TEKHN. FIZ. 42 (2), 391–4, 1972. (FOR ENGLISH TRANSLATION SEE TPRC NO. 73466)
73466	**STIMULATED EMISSION FROM A RUBY LASER EXPOSED TO COBALT-60 GAMMA-RAYS.** BEDILOV, M. R. KHAIDAROV, K. SOV. PHYS. TECH. PHYS. 17 (2), 311–3, 1972. (ENGLISH TRANSLATION OF ZH. TEKHN. FIZ., 42 (2), 391–4, 1972; FOR ORIGINAL SEE TPRC NO. 73465)
73503	**NEW DEVELOPMENT OF AN APPARATUS FOR THERMAL CONDUCTIVITY MEASUREMENT OF REFRACTORY BRICKS BY HOT WIRE METHOD.** HAYASHI, K. FUKUI, M. UEI, I. YOGYO KYOKAI SHI 82 (1), 13–21, 1974.
73516	**THERMAL CONDUCTIVITY OF CARBONATE ROCKS.** THOMAS, J., JR. FROST, R. R. HARVEY, R. D. ENG. GEOL. (AMSTERDAM) 7 (1), 3–12, 1973.
73520	**DILATOMETRIC DETERMINATION OF TRANSFORMATION POINT, DILATOMETRIC SOFTENING POINT AND THE COEFFICIENT OF EXPANSION IN THE SILICA-ALUMINA-SODIUM OXIDE-POTASSIUM OXIDE-CALCIUM OXIDE-MAGNESIA GLASS SYSTEM.** LAKATOS, T. JOHANSSON, L. G. SIMMINGSKOELD, B. GLASTEK. TIDSKR. 28 (5), 69–73, 1973.
73529	**VISIBLE AND NEAR INFRARED SPECTRA OF MINERALS AND ROCKS. VI. ADDITIONAL SILICATES.** HUNT, G. R. SALISBURY, J. W. LENHOFF, C. J. MOD. GEOL. 4 (2), 85–106, 1973.
73585	**AMPLIFICATION OF WAVES DURING REFLECTION FROM A ROTATING BLACK HOLE.** STAROBINSKII, A. A. ZH. EKSP. TEOR. FIZ. 64 (1), 48–57, 1973. (FOR ENGLISH TRANSLATION SEE TPRC NO. 73586)
73586	**AMPLIFICATION OF WAVES DURING REFLECTION FROM A ROTATING BLACK HOLE.** STAROBINSKII, A. A. SOV. PHYS. JETP 37 (1), 28–32, 1973. (ENGLISH TRANSLATION OF ZH. EKSP. TEOR. FIZ., 64 (1), 48–57, 1973; FOR ORIGINAL SEE TPRC NO. 73585)
73600	**NEUTRINO THERMAL CONDUCTIVITY IN COLLAPSING STARS.** IMSHENNIK, V. S. NADEZHIN, D. K. ZH. EKSP. TEOR. FIZ. 63 (5), 1548–61, 1972. (FOR ENGLISH TRANSLATION SEE TPRC NO. 73601)
73601	**NEUTRINO THERMAL CONDUCTIVITY IN COLLAPSING STARS.** IMSHENNIK, V. S. NADEZHIN, D. K. SOV. PHYS. JETP 36 (5), 821–7, 1973. (ENGLISH TRANSLATION OF ZH. EKSP. TEOR. FIZ., 63 (5), 1548–61, 1972; FOR ORIGINAL SEE TPRC NO. 73600)
73637	**EMISSION OF A RUBY LASER UNDER COBALT GAMMA RADIATION.** BEDILOV, M. R. KHAIDAROV, K. OPT. SPEKTROSK. 34 (4), 765–7, 1973. (FOR ENGLISH TRANSLATION SEE TPRC NO. 73638)
73638	**EMISSION OF A RUBY LASER UNDER COBALT GAMMA RADIATION.** BEDILOV, M. R. KHAIDAROV, K. OPT. SPECTROS., USSR 34 (4), 441–2, 1973. (ENGLISH TRANSLATION OF OPT. SPEKTROSK., 34 (4), 765–7, 1973; FOR ORIGINAL SEE TPRC NO. 73637)
73651	**HIGH-EFFICIENCY SCHEME OF INTRACAVITY GENERATION OF HARMONICS AND PARAMETRIC OSCILLATION.** VOLOSOV, V. D. KRYLOV, V. N. OPT. SPEKTROSK. 35 (1), 120–4, 1973. (FOR ENGLISH TRANSLATION SEE TPRC NO. 73652)

TPRC Number	Bibliographic Citation
73652	**HIGH-EFFICIENCY SCHEME OF INTRACAVITY GENERATION OF HARMONICS AND PARAMETRIC OSCILLATION.** VOLOSOV, V. D. KRYLOV, V. N. OPT. SPECTROS., USSR 35 (1), 69-71, 1973. (ENGLISH TRANSLATION OF OPT. SPEKTROSK., 35 (1), 120-4, 1973; FOR ORIGNAL SEE TPRC NO. 73651)
73659	**OPTICAL CONSTANTS OF HEMATITE IN THE INFRARED RANGE OF THE SPECTRUM.** POPOVA, S. I. TOLSTYKH, T. S. IVLEV, L. S. OPT. SPEKTROSK. 35 (5), 945-55, 1973. (FOR ENGLISH TRANSLATION SEE TPRC NO. 73660)
73660	**OPTICAL CONSTANTS OF HEMATITE IN THE INFRARED RANGE OF THE SPECTRUM.** POPOVA, S. I. TOLSTYKH, T. S. IVLEV, L. S. OPT. SPECTROS., USSR 35 (5), 551-2, 1973. (ENGLISH TRANSLATION OF OPT. SPEKTROSK., 35 (5), 945-55, 1973; FOR ORIGINAL SEE TPRC NO. 73659)
73665	**THERMAL CONDUCTIVITY OF VARIOUS MATERIALS.** TAYLOR, T. S. PHYS. REV. 13, 150-1, 1919.
73669	**HEAT TRANSFER IN A UNIFORM MIXTURE OF TWO DISPERSED MATERIALS.** GORBIS, Z. R. KNYAZEV, L. P. KUKLINSKII, V. V. INZH. FIZ. ZH. 18 (1), 45-51, 1970. (FOR ENGLISH TRANSLATION SEE TPRC NO. 73670)
73670	**HEAT TRANSFER IN A UNIFORM MIXTURE OF TWO DISPERSED MATERIALS.** GORBIS, Z. R. KNYAZEV, L. P. KUKLINSKII, V. V. J. ENG. PHYS. 18 (1), 33-7, 1970. (ENGLISH TRANSLATION OF INZH. FIZ. ZH., 18 (1), 45-51, 1970; FOR ORIGINAL SEE TPRC NO. 73669)
73679	**DETERMINATION OF TIME OF SOIL FREEZING BY METHOD OF CONCENTRIC ISOTHERMS.** SHAFEEV, M. N. METENIN, V. I. SHMAREV, A. T. INZH. FIZ. ZH. 18 (3), 499-507, 1970. (FOR ENGLISH TRANSLATION SEE TPRC NO. 73680)
73680	**DETERMINATION OF TIME OF SOIL FREEZING BY METHOD OF CONCENTRIC ISOTHERMS.** SHAFEEV, M. N. METENIN, V. I. SHMAREV, A. T. J. ENG. PHYS. 18 (3), 348-53, 1970. (ENGLISH TRANSLATION OF INZH. FIZ. ZH., 18 (3), 499-507, 1970; FOR ORIGINAL SEE TPRC NO. 73679)
73689	**EXPERIMENTAL RESEARCH TO DETERMINE THE EFFECTIVE THERMAL CONDUCTIVITIES OF A LAYER OF FURNACE CHARGE.** ABZALOV, YU. M. NEVSKII, A. S. INZH. FIZ. ZH. 19 (1), 42-6, 1970. (FOR ENGLISH TRANSLATION SEE TPRC NO. 73690)
73690	**EXPERIMENTAL RESEARCH TO DETERMINE THE EFFECTIVE THERMAL CONDUCTIVITIES OF A LAYER OF FURNACE CHARGE.** ABZALOV, YU. M. NEVSKII, A. S. J. ENG. PHYS. 19 (1), 827-30, 1970. (ENGLISH TRANSLATION OF INZH. FIZ. ZH., 19 (1), 42-6, 1970; FOR ORIGINAL SEE TPRC NO. 73689)
73730	**RADIANT AND MOLECULAR HEAT TRANSFER IN THERMALLY ISOLATED MATERIALS HAVING SMALL VOLUME DENSITIES.** KOSTYLEV, V. M. BELOSTOTSKAYA, V. YA. INZH. FIZ. ZH. 21 (2), 290-5, 1971. (FOR ENGLISH TRANSLATION SEE TPRC NO. 73731)
73731	**RADIANT AND MOLECULAR HEAT TRANSFER IN THERMALLY ISOLATED MATERIALS HAVING SMALL VOLUME DENSITIES.** KOSTYLEV, V. M. BELOSTOTSKAYA, V. YA. J. ENG. PHYS. 21 (2), 1019-22, 1971. (ENGLISH TRANSLATION OF INZH. FIZ. ZH., 21 (2), 290-5, 1971; FOR ORIGINAL SEE TPRC NO. 73730)
73732	**THERMAL CONDUCTIVITY OF POROUS MATERIALS AS A FUNCTION OF MOISTURE CONTENT.** NOVIKOV, P. A. MIKHNYUK, B. G. INZH. FIZ. ZH. 21 (2), 296-300, 1971. (FOR ENGLISH TRANSLATION SEE TPRC NO. 73733)
73733	**THERMAL CONDUCTIVITY OF POROUS MATERIALS AS A FUNCTION OF MOISTURE CONTENT.** NOVIKOV, P. A. MIKHNYUK, B. G. J. ENG. PHYS. 21 (2), 1023-6, 1971. (ENGLISH TRANSLATION OF INZH. FIZ. ZH., 21 (2), 296-300, 1971; FOR ORIGINAL SEE TPRC NO. 73732)
73753	**THE THERMAL CHARACTERISTICS OF THAWING SILTS OF GLACIAL GENESIS.** RASHKIN, A. V. SHUVALOV, N. G. VEDYAEV, YU. M. INZH. FIZ. ZH. 16 (2), 326-9, 1969. (FOR ENGLLISH TRANSLATION SEE TPRC NO. 73754)
73754	**THE THERMAL CHARACTERISTICS OF THAWING SILTS OF GLACIAL GENESIS.** RASHKIN, A. V. SHUVALOV, N. G. VEDYAEV, YU. M. J. ENG. PHYS. 16 (2), 221-3, 1969. (ENGLISH TRANSLATION OF INZH. FIZ. ZH., 16 (2), 326-9, 1969; FOR ORIGINAL SEE TPRC NO. 73753)
73763	**INVESTIGATING THE THERMAL INTERACTION BETWEEN A GAS WELL AND PERMAFROST SOILS.** KRIVOSHEIN, B. L. KOSHELEV, A. A. ROGOZHIN, KH. YA. SIDLER, L. E. INZH. FIZ. ZH. 16 (5), 872-7, 1969. (FOR ENGLISH TRANSLATION SEE TPRC NO. 73764)
73764	**INVESTIGATING THE THERMAL INTERACTION BETWEEN A GAS WELL AND PERMAFROST SOILS.** KRIVOSHEIN, B. L. KOSHELEV, A. A. ROGOZHIN, KH. YA. SIDLER, L. E. J. ENG. PHYS. 16 (5), 609-13, 1969. (ENGLISH TRANSLATION OF INZH. FIZ. ZH., 16 (5), 872-7, 1969; FOR ORIGINAL SEE TPRC NO. 73763)
73781	**DETERMINATION OF THERMOPNYSICAL PROPERTIES OF ROCKS UTILIZING MONOTONIC HEATING OF A TWIN-LAYERED PLATE.** MERZLYAKOV, E. I. RYZHENKO, I. A. HEAT TRANSFER-SOV. RES. 6 (1), 156-9, 1974.
73785	**LOW-TEMPERATURE SPECIFIC HEAT OF LITHIUM OXIDE - ALUMINUM OXIDE - SILICON OXIDE GLASSES AND KEATITE-PHASE GLASS CERAMICS.** BOHN, R. G. J. APPL. PHYS. 45 (5), 2133-7, 1974.
73810	**THERMAL CONDUCTIVITY OF PARTICULATE MATERIALS: A SUMMARY OF MEASUREMENTS TAKEN AT THE MARSHALL SPACE FLIGHT CENTER.** FOUNTAIN, J. A. MARSHALL SPACE FLIGHT CENTER, HUNTSVILLE, ALA. 72PP., 1973. (N73-27801, NASA-TM-X-64759)
73825	**THE TEMPERATURE OF PLANETS.** TSIOLKOVSKIY, K. YE. SOBRANIYE SOCHINENIY 4, 97-102, 1964. (FOR ENGLISH TRANSLATION, SEE TPRC NO. 73826)
73826	**THE TEMPERATURE OF PLANETS.** TSIOLKOVSKIY, K. YE. LINGUISTIC SYSTEMS, INC., CAMBRIDGE, MASS. 12PP., 1973. (ENGLISH TRANSLATION OF SOBRANIYE SOCHINENIY, 4, 97-10l, 1964; FOR ORIGINAL SEE T73825) (N73-33806, NASA-TT-F-15040)
73838	**DETERMINATION OF ABSORPTION AND SCATTERING COEFFICIENTS FOR NONHOMOGENEOUS MEDIA: II. EXPERIMENT.** EGAN, W. G. HILGEMAN, T. REICHMAN, J. GRUMAN AEROSPACE CORP., BETHPAGE, N. Y. 36PP., 1973. (AD-760203, RE-451J)
73840	**TRANSITION METAL ION SUBSTITUTION IN AN ALUMINOSILICATE GLASS.** DUNCAN, J. F. SITHARAMARAO, D. N. REV. CHIM. MINER. 9, 543-8, 1972. (AD-760247, AFOSR-TR-73-0755)
73843	**THERMAL CONDUCTIVITY MEASUREMENTS OF CRYOGENIC INSULATIONS AT HIGH PRESSURES.** TATE, K. W. WESTPHAL, R. C. NAVAL CIVIL ENGINEERING LAB., PORT HUENEME, CALIFORNIA 47PP., 1974. (AD-774470, NCEL-TN-1302)

TPRC Number	Bibliographic Citation
73845	**TEMPERATURE MEASUREMENT OF THE THERMAL CONDUCTIVITY OF CERAMICS.** KIRYUTIN, A. A. KISELEV, N. P. KOZITSYNA, I. V. IZMER. TEKH. (2), 51-2, 1972. (FOR ENGLISH TRANSLATION SEE TPRC NO. 73846)
73846	**TEMPERATURE MEASUREMENT OF THE THERMAL CONDUCTIVITY OF CERAMICS.** KIRYUTIN, A. A. KISELEV, N. P. KOZITSYNA, I. V. MEAS. TECH., USSR (2), 265-7, 1972. (ENGLISH TRANSLATION OF IZMER. TEKH., (2), 51-2, 1972; FOR ORIGINAL SEE TPRC NO. 73845)
73866	**A STUDY OF THE EMISSIVITY FACTOR OF ASH DEPOSITS AND CERTAIN REFRACTORY MATERIALS.** MITOR, V. V. KONOPELKO, I. N. TEPLOENERGETIKA 17 (10), 41-3, 1970. (FOR ENGLISH TRANSLATION SEE TPRC NO. 73867)
73867	**A STUDY OF THE EMISSIVITY FACTOR OF ASH DEPOSITS AND CERTAIN REFRACTORY MATERIALS.** MITOR, V. V. KONOPELKO, I. N. THERM. ENG., USSR 17 (10), 61-3, 1970. (ENGLISH TRANSLATION OF TEPLOENERGETIKA, 17 (10), 41-3, 1970; FOR ORIGINAL SEE TPRC NO. 73866)
73893	**VITROCERAMIC NUCLEAR FUEL COATING. II. COATINGS BASED ON ALUMINO-SILICATES.** MAXIM, I. V. APOSTOL, D. J. NUCL. MATER. 41 (1), 80-6, 1971.
73900	**THE THERMAL CONDUCTIVITY OF PLUTONIUM-URANIUM DIOXIDE AT TEMPERATURES UP TO 1273 K.** GOLDSMITH, L. A. DOUGLAS, J. A. M. J. NUCL. MATER. 43 (3), 225-33, 1972.
73901	**THERMAL DIFFUSIVITY OF MIXED THORIUM URANIUM OXIDES AND SOME MATERIALS TO BE USED AS REFERENCE IN THE RANGE 650 - 2700 K.** FERRO, C. PATIMO, C. PICONI, C. J. NUCL. MATER. 43 (3), 273-6, 1972.
73921	**APPLICATION OF A STRUCTURAL MODEL TO THE ELUCIDATION OF THERMAL CHARACTERISTICS OF DISPERSIVE SUBSTANCES.** KLUVANEC, D. ACTA PHYS. SLOVACA 23 (4), 234-45, 1973.
73927	**STUDIES OF STATE ANALYSIS WITH INSTRUMENTAL METHOD. IV. INFLUENCE OF GRINDING METHOD ON INFRARED SPECTRA OF QUARTZ POWDERS.** SATO, K. BUNSEKI KAGAKEE 22 (7), 824-31, 1973.
73939	**A COMPARISON OF THERMAL CONDUCTIVITY WITH ELASTIC PROPERTIES FOR SIX IGNEOUS ROCK TYPES.** JOHNSON, S. A. PURDUE UNIVERSITY, M. S. THESIS 104PP., 1974.
73943	**ULTRAVIOLET REFLECTIVITY AND GEOMETRICAL ALBEDO OF TITAN.** BARKER, E. S. TRAFTON, L. M. ICARUS 20 (4), 444-54, 1973.
73996	**COLOR AND ITS MEASUREMENT.** MEES, C. E. K. AMER. SOC. TEST. MATER., PROC. 30 (PT. II), 9-25, 1930.
74002	**ANALYSIS OF HEAT FLOW DATA IN SITU THERMAL CONDUCTIVITY MEASUREMENTS.** BECK, A. E. ANGLIN, F. M. SASS, J. H. CAN. J. EARTH SCI. 8 (1), 1-19, 1971.
74003	**GEOTHERMAL DATA FROM THE GRANDUC AREA, NORTHERN COAST MOUNTAINS OF BRITISH COLUMBIA.** MATHEWS, W. H. CAN. J. EARTH SCI. 9 (10), 1333-7, 1972.
74004	**RELATIONS BETWEEN THE THERMAL AND ELASTIC PROPERTIES OF ROCKS.** MEDVEDEV, R. V. FIZ. TEKH. PROBL. RAZRAB. POLEZ. ISKOP. (3), 30-6, 1967. (FOR ENGLISH TRANSLATION SEE T74005)
74005	**RELATIONS BETWEEN THE THERMAL AND ELASTIC PROPERTIES OF ROCKS.** MEDVEDEV, R. V. SOV. MIN. SCI. 1 (3), 238-42, 1967. (ENGLISH TRANSLATION OF FIZ. TEKH. PROBL. RAZRAB. POLEZ. ISKOP., (3), 30-6, 1967; FOR ORIGINAL SEE T74004)
74008	**THE THERMAL PROPERTIES OF ROCKS IN A TEMPERATURE FIELD.** DMITRIEV, A. P. DERBENEV, L. S. GONCHAROV, S. A. FIZ. TEKH. PROBL. RAZRAB. POLEZ. ISKOP. (2), 107-8, 1969. (FOR ENGLISH TRANSLATION SEE T74009)
74009	**THE THERMAL PROPERTIES OF ROCKS IN A TEMPERATURE FIELD.** DMITRIEV, A. P. DERBENEV, L. S. GONCHAROV, S. A. SOV. MIN. SCI. (2), 199-200, 1969. (ENGLISH TRANSLATION OF FIZ. TEKH. PROBL. RAZRAB. POLEZ. ISKOP., (2), 107-8, 1969; FOR ORIGINAL SEE T74008)
74010	**THERMAL CONDUCTIVITY OF PARTICULATE BASALT AS A FUNCTION OF DENSITY IN SIMULATED LUNAR AND MARTIAN ENVIRONMENTS.** FOUNTAIN, J. A. WEST, E. A. J. GEOPHYS. RES. 75 (20), 4063-9, 1970.
74011	**THE HEAT CAPACITY AND THERMAL CONDUCTIVITY OF APOLLO 11 LUNAR ROCKS 10017 AND 10046 AT LIQUID HELIUM TEMPERATURES.** MORRISON, J. A. NORTON, P. R. J. GEOPHYS. RES. 75 (32), 6553-7, 1970.
74012	**THERMAL CONDUCTIVITY OF ROCKS FROM MEASUREMENTS ON FRAGMENTS AND ITS APPLICATION TO HEAT FLOW DETERMINATIONS.** SASS, J. H. LACHENBRUCH, A. H. MUNROE, R. J. J. GEOPHYS. RES. 76 (14), 3391-3401, 1971.
74013	**HEAT FLOW NEAR MAJOR STRIKE SLIP FAULTS IN CALIFORNIA.** HENYEY, T. L. WASSERBURG, G. J. J. GEOPHYS. RES. 76 (32), 7924-46, 1971.
74014	**MEASUREMENT OF THE THERMAL CONDUCTIVITY AND THERMAL DIFFUSIVITY OF NATURAL ROCKS AT LOW TEMPERATURE.** BIA, P. COMBARNOUS, M. J. PHYS. SCI. INSTR. 3 E, 536-40, 1970.
74019	**EXPERIMENTAL STUDY RELATING THERMAL CONDUCTIVITY TO THERMAL PIERCING OF ROCKS.** MIRKOVICH, V. V. INT. J. ROCK MECH. MIN. SCI. 5 (3), 205-18, 1968.
74048	**THERMAL CONDUCTIVITY OF RAW MATERIALS DURING ROASTING.** MIKHAILOV, V. N. CHISTYAKOVA, A. A. TSEMENT (8), 12-13, 1973.
74070	**THERMAL AND THEMOMECHANICAL PROPERTIES OF LEAD SINTER.** KOVGAN, P. A. EVDOKIMENKO, A. I. TSVET. METAL 43 (2), 13-, 1970. (FOR ENGLISH TRANSLATION SEE TPRC NO. 74071)
74071	**THERMAL AND THERMOMECHANICAL PROPERTIES OF LEAD SINTER.** KOVGAN, P. A. EVDOKIMENKO, A. I. SOV. J. NON-FERROUS METALS 11 (2), 13-6, 1970. (ENGLISH TRANSLATION OF TSVET. METAL, 43 (2), 13-, 1970; FOR ORIGINAL SEE TPRC NO. 74070)
74074	**EFFECT OF FLUORIDE ADDITIVES ON THE SINTERING OF SLIME CHARGE.** NURMAGAMBETOV, KH. N. SHCHERBAN, S. A. NURKEEV, S. S. TSVET. MET. 43 (6), 33-, 1970. (FOR ENGLISH TRANSLATION SEE TPRC NO. 74075)
74075	**EFFECT OF FLUORIDE ADDITICES ON THE SINTERING OF SLIME CHARGE.** NURMAGAMBETOV, KH. N. SHCHERBAN, S. A. NURKEEV, S. S. SOV. J. NON-FERROUS METALS 11 (6), 37-9, 1970. (ENGLISH TRANSLATION OF TSVET. MET., 40 (6), 33-, 1970; FOR ORIGINAL SEE TPRC NO. 74074)

TPRC Number	Bibliographic Citation

74170 **EFFECT OF WEAKLY REFLECTING COMPONENTS ON THE COEFFICIENT OF REFLECTION OF MIXED POWDERS.**
RAICHENKO, A. I. NIKITIN, YU. I.
BONDARENKO, B. I. PRIMACHUK, V. L.
ZHENNI-MAISKAYA, L. O. STYSKIN, B. S.
POROSH. MET.
11 (2), 101-3, 1972.
(FOR ENGLISH TRANSLATION SEE T74171)

74171 **EFFECT OF WEAKLY REFLECTING COMPONENTS ON THE COEFFICIENT OF REFLECTION OF MIXED POWDERS.**
RAICHENKO, A. I. NIKITIN, YU. I.
BONDARENKO, B. I. PRIMACHUK, V. L.
ZHENNI-MAISKAYA, L. O. STYSKIN, B. S.
SOV. POWDER MET. METAL CERAM.
11 (2), 165-6, 1972.
(ENGLISH TRANSLATION OF POROSH. MET., 11 (2), 101-3, 1972; FOR ORIGINAL SEE T74170)

74222 **THERMAL AND ELECTRICAL CONDUCTIVITIES OF SANDSTONE ROCKS AND OCEAN SEDIMENTS.**
HUTT, J. R. BERG, J. W., JR.
GEOPHYSICS
33 (3), 489-500, 1968.

74223 **PROBLEM OF THE THERMAL CONDUCTIVITY OF COKING DINAS.**
STRELOV, K. K. PEREPELITSYN, V. A.
BICHURINA, A. A.
REFRACTORIES
(2), 116-21, 1973.
(ENGLISH TRANSLATION OF OGNEUPORY, (2), 48-52, 1973; FOR ORIGINAL SEE T70496)

74262 **FIELD SPECTROSCOPY FOR MULTISPECTRAL REMOTE SENSING: AN ANALYTICAL APPROACH.**
LONGSHAW, T. G.
APPL. OPT.
13 (6), 1487-93, 1974.

74265 **A THEORETICAL MODEL FOR LUNAR SURFACE MATERIAL THERMAL CONDUCTIVITY.**
KHADER, M. S. VACHON, R. I.
AMERICAN SOCIETY OF MECHANICAL ENGINEERS AND AMERICAN INSTITUTE OF CHEMICAL ENGINEERS, HEAT TRANSFER CONFERENCE, ATLANTA, GA.
35PP., 1973.
(ASME-ANNUAL MEETING-73-HT-35)

74269 **ABSORPTION SPECTRA OF LUNAR SECTIONS FROM DIFFERENT LUNAR AREAS.**
ANTIPOVA-KARATAEVA, I. I. STACHEEV, IU. I.
TARASOV, L. S.
SPACE RESEARCH XIII; PROC. OF THE FIFTEENTH PLENARY MEETING
2, 997-1000, 1973.

74283 **MEASUREMENT OF THE THERMAL DIFFUSIVITY OF URANIUM PLUTONIUM DI OXIDES. STUDIES AS A FUNCTION OF STOICHIOMETRY AND PLUTONIUM CONTENTS.**
WEILBACHER, J.-C.
PARIS UNIVERSITY, FRANCE, PH.D. THESIS
111PP., 1972.
(FRNC-TH-326, N74-13668)

74326 **THERMAL EXPANSION OF DIOPSIDE.**
DEGANELLO, S.
Z. KRISTALLOGR., KRISTALLGEOM., KRISTALLPHYS., KRISTALLCHEM.
137 (2-3), 127-31, 1973.

74339 **A CALORIMETRIC INVESTIGATION OF MOISTURE IN TEXTILE FIBERS.**
MAGNE, F. C. PORTAS, H. J. WAKEHAM, H.
J. AMER. CHEM. SOC.
69, 1896-1902, 1947.

74368 **EFFECT OF STOPPAGES ON THE THERMAL STATE OF THE LINING OF A 130 T OXYGEN-BLOWN CONVERTER.**
MARON, V. D. RYBALKO, L. G.
CHEREPANOV, K. A. KALASHINIKOVA, V. K.
IZV. VYSSH. UCHEB. ZAVED., CHERN. MET.
(4), 142-6, 1971.
(FOR ENGLISH TRANSLATION SEE TPRC NO. 74369)

74369 **EFFECT OF STOPPAGES ON THE THERMAL STATE OF THE LINING OF A 130 T OXYGEN-BLOWN CONVERTER.**
MARON, V. D. RYBALKO, L. G.
CHEREPANOV, K. A. KALASHINIKOVA, V. K.
STEEL IN THE USSR
1 (4), 279-81, 1971.
(ENGLISH TRANSLATION OF IZV. VYSSH. UCHEB. ZAVED., CHERN. MET., (4), 142-6, 1971; FOR ORIGINAL SEE TPRC NO. 74368)

74374 **HELIUM-3 QUASIPARTICLES IN THE COLLISIONLESS REGIME. I. DIRECT MEASUREMENT OF THE FERMI VELOCITY BY HEAT CONDUCTION.**
BETTS, D. S. BREWER, D. F. HAMILTON, R. S.
J. LOW TEMP. PHYS.
14 (3-4), 331-47, 1974.

74406 **ANISOTROPY OF SOME ANHARMONIC EFFECTS IN TOPAZ.**
GESHKO, E. I. MIKHAL'CHENKO, V. P.
SHARLAI, B. M.
UKR. FIZ. ZH. (RUSS. ED.)
19 (2), 329-31, 1974.

74420 **THERMAL EXPANSION AT LOW TEMPERATURES OF URANIUM DIOXIDE AND URANIUM DIOXIDE/THORIUM OXIDE.**
WHITE, G. K. SHEARD, F. W.
J. LOW TEMP. PHYS.
14 (5-6), 445-57, 1974.

74443 **FAR INFRARED DIELECTRIC DISPERSION IN ANTIMONY(III) SULFIDE, BISMUTH(III) SULFIDE, AND ANTIMONY(III) SELENIDE.**
PETZELT, J. GRIGAS, J.
FERROELECTRICS
5 (1-2), 59-68, 1973.

74495 **HIGH THERMAL CONDUCTIVITY MICROWAVE SUBSTRATES.**
POPE, B. J. HORTON, M. D. BOWMAN, L. S.
HALL, H. T. ADANIYA, H. N.
U. S. ARMY ELECTRONICS COMMAND, FORT MONMOUTH, N. J.
29PP., 1973.
(ECOM-0-89-1)

74498 **REFERENCE MATERIALS AT LOW TEMPERATURES.**
WHITE, G. K.
AIP (AMER. INST. PHYS.), CONF. PROC.
(17), 1-7, 1974.

74504 **THERMAL EXPANSION BEHAVIOR OF INTACT AND THERMALLY FRACTURED MINE ROCKS.**
THIRUMALAI, K. DEMOU, S. G.
AIP (AMER. INST. PHYS.), CONF. PROC.
(17), 60-71, 1974.

74507 **THERMAL EXPANSION OF A BOROSILICATE GLASS FROM 80 TO 680 K - STANDARD REFERENCE MATERIAL 731.**
HAHN, T. A. KIRBY, R. K.
AIP (AMER. INST. PHYS.), CONF. PROC.
(17), 93-101, 1974.

74511 **DIFFERENTIAL DILATOMETRY A POWERFUL TOOL.**
PLUMMER, W. A.
AIP (AMER. INST. PHYS.), CONF. PROC.
(17), 147-58, 1974.

74515 **INFLUENCE OF POISSON'S RATIO ON THE COMPOSITE EXPANSION OF LAMINATED PLATES AND CYLINDERS.**
GULATI, S. T. PLUMMER, W. A.
AIP (AMER. INST. PHYS.), CONF. PROC.
(17), 196-206, 1974.

74521 **THERMAL EXPANSION OF SOLIDS AT LOW TEMPERATURES.**
WHITE, G. K. COLLINS, J. G. SHEARD, F. W.
SMITH, T. F.
AIP (AMER. INST. PHYS.), CONF. PROC.
(17), 278-9, 1974.

74522 **DIMENSIONAL STABILITY OF FUSED SILICA AND SEVERAL ULTRALOW EXPANSION MATERIALS.**
JACOBS, S. F. NORTON, M. A.
AIP (AMER. INST. PHYS.), CONF. PROC.
(17), 280-96, 1974.

74526 **COORDINATION OF THE COBALT DIPOSITIVE IONS IN FINELY GRAINED SOLID SOLUTIONS COBALT(II) OXIDE-MAGNESIUM OXIDE.**
DYREK, K. SHVETS, V. A.
BULL. ACAD. POL. SCI., SER. SCI. CHIM.
22 (4), 315-19, 1974.

74535 **PROPERTIES AND STRUCTURE OF GLASSES IN THE SYSTEM LEAD MONOXIDE-GERMANIUM DIOXIDE-SILICON DIOXIDE.**
TOPPING, J. A. FUCHS, P. MURTHY, M. K.
J. AMER. CERAM. SOC.
57 (5), 205-8, 1974.

74536 **PROPERTIES AND STRUCTURE OF GLASSES IN THE SYSTEM LEAD MONOXIDE-GERMANIUM DIOXIDE.**
TOPPING, J. A. HARROWER, I. T. MURTHY, M. K.
J. AMER. CERAM. SOC.
57 (5), 209-12, 1974.

74548 **LOW TEMPERATURE DEFORMATION TEST FOR TAR-BONDED BASIC REFRACTORIES.**
CARNIGLIA, S. C. MARTINEK, C. A.
NEELY, J. E.
AMER. CERAM. SOC. BULL.
53 (7), 539-42, 547, 1974.

74549 **CERAMIC CUTTING TOOLS.**
WHITNEY, E. D.
POWDER MET. INTERN.
6 (2), 73-6, 1974.

74556 **X-RAY DIFFRACTION STUDY OF THE THERMAL EXPANSION OF KAOLINITE AND ILLITE AND THEIR INTERSALATED COMPLEXES WITH ALKALI ACETATES.**
BORA, M. N. HATIBARUA, J. MAHANTA, P. C.
HIGH TEMP. HIGH PRESSURES
5 (1), 47-53, 1973.

74558 **STUDY OF SOME THERMOPHYSICAL CHARACTERISTICS OF METALLURGICAL RAW MATERIALS.**
BRATCHIKOV, S. G.
HIGH TEMP. HIGH PRESSURES
5 (1), 85-90, 1973.

TPRC Number	Bibliographic Citation
74596	**THERMAL CONDUCTION OF DEPOSITS OF IRON OXIDES.** RASSOKHIN, N. G. KABANOV, L. P. TEVLIN, S. A. TERSIN, V. A. THERM. ENG., USSR 20 (9), 16–9, 1973. (ENGLISH TRANSLATION OF TEPLOENERGETIKA, 20 (9), 12–5, 1973; FOR ORIGINAL SEE TPRC NO. 72486)
74602	**ION DEPLETION OF GLASS AT A BLOCKING ANODE: II, PROPERTIES OF ION-DEPLETED GLASSES.** CARLSON, D. E. HANG, K. W. STOCKDALE, G. F. J. AMER. CERAM. SOC. 57 (7), 295–300, 1974.
74606	**TRANSIENT TEMPERATURE RESPONSE OF COMPOSITE SLABS.** SUGIYAMA, S. NISHIMURA, M. WATANABE, H. INT. J. HEAT MASS TRANSFER 17 (8), 875–83, 1974.
74621	**TEMPERATURE DEPENDENCE OF THE FORBIDDEN-BAND WIDTH OF BARIUM (0.50) STRONTIUM (0.50) NIOBIUM OXIDE SINGLE CRYSTALS IN THE PHASE TRANSITION REGION.** SAVITSKII, V. G. KOVTUN, R. N. ALEKSYUK, V. E. PROTSAKH, P. F. FIZ. TVERD. TELA 15 (12), 3696–8, 1973. (FOR ENGLISH TRANSLATION SEE TPRC NO. 74622)
74622	**TEMPERATURE DEPENDENCE OF THE FORBIDDEN-BAND WIDTH OF BARIUM (0.50) STRONTIUM (0.50) NIOBIUM OXIDE SINGLE CRYSTALS IN THE PHASE TRANSITION REGION.** SAVITSKII, V. G. KOVTUN, R. N. ALEKSYUK, V. E. PROTSAKH, P. F. SOV. PHYS. SOLID STATE 15 (12), 2464, 1974. (ENGLISH TRANSLATION OF FIZ. TVERD. TELA, 15 (12), 3696–8, 1973; FOR ORIGINAL SEE TPRC NO. 74621)
74674	**ABSORPTION SPECTRUM OF BENITOITE IN THE NEAR INFRARED.** BERAN, A. TSCHERMAKS MINERAL. PETROGR. MITT. 21 (1), 47–51, 1974.
74696	**CORRELATION FUNCTIONS FOR PREDICTING PROPERTIES OF HETEROGENEOUS MATERIALS. IV. EFFECTIVE THERMAL CONDUCTIVITY OF TWO-PHASE SOLIDS.** CORSON, P. B. J. APPL. PHYS. 45 (7), 3180–2, 1974.
74722	**THERMOPHYSICAL PROPERTIES OF APOLLO 14 FINES.** CREMERS, C. J. AMER. INST. AERONAUTICS ASTRONAUTICS, AEROSPACE SCI. MEET. 8PP., 1974. (AIAA-PAPER-74-116)
74724	**CARBON FIBER REINFORCED THERMOPLASTICS.** THEBERGE, J. ARKLES, B. ROBINSON, R. WIDE WORLD OF REINFORCED PLASTICS, PROC. ANN. CONF. 20-D, 1 TO 14PP., 1974.
74725	**THERMAL EXPANSION OF HIGH MODULUS GRAPHITE FIBER/EPOXY COMPOSITES.** KALNIN, I. L. WIDE WORLD OF REINFORCED PLASTICS, PROC. ANN. CONF. 21-C,1 TO 8,PP., 1974.
74727	**REGULAR NORMAL TRANSMITTANCE OF ROUGHENED DIELECTRIC INTERFACES - COMPARISON OF THEORY AND DATA.** SMITH, A. M. SNYDER, E. F. AMERICAN INSTITUE OF AERONAUTICS AND ASTRONAUTICS, AEROSPACE SCIENCES MEETING 9PP., 1974. (AIAA-PAPER-74-118)
74804	**THERMOPHYSICAL PROPERTIES OF STEEL SCALE AT DIFFERENT TEMPERATURES.** SEVERDENKO, V. P. MAKUSHOK, E. M. RAVIN, A. N. VESTSI AKAD. NAVUK BELARUS. SSR, SER. FIZ.-TEKH. NAVUK (1), 33–7, 1974.
74813	**CONVECTION IN THE EARTH'S MANTLE.** TURCOTTE, D. L. TORRANCE, K. E. HSUI, A. T. J. METHODS COMPUTATIONAL PHYS. 13, 431–54, 1973. (AD-776 852, AFOSR-TR-74-0408)
74816	**A STUDY OF SODIUM (20-TITANIUM (02)-SILICON (02)) GLASSES BY INFRARED SPECTROSCOPY.** MANGHNANI, M. H. HAWAII INST. OF GEOPHYSICS, HONOLULU, HAWAII 20PP., 1974. (AD-779 487)
74859	**THERMOPHYSICAL PROPERTIES OF HIGH-LEAD-CONTAINING GLASSES IN THE ANOMALOUS REGION.** RALKO, A. V. RODNIKOVA, V. V. PLEMYANNIKOV, N. N. GORODOV, V. S. SICKERT, G. DOPOV. AKAD. NAUK UKR. RSR, SER. 36 B (2), 147–50, 1974.
74868	**CHANGE IN THE ETHALPY AND SPECIFIC HEAT OF BLAST-FURNACE SLAGS DURING THEIR HEAT TREATMENT.** OSINOVSKIKH, L. L. BRATCHIKOV, S. G. YUR'EV, B. P. IZV. VYSSH. UCHEB. ZAVED., CHERN. MET. (2), 35–9, 1974.
74933	**REGULAR NORMAL TRANSMITTANCE OF ROUGHENED DIELECTRIC INTERFACES: COMPARISON OF THEORY AND DATA.** SMITH, A. M. ARNOLD ENGINEERING DEVELOPMENT CENTER, ARNOLD AIR FORCE STATION, TENNESSEE 23PP., 1974. (AEDC-TR-74-35)
74958	**THERMAL CONDUCTIVITY OF FUEL ELEMENTS. II. MIXED OXIDES.** CHAMERO FERRER, A. GISPERT BENACH, M. ENERG. NUCL. (MADRID) 17 (86), 427–35, 1973.
75009	**ENERGY TRANSMISSION THROUGH A MEDIUM WITH LOW OPTICAL DENSITY.** KOSTYLEV, V. M. KOMAROVSKAYA, N. V. J. ENG. PHYS. 22 (5), 636–9, 1974. (ENGLISH TRANSLATION OF INZH. FIZ. ZH., 22 (5), 907–12, 1972; FOR ORIGINAL SEE TPRC NO. 68895)
75013	**DETERMINATION OF THE SPECIFIC HEAT OF ORE-COAL PELLETS DURING METALLIZATION.** STOROZHEV, YU. I. TELEGIN, A. S. ZAVARZIN, V. P. STEEL USSR 1 (5), 331–2, 1971. (ENGLISH TRANSLATION OF IZV. VYSSH. UCHEB. ZAVED., CHERN. MET., 14 (5), 27–9, 1971; FOR ORIGINAL SEE TPRC NO. 62563)
75014	**THERMOPHYSICAL PROPERTIES OF A BED OF ORE-COAL PELLETS.** STOROZHEV, YU. I. TELEGIN, A. S. KUDRYAVTSEV, V. S. PCHELKIN, S. A. STEEL, USSR 1 (10), 763–4, 1974. (ENGLISH TRANSLATION OF IZV. VYSSH. UCHEB. ZAVED. FIZ., 14 (10), 30–2, 1971; FOR ORIGINAL SEE TPRC NO. 44543)
75083	**THERMAL CONDUCTIVITY OF MOIST BULK MATERIALS.** TANAYEVA, S. A. HEAT TRANSFER-SOV. RES. 6 (2), 107–11, 1974. (ENGLISH TRANSLATION OF NIZKOTEMP. ENERGETIKA I MASSOOBMEN VAKUUM, 141–7, 1973; FOR ORIGINAL SEE TPRC NO. 77333)
75084	**INVESTIGATION OF THE EFFECTIVE THERMAL CONDUCTIVITY OF POROUS MATERIALS.** TANAYEVA, S. A. HEAT TRANSFER-SOV. RES. 6 (2), 112–8, 1974. (ENGLISH TRANSLATION OF NIZKOTEMP. ENERGETIKA I MASSOOBMEN VAKUUM, 148–58, 1973; FOR ORIGINAL SEE TPRC NO. 77332)
75085	**OPTIMUM SHAPES AND DIMENSIONS OF SPECIMENS FOR DETERMINATION OF THERMOPHYSICAL PROPERTIES.** DOMOROD, L. S. HEAT TRANSFER-SOV. RES. 6 (2), 119–24, 1974. (ENGLISH TRANSLATION OF NIZKOTEMP. ENERGETIKA I MASSOOBMEN VAKUUM, 159–66, 1973; FOR ORIGINAL SEE TPRC NO. 77334)
75087	**EXAMPLES OF CHEMICALLY ACTIVATED SINTERING.** HEIMKE, G. POWDER MET. INTERN. 6 (3), 133–6, 1974.
75122	**DIRECTIONAL SOLIDIFICATION OF THE ZIRCONIUM OXIDE-MAGNESIUM OXIDE EUTECTIC.** KENNARD, F. L. BRADT, R. C. STUBICAN, V. S. J. AMER. CERAM. SOC. 57 (10), 428–31, 1974.
75123	**VISCOSITY AND THERMAL EXPANSION OF ALKALI GERMANATE GLASSES.** SHELBY, J. E. J. AMER. CERAM. SOC. 57 (10), 436–9, 1974.
75168	**CORRELATION BETWEEN THE IMPACT STRENGTH AND THERMAL EXPANSION OF CAST REFRACTORY BLOCKS.** ROMWALTER, A. VASSEL, K. R. GURUBI, L. FEMIP. KUT. INTEZ. KOZLEM. 9, 131–8, 1971.
75178	**SOLID SOLUTIONS IN THE HAFNIUM OXIDE-TITANIUM DIOXIDE SYSTEM.** KOCHEREGIN, S. B. KUZNETSOV, A. K. IZV. AKAD. NAUK SSSR, NEORG. MATER. 10 (2), 297–9, 1974. (FOR ENGLISH TRANSLATION SEE TPRC NO. 76900)

TPRC Number	Bibliographic Citation
75182	ANALYTICAL EXPRESSIONS FOR ENTHALPY AND HEAT CAPACITY FOR URANIUM (DIOXIDE)-PLUTONIUM DIOXIDE. GIBBY, R. L. LEIBOWITZ, L. KERRISK, J. F. CLIFTON, D. G. J. NUCL. MATER. 50 (2), 155-61, 1974.
75183	THERMAL CONDUCTION IN GLASSES AND POLYMERS AT LOW TEMPERATURES. MORGAN, G. J. SMITH, D. J. PHYS. 7 C (4), 649-64, 1974.
75189	STRUCTURE OF THE ABSORPTION SPECTRUM OF DIOPSIDES. BOKSHA, O. N. VARINA, T. M. KOSTYUKOVA, I. G. KRISTALLOGRAFIYA 19 (2), 392-4, 1974. (FOR ENGLISH TRANSLATION SEE TPRC NO. 76546)
75192	PROPERTIES OF SYNTHETIC MULLITE CERAMICS IN RELATION TO THEIR PURITY. BAKUNOV, V. S. VU, N. C. AGAFONOVA, M. I. RYABTSEV, K. I. POLUBOYARINOV, D. N. OGNEUPORY (4), 32-8, 1974. (FOR ENGLISH TRANSLATION SEE TPRC NO. 76872)
75201	THERMAL CONDUCTIVITY OF BORON OXIDE. LEZHENIN, F. F. KORNEICHUK, A. A. BERZHATYI, V. I. GRITSAENKO, V. P. TEPLOFIZ. TEPLOTEKH. 25, 94-6, 1973.
75213	OPTICAL PROPERTIES OF MAGNESIUM-ALUMINUM VERNEUIL-GROWN SPINEL IN THE INFRARED SPECTRAL REGION. RIEDE, V. SOBOTTA, H. EXP. TECH. PHYS. 22 (1), 17-23, 1974.
75232	REFLECTANCE AND OPTICAL CONSTANTS FOR CER-VIT FROM 250 TO 1050 A. OSANTOWSKI, J. F. J. OPT. SOC. AMER. 64 (6), 834-8, 1974.
75244	HEAT CAPACITY OF MINERAL IMPURITIES AND ASH OF COALS. AGROSKIN, A. A. GLEIBMAN, V. B. GONCHAROV, E. I. YAKUNIN, V. P. KOKS KHIM. (2), 3-4, 1974.
75274	SPECIFIC HEATS OF TYPICAL SOLID FINELY DIVIDED MATERIALS IN THE 123-293 K RANGE. DUSHCHENKO, V. P. ADRIANOV, V. M. KUCHERUK, I. M. GRISHCHENKO, E. V. KOSTYUK, N. S. TEPLOFIZ. TEPLOTEKH. 25, 30-3, 1973.
75284	FUSED BASIC REFRACTORIES IN THE ARGON-OXYGEN-DECARBURIZATION PROCESS. WHITWORTH, D. A. JACKSON, F. D. PATRICK, R. F. AMER. CERAM. SOC. BULL. 53 (11), 804-8, 1974.
75296	ANATASE OF TAPIRA (MINAS GERAIS, BRAZIL). CASSEDANNE, J. P. CASSEDANNE, J. O. BULL. SOC. FR. MINERAL. CRISTALLOGR. 96 (4-5), 316-18, 1973.
75300	NEW MAGNESIUM CARBONATE HYDRATE MINERAL, FROM YOSHIKAWA, AICHI PREFECTURE, JAPAN. SUZUKI, J. ITO, M. GANSEKI KOBUTSU KOSHO GAKKAISHI 68 (11), 353-61, 1973.
75330	IO: A SURFACE EVAPORITE DEPOSIT. FANALE, F. P. JOHNSON, T. V. MATSON, D. L. SCIENCE 186 (4167), 922-5, 1974.
75361	APPROXIMATION OF THE FUNCTIONAL DEPENDENCE OF THE HEAT CAPACITY OF PEAT COMBUSTION PRODUCTS ON SOME VARIABLE FACTORS. OPMAN, YA. S. VENZEL, E. F. IZV. VYSSH. UCHEB. ZAVED., ENERG. 17 (1), 145-7, 1974.
75441	THERMAL CONDUCTIVITIES OF POROUS SOLIDS. SAEGUSA, T. KAMATA, K. IIDA, Y. WAKAO, N. HEAT TRANSFER JAP. RES. 3 (2), 47-52, 1974.
75444	FABRICATION OF CHANNEL OPTICAL WAVEGUIDES IN GLASS BY CONTINUOUS WAVE LASER HEATING. PAVLOPOULOS, T. G. CRABTREE, K. J. APPL. PHYS. 45 (11), 4964-8, 1974.
75449	THERMODYNAMIC ANALYSIS OF REACTIONS OF SILICATE FORMATION IN THE SYSTEM LEAD OXIDE-SILICON OXIDE. MATVEEV, M. A. MATVEEV, G. M. EL'KIN, G. B. ZH. PRIKL. KHIM. 43 (2), 267-73, 1970. (FOR ENGLISH TRANSLATION SEE TPRC NO. 75450)
75450	THERMODYNAMIC ANALYSIS OF REACTIONS OF SILICATE FORMATION IN THE SYSTEM LEAD OXIDE-SILICON OXIDE. MATVEEV, M. A. MATVEEV, G. M. EL'KIN, G. B. J. APPL. CHEM., USSR 43 (2), 274-9, 1970. (ENGLISH TRANSLATION OF ZH. PRIKL. KHIM., 43 (2), 267-73, 1970; FOR ORIGINAL SEE TPRC NO. 75449)
75457	MAGNETIC PROPERTIES OF LEAD SILICATE GLASSES CONTAINING IRON OXIDES. VARGIN, V. V. ZARUBINA, T. V. STEPANOV, S. A. ZH. PRIKL. KHIM. 43 (6), 1225-9, 1970. (FOR ENGLISH TRANSLATION SEE TPRC NO. 75458)
75458	MAGNETIC PROPERTIES OF LEAD SILICATE GLASSES CONTAINING IRON OXIDES. VARGIN, V. V. ZARUBINA, T. V. STEPANOV, S. A. J. APPL. CHEM., USSR 43 (6), 1235-7, 1970. (ENGLISH TRANSLATION OF ZH. PRIKL. KHIM., 43 (6), 1225-9, 1970; FOR ORIGINAL SEE TPRC NO. 75457)
75466	INFLUENCE OF BORIC ANHYDRIDE ON THE STRUCTURE OF COLORING IONS IN WHITE PORTLAND CEMENT CLINKERS. GRACH'YAN, A. N. ROTYCH, N. V. J. APPL. CHEM., USSR 43 (9), 1920-2, 1970. (ENGLISH TRANSLATION OF ZH. PRIKL. KHIM., 43 (9), 1894-7, 1970; FOR ORIGINAL SEE TPRC NO. 75465)
75467	INFLUENCE OF MOISTURE ON THE ADSORPTION AND RHEOLOGICAL PROPERTIES OF ZINC YELLOW. IVANOVA, M. I. ERMILOV, P. I. ZH. PRIKL. KHIM. 43 (9), 1959-63, 1970. (FOR ENGLISH TRANSLATION SEE TPRC NO. 75468)
75468	INFLUENCE OF MOISTURE ON THE ADSORPTION AND RHEOLOGICAL PROPERTIES OF ZINC YELLOW. IVANOVA, M. I. ERMILOV, P. I. J. APPL. CHEM., USSR 43 (9), 1980-3, 1970. (ENGLISH TRANSLATION OF ZH. PRIKL. KHIM., 43 (9), 1959-63, 1970; FOR ORIGINAL SEE TPRC NO. 75467)
75483	STUDY OF THE PRODUCTS OF HYDROTHERMAL HARDENING OF ALUMINA-SILICA MIXTURES. VED', E. I. LITVINOVA, Z. S. ZH. PRIKL. KHIM. 44 (5), 1163-6, 1971. (FOR ENGLISH TRANSLATION SEE TPRC NO. 75484)
75484	STUDY OF THE PRODUCTS OF HYDROTHERMAL HARDENING OF ALUMINA-SILICA MIXTURES. VED', E. I. LITVINOVA, Z. S. J. APPL. CHEM., USSR 44 (5), 1177-80, 1971. (ENGLISH TRANSLATION OF ZH. PRIKL. KHIM., 44 (5), 1163-6, 1971; FOR ORIGINAL SEE TPRC NO. 75483)
75485	INFRARED SPECTROSCOPIC STUDY OF THE MECHANISM OF ACTION OF ADDED FLUORIDES DURING SINTERING OF ALUMINOUS MIXES. NURMAGAMBETOV, KH. N. SHCHERBAN, S. A. ZH. PRIKL. KHIM. 44 (7), 1467-73, 1971. (FOR ENGLISH TRANSLATION SEE TPRC NO. 75486)
75486	INFRARED SPECTROSCOPIC STUDY OF THE MECHANISM OF ACTION OF ADDED FLUORIDES DURING SINTERING OF ALUMINOUS MIXES. NURMAGAMBETOV, KH. N. SHCHERBAN, S. A. J. APPL. CHEM., USSR 44 (7), 1491-6, 1971. (ENGLISH TRANSLATION OF ZH. PRIKL. KHIM., 44 (7), 1467-73, 1971; FOR ORIGINAL SEE TPRC NO. 75485)
75495	STUDY OF THE THERMAL PORE-FORMATION PROCESS IN PERLITE BY QUANTITATIVE THERMOGRAPHY. ROKHVARGER, A. E. MATVEEV, G. M. ZH. PRIKL. KHIM. 44 (8), 1734-9, 1971. (FOR ENGLISH TRANSLATION SEE TPRC NO. 75496)
75496	STUDY OF THE THERMAL PORE-FORMATION PROCESS IN PERLITE BY QUANTITATIVE THERMOGRAPHY. ROKHVARGER, A. E. MATVEEV, G. M. J. APPL. CHEM., USSR 44 (8), 1756-61, 1971. (ENGLISH TRANSLATION OF ZH. PRIKL. KHIM., 44 (8), 1734-9, 1971; FOR ORIGINAL SEE TPRC NO. 75495)

TPRC Number	Bibliographic Citation
75511	**INFLUENCE OF THE SOLID PHASE-MELT TRANSITION ON THE EMISSION SPECTRA OF SLAGS IN THE REGION OF STRONG ABSORPTION.** KURBATOV, G. A. KALMYKOV, V. A. ZH. PRIKL. KHIM. 45 (1), 191-2, 1972. (FOR ENGLISH TRANSLATION SEE TPRC NO. 75512)
75512	**INFLUENCE OF THE SOLID PHASE-MELT TRANSITION ON THE EMISSION SPECTRA OF SLAGS IN THE REGION OF STRONG ABSORPTION.** KURBATOV, G. A. KALMYKOV, V. A. J. APPL. CHEM., USSR 45 (1), 181-2, 1972. (ENGLISH TRANSLATION OF ZH. PRIKL. KHIM., 45 (1), 191-2, 1972; FOR ORIGINAL SEE TPRC NO. 75511)
75615	**ESR SPECTRUM OF THE O-HOLE CENTER IN NATURAL QUARTZ.** SAMOILOVICH, M. I. NOVOZHILOV, A. I. TSINOBER, L. I. MALYSHEV, A. G. ZH. STRUKT. KHIM. 14 (3), 455-8, 1973. (FOR ENGLISH TRANSLATION SEE TPRC NO. 75616)
75616	**ESR SPECTRUM OF THE O-HOLE CENTER IN NATURAL QUARTZ.** SAMOILOVICH, M. I. NOVOZHILOV, A. I. TSINOBER, L. I. MALYSHEV, A. G. J. STRUCT. CHEM., USSR 14 (3), 416-9, 1973. (ENGLISH TRANSLATION OF ZH. STRUKT. KHIM., 14 (3), 455-8, 1973; FOR ORIGINAL SEE TPRC NO. 75615)
75623	**INFLUENCE OF ANNEALING ON DIELECTRIC PROPERTIES OF GLASS-BONDED CERAMIC COATINGS.** AVGUSTINIK, A. I. ALEKSEEV, V. P. ZHURAVLEV, G. I. ZH. PRIKL. KHIM. 45 (11), 2411-6, 1972. (FOR ENGLISH TRANSLATION SEE TPRC NO. 75624)
75624	**INFLUENCE OF ANNEALING ON DIELECTRIC PROPERTIES OF GLASS-BONDED CERAMIC COATINGS.** AVGUSTINIK, A. I. ALEKSEEV, V. P. ZHURAVLEV, G. I. J. APPL. CHEM., USSR 45 (11), 2527-31, 1972. (ENGLISH TRANSLATION OF ZH. PRIKL. KHIM., 45 (11), 2411-6, 1972; FOR ORIGINAL SEE TPRC NO. 75623)
75634	**INFLUENCE OF ADDITION OF NITRITE-NITRATE MIXTURES ON WHITENESS OF WHITE PORTLAND CEMENT CLINKER.** GRACH'YAN, A. N. KUDRYAVTSEV, YU. D. ROTYCH, N. V. TARARIN, V. K. ZH. PRIKL. KHIM. 46 (4), 724-8, 1973. (FOR ENGLISH TRANSLATION SEE TPRC NO. 75635)
75635	**INFLUENCE OF ADDITION ON NITRITE-NITRATE MIXTURES ON WHITENESS OF WHITE PORTLAND CEMENT CLINKER.** GRACH'YAN, A. N. KUDRYAVTSEV, YU. D. ROTYCH, N. V. TARARIN, V. K. J. APPL. CHEM., USSR 46 (4), 774-7, 1973. (ENGLISH TRANSLATION OF ZH. PRIKL. KHIM., 46 (4), 724-8, 1973; FOR ORIGINAL SEE TPRC NO. 75634)
75640	**RELATIONSHIP BETWEEN PHASE COMPOSITION AND PROPERTIES OF GLASS-CERAMIC COATINGS IN THE SYSTEM R(2)O-ZNO-P(2)O(5)-SIO(2).** PEVZNER, B. Z. ZH. PRIKL. KHIM. 46 (5), 976-80, 1973. (FOR ENGLISH TRANSLATION SEE TPRC NO. 75641)
75641	**RELATIONSHIP BETWEEN PHASE COMPOSITION AND PROPERTIES OF GLASS-CERAMIC COATINGS IN THE SYSTEM R(2)O-ZNO-P(2)O(5)-SIO(2).** PEVZNER, B. Z. J. APPL. CHEM., USSR 46 (5), 1039-43, 1973. (ENGLISH TRANSLATION OF ZH. PRIKL. KHIM., 46 (5), 976-80, 1973; FOR ORIGINAL SEE TPRC NO. 75640)
75648	**DIFFUSION PROCESSES IN REDUCTION OF TRICALCIUM PHOSPHATE BY HYDROGEN IN PRESENCE OF ALUMINA.** KUSHNIR, S. V. DUBROVSKII, V. P. ZH. PRIKL. KHIM. 46 (8) 1647-52, 1973. (FOR ENGLISH TRANSLATION SEE TPRC NO. 75649)
75649	**DIFFUSION PROCESSES IN REDUCTION OF TRICALCIUM PHOSPHATE BY HYDROGEN IN PRESENCE OF ALUMINA.** KUSHNIR, S. V. DUBROVSKII, V. P. J. APPL. CHEM., USSR 46 (8), 1755-9, 1973. (ENGLISH TRANSLATION OF ZH. PRIKL. KHIM., 46 (8), 1647-52, 1973; FOR ORIGINAL SEE TPRC NO. 75648)
75668	**ZINC GERMANATES.** KUZNETSOVA, G. N. TREGUBOVA, T. M. ZH. PRIKL. KHIM. 46 (12), 2763-5, 1973. (FOR ENGLISH TRANSLATION SEE TPRC NO. 75669)
75669	**ZINC GERMANATES.** KUZNETSOVA, G. N. TREGUBOVA, T. M. J. APPL. CHEM., USSR 46 (12), 2921-3, 1973. (ENGLISH TRANSLATION OF ZH. PRIKL. KHIM., 46 (12), 2763-5, 1973; FOR ORIGINAL SEE TPRC NO. 75668)
75753	**ANODIC PROTON INJECTION IN GLASSES.** CARLSON, D. E. J. AMER. CERAM. SOC. 57 (11), 461-6, 1974.
75779	**PYROCHLORES. VIII. STUDIES OF SOME 2-5 PYROCHLORES AND RELATED COMPOUNDS AND MINERALS.** BRISSE, F. STEWART, D. J. SEIDL, V. KNOP, O. CAN. J. CHEM. 50 (22), 3648-66, 1972.
75835	**MOLECULAR INTERPRETATION OF THE RELATIONSHIP BETWEEN THE MECHANICAL AND THERMAL PROPERTIES OF GLASSES AND THE GLASS TRANSITION TEMPERATURE.** SANDITOV, D. S. BARTENEV, G. M. ZH. FIZ. KHIM. 47 (9), 2231- , 1973. (FOR ENGLISH TRANSLATION SEE TPRC NO. 75836)
75836	**MOLECULAR INTERPRETATION OF THE RELATIONSHIP BETWEEN THE MECHANICAL AND THERMAL PROPERTIES OF GLASSES AND THE GLASS TRANSITION TEMPERATURE.** SANDITOV, D. S. BARTENEV, G. M. RUSS. J. PHYS. CHEM. 47 (9), 1261-3, 1973. (ENGLISH TRANSLATION OF ZH. FIZ. KHIM., 47 (9), 2231- , 1973; FOR ORIGINAL SEE TPRC NO. 75835)
75868	**REFLECTIVITY CHARACTERISTICS OF SOME ORE MINERALS AND METALLIC ELEMENTS.** NAKHLA, F. M. SALEH, S. A. ABDEL-AAL, O. Y. CHEM. ERDE 32 (4), 279-93, 1973.
75869	**TEKTITES. IRON-57 MOESSBAUER SPECTRA AND INFRARED SPECTRA OF AUSTRALITE AND MOLDAVITE.** SHIMA, M. OKADA, A. CHISHITSUGAKU ZASSHI 79 (12), 787-91, 1973.
75883	**INFRARED SPECTRA OF AUTOGENIC ISOTROPIC SILICA.** SENKOVSKII, YU. N. GEOL. GEOKHIM. GORYUCH. ISKOP. 35, 14-19, 1973.
75943	**MATHEMATICAL MODELS FOR DETERMINING THE CHEMICAL COMPOSITION OF GLASS.** SADOVSKI, A. STROIT. MATER. SILIKAT. PROM. 15 (1), 16-18, 1974.
75950	**LOW-TEMPERATURE HEAT CAPACITY AND VELOCITY AND ABSORPTION OF ULTRASOUND OF SILICATE GLASSES CONTAINING A TUNGSTEN MICROIMPURITY.** ZHILOVA, A. N. RATOBYL'SKAYA, V. A. BOZHKO, YU. A. SIMONOVA, L. A. TR. MOSK. KHIM.-TEKHNOL. INST. 71, 13-15, 1972.
75951	**EFFECTIVE THERMAL DIFFUSIVITY OF A MIXED LAYER OF SOLID PARTICLES.** KORTUNOVA, T. P. LEKAE, V. M. TR. MOSK. KHIM.-TEKHNOL. INST. 73, 163-7, 1973.
75982	**HEAT TRANSFER TO FLOWING GRANULAR MEDIA.** SULLIVAN, W. N. SABERSKY, R. H. INT. J. HEAT MASS TRANSFER 18 (1), 97-107, 1975.
75986	**THERMAL CONDUCTIVITY OF FIBROUS MATERIALS. POSSIBILITIES OF CALCULATION.** BROECKERHOFF, P. BER. KERNFORSCHUNGSANLAGE JULICH 49PP., 1973.
75987	**OPTICAL PROPERTIES OF CHERVETITE, A LEAD PYROVANADATE.** CERVELLE, B. CESBRON, F. BULL. SOC. FR. MINERAL. CRISTALLOGR. 96 (6), 391-2, 1973.
76037	**PROPERTIES AND STRUCTURE OF BORIC OXIDE-GERMANIUM OXIDE GLASSES.** SHELBY, J. E. J. APPL. PHYS. 45 (12), 5272-7, 1974.
76046	**RELATION OF THE STRUCTURE AND PROPERTIES OF DOUBLE-PHASE GLASS.** USHAKOV, D. V. GILEV, I. S. MILYUKOV, E. M. IZV. AKAD. NAUK SSSR, NEORG. MATER. 10 (5), 905-8, 1974. (FOR ENGLISH TRANSLATION SEE TPRC NO. 76906)

TPRC Number	Bibliographic Citation
76069	**LIQUID OPTICAL ELEMENT OF PERTURBED TOTAL INTERNAL REFLECTION WITH A VARIABLE NUMBER OF REFLECTIONS.** SAIDOV, G. V. YUDOVICH, M. E. OPT. SPEKTROSK. 36 (6), 1216–17, 1974. (FOR ENGLISH TRANSLATION SEE T76440)
76072	**INFLUENCE OF HEAT TREATMENT ON THE OPTICAL ABSORPTION OF GLASSES CONTAINING IRON AND TITANIUM OXIDES.** VAN DE GRAAF, M. A. C. G. DE VRIES, K. J. BURGGRAAF, A. J. PHYS. CHEM. GLASSES 14 (3), 53–9, 1973.
76073	**PROPERTIES OF GLASSES IN THE SYSTEM LEAD(II) OXIDE–INDIUM(III) OXIDE–GERMANIUM(IV) OXIDE.** MURTHY, M. K. PHYS. CHEM. GLASSES 15 (1), 32–3, 1974.
76074	**THERMAL EXPANSION OF STISHOVITE.** ITO, H. KAWADA, K. AKIMOTO, S. PHYS. EARTH PLANET. INTER. 8 (3), 277–81, 1974.
76085	**PORCELAIN GLAZES. V. PRODUCTION OF GLAZES ON PORCELAIN.** MATEJKA, J. SKLAR KERAM. 24 (1), 17–22, 1974.
76092	**PROPERTIES OF LOW–MELTING GLASSES IN THE SODIUM OXIDE–ZINC OXIDE–BORON OXIDE–SILICA SYSTEM.** LYUTSKANOV, ST. KARADZHOVA, N. STROIT. MATER. SILIKAT. PROM. 15 (2), 25–8, 1974.
76107	**PARARAMMELSBERGITE FROM THE BERIKUL'SKII DEPOSIT, THE FIRST FIND IN THE USSR.** VINOGRADOVA, R. A. BORISHANSKAYA, S. S. EREMIN, N. I. VYAL'SOV, L. N. ZAP. VSES. MINERAL. OBSHCHEST. 103 (1), 128–31, 1974.
76112	**ATOMIC VIBRATIONS AND THERMAL EXPANSION OF SILICATES AT HIGH TEMPERATURES.** DEGANELLO, S. Z. KRISTALLOGR., KRISTALLGEOM., KRISTALLPHYS., KRISTALLCHEM. 139 (3–5), 297–316, 1974.
76135	**THERMODYNAMIC PROPERTIES OF MINERALS OF THE EPIDOTE GROUP.** KISELEVA, I. A. TOPOR, N. D. ANDREENKO, E. D. GEOKHIMIYA (4), 543–53, 1974. (FOR ENGLISH TRANSLATION SEE T91262)
76145	**THERMAL DIFFUSIVITY OF BORATE GLASSES.** HATTORI, M. FUKUMOTO, T. JAP. J. APPL. PHYS. 13 (6), 1013–14, 1974.
76173	**ANISTROPY OF THERMAL CONDUCTIVITY IN MAGNETITE SINGLE CRYSTALS.** SHAKHSHAEV, G. M. KAMILOV, I. K. PRIKL. FIZ. TVERD. TELA 115–17, 1973.
76179	**PRODUCTION AND PROPERTIES OF HEAT–INSULATING CORDIERITE ARTICLES WITH BURNABLE ADDITIVES.** LEPKOVA, D. GERASIMOV, E. STROIT. MATER. SILIKAT. PROM. 15 (2), 15–17, 1974.
76210	**DIFFRACTION LOSSES ASSOCIATED WITH TUNGSTEN LAMPS IN ABSOLUTE RADIOMETRY.** BOIVIN, L. P. APPL. OPT. 14 (1), 197–200, 1975.
76238	**THERMAL PROPERTIES OF WET POROUS MATERIAL NEAR FREEZING POINT: A TRANSIENT SIMULTANEOUS MEASURING METHOD.** KATAYAMA, K. HATTORI, M. KASAHARA, K. KIMURA, I. OKADA, M. ASHRAE JOURNAL 15 (4), 56–61, 1973.
76265	**DETERMINATION OF MECHANICAL AND HEAT PROPERTIES OF KUNASHAK AND ELENOVKA METEORITES.** MEDVEDEV, R. V. METEORITIKA 33, 100–4, 1974.
76266	**REFLECTIVITY OF SULFIDES FROM THE HYDROTHERMALLY ALTERED ROCKS OF THE URUPSKAYA GROUP DEPOSITS.** GLUKHOV, YU. YU. MINERAL.–PETROGR. GEOKHIM. ISSLED. SEV. KAVKAZE DONBASSE 35–8, 1972.
76273	**TARGET–SIGNATURE MEASUREMENTS.** AIR FORCE – AVIONICS LAB. AIR FORCE AVIONICS LAB., WRIGHT–PATTERSON AIR FORCE BASE, OHIO 214PP., 1968. (AFAL–TR–68–198)
76274	**USE OF QUARTZ CERAMICS AS A STANDARD FOR DIFFUSE REFLECTION.** BORODAI, S. P. BORODAI, F. YA. OPT.–MEKH. PROMST. 41 (5), 45–7, 1974. (FOR ENGLISH TRANSLATION SEE TPRC NO. 77217)
76328	**A SPECTROGRAPHIC INTERPRETATION OF THE SHOCK–PRODUCED COLOR CHANGE IN RHODONITE (MANGANESE SILICON OXIDE): THE SHOCK–INDUCED REDUCTION OF MANGANESE (III) TO MANGANESE (II).** GIBBONS, R. V. AHRENS, T. J. ROSSMAN, G. R. AMER. MINERAL. 59, 177–82, 1974. (AD–786 295, CONTRIB–2337, AROD–8783.3–EN)
76329	**SPECTRAL REFLECTANCE MEASUREMENTS.** RAINES, G. L. LEE, K. PHOTOGRAMMETRIC ENGINEERING 547–50, 1974. (AD–784 540, AROD–9843.5–EN)
76340	**TARGET SIGNATURE MEASUREMENTS GROUND–BASED FIELD.** ANDRYCHUK, D. PALMER, W. TEXAS INSTRUMENTS, INC., NON–SERIAL REPT. 99PP., 1968.
76371	**POLARIZED EMITTANCE, VOLUME 1: POLARIZED BIDIRECTIONAL REFLECTANCE WITH LAMBERTIAN OR NON–LAMBERTIAN DIFFUSE COMPONENTS.** MAXWELL, J. R. WEINER, S. F. ENVIRONMENTAL RESEARCH INST. OF MICHIGAN, ANN ARBOR 126PP., 1974. (AD–782 178, BRL–CR–154, ERIM–192500–1–T)
76412	**THE ALBEDO OF TITAN.** YOUNKIN, R. L. ICARUS (21), (MAR.), 219–229, 1974.
76427	**A MODEL OF HEAT TRANSFER IN GAS FLUIDIZED BEDS.** KUBIE, J. BROUGHTON, J. INT. J. HEAT MASS TRANSFER 18 (2), 289–99, 1975.
76440	**LIQUID OPTICAL COMPONENT THAT USES ATTENUATED TOTAL INTERNAL REFLECTION WITH A VARIABLE NUMBER OF REFLECTIONS.** SAIDOV, G. V. YUDOVICH, M. E. OPT. SPECTROS., USSR 36 (6), 707–8, 1974. (ENGLISH TRANSLATION OF OPT. SPEKTROSK., 36 (6), 1216–7, 1974; FOR ORIGINAL SEE TPRC NO. 76069)
76473	**OPERATION OF THE SECTIONAL CERAMIC WALLS IN AN MHD CHANNEL.** BELOGLAZOV, A. A. BURENKOV, D. K. VYSOTSKII, D. A. ZALKIND, V. I. KIRILLOV, V. V. PANOVKO, M. YA. ROMANOV, A. I. SMIRNOVA, L. G. SOKOLOV, YU. N. SHUMYATSKII, B. YA. TEPLOFIZ. VYS. TEMP. 12 (3), 605–13, 1974. (FOR ENGLISH TRANSLATION SEE TPRC NO. 76474)
76474	**OPERATION OF THE SECTIONAL CERAMIC WALLS IN AN MHD CHANNEL.** BELOGLAZOV, A. A. BURENKOV, D. K. VYSOTSKII, D. A. ZALKIND, V. I. KIRILLOV, V. V. PANOVKO, M. YA. ROMANOV, A. I. SMIRNOVA, L. G. SOKOLOV, YU. N. SHUMYATSKII, B. YA. HIGH TEMP. 12 (3), 520–7, 1974. (ENGLISH TRANSLATION OF TEPLOFIZ. VYS. TEMP., 12 (3), 605–13, 1974; FOR ORIGINAL SEE TPRC NO. 76473)
76486	**A CALORIMETRIC INVESTIGATION OF MOISTURE IN TEXTILE FIBERS.** MAGNE, F. C. PORTAS, H. J. WAKEHAM, H. PHYS. REV. 69, 1896–1902, 1947.
76522	**SURFACE AND SIZE EFFECTS IN THE SPECIFIC HEAT OF SOLIDS.** NOVOTNY, V. J. UNIVERSITY OF TORONTO, PH.D. THESIS NO PP. GIVEN, 1972. (AVAIL. NATL. LIBR. CANADA, OTTAWA, ONT.)
76546	**STRUCTURE OF THE ABSORPTION SPECTRUM OF CHROMDIOPSIDES.** BOKSHA, O. N. VARINA, T. M. KOSTYUKOVA, I. G. SOV. PHYS. CRYSTALLOGR. 19 (2), 241–2, 1974. (ENGLISH TRANSLATION OF KRISTALLOGRAFIYA, 19 (2), 392–4, 1974; FOR ORIGINAL SEE TPRC NO. 75189)

TPRC Number	Bibliographic Citation
76565	**PROPERTIES AND STRUCTURE OF GLASSES IN THE SYSTEM BISMUTH OXIDE - SILICON OXIDE - GERMANIUM OXIDE.** TOPPING, J. A. CAMERON, N. MURTHY, M. K. J. AMER. CERAM. SOC. 57 (12), 519-21, 1974.
76616	**ACOUSTO-OPTIC PROPERTIES OF DENSE FLINT GLASSES.** ESCHLER, H. WEIDINGER, F. J. APPL. PHYS. 46 (1), 65-70, 1975.
76618	**THERMAL EXPANSION OF SOME DIAMONDLIKE CRYSTALS.** SLACK, G. A. BARTRAM, S. F. J. APPL. PHYS. 46 (1), 89-98, 1975.
76620	**THERMAL EXPANSION OF MIXED-ALKALI GERMANATE GLASSES.** SHELBY, J. E. J. APPL. PHYS. 46 (1), 193-6, 1975.
76621	**MODIFICATION OF THE PARTICAL-MODEL THEORY OF DIFFUSE REFLECTANCE PROPERTIES OF POWDERED SAMPLES.** SIMMONS, E. L. J. APPL. PHYS. 46 (1), 344-9, 1975.
76628	**IR-SPECTROSCOPIC STUDY OF PYROXENE GLASSES CONTAINING NEODYMIUM AND SAMARIUM IONS.** GULYAMOV, B. M. ISMATOV, A. A. SHEVYAKOV, A. M. AKAD. NAUK UZBEKSKOI SSR, DOKL. 31 (5), 34-6, 1974.
76640	**A CALORIMETER FOR THE TEMPERATURE REGION 1 TO 20 K—THE SPECIFIC HEAT OF SOME GRAPHITE SPECIMENS.** DE SORBO, W. NICHOLS, G. E. GENERAL ELECTRIC RESEARCH LABORATORY 35PP., 1957. (57-RL-1705)
76745	**ASTEROIDS. SURFACE COMPOSITION FROM REFLECTION SPECTROSCOPY.** MC CORD, T. B. GAFFEY, M. J. SCIENCE 186 (4161), 352-5, 1974.
76782	**METASTABLE MIXED CRYSTALS WITH QUARTZ STRUCTURE WITH THE OXIDE SYSTEM LITHIUM OXIDE - MAGNESIUM OXIDE - ZINC OXIDE - ALUMINUM OXIDE - SILICON OXIDE.** PETZOLDT, J. KANNER (LEO) ASSOCIATES, REDWOOD CITY, CALIF. 36PP., 1974. (ENGLISH TRANSLATION OF GLASTECH. BER., 40 (10), 385-96, 1967; FOR ORIGINAL SEE TPRC NO. 47477) (N74-16463, NASA-TT-F-15340)
76783	**THERMAL CONDUCTIVITY OF HETEROGENEOUS MIXTURES AND LUNAR SOILS.** VACHON, R. I. PRAKOURAS, A. G. CRANE, R. KHADER, M. S. THERMOSCIENCE GROUP, AUBURN UNIV., ALABAMA 389PP., 1973. (N74-19579, NASA-CR-120162)
76803	**TOWARDS INVISIBLE GLASS.** WARD, J. VACUUM 22 (9), 369-75, 1972.
76810	**SOME MEASUREMENTS OF INFRARED TRANSMISSION OF GLASS AND PLASTICS.** AUTHOR ANON. DEFENSE DOCUMENTATION CENTER 5PP., 1951. (AD-147 688)
76816	**STUDY OF HEAT TRANSFER OF CERAMIC MATERIALS.** SOXMAN, E. J. MONROE, J. E., JR. CRANDALL, W. B. ALFRED UNIVERSITY 83PP., 1957. (AD-134 314)
76818	**FAR INFRARED SPECTRA OF CRYSTALS AND SOLIDS IN A LARGE RANGE OF TEMPERATURES.** LECOMTE, J. HADNI, A. CAMBRIDGE RESEARCH LAB. 14PP., 1963. (AD-430 749, AFCRL-64-22)
76819	**TRANSMITTANCE OF INFRARED ENERGY. GLASSES, I INTERFERENCE FILTERS, SINGLE AND POLYCRYSTALLINE MATERIALS.** OLSEN, A. L. NICHOLS, L. W. REGELSON, E. U. S. NAVAL ORDNANCE TEST STATION, CHINA LAKE, CALIF. 65PP., 1957. (AD-150 475, NAVORD-5584)
76853	**THE DEVELOPMENT OF IRDOMES.** JENNESS, J. R., JR. U. S. NAVAL AIR DEVELOPMENT CENTER, JOHNSVILLE, PENNSYLVANIA 18PP., 1953. (AD-30 669, NADC-EL-53112)
76855	**EVALUATION OF OPTICAL PROPERTIES OF IRDOMES FOR SIDEWINDER 1C MISSILES.** OLSEN, A. L. U. S. NAVAL ORDNANCE TEST STATION, CHINA LAKE, CALIF. 22PP., 1963. (AD-345 739, NAVWEPS-8366, NOTS-TP-3265)
76856	**DUAL MODE DOME DESIGN PROBLEMS.** GRAY, B. C. TURNER, W. L. WRIGHT AIR DEVELOPMENT DIVISION, WRIGHT-PATTERSON AIR FORCE BASE, OHIO 43-56, 1960. (AD-321 784, WADD-TR-60-274, VOL. II)
76857	**DEVELOPMENT OF SPECIAL INFRARED TRANSMITTING GLASSES.** KING, B. W. DUCKWORTH, W. H. BATTELLE MEMORIAL INSTITUTE, COLUMBUS, OHIO 54PP., 1956. (AD-112 833)
76865	**DEVELOPMENT OF SPECIAL INFRARED TRANSMITTING GLASSES.** KING, B. W. DUCKWORTH, W. H. BATTELLE MEMORIAL INSTITUTE, COLUMBUS, OHIO 14PP., 1956. (AD-101 261)
76872	**RELATION BETWEEN THE PROPERTIES AND PURITY OF SYNTHETIC MULLITE CERAMIC.** BAKUNOV, V. S. VU, N. K. AGAFONOVA, M. I. RYABTSEV, K. I. POLUBOYARINOV, D. N. REFRACTORIES, USSR (4), 233-8, 1974. (ENGLISH TRANSLATION OF OGNEUPORY, (4), 32-8, 1974; FOR ORIGINAL SEE TPRC NO. 75192)
76886	**ULTRA-LOW-EXPANSION GLASSES AND THEIR STRUCTURE IN THE SILICON OXIDE - TITANIUM OXIDE SYSTEM.** SCHULTZ, P. C. SMYTH, H. T. PROC. INTERN. CONF. PHYS. NON-CRYSTALLINE SOLIDS, 3RD 453-61, 1972.
76887	**THEORETICAL STUDY OF THE CHOICE OF THE CONSTITUENT MATERIAL OF MIRRORS OF A TELESCOPE ON BOARD A SATELLITE.** PIEUCHARD, G. ESPIARD, J. REV. D'OPT. 47, (5-6), 197-205, 1968.
76900	**SOLID SOLUTIONS IN THE SYSTEM HAFNIUM OXIDE - TITANIUM DIOXIDE.** KOCHEREGIN, S. B. KUZNETSOV, A. K. INORG. MAT., USSR 10 (2), 254-6, 1974. (ENGLISH TRANSLATION OF IZV. AKAD. NAUK SSSR, NEORG. MATER., 10 (2), 297-9, 1974; FOR ORIGINAL SEE TPRC NO. 75178)
76906	**RELATION BETWEEN THE STRUCTURE AND PROPERTIES OF TWO-PHASE GLASS.** USHAKOV, D. F. GILEV, I. S. MILYUKOV, E. M. INORG. MAT., USSR 10 (5), 775-7, 1974. (ENGLISH TRANSLATION OF IZV. AKAD. NAUK SSSR, NEORG. MATER., 10 (5), 905-8, 1974; FOR ORIGINAL SEE TPRC NO. 76046)
76913	**INFRARED SPECTRA OF CRYSTALLINE LEAD GERMANATES IN THE COMPOSITION REGION BETWEEN 50 AND 100 MOLECULAR LEAD OXIDE.** MOROZOV, V. N. J. APPL. SPECTROSC., USSR 11 (6), 1431-4, 1969. (ENGLISH TRANSLATION OF ZH. PRIKL. SPEKTROSK., 11 (6), 986-90, 1969; FOR ORIGINAL SEE TPRC NO. 58350)
76916	**THE USE OF INFRARED SPECTROSCOPY FOR THE DETERMINATION OF THE NATURE OF HIDDEN INCLUSIONS IN MICA.** RESHETNIKOVA, A. K. J. APPL. SPECTROSC., USSR 12 (4), 566-7, 1970. (ENGLISH TRANSLATION OF ZH. PRIKL. SPEKTROSK., 12 (4), 749-51, 1970; FOR ORIGINAL SEE TPRC NO. 61231)
76917	**IR AND RAMAN SPECTRA IN RELATION TO THE STRUCTURE OF GLASSES IN THE SYSTEM SODIUM OXIDE - NIOBIUM PENTOXIDE - SILICON DIOXIDE.** SIDOROV, T. A. J. APPL. SPECTROSC., USSR 12 (6), 841-2, 1970. (ENGLISH TRANSLATION OF ZH. PRIKL. SPEKTROSK., 12 (6), 1124-5, 1970; FOR ORIGINAL SEE TPRC NO. 60925)

TPRC Number	Bibliographic Citation
76928	**STRUCTURE OF GLASSES BELONGING TO THE SODIUM OXIDE - BORON OXIDE - ALUMINUM OXIDE - SILICON DIOXIDE AND SODIUM OXIDE - BORON OXIDE - ALUMINUM OXIDE - ANTIMONY SESQUIOXIDE - SILICON DIOXIDE SYSTEMS AS REVEALED BY INFRARED SPECTROSCOPY, ELECTRON SPIN RESONANCE, AND ELECTRON ABSORPTION SPECTRA.** GOROBETS, F. T. SIDOROV, T. A. TYUL'KIN, V. A. J. APPL. SPECTROSC., USSR 16 (2), 234-8, 1972. (ENGLISH TRANSLATION OF ZH. PRIKL. SPEKTROSK., 16 (2), 313-9, 1972; FOR ORIGINAL SEE TPRC NO. 67510)
76930	**STUDY OF THE STRUCTURE OF GLASSES BELONGING TO THE CALCIUM OXIDE - ALUMINUM OXIDE - SILICON DIOXIDE SYSTEM BY INFRARED SPECTROSCOPY.** BOBKOVA, N. M. EVMENOVA, I. P. J. APPL. SPECTROSC., USSR 16 (3), 363-6, 1972. (ENGLISH TRANSLATION OF ZH. PRIKL. SPEKTROSK., 16 (3), 494-7, 1972; FOR ORIGINAL SEE TPRC NO. 67739)
76936	**THE INCLUSION OF WATER IN GLASSES. ULTRARED MEASUREMENTS ON ADDITIONAL GLASSES.** SCHOLZE, H. GLASTECH. BER. 32 (7), 278-81, 1959.
76957	**THE THERMOPHYSICAL PROPERTIES OF FIBER REINFORCED PLASTICS.** KIM, D. H. AIAA, ASME, AND SAE, STRUCTURES, STRUCTURAL DYNAMICS, AND MATERIALS CONF., 13TH 9PP., 1972. (AIAA-PAPER-72-366)
76986	**INFRARED SPECTRA OF DIALKALINE SILICATE GLASS.** CHEBOTAREV, T. YE. PROKIN, A. A. MOLCHANOV, V. S. FOREIGN TECHNOLOGY DIV., WRIGHT PATTERSON AFB, OHIO 18PP., 1968. (AD-675 574, N69-11985, FTD-MT-24-64)
76990	**PROPERTIES OF GLASSES IN SOME TERNARY SYSTEMS CONTAINING BARIUM OXIDE AND SILICON OXIDE.** CLEEK, G. W. BABCOCK, C. L. ARIZONA UNIV., TUCSON OPTICAL SCIENCES CENTER 45PP., 1973. (AD-777 138, AFOSR-TR-74-0480, NBS-MONOGRAPH-35, COM-73-50840)
76994	**DEVELOPMENT OF PORCELAIN ENAMEL PASSIVE THERMAL CONTROL COATINGS.** LEVIN, H. LENT, W. E. BUETTNER, D. H. HUGHES AIRCRAFT CO., CULVER CITY, CALIF. 82PP., 1973. (NASA-CR-124459, N73-33459)
76999	**AN EQUATION FOR THERMAL EXPANSION COEFFICIENT AT HIGH PRESSURES.** BONAFEDE, M. RIV. ITAL. DI GEOFISICA 21, 207-10, 1972.
77005	**STRENGTH OF INTERNALLY STRAINED BRITTLE MATRIX COMPOSITES.** YOUNG, C. A. UNIVERSITY OF CALIFORNIA, BERKELEY, M. S. THESIS 20PP., 1973. (UCRL-20395, N71-26987)
77025	**INVESTIGATION OF INFRARED SPECTRA OF REFLECTION OF VITREOUS LEAD BORATES.** KURTSINOVSKAYA, R. I. J. APPL. SPECTROSC., USSR 17 (5), 1441-5, 1972. (ENGLISH TRANSLATION OF ZH. PRIKL. SPEKTROSK., 17 (5), 834-9, 1972; FOR ORIGINAL SEE TPRC NO. 70002)
77029	**DEGREE OF BLACKENING OF DOPED QUARTZ GLASSES.** MEN', A. A. PROKHOROVA, T. I. SETTAROVA, Z. S. J. APPL. SPECTROSC., USSR 17 (2), 1075-7, 1972. (ENGLISH TRANSLATION OF ZH. PRIKL. SPEKTROSK., 17 (2), 325-8, 1972; FOR ORIGINAL SEE TPRC NO. 68812)
77037	**RADIOMETRIC IMAGES OF IR RESTSTRAHLEN EMISSION FROM ROCK SURFACES.** HOVIS, W. A. GODDARD SPACE FLIGHT CENTER, GREENBELT, MD. 4-1-4-13, 1972. (N72-29305, NASA-TM-X-68564)
77038	**SPECTRAL REFLECTANCE MEASUREMENTS OF PLANT-SOIL COMBINATIONS.** MAC LEOD, N. H. GODDARD SPACE FLIGHT CENTER, GREENBELT, MD. 6-1-6-11, 1972. (N72-29307, NASA-TM-X-68564)
77119	**THERMALLY DARKENABLE PHOTOCHROMIC GLASSES.** SEWARD, T. P., III J. APPL. PHYS. 46 (2), 689-94, 1975.
77128	**REFLECTION SPECTRA OF CRYSTALLINE AND VITREOUS MODIFICATIONS OF GERMANIUM DIOXIDE AND SILICON DIOXIDE IN THE 5-50 MU REGION.** VENEDIKTOV, A. A. MOROZOV, V. N. POLUKHIN, V. N. ZH. PRIKL. SPEKTROSK. 10 (6), 969-71, 1969. (FOR ENGLISH TRANSLATION SEE TPRC NO. 77129)
77129	**REFLECTION SPECTRA OF CRYSTALLINE AND VITREOUS MODIFICATIONS OF GERMANIUM DIOXIDE AND SILICON DIOXIDE IN THE 5-50 MU REGION.** VENEDIKTOV, A. A. MOROZOV, V. N. POLUKHIN, V. N. J. APPL. SPECTROS., USSR 10 (6), 656-7, 1969. (ENGLISH TRANSLATION OF ZH. PRIKL. SPEKTROSK., 10 (6), 969-71, 1969; FOR ORIGINAL SEE T77128)
77144	**PROPERTIES OF GLASS FIBERS AT ELEVATED TEMPERATURES.** OTTO, W. H. PROC. SAGAMORE ARMY MATER. RES. CONF., 6TH 277-303, 1960. (AD-233 158)
77166	**SIMPLIFIED RECORDING MICROSPECTROPHOTOMETER.** CASELLA, A. J. STROTHER, G. K. CONNOLLY, J. W. APPL. OPT. 14 (3), 771-7, 1975.
77206	**GLASSES IN THE INFRARED FIELD.** MEYER, H. ANN. CHIM. (PARIS) 9 (1), 111-7, 1974.
77217	**UTILIZATION OF QUARTZ CERAMICS AS A DIFFUSE REFLECTION STANDARD.** BORODAI, S. P. BORODAI, F. YA. SOV. J. OPT. TECHNOL. 41 (5), 279-80, 1974. (ENGLISH TRANSLATION OF OPT. MEKH. PROM., 41 (5), 45-7, 1974; FOR ORIGINAL SEE TPRC NO. 76274)
77232	**ANTIREFLECTION COATING OF CORUNDUM FOR SERVICE IN THE INFRARED.** KUDRYAVTSEVA, A. G. SUYKOVSKAYA, N. V. SOV. J. OPT. TECHNOL. 37, 334, 1970.
77237	**HIGH THERMAL CONDUCTIVITY MICROWAVE SUBSTRATES.** POPE, B. J. HORTON, M. D. BOWMAN, L. S. HALL, H. T. ADANIYA, H. N. U. S. ARMY ELECTRONICS COMMAND, FORTH MONMOUTH, NEW JERSEY 49PP., 1975. (ECOM-73-0289-F)
77264	**THERMAL EXPANSION OF CERMET COMPONENTS BY HIGH TEMPERATURE X-RAY DIFFRACTION.** MAUER, F. A. BOLZ, L. H. U. S. DEPT. OF COMMERCE, NATIONAL BUREAU OF STANDARDS 22PP., 1956. (AD-112 812, NBS-REPT.-4884)
77265	**CHEMICAL STRENGTHENING OF CERAMIC MATERIALS.** KIRCHNER, H. P. GRUVER, R. M. LINDEN LABORATORIES, INC., STATE COLLEGE, PENNSYLVANIA 41PP., 1964. (AD-452 185)
77266	**INORGANIC BINDERS FOR C CORES.** SEIDEL, J. WESTINGHOUSE ELECTRIC CORP., MATERIALS LAB., PITTSBURGH, PENNSYLVANIA 16PP., 1961. (AD-262 262)
77272	**LIGHTWEIGHT THERMAL PROTECTION SYSTEM DEVELOPMENT. VOLUME II: MATERIALS-EXISTING DATA AND RECOMMENDED DATA ACQUISITION.** LANGE, R. A. JONES, H. RYAN, J. M. ROSENTHAL, H. STITT, J. BALATA, J. F. AF MATERIALS LAB., RESEARCH AND TECHNOLOGY DIVISION, WRIGHT-PATTERSON AFB, OHIO 30PP., 1963. (AD-422 941, ASD-TDR-63-596-VOL II)
77274	**THE DEGRADATION OF DURESTOS COMPONENTS AT HIGH TEMPERATURES.** ROGERS, K. F. ROYAL AIRCRAFT ESTABLISHMENT, FARNBOROUGH 20PP., 1962. (AD-290 763, RAE-CHEM-1401)

TPRC Number	Bibliographic Citation
77282	**DEVELOPMENT OF REFRACTORY CERAMICS THAT CAN BE PROCESSED AT TEMPERATURES CONSIDERABLY LOWER THAN THEIR MAXIMUM USE OF TEMPERATURE.** SMOKE, E. J. SCHOOL OF CERAMICS, RUTGERS, THE STATE UNIV. 23PP., 1963. (AD-419 255)
77283	**METAL-CERAMIC SEAL TECHNOLOGY. C.V.D. RESEARCH PROJECT RP3-54.** WALLIS, D. R. THE GENERAL ELECTRIC COMPANY LIMITED, HIRST RESEARCH CENTRE, WEMBLEY, ENGLAND 23PP., 1964.
77294	**DEVELOPMENT OF COMPOSITE STRUCTURES FOR HIGH TEMPERATURE OPERATION.** LICCIARDELLO, M. R. OHNYSTY, B. BOLAND, J. SOLAR, DIVISION OF INTERNATIONAL HARVESTER COMP., SAN DIEGO, CALIF. 43PP., 1965. (AD-461 369)
77313	**DEVELOPMENT OF A 1200 F RADOME.** CHASE, V. A. COPELAND, R. L. BRUNSWICK CORP., DEFENSE PRODUCTS DIVISION, 60PP., 1964. (AD-438 134)
77314	**LITHIUM-ALUMINUM SILICATES AS CERAMIC MATERIALS.** MEHMEL, M. GLASS-EMAIL-KERAMO-TECH. 10 (9), 337-40, 1959. (FOR ENGLISH TRANSLATION SEE TPRC NO. 77315)
77315	**LITHIUM-ALUMINUM SILICATES AS CERAMIC MATERIALS.** MEHMEL, M. DEPT. OF RESEARCH AND DEVELOPMENT SERVICES, ADMIRALTY 11PP., 1963. (ENGLISH TRANSLATION OF GLASS-EMAIL-KERAMO-TECH., 10 (9), 337-40, 1959; FOR ORIGINAL SEE TPRC NO. 77314) (AD-406 341, A.C.S.I.L.-TRANS.-1396)
77316	**TARGET AND BACKGROUND CHARACTERISTICS. ANALYSIS AND APPLICATIONS.** ENVIROMENTAL RESEARCH INST. OF MICHIGAN AIR FORCE AVIONICS LAB., WRIGHT-PATTERSON AIR FORCE BASE, OHIO 90PP., 1974. (AFAL-TR-73-304, ERIM-196400-7-F)
77317	**DEVELOPMENT OF SPECIAL OPTICAL GLASSES.** CLEEK, G. W. HAMILTON, E. H. SCUDERI, T. G. NATIONAL BUREAU OF STANDARDS, WASHINGTON, D. C. 20PP., 1956. (NBS-REPT-4866)
77322	**DEVELOPMENT OF SPECIAL OPTICAL GLASSES.** CLEEK, G. W. HAMILTON, E. H. SCUDERI, T. G. NATIONAL BUREAU OF STANDARDS, WASHINGTON, D. C. 15PP., 1956. (NBS-REPT-4614)
77323	**DEVELOPMENT OF SPECIAL OPTICAL GLASSES.** CLEEK, G. W. SCUDERI, T. G. MOORE, J. A. NATIONAL BUREAU OF STANDARDS, WASHINGTON, D. C. 34PP., 1957. (NBS-REPT-5368)
77324	**DEVELOPMENT OF SPECIAL OPTICAL GLASSES.** HAMILTON, E. G. CLEEK, G. W. SCUDERI, T. G. NATIONAL BUREAU OF STANDARDS, WASHINGTON, D. C. 28PP., 1955. (NBS-REPT-4451)
77325	**DEVELOPMENT OF SPECIAL OPTICAL GLASSES.** CLEEK, G. W. HAMILTON, E. H. NATIONAL BUREAU OF STANDARDS, WASHINGTON, D. C. 18PP., 1955. (NBS-REPT-4337)
77326	**DEVELOPMENT OF SPECIAL OPTICAL GLASSES.** CLEEK, G. W. HAMILTON, E. H. NATIONAL BUREAU OF STANDRDS, WASHINGTON, D. C. 16PP., 1954. (NBS-REPT-3237)
77327	**DEVELOPMENT OF SPECIAL OPTICAL GLASSES.** CLEEK, G. W. HAMILTON, E. H. NATIONAL BUREAU OF STANDARDS, WASHINGTON, D. C. 7PP., 1953. (NBS-REPT-2985)
77328	**DEVELOPMENT OF SPECIAL OPTICAL GLASSES.** CLEEK, G. W. HAMILTON, E. H. NATIONAL BUREAU OF STANDARDS, WASHINGTON, D. C. 21PP., 1955. (NBS-REPT-4009)
77330	**TRANSMITTANCE OF INFRARED ENERGY BY GLASSES.** GLAZE, F. W. BULL. AMER. CERAM. SOC. 34 (9), 291-4, 1955.
77332	**INVESTIGATION OF THE EFFECTIVE THERMAL CONDUCTIVITY OF POROUS MATERIALS.** TANAYEVA, S. A. NIZKOTEMP. ENERGETIKA I MASSOOBMEN V VAKUUM 148-58, 1973. (FOR ENGLISH TRANSLATION SEE T75084)
77333	**THERMAL CONDUCTIVITY OF MOIST BULK MATERIALS.** TANAYEVA, S. A. NIZKOTEMP. ENERGETIKA I MASSOOBMEN V VAKUUM 141-7, 1973. (FOR ENGLISH TRANSLATION SEE TPRC NO. 75083)
77334	**OPTIMUM SHAPES AND DIMENSIONS OF SPECIMENS FOR DETERMINATION OF THERMOPHYSICAL PROPERTIES.** DOMOROD, L. S. NIZKOTEMP. ENERGETIKA I MASSOOBMEN V VAKUUM 159-66, 1973. (FOR ENGLISH TRANSLATION SEE TPRC NO. 75085)
77338	**INVESTIGATION OF DIELECTRIC OPTICAL COATINGS FOR LASERS.** WOOD, R. M. TAIT, J. M. ROUSE, R. L. GENERAL ELECTRIC CO., LTD., WEMBLEY, ENGLAND 16PP., 1971. (AD-891 768, DRIC-BR-28067)
77341	**THERMAL CONDUCTIVITY POTTING COMPOUNDS FOR 3-D WELDED ELECTRONICS.** JANSSON, R. M. INSTRUMENTATION LAB., MASSACHUSETTS INST. TECH., CAMBRIDGE, MASS. 10PP., 1960. (AD-605 679)
77342	**THE ABSORPTION AND REFLECTION CHARACTER OF INFRARED BANDS OF POWDERED INORGANIC MATERIALS.** PHILLIPPI, C. M. AIR FORCE MATERIALS LAB., WRIGHT-PATTERSON AIR FORCE BASE, OHIO 59PP., 1969. (AD-685 187, AFML-TR-68-317)
77361	**TARGET PHENOMENOLOGY: TASK III: EXPERIMENTAL DETERMINATION OF THE INFRARED SPECTRAL OPTICAL PROPERTIES OF BULK AND POWDERED ALUMINUM OXIDE.** STREED, E. R. CUNNINGTON, G. R. LIU, C. K. LOCKHEED PALO ALTO RESEARCH LAB., PALO ALTO, CALIF. 64PP., 1973. (AD-910 500, AFRPL-TR-73-3, LMSC-D313095)
77362	**VACUUM ULTRAVIOLET SURVEILLANCE TECHNIQUES.** MARMO, F. F. ENGELMAN, A. SCHULTZ, E. D. GCA CORPORATION, BEDFORD, MASS. 58PP., 1967. (AD-379 842, GCA-TR-67-2-A)
77371	**THERMAL CONDUCTIVITY OF CARBON AND GRAPHITE CLOTHS PHENOLIC AT 100 TO 1100 C.** BARTEL, E. H. NAVAL WEAPONS CENTER, CHINA LAKE, CALIF. 54PP., 1969. (AD-855 281, NWC-TP-4647)
77381	**PRELIMINARY EVALUATION OF CAPABILITIES FOR SPACE STUDIES IN THE FIELD OF MATERIALS AND PROCESSES.** TURNER, H. C. KELLER, E. E. FAULKENBERRY, B. H. GENERAL DYNAMICS, CONVAIR, SAN DIEGO, CALIF. 8PP., 1959. (AD-837 722, GDA-MP-59-291)
77418	**THERMAL CONDUCTIVITY OF ROCKS, ITS MEASUREMENT AND ROLE IN HEAT FLOW INVESTIGATION.** CERMAK, V. MINER. SLOVACA 5 (4), 507-13, 1973.
77420	**THERMAL DIFFUSIVITY OF DRY SUBBITUMINOUS COAL.** WALKER, I. K. MAC KENZIE, K. J. D. N. Z. J. SCI. 17 (1), 77-92, 1974K.
77445	**INFRARED ABSORPTION SPECTROSCOPY OF COPPER-CALCIUM-PHOSPHATE GLASSES.** MORIDI, G. R. HOGARTH, C. A. INT. J. ELECTRON. 37 (1), 141-3, 1974.
77502	**SURFACE ROUGHNESS EFFECTS ON BIDIRECTIONAL REFLECTANCE.** SMITH, T. F. UNIVERSITY OF ILLINOIS AT URBANA-CHAMPAIGN, PH.D. THESIS 123PP., 1972. (UNIV. MICROFILM NO. 73-17428, N73-33611)
77506	**TUBULAR COMPOSITE STRUCTURAL ELEMENTS HAVING ZERO THERMAL EXPANSION.** DOW, N. F. ROSEN, B. W. ISR. J. TECHNOL. 7 (3), 209-18, 1969.

TPRC Number	Bibliographic Citation
77511	**EMISSIVITIES OF ALUMINOSILICATE REFRACTORIES AT HIGH TEMPERATURES.** MICHAUD, M. COMPT. REND. 228, 1115–6, 1949.
77520	**OPTICAL ABSORPTION AND ELECTRON–MICROPROBE STUDIES OF SOME HIGH–TITANIUM ANDRADITES.** MANNING, P. G. HARRIS, D. C. CAN. MINERAL. 10 (PT. 2), 260–71, 1970.
77521	**REFLECTION OF RADIANT ENERGY FROM SOILS.** BOWERS, S. A. HANKS, R. J. SOIL SCIENCE 100 (2), 130–8, 1965.
77522	**THE SPECTRAL REFLECTANCE OF AMERICAN SOILS.** CONDIT, H. R. PHOTOGRAMMETRIC ENG. 36, 955–66, 1970.
77523	**ANISOTROPIC SOLAR REFLECTANCE OVER WHITE SAND, SNOW AND STRATUS CLOUDS.** SALOMONSON, V. V. MARLATT, W. E. J. APPL. METEOROL. 7, 475–83, 1968.
77529	**METAL CERAMIC SEAL TECHNOLOGY.** WALLIS, D. R. M–O VALVE CO., LTD., LONDON, ENGLAND 20PP., 1965. (AD–478 648)
77530	**SYNTHESIS OF INORGANIC PHTOTROPIC MATERIALS FOR HIGH DENSITY COMPUTER MEMORY APPLICATIONS.** RADLER, R. CHENOT, C. BRAY, R. DREYER, J., JR. MC DONALD, J. AIR FORCE AVIONICS LAB., WRIGHT PATTERSON AIR FORCE BASE, OHIO 73PP., 1964. (AD–608 426, AL–TDR–64–170)
77532	**MICROWAVE HIGH DIELECTRIC CONSTANT MATERIALS.** READEY, D. W. MAGUIRE, E. A. HARTWIG, C. P. MASSE, D. J. PALADINO, A. E. RAYTHEON COMP., RESEARCH DIVISION, WALTHAM, MASS. 27PP., 1970. (AD–872 558L, ECOM–0455–3)
77545	**EFFECT OF ALUMINUM ON THE INFRARED ABSORPTION SPECTRA OF NATURAL SORBENTS.** NADIROVA, G. A. ZH. FIZ. KHIM. 45 (5), 1117–20, 1971. (FOR ENGLISH TRANSLATION SEE TPRC NO. 77546)
77546	**EFFECT OF ALUMINUM ON THE INFRARED ABSORPTION SPECTRA OF NATURAL SORBENTS.** NADIROVA, G. A. RUSS. J. PHYS. CHEM. 45 (5), 626–8, 1971. (ENGLISH TRANSLATION OF ZH. FIZ. KHIM., 45 (5), 1117–20, 1971; FOR ORIGINAL SEE TPRC NO. 77545)
77552	**INTEGRAL GLASS COVERS FOR SILICON SOLAR CELLS.** RAUCH, H. W. ULRICH, D. R. SPACE SCIENCES LAB., GENERAL ELECTRIC COMP., PHILADELPHIA, PA. 55., 1974. (AD–A002 846, AFAPL–TR–74–14)
77553	**LOW–MELTING GLASSES.** PAVLUSHIN, N. M. LEGKOPLAVKIYE STEKLA MOSCOW, ENERGIYA 145PP., 1970. (FOR ENGLISH TRANSLATION SEE TPRC NO. 77554)
77554	**LOW MELTING GLASSES.** PAVLUSHIN, N. M. ARMY FORESIGN SCIENCE AND TECHNOLOGY CENTER, CHARLOTTESVILLE, VIRGINIA 250PP., 1974. (ENGLISH TRANSLATION OF LEGKOPLAVKIYE STEKLA MOSCOW, ENERGIYA, 145PP., 1970; FOR ORIGINAL SEE TPRC NO. 77553) (AD–A002 763, FSTC–HT–23–1869–73)
77556	**EVALUATION OF BERYLLIA CERAMICS FOR ELECTRONIC APPLICATIONS.** AUTHOR ANON BRITISH CERAMIC RESEARCH ASSOCIATION, STOKE–ON–TRENT 9PP., 1970. (AD–878 680, NSTIC–28424)
77562	**MICROWAVE HIGH DIELECTRIC CONSTANT MATERIALS.** READEY, D. W. MAGUIRE, E. A. HARTWIG, C. P. MASSE, D. J. RAYTHEON RESEARCH DIVISION, WALTHAM, MASS. 29PP., 1970. (AD–876 010L, ECOM–0455–4)
77589	**DIFFUSIONAL PRESTRESSING OF CERAMICS.** BICKELHAUPT, R. E. COX, H. P. HURST, R. D. SOUTHERN RESEARCH INST., BIRMINGTHAM, ALABAMA 76PP., 1967. (AD–819 014)
77590	**CHARACTERISTICS OF INITIAL STAGES OF SITALIZATION OF GLASSES NUCLEATED WITH ZIRCONIUM DIOXIDE.** PAVLUSHKIN, N. M. ELLERN, G. A. TR. MOSK. KHIM. TEKHNOL. INST. (55), 74–80, 1967. (FOR ENGLISH TRANSLATION SEE TPRC NO. 77591)
77591	**CHARACTERISTICS OF INITIAL STAGES OF SITALIZATION OF GLASSES NUCLEATED WITH ZIRCONIUM DIOXIDE.** PAVLUSHKIN, N. M. ELLERN, G. A. ARMY FOREIGN SCIENCE AND TECHNOLOGY CENTER, CHARLOTTESVILLE, VIRGINIA 13PP., 1974. (ENGLISH TRANSLATION OF TR. MOSK. KHIM. TEKHNOL., INST., (5), 74–80, 1967; FOR ORIGINAL SEE TPRC NO. 77590) (AD–A002 634, FSTC–HT–23–1456–73)
77596	**DEVELOPMENT OF NEW AND IMPROVED HIGH–TEMPERATURE SOLID FILM LUBRICANTS.** HOPKINS, V. LAVIK, M. T. CLOW, W. L. BOLZE, C. C. HUBBELL, R. D. MIDWEST RESEARCH INST., KANSAS CITY, MO. 266PP., 1965. (AD–466 660)
77603	**THE ANGULAR BACKSCATTERING CHARACTERISTICS OF PLANE SURFACES ILLUMINATED BY CARBON DIOXIDE LASER RADIATION.** SEANEY, R. J. ROYAL ARMAMENT RESEARCH AND DEVELOPMENT ESTABLISHMENT 45PP., 1969. (AD–850 443)
77616	**TEMPERATURE DEPENDENCE OF THE HEAT CONDUCTIVITY OF LUNAR MATTER FROM THE RESULTS OF OBSERVATIONS AT MILLIMETER WAVE FREWUENCIES.** KISLYAKOV, A. G. ASTRONOMICHESKII VESTNIK 8, 138–41, 1974.
77626	**DETERMINATION OF THE MECHANICAL AND THERMAL PROPERTIES OF THE KUNASHAK AND ELENOVKA METEORITES.** MEDVEDEV, R. V. METEORITKA (33), 100–4, 1974.
77670	**INFRARED VIBRATIONAL SPECTRUM OF CRYSTALLINE ARSENIC SULFIDE.** TREACY, D. TAYLOR, P. C. PHYS. REV. 11 B (8), 2941–7, 1975.
77746	**SPECTRAL REFLECTANCE MEASUREMENTS.** RAINES, G. L. LEE, K. PHOTOGRAMMETRIC ENG. 547–50, 1974. (AD–784 540)
77753	**MARS: COMPONENTS OF INFRARED SPECTRA AND THE COMPOSITION OF THE DUST CLOUD.** HUNT, G. R. LOGAN, L. M. SALISBURY, J. W. ICARUS 18, 459–69, 1973. (AD–764 161, AFCRL–TR–73–0333)
77754	**EFFECTS OF IMPURITIES AND SYNTHESIS ATMOSPHERE UPON THE OPTICAL PROPERTIES OF SOME GLASSES.** GINTHER, R. J. KIRK, R. D. NAVAL RESEARCH LAB., WASHINGTON, D. C. 1–34, 1973. (AD–763 338, NRL–MR–2590)
77755	**ULTRAVIOLET SPECTRA OF SILICATE GLASSES. A REVIEW OF SOME EXPERIMENTAL EVIDENCE.** SIGEL, G. H., JR. NAVAL RESEARCH LAB., WASHINGTON, D. C. 42–88, 1973. (AD–763 338, NRL–MR–2590)
77757	**VEGETATION CANOPY REFLECTANCE MODELS.** OLIVER, R. E. SMITH, J. A. FORT COLLINS COLLEGE OF FORESTRY, COLORADO STATE UNIVERSITY 82PP., 1973. (AD–763 222)
77763	**SEA FLOOR GEOTHERMAL MEASUREMENTS FROM VEMA CRUISE 25.** LANGSETH, M. G. MALONE, I. E. BOOKMAN, C. A. LAMONT–DOHERTY GEOLOGICAL OBSERVATORY, PALISADES, N. Y. 159PP., 1972. (AD–748 309)

TPRC Number	Bibliographic Citation
77765	**AIRBORNE MEASUREMENTS OF OPTICAL ATMOSPHERIC PROPERTIES, SUMMARY AND REVIEW.** DUNTLEY, S. Q. JOHNSON, R. W. GORDON, J. I. SCRIPPS INSTITUTION OF OCEANOGRAPHY, SAN DIEGO, CALIF. 79PP., 1972. (AD-754 898, AFCRL-72-0593, SIO-REF-72-82)
77769	**STRUCTURES AND KINETICS OF GLASS CORROSION.** SANDERS, D. M. HENCH, L. L. DEPT. OF MATERIALS SCIENCE, FLORIDA UNIV., GAINESVILLE 4-36, 1972. (AD-764 319)
77770	**QUANTITATIVE ANALYSIS OF GLASS STRUCTURE USING INFRARED REFLECTION SPECTRA.** SANDERS, D. M. PERSON, W. B. HENCH, L. L. DEPT. OF MATERIALS SCIENCE, FLORIDA UNIV., GAINESVILLE 37-65, 1972. (AD-764 319)
77778	**MANTLE CONVECTION AND THE NEW GLOBAL TECTONICS.** TURCOTTE, D. L. OXBURGH, E. R. ANN. REV. FLUID MECH. 4, 33-68, 1972. (AD-761 716, AFOSR-TR-73-1029)
77836	**ULTRAVIOLET PLEOCHROISM AND ABSORPTION IN CORUNDUM CRYSTALS.** GUM-GRZHIMAILO, S. V. RUDNITSKAIA, E. S. METEOROLOGICAL SOCIETY, BOSTON, MASS. 89-113, 1968. (AD-669 039, N68-27373)
77844	**REFLECTION SPECTRA OF SILICON, GALENA, AND PYRITE IN THE 0.62-5.48 EV REGION.** KORSUNSKII, M. I. MITRYAEVA, N. M. MURATOV, E. M. IZV. AKAD. NAUK KAZ. SSR, SER. GEOL. 26 (6), 11-22, 1969.
77861	**PUSH ROD DILATOMETER MEASUREMENTS UP TO 2,800 C OF REVERSIBLE AND IRREVERSIBLE LENGTH CHANGES IN POLYCRYSTALLINE CARBON AND COMPARISON WITH RESULTS OF X-RAY IN SITU MEASUREMENTS.** FITZER, E. WEISENBURGER, S. THERMOPHYSICAL PROPERTIES OF SOLIDS AT HIGH TEMP., EUROPEAN CONF., 4TH 69-73, 1974.
77865	**THERMAL CONDUCTIVITY OF MICA AS A FUNCTION OF TEMPERATURE AND DIRECTION OF HEAT FLOW.** MIRKOVICH, V. V. THERMOPHYSICAL PROPERTIES OF SOLIDS AT HIGH TEMP., EUROPEAN CONF., 4TH 106-9, 1974.
77908	**THERMAL PROPERTIES OF UNREINFORCED PPQ AND PHENOLIC POLYMERS.** ENGELKE, W. T. SOUTHERN RESEARCH INST., BIRMINGHAM, ALABAMA 49PP., 1974. (AD-919 087L, SORI-EAS-74-048-3069-3-F)
77935	**RESEARCH IN OPTICAL MATERIALS AND STRUCTURES FOR HIGH POWER LASERS.** HORRIGAN, F. A. DEUTSCH, T. F. RUDKO, R. I. LOTTI, R. WILSON, D. RAYTHEON RESEARCH DIV., WALTHAM, MASS. 60PP., 1970. (AD-875 339L)
77938	**THE GRUNEISEN PARAMETER AND THERMAL PROPERTIES OF POLYMERS.** URZENDOWSKI, R. GUENTHER, A. H. ALBUQUERQUE UNIVERSITY, NEW MEXICO 228PP., 1971. (AD-886 413L, AFWL-TR-71-6)
77941	**INFORMATION ON MECHANICAL PROPERTIES LEADING TO THE RECOMMENDATION OF NEW LIMITS FOR UPGRADING SPECIFICATION REQUIREMENTS FOR LAMINATED PLASTIC MATERIALS, MIL-P-8013, MIL-P-25421, AND MIL-P-25515.** VADALA, E. TH. NAVAL AIR SYSTEMS COMMAND, WASHINGTON, D. C. 7PP., 1967. (AD-814 504L, AML-2595)
77949	**MANUFACTURING METHODS FOR DIMENSIONALLY STABLE COMPOSITE MICROWAVE COMPONENTS.** KELLER, L. B. RAECH, H. AIR FORCE MATERIALS LAB., WRIGHT PATTERSON AFB, OHIO 574PP., 1974. (AD-920 253L, AFML-TR-74-70)
77957	**ABLATIVE MATERIALS FOR HIGH HEAT LOADS. PART III. INSULATIVE LAYERS OF CARBON/PHENOLIC HEATSHIELDS.** JUNEAU, P. W. FENTON, R. M. FEHL, C. METZGER, J. W. BRAZEL, J. AIR FORCE MATERIALS LAB., WRIGHT PATTERSON AFB, OHIO 128PP., 1973. (AD-912 718L, AFML-TR-70-95-PT. 3)
78023	**BERYLLIUM OXIDE CERAMICS FOR MICROWAVE APPLICATIONS.** HOLLADAY, C. O. LYNCH, J. F. BROWN, R. J. THE BRUSH BERYLLIUM COMP., ELMORE, OHIO 23PP., 1966. (AD-489 664L, ECOM 01242-4)
78045	**THERMAL CHARACTERISTICS FOR RADAR REFLECTING AND ABSORBING INORGANIC FOAMS AND COATING SYSTEMS.** KARDOS, K. P. HORIZONS INCORPORATED, CLEVELAND, OHIO 20PP., 1964. (AD-452 830L)
78047	**THERMAL CHARACTERISTICS FOR RADAR REFLECTING AND ABSORBING INORGANIC FOAMS AND COATING SYSTEMS.** KARDOS, K. P. HORIZONS INCORPORATED, CLEVELAND, OHIO 28PP., 1964. (AD-452 791L)
78202	**THERMAL CONDUCTUVIVITY OF BOROSILICATE GLASS.** ZAITLIN, M. P. ANDERSON, A. C. PHYS. REV. LETT. 33 (19), 1158-61, 1974.
78248	**SPECTRAL REFLECTIVITY OF CERTAIN MINERALS AND SIMILAR INORGANIC MATERIALS.** GIER, J. T. DUNKLE, R. V. UNIVERSITY OF CALIFORNIA, INST. OF ENGINEERING RESEARCH, BERKELEY 15PP., 1954. (AD-26 394L)
78249	**DEVELOPMENT OF NON-MELT DEPOSITION TECHNIQUES FOR OPTICAL COATING FOR RECONNAISSANCE.** SULZBACH, F. C. BARNUM, J. C. GARDNER, W. R. HALL, L. H. WIEMER, K. C. TEXAS INSTRUMENTS INC., DALLAS, TEXAS 100PP., 1972. (AD-905 989L, TI-DM-72-06-14, AFAL-TR-72-250)
78254	**DIRECTIONAL REFLECTANCE OF POLARIZED 1.06 MICRON WAVELENGTH LASER RADIATION FROM VARIOUS SURFACES.** WARSHAW, E. C. NAVAL WEAPONS LAB., DAHLGREN, VA. 83PP., 1972. (AD-903 820L, NWL-TR-2808)
78256	**RECONNAISSANCE WINDOW STUDY.** YOUNG, P. S. ROBERTS, F. E. STROUSE, E. A. AIR FORCE AVIONICS LAB., WRIGHT PATTERSON AIR FORCE BASE, OHIO 320PP., 1972. (AD-904 562L, AFAL-TR-72-193)
78263	**PARAMETERS FOR REFRACTORY MATERIALS FOR 1200 PSI MAIN PROPULSION BOILERS.** BURT, R. C. NAVAL SHIP ENGINEERING CENTER, PHILADELPHIA, PA. 23PP., 1973. (AD-913 942L)
78264	**OPTICAL AND THERMAL PROPERTIES OF INFRARED TRANSMITTING MATERIALS AS A FUNCTION OF TEMPERATURE.** HEDGE, J. C. OLSON, O. H. IIT RESEARCH INST., CHICAGO, ILL. 37PP., 1969. (AD-872 057L)
78266	**INFRARED DIFFUSE REFLECTOR COATING, PART II-CANDIDATE MATERIALS.** SCHMIDT, R. N. ZIMMER, P. B. BLAU, H. H. AIR FORCE MATERIALS LAB., WPAFB, OHIO 44PP., 1967. (AD-390 086L, AFML-TR-67-337, 12041-FR2)
78269	**PRODUCTION ENGINEERING MEASURE FOR OPTICAL INTEGRATORS, OPTICAL WINDOWS AND PROJECTION LENSES FOR LASER DIODE ARRAYS.** OKOOMIAN, H. J. WALLERSTEIN, E. P. VALPEY CORP., OPTICAL SUB-SYSTEMS DIV., HOLLISTON, MASS. 241PP., 1972. (AD-909 609L)
78275	**OPTICAL MATERIALS FOR THE ULTRAVIOLET AND INFRARED SPECTRAL RANGE.** SCHROEDER, H. NEUROTH, N. SCHOTT INFORMATION (4), 1-28, 1967. (FOR ENGLISH TRANSLATION SEE TPRC NO. 78276)
78276	**OPTICAL MATERIALS FOR THE ULTRAVIOLET AND INFRARED SPECTRAL RANGE.** SCHROEDER, H. NEUROTH, N. FOREIGN TECHNOLOGY DIVISION, WPAFB, OHIO 39PP., 1972. (ENGLISH TRANSLATION OF SCHOTT INFORMATION, (4), 1-28, 1967; FOR ORIGINAL SEE TPRC NO. 78275) (AD-901 700L, FTD-HC-1807-71)

TPRC Number	Bibliographic Citation
78278	**ADVANCED COMPOSITE APPLICATIONS FOR SPACECRAFT AND MISSILES–PHASE I FINAL REPORT, VOLUME II, MATERIAL DEVELOPMENT.** HERTZ, J. CHRISTIAN, J. L. VARLAS, M. AIR FORCE MATERIALS LAB., WRIGHT PATTERSON AIR FORCE BASE, OHIO 394PP., 1972. (AD–893 715L, AFML–TR–71–186–VOL–2)
78310	**AN INTEGRATING SPHERE FOR MEASURING DIFFUSE REFLECTANCE IN THE NEAR INFRARED.** JACQUEZ, J. A. MC KEEHAN, W. HUSS, J. DIMITROFF, J. M. KUPPENHEIM, H. F. ARMY MEDICAL RESEARCH LAB., FORT KNOX, KENTUCKY 11PP., 1955. (AD–59 874L, AMRL–188)
78327	**THERMAL AND MECHANICAL PROPERTIES OF ADVANCED HEATSHIELD RESINOUS (CP) AND CARBONACEOUS (CC) COMPOSITES. VOLUME I. TEST METHODS, COMPARATIVE DATA, RECOMMENDED INPUTS AND ANALYSIS.** LEGG, J. K. SANDERS, H. G. ENGELKE, W. T. PEARS, C. D. AIR FORCE MATERIALS LAB., WPAFB, OHIO 212PP., 1972. (AD–904 361L, AFML–TR–72–160–VOL–1)
78332	**NEW LABORATORY METHODS DETERMINATION OF BORON OXIDE CONTENT IN EUTAL GLASS USING THE ABSORPTION METHOD OF INFRARED SPECTROSCOPY.** NEMEC, L. SILIKATY 15 (2), 205–12, 1971. (FOR ENGLISH TRANSLATION SEE TPRC NO. 78333)
78333	**NEW LABORATORY METHODS DETERMINATION OF BORON OXIDE CONTENT IN EUTAL GLASS USING THE ABSORPTION METHOD OF INFRARED SPECTROSCOPY.** NEMEC, L. FOREIGN TECHNOLOGY DIVISION, WPAFB, OHIO 13PP., 1972. (ENGLISH TRANSLATION OF SILIKATY, 15 (2), 205–12, 1971; FOR ORIGINAL SEE TPRC NO. 78332) (AD–904 279L, FTD–HC–23–0986–72)
78340	**THERMAL ROCK BREAKER FOR HARD ROCK CRUSHING–A FEASIBILITY STUDY.** THIRUMALAI, K. CHUEUNG, J. B. CHEN, T. S. DEMOU, S. G. KRAWZA, W. G. BUREAU OF MINES, MINNEAPOLIS, MINN. 272PP., 1972. (AD–906 785L)
78350	**BERYLLIUM OXIDE CERAMICS FOR MICROWAVE APPLICATIONS.** HOLLADAY, C. O. U. S. ARMY ELECTRONICS COMMAND, FORT MONMOUTH, N. J. 16PP., 1967. (AD–809 315L, ECOM–01242–6)
78359	**THE MISUNDERSTOOD GLASS STATE.** TARTE, P. IND. CHIM. BELGE 27 (1), 4–15, 1962.
78497	**CARBON FIBRE REINFORCED CERAMICS.** BOWEN, D. H. PHILLIPS, D. C. SAMBELL, R. A. J. BRIGGS, A. PROC. INT. CONF. MECHANICAL BEHAVIOR MATERIALS, 1ST 5, 123–34, 1972.
78659	**INVESTIGATION OF THE CORRELATION BETWEEN THE HEAT CAPACITY AND DENSITY OF GLASSES OF VARIOUS TYPE.** RALKO, A. V. GORODOV, V. S. RODNIKOVA, V. V. ZIKERT, G. PLEMYANNIKOV, M. M. AKADEMIIA NAUK UKRAINS'KOI RSR, DOPOVIDI, GEOLOGIIA, GEOFIZIKA, KHIMIIA I BIOLOGIIA 36 B, 916–18, 1974.
78661	**THE OPTICAL PARAMETERS OF MARE AND HIGHLAND LUNAR SOILS.** ANTIPOVA–KARATAEVA, I. I. AKHMANOVA, M. V. DEMENT'EV, M. N. MARKOV, I. I. S. TARASOV, L. S. PROC. SIXTEENTH PLENARY MEETING, KONSTANZ, WEST GERMANY 621–4, 1974.
78683	**THERMAL CHARACTERISTICS OF APOLLO 16 LUNAR FINES.** CREMERS, C. J. HSIA, H. S. BIRKEBAK, R. C. PROC. FIFTH INTERN. CONF., TOKYO, JAPAN 128–32, 1974.
78685	**REMOTE SENSING OF ROCK TYPE IN THE VISIBLE AND NEAR–INFRARED.** SALISBURY, J. W. HUNT, G. R. INTERN. SYMP. REMOTE SENSING ENVIRONMENT, 9TH, ANN ARBOR, MICH. 1953–8, 1974.
78687	**EMISSIVITIES AND FORWARD SCATTERING OF NATURAL AND MAN–MADE MATERIAL AT THREE MILLIMETER WAVELENGTH.** SCHANDA, E. HOFER, R. INTERN. SYMP. REMOTE SENSING ENVIRONMENT, 9TH, ANN ARBOR, MICH. 1585–92, 1974.
78688	**USE OF VISIBLE, NEAR–INFRARED, AND THERMAL INFRARED REMOTE SENSING TO STUDY SOIL MOISTURE.** BLANCHARD, M. B. INTERN. SYMP. REMOTE SENSING ENVIRONMENT, 9TH, ANN ARBOR, MICH. 693–700, 1974.
78699	**MEASUREMENT OF THE THERMAL EXPANSION COEFFICIENTS OF MATERIALS WITH A LASER INTERFEROMETRIC DILATOMETER.** EGOROV, V. P. KVANTOVAIA ELEKTRONIKA (MOSCOW) 1, 2131–7, 1974.
78777	**THERMAL PROPERTIES OF FOOD MATERIALS.** RHA, C. THEORY, DETERMINATION AND CONTROL OF PHYSICAL PROPERTIES OF FOOD MATERIALS: SERIES IN FOOD MATERIAL SCIENCE 1, 311–55, 1975.
78785	**LOW EMISSIVITY COEFFICIENT AND HIGH ELECTRIC RESISTANCE LAYERS FOR ALTERNATOR CRYOSTAT.** COMPAGNIE GENERALE D'ELECTRICITE COMPAGNIE GENERALE D'ELECTRICITE, MARCOUSSIS (FRANCE) 25PP., 1973. (N74–23224)
78827	**THERMALLY INDUCED TRANSPORT PHENOMENA IN PARTIALLY SATURATED POROUS MEDIA.** MOORE, N. R. DEAVER, F. K. WOLF, H. HEAT TRANSFER, PROC. INT. HEAT TRANSFER CONF., 5TH 5, 103–7, 1974.
78841	**VITREOUS ALUMINOSILICATES.** MALMENDIER, J. W. PROC., SYMP. ART GLASSBLOWING 16, 12–20, 1971.
78860	**THERMAL DIFFUSIVITY OF SIX IGNEOUS ROCKS AT ELEVATED TEMPERATURES AND REDUCED PRESSURES.** LINDROTH, D. P. U. S., BUR. MINES, REP. INVEST. RI–7954, 38PP., 1974.
78899	**PYROELECTRIC VIDICON TARGET MATERIALS.** GARN, L. E. IEEE TRANS. PARTS, HYBRIDS PACKAG. PHP–10 (4), 208–21, 1974.
78950	**NATURE OF THE LUNAR GLOBE SURFACE ALBEDO.** SHEVCHENKO, V. V. ASTRON. ZH. 51 (5), 1064–71, 1974. (FOR ENGLISH TRANSLATION SEE T92421)
78981	**SURFACE COMPOSITION OF IO.** WAMSTEKER, W. KROES, R. L. FOUNTAIN, J. A. ICARUS 23 (3), 417–24, 1974.
78988	**TEMPERATURE DEPENDENCE OF THE THERMAL CONDUCTIVITY OF CERAMIC MATERIALS DURING GASEOUS MEDIUM RAREFACTION.** LITOVSKII, E. YA. ZBOROVSKII, I. D. USPENSKII, I. N. INZH.–FIZ. ZH. 27 (4), 648–54, 1974. (FOR ENGLISH TRANSLATION SEE T83663)
79014	**OPTICAL PROPERTIES, PHOTOGRAPHIC, AND HOLOGRAPHIC APPLICATIONS OF THIN PHOTOSENSITIVE LAYERS.** TUBBS, M. R. BOOTH, H. J. PHOT. SCI. 22 (2), 61–8, 1974.
79125	**COMPOSITE MATERIALS WITH A LOW COEFFICIENT OF LINEAR EXPANSION.** FRIDLYANDER, I. N. KLYAGINA, N. S. TYKACHINSKII, I. D. DAIN, E. P. GORDEEVA, G. D. METALLOVED. TERM. OBRAB. METAL. (6), 36–8, 1974. (FOR ENGLISH TRANSLATION SEE TPRC NO. 79405)
79147	**DETERMINATION OF THE HEAT CAPACITY OF REDUCED SINTERS BY THERMAL ANALYSIS.** GAVRILKO, S. A. POTEBNYA, YU. M. RIKHTER, R. G. IZV. VYSSH. UCHEBN. ZAVED., CHERN. METALL. (10), 23–5, 1974.
79241	**THERMAL CONDUCTIVITY OF THE ELEMENTS: A COMPREHENSIVE REVIEW.** HO, C. Y. POWELL, R. W. LILEY, P. E. J. PHYS. CHEM. REF. DATA 3 (SUPPL. NO. 1), 796PP., 1974.
79311	**PHYSICAL PROPERTIES OF GLASS–CERAMIC COMPOSITES. III. THERMAL EXPANSION AND PHASE TRANSITION OF ALKALI–ALUMINOSILICATES.** KUNUGI, M. SOGA, N. YAMAMOTO, J. ASAHI GARASU KOGYO GIJUTSU SHOREIKAI KENKYU HOKOKU 23, 129–40, 1973.

TPRC Number	Bibliographic Citation
79405	**COMPOSITE MATERIALS WITH A LOW COEFFICIENT OF THERMAL EXPANSION.** FRIDLYANDER, I. N. KLYAGINA, N. S. TYKACHINSKII, I. D. DAIN, E. P. GORDEEVA, G. D. METAL SCI. HEAT TREAT. METALS, USSR (6), 495–7, 1974. (ENGLISH TRANSLATION OF METALLOVED. TERM. OBRAB. METAL., (6), 36–8, 1974; FOR ORIGINAL SEE TPRC NO. 79125)
79531	**INFLUENCE OF AN ELECTRIC FIELD ON THE SPECTRUM OF INDUCED OPTICAL ABSORPTION IN A SODIUM – BORATE GLASS IRRADIATED WITH ELECTRONS.** VAKHROMEEV, V. G. ZINOV'EVA, S. P. IZV. VUZ FIZ. 11, 146–8, 1974. (FOR ENGLISH TRANSLATION SEE T83743)
79601	**THE THERMAL EXPANSION COEFFICIENTS OF SOME GLASS-REINFORCED PLASTICS AND THEIR COMPONENTS AT LOW AND HIGH TEMPERATURES.** KRITSUK, A. A. STRENGTH MATER. 4 (5), 611–15, 1972. (ENGLISH TRANSLATION OF PROBL. PROCHNOSTI, 4 (5), 98–102, 1972; FOR ORIGINAL SEE T067805)
79605	**THERMAL EXPANSIONS OF KOVAR AND CERAMVAR AND SEALS OF THESE MATERIALS TO ALUMINA.** BERTOLOTTI, R. L. SANDIA LABS., LIVERMORE, CALIFORNIA 15PP., 1974. (N75–16638, SAND–74–8003)
79799	**SURFACE TEMPERATURE AND EMISSIVITY OF MERCURY.** HANSEN, O. L. ASTROPHYS. J. 190 (3), (PT. 1), 715–17, 1974.
79806	**THERMAL EXPANSION OF NATURAL GRAPHITE IMBEDDED IN GLASSY CARBON.** KAMIYA, K. INAGAKI, M. CARBON 11 (4), 429–30, 1973.
79859	**MANUFACTURING METHODS, STRUCTURE AND PHYSICOCHEMICAL PROPERTIES OF OPTICAL CERAMICS.** VOLYNETS, F. K. OPT. MEKH. PROM. 40 (9), 48–61, 1973. (FOR ENGLISH TRANSLATION SEE T79860)
79860	**MANUFACTURING METHODS, STRUCTURE AND PHYSICOCHEMICAL PROPERTIES OF OPTICAL CERAMICS.** VOLYNETS, F. K. SOV. J. OPT. TECHNOL. 40 (9), 578–92, 1973. (ENGLISH TRANSLATION OF OPT. MEKH. PROM., 40 (9), 48–61, 1973; FOR ORIGINAL SEE T79859)
79915	**THE ULTRAVIOLET REFLECTIVITY OF THE MOON.** CARVER, J. H. HORTON, B. H. O'BRIEN, R. S. O'CONNOR, G. G. MOON 9 (3–4), 295–303, 1974.
79928	**HIGH TEMPERATURE X-RAY DIFFRACTION STUDIES ON CHROMIUM-DOPED VANADIUM OXIDE.** CHANDRASHEKHAR, G. V. SINHA, A. P. B. MATER. RES. BULL. 9 (6), 787–98, 1974.
80006	**THERMAL CONDUCTIVITY OF BORON OXIDE.** LEZHENIN, F. F. KORNEICHUK, A. A. BERZHATYI, V. I. GRITSAENKO, V. P. TEPLOFIZ. TEPLOTEKH. (USSR) (25) 94–6, 1973.
80319	**EFFECT OF ADDITIVES ON THE THERMAL CONDUCTIVITY OF BERYLLIUM OXIDE.** BELYAEV, R. A. SMOLYA, A. V. VAVILOV, YU. V. SHALIMOVA, I. F. PANTELEEVA, I. F. POPOVA, V. P. OGNEUPORY (2), 51–3, 1975. (FOR ENGLISH TRANSLATION SEE T81987)
80433	**THERMAL PROPERTIES OF THE SOIL OF LUNA-20 AND ITS TERRESTRIAL ANALOGS.** RZHEVSKII, V. V. DUKHOVSKOI, E. A. KRUGLOV, N. T. PETROCHENKOV, R. G. SILIN, A. A. SHVAREV, V. V. DOKL. AKAD. NAUK SSSR 218, 1043–5, 1974. (FOR ENGLISH TRANSLATION SEE T80434)
80434	**THERMAL PROPERTIES OF THE SOIL OF LUNA-20 AND ITS TERRESTRIAL ANALOGS.** RZHEVSKII, V. V. DUKHOVSKOI, E. A. KRUGLOV, N. T. PETROCHENKOV, R. G. SILIN, A. A. SHAVAREV, V. V. SOV. PHYS. DOKL. 19, 627–8, 1975. (ENGLISH TRANSLATION OF DOKL. AKAD. NAUK SSSR, 218, 1043–5, 1974; FOR ORIGINAL SEE T80433)
80464	**MEASUREMENT OF THERMAL CONDUCTIVITY IN COMPOSITE STRUCTURES.** CREMA, L. B. L AEROTECNICA – MISSILI E SPAZIO 54, 97–100, 1975.
80491	**EFFECTIVE THERMAL CONDUCTIVITY OF POROUS CATALYST SUPPORTS.** JIRATOVA, K. CHEM. PRUM. 24 (12), 611–14, 1974.
80507	**THERMAL CONDUCTIVITY AND DIFFUSIVITY OF GREEN RIVER OIL SHALES.** PRATS, M. O'BRIEN, S. M. J. PET. TECHNOL. 27 (1), 97–106, 1975.
80566	**SYMMETRY OF NORMAL OSCILLATIONS OF TOPAZ CRYSTAL.** KOVALEVA, L. T. ZH. PRIKL. SPEKTROSK. 22 (2), 311–16, 1975. (FOR ENGLISH TRANSLATION SEE T86759)
80596	**MEASUREMENT OF THERMAL PROPERTIES OF NUCLEAR MATERIALS BY THE LASER FLASH METHOD.** TAKAHASHI, Y. MURABAYASHI, M. J. NUCL. SCI. TECHNOL. 12 (3), 133–44, 1975.
80708	**UNKNOWN COBALT–RON SELENIDE.** DE MONTREUIL, L. A. ING. GEOL. 16, 65–70, 1974.
80711	**MANGANESE TAPIOLITE FROM NORTHERN MORAVIA, CZECHOSLOVAKIA.** CECH, F. ACTA UNIV. CAROL., GEOL. (1–2), 37–45, 1973.
80733	**HIGH–TEMPERATURE HEAT CAPACITY OF SAPPHIRINE.** KISELEVA, I. A. TOPOR, N. D. GEOKHIMIYA (2), 312–15, 1975. (FOR ENGLISH TRANSLATION SEE T91263)
80784	**ABSORPTION SPECTRA OF LUNAR SECTIONS FROM DIFFERENT LUNAR AREAS.** ANTIPOVA–KARATAEVA, I. I. STACHEEV, YU. I. TARASOV, L. S. SPACE RES. 13 (2), 997–1000, 1973.
80843	**PHONON THERMAL CONDUCTIVITY IN THE EARTH'S MANTLE.** PETRUNIN, G. I. YURCHAK, R. P. DOKL. AKAD. NAUK SSSR 220 (4), 833–6, 1975.
80845	**BELLIDOITE. NEW COPPER SELENIDE.** DE MONTREUIL, L. A. ECON. GEOL. BULL. SOC. ECON. GEOL. 70 (2), 384–7, 1975.
80884	**IN SITU ROCK REFLECTANCE.** RAINES, G. L. LEE, K. PHOTOGRAMM. ENG. 41 (2), 189–98, 1975.
80989	**SPECIFIC HEAT AND THERMAL CONDUCTIVITY MEASUREMENTS ON THIN FILMS WITH A PULSE METHOD.** FILLER, R. L. LINDENFELD, P. DEUTSCHER, G. REV. SCI. INSTRUM. 46 (4), 439–42, 1975.
81116	**LOW TEMPERATURE SPECIFIC HEAT OF (VANADIUM CHROMIUM) OXIDE AND (VANADIUM ALUMINUM) OXIDE.** WENGER, L. E. KEESOM, P. H. PHYS. REV. 12 B (11), 5288–96, 1975.
81213	**THE THERMAL CONDUCTIVITY OF SOME ROCKS FROM ROMANIA.** DEMETRESCU, C. STUD. CERCET. GEOLOGIA, GEOFIZICA GEOGRAFIE, SER. GEOFIZICA 11 (2), 255–62, 1973.
81230	**OPTICAL PROPERTIES OF NATURAL MAGNETITE IN THE REGION FROM 2.5 TO 50 MICROMETERS.** MOUSSOUROS, P. K. INFRARED PHYS. 15 (2), 69–72, 1975.
81256	**THERMAL CONDUCTIVITY OF MICA AS A FUNCTION OF TEMPERATURE AND DIRECTION OF HEAT FLOW.** MIRKOVICH, V. V. REV. INT. HAUTES TEMP. REFRACT. 12 (2), 106–9, 1975.
81297	**HIGH THERMAL CONDUCTIVITY MICROWAVE SUBSTRATES.** POPE, B. J. HORTON, M. D. HALL, H. T. U. S. ARMY ELECTRONICS COMMAND, FORT MONMOUTH, NEW JERSEY 13PP., 1975. (ECOM–75–1304–1)

TPRC Number	Bibliographic Citation
81307	**REFLECTIVITY MEASUREMENTS OF SOME ORE MINERALS.** ABDEL-AAL, O. Y. NAKHLA, F. M SALEH, S. A. J. GEOL. SOC. IRAQ 7, 35-49, 1974.
81682	**THERMAL CONDUCTIVITY OF ANHYDROUS BORAX, BORIC OXIDE, AND SODIUM SULFATE.** CREFFIELD, G. K. WICKENS, A. J. J. CHEM. ENG. DATA 20 (3), 223-5, 1975.
81726	**SEMICONDUCTING PROPERTIES OF NATURAL PYRITE. IV. OPTICAL TRANSITIONS.** HORITA, H. TOHOKU DAIGAKU SENKO SEIREN KENKYUSHO IHO 30 (2), 99-108, 1974.
81871	**CUPRIFEROUS TROILITE IN ORES OF THE OKTYABRSK AND TALNAKH DEPOSITS, NORILSK ORE DISTRICT.** KARPENKOV, A. M. MITENKOV, G. A. RUDASHEVSKII, N. S. SOKOLOVA, N. G. SHISHKIN, N. N. GEOL. RUDN. MESTOROZHD 17 (1), 64-6, 1975.
81872	**OPTICAL PROPERTIES OF ANISOTROPIC MINERAL ORES IN REFLECTED LIGHT.** VYALSOV, L. N. GEOL. RUDN. MESTOROZHD. 17(1), 47-57, 1975.
81874	**MICROWAVE MEASUREMENT OF THE TEMPERATURE COEFFICIENT OF PERMITTIVITY FOR SAPPHIRE AND ALUMINA.** AITKEN, J. E. LADBROOKE, P. H. POTOK, M. H. N. IEEE TRANS. MICROWAVE THEORY TECH. MTT23(6), 526-9, 1975.
81946	**THERMAL CONDUCTIVITY AND DIFFUSIVITY OF ENGINEERING CERAMICS.** MATTA, J. E. SIEBENECK, J. H. GANGULY, B. K. HASSELMAN, D. P. H. LEIHIGH UNIVERSITY CERAMICS RES. LAB., BETHLEHEM, PA. 84 PP., 1975. (AD-A015 717)
81954	**HANDBOOK OF THE INFRARED OPTICAL PROPERTIES OF ALUMINUM OXIDE, CARBON, MAGNESIUM OXIDE, AND ZIRCONIUM OXIDE. VOLUME I.** WHITSON, M. E. AEROSPACE CORP., EL SEGUNDO, CALIFORNIA 469PP., 1975. (AD-A013 722, SAMSCO-TR-75-131-VOL-1)
81966	**TWO NEW INTERESTING CONSTANTS FOR THE VISCOSITY AND THERMAL EXPANSION OF ENAMEL AND GLASS.** DEKKER, P. GLAS-EMAIL-KERAMO-TECH. (4), 127-34, (5), 161-4, 1972.
81987	**ADDITIVES AS A FACTOR IN THE THERMAL CONDUCTIVITY OF BERYLLIA.** BELYAEV, R. A. SMOLYA, A. V. VAVILOV, YU. V. SHALIMOVA, I. F. PANTELEEVA, I. F. POPOVA, V. P. REFRACTORIES, USSR (2), 120-2, 1975. (ENGLISH TRANSLATION OF OGNEUPORY, (2), 51-3, 1975; FOR ORIGINAL SEE T80319)
82018	**CHARACTERIZATION OF ROCK MELTS AND GLASSES FORMED BY EARTH-MELTING SUBTERRENES.** LUNDBERG, L. B. LOS ALAMOS SCIENTIFIC LAB., N. MEX. 13PP., 1975. (LA-5826-MS, N75-25271)
82019	**THERMAL CONDUCTIVITY OF A FROZEN HUBBARD LOAMY SAND.** GILLEY, J. E., JR. MINNESOTA UNIVERSITY, ST. PAUL, M. S. THESIS 91PP., 1974. (PB-240 025/7, N75-25286)
82040	**POROSITY-THERMAL CONDUCTIVITY CORRELATIONS FOR CERAMIC MATERIALS.** RHEE, S. K. MATER. SCI. ENG. 20 (1), 89-93, 1975.
82103	**USE OF SILICA-CONTAINING TAILINGS IN THE PRODUCTION OF ARTICLES FOR INDUSTRIAL HEAT-INSULATION.** TORGASHOV, V. V. TABACHISHINA, N. M. OLGINSKII, A. G. SPIRIN, YU. A. STROIT. MATER (3), 29-30, 1975.
82185	**DETERMINATION OF THERMAL CONDUCTIVITY OF REFRACTORY BRICKS BY THE TRANSIENT LINE SOURCE METHOD AND ITS RELATION WITH TEMPERATURE AND POROSITY.** CHAWLA, O. P. INDIAN J. TECHNOL. 13 (2), 80-4, 1975.
82320	**DIRECT MEASUREMENT OF THERMAL DIFFUSIVITY OF POWDERED MATERIALS BY TRANSIENT HEAT FLOW METHOD.** AMANCHARLA, J. UNIVERSITY OF TEXAS, ARLINGTON, M. S. THESIS 37 PP., 1974.
82363	**THERMAL CONDUCTIVITIES OF ARSENIC-SELENIUM GLASSES.** KURIYAMA, M. J. AM. CERAM. SOC. 58 (7-8), 302-3, 1975.
82557	**THERMAL CONDUCTIVITY AND DIFFUSIVITY OF SELECTED POROUS INSULATIONS BETWEEN 4 AND 300 K.** REINKER, R. P. TIMMERHAUS, K. D. KROPSCHOT, R. H. ADVAN. CRYOG. ENG. 20, 343-54, 1975.
82575	**THERMAL CONDUCTIVITY OF APOLLO 16 LUNAR FINES.** CREMERS, C. J. HSIA, H. S. PROC. LUNAR SCI. CONF., 5TH 3, 2703-8, 1974.
82684	**THERMAL EXPANSION OF NATURAL SPINEL, FERROAN GAHNITE, MAGNESIOCHROMITE, AND SYNTHETIC SPINEL.** SINGH, H. P. SIMMONS, G. MC FARLIN, P. F. ACTA CRYSTALLOGR. 31 A (6), 820-2, 1975.
82970	**THE HEAT CAPACITY OF MAGNETITE NEAR THE METAL-SEMICONDUCTOR TRANSITION POINT.** WOLF, J. R. THE AMERICAN UNIVERSITY, WASHINGTON, D. C., PH. D. THESIS 137PP., 1975. (UNIV. MICROFILM NO. 75-18571)
82998	**THE LOW TEMPERATURE THERMAL PROPERTIES OF GLASSES.** STEPHENS, R. B. CORNELL UNIV., PH.D. THESIS 1-307, 1974. (UNIV. MICROFILM NO. 75-4269, N75-21452)
83208	**SPECTROPHOTOMETRIC MEASUREMENT OF TRANSMISSION SPECTRA IN THE OPTICAL RANGE OF SILICATE GLASSES.** BOCHYNSKI, Z. SMYKALA, M. SZKLO CERAM. 26 (8-9), 232-4, 1975.
83260	**PHYSICAL PROPERTIES OF BASALTS FROM ARMENIA.** TOLSTOI, M. I. PRODAIVODA, G. T. MOLYAVKO, V. G. GEOFIZ. SB. (USSR) (61), 75-80, 1974.
83281	**THE ABSOLUTE DETERMINATION OF THE THERMAL CONDUCTIVITY OF SOME CERAMIC MATERIALS IN THE TEMPERATURE RANGE 100 TO 1000 DEGREES C.** JAKUBOWSKI, W. TRYKOZKO, R. JABLONSKI, W. PERYT, W. SZKLO AND CERAM. (POLAND) 26 (4), 108-11, 1975.
83343	**1. SPATIALLY RESOLVED ABSOLUTE SPECTRAL REFLECTIVITY OF JUPITER: 3390 - 8400 ANGSTROMS. 2. THE JOVIAN THERMAL STRUCTURE FROM PIONEER 10 INFRARED RADIOMETER DATA. 3. OBSERVATIONS AND ANALYSIS OF 8 - 14 MICRON MICRON THERMAL EMISSION OF JUPITER: A MODEL OF THERMAL STRUCTURE AND CLOUD PROPERTIES.** ORTON, G. S. CALIFORNIA INST. OF TECHNOLOGY, PH.D. THESIS 1-187, 1975. (UNIV. MICROFILM NO. 75-20,007)
83346	**INVESTIGATION OF A METHOD TO DETERMINE THE EFFECTS OF VARIABLES ON REFLECTANCE AND EMITTANCE DATA FROM NATURAL SURFACES UNDER NATURAL CONDITIONS.** HAGEWOOD, J. F. LOUISIANA STATE UNIV., PH.D. THESIS 1-380, 1974. (UNIV. MICROFILM NO. 75-14252, N75-28866)
83382	**LOW-TEMPERATURE AND LOW-EXPANSION GLASS-CRYSTAL COMPOSITES BY THE FORMATION OF PEROVSKITE LEAD TITANATE.** TUMMALA, R. R. J. MATER. SCI. 11 (1), 128-8, 1976.
83509	**RUTHENIUM, A NEW MINERAL FROM HOROKANAI, HOKKAIDO, JAPAN.** URASHIMA, Y. WAKABAYASHI, T. MASAKI, T. TERASAKI, Y. MINERAL. J., 7 (5), 438-44, 1974.
83559	**HEAT TRNNSFER MECHANISMS IN EVACUATED POWDER INSULATION.** GLASER, P. E. INTERN. HEAT TRANSFER CONF. 829-37, 1961.

TPRC Number	Bibliographic Citation
83577	**DEVELOPMENT OF SPACE STABLE THERMAL CONTROL COATINGS FOR USE ON LARGE SPACE VEHICLES.** GILLIGAN, J. E. HARADA, Y. IIT RESEARCH INSTITUTE, CHICAGO, ILLINOIS 464PP., 1976. (IITRI-C6233-57)
83581	**ANTIREFLECTION AND PROTECTIVE COATINGS OF COMPONENTS MADE FROM LEAD GERMANATE GLASSES.** SHIROKSHINA, Z. V. OPT. MEKH. PROM. 41, 39-41, 1974. (FOR ENGLISH TRANSLATION SEE T83582)
83582	**ANTIREFLECTION AND PROTECTIVE COATINGS OF COMPONENTS MADE FROM LEAD GERMANATE GLASSES.** SHIROKSHINA, Z. V. SOV. J. OPT. TECHNOL. 41 (9), 411-12, 1974. (ENGLISH TRANSLATION OF OPT. MEKH. PROM., 41, 39-41, 1974; FOR ORIGINAL SEE T83581)
83589	**SOLID SOLUTIONS AND PHASE TRANSITION IN THE SYSTEM STRONTIUM CERIUM OXIDE - STRONTIUM HAFNIUM OXIDE.** SOLOV'EVA, A. E. GAVRISH, A. M. IZV. AKAD. NAUK SSSR, NEORG. MATER. 10 (8), 1549-50, 1974. (FOR ENGLISH TRANSLATION SEE T83590)
83590	**SOLID SOLUTIONS AND PHASE TRANSITION IN THE SYSTEM STRONTIUM CERIUM OXIDE - STRONTIUM HAFNIUM OXIDE.** SOLOV'EVA, A. E. GAVRISH, A. M. INORG. MAT., USSR 10 (8), 1336-7, 1974. (ENGLISH TRANSLATION OF IZV. AKAD. NAUK SSSR, NEORG. MATER., 10 (8), 1549-50, 1974; FOR ORIGINAL SEE T83589)
83593	**REGION OF EXISTENCE OF FLUORITE-LIKE SOLID SOLUTIONS IN THE SYSTEM HAFNIUM OXIDE - MAGNESIUM OXIDE, AND THEIR PHYSICOCHEMICAL PROPERTIES.** KUZNETSOV, A. K. TIKHONOV, P. A. KRAVCHINSKAYA, M. V. IZV. AKAD. NAUK SSSR, NEORG. MATER. 10 (8), 1498-1501, 1974. (FOR ENGLISH TRANSLATION SEE T83594)
83594	**REGION OF EXISTENCE OF FLUORITE-LIKE SOLID SOLUTIONS IN THE SYSTEM HAFNIUM OXIDE - MAGNESIUM OXIDE, AND THEIR PHYSICOCHEMICAL PROPERTIES.** KUZNETSOV, A. K. TIKHONOV, P. A. KRAVCHINSKAYA, M. V. INORG. MAT., USSR 10 (8), 1290-3, 1974. (ENGLISH TRANSLATION OF IZV. AKAD. NAUK SSSR, NEORG. MATER., 10 (8), 1498-1501, 1974; FOR ORIGINAL SEE T83593)
83595	**THERMAL DIFFUSIVITY OF OLIVINE.** HOLT, J. B. EARTH PLANET. SCI. LETT., (NETHERLANDS) 27 (3), 404-8, 1975.
83663	**ON THE TEMPERATURE DEPENDENCE OF THE THERMAL CONDUCTIVITY OF CERAMIC MATERIALS WITH RAREFACTION OF THE GAS MEDIUM.** LITOVSKII, E. YA. ZBOROVSKII, I. D. USPENSKII, I. N. J. ENG. PHYS., USSR 27 (4), 1225-30, 1974. (ENGLISH TRANSLATION OF INZH. FIZ. ZH., 27 (4), 648-54, 1974; FOR ORIGINAL SEE T78988)
83743	**THE EFFECT OF AN ELECTRIC FIELD ON THE INDUCED OPTICAL ABSORPTION SPECTRA OF ELECTRON-IRRADIATED SODIUM BORATE GLASSES.** VAKHROMEEV, V. G. ZINOV'EVA, S. P. SOV. PHYS. J (11), 1612-3, 1974. (ENGLISH TRANSLATION OF IZV. VYSSH. UCHEB. ZAVED., FIZ., (11), 146-8, 1974; FOR ORIGINAL SEE T79531)
83746	**THERMOPHYSICAL PROPERTIES OF SELECTED AEROSPACE MATERIALS. PART I: THERMAL RADIATIVE PROPERTIES** TOULOUKIAN, Y. S. HO, C. Y. THERMOPHYSICAL AND ELECTRONIC PROPERTIES INFORMATION CENTER, PURDUE UNIVERSITY 1021PP., 1976.
83863	**JOAQUINITE: THE NATURE OF ITS WATER CONTENT AND THE QUESTION OF FOUR-COORDINATED FERROUS IRON.** ROSSMAN, G. R. AMER. MINERAL. 60 (5-6), 435-40, 1975.
83895	**INTERPRETATION OF VISIBLE AND NEAR-INFRARED DIFFUSE REFLECTANCE SPECTRA OF PYROXENES AND OTHER ROCK-FORMING MINERALS.** ADAMS, J. B. INFRARED RAMAN SPECTROSC. LUNAR TERR. MINER., 91-116, 1975.
83960	**REFLECTIVITY OF METALS AND THEIR CRYSTAL STRUCTURE.** NAKHLA, F. M. CHEM. ERDE 34 (2), 175-80, 1975.
84118	**INFRARED SPECTRA OF ACIDIC SOILS.** ALI, S. PANDEY, R. V. DE, S. K. J. INDIAN CHEM. SOC., 52 (9), 815-19, 1975.
84280	**OPTICAL PROPERTIES AND CHEMICAL COMPOSITIONS OF APOLLO 17 ARMALCOLITES.** WILLIAMS, K. L. TAYLOR, L. A. GEOLOGY 2 (1), 5-8, 1974.
84446	**NICKEL MINERALS FROM BARBERTON, SOUTH AFRICA. III. BONACCORDITE, THE NICKEL ANALOG OF LUDWIGITE.** ED WAAL, S. A. VILJOEN, E. A. CALK, L. C. TRANS. GEOL. SOX. S. AFR., 77 (3), 375, 1975.
84491	**THERMOPHYSICAL CHARACTERISTICS OF IRON-ORE PELLETS. COMMUNICATION 1.** BRATCHIKOV, S. G. STATNIKOV, B. SH. USOLTSEVA, G. I. IZV. VYSSH. UCHEBN. ZAVED., CHERN. METALL. (1), 32-5, 1976.
84492	**INTERFACIAL BONDING AND THERMAL EXPANSION OF FIBER-REINFORCED COMPOSITES.** MAROM, G. GERSHON, B. J. ADHES. 7 (3), 195-201, 1975.
84581	**STUDY OF THE PROPERTIES AND STRUCTURE OF PERICLASE-YTTRIA MATERIALS.** KAIBICHEVA, M. N. KAMENSKIKH, V. A. GILEV, YU. P. OGNEUPORY (1), 50-4, 1976. (FOR ENGLISH TRANSLATION SEE T91876)
84618	**RECENT DATA ON THE PHYSICAL PROPERTIES OF ROCK-FORMING MINERALS AND TRACE ELEMENTS OCCURRING IN THEM.** JAWORSKI, A. TECH. POSZUKIWAN GEOL. 14 (3), 11-16, 1975.
84691	**THERMAL, MECHANICAL, AND BRITTLE-TO-DUCTILE BEHAVIOR OF THORIATED TUNGSTEN.** IANNUZZI, F. A. STARRETT, H. S. PEARS, C. D. AIR FORCE MATERIALS LAB., WPAFB, OHIO, 135 PP., 1975. (AFML-TR-75-203)
84944	**HEAT EXTRACTION AROUND CORNERS IN CASTINGS.** APPADURAI, R. CHINNATHAMBI, K. PRABHAKAR, O. TRANS. INDIAN INST. MET. 28 (5), 411-18, 1975.
85031	**SYNTHETIC SAPPHIRE, AN INFRARED OPTICAL MATERIAL.** OLT, R. D. PROC. IRIS 3 (4), 141-53, 1958.
85032	**HIGH TEMPERATURE PROPERTIES OF INFRARED OPTICAL MATERIALS.** MC ALISTER, E. D. PROC. IRIS 4 (4), 139-45, 1959.
85093	**NEAR INFRARED ABSORPTION SPECTRA OF OPALS AND THE ROLE OF WATER IN THESE HYDRATED SILICON DIOXIDE MINERALS.** LANGER, K. FLOERKE, O. W. FORTSCHR. MINERAL. 52 (1), 17-51, 1974.
85197	**OPTICAL RADIATION MEASUREMENTS: DEVELOPMENT OF AN NATIONAL BUREAU OF STANDARDS REFERENCE SPECTROPHOTOMETER FOR DIFFUSE TRANSMITTANCE AND REFLECTANCE.** ENABLE, W. H., JR. HSIA, J. J. WEIDNER, V. R. NATIONAL BUREAU OF STANDARDS, WASHINGTON, D. C. 47PP., 1976. (NBS-TN-594-11)
85206	**TRANSPORT PROPERTIES OF DENSE MATTER.** FLOWERS, E. ITOH, N. ASTROPHYS. J. 206 1 (PT. 1), 218-42, 1976.
85218	**INDEXES OF REFRACTION (N) AND REFLECTIVITY (R) OF NATURAL GARNETS.** BISWAS, D. K. INDIAN MINERAL. 14, 74-9, 1975.
85230	**INFLUENCE OF WATER ON VISCOSITY AND THERMAL EXPANSION OF SODIUM TRISILICATE GLASSES.** SHELBY, J. E. MC VAY, G. L. J. NON-CRYST. SOLIDS 20 (3), 439-49, 1976.

TPRC Number	Bibliographic Citation

85232 **A VANADIUM PENTOXIDE PHOSPHATE FILTER GLASS WITH BANDPASS IN THE VISIBLE REGION.**
RES, M. A. KOK, C. J. BEDNARIK, J.
J. PHYS.
9 D (10), L119–L121, 1976.

85285 **REFLECTIVITY OF SPHALERITE AND ITS VARIATION WITH THE IRON CONTENT.**
BISHARA, W. W.
BULL. FAC. SCI., ASSIUT UNIV.
2 (2), 79–86, 1973.

85290 **DEVELOPMENTS IN THE TECHNIQUES OF COAL MICROSCOPY.**
DAVIS, A
MICROSTRUCT. SCI.
3 (PT. B), 973–90, 1975.

85313 **THERMODYNAMIC FUNCTIONS OF GEDRITE AND SOME EQUILIBRIUMS WITH ITS PARTICIPATION.**
KOLESNIK, YU. N. NOGTEVA, V. V.
NAUMOV, V. N. BUTENKO, V. I.
PAUKOV, I. E.
GEOKHIMIYA
(3), 441–7, 1976.
(FOR ENGLISH TRANSLATION SEE T91690)

85439 **LASER WINDOW TEST FACILITY.**
ZAR, J. L.
LASER INDUCED DAMAGE IN OPTICAL MATERIALS: 1975
175–88, 1976.
(NBS–SP–435, PB–252 186)

85445 **POWER DEPENDENCE OF REFLECTIVITY OF METALLIC FILMS.**
YEH, Y. C. STAFSUDD, O. M.
APPL. OPT.
15 (1), 127–31, 1976.

85499 **LOW TEMPERATURE THERMAL PROPERTIES OF GRANULAR MATERIALS.**
TAIT, R. H.
CORNELL UNIV., ITHACA, NY, PH.D. THESIS
1–191, 1975.
(UNIV. MICROFILM NO. 75–27864, N76–14954)

85507 **THE LOW TEMPERATURE HEAT CAPACITIES OF ANALBITE, LOW ALBITE, MICROCLINE, AND SANIDINE.**
OPENSHAW, R. E.
PRINCETON UNIVERSITY, PRINCETON, N. J., PH. D. THESIS
325PP., 1974.
(UNIV. MICROFILM NO. 75–23,230)

85517 **OPTICAL PROPERTIES AND ELECTRONIC STRUCTURE OF MANTLE SILICATES.**
NITSAN, U. SHANKLAND, T. J.
GEOPHYS. J. R. ASTRON. SOC.
45 (1), 59–87, 1976.

85563 **PRIMARY STATE STANDARD FOR THE THERMAL CONDUCTIVITY UNIT OF SOLIDS BETWEEN 90 AND 500 K.**
TATARASHVILI, D. A. SERGEEV, O. A.
CHISTYAKOV, YU. A.
IZMER. TEKH.
18 (4), 49–51, 1975.
(ENGLISH TRANSLATION SEE T85564)

85564 **PRIMARY STATE STANDARD FOR THE THERMAL CONDUCTIVITY UNIT OF SOLIDS BETWEEN 90 AND 500 K.**
TATARASHVILI, D. A. SERGEEV, O. A.
CHISTYAKOV, YU. A.
MEAS. TECH., USSR
18 (4), 552–5, 1975.
(ENGLISH TRANSLATION OF IZMER. TEKH., 18 (4), 49–51, 1975; FOR ORIGINAL SEE T85563)

85565 **TRANSPARENT HEAT MIRRORS FOR SOLAR-ENERGY APPLICATIONS.**
FAN, J. C. C. BACHNER, F. J.
APPL. OPT.
15 (4), 1012–17, 1976.

85566 **THIN–FILM CONDUCTING MICROGRIDS AS TRANSPARENT HEAT MIRRORS.**
FAN, J. C. C. BACHNER, F. J. MURPHY, R. A.
APPL. PHYS. LETT.
28 (8), 440–2, 1976.

85570 **AN IMPROVED METHOD OF COMPUTING THE THERMAL CONDUCTIVITY OF FLUID–FILLED SEDIMENTARY ROCKS.**
BECK, A. E.
GEOPHYSICS
41 (1), 133–44, 1976.

85580 **THERMAL CONDUCTIVITY MEASUREMENT BY AN IMPROVED AC HOT–WIRE METHOD.**
IRVING, J. B. JAMIESON, D. T.
NATIONAL ENGINEERING LAB., EAST KILBRIDE, SCOTLAND
46PP., 1976.
(NEL–REPT–609)

85637 **THERMODYNAMICS OF IRON(II)IRON(III) OXIDE SYSTEMS. II. ZINC– AND CADMIUM–DOPED FERRIFERROUS OXIDE AND CRYSTALLINE MAGNETITE.**
BARTEL, J. J. WESTRUM, E. F., JR.
J. CHEM. THERMODYN.
8 (6), 583–600, 1976.

85644 **THE THERMAL CONDUCTIVITY OF STABILIZED ZIRCONIA.**
GARVIE, R. C.
J. MATER. SCI.
11 (7), 1365–7, 1976.

85747 **OPTICAL STUDY OF CUPROSTIBITE.**
LOPEZ–SOLER, A. BOSCH–FIGUEROA, J. M.
KARUP–MOELLER, S. BESTEIRO, J.
FONT–ALTABA, M.
FORTSCHR. MINERAL.
52, 557–65, 1975.

85755 **ANISOTROPIC THERMAL EXPANSION CHARACTERISTICS OF WOLLASTONITE.**
WESTON, R. M. ROGERS, P. S.
MINERAL. MAG.
40 (314), 649–51, 1976.

85762 **THE LINEAR EXPANSION OF BOROSILICATE GLASSES IN RELATION TO THEIR STRUCTURE.**
KONIJNENDIJK, W. L. STEVELS, J. M.
VERRES REFRACT.
30 (3), 371–7, 1976.

85815 **THE THERMAL CONDUCTIVITY OF DIAMONDS.**
BERMAN, R. MARTINEZ, M.
THERMAL CONDUCTIVITY CONFERENCE, 14TH
3–10, 1976.

85837 **ANISOTROPIC LATTICE THERMAL CONDUCTIVITY OF ALPHA–QUARTZ AS A FUNCTION OF PRESSURE AND TEMPERATURE.**
DARBHA, D. M. SCHLOESSIN, H. H.
THERMAL CONDUCTIVITY CONFERENCE, 14TH
183–90, 1976.

85868 **A NEW TECHNIQUE FOR THE MEASUREMENT OF THERMAL CONDUCTIVITY OF THIN FILMS.**
BOYCE, T. C. CHUNG, Y. W.
THERMAL CONDUCTIVITY CONFERENCE, 14TH
499–506, 1976.

85954 **INFRARED REFLECTANCE: INDEPENDENT MEASUREMENTS YIELD GOOD AGREEMENT.**
WILLEY, R. R.
APPL. OPT.
15 (5), 1124, 1976.

86042 **THERMAL DIFFUSIVITY OF YTTRIA–STABILIZED ZIRCONIA.**
MIRKOVICH, V. V.
HIGH TEMP. HIGH PRESSURES
8 (2), 231–5, 1976.

86147 **EFFECTS OF NUCLEAR RADIATION ON THE OPTICAL PROPERTIES OF CERIUM–DOPED GLASS.**
MC GRATH, B. SCHOENBACHER, H.
VAN DE VOORDE, M.
NUCL. INSTRUM. METHODS
135 (1), 93–7, 1976.

86303 **THE THERMODYNAMICS AND KINETICS OF PHASE DECOMPOSITION IN THE SECOND ORDER PHASE TRANSITION OF A BOROSILICATE GLASS.**
COSTA, J. M. L.
CATHOLIC UNIV. OF AMERICA, WASHINGTON, D. C., PH. D. THESIS
169PP., 1975.
(N76–15312, UNIV. MICROFILM NO. 76–1179)

86304 **THE FAR ULTRAVIOLET ALBEDO OF THE MOON.**
LUCKE, R. L.
JOHNS HOPKINS UNIV., BALTIMORE, MD., PH.D. THESIS
1–163, 1975.
(UNIV. MICROFILM NO. 76–1573, N76–15963)

86706 **PHONON THERMAL TRANSPORT IN NON–CRYSTALLINE MATERIALS.**
ZAITLIN, M. P.
UNIVERSITY OF ILLINOIS AT URBANA–CHAMPAIGN, PH. D. THESIS
159PP., 1975.
(UNIV. MICROFILM NO. 76–7022)

86711 **INFRARED ASTRONOMICAL SPECTROSCOPY FROM ALTITUDE AIRCRAFT.**
SCHAACK, D. F.
CORNELL UNIV., N.Y., PH.D. THESIS
1–138, 1975.
(UNIV. MICROFILM NO. 76–12890, N76–28117)

86712 **PHOTOMETRIC STUDIES OF PHOBOS, DEIMOS, AND THE SATELLITES OF SATURN.**
NOLAND, M. C.
CORNELL UNIV., N.Y., PH.D. THESIS
1–275, 1975.
(UNIV. MICROFILM NO. 76–12881, N76–28118)

86759 **SYMMETRY OF NORMAL VIBRATIONS OF A TOPAZ CRYSTAL.**
KOVALEVA, L. T.
J. APPL. SPECTROS., USSR
22 (2), 237–41, 1975.
(ENGLISH TRANSLATION OF ZH. PRIKL. SPEKTROSK., 22 (2), 311–16, 1975; FOR ORIGINAL SEE T80566)

TPRC Number	Bibliographic Citation
86771	**MERCURY: SPECTRAL REFLECTANCE MEASUREMENTS (0.33–1.06 MICROMETER) 1974/75.** VILAS, F. MC CORD, T. B. ICARUS 28 (4), 593–9, 1976.
86877	**THERMAL CONDUCTIVITY OF MICA FROM 1.5 TO 4.2 K.** FALCO, C. M. J. APPL. PHYS. 47 (7), 3355–6, 1976.
87010	**THERMAL EXPANSION AT LOW TEMPERATURES OF GLASS–CERAMICS AND GLASSES.** WHITE, G. K. CRYOGENICS 16 (8), 487–90, 1976.
87015	**THERMAL EXPANSION OF EUROPIUM SESQUIOXIDE.** KELLY, B. T. MC CALLA, B. A. WOSTENHOLM, G. H. YATES, B. J. NUCL. MATER. 61 (2), 221–4, 1976.
87028	**SOIL HEAT FLUX DETERMINATION: TEMPERATURE GRADIENT METHOD WITH COMPUTED THERMAL CONDUCTIVITIES.** KIMBALL, B. A. JACKSON, R. D. NAKAYAMA, F. S. ISDO, S. B. REGINATO, R. J. SOIL SCI. SOC. AM., J. 40 (1), 25–8, 1976.
87131	**THE INTERACTIONS OF PAIRS OF DIFFERENT IONS IN GLASSES.** SKOROSPELOVA, V. I. STEPANOV, S. A. FIZ. TVERD. TELA 17 (9), 2810–12, 1975. (FOR ENGLISH TRANSLATION SEE T87132)
87132	**THE INTERACTIONS OF PAIRS OF DIFFERENT IONS IN GLASSES.** SKOROSPELOVA, V. I. STEPANOV, S. A. SOV. PHYS. SOLID STATE 17 (9), 1878–9, 1976. (ENGLISH TRANSLATION OF FIZ. TVERD. TELA, 17 (9), 2810–12, 1975; FOR ORIGINAL SEE T87131)
87185	**HIGH THERMAL CONDUCTIVITY MICROWAVE SUBSTRATES.** POPE, B. J. HORTON, M. D. HALL, H. T. U. S. ARMY ELECTRONIC COMMAND, FORT MANMOURTH, N. J. 29PP., 1977. (ECOM–75–1304–F)
87213	**SPECTROPHOTOMETER ATTACHMENT FOR MEASURING THE REFLECTION OF LIGHT FROM THE SURFACE OF SPECIMENS UP TO 300 MILLIMETER IN DIAMETER.** KUDRYAVTSEVA, A. G. FEDOTOVA, G. S. OPT. MEKH. PROM. 43, 73–4, 1976. (ENGLISH TRANSLATION SEE T87214)
87214	**SPECTROPHOTOMETER ATTACHMENT FOR MEASURING THE REFLECTION OF LIGHT FROM THE SURFACE OF SPECIMENS UP TO 300 MILLIMETER IN DIAMETER.** KUDRYAVTSEVA, A. G. FEDOTOVA, G. S. SOV. J. OPT. TECHNOL. 43 (2), 131–2, 1976. (ENGLISH TRANSLATION OF OPT. MEKH. PROM., 43, 73–4, 1976; FOR ORIGINAL SEE T87213)
87238	**AN EVALULTION OF WINDOW GLASS FOR AIR TRAFFIC CONTROL TOWER CABS.** PAUL, L. E. BRADLEY, J. R. MARTIN, D. A. NATIONAL AVIATION FACILITIES EXPERIMENTAL CENTER, ATLANTIC CITY, NEW JERSEY 45PP., 1976. (AD–A031 924, FAA–RD–76–105)
87277	**X–RAY DILATOMETRY OF NATURAL AND SYNTHETIC ZIRCON.** GAVRISH, A. M. ZOZ, E. I. SOLOV'EVA, A. E. IZV. AKAD. NAUK SSSR, NEORG. MATER. 12 (8), 1502–3, 1976. (FOR ENGLISH TRANSLATION SEE T89152)
87278	**DIELECTRIC AND THERMOPHYSICAL PROPERTIES OF TRANSITION METAL OXIDES.** SAMSONOV, G. V. FEN, E. K. MALAKHOV, YA. S. MALAKHOV, V. YA. IZV. AKAD. NAUK SSSR, NEORG. MATER. 12 (8), 1404–10, 1976. (FOR ENGLISH TRANSLATION SEE T89160)
87315	**X–RAY DETERMINATION OF THERMAL EXPANSION OF OLIVINES.** SINGH, H. P. SIMMONS, G. ACTA CRYSTALLOGR. 32 A (5), 771–3, 1976.
87316	**ANALYSIS OF HEAT FLOW DATA – CORRELATION OF THERMAL RESISTIVITY AND SHOCK METAMORPHIC GRADE AND ITS USE AS EVIDENCE FOR AN IMPACT ORIGIN OF THE BRENT CRATER.** BECK, A. E. HAMZA, V. M. CHANG, C. C. CAN. J. EARTH SCI. 13 (7), 929–36, 1976.
87322	**SYNTHESIS AND STUDY OF ALUMINA FOR A TRANSPARENT CERAMIC.** SOKOL, V. A. ROKHLENKO, D. A. KONONOVA, L. I. ZAVORUEVA, R. S. BYCHKOV, N. V. BROMBERG, A. V. IZV. AKAD. NAUK SSSR, NEORG. MATER. 12 (8), 1419–23, 1976. (FOR ENGLISH TRANSLATION SEE T89153)
87344	**PREPARATION OF ULTRA–THIN MICA WINDOWS.** MULDER, B. J. J. PHYS. 9 E (9), 724–5, 1976.
87429	**DIMENSIONAL STABILITY OF FUSED SILICA, INVAR, AND SEVERAL ULTRALOW THERMAL EXPANSION MATERIALS.** BERTHOLD, J. W., III. JACOBS, S. F. NORTON, M. A. APPL. OPT. 15 (8), 1898–9, 1976.
87488	**THERMAL PROPERTIES OF SOME MATERIALS USED IN IGNITERS.** ROBERTSON, M. M. SANDIA LABS., ALBUQUERQUE, N. MEX. 30PP., 1976. (SAND–76–0134)
87489	**APPLICATION OF LANDSAT IMAGERY TO GEOLOGIC MAPPING IN THE ICEFREE VALLEYS OF ANTARCTICA.** HOUSTON, R. S. MARRS, R. W. SMITHSON, S. B. WYOMING UNIVERSITY, LARAMIE, REMOTE SENSING LAB. 78PP., 1976. (E76–10345)
87491	**SPECTRAL ESTIMATES OF PLANET ALBEDO AND COMPARISON TO SKYLAB OBSERVATIONS.** RAINEY, D. A. MARLATT, W. E. COLORADO STATE UNIV., FORT COLLINS 22PP., 1975. (PB–253 550)
87561	**THERMAL EXPANSION OF MIXED–ALKALI SILICATE GLASSES.** SHELBY, J. E. J. APPL. PHYS. 47 (10), 4489–96, 1976.
87595	**SUBSURFACE WATER PARAMETERS: OPTIMIZATION APPROACH TO THEIR DETERMINATION FROM REMOTELY SENSED WATER COLOR DATA.** JAIN, S. C. MILLER, J. R. APPL. OPT. 15 (4), 886–90, 1976.
87604	**COMPARISON OF SPECTRAL REFLECTION COEFFICIENTS OF WHITE SURFACE SAMPLES MADE FROM MS–20 GLASS ACCORDING TO THE MEASUREMENT RESULTS IN THE GDR AND USSR.** KOENIG, H. NIKONOVA, E. I. POKROVSKAYA, G. V. YUSTOVA, E. N. IZMER. TEKH. (8), 21–3, 1976. (FOR ENGLISH TRANSLATION SEE T91194)
87714	**THERMAL CONDUCTIVITY MEASUREMENT AND PREDICTION FROM GEOPHYSICAL WELL LOG PARAMETERS WITH BOREHOLE APPLICATION.** GOSS, R. COMBS, J. PROC. UNITED NATIONS SYMP. ON THE DEVELOPMENT AND USE OF GEOTHERMAL RESOURCES, 2ND 2, 1019–27, 1976.
87952	**PHASE RELATIONS AND THERMAL EXPANSION IN THE HAFNIUM OXIDE – TITANIUM(4H) OXIDE SYSTEM.** RUH, R. HOLLENBERG, G. W. CHARLES, E. G. PATEL, V. A. J. AM. CERAM. SOC. 59 (11–12), 495–9, 1976.
87966	**THE PHYSICAL PROPERTIES OF OCEANIC BASEMENT ROCKS FROM DEEP DRILLING ON THE MID–ATLANTIC RIDGE.** HYNDMAN, R. D. DRURY, M. J. J. GEOPHYS. RES. 81 (23), 4042–52, 1976.
87993	**THERMAL CONDUCTIVITY OF FROZEN SOILS.** FUKUDA, M. SAKIKAWA, S. LOW TEMP. SCI. A, (33), 259–63, 1975.
88022	**SPECTRA OF DIFFUSE REFLECTION AS AN INFORMATION SOURCE ABOUT THE ABSORPTION SPECTRA OF ADSORBED MOLECULES.** KOTOV, Y. I. NAVAL INTELLIGENCE SUPPORT CTR., WASHINGTON, D. C., 9 PP., 1976. (ENGLISH TRANSLATION OF TEOR. PRIKL. PROBL. RASSEYAN. SVETA, 387–95, 1971; FOR ORIGINAL SEE T067728) (AD–A032 547 NISC–TRANS–3842)
88036	**THEORETICAL STUDIES OF MATERIALS FOR HIGH–POWER INFRARED COATINGS.** SPARKS, M. XONICS INC., VAN NUYS, CA 345PP., 1975. (AD–A031 948, N77–21943, AD–A034 087)

TPRC Number	Bibliographic Citation
88047	**MEASUREMENT OF FREE ELECTRON DENSITY AT THE ONSET OF LASER-INDUCED SURFACE DAMAGE IN BSC-2.** ALYASSINI, N. PARKS, J. H. LASER INDUCED DAMAGE IN OPTICAL MATERIALS: 1975 356-61, 1976. (NBS-SP-435, PB-252 186)
88185	**A TRANSIENT METHOD FOR RAPIDLY MEASURING THERMAL CONDUCTIVITY AND DIFFUSIVITY OF SALT MINE CORE SAMPLES (THE PLANE PROBE).** JURY, S. H. GODFREY, T. G. OAK RIDGE NATL. LAB., OAK RIDGE, TN 43PP., 1976. (ORNL-TM-4956)
88190	**STUDY OF THE PETROGRAPHY AND THERMAL CONDUCTIVITY OF SOME NATURAL MATERIALS UTILIZED IN CIVIL ENGINEERING.** DEVISME, J. M. DEVISME, V. MERIAUX, E. ANN. MINES BELG. (4), 353-64, 1976.
88199	**THERMAL EXPANSION - NONMETALLIC SOLIDS.** TOULOUKIAN, Y. S. KIRBY, R. K. TAYLOR, R. E. LEE, T. Y. R. THERMOPHYSICAL PROPERTIES OF MATTER - THE TPRC DATA SERIES 13, 1786PP., 1977.
88203	**THERMAL CONDUCTIVITIES OF LITHIUM, SODIUM, POTASSIUM, AND SILVER BETA-ALUMINA BELOW 300 K.** ANTHONY, P. J. ANDERSON, A. C. PHYS. REV. 14 B (12), 5198-204, 1976.
88314	**THE GLASS OF THE PRIMARY MIRROR OF THE LARGE AZIMUTHAL TELESCOPE.** BUZHINSKII, I. M. KORYAGINA, E. I. DMITRIEVA, Z. P. SOV. J. OPT. TECHNOL. 43 (11), 669-71, 1976. (FOR ENGLISH TRANSLATION SEE T91214)
88372	**HIGH-TEMPERATURE HEAT CAPACITIES OF SODIUM ZIRCONO- AND HAFNOSILICATES.** SHIBANOV, E. V. CHUKHLANTSEV, V. G. ZH. FIZ. KHIM. 50 (9), 2447-8, 1976. (FOR ENGLISH TRANSLATION SEE T90594)
88395	**APPLICATION OF OPTICAL FIBERS TO THE TRANSMISSION OF SOLAR RADIATION.** KATO, D. NAKAMURA, T. J. APPL. PHYS. 47 (10), 4528-31, 1976.
88411	**DIELECTRIC COATINGS ON METAL SUBSTRATE.** GLAROS, S. S. BAKER, P. MILAM, D. LASER INDUCED DAMAGE IN OPTICAL MATERIALS 331-7, 1976. (NBS-SP-462, PB-262 553/1SL)
88413	**OPTICAL CONSTANTS OF CARBONS AND COALS IN THE INFRARED.** FOSTER, F. J. HOWARTH, C. R. CARBON 6, 719-29, 1968.
88416	**CADMIUM STANNATE SELECTIVE OPTICAL FILMS FOR SOLAR ENERGY APPLICATIONS.** HAACKE, G. BURTON, L. C. CHEM. RES. DIV., AMER. CYANAMID CO., STANFORD, CONN. 63PP., 1976. (NASA-CR-143107 NSF/RANN/SE/AER73-07957/FR/76/2)
88455	**VISCOSITIES OF BASALT AND ANDESITE MELTS AT HIGH PRESSURES.** KUSHIRO, I. YODER, H. S., JR. MYSEN, B. O. J. GEOPHYS. RES. 81 (35), 3651-6, 1976.
88484	**STUDY OF THE THERMAL CONDUCTIVITY OF MOLTEN OXIDES.** SHURYGIN, P. M. BUZOVKIN, V. P. KALININA, K. TEPLOFIZ. VYS. TEMP. 14 (5), 1101-5, 1976. (FOR ENGLISH TRANSLATION SEE T89373)
88573	**REFLECTION SPECTRA OF TRIGONAL MERCURY(II) SULFIDE.** DONETSKIKH, V. I. SOBOLEV, V. V. OPT. SPEKTROSK. 42 (2), 401-3, 1977. (FOR ENGLISH TRANSLATION SEE T92560)
88621	**LOW-TEMPERATURE HEAT CAPACITY OF ALKALI-METAL AND SILVER BETA-ALUMINAS.** MC WHAN, D. B. VARMA, C. M. HSU, F. L. S. REMEIKA, J. P. PHYS. REV. 15 B (2), 553-60, 1977.
88682	**GLASSES DISIGNED FOR USE AS INTEGRAL SOLAR CELL COVERS.** RAUCH, H. W. ULRICH, D. R. GREEN, J. M. CONF. REC. IEEE PHOTOVOLTAIC SPEC. CONF., 10TH 182-7, 1974.
88782	**FORMATION AND COLOR OF THE SPINEL SOLID SOLUTION IN COBALT(II) OXIDE - ZINC OXIDE - ALUMINUM OXIDE - CHROMIUM(III) OXIDE - TITANIUM(IV) OXIDE SYSTEM.** OHTSUKA, A. KAZAMA, K. YOGYO KYOKAI SHI 84 (10), 457-69, 1976.
88798	**CALORIMETRIC INVESTIGATIONS OF THE MAGNETIC ALLOYS: DILUTE GOLD - IRON AND DILUTE COPPER MANGANESE; AND OF THE SESQUIOXIDES: (VANADIUM - CHROMIUM) OXIDE (VANADIUM - ALUMINUM) OXIDE AND (TITANIUM - VANADIUM) OXIDE.** WENGER, L. E. PURDUE UNIVERSITY, PH. D. THESIS 126PP., 1975. (UNIV. MICROFILM NO. 76-20,298)
88837	**EFFECT OF MAIN SUBSTITUTIONS ON OPTICAL PROPERTIES IN THE TENNANTITE-TETRAHEDRITE SERIES.** CHARLAT, M. LEVY, C. BULL. SOC. FR. MINERAL. CRISTALLOGR. 99 (1), 29-37, 1976.
88841	**PEROVSKITES AND GARNETS.** KHATTAK, C. P. WANG, F. F. Y. BROOKHAVEN NATL. LAB., UPTON, N. Y. 173PP., 1976. (BNL-21638)
88857	**MODELS FOR THERMAL AND VIBRATIONAL PROPERTIES OF GLASSES AT LOW TEMPERATURES.** LU, M. S. CORNELL UNIV., PH.D. THESIS 1-105, 1975. (UNIV. MICROFILM NO. 76-8147)
88872	**THERMAL CONDUCTIVITY AND DIFFUSIVITY OF ENGINEERING CERAMICS.** SIEBENECK, J. H. HASSELMAN, D. P. H. CERAMICS RES. LAB., LEHIGH UNIV., BETHLEHEM, PA. 104PP., 1976. (AD-A029 660, N77-18273)
88904	**AN ABSOLUTE INTERFEROMETRIC DILATOMETER.** BENNETT, S. J. J. PHYS. 10 E (5), 525-30, 1977.
88941	**THE SPECTRAL REFLECTIVITY OF ION-BOMBARDED SURFACES.** PRIMAK, W. J. APPL. PHYS. 48 (4), 1556-62, 1977.
88950	**THE THERMAL CONDUCTIVITY OF FIFTEEN FELDSPAR SPECIMENS.** SASS, J. H. J. GEOPHYS. RES. 70 (16), 4064-5, 1965.
88951	**OPTICAL PROPERTIES OF ZIRCONIUM DIOXIDE IN THE INFRARED REGION.** ALEKSANDROV, V. A. VASILEV, A. B. KALGIN, YU. A. KISLOVSKII, L. D. TATARINTSEV, V. M. OPT. SPEKTROSK. 40 (6), 1087-9, 1976. (FOR ENGLISH TRANSLATION SEE T88952)
88952	**OPTICAL PROPERTIES OF ZIRCONIUM DIOXIDE IN THE INFRARED REGION.** ALEKSANDROV, V. A. VASILEV, A. B. KALGIN, YU. A. KISLOVSKII, L. D. TATARINTSEV, V. M. OPT. SPECTROS., USSR 40 (6), 627-8, 1976. (ENGLISH TRANSLATION OF OPT. SPEKTROSK., 40 (6), 1087-9, 1976; FOR ORIGINAL SEE T88951)
88957	**THE OPTICAL CHARACTERISTICS OF NATURAL CRYSTALS OF ICELAND SPAR.** VOLKOVA, N. V. ROZENMAN, L. S. FOMINA, V. N. TSIRUL'NIK, P. N. OPT. MEKH. PROM. 43, 38-40, 1976. (ENGLISH TRANSLATION SEE T88958)
88958	**THE OPTICAL CHARACTERISTICS OF NATURAL CRYSTALS OF ICELAND SPAR.** VOLKOVA, N. V. ROZENMAN, L. S. FOMINA, V. N. TSIRUL'NIK, P. N. SOV. J. OPT. TECHNOL. 43 (6), 366-8, 1976. (ENGLISH TRANSLATION OF OPT. MEKH. PROM., 43, 38-40, 1976; FOR ORIGINAL SEE T88957)
89014	**INVESTIGATIONS ON THE THERMAL EXPANSION OF THICK-FILM MATERIALS AND CERAMIC SUBSTRATES.** JENTZSCH, J. OSTWALD, R. BOGENSCHUETZ, A. F. WISS. BER. AEG-TELEFUNKEN 49 (6), 229-34, 1976.

TPRC Number	Bibliographic Citation
89029	**THE FIRING REACTION OF BONE CHINA BODY IN THE USE OF CALCIUM HYDROGEN PHOSPHATE AS A BY-PRODUCT OF GELATIN PRODUCTION.** ICHIKO, T. ISONO, T. MOCHIZUKI, K. YOGYO KYOKAI SHI 85 (1), 1-8, 1977.
89033	**EFFECTS OF QUARTZ ADDITION ON FIRING PROCESS OF BONE CHINA BODY.** ICHIKO, T. ISONO, T. MOCHIZUKI, K. YOGYO-KYOKAI-SHI 85 (5), 24-31, 1977.
89035	**CHARACTERIZATION AND THERMAL CONDUCTIVITIES OF SOME SAMPLES OF CONASAUGA SHALE.** DELL'AMICO, J. J. CAPTAIN, F. K. CHANSKY, S. H. OAK RIDGE NATIONAL LAB., TENN. 48PP., 1967. (ORNL-MIT-20)
89038	**OBSERVATION ON CORDIERITE-NULLITE MATERIALS.** HOMBACH, K. RASCH, H. SILIK. J. 15 (9), 285-9, 291-2, 1976.
89070	**STUDIES ON BINARY PHOSPHATE GLASSES FREE FROM GLASS MODIFYING OXIDE - PROPERTIES AND STRUCTURE OF GLASSES IN THE ANTIMONY OXIDE - PHOSPHORUS OXIDE SYSTEM.** TAKAHASHI, K. MOCHIDA, N. MATSUI, H. AS1HI GARASU KOGYO GIJUTSU SHOREIKAI KENKYU HOKOKU 27, 41-54, 1975.
89071	**PROPERTIES AND STRUCTURE OF GLASSES IN THE SYSTEMS SILICON OXIDE - PHOSPHATE AND GERMANIUM OXIDE - PHOSPHATE.** TAKAHASHI, K. MOCHIDA, N. MATSUI, H. TAKEUCHI, S. GOHSHI, Y. YOGYO KYOKAI SHI 84 (10), 482-90, 1976.
89094	**ESTIMATION OF THERMAL DIFFUSIVITY FROM FIELD OBSERVATIONS OF TEMPERATURE AS A FUNCTION OF TIME AND DEPTH [LAKE SEDIMENTS].** ADAMS, W. M. WATTS, G. MASON, G. AM. MINERAL. 61 (7-8), 560-8, 1976.
89122	**THERMAL DIFFUSIVITY, CONDUCTIVITY AND THERMAL INERTIA OF APOLLO 11 LUNAR MATERIAL.** HORAI, K. SIMMONS, G. KANAMORI, H. WONES, D. PROC. OF THE APOLLO 11 LUNAR SCI. CONF. 3, 2243-9, 1970.
89123	**THERMAL DIFFUSIVITY OF LUNAR ROCK SAMPLE 12002,85.** HORAI, K. WINKLER, J., JR. PROC. LUNAR SCI. CONF., 6TH 3207-15, 1975.
89124	**THERMAL DIFFUSIVITY OF FOUR APOLLO 17 ROCK SAMPLES.** HORAI, K. WINKLER, J., JR. PROC. LUNAR SCI. CONF., 7TH 3183-3204, 1976.
89152	**X-RAY DILATOMETRY OF NATURAL AND SYNTHETIC ZIRCON.** GAVRISH, A. M. ZOZ, E. I. SOLOV'EVA, A. E. INORG. MAT., USSR 12 (8), 1232-3, 1976. (ENGLISH TRANSLATION OF IZV. AKAD. NAUK SSSR, NEORG. MATER., 12 (8), 1502-3, 1976; FOR ORIGINAL SEE T87277)
89153	**SYNTHESIS AND INVESTIGATION OF ALUMINUM OXIDE FOR TRANSPARENT CERAMIC.** SOKOL, V. A. ROKHLENKO, D. A. KONONOVA, L. I. ZAVORUEVA, R. S. BYCHKOV, N. V. BROMBERG, A. V. INORG. MAT., USSR 12 (8), 1170-3, 1976. (ENGLISH TRANSLATION OF IZV. AKAD. NAUK SSSR, NEORG. MATER., 12 (8), 1419-23, 1976; FOR ORIGINAL SEE T87322)
89160	**DIELECTRIC AND THERMOPHYSICAL PROPERTIES OF TRANSITION METAL OXIDES.** SAMSONOV, G. V. FEN, E. K. MALAKHOV, YA. S. MALAKHOV, V. YA. INORG. MAT., USSR 12 (8), 1159-63, 1976. (ENGLISH TRANSLATION OF IZV. AKAD. NAUK SSSR, NEORG. MATER., 12 (8), 1404-10, 1976; FOR ORIGINAL SEE T87278)
89182	**VISIBLE AND NEAR INFRARED SPECTRA OF MINERALS AND ROCKS. XI. SEDIMENTARY ROCKS.** HUNT, G. R. SALISBURY, J. W. MOD. GEOL. 5 (4), 211-17, 1976.
89183	**VISIBLE AND NEAR INFRARED SPECTRA OF MINERALS AND ROCKS. XII. METAMORPHIC ROCKS.** HUNT, G. R. SALISBURY, J. W. MOD. GEOL. 5 (4), 219-28, 1976.
89235	**COLORED FILTER GLASSES: AN INTERCOMPARISON OF GLASSES MADE BY DIFFERENT MANUFACTURERS.** DOBROWOLSKI, J. A. MARSH, G. E. CHARBONNEAU, D. G. ENG, J. JOSEPHY, P. D. APPL. OPT. 16 (6), 1491-512, 1977.
89267	**THE HEAT CAPACITIES OF CALORIMETRY CONFERENCE COPPER AND OF MUSCOVITE, PYROPHYLLITE, AND ILLITE, BETWEEN 15 AND 375 K AND THEIR STANDARD ENTROPIES AT 298.15 K.** ROBIE, R. A. HEMINGWAY, B. L. WILSON, W. H. J. RES. U. S. GEOL. SURV. 4 (6), 631-44, 1976.
89315	**MATERIALS DATA BOOK FOR ENGINEERS AND SCIENTISTS.** PARKER, E. R. MATERIALS DATA BOOK FOR ENGINEERS AND SCIENTISTS 72-3, 92-3, 120-3, 140-3, 158-61, 172-3, 188-91, 200-1, 206-7, 212-3, 220-1, 225, 278-81, 286-91, 322-9, 340-7, 1967.
89330	**THE PREPARATION OF MICA CYLINDERS FOR MEASUREMENT OF THERMAL CONDUCTIVITY.** MC DONALD, A. G. MIRKOVICH, V. V. CANADA CENTRE FOR MINERAL AND ENERGY TECHNOLOGY 12PP., 1975. (CANMET-REPT.-76-4)
89373	**INVESTIGATION OF THERMAL CONDUCTIVITY OF OXIDE MELTS.** SHURYGIN, P. M. BUZOVKIN, V. P. LEONOV, V. V. HIGH TEMP. 14 (5), 981-4, 1976. (ENGLISH TRANSLATION OF TEPLOFIZ. VYS. TEMP., 14 (5), 1101-5, 1976; FOR ORIGINAL SEE T88484)
89398	**SINGLE GLASS FILTER WITH SPECTRAL TRANSMITTANCE SHOWING V (WAVELENGTH) CHARACTERISTICS.** RES, M. A. KOK, C. J. J. PHYS., APPL. PHYS. 7 D (17), L196-L198, 1974.
89429	**THE INFLUENCE OF BASICITY AND RATE OF HEATING ON THE EFFECTIVE THERMO-PHYSICAL CHARACTERISTICS OF SLAGS (APPARATUS FOR SPECIFIC HEAT MEASUREMENT).** ABZALOV, V. M. YUR'EV, B. P. BRATCHIKOV, S. G. IZV. AKAD. NAUK. SSSR MET. (6), 10-17, 1976. (FOR ENGLISH TRANSLATION SEE T91205)
89460	**THERMAL CONDUCTIVITY OF MICA AT LOW TEMPERATURES.** GRAY, A. S. UHER, C. J. MATER. SCI. 12 (5), 959-65, 1977.
89480	**INFLUENCE OF FERROUS IONS ON THE THERMAL EXPANSION AND HEAT CAPACITY OF CUBIC ZINC SULFIDE.** SHEARD, F. W. SMITH, T. F. WHITE, G. K. BIRCH, J. A. J. PHYS. 10 C (5), 645-55, 1977.
89575	**ZONAL METHOD OF MEASURING THE INTEGRATED TRANSMISSION COEFFICIENT OF DARK FILTERS.** DEMKINA, L. V. OPT. MEKH. PROM. 43 (8), 71, 1976. (ENGLISH TRANSLATION SEE T89576)
89576	**ZONAL METHOD OF MEASURING THE INTEGRATED TRANSMISSION COEFFICIENT OF DARK FILTERS.** DEMKINA, L. V. SOV. J. OPT. TECHNOL. 43 (8), 517-18, 1976. (ENGLISH TRANSLATION OF OPT. MEKH. PROM., 43 (8), 517-18, 1976; FOR ORIGINAL SEE T89575)
89612	**AN ESTIMATE OF THE SURFACE FLOW OF HEAT IN THE WEST TEXAS PERMIAN BASIN.** BIRCH, F. CLARK, H. AMER. J. SCI. 243 A, 69-74, 1945.
89613	**A HEAT FLOW VALUE FOR A WELL IN CALIFORNIA.** BENFIELD, A. E. AMER. J. SCI. 245 A, 1-18, 1947.
89614	**TEMPERATURE, THERMAL CONDUCTIVITY, AND HEAT FLOW IN A DRILLED HOLE NEAR OAK RIDGE, TENNESSEE.** DIMENT, W. H. ROBERTSON, E. C. J. GEOPHYS. RES. 68 (17), 5035-47, 1963.
89615	**TERRESTRIAL HEAT FLOW NEAR WASHINGTON, D. C.** DIMENT, W. H. WERRE, R. W. J. GEOPHYS. RES. 69 (10), 2143-9, 1964.
89616	**SOME MEASUREMENTS OF HEAT FLOW THROUGH THE FLOOR OF THE NORTH ATLANTIC.** LISTER, C. R. B. REITZEL, J. S. J. GEOPHYS. RES. 69 (10), 2151-4, 1964.

TPRC Number	Bibliographic Citation
89617	**TERRESTRIAL HEAT FLOW NEAR ALBERTA, VIRGINIA.** DIMENT, W. H. RASPET, R. MAYHEW, M. A. WERRE, R. W. J. GEOPHYS. RES. 70 (4), 923-9, 1965.
89618	**HEAT FLOW DETERMINATIONS IN THE NORTHWESTERN UNITED STATES.** BLACKWELL, D. D. J. GEOPHYS. RES. 74 (4), 992-1007, 1969.
89619	**TERRESTRIAL HEAT FLOW DETERMINATIONS IN THE NORTH CENTRAL UNITED STATES.** COMBS, J. SIMMONS, G. J. GEOPHYS. RES. 78 (2), 441-61, 1973.
89620	**HEAT FLOW AT SPOR MOUNTAIN, JORDAN VALLEY, BINGHAM, AND LA SAL, UTAH.** COSTAIN, J. K. WRIGHT, P. M. J. GEOPHYS. RES. 78 (35), 8687-98, 1973.
89708	**THE OPTICAL PARAMETERS OF MARE AND HIGHLAND LUNAR SOILS.** ANTIPOVA-KARATAEVA, I. I. AKHMANOVA, M. V. DEMENT'EV, B. V. MARKOV, M. N. STAKHEEV, YU. I. TARASOV, L. S. SPACE RES. 14, 621-4, 1974.
89725	**THE SIGNIFICANCE OF THE FREQUENCY DEPENDENCE OF PHONON SCATTERING BY PLATELETS IN TYPE I(1) DIAMOND.** HUDSON, P. R. W. PHYS. STATUS SOLIDI 37 A (2), 645-51, 1976.
89727	**THERMAL CONDUCTIVITY OF HALITE USING A PULSED LASER.** SMITH, D. D. OAK RIDGE Y-12 PLANT, OAK RIDGE, TENN. 39PP., 1976. (Y/DA-7013, N77-28968)
89777	**EFFECT OF LIGHIUM AND SODIUM OXIDES ON THE THERMOOPTICAL PROPERTIES OF SILICATE AND PHOSPHATE GLASSES.** MOLEV, V. I. SHCHAVELEV, O. S. OPT.-MEKH. PROM-ST. 44 (4), 27-8, 1977. (FOR ENGLISH TRANSLATION SEE T91150)
89883	**MONOLITHIC LININGS FOR STEEL TEEMING LADLES.** TANTSYREV, O. V. AKSYUCHITS, N. N. GLAZYRIN, B. S. LYAPIN, YU. V. LARIONOV, E. D. STARKOV, YU. A. FLYAGIN, V. G. SOLODOVA, L. I. METALLURG (11), 24-6, 1976. (FOR ENGLISH TRANSLATION SEE T89884)
89884	**MONOLITHIC LININGS FOR STEEL TEEMING LADLES.** TANTSYREV, O. V. AKSYUCHITS, N. N. GLAZYRIN, B. S. LYAPIN, YU. V. LARIONOV, E. D. STARKOV, YU. A. FLYAGIN, V. G. SOLODOVA, L. I. METALLURGIST (11), 768-70, 1976. (ENGLISH TRANSLATION OF METALLURG, (11), 24-6, 1976; FOR ORIGINAL SEE T89883)
89894	**THERMAL CONDUCTIVITY OF S. E. NEW MEXICO ROCKSALT AND ANHYDRITE.** ACTON, R. U. SANDIA LABS., ALBUQUERQUE, N. MEXICO 24PP., 1977. (SAND-77-0962C)
90099	**MICROSTRUCTURE OF A BASE METAL THICK FILM SYSTEM.** MENTLEY, D. E. CALIFORNIA UNIV., BERKELEY, M.S. THESIS 1-36, 1976. (LBL-5176, N77-22220)
90120	**DEVICE FOR MEASURING THE THERMAL CONDUCTIVITY OF DIAMOND SINGLE CRYSTALS.** OSITINSKAYA, T. D. TSENDROVSKII, V. A. VISHNEVSKII, A. S. INZH.-FIZ. ZH. 32 (4), 620-4, 1977. (ENGLISH TRANSLATION SEE T92059)
90126	**PROPERTIES OF CARBON-CARBON COMPOSITE MATERIALS OF SYNTACTIC FOAM TYPE.** VIRGILEV, YU. S. VOLKOV, G. M. ZAKREVSKII, E. A. LOPATTO, YU. S. MAKARCHENKO, V. G. IZV. AKAD. NAUK SSSR, NEORG. MATER. 13 (6), 1009-12, 1977. (FOR ENGLISH TRANSLATION SEE T94999)
90178	**CHARACTERIZATION AND DISTRIBUTION OF LUNAR MARE BASALT TYPES USING REMOTE SENSING TECHNIQUES.** PIETERS, C. MASSACHUSETTS INST. TECH., CAMBRIDGE, PH.D. THESIS 1-350, 1977. (NASA-CR-153267, N77-27051)
90188	**METAMICTIZATION AND URANIUM - LEAD SYSTEMATICS - A STUDY BY INFRARED ABSORPTION SPECTROMETRY OF PRECAMBRIAN ZIRCONS.** DELIENS, M. DELHAL, J. TARTE, P. EARTH PLANET. SCI. LETT. 33 (3), 331-44, 1977.
90192	**THERMAL PROPERTIES OF CLAYS AND SHALES.** WEAVER, C. E. GEORGIA INST. OF TECHNOLOGY, ATLANTA 102PP., 1976. (Y/OWI/SUB-7009/1)
90208	**PHOTOCHROMIC PROPERTIES OF NATURAL SODALITE.** HASSIB, A. BECKMAN, O. ANNERSTEN, H. J. PHYS. APPL. PHYS. 10 D, 771-7, 1977.
90219	**TEMPERATURE, ITS MEASUREMENTS AND CONTROL IN SCIENCE AND INDUSTRY. APPENDIX.** ROESER, W. F. WENSEL, H. T. TEMP. ITS MEAS. CONTR. SCI. IND. 1293-1323, 1941.
90225	**OPTICAL PYROMETRY.** FORSYTHE, W. E. TEMP. ITS MEAS. CONTR. SCI. IND. 1115-31PP., 1941.
90241	**MEASUREMENT OF LINEAR THERMAL EXPANSION OF SOLIDS BY A CAPACITANCE METHOD.** RAO, K. V. MAITI, J. INDIAN J. PURE APPL. PHYS. 15 (6), 437-40, 1977.
90258	**FILTER GLASSES WITH BAND-PASS CHARACTERISTICS FOR THE YELLOW-ORANGE REGION OF THE VISIBLE SPECTRUM.** RES, M. A. BEDNARIK, J. KOK, C. J. KROGER, K. OPTIK 48 (4), 371-382, 1977.
90271	**OPTICAL ABSORPTION IN DISCONTINUOUS GOLD FILMS.** NORRMAN, S. ANDERSSON, T. GRANQVIST, C. G. HUNDERI, O. SOLID STATE COMMUN. 23 (4), 261-5, 1977.
90289	**POTENTIAL SYSTEMATIC ERROR WHEN MEASURING THE THERMAL CONDUCTIVITY OF POROUS ROCKS SATURATED WITH A LOW-CONDUCTIVITY FLUID.** BECK, A. E. TECTONOPHYSICS 41 (1-3), 9-16, 1977.
90290	**HEAT-FLOW MEASUREMENTS ON THE NORTHERN APENNINE ARC.** BOCCALETTI, M. FAZZUOLI, M. LODDO, M. MONGELLI, F. TECTONOPHYSICS 41 (1-3), 101-12, 1977.
90324	**EFFECT OF CRYSTALLIZATION OF THE THERMAL DIFFUSIVITY OF A MICA GLASS-CERAMIC.** SIEBENECK, H. J. CHYUNG, K. HASSELMAN, D. P. H. YOUNGBLOOD, G. E. J. AM. CERAM. SOC. 60 (7-8), 375-6, 1977.
90330	**THE ANGULAR REFLECTANCE OF SINGLE-LAYER GRADIENT REFRACTIVE-INDEX FILMS.** MINOT, M. J. J. OPT. SOC. AM. 67 (8), 1046-50, 1977.
90372	**A SPIN-ORBIT CONSTRAINT ON THE VISCOSITY OF A MERCURIAN LIQUID CORE.** PEALE, S. J. BOSS, A. P. J. GEOPHYS. RES. 82 (5), 743-9, 1977.
90381	**STUDY OF THE THERMAL EMISSION OF SOLIDS. APPLICATION TO MARBLE.** MARCHE, P. JOUVE, P. ANN. SCI. UNIV. REIMS ARERS 14, 9-15, 1976.
90383	**FABRICATION OF TRANSPARENT SPINEL CERAMICS BY REACTIVE HOT-PRESSING.** HAMANO, K. KANZAKI, S. YOGYO KYOKAI SHI 85 (5), 225-30, 1977.
90396	**REFLECTIVITY MEASUREMENTS WITH A HALLIMOND VISUAL MICROPHOTOMETER.** LEONARD, B. F. ECON. GEOL. 55 (6), 1306-12, 1960.

TPRC Number	Bibliographic Citation
90397	**GEOLOGICAL ENGINEERING PROPERTIES OF METAMORPHIC ROCKS FROM THE GYEONGGI METAMORPHIC COMPLEX.** SO, C. S. CHOI, B. R. NA, K. C. CHIJIL HAKHOE CHI 12 (2), 91-100, 1976.
90398	**ESTABLISHING A SCALE OF DIRECTIONAL HEMISPHERICAL REFLECTANCE FACTOR I: THE VAN DEN AKKER METHOD.** VENABLE, W. H., JR. HSIA, J. J. WEIDNER, V. R. J. RES. NAT. BUR. STAND. 82 (1), 29-55, 1977.
90440	**EFFECT OF PHYSICOCHEMICAL PROCESSES IN REFRACTORY MATERIALS ON THEIR THERMAL CONDUCTIVITY.** LITOVSKII, E. YA. KAPLAN, F. S. KLIMOVICH, A. V. INZH.-FIZ. ZH. 33 (1), 101-7, 1977. (FOR ENGLISH TRANSLATION SEE T95993)
90502	**A CRYSTAL MAXIMUM TEMPERATURE MEASURER (JMTK) FOR SPECIAL APPLICATIONS.** NIKOLAENKO, V. A. MOROSOV, V. A. KASIANOV, N. I. REV. INT. HAUTES TEMP. REFRACT. 13 (1), 17-20, 1976.
90594	**HIGH TEMPERATURE HEAT CAPACITIES OF SODIUM ZIRCONO- AND HAFNO-SILICATES.** SHIBANOV, E. V. CHUKHLANTSEV, V. G. RUSS. J. PHYS. CHEM. 50 (9), 1470, 1976. (ENGLISH TRANSLATION OF ZH. FIZ. KHIM., 50 (9), 2447-8, 1976; FOR ORIGINAL SEE T88372)
90666	**RESOLUTION OF INFRARED REFLECTION SPECTRA IN POTASSIUM OXIDE - SILICON DIOXIDE SYSTEM GLASSES.** TAKASHIMA, H. SAITO, H. NAGOYA KOGYO GIJUTSU SHIKENSHO HOKOKU 25 (8), 246-51, 1976.
90680	**FURTHER DATA ON THE SOUTH-WEST ENGLAND HEAT FLOW ANOMALY.** TAMMEMAGI, H. Y. WHEILDON, J. GEOPHYS. J. ROY. ASTRON. SOC. 49 (2), 531-9, 1977.
90681	**HEAT FLOW MEASURED IN FIVE HOLES IN EASTERN AND CENTRAL SLOVAKIA.** CERMAK, V. EARTH PLANETARY SCI. LETT. 34 (1), 67-70, 1977.
90682	**HIGH TEMPERATURE HDAT CONTENT AND HEAT CAPACITY OF SILICATE GLASSES: EXPERIMENTAL DETERMINATION AND A MODEL FOR CALCULATION.** BACON, C. R. AM. J. SCI. 277 (2), 109-35, 1977.
90683	**USE OF THERMAL RESISTIVITY LOGS IN STRATIGRAPHIC CORRELATION.** BECK, A. E. GEOPHYSICS 41 (2), 300-9, 1976.
90706	**LOCALIZED ACOUSTIC PHONONS IN GLASSES AND THEIR EFFECTS ON LOW TEMPERATURE THERMAL CONDUCTIVITY AND RAMAN SPECTRA.** BURGESS, S. SHEPHERD, I. W. CHEM. PHYS. LETT. 50 (1), 112-15, 1977.
90713	**CONDUCTIVITY OF POROUS BODIES.** ZORIN, F. I. KOTOSONOV, A. S. IZV. AKAD. NAUK SSSR, NEORG. MATER. 13 (8), 1533-4, 1977. (FOR ENGLISH TRANSLATION SEE T95009)
90745	**HEAT FLOW, HEAT PRODUCTION AND CRUSTAL DYNAMICS IN THE CENTRAL ALPS, SWITZERLAND.** RYBACH, L. WERNER, D. MUELLER, S. BERSET, G. TECTONOPHYSICS (1-3), 113-26, 1977.
90746	**ANALYSIS OF HEAT FLOW DATA: DETAILED OBSERVATIONS IN MANY HOLES IN A SMALL AREA.** LEWIS, T. J. BECK, A. E. TECTONOPHYSICS (1-3), 41-59, 1977.
90800	**THE SYSTEM ERBIA-ZIRCONIA.** DURAN, P. J. AM. CERAM. SOC. 60 (11-12), 510-13, 1977.
90853	**ELECTRICAL AND THERMAL INVESTIGATIONS OF ROCKS AND MINERALS UNDER EXTREME PRESSURE, TEMPERATURE-CONDITIONS.** STILLER, H. VOLLSTAEDT, H. SEIPOLD, U. WAESCH, R. PURE APPL. GEOPHYS. 114 (2), 263-72, 1976.
90886	**THE USE OF REFLECTIVITY STANDARDS IN AN IMAGE ANALYSIS SYSTEM.** HARRIS, L. A. MICROSTRUCT. SCI. 5, 303-9, 1977.
90890	**MEASUREMENT OF THERMAL DIFFUSIVITY USING THE CONSTANT-RATE HEATING METHOD. I. APPLICATION TO PYROPHYLLITE AND STAINLESS STEEL.** KOSAKA, M. ASAHINA, T. IKUTA, S. REP. GOV. IND. RES. INST. NAGOYA 26 (1), 11-14, 1977.
90913	**DENSITY MEASUREMENTS ON ROCK SALT CRYSTALS.** GEISS, W. NED. TIJDSCHR. NATUURK. PHYSICA 4 (8), 225-30, 1924. 4 (8), 225-30, 1924.
90953	**THERMAL CONDUCTIVITY OF NATURAL DIAMOND BETWEEN 320 AND 450 K.** BURGEMEISTER, E. A. PHYSICA 93 B, 165-79, 1978.
91150	**THE EFFECT OF LITHIUM AND SODIUM OXIDES ON THE THERMOOPTIC PROPERTIES OF SILICATE AND PHOSPHATE GLASSES.** MOLEV, V. I. SHCHAVELEV, O. S. SOV. J. OPT. TECHNOL. 44 (4), 212-3, 1977. (ENGLISH TRANSLATION OF OPT. MEKH. PROM., 44 (4), 27-8, 1977; FOR ORIGINAL SEE T89777)
91194	**COMPARISON OF THE SPECTRAL REFLECTION COEFFICIENTS OF A WHITE SURFACE SAMPLE MADE OF MS-20 GLASS FROM MEASUREMENTS PERFORMED IN THE GERMAN DEMOCRATIC REPUBLIC AND THE USSR.** KOENIG, H. NIKONOVA, E. I. POKROVSKAYA, G. V. YUSTOVA, E. N. MEAS. TECH., USSR (8), 1108-10, 1976. (ENGLISH TRANSLATION OF IZMER. TEKH., (8), 21-3, 1976; FOR ORIGINAL SEE T87604)
91205	**INFLUENCE OF BASICITY AND HEATING RATE ON THE APPARENT THERMAL PROPERTIES OF PELLETS.** ABZALOV, V. M. YUR'YEV, B. P. BRATCHIKOV, S. G. RUSS. MET. (METALLY) (6), 9-15, 1976. (ENGLISH TRANSLATION OF IZV. AKAD. NAUK SSSR, MET., (6), 10-17, 1976; FOR ORIGINAL SEE T89429)
91214	**THE GLASS OF THE PRIMARY MIRROR OF THE LARGE AZIMUTHAL TELESCOPE.** BUZHINSKII, I. M. KORYAGINA, E. I. DMITRIEVA, Z. P. SOV. J. OPT. TECHNOL. 43 (11), 669-71, 1976. (ENGLISH TRANSLATION OF OPT. MEKH. PROM., 43 (11), 34-6, 1976; FOR ORIGINAL SEE T88314)
91230	**THERMOPHYSICAL PROPERTIES OF BIABASE.** TKACH, G. F. IZV. AKAD. NAUK SSSR, FIZ. ZEMLI (10), 114-16, 1970. (ENGLISH TRANSLATION SEE T91231)
91231	**THERMOPHYSICAL PROPERTIES OF DIABASE.** TKACH, G. F. BULL. ACAD. SCI., USSR, EARTH PHYS. (10), 683-4, 1970. (ENGLISH TRANSLATION OF IZV. AKAD. NAUK SSSR, FIZ. ZEMLI, (10), 114-16, 1970; FOR ORIGINAL SEE T91230)
91236	**THE THERMAL EXPANSION OF TUGTUPITE.** HENDERSON, C. M. B. TAYLOR, D. MINERAL. MAG. 41 (317), 130-1, 1977.
91238	**THERMAL PARAMETERS OF SOME PRECAMBRIAN ROCKS IN SWEDEN.** PARASNIS, D. S. PURE APPL. GEOPHYS. 114 (2), 319, 1976.
91242	**GEOTHERMAL MEASUREMENTS IN PALAEOGENE, CRETACEOUS AND PERMO-CARBONIFEROUS SEDIMENTS IN NORTHERN BOHEMIA.** CERMAK, V. GEOPHYS. J. ROY. ASTRON. SOC. 48, 537-41, 1977.
91243	**ON THE ORIENTATION OF THE THERMAL AND COMPOSITIONAL STRAIN ELLIPSOIDS IN FELDSPARS.** WILLAIME, C. BROWN, W. L. PERUCAUD, M. C. AMER. MINERAL. 59 (5-6), 457-64, 1974.

TPRC Number	Bibliographic Citation
91244	**HEAT FLOW STUDIES IN STEAMBOAT MOUNTAIN-LEMEI ROCK AREA, SKAMANIA COUNTY, WASHINGTON.** SCHUSTER, J. E. BLACKWELL, D. D. HAMMOND, P. E. HUNTTING, M. T. AM. ASSOC. PET. GEOL. BULL. 60 (8), 1410, 1976.
91250	**CHEMICAL STRENGTHENING OF CERAMIC MATERIALS.** KIRCHNER, H. P. GRUVER, R. M. LINDEN LABS., INC., STATE COLLEGE, PA. 26PP., 1964. (AD-448 226)
91261	**THERMAL EXPANSION BEHAVIOR OF SOME SULFATE-CONTAINING LEAD GLASSES.** KARKHANAVALA, M. D. SHUKLA, B. S. CENT. GLASS CERAM. RES. INST. BULL. 23 (1), 33-7, 1976.
91262	**THERMODYNAMIC PARAMETERS OF MINERALS OF THE EPIDOTE GROUP.** KISELEVA, I. A. TOPOR, N. D. ANDREYENKO, E. D. GEOCHEMISTRY, USSR (4), 389-98, 1974. (ENGLISH TRANSLATION OF GEOKHIMIYA, (4), 543-53, 1974; FOR ORIGINAL SEE T76135)
91263	**THE HIGH TEMPERATURE SPECIFIC HEAT OF SAPPHIRINE.** KISELEVA, I. A. TOPOR, N. D. GEOCHEMISTRY, USSR 12 (1), 196-8, 1975. (ENGLISH TRANSLATION OF GEOKHIMIYA, (2), 312-5, 1975; FOR ORIGINAL SEE T80733)
91326	**THE ROLE OF POROSTIY IN THE ACCOMMODATION OF THERMAL EXPANSION IN GRAPHITE.** SUTTON, A. L. HOWARD, V. C. J. NUCL. MATER. 7 (1), 58-71, 1962.
91328	**THERMAL EXPANSION OF GRAPHITE OVER DIFFERENT TEMPERATURE RANGES.** MARTIN, W. H. ENTWISLE, F. J. NUCL. MATER. 10 (1), 1-7, 1963.
91329	**ON THE EXPANSION OF DIFFERENT KINDS OF STONE FROM AN INCREASE OF TEMPERATURE, WITH A DESCRIPTION OF THE PYROMETER USED IN MAKING THE EXPERIMENTS.** ADIE, A. J. TRANS. ROY. SOC., (EDINBURGH) 13, 354-72, 1836.
91331	**STRUCTURE SENSITIVE EXPANSIVITY IN SILICA.** WRIGHT, A. F. PHYS. NON-CRYST. SOLIDS, INT. CONF., 4TH 598-603, 1977.
91435	**DETERMINATION OF THE SPECIFIC HEAT CAPACITY OF SPHENE AND LOPARITE.** VASIN, S. K. MEL'NIKOVA, V. M. KHIM. KHIM. TEKHNOL. MINER. SYR'YA 90-4, 1975.
91436	**MEASUREMENT OF THE REFLECTIVITY OF VITRINITES IN PARTIALLY POLARIZED LIGHT.** DOBRONRAVOV, V. F. KHIM. TVERD. TOPL. (MOSCOW) (5), 61-9, 1976.
91438	**CHANGE IN THE INTERLAYER SPACING IN THE STRUCTURE OF CARBON MATERIALS AT A LOW TEMPERATURE.** SHMAKOVA, E. S. LEBEDEV, YU. N. NAGORNYI, V. G. KHIM. TVERD. TOPL. (MOSCOW) (1), 123-5, 1977.
91464	**ACCESSORY MINERALS OF NIOBIUM AND TANTALUM FROM THE PEGMATITES OF THE PRYAZOV REGION.** LITOVCHENKO, E. I. KUTS, V. P. MINERAL. SB. (LVOV) 29 (4), 32-41, 1975.
91482	**ENTHALPY AND HEAT CAPACITY OF DIAMONDS IN A WIDE RANGE OF TEMPERATURES.** VOLGA, V. I. BUCHNEV, L. M. MARKELOV, N. V. DYMOV, B. K. SINT. ALMAZY (3), 9-11, 1976.
91485	**BISMUTHINITE, EMPLECTITE, AND AIKINITE IN THE ZIDAROVO ORE FIELD.** KOVACHEV, V. SPIS. BULG. GEOL. DRUZH. 37 (1), 89-95, 1976.
91513	**RARE ANTIMONY MINERALS AND THEIR PARAGENESES IN YUZHNYI DEPOSIT ORES (TETYUKHE REGION, SOUTHERN MARITIME TERRITORY).** BORTNIKOV, N. S. BORODAEV, YU. S. VYALSOV, L. N. MOZGOVA, N. N. TR. MINERAL. MUZ., AKAD. NAUK SSSR 24, 3-13, 1975.
91530	**SPECTRAL STUDY OF THE COBALT(2+) ION ENVIRONMENT IN KAOLINITE.** TARASEVICH, YU. I. SIVALOV, E. G. UKR. KHIM. ZH. 43 (4), 367-70, 1977.
91533	**DEVELOPMENT OF GLASS FOR SOLDERING WITH IRON - NICKEL.** ISMATOV, A. A. FAZYLBEKOV, A. I. UZB. KHIM. ZH. (2), 23-4, 1977.
91551	**RUCKLIDGITE - A NEW MINERAL FROM THE GOLD ORE DEPOSITS ZOD AND KOCHKAR.** ZAVYALOV, E. N. BEGIZOV, V. D. ZAP. VSES. MINERAL. O-VA 106 (1), 62-8, 1977.
91585	**THE INFLUENCE OF VARIOUS OXIDES AND FLUORIDES ON THE INFRARED TRANSMITTANCE AND THERMAL EXPANSION OF GLASSES IN THE GERMANIUM DIOXIDE - LEAD MONOXIDE SYSTEM.** KROGER, K. RES, M., JR. GLASS TECHNOL. 18 (1), 5-6, 1977.
91623	**THE THERMAL EXPANSION OF TUGTUPITE.** HENDERSON, C. M. B. TAYLOR, D. MINERAL. MAG. 41 (317), 130-1, 1977.
91628	**ABSORPTION SPECTRA OF TRANSITION METAL-BEARING MINERALS AT HIGH PRESSURES.** ABU-EID, R. M. PHYS. CHEM. OF MINERALS AND ROCKS 641-75, 1976.
91632	**SPRAY DEPOSITION OF CADMIUM STANNATE FILMS.** HAACKE, G. ANDO, H. MEALMAKER, W. E. J. ELECTROCHEM. SOC. 124 (12), 1923-6, 1977.
91645	**ELECTRICAL AND THERMAL INVESTIGATIONS OF ROCKS AND MINERALS UNDER EXTREME PRESSURE, TEMPERATURE CONDITIONS.** STILLER, H. VOLLSTADT, H. SEIPOLD, U. WASCH, R. PAGEOPH 114, 263-72, 1976.
91646	**DEFECTS IN NATURAL TYPE IB DIAMOND.** HUDSON, P. R. W. PHAKEY, P. P. NATURE 269, 227-9, 1977.
91687	**SUBSURFACE TEMPERATURE DATA IN THE SOCORRO PEAK KGRA, NEW MEXICO.** REITER, M. SMITH, R. GEOTHERMAL ENERGY MAGAZINE 5 (10), 37-42, 1977.
91690	**THERMODYNAMIC FUNCTIONS OF GEDRITE AND SOME EQUILIBRIA IN WHICH IT PARTICIPATES.** KOLESNIK, YU. N. NOGTEVA, V. V. NAUMOV, V. N. BUTENKO, V. I. PAUKOV, I. E. GEOCHEM. INT. (2), 89-95, 1976. (ENGLISH TRANSLATION OF GEOKHIMIYA, (3), 441-7, 1976; FOR ORIGINAL SEE T85313)
91691	**THERMODYNAMIC PROPERTIES AND STABILITY OF PYROPE.** KISELEVA, I. A. GEOKHIMIYA (6), 845-54, 1976. (FOR ENGLISH TRANSLATION SEE T91692)
91692	**THERMODYNAMIC PROPERTIES AND STABILITY OF PYROPE.** KISELEVA, I. A. GEOCHEM. INT. (3), 139-46, 1976. (ENGLISH TRANSLATION OF GEOKHIMIYA, (6), 845-54, 1976; FOR ORIGINAL SEE T91691)
91693	**HEAT FLOW AT GRASS VALLEY, CALFORNIA.** CLARK, S. P., JR. TRANS. AMER. GEOPHYS. UNION 38 (2), 239-44, 1957.
91694	**QUANTITATIVE ESTIMATION OF TEMPERATURE AND GEOLOGICAL TIME AS FACTORS IN CARBONIFICATION OF DISSEMINATED COAL RESIDUES, AND POSSIBLE APPLICATIONS IN OIL GEOLOGY.** KARPOV, P. A. STEPANOVA, A. F. SOLOVYEVA, N. V. AGULOV, A. P. GOZHAYA, A. L. GOLIKOV, V. A. CHAITSKIY, V. P. INT. GEOL. REV. 18 (4), 397-405, 1976. (ENGLISH TRANSLATION OF IZV. AKAD. NAUK SSSR, SER. GEOL., (3), 103-13, 1975; FOR ORIGINAL SEE T92746)
91695	**HEAT FLOW IN WEST TEXAS AND EASTERN NEW MEXICO.** HERRIN, E. CLARK, S. P. GEOPHYSICS 21 (4), 1087-99, 1956.

TPRC Number	Bibliographic Citation
91696	**THE EFFECT OF CRACKS ON THE THERMAL EXPANSION OF ROCKS.** COOPER, H. W. SIMMONS, G. EARTH PLANET. SCI. LETT. 36 (3), 404–12, 1977.
91702	**THERMAL EXPANSION OF TEN MINERALS.** SKINNER, B. J. GEOLOGICAL SURVEY RES. D109–D112, 1962. (ARTICLE 152)
91731	**PLUMBIAN TENNANTITE FROM SARK, CHANNEL ISLANDS.** BISHOP, A. C. CRIDDLE, A. J. CLARK, A. M. MINERAL. MAG. 41 (317), 59–63, 1977.
91732	**NEW REFRACTORIES FOR TANK-TYPE GLASS-REFINING FURNACES IN GLASS FIBER PRODUCTION.** DEGTYAREVA, E. V. ORLOVA, I. G. KOLESOV, YU. I. PISTSOV, YU. N. OGNEUPORY (10), 57–62, 1977. (FOR ENGLISH TRANSLATION SEE T96251)
91785	**THERMAL CONDUCTIVITY AND DIFFUSIVITY OF ENGINEERING CERAMICS.** HASSELMAN, D. P. H. YOUNGBLOOD, G. E. MONTANA ENERGY AND MHD RES. DEV. INST. 114PP., 1977. (AD–A045 527, N78–13211)
91828	**MAGNETIC PROPERTIES OF APOLLO 11 LUNAR SAMPLES.** RUNCORN, S. K. COLLINSON, D. W. O'REILLY, W. BATTEY, M. H. STEPHENSON, A. JONES, J. M. MANSON, A. J. READMAN, P. W. PROC. APOLLO 11 LUNAR SCI. CONF. 3, 2369–87, 1970.
91831	**1976 REMEASUREMENT OF NBS SPECTROPHOTOMETERINTEGRATOR FILTERS.** ECKERLE, K. L. VENABLE, W. H., JR. COLOR RES. APPLICATION 2 (3), 137–41, 1977. (PB–274 515)
91838	**PLATINUM-GROUP MINERALS FROM ONVERWACHT. III. GENKINITE, PLATINUM PALLADIUM ANTIMONY, A NEW MINERAL.** CABRI, L. J. STEWART, J. M. LAFLAMME, J. H. G. SZYMANSKI, J. T. CAN. MINERAL. 15 (PT. 3), 389–92, 1977.
91876	**PROPERTIES AND STRUCTURE OF PERICLASE-YTTRIUM MATERIALS.** KAIBICHEVA, M. N. KAMENSKIKH, V. A. GILEV, YU. P. REFRACTORIES, USSR (1), 52–6, 1976. (ENGLISH TRANSLATION OF OGNEUPORY, (1), 50–4, 1976; FOR ORIGINAL SEE T84581)
91920	**CERTAIN PROPERTIES OF DIAMOND CERAMICS.** FEDOSEEV, D. V. SOKOLINA, G. A. BANTSEKOV, S. V. DERYAGIN, B. V. IZV. AKAD. NAUK SSSR, NEORG. MATER. 13 (10), 1904–5, 1977. (FOR ENGLISH TRANSLATION SEE T95298)
91966	**DOWNHOLE MEASUREMENTS OF THERMAL CONDUCTIVITY IN GEOTHERMAL RESERVOIRS.** MURPHY, H. D. LAWTON, R. G. J. PRESSURE VESSEL TECHNOL. 99, 607–11, 1977. (ASME PAPER–77–PET–23)
91972	**HEAT FLOW THROUGH THE SOUTHERN CALIFORNIA BORDERLAND.** LEE, T. C. HENYEY, T. L. J. GEOPHYS. RES. 80 (26), 3733–43, 1975.
91973	**ON SHALLOW-HOLE TEMPERATURE MEASUREMENTS — A TEST STUDY IN THE SALTON SEA GEOTHERMAL FIELD.** LEE, T. C. GEOPHYSICS 42 (3), 572–83, 1977. (PB–262 643)
91994	**SPECIFIC HEATS OF LITHIUM, SODIUM, POTASSIUM, AND SILVER BETA-ALUMINA BELOW 1 K.** ANTHONY, P. J. ANDERSON, A. C. PHYS. REV. 16 B (8), 3827–8, 1977.
92035	**SPECIAL REFRACTORY MATERIALS FOR USE IN GAS REFROMING.** MINER, R. R. PROC. MEET. - UNIDO-FAI INTER-REG. MEET. SAF. DES. OPER. AMMONIA PLANTS 18PP., 1976. (PAPER NO. VII)
92038	**HEAT FLOW MEASUREMENTS IN YELLOWSTONE LAKE AND THE THERMAL STRUCTURE OF THE YELLOWSTONE CALDERA.** MORGAN, P. BLACKWELL, D. D. SPAFFORD, R. E. SMITH, R. B. J. GEOPHYS. RES. 82 (26), 3719–32, 1977.
92040	**MEASUREMENT TECHNIQUES FOR LOW EXPANSION MATERIALS.** WOLFF, E. G. NATL. SAMPE TECH. CONF. 9, 57–72, 1977.
92045	**OPTIMIZATION OF COATINGS FOR FLAT PLATE SOLAR COLLECTORS. FINAL REPORT.** LIN, R. J. H. ZIMMER, P. B. MAR, H. Y. B. PETERSON, R. E. HONEYWELL INC., MINNEAPOLIS, MINN. 133PP., 1977. (COO–2930–12, COO–2930–10)
92059	**AN EXPERIMENTAL DEVICE FOR MEASUREMENT OF THERMAL CONDUCTIVITY OF DIAMOND MONOCRYSTALS.** OSITINSKAYA, T. D. TSENDROVSKII, V. A. VISHNEVSKII, A. S. J. ENG. PHYS., USSR 32 (4), 388–91, 1977. (ENGLISH TRANSLATION OF INZH. FIZ. ZH., 32 (4), 620–4, 1977; FOR ORIGINAL SEE T90120)
92103	**INTERPRETATIONS OF OPTICAL OBSERVATIONS OF MERCURY AND THE MOON.** HAPKE, B. PHYS. EARTH PLANET. INTER. (2–3), 264–74, 1977.
92117	**SPECTROSCOPIC MODE GRUNEISEN PARAMETERS FOR DIAMOND.** PARSONS, B. J. PROC. R. SOC. 352 A (1670), 397–417, 1977.
92134	**ROCK PROPERTIES FOR THERMAL ENERGY STORAGE SYSTEMS IN THE 0 TO 500 C RANGE.** PFANNKUCH, H. O. EDENS, M. H. PROC. ANNU. MEET. EM DASH AMER. SECT. INT. SOL. ENERGY SOC. 1, 5PP., 1977.
92136	**PHYSICAL PROPERTIES OF SAMPLES FROM THE JOIDES, LEG 37, DEEP SEA DRILLING PROJECT.** SCHLOESSIN, H. H. DVORAK, Z. D. INITIAL REP. DEEP SEAL DRILL. PROJ. 37, 403–15, 1977.
92161	**AGE-COLOR RELATIONSHIPS IN THE LUNAR HIGHLANDS.** CHARETTE, M. P. SODERBLOM, L. A. ADAMS, J. B. GAFFEY, M. J. MC CORD, T. B. GEOCHIM. COSMOCHIM. ACTA, SUPPL. 7, 2579–92, 1976.
92163	**ZERO AND LOW COEFFICIENT OF THERMAL EXPANSION POLYCRYSTALLINE OXIDES.** SKAGGS, S. R. LOS ALAMOS SCIENTIFIC LAB., N. MEX. 12PP., 1977. (LA–6918–MS, LA–UR–77–1307)
92180	**X-RAY DETERMINATION OF THE THERMAL EXPANSION OF NADORITE, LEAD ANTIMONY OXIDE CHLORIDE.** CHRISTIDIS, P. C. RENTZEPERIS, P. J. J. APPL. CRYSTALLOGR. 10 (6), 486, 1977.
92233	**METHODS FOR MEASURING THERMAL CONDUCTIVITY OF EARTH MATERIALS WITH APPLICATION TO ANISOTROPY IN BASALT AND HEAT FLUX THROUGH LAKE SEDIMENTS.** WATTS, G. P. UNIV. HAWAII, MANOA, M.S. THESIS 1–103, 1975.
92257	**MIXED-SEMICONDUCTOR EVAPORATED FILMS AS FILTERS WITH VERY SHARP CUTOFF AND SOME POSSIBLE APPLICATIONS IN ELECTROOPTICS.** KOTTLER, W. BONNET, D. GANSON, G. APPL. OPT. 17 (2), 164–5, 1978.
92321	**PALLADSEITE, A NEW MINERAL FROM ITABIRA, MINAS GERAIS, BRAZIL.** DAVIS, R. J. CLARK, A. M. CRIDDLE, A. J. MINERAL. MAG. 41 (317), 123, 1977.
92421	**ALBEDO CHARACTERISTICS OF THE LINAR GLOBE.** SHEVCHENKO, V. V. SOV. ASTRON. 18 (5), 628–32, 1975. (ENGLISH TRANSLATION OF ASTRON ZH., 51 (5), 1064–71, 1974; FOR ORIGINAL SEE T78950)
92443	**A HIGH PERFORMANCE GOLD/DIELECTRIC/RESISTOR MULTILAYER SYSTEM.** SPROULL, J. F. GERRY, D. J. BACHER, R. J. PROC. INT. MICROELECTRON. SYMP. 20–4, 1977.

TPRC Number	Bibliographic Citation
92471	**OPTICAL REFLECTANCE AND TRANSMISSION OF A TEXTURED SURFACE.** STEPHENS, R. B. CODY, G. D. THIN SOLID FILMS 45 (1), 19–29, 1977.
92523	**LOW-TEMPERATURE HEAT CAPACITY, STANDARD ENTROPY, AND ELASTIC PROPERTIES OF DATOLITE AND DANBURITE.** ZHDANOV, V. M. TURDAKIN, V. A. ARUTYUNOV, V. S. SEMENOV, YU. V. MALINKO, S. V. KHODAKOVSKII, I. L. GEOKHIMIYA (12), 1817–24, 1977.
92524	**HIGH-TEMPERATURE ENTHALPIES OF DATOLITE AND DANBURITE IN THE TEMPERATURE RANGE 298.15 TO 973.15 K.** AGOSHKOV, V. M. SEMENOV, YU. V. MALINKO, S. V. KHODAKOVSKII, I. L. GEOKHIMIYA (12), 1825–30, 1977.
92525	**CRYSTALLIZABILITY OF GLASSES OF THE COPPER(II) OXIDE - TELLURIUM DIOIXDE - VANADIUM(V) OXIDE SYSTEM.** IVANOVA, I. DIMITRIEV, YA. MARINOV, R. STAVRAKEVA, D. GOD. VISSH. KHIM.-TEKHNOL. INST., SOFIA 22 (4), 55–66, 1977.
92560	**REFLECTANCE SPECTRA OF TRIGONAL MERCURY SULFIDE.** DONETSKIKH, V. I. SOBOLEV, V. V. OPT. SPECTROS., USSR 42 (2), 224–6, 1977. (ENGLISH TRANSLATION OF OPT. SPEKTROSK., 42 (2), 401–3, 1977; FOR ORIGINAL SEE T88573)
92614	**CHANGE IN THE PROPERTIES OF GLAZE THROUGH FIRING AND ITS EFFECT ON THE STRESS IN GLAZE. STUDIES ON THE GLAZE FITNESS OF PORCELAIN WARE, NO. 4.** INADA, H. YOGYO KYOKAI SHI 86 (3), 108–14, 1978.
92618	**RELATIONS AMONG THE CRAZING TENDENCY OF VITREOUS CHINA, THE THERMAL EXPANSION COEFFICIENT OF THE BODY, AND ITS QUARTZ AND CRISTOBALITE CONTENT.** INADA, H. YOGYO KYOKAI SHI 85 (12), 580–6, 1977.
92619	**STUDIES ON THE GLAZE FITNESS OF PORCELAIN WARE. PART 3. CHANGE IN THE STRESS IN GLAZE WITH TEMPERATURES AND ITS RELATION TO THE DIFFERENCE OF THE THERMAL EXPANSION CURVES OF THE BODY AND THE GLAZE OF CHINAWARE.** INADA, H. YOGYO KYOKAI SHI 86 (2), 77–85, 1978.
92625	**THERMAL ENERGY STORAGE USING FLUORIDES OF ALKALI AND ALKALINE EARTH METALS.** SCHROEDER, J. ENERGY STORAGE 206–20, 1976.
92686	**PHYSICAL PROPERTIES OF RECOMPRESSED, EXPANDED GRAPHITE.** BERGER, D. MAIRE, J. MATER. SCI. ENG. 31, 335–9, 1977.
92721	**OPTICAL PROPERTIES OF CHEMICALLY SPRAYED CADMIUM SULFIDE / COPPER SULFIDE PHOTOVOLTAIC CELLS.** BERG, R. S. MYERS, T. M. SANDIA LABS., ALBUQUERQUE, N. MEX. 39PP., 1977. (SAND-76-0729, N77-33656)
92746	**QUANTITATIVE ESTIMATION OF TEMPERATURE AND GEOLOGICAL TIME AS FACTORS IN CARBONIFICATION OF DISSEMINATED COAL RESIDUES, AND POSSIBLE APPLICATIONS IN OIL GEOLOGY.** KARPOV, P. A. STEPANOVA, A. F. SOLOVYEVA, N. V. AGULOV, A. P. GOZHAYA, A. L. GOLIKOV, V. A. CHAITSKIY, V. P. IZV. AKAD. NAUK SSSR, SER. GEOL. (3), 103–13, 1975. (FOR ENGLISH TRANSLATION SEE T91694)
92758	**SPECIAL GLASSES FOR TECHNOLOGY - PRINCIPLES, CURRENT IMPORTANCE, TRENDS.** SCHROEDER, H. CHET, CHEM.: EXP. TECHNOL. 3 (10), 369–78, 1977.
92918	**THE THERMAL CONDUCTIVITY OF DIAMONDS.** BERMAN, R. MARTINEZ, M. DIAMOND RESEARCH 7–13, 1976.
92919	**TERRESTRIAL HEAT FLOW IN THE SWISS ALPS.** CLARK, S. P. NIBLETT, E. R. MON. NOT. ROY. ASTRON. SOC., GEOPHYS. SUPPL. 7, 176–95, 1956.

TPRC Number	Bibliographic Citation
92925	**POLYCARBONATES.** BONFIGLIO, G. POLIPLASTI PLAST. RINF. 22 (198), 6–15, 1974.
92927	**PROPERTIES OF HIGH ALUMINA MATERIALS OBTAINED FROM DIFFERENT (TECHNICAL-GRADE) ALUMINUM OXIDES.** WROBLEWSKA, G. SZKLO CERAM. 27 (6), 149–52, 1976.
92930	**PREDICTING THERMAL CONDUCTIVITIES OF FORMATIONS FROM OTHER KNOWN PROPERTIES.** ANAND, J. SOMERTON, W. H. GOMAA, E. SOC. PETROL. ENG., J. 13, 267–73, 1973.
92931	**HEAT FLOW STUDIES UNDER UPPER MANTLE PROJECT.** GUPTA, M. L. RAO, G. V. BULL. NATL. GEOPHYS. RES. INST. 8 (3–4), 87–112, 1970.
92934	**APPROXIMATION OF THE TRENDS OF ISOTHERMAL EXPANSION LINES OF CORDIERITE TYPE OF MATERIALS.** DE WYS, E. C. LIBYAN J. SCI. 3, 35–40, 1973.
92946	**MEASUREMENT OF THERMAL CONDUCTIVITY OF CONSOLIDATED DRY ROCKS AND WATER-SATURATED SANDS.** AMIRKHANOV, KH. I. GAIRBEKOV, KH. A. SUYETNOV, V. V. PROSVIROVA, T. A. GEOTERMICHESKIYE ISSLED. DAGESTANE I VOPROSY PRAKTICHESKOGO ISPOL. TEPLA ZEMLI 16–18PP., 1970.
92949	**DEPENDENCE OF THE THERMAL CONDUCTIVITY OF SANDY ROCKS ON THE MOISTURE CONTENT OF A SAMPLE USED FOR EXPERIMENTS.** GAIRBEKOV, KH. A. LEVKOVICH, R. A. GEOTERMICHESKIYE ISSLED. DAGESTANE I VOPROSY PRAKTICHESKOGO ISPOL. TEPLA ZEMLI 39–42PP., 1970.
92952	**THE MEASUREMENT OF THERMAL CONDUCTIVITY OF ROCKS BY MEANS OF AN AUTOMATIC APPARATUS.** KRESL, M. GEOELECTRIC AND GEOTHERMAL STUDIES 38–43, 1976.
92953	**THE MEASUREMENT OF THE PRESSURE-DEPENDENCE OF THERMAL DIFFUSIVITY IN ROCKS.** SEIPOLD, U. GEOELECTRIC AND GEOTHERMAL STUDIES 44–7, 1976.
92954	**CORRELATION BETWEEN THERMAL CONDUCTIVITY AND OTHER PHYSICAL PARAMETERS OF ROCKS.** PLEWA, S. GEOELECTRIC AND GEOTHERMAL STUDIES 48–52, 1976.
92956	**MICROSTRUCTURES, REFLECTIVITY, AND MICROHARDNESS OF SPHALERITES AND GALENAS OF ORE DEPOSITS IN THE ITALIAN EASTERN ALPS.** DI COLBERTALDO, D. VAGHETTI, A. INT. MINERAL. ASS., PAP. PROC. GEN. MEET., 7TH (1), 165–73, 1971.
92963	**DISTRIBUTION OF TEMPERATURES IN THE PROXIMITY OF A MAGMATIC INTRUSIVE SHEET DURING COOLING BY CONDUCTION.** RUBIA, J. DORIA, J. OSORIO, A. CALLEJA, J. M. ARANA, V. ESTUD. GEOL. 26 (3), 273–80, 1970.
92968	**THERMAL CONDUCTIVITY MEASUREMENTS OF SOME APULIAN LIMESTONES BY THE CUT CORE METHOD.** MONGELLI, F. ASS. GEOFIS. ITAL., ATTI CONV. ANN. (PART 1), 137–54, 1970.
93150	**THE REFLECTIVITY SPECTRA OF COAL VITRAINS IN THE VISIBLE AND THE ULTRAVIOLET.** GILBERT, L. A. FUEL 39, 393–400, 1960.
93151	**REFLECTANCE OF COALS, GRAPHITE, AND DIAMOND.** ERGUN, S. MC CARTNEY, J. T. FUEL 39, 449–54, 1960.
93152	**REFLECTIVITY SPECTRA OF COAL VITRAINS IN THE VISIBLE AND THE ULTRAVIOLET.** GILBERT, L. A. FUEL 40, 72–3, 1961.
93213	**BASIC PROPERTIES OF SOME SULPHUR-BOUND COMPOSITE MATERIALS.** SHRIVE, N. G. LOOV, R. E. GILLOTT, J. E. JORDAAN, I. J. MATER. SCI. ENG. 30 (1), 71–9, 1977.

TPRC Number	Bibliographic Citation
93218	TECHMOLOGICAL DEVELOPMENT AND ADOPTION OF A MAGNESITE REFRACTORY ON A SPINEL BINDER WITH DECREASED THERMAL CONDUCTIVITY. SHUBIN, V. I. SERGEEVA, V. M. MURASHOVA, M. V. PODDUBNYI, I. M. SIMONOV, K. V. BUGAEV, N. F. OGNEUPORY (2), 3-9, 1978. (FOR ENGLISH TRANSLATION SEE T98789)
93227	A THEORETICAL ANALYSIS OF THE OPTICAL PROPERTIES OF OF SPECTRAFLOAT GLASS. BAMFORD, C. R. PHYS. CHEM. GLASSES 17 (6), 209-13, 1976.
93254	EFFECTIVE THERMAL CONDUCTIVITY OF GLASS-CERAMICS IN VARIOUS STAGES OF HEAT TREATMENT. BOLSHAKOVA, N. V. MALTER, V. L. STEKLO KERAM. (2), 16-18, 1978. (FOR ENGLISH TRANSLATION SEE T98018)
93343	SOLAR ENERGY APPLICATIONS FOR HEAT-ABSORBING GLASS. DEMINET, C. PROC. ANNUAL MEET. AMER. SECT. INT. SOLAR ENERGY SOC. 1, 12.6-12.8, 1977. (CONF-770603)
93353	ON THE OPTICAL PROPERTIES OF IRRADIATED COBALT GLASS AND OTHER GLASSES. ONO, M. HIROKAWA, S. NAGAI, H. MEM. FAC. ENG., NAGOYA UNIV. 28 (2), 237-64, 1976.
93507	PHYSICAL PROPERTIES OF EUROPIUM SESQUIOXIDE. GILCHRIST, K. E. BROWN, R. G. PRESTON, S. D. J. NUCL. MATER. 68 (1), 39-47, 1977.
93565	THERMOPHYSICAL PROPERTIES RESEARCH ON REFRACTORY MATERIALS IN FRANCE. CABANNES, F. SYMP. THERMOPHYSICAL PROPERTIES, 7TH. 26-36, 1977.
93573	THERMAL CONDUCTIVITY OF GRANULAR MATERIALS. CRANE, R. A. VACHON, R. I. KHADER, M. S. SYMP. THERMOPHYSICAL PROPERTIES, 7TH. 109-23, 1977.
93594	EVALUATION OF A FOURIER TRANSFORM INFRARED SPECTROPHOTOMETER FOR MEASUREMENT OF DIFFUSE REFLECTANCE. WINN, R. A. DEWITT, D. P. SYMP. THERMOPHYSICAL PROPERTIES, 7TH. 285-94PP., 1977.
93606	THE THERMAL TRANSPORT PROPERTIES AT NORMAL AND ELEVATED TEMPERATURE OF EIGHT REPRESENTATIVE ROCKS. HANLEY, E. J. DEWITT, D. P. TAYLOR, R. E. SYMP. THERMOPHYSICAL PROPERTIES, 7TH. 386-91, 1977.
93608	ELECTRICAL AND THERMAL TRANSPORT PROPERTIES OF GREEN RIVER OIL SHELL HEATED IN LITROGEN. NOTTENBURG, R. RAJESHWAR, K. DUBOW, J. ROSENVOLD, R. SYMP. THERMOPHYSICAL PROPERTIES, 7TH. 396-403, 1977.
93609	ESTIMATING THE THERMAL CONDUCTIVITY OF ROCKS - AN AN1LYTICAL APPROACH. REDDY, M. S. SYMP. THERMOPHYSICAL PROPERTIES, 7TH. 404-11, 1977.
93643	THE CODATA TASK GROUP ON TRANSPORT PROPERTIES. MINGES, M. L. SYMP. THERMOPHYSICAL PROPERTIES, 7TH 943-8, 1977.
93644	STANDARD REFERENCE MATERIALS FOR THERMOPHYSICAL PROPERTIES. KIRBY, R. K. SYMP. THERMOPHYSICAL PROPERTIES, 7TH 949-65, 1977.
93700	OPTICAL PROPERTIES OF GRAPHITE. KWIECINSKA, B. J. MICROSC. 109 (PT. 3), 289-302, 1977.
93719	ADIABATIC POTENTIAL OF DIAMOND-LIKE SEMICONDUCTORS IN QUASIMOLECULAR MODEL. REZNIK, I. M. FIZ. TVERD. TELA 19 (2), 463-8, 1977. (FOR ENGLISH TRANSLATION SEE T93720)
93720	ADIABATIC POTENTIAL OF DIAMOND-LIKE SEMICONDUCTORS IN QUASIMOLECULAR MODEL. REZNIK, I. M. SOV. PHYS. SOLID STATE 19 (2), 266-9, 1977. (ENGLISH TRANSLATION OF FIZ. TVERD. TELA, 19 (2), 463-8, 1977; FOR ORIGINAL SEE T93719).
93760	INFRARED LATTICE VIBRATIONS AND DIELECTRIC DISPERSION IN ALPHA-IRON OXIDE. ONARI, S. ARAI, T. KUDO, K. PHYS. REV. 16 B (4), 1717-21, 1977.
93881	THERMAL CONDUCTIVITY OF REFRACTORY PASTEBOARD. AKSELROD, E. I. VISHNEVSKII, I. I. SORIN, M. N. YUTINA, A. S. TEPLOFIZ. VYS. TEMP. 16 (2), P.443, 1978. (FOR ENGLISH TRANSLATION SEE T95017)
94057	STUDY OF THE THERMAL PROPERTIES OF GRANULAR MEDIA. BERTAUX, M. G. BIENFAIT, G. JOLIVET, J. ANN. GEOPHYS. (FRANCE) 33 (3), 385-90, 1977.
94058	HIGH TEMPERATURE PROPERTIES OF ZIRCALOY - OXYGEN ALLOYS. BUNNELL, L. R. BATES, J. L. MELLINGER, G. B. HANN, C. R. ELECTRIC POWER RESEARCH INST., PALO ALTO, CALIFORNIA 207PP., 1977. (EPRI-NP-524)
94194	PHYSICAL PROPERTIES AND HEAT TRANSFER CHARACTERISTICS OF MATERIALS FOR KRYPTON-85 STORAGE. CHRISTENSEN, A. B. IDAHO CHEMICAL PROGRAMS OPERATIONS OFFICE, ALLIED CHEMICAL CORP. 48PP., 1977. (ICP-1128)
94196	TRANSPARENT CONDUCTING FILMS OF TIN OXIDE PREPARED BY THERMAL DECOMPOSITION OF TIN SALT OF ORGANIC ACID. KAWAMATA, E. KAMBE, S. OHSIMA, K. OYO BUTURI 46 (5), 490-9, 1977.
94200	THERMOPHYSICAL PROPERTIES OF SELECTED AEROSPACE MATERIALS. PART II: THERMOPHYSICAL PROPERTIES OF SEVEN MATERIALS. TOULOUKIAN, Y. S. HO, C. Y. CINDAS, PURDUE UNIVERSITY 230PP., 1977.
94212	THERMAL CONDUCTIVITY OF POWDER. WAKASHIMA, H. TERAOKA, S. ZAIRYO 26 (285), 565-9, 1977.
94260	PHYSICAL PROPERTIES OF SODIUM CHLORIDE IN CRYSTAL, LIQUID, GAS, AND AQUEOUS SOLUTION STATES. KAUFMAN, D. W. SODIUM CHLORIDE (CHAPT. 25), 587-625, 1960.
94271	ALUMINA CERAMICS. HOUSTON, A. M. MATER. ENG. 83 (6), 51-7, 1976.
94314	SPECIFIC HEAT OF SMOKY QUARTZ. SAINT-PAUL, M. PICOT, B. NAVA, R. PHYS. LETT. 66A (5), 389-91, 1978.
94376	NEW MINERALS OF THE GROUP OF IRON ANTIMONIDES AND ARSENIDES FROM SEINAJOKI DEPOSIT, FINLAND. MOZGOVA, N. N. BORODAEV, YU. S. OZEROVA, N. A. PAAKKONEN, V. BULL. GEOL. SOC. FINL. 49 (PT. 1), 47-52, 1977.
94378	CARBON REFRACTORIES. LACROIX, S. CESSID, CENT. ETUD. SUPER. SIDER. FR. 75-52, 15PP., 1975. (CESSID-75-52)
94406	PICOTPAULITE, A NEW MINERAL. JOHAN, Z. PIERROT, R. SCHUBNEL, H. J. PERMINGEAT, F. BULL. SOC. FR. MINERAL. CRISTALLOGR. 93 (5-6), 545-9, 1970.
94412	THERMAL BEHAVIOR OF THE CALCAREOUS STONES. SERSALE, R. REND. ACCAD. SCI. FIS. E MAT. 20, 245-56, 1953.
94419	THE THERMAL CONDUCTIVITY OF SOILS. KERSTEN, M. S. HIGHWAY RESEARCH BOARD, PROC. 28, 391-409, 1948.

TPRC Number	Bibliographic Citation
94422	**SPECIFIC HEAT OF COAL, ITS RELATION TO PRESENCE OF COMBINED WATER IN COAL SUBSTANCE.** PORTER, H. C. TAYLOR, G. B. J. IND. ENG. CHEM. 5, 289-93, 19 .
94424	**THE THERMAL DIFFUSIVITY AND THE WATER CONTENT OF THE SOIL.** TSUIJI, Y. J. MET. SOC. JAPAN 31, 1913.
94431	**MEASUREMENT OF THERMAL CONDUCTIVITY OF GRANITES AT HIGH TEMPERATURES FROM 100 TO 690 C.** KUMAGAI, N. ITO, H. MEM. COLL. SCI. UNIV. KYOTO 22 B, 264-80, 1955.
94434	**HIGH-TEMPERATURE THERMAL EXPANSION OF ROCKSALT.** MERRIAM, M. F. SMOLUCHOWSKI, R. WIEGAND, D. A. PHYS. REV. 125, 65-7, 1962.
94446	**RADIATIVE HEAT TRANSFER IN DENSE MEDIA AND ITS MAGNITUDE IN OLIVINES AND SOME OTHER FERROMAGNESIAN MINERALS UNDER TYPICAL UPPER MANTLE CONDITIONS.** PITT, G. D. TOZER, D. C. PHYS. EARTH PLANET. INTERIORS 2 (3), 189-99, 1973.
94447	**EFFECT OF PRESSURE AND TEMPERATURE ON THE MOLAR VOLUMES OF WUSTITE AND OF THREE IRON MAGNESIUM SILICATE SPINEL SOLID SOLUTIONS.** MAO, H. K. TAKAHASHI, T. BASSETT, W. A. WEAVER, J. S. AKIMOTO, S. J. GEOPHYS. RES. 74 (4), 1061-9, 1969.
94485	**MEASUREMENT OF THE COEFFICIENTS OF EXPANSION OF SODIUM CHLORIDE, LITHIUM FLUORIDE, POTASSIUM CHLORIDE, AND POTASSIUM BROMIDE BY THE FLOTATION METHOD.** KONSTANTINOV, B. P. EFREMOVA, Z. N. RYSKIN, G. YA. ZHUR. TEKH. FIZ. 28, 1740-7, 1958. (FOR ENGLISH TRANSLATION SEE T94486)
94486	**MEASUREMENT OF THE COEFFICIENTS OF EXPANSION OF SODIUM CHLORIDE, LITHIUM FLUORIDE, POTASSIUM CHLORIDE, AND POTASSIUM BROMIDE BY THE FLOTATION METHOD.** KONSTANTINOV, B. P. EFREMOVA, Z. N. RYSKIN, G. YA. SOV. PHYS. TECH. PHYS. 3, 1604-11, 1958. (ENGLISH TRANSLATION OF ZHUR. TEKH. FIZ., 28, 1740-7, 1958; FOR ORIGINAL SEE T94485)
94558	**MEASUREMENT OF THE THERMAL CONDUCTIVITY OF BOROSILICATE GLASSES BY LASER FLASH METHOD.** TERAI, R. HORI, M. OSAKA KOGYO GIJUTSU SHIKENSHO KIHO 28 (3), 219-24, 1977. (YOGYO-KYOKAI-SHI) (85 (3), 140-4, 1977)
94572	**ON A METHOD FOR DETERMINING THE THERMOPHYSICAL PROPERTIES OF ROCKS.** SOKOLOV, V. I. SHIVRIN, O. N. VINITI (MOWCOW) 12PP., 1976. (VINITI-DEPOSITED DOC.-1754-76)
94579	**THERMAL CONDUCTIVITY ENHANCEMENT OF EPOXIES BY THE USE OF FILLERS.** NIEBERLEIN, V. A. IEEE TRANS. COMPONENTS, HYBRIDS, MANUF. TECHNOL. CHMT-1 (2), 172-6, 1978.
94580	**THE THERMAL CONDUCTIVITY OF 2-PHASE URANIUM OXIDE.** BUYKX, W. J. PROC. - AUST. CERAM. CONF. 7, 4A/3, 6PP., 1976.
94615	**TEMPERATURE AND THERMAL CONDUCTIVITY OF GLAST-FURNACE SLAGS.** OSINOVSKIKH, L. L. YUREV, B. P. ORININSKII, N. V. BRATCHIKOV, S. G. KHOMUTININ, V. S. IZV. VYSSH. UCHEBN. SAVED., CHERN. METALL. (5), 36-40, 1977.
94616	**OPTICAL AND MICROHARDNESS STUDY OF SOME SILVER - COPPER - LEAD - BISMUTH SULFIDES.** VENDRELL-SAZ, M. KARUP-MOELLER, S. LOPEZ-SOLER, A. NEUES JAHRB. MINERAL., ABH. 132 (1), 101-12, 1978.
94662	**CYCLIC HEAT FLOW METER APPARATUS TO MEASURE THERMAL CONTACT RESISTANCE, CONDUCTIVITY AND DIFFUSIVITY.** SHIRTLIFFE, C. J. STEPHENSON, D. G. BROWN, W. C. THERMAL CONDUCTIVITY CONF., 15TH 169-76, 1978.
94672	**THERMAL CONDUCTIVITY OF S.E. NEW MEXICO ROCKSALT AND ANHYDRITE.** ACTON, R. U. THERMAL CONDUCTIVITY CONF., 15TH 263-76, 1978.
94673	**THERMAL RESPONSES IN UNDERGROUND EXPERIMENTS IN A DOME SALT FORMATION.** LLEWELLYN, G. H. THERMAL CONDUCTIVITY CONF., 15TH 277-88, 1978.
94674	**THERMAL CONDUCTIVITY OF CERTAIN ROCK TYPES AND ITS RELEVANCE TO THE STORAGE OF NUCLEAR WASTE.** MIRKOVICH, V. V. SOLES, J. A. THERMAL CONDUCTIVITY CONF., 15TH 297-304, 1978.
94675	**THERMAL DIFFUSIVITY AND THERMAL EXPANSION OF SOME ROCK TYPES IN RELATION TO STORAGE OF NUCLEAR WASTE.** MIRKOVICH, V. V. THERMAL CONDUCTIVITY CONF., 15TH 305-13, 1978.
94680	**THERMAL CONDUCTIVITY OF ROCKS ASSOCIATED WITH ENERGY EXTRACTION FROM HOT DRY ROCK GEOTHERMAL SYSTEMS.** SIBBITT, W. L. DODSON, J. G. TESTER, J. W. THERMAL CONDUCTIVITY CONF., 15TH 399-21, 1978.
94682	**HEAT TRANSFER THROUGH COALS AND OTHER NATURALLY OCCURRING CARBONACEOUS ROCKS.** VANDERBORGH, N. E. BERTINO, J. P. CORT, G. E. WAGNER, P. THERMAL CONDUCTIVITY CONF., 15TH 433-43, 1978.
94687	**DISCOVERY OF MENEGHINITE AND SEQUENCE OF FORMATION OF LEAD SULFOANTIMONITES IN ORES OF THE BESTOBE DEPOSIT (CENTRAL KAZAKHSTAN).** USPENSKAYA, M. E. SHCHIBRIK, V. I. VESTN. MOSK. UNIV., GEOL. 31 (1), 104-8, 1976.
94691	**APPLICATION OF ZEOLITE TUFFS. SPECIFIC HEAT MEASUREMENT OF ZEOLITE TUFFS.** TORII, K. HOTTA, M. ASAKA, M. TOHOKU KOGYO GIJUTSU SHIKENSHO HOKOKU 6, 68-70, 1975.
94710	**ELECTRICAL AND THERMAL CONDUCTIVITY OF ION-SELECTIVE MEMBRANES.** THORSLEY, J. D. UNIV. WEST ONTARIO, LONDON, ONT., PH.D. THESIS 1-, 1972.
94768	**FABRICATION EFFECTS ON THERMAL CONDUCTIVITY OF MIXED-OXIDE FUELS.** LAWRENCE, L. A. GIBBY, R. L. CHRISTENSEN, J. A. WEBER, E. T. TRANS. AMER. NUCL. SOC. 14 (2), 598-9, 1971.
94788	**THERMOPHYSICAL PROPERTIES OF THORIUM AND URANIUM SYSTEMS FOR USE IN REACTOR SAFETY ANALYSIS.** FINK, J. K. CHASANOV, M. G. LEIBOWITZ, L. ARGONNE NATIONAL LAB., ARGONNE, IL. 72PP., 1977. (ANL-CEN-RSD-77-1)
94818	**CHARACTERISTICS OF AN INDIUM - TIN OXIDE TRANSPARENT CONDUCTOR DEPOSITED FROM ORGANOMETALLIC COMPOSITIONS.** SIUTA, V. STEIN, S. J. IEEE TRANS. COMPONENTS, HYBRIDS, MANUF. TECHNOL. CHMT-1 (3), 237-41, 1978.
94823	**THERMAL EXPANSION OF BERYLLIUM ALUMINUM SILICATE.** GOWDA, K. A. GOWDA, S. S. S. KUCHELA, K. N. INDIAN J. PHYS. 52 A (4), 311-15, 1978. (PROC. NUCL. PHYS. SOLID STATE PHYS. SYMP., 19C, 266-8, 1976)
94862	**INFRARED SPECTRA OF SOME SELECTED MINERALS, ROCKS AND PRODUCTS.** GHOSH, S. N. J. MATER. SCI. 13 (9), 1877-86, 1978.

TPRC Number	Bibliographic Citation
94999	**PROPERTIES OF CARBON - CARBON COMPOSITE MATERIALS OF THE SYNTACTIC-FOAM TYPE.** VIRGILEV, YU. S. VOLKOV, G. M. ZAKREVSKII, E. A. LOPATTO, YU. S. MAKARCHENKO, V. G. INORG. MAT., USSR 13 (6), 821-4, 1977. (ENGLISH TRANSLATION OF IZV. AKAD. NAUK SSSR, NEORG. MATER., 13 (6), 1009-12, 1977; FOR ORIGINAL SEE T90126)
95009	**CONDUCTIVITY OF POROUS BODIES.** ZORIN, F. I. KOTOSONOV, A. S. INORG. MAT., USSR 13 (8), 1244-5, 1977. (ENGLISH TRANSLATION OF IZV. AKAD. NAUK SSSR, NEORG. MATER., 13 (8), 1533-4, 1977; FOR ORIGINAL SEE T90713)
95017	**THERMAL CONDUCTIVITY OF FIRE-RESISTANT CARDBOARDS.** AKSELROD, E. I. VISHNEVSKII, I. I. GLUSHKOVA, D. B. SORIN, M. N. YUTINA, A. S. HIGH TEMP., USSR 16 (2), 376, 1978. (ENGLISH TRANSLATION OF TEPLOFIZ. VSP. TEMP., 16 (2), 443, 1978; FOR ORIGINAL SEE T93881)
95041	**TRANSPARENT CONDUCTING FILMS OF TIN OXIDE PREPARED BY THE THERMAL DECOMPOSITION OF TIN SALTS OF ORGANIC ACIDS. II.** KAWAMATA, E. KAMBE, S. OHSHIMA, K. OYO BUTSURI 47 (1), 16-22, 1978.
95052	**WOLLASTONITE: SHORT-FIBER FILLER/REINFORCEMENT.** COPELAND, J. R. RUSH, O. W. PLASTICS COMPOUNDING 26-41, 1978.
95156	**INVESTIGATIONS OF THERMAL AND ELASTIC PROPERTIES OF ROCKS BY MEANS OF A CUBIC PRESS.** STILLER, H. VOLLSTAEDT, H. SEIPOLD, U. PHYS. EARTH PLANET. INTERIORS 17, 31-4, 1978.
95298	**SOME PROPERTIES OF DIAMOND CERAMIC.** FEDOSEEV, D. V. SOKOLINA, G. A. BANTSEKOV, S. V. DERYAGIN, B. V. INORG. MAT., USSR 13 (10), 1532-3, 1977. (ENGLISH TRANSLATION OF IZV. AKAD. NAUK SSSR, NEORG. MATER., 13 (10), 1904-5, 1977; FOR ORIGINAL SEE T91920)
95448	**INFRARED REFLECTIVITY, RAMAN SCATTERING, AND THE VIBRATIONAL DYNAMICS OF UNIAXIAL CRYSTALS AND MIXED CRYSTALS.** NICHOLS, H. F. UNIV. OKLAHOMA, NORMAN, OKLA., PH. D. THESIS 180PP., 1977. (UNIV. MICROFILM NO. 78-15375)
95474	**HEAT FLOW STUDIES IN THE STEAMBOAT MOUNTAIN-LEMEI ROCK AREA, SKAMANIA COUNTY, WASHINGTON.** SCHUSTER, J. E. BLACKWELL, D. D. HAMMOND, P. E. HUNTTING, M. T. WASHINGTON STATE DEPT. OF NATURAL RESOURCES 61PP., 1978. (PB-287 513, INFORMATION CIRCULAR-62)
95500	**SCHREYERITE, VANADIUM TITANIUM OXIDE, A NEW MINERAL.** MEDENBACH, O. SCHMETZER, K. AM. MINERAL. 63 (11-12), 1182-6, 1978.
95686	**THERMOPHYSICAL PROPERTIES OF BORATES TREATED WITH SULFURIC ACID.** PETROPAVLOVSKII, I. A. TOROCHESHNIKOV, N. S. KOSTYLKOV, I. G. ZH. PRIKL. KHIM. (LENINGRAD) 51 (10), 2164-7, 1978. (FOR ENGLISH TRANSLATION SEE T98588)
95767	**THE THERMAL EXPANSION OF AFWILLITE.** SHAW, R. ACTA CRYST. 6, 428-9, 1953.
95768	**REFLECTIVE POWER OF METALS AND DIELECTRICS IN THE ULTRAVIOLET.** LEWIS, E. P. HARDY, A. C. PHYS. REV. 14, 272-4, 1919.
95849	**THERMAL EXPANSION COEFFICIENTS OF UNIDIRECTIONAL COMPOSITES.** ISHIKAWA, T. J. COMPOS. MATER. 12, 153-68, 1978.
95858	**LATTICE CONDUCTIVITIES OF SINGLE-CRYSTAL AND POLYCRYSTALLINE MATERIALS AT MANTLE PRESSURES AND TEMPERATURES.** BECK, A. E. DARBHA, D. M. SCHLOESSIN, H. H. PHYS. EARTH PLANET. INTER. 17 (1), 35-53, 1978.
95916	**REFLECTANCE STUDIES OF THE GENOTYPES IN THE LIGNITE COAL OF THE MARITSA EAST BASIN.,** SISKOV, G. D., COMPT. REND., ACAD. BULG. SCI., 24 (1), 79-82, 1971.
95930	**1. SILICON VIDICON IMAGING OF JUPITER 4100-8300 ANGSTROM: SPECTRAL REFLECTIVITY, LIMB-DARKENING, AND ATMOSPHERIC STRUCTURE. 2. SIMULTANEOUS ULTRAVIOLET (0.36 MICROMETER) AND INFRARED (8-20 MICROMETER) IMAGING OF VENUS: PROPERTIES OF CLOUDS IN THE UPPER ATMOSPHERE.** DINER, D. J. CALIFORNIA INST. TECH., PASADENA, PH.D. THESIS 1-248, 1978. (UNIV. MICROFILM NO. 78-05,693)
95944	**LABORATORY STUDIES OF THE DIFFUSE REFLECTANCE SPECTRA OF FROSTS AND MINERALS OCCURRING ON ASTRONOMICAL OBJECTS.** GLASER, F. M., PAN AMERICAN UNIV., EDINBURG, TEX., 83 PP., 1978. (NASA-CR-157298, N78-28019)
95993	**EFFECT OF PHYSICOCHEMICAL PROCESSES IN FIREPROOF MATERIALS ON THEIR THERMAL CONDUCTIVITY.** LITOVSKII, E. YA. KAPLAN, F. S. KLIMOVICH. A. V. J. ENG. PHYS., USSR 33 (1), 813-18, 1977. (ENGLISH TRANSLATION OF INZH. FIZ. ZH., 33 (1), 101-7, 1977; FOR ORIGINAL SEE T90440)
96027	**STANDARD REFERENCE MATERIALS FOR THERMOPHYSICAL PROPERTIES.** HUST, J. G. KIRBY, R. K. ADVAN. CRYOGENIC ENG. 24, 232-9, 1978.
96083	**LOW TEMPERATURE THERMAL CONDUCTIVITY OF TWO NATURAL DIAMONDS: ANISOTROPIC HEAT CONDUCTION IN THE BOUNDARY SCATTERING REGIME.** VANDERSANDE, J. W. J. PHYS., COLLOQ. (ORSAY, FRANCE) 2 (6), 1017-18, 1978.
96140	**THE PHYSICAL BASIS OF MINERAL OPTICS. I. CLASSICAL THEORY.** STRENS, R. G. J. FREER, R. MINERAL. MAG. 42 (321), 19-30, 1978.
96156	**A STUDY OF ALTERATION ASSOCIATED WITH URANIUM OCCURRENCES IN SANDSTONE AND ITS DETECTION BY REMOTE SENSING METHODS.** CONEL, J. E. ABRAMS, M. J. GOETZ, A. F. H. JET PROPULSION LAB., CALIFORNIA INSTITUTE TECH. 1 AND 2, 1- PP., 1978. (NASA-CR-157600, NASA-CR-157601, N78-31512, N78-31511, JPL-PUB-78-66-VOL-1 AND VOL-2, GJBX-120(78)(VOL. 2))
96185	**APPLICATION OF HEAT-FLOW TEMPERATURE MODEL FOR REMOTELY ASSESSING NEAR SURFACE SOIL MOISTURE BY THERMOGRAPHY.** TUNHEIM, J. A. SOUTH DAKOTA STATE UNIV., BROOKINGS 50PP., 1977. (PB-279 616/7, N78-31518)
96251	**NEW REFRACTORIES FOR THE GLASS FURNACES OF THE GLASS-FIBER SECTOR.** DEGTYAREVA, E. V. ORLOVA, I. G. KOLESOV, YU. I. PISTSOV, YU. N. REFRACTORIES, USSR (10), 610-16, 1977. (ENGLISH TRANSLATION OF OGNEUPORY (10), 57-62, 1977; FOR ORIGINAL SEE T91732)
96284	**METEOROLOGICAL OBSERVATIONS.** FRANKLIN, B. THE WRITINGS OF BENJAMIN FRANKLIN 3, 186-8, 1905. (WRITTEN IN REPLY TO CADWALLADER COLDEN, NOV. 19, 1753 AND READ AT THE ROYAL SOC. LONDON, NOV. 4, 1756)
96401	**THE MECHANICAL PROPERTIES OF STRIPA GRANITE.** SWAN, G. LAWRENCE BERKELEY LAB., UNIV. CALIF., BERKELEY 30PP., 1978. (LBL-7074)

TPRC Number	Bibliographic Citation
96430	**LOW-TEMPERATURE INTERNAL STRESSES IN NIOBIUM FILMS ON DIELECTRIC SUBSTRATES.** SHERMERGOR, T. D. TUMANOVA, L. A. MAKHOV, V. I. DYUZHEV, N. A. FIZ. NIZK. TEMP. (KIEV) 4 (10), 1300–4, 1978. (FOR ENGLISH TRANSLATION SEE T98501)
96544	**PROPERTIES OF TIN-DOPED INDIUM OXIDE PREPARED BY HIGH RATE AND LOW TEMPERATURE RF SPUTTERING.,** ITOYAMA, K., JPN. J. APPL. PHYS., 17 (7), 1191–6, 1978.
96703	**MEASUREMENT OF THERMAL PROPERTIES OF ROCK USING A CARBON DIOXIDE LASER.** MURAHARA, M. HASHIMOTO, B. WASEDA DAIGAKU RIKOGAKU KENKYUSHO HOKOKU 80, 41–8, 1978.
96710	**LARGE SCALE PERMEABILITY TEST OF THE GRANITE IN THE STRIPA MINE AND THERMAL CONDUCTIVITY TEST.** LUNDSTROEM, L. STILLE, H. LAWRENCE BERKELEY LAB., CALIF. UNIV., BERKELEY 38PP., 1978. (LBL–7052)
96756	**INVESTIGATIONS OF INFRA-RED SPECTRA. APPENDIX V. WATER OF CRYSTALLIZATION.** COBLENTZ, W. W. CARNEGIE INSTITUTE OF WASHINGTON (PART I), (APPENDIX V), 129–31, 283–4, 1905. (CARNEGIE INST. WASHINGTON–PUBL.–35)
96757	**INVESTIGATIONS OF INFRA-RED SPECTRA. PART III. INFRA-RED TRANSMISSION SPECTRA.** COBLENTZ, W. W. CARNEGIE INSTITUTE OF WASHINGTON (PART III), 1–70, 1906. (CARNEGIE INST. WASHINGTON–PUBL.–65)
96758	**INVESTIGATIONS OF INFRA-RED SPECTRA. PART IV. INFRA-RED REFLECTION SPECTRA.** COBLENTZ, W. W. CARNEGIE INSTITUTE OF WASHINGTON (PART IV), 73–128, 1906. (CARNEGIE INST. WASHINGTON–PUBL.–65)
96759	**INVESTIGATIONS OF INFRA-RED SPECTRA. PART V. INFRA-RED REFLECTION SPECTRA.** COBLENTZ, W. W. CARNEGIE INSTITUTE OF WASHINGTON (PART V), 1–38, 1908. (CARNEGIE INST. WASHINGTON–PUBL.–97)
96760	**INVESTIGATIONS OF INFRA-RED SPECTRA. PART VI. INFRA-RED TRANSMISSION SPECTRA.** COBLENTZ, W. W. CARNEGIE INSTITUTE OF WASHINGTON (PART VI), 41–68, 1908. (CARNEGIE INST. WASHINGTON–PUBL.–97)
96789	**TUCEKITE, A NEW ANTIMONY ANALOGUE OF HAUCHECORNITE.** JUST, J. FEATHER, C. E. MINERAL. MAG. 42 (322), 278, M21–M22, 1978.
96807	**EMISSIVITY OF BASALTIC MELTS.** VAKULENKO, O. V. DZHIGIRIS, D. D. DEMYANENKO, YU. N. SHUTOV, B. M. STEKLO KERAM. (9), 14–16, 1978. (FOR ENGLISH TRANSLATION SEE T98853)
96877	**DETERMINING CONDUCTION THERMAL FLUX IN AN UNSATURATED SOIL FROM IN SITU MEASUREMENTS.** VAUCLIN, M. LAGOUARDE, J. P. THONY, J. L. HAMBURGER, J. CRAUSSE, P. SIXTH INTERNATIONAL HEAT TRANSFER CONFERENCE, PT. III. 13–18PP., 1978.
96889	**HIGH-TEMPERATURE HEAT CAPACITIES OF CORUNDUM, PERICLASE, ANORTHITE, CALCIUM ALUMINUM SILICATE GLASS, MUSCOVITE, PYROPHYLLITE, POTASSIUM ALUMINUM SILICATE GLASS, GROSSULAR, AND SODIUM ALUMINUM SILICATE GLASS.** KRUPKA, K. M. ROBIE, R. A. HEMINGWAY, B. S. AM. MINERAL. 64 (1–2), 86–101, 1979.
97008	**THERMAL CONDUCTIVITY OF GAMMA-IRRADIATED ALPHA-QUARTZ BELOW 10 K.** WASIM, S. NAVA, R. PHYS. STATUS SOLIDI 51 A (2), 359–66, 1979.
97010	**THE THERMAL EXPANSION OF REINFORCED NYLON 6 COMPOSITES THROUGH THE MATRIX GLASS TRANSITION TEMPERATURE.** SEGAL, L. POLYM. ENG. SCI. 19 (5), 365–72, 1979.
97029	**HEAT CAPACITY AND THERMAL CONDUCTIVITY OF SODIUM METAPHOSPHATE GLASS WITH DIFFERENT HEAT TREATMENT.** YACHMENEV, V. E. STROGANOV, E. F. DIMITROV, D. P. RABINOVICH, M. E. DEPOSITED DOCUMENT, VINITI 6PP., 1974. (VINITI–623–74)
97131	**EFFECT OF GAS COMPOSITION ON THERMAL CONDUCTIVITY OF CERAMIC FUELS.** MC CARTHY, W. H. TRANS. AM. NUCL. SOC. 30, 176–7, 1978.
97168	**EFFECT OF GAS COMPOSITION ON THERMAL CONDUCTIVITY OF CERAMIC FUELS.** MC CARTHY, W. H. ANS TRANS. 30, 176– , 1978. (ARSD–SP–007)
97338	**LATTICE THERMAL EXPANSION OF PYROCARBONS PRODUCED IN A FLUIDIZED BED.** PELLEGRINI, G. MOULAERT, M. ZUBANI, I. BIENN. CONF. CARBON, EXT. ABSTR. PROGRAM, 9TH 187–8, 1969. (AD–687 835)
97376	**VISCOSITY AND THERMAL EXPANSION OF LITHIUM ALUMINOSILICATE GLASSES.** SHELBY, J. E., J. APPL. PHYS., 49 (12), 5885–91, 1978.
97378	**HIGH-TEMPERATURE CRYSTAL CHEMISTRY OF HYDROUS MAGNESIUM- AND IRON-CORDIERITES.** HOCHELLA, M. F., JR. BROWN, G. E., JR. ROSS, F. K. GIBBS, G. V. AM. MINERAL. 64 (3–4), 337–51, 1979.
97392	**TRANSPARENT CONDUCTING INDIUM (III) OXIDE FILMS PREPARED BY THERMAL DECOMPOSITION OF INDIUM (III) CHLORIDE.** KAWAMATA, E. OHSHIMA, K. JPN. J. APPL. PHYS. 18 (1), 205–6, 1979.
97505	**THERMAL CONDUCTIVITY OF CRYSTALLINE ROCKS ASSOCIATED WITH ENERGY EXTRACTION FROM HOT DRY ROCK GEOTHERMAL SYSTEMS.** SIBBITT, W. L. DODSON, J. G. TESTER, J. W. J. GEOPHYS. RES. 84 B (3), 1117–24, 1979.
97516	**REFLECTIVITY AND EMISSIVITY OF TUNGSTEN.** HAMAKER, H. C. UNIVERSITY UTRECHT, PH. D. THESIS 79PP., 1934.
97519	**SOLAR OPTICAL PROPERTIES OF THIN FILMS OF SILVER, COPPER, GOLD, ALUMINUM, IRON, CHROMIUM AND NICKEL.** KARLSSON, B. RIBBING, C. G. UPPSALA UNIV., SWEDEN 22PP., 1977. (UPTEC–77–56–R)
97520	**HEAT TRANSPORT AND SOLAR TRANSMISSION THROUGH A WINDOW SYSTEM WITH LOW-EMITTING COATINGS.** KARLSSON, B. RIBBING, C. G. UPPSALA UNIV., SWEDEN 20PP., 1977. (UPTEC–77–57–R)
97563	**VISIBLE AND NEAR-INFRARED SPECTRA OF ROCKS FROM CHROMIUM-RICH AREAS.** HUNT, G. R. WYNN, J. C. GEOPHYSICS 44 (4), 820–5, 1979.
97572	**ANISOTROPY OF THE THERMAL EXPANSION OF ALUMINUM OXIDE-CHROMIC OXIDE SOLID SOLUTIONS.** ALAPIN, B. G. DEGTYAREVA, E. V. LYSAK, S. V. IZV. AKAD. NAUK SSSR, NEORG. MATER. 15 (2), 297–9, 1979. (FOR ENGLISH TRANSLATION SEE T100641)
97587	**USING BERYLLIUM [FOR MIRRORS] INSTEAD. PART I.** FULLERTON-BATTEN, R. C. BAYMOR, A. FAUSON, W. O. OPT. SPECTRA 13 (7), 49–52, 1979.
97659	**A STUDY OF SOME PROPERTIES OF IRON - CHROMIUM HIGH-TEMPERATURE SHIFT CONVERSION CATALYST.** POPA, O. ROTARU, P. BLEJOIU, S. I. PANDELE, L. BUNESCU, O. BRASOVEANU, I. MARCHIDAN, D. I. REV. ROUM. CHIM. 24 (1), 153–8, 1979.

TPRC Number — Bibliographic Citation

97669 POLYCRYSTALLINE HEAT-RESISTANT ELECTRIC INSULATING CERAMICS.
MASLENNIKOVA, G. N. FOMINA, N. P.
DAMIR, D. A. SOKOLINA, E. A.
STEKLO KERAM.
(11), 29-31, 1978.
(FOR ENGLISH TRANSLATION SEE T98577)

97680 THERMOPHYSICAL PROPERTIES OF OIL SHALE MINERALS.
WANG, Y. RAJESHWAR, K. NOTTENBURG, R. N.
DUBOW, J. B.
THERMOCHIM. ACTA
30 (1-2), 141-51, 1979.

97784 ANOMALOUS SPECIFIC HEAT OF PURE GRAPHITE AROUND 1.45 K IN MAGNETIC FIELDS UP TO 13.7 T.
KHATTAK, G. D. BOHM, H. V. KEESOM, P. H.
PHYS. REV. B: CONDENS. MATTER
18 (11), 6178-80, 1978.

97863 MACROSCOPIC AND MICROSCOPIC GRUENEISEN PARAMETERS OF GARNET, OLIVINE AND PYROXENE.
DIETRICH, P. ARNDT, J.
HIGH-PRESSURE SCI. TECHNOL., AIRAPT CONF., 6TH
2, 125-6, 1979.

97880 RADIATION EFFECTS ON THE TRANSMISSION OF VARIOUS OPTICAL GLASSES AND EPOXIES.
SHETTER, M. T. ABREU, V. J.
APPL. OPT.
18 (8), 1132-3, 1979.

97884 PREPARATION OF TIN-DOPED INDIUM OXIDE FILMS AT LOW DEPOSITION TEMPERATURES BY ION-BEAM SPUTTERING.
FAN, J. C. C.
APPL. PHYS. LETT.
34 (8), 515-17, 1979.

97889 THERMAL CONDUCTIVITY OF IRON OXIDE-(BORON OXIDE - LEAD OXIDE) GLASSES.
KELEMEN, F. ARDELEAN, I.
CZECH. J. PHYS.
B29 (6), 680-6, 1979.

97914 THERMAL EXPANSION OF ANDALUSITE.
SCHNEIDER, H.
J. AM. CERAM. SOC.
62 (5-6), 307, 1979.

97928 THE THERMAL EXPANSION OF CHRYSOTILE ASBESTOS AND ITS COMPOSITES.
HARWOOD, C. YATES, B. BADAMI, D. V.
J. MATER. SCI.
14 (5), 1126-40, 1979.

98018 EFFECTIVE THERMAL CONDUCTIVITY OF SITALLS AT VARIOUS STAGES OF HEAT TREATMENT.
BOLSHAKOVA, N. V. MALTER, V. P.
GLASS AND CERAMICS, USSR
(2), 85-8, 1978.
(ENGLISH TRANSLATION OF STEKLO KERAM., (2), 16-18, 1978; FOR ORIGINAL SEE T93254)

98069 THERMAL EXPANSION OF PYROPHYLLITE.
TAYLOR, L. A. BELL, P. M.
CARNEGIE INST. WASH., YEARB.
(69), 193-4, 1971.

98070 THERMAL CONDUCTIVITY, DENSITY, AND SONIC VELOCITY MEASUREMENTS OF SAMPLES OF ANHYDRITE AND HALITE FROM SITES 225 AND 227.
WHEILDON, J. EVANS, T. R. GIRDLER, R. W.
DEEP SEA DRILLING PROJ., INITIAL REPT.
23, 909-11, 1974.

98105 THERMAL CONDUCTIVITY OF IGNEOUS ROCKS.
NEUMANN, W.
BER. OESTERR. STUDIENGES. ATOMENERG.
14PP., 1979.
(SGAE-BER.-A0059)

98110 CHARACTERISTICS OF AN APPARATUS BASED ON CURVE-FITTING METHOD FOR MEASURING THERMAL PROPERTIES IN CRYOGENIC TEMPERATURES.
YOSHIWA, M. IWATA, A.
TEION KOGAKU
10 (3), 90-103, 1975.

98122 SOME PROPERTIES OF SINTERED DIAMOND.
HORTON, M. D. POPE, B. J. HORTON, L. B.
RADTKE, R. P.
HIGH-PRESSURE SCI. TECHNOL., AIRAPT CONF., 6TH
1, 923-30, 1979.

98161 THERMAL CONDUCTIVITY MODEL FOR (THORIUM, URANIUM) DIOXIDE TO MELTING.
HIMES, D. A.
TRANS. AM. NUCL. SOC.
30, 174-5, 1978.
(HEDL-SA-1590-S, CONF-781105-99)

98168 VACUUM ULTRAVIOLET SPECTRA OF CARBONACEOUS CHONDRITES.
COHEN, A. J. WAGNER, J. K. HAPKE, B. W.
PARTLOW, W. D.
METEORITICS
13 (4), 420-7, 1978.

98172 THERMOPHYSICAL PROPERTIES OF CONASAUGA SHALE.
SMITH, D. D.
OAK RIDGE Y-12 PLANT, OAK RIDGE, TENN.
28PP., 1978.
(Y-2161)

98173 PRESSURE EFFECTS ON THERMAL CONDUCTIVITY AND EXPANSION OF GEOLOGIC MATERIALS.
SWEET, J. N.
SANDIA LABS., ALBUQUERQUE, N. MEXICO
48PP., 1979.
(SAND-78-1991)

98199 GRUENEISEN PARAMETER AND SPECIFIC HEAT OF A PICRITE.
RAMANA, Y. V. SARMA, L. P.
ACOUST. LETT.
2, 74-7, 1978.

98240 THERMAL CONDUCTIVITY COMPARISON METHOD BASED ON SILICON DIODE THERMOMETRY.
LAWLESS, W. N. CLARK, C. F.
REV. SCI. INSTRUM.
50 (6), 787-9, 1979.

98301 CORRELATION BETWEEN THE CRYSTALLIZATION STATE AND THERMAL BEHAVIOR OF KAOLINITE.
MURAT, M. NEGRO, A.
GEOL. APPL. IDROGEOL.
12 (2), 79-87, 1977.

98302 THE MERCURY SOIL: PRESENCE OF IRON DIPOSITIVE ION.
MC CORD, T. B. CLARK, R. N.
LUNAR PLANET. SCI.
10 (PT. 2), 789-91, 1979.

98303 NEWLY DEFINED INFRARED ELECTRONIC ABSORPTION FEATURES IN TELESCOPIC REFLECTANCE SPECTRA: THE LUNAR CASE.
MC CORD, T. B. CLARK, R. N. MC FADDEN, L. A.
PIETERS, C.
LUNAR PLANET. SCI.
10 (PT. 2), 792-4, 1979.

98328 SPECTRAL REFLECTANCE IMAGES OF LUNAR CORES: NEW RESOURCES.
BUTLER, P. GOLDBERG, A. PIETERS, C.
WALTZ, S. NAGLE, S.
LUNAR PLANET. SCI.
10 (PT. 1), 175-7, 1979.

98331 UTILIZATION OF SPENT OIL SHALE IN THE PREPARATION OF GLASS-FIBER AND GLASS-CERAMICS.
HORIUCHI, T. MIZUNO, T. CHUNG, C. H.
MACKENZIE, J. D.
PROC. MINER. WASTE UTIL. SYMP.
6, 109-12, 1978.

98332 SINGLE CRYSTAL ULTRAVIOLET SPECTRA OF OLIVINES, (IRON, MAGNESIUM) SILICON OXIDE, UP TO 37,000 CENTIMETER AT ELEVATED PRESSURES.
ABU-EID, R. M. LANGER, K.
ANNU. RES. REP. - KUWAIT INST. SCI. RES.
78-81, 1977.

98462 THERMAL EXPANSION OF DIAMOND.
KRISHNAN, R. S.
PROC. INDIAN ACAD. SCI.
24 A, 33-44, 1946.

98501 LOW-TEMPERATURE INTERNAL STRESSES IN NIOBIUM FILMS ON DIELECTRIC SUBSTRATES.
SHERMERGOR, T. D. TUMANOVA, L. A.
MAKHOV, V. I. DYUZHEV, N. A.
SOV. J. LOW TEMP. PHYS.
4 (10), 613-15, 1978.
(ENGLISH TRANSLATION OF FIZ. NIZK. TEMP., 4 (10), 1300-3, 1978; FOR ORIGINAL SEE T96430)

98525 AN ANALYSIS OF THE DEPENDENCE OF THERMAL TRANSPORT PARAMETERS ON ORGANIC CONTENT FOR GREEN RIVER OIL SHALES.
WANG, Y. RAJESHWAR, K. DU BOW, J.
J. APPL. PHYS.
50 (4), 2776-81, 1979.

98577 POLYCRYSTALLINE, HEAT-RESISTANT, ELECTROINSULATING CERAMIC.
MASLENNIKOVA, G. N. FOMINA, N. P.
SOKOLINA, E. A. DAMIR, D. A.
GLASS CERAM., USSR
(11), 679-81, 1978.
(ENGLISH TRANSLATION OF STEKLO KERAM., (11), 29-31, 1978; FOR ORIGINAL SEE T97669)

TPRC Number	Bibliographic Citation

98588 THERMOPHYSICAL CHARACTERISTICS OF BORATES TREATED WITH SULFURIC ACID.
PETROPAVLOVSKII, I. A. TOROCHESHNIKOV, N. S.
KOSTYLKOV, I. G.
J. APPL. CHEM., USSR
51 (10), 2062–5, 1978.
(ENGLISH TRANSLATION OF ZH. PRIKL. KHIM., 51 (10), 51 (10), 2164–7, 1978; FOR ORIGINAL SEE T95686)

98629 SOLAR-CELL CHARACTERISTICS AND INTERFACIAL CHEMISTRY OF INDIUM-TIN-OXIDE/INDIUM PHOSPHIDE AND INDIUM-TIN-OXIDE/GALLIUM ARSENIDE JUNCTIONS.
BACHMANN, K. J. SCHREIBER, H.
SINCLAIR, W. R. SCHMIDT, P. H. THIEL, F. A.
SPENCER, E. G. PASTEUR, G. FELDMANN, W. L.
SREE HARSHA, K.
J. APPL. PHYS.
50 (5), 3441–6, 1979.

98632 THE COEFFICIENT OF SOLAR ABSORPTIVITY AND LOW TEMPERATURE EMISSIVITY OF VARIOUS MATERIALS. A REVIEW OF THE LITERATURE.
HOLDEN, T. S. GREENLAND, J. J.
COMMONWEALTH SCIENTIFIC AND INDUSTRIAL RES. ORGANIZATION, AUSTRALIA
26PP., 1951.
(CSIR-R-6)

98666 DENSITY AND VISCOSITY OF HYDROUS CALC-ALKALIC ANDESITE MAGMA AT HIGH PRESSURES.
KUSHIRO, I.
YEAR BOOK - CARNEGIE INST. WASHINGTON
675–7, 1978.

98673 SINGLE CRYSTAL UV-SPECTRA OF OLIVINES (IRON, MAGNESIUM) SILICON OXIDE UP TO 37,000 CM-1 AT ELEVATED PRESSURES.
ABU-EID, R. M. LANGER, K.
PETROLEUM, PETROCHEMICALS AND MATERIALS
78–81, 1977.

98785 THERMAL CONDUCTIVITY OF ROCK SALT FROM LOUISIANA SALT DOMES.
MORGAN, M. T.
OAK RIDGE NATIONAL LAB., OAK RIDGE, TENNESSEE
21PP., 1979.
(ORNL/TM-6809)

98789 SPINEL-BONDED MAGNESITE REFRACTORIES WITH REDUCED THERMAL CONDUCTIVITY.
SHUBIN, V. I. SERGEEVA, V. M.
MURASHOVA, M. V. PODDUBNYI, I. M.
SIMONOV, K. V. BUGAEV, N. F.
REFRACTORIES, USSR
(2), 70–5, 1978.
(ENGLISH TRANSLATION OF OGNEUPORY, (2), 3–9, 1978; FOR ORIGINAL SEE T93218)

98803 THERMAL CONDUCTIVITY OF SOLIDIFIED WASTE PRODUCTS.
NEUMANN, W.
INST. METALLURGIE, FORSCHUNGSZENTRUM SEIBERSDORF
33PP., 1979.

98853 EMMISIVITY OF BASALT MELTS.
VAKULENKO, O. V. DZHIGIRIS, D. D.
DEMYANENKO, YU. N. SHUTOV, B. M.
GLASS CERAMICS, USSR
(9), 530–2, 1978.
(ENGLISH TRANSLATION OF STEKLO KERAM., (9), 14–16, 1978; FOR ORIGINAL SEE T96807)

98902 THERMAL PROPERTIES OF GABLE MOUNTAIN BASALT CORES HANFORD NUCLEAR RESERVATION.
MARTINEZ-BAEZ, L. F. AMICK, C. H.
LAWRENCE BERKELEY LAB., CALIFORNIA UNIVERSITY
10 PP., 1978.
(LBL-7038)

98927 MEASUREMENT OF THERMAL CONDUCTIVITY OF GREEN RIVER (COLORADO) OIL SHALES BY A THERMAL COMPARATOR TECHNIQUE.
NOTTENBURG, R. RAJESHWAR, K.
ROSENVOLD, R. DUBOW, J.
FUEL
57 (12), 789–95, 1978.

98944 SPECIFIC HEAT ANOMALIES AND SPIN-SPIN INTERACTIONS IN CARBONS: A REVIEW.
MROZOWSKI, S.
J. LOW TEMP. PHYS.
35 (3–4), 231–98, 1979.

98998 THERMODYNAMIC PROPERTIES OF REFRACTORY INORGANIC MATERIALS AT HIGH TEMPERATURES.
COLWELL, J. H.
NATIONAL BUREAU OF STANDARDS, WASHINGTON, D. C.
103–31, 1979.
(NBSIR-79-1767, PB-297 465)

99103 TEMPERATURE-DEPENDENCE OF IRON - TITANIUM SPECTRA IN THE VISIBLE REGION: IMPLICATIONS TO MAPPING TITANIUM CONCENTRATIONS OF HOT PLANETARY SURFACES.
OSBORNE, M. D. PARKIN, K. M. BURNS, R. G.
PROC. LUNAR PLANET. SCI. CONF., 9TH
2949–60PP., 1978.

99162 SPECTRA OF ALTERED ROCKS IN THE VISIBLE AND NEAR INFRARED.
HUNT, G. R. ASHLEY, R. P.
ECONOMIC GEOLOGY
74, 1613–29, 1979.

99163 NEAR-INFRARED (1.3–2.4 MICROMETER) SPECTRA OF ALTERATION MINERALS - POTENTIAL FOR USE IN REMOTE SENSING.
HUNT, G. R.
GEOPHYSICS
44 (12), 1974–86, 1979.

99173 GROSSULAR HEAT CAPACITY UNDER THE 13–300 K TEMPERATURE INTERVAL AND THERMODYNAMICS OF SOLID SOLUTION PYROPE-GROSSULAR.
KOLESNIK, YU. N. NOGTEVA, V. V.
ARKHIPENKO, D. K. OREKHOV, B. A.
PAUKOV, I. E.
GEOKHIMIYA
(5), 713–21, 1979.

99174 TETRAHEDRITE HIGH IN SILVER, ZINC AND CADMIUM FROM JIHLAVA, CZECHOSLOVAKIA.
CECH, F. HAK, J.
CAS. MINERAL. GEOL.
24 (1), 83–7, 1979.

99206 PHYSICOCHEMICAL PROPERTIES AND STRUCTURE OF GERMANIUM OXIDE-SILICON DIOXIDE-PHOSPHORUS OXIDE SYSTEM GLASSES.
SEDMALIS, U. BOL'SHII, YA. YA.
KOZYUKOV, V. M.
LATV. PSR ZINAT. AKAD. VESTIS, KIM. SER.
(5), 525–30, 1978.

99207 GLASSES AND CRYSTAL PHASES IN BORON OXIDE-GERMANIUM DIOXIDE-SILICON DIOXIDE-PHOSPHORUS PENTOXIDE AND ALUMINUM OXIDE-GERMANIUM DIOXIDE-SILICON DIOXIDE-PHOSPHORUS PENTOXIDE SYSTEMS.
KOZYUKOV, V. M. BOL'SHII, YA. YA.
SEDMALIS, U.
LATV. PSR ZINAT. AKAD. VESTIS, KIM. SER.
(1), 23–6, 1979.

99265 THERMAL EXPANSION OF CERAMICS IN THE MAGNESIA-CALCIUM OXIDE SYSTEM.
DE ALMEIDA, M. BROOK, R. J.
CARRUTHERS, T. G.
J. MATER. SCI.
14 (9), 2191–4, 1979.

99278 NUKUNDAMITE, A NEW MINERAL, AND IDAITE.
RICE, C. M. ATKIN, D. BOWLES, J. F.
CRIDDLE, A. J.
MINERAL. MAG.
43 (326), 193–200, 1979.

99309 ANALYSIS OF THERMAL STRESS RESISTANCE OF MICROCRACKED BRITTLE CERAMICS.
HASSELMAN, D. P. SINGH, J. P.
AM. CERAM. SOC. BULL.
58 (9), 856–60, 1979.

99351 NEW DATA ON STANNITE AND RELATED TIN SULFIDE MINERALS.
KISSIN, S. A. OWENS, D. R.
CAN. MINERAL.
17 (1), 125–35, 1979.

99357 THERMAL CONDUCTIVITY TESTS ON HIGHLY COMPACTED BENTONITE.
KNUTSSON, S.
TEK. HOEGSK., LULEAA, SWEDEN
I.PP., U.MM.
(KBS-TR-72)

99431 PLAGIOCLASE FELDSPARS: VISIBLE AND NEAR INFRARED DIFFUSE REFLECTANCE SPECTRA AS APPLIED TO REMOTE SENSING.
ADAMS, J. B. GOULLAUD, L. H.
PROC. LUNAR PLANET. SCI. CONF., 9TH
2901–9, 1978.

99434 LONG-TERM GLAZING PERFORMANCE.
JORGENSEN, G. J.
SOLAR ENERGY RESEARCH INST., GOLDEN, CO
1–12, 1979.
(SERI/TP-31-193)

99523 NEW GLASSES PASS INFRARED, CONDUCT ELECTRICITY, ABSORB WATER.
AUTHOR ANON.
MATER. DESIGN ENG.
51 (5), 209–10, 1960.

99593 ZERO AND LOW COEFFICIENT OF THERMAL EXPANSION POLYCRYSTALLINE OXIDES.
SKAGGS, S. R.
REV. INT. HAUTES TEMP. REFRACT.
16 (2), 157–67, 1979.

TPRC Number	Bibliographic Citation
99652	**A STUDY OF LASER-INDUCED DAMAGE TO DIELECTRIC SURFACES AND THIN FILMS.** ALYASSINI, M. N. UNIV. OF SOUTHERN CALIFORNIA, LOS ANGELES, PH. D. THESIS 1- PP., 1975.
99672	**LOW-TEMPERATURE SPECIFIC HEAT OF A YTTRIUM OXIDE-DOPED ZIRCONIUM OXIDE OXYGEN CONDUCTOR.** LAWLESS, W. N. PHYS. REV. B: CONDENS. MATTER 21 (2), 585-8, 1980.
99698	**QUANTITATIVE RELATIONSHIPS OF SURFACE GEOLOGY AND SPECTRAL HABIT TO SATELLITE RADIOMETRIC DATA.** MARSH, S. E. STANFORD UNIV., PH.D. THESIS 1-290, 1979. (UNIV. MICROFILM NO. 79-17258)
99709	**DIDYMIUM GLASS FILTERS FOR CALIBRATING THE WAVELENGTH SCALE OF SPECTROPHOTOMETERS - SRM 2009, 2010, 2013, AND 2014.** VENABLE, W. H. ECKERLE, K. L. NATIONAL BUREAU STANDARDS, WASHINGTON, DC 1-89, 1979. (NBS-SP-260-66, PB-80-104961)
99742	**STORAGE AND DISPOSAL OF RADIOACTIVE WASTE AS GLASS IN CANISTERS.** MENDEL, J. E. BATTELLE PACIFIC NORTHWEST LAB., RICHLAND, WASH., 97 PP., 1978. (PNL-2764)
99744	**EXPERIMENTAL ENGINEERING SUPPORT OF IN SITU GASIFICATION PROCESSES.** WESTMORELAND, P. R. DICKERSON, L. S. THOMA, A. L. RODGERS, B. R. OAK RIDGE NATIONAL LAB., OAK RIDGE, TENN. 50-82, 1979. (ORNL-5487)
99817	**ON THE SECULAR INSTABILITY FROM THERMAL CONDUCTIVITY ROTATING STARS.** LINDBLOM, L. ASTROPHYS. J. 233 (3, PT. 1), 974-80, 1979.
99886	**IR SPECTRA OF INORGANIC COMPOUNDS. 4. IR SPECTRA OF INORGANIC COMPOUNDS AND MINERALS.** WADA, K. KANZEI CHUO BUNSEKISHOHO 19, 133-87, 1978.
99888	**QUANTITATIVE REFLECTIVITY MEASUREMENTS.** TANNER, B. K. PHYS. EDUC. 13 (2), 120-2, 1978.
99928	**MODEL DEVELOPMENT FOR IN SITU TEST RESULTS IN ARGILLACEOUS ROCK.** TYLER, L. D. SANDIA LABS., ALBUQUERQUE, N. MEX. 23PP., 1979. (SAND-79-1275)
99938	**SELECTED VALUES OF CHEMICAL THERMODYNAMIC PROPERTIES. PART 2. TABLES FOR THE ELEMENTS ELEMENTS TWENTY-THREE THROUGH THIRTY-TWO IN THE STANDARD ORDER OF ARRANGEMENT.** WAGMAN, D. D. EVANS, W. H. HALOW, I. PARKER, V. B. BAILEY, S. M. SCHUMM, R. H. NATIONAL BUREAU OF STANDARDS, WASHINGTON, D. C. (PT. 2), 62PP., 1966. (NBS-TN-270-2)
99957	**PROPERTIES OF IRON AT HIGH PRESSURES AND THE STATE OF THE CORE.** JEANLOZ, R. J. GEOPHYS. RES. 84 (B 11), 6059069, 1979.
99991	**PRELIMINARY REPORT ON PHYSICAL AND THERMAL PROPERTIES OF BASALT: DRILL HOLE DC-10 POMONA FLOW - GABLE MOUNTAIN.** DUVALL, W. I. MILLER, R. J. WANG, F. D. COLORADO SCHOOL OF MINES, GOLDEN, CO. 46PP., 1978. (RHO-BWI-C-11)
99992	**TERRESTRIAL HEAT FLOW IN EASTERN ARIZONAD A FIRST REPORT.** REITER, M. SHEARER, C. J. GEOPHYS. RES. 84 (11), 6115-20, 1979. (ALO-3721-12)
100051	**STRUCTURAL CHARACTERIZATION OF THE TERNARY SYSTEM COBALTOUS OXIDE-ZINC OXIDE-MAGNESIUM OXIDE SOLID SOLUTIONS.** PEPE, F. SCHIAVELLO, M. MINELLI, G. LO JACONO, M. Z. PHYS. CHEM. (WIESBADEN) 115 (1), 7-15, 1979.
100061	**MARS SECTION OF REMOTE SENSING OF BASALTS IN THE SOLAR SYSTEM. CHAPTER 2 OF BASALTIC VOLCANISM IN THE SOLAR SYSTEM.** MC CORD, T. B. INST. FOR ASTRONOMY, UNIVERSITY OF HAWAII, HONOLULU 22PP., 1978. (RSL-221)
100063	**A ROOM-TEMPERATURE PHASE TRANSITION IN MAXIMUM MICROCLINE. UNIT CELL PARAMETERS AND THERMAL EXPANSION.** OPENSHAW, R. E. HENDERSON, C. M. B. BROWN, W. B. PHYS. CHEM. MINER 5 (1), 95-104, 1979.
100083	**KEROGEN MATURATION OF THE MITI HAMAYUCHI CORE SAMPLES AND ITS IMPLICATION TO OIL GENERATION STUDY.** UJIIE, Y. SEKIYA GIJUTSU KYOKAISHI 44 (4), 175-82, 1979.
100105	**REINFORCED REACTION INJECTION MOLDING.** LEIS, D. G. J. ELASTOMERS PLAST. 11 (4), 301-6, 1979.
100115	**HIGH-TEMPERATURE THERMAL CONDUCTIVITY OF ELECTRON-IRRADIATED DIAMOND.** BURGEMEISTER, E. A. AMMERLAAN, C. A. J. PHYS. REV., B 21 (6), 2499-2505, 1980.
100149	**EVALUATION OF CELLULAR GLASSES FOR SOLAR MIRROR PANEL APPLICATIONS.** GIOVAN, M. ADAMS, M. JET PROPULSION LAB., PASADENA, CA 96 PP., 1979. (DOE-JPL-1060-24)
100206	**POTENTIAL APPLICATIONS OF TWO HIGH GRADE KAOLINITIC CLAYS AND A PYROPHYLLITE DEPOSIT.** BOWMAN, R. TAUBER, E. J. AUST. CERAM. SOC. 15 (1), 5-8, 1979.
100244	**THERMAL CONDUCTIVITY OF THE ROCKS IN THE BUREAU OF MINES STANDARD ROCK SUITE.** MORGAN, M. T. WEST, G. A. OAK RIDGE NATIONAL LAB., OAK RIDGE, TENN. 54PP., 1980. (ORNL-TM-7052)
100246	**MEASUREMENT OF THERMAL DIFFUSIVITY AND APPLICATION TO IN SITU COAL GASIFICATION.** WESTMORELAND, P. R. PROC. INTERSOC. ENERGY CONVERS. ENG. CONF., 14TH 1, 1018-21, 1979.
100311	**THE TEMPERATURE DEPENDENCE OF THE ELASTIC CONSTANTS OF A MACHINABLE GLASS-CERAMIC.** HARRINGTON, S. N. BREWER, J. A. PEDERSON, D. O. J. APPL. PHYS. 51 (4), 2043-6, 1980.
100317	**DEPENDENCE OF THERMAL DIFFUSIVITY ON ORGANIC CONTENT FOR GREEN RIVER OIL SHALES-EXTENSION OF THE MODIFIED CHENG-VACHON MODEL TO THE PARALLEL HEAT FLOW CASE.** WANG, Y. RAJESHWAR, K. DU BOW, J. J. APPL. PHYS. 51 (3), 1829-30, 1980.
100400	**SINTERED SUPERHARD MATERIALS.** WENTORF, R. H. DEVRIES, R. C. BUNDY, F. P. SCIENCE 208, 873-80, 1980.
100404	**EVALUATION TECHNIQUES FOR DETERMINING THE REFLECTIVITY, SPECULARITY, AND FIGURE OF SOLAR MIRRORS.** GRIFFIN, J. W. LIND, M. A. PHILLIPP, L. D. SOLAR ENERGY RESEARCH INST., GOLDEN, CO 1-110, 1980. (SERI-TR-98366-1)
100601	**A SUPERIOR BROADBAND ANTIREFLECTION COATING.** SUN, W-R. OPT. SPECTRA 14 (7), 57-60, 1980.
100605	**LABORATORY MEASURED MATERIAL PROPERTIES OF QUARTZ MONZONITE, CLIMAX STOCK, NEVADA TEST SITE.** PRATT, H. R. LINGLE, R. SCHRAUF, T. TERRA TEK, INC., SALT LAKE CITY, UT 1-24, 1979. (UCRL-15073)
100606	**GEOLOGICAL DISPOSAL INVESTIGATIONS IN GRANITE AT THE NEVADA TEST SITE.** BALLOU, L. B. LAWRENCE LIVERMORE LAB., CALIFORNIA UNIV., LIVERMORE 1-5, 1979. (UCRL-82284, CONF-790304-7)

TPRC Number	Bibliographic Citation
100616	**CONASAUGA NEAR-SURFACE HEATER EXPERIMENT.** KRUMHANSL, J. L. SANDIA LABS., ALBUQUERQUE, NM 1-98, 1979. (SAND-M.-U,KK)
100641	**THERMAL EXPANSION ANISOTROPY OF SOLID SOLUTIONS ALUMINUM OXIDE-CHROMIC OXIDE.** ALAPIN, B. G. DEGTYAREVA, E. V. SYSAK, S. V. INORG. MAT., USSR 15 (2), 234-6, 1979. (ENGLISH TRANSLATION OF IZV. AKAD. NAUK SSSR, NEORG. MAT., 15 (2), 297-9, 1979; FOR ORIGIANL SEE T097572)
100720	**DETERMINATION OF BASALT PHYSICAL AND THERMAL PROPERTIES AT VARYING TEMPERATURES, PRESSURES, AND MOISTURE CONTENTS.** MILLER, R. J. BISHOP, R. C. COLORADO SCHOOL OF MINES, GOLDEN, CO 1-83, 1979. (RHO-BWI-C-50)
101370	**POINT DEFECTS AND THE ELECTRICAL PROPERTIES OF APATITE.** SIEGEL, J. A. PRINCETON UNIV., PH.D. THESIS 1-180, 1970. (UNIV. MICROFILMS NO. 71-14,419)

Part D
AUTHOR INDEX

PART D

USE OF AUTHOR INDEX

The *Author Index* is given in two parts, namely, personal authors and coauthors followed by corporate authors. The numbers shown along with the name of each author refer to the TPRC serial numbers of the references cited in the *Bibliography* (Part C).

For simplification only the author's last name and his initials are used. Naturally, in the case of some of the more popular last names having the same initials, this practice will result in improperly crediting papers to similarly named authors. For this the Editors express their sincere apology. In such instances, the user of this *Author Index* should inspect the titles of the documents cited in the *Bibliography* to identify those technical papers which may have been improperly credited.

The term "corporate author" applies to industrial organizations, government agencies, colleges, universities, etc. In alphabetizing corporate names, the main listing is made under the key word of the parent company or organization's name, and names of divisions, laboratories, departments, etc., are cross referenced under the key words of their names.

In the case of universities, the word "university" always appears second in abbreviated form. In identification of corporate names, information is listed in the following sequence: name of organization, city, state, division, laboratory, etc.

PERSONAL AUTHOR INDEX

Author	Numbers
BELL P M	71054 98069
BELL R J	49293 59521
BELL R T	52892
BELL RAYMOND M	1557
BELL W C	48788
BELLE J	46952 71915
BELLIERE M	38471
BELLOTTI L H	58983
BELOGLAZOV A A	76473 76474
BELORGEOT C	32110
BELOSTOTSKAYA V YA	73730 73731
BELOUSOV M V	57308 57309 59430 59431 72646 72647
BELOUSOV V M	44133 44134
BELOUSOV V V	39422
BELOVA A N	56962
BELSER R B	33388
BELYAEV R A	80319 81987
BELYAEVA I D	50280 60144 60145 64014
BENDEROVICH A M	274
BENECKI W T	48014
BENEDICT ELISABETH	31809
BENFIELD A E	21536 89613
BENHAM C B	26766
BENN M	52855
BENNETT A J R	49087
BENNETT D G	52279 59516
BENNETT H E	45929
BENNETT H S	61349
BENNETT S J	88904
BENNEWITZ K	3421
BENZEL J F	52661
BERAN A	74674
BERDYEV B M	44372
BEREZHNOI A I	51980 51981 55387
BEREZHNOI A N	32198
BEREZHNOI A S	19538 55434 55435
BERG I V	57098 57099
BERG J W JR	74222
BERG M	52795
BERG R S	92721
BERG W T	48157
BERGER D	92686
BERGER D F	47426
BERGER E	37975
BERGERON C G	58728 64464
BERGGREN G	48431
BERGGREN W P	46820
BERKEY WM E	22494
BERKOWITZ J	29508
BERL E	10317 28664
BERLINCOURT D	24048
BERLINCOURT D A	45380
BERLOT R	44519 45483
BERMAN L U	50030
BERMAN R	56 266 917 1108 1161 1411 6702 7200 9398 16756 21176 25453 33801 85815 92918
BERMAN R M	46952 71915
BERNETT E C	24360 29364 29491 39516
BERNSTEIN J B	45698
BERREMAN D W	9395
BERRY CLIFFORD E	798
BERSAN J	32113
BERSET G	90745
BERSHAK V I	43143 45107 53327
BERSHOV L V	64691 64692
BERSHTEIN V A	69642 69643
BERTAUX M G	94057
BERTHIER G	42720
BERTHOLD J W III	59834 87429
BERTINO J P	94682
BERTOLOTTI R L	79605
BERTRAM W W	22766
BERZHATYI V I	75201 80006
BERZINS R	49689
BESPALKO V P	36028
BESPALOV YU I	56285
BESSONOV A F	50788 51130 51131 52347
BESSONOV M V	51623 66064
BESTEIRO J	85747
BETTA V	36904
BETTANY C	42209
BETTS D S	74374
BEUKEN L	26667
BEUKES G J	58992
BEUS M D	49052
BEUSSEM W R	30255
BEVANS J T	7350 7351 29596
BEYERSDORFF L E	42306
BEZBORODOV M A	26143
BEZNOSIKOVA A V	52116 52117
BEZSMERTNAYA Z G	57179 57180
BHANDARI R C	46332 48921 51330 51754 52361 55156
BHASKARA-RAO B	40403
BHATTACHARYYA B N	41301
BHATYE S V	42435
BHUSHAN B	60562
BHUTA V S	32869
BIA P	41374 74014
BIANCHERIA A	46259
BIBIK V P	27540
BICHURINA A A	58259 70496 74223
BICKELHAUPT R E	53097 77589
BIELANSKI A	57964
BIEN A S	25524
BIENERT W B	36468 57665
BIENFAIT G	94057
BIERMASZ TH	8245 9293 9294 11913
BIGELEISAN J	11047
BIGHLEY L D	47385
BIJL A	54992
BIJL D	26
BIL V S	36391 38075
BILLINGTON N S	12033
BILLMEYER F W JR	36348 54711
BILLMEYER FRED W JR	9450
BILTZ W	58390 60768
BINDER A B	66640
BING F G	6940
BING G F	6567
BIRCH F	16341 20780 28780 89612
BIRCH J A	38865 89480
BIRCH R E	60445
BIRGE G W	8882
BIRKEBAK R C	37601 56105 60704 65323 66830 68482 68484 68911 70817 73329 78683
BISCHOFF F	27579
BISHARA W W	85285
BISHAY A	18621
BISHAY A M	27407 41419
BISHOP A C	91731
BISHOP P H H	10723 16520
BISHOP R C	100720
BISHUI B M	10271 19768
BISMUTH G	65133
BISWAS D K	85218
BITSIANES G	44135
BIZETTE H	1213 54354
BLACHERE S T	47390
BLACK I A	41217 44628 50638
BLACK M H	39835
BLACK W E	43744
BLACKBURN A R	55621
BLACKWELL D D	89618 91244 92038 95474
BLACKWOOD J M	62381
BLAETTNER W	47316
BLAIN B J	43033
BLAINE L R	49290
BLAIR G E	18637
BLAIR G RICHARD	18630
BLAISSE B S	46615
BLAMONT J E	64445
BLANCHARD M B	78688
BLANCO E P	747 28264
BLASZYK P E	46030
BLATTNER W	44932
BLAU H H	78266
BLAU H H JR	32045 38312
BLAU HENRY H JR	7023

BRATSCHUN W R 54273
BRATT PETER R 26640
BRAUN G 1628
BRAUN H B 41478
BRAUN H G 41479
BRAUN W 64009
BRAVINSKII V G 53739
BRAY R 77530
BRAZEL J 77957
BRAZEL J P 71449
BRE M 56148
BRECKENRIDGE R G 55185
BREDEN C R 16712
BREEN C 52568
BREGER D 61725 65225
BREHAT F 40087 43975
BREKHOVSKIKH S M 38334 38846 49154 49155 51970 51971
BRENER A I 65854 73366
BRENNAN B 54960 57798
BRENNAN D 37606
BRESKER R I 55416 63075
BRESLICH F N 36144
BRESLICH F N JR 52574
BRETSZNAJDER S 40741
BRETT N H 46880
BREUCH R A 59196
BREWER D F 74374
BREWER J A 100311
BREWER R E 8882
BRIDENBAUGH P M 27989
BRIDGMAN P W 24551 39337
BRIGGS A 78497
BRIGGS F 70043
BRIL A 25906
BRINKMAN W F 69622
BRINN M S 40389
BRISBANE S M 46503
BRISSE F 75779
BROCK F H 17062
BROCK J C F 55170
BROCKES A 51581
BRODALE G E 47910 69939
BRODERICK A H T 7336
BROECKERHOFF P 75986
BROMBERG A V 87322 89153
BROMLEY L A 57380
BRON V A 1942 58259 64828 64829
BROOK R J 99265
BROOKS M H 19822
BROUGHTON J 76427
BROUKAL J 41958
BROVIKOV V N 62094
BROVKIN L A 36576 46500

BROWDER J S 47871 53989 54713 65401
BROWN F H JR 60458
BROWN G E JR 97378
BROWN J D 10184 11373
BROWN J J 46491
BROWN J J JR 34704
BROWN J R JR 72699
BROWN R 10682
BROWN R G 93507
BROWN R J 78023
BROWN R W 25424
BROWN S D 32616
BROWN W B 100063
BROWN W C 94662
BROWN W E 44136
BROWN W L 91243
BROWN W R 52662 52694
BROWNE C C 7001
BROWNELL W E 40427
BRUCE H D 29200
BRUECKNER R 41425
BRUESCH P 71147
BRUGEL W 33230
BRUMMAGE K G 182
BRUNEL R 34629 54251 59429 65152 68386 71786
BRUNNER C D 56188
BRUNSCHWEG M 20867
BRYDEN J G 68377
BRYDON J E 52038
BRYKSIN V V 64731 64732
BRYZGALOV I A 66874
BRYZHINA M F 50339 64018 64019
BUBIK I 44266
BUCHNEV L M 91482
BUCHROEDER R A 68344
BUCHVAROV S 47181
BUCKHAM J A 29258
BUCKNER O S 20589
BUDIN I K 46593 46594 57257 57258
BUDNIKOV P P 268 785 41085 41086 43656 44828 46096 46097 46428 46593 46594 47694 47695 48695 49332 49333 50076 50602 52086 52164 54560 57261 57262 59345 59346 59405 59406 60851 60852 61595 61596
BUDRIN D V 23645 27608 30775 31046 32761
BUESSEM W R 43625 52780 57977
BUETTNER D H 76994
BUETTNER K 10097 16344 18237
BUETTNER K J K 27850 34138
BUETTNERK J K 35414
BUFFINGTON R M 50877
BUGAEV N F 93218 98789
BUGAJSKI W 54559

BUKAREV A 47181
BULATOV M I 41075 41076
BULAVIN I A 46267 46268
BULJAN S T 68735
BULLARD E C 28787 28792
BULLARD SIR EDWARD 16331
BULLER F H 26766
BULLRICH K 44932 47316
BULMER R H 53618
BUNDY F P 42824 100400
BUNESCU O 97659
BUNNELL L R 94058
BUNTING E N 35768 35781 36061 36234 37113 37123 47957 52135 59584
BURCH D E 25207
BURCKHARDT G 19249
BURDG C E 45297
BURDICK M D 55856
BURDICK R B 31136 48050
BURENKOV D K 76473 76474
BURGEMEISTER E A 90953 100115
BURGER E E 40392 43056 57719
BURGER O K 58417
BURGESS G K 22559
BURGESS S 90706
BURGGRAAF A J 76072
BURGOYNE T 24364
BURK D L 10669 23953 40722
BURK MAKSYMILIAN 20313
BURKART-BAUMANN I 64343
BURKIG V W 44403
BURKS T L 22272 37398
BURNETT S J 55310
BURNHAM C W 71982
BURNS E A 21660 21932
BURNS R G 60195 70057 99103
BUROV A B 50610
BUROV G V 51130 51131
BURRELL R W 26766
BURSON J H III 23525
BURSTEIN E 33156
BURT R C 78263
BURTON L C 88416
BUSBY T S 27783
BUSER R G 36465
BUSH E A 30255 43564
BUSHUEVA E B 65920
BUTENKO V I 85313 91690
BUTLER C P 39333
BUTLER C T 29429
BUTLER P 98328
BUTTERWORTH A V 6283
BUTTERWORTH G J 57997
BUTTON D D 52576

BUTZLER E W	21969		
BUYCO E H	61236		
BUYKX W J	94580		
BUZHINSKII I M		54919	88314
91214			
BUZOVKIN V P	88484	89373	
BUZOVKINA T B	70445	71677	
BYCHKOV N V	87322	89153	
BYERS H G	14188		
BYK S I	69533		
BYKOV I I	48703	49415	61898
72171			
BYKOV V N	69818	69819	
BYKOV V P	62413		
BYRNE R F	32388		
BYSTRIKOV A S	61002	61003	
BYURGANOVSKAYA G V		63162	63163
CABANNES F	23879	60595	67713
72419 72822	93565		
CABELLO A S	58630		
CABRI L J	45486	58974	91838
CADY F E	47965		
CAHOUR J A	46297		
CAILLET M	49826		
CAIN D C	4438		
CALDWELL C S	47497	48428	
CALDWELL J J	69810	70976	
CALK L C	84446		
CALLAHAN W R	33774	42772	49290
CALLEJA J M	92963		
CALVERT T A	54305		
CAMAC M	29193		
CAMERON E N	56101		
CAMERON N	76565		
CAMMERER C	37646		
CAMMERER J S	1070	1444	6215
CAMMERER W F	26150	28817	
CAMPAN T	51235		
CAMPAN T I	14148	16208	20061
CAMPBELL A N	38131	48550	
CAMPBELL D E	27922		
CAMPBELL JOHN M		29285	
CAMPBELL M J	56102		
CAMPBELL W J	52754		
CANNING R G	10306		
CAPTAIN F K	89035		
CARITHERS M D	33388		
CARLSON D E	74602	75753	
CARLSON W G	45272		
CARMAN A P	21495		
CARNIGLIA S C	74548		
CARPENTER H C	51892		
CARPENTER L G	37819		
CARPENTER R H	71157		
CARRARD A	48024		
CARRELL M A	44408		

CARROLL W	41149		
CARRUTHERS T G		25498	99265
CARSON J E	24910		
CARSTENS J P	70719		
CARTE A E	32862	35118	
CARTER CLYDE L		6227	
CARTWRIGHT C HAWLEY		33142	
CARVER J H	79915		
CARWILE L C K	26069	42676	
CASELLA A J	77166		
CASERTANO L	56973		
CASIMIR H B G	8327		
CASON J L JR	47036	53424	
CASSAN HENRY	6481		
CASSEDANNE J O		75296	
CASSEDANNE J P		75296	
CATANA V	71321		
CAVALLARO Y	20510		
CAVE E F	56540		
CECH F	80711	99174	
CERMAK V	50428	77418	90681
91242			
CERVELLE B	49675	66864	69877
75987			
CERVELLE B D	63774		
CESBRON F	75987		
CHABANOVA S	57518		
CHABERSKI A Z	20774		
CHACHANIDZE G D		37038	51368
CHADEYRON A A	59052		
CHAITSKIY V P	91694	92746	
CHAKLADER A C	25498		
CHALLIS L J	67775		
CHALLONER A R	8097	9350	
CHAMERO FERRER A		74958	
CHAMNESS L	30621		
CHAMPION F C	9433		
CHAMPOUSSIN J C		57876	
CHAN S	51146		
CHANDRA D	35594		
CHANDRASEKHARAN V		59793	
CHANDRASHEKHAR G V		79928	
CHANG C C	87316		
CHANG G K	26836		
CHANG J H	10667		
CHANG L	42979		
CHANG S C	6309	39860	55981
CHANG TETSE	9395		
CHANSKY S H	89035		
CHANT C A	27306		
CHANUSSOT G	54531		
CHANUSSOT GUY	54531		
CHAO C	52418		
CHAPMAN C R	71987		
CHAPMAN I D	32319		

CHARAK I	16712		
CHARBONNEAU D G		89235	
CHARBONNIER M	44166		
CHARETTE M P	65340	69103	92161
CHARI M S R	48890		
CHARLAT M	88837		
CHARLES E G	87952		
CHARLESWORTH D H		10898	
CHARNOCK H	18317	29365	
CHARTERS M L	45668		
CHASANOV M G	94788		
CHASE V A	36084	37509	77313
CHAUDHARY D R	46332	48921	51330
52361 55156			
CHAUDRON G	59929		
CHAWLA O P	82185		
CHEBOTAREV T YE		76986	
CHEBOTAREVA T E		61201	
CHECHERNIKOV V I		43289	
CHECHINA A F	49853		
CHEKHOVSKOI V YA		27546	28567
28597 44062	44063	56210	56605
70092			
CHELKO A J	58318		
CHELTON D B	16875	16895	
CHEN B H	67445		
CHEN J H	35227	61518	
CHEN K H	52209		
CHEN R Y	70966		
CHEN R Y S	59444		
CHEN S N C	16740		
CHEN T	47302		
CHEN T S	78340		
CHENG S C	58368		
CHENG T Y	48527		
CHENOT C	77530		
CHEPIZHNYI K I		56269	
CHEPNITSOVA N M		66874	
CHEREPANINA L I		56253	
CHEREPANOV A M		43656	
CHEREPANOV K A		74368	74369
CHERKASHCHENKO A YA		36019	46353
CHERKASOV YU A		61103	
CHERKOV D A	32646		
CHERNEVA E F	47474	47475	58575
58576			
CHERNIK L N	44608		
CHERNOCK W P	24962	48904	54697
54698			
CHERNOPLEKOV N A		31165	31166
CHERNOVA V I	57536	59464	
CHERNYAK YA N	16452		
CHERNYAVSKII V L		63748	
CHERNYKH V I	63801		
CHERNYSHOVA O V		45866	
CHERRY J R	47497		
CHESLEY F G	55185		

DECKER J A JR 70824
DECROLY CLAUD 7703
DEEG E 47364
DEEM H W 6567 6940 10947 17039 24748 25558 35917 53006
DEEV A N 45522
DEGANELLO S 74326 76112
DEGTYAREVA E V 52350 52351 61531 61532 64822 64823 91732 96251 97572 100641
DEJONG J 49783
DEKHTYAREV V L 69818 69819
DEKKER P 56820 81966
DEL MAR WILLIAM A 26766
DELANEY A 72674
DELHAL J 90188
DELIENS M 71252 90188
DELLAMICO J J 89035
DELLE CANNE GIUSEPPE 6752
DEMASHINA A I 35610
DEMCHENKO N S 47640
DEMENTEV B V 70141 89708
DEMENTEV M N 78661
DEMESHINA A I 38592
DEMETRESCU C 81213
DEMIDOV L G 71345
DEMIDOVA G A 40891 40892
DEMINET C 93343
DEMIRSOY S 55278
DEMKINA L V 89575 89576
DEMONIS I M 49621 55432 55433
DEMOU S G 60293 67289 74504 78340
DEMYANENKO YU N 96807 98853
DEMYANOV YU A 40459
DEN HARTOG H 72814
DENEKA C W 45628 47870
DENISOV YU V 51970 51971
DENISOVA I A 54757
DENNIS M D 61737
DENNY R C 39856
DEPORTES C 49826 57614
DERBENEV L S 74008 74009
DEREVNINA T O 48545
DERIBERE M 6279
DERIE R 55539
DERMAN A S 59198 60944 60945 62295
DERTEV N K 57203 57204
DERYAGIN B V 91920 95298
DESBOROUGH G A 55641
DESHAZER L G 58799
DESNOYERS J E 9820
DESPRETZ C 16209
DEUSS E 24550
DEUSS ELSA 22607
DEUSTACHIO D 6243 6938 24201

DEUTSCH T F 77935
DEUTSCHER G 80989
DEVEREUX R J 19362
DEVINE JANET 23542
DEVISME J M 88190
DEVISME V 88190
DEVRIES R C 100400
DEVYATKOVA E D 20183 26060
DEWAR J 40240
DEWITT D P 93594 93606
DEZSO E 43202
DHAR P R 23287
DHAR R N 34399
DHONT M 10394
DI COLBERTALDO D 92956
DI MARCELLO F V 47117
DIAL R 52112
DIAMOND A E 57380
DICKERSON L S 99744
DICKERSON R F 45513
DICKERSON R W JR 54081
DICKINSON H C 48034
DICKSON G 48237
DIDRIKIL L N 63800
DIEMKE H 9693 31888
DIESPEROVA M I 64828 64829
DIETERT H W 60425
DIETRICH P 97863
DIETRICH W 45262
DIETZ U 62306
DIETZEL A 14762 58735
DIKERMANN N 25525
DILKS B H JR 41699
DILLINGER J R 61376
DIMAROVA E N 34098 35752 36035 48739 67665
DIMBLEBY V 37634 37974
DIMENT W H 46855 89614 89615 89617
DIMITRIEV YA 92525
DIMITROFF J M 78310
DIMITROV D P 97029
DINA ALBERTO 20655
DINER D J 95930
DINGER C 14762
DINKLER W 22655
DIVITA S 44329
DIXON M 55946
DIXON W P 53595
DMITRIEV A P 69768 69769 74008 74009
DMITRIEV G I 62953 62954
DMITRIEVA V I 56889
DMITRIEVA Z P 88314 91214
DOBORZYNSKI D W 36300 48880
DOBRIN M 57568

DOBRONRAVOV V F 91436
DOBROVOLSKII G B 54917 57259 57260 57266 64834 64835
DOBROWOLSKI J A 89235
DODD A E 1719 60559
DODD C M 47675
DODGE M J 62634
DODSON J G 94680 97505
DOERNER E C 8818
DOHM K D 45067 53723
DOLCH W L 23406
DOLGINA G Z 46579 46580 57271 57272
DOLKART F Z 28492
DOLLFUS A 56103
DOLLING G 39188
DOMASHNEV I A 72260 72261
DOMBROVSKAYA V K 49689 56245
DOMNICH A I 20043
DOMNIKOVA G I 64346 64851
DOMOROD L S 65443 68674 75085 77334
DONCHEV S 71516
DONCHEV ST 68514
DONETSKIKH V I 66906 88573 92560
DONEY L M 46789
DONOVAN R J 47761
DORF A 43625
DORIA J 92963
DOROKHINA YU M 51010 65523
DORSCHNER J 64194
DORSEY H G 40328 49396
DOUGILL G 31868
DOUGLAS J A M 73900
DOUGLAS N J 59647
DOUGLAS R W 1713 39945 43033
DOUGLAS T B 10583 17036 39351 40494
DOW N F 77506
DOWNS L E 29200
DOYLE W P 37046
DOZIER J B JR 46042
DRABINA J 44322
DRACASS J 30293
DRAGOMIR I 61774
DRAKE E M 63519
DRAPER A L L 45538
DREYER J JR 77530
DREYFUS B 23210 27738 56698
DRIFFOED MAURICE 33568
DRIZHERUK M E 54779 54780
DRONOV A V 48802 49364
DROZDOVA T A 57817 57818
DRUMMOND A J 17182 24943
DRUMMOND D G 59958
DRUMMOND W W 10933

Name			
MC FARLIN P F	82684		
MC GARVEY J W	68251		
MC GEE T D	47675	61383	
MC GRATH B	86147		
MC HENRY R E	46877		
MC ILVRIED K E		46537	
MC KEEHAN W	78310		
MC KELLAR L A	24746		
MC KINSTRY H A		44960	46637
52780 60010			
MC LINDEN H G	29377	36486	48370
MC MAHON H O	9560	20468	41918
MC MASTER D G	38320		
MC MILLAN P W	37530	41584	49644
MC MURDO A	59921		
MC MURTRY C	52112		
MC NAMARA E P JR		56454	
MC NAUGHTAN I I		36327	
MC NEILLY C E	50485	59685	
MC NICHOLAS H J		40591	
MC QUARRIE M	16826		
MC QUARRIE M C		10083	16820
16821			
MC QUISTAN R B		55041	
MC SWAIN B D	30814	39896	
MC VAY G L	85230		
MC VAY T N	38130	42418	52479
MC WHAN D B	69622	88621	
MCGRAW D A	18322		
MCHEDLOV-PETROSYAN O P			19586
63748			
MCMAHON H O	31084		
MCMULLEN J C	14673		
MCNAMARA E P	10682		
MCQUARRIE M C	7156		
MCQUARRIE MALCOLM		196	
MEALMAKER W E	91632		
MECKE R	33451	33453	
MEDENBACH O	95500		
MEDLIN W L	38654		
MEDVEDEV N N	15835	49504	50070
64177 65123			
MEDVEDEV R V	58631	74004	74005
76265 77626			
MEDVEDOVSKAYA E I		41896	46267
46268 61595	61596	63068	
MEEKS D	61710		
MEES C E K	73996		
MEGAW H D	60348		
MEGLA G K	39012		
MEGLES J E JR	46742		
MEHMEL M	77314	77315	
MEINKEN R H	27989		
MEIRING G L	39168		
MEISSNER A	31878	31927	
MEISSNER W	6748		
MEITZNER K	58456		

Name			
MEKER GEORGES	21563		
MELAMED R IA	32675		
MELCHAKOVA L V		65249	66891
71078 71882			
MELDE G F	68234		
MELLINGER G B	94058		
MELLOR J W	55983		
MELLOR M	61727		
MELNICHENKO V S		71780	
MELNIK M T	26143		
MELNIKOV F I	8886		
MELNIKOV O K	34118	60093	
MELNIKOVA O V	54930		
MELNIKOVA V M	91435		
MELONAS J V	7695		
MELTSANSKAYA T N		64215	
MEN A A	68812	77029	
MENDEL J E	99742		
MENDELEEV D I	20683	31150	
MENDELENKO A K		57271	57272
MENTH A	69622		
MENTLEY D E	90099		
MENZHULINA F M		66437	66438
MERGERIAN D	10967		
MERIAUX E	88190		
MERILL R B	56537		
MERKER L	27646		
MERKULOVA E V	64840	64841	
MERLIN M	35068		
MERRIAM M F	94434		
MERRILL E W	10721		
MERRILL R	26087		
MERRILL R B	51904	52300	
MERRITT E	30195		
MERRITT G E	55980	60434	
MERZ K M	45778	52652	52653
52694			
MERZLYAKOV E I		73781	
MESECAR R S	68244		
MESSMER J H	19024	20265	44339
45653			
METCALFE A G	52286		
METENIN V I	73679	73680	
METSIK M S	38189		
METZ A	62289		
METZGER C	36685	36690	44351
METZGER J W	77957		
METZIER N	73315		
METZLER J	67454		
MEUTHEN A	32447		
MEWES RUDOLF	23176		
MEYER H	48685	77206	
MEYER W W	42478	44727	
MICHAUD M	4834	77511	
MICHAUD R	69660		
MICHERON F	65133		

Name			
MICHON M	53706		
MIDDLETON W E K		39842	
MIEHR W	19685		
MIELENZ K D	72581		
MIGLYACHENKO A F		20247	
MIGONADZHIEV A S		44218	44219
MIHAILOVIC Z	27691	61396	
MIHALIK P	45890		
MIKAILOFF H	56968		
MIKAMI H M	44136		
MIKASHIMA H	192		
MIKHAILOV O M	59391	59392	
MIKHAILOV P E	60936	60937	
MIKHAILOV V N	74048		
MIKHAILOVA K A		58852	58853
MIKHALCHENKO R S		49921	65125
MIKHALCHENKO V P		74406	
MIKHALEVA L A	67298		
MIKHAYLOV A R	45831	45832	
MIKHEEV V N	45866	62094	
MIKHLYUKOV E I		21323	
MIKHNYUK B G	71835	73732	73733
MILAM D	88411		
MILES JOHN L	7023		
MILFORD F J	30875		
MILGRAM A A	43437		
MILIN V B	22225		
MILLEDGE H J	60221		
MILLER A R	40665		
MILLER C D	33458		
MILLER C F	10853		
MILLER E	41149		
MILLER F	70283		
MILLER J E	71248		
MILLER J R	87595		
MILLER R J	99991	100720	
MILLER W A	37802	49179	
MILLER W E	37896		
MILLERS T	56245		
MILLS R W	60590		
MILOVIDOVA T I		53880	53881
MILSHENKO R C	55280	57270	
MILSHENKO R S	58848	58849	59046
60836			
MILVIDSKII M G		35669	
MILYUKOV E M	76046	76906	
MIMELMAN E L	30309		
MINAGAWA K	45644		
MINAKAMI T	14655		
MINCHER A L	1945		
MINELLI G	100051		
MINER R R	92035		
MING-CHONG L	67398		
MINGES M L	93643		

SILVERMAN S M 341

SILVESTROVICH S I 20917 33745 33746 53157 60905

SIMIONESCU A 14148 51235

SIMMINGSKOELD B 73520

SIMMONS E L 70035 76621

SIMMONS G 55729 56107 57604 68734 70283 82684 87315 89122 89619 91696

SIMMS R J 36758 37430

SIMON F 11805 21397 31607

SIMON F E 266 7200 25453

SIMON I 4515 29508 30221 31084 41918 46250

SIMONATO J 60595

SIMONE R 44924

SIMONIS M 24397

SIMONOV K V 46604 46605 93218 98789

SIMONOVA L A 75950

SIMPSON A W 73134

SIMPSON F H 22261 52417

SIMPSON H E 52523 52524 60009

SIMPSON O C 19419

SIMPSON P R 56100

SINCLAIR D H 46217

SINCLAIR W A 26766

SINCLAIR W R 98629

SINELHIKOV N N 9254

SINELNIKOV N N 619 8506 15501

SINGER F 24372 25148

SINGH B D 66020

SINGH D S 42064

SINGH H P 82684 87315

SINGH J P 99309

SINGH S 73297

SINHA A P B 79928

SINKE G C 8282

SINNATT F S 21665

SINTON W M 40784 40789

SINTSOVA I T 46274 46275

SIPLE G E 46855

SIRAZHIDDINOV N A 42125 51253 51254 55146

SIRCAR N R 53652

SIRKIS M D 52665

SIROTENKO D YA 15129

SIROTKIN G D 45866

SISKOV G D 95916

SISMAN O 15754 25109 35437

SITHARAMARAO D N 73840

SITNIKOVA A YA 52569 52570 58287

SIUTA V 94818

SIVALOV E G 91530

SIVCHIKOVA M G 52538 52539 58769 62788

SIVKO A P 65218 66790

SIZOV I D 46573 46574 52348 52349

SJOSTRAND M E 64004 65865 70202

SJOSTRAND M E C 71391

SKAGGS R W 45843

SKAGGS S R 92163 99593

SKANAVI G I 790 14540

SKAUPY F 31928

SKEIB G A 23243

SKINNER B J 42284 43391 52509 52532 91702

SKINNER D R 34852

SKLAREW S 15666 23145

SKOGEN H S 901 27989

SKOROSPELOVA V I 87131 87132

SKRABEK E A 36468 57665

SKRAMTAEV B G 20043

SKRIPAK V N 28591 30257 34784 34785 42464 44369 44488 46599 47476 49405 53442 54102 61097 64309 64310 64345 64850 67215

SKRYABIN A K 11886

SKULSKA I M 21511

SKURIDIN V A 71302

SLACK G A 64 29253 33123 33967 45417 62719 67372 68190 70467 76618

SLADE M L 62140

SLADEK T A 58484

SLATTERY J C 52470

SLAUGHTER C W 56355

SLAVSKAYA A I 59194 64869

SLOBODKIN L S 58230 64246

SMAKULA A 61603

SMELTZER W W 63145

SMELYANSKII I S 5009

SMIRNOV I A 20183 26060 34423 47303 47358 53478 59436 59437 64741 64742 67345 69632 69633

SMIRNOV S N 41750 41751

SMIRNOV V I 42066

SMIRNOVA E A 38598

SMIRNOVA E V 34183 38599 50057 50058 52757 61203

SMIRNOVA L G 76473 76474

SMIRNOVA M N 53327

SMIRNOVA S A 72394

SMIRNOVA T A 66892

SMITH A M 61369 74727 74933

SMITH A R 6940

SMITH C F 48050

SMITH D 75183

SMITH D D 89727 98172

SMITH D T 56479 57504 59551 59820

SMITH D W 46030

SMITH E M 68980

SMITH G S 7365

SMITH G T 46217

SMITH H L 33172

SMITH J A 77757

SMITH J M 17349 25813 28710 28956 30401 32632 39858

SMITH J S 44936

SMITH J S II 35229

SMITH J W 40398

SMITH M S 72795 72869

SMITH P L 45258

SMITH R 91687

SMITH R A 41709 56348

SMITH R B 92038

SMITH R L 32896

SMITH R T 55736

SMITH S D 53509

SMITH SIDNEY LOUIS 29043

SMITH T 8831

SMITH T F 74521 77502 89480

SMITH W O 4561 5200 14188

SMITHSON S B 87489

SMOKE E J 1132 6939 7036 7067 7102 7104 10127 10272 10518 30225 31538 35277 43395 44149 52405 52418 52568 52665 52670 52854 56215 58115 58145 77282

SMOLAK G R 46217

SMOLENSKII G A 36019 46353

SMOLUCHOWSKI M 24647

SMOLUCHOWSKI R 70325 94434

SMOLYA A V 80319 81987

SMOLYANINOV N P 64915 64916

SMOOT T W 52481

SMUDA P A 47768

SMYKALA M 83208

SMYTH H T 27989 45778 52662 67656 76886

SNIDER C R 56455

SNITZER E 39603 63697 70702

SNODDY W 26087

SNYDER D G 66852

SNYDER E F 74727

SNYDER F T 35241

SNYDER N H 7036 7067 10127 10518 30225 31538 35277

SO C S 90397

SOBOLEV R N 65249 71078

SOBOLEV V V 66906 88573 92560

SOBOLEVA P A 60073 60074

SOBOTTA H 75213

SOCHAVA I V 9440 17360

SOCHNEVA E G 70800

SODERBLOM L A 92161

SOFER G 52112

SOGA N 43453 68349 79311

SOGAWA S 27213

SOHNS H W 51892

SOKASKI MICHAEL 17994

CORPORATE AUTHOR INDEX

CONDENSED MATERIALS INDEX TO SEVEN-VOLUME RETRIEVAL GUIDE

Volume 1: Elements
Volume 2: Inorganic Compounds
Volume 3: Organic Compounds and Polymeric Materials
Volume 4: Alloys, Intermetallic Compounds, and Cermets
Volume 5: Oxide Mixtures and Minerals
Volume 6: Mixtures and Solutions
Volume 7: Coatings, Systems, Composites, Foods, Animal and Vegetable Products

The above volumes supersede and expand on the earlier publications of the Retrieval Guide:

Basic Edition (Plenum, 1967)
Supplement I, 1964–1970 (IFI/Plenum, 1973)
Supplement II, 1971–1977 (IFI/Plenum, 1979)